Voisinage Arrangement

Pádraig Ó Tuama

Two tongues
of water
lick the
recent
borders
of our
shoreline.

Neighbours,
it seems,
we share
unshared
tidelines.
Where the
water rises,
conflict
rises too.

Seems we
are
used to
the stasis
of salt
and rock
and disagreement.

Sands still
shift
making
molehills
out of
underwater
mountains.

And people
swim
in waters
that divide
them, in
histories
that deny
them their
shared
longing,
held up
by salt
that licks
the cuts
that we've
inflicted
on our
living
here.

Praise for *The Coastal Atlas of Ireland*

Ireland's ancient relationship with the waters that surround our island is one that has shaped so much of what we are. The sea has played a fundamental part in all periods of our social and economic history. This important *Atlas*, a celebration of Ireland's coastal and marine endowments, takes the reader on a voyage from the distant geological past, by way of the prehistoric era, with a focus on the island's physical environments. An important contribution, it moves through the human colonisation of Ireland to the complex cultural and economic landscapes of more recent times. A consideration of our coastal and marine environments is essential to our understanding of the history of Ireland and its people and of our present circumstances. It is also vital, if we hope to identify future challenges and opportunities as we look seaward and seek to create a truly sustainable future for ourselves and the marine ecosystems that surround us.

MICHAEL D. HIGGINS, Uachtarán na hÉireann, President of Ireland

Clearly, an island that has no less than sixty-eight maritime museums and heritage centres is an island whose people have a deep connection to its seas. *The Coastal Atlas of Ireland* illustrates the wealth and depth of knowledge we have of the Irish coastline. The result of collaborations by many experts from diverse fields of study, the *Atlas* throws light on the multiple ways we experience our coasts. Whether stunning seascapes, invasions (e.g. Viking, Norman, or tourist), or happy childhood memories of sandcastles, this Atlas has something for everyone. Beautifully and vividly illustrated, it takes the reader on a journey from the deep past to the present and towards our future coast, under pressures from climate change and development. The editors and authors are to be commended on the magnitude of this highly accessible work. It stands as the definitive published collation of knowledge for Irish coasts.

MARY M. BOURKE, Associate Professor of Geography, Trinity College, Dublin

This *Atlas* represents a fascinating picture of Ireland. Because Ireland is an island, the coastal focus covers a vast area of our natural environments, economic and social life, and history and are all presented in an accessible and readable way. I am always a sucker for maps and, as one would expect, there is a huge range, illustrating everything from the distribution of different types of whales to where the best wind resource is to be found. As well as a valuable tool for policy makers and researchers, the *Atlas* is great fun to read, while the illustrations, photographs and maps make it visually very attractive. Even if many of those who will use this *Atlas* may not read it from cover to cover, everyone will find something in it that intrigues or inspires.

JOHN D. FITZGERALD, Research Affiliate and Former Research Professor, ESRI

This *Atlas* is a rare and extraordinary fusion of land and seascapes, scientific inquiry and artistic talent, a view of the Irish coastline as you have never seen it before. It filled this exile with gratitude –for the coast itself and for the makers of this exemplary work.

FERGAL KEANE, BBC

This *Atlas* is a survival book for you and your children. It describes the inextricable link between Irish people and the sea over the course of history, and examines the challenges facing coastal communities into the future. The very reason we, as a species, are alive is because of the manner in which the sea drives the global ecosystem. Our survival will depend on how we harness our intelligence to improve our understanding of the complexity of our relationship with our oceans and coasts. *The Coastal Atlas of Ireland* is a stand-out source of the valuable knowledge we need to gather for sustainability.

MARK MELLETT, Vice-Admiral, Chief of Staff, Irish Defence Forces (retired)

Like winkle collectors, forensically searching the shoreline for what the outgoing tide might reveal, this *Atlas* leaves no stone unturned in telling the story of Ireland's coastline and our complicated relationship with the sea. If there is a truth in the lament that as a nation we have turned our backs on the sea, this *Atlas*, through in-depth analyses of the forces, natural and human, that shape the coastline, will go a long way to reintroducing us to our constant companion, its history and heritage, and the miracle of its ecosystems. It also charts a course to a more sustainable future for the planet based on the resources of the seas. Meticulously researched and beautifully illustrated, *The Coastal Atlas of Ireland* is a triumph!

CONOR NEWMAN, Former Chair of the Heritage Council

The Coastal Atlas of Ireland is the definitive examination of Ireland's unique relationship to the sea. A rare combination of the historical and the natural, the *Atlas* is as comprehensive as it is beautiful and accessible.

GRAHAM NORTON, writer and broadcaster

This fantastic *Atlas* gives an in-depth look into Ireland's past and present relationship with our fascinating coastal environments, which are vital to our existence. Not only is it all-embracing educationally on how magical nature is, but it also gives us much food for thought for the future role of our environment.

ALICIA JOY O'SULLIVAN, Climate Activist, Skibbereen

This beautifully researched and illustrated *Atlas* is a treasure trove of information on Ireland's stunning coastline. Just like a dip into a rock pool, it allows the reader to explore the many facets of our coastline in an accessible manner.

LYNN SCARFF, Director, National Museum of Ireland

This is so much more than an atlas. It tells the story of our coastline – where the land meets the sea and people, nature and industry gather to reap the benefits of a diverse mix of resources. The story is told by experts with different insights and perspectives to give a rounded and complete view of the Irish coast, looking back in time and forward to a promising future.

TARA SHINE, Director of Change by Degrees, Ireland

In celebrating Ireland's coastal and maritime spaces, *The Coastal Atlas of Ireland* also establishes a firm basis for exploring their potential within developing futures. I particularly enjoyed the evidencing of these spaces and the supportive case studies depicting the strength of relationships of its coastal spaces with Ireland's social, economic, and environmental well-being across time and space.

ROGER STREET, Research Associate, Environmental Change Institute, University of Oxford

Where the land meets the sea, and the sea meets the land provides a multi-disciplinary context for the portrayal of nature and life in this grand *Coastal Atlas*. With its vast array of coastal landscapes that have evolved at both geological and human time scales, Ireland's coast offers an ideal environment in which natural and cultural forces interact over time. There are other places where such forces can be discerned, but a wealth of research and availability of information makes this *Atlas* possible at a scale that can be appreciated and utilised. As a result, *The Coastal Atlas of Ireland* offers a unique and beautiful overview of a country that, in the words of the editors, is 'all-embracing and multi-faceted'. I pay tribute to all involved in the production of this grand work. It opens the way for other nations to look more closely at adopting the Irish approach to communicating the values and character of their coastal lands and seas.

BRUCE THOM, Emeritus Professor in Geosciences, University of Sydney

Ireland has made an indelible imprint on the world, not only through its music, literature, folklore and breweries, but also through its leadership in the study of ocean systems, resources, and maritime industries. This comprehensive *Atlas* is another great contribution; richly compelling by way of its scientific rigor, its many journeys through Irish history, and its cautionary tales for a sustainable future.

DAWN J. WRIGHT, Professor of Geography and Oceanography, Oregon State University

THE COASTAL ATLAS
OF IRELAND

For future generations, and their relationships with the coast. [Source: Robert Devoy]

THE COASTAL ATLAS
OF IRELAND

EDITORS: ROBERT DEVOY, VAL CUMMINS,
BARRY BRUNT, DARIUS BARTLETT
AND SARAH KANDROT

CORK UNIVERSITY PRESS

First published in 2021 by
Cork University Press
The Boole Library
University College Cork
Cork, T12 ND89
Ireland

Reprinted 2021

Library of Congress Control Number: 2020945092
Distribution in the USA: Longleaf Services, Chapel Hill, NC, USA

The authors have asserted their moral rights in this work
ISBN: 9781782054511

Print origination and design: Maria O'Donovan, Cork University Press

Printed in Italy by Printer Trento

CONTENTS

DEDICATION

To the memories of Bill Carter and John de Courcy Ireland. The lives and work of these two engaging and larger-than-life personalities have influenced greatly our knowledge, understanding and appreciation of Ireland's coasts; of their histories, environments and their links with the wider maritime world.

RICHARD WILLIAM GALE CARTER (Bill Carter) (1946–1993) came to live in Coleraine in the north of Ireland in 1968. Originally from the region of Bristol in south-west England, he took up a research studentship to study coastal geomorphology with Professor Frank Oldfield in the newly formed University of Ulster. He continued to work there until his early death from cancer, becoming Professor and head of the expanding Department of Environmental Sciences. His research and teaching on many different aspects of coastal science, management and community engagement with coastal matters was seminal and inspired countless people in different walks of life to take a greater interest in the world's coasts. His work continues to inform contemporary approaches in the study and use of coastal environments, both within Ireland and internationally.

JOHN DE COURCY IRELAND (1911–2006) was born of Irish parentage in Lucknow, India, coming to study for a doctorate at Trinity College Dublin. The title of his thesis, The Influence of the Sea on Civilisation (1951), in a way, says it all about his subsequent life, interests and involvement in matters marine. For many growing up in Ireland in the period from the 1950s and on to the 1980s, his name was almost synonymous with maritime history, folklore of the sea and seafaring in Ireland. His career, primarily as a school teacher, developed into that of a well-known national figure in maritime history, as well as a writer, broadcaster, politician and political thinker. His close involvement with all things marine did much to help improve and restore the importance of Ireland's sea and its coasts in people's imaginations and lives, as places to value and enjoy; a notion lost since the times of the nineteenth-century famines.

IN MEMORIAM

Sadly, the following contributors to this work have died since the first development, writing and later production stages of the *Atlas* began, and the editors wish to give a special acknowledgement of their work.

SUSAN LETTICE
MATTHEW PARKES
TIM ROBINSON
TIM SEVERIN
PETER WOODMAN

LIST OF CONTRIBUTORS

MICHAEL BARRY is a family historian and biologist with a life-long interest in sailing and Irish marine resources.

GREG BEECHINOR is an Inspector at the Environmental Licensing Programme, Environmental Protection Agency.

JAMES BELL is a Professor at the School of Biological Sciences, Victoria University of Wellington, NZ.

JOHN BORGONOVO is a Lecturer in the School of History, University College Cork.

MARY BOURKE is an Associate Professor in the Department of Geography, Trinity College, Dublin.

KARL BRADY is an Underwater Archaeologist at the National Monuments Service, Department of Housing, Local Government and Heritage, Dublin.

NIALL BRADY is Director of the Archaeological Diving Company Ltd (ADCO), Bray, County Wicklow.

WILLIAM BRADY is a Lecturer in the Centre for Planning Education and Research, University College Cork.

COLIN BREEN is Reader in the School of Geography and Environmental Sciences, Ulster University, Coleraine.

DAIRE BRUNICARDI is a Master Mariner and Senior Lecturer (retired), National Maritime College of Ireland, Cork.

BRIAN BURKE is a member of the Surveys and Monitoring Team, Birdwatch Ireland, Bullford Business Campus, Kilcoole, County Wicklow.

FIONA CAWKWELL is a Lecturer in the Department of Geography, University College Cork.

DEBORAH CHAPMAN is Director of the UNEP GEMS/Water Capacity Development Centre, School of Biology, Ecology and Earth Sciences and Environmental Research Institute, University College Cork.

LIAM COAKLEY is a Lecturer in the Department of Geography, University College Cork.

NIALL COLFER is Assistant City Archaeologist for Dublin City Council.

STEPHEN CONLON is a Design and Marketing Management specialist, based at Beginish, Bray, County Wicklow.

CLAIRE CONNOLLY is a Professor and Head of the School of English and Digital Humanities at University College Cork

ANDREW COOPER is the Professor of Coastal Studies, School of Geography and Environmental Sciences, Ulster University, Coleraine.

GRACE COTT is an Assistant Professor, School of Biology and Environmental Science, University College Dublin.

KEN COTTER is a writer and a musician, Cork.

RÓNADH COX is a Professor of Geosciences at Williams College, Massachusetts, USA.

MARGOT CRONIN is an Oceanographic Researcher, Marine Institute.

MICHELLE CRONIN is a Research Fellow (retired) with the School of Biological, Earth and Environmental Sciences, University College Cork.

JOHN CROWLEY is a Lecturer in the Department of Geography, University College Cork.

NIAMH CULLEN is a Postdoctoral Researcher, Department of Geography, Trinity College, Dublin.

SEAN CULLEN is Head of Marine and Coastal Unit (MCU) and INFOMAR Programme Manager (Joint with MI).

JOHN DAVENPORT is Emeritus Professor of Zoology, School of Biological, Earth and Environmental Sciences, University College Cork.

AOIFE DELANEY is a Coastal Specialist with the National Parks and Wildlife Service, Dublin.

HERBIE 'JOHN' DENIS works in Aquaculture Data Management for Bord Iascaigh Mhara, Clonakilty, County Cork.

FIONA DEVOY MCAULIFFE is a Research Fellow at MaREI: the SFI Centre for Energy, Climate and Marine, Environmental Research Institute, University College Cork.

EAMON DOYLE is a Geologist for Clare County Council.

TOM DOYLE is a Lecturer in Zoology, School of Biological, Earth and Environmental Sciences, University College Cork.

PAUL DUNLOP is Professor, School of Geography and Environmental Sciences, Ulster University, Coleraine.

ROBIN EDWARDS is Associate Professor, School of Natural Sciences, Trinity College, Dublin.

EOIN FANNON is a Barrister, specialising in maritime matters, Dublin.

EUGENE FARRELL is a Lecturer, Discipline of Geography, National University of Ireland, Galway.

KYLE FAWKES is a former Research Assistant with the Environmental Research Institute, University College Cork.

BRIAN FITZGERALD is Director of External Affairs and Stakeholder Liaison, Simply Blue Group and former captain (retired) in the Irish Naval Service.

LINDA FITZPATRICK is Press Officer at the National Space Centre Ltd, Elfordstown Earthstation, Cork.

MIKE FITZPATRICK is Managing Director, Irish Observer Network Ltd, Cork.

RONAN FOLEY is Associate Professor, Department of Geography, Maynooth University.

MARITA FOSTER is Deputy Director, International Office, University College Cork.

JOÃO FRIAS is a Marine Litter and Microplastic Researcher, Marine and Freshwater Research Centre, Galway-Mayo Institute of Technology.

CORMAC GEBRUERS is Head of College, National Maritime College of Ireland, Cork.

MICHELLE GILTRAP is a Lecturer, School of Food Science and Environmental Health, Technological University Dublin, City Campus.

ROB GOODBODY is an Historic Building Consultant based in Dublin.

ROY GRIFFIN is Policy Officer at the Dept of Agriculture, Environment and Rural Affairs, Downpatrick, County Down.

DAMIEN HABERLIN is a Research Scientist at MaREI: the SFI Centre for Energy, Climate and Marine, Environmental Research Institute, University College Cork.

JOHN HEGARTY is a Consultant Architect at Fourem, Cork.

KIERAN HICKEY is a Senior Lecturer, Department of Geography, University College Cork.

MADELINE HUTCHINS is a Researcher, writer and event organiser for the Ellen Hutchins Festival, Bantry, County Cork.

DEREK JACKSON is the Professor of Coastal Geomorphology, School of Geography and Environmental Sciences, Ulster University; and School of Agriculture, Environment and Earth Sciences, University of KwaZulu-Natal, Durban, South Africa.

EMMET JACKSON is an Economist, Seafood Economic and Strategic Services Unit, Bord Iascaigh Mhara, Limerick.

MARK JESSOPP is a Lecturer in Zoology, School of Biological, Earth and Environmental Sciences, University College Cork.

NORMAN KEAN is Editor of Irish Cruising Club Sailing Directions, Courtmacsherry, County Cork.

MICHAEL KEANE is a Senior Lecturer (retired), Department of Food Business and Development, University College Cork.

CONNIE KELLEHER is an Underwater Archaeologist, National Monuments Service, Department of Housing, Local Government and Heritage, Killarney, County Kerry.

BEATRICE KELLY is Head of Policy and Research for the Heritage Council, Kilkenny.

SEÁN KELLY is an Ecologist at the National Parks and Wildlife Service, Dublin.

THOMAS KELLY is Lecturer Emeritus in Zoology, School of Biological, Earth and Environmental Sciences, University College Cork.

GRÁINNE KILCOYNE is a member of the Wild Atlantic Way Team, Fáilte Ireland.

KATHRIN KOPKE is a Research Scientist at MaREI: the SFI Centre for Energy, Climate and Marine, Environmental Research Institute, University College Cork.

MAXIM KOZACHENKO is a Coastal and Marine Scientist, GEOCOAST, County Cork.

MARCUS LANGE is Research and Projects Officer at Helmholtz-Zentrum hereon GmbH, Germany.

ADAM LEADBETTER is Team Leader for Data Management, Marine Institute, Oranmore, County Galway.

SUSAN LETTICE (1985–2017) was a Postdoctoral Researcher, School of Biological, Earth and Environmental Sciences, University College Cork.

LESLEY JANE LEWIS is an Ecologist, BirdWatch Ireland/Limosa Environmental.

TONY LEWIS is the Emeritus Beaufort Professor at Mthe SFI Research Centre for Energy, Climate and Marine research and innovation, Environmental Research Institute (ERI), University College Cork.

AARON LIM is a Marine Geoscientist, Green Rebel Marine Ltd., Cork, Ireland.

ALASTAIR LINGS is an Environmental Scientist, Galashiels, Scotland.

COLIN LITTLE is a Senior Lecturer (retired), School of Biological Sciences, University of Bristol, UK.

JAMES LYTTLETON is an Archaeologist at AECOM, Bristol, UK.

TRISTAN MacCANA is a Freelance Surfing Instructor.

CRÍOSTÓIR MAC CÁRTHAIGH is Director of the National Folklore Collection, University College Dublin.

EOIN MacCRAITH is a Marine Geologist, INFOMAR, Geological Survey Ireland, Dublin.

PIARAS MacÉINRÍ is a Lecturer, Department of Geography, University College Cork.

DONAL MAGUIRE is the Director of Aquaculture Development Services, Bord Iascaigh Mhara.

GERLANDA MANIGLIA is Local History Librarian, Wicklow Library Service.

ANNALEIGH MARGEY is a Lecturer in History, Department of Humanities, Dundalk Institute of Technology.

ALICIA MATEOS-CÁRDENAS is a Postdoctoral Researcher, School of Biological, Earth and Environmental Sciences, University College Cork.

ROB McALLEN is the Professor of Marine Biology, School of Biological, Earth and Environmental Sciences, University College Cork.

DES McCAFFERTY is the Professor of Geography, Department of Geography, Mary Immaculate College, University of Limerick.

STEPHEN McCARRON is an Associate Professor, Department of Geography, Maynooth University.

BERNIE McCARTHY is Curator of Dún na Seád Castle, Baltimore, County Cork.

EVIN McGOVERN is a Senior Chemist, Marine Institute, Oranmore, County Galway.

TRIONA McGRATH is in Earth and Ocean Sciences, School of Natural Sciences, National University of Ireland, Galway.

BRENDAN McHUGH is a Marine Chemist, Marine Institute, Oranmore, County Galway.

CHARISE MCKEON is a Geologist, Geological Survey Ireland, Dublin.

DAVID MCMYLER is Operations and Training Officer, Irish Coast Guard, Dublin.

MARIA MCNAMARA is the Professor of Palaeontology, School of Biological, Earth and Environmental Sciences, University College Cork.

PAT MEERE is Head of Geology, School of Biological, Earth and Environmental Sciences / Environmental Research Institute, University College Cork.

DAVID MEREDITH is a Senior Research Officer, Rural Economy Development Programme, Teagasc, Ashtown, Dublin 15.

MICK MONK is a Lecturer (retired), Department of Archaeology University College Cork.

XAVIER MONTEYS is a Senior Geologist, Marine and Coastal Unit, Geological Survey Ireland, Dublin.

FIONNBARR MOORE is a Senior Archaeologist, Underwater Archaeology Unit, National Monuments Service.

HIRAM MORGAN is a Senior Lecturer, School of History, University College Cork.

BREDA MORIARTY is a Community Water Officer, Local Authority Waters Programme, Environmental Services, Kerry County Council

JIMMY MURPHY is a Senior Lecturer, School of Engineering, University College Cork.

MATT MURPHY is the Director, Sherkin Island Marine Station, County Cork.

RACHEL MURPHY is a Lecturer, Department of History, University of Limerick.

RÓISÍN NASH is a Lecturer, Department of Natural Sciences, Galway-Mayo Institute of Technology.

DAVID NAYLOR is an Adjunct Professor, School of Earth Sciences, University College Dublin.

VERNEY NAYLOR is a Garden Designer based at Durrus, Bantry, County Cork.

MUIREANN NÍ CHEALLACHÁIN is Excavation site director, Irish Archaeological Consultancy.

TREASA NÍ GEARRAIGH is Community Development Manager, Comhar Dún Chaocháin Teo, County Mayo.

SÍORCHA NÍ LONGHUIRT is a Marine Scientist, Environmental Protection Agency, Ireland.

ROKSANA NIEWADZISZ (Nievadis) is an Artist-researcher, University College Cork.

JULIA NUNN was formerly a Marine Biologist for the Centre for Environmental Data and Recording (CEDaR), National Museums Northern Ireland.

SHANE O'BOYLE is a Senior Scientist, Environmental Protection Agency, Ireland.

CLÍONA O'CARROLL is a Lecturer in Béaloideas/Folklore and Ethnology, University College Cork, and Research Director of the Cork Folklore Project.

RAY O'CONNOR is a Lecturer, Department of Geography, University College Cork.

BARRA Ó DONNABHÁIN is a Lecturer, Department of Archaeology, University College Cork.

NIAMH O'DONOGHUE is a PhD candidate, Department of Geography, University College Cork.

ELAINE O'DRISCOLL-ADAM is a marine cultural geographer with a keen interest in people and place, food and the environment.

PATRICK O'FLANAGAN is Emeritus Professor of Geography, University College Cork.

EOIN O'GRADY is Information and Services Development Manager, Marine Institute, Oranmore, County Galway.

ANNE MARIE O'HAGAN is a Senior Research Fellow, MaREI: the SFI Centre for Energy, Climate and Marine, Environmental Research Institute, University College Cork.

EUNAN O'HALPIN is Emeritus Professor of Contemporary Irish History, Trinity College Dublin.

EIMEAR O'KEEFFE is a Scientific and Technical Officer, Marine Institute, Galway.

CATHAL O'MAHONY is EU Grant Manager, MaREI: the SFI Centre for Energy, Climate and Marine, Environmental Research Institute, University College Cork.

RUTH O'RIORDAN is the Professor in Zoology, School of Biological, Earth and Environmental Sciences/Environmental Research Institute and Dean of Graduate Studies, University College Cork.

MICHAEL O'SHEA is a Research Fellow, MaREI: the SFI Centre for Energy, Climate and Marine, Environmental Research Institute, University College Cork.

BRENDAN O'SULLIVAN is a Senior Lecturer, Centre for Planning Education & Research, University College Cork.

DAVID O'SULLIVAN is Value Added Programme STO, INFOMAR, Geological Survey Ireland, Dublin.

RONAN O'TOOLE is Senior Geologist, Geological Survey Ireland, Dublin.

JULIAN ORFORD is Emeritus Professor, School of Natural and Built Environment, Queen's University Belfast.

MATTHEW PARKES (1961–2020) was Geological Curator, Natural History Museum, National Museum of Ireland, Dublin.

ORLA PEACH-POWER is a Research Assistant at MaREI: the SFI Centre for Energy, Climate and Marine, Environmental Research Institute, University College Cork.

MICHAEL POTTERTON is a Lecturer, Department of History, Maynooth University.

SOPHIE POWER is a Research Assistant at MaREI: the SFI Centre for Energy, Climate and Marine, Environmental Research Institute, University College Cork.

JOHN QUINN is a Professor in Zoology, School of Biological, Earth and Environmental Sciences, University College Cork.

KAREN RAY is a Lecturer, Centre for Planning Education and Research, University College Cork.

DICK ROBINSON is a writer and long-time member of Valencia RNLI.

NINI RODGERS is Honorary Senior Research Fellow, School of History, Anthropology, Philosophy and Politics, Queen's University Belfast.

STEPHEN ROYLE is Emeritus Professor of Island Geography, School of Natural and Built Environment, Queen's University Belfast.

ANNA RYAN is a Lecturer in Architecture, School of Architecture, University of Limerick.

COLIN RYNNE is a Senior Lecturer, Department of Archaeology, University College Cork.

HOLGER SCHWEITZER is an Underwater Archaeologist, Würzburg, Germany.

GILL SCOTT is a Marine Geologist, INFOMAR, Geological Survey of Ireland, Dublin.

RICHARD SCRIVEN is a Lecturer, Department of Geography, University College Cork.

REGINA SEXTON is a food and culinary historian; a Lecturer, Adult Continuing Education and School of History; and Programme Manager of the Postgraduate Diploma in Irish Food Culture, University College Cork.

JOHN SHEEHAN is a Senior Lecturer, Department of Archaeology, University College Cork.

JOHN SIMPSON is an artist, educator and fine art conservator, west Cork.

GERRY SUTTON is a Senior Research Fellow, at MaREI: the SFI Centre for Energy, Climate and Marine, Environmental Research Institute, University College Cork.

CHRISTINA TLUSTOS is Chief Specialist - Chemical Safety, Food Safety Authority of Ireland, Dublin.

CYNTHIA TROWBRIDGE is a Senior Research Associate, Oregon Institute of Marine Biology, USA.

MICHAEL WALDRON is an Assistant Curator of Collections and Special Projects, Crawford Art Gallery, Cork.

ANDREW WHEELER is Head of the School of Biological, Earth and Environmental Sciences, and Professor of Geology, University College Cork.

ROBERT WILKES is a Scientific Officer, Environmental Protection Agency.

PETER WOODMAN (1943–2017) was Emeritus Professor, Department of Archaeology, University College Cork.

FOREWORD

A team of editors and numerous contributors has unravelled the charm of the Irish coast, revealed in this remarkable and authoritative atlas.

The ocean covers more of the Earth than land does; and the coast, where sea meets land, has been disproportionately significant in human history. It remains crucial to our future endeavours, economic, cultural and recreational. However, despite our familiarity with coasts, widespead imagery of the mapping of the surface of the Moon and Mars means these are known in greater detail than most of the ocean floor.

Ireland is the quintessential island. Its relatively small size belies the enormity of its offshore marine environments. Ireland, in common with Australia, is surrounded by continental shelf and an Exclusive Economic Zone (EEZ) vastly exceeding the area of the island itself. Protruding defiantly into the Atlantic ocean, detailed seabed surveys over recent decades have revealed a complex topography with canyons, cliffs and submarine landslides, a landscape that was submerged as sea level rose after the last Ice Age. Bathed by the relatively warm waters of the Gulf Stream, these extensive underwater habitats support substantial biodiversity. The relatively clean and healthy seas are home to seabirds and marine mammals; they offer great potential for fisheries and aquaculture.

Ireland was an inhospitable place, largely covered by ice, 20,000 years ago. As the ice retreated and seas rose, plants and then animals re-occupied the landscape; the earliest people, exploiting shellfish along the coast, arrived perhaps some 30,000 years ago. It is not known whether they entered across the impressive threshold of the Giant's Causeway, as the myth says Finn MacCool did, or via some other route. The coast, then, as now, was attractive; initially for its resources and for its recreational values. Shipwrecks tell of the importance of sea transport. Numerous headland forts and castles, each with its essentially Irish charm, are evidence of the need for defence against raiding parties. Today, much of the population is focused along the coast, primarily for industrial or transport purposes.

It is important to facilitate exchange of knowledge about coastlines of the world, their functioning, history, management and governance, in order to better manage and protect them. This magnificent volume is an outstanding example of the richness and depth of coastal information. Beautifully illustrated, and meticulously edited, *The Coastal Atlas of Ireland*, with its numerous tantalising vignettes by a wide range of authors, reveals the hidden stories of the land and its peoples. It explains the appeal of the rugged rocky western coast, including the Wild Atlantic Way and the iconic Aran Islands. It describes the wetlands and numerous estuaries that support invertebrate communities and waterbirds. It describes the broader significance of beaches to residents and tourists; the appeal of fishing, surfing, swimming, walking and sailing; as well as the deeper spiritual values and the rich tapestry of cultural heritage inspired by the coast and expressed in song, music, drama, literature and the visual arts.

As Chair of the Commission on Coastal Systems, I have watched with admiration the significant role that Irish researchers have played in the broader development of coastal and marine studies in an international context. University College Cork, in particular, has been a global leader in the development and dissemination of increasingly sophisticated coastal surveys. They have played a pivotal role in pioneering the application of digital technologies to study of the coast, development of computer mapping and geographical information systems, and compilation into digital atlases and online internet resources. Collection of data, but more particularly its dissemination, is crucial in order to enable society to reduce its negative impacts on coastal and marine environments and as a basis for Maritime Spatial Planning. Dedicated projects, such as that supported by INFOMAR, have revealed an underwater world that could not otherwise have been imagined. Just as important has been the development of repositories, translating observations into computer mapping and geographical information systems, and compilation into digital atlases, or internet portals, in order to make information available for the widest possible use.

My visits to Ireland have been too short and too infrequent, but this magnificent volume gives me, here in Wollongong almost as far away from Ireland as one can be, a vivid insight into the splendour and diversity of its shorelines. It offers the opportunity to explore this 'paradise on earth', and to see that the often cold and grey spaces of the Irish coast contain an astonishing mix of wonder, mystery and charm.

PROFESSOR COLIN WOODROFFE
Chair, Commission on Coastal Systems (International Geographical Union)
University of Wollongong, Australia

We all need water to drink and food to eat, and we need to know that we belong to something bigger than ourselves. Growing up in a small coastal village in Norway, I always felt that the ocean represented this notion of something bigger than myself. The ocean represented the universe and was essential to life. The ocean could take me to new foreign shores and people. The ocean connected my small village with the rest of the world. In my area of Norway, the south-west, we have always travelled west, in the direction of Iceland, Orkneys, Shetland, Scotland and Ireland. My home region was where the first Vikings who travelled west to Ireland came from. Maybe the salted sea is in our DNA.

The ocean is essential to all life, since it is the largest ecosystem on Earth, and it is this planet's life support system. Oceans generate half of the oxygen we breathe and contain most of the world's water. Oceans provide at least a sixth of the animal protein we eat. Our security, our economy, our very survival depends on the ocean.

There is increasing international understanding and cooperation towards enhancing the conservation and sustainable use of our interconnected oceans, seas and coasts. *The United Nations 2030 Agenda for Sustainable Development* includes a specific goal about the importance of healthy oceans. All the nations of the world have agreed to prevent and significantly reduce marine pollution of all kinds and to sustainably manage and protect marine and coastal ecosystems. Increasing scientific knowledge, developing research capacity and involving all parts of society are important in the follow-up.

The publishing of *The Coastal Atlas of Ireland* is an example of an Irish commitment to the United Nations Sustainable Development Goal (SDG) for a healthy ocean. Ireland has one of the longest coastlines in Europe, and the Irish continental shelf is one of the largest seabed areas in the EU, with a seabed ten times bigger than the landmass. The Irish ocean and coastal resources have huge potential to support a diverse marine economy for seafood, tourism, energy and new applications for health, medicine and technology.

The ocean has always connected Ireland and Norway. The earliest Viking raids on the Irish coast occurred around the mid 790s AD. The Norse were not merely raiding expeditions. They were motivated by the quest for trade, tribute and land. Seaborne contact between the Norse world and Ireland intensified from the tenth century onwards, and Viking-style ships were built in Ireland. The Irish were of course connected to the sea long before the Norse came, but the new contact shifted the Irish economic power from the inland to the coastline. The Norse made Dublin the most important trading city in the west Atlantic ocean area. Many place names along the Irish coast have a Norse origin.

For many hundreds of years, Ireland and Norway were among the poorest regions on the outskirts of Europe and separated by the sea. Now we are united again to work for the sustainable development of our oceans. I will always regard myself as a Foreign Gael, Norse influenced and part of two cultures and two countries: Ireland and Norway.

Like my Viking ancestors who settled and became integrated in the Irish culture and society, we Foreign Gaels are in between the two cultures. The love for the sea and the need for a sustainable use of marine and coastal resources unite us.

Congratulations on the publication of the first ever *Coastal Atlas of Ireland*. This is a historic event.

ELSE BERIT EIKELAND
Former Norwegian Ambassador to Ireland, 2016–2019

ACKNOWLEDGEMENTS

INSTITUTIONAL SUPPORT

The Coastal Atlas of Ireland would not have been possible without the support of University College Cork (UCC) in aiding and promoting the development of this work The editors wish to thank the members of the Cork University Foundation and the staff of the Boole Library, UCC, with especial recognition of John Fitzgerald, Jean van Sinderen-Law and Cal Healy, for facilitating the essential financial support that helped underwrite the costs incurred, especially in the early stages of the work.

We would like to express particularly our appreciation to our families, who showed admirable forbearance and provided constant encouragement during this project.

PROJECT PARTNERS

The preparation of this book has been a very significant financial and logistical undertaking. Our thanks for the generous support given by the following individuals and institutions: at UCC the School of Biological, Earth and Environmental Sciences (BEES), for hosting the editorial support team and for financial assistance; the Department of Geography, for its support of the project in its initial stages of set up and conceptual development; the Environmental Research Institute, for also helping in hosting the editorial team; the Science Foundation Ireland (SFI) Research Centre for Energy, Climate and Marine (MaREI), for its interest, promotion and sponsorship. We also thank Peter Barrett; Dr. Tony and Delia Barry; the Environmental Protection Agency (www.epa.ie); the INFOMAR Programme (www.infomar.ie) co-managed by Geological Survey Ireland and the Marine Institute, and funded by the Department of Communications, Climate Action and Environment, (now the Department of Environment, Climate and Communications); Simply Blue Group; Green Rebel; the Petroleum Affairs Division, DECC, who also supplied data; and the Irish Shelf Petroleum Studies Group (ISPSG) of the Irish Petroleum Infrastructure Programme, especially Nick O'Neill, Project Manager. The ISPSG members include: AzEire Petroleum Ltd, BP Exploration Operating Company Ltd, Cairn Energy Plc, Chevron North Sea Limited, CNOOC Petroleum Europe Ltd, ENI Ireland BV, Equinor Energy Ireland Ltd, Europa Oil & Gas Plc, ExxonMobil E&P Ireland (Offshore) Ltd, Petroleum Affairs Division, DECC, Providence Resources plc, Repsol Exploración SA, Serica Energy Plc, Sosina Exploration Ltd, Total EP, Tullow Oil Plc and Woodside Energy (Ireland) Pty Ltd.

CONTRIBUTORS

The editors wish to thank all those who were directly involved in the production of this *Atlas*. To our many contributors, who have given an enormous deal of support and patience in navigating what has been a challenging process, we are grateful. It is these contributions, driven by collaboration, expertise in a diverse range of themes, and a shared love of and interest in the Irish coast, that have brought together this publication. Our thanks are due also to Presidents Michael Murphy and Patrick O'Shea (UCC) for their interest in the project as it took shape and to Jeremy Gault, MaREI's coordinator for Coastal and Marine Systems, for his encouragement.

A number of contributors helped in additional ways, these included Manon Haag and Becky Bartlett (Film and Television Studies, University of Glasgow), Des McCafferty (Mary Immaculate College, Limerick), Margot Cronin (Marine Institute), Bill Dore (Marine Institute), Brendan McHugh (Marine Institute), and David McMyler. The editors sincerely thank Colin Woodroffe and Else Eikeland, for contributing an eloquent foreword. We are indebted to Pádraig Ó Tuama and Theo Dorgan for their inspirational poetry, which form the endpages of the *Atlas*.

ATLAS EDITORIAL TEAM

The editors wish to warmly thank Kyle Fawkes and Zoë O'Hanlon, as principal supporters of the editorial team, without whom this publication could not have come to fruition. Their diligence, care and competence were essential to the editorial process and ultimately the publication of this *Atlas*. Thanks are also due to Aisling O'Grady and Ken Cotter for their assistance in gathering imagery and helping with administration.

Special thanks to Maxim Kozachenko, a member of the original team and one of the instigators of the idea of a *Coastal Atlas of Ireland*. Dr Maxim Kozachenko is a coastal and marine scientist, and a former lecturer in physical geography and Geographical Information Systems (GIS) at University College Cork. For over twenty years he has been exploring and mapping Ireland's coastal and marine environments, and is a founder of the GEOCOAST YouTube channel, which aims to increase people's awareness, understanding and appreciation of Ireland's coastal and marine environments. In addition to contributing directly to several chapters, Max has produced a number of video interviews, which provide further insight into subject areas addressed within the *Atlas* (see Chapter 1 for further details).

IMAGERY/DATA

We acknowledge the many institutions and members of the public for the wonderful support they have given to the *Atlas*, in providing much of the imagery and data that helped bring this publication to life.

A special appreciation is due to photographers Robbie Murphy, Gordon Dunn, Geraldine Hennigan and Norman Kean, for their significant contributions of stunning imagery within this *Atlas*. The team also wish to sincerely thank additional contributors, including all the open call participants; Zoë Devlin from Wildflowers of Ireland; John Cunningham; David Jones, Irish Naval Service; Donegal County Council; Douglas Cecil; and a wide variety of talented artists, including John Simpson, Brian Cleare, Gerard Byrne.

We appreciate particularly the assistance of Bernadette Metcalfe and the staff in Special Collections and Digital Collections at the National Library of Ireland for their continuous support and insight in sourcing imagery; Sara Mackeown at Port of Cork; and Connie Kelleher and colleagues at the National Monuments Service. We further thank the National Museum of Ireland, National Museums Northern Ireland, the Irish Air Corps, Tourism Northern Ireland, the Board of Trinity College Dublin and the National Folklore Collection, University College Dublin. We also recognise the important role contributors to open source platforms have played, in providing some of the images.

Data processing and map production was led by Sarah Kandrot, in the editorial team. Sarah's passion for cartography, attention to detail and capacity for processing a huge volume of data and information, was key to the production of the original and recreated maps and graphics in the *Atlas*.

CORK UNIVERSITY PRESS

The editors wish to thank the editorial board of Cork University Press for their continued support and encouragement in advancing and bringing the *Atlas* to completion. We would especially like to acknowledge publications director Mike Collins, for commissioning this book, and Editor Maria O'Donovan who has played a significant role in supporting the editorial team throughout the publication process. Thank you also to Eileen O' Carroll for her diligence and skills in copyediting this large and complex publication.

DISCLAIMER

All opinions expressed in this *Atlas* are those of the contributors. It is important also to recognise that the *Atlas* was developed over a number of years, up to and including the early part of 2021: any data or information presented may be subject to changing political, economic and/or other conditions that might arise. The presentation of any boundaries, names or other data shown in the maps, and/or included elsewhere in the *Atlas*, does not imply the expression of any opinion whatsoever on the part of the Editors or the publishers, concerning the legal or political status or the borders of the countries, territories or cities concerned. While every effort has been made to ensure the maps are free from errors, there is no warranty that they are either spatially or temporally accurate, or fit for any particular use. Especially in offshore areas, the mapping should not be used for navigational purposes.

The *Atlas* text was finalised before the ratification of Brexit and the appearance of the COVID-19 pandemic in 2020. Both events have implications, especially in Sections 4 and 5, for chapters which focus on the impact of contemporary developments on the coast. Brexit and COVID issues are intertwined with the fundamental need for the sustainable development of our coasts and seas.

ACRONYMS

AIS	Automatic (ship) Identification System	ICES	International Council for the Exploration of the Sea
BCE	Before the Christian Era	ICZM	Integrated Coastal Zone Management
BIM	Bord Iascaigh Mhara	IFA	Irish Farmers' Association
BP	Before Present	IFSC	International Financial Services Centre (Dublin)
CAP	(The European Union's) Common Agricultural Policy	IHO	International Hydrographic Office
CAS	Census Area Statistics	IMO	International Maritime Organisation
CCRS	Coast and Cliff Rescue Service	INFOMAR	Integrated Mapping For the Sustainable Development of
CDB	Congested Districts Board		Ireland's Marine Resource
CE	Christian Era	INSS	Irish National Seabed Survey
CFP	(The European Union's) Common Fisheries Policy	IODE	International Oceanographic Data Exchange
CHDDA	Custom House Dock Development Authority (Dublin)	IPCC	Irish Peatland Conservation Council
CIL	Commissioners of Irish Lights	IPCC	Intergovernmental Panel on Climate Change
CIT	Cork Institute of Technology	IRCG	Irish Coast Guard
CLCS	Commission on the Limits of the Continental Shelf	IUCN	International Union for Nature Conservation
CMRC	Coastal and Marine Research Centre	IWDG	Irish Whale and Dolphin Group
CRBI	Community Rescue Boats Ireland	LAT	Lowest Astronomical Tide
CSO	Central Statistics Office	LCA	Landscape Character Assessment
CWA	Coastal Web Atlas	LGM	Late-Glacial Maximum
CZM	Coastal Zone Management	LNG	Liquified Natural Gas
CZMD	Coastal Zone Management Division [of the Department of	lo-lo	Load on - load off
	Marine and Natural Resources]	LPG	Liquified Petroleum Gas
DAERA	Department of Agriculture Environment and Rural Affairs (NI)	LWST	Low Water mark of Spring Tide
DAFM	Department of Agriculture, Food and Marine	MaREI	The SFI Research Centre for Energy, Climate and Marine (UCC)
DCCAE	Department of Communications, Climate Action and	MBES	MultiBeam Echo Sounder
	Environment [now the Department of Environment, Climate	MHWN	Mean High Water mark of Neap (tide)
	and Communications]	MIDA	Marine Irish Digital Atlas
DCU	Dublin City University	MIS	Marine Isotope Stage
DDDA	Dublin Docklands Development Authority	MPA	Marine Protected Area
DHPLG	Department of Housing, Planning and Local Government	MSC	Marine Stewardship Council
DOMNR	Department of Marine and Natural Resources	MSFD	(The European Union's) Marine Strategy Framework Directive
DWT	Dead Weight Tonnes	MSL	Mean Sea Level
ECDIS	Electronic Chart Display and Information System	MSP	Maritime Spatial Planning (and/or Marine Spatial Plan)
ECS	Electronic Charting System	NFC	National Folklore Collection
ED	Electoral Division	NHA	National Heritage Area
EEC	European Economic Community	NI	Northern Ireland
EEZ	Exclusive Economic Zone	NISRA	Northern Ireland Statistics and Research Agency
EFSA	European Food Safety Authority	NMCI	National Maritime College of Ireland
EFTA	European Free Trade Association	NMPF	National Marine Planning Framework
EIA	Environmental Impact Assessment	NMS	National Monuments Service
EMEC	European Marine Energy Centre	NMWS	National Monuments and Wildlife Service
EMSA	European Maritime Safety Agency	NNR	National Nature Reserve
EPA	(Irish) Environmental Protection Agency	NPWS	National Parks and Wildlife Service
ESA	European Space Agency	NUIG	National University of Ireland, Galway
ESRI	Economic and Social Research Institute	NUIM	National University of Ireland, Maynooth
FLAG	Fisheries Local Action Group	ONB	Area of] Outstanding Natural Beauty
FSAI	Food Safety Authority of Ireland	OPRC	Oil Pollution Preparedness, Response and Co-operation
GHG	Greenhouse Gas		Convention
GIS	Geographical Information System	OPW	Office of Public Works
GPS	Global Positioning System	ORE	Ocean Renewable Energy
GRT	Gross Register Tonnage	OSi	Ordnance Survey of Ireland
GSI	Geological Survey of Ireland	OSPAR	OSlo-PARis Convention for the Protection of the Marine
HAB	Harmful Algal Bloom		Environment of the North-East Atlantic
HAT	Highest Astronomical Tide	OWC	Oscillating Water Column
HWM	High Water Mark	QUB	Queen's University Belfast
HWST	High Water mark of Spring Tide	RBMP	River Basin Management Plan
ICAN	International Coastal Atlas Network	RCP	Representative Concentration Pathway

RNLI   Royal National Lifeboat Institute
ro-ro   Roll on - roll off
ROI   Republic of Ireland
ROV   Remotely Operated Vehicle
RSL   Relative Sea-level
SA   Small Area
SAC   Special Area of Conservation
SAR   Search and Rescue
SDZ   Strategic Development Zone
SLC   Sea-level Change
SLR   Sea-level Rise
SOLAS   Safety Of Life At Sea (Convention)
SPA   Special Protection Area
SSI   Site of Special Scientific Interest
SSP   Shared Socioeconomic Pathway
TAC   Total Allowable Catch

TCD   Trinity College Dublin (Dublin University)
TLS   Terrestrial Laser Scanning
TNC   TransNational Corporation
UAU   Underwater Archaeology Unit
UAV   Unmanned Aerial Vehicle
UCC   University College Cork
UCD   University College Dublin
UHO   Underwater Heritage Order
UKHO   United Kingdom Hydrographic Office
ULCC   Ultra Large Crude Carrier
UNCLOS United Nations Convention on the Law of the Sea
UNESCO United Nations Educational, Scientific and Cultural Organisation
VMS   Vessel Monitoring Systems
WGS84   World Geodetic System (1984 revision)
WHS   World Heritage Site

SECTION 1
THE PHYSICAL, BIOLOGICAL AND HUMAN SETTINGS

Basalt cliffs of the Causeway Coast, County Antrim
[Source: Gordon Dunn]

IRELAND'S COASTS: SETTING THE SCENE

Darius Bartlett, Barry Brunt, Robert Devoy, Val Cummins and Sarah Kandrot

WHERE THE SEA MEETS THE LAND. The coastline of County Derry from Castlerock and the mouth of the River Bann, with its training walls or 'moles', looking westward to the spit of Magilligan Strand at the mouth of Lough Foyle and the mountains of Donegal in the far distance. [Source: Gordon Dunn]

The real Ireland is coastal Ireland: to one side is the sea; to the other is the land! The seas that surround Ireland affect everyone on the island, no matter how far from the coast. As a result, almost every aspect of life on the island has to be viewed through the lens of 'the coast'; even though issues in the history and culture of the island have often combined to suppress the coastal element and character of Ireland's population. The coast has had a direct role in shaping Ireland's history and its evolving, multi-layered identities, and continues to play a major role in influencing how the island functions environmentally, economically, socially, culturally and politically. While the sea connects us with Europe and the rest of the world, it also separates us, and the resulting relationships will affect all of our futures. Ireland's coastal environments are embedded in our culture and our imaginings, as much as in the daily practicalities of our lives. This *Atlas* presents the story of these coasts, the processes that formed and continue to shape them, the people who live along them and their evolving interrelationships with coastal resources (Figure 1.1).[1]

Although the coast, as an entity, is something that virtually everyone can recognise intuitively, arriving at a precise and universally agreed definition of what constitutes this environment is very challenging. While the coast is a distinct and critical element in the geography of Ireland, it is almost impossible to recognise where it begins and ends: how far inland do you have to go, and how far offshore, until you are no longer in the coastal zone? For geographers, anywhere on land that lies within reach of the influence of the sea, and anywhere offshore that is affected by what happens on land, may be considered part of the coastal zone.

The coastlands of the world are home to increasing numbers of people. At least 40 per cent of the global human population lives within 100km of the sea and the proportion of coastal dwellers is increasing much more rapidly than that of people living further inland. This is as true on the island of Ireland as it is elsewhere, with the increased urbanisation of the coast forming a marked trend in meeting the needs of this expanding population. The coastal margins of Ireland, however, provide more than simply living space; they also provide a wealth of other tangible and intangible benefits, including minerals and sources of energy, fish and other foodstuffs, trade links and transportation routes, flat land for agricultural and industrial development, space for leisure and recreation, and a host of irreplaceable and essential ecosystem services (Figure 1.2).

This multiplicity of uses and perspectives on coastal spaces means that the study of the coast transcends traditional scholarly divisions between the natural sciences, humanities and arts, as well as our human tendency to compartmentalise and define sub-disciplines within these larger categories. We can only properly begin to appreciate and understand the coast through a profoundly multidisciplinary approach. Furthermore, people's awareness and perception of coasts vary enormously, especially given the great diversity and changeability of features – such as colour, light and shade, storminess and tranquility – that can be found in coastal areas. The emotional responses generated by such seascapes and skyscapes can often have subjective and spiritual value, and help explain the attraction of coastal regions to people throughout the world. Thus, in this *Atlas* you will find contributions that draw on the diverse insights of archaeologists and architects, artists, biologists, botanists, cartographers (ancient and modern), conservationists, dancers, developers, ecologists, economists, engineers, entrepreneurs, explorers, farmers, filmmakers, gardeners, geographers, geologists, historians, industrialists, islanders, marine scientists, naval personnel, navigators, oceanographers, photographers, planners, poets, sailors, seanchaithe, sociologists, tourism and hospitality professionals, urban dwellers, yachtsmen

Fig. 1.1 IRELAND'S COASTS AND SEAS. a. Bray, County Wicklow, and b. the west coast of Achill Island, County Mayo. The seas surrounding Ireland, a relatively small island located off the western seaboard of Europe, have long defined its natural and cultural landscapes. a. Around the eastern and southern coasts in particular, they have provided the principal means of connectivity to link Ireland to its neighbouring island of Britain, as well as to Europe and the world. Such well-established links exposed these coastal regions to strong external influences and, together with their more benign environmental conditions, encouraged the development of relatively prosperous and outward-looking rural and urban communities. This is reflected in the shoreline at Bray in County Wicklow, which has developed as a tourist resort and commuter town for Dublin, located 20km to the north. [Source: Mark and Dave Murphy]; b. In contrast, more extreme weather conditions, stormy seas and a rugged coastline typify much of Ireland's western seaboard. This, together with the vast expanse of the Atlantic ocean to the west, contributed to a perception of the seas holding the island and its traditional culture in isolation from the rest of the world. The western coastline of Achill Island evokes this 'separateness', although now it is precisely this sense of being a 'world-apart' that sees such areas becoming more globally integrated, via international tourism. [Source: Brigid Bates Furlong © 2018]

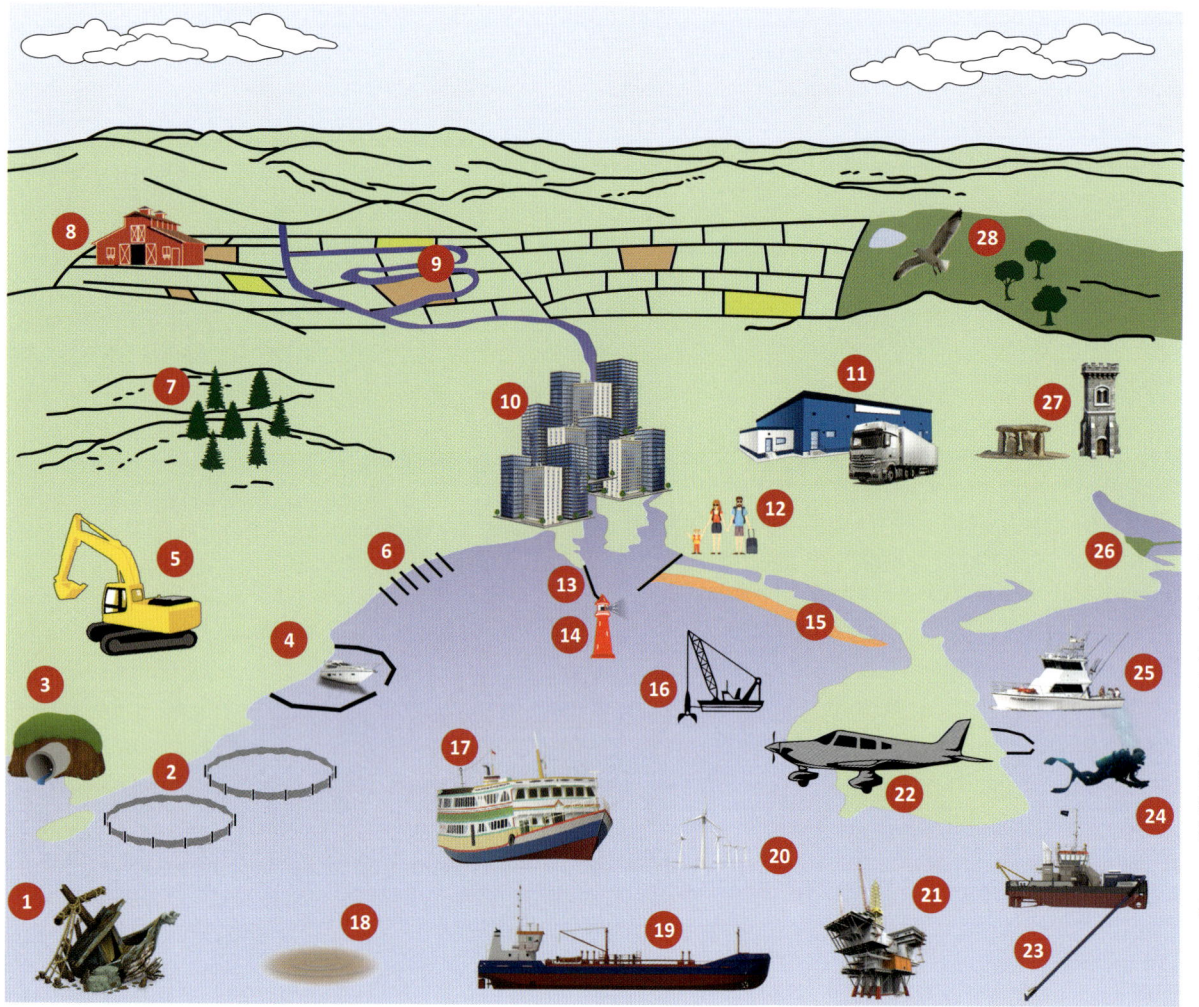

1. Shipwrecks
2. Aquaculture
3. Sewage outfall
4. Marina
5. Quarry
6. Erosion control
7. Forestry
8. Agriculture
9. Irrigation
10. Urbanisation
11. Industry
12. Tourism
13. Seawalls
14. Hazard warning
15. Beaches
16. Dredging
17. Ferries
18. Waste dumping
19. Oil tanker movements
20. Offshore renewable energy
21. Oil and gas extraction
22. Military activities
23. Sand and gravel extraction
24. Diving
25. Inshore fishing
26. Land reclamation
27. Cultural and archaeological heritage
28. Nature conservation

Fig. 1.2 THE COASTAL ZONE IS SUBJECT TO MULTIPLE, AND OFTEN CONFLICTING, USES, STAKEHOLDERS AND INTERESTS.

and women, and zoologists, all of whom bring to these pages, and share with us, their unique ways of looking at the coast.

Beyond the line of the shore extends the ocean. The marine realm is an essential part of Ireland's story and it, too, is a major focus of this *Atlas*. At the planetary level, the oceans remain largely unknown and poorly understood. It is commonly stated that we have mapped more of the surface of our Moon and nearer planetary neighbours, and in greater detail, than we have mapped the floor of our oceans. However, due to recent developments, such as the INSS/INFOMAR programmes (presented in detail in later chapters), Ireland is in many ways an exception to this rule. The country has often emerged as a global leader in its many engagements with the sea, including in marine and coastal science, the pursuit of a 'blue' (and green) economy, and in the development of sustainable marine renewable-energy resources. The *Atlas* celebrates these achievements, while pointing the way for future research and explorations that build on these foundations.

To the best of our knowledge, nowhere else in the world has such an all-embracing and multifaceted exploration of a nation's, or an island's, coast been undertaken. In preparing the ground for this *Atlas*, we acknowledge a classic coastal dichotomy: on the one hand, the coast represents a natural boundary between land and sea, and does not recognise political divisions; while on the other, socio-economic realities have to be accepted, and policies enacted in one jurisdiction may have profound implications for a neighbouring territory. Furthermore, in a globalised world, a small island like Ireland is inevitably exposed to events and policies enacted on much larger geographical scales. We have, therefore, taken an explicitly all-island approach to the contents and presentation throughout the chapters that follow and, where appropriate to do so, have endeavoured to place Ireland's marine and coastal stories in their wider regional and planetary settings.

The coasts of Ireland can also be viewed as a laboratory and baseline for the study of key processes and patterns of Earth System functioning, as well as for developing and testing methods or technologies, in areas such as renewable energy or aquaculture, that might contribute to the future sustainable development of the world's societies. The island is situated on the Atlantic and European continental margins, and therefore receives the first signals of ocean and atmosphere responses to variations in their functioning. Consequently, the complex of physical and human themes developed in this *Atlas* has international relevance for coastal communities worldwide, but especially those located in mid-latitudes. Given the island's strategic location and the array of coastal and marine research expertise available in Ireland, further investment in such research has real potential to contribute even more to our understanding of coastal issues of global concern.

HOW THE Atlas IS ORGANISED

To present a comprehensive approach to unravelling the diversity and fascination of Ireland's coastline, the *Atlas* is organised into six sections. These focus on what we consider to be the fundamental ingredients necessary to understand the evolution, variety and vitality of Ireland's coastal landscapes, and which help define the island's past and present, as well as its future. The sections are built up around a number of key chapters, each of which provides a description and analysis of a particular subject area, and one considered vital to an understanding of the overarching theme of the section. The chapters have a wealth of illustrations – including maps, diagrams, graphs and photographs – all designed to further the reader's appreciation of the complexities and beauty of the island's coastal landscapes. While each chapter can be read as a self-contained piece of work, and one that may be of particular interest to a reader, they also form part of an evolving sequence in which each chapter can be considered to be a 'building block' designed to take readers in a logical progression to a fuller appreciation of a section's central theme.

Within the chapters, in addition to the core text, you will find a series of featured subjects and case studies that provide greater-depth explorations of particular topics or examples related to the central theme. In addition, the maps, photos and other illustrations that accompany the text have been provided with self-contained captions that may also be browsed to gain an appreciation of the chapter's subject matter, before a more immersive reading is undertaken. Thus, the *Atlas* is equally suited to being read in progression or, if preferred, it can be dipped into and navigated according to the specific interests of the reader.

The six sections also adopt an evolutionary approach, taking the reader from the distant geological past, by way of the prehistoric era and a principal focus on the island's physical environments, through time and the human colonisation of Ireland, to the complex cultural and economic landscapes of the present and concerns for the sustainability of its future coastal landscapes. Although not claiming to be encyclopaedic, when read in its entirety the *Atlas* will provide readers with a fascinating and comprehensive excursion through time and space along Ireland's coastline. Furthermore, it provides critical insights into the importance of coastal environments in understanding the island's past, appreciating the present and contemplating future opportunities and challenges.

Section 1 (Chapters 1–9) begins by providing an overview of the key characteristics and processes that have helped shape and design Ireland's coastal environments. It begins by introducing the roles of core physical processes, such as relief, climate and tidal movements, together with marine biology, and shows how these created the natural landscape that humans came to colonise (Figure 1.3). The human presence, and subsequent transformation of Ireland's coastal environment, is viewed initially through the introduction and subsequent development of agriculture. Even today, the spectacle of Ireland's coastline is epitomised for many, and especially for international tourists, in the combination of the 'natural' environment and the superimposition of a cultural landscape comprising traditional family farms, small field patterns

Fig. 1.3 TWO OF THE MOST PROMINENT PHYSICAL SETTINGS FOUND ON IRELAND'S COASTS: SANDY BEACHES AND HARD ROCK WITH CLIFFS. a. Tullan Strand, County Donegal. Situated north of the seaside town of Bundoran (visible top centre), the strand is part of this coast's complex of sandy beaches, and is linked to a prominent sandspit feature (centre). The dunes visible behind the beaches are heavily degraded from use by people, particularly from the recent impacts of beach tourism and leisure. The sand was originally deposited on the continental shelf by retreating glaciers at the end of the Ice Age, and moved onshore under Atlantic waves and storms. Today, the beaches are linked to the dunes as sources of sand. [Source: Gordon Dunn]; b. The glaciated coast of the Iveragh Peninsula, south-west Ireland. Valencia Island lighthouse at Cromwell Point, with Beginish Island behind it, was built as a warning to sailors of the dangers of this rugged and indented rocky coast. The hard sandstone and quartzite rocks that form this coastline are resistant to the high-energy Atlantic wave action that characterises the shores of western Ireland. The results are steep and cliffed coastal landscapes, with the cliffs commonly 10–30m high. These coasts have been heavily affected by ice action in the recent past, with ice cover melting back from the region only *c.*18,000 years ago. Beginish Island shows the distinct, smoothed rock surfaces that indicate ice action, with the steeper slopes around the upland of Cloghanelinagan, MacGillycuddy's Reeks, in the background. The seaward coastal areas were subsequently flooded by rising sea levels as the ice retreated, creating the deep-water harbours, embayments and extensive estuaries that punctuate the line of Ireland's coasts today. [Source: Robert Devoy]

and the rearing of cattle and/or sheep. More detailed accounts are then provided of the island's formative processes, including the laying of its geological foundations and the impacts of glaciation and past sea-level changes around the coast. The section concludes by reviewing various ways in which people have interpreted the coast, both onshore and offshore, through different techniques of mapping and visualisation.

Fig. 1.4 CLIFFS OF MOHER. Standing as a bulwark to the full force of Atlantic waves, the Cliffs of Moher rise almost vertically to a height of 214m above the sea, and extend for more than 14km along the County Clare coastline. Today, the cliffs are one of the top tourist attractions in Ireland, receiving more than 1.5 million paying visitors annually. They are formed of shale and sandstone, laid down in the delta of a massive river that flowed through the area around 320 million years ago, and are gradually retreating landward as waves attack and undermine their base, causing layers of rock above to fall into the sea under their own weight. Sea stacks, such as the one seen here, mark previous positions of the cliffs, but these will be worn down and disappear in due course. As well as attracting human visitors, the cliffs are also home to a large and varied population of seabirds, many of which nest on ledges on the cliffs, while seals, basking sharks, dolphins and porpoises may often be seen offshore. Below the surface of the sea, a shallow reef extends several kilometres seaward from the base of the cliffs, causing refraction (bending) and focusing of incoming waves. Particularly after Atlantic storms and at certain states of the tide, this results in the creation, some 3km seaward of the cliffs themselves, of some of the most celebrated surfing waves in the world. This includes the wave known as Aileen's (after the original Irish name Aill na Searrach, 'Cliff of the Foals'), which can rise as a giant plunging breaker, sometimes more than 12m in height. [Source: Bernard Niess, flickr, CC BY 4.0]

The four chapters (Chapters 10–13) that constitute Section 2 focus on the physical expressions of Ireland's natural coastal environments. Readers are exposed to the great diversity of the island's shoreline, such as its rocky coasts, which display some of Ireland's most iconic seascapes backed by steep cliffs plunging into the surrounding seas (Figure 1.4). Ireland's wealth of beach and dune landscapes, wetlands and estuarine environments are also explored. In aggregate, this range of what are perceived to be spectacular coastal environments is of fundamental importance to the island's large and growing tourist industry. Such landscapes, however, are coming under increasing threats from over-exploitation and ever-expanding urbanisation. Careful planning will be required to preserve the quality and vitality of these environments.

The impact of people on Ireland's coastline since they first colonised the island some 9,000 years ago is the focus of Section 3 (Chapters 14–23). The section opens and closes with chapters that centre on how Irish people have imagined – and continue to imagine – the coast as a component of their

Fig. 1.5 DUNLUCE CASTLE, COUNTY ANTRIM. The human presence in Ireland can be traced back some 9,000 years, and involved migration and colonisation of the island from the east. From its east coast, people spread to occupy the whole island. More peripheral and isolated areas along the western seaboard, in particular, exhibit excellent examples of ancient settlement structures, such as beehive huts and ring forts. Over the last millennium, however, the colonial imprint became more deeply embedded along the north, east and south coasts. Vikings and (especially) Norman invaders, as well as later British colonists, developed most of their sites of control at strategic locations along these coasts, which allowed ease of access to Britain. These include historic landmarks such as castles, plantation estates and historic ports and towns that trace their origins to the Viking and Norman eras. Dunluce Castle provides an excellent example of the historic importance of strategic coastal sites. Replacing an earlier Irish fort from the thirteenth century, Dunluce Castle was built in the early 1500s on a basaltic outlier, with steep cliffs plunging to the sea. It became the control centre for the MacDonnell clan, who held extensive territories in Scotland and the north of Ireland, and became the seat of the earls of Antrim in the seventeenth century. Today, these iconic ruins are a tourist 'hotspot', not only for the spectacular views, but also due to their use as a location for the television series *Game of Thrones*. [Source: Colin Breen]

Fig. 1.6 DUBLIN. Some 40 per cent of Ireland's population resides within 5km of the sea. An increasing dominance of Ireland's coast as the preferred location of residential and economic activities has become particularly apparent since the 1960s. From this decade, the island experienced modern economic development associated with successful engagement in global trade and accession to the European Union. Emblematic of these endeavours is Dublin, the island's largest city (2016: 1.3 million population in the city and its suburbs, which extend over the former County Dublin) and capital of the Republic. This historic city, which traces its urban foundation to the Vikings, has long served as the principal gateway into the island, and departure point, serving Britain and the rest of the world. The photograph shows the port (top left) through which flows almost one-third of the island's trade. Extensive areas of the historic quays have also experienced urban renewal and now accommodate a highly successful and expanding hub of offices involved in international financial services and high-tech industries. Dublin Airport, the island's largest hub for air freight and passenger traffic (30 million passengers in 2017) is only 10km from the city. Centre-left in the photograph is Heuston railway station, one of two primary rail termini located in Dublin. These, together with the motorway system that is also focused on Dublin, extend its effective hinterland to cover most of the island. This strategic gateway city has generated large-scale urban sprawl, which has extended into the adjacent counties of Meath, Kildare and Wicklow. This extended urban area is termed the Greater Dublin Area. In 2016 its population was 1.9 million: 40 per cent of the Republic's population, and 29 per cent of the island total. Such growth and polarisation of development has placed increasing pressure on adjacent coastal areas, as well as deflecting growth from elsewhere on the island. [Source: Irish Air Corps]

identity. Despite what many Irish people would feel to be a stronger attachment to 'the land', in the shape of its rural economy and culture, Ireland possesses a deep maritime heritage and traditions (Figure 1.5). These take the form, for example, of many historic structures and ancient ruins located along the shoreline, including tower houses and Martello towers, as well as long-established marine institutions. Offshore lie the wrecks of many ships that foundered in the stormy seas that can pound the shoreline. Taken together, they continue to define coastal communities, whether established settlers or more recent 'blow-ins'. The rest of this section traces the colonisation of Ireland from its beginnings, onwards through the Vikings, Normans, the plantation system and into the nineteenth century with its elements of modernisation. This latter century, however, also embraced the tragedy of the Great Famine, with consequences that continue to define much of the country's landscape and culture, most especially along the depopulated western seaboard.

Section 4 (Chapters 24–29) takes the reader into the twentieth century and to the present. Focus is placed on the coastline's contemporary resources, which became increasingly the foundation of much of the island's economic and social development since the 1960s. Central to this growth in prosperity and renewed optimism was the surge in international trade through Ireland's port

infrastructure and the linked growth of population. This development, however, polarised increasingly in and around the larger city-ports located on the east coast, and was reflected especially in the urbanisation of large areas of the coastline and adjacent inland areas (Figure 1.6). Coastal developments associated with more traditional industries such as fishing and tourism are explored, as are more modern growth sectors, involving mining and the offshore exploitation of renewable energies.

Arising from the rapidity and uncontrolled nature of much of Ireland's coastal development, these areas have experienced a significant increase in environmental impacts, from pollution and many other stresses that arise when expanding human populations and their infrastructures strive to occupy finite areas of coastal and estuarine land, a process sometimes known as 'coastal squeeze' (Figure 1.7). In addition, while Ireland benefited significantly from its growing involvement in processes of globalisation and membership of the European Union (through, for example, inward investment, increased employment and economic prosperity), the country was exposed increasingly to external threats over which it had comparatively little control, including the hypermobility of capital and rising sea levels linked to climate change. While Ireland, as a small and an increasingly open economy and society, has limited powers to control external

Fig. 1.7 WATERFORD CITY. Founded by the Vikings in AD914, Waterford is reputed to be Ireland's oldest city, and today is the fifth most populous settlement in the Republic (2016: city population of 53,504). Despite being almost 30km inland from the open sea, its location on the navigable River Suir enabled it to become an important trading centre and benefit from being Ireland's closest deep-water port to the continent of Europe. This, together with ease of access to a productive and prosperous agricultural hinterland, allowed the city to become the major commercial and administrative hub for the south-east. The brown colour of the Suir, seen in the photograph, indicates its heavy load of suspended soil and silt that comes downriver from the hinterland at times of heavy rainfall. This supplies sediment to the marshes and mudflats of the estuary and the open coasts nearby, but also leads to the need for periodic maintenance dredging to keep channels open for shipping. Following marked prosperity in the eighteenth century, Waterford entered a lengthy period of decline. In the 1980s, however, urban renewal began, linked to national development strategies and the availability of extensive areas of waterside land allowing the city's port functions to be transferred downstream in 1992 to a new outport at Belview. Regeneration of the vacated south quays (left side of photograph) and of the adjacent buildings became the initial key objective. Its success has transformed and revitalised this historic area through a new marina, amenity space and refurbished waterfront properties. The long-unused, north inner-city quays are now designated a Strategic Development Zone and will involve the planned construction of hotel, office and residential properties, together with a new transport hub for the city. Integration of this expanded central area with the existing city core to the south of the river will be achieved through a new link bridge to relieve heavy cross-river traffic over Rice Bridge. Furthermore, to reduce large volumes of through traffic and enhance quality of life within Waterford, a ring road and cable-stayed bridge across the Suir (top-right of photograph) have been completed to the north of the city. [Source: Geraldine Heningan and Norman Kean]

threats of this nature, planning can assist in developing an integrated response that better reflects the aims of more sustainable human and physical environments. These crucial issues for Ireland's coastal areas and communities are explored in Section 5 (Chapters 30–32), which offers some potential lessons that may benefit coastal communities located in other parts of the world.

Finally, Section 6 consists of a single chapter (Chapter 33), in which conclusions are outlined and some thoughts on future coastal problems and solutions are presented. Ireland is currently having to confront an era of increasing precariousness in terms of issues such as climate change and future uncertainties, shifting patterns of global and local politics, the consequences of Brexit, restructuring of the world's energy supplies, food security, globalisation and emerging counter-globalisation sentiments in the wake of COVID-19 (Figure 1.8). All these issues have implications for Ireland's coastal and marine spaces, as well as for its ability to manage them for present and, especially, future generations. This section confronts these issues and outlines some of the structural changes that need to be made in our political, economic and social

systems, if Ireland is to increase its resilience and capacity to overcome the adverse consequences of these trends. From the emergence of purely local, grassroots campaigning groups, to the 'greening' of policies and practice at national and international levels, we recognise and celebrate a growing public awareness that humanity's stewardship of the planet must be placed on a much more sustainable footing. Although the challenges involved are significant and becoming increasingly urgent, we nonetheless end on an optimistic note.

The Government of Ireland INFOMAR programme, with a primary aim to map the physical, chemical and biological features of Ireland's seabed, has contributed to *The Coastal Atlas of Ireland* through a chapter in the publication. It has also supported the production of online videos and a link to this material can be found on the Atlas' webpage at https://www.ucc.ie/en/coastal-atlas/. These videos complement the printed Atlas by presenting interviews with people involved in writing parts of the text and provide additional insights into the subject areas dealt with in the *Atlas*.

THE DIGITAL STORYMAP

A comprehensive suite of supporting digital materials has been compiled to complement and supplement the contents of this published Atlas. This material is presented as a StoryMap, a web-based application that contains interactive digital maps in the context of narrative text that aligns with the themes covered in the printed *Atlas*. It can also be accessed on the *Atlas* webpage at https://www.ucc.ie/en/coastalatlas/.

Much of the data for the maps produced in the *Atlas* were derived from freely accessible digital sources, such as INFOMAR, Ireland's Marine Atlas, the Central Statistics Office, NISRA and elsewhere. A selection of these data is presented within individual web maps contained in our StoryMap, and visualised via Web Map Services (WMS), an international standard protocol for the delivery of map images over the internet. The data delivered by the WMS are housed in geographical information system (GIS) databases (see Chapter 8: Monitoring and Visualising, the Coast),

and are curated by the various agencies responsible for their collection. The *Atlas* StoryMap was built on the ArcGIS software platform, developed by the US company, Esri (www.esri.com). Other digital materials about Ireland's coasts (e.g., video/ film) can be found at the GEOCOAST link in (https://www.ucc.ie/en/coastal-atlas/).

For each of the six sections of the *Atlas*, our StoryMap highlights a selection of topics from the main text. The interactive maps allow readers to probe more deeply into themes and issues that are introduced in the main *Atlas*. Thanks to compliance with international standards for data organisation and exchange, the StoryMap also lets users discover and explore a wide range of datasets relevant to the coast of Ireland, and to download many of these for their own analyses and projects, should they wish. We hope that this will enhance our readers' engagement with the *Atlas* contents and motivate them to explore, in more detail and at their leisure, the ever-expanding range of available coastal and marine data.

Fig 1.8 KINSALE, COUNTY CORK. This picturesque town is located at the mouth of the Bandon River, and looks out into the broad and sheltered Kinsale Harbour. While historically an important fishing centre and port, guarded by the impressive Charles Fort, Kinsale declined following the formation of the Irish Free State. By 1961, as with many other small coastal communities, the town was in a relatively depressed condition and prospects of economic development for the then population of 1,587 were limited. Since the 1960s, however, the prosperity of many coastal towns has improved, especially within the context of Ireland's Celtic Tiger. Located 25km from Cork city, Kinsale's attractive coastal environment, a growing range of amenities in the town – including a yachting marina and high-quality restaurants – and improved road infrastructure not only encouraged the growth of a tourist sector, but also increased its role as a commuter town for Cork. The result has been significant urban renewal, population growth (2016: 5,281 population) and increasing pressures on the natural environment and social infrastructures of the town, especially in the peak summer tourist season. Population growth has now spread into smaller communities adjacent to Kinsale with similar consequences. For coastal communities, such as those of Kinsale, it is a matter of urgency that more sustainable and integrated planning systems are put in place, to preserve the quality of life offered by these towns, and to protect their diverse and distinctive natural environments for the enjoyment of future generations. [Source: Geraldine Hennigan and Norman Kean]

The Coastal Environment: Physical systems, processes and patterns

Robert Devoy, Andrew J. Wheeler, Barry Brunt and Kieran Hickey

GOLDEN STRAND, ACHILL ISLAND, COUNTY MAYO. The strand here is just one of the many highly attractive scenic beaches found along Ireland's sediment-dominated coasts. This coastline is composed of beaches that are often linked with large dunes and sand gravel barriers, as well as with other soft, sedimentary estuary and wetlands environments. This aerial view over Golden Strand shows the operation of the key, common coastal controls of wave action, tides, wind and across-beach stream and other freshwater discharges. Together, these hydrodynamic flows are responsible for moving the mainly sand sediments at this site into the patterned shapes shapes that can be seen here, as well as on all such beaches. [Source: Eugene Farrell]

IRELAND'S COASTS

The coast, which forms the edge zone of the Earth's continents (over 1 million km in length), is a palimpsest of landscapes and environments (for example, rocky shores, sand dunes, beaches): the coastal zone. As a well-established definition, the coast is, at its simplest, the three-way boundary zone of land, ocean and atmosphere. One of these components, the land, is relatively stable, at least in human terms. The other two are extremely dynamic. Coastal changes also occur at a variety of spatial scales – from the microscopic to the national and beyond – as well as at a range of temporal scales, measured from microseconds to thousands and millions of years.

Most significant for the day-to-day functioning of the coastal zone are the controls of waves and storms, as is the case for Ireland, western Scotland and the Western Isles; tides, as shown in the large tidal ranges experienced in the Bristol Channel, south-west England and south Wales, or the Bay of Fundy, north-east Canada, in contrast to the minimal tides of the Sea of Japan and the Mediterranean's coasts; and sediments, the lack or abundance of them. The role of sea-level change as an additional control to coastal behaviour is discussed separately (see Chapter 7: Ancient Shorelines and Sea-level Changes). The margins (vertical and horizontal boundaries) of coastal environments (or systems) operating under these controls overlap each other and share flows of mass and energy over place and space and through time; these are the primary ingredients of all Earth systems.[1]

Ireland's coasts are made up of many different types of generic systems. They comprise rock-dominated and, generally, wave-exposed settings (>40 per cent by length), as on the western shores of the island. Of equal importance are its coasts of lower wave-energy situations, such as estuaries. These environments are characterised more by elements of sediment movements and tides, by wide sand beaches and, as is often found, the occurrence of extensive sand dunes. Other important coasts are those of low height (<10–30m), soft cliffs type, composed of sediments (e.g., glacigenic sediments) derived from the impacts of the not-so-time

Fig. 2.1 SOME OF IRELAND'S MANY COASTAL LANDSCAPES, OF VERTICAL CLIFFS, ROCKS AND WAVES, WETLANDS AND SANDY ESTUARIES a. The steeply cliffed coasts of Tory Island, County Donegal, are formed of the ancient Pre-Cambrian acid intrusive rocks (such as granites) of the region. The island is exposed to frequent high-energy Atlantic storms, creating these rocky and eroding coastlines; showing offshore sea stacks indicate coastal retreat over time. [Source: Donegal County Council]; b. The crenulate, cliffed and rocky beach of Beenbawn, Dingle, County Kerry, at the entrance to Dingle Harbour. Here, the beach and embayment is exposed to Atlantic swell waves, which can be seen breaking on the mid-Devonian age sandstones of this coastline. [Source: Noel O'Neill]; c. Extensive areas of marshland (containing salt- to freshwater-dominated environments) are located along Ireland's c.7,000km of coast. The view is of the wetlands at Whiddy Island, Bantry Bay, County Cork, showing the different vegetation communities developed here, together with associated open-water lagoons. [Source: Robbie Murphy]; d. White Strand, Corragaun (south of Clew Bay), County Mayo. A complex of coastal environments are shown in this panoramic view, which contains the sandflats and beach of the open coast (photograph centre), framed by low rock coastlines (for example, bottom left). The visible sandy shorelines are backed landwards by low, vegetation-covered sand-sheets and sandhills, forming machair-type environments, which lead into (centre and right) a lagoon lake and stream-outlet system. Such complexes are developed particularly along the western seaboard of the island. [Source: Gordon Dunn]

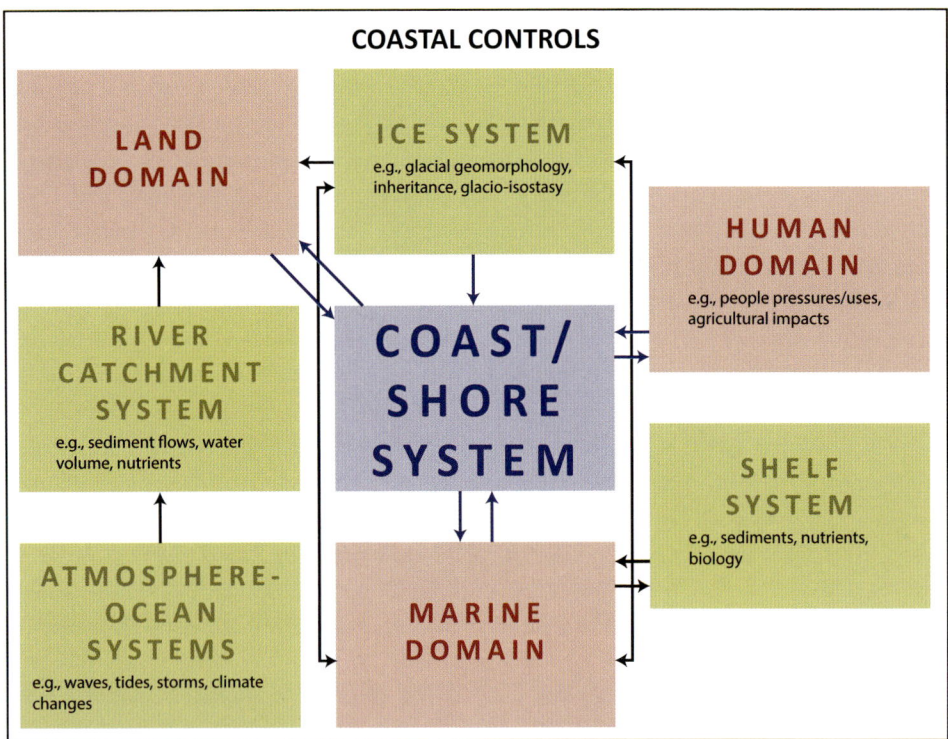

COASTAL CONTROLS

LAND DOMAIN

ICE SYSTEM
e.g., glacial geomorphology, inheritance, glacio-isostasy

HUMAN DOMAIN
e.g., people pressures/uses, agricultural impacts

RIVER CATCHMENT SYSTEM
e.g., sediment flows, water volume, nutrients

COAST/ SHORE SYSTEM

SHELF SYSTEM
e.g., sediments, nutrients, biology

ATMOSPHERE-OCEAN SYSTEMS
e.g., waves, tides, storms, climate changes

MARINE DOMAIN

Fig. 2.2 MODEL OF THE MAIN MASS AND ENERGY CONTROLS THAT CREATE AND STRUCTURE IRELAND'S COASTS. The model centres on the coast/shore and shows the main systems and domains that affect it. Within these, the driving controls on coasts – such as waves and tides for energy (as in Atmosphere-Ocean Systems) and sediments for mass (as in Shelf System) – feed into the Coast/Shore System and have primary interconnections with each other (links and feedbacks), as shown by the arrows. The controls are part of, and operate through, regulating environmental systems (for example, Atmosphere-Ocean), which together drive (affect) coastal functioning. The systems are, in turn, linked closely with the domains (sources or homes) of the different control factors. (Every link and feedback possible is not shown.)

distant last Ice Age. To these can be added more particular types of coastal environment, such as distinctive long stretches of gravel beaches and barriers, the pockets of machair coast (sandy, fertile coastal plain) of mid- to north-west western Ireland, saltmarshes, the mud- and sandflats of the many estuaries and bays, lagoons and extensive fresh-to brackish-water wetlands. Collectively, these different coastal systems occupy >14,000km^2 of Ireland's c.84,400km^2 total land and seawater margins (Figure 2.1) (see Chapters 10–13, on rock, sand beach, wetlands, estuary and lagoon coasts).

The primary geomorphological processes that control coasts as Earth systems are rock weathering (rock decomposition), erosion (rock and sediment removal) and transport (or transfers) of energy and rock materials to new locations. The way these operate on the island and their role together in balancing the controlling ingredients – of water (waves, tides, currents), rock and sediments and flows of chemical

Fig. 2.3 FIVE FINGER STRAND, TRAWBREAGA BAY, INISHOWEN, COUNTY DONEGAL. This shows the beaches, dunes and waves entering the outer bay, with a view southwards along the channel of the connected estuary, with its feeder streams into the bay. This area forms a good illustration of a functioning coastal system and its subsystems, with their interconnected flows of mass and energy. Water can be seen flowing into this bay and estuary system from the landward end and from seaward by the tides, waves and coastal currents. Together, these move and mould the available rock materials (sediments) into different physical shapes (landforms), such as spits, beaches, sandbars, mud- and sandflats and saltmarshes. Each of these environments can operate as a subsystem of the whole (of the bay and estuary). Together, they form part of the bigger system: the whole of the coast itself. [Source: Geraldine Hennigan and Norman Kean]

nutrients for plants and animals, as well as other elements – give Ireland's coast its distinct and unique character.

This chapter explores Ireland's coastal environment as a physical system, the product of geological, climatic, oceanographic and human forces, or drivers. It recounts the story of how today's coast came to be, based on the best available knowledge to date. While Ireland's coast shares many characteristics with other Atlantic and global coastal environments, it is also unique in its character and such distinctiveness inspired the production of this *Atlas*. Although there is still much to discover and understand, we begin here with a geological history of Ireland's seabed, continuing with a review of the influences of climate and weather, waves and tides, and people, and how these have shaped and moulded Ireland's contemporary coastal landscape (Figures 2.2 and 2.3).

OFFSHORE ENVIRONMENTS

Where exactly does the coastal zone begin and end, and how does it link to the offshore zone? These can be difficult questions to answer. Regarding the coastal zone, the question centres fundamentally on where the sea and the maritime influence cease to be significant driving forces in defining Earth's environmental character and functioning. Answers to this may mean that on islands like Ireland, everywhere on the island is part of the coast. For the offshore zone, the answer is simpler.

Ireland's offshore environment extends from its landward margin, across the continental shelf, and into the deep ocean of the North Atlantic (Figure 2.4). By definition, this includes all seabed areas around our shores down to water depths of *c*.200m along the continental shelf edge and slope, as well as the distinctive shelf sea areas of the Irish Sea, Celtic Sea and the Malin Shelf Sea. This zone reaches an effective boundary, in physical, economic and legal terms, of *c*.200km from Ireland's western coastline. The offshore environment hosts a diverse range of ecosystems and plant and animal species, forming a rich and highly productive food environment (Figure 2.5) (see Chapters 26: Coastal Fisheries and Aquaculture and 32: Management and Planning).[2]

IRELAND'S CONTINENTAL SHELF

During the last major glacial cold stage (*c*.16,000–30,000 years ago), and also at periods earlier in the Quaternary, sea level was much lower than it is at present and in some areas Ireland's shorelines lay close to the edge of the continental shelf. Beyond the shelf lies the deeper-water environments of the Atlantic Continental Margin. This is formed primarily by the continental slope (shelf margin) that connects the plateau-like surface of the shelf (at 100–200m water depths) to the deep Abyssal Plain, thousands of metres below the ocean surface. In general terms, this slope area has been built up over the long term (millions of years), by sediments coming off the land via coastal systems and transiting across the shelf and down into deeper ocean areas (Abyssal Plain). This material may be 'wafted off the shelf' as suspended plumes of sediments, which settle slowly onto the seabed and beyond. These sediments may also become entrained in the many submarine canyons and other channels cut into the slope, and then fast-tracked into the deep ocean beyond. Thick accumulations of the slope sediments (for example, fan structures, such as the Donegal Fan) may become subject to collapse. This can result in submarine landslides and avalanches (called turbidity currents), or may become reworked by

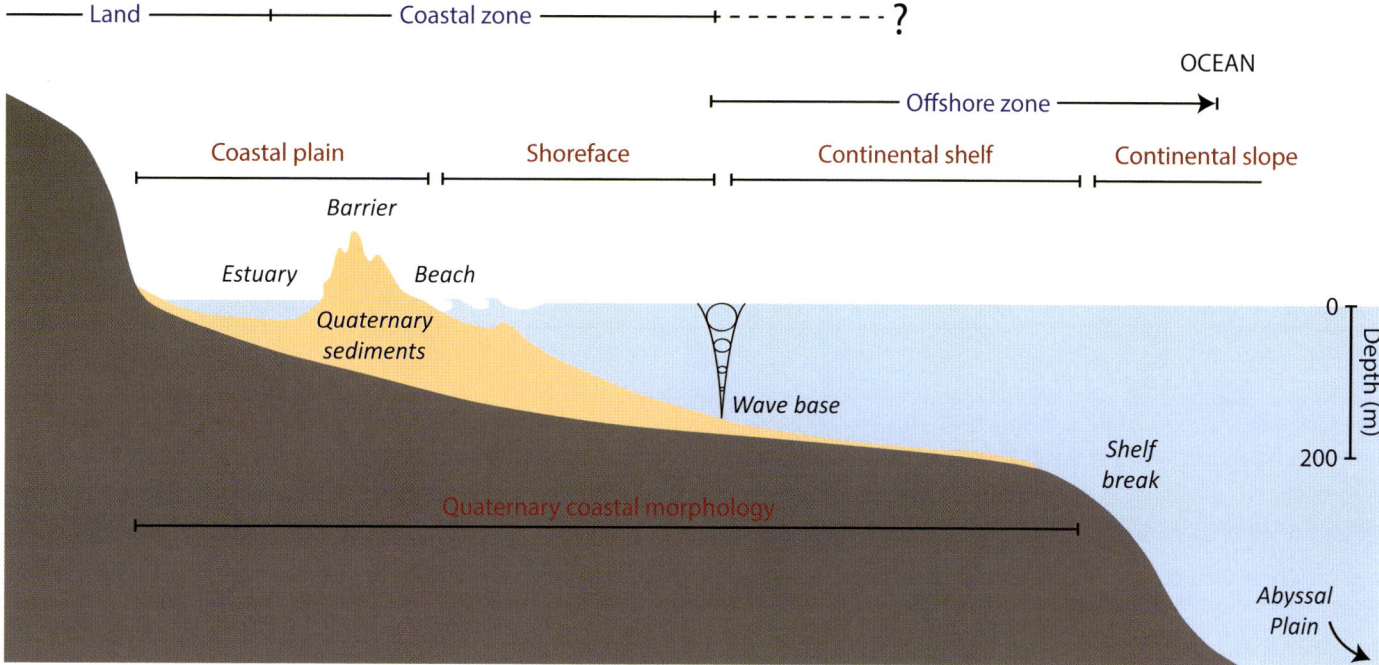

Fig. 2.4 COAST TO OCEAN: KEY AREAS WITHIN THE BOUNDARY ZONE BETWEEN THE COAST AND OFFSHORE ENVIRONMENTS. The cross-section is for Ireland's location (not to scale), of the Earth's land surface and the ocean margin. The relative positions and extents of key areas within this boundary zone are shown, together with illustration of some coastal features and terms used commonly in association with this margin. In particular, the wave base describes the point vertically in the water column, below the sea surface, where orbital wave motion ceases. Where this point (at water depths of *c*.180m) intersects with the seabed then sediment entrainment (capture and suspension) and movement toward the coast begins. The distance the coastal zone extends into the ocean is subject to some functional variability. It is taken generally as the distance to the point of shelf break on most ocean margins, but may be even greater in some coast–ocean settings. For Ireland, this is a maximum distance on its western coasts of *c*.200km.

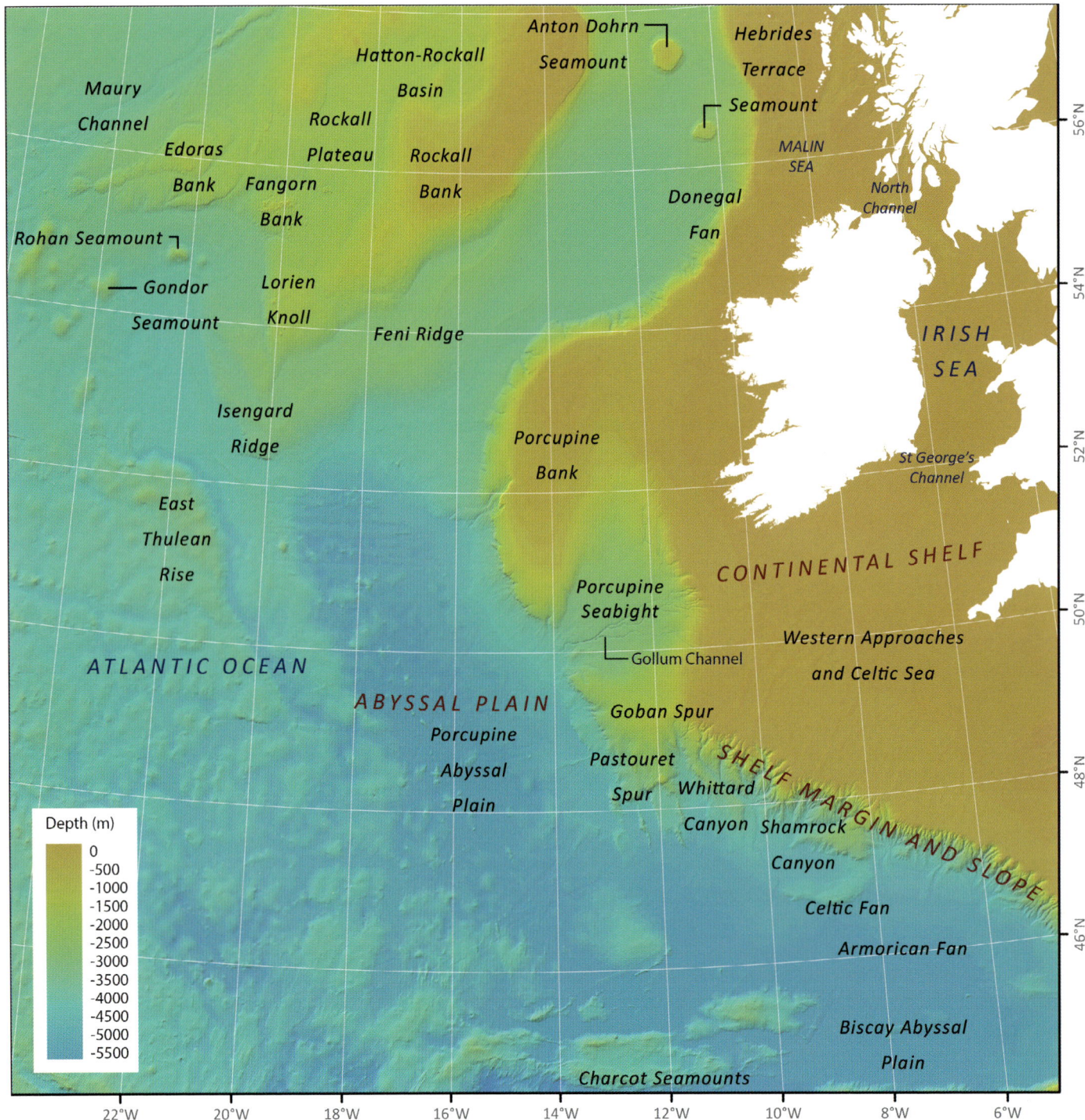

Fig. 2.5 WATER DEPTHS AND THE EXTENT OF IRELAND'S CONTINENTAL SHELF MARGINS AND OFFSHORE ENVIRONMENTS. The continental shelf, which includes the inner shelf regions of the Celtic and Irish seas, extends over 200km west of Ireland into the North Atlantic ocean. To the north-west of the main shelf lies a large zone of separated shelf and elevated submarine seafloor. Most of these areas are part of Ireland's territorial waters, which have a total seabed coverage of approximately 900,000km². The surface of the continental shelf maintains a low angle of <1° and reaches a maximum mean depth along the shelf margin of *c*.200m. Here, the seabed gradient steepens to form the continental slope, before descending further into the deep ocean (to areas of the Abyssal Plain). The surfaces of the shelf (seabed) are cut in places, particularly within the Celtic Sea, by deep palaeo-channels. These were formed as part of former river and glacial meltwater systems that once crossed the shelf at times of low sea level. The seabed itself is composed of both bare rock surfaces, swept clear by seabed currents, as well as extensive areas of sands and finer sediment deposits. The location of some of the main geological structural and geomorphological seabed features are also shown on the map, for example, the Porcupine Seabight sedimentary basin. This is a primary source region of potentially viable oil and gas resources. The shelf margin is characterised by large and deeply incised submarine canyon and gully systems. Examples are of those linked to the Gollum Channel within the Porcupine Seabight, but there are many such features along the entire shelf edge, particularly southwards. Some 280km from Ireland's south coast occur the large Whittard and Shamrock canyons. Both are associated with the submarine sediment fans of deeper ocean areas, such as the Celtic Fan and Armorican Fan. [Map data: GEBCO Compilation Group 2019. GEBCO 2019 Grid (doi:10.5285/836f016a-33be-6ddc-e053-6c86abc0788e)https://www.gebco.net/data_and_products/gridded_bathymetry_data/#global]

strong margin-hugging currents.

The Irish continental slope borders the western and southern Atlantic shelves and is vast. It also encompasses the Porcupine

Seabight embayment and includes an isolated area surrounding the (Irish sector of) the Rockall Bank. The angle of these slope areas is slight (generally the continental slopes are between angles of 1–

10 degrees with gradients of *c*.1:6) and, unlike slopes on land, continental slopes extend over considerable distances. For example, due west of Galway, the margin, for a distance of *c*.80km, descends from 170m to 2,500m water depth; that is an average slope of only 1.7 degrees but a fall of over twice the height of Carrauntoohil (Ireland's highest mountain).

Not surprisingly, such an extensive environment contains many features, several of which have only recently been discovered and about which we know very little. Within a European context, the Irish continental margin is special, in that it straddles the southern limit reached by the European ice sheets and glaciers during the last Ice Age. Glaciers are incredibly erosive and the continental margins of northern Europe are mantled by thick accumulations of glaciomarine sediments that were dumped by the ice as it extended to the edge of the continental shelf, and sometimes extending beyond the continental margin as floating ice (ice shelves). These accumulations of glacially derived sediment can over-steepen the margin, causing it to collapse. The large submarine slides and avalanches that have resulted from this dominate the margin topography. To the south of these areas, beyond the maximum reach of the ice, the southern European continental margin is characterised by incised submarine canyons, in many cases fed directly by present-day river systems or ancient glacial meltwater-fed rivers. Ireland's continental margin contains both these types of features and many others besides (see Chapter 6: Glaciation and Ireland's Arctic Inheritance).[3]

As far as we know, most submarine slides on the Irish continental margin occurred before the end of the last glaciation. The rise in global sea level, due to the melting of the ice, increased water pressure on the slopes (hydro-isostasy). This may have been accompanied by earthquakes, as Earth's crust readjusted to the removal of glacier ice and the change to renewed water loading, causing parts of the margin to collapse. The resulting submarine avalanches were huge, with hundreds of cubic kilometres of sediment moving down-slope. These rapid collapses almost certainly caused large tsunami to occur that would have inundated Irish coastal areas. Today, we see huge scallop-shaped indentations in the margin, representing the headwalls of these slides, and we can trace displaced sediment avalanching nearly 200km into the deep basins. Evidence is becoming available that these may still contribute to present-day tsunami waves that continue periodically to impact Ireland's coasts (see Chapter 7: Ancient Shorelines and Sea-level Changes).

The submarine canyons of the margin (for example, within the Porcupine Seabight) and further south erode multi-kilometre-wide chasms into the margin slopes. Ireland hosts the largest submarine canyon in Europe, the Gollum Channel, extending hundreds of kilometres across the Porcupine Seabight (Figure 2.6). These canyons are conduits through which sediment is transported rapidly into the deep oceans and play a major role in connecting surface and deep basal ocean waters. Nutrient flux through the canyons supports abundant life, but this is also a dynamic and violent environment. Episodic flushes of dense sediment-laden waters shape the canyon floor topography, forming thick layers of sediments as they settle (turbidites).[4]

Deep erosion of these canyons also occurs and can generate precipitous rock cliffs as part of the canyon systems. Recent explorations have revealed that these cliffs can be vertically extensive in area and support vibrant habitats, colonised by corals and sponges. Ireland's world-famous cold-water coral carbonate mounds are also prominent features of the continental margin. Additional erosive features found on parts of the margin are due to the impacts of icebergs. As these have grounded, they have dragged their keels through the seabed muds, particularly during the times of the last glaciation, forming extensive areas of criss-cross patterned furrows. These so-called iceberg plough marks can be tens of metres deep and extend for distances of kilometres (see Chapter 6: Glaciation and Ireland's Arctic Inheritance).

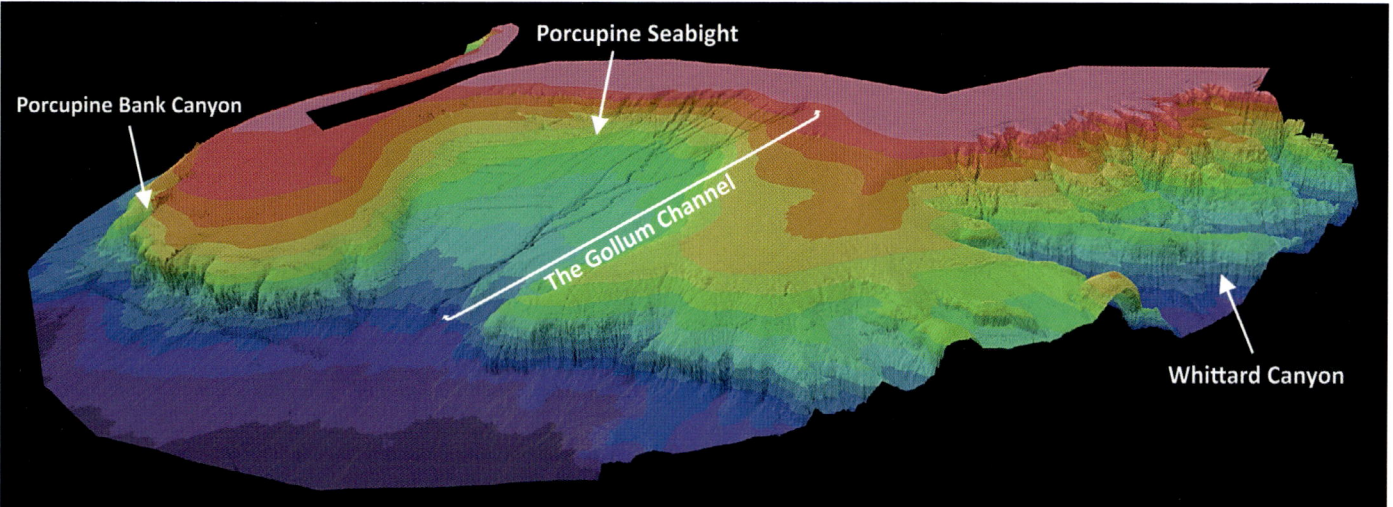

Fig. 2.6 THREE-DIMENSIONAL VIEW OF IRELAND'S SOUTHERN CONTINENTAL MARGIN. This colour-coded multibeam bathymetry digital terrain model covers a small part of Ireland's western to southern continental margins. The model is structured along the shelf edge and taken between the Porcupine Bank on the left (orientation to the north-west) and the heavily incised feature of the Whittard Canyon on the right. The middle section of the model shows the Porcupine Seabight, with Europe's longest submarine canyon, the Gollum Channel. This stretches from the shelf break (in pink) and down the continental slope to the continental rise (in blues). The colours in the model mark depth zonations, with the shallowest pink-red boundary at 400m water depth and the deepest purple/dark purple boundary at 4,000m water depth. [Data source: bathymetry data is courtesy of the Geological Survey Ireland and Marine Institute INFOMAR programme, www.infomar.ie]

COLD-WATER CORALS, REEFS AND CARBONATE MOUNDS

Andrew J. Wheeler and Aaron Lim

Contrary to popular belief, corals are not exclusively tropical or restricted to the shallow-water photic zone where light penetrates down to the seabed. In fact, the majority of coral species exist in deeper, colder waters (about 5,100 documented species) and occur as far north as the Arctic. These are mainly soft-bodied corals, but calcareous corals also form spectacular 'reefs of the deep'. The corals feed generally on dead organic remains that rain down from productive surface waters, as well as live prey, mainly small shrimp-like copepods. Ireland's continental margin is famous for its cold-water coral reefs and giant coral carbonate mounds (Figure 2.7).

Ireland's cold-water corals inhabit a depth range of about 500 to 1,200m below sea level on the continental margin where water temperatures are decidedly untropical and permanently dark. Reefs thrive at the boundary between surface East North Atlantic Water and the more dense water masses of Mediterranean Outflow Water and North Atlantic Deep Water. Here, organic food particles – such as dead plankton – which normally settle through the ocean, stall temporarily, thereby creating concentrations of food for the corals. Fast-flowing currents and suspended sand particles also keep the corals clean from clogging muds, while mobile sands can infill the coral framework bases. This protects them from predation and adds stability to the coral framework, thereby allowing reefs to reach impressive heights. For these reasons, certain parts of the margin are particularly hospitable to cold-water corals, allowing cold-water coral reef provinces to be defined (Figure 2.8).[5]

Although reefs thrive today, they are susceptible to climate change. Perhaps the biggest threat to cold-water corals globally is ocean

Fig. 2.7 CORALS OF THE COLD-WATER REEF ENVIRONMENTS OF THE IRISH CONTINENTAL SLOPE MARGINS. This image is from the Piddington Mound, at just under 1,000m water depth, south-west of Ireland. Cold-water coral reefs provide important habitats for many organisms. Although many dead coral frameworks (grey) are visible, on closer inspection the tips of these colonies have been found to be alive and growing (white and pink). There are several coral species present, the large frameworks being composed of *Desmophyllum pertusum* and *Madrepora oculata*. The pale yellow upright coral on the left is *Acanthorgorgia armata*. Two much smaller and barely visible corals are labelled (1) *Zoanthidea* sp. and (2) *Pliobothrus symmetricus*. Conspicuous white glass sponges (*Aocallphristes beatrix*) and less conspicuous bright yellow glass sponges (Hexadella dendritifera) are also common. Other organisms found here include (3) squid, (4) anemones, (5) sea urchins (*Echinus* sp.), (6) squat lobsters, (7) starfish (*Porania* sp.), as well as others living within the coral framework. Many of these are hiding from the bright lights of the cameras that have intruded on this world of perpetual darkness. [Source: courtesy of University College Cork and the Marine Institute]

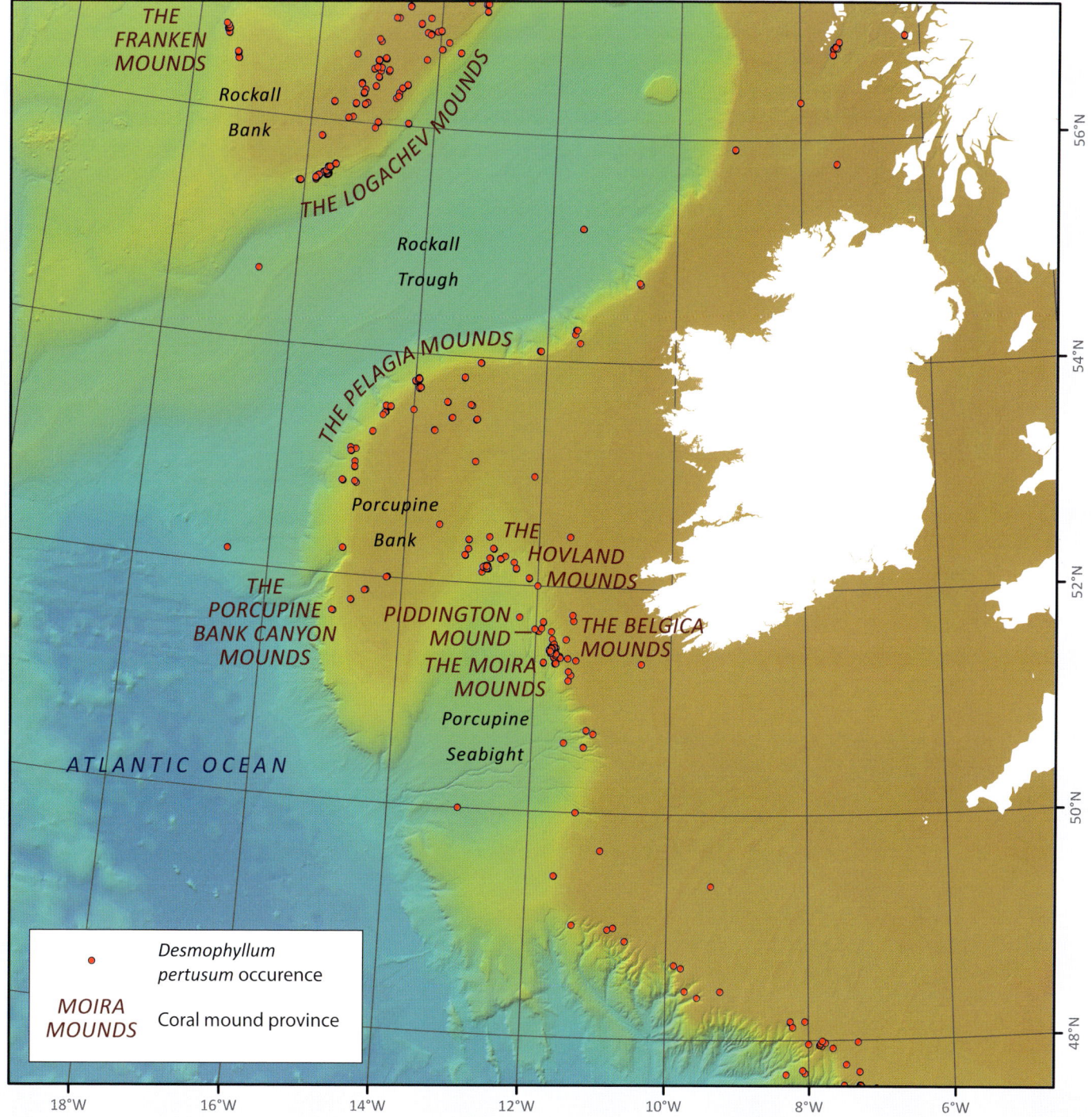

Fig. 2.8 DISTRIBUTION OF THE SEVEN MAIN PROVINCES OF COLD-WATER CORAL REEFS AND MOUNDS ON IRELAND'S CONTINENTAL SHELF. Confirmed observations of the principal Irish reef-forming cold-water coral, *Desmophyllum pertusum*, show a distribution along the upper continental margin. In some cases, these are isolated colonies, whereas elsewhere reefs, coral gardens or giant carbonate mounds occur. The seven main coral mound provinces, and others, have developed as significant topographic structures over thousands to millions of years. The Moira Mounds are perhaps the best studied of these, of which over 250 have been identified. Despite the Geological Survey of Ireland's mapping of most of Ireland's deep-water seabed, the smaller, cold-water coral patch reefs and mounds are hard to find. The Piddington Mound, within the Porcupine Seabight, is possibly the most impressive example, and supports communities of corals, sponges, crustaceans and fish. It is 50m across and shows a clear zonation of corals, reflecting a reef-generated micro-climate of more coral-friendly environments in the centre. [Data source: Freiwald et al, 2017. Global Distribution of Cold-water Corals (version 3.0). Second update to the dataset in Freiwald et al. (2004) by UNEP-WCMC, in collaboration with Andre Freiwald and John Guinotte. Cambridge: UNEP World Conservation Monitoring Centre. Available at: http://data.unep-wcmc.org/datasets/3]

acidification. This is the decrease in ocean pH due to its absorption of atmospheric CO_2 and the consequent formation of carbonic acid. Calcareous animals, like cold-water corals, cannot secrete carbonate in low acidity conditions and, therefore, literally dissolve with increased acidification. Past ocean acidity events have resulted in the mass extinction of many coral species, such as during the end of the Permian, 250 million years ago. Compared with most of the world, Irish waters have a deep aragonite compensation depth: the depth at which coral becomes fatally stressed by pH. In a worst-case, climate-disaster endgame scenario, Ireland may support some of the last remaining cold-water corals, which may avoid extinction, when and if oceanic conditions improve. Continued protection and understanding of these hidden and poorly understood habitats is, therefore, of major importance to the biological future of our planet (see chapters, 31: Pollution and 33: Climate Change and Coastal Futures).

As well as contemporary reefs, Ireland is also famous for its giant coral carbonate mounds. These can reach up to 350m in height and cover several kilometres across mounds of coral debris formed by successive reef development over the last 2.6 million years. As the coral reefs forming the carbonate mounds grow, they trap sediment and thereby preserve an archive of the changing ocean that can be interpreted by geologists. The study of sediments associated with these reefs shows that they were dropped from floating icebergs. These had melted and deposited sediment onto the reef over two million years ago, and had originated from glaciers in Ireland. This is remarkable, as it was previously thought that glaciation was only confined to the Arctic at this time.[6]

Ireland's shelf seas

The continental shelf is the seabed area that exists immediately seaward off the coast as far as the shelf break and continental slope, where the seabed plunges to the deep ocean (Figure 2.4). Shelf seas are, therefore, relatively shallow, typically between 50 and 200m water depth; and such seas surround the entire island of Ireland. They include the Irish Sea to the east, the Celtic Sea to the south and the North Atlantic Irish Shelf to the west and north. It is best to consider the shelf seabed areas as presently submerged land since, during the peak of the last glaciation, global mean sea level was over 100m lower than it is today. This means that nearly all of the Irish continental shelf was land, but was far from being dry land. It would have been covered variably over these and earlier times of cold climate by glaciers, meltwater-fed braided rivers, tundra vegetation, wetland environments and walked across by the iconic Irish giant elk (*Megaloceros giganteus*). With the melting of the ice caps and the subsequent rise in sea level, these shallow areas became submerged (see chapters 7: Ancient Shorelines and Sea-level Changes and 16: The Inhabitants of Ireland's Early Coastal Landscapes).

The inundation of these shelf areas by the sea at the end of the last glaciation was incredibly destructive. It is useful to think, in this context, about today's coastline and the impacts of coastal erosion, as postglacial sea level continues to rise into the modern era. The operation of coastal processes, particularly through the actions of storm waves, is constantly eroding the coasts. Beaches continue to march inexorably landward, destroying fences and fields and even concrete structures that stand in their way. Similarly, these destructive geomorphological forces accompanied the early rises of sea level on the now offshore areas of the shelf. The relentless progress of the coast landward planed off and removed exposed land surfaces, sometimes leaving only a platform of rock. Today, the seabed of the Irish shelf shows a variety of environments, reflecting both relict glacial features and contemporary conditions. The composition of the seabed sediment is largely a complex mixture of eroded rocks, first scoured and moved by glaciers and then subsequently re-mobilised and mixed up in the sea, with an addition of shells and other remains from marine organisms (see Chapters 11: Beaches and Barriers and 30: Engineering for Vulnerable Coastlines).

The western and northern areas of the Irish shelf seabed are exposed to the full forces of westerly gales and are worked by strong currents generated by large swell waves. For this reason, the seabed, particularly of the western continental shelf, has the most extensive areas of gravels, as smaller sediment grains are swept away. Despite the erosive energy regime, long ridges run across the shelf sub-parallel to the shoreline. These are terminal moraines, formed at the front of the Irish ice sheets that once spread onto the shelf when sea levels were lower. The outermost moraines mark the ultimate extent of the last ice sheet (Figure 2.9).[7]

Off the south coast, the seabed energy regime is less extreme: barren rock, coarse sand and shell make up most of the seabed. Strong tidal flows mobilise this sediment and, in places, form trains of submarine dunes and sediment waves. These areas of mobile seabed form a distinct habitat, but also indicate to geologists the strength and direction of tidal flows. The Celtic Sea shelf, off the south coast,

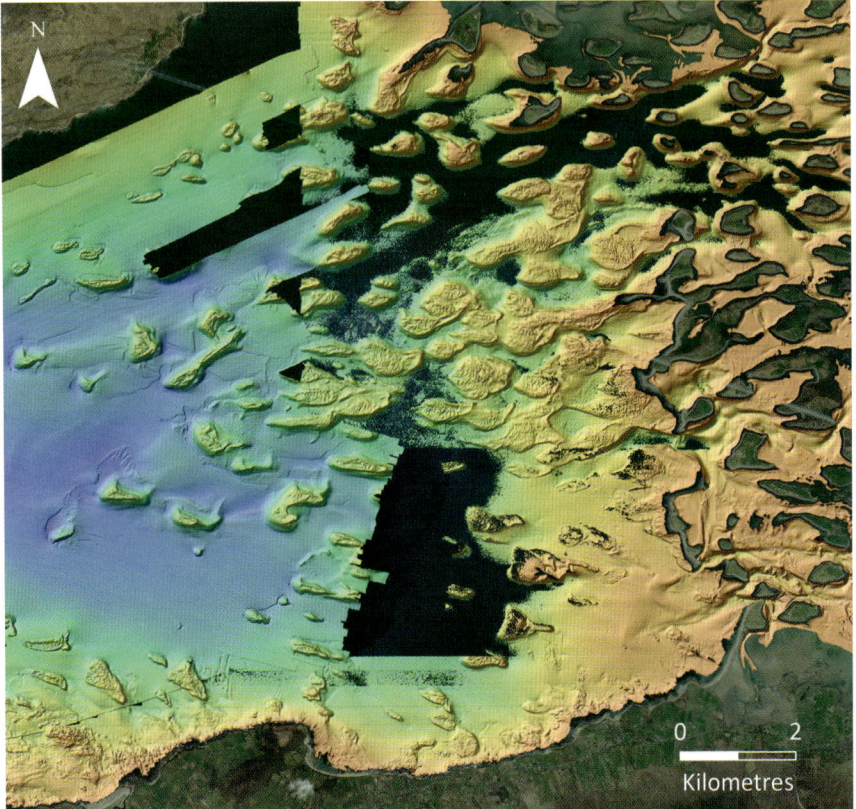

Fig. 2.9 SWATH BATHYMETRY MULTIBEAM DATA MERGED WITH A SATELLITE IMAGE SHOWING CLEW BAY DRUMLINS. The islands in Clew Bay, County Mayo, are drumlins, or hills of sub-glacial deposits formed under the retreating ice sheets towards the end of the last glaciation. The bathymetry data shows their true extent, preserved on the seabed as part of a submerged glacial landscape. Gaps in the bathymetry are dark blue, browns are shallow water and blues are deeper. [Data and imagery: courtesy of the INSS/ INFOMAR programme, Geological Survey of Ireland and Marine Institute (www.infomar.ie) and Siobhán Burke, University College Cork]

0   2
Kilometres

is Ireland's most extensive, stretching to the Whittard submarine canyon beneath the Southwest Approaches (see Figure 2.5). This large shelf witnessed the surge of an enormous glacier (the Irish Sea Ice Stream) that carved out a broad, deep trough called the St George's Channel. As a result of the Irish Sea Ice Stream, the sub-seabed structure of the Irish Sea is complex and its seabed exhibits several linear, deep channels that are believed to have been cut by subglacial meltwater channels. In the southern Irish Sea, there are a number of coast-parallel banks, some of which have developed on ice-marginal, morainic ridges, but are now maintained by strong tidal currents and waves.[8]

THE COAST: SYSTEMS, PROCESSES AND PATTERNS

The shaping of Ireland's coastal environments is the product of many linked processes and controls. The driving factors include the geology, climate, weather, and tides, as well as people (see Figure 2.2). They are contained within and operate as part of Earth's systems, the fundamental web of mass and energy flows on the Earth and wider.

Coastal topography

Coastal topography, or relief, refers to the physical-environmental factor of shape, and is the common denominator bringing together the processes and controls that create coastal landscapes. Ireland, with its varied coasts, has a diversity of landscapes that can focus and encapsulate people's feelings, perceptions and understandings of coasts (see Chapters 14: Imagining Coasts and 15: Coastal Heritage) (Figure 2.10). There are coastal landscapes that are centred on their vertical scale and dimensions, such as mountains. Examples of these could be the sharp and stark basalt shoreline edges of County Antrim (see Figure 10.1), the bare limestones of the Burren, or the softer and distanced backdrop slopes of the Comeraghs to the coast of the south-east. In contrast to these spectacles are the low, or even flat, and often dull grey-green landscapes of Ireland's coastal wetlands (bogs and saltmarshes), which, when the sun shines, can suddenly appear to sparkle and become magical. Then, there are the almost countless sandy, dunes-backed beaches, of all shapes and dimensions, that are beloved by photographers, surfers, sun bathers and dog-walkers alike (see Figure 2.1).

The classic view of the island's topography, part of the fundamental knowledge from school geography, is of the island's saucer-like 'rim' of mountains, geologically constructed around its lowland interior (Figure 2.11). This coastal mountain rim is formed by many, but individually distinctive, upland areas, such as those of Connemara in western Ireland, or of the south-western peninsulas along the

coastline of counties Kerry and Cork, the Mourne Mountains of County Down and the Wicklow Mountains, each with their own character (Figure 1.4).

The geological core of Ireland is composed of ancient rocks which, over millions of years, have been exposed to phases of

Fig 2.10 LANDSCAPES AND ENVIRONMENTAL IMPACTS OF IRELAND'S RIM OF UPLANDS AND MOUNTAINS. a. The glaciated coastal uplands of the central MacGillycuddy's Reeks, County Kerry, showing the pyramidal-shaped summit of Carrauntoohil mountain, to the top-right of the photograph, Ireland's highest peak at 1,039m, with the view beyond to the Kenmare River coast. Former ice basins (corries), associated lakes and glaciated troughs of this landscape can be seen in the foreground at the base of the mountain peaks. Valleys leading away from these central areas have been broadened and their slopes lowered in postglacial time by geomorphological processes and actions of rainfall and gravity (mass movements). Glaciated landscapes similar to these can be found throughout the island's uplands rim, which have been a constant source of sediments and water, via streams and rivers, to the coastal and offshore zones. [Source: Valerie O'Sullivan]; b. The postglacial, humid-climate landscape of Ballydonegan Bay, Allihies on the Beara Peninsula, County Cork, showing the lower-angled slopes of the coastal lowland strip at the foot of steep upland areas. Coastal lowlands such as these are of very varied widths, dependent on the proximity of their backing upland slopes. They are found in many of Ireland's coastal regions and have formed valuable settlement locations since the first arrivals of people on the island. Their long-settled character is indicated in this example by the nineteenth-century field patterns and the widespread development of farms, houses and other buildings. Agriculture, together with fishing, has traditionally formed the primary economic activity of these coastal communities; though leisure pursuits and tourism have drawn people to their coastlines and particularly the beaches, such as those seen in the centre of the photograph. Tourism now often rivals the former traditional rural activities in economic importance, both locally and nationally. [Source: Barry Brunt]

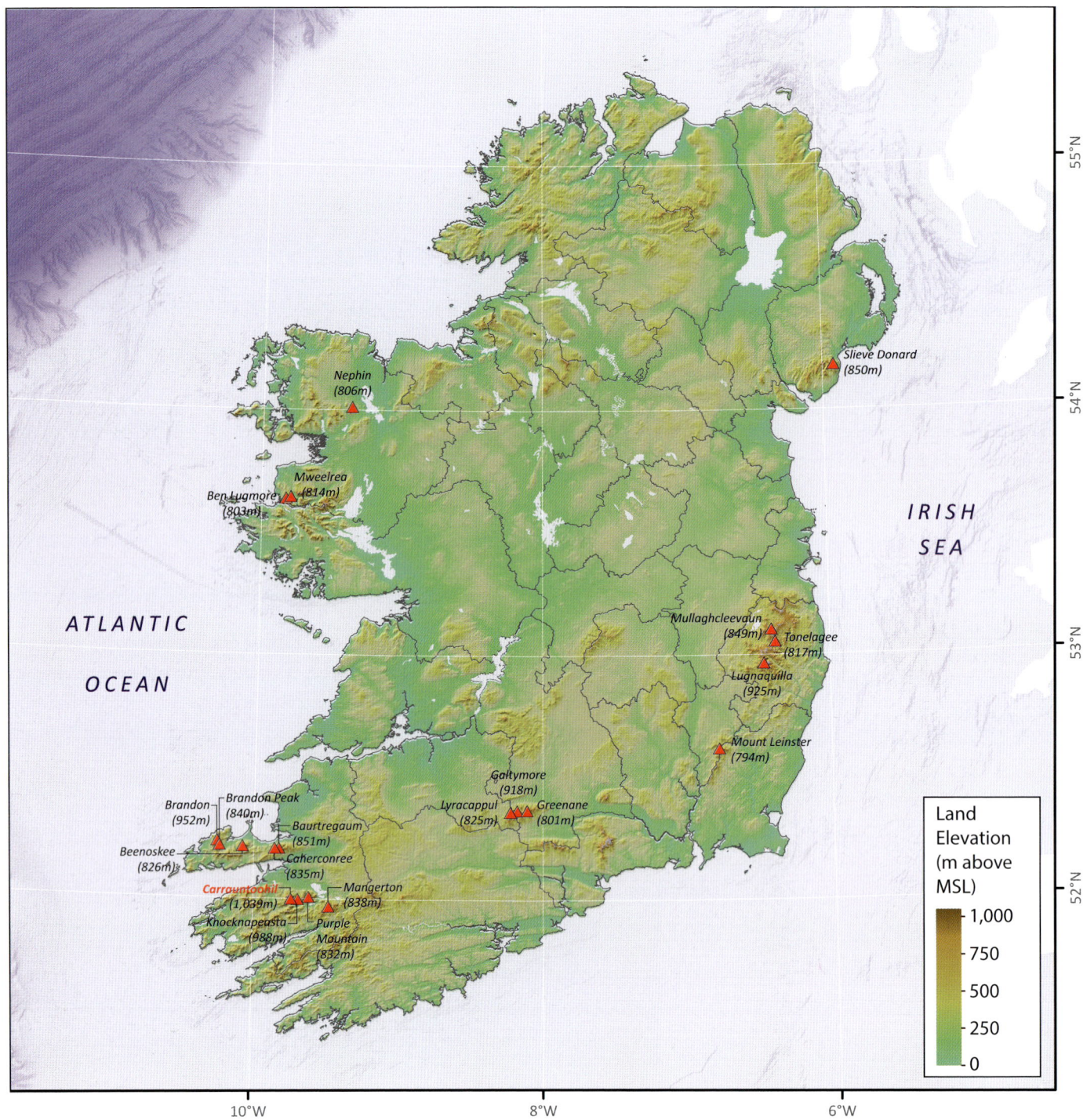

Fig. 2.11 THE TOPOGRAPHY OF THE ISLAND OF IRELAND, SHOWING THE HEIGHT VARIATIONS IN RELIEF. Ireland is characterised by a mountainous or upland rim, with river systems providing access to a more lowland interior. The western and northern seaboard is dominated by a series of mountain ranges that restrict lowlands to a relatively narrow coastal zone. Of the twenty highest mountains, nine form part of the five peninsulas located in the south-west of the country. Eastern and southern coasts are less dominated by high relief, apart from the mountains centred in County Wicklow and the Mourne Mountains of County Down. The green shading shows lowland areas below 250m elevation and emphasises the narrow width of coastal lands around most of the island, where the bulk of the population reside. The extent of river-dominated catchment areas, containing their coastal-end estuary environments, is also illustrated, as for example the rivers Shannon, Liffey and Bann. Contrasts in relief, especially between the higher and more difficult terrain of the west, and the lower and more fertile lands of the east, have played critical roles in shaping the island's human landscapes. While the west was able to support a high population until the Great Famine, today it is characterised by a low population density and is relatively underdeveloped. In contrast, the more benign relief and environmental conditions of the eastern and southern coastline support well-developed urban and rural communities. [Data source: NASA SRTM; bathymetry: GEBCO Compilation Group, 2019. GEBCO 2019 Grid (doi:10.5285/836f016a-33be-6ddc-e053-6c86abc0788e)]

mountain-building (orogenesis) and processes of denudation. Most of the island's mountains are linked to the Caledonian orogeny (490–390 million years ago). This resulted initially in the formation of a high chain of fold mountains, which today runs in a north-east–south-west direction and spans much of the north-west fringe of Europe. The southern part of this forms Ireland's mid- to north-western coastal uplands of County Donegal and Connemara. The mountains of the southern coasts, however, are younger and possess the distinctive east–west grain associated with the Hercynian or Variscan orogeny (370–290 million years ago) of

Fig. 2.12 SLIEVE LEAGUE, COUNTY DONEGAL. The mountains that form the coastline of north-west Ireland, from Connemara to County Donegal, are remnants of fold mountains formed during the Caledonian orogeny. Composed of hard, resistant quartzite rock, they were subjected to intense glacial erosion and, more recently, erosion linked to the high levels of precipitation that fall in these uplands. This has left a contemporary landscape of rounded mountaintops and large areas where the soil has been stripped away, leaving bare rock surfaces, glacial lakes, extensive bogland and a marked absence of human settlement. Where their western slopes plunge to the sea, they give rise to some of the highest and most spectacular coastal cliffs on the island. Here, at Slieve League, we see the steep uplands around Carrigan Head, which rise to a height of over 600m and drop almost vertically to form the coastline. [Source: Tony Webster, Wikimedia Commons, CC BY 4.0]

continental Europe.

In addition to this primary role of geology in relief, Ireland has been covered repeatedly by extensive ice sheets and valley glaciers. During the last Ice Age, and also earlier, through much of the mid- to late Quaternary (c.0.78 million years ago), all of the country's contemporary relief was moulded by ice action. In mountain areas, erosion has been the dominant process. Here, ice action scraped away earlier Tertiary sediments and soils and excavated the geology, leaving mainly bare rock surfaces, steep slopes and new, thin mineral soils. Prominent relief features created by the ice – such as ice basins

Fig. 2.13 MUSSENDEN TEMPLE, NEAR DOWNHILL, COUNTY DERRY, VIEWING WEST FROM CASTLEROCK TOWARD MAGILLIGAN FORELAND. Basalt rock forms the upland surface on this coast, providing a very narrow strip of immediate and usable ground at the cliff base, which has been colonised for limited settlement and the coastal rail line from Coleraine to Derry. Rockfalls and debris slides have formed the source of the material for this area between the cliff and the sea, which is a concern for the stability of this coast and the prudence in using spaces like this for buildings and infrastructural developments. These are issues that are raised frequently today on coasts where people are creating continuing pressures for the provision of new living space. The field patterns of the farming in this otherwise rural landscape contrast strongly with the leisure use of the spectacular beaches along this topographically challenging coastal zone. [Source: Gordon Dunn]

(corries), U-shaped glacial troughs and glacial lakes – can be found throughout the different mountain areas (Figure 2.10a).

In the later, warmer postglacial climate, erosion associated with rainfall and concentrated water flows in streams and rivers has added further to the work of ice. The result has been the production of separate, distinctive, humid coastal landscapes, showing lower-angled slopes around the margins of the uplands and with deeper soils and vegetation cover. Over the uplands and their steeper wind-swept slopes, present-day vegetation is dominated by shrub-woodlands, heather moors, grasslands (including developed pasturelands) and extensive blanket boglands (Figure 2.10b). This long history of rock denudation has led to the progressive rounding and lowering of these remnant mountains, such that few now exceed 1,000m.[9]

At averaged heights inland of c.500m, coastal uplands characteristically drop steeply over short distances, generally of only a few kilometres, to a relatively narrow coastal lowland strip and its shoreline. This is especially apparent along the coasts of counties Cork and Kerry, where Hercynian-age folding created east–west trending mountain spines that now stretch several kilometres into the Atlantic ocean. Separating these 'spines' are glacially overdeepened valleys, which were flooded by the sea to create five distinctive and spectacular peninsulas that attract a huge influx of tourists annually. Elsewhere, cliffs plunge directly into the sea, as along the coastline of County Donegal, where remnants of the Caledonian fold mountains, composed of hard quartzite rocks,

provide some of the most spectacular cliff coastlines in Ireland (Figures 2.12 and 2.13). Such dramatic profiles, which can plunge from peak elevations of over 800m to sea level, serve to enhance their iconic and integral role in defining the island's coastal zone. These uplands, such as those of 'the Mountains of Mourne, which sweep down to the sea', when viewed from the coast can seem to rival even the Alps!

In the land corridors between shoreline and upland peaks, the environmental influence of the sea diminishes rapidly inland. This can be seen, for example, in changes in the vegetation and soils, the declining availability of flat land usable for living space and in agricultural land use (cropping and animal stocking patterns) (Figure 2.10b). The narrowness of many of these corridors, which are pronounced especially along the western seaboard, has therefore influenced their development potential for people. Not only is lowland restricted, but the many fast-flowing streams coming off the upland slopes rarely coalesce to form significant rivers. As a result, opportunities for sizable settlements to develop around ports or harbours where rivers enter the sea are limited. This is well-illustrated in the Burren, County Clare. Here, bare limestone rock surfaces and the absence of any significant coastal lowland are powerful control factors on the development of settlement and on the provision of transport links to meet the demands of tourism (Figure 2.14) (see Chapter 27: Tourism and Leisure).

Separating the uplands and connecting deep into the island's

Fig. 2.14 THE BARE AND UNYIELDING ROCK SURFACES OF THE BURREN'S COAST. The Burren is an extraordinary landscape that occupies an area of between 250 and 560km² and forms the coast of north County Clare in the west of Ireland. The region is composed of an extensive surface of bare limestone, which was laid down in the Carboniferous era and can extend to some 800m in thickness. These rocks were subject to intensive glacial action in the past and now provide one of the finest examples of glacio-karst landscape in the world. Spectacular seascapes and landscapes are combined with the rich and unusual botany of the area, which includes both alpine flora – such as gentians – and sub-tropical orchids. These ingredients have made the Burren a tourist 'hotspot'. The key tourist road that parallels the sea margin, and which is seen in the photograph, however, is restricted by the topography to quite a narrow artery. In an age of large-volume visitor tourism passing through the area, this route is often unable to accommodate the high car and coach transport loads being placed on it. To engineer a way out of this problem is prohibitively expensive! [Source: David Hodgeson]

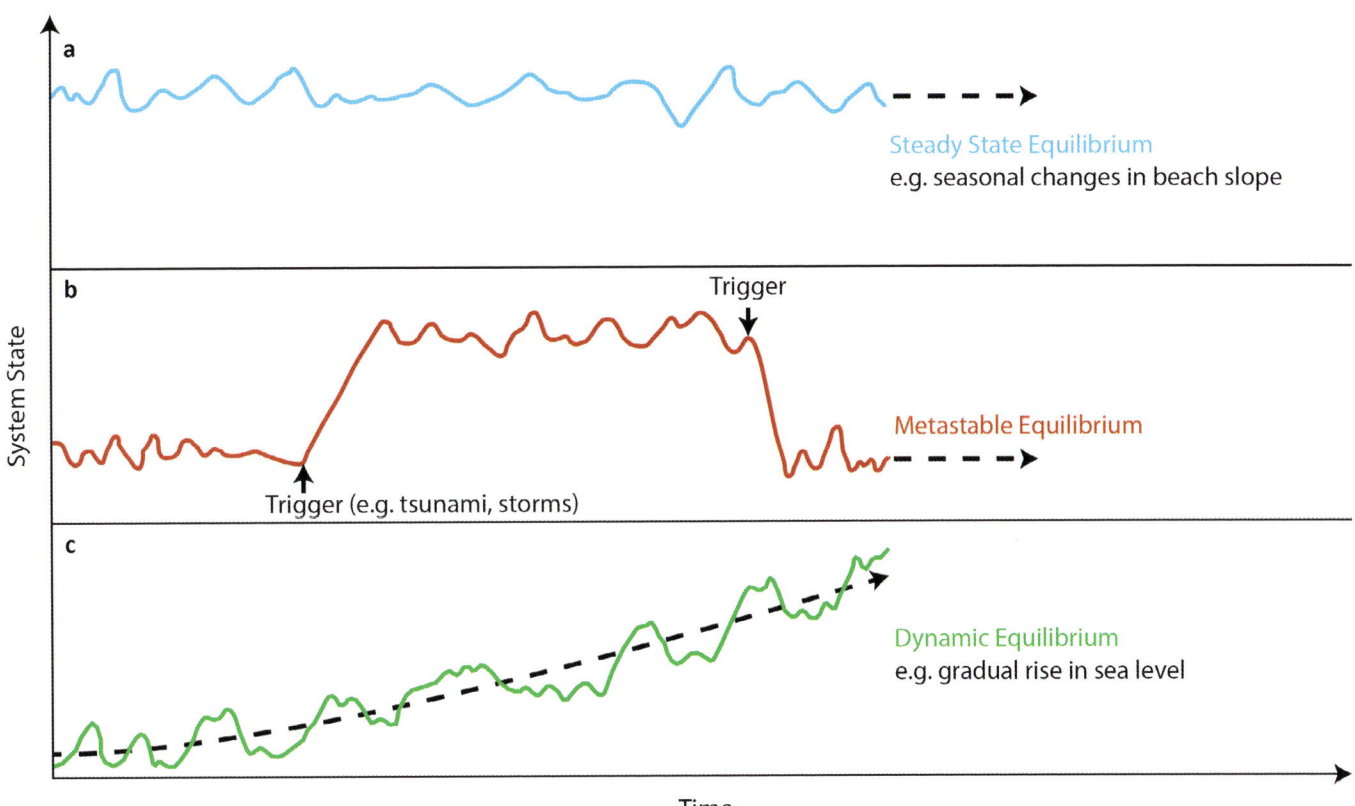

System State (vertical axis label)

a Steady State Equilibrium
e.g. seasonal changes in beach slope

b Trigger
Metastable Equilibrium
Trigger (e.g. tsunami, storms)

c Dynamic Equilibrium
e.g. gradual rise in sea level

Time

Fig. 2.15 PATTERNS OF COASTAL-SYSTEMS FUNCTIONING, SHOWING CRITICAL THRESHOLD AND STATE, OR STAGE, CHANGES THROUGH TIME. Note the time and system state axes are both linear. The diagram illustrates three key and linked ideas in the functioning of systems (as in those of the coast). Graph a. describes an apparently 'unending' and balanced system through time, in its operation, appearance and characteristics; b. shows more the reality of systems: that they are subject to repeated and major changes in state (that is, they are dynamic). Sometimes changes are so large and instantaneous that a system's character and appearance alter completely. The causes of such changes in the state of a coastal system (a trigger event) may be, for example, an extreme storm. Here, the ability of the system to resist and/or absorb the pressure to change is exceeded and a system threshold limit is passed. Post the event, the new system state of balance is maintained at a different level (state) for a period of time, before the next event as trigger to change occurs; c. indicates a different level of system reality, in which a. and b. are combined, to develop the system in a series of cyclical stages along a trajectory of say, energy or mass availability. For example, since the end of the last Ice Age, global mean sea levels have been rising incrementally as the available water mass of the oceans increases. Consequently, the coastlines of the world have moved inexorably onland over the last 18,000 years. While the mean trajectory of this change is always upward over this time, the pattern of 'small' or 'large' changes (system variability) has been cyclical. The cycles are driven by the trigger events and system boundaries and thresholds (e.g., ocean basin capacity), influencing the timing and scale of change.

interior are another type of intervening lowland, different from those of the shore-aligned coastal corridors. These dissect upland areas with relatively small river-dominated landscapes, which can develop into numerous estuaries along coasts. This is most apparent along the eastern and southern coastlines, where the majority of the island's important rivers – such as the Lagan, Liffey, Suir and Lee – enter the sea, providing attractive sites for large-scale urban developments. The eastern coast of the island, north of Dublin, in particular, constitutes a significant low-lying zone of access to the interior lowlands basin of the island. Landscapes such as these are different again from the more plains-like areas of the larger estuary coastal systems, such as that of the River Shannon region, the island's largest estuary zone; and different again from the smaller and more topographically confined ones, like that of the River Foyle in Northern Ireland. Overall, the different uplands and adjacent lowlands provide their coasts with distinctive local characteristics, resulting from their varied geological and geomorphological histories, soils and vegetation patterns and their exploitation for settlement, themes expanded on in later chapters of the Atlas (see, for example, Chapters 5, 6 and 13).

Perhaps one of the most significant aspects of these topographic differences lies in the importance of the interactions of the uplands with the movement of Atlantic air masses over the island. These air movements impact all aspects of Ireland's climate, as well as key energy and mass transfers (for example, heat, water, winds). The influence of the western mountain regions is to force Atlantic air to rise (orographic effect), which results in it being cooled and, in outcome, producing high levels of rainfall for the west of Ireland, with significant reductions eastwards. The consequences of this are profound, for example in the volumes of water and suspended sediment flows in rivers, soil type and land capability, as well as agricultural patterns and productivity (see Chapter 4: People, Agriculture and the Coast). In terms, for example, of the natural vegetation, biodiversity and plant and animal ecology, there are distinct west coast (wet) to east coast (dry) differences. A notable one concerns the peatlands that come down from the uplands onto the coast in many western areas of Ireland, but are absent in the east. These peatlands form the unique, organic-dominated, flat-surfaced saltmarsh type shorelines of western Ireland (see Chapter 12: Coastal Wetlands).

SYSTEMS AND COASTAL BEHAVIOUR

The importance of viewing the coast as a system, as is now common practice, is that it allows us to focus on how its different ingredients (including rock, sediment, water, plants, animals, people) are interconnected and function on a day-to-day basis. In real terms, this image (conceptual model) of the coast as a system is too simple. Earth systems are more complex, as many other factors can distort the model, creating special and limiting cases in its functioning. For example, storm and tsunami events can dramatically alter, almost instantaneously, the physical appearance and state of the coast, such as a beach (Figure 2.15; also Figure 2.2). However, after the event, if we observe the beach for long enough, we would see how the shape of the beach, its levels and appearance, settle and stabilise again. Effectively, the beach system self-regulates, until the next critical threshold-changing event. In studying coasts in this way, their functioning can be quantified, numerically modelled, accurate projections of future changes made and thus managed intelligently by people.[10]

COASTS AS SYSTEMS

Darius Bartlett

The idea of 'systems' first entered scientific and technical literature in the 1930s with the work of German biologist, Ludwig von Bertalanffy, as a way of describing and explaining what he saw as 'an amazing order, organisation, maintenance in continuous change [and] regulation' in biological organisms he was studying.[11] Over subsequent years, and especially in the 1970s and 1980s, systems thinking has increasingly made its way into many aspects of our vocabulary and daily lives. We talk, for example, of information systems, economic systems and social systems.

Any system comprises an ordered, interrelated set of objects that have more in common with each other than they do with their external environment. A system thus has three component parts: (i.) attributes and structure (its contents, plus the properties that describe

Fig. 2.16 ROSSBEHY SPIT, DINGLE BAY, COUNTY KERRY, AS AN EXAMPLE OF A COASTAL SYSTEM, WITHIN A COMPLEX OF OTHER SEPARATE BUT INTERLINKING SYSTEMS. This panoramic view over Castlemaine Harbour shows Rossbehy Spit (left) as one of the two fronting dune and sand-barrier physical systems that form the western boundary to the harbour environments, some of which (for example, saltmarshes, sandflats) can also be seen. The dunes at Rossbehy are 10–20m in height. The other spit feature of Inch is just visible in the far distance, in front of the backing uplands of the Dingle Peninsula. Seen as a system, the Rossbehy Spit has structure: we can describe it in terms of its length, width and height, its position within the Dingle Bay–Castlemaine Harbour environment, and its orientation with respect to the dominant waves and wind. The spit has state: the tide can be seen in the photograph at a particular point in its diurnal rise and fall. Similarly, the quantity of sand that makes up the dunes, the parameters of wind, waves, currents and weather can all be described, measured and defined. The spit is also subject to the the operation of processes. Change the state of the tide, for example, and the way that wave energy impinges on the beach will also change; remove the surface cover of vegetation, perhaps through trampling by visitors, and the exposed sand will be more subject to erosion. In the storms of 2008 and 2014, the sand-barrier was breached by wave action. This change in the structure of the system has led to the spit adjusting to a new and quite different state. These changes may further result in longer-term, knock-on changes to the hydrology and sediment dynamics of the inner Castlemaine Harbour system. [Source: Darius Bartlett]

the contents and their interrelationships); (ii.) a state (whether it is static or dynamic); and (iii.) processes (the controls in its functioning and its outcomes).

Good examples of coastal systems showing these three parts are those of the sandy coasts of the Inishowen Peninsula, County Donegal (Figure 2.3) and the more extensive dune and beach-barrier systems of Rossbehy and Inch in Castlemaine Harbour, County Kerry (Figure 2.16). This latter sand system has been studied now for over thirty years as part of EU-funded research projects designed to help the understanding of coastal processes and their links to practical measures for future coastal management (see Chapters 30: Engineering for Vulnerable Coastlines and 33: Climate Change and Coastal Futures).[12]

For scientists and engineers, in particular, the systems concept is both an intellectual construct and a valuable tool for management. There are a number of ways in which an intervention (whether deliberate or accidental) can change, for example, the nature and workings of a harbour, beach or other stretch of coastline. The change may be linear, with a simple and progressive one-way transition, from the initial to the final state; or it may be cyclical with, for example, sand being eroded from dunes during storms, and then gradually returned to the dunes when calmer conditions prevail. As the Rossbehy and Five Finger Strand coastal system examples indicate, thinking of the coast in terms of a nested and interrelated set of systems is of huge importance for anyone wishing to examine and understand its workings.

CLIMATE AND WEATHER

Central to understanding Ireland's climate and weather is the role of its surrounding seas. Also, an understanding of Ireland's coasts requires accurate knowledge of its climate and weather. Nowhere on the island is far enough from these seas, including estuaries, not to be considered part of its maritime climate. There are, however, differences between coastal zones and inland areas, especially due to a rain-shadow effect to the lee of the mountain ranges along its western seaboard.

Ireland's mid-latitude location on the western coastline of the Eurasian continent determines its climate as a Temperate Oceanic Climate (Cbf), as indicated under the Köppen Climate Classification. This climate (that is, Cbf class) can be described as 'equable', as it is characterised throughout the year by relatively mild temperatures (with no extremes) and abundant rainfall. In addition, perhaps its most significant descriptor is the changeability of its weather patterns through the year. This promotes a belief that 'Ireland does not have a climate, only weather'!

The major factor shaping the climate of Ireland is the dominant westerly to south-westerly (onshore) winds associated with the global atmospheric circulation found in mid-latitudes. Embedded in this low pressure belt are frontal weather systems (cyclones), which track from south-west to north-east across the Atlantic ocean and bring moisture-laden winds to the island throughout the year. Second, being surrounded by the Atlantic ocean and, to a lesser extent, the Irish Sea, maritime influences are emphasised in Ireland, providing for small temperature ranges and abundant, year-round precipitation. Finally, the North Atlantic Drift – part of a clockwise-rotating ocean current system that, under the influence of prevailing winds, transfers a large flow of relatively warm water in a north-easterly direction across the Atlantic to the shores of north-west Europe – keeps Ireland warmer than its latitudinal position would indicate.[13]

Ireland has a small temperature range from summer to winter and shows few extremes throughout the year. Winter temperatures rarely fall below freezing point, with average daily air temperatures ranging between 4 and 8°C . Summer months remain relatively cool, rising to an average of only 17–19°C in July–August. However, in spite of the island's small size, there are important regional variations. In particular, when compared to inland sites, coastal area temperatures are generally milder in winter and cooler in summer, with distinctive coastal microclimates developing, as in counties Cork and Kerry (Figure 2.17) (see Chapter 27: Tourism and Leisure). This is due to the surrounding seawaters warming up more slowly than land surfaces in summer months, while in winter the sea is better able to retain heat in its surface layers as opposed to the land (Figure 2.18). Such surface sea water temperatures are transferred to adjacent coastal areas by prevailing onshore winds. The presence of the warm waters of the North Atlantic Drift contributes to this moderating effect, especially along the coast of south-west Ireland.[14] The difference in temperature between Ireland's southern and northern coasts is due primarily to the influence of latitude.

Valencia Island, off the coast of County Kerry, displays particularly mild temperature conditions, which allow for the growth of sub-tropical plant species. Here, January temperatures average c.7°C, while in July they rise to c.15.5°C. In contrast, seasonal temperatures along the east coast are less mild, as prevailing winds passing across the island's interior blow offshore by the time they reach eastern areas. This reduces the modifying influence of the Irish Sea. Dublin's January and July temperatures, therefore, average c.2°C and 20°C respectively.

Rainfall patterns in Ireland have a strong west-to-east gradient (Figure 2.17b). Most rain is associated with Atlantic depressions and their frontal systems which, in winter especially, track in an almost relentless sequence north-easterly across the island. Where these onshore and moisture-laden winds are confronted by high mountains, which parallel the west coast, they are forced to rise and, in so doing, produce much cloud and rainfall totals that can reach 4,000mm annually (Figure 2.19). As these winds descend to the east of the mountains, they become warmer and drier, creating a rain-shadow effect. This results in lower rainfall totals, especially over lowland areas, which are more common in the interior and along the island's east coast. Here, annual rainfall can be four times less than in the high mountains of the west coast. The only east coast exceptions to this are linked to the Mourne and Wicklow mountains and the Antrim Plateau. During winter, low temperatures associated with high altitudes can result in snowfall occurring on Ireland's mountain ranges. Ireland's map of average rainfall totals correlates strongly with a topographical map of the island (see Figures 2.10 and 2.17b).

Associated with relatively high totals of rainfall and cloud cover, are the comparatively low levels of sunshine that characterise much of the country. May and June are the sunniest months,

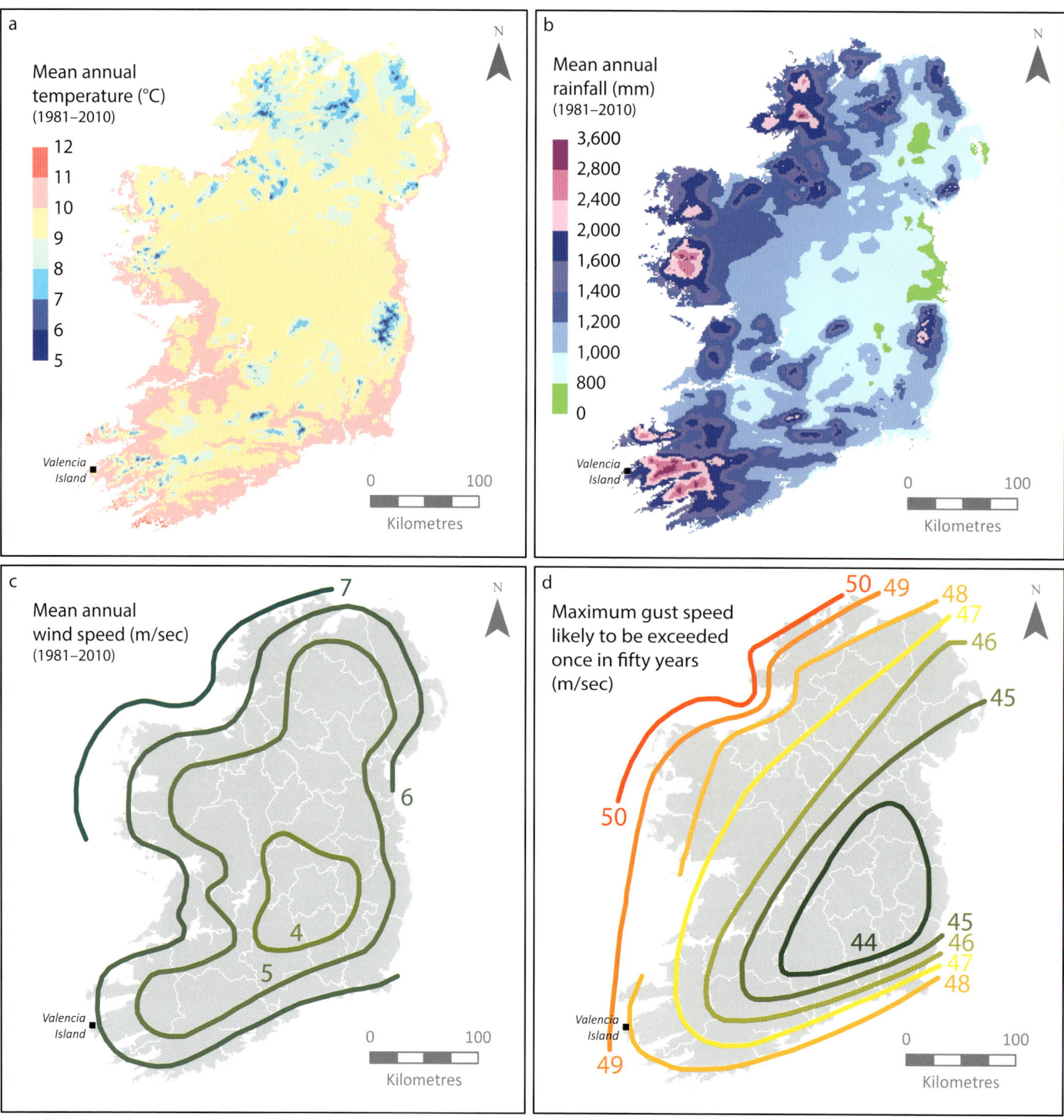

Fig. 2.17 IRELAND'S CLIMATIC PATTERNS AND THE LINKS TO COASTS. Isoline distribution diagrams of a. Mean annual temperature (1981–2010). The main differences occur between inland and coastal areas, and also from the south-west to the north-east of the island. Moderating influences from the surrounding seas, together with prevailing, onshore south-westerly winds, allow the coasts to benefit from higher year-round temperature over inland areas. The south-west to north-east gradient in temperatures also reflects the influence of latitude. As temperatures decline with altitude, lowest temperatures occur in mountainous areas, especially to the north-west of the island; b. Mean annual rainfall (1981–2010). Ireland is characterised by moist conditions and year-round rainfall. Depressions and frontal systems from the Atlantic bring rain-bearing winds to the island, especially in winter. These, together with relief-induced rainfall caused by winds rising over coastal mountains, result in most areas along the western seaboard receiving between 1,000 and 1,500mm of rain annually. In contrast, the rain-shadow effect of these mountains causes lowland areas along the east coast to receive much lower levels. Highest annual rainfall totals of some 4,000mm occur on the high mountains ranges located on the peninsulas of the north-west and south-west. In contrast, lowlands around Lough Neagh receive less than 800mm, Dublin only 760mm; c. Mean annual wind speed (1981–2010); and d. Maximum gust speed likely to be exceeded once in fifty years. In both c. and d. above, wind speeds display a north-west to south-east gradient across the island, related to prevailing winds which reach the western coast from the Atlantic. As these winds flow across the island, the frictional drag of land surfaces reduces both average and maximum gust speeds, especially in the south-east of Ireland. Higher and more extreme wind speeds in the north-west are linked especially to winds from the north-west, which sweep in behind deep low pressures that track across the country. Northern Ireland is considered to be one of the windiest areas in Britain and Ireland. [Map data source: Met Éireann]

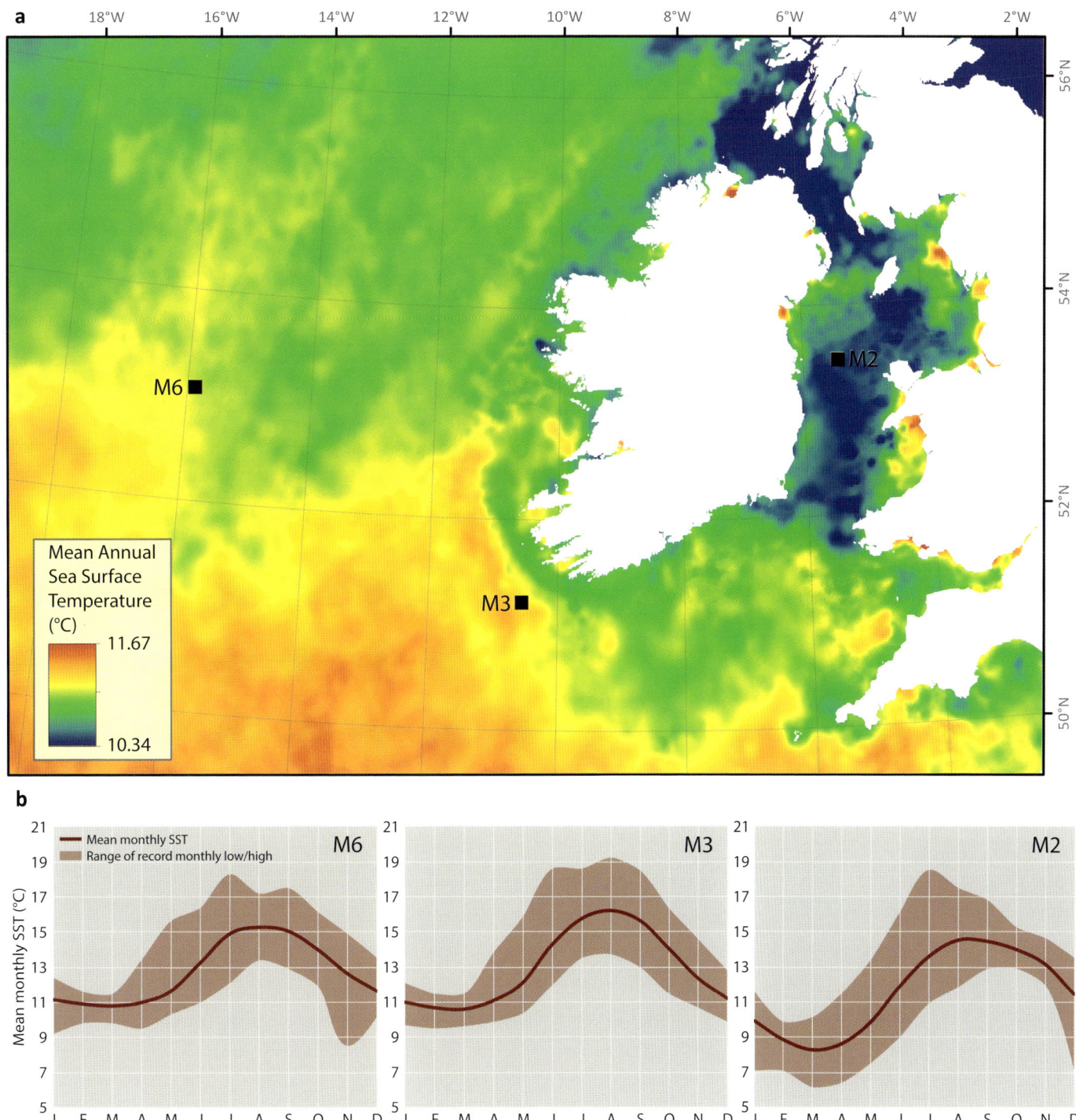

Fig. 2.18 THE MEAN ANNUAL AND MONTHLY SEA SURFACE TEMPERATURES (SSTs) FOR THE EASTERN ATLANTIC OCEAN AND IRELAND'S SHELF SEAS. a. The pattern of mean annual Sea Surface Temperatures around Ireland, based on satellite measurements (1981–2016). These data show the relatively cooler waters of the Irish Sea region (area of M2 Buoy, in 2.18b) and the overall south-west to north-east gradient of cooling in these surface waters. The background and long-term warming effect of the Gulf Stream's North Atlantic Drift on south-west Ireland is marked. Overall, the SSTs of these waters around Ireland have increased by *c.*1°C since 1980. [Map data source: Good et al., 2019. ESA Sea Surface Temperature Climate Change Initiative (SST_cci): Level 4 Analysis Climate Data Record, version 2.1. Centre for Environmental Data Analysis, 22 August 2019. (doi:10.5285/62c0f97b1eac4e0197a674870afe1ee6]; b. A more detailed view of the mean monthly SST data from three recording buoys situated in Ireland's offshore zone (the buoy positions are shown in 2.18a). The data used covers the periods: M6: 2006–2020; M3: 2002–2020; M2: 2001–2020. [Data source: Generated using EU Copernicus Marine Service Information and the Marine Institute Weather Buoy Network Temperature Salinity, data licensed under CC BY 4.0]

averaging 5–6.5 hours per day, but in December these averages fall to only one hour in the north and some two hours in the south-east. While coastal areas generally receive more sunshine than inland areas, this is most apparent along the south-east coast. This area receives some 1,600 hours of sunshine a year and is advertised, especially for tourism, as 'the sunny south-east' (see Chapter 27: Tourism and Leisure). This contrasts, for example, with the County Down coast in Northern Ireland, which receives an annual average of approximately 1,250 hours of sunshine.

The combination and spatial patterns of temperature, rain, sunshine and wind have profound implications for human settlement and land use. In particular, such physical inputs are vital

Fig. 2.19 VIEW WEST TOWARDS BALLINSKELLIGS BAY OVER THE WESTERN END OF THE BLACK VALLEY, MACGILLYCUDDY'S REEKS, COUNTY KERRY. The mountains that parallel the coast of the west of Ireland have a major influence on rainfall patterns across the island. Here, onshore south-westerly and moisture-laden winds coming off the Atlantic ocean are forced to rise over the relief barrier (the orographic effect). In doing so, the ability of the air to hold moisture in its gaseous form declines. The result is the build-up of clouds, and frequent and often high levels of precipitation. In addition, the combination of moisture-laden air and the relatively mild temperatures that typify these coasts can give rise to sea mists and reduced levels of visibility. Such weather conditions can be seen in this photograph, which shows the landscape of the mountainous peninsulas along the south-west coast when under the influence of moisture-laden air and clouds. These mountains can receive up to some 4,000mm annually, while the adjacent lowlands receive less, for example, the recording station at Valencia, County Kerry, receives 1,550mm. [Source: Robert Devoy]

for agricultural activities and emphasise the marked tradition of extensive pastoral farming that dominates much of the country. This is most apparent along the mountainous western seaboard, where the raising of sheep and cattle is best suited to withstanding the relative low temperatures, high rainfall and limited sunshine. In contrast, more benign weather conditions and relief along the eastern and north-eastern coasts can support more intensive arable farming and denser patterns of population and settlement (see Chapters 4: People, Agriculture and the Coast and 25: Urbanisation of Ireland's Coast).

Winds and storms

A major factor shaping Ireland's climate and a key feature of its coastal climate, is wind. Westerly and south-westerly winds dominate, associated with the island's location within the global system of mid-latitude westerly airstreams. Furthermore, the extensive Atlantic ocean to the west means that there are no effective obstructions to act as a frictional drag on winds before they reach Ireland's western seaboard. The west, and especially the north-west coasts are, therefore, exceptionally windy by European standards, with Malin Head, County Donegal, identified as the windiest location in Ireland. Such strong, persistent onshore winds have also a significant influence on the wave-energy systems that break on the country's shores, thereby helping shape its morphology (see Chapters 10: Rocky Coasts and 11: Beaches and Barriers). Calm spells of more than a few hours are rare along the coastline. However, there is a notable wind gradient from the coast to the interior, with wind speeds being reduced significantly. Windward slopes of coastal mountains are, therefore, often stormy and

windswept, while leeward slopes and inland areas are more sheltered and considerably less windy (see Figure 2.17c and d.).

Not only does Ireland experience persistent winds but on many occasions, especially in the winter months, depressions forming in the Atlantic undergo a significant fall in their central pressure. Related to this deepening is a marked tightening of isobars rotating around the core of the cell. This generates strong winds, which can increase to storm-force conditions with gale- (storm-) force winds, generally taken as wind speeds > 61km per hour (c.17m/sec). Such storms (and the linked generation of increased water and wave heights, as storm surges) have a long history in shaping the morphology of the Irish coastline and the collective memories of coastal communities (see Night of the Big Wind, pp.32–33).

North Atlantic storms are most common during winter, although they can occur in any month. Three types of storm affect Ireland, with the majority evolving as deepening cells of low pressure (cyclones), which form within the mid-latitude westerly wind belt. Second, less frequent, but often intense, storms reach the Irish coast as the tail-end of hurricanes or tropical storms. These have usually tracked across the Atlantic from warmer waters off the coast of the southern USA. Finally, small-scale tornadoes and waterspouts can occur as localised events of extreme weather. Most produce weather conditions that are to be expected during winter and have little impact on the island's physical and human environments. However, in any given year, two or more stronger and more damaging storms can occur. These typically cause power outages across many parts of Ireland, structural damage to buildings and the downing of trees, disrupting travel. The impacts are often most intense in coastal and adjacent areas, which become subject to high tides, along with storm surges and high waves. These raise sea levels and often result in coastal flooding, erosion of shorelines and disruption to shipping (Figure 2.20) (see Chapters 30: Engineering for Vulnerable Coastlines and 33: Climate Change and Coastal Futures).[15]

The exceptionally stormy winters of 2013–2014 and 2019–2020, and the tail-end of Hurricane Ophelia in 2017 (the strongest eastern Atlantic storm to affect Ireland in some 150 years), are reminders of how vulnerable Ireland is to such stormy conditions, which centre on destructive winds and intense rainfall. This is apparent especially along the west coast, which usually receives the initial and full impact of Atlantic storms as they reach landfall. The winter of 2013–2014 is considered to be Ireland's stormiest since systematic daily synoptic charts were first plotted in 1877. The storm events of these years can be seen as a part of the prelude of the present day to times of even more severe future coastal storm events.[16]

Storms in past times have shown little variation in their tracking, generating winds often in excess of 100km/h. One notable example was Storm Darwin, which made landfall in Ireland on 12 February 2014. A maximum gust of 160km/h was recorded at Shannon Airport, while a maximum wave height of 25m was recorded by the Kinsale Energy Platform off the south coast. The storm caused extensive damage, including power outages affecting 215,000 people, 7.5 million trees being blown down causing considerable disruption to transport systems and many buildings suffering structural damage. The impact on the coastline was enormous, with high tides causing extensive flooding, affecting many coastal towns and villages. Major coastal erosion occurred, involving the destruction of some coastal roads.

Occasionally, Ireland is affected by the tail-ends of hurricanes, or tropical storms, that have tracked across the North Atlantic. The hurricane season for the USA occurs from late August to October and Ireland can experience weather conditions associated with such storms during these months. This time frame is outside Ireland's 'normal' winter storm season, but in some years one or more events can occur, bringing very strong winds and intense rainfall to the country. Storm Ophelia is a recent example of this type of extreme weather event, affecting Ireland for a number of days in October 2017 (Figure 2.21). Such hurricane-scale events may increase in the future, given the climate projections for extreme weather events linked to global warming.[17]

Tornadoes, waterspouts and funnel clouds also occur on the island of Ireland and, in the case of waterspouts and funnel clouds, are particularly associated with coastal waters and the bigger lakes. Collectively, these are generally relatively small and short-lived events, rarely causing any significant damage. However, destruction of coastal property from these events can happen, as shown at sites such as Youghal, along the southern coast.[18]

While severe storms have a long history in influencing Ireland and its coastline, recent years have seen what appears to be an increasing frequency of these extreme weather events. This is linked to issues associated wth climate change, with problems of prolonged summer droughts, destructive winter storms and the inundation of coastal lands likely to test the current and future resilience of the country's physical and human infrastructures. In this context, effective planning to adapt to Ireland's evolving and more extreme weather events is a matter of increasing national importance (see Chapter 33: Climate Change and Coastal Futures).

Fig. 2.20 STORM SWELL WAVES BREAKING ON COAST OF RATHLIN ISLAND, COUNTY ANTRIM. Ireland is located within the mid-latitude, low-pressure belt of the world's system of atmospheric circulation. This means that the island's prevailing winds flow principally from the south-west and off the Atlantic Ocean. Embedded within these winds are storms, which evolve most commonly from deepening cells of low pressure in the mid-latitudes, but less frequently are the tail-ends of hurricanes and tropical storms. When these storms gather power as they pass over the relatively warm waters of the Atlantic, and their central pressure falls, wind speeds increase and sometimes reach gale-force conditions. The result is that Ireland's northern, western and southern coastlines are often exposed to strong winds and the coastline is pounded by high waves and storm surges. Coastal erosion, flooding and damage to infrastructures, such as promenades and piers, can occur with costly consequences for coastal communities. Here, storm waves are seen pounding the steep cliff face of the north-eastern headland of Rathlin Island, County Antrim, while dark grey and heavy storm clouds are gathered above the island's East Lighthouse. [Source: Douglas Cecil]

Fig. 2.21 HURRICANE OPHELIA. This tropical storm developed, almost uniquely, out of a powerful hurricane that formed in the eastern North Atlantic and, from the Azores, followed a north-easterly track before making landfall in Ireland. Cloud banks associated with its intense circulatory system are clearly shown in this image. Extreme winds, which rotate in an anticlockwise direction around its deep, low pressure centre, were recorded at the Fastnet Rock on the extreme tip of south-west Ireland, with gusts of 191km/h. These are the highest wind speeds recorded in Ireland. High waves pounded the southern and western coastline, while extreme winds caused extensive damage throughout the island, but especially for communities adjacent to the coastline. Concerns over the threat to life from extreme weather led to the closure of education facilities and businesses for two days, and some 385,000 homes experienced a loss of electricity supply. Despite such preparations, five fatalities were linked to the storm, while total costs were estimated at close to €1 billion, due primarily to a loss of output from the closure of businesses. Structural damage was estimated at €70 million. [Source: NOAA National Environmental Satellite, Data, and Information Service (NESDIS)]

Night of the Big Wind 1839

Kieran Hickey

This major storm affected Ireland and Britain from late on the night of 6 January 1839 to the following day, and is still considered Ireland's greatest storm – in terms of fatalities and damage – in the last 500 years. It was an exceptional mid-latitude storm, with an extremely low central pressure (estimated to be as deep as in the 930s hPa) when it directly impacted Ireland and Britain (Figure 2.22). The consequences of this extreme weather event were enormous, with large-scale damage reported all over the country. Hundreds of fatalities were recorded, along with numerous injuries, caused mostly by the collapse of buildings and other structures, along with flying debris, such as slates. Many fires were started by chimneys being blown down and sparks setting fire to roofing material, especially thatch, which then spread easily to adjacent buildings.

Fig. 2.22 ARTIST'S REPRESENTATION OF THE NIGHT OF THE BIG WIND STORM AT SEA. The centre of this deep, low-pressure system travelled in a north-east direction off the north coast of Ireland, producing very strong pressure gradients and exceptionally high winds across the island. During the storm, sustained winds of 112km/h were recorded, with gusts reaching 178km/h. Curiously, contemporary reports of the devastation caused by the storm rarely mention its direct impact on coastal areas, whether from flooding, shoreline erosion or sand-blown events, but these must have been both significant and extensive. Exceptionally high seas would have been associated with this event, given the number and size of vessels recorded as having been carried inland. At Rossnowlagh, County Donegal, it was claimed that the sea rose to such a height that the poor inhabitants thought it was the end of the world! Along with extreme winds, heavy rains caused extensive flooding in the catchment areas of many rivers, while numerous small tornadoes were recorded. Such was the strength of the storm's winds, it was reported that adults, at times, could not stand up and young children were placed in all sorts of containers in order to try to save them from being blown off their feet. [Source: BL/EP/B/3304,I, Bantry Estate Collection, UCC Library Archives Service]

Storm-force conditions at sea were particularly destructive, with exceptionally high seas pounding the coastline in the form of storm surges and high waves, resulting in extensive coastal erosion with sand and numerous boats being blown significant distances inland. At least 200 vessels were also shipwrecked, lost at sea, torn from their moorings or anchorage and destroyed or seriously damaged. Some 213 fatalities were related to these events, although the true figure was probably higher as records were incomplete. The biggest loss of life occurred when nine fishing boats from Skerries, County Dublin, were lost, each with a crew of nine or ten men. The 1839 storm has yet to be exceeded, but with changing patterns of storminess associated with climate change, this may not be the case for much longer (see Chapter 33: Climate Change and Coastal Futures).[19]

WAVES AND TIDES

Waves and currents (for example, ocean, littoral zone types) are two of the main means by which energy from the Sun flows from the atmosphere (via air mass movements, or 'wind') into ocean water and from there to our shorelines. A third source of energy reaching coasts is that of the daily cycling of tides. These long-period types of wave, affecting both the ocean and the land, are driven by the movements and consequent variations in gravitational attraction of the Earth and Moon with the Sun. Of these three sources of energy, wind-generated waves form the most significant link to coastal processes and functioning, accounting for more than 80 per cent of the energy available to coasts. Wind waves at the sea surface

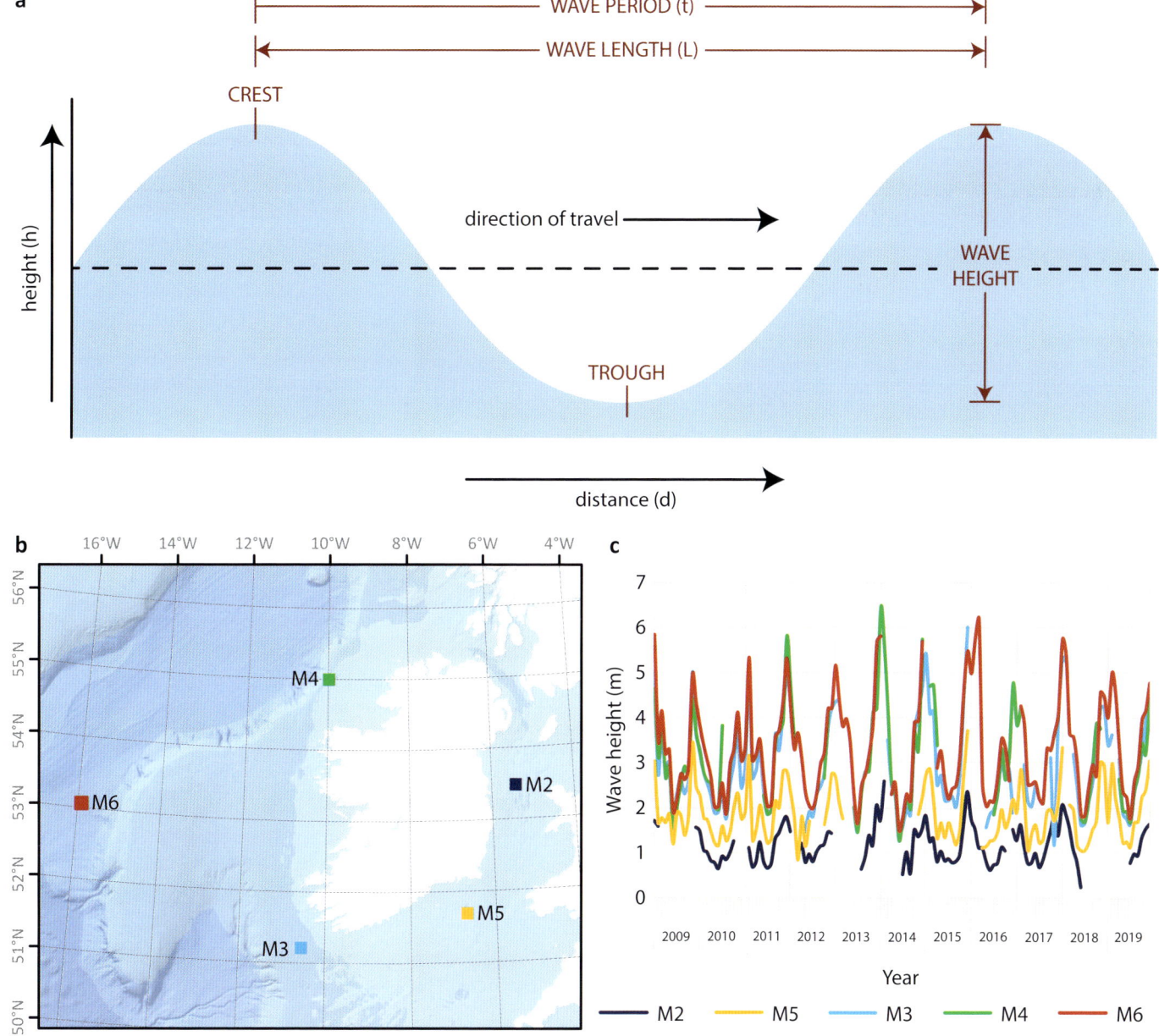

Fig. 2.23 SEA SURFACE WAVE SHAPE AND HEIGHT. a. The main elements of the shape and motion of a swell, or sea surface gravity wave. Waves of this form oscillate vertically around their long axis (the axis perpendicular to the wave shape) and energy is transmitted through this motion in the direction of wave travel. Waves vary greatly in both shape and size, with wave height changing under wind stress day-to-day and even hourly; b. and c. Mean monthly wave heights around Ireland (2009–2019). Swell waves developed within the North Atlantic and approaching inshore on Ireland's western coasts have generally a mean annual height of <7m, reducing to values of 1–2m within Irish Sea areas. Under the influence of storm conditions, these waves commonly reach heights in offshore deep waters of the Atlantic Shelf, Celtic and Malin seas of 12–20m, again diminishing in height eastwards into the Irish Sea. [Data source: Met Éireann licensed under CC BY 4.0]

a

constructive (flat) waves

strong wash; much water is lost through percolation; sand carried up the beach and forms a berm

relatively flat and gentle waves

berm

original beach profile

new beach profile

smaller longshore (breakpoint) bar

weak backrush; little material is returned to the beach

b

destructive (steep) waves

high, steep waves

some large material forming a storm beach

original beach profile

new beach profile

larger longshore (breakpoint) bar

weak swash

gradient decreases down the beach

little water lost through percolation; most of material comes down beach by backwash

Fig. 2.24 THE WORK OF BREAKING WAVES. On approaching the coastline and entering shallow water the frictional drag of the seabed causes waves to slow down, to increase in height, change shape and finally to break on the beach. Many ingredients are involved in creating the active conditions for wave breaking. These include the angle of wave approach to the shore; water depth changes; seabed and shore gradients; sediment availability, sizes and volume; seasonality (summer to winter) and coastline shape. A useful idea in describing the result of wave breaking and the resulting movement of sediments is that of constructive a. and destructive b. waves. For a., particularly where sandy and wide-beach systems occur, waves tend to be low in height (flatter shapes) and move sediments onshore to build up the beach. In b., seabed gradients may be high (>4°) and/or beach sediments may be coarse (for example, cobbles). These factors result in a steep beach angle, with deeper water depths occurring inshore. Consequently, the wave shape is also steep and waves can approach closer to the shore before breaking. This process then causes sediments to be moved both onto the beach and then back down (wave backwash), often with the removal of beach sediments offshore, or further alongshore. [Source: adapted from Quizlet (https://quizlet.com/221821828/constructive-and-destructive-waves-diagram/)]

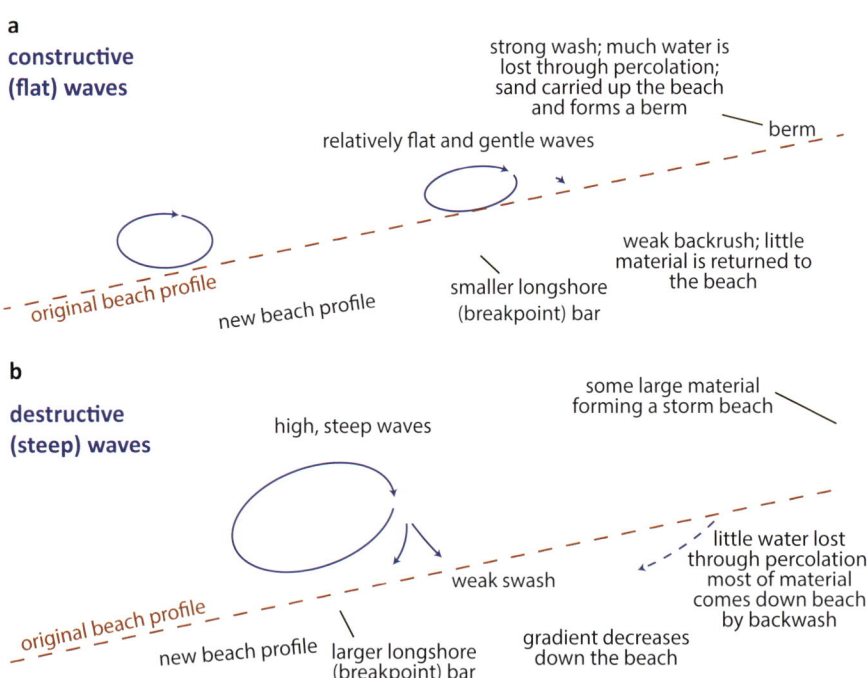

are caused by storm events and result from the frictional force of winds (as stress, or drag) blowing across water surfaces. Energy in the wind is transmitted (lost) to the water through this motion, which is stored and subsequently moved across ocean surface waters in the shape (form) and vertical oscillatory motion of waves (Figure 2.23a–c).[20]

Once developed and regularised in the oceans, waves (as swell waves) move progressively toward the continents (land) under the influences of continued wind action (such as from storms) and the Earth's gravity. As waves approach the shallowing waters of the land margins, their stored energy is lost. Energy is dissipated within the water column and to the

Fig. 2.25 WAVES AT THE COAST. a. Breaking Atlantic swell waves at Dawros Bay, County Donegal. The photograph captures the waves increasing in height as they move into the shallow waters of the shore zone. The wave closest to the camera has reached the point of peaking, curling over and collapsing. The waves behind will follow a similar pattern, though no two waves are exactly the same. Offshore, wave crests approaching the shore are being blown off as sea spray and the form of the wave is being changed under the influence of strong offshore winds. [Source: Ciaran Lee, flickr, CC BY-SA 4.0]; b. An aerial view at Whitestrand, County Mayo, of the patterns of breaking and refracting waves forming as the series of Atlantic swell waves (wave train) approaches the three-dimensional topography of the coast. The shallow water areas of the beach and offshore bar (left) and sand shoals (right) show collapsing waves, breakers and areas of white water. The deeper waters in between these (centre) show the approaching swell wave crests continuing onshore, bending or spreading in a circular pattern (wave refraction) to fit the shape of the beach. The waves bring with them the shelly sands of this coast, which form – with wave and wind action – the dunes and machair environments of this high-energy coastline. [Source: Gordon Dunn]

seabed through frictional drag and, finally, as breaking waves on the seashore (Figure 2.24). In these different energy flows work is being done by wave action. Rock and sediments are eroded from the seabed and shorelines, as well as being transported to new locations along coasts (for example, to beaches, sandbanks, estuaries, marshes). This wind-wave activity at the coast is added to and modified by the actions of other water motions – such as tides and currents, joined now by river and other freshwater outflows – and energy sources (such as Earth heat losses).

The size (magnitude), number and frequency of storms from the North Atlantic are the most important control on the amount of wave energy reaching Ireland's shores and thus on the coastal processes operating. In short, waves and storms together drive our coasts, as they do most mid-latitude coasts in the northern and southern hemispheres. Around Ireland, maximum wave energy is received on its storm-exposed west coasts, with progressively

reducing levels of wave energy recorded on shorelines eastwards into the Irish Sea (see Figure 2.23b and c).[21] This effective 'wave-stilling' and sheltering effect of Ireland for land areas further east (the coasts of Wales and north-west England particularly) has produced the observation of the island as being 'one of the world's largest natural breakwaters' (see Chapter 30: Engineering for Vulnerable Coastlines).

Waves, as they enter shallow waters, begin to move sediments from offshore to the beaches. Wave crests begin to increase in height and bend, to mirror (in the process of wave refraction) the different shapes of the land (such as headlands and bays) and to break eventually (collapse) (Figure 2.25). This wave action produces a wide range of shoreline responses over time, not only of rock and sediment erosion, but also, importantly, the transport of sediments across and along our beaches and other coastal types.

TIDES AND THEIR IMPACTS ON IRELAND'S COASTS

Eugene Farrell

Tides, as observed in the Earth's oceans, form the periodic rise and fall of the sea surface, measured against the land (coasts). For coasts in Ireland, and most globally, this rhythmic oscillation of ocean waters occurs as a change in the sea's surface (tide height) between high-tide and low-tide positions, approximately every twelve hours and twenty-five minutes. This changing level exposes an intertidal area of the shore zone to the air twice daily. In some parts of the world, 'double' tides occur; in others, none. These tidal changes are caused primarily by variations in the gravitational pull of the Moon and the Sun on Earth's surface, as the Moon rotates around the Earth in its monthly lunar cycle (c.29.5 days) (Figure 2.26).

This wetted edge of the land exposed by the tidal cycle forms the shore zone. If we consider the position of mean sea level to be constant over short time periods, then there are three primary variables driving coastal functioning and evolution: (i) waves and storms

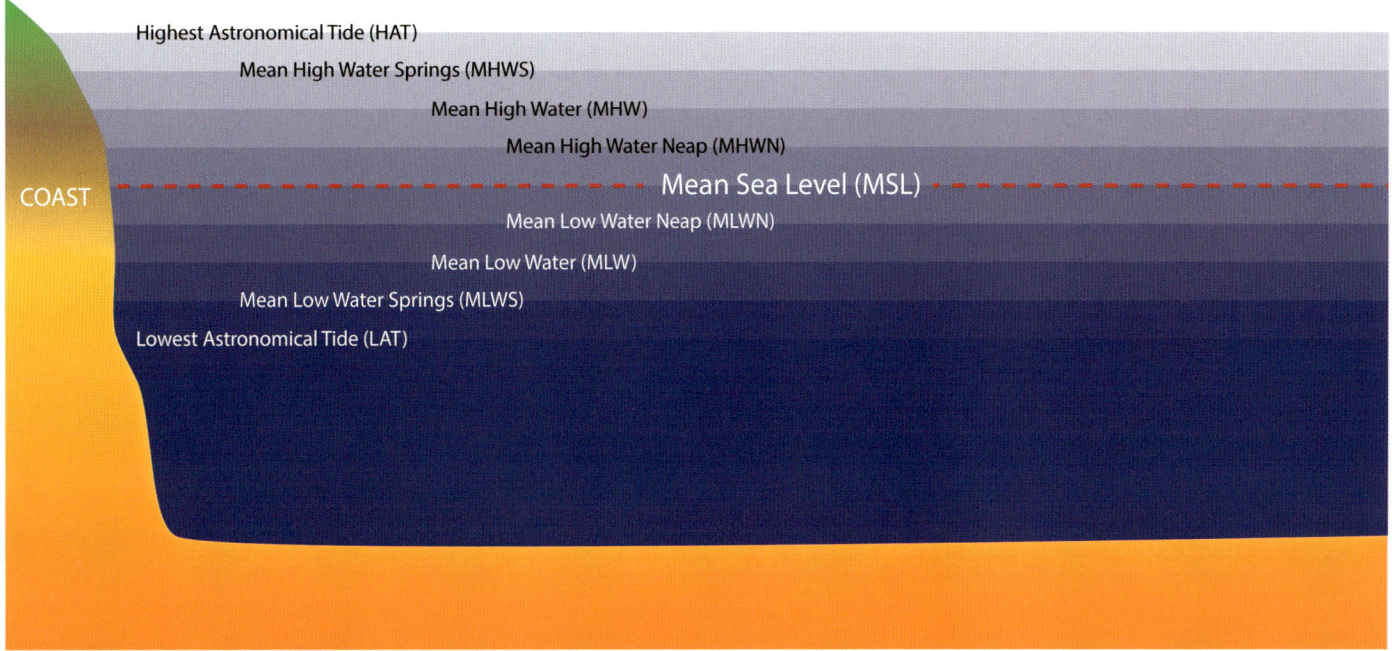

Fig. 2.26 TIDE POSITIONS ON A SHORELINE, FROM THE LEVELS OF HIGH WATER TO LOW WATER. This regular cycling of changes in tidal height and position on a coast operate between maximum upper (high tide) and lower (low tide) limits on the shore, as twice-monthly periods of spring tides and, conversely, the alternating phases of minimum upper and lower tide levels, known as neap tides. The diagram shows these different high-water and low-water positions. The most extreme of these water levels are referred to as astronomical tides and record the effect on sea level of the maximum gravitational pulls on the Earth.

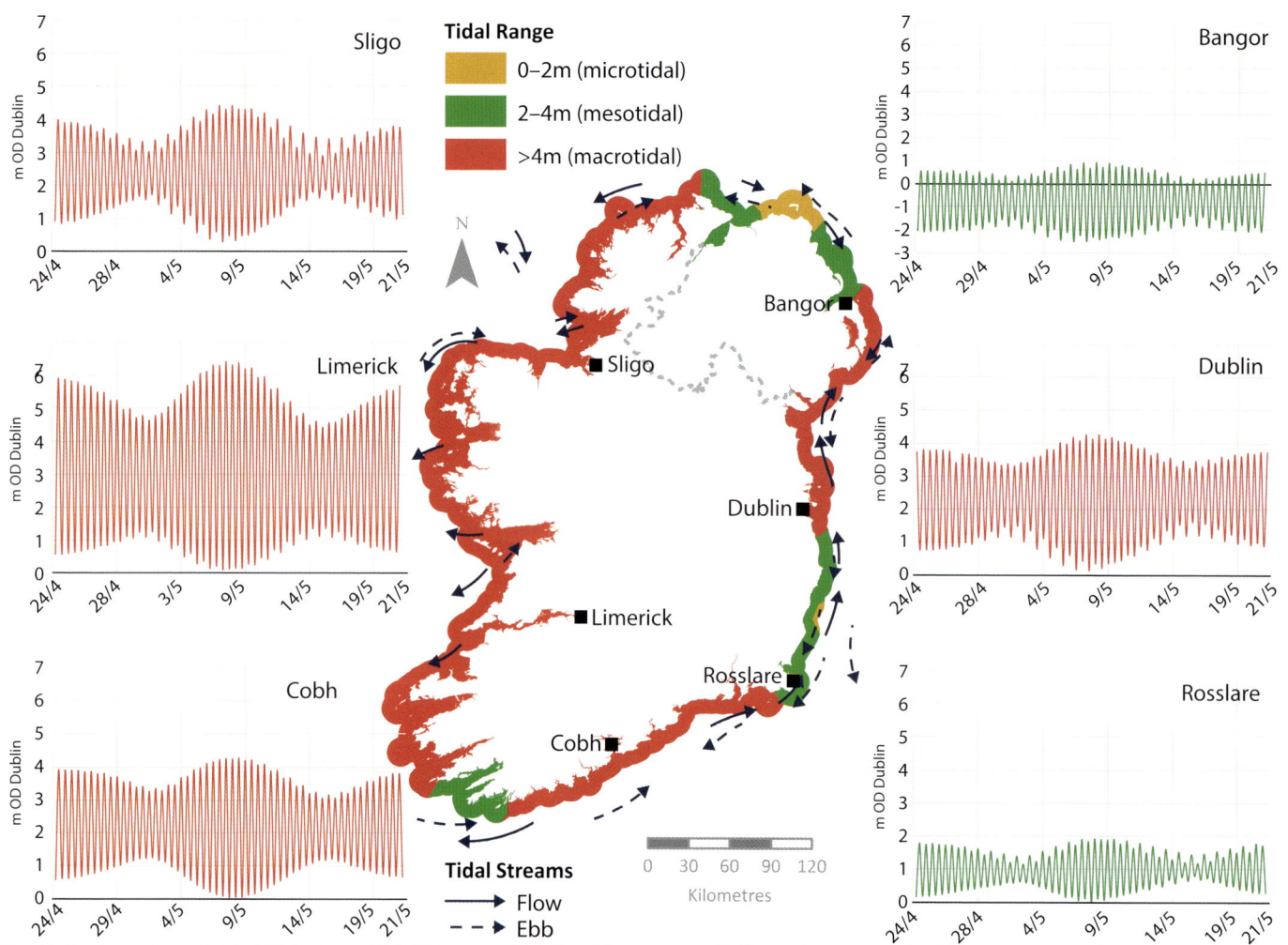

Fig. 2.27 TIDAL RANGES AROUND IRELAND, SHOWING THE VARIATION FROM MACRO- TO MICROTIDAL RANGE COASTS. The map indicates the main tidal ranges found on Ireland's coasts. The graphs show predicted tide data for a selected one-month period (24 April–21 May 2020), and illustrate the daily to weekly variations in tide heights and ranges for six standard, or tide-recording, ports around the island. It is worth noting that truly 'microtidal coasts' are really those as found in, for example, the Mediterranean and Sea of Japan, which experience tidal ranges of <1m. These very small ranges are not found on Ireland's wind- and storm-dominated coasts. [Map data: VORF data supplied by the UKHO and reproduced under licence. © Crown Copyright 2019; tide data: National Oceanography Centre (NOC), 2020]

(which have been considered briefly already), (ii) tides with long-term relative sea-level changes and (iii) sediment availability. Most centrally, much of the character of shorelines, both in Ireland and globally, results from the combined working of tides and waves, operating on a platform of changing sea levels. Their interaction impacts a wide range of coastal applications, including, but not limited to, coastal sediment budgets, shoreline morphodynamics, computation of the long-term stability of beaches, modelling coastal evolution and assessing the impact of sea-level rise (SLR) on coasts. The role of tides is integral to understanding the operation of all these coastal processes, and more (see Chapter 7: Ancient Shorelines and Sea-level Changes).[22]

Tidal characteristics have been used in global coastal classifications to characterise different shorelines in terms of their physical properties, functioning, evolution, or geographic location. In Ireland, we can use the spatial distribution of tides and their classification to further understand the distribution and functioning of other coastal environmental controls, such as variability in wave climates or the legacy on coasts of inherited geomorphological factors, especially the impacts of past glaciations (see Chapter 6: Glaciation and Ireland's Arctic Inheritance).[23]

Depositional coasts – such as beaches, estuaries and lagoons – have morphologies that are tied strongly to wave energy and tidal currents. Tidal range (the height difference between consecutive high and low waters) has been used to subdivide shorelines in Ireland, and worldwide, into three main classes: micro- (0–2m), meso- (2–4m) and macrotidal (>4m) environments. This focus is justified because tidal range controls the length of time waves act on any portion of the shore (Figure 2.27). The extent to which tide-related processes affect coastal evolution increases with tidal range. Conversely, the effectiveness of wave energy decreases with increasing tidal range. The type, distribution and frequency of shoreline features (for example, erosion-dominated coasts) has also been applied to this model of tidal classes, to make a generalised link between tidal range and coastal morphology (see Figure 30.8).

Based on this approach, three major types of depositional coastlines have been defined: wave-dominated, mixed-energy and tide-dominated. More recent coastal classification models have incorporated tidal dynamics. These show the influence of (i) tidal stage, which determines where nearshore processes operate on the shore; (ii) tidal translation rates, which determine how long tidal processes act on the seabed and affect beach morphology and (iii) tidal cycles (tidal amplitude and duration). This understanding of the way tides influence other coastal processes and functioning links to the late twentieth-century, revolutionary morphodynamics approach to the study of beaches and other coastal environments.[24]

A threefold classification of Irish estuaries has been established (drowned river valley estuary, river-dominated estuary and bar-built estuary), based on the genetic characteristics (mode of origin) of coasts. However, the classification indicates that the diverse range of estuary forms in Ireland is a response to a combination of both contemporary dynamic conditions (for example, wave-dominated or tidal-dominated coastline) and genetic, inherited factors (for example, glacial valleys or early river valleys), and that tidal range had a key role on each. For example, beach and dune back-barrier type coasts often have seepage lagoons and other wetland environments behind them, as at Carrownisky Strand, County Mayo, and Rossbehy Spit, County Kerry. Most of these lagoons and low-salinity wetlands are cut off from the sea by barriers, so that tides and waves have limited influence on them, except during storm surge events (as at Lady's Island Lake, County Wexford), while many are only flooded by the sea at high tides. Lagoons on macrotidal coasts have developed strong currents due to the greater tidal range and, in these, tidal channels are better developed (for example, at Rusheen Bay, County Galway). In other coastal settings, the thickness of tidal flat deposits is controlled by the tidal range and depositional rates on saltmarshes generally increase with tidal range (see Chapters 11: Beaches and Barriers and 13: Estuaries and Mudflats).[25]

As noted earlier, most Irish coasts have semi-diurnal tidal variations. This results in different parts of the shoreline and beach area being exposed to wave energy each day, with fundamental impacts on coastal biology (see Chapter 10: Rocky Coasts). On macrotidal coasts, the lateral movement of the shoreline, between high and low water, can extend over several hundred metres or more, as experienced in particular on exposed western shores. Where sediment is available, as in the Shannon Estuary, tidal currents are able, potentially, to move large volumes of fine-sized sediment. This mechanism is important in the geomorphological and ecological functioning of tidal mudflats and sandflats, saltmarshes and lagoons. These different environments all host critical habitats that are legally protected under the EU Habitats and Birds Directives (Natura 2000) and the Ramsar Convention. Unfortunately, many of these tidally dominated and protected environments have been poorly studied and monitored. These areas often lack sufficient data to assess their overall conservation status, or their likely long-term response to pressures from aquaculture, fishing, coastal development, water pollution and climate change (see Chapter 32: Coastal Management and Planning).[26]

Tides interact in complex ways with coastal topography, river flows, seawater density and salinity gradients when entering shallow-water coastal areas. Tidal dynamics are greatly distorted as they move into estuaries, bays and tidal inlets, resulting in very local tidal regimes. In locations with prominent flood (rising tide) and ebb (falling) tidal flows, complex networks of channels and bars, associated with asymmetrical patterns of discharge (volume, speed, duration) and sediment transport pathways, are present. The mixing of freshwater and seawater in estuaries can reduce tidal current flows and lead to the deposition of fine sediments. These often form extensive intertidal mudflats and sandflats (for example, Lough Swilly, County Donegal, and Courtmacsherry Bay, County Cork) and saltmarshes (for example, Tralee Bay, County Kerry) (see Chapter 13: Estuaries and Lagoons).

In coastal regions around the Irish Sea, tidal flows are also complex. As the gravitational attraction of the Moon pulls the oceans, water flows from the west around the island of Ireland and into the Irish Sea from both the north and south. A circulatory tidal flow is generated, with the unusual phenomenon developed at Courtown, County Wexford, of four weak daily tides (as opposed to the more usual two), as incoming and outgoing tides from each direction meet and partly cancel each other out. Tidal flows in the south Irish Sea are strong and ebb and flow tides are of near-equal strength. These have produced large symmetrical sediment waves, where sands and gravels have heaped up on the seabed. In the north-east of the Irish Sea, tidal flows are slow and large areas of silts and muds have accumulated that are heavily burrowed by langoustines or Dublin Bay prawns.

More widely, shoreline configuration (3-D shape changes) can enhance the observed tidal range on coasts, due to funnelling effects found in many bays and larger estuaries, such as that found in the rivers Shannon (County Limerick), Bandon and Bride (County Cork), Foyle (counties Derry and Donegal), Barrow and Slaney (County Wexford) estuaries. This explains why the average tide range half-way up the Shannon Estuary is c.4.5m, but at its head, east of Limerick City, it is almost 1m larger (Figure 2.27). Further, the delicate balance between the times and extents of flood and ebb tides is also a major influence on the location of our coastal environments. Saltmarshes, for example, are found typically between the upper limits of neap and spring tides on sheltered coastlines. In locations where saltmarshes overlay freshwater peat, we can infer that the sea invaded the land through marine flooding mechanisms as at, for example, Castlemaine Harbour, County Kerry, and that tidal changes in such areas occurred there subsequently, linked to the alteration of coastal heights and geometry (see Chapters 11: Beaches and Barriers and 30: Engineering for Vulnerable Coastlines).

Tides play a critical role in potentially increasing the impact of storm surges along the island's shorelines. This can occur if high tides combine with the peak of a storm surge. Storm surges occur where a combination of wind, atmospheric pressure and wave set-up lead to an increase in sea levels over a number of hours, or even days. For instance, Storm Eleanor, only a Category 2 storm in Irish waters (storm range scale, Categories 1–5), caused severe flooding in Galway city on 2 January 2018. Landfall of the storm coincided with a high spring tide. Together, this caused the highest water levels (4.7m) ever recorded at Wolfe Tone Bridge on the River Corrib, located at the river mouth. The previous record of 4.5m was from winter storms in February 2014. Extreme water levels of this type can cause significant morphological changes to coasts, as well as causing widespread flooding.[27]

IMPACTS AND IMPLICATIONS OF TSUNAMI ON IRELAND

Robert Devoy

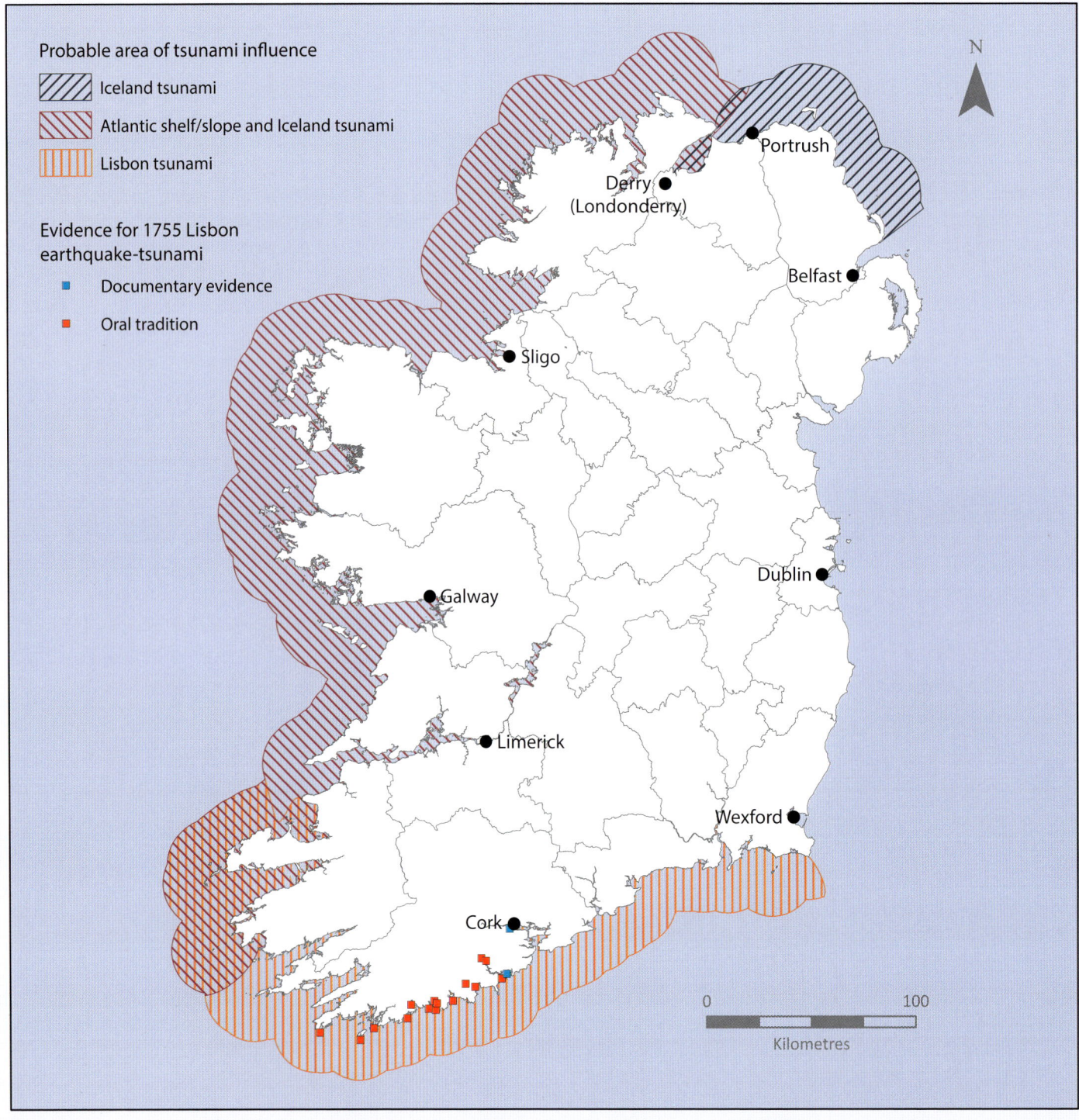

Fig. 2.28 SITE LOCATIONS OF THE 1755 LISBON TSUNAMI AND THE COASTS LIKELY TO HAVE RECORDED TSUNAMI FROM DIFFERENT SOURCES IN THE NORTH-EAST ATLANTIC. In addition to recorded impacts on the south-west and south coasts of Ireland of the 1755 Lisbon earthquake and tsunami, two other possible sources of tsunami affecting Ireland exist: first, from the Atlantic shelf edge, as generated by submarine sedimentary fan slides. These would have resulted from the build-up of material on the continental shelf margin, as in the Donegal Fan. Tsunami resulting from sediment-slide sources such as this could have caused events similar in character, though probably not as large, as the catastrophic Storegga slide and tsunami in the North Sea region *c.*8,000 years ago. In spite of some detailed studies, to date no good evidence for such events in the past has not yet been found. Ireland's shelf-edge areas are large, as are associated technical survey problems. However, it remains likely that such events have occurred, most probably triggered and compounded by generating sources in eastern Canada. A second potential source region for tsunami impacts on Ireland's coasts is Iceland. These would be caused by catastrophic ice-melt origins (jökulhlaups), or combined with shallow crustal earthquakes and volcanic eruptions. [Map data source for 1755 Lisbon earthquake sites: Anthony Beese]

While historical records of tsunami events (a series of fast-travelling large sea waves) impacting the Irish coast in the recent past exist, until relatively recently, few tsunami have been recognised. The most abundant historical (as maps and archives) and physical evidence (sedimentary and geomorphological) is for south-west Ireland, in counties Cork, Kerry and possibly Clare. These data are associated, invariably, with the 1755 Lisbon earthquake-generated tsunami, though there may have been earlier events on these coasts that have gone unrecorded. The subject field is controversial (Figure 2.28).[28]

The Lisbon event hit Ireland's shores on All Saints Day, 1 November 1755, some hours after the earthquake and tsunami had destroyed much of the prosperous coastal city of Lisbon. In Ireland, significant changes to the shape and the geomorphology of many south-western coasts occurred, as found at Rosscarbery, Owenahincha, Longstrand, Kinsale, Clear Island and in neighbouring areas of Baltimore, County Cork.[29] Fortunately, no major loss of life is known to have resulted in Ireland, though deaths may have happened, particularly in rural areas, that went unrecorded. The major change to coasts here involved the immediate to longer-term reorganisation of nearshore sediments, with sands, gravels and boulders being moved both onshore and alongshore (Figure 2.29). This resulted in the complete alteration of the coastal 3-D shape of some areas. For example, reconstruction of the event at Tranabo Cove

Fig. 2.29 HIGH-LEVEL MARINE SANDS AT THE LINKED SITES OF TRANABO COVE AND TRALISPEAN BAY, EAST OF LOUGH HYNE, COUNTY CORK. These marine sands are situated at *c.*8–10m above the present cobble beach at Tranabo Cove. The sediments were most likely deposited over underlying glacigenic materials and rock by the 1755 Lisbon tsunami event. Boulders found within the marine sands (small boulder shown in photograph foreground), derived from Tranabo Cove and moved across the rock headland into Tralispean Bay, were deposited with the sands by the tsunami wave in a thick layer of water-sorted sediments across the area. [Source: Robert Devoy]

and Tralispean Bay, near Lough Hyne, Skibbereen, County Cork, showed that the impact of the tsunami wave was to cause extensive cliffing, erosion and scouring, destroying all in its path as it moved across the neighbouring headlands of the bay and generally left only bare, washed rock surfaces in its aftermath. Any coastal dwelling in its pathway would have been washed away. The wave also transported large boulders (megaclasts, >1–3m diameter) onshore, together with thick sequences of marine sands, leaving these deposited many metres above the contemporary shoreline. Elsewhere, as in Kinsale Harbour and Clear Island, coastal cliffing and landslides took place, some of which are still visible. This may have occurred on many other exposed coasts in the region.

The height of the main tsunami wave varied along the coastline. On Clear Island, the wave probably washed completely over a corridor of low ground at its western, Atlantic end, at levels of *c.*40–45m above sea level (Ordnance Datum). Further east, in the Tralispean Bay area, it washed up to heights of 18–20m, but had fallen further east, at Harbour View (Courtmacsherry), County Cork, to levels of 11–14m. This apparent eastward, along-coast, reduction in wave height was most likely the result of frictional effects of seabed and coastal geomorphology as the wave travelled further into the Celtic Sea. Direct observation of the wave in Cork Harbour, as recorded in ship's logs, shows that it had dropped to only a few metres in height, which caused ships anchored in the outer harbour to yaw heavily on their moorings, while some coastal flooding occurred. In other low-lying areas, the tsunami travelled substantial distances inland, as in the Carbery River estuary (at Rosscarbery), while at Kinsale, the wave travelled up the Bandon River from the open sea for *c.*17km to Innishannon.

Good evidence of other tsunami events are not known from Ireland's coasts, though they may well have occurred, particularly in prehistory, rather like the story of Doggerland in the North Sea (see Chapter 16: The Inhabitants of Ireland's Early Coastal Landscapes). The 1755 earthquake in Lisbon, which caused the tsunami, was part of long-term regional Earth crustal faulting and plate tectonic movements. These are relatively rare as large-scale events on these passive, eastern Atlantic continental margins (that is, from Iceland to the western Mediterranean), though small-scale earthquakes and tremors are common and are recorded regularly in Ireland. It is improbable though that earlier versions of Lisbon 1755 had not impacted Ireland at some point in the past from this southern European origin (see Figure 2.28).[30]

As with severe storm surges of lower magnitude, but similarly large coastal impacts, tsunami have been shown in other parts of the world to throw megaclasts (some weighing >100 tonnes) over cliff tops tens of metres in height above the shoreline. The possibility that this happened along Ireland's western coasts, in particular, has been advanced since the early twentieth century. More recently, however, studies of cliff area megaclasts – for example, in the Iveragh Peninsula, County Kerry, the coasts of County Donegal, the Aran Islands and elsewhere in County Clare – have swung in favour of their emplacement by storms (see Chapter 10: Rocky Coasts).

The significance of the Lisbon event in understanding the operation of coastal processes in Ireland is that it shows that a tsunami can happen here, even though the evidence for them may not be recognised easily. This event, one of the largest known to have occurred within European waters, demonstrated how such waves can change instantaneously the appearance, and even the functioning, of an entire coastline. On many tsunami-prone coasts around the world (such as Japan, New Zealand, Portugal, Sri Lanka, Chile), anyone living there is very vulnerable and the loss of life from a tsunami is often large (numbering tens of thousands or more, as in the Indian ocean tsunami, 2004).

Equally, damage to coastal infrastructures and urban areas is massive. It is the catastrophic character and scale of a tsunami, as well as similarly scaled megastorms that Ireland experiences (for example, the Big Wind, hurricanes Debbie and Ophelia), that is most significant.

Ireland's coastal population is in excess of 55 per cent of the total, and although the majority resides on the wave-protected Irish Sea coasts, the distinct possibility of another 1755-type event occurring, and the certainty of megastorms, are both cause for concern (see Chapter 25: Urbanisation of Ireland's Coast). Further, there is a lesson regarding slower but equally large, or larger-scale, coastal changes to be drawn from such megawave events. Global to regional sea-level rise (SLR) has been relatively slow (mm/year), but is now accelerating and will change even faster in future, with total vertical mean SLR of at least 1–2m by 2200. Unlike a tsunami, or a storm, this water level never retreats; as with the tsunami, the coast has to change irrevocably. The intelligent response of people to this is to be aware and be prepared for coastal change.

SEDIMENTS

Knowledge of how sediments are supplied to coasts, and where this rock material goes, is essential in understanding how coasts respond

Fig. 2.30 a. BEACH SEDIMENT VARIATION IN SPACE AND TIME. The reappearance of the sandy beach at Dooagh, on the southern shore of Achill Island, County Mayo, after an absence of thirty years attracted much international media attention, with great Facebook and Twitter interest. This photograph, taken on 28 April 2017, shows the beach area after this overnight deposition of hundreds of thousands of tonnes of sand. The loss of the earlier wide and scenically attractive sandy beach in the winter storms of 1984 left behind a much narrower and gravelly environment. This apparent erosive action created concerns for local residents for the future erosion of these coasts and, most particularly, for the impacts that the disappearance of the beach would have on a developing local tourism trade. Its subsequent disappearance again in the major storms of 2018/2019 emphasises that sediments rarely remain stable on coasts. They are moving constantly under daily, seasonal and longer term waves, tide, current and storm actions. [Source: Sean Molloy]; b. Gravel and cobble beach, Kells Bay, County Kerry. This small coastal embayment has a gravel, cobble and boulder beach, with a noticeable lack of sand. Such beaches have high scenic attractiveness for coastal tourism. The steep face of the beach results from the large sizes of these coarse sediments, as well as the high-energy wave conditions experienced on this Atlantic coast, which remove the finer sandy sediments and concentrate the coarser fractions. The sources of these materials are generally from local glacigenic deposits, which can be seen in the vertical, soft cliff face on the right of the photograph. Due to the extensive coverage of Ireland's coastal areas with these glacial sediments (for example, boulder clays), gravel beaches and barriers can be found in most of the island's coastal regions. Their prominence has been noted internationally as part of high-latitude former glacial environments (paraglacial coasts), as found here and in north-east Canada. [Source: Robert Devoy]

to environmental pressures to change (that is, coastal resilience) and how people can manage them effectively. Sources of sediment supply to Ireland's different beach and other coastal systems include: (i) the offshore zone, where sediments inherited from glaciation and other mainly past geomorphological changes have been deposited, and (ii) the erosion of the present littoral/nearshore areas, with new sediment generated from the rocks and sedimentary cliffs on land.[31]

The transfers of sediments on the island's coasts is a complex process. It depends on the constant interplay of factors of sediment availability, effects of coastal shape, proximity to sources and of hydrodynamics (waves, wind, tides and currents). The dominant patterns of these transfers are both onshore and offshore sediment transport, but, importantly, along the shore as well (Figures 2.3 and 2.16). These movements are driven by daily and seasonal factors. During the storminess of winter (September–March), sediments move predominantly offshore, characterised by the distinctive combing-down and steepened angle of beaches, together with the generally temporary transport of finer sediments offshore. In summer (April–September), onshore sediment transport becomes dominant, with the building up of beaches and the lowering of beach angles. Sediments (mainly sands) that have accumulated offshore over the winter in deeper waters, now move onshore under the lower-energy wave conditions of summer, as sandbars and shoals.

Sediments that have moved offshore during the winter may not always reappear the following year, or for even longer periods. These sediments may, over time, become the supplies to neighbouring sections of coast, possibly where erosion has begun or been renewed (from whatever cause: human or natural), and now these coasts require more sediment. Alternatively, the longer-term operation of waves, particularly at times of high storm surges, may override sediment storage thresholds and bring the materials back, after years of absence. A good example of such a pattern, and one that has caught people's attention, is the reappearing beach at Dooagh, Achill Island, County Mayo (Figure 2.30a).

The work of waves in the erosion, transport, sorting and re-supply of sediments is an essential dynamic for our coasts. It has also been a critical factor in the creation of distinctive beach characters and even whole coastal landscapes, such as the development of Ireland's gravel-dominated beach systems, as found particularly on the shores of south and south-east Ireland (in counties Cork, Waterford and Wexford) and County Donegal. The material source here is the coarse debris of Ireland's widely distributed glacial sediments, found on almost all of its coasts

(Figure 2.30b). Another almost iconic coastal landscape resulting from this interaction of sediment source and wave action is the coarse boulder deposits of Ireland's storm-dominated mid- to north-west coasts. On Inishmore, one of the Aran Islands, large boulder deposits accumulate along the cliff tops, tens of metres above the sea surface, thrown there by storm waves. Wave action here is able to move rock blocks weighing over 800 tonnes (see Figure 10.15 and Chapter 11: Beaches and Barriers).

Over the long term (that is, at decadal scales or longer), coasts receive progressive onshore movement of sediments, particularly as part of coastal changes linked to the postglacial rise of sea levels (see Chapters 7: Ancient Shorelines and Sea-level Changes and 30: Engineering for Vulnerable Coastlines). Most sediment supply to coasts from shelf and offshore sources overall appears now to have reduced significantly, or even stopped. However, increasing rates of SLR in future will lead to the rejuvenation of this supply process. Currently, sediments are recruited primarily from areas of the nearshore and cliff line through wave erosion. Alongshore transport of these materials is now critical in moving the available sediments between different beach and bay systems.

Around the island, at the larger scale and regional level, sedimentary movements from the central Irish Sea mud basins toward coasts in the Irish Sea region are divided between those predominantly to the north (toward the North Channel) and those to the south and out into the Celtic Sea. However, actual nearshore coastal sediment movements in the region are mainly southward, from areas of Dundalk Bay toward Dublin and, separately, south from Wicklow Head to Curracloe beach and Wexford Harbour. On the western coasts of Ireland transfers are northward, under Atlantic ocean current, tide and cyclone-powered wave directions, while on the north coast they are predominantly eastward. Along southern coasts, the main drift direction is eastwards, again due to the dominant westerly wind and wave directions. Overall, the larger-scale patterns of sedimentary flows and supply are modified continually by the local actions of waves and tidal currents, activity that may even appear for a time to reverse the dominant transfer directions. This is seen on some coasts in south-east Ireland where a westward drift element does occur closer to the shoreline, under the control of local coastal embayment and sediment-cell processes (see Figure 11.11).[32]

FROM SOURCE TO SINK: STUDYING A COASTAL CATCHMENT

Eugene Farrell and Robert Devoy

The coast of Golden Strand, Achill Island, County Mayo, provides a good analogue for many of Ireland's sedimentary coasts (Figure 2.31). This site formed the location for a novel series of integrated coast and river catchment field studies, undertaken between 2015 and 2017.[33] These involved the multidisciplinary monitoring of this coastal environment during both fair-weather and storm conditions. Golden Strand forms a closed beach system, which is both morphologically and geologically controlled, and is bounded by two prominent headlands that compartmentalise it. Sediments are supplied from both the offshore zone and, possibly, the adjacent glacial sedimentary (till) cliffs.

Observations from the sediment budget work carried out showed that this coast is tuned to high-energy wave conditions. However, it seems to have been largely insensitive to the big storms that occurred during the study period. This result supports findings of earlier research from elsewhere in western Ireland, which show that beaches and dunes exposed to high-energy wave regimes often require extreme storm events to cause

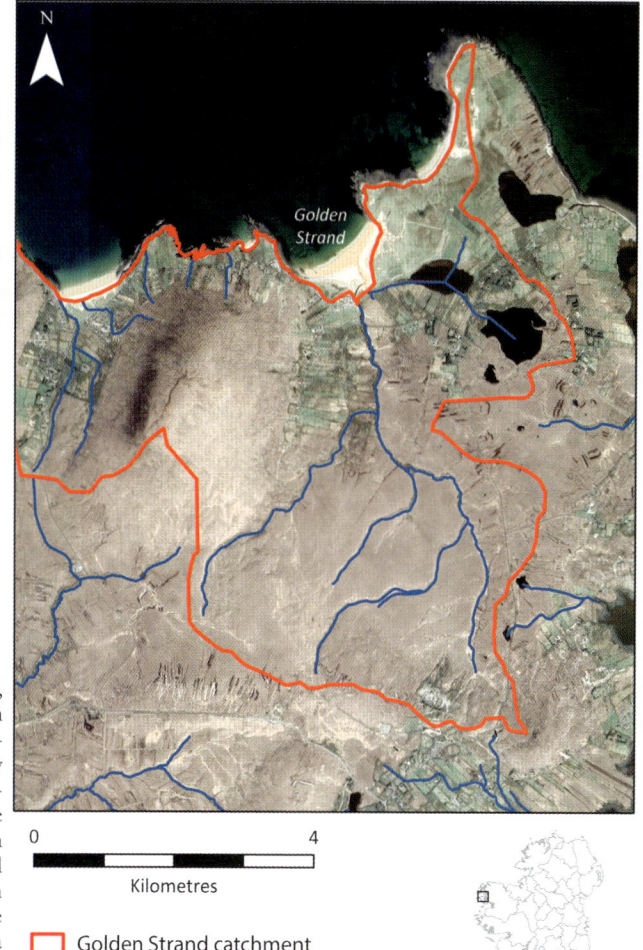

Fig. 2.31 BEACHES AND THE RIVER CATCHMENT FEEDING INTO GOLDEN STRAND, NORTH ACHILL ISLAND, COUNTY MAYO. The photographic map shows the main high-ground terrain that frames the sand-dominated beach, dune and machair environments found at Golden Strand and more widely along these coasts of County Mayo and further north into County Donegal. Three coastal, fresh-to-slightly brackish water lakes can be seen as forming part of this environment, two of which lie within the main river and tributary streams catchment (river basin) that focuses on this strand. Most of the slopes in the catchment are covered by peat and associated bogland plant communities. The immediate coastal lowlands, as around Golden Strand, form part of the sand machair and are farmed or covered by grassland type vegetation. [Source: Farrell et al., 2021. From Source to Sinks: Responses of a coastal catchment area to large-scale changes (Golden Strand Catchment, Achill Island, County Mayo), EPA Report 376, 84pp. Dublin: EPA; Map data: Google, CNES/Airbus, 2020]

0   4

Kilometres

□ Golden Strand catchment
— Rivers

Fig. 2.32 COASTAL PROCESSES MONITORING AND SURVEY WORK AT GOLDEN STRAND, ACHILL ISLAND, COUNTY MAYO. An important component of the study of this sand-dominated coastal system is the collection of a detailed time series of terrain, sediments and wind movement information: a. shows the recording of wind and other atmospheric data, from ground level to *c*.5m into the air-stream above using anemometer tower equipment; b. surveying with high-precision GPS techniques the ground levels on the dune and machair terrain at the site. The machair sand system forms a significant type of coastal landscape, which is developed more fully further north in Donegal and on to the Hebrides (the Western Isles) of Scotland. Machair is an inherently dynamic geomorphological environment, with high sensitivity not only to wind and wave sediment transport, but also to freshwater fluxes through the system (groundwater and near-surface water flows) and to shifts in climate (such as winds, temperature and rainfall). Monitoring this area at Golden Strand has indicated that it shows evidence of moving now into a new state; one more representative of a fixed dune and wet machair. In light of this, it is suggested that the current grazing regime of this sensitive environment be altered, to change from sheep to cattle. To reduce soil compacting and enhance machair stability and longevity, farming practices such as the use of heavy machinery should also be avoided. [Source: Eugene Farrell]

significant morphological change.[34] It remains uncertain how much time is needed, or what combination of environmental conditions (coastal controls) are required, for a substantial volume of beach sediment at the Golden Strand site to be 'rolled-over' and/or pushed (bulldozed) landward. What is apparent from monitoring is that this beach system is increasingly likely to move inland soon, if the number of storms and their intensity continues to rise, as is projected for these western coasts. The extent to which Golden Strand will undergo significant process and domain changes over time (Figure 2.31), to break its current cyclical (seasonal) operation (that is, across shore sediment exchanges), is still difficult to determine. This is due to the uncertainty about the nature of the size of the storms required to generate intransitive system changes (Figure 2.15).

One of the important outcomes of the study is recognition of the need to monitor many more such coastal catchment systems (Figure 2.32). This should be carried out now, both in Ireland and abroad, in order to establish robust numerical models of their functioning and behaviour, enabling more reliable projections of beach system changes to be made. For example, Irish coasts vary significantly in terms of both their static boundary conditions for such changes (the geology, orientation, sediment sizes and abundance) and their dynamics (wave-climates, tides, winds and climate). This has resulted in Ireland's rich diversity of coastal environments. One model, however, does not fit all situations or individual sites and many small-area studies are required.[35]

PEOPLE

The majority of the island's population live in or around its coasts (approximately 55 per cent of the total), concentrated mainly within urban and employment hubs, such as Dublin, Belfast, Derry, Cork, Galway and Limerick. The remainder of the population is more dispersed and rural in character (Figure 2.33a and b). Many of this rural population live within some 50km of coastlines and are engaged in coast-related activities, ranging through agriculture; tourism and leisure; small, high-technology businesses; retirement and second-home investment. In common with all other coasts of the world, these human activities impact every aspect of the functioning of coastal systems (atmosphere, water and flows of mass and energy) (Figure 2.2). Today, the input of human agency is one of the most important coastal system controls and one that will have long-term consequences for the island's coastal landscapes and environmental qualities. To understand fully how these environments will develop in future, the dynamics of people have

to be understood as an integral part of the system, in much the same way as that of, for example, waves and sediments.

People have been making their presence felt in Ireland's coastal environments since the time of their first arrival, over 9,000 years ago (see Chapter 16: The Inhabitants of Ireland's Early Coastal Landscapes). Every part of the coastline has a harbour or port, which shows the significance of trade and coastal activities to its inhabitants. Furthermore, the diversity of Ireland's coastal environments have long-held, important implications for human settlement and development. In particular, the presence or absence of available resources presented significant opportunities, as well as challenges, for humans to exploit. Through the adoption and adaptation of particular forms of land use, Ireland's populations have sought to improve the quality and security of their lives. In doing so, they have also greatly altered the coastal environment.

As an illustration of earlier, historical use of coastal lands, locations in counties Donegal, Galway, Kerry and Cork provide ample historical records of human activity. Many ancient, built coastal structures (such as fences, sea walls, jetties and embankments) have been found in these areas, positioned on low-lying mud- and sand-flats, marshes and lagoons. In some areas, such as the Shannon Estuary, reclamation practices may have begun as early as the tenth or eleventh centuries (see Chapter 17: The Vikings and Normans: Coastal invaders and settlers). Such works indicate that these coastal lands have been used intensively for a long time. This activity further increased in the late eighteenth and nineteenth centuries, with the onset of industrialisation and high population pressures, for example, during the period leading to the Great Famine (see Chapters 18: Era of Settlement: Trade, plantation and piracy; 19: Changing Coastal Landscapes and 20: The Great Famine).

In the twentieth century, until the late 1980s, there had been a noted lack of investment in the maintenance and improvements of the built coast and linked river catchment infrastructures. This has probably compounded the increasing impacts of storms, flooding and relative sea-level rise in these areas. On many coasts and perhaps most noticeably in harbours, human-driven sediment silting has also occurred. This, however, is a long-established impact pattern and is linked to people's earliest urban and agricultural uses of the land. These sediment-change problems continue to require major ports in particular to engage in constant dredging to keep them open. Interventions, such as artificial embankments for flood protection and sea defences, have also altered significantly the tidal flow and the hydrodynamics of estuary areas. In consequence, during river floods and other like events, the water storage capacity of adjoining low-lying areas that were once tidal wetlands has been reduced significantly. Observations such as these make it clear that no parts of the island's coastal zone have remained pristine and without human impacts (see Chapters 30: Engineering for Vulnerable Coastlines, 31: Pollution and 32: Coastal Management and Planning).[36]

For much of history, occupancy of the island has been linked directly to the pursuit of the primary activities of agriculture, forestry and fishing. These depend to a very large degree on inputs from the physical environment, which essentially define the success or failure of settlement. Along the western seaboard, the dominance of uplands and the paucity of soil cover, combined with relatively wet and steep wind-swept slopes, gave rise to a pastoral rural economy best suited to the limited carrying capacity of these areas. This pattern continues to characterise much of the western seaboard today and signifies forces of inertia that have helped

Fig. 2.33 PEOPLE AS PROCESS. a. A pastoral landscape on the Iveragh Peninsula, County Kerry. Ireland's western coast is dominated by uplands, characterised by relatively high annual rainfall and slopes that often plunge steeply to the sea. These environments provide a limited resource base to sustain intensive farm practices. Despite such constraints, and until the mid-nineteenth century, these uplands and their relatively narrow coastal lowlands supported a large rural population. Consequences of the Great Famine and a subsequent history of large-scale emigration from these areas, however, have produced a contemporary landscape that combines many features from the past with current land-use practices. This is reflected on the slopes of the Iveragh Peninsula, which show evidence of depopulation, with abandoned field patterns and run-down, stone-wall field boundaries. The rough pasture that dominates today supports extensive pastoral activities and a low-density rural population. Such environments – with their traditional culture and spectacular land and seascapes – now attract large numbers of domestic and international visitors. This is breathing new life into these areas, centred in particular on their small coastal settlements. [Source: John Crowley]; b. Intensive land use along the coastal strip of Castlerock, Downhill and the River Bann estuary, County Derry. (Mussenden Temple, Figure 2.13, is visible on the cliff top, centre left.) Northern Ireland's early engagement with the industrial revolution in the nineteenth century created a growing urban population and increased demands for food. This encouraged farmers to pursue more intensive forms of agricultural activities, especially as improvements in transport allowed them easy access to both local and regional markets. Such well-developed transport links also encouraged an emerging middle class, but also a rapidly growing working-class population, to seek recreational outlets in the many tourist resorts that were evolving from small fishing ports/market centres along the coastline. These long-established demands are reflected in the current land uses along the coastal strip near Castlerock, County Derry. Here, a sharp cliff-line separates the basalt plateau from the shoreline. On the plateau, fertile soils encourage a farm landscape, typified by large fields and intensive arable and pastoral land uses. At the base of the cliff, both road and railway provide efficient transport to link coastal communities and provide ease of access to the splendid beaches found along the Northern Ireland coastline. In the middle-background is the commuter/tourist town of Castlerock, located at the mouth of the River Bann. Further along the coast is the larger tourist centre of Portstewart. [Source: Gordon Dunn]

maintain the iconic and picturesque landscapes of its coasts, beloved by tourists. These are characterised by sweeping upland landscapes, extensive areas of boglands and rough pastures, the absence of tree cover and dispersed, small-farming rural communities (Figure 2.33.a) (see Chapters 4: People, Agriculture and the Coast; 19: Changing Coastal Landscapes and 27: Tourism and Leisure).

In marked contrast is the eastern coast, but extending to include significant areas of the southern and northern coastal zone. Here, benign environmental conditions have encouraged more productive and profitable farming, based on intensive pastoral and/or arable land uses. To raise productivity further, farmers focus on creating larger farms, replacing smaller with larger field patterns, to facilitate mechanisation and the replacement of labour by an intensive use of capital inputs, such as machinery and chemical fertilizers. Such rural landscapes have come under increasing pressure from pollution, soil and sediment losses and processes of modernisation. This has resulted in the loss of traditional field boundaries and farm practices, as well as growing threats to biodiversity (Figure 2.33a).

Since the 1960s, the most important and all-embracing impact of people on coastal areas has been urbanisation (see Chapters 24: Ports and Shipping and 25: Urbanisation of Ireland's Coasts) (Figure 2.33b). Until that decade, the number and scale of towns throughout Ireland was limited and growth confined essentially to ports and their hinterlands along the eastern and northern coasts. A modern industrial revolution was to change this, especially in the Republic, but has since extended into Northern Ireland following the cessation of the 'Troubles'. Rapid and large-scale expansion of trade through the island's ports, together with a preference of modern growth industries to concentrate in and around Ireland's larger coastal cities and towns, resulted in a surge in their population totals. The consequence has been an almost uncontrolled urban sprawl into the rural hinterlands of these coastal centres. Furthermore, as places of residence became increasingly separated from workplaces, a dense transport infrastructure has grown. This has been built to integrate cities and larger towns with the increasing number of commuter towns and attractive tourist resorts that now typify these highly developed zones, such as along the Dublin to Belfast Corridor.

As so many people now live as stakeholders on coasts, with varied and different uses for and perspectives on this space, there is a need to examine people's behaviours and motivations in how they interact with coasts. This information is essential in determining answers to the questions of: 'Why do we use coasts?'; 'What is being built there?' and 'How do we live in the coastal environment in the twenty-first century?' Critical elements in developing some understanding are the workings of politics and governance (broadly, planning and management) as they apply to coastal spaces. These activities have become critical components of studies, in the late twentieth century, of Integrated Coastal Zone Management (ICZM or CZM) and Maritime Spatial Planning (MSP). Work on Ireland's coastal zone has made significant contributions to both of these study areas and has helped in the decision-making processes of management, as well in developing the science of the role of people as a coastal control.[37]

Ireland's coastal zone is now a highly contested space. The consequences of this focus on human growth, however, have been increasing levels of pollution and other negative impacts on coasts. These have caused real and growing concern regarding the degradation of a fragile environment and unsustainable social and economic pressures on coastal communities. Preserving the diversity and qualities of human and physical coastal environments and landscapes, while accommodating the preference of Ireland's changing population to live and work on or near the coast, and access its recreational attributes, have become major issues for government. Effective and sustainable planning is required to achieve the long-term viability of these coastal areas (see Chapter 33: Climate Change and Coastal Futures).

SEA STACKS AT NOHOVAL COVE, COUNTY CORK. [Source: Alix Scullion]

COASTAL WATERS: MARINE BIOLOGY AND ECOLOGY

Mark Jessopp and Michelle Cronin

ONE OF IRELAND'S MOST COMMON SEABIRDS, THE CORMORANT, DRYING ITS WINGS. The charismatic pose of the cormorant, with its wings spread, is a majestic and common sight around the coast. The reason for this characteristic display is that cormorants lack the waterproofing oils of other diving birds. Hence they spread their wings in order to dry them. The cormorant, a large waterbird with a wingspan of up to 160cm, has a primitive appearance, with a long neck and dark plumage. It is a supreme fisher, can dive for up to seventy seconds and feeds primarily on fish, but also on crustaceans, insects and amphibians. Its efficacy as a fisher has meant that cormorants are perceived as a nuisance by anglers and fish farmers, although there is little scientific evidence to indicate that they have a significant negative impact on fisheries, including salmon smolts. The need for a cull on cormorants in response to this threat is an ongoing subject of debate. Historically, the cormorant was hunted in Europe to the verge of extinction. Breeding primarily along coastal cliffs, numbers of cormorants equate to approximately 5,000 breeding pairs. [Source: Shay Connolly, Courtesy of BirdWatch Ireland]

Ireland's offshore marine environment is vast (some 900,000km²), and is approximately three times the size of Germany. While to the casual observer it may appear to be a featureless expanse of the North Atlantic, it is a highly complex, three-dimensional dynamic environment, brimming with a multitude of resources. These include abundant fisheries, plants, mineral deposits, oil and gas reserves and almost unrivalled renewable energy generation potential through wind, wave and tidal currents (see Chapters 26: Coastal Fisheries and Aquaculture, 28: Renewable Energies: Wind, wave and tidal power and 29: Coastal Mining, Quarrying and Hydrocarbon Exploration). Ireland's marine waters also support an impressive range of habitats and associated biodiversity, including important spawning areas for fish, internationally important numbers of breeding seabirds and a high diversity of whales and dolphins (Figure 3.1).

In recent years, offshore development of marine resources has gained considerable interest and is of clear importance to Ireland's future. However, Irish coastal waters are probably the least understood of all Ireland's habitats, necessitating a cautious approach to development and dedicated research into the distribution and ecology of marine species. Under an Ecosystems-Based Approach to resource management, economic benefit must be balanced with environmental impact, in order to ensure the continued health of coastal and marine ecosystems for future generations.

This chapter begins by introducing the unique set of conditions that enables Irish coastal waters to support a large and diverse range of marine species. The following sections discuss ongoing research into how marine and coastal biodiversity is distributed and how animals and plants interact with human activities, with a focus on predators at the top of the food chain. A conclusion offers some insights into how humans and marine biodiversity can effectively co-exist.

ENVIRONMENTAL DRIVERS OF BIODIVERSITY

Ireland's coast is influenced by the North Atlantic Current (Drift), a north-easterly directed component of the Gulf Stream, which moves heat from the equator to the Arctic (Figure 3.2). The meeting of this warm water with the western European continental shelf results in much warmer conditions than would be expected for Ireland's northerly location. Coupled with this is the role played by the steep break associated with the continental shelf. Here, the disturbances caused by ocean currents allow them to interact with the seabed to force deep, nutrient-rich waters upwards. When these waters reach shallower depths where light penetrates, the combination of nutrients, sunlight and warm ocean temperatures stimulates the rapid growth of small, free-floating organisms (phytoplankton). These microscopic plants form the base of the

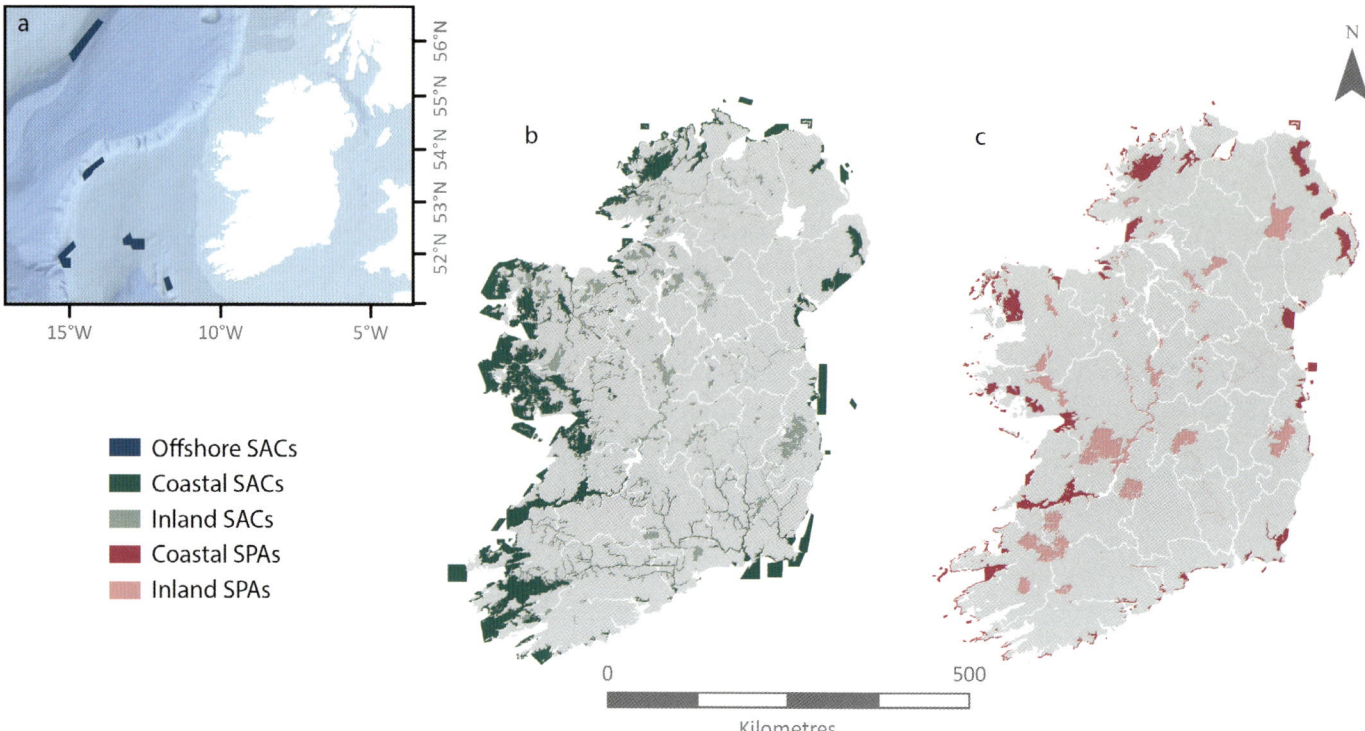

Offshore SACs
Coastal SACs
Inland SACs
Coastal SPAs
Inland SPAs

Fig. 3.1 AREAS PROTECTED FOR CONSERVATION OF HABITATS AND SPECIES AROUND THE COAST AND FURTHER OFFSHORE. An impressive range of important natural habitats and associated biodiversity in the coastal zone are designated for protection as a. Offshore Special Areas of Conservation, b. Special Areas of Conservation, and c. Special Protection Areas. Often referred to as SACs or SPAs, these designations form part of the European Union's Natura network of conservation areas and stem from directives that restrict damaging activities and promote best practice in monitoring and protection of plants and animals. As the maps show, these networks occur all around the coast and there is hardly a 50km stretch of coast without some form of conservation designation. These include wetlands habitats, sand dunes, machair, estuaries and bays. In addition, species such as salmon, otters, freshwater pearl mussels and others, because of their importance at Irish and European levels, are afforded protection. SPAs, designed for the protection of birds, cover productive intertidal zones, bays, estuaries and wetlands that provide food resources for waterbirds, seabirds, ducks, divers and grebes. Every summer, several species of seabird breed along the coast on cliffs and offshore rocks and islands, which provide access to the rich waters of the continental shelf. Approximately 80 per cent of Ireland's SPAs are designated for breeding seabirds and overwintering waterbirds. [Data source: Republic of Ireland SACs and SPAs map data from NPWS; Northern Ireland SAC map data from Joint Nature Conservation Committee, Department of Environment Food, and Rural Affairs; Northern Ireland SPA map data from Department of Agriculture, Environment and Rural Affairs]

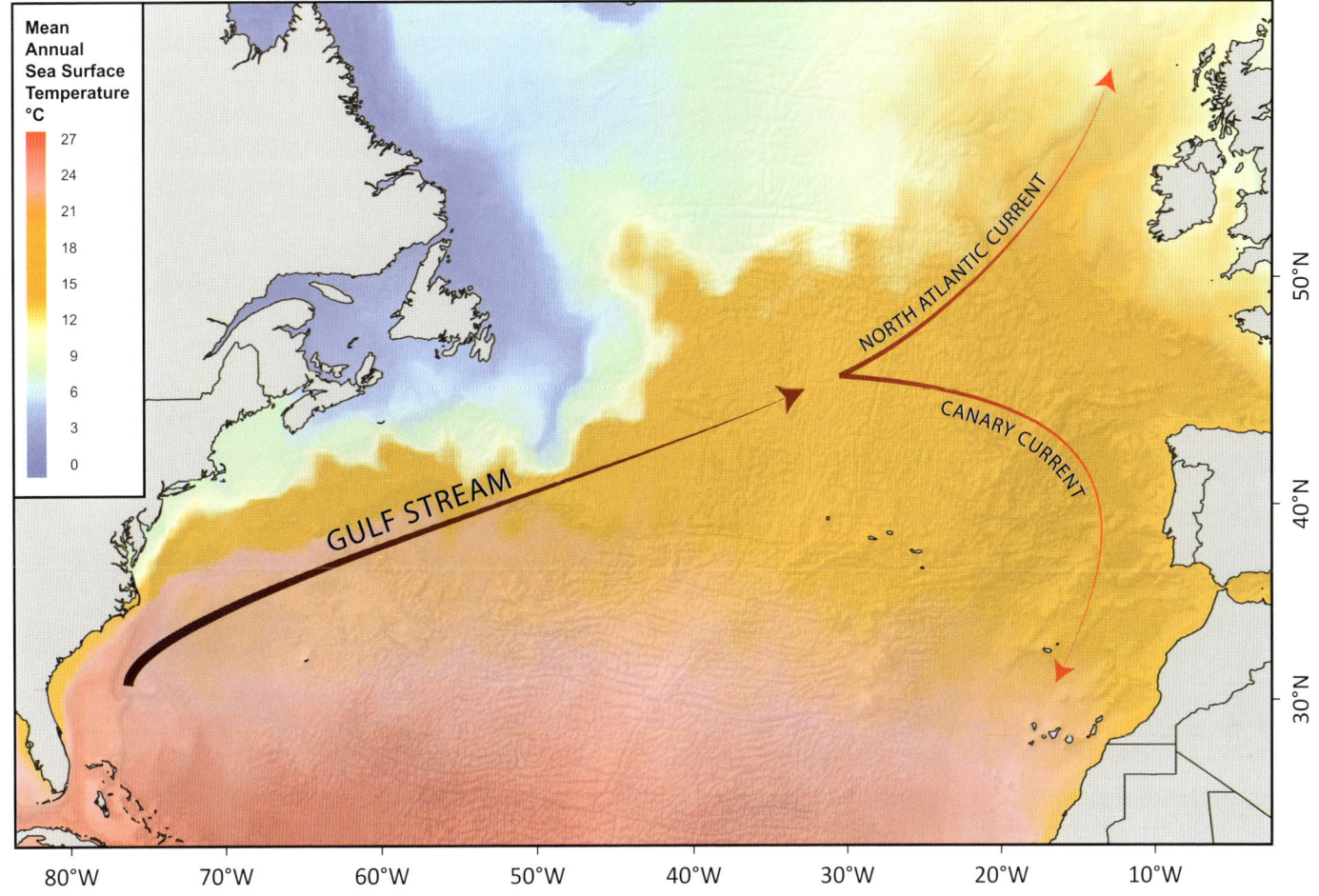

Mean
Annual
Sea Surface
Temperature
°C

27
24
21
18
15
12
9
6
3
0

GULF STREAM

NORTH ATLANTIC CURRENT

CANARY CURRENT

50°N

40°N

30°N

80°W   70°W   60°W   50°W   40°W   30°W   20°W   10°W

Fig. 3.2 THE INFLUENCE OF THE GULF STREAM ON MARINE BIODIVERSITY IN IRELAND. Ireland's location at the edge of the continental shelf means that the country is influenced heavily by the Gulf Stream, which creates warmer ocean temperatures than at comparable latitudes. This has significant implications on the distribution and abundance of marine life, particularly off the west coast, making Ireland's coastal waters and seas among some of the richest temperate areas for marine plants and animals, from microscopic phytoplankton at the bottom of the food web, to charismatic marine megafauna such as whales and dolphins. The North Atlantic ocean's circulation system, which includes the Gulf Stream, is weakening as a result of increased temperatures associated with climate change. This could trigger a sudden change in Ireland's climate, which could have major implications for marine biology as we currently know and understand it.

marine food chain. Large phytoplankton blooms that are visible from satellites occur regularly off Ireland's south-west coast, providing a feast for zooplankton (free-floating microscopic animal life) and numerous small fish species (Figure 3.3). These, in turn,

Fig. 3.3 (right) EXTENSIVE PHYTOPLANKTON BLOOM OFF THE WEST COAST OF IRELAND. A dramatic and extensive blue-green phytoplankton bloom is evident along the Irish continental shelf in the Atlantic ocean. Marine phytoplankton are composed mainly of microscopic marine algae, known as dinoflagellates and diatoms, and these form the basis of the marine food chain. A well-balanced ecosystem depends on these primary producers as a source of food for everything from zooplankton to small fish, invertebrates and even whales. Phytoplankton live near the surface of the sea as they require sunlight to grow. Blooms, which are high concentrations of phytoplankton, occur when conditions for their growth are optimal, including the availability of sunlight and inorganic nutrients such as nitrogen and phosphorous. Blooms can restrict the penetration of light lower in the water column, causing adverse effects for other marine species. They have been known, for example, to create mass mortality by clogging the gills of fish. However, not all phytoplankton blooms are harmful. As well as being vital to the food web, phytoplankton play an important role in the carbon cycle. Through the process of photosynthesis, phytoplankton absorb and store carbon dioxide from the atmosphere and release oxygen. Rainforests are responsible for one-third of the Earth's oxygen, while over two-thirds of the oxygen in the atmosphere is produced by marine plants, including phytoplankton. The image was acquired by the Moderate Resolution Imaging Spectroradiometer (MODIS) on NASA's Aqua satellite on 2 June 2006. [Source: NASA image courtesy Jeff Schmaltz, MODIS Land Rapid Response Team at NASA GSFC. Public domain, via Wikimedia Commons]

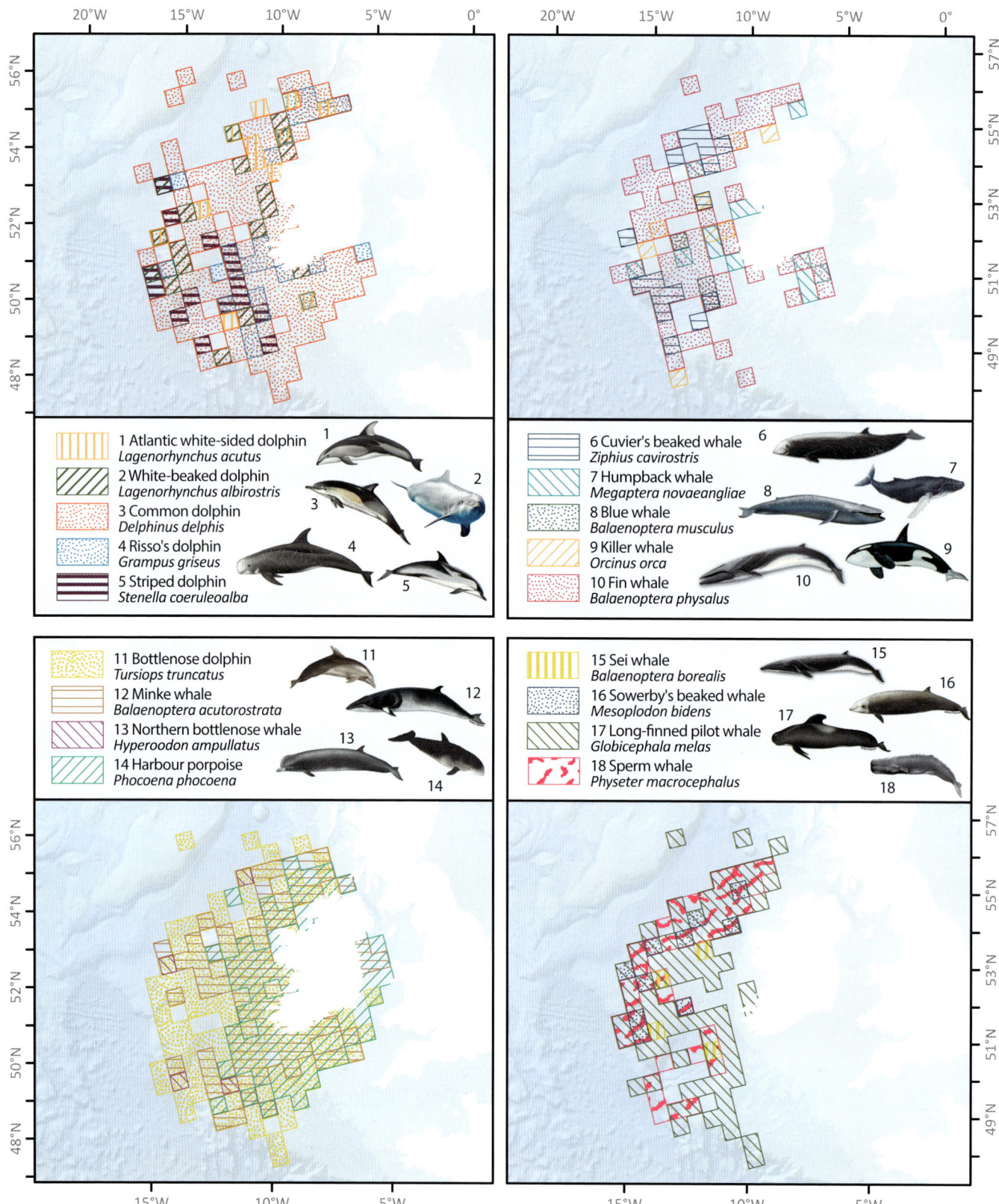

Fig. 3.4 IRISH WHALE AND DOLPHIN SANCTUARY BOUNDARY AND DISTRIBUTION OF SPECIES. To date, twenty-five species of cetaceans have been identified in Irish waters, including the larger whales, such as fin whales. In 1991, the government of Ireland declared Irish waters as a whale and dolphin sanctuary, through which the hunting of all whale species, including dolphins and porpoises, was banned to 200 nautical miles from the coast. This is significant as Ireland is one of the most important places in Europe for cetaceans (whales, dolphins and porpoises). The coastal zone provides a breeding and feeding ground for dolphins, such as the common dolphin and the bottlenose dolphin, and harbour porpoises, as well as for some smaller whales such as the minke whale. Larger whales, such as the fin whale, can also be observed relatively easily from headlands during settled conditions. The maps depict eighteen different species observed from sighting records collected by the National Parks and Wildlife Service to comply with statutory reporting requirements for Annex II and Annex IV species under the European Habitats Directive. Distribution of individual species is mapped according to 50km grid cells, from data recorded during national fisheries surveys, dedicated cetacean surveys, or during seismic surveys. Data used to generate these distribution maps does not include shore-based sightings. The spatial distribution of cetaceans arising from these data shows that a relatively limited number of species make their way to the Irish Sea. This may be a feature of migration patterns and feeding grounds. However, the bottlenose dolphin, the minke whale, the northern bottlenose whale and the harbour porpoise have all been recorded in Ireland's coastal waters. [Data source: National Parks and Wildlife Service]

Fig. 3.5 THE ELUSIVE OTTER. One species of otter occurs in Ireland (*Lutra lutra*). Although widespread along the coast and in rivers and lakes, the otter is one of the most elusive of our native mammals. This shy and charismatic mammal preys on a variety of fish, eels, frogs and crayfish, as well as shellfish. They live where they can access a good source of food. Otters also need access to freshwater to wash their coats, which means that they tend to occupy parts of the coast with a freshwater stream closeby. Their habitats include 'couches', which are secluded resting places made amongst dense scrub or reed beds. They also occupy a network of 'holts', which are below ground in tunnels, caves and amongst rocks. Survey data indicates a decline in otter numbers since the first national survey in the 1980s. This trend could be the result of land drainage and water pollution. The population is estimated at about 12,000 animals. [Source: Stephen Dunbar]

are eaten by larger predatory fish, marine mammals, seabirds and even jellyfish. While an estimated 1.4–1.6 million species inhabit the open sea worldwide, the exact number of species in the Irish context is unknown.[1] Irrespective of this, Ireland's unique marine environment is known to support a remarkable diversity of plant and animal life.

An estimated 375 species of fish live in Irish coastal waters, which are the preferred food choice of a number of predators, such as seabirds and marine mammals.[2] At least forty-five species of seabirds have been recorded during ship-based surveys with twenty-three of these breeding regularly around the coast.[3] In addition to two species of seal – grey seal (*Halichoerus grypus*) and the harbour or common seal (*Phoca vitulina*) – twenty-five cetacean (whale and dolphin) and nineteen species of shark occur. The importance of

these waters was recognised in their designation as a sanctuary for whales and dolphins in 1991, the first of its kind in Europe (Figure 3.4). Five species of turtle have also been identified around the coast and, although all can be classified as migratory species that breed in tropical waters, they are often recorded as they stray close to the west coast of Ireland during their migration. In particular, adult leatherback turtles (*Dermochelys coriacea*) may well spend more time foraging in cool temperate waters, where their preferred jellyfish prey are more abundant, than in tropical waters. Irish coastal wetlands are also a vital resource for over fifty species of waterbirds that migrate annually to Ireland, providing food and protection for over three-quarters of a million of these waterbirds each year. The otter (*Lutra lutra*), an elusive and charismatic mammal, also occurs in rivers and along the coast (Figure 3.5).

Case Study: Waterbirds in Irish Coastal Areas

John Quinn, Brian Burke and Seán Kelly

Ireland's isolation on the edge of the Atlantic is no barrier to the migratory populations of waterbirds found along its extensive coastline (Figure 3.6). More than any other avian group, it is perhaps the waterbirds for which Ireland is especially important in terms of international populations.

One of the most striking features of waterbirds is the diversity of niches that they occupy (Figure 3.7). In Ireland, they are found on sandflats, mudflats, open estuarine or coastal waters, coastal lagoons (artificial and natural) and saltmarshes. Some species also occur in substantial numbers across inland wetland and terrestrial habitats. Waterbirds feed on a diverse range of food types – vertebrates, invertebrates and plant species – and these occur at different depths in the soil or water throughout these habitats. Over evolutionary time, this marked variation in food diversity has led to dramatic adaptive features in these birds. This is most apparent in two key anatomical features – their legs and bills – which play a key role in determining in which habitats waterbirds live.

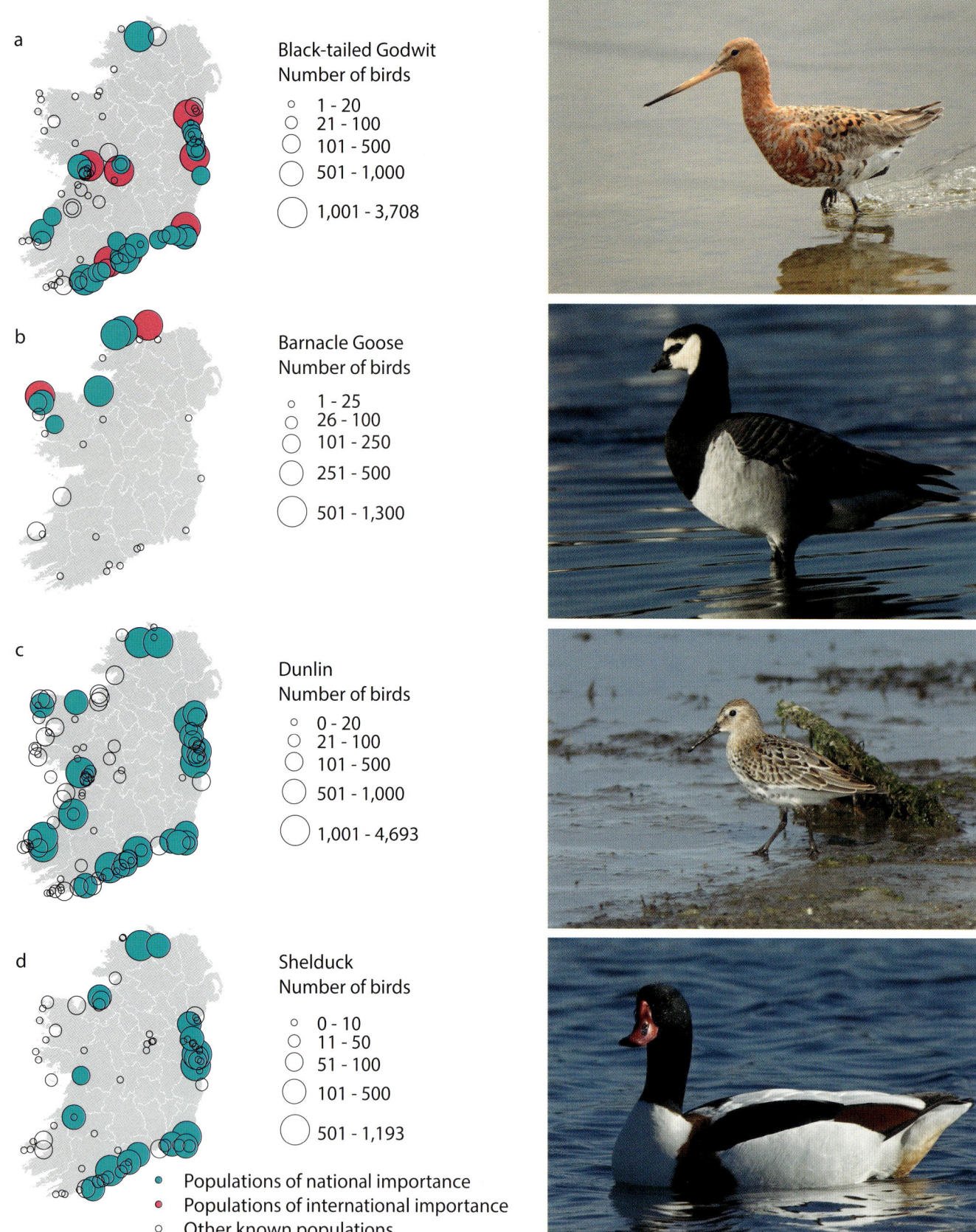

a Black-tailed Godwit
Number of birds

○ 1 - 20
○ 21 - 100
○ 101 - 500
○ 501 - 1,000
○ 1,001 - 3,708

b Barnacle Goose
Number of birds

○ 1 - 25
○ 26 - 100
○ 101 - 250
○ 251 - 500
○ 501 - 1,300

c Dunlin
Number of birds

○ 0 - 20
○ 21 - 100
○ 101 - 500
○ 501 - 1,000
○ 1,001 - 4,693

d Shelduck
Number of birds

○ 0 - 10
○ 11 - 50
○ 51 - 100
○ 101 - 500
○ 501 - 1,193

● Populations of national importance
● Populations of international importance
○ Other known populations

Fig. 3.6 THE DISTRIBUTION OF FOUR COMMON WATERBIRDS AROUND THE COAST OF IRELAND. Black-tailed godwit, barnacle goose, dunlin and shelduck are among the most common waterbirds of Ireland. The maps show the distribution of each species around the coast from data available for the Republic of Ireland: a. black-tailed godwit in breeding plumage [Source: Liam Kane, courtesy of BirdWatch Ireland]; b. barnacle goose [Source: Clive Timmons, courtesy of BirdWatch Ireland]; c. juvenile dunlin, Tacumshin, County Wexford [Source: Tom Shevlin]; d. shelduck [Source: Clive Timmons, courtesy of BirdWatch Ireland]. For each species, populations of national importance (indicated by the blue circles) can be found at specific coastal sites. For barnacle geese, significant numbers of birds tend to be concentrated in the north-west of the country, whereas for other species, significant populations are more widely distributed. Barnacle geese and black-tailed godwit occur on some sites in internationally significant numbers (pink circles). For example, the black-tailed godwit can be seen in large numbers wintering in locations such as Wexford Harbour and Slobs, Dublin Bay and Dundalk Bay. [Map data source: The Irish Wetland Bird Survey (I-WeBS), a scheme funded by the National Parks and Wildlife Service (NPWS), and coordinated by BirdWatch Ireland]

Herons and egrets (from the family *Ardeidae*) have long legs that allow them to wade in relatively deep water, while their pointed bills minimise splash when stabbing at prey in water (Figure 3.8). Cormorants (*Phalacrocoracidae* sp.) and divers (*Gavia* sp.) – for example, the great northern diver (*Gavia immer* – are predominantly piscivorous and pursue their prey underwater in tidal channels and coastal lagoons, using their webbed feet for propulsion and their sharply pointed or hooked bills to catch prey. Members of the rail family are not known for their strong swimming ability. However, one of their best-known members, the coot (*Fulica* sp.), feeds on sometimes deeply submerged aquatic vegetation. They also graze like wildfowl and are readily seen in large flocks on grass swards. Two other species of rails, the common moorhen (*Gallinula chloropus*) and water rail (*Rallus aquaticus*), are much more secretive and solitary, confining themselves to foraging amongst reedbeds, most commonly alone.

Leg length among the true 'waders' – including oystercatchers (*Haematopus ostralegus*), plovers (*Charadriinae* sp.) and sandpipers (*Scolopacidae* sp.) – varies even more dramatically than in herons, allowing some species, such as the black-tailed godwit (*Limosa limosa*) (Figure 3.6a), to wade in quite deep water. Others like the dunlin (*Calidris alpina*) (Figure 3.6c) and sanderling (*Calidris alba*) straddle the tideline to feed on the wet mud, venturing only briefly into very shallow water. Yet others, including the ringed plover (*Charadrius hiaticula*) and the grey plover (*Pluvialis squatarola*), feed primarily on mudflats and avoid the water's edge.

Differences in bill shape also occur within waders. The redshank (*Tringa totanus*) and dunlin (Figure 3.6c) have relatively small bills and feed on small surface invertebrates. In contrast, the long-billed bar-tailed godwit (*Limosa lapponica*) can feed on invertebrates located relatively deep in the mud or sand, including the sand mason worm.

Bill shape also varies somewhat among waders. Although most are straight and pointed, allowing

Fig. 3.7 VARIABILITY IN WATERBIRD HABITATS, FROM SANDFLATS TO FEEDING GROUNDS AMIDST COASTAL GRASSLANDS. One of the most striking features of waterbirds is the diversity of niches that they occupy. In Ireland they tend to be found on sandflats, mudflats, open estuarine or coastal waters, coastal lagoons (artificial and natural) and saltmarshes. The image shows a. oystercatchers resting on the shores of Cow Strand, County Cork; [Source: Robbie Murphy], b. barnacle geese feeding in coastal fields on the Inishkea Islands, County Mayo. [Source: David Cabot].

easy penetration of the substrate, others are subtly or more obviously curved. The downward-curved bill of the curlew (*Numenius arquata*) is thought to allow them to extract lugworms from their U-shaped burrows. In contrast, the upward-curved bill of the greenshank (*Tringa nebularia*) facilitates catching fish and crustaceans as they swish their bills back and forwards in the shallows.

Surprisingly, bill shape can display major differences within species too. In Dublin Bay and across Europe in general, individual oystercatchers with blunt-ended bills tend to hammer open cockles and mussels, while those that have pointed bills stab open the cockles through the gap between the two shells. Wintering oystercatchers feed predominantly on estuaries, but switch to terrestrial habitats when midday high tides limit their feeding time on intertidal mudflats, or when rain makes grassland soils more penetrable and brings earthworms closer to the surface (Figure 3.7a). Although the majority of waders spend most of their time on the coast, that is not true of all species. The lapwing (*Vanellinae*) and golden plover (*Pluvialis apricaria*) feed primarily on grasslands and use coastal estuaries mainly as roosting areas.

With their relatively short, robust bills and their strong legs for walking, many wildfowl species can be found grazing or grubbing on Ireland's coastal grasslands, including whooper swans (*Cygnus cygnus*) and geese from the family *Anatidae*. The barnacle goose (*Branta leucopsis*) (Figure 3.6b) feeds along the western seaboard on natural machair, saltmarsh and agricultural coastal grasslands, especially on the Inishkea

Fig. 3.8 GREY HERON IN KINISH HARBOUR, SHERKIN ISLAND, COUNTY CORK. The grey heron is one of the most distinctive birds in Ireland, with its long neck and legs, razor-sharp bill and head crest. It can be seen gracefully stalking prey in shallow waters, poised almost motionless, until it strikes rapidly to catch fish, frogs and other amphibians. It will also feed on other small birds and small rodents. The grey heron is distributed widely in coastal waters, among wetland habitats and estuaries. They are often encountered as solitary birds, but may also occur in pairs and in breeding colonies in groups of up to fifty birds. [Source: Robbie Murphy]

Islands off County Mayo (Figure 3.7b). So too does the brent goose (*Branta bernicla*), which is also well known for feeding on the eelgrass beds and algae exposed at low tide on some of Ireland's estuaries. This is most apparent on Strangford Lough, where they spend several weeks feeding when they first arrive in autumn, before dispersing around the Irish coast for the rest of the winter.

Ducks are generally omnivorous and dabble for invertebrates and seeds in muddy substrates. The wigeon (*Mareca*) is an exception and on the coast grazes primarily on saltmarsh graminoids, or on algae and eelgrass on mudflats. Although most ducks use inland and coastal habitats equally, others such as the shelduck (*Tadorna*) (Figure 3.6d) are tied more strictly to mudflats, collecting enormous numbers of the tiny hydrobia as one of its main food items. Other coastal ducks include the eider duck *(Somateria mollissma)* and the greater scaup (*Aythya marila*). These species' legs are set far back on the body and are poor for walking. Instead, they dive for their main food, benthic molluscs. As a collective, therefore, waterbirds are a diverse group of species that are well adapted to exploiting Ireland's diverse coastal habitats.

THE ORIGIN OF IRELAND'S WATERBIRDS

Some of Ireland's waterbirds occur in coastal areas only during winter months, spending the summer nesting usually away from the coast on tundra landscapes in northern latitudes. Many of these – such as the black-tailed godwit (Figure 3.6a), greylag goose (*Anser anser*) and dunlin (Figure 3.6c) – breed in Iceland. Others head as far as Greenland and the Canadian Arctic, including brent geese (Figure 3.9), barnacle geese (Figure 3.6b), knot (*Calidris canutis*) and grey plover. To the east, many breed in northern Scandinavia, especially wildfowl, including the wigeon and teal (*Anas crecca*) and several of our waders. Others, such as bar-tailed godwits and Bewick's swans (*Cygnus colombianus bewickii*), travel as far as the Russian Arctic. The sound of the whimbrel's (*Numenius phaeopus*) urgent and distinctive call on our coasts is a sure sign that the seasons are changing. Whimbrel pass through Ireland in the spring and autumn on their way from Africa to breed in Iceland.

Understanding where Ireland's waterbirds originate is often complicated by the fact that many independent populations or races of the same species are found on our coastal wetlands at different times of the year. Eurasian curlews (*N. Arquata*), for example, come from two distinct populations: those from Scandinavia and Iceland, and those that now breed in very small numbers within Ireland. The same is true for several of Ireland's smaller wader species, such as the ringed plover, while our wintering sanderling originate in Siberia. Sanderlings are both a passage migrant and a winter visitor, with a small number of birds choosing to stay in Ireland on the passage to or from West Africa. The dunlin (Figure 3.6c) has a particularly complex biogeography: those from the *alpina* population breed in Scandinavia/Russia and winter in Ireland; those from the *schinzii* breed in Ireland and the UK, migrating to southern Europe and north-west Africa for winter; while *arctica* breed in north-east Greenland, passing through Ireland en route to wintering grounds in Africa. Thus, our coastal waterbirds are a diverse mix of species, not just in terms of their taxonomy and ecology, but also in terms of their origins and migratory strategies. This makes them a fascinating but difficult group of birds to conserve.

DISTRIBUTION AND IMPORTANT COASTAL WETLANDS

Waterbirds are distributed all around the Irish coast and are monitored each winter, from September to March, on coastal (and inland) wetlands (the Irish Wetland Bird Survey or 'I-WeBS').[4] They are also surveyed occasionally along the coast itself, on rocky shores, beaches and in shallow marine waters (Figure 3.7a) (the non-Estuarine Coastal Waterbird Survey or 'NEWS').[5] Most waterbirds are found on estuaries and lagoons, but the rocky shorelines, sandy beaches and shallow coastal waters hold important numbers of divers, cormorants and several species of wader, especially purple sandpiper (*Calidris maritima*), sanderlings, ringed plovers, turnstones (*Arenaria* sp.) and oystercatchers. The largest concentrations of waterbirds occur on the east and south coasts of the country (Figure 3.6), mainly because estuaries here are more extensive and more productive, due to higher sediment availability, higher nutrient input from extensive, fertile catchment areas and from partially treated sewage emanating from large settlements of human populations.

Many waterbirds use saltmarshes, but Merne and Robinson[6] suggest that of an inventory of 259 saltmarshes made in 1998,[7] only a small proportion are important for waterbirds. These include North Bull in Dublin and Bannow Bay in County Wexford, with a few others on the south and north coasts. Few saltmarshes on the relatively exposed and less productive west coast are good for waterbirds, notable exceptions being in Tralee Bay and on parts of the Shannon. Of the one hundred coastal lagoons around Ireland's coast, three are particularly important for waterbirds: Tacumshin and Lady's Island lakes in County Wexford and Inch in County Donegal, the latter two being especially important for nesting terns in summer, notably roseate terns (*Sterna dougallii*) in Lady's Island.

Important sites for waterbirds are identified using criteria set out by an international treaty for the protection and sustainable use of wetlands, the Ramsar Convention. These are used by Wetlands International to identify Important Bird Areas (IBAs) and guide EU legislation (see Chapter 12: Coastal Wetlands). One criteria for identifying an IBA is that it regularly holds more than 20,000 migratory waterbirds (or 10,000 pairs of seabirds); in Ireland, twenty-two estuaries and coastal lagoons meet these criteria.[8] In the Republic of Ireland, however, the number of designated IBA sites has been declining for a number of years and now only five sites remain. Almost half of Ireland's sites that hold more than 20,000 wildfowl are important primarily for the light-bellied Brent goose, a population that breeds in the high Canadian Arctic and winters almost exclusively in Ireland (Figure 3.9). Six others – Strangford Lough, Dundalk Bay, Wexford Harbour and Slobs, Lough Foyle, Dublin Bay and Lough Swilly – are internationally important for a range of species. Locations such as the Wexford Harbour Slobs provide visitors with unique birdwatching experiences. In this case, thousands of white-fronted geese (*Anser albifrons*) can be observed moving in huge flocks, from October to April, to and from the sandbanks in the harbour to roost.

Another criteria for identifying important areas is when a site holds 1 per cent of the international biogeographic or 'flyway' population of a single species. Flyways are defined by the African-Eurasian Migratory Waterbird Agreement (AEWA), and embrace two key features of waterbird ecology. One is that they ignore national borders. The other is that populations that are essentially independent should be treated as distinct conservation units, deserving protection in their own right. The 1 per cent principle for identifying important

Fig. 3.9 LIGHT-BELLIED BRENT GEESE IN FLIGHT. Light-bellied Brent geese are a winter migrant from High-Arctic Canada, arriving in Ireland typically between the months of October and April. The majority of these wintering migrants can be found in areas such as Strangford Lough and Lough Foyle, Northern Ireland. They differ from the dark-bellied Brent goose in their features, breeding and migratory paths. These geese feature dark underbellies, as the name suggests, and breed in northern Russia, from where they migrate to southern and eastern England during the wintering season. Brent geese migrate in family groups and are seldom seen flying in the characteristic 'v'-shaped formation as with other species. Light-bellied Brent geese are found typically on coastal estuaries where they feed primarily on eelgrass, but when such sites become depleted, they often relocate to grasslands from mid-winter until their departure. Following their departure in April, light-bellied Brent geese fly to Iceland, where they increase their weight by up to 40 per cent, which prepares them for the 3,000km journey to their Canadian breeding grounds. In terms of species conservation, the light-bellied Brent goose is a protected species listed as amber among the birds of conservation concern in Ireland. [Source: Richard Mills, Courtesy of BirdWatch Ireland]

sites for a species is, furthermore, independent of the 20,000 bird threshold. So, for example, sites that hold fifty great northern divers are considered to be internationally important because only 5,000–6,000 are thought to occur in Europe. For the well-known mute swan (*Cygnus olor*), the Irish national population is the flyway population because mute swans in Ireland are sedentary. In contrast, the threshold for the lapwing (*Vanellus* sp.) is 72,300 and there are no sites in Ireland that hold this number in winter.

Though the highest concentrations of waterbirds occur in a few estuaries around the east, north and south coasts, most of our waterbird species are scattered around the entire Irish coast, with some notable exceptions. Two of the three main sea ducks, the greater scaup and common eider, are restricted primarily to the north of the country. In contrast, barnacle geese (Figure 3.6b) are found mainly to the west and north-west, while white-fronted geese and Icelandic greylag geese are restricted to a handful of widely scattered sites around the country, even though 'feral' greylag geese are found more widely. Bewick's swans now use only two sites in Ireland. Both occur in County Wexford and are used by the same individuals, pointing to their precarious existence in this country.

POPULATION TRENDS AND CONSERVATION ISSUES

Waterbird population trends at the international flyway level are known only partially for many species. Nevertheless, it appears that about twenty-one species are showing favourable trends (stable or increasing), although another twenty-four have unfavourable trends (decreasing). In Ireland we can be more confident of trends because a network of volunteers, as well as staff from both the National Parks and Wildlife Service and BirdWatch Ireland, undertake diligent counts of waterbird populations in most of our coastal areas (and inland sites). This operation has occurred since the 1990s, funded by the National Parks and Wildlife Service and coordinated by BirdWatch Ireland. Recent estimates suggest that from the mid-1990s to the period ending in winter 2015/16, four species are stable, while a further seven are increasing in numbers. In contrast, an enormous thirty-one species have shown declines of 5 per cent or more, twenty-three of which exhibit rates of decline exceeding 30 per cent.[9] Estimates for a further fifteen species are conservative because, for various reasons, they are under-recorded.

Common threats facing our waterbird populations include development, recreational disturbance, mariculture and invasive plants on saltmarshes. Evidence suggests that climate change is also playing a major role, with many species that normally migrate to western Europe 'short-stopping', or choosing to winter further east as European winters become milder. This has been marked in particular for the Bewick's swan. The challenge, however, is to distinguish between loss of birds from an area due to international shifts in winter distribution and declines in flyway population numbers caused by lower survival rates or poor breeding productivity on the Arctic breeding grounds. It remains unclear whether rising sea levels caused by climate change will lead to loss of habitat due to 'coastal squeeze'.[10] Nevertheless, seventy-one low-lying sites of importance are now thought to be vulnerable to increasing sea levels,[11] while climate change has been cited as a 'medium-level' threat to eleven wader and two wildfowl species in Ireland's Article 12 report to the EU commission.[12]

Another potential issue facing waterbirds in coastal Ireland is the effect of the EU Water Improvement Framework on improved wastewater treatment. Ironically, poor treatment in the past has led to increased nutrient input and productivity in estuaries. Although this

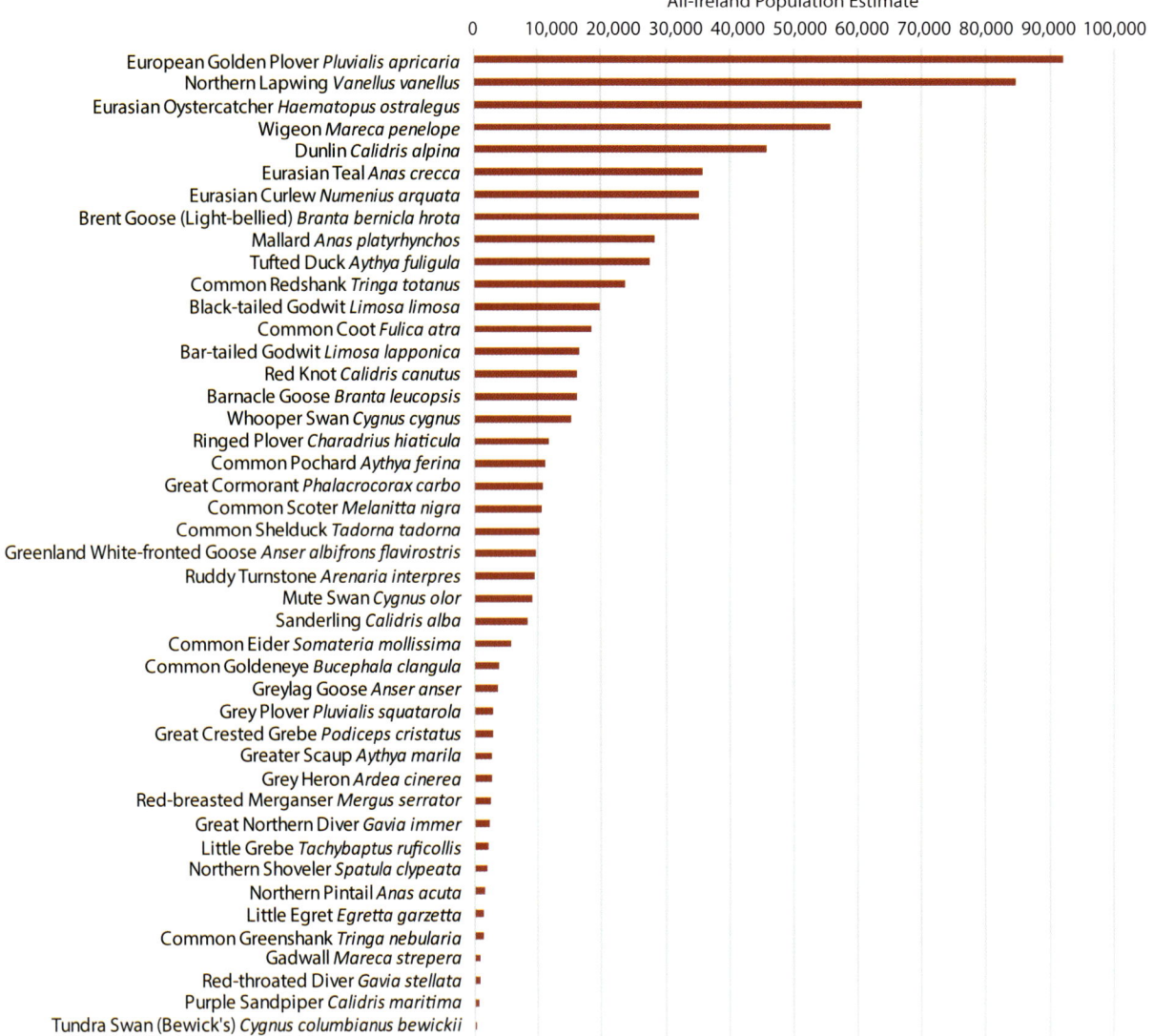

Fig. 3.10 POPULATION ESTIMATES OF IRISH WATERBIRD SPECIES. More than any other avian group, it is perhaps the waterbirds for which Ireland is especially important in terms of significant populations. The highest concentrations of waterbirds occur in a few estuaries around the eastern, northern and southern coasts. One such location, Wexford Harbour's North Slob, is one of the most renowned wildfowl reserves in the country. The arrival of 10,000 Greenland white-fronted geese each autumn (about one-third of the world population of this species) is an annual spectacle. Nevertheless, most waterbird species are scattered around the entire Irish coast. Ireland's coastline supports a tremendous diversity of waterbirds, some of them in internationally important numbers. High population densities are noted for species such as the European golden plover, the northern lapwing and the eurasian oystercatcher. Throughout the winter, the golden plover can be seen regularly in large, dense flocks, distributed all around the coast. Other species occur in very small numbers, such as the Bewick's swan, the purple sandpiper or the red-throated diver. Numbers of the Bewick's swan (the smallest of the three white swan species occurring in Ireland), have declined from over 1,000 to just 300 in the last swan census in 2005. [Data source: Burke et al., 2018]

has benefited many waterfowl, it remains unclear whether the net impact of improving quality is necessarily negative because waterbirds may compensate by switching to other prey items in the system.

Ireland's coastline supports a tremendous diversity of waterbirds, some of them in internationally important numbers (Figure 3.10). Although we have some cause to be optimistic because of the protection afforded by EU legislation, clearly their populations face many threats, some of which will be extremely difficult to control.

DISTRIBUTION OF ANIMALS AND PLANTS IN COASTAL WATERS

Understanding the distribution of habitats and species in Ireland's coastal waters is challenging. The large area covered by coastal waters makes surveying – such as the national bird counts undertaken by BirdWatch Ireland – logistically demanding and potentially expensive. In addition, being located on the Atlantic margin, the Irish coast is subject to extremes of weather, particularly in winter months, including high winds and waves, that often prevent surveys from being undertaken at all. A variety of methods have, therefore, been employed to determine the distribution of species in coastal and oceanic waters, focusing especially on land-based observations,

large-scale ship or aerial surveys and a combination of tracking and species distribution modelling studies. The data obtained through these methods are important for informing conservation legislation as well as marine industry licensing requirements (see Chapter 32: Coastal Management and Planning).

Early understanding of the occurrence of many species was influenced primarily by visual observations from land-based vantage points. Pioneering Irish scientists, such as Maude Delap and Ellen Hutchins, used their passion for natural history to develop new knowledge of jellyfish and marine plants respectively from their observations at the coast.

Maude Delap

Damien Haberlin

Maude Jane Delap (1866–1953) was born in Templecrone, County Donegal, and moved to Valencia Island, County Kerry, at the age of eight (Figure 3.11). She and her sisters became avid collectors of all marine animals, but it was in the area of jellyfish that Maude left an indelible legacy in the world of marine biology. She was the first person to culture jellyfish in aquaria and to describe fully the complex lifecycles of several species of jellyfish. In 1928, the rare sea anemone *Edwardsia delapiae* was named in honour of Delap and, in further recognition of her research, she was made an associate of the Linnean Society of London in 1936. Delap worked throughout her life, publishing her last article in the *Irish Naturalist* in 1924. Some of the specimens she collected can still be seen in the Natural History Museum in Dublin.

Fig. 3.11 MAUDE DELAP (FAR RIGHT) AND HER SISTERS, MARY AND CONNIE, AT HOME IN REENELLEN, KNIGHTSTOWN, VALENCIA ISLAND, COUNTY KERRY. [Source: Nessa Cronin, courtesy of Valencia Island Heritage Centre]

ELLEN HUTCHINS: IRELAND'S FIRST FEMALE BOTANIST

Madeline Hutchins

If you walk the shores of inner-Bantry Bay and Whiddy Island, you will be following in the footsteps of Ellen Hutchins (1785–1815), Ireland's first female botanist. It was during an eight-year period (1805–1812), however, that she was most active as a botanist and during which she would make a significant contribution to the scientific knowledge of Ireland's non-flowering plants. She lived most of her life in Ballylickey, on the north coast of inner-Bantry Bay, a botanically rich corner of Ireland, but at a time when it had been little explored. Ellen was a remarkable young woman. Her short life was complicated by illness and family circumstances, but she was driven by a desire to be useful and, living remotely, she made connections with people through plants and letters.

Fig. 3.12 WATERCOLOUR DRAWING OF SEAWEED *Fucus asparagoides* (1811) BY ELLEN HUTCHINS, IRELAND'S FIRST FEMALE BOTANIST. Issued for the Ellen Hutchins Festival in Bantry Bay Heritage Week (August 2015), Ellen's drawings are exquisite, with an incredible level of detail and great accuracy. This is a high-quality print with careful colour balancing that does justice to the original. This print is taken from the single drawing owned by the Hutchins family. It was one of the images displayed in Bantry House in the first-ever exhibition of Ellen Hutchins' drawings in the Republic of Ireland (August 2015). The original drawing has been shown in the pop-up exhibitions during the Ellen Hutchins festivals, and at the Boole Library, University College Cork. The print edition is limited to 200, to reflect the 200th anniversary of Ellen's death in 1815. On Ellen's inventory of over 1,100 plants she identified and listed in the Bantry Bay area, she noted that she found *Fucus asparagoides* at Ardnagashel. [Source: Madeline Hutchins]

When James Mackay, botanist in charge of Trinity College Dublin's Botanic Garden, travelled to west Cork in July 1805 on a plant-hunting expedition, he spent a few days with Ellen at Ballylickey. She was delighted to find someone who shared her 'pleasure in plants' and impressed him with her skills and knowledge. He suggested she study seaweeds, showing her how to spread them onto paper as specimens.

James Mackay forwarded Ellen's specimens to botanists making specialist studies of each species and, if new, they described and published them. Within two years, Ellen had found at least seven species new to science or new to Ireland, and went on to find more seaweeds but also mosses, liverworts and lichens. Some plants Ellen found were named *hutchinsiae* by leading botanists as recognition of her contribution to botany. Her specimens, held in herbaria in Ireland, the UK and the USA, are still used for research purposes Her frequent ill health and caring responsibilities for her mother, however, often prevented her from getting out to collect plants and, sometimes for months on end, she would be too weak even to write. She wrote 'working for oneself is very dull, but to doing anything for another gives one spirit to proceed'.[13]

From 1808 onwards, Ellen made watercolour drawings of the seaweeds she found (Figure 3.12). These drawings, which are wonderfully detailed and accurate, were greatly valued by the UK botanist, Dawson Turner, who used some as plates in his book and provided others to fellow botanists for their publications.

Ellen had a great appreciation of the beauty in the plants she studied, calling them 'treasures' and 'exquisite little beauties', and also in the coastal landscape of Bantry Bay and its surrounding mountains. For example, she praised the views from her home:

> There are many pretty peaceful looking spots, few more so than the little place we inhabit. The view from our door is very soft and pleasing particularly at evening when the sun retiring behind a beautiful mountain, Sugarloaf, just gilds the high ground and leaves the little cove before us in deep soft shade, or when the moon shines bright upon it.[14]

She delighted in details seen under her microscope, sometimes spending days admiring the profile of a single seaweed. Ellen died in 1815 at the age of twenty-nine, following a long, final illness. She was buried in an unmarked grave in the old Garryvurcha churchyard in Bantry, where a plaque has now been erected. The annual Ellen Hutchins Festival, held in August in Bantry, celebrates her botany and botanical art.

Historic and contemporary observations by researchers over decades have helped to build a picture of marine ecosystem functioning. For example, observations from strategic locations, such Cape Clear Island, revealed a major migration route of seabirds off the south-west coast. In addition, over 13,000 sightings of marine mammals have been made through the Irish Whale and

Fig. 3.13 MARINE MAMMAL OBSERVERS ON BOARD THE *Celtic Mist*. Our understanding of the distribution of the various plants and animals around the Irish coast is dependent on multiple forms of observations. Some forms are autonomous, but others depend on human observers, such as these marine mammal observer volunteers on board the *Celtic Mist*, who are trained to identify cetaceans in Irish waters by the Irish Whale and Dolphin Group (IWDG). Impressively, over 13,000 sightings of marine mammals have been made through the IWDG since it was established in 1990. The yacht owned previously by former taoiseach, Charles Haughey, the *Celtic Mist*, serves as an excellent platform for cetacean spotting, as wind propulsion mitigates against the acoustic impacts caused by boat engines. As well as research, the IWDG, as a Non-Governmental Organisation (NGO), has advocated actively for the conservation of cetaceans. One of the first major achievements of the fledgling organisation was the declaration of an Irish Whale and Dolphin Sanctuary by the government in 1991. This designation aims to protect the twenty-five cetacean species present in Irish waters, from the elusive harbour porpoise to the largest animal on the planet, the blue whale. [Source: Robbie Murphy]

Dolphin Group reporting scheme (Figure 3.13). This has identified seasonal occurrence and migrations of species, such as humpback whales (*Megaptera novaeangliae*) and fin whales (*Balaenoptera physalus*), along the south coast. The distribution of these species during the non-breeding season is, however, particularly difficult to determine, especially away from inshore waters.

Dedicated ship-based surveys of seabirds and marine mammals were first conducted in Ireland's Atlantic region in 1978, and off the south and west coasts in 1980. These surveys have also been used to record the occurrence of other large marine animals, including sharks and turtles. Also, given the increasing importance attached to understanding marine ecology, the Irish government funded a series of extensive aerial and acoustic surveys (the ObSERVE programme, 2015–2018) covering Irish territorial waters, to assess the abundance and distribution of marine mammals and seabirds. In addition to recording marine mammals and seabirds, the aerial surveys recorded sharks, tuna, jellyfish, turtles and ocean sunfish, providing important information on the distribution and abundance of other marine predators. Such efforts are helping to bridge the gaps in our knowledge of the distribution and habitat use of important top predator species in Irish waters, many of which are considered to be key indicators of ecosystem change.

Information gathered over the past two decades of visual surveys in Irish waters suggests that large, local occurrences (hotspots) for top marine predators occur notably on the edge of the continental shelf, but also in coastal waters off south-west Ireland. The largest of Ireland's marine species – the blue whale (*Balaenoptera musculus*) (Figure 3.14), fin whale and humpback whale – have all been sighted in waters off south and south-west Ireland.

These areas are believed to provide rich foraging grounds, with early autumn herring (*Clupea harengus*) spawning and sprat (*Sprattus sprattus*) shoaling along the south coast, attracting the whales. The coastline around Ireland also supports numerous colonies of seabirds (Figure 3.15). The south-western coast hosts the largest northern gannet (*Morus bassanus*) (Figure 3.16) colony in Ireland (on the island of Little Skellig, County Kerry, with an estimated 35,000 breeding pairs) as well as significant breeding populations of Manx shearwater (*Puffinus puffinus*), European storm petrel (*Hydrobates pelagicus*) and Atlantic puffin (*Fratercula arctica*) (Figure 3.17). This area has also been identified as a hotspot for the ocean sunfish (*Mola mola*).

Fig. 3.14 BLUE WHALES IN IRISH WATERS OFF THE SOUTH COAST. These two blue whales were spotted 80 to 100 nautical miles off Ireland's south coast by the Irish Air Corps. In general, blue whales have been observed off the coast in the summer and autumn, migrating along the continental shelf, often in waters of approximately 700m deep. At almost 30m in length, the adult blue whale is the largest animal known to have existed. Populations of these impressive marine mammals are now recovering slowly following their decimation by whaling over previous centuries. Current anthropogenic threats include ship strikes, entanglement in commercial fishing gear, ocean noise, pollution and plastics and microplastics, although little is known, as yet, about the latter. [Source: Irish Air Corps]

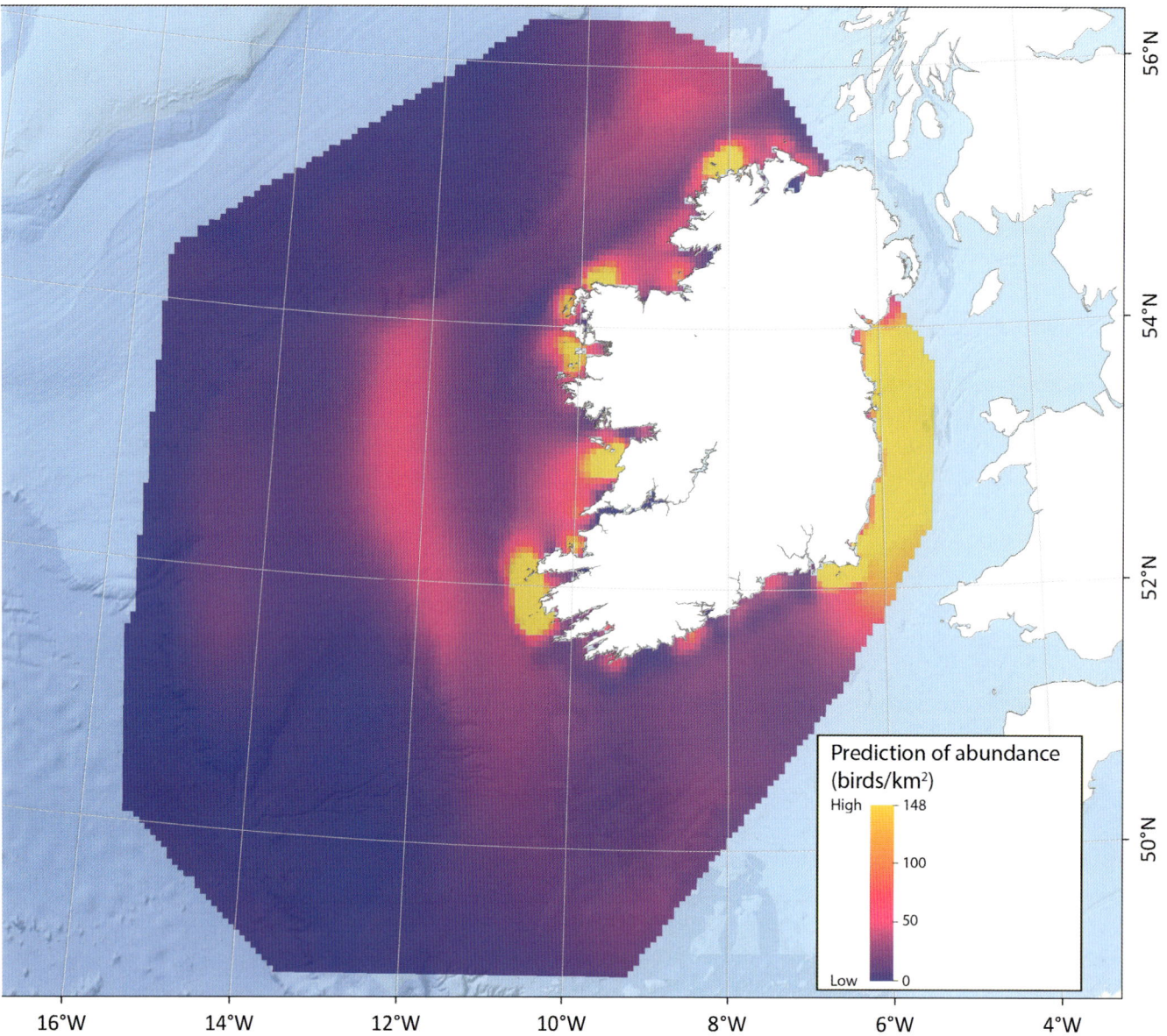

Fig. 3.15 PREDICTED DISTRIBUTION OF ALL SPECIES OF SEABIRDS IN IRISH WATERS. The coastline around Ireland supports numerous colonies of seabirds, by offering an ideal base for large aggregations, with safe nesting sites, which are vital for survival. The coastline also provides close proximity to dependable food supplies for local breeding and non-breeding seabirds, as well as for migrating seabirds. Survival, breeding success and chick growth are related closely to the availability and distribution of prey. Coastal upwelling and frontal systems around the coast enhance primary productivity. This yields a rich source of nutrition, leading to ideal feeding grounds for an abundance of seabirds, as shown along the coastline of the Irish Sea, as well as clusters off the south-west and western coast. For example, the south-west hosts large breeding populations of Manx shearwater, European storm petrel, Atlantic puffin and the northern gannet. The map has been produced based on data collected during the ObSERVE seabird surveys from 2015 to 2016, under the ObSERVE programme promoted by the Department of Community, Climate Action and Environment. This builds on colony data for the predicted foraging radius distributions for seabirds in coastal areas (Seabird 2000 survey data held by the Joint Nature Conservation Council, UK) and surveys conducted between 2015 and 2018 by BirdWatch Ireland, the National Parks and Wildlife Service and University College Cork.

While information on the broad-scale distribution of large marine species can be achieved through visual surveys, smaller species, as well as those that rarely come to the surface to be observed, must be surveyed more directly through the use of net sampling or acoustic techniques. These have shown Irish coastal waters to contain a rich biodiversity. Of note are large populations of fish, the most abundant being the Norway pout (*Trisopterus esmarkii*), a small species (usually 19–20cm in length) and member of the cod family that is used most commonly for fishmeal. Important spawning aggregations of haddock (*Melanogrammus aeglefinus*), herring (*C. harengus*) and whiting (*Merlangius merlangus*) occur in nearshore coastal waters, supporting the livelihoods of many coastal

communities (see Chapter 26: Coastal Fisheries and Aquaculture).

Advances in technology have opened up an exciting field of research in animal biotelemetry. This involves attaching data loggers to animals to collect information on their patterns of movement, feeding and resting; how deep they dive for prey and a range of other ancillary data, including temperature and conductivity of the water. This enhances our ability to map the distribution of animals, their habitat use, the identification of important life-history events such as migration and breeding, and the investigation of interactions between different species and human activities.

Numerous species have been tracked in Irish waters, including both species of seal and many seabirds, fish, turtles and jellyfish.

Fig. 3.16 NORTHERN GANNET DIVING IN WATERS BETWEEN SHERKIN ISLAND AND CAPE CLEAR. Northern gannets (*Morus bassanus*) are one of Ireland's most easily recognised seabirds, being the largest seabird in the North Atlantic, and are seen often performing thrilling plunge dives to catch fish at depths of up to 20m. Ireland hosts internationally important numbers of gannets, with significant colonies on the south-west and south-east coasts. The largest colony, on Little Skellig, County Kerry, holds over 35,000 breeding pairs, which is approximately 10 per cent of the world population. Gannets are relatively large (weighing over 3kg), which facilitates the attachment of data loggers to their feathers without overburdening them and affecting their behaviour or ability to capture prey. They return regularly to breeding colonies over the summer to feed their chicks, which facilitates the retrieval of loggers. As a result, detailed information, which otherwise would be impossible to collect, can be assembled to better understand their location and behaviour patterns at sea. This makes them a model species for investigating interactions between seabirds and human activities in coastal waters, such as the potential risk of collision and displacement caused by offshore wind farms. The extent to which this may become an issue as offshore wind farms are developed is, however, questionable, as studies in other jurisdictions show that gannets have an ability to adapt their behaviour to avoid such obstructions. [Source: Robbie Murphy]

These studies demonstrate that many species are wide-ranging and are found well beyond the coastal waters. For example, Atlantic puffins tracked from Skellig Michael, County Kerry, crossed the North Atlantic during the non-breeding season (Figure 3.17);[15] grey seals tagged on the Blasket Islands, County Kerry, swam to the Hebrides, Scotland (Figure 3.18 and Figure 3.25)[16] while blue sharks (*Prionace glauca*) tagged off the coast of County Cork were tracked to the Bay of Biscay.[17] However, tracking studies also show that species often stay nearshore, particularly during the summer months. For example, tracked gannets remain in coastal waters when provisioning their chicks while European sea bass (*Dicentrarchus labrax*), tracked using a network of acoustic receivers in Cork Harbour, remain localised within the harbour during the summer before migrating out to sea in winter.[18]

HUMAN INTERACTIONS

It is inevitable that most marine species come into contact with humans and there are numerous examples where human activities in coastal regions have impacted negatively on marine organisms. Examples include overfishing, with large numbers of seabirds and marine mammals being recorded as an accidental bycatch. The impacts of underwater noise from shipping, especially along major maritime trading routes, also act to disturb the natural environmental conditions of many marine species. Significant habitat modification is associated especially with shore-based infrastructure development (for example, piers and marinas), but also with the construction of offshore oil and gas platforms and other installations. These can change the distribution of both prey and predators, as well as having adverse effects on the numbers and wellbeing of stocks of fish,

Fig. 3.17 A TAGGED ATLANTIC PUFFIN ON SKELLIG MICHAEL, COUNTY KERRY. Puffins are an iconic seabird, but are currently listed as endangered in Europe due to declining populations. Irish puffins fly approximately 40km from the coast, diving to around 12m deep to feed on their preferred prey of sand eels. They can carry up to twenty sand eels in their beak while transporting them back to the breeding colonies to feed their chicks. Irish puffins migrate as far as Canada during the non-breeding season, although this may also affect their ability to breed in the following breeding season. The leg-mounted tracking device shown in the image weighs as little as 1.5g and provides important information on the broad-scale distribution of puffins over winter to better understand their ecology and risks to the population. Data recovered from these tags show puffins travelling as far as Canada between breeding seasons and spending the majority of the winter in the inhospitable seas of the central North Atlantic. [Source: Mark Jessopp]

Fig. 3.18 TAGGED SEAL AND SEALS HAULED OUT ON THE BLASKET ISLANDS, COUNTY KERRY. a. A tagged grey seal on the Blasket Islands, County Kerry. Sophisticated tags, which relay data to researchers via the mobile phone network, can be attached directly to the fur to provide detailed information on where seals are going and how deep they are diving. This information is used to show how seals are interacting with Irish fisheries, leading to better management of marine resources. The tags drop off naturally when the seal moults its coat annually; [Source: Michelle Cronin] b. Seals on the shore of the Blasket Islands, County Kerry. [Source: Geraldine Hennigan and Norman Kean]

Fig. 3.19 MATT MURPHY, FOUNDER AND DIRECTOR OF THE SHERKIN ISLAND MARINE STATION. Founder of the Sherkin Island Marine Station, Matt Murphy is a pioneering figure whose passion for marine observation has resulted in some of the most comprehensive databases of plankton and seashore ecology, not just in Ireland, but worldwide. Sherkin Island Marine Station, a privately run marine station, was founded in 1975 by Matt Murphy and his late wife, Eileen. Staffed by volunteer scientists from April to November each year up to 2015, baseline data were collected on plankton, rocky shore ecology, otters, birds, underwater macrofauna, rockpools, insects and plants in Roaringwater Bay and along the west Cork coast. Matt Murphy has been an inspiration to the hundreds of marine biologists who had the opportunity to volunteer at the unique marine station over the decades. The Sherkin Island Marine Station has also been extremely effective in raising awareness of the marine environment and the potential of the sea to create jobs, long before 'Blue Growth' became fashionable as a marine development strategy. [Source: roaringwaterjournal.com]

seabirds and marine mammals. In addition, they can create barriers to migration or movement between feeding and breeding areas, and change the geography of food availability.

Conversely, some marine animals have significant impacts on humans. Seals, for example, are known to cause considerable damage to fish catches through the raiding of fishing nets, while jellyfish blooms can cause the closure of beaches due to the threat of stings, and have significant economic impacts on coastal

aquaculture (see Chapter 26: Coastal Fisheries and Aquaculture: Case study: Jellyfish and Aquaculture). Even microscopic marine plants, known as phytoplankton, can be toxic to humans and to other animals, when Harmful Algal Blooms (HABs) occur. Phytoplankton, unseen to the naked eye, yet a vital component of the food chain, have been observed from the remote island of Sherkin Island in west Cork over a unique programme that has spanned forty years. (Figure 3.19)

Case Study: Jellyfish in Irish Coastal Waters

Tom Doyle

Jellyfish are distinctly recognisable lifeforms with their transparent bodies, trailing tentacles and round symmetry. Many of them are large and form spectacular blooms (or aggregations) that wash up along Ireland's shores from mid-May to September. Together, these attributes

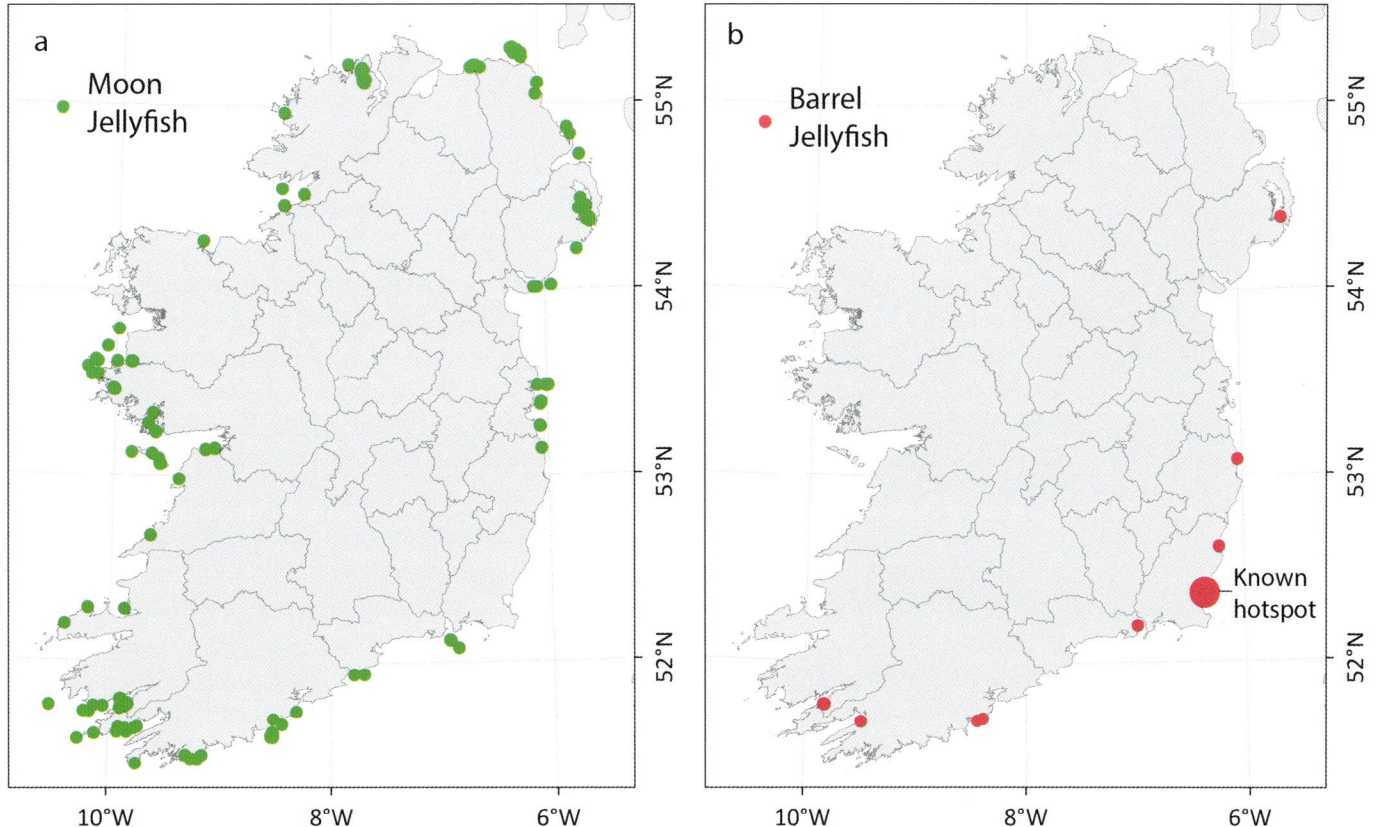

Fig. 3.20 JELLYFISH DISTRIBUTION. The maps show the distribution of two native jellyfish in Irish waters: a. moon jellyfish; and b. barrel jellyfish. As can be seen, the moon jellyfish is the most common species and can be found all around the coast. The barrel jellyfish is normally confined to a few localised places, most notably a hotspot in the coastal waters between Rosslare Harbour and Curracloe, County Wexford, but individuals wash up occasionally on other beaches on the east and south coasts. [Source: retrieved and referenced from the National Biodiversity Data Centre]

make them one of the most conspicuous animals in Ireland's coastal seas. Anyone who has spent some time on an Irish beach is familiar with the common jellyfish, a saucer-sized disc of jelly with four pink rings.

DIVERSITY AND DISTRIBUTION OF IRISH JELLYFISH

True jellyfish are the stereotypical jellyfish with which we are most familiar, and six of these occur around the Irish coastline. The most abundant species is the common or moon jellyfish (*Aurelia aurita*) and is found in all Irish coastal waters (Figure 3.20a). It has hundreds of very fine tentacles around the perimeter of its bell and swims continuously to create a vortex of water through its tentacles to help capture food. Furthermore, they can often form spectacular coastal blooms as part of their lifecycle.

The blue jellyfish (*Cyanea lamarckii*) is probably the least abundant jellyfish that occurs in these coastal seas, and is found typically in only ones or twos. As the name suggests, many have a striking blue colouration, although some individuals are transparent or pale yellow in colour. A closely related species is the lion's mane jellyfish (*Cyanea capillata*) (Figure 3.21a), which can measure up to one metre in diameter and is of a dark red colour. The bell has eight distinctive lobes, each bearing hundreds of long tentacles that are known to inflict serious injuries on bathers. Unlike blue jellyfish, the lion's mane jellyfish has a northern distribution, which extends typically north of Wicklow town, around the northern coasts of Ireland and down the west coast as far as County Galway. Given the high density and large numbers of people that reside in Dublin and its hinterland, it is unfortunate that the Forty-Foot bathing spot in Dún Laoghaire is a known hotspot for the lion's mane jellyfish; so too are other beaches along the Dublin and Meath coastline.

Uniquely among Irish jellyfish, the barrel jellyfish (*Rhizostoma pulmo*) is confined to a few localised places, most notably the coastal waters between Rosslare Harbour and Curracloe beach, County Wexford (Figure 3.20b). The barrel jellyfish is the most impressive in terms of size. A single individual can weigh up to 30kg and can be close to one metre in diameter (Figure 3.22). They have a white colour and, rather than possessing tentacles, have eight large oral arms. It is thought that these jellyfish overwinter on the seabed as, in early April, large individuals (often more than 70cm in diameter) are often seen swimming and feeding on the spring phytoplankton bloom.

One jellyfish-like animal, the Portuguese man of war (*Physalia physalis*) has evolved a very unique way of exploiting the oceans (Figure 3.21b). They have a large surface float that captures the wind like a sail, and which propels them across ocean waters. Remarkably, these floats are just one individual of a colony of polyps that make up a composite (colonial) animal. The remaining polyps are attached to the base of the float and include very long tentacles that are used to capture food. Of all the Irish jellyfish and jellyfish-like animals, the Portuguese man of war is by far the most venomous.

Fig. 3.22 (above) BARREL JELLYFISH. The barrel jellyfish occurs notably in large blooms every summer off Rosslare and Wexford harbours, and is one of the largest jellyfish found in Irish waters. It is distinguished by its large size and its frilly appearance, with eight cauliflower-like extensions. Despite its relatively large size, its sting is relatively mild, but prolonged exposure could trigger an allergic reaction in humans. The barrel jellyfish is preyed upon by leatherback turtles, which can attract these charismatic marine animals to the Irish Sea. [Source: Damien Haberlin]

Fig. 3.21 a. A LION'S MANE JELLYFISH BEING TAGGED WITH AN ACOUSTIC TAG IN DUBLIN BAY (2010); b. PORTUGUESE MAN OF WAR WASHED UP ON DERRYNANE BEACH, COUNTY KERRY. The lion's mane jellyfish a. is the largest jellyfish found in Irish waters and has over a thousand tentacles stretching out from under its bell shape. Despite being one of the least abundant jellyfish in Irish coastal waters, it has received considerable attention, as a result of its severe sting. Regular occurrences of lion's mane jellyfish in popular swimming locations, such as the Forty Foot diving area in Dún Laoghaire, have triggered media attention and heightened public awareness among sea swimmers. Scientists working to gain an enhanced understanding of jellyfish biology utilise a range of measures, from remote observations to tagging. [Source: Damien Haberlin]; b. The Portuguese man of war also receives media attention, as it is infamous for its extremely painful sting, which can be fatal for humans. Despite appearances, the Portuguese man of war is not a jellyfish. It is, in fact, a siphonophore, being made up of a colony of animals working together. It occurs rarely in Irish waters, but since 2016, significant numbers have stranded on Irish beaches every October. Scientists are not entirely sure of the cause of these episodes, but they could be a consequence of storm patterns. They are sometimes referred to as 'armadas' as a result of large clusters occurring together at sea. [Source: Vincent Hyland]

The compass jellyfish (*Chrysaora hysoscella*) is a southern species, found mainly around the south and west coast of Ireland. This is a large jellyfish with twenty-four long tentacles dangling from the margin of the bell and four frilly mouth-arms (Figure 3.23). It also has many distinctive brown, v-shaped markings on the bell that are reminiscent of the pointers on a compass rose. At times, this species can be extremely abundant throughout the entire Celtic Sea, but in some years it can be very rare.

A closely related species to the compass jellyfish is the mauve stinger (*Pelagia noctiluca*) (see Chapter 26: Coastal Fisheries and Aquaculture, Figure 26.27a). Unlike all other jellyfish, the mauve stinger is an oceanic species that is swept into Ireland's coastal waters by onshore winds and coastal currents. It is a small jellyfish, being no larger than a closed fist and, while juveniles are coloured a golden-brown, it becomes a mauve-pink in colouration when fully mature. This species occurs most frequently off the west and north-west coasts of Ireland.

There are at least twenty other species of jellyfish-like animals found in Ireland's coastal waters. Some are very closely related to true jellyfish and share some common characteristics: they are largely gelatinous (>90 per cent water), transparent, fragile and have a planktonic existence.

Since the end of the twentieth century, increasing levels of concern have been voiced regarding the extent to which jellyfish are increasing in our oceans as a result of climate change, eutrophication, overfishing and translocations. However, while many cases of jellyfish proliferation have emerged, many media reports involve exaggeration and/or are purely speculative. In fact, there is no robust evidence confirming a global increase in jellyfish populations. This is because there are few long-term jellyfish datasets (>30 years of data) to examine trends in jellyfish abundance. The best available data for Irish waters, for example, is a sixteen-year time series (1994–2009) collected from the Irish Sea. This study found that there was a slight upward

Fig. 3.23 (left) COMPASS JELLYFISH. The compass jellyfish is seen frequently over the summer months, at sea or washed up on the shoreline, mainly on the south and west coasts. Its brown radial pattern, which resembles a compass, is a distinguishing feature. These jellyfish are carnivores, preying on plankton, by stinging them with their tentacles. The compass jellyfish is one of six true jellyfish found in Ireland (including the barrel jellyfish, blue jellyfish, common jellyfish, mauve stinger and lion's mane jellyfish). [Source: Damien Haberlin]

trend in jellyfish abundance during this time period and that the strongest driver in jellyfish abundance was climatic variation (for example, changes in temperature and precipitation).

Interest levels in trying to ascertain trends in abundance levels of jellyfish are related to the well-established fact that jellyfish play important roles in these coastal waters. For example, leatherback sea turtles (*Dermochelys coriacea*) can eat up to several hundred lion's mane jellyfish in a day, and hundreds of fish species are now known to consume jellyfish routinely, or as a part of their diets. Jellyfish are also important predators of plankton (that is, crustaceans and fish eggs) and in some situations can control the populations of other zooplankton. Furthermore, in a pelagic world devoid of physical surfaces, jellyfish can provide a floating habitat with 'nooks and crannies' available for other organisms to exploit. For example, some juvenile fish species are known to hide in amongst the tentacles of large jellyfish during the day to avoid being eaten by larger fish. However, jellyfish are also known to impact negatively on coastal activities, as in 2007, when a swarm of mauve stingers swept through a salmon farm in Northern Ireland, killing all 200,000 fish (see Chapter 26: Coastal Fisheries and Aquaculture). In recent years, many open water swimmers have also been stung badly by the lion's mane jellyfish in the Irish Sea. While there is no evidence for an increase in such incidents, such negative interactions are likely to rise as we increase our activities in the sea, regardless of whether we see an increase in jellyfish or not.

LESSONS LEARNED FROM LONG-TERM PHYTOPLANKTON MONITORING AT SHERKIN ISLAND, WEST CORK

Matt Murphy

The Sherkin Island Marine Station regularly recorded plankton from water samples taken at a series of nearby coastal and offshore sites between 1978 and 2015. This provides one of the few available and long-term phytoplankton records in existence worldwide. The data include known toxic and nuisance species associated with Harmful Algal Blooms (HABs). These occur when algae grow very fast and accumulate into dense patches at the surface of the sea to depths of around 50m. HABs, which refer to bloom phenomenon that contain species that produce neurotoxins, warrant serious consideration as they can have consequences for human health, wealth and wellbeing. For example, toxic shellfish poisoning can result in humans who consume shellfish contaminated by HABs. Research into HABs at the Sherkin Island Marine Station provided an early warning system for the local aquaculture industry, as some phytoplankton contain toxins that can become concentrated in filter-feeding animals, such as mussels.

Analysis of the long-term data from Sherkin Island Marine Station shows greater variation of recorded species in time and space compared to other international studies, suggesting that ecological complexity may be critically misrepresented in scientific investigations and models, which typically have far less comprehensive sampling than efforts undertaken at Sherkin.[19] The Sherkin Island Marine Station Phytoplankton Dataset revealed that differences can occur in concentration levels of individual phytoplankton species in water-depth intervals of 2–5m at any one sampling station, as well as inter-annual variability within and between different species.

This has implications for modelling and long-term prediction of HABs. Current modelling efforts tend to be limited by the fact that they are designed to extrapolate and interpret data based on sparse observations. The work at Sherkin Island Marine Station shows the value of long-term monitoring, which is crucial in determining, in future, which species may bloom, where, and when. This becomes increasingly important in light of climate change, as scientists strive to understand the implications of increases in sea surface temperature on HAB events. The data collected as part of the plankton monitoring programme at Sherkin Island Marine Station included records of additional vectors, such as the presence of zooplankton, seawater temperature and salinity, as well as hydrological and meteorological conditions, which are of immense value in analysis of ecosystem change.

The Sherkin Island Marine Station Annual Rocky Shore Survey Record is another internationally significant longitudinal record of 147 coastal sites from Cork Harbour to Bantry Bay, surveyed, in the main, in the period from 1981 to 2015. The methodology involved the surveying of each site during a spring tide and occurred at approximately the same date each year along measured transects. On occasions, when tidal conditions were particularly favourable,

Fig. 3.24 SHORE CRAB ON SHERKIN ISLAND. The unique scientific value of the Sherkin Island Marine Station Annual Rocky Shore Survey Record is the spatial and temporal dimension of the records, covering data collected from 147 coastal sites from Cork Harbour to Bantry Bay over a thirty-year period. The records contain more than 230 species of plant and animal life, from rare seaweeds to the common shore crab. [Source: Robbie Murphy]

extreme low tides exposed the sublittoral zone to allow unique opportunities to record species seen less frequently in the intertidal area. This led to a unique record of well over 200 different species. Initial systematic analysis of the data, published in *The Ecology of the Rocky Shores of Sherkin Island,* shows shifts in population dynamics over the duration of the study[20] Observing such systemic changes in the distribution and abundance of plants and animals living on rocky shores is the first step in understanding ecosystem functioning in the context of climate and other anthropogenic stressors (Figure 3.24).

Telemetry studies also provide insights into interactions between humans and marine life, by supplying alternative sources of data to address how human impacts on marine biodiversity might be reduced. For example, tracking seabirds to identify important feeding areas occurring in intensively fished areas can lead to the introduction of simple management measures. One such measure involves the weighting of bait lines to limit the amount of time that bait is available at the surface, which can reduce significantly the extent of seabird bycatch. Tagging studies show that grey seals may target inshore passive gear fisheries, suggesting that management interventions should focus on minimising interaction at nets (Figure 3.25).[21] Other work shows how marine species may benefit from interactions with human activities, such as increased food availability for seals around wind farm turbines, or gannets taking advantage of additional food in the form of discarded fisheries waste.[22]

At the apex of Ireland's rich biodiversity are the seabirds, marine mammals and large fish that play a key role in the functioning of a healthy marine ecosystem. Many of these animals are well known and highly valued by coastal communities, where they underpin tourism and leisure opportunities, such as recreational fishing, whale watching and sightseeing trips. Locations like the Cliffs of Moher and Skelligs are internationally renowned

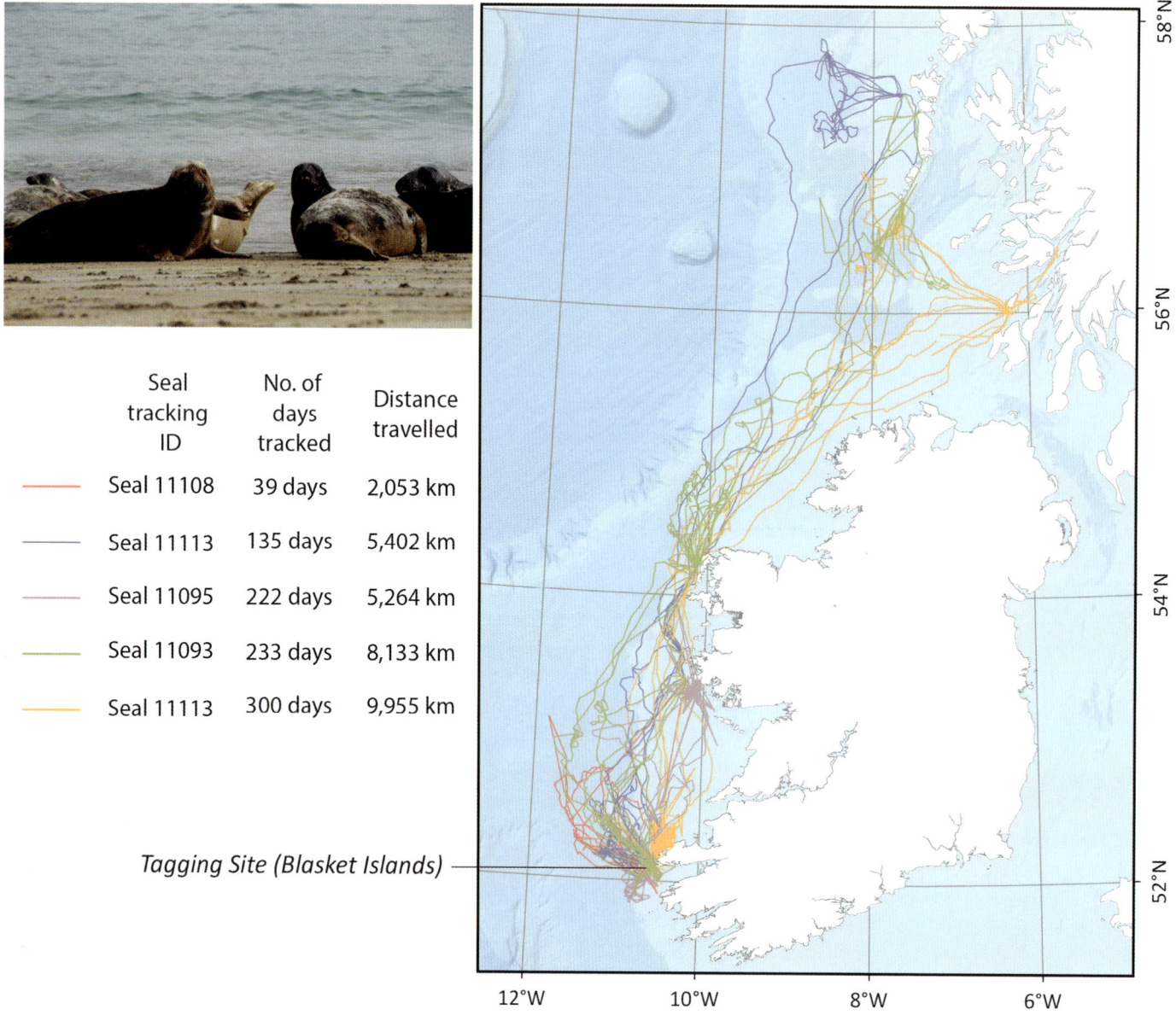

Seal tracking ID	No. of days tracked	Distance travelled
Seal 11108	39 days	2,053 km
Seal 11113	135 days	5,402 km
Seal 11095	222 days	5,264 km
Seal 11093	233 days	8,133 km
Seal 11113	300 days	9,955 km

Tagging Site (Blasket Islands)

Fig. 3.25 SEAL MOVEMENT MAP. Five grey seals, tagged and tracked from the Blasket Islands, were shown to have a broad range, with one animal extending as far as 9,955km in 300 days. This research is vital to help understand the movements of seals relative to fishing activities, bearing in mind that seals and humans both target fish, leading to competition for resources. [Map data source: subset of seal tracks from Jessopp, M., Cronin, M., and Hart, T., 2013. Habitat-mediated Dive Behavior in Free-ranging Grey Seals. *PLOS ONE* (5), e637201-8; Photo: David Thompson]

Fig. 3.26 FUNGIE THE DINGLE DOLPHIN. Fungie, the Dingle Dolphin, was a male Atlantic bottlenose dolphin that attracted visitors to Dingle since first sighted in the bay in 1983. Since then, Fungie has charmed locals and visitors alike with frequent interactions with boats and swimmers. As a result, a significant ecotourism industry developed locally around Fungie, as boat operators and tour guides took visitors to see him. This is a classic example of how public appreciation of biodiversity can help to create and sustain jobs in peripheral coastal locations. [Source: Paul Britten, courtesy of www.dingledolphin.com]

Fig. 3.27 DYNAMIC INTERACTION OF HUMANS, FISH AND BIRDS. Ireland's internationally renowned marine biodiversity has been appreciated and valued for generations. However, the myriad threats to the sustainability of marine life, from the blue whale to the humble shore crab, place a critical emphasis on society to invest in science, management and innovative responses to protect these precious resources. The bounty of coastal waters has dropped. Marine biodiversity has been diminished. Reversing this trend requires an understanding of the complex interplay between humans, marine life and ecosystem functioning. Acclaimed marine biologist Daniel Pauly coined the expression 'shifting baselines' to describe a phenomenon of lowered expectations, whereby each generation normalises a progressively poorer natural world. Scientific monitoring has an important role to play in mapping baselines and informing the decision-making process. Advances in marine biology, with techniques such as tagging, are extremely progressive. However, there needs to be equal, if not more, pressure to transfer this knowledge into meaningful action for sustainability. [Source: John Cunningham]

Fig. 3.28 NURTURING THE NEXT GENERATION OF MARINE BIOLOGISTS. a. The next generation of marine biologists are encouraged to get involved in marine conservation through an increasing number of citizen science projects. These sources of data and information can add significant value to data collected from formal scientific investigations and sophisticated, and often expensive, technology. A Coastwatch Survey, operating since 1987, mobilises local groups and individuals around the coast to make observations each year of 500m stretches of shoreline. The Environmental Protection Agency also supports the Seashore Spotter Survey, as an entry-level survey designed to capture records of marine species along the shore, from periwinkles to dogwhelks. A simple online form or phone app can be used to capture data, and digital photographs are used to validate efforts to identify different species by members of the public. The National Biodiversity Data Centre also plays an increasingly important role in managing and sharing datasets on marine biology. b. The rocky shore provides an ideal living laboratory as a result of the diversity of plant and animal life found there. Rocky shore ecology features winkles, whelks and anemones and sponges among a myriad of species that adapt to the rhythm of the tides and exposure to seawater. [Source: Ailbhe Cotter]

for their incredible array of nesting seabirds, while the Dingle Peninsula remains famous for its dolphin, Fungie (Figure 3.26). Marine and coastal tourism and leisure contributed approximately €648 million to the Irish economy in 2018.[23] This is a testament to the value of Ireland's marine biodiversity.

Arising from existing benefits and the significant potential that has been identified for further exploitation of other marine resources, there is a clear need for development to be sustainable (Figure 3.27). This is to ensure that it does not impact on the environmental, cultural and societal benefits already enjoyed, and to safeguard the long-term health of these waters and marine biodiversity for future generations. In this context, ongoing surveys in coastal waters remain essential in providing critical data on the distribution and habitat use of marine species, as well as helping to identify key areas of development. Telemetry studies, in particular, provide insights into key factors that underpin species distributions

and interactions with coastal activities. Technological advances such as these are complemented increasingly by the power of data collected through citizen science (Figure 3.28) and distributed via social media.

In conclusion, the collection and interpretation of information from a variety of sources can help to achieve a much better understanding of Ireland's coastal biodiversity and, thereby, enable society to reduce any negative impacts on coastal and marine environments. Technology has been shown to play a pivotal role in this pursuit. Juxtaposed with this is the passion of the Irish men and women who have dedicated their careers to improving our understanding of marine biology in the Irish coastal zone, from the efforts of the inspiring individuals who have been featured here, to the collective actions of the individuals and institutions that support capacity building, through a wide range of biological survey and observation work.

People, Agriculture and the Coast

Barry Brunt, Michael Keane and David Meredith

The agricultural landscape of Lurigethan, the Glens of Antrim, looking toward the Causeway Coastal Route and beyond to Scotland. Located on the north-eastern coast of Ireland, the Antrim Plateau sweeps down to the coast in steep cliffs. This farmed landscape highlights the dominance of the grasslands and pastoral enterprises that typify most of the country's coastal areas. Extensive rearing of cattle and sheep have long dominated the upland areas, characterised by common grazing and rough pasture on the poorly drained and shallow soils on the basalts of the Antrim Plateau. Here, low returns on these efforts have often resulted in the abandonment of farms and field patterns. This contrasts with the well-managed and intensive pastoral enterprises that occur on the more fertile, drift-covered lowlands, where hedgerow-bounded fields enclose improved pasturelands attached to specific private landholdings.
[Source: © Tourism Northern Ireland]

From the earliest recorded presence of humans in Ireland at least 9,000 years ago until well into the twentieth century, the evolution of the island's culture, society and economy has been defined strongly by agriculture and its associated rural landscapes. While Northern Ireland experienced an urban-industrial revolution in the nineteenth century, such a transformation over the rest of the island awaited an economic 'take-off' in the 1960s (Figure 4.1). Although the absolute and relative roles of agriculture in shaping contemporary Ireland have declined, this primary sector continues to exert a disproportionate influence on the island's identity. This is especially the case in coastal areas.

Pivotal to the development of agriculture has been the ability of humans to adapt successfully to the opportunities and challenges presented by Ireland's diverse natural environments. Such interactions between people and their immediate environment, undertaken in order to achieve basic requirements for food and to meet expectations for improved standards of living, have yielded the rural landscapes that now typify much of the country. The roles played by evolving historical and contemporary processes of economic and political governance have also been, and remain, important in shaping agricultural land use and management of coastal environments. In this context, Ireland's peripheral location off the western seaboard of Europe, together with an historic underdevelopment linked directly to political and economic dominance under its nearest neighbour, Britain, reduced the introduction of innovative practices (see Chapters 17: The Vikings

and Normans: Coastal invaders and settlers and 18: Era of Settlement: Trade, plantation and piracy). As a result, forces of inertia and expressions of traditional farm practices have tended to dominate rural landscapes. Although innovation and change has gathered pace since the 1960s, many of the island's agricultural and rural landscapes remain conditioned by past traditions, as well as dynamic contemporary processes, such as the Common Agricultural Policy of the European Union.[1]

As a comparatively small island of some 83,000km^2, the colonisation and subsequent development of Ireland focused primarily on its limited coastal lowlands. From the earliest times, agriculture and fishing together played key roles in supporting most coastal communities (see Chapters 16: The Inhabitants of Ireland's Early Coastal Landscapes and 26: Coastal Fisheries and Aquaculture). In addition, other activities that have become key characteristics in defining modern Ireland can also trace their initial presence to the country's coastline and to areas where development was shaped strongly by relationships with agriculture. Although its direct role has declined, the sector continues to retain significant relationships with other components of Ireland's economy (Figure 4.2). For example, the richer agricultural land uses found along the eastern and southern coastline and major river valleys were vital factors in determining Ireland's evolving pattern of urbanisation, with almost all principal towns sited on the coast (see Chapter 25: Urbanisation of Ireland's Coast). Output from agriculture has also been a key input for industrial and economic development in other

Fig. 4.1 FARMING ON MWEENISH ISLAND, COUNTY GALWAY, IN THE EARLY TWENTIETH CENTURY. Although processes of urbanisation and industrialisation were well underway in Northern Ireland, this was a relatively localised transformation. By the start of the twentieth century, some fifty years after the Great Famine, large numbers of the island's agricultural communities continued to live in abject poverty, as exemplified in this historic image of a farm community on Mweenish Island off the Connemara coast of County Galway. Here, the harsh environment – dominated by bogs and thin, stony soils and exposed to the strong, rain-bearing winds that sweep in from the Atlantic ocean – presented few opportunities for productive farming. The poor carrying capacity of such land, littered with small enclosed fields, combined with a low average farm size, supported relatively few cattle and sheep, the principal enterprise. Large-scale emigration typifies such areas, with Mweenish Island's 2011 population of 145 contrasting with 376 in 1841. Most of the current population are no longer full-time farmers, but include many part-time farmers, residents involved in tourism, retirees and commuters travelling to employment on the mainland. [Source: National Library of Ireland, call no.: CDB82]

Fig. 4.2 CORK CITY AND ITS FARMING HINTERLAND. The distribution and development of Ireland's towns and cities has long been associated with agriculture. Most emerged as market centres for their rural hinterlands, while the growth of the island's ports was conditioned to a large extent by their role in the export of farm-related produce. Cork city provides a good example of this relationship. Located at the lowest bridging point on the River Lee, where the river flows into Cork Harbour on Ireland's southern coast, its environs are composed of low and undulating hills, relatively rich soils and mild climatic conditions that favour the year-round growth of grass. The area has long been associated with dairying, but from the eighteenth century the intensity and specialisation of this land use increased markedly. This was due to growing demands for butter from the developing North Atlantic provisions trade, which encouraged the commercialisation of dairy-ing in Cork and Munster. Given its strategic location, the city established itself as the most important centre for trade in butter, but also in beef, on the North Atlantic seaboard. The result was a growth in the scale and prosperity of the city and port, which continues to act as a focus for agricultural trade and a buoyant agri-food sector. Although developments since the 1960s have been conditioned more by the city's attraction for manufacturing and service activities, Cork retains an intimate relationship with its agricultural communities. This is reflected in the well-organised pastoral landscape, composed of large fields and prosperous farms, that surrounds the city. Pressures on that rural environment are increasing, however, as seen in the spread of suburban housing and industrial estates into these areas, as well as around the margins of Cork Harbour. [Source: Maxim Kozachenko]

sectors (for example the linen industry in Northern Ireland and modern agri-food industries in the Republic). This, in turn, encouraged the island's long-established traditions of export trade and port development (see Chapter 24: Ports and Shipping). Finally, cultural landscapes associated with traditional farming practices and rural environments, especially along the western seaboard, are central elements in the attractions of coastal areas for large-scale inflows of tourists. Such relationships, however, have also placed sensitive rural landscapes in coastal areas under increasing pressures, such as urban sprawl and degradation of the natural environment. This points to the need for effective integrated planning (see Chapter 32: Coastal Management and Planning).[2]

Along with agriculture, many other developments and signs of encroaching modernity (especially the introduction of new methods and technologies, including the appearance of tractors in the early twentieth century) originated along Ireland's coast, from where they migrated inland. In doing so, they became modified to some extent. However, such is the limited area of the island that differences between coastal identities and those expressed in the interior are often marginal. In effect most, if not all, of the island of Ireland can be considered to be coastal!

Yet, coastal agriculture does have some unique characteristics when compared to inland farming. For example, prior to the modern era of chemical fertilisers, sand and seaweed played major roles in enhancing fertility levels in those areas of the coast and adjacent uplands where soils were of poor quality. Milder climates

around the coast, especially in the south, facilitate an earlier and extended season for grass growth, as well as opportunities for tillage and the raising of specialist crops – such as potatoes, strawberries and flowers – along the east and south-east coasts. Many areas along the western seaboard, however, are also more exposed to severe weather, which restricts the livestock carrying capacity of land as well as causing some coastal erosion and loss of farmland.

Given agriculture's deep historic and contemporary roles within Ireland, this chapter presents an opportunity to assess its significance for national development. This review of farming and the island's rural environments also provides a link to some other significant human components of the coast that are treated in later chapters. The chapter opens with a brief review of the importance of the island's diverse natural environments for coastal agriculture, before reviewing the historical context and influences that contributed to Ireland's agricultural economy and rural landscapes around its coastal zone. The chapter concludes with a detailed description of the patterns of contemporary coastal agriculture, together with some of the pressures faced and the planning required to sustain its role within Ireland's sensitive coastal environments.

NATURAL ENVIRONMENTS AND AGRICULTURE

Although Ireland's coastal agricultural landscapes are essentially the product of human endeavour, they are also shaped strongly by the natural environments (see Chapter 2: The Coastal Environment:

Fig. 4.3 THE IVERAGH PENINSULA, COUNTY KERRY. The Iveragh Peninsula is the largest of six east–west trending peninsulas that protrude into the Atlantic ocean in the south-west of Ireland. Composed essentially of sandstone, the spine of this peninsula is made up of the Macgillycuddy's Reeks, which include the highest mountain on the island. Over recent geological periods, intense denudation by ice and water has rounded the appearance of these upland areas, while exposure to high levels of precipitation and the clearance of forests in historic times have combined to define its current treeless landscape, dominated by extensive coverage of blanket bog. The photograph displays such a landscape at Bolus Head, located near the western extremity of the Iveragh Peninsula. As elsewhere along the peninsula's limited coastal lowlands and lower slopes, rough grazing of cattle and sheep dominates farm enterprise. Although this extensive form of agriculture persists, many of the walled fields that characterise the area's more densely populated past have now been abandoned. This reflects low income returns, which have encouraged many farmers to cease full-time farming. The spectacular coastal scenery and deep cultural traditions vested in its landscape, however, attract large numbers of tourists to the 175km-long Ring of Kerry coastal route that surrounds the Iveragh Peninsula and which, with the Dingle Peninsula located immediately to the north, is one of the island's principal tourist destinations. [Source: John Crowley]

Physical systems processes and patterns). Relief, climate and soil types combine, in particular, to influence the livestock carrying capacity of the land for farming and the type of farm enterprise that occurs.

As will be discussed in detail in the following two chapters, almost three-quarters of the island is classified as lowland, being dominated by the broad Central Lowlands composed of carboniferous limestone, but overlain by extensive deposits of glacial drift and raised bogs (see Chapters 5: Geological Foundations and 6: Glaciation and Ireland's Arctic Inheritance). Surrounding these lowlands are a series of distinct uplands/mountains, composed of more resistant rock types which, over time, have been subjected to significant denudation, including glaciation. This has created the spectacular scenery that attracts tourists to the country's narrow coastal zone, especially in the west, with its fringe of islands and special landscapes (see Chapter 21: Ireland's Islands). In terms of agriculture, however, these areas of mountainous relief, more extreme climatic conditions and a paucity of soil cover, are

inhospitable for settlement and most types of farming other than rough grazing of sheep and cattle (Figure 4.3).

Variations in levels and seasonality of precipitation, temperature, sunshine and exposure to storms are also important in shaping farm practices. Ireland's mid-latitude location off Europe's western seaboard ensures that it experiences a temperate oceanic climate. This means it benefits from relatively mild temperatures and high levels of precipitation throughout the year, which favour pastoral farm enterprises. Additionally, the mountains of western Ireland receive comparatively higher amounts of precipitation, while the lowlands along the east coast experience a 'rain-shadow' effect. This, in particular, raises the number of sunshine hours in the 'sunny south-east', which, in conjunction with deep, well-drained soils, has been critical in this region's greater commitment to tillage farming. Against this backdrop of climate and the dependent pattern of agriculture, significant landscape changes have developed as rural transformations occurred from the eighteenth century, particularly as a result of technical innovation, such as windmills. The roles of

WINDMILLS

Robert Devoy

Many of the sixty or so windmills that were built across Ireland in the eighteenth and nineteenth centuries now have only their bases or towers visible. A few of these had their origin in the sixteenth and seventeenth centuries, while others were combined with the much older use of water in coastal, tidal and river locations for milling (see Chapter 17: The Vikings and Normans: Coastal invaders and settlers). Only seven still have their sails and these are found mainly in coastal areas of the island. They were built primarily for the milling of grain-type crops (for example, wheat, barley, rye and oats) and also flax but, following the end of the Napoleonic War (1815), cereal growing and milling in Ireland became far less profitable. As a result, many windmills, as well as reclaimed coastal and other lands, fell into disuse (see Chapter 30: Engineering for Vulnerable Coastlines). In spite of this, these buildings had established themselves as part of the local landscape and, in some cases, are now important attractions for tourism.

In other European countries, such as the Netherlands or the eastern counties of England, windmills have become iconic elements of the landscape, being used essentially as power sources to aid in the drainage and reclamation of land as much as for grain milling. This is less the case in Ireland, where they have been used rarely in drainage works. Nonetheless, they provide a valuable illustration of the innovation of a new technology as an instigator of landscape transformation. If the appearance of the windmill was ever controversial for people, they are accepted now as part of the landscape and highly valued. Today, the appearance of wind turbines – which are springing up like mushrooms across Ireland's uplands, boglands, coasts and now offshore – is a development that resonates with the much earlier

Fig. 4.4 COASTAL WINDMILLS. a. Ballycopeland, County Antrim; b. Elphin, County Roscommon; c. Blennerville, County Kerry; d. Tacumshane, County Wexford. [Source: Ian Johnson]

appearance of windmills on rural landscapes. In terms of their aesthetics and environmental impacts, however, wind turbines (the new windmills) are a source of dispute and even conflict for many people. In spite of this, their technology is innovative, perhaps more so than the windmills, and may prove to be essential to Ireland in the provision of future renewable, electric power.

Represented in Figure 4.4 is: a. Ballycopeland mill, a brick, whitewashed tower mill, with four sails and a black cap, not unlike the mills in Lancashire (Lytham St Annes and Blackpool); a store cupboard has been attached to the base of the tower. b. Elphin windmill is an eighteenth-century, circular, three-stage tower mill, with open-trestle sails and, unusually, the rotating conical-cap roof is thatched, originally in rye. It was built c.1730 and provided corn meal locally, as well as milling flax for almost a hundred years. The recessionary environment for farming immediately after the Napoleonic Wars, however, meant that the Elphin mill was in ruins by the 1830s. Recently, the windmill has been restored as part of a National Millennium Committee Project. The National Inventory of Architectural Heritage for Ireland includes the mill as an important piece of industrial architectural heritage and possibly its oldest operational windmill. c. Blennerville windmill was built in 1800 and is 21.3m high, forming a prominent feature in the local landscape, set against the coastal backdrop of Tralee Bay. It is situated near the magnificent bridge across the seaward end of the Tralee Ship Canal, which was opened in 1846. Blennerville is a technically sophisticated round-capped tower windmill, with four sails, five floors and a gallery from which the sails can be maintained. The mill was used for the local milling of grain, supplied in part by arable farmers occupying reclaimed coastal marshlands in the region. Much of the product was exported to Britain from Blennerville Harbour. The innovation of steam power, the silting up of Blennerville Harbour, the opening of the canal to Tralee, and later the port at Fenit in 1880, were amongst the factors that led to the demise of the mill in the late nineteenth century. The mill was restored, however, by Tralee Urban District Council and opened officially as a centre for tourism in 1990. It provides a range of attractions, including visitor and craft centres, an art gallery, model railway and cafe. d. Tacumshin (colloquially known as Tacumshane) windmill was built in 1846, has four sails and operated commercially as a mill until 1936. Restoration of the building was undertaken in the 1950s.

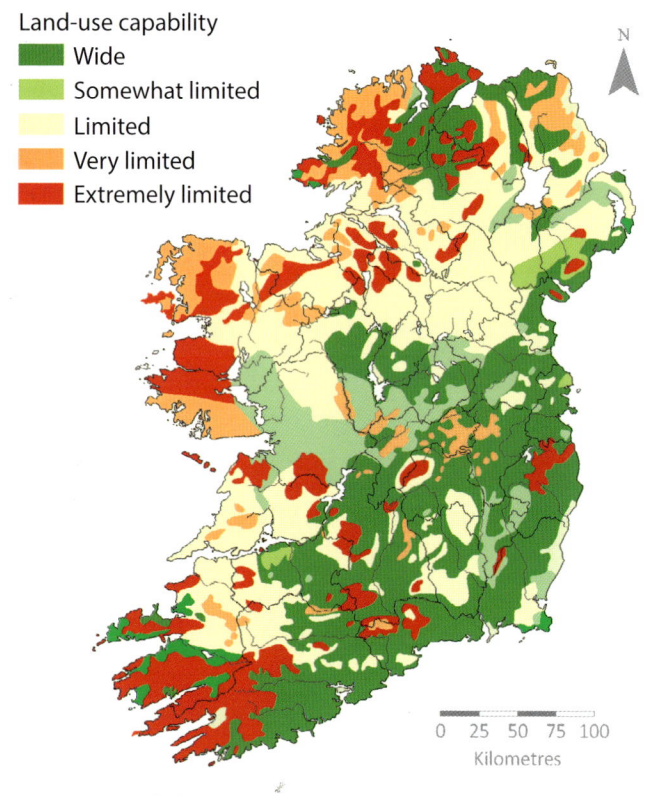

Land-use capability
- Wide
- Somewhat limited
- Limited
- Very limited
- Extremely limited

0 25 50 75 100
Kilometres

Fig. 4.5 LAND-USE CAPABILITY FOR AGRICULTURE. Within Ireland, there are significant variations in both the opportunities and difficulties that confront agriculture. Based on physical factor inputs – such as climate, drainage, geology, relief and soil structure – five categories of land-use capability can be outlined. In general terms, these display a west/north-west to south/south-east gradient, which divides the island into two contrasting regions. As much of the western seaboard is dominated by upland relief, relatively high precipitation, extensive areas of bogland and soils with low levels of fertility, this area possesses either very limited or extremely limited capacity for agricultural enterprises. In contrast, the eastern and southern coastlines have larger areas that offer a wider range of capabilities. The result is a more diversified and prosperous system of farm enterprises than is found to the west. [Data source: Aalen, F.H.A., Whelan K. and Stout, M., eds., 2011. *Atlas of the Irish Rural Landscape*. 2nd ed. Cork: Cork University Press]

geology, relief and climate are central to the formation of soil. In general, all of the island enjoys relatively high levels of precipitation, which leads to moist ground conditions where leaching of nutrients is the dominant process. The result is that relatively impoverished podsol-type soils and peat boglands are found across much of Ireland, particularly in the higher upland areas of the west. On lowlands benefiting from layers of glacial drift, more fertile soils such as brown earths occur. These support higher levels of land-use capability, as illustrated along the eastern seaboard, and contrast strongly with far more limited options identified along most of the western seaboard (Figure 4.5).

The combination of physical factors has had a critical role in influencing how Ireland's farming communities have shaped their agricultural landscape. This is summarised well by Davies and Stephens, who highlight fundamental differences in geomorphology between different segments of the coastline, especially between the rugged Atlantic coast and the mainly drift-clad Irish Sea lowlands.[3] In effect, this provides an essential explanation for differences in the types of contemporary coastal farming found on the island. Dry-stock enterprises, involving extensive rearing of sheep or beef cattle, predominate along the northern and western coasts, characterised by poorly drained peats and gley soils (Figure 4.6). In contrast, intensive dairy and tillage enterprises are more common on the lowlands possessing fertile soils that are prevalent along the southern and eastern coasts (Figure 4.7).

Due to difficult natural conditions that typify many coastal and upland areas, farmers have often been induced to engage in the long-standing practice of combining agriculture with other activities linked to their immediate environments in an effort to sustain both themselves and their families (Figure 4.8). Ireland's island communities, in particular, fished into offshore waters to supplement their meagre diets and sources of income. Sexton's case study on sea and shore foods highlights this practice in the Great

Fig. 4.6 SHEEP REARING IN THE SPERRIN MOUNTAINS, NORTHERN IRELAND. The Sperrin mountain range forms one of the largest upland areas in Ireland. Straddling counties Down and Tyrone, it separates the more intensive farming areas associated with the lowlands around Lough Neagh and Lough Foyle respectively. Its highest peak rises to 678m and, with its northerly latitude, weather conditions can be relatively severe, with snow cover common in winter. Extensive areas of bog and moorland often restrict farming to a specialisation in sheep-rearing, which involves grazing the animals on the rough pastures that cover much of the uplands. In winter, the sheep are often brought to lower slopes where pasture and grazing is better. The Sperrins are now classified as an Area of Outstanding Natural Beauty and this has imposed limits on what farmers can do to modernise their land use. While this may benefit tourists and daytrippers from urban areas such as Belfast and Londonderry, it can act as a further constraint on farmers seeking to improve income levels from a more productive use of their land. [Source: Copyright Tourism Northern Ireland]

Fig. 4.7 AGRICULTURAL LANDSCAPE ALONG THE COUNTY LOUTH COAST. Between Dundalk and Dublin, the Central Lowlands reach the coast to provide an extensive stretch of low-lying land, uninterrupted by uplands. Glacial drift deposited in this area is deep and provides the foundation for relatively fertile soils, while the east-coast location provides benign weather conditions. In combination, these natural attributes allow the area to offer a wide range of land use capabilities. The above-average farm size and the well-managed and large field patterns are also both a cause and an effect of the prosperous and innovative rural communities that reside in this historic core region of Ireland. The result is an agricultural landscape characterised by a specialisation in intensive arable and pastoral enterprises. It is not, however, a monoculture as farmers diversify land use by responding to changes in the market environment. Such a landscape is highlighted above, which shows a portion of the County Louth coast looking south to Clogher Head. This rural landscape contrasts strongly with the small-scale and extensive form of pastoral farming that characterises the west coast. [Source: Gordon Dunn]

Fig. 4.8 FISHING COMMUNITY ON ACHILL ISLAND, COUNTY MAYO, IN THE EARLY TWENTIETH CENTURY. Achill Island is Ireland's largest island and is located off the west coast of County Mayo. In 1841, it supported a population of 4,900. Despite the ravages of famine, this increased to 5,260 by 1911. Most of the island is covered in bogland and agriculture was confined essentially to the rearing of sheep. This enterprise could not sustain such a large population and many islanders, therefore, owned boats and became skilled fishermen. In contrast to the low carrying capacity of the land, offshore waters are rich in many varieties of fish. Using their traditional boats, or currachs (as shown above), large quantities of fish could be caught to supplement their diet but also to provide a source of additional income. Herring, in particular, was salted and stored in barrels, surplus amounts being sold to buyers from the mainland. The island's population has declined consistently since 1911 to 2,440 in 2016. Both agriculture and fishing now provide few jobs. Although a limited amount of pastoral farming persists, tourism is the dominant source of income on the island and some families possess a licence to take visiting tourists offshore to fish the rich fishing grounds. [Source: National Library of Ireland, call no.: L_IMP_1289]

Blasket Island and elsewhere (see below). Another tradition for some coastal communities located along the western seaboard involved the gathering, drying and burning of kelp seaweed to extract iodine for sale. Here, the plentiful and renewable resource of seaweed was also harvested to spread over the land to add nutrients to the thin and unproductive soils that dominated much of the landscape. The ubiquitous lazy beds of the nineteenth century, which yielded large crops of potatoes and supported a high density of population prior to the Famine, highlight this practice.

In contemporary times, the cultural and natural landscapes along Ireland's coastal areas provide valuable additional income to farming communities, as large numbers of tourists flock to the area. These offer opportunities for income to be derived from, for example, farmhouse bed-and-breakfast. In other locations, boglands – which are perhaps the most iconic feature of the natural environments of upland coastal areas – not only restrict the principal form of land use to extensive pastoral activities, but also provide for an alternative, farm-related activity. This involves the cutting of peat or turf from extensive areas of mainly blanket bog to provide a domestic source of fuel for rural communities.

Case Study: Blanket Bogs and the Cutting of Peat/Turf

Barry Brunt

Boglands, which cover some one-sixth (c.1.2 million hectares) of the island, constitute one of Ireland's most characteristic rural landscapes (Figure 4.9). These bogs are essentially waterlogged environments in which peat has accumulated during the postglacial period, and especially over the past 5,000 years. The peat is composed of partially decomposed plant material, such as the mosses, coarse grasses and heather that typify the surface vegetation of these areas. When cut and dried, the peat (called turf when dried) can be burned as a fuel. Since the seventeenth century, when most of Ireland's forest cover had been removed to create additional land for farming, the practice of digging out or harvesting peat gained considerable momentum, as the rural population looked to turf as their principal source of fuel for heating

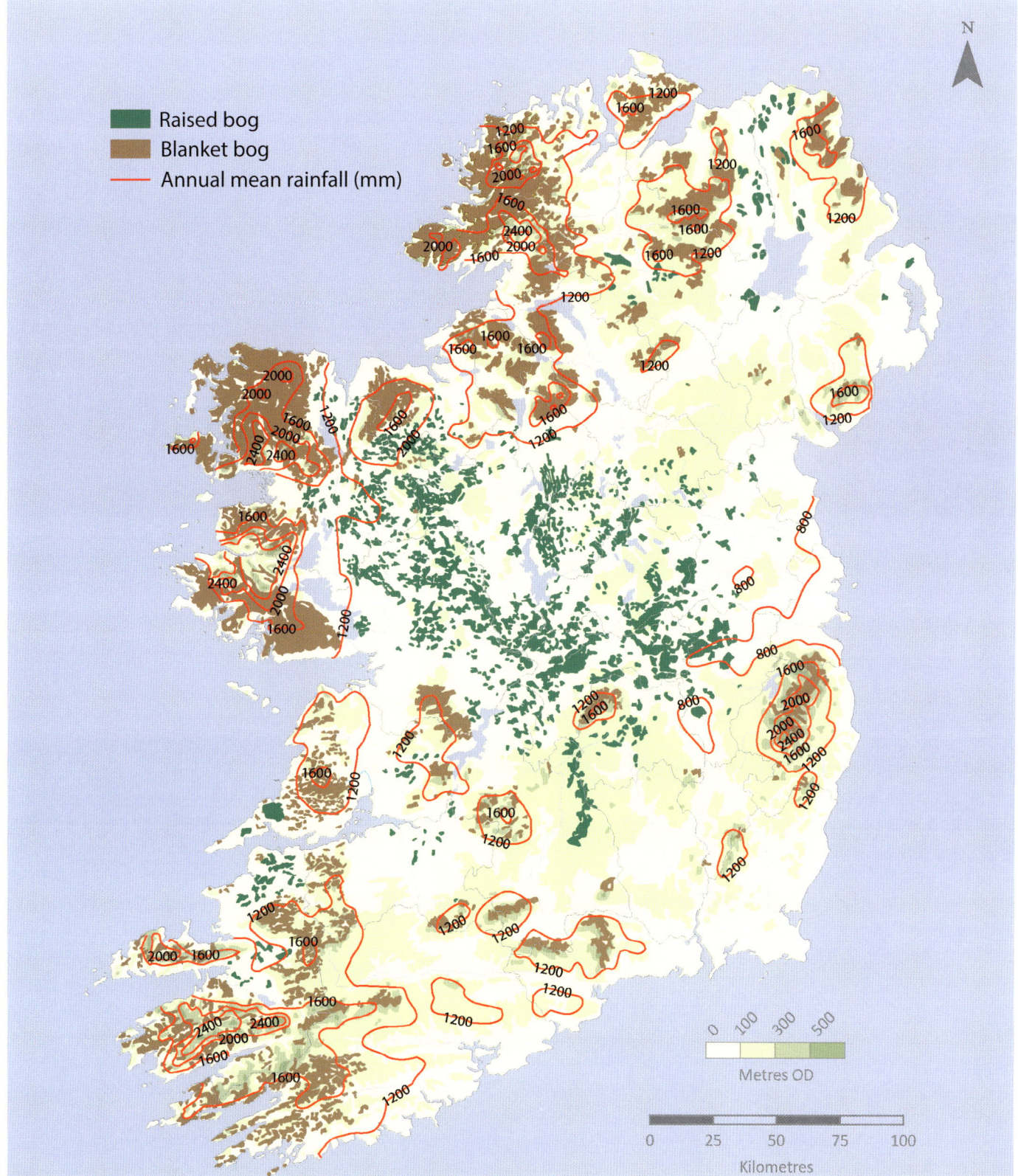

Fig. 4.9 DISTRIBUTION OF BOGLANDS IN IRELAND. Boglands form one-sixth of Ireland's surface area and constitute an important component of its rural land-scapes. Smaller, more fragmented, but deeper layers of raised bog are confined essentially to the the Central Lowlands. Since the end of the Second World War, these have been subjected to large-scale commercial exploitation by the semi-state company, Bord na Móna, to supply turf to power stations for the production of electricity. Exhaustion of supplies and concerns over the detrimental environmental effects of burning peat will see such operations cease in the 2020s. In contrast, blanket bogs occupy far more extensive areas of the island, especially in uplands along the western seaboard, but with important outliers to the east, such as the Wicklow Mountains. Here, harvesting of peat is confined largely to individuals working their own area of bog. High levels of precipitation (>1,200mm) in these uplands are important for the formation and retention of peat. In addition, widespread deforestation, commencing some 4,000 years ago by farmers seeking to extend pasture land for cattle rearing, was also a vital formative factor. This resulted in rainwater from the heavy precipitation that would previously have been taken up by trees remaining in the ground. Here, it caused leaching of soil nutrients and the creation of a 'hardpan' in the lower layer of the soil, which impeded drainage. Furthermore, a surface vegetation cover of rough grasses, shrubs and mosses was able only to decompose partially in such acidic and waterlogged environments. The result was the build-up of layers of peat and the creation of blanket bog. For over 400 years, farmers occupying areas of blanket bog have cut into these bogs to harvest peat as a source of domestic fuel. [Data source: Aalen, F.H.A., Whelan K. and Stout, M., eds, 2011. *Atlas of the Irish Rural Landscape*. 2nd ed. Cork: Cork University Press]

Fig. 4.10 TRADITIONAL HARVESTING OF PEAT ON BLANKET BOGS. The cutting of peat to provide turf from extensive areas of blanket bog in upland areas along Ireland's western coastline has a long tradition. With the loss of most of the country's forest cover by the 1700s, layers of peat, partially decomposed plant material that builds up in the acidic waterlogged environment of the bog, offered a cheap and effective alternative source of fuel. Traditionally, peat was removed by cutting into the deposits using a specialised spade known as a sleán. The water-saturated sods of peat had to be lain on the bog surface to be dried by the wind. When dried, the peat, or turf, was then usually stacked in larger quantities to be transferred to farmstead, by donkey and cart where possible, and sometimes simply in baskets or creels carried on peoples' backs. Although the individual amounts harvested were limited, over time the practice has reduced considerably the area and quality of remaining bogland. The labour-intensity, time expended and hard work involved has weakened this tradition, especially as alternative fuels have become available and environmental restrictions have been placed on cutting blanket bogs. [Source: National Library of Ireland, call no: VAL 8284]

and cooking. By the nineteenth century, access to turf could be considered a necessity of life for a poverty-stricken peasantry. The historic rights to cut peat for fuel from a particular bog (called turbary) are regarded as a well-established tradition in rural Ireland and are jealously guarded by those farmers whose lands include these bogs.

There are two main types of bog in Ireland, raised and blanket bogs, with the latter accounting for some 80 per cent of all bogland.[4] Blanket bogs are also the most characteristic vegetation cover to be found in upland areas adjacent to the western seaboard, where annual precipitation levels in excess of 1,200mm contribute fundamentally to its formation and retention. Although the depths of peat formation in blanket bogs are much less than those found in the raised bogs that occupy the interior of the country, layers of between 2 and 5 metres in thickness have long provided a significant resource for local farming communities to exploit. As a result, the area of blanket bog has been depleted severely: by some 15 per cent in the Republic and by over 50 per cent in Northern Ireland.

While yielding a harvest of turf, until recently these areas of blanket bog were considered generally to be wilderness areas suitable, at best, for extensive grazing of cattle and sheep. In effect, they came to symbolise the paucity of Ireland's rural landscapes and the poverty of its farmers (Figure 4.10). Although this generality continues to hold some validity, views on the roles and importance of blanket bogs changed considerably in the latter half of the twentieth century. During this time, a more positive attitude emerged with regard to 'farming' these areas. With financial encouragement from the European Union to raise farm output, drainage schemes were introduced in many bogs to create improved grassland to accommodate higher stocking rates of cattle and sheep. Also, since Ireland has the lowest percentage of forested land of any member state of the European Union, incentives were offered to plant coniferous trees on the boglands.

These developments resulted in a more productive use of the bogs, but had significant implications for rural landscapes along the western seaboard (Figure 4.11). Draining these bogs, for example, influences the rate for surface run-off of water. This, in turn, raises concerns over flood control in the drainage basins into which they feed and, therefore, the vulnerability to flooding of many coastal (and other) settlements. Drainage has also accelerated the loss of what are recognised increasingly as unique natural environments that hold important scientific and cultural value. Within the European Union, apart from Ireland and Finland, few natural boglands remain. The Irish Peatland Conservation Council (IPCC), for example, estimates that some 8 per cent of the world's blanket bogs, which remain in good condition, are to be found in Ireland.[5] Furthermore, with a growing recognition that removal of boglands equates to a loss of areas of unique scientific interest, biodiversity and cultural and aesthetic landscapes that hold tourist potential, legislation has been passed by the European Union that demands the protection of such environments for future generations. Financial incentives to drain boglands have been removed and restrictions placed on changing the land use in bogs designated as special areas of conservation. This is required given that the IPCC, which acts as a 'watchdog' for over 180 blanket bogs, suggests that over 60 per cent of them are under threat from a number of

Fig. 4.11 BADLY DEGRADED BOG NEAR SLIEVE LEAGUE, COUNTY DONEGAL. Many of Ireland's coastal uplands, especially in the west but also in the Wicklow Mountains and the Mountains of Mourne to the east, are mantled in layers of peat, formed of sphagnum moss, heathers and other acid-loving and water-tolerant plants. Poor in nutrients and waterlogged for much of the year, these blanket bog landscapes constitute fragile and unique habitats, now recognised as being of international significance for wildlife conservation. In economic terms, these uplands have long been considered of marginal and limited economic value, mostly used for grazing sheep and goats and the extraction of peat for domestic fireplaces. In recent years, however, they have come under pressure from overgrazing, trampling by increasing numbers of hillwalkers and mountain bike enthusiasts, construction of wind farms and conversion of the land to upland forestry. As well as impacting wildlife and habitats, the degradation of peat bogs also releases large amounts of carbon dioxide and methane into the atmosphere, contributing to climate warming. Conversely, intact, preserved and restored peatlands have important potential as carbon sinks, helping Ireland meet its commitments under international protocols to limit or reduce greenhouse gases and global climate change. [Source: Kyle Fawkes]

quarters, including continued cutting of turf, overgrazing (especially from sheep), trampling and rutting by hillwalkers as well as, in recent years, degradation by mountain bikers and the promotion of alternative land uses, such as forestry and schemes for the generation of wind energy.

Environmentalists regard legislation to protect and preserve blanket bogs as essential. However, many farmers, especially those with rights of turbary, often view restrictions as infringing their ways of life and historic interactions with the boglands. Such conflicts of interest, involving the delicate interface between oftentimes marginal rural communities and physical environments that are sensitive to change, require careful and integrated planning. The absence of such an approach threatens the sustainability of traditional rural communities as well as a critical component of their cultural and physical environments.

FARMING IN COASTAL IRELAND THROUGH THE AGES

While there are some unique elements to Ireland's coastal farm practices, as outlined above, in general terms they have mirrored broadly those that operated through the ages for the island as a whole. Hence, their evolution has to be discussed in the context of overall changes in the Irish agricultural sector.

Initial habitation of Ireland from around 9,000 years ago involved a hunter-gatherer existence rather than farming. This pre-farming era originated on the coastline but spread gradually inland along the island's estuaries and rivers.[6] Evidence of a transition to recognisable farming occurred from the Mesolithic through to the Neolithic eras, as represented compellingly by the discovery of the Céide Fields adjacent to the coastline of north County Mayo. This remarkable archaeological find, which extends to hundreds of acres, uncovered a landscape of stone-walled fields, houses and megalithic tombs from around 6,000 years ago, preserved under blanket bog (see Chapter 16: The Inhabitants of Ireland's Early Coastal Landscapes, Fig. 16.25). It revealed a seemingly organised community of farmers who had replaced forestry with farmland and regular field systems. The dominant activity was cattle rearing, although artefacts indicate that these ancient farmers may also have used wooden ploughs, with a stone-cut edge and drawn by oxen, for cultivation of fields. Study of other Neolithic settlements, such as Newgrange in the east of Ireland, confirm that farming at that time was primarily pastoral and dominated by cattle rearing for beef consumption.

Through the Bronze Age and into the Iron Age, evidence emerges of some growth in arable farming. Despite the appearance of what could be described as a more balanced system of sustainable husbandry, the inherent suitability of Ireland's natural environment for pastoral activities was difficult to ignore.[7] As a result, the rearing of livestock and dairying to produce beef and butter became predominant farm enterprises for much of the country throughout the early- and late-medieval periods. On a regional basis, however, tillage retained an importance along the eastern seaboard, especially around Dublin, where crops included wheat, barley, oats and legumes. This mix of farm enterprise matched more closely that found in Britain and most of continental Europe, where tillage played a more dominant role in land use. In spite of this, the disproportionate dependency of pre-modern Irish

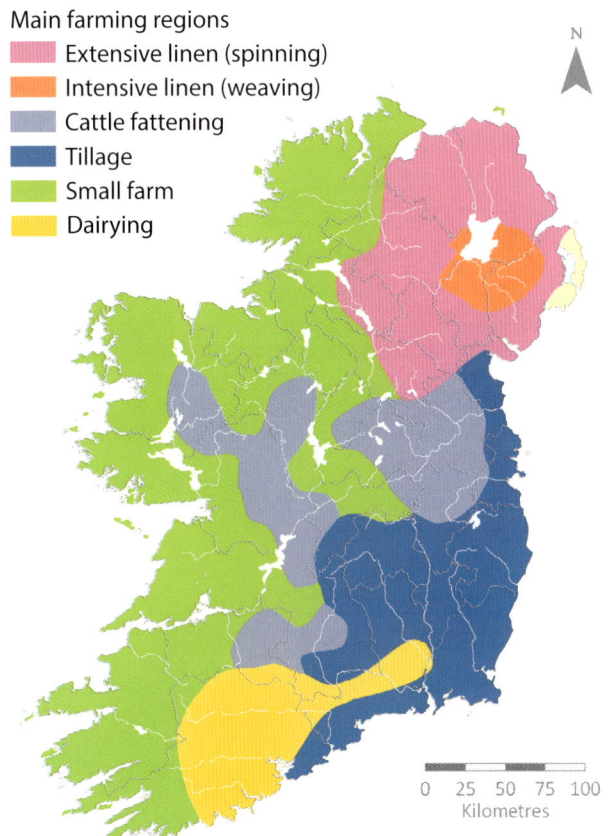

Main farming regions
- Extensive linen (spinning)
- Intensive linen (weaving)
- Cattle fattening
- Tillage
- Small farm
- Dairying

N

0 25 50 75 100
Kilometres

Fig. 4.12 IRELAND'S FARMING REGIONS IN THE MID-EIGHTEENTH CENTURY. The introduction of the plantation system and a rent-paying tenancy, combined with the island's diverse natural environments, gave rise to changes in its agricultural landscape. By the mid-eighteenth century, five distinct farming regions could be discerned. In the north, a proto-industrialisation encouraged tenant farmers to grow increasing quantities of flax for the emerging linen industry, as well as raising demand for its traditional food products. Along the east coast, commitments to tillage were intensified to respond to the twin demands of a growing domestic population and increasing overseas trade. Serviced by an improving transport network, which linked this region's prosperous and dense network of rural towns and villages to the main east coast ports and which facilitated the rapid transport of perishable goods to market, its status as the island's core farming region was confirmed. Along the south coast and its hinterland, environmental qualities that had long favoured intensive dairying enterprises were now amplified by the emergence of Cork as the most important centre in the country for butter production, as well as it being a key supply port for the booming North Atlantic trade in provisions. The entire western seaboard, however, showed little, if any, change from its historic landscape dominated by small farms. [Data source: Aalen, F.H.A., Whelan K. and Stout, M., eds, 2011. *Atlas of the Irish Rural Landscape.* 2nd ed. Cork: Cork University Press]

agriculture on pasture and livestock created an agricultural system that differed fundamentally from that which evolved in the rest of Europe. This difference persists to the present day.

Early social structures in Ireland also developed progressively into a widespread clan system that involved land-based tenure and communal ownership of land. The clan chieftain held private ownership of some land, from which revenues could be derived to help finance the support of soldiers and fortifications needed for protection and/or expansion of clan territory. All clan members, however, held some common rights and this encouraged a strong sense of identity with the land, together with the importance of its use for the wellbeing of the community. The clan structure and attendant land-use patterns with a focus on pastoralism prevailed until the arrival of the plantation system in the mid-sixteenth century.

This stability of social structures and a sympathetic relationship between farmers and their natural environments encouraged a period of prosperity. In many coastal communities, the combination of productive pastoral pursuits, opportunities to fish rich offshore waters and growth of trade with Britain and the rest of Europe became notable characteristics. To protect and nurture such wealth, a remarkable array of approximately 3,500 tower houses were constructed throughout Ireland, many located along the coast between counties Waterford and Galway (see Chapter 18: Era of Settlement: Trade, plantation and piracy). Wealth was also reflected in the appearance in the countryside of many new monastic orders and monasteries (see Chapter 17: The Vikings and Normans: Coastal invaders and settlers).

PLANTATIONS: A TRANSFORMATIVE PERIOD FOR AGRICULTURE

For several millennia, Ireland's agriculture had evolved slowly as farmers adapted to their natural environments. By the late sixteenth century, the island was a pastoral-based economy, organised essentially under traditional forms of communal land ownership and a dispersed pattern of small settlements. The mid-1500s to the late 1600s, however, was a period of major conflict with English rule, such as the revolt by the Earl of Desmond (1579–1583), the conquest of Ulster by English forces leading to the Flight of the Earls (1607), Cromwellian suppression (1649–1653) and the Williamite Wars (1688–1691). During this time, both economy and population declined and the traditional Gaelic way of life was marginalised, as Ireland became integrated into the English state.[8] This would expose the island to new and powerful external forces of change, which resulted in some fundamental transformations of its agriculture and rural landscapes.

Central to the imposition of English rule and land practices was the introduction of the plantation system. This involved the confiscation of Gaelic lands and the transferral of ownership of these mainly to English and Scottish landlords, who were usually Protestant.[9] Along with a new landlord class, more than 100,000 English, Scottish and Welsh settlers were brought to Ireland, and especially to Ulster, to work on the new estates and to consolidate the system. At least a quarter of farmland became owned by absentee British landlords and, by 1700, less than 10 per cent of land remained in Catholic possession.

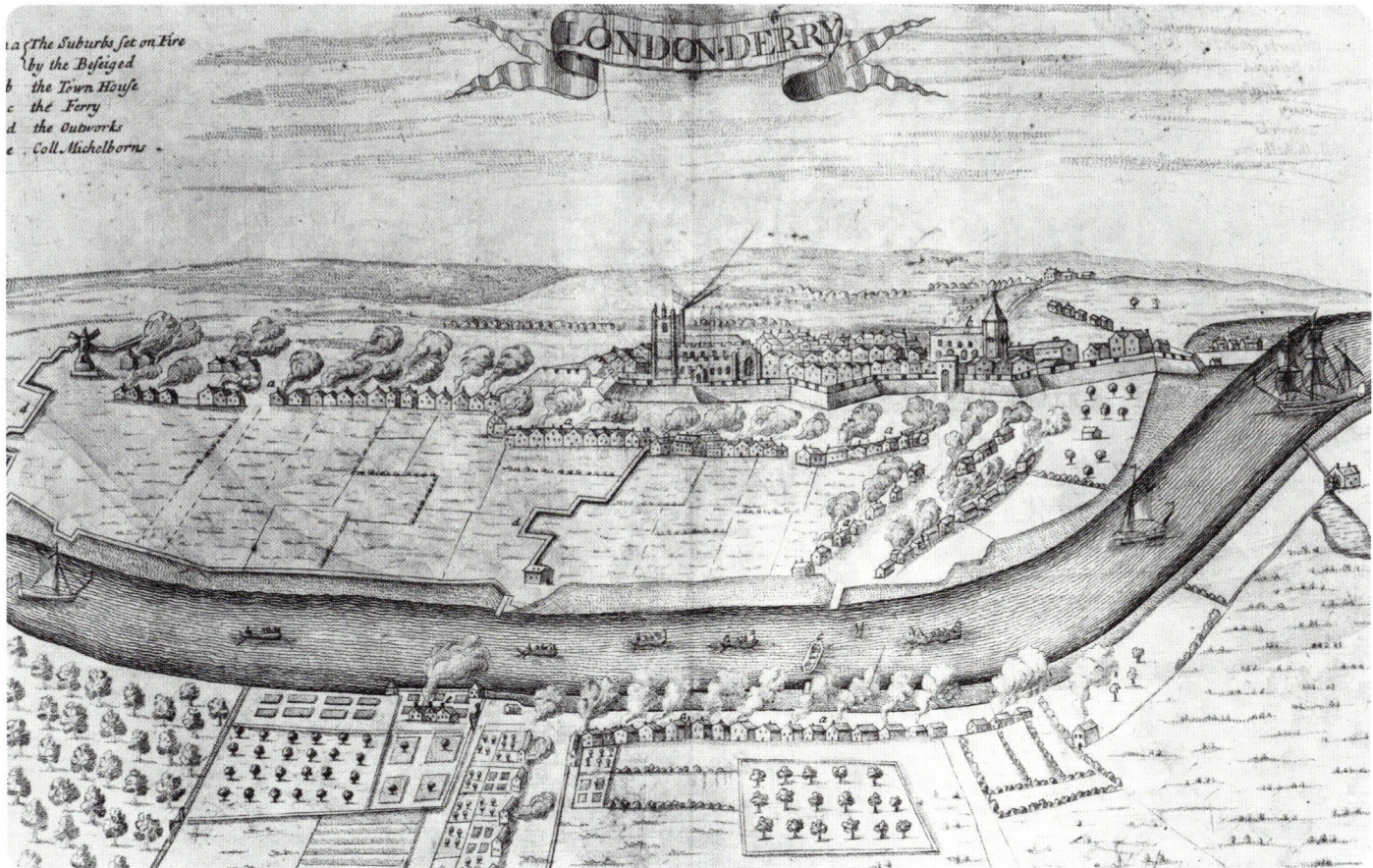

a The Suburbs set on Fire
  by the Besieged
b the Town House
c the Ferry
d the Outworks
e Coll. Michelborns

LONDON-DERRY

Fig. 4.13 THE PLANTATION TOWN OF LONDONDERRY, (c.1689), ENGRAVING BY JOHN MITCHELBURN. The suppression of the Ulster Rebellion in the early seventeenth century initiated a period of intense plantation within the province. This involved the confiscation of lands owned by Gaelic lords, which were then granted to a new class of English, Scottish and Welsh landlords. To work and settle the new estates, an estimated 80,000 or more settlers from England, Scotland and Wales had arrived by the 1630s to create areas of Protestant majority populations in several key, strategic areas. The Foyle River Valley was one such area and in 1613 the first planned city in Ireland was commenced at Londonderry, located on its banks. Here, a pre-existing town was rebuilt, with impressive walls designed to protect the growing settler population and to control an increasing trade through its port. Central to the success of this and other planter towns was the reorganisation of land use on the estates. As seen above, by the 1680s Londonderry was surrounded by a rural landscape arranged in a well-ordered pattern of fields. This represented the emergence of intensive pastoral, but especially arable, farming and was suggestive of a prosperous community. The pattern essentially replaced the traditional but less productive Gaelic system of land use, centred more on community ownership of land and more extensive pastoral enterprises. [Source: Thomas, A., 2005. *Irish Historic Towns Atlas*, No. 15, Derry~Londonderry. Dublin: RIA]

The estate system established under plantation introduced a more commercial approach to farming, especially in areas where the carrying capacity of the natural environment favoured a productive land use. By the mid-eighteenth century, this externally imposed system appeared to be firmly in place and defined a new map of specialised farming regions (Figure 4.12). This was encouraged by Ireland's location, which had acquired a pivotal position for England as it sought to extend control over opportunities presented with the opening up of trade across the North Atlantic. Many landlords and large-farm tenants moved away from the rearing of animals to focus instead on producing commodities, such as beef and butter, for the North Atlantic provisions trade. Further stimuli were presented by the demands and high prices for food from Britain's emerging urban-industrial workforce and, later, from the Napoleonic Wars. In contrast to Britain, however, where major gains in productivity were achieved from its agricultural revolution, progress in Ireland was slower due to inherent constraints and the stultifying effects that characterised Ireland's landlord and tenant system.

A key dimension of the new estate system in Ireland was the introduction of rent-paying tenancies, which pressurised tenants to become more responsive to prices and market demands in order for them to meet their obligations to landlords. With high prices and returns from rents rising up to four-fold in the fifty years prior to the end of the Napoleonic War in 1815, successful landlords were able to invest in the promotion of new towns, villages and improved transport networks to facilitate a more commercialised system of farming (Figure 4.13). In addition to such infrastructure, many newer farm enterprises, especially tillage, were more labour intensive than traditional forms of extensive rearing of cattle and sheep. This stimulated additional demands for farm labour. Such a period of political stability and relative prosperity would see total population on the island increasing from some three million in 1700 to about seven million by 1821, which placed additional pressures on available land.[10]

The tillage areas of the east coast of Ireland proved particularly well suited to increasing the growth and productivity of cereal crops. This led, however, to an increased demand from established and relatively large-scale tenant farmers for additional agricultural labourers to work the arable landscape.[11] In many cases, this involved tenant farmers using unpaid labour or 'cottiers' in exchange for the provision of a small plot of land on which potatoes could be grown (a 'potato wage') and a one-room cottage built. The contrasts between landowning classes and the majority of the rural population were immense. However, although the workers lived in conditions of abject poverty, the introduction of the potato

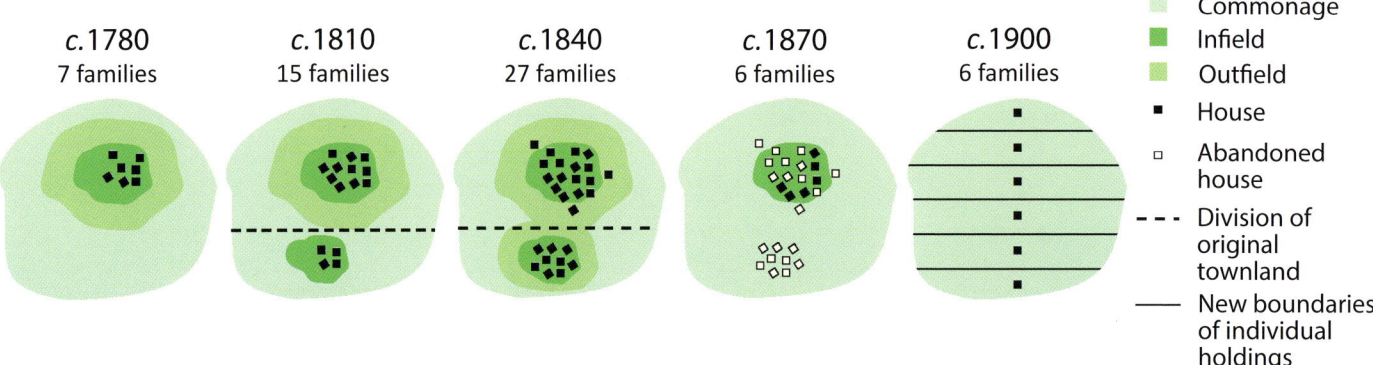

	Commonage
	Infield
	Outfield
■	House
□	Abandoned house
- - -	Division of original townland
——	New boundaries of individual holdings

*c.*1780 — 7 families   *c.*1810 — 15 families   *c.*1840 — 27 families   *c.*1870 — 6 families   *c.*1900 — 6 families

Fig. 4.14 A SCHEMATIC EVOLUTION OF THE CLACHAN AND RUNDALE SYSTEM OF LAND USE. By the late eighteenth century, clusters of family farm dwellings (clachans), often linked by kinship, had become a common feature in what had formerly been unsettled and marginal lands, typical of townlands located in the upland slopes in the west of Ireland. As distance from a clachan increased, the leased and shared land was worked more extensively. A small infield adjacent to the dwellings was given over to intense cultivation of crops (especially potatoes), with families boosting its carrying capacity by application of fertilising materials, such as seaweed. The larger outfield offered grazing opportunities for animals as well as the extension of lazy beds for cultivation of more potato crops. The most marginal land at the periphery of the leased area was commonage, on which limited numbers of cattle could graze. As population grew into the nineteenth century, both within an existing clachan as well as through migration from elsewhere in Ireland, available land was subdivided increasingly amongst a larger number of people. The division of townlands also allowed for new settlements, with their associated rundale system (broadly similar to the run-rig system of the Highlands of Scotland, from which it may have been partially derived, and with which it shared both its strengths and, ultimately, its weaknesses) to be extended into more marginal commonage. The Famine emphasised the unsustainability of such population pressures and the rundale system collapsed, with the abandonment of many individual farmsteads and villages that occupied the most marginal land. By the late nineteenth century, the rundale system was replaced, with individual farmers being granted personal ownership of land, organised usually into rectangular strips, or ladder farms. [Data source: Aalen, F.H.A., Whelan K. and Stout, M., eds, 2011. *Atlas of the Irish Rural Landscape*. 2nd ed. Cork: Cork University Press]

to Ireland provided them with a basic food supply. It also encouraged the continued and large-scale increase in population. By the early nineteenth century, almost one-in-two of the population residing in these tillage areas were cottiers.

While much of the country was transformed under the estate system, a more traditional and communal form of land use and small farms continued to dominate large areas of the more marginal environments along the western seaboard. Here, families, often united by kinship, came to reside in relatively small nucleated groups of farmhouses, or clachans, and worked their leased land through a rundale system (Figure 4.14). This involved sharing a small area of improved land (the infield) located adjacent to the clachan, which was used for the cultivation of crops. Surrounding this was a larger outfield, composed of more marginal land, on which families reared animals and/or extended the cultivation of potato crops as well as providing a source of turf. Beyond this was commonage of rough pasture, which provided grazing rights to the families and was controlled to avoid overgrazing. In many respects, it was an effective adaptation by farming communities to local environmental constraints. This intimate relationship between local environments and mostly Irish-speaking communities also found expression in a rich tradition of naming landscape features to give them a further identity and meaning within their culture.

Case Study: Deep Geography: Memory, community and continuity of coastal place names

Patrick O'Flanagan

By the early nineteenth century, Ireland's western coastal zone, including its many islands, had become one of the most populous and densely named rural areas in Ireland. Prior to the 1840s, high birth rates within local communities, as well as migration from elsewhere in Ireland of people attracted by the multiple resources of the coastal area, contributed to this surge in population. The relative isolation of the area also served to shield this traditional Irish-speaking society from increasing forces of anglicisation, with their cores located outside of Ireland. Following the Great Famine, however, a population adjustment occurred that involved high losses in population. This trend would persist into the latter half of the twentieth century as processes of continuous outmigration left these areas severely depopulated and, in some cases, even abandoned.

Population losses on such a scale have had major implications for the long-term economic and social vitality of the west of Ireland. They also resulted in a deep cultural impact, reflected in a linguistic transformation of many parts of the region over the last two centuries. In effect, a strong Irish-speaking tradition has been diluted severely by external forces which have increased the dominance of the English

Fig. 4.15 THE CULTURAL LANDSCAPE OF KILGALLIGAN, COUNTY MAYO. The townland of Kilgalligan, located on the northern shore of Broadhaven Bay in north County Mayo, remains a mainly Irish-speaking area. Its landscape is characterised by extensive areas of blanket bog and upland slopes, providing mostly rough grazing for limited numbers of cattle and sheep. Part of the outfield is illustrated in the foreground, which is now essentially cut-away bogland. The remains of what was once the more productive and intensively used infield pattern occurs in the background as a more regular arrangement of fields trending downslope from the settlement. Here, and elsewhere in western Ireland, from the late nineteenth century the Congested Districts Board and later the Land Commission reorganised many former rundale arrangements of landholdings. The result was usually the creation of a ladder-like organisation of fields, with farmhouses built on each newly created strip of land. [Source: Patrick O'Flanagan]

language. Contributing to these changes were major declines in the small-scale, mixed farming and local fishing activities that had been the mainstay of most communities. These localised activities have been replaced, to a large extent, by externally driven leisure and tourist pursuits. This is exemplified by the sale of building sites and the erection of second homes, together with the large numbers of tourists that pass through the area. In essence, Irish-speaking populations have declined significantly.

Abandoned areas and other places that have become more anglicised have witnessed a large-scale loss of unrecorded place names. In contrast, there are some relatively small areas where the Irish language remains strong and acts as the lingua franca of local communities (see Chapter 20: The Great Famine, Fig. 20.19). Most of these are located along the western seaboard and are known collectively as the Gaeltacht. These relatively remote and predominantly rural areas remain rich repositories of oral Irish culture, exemplified in speech, the conservation of significant volumes of place names and traditional music. Although frequently changed, and often politicised, place names can also illustrate the strong relationships and identities that traditional communities hold with their local surroundings. Research into these oral sources allows for a better understanding of how place names function within local communities.

Most major physical features along Ireland's coastline, visible from the sea, were recorded by maritime cartographers and chart-makers visiting the island. Over time, different visitors to Ireland's coast – such as Basque, Galician and Portuguese mariners – compiled their own shoreline compendiums of place names.[12] From 1800 onwards, many anglicised coastal place names were also inserted on Admiralty charts and the work of O'Donovan and O'Curry, agents of the Ordnance Survey, rescued many such place names from oblivion. The principal agents for the naming of places, however, were local communities.

Research in County Mayo has illustrated the extraordinary diversity and richness of coastal place names still evident in a few Gaeltacht areas. Focusing on the townland of Kilgalligan and its adjacent Irish-speaking area of Dún Chaocháin to the north of Belmullet, research in the 1970s indicated the continued presence of a rich harvest of historic Irish place names. It furthermore confirms that in some Irish-speaking farming communities, place names discharge many different functions for a largely non-literate community. Their maintenance in such contexts owes more to their past and contemporary functions, rather than their role as place identifiers. In effect, they serve as archives for local communities and help evoke strong linkages with their surroundings, in both a metaphysical and physical sense (Figure 4.15).[13]

Irish place names in Kilgalligan and elsewhere can be classified into four principal groups. Toponyms represent physical features, such as *droim* (hill), *fothair* (cliff), *rinn* (point), *sliabh* (mountain), *trá* (beach) and *béal* (mouth). Second, hydronyms are water-related features, such

FIG. 4.16 THE SEANCHAÍ, JOHN HENRY (*c.*1975). An oral tradition has long been central to the understanding of the large numbers of Irish place names that occur in Gaeltacht areas along the western coastline of Ireland. As local populations collapsed with the exodus of their youth and the death of older residents, the meanings of places and place names have generally faded from collective memory. This is a major loss for the cultural expression of such communities. Fortunately, many of these historic place names were archived orally within local communities by leading seanchaí and spokespersons, such as the late John Henry (*c.*1915–1998). This man was a monoglot Irish-speaker who lived his life at Kilgalligan, County Mayo. As a gifted and highly knowledgeable seanchaí or traditional storyteller and historian, he acted as a major source in the research conducted by Séamus Ó Catháin and Patrick O'Flanagan in the early 1970s, which is cited in this case study. The importance of understanding and preserving the store of place names in such places has been continued by contemporary researchers, such as Treasa Ní Ghearraigh, Uinsíonn Mac Graith, Séamus Ó Mongain and through the maps and writings of the English scholar, the late Tim Robinson. [Source: Professor Séamus Ó Catháin]

as *cuan* (harbour), *inbhear* (inlet) and *tobar* (well). Third, bionyms relate to plants and animals, such as *bogach* (bog), *machaire* (mahair) and *nead* (nest). The final group involve anthroponyms, or names that relate to human activities or particular events; examples include *tamhnaigh* (a fallow enclosure), *straidhp* (a strip of land) and *geata* (gate). The majority of place names recorded at Kilgalligan in the early 1970s were either toponyms or anthroponyms.

Talamh an Fhir Dhuibh (the dark man's holding), *An Fiodán Tuileach* (the stream in spate), *Ruball an Lachán Chaol* (the tail of the thin duck), and *Garraí Micín Larry* (the garden of Micín Larry) are just some of the place names of Kilgalligan that illustrate the ingenuity and inventiveness of their makers (Figure 4.16). Some 800 hitherto unrecorded place names were also discovered during research, most of which were known only to, or were used only by, the local townland communities. Such 'in-group' names had virtually no currency outside the townland and only a handful were ever recorded and mapped by the Ordnance Survey.

Most of Ireland's coastal place names were nurtured in contexts of small and often remote rural communities, where oral practice and tradition supersede the written word. As a result, place names evolved in a context where they served various functions, the most important of which was the endowment of local spaces with names that embodied elements of collective memory and beliefs for the community in which they were used. With large-scale population losses through migration and the demise of the Irish language, collective memories that are central to appreciating the roles of place names have also been eroded. The research was able to show, therefore, that even when most place names retained a general currency in Kilgalligan, not all were known or appreciated by the townland's contemporary residents. More recent work in this and other townlands has furthered knowledge of the rich lexicon of place names that occurs along the western seaboard.[14]

THE DÚN CHAOCHÁIN PLACENAME COLLECTION PROJECT

Treasa Ní Ghearraigh agus Uinsíonn Mac Graith

The Gaeltacht community of Dún Chaocháin, Erris, County Mayo, identifies easily with their environment because they live in a landscape that is littered with Irish place names. These provide meaning to features of that landscape and give them a place in the lives of the people. Along this coast, every rock, inlet, cave, reef, stack and headland is imbued with its own special name. Inspired by earlier research on Irish place names undertaken at Kilgalligan, Uinsíonn Mac Graith commenced work in July 1998 on the Dún Chaocháin Placename Collection Project, which resulted in the collection and mapping of over 1,500 place names by December 1999.[15]

Some 2.5km from the mainland of Ceathrú na gCloch townland is a collection of five impressive sea stacks known to seafarers as the Stags of Broadhaven and referred to by local fishermen as *Na Stácaí* (Figure 4.17). From north-west to south-east these rugged, steeple-like rocks are known as *An tOigheann* (the cauldron, 78m high), *Carraig na Faoileoga* (rock of the seagull), *Teach Dhónaill Uí Chléirigh* (Dónall Uí Chléirigh's house, 97m), *An Teach Beag* (the small house) and *An Teach Mór* (the big house, 94m).

Composed mainly of schist and originating as sandy sediments on a shallow seafloor, they are among the oldest rocks in the country. With their counterparts in Nova Scotia and Newfoundland, they have stood proud against the relentless weathering of the Atlantic waves. Through the schist are bands of calcium silicate which have left the impression of a white horse on *An Teach Mór*, referred to locally as *Tóin a' Ghiorráin Bháin* (the backside of the white gelding).

Solas Theach Dhónaill, a gap beside *Teach Dhónaill Uí Chléirigh*, is a navigation aid to sailors. To the north-east, and proximate to *Na Stácaí*, is *An Bád Bréige* (the false boat), a small rock often mistaken for a boat when seen from a distance.

An tOigheann

Carraig na Faoileoga

Teach Dhónaill Uí Chléirigh

An Teach Beag

An Bád Bréige

An Teach Mór

Fig. 4.17 *Na Stácaí* (THE STAGS), BROADHAVEN BAY, COUNTY MAYO. A set of prominent sea stacks are located off the coast at Benwee Head, Broadhaven Bay, County Mayo. Each stack has its own genealogy crystallised in its name. The stacks provide a good breeding habitat for seabirds and support a diverse offshore marine ecosystem. They are a popular site for divers and sea kayakers. [Source: Paddy McGuirk, courtesy of Uinsíonn Mac Graith and Treasa Ní Ghearraigh]

Formerly, *An Teach Mór, Teach Dhónaill Uí Chléirigh* and *An Teach Beag* provided grazing for eight, six and two sheep respectively. These were transported by currach to the stacks in May and, despite the difficult environment, they were collected and returned to the mainland by late September, often in a thriving condition.

The sea cliffs and islands provide an excellent habitat for breeding seabirds. Some are protected species, such as the rare Leach's petrel, which nests on the Stags, one of its two nesting sites in Ireland. Other notable species include the storm petrel, puffin, fulmar and kittiwake and a more recent visitor, the great skua.

Na Stácaí (The Stags) are now a popular site for divers and kayakers to explore the sea arches that have been gouged out, as well as the tunnel-like cave running straight through *An Teach Beag*.

Fig. 4.18 INIS OÍRR (INISHEER), ARAN ISLANDS, COUNTY GALWAY. The search for available land by Ireland's growing population had, by the mid-nineteenth century, resulted in settlement being extended to some of the most peripheral and seemingly inhospitable landscapes. This is illustrated in the now abandoned dwellings and fields enclosed by dry stone walls that typify much of the landscape of the Atlantic seaboard and offshore islands. Inis Oírr, the smallest of the three Aran Islands and located 8km off the coast of County Galway, presents such a relic landscape. These islands are a western extension of the Burren and their topography is, therefore, composed of relatively flat pavements of rock-strewn limestone with only limited soil cover. To support a growing population the land needed to be cleared of stones. The most effective way of doing this was to remove the stones by hand to create dry stone walls, which would designate individual fields on which soil could be improved, and which also provided some shelter from strong winds for cattle and sheep. On the three Aran Islands there remains an estimated 1,600km of dry stone walls, which reflects the intensity of effort required to settle such areas. In 1841, such a farm landscape, plus offshore fishing, supported 456 islanders on Inis Oírr. Although the promotion of a craft-based industry in the late nineteenth century (Aran woollen goods linked to local sheep rearing) helped to offset population loss (1911: 480 people), decline typified much of the twentieth century. In 1981, only 239 islanders remained on the island, as lack of employment encouraged migration. Since that date, a developing tourist industry has brought new prosperity to the island and has helped stabilise the population (2016: 260). These tourists access Inis Oírr via ferry boats from the mainland and visit the island to see the iconic landscape of dry stone walls, the ruins of the fourteenth-century O'Brien's Castle (used to collect tolls from vessels accessing Galway port), and also to experience the culture of this Gaeltacht community. Here, the difficulties of the past have been harnessed as a heritage-based resource, upon which to build the island's future prosperity. [Source: Irish Air Corps]

As population pressures in Ireland increased through the eighteenth and early nineteenth centuries, the rundale system came under intense pressure as a sustainable form of land-use management. Growth of family size within the clachan and its system of partible inheritance saw increasing subdivision of land shares. In addition,

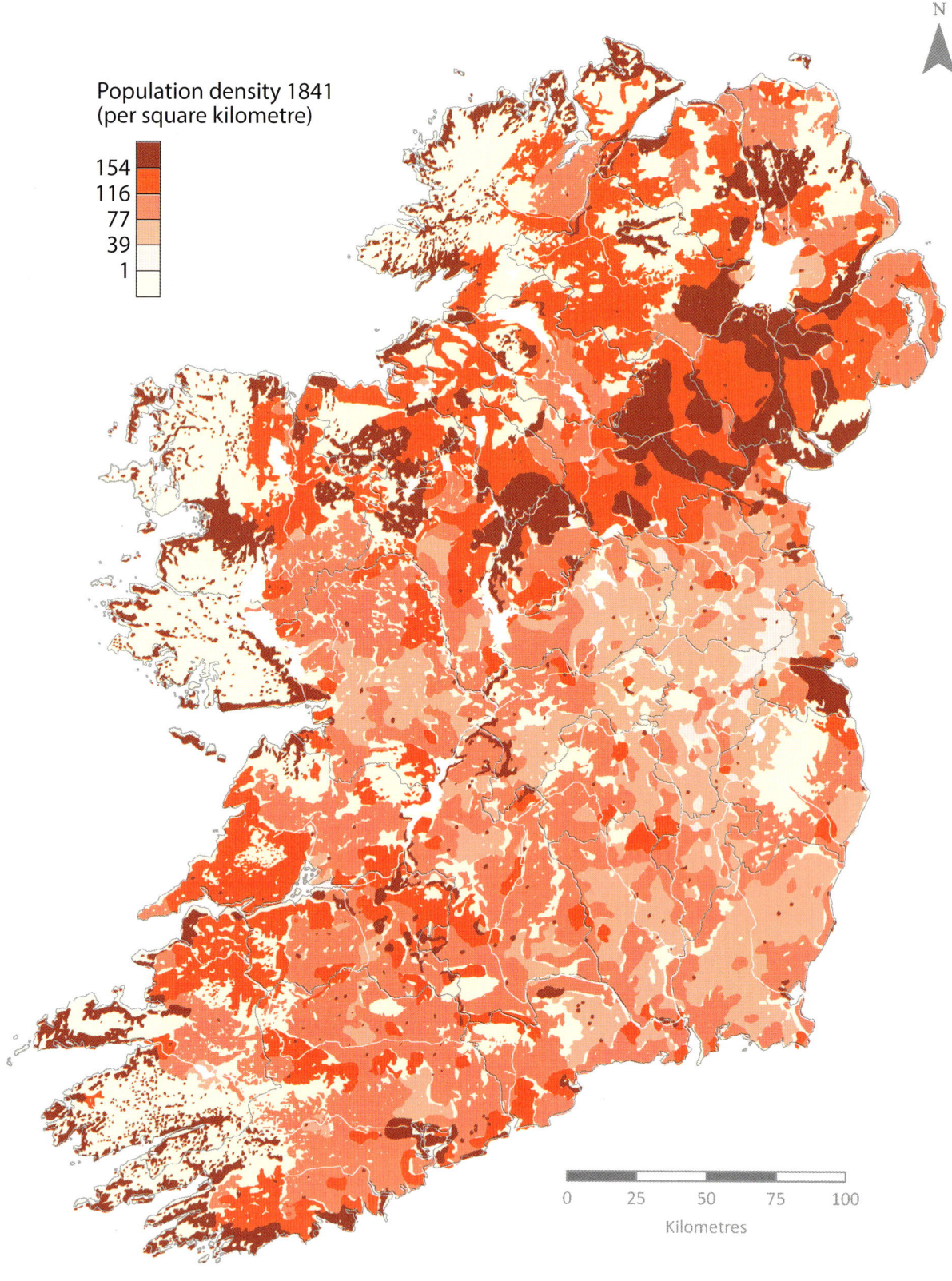

Population density 1841
(per square kilometre)

154
116
77
39
1

N

0   25   50   75   100
Kilometres

Fig. 4.19 DENSITY OF POPULATION IN IRELAND (1841).While the population of Ireland doubled to 8.2 million between 1780 and 1841, there were significant region-al differences in rates of growth as reflected in this map of population density for 1841. The largest area of highest densities relates to north-east Ulster, with some extension into the north Midlands, where proto-industrialisation supported a large growth in population. Areas along the eastern and southern coasts, where the environment possessed a high carrying capacity for agriculture, display relatively low densities. This relates to the dominance of larger and more prosperous farms, which were engaging in less labour-intensive enterprises, as well as lower birth rates associated with their significant Protestant communities. The growth of Dublin, Cork and especially Belfast, with its expanding industrial hinterland, is apparent. Along the western seaboard, extensive areas of upland remain virtually uninhab-ited. However, along its coastline and adjacent lower slopes of uplands, densities often in excess of 150 per square kilometre were recorded. This reflected the extent of colonisation that occurred in these more marginal environments, enabled by the availability of the potato to sustain large numbers of people. The pat-tern is highlighted around the peninsulas of counties Cork and Kerry and along the Connemara coast from Clew Bay to Killala. [Data source: Aalen, F.H.A., Whe-lan K. and Stout, M., eds, 2011. *Atlas of the Irish Rural Landscape*. Cork: Cork University Press, adapted from T.W. Freeman]

to accommodate increasing numbers of people seeking land, landlords seized the opportunity to extend farming into areas of commonage that were the least supportive of permanent settlement. Initially using lazy beds to facilitate the cultivation of high-yielding crops of potato, large numbers of farm families were able to occupy the increasingly marginal slopes of the uplands in western coastal regions and the offshore islands (Figure 4.18). This process also contributed strongly to a shift in Ireland's population away from the more benign farm environments of the east. By 1841, some of the densest areas of population were to be found along the limited lowlands and adjacent upland slopes of the western seaboard (Figure 4.19). Such a perverse relationship, in which areas with the least carrying capacity for settlement supported the highest densities of people, created the context for the Great Famine, which would soon devastate the island.

FROM FAMINE TO THE EUROPEAN UNION

The thirty years that followed the end of the Napoleonic Wars was a time of great distress in Ireland's agricultural sector. Less buoyant market conditions saw prices for its main farm outputs fall by some 30–50 per cent. This encouraged a shift away from tillage to more traditional but less labour-intensive pastoral farming, with one result being the release of large numbers of agricultural labourers who went in search of land. At the same time, farmland ownership and tenancy practices had become increasingly dysfunctional. Large numbers of absentee landlords, together with an increasing practice of multiple sub-tenancies, no fixity of tenure and high rents combined to restrict investment and innovative practices on the land. In such circumstances and with the vast majority of the island's poverty-stricken rural population depending almost exclusively on the potato for survival, the appearance of the potato blight in 1845 proved disastrous.[16] This caused the almost complete failure of the crop and a series of devastating famines, which collectively are known as the Great Famine (see Chapter 20: The Great Famine). These land and farming conditions also nurtured the rise of a period of progressive political instability and popular movements for independence, as illustrated in the life and work of Daniel O'Connell.

Daniel O'Connell and Derrynane: The coastal connection

Robert Devoy

Darrynane Abbey (known now as Derrynane House) is situated close to Derrynane Bay on the south-western edge of the Iveragh Peninsula (Figure 4.20). The location within the bay clearly forms the landscape background to the painting, with its small islands, including Abbey Island and the ruins of the abbey that was founded originally around the sixth-century. Multiple sand bars, small sand spits and a fronting dune and beach-barrier sedimentary system comprise the physical features within the bay itself, all of which are products of the high-energy wave conditions of the Atlantic ocean to the west. The house and coast, together with their historic associations, form a widely known heritage site and scenic attraction for tourists, part of the Wild Atlantic Way. The house was the family home of Daniel O'Connell (1775–1847), one of Ireland's most notable political leaders and advocate for fundamental reforms to improve the living conditions of the country's Catholic and poverty-stricken population in the first half of the nineteenth century.

O'Connell was born and grew up in the area of Carhan, close to the town

Fig 4.20 DARRYNANE ABBEY. Attributed to Robert Havell the Elder (fl. 1800–1840) and Robert Havell the Younger (fl. 1820–1850), after an original watercolour of c.1831 by John Fogarty. Darrynane Abbey, now known as Derrynane House, was the family home of Daniel O'Connell (1775–1847), who inherited it from his uncle, Maurice O'Connell, in 1825. It was a favourite rural residence for Daniel O'Connell, who was one of Ireland's leading political figures in the first half of the nineteenth century and a champion of the country's landless masses. Here, with his wife Mary O'Connell (1778–1836) and large family, he was able to retreat and find some tranquility, especially in the summer months and in the latter part of his life, from the turmoil and controversies that accompanied his political work. The death of his wife affected him greatly, and she was buried in the O'Connell family tomb on Abbey Island (shown in the bay behind the house in the picture). [Source: Image courtesy of the Office of Public Works/the Department of Arts, Heritage and the Gaeltacht]

of Cahersiveen and Valencia Island, until being sent to school at the age of thirteen with his brother, Maurice, first to Queenstown (Cobh) and then to France in 1790. Though later to become wealthy, through his career as a barrister and then by family inheritance, O'Connell's early life was rooted within the farming communities of these coastal districts of County Kerry. He was an Irish speaker from childhood and knew well the hardships of rural life and of being poor. As O'Connell's legal and political career grew, this brilliant orator became increasingly known as a champion and voice of the people of Ireland, particularly the landless masses.

During the early nineteenth century increasing numbers of Irish people became landless or displaced and were forced to live in agriculturally marginal coastal and upland areas. Caused initially by a complex of engrained historical and political factors and patterns of economics and trade, these degradations in living space were added to by more specific issues. Pressure from landowners on landholding and farm tenancy, as well as from land reclamation and agricultural improvement, further contributed to the problems of the rural poor. The resulting deterioration in people's lives was not ameliorated by the introduction of the potato as a monocrop to provide a staple food for the growing population on the island. The arrival of potato blight in Ireland in 1845 would confirm the problems of excessive dependency on this crop and led to the Great Famine (1845–1849). O'Connell's legal and political campaigns for Catholic emancipation, support of people's rights and an end to the political union with Britain were set against this backdrop.

The effects of the Famine were both severe and immediate, with some one million people dying of starvation and disease, and a further two million emigrating overseas. The number of landholdings of less than five acres (about 2.03 ha) fell nationally by 70 per cent, from an estimated 442,000 to 126,000 in the decade 1841–1851. These holdings were most prevalent along the western coastline and, with their dependency on potatoes being almost complete, many areas experienced losses in rural populations well in excess of 50 per cent. In contrast, the more diversified economies along the eastern and northern coasts suffered less severe losses. However, the implications of the Famine for Ireland's agriculture and rural development, especially in coastal areas, extended well beyond the traumatic decade of the 1840s.

First, the move of farm enterprises away from tillage to a reassertion of pasture-based cattle rearing gathered further momentum. Numbers of drystock, for example, almost doubled in Ireland in the fifty years prior to the First World War, while arable land use was halved to 0.9 million hectares by 1911.[17] New relationships based on land use devoted to cattle were also being forged. In particular, western areas became focused increasingly on the rearing and supply of young cattle, which would then be transported for finishing on the richer grazing lands around Dublin and the east coast.

Second, and arguably the most notable transformation in agriculture that occurred in post-Famine Ireland was the fundamental change in farm ownership. Landlords had come to

Fig. 4.21 LADDER FARMS ALONG THE WESTERN SEABOARD. The break up of the rundale system and large landlord estates allowed for a new organisation of land use. Through the work of the Congested Districts Board, individual farmers would be given ownership of land, on which their farm buildings would be located. In western, coastal areas, farms were reorganised generally through the creation of relatively large and usually rectangular-shaped fields, to encourage effective management. Individual farm dwellings would be located close to each other along a road, with fields arranged at right-angles to the road and running parallel to those of adjacent farms. The regularity of the field pattern gave rise to the term 'ladder farms'. These are common along the western coastline, as illustrated above, at Gowlaun in Connemara, County Galway. Here, the individual farms consist essentially of large, rectangular-shaped fields that stretch from the water's edge to the lower slopes of the surrounding uplands. This follows the principles of rundale in that it provides farmers with different qualities of land on their individual holding. Pastoral farming continues to dominate in these areas, although at the present time many fields are underutilised or have been abandoned for active farming. Most farms are located on or adjacent to the coastal road, which helps to improve accessibility within the rural community. Proximity to the waterfront and beach has also resulted in many farms being sold to 'outsiders' as second homes, or they have become involved in farmhouse B&B to supplement incomes. [Source: Irish Air Corps]

Fig. 4.22 (right) POPULATION CHANGES IN COASTAL COUNTIES (1841–1911). The devastating effects of the Famine on the island's economy and society led directly to a halving of the total population from 1841 to 1911. This was the almost universal experience in all coastal counties, apart from County Antrim where the growth of a powerful industrial base, centred on Belfast and its hinterland, allowed population to increase by over 60 per cent. One result of this was that the rate of population loss in the six counties of Northern Ireland (24 per cent: 390,000) was significantly lower than that experienced in the rest of the island (52 per cent: 3.4 million). Only County Dublin in the south was able to offset partially a large-scale decline, due to its strengthening urban and economic functions. In 1911, over 3.1 million people (72 per cent) of the island's total population of 4.3 million, continued to reside in coastal counties. [Data source: Census of population, 1841 and 1911]

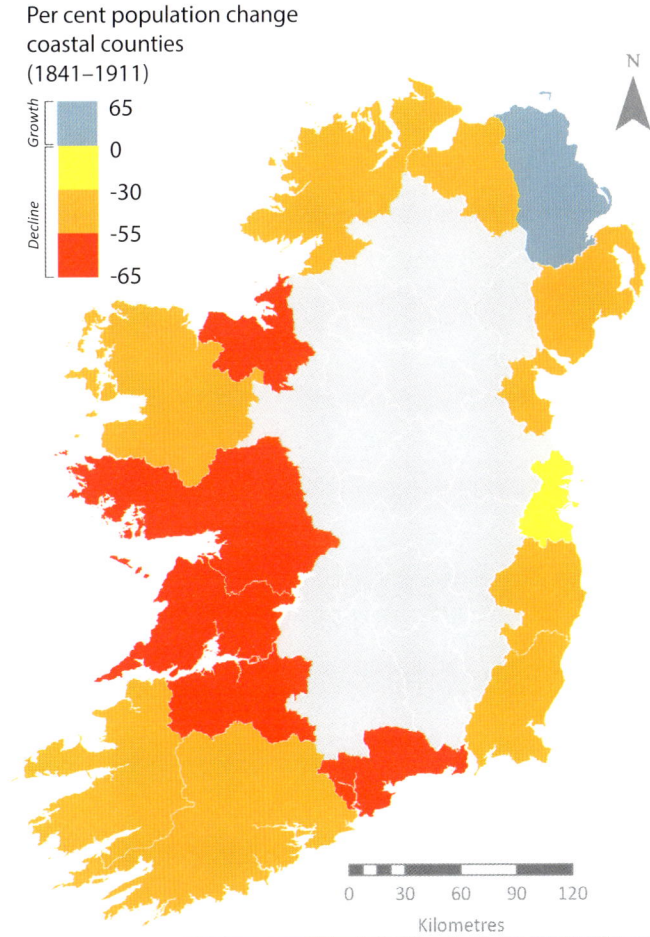

Per cent population change
coastal counties
(1841–1911)

Growth	65
	0
	-30
Decline	-55
	-65

0   30   60   90   120
Kilometres

recognise that the maintenance of small tenant farms imposed more costs than benefits, while the focus on subsistence within the rundale system had encouraged inertia rather than the innovation that was necessary to increase productivity and returns from land. To redress such economic concerns and also to satisfy a growing political movement in Ireland dedicated to resolving the rural population's historic demands for land, a transition from tenancy to owner-occupancy occurred.

Central to the process was the Irish Land League, founded by Michael Davitt in 1878, while the establishment of a Land Commission (1881) facilitated the division of many large estates to create individually owned farms for former tenants. As well as establishing a new social class of peasant proprietors, the farming landscape also underwent significant change. In this, the generally preferred spatial arrangement involved creating square-shaped farms in which the dwelling was centred within the holding. While facilitating the efficient use of land, it tended to isolate farm families and thereby worked against a long-established practice focused around social interaction. However, in western coastal and upland areas, where the rundale system was strongly represented, a variant of the rearrangement of land organisation appeared. Described as 'ladder farms', these involved locating farm dwellings close to each other along a road, with their associated landholdings organised in large, rectangular-shaped fields, running parallel to each other and perpendicular to the road. In coastal areas these rectangular-shaped fields provided farmers with a variety of environmental resources as they stretched from the seashore to lower mountain slopes (Figure 4.21). While helping to resolve a political issue, this reorganisation of landholdings failed to address longer-term economic problems. This was due in part to the scale of demand for farm ownership, which resulted in the continued dominance of relatively small-scale land units. In 1911, therefore, farms of less than thirty acres accounted for two-thirds of all farms in the twenty-six counties of southern Ireland.

Fig. 4.23 DESERTED VILLAGE ON GREAT BLASKET ISLAND, COUNTY KERRY. Located some 2km offshore from the Dingle Peninsula in County Kerry, the now-deserted village on the Great Blasket Island was once Ireland's most westerly settlement. In the 1841 Census, 153 people were recorded as being resident on the island, the majority in the village located on the more sheltered, north-eastern coast and facing the mainland. Most were involved in fishing but also raised cattle and sheep to graze rough pastures on the windswept slopes that dominate the landscape. In spite of difficult living conditions, the island spawned a rich literary and linguistic tradition, epitomised by several key writers, such as Peig Sayers. While the island's population remained resilient until the First World War, thereafter difficulties of working in such exposed environments, low living standards and isolation from services, combined to cause significant depopulation. In 1953, the last residents left the island, leaving the village deserted. The island is now a national park, with the objective of conserving and preserving its cultural landscape. Several houses in the deserted village have, therefore, been renovated to help 'bring-to-life' the island's cultural traditions for tourists who access the island by ferry service from Dunquin or Dingle. [Source: Barry Brunt]

Such a size distribution, especially in marginal environments, presented few opportunities for farmers to innovate and thereby increase levels of production.

Third, although the creation of individual family farms met some of the expectations within rural Ireland, it could not satisfy all of the needs of its large total population. The result was the continuation of significant declines in population, as those who failed to access land, or could not find alternative employment, continued to emigrate in substantial numbers. Between 1841 and 1911, the island's population was almost halved to 4.4 million, with 90 per cent of that loss occurring in the twenty-six counties that would later become the Republic. Much of this collapse in population was focused in coastal counties, especially along the western seaboard (Figure 4.22). Here, some villages and communities virtually disappeared, while the retreat of active farming from higher slopes and bog margins resulted in a landscape littered with abandoned cottages, fields and lazy beds (Figures 4.23, 4.24). The process of decline did not end with the First World War, but rather continued into the post-war period. Such a relic landscape is a reminder of a region that was once more densely settled and actively farmed.

The combination of the First World War, Ireland's War of Independence and the subsequent civil war had profound consequences for Ireland, not least in political terms with the emergence of two separate jurisdictions on the island.[18] Six counties in Northern Ireland remained a region within the United Kingdom, while twenty-six counties became an independent state. Reflecting historical differences in cultural and economic trends, this partition highlighted and intensified major contrasts in the degree of dependency on agriculture for employment between the two states. By 1926, Northern Ireland's successful engagement with industrialisation meant that only one-in-three of its workforce remained in agriculture. In marked contrast, Ireland's continued role as an underdeveloped rural economy was highlighted by the fact that 53 per cent of total employment was engaged in farming activities. Levels of dependency were particularly high along the western seaboard (Figure 4.25). Although an increased demand for food products in both world wars provided short-term gains for farming communities, longer-established negative trends continued to characterise agriculture in both jurisdictions.

Having become a regional economy within the United Kingdom, farmers in Northern Ireland were less impacted since they benefited from subsidies and uninterrupted access to the large British market for food, and were thus able to offset the scale of decline. In contrast, the problems and their effects were more severe in the south. Following on from the global economic crash (1929–1932), the Irish government decided to embrace a policy of protectionism to favour self-sufficiency. This was complemented by the state's commitment to emphasise its rural-based society and Gaelic traditions in order to distinguish itself from the anglicised,

Fig. 4.24 SOUTH-FACING SLOPE OF THE DINGLE PENINSULA AT COUMEENOOLE, COUNTY KERRY. The Dingle Peninsula hosted a large rural population throughout the nineteenth century, even though this was depleted severely in the aftermath of the Famine. Here, the southern slopes of the uplands that protrude into the Atlantic ocean carry the imprint of such dense populations. Large numbers of individual fields, marked by well-preserved boundaries of stone walls, extend far up the slopes as farmers strove to increase the availability of pasture for their cattle and sheep. The majority of these are now relic features in a landscape that is becoming more defined by tourism. Evidence of this is clear from the modern structures built along the coastal road, to the left of older and more traditional stone-built houses, some of which are abandoned, as well as the adjacent pattern of small, enclosed fields. [Source: Barry Brunt]

urban-industrial system of its former colonial ruler. In this cultural and political context, land was treated more as a social factor of production, which would give its owner identity and meaning, rather than as an economic factor of production designed to maximise profit. The family attachment to and ownership of land also discouraged its sale. This limited the opportunities for younger and/or enterprising farmers to become more competitive through purchase of land to consolidate and increase farm size. The result was a significant absence of innovation, leading to a fossilisation of the Republic's farm landscape and a detachment from global market forces.

Between the Second World War and 1960, agricultural production in the Republic increased by only 8 per cent. This contrasts with an average of 67 per cent in the Organisation for Economic Cooperation and Development, of which Ireland was a founding member. Furthermore, this growth, limited as it was, remained focused on the larger dairy, beef cattle and tillage farmers located along the eastern and southern seaboard. Differences between this relatively prosperous core and the underdeveloped western coastline – with its historic focus on small-farm units and extensive pastoral activities – were therefore amplified rather than reduced.[19] One consequence was the much higher rates of emigration and population loss that characterised this peripheral western seaboard between 1926 and 1961 (see Chapter 25: Urbanisation of Ireland's Coast). This reflected powerful inefficiencies within the state's agricultural system and strengthened the need for modernisation.

Crucial to the beginning of this modernisation in the Republic was the launching of the government's First Programme for Economic Expansion in 1958. This recognised the need to reject protectionism and embrace an export-based strategy to stimulate long-term, sustainable development.[20] While the attraction of new, foreign-based manufacturing activities would be the central component of this strategy, modernisation of farming – to raise productivity and international competitiveness – was also recognised as vital for success. This is apparent from the fact that, in spite of protection, the Republic's workforce in agriculture declined by over 380,000 jobs (58 per cent) from 1926 to 1971; heavy losses (totalling almost 245,000) were recorded in all of its coastal counties. Rates of decline were as severe for the three coastal counties of Northern Ireland. Here, its Agricultural Census recorded a loss of over 68,000 (65 per cent) in the agricultural workforce from 1926 to 1974. By 1971, only 8 per cent of this region's workforce remained in agriculture as alternative prospects in more urban-centred employment attracted an increasing proportion of the labour market. This compared with a 25 per cent dependency for the Republic, where the benefits of its policy to attract industrial investment had yet to create a more diversified economic base (Figure 4.26).

The 'treadmill' effect of pressure to achieve economic viability on small-farm operations also resulted in a decline in their number. By the 1960s, fewer than one-half of landholdings in the Republic were less than thirty acres (12 ha), compared with 70 per cent in 1911. Most intense losses continued to occur in western, coastal

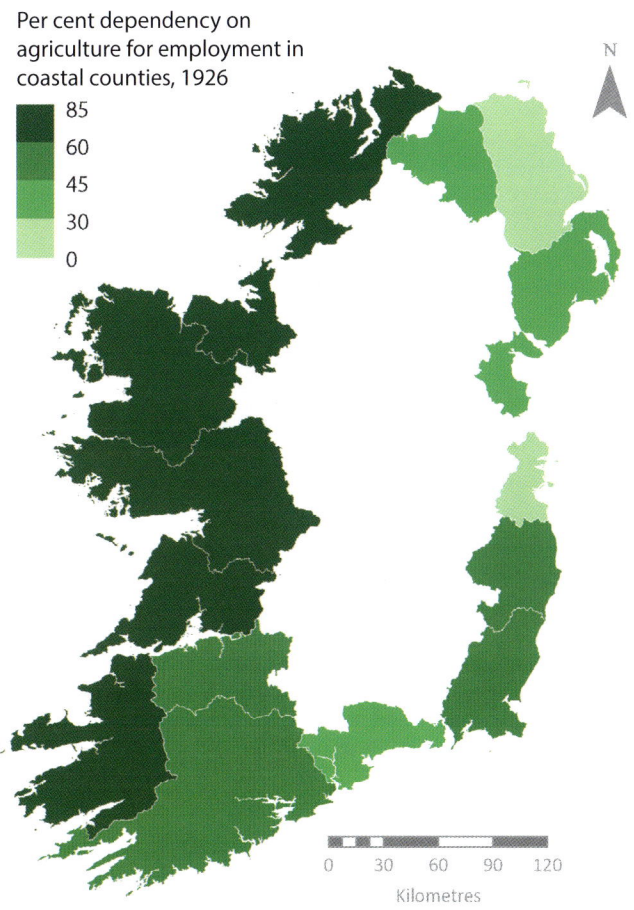

Per cent dependency on agriculture for employment in coastal counties, 1926

85
60
45
30
0

Fig. 4.25 DEPENDENCY ON AGRICULTURE FOR EMPLOYMENT IN COASTAL COUNTIES OF IRELAND (1926). In 1926, total employment in agriculture on the island amounted to 0.85 million, almost 80 per cent of which was in the twenty-six counties of the new state. The absence of a successful industrial revolution in the South meant, furthermore, that dependency on agriculture was significantly higher (53 per cent) than in the North (35 per cent). Dependency levels were especially high along the western seaboard, where counties north of Limerick recorded in excess of 70 per cent, with Mayo the highest at 81 per cent. Alternative employment prospects to agriculture in counties possessing important port cities along the southern and eastern coasts help explain their lower levels, most occupying a narrow range of between 46 and 50 per cent. The lowest dependency (5 per cent) occurred in County Dublin, which hosts the capital city. Successful industrialisation in the North translated into relatively low dependency rates in coastal counties, but especially in Antrim where most of Belfast's administrative area is located. [Data source: Census returns for Republic and Northern Ireland, 1926]

counties. Modernisation, however, was necessary if Ireland was to prepare effectively for its commitment to gain accession to the European Economic Community (EEC), the precursor of the present-day European Union.

When Ireland and the United Kingdom (along with Denmark) joined the EEC in 1973, both states were provided with substantial economic benefits. Agriculture gained unlimited access to a major market for its principal products, and at high guaranteed prices, benefits that had been yearned for by all Irish farmers for many years. Over the last fifty years, therefore, the implications of the EU's Common Agricultural Policy (CAP) have been fundamental in helping to transform Ireland's agriculture.[21]

The high and guaranteed price regime of the CAP resulted in an immediate and positive production response for most farmers around Ireland's coastline. Output of beef was encouraged and,

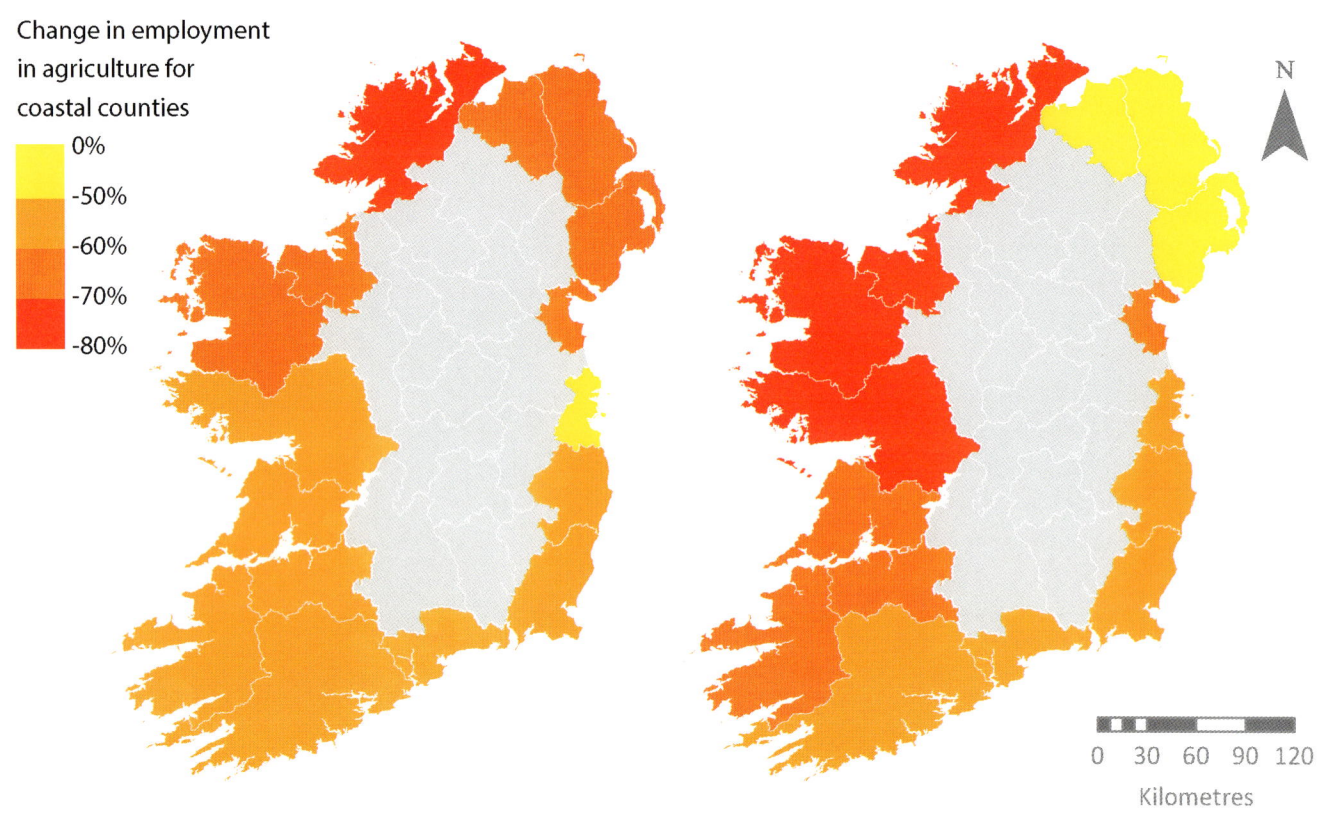

Change in employment
in agriculture for
coastal counties

0%
-50%
-60%
-70%
-80%

0 30 60 90 120
Kilometres

a 1926–17 (ROI) / 1926–74 (NI)   b 1971–2011 (ROI) / 1974–2011 (NI)

Fig. 4.26 TRENDS IN AGRICULTURAL EMPLOYMENT FOR COASTAL COUNTIES OF IRELAND, 1926–1971 and 19712011. a. Between 1926 and 1971, total employment in agriculture on the island fell by over 0.5 million jobs (60 per cent). Most of this decline was concentrated in coastal counties (312,000) and especially in those located in the Republic (245,000). Given the paucity of alternative jobs in these areas, the extent of such losses was a major problem for the state during its formative decades. Apart from Dublin, where less than 2 per cent of its workforce remained in farming (25 per cent nationally), all other coastal counties recorded declines of over 50 per cent in their agricultural workforce, rising to 70 per cent for County Donegal. In spite of the scale of its losses, dependency rates remained high, exceeding 40 per cent in all west-coast counties, except Limerick (26 per cent). In Northern Ireland, rates of decline were also high, but the region's more diversified economy was better able to compensate for the losses. By 1974, therefore, only 7 per cent of this region's total employment remained in agriculture. b. Despite the objective of the CAP (Common Agricultural Policy) to maintain the presence of farm workers on the land, the following four decades witnessed the continuation of extremely high rates of employment loss. In total, this amounted to the loss of almost 200,000 (60 per cent) employed in Irish agriculture, with 90 per cent occurring in the Republic. Declines in excess of 60 per cent in agricultural employment typified many coastal counties of the Republic, with highest levels occurring along the entire western seaboard. This, together with government policy to disperse employment opportunities throughout the country, caused the national dependency level on agriculture to fall significantly to some 5 per cent. Although all western counties exceeded this level, only counties Mayo and Kerry approached the level of 10 per cent. In Northern Ireland, rates of employment loss for agriculture were generally much lower, as much of what could be considered 'surplus' or inefficient farm units had been largely removed in the previous period. By 2011, the region's dependency level on agriculture was a little over 2 per cent. [Data source: Census of population CSO and NISRA]

even though their average herd size was quite low, the higher prices brought some benefits to the many small-scale farm enterprises engaged in the rearing of cattle along the western seaboard. In contrast, those west coast farmers who specialised in sheep rearing experienced initially a fall in their flock sizes, as this farm enterprise was excluded from market price support. Since the 1980s, support under the CAP was extended to sheepmeat, to the benefit of coastal and upland sheep farmers. One negative consequence of increasing numbers of sheep, however, was overgrazing of the rough pasture available on the bogs and uplands located along Ireland's west coast. This, in turn, created problems of soil erosion in these sensitive environments.

While some increase in the profitability of these small-scale and extensive farm enterprises, located disproportionately along the western and northern shoreline, led to an advancement in farmers' standards of living, it would be the larger-scale and more intensive farm enterprises involved in tillage and dairying that benefited the most. Milk production, for example, doubled in the decade

following accession, emphasising the prosperity of the specialised dairying area along the Republic's south coast. In addition, the specialised tillage enterprises of the eastern coastline responded significantly to higher prices and consolidated their status as a core farming area. In effect, the combination of strong market supports and natural environmental advantages for pastoral farm enterprises meant that the Republic, in particular, become a major supplier of beef and dairy products to the European Union.

The above trends were clear responses to the fundamental economic principle that guided the initial CAP: the principle of comparative advantage. This works to encourage both intensification as well as specialisation to occur in regions best able to produce an output most efficiently. In Ireland, therefore, it was, almost inevitably the larger farm enterprises, but especially those engaged in dairying, tillage and beef-fattening along the southern and eastern seaboard, that benefited most. Not only is this area favoured by supportive environments that help reduce input costs, but it also enjoys above-average farm incomes which encourage investment, as well as an innovative culture that in turn raises productivity and levels of output.

In contrast, farm units located along the western and northern coasts have been less able to respond positively to incentives to increase production. Here, difficult natural environments, small farm size and the legacy of a traditional form of extensive pastoral farming has meant that their low levels of output yielded poor returns. In these circumstances, many farmers came to depend increasingly on income support schemes that were designed to maintain living standards and enable them to remain on their land. While helping to augment income levels, one of the 'downsides' of such support was that it discouraged some farmers, and especially the more elderly, from adopting innovative practices or else retiring and passing on the land to younger farmers. In effect, the rich were getting richer, while the poor were becoming poorer.

By the 1990s, the CAP had been revised in response to criticisms that inefficiencies, and their role in encouraging surpluses of output, had raised budget costs to high and unsustainable levels. Furthermore, the initial focus on raising production was identified as having significant and negative consequences for the cultural and natural environments of Europe. These were apparent especially in core farming regions, such as those along Ireland's eastern and southern coasts and hinterlands. Here, intensification in land use was accompanied usually by removal of historic field boundaries, such as hedgerows and stone walls, to create large, uniform fields that facilitated mechanisation. It also led to increased pollution of soil and groundwater by overuse of fertilisers and loss of vernacular architecture, including traditional farm buildings, and their replacement by, for example, continental-styled bungalows and stainless steel grain silos. To help redress these problems, the revised

CAP not only retains a commitment to encourage a competitive agricultural sector, but now also stresses farm practices that are environmentally sustainable. To achieve this, income supports to farmers are linked strongly to the maintenance and improvement of their cultural and physical environments.

Such modifications to the CAP have been well accommodated in the larger-scale and more intensive farm enterprises that dominate the eastern and southern coasts and their immediate hinterlands. The high demand for efficiently produced food and a capacity to innovate in response to changing market conditions have meant that these remain specialised and core agricultural areas, with a farm workforce engaged primarily on a full-time basis. Pressures on these rural landscapes are, however, increasing through urban sprawl around the island's principal cities, which are also located primarily in the east and south of the island. This is reflected in extensive commuter belts and the profusion of bungalows (sometimes referred to, disparagingly, as 'bungalow blitz') that have inflated rural land values and led to the loss of available farmland, especially in the hinterlands of Belfast and Dublin.

The increased focus on the environment, however, offers important benefits to farming communities in peripheral areas. Under the revised CAP, farmers are encouraged to diversify their functions to embrace a heightened demand for environmental goods and services. This is linked directly to Europe's growing urban populations and tourist trade, which look increasingly for leisure and recreational opportunities in rural areas. In this context, Ireland's western seaboard is the region that best preserves the island's historic cultural and natural landscapes. As a result, in addition to supporting traditional agricultural production, many farmers have opted to become involved more directly in the

Fig. 4.27 BLANKET BOG AND PARTLY CLEARED FORESTRY ON THE LOWER SLOPES OF STRADBALLY MOUNTAIN, COUNTY KERRY, WITH THE MAHAREES PENINSULA VISIBLE IN THE DISTANCE. Since its launch in 1990, the Irish government's forestry programme has provided over €2.5 billion in financial incentives to landowners to invest in farm forestry schemes, with more than 300,000 ha of land being planted with trees. There is a preference, where conditions allow, for the planting of native broadleaf trees to encourage biodiversity, but extensive plantations of fast-growing conifer species, including sitka spruce and lodgepole pines, have appeared on many coastal hillsides, particularly in the north and west. Planting forests on marginal land can allow farmers to concentrate on working more productive and profitable areas of their farms, while also providing a welcome pension fund, to be drawn down when the trees mature after thirty or forty years of growth. [Source: Darius Bartlett]

growing coastal tourist trade. This is expressed in the almost-ubiquitous signs for farmhouse bed-and-breakfast that are displayed along scenic, coastal roads, and the designation of branded routes and tourist regions, such as the Wild Atlantic Way (see Chapter 27: Coastal and Marine Tourism: Building opportunities in Blue Growth). Furthermore, a concentration of tourist facilities in small towns and villages allows the retention of key services required by local populations, as well as offering expanded employment prospects.

Areas of new forestry have also appeared on the landscape, as improved subsidies have encouraged farmers to plant fast-growing coniferous trees on their more marginal land, rather than persisting exclusively with rough grazing of cattle and sheep (Figure 4.27). The most recent 'intrusion' onto rural landscapes has involved the construction of wind farms, as Ireland seeks to meet its obligations to increase the use of renewable forms of energy (see Chapter 28: Renewable Energies: Wind, wave and tidal power). These are located especially in marginal upland areas to take advantage of stronger and more consistent wind flows. The net result of a more multifunctional approach is not only an increase in part-time farming, but also the creation of a new regional dynamic and improved income potential.

Despite the opportunities and revisions of the CAP, modernisation within agriculture, together with greater prospects for higher standards of living outside of farming, have translated into an accelerated loss of employment. For the island as a whole, by 2011 the sector had lost 60 per cent (c.200,000) of its 1971 workforce. In Northern Ireland, approximately 2 per cent of its workforce remained in agriculture while in the Republic, the success

of its programme of industrialisation and movement into a post-industrial economy and society resulted in a similar collapse in dependency. By 2011, only one-in-twenty of its workforce remained in agriculture, compared to some 25 per cent at the time of accession to the EU. This represents an erosion of some 175,000 jobs for the forty year period, with heavy losses typifying all coastal counties (Figure 4.26). Unlike in previous periods, however, significant gains in industrial and service sectors have compensated more easily for these losses. As a result, total employment and population have increased for all coastal counties in Ireland and dependency levels on the primary sector are now generally less than 10 per cent of the workforce. The overall decline in numbers of farmers, however, released land onto the market. This facilitated an increase in average farm size in the Republic and Northern Ireland to 33 and 40ha respectively, with beneficial implications for farm productivity.

CURRENT COASTAL FARM ENTERPRISES

A detailed micro-level review of the spatial distribution of different farm enterprises in the Republic of Ireland demonstrates the wide variations as well as specialisations that occur along its coastline.[22] For the purposes of the analysis presented here, 'coastal' is defined as the geographic area encompassed by Electoral Divisions (EDs) that form part of the coastline. In broad terms, coastal areas throughout Ireland continue to be dominated by pastoral farm activities.

Rearing beef cattle is clearly the single most important farm activity around Ireland's coastline. With the inclusion of dairy cattle and sheep, pastoral farming accounts for almost three-quarters of

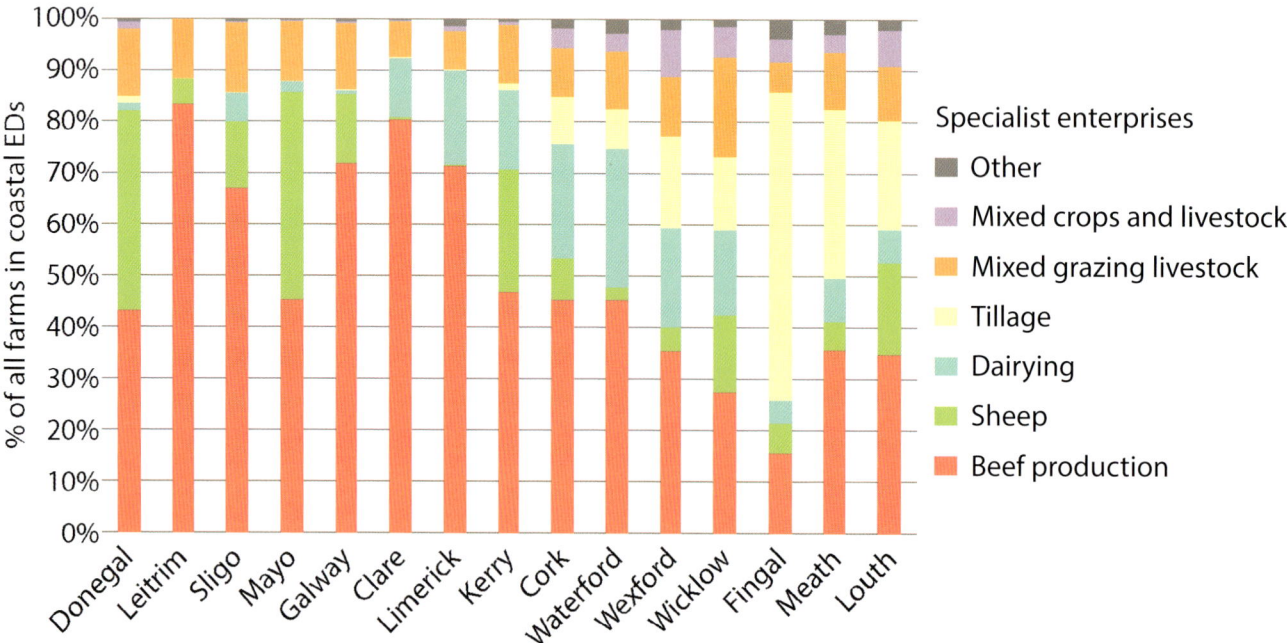

Fig. 4.28 COMPOSITION OF FARM ENTERPRISES IN COASTAL AREAS (2010). Pastoral farming dominates agriculture around the coastline of Ireland, with most counties having at least three-quarters of their coastal farm enterprises involved in the rearing of beef and dairy cattle and sheep. Raising of beef cattle is, apart from Fingal (Dublin), the predominant farm activity, especially along the entire western seaboard. Concentrations of specialist sheep farms occur along the more rugged stretches of coastline in counties Donegal, Mayo and Kerry, but with important outliers in counties Wicklow and Louth on the east coast. Dairying is more common in the coastal lowlands of counties Limerick, Cork and Waterford. Here, environmental conditions are more favourable and support the most diversified range of agricultural enterprises to be found along the Republic's coastline. Arable farming (tillage) is confined largely to Dublin (Fingal) and its adjacent counties. The remaining specialised enterprises occupy generally less than 20 per cent of coastal farmland. [Data source: Author's calculations based on analysis of the CSO (2012). Census of Agriculture – 2010 Final Results. Dublin: Stationery Office]

all farm enterprises in the majority of coastal counties (Figure 4.28). Arable (tillage) farm enterprises are more restricted and occur in relatively small coastal areas of specialisation. Fingal, as part of the Greater Dublin region, is the only coastal county to show a specialisation of arable/tillage. Here, intense pressure on land for building, and its high cost, preclude extensive forms of pastoral activities as a viable option for farmers.

Within these generalisations, significant variations in specialisation can be observed at the Electoral Division level both within and between coastal counties (Figure 4.29). The dominance of rearing beef cattle in the Republic is most apparent along the western coast, between counties Donegal and Kerry, with some EDs showing a dependency that exceeds 80 per cent of all farm enterprises. In coastal counties characterised by mountainous terrain, however, the rearing of sheep is an important activity. Counties Mayo and Donegal, for example, have some coastal EDs in which the vast majority of farms are specialist sheep enterprises. The peninsulas of County Kerry are also associated strongly with this farm enterprise, as are outliers of higher ground along the east coast, including the Wicklow Mountains.

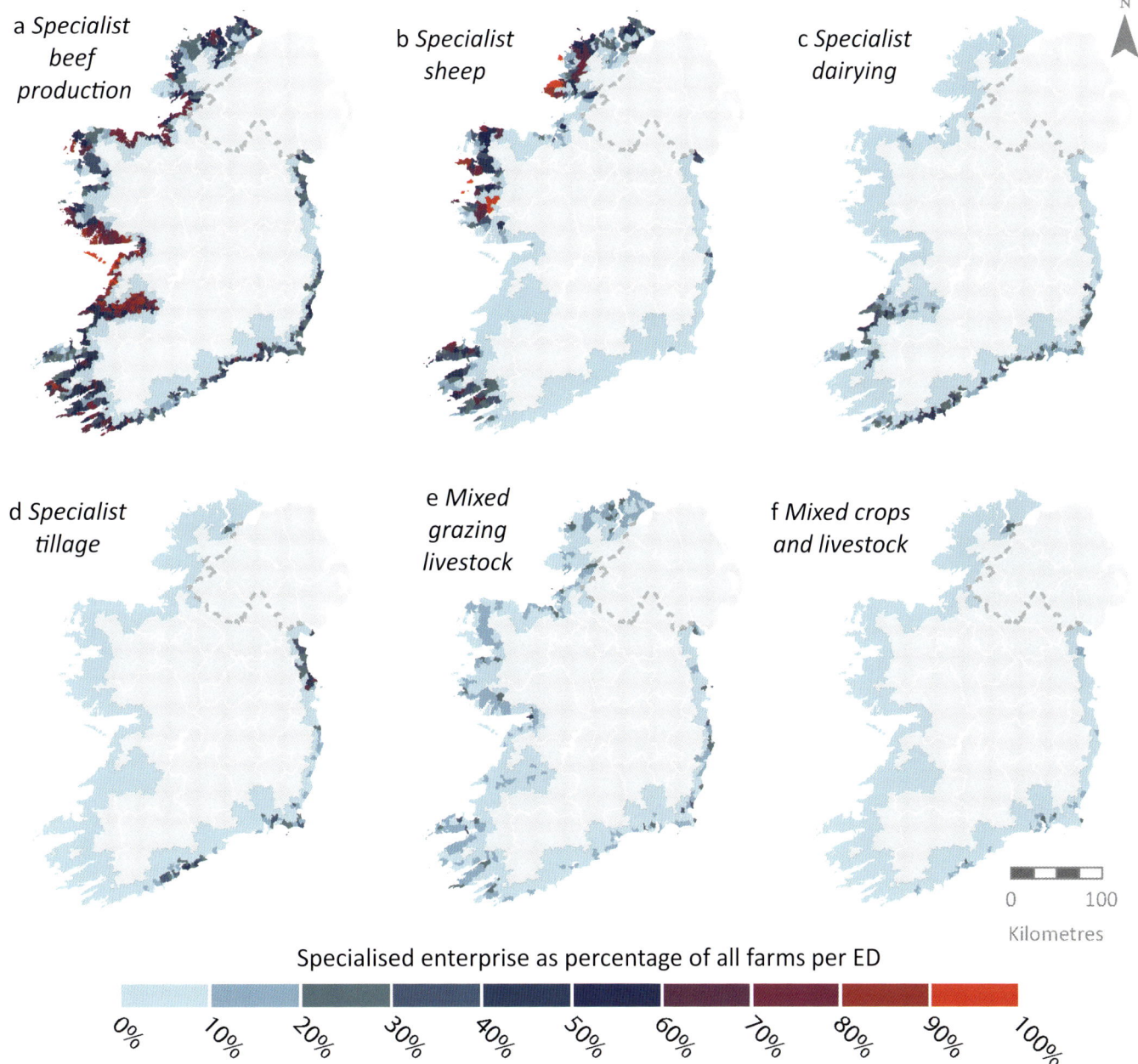

Specialised enterprise as percentage of all farms per ED

0% 10% 20% 30% 40% 50% 60% 70% 80% 90% 100%

Fig. 4.29 FARM ENTERPRISE SPECIALISATION AROUND THE COASTLINE OF THE IRISH REPUBLIC (2010). Using the relatively small geographic unit of Electoral Districts to define coastal areas, a. the dominance of farm enterprises engaged in specialist beef production is apparent, especially along the entire western seaboard of the Republic. Some coastal areas in Connemara, and in counties Donegal and Kerry, for example, have at least four out of every five farms engaged in this activity; b. Specialised sheep farming is also confined largely to the western coast, although this is focused most particularly in the more mountainous areas of Connemara, Donegal and the peninsulas of the south-west; c. In contrast, intensive dairying is confined especially to the southern coastline, reflecting the more benign environmental conditions which favour the year-round growth of grass, and which stretches from the Shannon Estuary to County Wexford; d. Environmental conditions in Ireland are not especially favourable to tillage and this enterprise is restricted essentially to relatively small lowland areas that possess fertile soils. These conditions occur in some parts of the eastern and southern coastline, but pressures from urban expansion in these areas places considerable pressure on this, as well as other types of farm activity; e–f. The significance of animals in the farm landscape around Ireland's coasts is further amplified in mixed-farm enterprises, which combine livestock with mixed grazing and field crops. [Data source: Author's calculations based on analysis of CSO (2012). Census of Agriculture – 2010 Final Results. Dublin: Stationery Office]

The association between topography and farm enterprise also extends to the coastal areas of Northern Ireland. Data from the Census of Agriculture 2018 highlight the classification of much of this area – that is, the upland areas of the Mourne Mountains in south County Down, virtually all of the County Antrim coast and western parts of County Derry (Londonderry) – as 'severely disadvantaged' or 'disadvantaged'.[23] Assessment of the mix of farm enterprises in these areas points to the dominance of 'Cattle and Sheep', a single classification used for the reporting of agricultural statistics in Northern Ireland. In summary, for the eleven rural districts that cover the Northern Ireland coastal zone, 'Cattle and Sheep' comprises, on average, 77 per cent of all farm enterprises.

Overall, extensive rearing of beef cattle and sheep appears well suited to western and northern coastal areas, as these animals are better able to survive in rugged and difficult upland environments where grazing is generally poor and weather conditions more extreme. Alternative farm enterprises would not be a viable option in these conditions. Though not as prominent as it once was, large numbers of cattle continue to be raised in west-coast upland areas before being transported to be fattened on improved grasslands located in lowland areas of the east.

Dairying is more common in an arc of coastal EDs from counties Limerick to Wicklow. This reflects a topography characterised by greater areas of lowlands, deeper and well-drained soils, more favourable weather conditions for year-round growth of grass and proximity to large urban markets. In combination, these factors facilitate a greater carrying capacity of the land for more intensive farm practices. Dairying is prominent particularly in coastal areas of counties Cork, Waterford, Wexford and Limerick, where in some coastal EDs it can account for approximately one-half of all farm enterprises.

The more limited areas committed to tillage enterprises occur primarily on the east coast, which comes under the influence of the large Dublin conurbation, especially in counties Louth and Wexford. Some coastal EDs in east Cork are also of significance. In such areas, competition for land from alternative urban-based activities, together with an advantageous natural environment, have encouraged farmers to work their land intensively and, where possible, engage in multiple cropping to maximise returns to remain economically viable.

In summary, supportive environmental conditions for farming, together with proximity to large urban markets for food, have given rise to a more diversified and intensive range of farm enterprises along the southern and eastern coastlines. In contrast, the west coast is given over primarily to the extensive rearing of beef cattle and sheep. There are exceptions, however, to such a broad generalisation: for example, the presence of specialist sheep enterprises occurring in some areas of Wicklow and Louth, which reflect the influence of local coastal uplands, that is, the Wicklow and Cooley mountains respectively.

Income levels for the six main farm enterprise types in the Republic are provided annually by the National Farm Survey published by Teagasc.[24] The three types that involved cattle and sheep all had income levels in 2018 significantly below the national

a Percentage of family farms

b Average family farm income by enterprise

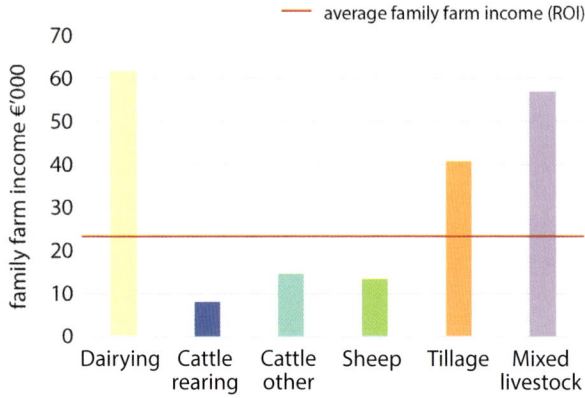

c National farm output by enterprise
Total value €2.1 billion

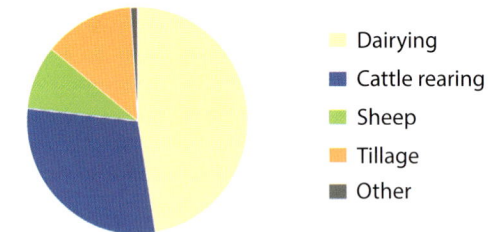

Fig. 4.30 FARM SURVEY RESULTS BY ENTERPRISE TYPE IN THE REPUBLIC (2018). The estimated average farm income of €23,333 contrasts with the average 2018 wage for workers in the Republic of €38,878, and highlights the income difficulties faced by many farming communities. a. The three enterprises – cattle (rearing and other) and sheep – that account for almost three-quarters of all farms, generate income levels far below the national average. b. Cattle rearing, for example, is the least rewarding, with incomes averaging only €8,300. Despite this, over one-in-four farmers are engaged in this extensive pastoral activity. When other forms of cattle farming are included (for example, fattening), almost 60 per cent of the country's 92,720 farm enterprises are linked to cattle. In contrast, the 17 per cent of farmers engaged in dairying receive the highest income, over seven times that earned from rearing beef cattle. c. In addition, despite their relatively smaller numbers, the capital intensity and high productivity of dairying contributes a disproportionate amount (46 per cent: €1 billion) to the total value (€2.1 billion) of Irish farm output. This is in marked contrast to all cattle and sheep enterprises which combine to account for 38 per cent (€0.8 billion) of national farm output. The remaining value is provided mainly by the 13 per cent of farmers engaged in tillage. The domination of low-income farm activities along the western seaboard helps explain its less developed regional economy, when compared to areas along the eastern and southern coasts. [Data source: Dillon et al., 2018. Teagasc National Farm Survey. Athenry: Teagasc]

average farm income of €23,333 (Figure 4.30). This was down by 26 per cent against the 2017 figures, due primarily to higher costs imposed by difficult weather conditions, and highlights both the continued importance of the natural environment for Irish agriculture and the precariousness for farming that it imposes. Despite such poor returns, however, these categories accounted for 73 per cent of the state's total number of farms, while in Northern

Ireland they represent 77 per cent of the total. In contrast, the three more intensive forms of farming generated much higher returns, with dairying providing an average income of just over €61,000, some seven times higher than cattle rearing. The significance of dairying is emphasised further in that, although representing only 17 per cent of farms, it generates almost one-half of the Republic's total value of family farm output.

Returns from farming are not, however, determined simply by enterprise type. Other key variables include location, conditions of the natural environment, the size and scale of farming and the demographic structure of the farming community. A comparison of Figures 4.29 and 4.30 allows an interpretation of such variables, together with their likely implications for farm income levels and the overall status of farm development around the island's margins.

The coastal areas of the Republic that experience the lowest levels of average farm income occur primarily along the west and north-west coast between counties Clare and Donegal. Low incomes here can be linked directly to the predominance in these areas of the rearing of beef cattle and sheep, and the difficult environmental conditions that reduce opportunities for more productive farming (Figure 4.31a and b). In addition, some 40 per cent of farms in this area are less than 12ha in size which, with the low carrying capacity of the land, mean generally low stocking rates and herd size. The low returns available in such circumstances and the relative isolation of such areas from urban centres have also encouraged many younger people to opt out of farming or migrate. These communities are characterised, therefore, by much higher-than-average numbers of farmers over sixty-five years of age. This age profile, often compounded by the lack of heirs on farms owned by bachelor farmers, removes much of the incentive to innovate in order to promote more productive enterprises. Many farms along the west coast are also worked on a part-time basis, with income

Fig. 4.31 MAIN FARM ENTERPRISE TYPES AROUND IRELAND'S COASTLINE. a. The rearing of beef cattle is a common farm enterprise along Ireland's western coastline, as here on rough pasture land near Doolin in County Clare. The extensive nature of this type of enterprise yields very low returns to farmers. [Source: Barry Brunt]; b. The steep terrain, exposure to extreme weather conditions and low carrying capacity of the landscape characteristic of mountainous areas in the north-west of Ireland provide few opportunities for farming, other than extensive sheep rearing. This is typified by the slopes of Slieve League in County Donegal, where outcropping bare rock surfaces and vegetation dominated by heather provide limited grazing for the sheep that roam these areas. [Source: Kyle Fawkes]; c. Intensive dairying is the dominant farm enterprise along much of Ireland's southern coastline. Here, overlooking Dungarvan Bay and the town of Dungarvan in County Waterford, a rolling landscape of large fields and rich pasture land can be grazed by sizable herds of dairy cattle on an almost year-round basis. This type of enterprise yields the highest average returns to family farms in the country. [Source: Michelle Byrne Photography]; d. Arable or tillage farming is restricted in Ireland by its requirements for environmental conditions that include level topography, relatively rich soil and a favourable climate that combines both rainfall and ample sunshine to ripen the crops. Such conditions occur along the eastern coastline, as here around Dunany Point in County Louth. The large and well-managed field patterns facilitate the mechanised and intensive farm practices that define this prosperous farming area. [Source: Gordon Dunn]

supplemented by off-farm employment and/or additional on-farm ventures. In more visually attractive coastal areas, farmers may also sell off plots of land for holiday homes as a once-off means of raising income. In doing so, farmers reduce land available for raising animals, further limiting future prospects for generating farm income.

In contrast to the low returns from farming around the coastline of west and north-west Ireland, but also extending to include the mountainous peninsulas of west Cork and Kerry, coastal areas in the rest of the Republic are characterised by higher levels of farm income. For example, where coastal lowlands occur from counties Limerick to Waterford, the long seasonal growth of grass, linked to the relatively mild year-round weather conditions associated with onshore south-westerly airstreams, encourages specialisation in dairy farming (Figure 4.31c). This enterprise, as well as providing the highest returns, also supports a small number of significant agri-food producers that have emerged through the consolidation of a large number of creameries and the conversion of farmer co-operatives to publicly traded companies. Proximity to the large coastal cities of Limerick, Cork and Waterford further-more provide good opportunities for farm families to diversify their income levels with off-farm employment. The result is the presence around Ireland's southern coastline of some relatively prosperous rural communities.

High farm incomes are also apparent along the eastern coastline. Here, rich soils, the availability of lowland pastures and proximity to urban markets encourage a diversified range of farm enterprises as well as more intensive farm practices (Figure 4.31d). Tillage, mixed grazing with livestock, as well as dairying, all provide family farm returns that are above the national average wage. The focus on competitive and productive farm practice is aided further by a younger age profile of farmers. Relatively high farm incomes and access to social and recreational facilities in the many urban centres located in these areas allow for a good quality of life and help to reduce the attrition of younger and potentially more innovative farmers from the sector. In general, although the eastern coastline is experiencing increasing rates of urbanisation, the high levels of returns from specialised and intensive farm enterprises help to maintain the presence of relatively vibrant farming communities within the Republic's economic core.

COASTAL AGRICULTURE IN THE LONGER TERM

One of the lessons of history of relevance to the farming communities of Ireland involves the need to adapt to what is referred to as a long-term income treadmill, in order to survive on the land. As other sectors of the economies in both parts of Ireland grew in significance – especially following the industrial revolution of the nineteenth century – and through the emergence of more competitive global forces – from colonialism and other factors – agriculture has experienced both relative and absolute decline in terms of share of the economy, employment and the number of households deriving a livelihood from the sector. It is important to note, however, that agricultural productivity and its total volume and value of production have experienced growth over time. In this context, the treadmill analogy refers to a range of intersecting processes and associated policies (as witnessed in the CAP for example), which are driving the need for farmers, food processors and rural communities to specialise and achieve economies of scale. In turn, this has demanded greater investment in land, technology and labour to remain relevant and sustainable. This has not been easy for Ireland and problems of adjustment will persist for farmers.

Many coastal areas face an uncertain future without further policy support, especially where rugged and hilly terrain dictate that only extensive rearing of sheep and cattle are possible. Such areas already have the lowest levels of family farm income and have become almost totally dependent on direct payments under the CAP. Farmers in these 'at risk' areas also have very high age profiles and, in many cases, no obvious successor for their farm enterprise. This pattern represents major and long-term challenges to the sustainability of agriculture in these marginal farming areas and, by extension, to the viability of the communities in which they are embedded.

For both the Irish and EU governments, a major challenge is the need to provide more effective and ongoing support to prevent 'at risk' coastal regions becoming denuded of population and returning to non-farmed habitats. Central to this will be effective incentive schemes, designed to sustain the small-scale farming and associated land-use practices that are critical for the maintenance of the island's coastal landscapes and seascapes. In addition, these culturally defined and scenic landscapes are also the basis for a growing and vitally important local and national tourism industry, highlighted for example in the attraction and success of the Wild Atlantic Way (see Chapter 27: Coastal and Marine Tourism: Building opportunities in Blue Growth). A combination, therefore, of part-time farming with off-farm involvement in sectors such as tourism and maritime activities, supported by benign pro-agricultural policies, offers the best prospect for longer-term viability for many of these coastal farms and local communities.

For those coastal areas characterised by larger-scale and more intensive farming practices, as along the eastern and southern seaboard, the challenges will be different and alternative policy options will need to be tailored to meet their needs. Growing international demand for food products present opportunities as well as challenges. The successful expansion of intensive dairy, for example, is likely to be conditioned by establishing more effective linkages with related farm enterprises engaged in contract rearing and specialist fodder production. Such synergies in the production chain of specialist enterprises should help support broader-scale economic development in rural areas. However, the continuing expansion of Ireland's large coastal cities – especially Dublin, Belfast, Cork, Limerick, Galway and Waterford – will exert considerable pressure on both the price and availability of land, as well as on the viability of farming in their immediate hinterlands. In overall terms, territorially tailored government responses to meet the diverse farming needs of the different coastal regions will be of much benefit in helping to maintain the sustainability of coastal farms in the longer term.

Case Study: Sea and Shore Food

Regina Sexton

In a flight of fancy, food writer and food historian, Alan Davidson, invites us to imagine what might happen to seafood consumption rates and seafood cookery along the Irish coasts 'if the people of say, Singapore were suddenly transported thither'.[25] He posits the idea in an attempt to explain the 'reserved attitude to most species of fish' in Ireland, which proved problematic for him when he struggled to find fish recipes that could be labelled as distinctively Irish, or long-lived and inherited in a traditional sense. An Irish collective might suggest salt cod/ling/wrasse, salmon and smoked salmon, but beyond these suggestions, which are shaped largely by more recent food preferences, Ireland's contribution to seafood cookery is frail and underdeveloped. In a similar way of thinking, JP McMahon, Michelin Star chef at Aniar in Galway, compares the outward-looking Japanese and their extensive and often elaborate fish cooking with the more modest Irish approach to fish: 'We look inland', he says, 'we're a land people'.[26] This reluctance to look outward to the sea explains something of the complex Irish relationship with marine foods. With meat held in higher esteem, fish is often considered less desirable and less substantial, with satiety associated with the consumption of animal proteins. Long-established barriers to consumption carry forward from the past, to shape a distinctive Irish attitude and culinary approach to sea and shore foods. At the core of this complicated relationship with seafood is the nexus between land and sea, and how the tensions between these two facets of Irish life can be seen to impact on the ways in which value and meaning are assigned to fish, shellfish and shore foods.

The investment of energy and resources in farming, at the expense of developing fisheries, was ingrained from the prehistoric period, and was marked by the transition from a Mesolithic food-supply system of fishing/hunting/gathering to a land-based means of producing animal and plant foods from the Neolithic period onwards. Over time, the emergence of a farming society, in which patterns of production were in the main predictable, resulted in distinctive differences in the lifestyles of inland and coastal communities and their associated dietary patterns. McKenna extends the divide further in his exploration of food on the Great Blasket Island between 1850 and 1950.[27] While mainlanders were farmers first and then fishermen, McKenna saw the islanders as predominantly fisher-hunters, who secured food with a mix of skill, chance and, at times, bravado:

> … island men spend a great deal of time fishing. In spite of this, it would not be accurate to describe the island men simply as 'fishermen', as this term doesn't convey adequately the range of activities they engaged in to supply themselves with food. Hunting for seals, porpoises, rabbit and sea-birds [and their eggs] was extremely important to the Islanders, as their basic diet of fish had to be supplemented in some way. In this sense, they were as much hunters as fishermen. Though fishing and hunting could provide large quantities of food – a single boat could catch thousands of mackerel in a night – there was no such quick and copious rewards for those who worked the land … the young men of the island as soon as they were old enough, set their sights on the sea rather than the land.[28]

This threefold distinction between fisher-hunters, farmer-fishers and farmers brought difference to the respective diets of those they fed, with coastal and inland patterns determined by the extent to which marine resources were integrated into daily patterns. In particular, it was the variety of such foods and how they appeared in the diet – either fresh or cured – that determined inland or coastal culinary styles and, in turn, directed how people engaged with seafood on a sensory and, indeed, a psychological level. The importance of marine resources for the Blasket islanders, as described by McKenna, is a good illustration of the extent to which locally available, non-commercial foods were exploited to bring diversity and seasonal variance to their diet. In addition, the assignment of special status to foods such as puffins and fledgling puffins, relished by the islanders as the 'chickens of the sea', and their indifference to the gastronomic qualities of lobsters[29] are suggestive of a local value-system that was based on fundamental factors of taste, place and preference, rather than being shaped by received and socially constructed notions of luxury.

Coastal and island fishermen, whose resources and skills extended to the acquisition and upkeep of boats and fishing tackle, had access to a wide variety of fresh seafood. While much of the catch was often destined for the market, any surplus was consumed fresh or, if taken in glut, could be salted down to sell on or to be kept for domestic consumption (Figure 4.32). In particular, stocks of salted white fish, herring or mackerel brought a sense of food security to households in the lean winter months and through periods when the fishing was slow. As late as the 1930s, a Donegal informant to the Schools' Collection describes how fishing communities in Donegal managed a perishable commodity in precarious market conditions:

Fig. 4.32 FISH CURING AT DOWNINGS PIER, DONEGAL (1906–1914). The oftentimes difficult terrain and poor soil conditions around much of Ireland's coast-line meant that many coastal communities looked to the sea for their principal livelihood. In contrast to the land, offshore waters provided abundant resources. Fish not only supplemented the relatively meagre supplies of food obtained from farming, but also any surplus catch was available to sell to markets and, there-fore, provided a key source of income. The fisheries offered a diverse range of fresh seasonal fish and shellfish to markets close to the coast. However, fish destined for inland markets and export was first prepared and cured, usually at the site where the catch was landed. The process involved gutting and generally preparing the fish for salting before packing it in barrels to ensure preservation while it was transported to markets. Such activities offered additional employ-ment opportunities, especially for women. This image illustrates the gutting of fish prior to salting in a small coastal community in County Donegal. The barrels into which the fish would be encased are seen adjacent to where the labour-intensive work is conducted on the quayside. [Source: National Library of Ireland, call no.: L_ROY_04566]

> Fish of all kinds were cured at home as the immediate sale of fresh fish was poor. When cured these fish were packed in the side of the kitchen layer by layer until they constituted a 'stack' perhaps eight feet long by four feet wide by four feet high. Persons from inland places came and bought a cart load of these cured fish and sold them from house to house on their way.[30]

Beneath this description of catching and curing fish to sell on for further distribution lay a network of interconnected factors that shaped the nature of fish-consumption patterns in Ireland. Until the nineteenth century, Ireland's poor infrastructure hampered delivery of fresh fish and shellfish to inland areas. This, coupled with the prescription of the Catholic Church to abstain from meat and dairy produce on fast days, made it expedient to produce salt white fish and salted and smoked herring. It is difficult to overemphasise the ubiquity of salted and cured fish in the diet of the faithful during Lent, Advent and the vigil of holy days. Before refrigeration, hard-curing of ling and cod produced an extremely salty and dehydrated product that had to be desalinated, often soaked, strained and soaked again, before it was ready for boiling. Fish for inland communities tasted of salt first and then of fish or, in the case of herrings, salt and smoke first and then fish. For inland communities, salt fish dishes were not only insipid but also the act of consumption was held by many as a performance of penance. With price and distribution problems as deterrents to the consumption of fresh fish in inland regions, fish that cured well and were caught in quantity were able to service fast-day obligations.

Domestic and commercially produced salted white fish and salted and smoked herring proved popular until well into the twentieth century. In coastal areas, for example, herring continued to be home-cured and chimney-smoked and were roasted over the open hearth using tongs or placed on a grid-iron for broiling. They were also boiled and, like salt white fish, served in a white sauce of milk and onions called 'squint'. Cooked and flaked herring was also added with oatmeal to mashed potatoes. Amongst the poor, cured herring served as a 'kitchen' or a flavour enhancer to the otherwise bland potato and milk diet, while their high fat content, together with the salt and smoke elements in cured fish, turned the fish-cooking liquor into a tasty relish for potatoes. The familiar 'Dip the dip and leave the herring for your father' anecdote indicates not only the customary use of the cooking liquor of herring as a thin sauce for potatoes, but also the ritualised eating of the components of the meal – fish, potatoes and liquor – confirms preferential treatment given to the male head of the family unit. It also highlights how status can be affirmed through the simple and few ingredients involved in a poor food economy.

A poor status was also assigned to shore foods, like mussels, cockles, clams, limpets, periwinkles, and seaweeds such as sloke, dulse and sea lettuce. These items were picked traditionally by women and children, while sand was being raked for razor clams and in summer for sand eels (Figure 4.33). Shellfish-gathering began in spring when they were considered good for eating. This tied in conveniently with Lent, and the black fast on Good Friday was marked in coastal areas with the collection and consumption of shellfish and seaweeds. Depending on the take, simple dishes of mixed shellfish and seaweed boiled with sea water made a tasty relish for the potato meal, especially on fast days when milk and butter were prohibited and at times when milk was scarce. Stigmatised as free foods, molluscs and seaweeds characterised a food culture of the rural coastal poor.

As free and perishable foods, shellfish gathered at the shore remained peripheral to a market economy that gave value and esteem to

Fig. 4.33 COCKLE PICKERS (1890–1910), LOCATION UNKNOWN. For many coastal communities, the chore of trying to provide sufficient food to feed families was a constant struggle. The difficult environmental conditions made farming an activity that functioned at little more than subsistence level, while adverse weather often made the reliability of the fish catch questionable. Availability of shore foods such as shellfish were, therefore, an important element in supplementing the diet of poor coastal communities. Scouring the shoreline for such foods was laborious, time-consuming and often dangerous work and generally the task of women. Here a group of women carry the baskets into which they place the cockles collected from the shoreline. In addition to supplementing family diet, the cockles, periwinkles or other shellfish could also be bartered in coastal towns and villages in exchange for commercial food items, such as bread. [Source: National Library of Ireland, call no.: CLAR116]

product only in terms of its monetary worth. On fast days and in coastal areas, however, men and women often hawked from house to house with sack-loads of cockles or other shellfish, exchanging them for commodities like bread, flour or tea. This served to circumvent the increasingly important commercial food system, in which the acquisition and consumption of produce bought in shops shaped the economic standing of most people into the second half of the twentieth century. It also determined how people made their choices of food and undermined the value of free, unprocessed and locally produced items. The play-off between the local and the global was particularly detrimental to the status of shellfish and seaweeds. Their low status was compounded further by their links to the efforts of destitute and starving coastal communities to feed themselves during the Great Famine. Undervalued at the best of times, the association in folk memory of shore food with deprivation moved perception of these foods beyond their role as stigma-markers and closer toward taboo status.

Despite the low status of shellfish, collectively termed *bia bocht* (poor food or poor people's food), a hierarchy of preference recognised the superior qualities of certain fish above others. Limpets and periwinkles ranked lowest because of their tough texture and close association with consumption patterns of the coastal poor. McKenna illustrates their low value with reference to Peig Sayers' comment: '*seachain an teach tábhairne nó is bairnigh is beatha duit*' ('stay away from the pub, or you'll end up living on limpets').[31] On the other hand, scallops and razor fish were held in high esteem due to their tender and sweet flesh, although native oysters were the most prized and sought-after. By the eighteenth century, natural and artificial oyster beds satisfied demand, with numerous coastal areas noted for the quality of their oysters. Rutty's description of the beds in Arklow and at Poolbeg, Howth, Rush and Skerries in eighteenth-century Dublin is an indication of the popularity and extensive cultivation of oysters, and also details the variations in quality that determined their culinary uses. Oysters were consumed fresh and in pickled form, and cooking oysters were widely used in meat dishes, in soups, pies and sauces. The presence of oyster rooms and saloons in many Irish cities until the late nineteenth century and beyond is evidence of their continued appeal. Excessive demands of both home and export markets, however, brought native stocks close to collapse, encouraging the introduction in 1972 of the Pacific Oyster (Figure 4.34).

Notwithstanding Ireland's traditionally strained relationship with sea and shore foods, consumption of fish increased significantly from the second half of the twentieth century. Between 1963 and 1987, per capita consumption of fish doubled from 3.4kg (9.5lbs.) to 6.6kg (15lbs), and by 2016 had more than tripled to 23kg (50.7lbs).[32] While this rate of increase is notable, it is not particular to Ireland but is rather the Irish expression of a general pattern of increased protein consumption common to the western world, and one that is facilitated by the globalisation of food industries. Improvements in fishing techniques and transportation brought a greater diversity of fresh fish to markets and made possible the importation of fresh fish and shellfish.[33] This also allowed for the appearance of a wider variety of value-

Fig. 4.34 OYSTER VENDER, CARLINGFORD (c.1950). Consumption of oysters has a long history in Ireland and is reflective of the appreciation coastal communities held for their taste and diversity of use in cooking. Throughout the eighteenth century, if not before, oysters were shucked by hawkers in coastal cities and towns and offered for sale for immediate on-street consumption. Considerable quantities were eaten cooked and pickled as evidenced in eighteenth- and nineteenth-century oyster-cooking styles. The tradition of eating fresh and locally harvested oysters – here from the oyster beds in Carlingford Lough – extended into the second-half of the twentieth century. Such a practice is now far less common, as local oyster beds have been depleted and/or the available harvests sold directly to the retail sector. In the photograph above, fresh oysters are served with a little dash of brown vinegar, possibly with the aim of adding a little piquancy in the absence of lemons. [Source: C050.15.00009; Photographer: Maurice Curtin. © National Folklore Collection UCD]

added fish products on the shelves of supermarkets, thereby presenting consumers with an illusion of great choice. These developments, combined with a range of socio-cultural changes – increased foreign travel, the influence of celebrity chef culture and a desire by consumers to engage with food in meaningful ways (be they environmental, ethical or health-based) – impacted on Ireland's changing relationship with fish. However, consideration of the nature of that change reveals that the conservative approach to fish persists. In particular, this is centered on the eating of a narrow range of fish and shellfish, with most consumers' knowledge of a diversity of seafood products low to the point of embarrassment. Traditional favourites like salmon, cod, hake, mackerel, whiting, trout and prawns are the varieties purchased with the highest frequency. This high demand for a limited variety of familiar fish has impacted severely on depleting native stocks, such that demand is now satisfied by the importation of cod and the expansion of aquaculture, especially salmon farm-fisheries.

A reluctance to eat outside the familiar range not only undermines sustainable fishing practice, but a limited engagement with seafood, particularly in domestic kitchens, also stifles culinary creativity and inhibits the development of a gastronomic appreciation of fish and shellfish. Not surprisingly, consumers still approach fish with a sense of unease. A 2006 Safefood report isolated worries about freshness and uncertainty about cooking fish as the main barriers to fish consumption, alongside concerns over food safety, price and sensory aspects such as smell, taste and the presence of bones, heads and tails.[34] While concern about freshness may be a remnant of an older relationship with seafood, the issue has relevance in the context of present-day food distribution and retailing systems. Consumers' perception of what constitutes freshness is influenced by the time it takes to get fish from the sea to the table. There is a considerable difference between day-boat fish, landed within hours of catching from small boats with a small trawl, and fish caught by larger fleets and placed on ice for days before it goes to market.

The freshest of fish requires extremely simple cooking. Fish and shellfish are delicate proteins and the refined taste and texture of fresh produce are characteristics that fade quickly. The gastronomic qualities of iced, days-old fish, compared to day-old fish, differ considerably and determine consumers' affective responses to and appreciation of seafood. In addition, the freshest fish is the quintessential fast food. Once the protein sets, the translucent flesh turns opaque and the fish is cooked. Yet, despite the simple nature of fresh fish cookery, the culture of Irish seafood cooking is fraught with problems. For many Irish people, fish and chip shops are the only point of contact with fish cooking and consumption. Here, heavily battered and deep-fried cod or white fish disguises, if not destroys, the flavour and texture of the fish, in the same way that salt overwhelmed cured cod and ling. In home kitchens, the division between meat and fish cookery is usually unclear, and home cooks tend to apply meat-cooking methods to fish, thereby destroying the delicate taste and mouth-feel through overcooking. The tacit skills required to prepare fish and shellfish are not part of the repertoire of many home cooks, while the complication of shells, claws and tails deters the consumption of molluscs and crustaceans. Little wonder that chefs and restaurant owners have no end of anecdotes concerning diners and their first taste of a mussel, or first encounter with a raw oyster.

In spite of this lingering conservative approach to seafood consumption, there is a community of young chefs, small fishmongers, day-boat fishermen, artisan producers and shore foragers who are attempting to reset the Irish relationship with sea and shore foods. Operating outside mainstream distribution and retail practice and guided by principles of using local, seasonal and traditional produce, they use their restaurants, shop floors and market stalls as educational spaces to encourage more informed and adventurous eating (Figure 4.35). In addition, this movement is also active in applying the snout-to-tail philosophy of meat cookery to fish cookery. The practice of fin-to-gill eating makes use of all parts of the fish in an attempt to reduce waste and to encourage sustainable use and consumption of seafood. This alternative seafood community takes impetus, in part, from outside trends and movements, such as Slow Food and, more recently, New Nordic Cuisine. However, pre-empting these developments was the work of the late Myrtle Allen of Ballymaloe House in east County Cork, particularly in the early, formative years of the house restaurant in the 1960s and 1970s. By choosing to profile under-valued traditional produce, like cockles, mussels, herring, eel, mackerel and carragheen moss in a restaurant context, Allen encouraged diners to appreciate the intrinsic value of local fresh seafood and seaweed. The simplicity of her cooking style maintained the integrity of the produce that she sourced from fishermen in nearby Ballycotton. She earned a Michelin Star in 1975, and this external validation encouraged a new generation

Fig. 4.35 FISH STALL, ENGLISH MARKET, CORK CITY. The increasing levels of fish consumption in Ireland since the latter part of the twentieth century are linked strongly to changing attitudes to fish as a principal component in the national diet. However, consumption patterns are often limited to a narrow range of fresh fish and processed breaded fish products. Nonetheless, many urban markets and small fishmongers are active in encouraging customers to consume a greater diversity of fish species. This is also reflected in the more sophisticated and consumer-friendly way in which fish is now marketed. Many fishmongers are also highlighting that their stocks originate from local Irish waters, rather than coming from imported supplies, and that day-boat fishermen supply the freshest of fish that has been handled and iced with care en route to market. Complementing this more urban expression of retailing are the growing numbers of fresh-fish stalls that are appearing in many local country markets, which extend the market for fish into smaller towns and rural areas. [Source: Barry Brunt]

of chefs and producers to reconsider how they engaged with the sea in developing an Irish signature to seafood cooking.

Contemporary advocates of this counter-culture include individuals like Prannie Rhatigan and Sally McKenna, who work to promote the gastronomic values of Irish seaweeds, while a first generation of artisan fish and shellfish smokers now distribute a variety of ethically sourced local products, based on sea and shore foods, through the farmers' markets networks. In the context of restaurants, a new style of lighter seafood cookery has emerged that has a distinctive Irish signature and which is built on local produce. The menus of many high-profile Irish chefs now display sea and shore foods prominently; these can include often-undervalued and less well-known varieties of fish, such as ray and pollock, as well as seaweeds and shore plants such as sea radish (*Raphanus maritimus*), marsh and rock samphire (*Salicornia europaea* and *Crithmum maritimum*) and sea beet (*Beta vulgaris subsp. maritima*). Such an approach, furthermore, has demonstrated successfully the diversity, versatility and exceptional quality of locally sourced produce and has challenged the hegemony of French classic cuisine (Figure 4.36).

Collectively, this alternative sea-food community is working to develop a more nuanced and respectful attitude to ingredients from the sea and shore, so that informed consumers can bring more meaning to their food choices and dining experiences. Young chefs, unburdened by the idiosyncratic culture of seafood consumption common to older generations, are designing a contemporary Irish seafood cuisine that brings with it a tangible sense of belonging to place and people. These efforts have succeeded in creating a positive identity for Irish sea and shore foods that satisfies the curiosities and expectations of domestic and, more especially, international tourists. Furthermore, initiatives that encourage food and culinary tourism – such as the Wild Atlantic Way's 'Taste the Atlantic: A Seafood Journey' – together with numerous other seafood trails and seafood festivals, bring economic benefit to coastal regions. More significantly, this developing appreciation of local marine resources unsettles the inherited tendency to assign greater value to products of a global food system, although, in light of the enigmatic Irish relationship with seafood, it is difficult to predict how this alternative approach may impact on overall consumer behaviour, mainstream dietary practice and majority thinking. Reforming attitudes to seafood may be just one aspect in the complex process of how Irish people navigate the tensions between local and global food systems, particularly in terms of sustainable practice.

Fig. 4.36 JP MCMAHON'S OYSTER AND SEAWEED. Dooncastle oyster, pickled channelled wrack, sea purslane, seaweed gel. This dish from Aniar Restaurant, Galway, is evidence of a contemporary food movement, which supports and celebrates local produce in season. With the emphasis on quality shellfish, the refined nature of the dish stands in contrast to the fish and 'two-veg'-type dishes that were traditionally prepared in Irish homes and in some catering establishments. The inclusion of seaweed and shore plants also marks a growing interest in foraging and an increase in the use of Irish seaweeds in cooking. [Source: © Anita Murphy 2018]

GEOLOGICAL FOUNDATIONS

Patrick A. Meere

UPPER SILURIAN TO MID-DEVONIAN AGE ROCKS AT FERRITER'S COVE, COUNTY KERRY. This shows the effects of geological structures – such as joints, faults and rock cleavage – in helping the process of rock shattering and erosion by the sea. [Source: Sarah Kandrot]

The geomorphological characteristics (that is, the Earth surface processes) of coastal zones in tectonically quiescent areas of the planet, such as Ireland, are determined by the complex interaction of marine coastal processes with the geology of the land and neighbouring offshore areas. Coastal processes are amongst the most dynamic on our planet, allowing significant changes in the morphology of our coasts to take place over extremely short geological timescales (see Chapters 2: The Coastal Environment: Physical systems, processes, and patterns and 30: Engineering for Vulnerable Coastlines). In this chapter we will explore, at different scales, the role of bedrock geology in shaping Ireland's coastal and linked offshore environments.

COASTAL GEOLOGICAL HERITAGE

Historically, Ireland has provided coastal localities that were, and still are, deemed critical in the conceptual evolution of the science of geology. In the late eighteenth to early nineteenth centuries the pioneers of the modern science of geology were divided into two rival camps, based on their respective theories for the formation of the planet. The Neptunists' Earth origin theory postulated that all rocks were a product of a process of sedimentation that involved crystallisation of minerals out of the Earth's early oceans.

The rival theory, Plutonism, was based on the belief that the planet's rocks were simply a product of volcanic activity, producing igneous-type rocks (such as basalt and granite). Portrush, on the north Antrim coast, was for a time at the very centre of this worldwide debate. In 1799 the Rev. William Richardson, in examining the rock sequences in this area, noted the presence of fossil ammonites (Figure 5.1) in what he identified as igneous rocks, so posing an intriguing puzzle in this debate. Had this rock identification indeed been the case, it would have provided very powerful evidence supporting the Neptunist view. However, in 1802 John Playfair correctly identified these supposed 'basalts' as 'stratified stone, which had acquired a high degree of induration'.[1] The fossils are, in fact, sitting in a late Jurassic mudstone that had been baked by a later igneous rock intrusion to form a dark contact metamorphic rock.

The survey and mapping work by the Geological Survey of Ireland (GSI), established in 1845, provided an objective recording of the different rock sequences found on the island and, progressively, brought clarity to these earlier science debates. The work of the survey, based often on coastal rock exposures, was undertaken through the nineteenth century by teams of field scientists, such as George Du Noyer.

Fig. 5.1 AMMONITES FOSSILISED IN COASTAL JURASSIC SEDIMENTARY ROCKS, LANDSDOWNE ROAD, RAMORE HEAD, PORTRUSH, COUNTY ANTRIM. Ammonites are a now extinct marine mollusc; beginning life in the Devonian period of geological time and flourishing in the Earth's oceans from the Permian to Cretaceous period (250–65 million years ago). They had shells with multiple chambers and belong to the group of sea creatures known as cephalopods, with their closest living relatives today being the octopus, squid and nautilus. They lived in the shelf environments of shallow, warm, tropical seas. The ammonites at Landsdowne Road can be seen on the foreshore at this publicly accessible site, though they are covered periodically by the tides. To ensure their preservation, these fossils have legal protection and must not be removed. [Source: Mike Simms, National Museums Northern Ireland]

The Coast Through the Eyes of a Geologist

Robert Devoy

George Du Noyer was a Dubliner of Huguenot descent. Known today as a celebrated artist, Du Noyer was an able geologist, field scientist and antiquarian and became a member of the Royal Irish Academy. His artwork frequently depicts landscape scenes across Ireland, particularly of coastal areas, as well as of plants and animals, historic buildings and rural life. The subject material of most of this work is focused on the natural world, but often includes acute observation of people and the life of the time. Equally, the aesthetic and emotional dimensions of landscapes and their links to nature were appreciated by him.

Much of his everyday working life was spent with the Irish Ordnance Survey and later the Geological Survey of Ireland. He was one of many similar characters working in the field of geology in Ireland and Britain in the nineteenth century who brought their acute observational skills to the study of the natural sciences. The result was not only a rigorous recording of the worlds of geology, geomorphology, botany, zoology and archaeology, but also the development of visionary concepts in the understanding of Earth environmental processes. Du Noyer, along with others, showed an awareness of the integrative nature of the Earth's key environmental spheres. This was a subject that was largely ignored in the sciences until the formal development of Earth systems science in the late twentieth century.

Du Noyer's work in helping map the geology of Ireland often took him to the coast, where rock sections, structures and sedimentary profiles were readily accessible (Figure 5.2). His eye for detail and his recognition of Earth processes helped him see intuitively the connections between different types of environmental data. For Du Noyer, and his contemporaries, there were obvious links between what became the separated disciplines of geology, geomorphology, botany and the marine sciences. He was a prolific artist, producing over 5,000 watercolours, paintings and line drawings; most of this artwork is held by the Geological Survey of Ireland. His life has been celebrated in recent years in a BBC documentary series (2015), in books and in exhibitions of his artwork.[2]

Fig. 5.2 SANDBANKS NORTH OF ARDMORE POINT, COUNTY WATERFORD (1850) BY GEORGE VICTOR DU NOYER. [Source: Geological Survey Ireland]

The island of Ireland is endowed with an incredible variety of rock types that span a two-billion-year history and which records most types of known geological settings, including sedimentary basins, high mountain ranges and active volcanoes (Figure 5.3 and Figure 5.4).[3] This rich and varied geology is best exposed along the coastline, as found in the early work of the GSI, where the geology exerts a fundamental control in defining the shape of the coast.

The oldest known rocks in Ireland are metamorphic, that is,

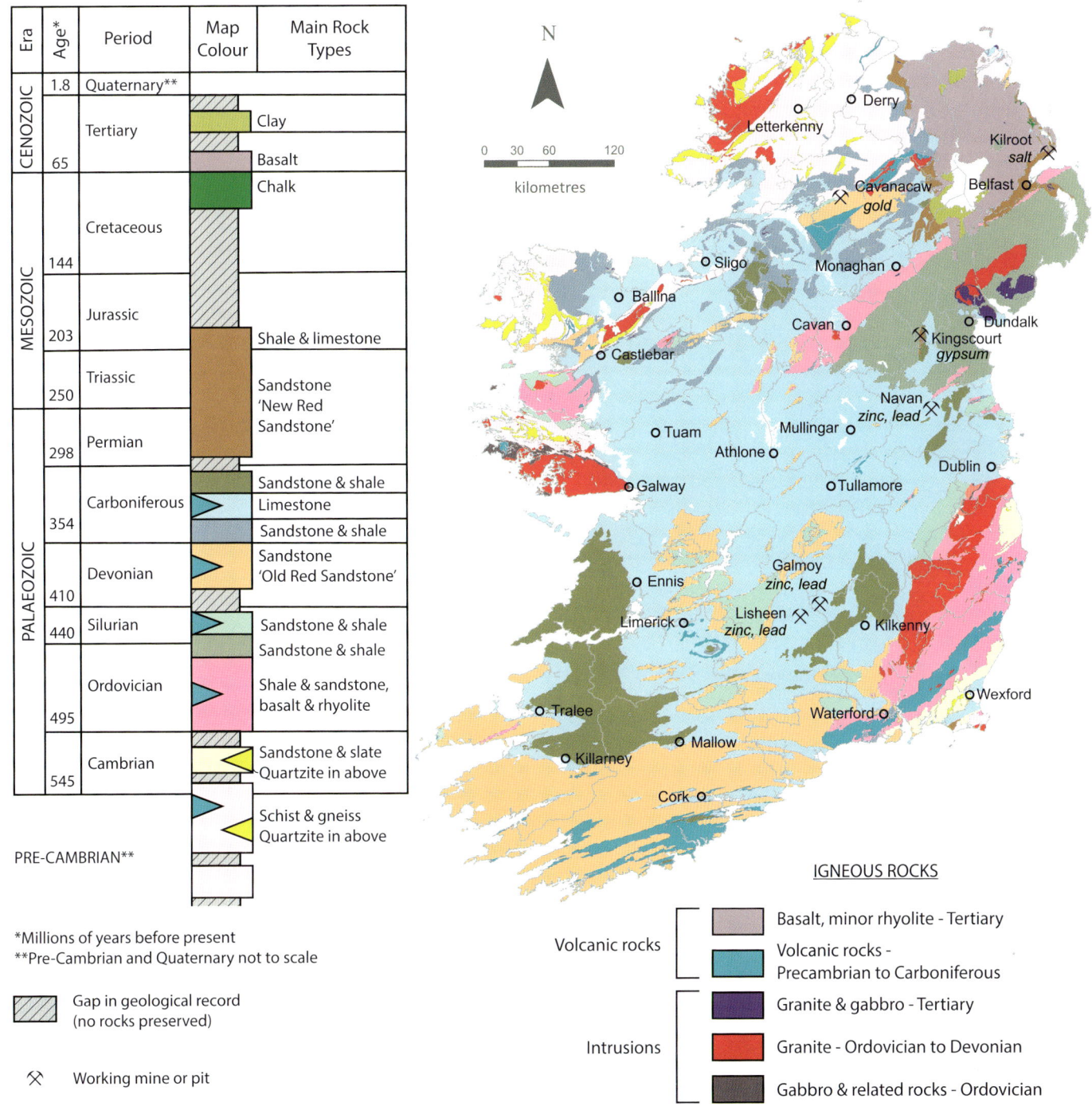

Era	Age*	Period	Map Colour	Main Rock Types
CENOZOIC	1.8	Quaternary**		
		Tertiary		Clay
	65			Basalt
		Cretaceous		Chalk
MESOZOIC	144			
		Jurassic		
	203			Shale & limestone
		Triassic		Sandstone 'New Red Sandstone'
	250			
	298	Permian		
		Carboniferous		Sandstone & shale
	354			Limestone
				Sandstone & shale
PALAEOZOIC		Devonian		Sandstone 'Old Red Sandstone'
	410			
	440	Silurian		Sandstone & shale
				Sandstone & shale
		Ordovician		Shale & sandstone, basalt & rhyolite
	495			
		Cambrian		Sandstone & slate Quartzite in above
	545			
		PRE-CAMBRIAN**		Schist & gneiss Quartzite in above

*Millions of years before present
**Pre-Cambrian and Quaternary not to scale

Gap in geological record (no rocks preserved)

⚒ Working mine or pit

IGNEOUS ROCKS

Volcanic rocks
- Basalt, minor rhyolite - Tertiary
- Volcanic rocks - Precambrian to Carboniferous

Intrusions
- Granite & gabbro - Tertiary
- Granite - Ordovician to Devonian
- Gabbro & related rocks - Ordovician

Fig. 5.3 THE BEDROCK GEOLOGY OF IRELAND. [Data source: Geological Survey of Ireland]

they were created by the alteration of earlier-formed rocks, through heat and pressure over long periods of geological time. They are c.1.7 billion years old and are located in Inishtrahull, County Donegal. The old geological terranes of Donegal, Connemara and south-west Leinster are also composed predominantly of very old medium to high grade metamorphic rocks. Both extrusive and intrusive igneous rocks – those that form above and below the Earth's surface, respectively – can also be found along the Irish coast. Extrusive igneous rocks are most common in the north-east, where basaltic lavas have given rise to the Antrim–Derry plateau. In contrast, intrusive igneous rocks are located in counties such as Wicklow, Galway and Donegal, where granite rock is most prevalent. Sedimentary rocks, which

are those formed by the accumulation of sedimentary materials on the Earth's surface, are widespread throughout Ireland. Of these sedimentary rocks the Old Red Sandstone sequence – consisting of Upper Devonian terrestrial sandstones, siltstones and mudstones – is common in the 'ridge and valley' landscape of much of south-west Ireland.[4] In contrast, limestone, a carbonate sedimentary rock most notable in Ireland in its karst landscape, covers the majority of the Midlands and areas such as the Burren, County Clare. Continuous and freshly eroded coastal sections expose the variety of geology Ireland has to offer, and provide excellent opportunities for study and research. They make it easy for geologists to acquire high-quality data and interpret the history of these rocks (Figure 5.5).

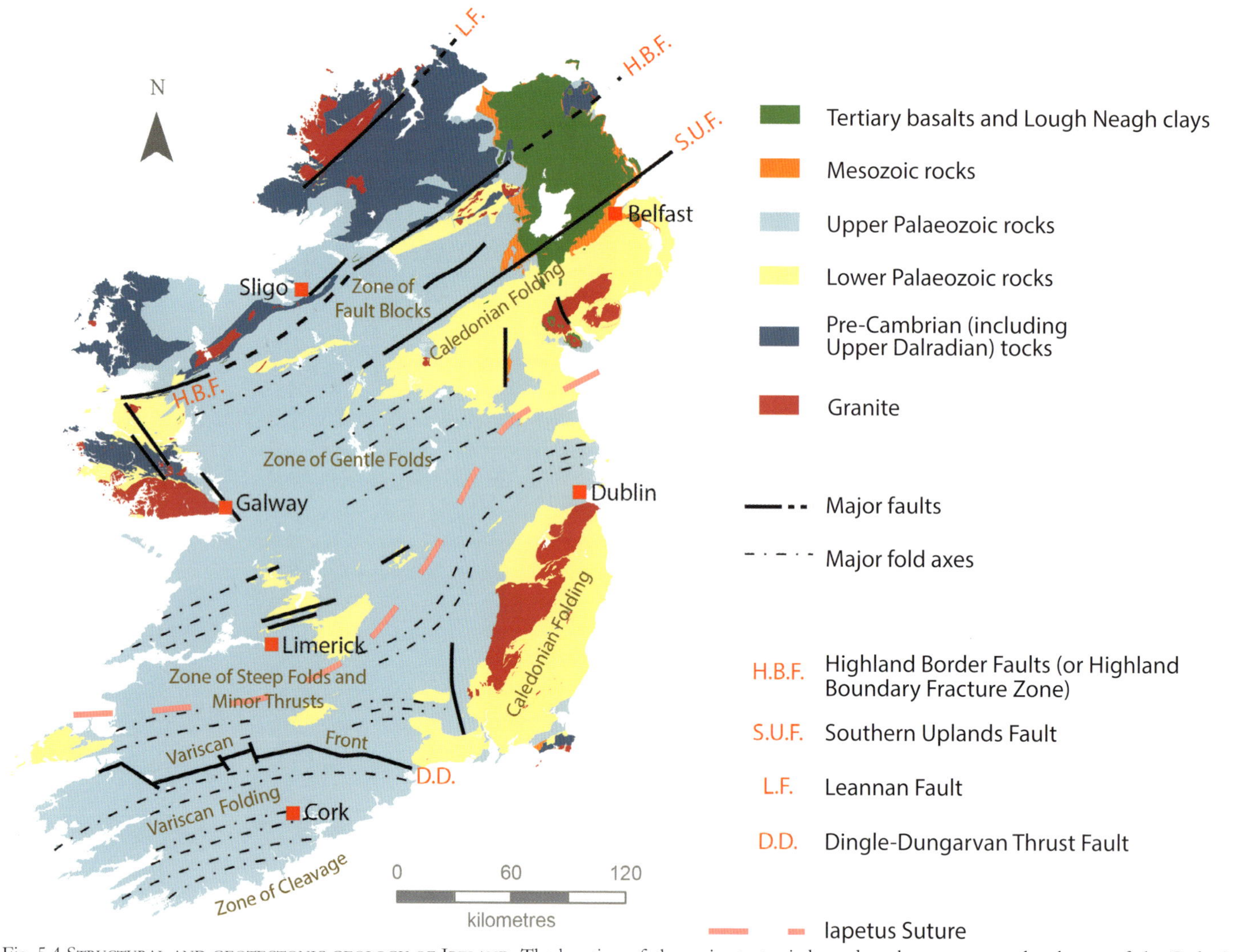

N

Tertiary basalts and Lough Neagh clays

Mesozoic rocks

Upper Palaeozoic rocks

Lower Palaeozoic rocks

Pre-Cambrian (including Upper Dalradian) tocks

Granite

L.F.

H.B.F.

S.U.F.

Belfast

Sligo

Zone of Fault Blocks

Caledonian Folding

H.B.F.

Zone of Gentle Folds

Galway

Dublin

Caledonian Folding

Limerick

Zone of Steep Folds and Minor Thrusts

Variscan   Front

Variscan Folding

D.D.

Cork

Zone of Cleavage

— · — Major faults

— · · — Major fold axes

H.B.F. Highland Border Faults (or Highland Boundary Fracture Zone)

S.U.F. Southern Uplands Fault

L.F. Leannan Fault

D.D. Dingle-Dungarvan Thrust Fault

Iapetus Suture

| 0 | 60 | 120 |

kilometres

Fig. 5.4 STRUCTURAL AND GEOTECTONIC GEOLOGY OF IRELAND. The location of the major tectonic boundary that represents the closure of the Ordovician and Silurian Iapetus Ocean, at the end of the Silurian period, runs across the island from just south of the Shannon to north of Dublin. [Data source: adapted from: Geological Survey Ireland and Whittow, J.B., 1978. *Geology and Scenery in Ireland.* London: Penguin, p. 301]

Fig. 5.5 STUDENTS ON A COASTAL GEOLOGICAL FIELD TRIP AT KILKEE, COUNTY CLARE. The group is sitting on the layered and dipping beds of the grey Carboniferous limestone (aged between *c.*313 and 331 million years old) that forms the low cliffs along this coast. [Source: Zoë O'Hanlon]

Tetrapod Trackway, Valencia Island, County Kerry

Kenneth T. Higgs

The Valencia Island tetrapod trackway located at Dohilla on the north coast of Valencia Island is of international geological importance (Figure 5.6). It records the oldest reliably-dated evidence in the world of four-legged vertebrates (amphibians) first walking on land. The trackway occurs in the Middle Devonian Valencia Slate Formation, which is the oldest rock formation in the Iveragh Peninsula. Volcanic rocks within the Valencia Slate Formation have given an isotopic age of 385 million years for the trackway.[5] The tetrapod tracks were first discovered and described by Ivan Stossel, a Swiss geology student carrying out an undergraduate mapping project on the island in 1993. He found 150 footprints arranged in a parallel trackway that meanders across a large sandstone bedding plane surface. Since this initial discovery a further eight trackways have been discovered on the north coast of the island and some of these trackways record body and tail drag impressions.[6] The rocks of the Valencia Slate Formation have been subjected to strong tectonic deformation and compression, which has shortened the original distance between the footprints. However, using strain analysis techniques it has been possible to un-strain the rocks and calculate the original body size of the primitive amphibian at approximately 1m in length and 0.5m in width. Sedimentological studies of the rocks associated with the trackways show that tetrapods lived adjacent to river channels that crossed the wide alluvial plains on the southern side of the Old Red Sandstone continent. The Valencia Island trackways provide unique insights into the biology and ecology of these early vertebrates and the timing of the transition of animal life from water to land, which was one of the most momentous steps in evolution.

Fig. 5.6 a. VIEW OF THE UPPER DEVONIAN TETRAPOD TRACKWAY AT VALENCIA ISLAND, COUNTY KERRY; b. Close-up view of the trackway footprints. [Source: Ken Higgs]

GEOLOGICAL CONTROLS ON COASTAL GEOMORPHOLOGY

Coastal geomorphology can be defined as 'the study of the morphological development and evolution of the coast as it acts under the influence of winds, waves, currents, and sea-level changes' (see Chapter 2: The Coastal Environment: Physical systems processes and patterns).[7] When the influence of geology in the geomorphological expression of any given section of coast is considered, primarily two overlapping aspects of geology are of importance: the compositional strength of lithologies (rocks) exposed at the coast; and the geological structures exhibited within these lithologies.

Lithological controls

'Lithology' defines the physical characteristics of a rock, including its colour, composition and texture. The variable mechanical strength of rocks (that is, the ability of a rock to withstand exposure to stress exerted by physical forces, as based on the composition of different lithological types, be they igneous, sedimentary or metamorphic in origin) plays a fundamental role in determining the shape of erosional coastlines. Rocks that are susceptible to erosion tend to form coastal bays, while those more resistant to erosion tend to form headlands and promontories. The strength of rock in the first instance is determined by the mineralogical composition and shape of grains that make up these rocks and their possible modification by heat or pressure. Secondary processes, such as weathering, can also play a significant role in determining the strength of a given rock mass (see Chapter 10: Rocky Coasts).

This is particularly evident along Ireland's western seaboard, where exposure to the higher wave energy of the North Atlantic results in the more active erosion of the lithologically weaker Upper Palaeozoic sedimentary rocks. On a smaller scale, the strength of rock can have a direct influence on the development of bays and promontories. A particularly dramatic example of this can be seen at Tramore Bay, County Waterford, where the regular rectilinear geometry of the bay is a product of the selective marine erosion of the soft mudstones and siltstones of the Cambrian Booley Bay and Ordovician Tramore Shale Formations (Figure 5.7).

Dublin Bay is another example of a significant coastal geomorphological feature that has been formed primarily by the process of preferential rock erosion: of the 'Calp' limestone of the Lower Carboniferous Lucan Formation (Figure 5.8). This deep water rock formation consists typically of thin (30–50cm thick) beds of limestone, which are interbedded with black organic rich shales. The limestone beds in themselves are mechanically quite strong. The weak interbedded shales, however, are particularly susceptible to wave erosion and, as a consequence, result in a

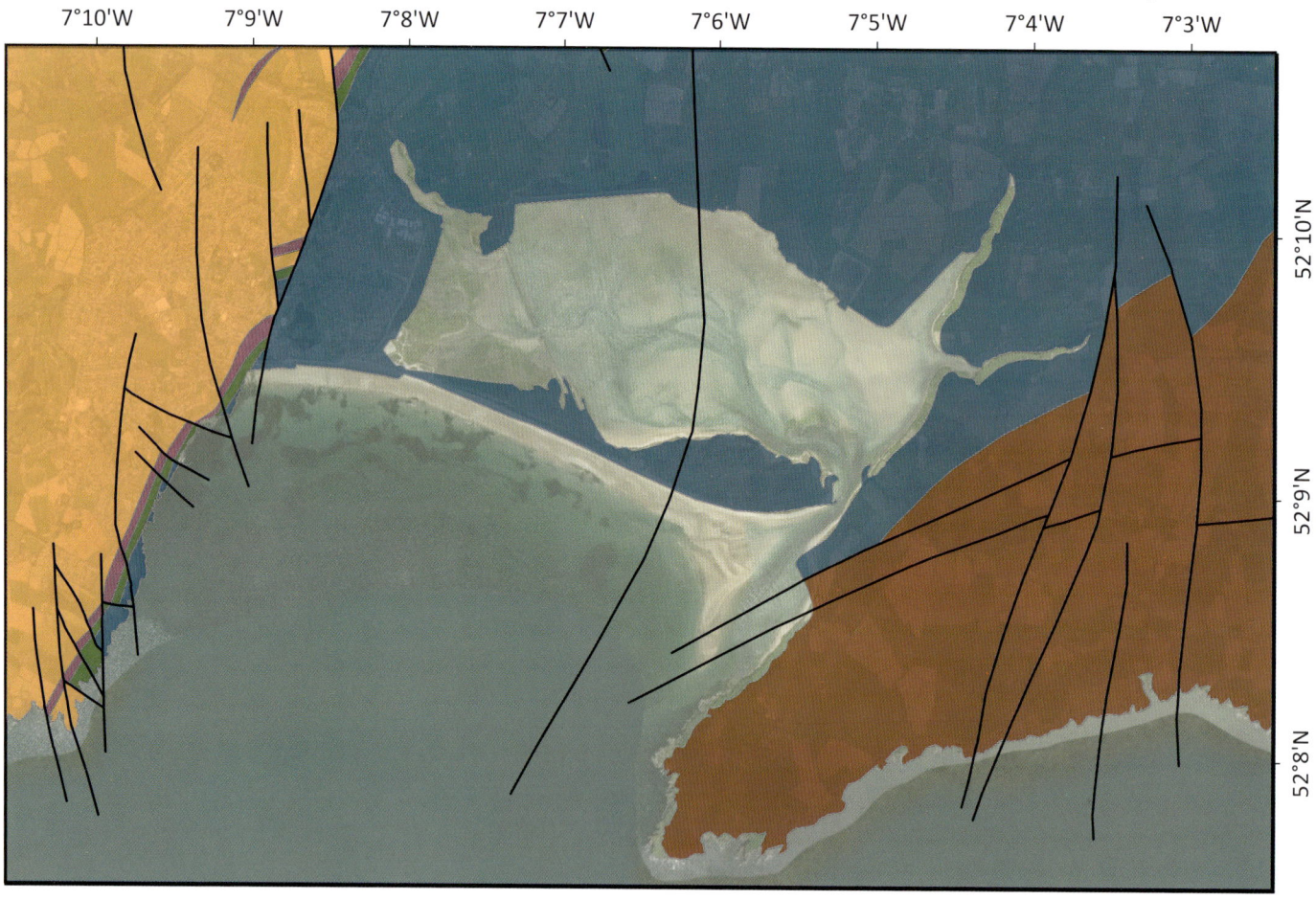

Bedrock Rock Units

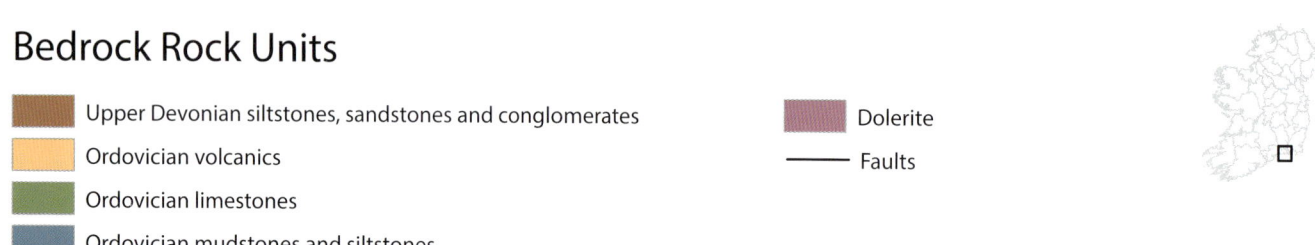

■ Upper Devonian siltstones, sandstones and conglomerates

□ Ordovician volcanics

■ Ordovician limestones

■ Ordovician mudstones and siltstones

■ Dolerite

—— Faults

Fig. 5.7 ONSHORE GEOLOGY OF THE TRAMORE BAY AREA, COUNTY WATERFORD. The coastal geology is shown around the central sandspit and intertidal mudflats in the bay (white and grey colours). [Data source: Geological Survey of Ireland, 1:100,000 data; Google, DigitalGlobe]

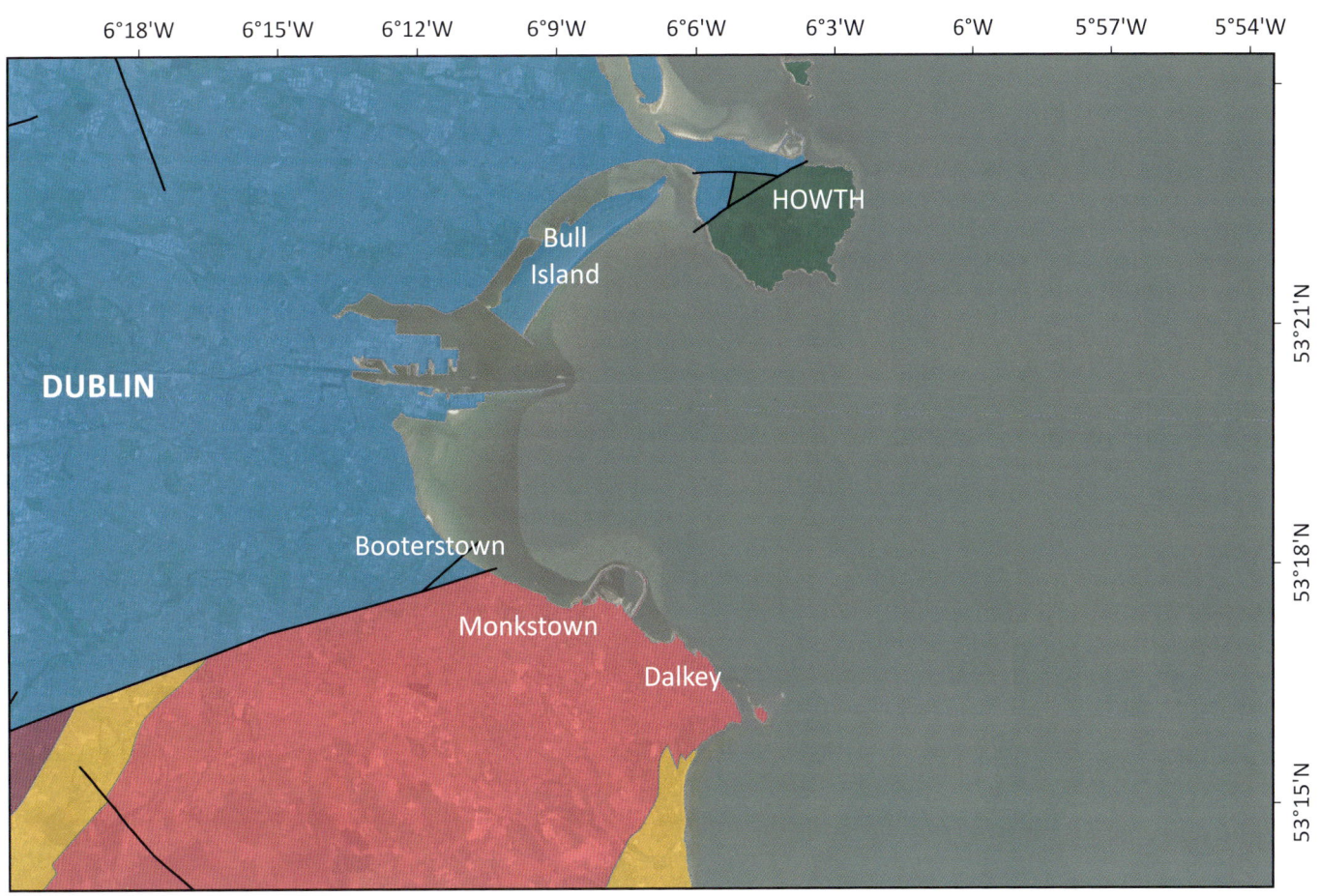

6°18'W 6°15'W 6°12'W 6°9'W 6°6'W 6°3'W 6°W 5°57'W 5°54'W

HOWTH

Bull
Island

DUBLIN

53°21'N

Booterstown

53°18'N

Monkstown

Dalkey

53°15'N

Bedrock Rock Units

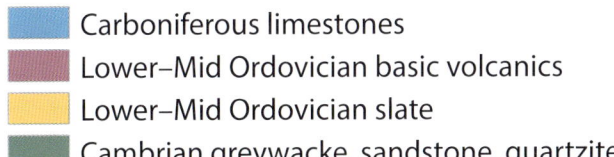 Carboniferous limestones
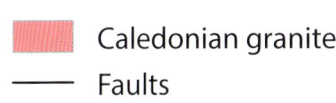 Caledonian granite
Lower–Mid Ordovician basic volcanics
— Faults
Lower–Mid Ordovician slate
Cambrian greywacke, sandstone, quartzite

Fig. 5.8 ONSHORE GEOLOGY OF THE DUBLIN BAY AREA. [Data source: Geological Survey of Ireland, 1:500,000 data; Google, Digital Globe]

relatively higher degree of shoreline retreat. The width of the bay is further constrained to the north by hard (competent) Cambrian quartzites and sandstones in the Howth area and to the south by the northern extent of the Upper Devonian Leinster Granite.

The slope angles of coastal profiles are also heavily influenced by the mechanical resistance of the rocks and recent sediments to erosion. Coastal cliffs are developed where significant areas of elevated topography intersect the coastline. Those rocks more resistant to erosion tend to form steeper slopes and vertical cliff faces (Figure 5.9a), while those less resistant tend to produce lower-angle (10°–40°) slope profiles (Figure 5.9b) (see Chapter 10: Rocky Coasts).[8]

Fig. 5.9 (right) COASTAL CLIFFS AND SLOPES, RESULTING FROM DIFFERENCES IN ROCK AND SEDIMENT TYPES. a. Hard Pre-Cambrian quartzites comprise the rocks of the almost vertical sea cliffs at Croaghaun, west Achill Island, County Mayo. At 688m high, these cliffs are the second highest in Europe. [Source: Robert Burns]; b. Clew Bay, County Mayo coast, which is characterised by the low-angled slopes of the eroded remains of numerous drumlins, developed in relatively soft, glacial till. Many such glacial features in the bay have been overtaken by the postglacial rise of sea level and now lie submerged offshore. [Source: Geraldine Hennigan and Norman Kean]

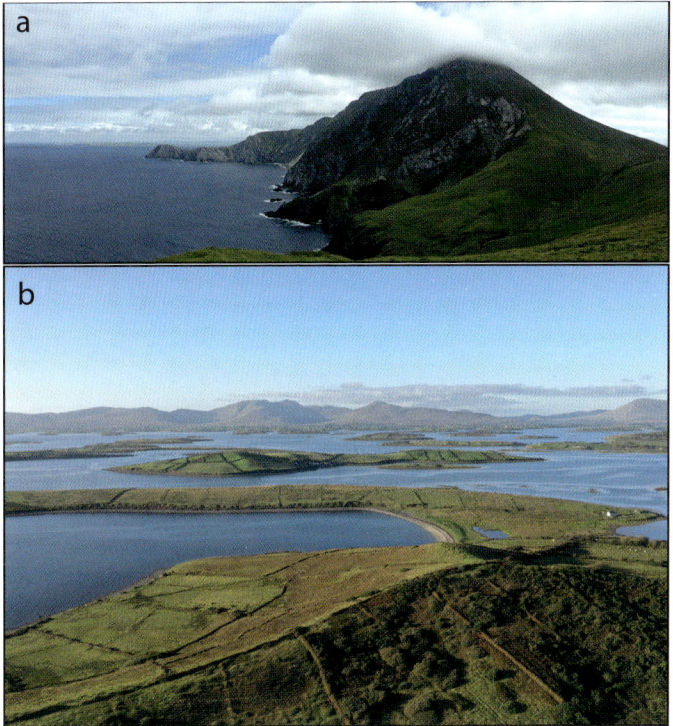

a

b

Fig. 5.10 GEOLOGICAL FOLDS DEVELOPED DURING VARISCAN (END CARBONIFEROUS) MOUNTAIN BUILDING AT LOUGHSHINNY, COUNTY DUBLIN. [Source: Siim Sepp, Wikimedia Commons, CC BY-SA 3.0]

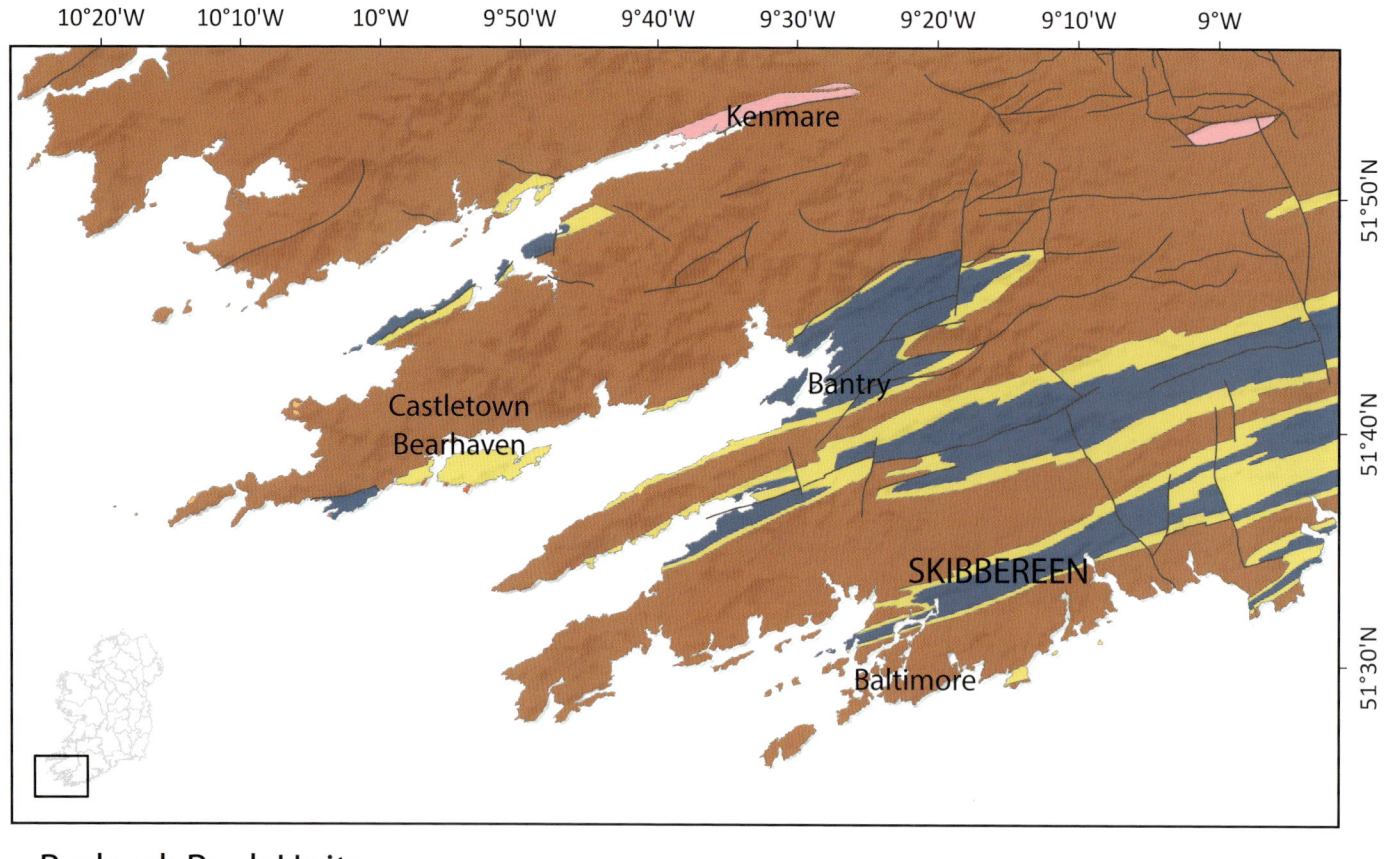

Bedrock Rock Units

- Upper Devonian Old Red Sandstone
- Carboniferous marine mudstones and sandstones
- Upper Devonian marine sandstone
- Carboniferous limestones
- Carboniferous volcanics and minor intrusion
- Devonian basic volcanics, minor intrusion
- —— Fault

Fig. 5.11 REGIONAL GEOLOGY OF SOUTH-WEST IRELAND. [Data source: Geological Survey of Ireland, 1:500,000 data]

Structural controls

Plate tectonics, as a unifying theory of the Earth's geophysical and geological functioning, has impacted in a very profound manner in helping the understanding of how the planet operates. This paradigm shows the planet as consisting of an outer, relatively rigid layer, the lithosphere (the Earth's crust), which is broken into approximately twenty distinct tectonic plates. The lithosphere is underlain by a relatively deformable and ductile asthenosphere that facilitates movement of the tectonic plates across the outer face of the planet. The lithospheric plate movements can result in the large-scale deformation of Earth crustal rocks at plate boundaries and the development of distinctive geological structures. In the context of coastal geomorphology, these structures include features in coastal rock sequences of principally geological folding and fracturing (Figure 5.10).[9]

Geological folding involves the bending of stratified 'layered' rocks, usually as a response to tectonic compressional stress. It is a very effective mechanism for shortening a given sequence of rocks as part of a deformation process related to tectonic plate collision and mountain building. The coastal geomorphology of south-west Ireland is fundamentally controlled by the presence of regional-scale north-east to south-west aligned geological folds that were developed during the Variscan mountain-building event at the end of the Carboniferous period (Figure 5.11).

The distinctive headlands and peninsulas of south-west Ireland (for example, Mizen Head, Sheep's Head, the Beara and the Iveragh peninsulas) correspond to upward-closing anticlinal ridges, while the intervening bays (Dunmanus, Bantry and Kenmare river bays) correspond to downward-closing synclinal troughs. This large-scale rock architecture is further defined by the preservation of mechanically weaker lithologies in the core of the synclines, which facilitate more rapid shoreline retreat along their axes. On a more local scale, the presence of geological folds are responsible for the development of coastal erosional features, such as sea arches and sea stacks (Figure 5.12).[10]

Geological fractures, forming discrete breaks in a rock, impart an inherent structural heterogeneity to a given rock mass and consequently have the potential to greatly influence the geomorphology of actively eroding coastlines. Joint fractures are the most ubiquitous and obvious structures seen in coastal outcrops. These represent simple opening fractures that exhibit no observable simple shear strain, that is, rock slip parallel to the fracture surface. As a consequence, they are discrete features that exhibit little or no damage to the host rock either side of the fracture. Typically, the orientation of joint fractures is very systematic, with a few fracture sets having the same orientation and spacing developed in a given area. This regular patterning inherent in the Earth's

rocks is displayed well along Ireland's coasts. For example, the coastal geology of south-west Ireland is overprinted with a regional north-north-west–south-south-west striking joint fracture set that is particularly well exposed at the western end of the Beara Peninsula. The presence of this mesoscopic scale fracture set imparts a very rectilinear pattern to the coastal profile in this region (Figure 5.13).[11]

Geological faults are fractures across which there is observable simple shear failure and displacement, indicating vertical and horizontal movements of the rocks. As a consequence these structures have usually an associated damage zone, formed either side of the fracture surface. In some cases, a given fault, can consist of multiple fracture surfaces defining a fault zone. The rocks hosting faults can, therefore, be modified

Fig. 5.12 a. AERIAL VIEW OF SEA ARCH AT THE BRIDGES OF ROSS, LOOP HEAD, SOUTH-WEST COUNTY CLARE, CREATED BY SEA EROSION. [Source: David Hodgson]; b. Ground view of the sea arch. The development of this feature is controlled by the presence of a Variscan age folding in the well-bedded Upper Carboniferous siltstones and sandstones. Some of the sea arches known from the site have now fallen into the sea. [Source: Zoë O'Hanlon]

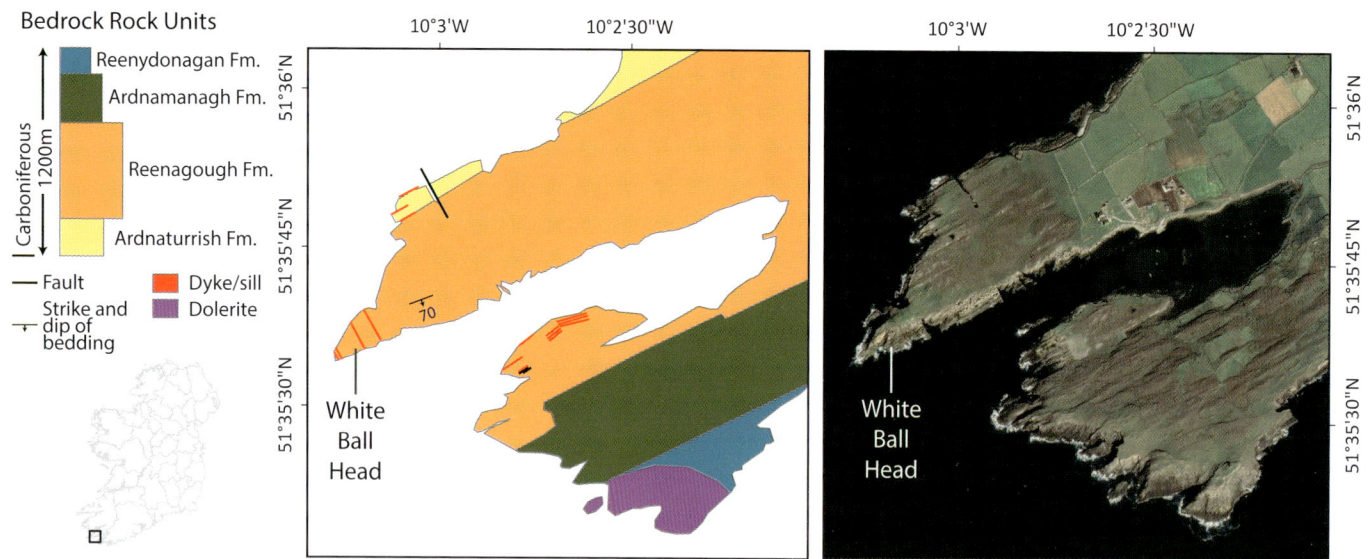

Bedrock Rock Units

Reenydonagan Fm.
Ardnamanagh Fm.
Reenagough Fm.
Ardnaturrish Fm.

Carboniferous 1200m

— Fault
Strike and dip of bedding
Dyke/sill
Dolerite

White Ball Head

White Ball Head

FIG. 5.13 SIMPLIFIED GEOLOGY OF THE WHITE BALL HEAD AREA, ON THE SOUTH-WEST SIDE OF THE BEARA PENINSULA, COUNTY CORK. Rectilinear pattern visible in the rocks of this area, resulting from the combination of local to larger-scale rock fractures and faults. [Data source: Quinn, D., Meere, P.A. and Wartho, J.A., 2005. A Chronology of Foreland Deformation: Ultra-violet laser 40Ar/39Ar dating of syn/late-orogenic intrusions from the Variscides of southwest Ireland. *Journal of Structural Geology*, 27(8), pp. 1,413–1,425; Google, CNES/Airbus, 2019; Geological Survey Ireland 1:100,000 data. Google, DigitalGlobe]

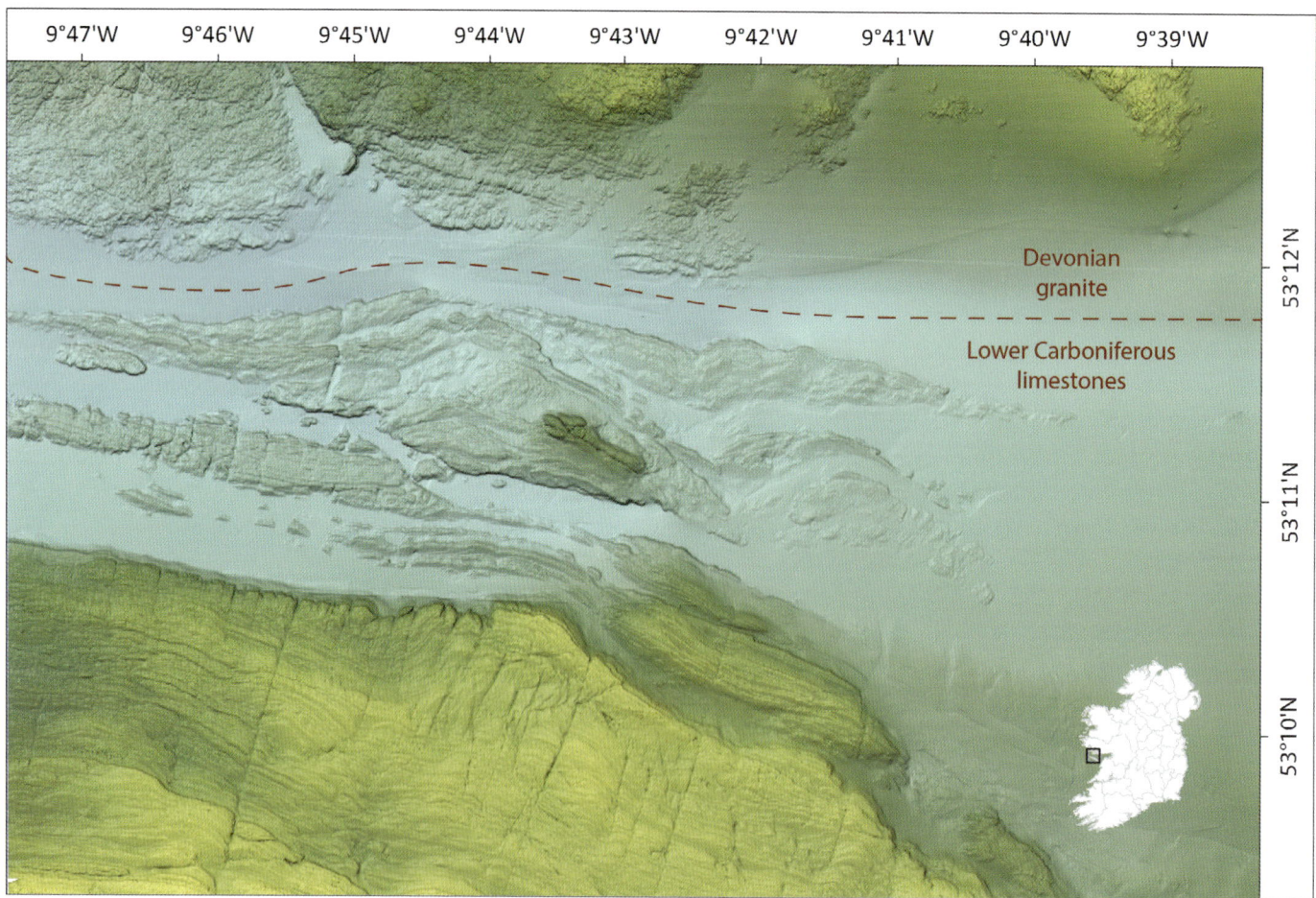

Devonian granite

Lower Carboniferous limestones

Fig. 5.14 BATHYMETRY AND GEOLOGY OF THE NORTH PASSAGE AREA OF GALWAY BAY. The boundary between the Galway granite and Carboniferous limestones is marked by a dashed line. Based on the known onshore geology, this boundary between a sequence of Lower, well-bedded limestones in north County Clare and the Lower Devonian Galway granite, of south County Galway, must lie somewhere within this area of Galway Bay. Examination of the INFOMAR bathymetry of the area reveals the boundary between two distinct seafloor morphologies that represent the two very different geological units. The submarine limestone outcrops to the south are characterised by a distinctly uniform surface texture, with clear primary layering (bedding) and north–south orientated cross-cutting joint fractures that correspond to these we see onshore. On the other hand, the submarine granite outcrops have a distinct rough surface texture and lack any distinct layering or systematic fracture set. [Source: INFOMAR]

radically by prolonged histories of shear displacement across fracture surfaces, which produce very distinct fault rocks, such as fault gouge and breccia. Under low pressure and temperature deformation conditions, these different fault rock lithologies form relatively weak, incohesive materials that are particularly susceptible to preferential coastal erosion. An example of this type of erosion is illustrated beautifully by the presence of a deeply incised inlet, containing two aligned 'blow holes', that corresponds to the trace of a north-west to south-east orientated fault plane on the north shoreline of White Ball Head (see Chapter 10: Rocky Coasts).

Faults occur on all scales on planet Earth, from regional kilometre-scale tectonic plate boundaries, to local metre-scale features, to micron-scale microstructural discontinuities. When focusing on coastal geomorphology we are principally concerned with the upper end of this size scale range. The island of Ireland is transected by a number of major tectonic suture, or fault zones. Arguably, the most important of these is the Iapetus Suture Zone, extending from the Shannon Estuary in the west to County Louth in the east. This tectonic boundary represents the closure of the Ordovician and Silurian Iapetus Ocean at the end of the Silurian period, which represents the final significant coming together of the Earth tectonic plates that make up the island of Ireland (see Figure 5.4).[12]

Offshore geology

The rocks visible at the coast represent a snapshot of the wider geological picture, which is, of course, developed both landwards and offshore. On land, the rocks and associated structures can be traced relatively easily, for example, in naturally occurring cliff sections, quarries, river valleys and other exposures, but offshore, with much greater difficulty. In the past, marine geology and geophysics were based largely on seabed rock and sediment coring and on the use of a limited range of seismic and other remotely sensed survey techniques. The last two decades, however, have seen an unprecedented increase in the range and availability of high-resolution survey techniques, as well as in the mapping of the seabed of Ireland.

The Irish National Seabed Survey (INSS) and its successor, the Integrated Mapping For the Sustainable Development of Ireland's Marine Resource (INFOMAR) programme, have provided a rich database to help geoscientists reveal the physical and chemical features of the seabed in the nearshore area (see Chapter 9: Underwater Surveys: The INFOMAR project). This work, which is a joint venture between the GSI and the Marine Institute (MI), uses a combination of cutting-edge geophysical mapping techniques to determine the bathymetry (water depth) of the survey areas. These include single and multibeam echo sounding, side scan sonar, light detection and ranging (LiDAR) and shallow seismic profiling. This suite of techniques, in association with the geological sampling methods outlined below, have also proven to be important in characterising the nature of sediments and hard rocks on and below the seabed (see Chapter 8: Monitoring and Visualising the Coast). The resulting data sets, all of which are available on the GSI website (www.gsi.ie), include detailed maps of the seafloor morphology. Shaded relief maps are particularly useful in characterising the bedrock geology of the seafloor. The morphology of any rock outcrop is principally controlled by the lithology type and associated geological structures. A good example of this can be seen from the bathymetry map of the North Passage area, Galway Bay (Figure 5.14).

THE COLLECTION OF GEOLOGICAL DATA FROM SHELF AND COASTAL WATERS

Aaron Lim

Approximately 90 per cent of Ireland's territory and landscape is submerged, deep beneath the ocean (see Chapter 7: Ancient Shorelines and Sea-level Changes and Chapter 32: Management and Planning, Figure 32.3). To study Irish geology effectively, it is essential to collect data from each part of the Irish landscape. Given the time, cost and weather dependency associated with marine data acquisition, collecting rock and sediment samples from beneath hundreds to thousands of metres of water is an extremely difficult task. To do this, geoscientists utilise a number of different methods to collect these samples (Figure 5.15).

A relatively simple method for retrieving geological samples from the seabed can be completed using a rock dredge, and other grab-type samplers. A rock dredge is a circular, steel-jawed frame that is used to break off pieces of bare rock from the seabed as it is towed behind a research vessel. Samples are then trapped within a chain mail bag and can be retrieved once the dredge is brought back onboard. This is a relatively quick and low-cost method that can be used in any water depth, as there are no electronic and motorised parts on the dredge. However, using the dredge alone does not provide information on exactly where the sample is from. As such, it is assumed that the sample is collected along the transit line of the vessel.

Irish geoscientists have been developing a new way to collect samples from the ocean floor that allows the user to see exactly what is being sampled. To do this, a rock drill is mounted to the front of a Remotely Operated Vehicle (ROV). The ROV operator onboard the ship can guide the equipment using a high-definition video camera system, to reach precise locations and hard-to-reach rock outcrops.

Fig. 5.15 GEOLOGICAL SAMPLING TECHNIQUES OFFSHORE. a. The *Holland* 1 ROV on board the RV *Celtic Explorer*. [Source: Zoë O'Hanlon]; b. Day grab being deployed on board the RV *Celtic Voyager*. A day grab is typically used for 'ground truthing' and sampling of sandy and muddy sediments. [Source: Darius Bartlet]; c. Vibrocore in use onboard the RV *Celtic Explorer* [Source: Dr Aaron Lim]; d. CTD deployment on a SMART Seaschool training programme. This probe measures the salinity, temperature and depth of the water column and is rated to 3,000m. [Source: Strategic Marine Alliance for Research and Training]; e. Analysing sediment from a core. [Source: Dr Aaron Lim]

When a suitable outcrop is identified, the ROV guides the rock drill into the outcrop where it can take a 20cm-long sample. This affords scientists the ability to retrieve samples precisely from 10m to 3,000m water depth with confidence. However, this method requires relatively calm sea states, prior knowledge of the study site and is relatively expensive.

Marine geoscientists are also interested in the character of the sediments deposited on the seabed. When analysed, these samples can provide information on current speeds and the benthic environmental conditions in space and over time. To take sediment samples from the surface of the seabed, a day grab-type sampler is commonly used. This is an open-jawed, bucket-type apparatus that is lowered through the overlying water to the seabed from a research vessel. On reaching the seabed, it is triggered to take a scoop of sediment. The sample is stored within the day grab and brought back on board, where it is washed through several sieves to determine the particle sizes and composition of the sediments.

These types of surface sediment samplers only provide information about contemporary seabed processes. Marine scientists are also interested in how seabed areas have changed over time and in the patterns of sediment deposition beneath the seabed surface. To retrieve a longer sedimentary record, coring systems are used. There are many different types of sediment corers. Given the thick sequences of soft marine, glacial and other sediments that surround Ireland, a device used commonly in Irish waters is a vibrocorer. This is a 6m-long cylinder with a spoke-like stabiliser frame at the base and a motor at the top. The corer is deployed from the research vessel onto the seabed and the motor vibrates and pushes the cylinder into the underlying sediment, which is retained within the vibrocorer in a core liner. The vibrocorer is then recovered and the sediment core extruded from the cylinder's liner on the research vessel. The cores are generally brought to a laboratory, where they can be split open along the length of the core and the sediment described. Results from core studies are often a continuous description of seabed processes through time, providing detailed understanding of seabed development.

Preserving coastal heritage

Ireland's unique geological heritage attracts visitors to the most remote areas of the coastline due to its stunning geology and geomorphology (Figure 5.16). Key programmes with a role in the protection and promotion of these important geological sites seek to preserve this heritage (see Chapter 15: Coastal Heritage). An example is the Geoheritage programme, run by the GSI, which identifies sites – such as Lambay Island, Fingal, north County Dublin – as county geological sites, affording them protection from inclusion in the planning system. These sites are included in Ireland's County Development Plans and County Heritage Plans, ensuring that due regard to such sites is given. Locations of significant importance are recommended to the National Parks and Wildlife Service for future potential designation as Natural Heritage Areas, providing them statutory levels of protection.[13]

The Giant's Causeway, on the north Antrim Coast, is an example of other coastal areas of importance in Ireland's geological heritage, in this case at an international level. The significance of the Giant's Causeway in understanding Earth's geological history was recognised in its designation as a UNESCO World Heritage Site in 1986 (see Chapter 27: Tourism and Leisure).[14] Most of the property in this coastal region has been designated a National Nature Reserve and area of Special Scientific Interest, with the area attracting up to one million tourists each year. Additional

locations, including two European geoparks – at Waterford's Copper Coast and at the west Clare Burren and Cliffs of Moher Geopark – also contribute to the development and enhancement of high-quality and high-impact eco-tourism. These geoparks bring socio-economic benefits to the coastal communities in these regions, especially outside the traditional peak tourism seasons.

Ireland's geology has given a distinctive character to its coasts,

Fig. 5.16 THE COAST OF IRELAND CONTAINS SOME SPECTACULAR EXAMPLES OF GEOLOGICAL AND GEOMORPHOLOGICAL FEATURES. a. Kinsale, County Cork [Source: Shane O'Connor]; b. Whiterocks Beach, Portrush [Source: © Tourism Northern Ireland]; c. Tearaght Island, west of the Dingle Peninsula [Source: Geraldine Hennigan and Norman Kean]; d. Loop Head Cliffs, County Clare [Source: Zoë O'Hanlon]; e. Inishtooskert, County Kerry [Source: Geraldine Hennigan and Norman Kean]; f. Blind Harbour, Dungarvan, County Waterford. [Source: Geraldine Hennigan and Norman Kean]

through both the onshore and offshore geology and the operation of the contemporary agents of coastal change: water, waves and wind. At one level this geology is simple, with the rocks dominated by sandstones, limestones, granite, basalt and soft glacial sediments. In contrast, the geology is old and structurally complex, with the island's tectonic history stretching from before the Pre-Cambrian to the present. This structural geology has resulted in the creation of an upland coastal rim of mountains around the island, together with high cliffs, long indented valleys, wide embayments, rock headlands and numerous sandy beaches. These unique features have imparted a high level of attractiveness to Ireland's landscapes, with land surfaces often rising from sea level to over 500m within 5-10km of the coastline. At another level, the geology provides the island with economically valuable resources. The coastal zone, whilst not extremely rich in mineral resources, does have wide and shallow water shelf seas. These areas form a good platform for wind- and wave-energy development, as well as containing the now less desirable resources of oil and gas. This geological inheritance also provides a rich legacy to the island's coasts for trade, which includes its many deep-water harbours and ports for shipping, as well as being a valuable asset for recreation and tourism. The rich natural heritage of Ireland's coastline will continue to provide the Irish professional and amateur geoscience community with an incredibly varied natural laboratory to explore and study.

LOOP HEAD 'THE ROSS SLIDE' COASTAL GEOLOGY, COUNTY CLARE (Source: Zoë O'Hanlon)

GLACIATION AND IRELAND'S ARCTIC INHERITANCE

Paul Dunlop

KILLARY HARBOUR AND BUNDORRAGHA RIVER, LEENANE, COUNTY MAYO. Killary Harbour, in the foreground of this picture, is a sinuous, glacially overdeepened valley, carved out by the repeated advance and retreat of ice during the Quaternary. It is commonly known as a fjord, even though it does not meet all the criteria normally associated with true fjords – but what's in a name? Bundorragha River, seen here at its mouth, links the waters of the harbour to the glaciated plateau uplands to the north. [Source: Geraldine Hennigan and Norman Kean]

When you walk along the Irish coastline today it can perhaps be hard to comprehend that most of what you see has been influenced in some way by glacial geomorphological processes that took place during the last Ice Age. To understand how this is so, one needs to have an understanding of the last glacial period and also how ice sheets erode and transport sediments to create unique glacial landscapes. Ice sheets and glaciers are powerful systems that sculpt the topography they flow across. They are able to grind down into solid bedrock, creating new sediments that are then transported over sometimes large distances, before being deposited and moulded into new landforms. In Ireland, these can extend as far as the edge of the continental shelf.[1]

The landscape we see today, as well the coastline and its continental shelf, have all been directly influenced by the growth, flow and decay of large ice sheets during the last 2.6 million years, a geologic period of time known as the Quaternary (Figure 6.1). The Quaternary is divided by geologists into two epochs: the Pleistocene, which stretched from 2.6 million years ago to approximately 11,700 years ago, and the Holocene. The latter marks the end of the last glacial cold stage and the beginning of the current warm period of earth's climate (the postglacial, beginning fully from c.10,500 years ago).

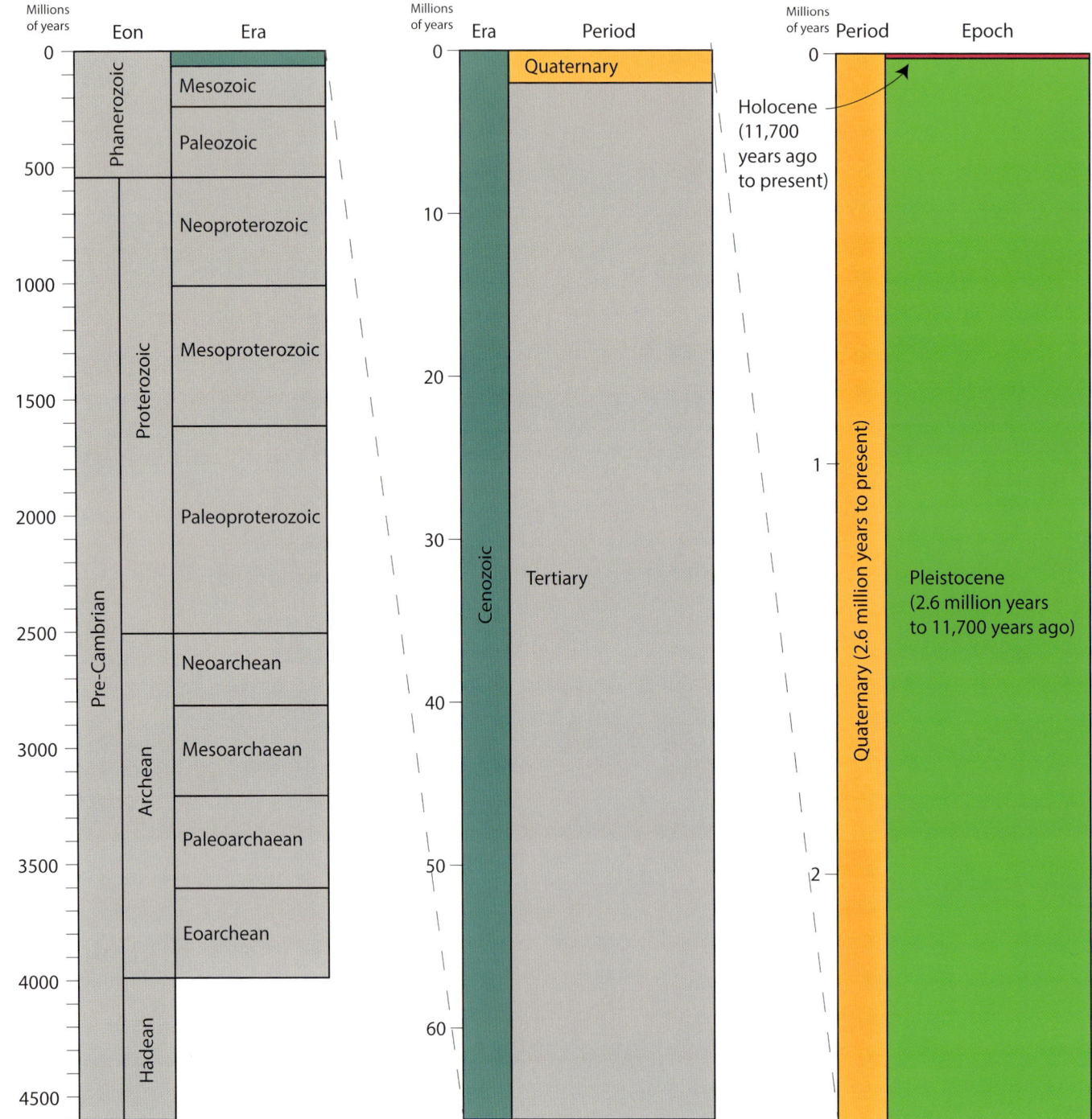

Fig. 6.1 QUATERNARY PERIOD AND EPOCHS IN THE CONTEXT OF GEOLOGIC TIME. Part of the Cenozoic era, the Quaternary period covers the last 2.6 million years of the Earth's c.4.6 billion-year history. It comprises two epochs – the Pleistocene and the Holocene. The Pleistocene covers the period from 2.6 million years to 11,700 years before present and is characterised by major climatic changes that saw the growth and decay of large ice sheets that covered parts of North America, Eurasia, Britain and Ireland. The Irish landscape that we are familiar with today has been shaped largely by Quaternary glaciations. The current warm period, which began approximately 11,700 years ago, is known as the Holocene.

118

The defining feature of the Pleistocene has been fluctuations of the global climate system, which produced long periods of extremely cold temperatures known as glacials. Cores of sediments taken from the deep ocean, as well as from ice in Antarctica and Greenland, tell us that there were at least eight of these glacial episodes in the past three quarters of a million years, though in many cases the more recent events obliterated any evidence on land for the previous ones. In between the glacials, periods of warmer climate occurred, some of them with temperatures at or greater than those seen today; these are known as interglacials.

During the glacials, enormous thicknesses of ice grew and expanded over much of the northern hemisphere, covering large parts of North America, Eurasia, Britain and Ireland, and spreading across the coasts onto the continental shelf where, in places, they merged with other ice sheets from adjacent landmasses. At its maximum extent, the ice sheet that covered most of Britain spread across the North Channel, south-west from Scotland and across the Irish Sea, and coalesced with the Irish Ice Sheet to form one large ice mass, known as the British-Irish Ice Sheet. The interaction of this ice sheet with the underlying topography has left a rich geological record, which can be used to decipher the evolution of this ice sheet, the coastline and the continental shelf through time. This chapter will review, with particular focus upon the evidence from north and mid-western Ireland, the different types of landforms and sediments that an ice sheet creates. It will outline how successive glacial episodes, and particularly the last such event, have helped shape Ireland's coastal and submarine landscapes (Figure 6.2).[2]

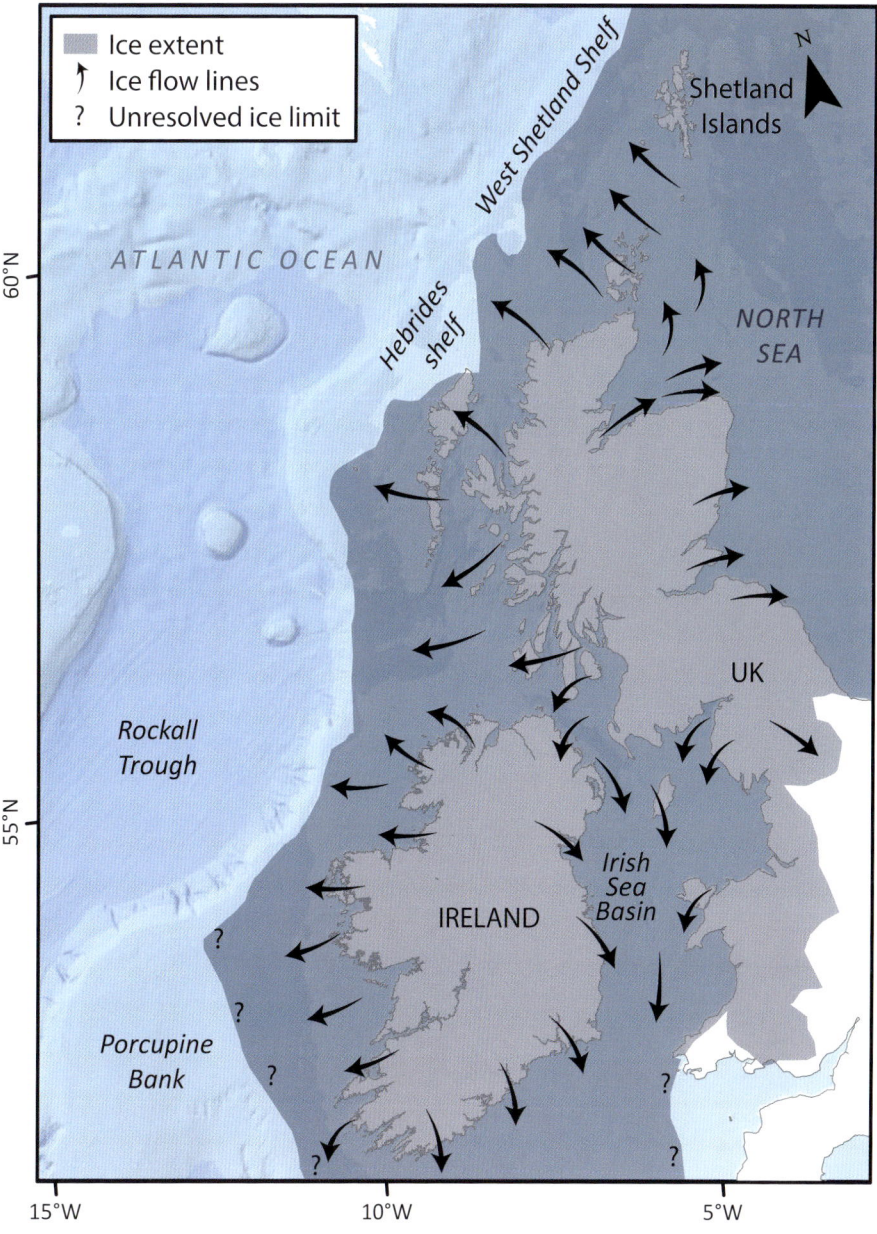

Fig. 6.2 THE MAXIMUM KNOWN EXTENT OF THE FORMER BRITISH-IRISH ICE SHEET DURING THE LAST GLACIAL PERIOD. Arrows show the approximate ice limits and the generalised ice-flow pattern at maximum extent, between approximately 20,000 and 40,000 years ago. At its full extent, all of Ireland and large parts of the Irish continental shelf were covered by ice, in places over 1km thick, which flowed across the topography and the present-day seafloor. At its margins, and especially as conditions started to ameliorate, the ice became detached from the seabed and formed extensive floating sea ice.

Ailsa Craig

Darius Bartlett

Ailsa Craig is a small, now-uninhabited island in the Firth of Clyde on the west coast of Scotland, approximately 15km from the town of Girvan (see Figure 6.3a). In geological terms it represents the remains of a volcano that erupted *c.*60 million years ago, during the opening of what is now the Atlantic ocean. The microgranites that make up the plug (the solidified magma neck) of the volcano have a distinctive composition and crystalline structure, found only in stone from this one location. This makes Ailsa Craig granite easy to recognise and, hence, an ideal marker for tracing the passage of Scottish ice outwards and away from its centre of accumulation in the Highlands to the north. Pebbles and boulders of Ailsa Craig granite can be found as glacially derived erratics on many shorelines on the

119

Fig. 6.3 (left) AILSA CRAIG GRANITE IN IRELAND. a. The island of Ailsa Craig in the outer Firth of Clyde. [Source: Mary and Angus Hogg, https://www.geograph.org.uk/]; b. The beach at Killiney, a classic Irish Sea location for finding pebbles of Ailsa Craig granite. [Source: William Murphy, CC BY-SA 4.0, flickr]; c. (inset photograph) Pebble of Ailsa Craig granite, found as an erratic in the cliffs at Ballycotton Bay, County Cork, in till derived from Irish Sea ice. [Source: Robert Devoy]

east coast of Ireland, including sites at Killiney in County Dublin (Figures 6.3b), and round Wicklow Head at least as far west as Ballycotton Bay in County Cork (Figure 6.3c). Distribution of this granite, as well as large amounts of linked glacial sedimentary evidence, show how an important ice stream from the British-Irish Ice Sheet made its way down the Irish Sea, and out and westward into the Celtic Sea. The glacial tills at the base of cliffs to the rear of Ardnahinch and Shanagarry beaches in Ballycotton Bay also contain clasts of Ailsa Craig granite; these tills are overlain by more recent ones that were deposited by ice that travelled eastward down the Lee Valley from the Kerry mountains in the west, as part of a shorter-lived resurgence of glacial conditions towards the close of the last Ice Age, around 20,000 years ago.

QUATERNARY SEDIMENTS AND LANDFORMS

The legacy of Quaternary glaciation can be understood by careful examination of the landforms and sediments seen in the present day (Figure 6.4a and b). These have been investigated in Ireland for over 150 years, with Ireland being one of the first places in Europe where intensive scientific investigations of glacial sediments and landforms were undertaken. Our understanding of the last ice sheet that covered much of Ireland, and of its several predecessors, has become increasingly more complex as new lines of evidence have been presented. One of the longest-studied of the landforms that resulted from glacier action and deposition of ice-eroded debris has been the pattern of subglacial bedforms (for example, drumlins). Different assemblages of these bedforms cover large swaths of the country, from the interior of the island to the present-day coast. Increasingly, as technologies in geophysical and remote sensing investigations have allowed, these bedforms and the associated glacial sediments (moraine) found off these coasts and further onto the continental shelf have also become mapped and studied intensively.[3]

Understanding of these bedforms is important for both the reconstruction of the Quaternary ice patterns that developed over Ireland as well as the interactions of this ice with the underlying bedrock. The land topographies, associated sediments and landforms that have resulted from bedform development now underpin many of Ireland's coastal areas and shelf environments. Subglacial bedforms are either longitudinal or transverse accumulations of sediment that were formed at the base of an actively flowing ice sheet. The longitudinal forms are collectively termed glacial lineations and in forming were streamlined parallel to ice-flow direction. Depending on their size, they are further subdivided into flutes, drumlins or mega-scale glacial lineations. These can increase in length from commonly a few hundred metres, to reach over hundreds of kilometres (especially in continental environments, as in Scandinavia and Canada). The most significant of the transverse bedforms are known as ribbed moraines, which in Ireland can be up to 1.5km wide and 17km in length (Figure 6.5). The largest concentration of subglacial bedforms is located in the central and northern half of the island, where they comprise mainly ribbed moraines. These bedforms have been partially overprinted by drumlins, forming distinctive landscapes, particularly in coastal areas.[4]

Because of the complexity of these landform arrangements, they were overlooked or misunderstood by early researchers who, as a result, arrived at relatively simple ice-sheet reconstructions that depicted the ice sheet as one large dome, radiating ice away from central Ireland towards the coast.[5] However, this simple radial ice-flow pattern has since been rejected because it does not explain more recently identified bedforms that reveal the superimposition of landforms and sediments, or more than one direction of ice movement. Studies indicate that a more complex ice sheet existed on the island, which at times crossed over Ireland's present coasts and flowed out onto the continental shelf. We now know that the last ice sheet in Ireland consisted of several large, semi-independent ice domes. These coalesced gradually and had distinctive ice-flow corridors of varying strength, that switched periodically on and off, and that converged with ice flowing from the other domes, including those from northern Britain at times. The cross-cutting landform patterns of the resulting landscapes were created by these changing

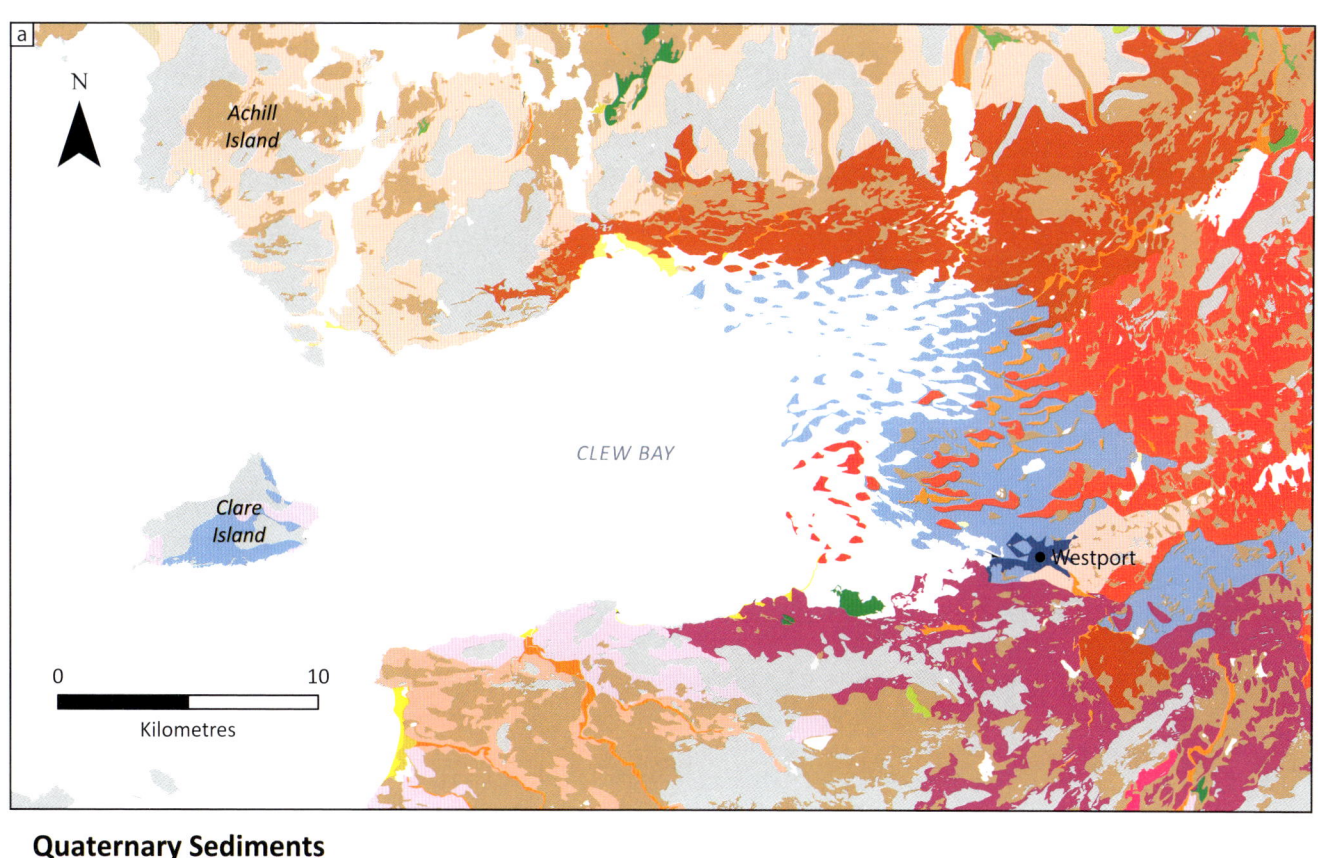

Quaternary Sediments

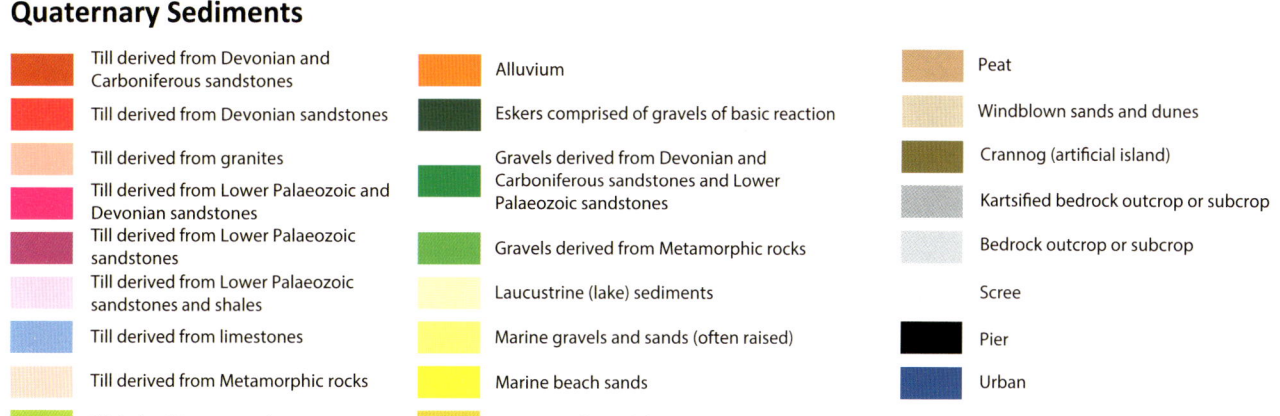

- Till derived from Devonian and Carboniferous sandstones
- Till derived from Devonian sandstones
- Till derived from granites
- Till derived from Lower Palaeozoic and Devonian sandstones
- Till derived from Lower Palaeozoic sandstones
- Till derived from Lower Palaeozoic sandstones and shales
- Till derived from limestones
- Till derived from Metamorphic rocks
- Till derived from quartzites

- Alluvium
- Eskers comprised of gravels of basic reaction
- Gravels derived from Devonian and Carboniferous sandstones and Lower Palaeozoic sandstones
- Gravels derived from Metamorphic rocks
- Laucustrine (lake) sediments
- Marine gravels and sands (often raised)
- Marine beach sands
- Estuarine silts and clays

- Peat
- Windblown sands and dunes
- Crannog (artificial island)
- Kartsified bedrock outcrop or subcrop
- Bedrock outcrop or subcrop
- Scree
- Pier
- Urban

Fig. 6.4 QUATERNARY SEDIMENTS AND LANDFORMS AROUND CLEW BAY. a. These excerpts from the Geological Survey of Ireland's Quaternary geology mapping programme show Quaternary sediments and landforms around Clew Bay, County Mayo. The study of such physical features can help us reconstruct past environments (known as palaeoenvironments), so that we can understand, for example, ice-flow patterns and subglacial and periglacial (marginal ice) processes that once operated here. Tills, for example, are deposited directly by glacial ice, whereas alluvium is deposited by meltwater. The composition of the sediments can tell us about their source (or provenance). b. (overleaf) These sediments form the basis of the depositional landforms shown on the geomorphology map, such as moraines, drumlins and eskers. The landforms found here are characteristic features of periglacial environments. Their distribution, composition and orientation provide important evidence about the ice extent, flow and timing of deglaciation. The data on which these maps are based are free to view or download for the whole island of Ireland from the GSI website. [Data source: Ordnance Survey Ireland, CC-BY 4.0; Geological Survey Ireland. [online] Available at: https://dcenr.maps.arcgis.com/apps/webappviewer/index.html?id=de7012a99d2748ea9106e7ee1b6ab8d5&scale=0]

ice flows and by large migrations of the centres of ice dispersal to different parts of Ireland, as they grew and decayed in response to climate change (see Figure 6.2).[6]

In terms of the coastal influence of the ice sheet, the entire Irish coastline would have been shaped extensively by glacial erosion and landform development during the last glacial period. It is important to remember though, that as global and also local sea levels around Ireland were significantly lower at this time (by over 100m or so), the coastline itself would have been situated far to the west and south of the island. The contemporary, and at times ice-marginal, coast would have been out on the present-day continental shelf, extending as far as the continental slope in some places. Part of the evidence for this comes from the studies of these extensive ice-depositional bedforms. Two of the most spectacular coastal examples of this landform pattern can be seen on the fringes of Donegal town and further south in Clew Bay in County Mayo (Figure 6.6), where several hundred ribbed moraines, drumlins and other bedforms converge into the bays and extend offshore, recording the former flow pathway of the ice sheet as it streamed through the bay, past Clare Island and onto the outer continental shelf (Figure 6.7).

IRELAND'S GLACIATED CONTINENTAL SHELF

Recent breakthroughs in marine acoustic scanning techniques have

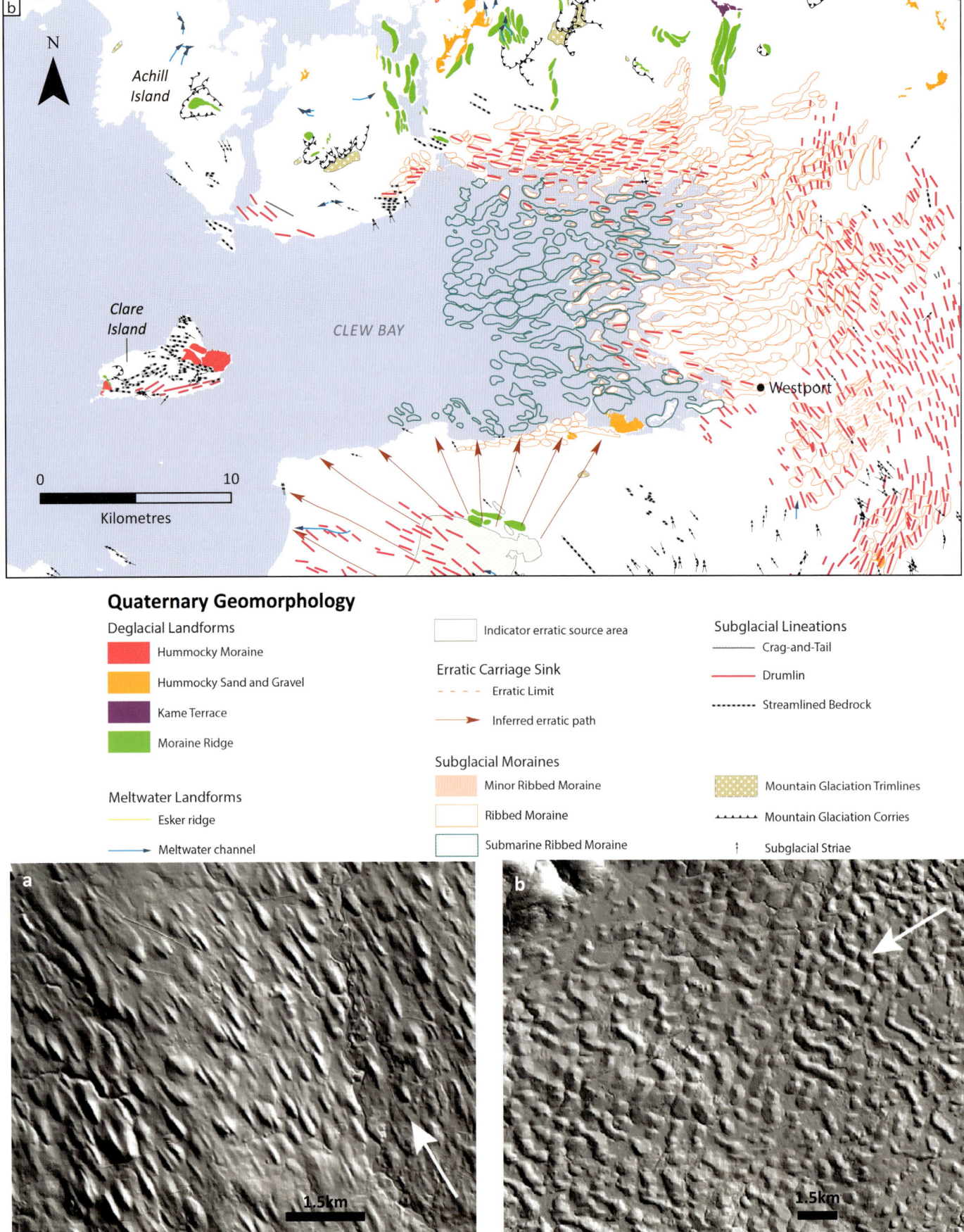

Quaternary Geomorphology

Deglacial Landforms

- Hummocky Moraine
- Hummocky Sand and Gravel
- Kame Terrace
- Moraine Ridge

Meltwater Landforms

- Esker ridge
- Meltwater channel

Indicator erratic source area

Erratic Carriage Sink

- Erratic Limit
- Inferred erratic path

Subglacial Moraines

- Minor Ribbed Moraine
- Ribbed Moraine
- Submarine Ribbed Moraine

Subglacial Lineations

- Crag-and-Tail
- Drumlin
- Streamlined Bedrock

- Mountain Glaciation Trimlines
- Mountain Glaciation Corries
- Subglacial Striae

Fig. 6.5 SUBGLACIAL BEDFORMS IN THE BANN AND OMAGH BASINS, NORTHERN IRELAND. These enhanced digital terrain models show patterns of subglacial bedforms on the landscape of Northern Ireland. The arrows show the direction of ice flow. a. Drumlins in the Bann Valley, Northern Ireland, give the landscape a streamlined appearance. Note how the long axis of each drumlin is oriented in the same direction. Here they record ice flowing from Lough Neagh towards the north Antrim Coast. b. Ribbed moraines in the Omagh Basin lie transverse to the former ice-flow direction and appear like large kilometre-scale ripples on the landscape. [Source: Geological Survey of Northern Ireland 10m Enhanced DTM]

made it possible to image the seabed at resolutions that can reveal the shape and distribution of landforms on Ireland's continental made it possible to image the seabed at resolutions that can reveal the shape and distribution of landforms on Ireland's continental shelf (see Chapter 9: Underwater Surveys: the INFOMAR project). Multibeam sonar systems, in particular, are extremely important in seabed research because they produce high-resolution, 3-D images of the seabed, making it possible to map submarine geomorphology in detail. Recently, this imagery has been used to investigate the glacial record on the continental shelf west of Ireland. A number of scientific cruises have also retrieved sediment cores from glacial deposits on the shelf. Examination of these has made it possible to date the glacial history of the ice sheet in this region. The discussion in the following section reviews what has become better understood about Ireland's glaciated continental shelf and coastal zone.[7]

While drumlins are common on the Irish landscape, multibeam data have also revealed their presence on the seafloor of the continental shelf at three locations. This has been an important find because it confirms that the last ice sheet was not floating on the sea, but was actually grounded over large areas of this eastern part of the North Atlantic continental shelf. It also shows that ice was moving actively across what is now the present-day seabed. In all cases, the orientation of these drumlins confirms that the ice sheet flowed across the shelf towards the shelf edge. The western part of the Malin

Fig. 6.6 CLEW BAY DROWNED DRUMLINS, COUNTY MAYO. The drumlins of Clew Bay were formed when the sea level was much lower than it is today. The pattern of these subglacial bedforms, which are comprised of glacial till, shows that ice flowed from east to west across the present-day coastal fringes and out onto the shelf. This happened probably as part of the rapid collapse and downwasting of the ice mass (the regional ice sheet) over Ireland. As regional sea level rose subsequently, during the lateglacial and Holocene times, the drumlins became submerged (or 'drowned'), creating the hundreds of small islands we see today. [Source: Geraldine Hennigan and Norman Kean]

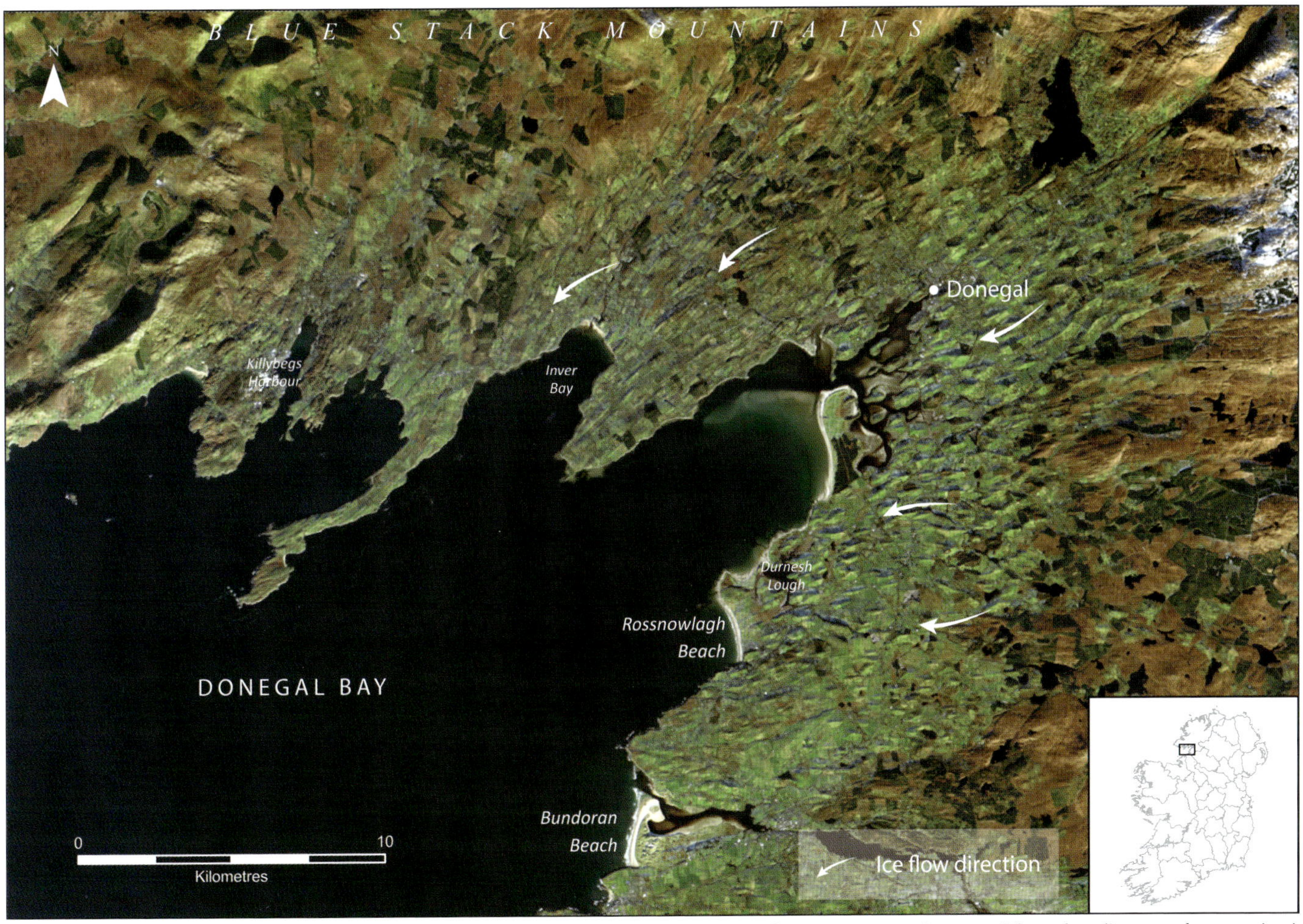

Shelf, situated north of Inishowen and 80km north-west of Malin Head in County Donegal, contains a large drumlin swarm. Here, the drumlins record a north-westerly ice flow across the shelf. The ice came from terrestrial Ireland sources, and merged with ice flowing across the Malin Shelf from south-west Scotland.[8] Further south, smaller drumlin swarms are located on the seabed north-west of Tory Island and Arranmore Island (Figure 6.8). These drumlins all have the same north-west–south-east alignment and provide clear additional

Fig. 6.7 A LANDSAT-8 SATELLITE IMAGE, SHOWING THE DRUMLIN SWARM ENTERING DONEGAL BAY, NORTH-WEST IRELAND. These drumlins record converging ice flow, from the land out into the bay, during the last glacial period. Drumlin fields are classically referred to in physical geography school texts as 'basket of eggs' topography, due to their resemblance to half-buried eggs. Here, the drumlins trend generally north-west to south-east, indicating the direction of ice movement in the late Quaternary. [Source: Landsat-8 image courtesy of the US Geological Survey, 2016]

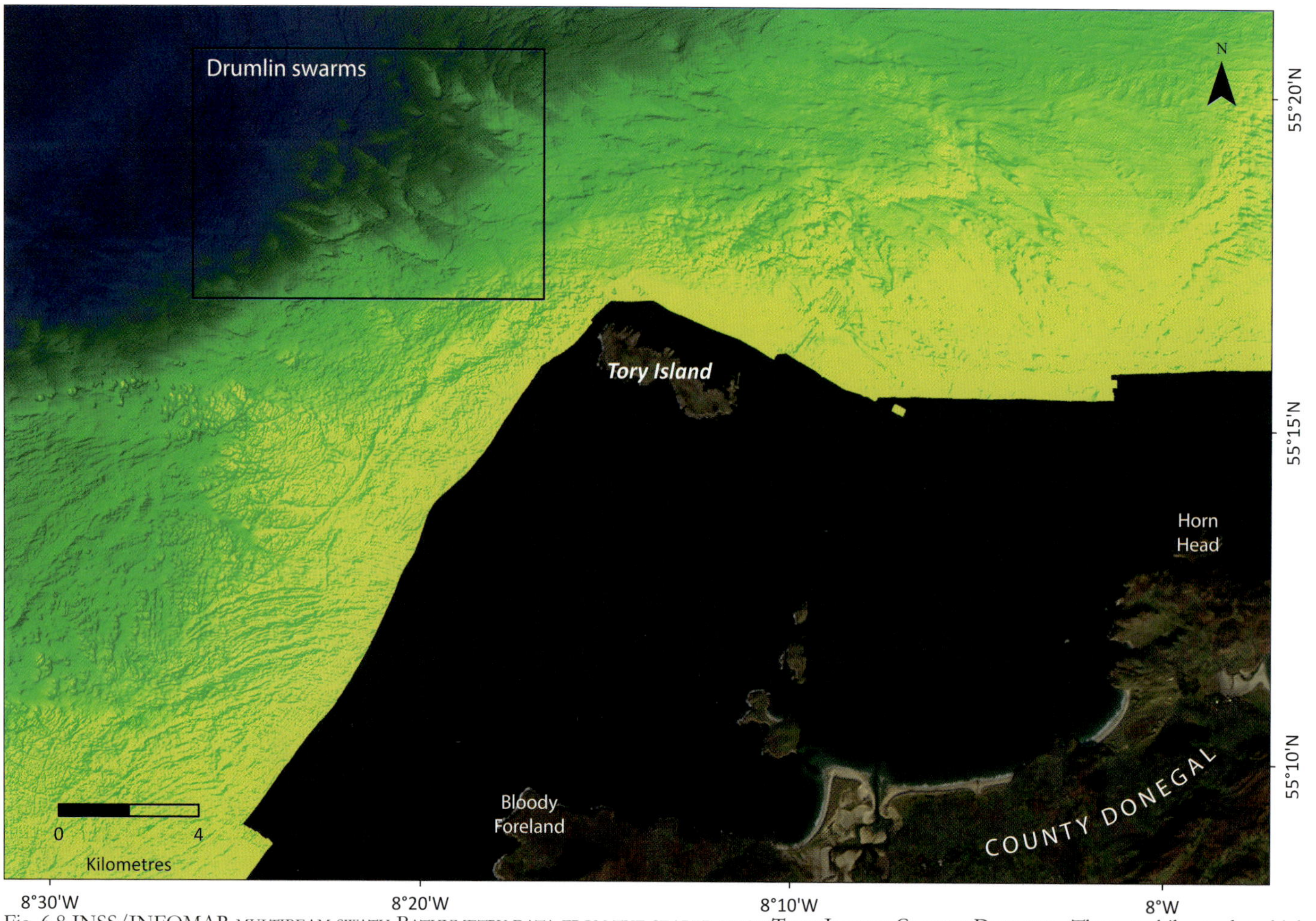

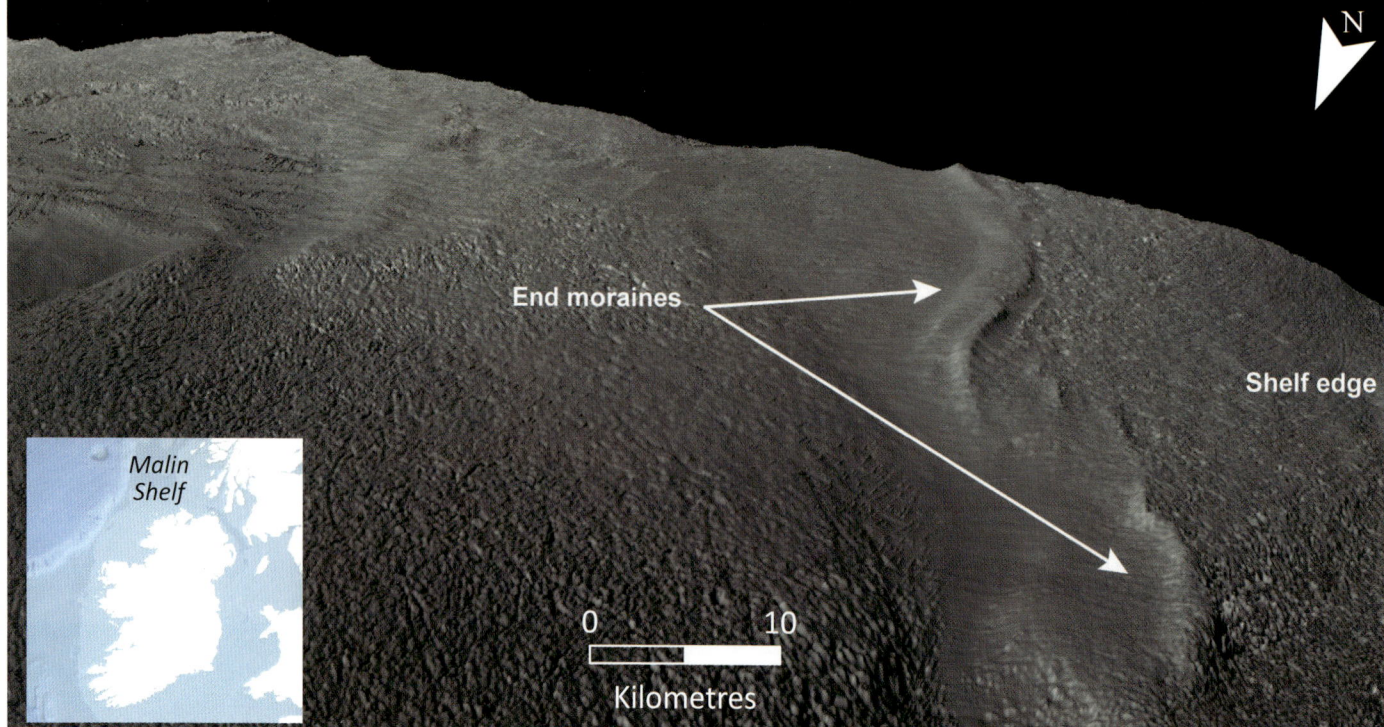

Fig. 6.8 INSS/INFOMAR MULTIBEAM SWATH BATHYMETRY DATA FROM THE SEABED NEAR TORY ISLAND, COUNTY DONEGAL. These multibeam data, highlighting the seabed morphology, are merged here with a Landsat satellite image to show the mainland northern Donegal coast and islands. Note the strongly streamlined appearance of the seabed bathymetry (inset black box), north-west of Tory Island. This has resulted primarily from the formation of drumlins by ice streaming. The presence of these drumlin landforms confirms that the ice sheet was once grounded in this location and slid over this present-day seafloor area during the last glaciation. [Source: INFOMAR; Landsat-8 image courtesy of the US Geological Survey, 2016]

Fig. 6.9 INNS/INFOMAR MULTIBEAM DATA FROM THE MALIN SEA SHELF, SITUATED NORTH OF MALIN HEAD. The large arcuate end moraine ridges shown in this image by these data mark the end (terminal) position of the last ice sheet to cover the area. The moraines were created in a zone of ice confluence, where ice flowing across the shelf from Scotland merged with that flowing onto the shelf from Ireland. The two ice streams continued on together, reaching as far as the shelf edge. This occurred when the British-Irish Ice Sheet was at its maximum extent. [Source: INFOMAR]

124

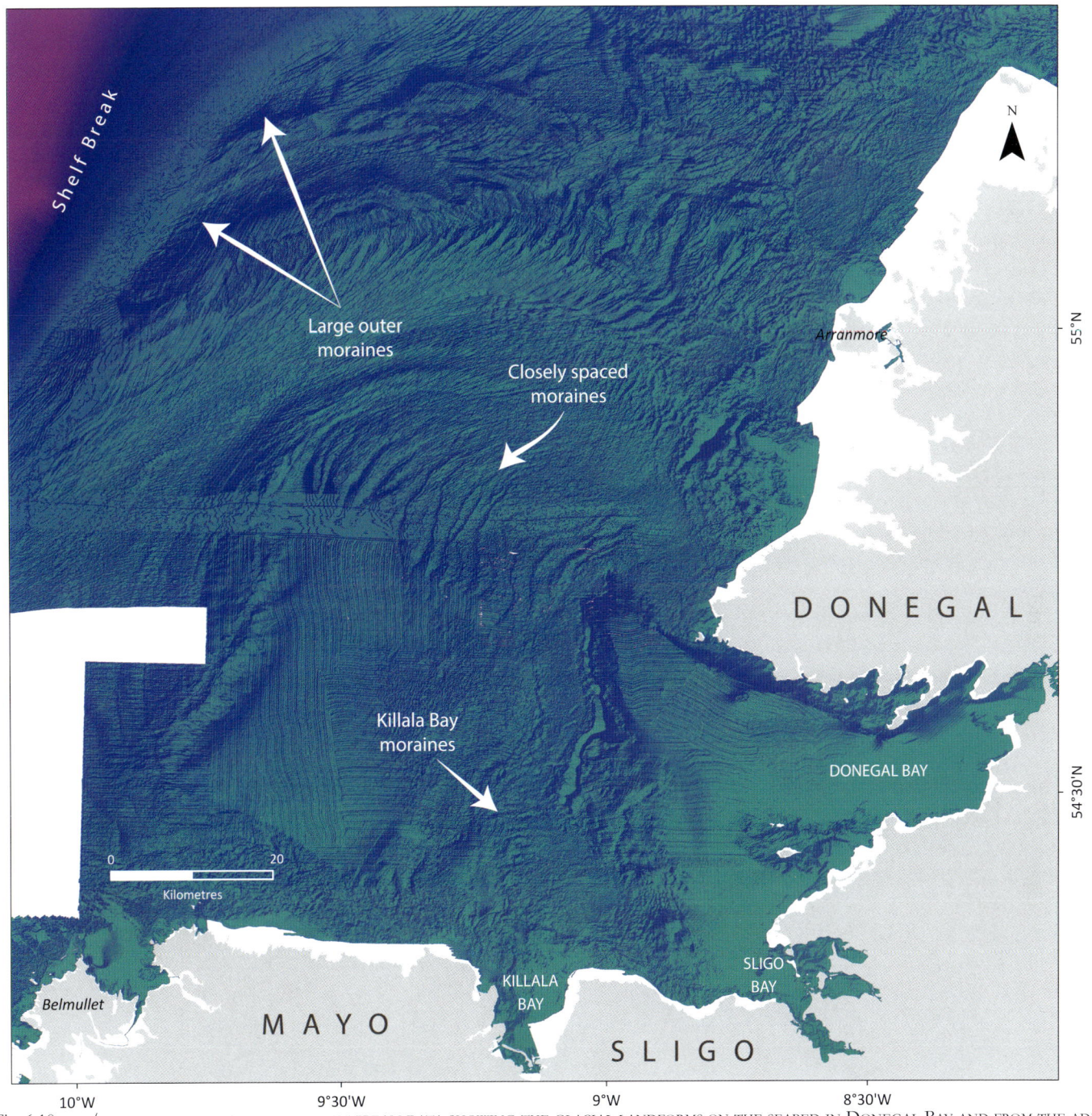

Fig. 6.10 INSS/INFOMAR SWATH BATHYMETRY, MULTIBEAM DATA SHOWING THE GLACIAL LANDFORMS ON THE SEABED IN DONEGAL BAY AND FROM THE ADJACENT NORTH-WEST CONTINENTAL SHELF AREAS. The seabed in this region shows the pattern of end moraines that record the westward movement and extent of the Irish ice sheets across this North Atlantic shelf area. This west-to-east aligned series of arcuate moraines indicated link back from the former shelf-edge ice margin, toward the present-day coastal zone and the mainland. The multibeam data record graphically the maximum extent of the ice, together with its subsequent retreat back towards the present-day Donegal and north Mayo coasts. [Source: INFOMAR]

evidence of a north-westerly ice flow towards the shelf break.

The extent of the outer edges of this ice sheet is recorded by sequences of large seabed moraine ridges that straddle the shelf at various locations, from the Malin Shelf in the north and as far south as the shelf west of County Kerry.[9] These ridges are by far the most striking glacial landforms that have been identified on the Irish shelf. These moraines mark the outer reaches of the former ice sheet and its subsequent pattern of retreat back on shore as the ice sheet melted. On the Malin Shelf, prominent end moraines are located on the western end, close to the shelf edge, where they form two discontinuous arcs on the seabed (Figure 6.9). Their orientation shows that they were formed by ice that came from north-west

Ireland. This ice merged in an extensive confluence zone with a separate large ice sheet that flowed from south-west Scotland. This combined ice sheet subsequently retreated back onshore to Ireland and then eastwards across the Malin Shelf towards Scotland.

The most striking pattern of moraines, however, can be seen on the shelf south of Malin, stretching from the centre of Donegal Bay to the shelf break approximately 90km off the north-west coast of Donegal (Figure 6.10). The largest moraine in this sequence extends for some 125km along the edge of the shelf, forming a prominent arcuate bedform on the outer shelf. Further to the south-east, the moraines become spaced closely together in a sequence across the shelf towards Donegal Bay and are aligned in a north-east to south-west fashion. The largest ridge in this sequence measures about 35km in length and extends across the mouth of Donegal Bay. The moraines here show the ice sheet retreat from its outermost extent west of Donegal, each one recording a position of where the ice front stopped for a while. The ice lobe responsible for these moraines was large, extending over 80km from the mouth of Donegal Bay to the shelf edge, and was about 120km across at its widest point. It would have been fed by ice that was centred over Donegal in the Bluestack Mountains and further inland in the Omagh Basin (see Figure 6.5).[10]

In southern Donegal Bay, there is an interesting further set of small moraine ridges. These have a general east–west alignment, extending southwards in a closely nested fashion from the centre of Donegal Bay towards the mouth of Killala Bay, close to the north County Mayo coast (c.8km offshore). They are superimposed on top of the prominent moraine ridge opposite Donegal Bay and must have been deposited after the formation of the larger moraine feature. The most likely explanation for this is that the Killala Bay moraines represent a late-stage re-advance into the southern part of Donegal Bay from ice centered in County Mayo. This presumed re-advance also created prominent sharp-crested ridges called thrust moraines, formed when the ice stacked sheets of sediment on top of one another at its base as it advanced onto the shelf. They are easily seen along the coastline between the towns of Killala and Inniscrone, County Mayo. The pattern of retreat in all cases appears to have been in the form of three independently operating ice lobes, each occupying different parts of the continental shelf off the coast of Donegal and North Mayo.[11]

South of Donegal Bay, and central to Ireland's western coasts, INSS/INFOMAR data for the shelf areas offshore are lacking. However, another dataset, Olex bathymetric data, has been compiled from stitching together seabed images captured mostly by fishing vessels.[12] This has been used to map the glacial landform record where there are gaps in the available INSS/INFOMAR data.

Research using these data on the shelf west of Mayo, Galway, Clare and Kerry has again located large moraine bedforms, similar to those described for west of Donegal. Sediment cores retrieved from these provide more detail of the sedimentary structure of the ridges, while shells found within these sediments have been radiocarbon dated, allowing determination of the timing of the ice sheet's advance and retreat on this part of the shelf.

The outermost moraine ridge here, known as the West Ireland Moraine, is located 20km from the shelf edge (Figure 6.11). It measures c.80km in length, is made of subglacial till sediment (boulder clays) and is capped by glaciomarine sediments. The presence of till at the base of the ridge confirms that the ice sheet here was also grounded on the seabed. Dating this till places the minimum westward extension of the ice sheet to at least c.24,000 years ago. Subsequently, the ice sheet retreated to approximately the position of the modern day's 200m water-depth contour, where it became grounded on the seafloor during a period of stability. Sometime between 21,200 and 18,500 years ago the ice front stopped retreating and created a ridge approximately 150km long, 22km wide and 20m high (the Galway Lobe Grounding Zone Wedge).[13]

As well as impacting on the inner and mid-shelf, the outermost part of the continental shelf and the upper slope down to 500m water depth are also characterised by large furrows that are incised into the seabed. These cut across one another, range from a few hundred metres to 7km in length, and are c.15m in width and a few metres deep (Figure 6.12). They are interpreted as the seabed scour-marks from the movement of icebergs, which ploughed across the newly submerging shelf margin, forming as sea levels rose and the ice sheet broke up during deglaciation. They can be seen stretching along the length of the outer shelf close to the shelf break, from the Malin Shelf in the north to the Porcupine Bank west of Ireland, which is also heavily scoured by criss-crossing grooves on the seabed.[14]

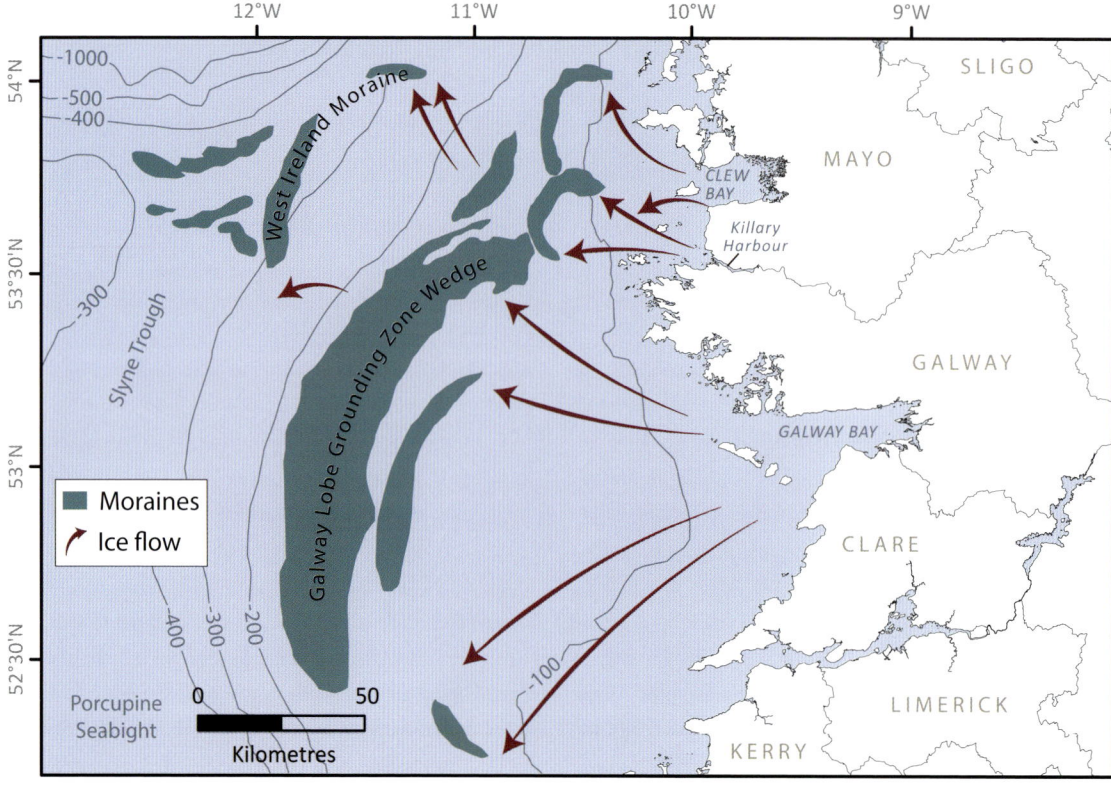

TIDEWATER GLACIAL SEDIMENTATION IN IRELAND: IDENTIFICATION AND SIGNIFICANCE

Stephen McCarron

Glacial sedimentary deposits found in Ireland today reveal information about the characteristics of the glaciers that once existed during the last Ice Age. Layers of sediment with different characteristics, called 'facies', indicate the former presence of a particular type of glacier called a tidewater glacier environment. Tidewater glaciers terminate in the ocean and have been studied from many contemporary ice sheet margins (such as Arctic Canada, western Greenland and from the West Antarctic ice sheet). On the outer edges of ice masses (including glaciers), where they make contact with the sea, submarine melting would have occurred (both seasonally and periodically for longer) due to the relative warmth of the ocean waterbody. When combined with the water's buoyancy and possibly amplified 'glacial supertides', this results in a steep, rapidly eroding (calving) ice front, with little or no fringing ice shelf (Figure 6.13).

Icebergs, sediment and meltwater produced at tidewater glacier margins enter the global ocean system, as evidenced on Ireland's Atlantic continental shelf margin (see Figure 6.12) (see Chapter 7: Ancient Shorelines and Sea-level changes). This can have, and continues to have, a large-scale impact on the Earth's climatic system, affecting the planet's meridional overturning seawater circulation (Thermohaline Ocean Currents). This circulation is driven by the ocean surface to bottom temperature and salinity (and consequently, density) gradients of seawater and, hence, it is affected by the behaviour of tidewater glaciers.

Former climatic episodes in which extensive iceberg discharges occurred, releasing large amounts of meltwater into the ocean, are called 'Heinrich Events'. In the past, such events in the North Atlantic, sometimes known as binge-purge cycles, punctuated the last glacial marine record. They were not necessarily linked to times of climate warming, but probably more to ice-sheet dynamics. These cycles have been associated with repeated millennial-scale catastrophic collapses of the former Laurentide (eastern North American) ice sheet and, thus, regional ice advances and retreats. Changes in the Laurentide ice sheet were important, together with Scandinavian and other Arctic ice masses, in controlling Quaternary climate, ice volumes and sea levels in the North Atlantic region, which of course impacted Ireland.[15]

A possible enabling mechanism of this dynamic of binge-purge ice behaviour is the 'Jakobshavn Effect'. This suggests that glacier ice flows accelerate towards tidewater calving margins and act to lower the ice-sheet profile, a process called dynamic thinning (or ice downdraw). This increases the ice-loss (ablation) area and encourages stagnation of ice masses (that is, ice sheets, glaciers). The portion of the glacier that moves faster than the surrounding ice is known as an ice stream. One current cause for concern is the role of modern tidewater margins in the initiation and maintenance of ice streams, particularly in Greenland and Antarctica. These ice streams account for the losses of the majority of ice from present-day polar ice sheets (for example, Jakobshavn Isbræ in south-west Greenland; Lambert Glacier in East Antarctica) and may be involved in the drawdown of the West Antarctic ice sheet. This will lead to rapid rises in global-to-regional sea levels in future.[16]

TIDEWATER EVIDENCE IN IRELAND

Modern sedimentary environments associated with tidewater glaciers are relatively inaccessible and observational data are unfortunately limited. Key models for these have, however, been produced for high-Arctic fjord glaciers, from the study of exposures of recently deposited sediments (Figure 6.14).[17] In Ireland, glaciomarine models that link the past coastal depositional environments from locations around the Irish Sea Basin, and wider, to an ice stream-dominated tidewater system have been produced. These have been important in developing the understanding of late Quaternary patterns of ice-sheet movements for the island.[18]

The sedimentary facies associated with tidewater environments include both subglacial (deposited under the ice) and subaqueous (deposited on the seafloor) sediments. Where the seabed has been previously scoured by ice, glaciomarine sediments infill the troughs. These are overlain discontinuously by other sedimentary layers of varying characteristics, which depend on proximity to the ice margin, the meltwater source and the degree of energy associated with their transport and deposition. Within tidewater models, subaqueous sediments

Fig. 6.11 (opposite) THE MAIN GLACIAL MORAINE-TYPE LANDFORMS FOUND ON THE CONTINENTAL SHELF OF WESTERN IRELAND. At its maximum extent, *c.*24,000 years ago, ice from Ireland extended westward, across the continental shelf as far as the edge of the Slyne Trough, although different interpretations of the data and events exist. Large moraine ridges and sedimentary evidence mark the terminal positions of this ice sheet. The ridge shown as the Galway Lobe Grounding Zone Wedge was created during a period of ice still-stand in the retreat of the ice sheet back to its source regions inland. The main ice masses had probably disappeared from Ireland by *c.*18,000 years ago. In later stages of deglaciation, large ice lobes withdrew back into Clew Bay, Killary Harbour and Galway Bay and from there onshore, leaving behind sediments and other traces of their passage as they went. [Adapted from Peters, J.L., Benetti, S., Dunlop, P., O'Cofaigh, C., Moreton, S.G., Wheeler, A.J. and Clark, C.D., 2016. Sedimentology and Chronology of the Advance and Retreat of the Last British-Irish Ice Sheet on the Continental Shelf West of Ireland. *Quaternary Science Reviews*, 140, pp. 101–124.]

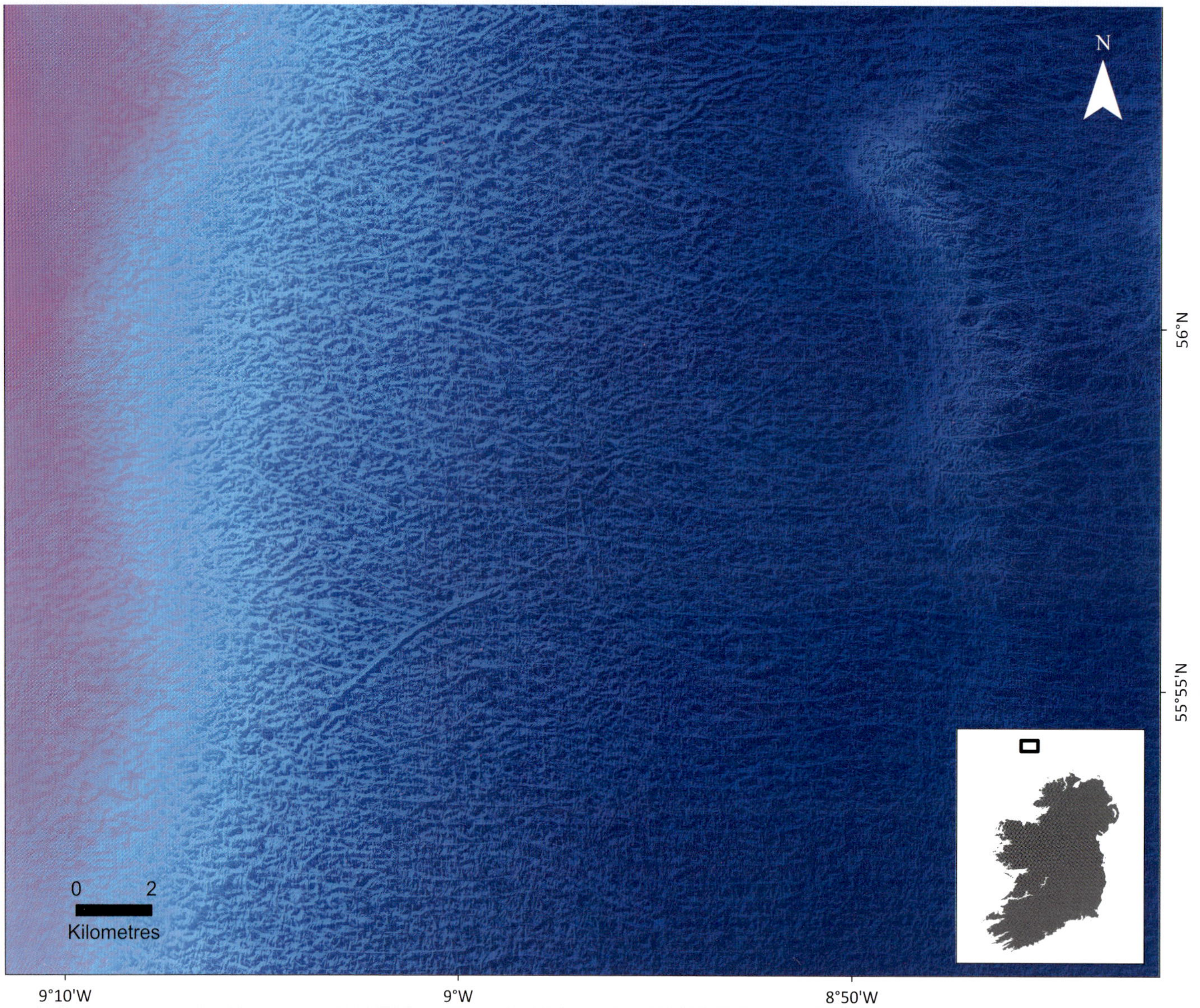

N

0   2
Kilometres

Fig. 6.12 ICEBERG PLOUGH MARKS ON THE MALIN SHELF. At the start of deglaciation, large icebergs broke off from the front of the ice sheet which, at the time, was situated some distance offshore to the west of Ireland. These bergs sometimes grounded on the seabed, where they were subject to movement by waves, wind and tide, causing their keels to plough furrows through the sediments, leaving a criss-cross morphological pattern on the seafloor. This pattern can still be seen clearly today, as shown in this image created from INSS/INFOMAR swath bathymetry multibeam data. [Source: INFOMAR]

Fig. 6.13 (left) CALVING ICE FRONT, HUBBARD GLACIER, ALASKA, USA. Tidewater glaciers like this one terminate at the boundary between land-derived ice and the ocean. Coastal and seabed sedimentary evidence suggests that these environments, similar to those found today in Alaska and northern Canada, existed in Ireland during the late Quaternary and gave rise to some of its coastal glacigenic cliff sediments. When chunks of ice break from the edge of a glacier, a process called 'calving', they carry with them large amounts of sediment. As the ice melts, the sediment is deposited on the seafloor. In addition, sediment carried by subglacial meltwater also gets deposited on the seafloor, though in this situation it tends to be reworked and sorted by the water transport mechanisms. Sediments arising from these processes can also be found in Ireland, including at Cooley Point, in Dundalk Bay, County Louth. [Source: Bernard Spragg, public domain]

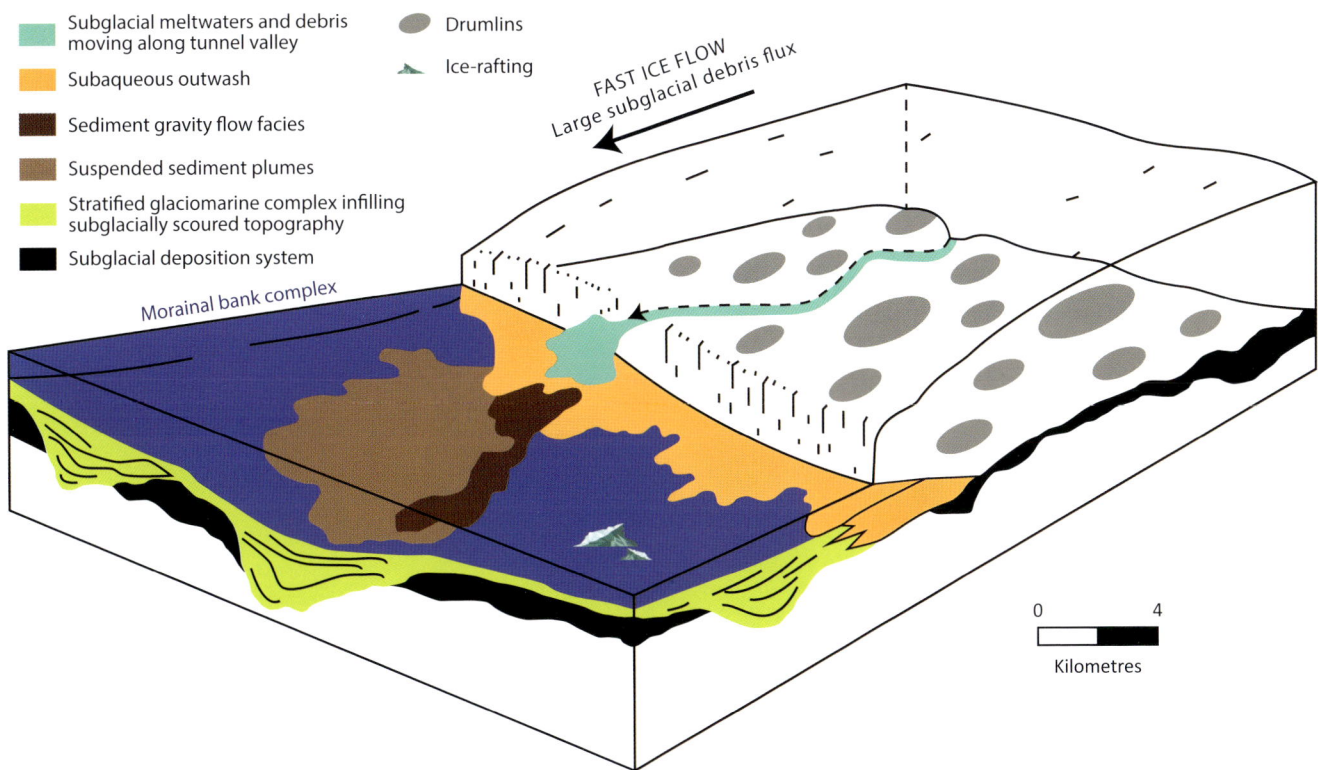

Legend:
- Subglacial meltwaters and debris moving along tunnel valley
- Subaqueous outwash
- Sediment gravity flow facies
- Suspended sediment plumes
- Stratified glaciomarine complex infilling subglacially scoured topography
- Subglacial deposition system
- Drumlins
- Ice-rafting

FAST ICE FLOW
Large subglacial debris flux

Morainal bank complex

0   4
Kilometres

Fig. 6.14 A GENERALISED CONCEPTUAL MODEL OF SEDIMENTATION PATTERNS AND FACIES (THAT IS, SEDIMENTARY ENVIRONMENTS) DEVELOPED ALONG A TIDE-WATER ICE MARGIN. This glaciomarine model links subaqueous sediments, deposited in morainal banks and associated facies, with sediment supply from subglacial erosion at the bed of a fast-moving ice stream. [Adapted from McCabe, M. and Dunlop, P., 2006. The Last Glacial Termination in Northern Ireland. Geological Survey of Northern Ireland]

derived from subglacial ice are shown to give way laterally to more extensive spreads of stratified, progressively finer-grained sediment layers. These sediments are then deposited as outwash, and/or as gravity flows, near the mouth of subglacial drainage channels.

Studies of former tidewater glaciers in Ireland have sought to link deglacial tidewater sedimentation during the late Quaternary with sediment supply from the drumlin coasts of Ireland and western Britain, for example, of Donegal Bay, Dundalk Bay, Galway Bay, Clew Bay, Ards Peninsula, Cumbria and the Kintyre peninsula in Scotland. At Cooley Point in Dundalk Bay, County Louth, on the north-west Irish Sea coast, a layer of boulder-sized clasts (rocks) is associated with a striated rock pavement created by ice action. This is identical with such features found today within intertidal zones on high Arctic coasts (Figure 6.15).[19] The Cooley Point pavement overlies massive to crudely stratified glaciomarine muds, dated to the last deglaciation. This provides an explicit indication of the existence of a former sea level, during a phase of coastline emergence and ice retreat from further east in the Irish Sea. Based on this type of sedimentary evidence, ice process links can be made between basin topography, the position of sea level and the formation of tidewater sediment accumulations at many locations along Irish coastlines (e.g., Courtmacsherry Bay, County Cork). The study of tidewater sediments in Ireland has allowed

scientists to better understand the sensitivity of the last glacial ice sheet to millennial-scale palaeoclimatic and palaeoceanic events in the circum-North Atlantic. This is important today because it can help us to understand the response of modern ice sheets to contemporary climate warming.[20]

Fig. 6.15 COASTAL CLIFF SECTION AND PROFILE, COMPRISING TIDEWATER GLACIER SEDIMENTARY FACIES AT COOLEY POINT, DUNDALK BAY, COUNTY LOUTH. Grey, glaciomarine muds (a.) dated to about 18,700 years before present, are capped by a thin layer of inter-tidal boulder pavement (b.) Although these now sit 2–5m above contemporary mean sealevel, at the time of deposition these muds would have been sub-marine when formed. The muds contain iceberg-rafted 'dropstones' (that is, isolated rock fragments deposited by melting icebergs that once floated on the sea surface). The muds also contain a well-preserved calcareous marine fauna. This is dominated by ice-proximal Arctic foraminifera species, such as *Elphidium excavatum*, a micro- to macroscopic (*c.*1mm in diameter) organism, which is abundant and found from the coast to the shelf edge waters. The boulder pavement was formed by intertidal ice stranded at the former sea-level, as occurs in Polar paraglacial coastal settings today. The lower sediment layers (dark grey-purple in colour) are overlain by subaerial outwash sand and gravel (c.) These were transported and deposited by meltwater. [Source: Stephen McCarron]

Fig. 6.16 GLACIAL TILL CLIFFS ON THE EASTERN SHORELINE OF HOG ISLAND, SHANNON ESTUARY Glacial till cliffs, like this one at Hog Island, comprise long stretches of the Irish coast. In this view of such a coast, a large glacial erratic boulder is visible on the foreshore in front of the eroding cliffs moved and left as part of the till by the ice and now exposed by wave erosion. The till was deposited by south and westward ice movements across the area in late Quaternary time. These soft glacial sediments are now subject to active erosion by waves and tides, as well as undergoing slumping and other mass movements, due to gravity. These processes are assisted by lubrication from percolating groundwater. This erosion exposes vertical cliff sections, as well as delivering sediments to beaches along the coast at the same time. [Source: Robert Devoy]

THE FINAL DEGLACIATION OF IRELAND AND ITS COASTAL LEGACY

The final stages of ice decay during deglaciation in Ireland (and in the wider North Atlantic region) were punctuated with standstills and minor re-advances of the ice sheet. The most studied of these ice re-advances occurred during a cold climatic period known as the Killard Point Stadial, which reached its peak *c*.14,000 years ago. This event was associated with the earlier pulses of meltwater and icebergs discharging into the North Atlantic from the retreating Laurentide Ice Sheet in North America, around 16,500 to 14,700 years ago (Heinrich Event H1). (This was the last major Heinrich event, which operated in the late Quaternary on a cyclical time series of 5,000–10,000 year cycles.) These iceberg pulses are known to have shut down the North Atlantic Thermohaline Ocean current, which contributes to bringing warm oceanic water into the world's oceans. During the Killard Point Stadial the climate cooled and the ice sheet re-advanced. In the north-west of Ireland, the most recent sets of drumlin realignments record ice flow westward onto the coastal fringe of western Ireland, as well as again offshore onto the shelf, with the drumlins forming a landscape signature of this event (see Figure 6.6). Further north, ice from the relatively small Donegal ice cap barely reached the coast during this time. Its terminus is marked by small end moraines at Arduns near Gweedore and Ballycrampsy in Inishowen. The Killard Point Stadial was the last significant climatic event that is recorded in Ireland's glacial record.

After this time the ice sheet finally thinned, stagnated and eventually disappeared, leaving behind a landform and sedimentary history that straddles the Irish coast today.[21]

Most of the sand and gravel deposits that we find on Ireland's beaches today have their origins in the last Ice Age. The ice brought vast amounts of glacial sediment onto the continental shelf and moulded the landscape it flowed over. As the ice sheets retreated, the sea level rose and the environment of the succeeding postglacial period was established. During this time of 'liquid' water, rather than ice, the combined actions of ocean and coastal currents, with waves and storms, have together brought shelf sediments ashore, helping to create the characteristic sand and shingle beaches of the contemporary Irish coast. Modern tidal regimes, currents and waves now rework continually the unconsolidated glacial till cliffs that line large sections of the coast (Figure 6.16). The erosion and slumping of these cliffs is an important lifeline for the nourishment of beach systems, as new sediments are supplied to the shore and reworked by coastal processes (see Chapters 2: The Coastal Environment: Physical systems, processes, and patterns and 11: Beaches and Barriers). Whilst it might be easy to think of beaches and the coastline as being the product of modern-day processes, former ice sheets have played a major role in the controlling and shaping of Ireland's coast through time.

ANCIENT SHORELINES AND SEA-LEVEL CHANGES

Robin Edwards and Robert Devoy

'RAISED BEACHES' AT BALLYHILLIN, MALIN HEAD, THE INISHOWEN PENINSULA, COUNTY DONEGAL. A 'staircase' of late- and postglacial raised shorelines can be seen in this photograph, developed landwards of the modern wave-reflective beach. The highest and oldest of these former coastlines is formed in the area of the line of cottages, in the upper right of the image. This bench-like beach landform is fronted by steep slopes and an early coastal 'cliff-line', which has been modified subsequently by postglacial slope development. The green, flat area below, now divided into farmers' fields, was the intertidal zone of the raised beach, formed approximately 9,000–11,000 years ago. A second and younger raised shoreline (centre of photograph) is associated also with small, former cliff exposures. Finally, we have the present-day beach, composed mostly of sands and shingle (gravels) that are themselves derived from glacial sediments deposited initally on the seabed. The finer particles from these sediments were brought back onshore as sea levels rose at the close of the last glaciation and are now being further reworked by wind, waves and tide. [Source: Andreas F. Borchert, CC BY-SA 3.0 DE, Wikimedia Commons]

Ireland boasts a wide diversity of coastline types, from the towering rocky cliffs of its Atlantic façade, assaulted by giant breaking waves, to its low-lying estuaries and saltmarshes, where much lower wave and tidal energies prevail. This diversity arises from the interplay of form (shape) and process, both of which dictate that all coasts are, almost by definition, dynamic and constantly 'evolving'. Further, shoreline forms reflect the nature of the material from which they have been sculpted (for example, from 'hard' rock or loose accumulations of sands and gravels), as well as the physical agencies acting on them (principally wind, waves and tides) (see Chapter 2: Coastal Environment: Physical systems processes and patterns). Importantly, the current coastline is not simply the product of the processes and controls operating today. It also contains within it the historic legacy of its ancient former shorelines, which continue to exert an influence on present-day coastal shapes (that is, the factor of coastal inheritance). In Ireland, this inheritance arose primarily from Earth processes connected with the growth and decay of its ice masses (ice sheets and glaciers) and by associated multiple episodes of relative sea level (RSL) change. As we shall see, these two phenomena are themselves intimately linked in their operation.

The previous chapter outlined the role of ice in shaping Ireland's coastal landscapes and the adjacent continental shelf. This study looks now at the evidence used to uncover more of the nature and impacts of this glacial inheritance. The discussion outlines the evolving picture of Irish relative sea level changes that have emerged from geomorphological and other studies within the Earth sciences.[1]

Over the past two million years (the geological time known as the Quaternary), Ireland's climate has been characterised by pronounced swings between interglacial warmth – during which the mean temperature was considerably higher than today – and times of glacial cold (see Chapter 6: Glaciation and Ireland's Arctic Inheritance). During cold phases, terrestrial ice sheets, ice caps and smaller valley-type glaciers expanded over large parts of northern Europe with, at times, ice sheets developing independently in Ireland and extending across the island on several occasions. The Irish landscape bears the clear imprint of this moving ice, with the uplands carved by glaciers and lowlands areas mantled in the eroded detritus they deposited.[2] The coasts also show the legacy of these events in a number of ways. Much of the sand and gravel contained within Ireland's 'soft' coastal cliffs, or distributed along beaches and dunes, is the product of this glacial past. Similarly, many modern coastal landforms have evolved from Ice-Age antecedents, such as the shingle spit at Cromane, which still marks the ancient margin of

Fig. 7.1 THE SHINGLE SPIT AT CROMANE IN CASTLEMAINE HARBOUR, COUNTY KERRY. The sediments underlying this coastal spit-like structure are glacial in origin and have been (and continue to be) reworked by coastal processes during the postglacial (Holocene) rise of sea level. The spit features of inner Dingle Bay and Castlemaine Harbour (Inch, Rossbehy, Cromane) have complex origins and probably developed together initially, as part of a local to regional ice front advance-and-retreat complex, developed by ice moving westward in the late Quaternary through Castlemaine Harbour and the Lough Caragh region of the Iveragh uplands. Cromane, and probably Inch and Rossbehy, are underlain by glacial till (moraine or boulder clay), which was pushed into ridges at the front of glacier advance. Changes in sea level during the Holocene have resulted in the progressive drowning of these moraines, with wave and tidal processes sorting and redistributing the glacial sediments to create the coastal sand-gravel barrier landforms we see today, including the contemporary beaches, dunes and fringing saltmarshes. [Source: Tadhg Hayes]

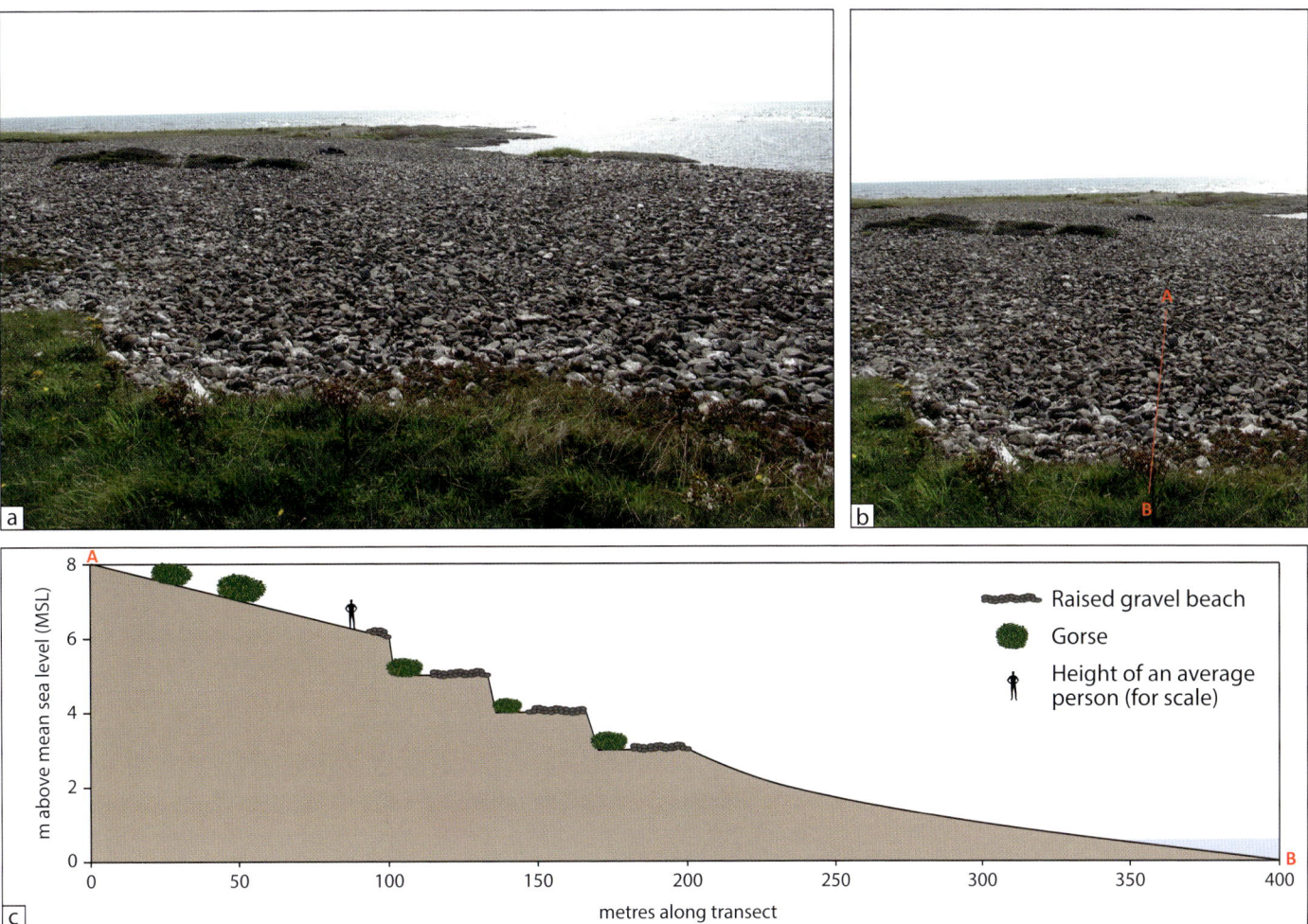

Fig. 7.2 THE RAISED GRAVEL BEACH AT BALLYQUINTAN POINT, OUTER STRANGFORD LOUGH, COUNTY DOWN. a. View of the contemporary beach at Ballyquintan coastal headland, showing the extensive area of gravel and cobble-sized sediments, which dip down toward the sea. b. An aerial image of the area shows the remnants of past coastal landscapes on the southern tip of this headland and the broader coastal context of these ancient shorelines. A transect-line through the raised beaches is marked as the red line A–B on the photograph; c. A cross-section of the preserved raised beaches taken along this transect. Three raised shorelines have been identified, forming a benched, or terraced, coastal profile. The beaches reach an elevation of *c*.5–6m above modern-day mean sea level. These raised shorelines were formed when relative sea level (RSL, the position of the sea relative to land) was higher than it is today. At the end of the last Ice Age, unloading (the melting) of huge ice masses from the Earth's land surfaces caused the underlying crust to rebound, or rise. In the north of Ireland this occurred relatively faster than the return of water to the oceans and, consequently, was more rapid than the rate of water-level rise against the land (the continental freeboard). As a result, beach environments such as this at Ballyquintan Point appear raised high above modern coastal levels. [Photograph: Eric Jones, CC BY-SA 4.0; Map data: Google, Maxar Technologies, 2003]

the glacier that once occupied Castlemaine Harbour (Figure 7.1) (see Chapter 11: Beaches and Barriers).[3]

The waxing and waning of ice sheets over Scandinavia and northern Europe, together with upland ice masses – as in the Alps and other mountain ranges further south in Europe, the Middle East and beyond (for example, valley-type glaciers) – occurred on at least six occasions in the last 500,000 years. Elsewhere, the ice build-up over Europe was mirrored by the growth and decay of larger ice sheets in North America and Siberia, increased ice cover of Antarctica and the expansion of glaciers in the other continents of the southern hemisphere. Each time these different ice masses built up, enormous amounts of water were abstracted from the world's oceans, to be deposited on land as snow. In the final stages of the last Ice Age (post-*c*.35,000 years ago), this abstraction lowered the world's relative sea levels by around 120m, leaving much of the present-day north-eastern Atlantic continental shelf as exposed land and connecting Ireland and Britain to continental Europe (see Chapter 6: Glaciation and Ireland's Arctic Inheritance).

Then, when the climate ameliorated, the ice sheets melted, the abstracted water was released back into the ocean and sea levels returned to positions close to those of the present day. As well as changes arising from these variations in the amount (volume) of water in the ocean, the landmass of Ireland itself was pressed downwards under the enormous weight of the overlying ice, a process known as isostasy. Once that pressure was removed the land slowly recovered. In Ireland, as in other regions adjacent to large ice sheets, the extent, thickness and duration of local ice cover played a pivotal role in controlling the height of the land relative to the sea. Further, as a result of these two processes (glacio-isostasy), the coastlines of Ireland advanced and retreated laterally, sometimes over many kilometres. It also rose and fell vertically several tens of metres, with these movements being repeated many times and over a range of timescales (see Chapter 16: The Inhabitants of Ireland's Early Coastal Landscapes).

The resulting histories of RSL change are complex and spatially variable in Ireland. All round the coasts of the island, relics and fragments of past shorelines may be seen, as also in many other parts of the world. An example of this can be found at Ballyquintan, County Down, where modern beach gravels are trapped within a rock basement on an exposed headland, while remnants of older gravel beaches, outlined by strips of flowering gorse or whin bushes, are stranded at higher elevations (*c.*6m above contemporary mean sea level) (Figure 7.2).[4]

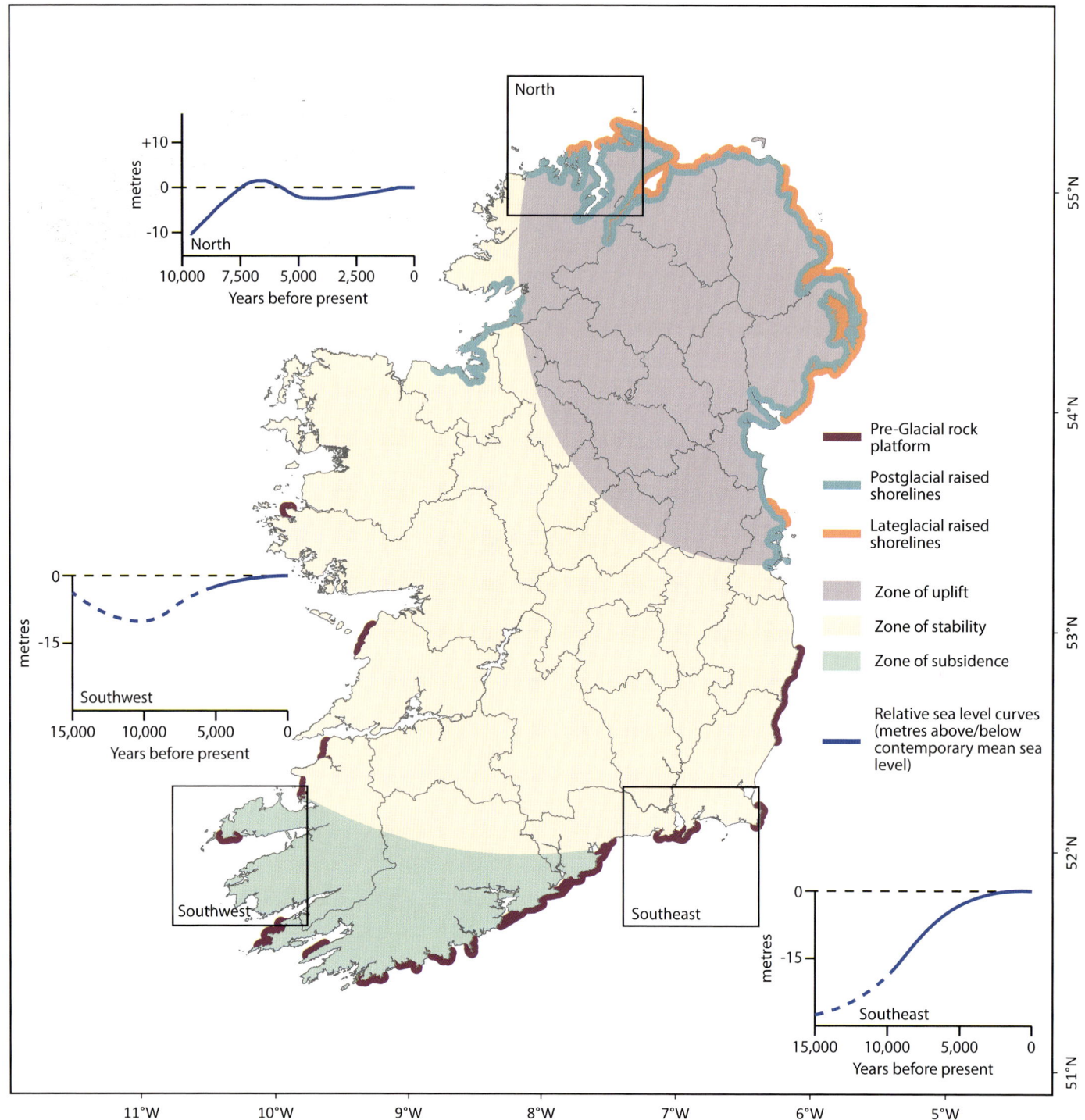

Fig. 7.3 RELATIVE SEA-LEVEL CHANGES AND IRELAND'S EARLIER SHORELINES. Sea-level changes during postglacial time (the Holocene) did not occur uniformly across Ireland. The three graphs here illustrate the broad regional patterns of rise and fall of RSL around the island's coasts since the end of the last glaciation, based on numerical model simulations (projections) of RSL. They have been validated against field evidence and observed sea-level data (precise heights of sedimentary and other shoreline positional information), and are shown in the form of generalised trends (curves) of sea-level change. The RSL positions, plotted against time, are shown for the northern, south-western and south-eastern regions of Ireland during the last 10,000–15,000 years. In the south and south-west, RSL has risen progressively over the land through the Holocene, continually submerging coastal areas. In contrast, in most northern parts of Ireland, RSL can be seen as having fallen periodically, relative to the land. This has led to shorelines in this region appearing as stable or even building seawards. These differences in RSL behaviour are due to glacio-isostatic adjustment, a process by which the Earth's crust changes vertically in height (rises and falls) due to the removal of downward pressure caused by the weight of former massive ice sheets. The long-term trend of relative lowering of sea level in the north of Ireland is why we can now see raised beaches preserved along its coasts. The map in this illustration shows the spatial distribution of raised shorelines of lateglacial and postglacial ages. Also shown is the extent of the earlier, interglacial-age, Pre-Glacial rock platform. The three zones marked show the general pattern of isostatic crustal movements relative to contemporary sea level. [Adapted from: Devoy, R.J.N., et al., 1996. Coastal Stratigraphies as Indicators of Environmental Changes upon European Atlantic Coasts in the Late Holocene. *Journal of Coastal Research*, pp. 564–588 and Brooks et al., 2008. Postglacial Relative Sea-level Observations from Ireland and their Role in Glacial Rebound Modelling. *Journal of Quaternary Science*, 23, pp. 175–192]

Ireland's ancient shorelines have been investigated by scientists since at least the middle of the nineteenth century and are still the subject of active current research and debate. Yet, many elements of the stories that they tell remain imperfectly known and understood. Nevertheless, detailed examination of these fragments in the field, along with field and laboratory study of sediments preserved in key locations, as well as comparison with similar evidence from other parts of the world, is gradually allowing scientists to build up a history of the changes that they represent (Figure 7.3).[5]

SEA LEVELS AND IRELAND'S ANCIENT SEABEDS

Andrew J. Wheeler

Sea-level fluctuations on the scale of tens to hundreds of thousands of years, changing in response to climate-controlled global ice budgets, have many times moved Europe's Atlantic coastlines back and forth across the continental shelf. On timescales ranging from tens to hundreds of millions of years, more fundamental changes have occurred that are driven by the deeper geological and geophysical processes of plate tectonics. When we look at the bedrock geology preserved off Ireland's shores, the depositional histories of the sedimentary

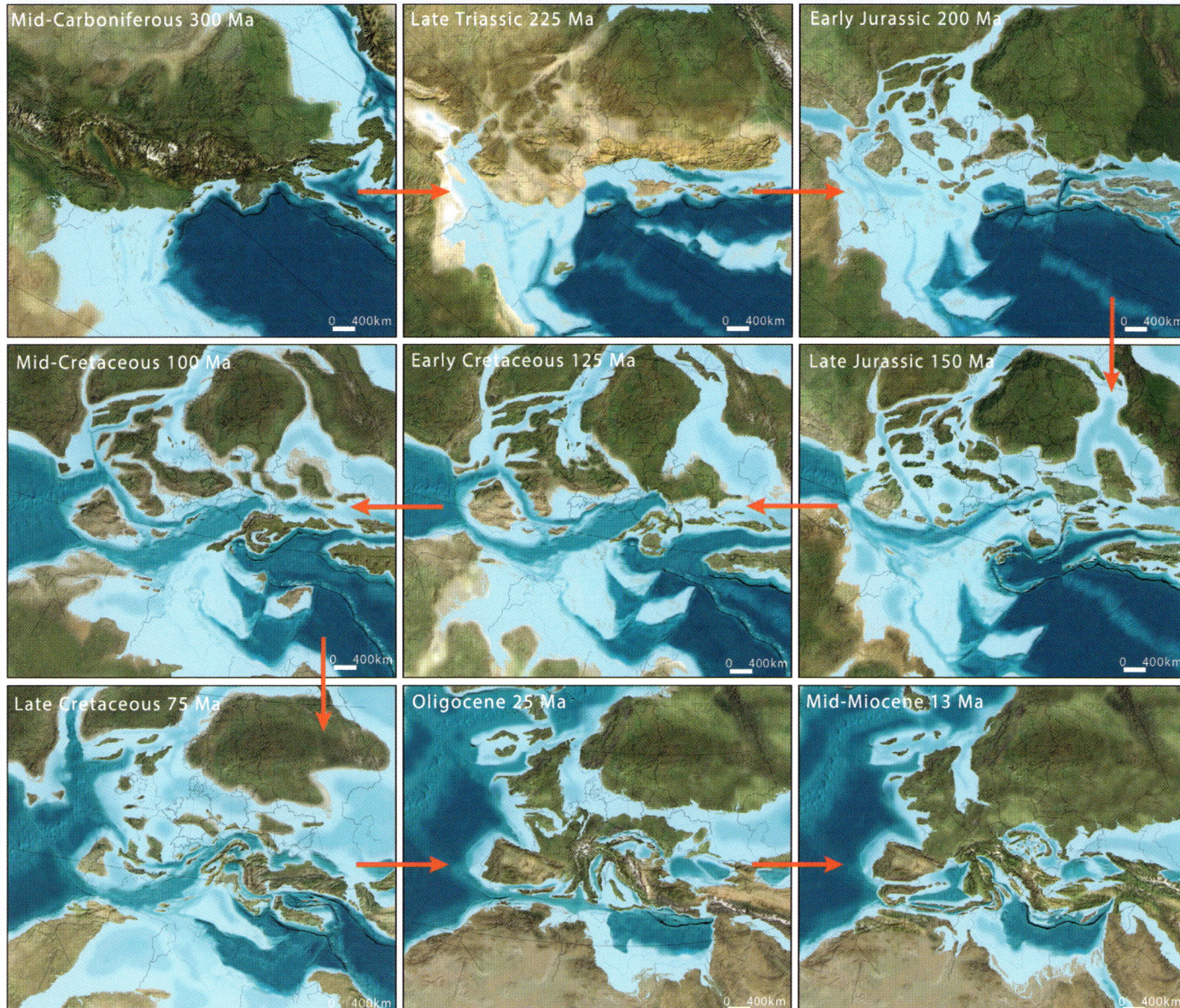

Fig. 7.4 PALAEOGEOGRAPHIC RECONSTRUCTION OF THE FORMATION OF THE NORTH ATLANTIC OCEAN, WITH THE EUROPEAN CONTINENTAL SHELF MARGIN AND MEDITERRANEAN BASIN, UP TO AROUND THIRTEEN MILLION YEARS AGO. In this palaeogeographic reconstruction of the configuration of the land and sea, as the North Atlantic ocean developed over the last 300 million years, the contemporary coastline is superimposed in dark grey on each panel. The sequence of panels shows the breaking up of the Pangea supercontinent and the opening of the Atlantic. Ireland remains as one of the many islands in the early Atlantic, at times semi-submerged. [Source: © 2011 Colorado Plateau Geosystems Inc. The map(s) are subject to copyright protection under licence no. 40520 and cannot be reproduced without prior consent from Colorado Plateau Geosystems Inc.]

sequences that we see provide a record of our Earth through time. These depict radical changes in coastal positions, relative sea levels, past environmental conditions and climates, tectonic upheavals and subsidence.

Some 250 million years ago, the continental crust that now underlies Ireland was in the middle of a large supercontinent that geologists call Pangea, with the coastline postulated to be about 1,000km distant (Figure 7.4). By 225 million years ago, that Pangean crust started to stretch and a rift valley formed. This deepened and was eventually flooded by the sea and, by 200 million years ago, Pangea had separated into what is now North America/Greenland on one side and Europe on the other, with the resultant shallow seaway between the two forming the proto-Atlantic, containing several large islands, including one that was to become Ireland. By fifty million years ago, the Atlantic was a well-established ocean and the European coastline beginning to look familiar. Throughout the last 200 million years of Earth's history, the Irish landmass has remained relatively consistently above sea level, feeding eroded sediments into the adjacent offshore basins.[6]

The rifted opening of the proto-Atlantic was complex and several rift basins formed, downthrowing the crust within them by over several kilometres. As the crust continued to thin, the lower crust melted and volcanic provinces developed, emplacing large bodies of magma into the thinning crust and producing eruptive volcanic edifices. Eventually, the crust failed fundamentally in one of these basins and volcanic eruptions created a new oceanic crust at the site of the current Mid-Atlantic Ridge. Consequently, extensional stress was relieved from the other rifted basins and these became less tectonically active (Figure 7.5).

These rifted basins contain up to 28km thicknesses of sedimentary deposits, representing offshore- and land-sourced deposition from 300 million years ago to the present day. The sedimentary sequence consists of a full range of rock types, although we don't have a detailed knowledge of the entire rock sequence, since only a limited number of boreholes have been drilled through the entire basin. However, we have collected geophysical imagery or seismics from thousands of kilometres of survey lines across them, so we do know how the layers of rock are arranged, but not necessarily what specific types of rocks make up the layers or how old they are. Nevertheless, from studying rock outcrops exposed on land, by analogy we can assume that these rocks contain evidence of dramatic changes in landscapes and climates, including earthquakes and other natural hazards, and a fossil record showing the evolution and extinction of many organisms, including the dinosaurs.

The Irish offshore rock sequences also contain oil and gas resources, which result from long-term, large-scale relative changes in global- to regional-scale sea levels.[7] These hydrocarbon resources date primarily to the Jurassic and Cretaceous periods of time, post-c.180 million years ago. The shallow sub-tropical seas of the Jurassic were biologically fertile, and the seabed sediments often accumulated significant concentrations of organic material that did not decompose, but which was buried by further sedimentation and preserved. This organic material, when buried to depths of between 2 and 5km (the so-called hydrocarbon window), becomes 'cooked' over millions of years and transformed into natural gas and crude oil. Hard to find and then extract, much of this oil and gas still remains trapped within the rock; it may become a future source of energy and wealth for Ireland, but we need to be mindful that it is also an immense reservoir of carbon stored safely underground, and perhaps better to let it remain there than be transferred to our atmosphere through combustion (see Chapter 29: Coastal Mining, Quarrying and Hydrocarbon Exploration).

Fig. 7.5 A simplified cross-section through the Earth's crust underlying the shelf-edge region of the Porcupine Seabight, showing the configuration of sedimentary units. During the early opening of the Atlantic (225 to 145 million years ago) this portion of the Earth's crust was stretched and rifted, forming a hyper-extended basin. Rotational fault blocks formed, downthrowing rocks of Upper Palaeozoic to Jurassic age, and causing magma (now crystallised into gabbros and basalts) to pond just below the crust. As the amount of crustal stretching eased, from 145 million years onwards, the basin continued to be infilled, producing large, near-continuous layers of faulted Cretaceous- to Cenozoic-age sediments. The plate tectonic collision of Africa into Europe (producing the Alps) saw some compression of the crust here, and reactivated some major rock faults. These processes buried organic-rich sediments, forming hydrocarbons, that have risen toward the seabed and become trapped in porous rock reservoirs hidden within this sequence of rocks.

IDENTIFYING IRELAND'S RAISED SHORELINES

The most visible of Ireland's ancient shorelines are those that exist above modern sea level (that is, raised shorelines). Examples of these can be found at many sites along Ireland's eastern coasts, north of Dublin Bay, for example, around Drogheda and the Boyne Valley coasts, County Meath; Carlingford Lough and Strangford Lough, County Down; as well as across much of the northern coasts of Ireland, from County Antrim to County Donegal (see Chapter 16: The Inhabitants of Ireland's Early Coastal Landscapes).[8]

The recognition of these earlier positions of the coast relies on identifying the geomorphological (shape) or sedimentological (composition) characteristics of a shoreline, and is based on the well-established geological principle that 'the present is the key to

Fig. 7.6 RAISED BEACH, FORMER SEA STACKS AND SEA ARCH, WEST OF BALLINTOY HARBOUR, NORTH COAST, COUNTY ANTRIM. These landforms are clearly of coastal and marine origin and resemble closely ones that can be found on the modern shoreline, but these now lie well above the reach of today's waves and tides. The raised shoreline here shows a former sea cliff (left), with a raised beach/intertidal area in front of it, which slopes down to grade into the present-day beach (bottom right). Coastal erosion features of sea stacks and a sea arch cut in rock, as a former coastal promontory, are associated with this raised shoreline. By identifying and mapping relict coastal landforms such as these, reconstruction of former shorelines can be undertaken, together with determination of vertical and horizontal changes in the line of the coast and relative sea-level changes. This information about past trends and events is of value in helping scientists and policy-makers better understand the implications of current climate changes and sea-level rise into the foreseeable future. [Source: Colin Park, CC BY-SA 4.0, image contrast increased]

the past'. In other words, using the landforms and sediments of present-day coasts as exemplars, we can make comparisons with possible traces of ancient ones, for example from the coastal features created by erosion, such as cliffs and wave notches, sea arches, sea stacks, caves, rock platforms and terraces. Where these features survive, they may exist in positions that are no longer attacked directly by the sea. These may be used to identify and mark the relict cliff-line and positions of former coasts (Figure 7.6) (see Chapter 10: Rocky Coasts).

In some instances, as occurs commonly on many raised rock-cut shores in Northern Ireland and Scotland, the upper limit of sea action may be inferred from the absence of sediments associated with such features. This is evidenced by exposures of bare relict coastal rock surfaces, which have been washed clean of any original overlying beach materials (that is, shoreline washing limits). In other cases, evidence may take the form of coastal sediments, the characteristics of which reflect their original deposition in coastal and nearshore environments. As with the features created originally by erosion, these coastal sediments are found today some distance from the reach of waves. Such sediments include raised beach deposits composed of sands and gravels and of finer-sized sediments, such as former estuarine and saltmarsh muds. These sediments frequently contain marine shells and other brackish- or marine-water flora and fauna.[9]

One of the main challenges in trying to identify an ancient shoreline is that, in contrast to its modern counterpart, its physical expression is rarely continuous. Instead, its reconstruction requires piecing together the scattered and disparate fragments of the shoreline that have escaped denudation by the Earth processes

operating after their formation. Whilst the size and resistance of a feature may assist in its survival, serendipity is perhaps the ultimate determinant for entry into the geological record. Assembling the incomplete shoreline jigsaw puzzle requires accurate correlation between these degraded remnants, in order to 'connect the dots' and reconstruct the wider picture of the shoreline's genesis and subsequent history. In most instances, this correlation is based principally on spatial relationships, such as the height of the relict shoreline pieces above present-day sea level. However, even on present-day coasts, different features associated with the same shoreline may develop across a considerable height range, and this is equally true of the preserved traces of past shorelines, especially where the precise nature of the evidence is unclear.

Whilst more resistant features, such as rock-cut platforms, have higher preservation potential, their longevity increases the likelihood that their form is the product of repeated phases of modification by a range of different processes operating over a long period of time. Furthermore, recent scientific literature on the rock platforms around Ireland's coasts now broadly acknowledges the possibility that they may be the cumulative product of more than one phase of erosion, each of which possibly having been separated by many thousands of years (see Chapter 10: Rocky Coasts).

Ultimately, reliable correlations are best developed in the presence of chronological data that can help anchor inferred connections and associations along a firm timeline, so that phenomena such as inheritance (where a single shoreline feature is the product of more than one process or phase of development) or the time-transgressive development of a shoreline can be quantified. Recent advances in geochronology are beginning to provide these tools and promise to aid greatly the task of quantifying the age of ancient shorelines, although the application of these new techniques is a work in progress. In some instances the newly-emerging data support and refine the qualitative inferences drawn by early workers; in other cases, the dates are forcing revision and re-appraisal of traditionally held views.[10]

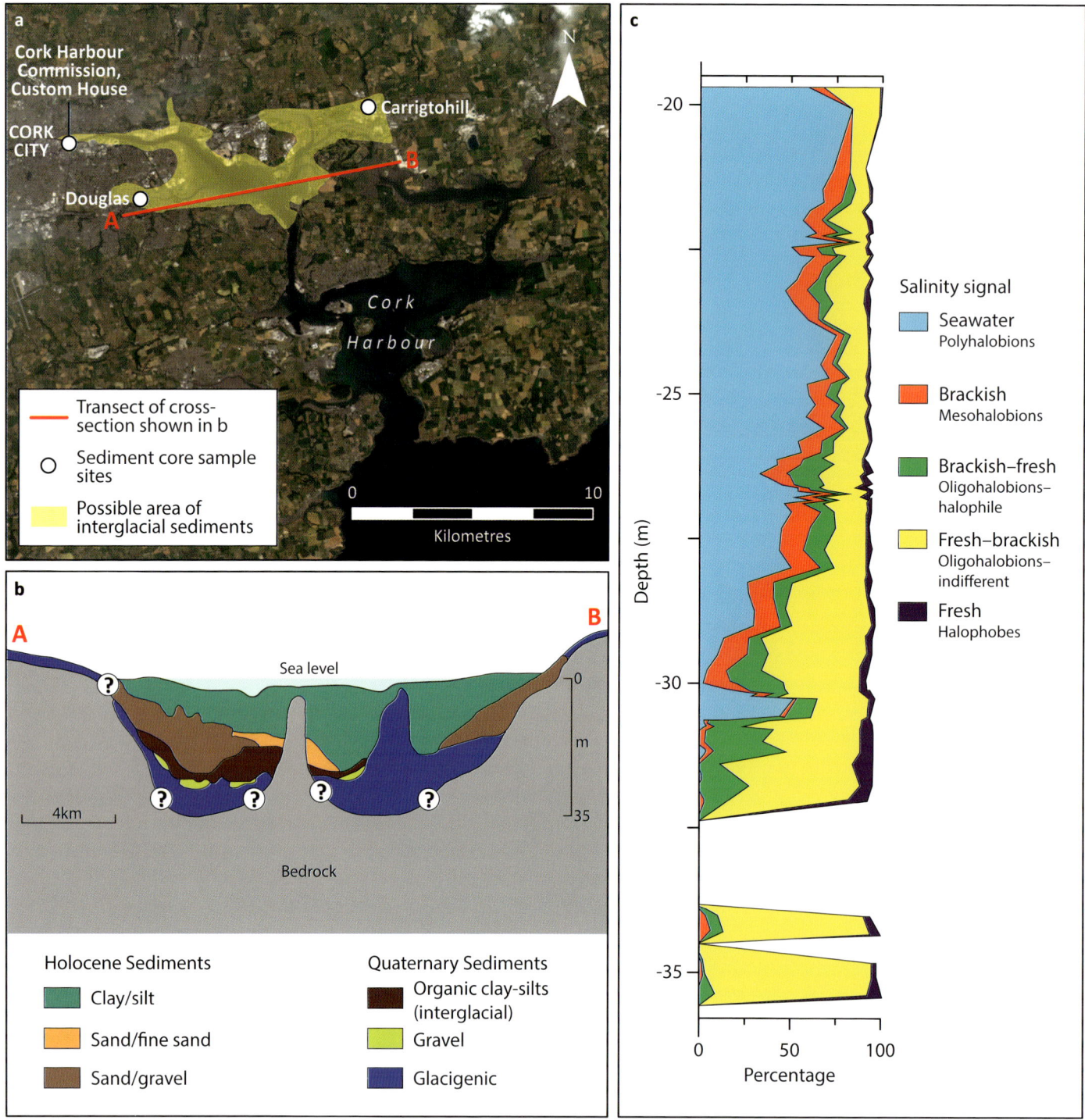

Fig. 7.7 INTERGLACIAL-AGE MARINE AND ESTUARINE SEDIMENTS OF CORK HARBOUR, COUNTY CORK. a. The Cork Harbour region, showing the known extent (based on borehole and core data) of interglacial-age marine and brackish water sediments that contain a record of past RSL changes. The red line on the satellite image marks the position of the cross-section through these sediments (line A-B, in 7.7b.). Post their accumulation in this heavily rock-faulted, deep-water harbour basin, the interglacial sediments have undergone differential warping in height westwards (of up to 13m height difference) between the eastern and western sides of the region. [Map data source: Landsat-8 image courtesy of the US Geological Survey, 2018]; b. A simplified cross-section of the late-Quaternary sedimentary stratigraphy overlying the basal limestone rock surfaces of the basin. The organic-rich, interglacial marine sediments (brown colour shading) are dominated by silts and clays deposited in an estuary. These are sandwiched between underlying river (fluvial) gravels and superimposed glaciofluvial sands and gravels, which record the melting of ice in the River Lee valley and wider region during the last ice stage. [Adapted from Coxon, P. and McCarron, S.G., 2009. Cenozoic: Tertiary and Quaternary (until 11,700 years before 2000). In: Holland, C.H. and Sanders, I.S., eds, 2009. *The Geology of Ireland*. Edinburgh: Dunedin Press, pp. 354–398]; c. A diatom diagram, showing the changes in the frequency of this microscopic algae within this sedimentary sequence. Diatoms are particularly adjusted in their habitat to the salinity of water and form an excellent indicator of changes from marine through to freshwater environments. The diagram records the dominance of the fresh to brackish salinity conditions of an estuary during the early- to mid-interglacial time (depths of c.31–36m). Above 31m sediment depths, an increasing marine water influence is shown, indicating the rapid invasion of seawater into the relatively fresher waters of the inner estuary at times of lower RSL. This represents the later phases of the interglacial, before full cold conditions are re-established. [Data source: Robert Devoy and Anne Synnott; diagram based on Devoy, in Crowley et al., eds., 2005. *The Atlas of Cork City*. Cork. Cork University Press].

PRE-GLACIAL RAISED SHORELINES

The process of identifying, describing, correlating, dating and re-interpreting ancient raised shorelines is exemplified in the quest to understand the prominent coastal landforms of the Pre-Glacial shoreline of Ireland (the term Pre-Glacial refers to the last ice and cold climate period). Evidence and understanding of the far less visible and often-submerged ancient depositional coastal environments – such as early estuaries, saltmarshes and lagoons – has been difficult to find. Some insight to these environments, as part of Ireland's early shore zones, has occurred in Cork Harbour. Three main sites across the harbour area have been cored, establishing the existence in some places of over 15m of interglacial to early cold stage sediments. Together, the different cores provide one of the longest sequences of marine and estuarine sediments in western Europe. The sediments record the progress of RSL changes on the south coast region during probably the last interglacial warm period (c.110,000 years ago) and possibly even earlier in the Quaternary, though precise dating of the sediments has been difficult (Figure 7.7).[11]

In contrast, the Pre-Glacial raised shore platform provides very visible evidence of early coasts in Ireland. This is a feature that has been known to science since the middle of the nineteenth century. It is similar in its form and heights to shorelines found along the coasts of western Britain and across the Celtic Sea to Brittany. In the field, the shoreline is seen most commonly as a relatively smooth, rock-cut, bench-like surface (a former wave-cut platform), which is approximately level to sub-horizontal in form, though generally dipping seaward in a series of steps. It is positioned approximately 2–7m above the modern high water mark. The platform can be seen occurring intermittently at these heights on the southern coast of Ireland, particularly on the wave-sheltered western side of coastal embayments, and runs along the coast for distances of over 250km from County Waterford to County Kerry. Probably the same surface can be traced northwards at lower levels into County Clare in the west and Wexford on the Irish Sea coasts. Importantly, in helping reconstruct ice and sea-level histories, the platform is backed (as a cliff-line) and overlain in some locations by bedded yellow-orange sand and gravels. These sediments were interpreted in the past as a raised beach. The best expressions of the complete fossil shoreline (platform and 'beach') can be found at sites in and close to Courtmacsherry Bay, west Cork. Here, the shoreline is often referred to as the Courtmacsherry raised beach (also the Courtmacsherry Formation raised beach) and rock platform, as this was the area where the platform and the overlying sedimentary sequence was first described in detail in the early years of the twentieth century (Figure 7.8) (see Chapter 10: Rocky Coasts, Figure 10.3).[12]

Fig. 7.8 THE COURTMACSHERRY ROCK PLATFORM AND RAISED SHORELINE, LOOKING WESTWARD ALONG THE COAST FROM THE OLD HEAD OF KINSALE, COUNTY CORK. Fragments of this distinctive ancient coastal feature, also known as the Pre-Glacial rock platform, or the South of Ireland Shore Platform (SISP), are found preserved intermittently along much of the south coast of Ireland. The platform and, at some sites, an overlying sequence of bedded sands and gravels that was originally thought to be a raised beach, was first described in detail in the early years of the twentieth century. This prominent shoreline feature has been the subject of research and controversy ever since, as geologists have tried to unravel its origins and spatial extent. Its height above present-day reach of high tide and all but the strongest of storm wave conditions, implies that it was formed at a time of higher RSL than at present. According to current scientific thinking, the platform pre-dates the last glacial epoch but, beyond this, its age and the precise manner of its formation remain uncertain. It may even be the cumulative product of separate, successive phases of erosion in previous times of warm climate (that is, Quaternary-time interglacials), or it may have been shaped as a shore platform, at least in part, by ice processes. It has certainly been modified by people, who have used it in some places as a roadway to the beach, as at Maloney's Strand, Dunworley, west Cork. [Source: Karl Grabe]

Before *c*.1970, the absence of suitable dating techniques prevented the fixing of a definitive age for either the eroded rock platform or the overlying raised beach-type sediments. However, both features were found to underlie cold-climate, periglacial sediments and this fact, coupled with evidence that the rock platform had been scoured by ice, led to the conclusion that the shoreline pre-dated the last time Ireland was extensively glaciated (35,000–20,000 years ago) (see Chapter 6: Glaciation and Ireland's Arctic Inheritance). Subsequent research has tried to correlate the Courtmacsherry rock platform with similar platform fragments found along the western, eastern and (more tentatively) the northern coasts (Figure 7.3), but none of these suggested equivalences have yet proved conclusive. Despite more than a century of investigations, the ages of both the Courtmacsherry raised beach sediments and the rock platform on which it rests still remain uncertain, as do the physical processes and environmental conditions in the Quaternary responsible for their formation.

The development of independently dated chronologies for platform and 'beach' formation will be important in ultimately resolving this debate and improving the reliability of correlations with similar features found elsewhere. In recent years, the application of new geochronological tools, such as radiometric Uranium-Thorium dating (UTD) and Optically Stimulated Luminescence (OSL) type techniques, have started to provide some dating evidence and age constraints. For the raised beach, its depositional settings are technically challenging environments in which to work. Because of this, any attempted chronology, and interpretations of the 'beach' sediments based on these new techniques, must remain open to question regarding their reliability. However, radiometric dating of the sands and gravels found above the platform at three sites around Courtmacsherry Bay (using the OSL technique) has returned ages ranging from between *c*.30,000 and 90,000 years ago. This constrains the accumulation of these sediments to a known period of climatically cold conditions that occurred after the last interglacial (that is, pre *c*.115,000 years ago), but prior to the maximum build-up of ice at the height of the last glacial.[13]

Interestingly, as part of this work, cross-stratified sands immediately overlying these 'beach' deposits have been re-evaluated as being of shallow marine origin, forming in the ice-front waters of tidewater glaciers, rather than being wind-blown (terrestrial) in origin, as in earlier interpretations. If this assessment is correct, it presents an interesting paradox and conundrum. It requires that RSL was several metres above present-day levels, during a cold, ice-dominated climate stage, in which sea levels globally were several tens of metres below their current level. In turn, if this scenario is correct, it would suggest that both ice loading over this region of Ireland, and the attendant isostatic depression of the land surface, were much greater than envisaged by the current generation of sea-level models.[14]

Further along the south coast, at Fethard in County Wexford, a sequence of bedded sands and gravels is to be found, resting on a rock platform. Original attempts to date these sediments, using a different chronological technique (Infra-red Stimulated Luminescence, or IRSL), produced approximate ages of between around 110,000 and 180,000 years ago, suggesting they were deposited sometime during the last interglacial, or perhaps even as early as the preceding cold stage (see Chapter 6: Glaciation and Ireland's Arctic Inheritance). More recent re-dating of these deposits, using the same technique as employed at Courtmacsherry (the OSL technique), resulted in a revised age of between approximately 40,000 and 60,000 years ago, which would suggest they are broadly contemporaneous with the sediments further to the west.

The only other age data for Ireland's ancient shorelines have come from pioneering work on the west coast during the early 1990s, which focused on peat horizons found in association with raised beach sediments exposed around the coast of Tralee Bay in County Kerry. These peats returned ages of between 114,000 and 123,000 years ago, indicative of accumulation during the cool temperate interval that occurred toward the close of the last interglacial. The organic deposits are found beneath the raised beach in the eastern end of the section and so could be compatible with either of the date ranges reported at Fethard.[15]

One intriguing feature of the Pre-Glacial rock platform is its apparent lack of tilt. This is unusual, because elsewhere, as in Scotland for example, the isostatic loading of the land by ice, and its subsequent recovery, is usually greatest at the centre of ice accumulation and decreases with distance outwards (Figure 7.9a and b).

This lack of tilt has led some investigators to interpret this as evidence that, at the time the platform was forming, relative sea levels around the Irish coast were similar to today's. Such a scenario is plausible if the platform was cut during the latter part of a warm interval (an interglacial), by which time vertical land movement caused by ice loading had effectively ceased. If this was not the case, and if the platform formed during a cold (glacial, or stadial) phase, it would imply that the land surface must have experienced equal amounts of depression (for example, from the effects of ice loading on this south coast region) along the entire length of the shoreline. This would indicate that our understanding of former ice extents and thickness in Ireland requires revision.

Alternatively, the rock platform may, in fact, vary in age along its length, and/or be the cumulative product of successive phases of erosion, each of which could have been separated from the other by up to several thousands of years. It is also important to acknowledge that similarity of height is simply one of several criteria by which the discontinuous features constituting the 'shoreline' are correlated, and so the absence of an apparent tilt may have less to do with how the shoreline formed and more to do with how its fragments have been reconstructed. Ultimately, more dating and related field evidence will be required to establish with greater certainty the nature of Ireland's Pre-Glacial raised shoreline(s); to underpin the reliability of existing age estimates and also the significance, if any, of these shorelines in the regional to island-wide patterns of glaciation and the impacts of glacio-isostasy.

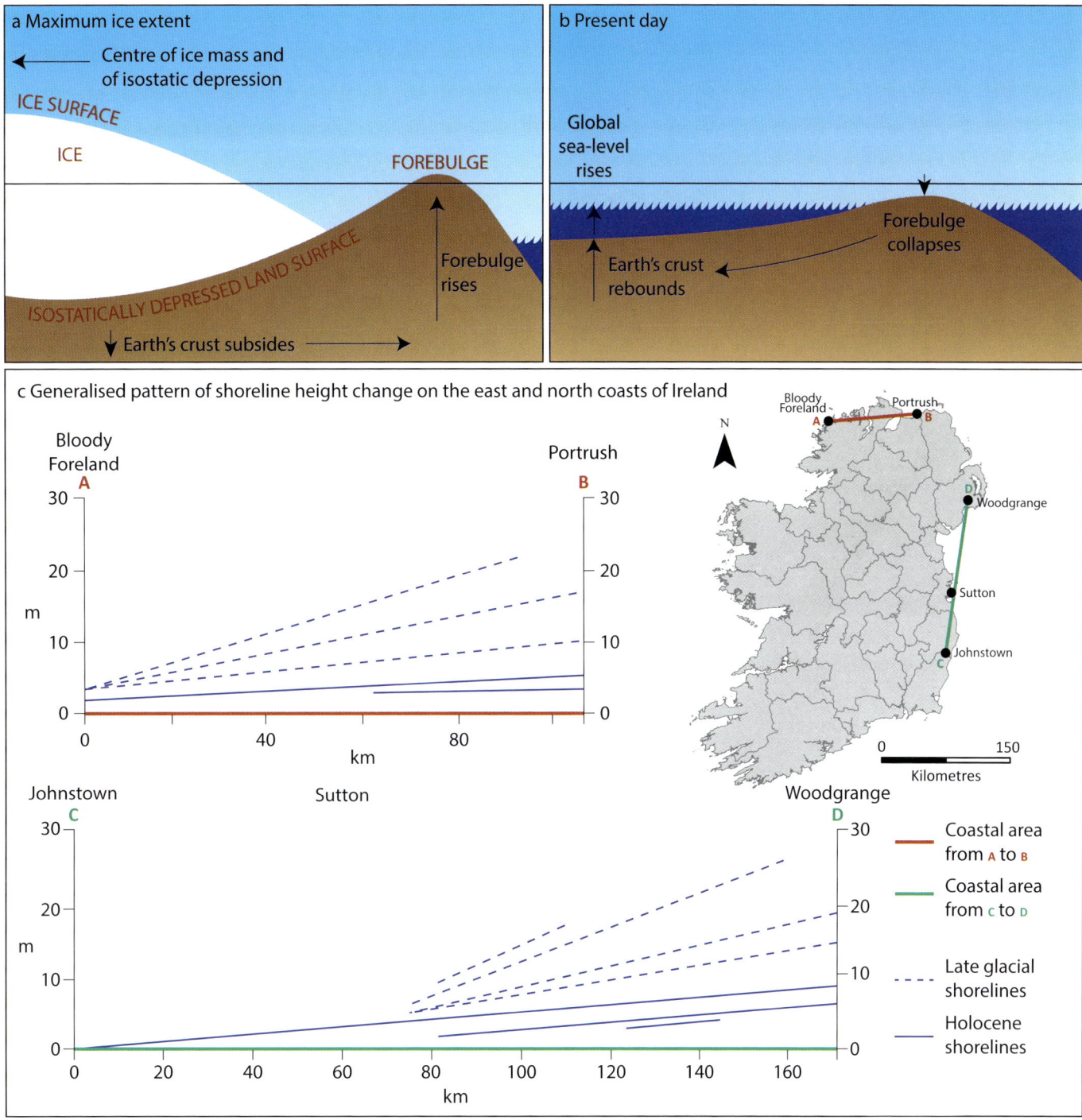

Fig. 7.9 GLACIAL ISOSTASY AND THE MAIN RAISED SHORELINES IN IRELAND. a. Conceptual diagram of Earth crustal behaviour under ice loading through the process of glacio-isostasy. The development of a raised crustal forebulge region, as part of this process, is also indicated (right). This feature is formed at distances proportionate to the thickness of the ice mass loading in its central area. The outcomes of glacio-isostasy and linked forebulge behaviour, where this occurs, are recorded on coasts in the preservation, at varying heights above or below present-day sea level, of former shorelines. Where the dominant crustal signal in response to the ice unloading has been land uplift, then raised shorelines are found. [Adapted from: Lowe, J.J. and Walker, M.J.C., 1997. *Reconstructing Quaternary Environments*. New York: Pearson]; b. The different mechanisms involved in the operation of glacio-isostasy and its corollary, hydro-isostasy, are complex and lead to spatially varied effects in crustal rocks. This diagram shows the movement and subsidence of a forebulge area toward the centre of former ice loading of the crust as deglaciation occurs, with progressive ice melting and retreat. This process leads to relative seafloor changes, involving an initial elevation, with forebulge migration and subsequent potential sinking of the crust as the forebulge passes through. [Based on: Weerts, H.J., 2013. Holocene Sea-level Change, Sedimentation, Coastal Change and Palaeogeography in the Southern North Sea Lowlands: A 2012 geological literature overview. In: Thoen, E., Boeger, G.J., de Kraker, A.M.J. et al., eds. *Landscapes or Seascapes? The history of the coastal environment in the North Sea area reconsidered*. Turnhout: Brepols, pp. 145–173]; c. Patterns of broadly north-to-south and east-to-west Earth surface and coastal tilt directions in Ireland, based on the height data of relict raised shorelines. The generalised ages and heights of the principal raised shorelines found on these eastern and northern coasts is shown. Across the north of Ireland, differences exist in the heights, timing and pattern of RSL rise as recorded by the shorelines of the east coast (for example, at Woodgrange, Killard, Belfast Lough and Ballyquintan) when compared to those of the north-coast sites (for example, Portrush, Magilligan, Rathlin Island). These differences have been interpreted broadly as resulting from the distance of these coasts and shorelines from the former centre of ice loading in the region.

LATE- AND POSTGLACIAL RAISED SHORELINES

In contrast to their older counterpart, the raised shorelines developed during and after the melting of Ireland's last ice sheet are characterised by marked differences in their height at different points around the coast, primarily reflecting the differential effects of glacio-isostatic adjustment. A general increase in the height of

Fig. 7.10 (left) RAISED GLACIOMARINE 'RED' LATEGLACIAL MUDS FOUND AT KILKEEL STEPS, KILKEEL, COUNTY DOWN. Sites similar to this, recording the early deposition under RSL rise of sediments dominated by clay and silts and often containing interbedded sands (as shown here close to the trowel), can also be found elsewhere along the north-east coasts of Ireland (for example, Ballyquintan Point, Killard Point and Roddan's Port, all three in County Down). These sediments evidence the formation of deeper-water marine facies, to the seaward of coarser sediments (sands and gravels) that compose the neighbouring and higher environments of the raised beach shorelines. The 'red clays', which often contain shells and other micro- and macrofossils, were deposited in marine, but low salinity, cold waters. This water occurred at the margins of the retreating ice fronts, as the sea flooded over the ice-depressed land surface areas and – as at Woodgrange, west of Strangford Lough, County Down – the marine waters filled the land surface hollows between the now-exposed glacial drumlin landforms. [Source: Jasper Knight]

these raised shorelines with distance north from Dublin was recorded in the nineteenth century. The subsequent mapping and measurement of the height and tilt of these shoreline fragments have produced 'shoreline correlation diagrams' (Figure 7.9c). These

have been used to infer the degree and pattern of isostatic recovery following the removal of ice loading over the island. General patterns have emerged from these diagrams, such as the reduction in shoreline height with increasing distance towards southern and western coasts. According to generally accepted explanations, this tilt is due to increasing distance from the centre of ice accumulation, while a distinction can also be made between lower-elevation postglacial raised beaches, composed of sand, gravel and shell, and higher-elevation lateglacial shorelines that occur above them. These latter shorelines have been found commonly to be linked with, or incised into, cold-climate deposits of sediment.[16]

In addition to these morphological features, sediments indicative of marine conditions have been studied at several locations around the Ireland's coasts, with a particular concentration

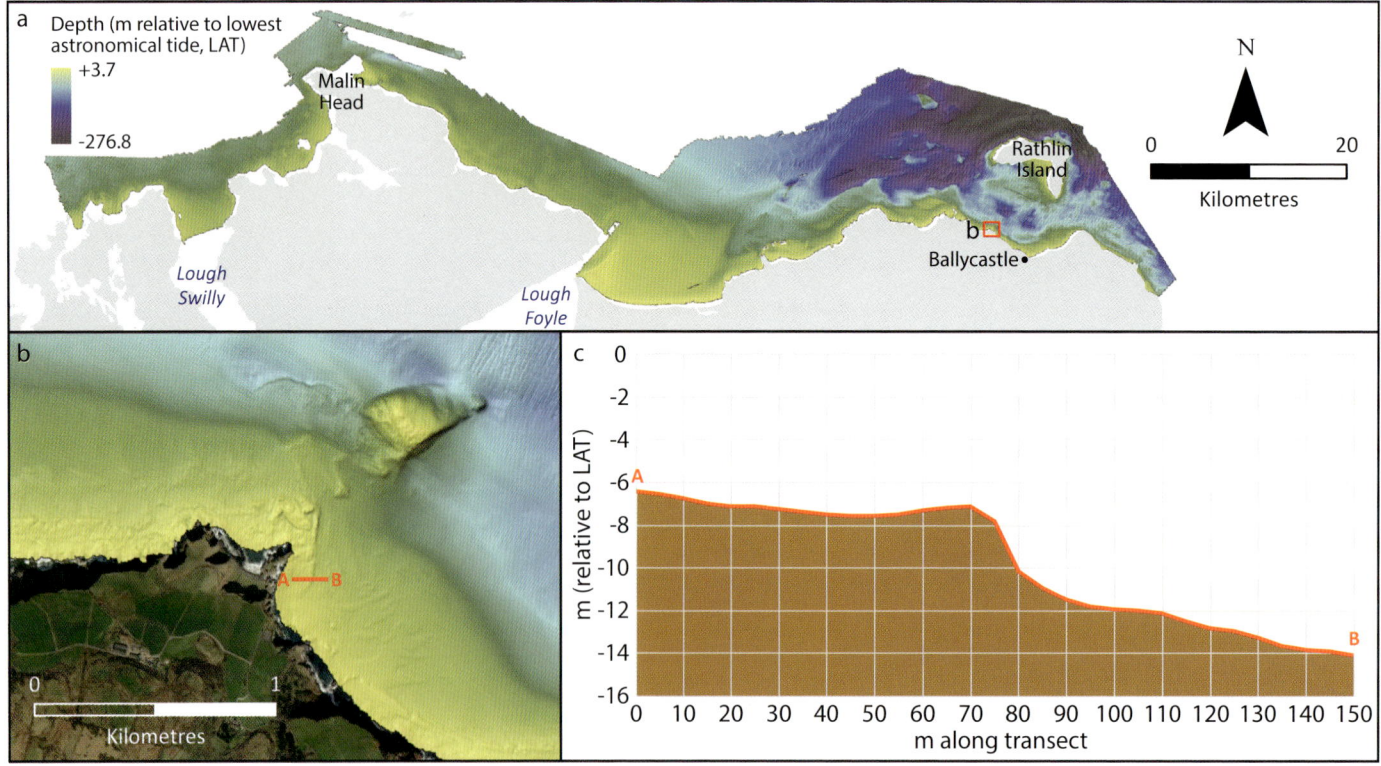

Fig. 7.11 A SUBMERGED SHORE PLATFORM NEAR BALLYCASTLE, COUNTY ANTRIM. a. Analysis of offshore bathymetry data, collected from the Joint Irish Bathymetric Survey (JIBS), has enabled the identification of a suite of submerged rock platforms, extending along the Northern Ireland coast from Rathlin Island to the mouth of Lough Swilly. These platforms were created by marine erosion, at a time when the relative sea level along the north coast of Ireland was lower than that of the present day. b. This visualisation shows one such platform near Kinbane Castle, approximately 5km north-west of Ballycastle, County Antrim. c. The cross-section helps to illustrate the profile of the platform. The platform is relatively flat, terminating at its seaward margin with a marked cliff that drops into the sea. This type of platform is known as a 'Type B' platform (as opposed to 'Type A' platforms, which grade gently into the sea). The JIBS Project (EU INTERREG IIIA Programme) was co-ordinated by the Department of the Environment for Northern Ireland and led by the Maritime and Coastguard Agency in partnership with the Marine Institute of Ireland. [Map data: JIBS]

in the north-east of the country, where the evidence for Lateglacial raised shorelines is most abundant. For example, at Kilkeel Steps in County Down, thick sequences of laminated mud containing marine microfossils infilling channels cut into glacial sediments can be found. The association of these sediments (as marine in origin) with neighbouring raised shoreline features has been used to infer a complex pattern of RSL change in this area (Figure 7.10).

Whilst the interpretation of these deposits is contested, the raised marine muds are particularly important since microfossils obtained from them have been radiocarbon dated to around 19,000 years ago, providing rare chronological and chronostratigraphic information for this period. Where it can be demonstrated that the dated material is in its original position of deposition, and has not been reworked subsequently, sedimentary sequences of this kind provide evidence that RSL at that time was effectively higher than its present-day relative position. Since this situation also coincided with a time when global sea levels were lower than those of the present day (due to the abstraction of water from the sea to create the ice sheets and glaciers present on land), then the land surface in this area must have been depressed to an even greater extent. Further, this in turn indicates that there was a significant load of ice in this area and the wider region at the time the now-raised shoreline was forming.

BURIED AND SUBMERGED COASTS

Whilst raised shorelines provide striking evidence for past RSL, their use is limited by definition to elucidating intervals when relative sea level was higher than present. In contrast, any ancient shorelines that formed during lower phases of RSL will now be located offshore, or else have been buried beneath more recent accumulations of sediment. In both instances, direct observation and logging of exposures is more challenging, and the volume of available data is correspondingly more restricted. In recent years, however, advances in marine technology have made it possible to image the seafloor and probe the nature of the sediments beneath it.

Since the start of the present millennium, the Geological Survey of Ireland and the Marine Institute have undertaken extensive mapping of Ireland's seafloor under the auspices of the Irish National Seabed Survey (INSS) and the subsequent INFOMAR programmes (see Chapter 9: Underwater Surveys: The INFOMAR Project). Although the nearshore areas that are of most interest from a RSL perspective are among the last to be mapped, due to the logistical challenges of surveying in very shallow water, nonetheless several important datasets are now available. Where detailed mapping of seafloor morphology has been possible, many topographic features that resemble possible ancient shoreline fragments have been revealed. For example, along the northern coastline, a suite of submerged rock platforms has been identified, extending between Rathlin Island and Lough Swilly (Figure 7.11), that would have been formed during a period of lower relative sea level than at present. Similar rock-cut features are also apparent offshore in other parts of the country, while drowned palaeovalley systems, that may be linked to phases of lower RSL, have been reported off the present-day coasts of counties Cork and

Waterford. Whilst these kinds of evidence provide some qualitative information on the position of RSL, difficulties in determining the age of these features currently limits their value in reconstructing a reliable history of change.[17]

The offshore evidence is not limited to erosional features and in some places accumulations of soft sediment, that may reflect ancient beach, barrier or spit-like morphologies, have also been identified. The preservation potential of soft sedimentary features is generally low, due to the high wave energies and current velocities that typify much of the Irish continental shelf. However, in sheltered, lower-energy settings, or where sediments accumulate rapidly, there is a possibility that some traces may be preserved, even if these now lie buried beneath more recent sediment cover. For

Fig. 7.12 a. WAVE-ROLLED NODULES OF PEAT (MOORLOGS), AND b. WOOD FROM A FOSSIL PINE TREE, BOTH PART OF THE PEAT (ORGANIC RICH SEDIMENTS) EXPOSED ON THE BEACH AT SHANAGARRY STRAND, EAST COUNTY CORK. 'Submerged forests', such as this, containing the remains of pines, oaks, willows and other trees, as well as layers of wood and monocot-rich peats, are to be found at many locations around Ireland's coasts, including Wexford Harbour, Ballinskelligs Bay, Galway Bay and Magilligan Foreland. Detailed observation of these intertidal zone peat exposures, and their significance for RSL changes in Europe and further afield, started in the early twentieth century in Ireland with the work of the geologist and natural scientist, Clement Reid. Subsequent studies have shown that most of the peats found on present-day shorelines, together with other sediments of freshwater and brackish-water environments, formed originally behind coastal sand-gravel barriers. In most cases they date to the period of around 4,000–5,000 years ago, when the rate of earlier postglacial RSL in European waters had decreased significantly and sediment supply to coasts had risen sharply under human influences. [Source: Darius Bartlett]

example, some studies have used a combination of geophysical techniques and evidence from boreholes to identify submerged and buried land-surfaces near Portrush in County Antrim. They then combined these data with simulations of RSL change to analyse how the palaeogeography of the area is likely to have evolved over time, and to assess the implications this has for the archaeological record of the area. This kind of offshore work is in its infancy, but has significant potential to further our understanding of past RSL and is likely to be an area of significant development in the future.[18]

Coastal peats

At a number of locations around Ireland's coasts, the falling tide exposes beds of peat on the foreshore, many of which contain abundant remains of pines, oaks, willows and other trees (Figure 7.12). Since the plant species they contain show that these 'submerged forest' beds are clearly indicative of freshwater conditions, they constrain the position of RSL and tell us that the trees must have been growing at some (unknown) height above sea level at the time. Similarly, where interbedded deposits of now-buried freshwater or saltmarsh peat and estuarine sediments are revealed in exposed sections, or seen in cores extracted from drilled boreholes, this can also be used to help determine the position of past RSL (Figure 7.13a and b) (see Chapter 16: The Inhabitants of Ireland's Early Coastal Landscapes). Significantly, the organic-rich nature of these types of material often means that radiocarbon dating can be undertaken, permitting the changing position of RSL to be tracked through time, while fossil pollen and diatoms contained within the sediment can give additional information about how the environment gradually changed from freshwater, through brackish to fully marine. The assembly of such data from locations

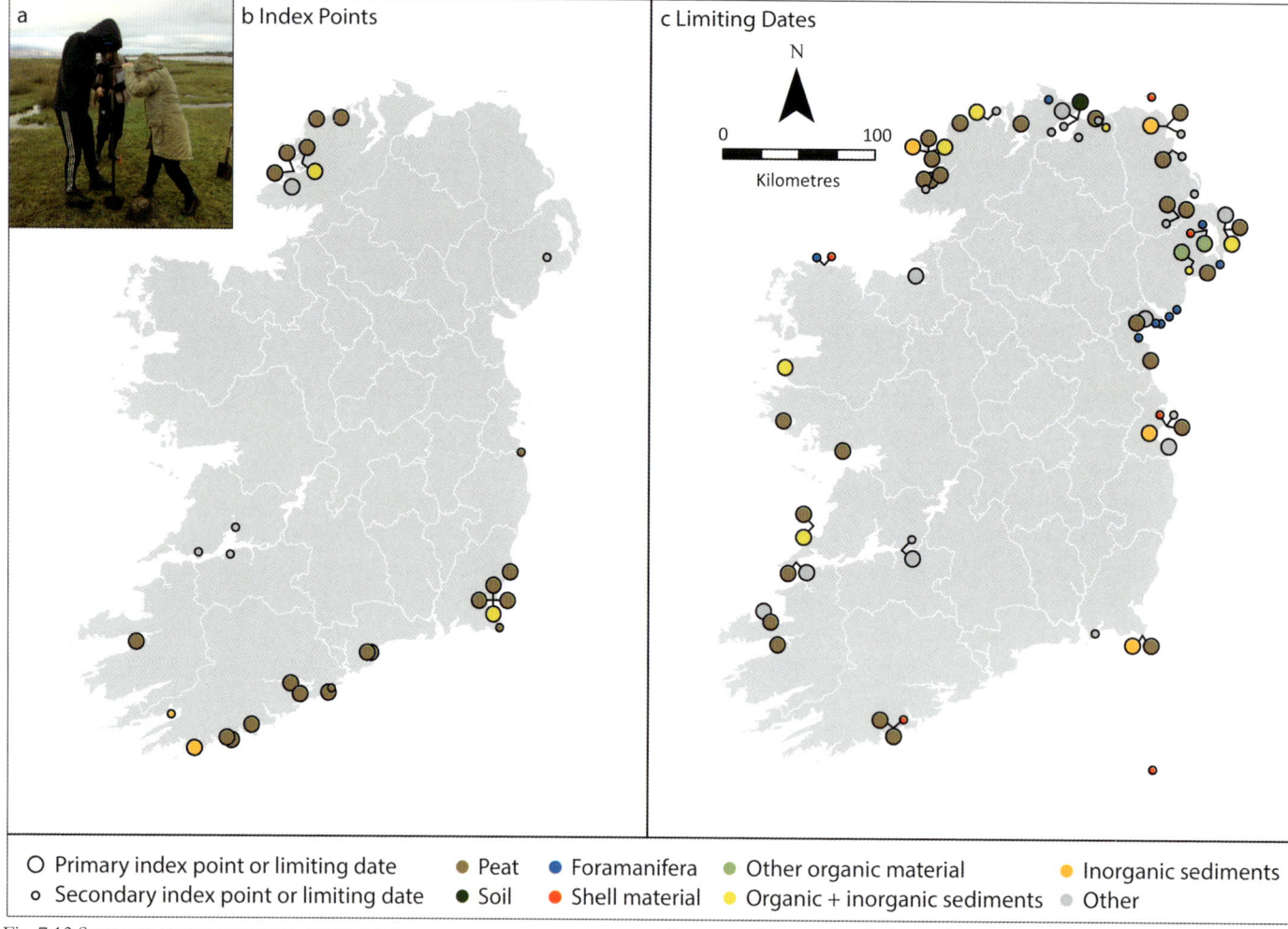

○ Primary index point or limiting date   ● Peat   ● Foramanifera   ● Other organic material   ● Inorganic sediments
∘ Secondary index point or limiting date   ● Soil   ● Shell material   ● Organic + inorganic sediments   ● Other

Fig. 7.13 SIGNIFICANT RELATIVE SEA-LEVEL DATA POINTS AROUND THE COASTS OF IRELAND. Changes in relative sea level (RSL) can be reconstructed by examining the characteristics of sedimentary layers formed in coastal environments (in estuaries, lagoons and saltmarshes). These can be sampled using sediment corers (a cylindrical, metal tube device, which is inserted into the layers of sediments and then used to take a sample of the sedimentary sequence). a. A sediment core being extracted from a saltmarsh at Rossbehy, County Kerry. RSL change can be inferred from examining the zones of contact between different sedimentary layers, for example between terrestrial or freshwater sediments (such as peats) and estuarine or marine sediments (typically represented by clay, silts and sands). The timing of the onset or removal of marine conditions can be determined by the radiometric dating of these sediments, either directly from the sediment's composition (for example, woody peat, allowing a radiocarbon date to be taken) and/or from their fossil contents (for example, from shells, foraminifera). [Source: Sarah Kandrot]; b. In Ireland, many such palaeoenvironmental reconstruction studies have been undertaken. Brooks and Edwards (2006) have collated the results of this work to produce a sea-level database for Ireland. This consists of sea-level index points and limiting dates, as shown here. Index points are localities where sedimentary data have been found that delimit the past altitude of a sea-level position in time and space. c. Limiting dates are derived from samples that are of known age and are associated with a known environment, but that have no quantifiable relationship with sea level. As a consequence, it is only possible to infer from this latter type of sample whether sea level was above or below a certain altitude at a given time. They are further subdivided, based on the quality of the data, with primary data being of higher quality, or associated with more certainty than secondary data. While the Brooks database is extremely useful, it is now more than a decade old and new data have become available since it was compiled. [Map data source: Brooks, A. and Edwards, R., 2006. The Development of a Sea-level Database for Ireland. *Irish Journal of Earth Sciences*, pp. 13–27]

around the Irish coast allows a more quantitative view of past RSL change to be elucidated, and provides essential data and constraints for the development of numerical sea-level models. These models, in turn, allow scientists to infer RSL histories for those parts of the coastline where no reliable field data exist currently.[19]

CHANGES IN IRELAND'S RELATIVE SEA LEVELS

Tide gauges and other instruments are designed to provide direct measures of sea level. In contrast, the various kinds of geological evidence outlined in this chapter should be seen as indirect indicators, and these must be interpreted if they are to be used to produce reconstructions of past RSL. In many instances, deposits only permit qualitative inferences to be drawn regarding past RSL position. Hence, freshwater peat or marine clays simply constrain the possible upper and lower limits of RSL at the time of their formation. However, just as modern shorelines form under the influence of multiple controlling factors, of which RSL is only one, interpreting the past position of RSL from ancient shorelines is not straightforward. This situation is compounded when dateable material is absent or potentially reworked, such that age estimates are poorly constrained and potentially erroneous (see Chapter 2: The Coastal Environment: Physical systems processes and patterns) (Figure 7.13c).

As a consequence, sea-level scientists have placed particular emphasis on the collection of material from low-energy, intertidal deposits (for example, in situ shells and peats from estuaries, coastal embayments). These sediments can usually be dated reliably and quantitatively related to a defined tide level. Although these kinds of data permit precise reconstruction of past RSL, they are comparatively few in number and distributed unevenly in space and over time. Because of this, whilst the general pattern of RSL change around the Irish coast has been known for many decades, there are still gaps in detail, particularly for certain parts of the country and during certain intervals of time. Since RSL is highly variable in time and space, attempts to fill data gaps by compiling field evidence from different locations into a single record is prone to significant error. To overcome this, a framework that accounts for regional differences in vertical land movement around the Irish coast is needed, to assist in developing a picture of RSL change. This framework is provided by the Glacial Isostatic Adjustment (GIA) numerical models for ice-loading patterns and the consequent RSL changes. These models are the subject of ongoing refinement and apparent misfits between simulated and inferred RSL are valuable signposts that point to areas of incomplete understanding in Earth crustal and also coastal behaviour.[20]

Synopsis of RSL changes

The pattern of RSL change simulated for Ireland by the ice-loading models generally now show excellent agreement with the postglacial-age shoreline data. The model simulations of RSL change are consistent with the broad picture shown by these data for sea-level and coastal changes. Field data, from the different sources of evidence, are most abundant on the southern coasts of Ireland. In this region, stretching northwards to approximately a line between the Shannon and Dublin Bay, many suitable environments for sediment accumulation exist. The many sheltered estuaries and embayments of these coasts have facilitated the accumulation of low-energy deposits under rising RSL. Along the coasts from Kerry to Wexford, the field data and the simulated GIA model curves together show RSL rising throughout the Holocene. Relative sea level is shown to reach within a couple of metres of present-day sea level by around 5,000–6,000 years ago. Precise field data are, however, absent for the lateglacial. The models simulate a low-stand of RSL of 60m below that of the present day, following deglaciation in the region. Sea levels at the Late-Glacial Maximum for ice build-up were possibly slightly lower still, reflecting the peak draw-down of global ocean water volume (Figures 7.3 and 7.13).[21]

Reliable field data are more limited for central Ireland, though the comparative lack of RSL evidence for regions here may in part reflect the relative lack of high-quality studies undertaken in the past. On western coasts, data absence is also likely to reflect, in particular, the prevalence of high-energy wave conditions and storminess. These will have limited the opportunity for the accumulation and preservation of suitable sedimentary deposits. Further, persistent onshore movements of sand dunes and other coastal barrier systems under SLR dominate these coasts. The result has been the constant erosion and reworking of shoreline sedimentary environments as evidenced, for example, in Ballinskelligs Bay and Dingle Bay, Galway Bay and Achill Island (see Chapter 30: Engineering for Vulnerable Coastlines). The absence of good quality sea-level data along eastern coasts may reflect these west coast factors too, as well as a general lack of coastal accommodation space (for example, deep bays and estuaries). This is coupled with enhanced retreat rates of the soft-cliff glacial sediments found on these coasts, which has resulted in both the submergence and the erosion of any suitable deposits that may have formed in the past.

In the north and north-east, evidence for ancient sea levels seen in the field typically comprise deposits that only limit the upper and lower bounds of RSL and that date predominantly to the early Holocene. This different distribution is explained by the fact that, in contrast to the southern sites, RSL in the late- and postglacial periods in this region was higher than that seen at the present day (Figure 7.3). In fact, the simulated RSL curves for the north-east reveal a complex pattern of change, in which RSL rises, falls and rises again in response to the shifting balance between the rates of isostatic recovery of the land and that of SLR as water was returned to the oceans.

It is against this backdrop of complexity that the lateglacial raised marine muds exposed at several locations on these coasts must be interpreted (Figure 7.10). To date, their significance as indicators of higher-than-present RSL has been the source of some controversy since they require considerable glacio-isostatic depression, in excess of that which could be produced by early GIA model simulations. A partial answer to this paradox may be found if we are willing to consider that the ice sheet covering Ireland was thicker than has traditionally been envisaged. There is growing

-200 -100 -50 -30 -20 -10 0 10 20 30 50 100

Elevation (m)

Fig. 7.14 PALAEOGEOGRAPHIC MAPS OF THE CHANGING SEA AREAS AND COASTS AROUND IRELAND DURING THE LATEGLACIAL. Large-scale (First Order) palaeogeographic reconstructions for Ireland and its surrounding seas from around 20,000 to 14,000 years ago: a. 20,000; b. 18,000; c. 16,000; d. 14,000 years ago. The maps provide an initial lateglacial time series, showing the evolving land and sea areas, which resulted from the regional- to global-scale RSL changes that followed the melting of the ice. The ice modelling for the region has since been modified, but the current models remain similar to those used here, which were first produced around 2006. The map reconstructions for the changing ice dimensions and limits, together with land and sea elevations (bathymetry), show evidence of the areas that existed formerly across the Celtic Sea and Irish Sea as potential landbridges with Britain, and possibly more directly with continental Europe. However, any landbridge connections had gone by c.14,000–16,000 years ago. [Source: After Edwards, R.J. and Brooks, A. 2008. The Island of Ireland: Drowning the myth of an Irish landbridge? In: Davenport, J.J., Sleeman, D.P. and Woodman, P.C., eds. Mind the Gap: Postglacial colonisation of Ireland. Special Supplement to the *Irish Naturalists' Journal,* pp. 19–34]

IRELAND'S EVOLVING COASTLINE

Whilst the focus of this discussion of past sea level has been on tracing the vertical movements of RSL around Ireland's coasts, of equal importance have been the lateral shifts in the coastline that such oscillations cause. One way to examine these large-scale, regional patterns of palaeogeographic change is to combine topographic and bathymetric data with the simulations of RSL. The result is a series of maps charting the shifting balance between land and sea (Figure 7.14). In practice, this approach examines the impact of the raising or lowering sea level over a land surface of constant morphology. The maps are most accurate, therefore, in areas where the erosion and deposition of coastal and offshore sediments have been minimal.

The simulations indicate that around 20,000 years ago, during the last glacial period, south-eastern Ireland was joined to south-west Britain by a low-lying isthmus of land. Similarly, the island of Britain also remained joined to mainland Europe at this time. The consequent landbridges facilitated, for a brief window of opportunity, the migration of plants, animals and humans into these later islands. Britain and Ireland also continued to be linked by their shared ice sheet, which was likely fringed by large expanses of sea ice. Sometime between 20,000 and 16,000 years ago, as the ice covering Ireland melted, rising global sea levels outpaced the rebounding land surface, flooding the shelf and severing the connection with Britain. By the start of the Holocene (c.10,500 years ago), the model shows that Ireland was separated from Britain by water of c.50m or more in depth.

The bathymetry around the modern Irish coast means that RSL rise did not usually translate into the kinds of land loss experienced in other parts of the region (for example, the southern North Sea). Nevertheless, in some areas during the early Holocene, substantial changes in the configuration of the coast are indicated, which would have had profound impacts

evidence that this may have been the case (see Chapter 6: Glaciation and Ireland's Arctic Inheritance).[22]

on the landscape and its inhabitants. For example, the modelling indicates that around 8,000 years ago, the Aran Islands and Clare Island were joined to the Irish mainland. Further, present-day shallow coastal embayments, such as Clew Bay and Galway Bay, most likely contained considerable expanses of terrestrial and/or intertidal environments. These kinds of palaeogeographic reconstruction can be useful in the assessment of archaeological potential, long-term trajectories of coastal change and the potential occurrence of sedimentary contexts where RSL data may be collected (see Chapter 16: The Inhabitants of Ireland's Early Coastal Landscapes).[23]

THE FUTURE OF IRELAND'S ANCIENT SHORELINES

As the above snapshot of coastal palaeogeographies and the earlier brief discussion of Ireland's varying complex of past sea-level changes have demonstrated, most of Ireland's coasts today are eroding and retreating. This observation is true both for those regions on the island that have former raised shorelines preserved, as well as those coasts that are submerging under RSL. The processes of coastal functioning continue, of course, to create areas of sediment transfer and build-up, giving rise to apparent, but temporary (decadal scales) coastal land progradation on many coasts. Considering the past patterns of relative sea- and land-level changes, a now well-established picture for Ireland has emerged. In essence, northern coasts are still rising relatively, in response to the ending of the last Ice Age. The broad outcome of this is that many

sections of early former coasts are still visible and are preserved as raised shorelines above the present-day intertidal zone. To the south of the island, along a broad sout-west–north-east aligned hinge zone (axis of movement), from the Shannon Estuary to around Dublin Bay, the Earth's crustal surface (land level) is in a quasi-stable state. Submergence of these southern coastal regions has also been occurring, at slow and declining rates, for the last 4,000–5,000 years, under the continuance of postglacial sea-level rise.

Post-2000 and into the twenty-first century, the impact of long-term ice melt from the last glaciation is being added to by the outcomes of global warming and enhanced ice melt. These factors together are rapidly raising relative sea levels on Ireland's coasts and worldwide. Current regional mean rates of SLR are c.3–4mm/year, which represents a doubling of the rates of change seen over the last c.2,000 years. These levels of change are projected to rise to above 5–6mm/year by 2040. Along the coasts of the north of Ireland, the combined impacts of climate change and this accelerating sea-level rise mean that the coastline is mostly eroding, and so will effectively retreat despite continuing isostatic emergence, as the twenty-first century progresses. The implications and future significance of these relative sea-level changes for the coasts of Ireland will be returned to later in this *Atlas*, particularly in Chapter 33: Climate Change and Coastal Futures.

INTERGLACIAL AGE RAISED SHORELINE, DUNWORLEY BAY, COUNTY CORK. (Source: Robert Devoy)

MONITORING AND VISUALISING THE COAST

Darius Bartlett

CONNEMARA AND CLEW BAY FROM THE WEST, IMAGE TAKEN FROM THE INTERNATIONAL SPACE STATION ON 26 FEBRUARY 2018 [Source: NASA].

Fig. 8.1 CASPAR PLAUTIUS' ILLUSTRATION OF ST BRENDAN'S SHIP ON THE BACK OF A WHALE. With typical mordant wit, the Anglo-Irish writer and Dean of St Patrick's Cathedral in Dublin, Jonathan Swift (1667–1745), claimed that 'So geographers, in Afric maps, With savage pictures fill their gaps, And o'er unhabitable downs, Place elephants for want of towns.' [Swift, J., 1733. *On Poetry: A Rapsody*. Lines 179–182]. This alleged early practice of using illustrations and other elements to mask areas of *terra incognita* on maps was not confined to land-based cartography. In the example seen here, Spain, north-west Africa, the Canary Islands and the entrance to the Mediterranean are combined with imaginative and possibly allegorical elements to illustrate St Brendan the Navigator's travels in the North Atlantic. The book from which this illustration comes was published in honour of, and may perhaps have been actually written by, Caspar Plautius, Abbot of the Benedictine monastery of Seitenstetten (in present-day Austria) in 1621, as an account of the role of Benedictine missionaries in the exploration of the Americas. [Source: Plautius, C., 1621. *Nova typis transacta navigatio*. Courtesy of Houghton Library, Harvard University, Wikimedia Commons, Public Domain]

For our coastal-dwelling ancestors, the ocean was a potentially hazardous realm, full of unknown dangers, ranging from submerged rocks and terrifying storms to strange and hostile animals (Figure 8.1). Pirates, raiding parties and invading armies arrived by sea to terrorise shore-based populations. The oceans were also immeasurably vast and, once out of sight of land, navigation

Fig. 8.2 LITTLE SAMPHIRE LIGHTHOUSE. Little Samphire Lighthouse was built in 1854 and positioned outside Fenit near Tralee, County Kerry. This attractive representation by marine artist, Kenneth King (1939–2019), shows little alteration to the light from the time it was first erected. It also stresses how this minute lighthouse transforms the vista of this section of Tralee Bay. [Source: by kind permission of Kenneth King, Glencolmcille, County Donegal. See also, O'Kane Boal, M., 2013. *Kenneth King: Life and works*. Mount Charles, County Donegal: Bad Toy Books]

Fig. 8.3 POSTCARD OF THATCHED COTTAGE, COUNTY GALWAY. Particularly during the latter part of the twentieth century, no holiday was considered complete without the almost-obligatory picture postcards sent to friends and family at home. In Ireland from the 1950s onward, John Hinde's distinctive and brightly coloured depictions of rural and urban scenes rapidly became iconic. Often featuring enhanced colours and carefully staged composition, as well as stereotypes such as freckled children, donkeys and the traditional thatched cottage seen in the example here, Hinde's postcards played no small part in presenting a romanticised vision of Ireland to the wider world. [Source: © The John Hinde Archive/Mary Evans Picture Library]

wide variety of means, and this has long been an activity of importance to maritime states. This has been as much the case in Ireland as elsewhere.

Any graphic or pictorial representation of a country or a place is a visualisation; it enables information about that location to be captured, stored for posterity, and shared with an audience. Sometimes, the visualisation is created purely for artistic reasons, to communicate the creator's emotional or other subjective responses to a place (Figure 8.2). Today, many people like to take a photograph of the beach that they visit on holiday, as a souvenir, to show to their friends or, increasingly, to post on social media. These are also visualisations of the coast, as are the picture postcards that were, and often still are, sold in seaside resorts and other tourist locations (Figure 8.3).

However, visualisations can have the more objective, normative purpose of recording, as faithfully as possible, what is present. In this context, throughout much of recorded history, maps and charts have been particularly important and effective tools for visualising and communicating the geography of Irish coastal landscapes and places. Often combining science and artistry, it is no surprise that many of these historical maps are prized today for their aesthetic and decorative qualities, as much as for the story they tell about how our ancestors understood and shared information about the world they lived in.

presented many challenges. And yet, the oceans were also the source of fish and other important resources; they were often the safest means of getting from one place to another (for example, a sea passage from Cork city to Dublin was often preferred over the more hazardous and less comfortable journey by land); and they offered access to far-off lands for exploration, trade and, for some maritime nations, conquest. For these and many other reasons, human beings have been observing, recording, communicating and visualising the geography of the world's coasts and oceans for millennia, through a

Case Study: Ptolemy's Inventory of Ireland: Geographical features and places

Mick Monk

Claudius Ptolemaeus (Ptolemy) (*c.*90–168AD) was a Graeco-Roman astronomer and mathematician, who worked in the celebrated library of Alexandria in Egypt during the second century AD. He wrote many treatises on a range of subjects, from astrology to music, astronomy, philosophy and mathematics. Of these, his 'Handbook of Geography' (the *Geographica*), is perhaps the best known. This was written around 150AD and, in its time and well beyond, it was considered the most thorough, accurate and authoritative geographical text of the known world. It continued to form the basis for map-making in Europe well into medieval times. Unfortunately, Ptolemy's original version of the *Geographica* is long lost and no copy earlier than the thirteenth century AD is known to exist; there are, however, many manuscript copies that post-date that time and there is relative consistency between them.

The *Geographica* consisted of a list of peoples, places, settlements, rivers and coastal promontories of Europe and parts of Asia and North Africa. It includes Ireland and other places that were outside the limits of the Roman empire. The latitude and longitude of named places were also provided, based on Ptolemy's own calculations. Ptolemy was primarily a mathematician and geographer, so he relied on others for the detail of the names of the peoples, places, etc. and their relative position. While it is likely that he obtained some of his information from his contemporaries, most authorities agree that much of the information is at least two or three stages removed from the actual travellers who visited and experienced the places listed. These earlier sources included, in particular, Marinus of Tyre, who wrote in the late first and early second centuries AD and who, in turn, derived his information from the much earlier first-century AD geographer, Philemon, about whom little is known.[1] In many cases, these travellers would have recorded the information in their own language (whether it be another form of Celtic, Gaulish, Greek or Early Latin) and passed it on to others, perhaps some years after their trips. This information was then committed to Greek by Ptolemy.

From the early seventeenth century onwards, numerous scholars have attempted to locate the places listed in Ptolemy's inventory and analyse the etymology of the names, and the language roots from which they came. More recently, and especially since the beginning of the twentieth century, archaeological information has also helped identify some of the places listed, while more exacting methods have been developed and applied to reappraise the earlier historical and linguistic evidence. Since 2000, there has been renewed interest in Ptolemy's *Geographica* from geographers, and in two cases this has included the application of Geographical Information Systems (GIS) software and Google Earth technology to a digital reconstruction of Ptolemy's information for Ireland.[2]

Today, most researchers would agree that Ptolemy's inventory was quite accurate for its time and provides a valuable window on Ireland and other regions (Figure 8.4). As such, it provides historians, geographers and archaeologists with an important baseline of knowledge on which to build more contemporary work.[3]

PTOLEMY'S IRELAND

By comparison with other sections of the *Geographica*, particularly those for the Mediterranean region, fewer places, peoples and geographical features are listed for Ireland. The list is organised by the north, east, south and west coasts. The sections for each coast have separate lists

Fig. 8.4 PTOLEMY'S MAP OF ANGLIA AND HIBERNIA (IRLANDA). A late fifteenth-century AD map of Ireland and Britain based on Ptolemy's inventory of Places. Names are in Latin. Claudius Ptolemy lived and worked in the Graeco-Egyptian city of Alexandria during the second century AD, from where he wrote on numerous scientific subjects, including astronomy and geography. He was particularly notable for his *Geographica*, an encyclopaedic listing that gave the latitude and longitude of known places around the world, extending from the Atlantic coasts of Europe as far east as present-day Malaysia, and from the Equator northward to approximately the Arctic Circle. Ptolemy also explored ways of projecting the curved surface of the world onto flat sheets of paper and, as seen in this illustration, used a system of geographical co-ordinates that more easily allowed scholars and explorers who came after him to correct and add to his initial recordings. Sadly, no surviving examples of Ptolemy's original maps still exist and the earliest known manuscripts based on his work date from the late twelfth or early thirteenth century. However, his lists became especially significant during the fifteenth century for three primary reasons: the translation of the *Geographica* into Latin from its original Greek; a rapid expansion of European voyages of discovery, initially along and round the African coast and subsequently to the New World and beyond; and, arising largely from these, a growth in the philosophical interest in theoretical geography. The fifteenth century also saw the emergence of early printed editions of maps based on Ptolemy's data, which facilitated greatly their proliferation and diffusion around Europe. [Source: Royal Irish Academy]

for promontories, river mouths, islands, names of peoples and settlements, presented in the same order. Unsurprisingly, there is a greater abundance of place names recorded for the east coast, since this is the coast that informants would have visited most frequently. While the list incorporates the longitude and latitude of settlements, which Ptolemy terms *polis* (towns), these were likely to be merely areas of denser population or of regular assemblies, and not towns in the sense we understand them today. In total, there are fifteen named and positioned coastal river mouths, five promontories, ten or eleven towns and nine islands. In addition, Ptolemy indicates the relative position of sixteen tribes or peoples, and names the four oceans that surround Ireland. Later scholarship, using positioning and drawing from etymology, has attempted to suggest the identity of many of these places and the association of some of the peoples listed, but there is much disagreement amongst scholars regarding these interpretations. Overall, except where access is by a navigable river, it seems that the places in Ptolemy's inventory that are furthest from the coasts have the least reliable associations and, hence, identification.

THE SIGNIFICANCE OF PTOLEMY'S INVENTORY/'MAP'

Despite the many debatable interpretations of the names and locations of places and peoples recorded in the *Geographica*, and the fact that his original manuscript has not survived, Ptolemy's inventory remains a very valuable source of information for a time when there are few extant references to Ireland. It provides the earliest written view of Ireland's salient features and occupants and has been described as a 'window' on the later Irish Iron Age. It marks the very beginnings of Ireland's emergence into history, even if it is seen through the eyes of outsiders, whom Ptolemy never met and who transliterated the names of peoples and places into their own languages. The inventory/'map' provides a starting point from which an attempt can be made to begin to understand the social and political dynamics of human groupings and their activities during the transitional period from the later Iron Age into the beginnings of the early medieval period. With focused studies of the archaeological evidence that has emerged over the last few years, it may be possible to add more substance to our knowledge of the places and peoples listed in this very valuable source.

Today, visualisations of the coast can be performed for a variety of reasons, including:

• for artistic and creative purposes, to serve as expressions of the creator's subjective impressions or interpretations of the coastal landscape (see Chapter 14: Imagining Coasts);
• to convey information about a coastal location to a wider audience, for example in journalism, to provide context for a news item;
• to make it easier to understand the location and distribution of objects or phenomena that lie under water and hence cannot normally be easily seen from shore, such as shipwrecks or coral reefs (see Chapter 9: Underwater Surveys: The INFOMAR project);
• to support pure and applied scientific research, including in the fields of oceanography, meteorology, marine and coastal engineering, marine biology, and many other disciplines;
• in support of decision making, for example, in coastal zone management and marine spatial planning, for safe navigation of ships at sea, or to anticipate and respond to hazards such as oil spills, erosion, coastal flooding and storm impact.

Visualisations can also help us see the 'bigger picture' of marine and coastal phenomena that operate at regional, national or even global scales, in particular those that have also a significant dynamic component; for example, by recording the state of weather systems, ocean waves, water temperature and other variables at a number of locations, a much more complete understanding can be obtained of how these parameters operate across a wider area (Figure 8.5). Similarly, a single photograph or picture, taken at a given location, gives a snapshot of how that place looks at any one time. However, if you continue returning to the same location and take a photograph of the same feature on a repeated basis, then the resulting images can enable the viewer to see and understand how the object evolves through time; this is the basis of monitoring coastal phenomena and is usually related to solving scientific or management challenges, such as trying to understand the reasons for and patterns of coastal erosion at a given location (which is science), or to assist in decision-making regarding design and construction of the potential coastal protection measures that might respond appropriately to this (which is part of coastal management).

DIGITAL TECHNOLOGIES AT THE COAST

For most of the past 2,000 years or more, paintings, drawings and maps were the only means available to visualise the coast. However, from the 1950s onwards, and particularly into the twenty-first century, these traditional, analogue tools and methods have been supplemented, and in many cases replaced, by a growing array of visualisation and monitoring techniques based on digital computer technologies.

At least 70 per cent of all information collected and used in the public and private sectors has a geographic component and this is especially the case regarding data relevant to marine and coastal areas. This geography – based on spatial relationships and concepts such as position, direction, distance, proximity, adjacency and trajectory – is essential in most commercial, managerial and scientific operations at the coast, including planning and decision-making, navigation, resource exploration and extraction, protection of environments and ecosystems, maintenance of security and much more.

By the 1970s and 1980s, computer software that harnessed these geographical concepts had entered widespread use in many areas of coastal science and management – particularly in the fields of biological, chemical and physical oceanography, geoscience, navigation and charting – and was starting to be used in Ireland. Pioneering adopters in Ireland during the early 1980s included geographers at what was then the New University of Ulster in

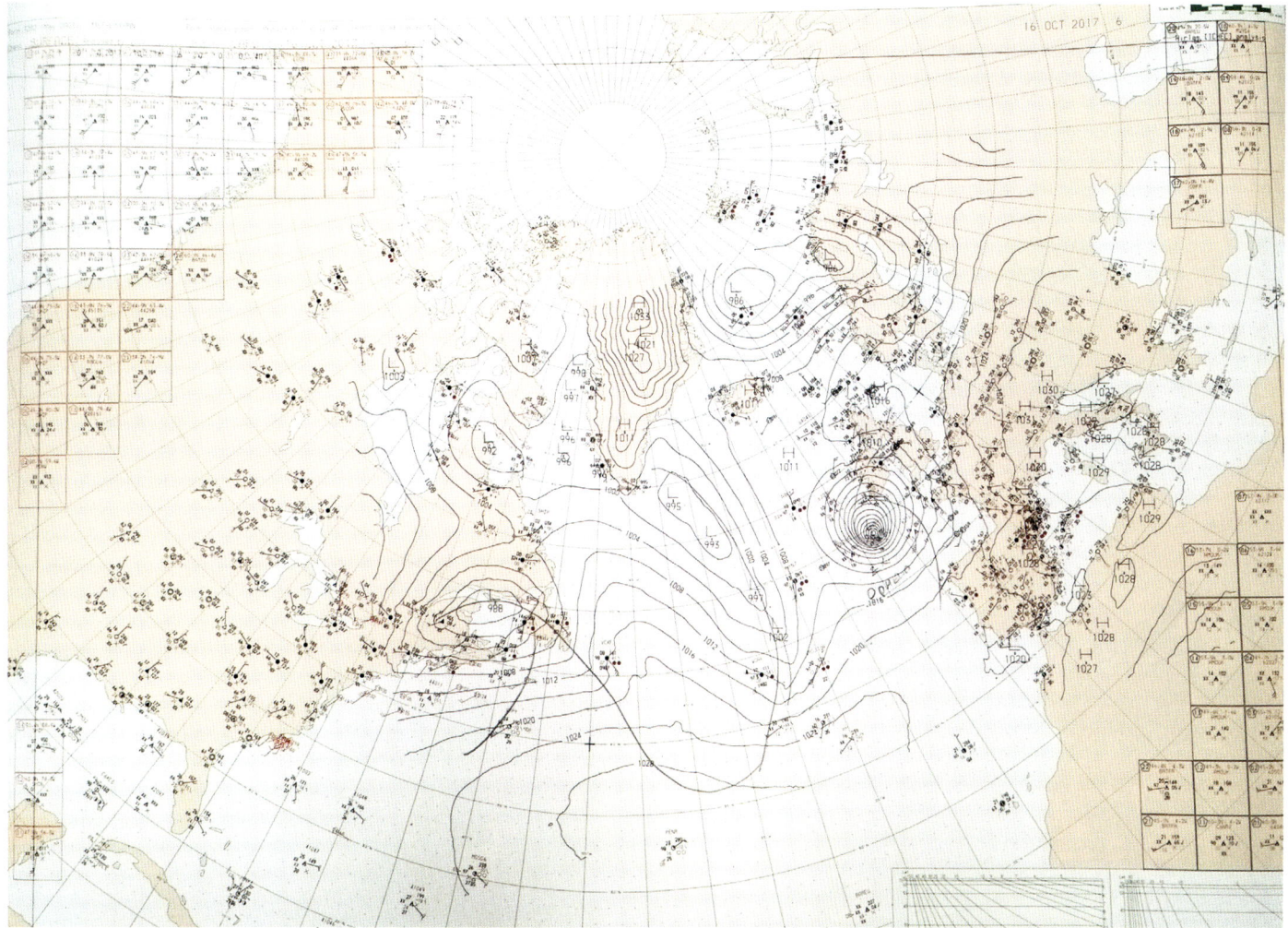

Fig. 8.5 SYNOPTIC WEATHER CHART OF THE NORTH ATLANTIC. For at least two millennia, human society has been observing and recording atmospheric and meteorological conditions, mostly to help monitor seasonal and other changes in the weather. While many of these observations were remarkably precise, it was only in the European Renaissance of the fifteenth century onwards that scientific instruments started to add precision to measurements of temperature, precipitation and other weather phenomena. During the seventeenth and, especially, the middle of the eighteenth century, increasing numbers of individuals were recording the workings of the atmosphere at scattered locations around the world. With the advent of the electric telegraph in the nineteenth century, and hence the means of rapidly transmitting and centralising data collected at these far-flung locations, it became possible to study the workings of weather systems at regional and wider scales. This catalysed the development of the first meteorological charts and, by the 1860s, the birth of the science of synoptic weather forecasting, based on the integration and analysis of large numbers of observations taken simultaneously over a wide area. Today, meteorological charts such as the one shown here are routinely used in weather forecasting. This particular example, compiled by scientists at Met Éireann in Dublin, shows the meteorological situation at 6p.m. on 16 October 2017, with the intense low-pressure system of Hurricane Ophelia lying out in the Atlantic to the immediate west of Ireland and about to arrive onshore. [Source: Met Éireann]

Coleraine, who used a computer program developed by academics at Florida State University in the USA to investigate the behaviour of sediment cells on the coast of County Wexford, and at Queen's University, Belfast, who applied the same program to study coastal longshore wave power and variations in the distribution of sediment along the Ards coast of County Down.

Initially, the use of tools such as these was limited by the size and expense of computers, by the limited storage, processing and graphic output capabilities of the machines available at the time and by the challenges posed in sharing data between different stand-alone programmes (Figure 8.6). By the 1980s, collections of integrated programmes for handling spatial data were being brought together, packaged and commercialised in the form of off-the-shelf, general-purpose Geographical Information System (GIS) toolboxes. Although they were aimed initially at terrestrial applications and their users, the potential of these new software tools to address coastal and marine information-handling needs

soon became recognised and exploited.[4]

Strategic recruitment at the end of the 1980s allowed the Department of Geography at University College Cork (UCC) to position itself as one of the first university departments in Ireland to put GIS on the teaching curriculum. The department had already established a national and international reputation in the field of coastal zone science and, by now adding GIS expertise to its portfolio, Cork soon emerged as a pioneering centre in Europe and beyond, in the application of GIS to coastal zone matters. This led to the establishment at UCC of a specialist research unit, the Coastal Resources Centre, which expanded gradually to become the Coastal and Marine Research Centre (CMRC) and was eventually subsumed within the current MaREI research facility at Ringaskiddy on the shores of Cork Harbour. From the outset, GIS and other geoinformation technologies featured strongly in the many applied research projects undertaken successfully by this group.

Also of significance was the launch at UCC, in 2004, of a

taught Masters degree programme in Coastal Zone Management and GIS, one of the first to be developed and offered anywhere in the world. The course ran for several years and, over its lifetime, was instrumental in training a new and influential generation of coastal zone managers, who were also proficient in the use of computer technologies to support their work (see Chapter 30: Engineering for Vulnerable Coastlines).

Today, spatial information technologies are used extensively in Ireland and worldwide to support coastal and marine operations across a wide range of public and private sectors. They involve increasingly the convergence and integration of a wide range of components and once-separate technologies, including GIS, remote sensing, digital mapping and charting, global satellite positioning systems, spatial database systems, laser technologies, mobile telephony, sensors and autonomous data-collecting devices, and others. Collectively, the term 'geoinformatics' refers to these tools, their applications and the scientific principles on which they are based.[5]

GEOGRAPHICAL INFORMATION SYSTEMS

Darius Bartlett

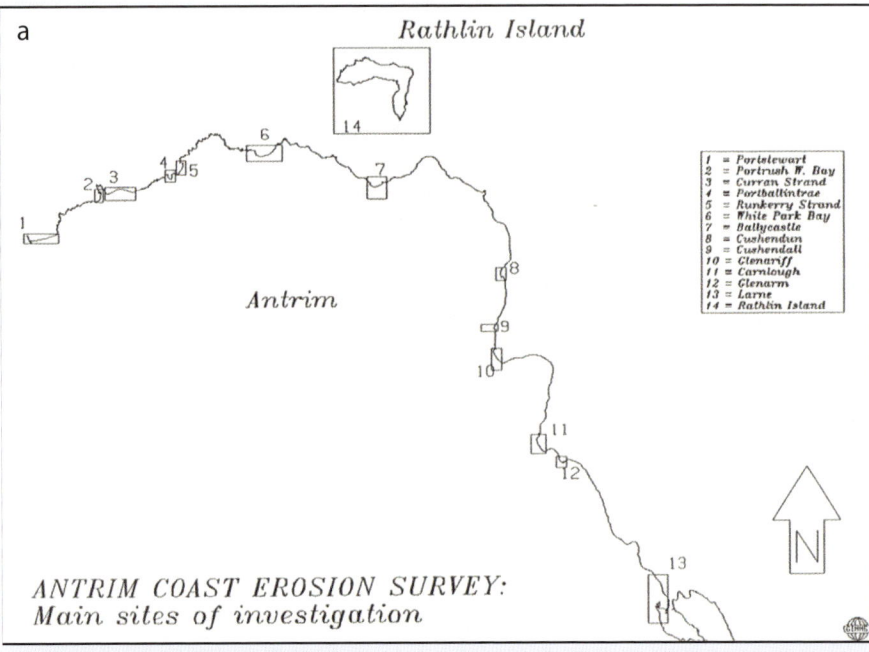

ANTRIM COAST EROSION SURVEY:
Main sites of investigation

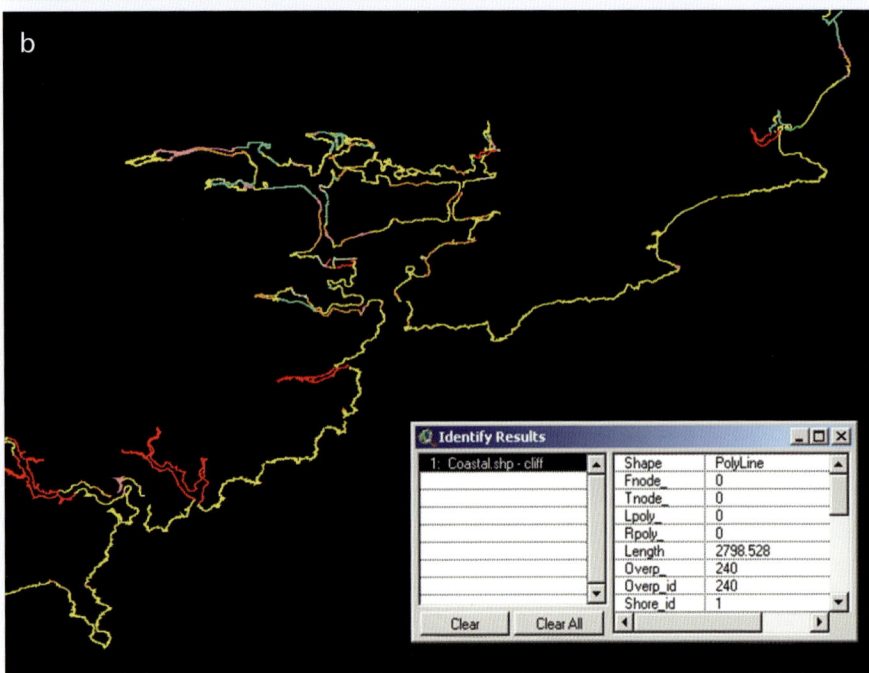

Geographical Information Systems (GIS) are integrated computer software packages that combine computer databases for storing and

Fig. 8.6 a. MAPPING AND ANALYSING EROSION ON THE ANTRIM COAST WITH DIGITAL TECHNOLOGIES IN THE MID-1980S. This pioneering study, undertaken at the University of Ulster at Coleraine in the mid-1980s, explored the use of computer mapping technology to support research into the distribution and causes of coastal erosion on the Antrim coast between the mouth of the River Bann and Larne in the east. In the absence of integrated software for the purpose, the maps required for the study were created by retrieving coastal data stored locally on a desktop PC, which were then transferred to the university's central mainframe computer, where they were reformatted and incorporated into a one-off set of instructions for a stand-alone computer mapping programme. The whole process usually took one or two days to complete and, if the resulting map had any errors, or turned out not to be what was expected, the whole sequence of steps had to be repeated once more. Nonetheless, despite the effort and considerable potential delays in the workflow as well as, by today's standards, the very primitive nature of the maps produced, it was still seen to be a much quicker and more effective way of working than the previous manual process. As one of the earliest projects of this kind to be undertaken anywhere in the world, it also laid important conceptual and methodological foundations to be followed and built on subsequently. [Source: Darius Bartlett]; b. Dynamic segmentation of the east Cork coastline. In the 1980s and early 1990s, one of the major challenges for modelling coastal phenomena in emerging GIS technologies was the difficulty of linking thematic data (properties and attributes such as geology, sediment type and presence or absence of coastal defence structures) to the stored co-ordinates that represent the geometry of the shoreline. In a European Union-funded project to assess the vulnerability of the Irish south coast to sea-level rise, researchers at University College Cork adapted a technique known as dynamic segmentation, developed originally in the US and Canada for modelling traffic flows in urban settings, to divide the line of the shore according to its various attributes. This method enabled for the first time multiple thematic properties of the coast – in the example shown here, the presence and height of cliffs, shown in different colours – to be captured, stored, integrated and analysed within the computer system, all linked to a single set of co-ordinates that described the geometry of the shore. [Source: Darius Bartlett]

Fig. 8.7 ASSESSMENT OF COASTAL EROSION RISK BY MEANS OF A MODERN WEB-ENABLED GIS. This GIS application was created at MaREI, the SFI Research Centre for Energy, Climate and Marine at University College Cork, for a coastal erosion policy and practice study conducted in 2017 on behalf of Ireland's coastal local authorities. It contains information about the physical characteristics of the coast and sites known to be at risk of coastal erosion. Implemented as a fully functional Web GIS, created on ArcGIS software from the US-based company ESRI, this platform functions in much the same way as a desktop GIS would, but allows the data storage, retrieval and analysis functions to be shared and accessed by multiple users via the internet. Some of its key features are highlighted here. The left panel (a.) lists the contents of the GIS, or the geographical layers contained in the database. In this case, this includes various different vector layers. Vector layers represent geographical features as discrete points, lines or polygons (shapes). In this case, the features in the roads layer (b.) are represented as line features. Layers can be turned off or on in the map view c. by ticking or unticking boxes in the table of contents. In this way, the two panels are dynamically linked. A GIS contains not only information about the location of features in space, but also additional information about those features. This is known as thematic or attribute information, and can be found in the table view panel (d.) Each row (e.) of data in the attribute table is linked to a specific geographical feature in the map view. The columns contain different types of attribute information about each individual feature, for example, (f.) the Road ID (its unique identifier). There are a number of ways to search, filter and select features, based either on their attribute information or geographical location. This is one of the many facets that makes a GIS so powerful.

retrieving information, tools for creating and displaying both static and dynamic maps, and a range of statistical, modelling and analytical tools that allow the geographical relationships between phenomena to be studied.

Particularly since the 1980s, GIS has become an essential tool for managing, visualising and analysing spatial data used by a wide range of agencies and commercial enterprises concerned with coastal and marine matters in Ireland. The benefits of applying GIS to these tasks include the ability to work with, integrate and analyse larger, richer and more comprehensive databases; improved sharing of information between communities, sectors, professions and stakeholders; greater control over the consistency and quality of input data and information products derived from these data; the ability to model and simulate coastal processes; non-invasive scenario testing and validation of alternative management options; enhanced, evidence-supported decision-making, not least when a rapid response is required in the face of real or potential disaster; and improved visualisation and communication of concepts, information and ideas to a greater variety of audiences.

Originally, these systems were installed on stand-alone computers and required intensive training in their use, but in recent years they have become much more accessible and user-friendly, while cloud computing and the internet has led to the emergence of a whole new generation of mobile and web-enabled GIS technologies. These latter systems make it much easier for researchers to collect and work with data in real time – often while still in the field or at sea – and they also provide enhanced tools for online publishing of map-based and other information in the form of digital web atlases, so facilitating much greater public participation in the decision-making processes.

Typical marine and coastal applications of GIS include natural and human resource management, route planning and navigation, optimising the location of commercial activities, environmental protection, marine spatial planning, heritage management, fisheries and related activities, exploration and exploitation of marine resources, emergency preparation and response, policy-making and decision support, defence, law enforcement and pure and applied scientific research. For example, the state agency with responsibility for developing Ireland's sea fishing and aquaculture industries, Bord Iascaigh Mhara, uses GIS to identify and evaluate potential locations for fish farming, including assessment of the environmental impacts of this activity, as well as to monitor the location and movement of fish stocks; the Underwater Archaeology unit of the National Monuments and Wildlife Service (NMWS) at the Department of Culture, Heritage and the Gaeltacht uses GIS to maintain a dynamic map and database of shipwrecks in Irish coastal waters (viewable online at https://www.archaeology.ie/underwater-archaeology/wreck-viewer); the Irish Naval Service uses GIS to support its fisheries protection and related duties in Irish waters; scientists and engineers at MaREI use GIS to help plan the development of marine renewable energy resources; and most local authorities across the island of Ireland now use GIS as an integral part of their information infrastructure, for a diversity of tasks, including those relating to coastal planning and management (Figure 8.7). The maps created for this *Atlas* were compiled using GIS software and methods.

Case Study: The Impact of Coastal Web Atlas Development

Kathrin Kopke, Sophie Power, Adam Leadbetter and Eoin O'Grady

The advent of Geographical Information Systems (GIS) has transformed the way we measure, research and visualise coastal and marine information. Development of web-based technologies, such as Google Maps, as well as those based on more powerful analytical technologies, such as the industry-leading ArcGIS Online, has improved adoption and use of GIS around the world. Collectively known as web-GIS, these technologies combine the advantages presented by GIS with those of the internet, for use by spatial information specialists and non-specialists alike. This technology also provides much-improved access to coastal and marine data and information, and has been of particular value in the development of Integrated Coastal Zone Management projects. It supports decision-makers on issues such as national sovereignty, resource management and maritime safety and hazards, as well as raising awareness of coastal resources and environments.[6]

While the quantity and quality of spatial data relating to coastal and marine matters has increased enormously during and subsequent to the latter part of the twentieth century, for a long time these data holdings were dispersed across multiple organisations, were stored in many different locations and lacked standardisation. These, and other obstacles, made it very difficult at times for potential users of these data to find out what was available, who to contact to obtain access to the data and what would be required to integrate the data into existing or new workflows and analyses.

Coastal Web Atlases (CWAs) seek to resolve these challenges, in the form of a particular application of web-GIS. A coastal web atlas may be defined as a 'collection of maps and datasets with supplementary tables, illustrations and information that systematically illustrate the coast, oftentimes with cartographic and decision support tools, and all of which are accessible via the Internet'.[7]

An early example of a CWA was the Marine Irish Digital Atlas (MIDA). Developed in Cork and launched in 2006, this pioneering initiative further strengthened Ireland's position as a world leader in the application of GIS and web mapping technology to coastal zones. International collaboration between the MIDA team at University College Cork and the Oregon Coastal Atlas team at Oregon State University (USA) led to the founding in 2008 of the International Coastal Atlas Network (ICAN), which seeks to promote technical solutions to coastal web atlas development, overcome barriers to interoperability, encourage the exchange and sharing of coastal or marine datasets and improve the ways in which end-users of coastal atlases can search for and interact with data (Figure 8.8).

ICAN also encourages the development of common standards to be applied across different coastal atlas products. An ICAN CWA contains typically a map area, legend and tools that allow the display and query of coastal and marine data. Crucially, it also provides access to comprehensive metadata (or cataloguing information) about the geospatial datasets, including information about the owner of the dataset, the publishing date and how the data were created. Multimedia elements and other non-spatial features are used within CWAs to help the user understand and contextualise this coastal and marine geospatial information. ICAN has identified four broad categories of CWA user, namely the scientific community, policy- and decision-makers, the general public and the education sector, each of which have specific requirements but also some common end-user expectations.

The potential role of CWAs was recognised by the EU as early as 2006, in a Green Paper on Future Maritime Policy. In 2008, the European Commision published its vision for an Integrated Maritime

Fig. 8.8 ICAN STEERING GROUP, AND KEYNOTE SPEAKER FROM INTERGOV-ERNMENTAL OCEANOGRAPHIC COMMISSION UNESCO, FRANCESCA SAN-TORO, AT THE ICAN 8 WORKSHOP: COASTAL WEB ATLASES – ENHANCING OCEAN LITERACY, WHICH TOOK PLACE IN SANTA MARTA, COLUMBIA, IN SEPTEMBER 2017. The International Coastal Atlas Network (ICAN) was established in 2008, and grew out of earlier research collaboration between scientists at University College Cork and at Oregon State University in the USA. In 2013 ICAN was adopted as an official project of the International Oceanographic Data and Information Exchange (IODE) Programme of UNESCO. The long-term strategic goal of the project is to encourage and help facilitate the development of digital atlases of the global coast, based on distributed, high-quality data and information. These atlases can be local, regional, national and international in scale. In little more than a decade, ICAN has hosted training events and workshops on three continents, and the network has grown to over seventy member organisations in fifteen countries. A typical coastal atlas is an interactive tool, built on Web GIS technologies, that allows the user to explore, visualise and query geospatial data directly via a standard web browser. In addition, many coastal atlases are designed to let the user discover, view and download coastal data for use in their own applications and analyses. [Source: INVEMAR – The Marine and Coastal Research Institute of Colombia]

Policy (IMP), which included initiatives like the European Atlas of the Seas, seabed habitat mapping and promotion of maritime spatial planning, all of which underpinned recognition of the role of CWAs as supporting tools within an EU legal and policy context. The significance of CWAs was also recognised by UNESCO, when ICAN became part of the International Oceanographic Data and Information Exchange (IODE) Programme in 2013.

Ireland continues to play a pivotal role within ICAN. MaREI-UCC co-chairs the network, together with the Oregon Coastal Management Program (USA) and two Irish CWAs – the Marine Irish Digital Atlas (MIDA) hosted in UCC and Ireland's Marine Atlas – are represented within the network.[8] This latter atlas (http://atlas.marine.ie) was developed by the Marine Institute as a decision support tool and provides a spatial data platform that allows viewing and download of marine environmental data in support of reporting requirements under the Marine Strategy Framework Directive (MSFD) (Directive 2008/56/EU). Since its foundation, ICAN has hosted member workshops on three continents and the network has grown to over seventy member organisations in fifteen countries. The wide-ranging and varying expertise within ICAN allows for sharing of experiences and lessons learned, which allows developers of new atlases to avoid pitfalls that their predecessors may have encountered and to learn from successes in order to enhance a CWA. Experts within the network also came together to produce a peer-reviewed book on web atlas design and implementation.[9] Training is a key activity of ICAN and specific events have taken place at a number of international venues to provide fundamental knowledge and guidance on atlas tools and their usage, for example in Latin America and in Africa. In addition, ICAN has produced a number of 'best-practice' guides concerning CWA formats, software, design and user-interactions, as well as facilitating researcher placements.

CWA user interactions have been a key component of the ICAN project and members, fostering a culture of information sharing. This has allowed the wealth of experience gained by network members to be collated and documented within ICAN publications, in order to allow both new and established CWA developers and hosts to benefit from best-practice examples.[10] ICAN started off as a series of informal meetings of organisations that shared a common interest in CWA development and has since grown into an international network that successfully facilitates sharing of software, know-how, experiences and technical expertise. The interest in ICAN activities and the diversity of CWAs represented within the network has not only allowed ICAN to established itself as a global reference in CWA development, but also demonstrates the value and positive impact that CWAs can have at local, regional, national or continental level and their ability to support specific or multiple purposes as well as catering for general interests.

Case Study: Deep Maps: West Cork Coastal Cultures

Claire Connolly, Rachel Murphy, Breda Moriarty, Orla-Peach Power, Michael Waldron and Rob McAllen

The cultural heritage of our coastlines is rich but elusive, a record of human encounters with land and sea that is deeply felt and experienced, yet often described as intangible. Inherently intricate, fluid and changeable, coasts are often understood in disjointed ways and addressed differently by researchers in the sciences and the humanities.

'Deep Maps: West Cork Coastal Cultures' was an interdisciplinary project, funded by the Irish Research Council, that ran from 2015 to 2019. It aimed to bring to life the rich physical, cultural and symbolic heritage of the south-west Cork coastline by mapping and documenting the biological, cultural and historical contexts of the area from 1700 to 1920, using modern geoinformatics methods and technologies. Additional objectives of the project were to promote a more complete public understanding of current environmental priorities and to communicate a sense of shared identity and ownership of the maritime heritage that it documented.

The project focused on the stretch of coast from Clonakilty to Bantry Bay, as it is shaped by sea and land and as it was imagined in poems and drawings from the eighteenth and nineteenth centuries. This location was selected because there exists an important record of cultural engagements with Ireland's south-west coast from the early eighteenth century onwards, in the form of responses in different cultural media by individuals drawn to the unique cultural and biological aspects of a unique maritime environment. The period chosen began with antiquarian enquiries and poetic responses and ends with the start of scientific field research in this area, when Professor Louis Renouf of University College Cork began visiting Lough Hyne, near Skibbereen, from the 1920s onward (Figure 8.9).

The project was both extensive and intensive; it selected a biologically meaningful sweep of coastline that makes sense as a cultural unit and then developed digital representations of the cultural and historical layers of meaning that shape these encounters with sea and land. The concept of 'place' connects the three main parts of the project: scientific knowledge, historical representation and community perception of dangers to the marine environment.

157

Fig. 8.9 LOUGH HYNE, COUNTY CORK, LOOKING INLAND FROM BARLOGE CREEK. The lough is a very sheltered, fully marine basin, connected to the open sea by the narrow inlet of Barloge Creek, seen in the bottom right of the photograph, with its famous 'reversing waterfall' at the point of outflow from the lake. Scientific research at Lough Hyne has been undertaken for well over a century, and UCC maintains currently three research and teaching laboratories at the site (two of which are just visible in the photograph, one on either side of the creek). In 1981, Lough Hyne was declared Europe's first statutory Marine Nature Reserve. [Source: Geraldine Hennigan and Norman Kean]

DEEP MAPPING

Physical and digital maps have become ubiquitous in contemporary life. In an age of GPS and location-enabled devices, we are perhaps more conscious than ever of our own geographic position and more reliant on new technologies to orientate ourselves within a globalised world. While containing essential information, such maps often remain basic, flat, or passive and fail to communicate a deeper knowledge of place.

Fig. 8.10 SKETCH SHOWING THE WEST CORK COAST WITH ALLIHIES MINES. The West Cork Coastal Cultures project brings together a number and variety of historical images, topographic sketches and other documents, organised into a spatially referenced database for viewing and interrogating via an interactive web-enabled map interface. [Source: BL/EP/B/3318, Bantry Estate Collection, University College Cork Library Archives Service]

The concept of deep mapping seeks to address these limitations. A deep map departs from the strict objectivity of scientific cartography and seeks also to incorporate the subjective and lived experiences of place, by harnessing the power of multimedia to depict a place in ways that go beyond the strictly tangible. Susan Naramore Maher describes the deep map as representing 'the multiple histories of place, the cross-sectional stories of natural and human history as traced through eons and generations'.[11] This innovative mode of mapping demands a form of 'layered storytelling' to express the complex multiple narratives of space and time effectively, and lends itself very well to implementation through the capacities of digital environments in general and geoinformatics specifically.

Fig. 8.11 Net-making in Baltimore School, County Cork. The concept and technique of deep mapping seeks to combine subjective and often nuanced information, drawn from social geographies and the humanities, with observations drawn from the natural and physical sciences. It also seeks to link the present with the past, through the incorporation of both historical and contemporary data and information. The aim of the approach is to enable an holistic, multi-dimensional and cross-disciplinary understanding of a place and its transformations over time. The image shown here is one of many that were included in the deep map produced for the West Cork Coastal Cultures project. It illustrates an aspect of the local economy of Roaringwater Bay that has long been relegated to history, but which remains an important element of the cultural heritage of the area. [Source: National Library of Ireland, call no.: L_ROY_02595]

Methods

The West Cork Coastal Cultures project integrated scientific, stakeholder and cultural data and knowledge about coastal environments. The research methods included a review of relevant literature, including scientific publications and reports on key threats to the marine environment; questionnaires and workshops to obtain more detailed views and perspectives from targeted groups of stakeholders in the areas of tourism, heritage, science, the fishing industry, the general public and policy-makers; and in-depth interviews with selected individuals from the local communities. Along with the interviews, school visits were also conducted, to engage primary school children, determine what they value about the coast and what they view as the problems in their locality. The information gathered via these consultation exercises was used to establish stakeholder priorities, which in turn helped to shape the project visualisations.

Archival research was also undertaken, to gain a historical perspective on the west Cork coastline. Material consulted included estate papers, eighteenth- and nineteenth-century Irish poetry, historic newspapers, topographical drawings and photographs (Figures 8.10 and 8.11).

The scientific, stakeholder and cultural data were organised in a digital database and visualised using a range of geoinformatics techniques. These visualisations combine traditional, chronological approaches to landscape studies (including analysis of geology, climate and settlement patterns) with a consideration of the landscape as symbolic and socially formed (as evidenced through oral accounts, literature, poetry and art from a wide range of sources).

Visualisations

A fully searchable, open-access project website was created (www.deepmapscork.ie). This incorporates research on environmental priorities,

Legend
Art
Biology
Environment
Folklore
History
Literature

weed'. It was first noted in Ireland in 1995 at Strangford Lough. Since then it has been observed near Lough Hyne.
Wireweed may be brought in by aquaculture (oyster-growers acquiring half-grown oysters from other geographical locations). It can also be introduced and spread accidentally via boats and other marine vessels.

Zoom to

Lough Hyne: A Deep Map

Fig. 8.12 EXTRACT FROM MULTI-LAYERED DIGITAL MAP OF LOUGH HYNE, COUNTY CORK. The database developed for the West Cork Coastal Cultures project has been implemented through digital geoinformatics technologies. The web-enabled GIS interface shown here lets the user scan and navigate across the project study area interactively, turning on or off different layers of information, and selecting specific features, images, video clips and sound files as they go. This multimedia approach allows a much richer and more immersive understanding of a place to be communicated to the visitor than can be achieved by more conventional mapping.

exhibitions linking past and present issues, maps and timelines and project resources. The maps and timelines are original, research-based digital representations of aspects of the coastal area and reflect the diverse and layered approach of deep mapping (Figure 8.12). The technologies used to create and publish the visualisations were selected with longevity, cross-platform integration and ease of dissemination in mind. These included a project website, a GIS, open-access tools and software and a number of social media outlets and platforms including Facebook, Instagram and Twitter.

LEGACY AND CONCLUSIONS

One of the risks associated with any digital project is the rapidly changing nature of the technologies. With longevity in mind, the team worked with library scientists at UCC and the Digital Repository of Ireland (DRI) to compile a records management strategy for the project, including the website and visualisations. Resources developed for the project are being incorporated into the DRI in preservation-friendly formats, so that the results of the research will be easily accessible to others in the future. Along with these resources, the project's underlying interdisciplinary methodology can also be shared, applied and scaled to other coastal environments.

DIGITAL MAPPING AND CHARTING

Darius Bartlett

Instead of cabinets full of paper charts that are easily damaged, become rapidly out of date, require lots of space for storage and are particularly difficult to read in poor visibility (Figure 8.13), the safe operation of most ships today is based on the use of a range of digital spatial data products – built largely on GIS foundations – that meet the specialised needs of mariners. Electronic Charting Systems (ECS), the marine equivalent of the satellite navigation systems that are now commonplace in cars and other terrestrial transport systems, are

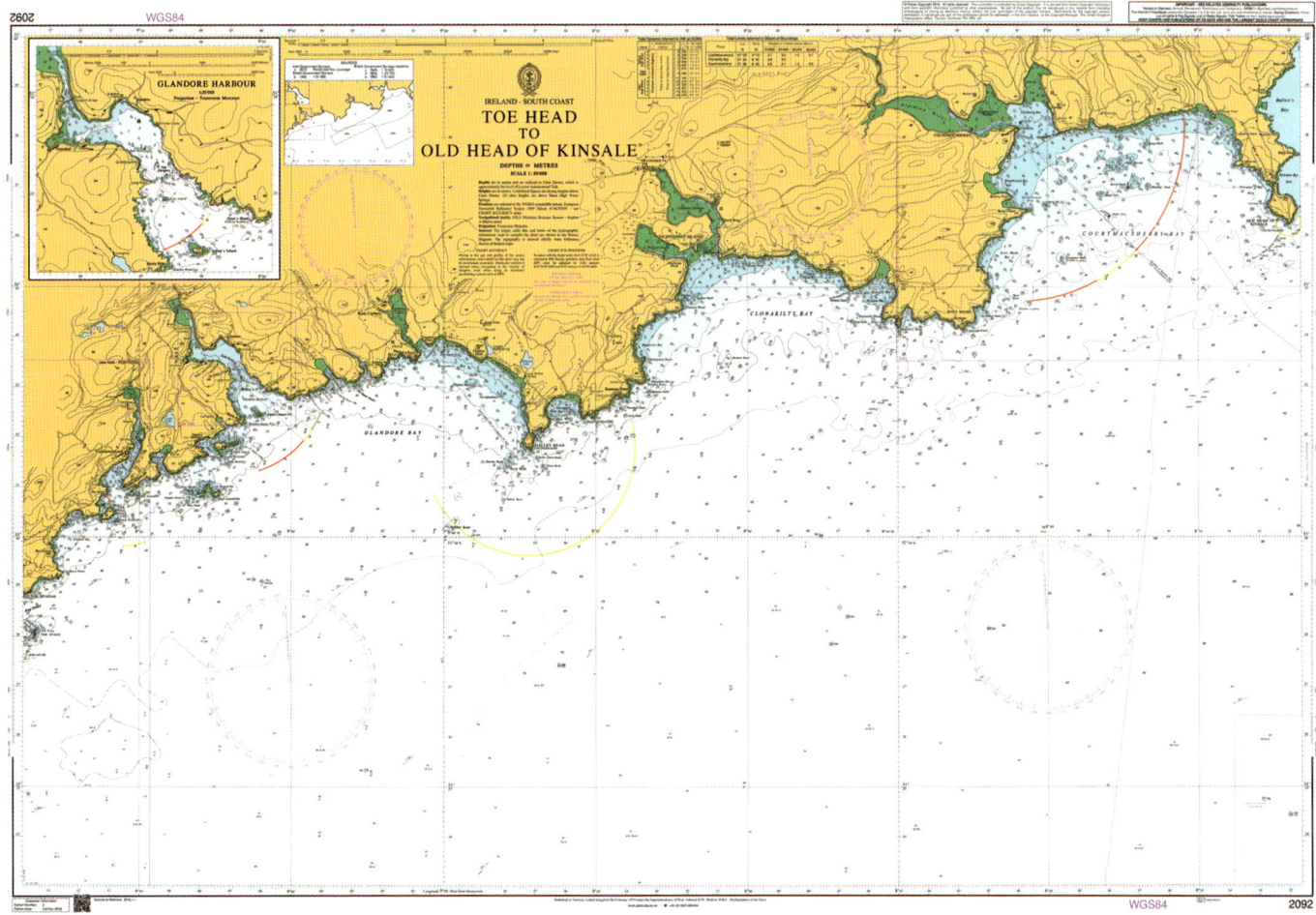

Fig. 8.13 A TRADITIONAL ADMIRALTY CHART OF IRISH COASTAL WATERS. Before the advent of digital technologies, mariners the world over relied on published seacharts for navigating at sea. These specialist maps typically show coastlines and river mouths, harbours and sea defences, navigation aids such as buoys and lighthouses, and known hazards, including reefs and shipwrecks. They also normally include information about water depths, essential for the safe passage of shipping in potentially shallow water. Today, official navigation charts are issued by the national hydrographic survey offices of many countries. Although some maritime countries produce 'local' charts, the design and production of most official charting around the world is co-ordinated by the International Hydrographic Office in Monaco as part of a global, collaborative and standardised 'international' hydrographic chart series. Ireland does not have its own hydrographic office but, instead, navigation charts for Irish waters are currently produced on behalf of the state by the UK Hydrographic Office in Taunton, Somerset, using data obtained by the Geological Survey of Ireland and the Irish Marine Institute under the collaborative INFOMAR project. [Source: © Crown Copyright and/or database rights. Reproduced by permission of the Controller of Her Majesty's Stationery Office and the UK Hydrographic Office (www.GOV.uk/UKHO)]

perhaps the most important of these products. Tailored for the navigation of shipping in marine and inland waterways, they use computers and other electronic systems, along with digital chart data, to plot and track a vessel's position. They can also incorporate other information, such as the location and movements of other shipping in the immediate vicinity, the existence of underwater or other hazards that might put the ship at risk, the position and identifying characteristics of lights and other navigational aids and the optimum direction and bearing for efficient route-finding.[12] In most ECS, the display of chart data on the screen can be adjusted to ensure easy visibility, even under the extreme wave and weather conditions that may sometimes be found at sea, as well as at night or at times of restricted visibility.

One of the most important categories of ECS is known as an Electronic Chart Display and Information System (ECDIS). As well as displaying information from electronic navigation charts, ECDIS also integrates information from GPS and other satellite positioning systems, navigation sensors such as radar and automatic ship identification systems (AIS). It may also display additional information, such as sailing directions, tide tables and, in polar waters, ice conditions. These systems are designed to be used worldwide and adhere to the requirements of the internationally agreed Safety of Life at Sea (SOLAS) convention. Because of this, ECDIS equipment must conform to strict standards, developed and maintained by the International Hydrographic Office (IHO) in Monaco, to ensure that they all look and behave in a similar manner, irrespective of where they are being used, the nationality of the ship on which they are being used, or the originating country of the chart data. All commercial shipping operating in Irish and international waters, including passenger ships greater than 500 tonnes (gross tonnage), tankers larger than 3,000 tonnes and cargo vessels greater than 10,000 tonnes are required to use ECDIS (Figure 8.14). To support the need for crew suitably qualified in the use of ECDIS, merchant mariners receive instruction in the technology as part of their training at the National Maritime College of Ireland in Ringaskiddy, County Cork.

As well as official ECDIS charting systems, a wide range of non-compliant ECSs are also in widespread use. These products are typically smaller, less sophisticated and hence usually much less expensive than the fully compliant ECDIS, and usually use data from one or more commercial company rather than from official national hydrographic offices. These non-compliant ECS are especially popular

Fig. 8.14 MARINER MICHAL JAGIELSKI USING AN ELECTRONIC CHART DISPLAY AND INFORMATION SYSTEM (ECDIS) ON THE BRIDGE OF THE IRISH-OWNED AND FLAGGED VESSEL *Arklow Falcon*. Under the terms of the international Safety of Life at Sea (SOLAS) convention, systems such as these are obligatory equipment on all commercial shipping operating in Irish and international waters, including passenger ships greater than 500 tonnes, tankers larger than 3,000 tonnes and cargo vessels greater than 10,000gt. An ECDIS combines electronic charts with additional information essential for safe navigation of ships in coastal waters, including GPS, radar and Automatic Identification Signals (AIS). The display can be adjusted for daylight or nighttime use, and the system is designed to be usable even at times of stormy sea conditions and poor visibility. [Source: Captain Aidan Fleming, courtesy of Arklow Shipping]

within the leisure boating community and are in widespread use on yachts and cabin cruisers in Irish inland and marine waters. However, these smaller systems are less likely to conform to internationally agreed standards and cannot legally be used on board commercial shipping as a substitute for paper charts, whereas a fully compliant ECDIS can.

TEXTUAL AND PHOTOGRAPHIC DESCRIPTIONS OF THE COAST

Norman Kean

Nautical charts, whether paper or electronic, are among the most essential tools at the marine navigator's disposal. However, written and photographic materials also play an important role. Standard practice is to present these items in the form of Sailing Directions or 'Pilot Books'. The latter term refers to their use in pilotage, the art and science of navigation close to land, the purpose of which is to enable a vessel to find her way into, out of and between harbours in a safe, efficient and seamanlike manner. In respect of this type of information, the coasts of Ireland are among the most thoroughly documented in the world.

Pilot books designed principally for large vessels are published by the hydrographic offices of many nations and are considered essential aboard ships subject to the international SOLAS (Safety of Life at Sea) convention. The British Admiralty's *Irish Coast Pilot* can trace its origin to 1866 – the first compiler was Richard Hoskyn, Master RN, a Cornishman who played a central role in the mid-nineteenth-century

Fig. 8.15 SOUTHERN ENTRANCE TO ACHILL SOUND. Achill Sound is a narrow, winding waterway that separates Achill Island from the mainland of Ireland. It dries completely in places at low tide. While the bridge to Achill Island is capable of opening, the tidal and operational restrictions involved prevent the use of the Sound as a routine shortcut for small-craft sailors who might wish to avoid the passage round Achill Head. However, both ends of the sound provide useful havens on this isolated coast. The narrows pictured here, between Darby's Point (Achill Island) and Gubnacliffamore (mainland), form the entrance to the South Sound. Tidal currents through the narrows can exceed four knots and the sandbanks are subject to movement. While the journey into this maritime refuge is tricky, negotiating challenging waterways to reach isolated harbours is essential for small craft along the west coast of Ireland. [Source: Geraldine Hennigan and Norman Kean]

century charting of the coast – and in 2019 was in its twenty-first edition.[13] It describes the coast in considerable detail, beginning with its maritime topography, its currents and tidal streams, its climate and weather and through routes from the Atlantic to the Irish Sea. There follows a more detailed section-by-section description, with routeing advice for shipping and port, harbour and anchorage information. This includes such things as bunkering facilities, the depths alongside commercial quaysides, maximum sizes of vessels accommodated, the availability of tugs and the capacity of dockside cranes.

While the basic principles of safe navigation are the same for all, the needs of small craft, particularly leisure vessels, are slightly different from those of larger ships. They have no need for tugs and commercial quayside cranes. They can navigate very close to shore and their definitions of 'harbour' and 'anchorage' extend to places where the masters of SOLAS ships would never dream of going (Figure 8.15). Facilities for them are often purpose-built and segregated, such as marinas, although on some stretches of the Irish coast these are few and far between (see Chapter 27: Tourism and Leisure). While the basic headings under 'pilotage information' are the same as for large ships, the detail and focus are different and pilot books for small craft tend to reflect a much more intimate view of the coast. They are provided, in Ireland as elsewhere, by commercial publishers and by voluntary organisations, such as clubs. *Reed's Nautical Almanac*, covering the Atlantic coast from Denmark to Gibraltar and appearing annually since 1932, is a leading reference work, commercially edited and published but kept up to date by a network of volunteer correspondents around the coast, in Ireland as elsewhere.[14]

Sometime around 1910, Henry Donegan, a lawyer from Cork and a keen yachtsman, began gathering pilotage information on the south and south-west coasts. In 1930, as a founder member of the newly formed Irish Cruising Club, he had this information published for the use of club members. The same was done for the east coast by his fellow-member, the Dublin dentist, Billy Mooney. In 1946, second editions were published and made available to the public. By 1962, the Irish Cruising Club *Sailing Directions* covered the whole coast and rapidly became accepted as the standard work on the small-craft pilotage of Ireland. The two volumes are now in their thirteenth and fifteenth editions respectively.[15] Since 2005, other publishers have offered alternatives for the south, south-west and east coasts.[16]

The successive authors and compilers of these books have contributed to our knowledge of the coast, often finding uncharted hazards or conducting amateur surveys of out-of-the-way places that have a low priority in professional surveying. The Irish Cruising Club collaborates closely with, and is respected by, the UKHO, the Commissioners of Irish Lights and official survey organisations. Its publications are carried by specialised larger ships with a particular interest in pilotage close inshore, such as the Irish Lights Vessel, the survey vessels of the GSI and the ships of the Irish Naval Service. They have also been used as the basis of local knowledge manuals by several RNLI lifeboat stations.

The development of digital, aerial and now drone photography, desktop publishing software, the internet and affordable high-quality, four-colour printing has made good pilotage information much more accessible. Meanwhile, the advent of global satellite positioning has enabled pilotage advice for small craft to be published for many places in Ireland where, formerly, it had to be advised that local knowledge was essential.

Written descriptions, illustrative plans and aerial and sea-level photographs in pilot books of all types provide a view of the coast that

Bofin Harbour from the NE: the inner quay, centre, and the ferry pier, R

Entering Bofin Harbour: (above) the leading daymarks in line, R, and the PEL (showing red), centre

(L) Gun Island tower: the leading daymarks, extreme L; the light beacon is on the slim pole, R

Lights and Marks - Bofin Harbour

Gun Island, white tower, with adjacent bn Fl(2) 6s 8m 4M
Bofin Harbour, leading daymarks 032°, white towers, unlit
Bofin Harbour Port Entry Light, Dir Oc WRG 6s 22m 11M. Close W of the front leading mark, shows a narrow white sector centred on 021° with 5° sectors either side, red to W, green to E
Inishbofin Main Pier, Fl(2) R 8s
Inishbofin Inner Pier leading beacons, triangles on poles, L Fl 6s
Inishbofin Inner Pier, Fl R 5s

Fig. 8.16 EXCERPT FROM IRISH CRUISING CLUB'S 15TH EDITION OF *South and West Coasts of Ireland Sailing Directions*. The main text (not reproduced here) provides a written description and guidance. There are also boxes describing tides and tidal streams and listing particular dangers, and sketch charts to illustrate the text. [Source: Kean, N., 2020. *South and West Coasts of Ireland Sailing Directions*. 15th ed. Kilbrittain: Irish Cruising Club Publications CLG]

complements navigational charts (Figure 8.16). A written and illustrated description can add much in the way of useful detail to a chart symbol denoting, for example, a tide race or a navigational aid. A list of dangers may draw attention to a hazard that might otherwise be overlooked, especially on a vector chart. An overhead bridge clearance marked on the chart is essential, but a photograph of the bridge is helpful and reassuring. Photographs of visual transits can be very useful. The use of a drone provides informative low-level aerial pictures, and the drone – in contrast to a fixed-wing aircraft – can be flown so as to capture the required picture in the optimum conditions of weather, lighting and tidal height. The combination of charts and plans with text and photographs enables increased spatial awareness and also addresses the fact that people have different aptitudes and different ways of absorbing information.

For small craft, in particular, the paper pilot books for Ireland remain a popular format. A paper product has well-defined advantages over an electronic publication: it can stand a great deal of abuse and it can be used in all but the most extreme conditions. It is easy to refer back and forwards and a book doesn't need instructions, batteries or warm dry hands to operate it. It will not stop working and if it gets damaged it can be repaired, if necessary, on board.

Over 90 per cent of the copies of the Irish Cruising Club's *Sailing Directions* are bought by intending visitors to Ireland from overseas and they are shipped all over the world. For a yacht's crew, the coastline of Ireland has been described recently as one of the most fascinating in Europe. The north and west coasts, fully exposed to the Atlantic, with formidable headlands – such as Slyne Head, Achill Head, Erris Head and Malin Head – offer particular challenges to the adventurous sailor. Cruising these coasts in a small vessel demands a considerable degree of self-sufficiency and has been variously and aptly described as 'wilderness sailing' and (by analogy with one of Ireland's favourite sports) 'senior-league hurling'. Places such as Strangford Lough and west Cork provide an extremely attractive, safe and uncrowded environment for leisure sailing and are justly popular.

A growing number of websites also provide small-craft pilotage information, including for Ireland, and as appropriate signal coverage becomes widely available within sight of land, these are becoming accessible at sea. Some apps, available on tablets and smartphones, are simply a means of accessing website pilotage reports. Websites have the advantages that they are not constrained by page count and (given signal) they are instantly available and normally free to the user. Some of them are compilations of individual reports and descriptive material, often written by experts with local knowledge, such as harbourmasters. The reports may be excellent, but unless they are centrally edited, they will inevitably display inconsistencies of terminology. 'Deep' and 'shallow' can be quantified, but many adjectives cannot. One writer's 'challenging' may be another's 'dangerous'. Some websites are confined to describing ports and harbours, with little or no information on passagemaking. Port authority websites, such as those of Dublin, Belfast and Cork, include regulations and guidance for the conduct of small craft within the port areas. Some websites are commercially driven and publish only paid-for content; some attempt to provide coverage well beyond their home base, possibly beyond their ability to check information and put it into a local context. Other websites are accumulations of reports from single visits, which may be posted either on open forums, hosted by clubs, or on individual blogs of single voyages. Websites are often a very good source of port information, but in general they lack the defining preliminary sections published in traditional pilot books.

Distinct from those that simply access websites, there are also apps that are usually very specific in their purpose, including such things as:
- harbour-entry guidance, usually in places with mobile sandbanks and variable channels (for example, Wexford);
- webcams trained from landward onto hazardous rivermouth bars, such as is in place for the River Bann;
- real-time weather data tweeted from navigation buoys.

These facilities are maintained by the relevant harbour authorities and (in the last case) the Commissioners of Irish Lights.

VESSEL MONITORING, IDENTIFICATION AND TRACKING SYSTEMS

Darius Bartlett

Automatic Identification Systems (AIS), the Long-Range Identification and Tracking (LRIT) system and Vessel Monitoring Systems (VMS) are all ship-board systems that transmit data about the ship's identity, type, position, course, speed, navigational status and other safety-related information to shore-based stations as well as to other suitably equipped shipping (Figure 8.17). These systems are now widely used for collision avoidance and other safety-related purposes, Search and Rescue (SAR), port and harbour management, environmental protection and, increasingly, for security and law enforcement purposes, including fisheries protection, the prevention and detection of smuggling and responding to terrorism threats. Information received from these systems may also be incorporated into electronic charting systems, so that the crew of one vessel can easily keep track of the movement of other shipping, particularly in harbours or busy shipping channels.

AIS uses VHF radio links to send information, and has a range of approximately 40 nautical miles. It is required under SOLAS regulations for all vessels of 300 tonnes and over that are on international voyages, all passenger ships and all vessels of 500 tonnes and over whether or not on an international voyage.[17] In Ireland, the National Space Centre outside Midleton, County Cork, is a receiving station for AIS transmissions, which are combined with satellite observation for improved monitoring and tracking of shipping in North Atlantic waters.

Fig. 8.17 AVERAGE SHIPPING TRAFFIC DENSITY IN EUROPEAN WATERS (2017). Based on data derived from Automated Identification Systems (AIS) transmissions, this map shows shipping density in 1km x 1km cells of a grid covering all EU waters and some neighbouring areas. Density is expressed in hours per square kilometre per month, and shows clearly that the majority of shipping stays relatively close to shore, while the waters around Ireland, the UK, northern France and within the North Sea are particularly heavily trafficked. Shown too are longer-distance trade routes, including those within and through the Mediterranean and down the coast of Africa. Perhaps surprisingly, there is comparatively less maritime traffic crossing the Atlantic between Europe and North America. [Data source: European Marine Observation and Data Network (EMODnet), 2017, courtesy of Cogea, 2019]

LRIT was adopted by the International Maritime Organisation (IMO) in 2006 as an amendment to SOLAS,[18] and became operational in 2009. It is a satellite-based tracking system, for vessels over 300 tonnes, high-speed craft, all passenger vessels, and offshore Mobile Drilling Units (MDUs). Under the LRIT system, a ship is required to transmit a minimum of four positions per day (once every six hours). In Europe, the EU Long Range Identification and Tracking of Ships Cooperative Data Centre (EU CDC), based in Lisbon in Portugal, is one of the largest data centres in the whole LRIT system. It tracks over 8,500 ships, estimated to account for 25 per cent of the world fleet subject to LRIT, on behalf of thirty-seven member countries, including Ireland, as well as EEA countries and Overseas Territories.

Information collected and stored by LRIT data centres is made available to eligible users, including flag states, which may request information on the location of their own vessels around the world; coastal states, who may request information on ships up to 1,000 nautical miles from their coasts irrespective of their flag; and port states, who may request information on those ships that have declared one of their ports as destination, irrespective of their location or flag. In addition, LRIT information may also be made available to search and rescue authorities, as well as to defence agencies including, in Ireland, the Irish Naval Service.

The Vessel Monitoring System (VMS) is designed specifically for purposes of fisheries monitoring and control worldwide. It is a satellite-based system, which collects data on the location, course and speed of fishing vessels, and returns these data to the relevant fisheries authorities. Under European Union legislation, all fishing vessels greater than 15m in length, operating in EU waters, are required to be fitted with VMS and all coastal EU countries are obliged to implement systems that are compatible with each other, to facilitate the sharing of data as well to make it easier for the European Commission to monitor adherence to fisheries rules and regulations (Figure 8.18).

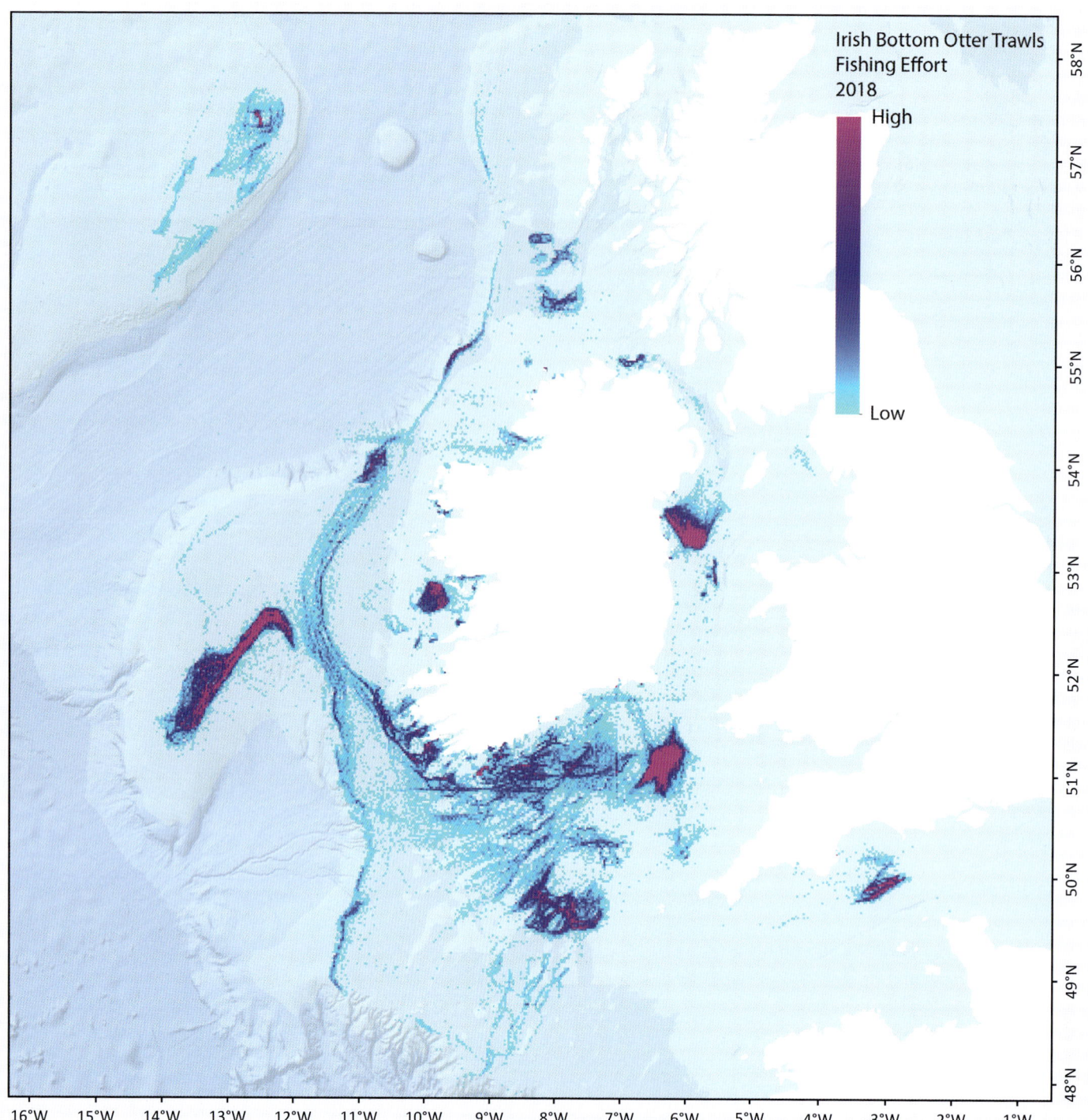

Fig 8.18 BOTTOM OTTER TRAWLS (2018) FISHING EFFORT FROM VMS DATA. A bottom otter trawl is a type of net used commonly onboard fishing vessels operating in Irish waters. The trawl is cone-shaped, with two rigid boards, known as otters, designed to keep the mouth of the net open while it is dragged along the seabed. Based on data derived from Vessel Monitoring Systems (VMS) transmissions, this map shows fishing effort for bottom otter trawls ≥15m around Ireland, in 0.03° longitude x 0.02° latitude cells, during 2018. Density is expressed in hours per square kilometre per year, and the map shows clearly the main 'hot-spots' of this particular fishing activity are to be found in the north Irish Sea around Dundalk Bay; in the south Irish and Celtic seas; along the south Cork and Waterford coasts; and on the shallower parts of the Irish continental shelf, and particularly the Porcupine Bank, in the Atlantic to the west. [Data source: Marine Institute, 2018]

SATELLITE REMOTE SENSING OF THE COASTAL REGIONS OF IRELAND

Fiona Cawkwell

Measurements made by an instrument that is not directly in contact with the surface that is being studied are referred to as remote sensing. This broad term can refer to an X-ray in a medical context, or the use of a telescope for astronomy; satellite sensors have been used for

Fig. 8.19 a. TRUE COLOUR COMPOSITE SATELLITE IMAGE AND b. FALSE COLOUR COMPOSITE IMAGE OF SHANNON ESTUARY, WEST OF THE AIRPORT. Many satellite sensors record the amount of energy reflected from the surface in both the visible range of the electromagnetic spectrum and at longer wavelengths, which our eyes cannot detect. To display this information, the reflectance from three different wavelengths can be shown as a colour composite. A true colour composite displays the blue, green and red reflectance amounts in their correct colours, to generate an image very similar to that which our eyes view. A false colour composite displays any other reflectance amounts, including those that are not in the visible region of the electromagnetic spectrum, in red, green and blue shades. a. Shows a true colour composite of part of the Shannon Estuary (with Shannon airport in the bottom right corner) as acquired by the European Space Agency Sentinel-2 sensor on 28 June 2018. b. Shows a false colour composite of the same area at the same time from the same sensor, but with the near-infrared reflectance shown in the red colour, as well as the red and green reflectance in green and blue colours respectively. The red areas indicate vegetation, which reflects the majority of near-infrared radiation incident upon it, with the different shades of red corresponding to different species or phenological stages, while urban areas are shown in turquoise and white. Some features, such as the airport runways, are much more apparent in the false colour, while others, for example the sediment in the river, are more clearly seen in the true colour image. [Source: European Space Agency sentinel-2 Sensor, 28 June 2018]

military purposes since the late 1950s. The first civilian Earth-observing satellites were launched in the 1970s and, since then, their potential for viewing and monitoring national and global coastal and marine environments has been increasingly valued. The earliest satellite sensors, such as those on the American Landsat family of satellites launched from 1972 onwards, were only able to record the amount of sunlight reflected directly from the Earth below. This, along with the relatively coarse spatial resolution of the images obtained, limited their use for studying the Earth and sea surface to large-scale phenomena and cloud-free daylight hours. More recent instruments, however, are capable of much more detailed measurements, at spatial scales of less than one metre. They also use a much wider range of radiation, including in some cases microwave (radar) energy that is able to penetrate through cloud cover and is not limited to daytime use, thus significantly increasing the applications of the imagery. In addition to satellites, manned and unmanned aerial platforms are also used to collect a range of information that is typically of a scale of centimetres to tens of centimetres. However, unlike satellites, these tend to be one-off missions that do not have an extensive archive of historic measurements.

Every surface type reflects and absorbs wavelengths of electromagnetic radiation differently, to create a unique spectral signature for that surface. For example water absorbs almost all incoming near-infrared radiation from the sun, while vegetation, soil and rock reflect large amounts of the same wavelengths, so the shoreline can be readily discriminated on a satellite image by comparing the amount and type of light energy received by the sensor in different parts of the scene. These and other features of the coast can be accentuated in the resulting images through various techniques for processing the data received from the satellite (Figure 8.19). Variations of this type allow different coastal and marine habitats and land- or sea-surface characteristics to be identified and analysed, while changes in dynamic coastal environments – such as the erosion of a sandbank following a storm surge or the accretion of a spit due to longshore drift – can be distinguished easily. Where several satellite images of the same area are available, taken over an extended time period, the ability to detect change at both short and long timescales is possible, providing the coastal phenomena being examined are greater in size than the spatial resolution of the satellite image. An archive of images from the Landsat family of sensors, dating back to the early 1970s, is particularly important in this regard, despite their comparatively coarse resolution. Notably, however, many of the changes in Irish coastlines are of the order of centimetres to metres on an annual basis, thus making them evident only on the very high spatial resolution imagery acquired in more recent years.

In addition to distinguishing between water and non-water environments, the multiple wavelengths recorded by satellite sensors enable different coastal ecosystems to be studied. Although individual plants cannot be identified from space, the assemblages of species that make up particular habitats can be separated. Although the vast majority of animal species cannot be viewed directly from space, vegetative habitats can act as proxy indicators for the potential presence of mammals, birds and insects. With the development of both higher spatial resolution sensors and those that record reflectance in a larger number of wavelengths, this task becomes easier; while it cannot entirely replace the field identification of species, the use of satellite imagery does assist in identification of areas of change. The very high spatial resolution offered by some satellite sensors does, however, permit the identification of large marine mammals such as whales and, while a single snapshot may not capture all members of a pod, it can reveal individuals at the surface as well as those stranded onshore.

Offshore, the large concentrations of coloured phytoplankton found in algal blooms are also readily identifiable from space, especially as they can extend over tens to hundreds of kilometres in some instances and can have a spectral signature that is markedly different from the water in which they occur. The growth of these algal blooms can provide important information on the nature of water masses, as they are typically found when nutrient and temperature levels rise (see Chapter 3: Coastal Waters: Marine biology and ecology, Figure 3.3, acquired by the NASA MODIS sensor on the Aqua satellite on 2 June 2006, where the bloom highlights upwelling near the continental shelf). Many blooms are short-lived, lasting only hours or days; therefore to be seen from space requires a satellite sensor to pass overhead frequently and to track their movement requires multiple sensors. Algal blooms can pose major hazards for marine ecosystems, as well as for aquaculture and other commercial activities, so monitoring the progress of these masses from satellite imagery, along with numerical modelling and aerial surveillance, can provide essential early warning of potential threats.

Other coastal water-quality issues that can be detected and monitored from space include surface oil slicks, which are most readily seen on microwave imagery because oil smooths the rough water surface, reducing the amount of energy scattered back to the sensor. These slicks, which may occur naturally due to oil seeping from beneath the sea bed, or can arise from an accidental or deliberate spillage, are equally harmful to coastal habitats and therefore monitoring their movement can facilitate protection of vulnerable areas.

Many commercial companies and governments – including the Irish government, through its membership of the European Space Agency – are committed to the launch and maintenance of further satellite sensors over the coming decades. With an increasing number of sensors and the ability to detect radiation in a greater range of wavelengths, the potential for monitoring events as they happen will increase and the extension of existing datasets will enable change over longer time periods to be evaluated. A recent development has been the growth of commercial companies launching constellations of tens to hundreds of identical satellite platforms, carrying relatively small and cheap instruments, which increases the frequency with which the same region of the Earth's surface is viewed, thus enabling short-lived features, and often cloud-covered regions, to be monitored in more detail. The use of satellite imaging will, therefore, continue to be a growing source of beneficial data for those who live and work within the Irish coastal region for many years to come.

LASER TECHNOLOGIES

Sarah Kandrot

Laser scanning is an active remote sensing technology that uses either a reflected laser pulse or, less commonly, differences in phase from a continuous beam, to measure the distance to an object or surface. There are two types of laser scanning currently in widespread operation: first, airborne laser scanning, often known as Light Detection and Ranging (LiDAR), in which the laser beam is fired from an instrument carried in an aircraft or, in recent years, a remotely operated drone, and is used to detect changes in the surface elevation of the ground beneath (Figure 8.21); and second, Terrestrial Laser Scanning (TLS), where a portable instrument is mounted on a tripod and can be used to measure vertical phenomena such as cliffs and coastal dunes (Figure 8.20). Irrespective of the platform used, laser scanning provides a rapid, cost-effective and efficient surveying technology that is increasingly being employed in coastal mapping and monitoring practices.

Fig. 8.20 TERRESTRIAL LASER SCANNING. a. Terrestrial laser scanned data are represented by what is known as a point cloud – a densely packed collection of 3-dimensional (x, y, z) co-ordinates in space. A typical point cloud may contain many million individual points, which can each be colour-coded to show its distance from a reference point, as in the example above, or can be keyed to represent some property of the surface that has been scanned, such as its texture. The graphic above shows a point cloud merged with an image of coastal dunes at Inch, County Kerry. The resolution of the data is 2.5cm. [Source: Sarah Kandrot]; b. A terrestrial laser scanning workstation, set up on the foreshore to survey dunes at Rossbeigh Strand, County Kerry. This instrument, a Leica ScanStation, is powered by a battery (the red block mounted on the tripod) and is operated via a laptop computer, connected via an ethernet cable. Many newer models contain an in-built GPS, smaller, longer-lasting batteries, on-board data storage and touch screens for their operation. [Source: Sarah Kandrot]

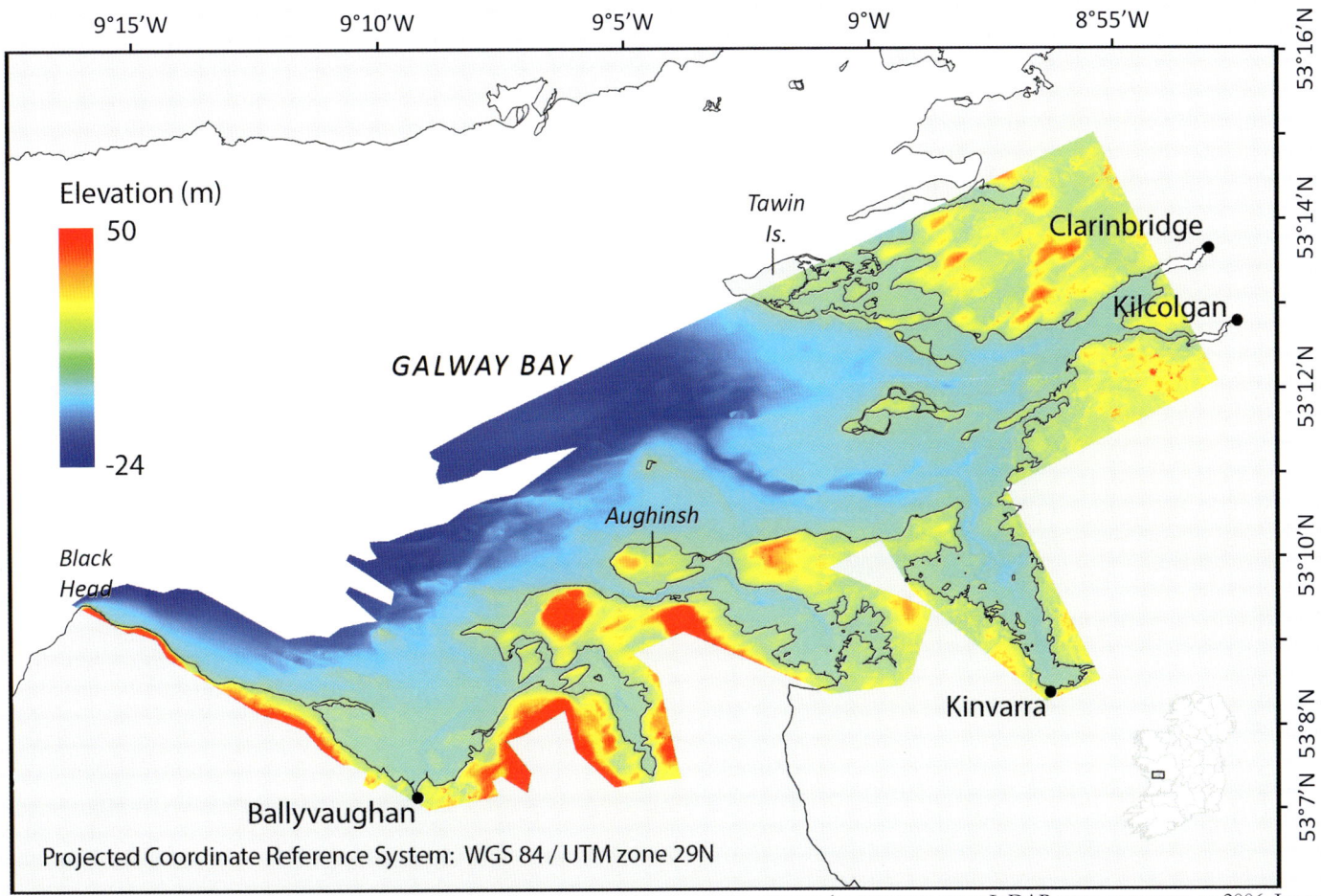

Fig. 8.21 COASTAL ELEVATION MODEL OF SOUTH GALWAY BAY, WITH A GROUND RESOLUTION OF 6M, DERIVED FROM LiDAR DATA CAPTURED IN 2006. Invented in the early 1960s as a means of tracking orbiting satellites, Light Detection and Ranging (LiDAR) technology is widely used today as a means of producing very high-resolution, high-precision maps of terrain, such as the example shown here. Although used mostly for surveying the land surface, some wavelengths of light used in LiDAR are also able to penetrate shallow water. Particularly when airborne LiDAR surveys are undertaken at times of low tide, this allows the use of the technology to accurately map those parts of the shoreline that would otherwise be inaccessible to survey ships as well as from land. [Source: Geological Survey Ireland]

Laser scanners capture ultra-high resolution information about the geometry of solid objects or surfaces in the form of a 'point cloud', a densely packed collection of very precisely measured, three-dimensional co-ordinates in space. Data from these surveys can form the basis of Digital Elevation Models (DEMs), which are a basic requirement of a wide range of coastal applications.

Coasts are perhaps the most active of all geomorphic environments and are subject to a variety of complex and dynamic processes that operate over various space and time scales. From a management point of view, effective adaptation to this dynamism necessitates a scientific understanding of these processes, underpinned by information gathered from field-monitoring campaigns at relevant scales. Depending on the application, high-resolution 3-D data (on the order of millimetre to metre) may be required. While other, related, 3-D surveying technologies exist – particularly for ground-based operations – they may not be capable of capturing the required spatial resolution or their cost may be prohibitive, especially if regular surveying is necessary or if surveying is required on demand, for example in the immediate aftermath of a storm.

Terrestrial laser scanners are generally capable of capturing data at centimetre to millimetre resolution, and at ranges of 100m to sometimes upwards of 1,000m; airborne LiDAR can achieve ground resolutions of better than half a metre. In the past, the rugged TLS models most suited to outdoor use (for example, in coastal environments) were heavy and unwieldy, but newer models are smaller, lighter and easier to transport and operate. In addition, many new models have in-built Global Positioning Systems (GPS), so georeferencing the data (adding real-world ground co-ordinates) is now easier than ever. Depending on the specifications, terrestrial laser scanners cost typically between €40,000 and €150,000, with mobile TLS systems costing around €300,000. They can also be rented at a much more modest cost.

In a research setting, terrestrial laser scanning has been employed successfully as a monitoring tool in coastal environments in a number of cases; for example, work by Kandrot demonstrated the usefulness of TLS in assessing the impacts of storm events on the beach and dune systems of Rossbeigh in County Kerry.[19] TLS surveys were undertaken regularly over the course of a two-year monitoring campaign. Volumetric change analyses performed on the data revealed significant changes to the dune system in the aftermath of winter storms, particularly during the 2013/2014 storm season. Such assessments are required so that coastal planners and administrators can make informed management decisions in the aftermath of hurricanes and other natural disasters, and can also feed into scientific studies of the impacts of climate change and sea-level rise. Elsewhere, data derived from both LiDAR and TLS are increasingly being used by civil engineers to design, build and monitor the performance of coastal defence works, bridges and other civil engineering structures. Finally, a number of studies have found TLS to be a useful tool in examining the evolution of hard and soft sea cliffs, especially in places of limited accessibility. TLS offers an advantage over aerial LiDAR in that the instrument can be orientated in front of sea cliffs during surveying, rather than above them, allowing for a greater coverage area. Information from TLS surveys can help coastal managers better understand the risks of coastal hazards associated with sea cliff erosion, which would ultimately benefit landowners and other stakeholders.

While laser scanning is in the process of emerging as a useful coastal monitoring tool, there are some factors that should be considered prior to its usage. These include issues such as the storage and processing of large datasets, vegetation removal, DEM generation and error propagation. An excellent resource that addresses these and other issues is found in Heritage and Large (2009).[20]

Sensors and Autonomous Data Collecting Devices

Darius Bartlett

Collecting data about the marine environment presents many challenges. Traditionally, these data were usually collected by scientists working on board ships at sea, using conventional sampling techniques and subsequent laboratory analysis. As well as often being costly to undertake, data collected in this manner were often inconsistent, prone to error or uncertainty and poorly suited for drawing broader conclusions

Fig. 8.22 SCHEMATIC OF THE SMARTBAY MARINE AND RENEWABLE ENERGY TEST SITE IN GALWAY BAY. The SmartBay test site is located 1.5km off the coast of Spiddal, County Galway, and is delineated at sea by four cardinal marks for navigation purposes. The test site has an average water depth of 23m and contains three berths for testing prototype marine renewable energy devices, including wave energy devices, and a further berth that can be used for floating offshore wind-energy prototypes. The facility also has an underwater cabled observatory, connected to a shore station in Spiddal by 4.8km of fibre optical cable. This cable also provides low-level power to the observatory for testing underwater sensors, lights, cameras, hydrophones and various other types of equipment. The test site also has a floating platform or buoy which can host a range of prototype equipment, for example components, materials and sensors. All data from the observatory and buoy platform are made available to users via a dedicated web portal. Users of the facility can avail of a range of additional on-site and local facilities including vessels, ROVs, data buoys, engineering support and 4G communications to support testing programmes. A number of existing or proposed instruments that make up the SmartBay cluster are shown here, including: a. prototype floating offshore wind device; b. proposed SeaLab facility, to provide additional power to, and dissipate power generated by, Ocean Energy devices (not currently available on the test site); c. prototype wave-energy device; d. floating platforms for testing and validating a range of novel sensors and efficient gathering of metocean time-series data; e. an array of subsea sensor equipment; f. fibre optic data and 400v power cable; g. subsea frame with high-definition camera and subsea lighting on top and Conductivity, Temperature and Depth (CTD) sensor on the bottom of the frame; h. subsea cabled node, which operates as a sensor platform, to provide power to and collect data from a host of sensors and equipment that can be tested and demonstrated in near real-time; i. subsea static cable; j. high-frequency hydrophone; k. acoustic doppler current profiler. [Source: Image courtesy of SmartBay Ireland]

regarding the geographic distribution of the phenomenon of interest or its variation over time. The availability of remote sensing imagery has greatly improved the situation, but the images collected by satellite or other means are limited in their ability to penetrate the water column, and so are mostly of use in examining phenomena at or near the sea surface. Particularly since the 1990s, a growing range and variety of sensors have been developed which, when deployed in the marine environment, can help to reduce significantly the costs of data collection, are able to measure and monitor changes taking place on the seabed and within the water column, and provide much more useful and accurate monitoring of conditions over space and time.

A sensor is a device that is designed to collect data relating to events or changes in its environment, either completely or semi-autonomously, and send these data via wireless links to a receiving station for processing. Each individual sensor contains a transducer that interacts physically with some aspect of its environment (temperature, pressure, sound, the presence or amount of particular chemicals, etc.), and converts this into an electrical signal.[21] The sensor will also have its own individual power supply, or the means of generating electricity by, for example, solar cells, and either a means of transmitting the data to the base station in real time, or some form of onboard storage for the measurements, which can be downloaded when the sensor is recovered at a later time.

In some cases a single sensor may have practical use, for example a sensor calibrated to record the state of the waves or the tide and mounted on a buoy moored offshore. The sensor may also be part of a tracking device fitted to a whale, seal or other animal to monitor its location and help scientists learn more about its behaviour. This latter technique, sometimes referred to as telemetry, has been used in

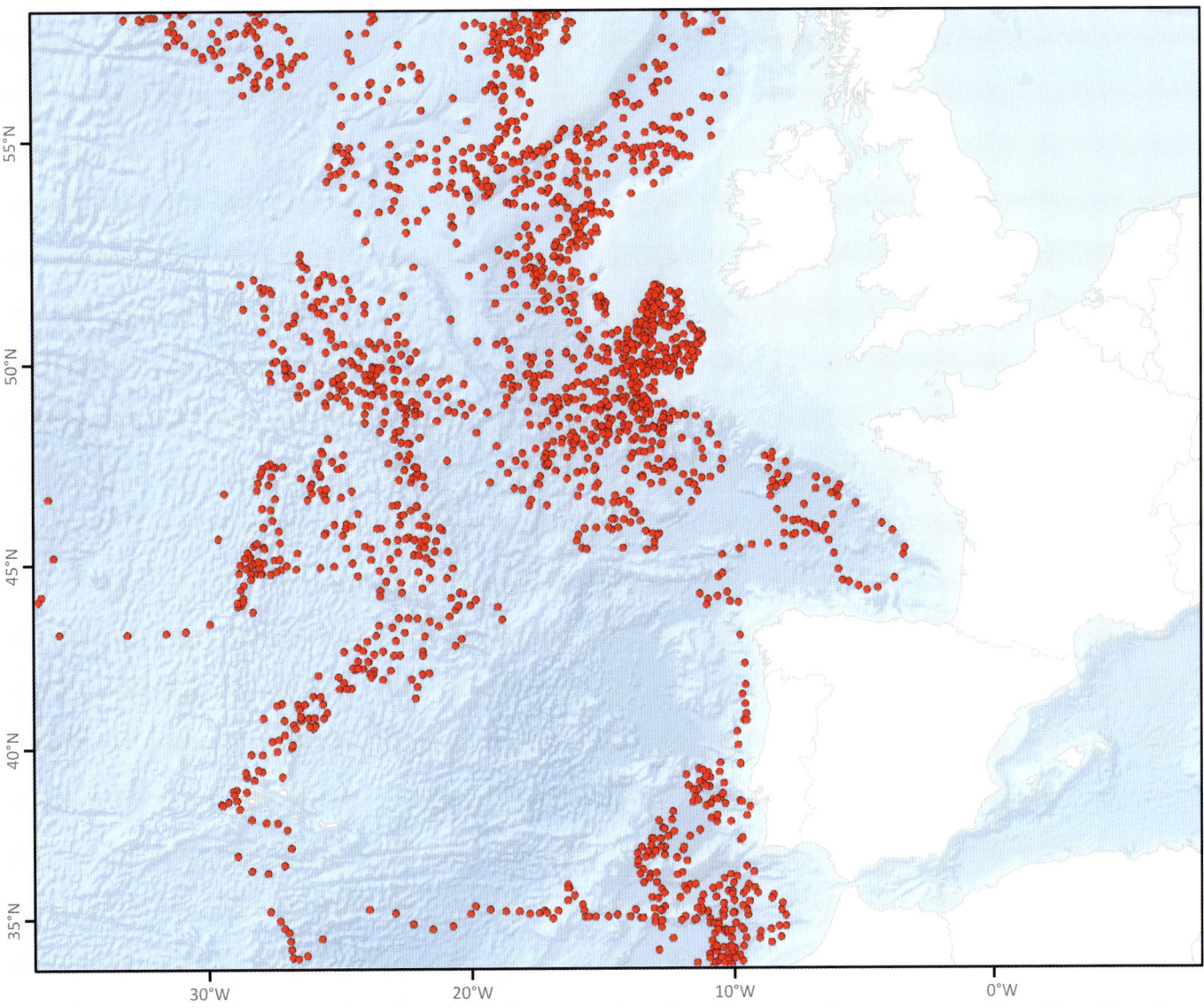

Fig. 8.23 Argo FLOAT DISTRIBUTION IN THE NORTH ATLANTIC. Argo is a global network of autonomous data collecting devices known as floats, deployed as part of a worldwide scientific ocean observing and monitoring system, to capture data on ocean salinity, temperature and other key parameters. Currently, over 3,800 floats are in operation, in all the seas and oceans of the world, including the Arctic Ocean, the Baltic and the Black Sea, and over 800 more are added to the network every year. A primary focus of Argo is to document seasonal-to-decadal climate variability and to aid our understanding of its predictability, while other objectives include providing key information about the chemical and physical properties of seawater, which can be used to fine-tune satellite-derived ocean surface elevation data; and provision of data to support the testing of ocean and climate predictive models. As well as providing essential information about the state of the world's oceans, Argo has also had an important role in helping the scientific community develop better ways of collaborating internationally and in the development of more effective systems for data collection, management and exchange. Ireland is a contributing participant in the Argo programme and, since 2016, has contributed three floats to the network. [Source: Marine Institute, 2017]

Irish coastal waters to study the behaviour of seabirds and seals, and their interactions with the fishing industry and with offshore wind farms (see Chapter 3: Coastal Waters: Marine biology and ecology).

The greatest value of these sensors becomes apparent when they are deployed in arrays, rather than singly. This has given rise to the concept of the 'Sensor Web',[22] a network of wireless, intercommunicating monitoring devices, distributed over a wide geographical area, in order to explore and monitor changing environments on a more regional, or even global, scale. The Sensor Web has been subsumed within the even more extensive and ambitious Internet of Things (IoT), which allows data to be harvested from a vast number of varying but inter-operating interconnected devices, and integrated for richer and more complete analyses of changing conditions. A number of sensor-based environmental monitoring systems are now in operation globally, as well as at European and Irish national levels. Many of these involve the integration of data collected by sensors with observations obtained from other platforms, such as satellites, ships or ground stations located onshore.

In Ireland, a major initiative to research and develop a comprehensive, sensor-based infrastructure for studying and monitoring the marine environment was launched in 2010. Called SmartBay, this project is a collaboration between scientists at Dublin City University (DCU), National University of Ireland Galway (NUIG), Maynooth University and University College Dublin (UCD), with support from Intel and IBM. SmartBay seeks to develop a national test and research infrastructure, to support the growth of the Blue Economy in Ireland, through the testing and development of materials, components, coatings, sensors and ocean energy devices. To further these aims, an underwater 'cabled observatory' has been established in Galway Bay, where a diverse array of instrumented buoys, sensors, high-speed optical fibre communications networks, power sources and supporting transport vessels have been deployed (Figure 8.22). The observatory includes instruments installed on a permanent basis to record and monitor water conductivity, temperature and depth; dissolved oxygen; turbidity and fluorescence (these last as indicators of water clarity and the possible presence of bacteria and other microorganisms); an acoustic doppler current profiler to measure the speed of the currents; a high-frequency hydrophone to facilitate detection and monitoring of cetaceans and other marine animals; and an acoustic fish tag detector.[23]

A growing number of marine sensor networks also exist or are under development at the global scale, of which the Argo project[24] is perhaps the most significant. Argo is based on the deployment of an array of autonomous, floating sensors, which collect high-quality temperature and salinity profile data from the upper 2km of the water column in oceans and seas around the world. Deployment of Argo floats began in 2000, and continues today with about 800 new launches per year. Argo is a truly international undertaking, with more than thirty countries taking part, including Ireland, as well as the European Union and the United Nations, both of which participate in Argo in their own right. As of June 2018, approximately 4,000 floats were in operation in seas around the world (Figure 8.23).

The floats spend most of their time drifting at around 1,000m depth within the water column but, at typically ten-day intervals, they are programmed to sink to 2,000m and then rise slowly to the surface while measuring a continuous vertical temperature and salinity profile, by means of their onboard sensors. When the float reaches the surface, its position is recorded using satellites or GPS, and the float transmits the data it has been collecting to satellites. The transmitted data are received initially at the Argo Information Centre in France, and also at one or more national data centres operated by countries that participate in the Argo project, where they are examined for quality. Once any errors have been dealt with, the data are then sent to Global Data Assembly Centres – one in Brest, France, and the other in Monterey, California – where they are integrated with previous data holdings. In most cases, the data are available to the analyst within twenty-four hours of having been transmitted by the float.

Argo data are freely available for download via the internet. Examples of the use of these data in Irish-led studies (in conjunction with GIS and other geoinformatics tools) include applications for ocean surface current mapping, marine biological production,[25] and the study of water mass circulation at intermediate depths at the entrance to the Rockall Trough.[26]

ELFORDSTOWN EARTHSTATION: IRELAND'S STRATEGIC LINK

Linda Fitzpatrick

Elfordstown Earthstation is located near Midleton, County Cork, in the south of Ireland (lat/long 51° 57.192°N 8° 10.55W), and since the 1980s has played an important role in global communications and Earth-observation infrastructure networks (Figure 8.24).

Elfordstown originally entered service in 1984, commissioned by Bord Telecom in association with Eutelsat as one of the first state-of-the-art broadcast sites in Europe. It remains the closest westerly positioned teleport for connecting to the USA.

Eutelsat had been established seven years earlier, in 1977, to build a European satellite industry and to operate the first generation of communications satellites ordered by the European Space Agency (ESA). Climate, site size and its southerly aspect were important factors in the Midleton site's original selection as part of Europe's largest telecom infrastructure project of the era.

Elfordstown Earthstation opened its doors on 11 May 1984, hailed as the new face of Irish satellite communications at the official

Fig. 8.24 NATIONAL SPACE CENTRE, ELFORDSTOWN, MIDLETON, COUNTY CORK. In existence since 1984, the Elfordstown Earthstation was initially developed as the most westerly satellite teleport in Europe. It was the main hub for all satellite transmissions to and from Ireland, and could simultaneously handle 300 telephone calls between the USA and Ireland. However, by 2007, transatlantic fibre optic cables had replaced satellites and the facility was wound down. It was reopened in January 2010, and redeveloped by National Space Centre Ltd (NSC) to provide internet services, as well as supporting ship tracking, maritime surveillance operations, ship-to-shore communications and a number of other satellite-based functions. [Source: Darius Bartlett]

opening by the then Bord Telecom chairman, Michael Smurfit, and government Junior Minister Ted Nealon. The Elfordstown station cost IR£8 million to build and consisted of a 32m C band antenna resting on a 300 tonne concrete base, an 11m C band antenna and a microwave tower. It was the main hub for all satellite transmissions to and from Ireland and could simultaneously handle 300 telephone calls between the USA and Ireland.

A 13.1m KU band antenna was added in 1992 by Eutelsat to take TDMA (Time Division Multiple Access) traffic from Cyprus to Ireland and on to America, as part of an earthstation network that extended across nineteen European countries. This antenna was used for broadcasting the European Union's Eurovision and Euroradio transmissions. (TDMA allows several users to share the same frequency channel by dividing the signal into different time slots.)

By 1997, transatlantic fibre optic cables had replaced satellite-based communication systems and the satellite-based system was shut down. Fibre optic cables had the advantage of being able to carry a much larger volume of information without the time delay associated with satellite communications. As operations wound down, the site was maintained by Eircom and was used as a local depot until 2010.

In January 2010 Elfordstown Earthstation was reopened and redeveloped as a teleport by National Space Centre Ltd (NSC), to provide Eutelsat/Skylogic Tooway Satellite Broadband Internet from a new 9.1m KA band antenna. It opened as part of a ten-gateway European teleport network, to provide internet services, communicating with the Eutelsat KA-SAT satellite, to users in areas where fast ASDL or fibre broadband is not available. The service went live across Europe on 31 May 2011. From 2010 to date, the development of services at Elfordstown continued with the construction of several more antennae for various purposes.

Partnering with satellite operators, National Space Centre Ltd has delivered a number of projects in the maritime domain, with a new AIS tracking antenna linking with Low Earth Orbit (LEO) satellites transmitting Earth-observation data to assist in the area of maritime and ship tracking. Other maritime-based projects included participation in a pan-European Recognised Marine Picture (RMP) project, and management of satellite-based maritime communications and associated data management.

The location criteria that were key to the original selection of the Elfordstown site as a teleport location remain as relevant today as they were back in the 1980s. As Europe's most westerly teleport, and with access to key infrastructure to support power and connectivity, including the significant transatlantic cable infrastructure passing just south of Cork Harbour, the Elfordstown site continues to evolve to serve the communications and data needs of the ever-changing global telecommunications and broadcast network.

Underwater Surveys: The INFOMAR project

Eoin MacCraith, Sean Cullen, Charise McKeon, Eimear O'Keeffe, David O'Sullivan, Ronan O'Toole, Gill Scott and Xavier Monteys

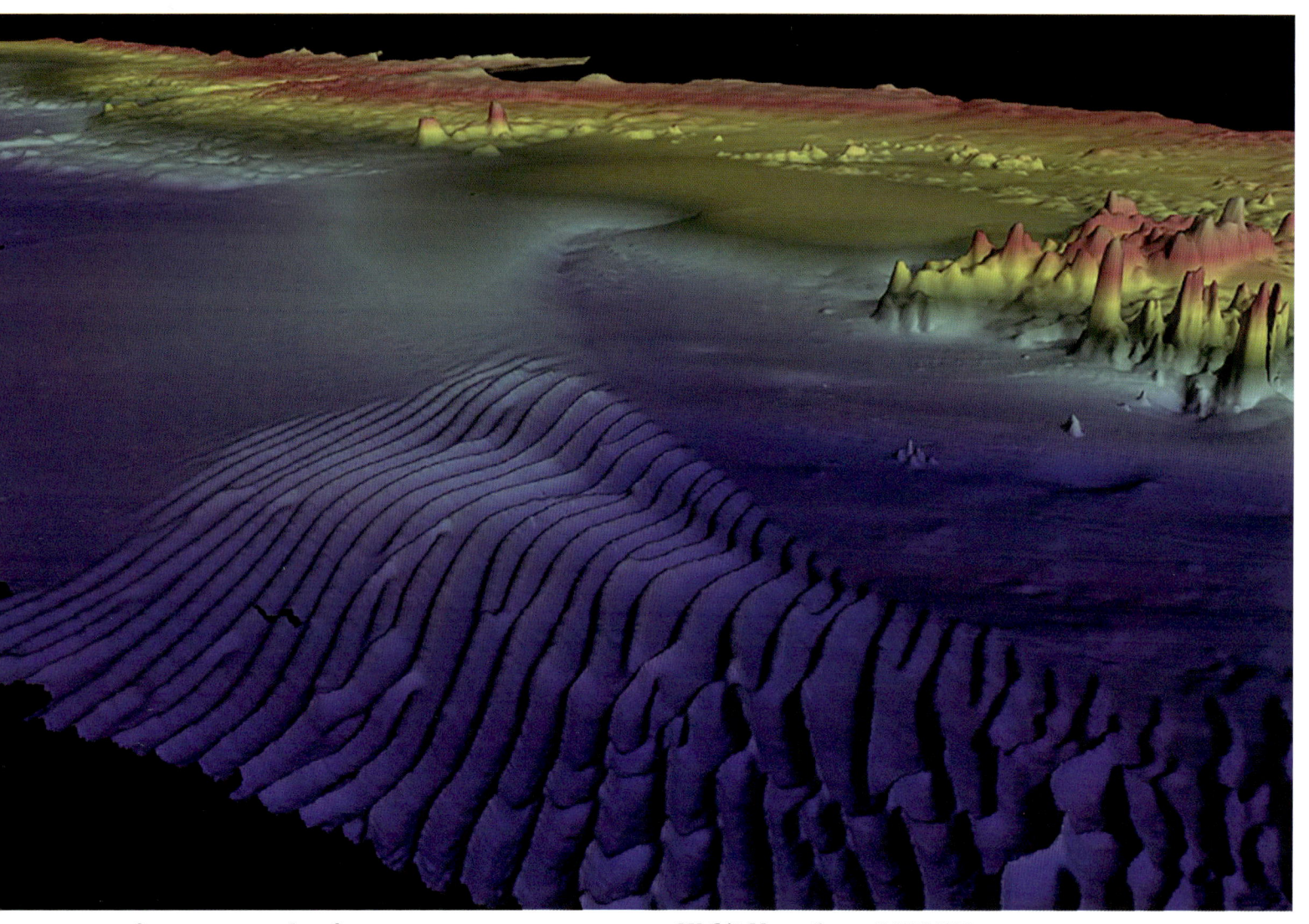

Sand waves in the Irish Sea, mapped by researchers on board the RV *Celtic Voyager*. [Source: INFOMAR]

Humankind has long been fascinated with what lies beneath the sea. When we look at the ocean, we see its turbulent surface, but the landscape beneath it is hidden from our view. For centuries we have coarsely mapped the seabed's depths, using data obtained mostly by lowering weighted lines from the decks of ships, to provide some level of safe navigation. However, it was generally believed that the deeper ocean floor was a featureless plain, and we were unaware of the details of its submarine canyons, rock outcrops and sedimentary forms, an entire submarine world of complex features and biological habitats (see Chapter 6: Glaciation and Ireland's Arctic Inheritance). As Robert Kunzig has remarked, 'the ocean remains largely unexplored – and yet oceanographic exploration in the past half century has completely transformed our view of it'.[1] Nowhere is this better seen than in Irish offshore and coastal waters.

From the birth of the state in 1922 until quite recently, bathymetric surveying of Irish waters had been very limited. For example, even as late as 2003, charts for Dublin Bay were still based on data collected in the nineteenth century by Captain William Bligh of 'Mutiny on the Bounty' fame. However, the Irish offshore mapping programme, the Irish National Seabed Survey (1999–2005), together with the subsequent inshore mapping programme, INFOMAR (2006–2026), has completely transformed this position and Irish waters are now among the most extensively surveyed at high resolution of any worldwide.

While detailed mapping of Ireland's deep offshore areas was completed in 2002, from the edge of the continental shelf at 200m depth down to the Abyssal Plain at 4,800m depth, the surveying of the inshore continental shelf is still underway and presents a number of unique challenges. In a non-intuitive turn of events, mapping shallower water with modern sonar systems is actually more time-consuming than mapping deep water. This is due to the fact that sonar has a 'footprint', much like a torch shining at a wall, which shrinks as it approaches the surface being lit. In the inner shelf, as a survey vessel makes successive passes over a particular area of seabed, each pass collects an increasingly narrower strip of data as the vessel moves into shallower water. The resulting maps have a much higher resolution than those further offshore, but with a significant increase in survey time and effort.

To quantify this, take for example a vessel surveying in water that is 1km deep. The slice of seabed that the sonar instrument is able to map on either side of the vessel is usually four times wider than the water depth. Therefore, the vessel will collect detailed depth data about the seabed across a 4km-wide strip (known as a 'swath') as it travels along its survey line. By contrast, a vessel in 10m water depth will only map a 40m-wide strip with one pass. Therefore, many more passes will be required to map an equivalent area in shallower water. To visualise this law of diminishing returns, imagine mowing a lawn with the lawnmower shrinking in size as one approaches the edge of the garden. See Dorschel et al. (2010) for a more detailed explanation.[2]

In addition, there are the risks and disruptions posed by other shipping in busy coastal areas, while working in very shallow depths close to shore also carries a significant potential hazard of striking uncharted, shallow rocks and sandbanks. The latter is a limiting factor in coastal surveys and has contributed to the so-called 'white ribbon', a zone very close to the shore lacking detailed data, that has often been shown on a map of Ireland as a blank strip running parallel to the coastline. This white ribbon also results in some potentially dangerous shoals and smooth mudflats being absent from our maps, as the diminishing returns due to the shrinking sonar footprint and difficulty of access means that some very shallow areas would take prohibitively long to survey despite their relatively small extents. Regardless of the challenges, inshore mapping is an essential activity and is fundamental both for marine development and for protection. As technology continues to be developed, the solutions to mapping the nearshore are becoming available in various forms.

Like many countries worldwide, Ireland's greatest population densities are found along the seaboard, giving rise to heavy marine traffic. Therefore, one of the primary goals of INFOMAR is the detection of hazards to navigation such as shoals (shallow rocks and sandbanks) and shipwrecks. However, the impetus for mapping is much broader than this. For example, the Irish coast is under increasing threat from climate-change induced sea-level rise, episodic flooding and coastal erosion due to increased storminess (see Chapters 30: Engineering for Vulnerable Coastlines and 33: Climate Change and Coastal Futures). Accurate coastal seabed mapping provides a baseline against which future changes can be measured and is essential to predictive modelling, so that government policies and actions can provide mitigation.

In addition, while coastal waters have always been important as a source of food and as a conduit for trade, there has recently been a technology-driven acceleration in efforts to develop the marine commercially while keeping in mind the need for sustainability, for example in the quest for offshore renewable energy (see Chapter 28: Renewable Energies: Wind, wave and tidal power). As an island nation, this 'blue growth' is very welcome and clearly requires accurate seabed information to support it, but there is also an onus on us to protect those marine ecosystem services from which we already benefit. Therefore Ireland, like many countries, is adopting an ecosystem-based management approach to marine spatial planning (see Chapter 32: Coastal Management and Planning). This is largely driven by EU legislation (EU 2014\89),[3] which obliges member states to develop a national marine strategy, including the establishment of a monitoring programme. There is a need for accurate and repeated marine mapping to underpin this.

SONAR

Darius Bartlett

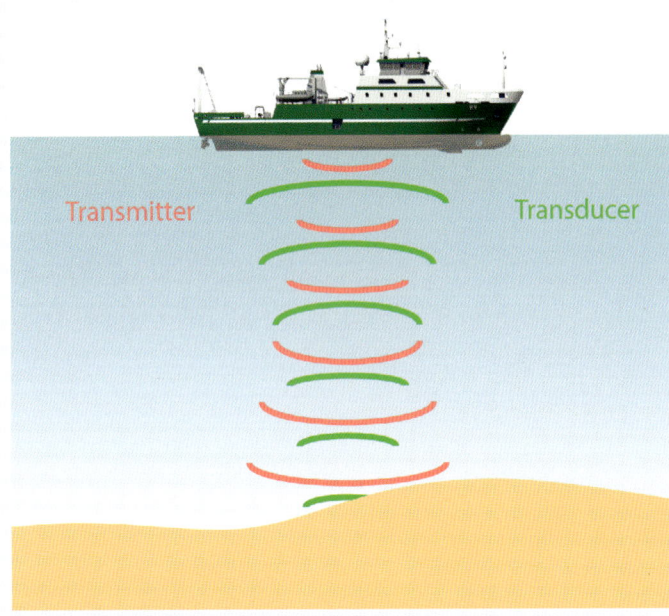

Transmitter   Transducer

Fig. 9.1 THE PRINCIPLES OF SONAR. sonar (Sound Navigation And Ranging) is a method of acoustic remote sensing, developed originally in the United States during the Second World War, and commercialised for civilian use in the 1960s. It is based on the measurement of time taken for a continuous pulse of sound to travel from a transmitter to a reflective surface (for example, the seabed, a shoal of fish, a submerged structure or even a sharp boundary between water bodies of different salinity or temperature) and back to a receiver. By also measuring differences in the attenuation of the transmitted and received sound signal, information can also be obtained regarding the roughness and density of the reflecting surface; from this its structure and composition may be inferred. Sonar is now a widely used technology in a number of applications, including seabed mapping, marine geology and archaeology, detecting obstructions, monitoring the condition of underwater pipelines and other structures, and fisheries research and management.

One of the biggest limitations of satellite or airborne remote sensing for marine applications is the inability of most forms of electromagnetic radiation, including in the optical and microwave parts of the spectrum, to penetrate any significant depth of seawater. In contrast, sound waves travel easily through the water column. Sonar (an acronym for SOund Navigation And Ranging) uses the transmission of sound waves through the water column to provide a specialised form of acoustic remote sensing, specifically suited to a number of applications in the marine sector (Figure 9.1).

The basic principle of sonar is that a stream of sound wave pulses is emitted from a transmitter located either on a ship, or else from a device towed behind the vessel (Figure 9.2). The sound is reflected off the seabed or off certain other objects within the water column itself (such as a submerged submarine, a shoal of fish, or even a sharp boundary between water bodies of different temperature or density) and the return signal is picked up by hydrophones.

Side scan sonar, as the name suggests, casts the beam of sound waves from the transmitter outwards and sideways into the water column, so that it strikes the target at an angle. If the sound waves strike an object that is raised above the seafloor, or suspended in the

Fig. 9.2 SIDE SCAN SONAR BEING DEPLOYED OFF THE STERN OF THE RV *Keary* IN WATERFORD ESTUARY. [Source: INFOMAR]

water column, it results in a shadow being cast behind the reflector. This makes the side scan a successful tool in object detection, and it has found applications in seabed mapping and geological investigations of the Irish continental margin, searching for shipwrecks, monitoring the condition of pipelines and other installations, underwater archaeology, assessing the impacts of trawl fishing on seafloor biological communities and the forensic investigation of certain crimes at sea.[4]

Multibeam echosounders (MBES) provide the other main form of sonar technology and are well suited to bathymetric surveying, providing 100 per cent coverage of the seafloor to better than International Hydrographic Organisation (IHO) specifications.[5] This was the primary surveying technology employed in the Irish National Seabed Survey (INSS), established in 1999 and, since 2006, the INFOMAR programme. In addition to providing information about the depth of water and, hence, the submarine topography, the intensity of the returning signal (known as backscatter) can also give valuable information about the nature, texture and composition of the seabed itself, as well as the structure of the water column and objects such as fish shoals suspended within it.

THE INFOMAR SURVEY FLEET

It is necessary for INFOMAR to maintain a fleet of survey vessels of differing sizes. Each vessel is particularly suited to a particular range of water depths, sea states and operational distance from shore. All of these aspects come into play when selecting the combination of survey vessels needed to map a given area of seabed. While smaller vessels, having shallower draughts and tighter turning circles, are most appropriate to nearshore areas, larger vessels have the advantage of greater weather and sea state endurance (Figure 9.3).[6] In practice, however, the smaller INFOMAR inshore vessels tend to work in association with the largest, such that all of the target coastal area, from deepest to shallowest, is surveyed simultaneously. In addition, the larger vessels provide survey support, for example by supplying accommodation and data processing facilities. Each survey vessel is a self-contained scientific platform that uses advanced mapping systems in conjunction with highly accurate satellite-based positioning to pinpoint depth soundings on the seafloor. Additionally, some vessels carry geophysical equipment, remotely operated vehicles and mechanical sampling gear for retrieving sediment. Finally, and most important, are the teams of dedicated marine scientists and crew that operate these vessels safely.

The RV *Keary* is a highly versatile inshore survey vessel. At 15m in length but with a draught of just 2.1m with sonar equipment deployed, this aluminium catamaran is used to survey nearshore waters deeper than 5m. The boat can sleep up to four crew and routinely provides support to the smaller vessels in the fleet, which include the RV *Tonn*, a 7.9m Cheetah catamaran, and the RV *Geo*, a 7.4m Redbay Stormforce RIB (Rigid Inflatable Boat). With draughts of 0.7m (with survey equipment deployed) and 1.3m respectively, these smaller vessels can survey in all but the shallowest of waters and can pass close to the shoals and rocky outcrops that are essential to map, but which are too close to the surface for vessels such as the RV *Keary*. All vessels are operational throughout the summer and autumn months.

For the deeper shelf waters, INFOMAR utilises the Marine Institute's RV *Celtic Voyager* and RV *Celtic Explorer*. At 31.4m in length, the *Celtic Voyager* can sleep up to eight survey crew in addition to ship's personnel. Less hindered by shoals in its deeper operational areas, the vessel can survey continuously on a twenty-four-hour basis, providing extensive coverage in a relatively short period of time. The *Celtic Explorer* is a 65.6m long, multi-purpose

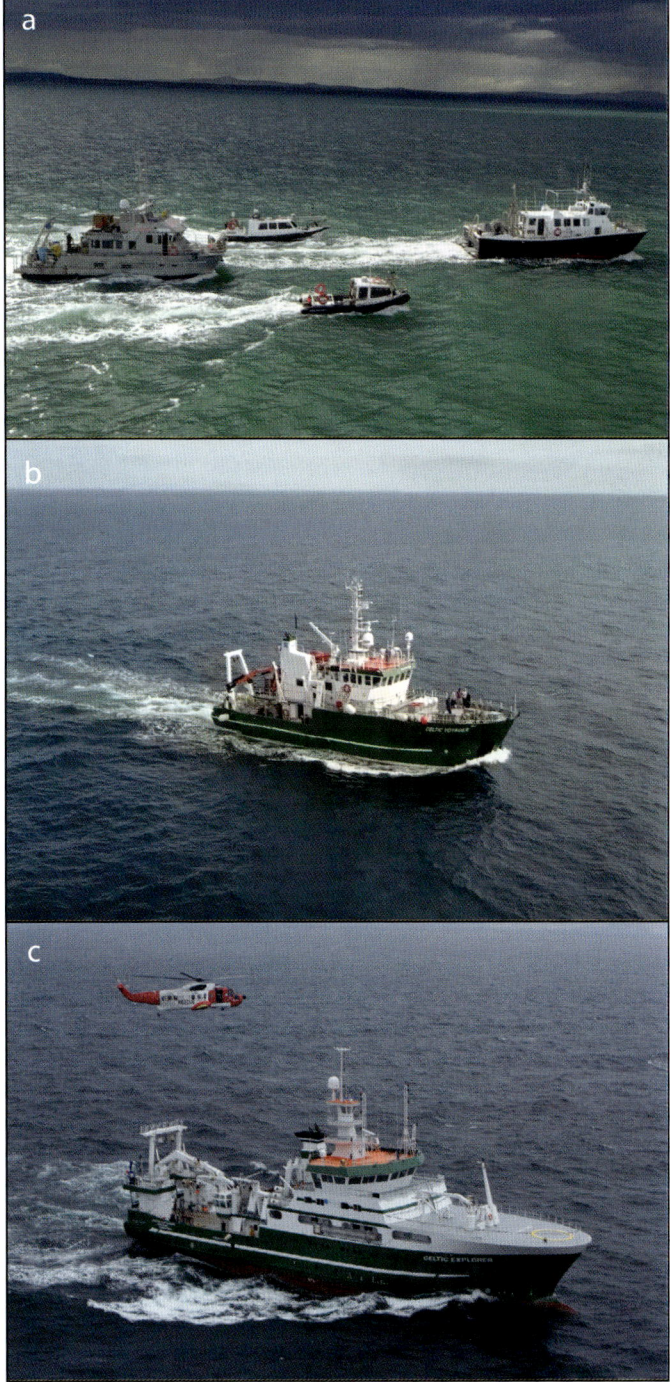

Fig. 9.3 THE INFOMAR FLEET. a. left to right: RV *Keary*, RV *Lir*, RV *Geo* and RV *Mallet*, operated inshore by Geological Survey Ireland; b. RV *Celtic Voyager*, operated by the Marine Institute; c. RV *Celtic Explorer*, operated by the Marine Institute. [Source: INFOMAR]

research vessel and can accommodate thirty-five personnel, including up to twenty-two scientists. This vessel has a maximum endurance of thirty-five days and is well suited to surveying in Ireland's farthest offshore regions.

Two new inshore survey vessels, commissioned by Geological Survey Ireland, have been available to the INFOMAR programme since early 2017 – the RV *Mallet*, an 18m fibreglass catamaran, and the RV *Lir*, an 11m RIB. Together with the existing vessels, these will significantly broaden the mapping effort in Irish coastal waters. Like the *Keary*, the *Mallet* can provide accommodation and data processing facilities both for its own crew and for those of the smaller vessels, along with offering an additional survey platform for working that bit further from the coast with greater endurance. The *Lir* is a particularly versatile craft, being able to survey in extremely shallow waters as well as further offshore when needed. Its ability to transit at high speed, combined with a secure, enclosed and spacious cabin area, places it between the *Keary* and the *Geo* in terms of sea-keeping and endurance.

CONDUCTING A SEAFLOOR SURVEY

The start of the survey season for INFOMAR is always preceded by a period of planning and preparation, during which system checks, maintenance and trouble-shooting are carried out. The INFOMAR team strives to keep abreast of any technological advances that can increase the effectiveness of the survey fleet and so, during the winter, sonar and positioning systems are often upgraded and replaced. The vessels are sometimes brought into dry dock to provide better access to the sonar equipment, which hang usually below the keel, along with receiving a scrubbing and repainting of the hull to remove bio-fouling.

In addition, for each survey area, an advance team must install an onshore GPS base station to increase the accuracy of the vessels' positioning, along with a tide gauge to determine the level of the lowest astronomical tide (see Chapter 2: The Coastal Environment: Physical systems processes and patterns, box on tides). When this onshore infrastructure is in place, the vessel or fleet can make its way to the survey area. It is only once on-site that the complexities of the coastline can be properly assessed and the survey lines finalised. For the smaller boats, which must negotiate the nearshore rocks and shoals, these lines are often challenging and technical, but they are relished by the crew, who get to see a unique aspect of Ireland's coastline from close quarters.

Survey days start early and finish late when conditions are good. As for all mariners, the crew's ability to read the signs of on-coming weather and sea states is crucial. Survey plans need to be flexible in order to take advantage of more sheltered conditions on one side of a bay or behind a headland, for example. Even in perfect conditions it may be necessary to deviate from the planned course to accommodate other marine traffic or to avoid deployed fishing gear.

The coastal areas around Ireland receive a lot of freshwater run-off, which reduces seawater salinity and shallow waters are also more susceptible to heating and cooling effects. These variations result in refraction of the sound waves from the mapping sonar,

bending them and potentially degrading the quality of the depth data received. To correct for this effect, regular vertical sound velocity profiles (SVP) are recorded by lowering a probe to the seabed. In areas of high variability, such as in estuaries, it may be necessary to take thirty or more of these profiles in a single day; in contrast, offshore work can often require only one or two profiles per day. The sound velocity information from these profiles is incorporated into the processing of the bathymetric data, effectively countering the refraction effects.

A small overlap is maintained between successive survey lines, to ensure comparability of data. The amount of overlap is increased in very shallow waters or near treacherous coastlines, to ensure that the deepest part of the vessel stays within the area that has already been mapped on the previous pass. Once significant shoals have been identified, additional lines are run across the area to ensure that their shallowest point has been detected. In addition, following completion of the mapping a set of cross-lines are run, perpendicular to the main grid of survey lines and at a much wider spacing. These, together with regular SVPs, are essential to ensure the integrity of measurements and are required for compliance with the International Hydrographic Organisation standards to which the INFOMAR survey data adheres.[7]

FROM INSTRUMENT TO MAP

The main instrumentation deployed during a survey is the multibeam echosounder (MBES). This emits acoustic energy which is then reflected back from the seabed. The time taken for the acoustic return is directly proportional to the travel distance and can be converted to water depth; additionally, the strength of the return, known as 'backscatter', is a function of the seafloor's hardness and can be used to extract information such as sediment type. However, to do this the backscatter data must be validated with 'ground truthing', or direct seabed samples for a selection of locations. These samples can be collected physically using a sampler deployed from the vessel, or they can take the form of imagery collected by underwater video mounted to a sledge which is dropped from the vessel and dragged across the seafloor for a short distance (Figure 9.4).

Processing of MBES data is commonly the most challenging and time-consuming step in bathymetric mapping, depending on the MBES systems used, the water depths mapped, the planned uses of the data and on issues encountered during acquisition. Processing is initially carried out on board the vessel, but can continue long after the survey has finished. The processing workflows generally comprise the following steps: correction for vessel motion; conversion of raw readings to depth values; correction for water column effects based on the SVP readings; vertical corrections for tide and for changes in vessel dynamics; and reduction of depths relative to a specified vertical datum (usually lowest astronomical tide). The final and most time-consuming step – data cleaning to remove outliers and anomalous data points – is usually achieved using a combination of manual and statistical methods, along with a range of visualisation techniques. One of the most common causes of data error, for example, stems from the

Fig 9.4 'GROUND TRUTHING' VIDEO STILL, CAPTURED WITH A DROP CAMERA IN MULROY BAY, COUNTY DONEGAL. [Source: INFOMAR]

presence of kelp on a rocky seabed; the acoustic reflections caused by the kelp cover need to be cleaned and removed from the data, in order to reveal the actual seabed. Shoals of fish can also cause false returns. Once the data are cleaned, digital elevation models can be produced by interpolating and gridding to the optimal spacing. Finally, INFOMAR then produces a wide range of seabed maps and map products which are made freely available to download from a number of web portals.[8]

Shallow seismic data are often acquired alongside multibeam data. Seismic instrumentation also emits acoustic pulses, but at lower frequencies than the MBES, and these can penetrate up to 40m below the seafloor, providing an image of the shallow subsurface sediments and sometimes depth to bedrock. The shallow seismic data are also processed, before being merged with navigation-based information. The final outputs are a suite of seabed and subsurface datasets that, once subject to expert interpretation, can be used to compile integrated seabed geology maps (Figure 9.5).

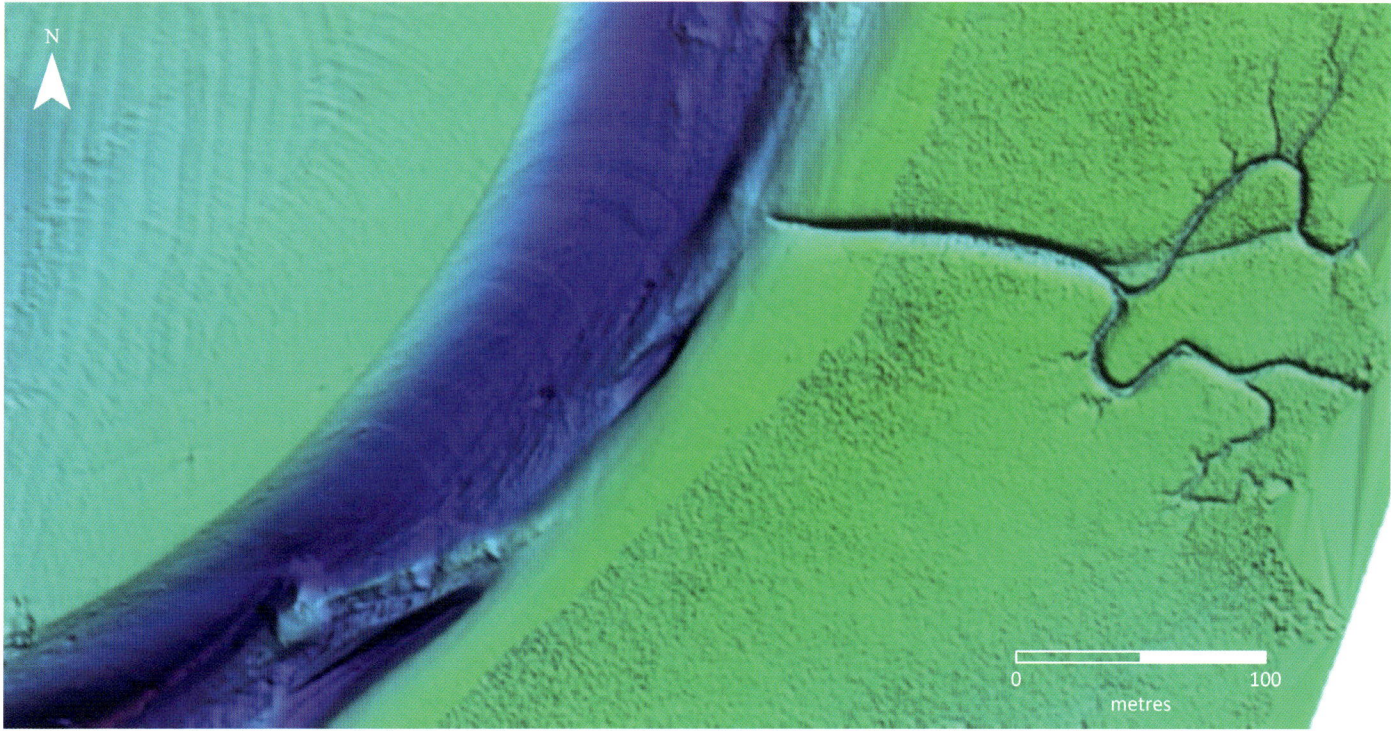

Fig. 9.5 INTEGRATED MBES BATHYMETRY (DEEP CHANNEL ON THE LEFT) AND UAV TOPOGRAPHY DEM (MEANDERING CHANNELS IN THE INTERTIDAL ZONE TO THE RIGHT): THE 'WHITE RIBBON'. [Source: INFOMAR]

Fig. 9.6 VERTICAL AERIAL PHOTOGRAPH (ORTHOPHOTOGRAPH), SHOWING OMEY ISLAND AND ADJACENT COASTLINE, CONNEMARA. The idea of taking photographs from the air dates back to the mid-1800s, but flourished particularly during the First World War, when it was practised by both sides in the conflict as an aid to reconnaissance and rapid updating of maps. This led, after the war, to the emergence of commercial and civilian air photography, and the needs of high-accuracy peacetime surveying and cartography. For these purposes the captured air photographs needed to be rectified to correct for distortions caused by the angle and motion of the aircraft and variations in the distance between airplane and ground due to the rise and fall of the terrain. These rectified photographs are known as orthophotographs. They have a uniform scale across all areas of the image and fit a standard coordinate system (Irish National Grid in the case of air photography produced in Ireland). Mapping agencies such as the Ordnance Survey of Ireland (OSI) make extensive use of orthophotography, today in digital form, for the production and updating of their various map products. OSI has also been capturing aerial photography of Ireland for several decades, for sale to third-party customers. It operates its own dedicated survey aircraft, based at Shannon Airport, to acquire high-resolution digital imagery for all twenty-six counties of the Republic, with complete update and revision at least once every three years. The image shown here was captured with a pre-digital, large-format film camera. It is oriented facing south, and shows Omey Island (along the bottom), Knightstown Bay and Streamstown Bay (on the left) and Inishturk and Turbot islands (near top right). It was captured by the French national mapping agency, IGN, under contract to the Geological Survey of Ireland, on 25 April 1973. [Source: © Ordnance Survey Ireland/Government of Ireland Copyright Permit No. MP 000620]

MAPPING FROM AIR AND SPACE

Aerial mapping is particularly important in mapping the 'white ribbon' between land and deeper water, given the difficulties posed by shoals and other navigational hazards to traditional, ship-based methods. Airborne mapping was first carried out in Ireland in the 1950s, when the first campaign to gather aerial orthophotography of the country was run by the Irish Air Corps. Between 1973 and 1977, Geological Survey Ireland commissioned the first complete aerial photographic survey of the Republic of Ireland, primarily for photo-geological interpretation. The survey was flown at a height of 15,000ft to give a nominal scale of 1:30,000. The Ordnance Survey of Ireland (OSI) has also carried out aerial mapping for several decades and maintains its own aircraft and crew, based at Shannon Airport, for this purpose. The early photographic surveys were carried out using large-format film cameras, mounted in the fuselage of dedicated aircraft (Figure 9.6). Small amounts of overlap between successive frames ensured complete coverage and also allowed the use of the photographs as stereoscopic pairs for extracting elevation information, a technique known as photogrammetry. Appropriate georectification techniques ensured that each photograph was positioned accurately in terms of map coordinates. Today, digital cameras and GPS are used, yielding data that can be readily worked with using standard image processing technologies. Whether from film or from digital sources, historic and present-day photographs can provide an excellent resource from which to map Irish land cover, including coastal environments and the intertidal zone.[9]

Another aerial mapping technique, airborne LiDAR (Light Detection and Ranging), was developed in the 1980s and became available widely from the 2000s. LiDAR is a remote sensing method that uses light in the form of a pulsed laser to measure distance to a target surface, generating precise elevation maps (see Chapter 8: Monitoring and Visualising the Coast, box on laser technologies), and LiDAR datasets collected in Ireland by the OSI and the Office of Public Works (OPW) have been used by INFOMAR to delineate the coastline and the exposed intertidal zone. In addition, INFOMAR has commissioned airborne marine LiDAR for bathymetric mapping. Marine LiDAR uses laser light that is generated simultaneously at two different wavelengths, and hence colours: green/blue laser light is able to penetrate shallow water depths, while red light does not and is reflected directly from the sea surface, so that the difference in return time between the two beams can be used to calculate the water depth. Such systems have proven successful in surveying shallow depths (0–30m) in clear coastal waters,

given the right environmental conditions. However, water penetration by LiDAR is impacted by turbidity and cavitation (aeration of the water caused by breaking waves or strong currents), which is commonplace in the coastal zone, especially near rivers and the utility of the technique is also impaired by cloud cover and precipitation. In general, LiDAR surveys have proved to be more successful on the west coast of Ireland, where water clarity is significantly higher than in the Irish Sea.

Satellite-derived bathymetry in coastal areas is based on multispectral data from the optical part of the light spectrum (see Chapter 8: Monitoring and Visualising the Coast, box on satellite remote sensing of the coastal regions of Ireland). This can provide information on the seabed to a maximum of 30m water depth. The technique is based on the principle that as light passes through water it becomes attenuated to differing degrees, depending on the wavelength of the light, so that shallower areas appear brighter and deeper areas darker. Sun glint and atmospheric corrections need to be applied to the data and, like LiDAR, water turbidity can interfere with light penetration.

The utility of optical satellite imagery for predicting water depth in Irish coastal waters was demonstrated by INFOMAR geoscientists who applied spatial statistics to satellite imagery from Dublin Bay to produce water depth predictions.[10] While the level of uncertainty contained in these predictions means the method is unsuitable for producing navigation charts, the method does have useful applications in environmental monitoring, seabed mapping and coastal zone management. Work on this is continuing, with the aim of developing better models and improving predictions, but while these are unlikely to exceed sub-metre accuracy, the advantage of satellite bathymetry is that it is timely, cost effective and can be repeated at regular intervals. As a result, the INFOMAR programme is exploring the potential of time-series satellite data for monitoring coastal change.

The potential of Unmanned Aerial Vehicles (UAV) or drones to map the white ribbon is another area of active development by INFOMAR. With careful planning of survey operations to coincide with low spring tides, coastal zone areas can be mapped to absolute accuracy of a few centimetres, using photogrammetry alongside standard red-green-blue photographs taken by Single Lens Reflex (SLR) cameras mounted on a UAV. The use of UAVs for coastal mapping has verified applications for quantifying changes along the coastline through repeat surveys. Other applications in development are the validation of Earth Observation data products and bathymetry derived from photogrammetry.

UAVS FOR COASTAL ZONE MAPPING

Ronan O'Toole

Recent advances in Unmanned Aerial Vehicles (UAVs), also known as Remote Piloted Aerial Systems (RPAS), Small Unmanned Aircraft

(SUA), Unmanned Aerial Systems (UAS) or simply 'drones', have already made disruptive impacts on activities across many sectors and industries, with applications as varied as agriculture and film-making; and UAVs are also starting to revolutionise the surveying and land-mapping industries (Figure 9.7). Several inherent benefits of UAVs are contributing to a change in perception of these devices and their growing acceptance as powerful data collection tools for the professional. These are related largely to availability and cost of deployment, alongside developments in automation, onboard position and motion sensing and ever-improving computer software, hardware and data processing solutions, all of which have allowed their advocates to make compelling cases for their use.

Ireland's National Seabed Mapping programme, INFOMAR, has taken a proactive approach to this new technology, where its use is being investigated for mapping the 'white ribbon' of Ireland's intertidal zone, thereby providing a much-needed means of integrating terrestrial topographic data with INFOMAR's bathymetric (below sea-level) datasets acquired using vessel-borne sonar. A series of standard operating procedures have now been developed for use in four UAVs (two fixed-wing and two multi-rotor aircraft) operated by Geological Survey Ireland, which encompass all aspects of aerial operations from training, regulations and licensing, to best practice in field procedures and data processing methodologies. This has led directly to a GSI/INFOMAR capacity for carrying out airborne mapping operations.

The datasets resulting from the use of these drones are impressive. High-resolution photorealistic three-dimensional (xyz) point clouds of 5cm accuracy can be generated repeatedly and reliably. One twenty-five minute flight will result in the mapping of more than one square kilometre at this high level of detail, which produces corresponding benefits in terms of information that may be extracted from the Digital Terrain Model (DTM) and orthomosaic imagery (Figure 9.8). As well as producing intuitive and realistic terrain models for effective visualisation, the level of accuracy makes it possible for GIS (Geographic Information Systems) analysts to place important quantitative values on geomorphological features such as dunes, drumlins and slopes at risk of failure. These capabilities are being harnessed and evaluated as part of a European-funded research project called CHERISH.

CHERISH (Climate, Heritage and Environments of Reefs, Islands and Headlands) brings together researchers from the Royal

Fig. 9.7 Unmanned Aerial Vehicle (UAV or drone) and operator near the Maharees, County Kerry. Developed initially for military uses from the early years of the twentieth century onwards, recent rapid advances in miniaturisation of the technology, accompanied by falling production costs, have allowed Unmanned Aerial Vehicles (drones) to enter widespread civilian use by both professionals and hobbyists. The first reported use of drones for commercial purposes occurred in Japan in the 1980s and a report published in November 2018 by management consultants, PwC, in conjunction with the Engineering Council of India, predicted that by 2021 the global UAV market would reach US$21.47 billion (PwC and Engineering Council of India, 2018. *Flying High: Drones to drive jobs in the construction sector*. Available online at https://www.pwc.in/assets/pdfs/publications/2018/flying-high.pdf)). Today, drones are finding application within a growing range of industries and professions, including agriculture, transport, security and law enforcement, media and entertainment, mining, telecommunications, insurance and commerce, many of which extend their use of these devices into the coastal and maritime domains. Several countries are putting in place policies and regulations to ensure responsible use of drones, or have already done so. In Ireland, the Irish Aviation Authority (IAA) currently requires all UAVs greater than 1kg to be registered and operation of a UAV heavier than 4kg requires a licence issued by the IAA. For coastal scientists and managers, UAVs greatly facilitate the examination and recording of sections of shore that would otherwise be inaccessible, as well as enabling repeated, rapid and cost-effective surveying, monitoring and mapping of areas of interest. [Source: Eugene Farrell]

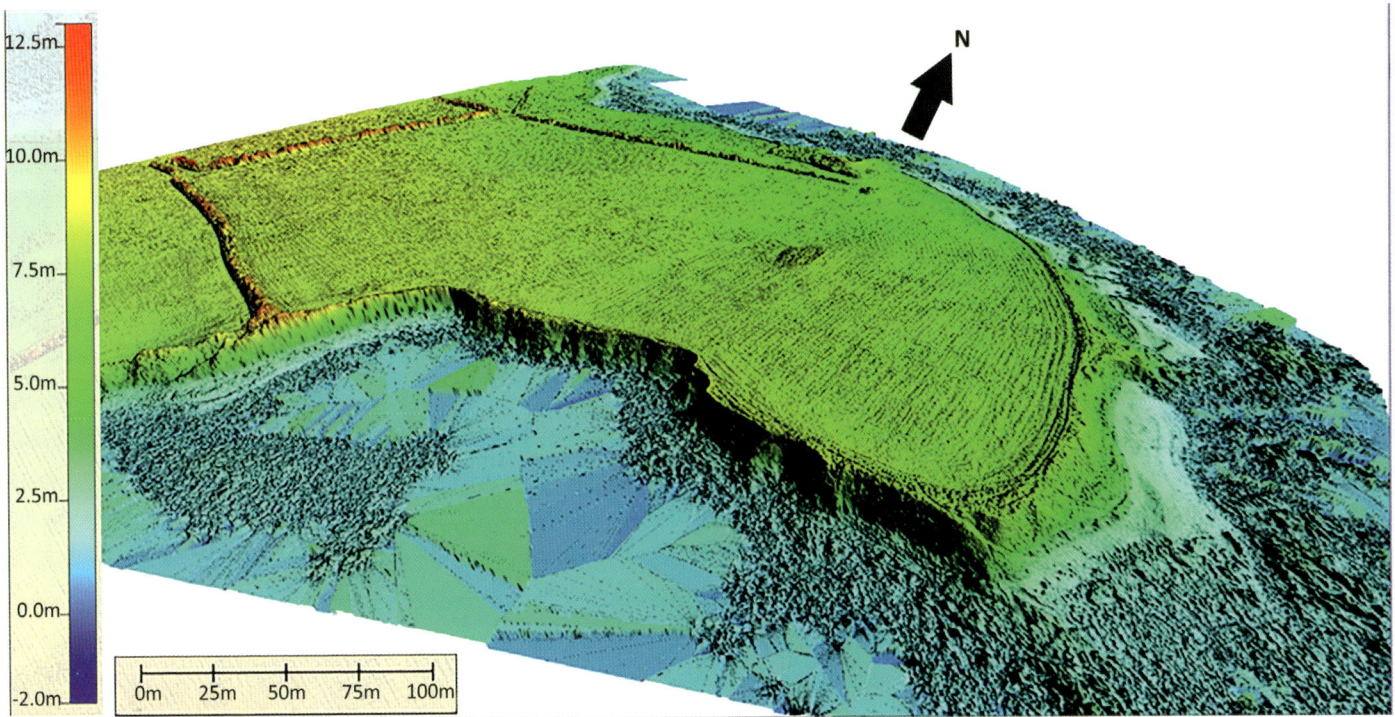

Fig. 9.8 BRAEMORE HEAD, COUNTY DUBLIN, MODELLED USING A COMBINATION OF BATHYMETRIC DATA OBTAINED USING SHIP-BASED SONAR AND TERRESTRI-
AL DATA CAPTURED BY MEANS OF AN UNMANNED AERIAL VEHICLE (UAV OR 'DRONE'). The image shows the transition from terrestrial headland (green) into
intertidal and offshore rock outcrops (blue), and demonstrates how the integration of data captured through these two separate but complementary technologies can be used to map the problematic 'white ribbon' between land and sea. [Source: INFOMAR]

Commission on the Ancient and Historical Monuments of Wales, the Discovery Programme's Centre for Archaeology and Innovation
Ireland, Aberystwyth University's Department of Geography and Earth Sciences and Geological Survey Ireland. It aims to raise awareness
and understanding of the past, present and near future impacts of climate change, storminess and extreme weather events on the rich
cultural heritage of the Irish and Welsh regional seas and coasts.

Building on INFOMAR's capacity, the CHERISH team at Geological Survey Ireland have put UAVs to great use in mapping the coastal
zone in areas around Ireland. These UAV data are used to establish baseline monitoring of coastal geomorphology and, through repeat
surveys, are being used to estimate volume loss of material at sites that are vulnerable to coastal erosion. In addition, the team have deployed
a tripod-mounted laser scanner in the field (see Chapter 8: Monitoring and Visualising the Coast, box on laser technologies) and merged
the data with that of the UAV to get full three-dimensional coverage of coastal features.

The experience gained from the CHERISH project, as well as from other exploratory initiatives, has shown that UAVs can be deployed
for coastal monitoring purposes successfully, rapidly and inexpensively. Coastal areas most at risk from erosion, pollution and environmental
change may therefore be targeted, mapped and monitored more effectively, and repeat surveys can be conducted as often as necessary. The
unprecedented level of detail available in the derived imagery (typically 1.5–3cm), and advances in automated image segmentation, mean
that small-scale features, such as seaweed and even plant species type, may be mapped at resolutions that were previously impractical and
for a fraction of the cost of traditional airborne techniques. The technology also holds significant potential as a means of validating and
ground truthing coarser regional-scale maps and models. The incorporation of UAV workflows into pre-existing Earth Observation
methods such as satellite and aircraft-based mapping is an area for future development. Successful management strategies will require good-
quality information in key areas of the coast, in order to underpin effective decision-making. Datasets of high resolution and accuracy, such
as those now available through the INFOMAR programme, will undoubtedly prove useful to individuals and organisations involved in the
observation, management and protection of Ireland's coastal areas and UAV-generated data are now part of this useful resource.

Marine Spatial Planning (MSP) has been widely adopted globally to
manage oceans and coastal regions in an efficient, safe and sustainable
manner (see Chapter 32: Coastal Management and Planning). The
overlap between human activities and the pressure to manage and
protect the coastal environment is one of the most serious societal
challenges we face during the next decade, especially in the light
of climate change and sea-level rise scenarios (see Chapter 30:
Engineering for Vulnerable Coastlines). Precise navigation is also of
the utmost importance in maintaining safety at sea. Modern marine
surveying and aerial techniques are fundamental to map our coastal
regions effectively and the wealth of data collected from the
INFOMAR programme has supported a wide and growing range of
applications, covering studies of geology, marine eco-systems,
shipwrecks and safety at sea in Irish coastal and marine territories, as
well as new technologies in addressing climate change. Some of these
applications are described in further detail in the accompanying
featured sections.

INVESTIGATING THE WRECK OF THE GUINNESS SHIP, THE SS *W.M. Barkley*

Charise McKeon

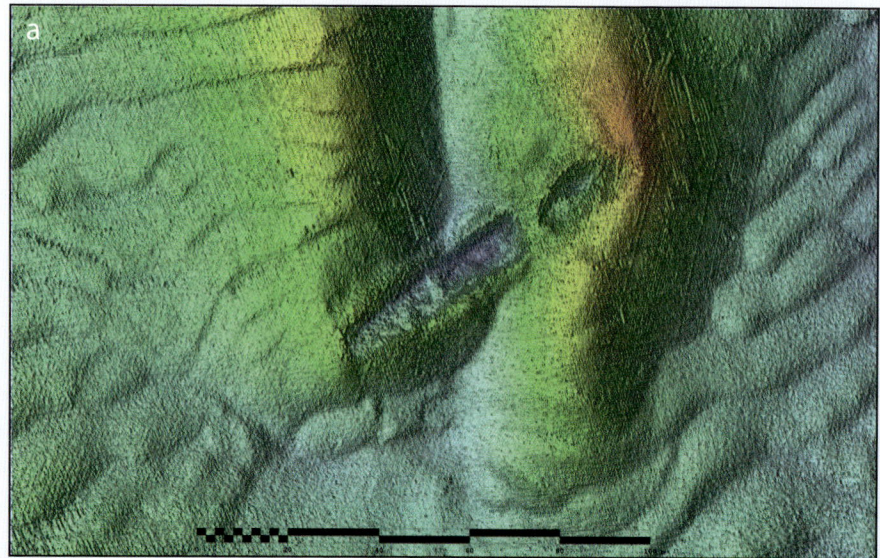

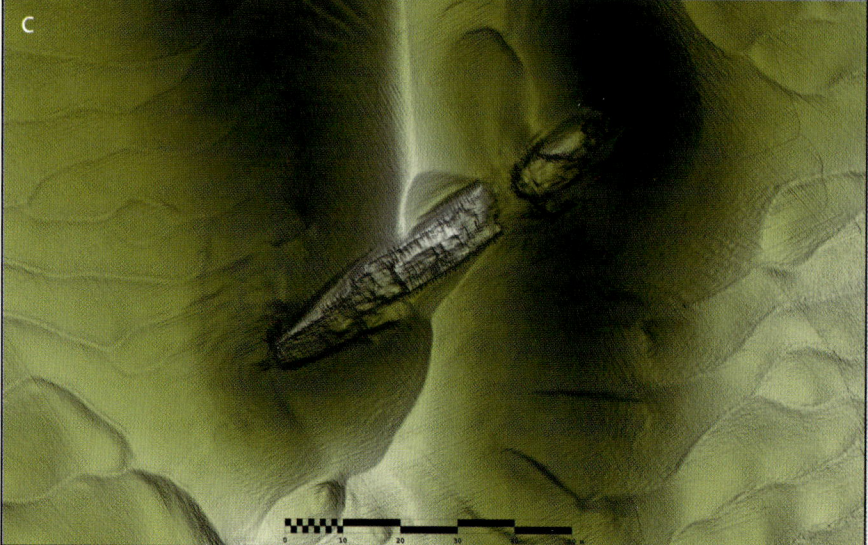

Ireland has a rich underwater cultural heritage (see Chapter 15: Coastal Heritage) and – according to the Shipwreck Inventory of Ireland, maintained by the Underwater Archaeology Unit at the National Monuments Service – almost 15,000 shipwrecks are believed to lie in Irish marine, coastal and inland waterways. Locating, mapping and visualising these shipwrecks is an integral part of the INFOMAR programme and, to date, over 300 have been recorded. This provides a vital dataset for users as diverse as the UK Hydrographic Office, for the production of navigational charts (see Chapter 8: Monitoring and Visualising the Coast, box on digital mapping and charting); for state archaeologists and their archives; for the diving community; and, most importantly, for those who are interested in what lies beneath our seas.

Twenty-six kilometres east of Howth Head, the wreck of the SS *W.M. Barkley* lies at a general water depth of 56m (Figure 9.10). This steamship had been launched as a coaster (cargo vessel) in June 1898, by the Ailsa Shipbuilding Company in Troon, Scotland, and, in December 1913, became the first steamship owned by the Dublin brewery company, Arthur Guinness and Sons Ltd. During most of the First World War the vessel, along with other Guinness ships, was commandeered by the British Admiralty and used to transport cargo, mainly to France. Having been returned to the Dublin brewery, on 12 October 1917 en route from Dublin to Liverpool carrying a delivery of stout, the SS *W.M. Barkley* was torpedoed without warning by German submarine, UC-75, *c.*11km east of the Kish Light Vessel (which was in place and operational throughout the war). The cargo of barrels helped keep the vessel afloat just long

Fig. 9.9 MULTIBEAM TIME-SERIES IMAGES OF THE SS W.M. *Barkley* SHOWING HOW SUCCESSIVE SURVEYS HAVE RESULTED IN IMPROVED DETAILED IMAGERY. a. 2010 (EM3002 multibeam survey by INFOMAR, using the EM3002 sounder manufactured by Konsberg. This sounder works at only one frequency and is used to survey depths in the range 1m to 350m approximately); b. 2011 multibeam re-survey by INFOMAR, using the EM3002 sounder; and c. 2015 multibeam survey by F. Sacchetti/R. Plets from the University of Ulster and INFOMAR, using Konberg's EM2040 Sounder. This improved technology works at three frequencies, and produces high-resolution imagery in shallow to medium water depths (1–500m). [Source: INFOMAR]

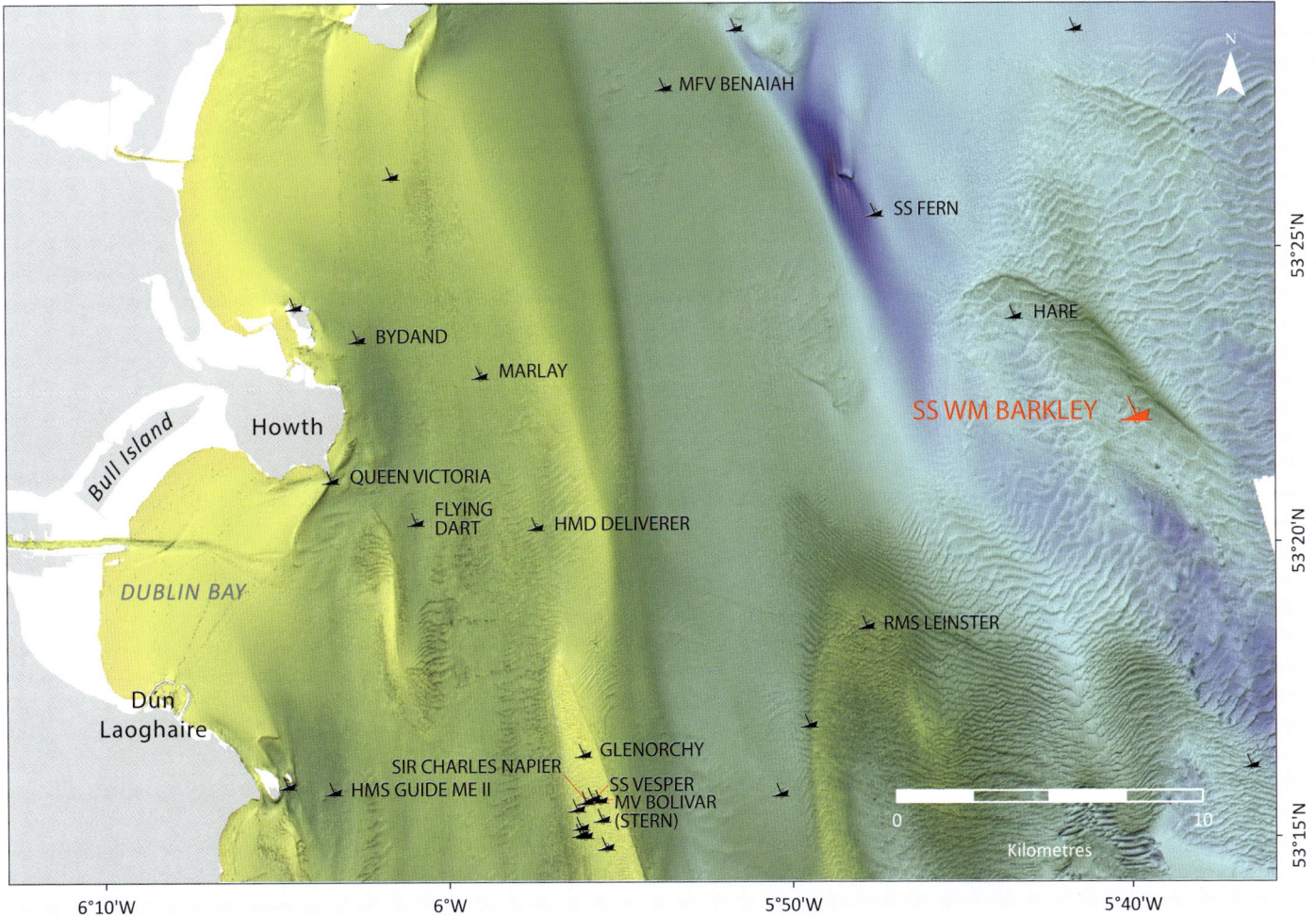

Fig. 9.10 LOCATION OF THE WRECK OF THE SS *W.M. Barkley* AND OTHER KNOWN SHIPWRECKS IN THE VICINITY, WITH SHADED RELIEF BATHYMETRY ALSO SHOWN. [Data source: INFOMAR]

enough to enable some of the crew to launch a lifeboat and escape, but the steamship quickly sank with the loss of four lives, including the ship's master. In the days that followed, casks of Guinness washed ashore along the Welsh and Irish coasts.[11] A model of the SS *W.M. Barkley* is on display and can be viewed in the Guinness Storehouse in Dublin.

The wreck of the SS *W.M. Barkley* is protected under the National Monuments (Amendment) Acts 1987 and 1994. Diving on the wreck requires a licence and can be challenging as visibility is usually quite poor. It measures 69m long and 9m wide, and is orientated in a north-east–south-west direction. In 2010, a full site survey of the wreck was acquired for the first time, using multibeam sonar onboard the RV *Celtic Voyager*, as part of a wider INFOMAR seabed survey. The site of the SS *W.M. Barkley* was subject to a second, targeted, survey by INFOMAR in 2011 and again in 2015 by a team of marine scientists from the University of Ulster and INFOMAR, as part of a project surveying First World War wrecks lost in the Irish Sea. The 2015 survey used a higher-resolution multibeam system that had only recently been fitted to the Voyager and extra lines were run across the wreck at slower speeds, producing a far denser dataset and capturing much more detail (Figure 9.10).

The resulting imagery is the best achievable with current technology and has enabled better analysis and interpretation of the wreck. From this it could be determined that the stern of the vessel is still intact, though the deck plating has corroded away, while the mid-ships superstructure, with the hole for the funnel in the middle, is clearly discernible. The wooden decking has largely collapsed along the length of the wreck, exposing the ship's holds below. Forward, the wreck is broken up in the vicinity of the torpedo strike and the bow is essentially separate from the main body of wreckage. A sediment scour has developed at the north-east end of the vessel, exposing its crippled bow.

MAPPING HERRING SPAWNING BEDS WITH REPORTED FISHERIES AND BACKSCATTER DATA

David O'Sullivan

The Atlantic herring, *Clupea harengus*, is an important commercial species in Irish waters (see Chapters 3: Coastal Waters: Marine biology and ecology and 26: Coastal Fisheries and Aquaculture).[12] The species is also ecologically important, as it occupies a central position in the pelagic marine food-web. Spawning takes place in coastal waters, mostly between September and February in high-energy environments, and usually at the mouths of bays and estuaries. Herring are benthic spawners and require specific substrates on which to lay their eggs, primarily gravel, stones, broken mussel shell and/or flat rock.[13] They often form dense spawning aggregations, that can be prone to high mortality from even relatively light fishing activities. Spawning grounds are also vulnerable to disturbance from deposits of dredge spoil and from the construction of structures on or over the seabed. Given this susceptibility to anthropogenic activities, mapping of spawning grounds is important, as it ensures that such information is in the public domain and available for decision-making in the context of marine spatial planning and fisheries management.

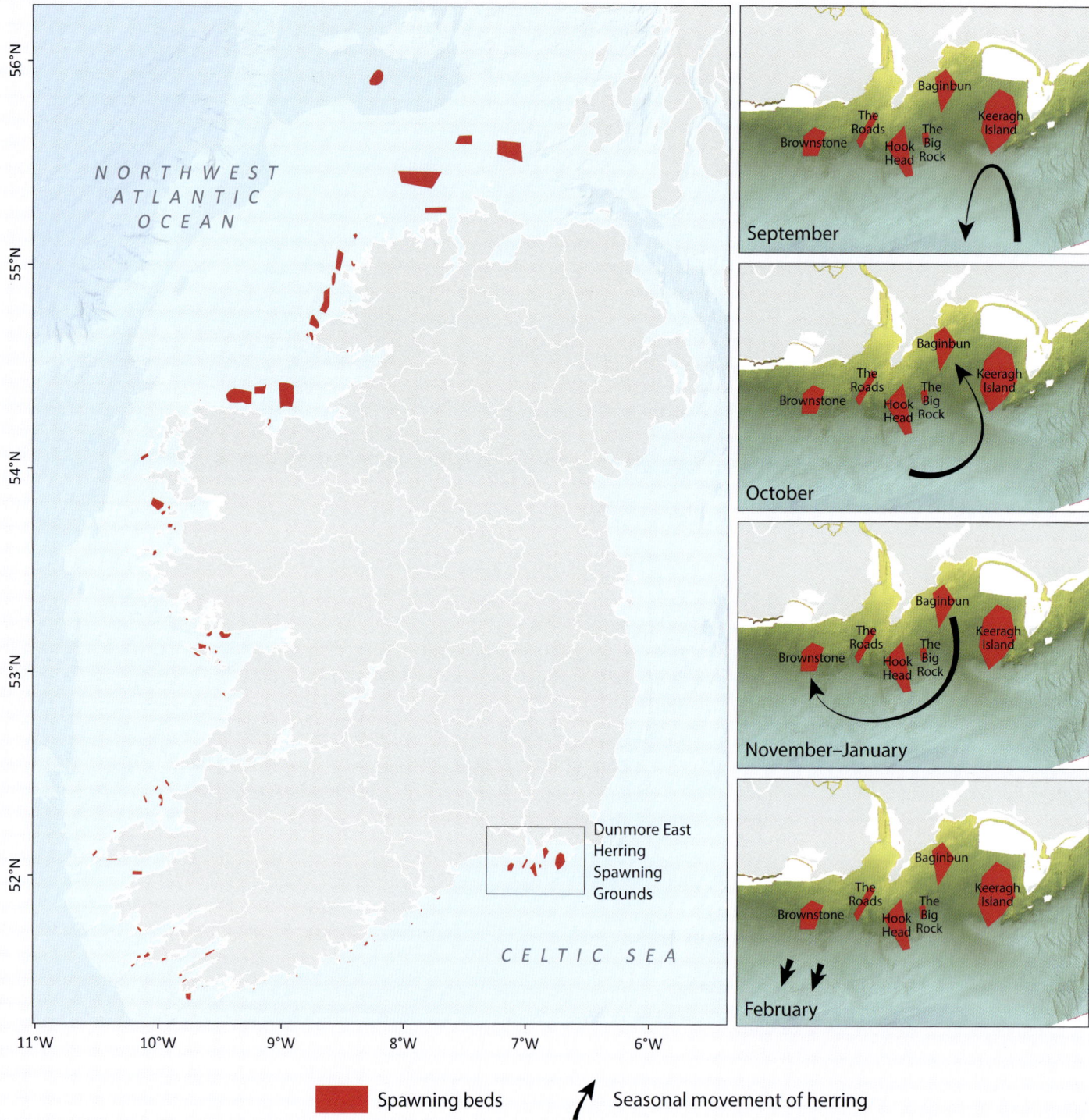

Fig. 9.11 The location of herring spawning beds and grounds around Ireland, identified from INFOMAR seabed data combined with knowledge obtained from workers in the fishing industry. This enables spatial and temporal analysis of herring spawning patterns, as seen in the right-hand column, which shows the seasonal movement of herring off the coast of Dunmore East, a commercially important area. [Data source: Information supplied by the Irish fishing industry with the support of the Celtic Sea Herring Management Advisory Committee and verified using INFOMAR seabed classification data]

Using multibeam echosounder data, the INFOMAR project has produced a series of substrate maps that classify seabed sediments using the amplitudes and statistical properties of backscatter images. This information on the bathymetry and the nature of the seabed can then be used to interpret the substrate type present; physical sampling was used to ground truth and verify the substrate maps produced. Sediment samples were subject to particle size analysis to determine sediment grain sizes and, from the resulting data, the sediments were grouped into eight physical habitat classes: rock (encompassing small stones, pebbles, boulders and bedrock), coarse sediment, mixed sediment, sand, muddy sand, sandy mud, mud and seaweed-dominated sediment.

To identify herring spawning beds, the seabed classification data were consulted and potential habitats identified. This was compared with data from larval surveys undertaken by the Marine Institute and with a comprehensive fishing industry survey conducted specifically for the study. Finally, a detailed inventory of individual herring spawning beds, grounds and areas around the Irish coast was produced.

The maps confirmed that all spawning beds reported by fishermen were composed of gravel/broken rock substrates, excepting one small sandy area in Baltimore harbour. The beds ranged in depth from 7m to approximately 90m, but on the north coast they are generally found in the deeper waters, up to 80m, and located further offshore. In the Celtic Sea, by contrast, the average spawning bed depth was at approximately 30m, and the grounds were found no more than 3km from the coast and often located within bays and estuaries (Figure 9.11).

A key finding was that actual spawning grounds occupy only a fraction of the potential habitats that are available, especially in the north-west. The reasons for this are unclear, but there are probably habitat cues used by herring other than substrate that are not captured by the acoustic backscatter data.

The maps produced from this study are based on the extensive knowledge of the fishing industry, combined with and validated by seabed data from INFOMAR. They are essential for the purposes of marine spatial planning (see Chapter 32: Coastal Management and Planning), and especially to avoid negative impacts on herring spawning grounds. The importance of herring as a forage fish and as a commercial resource relies on the ability of stocks to replenish, so the loss of spawning beds should be avoided. Taking the precautionary approach, this means that all known beds should be afforded maximum protection.

MAPPING THE SEABED GEOLOGY OF INISHBOFIN, COUNTY GALWAY, WITH BATHYMETRIC DATA

Eoin Mac Craith

The rocky Irish shoreline continues beneath the sea as a fascinating, submerged landscape. However, until the advent of high-resolution bathymetric mapping by multibeam echosounder, it was difficult to infer the extent and detail of seabed rock outcrop without significant guesswork.

While it is difficult to determine rock types using multibeam imagery alone, to an extent the onshore geology can be extrapolated offshore, based on the fabric and morphology of the seabed outcrops. Furthermore, the seabed is an ideal place in which to study geological structures and patterns in the rock, since it is particularly well exposed, with little or no large-scale vegetation cover and few human constructions or artefacts.

While seabed mapping is usually carried out with an intended use (for example, navigation), the data collected can be used for a variety of other purposes, including the mapping of seabed geology.

In 2012, the survey vessels RV *Keary* and RV *Geo* conducted a seabed mapping exercise around the island of Inishbofin, County Galway. The finished map, entitled the 'Real Map of Inishbofin' (Figure 9.12), potentially supports a range of activities in the area, from fisheries and habitat mapping to recreational diving and tourism. Its aim is to highlight the continuity of Inishbofin's landscape beneath the sea surface, and show the host of interesting features that can be found there.

The map reveals the complex geology of the submerged bedrock surrounding the island, and shows the rock outcrops to be heavily fractured and without clear sedimentary bedding. These same characteristics are found in the rock of the island, which is primarily ancient, deformed and metamorphic, suggesting that the offshore geology belongs to the same group, known as the Dalradian. The creation of these rocks began with the breakup of the supercontinent Rodinia, approximately 750 million years ago. The rocks were formed from sediments laid down in a marine basin, as the continental fragments spread apart and the crust between them thinned and subsided. Later, they were deformed and metamorphosed as the basin and its sedimentary contents were compressed between converging plates.

Potential large faults can be traced in the seabed imagery. These weak points in the rock have been preferentially eroded over time,

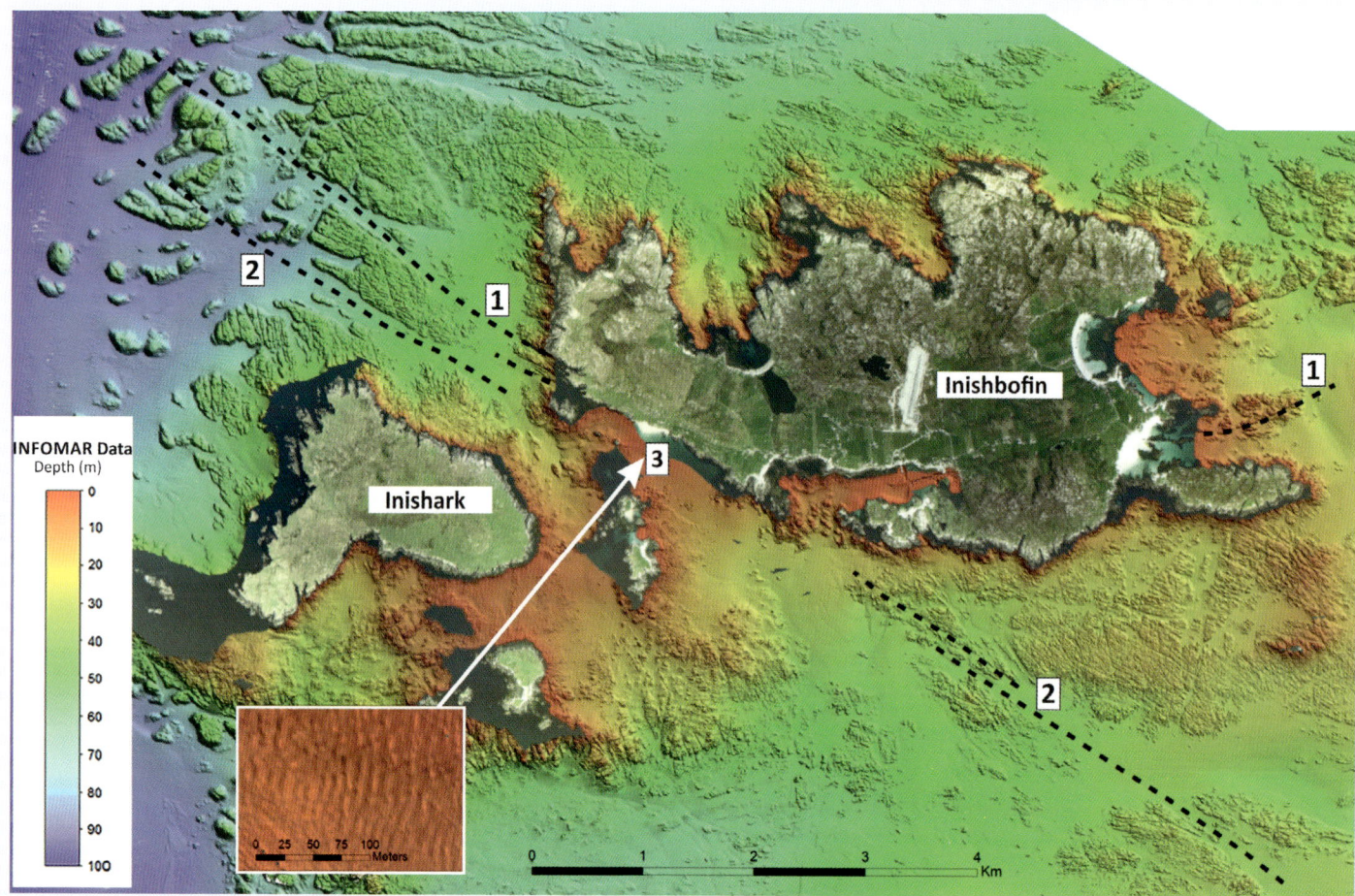

Fig. 9.12 THE SEABED AROUND INISHBOFIN, COUNTY GALWAY. Data vertically exaggerated four times to highlight features. Feature 1–1 is the potential marine expression of the Renvyle-Bofin Slide, while Feature 2–2 may be a large fault emanating from the mainland. The inset (Feature 3) highlights a field of sand ripples, found just off the west coast of Inishbofin island. [Source: INFOMAR; Background imagery: aerial photographs provided by Ordnance Survey Ireland – Licence No. EN 0047206]

rendering them visible as splits in the seafloor outcrops. One of these, marked as '1' in Figure 9.12, may be the submerged continuation of a large fault, known as the Renvyle-Bofin Slide, which cuts through Inishbofin, separating two geological groups within the Dalradian succession. This fault is also seen on the mainland, coming ashore at Renvyle Point.

Other large structures are also visible. A north-west to south-east trending feature, marked as '2' in Figure 9.12, lies between Inishark and Inishbofin and may have created the topographic low that separates the two islands. Based on geological land mapping, this may be the same feature found on the mainland at Cleggan, the local ferry port. On the seabed these features take the form of troughs running between the outcrops, perhaps having been eroded and widened during periods of lower sea level. One possible interpretation is that they represent landscape features that were once exposed to the air at a time of lower sea levels and were shaped by the erosional processes of that environment.

This map also reveals patterns in the soft sediment, which can provide useful information on the local current regime. Close to a beach on Inishbofin's west coast, a field of sand ripples (inset to Figure 9.12) is indicative of strong currents, with the direction of flow being broadly perpendicular to the crest-lines of the ripples.

The area around Inishbofin displayed here represents an excellent example of a rocky seabed, the details of which are clearly revealed through seabed mapping with multibeam sonar. It is sited within a range of water depths (approximately 0–100m), with complex topography and a bedrock structure that depicts a long and dramatic geological history.

Habitat Mapping of Kenmare River Using Multibeam Echosounder Data

Eimear O'Keeffe

The waters around Ireland are teeming with marine life, with different seabed habitats supporting different communities of animals. Numerous factors contribute to this, but two of the most important are water depth and seabed type.[14] The high-resolution information on bathymetry (depth) and backscatter (seabed hardness) provided by multibeam echosounders can be used to delineate accurately these different physical habitats.

Although the term habitat mapping is relatively new, marine biologists have been documenting coastal and marine flora and fauna for over a century. These surveys have evolved from early studies that recorded species collected by nets, trawls and dredges, while in situ sampling of the undersea world by diving became popular in the 1950s. Current technology makes it possible to view live footage of the seabed using Remotely Operated Vehicles (ROVs) equipped with high-resolution cameras from the comfort of a research vessel. However, despite all these advances in habitat surveying, detailed marine habitat maps for Irish coastal waters are still lacking.

In response to this need, INFOMAR has developed a methodology to create habitat maps using backscatter and bathymetric data, combined with sediment samples and video footage of the seabed. One example of where this methodology has been applied was in a survey of Kenmare River, County Kerry, conducted by INFOMAR aboard the RV *Celtic Voyager* in August 2011, using multibeam echosounders. The bay was considered to be an appropriate location for the study, because it is designated as a Special Area of Conservation (SAC) due to the presence of reefs, seals and otters.

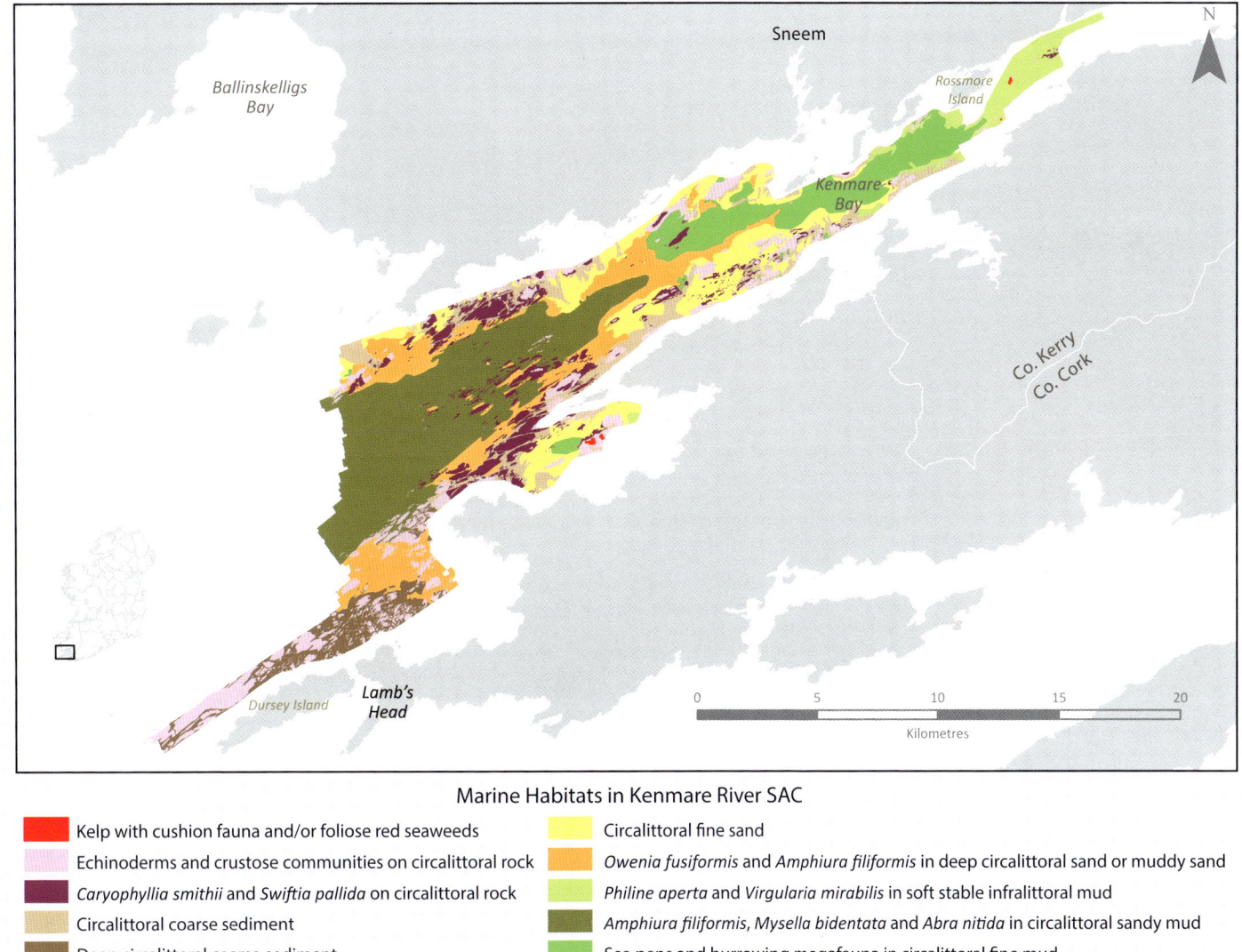

Marine Habitats in Kenmare River SAC

- Kelp with cushion fauna and/or foliose red seaweeds
- Echinoderms and crustose communities on circalittoral rock
- *Caryophyllia smithii* and *Swiftia pallida* on circalittoral rock
- Circalittoral coarse sediment
- Deep circalittoral coarse sediment
- Circalittoral fine sand
- *Owenia fusiformis* and *Amphiura filiformis* in deep circalittoral sand or muddy sand
- *Philine aperta* and *Virgularia mirabilis* in soft stable infralittoral mud
- *Amphiura filiformis*, *Mysella bidentata* and *Abra nitida* in circalittoral sandy mud
- Sea pens and burrowing megafauna in circalittoral fine mud

Fig. 9.13 Kenmare River marine habitat map, comprising three reef habitats and seven sediment habitats. [Data source: INFOMAR]

Fig. 9.14 MARINE FLORA AND FAUNA OBSERVED IN KENMARE RIVER. a. Brittle star (*Amphiura filiformis*) [Source: Bernard Picton]; b. sea pen (*Virgularia mirabilis*) [Source: Ben James, Scottish Natural Heritage]; c. Dublin Bay prawn (*Nephrops norvegicus*) [Source: Nicolas Chopin, Bord Iascaigh Mhara]; d. kelp and red seaweeds [Source: Nicolas Chopin, Bord Iascaigh Mhara]; e. urchin (*Echinus esculentus*) [Source: Nicolas Chopin, Bord Iascaigh Mhara]; and f. jewel anemones (*Corynactis viridis*). [Source: Nicolas Chopin, Bord Iascaigh Mhara]

A habitat map is created in stages. Reefs are delineated from shaded relief images of the bathymetric data. Areas with a similar backscatter response are grouped into acoustic classes. Sediment samples are obtained to identify or 'ground truth' the acoustic classes into sediment types (for example, mud, sand). Faunal assemblages present in the samples are identified and used to describe the different habitats supported by the different sediment types. Video footage is then used to document the species occurring on the reefs. All of these data are combined to produce a habitat map (Figure 9.13), which displays the diverse array of species living on the seabed.

Fine-grained sediments, such as mud or sand, or a mixture of both, allow animals to burrow into the substrate where they can seek refuge. While this habitat can appear desolate in comparison to the reefs, burrowing megafauna in addition to epifaunal species of sea pens and brittle stars are often in abundance. The deeper muds support clam species (*Abra nitida*), whereas muddy habitats at the shallower end of the bay support communities of sea pens (*Virgularia mirabilis*) and the burrowing Dublin Bay prawn (*Nephrops norvegicus*) (Figure 9.14).

The rocky areas of seabed support a more colourful array of species. Kelp, seaweeds, sponges, urchins and anemones all cling to the sandstone reef. The characteristic changes that occur in community composition with depth are evident: kelp and red seaweeds dominate the surface waters where there is more light; filter-feeding species such as sea fans prefer the deeper, darker waters where they rely on current-driven nutrients for sustenance.

Kenmare River is one of a number of marine protected areas around our coast. The availability of detailed habitat maps greatly facilitates our ability to protect, conserve and sustainably manage these areas. The INFOMAR project has surveyed twenty-six bays around the coast, many of which are designated SACs. These multibeam echosounder data can be used to deliver standardised and detailed marine habitat maps for the entire coastal area.

TANKER ROCK: A 'RARE EVENT' JUSTIFICATION FOR THE INSHORE MAPPING PROGRAMME

Sean Cullen

Imagine the relief of the captain of an oil tanker with engine trouble, on taking shelter in the lee of Slyne Head during a south-easterly

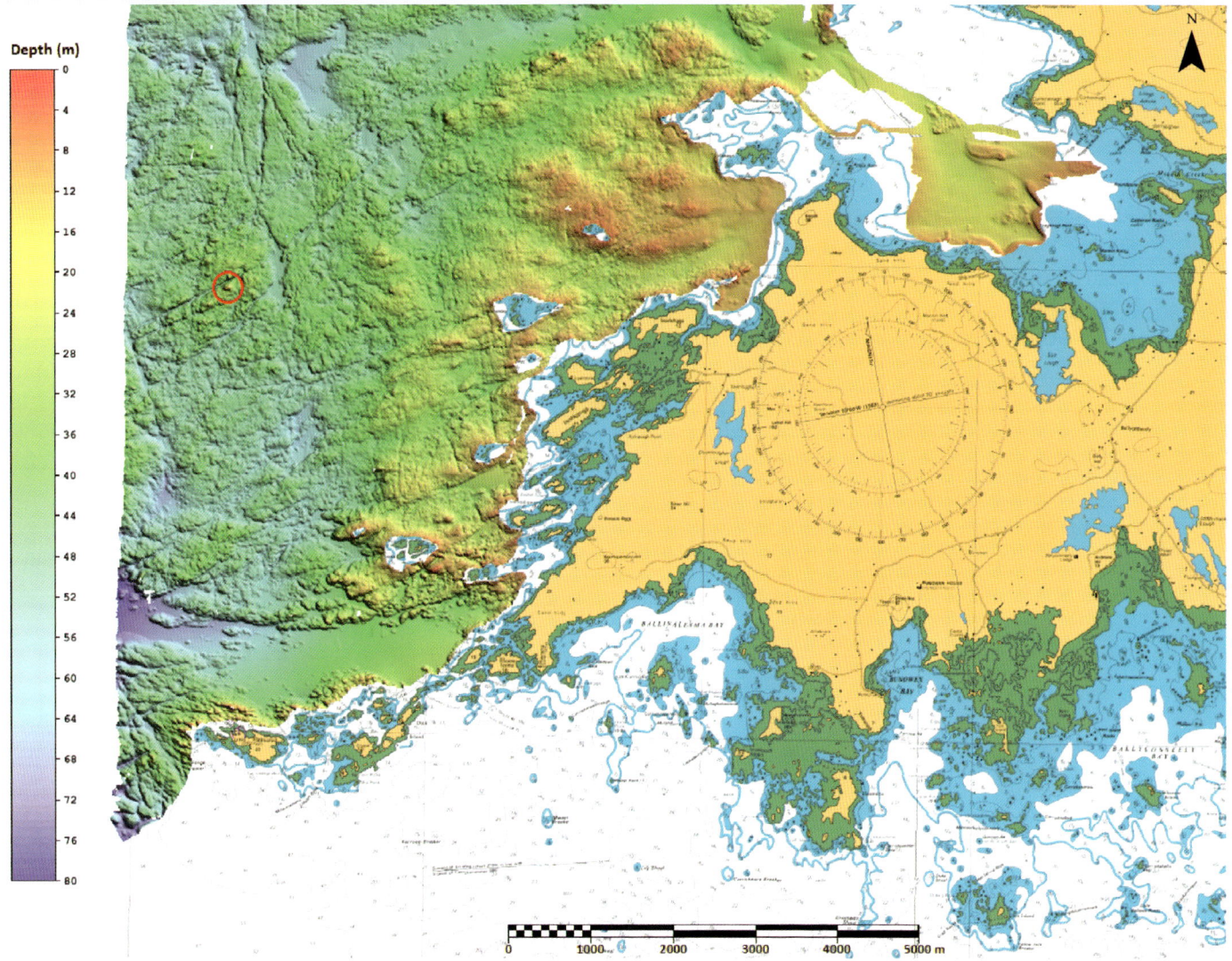

FIG. 9.15 THE TANKER ROCK. The previously uncharted shoal, of which Tanker Rock forms part, lies to the west of Mannin Bay in County Galway. Long known to the crews of fishing vessels, but unrecorded on official maps and charts until it was discovered by the RV KEARY as part of the 2010 INFOMAR survey, this rock rises to give just over 12m of clearance, and poses a serious potential hazard to shipping and, in particular, heavily laden oil tankers rounding the nearby Slyne Head. [Data source: bathymetry provided by INFOMAR. Chart data © Crown Copyright and/or database rights. Reproduced by permission of the Controller of Her Majesty's Stationery Office and the UK Hydrographic Office (www.GOV.uk/UKHO)]

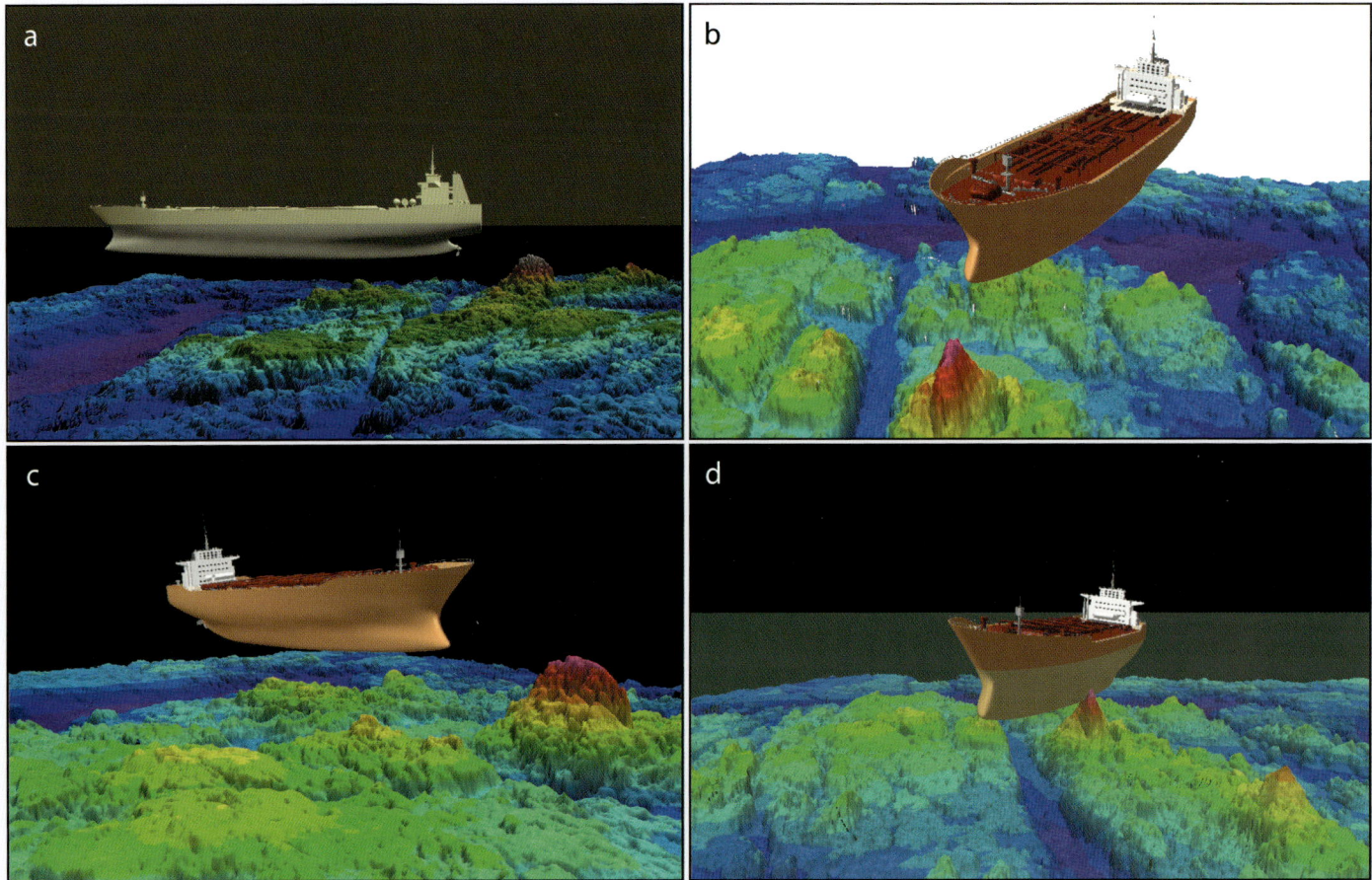

Fig 9.16 TANKER ROCK: A POTENTIAL HAZARD TO SHIPPING. In this assemblage of images, the shoal west of Mannin Bay, County Galway, has been rendered and displayed in three dimensions, with a virtual tanker model added to show the hazard it presents to shipping. The tanker is shown with a typical depth of hull for a ship of that size, indicating that the shoal rises to a point shallower than its keel. [Data source: INFOMAR]

storm raging over Connemara, County Galway. Then, as the wind abates and veers, the ship swings and the sudden change of motion is accompanied by the terrifying sound of rendered metal, confirming the seafarer's worst nightmare – a grounding. Crude oil spills onto the pristine coastline across a black, flat sea … relief instantly transformed to horror!

The chart for the sea area north of Slyne Head had always shown a safe anchorage depth of 30–50m outside of Mannin Bay. However, in 2010, while surveying in the area, the RV *Keary* identified a previously uncharted hard shoal, rising to give 12.2m of water between the top of the rock and the sea surface (Figure 9.15). The survey team immediately dubbed it 'Tanker Rock' in recognition of its significance and the danger it could pose to a tanker with a draft of 12m. An H-Note, the established mechanism for reporting navigational hazards, was sent to the UK Hydrographic Office (UKHO) in Taunton, which then generated a radio 'notice to mariners' to update charts. The UKHO has a bilateral agreement with the Irish Department of Transport for charting Irish waters and informing mariners. While listening to the notice over the coast guard VHF channel, local fishermen told the survey crew that they had long known of the shoal, a favourite lobster 'reef'. Nonetheless, it had never before been officially recorded.

Slyne Head sees significant oil tanker traffic, though usually this stays well out to sea. However, it is not inconceivable that it might offer shelter from southerly to easterly winds if required by a tanker restricted in its ability to manoeuvre. While the seabed is unsuitable for anchoring, such concerns would be overridden by the need for shelter for a vessel in difficulty. The scenario is, therefore, a real possibility (Figure 9.16).

When the *Exxon Valdez* hit the Bligh Reef in Prince William Sound, Alaska, on 24 March 1984, 11 million gallons of crude oil were released, leading to the catastrophic pollution of large areas of conservationally important coastline, and ultimately costing Exxon billions of dollars to clean up. The financial costs in this instance were borne by the company, but if it had been shown that the state was negligent in reporting the depth of the Bligh Reef, then the onus would have fallen fully or partially on the state. However, monetary compensation does not ultimately address the socio-economic and environmental costs of a large oil spill along a coastline, which are ultimately borne by the state and its citizens. The International Convention on Civil Liability for Oil Pollution Damage (1969) states that the ship owner will pay all clean-up costs, but that there is an exception under Article III 2 (c) that allows for the failure by the state to properly maintain navigational aids, which have been shown to include navigational charts.[15] This was enacted in Ireland under the Oil Pollution of the Sea Act 1988, Section 8 (c).

In 2008 the INFOMAR programme commissioned PwC consultancy to undertake a cost–benefit analysis for the inshore mapping programme. This put the return on investment at between four and six times the running costs.[16] Similar cost–benefit analyses in other countries have indicated even higher returns because they also account for mitigation against possible disasters such as oil spills, but PwC were expressly briefed not to include such rare events in their calculations. Nonetheless, if a large spill is averted through accurate mapping of shoal hazards, then the justification for the programme costs are self-evident.

The INFOMAR survey teams have always held the hydrographers of yesteryear in high esteem, when comparing the good agreement between modern surveys and old Admiralty charts that have not been updated since the use of leadlines and horizontal sextants. Modern multibeam echosounders can now scan an entire seabed area, whereas leadline surveys were done in rows of discrete soundings spaced widely apart, which inevitably led to some large rock shoals being totally missed. 'Tanker Rock' will be properly named at some stage, but in the meantime the updated Chart No. 2708 will give tanker captains pause for thought before venturing for shelter behind Slyne Head.

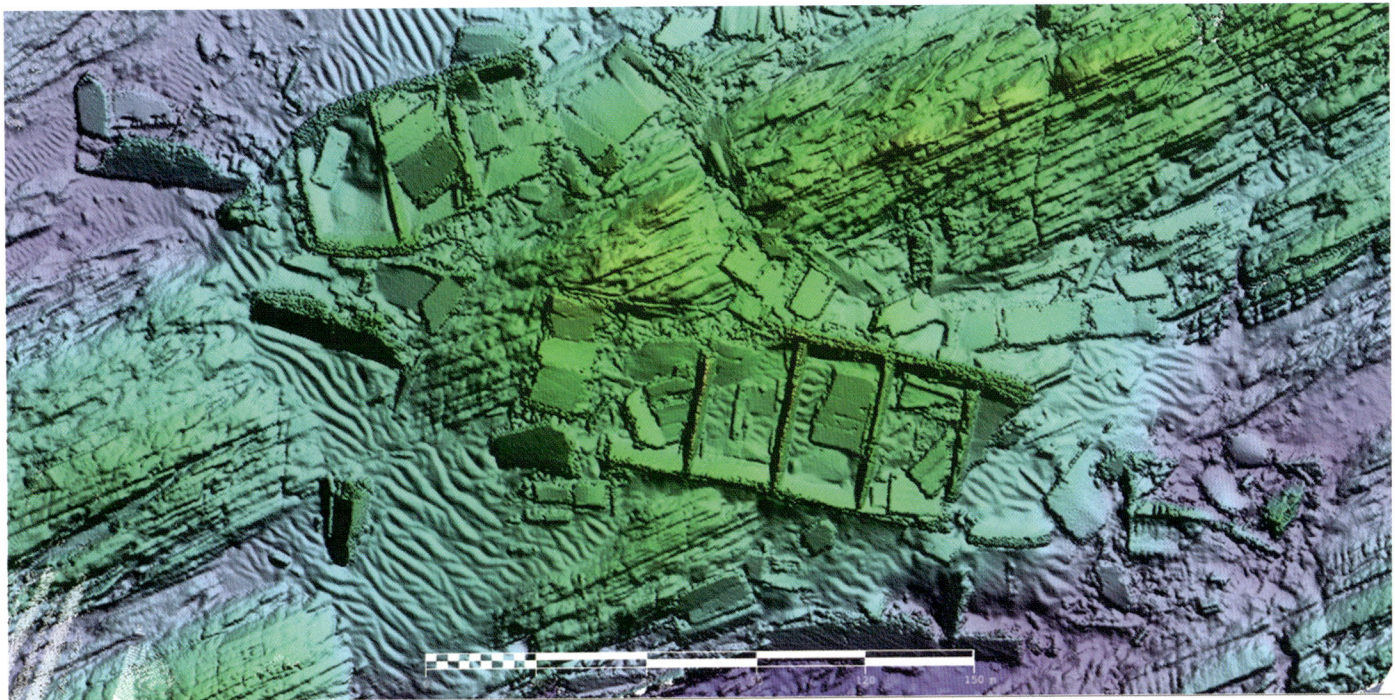

KOWLOON BRIDGE SHIPWRECK [Source: INFOMAR]

SECTION 2
NATURAL COASTAL
ENVIRONMENTS

Bays and Headland Coast, Derrynane, County Kerry [Source: Geraldine Hennigan and Norman Kean]

ROCKY COASTS

Maxim Kozachenko, Ruth M. O'Riordan, Rob McAllen and Robert Devoy

MIZEN HEAD, COUNTY CORK. Rocks, sea stacks and storms – just some of the ingredients that control the spectacular rocky coasts of Ireland. The island has numerous peninsulas and headlands like this, which have developed on its many types of hard rock coasts. These have been eroded and changed quite slowly under the action of waves and storms over the last 2,000 or so years. The photograph gives a flavour of the scenery of such coastlines, showing the multiple rock pinnacles (or sea stacks) developed in the folded and faulted, hard sandstone rocks of the Mizen. [Source: Michelle Byrne]

The coastlines of Ireland developed during the Cenozoic era of geological time (formerly the Tertiary, that is, since *c.*65 million years ago) and result from two main factors. Firstly, the process whereby the long-term relative rise of global mean sea levels over this time against the Earth's continental land margins – as water was redistributed in the world's ocean basins through earth crustal changes – eventually created the islands of Ireland and Britain, with the submergence of the continental crustal margins of north-west Europe. Secondly, more frequent oscillations of sea level around this baseline position occurred, resulting from the repeated glaciations and warmer interglacials of the last one to two million years (the Quaternary period, that is, since *c.*2.6 million years ago until now, the Anthropocene), with times of significant falls and then balancing rises of sea level. These oscillations, and the earlier changes of the sea's surface, have cut into Ireland's bedrocks to create the present coastline, including its rocky coasts (see Chapter 6: Glaciation and Ireland's Arctic Inheritance and 7: Ancient Shorelines and Sea-level Changes).

The different nature and special regional characters of these coasts have been addressed and detailed by geomorphologists, geologists and various other authors since the nineteenth century (see Chapter 2: The Coastal Environment: Physical Systems' Processes and Patterns).[1] The rocks, and their geology, have imparted much of the distinctive character now observed on these coasts. During the Cenozoic, a rim of uplands (for example, the Wicklow, Comeragh and Mourne mountains) was formed around the island, through plate tectonics and linked geophysical rock structural mechanisms, causing the differential uplift and downwarping (diastrophism) of Ireland's rocks, as exemplified by those of Lough Hyne, County Cork (see Chapter 5: Geological Foundations). Rock-dominated and cliffed coasts, or rocky coasts, represent arguably the most spectacular and dramatic of Ireland's coastal landscapes, with some of these – most notably the Giant's Causeway, County Antrim, and the Cliffs of Moher, County Clare – attracting many tens of thousands of tourists each year (see Chapter 27: Coastal and Marine Tourism). Erosional, rocky coasts account for over 40 per cent of the total length of Ireland's coastline. In terms of their geomorphology, these environments are linked most with erosional types of landforms, as opposed to sedimentary and depositional systems, as represented, for example, by sand beaches, barriers, spits and dunes (see Chapter 11: Beaches and Barriers). Rocky coasts are controlled particularly by waves and storms and form part of the recognised wave-dominated coastal systems of the world. Tides, as a separate control on rocky coasts, are also important, but less so for Ireland's coasts, which fall within the meso- to macro- ranges of tides. The result is synergistic, simply amplifying the dominant impacts of wind-generated waves rather than forming a distinctive control mechanism, except perhaps for the biota and rocky shore ecology.

The hard rocks forming the island's coasts range in geological age from *c.*1,800 million years old, in south-east and north-west Ireland, to *c.*65 million years, in the Upper Cretaceous chalk and

Fig. 10.1 RATHLIN ISLAND, NORTHERN IRELAND, SHOWING THE DISTINCTIVE CHALK AND BASALT, 30–40M HIGH, SEA CLIFFS OF THE COUNTY ANTRIM COASTLINE. The chalk, as a sedimentary limestone rock, developed first in the tropical, shallow seas of the Cretaceous geological period. At this time, the position of the Earth's crust (beneath Ireland) was much closer to the Equator (the process of Continental Drift having moved it progressively northward over the last 100 million years). This rock is composed of countless sea creatures (for example, microalgae, such as diatoms, zooplankton, such as radiolaria and a diverse shell fauna), as well as quartz and other minerogenic sediments, which have formed in sedimentary layers (rock strata). This composition probably gave the chalk here originally a much lighter, white colour, similar to the chalks of southern England and northern France. Later Tertiary age volcanic activity and the outpouring of molten basaltic lavas across the Antrim region subsequently baked and altered the chalk to a dirty grey colour and into a harder rock; the basalts are also a dark grey. The resulting thick and massive rocks form steep cliffs along this coastline, their hardness allowing for a high resistance to erosion. Structural rock weaknesses lead, however, to the persistent occurrence of rock falls and slides, which, with wave activity, does enable cliff retreat. [Source: Geraldine Hennigan and Norman Kean]

Tertiary basalt cliffs of County Antrim (Figure 10.1). The morphology (that is, the shape and appearance) and types of rocky coasts are, as one expects, influenced strongly by the bedrock geology of a particular location (see Chapter 5: Geological Foundations). An important factor in developing such environments is, therefore, rock hardness, linked with the ability to withstand erosion by the sea. Rock structures (for example, bedding, or layering of the rocks, joints, folds, faults and fractures) also contribute towards establishing the morphology of these coasts (Figure 10.2).

In contrast, contemporary coastal features developed in the wide range of rock types are generally much younger and owe much of their present forms to the later glacial and interglacial episodes of the last 0.5–1 million years (the Quaternary), a time of rapid changes in Earth's climate and sea-level positions. Many coastal cliffs and some rock platform features began their formation during these interglacial periods, at times of predominantly rising sea levels and later, during shorter periods of stable maximum/high sea levels. This sort of coast is shown well in the exposures of the South of

Ireland Shore Platform (SISP), displayed today as a prominent rock-cut feature along Ireland's southern and western shorelines (Figure 10.3). Others have been shaped during glacial stages by contact with ice and erosion, as well as by freeze–thaw and low sea-temperature-based erosional mechanisms. Some rock coastal landforms, however, are older than the Quaternary, having their origin in the humid and 'river-driven' geomorphological processes operating during the late Cenozoic (for example, the Aran Islands, the Burren and Cork Harbour coastal landscapes).[2]

It is often the case that proto beaches, formed of large-sized and angular local boulders, together with finer material derived from cliff rock falls, accumulate at the base and in front of cliffs. This material is part of the long-term operation of gravity-driven mass movement processes working on rocky coasts combined with the structurally linked mechanisms of rock rotation, slip and slide. Together, these features of effective rock accumulation, derived from cliff rock falls, coexist with erosional rock shoreline landforms, such as shore platforms, coastal boulder deposits (CBDs), cliffs, caves and sea stacks. On a regional scale, rock headlands (peninsulas)

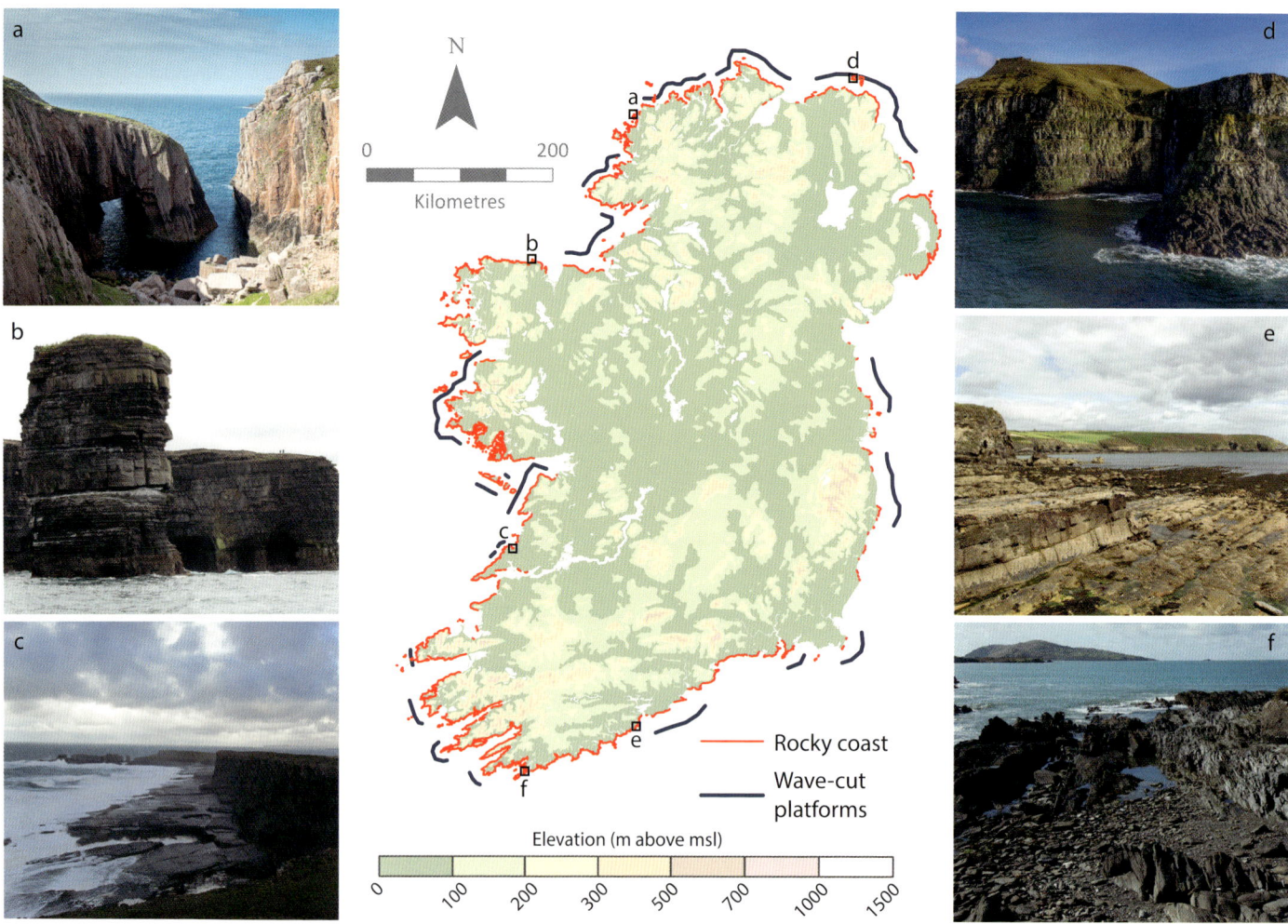

Fig. 10.2 ROCKY COASTS OF IRELAND. Distinctive coastal landforms and landscapes found around Ireland include, a. a sea arch, Gola, County Donegal [Source: Donegal County Council]; b. a sea stack, Downpatrick Head, County Mayo [Source: Geraldine Hennigan and Norman Kean]; c. wave-cut platform, backed by a mudstone, siltstone and sandstone rock cliff, at Ballard Bay south-west of Doonbeg, County Clare [Source: Niamh Cullen]; d. cliffs and embayment with coastal cave, Rathlin Island, County Antrim [Source: Douglas Cecil]; e. an irregularly shaped rock coast, due to wave erosion of the dipping rock layers, Rocky Bay, County Cork [Source: Robbie Murphy]; f. low rocky shore, Sherkin Island, County Cork [Source: Robbie Murphy]. Ireland's rock-dominated coasts, often visualised as cliffs, are closely linked also with areas of wave and storm-driven erosion, but they are not synonymous with this important coastal process of geomorphological change. Many such coasts are, in fact, relatively stable in shape and position, only changing slowly. In contrast, long sections of the coast are not rocky (the areas not marked by a red line). These are most commonly composed of soft sediments, of glacigenic origins, and also of sands (for example, dunes and beach geomorphological systems). Many of these coasts also develop cliffs when exposed to wave attack and many are now retreating rapidly under rising sea levels.

coincide quite frequently with anticlinal geological folds (ridges, or rock up-folds), while embayments, or bays, coincide with synclinal geological folds (valleys, or rock down-folds) (Figure 10.4).[3]

The prominent character of Ireland's rock coastlines has a powerful attraction for many people to view their iconic features of, for example, cliffs, sea stacks and caves (Figure 10.2). These, and other coastal landforms, are the result of geomorphological process variations which control the way water (physics, chemistry and hydrodynamics) and the mechanisms of wave action, storms, tides, sea-level changes and biology, work with the different coastal rock types (their compositions and structures). Further, rock-dominated coasts are important in the context of Integrated Coastal Zone Management (ICZM), as they represent one factor in the natural forms of protection in coastal environments, forming buffer zones to marine flooding of areas inland (see Chapter 30: Engineering for Vulnerable Coastlines).

Cliffs, as landforms, represent the default, or 'parent' form of most types of features that comprise rocky coasts. This chapter examines some of these features as a means of viewing this important physical theme of coastal appearance, which has often been used as the evocative source material of photographs in many 'coffee-table' books! In addition, the distribution and ecology of marine plant and animal life found on rocky coasts is presented.

CLIFFS

Cliffs, as coastal landforms, are simply slopes cut into the bedrock. They develop different shapes under the primary control of the

Fig. 10.3 COASTAL ROCK PLATFORM, THE SOUTH OF IRELAND SHORE PLATFORM (SISP), SIMON'S COVE, COUNTY CORK. This raised interglacial shoreline, cut by sea and ice action in the late Quaternary first formed as a coastal feature at least 110,000 years ago, if not earlier. The platform is developed as a sub-horizontal and bevelled surface in Devonian age quartzites and sandstones of the West Cork Sandstone Series, which dominate the coastal rocks of south-west Ireland. More recent glacial action is evidenced in numerous ice erosion features etched into this platform as gouge marks, which can measure several metres in length. Glacial sediments deposited from tidewater Arctic glacier margin-type environments overlie the platform and are dated to post *c*.24,000 years ago. [Source: Robert Devoy]

Earth's gravity (the geomorphological process of mass movement). Many of Ireland's coastal cliffs appear as near vertical (90°) in form, although in reality their averaged and long-term maintained slope angle is much lower, but generally >45°. Any coastal slope of

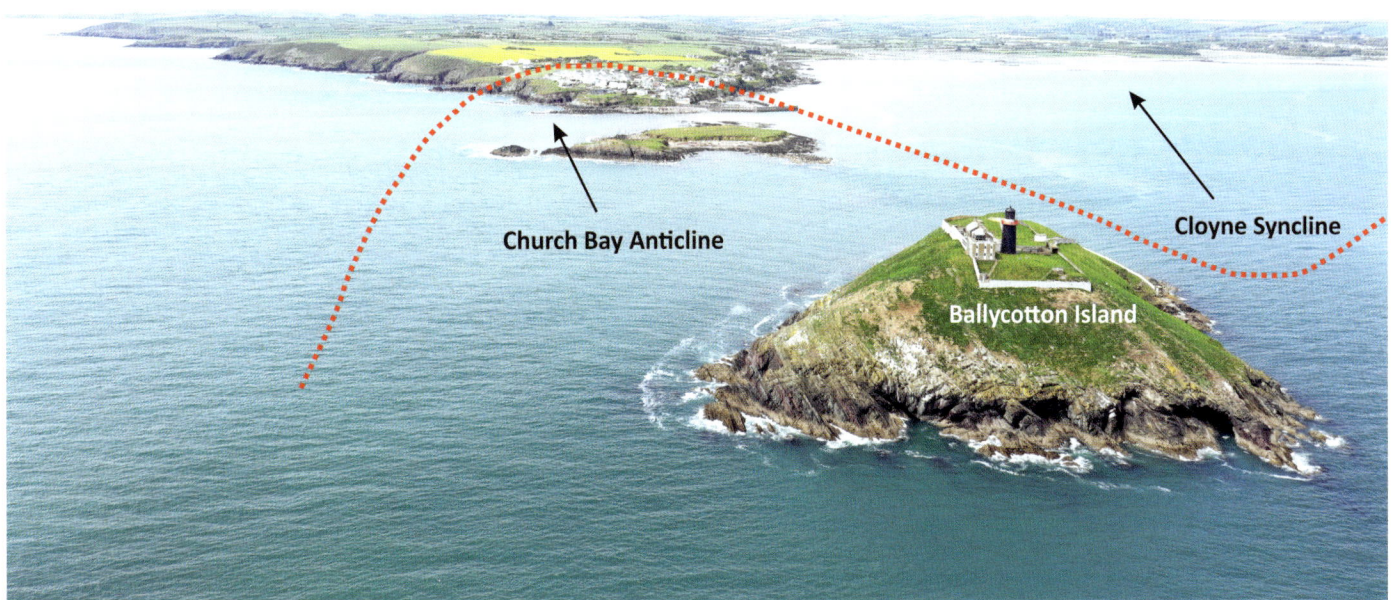

Fig. 10.4 BALLYCOTTON ISLAND, COUNTY CORK, ILLUSTRATING THE LOCAL COASTAL AREA AND REGIONAL-SCALE IMPACT OF PAST ROCK FOLDING IN THE DEVELOPMENT OF COASTAL CLIFFS AND SEDIMENTARY DEPOSITIONAL EMBAYMENTS. This aerial image depicts Ballycotton Island, Ballycotton cliffs (left) and the western embayed side of Ballycotton Bay (centre to right). The Bay and the associated low-angle beaches and wide intertidal zone are formed in the Cloyne Syncline (down-folded Carboniferous age rocks). The southerly wave-exposed rock headland, together with Ballycotton Island, have been cut in the older and hard sandstones and quartzite rocks of the Devonian and coincide with the Church Bay Anticline (rock up-fold). Conceptually, the direction and broad shape of these now heavily eroded rock folds are illustrated by the red dashed line. Ballycotton Island forms a good example of the long-term impacts of past relative sea-level changes, together with wave action, in causing coastal retreat and the removal of headland environments. Such processes can leave progressively isolated rock pinnacles (such as sea stacks) and larger offshore islands. Other examples of the operation of similar geological processes contributing to the development of Ireland's hard rock coastal geomorphology are found in the peninsulas of south-west Ireland. These are coincident with the major regional anticline structures (for example, Mizen Head, Sheep's Head, Dursey Head and the Iveragh Peninsula). The bays separating the peninsulas (for example, the Kenmare River) are the drowned, offshore zones of earlier rivers and glaciated valleys, formed following postglacial sea-level rise. Coastal features like these are often referred to as rias. [Source: Maxim Kozachenko]

between 40° and 90° and even appearing to 'overhang' (that is, greater than 90°), can be termed a cliff (Figure 10.5). As with all erosional landforms, cliff development is an irreversible process and, in the geological short term (that is, <1 million years), creates a landform that appears to last. This is, of course, an illusory human viewpoint; cliffs do change their shape over time and disappear as land surfaces 'evolve'.[4]

In describing the mechanisms and processes involved in cliff formation, a distinction should be noted between the terms cliff erosion (as a process) and recession (the response of the land area, as seen in the form of the cliff-line, to the operation of erosion). Erosion relates to the mass, or volume, of rock and sedimentary materials being removed from the cliff face and from adjoining coastal areas. In contrast, recession describes more the overall changes through time in the positions of coasts, as seen by the progressive landwards movements of a cliff-line. This is measured as a rate of coastal retreat in metres per year. Cliff retreat can be seen most clearly in vertical (plan) view of a time series of aerial photographs and plotted as changes in coastal positions on maps (see Chapter 8: Visualising, Mapping and Monitoring Coasts).

The rate and nature (processes and mechanisms) of cliff recession and erosion are controlled by many factors: by the rock type, geological structures, the flows of coastal water (hydrodynamics) and by the climate. Hydrodynamics are associated closely with ocean and offshore current movements, onshore wind directions, the operation particularly of wave climate and storminess patterns and of seabed morphology. The seabed's shape and structures – such as submerged rock outcrops, reefs, shore platforms and sand banks and/or gravel ridges – act importantly as natural wave breakers and provide dynamic feedback into the coastal water movements. The effect of changes in seabed shape and gradient is to alter locally the power (available energy) of waves approaching the coast, or to change the direction and intensity of nearshore currents. A decrease in water depth leads to a reduction in the ability of waves to power the process of erosion and, thus, cliff development (Figures 10.6 and 10.7).

Ireland's cliffs are formed mostly by sedimentary rocks, deposited originally in horizontal layers. Therefore, the morphology of cliffs will be influenced particularly by the angle of bedding, dip and orientation (strike) of these rocks, in relation to the sea levels operating at particular locations (mean low water to mean high water) and the dominant wave directions. Cliffs cut into massive, igneous rocks, such as granite, basalt and other volcanics, also occur commonly, for example, on the Copper Coast, County Waterford, and at Carnsore Point, County Wexford, and Killiney, County Dublin.

A cliff is a three-dimensional structure and its morphology can be described by both its vertical profile and its plan view. The vertical profile is controlled by differences in the hardness of the rocks composing the cliff face (the result of rock composition and

Fig. 10.5 'KERRY CLIFFS', PORTMAGEE, VALENCIA, COUNTY KERRY. The cliffs near Portmagee are now referred to on maps presented by the Wild Atlantic Way (Fáilte Ireland) as the 'Kerry Cliffs'. The rocks here are composed of hard, sub-horizontally bedded Devonian age sandstones and quartzites. These cliffs are controlled strongly by geological structures (local fractures and faults), which cut vertically through the rock layers. Many of these structures are linked with rock falls visible along this coastal cliff section. Some of these falls consist simply of large pieces of rock that have broken free from the cliff face, creating near-vertical sections of cliff behind them (indicating the underlying control of gravity in cliff development). The overall stepped-like morphology of some cliff sections here results also from the geological bedding (rock layers). Cliff sections similar to this are common throughout Ireland. The most well-known of Ireland's many rock cliffs are probably the Cliffs of Moher, County Clare, which rise to over *c.*200m above sea level (see Chapter 27: Coastal and Marine Tourism, Figure 6). The highest cliffs are located, however, on Achill Island at Croaghaun (688m) and those of Slieve League, County Donegal (highest mountain point, 596m). [Source: Maxim Kozachenko]

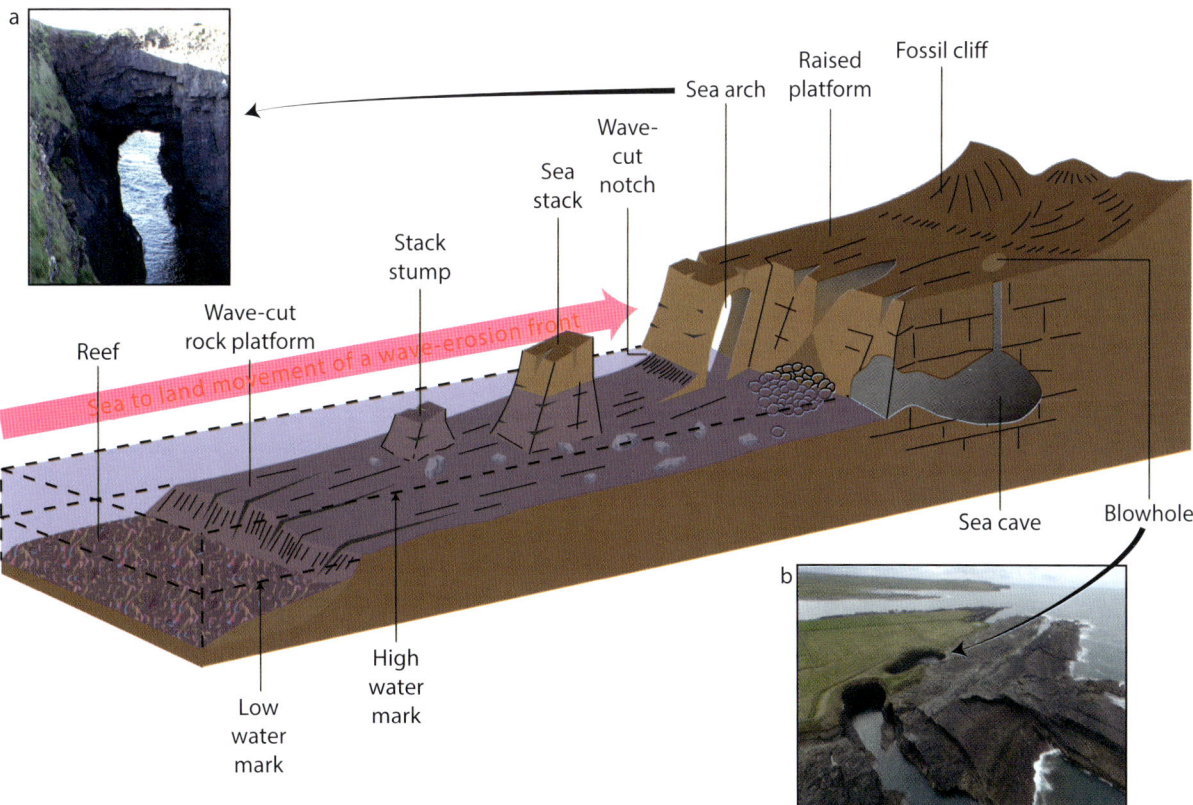

Fig. 10.6 A DESCRIPTIVE MODEL OF LANDFORM FEATURES TYPICAL OF THE GEOMORPHOLOGY OF ROCKY COASTS, AS LINKED TO THE PRIMARY CLIFF COASTAL FORM (SEE FIGURE 10.7). It indicates the spatial and quasi-genetic, developmental sequencing over time of such features (the oldest features are shown on left of the diagram) features. The dominant controls in their development are the regular tidal changes in the water levels (high and low water) and exposure to the landwards movement of the sea and of a linked wave-erosion front. These coastal landforms can also be found at numerous similar rock-cliff locations around Ireland's shorelines. Inset photographs show: a. a sea arch eroded in Devonian age sandstones at Coosfadda, Canglass Point, County Kerry [Source: Robert Devoy]; b. sea arches and former blowholes in the limestone coast of the Bridges of Ross, County Clare [Source: David Hodgeson]. [Diagram adapted from Briggs, D., Smithson, P., Addison, K. and Atkinson, K., 1997. *Fundamentals of the Physical Environment*. Second edition. New York: Routledge, p. 557]

Fig. 10.7 CLIFF DEVELOPMENT AND ROCKY SHORE LANDFORMS, OLD HEAD OF KINSALE, COUNTY CORK. The formation of erosional rock landforms is shown well on the south coast of Ireland at the Old Head of Kinsale. This location, with its stunning coastal views into the Celtic Sea, is the first signage point of the Wild Atlantic Way coastal tourist route (see Chapter 27: Coastal and Marine Tourism). Included in its attraction as a physical coastal landscape, as well as a heritage and popular tourism site, is its historical link with the torpedoing on 7 May 1915 during World War I of the liner RMS *Lusitania*, with the loss of over 1,100 people. The position of the sinking lies *c.*18km south of the lighthouse at the Old Head of Kinsale. Many of the casualties were first landed at harbours and other sites in this area. [Source: Maxim Kozachenko]

201

mineralogy), as well as its geological structures (Figure 10.8). Averaged rates of cliff erosion on Ireland's coasts are generally low (<0.01m/century) in hard rocks, such as those of the Precambrian and Cambrian age rocks of Donegal and Connemara, the volcanics and basalt of County Antrim and the granite and the Cambrian age rocks of the coasts from Dún Laoghaire to Wicklow (Figure 10.9). In contrast, rates of erosion for cliff-lines formed in soft glacigenic sediments often reach values of 0.5–1.0m/year (see Chapter 30: Engineering for Vulnerable Coastlines).[5]

Measurement of the rate of rock surface erosion is often undertaken today using micro-erosion meter techniques. In contrast, the calculation of the total volumes of coastal rock removal, as linked to the averaged coastline rates of retreat and loss of cliffs, is measured using archived historical maps, photographs and other imagery (including satellite images), together with many direct monitoring measurements (such as LiDAR and UAV/drone surveys and erosion pins positioned on cliff/coast edges) (see Chapter 8: Visualising, Mapping and Monitoring Coasts).

Elements of rock composition and coastal shape have frequently provided the basis for characterising coastal cliffs. Coasts with geological structures such as synclines and anticlines, which contain faults and heavily rock-fractured zones, develop commonly as rugged cliff forms (see Figure 10.4). These coasts are described as crenulate (indented) in appearance and, at the large to regional scale, form as bay and headland coastlines. Geomorphological features associated with these coasts include sea stacks and caves, developed on the promontory headlands separating the bays.

Fig. 10.8 A NARROW GORGE TYPE OF COASTAL FEATURE, OR 'GEO', TULLIG, COUNTY CLARE. A gorge, or narrow rock corridor-type coastal indentation, created by two cliffs facing each other and developed in sub-horizontally bedded weak limestones and shales, along the line of a geological fault. It has been formed, probably through multiple cycles of coastal erosion over two or three interglacial times of high sea level, by the removal of rocks weakened by high-energy wave action. Many similar geo-type forms occur on Ireland's rock coasts, conditioned by long-term exploitation by the combined action of waves, storm-surges and sea-level changes on the pre-existing geological structures. Waves eroding and impacting these high-energy Atlantic ocean coasts of western Ireland regularly reach elevations on cliff-lines of above 7–10m above HWM during storm conditions (see Cronin, A., Devoy, R., Bartlett, D., Nuyts, S. and O'Dwyer, B., 2018. Investigation of an Elevated Sands Unit at Tralispean Bay, South-west Ireland: Potential high-energy marine event. *Irish Geography*, 51, pp. 229–260). [Source: Zoë O'Hanlon]

Fig. 10.9 CLIFFS OF CAMBRIAN AND ORDOVICIAN AGE ROCKS ON THE BRAY TO WICKLOW COAST, COUNTY WICKLOW, EAST COAST OF IRELAND. The railway here is perched on this steeply cliffed coast and in many locations has been constructed by tunnelling through these hard, highly fractured and unstable coastal rocks. Rates of cliff retreat have been slow, but observable, since the building of the railway line in the nineteenth century. In some places along this coast the railway line has subsequently (particularly since the 1980s) had to be re-positioned further inland, to offset the immediate impacts of this erosion. Such impacts will continue to be experienced by infrastructures such as railways and roads on this east coast. [Source: Iarnród Éireann Irish Rail]

Examples are those of Slea Head, County Kerry, and the Old Head of Kinsale, County Cork (Figure 10.10a and b). Coastal cliffs created in rocks with varied rock properties and structures (such as sub-vertical rock bedding and cleavage) can create an irregular, serrated vertical profile. Elsewhere, sea erosion operating along the line of a geological fault, or by the erosion of rocks with uniform properties, have commonly a comparatively straight plan profile form. Similarly, such steep vertical cliff-shape lines are formed where processes of sudden tectonic uplift, or tsunami, have occurred, both of which have been relatively rare events in Ireland's recent coastal history (see Chapter 2: The Coastal Environment: Physical Systems, Processes and Patterns).

Some cliffs, particularly those with thick sedimentary cover above a bedrock of glacigenic sediments, can develop a so-called 'slope-over-wall' type feature, as a seawards overhanging vertical profile. Such features often result from these sediments moving down-slope toward the sea edge, under gravity and saturated groundwater conditions. This situation occurred frequently in Ireland during the times of cold climate (the Ice Ages) and results from periglacial freeze–thaw action, when water gets into the cracks in the rock (at micro- to macro-scales) and, when turned into ice, expands and loosens material from the upper part of the cliffs. Direct loading of coastal rocks by ice and plastic deformation of the bedrock also occurs during times of glaciation and may add to this process. However, in some locations, a similar cliff appearance could be derived more from the primary geological structures and represent, for example, the seaward slope of an anticline – the phenomenon of equifinality in geomorphology.

Rock material fallen under gravity from cliff surfaces during storms often accumulates as a steep beach at the cliff foot and forms a protection from further sea erosion.

Alternatively, and over time, this material will be carried alongshore, as well as offshore, by waves and currents. At other times, or even synchronously, rock will be lifted into suspension in coastal waters, thus increasing the erosional power of waves against the cliff face

Fig. 10.10 THE TOWERING SANDSTONE CLIFFS OF THE SOUTH-WEST DINGLE PENINSULA AND BLASKET ISLANDS, COUNTY KERRY. a. This crenulate and rugged rocky shore is located at Ballymacadoyle Hill, just west of the mouth of Dingle Harbour. The cliffs here are characteristic of many such coasts of the south-west of Ireland. The Eask Tower, set above the clifftop and the adjoining steep slopes at a height of *c.*180m, can be used to vertically scale this coast. [Source: Barry Brunt]; b. The Cathedral Rocks, on Inis na Bró, Blasket Islands, County Kerry, are the 'spectacular' rock formations at the north-east point of one of the seven islands that form the archipelago of the Blasket Islands. These rock cliffs have been likened to a Gothic cathedral, hence the name. [Source: Barry Brunt]

(abrasion). The direct wetting of the rock by sea water and spray adds to this action (that is, the process of geomorphological weathering and linked chemical corrosion). Apart from this removal of material from the lower cliff sections, there are other drivers of rock breakdown and cliff development. For example, air trapped between the water and rock surfaces, especially in cracks, rock-joints and other fissures is compressed by the hydraulic action of waves, resulting in increased pressure in these fissures, with consequent fracturing and subsequent rock removal.

These different elements of cliff formation provide the basis for a very much simplified three-stage descriptive model of cliff development. The stages are: 1. formation of an erosional and rock-weathered notch (concave, or c-shaped rock surface) in the lower part of the cliff, in the zone of regular tidal and wave action (this feature is not formed or observed easily everywhere); 2. as this feature is enlarged through seawater and wave action, a consequent collapse occurs under gravity of overhanging parts of cliff areas above the notch; 3. erosion of the air-exposed (subaerial) part of the cliff under the influence of local climate conditions.

In locations where different rock hardnesses and lithologies comprise the lower parts of cliffs, a notch will generally grow faster when developed on softer, rather than harder, rocks. This differential exploitation of rock characteristics can lead to the formation of caves and sea-arch type landforms, which are the outcome of a continuum of mechanisms and processes involved in cliff recession and the alteration of coastal shape.[6]

LOW-RELIEF ROCKY SHORES

Many sections of Ireland's coasts exist as relatively low sedimentary cliffs, formed in glacigenic sediments. Cliffs in these 'soft' rock materials can reach >20–30m in height (though more commonly lower), particularly along eastern coasts south of Dundalk, County Louth, which have thick and extensive sequences of these sediments (see Chapter 6: Glaciation and Ireland's Arctic Inheritance). These coasts will be underlain by bedrock, which generally emerges in the intertidal zone unless covered by beach sands and gravels, as found in most coastal embayments. Where the bedrock does appear in the intertidal zone, it forms areas of irregular, low-relief accompanied by rock platform type surfaces. Closer to the high-tide line it may form topographically as low height and rounded rock outcrops (<5–10m height), which some have termed loosely as coastal bluffs. In areas of soft sediments, the cliff-line is maintained by wave erosion, with sea-level changes and storminess principally controlling the rates of retreat of these coasts. As suggested by recent coastal monitoring and historical mapping, these rates are reaching levels of >1m/ year on some exposed southern and western coasts, and even within the more wave-sheltered greater Dublin Bay region.[7]

In many formerly glaciated coastal areas – such as Ballycotton Bay, Dingle Bay, Clew Bay and Donegal Bay, or the coastlines of County Antrim – prominent areas of loose rock pavement exist across the foreshore. Under earlier periods of freezing conditions, which developed during glaciations, periglacial action and permafrost conditions became established on exposed glacial sedimentary surfaces. This freezing regime imparted an increased hardness to these sediments, with the ice and freezing process concentrating the large-sized rock debris (>large gravel to boulder sizes) into extensive sub-horizontal levels. Postglacial wave erosion has exposed these former surfaces along coasts as rocky intertidal pavements, full of rock pools (Figure 10.11) These pools have become important habitats for seaweeds and many other group of rocky shoreline biota.

Fig. 10.11 PERIGLACIAL STONE PAVEMENT EXPOSED AT LOW TIDE AT GARRYVOE, BALLYCOTTON BAY, COUNTY CORK, WHICH FORMS A DISTINCTIVE TYPE OF LOW ROCKY SHORELINE. Periglacial processes operating in Ireland's glacial past (10,000 to 20,000 years ago) have left extensive areas of hard, coarse stone surfaces (pavements) on coasts developed in these glacigenic sediments. The pavement was created by seasonal frost-shattering of this former periglacial surface, now part of a coastal intertidal zone. These rocky shorelines are also often structured by the earlier freezing mechanisms into distinctive shapes, such as stone stripes and polygons, and are still in evidence on many similar shorelines formed from periglacial sediments found around Ireland's coasts. The inset photograph shows the coarse and angular rock fragments of a wave-eroded stone polygon at Garryvoe. The rock is generally hardened by a process of iron cementation, giving it resistance to wave erosion. These surfaces often contain extensive areas of rock pools and are particularly important as habitats for biota, including many different macroalgae (for example, seaweeds). [Source: Maxim Kozachenko]

SHORE PLATFORMS

Niamh Cullen and Mary Bourke

Shore platforms are primary erosional geomorphological features associated with rocky coastlines. These landforms have many names, which are often based upon the inferred process involved in their formation. Examples include wave-cut platforms, abrasion platforms, intertidal platforms and wave benches. Many of these features are backed by cliffs, although they may also occur on low-lying rock shorelines or in association with beaches and sea walls. In some places, boulder banks and the accumulation of other sediments may obscure them from view, even at low tide. Globally, the percentage of coastlines fronted by shore platforms is unknown. In Ireland, however, mapping has revealed their widespread distribution along its rock-dominated coastlines.[8]

These platforms have been classified into three main types but this simple and widely used designation belies the diversity of platform morphologies that exist (Figure 10.12).[9] Shore platforms along the Irish coast occur in a variety of rock types, including limestone, sandstone, mudstone and hard igneous rocks. How and when these features formed is still under debate, but the huge variability in their morphology suggests that a combination of inheritance factors – such as rock type, geological, climatic and sea-level history – all play a significant role in their formation; they are not formed simply by contemporary processes (Figure 10.13a–e).

Present-day processes that influence the weathering and erosion of these extensive shoreline surfaces, however, include marine (for example, hydraulic action) and subaerial (for example, thermal expansion and contraction and salt weathering) processes. In addition to these, biological and chemical action also contribute to the rates and styles of rock breakdown. In Ireland, chemical weathering is most evident on the Carboniferous limestone shore platforms located along western coasts, as found in County Clare and County Sligo, and results in distinct morphologies. These platforms are controlled particularly by the angle of the rock bedding and thickness, with many distinctive rock benches exposed above the active shoreline platform. In contrast, bioerosion occurs more frequently in soft rock lithologies, such as siltstones and mudstones. These rock types are widespread in occurrence on Ireland's coasts, including in areas dominated by limestones, as in County Clare, or the sandstones of southern coasts (Figure 10.14a and b).

The rate of platform erosion along Ireland's coast is currently unknown. Globally, mean annual rates of vertical erosion (downwearing) on shore platforms, measured using a micro-erosion meter, are 0.397mm/yr, 1.282mm/yr and 0.625mm/yr for igneous, sedimentary and metamorphic rocks respectively. Although measurements of shore platform erosion have tended to focus on the micro-scale (mm/yr to cm/yr), geological controls, such as the spacing and orientation of joints and fractures, have been shown to strongly influence erosion of

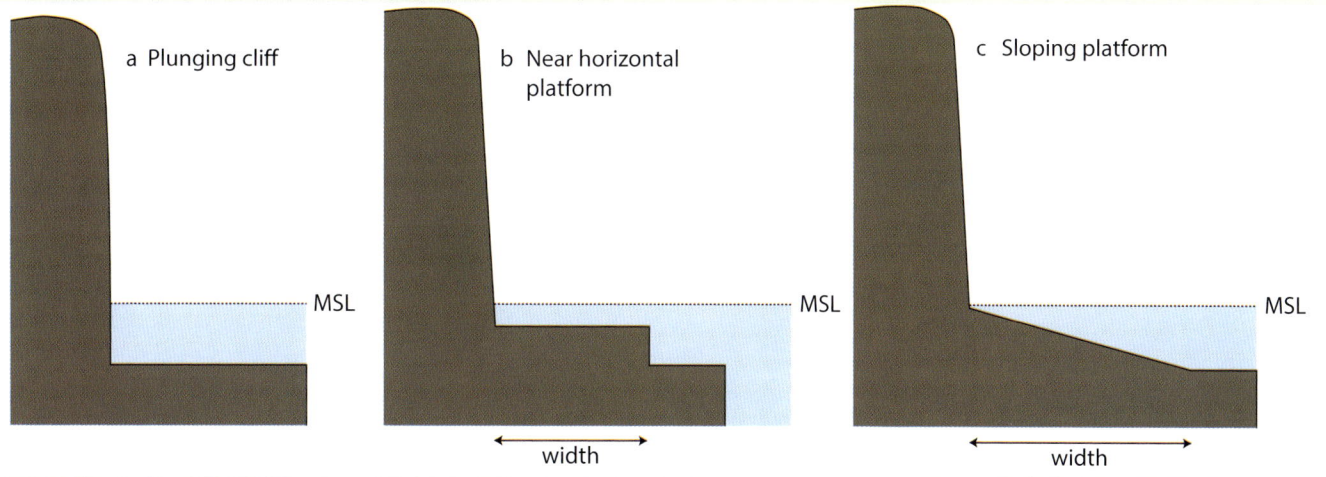

Fig. 10.12 COASTAL PLATFORM TYPES. Three main types of platform are recognised, based on their relationship to mean sea level and shape: a. plunging, b. horizontal, c. sloping. Shore rock platforms are generally exposed during some stage of a tidal cycle, but also exist completely submerged below water level at all times. If the shore platform is never exposed, even at low tide, and the adjacent cliff dives straight into deep water, it is termed 'plunging', as illustrated in a. Many examples of plunging cliffs and linked rock platforms occur along the west coast of Ireland. Most commonly this type of platform results from the dominant influence of the bedrock characteristics (that is, where the rock bedding is horizontal) and/or possibly from former sea-level changes. Shore platforms, also known as wave abrasion surfaces, develop usually at the toe of a cliff face, through wave erosion of the bedrock. They extend seaward from the cliff to the offshore limit of wave erosion, or to the seaward extent of the rock outcrop. In some places, the present-day shore platform is truncated offshore by another cliff, as shown in b, which may have been formed during an episode of lower sea levels. Platforms generally are either horizontal in shape (b), or sloping (c), at an angle of *c*.1–5 degrees in the seaward direction. This basic morphology is influenced strongly by the factors of bedrock geology and the tidal height range, over which the local spectrum wave action operates. The seaward dipping platforms are characteristic of wave-and-tide-dominated coasts. Comparatively flat platforms can occur in sedimentary rocks possessing horizontal bedding. Their shape and width, often forming as narrow platforms, can also be the result of the control of a microtidal range and/or exposure to processes of water freezing and are developed on ice dominated coasts, known as paraglacial. These are found today in high-latitude, Arctic-type regions, or in formerly glaciated environments, such as those of Ireland. [Source: adapted from Sunamura T., 1992. *Geomorphology of Rocky Coasts*. Chichester: John Wiley]

Fig. 10.13 (above) SHORELINE PLATFORMS IN IRELAND. a. Steeply dipping sandstone platform at Corraun, County Mayo [Source: Niamh Cullen]; b. washboard-type morphology in sandstone at Myrtleville, County Cork [Source: Niamh Cullen]; c. sub-horizontal limestone creating rock-benched cliffs and a lower shore platform, near the Bridges of Ross, County Clare [Source: David Hodgeson]; d. cliff overhangs and horizontal intertidal to subtidal platform (plunging type), developed in hard, interbedded sandstones and thin mudstones of Precambrian age at Muckross Head, County Donegal [Source: Liam Carr]; e. gently sloping platform backed by sea wall and unconsolidated Quaternary cliffs at Spanish Point, County Clare. [Source: Niamh Cullen]

Fig. 10.14 (left) SHORE PLATFORMS AND ROCK BENCHES ON INIS MÓR, THE ARAN ISLANDS, SHOWING THE RESULT OF THE CHEMICAL AND BIO-WEATHERING OF THESE SURFACES. a. The bedrock-controlled rock benches exposed on the coast close to Dún Dúchathair (Black Fort), Killeany, Inis Mór. These benches occur within the present intertidal zone and have developed as the contemporary shore platform under wave and tide influences. Storms often bring the higher benches, seen at this site, under the effect of sea and salt spray. Consequently, these limestone surfaces, particularly those forming the active shore platform, are controlled by chemical weathering and, at times, the erosive action of waves and seawater. This creates surfaces showing extensive microscale rock pitting, with multiple larger shallow depressions. On the benches close to sea level, these develop as rock pools, with distinctive marine plant and faunal communities. Many features of normal karst weathering can occur on these surfaces. Sea lettuce (Ulva) and other green algae can be seen in the photograph, covering the lowest bench closest to the sea. Rock pools formed on higher benches are still within the intertidal zone and show white salt crusting around their edges as drying-out occurs between high tides. [Source: Robert Devoy]; b. Close-up photograph of the chemical and associated bio-weathering morphology in this limestone platform, showing the rock pools formed at the shoreline. The benthic flora and fauna colonising these shore platforms often provide a form of natural protection. The covering of biota helps to slow down the process of vertical wave erosion and, in a way, contributes towards overall coastal protection. A similar protecting role can be played by the wider supratidal ecology of these shores, as well as by any loose sediments covering the platform. In many places cliff erosion supplies boulder-size material, resulting in the formation of storm beaches with a steep/high gradient. These can also act as natural revetments, forming effectively a coastal protection measure. [Source: Niamh Cullen]

shore platforms at meso-scale (cm/yr to m/yr) by facilitating the removal of blocks via wave quarrying.[10]

Emerging evidence suggests that shore platforms may play an important role in protecting the coastline from erosion, by attenuating wave energy before it reaches the hinterland. Shore platforms also provide habitats for a range of intertidal organisms, such as limpets, mussels and micro- and macro-algae, which may be at risk from the effects of 'coastal squeeze', due to rising sea levels associated with climate change. Despite the intrinsic value of shore platforms and their geodiversity, until recent decades they have received relatively little attention as coastal environments, especially compared to the recognition given to the uses and value of soft sedimentary coasts with their beaches and dunes. In addition to ecosystem services provided by shore platforms, and of rocky coasts in general, the cultural value (for example, tourism and social amenity provision) of these coastal landforms warrants much greater attention (see Chapters 27: Tourism and Leisure and 33: Climate Change and Coastal Futures).[11]

COASTAL BOULDER DEPOSITS ON THE ARAN ISLANDS

Ronadh Cox

Ireland's western coasts, imaged by the media most commonly as steep bedrock cliffs with open-ocean exposure, host some of the world's most spectacular coastal boulder deposits (CBD). These landforms easily rival in dimensions and scale those CBD found, for example, on the hurricane-driven coastal systems of Pacific ocean islands, including Hawaii and coasts of the Pacific shoreline rim, including those found in New Zealand, Chile or those of Atlantic Canada. CBD are generally perched above the highest high-tide levels and preserve a partial record of the history of these coasts' major storm events. Storms and waves are capable of transporting large rocks against gravity, depending on their shape characteristics, with flatter boulders able to be moved preferentially much further. As a result, very large boulder-sized blocks, and bigger, can be moved both upward and inland, to sit stranded for years or even centuries, as CDB, before the next big storm event, or even a tsunami (see Chapter 2: The Coastal Environment: Physical Systems, Processes and Patterns).

The highest deposits, situated *c.*45m above high water, are on Erris Head in County Mayo, and near Cathaoir Synge on the western coast of Inis Meáin. Boulders at those elevations are relatively small, mostly weighing less than 1 tonne, but have been moved and assembled in heaps at such sites by wave

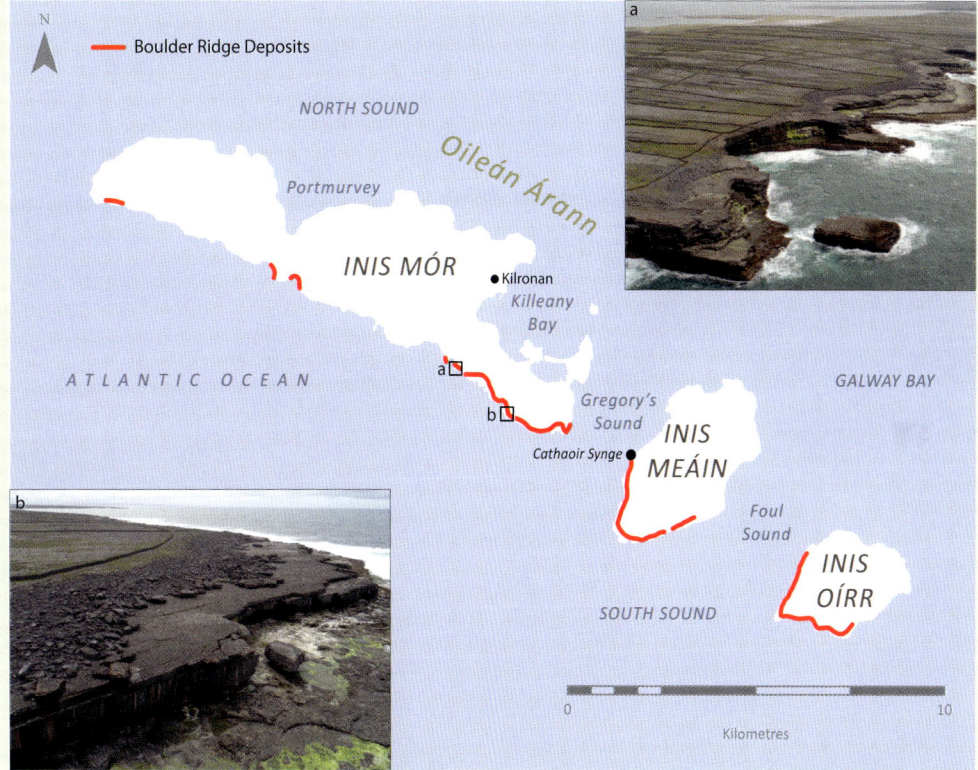

Fig. 10.15 COASTAL BOULDER DEPOSITS (CBD) ON THE ATLANTIC-FACING COASTS OF THE ARAN ISLANDS (OILEÁN ÁRANN), COUNTY GALWAY. a. View along the south-west coast of Inis Mór, east toward the Burren and the other Aran Islands, showing the extent and scale of these limestone-dominated coasts. The prominent cliff-edge rock benches, shore platforms and the linked CBD are visible. The heights and widths of the rock benches, particularly of the upper surfaces, are controlled by the bedrock geology. The effects of former Quaternary (or even earlier) sea-level changes on these geological features, and the subsequent development of shore platforms through the operation of coastal processes, impact only the rock benches closest to the sea. Extensive vegetation is largely absent from the cliff top, upper surface, on which the main spread of boulders and other loose rocks occurs. This surface has also been stripped of former soil cover, both by glaciations and, more recently, by the processes of erosion together with subaerial weathering and the impacts of people, from the time of the first Neolithic farmers onward (see Chapter 16: The Inhabitants of Ireland's Early Coastal Landscapes). The linear boulder spread over this surface (that is, the length and width of the main accumulation of boulders) is visible between the cliff edge and the field boundary wall. The boulders, once broken away from the eroding cliff edges, have been emplaced in their present positions by storm surge events. b. A closer view of the CBD on this coast, showing the range in size of the mega-clast boulders. A few of the limestone boulders can weigh >500 tonnes. Examples of some of these can be seen on both the upper and lower rock benches. Observation and monitoring show that the boulders are moving on the shore platform through contemporary wave and storm actions. [Photo source: Peter Cox; map data source: Cox, R., Zentner, D.B., Kirchner, B.J. and Cook, M.S., 2012. Boulder Ridges on the Aran Islands (Ireland): Recent movements caused by storm waves, not tsunamis. *The Journal of Geology*, 120(3), pp. 249–272]

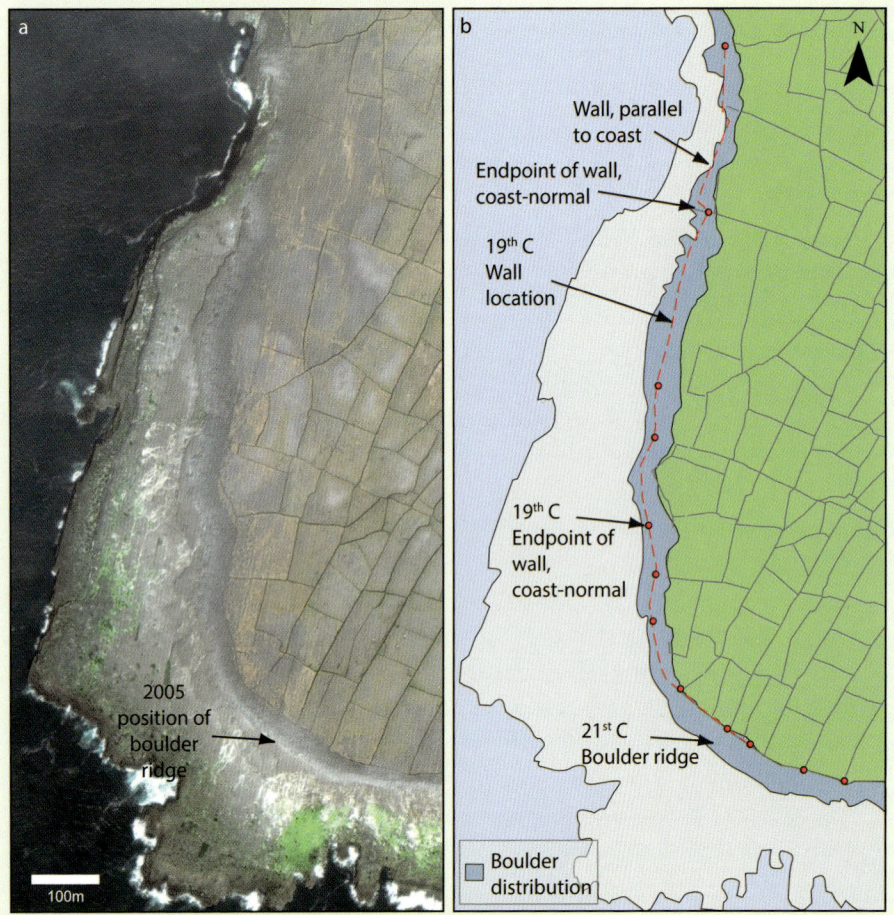

Fig. 10.16a. STORM BOULDER RIDGE ON INIS MEÁIN, ARAN ISLANDS, COUNTY GALWAY. a. Aerial photograph from 2015 of the exposed Atlantic coast of the island [Adapted from: Google, CNES/Airbus, 2015]; b. the site and historic locational information of this prominent boulder ridge on Inis Meáin, showing the position of the nineteenth-century field boundary at the coast in relation to the ridge extent. The ridge has been built from the storm boulder blocks and other slabs of the clifftop Carboniferous limestone. The map indicates the increase in the boulder ridge extent (blue shading) caused by storm waves. Coastal boulder deposits (CBD) do not change location very often; it takes extraordinary energy to dislocate massive rock blocks near sea level, or smaller boulders at high elevations. Research studies to show with certainty that these were storm deposits have taken time. Comparison of nineteenth-century maps with modern orthophotos has demonstrated that the boulder deposits have migrated inland in several places, overriding old field walls in the process. The strong winter storms of 2013–2014 caused significant changes in the distribution of the boulders, which has been documented with before-and-after photographs. [Adapted from: Cox, R., Zentner, D.B., Kirchner, B.J. and Cook, M.S., 2012. Boulder Ridges on the Aran Islands (Ireland): Recent movements caused by storm waves, not tsunamis. *The Journal of Geology*, 120(3), pp. 249–272]

action. The biggest piles, forming boulder accumulations up to 7m high, 40–50m wide and incorporating individual boulders weighing up to 100 tonnes, occur at 10–20m elevation, lining the inland edges of wide bedrock platforms. Notable examples occur at the southern tip of Inis Oírr, along the south-western coast of Inis Meáin, on the south-western side of Inis Mór, on the Inishkea Islands, at Annagh Head in County Mayo and on the County Clare coast between Doolin and Fanore (Figure 10.15). The largest individual boulders, of 100–700 tonne megaclasts, occur in isolation on low-elevation shore faces, located a few metres above the high-tide mark and several tens of metres inland. Impressive examples of these are to be found on Inis Mór and on the coasts of County Clare. Boulders in all of these categories have been moved by storms in recent years.

A common misconception about CBD is that because they are transported by waves, the boulders must have come from the ocean. In most cases, however, they have been excavated quite close to the sites where they are currently located. For cliff-top deposits, boulders have been broken free by mass movement (gravity-driven) mechanisms from the lip of the cliff, while boulders situated towards the back of shore platforms have been removed by storm surges (that is, hydrostatic mechanisms) from their subjacent bedrock, as waves exploit fractures and bedding plane weaknesses. Isolated mega-clasts (large boulders) form when sections of cliff detach as coherent blocks, which may then be shunted along the platform by extreme waves.[12]

CBD were first mentioned in science literature by the Ordnance Survey mappers of the Aran Islands, who reported briefly that, 'Great quarrying seems to be going on here during the gales. Blocks 30 x 15 x 4 feet tossed and tumbled about.'[13] Ireland's CBD remained largely unstudied for more than a century until detailed studies in the twenty-first century inserted them into international debates about whether boulders of such size could be moved by storm waves, or whether their transport required tsunami, or are even the result of earlier ice action (Figure 10.16a and b).[14]

CAVES, BLOWHOLES, SEA STACKS AND ARCHES

All of these cliff-related features are the classic, school textbook landforms of the rocky coast (Figures 10.6 and 10.7).[15] True sea caves are not very common on Ireland's long coastline of over 7,000km, in spite of the extensive occurrences of limestone caves and passages (karst) across the island, with long stretches of coastal limestones. In the Burren, deep-level cave systems are known to intersect with the coast, but their outlets (as freshwater springs) occur below sea level and generally the groundwater emerges well offshore. The origin of these limestone passages dates to the late Cenozoic and the much more humid (climatically warm and wet) landscapes developed at that time on Ireland's rocks.

If sea caves are defined more simply, however, as a rock-covered coastal fissure, then they are to be found everywhere on our coast! These caves are initiated in the lower part of cliffs, forming through the collective interactions of seawater, groundwater, tides and abrasive wave action. They are often created in those parts of cliffs composed of soft and/or fractured rock material, which erode faster than surrounding areas of relatively harder rock. Alternatively, or in combination, they can form where the rocks have structural weaknesses resulting from, for example, rock fracturing, faulting, bedding and folding. These conditions are added to in caves developing within limestone sea cliffs, where mechanisms of mechanical erosion are coupled with the chemical dissolution of rocks by water and other forms of weathering (Figure 10.14).

A sea cave often starts in association with the growth of an erosional notch. The mechanism is similar, therefore, to that operating in primary cliff-line development; the only difference is that erosion is concentrated now in a small cliff zone of comparatively weaker rocks (an area of rock fissures). As an irregular-shaped hollow is initially enlarged, it may result in increased rock release and, over time, the size of this proto-cave expands, transforming eventually into a recognisable sea cave (that is, increasing progressively in height, width and depth). Additionally, coastal sediments in transport (such as sands), and eroded larger rock fragments from recent cliff falls, contribute towards this process of enlargement within the cave.

If visiting these caves, a note of caution should be heeded. The floor of a cave is normally coincident approximately with the level of an adjacent shore platform, or a beach. Consequently, many caves can be accessed from along the shoreline on apparently 'dry' land. But the timing and operation of the local tides should be taken into account in any visit, as some caves may be exposed only at low tide and are deceptively accessible. On some coasts, people can be cut off easily by the rising tide. Other caves, however, can only be explored from the sea, through sailing or kayaking, or with the use of modern technology, such as a drone.

Blowholes form commonly where fractures and fissure development in cliff-face rocks are propagated landwards. Such features are generated through the oscillation of hydrostatic pressures from wave action and as compressive forces within these fissures. These conditions operate effectively through the air that is trapped between the oscillating waveform and the rocks, especially in microcracks, and can speed up cave formation. Within these cave-fissure systems, blowholes can form spectacular features that develop at the far end of sea caves, and where the cave roof partly collapses. These features connect to the land surface above from, for example, the cave passage and through an irregular to semi-circular shaped vertical shaft, often tens to hundreds of metres inland, behind the cliff edge. Wave pressure build-up in the rock system, particularly during times of storm surge, can result in an explosive release of air and water up the shaft, with a geyser-like fountaining of water from the blowhole travelling tens of metres into the air. The pressure built up in these systems is such that loose rock material weighing many tonnes can be ejected and spread great distances around the shaft and along the neighbouring cliff.

Examples of these landforms are widespread and can be found on Inishbofin Island, County Galway; Raghly in Sligo Bay; Coosfadda, Canglass Point, County Kerry; Poll na Scantoine, Ballycastle, Downpatrick Head, County Mayo and Donegal Point, Baltard, County Clare (Figure 10.17a–c). These features are less common on the coasts of Northern Ireland.

Sea arches are developmentally related landforms to caves and blowholes representing, together with sea stacks and rock pillars, one of the most photogenic results of cliff erosion. They form in places where a sea cave passage eventually cuts through a narrow rock headland, as situated between two coastal geo-indents or embayments (Figures 10.2a and 10.6).

Sea stacks are stand-alone rocks rising above sea level; they may have pointed or 'flat' tops depending on the parent geology (Figures 10.2b and 10.7). A simple rock erosion story attaches to such iconic landforms. Most sea stacks were once part of promontory headland-type cliff settings and are linked in their development to

Fig. 10.17 BLOWHOLE SEA EROSION LANDFORMS. a. Rock-structured geo and a linked blowhole at Raghly, Sligo Bay. Inland of the cliff edge, and situated along the line of a fault-guided geo, a blowhole feature has formed in the highly fractured Visean limestone rocks of the area. These have been further exploited by wave hydrostatic pressure to create this blowhole behind the low (<40m high) cliffed coast. [Source: Sally Siggins]; b. Atlantic storm structured promontory cliff headland at Coosfadda, Canglass Point, County Kerry. A geo-linked blowhole has been developed at the head of a rock fracture line. The steep cliffs here are formed in the regional Devonian sandstones and reach elevations well above 50m. [Source: Robert Devoy]; c. rock debris (boulder sized, some >0.5m in diameter) can be seen lying around the blowhole shaft, emitted above the cliff-line by explosive wave action during storm events, when water and spray may rise over 20m over the ground. [Source: Robert Devoy]

the erosional continuum of: sea caves–passages–blowholes–sea arches–sea stacks. In these cases the sea stack may be the outcome of initial sea-cave system development and subsequent progressive rock collapse. In other locations there may be no such cave–sea arch linkage, but simply situations of coastal headlands or other cliff areas, once part of the mainland, being eroded by the sea. This commonest of settings happens most frequently in places coincident with local geological faults or fracture zones. If the top part of a sea stack becomes worn down due to weathering and erosional processes, it will eventually become a lower-profile sea stump, a standalone rock with a relatively flat top, rising just above sea level. The shape and size of sea stacks depends on the height and geology of the adjacent cliffs. Sea stacks are often home to numerous colonies of seabirds as they provide a natural living environment with minimal disturbance from human activities (see Chapter 3: Coastal Waters: Marine Biology and Ecology).

THE ECOLOGY OF IRELAND'S ROCKY SHORES

Rocky shorelines occur wherever the geology and hydrographic conditions are suitable and can include those formed of both solid (the bedrock) and loose rocks (cobble to boulder-sized materials). Unlike the different habitats of soft sedimentary coasts (for example, estuaries, mudflats and saltmarshes), those of rocky shorelines are exposed to a wide spectrum of coastal wave activity. These extend from very sheltered environments, to those of high wave-energy coasts, which receive the full impact of storm surges.

In addition to these factors, coastal geology and geomorphology, particularly the slope and surface shapes of coasts, also play an important part in the heterogeneity of shoreline habitats. For instance, some shores will have more rock pools, cracks and crevices than others. These can act as important refugia for marine biota. Cracks and crevices, especially on more sheltered shores, can harbour a large range of small fauna. Further, the colour of the rock, as linked to its capability to reflect and absorb heat, together with rock softness and hardness, which influence the

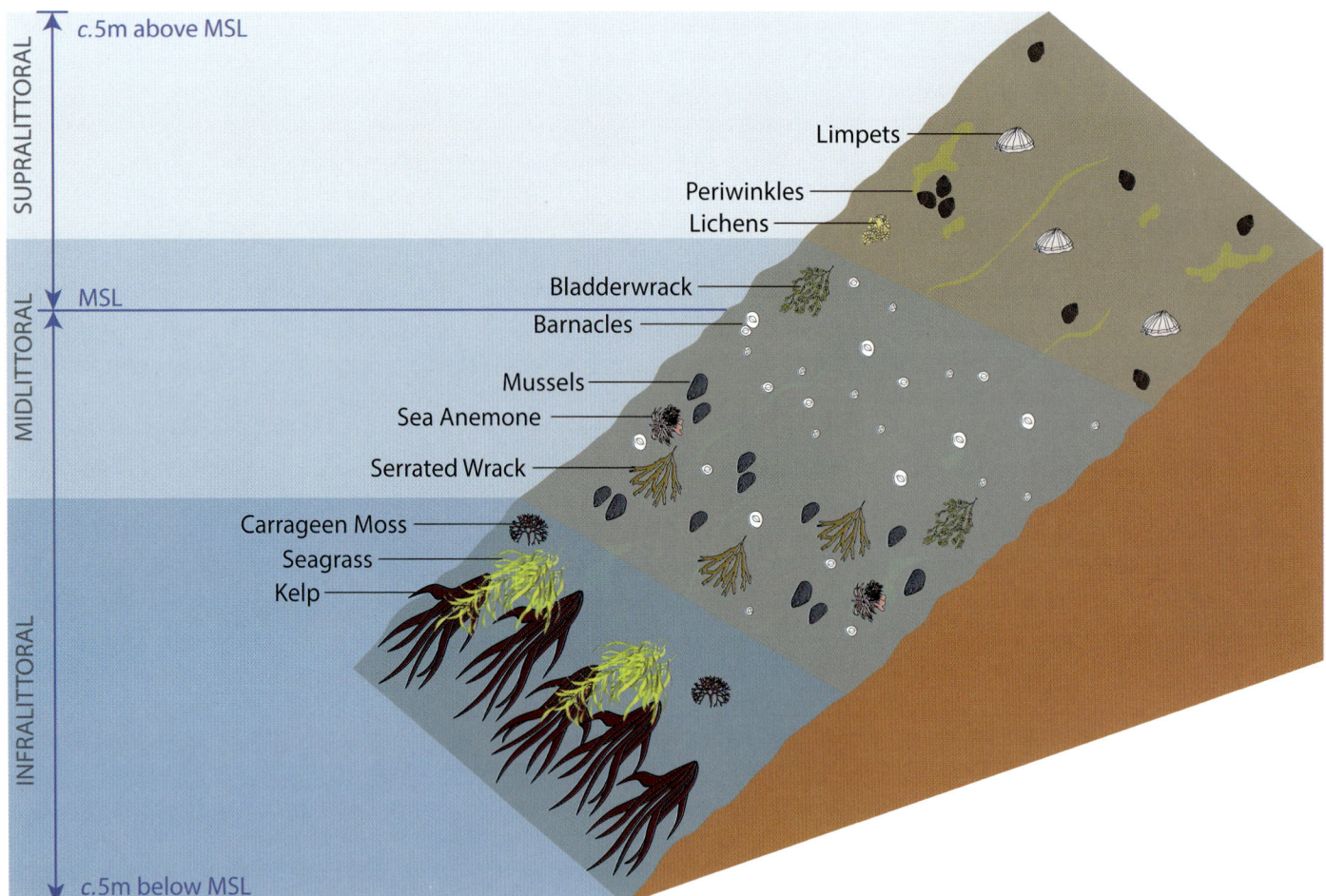

Fig. 10.18 ECOLOGICAL ZONATION ON ROCKY SHORES. Ireland's shores span vertically the areas from supralittoral (upper intertidal) environments, the upper limits generally between 3–5m above the high water mark (HWM), through the midlittoral and down to those of the infralittoral (neritic). Below these areas occur the deeper waters of the shelf. These cover the environments beyond the low-tide mark and, around Ireland, reach depths of 150–300m below sea level. The vertical zone nearest the land, the supralittoral, is termed the splash zone. As its name suggests, this is affected by seawater spray, waves in rough weather and high spring tides. The uppermost part of the supralittoral is defined by the highest limits of periwinkles, while its lower limit reflects the upper limit for barnacles, which are filter-feeding crustaceans. Lichens also characterise the supralittoral fringe. The diversity of species is low in this zone, although marine organisms that can survive within this near-terrestrial environment can occur in abundance. The midlittoral zone is mostly covered and uncovered on a daily basis, at both spring and neap tides. The top of the midlittoral zone occurs where barnacles start to become more abundant, while its lower limit is defined by the upper limit for kelps (laminarians). Below this is the infralittoral fringe, the lowest limit being uncovered only during the extreme low-water conditions of spring tides. The biodiversity in the infralittoral fringe can be very high, but concomitantly there may be great competition for resources, including space. Below the infralittoral fringe are the infralittoral to offshore zones, which are not usually uncovered, even during spring tides.

roughness (rugosity) of surfaces, both affect the composition of the biota. These and other physical environmental factors controlling Ireland's coasts have created a distinctive rocky shore ecology on the island. This study provides a brief look at the key elements of the flora and fauna to be found on this type of shoreline.

Rocky shores and their biota

Rocky shores form over 40 per cent of Ireland's coastline and are the island's most dominant type of intertidal shoreline surface (substrate). Such surfaces exhibit the whole spectrum of rocky shore environments, ranging from those that are relatively low-energy and sheltered (such as enclosed bays), to those of exposed and high wave-energy conditions. This diversity of rocky coasts is reflected also in the composition of their respective plant and animal communities.[16]

Ireland's Atlantic west coast has some of the most exposed and high-energy rocky shores in Europe (Figures 10.7 and 10.10 and also 30.9). In contrast, other rocky shorelines around the island can receive some protection from the full effects of such wave action by the presence of sheltering headlands and of the calmer waters of linked indented bay environments. For eastern coasts, the island's landmass itself provides protection from the severe Atlantic ocean wind and wave regime. Ireland has even been described as Europe's (if not the world's) biggest natural breakwater.[17] This variation and diversity in the wave regimes of Ireland's rocky coasts is reflected in the composition of different biotic communities. In addition, these coasts commonly have large tidal ranges (low-water to high-water marks), of up to ~5.0–5.5m during spring tides (from meso- to macro-coastal tides). This results in a broad and distinctive vertical zonation of organisms within the shoreline environments. Other factors – including water temperature, salinity and geology – also drive a horizontal, or alongshore, zonation in the composition of these biota. Interestingly, an exception occurs within Lough Hyne, west County Cork, where the vertical zonation of algae and fauna is compressed, due to its reduced tidal range (microtidal of only 0.7–1.0m). Many other low-tidal amplitude loughs – such as Strangford Lough, Carlingford Lough and Lough Swilly – also have unique aspects to their ecology (Figure 10.18).

The horizontal and vertical zonation of shorelines involves the organisation of organisms into distinct bands, or strata, of the biota. These occur at different heights and widths along a shoreline, extending from the land to the water's edge and beyond. This structuring reflects the ability of the biota to withstand different levels of, for example, wave action, salinity and duration of exposure to the air. Organisms show their biological responses to these different physico-chemical parameters, as well as to biological factors (such as competition, grazing, predation or differences in growth rates). Biota can show physiological, morphological and/or behavioural adaptations within these zones. Many species living in these zones are sessile, being attached to rock surfaces, while others are mobile and can move up and down the shore, even into infralittoral areas.

Another important driver of some of these biota is that of Ireland's location, which straddles a significant biogeographic boundary on the north-eastern Atlantic ocean margin. This location is influenced strongly by the warmed currents of the North Atlantic Drift/Gulf Stream. For many cold-adapted (arctic-boreal) rocky-shore species, these coastal waters mark their southern limits, while warm-adapted (Lusitanian) species are reaching their northern limits. Hence, the composition of the island's rocky coast communities of biota is a mixture of cold-water, warm-water and broadly distributed species (with ranges that extend from Norway to North Africa or the Mediterranean) (see Chapter 2: The Coastal Environment: Physical Systems, Processes and Patterns and 33: Climate Change and Coastal Futures).[18]

In general, Ireland's rocky shores are dominated by seaweeds and invertebrates. These are mainly marine in origin, so the most significant environmental stress for these organisms is their emergence out of seawater (for example, in supratidal settings). This situation contrasts with areas of seawater immersion (submergence), where the organisms are bathed in a nutrient broth of relatively constant temperature and salinity (intertidal settings). When these environments are uncovered during low tides, they have to cope with varying salinity extremes (perhaps due to rainfall or water evaporation), aerial temperatures, as well as greater variations in CO_2, O_2 and pH values. The impact of this is reflected in the resulting shoreline plant and animal species composition and diversity.

THE SUPRALITTORAL

This area of Irish shores is adjacent to freshwater communities on the land and is known as the splash or spray zone (that is, the upper intertidal and areas above the high water mark of spring tides) (see Figure 10.18). It is characterised by terrestrial plants that can tolerate seawater spray. These commonly include thrift (*Armeria maritima*), sea plantain (*Plantago maritima*) and on some rocky shores, such as those of the Aran Islands, sea campion (*Silene vulgaris subsp. maritima*), rock samphire (*Crithmum maritimum*), sea kale (*Crambe maritima*) and sea beet (*Beta vulgaris subsp. maritima*), as well patches of lichens (see Figure 27.30). On shores where there is a freshwater input and plenty of nutrients present (for example, due to water run off from the land), the distinctively green gutweed (*Ulva intestinalis*) may be abundant, though this is often bleached white during hot weather (see Figure 10.14a). The main lichen genera present in the supralittoral are ramalina, xanthoria, caloplaca, lecanora, verrucaria and the black lichen (*Lichina pygmaea*), amongst which may be found the red bivalve (*Lasaea rubra*).[19]

Amongst the shell fauna, the small periwinkle (*Melarhaphe neritoides*) is most characteristic of this zone. It can survive here due to a number of adaptations. These include its capability to take oxygen from the atmosphere, its production of uric acid (instead of ammonia or urea) to conserve water and the retention of water inside its shell. It feeds on diatoms and fine algae, and 'on some exposed cliffs it has been found 10m above the reach of the spring tides'.[20] Toward the lower part of the supralittoral, this species may be joined by the rough periwinkle (*Littorina saxatilis*) and also *Littorina arcana*. These winkle species occur generally lower down the shore, but adaptations allow them to exploit the supralittoral fringe,

Fig. 10.19 ALGAE AND INVERTEBRATES COMMONLY FOUND AS ELEMENTS OF THE BIOTA ON MANY ROCKY SHORES AROUND IRELAND. a. Macro-brown algae, including bladder wrack and other fucoid species, cover extensive areas of the intertidal areas along Ireland's shoreline. This is illustrated here at Kilkieran Bay, Connemara, County Galway, where a survey using a drone was used to measure the presence and abundance of biota along this intertidal area. The 'white taped' squares visible in the photograph mark the positions of quadrat-type survey points of the biota. [Source: Eugene Farrell]; b. egg cases of the dogwhelk (*Nucella lapillus*) along with some dogwhelk individuals on the low shore [Source: Elizabeth Cotter]; c. underside of a low-shore intertidal boulder with two specimens of the common starfish (*Asterias rubens*), tubeworm cases of *Pomatoceros lamarckii* and empty barnacle shells probably predated on by the dogwhelk (*Nucella lapillus*). [Source: Elizabeth Cotter]; d. intertidal seaweed (*Ascophylum nodosum*) with single egg-shaped air bladders along with individuals of *Littorina mariae* present. [Source: Cynthia Trowbridge]

where environmental conditions are too challenging for many other intertidal molluscan faunas.

Low numbers of barnacles also begin to appear in this area. On more sheltered shores, the channelled wrack (*Pelvetia canaliculata*) and the spiral wrack (*Fucus spiralis*) are the algae characteristic of lower parts of the supralittoral fringe (Figure 10.19a–d). These species can tolerate a lot of desiccation, despite appearing black and shrivelled at times. On wave-exposed shores, the dominant macroalgal species present in the supralittoral is the red alga (*Porphyra umbilicalis*), known as laver. Underneath these, and on the rock surface, is a coating of biofilm, which may include blue-green algae, diatoms and Ralfsia-like species (brown seaweeds).[21]

LOUGH HYNE: A MARINE RESERVE IN CRISIS

Rob McAllen, Cynthia Trowbridge, James Bell, Julia Nunn and Colin Little

Lough Hyne is the Republic of Ireland's only (and Europe's first) Statutory Marine Reserve. Designated in 1981, it represents the 'jewel in the crown' of the state's marine conservation strategy (Figure 10.20). The Lough is an area of long-term, local to regional, interests in marine

Fig. 10.20 LOUGH HYNE, COUNTY CORK. A panoramic view across Lough Hyne and this south-west County Cork coast to the Celtic Sea beyond. The top left of the photograph shows Tranabo Cove and Tralispean Bay. Both these sites record the 1755 Lisbon tsunami that struck this part of Cork's open, crenulated and cliffed coastline. The Lough was formed as a rock basin during the early Neogene period (probably between 17–23 million years ago), by earth crustal movements. These led to rock subsidence and differential faulting, causing topographic height differences of *c*.300m across the area. The basin was later filled by marine water as sea levels rose and further coastal erosion occurred. [Source: Superbass, Wikimedia Commons, CC BY-SA 4.0]

environmental matters and heritage (see Chapter 8: Visualising, Mapping and Monitoring Coasts). It is a semi-enclosed marine lake, or lagoon, situated *c*.5km west of Skibbereen and some 80km from Cork city in south-west Ireland. Measuring just 0.8km by 0.6km, it is believed that the Lough was a freshwater lake up to *c*.4,000 years ago, when a rise in sea levels joined it with the sea. As a result, it is now a highly sheltered, seawater basin connected to the North Atlantic ocean via a narrow inlet, called Barloge Creek. All tidal exchange goes through a narrow, shallow channel, known as the Rapids. This constriction causes a highly variable current flow regime at different points in the Lough and is the principal reason for the rich biodiversity occurring within its depths (Figure 10.21).[22]

Regrettably, the reserve is in crisis, with the rapidly declining environmental quality of the Lough leading to altered biological communities. The grounds for concern and the actions needed to remedy this situation are presented in this short discussion.

LOUGH ECOLOGY

Lough Hyne is recognised globally as a biodiversity hotspot and has been the focus of intense scientific interest, beginning in the nineteenth century but expanding rapidly from the 1950s onwards. It has been the site of ground-breaking, field-based experiments and is associated with more than 460 scientific publications. As a result, Lough Hyne is one of the most intensively studied coastal sites in the world, with in excess of 1,800 marine species recorded within its waters. The environmental decline of such a site should send shock waves through the scientific community and be a serious concern for local and national stakeholders.[23]

The midlittoral (intertidal) zone

The rocky shoreline exposed at low tide has not changed in overall appearance in recent time. There are still normal cycles in which barnacles and seaweeds are dominant and can easily be recognised. Unfortunately, this may have led to a local perception that the Lough remains 'pristine'. This is far from the truth. Most of the barnacle population is now made up of an invasive Australasian species, an acorn-type barnacle, *Austrominius modestus*. Smaller and more delicate organisms, such as bryozoans (sea-mats), have declined in abundance since 2010, in parallel with a decline in available dissolved oxygen. In addition, since the late 1990s, there has been a rise in mussel populations, suggesting an increase in phytoplankton food, linked to rising nutrient levels.

The shallow sublittoral (infralittoral) zone

Changes in the shallow waters around the edge of the Lough have also been profound and obvious. From the 1920s to the early 2000s, much of the shallow subtidal zone was dominated by the 'boom-and-bust' purple sea urchin (*Paracentrotus lividus*), which was numbered in tens of thousands at its peak. This ecologically significant urchin grazed ephemeral algae, helping to maintain a diverse ecosystem of invertebrates. In the early 2000s, however, it declined massively, resulting in the excessive overgrowth of these algae. The process was enhanced further by rising nutrient levels, which led to declines in water quality and available oxygen, contributing directly to the disappearance of many invertebrates (Figure 10.22). Causes of the urchin's decline are probably diverse and include a population explosion of carnivorous spiny starfish (*Marthasterias glacialis*), as well as the over-collection of adult urchins and an increase in pathogens. The result has been a profound community change in the biota, particularly that of reduced species diversity and increased abundance of short-lived, opportunistic species.

The deeper sublittoral zones

Lough Hyne is known globally for its abundance and biodiversity of sponges, with over one hundred different species forming extensive sponge gardens. However, between 1999 and 2017, the deeper areas of the Lough, particularly its sublittoral cliffs, have changed dramatically. Sponge abundance declined to 10–15 per cent of its original level across nearly all species and large, decades-old sponges have disappeared with no evidence of new individuals. Once abundant sponge species are now locally extinct. The presence and abundance values across other taxonomic groups of species mirror this decline. For example, cliff areas of the Lough, excluding Whirlpool Cliff, now exhibit either bare silty rock, or are occupied primarily by large sea squirts. Almost all upward-facing surfaces are devoid of algae, while other species have not been seen for many years, including the Dublin Bay prawn (*Nephrops norvegicus*). Observations by divers indicate that the marked decline occurred after 2010, but before 2012, when it was first reported. This suggests either some larger-scale environmental impact, such as changes in nutrient or oxygen levels to the Lough water mass reaching a tipping point, or possibly a toxic event.

CAUSES OF DECLINE

Overall, the changes in littoral and sublittoral communities demonstrate a massive and unacceptable decline in the environmental quality of

Fig. 10.21 THE RAPIDS AREA OF LOUGH HYNE MARINE NATURE RESERVE. A mussel bed, principally of the blue mussel (*Mytilus edulis*), is shown here, although other species of mussel live in the warm waters of the edge Lough, especially the Mediterranean mussel (*Mytilus galloprovincialis*). In addition, stipes (the stem-like part of seaweeds) and fronds are shown of kelp species that grow in these, at times, fast-moving tidal channel waters of the Lough. [Source: Rob McAllen]

these coastal waters. There are probably several drivers of this decline. These include the increased use of the Lough as a tourist site, the arrival of invasive and non-native species and climate change. The temperature of Ireland's coastal waters has increased by $c.1°C$ since 1990, as mirrored in the water temperature within the Lough.[24]

The most significant influence, however, appears to be linked to increasing nutrient levels. Ireland's Environmental Protection Agency's STRIVE-funded project, NEIDIN, showed that nitrogen levels had increased from $210mg/m^3$ total nitrogen in the 1990s to $720mg/m^3$ in the late 2000s. The EPA standard level for waters to be considered eutrophic is $>350mg/m^3$ for total nitrogen. Locally, the primary source of this nitrogen is from outside the

Fig. 10.22 THE GROWTH OF GREEN, EPHEMERAL ALGAE IN LOUGH HYNE. The increased and abundant growth of different species of benthic green algae in the warm-water environments found along the west shore of Lough Hyne Marine Nature Reserve is an indication of the decline in numbers of purple sea urchin since around 2000. These changes in the Lough ecology are coupled probably with increasing eutrophication of the Lough and, more widely, of other coastal waters in south-west Ireland. [Source: Rob McAllen]

Lough and is derived from neighbouring rivers, estuaries and shorelines along the south-west coast of Ireland. The hydrography and geography of the Lough may, however, act to amplify the potential effects of this eutrophication on biological communities.[25]

Active restoration action is still not widespread in marine environments. However, due to the Lough's isolation, this is likely to be required to restore Lough Hyne to its condition prior to 2000. Actions to investigate and halt or reverse the decline include:

- reduction of the nitrogen runoff into rivers and from shores within regional water catchments;
- extending regular and wider regional monitoring of nitrogen levels to cover the Lough Hyne Special Area of Conservation (SAC);
- investigating direct and indirect effects of nitrogen on marine communities inside and outside the Lough, but within the designated SAC;
- determining the stress tolerance thresholds for key organisms, to enable environmentally relevant thresholds to be established that may be lower than current 'good ecological status levels'.

Recommendations for management

A number of actions are proposed, and need to involve multiple government agencies to help enhance the management of Lough Hyne. These include:

- producing a specific management plan for the Lough;
- developing and implementing a long-term monitoring plan for the subtidal cliffs in the Lough;
- establishing a recovery and restoration plan for the Lough (after the water quality has improved).[26]

THE MIDLITTORAL

The midlittoral is known commonly as the intertidal zone or the foreshore (Figure 10.18). On some very exposed Irish rocky shores, acorn barnacles (*Semibalanus balanoides*) are often the dominant fauna and can be virtually the only sessile (immobile) crustacean found on open rock surfaces in this zone. Along more sheltered shores, the midlittoral may become dominated by brown seaweeds or there may

be sub-zones with bands of barnacles, mussels and mobile gastropods. Found commonly among the barnacles in this zone is the small periwinkle gastropod, which can reach high densities of up to $50,000/m^2$. An additional element in the fauna here is the beadlet anemone (*Actinia equina*), which is able to survive on open rock in this zone as it is covered by a mucus layer, as well as being able to withdraw its tentacles completely. It is also found in rock pools, whereas other species of anemone, such as the snakelocks anemone (*Anemonia viridis*), which cannot retract its tentacles, are confined entirely to rock pools.

The majority of molluscs occurring on the midlittoral are different species of gastropods (including top shells, winkles, dogwhelk and limpets). Just a few species of bivalves (for example, mussels, oysters, cockles) occur, together with chitons, nudibranchs (sea slugs) and sea hares. Two species of mussel (*Mytilus spp.*) are common here: the edible blue mussel (*Mytilus edulis*) and the warm-water Mediterranean mussel (*Mytilus galloprovincialis*). At Lough Hyne, the distribution and abundance of both these species of mussels, and their hybrids, have been shown to be controlled by predatory crabs and starfish. The widely distributed volcano-shaped common limpet (*Patella vulgata*) also occurs on these shorelines and can span the whole range of wave exposure (Figure 10.23).[27]

Seaweeds in the midlittoral are characterised generally by a canopy of brown algae, in particular by wracks (fucoids), which show distinctive vertical and horizontal zonations. These include the spiral wrack (*Fucus spiralis*), the bladder wrack (*Fucus vesiculosus*) and its hybrids, the toothed or serrated wrack (*Fucus serratus*), as well as the egg wrack (*Ascophyllum nodosum*), with its red filamentous algal epiphyte, *Vertebrata lanosa*. These seaweeds may be joined on more sheltered coasts by green algae, such as sea lettuce (*Ulva lactuca*) and *Codium* species. Often, some algae and fauna on these shores only become apparent when the macroalgae are moved aside. In contrast, on exposed shores, where there is more wave splash, the dominant macroalga present is the red false Irish moss (*Mastocarpus stellatus*). Here, an underlying coating of biofilm, composed principally of microalgae (for example, diatoms), is often apparent on the rock surfaces. Many species of invertebrates benefit from the moist sheltered environment under this midlittoral canopy of seaweeds, which forms an important foodstuff for biota. Further, these seaweeds provide a substrate on which to live, for example, for the yellow and flat winkles – which imitate the bladders of the seaweeds – as well as for the spiral calcareous coils of the filter-feeding tubeworms (*Spirorbis*), for colonial bryozoans (sea-mats) and hydroids (Figure 10.19).

THE INFRALITTORAL

The infralittoral zone is subtidal (formed below low water) and can extend up to *c*.5m below the low water mark of spring tides. Further, beyond this zone into deeper water is the sublittoral zone (that is, neritic or offshore environments). The term sublittoral is used to refer to the rock and sedimentary substrata of the continental shelf, which reach to depths of 150–300m below sea level (see Chapter 2: The Coastal Environment: Physical Systems, Processes and Patterns).

As with the midlittoral, this zone is also commonly dominated by algae. Species diversity here can be very high, but challenges from physical factors remain, such as increasing wave exposure into shallower waters, as well as from biological controls. The biota of this zone are characterised by large brown algae – for example, kelp species (*Laminariales*) – with an understorey of small red algae among their holdfasts (the anchoring, root-like structures of some

macroalgae, such as kelps). Most of the soft and more delicate red algal species do not have well-developed adaptations to cope, for example, with desiccation and direct sunlight. Consequently, they occur only where they will be submerged in a semi-permanent and protective layer of seawater (that is, in the infralittoral or in pools), or under a moist canopy of other algae.

At the lower limits of the infralittoral fringe, during a spring tide, subtidal barnacles can be seen, together with calcareous tube worms (*Spirobranchus*) (Figure 10.19c). This fauna is joined by the first appearance of sponges on rocky shores, other than in rock pools.: for example, the breadcrumb sponge (*Halichondria panicea*), sometimes the solitary cup coral (*Caryophyllia sp.*), colonial sea squirts, the common starfish (*Asterias rubens*) and cushion stars (for example, *Asterina gibbosa*). A number of genera of small gastropod molluscs (for example, *Tritia, Rissoa, Trivia*) can also occur in this zone, as well as the open rock limpets, *Patella vulgata* and *Patella ulyssiponensis*.[28]

BETWEEN THE ZONES

In addition to the fauna found at times of low tides on rocky shores, numerous species of mobile marine invertebrates and vertebrates reside between the zones. Most of these will be apparent only when the tide is in, since they do not have the adaptations to tolerate being emerged. Many species, however, move up and down the shore zones as the tide flows and ebbs. Examples of this biota include numerous species of fish, such as wrasse. A range of crab, jellyfish and echinoderms, such as the common starfish (*Asterias rubens*), can be seen in the water when the tide is in, or are left stranded in the midlittoral rock pools as the tide recedes (Figure 10.19c). In recent years, increasing numbers of jellyfish, of different species, have been recorded along Ireland's inshore waters and coasts (see Chapter 3: Marine Biology and Ecology).

Mammals can commonly be found living in the between-zone areas, such as seals and otters (*Lutra lutra*), particularly when a freshwater source nearby is available. Terrestrial vertebrates also visit and live on rocky shores, using them as part of their habitat. Included in these are badgers (*Meles meles*), which may forage on amphipods left on the tidelines (strandline), while rats (*Rattus norvegicus*), foxes (*Vulpes vulpes*) and stoats (*Mustela erminea*) can be found feeding along the seashore.[29]

Reefs and rockpools

Reefs developed in the littoral environment (mid- to sublittoral) and further offshore can be of both bedrock and sedimentary origins, or a combination of these. They represent areas of relatively shallow water and, in the coastal zone particularly, areas of potential wave breaking. Consequently, they can form features for naturally protecting coasts from wave erosion. In the context of rocky shore ecology, the role of the biota can be important in the development of such reef environments (Figure 10.24).

In contrast, rockpools represent periodically air-exposed environments, found particularly in the infralittoral and midlittoral zones (see Figure 3.28). The biota of these areas, especially of large rockpools, may be similar to those of neighbouring open rock surfaces, due to the prolonged submergence of these pools by the

Fig. 10.23 SOME BIOTA OF ROCKY SEASHORES. Many species of plants (particularly algae) and animals (for example, gastropods) live in the different habitats provided by the rocky shorelines that have developed around Ireland's coasts. Examples of these include: a. the European painted top shell (*Calliostoma zizyphinum*) [Source: Elizabeth Cotter]; b. honeycomb worms (*Sabellaria alveolata*), shown here from the lower shore zone at Garrettstown, County Cork [Source: Elizabeth Cotter]; c. snakelocks anemone (*Anemonia viridis*) together with the shells of limpets (*Patella sp.*) at Tranabo Bay, West Cork [Source: Cynthia Trowbridge]; d. example of a subtidal reef habitat, typical of the mid- and infralittoral zones and showing different species of macroalgae, including kelps. [Source: MERC Consultants Ltd and the National Parks and Wildlife Service]

sea. Small fish, such as gobies and blennies, can be found in pools, along with a range of crustaceans, such as shrimps, hermit crabs and other crab species. Further, a characteristic species seen floating on the surface of many midlittoral pools is the small, black-coloured insect *Lipura maritima*. Other biota in these environments, often occurring in higher numbers sub-littorally, include the warm-water purple sea urchin. This urchin occurs primarily on western and southern Irish shores and can be found burrowed into the rock in rock pools on the middle-lower part of the shore.

Rock pools that occur in the upper shore, or supralittoral zone, but especially those that are small in size, are a challenging environment for marine organisms' survival. This is due to the extreme environmental variations that occur here. As a result, these pools contain only a limited number of species, such as the harpacticoid copepod (*Tigriopus brevicornis*) and midge larvae (*Halocladius*). Young mullet and prawns may also be found in high-shore rock pools.

The rocky surfaces of many pools often appear to be stained pink, due to the growth of calcareous red algae. These species may be bleached white at the edges of these environments during the summer, when seawater in the pools evaporates. As observed by McGrath (1984), other refuges for species associated with such

rock pool habitats are the 'damp, dark recesses beneath rocks'.[30] The array of organisms that might be seen underneath a rock, as for example on the limestone coast of County Clare, could include purple sea urchins, starfish, cushion stars and brittle stars, along with a range of fish, such as butterfish, rocklings, scorpion fish, blennies and pipefish, as well as perhaps a bootlace worm. Six different crab species might be observed, along with squat lobsters, cowries, sea slugs, sea squirts, sea mats, sea firs and sponges, while bivalve molluscs may have bored into the limestone rock. In this setting, 'boulders and loose rock increase the degree of shelter available to animals and plants and hence the diversity of the communities'.[31]

FUTURE ECOLOGY

Shorelines occur at the environmentally sensitive interface between land and sea. Climate changes into the future will affect the biota of Ireland's rocky shore and other coastal environments due to sea-level rise, as well as increases in all, or any, of the following: air and seawater temperatures, ultraviolet light penetration, storminess and wave action.[32] Since the islands of Ireland and Britain are located in a critical transition area between northern cold-water and southerly warmer-water species, changes in populations of certain

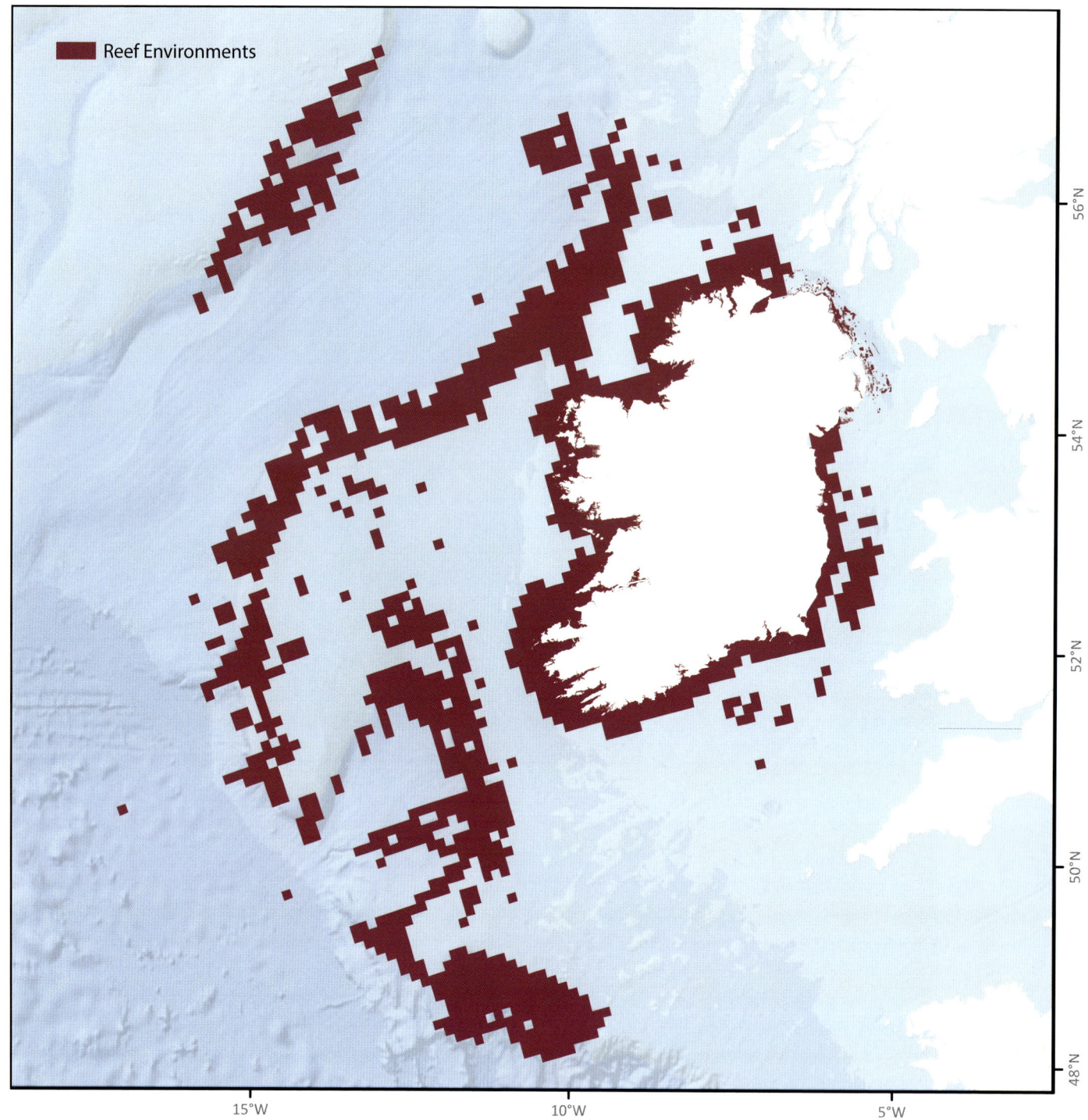

Reef Environments

Fig. 10.24 ROCK REEF ENVIRONMENTS AROUND IRELAND'S COASTS AND OFFSHORE. The mapping of reef environments from Irish (RoI) and offshore deeper shelf waters has been undertaken using a 10km x 10km grid. Data for Northern Ireland are limited to coastal waters and have been collected at a higher resolution. Two species of reef-building worms occur on Irish coasts, the honeycomb worm (*Sabellaria alveolata*) and the Ross worm (*Sabellaria spinulosa*). These worms develop colonies only in areas where sand occurs next to a hard rocky substrate, since the worms create their tubes using the sand and other available marine debris. The shape of the colonies varies from small, single hummocks to reefs. Similar to kelp 'forests', reefs are an important habitat for a range of other organisms. Honeycomb worms are protected under the EU Habitats Directive and are predicted to increase their areas of coverage with rising water temperature. [Data source RoI: National Parks and Wildlife Service, 2019; NI data source: Department of Agriculture, Environment and Rural Affairs and Joint Nature Conservation Committee]

key rocky shore invertebrates and algae can be used as indicators of climate change. There have been many surveys of Irish rocky shores and thirteen of these provide whole-community studies of the biota. During the 1950s, fifty-three species were surveyed from 205 rocky shore sites around Ireland, while in 2003, sixty-three of these sites were resurveyed as part of the MarClim project. In addition, some rocky shores on Ireland's offshore islands – including those of Sherkin Island, County Cork, and Clare Island,

County Mayo – also attracted long-term research, comparing their biota to shores on the mainland.[33] These surveys provide a good basis for assessing the future impacts of climate change on rocky shore ecology.

Comparison of data from these surveys shows that climate-induced changes have occurred in species' distributions (both abundance and range). For example, a southern species of black-footed limpet (*Patella depressa*), whose most northerly limit is currently in south-west Britain, may spread to Ireland's shores. Further, there may be increases in the spread and distribution of some warm-water species of top shells already found in Ireland, such as the flat top shell (*Steromphala umbilicalis*) and the lined top shell (*Phorcus lineatus*), and which are currently absent on parts of the east coast.

It is not known how warmer waters may affect interspecific interactions, but one concern is that they may allow increases in some non-native species that are currently present, as well as for the introduction and establishment of new non-natives (Figure 10.25). For example, the non-native alga Japanese wireweed (*Sargassum muticum*), which is endemic to Japan, has been present on Irish shores since around the late 1990s, and is of concern due to its impacts on native species, especially in areas such as Lough Hyne. The Asian leathery sea squirt (*Styela clava*) was also first recorded on Irish rocky shores in 1972, in Cork Harbour, but has been found subsequently in Dublin Bay and the waters around Dingle, County Kerry. The adults of this species become attached to substrates in the lower parts of the midlittoral, as well as subtidally. However, as it needs seawater temperatures of greater than 15°C to spawn, this currently restricts its period of reproduction in Irish waters. To date, therefore, it has not caused too much concern, except for fouling on commercial oysters. Therefore, should seawater temperatures increase as predicted, this species will be of growing concern, since elsewhere in its invasive range it has been shown to foul mooring buoys and lines, as well as to compete trophically and spatially with native species.[34]

In addition to the probable influences of climate change, increasing levels of pollution also present a significant challenge for species living on Ireland's coasts. On rocky shorelines, the gastropod species *Nucella lapillus* and *Littorina littorea* were affected badly by the pollution of Tributyltin compounds (TBTs), through their effects on reproduction (see Boats, Paint and Transgender Snails, in Chapter 31: Pollution). TBTs were contained in paints used to prevent biotic fouling on boat and ship hulls, as well as, for example, on aquaculture installations. A more recent concern is the discovery of microplastics in some rocky shore invertebrates, but the extent of this problem, and whether microplastics affect the long-term biology of these organisms, will not be known until there is further research on the many different fauna found on these coasts.

Ireland's rocky coasts represent a vital component of the coastline. They attract people who visit and admire their natural beauty, walk the cliff tops, engage in rock fishing, or explore the coastline while sailing or kayaking. Tourists particularly enjoy visiting the many small bays, coves and pocket beaches, which form hidden and cosy oases enclosed in the rugged coastline. A lot of the landscapes formed by these coasts are of significant natural heritage value and can be more popular, and bring in more revenue, than many museums and historic buildings. The Cliffs of Moher in County Clare, for example, attract over one million visitors each year, making a considerable contribution to local and regional economies. Other places of great heritage value include: the Copper Coast, County Waterford; the Old Head of Kinsale and Mizen Head, County Cork; Loop Head and the Bridges of Ross, County Clare; Downpatrick Head, County Mayo, and the

FIG. 10.25 THE ASIAN LEATHERY, OR STALKED, SEA SQUIRT (*Styela clava*). This was first recorded on Irish rocky shores in 1972, in Cork Harbour. This animal is hardy, living in seawater temperatures and salinities as low as -2°C and 10ppt. They inhabit the surfaces of rocks and other hard materials (e.g., mussel shells, piers), and sometimes seaweeds, developing dense colonies of as many as 1,500/m². This sea squirt originated on Asian Pacific ocean coasts, but had found its way into European waters by the early 1950s, primarily through human agencies. [Source: Matthieu Sontag, Wikimedia Commons, CC BY-SA 3.0]

Giant's Causeway, County Antrim.

From a scientific point of view, cliffs, caves and sea stacks allow direct access to rock layers formed millions of years ago and provide natural outdoor laboratories in which people can investigate and reconstruct the geological and environmental history of our planet. Given that most of Ireland's rocky coasts respond only slowly to wave and storm-driven erosion, these parts of the island are likely to keep their position on the map far into the future, regardless of any increased storm sizes that may develop under climate change. Further, these wave-resistant coasts serve a valuable role benefiting the inland areas behind, in protecting them from storms and the direct effects of marine flooding. These environments can, however, prove to be expensive. Some sections of rocky coast may need artificial protection measures, mainly those with roads and railways close to cliff margins, and can often require costly engineering works to be carried out (see Chapters 30: Engineering for Vulnerable Coastlines and 32: Coastal Management and Planning). In spite of this potential coastal management issue, the island's rocky coastlines are a valuable national asset, both scenically and monetarily and, perhaps most significantly, an important ecological habitat.

BEACHES AND BARRIERS

Julian Orford

BARTRAGH ISLAND IN KILLALA BAY, COUNTY MAYO. This long, narrow sand barrier island hosts one of the most biologically important coastal sand dune systems in north-west Ireland. [Source: Geraldine Hennigan and Norman Kean]

BEACH PROCESSES: OVERVIEW

The Irish coastline and its constituent elements (headlands, waves, beaches, dunes and estuaries) are a continuing theme for actual and imaginary visits. As such, they define a developmental backcloth for many past and likely future generations. Central to the idea of coastlines are 'beaches', in all their diversity of site, form and constituent rock and mineral materials, the latter, together, known more generally as 'sediment'. This chapter presents the range and structure of Ireland's sediment-dominated coasts, primarily its beaches and sand systems. This subject field is technically complex and demanding, but the material is discussed here as clearly as possible.

Coastal sediment is seen most obviously above the lowest tidal level, but also has continuity with sediment found below low tide in the marine environment. The area between low and high tide is colloquially known as the 'beach' and is most commonly composed of sands. It is shaped and structured principally by breaking waves and, to some extent, by water currents driven by both waves and tides. Landward of a sand beach, above the reach of waves and tides, is an area dominated by wind-blown sediment, where sand dunes can develop, fed by the beach sediments. Coastal sediments have a continuing erosional history at the shoreline, mostly being worn-down by attrition through breaking wave activity.[1]

Ireland has a rich diversity of beach types and forms, essentially differentiated on the basis of the wave energy experienced (known as the wave climate), and the range of sediment sizes available for waves and currents to move. Shoreline sediment is texturally very diverse and ranges in size from boulders to cobbles to gravel to grit to sand and, finally, down to mud (Table 11.1). Sediment size presence on the shoreline is approximately related to the wave energy experienced. The highest wave energies typically occur at rock headlands, where wave energy is concentrated due to refraction. Here, waves can erode and move sediment the size of boulders. The smallest particles (muds) are associated with areas that experience minimal wave energies and high tidal currents, such as estuaries. Sand, a typical constituent of beaches, is found across a range of wave energies. Sediment presence on beaches poses two questions: where do these sediments come from and why are particular sediment sizes found on different beaches?

Much of Ireland's landscape has been shaped by the effects of several glacial episodes over the last 120,000 years (see Chapter 6: Glaciation and Ireland's Arctic Inheritance). The direct erosion by glacier ice, plus freeze-thaw action on bare rock (in the periglacial environments), generated most of the unconsolidated debris that now covers Ireland and its coasts. The umbrella term for this type of debris is glacigenic drift. This material is characterised by a heterogeneous mix of sediment sizes, from boulders to clays; hence, it has been referred to colloquially, since the term's nineteenth-century origin, as 'boulder clay'. Today, it is more formally known as 'diamict' or 'till'.

As the glaciers retreated and sea level started to rise, these deposits were quickly eroded, allowing a cascade of sediment sizes to appear on the newly developed shorelines. Likewise, as the ice margins retreated over land, glacial meltwater selectively transported different sediment sizes and also preferentially washed these across the newly emerging landscapes and into the rising seas. This process allowed further available sources of sorted sediment to be reworked and redistributed along the coast by the rising seas.[2]

The hard rock, geological character of the Irish landscape also exerts an important control on coastal sedimentary characteristics. Many interbedded layers of resistant and erodible rock lithologies have been folded and faulted into discordant headlands along the coast. The geographical distribution of these and other local geology factors controls the type of sediment derived from this source (see Chapter 5: Geological Foundations). Over the 15,000 years since the end of the last major Irish ice cover, even the hardest rock headlands – granitic and metamorphosed as in Donegal, or basaltic lavas in Antrim – have provided a supply of sediment to populate nearby beaches. Generally, a beach's sediment can be related to both local and reworked glacial debris. For example, sediment derived from Scottish granites, present along the Antrim and Down coasts, is evidence of faraway sediment sources from which sediment was transported by past ice-streams.[3] A final sediment source is the calcium carbonate fragments derived from the shells and coralline red algae (maërl beds) found on all Irish coasts, but peaking on western Irish beaches. Such calcareous sediments form the basis of some of Ireland's most unique and special sedimentary systems, such as machair.

Class	Grain size
boulders	>1m
cobbles	256–128mm
gravel	128–2mm
grit	2–1mm
sand	1–0.125mm
mud	<0.063mm

TABLE 11.1 SEDIMENT GRAIN SIZE CLASSIFICATION.

Machair

Derek Jackson

Machair systems are complex sand dune habitats, found on flat coastal sand plains with a history of grazing. Their rarity and uniqueness have led to them being classified as priority habitats, as defined by the EU Habitats Directive. In Ireland, machair are generally confined to the north-west coast, stretching from Galway Bay to Malin Head in Donegal (Figure 11.1), though some coasts further south (for example, Castlegregory, Brandon Bay, County Kerry) arguably have sand systems with some machair-like characteristics. The National Parks and Wildlife Service (NPWS) lists a total of fifty-nine sites, representing over 2,750ha (though other sites exist). Unique to Scotland

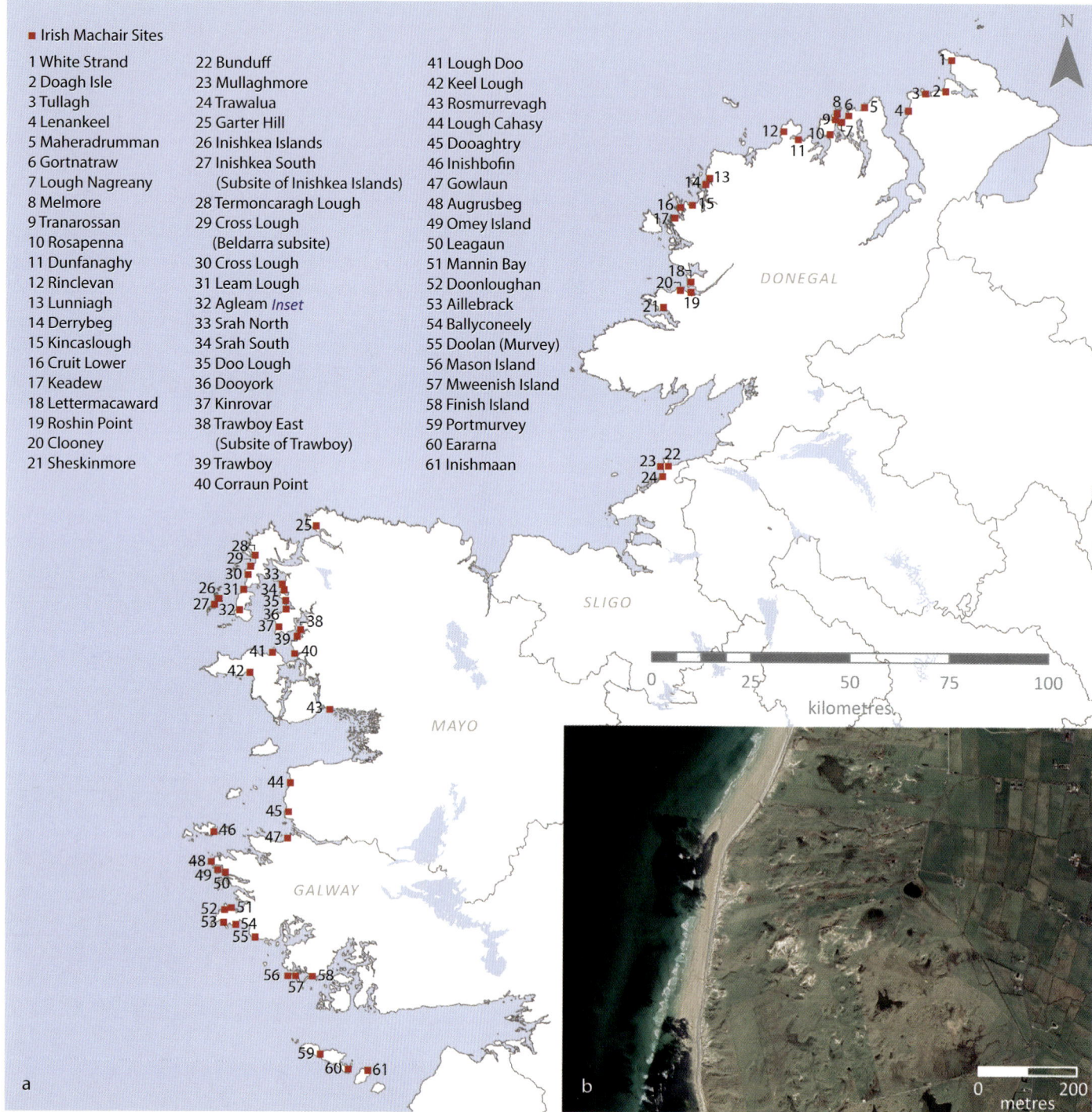

■ Irish Machair Sites

1 White Strand	22 Bunduff	41 Lough Doo
2 Doagh Isle	23 Mullaghmore	42 Keel Lough
3 Tullagh	24 Trawalua	43 Rosmurrevagh
4 Lenankeel	25 Garter Hill	44 Lough Cahasy
5 Maheradrumman	26 Inishkea Islands	45 Dooaghtry
6 Gortnatraw	27 Inishkea South	46 Inishbofin
7 Lough Nagreany	(Subsite of Inishkea Islands)	47 Gowlaun
8 Melmore	28 Termoncaragh Lough	48 Augrusbeg
9 Tranarossan	29 Cross Lough	49 Omey Island
10 Rosapenna	(Beldarra subsite)	50 Leagaun
11 Dunfanaghy	30 Cross Lough	51 Mannin Bay
12 Rinclevan	31 Leam Lough	52 Doonloughan
13 Lunniagh	32 Agleam *Inset*	53 Aillebrack
14 Derrybeg	33 Srah North	54 Ballyconeely
15 Kincaslough	34 Srah South	55 Doolan (Murvey)
16 Cruit Lower	35 Doo Lough	56 Mason Island
17 Keadew	36 Dooyork	57 Mweenish Island
18 Lettermacaward	37 Kinrovar	58 Finish Island
19 Roshin Point	38 Trawboy East	59 Portmurvey
20 Clooney	(Subsite of Trawboy)	60 Eararna
21 Sheskinmore	39 Trawboy	61 Inishmaan
	40 Corraun Point	

Fig. 11.1 IRELAND'S MACHAIR COASTS. a. Irish machair sites; b. A site of the machair coast from the Belmullet Peninsula, north County Mayo, which was not included as such in the National Parks and Wildlife Service inventory. The vegetation cover of the underlying machair sands sheet here can be seen extending several hundred metres behind the shoreline. [Source: map: National Parks and Wildlife Service, 2012; inset map data: Google, CNES/Airbus]

and Ireland, the genesis of machair (from the Gaelic word meaning 'fertile plain') is believed to be a result of beach-supplied calcareous sand feeding flatter or gently undulating plains in the lee of the main coastal dune areas.[4]

Anthropogenic agricultural inputs – such as seaweed spreading, grazing and dung – are also thought to add to the formation of these habitats, beginning possibly as early as Bronze Age times (see Chapter 16: The Inhabitants of Ireland's Early Coastal Landscapes). For example, grazing of herbaceous vegetation encourages low densities of typical sand-binding vegetation, such as marram grass (*Ammophila arenaria*), in the lee of dune systems. Machair forms part of a larger land-vegetation system that usually has a mosaic of wet and dry grasslands. In terms of coastal geomorphology, they could be viewed as sediment leakage zones from the main beach/dune system and effectively as the inland flattening of the coastal dune area. Human interactions, however, complicates an understanding of their role in dune dynamics. The machair on the Belmullet Peninsula (Figure 11.1) exemplifies the traditional way that the field patterns were initially divided into long field sections formed at right angles to the shoreline: constructed to include a proportional share of all of the varying dune-machair conditions.

MAËRL

Eugene Farrell

Maërl is a collective term for coralline red algae, made up of mostly *Phymatolithon purpureum* and *Lithothamnion coralloides*, found near the seabed in clear, shallow waters, particularly on the western coasts of Ireland. It is a type of red seaweed in the order *Corralines*, because its branch-shaped, hard calcified body structure resembles coral. Historically, the term maërl relates to the Breton (north-west France) word for pearl, as both species have skeletons comprising calcium carbonate.

Similar to all seaweeds and plants, maërl requires light to photosynthesise. It is therefore depth limited, occurring from the lower shore zone to depths of 40m or more. It can form extensive beds in shallow water on the open coast or in sheltered bays in preferential ocean conditions that are highly oxygenated, have reduced turbidity and strong current flow. Maërl beds are not temperature restricted and are found commonly along the Atlantic coasts of northern and western Europe, the Mediterranean and other temperate regions around the world. In Ireland, the majority (65–70 per cent) of maërl beds are found in the Galway-Connemara region (for example, Clew Bay, Greatmans Bay, Kilkieran Bay and Mannin Bay); and also in the south-west (20–25 per cent), for example, in Roaringwater Bay, Bantry Bay and Kenmare Bay. Other less extensive records have been established in other parts of the country.[5] Radiocarbon dating has established that maërl can live for hundreds of years, growing concentrically approximately 1mm per year. In this way, they are similar to tree rings and can record past climate conditions within their structure.

When maërl dies, it loses its red-purple-pink colour and breaks up into hard grey, white and brown pieces that are often mistaken for coral. Banks of dead maërl (or debris) can occur and be washed up on beaches (often mislabelled as coral strands) due to the action of storm waves. Two classic examples of maërl debris beaches are to be found at Carraroe and Mannin Bay in County Galway (Figure 11.2).

Maërl beds are benthic habitats supporting high associated invertebrate and algal biodiversity, which are of great conservation

Fig. 11.2 MAËRL BEDS AND SANDS. Maërl beds are some of the rarest, and least understood, coastal environments in the world. a. Live maërl has a distinctive red-purple-pink colour, as shown in this maërl bed in Falmouth, Cornwall, UK; b. at Trá na Dóilín (Coral Strand), County Galway, dead maërl fragments can be seen scattered across the strand. After becoming crushed by incoming wave action and bleached by the sun, maërl develops a rich calcareous type of sand-sized beach sediment. [Source: a. Mark Milburn/Atlantic Scuba; b. Eugene Farrell]

significance. The calcified nature of maërl beds makes them 'ecosystem engineers' that build habitat or, in some extreme cases, biogenic reefs, providing shelter and substrate for a host of marine species, such as anemones, sea cucumbers, sea grasses and bivalves, including scallops, oysters and mussels. Because of the abundance of potential prey at these sites, small fish and juveniles are attracted to maërl beds, establishing them as nursery areas for a range of commercially important fish species, such as cod and pollock.

In Ireland, maërl beds are protected as Special Areas of Conservation under the European Habitats Directive. Both Ireland and the EU recognise maërl as a keystone habitat, containing up to six different species of coralline algae, that is different from its surroundings. Maërl beds are also recognised as biodiversity hotspots. Multiple species are endemic to these habitats and depend on their biological and structural characteristics.[6]

In spite of their highly protected status in European and Irish conservation legislation, maërl beds are subject to a variety of natural and anthropogenic pressures. The Convention for the Protection of the Marine Environment of the north-east Atlantic (the 'OSPAR Convention') lists maërl beds as 'threatened and/or declining species and habitats', based on an assessment of their distribution, extent and condition. Their unique status has not always been recognised. For example, traditionally, maërl beds were mined for fertiliser on acid soils, although this practice has largely been abandoned. Maërl has also been used as a food additive for animal feed, as a component in water-filtering systems and as a replacement for lime as a soil conditioner. In Connemara, maërl is still widely used on graves in the local cemeteries, a practice dating as far back as the 1930s. It is also still used in landscaping as ornamentation on paths and walls. In an effort to protect these important habitats from anthropogenic threats, there are now 'no go' buffer areas around maërl beds that restrict aquaculture and fishing practices.

More recent concerns have highlighted the impacts of climate change on the condition and extent of maërl in Europe. Ocean temperature and carbon dioxide levels are predicted to increase in the north-east Atlantic during the next decades. It is projected that algae and animals that need abundant calcium carbonate will be less abundant in acidified water and will have weaker skeleton structures. This will make maërl beds more vulnerable to storm damage and less stable habitats (see Chapter 33: Climate Change and Coastal Futures).

In addition to particle size and lithology, waves and currents play a key role in moving and reworking coastal sediments. Waves are driven by wind, with the westerly Atlantic storms generating the majority of waves around the Irish coast. Generally, the west and south coasts experience the highest inshore wave energy (median offshore wave height: >3m), while east coasts experience lower wave energy (median offshore wave height: <2m). On the west coast, this energy is about two orders of magnitude higher than the wave energy of the eastern coasts. This spatial wave energy difference is exemplified in the different coastal forms characteristic of the two regions. Variable waves operate in conjunction with the predictable beat of the rise and fall of tidal levels.[7]

The pattern of variation in tidal range around the coast is affected by Ireland's relative position to Britain, such that the lowest extremes of these levels (0.5–5m range) are experienced along the east coast, which is enclosed by the Irish Sea. Tidal ranges around the rest of Ireland, bordering on the open ocean, generally fall between 2–4m, but usually increase where the tide is funneled by bays and estuaries (see Tides, in Chapter 2: The Coastal Environment: Physical Systems, Processes and Patterns). Tidal range, as it gets larger, is important in physically increasing the width of beaches, as well as driving currents sufficient to move finer sediments in protected coasts, such as estuaries, where high-energy waves fail to penetrate. Waves and currents combine, but the relative transport power on beaches is wave dominated.

As waves move from open water onshore, they adjust their three-dimensional shape to reflect the reducing water depths associated with the shallowing bathymetry. In plan view, this change in wave crest form is seen in an increasing crest adjustment to the overall shape of the shoreline, a process known as wave refraction (Figure 11.3). This is an important mechanism by which different offshore slopes impose changes both in wave energy (proportional to variable wave height squared) and wave direction, both are experienced variably alongshore. It is these longshore differences in breaking waves that drive sediment variations and movements through differential transport capacity. Wave crests that break at an angle to the shoreline have longshore sediment transport potential proportional to this angle, while waves breaking parallel to the shoreline have negligible longshore transport potential.[8]

Over time these transport differences pick out longshore sediment size variation as a function of sediment source areas (cliffs), transport corridors (beaches) and then, finally, sediment sinks (stores of sediment or depositional areas), where transport potential is insufficient to move the sediment. The source–corridor–sink idea constitutes the concept of a sediment pathway. Often headlands and estuaries prove to be boundaries to longshore sediment pathways. Consequently, highly discordant shorelines, such as the Irish west coast, will have numerous longshore-segmented pathways. Each of these pathways will be part of a separate wave-generated sediment cell (a functional physical coastal unit, or system). The direction of transport reflects the dominant longshore direction of the breaking wave. Changing the wave approach may also shift the transport direction of the cell, though the general uniformity of westerly Atlantic depressions means that most Irish sediment cells show overall directional consistency at present, though they may fluctuate over decades to centuries with climatic shifts.[9]

Beaches are a major component of sediment pathways and usually reflect either the transport corridor stage or the final sediment sink stage. Sediment availability is a major control on a pathway, as without sediment the pathway is unrealised. Once a source area's sediment diminishes – for example, from an eroding cliff – the corridor starts to reduce until all its sediment is moved to the final sink. Importantly, for Ireland's coasts, many southern

Fig. 11.3 WAVE REFRACTION AT SILVERSTRAND, COUNTY MAYO. This shows the principal swell waves from the ocean approaching the coast. Refraction is the bending of wave crests approaching the shoreline as a result of variation in the underlying bathymetry. This occurs because the portion of the wave in shallower water moves slower than the portion in deeper water. The maximum focus of wave energy is concentrated on headlands. As waves move into shallow water, the wave crest becomes more defined and adjusts, in plan view, to the sandy shoreline, with the wave crests eventually parallelling the beachline. Wave energy is lost in this process and the remaining energy spread over the length of the beach areas (bays). Where there is insufficient time to adjust, then the breaking wave meets the shoreline obliquely. The angle between the crest and the depth contour defines the longshore sediment transport potential. Where waves break parallel to the shoreline, there is limited sediment transport potential (longshore drift). [Map data: Google, Digital Globe, Maxar Technologies]

and western Irish beaches appear as sinks, as contemporary sediment supply has become scarce or negligible.

Pathways are set up by the interaction of a wave climate with a coastal landscape and the provision of transportable sediment. These processes, and their outcomes on the Irish coastline, are similar to any other shoreline around the world. However, the idiosyncrasies of Irish geology and landscape evolution allow scope for differences in detail. Such differences can also be seen in the variation around the Irish shoreline of environmental controls, such as sea level, storms, wave refraction and sediment supply.

Why Do Beaches Erode?

Andrew Cooper

At the simplest level, beaches are accumulations of loose grains of sand, gravel or even boulders. They are able to change their shape easily in response to the amount of sediment they contain and the nature and levels of energy to which they are subject, within the constraints imposed by the surrounding geology and sea level. This ability to adapt enables them to survive hostile conditions during storms that can cause sea walls and cliffs to collapse.

Fig. 11.4 ERODING CLIFFS AND BEACH AT BALLYCONNIGAR, COUNTY WEXFORD. Wave approach to the shoreline allows the rapid removal of the soft glacigenic sediments (land/cliff area) on this coast, which are then removed in the dominant offshore sediments drift southwards, toward Carnsore Point. Both fine sands through to cobble-sized sediments are released from the cliffs, with the coarse material remaining on the beach at the foot of the cliff as a beach berm. Erosion rates on these eastern glacigenic sedimentary coasts of Ireland are high, reaching values of 0.5–1.0m per year. [Source: Maxim Kozachenko]

In adapting to changing energy and sand volumes, beaches alter their shape and can move landwards and seawards easily. High wave energy during storms normally causes the beach to change and assume a more gentle gradient, breaking the wave energy across a wide gentle slope. This reduction in gradient causes the high water mark to move landwards; where dunes occur these may be eroded in this process. The extra sediment brought in from eroded dunes, however, will also help the beach break the incoming wave energy, with the sediment often returned to the dunes in calm weather conditions. Abundance or scarcity of sediment at any given time will also cause a beach to either advance or retreat. All these are normal mechanisms of beach adjustment to changing conditions and its action as a natural wave energy buffer. Many Irish beaches derive their sediment from finite sources. Slow leakage of sand can lead to long-term natural erosion of these beaches, as at Ballyconnigar, County Wexford (Figure 11.4) (see Chapter 30: Engineering for Vulnerable Coastlines).[10]

Humans can, however, also cause erosion by altering the amount of sand, or by adjusting the surrounding constraints.[11] Such activities include beach mining, building piers, sea walls, or groynes and landfilling. Sea walls are particularly damaging to beaches. They protect landward property, but they limit the space for wave energy reduction during storms and, in so doing, they cause direct erosion of beaches, whereby waves reflect off sea walls and carry sand away. Even well-designed sea walls that attempt to break wave energy prevent beaches from adjusting to changing conditions and eventually lead to their demise.

The southern coasts of Ireland have a history of rising sea level over the last 8,000 years (see Chapter 7: Ancient Shorelines and Sea-level Changes), which means that beaches are generally pushed onshore, assuming there is landward accommodation space available, or space for the sediment to accumulate onshore. The northern half of the island, on the other hand, has experienced falling sea levels for the last 5,000 years, causing coasts to migrate seawards. The accommodation spaces created by this movement are sometimes filled by dunes. Twenty-first-century accelerated sea-level rise will bring new conditions for the landward movement of most beaches, though the speed of this change is likely to be three-to-five times faster than previously experienced.

Why are Dunes at the Coast?

Derek Jackson

Coastal dunes will accumulate when particular environmental conditions are present at the coast.[12] Adequate sediment supply, sufficient exposure at low tides and consistently strong onshore winds all promote dune formation. The delivery of sand from nearshore wave and current action feeds the main beach, resulting in a build-up of available sand. As the sand dries out, it becomes transportable by wind action (Figure 11.5). Under certain conditions of wind strength and direction, sand is picked up from the surface and transported further up the

Fig. 11.5 COASTAL SAND DUNE PROCESSES. a. Aeolian (wind-driven) sediment being transported along the beach at Rossnowlagh, County Donegal; b. fore-dunes at Back Strand, Inishowen Peninsula, County Donegal. The sharp wave-erosion front, creating steep and unstable cliffs in these dunes, is noteworthy; it is a common feature today in many Irish coastal dune situations. [Source: a. Karen Skelly; b. Oliver Dixon, CC-BY-SA 4.0]

beach. Obstacles such as seaweed line debris, or other topographic highs (and lows), will help retard the sand-laden flow and result in sand accumulation. These dry sand deposits may or may not form sand dunes, depending on their location relative to wave and tidal action; too close and they quickly disappear. Once the initial sand deposits accrue to sufficient levels they may then become more fixed through vegetation growth, binding the sediment together and forming new dune landforms at the back of the beach, known as foredunes. Older coastal dunes are good markers of former environmental conditions at the coast. We can learn about storms and other past environmental conditions that have occurred at coastal dune sites from their internal structures.

COASTAL DUNES

Derek Jackson

Sand, deposited on beaches able to absorb large amounts of wave energy (dissipative), when given adequate drying conditions and supply volumes, along with transporting winds, can result in the development of large sand dunes, acting as coastal sediment sinks. Around 14,300ha of dune fields exist along the Irish coastal fringe, represented by almost 200 individual sites (Figure 11.6). The largest accumulations cluster on the western seaboard of Ireland, where sediment availability and strong onshore winds have stimulated the proliferation of significant dune sites, as exemplified by Five-fingers Strand (County Donegal).

Most Irish dunes are readily stabilised with vegetation, which thrives due to the high levels of coastal precipitation characteristic of Ireland's climate. Once established, coastal dune fields represent responsive geomorphological landscapes. These can react rapidly and closely with changes in the forcing parameters, such as, sediment supply, wind and storminess. Dunes are sensitive to changing climatic conditions, which, along with animal and human trampling, stress the vegetation and hasten the reworking of the dune sediment by the wind. The relatively high relief of many Irish dunes acts as a barrier to further inland sediment transfer, particularly where heavily vegetated forms exist. This also effectively halts any major overwash processes along most of the dune-fringed coastal stretches, though there are exceptions, for example, the Tacumshin–Ballyteige Burrow coast, County Wexford (see Chapter 12: Coastal Wetlands).[13] The sediment storage function of coastal dunes becomes important when coastal storms predominate. Dunes located on the west coast play significant roles in sediment redelivery (dune erosion) to the beach face during high-magnitude events.

As is the case with the majority of European dunes, most Irish dunes are considered relict features; that is, they were formed from processes that are no longer in operation, with the current limited nearshore sediment supply no longer delivering significant quantities of fresh sand to foredune areas. The oldest dune deposits in Ireland are approximately 5,000 to 6,000 years old. It is suspected that dune development correlates with past major climatic shifts. For example, increases in wind velocities may degrade dunes, resulting in onshore blowouts. Major dune reworking on the west coast has been linked with a stronger wind climate during the Little Ice Age (LIA), which occurred between the fourteenth and early nineteenth centuries. During this time, stronger than present south-west to west winds resulted in extensive restructuring of dune surfaces into major blowouts that can run across the original dune area, for example, at Inch, County

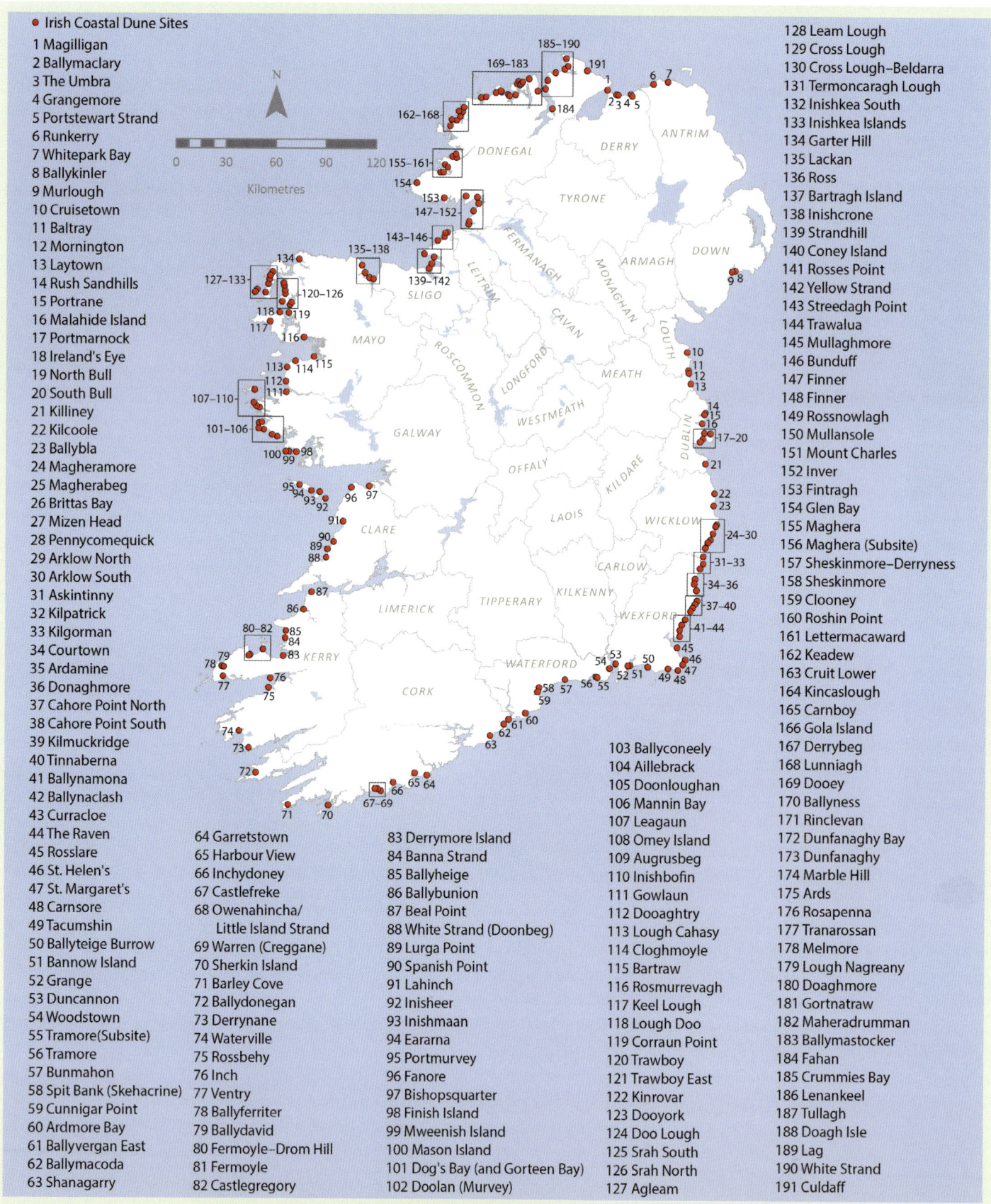

Irish Coastal Dune Sites

1 Magilligan
2 Ballymaclary
3 The Umbra
4 Grangemore
5 Portstewart Strand
6 Runkerry
7 Whitepark Bay
8 Ballykinler
9 Murlough
10 Cruisetown
11 Baltray
12 Mornington
13 Laytown
14 Rush Sandhills
15 Portrane
16 Malahide Island
17 Portmarnock
18 Ireland's Eye
19 North Bull
20 South Bull
21 Killiney
22 Kilcoole
23 Ballybla
24 Magheramore
25 Magherabeg
26 Brittas Bay
27 Mizen Head
28 Pennycomequick
29 Arklow North
30 Arklow South
31 Askintinny
32 Kilpatrick
33 Kilgorman
34 Courtown
35 Ardamine
36 Donaghmore
37 Cahore Point North
38 Cahore Point South
39 Kilmuckridge
40 Tinnaberna
41 Ballynamona
42 Ballynaclash
43 Curracloe
44 The Raven
45 Rosslare
46 St. Helen's
47 St. Margaret's
48 Carnsore
49 Tacumshin
50 Ballyteige Burrow
51 Bannow Island
52 Grange
53 Duncannon
54 Woodstown
55 Tramore(Subsite)
56 Tramore
57 Bunmahon
58 Spit Bank (Skehacrine)
59 Cunnigar Point
60 Ardmore Bay
61 Ballyvergan East
62 Ballymacoda
63 Shanagarry

64 Garretstown
65 Harbour View
66 Inchydoney
67 Castlefreke
68 Owenahincha/
  Little Island Strand
69 Warren (Creggane)
70 Sherkin Island
71 Barley Cove
72 Ballydonegan
73 Derrynane
74 Waterville
75 Rossbehy
76 Inch
77 Ventry
78 Ballyferriter
79 Ballydavid
80 Fermoyle–Drom Hill
81 Fermoyle
82 Castlegregory

83 Derrymore Island
84 Banna Strand
85 Ballyheige
86 Ballybunion
87 Beal Point
88 White Strand (Doonbeg)
89 Lurga Point
90 Spanish Point
91 Lahinch
92 Inisheer
93 Inishmaan
94 Eararna
95 Portmurvey
96 Fanore
97 Bishopsquarter
98 Finish Island
99 Mweenish Island
100 Mason Island
101 Dog's Bay (and Gorteen Bay)
102 Doolan (Murvey)

103 Ballyconeely
104 Aillebrack
105 Doonloughan
106 Mannin Bay
107 Leagaun
108 Omey Island
109 Augrusbeg
110 Inishbofin
111 Gowlaun
112 Dooaghtry
113 Lough Cahasy
114 Cloghmoyle
115 Bartraw
116 Rosmurrevagh
117 Keel Lough
118 Lough Doo
119 Corraun Point
120 Trawboy
121 Trawboy East
122 Kinrovar
123 Dooyork
124 Doo Lough
125 Srah South
126 Srah North
127 Agleam

128 Leam Lough
129 Cross Lough
130 Cross Lough–Beldarra
131 Termoncaragh Lough
132 Inishkea South
133 Inishkea Islands
134 Garter Hill
135 Lackan
136 Ross
137 Bartragh Island
138 Inishcrone
139 Strandhill
140 Coney Island
141 Rosses Point
142 Yellow Strand
143 Streedagh Point
144 Trawalua
145 Mullaghmore
146 Bunduff
147 Finner
148 Finner
149 Rossnowlagh
150 Mullansole
151 Mount Charles
152 Inver
153 Fintragh
154 Glen Bay
155 Maghera
156 Maghera (Subsite)
157 Sheskinmore–Derryness
158 Sheskinmore
159 Clooney
160 Roshin Point
161 Lettermacaward
162 Keadew
163 Cruit Lower
164 Kincaslough
165 Carnboy
166 Gola Island
167 Derrybeg
168 Lunniagh
169 Dooey
170 Ballyness
171 Rinclevan
172 Dunfanaghy Bay
173 Dunfanaghy
174 Marble Hill
175 Ards
176 Rosapenna
177 Tranarossan
178 Melmore
179 Lough Nagreany
180 Doaghmore
181 Gortnatraw
182 Maheradrumman
183 Ballymastocker
184 Fahan
185 Crummies Bay
186 Lenankeel
187 Tullagh
188 Doagh Isle
189 Lag
190 White Strand
191 Culdaff

Fig. 11.6 COASTAL SAND DUNE SITES AROUND IRELAND. These mapped sites represent a general overview of recognised coastal dune sites at an island-wide level. While based on the best available data for sand dune distribution in Ireland, it is possible that there are other dune sites around the coast that are not identified on the map. No complete dataset of coastal dune distribution and characterisation for the island has been compiled to date. [Source: National Parks and Wildlife Service, 2012; DAERA (NI); Doody, J.P., ed. 2008. *Sand Dune Inventory of Europe*, 2nd Edition. National Coastal Consultants and EUCC - The Coastal Union, in association with the IGU Coastal Commission]

Kerry.[14] Many of these features have still to recover their vegetation, as their size encourages sand mobility sufficient to resist vegetation growth. General beach sediment scarcity means that any new aeolian (wind-driven) growth is now generally restricted to small-scale reworking of the back-beach areas and/or limited dune blowout activity.

ECOLOGY OF SAND DUNE HABITATS IN IRELAND

Aoife Delaney

Ireland's coastal dunes host rich and biodiverse ecosystems (Figure 11.7). Dune habitats span a steep environmental gradient, from the exposed shoreline to the relatively sheltered and stable landward coastal boundary. This gradient is reflected in plant communities, so that a predictable suite of habitats can be found in many sand dune systems. Closest to the shore, orache (*Atriplex species*), sea rocket (*Cakile maritima*) and prickly saltwort (*Salsola kali*) colonise sand and gravel along the drift line where it is enriched by fragments of decaying organic debris.[15]

These plants are well adapted to deal with wind and salt spray and they can colonise quickly after disturbance. The drift line acts as a barrier and windbreak, allowing low deposits of sand, called embryonic dunes, to grow. These embryonic dunes are located just inshore of the harshest beach conditions and the sparse vegetation comprises sand couch (*Elytrigia juncea*) and lyme grass (*Leymus arenarius*) in addition to drift-line species. Both drift-line vegetation and embryonic dunes are ephemeral habitats that may appear and disappear according to weather and sea level, or become more stabilised and succeed to another habitat over time.

On the landward side of embryonic dunes' conditions are suitable for the growth of marram grass (*Ammophila arenaria*). Marram traps blown sand and grows rapidly through accumulating sand dunes. As a result, marram dunes are higher and steeper than embryonic dunes. The sand here is still mobile but salt input is lower and there is a greater range of plant species. Dandelion (*Taraxacum officinale agg.*), sand

Fig. 11.7 ECOLOGY OF SAND DUNES. a. Dune Slacks, Umbra Nature Reserve, County Londonderry [Source: Ulster Wildlife]; b. embryonic shifting sand dunes in Dooey, County Donegal [Source: Aoife Delaney, Courtesy of National Parks and Wildlife Service]; c. decalcified dune heath near Maghera, County Donegal [Source: Jim Martin, Courtesy of National Parks and Wildlife Service]; d. mobile dunes dominated by marram, County Wexford [Source: Mary and Angus Hogg, www.geograph.org.uk]

sedge (*Carex arenaria*), sea spurge (*Euphorbia paralias*) and sea holly (*Eryngium maritimum*) are scattered between marram tussocks, and are increasingly frequent moving away from the sea. As bare sand becomes less dominant, mosses and fine grasses appear and an increase in plant diversity and cover marks the transition to fixed-dune vegetation.

Fixed dunes were traditionally managed as pasture or as rabbit warrens in Ireland and this maintained the herbaceous vegetation that is our most common fixed-dune vegetation type (see Chapter 17: The Vikings and Normans: Coastal Invaders and Settlers). Because many Irish dunes are composed of sand with a high limestone or seashell content, they are frequently rich in calcareous material. These habitats tend to be species rich, with bird's-foot trefoil (*Lotus corniculatus*), kidney vetch (*Anthyllis vulneraria*), lady's bedstraw (*Galium verum*) and yarrow (*Achillea millefolium*) common among grass species (*Festuca rubra, Agrostis* sp., *Holcus lanatus, Poa humilis*) (Figure 11.8). There is notable moss and lichen cover and fixed dunes with herbaceous vegetation are sometimes called 'grey dunes', because of the abundance of lichen in the genus *Cladonia*. Fixed dunes can support a variety of orchid species, such as pyramidal orchid (*Anacamptis pyramidalis*) and bee orchid (*Ophrys apifera*). If grazing ceases, woody species can become

Fig. 11.8 FLORA FOUND WITHIN IRISH SAND DUNE HABITATS. a. Bee orchid (*Ophrys apifera*) near Ballycullane, County Wexford, 2019; b. kidney vetch (*Anthyllis vulneraria*) near Saltmills, County Wexford, 2017; c. lady's bedstraw (*Galium verum*) near Belmullet, County Mayo, 2015; d. sea spurge (*Euphorbia paralias*) interspersed with daisy (*Bellis perennis*) at Rossbehy, County Kerry; e. silverweed (*Potentilla anserina*) near Ballinesker, County Wexford, 2017; f. water mint (*Mentha aquatica*) near Gibletstown, County Wexford, 2015. [Source: a, b, c, e and f: © Z. Devlin, 2019. *Wildflowers of Ireland*. Cork: The Collins Press; d: Sandra Kandrot]

more prominent and in some parts of Ireland scrubby and woody species have become established on fixed dunes. Over time, dunes can lose their calcium content, leading to a change in the vegetation. Dune heath can develop in decalcified sites with little grazing pressure. Decalcified dune heath is characterised by heather (*Calluna vulgaris*), bell heather (*Erica* sp.) and gorse (*Ulex* sp.), frequently interspersed with more typical herbaceous dune species including grasses, sedges and lichens.

Wetland habitats called dune slacks can occur in valleys between fixed-dune ridges where the water table lies close to the surface. They are fed by groundwater and may be flooded for months at a time every year. Dune slacks form when a beach is cut off from the sea by a new dune ridge, or as the result of erosion in fixed dunes. Their characteristic vegetation includes rushes and sedges (*Juncus* sp., *Carex* sp.), marsh pennywort (*Hydrocotyle vulgaris*), water mint (*Mentha aquatica*), silverweed (*Potentilla anserina*) and orchids such as marsh helleborine (*Epipactis palustris*), butterfly orchid (*Platanthera* sp.) and marsh orchids (*Dactylorhiza* sp.). Over time, dune slacks tend to dry out as blown sand covers the slack floor. During this process the dune slack vegetation changes slowly as the influence of groundwater wanes. Creeping willow (*Salix repens*) commonly becomes dense as dune slacks dry out and the increase in evapotranspiration hastens succession away from dune slack vegetation. Eventually, the dune slack will come to resemble the surrounding grey dunes.

In western counties, machair plains host a variety of species. The soil in machair plains contains a high quantity of shell sand and they often comprise a mosaic of wet and dry patches as the water table is not far below the surface. Machairs are characterised by a large proportion of broadleaved herbs and share many species with fixed dunes, dune slacks and calcareous grasslands, for example lady's bedstraw (*Galium verum*), bird's foot trefoil (*Lotus corniculatus*), white clover (*Trifolium repens*) and daisy (*Bellis perennis*).

Sand dunes are also home to a wide range of fauna. Insects benefit from the small-scale variation in dunes, where there are warm, south-facing slopes, open water, flowering plants and a heterogeneous vegetation structure within a small area. Sand dune specialists (species adapted to the prevailing environment) include the solitary bee (*Osmia aurulenta*), which nests in empty snail shells. Some invertebrate species, such as the heath snail *Helicella itala*, were previously widespread, but they now have a strongly coastal distribution due to loss of suitable habitats inland. Among the vertebrates, the natterjack toad (*Epidalea calamita*) is heavily dependent on dune slacks in southern Ireland (Figure 11.9).[16] Dune slacks flood for long enough to allow tadpoles to mature, but because of their annual dry phase and lack of connection to permanent water bodies, they contain few predators. While some species can complete their lifecycle within the confines of dune habitats, for others, such as choughs, sand dunes form part of a range of habitats upon which they depend. Choughs tend to nest on coastal cliffs but can be seen foraging in the low grass on sand dunes in the west of Ireland. Machairs are also important breeding locations for wader species, including lapwing and redshank – which are

Fig. 11.9 THE NATTERJACK TOAD. The natterjack toad is a sand dune specialist, a species adapted to the sand dune environments. Sheltered dune slacks provide the perfect nursery grounds for tadpoles. Here, there are few predators due to their ephemeral nature and the lack of connection to permanent water bodies. [Source: Anthony Dawson, Irish Wildlife Trust, Kerry Branch]

International Union for Conservation of Nature (IUCN) red-listed species – and dunlin, a species listed in Annex I of the Birds Directive (see Waterbirds in Irish Coastal Areas, in Chapter 3: Marine Biology and Ecology).

Sand dunes are susceptible to changes in land management. Lack of grazing in fixed dunes leads to depleted plant and structural diversity, reducing the value of the dunes for wildlife. Destabilising activities – such as overgrazing, excessive cutting of vegetation, sand extraction, trampling and driving on dunes – can have an impact on their structural integrity, so that they become vulnerable to erosion events during storms. Conversely, coastal stabilisation measures such as groynes and rock armour can have unpredictable effects on erosion and deposition of sand, so that areas not directly linked to the stabilisation works may change unexpectedly as a result.

A number of non-native species have become invasive in Irish sand dunes. Probably the most visible of these is the sea buckthorn (*Hippophae rhamnoides*) (see Maherees Conservation Association, in Chapter 30: Engineering for Vulnerable Coastlines). Forming a dense thicket, sea buckthorn reduces the diversity of dunes and increases the rate of evapotranspiration, reducing water levels in the dunes. Planting dunes with conifers has a similar effect on their hydrology and changes the flora and fauna composition. Groundwater extraction for amenities at campsites, golf courses and holiday homes can also lower the water table. When groundwater levels decrease, dune slacks dry out and dunes are more vulnerable to erosion. The disposal of waste water may also be problematic if septic tanks present in the dune system are poorly maintained as this can lead to the contamination of groundwater entering dune slacks.

Sand dunes buffer human habitations and agricultural land from the destructive forces of the sea and provide a refuge for wildlife. Their importance in conserving biodiversity and ecosystem functioning has been recognised within the European Union under the Habitats Directive. As a result, large areas of sand dunes in Ireland now form part of the Natura 2000 network, a network of protected areas designed to prevent degradation of important natural habitats in Europe. Monitoring surveys indicate that effective management solutions must be put into place if sand dune habitats in Ireland are to achieve good conservation status (see Chapter 32: Coastal Management and Planning).

Beach functioning and form

All beaches experience extreme events where combinations of extreme wave heights, tidal elevations and storm surge can lead to dramatic changes in beach volume and shape. Western Ireland is likely to experience large extreme wave effects, while eastern Ireland is affected more by extreme surges. Some scientists suggest that it is these extreme events (of waves and surges) that leave the dominant 'fingerprints' shown by long-term beach structure.[17]

Ireland's bathymetric changes account for inshore wave refraction differences that control the physical setting of beaches. This factor, plus the availability of longshore sediment supply, defines the position of beaches as part of a sediment transport corridor. The corridor forms a continuing sediment supply stream to beach areas, or where there is no supply, it becomes a diminishing sink where sediments are deposited offshore. Beach shape alters to accommodate changing conditions. Storms will comb finer sediment offshore, while fine weather with swell waves will add sand volume to beaches. In the longer term, reducing longshore sediment supply means that beaches will start to cannibalise themselves (that is, they will lose sediment alongshore). This emphasises two forces: cross-beach transport under storms and the inevitable retreat of beaches; both of these force refracting waves to break parallel to the coast and, hence, further reduce longshore transport (swash-alignment).

This move to 'swash-alignment' (sediment transport orientated onshore, perpendicular to the direction of the incoming waves) is indicative of sediment pathway breakdown. Its form is in contrast to where high longshore supply builds up the beach face (leading to reducing nearshore depths and again, through refraction, finally reducing the waves' ability to transport longshore). This means the beach is aggrading, or appearing as 'drift-aligned'. That these two beach types are often seen adjoining – for example, at Rossbehy Spit, County Kerry – is not by chance (see Chapter 30: Engineering for Vulnerable Coastlines). The diminishing sediment supply due to swash-alignment is often the end product of exporting a sediment surplus to an adjacent down-drift, drift-aligned beach.[18]

While headlands provide the dominating control on sediment pathway starts, estuaries often provide the final boundary, or pathway termination. However, while beaches indicate the sand-sized elements of pathway load, and hence the most visual expression of a pathway, it should be remembered that pathway muds suspended by wave action can be transported into the estuaries where tidal forces will continue to exert depositional control on them (inter- and subtidal mud banks) (see Chapter 13: Estuaries and Lagoons).[19]

Although the plan form of beaches may show stability over time (years), the cross-beach elevation profile may show distinct seasonal (annual) changes. During winter, strong incident wave conditions during storms move sand down the beach, though gravel can move up-beach, under storms, creating a platform called a 'storm-berm'. Under fair-weather conditions, sediment is gradually moved onshore, allowing the beach to rebuild. There is a continuum of possible beach profile changes, though many beaches are limited to a narrow range of profiles, called domains.

The domain capacity of any beach relates to the range of wave conditions relative to the overall slope of the beach and its offshore extension, plus the mean beach sediment size. This balance defines the beaches' morphodynamic status (shape and process). It indicates the ability of energy in the waves to generate water currents that can move sand on- or offshore, as well as alongshore. Very low-angle beaches (gradient 1:100, or angle <1 degrees) show wide intertidal areas, dominated by sand, and are at one end of the beach shape (morphodynamic) continuum. These are called dissipative beaches. On these beaches, wave-generated currents have a tendency to produce intertidal sandbed forms, such as bars and troughs. Dissipative beaches are often associated with sand dunes developed landward of the high-tide shoreline.

At the other end of the continuum are reflective beaches. These have steeper gradients (1:<30 gradients, or angle >3 degrees) and are often gravel sediment dominated with minor beach morphology (cusps) and usually no dunes. Beach domains tend to be temporarily set, though sudden incursions of sediment along the pathway can reset the profile into a central position on the continuum. These beaches are considered 'intermediate'. Despite this possibility, it appears that Irish beaches are more set by limited wave range, beach slope and sediment variation so that 'end-of-continuum' positions dominate.[20]

IRELAND'S BEACH TYPES

Ireland's beaches might be expected to show a great diversity of form variation, given the range of coastal landscapes, waves and tides and the sediment supply variations experienced. However, its beach forms can be reduced to a few basic types.

—— Sediment source areas   —— Transport corridors or beaches   —— Sediment sinks or beaches

Fig. 11.10 COASTAL CELLS. Coastal cells are self-contained physical environmental units within which sediment circulates. These units can be thought of as quasi-closed systems. When the cell is in equilibrium, there is no net transport into, or out of, the system. The boundaries of coastal cells are usually defined by the land topography and shape of the coastline, although these boundaries may not necessarily be fixed. a. Merva Bay, County Donegal; b. Lehinch Bay, County Clare. In these two examples, the coastal cells are divided into sediment source areas (red), transport corridors or beaches (blue), and sediment sinks or beaches (yellow). Arrows show the net sediment transport direction, while broken lines indicate zones that can show combined functions. [Map data: Google, Maxar Technologies]

The coastal wave-sediment cell

The longshore wave-sediment cell is typical of the Irish coast (Figure 11.10), as illustrated by two example sites. The first, at Merva, County Donegal, shows the impact of a dominant south-to-north transport pathway, driven by south-west to west waves. On the southern coast of this site, a thin cover of glacigenic sediment overlying a rock platform provides the sediment source area for gravel-sand-muds. Today, the transport corridor is segmented in the small southern bays of the site. In the past, surplus sediment was driven by south-west longshore-directed breaking waves into a drift-aligned sink in the outer estuary. This was controlled by island remnants of the glacigenic cover.

As the sediment sink of the drift-aligned beach grew, the inner estuary was increasingly shut off from wave action. This created a sink area for pathway muds controlled by estuary tidal currents behind the beach. Over time, sequential drift aggradation increased the sink area northwards, into the area underlying the modern-day golf course. At the same time, a reduction in source area sediment allowed cannibalisation of the older western shoreline and a reworking of sediments into the contemporary realigned (north-to-south) sink area. This area has provided a sediment source for the sand dunes on which the golflinks is situated. The new alignment of the beach has brought sediment transport at the coast to near zero, though the estuary entrance is still transporting finer sediment. As this type of coastal cell evolves, sediment supply reduces and source areas expand alongshore, as available beach sediment moves down the corridor. Ultimately, sediment volume diminishes to a point where beaches disappear.[21]

The second site of Lehinch Bay, County Clare, is a wave-exposed coastal cell driven by Atlantic waves (Figure 11.10b). Lehinch village grew behind a high gravel barrier that was part of the coastal sediment transport corridor (see Chapter 30: Engineering for Vulnerable Coasts). The southerly promontory headland provided the source for the dominant gravel sizes of the modern-day beach sediment, derived from glacigenic sedimentary cover. Over time, finer sediments progressed northwards into a sink and nearly shut down the small estuary of the Deelagh and Cullenagh Rivers. This was unable to move further north, as it is trapped against the Carboniferous rocks that form the southern edge to the Cliffs of Moher.

The scarcity of new beach sediment at this site is exemplified by the need for anthropogenic intervention, in the form of rock boulders (rip-rap), to prevent the onshore movement of the gravel ridge. If left unchecked, the natural tendency towards swash-alignment would leave Lehinch vulnerable to storm waves. The installation of coastal protection measures here to date, however (Figure 11.11), has only served to push the focus of the south-westerly driven wave erosion further north. Here, wave cannibalisation of the sand sink (the dunes of another golflinks) have been deflected, by extending the rip-rap boulder protection

Fig. 11.11 LEHINCH BAY, BEACH AND WATERFRONT, COUNTY CLARE. The photograph shows the wave-scoured surface of the intertidal rock platform in front of the town of Lehinch. The rock platform is covered by a relatively thin veneer of sand, gravel and boulder-sized sediments. Lehinch Bay forms part of the larger Liscannor Bay and together they record the earlier movement onshore of a former large offshore gravel barrier, which once formed the sea-ward limit to these bays. The present-day sea wall and other coastal protection structures have been built to help stop erosion on this south side of Lehinch Bay. Regrettably, the defences have only served to shift the erosional focus of wave activity northward, into the dune-barrier coast beyond (visible, top of photograph). These dunes have been developed as an international and economically valuable golf links. This has required the building of separate gabion and boulder rip-rap defences. In the face of known future sea-level rises and storm impacts on these coasts under climate change, debate has begun on what to do next to defend Lehinch. Many argue for increased expenditure on continuing coastal protection works, including the construction of submerged breakwaters offshore, to cause reduction in wave heights and energy. Others argue for a more radical policy of coastal accommodation, through phased retreat from the present shoreline areas. (Source: Julian Orford)

another 0.5km. Longshore sediment transport potential is further reduced by the effects of oblique south-westerly directed wave crests and the wave refraction effects of nearly parallel shore wave breaking along the consequent drift-aligned sink.[22]

Drumlin coasts

One of the most distinctive responses of a retreating glacier is the formation of small hills, or 'drumlins', developed from glacial material. This is generally plastered in a streamlined fashion around an interior rock core. Drumlins tend to occur as multiple interconnected hills, whose orientation parallels the direction of a retreating ice front. Where such drumlin fields intersect with the coast, there is potential for a small archipelago of drumlin islands to appear. Archipelagos are most likely to occur where the overall seaward coastal slope is low and the sites are effectively protected from major ocean waves. The erosion of drumlins (during storms for example) can provide a sediment source for adjacent coastal cells, such as at Knocknagoneen, near Silverstrand Beach, County Galway (Figure 11.12).

Ireland has two splendid examples of drumlin archipelagos: in Clew Bay, County Sligo, and Strangford Lough, County Down (see Chapter 6: Glaciation and Ireland's Arctic Inheritance). Both sites have differing levels of protection from ocean waves, with Clare Island dominating the entrance to Clew Bay and narrowing the effective window of onshore Atlantic waves westwards into the southern and central sides of the Bay. Strangford Lough's virtually enclosed status means only local, westerly to southerly wind waves impact on the drumlins. The narrowness and consistency of wave vectors means that the islands here show a variation in coastal cell development that is dependent on drumlin position and spacing.

The Clew Bay drumlins have a west–east orientation. The outer, western drumlins show the most erosion, with cliff headlands forming sediment sources and the locations for drumlin-flanked beaches. The shallowness of the sea areas between the drumlin, and the close drumlin spacing, means that wave refraction has developed beaches that interlink and overlap from differing western headlands. These sink-type beaches protect the drumlins further from major wave attack. The outer drumlins are now limited in their sediment

Fig. 11.12 AN ERODED DRUMLIN (KNOCKNAGONEEN) NEAR SILVERSTRAND BEACH IN COUNTY GALWAY. The western face of this drumlin at Silverstrand is exposed to extreme wave action, which has developed a steep cliff face and boulder strewn beach-platform. Due to their composition and geography, drumlins at the coast often provide a sediment source for adjacent coastal cells. The Clew Bay drumlins (see main text) show a west-to-east alteration by coastal erosion. The outer drumlins at Clew Bay are more affected by erosion, which has formed cliffed headlands and associated drumlin flank beaches, similar to the features shown here at Silverstrand. [Source: Eugene Farrell]

Fig. 11.13 GRAVEL BEACHES ON THE GORTNAMULLAN COAST, COUNTY DONEGAL. Wave refraction around resistant hard rock outcrops has resulted in the development of gravel beaches in this area, as illustrated further in the inset photographs: a. narrower across the small promontory headland; b. wider beaches in the inter-promontory embayments. The original glacial sedimentary source areas on this coast lay offshore and the beach sediments were concentrated onshore by postglacial sea-level rise. Though surface sand is not evident, the beaches' exposure to high-energy waves means that any sand that moves along the transport corridor is washed into the offshore zone by storm backwash. Many gravel beaches commonly appear to be without surface sand. However, sand can often be found washed down into interstitial spaces between the gravel and found at depth. On the beach surface, gravel often display a size-shape grading. This is because both their size and shape act as controls on their transport potential. [Map data: Google, Maxar Technologies]

availability. As drumlin sediment sources finally fail, through erosion, their associated beaches are reworked to new positions. Any outer drumlin may show a history of episodic erosion as reworking of lateral beaches may create new, albeit time-limited, protection. As these beaches change they may reopen lines of wave access to the rear-drumlins and reactive small points of cliff erosion, feeding minor beaches.

The lack of major wave intrusion in the north of Clew Bay, with its shallower water depths, shows how individual drumlins have insufficient sediment transport potential for lateral connections through wave refraction. This accounts for more linear connections to drumlins positioned to the rear. In short, the pattern of the Clew drumlin beaches is accounted for by wave refraction of limited Atlantic wave penetration. This spatial sequence of drumlin change is akin to an erosional front working itself across the archipelago; its tempo of change is set by the rate of sea-level rise and liable to twenty-first-century future acceleration.[23]

Unlike the drumlins in Clew Bay, the drumlins in Strangford Lough are virtually enclosed, exposed only to limited sea waves via a southern, tidal-dominated strait called the narrows. The Lough has an overall north–south orientation, with drumlins showing an orientation of north-east to south-west, crossing the coastal area. Local short-period storm waves generated under the prevailing west-to-south wind field account for maximum erosion along the eastern shore of the Lough, where only the core remnants of the drumlins remain. Their reworking is shown by linear ridges formed from the largest boulders and gravel (sinks), similar to comet tails, aligned with wave travelling direction.

Gravel beaches

The term 'shingle' is a colloquialism used during the nineteenth to twentieth centuries by coastal scientists to describe beaches dominated by coarser than sand-sized sediment, ranging between 4 and 120mm. This is often described by the generic term 'gravel'. However, gravel beaches can include a wide sediment size range, from boulders to grit.

Beach sediment coarser than sand often indicates a sediment source that was glacial in origin, or near to glaciers, where freezing temperatures shattered bare rock into gravel sized-fragments, a process called periglacial action. Such processes were typical of northern to mid-latitudes during the last major glaciation, so the presence of concentrations of coastal gravel along Irish beaches is not unusual.

The size range of beach gravels (see Table 11.1) means that there is potential size variation along the transport corridors graded by longshore wave energy. This is highly likely, where refraction occurs around resistant hard rock outcrops, especially where there exists folded rock, such as in Donegal. Where this occurs, small coastal cells may occur, characterised by boulders at the mini-headlands and smaller cobble and gravel sediments in the intervening embayments. An example of this is shown at Gortnamullan, west County Donegal (Figure 11.13).[24]

Is Sediment Size the Only Determinant of Transport Potential?

Julian Orford

Sediment particle size is an important control on the transport potential of sediment on Irish beaches. However, it is not just size alone that controls the movement of sediment. All solid sediments have a 3-D nature or, more simply, a shape, characterised by three size dimensions known as the A, B and C axes (Figure 11.14).

Larger particles have an increased 3-D form, which starts to affect their potential to be transported by the swash flow. Furthermore, once entrained, the particle form influences how far the sediment might be moved, for example, how buoyant the particle might be in turbulent swash. There is a long history of shape being defined by relative axial ratios, but though shape indices are easily defined, they have proved inefficient in discriminating how shape affects the hydrodynamic properties of sediments. Nevertheless, particle shape can be an important determinant, in combination with size, in explaining how gravel beaches can show spatial particle variation under swash action. Generally, discoid and platey particles tend to be moved up beach by swash action, which lifts this more easily suspended material, while rollers and spheres tend to be more easily rolled down the beach slope by backwash. Different rock types can show a preferred particle shape potential that can lead to spatial discrimination in gravel beaches; sedimentary and metamorphic rocks tend to take disc and plate forms, while igneous rocks tend to spheres.

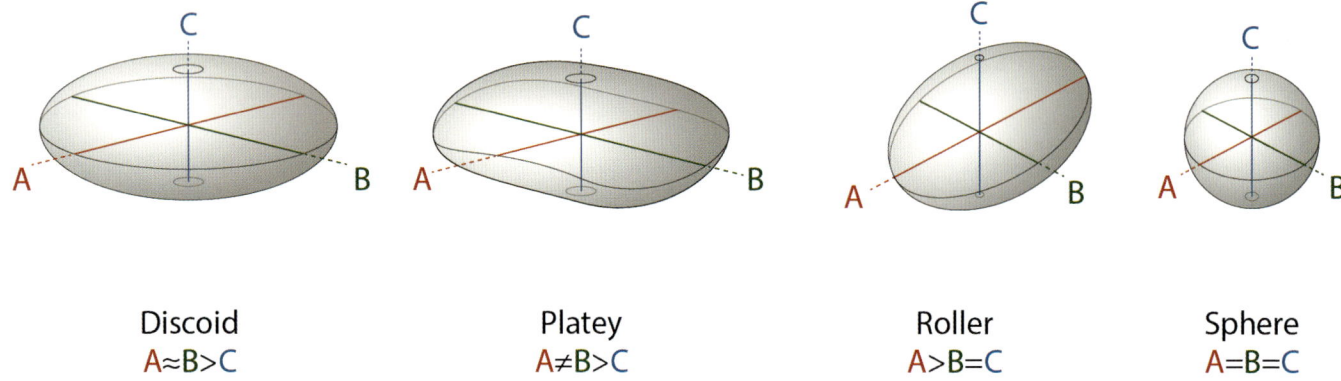

Discoid	Platey	Roller	Sphere
A≈B>C	A≠B>C	A>B=C	A=B=C

Fig. 11.14 SEDIMENT PARTICLE MORPHOLOGIES. Characteristics of sedimentary particles (or clasts), including size and shape, are important determinants of their transport potential. Generally discoid particles (similar A and B, but both larger compared to smaller C) and platey particles (unequal A and B, but both larger than C) tend to be moved up-beach by swash action, which lifts these more easily suspended particles. In comparison, rollers (large A compared to equal smaller B and C) and spheres (smaller but equal A, B and C) tend to be more easily rolled down the beach slope by backwash.

Gravels are usually associated with steep overall beach slope (30 degrees), which is indicative of a reflective morphodynamic regime (Figure 11.15). A distinctive feature of reflective beaches are beach cusps, or arc-shaped sedimentary formations on the shore. These are associated with rhythmic longshore variation in the run-up of swash. At many rock headlands where there is a history of falling sea levels – as in the north-east of Ireland – it is not uncommon to find a succession of gravel beach ridges similar to steps in the landscape. These are remnants of past coastal landscapes, called raised beaches, where the land surface was once at sea level before being raised due to glacio-isostatic effects (see Chapter 7: Ancient Shorelines and Sea-level Changes). Raised beaches are common along the Down and Antrim coasts. Their contemporary height above sea level can be used to help reconstruct past environments. As such, they are often the subject of geomorphic enquiry.

Fig. 11.15. A GRAVEL BEACH, AT GORTNAMULLAN, COUNTY DONEGAL (LOOKING SOUTHWARDS). Particle size and shape varies spatially across this steeply sloping gravel beach, where the reflective morphodynamic regime facilitates the development of beach cusps, remnants of which are visible on the lower foreshore areas. [Source: Julian Orford]

How High Can Beaches Reach?

Julian Orford

This may seem a strange question, but it is one that has caused coastal scientists concern in the last decade. Beaches are formed by waves that operate within the vertical framework of the tidal range. The beach may heighten when sediment is transported to higher elevations through the uprush of turbulent water across the beach, called swash, from waves breaking on the beach face. The returning backwash can also transport sediment, resulting in a lowering of beach height. Platforms on the upper beach level, called beach ridges, mark the highest elevations of the backshore area (Figure 11.16). These occur where the swash can reach upwards beyond where the waves break at high tide. During storms, waves and swash can be lifted vertically to higher than normal heights. This is partly due to the inverse barometric pressure effect (the raising of mean water level due to the low atmospheric pressure characteristic of storms). Water levels may be further raised by the geostrophic action of storms moving over the sea (a form of frictional drag), forming storm surges, as occur particularly on Ireland's western coasts.[25]

Pressure and geostrophic effects define the surge, or excess water level, beyond what is tidally predicted. This surge can lift accompanying swash to higher than normal levels. In severe storms this might amount to +1–2m above normal. There has always been an interest in the elevation that storm waves might move sediment, especially big boulders on rock-cliffed coastlines, such as on the Aran Islands. Previously, high elevated sediments were thought to be caused by past higher-than-present sea levels. In the last decade, though, the observed vertical reach of major tsunami flow has raised the possibility of ancient or historic tsunamis being responsible for some high level sedimentation on the Irish coast. Evidence is also accruing that suggests science has underestimated the power of severe storms in placing sediment at high elevations, so this elevation debate is still far from being resolved (see Chapters 2: The Coastal Environment: Physical Systems Processes and Patterns and Chapter 10: Rocky Coasts).

Fig. 11.16 GRAVEL BEACH RIDGE. Beach ridges, like this one at Inch Strand, County Kerry, mark the upper reach of wave movement. During storms, waves can reach heights higher than normal, leaving behind these deposits. The steep cliff of the dunes behind the gravel indicates that some trimming and erosion of the dune sands may also occur during these times of storms. [Source: Sarah Kandrot]

237

Gravel barriers

Beaches are essentially fringing structures that have no rear landward slope. Where there is a back slope to a beach, then technically the beach forms a barrier. In the case of a gravel barrier, the back slope is a result of a gravel crest developing landward of the high-tide position. This is caused by extreme storm waves pushing their post-breaking swash water landward and beyond the high-tide limit. In so doing, waves transport gravel to the swash limit above high-tide positions. Shorter swash events build the upper beach forward (aggradation), while middle-length swash flows may add material to a growing crest elevation (overtopping). Over time the succession of extreme events may add more and more material to the swash limit. Under really extreme events, swash may override the crest and transport gravel beyond the crest (overwashing). Assuming a constant sea level, an approximate equilibrium develops, whereby the majority of swash flows only reach seaward of the crest.

Approximately 2 per cent of swash flows reach the top of the

Fig. 11.17 OVERWASH (WASHOVER) SEDIMENTARY FANS DEVELOPED ON GRAVEL BARRIERS. a. A modern view of the gravel barrier at Tacumshin, County Wexford, showing large overwash fans crossing the barrier and their characteristic fan-shaped splays at the rear of the barrier [Map data: Google, Maxar Technologies]; b. the overwash fans shown in a. were deposited primarily by the storm of 24 December 1977, which generated major surge conditions and sediment overwashing of the gravel barrier. Although they have lain mainly dormant, and subsequently been covered with vegetation, the fans have been reactivated in recent time, probably as a result of the storms in the winters of 2013/2014 and 2014/2015. [Source: Julian Orford]

crest, thereby building up the crest. Only a very, very small percentage of that 2 per cent are of sufficient magnitude, as overwash events, to flow down the developing back slope of the barrier, transporting material from the crest and back slope beyond the edge of the back barrier. These latter events are associated only with the most extreme storms. On gravel beaches, overwashing swash is usually discontinuous in position along the crest, such that discrete swash channels, called throats, form at the rear of the crest (Figure 11.17a and b). These merge into fan-shaped deposits, formed as the overwash flows rapidly decelerate and come to a halt as the flow reaches the flat back-barrier area. These overwash fans can leave a very distinctive plan-view morphology, with apparent 'headland and bay' type relief. The finer the gravel size, the more likely it is that these longshore fans may merge sideways with adjacent fans.[26]

As sea level rises, the spatial template of swash extent is moved onshore, so that overwashing flows increase. As a consequence, more gravel is moved over the crest and down the barrier back slope. The dominant beach-face swash flows push more sediment towards the crest to replace landward-directed crest loss. As a result, the beach barrier crest is pushed onshore in what is termed 'barrier rollover'. It is important to realise that a shoreward moving gravel barrier is normal behavior, more so when sea level has been rising, as experienced on the southern and western coasts of Ireland over the last several millennia. Today, it is common practice to attempt to stabilise retreating gravel barriers, for example, by artificially raising the beach crest with new gravel, so that they act as flood defence structures. This creates a new problem, though. The prevention of overwash flows tends to maximise gravel losses from the barrier's beach face by longshore transport, so less sediment is re-included into the barrier by rollover. As a result, the shoreline is vulnerable to retreat.

The dominance of cross-shore transport on barriers is usually associated with barriers that have moved into swash-alignment. This process is enhanced where there is a longshore sediment deficit or scarcity, resulting in a thinning of the transport corridor. On such beaches, most waves break parallel to the coast, allowing minimal longshore movement and maximal onshore movement during storms.[27]

Barriers often have an initial history of dominant longshore activity. As their sediment source is reduced or exhausted, for example, where there is limited glacial debris sitting on a rock basement, a re-orientation of down-drift beaches from drift to swash alignment inevitably occurs. This process is called cannibalisation. Waves breaking closer to the thinning barrier produce more swash flows at the beach crest, which generate cross-beach movement, a prerequisite of barrier rollover. This in turn emphasises the movement of the overall beach/barrier into swash-alignment.

An example of a gravel-dominated barrier can be found between Kilmore Quay and the granite headland of Carnsore Point in County Wexford. This is a barrier that was initially fed from weathered Carnsore granite and several glacigenic sources around the bay. Sediment movement along the transport corridor is dominated by both Atlantic south-westerly waves and local south-easterly storm waves, which together have re-sorted the gravels into a generally fine–coarse trend from west to east. Over the last 5,000 years, rising sea level has pushed the developing gravel barrier onshore. This, combined with a reduced sediment supply, has encouraged a single major swash alignment bay from Kilmore to Carnsore. The kink in the bay just west of Carnsore Point is indicative of the resistant granite outcrop edge.

Along this coast, the barrier separates two lakes (Lady's Island Lake and Tacumshin) from the sea. The lakes exemplify how onshore barrier migration, along with longshore sediment transport, has shut off any sea outlet for freshwater drainage (see Chapter 12: Coastal Wetlands). Flooding around both lakes has encouraged human intervention to induce lake drainage. For example, there is a history of breaching by 'hand and donkey' at Lady's Island and building sluice gate systems at Tacumshin. Gravel extraction from the barrier by building contractors, now forbidden by law, threatened the best gravel sections around Carnsore, and certainly reduced the overall gravel budget. As a result, some sections now seem to be dominated by sand beaches. This is an illusion, though, in that the barrier is still gravel-dominated, if lost at times under sand cover.[28]

Another notable gravel barrier exists along the 17km stretch of coast between Wicklow and Greystones. This barrier has been incorporated into the Dublin–Rosslare railway foundations. The relative lack of recent landward migration and reworking of the sandier beach face by dominant south-easterly storm waves is evident in this area. The barrier has been prevented from landward migration since the late nineteenth-century engineering intervention, assisted by large boulders (rip-rap) placed to protect the seaward face of the barrier. The barrier's future is of considerable concern for this rail connection, as contemporary rising sea level has pushed the beach face into deficit sediment mode, thereby steepening the beach and allowing bigger breaking waves closer to the defended gravel barrier (see Chapter 30: Engineering for Vulnerable Coastline).

Sand barriers

In comparison to gravel barriers, there is a discernible lack of sand barriers in Ireland. This is not because of a lack of sandy coastal systems, but more due to an aspect of coastal sand transport that is not likely to affect gravel barriers: that of wind, and its ability to move finer sand onshore, especially in western Ireland. According to definition (as above), barriers should possess a rear (landward) slope, to any crestal build-up of sediment, positioned beyond the reach of all but the most extreme swash flows. Both the swash and wind together transport sand off the beach face and over the potential wave-built barrier crest, which means that dunes can start to develop once out of the reach of swash and wave action. Finding a distinctive backslope, however, to a sand beach in Ireland is rare. This is due to the high-energy, mid-latitude strong wind regime, the wind forming an ever-present transporting system that moves sand from the beach corridor and backslope areas to new landward dune sinks, thus limiting accumulation.

There are other requirements for dune formation, but sand

Fig. 11.18 CARRICKFINN SAND BARRIER, COUNTY DONEGAL. a. Carrickfinn has a wide dissipative beach system and limited rear-dune development. This is due to the absence of longshore sediment transport corridors, as evidenced by the scarcity of sand along the adjacent rock headlands. If active, the transport corridors would deliver source sediment to the dunes. Donegal's airport is strategically positioned on these low dunes. [Map data: Google, CNES/Airbus]; b. aground view at Carrickfinn, looking north, shows the wide beach and the densely vegetated dunes. [Source: Derek Jackson]

availability is a major factor. Where beach sand size is dominated by coarser sand (1–0.5mm), aeolian (wind) transport is limited and a beach crest is more likely to form. As with gravel barriers, the potential exists for extreme overwash to move sediment down a definite backslope. A major difference in morphology between sand and gravel barriers relates to their fans. With sand-dominated washover, there is more likely to be larger alongshore coalescing fans, so that the diagnostic rhythmic bays and headlands of gravel overwash tend to be missing. Another difference between the two types of barriers, at least in Ireland, relates to the temporal spacing of storm events causing overwashing. Sand overwash events appear to occur at larger intervals (over hundreds of years) than equivalent gravel overwashing. Silverstrand, County Mayo, is an example of a coarse-sand barrier that has experienced relatively few (approximately five major overwash events over the last several thousand years. This low number is possibly related to past low rates of sea-level change.

At Silverstand, there is an absence of major dunes, and the sandy overwash feeds extensive spreads of sand over the landward low-lying ground. This can be an issue for farming in the back barrier area, where farmers have had to deal with these occasional sand and saltwater incursions. In areas where sand supply is plentiful, the barrier morphology may get lost under the feed of blown sand to rear dunes. For example, Figure 11.18 shows this type of structure at Carrickfinn, County Donegal. The dunes here are not substantial, with growth predominantly confined to the dune's seaward edge. This is because sand supply to the dunes is now reduced, relative to the past. This is evidenced by the dominant vegetation and lack of bare sand in the dunes. Sand barriers, such as this one, are considered sediment sinks. They occur where there is minimal longshore transport and near-parallel wave breaking on the shore.[29]

Mixed sand and gravel barriers

Apart from sand- or gravel-dominated barriers, there is an intermediate situation where both sand and gravel are present in the same barrier. This may appear somewhat unlikely, as energy regimes for such size differences suggest they are not likely to be found at the same site. One reason this might be possible is that conditions favouring one or the other come to dominate at different times, for example, as a function of a changing dominant source of sediment supply or, more likely, a shift in the proportion of sand to gravel in the transport corridor and sink. For instance, as an initial dominant sand element is blown inshore, it may leave a gravel element to be consolidated as a gravel beach fringing a dune area. This shift in relative sediment proportions may be repeated over time as a result of changes in the sediment source, and/or as a result of changes in sea level. Streedagh Strand and Ballyliffin, County Donegal, form examples of these mixed sediment situations (Figure 11.19).[30]

Fig. 11.19 MIXED SAND AND GRAVEL BARRIER AT STREEDAGH STRAND, COUNTY DONEGAL. Mixed sand and gravel barriers occur as a result of changing conditions, for example, changes in sediment supply. The gravel component here provided the basis for the initial development of semi-rhythmic headland and bays of overwash fans, formed along the inner estuary side of the narrow barrier. Subsequent dune activity resulted in the later burial of the gravel. Although the sand component was once sufficient to cover the barrier back-slope, its dominance has since waned and gravel can now be seen breaking through the sand cover. [Source: Andrew Short]

Spits

Spits are depositional coastal landforms that appear, in plan view, as projections of a beach corridor into the sea from the mainland. They develop where there is a sudden change in the shape of the coastline, for example, at major headlands or estuary mouths. The refraction of waves around the up-drift corner of a coast results in a reduction in wave energy and, therefore, deposition of sediment. Sediment accumulation at this corner initiates spit growth. This will continue, either until the point where currents are sufficiently strong to limit sediment transport, or until the sediment supply runs out.

Although probably the most easily recognised coastal form, spits are not abundant around the Irish coast. Irish spits can be found, for example, at Tramore and Dungarvan, County Waterford; Inch and Rossbehy, County Kerry; Donabate, County Dublin, and at Magilligan, County Derry. Many of these features have behaved historically, morphologically speaking, in similar ways. For example, at the beginning of spit growth, the proximal (near land) end of a proto-spit often acts as a temporary sediment sink, forcing the transport corridor to deflect its focus away from the land. This movement in turn adjusts incoming wave refraction, such that the initial sink starts to expand seawards, with the set-up of a new transport corridor. This corridor terminates in a deepwater sink, which becomes the basis for the later seaward extension of the beach (see Chapter 30: Engineering for Vulnerable Coastline).[31]

There is a competition between the transport potential and the sink tendency as the spit extends seawards, such that longshore sediment supply becomes the key to spit extension and growth. When the transport rate slows, refracting waves curve around the end of the spit. This forces sediment into a spiraling pattern, forming a recurve (hook-like end) to the spit, which is characteristic of this landform. If the longshore sediment rate then increases, the spit can be re-energised and again extend seaward from the old terminus.

Spits usually show past recurves projecting out from the rear slope. These are truncated as the spit rolls back. The roll-back is usually aided by the cannibalisation of sediments from the up-drift shore zone to feed the down-drift spit extension. In this way the landform seems to rotate from its initial acute angle of departure from the coast, to become a later near right-angle to the mainland trend. The Tramore spit, for example, shows these signs of rotation. The Dungarvan spit is unusual, as it shows no sign of recurves or rotation (Figure 11.20). Spits rarely join to the other side of estuary mouths, as the tidal flow of estuary water maintains the channel at the spit end and prevents closure.

Fig. 11.20 THE CUNNIGAR SPIT, DUNGARVAN, COUNTY WATERFORD. The spit, shown here at low tide, extends north across the mouths of the River Brickey and River Colligan estuaries at Dungarvan Harbour. In spite of the extensive back-spit estuary area, with its accumulation of fine-sized sediments, there is no evidence of recurves or rotation of the spit. Many other Irish spits have these coastal shape features. Seaward of the Cunnigar, the dissipative, low-energy intertidal zone provides ideal conditions for a variety of fine-sands and muddy-shore shellfish species to thrive, such as the common cockle (*Cerastoderma edule*) and other clam species. Conditions are also favourable for the farming of oysters, with the rectangular arrays of oyster trestles visible at the low water edge. [Map data: Google, Maxar Technologies]

Forelands or nesses

Major changes of coastline direction occur around Ireland's many estuaries and sea loughs. Those with very wide, deep mouths and dominant longshore sediment supply can be associated with a distorted spit-like development, known as a foreland or ness. Typically, such features are located in the wake of two opposing wave directions, usually non-balancing. Where this occurs around a spit, there is a directional push to one side of the spit, so that waves from one side dominate the transport around the spit terminus. Here, the lesser opposing waves rework and push sediment onshore to help build (aggrade) the rear of the spit. Over time the spit grows broader at its coastal base, while still tapering towards its deep-water terminus, where strong tidal currents prevent any further spit extension. Long-term variation in sediment supply allows pulses of erosion and deposition along the up-drift dominant beach corridor, which then supply the sinks in the spit rear. Often the reduced energy area on the lee side of spits (that is, downwind) provides fertile ground for saltmarshes, and their growth helps to broaden the spit base. The end product of this spit transformation is a foreland, or ness.

In County Kerry, Derrymore Strand exemplifies the bidirectional influence of waves on coastal sediments, formed from the erosion of local glacigenic sedimentary sources. Here, the dominant wave-driven beach transport occurs on the western Atlantic side, with the secondary forcing from tidal power and local easterly waves in Tralee Bay. Periods of Atlantic wave reworking and a seaward extension of the tip of Derrymore have developed a foreland. The neck of the foreland (where its base meets the coast) has been expanded by substantial saltmarsh development.

A more conventional foreland is shown at Magilligan foreland, on the eastern side of the entrance to Lough Foyle (Figure 11.21). On this north Ulster coast, there is a dominant east-to-west coastal longshore sediment supply from the River Bann estuary. In addition, coastal erosion of boulder clay provides sand and gravel from open coasts to the north of the site. This, together with probable beach face onshore sand movement, has fed the development of a series of prograding beach ridges at Magilligan, each one nearly a small spit. Westerly waves in the Foyle have successfully trimmed and reworked the sequential ridges, forcing the two coasts to meet as a foreland tip. The increasing power of the tidal currents, squeezed into the narrowing entrance to the Lough, prevents the ness from ever closing off Lough Foyle. Progradation of the ness (its growth into the sea), has been assisted by falling sea level over the last 5,000 years.[32]

Bay-head beaches

The heavily indented coastline of Ireland, from Antrim round the west coast to Wexford, is based upon major resistant headlands of folded and faulted ancient rocks. Many adjacent headlands have enclosed embayments (bays) between them, with their entrances often directed into the face of Atlantic waves. These bays now lack beach sediments such that, with rising sea level, any sediment

Fig. 11.21 OBLIQUE AERIAL VIEW OF A FORELAND (OR NESS) AT MAGILLIGAN, COUNTY DERRY. Coastal forelands, or nesses, are nose-like features that form where spit type formation is influenced by bidirectional waves. At Magilligan foreland, sediment eroded from the beach and cliffs on coasts to the north have fed the development of the ness over the last 4,000 years or so; as shown by the age of peats formed in wetland hollows (slacks) trapped between the prograding series of spit, beach and accompanying dune ridges. Successive phases of beach and dune ridge development are visible on the left side of the photograph, as is the pattern of opposing wave crest directions (herringbone-like appearance) approaching the foreland (left and centre). [Source: Cinematic Sky, Courtesy of Scenic Lough Foyle Ferry; inset map data: Google, CNES/Airbus and Maxar Technologies]

Fig. 11.22 BAY-HEAD BEACHES AT a. BALLINSKELLIGS BAY (WATERVILLE BAY), COUNTY KERRY, AND b. CLEGGAN BAY, COUNTY GALWAY. Bay-head beaches occur where refracting waves, breaking onshore at an oblique angle, drive sediment into small embayments set within the larger bay. In some cases their existence is dependent on wave shadows, favourable for sediment deposition, created by small rock islands, as occurs in Ballinskelligs Bay. Elsewhere, more opportunistic bay-head beaches can occur, as at Cleggan, County Galway. Here, the weathering of the local Connemara granites has produced an especially coarse quartz beach sand. This resists wind transport, such that the beaches are trapped against the shore and do not leak landward as dunes. Embayments that are elongated and narrow, often fault-guided, have little potential for beach formation along their side walls, although they do exist. [Source and map data: Google, CNES/Airbus]

available from the former submerging shorelines has been rolled up into a few beach sinks. These cling almost opportunistically to the side walls of these embayments. The semicircular nature of embayments encourages the sideways expansion of waves refracting into the bays (through diffraction). Consequently, final breaking waves appear at the same time, almost, around the enclosed bays. This is what tends to bring and hold any residual sediment into some locations, often in the smaller bays inset within the coastal embayments.[33] One example of this can be found in Ballinskelligs Bay, Waterville, County Kerry, where the presence of several beach sinks within subset, smaller embayments within the bay can be observed. Generally, this is a coastline with scarce beach sediment availability. Often coastal defences have been installed to control the failing beach sediment volume (see A Recent History of Coastal Engineering in Waterville, County Kerry, in Chapter 30: Engineering for Vulnerable Coastlines)(Figure 11.22).

Estuary mouth and associated flanking beaches

Major embayments that are associated with estuaries are also the setting for a further coastal system, one that is mutually dependent on waves driving spits, and/or barriers, up-estuary into the estuary's tidal channel. As this channel narrows, water flows increase, holding open any existing barriers and also limiting the growth of estuarine spits. This 'holding situation' is often semi-balanced. In the case of spits, storms cause sediment to move around the distal (outer) end of the spit, thus narrowing the tidal channel. The increased tidal velocity associated with the narrowing results in the reworking of the storm deposits. As the channel width widens, due to erosion of the spit's end, the tidal current velocity will reduce and quasi-stability returns after the storm. The eroded sediment released by this process can move into the estuary, where sudden diminishing tidal currents (out

of the channel) allow subtidal deposition in the form of a flood-delta. Alternatively, sediment can move offshore to form an ebb-delta.

An Irish example of an ebb-delta system can be found between the Inch sand barrier and Rossbehy spit, in Dingle Bay, County Kerry. Here, the exchange of tidal flow occurs between the back barrier estuary – formed by Castlemaine Harbour – and the open ocean. In fair-weather conditions, it is likely that sediment from the ebb-delta is reworked onto the two adjacent flanking barriers, waiting for the next storm to drive its recirculation in and out of the estuary.[34] Elsewhere, an example of a flood-delta can be found at Falcarragh, County Donegal. Here, the delta is located on the estuarine (as opposed to the seaward) side of the barriers (Figure 11.23).

Emergence of either an ebb- or a flood-delta depends on the asymmetry (the balance) of the tidal flows in their semi-diurnal timing. Short-term peak ebb-currents (seaward flows) and longer-term reduced flood-currents (flows into the estuary) together create an ebb-delta and vice versa. What causes the tidal cycle asymmetry depends often on the flooding resistance, compared to the ebbing resistance, of the sedimentary environments present on an estuary's intertidal margins. Where extensive mudflats or saltmarshes have formed, this asymmetry may be more pronounced and if their growth persists, this can alter the stability of local spits. For example, nineteenth-century land reclamation activities at numerous Irish estuaries have affected many entrance and flanking spits. Such activities reduce the intertidal water volume in the estuary (called the tidal prism), which changes the inlet/outlet tidal asymmetry (see Chapter 4: People, Agriculture and the Coast).

The operation of this is shown on the coast at Wexford Harbour, where the remnants of the Rosslare spit can be seen. During the mid-nineteenth century, half of Wexford Harbour was reclaimed for agriculture. This generated a sudden and dramatic

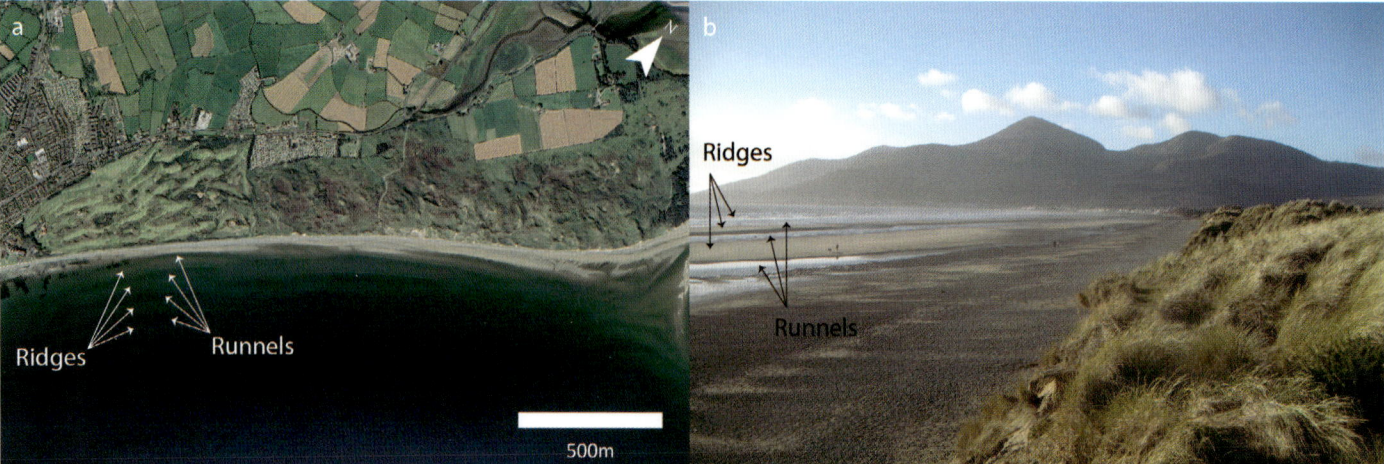

Fig. 11.23 THE ESTUARY MOUTH AND FLANKING BEACHES AT FALCARRAGH, COUNTY DONEGAL. The Dooey spit-barrier and Ballyness barrier are part of a flood delta system. Here, sediment accumulates as intertidal sand shoals behind the two barriers, but not on their seaward side as would be characteristic of an ebb delta system. The sediment is delivered into the area via primarily the main inlet channel, through which powerful tidal currents carry sediment. As the inlet channel widens suddenly to the south, the diminishing tidal currents allow sediment to be deposited, where it accumulates in the form of the intertidal shoals, shown here at low tide. [Map data: Google, Maxar Technologies]

change in tidal asymmetry, causing a dominant ebb-tidal flow that prevented the southward drift of beach sediment round the delta front and onto Rosslare spit. The loss of this sediment supply meant that by the 1930s, Rosslare spit had lost much of its sediment volume, causing the spit to thin and allowing major south-easterly storms to overwash the spit. The spit was finally beheaded near its present terminus, leaving some remnants of the original spit rolled over into the estuary as sand shoals. This shows how well-meant but

Fig. 11.24 RIDGE AND RUNNEL SYSTEM AT DUNDRUM, COUNTY DOWN. a. This satellite and accompanying ground-level view show the parallel intertidal bars and separating waterlogged beach runnels. The beach system operates under negligible wave energy at high tide. b. The ground view, looking south-west to the Mourne Mountains, was taken the day after the major tidal surge of 6 January 2014, which caused erosion to the beach and dunes at the site. Although lowered, the bars are still prominent. Three bars are exposed but the fourth (seen in a.) is identified by wave breaking in the surf zone. [Source: a. map data: Google, Maxar Technologies; b. Julian Orford]

uninformed alteration of coastal systems can cause major problems decades later for coastal inhabitants.

Intertidal bars (ridge and runnel) on sandy beaches

Along the eastern Irish coast, for example, from Malahide southwards, there are a number of spit and dune systems. The major difference between these and other Irish coast examples discussed in this chapter, is the presence of substantial intertidal sand bars fronting the spits. These extend along most of the sandy coast between Dublin and Down. The bars are actually ridges, which tend to have a steeper landward face and a gentle seaward slope merging into a runnel or ditch between the ridges. There are between two and four ridges on these beaches, with the seaward one lying just below the low-tide position. Ridges are essentially parallel with the shoreline and usually there is no major rhythmic interruption, hence the capacity

for such a beach to be used as an alongshore horse racecourse, as at Laytown, County Meath. Cross-ridge breaches, where water can drain from one runnel into the adjacent runnel, also occur. As the tidal range varies along the eastern Irish coast (for example, 0.9m at Courtown, County Wexford, to 5m at Dundalk, County Louth), then the number of ridges varies. Figure 11.24a shows a four-bar system in Dundrum Bay (County Down). These east coast sand beaches tend to have very low gradients (c.1:70 to 1:100), fine sand sizes and low median wave height. As such, the beaches are classified as morphodynamically 'very dissipative' (absorbing wave energy). The ridges are persistent through the seasons, although they become flatter during winter storms (Figure 11.24b), while building up in summer fair-weather conditions. Some ridges appear to migrate slowly onshore, adding sand to the backing dunes in the process, though this has not been reported along the Irish coast.[35]

BEACHES AND THE PROBLEM OF COASTAL DEFENCES

Andrew Cooper

Sandy beaches and linked coastal systems are inherently mobile. By their mobility they can accommodate naturally the effects of storms and so survive for millennia, forming excellent sea defences. This mobility, however, is seldom appreciated by contemporary society. Instead, beach mobility, which often threatens or appears to threaten badly located property, is seen as something to be controlled. After any major storm there are demands from homeowners for help in defending their properties. Almost always in Ireland, property is seen as more valuable than the beach. This attitude has led to badly placed development that suffers major damage during storms. Worse, from a beach perspective, it has led to the construction of a variety of structures built to protect roads, houses, golf courses, hotels and even agricultural fields (Figure 11.25). The alternative, of relocating these activities and built structures (such as roads, railways and houses) through coastal set-back or realignment, has yet to receive widespread consideration. Often, demands for coastal defence are based on a lack of understanding of erosion as a natural process. The coast is littered with useless sea walls that protect undeveloped dunes for no apparent reason, other than a perception that erosion must be stopped. At best, such structures are benign, but many cut the link between dune and beach and prevent the beach's ability to adjust during subsequent storms (see Chapters 27: Tourism and Leisure and 30: Engineering for Vulnerable Coastlines).

One of the best-documented illustrations of the misguided nature of coastal defences in Ireland is at Portballintrae, County Antrim (Figure 11.26).[36] Here, construction of a pier in 1895 caused a change in how the wave energy was spread around the bay. Over several decades, the changed conditions led to loss of sand from the beach. Rip-currents in the middle of the bay facilitated the movement of sand out of the bay. The progressive loss of sand can easily be tracked from photographs through time of the formerly popular beach. Not convinced of the reasons for the erosion, the local authority has spent a great deal of money on groynes to trap the non-existent sand, drainage to stabilise the now-exposed cliffs of soft glacial sediment and sea walls to prevent further wave erosion of the base of the cliffs. This situation is mirrored in many other coastal settings around Ireland.

An interesting and somewhat ironic situation exists near Buncrana on Lough Swilly (County Donegal). Here, sand is moved alongshore from north to south by Atlantic waves that penetrate far into the Lough. A dune system in the north is occupied by the Lisfannon golf club. Natural erosion of those dunes as sand moves southward caused the golf club to rock armour its entire perimeter. This cut the sand supply to the recreational beaches at Lisfannon and they have consequently begun to erode. As they erode, their sand is carried further south and

Fig. 11.25 LONGSHORE, GABION DEFENCES IN FRONT OF A HOTEL AT BETTYSTOWN, COUNTY MEATH. Poor planning and a general lack of awareness around the natural mobility of beaches has resulted in inappropriate development on the coast, both in Ireland and abroad. When properties become threatened by erosion and/or flooding, the response is often to construct defences, such as gabions, like those pictured here. These are often expensive to maintain and, worse, ineffective. Despite this, the engineered response remains the preferred, if ill-advised, action. [Source: Andrew Cooper]

Fig. 11.26 REMNANT BEACHES AND COASTAL ENGINEERING WORKS AT PORTBALLINTRAE, COUNTY ANTRIM. This site exemplifies the unintended consequences of coastal construction activities on or near sandy beaches. The construction of the pier at the mouth of the bay in the late nineteenth century affected local wave conditions, thus leading to the progressive loss of sand from within the bay. What had once been a wide sandy beach (extending over the area now occupied by groynes), on which Victorian ladies laid out tennis courts, has since been reduced to a narrow gravel ridge. Significant sums of money have been spent on protective measures, including groynes, wooden revetment and rock armouring (inset). All of these built works are unable to stem the cliff erosion, or beach sediment losses. [Map data: Google, Maxar Technologies; inset: Derek Jackson]

deposited in a newly constructed marina. The marina has now to be dredged, because there is too much sand, while the golf course has been protected to save its diminishing sand.

Until recently, much of Ireland's coast has been saved from the depredations of sea defences, because it has been relatively undeveloped. Further, money to both develop the coast and to build defences has been scarce. During the Celtic Tiger economic years, however, much inappropriate development was built on sand dunes and other vulnerable coastal locations. This has been accompanied by the construction of ad hoc sea defences that will continue to threaten beaches. Regrettably, the existing coastal management systems operating in Ireland are unable to accommodate more sustainable and less damaging alternatives, with the extent of sea defences ever increasing.

Ireland's sediment-dominated coasts are very important for both people and wildlife and account for some half of Ireland's coasts. Also, they are fascinating in their own right, in their diversity and functioning. This chapter has considered the physical attributes of these sandy coasts: from their types, sizes and shapes, to the critical influences of geology, glaciation, sea level, waves and, most particularly, the sediments. Further, the chapter has examined, albeit briefly, some of the significant issues that result from the interaction of the physical processes with people that have moulded its characteristic features, of beach, barrier and coastal spit.

In no small measure, it is due to the work of coastal scientists over many years that people are able to understand something of the complex of processes that operate across these culturally and biologically important environments. Yet, human interactions with the coast remain complicated. Where erosion and flooding threaten development, the current system of coastal management is inadequate, often causing more harm than intended. Now, under rising sea levels, for example, due to climate change, the significance of such problems is likely to worsen. As such, the need for refreshed long-term coastal management strategies is more pressing than ever before, as is the continued study of the morphodynamic behaviour of these captivating and and much-visited Irish beaches.

Coastal Wetlands

Deborah Chapman

Timoleague high marsh and saltmarsh, County Cork. This area of mudflat and saltmarsh wetlands lies on the margins of the Argideen Estuary, where it leads out to sea and into Coolmain and Courtmacsherry bays. The head of the Estuary here occurs close to the medieval settlement and Franciscan Friary of Timoleague. The wetlands were embanked in the early nineteenth century and partly reclaimed for agriculture, as was the case in many other coastal marshes around Ireland, but fell rapidly into disuse. The remains of stone embanking walls are visible on the right of the photograph and these still form a sharp division between the lower saltmarsh, and the mud- and sandflats of the Estuary itself. The walls are breached and the sea accesses these areas as part of the regular tidal cycle, maintaining the tidal creeks and marshes. The earlier nineteenth-century plough and drainage lines can still be seen reflected in the marsh vegetation cover, together with other more obvious human impacts on these marshes (such as modern roads and drainage ditches). [Source: Sarah Kandrot]

Fig. 12.1 THE SALTMARSHES AT CARRIG ISLAND IN THE SHANNON ESTUARY, COUNTY KERRY, SHOWING ALSO CARRIGFOYLE CASTLE IN THE CENTRE OF THE PHO-TOGRAPH. Wetlands are typically undrained areas that are periodically inundated and support water-loving plants. Wetland habitat types include marshes, lagoons, estuaries, saltmarshes, sand- and mudflats, turloughs, bogs, fens, swamps and wet woodland. Saltmarshes are among the wetland habitat types found commonly on the Irish coast. They consist of stands of vegetation that occur in marine and brackish waters. The dendritic (tree root-like) shapes of the tidal creek channels visible on the saltmarshes in this area are a distinctive morphological feature found in saltmarsh environments. They are formed in response to the daily tidal action and are an important control in the functioning of marshlands. They are particularly pronounced here, due to the large tidal range (high- to low-water conditions) experienced on the meso–macro coasts of the Shannon Estuary. [Source: Geraldine Hennigan and Norman Kean]

Our wetlands are places of immense beauty, distinct elements of our natural environment, to be celebrated and protected. They can be stark and desolate in the winter and alive with birds, insects and flowers in the summer. Boglands are perhaps amongst the most well-known of Irish wetlands. However, there is a wide variety of wetland environments, both terrestrial and marine, and each type supports a special ecosystem. Coastal wetlands are distinguished by the way they are flooded and saturated by seawater, creating vegetative conditions suited to particular aquatic plants that can tolerate waterlogged, hydric soils.

This chapter will provide a broad view of the diversity of coastal wetlands, including estuaries, lagoons, saltmarshes, sandflats, mudflats, tidal marshes and wet machair. (The specific aspects of Irish estuarine ecology, including mudflats, lagoons and aspects of

Fig. 12.2 ESTUARINE AND MARINE WETLAND ENVIRONMENTS. a. The coastal zone supports both estuarine and marine wetlands. The estuarine wetland shown here is from the Shannon Estuary. The relatively sheltered aspect of estuarine environments allows sediment to fall out of suspension in the water and accumulate in the form of mudflats and sandflats. The Shannon Estuary is the largest estuary in Ireland and is connected inland to a drainage system with a dense network of rivers, streams, creeks, arterial drainage channels and sluices. The mudflats of the estuary have been reclaimed for agricultural, industrial and commercial purposes over centuries. b. A marine wetland from Blackrock, Dundalk Bay, County Louth (with the Cooley mountains in the distance). This open sea bay is the outlet for four estuaries. As a result of the sediment dynamics associated with these features, it has extensive saltmarsh-es, intertidal sand- and mudflats. The marine wetlands of Dundalk Bay sup-port an abundance of crustaceans, molluscs and marine worms, which in turn provide a valuable food source for tens of thousands of water birds. The area is designated as a wetland of international significance under the European Ramsar Convention. This obliges the government, on behalf of the people of Ireland, to ensure the wise use of wetlands, their conservation, vigilance in planning decisions and the use of shared best international practice in their management. [Source: a. Simon Barron; b. Breffni Martin]

estuarine management, are outlined in more detail in Chapter 13: Estuaries and Lagoons.) Saltmarshes are highlighted later in this chapter as a case study (Figure 12.1). Wetlands in general share common challenges to thrive and survive in the face of increasing human impacts. These challenges, including climate change, are explored towards the end of the chapter, but first, the focus is on what makes coastal wetlands so special as a natural asset.

WHAT ARE COASTAL WETLANDS AND WHY ARE THEY SO SPECIAL?

There are many different definitions of wetlands, but the general consensus is that wetlands are comprised of habitats that are permanently, or occasionally, submerged with fresh, or saltwater. In the Irish coastal zone, wetlands are most commonly associated with shallow water marine areas (for examples, lagoons and seagrass beds), open and low-lying coastal environments subject to frequent seasonal flooding (such as fen and reedswamp), or estuaries (including tidal marshes and mudflats) (Figures 12.2a and b). Wetlands can occur in natural hydrological settings, such as rivers and estuaries, or as a result of modifications to existing environments through the building of dams, causeways and flood defences.[1]

In a guide produced by the Irish Ramsar Wetlands Committee, saltmarshes, dune slacks and wet machair (see Chapter 11: Beaches and Barriers), transitional waters (such as lagoons) and intertidal or subtidal habitats are all identified as coastal wetlands in Ireland.[2] Saltmarshes occur on flat areas between the neap and spring tides

in sheltered bays and comprise salt-tolerant vegetation growing on mud, sand or submerged peat in the intertidal or near-tidal zones. They are often used for grazing sheep and cattle, as for example in the Magilligan Foreland area, County Londonderry, where the meat from its saltmarsh-grazed sheep is highly prized (see Chapter 4: People, Agriculture and the Coast). Wet machair is a type of grassland growing on calcareous sand that has been blown behind a narrow band of sand dunes and where the water table is close to the surface. Within Europe, wet machair only occurs in Ireland and Scotland; within Ireland, it only occurs on the west coast from Galway Bay to Malin Head (Figure 12.3). It is characterised by fine-leaved grasses and small sedges (see Chapter 11: Beaches and Barriers).

Coastal lagoons are shallow brackish lakes with a limited exchange of water with the sea (see Chapter 13: Estuaries and Lagoons). Some are fed by freshwater streams or rivers and others are fed mainly from rainfall runoff from the surrounding land. Natural lagoonal wetlands are usually separated from the sea by a shingle bank, or sand dune or rock barrier. Most have a single, narrow exit and entrance, but some have multiple narrow channels, as in the Tacumshin and Lady's Island system, County Wexford. The lagoon channel controls the amount of saltwater entering and the brackish water leaving the lagoon (see Chapter 11: Beaches and Barriers). Some shallow estuaries have become lagoons as a result of the construction of causeways to allow the direct crossing of

Fig. 12.3 MACHAIR AT LOUGH DOO, COUNTY MAYO. The machair at Lough Doo, County Mayo, is a good example of a machair habitat. The word machair indicates a sandy plain. The image shows machair that is rich in flowering plants, gently undulating and not over-grazed. The tufts of longer vegetation are marram grass, which is found exclusively in coastal areas of Ireland and forms low, mobile dunes as seen on the left side of the background. In general, only a small portion of most machair plains lie close to the sea, and are often separated by an area of foredunes or saltmarsh. This emphasises how the influence of the marine environment can extend beyond what might be perceived as the immediate coastline. Even though the sea cannot be seen in the case of the Lough Doo example, this remains a protected coastal habitat. [Source: Aoife Delaney, Courtesy of National Parks and Wildlife Service]

vehicles. In some of these artificial lagoons, the exchange of seawater is controlled by the installation of automatic or manually operated gates or tidally operated sluice gates (see Chapter 25: Urbanisation of Ireland's Coasts).

Coastal wetlands in Ireland have not all been well studied and only within recent decades have inventories been compiled for these and related environments, partly as a response to European legislation.[3] The long, and in places inaccessible, coastline of Ireland means that new areas of saltmarsh are only recently being discovered and surveyed. Recent research on saltmarshes in the south-west of Ireland has identified a rare form of saltmarsh that develops on tidally inundated peat substrates formed originally in freshwater bogs.[4]

Wetlands provide many important hydrological and biological functions and support many essential services to human communities. For example, wetlands can play an important role in protecting the coast from the worst damages of storm surges and floods. They are also very effective at attenuating phosphorus,

Fig. 12.4 WETLANDS FAUNA. Wetland environments in Ireland support a diversity of fauna, including rare and endangered species. Pictured here are a. the common frog (*Rana temporaria*) [Source: Andrew Lynch]; b. common blue butterfly (*Polyommatus icarus*) [Source: Darren Ellis]; c. viviparous lizard (*Zootoca vivipara*) [Source: Shay Connolly]; d. Kingfisher (*Alcedo atthis*) [Source: Michael Finn]; e. emperor dragonfly (*Anax imperator*). [Source: Michael O'Donnell; all images here courtesy of BirdWatch Ireland]"

Fig. 12.5 COASTAL REEDSWAMP WETLAND AT TRALISPEAN BAY, COUNTY CORK. Examples of extensive and vegetatively important plant wetlands are very common throughout coastal Ireland. The site at Tralispean Bay is dominated by grasses, sedges and rushes, formed in a glaciated meltwater valley that exits into the Bay. The coastal features at Tralispean Bay record the occurrence and impact of the 1755 Lisbon tsunami in this region of south-west Ireland and, possibly, earlier events. The reedswamp, formed behind a former dune barrier at the nearby coast, is maintained by the ponding of freshwater behind the barrier and by locally high freshwater tables. The entire system is partly regulated here, as in all this type of coastal wetland, by the rise and fall of the tides. Traditionally, reedswamps would have provided raw material for thatching. [Source: Robert Devoy]

which can mitigate against excess nutrient loads from diffuse sources of pollution, such as agriculture. With regards to biological functioning, wetlands can nurture fish and shellfish, provide overwintering grounds for waterfowl and facilitate a diversity of species, including rare and endangered species (Figure 12.4). Beyond that, wetlands provide raw materials, including reeds for thatching, and also space for recreation and ecotourism, such as birdwatching and angling (Figure 12.5).

Different wetland functions arise according to the type of wetland, its size and location. For example, large coastal lagoon systems can support important commercial fisheries, whereas estuarine marshes support rich and diverse populations of invertebrates and wading birds. It has been estimated that approximately 75 per cent of commercially harvested fish and shellfish in the USA are dependent on estuaries and their associated wetlands.[5] Despite their importance, marine and coastal wetlands worldwide declined by approximately 35 per cent between 1970 and 2016.[6]

Case Study: Saltmarshes

Grace Cott

Saltmarshes are one of the most commonly occurring natural wetland habitats globally, comprised of dense stands of salt-tolerant vegetation. Located in the coastal intertidal zone between land and open saltwater, or on the margin of brackish water bodies, the marsh surfaces are flooded regularly (generally twice daily) by the tide (see Chapter 2: The Coastal Environment: Physical Systems, Processes and Patterns). Saltmarshes are formed usually on low- to meso-tidal range and low-energy coasts, as found in estuary environments, and are developed by the tidal accretion of mainly mineral sediments. The upper elevation limit of saltmarshes is approximately that of the Highest Astronomical Tide (HAT) while the lower limit is rarely below Mean High Water Neap (MHWN) tide level (Figure 12.6). The marsh vegetation can grow on a range of different underlying sediments (substrates), from fine silt and clays to coarse sand and also on peat. Large saltmarshes, particularly where they develop on fine sediments, include features such as tidal creeks, channels and saltpan pools. A complete inventory of Irish saltmarshes in the Republic of Ireland was carried out by Curtis and Sheehy Skeffington (1998), who classified marshes

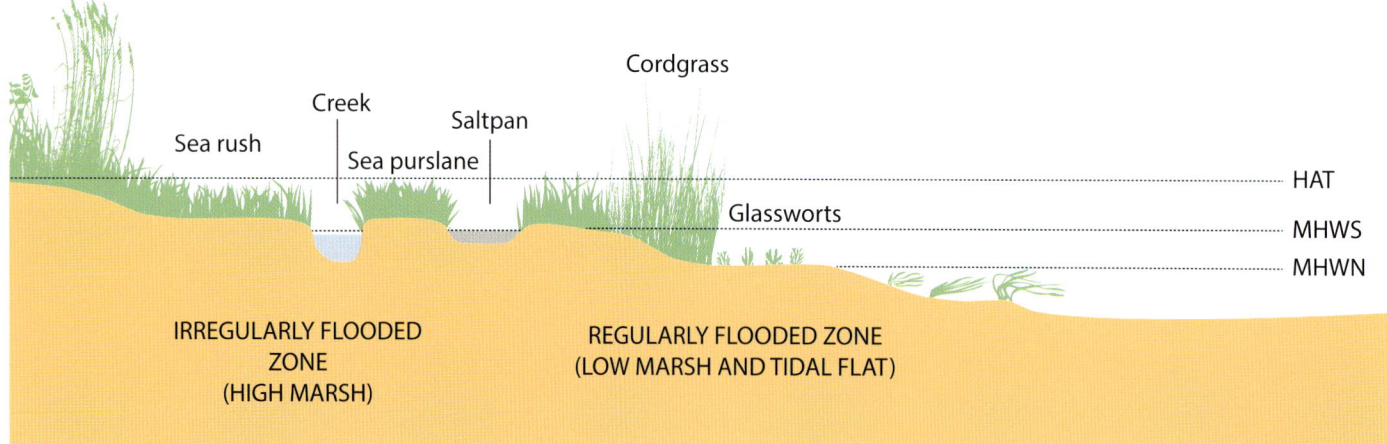

Fig. 12.6 SALTMARSH MORPHOLOGY. The cross-section profile show the morphological features of a saltmarsh and indicates the typical zonation of the high marsh and low marsh areas above and below the Mean High Water Spring (MHWS) tide level. The upper elevation limit of saltmarshes is approximately that of the Highest Astronomical Tide (HAT), while the lower limit is rarely below Mean High Water Neap (MHWN) tide level. Lower limits of plant zonation are usually set by environmental tolerances. Plants that thrive in the low marsh include cordgrass and glasswort. The appearance of plants, such as sea rush, in the upper marsh limits is mainly the result of interspecific competition between different plant species. [Source: adapted from Tiner, R.W. and Milton G.R., 2016. Estuarine Marsh: An overview. In: Finlayson, C.M., Milton, G.R., Pretince, C. and Davidson, N.C., eds. *The Wetland Book: II: Distribution, description and conservation.* Haarlem: Springer Netherlands]

according to their morphology and nature of the substrate.[7] They identified five basic types of saltmarsh: estuary, bay, sandflat, lagoon and fringe (Figures 12.7 and 12.8). The estuary type occurs at the mouths of medium to large rivers and is well represented in counties Dublin, Cork, Limerick and Clare. Bay-type saltmarshes form in sheltered bays where freshwater input is minimal and occur mainly on western coasts, principally around the coasts of Donegal, Clew Bay and Galway Bay. Both estuary and bay types form on silt and clay substrates. Sandflat saltmarshes typically form in association with dune systems and can develop as extensive seaward extensions of machair in the west of Ireland (Figure 12.7). The lagoon type, the rarest of the saltmarsh types, forms behind shingle or sand barriers and more rarely on peat.

The fringe (peat fringe) type overlies peat substrates and is found fringing sheltered rocky bays, or it develops as a narrow band between the sea and bog, or heath-dominated hinterland. These saltmarshes often link to more extensive areas of adjacent freshwater wetlands, characterised by reedswamp vegetation (Figures 12.9a and b). This peat fringe type is especially represented in counties Mayo, Galway and Kerry. These habitats have formed in a different manner to the other saltmarsh types, primarily in response to the abundant rainfall of western Ireland and are associated with high fresh groundwater tables. In the mid- to late postglacial period, an increasingly wet climate resulted in the

Fig. 12.7 SANDFLAT SALTMARSH AT BARLEYCOVE, COUNTY CORK. The habitat pictured here is an example of a sandflat saltmarsh. The fixed-dune habitat that occurs at Barleycove merges inland with a substantial area of saltmarsh and linked wetland areas, which supports both Atlantic and Mediterranean salt meadows. A fringe of glasswort also occurs at the lowermost part of this saltmarsh. [Source: John Devaney]

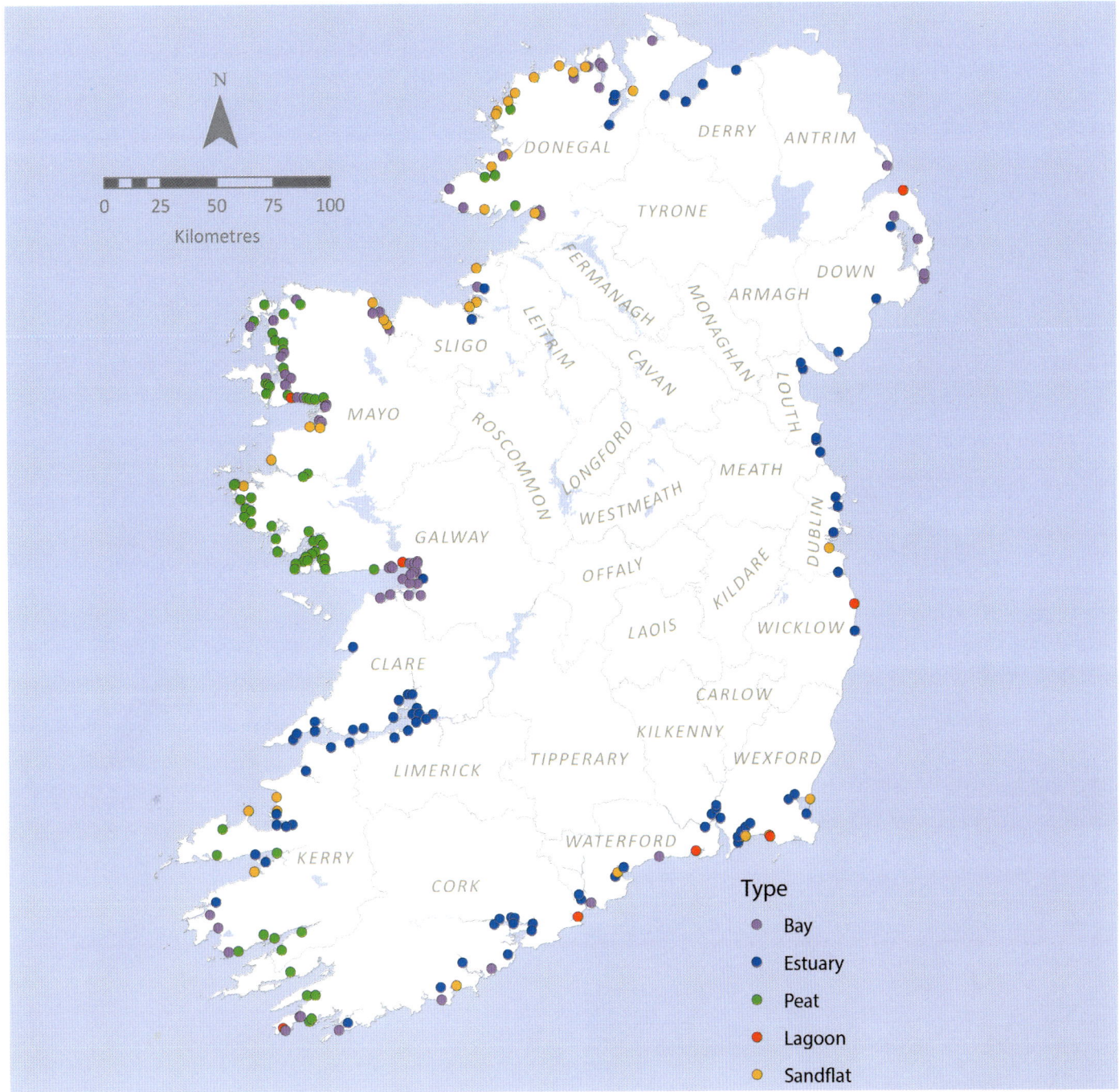

Figure 12.8 SPATIAL DISTRIBUTION OF SALTMARSH TYPES IN IRELAND. A complete inventory of saltmarshes in the Republic of Ireland was undertaken in 1998, in which these wetlands were classified according to their morphology and the nature of the marsh substrate (surface sediments). Five basic types of saltmarsh were identified: estuary, bay, sandflat, lagoon and peat. The map shows the distribution of these types around the coast of the Republic. Bay, estuary and sand-flat types are fairly well distributed around the coast. However, peat (peat fringe) types are only found on the west coast. Lagoon types are the lowest in frequency, but can be found on the east, south and west coasts. Subsequently, the saltmarsh monitoring project, undertaken on behalf of the National Parks and Wildlife Survey in 2008–2009, was based on new vegetation surveys and an assessment of threats and management practices. Similar data on the saltmarshes of Northern Ireland are available from the Department of Agriculture, Environment and Rural Affairs. [Source: adapted from Curtis, T.G.F. and Sheehy Skeffington, M., 1998. The Salt Marshes of Ireland: An inventory and account of their geographical variation. *Biology and Environment* 98B, pp. 87–104]

growth of large coastal boglands, with the development of freshwater peats. Sea-level rise *c*.4,000 years ago, together with phases of marine flooding, allowed the later colonisation of the seaward margins of these boglands by salt adapted (halophytic) vegetation. This resulted in the extensive development of this marsh type and of linked wetlands along Ireland's North Atlantic seaboard (see Chapters 7: Ancient Shorelines and Sea-level Changes and 16: The Inhabitants of Ireland's Early Coastal Landscapes).[8] These peat fringe type saltmarshes are probably unique to Ireland and to parts of western Britain (that is, Irish Sea coasts and western Scotland). All five saltmarsh types overlap to some degree.

Value of saltmarshes

Saltmarshes provide a number of valuable ecosystem services.[9] They function as essential habitats for overwintering and migratory birds, which use saltmarshes for foraging, nesting and roosting. They provide spawning sites and nursery grounds for a range of fish species. Saltmarshes provide a vital role in protecting the lands behind them from coastal flooding. They act as natural land buffers, dissipating wave

Fig. 12.9 (left) a. TIDAL PEAT SALTMARSH NEAR CARNA, COUNTY GALWAY. Salt-marshes develop typically on low-energy coasts under the influence of the tide. However, saltmarshes on peat substrates are formed in an inherently different way, unique to the saltmarshes of the west of Ireland and western Britain. The peat saltmarsh, pictured here in Galway Bay, is the remnant of a blanket bog which has ended up below sea level. The peat here formed in postglacial times when climatic conditions and former sea levels were conducive to the creation of boglands along the western Atlantic coast of Ireland. A subsequent phase of marine inundation of the coasts, associated with continued rises of sea level, created a shift from freshwater to saline conditions, hence these bogs now support saltmarsh vegetation. [Source: Jonathan Wilkins]; b. Freshwater peats of former coastal wetlands emerging from beneath the eroding sand dunes on the foreshore at Rossbehy, Dingle Bay, County Kerry. This peat outcrop is formed of fresh- and brackish-water peat sediments, which were laid down during the mid-Holocene, at least 5,000 years ago. At that time, sea level was lower than it is today. The shoreline would have been further seaward and this site would have been a terrestrial (as opposed to marine) environment. In the wider Castlemaine Harbour area (here and to the east of this site), rising post-glacial sea level facilitated the extensive development of reedswamp, saltmarsh and bog environments, particularly after c.4,000 years before present. These were first developed as extensive coastal wetlands, now preserved as peats, which formed behind seaward fringing sand dune barriers. Sedimentary evidence suggests that these sand dune barriers may have once extended uninterrupted across Dingle Bay, until they were breached around 3,000 years ago. This breaching is thought, by some, to be responsible for separating the modern-day Inch Spit and Rossbehy Spit from one another. The breaching event allowed marine invasion of the inner harbour areas and the progressive replacement of the freshwater vegetation by more brackish-water types of wetland and later by marine sand. As the sea level continued to rise, shorelines migrated landward and terrestrial environments were replaced by coastal sand barriers. This pattern of coastal behaviour is called barrier rollover and can be seen in the sedimentary record where organic peats are buried by marine sands. Sometimes, as at Rossbehy, modern-day beach erosion exposes these buried peats. Sedimentary evidence of barrier rollover is common on the Atlantic coasts of Ireland (for example, at Ballinskelligs Bay, Ballyferriter, Galway Bay, Clew Bay and Donegal Bay). Today's coastal peat wetlands are under threat from twenty-first-century sea-level rise. [Source: Robert Devoy]

energy and absorbing storm surges. These habitats also function to purify water by acting as filters and accumulating a wide variety of pollutants. Most importantly, saltmarshes store large amounts of carbon and play a vital role in the global carbon cycle.

Saltmarsh ecology

A distinct feature of saltmarsh habitats is the zonation of plant species (see Figure 12.6). Lower limits of plant zonation are usually set by environmental tolerances, while upper limits are mainly the result of competition between different plant species. At the lower, seaward side, saltmarshes are colonised by species that can tolerate high levels of salinity and waterlogged conditions, including cordgrasses (*Spartina* sp.) and glassworts (*Salicornia* sp.). The mid-shore is colonised by plants such as sea aster (*Tripolium pannonicum*), sea plantain (*Plantago maritima*) and sea arrowgrass (*Triglochin maritima*) (Figures 12.6 and 12.10). This transitions to the upper shore on the landward side, where species are less frequently inundated by the tide. The plant community here includes rushes, such as saltmarsh rush (*Juncus gerardii*) and grasses such as red fescue (*Festuca rubra*). These zonations may be influenced by regional environmental characteristics, such as those of climate, hydrology, geology and marsh surface shapes, which can result in vegetation mosaics developing rather than simple zonations.

Fig. 12.10 9 (above) SEA ASTER GROWING ON SHERKIN ISLAND, WEST CORK. This pretty, native plant grows in coastal saltmarshes, estuaries and on coastal cliffs. It displays purple-blue petals around a yellow floret when it blooms from July to October. Its distribution on salt marshes complies with levels of exposure to saltwater, a factor that differentiates plants to be found on the upper, mid- and lower shore within a saltmarsh. [Source: Robbie Murphy]

The vegetation composition of saltmarshes in Ireland differs significantly between the various marsh types, a characteristic which is also unique to Ireland. Both the diversity and species richness are highest on peat saltmarshes. A striking feature of these saltmarshes is the absence of otherwise common saltmarsh species, such as sea purslane and cordgrasses. The distribution of sea purslane in Ireland may be related to the intensity of animal grazing, as the plant is known to be sensitive to this activity. It is more likely, however, that the absence of

Fig. 12.11 HUMAN IMPACTS ON WETLANDS IN THE AVOCA ESTUARY, ARKLOW, COUNTY WICKLOW. Land reclamation of the wetlands in the Avoca Estuary has been considerable over the past couple of hundred years. The changing shoreline can be seen in the image, where the blue line shows former nineteenth-century channels and the yellow line shows the former location of the high water mark. These features were digitised from historic six-inch Ordnance Survey maps from c.1824–1841. For thousands of years prior to that, the sandflats at the rivermouth shifted course naturally. Erosion and accretion affect part of the processes of saltmarsh and sandflat development (ontogeny). When these are the result of natural processes, a coastal wetland will adapt, or reach equilibrium, in response to climatic and local changes. The historic position of the channels show the nature of the mouth of the Avoca river in the early nineteenth century. Human settlement in the town of Arklow was concentrated along its southern shores. In the early 1800s engineering works commenced to improve Arklow Port. This constrained the natural dynamics of the river, causing sandbanks to accumulate, predominantly deposited to the north of the town, which led to the creation of an extensive saltmarsh ecosystem. However, this was eventually infilled in 1872, as industrial development took hold. As shown in the Google Earth image, the earlier situation is altered dramatically today, as a consequence of the infill and reclamation of the wetlands to the north of the river. Port development at the mouth of the river to the south can also be seen. Arklow town had long been a fishing port, but the port improvement scheme in the nineteenth century facilitated the shipment of massive amounts of copper, as well as iron and silver, and to a lesser extent gold, from the Avoca mines. A considerable amount of trade between Arklow and Dublin went by sea due to a poor road network. [Source: data from Google, Maxar Technologies, 2018 and the OSI six-inch maps, 1824–1841]

sea purslane on peat saltmarshes is due to the plant's intolerance of the high moisture content of the peat-based habitats. Peat-based saltmarsh surfaces develop commonly as uniform in gradient (that is, they are flat) and occur at slightly higher elevations than other saltmarsh types, also making them less suitable for cordgrass species.

Threats to saltmarshes and their conservation

Wetland degradation and loss has substantial and lasting effects, including the loss of the ecosystem services provided by saltmarshes described previously. The conservation status of saltmarshes in Ireland is based on Annex I of the European Habitats Directive (92/43/EEC). Under this directive some 87 per cent of Ireland's saltmarshes have been designated as candidate Special Areas of Conservation (cSAC) (see Chapters 3: Marine Biology and Ecology and 32: Coastal Management and Planning). Within these cSACs, much of the land is in private ownership. As a result, in spite of this designation, damage by land infilling, reclamation and overgrazing occurs. The National Parks and Wildlife Service (NPWS) of Ireland has identified grazing as the biggest threat to Irish saltmarshes. Overgrazing by cattle and sheep together has affected about 20 per cent of saltmarsh surface areas. Infilling and land reclamation also poses a threat to saltmarsh habitats, particularly of the estuary type (Figure 12.11).

Saltmarshes and Global Climate Change: Blue carbon

Grace Cott

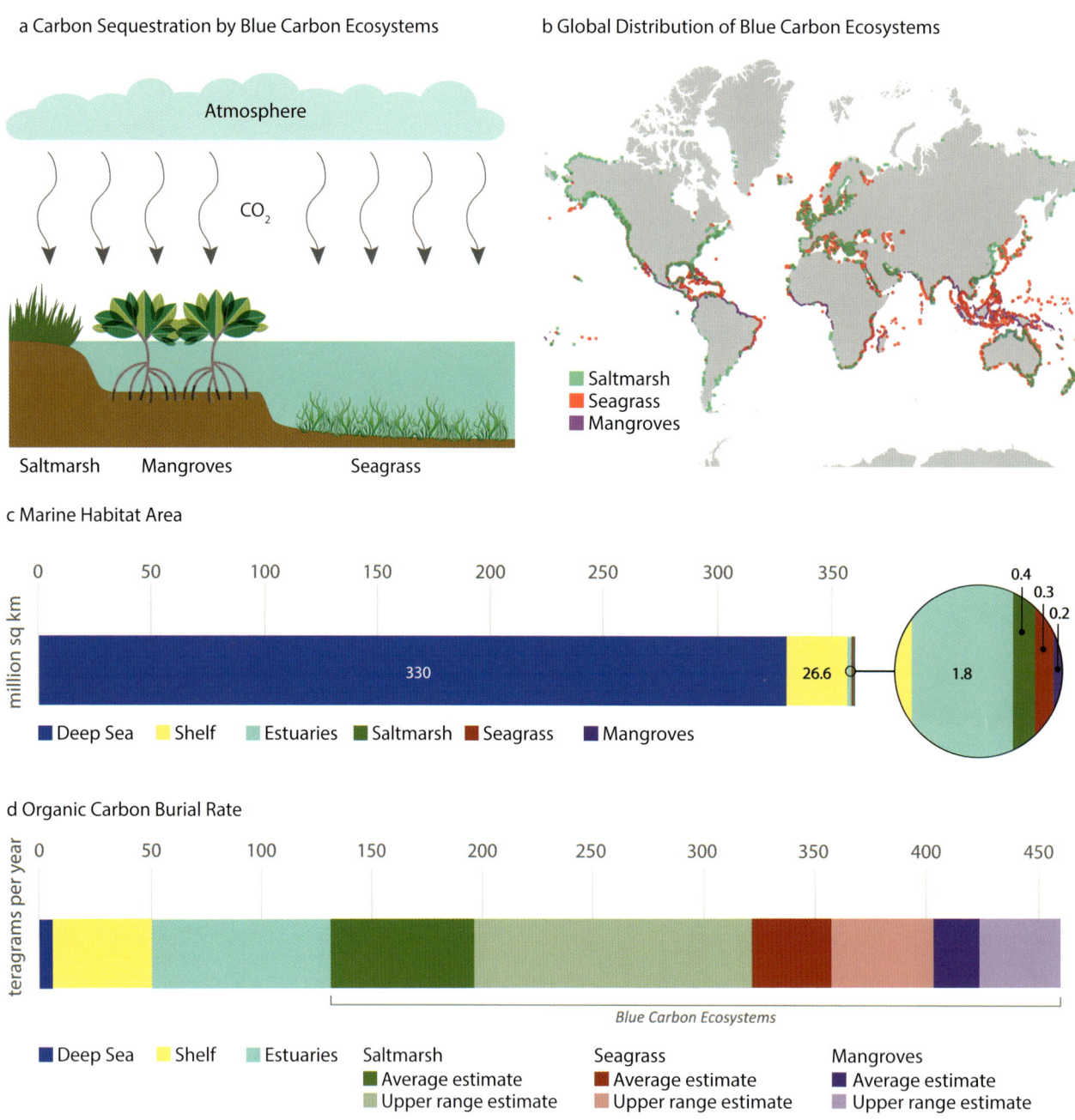

Fig. 12.12 a. CARBON SEQUESTRATION BY BLUE CARBON ECOSYSTEMS. The carbon sequestered in vegetated coastal habitats, specifically saltmarshes, seagrass beds and mangrove forests, is termed 'blue carbon'. Mangroves, saltmarshes and seagrass beds are all vital ecosystems that are under threat from human impacts. [Data source: adapted from PEMSE, http://pemsea.org/publications/brochures-and-infographics/infographics/infographic-coastal-blue-carbon]; b. Global distribution of blue carbon ecosystems. In Ireland, the predominant coastal habitat type is the saltmarsh, although seagrass beds occur in approximately twenty bays and estuaries around the coast. By comparison, mangroves are tropical ecosystems, which, of course, do not occur in Ireland. The degradation of these ecosystems negatively impacts on our ability to deal with climate change, as well as exacerbating coastal vulnerability to storminess and sea-level rise. [Data sources: Adapted from: UNEP World Conservation Monitoring Centre]; c. Marine habitat area. Although their global footprint relative to other marine habitats is relatively small, the efficiency and effectiveness of mangroves, seagrass beds and saltmarshes in sequestering carbon, means they punch above their weight in terms of their impact in this role; d. Carbon sequestration by blue carbon ecosystems. The combination of rapid plant growth, water-logged conditions and a distinctive efficiency for trapping carbon-rich sediments enables these ecosystems to store vast quantities of carbon for millennia. Annual organic carbon burial rates are shown. Total burial rates of organic carbon in estuarine and shelf sediments and deep-sea sediments are provided for comparison. [Data sources: c. and d. adapted from: Nellemann, C., Corcoran, E., Duarte, C.M., Valdés, L., De Young, C., Fonseca, L., Grimsditch, G., eds, 2009. *Blue Carbon. A Rapid Response Assessment*. United Nations Environment Programme, GRID-Arendal, www.grida.no]

Saltmarshes have an important role in climate mitigation and adaptation, as they are highly efficient carbon sinks (see Chapter 33: Climate Change and Coastal Futures). The carbon sequestered in vegetated coastal habitats, specifically saltmarshes, seagrass beds and mangrove forests, has been termed 'blue carbon' (Figure 12.12a).[10] The combination of rapid plant growth, waterlogged conditions and a distinctive efficiency in trapping carbon-rich sediments, enables these ecosystems to store carbon for millennia (Figure 12.12d). Their contribution per unit area to long-term carbon storage can be up to ten times greater than that of terrestrial forests. However, when these ecosystems are degraded, damaged or destroyed, this carbon can be released back to the atmosphere and ocean within a matter of years.[11]

Saltmarsh ecosystems can also contribute to the adaptation of coastal areas to sea-level rise, as densely vegetated saltmarshes dissipate the impact of wave energy. Ireland has approximately 100km^2 of saltmarsh habitat. If these areas are appropriately managed and protected they can adapt naturally to moderate future rises in sea level and hence ensure that their benefits, including that of carbon sequestration, are maintained into the future..

Spartina IN IRELAND

Grace Cott

Common cordgrass (*Spartina anglica*) is a perennial, rhizomatous grass that grows mostly on the mud and sandy substrates of intertidal habitats (Figure 12.13). This species has an unusual origin, in that it is a hybrid of *Spartina alterniflora* – a species native to the east coast of America believed to be introduced into Southampton, UK, in 1829 – and *Spartina maritima*, a long-recognised native species of Europe.[12] The natural hybridisation gave rise to the development of the common cordgrass, which became known for its vigour, its ability to withstand tidal emergence and its rapid accretion rates of sediments within its stands. These characteristics were reasons for it to be widely planted in estuaries, to stabilise sediments adjacent to shipping channels, reduce coastal erosion, aid in the reclamation of intertidal land and, on occasion, for use as animal fodder.

Fig. 12.13 *Spartina* ON BULL ISLAND, COUNTY DUBLIN. Common cordgrass propagates on mudflats, sandflats and saltmarshes and is listed on the national invasive species list. Monitoring of common cordgrass (also frequently referred to as *Spartina*) has been an ongoing management issue at Bull Island since it originally established itself on the mudflats north of the causeway in the late 1960s. *Spartina* is now distributed along most of this saltmarsh, as well as almost everywhere else on Irish coasts. This is in spite of attempts by the local authority to control extensive swarths of spartina on the lowest part of the saltmarsh, between 1975 and 1990. Its encroachment into mudflats can limit the extent of the feeding area available to wintering waders and wildfowl. It is not considered to have any adverse impacts on avian fauna locally, nor is it a significant threat to the Atlantic salt meadow habitat, though it may have an impact on the glasswort flats. In order to better understand the need for management interventions, improved monitoring to assess the potential spread of common cordgrass is required into the future. [Source: Grace Cott]

The common cordgrass, commonly referred to as simply spartina, is now a major low-marsh species in Northern Europe. It was first introduced to Ireland in Little Island, Cork Harbour, in 1925, with subsequent plantings in Baldoyle Estuary, County Dublin; Fergus Estuary, County Meath, and Belfast Lough, County Antrim.[13] Spartina colonises soft substrates, primarily muds and sands, and can form extensive communities in the low marsh. It is known as an 'ecosystem engineer' that can essentially build its own habitat by slowing down waves and currents and enhancing sediment deposition, in some coastal locations in Ireland by several centimetres of sediment per year.

Spartina can tolerate longer periods of inundation by the tide and lower levels of oxygen in the substrate than other perennial saltmarsh species, giving it an advantage in stressful wetland conditions. The grass can spread both vegetatively and by seed, easily dispersed by waves and tides. A heavy seed setting, when conditions are suitable, allows this species to colonise bare mud in channels and pans (shallow depressions) in the mid- and upper shores. *Spartina* is now prevalent in the east and south-east of Ireland, with some marshes colonised in County Kerry and in the Shannon Estuary. It has not been established in the saltmarshes of County Mayo and County Galway, which tend to have primarily a peat substrate and are higher in the tidal frame.[14] In general, the plant has been considered of low intrinsic value as it creates monospecific stands and lowers plant diversity. However, monospecific stands are an inherent characteristic of many coastal wetlands around the world. In the case of North Bull Island, County Dublin (Figure 12.13), there have been concerns that stands of spartina were lowering macrofauna diversity. However, research in Irish wetlands indicates that clumps of Spartina can provide a habitat for macro-invertebrate infauna as abundant, and species-rich, as areas vegetated by adjacent glassworts (*Salicornia* sp.) (Figure 12.14).[15]

In the context of climate change and rising sea levels, spartina is an extremely productive plant that has the potential to sequester large amounts of carbon from the atmosphere along with an ability to trap sediments, leading to the increased build-up of soil carbon. Stands of spartina have the potential to absorb wave energy, protecting the coast from storm surge. Because this species is an ecosystem engineer it can build habitat in response to rising sea levels, further protecting the coast. A challenge for managers is to balance conservation goals, taking into consideration the marine biodiversity and habitat functions in coastal systems, with climate action. The management of spartina

Fig. 12.14 GLASSWORT (*Salicornia*) MUDFLATS AT RATHMELTON, COUNTY DONEGAL. In the sixteenth century the term 'glasswort' was coined in England to describe plants that could be used for making soda-based glass. High in sodium carbonate content, one strain of the *Salicornia* plant, known as marsh samphire, was harvested and burned for this purpose. Its ash was infused with sand in glass production. Most of our saltmarshes contain less than 0.01ha of glasswort habitat. As it is found on the seaward side of saltmarshes, there tends to be less pressure for drainage, development and grazing. However, common cordgrass has the capacity to act as an invasive species by spreading into the *Salicornia* habitat and reducing its coverage. An annual plant, *Salicornia* produces minute flowers in August and September. Marsh samphire (also known by some as poor man's asparagus) has become popular in recent years for its nutritional and culinary value. It is also considered to have potential as a plant biofilter to degrade pollutants arising from marine aquaculture. [Source: Kristi Leyden, Courtesy of National Parks and Wildlife Service] The inset image shows a close-up of glasswort taken in June 2007 between Rosscarbery and Warren Beach, County Cork. [Source: Jenny Seawright, www.irishwildflowers.ie]

N

■ Coastal Wetlands
░ Inland Wetlands

Trawbreaga Bay
Lough Foyle
Garron Plateau
Larne Lough
Belfast Lough
Outer Ards
Strangford Lough
Killough Bay
Carlingford Lough
Dundalk Bay

Cummeen Strand
Killala Bay/Moy Estuary
Blacksod Bay and Broadhaven
Owenduff catchment

Rogerstown Estuary
Broadmeadow Estuary
North Bull Island
Sandymount Strand/Tolka Estuary

Inner Galway Bay

The Raven
Bannow Bay
Tramore Backstrand
Tralee Bay
Dungarvan Harbour
Castlemaine Harbour
Blackwater Estuary
Ballymacoda
Ballycotton
Cork Harbour

0   200
Kilometres

Fig. 12.15 DISTRIBUTION OF RAMSAR WETLAND HABITATS IN IRELAND. A total of forty-five sites have been designated as international Ramsar sites in Ireland to date. The objective is the conservation of wetland habitats, especially for waterfowl. Over half of the current Ramsar locations are coastal wetlands of significant value for nature. These sites have been selected on the basis of their distinct ecological, botanical, zoological, limnological and/or hydrological criteria. Irish Ramsar designations have no intrinsic legal protection. As a result, these sites are also afforded protection under the transposition of European Directives for conservation. [Source: ROI, wetlands.org; NI, DOENI]

plant stands in wider marsh and intertidal settings should be undertaken with great care, as exemplified in the case of the problems encountered with farmed oyster rafts in Strangford Lough. The use of herbicides and attempts to eradicate spartina here led to major releases of trapped sediments and the burying of the oysters.

PROTECTING COASTAL WETLANDS

Understanding the physical, chemical and biological interactions in coastal wetlands is essential to their conservation and effective management. Despite this, there has been little research on contemporary coastal wetlands and management has often been driven by a perceived need to preserve a particular aspect of their use or biodiversity. For example, coastal wetlands, particularly the intertidal zones of estuaries, and lagoons are important feeding grounds for wading birds (see Waterbirds in Irish Coastal Areas, in Chapter 3: Marine Biology and Ecoology) and some coastal lagoons are deliberately managed to provide shallow margins to encourage these birds. Such management may not take into consideration the other plant and animal groups that are unique to saline lagoons and which may be adversely affected by manipulation of water levels.

Wetlands are dynamic ecosystems that would normally adapt slowly to changing environmental conditions, such as precipitation and sea level. Many wetlands undergo a natural succession where some areas ultimately become terrestrial systems, while other new wetland areas can also be created. Coastal wetlands are influenced by the transport of sediment down rivers and by the erosion and redeposition of sediment with tidal influxes. Further, storms that lead to increased river flows, higher tidal surges and, together, the related occurrences of coastal marine flooding, can change the rate and enhance the impacts of these processes. Anticipated increases in storminess for Ireland, associated with climate change, are likely to result in the impacts of these factors increasing in the future. Some coastal wetlands, especially low-lying saltmarshes and mudflats, may be lost, or their plant and animal communities altered due to changes in salinity regimes. Evidence exists already for changes in the distribution of estuarine birds in the UK and of the associated invertebrates that are the main food source for the estuarine birds, as a result of sea-level rise along the coasts altering estuarine morphology.[16] These areas may also be at direct additional risk from sea-level rise itself, with increasing future marine inundation, combined with other human-related pressures (for example, buildings), resulting in the loss of 70 per cent of coastal wetlands globally by the 2080s and of the order of 30 per cent of those in Ireland (see Chapters 7: Ancient Shorelines and Sea-level Changes and 33: Climate Change and Coastal Futures).[17]

In the late 1960s, recognition of the importance of wetlands and the need to protect and restore them led to the signing of the intergovernmental Ramsar Convention on Wetlands in 1971, which came into force in 1975. The purpose of the Convention was to draw attention to the rate of degradation of wetlands. Signatories to the Convention demonstrate their willingness to support the reversal of wetland loss and degradation; by 2018, 170 countries had signed. Ireland signed the Convention in 1985 and there are currently forty-five Ramsar sites listed for Ireland (Figure 12.15). Despite this and other international efforts, such as the Convention on Biological Diversity, the trajectory of global wetland sustainability is one of persistent decline (Figure 12.16).

At the European level, there are two main directives that include provisions for the protection of wetlands: the Habitats Directive (CEC, 1992) and the Water Framework Directive (2000/60/EC). Under Annex I of the Habitats Directive, special habitats such as lagoons are designated as priority habitats and protected under the Natura 2000 network (see Chapter 32: Coastal Management and Planning). The Water Framework Directive requires all inland and coastal waters to reach good ecological status through the implementation of river basin management plans. The complexity of the many types of wetlands and their potential involvement in several different types of legislation led the European Commission to produce a guidance document to assist countries with the roles of the different directives that relate to wetlands.[18]

Instruments for the defence and protection of wetlands at the national level include planning and wildlife legislation. However, despite legislative provisions, we are still draining and destroying coastal wetlands at an alarming rate. Current policy approaches in climate-action planning aim to establish a baseline of wetland conditions with the stated objective of informing best practice for wetland management. However, the emphasis appears to be on peatlands currently exploited for peat extraction, with no vision for blue carbon or coastal wetlands specifically. Further, policies in the overarching management of coastal ecology need to return to those concerned with conservation principles and linked to allowing adaptation, rather than protection.[19]

FUTURE OUTLOOK FOR COASTAL WETLANDS IN IRELAND

It has been estimated that the biodiversity of all Irish wetlands (inland and coastal) brings €385 million each year to the country's economy generally, while wetlands also make up a component of the €330 million contributed by Irish habitats overall to the nature and tourism sectors.[20] Despite this, coastal wetlands suffer from a lack of appreciation of their value as ecosystems. Often perceived as of

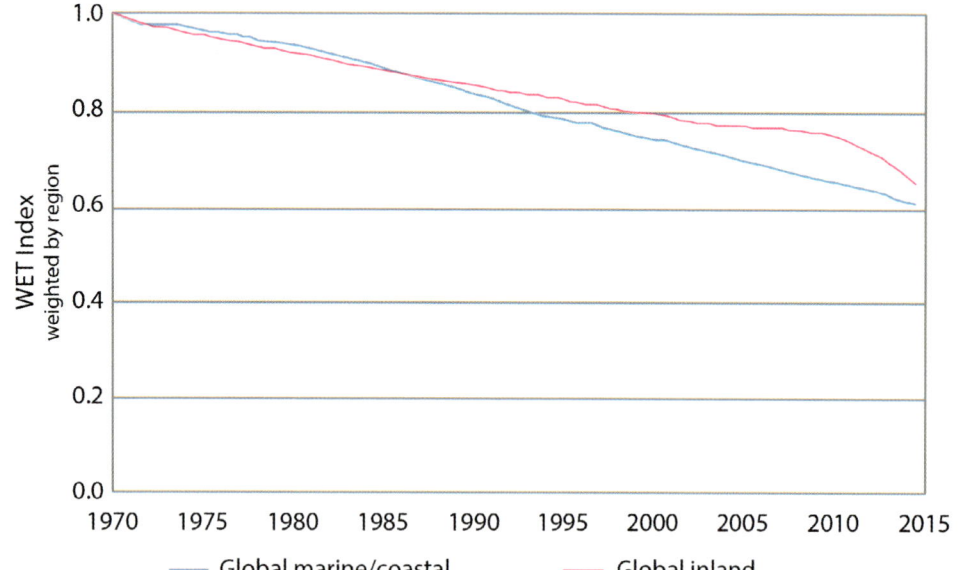

Fig. 12.16 WORLDWIDE TRENDS IN DECLINING WETLAND AREAS. In spite of their importance, marine and coastal wetlands worldwide declined by approximately 35 per cent between 1970 and 2016. The blue line shows the time trend in coastal and marine habitats from 1970, against inland wetlands shown in pink, as measured by the Wetland Extent Trends (WET) index. The index is an international tool for monitoring natural and artificially created habitats. As wetlands exist on the boundary of land and ocean, they are sensitive to climate change impacts (such as sea-level rise, temperature change and disease) and among the first indicators to respond to these environmental pressures. The data show well that countries are not on track in tackling internationally accepted global targets to reverse wetland habitat loss. [Source: adapted from Dixon, M.J.R., Loh, J., Davidson, N.C., Beltrame, C., Freeman, R. and Walpole, M., 2016. Tracking Global Change in Ecosystem Area: The Wetland Extent Trends index. *Biological Conservation*, 193, pp. 27–35 and UN World Conservation Monitoring Centre, 2017. Wetland Extent Trends [WET] Index. Cambridge, UK]

Fig. 12.17 WETLANDS ACTIVITY CENTRE, TRALEE BAY, COUNTY KERRY. Protected by a rocky promontory and sand spits, Tralee Bay hosts a complex of sand dunes and extensive freshwater marsh, saltmarsh and mudflat habitats. The environment provides a rich resource for Brent geese and other overwintering birds, such as sanderling, godwit, dunlin, plover and lapwing. As a result, it is designated as a Ramsar site, a Special Protection Area and a wildfowl sanctuary. The inner bay encompasses the estuary of the River Lee which flows through Tralee town. The wetlands associated with the estuary host an important expanse of seagrass. The Tralee Bay Wetlands and Activity Centre is a nine-hectare (twenty-two acre) site, which has been designed as a microcosm of the range of coastal wetland habitats in Tralee Bay. This ecotourism initiative provides visitors with an interpretation of the value of the protected area, including a 20m viewing tower. This type of interpretation is important in improving understanding of the importance of wetland habitats. [Source: Tralee Bay Wetlands Activity Centre]

little value, estuarine marshes and shallow lagoons have traditionally been drained to facilitate land reclamation for agriculture (see Chapters 4: People, Agriculture and the Coast and 30: Engineering for Vulnerable Coastlines). Placing an economic value on the benefits of wetland ecosystems can strengthen the case for conservation and even restoration, and makes the need for such actions more apparent to managers and the public. In the Irish context, coastal wetlands are often very small and fragmented, making it difficult to evaluate their benefit other than to small groups of interested users, such as wildlife conservation groups, who do not have the financial resources to invest in their management and remediation. One such group is BirdWatch Ireland, which produced an action plan in 2011 to protect shore and lagoon birds.[21]

In their report on the status of protected habitats in 2013, the National Parks and Wildlife Service assessed the overall status of Irish lagoons and machair as 'bad', with no appreciable change since the assessment of 2007; estuaries and mudflats as 'favourable' and improving, whereas the status of saltmarshes/meadows is 'inadequate', with no substantial changes.[22] It is clear that we have had very little appreciation of the value of our wetlands in Ireland, even though they are a future carbon sink and provide a habitat that can be enriched. Conservation, interpretation and restoration projects are necessary to support the important habitats provided by coastal wetlands (Figure 12.17). However, such projects require an understanding of the functioning of these ecosystems, together with financial and technical resources and stakeholder and community participation. Restoration projects can be extremely successful. There are many examples of good practice cited under the Ramsar initiative and they can lead to increased social and economic benefits for local communities. Greater awareness of the imperative for action in response to the biodiversity and climate crisis may concentrate attention on the value of the ecosystem services provided by our coastal wetlands, before it is too late.

ESTUARIES AND LAGOONS

Sorcha Ní Longphuirt and Robert Devoy

WARRENPOINT AND SURROUNDING COASTLINE FROM FLAGSTAFF VIEWPOINT. The view encompasses a tributary valley scene of the waters and estuary zone leading into Carlingford Lough, looking south-west to the inner Lough area of Warrenpoint, the Irish Sea beyond and the glaciated coastal landscapes of the Mourne Mountains. Although the inshore marine environment here is relatively deep water, it is an important area for inshore fisheries (such as oysters and mussels), whilst human impacts and urban developments around the shorelines are prominent in this otherwise rural and farming coastal landscape. [Source: © Tourism Northern Ireland]

Estuaries are formed in coastal areas where freshwater runoff from the land (for example, as groundwater, streams and rivers) meets the salt-rich sea, creating generally brackish-water (semi-saline) environments. In Ireland these transitional waters occupy an area of 844km². Estuaries play an integral role in the hydrological cycle, as well as in the broader links in the land-to-ocean continuum, and are considered to be some of the most biologically productive areas in the world (Figure 13.1). Although subject to tidal forcings and river flows, they are generally protected from the full force of coastal waves and storms due to their indented and often funnel-like shape

Fig. 13.1 MAJOR ESTUARIES OF IRELAND AND ESTUARY CATCHMENTS. It is hardly surprising that the major estuaries around the coast of Ireland correspond with important ports and cities, such as Belfast, Limerick, Cork and Dublin. These estuaries, where the Lagan, the Shannon, the Lee and the Liffey Rivers, respectively, enter the sea, have been ideal locations for human settlement and for port and harbour activities over centuries. Waterford Harbour, Galway Bay, Donegal Bay and Lough Foyle also form key areas in the drainage of each of their adjoining river systems, as well as forming the location for important coastal settlements. The island of Ireland is divided into forty hydrometric areas for the purpose of managing amalgamations of river basins, including estuaries downstream. Multiple agencies are involved in the monitoring of water quality and flood management on both sides of the border, including organisations such as local authorities, environmental authorities and utilities. Estuarine waters are particularly vulnerable to anthropogenic pressures as they tend to be surrounded by industrialised centres of population. Different approaches to estuarine management, such as hydrometric zonation, help to ensure that estuaries are maintained as vital arteries in support of social and ecological functioning.

(estuary morphology). This allows them to form a haven for unique plant and animal species, with the development of diverse habitats which offer important ecosystem services (see Chapters 11: Beaches and Barriers; 12: Coastal Wetlands and 30: Engineering for Vulnerable Coastlines). The sheltered nature of estuaries also makes them ideal living sites for people (towns and cities) and particularly

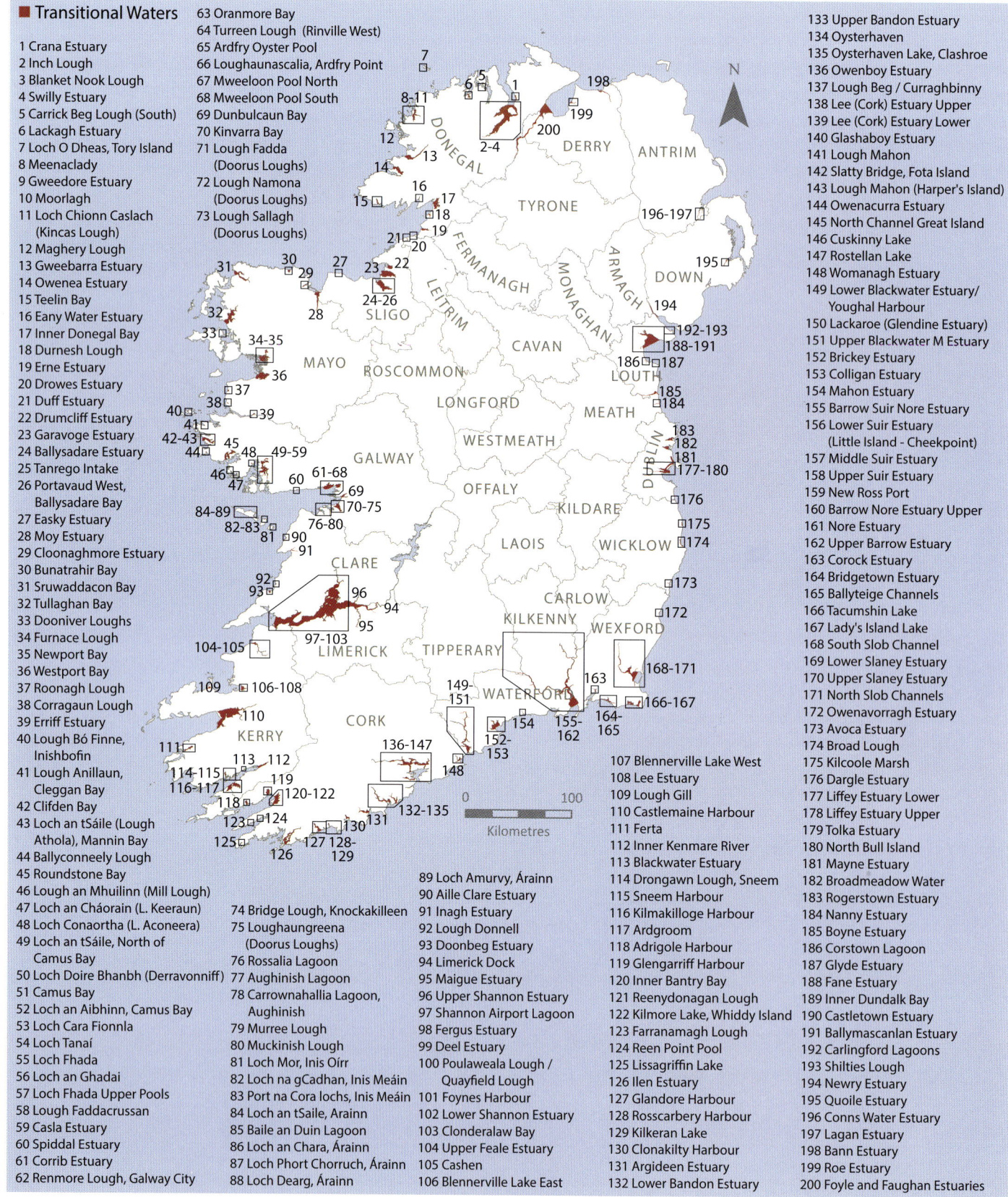

■ **Transitional Waters**

1 Crana Estuary
2 Inch Lough
3 Blanket Nook Lough
4 Swilly Estuary
5 Carrick Beg Lough (South)
6 Lackagh Estuary
7 Loch O Dheas, Tory Island
8 Meenaclady
9 Gweedore Estuary
10 Moorlagh
11 Loch Chionn Caslach (Kincas Lough)
12 Maghery Lough
13 Gweebarra Estuary
14 Owenea Estuary
15 Teelin Bay
16 Eany Water Estuary
17 Inner Donegal Bay
18 Durnesh Lough
19 Erne Estuary
20 Drowes Estuary
21 Duff Estuary
22 Drumcliff Estuary
23 Garavoge Estuary
24 Ballysadare Estuary
25 Tanrego Intake
26 Portavaud West, Ballysadare Bay
27 Easky Estuary
28 Moy Estuary
29 Cloonaghmore Estuary
30 Bunatrahir Bay
31 Sruwaddacon Bay
32 Tullaghan Bay
33 Dooniver Loughs
34 Furnace Lough
35 Newport Bay
36 Westport Bay
37 Roonagh Lough
38 Corragaun Lough
39 Erriff Estuary
40 Lough Bó Finne, Inishbofin
41 Lough Anillaun, Cleggan Bay
42 Clifden Bay
43 Loch an tSáile (Lough Athola), Mannin Bay
44 Ballyconneely Lough
45 Roundstone Bay
46 Lough an Mhuilinn (Mill Lough)
47 Loch an Cháorain (L. Keeraun)
48 Loch Conaortha (L. Aconeera)
49 Loch an tSáile, North of Camus Bay
50 Loch Doire Bhanbh (Derravonniff)
51 Camus Bay
52 Loch an Aibhinn, Camus Bay
53 Loch Cara Fionnla
54 Loch Tanaí
55 Loch Fhada
56 Loch an Ghadai
57 Loch Fhada Upper Pools
58 Lough Faddacrussan
59 Casla Estuary
60 Spiddal Estuary
61 Corrib Estuary
62 Renmore Lough, Galway City

63 Oranmore Bay
64 Turreen Lough (Rinville West)
65 Ardfry Oyster Pool
66 Loughaunascalia, Ardfry Point
67 Mweeloon Pool North
68 Mweeloon Pool South
69 Dunbulcaun Bay
70 Kinvarra Bay
71 Lough Fadda (Doorus Loughs)
72 Lough Namona (Doorus Loughs)
73 Lough Sallagh (Doorus Loughs)

74 Bridge Lough, Knockakilleen
75 Loughaungreena (Doorus Loughs)
76 Rossalia Lagoon
77 Aughinish Lagoon
78 Carrownahallia Lagoon, Aughinish
79 Murree Lough
80 Muckinish Lough
81 Loch Mor, Inis Oírr
82 Loch na gCadhan, Inis Meáin
83 Port na Cora lochs, Inis Meáin
84 Loch an tSaile, Arainn
85 Baile an Duin Lagoon
86 Loch an Chara, Árainn
87 Loch Phort Chorruch, Árainn
88 Loch Dearg, Árainn
89 Loch Amurvy, Árainn
90 Aille Clare Estuary
91 Inagh Estuary
92 Lough Donnell
93 Doonbeg Estuary
94 Limerick Dock
95 Maigue Estuary
96 Upper Shannon Estuary
97 Shannon Airport Lagoon
98 Fergus Estuary
99 Deel Estuary
100 Poulaweala Lough / Quayfield Lough
101 Foynes Harbour
102 Lower Shannon Estuary
103 Clonderalaw Bay
104 Upper Feale Estuary
105 Cashen
106 Blennerville Lake East
107 Blennerville Lake West
108 Lee Estuary
109 Lough Gill
110 Castlemaine Harbour
111 Ferta
112 Inner Kenmare River
113 Blackwater Estuary
114 Drongawn Lough, Sneem
115 Sneem Harbour
116 Kilmakilloge Harbour
117 Ardgroom
118 Adrigole Harbour
119 Glengarriff Harbour
120 Inner Bantry Bay
121 Reenydonagan Lough
122 Kilmore Lake, Whiddy Island
123 Farranamagh Lough
124 Reen Point Pool
125 Lissagriffin Lake
126 Ilen Estuary
127 Glandore Harbour
128 Rosscarbery Harbour
129 Kilkeran Lake
130 Clonakilty Harbour
131 Argideen Estuary
132 Lower Bandon Estuary

133 Upper Bandon Estuary
134 Oysterhaven
135 Oysterhaven Lake, Clashroe
136 Owenboy Estuary
137 Lough Beg / Curraghbinny
138 Lee (Cork) Estuary Upper
139 Lee (Cork) Estuary Lower
140 Glashaboy Estuary
141 Lough Mahon
142 Slatty Bridge, Fota Island
143 Lough Mahon (Harper's Island)
144 Owenacurra Estuary
145 North Channel Great Island
146 Cuskinny Lake
147 Rostellan Lake
148 Womanagh Estuary
149 Lower Blackwater Estuary/ Youghal Harbour
150 Lackaroe (Glendine Estuary)
151 Upper Blackwater M Estuary
152 Brickey Estuary
153 Colligan Estuary
154 Mahon Estuary
155 Barrow Suir Nore Estuary
156 Lower Suir Estuary (Little Island - Cheekpoint)
157 Middle Suir Estuary
158 Upper Suir Estuary
159 New Ross Port
160 Barrow Nore Estuary Upper
161 Nore Estuary
162 Upper Barrow Estuary
163 Corock Estuary
164 Bridgetown Estuary
165 Ballyteige Channels
166 Tacumshin Lake
167 Lady's Island Lake
168 South Slob Channel
169 Lower Slaney Estuary
170 Upper Slaney Estuary
171 North Slob Channels
172 Owenavorragh Estuary
173 Avoca Estuary
174 Broad Lough
175 Kilcoole Marsh
176 Dargle Estuary
177 Liffey Estuary Lower
178 Liffey Estuary Upper
179 Tolka Estuary
180 North Bull Island
181 Mayne Estuary
182 Broadmeadow Water
183 Rogerstown Estuary
184 Nanny Estuary
185 Boyne Estuary
186 Corstown Lagoon
187 Glyde Estuary
188 Fane Estuary
189 Inner Dundalk Bay
190 Castletown Estuary
191 Ballymascanlan Estuary
192 Carlingford Lagoons
193 Shilties Lough
194 Newry Estuary
195 Quoile Estuary
196 Conns Water Estuary
197 Lagan Estuary
198 Bann Estuary
199 Roe Estuary
200 Foyle and Faughan Estuaries

Fig. 13.2 IRELAND'S TRANSITIONAL COASTAL WATERS. Under the European Union's Water Framework Directive, estuarine waters are known as 'transitional waters'. Estuaries in Ireland can be split into two broad types, forming either in drowned river and glaciated valleys, or as sedimentary, bar-built systems and lagoons. Ireland's transitional coastal waters include both the estuary and lagoon type of estuarine environments. As seen in the map, 200 different transitional water bodies have been defined around the entire coastline. This features all of the major estuaries of Ireland, as well as the abundance of smaller but locally significant estuaries, from the Womanagh Estuary in East Cork, to Gweedore Estuary in County Donegal. [Data source: Environmental Protection Agency, 2017; Department of Agriculture, Environment and Rural Affairs, 2016]

Fig. 13.3 HUMAN IMPACTS ON ESTUARIES. Arguably, estuaries support the most diverse range of human activities to be found along the coast of Ireland. The image illustrates just some of these activities, all of which alter the natural functioning of the estuarine system to a greater or lesser extent: a. fish farms in Mulroy Bay, County Donegal [Source: Geraldine Hennigan and Norman Kean]; b. farming activity near Kildysart, County Clare [Source: Geraldine Hennigan and Norman Kean]; c. Shannon Airport on the Shannon Estuary, County Clare – the artificial lagoon in the background, created following the construction of the airport, has subsequently attracted a number of wetland species to it [Source: Irish Air Corps]; d. the wide and shallow Slaney Estuary, County Wexford, acquires inputs from towns and villages such as Baltinglass, Tullow and Bunclody before entering the sea at Wexford Harbour [Source: Wexford County Council]; e. the highly urbanised Liffey Estuary, County Dublin, featuring port infrastructure and shipping traffic. [Source: Irish Air Corps]

for the location of harbours and ports. Most coastal cities are situated on or around estuaries, forming vital coastal waters for shipping and transportation, as well as for recreation. As a consequence, they are some of the most heavily used and threatened ecosystems in Ireland (Figures 13.2 and 13.3) (see Chapters 24: Ports and Shipping and 25: Urbanisation of Ireland's Coasts).[1]

The formation of Ireland's coastal areas, and hence its estuaries, is determined by three important factors: those of environmental inheritance (such as, regional geology, postglacial sea-level rise),

hydrodynamics (including exposure to wave energy, tidal range, river discharge) and chemical exchanges (for example, salinity and nutrient changes). Estuaries in Ireland can be split into two broad types, forming in drowned river and glaciated valleys, or as sedimentary, bar-built systems and lagoons. Further complexities in these types will be influenced by the local operation of the key inherited, dynamic and chemical variables. Within the two main types a range of estuarine forms can be found.[2] This chapter provides a brief overview and insight into these important and often strategically critical coastal water bodies in Ireland (Figure 13.2).

ESTUARY TYPES

Sorcha Ní Longphuirt and Robert Devoy

Drowned river valleys, coastal plain estuaries, bar-built estuaries and lagoons, and complex estuaries, are useful typologies for understanding the main types of estuary found on Ireland's coasts (Figure 13.4).

Drowned river valley estuaries

In the early Holocene (after c.10,000 years ago) (See Chapter 16: The Inhabitants of Ireland's Early Coastal Landscapes and Figure 16.1 for an outline of the main archaeological time periods from the Palaeolithic to the Iron Age), the rapid postglacial rise of sea levels caused the re-submergence of continental shelf surfaces that had been exposed during the Ice Age, as well as the loss of the early coastal environments developing around Ireland. The pre-existing rivers and glaciated valleys on land were flooded and submerged rapidly by the rising seas, to

Fig. 13.4 DIFFERENT ESTUARY TYPES FOUND IN IRELAND. There are many different ways of classifying estuary systems: for example, by their shape (morphology), as in this figure; or importantly, by their hydrodynamics (including tidal behaviour); or their salinity and chemistry. a. Killary Harbour, County Galway, provides an example of an estuary environment developed in a drowned river valley, which has been heavily modified by glacier ice, creating a valley-trough estuary. Killary is sometimes referred to as Ireland's fjord. Another example of this type in Ireland is Lough Swilly, County Donegal. [Source: Geraldine Hennigan and Norman Kean]; b. the confluence area of the River Shannon estuary and that of its tributary, River Fergus (County Kerry and County Clare), is a coastal plain type estuary. These are generally characterised by large water discharges and sediment supply type estuary systems. [Source: Geraldine Hennigan and Norman Kean]; c. Ards Bay, County Donegal, a bar-built estuary. Another example of a bar-built estuary is Broadmeadow, Swords, County Dublin. [Source: Geraldine Hennigan and Norman Kean]; d. the River Lee estuary and Cork Harbour, County Cork, is a geologically controlled and complex type estuary. The role of earlier north–south aligned, Pleistocene age/Cenozoic period drainage patterns, together with regional rock folding and fault structures, have been important here in creating a naturally deep harbour and estuary area. [Source: Google, Landsat/Copernicus, 2015]; e. Kenmare River, County Kerry, represents an example of another distinctive form of drowned river valley type estuary, often referred to as a ria. The term ria relates to the development of coastal features found in northern Spain and Portugal. This type of estuary results from the combination of sea-level rise, often inherited earlier drainage patterns, together with strong geological structural controls. In the case of Kenmare River this is influenced by the regional rock folding, as part of the broadly SW–NE trending ridge and valley geological province of south Munster. [Source: Google, Landsat/Copernicus, 2015]

form the distinctive coastal embayments and estuaries of the present day. The depth and overall shape of these river valleys was generally heavily modified by their glacial history (see Chapter 7: Ancient Shorelines and Sea-level Changes).

The whole of Ireland was glaciated pre-*c.*18,000 years ago, with the meltwaters from the ice developing deep-channel systems across the continental shelf, away from the glacier sources and margins on the island. On land, where deep glaciated-valley troughs intersected the coast, new overdeepened river valleys and their estuaries were formed under the rising postglacial sea levels. Such glaciated coastal valleys are prominent today in north-west Ireland, due to the upland nature of this region, and include the long and sinuous estuary of Lough Swilly (County Donegal) and also Killary Harbour (County Galway/Mayo)(see Chapter 6: Glaciation and Ireland's Arctic Inheritance).

In south-west Ireland, older Cenozoic (Tertiary) age river valleys, often influenced by the underlying and controlling regional geological structures, as well as by glaciation, form another distinctive type of drowned river valley and estuary setting. These are often still referred to as 'ria' coasts, as exemplified by the Kenmare River, County Kerry, and Bantry Bay, County Cork. Due to the sediment-poor nature of these systems they are mainly marine dominated, with only the upper regions being diluted by freshwater. On the eastern coasts of Ireland the movement of land-based ice offshore was limited by the dominant south and westward flow of ice in the Irish Sea during the times of major glaciation (over the last *c.*200,000 years). Consequently, the existence of deep drowned river valleys is less apparent today in this region; they occur only offshore as, primarily, palaeo-river valley and glacial meltwater systems.

Some part-drowned coastal valleys are linked also in their development to the building of blocking-type sedimentary structures at their seaward ends. These estuaries are distinctively shallow in character and protected from the ocean by, for example, large sandbars, barrier islands, or sometimes by a tidal flow and ebb delta. These different types of barrier system result from the availability close by of large volumes of sandy sediments. As the barrier develops it creates a tidal asymmetry, which allows for further sediment infilling of the valley, and/or larger embayment, by marine shelf sands. Examples of this drowned estuary form are represented by the River Foyle, ending at the coast in Lough Foyle, County Londonderry/County Donegal, and the Rivers Maine and Laune together with the Castlemaine Harbour

Fig. 13.5 MUDFLATS DEVELOPED FROM THE TRIBUTARIES OF THE RIVER SHANNON. Estuaries are major sites for the accumulation of sediments along the coastline. The estuary of the River Fergus catchment feeds into the river Shannon and the resulting extensive mudflats (pictured here) show the large volume of sediment deposited at the mouth of this tributary river. The Shannon Estuary is an impressive coastal plain estuary, influenced by its feeder river catchments, which facilitate land drainage. As freshwater is less dense than seawater, when sediment-laden flood water enters an estuary, fine particles in suspension can be flushed out to sea. However, heavier particles sink to the bottom when the freshwater meets seawater. This explains why sedimentation is most pronounced in the upper reaches of the estuary. The transport and deposition of sediment in estuaries is a natural process that can be accelerated by human activities, such as the infilling of river channels. [Source: Alan Lauder]

system, County Kerry (see Chapters 11: Beaches and Barriers and 30: Engineering for Vulnerable Coastlines).

Coastal plain estuaries

These are considered as related to the drowned river valley types of estuary. Their structure and evolution, however, are dominated by water flow from large river-catchments systems. Unlike the glaciated valleys and rias, described above, the fluvial (freshwater) and marine waters of these systems are well mixed and are usually less than 30m in depth. They often have extensive intertidal mud- and sandflats in their upper reaches, due to high sediment supply and deposition from their feeder river catchments (Figures 13.5 and 13.6). The River Shannon system has Ireland's largest coastal plain estuary; others include the River Slaney, with Wexford Harbour, County Wexford.

Bar-built estuaries and lagoons

The bar-built estuary is one of the most common estuary forms found around Ireland's coasts and often constitutes the transitional and linking

13.6 FALSE COLOUR IMAGE OF SHANNON ESTUARY AND MUDFLATS. An alternative view of the upper reaches of the Shannon Estuary, shown here, further demonstrates the extensive sedimentation processes that give rise to mudflats. The mudflats in the false colour satellite image appear with the dendritic (root-like) patterns of the tidal channels and creeks (centre view). This perspective also shows the areas of saltmarsh vegetation (dull red shades), sandwiched between the vegetated farmlands on land (darker red and purple, with rectilinear field boundaries also visible) and, seaward, the tidally water-covered mud- and sandflat environments (blue grey in colour). These are characteristic of all estuaries globally and allow tidal runoff on the ebb tide, draining the fringing saltmarsh and mudflat surfaces. These soft sedimentary environments are highly bioproductive, providing important habitats for fish larvae, diatoms and micro-crustacea. [Source: European Space Agency Sentinel-2 sensor, 28 June 2018]

Fig. 13.7 SMALL BARRIER (BAR-BLOCKED) ESTUARY AT LOUGH DURNESH, COUNTY DONEGAL. Lough Durnesh is an example of a small barrier (bar-blocked) and percolation-controlled type of estuary, with accompanying lagoon system. The upper part of the estuary, featured here, hosts reed beds and is an important habitat for water birds. Lough Durnesh is separated from the sea by sand dunes and drumlins. An artificial channel connects the Lough to the sea and is the source by which inundation of seawater occurs. These estuary and lagoon systems are not always obvious as estuaries. They generally have their origin on coasts where small rivers and streams, or even formerly glacial spillway channels, reach the sea directly, as also seen at Kilkerran Lake, County Cork. [Source: Jonathan Wilkins]

environment from which many coastal lagoons develop. They are created where the inflowing rivers are relatively small, but have an abundant sediment supply from which the estuary-blocking barriers are constructed at their seaward ends. The sediment sources vary locally, but are derived typically from eroding glacigenic sediments, such as soft glacial cliffs at the coast.[3] These estuary systems are characteristically small, with low volumes and accompanying accommodation spaces. They can be split into two groups: firstly, systems that contain barriers cut by sea inlets, as exemplified by the Tramore River, County Waterford, or the Rogerstown Estuary, County Dublin; and secondly, systems that are considered closed and have no natural tidal inlet channel. Closed systems have either marine water exchange by percolation through the barrier, or by seawater overwashing of the blocking barrier. Examples of this type include Our Lady's Island Lake, County Wexford and Lough Gill, County Kerry. In some cases, an artificially cut channel, or sometimes a sluice-controlled outlet, has been created in these closed systems to allow water escape during times of flood, either from land-derived freshwater inputs or by sea overtopping of the coastal barrier, as seen at, for example, Lough Durnesh, County Donegal, and Kilkerran Lake, County Cork (Figure 13.7) (see Chapter 11: Beaches and Barriers).

Complex estuaries

Linked sometimes to drowned river valley type estuaries, the complex form of estuary is controlled particularly by the geological setting, as defined by factors of rock composition, folding, strike and faulting. An example of this rarer type is Cork Harbour (see Figure 13.4d). Here, the Devonian age rocks of the West Cork Sandstones comprise the upland anticlinal ridges, which define the areas of the now-submerged valley areas in between, formed in the softer Carboniferous limestones and shales. This geology is oriented from east to west, with north–south faults cutting through the bedrock. Proclaimed as one of largest natural harbours in the world, Cork Harbour is fed mainly by the River Lee, County Cork, but also by the smaller rivers Owennacurra, Glashaboy and Owenabue Rivers.[4]

THE PHYSICAL DYNAMICS OF IRISH ESTUARIES

In their earlier stages of development, estuaries are traps for sediments that flow downstream along the land-river-estuarine continuum. A highly turbid zone of suspended sediments and fast water flows is often found at the head of an estuary environment, with sedimentation occurring as water velocities decrease toward the estuary margins and the lower (seaward) regions of the system. Tidal amplitude changes, tributary stream inputs, together with storm impacts may distort this broad pattern. Estuarine wetlands and mudflats act to filter and store these sediments before they reach the outer estuary zone and contiguous open coastal areas. These environments, together with saltmarshes and sandbars, forming within the estuary, also serve as natural protecting sedimentary buffers from coastal storms and flood damage to inland habitats and adjoining urban areas. Estuaries also form sinks for the onshore movements of sediments of marine origin, which are deposited towards the mouth of the estuary under each tidal cycle. As estuaries are progressively infilled, a balance is found between the sedimentary inputs, the shape and depth of the estuary and the scouring action of tides and currents.[5]

There is commonly a mix of fresh- and saltwaters found in many estuaries, resulting in a range of salinity values and patterns, which change both spatially within the estuary and over time. Salinity (often denoted as PSU: practical salinity units) is a unit based on the properties of seawater conductivity and equivalent to the salinity of water in parts per thousand, or ppt). This can be classified as oligohaline or fresh (<0.5ppt), mesohaline (0.5–17ppt), polyhaline (18–30ppt), mixoeuhaline (>30ppt). The importance of the saline intrusion, and hence vertical and horizontal salinity gradients, will be constrained by the morphology of the estuarine basin, the overall strength of the river flow and seasonal oscillations in freshwater inputs. During wetter winter periods, flow will increase and overall salinity will decrease, while in the dryer summer season the saline intrusion, and hence salinity, will be greater (Figure 13.8).[6]

Though estuarine systems can be classified and characterised in a variety of ways, the structure and operation of changes in estuary salinity is often used.[7] Salinity types include highly stratified estuaries (for example salt-wedge and fjord type), which have a distinct halocline (a sharp salt to fresh boundary, or strong vertical salinity gradient within a body of water); partially mixed estuaries, where there is continuous mixing between the fresh and saline waters, with undiluted freshwater occurring only at the head of the system; and vertically homogenous estuaries, where salinity levels are almost the same from top to bottom at any given point in the estuary (see Chapter 2: The Coastal Environment: Physical Systems, Processes and Patterns). Most Irish systems can be considered as partially

mixed estuaries, though the upper sections of many of these are essentially tidal 'freshwater'. Here, the water depth will oscillate with the semi-diurnal tide (that is, twice daily, as is common in most European estuaries), whilst staying completely fresh. The River Blackwater estuary, County Cork, and the River Slaney estuary, County Wexford, for example, have large tidal freshwater regions.

The tidal range of estuarine systems around Ireland differs between microtidal (<2m), mesotidal (2–4m) and macrotidal (>4m) systems. Tidal range can affect the hydrology of transitional and coastal waters through its influence on water movements (for example, turbulence and changing current velocities), residence time and the degree of mixing. Mesotidal estuaries are the most common in the country and can be found on all coasts. Microtidal systems are found only on the east coast (such as the River Avoca estuary, County Wicklow, and Bridgetown Ballyteigue Bay coastal system, County Wexford). In contrast, most macrotidal systems are found mainly on western coasts (for example, River Shannon estuary, County Limerick, and River Glenhest, Newport, County Mayo), with some also occurring in north-east Ireland. The tidal regime will also alter the patterns of sedimentary composition (the substrate) and behaviour within an estuary, through changes in bed stresses, changing water-current directions and strengths, as well as the operation of sediment erosion and deposition regimes. The estuarine substrate is often dominated by soft muds (that is, composed of a combination of silts and clay) and is rich in organic matter (as comprised of algae and the macro-remains of other plants, shells and marine or brackish water animals). In contrast, in areas where strong currents occur the substrate will be dominated by sands. Conversely, if water-current velocities are low, as at estuary channel and creek margins, or close to the land-estuary edges, then fine silt will settle out and dominate the sediments (see Chapter 12: Coastal Wetlands). The type of substrate will determine the flora and fauna that can live and thrive on the estuary bottom (benthic habitat).

Along the east coast of Ireland many estuaries have been subject to major changes in shape. Causes of this include, in particular, dredging and channelisation for shipping access, port and

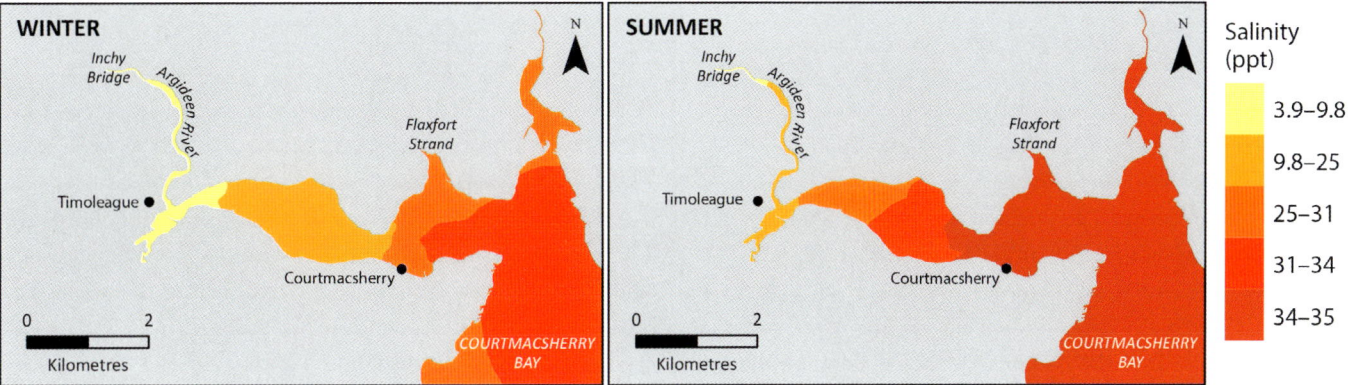

Fig. 13.8 SEASONAL VARIATION IN SALINITY LEVELS IN THE ARGIDEEN ESTUARY, COUNTY CORK. The mix of fresh and salt waters found in estuaries results in a range of salinity values and patterns, which change over space and time. Salinity levels in the Argideen Estuary showing the areas of median salinity (measured in parts per thousand, ppt) are based on data taken during the winter period (September to April) and bio-productive summer period (May to September) of 2010–2012. During wetter winter periods, water flow increases and overall salinity decreases; while in the dryer summer season the saline intrusion and hence salinity is greater. These types of salinity changes affect the chemical composition within an estuary. Organisms that live in estuaries are adapted to variable salinity, from daily changes with the tides, to the type of seasonal changes observed in the Argideen Estuary. This typically results in gradients in species that prefer freshwater upstream to saltier water further downstream. Increased salinity impacts on the ability of oxygen to dissolve in water. Higher salinity levels, or the saltiness of water, can stress or harm plants and animals. Climate change is expected to impact on the chemical composition of estuaries through altered river flows and the impacts of sea-level rise. [Source: EPA Ireland]

269

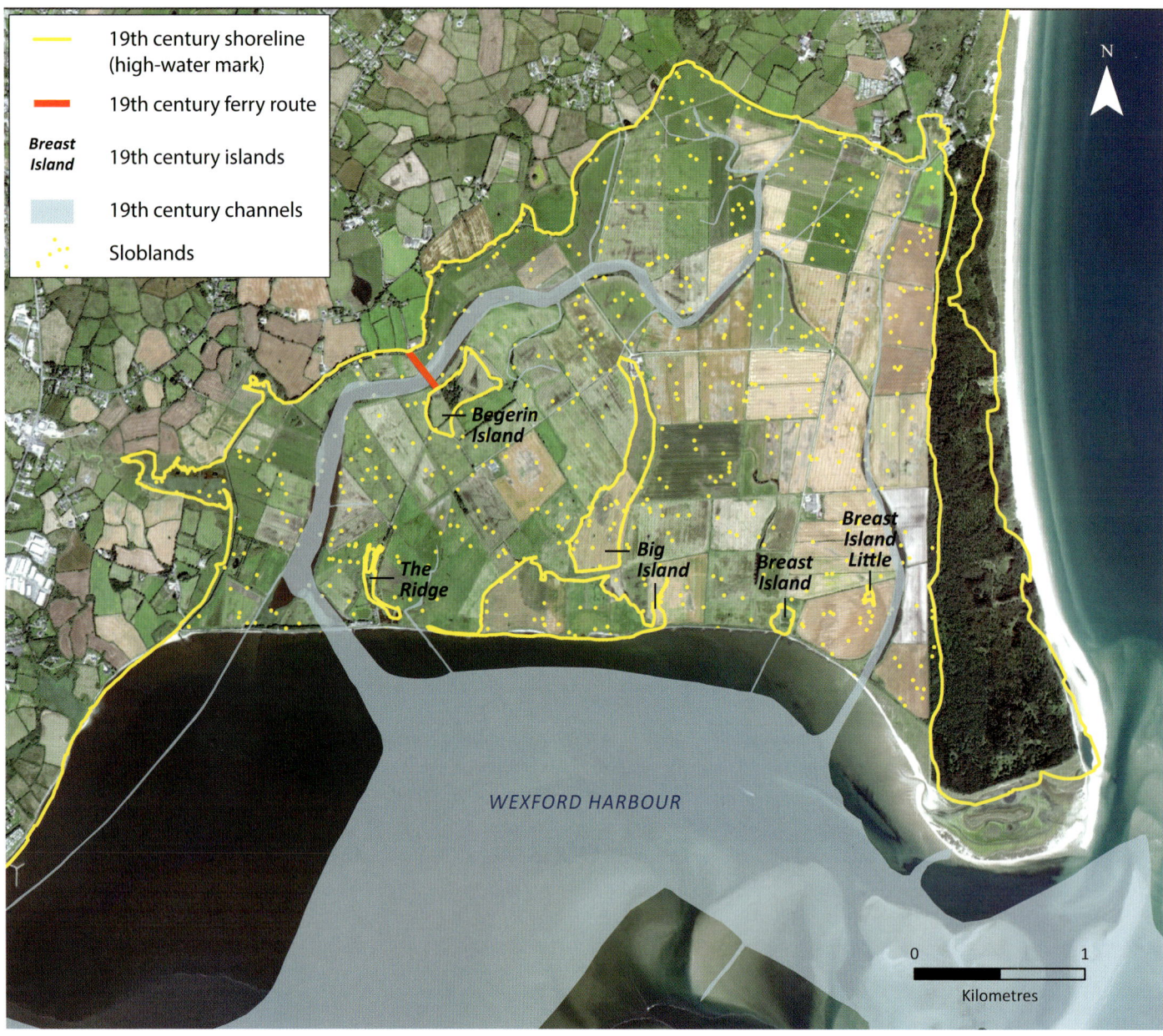

Legend:
- — 19th century shoreline (high-water mark)
- — 19th century ferry route
- *Breast Island* 19th century islands
- 19th century channels
- Sloblands

Begerin Island

The Ridge

Big Island

Breast Island

Breast Island Little

WEXFORD HARBOUR

N

0 — 1
Kilometres

Fig. 13.9 RECLAIMED SLOBLANDS IN WEXFORD HARBOUR. The evolution of sediment mechanics, together with structural changes arising from the reclamation of the wetland system, have generated major changes to the coastline where the River Slaney enters the sea at Wexford Harbour, over the past two hundred years. The extent of reclaimed land can be seen in the image, where features that were once connected to the coast by virtue of a ferry (for example, Begerin Island) are now part of a coastal hinterland. Dykes constructed in 1847 led to the previous extensive mudflats and arrays of small islands being transformed into what are now the north and south sloblands. The sloblands in Wexford Harbour – which are those areas that lie below sea level – are an important reserve for over 10,000 Greenland white fronted geese in winter. [Source: Map Data from Google, Maxar Technologies, 2014 and the OSI six-inch maps, 1824–1841]

linked urban developments (for example, River Liffey and the coast of Dublin Bay, or the River Boyne estuary, County Meath); alteration of river discharges, due to the damming of upstream areas (for example, River Lee estuary on the south coast), or the reclamation of wetlands (such as at Wexford Harbour) (Figure 13.9) (see Chapter 24: Ports and Shipping and 30: Engineering for Vulnerable Coastline). These structural changes can alter the ecological functioning of the system and the evolution of sediment mechanics in the estuary itself and the surrounding coastal zone.[8]

WATER CHEMISTRY AND PRIMARY PRODUCTION

In estuarine ecosystems, nitrogen and phosphorus provide the main building blocks for primary producers, such as phytoplankton, macroalgae and seagrasses (for example, in the North Atlantic region, including Ireland, *Zosteraceae* or eelgrasses). In Ireland, loads

of nitrogen and phosphorus that enter estuarine systems can be attributed to diffuse (for example, agricultural and forestry activities) and point sources (such as from urban areas, domestic wastewaters and industry) in the upstream zones of river-estuary catchments. The magnitude and apportionment of these loads will depend on land uses within catchments and, of course, on the proximity of urban agglomerations and industry to each estuarine system. In the cases, for example, of the River Liffey and River Tolka (County Dublin) and River Lee (County Cork) estuaries, phosphorus loads are largely attributed to wastewater treatment outfalls, while nitrogen originates from both downstream point sources as well as diffuse catchment sources. In more rural estuaries, such as the River Garavogue, County Sligo, and the River Deel, County Limerick, diffuse sources play a more significant role.

The catchment land uses will have major impacts on the

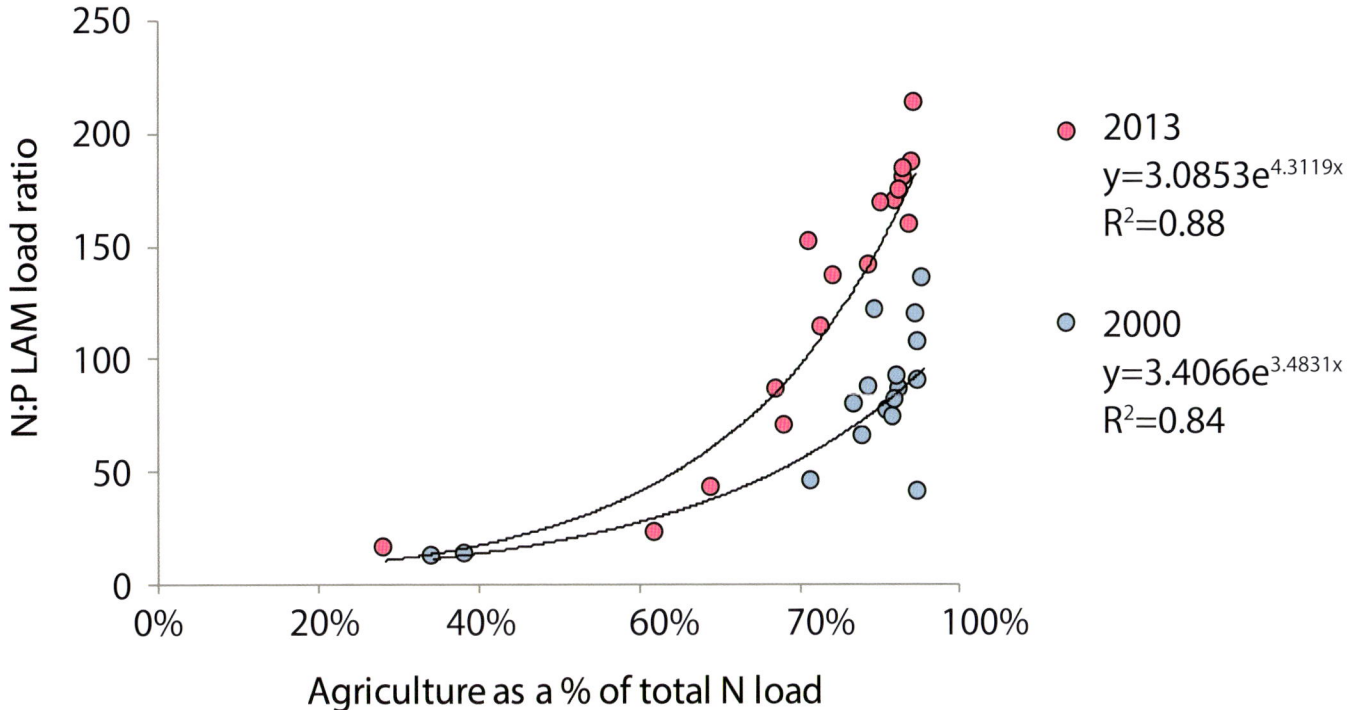

Fig. 13.10 NUTRIENT LOADING IN IRISH ESTUARIES. Nitrogen (N) and phosphorus (P) are nutrients that are natural parts of aquatic ecosystems. They support primary productivity, which underpins the health of the overall food web. Agricultural production interferes with natural cycles of nutrient elements, N and P, in Irish estuaries. Studies by the EPA show the relationship between the importance of agriculture in the overall nutrient load of the catchment, as well as the N:P load ratio. Blue dots represent the year 2000; red dots represent the year 2013. As agriculture increases the sources of nitrogen into the system through the spreading of fertiliser, the N:P load ratio entering estuary systems has also increased. Higher N:P ratios can alter phytoplankton biodiversity, and in an Irish context, have been shown to alter the size structure of phytoplankton communities. This can have implications for the health of estuarine ecosystems. [Source: EPA Ireland]

nutrient landscape of the downstream estuary, in terms of the magnitude of nutrient loads and their relative ratios to each other. This in turn will affect the ecological functioning of the estuaries into which they arrive. Generally, in Irish catchments, as the importance of agriculture as a source of nitrogen increases, so too does the ratio of nitrogen to phosphorus (N:P) delivered to the receiving estuarine system (Figure 13.10). This increase in load ratio can be linked to an increase in N:P ratios in the upper mesohaline areas of the estuary. Higher N:P ratios can alter phytoplankton biodiversity and, in an Irish context, have been shown to alter the size structure of phytoplankton communities. This change in species composition can have repercussions throughout the food web.[9]

The high productivity in estuarine systems can serve to remove nutrients from the water column before they reach the outer coastal zone and, hence, estuaries are often referred to as 'nature's natural filter'. This filtration capacity is dependent not only on the concentrations of nutrients themselves, but also the physical constraints on primary production. A further related factor includes water turbidity. In Ireland, many estuaries can be considered highly turbid (that is, high quantities of sediments in suspension in the water, of organic and mineral origins), for example, the River Suir estuary, the Upper Barrow Estuary, and in the Shannon Estuary. As a result, light penetration into the water column is insufficient for large amounts of phytoplankton to grow, even when nutrient concentrations are high. In contrast, in other estuary systems water flows are often rapid (high), resulting in the residence time for water and nutrient loads being relatively short (for example, Upper Slaney,

Tolka Estuary, Upper Blackwater). This timescale is too short to allow the phytoplankton to respond to high nutrient concentrations.

Oxygen concentration is a major factor dominating estuarine and lagoonal systems. Oxygen can be increased through processes of photosynthesis by primary producers, such as phytoplankton and macroalgae, and by physical mixing of estuary waters through currents, tides and turbulence. Oxygen can be reduced by the respiration of animals, but also by bacteria, which break down or re-mineralise organic material in the water column and the benthos. This organic material is created either within the system (for example, by the primary production of algae), or derived from organic matter from the catchment areas (detrital materials). If respiration, or oxygen demand, is greater than production then systems can become anoxic, making it difficult for plants and animals to survive. In Ireland, very low oxygen saturation has been found in the bottom waters of the Lee Estuary, County Cork (at 50 per cent saturation), indicating the probable re-mineralisation of organic material. Conversely, extremely high oxygen levels have been observed commonly in the Upper River Bandon, County Cork, the Broadmeadow estuary, County Dublin, and the Maigue Estuary, County Limerick, in the bio-productive summer period. This is mainly due to the photosynthesis of phytoplankton.

In some systems – for example coastal lagoons, such as Lough Hyne, County Cork, and Lough Furnace, County Mayo – low oxygen or deoxygenated conditions in deep waters may be a natural event. In Lough Hyne, a thermocline (a vertical temperature gradient) occurs naturally in the summer months, with an associated oxycline. In Lough Furnace, a halocline (vertical salinity gradient) exists due to the physical constraints of the system, leading to the deeper waters being deoxygenated (see Chapter 8: Visualising, Mapping and Monitoring Coasts).

ECOLOGY

The flora and fauna living in estuarine systems are unique, as they are adapted to a number of physico-chemical gradients, which often change with the tidal cycle. These include changes in salinity, water depth, exposure and inundation in intertidal areas, temperature, nutrient concentration and turbidity.

The dominant primary producers in estuarine systems will be determined by an estuary's structure and the biomes contained within the system, and distinct differences occur. For example, coastal plain estuaries, which have extensive mudflats, will have a large bioproduction generated by the microphytobenthos (benthic phytoplankton, for example, diatoms, cyanobacteria, flagellates and green algae). These may be responsible for up to 50 per cent of the primary production. Barred, drowned river valleys may contain extensive seagrass beds in their shallow intertidal areas. In rias and deep, drowned river valleys, primary production will be dominated by the phytoplankton that live free in the water column and also by the macroalgae, which live on the shorelines and offshore (subtidally) in the large kelp beds, such as those found in Bantry Bay. Bar-built estuarine systems often have extensive sandy and mudflat intertidal areas which, due to eutrophication issues, are commonly dominated by opportunistic algae (see Chapter 3: Marine Biology and Ecology).[10]

In each case, the predominant primary producers become the building blocks for the food webs in these different types of estuary. This production will only occur in the photic, or lightened zone of the estuary and, therefore, water depth gradients will often be very important to the biodiversity within an estuary. Excluding microphytobenthos and phytoplankton, primary producers (such as the macroalgae and seagrass) are usually a food source only after they have died and decomposed.

The benthic faunal communities of estuarine systems are dependent generally on the environmental conditions and sediment make-up. Crustaceans (such as crabs and shrimp), as well as polychaete and oligochaete worms, dominate mud to fine-sized sediment surfaces. Bivalves (for example, mussels, oysters) and polychaetes are more common in fine-sand to sand areas. These communities can be separated into: filter feeders (oysters, mussels and other bivalves), which are sessile (immobile) filter feeders; and deposit feeders (polychaetes, oligochaetes), which move through the sediment and feed in a fashion similar to worms in a terrestrial setting. The flow of food, or energy, within estuarine systems will travel from the water column to the bottom, as well as in the opposite direction. Bottom-dwellers feed on phytoplankton through filtering the water, or when their food drops to the estuary bed. In turn, bottom-living fauna and flora are fed on by a host of pelagic (upper and ocean) species, including, fish, birds and mammals, such as harbour porpoises, seals and otters. In spring and summer, when production is high, basking sharks can be found on the outer reaches of estuarine systems filter-feeding on plankton, while minke whales will come to feed on the shoals of mackerel, herring and sprat that are in turn drawn to these productive waters (see Chapter 3: Marine Biology and Ecology).[11]

Estuaries can be regarded as the 'nurseries of the ocean'. Irish estuaries are a spawning and nursery ground for many ecologically and commercially important fish species, such as cod, sole, herring and plaice. The fish communities living in these environments can also include more brackish and freshwater species, which inhabit more the freshwater reaches of the system. They can also include species such as salmon and lamprey, which pass through the estuary to reach freshwater spawning grounds (see Chapter 26: Coastal Fisheries and Aquaculture).

Case Study: The Ecology of Mudflats: Clonakilty Harbour

John Davenport, Lesley J. Lewis and Thomas C. Kelly

Clonakilty Harbour and Bay is a small estuary (roughly 1.9km²) on the south coast of Ireland (Figures 13.11 and 13.12). Clonakilty forms an example of a drowned river valley type estuary and is characteristic of many small estuary or embayment settings on the southern coasts of Ireland. Its ecology is broadly representative of other such estuaries in this region, with the sediments composed of muds and fine-sized sands. Intensively farmed land on the slopes of the associated valleys around Clonakilty, plus human sewage inputs, causes eutrophication and enrichment of these sediments and of neighbouring coastal waters. The estuary is less representative of those found on eastern coasts, such as Dublin Bay and Dundalk Bay, which are sandier and have higher densities of some wading bird species, such as the oystercatcher (*Haematopus ostralegus*).

The estuary adjoins the town of Clonakilty and has been much modified by human activities, notably by canalisation generated by wall- and road-building. These developments have been added to in recent times by the upgrading of sea- and flood-defence works. Originally the estuary, which discharges into Clonakilty Bay, had an island (Inchydoney) at its mouth. In the nineteenth and early twentieth centuries land reclamation and causeway construction transformed it, however, into a peninsula and separated the western portion of the estuary from the rest of the Harbour (see Chapters 4: People, Agriculture and the Coast and 30: Engineering for Vulnerable Coastlines). The estuary is fed at the landward end by the Feagle River, which flows through Clonakilty, but also by numerous streams and drains that supply freshwater into the area as a whole. At low water spring tide (LWST), the estuary empties almost completely into Clonakilty Bay, exposing large areas of sand- and mudflats that are roughly 1m above LWST.

There is a gradient of sediment type, from fine glutinous mud at the head of the estuary, via muddy sands, to pure sand close to the mouth (see Figure 13.12). As high water spring tide (HWST) is *c*.4m above LWST, it may be calculated that approximately 5.5–6.5 million tonnes of seawater flows into the Harbour on the spring flood tide and leaves again on the ebb. At HWST all sediments are flooded by seawater (salinity: 32–33psu), which displaces any lower-salinity water present, because the latter has a lower density. Salinity levels have been measured extensively (2000–2011) along the estuary on falling tides.[12] Broadly speaking, salinities are lower in wet weather, when the river flow and freshwater runoff from agricultural land around the estuary is greater, and on neap tides, when volumes of incoming seawater are less. At the head of the estuary, where the sediments are dominated by fine muds and the water table is at the surface, the salinity in the sediments at 10–15cm depth remains high (typically 25–28psu). This environmental condition persists,

Fig. 13.11 CLONAKILTY BAY, ESTUARY ZONE AND FRONTING COASTAL DUNES BARRIER. The aerial photograph shows the former Inchydoney Island (photograph centre), now joined to the Clonakilty hinterland by the building, in the nineteenth century, of protecting coastal embankments, which blocked off the tidal channel to the north of the island. This led subsequently to the shallowing of the Bay (now known as the Inchydoney estuary; about 1km² in area) and its progressive infilling by sands. This human impact was added to in the late twentieth century by, in part, the effects of onshore movements of sands under tidal action and storms and, increasingly, by the outcomes of climate change. This experience of onshore sands movement at this site is reflected in many estuaries and embayments along the south coast of Ireland, such as along the coast in Rosscarbery. In the late 1970s this area would have retained at least a 1m depth of water in most of its area at low tides; now it is almost dry. [Source: Irish Air Corps]

even if the weather is wet and the salinity of the main river channel is only 1–5psu. However, pore water in the top 1cm of the mud can be brackish (typically 2–15psu). In the middle part of the estuary, where sediments are mud plus fine sand, the water table at low water is 10–20cm below the surface; salinities at a depth of 20–30cm in the sediment are fully marine (32–33psu), though may be lower than this close to the main channel.

The fauna of fine muds at the head of Clonakilty Harbour is dominated by the polychaete ragworm, *Hediste diversicolor*, which lives at and near the surface of the mud (top 10–20cm) and by the peppery furrow shell, *Scobicularia plana,* which lives deep in the sediment (Figure 13.13). The ragworm lives in semi-permanent burrows and uses a net of mucus to trap fine material (plankton, detritus), but also leaves the burrow to prey upon small invertebrates, or scavenge on carrion. It is physiologically tolerant, being an osmoregulator that can pump salts inwards when exposed to low salinities. The furrow shell is not tolerant of low salinities and so remains inactive on falling and low tides. When the tide returns it feeds by projecting its inhalant siphon above the sediment surface and curving it down to hoover up the fine, organic-rich material (including benthic diatoms) on the surface of the mud. At low water the fringes of the mudflats feature very large numbers of gammarid amphipods under stones and weed debris. These are small crustaceans that feed on surface detritus, algae, carrion and small animals; they move out onto the mudflats as the tide rises.

In the middle of the estuary the muddy sands contain the ragworm, other polychaete worms (for example, *Phyllodoce maculate* and *Pygospio elegans*), as well as the peppery furrow shell. These are joined by the polychaete lugworm (*Arenicola marina*), adults of which burrow to 20–40cm depth. Lugworms are inactive at the bottom of their mucus-lined, U-shaped burrows when the tide is out, so only encounter fully marine conditions. When the tide is in they consume surface sediment laden with micro-organisms (for example, bacteria, fungi), benthic diatoms, meiofauna and detritus, turning over the top layers of the sediments like earthworms. The mudflats of the harbour also feature two small, surface-dwelling invertebrates: the gastropod laver spire shell, or mudsnail (*Hydrobia ulvae*) (<5mm shell height) and the amphipod mud shrimp (*Corophium volutator*) (<11mm long). Both species occur in huge numbers (100s to 1,000s per m²) and eat microbial layers and associated organic material at the mud surface. The laver spire shell does this by constant movement and browsing with its radula; the mud shrimp by using its limbs to scrape material to its mouth. Both species are euryhaline (tolerant of a wide range of salinities) and become active as soon as the sediments are covered by the incoming tides. Again, gammarid amphipods are common at the edges of the mudflats.

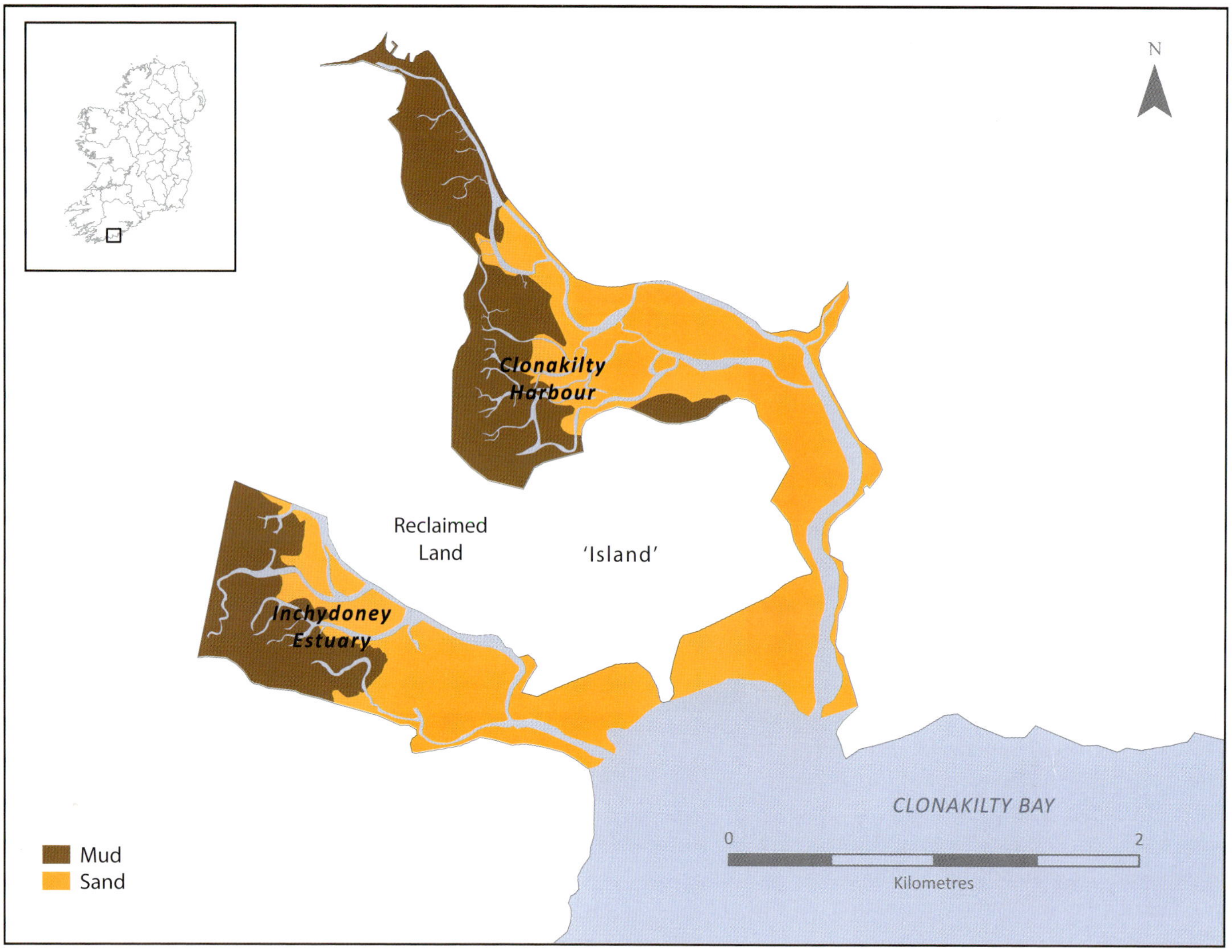

Fig. 13.12 CLONAKILTY BAY, CLONAKILTY HARBOUR AND THE INCHYDONEY ESTUARY AREAS AT LOW WATER. A gradient of sediment types, from fine glutinous mud at the head of the estuary, via muddy sands, to pure sand close to the mouth, can be observed at low water in the Inchydoney estuary and in Clonakilty Harbour. The coarser sand flats nearer to the estuarine mouth drain extensively and are 'damp' rather than wet at low water. This variability of sediment type supports a diversity of physiologically tolerant estuarine life, as shown in Figures 13.13 and 13.14. [Source: Digitised from a 2018 digital globe satellite image]

The great densities of mudflat invertebrates at Clonakilty Harbour support two groups of predators: those that feed when the tide is in; and those that feed when the tide is out. The former group has been poorly studied, so knowledge is limited to comparisons with other estuaries, plus observations of animals found in refuges at the edges of the estuary, in the main channel and in the latter's small tributaries at low water. When the tide rises, aquatic predators spread out over the mudflats, especially at night, when they are less vulnerable to predation themselves. These include shore crabs (*Carcinus maenas*) and shrimp (Crangon crangon) and a variety of flatfish

Fig. 13.13 INVERTEBRATE DETRITIVORES OF CLONAKILTY HARBOUR: ORGANISMS LIVING IN THE MUDFLATS. a. Ragworm (*Hediste diversicolor*); b. mud 'shrimp' (*Corophium volutator*); c. laver spire shells (*Hydrobia ulvae*) on a patch of *Ulva* sp.; d. peppery furrow shell (*Scobicularia plana*); e. casts (faeces) of the lugworm (*Arenicola marina*) on muddy sand. [Sources: a and e: Keith Hiscock; b, c and d: David Fenwich Sr]

Fig. 13.14 UNDERWATER PREDATORS OF CLONAKILTY HARBOUR. a. Shore crab (*Carcinus maenas*) [Source: John Davenport]; b. flounder (*Platichthys flesus*) [Source: Keith Hiscock]; c. common shrimp (*Crangon crangon*). [Source: David Fenwick Sr]

(for example, flounder, *Platichthys flesus*) (Figure 13.14). They are all generalists, often exploiting different prey items in different parts of the estuary. Flounders, for example, consume ragworm, mud shrimp and mudsnails; they will also attack shrimps and small crabs. Flatfish, in general, gain some energy intake, not in the form of whole prey items, but by biting off pieces of polychaetes or ends of bivalves' siphons ('cropping').

The other group of predators is much more conspicuous and better-studied. Clonakilty Harbour has many resident and migratory waterbird species (Figure 13.15) (see Chapter 3: Marine Biology and Ecology). The area is especially important nationally and internationally because it supports substantial populations of overwintering and migratory wading birds, notably black-tailed godwits (*Limosa limosa islandica*) and redshank (*Tringa totanus*). Black-tailed godwits breed in Iceland; Clonakilty mud-dwelling invertebrates enable them to survive the winter and, importantly, fuel their northwards migrations.[13] One bird species, the great cormorant (*Phalacrocorax carbo*) is a conspicuous predator when the tide is in and fish are present. These birds, usually around twenty in number, take both round and flat fish. They eat about a third of their body weight per day, so together they consume some 20kg of fish everyday.

Wading birds have a variety of bill lengths and shapes and this has sometimes been uncritically correlated with preferred prey. Although the curlew (*Numenius arquata*) can probe deeper into mud than other species, it is also capable of catching crabs at the mud surface. Conversely, because young furrow shells and lugworms live in much shallower burrows than large adults, they are accessible to short-billed birds. Hence, most waders have a relatively varied diet, but show some preferences: godwits at Clonakilty feed especially on furrow shells and the abundant ragworm; redshanks take ragworms, mud shrimp and mud snails. Many waders feed equally efficiently at low tide during day and night, while many also move to pastures around the estuary and take terrestrial prey, such as earthworms.

Besides overwintering birds, Clonakilty Harbour also has resident species that forage year-round. Grey herons (*Ardea cinerea*) and little egrets (*Egretta garzetta*) take fish and shrimp from the central channel, where they are concentrated when the tide is low. Little egrets are recent colonists of the UK and Ireland; they catch shrimps and consequently compete with wintering greenshank (*Tringa nebularia*), which have probably used the estuary for millennia. Shelduck (*Tadorna tadorna*) occur in small numbers; they are specialised feeders on mud snails, sweeping their bills to-and-fro to take in surface mud and sieve out large numbers of them. Omnivorous gulls and corvids are common;

Fig. 13.15 PREDATORY BIRD SPECIES OF CLONAKILTY HARBOUR. a. Black-tailed godwit (*Limosa limosa*); b. redshank (*Tringa totanus*); c. grey heron (*Ardea cinerea*); d. little egret (*Egretta garzetta*); e. black-headed gulls (*Chroicocephalus ridibundus*), seen in eclipse, foraging on *Hydrobia* and *Corophium*, f. shelduck (*Tadorna tadorna*) foraging on mudflat. [Sources: a, b, c, d and f: John Fox; e: John Davenport]

black-headed gulls (*Chroicocephalus ridibundus*) often take mud snails individually.

Since around 1990, Clonakilty Harbour has been afflicted by blooms of the green algae sea lettuce (of the genus *Ulva*) in spring and summer (Figure 13.16a) (see Chapter 31: Pollution). These seaweeds have thin, flat fronds and are either weakly attached to the sediment or free-floating. By 2000 it was common for drifts of sea lettuce, 10–20cm deep, to accumulate at the edges of the estuary and for a complex mosaic of covered and uncovered mud areas to develop in the summer. Such blooms of sea lettuce have been attributed to heightened nitrate levels (eutrophication), which has been proved to be the case elsewhere; for example, in the pioneering work in the Ythan Estuary, Scotland, during the 1980s and 1990s, where blooms of sea lettuce were often caused by unrestricted use of fertilisers. Untreated sewage (human and animal) is another potential source of nitrates.

Because the blooms blanket the substratum, they inhibit most burrowing mud dwellers. They also cause hypoxic or anoxic conditions

Fig. 13.16 BLOOMS OF THE GREEN ALGAE, SEA LETTUCE (OF THE GENUS *Ulva*) IN CLONAKILTY HARBOUR. a. The middle to inner sandflats in Clonakilty Harbour, which have been almost completely infilled by sands in recent years, showing brown and green algal growths, including extensive areas covered by sea lettuce (*Ulva spp.*). This organic enrichment, often caused by unrestricted use of fertilisers on land, became such a source of public contention in 2010, that the local authority intervened to clear the green algae. Detailed study of these intertidal plant and animal communities was completed in 2010–2011. Since that time changes of the mud- and sandflats have continued, reflecting the earlier major alterations in the Harbour and estuary that occurred over the preceding forty or so years, in sediment patterns, water flows and the dependent plants and animals. [Source: Dr Liam Morrison, National University of Ireland, Galway]; b. the main change recently has been the appearance in the system of the invasive red algal species, Papenfuss (*Agarophyton vermiculophyllum*) (Ohmi), which produces extensive blooms. This was first recorded in this area in 2014 by the Environmental Protection Agency. Papenfuss is probably originally native to the north-west Pacific ocean region. It has been introduced into the Atlantic ocean and forms a very invasive species, rapidly outcompeting native algae in estuaries. [Source: Stacy A. Krueger-Hadfield (*Agarophyton* thallus with *Ulva* and a horseshoe crab in the background at Savages Ditch Rd in Delaware)]

at the mud-to-weed interface, sometimes generating methane or hydrogen sulphide. Bacterial breakdown of blooms further enriches the sediments, creating a positive feedback loop. Studies at Clonakilty in the early 2000s revealed that *Ulva*-covered mud patches had reduced populations of mud snails and mud shrimp by comparison with bare patches. Redshank foraging was significantly less efficient in *Ulva* patches than on bare mud. Godwits were less affected. By 2008/09 blooms were even more extensive and it was found that the *Ulva* fronds supported huge populations of mud snails, which grazed on the fronds themselves. In 2010, quantities of green algae carried out to the Atlantic on the ebb tide were so copious that they were washing up on the surfing beaches of Clonakilty Bay, outside the estuary. They were cleared by civic authorities, who contemplated harvesting the algae for use as fertiliser. Blooms appear to be promoted by wet weather and enhanced runoff. Surveys in 2009–2011 failed to find mud shrimp species, including the common mud scud (*Corophium volutator*), on Clonakilty Harbour mudflats. While this is bad news for the wading birds that like to feed on them, there could be more far-reaching consequences. Canadian studies in the Bay of Fundy indicate that mud shrimp is a keystone species, essential to ecosystem functioning. In this way, organic enrichment acts as a double-edged sword with respect to the foraging success of wading birds. While birds indirectly benefit from mild organic enrichment because it fuels abundant densities of invertebrate prey, dense annual algal blooms can have negative effects upon foraging success for varying lengths of time.

HOW HEALTHY ARE OUR ESTUARIES?

Estuaries, together with other transitional coastal waters, provide significant ecosystem services (such as, coastal protection, nutrient control, fisheries), and many other benefits, including fisheries and recreation. However, multiple pressures threaten the ecological and economic stability of these environments in Ireland, as well as worldwide. Human actions and related pressures can cause their degradation, through people's physical uses (in, for example, the building of ports, dredging or channelling), bio-chemical (for example, wastewater outflows, industry, forestry and agriculture) and biological alteration (for example, fisheries and aquaculture) of these systems. In addition, global pressures – such as those arising from the impacts of invasive species and climate change – threaten the fundamental health of estuaries (see Chapter 33: Climate Change and Coastal Futures).

An assessment of the ecological status of our transitional water is carried out every six years under the EU's Water Framework Directive (WFD). This includes an assessment of the physico-chemical (for example, nutrients, oxygen), biological (for example, phytoplankton, macroalgae, seagrass, fish, benthic invertebrates), chemical (for example, metals, pesticides and industrial chemicals) and hydro-morphological elements. Hydromorphology considers the water flow and substrate available for physical habitation of flora and fauna. Physical structures can alter these elements of an ecosystem (through, for example, abstraction, dams and weirs, dredging, embankment, sea walls, piers and harbours) and put pressure on the ecological functioning of an estuary.

The saline waters of Ireland include 200 transitional estuaries and lagoons and 110 near coastal waters, 40 per cent of which were monitored in the period 2013–2018. Only 38 per cent of Irish

Fig. 13.17. ALGAL BLOOM IN THE TOLKA RIVER ESTUARY, COUNTY DUBLIN. One of the main contemporary issues for transitional water systems is eutrophication. This does occur naturally, as part of ecosystem functioning. But problems result particularly when people generate large nutrients loads in the environment, which enter a system and cause excessive growth of algae (blooms). This may lead to oxygen depletion of the water body and can cause harm to other flora and fauna in the system. Blooms can occur when conditions of light, temperature and residence time coincide to produce ideal conditions for the growth of algae. These blooms, particularly of microscopic phytoplankton (e.g., diatoms, dinoflagellates, green algae) can serve to not only degrade oxygen conditions but can also be harmful to fauna, such as mussels and oysters, due to the toxins they can produce. Blooms of green macroalgae also often occur in shallow estuaries with large sandy and muddy intertidal areas, as in the Argideen Estuary and Rogerstown and Tolka Estuaries. These can have negative impacts on fauna, clog the nets of fishermen and affect tourism in the area. [Source: Dublin City Council and the UNESCO Dublin Bay Biosphere Partnership]

estuarine systems are of good, or high environmental quality. In the case of near coastal waters, 80 per cent are at an acceptable status.[14] Agriculture is the main pressure affecting transitional waters. The impacts are most evident in the south-east of Ireland, where nitrogen levels in estuaries are the highest and the status of these water bodies (estuaries, bays and lagoons) is often only 'moderate', or even lower. Urban wastewater can also be a significant pressure on estuarine systems. This includes licensed outflows from treatment plants, but also storm water overflows. With climate warming impacts, this is becoming an increasingly serious problem for estuaries in urban settings, particularly in Dublin, Cork and Limerick (Figure 13.17).

Whether through sea-level rise, increased rainfall and freshwater discharges, or more frequent storms and droughts, global climate change processes are already having, and will continue to have, significant impacts on estuarine systems in Ireland. These will affect particularly the transport, dilution and the fate of nutrients and harmful chemicals in estuarine systems, thus amplifying their effects on flora and fauna. Other impacts will include the continued rise in the presence of invasive/new species in estuaries, changes in water temperatures, sedimentation and the levels of freshwater inflows, with significant seasonal differences. Management approaches are required that consider holistically the interaction of these present and future pressures within the estuary and catchment system, together with those of the adjacent coastal and marine areas. The focus and vision, hopefully, of such management and linked governance structures, is to retain the sustainability of these vulnerable systems and to protect and improve their ecological status. This will benefit not only the diverse flora and fauna in the ecosystems themselves, but also the communities in close proximity to estuaries who benefit from the many ecosystem services they have to offer.[15]

COASTAL LAGOONS: A BARRIER TO THE TERRESTRIAL ENVIRONMENT AND A FILTER FOR THE MARINE ENVIRONMENT

Susan Lettice, Greg Beechinor and Deborah Chapman

Worldwide, coastal lagoons are unique and valuable ecosystems that provide many ecosystem services. They are typically shallow, brackish-water lakes, commonly having a restricted water exchange with the adjacent sea and separated from it, either partially, or completely, by rock or sand and gravel barriers. The lagoon is supplied with saline water via both percolation and infiltration through the sedimentary barrier, or often by direct water flows through a single, narrow connecting channel. In some locations, lagoon systems play an important natural coastal defence role. They may provide protection for the hinterlands of coastal zones during storm events. Without this many cities and other coastal settlements, such as Venice, would not exist (see Chapter 11: Beaches and Barriers). Like estuaries, coastal lagoons are complex ecosystems, influenced by varying levels of marine and freshwater exchanges, nutrient fluxes and sediment flows, nutrient cycling and the decomposition and transformation of organic matter. Lagoons have often been managed to provide resources for nearby communities rather than to protect them as unique ecosystems (Figure 13.18).[16]

Compared with other ecosystems, little is known about coastal lagoons in Ireland, or even globally. They are difficult to define because of the temporal and spatial variability of the marine, freshwater and terrestrial influences to which they are subjected, which are often specific to each lagoon. They differ in terms of their size, depth and water dynamics, such as tidal exchange and geomorphology. Although considered shallow, lagoon depths can vary significantly, but are often 1–3m, reaching depths of 5m in channels, or in isolated hollows.

Around the world, lagoon sizes can range from less than 1ha up to 10,200km², as in the case of Lagos dos Patos in Brazil. Given the extent of these differences, it is unsurprising that much debate has occurred regarding the definitions and classifications of coastal lagoons. The most commonly used is based on the connection with the sea: 'a shallow coastal water body separated from the ocean by a barrier, connected at least intermittently to the ocean by one or more restricted inlets, and usually orientated shore-parallel'. A more comprehensive definition was outlined in the European Habitats Directive, which takes account of the vegetative state of the lagoon.[17]

Coastal lagoons occupy 13 per cent of continental coastlines worldwide and 5.6 per cent in Europe. On the island of Ireland, 118 lagoons have been recognised (Figure 13.19) and Irish lagoons are considered to be unique. They are relatively unusual as systems in a North Atlantic context because of their macrotidal (>4m) location on high energy coastlines. Despite this unique characteristic, the overall conservation status of these environments was classified as 'favourable' to 'bad'.[18]

Figure 13.18. RUSHEENDUFF LOUGH, SITUATED NEAR TULLY CROSS IN NORTH CONNEMARA, COUNTY GALWAY. A former brackish-water lagoon and connected to the sea, Rusheenduff Lough is now a shallow freshwater lake environment separated over time by a longshore and onshore gravel bar and beach sedimentary movements. The site is a Special Area of Conservation (SAC), representing important habitats and homes for plant and animal species listed on Annexes I and II of the EU Habitats Directive. The Lough is an oligotrophic to mesotrophic lake, which supports a range of aquatic plant species and several rarities. These include the slender naiad (*Najas flexilis*), an inhabitant of shallow bodies of brackish- and freshwater such as lakes and bays, as well as a population of the aquatic plant water thyme (*Hydrilla verticillata*). This is a very rare species in Ireland, being known from only one other site, a recently discovered population in another west Galway lake. Offshore, rock reefs formed in the predominant Gneiss rock in the area, as shown in the photograph, are created by the repeated impacts of high onshore wave energy in Atlantic storms. Breaches of the coarse sands and gravel (shingle) beaches by storm over-topping can cause periodic seasonal changes in the salinity of the Lough. These can threaten the survival of particularly the rare vegetation communities. Further, eutrophication of the Lough waters, caused by runoff from the surrounding farmland, or through the discharge of domestic sewage, may also threaten this sensitive environment. (National Parks and Wildlife Service, 2015. Site Synopsis. Department of Arts, Heritage and the Gaeltacht, Dublin. [online] Available at: https://www.npws.ie/sites/default/files/protected-sites/synopsis/SY001311.pdf). [Source: Irish Air Corps]

Oliver and Lettice have provided some of the earliest studies characterising lagoons in Ireland based on their environmental and biological characteristics (Figure 13.20).[19] Three coastal lagoon systems with similar meteorological influences were studied, at Cuskinny, Farranamanagh and Toormore, in County Cork, along the south-west coast of Ireland. Interestingly, precipitation was found to be the main driving force of the different salinity regimes in these locations. Seasonal nutrient patterns were also evident in the three systems, with chlorophyll a at its highest during late spring/early summer and at a minimum during the winter months. The recommendation for informed management arising from the study was that the monitoring of coastal lagoons in future should include, at a minimum, salinity and chlorophyll a measurements, as well as orthophosphate and nitrate concentrations.[20]

THE FLORA AND FAUNA OF COASTAL LAGOONS

Few studies have been carried out on the ecological communities in Irish coastal lagoons. In studies that have been done, in Lady's Island Lake, County Wexford, wide variations over a seventeen-year period were found in salinity and water levels, due to frequent breaking of the barrier. Observations during this time revealed changes in the salinity regime every two to four years. Vegetation was also found to vary from year to year, as the vegetation either grew abundantly, or was nearly absent. When vegetation was present it was dominated by aquatic plants, for example spiral tasselweed (*Ruppia cirrhosa*) and fennel pondweed (*Potamogeton pectinatus*), depending on the salinity. The faunal species were also found to alter when the salinity changed. For example, the opossum shrimp species (*Neomysis integer*) replaced another species of opossum shrimp, the chameleon shrimp (*Praunus flexuosus*), following the decline in salinity. Dramatic changes in the ecological community of Lady's Island Lake from one year to the next were also recorded. However, the lagoon system may be considered exceptional, as many of the changes found could have been the result of attempts to manage marine breaching of the barrier (see Chapter 11: Beaches and Barriers).[21]

Coastal lagoons support unique assemblages of fauna, in particular aquatic invertebrates. These organisms, particularly a group known as shredders (due to the manner in which they break up detrital material), play an important role in the decomposition of detritus. Their activity increases the rate of biomass decomposition and their absence delays the decomposition process in aquatic ecosystems. The most

Coastal Lagoons

1 Gransha
2 Black Brae
3 Donnybrewer
4 Longfield
5 Ballykelly
6 Myroe
7 Ballyaghran
8 Larne
9 Glynn A
10 Glynn B
11 Oldmill
12 Ballycarry
13 Whitehouse
14 Victoria Park
15 Belfast Harbour Lagoons
16 Castle Espie
17 Anne's Point
18 Rosemount
19 Mahee Point
20 Cadew Point
21 Quarterland
22 Rathgorman
23 The Dorn
24 East Down Yacht Club A
25 East Down Yacht Club B
26 Castleward
27 Blackcauseway
28 Granagh
29 Strand Lough
30 Dundrum South
31 Greenore Golf Course
32 Broadmeadow
33 Kilcoole
34 North Slob channel
35 South Slob channel
36 Lady's Island Lake
37 Coornagillagh
38 Ballyteige channels
39 Rostellan Lake
40 Baalyvodock lagoon
41 Cuskinny Lake
42 Bessborough Pond, Blackrock
43 Raffeen Lake, Shanbally
44 Lough Beg, Curraghbinny
45 Oysterhaven Lake, Clashroe

46 Commoge Marsh, Kinsale
47 Inchydoney
48 Clogheen/White's Marsh
49 Kilkeran Lake
50 Rosscarbery Lake
51 Toormore lagoon
52 Lissagriffin Lake
53 Farranamanagh Lake
54 Reen Point Pools
55 Kilmore Lake
56 Reenydonegan Lake
57 Lauragh
58 Drongawn Lake
59 Lough Gill
60 Blennerville lakes
61 Quayfield/ Poulaweala
62 Shannon Airport lagoon
63 Scattery lagoon
64 Cloonconeen Pool
65 Lough Donnell
66 Loch Mór, Inis Oírr
67 Port na Cora, Inis Meáin
68 Loch an tSaile, Árainn
69 Loch an Chara, Árainn
70 Loch Phort Chorruch, Árainn
71 Loch Dearg, Árainn
72 Muckinish Lake
73 Lough Murree
74 Aughinish
75 Rossalia
76 Bridge Lough, Knockakillee
77 Doorus Lakes
78 Rincarna pools
79 Mweeloon pools
80 Ardfry Oyster pond
81 Turreen Lough (Rinvile)
82 Renmore Lough
83 Lough Atalia
84 Lettermullen
85 Loch Fhada upper pools
86 Loch an Ghadaí
87 Loch Fhada
88 Loch an Aibhnin
89 Loch Tanaí
90 Loch Cara Fionnla
91 Loch Cara na gCaorach
92 Loch Doire Bhanbh
93 Loch an tSáile (Lough Atalia)
94 Loch Conaorcha (Aconeera)

95 Loch A' Chaorainn
96 Lough Ateesky
97 Loch an Mhuilinn (Mill Lough)
98 Loch Ballyconneely
99 Lough Athola
100 Lough Anillaun
101 Lough Bofin
102 Corragaun Lough
103 Roonah Lough

104 Claggan lagoon
105 Furnace Lough
106 Dooniver Lough, Achill Island
107 Cartoon Lough, Killala Bay
108 Portavaud, Ballysadare Bay
109 Tanrego
110 Durnesh Lake
111 Maghery Lough
112 Sally's Lough

113 Kincas Lough
114 Moorlagh
115 Loch Ó Dheas, Tory Island
116 Carrick Beg Lough
117 Blanket Nook Lough
118 Inch Lough

Fig. 13.19 COASTAL LAGOONS IN IRELAND. Prior to the introduction of the Habitats Directive, which provided a focus for the identification of diverse habitat types, only four lagoons were well known and studied in Ireland. These were Lady's Island Lake and Tacumshin Lake, County Wexford; Lough Murree, County Clare; and Furnace Lough, County Mayo. Subsequent surveys revealed a total of 118 coastal lagoons on the island of Ireland. Coastal lagoons are considered to be relatively unusual in a North Atlantic context, because of their macro tidal location in high-energy coastlines. [Data source: National Parks and Wildlife Survey, 2013; Department of Agriculture, Environment and Rural Affairs, 2019; Joint Nature Conservation Committee]

common and abundant shredders recorded in Irish lagoons include the shrimp *Neomysis integer*, *Gammarus spp.* (scuds), *Palaemonetes varians* (river shrimp) and *Lekanesphaera hookeri* (a pill isopod). It has been found, however, that the majority of invertebrate species in these coastal lagoons were only temporary residents; they colonised under favourable conditions, but could disappear due to seasonal, systematic or unpredicted changes in salinity.[22]

LAGOONS IN A CHANGING WORLD

The effect of global climate change on the physical structure, ecological characteristics and social values associated with lagoons is largely unknown. Sea-level rise (SLR) presents one of the biggest threats to these ecosystems, with the coastal lagoons that we know today possibly disappearing completely in the future under the impacts of future SLR; the saline regimes of many lagoons are also changing significantly. Many barrier-lagoon systems respond naturally to SLR by migrating landward along undeveloped, gently sloping coastlines. As the lagoon barrier retreats, future accelerations in SLR will lead to steeper and narrower barrier profiles, resulting in water retention behind the barrier and greater water depth. Provided the barrier is not broken, the biological communities will change due to reduced light reaching the more deeply submerged aquatic vegetation. The resultant reduction in the photosynthetic potential of the primary producers will change nutrient dynamics, making the lagoon more susceptible to eutrophication.

Both intense precipitation and high winds can be severe during major weather events, but their impacts on lagoons may be different in the context of climate change. Changes in precipitation patterns can affect the physical and ecological characteristics of these environments, through the alteration of freshwater inputs and associated changes in the salinity and dissolved oxygen concentration of the

Fig. 13.20 EXAMPLES OF LAGOON SYSTEMS IN IRELAND. a. Coastal lagoon on Inishirrer, County Donegal, forming on this coast behind a cobble barrier. It is classified as an isolated sedimentary lagoon. [Source: Donegal County Council]; b. Lough Furnace, a tidally influenced saline lagoon in County Mayo, is a well-known and relatively well-studied lagoon system in Ireland; [Source: Martin O'Grady, Courtesy of Marine Institute]; c. Bullogfemule, Drawngan Lough, County Donegal. [Source: Gordon Dunn]; d. Loughs on Inishbiggle, County Mayo, with Achill Sound and Island in the background. [Source: Geraldine Hennigan and Norman Kean]

lagoon. By contrast, storm events can lead to overtopping the barrier and erosion of the barrier by wind and wave action. Many of the unique coastal lagoons are, therefore, potentially vulnerable to climate change and more understanding of their dynamics is needed in order to determine the most appropriate management measures to protect them.

Coastal lagoons have been described as a 'neglected habitat'. However, they have gained a considerable amount of attention in Europe and Ireland since around the 1990s, primarily as a result of the EU's Habitats Directive. Although coastal lagoons are a rare habitat in Europe, their unique features have attracted significant research attention, due, for example, to them being home to rare or isolated species, or because they are environments that need protection from the impacts of people, or because of their role as 'first responders' in the environment to the influences of climate change. In spite of this, there are many features of their ecological functioning that are still poorly understood or remain a mystery. The Lagoons for Life initiative launched in University College Cork in 2017 recognises the need to bring scientists together with key stakeholders to improve understanding of how coastal lagoons respond to change at local, regional and global levels.

Ireland's coastal transitional waters, of estuaries and lagoons, represent, literally, the marginal landscapes of the island, located as they are on the 'edge' (see Chapter 14: Imagining Coasts). These waters and the neighbouring low coastal embayments were invariably the zones of contact with Ireland for the first settlers and, thereafter, the more driven and focused invaders of the island. Archaeological and historical evidence abounds for estuarine environments, especially, providing the attraction for core settlement and subsequent expansion (see Chapters 16: The Inhabitants of Ireland's Early Coastal Landscapes and 17: The Vikings and Normans: Coastal Invaders and Settlers). Today, these waters remain vitally attractive to people as living spaces, and as the location of economic activity and leisure activities also. Perhaps their most fundamental value is as natural environments, 'wild places', as reservoirs and habitats for plants and animals as well as people (see Chapter 27: Coastal and Marine Tourism). At a more human and emotive level, our perception of these environments is linked, perhaps, to how we see them. On a wet and cloudy day, as is so often the case in Ireland, they can appear to many as unattractive and bleak places. On a sunny day, they bring the coast alive.

SECTION 3
PEOPLE AND THE COAST

Monastic site of Illauntannig, Maharees Islands, County Kerry
[Source: Geraldine Hennigan and Norman Kean]

IMAGINING COASTS

Ronan Foley and Anna Ryan

MULLAGHMORE, COUNTY SLIGO, APRIL 2019. [Source: Mary-Therese O'Neill]

Fig. 14.1 MALONEY'S STRAND, DUNWORLEY BAY, COUNTY CORK. At low tide the exposed sandy intertidal zone (the littoral, or liminal edge of the coast) of the beach shows a zebra-patterned series of sand ripples and low-profile water-filled channels, developed as the tide falls on the ebb, the natural artwork of the sea. [Source: Robert Devoy]

From time immemorial, the sea has exerted a powerful influence over people's imaginations. The oceans were assigned their own gods: Poseidon for the Greeks; Neptune for the Romans; the Norse had Ægir, a *jötunn* or troll-like being who was particularly associated with the sea. In pre-Christian Ireland, the hero Manannán mac Lir ('son of the sea') was associated with the Tuatha Dé Danann, and travelled in a sea-borne chariot named *Scuabtuinne* (wave sweeper), drawn by the horse Enbarr (water foam).

Irish mythology relates how the celebrated hero of the Fenian cycle, Fionn Mac Cumhaill, or Finn MacCool, built a causeway to Scotland to fight his rival, the giant Benandonner. The latter crossed to Ireland to engage with his foe, but fled in terror when he saw Mac Cumhaill (dressed as a baby) ripping up the causeway and leaving remnants behind (namely, the Giant's Causeway on the coast of Antrim and Fingal's Cave on the island of Staffa in Scotland) (see Chapter 27: Tourism and Leisure).

Perhaps one of the most famous of Irish legends associated with the sea is that of the Children of Lir, sometimes claimed as having provided the basis for Tchaikovsky's celebrated ballet *Swan Lake*. According to this legend, the four children of King Lir were transformed into swans by their jealous stepmother. For 900 years they lived as swans by daylight, only able to resume human form in the light of a full moon, on Lake Davra, the sea of Moyle and the lake island of Glora in Mayo, before their spell was broken with the arrival of Christianity to Ireland.

Imaginings inspired by the seas around Ireland are not only found in legend, mythology and folklore; they extend well into more modern eras and occur in myriad other creative forms, including song, music, drama, dance, literature, the visual arts, popular entertainment and even architecture and design. Engagements with the marine realm, whether daily, seasonal or simply occasional, have shaped the collective psyche of the Irish as a people and the character of the nation. They help frame relationships between the 'local' and the 'visitor', no matter how near or far the latter has travelled to reach the coast, and imprint themselves on our memories in multiple layers of recollection, extending from childhood to old age. This chapter offers some of these more subjective perspectives on the sea and will, hopefully, encourage readers to bring their own dreams and creativity to their interactions with the coast.

COIS FARRAIGE – BESIDE THE SEA

Imagine … a small boy, barefoot, laughs as he runs across a beach of hard wet sand in the late-November sunshine splashing through a rivulet left behind by the recently ebbed tide. An elderly woman leans against a promenade wall, her cheeks red as the wind buffets strands of hair peeking from behind the scarf wrapped around her head. A wiry young man stretches, moving one hand and then, one foot, his body in conversation with the rocky cliffside to the sound of rolling waves below. Sliced pan sandwiches of butter and ham, warm in the heat of June, are eaten, slow bite after slow bite, by a couple sitting on the bonnet of their car and looking out to sea, all middle-aged softness and freckles. In Ireland, the sea is our daily neighbour; the coast is for us all.

Coasts are, and always have been, attractive aspects of the world we live in and have a sustained meaning in many people's lives, both as places to inhabit and to visit. For centuries and across multiple cultures, people have been drawn to the coast and this is no different for Irish people, living on a small island nation that is always proximate to its blue edge. As authors, our backgrounds as geographer and architect-geographer, with parallel interests in history, culture and different aspects of representation, shape how this chapter describes that engagement. Other writers with other perspectives might equally document a wholly different set of engagements. It is part of the generosity of the coast that it leaves space for this multiplicity of human responses.

Fig. 14.2 ROCKY BAY, COUNTY CORK. a. Even the same stretch of coast can present many faces to the visitor. At times it can be angry, hostile and dynamic, such as here, with the waves of Storm Ophelia (16 October 2017) beating against the shore. [Source: Louise Hill]; b. on other occasions it can be calming, restful and relaxed, inviting picnics, dog walks and quiet contemplation. [Source: Zoë O'Hanlon]

Coastal landscapes have the power to exist as 'in-between', mobile, liminal, edge spaces; this is a feature of both their historical development and our contemporary cultural understanding (Figure 14.1). Our books document coastal landscapes from both such perspectives and draw on an approach to the coast as a relational blue space.[1] According to this way of thinking, the coastal landscape can be described in ways that suggest different but well-understood overlapping meanings, such as the coast, the beach, the sea, the shore, the ocean, the water: *cois farraige*. The hybrid interplay of contrasting characteristics – local/visitor, winter/summer, bright/dark, calm/wild, safe/risky – emerge out of coastal landscapes, but equally shape them (Figure 14.2). New research in cultural geography has begun to talk increasingly about 'wet geographies' of ocean, sea and blue space, all identifying the coast as a 'zeitgeist' setting that reflects the spirit of the times.[2] At the coast, new lifestyle and leisure practices (for example windsurfing, kayaking, surfing, coasteering) interact with long-established cultural practices (such as swimming and bathing, fishing, horse-riding, walking and sailing) (see Chapter 27: Tourism and Leisure), but all are mixed up in the identity of coastal spaces and people's perceptions and understandings of them. Equally, the ongoing capacity of the coast as a space for contemplation, leisure, escape and recovery must also be set alongside the built coast, its inhabitations, livelihoods and place-attachments.[3] Here, one must always balance a romantic and intermittent engagement with the coast against the practicalities of making a living in this often harsh space.[4] It is this interplay between active, passive and contested dimensions of coastal landscapes that provides the basis for this chapter, though it is one that always recognises an emotional affinity with the coast and its many hybrid spaces.

RELATIONAL GEOGRAPHIES OF THE COAST

The idea of a relational geography is a popular one in current human geographical research, inspired in part by the description of a place as a mix of location, locale and sense of place. Relational thinking proposes that any single place acts as a node or link, in relation to other places, but also in relation to the people who occupy it.[5] Relational geographies of the coast have these same physical, material, inhabited and imaginative dimensions. The coast is also, as reflected in other elements of relational thinking, mobile, connective and, above all, fluid in its relations with other spaces and places. Think of the ways in which we place the coast as a contrast to areas inland, or the shifting lines identifiable in estuaries and cliffs, produced by deposition and erosion. Coasts are both topographical and topological at one and the same time. The explorations of west Cork by Claire Conolly, Rachel Murphy and colleagues, using deep mapping techniques is an example of this (see Chapter 8: Monitoring and Visualising the Coast), where they draw from a range of literary and other cultural narratives to place the coast as a rugged, fractal and fluid space of engagement, and to then represent these elements through the medium of digital technologies. Equally, recent explorations of Dún Laoghaire Harbour (Figure 14.3) and Dublin Port by artist, Silvia Löffler, extend these metaphors around protection, reclamation and safety and the always uncertain struggle between humans and the sea in urban settings.[6]

The sea and its coast have a tidal relationship; this applies to its specific physical interactions as well as to its human inhabitations (see Chapter 2: The Coastal Environment: Physical systems, processes and patterns). If we think about our relationships with the coast, especially those of us who grew up inland, we recognise that these relationships ebb and flow across the course of people's lives. Many families have old photographs of babies and their grandparents playing in the sand, eating ice-cream, or relaxing on a deckchair, in an inter-generational inhabitation that is passed on almost by osmosis. In his memoir, novelist Colm Tóibín captures this well:

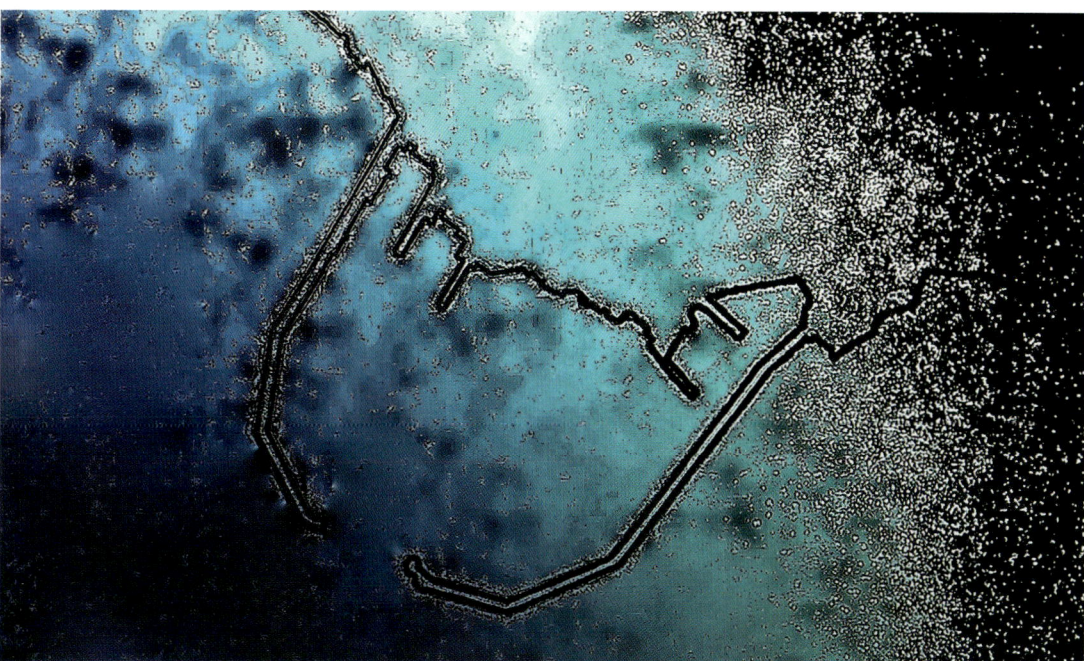

Fig. 14.3 SILVIA LÖFFLER'S DEEP MAP OF DUBLIN/DÚN LAOGHAIRE. Dr Silvia Löffler is an artist, researcher and educator in visual culture based in Dublin. She is interested in the emotional, social and cultural mapping of spaces – particularly at the coast – and much of her work combines the scientific precision of cartography with the fluidity and emotional subjectivity that has its inspiration in the imaginative use of colour and texture. The image shown here comes from an interdisciplinary participatory arts project, *Glas Journal: A deep mapping of Dún Laoghaire Harbour*, during which she created fourteen handmade artist's books representing sequential harbour locations bordering the sea between the West and the East Pier of Dún Laoghaire Harbour. Her work from this and other projects has been widely exhibited, including in the Terminal 1 building of Dublin Port. [Source: © Silvia Löffler]

> Recently, I have been going through the photographs that my mother kept beside her bed close to the poems and the missal and the list of important telephone numbers. So many of them were taken on those beaches on the Wexford coast, black and white photographs of moments when she and her sister and their friends were young or when their children were young … They were all as happy that summer, let us suppose, as they ever would be. They were on holidays. At night, they went to the bar of the Strand Hotel; during the day they lay on the beach and swam and watched the sky in case the sun might come out once more. The women wore slacks. There was a lot of laughter.[7]

Fig. 14.4 A LITTLE PLACE TO STAY BESIDE THE SEA. a. Aran Island Camping and Glamping, Inish Mór [Source: Aran Islands Camping and Glamping]; b. Cushendall, County Antrim [Source: © Tourism Northern Ireland]. There are at least 200 camping and caravan parks around the coasts of Ireland, ranging from small sites that cater primarily to the touring holiday-maker to enormous parks that can contain 200 or more chalets and mobile homes. Many of these originated, during and soon after the Second World War, as unplanned and sometimes ramshackle 'hutment' settlements, often created from old railway carriages and converted coach bodies, that sprung up to provide cheap seaside holiday accommodation for predominantly working-class urban families. Today's camping and caravan sites bear little resemblance to their predecessors and are now usually carefully landscaped, managed and enhanced with supporting infrastructure that may include shops, playgrounds, showers and laundry facilities. Rather than being simply a low-cost and generally spartan holiday accommodation option for the less-affluent, camping at the coast today can be seen as an expression of a wider emergent 'slow tourism' movement. This seeks to promote ideas of environmental sustainability, personal and communal wellbeing and enhanced social relationships, as a counter-balance to the stresses and pressures of modern, fast-paced capitalist society. [Source: © Tourism Northern Ireland]

The description of the beach as 'paradise on earth' applies even in the often cold and grey spaces of the Irish coast.[8] In other cultures and places, the little beach shack remains a precious node in that networked relationship over time, best seen in the New Zealand *bach*, the English beach-hut, or in the Irish coastal mobile home (Figure 14.4). Reflecting the idea of a mobile force moving us across space, the ebb and flow is also reflected in the 'draw' of the coast and its position as an edge, or limen, to the more central lived-spaces of everyday life. The intertidal zone is perhaps the ultimate liminal space, being never entirely land or sea and, according to Shields, this has always framed the seaside traditionally, as a sort of in-between place, where many of the rules of normal living might not apply.[9] To subvert the ban on mass during the penal times in eighteenth- to early nineteenth-century County Clare, the famous Little Ark of Kilbaha (Figure 14.5) was moved as a temporary church to the intertidal zone, where the laws of the land did not apply.

Permanent coastal dwellers have a very particular lived experience, quite different from that for whom the coast acts as a temporary or intermittent space in relation to home. Relatedly, the delicate placing of the coast, between remoteness and desirability, has always created an additional space to which artists and writers are irresistibly drawn. But, equally, tensions can sometimes exist between those who live by the sea all year round and the summer or occasional visitor.[10] Aspects of belonging, affordability and the making of a home by the coast – with its often seasonal jobs – scratch up against the presence of seemingly ephemeral holiday homes and residential dwellings, sitting cold and empty for much of the year; the 'feeling for the coast' can be a double-edged sword.[11] GIS mapping by the All-Ireland Research Observatory (AIRO) from

Fig. 14.5 LITTLE ARK OF KILBAHA. In the mid-nineteenth century, more than a quarter of a century after Catholic Emancipation had passed into law in Ireland (1829), there were still parishes where it was difficult, or impossible, in practice for priests to celebrate mass, conduct baptisms or give last sacraments to the many victims of the Famine. One of these was the parish of Carrigaholt in County Kerry, where the local landlord refused to allow a church to be built on his land. Such was the situation in 1839, when a new priest, Fr Michael Meehan, was appointed to the parish and arrived at an imaginative solution to the problem. Inspired by a bathing box he saw in use on the beach at Kilkee, Fr Meehan organised the construction of a small 'church on wheels'. This consisted of a timber frame covered in canvas, containing a low altar with a statue of the Sacred Heart and a crucifix. For five years, the Little Ark of Kilbaha, as it rapidly became known, was wheeled down to the beach and mass was conducted in all weathers, in the open air, between the hig-h and low-tide lines, where the law of the land no longer applied. Word of this novel arrangement soon spread, attracting visitors and wider attention and putting pressure on the landlord to release land for the construction of a more permanent church, Our Lady, Star of the Sea, at Moneen, a mile from the beach at Kilbaha. The first stone for this new church was laid in July 1857, with final dedication taking place in October of the following year. The ark is still preserved in the church at Moneen today and is celebrated in this fine stained-glass window. [Source: Krystyna Pomeroy]

the 2016 census reveals the ongoing clustering of second homes along the coast (Figure 14.6), many empty for eleven months of the year (see Chapter 27: Tourism and Leisure). This is being exacerbated by new geographies of coastal homelessness in Ireland. Contemporary anecdotal evidence from the Dingle Peninsula reports reduced access to housing (rental and owned) for local residents, caused by the issue of second homes, as well as a hybrid set of global and fast-growing forces linked to the needs of tourism and accommodation for seasonal workers. These pressures suggest the continuing fluidity of coastal landscapes and the potential erosion and hollowing out of both social and natural assets and resources.

HEALTHY BLUE SPACE: THE COAST AS A THERAPEUTIC LANDSCAPE

The draw of the coast and the flow of bodies towards it, especially for health and leisure purposes, have very old histories in Ireland. The first records of the use of the seaside for medical purposes date from very early in the eighteenth century, initially from the east coast north of Dublin in 1709; they note that people were copying the new British fashions for seawater drinking and bathing.[12] By the early decades of the nineteenth century there was already a thriving holiday trade from Limerick to Kilkee, where passengers were carried down the Shannon Estuary to Kilrush on steamers, with the journey being then completed by horse and cart. By the same period there were well-developed bathing cultures in Malahide and Sandymount Strand, records exist of complaints by local female 'dippers' (people who assisted bathers, of the same gender, in and out of the sea) at Blackrock in the 1830s about the disruption to their work, caused by the construction of the new Dublin–Kingstown railway. More broadly, across Ireland and Britain, the rapid development of the railway networks in the 1840s and 1850s revolutionised access to the seaside, especially for the less wealthy (see Chapter 19: Changing Coastal Landscapes). Towns like Bray were invaded by cheap day-trippers, driving the original posher denizens out to other locations; Greystones received the nickname 'Rathmines-super-Mare' in the 1860s and 1870s.

The ways in which coastal landscapes were developed as social spaces, on the back of these foundational health and leisure performances, threw up some surprising examples. While many seaside resorts before the middle of the nineteenth century had primarily Protestant middle-class histories, older leisure uses of the seaside were occasionally recorded. Wood-Martin recounts the amazement of Inglis in the 1830s at meeting a group of poor

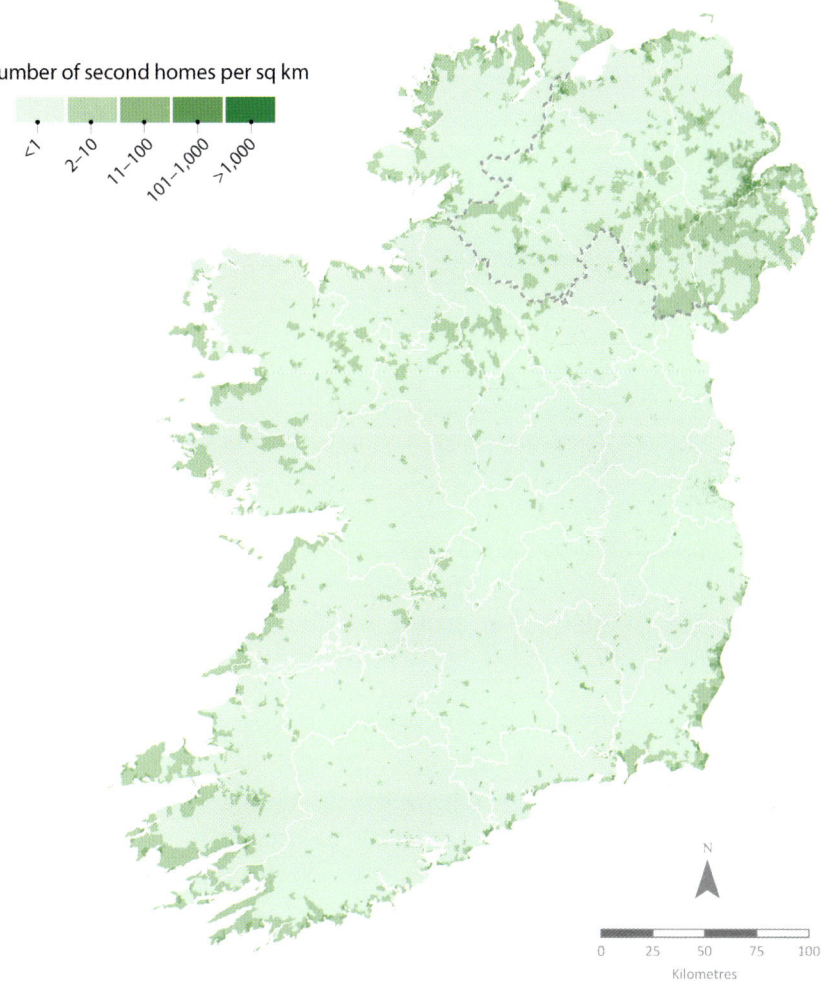

Number of second homes per sq km

<1 2–10 11–100 101–1,000 >1,000

N

0 25 50 75 100
Kilometres

Fig. 14.6 SECOND HOMES IN IRELAND. Holiday and second homes constitute a significant, and sometimes highly contentious, proportion of the overall housing stock of many coastal communities, particularly on the western and northern seaboards. Often lying empty for eleven months of the year, holiday homes frequently contribute to an overall 'gentrification' of outlying and rural settlements. In addition, they can exacerbate the highly seasonal viability of shops, pubs and other facilities, upon which the permanent residents depend, as well as inflating property prices beyond the reach of young first-time buyers from within the coastal communities themselves. Properties with a view of the sea, in particular, will often attract a higher price when sold as holiday accommodation. Due to differences in definition and interpretation adopted in the surveys, this map shows the distribution of 'unoccupied holiday homes' from the 2016 census in the Republic of Ireland and, for Northern Ireland, 'household spaces without usual residents' from the 2011 census. These latter include a slightly broader category of properties, including vacant and second homes, as well as holiday homes. [Data sources: Central Statistics Office, 2016 and Northern Ireland Statistics and Research Agency, 2011]

country-dwellers (known locally as 'sea-pikes') on their way to the seaside near Strandhill:

> I was surprised to meet every few hundred yards on the road (i.e. from Bole to Sligo) carts heavily laden with country people, many of them of the lowest order, and with different articles of furniture piled up, or attached to the carts; and I learned with some astonishment that all these individuals were on their way to sea-bathing. This is a universal practice over these parts of Ireland. A few weeks passed at the seaside is looked upon to be absolutely necessary for the preservation of health; and persons of all classes migrate thither with their families. On my way to Boyle, I met upwards of twenty carts laden with women, children and boys.[13]

There were also older histories of post-harvest visits to the seaside, identified as being a characteristic practice of western Atlantic coasts from Scotland to San Sebastian,[14] the above reference being a prime example. Descriptions from 1940s' Ballybunion recount similar post-harvest visits by farming families from inland Munster, where ladies' petticoats were recorded as billowing in the windy waves, leading to loud shrieking and hilarity.[15] That sense of release and letting go seems an ongoing capacity of the coast, as will be considered below. The impacts of the influx of wealthy summer visitors – often from the wealthier Anglo-Irish and middle-class Catholic families – were also recorded by nineteenth-century visitors. In Kilkee, various travellers' accounts noted that 'almost all of the houses in the village were given up for rent, with the locals bunking in with family and neighbours in the surrounding countryside'. Other travellers made explicit mention of poor local people putting string across the road north of the town towards Doonbeg, so they could stop carriages and beg for help.[16]

Yet throughout their histories (and even surviving today), the social elements that made the seaside town are almost stereotypical: funfairs, amusements, dance-halls, fish-and-chip shops, races, swimming, golf and donkeys (Figure 14.7). Most towns had their own specific catchments: Limerick for Kilkee, Cork for Youghal, Waterford for Tramore, Belfast for Bangor and so on. There is even a part of the riverside in Limerick city known as Little Kilkee, for those too poor to get to the coast. The design and management of seaside towns was developed across the Victorian era and generally made an attempt to instil some sort of order on settings that were instinctively designed to be disorderly and free. There was an odd obsession with Brighton – probably because of the royal patronage of this seaside resort – with many of the different towns labelling themselves the 'Brighton of the North' (Portrush), 'East' (Bray) or 'South' (Tramore). Bray took the comparison so seriously that it copied the Brighton promenade by-laws and made several attempts to build a pier.[17]

The ways in which gender separations were managed at the seaside tells us much about the clash of order and disorder, but equally provides an insight into the often complex Irish relationship with the body. In the early years of seaside resorts, men tended to bathe naked. As more children and families began to use beaches, men were then either confined to very early morning swimming or moved to 'men-only' swimming spots like the Guillamene in Tramore or Burns' Cove in Kilkee (see Chapter 19: Changing Coastal Landscapes). Different parts of seafronts were reserved for women at different points and there was even a nun's beach at Bray. These were regulated by fines, paid by over-inquisitive young men. By the 1920s, however, most of these gendered spatial constraints had been removed. In fairness to the young women of Tramore, this was the only resort at which females were specifically chastised for frequenting the male spaces.[18]

More recently, research in health geography has begun to explore the relationships between natural spaces and human health in a broader sense, as part of an interest in 'healthy nature' and being outdoors, something that is also at the heart of much

Fig. 14.7 DONKEY RIDES SOMETIME BETWEEN 1860 AND 1883, KILKEE, COUNTY CLARE. In the days before the advent of the motor car, donkeys were commonly used in seaside resorts to help transport people and their luggage from the train station to their holiday accommodation. In time, with the increasing availability of buses and, eventually, the private motor car, donkeys were no longer needed for their original purpose though, happily, many remained on the beach as a tourist attraction instead. For generations of children and their parents, donkey rides along the beach were an essential part of a seaside summer holiday, along with Punch and Judy shows, fairgrounds, fish and chips and ice cream. [Source: National Library of Ireland, call no.: STP_1660]

Fig. 14.8 THE SEASIDE AND HEALTH. Led by a team of medical and marketing professionals, social scientists and innovators, environmental scientists and ecologists from NUI Galway, the research project Nature and Environment to Restore Health (NEARHealth) explores, identifies and promotes the health-giving benefits of contact with and immersion in natural spaces. These include, as seen here at the beach and dunes, Kilshannig (Maharees Beach), County Kerry, the use of beaches and other coastal locations in different recreational pursuits. The project was jointly funded by the Environmental Protection Agency (EPA) and the Health Service Executive to support the implementation of the EPA Strategic Plan 2016–2020, Our Environment: Our wellbeing and the Healthy Ireland national framework for action to improve the health and wellbeing of the people of Ireland. [Source: Gesche Kindermann]

contemporary public health policy.[19] This has also involved moving beyond the traditional urban focus on green space to think about the notion of 'blue space'. Defined as 'health-enabling places and spaces, where water is at the centre of a range of environments with identifiable potential for the promotion of human wellbeing',[20] new research on blue space has exploded in the past decade.[21] Describing the coast as a 'blue gym', researchers in Cornwall (which has a very similar environment to Ireland) are beginning to use a range of psychological instruments, local and national surveys and new technological measures of physical activity (accelerometers, GPS, fitness trackers) to produce strong evidence for the benefits to human health of living by or using the coast.[22] This represents a new direction in public health policy, linked to concerns around reductions in physical activity and a rise in obesity and sedentary indoor lifestyles. Recent work by NUI Galway's NEARHealth project has identified multiple ways in which being by and interacting with coastal blue space has positive health benefits (Figure 14.8). For example, people who engage in beach clean-ups show enhanced psychological health, while the coastal environment also benefits; caring for the beach requires an act of caring that is equally beneficial to the care-giver.

LIVED EXPERIENCES OF THE IRISH COAST

In the Ireland of today, people particularly embrace the coast for leisure, wellbeing, thrill-seeking and health. Coasts have, for many people, become linked in their imaginations to the numerous outputs of cinema and television that feature coastal landscapes, either as emotive backdrops or as an integral part of the story. Additionally, coastal livelihoods from often seasonal economies also join hands with more traditional forms of coastal occupations, such as fishing and coastal farming, to generate opportunities for day-to-day living from and by the sea. Previous generations relied on the sea for survival but accompanying this was the natural fear of drowning. People purposely did not learn to swim as it was felt to be simpler to drown without a struggle. Nowadays, the lifestyle draw of the coast has transformed such fears; the action and image of surfing in Ireland has resulted in people being able to buy wetsuits with their groceries and body boards with their hardware (see Chapter 27: Tourism and Leisure). Beaches that were once covered in picnic blankets in the 1950s are now covered in black figures tumbling and gliding in the low waves and children can spend hours on their knees making sandcastles, while remaining relatively warm. The wetsuit has transformed the 'day out' at the Irish coast.

Walking, fishing, surfing, swimming and sailing are all common activities at the Irish coast. These activities involve the entire body in movement and as they are repeated across the seasons and over many years, they build up an embodied knowledge of the space where land meets sea. The repetitive nature of placing one foot in front of another on a coastal walk, the pause generated while holding a fishing rod at the end of a pier, the moment of adrenaline rush when the surge of the wave catches the surfboard with the weight of its carefully placed rider, or the rising and falling of the

Fig. 14.9 CORK SIMON COMMUNITY'S ANNUAL CHRISTMAS DAY SWIM AT WHITE BAY, CORK HARBOUR. In the summer months, a dip in the ocean can bring welcome, cooling respite from the heat of the sun. In the heart of winter, it can be a somewhat more daunting experience, bringing the joyful terror of running, shrieking and shivering, into the anticipated, breath-catching iciness of the sea. At several points around Ireland's coasts, groups and individuals gather at Christmas and new year – come sun, rain or even sleet and snow – to enter the sea en masse and raise much-needed funds to support good causes. One such event is the annual sponsored Christmas Day Swim at White Bay, on the east side of Cork Harbour. This has been taking place for more than twenty-five years and has raised over €250,000 in that time, to help the Cork Simon community provide essential services for homeless men and women in the city. [Source: Diane Cusack]

Fig. 14.10 *The Bathers*, JOHN LUKE (1929), OIL ON CANVAS. The beach is a place where many of the conventions of daily life can be put on hold. It is a place for socialising, relaxing, having fun and being seen to be having fun; enjoying the feel of sun and wind upon one's face and limbs, and even for trading the restrictions of formal daily clothing for the more colourful and exuberant fashions of the beach. The painting epitomises the ingredients of a beach holiday in Ireland, and also foretells the popularity of the beach later in the twentieth century and to the present day, as a worldwide destination for total recreation and commerce. [Source: Michael Tropea, courtesy of the O'Brien Collection]

ocean's depth felt while sea-swimming, all place the individual in a closely entwined relationship with the elements of the coast: land, sea, sky. These visceral practices are creative in nature and can also be moments where emotion comes to the fore. The coast can be experienced as restorative, a place to 'breathe'.

Swimming, and the places where people swim at the coast, also have strongly accretive and relational effects. Many hardy sea-swimmers have entered the water daily all of their adult lives (Figure 14.9). Many do so because their parents and grandparents did, and they in turn teach their own children and grandchildren to do so in those same shared spaces.[23] For the older and less able, the sea provides a space where unhealthy land-bodies become healthy sea-bodies. As one swimmer said, 'the only time I don't limp is when I am in the water'. Coastal swimming experiences, as noted earlier, trigger memories of different places, times and objects, as well as of people (Figure 14.10). From the eighteenth century onwards, in numerous towns around the coast, now-historic sea baths provided safe communal locations – at the shore but separated from the open sea – where many of these experiences were first triggered (Figure 14.11). We have suggested elsewhere in this chapter that improved health and wellbeing emerge from swimming as an accretive practice within blue space, particularly through repeated affective and emotional encounters,[24] but sometimes one's need to swim, in the company of other swimmers, is based on more immediate needs. Ruth Fitzmaurice's powerful memoir, *I Found my Tribe*, recounts the value of swimming with other women in Greystones, County Wicklow, as a means of coping with her husband's ultimately fatal illness.[25]

Fig. 14.11 BATHS AT a. ENNISCRONE, COUNTY SLIGO AND b. BRAY HEAD. Throughout history, from the thermae and spas of ancient Greece and Rome, to the sweat houses of the Vikings and several North American peoples, the hammams of the Islamic world, and the holy wells of Ireland, the therapeutic three-way relationship between water, health and place has featured prominently in many cultures and contexts. While informal sea swimming has probably taken place as long as people have occupied the coast, the idea of designating official swimming places and constructing facilities to support the activity came much later. Perhaps the earliest of these endeavours in Ireland was the proposal, put forward in 1754, to build a bathing place in Blackrock, County Dublin. It took almost a century, and the completion of the Dublin to Kingstown Railway in 1834, before the project took shape, taking advantage of the nearby Blackrock Railway Station. Other baths along the coast soon followed, including at Dún Laoghaire and Bray Head, as well as at Salthill in Galway and Portrush and Portstewart in County Antrim. Many of these offered a range of services and options for a modest charge, including sea and fresh-water bathing, heated as well as cold-water swimming, medicinal and seaweed baths, children's paddling pools and tearooms. [Source: a. Mark Waters, ccBy4.0By4.0; b. National Library of Ireland, call no.: VAL 1993]

THE COAST OF IRELAND ON SCREEN

Darius Bartlett

Production of screen media (including feature films, documentaries, high-end television drama, television animation and computer games) is one of the fastest growing industries in the world. In Ireland the audiovisual sector generated €1.05 billion in gross added value in 2016 and supported almost 17,000 full-time equivalent jobs.[26] Screen Ireland (the national development agency for the industry), the Irish Film and Television Network and Northern Ireland Screen all offer extensive support services and resources to help attract potential filmmakers to Ireland, including comprehensive guides and databases to assist in identifying suitable filming locations for their projects. Furthermore, in the Republic, the industry benefits from important government support, including a competitive tax incentive known as Section 481, which offers tax credits of up to 32 per cent of eligible Irish expenditure.

The sector is also extremely important in helping showcase Ireland's landscapes and cultures, and film tourism brings an increasingly important dimension to the country's wider tourism industry. Many coastal locations feature strongly in this regard (Fig 14.12). The trend probably started with the 'phenomenal influx of tourists'[27] that is still regularly drawn to sites around Cong in County Mayo, where some scenes from *The Quiet Man* (1952),[28] starring John Wayne and Maureen O'Hara, were filmed (Fig 14.13). These include Lettergesh Beach, 50km west of Cong, where Ford's iconic Inisfree horse-race scene was shot. Other notable film locations around the coast of Ireland include Inch Strand and Coumeenoole Beach, both in County Kerry, which featured in David Lean's *Ryan's Daughter*;[29] Curracloe Beach in County Wexford, which was used as a surrogate for Omaha Beach in Normandy in *Saving Private Ryan*,[30] starring Tom Hanks and which also featured in John Crowley's 2015 film, *Brooklyn*,[31] starring Saoirse Ronan; and Skellig Michael, which substituted for the planet Ahch-To in *Star Wars*

293

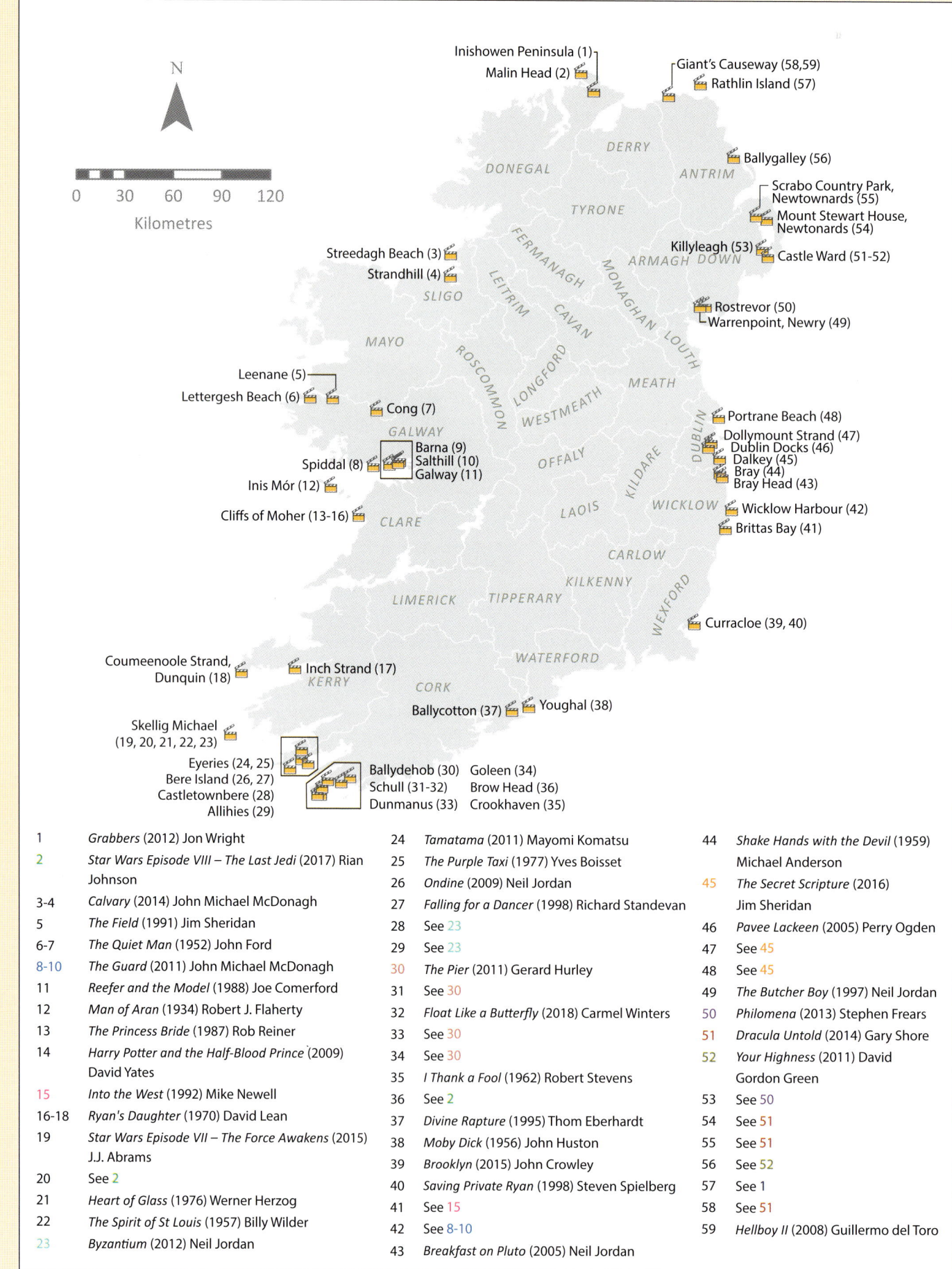

N

0 30 60 90 120
Kilometres

Inishowen Peninsula (1)
Malin Head (2)
Giant's Causeway (58,59)
Rathlin Island (57)

DERRY
DONEGAL
ANTRIM
TYRONE
Ballygalley (56)

Scrabo Country Park, Newtownards (55)
Mount Stewart House, Newtonards (54)

Streedagh Beach (3)
Strandhill (4)

FERMANAGH
MONAGHAN
ARMAGH DOWN
Killyleagh (53)
Castle Ward (51-52)

SLIGO
CAVAN
LOUTH
Rostrevor (50)
Warrenpoint, Newry (49)

MAYO
LEITRIM
ROSCOMMON
LONGFORD
MEATH

Leenane (5)
Lettergesh Beach (6)
Cong (7)

WESTMEATH

GALWAY
Barna (9)
Salthill (10)
Galway (11)
Spiddal (8)
OFFALY

DUBLIN
Portrane Beach (48)
Dollymount Strand (47)
Dublin Docks (46)
Dalkey (45)
Bray (44)
Bray Head (43)

Inis Mór (12)
Cliffs of Moher (13-16)
CLARE
LAOIS
KILDARE
WICKLOW
Wicklow Harbour (42)
Brittas Bay (41)

CARLOW
KILKENNY
LIMERICK TIPPERARY
WEXFORD
Curracloe (39, 40)

Coumeenoole Strand, Dunquin (18)
Inch Strand (17)
WATERFORD
KERRY
CORK

Ballycotton (37) Youghal (38)

Skellig Michael (19, 20, 21, 22, 23)
Eyeries (24, 25)
Bere Island (26, 27)
Castletownbere (28)
Allihies (29)
Ballydehob (30) Goleen (34)
Schull (31-32) Brow Head (36)
Dunmanus (33) Crookhaven (35)

1	*Grabbers* (2012) Jon Wright	
2	*Star Wars Episode VIII – The Last Jedi* (2017) Rian Johnson	
3-4	*Calvary* (2014) John Michael McDonagh	
5	*The Field* (1991) Jim Sheridan	
6-7	*The Quiet Man* (1952) John Ford	
8-10	*The Guard* (2011) John Michael McDonagh	
11	*Reefer and the Model* (1988) Joe Comerford	
12	*Man of Aran* (1934) Robert J. Flaherty	
13	*The Princess Bride* (1987) Rob Reiner	
14	*Harry Potter and the Half-Blood Prince* (2009) David Yates	
15	*Into the West* (1992) Mike Newell	
16-18	*Ryan's Daughter* (1970) David Lean	
19	*Star Wars Episode VII – The Force Awakens* (2015) J.J. Abrams	
20	See 2	
21	*Heart of Glass* (1976) Werner Herzog	
22	*The Spirit of St Louis* (1957) Billy Wilder	
23	*Byzantium* (2012) Neil Jordan	
24	*Tamatama* (2011) Mayomi Komatsu	
25	*The Purple Taxi* (1977) Yves Boisset	
26	*Ondine* (2009) Neil Jordan	
27	*Falling for a Dancer* (1998) Richard Standevan	
28	See 23	
29	See 23	
30	*The Pier* (2011) Gerard Hurley	
31	See 30	
32	*Float Like a Butterfly* (2018) Carmel Winters	
33	See 30	
34	See 30	
35	*I Thank a Fool* (1962) Robert Stevens	
36	See 2	
37	*Divine Rapture* (1995) Thom Eberhardt	
38	*Moby Dick* (1956) John Huston	
39	*Brooklyn* (2015) John Crowley	
40	*Saving Private Ryan* (1998) Steven Spielberg	
41	See 15	
42	See 8-10	
43	*Breakfast on Pluto* (2005) Neil Jordan	
44	*Shake Hands with the Devil* (1959) Michael Anderson	
45	*The Secret Scripture* (2016) Jim Sheridan	
46	*Pavee Lackeen* (2005) Perry Ogden	
47	See 45	
48	See 45	
49	*The Butcher Boy* (1997) Neil Jordan	
50	*Philomena* (2013) Stephen Frears	
51	*Dracula Untold* (2014) Gary Shore	
52	*Your Highness* (2011) David Gordon Green	
53	See 50	
54	See 51	
55	See 51	
56	See 52	
57	See 1	
58	See 51	
59	*Hellboy II* (2008) Guillermo del Toro	

Fig. 14.12 THE COAST OF IRELAND ON SCREEN. For more than half a century, the coast of Ireland has provided sought-after locations for both the domestic and the international film industry. This brings many financial and employment benefits, particularly in the technical and hospitality sectors, as well as bringing the attractiveness of the island's scenery to a worldwide audience, thus helping promote Ireland as a tourism destination.

Episode VII – The Force Awakens[32] and *Star Wars Episode VIII – The Last Jedi*.[33]

Through the medium of film, Ireland's coasts are frequently represented on screen as a signifier of the country's real-world natural rugged beauty, but in many cases the landscape is recontextualised and used as a stand-in for the otherworldly. Thus, audiences are alternately encouraged to recognise the landscape as being part of Ireland, but also to not recognise it as 'Ireland' but to appreciate its qualities regardless. In both situations, the country benefits from subsequent film tourism (in the case of the latter, often capitalising on existing fan cultures that surround popular film franchises, such as the *Star Wars* series, thereby offering fans an opportunity to visit these 'alien' planets).

More recently, and on the smaller screen, several locations around the coast of Northern Ireland – including Castle Ward in County Down and Ballintoy Harbour and Portstewart Strand on the north coast – were used as locations for the hugely successful *Game of Thrones* television series (Figure 14.15).[34] This has led to significant numbers of tourists visiting the province specifically to tour sites where filming of scenes from the

Fig 14.13 *The Quiet Man*. John Ford's 1952 film *The Quiet Man*, based on Kerry writer Maurice Walsh's short story 'Green Rushes', won Academy Awards for Best Director and Best Cinematography, and remains one of the most enduring films of all time. Shot in and around Cong in County Mayo, and set in the fictional coastal village of Inisfree, the film tells the story of the developing romance between villager Mary Kate (played by Irish-born actor Maureen O'Hara) and retired Irish-American boxer, Sean 'Trooper Thorn' Thornton (John Wayne), who was also born in the village but who grew up and spent most of his life in the United States. For more than sixty years since *The Quiet Man* was released, hoards of tourists have flocked to Cong to view locations immortalised in the film for themselves, including many of the houses where the action took place and the marvellous Lettergesh Beach, 50km to the west, where the Innisfree horse race was filmed. [Source: public domain, Wikimedia Commons]

Fig 14.14 THE DEMESNE AT CASTLE WARD, ON THE SHORES OF STRANGFORD LOUGH IN COUNTY DOWN. Owned by the National Trust, the courtyard and tower of the demesne became the Castle of Winterfell in the HBO television series *Game of Thrones*. This has encouraged an active and diverse tourism spin-off industry aimed at fans of the series, at Castle Ward and other sites around the Northern Ireland coast. This includes tours led by guides in theme-relevant costumes and gift shops filled with a prolific range of *Game of Thrones* merchandise. [Source: © Tourism Northern Ireland]

Portstewart Strand
Where Jaime and Bronn were captured by Dornish soldiers on the coast of Sunspear

Mussenden Temple
Dragonstone, where Melisandre convinces Stannis Baratheon to reject the seven old gods of Westeros

Binevenagh, Limavady
Where Daenerys and her dragon Drogon find refuge in the Dothraki Grasslands after they flee the fighting pits of Meereen

Giant's Causeway

Ballintoy Harbour Pyke, the home of Theon Greyjoy and the Ironborn

Fair Head Dragonstone Cliffs, where Jon Snow encounters Daenerys and her dragon

Murlough Bay Slaver's Bay

Dark Hedges, Ballymoney Part of the Kingsroad

Cushendun Caves Where Melisandre gives birth to a shadow demon

The Hidden Village of Galboly Runestone

Shillanavogy Valley The Dothraki Sea

Magheramorne Quarry

Quintin Castle, Portaferry Castle Stokesworth

Castle Ward Winterfell

Inch Abbey, Downpatrick Where the **War of the Five Kings** began

Tollymore Forest Park The Haunted Forest in Westeros

LOUGH FOYLE · Coleraine · Ballymoney

DERRY · Derry/Londonderry

ANTRIM

DONEGAL · Strabane

Ballymena · Larne

TYRONE · Antrim · Carrickfergus · Bangor

· Omagh

LOUGH NEAGH · BELFAST ★

Lisburn

STRANGFORD LOUGH

· Portadown

DOWN

FERMANAGH · Enniskillen

ARMAGH · Armagh

MONAGHAN

· Newry

CAVAN

LOUTH

N

0 10 20 30 40
Kilometres

Fig 14.15 Coastal locations in Northern Ireland that feature in the HBO series, *Game of Thrones*. The Dothraki Sea, Dragonstone, Winterfell, Pyke and Slaver's Bay are not to be found in any regular gazetteer or list of place names of Ireland. For fans of the enormously successful HBO television series, *Game of Thrones*, however, they are immediately recognisable as locations in the fictional land of Westeros, where the protagonists' struggles to claim the Iron Throne unfold. As shown in this map, many of the key settings for events in the series were shot in locations on and around the coast of Northern Ireland; from the mouth of Lough Foyle, via counties Antrim and Down, to the Mountains of Mourne. However, no doubt to the disappointment of *Game of Thrones* fans, Daenerys Targaryen's dragons, seen flying above the Dothraki Sea and elsewhere in the series, were added via computer-generated imagery in post-production, and are not part of the native coastal fauna of Ireland.

series took place (Figure 14.14). A 2017 report by Northern Ireland Screen anticipated that '[It] is possible that *Game of Thrones* will deliver the widest media exposure Northern Ireland has ever received outside of politics and the troubles'.[35] Furthermore, we are now seeing a welcome shift in the portrayal of Northern Ireland, away from the more traditional urban settings, rooted in gritty realism and history, towards greater use of rural settings and landscapes, in which the coast often plays a major starring role. Admittedly, in the case of *Game of Thrones* these are recast as fantasy locations, but nonetheless, this new onscreen image is one that is particularly encouraged by both the tourist and the film industries of Northern Ireland.

The coast is a fluid space, both physically and socially. The stuff of the coast, its matter, is always moving. The sea itself rises and falls with the daily tides, ebbing and flowing according to the springs or neaps. This moving water reveals and conceals, and reveals again in a dance, an interchange with sand, grass, rock, earth and wind, sometimes with force and drama and sometimes so gradually that the interplay is almost imperceptible (Figure 14.16). This flow of materials, our moving world, becomes very apparent to us at the coast. Our feet get wet, salt spray dries on our cheeks, sand exfoliates our skin. We are exposed. The sticky slipperiness of seaweed underfoot, the rhythm of the waves rushing and withdrawing, the sink, sink, sinking of the sun into the horizon, the prickle of thistle through the picnic mat. Our senses are heightened. We become more aware of our bodies: our

Fig. 14.16 WAVES BREAKING ON INISH MÓR, ARAN ISLANDS. This photograph provides an aerial view of the shoreline, with waves breaking as they hit land, giving an impression of the depths and ranges of light, texture and colour to be found where the sea meets the land. [Source: Unukorno, flickr, CC BY 4.0]

verticality in relation to the horizontality of the sea; our size in relation to its expanse; our vulnerability in relation to its strength. By its nature, the coast encourages a greater awareness of spatial sensations of our surroundings. These encounters between our bodies and our physical surroundings are often taken for granted. If we were to bring them more overtly to our attention on a regular basis, to become more actively aware of how we, as embodied human beings, engage with our more-than-human and multi-dimensional surroundings at the coast, we might make different decisions about its future. We might better recognise how so very

fluid the relationship between land and sea at the coast really is, how cyclical and transitional our interaction with it is and how insubstantial policies of 'coastal protection' are, when considered in the context of the inexorable operational forces of natural coastal systems. What is often considered fragile, the coast itself, is actually where the power lies; its fluidity and ever-changing nature is its strength, and it is more and more important that we recognise this. In becoming more aware of our embodied spatial sensations when at the coast, we better understand our insignificance, our 'small' place in the long-time of the coast.

ARCHITECTURE OF COASTAL ESSENCES: VICO, DUBLIN BAY

Anna Ryan

As a lecturer in architecture, I am fascinated by coastal spaces, their landscapes and possibilities. I have chosen to focus in this short piece on coastal architecture on one jewel from the necklace of constructed swimming places that texture the southern sweep of Dublin Bay. Their locations extend from the bend of Dublin Bay, at Vico in Dalkey, through Sandycove's Forty Foot, the remains of the Dún Laoghaire and now dismantled Blackrock Baths, the low enclosure of the Victorian pool on Sandymount Strand to the active Half Moon club on the South Wall, where Dublin's River Liffey meets her sea (Figure 14.17). Elemental in presence and simple in use, these bathing places epitomise the most significant qualities that architecture can provide for people. What draws me to these spaces as exemplars of coastal architecture is their responsiveness to the essence of the coastal experience. Each responds to the immediacy of its physical conditions at the land–sea edge (Figure 14.18). Each presents the role of architecture at its best: architecture as a careful observation of its context of place; of highlighting and augmenting that context through a specificity of construction, which offers the opportunity for new forms of social interaction.

In facilitating sea-swimming and access to the tidal waters, these six bathing places around Dublin Bay are deeply sensitive to the variety of dimensions and geometries to be found in this coastal area, the materiality that this stretch of the city's edge presents. At Vico baths, overlooking Killiney Bay and towards the undulating horizon of Bray and the Wicklow Mountains beyond (see Chapter 19: Changing

Fig. 14.17 ROYAL VICTORIA BATHS, NEAR DÚN LAOGHAIRE (AUGUST 2017). These baths are typical of the seaside swimming pools and bathing places constructed along the coast of Dublin Bay, and more widely. Built in 1843 and in use for over 150 years, the Royal Victoria Baths finally closed for good in the mid-1990s and stood disused and forlorn, with crumbling concrete, wave-battered sea walls and empty spaces where the doors and windows once were. As a built space, the mirror stillness of the water in the pools reflects the light of the sky and contrasts with the darker, urgent restlessness of the sea. The smooth contours of the concrete stands in juxtaposition to the roughness of the unadorned rocks, while the steps up to the baths provide a physical and metaphorical connection between the tamed enclosure and the wilder, wider openness. Graffiti on the walls tells that a new generation of youths claimed ownership of this space, endowing it with fresh meaning and an alternative cultural legitimacy. However, in March 2015, permission was given by the Department of the Environment for the baths to be filled in. The Victorian buildings were refurbished and the area landscaped and redeveloped as a complex of artists' studios, cafés and a gallery. As well as hosting countless families over the last century, these coastal bathing places to the south of the Liffey were the favoured haunts of Dublin's literary and cultural luminaries. These included James Joyce, Oscar Wilde, Brendan Behan and Samuel Beckett. [Source: Mary Therese O'Neill]

Fig. 14.18 CONTEMPORARY VIEW OF THE COAST AT BRAY, COUNTY WICKLOW. Coastal landscapes, such as this one, are increasingly dominated by the concrete and steel of sea defences, beach tourism and the designs of modern buildings, most of which could be found in almost any part of the world. The large family residences and more atmospheric buildings of bygone days are being absorbed into this uniformity of international architecture and the needs of present-day living and communications. [Source: Mark and Dave Murphy]

Coastal Landscapes) (Figure 14.19), the approach to the water is made from a height: through a gap in a wall, along a narrow, often-overgrown path above the sheer railway cutting, across a bridge, along a fine iron handrail, down a steeply paced set of steps, around and along another stretch of path bordered by high tumbling grasses, and doubling-back around a final corner, where the destination becomes finally visible, lodged on the rocks at the bottom of a cascade of steps. Descending these steps, some additional protection is offered from the vertical openness by a row of old surfboards strapped to the now azure-blue painted iron handrail, its blueness standing apart from the shifting tones of the

Fig. 14.19 VICO BATHING PLACE. Nestled below the villas and gated communities of Killiney Hill, between the railway and the sea and with views across to Dalkey Island, Vico bathing place has been a key social meeting-point on the coast of south Dublin as much as it has been a place for swimming. The sea runs deep but cold right to the foot of the rocks, encouraging adventurous bathers to leap and hang briefly in the air before plunging into the water below, while concrete stairs and iron railings are available to the less intrepid, and facilitate return to land. Traditionally, this was one of several locations along the coast where naked bathing was tolerated and, while less common than in earlier days, the practice does continue to this day. [Source: Brian Kenny/Outdoor Swimming Ireland]

frequently misty-grey shimmer of the sea below. Metal bolted onto rock and concrete, metal plunging into the sea, the delicate form of its thinness is highlighted by its surroundings. From above, the sea's moving energy contrasts with the crisp flatness of the concrete surfaces, while the rock acts as a material intermediary: once fluid liquid, now solidified, the grain of its angled strata curved in places, in part by the cyclical working of the water.

Concrete is poured onto granite and textured for grip, exfoliating the bare soles of the swimmers when dry, or puddled with their drip-water or the rain or spray from the sea. From the side, the constructed nature of the place is apparent, layers of concrete over concrete, the repairs of the generations demonstrating the longevity of use and care. The changing shelter sits, with its door-space and two window openings, nestled at the foot of the cliff. Beyond it and below lies the enclosure of a pool, bounded in part by natural outcrops of granite and infilled with concrete. In calmer weather, the pool holds still water, ruffled only by the wind and the movements of its bathers; on a stormy day or with higher tides, the pool is overwhelmed by the inundations of a swelling sea. Gaps between the larger rocks hold passageways to the pool, or other informal places to change or to shelter: the architecture of the rock. No longer the sole remit of male-only bathers, on busy days bodies sit here and there, leaning against the wall of the hut, or against an accommodating slope of rock, gathering to swim, sunbathe and talk, a regular habit in the local landscape of lives.

To enter and exit the rolling water can be a real feat. Walking down the steps, one's left hand gripping the metal rail or descending the ladder, one's back to the sea, then push-off and into the experience of the weightless depths. Once immersed, one's spatial experience continues; one is aware of dimension and time, the fast-rising hill of the rugged coastline fracturing the landing of the morning sun, or hiding itself in the evening shade. The swelling surge of the seawater lifts and rolls the body closer to and from the shifting granite edge. Then, when in a big sea, the careful timing needed to reach out, grasp the iron, regain the weight of the body onto ladder or steps and rise from the water, returning to the seeming stability of land. There is a great spatial intensity to this material experience of the body in this location, skin moving from a height of stone into the sea and back again. Vico: a composition of four elements – rock, metal, concrete and seawater – wrapped by the sky and experienced as an architecture of coastal essences by the moving body.

WHERE LAND MEETS SEA: AN EXPLORATION OF COASTAL LANDSCAPES

Anna Ryan

People are drawn to the coast, to its ever-moving rhythms and cycles. This short section presents some extracts from a large body of work carried out at two coastal locations in Ireland: the Maharees Peninsula, County Kerry, and at the South Wall in Dublin Bay.[36] The studies are based on conversations with people at these locations. The research focused on generating an awareness of the relationships forged between one's self and one's surroundings. Attention was drawn to people's embodied knowledge: the knowledge our body generates itself by being, moving and acting in a place, how, through our body, we feel and respond to what's around us, creating knowledge of the world that we tend to take for granted, that we don't often put into words.

Sixty-two participants in Kerry and Dublin were engaged in photography and drawing to enable this exploration of a coastal spatial experience. The idea was that as we become more and more aware of innate spatial knowledge, our bodies, in relationship with the space of the coast, together with our sense of being fully part of our ever-emerging world, become heightened. We become empowered by these awarenesses. This awareness can generate new ways of presenting and representing a spatial knowledge and environmental experience to ourselves. Such understanding puts people and communities in a better position to make decisions about coastal spaces, of where land meets sea: the 'coastal edge' (Figure 14.20).

Participants in the different studies used their bodies as the means by which they instinctively measure their surroundings. Figure 14.21

Fig. 14.20 THE COASTAL EDGE. This space, between the marine and land environments, is dominated by low-angle surfaces that offer interest at scales from the global, to the immediate, to the microscopic: a. of the beach and the sea; b. of rock, sand, seaweed and sky – broad, open panoramas and promises of lands over the horizon, contrasting with curious and intricate details and patterns close to hand. It is a space that changes constantly, and where our imaginations can soar. [Source: John Sunderland]

shows one participant facing the water and, in their words, imagining 'the big waves and the mountains and just the sheer force of nature really, and just the powerful feeling of it'. Another participant spoke of 'feeling tiny and insignificant' in a 'big world', because 'the sea is so vast and you see land so far away'. The bodily feelings go beyond the sense of vision, beyond what can be seen with the eyes. Talking about her drawing (Figure 14.22) of being on the Maharees Peninsula, this participant said, 'It's free, or gentle and soft … You just feel like you belong to the earth there.' Another participant felt 'safe, with the whole protective sort of nature of the mountains and living amongst them and the bay … it's kind of calming in a way'. Similarly, a participant on the South Wall said that walking there 'gives you that real feeling of, sense of … belonging and free … and nature'.

Fig. 14.21 DRAWING OF PERSON FACING THE WATER AND WAVES. [Source: Drawing by research participant no. 59, courtesy of Anna Ryan]

The moving body, through walking and many other types of physical recreation in coastal areas, can become a catalyst for deep engagements with the world. Another study participant on the South Wall described her photograph of a shadow (Figure 14.23): 'many times that I've walked there, feeling an awful crisis, or in huge dilemmas, or really, really happy and in great form and you meet all those other people walking in and out … You see the outside of the person, but you really don't know what's going on inside them … and you're always aware of that especially with some walkers.' Another participant said, 'it's kind of as you come out [the South Wall] you realise that it is, again it's like a bigger world … that whole idea of being … We're only just part of what's kind of going out there, you know the real sense of … hope.'

Coastal walking generates mind space through which moments of creativity and attentiveness emerge. One participant in the studies drew a picture of somebody nearly floating on air: 'there's no pressure on them and a chance to just think about things that you wouldn't normally think about'. Another said, 'it is the perfect place to actually just let your mind go totally clear … where you just walk along and just kind of go [the participant exhales and smiles], "ahh", you know, and nothing in your head'. Another said, 'I would seek out a beach some place that I could walk … and just simply, sometimes, something would shape itself … and that you would trust in leaving this kind of inner space and personal space: that you know something is being processed, almost unknown to oneself,

Fig. 14.22 LINE DRAWING, 'BEING ON THE MAHAREES'. [Source: drawing by research participant no. 29, courtesy of Anna Ryan]

Fig. 14.23 'SHADOW ON A BEACH'. [Source: photograph by research participant no. 27, courtesy of Anna Ryan]

which will bear fruit in some shape or other [the participant smiling!]. It's kind of stuff … percolating in some shape or other, depending on what's going on.' For some people, coastal walking allows the opportunity to be fully attentive to others. For others, the deep sense of attentiveness to the now is demanded by the place itself: 'it's just a place to be and, as you're focusing on the nature that's around you, if you're taking it in, whether it be just the stones or, whether you notice something about the sea, for that moment that you're noticing it, you're just simply being in that moment and your problems and your fears and your doubts and your whatever are somewhere else'.

The physical nature of the coastal edge offers many other senses of wellbeing. In the words of one participant, 'you know that's the openness, that's how you breathe … that's why people come, that's why people stop their car before they go working into the IFSC and have a coffee in front of the sea, or in front of the open land because it's a big breezy experience, there's wind and it's nice, it feels sane, it feels open'. Many participants reveal how moving physically through a coastal place is often paralleled by emotional transitions and awarenesses. Various participants refer to their coastal walks as, 'restorative', as being 'good for your soul' and as 'a great relaxation'. One participant said the beach lets 'one's soul rest a bit', another, that the beach gives 'just the feeling of freedom … it's just a really warm feeling', while another said of the beach, 'oh it's heaven itself to be honest with you'. These coastal places offer the feeling of 'being away from it all', or being 'far out!' These coastal surroundings can be tied deeply into how a person feels:

Fig. 14.24 'SEA, SKY AND CLOUDS'. [Source: photograph by research participant no. 35, courtesy of Anna Ryan]

And I mean, I have cried at the end of that pier [the South Wall] … it's a place that if I was really upset about something … I can just be out there, and just be very sad out there … And you know you can be any kind of feeling down there and I think even simply if you look at the sea and how much is reflected in the sky I think it can be a reflection of your emotions as well, 'cause it's … You know so much is going on there and you can be happy and sad all at the same time … I think the sky really just reflects that. [Figure 14:24]

The different participants' contributions emphasise the importance of emotion in accessing deep spatial sensibilities. The presence of the place is deeply embedded in the body. Similarly, they demonstrate how the social is intertwined with the physical nature of surroundings, how our sense of connectivity, and connectivity to others, is deeply part of our spatial entanglement with the world. The idea of 'where land meets sea' offers a very appropriate physical setting through which to explore these natures of spatial experience, these rhythms and interactions of daily life at the coast.

'Seal Woman Story'

Roksana Niewadzisz (Nievadis)

'Seal Woman Story' is about transformation and fluctuation, about betweenness; its inspiration comes from the sea and coasts (Figure 14.25). It is composed of three multilingual site-specific stage pieces that have been performed by the author in three different places: outdoors, for Global Water Dances, (2019), Cork; in the Theatre Laboratory, in the Department of Theatre, UCC, during the conference, 'Performing Translation: Translatorship in the twenty-first century'; and as an immersive piece in Graffiti Theatre, Cork.

Everyone is coming from somewhere. We all live between; between something or nothing, birth and death. We all live in translations, between one and the other, between water and land. The Seal Woman is a symbol of this betweenness; she is fluctuating from one state to another, from one place to another, from one skin to another but is always being trapped even in this fluctuation. In some way we can't escape and, no matter the country, no matter the language, there is always present a foreignness, an otherness. Who are we? Where are we going?

Seal Woman has regained her seal skin and she is heading to the sea. She crosses waste lands, finds shredded skins and thirsty souls. In order to revive them, she leads them to the shore, while telling them her story.

It was on the shore where, once upon a time, she took her seal skin off to dance freely in a human form and where her skin was stolen from her; what doomed her to foreignness, otherness and vulnerability on dry land. Now, on the shore, she will return, in her animal form, to the sea. The shore is the place of her transformations, because it is the quintessence of the transformation itself. It is a liminal space where water becomes land, and vice versa. It is located between and within both elements and only exists when they collide; sometimes through gentle caresses, sometimes stormy fights, which mark its skin.

The shore is a border, a place of separation and access. It is like skin, itself, which being a container and protector of a creature's interior, carries out an effective exchange with its environment, often paying for that with bruises and scars. By

Fig. 14.25 'SEAL WOMAN STORY'. [Source: photographs by Marcin Lewandowski, soundofphotography.com]

303

stealing a seal skin, a Man, 'the one', domesticated the Seal Woman, 'the other', through depriving her of her duality. Controlling the shore, he exploits the ocean and tames its exchangeable and migratory nature. The injuries of the 'capitalocene' are visible as much in damaged, industrialised and contaminated shores, as in human bodies that have succumbed to civilisation's diseases.

The shore is a theatre where the stories of these scars are told, whether they are inscribed on rocks, or brought from the farthest parts of the world by the incessant flow of water and from the land by the unceasing surge of humans. Theatre in turn, like a shore, can revive and transform them. Theatre, like a shore, is liminal, fluctuating between and within real and imaginary, told and silenced, seen and intuitive, understanding and not-understanding.

Case Study: Between the Tides: The influence of the coast on the life and work of a painter

John Simpson

Fig 14.26 'Flow' (2004). Oil on canvas, size: 51cm x 40cm, private collection. 'It is very difficult to run against the tide. Little or no progress is made while expending huge amounts of time and effort. We go with the flow. There is no point in doing otherwise. My paintings begin with a theme, or a notion, rather than with a fixed idea or an end result in mind. Often I work simultaneously on a series of canvasses where images can flow from one to another. There are always 'healthy struggles' which have to be worked through but are often resolved by 'letting go'. The work finds its direction as it goes through processes of change – applying paint, removing paint, covering, exposing, scraping back, layering, moving it around. It's all about the ebbing and flowing of the tides.' [Source: John Simpson]

My world is the space between the tides.
That place where the sea takes away and gives back.
The edge of No-Man's Land between creation and destruction.
That portal of arrival and departure.

Where the great tragedies and comedies are acted out.
Where the shells of the dead are washed up and the spirits of the living run and play to the sea.
Our living space between the highs and lows, our to-ing and fro-ing, our coming and going, our linking and parting, our being and ending.
That place that's neither here nor there, but at the same time everywhere.

That's where you'll find me – between the tides.

There is a certain 'place' where a painter needs to be in the process of making paintings. It lies somewhere around and between chaos and order, around and between intuition and logical thought. There is a need to take on board the concepts of creation and destruction. The blank canvas is destroyed in order to create something else. Painting is a physical act as well as being about feeling, using intelligence and the visual sense. The making of paintings involves change, development, putting paint on, taking it off, moving it around, making, destroying, thinking, feeling, acting, looking, until that 'place' is found. The 'place' is like the space between the tides, where many aspects of and metaphors for life are found (Figure 14.26).

The coastline is a place where earth, air, water and fire meet and interact. This meeting can be tranquil: the fires of the sun warming the glassy, lapping water; the soft, airy breezes and the rocks and sand. It can, at the other extreme, be wild, with sea, wind and rain ripping at earth and rock. There are, of course, many meetings in between, but there is always movement and rhythm. The spirals of shells echo the shape of the turning wave, as do the strata of the rock, the movement

Fig 14.27 *Seas Between Us* (2006). Oil on canvas, size: 61cm x 76cm. 'All the senses are heightened while observing and absorbing the coastal realm – sea smells on the air, tastes of sea vegetation and molluscs, the textures of rock, sand and seaweed on the hands and feet, the qualities of water, sea and rain on the skin, heat from the fiery sun in summer. There are symphonies: waves crashing over rock and sand, the tide rattling pebbles, the gentle lapping of quiet water, the buffeting of wind, the various musical sounds of gulls, oystercatchers and curlews. Then there is the overwhelming variety of visual stimuli – the changing light, shadows, colours, shapes, textures, spaces, moods and all the natural phenomena of this particular world. The tide comes and goes, and with it change – new patterns on the sand, new tidal marks, new things being cast ashore, old things being washed out to sea. It is the same beach, but the beach is not the same.' [Source: John Simpson]

Fig 14.28 *Leaving (2003).* Oil on canvas, size: 122cm x 122cm, private collection. 'You start the voyage at the stern of a ferry boat looking back to where you came from, having left the farewells, the embraces and the tears. The travelling, according to Robert Louis Stevenson, is best to be done 'hopefully'. You are standing there in the present, but, strangely enough, the future is behind you as you travel forward. That future and its outcomes are as yet unknown, but you are full of anticipation and excitement. The painting process can echo this experience. You leave where you were to go and discover something new, carrying your past experiences and history with you.' [Source: John Simpson]

of the wind blowing spirals of sand and the air forming shapes of cloud in the sky. The environment leads you to ask questions. How do you paint the wind? How do you paint the movement of water, the rhythms and patterns of waves? How do you paint water crashing, breaking and spraying over rock? How do you paint solid rock and the structure of its strata? The answers, if they are to be found, are there on the shore, to be observed and studied in relation to the elements of painting: point, line, plane, shape, three-dimensional form, tone, colour, texture, paint quality, and their diverse combinations and arrangements (Figure 14.27).

The artist discovers links and connections between elements that appear to be opposites. Contrasting light against dark, hard against soft, colour against colour. Reconciling differences within the four edges of the rectangle of the painting, employing geometry to order and offset chaos. The horizontal and vertical are there to hold the instability of the diagonal. The circle is set against the square and the triangle. Offsetting spontaneous marks against rigid grids to find the dynamic against the static. The artist finds proportion by intuition, or by applying the 'golden section', or Fibonacci's number sequences, which are rooted in nature. When order doesn't work, he or she uses chaos and chance, and throws paint at the canvas to let it mix and mingle and have its own say, like, 'the wave breaking on rock, taking its own form' (Figure 14.28).

Fig 14.29 (Right) *Neptune (2016)*. Oil on canvas, size: 103cm x 93cm, collection of the artist. 'Looking at harbour walls with the rising and falling of the tide is a fascinating experience. Even the unforgiving concrete shows variations of greys made by rain and the splashing sea. However, at low tide, there are living, growing and glowing arrangements of clusters of crustaceans and seaweeds – various browns, ochres and vivid greens, that become landscapes made by the sea, and shaped and scraped by the movement of boat moorings and fenders. At high tide there is an unseen submerged world with its forests of vegetation and communities of sea creatures, where in the world of painting, the invisible and the imaginative can meet the world of reality.' [Source: John Simpson]

There are times and stages in the making of paintings when the artist must destroy their own precious image in order to make progress. Removing parts by scraping back, or rubbing off with solvents, then overpainting to find new and fresh forms, surfaces and images. The sea does this constantly, eroding the land, grinding down pebbles and shells to sand, then moving the sand from one place to another. The space between the tides is always changing. The same foreshore is different on every visit. It is never the same shore (Figure 14.29). This is what is exciting about painting.

Furthermore, the work in progress is never static. It develops in a state of flux. Painting is not an easy process, like a leisure activity for relaxation. It is an all-encompassing physical, mental and emotional activity. All one's faculties are employed working in a state of unknowing. The painting is never 'finished', but at some point it gels into some kind of unity. What works for one painting does not necessarily work for the next. The artist works on a series of paintings for a period of time, months, years. Then, like the 'turnings-of-the-tide', he or she has to find new beginnings, that are often new ways of expressing the same thing. There is often repetitiveness, but the repeating patterns are never quite the same.

The tide has gone out and left a new, clear, open sandy beach – a fresh canvas – the start of another exciting adventure into the unknown, the space between the tides.

SAND SCULPTING: MAKING SHAPES OUT OF SAND

Kyle Fawkes

Sand sculpting is an ancient art form that involves manipulating sand into artistic shapes or designs (Figure 14.30). While its exact origins are unknown, it is possible the practice dates back to the Ancient Egyptian time period. Anecdotes and legends suggest that it may have historical roots in several cultures and prominence in forms of religious practice, artistic expression and play. By the nineteenth century, illustrations show that competitive sand-sculpting activities had become common along beaches in the United States and the British Isles. Today, sand sculpting is practised as both a leisurely beachside activity as well as a serious artistic endeavour. World championship events are now held across the world and have helped to develop the practice into a profession. Huge teams of designers, sculptors and architects will embark on large-scale design projects that push the historic limits of sand sculpting. Competitions around Ireland are generally smaller, but are often associated with maritime festivals and additional beachside activities, which draw substantial crowds to the seaside (Figure 14.31).

Despite the sculptures being predominantly a work of visual art, the experience of creating the sculpture may have important tactile and performative elements.[37] This experience is artistically unique in that it is deeply embedded within the context of the beach and its natural rhythms. The canvas is temporally constrained and forever changing with its surrounding environment; the medium is constantly on the verge of disintegration, threatened by the elements. Yet, this is the nature of sand sculpting. The work is ephemeral and subject to an expiry imposed by the forces of nature and, at the beach, is predominantly controlled by the tide. Although it may be the thrill of many children and even adults to battle against the incoming tide, these struggles almost inevitably end in forfeit. For some participants, the dynamic nature of the environment motivates the performative experience of sand sculpting. For others, understanding the sculpture's eventual

Fig. 14.30 SAND DESIGNS NEAR BUNMAHON, COUNTY WATERFORD. Waterford artist Sean Corcoran is revolutionizing traditional sand and beach art in Ireland. With an innovative raking technique, he moves beyond the traditional three-dimensional sculpting practice to a two-dimensional engraving form. His magnificent works, which include abstract designs, inspirational messages and animal sketches, often feature on social media and television. [Source: Michael Faulkner]

Fig. 14.31 SAND SCULPTURES FROM THE 2019 NATIONAL SANDCASTLE AND SCULPTURE COMPETITION, BETTYSTOWN, COUNTY MEATH. a. 'Horse' by Greg Szmigier in the senior sculpture category [Source: Pat Wogan]; b. 'Mermaid' by Amy Collins in the junior sculpture category [Source: Pat Wogan]; c. 'Beached Whale' by Connor Duff in the senior sculpture category and 'Bear' by Leva, Krista & Olie in the senior sculpture category [Source: Pat Wogan]; and d. 'Terry The Turtle' by Mags Kilcooley in the senior sculpture category [Source: Pat Wogan]. The National Sandcastle and Sculpture Competition was originally established by local Bettystown resident Dick O'Reilly in 2003 as a way to promote the natural amenity of coastal Meath. The competition is open to amateur artists and offers several thematic categories and age brackets. The sculptures and castles may only be designed using sand, driftwood, shells, seaweed and rocks found on the beach on the day of the contest. The 2019 competition had a strong emphasis on climate change and nature, with designs including lions, hippos, crocodiles, polar bears, mermaids, whales, turtles and every variation of sand castle imaginable. The 150 sculptors from Spain, France, Australia, Poland, Slovakia, Zimbabwe, Wales and Ireland attracted a crowd of more than 3,000.

termination may inspire certain design features.

As well as an art, sand sculpting is also a science. For instance, a sculpture's ability to hold shape is highly dependent on physical factors, such as the moisture content of the sand. Too wet and the sand grains become separated and flow freely; too dry and gravity takes over, pulling the sand grains apart. However, with a precise balance of water to sand – approximately a 1:100 ratio – adhesive forces called capillary bridges form between the sand grains and sculpting becomes possible.[38] The size and shape of sand grains are also important factors; generally, finer grains with a diverse range of shapes hold together better than larger and smoother grains. Both size and shape are determined by the physical environment. Typically, high-energy beaches with large waves produce smoother sand grains, whereas beaches subject to riverine sediment deposition may consist of smaller and more jagged sand particles. In many high-profile sand-sculpting competitions, organisers or sculptors will transport this highly desirable riverine sand to the competition's location. From here, artists may deploy a wide range of sculpting techniques to shape the sand. These may include carving, moulding, scraping, packing, brushing and painting. Another technique, which is especially common amongst novice artists, is dribbling. Dribbling is generally performed later in the design process, often when the general structure of the sculpture has been developed. The technique involves carefully dripping a watery sand mixture over selected features of the sculpture. As this mixture flows down over these features, it dries and layers build up, forming intricate stalagmite structures. These design techniques may be supplemented by using other materials such as seaweed, driftwood, rocks and shells to accessorise the sculpture.

Sand sculptures may encompass a wide range of designs and often involve eccentric themes, ranging from the abstract to fantasy. Perhaps the most recognisable of these designs is the classic 'sand castle', but newer and more imaginative subjects, themes and methods are constantly evolving and developing.

COASTAL HERITAGE

Beatrice Kelly, Gerlanda Maniglia and Val Cummins

MUSSENDEN TEMPLE, COUNTY LONDONDERRY. Mussenden Temple is located in the surrounds of Downhill Demesne, near Castlerock. It sits atop a dramatic 36m cliff, overlooking Downhill Strand. Taking inspiration from Italian architecture, it was built by Earl Bishop in 1785 as a summer library. It has endured as one of the most iconic landmarks along the coast. The site is managed today by the National Trust, which plays an important role in preserving and promoting coastal heritage along the coast of Northern Ireland. In 1997, the National Trust invested in cliff stabilisation work to counter the worst impacts of localised coastal erosion, and to avoid the loss of the building to the sea. [Source: © Tourism Northern Ireland]

The heritage of Ireland's coast is the outcome of the dynamic interaction between people and their coastal environment. Over thousands of years, Ireland's coast has provided its inhabitants with fish, shellfish, seaweed, stones, sand and other materials. Those inhabitants have explored, traded, invaded, defended, prayed and played along the coast. Physical reminders of these activities are to be found throughout the length of the Irish coast. These include lighthouses and navigation markers, ports and seaside towns, and harbours sheltering boats designed for many different uses. Documents, memorabilia and equipment associated with coastal and seagoing activity can be seen in museums, collections and archives around the island. Many events and activities have, however, left no physical trace over the centuries; instead we must decipher the past through clues in coastal place names, stories and music.

This chapter presents a tapestry of heritage insights and initiatives, for the purpose of illustrating this integrated, broad-based view of coastal heritage. It starts with an exploration of the rich cultural lore of the coast and sea. The role of the coast in influencing politics and sense of place has been highlighted through the lens of the Irish Revolution. The chapter portrays the natural heritage of the coast, in all its beauty, before examining aspects of built coastal heritage, combining influences from coastal defence (such as Martello towers) and maritime transport (such as shipwrecks, lighthouses and traditional boats). The enduring impact of key historical events is also explored using the examples of the Spanish Armada and the sinking of the *Lusitania*, to highlight the link between past events, remembered or commemorated to this day. The final section of this chapter focuses on the importance of recording, preserving and promoting coastal heritage, and the effectiveness of increased public engagement through voluntary and community initiatives to achieve these aims.

Heritage themes and stories are embedded in a number of chapters throughout the *Atlas*. However, the subject warrants a dedicated space to integrate the many facets of Ireland's coastal heritage. Further detail on some of the heritage features presented here can be found in other chapters. For example, as well as shaping our identity, heritage is at the heart of the Irish tourism industry, where international visitors come to appreciate heritage in all its forms, particularly around the coast. This is touched on here, but presented in further detail in Chapter 27: Tourism and Leisure. (Other chapters related to themes examined here include Chapter 3 on natural heritage; Chapter 22 on underwater heritage and Chapter 23 on maritime and nautical traditions.)

Ireland's coastal heritage evokes a sense of place, coloured by people's ongoing relationships with the seas and oceans around this island, with eras of maritime influences, with politics and power struggles over centuries, and with the fortitude, resilience and innovation of coastal communities. The Heritage Council of Ireland describes our heritage as our inheritance, what we have received from previous generations and those things that we choose, by accident and design, to pass on to future generations. The heritage of the Irish coasts blends many aspects of this inheritance: built, cultural and natural, including tangible (that is, built heritage, such as piers, or natural heritage such as marine life and habitats) and intangible aspects (including customs, skills and knowledge, such as currach building). The opening image of this chapter, from Northern Ireland, captures some of this diversity, including a stunning seascape and a contemporary coastal railway, which provides a unique experience for travellers. Built heritage is visible in the background, in the form of Mussenden Temple, perched dramatically on a receding clifftop 36m above the Atlantic ocean and which has undergone adaptive works to protect it from increasing erosion.

THE CULTURAL HERITAGE OF THE COAST

One of the explicit aims of the highly successful 2018 European Year of Cultural Heritage was to encourage people to discover and engage with heritage across Europe, to reinforce a shared sense of place and belonging. This work is being continued through the European Framework for action on cultural heritage,[1] thus providing an indicator of the importance placed by policy-makers on the role of cultural heritage in influencing public perception. Interventions are designed to celebrate cultural diversity, while bringing a common interest in heritage to the fore. This modern-day approach to heritage appreciation builds on millennia of storytelling, passed from generation to generation, reflecting a sense of place experienced by coastal communities. The relationship between coastal dwellers and their environment in Ireland is represented by the rich and extensive expression in folklore: in material culture, folk belief, and narrative and song. The following provides the merest taste of the wealth of cultural and material forms born from this creative relationship between people and place.

LORE OF THE SHORE

Clíona O'Carroll

Throughout history the coast has provided an abundance of resources to sustain livelihoods, allowing coastal dwellers to develop a range of related knowledge and skills. Traditionally, seaweed and shellfish were gathered from the shore for fertiliser and food; kelp was collected to be burned and sold as vegetable ash; and valuable timber and other kinds of 'wreck' washed up along the coast were also put to good use. 'Wreck' timber played an important role in the building of houses and furniture in times of timber scarcity. A range of other valuable

Fig 15.1 GREY SEAL, INIS GÉ THUAIDH, COUNTY MAYO. Folklore abounds with stories of tragedy, mystery and fantasy pertaining to the relationship between people and the sea. Of particular note was the folk belief that some people were descendants of seals, or the notion of the seal wife whose sealskin would be hidden by her husband, until the day she could reclaim her skin and return to the ocean. [Source: Oliver Ó Cadhla]

flotsam, such as rubber, wax, barrelled foodstuffs and beverages, were also recovered occasionally from the sea, to be used or sold, particularly during the wars of the twentieth century. A range of boat types were used for fishing, trading, smuggling and marine transport. Sailing and rowing boats were constructed according to local and imported designs and materials. These traditional coastal activities all had salient social dimensions, from the gendered division of labour in seashore gathering and the respect accorded to highly skilled *bádóirí* (boatmen) and crews, to the unwritten and often unspoken rules regarding the claiming of wreck through signs inscribed on the landscape (such as heaps of stones) or wreck items dragged above the high tide mark.

It is no wonder that the mystery and danger of the sea – which could provide such bounty, yet occasion sudden terrible human loss – is reflected strongly in folk belief and the narrative tradition in Ireland. The supernatural and the otherworld are close to the lore of all parts of the country, but this closeness is particularly apparent in coastal lore. It was believed that everything that existed on the land, existed in the sea also. Stories abound of sea cows and horses and of encounters (sometimes pleasant, sometimes less so) with sea men and women and settlements under the waves. Certain inhabitants of the sea were to be avoided, such as the sea woman

with glowing eyes, whose main desire was to sink boats. Others had long-term relationships with humans, such as the seal wife, whose sealskin would be hidden by her husband, until the day she could reclaim her skin and return home (Figure 15.1). Certain families, including the Conneellys and O'Dowds in the north-west, were considered to be blood relations of the seal population. Enchanted islands appearing off the coast, storm warnings from sea people or drowned relatives, ghost boats and hauls full of rats feature among the stories told.

Considering the danger and unpredictability of the sea, avoiding bad luck was a preoccupation. Boatmen patterned their behaviour to avoid mention of priests, foxes, salmon, pigs and anything russet; carried out movements in a sunwise direction; and dipped sails or beat oars off the water, out of respect, when passing certain holy sites. Personal testimony, riddles, proverbs, songs praising boats and mourning losses, and stories of wondrous happenings and terrible hardships all abound in the lore of the shore, and can be explored and enjoyed further using the sources discussed in this chapter, as a starting point.

COMMUNITY ENGAGEMENT

The changing role of heritage is characterised by the acknowledgement of its role at the core of Irish tourism, as well as its potential to help build social cohesion. Management of heritage has shifted from being solely the domain of experts to emphasising the importance of involving local communities and voluntary groups.[2] In many cases, management of coastal heritage sites is dependent entirely on the hands-on experience of community groups that put heritage at the centre of local initiatives. Community projects cover interests such as the recording of coastal place names, boats and boat building, coastal habitats and species, and the reuse of redundant buildings. The Friends of the Murrough and Meitheal Mara initiatives are just two of hundreds of coastal partnership and heritage projects now dotted around the entire coast.

FRIENDS OF THE MURROUGH

Gerlanda Maniglia

Wicklow town is a vibrant coastal town located south of County Dublin. The coastline extends from Wicklow Head to the south and northwards along the Murrough wetlands towards Kilcoole and Greystones. This stretch of coast supports a diverse range of habitats and associated flora and fauna, providing important wintering habitats and nesting sites for birds and an important breeding site for seals (Figure 15.2). A number of archaeological features and sites attest to the integral role the coast has played in the economic, political and spiritual development of the town.

The town is endowed with several traditional coastal paths that have been patronised by the community for generations. One path extends from the twelfth-century Anglo-Norman castle in the harbour to Wicklow Head, and the distinctive eighteenth-century lighthouse to the south, while a second path runs along the shingle ridge of the Murrough and inland to the large intertidal lagoon at Broad Lough.

Fig. 15.2 WICKLOW TOWN AND THE MURROUGH WETLANDS. The Murrough is a 15km stretch of coastline, extending south from Greystones to the north of Wicklow town. The coastline is dominated by a long shingle beach, which forms a natural barrier to the sea. Inland is the largest coastal wetland complex on the east coast. This coastal wetland ecosystem provides unique habitats – such as saltmarshes, mudflats, alkaline fen and reed beds – each of which supports an array of flora and fauna. Its designation as a Natura 2000 site means that the Murrough is recognised and protected for its biodiversity under EU legislation. The Murrough is flanked by Greystones and Wicklow towns, each of which has a strong maritime history, dating back to the arrival of the Normans in 1169. The ruin of Black Castle provides a vantage point over Wicklow town and northwards towards the Murrough, and forms part of the built heritage to be found along the coast. The coastal railway from Dublin to Rosslare transits through the Murrough, on the extensive shingle ridge that divides the seaward from the landward habitats. [Source: Fintan Clarke, courtesy of Friends of the Murrough]

Outstanding scenery, heritage and proximity to Dublin means that this stretch of coastline is an important recreational, educational and tourism amenity for both locals and visitors. Like many Irish coastal areas, Wicklow town has undergone significant expansion during the past thirty years, placing increasing pressure on the visual and environmental characteristics of the coastline.

Friends of the Murrough is a community group established in the 1990s in response to these pressures. One of the main aims of the group is to protect and enhance coastal paths and create new recreational amenities for the public. Friends of the Murrough have been instrumental in raising awareness of Wicklow's rich coastal heritage, highlighting issues and providing a platform for people to voice their support and concerns (Figure 15.3). In particular, they have played a vital role in preserving and enhancing public access to traditional coastal walks, especially in the aftermath of severe storm events, such as in the winter of 2015/2016. As well as promoting and lobbying for best environmental practice, the group has commissioned, using their own resources, a number of independent coastal and environmental studies, with a view to informing policy and planning. They organise regular clean-ups of coastal paths and the lough, and run talks and educational walks for adults and children. The walks have become so popular with locals and visitors that they have become an annual Heritage Week event. A comprehensive list of their activities and reports can be found on their website and Facebook page.

Local groups, such as Friends of the Murrough, possess invaluable knowledge and a long-term local perspective. Combined with an action-focused approach, they represent a willing and enthusiastic work pool. In many cases they are the only vehicle of communication for individuals and communities who feel they may not have the 'language' to make their voice or their concerns heard. The challenge for coastal management in Ireland is to recognise the vital role these groups play in the management, enjoyment and preservation of coastal heritage, and find ways to engage with and strengthen their visibility in the planning and decision-making process. Structures such as local authority Public Participation Networks offer potential for support.

Fig. 15.3 FRIENDS OF THE MURROUGH EDUCATIONAL WALK. Friends of the Murrough is a community group dedicated to the protection and appreciation of the Murrough. The group has been particularly effective in promoting awareness of the unique habitats within the Murrough, via educational walks, clean-up activities and the creation of coastal paths. Extensive pathways connect walkers to a variety of special ecological features, including Broad Lough, Kilcoole Marshes and the East Coast Nature Reserve, the latter of which is managed by BirdWatch Ireland. Depending on the season, visitors can see little terns and skylarks, or spot a kingfisher, brent or greylag geese, lapwings, golden plover and other birds. Otters, dragonflies, swans and other waterfowl can also be seen, in addition to beautiful wildflowers, including pyramidal orchids. Friends of the Murrough is an excellent example of how local communities can develop an abundance of knowledge and create connections with their coastal heritage. [Source: Fintan Clarke, courtesy of Friends of the Murrough]

MEITHEAL MARA

Val Cummins

Fig. 15.4 OCEAN TO CITY RACE. The Ocean to City rowing race is an annual spectacle in Cork Harbour. Run by Meitheal Mara, the race is inclusive of all forms of rowing and paddling, embracing everything from traditional wooden working boats, currachs, skiffs, gigs and longboats, to modern kayaks, canoes and even stand-up paddle boards. Spanning a distance of up to 28km, the race is an endurance test that inspires and motivates hundreds of participants. [Source: Clare Keogh, courtesy of Meitheal Mara]

While boatbuilding projects have grown in popularity over the past twenty-five years, one of the longest running is Meitheal Mara, a community boatyard in Cork city on the River Lee. Meitheal Mara translates as 'community or workers of the sea'. Founded in 1993, Meitheal Mara demonstrates the full social benefits of community-based, heritage-centred initiatives. Its boatbuilding programmes, normally of currachs, are aimed at helping people of all ages to progress. Its programmes assist the long-term unemployed, early school leavers and homeless youths, by developing their social and educational potential. Similar provision is made for people with physical challenges and mental health difficulties. Meitheal Mara conducts research and survey work too, most notably the County Galway boat audit between 2009 and 2012. Seamanship is developed and promoted through rowing and sailing on the Lee, and the 'Corkumnavigation' through the channels around Cork city. The Ocean to City race, organised by Meitheal Mara since 2005, is now an event of international attraction and the highlight of the week-long Cork Harbour Maritime Festival. The importance of boatbuilding and currach-building skills as part of Ireland's heritage was officially recognised in summer 2019 when sea-currach building was inscribed on the National Inventory of Intangible Cultural Heritage.[3] The annual Ocean to City race, from outside of Cork Harbour, to Cork city centre, covers a distance of 28km (Figure 15.4). It is Ireland's largest multicraft rowing and paddling race, and features everything from wooden boats, to currachs, kayaks and even Chinese dragon boats. With around 250 boats, the race creates a vibrant spectacle on the water and increasingly attracts overseas crews, from around Europe and the United States. Run by Meitheal Mara, since its inception in 2005, the race provides an opportunity to celebrate a diverse rowing heritage.

Several community-based place-name collecting initiatives have taken place around the coast, for example, on the Hook Peninsula in County Wexford, where the local placenames convey vividly the range of activities carried out around the shore, as well as the rhythms of past ways of life, from fishing to quarrying millstones (see Chapter 29: Coastal Mining, Quarrying and Hydrocarbon Exploration).[4] The Eachléim area of the Belmullet peninsula (County Mayo) and Oileán Cléire (County Cork) have also been the focus of similar interest and activity.[5]

The network of coastal forts around the coast – such as in Cork Harbour and Berehaven in County Cork and Lough Swilly in County Donegal – provides a focus for several voluntary-led heritage initiatives that combine many heritage interests, such as historic collections, events, wildlife and seascapes. Camden Fort Meagher at the mouth of Cork Harbour is tended to by a team of skilled volunteers in partnership with Cork County Council.[6] At the other end of the country, on Lough Swilly, Fort Dunree hosts a range of activities, facilitated by the community, covering military history, wildlife observation and coastal paths (Figure 15.5). The Dunree group receives some support from the local authority, Donegal County Council, which owns the site.

Previously, Donegal County Council also gave assistance to the voluntary-run Flight of the Earls Heritage Centre at Rathmullan Battery; however, this centre had to be closed as community resourcing was stretched beyond its limit. This reflects the challenges in funding cycles for local heritage projects. More recently, funding routed through European support for maritime and fisheries funds is having a positive impact on a diverse range of coastal community initiatives. The Fisheries Local Area Group (FLAG) scheme aims to support diversification within coastal communities and promote social wellbeing and cultural heritage in fisheries and aquaculture areas. Tangible outcomes of this programme are to be seen all around the coast. For example, on Achill Island in County Mayo, local fishermen supply the Achill Experience, a local heritage and interpretation centre supported by FLAG, with a variety of fish for the display tanks.

On Bere Island, the decaying state of the military batteries and fort structures motivated the islanders to take part in an innovative, heritage-led community process, resulting in the production of the Bere Island Conservation Plan (2002), developed in partnership with Cork County Council and the Heritage Council (Figure 15.6). Under the aegis of the plan,

Fig. 15.5 FORT DUNREE, COUNTY DONEGAL. Fort Dunree is an example of the historic coastal defence architecture that is dotted all around the coast of Ireland, much of it a legacy of the British military. Several forts have been renovated as visitor attractions, in recognition of the value of their inherent built heritage, but also as a result of their prime locations, typically enjoying expansive seaward views. Fort Dunree is acclaimed for its excellent collection of coastal artillery guns, including the Rockhill Collection, which has significant appeal to military history enthusiasts. However, it is also renowned for its stunning location on the Inishowen Peninsula and for the natural heritage that can be appreciated from its vantage point, including seabirds and cetaceans. Local volunteers have made a significant contribution to the success of the heritage centre since it opened its doors to the public in 1986. Partnership between the local community and Donegal County Council, as well as other relevant government departments and agencies, has been a key factor in maintaining Fort Dunree as a vibrant heritage resource and destination. [Source: Brendan Diver, courtesy of Donegal County Council]

Lonehort Battery at the east end of the island has been leased from the Department of Defence, and has been opened up for walking tours and artistic events. Phase one of its restoration was launched in July 2019. One of the two extant Martello towers on the island has been conserved and the island's multi-purpose heritage centre hosts an exhibition on the cultural and military history of the island. Such is the success of the conservation plan for the island that the Bere Island Projects Group is administering community schemes on other west Cork islands.[7]

While heritage centres provide an important focal point for the interpretation of local heritage, the entire coastline is akin to one large, living heritage resource. The case study of the heritage of the Irish Revolution, outlined below, describes how that period in more recent Irish history has left a legacy of evidence and stories that pock-mark the landscape, such as the remains of coast guard stations destroyed by the IRA. The extent to which the Irish Revolution was shaped by the marine landscape is seldom conveyed, although revolutionary acts at sea and along the coast have long endured in memory. This heritage is also kept alive through community groups, particularly by local history associations, who maintain and publish important local records and contribute to a sense of place at the local level.

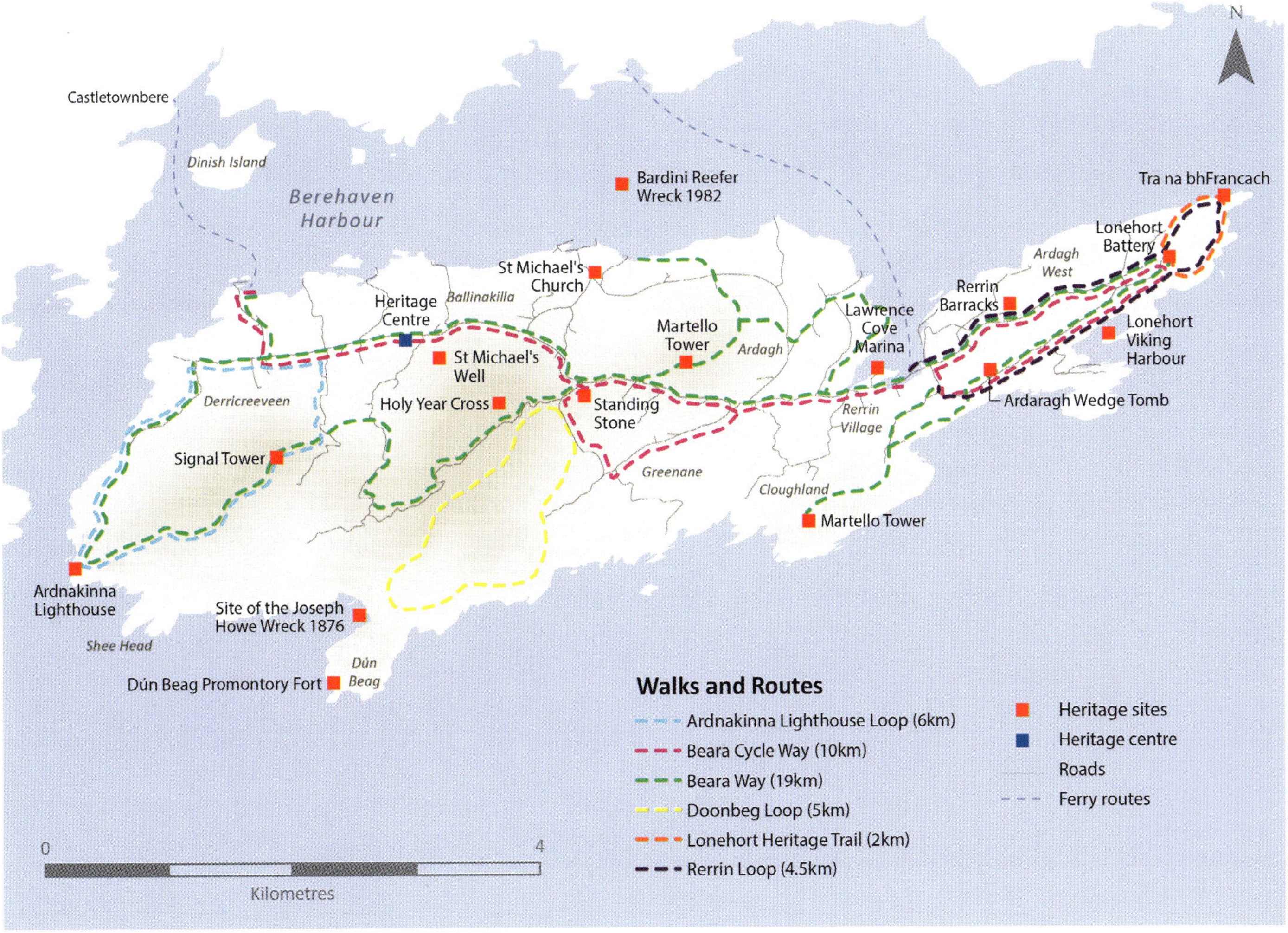

Walks and Routes

- – – Ardnakinna Lighthouse Loop (6km)
- – – Beara Cycle Way (10km)
- – – Beara Way (19km)
- – – Doonbeg Loop (5km)
- – – Lonehort Heritage Trail (2km)
- – – Rerrin Loop (4.5km)

■ Heritage sites
■ Heritage centre
— Roads
- – – Ferry routes

Fig 15.6 HERITAGE FEATURES OF BERE ISLAND. Bere islanders have worked together, and in partnership with government agencies and departments, to actively develop and promote the heritage of the island. The results have led to tangible impacts for the local community and visitors alike, with access to walking routes, Martello towers, Lonehort Battery, Ardnakinna Lighthouse and other vantage points and historic sites across the island. [Source: Bere Island Heritage Centre]

Case Study: The Heritage of the Revolution: Coastal legacies

John Borgonovo

The trajectory of the Irish Revolution (1914–1923) was often shaped by its relationship with the sea. Following the British parliament's approval of 'Home Rule' for Ireland in 1914, Ulster unionists threatened armed resistance. They formed their own parliamentary organisation, a step that was followed by Irish nationalists. Two arms importations raised tensions dramatically. During a meticulous operation by the (unionist) Ulster Volunteer Force, the steamer *Clyde Valley* successfully landed *c.*25,000 rifles at Larne, north of Belfast on 14 April.[8] Three months later, in July 1914, the (nationalist) Irish Volunteers imported 900 rifles via Howth, County Dublin, carried by the rigged yacht, *Asgard*, piloted by Erskine Childers, author of the popular maritime thriller, The *Riddle of the Sands* (Figure 15.7). These arms landings became catalysts in the escalating crisis, which was only stalled by the outbreak of the First World War in August 1914.[9] Two years later, Irish republicans staged an armed insurrection, the 'Easter Rising', in April 1916. The Rising was largely confined to Dublin because of a failed German arms landing on the Kerry coast a few days before, as featured below.

The Royal Navy's Irish fleet was comprised largely of slow sloops, trawlers and armed yachts.[10] They could not prevent a dramatic escalation of the German submarine campaign in 1917, which sank scores of vessels in Irish waters. Losses included most of Ireland's cross-channel mail boats, such as RMS *Leinster*. The tide only turned after the arrival of a major US Navy fleet (between thirty and fifty warships) to Cobh and Berehaven in County Cork, which allowed for the adoption of a highly-effective convoy system. The Irish naval

Fig. 15.7 THE 1914 HOWTH GUN-RUNNING. The delivery of 900 rifles and 29,000 rounds of ammunition in broad daylight alongside the pier in Howth in July 1914 represented one of the most daring gun-running missions in modern history. On board the yacht *Asgard*, commissioned as their wedding present, were Molly and Erskine Childers. Motivated by the landing of a shipment of *c.*25,000 rifles at Larne by the Ulster Volunteers, the Childers were part of a small group of aristocrats who hatched a plan to foil British intelligence and to arm the Irish Volunteers. It took a mere twenty minutes to offload and dispatch the cargo at Howth, as 800 Volunteers had been organised to meet the vessel upon its arrival. By the time authorities in Dublin Castle became aware of what was happening, the Volunteers has dispersed. [Source: The Board of Trinity College Dublin]

314

Fig. 15.8 IRA ATTACKS ON LIGHTHOUSES AND COAST GUARD STATIONS. Throughout the War of Independence, British-held lighthouses and coast guard stations were frequently attacked by the IRA. A total of ninety-seven individual raids were reported to the British cabinet. The map shows the distribution of raids and attacks around the coast, including the level of destruction that led to buildings being destroyed by fire. A proliferation of attacks centred on the south and south-west coasts, hotspots of dissident activity during the Irish Revolution. Armed raiders sought military artillery to help with the Volunteer effort.

bases were critical in the defeat of the German submarine menace, which in turn created the conditions for the final Allied victory over the Central Powers in 1918 (see Chapter 23: Maritime Traditions and Institutions).[11]

Meanwhile, Ireland's independence movement swept away constitutional nationalism in 1917 and 1918. The new Sinn Féin party espoused economic nationalism, advocating a strong marine policy of international trade and the exploitation of natural resources. After self-declaring an Irish Republic, separatists issued a 'Message to Free Nations of the World', which declared: 'Internationally, Ireland is the gateway of the Atlantic. Ireland is the last outpost of Europe towards the West: Ireland is the point upon which great trade routes between East and West converge: her independence is demanded by the Freedom of the Seas: her great harbours must be open to all nations, instead of being the monopoly of England.'[12] Republican propaganda emphasised the hydropower potential of Irish rivers, especially the lower Shannon.[13] A national loan handbill promised to 'put her flag on every sea … send her ships to every port … garner the harvest of the seas'.[14]

Fig. 15.9 TEELIN COAST GUARD STATION, COUNTY DONEGAL (1962). The IRA burned out numerous coast guard installations, such as this one in Teelin, in order to clear areas from possible British surveillance or to seize any rifles or revolvers held there. The Teelin Coast Guard Station, built in 1871, replaced an earlier building constructed in the 1820s. It was designed by English architect, Enoch Trevor Owen, who designed over thirty coast guard stations in Ireland during the course of his tenure in the Board of Works. The station was attacked by about fifty IRA members in 1921, resulting in the death of Coast Guard William Kennington. In 1923 the Irish Free State army apprehended several local residents, prompting retaliation from the IRA, who set the coast guard station ablaze. [Source: National Library of Ireland, call no: NPA TYN1315]

Sinn Féin's failure to secure Irish independence peacefully set the stage for its military wing, the Irish Republican Army (IRA), to launch a guerrilla campaign known as the Irish War of Independence (1919–1921). The IRA was a mass movement with members located across the island, including on its coastal exterior. Throughout the War of Independence, the IRA attacked lighthouses and coastguard stations, with a total of ninety-seven individual raids reported to the British cabinet (Figures 15.8 and 15.9).[15] The IRA typically seized from the lighthouses guncotton and rockets used to signal ships, with the explosives repurposed for bombs and mines.[16] Armed raiders often took anything else with a possible military utility, including signalling lamps and telescopes, which were essential to lighthouse-keeping duties. The Commissioners of Irish Lights issued a number of public appeals, warning of the risk to shipping, should stations continue to be deprived of signalling charges and equipment.[17] Fortunately, no shipwrecks appear to have been attributed to the IRA raids. The IRA also burnt out numerous coast guard installations, in order to clear areas from possible British surveillance, or to seize any rifles or revolvers held there (see Chapter 23: Maritime Traditions and Institutions).

Strong IRA units could be found in coastal communities with long traditions of service in the Royal Navy, such as Dungarvan, Cobh, Castletownbere, Dingle, Clifden and Westport. IRA activists exploited their marine landscape. Fugitives and couriers maintained caches of boats to covertly cross bays, inlets and estuaries, to avoid the crown forces that patrolled the roads. Republicans hid in coastal caves and used small islands to conceal arms or imprison civilians who fell foul of republican authorities.[18] Some of the populated islands without police barracks became IRA sanctuaries, such as Inish Mór in the Aran Islands, which was raided by a large search party of troops in 1920.[19] No major arms importations were undertaken during the War of Independence, but small amounts of arms and explosives trickled into Dublin and Cork harbours aboard merchant and passenger vessels. Republicans created transnational networks of sympathetic dockers and sailors (primarily Irish) to smuggle materials and messages, usually via Liverpool, Glasgow, Hamburg and New York. In response, the British government organised a special unit of the Auxiliary Division of the Royal Irish Constabulary (popularly known as Black and Tans) for service on the Dublin docks, but they did not stop the covert traffic.[20]

Despite its preeminence at the time, the Royal Navy was only peripheral to the Anglo-Irish guerrilla war. A small fleet of destroyers and sloops ferried soldiers and war materials as needed. Warships transported republican prisoners to prison camps in Britain and Ireland, and supplied military prisons on Spike Island (Cork Harbour) and Bere Island (County Cork). During British army sweeps of parts of counties Donegal, Mayo and Kerry, the Royal Navy landed troops at unexpected parts of the coast to surprise the IRA.[21] Warships occasionally assisted military or police posts attacked by the IRA, usually with spotlights or landing parties. Small detachments of Royal Marines were scattered across the coastline to defend exposed coast guard stations, and they sometimes exchanged fire with guerrilla fighters. There was less naval penetration of the inland waterways, with the exception of two patrol boats called the 'Shannon Flotilla', which patrolled the Shannon Estuary. Police and army units utilising commandeered boats were seldom effective.[22]

The War of Independence ended with a truce in July 1921, followed by negotiations and the passage of the Anglo-Irish Treaty by Dáil Éireann in January 1922. The Treaty created the Irish Free State, a partitioned dominion within the British Empire. Partition followed sustained and forceful rejection of Irish self-government by Ulster unionists, reinforced by intercommunal sectarian hostilities. A flashpoint for these tensions was the Belfast shipyards, which in July 1920 saw the forceful expulsion by Protestant unionists of roughly 7,000 Catholic workers and left-wing workers considered not sufficiently anti-Catholic. Partition, considered initially a temporary measure, soon hardened into seeming permanency.[23] The Anglo-Irish Treaty also granted the British three naval bases/anchorages in the Irish Free State, at Lough Swilly (County Donegal) and Cobh and Berehaven (County Cork). These 'treaty ports', as they became known, remained in British hands until 1938.

The Anglo-Irish Treaty was followed by a civil war between moderate (pro-Treaty) and militant (anti-Treaty) nationalists across the south of Ireland, which lasted from June 1922 to May 1923. Anti-Treaty (also called Republican) IRA and Cumann na mBan (the women's republican organisation) units, controlled initially most of the provinces of Munster and Connacht, but lost territory steadily to the pro-Treaty National Army (Figure 15.10). During this opening 'conventional' phase of the civil war, anti-Treaty forces tried to defend their exposed coastal flanks by occupying coast guard stations and mining landing piers. However, owing to the high number of possible landing

Fig. 15.10 DREDGE AND STEAMER SCUTTLED BY THE IRA IN THE CHANNEL AROUND LOUGH MAHON, DURING THE FREE STATE ARMY LANDINGS AT PASSAGE WEST, COUNTY CORK, IN AUGUST 1922. A series of National Army amphibious landings behind Republican lines in late July and early August 1922 were a turning point in the Irish Civil War. In Cork Harbour, the Free State landing was heavily contested during a three-day battle in the Rochestown/Douglas area. During the fighting, the anti-Treaty IRA sought to block the River Lee channel to prevent a seaborne attack on Cork city itself. The Republicans commandeered and scuttled in Lough Mahon the steamer *Gorilla*, owned by G. & J. Burns of Glasgow (in the foreground) and the Port of Cork dredge, *No. 1 Hopper*, referred to by some Corkonians as the 'republican dreadnought'. Despite these efforts, the National Army drove the IRA from Cork city, which was reopened to shipping a few weeks later. [Source: National Library of Ireland, call no: HOGW 4]

points and weapons and manpower shortages, these defences were quite flimsy. From mid-July to early August, the military campaign tipped decisively to the National Army when it staged a series of amphibious landings at Buncrana (County Donegal), Westport (County Mayo), Clifden (County Galway), Fenit and Kenmare (County Kerry), and Union Hall, Youghal and Passage West (County Cork).[24] Commandeered cross-channel mail boats ferried pro-Treaty troops, artillery and armoured cars, which overcame ineffective republican resistance. The anti-Treaty forces evacuated urban areas and retreated into the countryside.[25]

As the civil war evolved into a guerilla conflict, the IRA struggled to supply itself. In parts of counties Kerry, Mayo and Sligo, IRA units in commandeered boats ventured offshore and acted essentially as pirates. They seized a number of passing freight steamers and stripped them of usable cargo. Elsewhere, ship-born IRA fighters sniped occasionally at National Army shore posts from the water. The Free State lacked a navy, but the National Army did maintain a few steamers to patrol the shoreline, exchanging fire occasionally with IRA fighters ashore. Additional assistance came from the Royal Navy, which inspected inbound vessels for IRA arms and covertly assisted National Army units by relaying radio messages and illuminating positions with search lights, flares and star shells.[26]

Ultimately, the Irish Civil War ended in victory for the pro-Treaty cause. The Irish Free State that emerged was impoverished and increasingly both politically and culturally conservative. Among the many things it failed to address was how to enable the marine project its founders had envisioned. The result was that the state turned its back to the ocean. While this has been somewhat addressed in recent years, through advances in marine planning and policy, the 'seablindness' that became part of Irish cultural heritage, ultimately had implications for the institutionalisation of the state's relationship with the sea, as described in Chapter 23.

ROGER CASEMENT, 1916 AND THE USE OF THE COAST IN THE STRUGGLE FOR INDEPENDENCE

Fiona Devoy McAuliffe

The Irish coastline played a key role in the 1916 Rising, particularly in efforts to smuggle in guns to support the rebels. While some gun-running attempts were successful (for example, the guns landed in Howth in 1914 from the *Asgard*), the efforts of Roger Casement and the sinking of the *Aud* had unsuccessful outcomes, ending in arrests and a shipwreck respectively. Events such as these mark important milestones in Irish cultural heritage.

Plans for rebellion were fuelled by the First World War, which offered a chance for radical nationalists to make 'England's difficulty, Ireland's opportunity'. Seeking support for an insurrection, Fenian and Clan na nGael leader, John Devoy, met the German ambassador in New York in August 1914 and became the key contact between Ireland and Germany. With his assistance, Roger Casement travelled to Berlin where he planned to raise an Irish brigade made up of prisoners of war and to negotiate support for a rising. Born in Sandycove, Dublin, Casement was a former British diplomat. Disillusioned with British imperialism, he turned to radical nationalism and strongly advocated an Irish–German alliance to break British rule. Casement met with high-ranking German officials, but they were dubious of the Irish rebel group's ability to execute a revolt.

As plans took hold for the Easter Rising in 1916, as a token gesture, the German government agreed to send 20,000 captured Russian rifles, one million rounds of ammunition, ten machine guns and explosives. The shipment was dispatched for Kerry on 9 April aboard the SMS *Libau*, disguised as a Norwegian ship named the *Aud*. The mission aboard the *Aud* was led by Captain Karl Spindler. Roger Casement accompanied the consignment in a German submarine, the *U-19*. On 20 April, the *Aud* arrived off the island of Inishtooskert, County Kerry. Although the plan was to rendezvous with the submarine at the Maharees Islands, due to navigational errors, the U-boat and the *Aud* were unable to locate each other. The submarine captain set Casement and his companions – Robert Monteith and Daniel Beverley – ashore at Banna Strand in Tralee Bay, but the Royal Irish Constabulary (RIC) found their boat and arrested Casement and his companions shortly after. Roger Casement was convicted of treason and executed on 3 August 1916.

In the meantime, failing to find the submarine, the *Aud* progressed to Tralee Bay and signalled to the coastal village of Fenit expecting a response from shore from the local Irish Volunteers. However, the Volunteers were not expecting the arrival of the *Aud* for another couple of days. Last-minute plans to delay the landing from the *Aud*, for fear of discovery, were not conveyed, as the *Aud* had already left Germany and had no radio on board. With no answer from Fenit, the *Aud* moved back towards the rendezvous at the mouth of the bay.

The next morning the *Aud* was intercepted by an armed British trawler, following up reports of a vessel flashing lights. It is believed that British intelligence knew of the expected shipment and had ordered all vessels stationed nearby to watch the coastline for suspicious activity. However, Spindler convinced the interrogating British captain that his engine had broken down and showed him part of the cargo of pots and pans, which were catalogued in the ship's manifest to disguise the real shipment. With this narrow escape, Spindler decided to wait for darkness and to head for Lisbon, but later in the day, another armed trawler approached.

This time Spindler attempted to evade the trawler, the *Lord Heneage*, but the *Lord Heneage* radioed sloops, the *Zinnia* and *Bluebell*, for

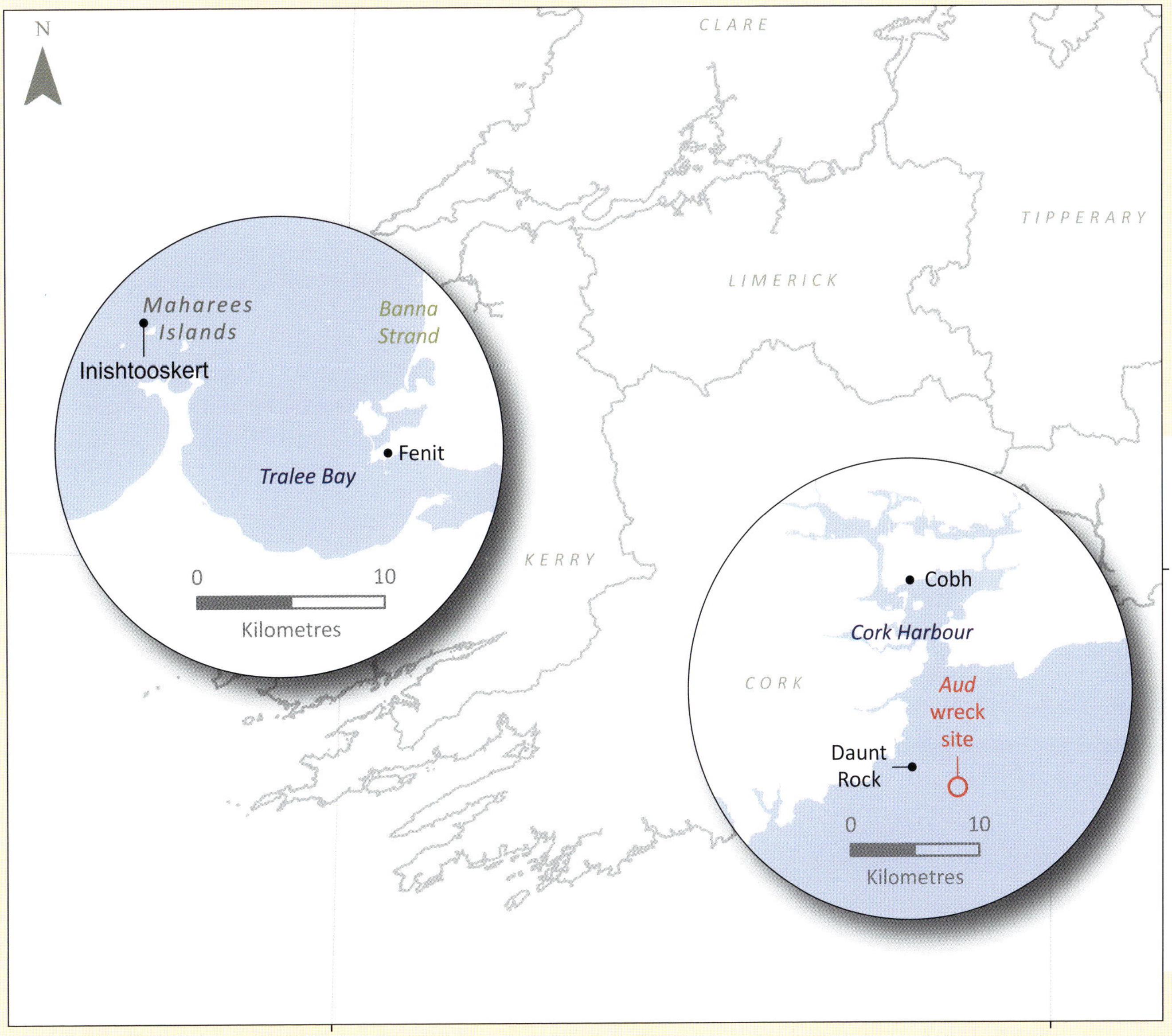

Fig. 15.11 KEY LOCATIONS IN THE STORY OF THE SINKING OF THE GUN-RUNNING SHIP, THE *Aud*. On 20 April 1916, the *Aud* arrived off the island of Inish-tooskert, County Kerry. Although the plan was to rendezvous with a German submarine at the Maharees Islands, due to navigational errors, the U-boat and the *Aud* were unable to locate each other. When the ship further failed to solicit a response from Volunteers ashore at the coastal village of Fenit, her captain decided to retreat. The ship was intercepted by the British navy off the Cork coast. Whilst being escorted to Cork Harbour, the *Aud* was abandoned by her crew and scuttled close to Daunt Rock.

additional support. Intercepting the *Aud*, the *Zinnia* ordered Spindler to sail to Queenstown (Cobh), escorted by the *Bluebell*. At the entrance to the harbour, the *Aud* halted near Daunt's Rock and began lowering lifeboats of crew, now in German naval uniforms. The crew surrendered, but explosives were set off to sink the vessel and its cargo rather than hand them over to the British (Figure 15.11). With them went republican hopes for an all-island insurrection.[27] The sinking of the *Aud* and the arrest of Roger Casement were key events in the 1916 Rising and exemplify the intrinsic link between Irish history and the coastline. The wreck of the *Aud* remains a permanent fixture. It was surveyed in 1997 and two of its anchors were recovered in 2012 and put on display – in Cobh, County Cork, and Fenit, County Kerry – following extensive restoration (Figure 15.12).

Fig. 15.12 (right) THE *Aud* ANCHOR. Two anchors from the 1916 gun-running ship, the *Aud*, were raised from the seabed in Cork Harbour in 2012. The *Aud* – a German vessel with a munitions shipment, disguised as a Norwegian vessel – was intercepted by the British navy off the Cork coast. A mix-up in communications meant that the *Aud* failed to land its illicit cargo, as originally planned, in County Kerry. It was intercepted shortly after commencing its return voyage to Germany. The captain gave the order to abandon and scuttle the ship as it was being escorted to Cork Harbour. The anchors underwent extensive restoration and are on display in Fenit, County Kerry, and Cobh, County Cork. [Source: Laurence Dunne Archaeology]

NATURAL HERITAGE AND SEASCAPES

The coast is home to many diverse habitats and species of birds, marine mammals and plants. These form part of the cultural and natural heritage, alongside the much celebrated seascape environment (see Chapter 3: Coastal Waters: Marine biology and ecology). Areas designated as Natura 2000 sites, under the EU Birds and Habitats directives, provide networks of conservation areas, essential to the protection of natural heritage at the coast (Fig 15.13). Non-Governmental Organisations (NGOs), such as the Irish Whale and Dolphin Group (IWDG) and BirdWatch Ireland, have been extremely effective in helping to preserve these valued natural heritage assets (Figure 15.14).

Ireland's coastal landscapes, often referred to as 'seascapes', are moulded by their landforms, habitats, cultural associations and uses (past and current) (Figure 15.15). Seascapes are now valued for many reasons, not least their contribution to community or personal identity, as well as the benefits that people gain from them. In the past, seascape features were put to practical use. For example, headlands served as strategic vantage points for navigation, defence, fishing and religious purposes, as evidenced by the existence of beacons, signal towers, military batteries and early Christian oratories. Nowadays, seascapes often provide the main attraction and backdrop for many recreational and tourism activities around the Irish coast (Figure 15.16). Coastal

Fig. 15.14 TERN WEIGHING IN THE MURROUGH, COUNTY WICKLOW. Bird-Watch Ireland is the largest independent conservation organisation in Ireland, with over thirty branches around the country. Non-Governmental Organisations such as BirdWatch Ireland are extremely effective, working in partnership with government and local communities, to achieve conservation objectives that preserve and protect wildlife. BirdWatch Ireland manages a network of nature reserves, including reserves along the coast, such as the East Coast Nature Reserve, encompassing the Murrough in County Wicklow. Habitat management in these nature reserves ensures an approach to land-use planning and development that supports healthy and threatened bird populations. The work of BirdWatch Ireland has ensured the recovery of seabirds such as roseate and little terns. [Source: BirdWatch Ireland]

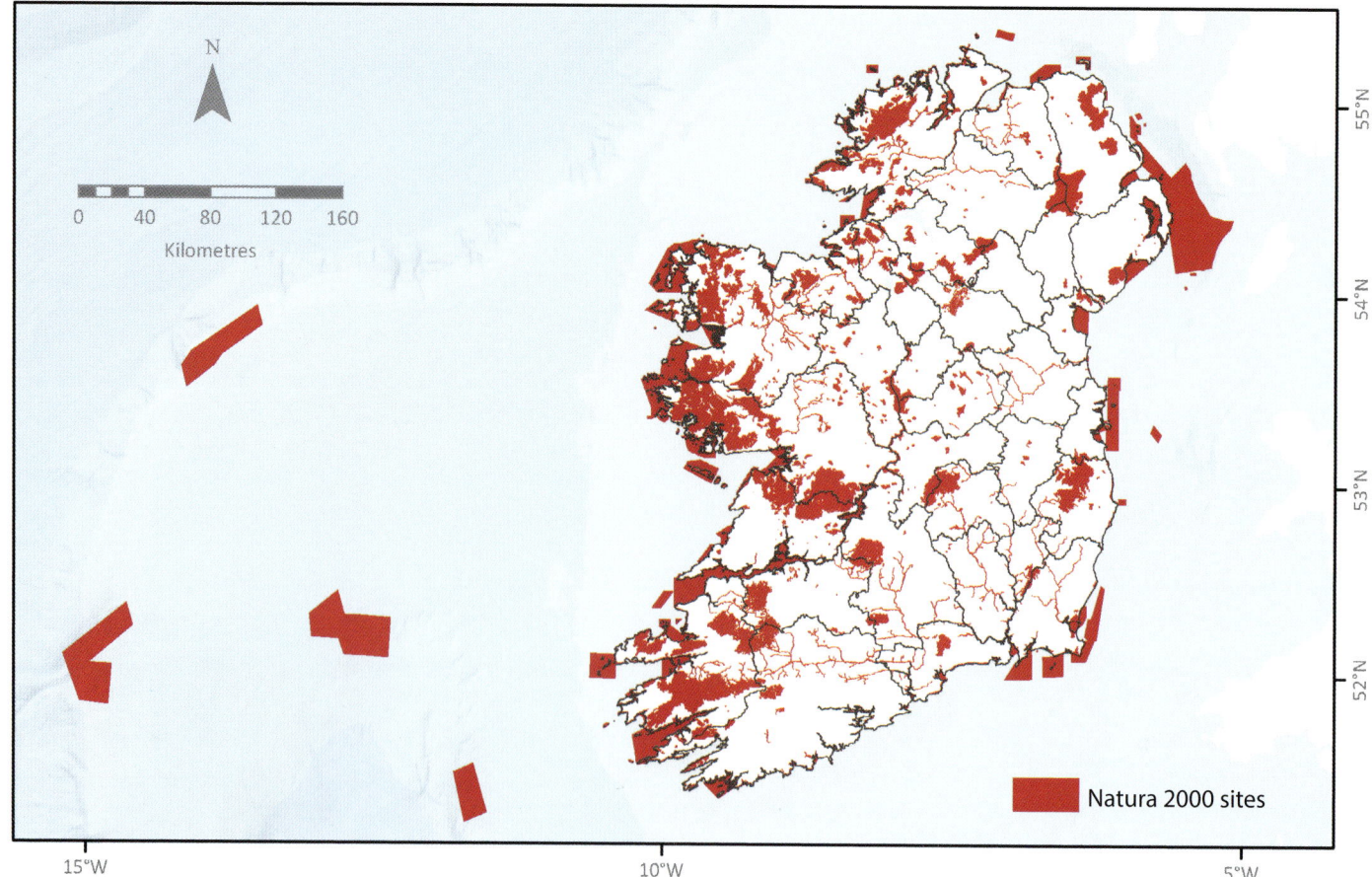

Fig. 15.13 NATURA 2000 SITES IN IRELAND AND NORTHERN IRELAND. Natura 2000 is a network of ecologically important sites designated for protection under European legislation. The National Parks and Wildlife Service (NPWS) is responsible for designating these conservation sites in Ireland. The map shows the strong coastal orientation of sites designated to date, reflecting the biological diversity of these coastal and marine waters, particularly on the west coast. The NPWS plays an important role in the ongoing management and monitoring of sites, including public access provisions and interpretation of natural heritage. [Data source: Directorate-General for Environment, European Environment Agency]

Fig. 15.15 (above) CEANN SIBÉAL (SYBIL HEAD), COUNTY KERRY. Sybil Head on the Dingle Peninsula in County Kerry is of stunning scenic value, appreciated by coastal walkers from near and far. With views of the Three Sisters, Smerwick Harbour and Mount Brandon as a backdrop, this south-west coast location provides visitors with a strong sense of place. The headland co-exists with the elements that have shaped the surrounding seascape over millennia. In the past, such locations may have provided important vantage points, or served as strategic locations for aids to navigation. Nowadays, the value attributed to seascapes extends to less tangible benefits, such as the impact of place on a sense of human wellbeing. Seascapes are an intrinsic part of our natural heritage and need to be considered as part of a forward-planning approach in the management of the coast. [Source: Irish Air Corps]

management has tended to focus on environmental processes, biodiversity, fisheries and aquaculture, and less so on issues of visual impact, built heritage and cultural significance. However, seascape character assessment is increasingly valued as a tool in the context of Marine Spatial Planning (Figure 15.17). Visual impact tends to

Fig. 15.16 CUSHENDUN SEASCAPE, COUNTY ANTRIM. Cushendun is a small coastal village, located along the stunning coastline of County Antrim. Its picturesque setting, as one of the nine Glens of Antrim, has led to it being designated as an Area of Outstanding Natural Beauty. A sheltered harbour at the mouth of the River Dun forms the heart of this coastal settlement. Its architectural heritage is also protected as a Conservation Area. The coastline features sandy beaches and dramatic cliffs, as well as built heritage features, such as the ruins of Carra Castle. The grounds of the fourteenth-century castle were used as a children's cemetery in medieval times. The Cushendun Caves, formed by 400 million years of weathering by the sea, formed the backdrop to filming in the renowned television series, *Game of Thrones*. [Source: © Tourism Northern Ireland]

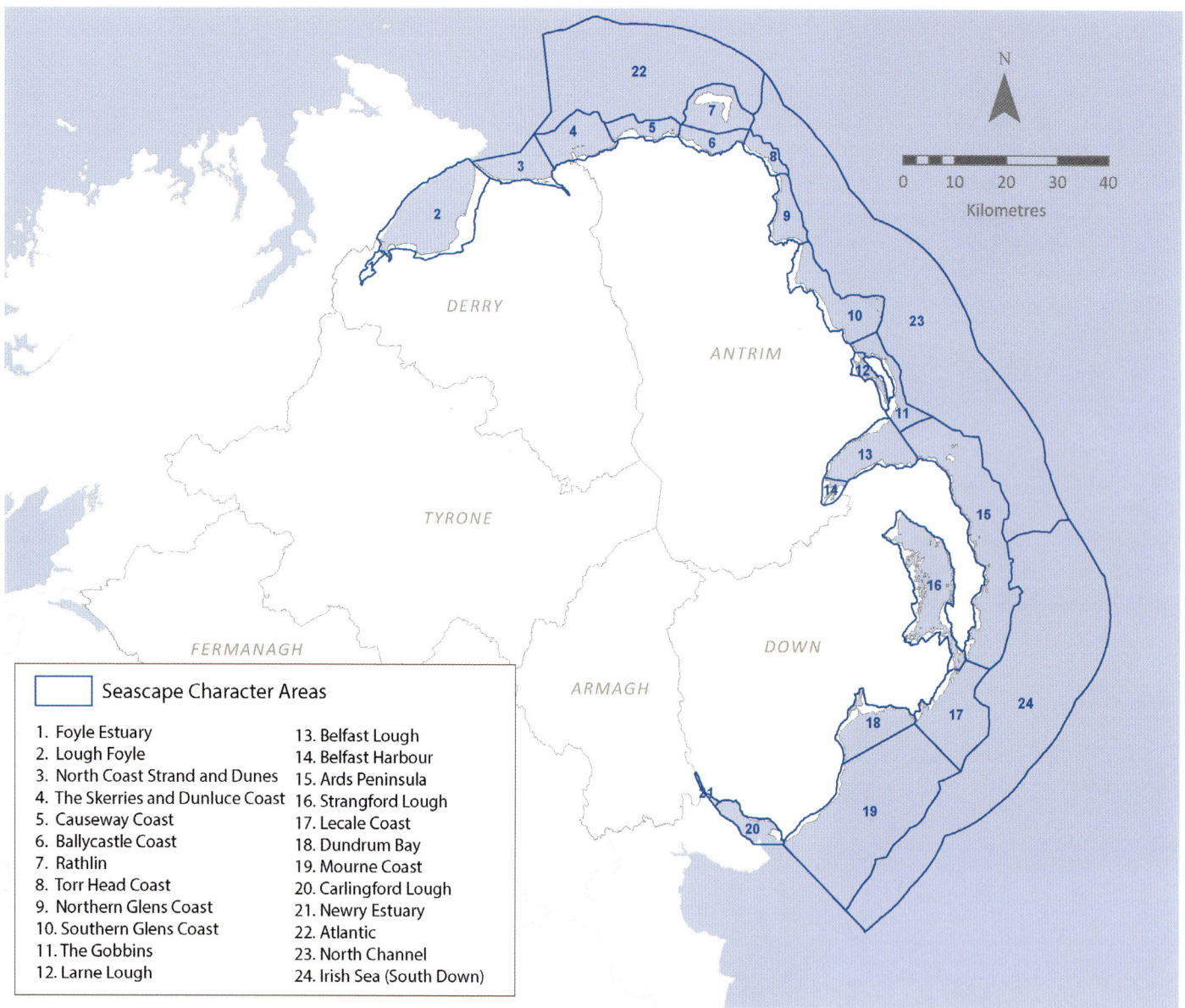

Fig. 15.17 SEASCAPE CHARACTER AREA DESIGNATIONS FROM THE NORTHERN IRELAND REGIONAL SEASCAPE CHARACTER ASSESSMENT. Seascape character assessment is a potentially valuable, integrative tool, which has generally been underutilised in the context of planning and management to date. However, a proactive approach to characterising and mapping seascape was undertaken by the Northern Ireland Environment Agency in 2014, which deals with over 650km of contrasting coastline, with a rich variety of natural, cultural and built heritage. The process defined twenty-four Seascape Character Areas (SCAs), which are shown in the map. The delineation of these areas is based on the weighting of multiple criteria, such as geology, elevation, landscape, natural heritage, cultural heritage, hydrology and marine uses. The boundaries represent areas of transition from one seascape type to another, as opposed to abrupt changes in seascape character. For example, areas 22, 23 and 24 reflect the transition from the Atlantic coast (22), to North Channel (23) to the Irish Sea (24). [Data source: Department of Agriculture, Environment and Rural Affairs, 2019]

be assessed on a case-by-case basis, for those landward parts of the coast, subject to building development. Drivers for equal treatment of the seascape include marine renewable energy developments, aquaculture, tourism and climate change. The Marine Institute is responding to this through a national seascape character assessment initiative.

BUILT COASTAL HERITAGE

Many natural habitats have been modified or even created by human activity. For example, the construction of piers and harbours has affected sediment distribution along stretches of coastline, while land reclamation has impacted on natural patterns of erosion and deposition elsewhere. Dramatic examples include the build-up of Bull Island in Dublin Bay during the nineteenth

century, following the construction of the North Bull Wall (Figure 15.18), and the erosion of Rosslare Point, County Wexford, as a result of land reclamation in Wexford Harbour in the eighteenth and nineteenth-centuries and the construction of Rosslare Harbour in the 1860s .[28]

In addition to coastal engineering and transport structures, built coastal heritage also includes architectural heritage, embracing buildings of architectural importance ranging from the Custom House in Dublin to small vernacular houses on the coast of County Donegal. Planning authorities are required to maintain a Record of Protected Structures for the statutory protection of architectural heritage. These can include lighthouses (such as Fastnet Lighthouse) and keepers' residences (Mizen Head), coast guard stations (Clifden), Martello towers (the Joyce Tower in

Fig. 15.18 SHORELINE CHANGE AT BULL ISLAND, DUBLIN BAY. Bull Island is a low-lying, sandy island in Dublin Bay, formed over 200 years ago as a consequence of the development of Dublin port. Historically, Dublin Bay had a problem with silting at the mouth of the River Liffey, causing problems for navigation and access to the port. The construction of the North and South Bull walls created a natural scour effect, which deepened the river channel. The deposition of sediment scoured from the river led to the gradual development of Bull Island. The change to the shoreline over time can be seen by contrasting the DigitalGlobe image with the mean high water mark recorded in the 1843 (first edition) Ordnance Survey map of Ireland. By the end of the nineteenth century the island had become a popular destination for beach-goers, with Dollymount Strand serving as an attraction to this day. It is also a destination for wildlife observers. In 1981 the island was designated as a UNESCO biosphere reserve, as a result of its unique biodiversity, which includes saltmarsh habitats and overwintering birds. For these reasons, Bull Island is an important part of the coastal heritage of Dublin Bay. [Data source: Google, DigitalGlobe, 2018]

Sandycove), bathing shelters (for example, the Forty Foot, Dublin), sea baths (Dún Laoghaire), dry docks (Cork Dockyard), coastal railways (West Clare) and archaeological remains (such as wrecks) (Figure 15.19).

KEY MARITIME EVENTS AND THEIR HERITAGE

A good example of a major historic maritime event being woven into the fabric of a local coastal community is the work of Cobh Town Council over a number of years to remember respectfully the sinking of the *Lusitania* during the First World War. The story of the Spanish Armada has been a significant, yet neglected, aspect of coastal heritage until relatively recently, but the significance of the event has been brought to life by a local heritage community group in Sligo. These two events are documented here to reflect how times past influence remembrance today.

Fig. 15.19 ARDAGH MARTELLO TOWER, BERE ISLAND. Martello towers represent an important feature of the built heritage of the coast of Ireland. Constructed in the early part of the nineteenth century, these small, defensive, round towers were built by the British government across Britain and Ireland in response to the failed French invasion of 1796. In total, fifty towers were built on the island of Ireland; with clusters of towers designed to be within line-of-sight of each other. The Martello tower in Sandycove has been made famous through the publication of James Joyce's novel, *Ulysses*. The image of Ardagh Martello tower shows one of a pair of such forts on Bere Island in west Cork. The tower was renovated to provide a vantage point to visitors over Berehaven and serves as an important local heritage attraction for the island community. [Source: Robbie Murphy]

THE SPANISH ARMADA IN IRELAND

Hiram Morgan

The shipwrecks of the Spanish Armada in September 1588 is a significant, yet neglected, aspect of Ireland's coastal heritage. At the time, Ireland had been considered in some early Spanish plans for an attack on England, but in the end, a preliminary landing in Ireland was rejected in favour of a direct invasion of England. Nor was Ireland included in any Spanish back-up plans, should their advance up the Channel fail to make landfall; rather, the fleet was ordered to take a wide berth of the island on an Atlantic return journey via the North Sea and Scotland. Along a similar vein of poor planning, the English state had left Ireland weakly defended at the time, with only 750 men in garrisons, though emergency plans to rush in reinforcements had been developed.

Battle-damaged Spanish ships, with injured, hungry and thirsty crews, were forced to seek a safe haven in Ireland when they were suddenly overwhelmed by a violent storm. Though some managed to resupply and sail home, at least twenty-one ships were wrecked in locations between the Dingle Peninsula and north Antrim. It is estimated that these carried about 6,000 soldiers and sailors; of these, some 3,750 died or were drowned, 1,500 were killed by the English and their Irish collaborators and 750 survived.

The biggest losses of life were at Streedagh Strand in Sligo, where three ships – *La Juliana*, *La Lavia*, and *Santa María de Visón* – were wrecked with 1,000 drowned (Figure 15.20) and then subsequently at Lacada Point in north Antrim, where *La Girona*, after gathering a number of survivors in west Donegal in addition to its own crew, suffered a similar-scale catastrophe.

Fig. 15.21 PORTRAIT OF SIR RICHARD BINGHAM. Sir Richard Bingham, governor of Connacht from 1584 to 1596, was an experienced military man, who fought at Lepanto against the Turks in 1571. Already having a ruthless reputation amongst the Irish, he was the principal executioner of the Armada survivors in Ireland. The incidents around the Spanish Armada show how the west coast of Ireland became a theatre for the political events being played out on the greater European stage. [Source: portrait courtesy of Lord Charles Spencer, Althorp Estate]

The killing of bedraggled and defenceless men coming ashore is a vexing issue and may amount to a war crime. Certainly, the English and their Irish allies had no means of keeping large numbers of prisoners and had no way of knowing whether or not these men were part of a more coordinated, disciplined landing force. There appears to have been no clear orders given regarding summary execution. Rather, expedient decisions were made locally as the situation developed, the basic premise being to rob, strip and kill ordinary Spaniards and to save richer and more important ones for interrogation and ransom.

Officials like sheriffs Denny in County Kerry and Clancy in County Clare took rapid action locally, but the most notable, Sir Richard Bingham, governor of Connacht, had martial law powers (Figure 15.21). He was infamous as the principal executor of the Armada survivors arriving in Ireland. In Inishowen, 300 men surrendering off *La Trinidad Valencera* were executed. An elite mercenary warrior in Mayo, Melaghlin McCabe, dispatched eighty unfortunates.

Where the Spaniards landed in Irish-controlled areas or could readily escape to them, they generally received assistance. Several hundred Spaniards were openly aided to reach Scotland and were subsequently repatriated to the Spanish Netherlands. A further group of Spanish officers, who were already in custody, escaped by taking control of the boat in which they were being transported across the Irish Sea. A number of these survivors provided accounts of their adventures when they were debriefed by the Spanish authorities. The most famous is that of Captain Francisco de Cuéllar, who wrote up his story for private circulation. He survived the foundering of *La*

Fig. 15.20 (left) SHIPWRECKS OF *La Lavia*, *La Juliana* and *La Santa Maria de Visón* LYING OFF STREEDAGH STRAND, COUNTY SLIGO IN A CONTEMPORARY MAP MADE FOR SIR RICHARD BINGHAM. It is estimated that twenty-one ships from the Spanish Armada, carrying some 6,000 soldiers and sailors, were wrecked off the coast of Ireland. Of these, some 3,750 died or were drowned; 1,500 were killed by the English and their Irish collaborators and 750 survived. The biggest losses of life were at Streedagh Strand in Sligo. Here, three ships – *La Juliana*, *La Lavia*, and *Santa Maria de Visón* – were wrecked with 1,000 drowned. Sir Richard Bingham played a notorious role in the fate of those that survived the perils of the sea. His map of Streedagh Strand, from 1589, shows the indicative locations of the wrecks. [Source: National Archives, ref. MPF1/91]

Lavia at Streedagh Strand, avoided the marauding English soldiers on the beach, was aided by the MacClancys and O'Rourkes in Leitrim, decided against boarding *La Girona* and eventually made it to Antwerp, via Scotland, a year later.

Modern research shows that the alleged widespread genetic inheritance from Armada survivors in the west of Ireland is a myth. However, the Spanish Armada's shipwreck had a very significant political impact at the time. Lord Deputy Fitzwilliam's march into the north-west – ostensibly a mopping-up operation – was a barely concealed search for treasure that ended up antagonising the local elite. Furthermore, Irish assistance rendered to the survivors was perceived as a strategic weakness and this increased the English drive against the O'Flahertys and Burkes in the west and against O'Rourke and Catholic bishops in the north-west.

By the same token, the Armada naturally opened the region to an influence from Spain, which hitherto had not penetrated further north than Galway. Some of the Spanish survivors, including de Cuéllar himself for a time, were already helping to introduce Spanish military methods and Pedro Blanco, a survivor of *La Juliana*, became a life-long manservant of Hugh O'Neill. At the time of the shipwrecks, the government in Dublin was worried about O'Neill's loyalty, declaring: 'We have a special distrust in the Earl of Tyrone'. Thereafter, he began to seek Spanish military support. After a formal alliance in 1596, Spain sent the Irish Confederates arms and money, but no actual military assistance, until the small Armada dispatched to Kinsale in 1601. Defeat there ended the important role Ireland had assumed in the Anglo-Spanish war.

The living memory of this infamous event is maintained in heritage centres, such as the old courthouse in Grange, County Sligo, which now serves as a Spanish Armada interpretive and visitor centre, kept open throughout the summer months by local volunteers. The Spanish Armada remembrance is held annually in Grange and plans are underway to develop a Spanish Armada trail.

RMS *Lusitania*: HISTORY OF A LOST LINER

Eunan O'Halpin

The Cunard passenger liner, RMS *Lusitania*, en route from New York to Liverpool, was sunk 11.5 nautical miles off the Old Head of Kinsale by the German submarine, *U-20*, on 7 May 1915. A reported 1,198 passengers and crew were lost, with 764 people rescued after hours adrift in lifeboats or clinging to wreckage and brought into Irish ports (Figure 15.22). The ship had moved closer to the Irish coast on foot of confused reports about German submarine activity off the south coast. Poor communications and misunderstandings between the ship and shore radio stations, and between the Admiralty and the ship's owners, meant that up-to-date information about where the greatest danger lay was not conveyed. The result was that the liner sailed into the sights of the German submarine.

Germany had given clear warnings that British and French civilian ships would be considered legitimate targets, arguing that this policy was in essence no different to that adopted by the Allies in their maritime blockade of Germany's seaborne trade. Furthermore, the German Embassy in the United States had placed advertisements in the local press warning prospective passengers of the dangers of seaborne travel. Though a civilian liner, the *Lusitania*'s cargo manifest showed that she was carrying some war material and conspiracy theorists believe she was carrying far more than had been declared.

The nature of her cargo is relevant because, although the captain of the *U-20*, Walther Schwieger, reported that it had fired only one torpedo, the strike was followed by a second explosion. Debate continues as to

Fig. 15.22 *Lusitania* VICTIM BURIALS, QUEENSTOWN (1915). When the *Lusitania* was torpedoed off the Old Head of Kinsale, County Cork, by a German submarine on 7 May 1915, it sank in just eighteen minutes, with the loss of 1,198 lives. Most of the 764 survivors, as well as the bodies recovered from the sea, were landed at Queenstown (Cobh). Of the 289 bodies landed, 169 were buried three days later in three mass graves in the Old Church cemetery on the outskirts of the town. Forty-five bodies remained unidentified and their graves are marked merely by numbers. The entire town was impacted by the horror of the event. Carts were assembled to ferry the coffins containing the dead to their final resting place. Naval personnel and others were involved in helping the survivors, as well as providing the dead with a dignified burial, as captured by historic photographs. [Source: National Library of Ireland, call no.: POOLED 2749/1]

whether this second explosion could have been caused by some undeclared portion of her cargo, rather than by one of her boilers. A British board of inquiry naturally laid the blame for the sinking entirely upon Germany. There are indications that the Admiralty was anxious about the suggestion that the ship had been carrying contraband and contrived to have this matter omitted from the board's conclusions. The board concluded that the ship had been hit by two torpedoes, but evidence suggests that *U-20* fired only one torpedo. A more probable cause was that one of the ship's boilers, as it flooded with seawater, led to a massive second explosion.

The sinking caused a sensation across the world and had a particular impact in the United States, with 128 of its citizens lost. The German government's argument that the *Lusitania* was a British-flagged ship travelling in a warzone, and therefore a legitimate target, outraged rather than persuaded neutral opinion. There was little sympathy for Germany's own difficulties arising from the Allied blockade of her maritime trade, which she argued constituted an unfair attack on her civilian population. Though mooted that the sinking was the catalyst that brought the United States into the war, other indications suggest that this was not the case. Just two months later, for example, the British ambassador in Washington wrote that the political situation in the United States was tilting against Britain and her allies. Those against the Allied blockade of Germany were being manipulated by Germany 'with all her skill'.[29]

The sinking underlined Germany's determination to prosecute the war against Britain through all available means. It upped the stakes in the Atlantic, obliging the Royal Navy to devote additional resources to anti-submarine patrols, although it was not until 1917 that Prime Minister David Lloyd George forced the Admiralty to adopt the escorted convoy system for merchant shipping, a ruling that saw a marked reduction in losses to enemy action.

In May 2015 the centenary of the sinking of the RMS *Lusitania* took place (Figure 15.23). The tragic loss of the ship shocked the world at the time and the wreck itself remains a fascination to all; the cause of the second explosion continues to be the subject of controversial debate and the focus for exploration and investigation. Perhaps one day the *Lusitania* will reveal her secrets.

Fig. 15.23 *Lusitania* CENTENARY COMMEMORATION. In May 2015 the centenary of the sinking of the RMS *Lusitania* took place. The tragic loss of the ship shocked the world at the time. The event is respectfully commemorated by the Lusitania Monument, by sculptor Jerome Connor, located in the town centre, in Cobh. [Source: Óglaigh na hÉireann, flickr, CC BY 4.0]

MANAGING COASTAL HERITAGE

The value of coastal heritage

As mentioned earlier, Ireland's seascapes provide a vibrant backdrop for recreational and tourism activities. Seascapes can be experienced visually, but also through hearing, smell, touch, memory and association.[30] Opportunities for these sensory experiences are dependent, however, on access to beaches, coastal paths, cliff tops and headlands. Coastal paths have been enjoyed by all generations and groups around the country have been involved for many years in ensuring access to these paths is maintained and their historic resonance still understood. Examples can be found around the coast linked to specific sites and places, including natural habitats, military forts and early Christian monastic sites (Figures 15.24 and 15.25).

Increasingly, new methods of analysis are being developed to understand the non-direct economic value of ecosystem goods and services. In this context, we refer to ways of measuring the value of coastal heritage where the sense of space, connection, spirituality, physical exercise and enjoyment to be obtained at the coast can have positive implications and beneficial economic consequences for human wellbeing. Coastal ecosystem services can be classified as provisioning (tangible resources such as fish stocks, which can be valued according to conventional economics) and regulatory (the regulatory functions of natural coastal processes, such as the benefits derived from saltmarshes as natural coastal defence structures). The valuation of regulatory services often depends on proxy values (such as the equivalent cost of a hard-engineering response to coastal defence) and cultural services (such as the physical, psychological and spiritual benefits from human interactions with nature). These latter intrinsic values are

Fig. 15.24 WALKING ON BLOODY FORELAND. The Bloody Foreland Walk is the coastal section of the Slí an Earagail–Slí Dhún na nGall trail. Situated in the remote north-west corner of Ireland, the name did not, as may be expected, derive from an ancient Celtic bloody battle. Rather, the name 'Bloody Foreland' relates to the intense red hue of the rocks at sunset on this part of the Atlantic coast. Walkers can enjoy moderate rambles across Bloody Foreland, on rough tracks and old bog roads. The headland as a Discovery Point on the Wild Atlantic Way. [Source: Seamus Doohan courtesy of WalkingDonegal.net]

Fig. 15.25 SELECTION OF MANAGED COASTAL WALKING TRAILS. Coastal cliff walks and hiking trails are a major attraction for visitors to the island of Ireland, as well as valuable recreational amenities for coastal residents. The map depicts Sport Ireland Trails in the Republic of Ireland and the Coastal Causeway Route in Northern Ireland. These are just a selection of the many trails around the country. These trails are distinguished by the fact that they have an information board, a trail map, way marking, and are managed and maintained by the aforementioned groups. Importantly, it also means that they have agreement from the landowners whose land is crossed by the trail and when on private land there is an insurance policy in place that indemnifies private landowners. Iconic walks include the Cliffs of Moher and the Giant's Causeway coastal route. Other coastal walks may consist of traditional routes that people have walked over the years, which may never have been formally developed as a trail, or may be under the management of different authorities, such as local councils. The Ballycotton, and the Bray to Greystones cliff walks, are two such examples of other, well-established and promoted cliff walks that fall outside the criteria for walks depicted on the map. [Data source: Sport Ireland; Causeway Coast and Glens Burrough Council]

challenging to integrate into traditional economic models; however, they are increasingly regarded as important to quantify in order to demonstrate, in economic terms, the benefits of heritage and biodiversity.

Of course, the link between the direct economic value of heritage and tourism, in terms of enterprise development, is relatively well established by comparison. Irish seascapes are dotted with numerous buildings and structures, which provide evidence of past coastal activities and livelihoods – for example boatyards, warehouses, lighthouses, coast guard and fishermen's cottages – many of which are now abandoned and neglected. Increasingly, they are being considered as important parts of local

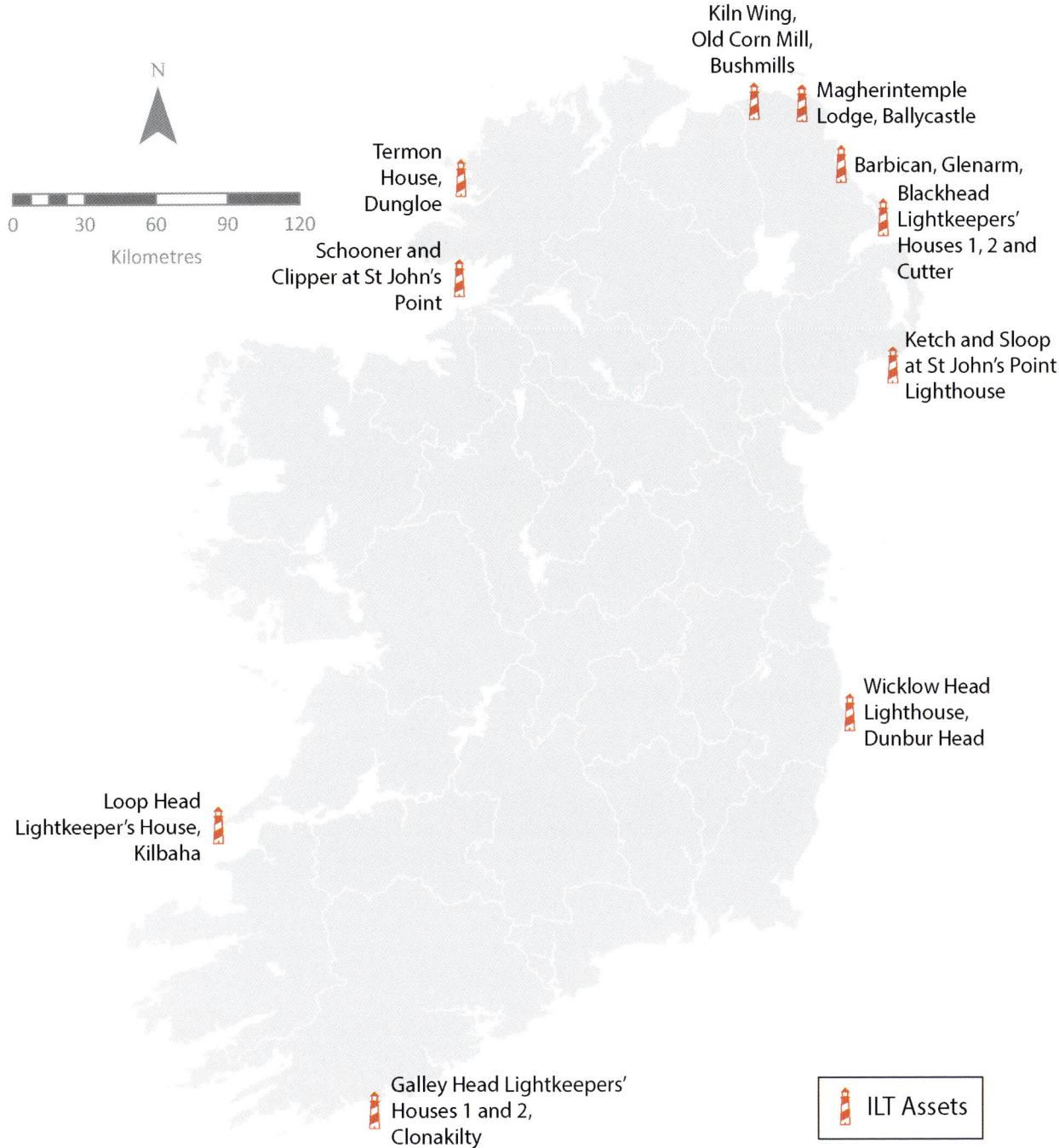

Kiln Wing,
Old Corn Mill,
Bushmills

Magherintemple
Lodge, Ballycastle

Termon
House,
Dungloe

Barbican, Glenarm,
Blackhead
Lightkeepers'
Houses 1, 2 and
Cutter

Schooner and
Clipper at St John's
Point

Ketch and Sloop
at St John's Point
Lighthouse

Wicklow Head
Lighthouse,
Dunbur Head

Loop Head
Lightkeeper's House,
Kilbaha

Galley Head Lightkeepers'
Houses 1 and 2,
Clonakilty

ILT Assets

Fig. 15.26 IRISH LANDMARK TRUST ASSETS AT THE COAST. Lighthouses are probably the most recognisable structures of the coast. Modern communications technology means that full time human presence in lighthouses is no longer needed so ancillary buildings such as keepers' cottages have lost their original purpose. New uses for these important structures have evolved through long-term partnerships between their owners, the Commissioners of Irish Lights (CIL) and the Irish Landmark Trust (ILT), and latterly with community groups at Hook Head, County Wexford, Valencia, County Kerry, and Fanad, County Donegal. Irish Landmark Trust has a long-standing interest in promoting access to the coast from its establishment in 1992. Collaboration with CIL has allowed the trust to realise this objective by providing holiday accommodation at lighthouse sites around the island of Ireland, as shown on the map. CIL is also working with community organisations at Hook Head, Valencia and Fanad where the lighthouse complexes form the hub of a range of activities from visits of the lighthouse to hosting music and cultural activities and wildlife spotting. Experiences from the ILT are also passed on to community organisations through these networks, which contribute to making these sites significant destinations for visitors and locals alike. [Source: Irish Landmark Trust GLG]

heritage and identity and exploited as alternative sources of employment and revenue. Many of these buildings now form key parts of Ireland's heritage tourism, as holiday accommodation and exhibition spaces, providing opportunities in the immediate locality for employment and as the focus of local interest and pride. Initiatives to promote lighthouse heritage tourism by the Commissioners of Irish Lights and the Irish Landmark Trust are excellent examples of this (Figures 15.26 and 15.27). Another excellent example of a unique aspect of the Irish coast being used to promote tourism is revealed at the Foynes Flying Boat and Maritime Museum (Figure 15.28).

Appreciation of the natural coastal heritage, by visitors and

Fig. 15.27 WICKLOW HEAD LIGHTHOUSE. (Inset photograph of the lighthouse interior.) The Irish Landmark Trust acquired Wicklow Head Lighthouse in 1996, when the trust embarked on a programme of restoration. Fast forward to today, and this remarkable structure, with commanding views over the Irish Sea, has been beautifully appointed to let for holiday makers. Six octagonal rooms were developed within the void that existed within the tower when it was taken up by ILT. The kitchen is at the top of the tower, following a climb of 109 steps! [Source: Steve Foster, courtesy of Irish Landmark Trust GLG; inset: Irish Landmark Trust GLG]

locals alike, has also created significant economic value. The marketing of the Wild Atlantic Way is acclaimed as an impactful approach to packaging and promoting the heritage of the west coast of Ireland. Numerous businesses have grown to depend on the natural heritage of the coastal environment, such as whale watching, diving and boat-based eco-tours (see Chapter 27: Tourism and Leisure) (Figure 15.29).

Impacts on coastal heritage

Much of Ireland's coastal heritage is not as well known as other

Fig. 15.28 (left) FOYNES FLYING BOAT AND MARITIME MUSEUM. Foynes Flying Boat and Maritime Museum is located in the former terminal building of the Port of Foynes. When the building was purchased in 2001, the archives of Foynes Harbour Trustees were donated to the museum. These archives, along with an additional loan from the Shannon Foynes Port Company, have assisted the expansion of the maritime exhibition floors, opened in 2013. Most independent museums run by voluntary groups in Ireland tend to struggle as there are so few areas of support, and the demands of caring for a collection are high. However, Foynes Flying Boat and Maritime Museum has demonstrated that collaboration with another agency, such as a port authority, can allow exhibition and activities on site to diversify, thus widening its potential range of interest for visitors in the longer term. The Hollywood actress Maureen O'Hara was a dedicated patron of the museum, up until her death in 2015. She had been married to Brigadier General Charles Blair, who had a distinguished flying career, and flew flying boats into Foynes between 1942 and 1945. The museum has become a beneficiary of many of the Hollywood legend's personal belongings, which feature in the Maureen O'Hara exhibition, thus diversifying the attractions available to visitors at the museum. [Source: Foynes Flying Boat & Maritime Museum (www.flying-boatmuseum.com)]

aspects of Ireland's heritage. Globalisation and technological advances have wrought long-term change on maritime and coastal culture in Ireland. The impacts of European common fisheries and agricultural policies have escalated the decline of traditional livelihoods along Ireland's coasts, as elsewhere. Many activities along the coast were once centred on commercial or individual exploitation of natural resources. When the activity became uneconomic and ceased, associated buildings and equipment fell redundant. As industries declined – for example mackerel fishing and processing, or the harvesting of seaweed for fertiliser – the implements and places associated with these activities (tools, buildings, skills, outdoor spaces) disappeared, decayed or were put to other uses. For this reason, place names are often the only surviving evidence of previous events and productive activities that formerly took place along the coast, and they are sometimes linked to local folklore, wildlife and the rhythms of daily life in the past.

Preserving coastal heritage

The dynamic nature of the coastal zone brings to the fore challenges and tensions faced by all aspects of heritage, but which are more extreme or acute along the coasts than in other locations. These can include erosion, storm damage, flooding, pressure from excessive visitor numbers, coastal squeeze and demands for new infrastructure. If we can find solutions that balance the varied and, at times, conflicting demands on heritage in coastal areas, we will be able to find solutions for heritage practice elsewhere. The Department of Culture, Heritage and the Gaeltacht is responsible for leading the Heritage Ireland 2030 strategy. Through the stakeholder consultation process, it is apparent that there is a great interest in the heritage of these coasts and seas. The department also oversees the sectoral climate adaptation planning for biodiversity, built and archaeological heritage. These, and other sectoral adaptation plans (for seafood and flooding sectors), are extremely relevant to the preservation of coastal heritage. These plans aim to build capacity to cope with climate change, including improved coordination with national and local government.

Planning for the coast, including the preservation of coastal heritage, is contingent on the availability of relevant data. Many sections of these coasts and upland areas were historically under-surveyed by the Ordnance Survey due to perceived lack of biological productivity.[31] Certain other aspects of heritage, such as seascapes and intertidal archaeology, have not been systematically surveyed due to lack of resourcing and difficulties with accessing remote areas. Furthermore, many coastal activities and industries have rarely been the subject of detailed research.[32]

The dynamic interaction of the sea and its place of landfall means that much of Ireland's coast is exposed to natural processes of storm damage and erosion. In addition, the coast is under pressure from many development demands, including port expansion, energy generation, tourism and recreational activities, urban/suburban development and the intensification of agriculture and aquaculture. As we lack baseline data on much of the coastal heritage resource, we have no clear idea of the extent to which it is under threat of damage, or at risk of being lost, nor do we have a

Fig. 15.29 APPRECIATING NATURAL HERITAGE: WHALE WATCHING OFF THE COAST. Irish waters were declared a whale and dolphin sanctuary in 1991. This was the first designation of its kind in Europe. Ireland's marine environment is akin to a biological super highway for cetaceans, who trawl the seas in search of food. Twenty-four different species of whales and dolphins are known to inhabit or transverse the waters around the coast, from majestic baleen whales, such as fin whales, to common dolphins and shy porpoises. This natural heritage provides a unique experience to tourists, who avail of boat based eco-tours in the anticipation of spotting the elusive wildlife. These trips also provide educational opportunities for novices and enthusiasts alike, of all ages. Data and information collected by the Irish Whale and Dolphin Group (IWDG) helps to promote conservation and understanding of whales, dolphins and porpoises within Irish waters. Established in 1990, the IWDG plays an important role in managing sightings and stranding data, as well as providing training for marine mammal observers and recommendations on best practice to eco-tour operators. [Source: Debs Allbrook]

context in which to assess its importance. To illustrate how easily sites and structures can move beyond recall, a shoreline survey by Dún Laoghaire Rathdown County Council in 2007–2008 found the remains of forgotten early nineteenth-century fortifications, in this densely populated and widely used stretch of coast (Figure 15.30).[33] The adaptation strategies for built and archaeological heritage include several actions to assess the types of sites and locations that are most at risk. This is an important advance in

Fig. 15.30 AERIAL INFRA-RED PHOTO OF BATTERY NO. 5 ON THE SHORES OF KILLINEY BAY. During a coastal survey in 2007–2008, the remains of hitherto forgotten early nineteenth-century fortifications were found. The image is an infra-red photograph, taken on the shores of Killiney Bay, which shows the remains of the fort, including fragments of the rear wall, battery, officers' and soldiers' quarters. The hedges indicate the road that ran to the fort. The fort was built 1804–1805, but the defensive walls were eroded and it was abandoned by 1812. The coast stabilised subsequently, with little erosion over the last 200 years. [Source: Compass Informatics, Marine Institute and Enterprise Ireland]

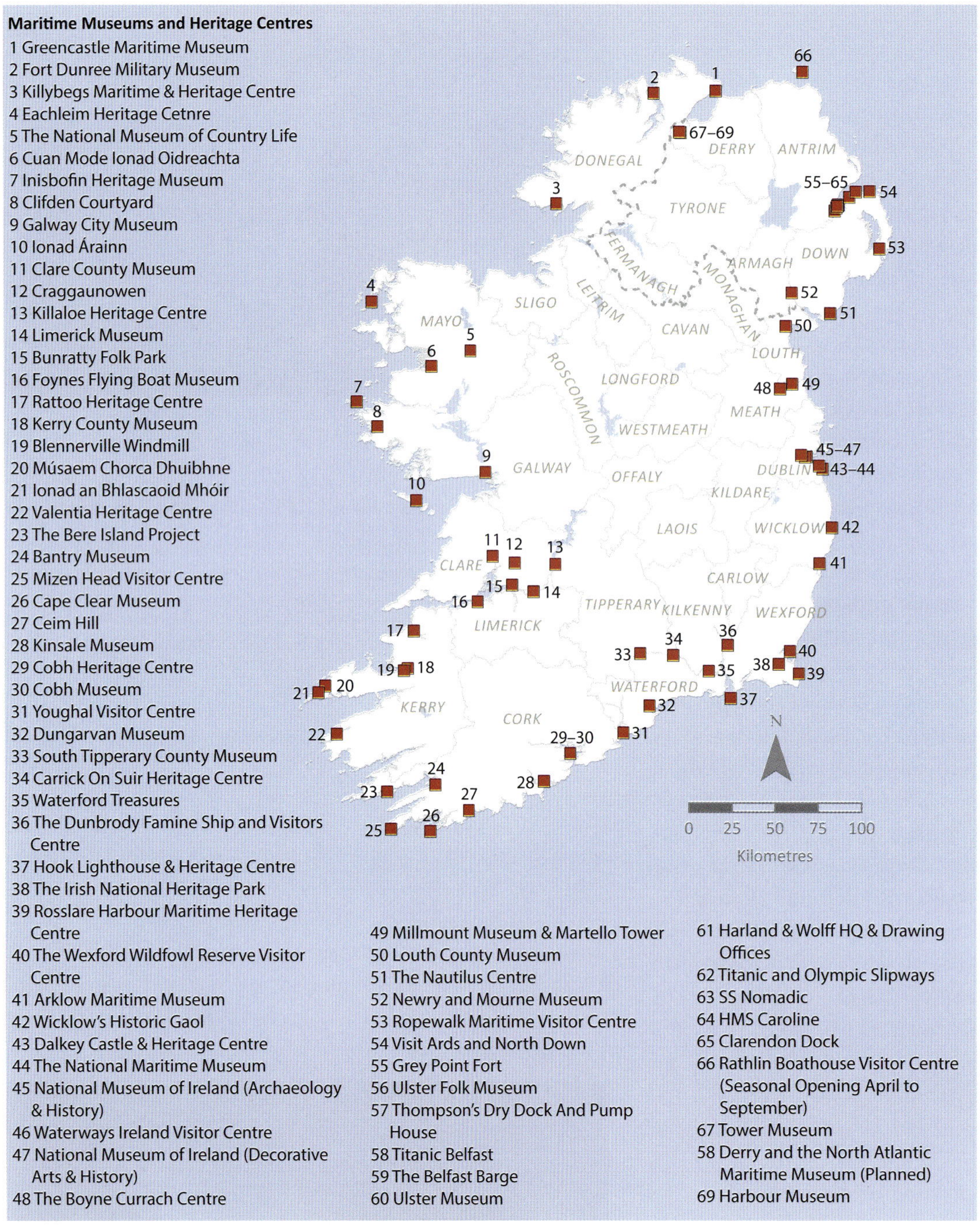
Maritime Museums and Heritage Centres
1 Greencastle Maritime Museum
2 Fort Dunree Military Museum
3 Killybegs Maritime & Heritage Centre
4 Eachleim Heritage Cetnre
5 The National Museum of Country Life
6 Cuan Mode Ionad Oidreachta
7 Inisbofin Heritage Museum
8 Clifden Courtyard
9 Galway City Museum
10 Ionad Árainn
11 Clare County Museum
12 Craggaunowen
13 Killaloe Heritage Centre
14 Limerick Museum
15 Bunratty Folk Park
16 Foynes Flying Boat Museum
17 Rattoo Heritage Centre
18 Kerry County Museum
19 Blennerville Windmill
20 Músaem Chorca Dhuibhne
21 Ionad an Bhlascaoid Mhóir
22 Valentia Heritage Centre
23 The Bere Island Project
24 Bantry Museum
25 Mizen Head Visitor Centre
26 Cape Clear Museum
27 Ceim Hill
28 Kinsale Museum
29 Cobh Heritage Centre
30 Cobh Museum
31 Youghal Visitor Centre
32 Dungarvan Museum
33 South Tipperary County Museum
34 Carrick On Suir Heritage Centre
35 Waterford Treasures
36 The Dunbrody Famine Ship and Visitors Centre
37 Hook Lighthouse & Heritage Centre
38 The Irish National Heritage Park
39 Rosslare Harbour Maritime Heritage Centre
40 The Wexford Wildfowl Reserve Visitor Centre
41 Arklow Maritime Museum
42 Wicklow's Historic Gaol
43 Dalkey Castle & Heritage Centre
44 The National Maritime Museum
45 National Museum of Ireland (Archaeology & History)
46 Waterways Ireland Visitor Centre
47 National Museum of Ireland (Decorative Arts & History)
48 The Boyne Currach Centre
49 Millmount Museum & Martello Tower
50 Louth County Museum
51 The Nautilus Centre
52 Newry and Mourne Museum
53 Ropewalk Maritime Visitor Centre
54 Visit Ards and North Down
55 Grey Point Fort
56 Ulster Folk Museum
57 Thompson's Dry Dock And Pump House
58 Titanic Belfast
59 The Belfast Barge
60 Ulster Museum
61 Harland & Wolff HQ & Drawing Offices
62 Titanic and Olympic Slipways
63 SS Nomadic
64 HMS Caroline
65 Clarendon Dock
66 Rathlin Boathouse Visitor Centre (Seasonal Opening April to September)
67 Tower Museum
58 Derry and the North Atlantic Maritime Museum (Planned)
69 Harbour Museum
DONEGAL
DERRY
ANTRIM
TYRONE
FERMANAGH
ARMAGH
DOWN
SLIGO
LEITRIM
MONAGHAN
MAYO
CAVAN
LOUTH
ROSCOMMON
LONGFORD
MEATH
WESTMEATH
GALWAY
OFFALY
DUBLIN
KILDARE
LAOIS
WICKLOW
CLARE
CARLOW
TIPPERARY
KILKENNY
WEXFORD
LIMERICK
WATERFORD
KERRY
CORK
N
0 25 50 75 100
Kilometres

Fig. 15.31 Maritime museums and heritage centres. A plethora of maritime museums and centres around the coast promote the conservation and interpretation of national to locally significant marine artefacts, biological resources and creative displays. The map shows the distribution of key maritime museums and heritage centres around the coast, but it cannot be said to be an exhaustive list, as many temporary displays are produced, for example, to run over the summer months in remote coastal villages. Other significant resources are in private ownership, such as the museum holdings of Sherkin Island Marine Station. [Data source: Heritage Council; Tourism Northern Ireland; Northern Ireland Museums Council; Department for Communities Northern Ireland]

managing coastal heritage.

On the other hand, a significant array of objects and documents associated with coastal and maritime history and heritage have survived and are held in public and private collections around the country. An overview of key centralised coastal heritage collections is provided in BiblioMara, the National Folklore Collection Schools' Collection and the Traditional Boats record. In a study carried out for the Heritage Council in 2005, 160 maritime collections were identified, of which only about 10 per cent are owned by local authorities or a statutory body. The vast majority are run by community-based and voluntary organisations.[34] Many of these collections contribute to local distinctiveness and as a result are central to local tourism. The National Maritime Museum in Dún Laoghaire, County Dublin, is the largest collection of this kind; it is owned and run

OXO
ELSInore

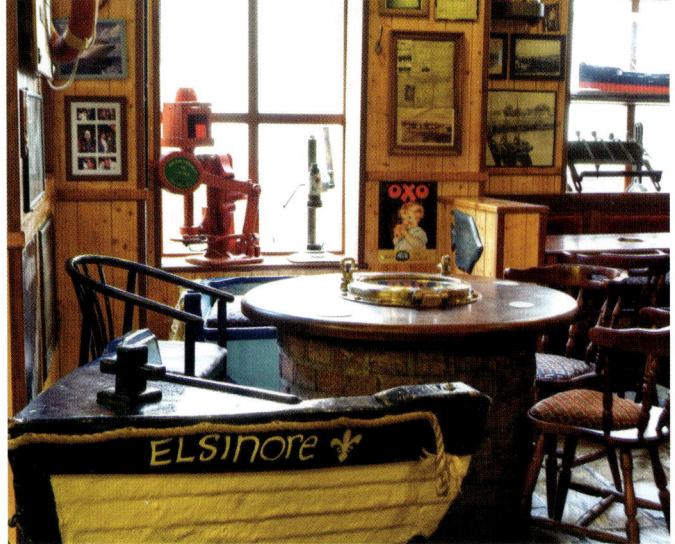

Fig. 15.32 Local maritime heritage on display in Harry's Bar Rosses Point, County Sligo. Displays of objects salvaged from the sea or recovered from the shoreline are often found in bars and hotels, juxtaposed with artefacts and memorabilia that evoke associations with local sailing, fishing, lifeboats and/or shipping activities. This image from Harry's Bar, Rosses Point, County Sligo, is indicative of how local maritime heritage can be appreciated and promoted. Harry's Bar was listed in the 2006 maritime audit undertaken by the Heritage Council, such is the significance of the artefacts on display in the public house. These include log books, a bow thruster converted into a table, a name plate and starboard telegraph from the Russian vessel T.R. *Wevenhko*, and a stunning photographic collection. [Source: Frances Muldoon]

on a voluntary basis by the Maritime Institute. Established in 1941, the institute was set up to support Ireland's maritime interests, including the promotion of awareness of maritime heritage. Its main activities today are the management of the National Maritime Museum and maintaining a library of *c*.5,000 books on maritime affairs. Other notable maritime museums are to be found at Greencastle, County Donegal, and Foynes, County Clare; other important collections are located in Arklow, County Wicklow, and Cobh, County Cork. Maritime material forms a sub-stantial part of the holdings of the National Museum of Ireland – Country Life, County Mayo, and several county museums (Figure 15.31). In some cases, it is a core theme in the main exhibition; Galway City Museum explores the maritime traditions of the Claddagh

basin and of Galway Bay. The widespread interest in maritime heritage is also informally acknow-ledged through many displays of objects salvaged from the sea and shoreline, often found in bars and hotels, such as Bushe's Bar, Baltimore, County Cork, and Harry's Bar, Rosses Point, County Sligo (Figure 15.32).

In addition to interest in objects and material relating to the coast, public interest in publications on coastal and maritime heritage is high, as the Heritage Council has discovered through the sustained demand for its series of leaflets on diverse topics, such as coastal geology and archaeology, sharks, jellyfish and seaweed and also in the quantity of new publications on maritime themes published with its support in the decade 2000–2010 (Figure 15.33).[35]

HERITAGE COLLECTIONS: SOURCES OF LORE FOR RESEARCH AND ENJOYMENT

Clíona O'Carroll

Most of the following resources focus specifically on matters maritime, but any collection of lore from a coastal region will reflect the centrality of the sea and shore as a physical and imaginative resource for the community.

A good starting point for enquiry is BiblioMara, a free annotated bibliography of works pertaining to the cultural and built maritime heritage of the island of Ireland. This bibliography gives a short description of each resource and contains 2,964 references to works in the English and Irish languages published from 1772 to 2004. These include: 1,195 book references, ninety book sections, 1,411 journal articles, eleven reports, five articles from serials and 252 postgraduate theses (published from 1934 to 2003). Browsing or carrying out word searches in BiblioMara can start myriad pathways of enquiry.

The National Folklore Collection (NFC), the bulk of which was collected by the Irish Folklore Commission between 1935 and 1971, is housed in

Fig. 15.33 HERITAGE COUNCIL COASTAL HERITAGE LEAFLETS. Interpretive pamphlets, such as these leaflets produced by the Heritage Council, contribute to understanding of natural and built marine heritage around the coast. The leaflets cover a range of diverse topics such as maritime archaeology, jellyfish in these coastal seas and seaweed. [Source: The Heritage Council]

the Delargy Centre for Irish Folklore and the National Folklore Collection, University College Dublin. This is a very rich source of maritime lore in Irish and English, including folk belief, personal memory, accounts of fishing and shore gathering, descriptions of boat designs and use, as well as local legends of supernatural encounters, tragedies and close escapes. *A Handbook of Irish Folklore* by Seán Ó Súilleabháin, now available as an electronic publication, provides an insight into the Irish folk tradition and serves as a guide to NFC materials. Tom Munnelly's two articles in the journal *Béaloideas* give a good orientation with regard to songs connected with the sea and shore in Ireland and outline their representation in published sources and in the NFC. Ó hEochaidh's ninety-seven-page 1965 *Béaloideas* article compiles sea-related lore collected in Irish in south-west and north-west Donegal into a collection that is representative of such lore countrywide and includes summaries and notes in the English language. O'Carroll's article draws upon lore from the NFC relating to supernatural encounters, and good and bad luck on the water, and lists relevant publications in the field of folklore (Figure 15.34).

The Schools' Collection, a subcollection of the NFC carried out in 1937–1938 by schoolchildren, is now available online at dúchas.ie, with a text search facility. Users can access (and transcribe) descriptions of coastal life, songs of the sea and coast, and stories of gold from the sea, shipwrecks, fairy boats and sea cows.

Fig. 15.34 (left) MEMBERS OF THE IRISH FAMILY HISTORY SOCIETY VISITING THE NATIONAL FOLKLORE COLLECTION IN 2017. Heritage collections provide important sources of material for the enjoyment, preservation and enhanced understanding of our shared coastal heritage. The National Folklore Collection contains a treasure trove of information on maritime lore, coastal lifestyles, songs and stories that reflect the dual realities, of both the vibrancy and the hardship, of living by the sea. Public engagement with such resources have been significantly enhanced with the advent of online access to heritage databases, such as BiblioMara. Informal collections, such as those displayed in local history centres and public houses, also provide huge value in their interpretation of local coastal heritage attributes. [Source: National Folklore Collection UCD]

Traditional Boats of Ireland: History, Folklore and Construction by Criostóir MacCárthaigh is the result of long-term collaboration among maritime and traditional boat scholars and practitioners, and makes a substantial and authoritative contribution to the documentation of Irish working vessels and their use countrywide (Figure 15.35).

Fig. 15.35 BALTIMORE TRADITIONAL BOATS REGATTA. Traditional boats and boat festivals are amongst the most popular aspects of coastal heritage. As a result of its history and varied coastline, Ireland is fortunate to have a wide range of boat types. Steps have been taken by many individuals and voluntary groups around the coast to protect and promote their local boating heritage. Such activities include boat repair, the building of replicas, the collection of lore and skills from the surviving boat builders, and festivals and regattas. This work is of great significance, as traditional boats and surviving nineteenth-century boats afloat are not protected by legislation unless they are classified as a wreck. Many regional and local boats are now redundant as their original uses for fishing, trade and transport have been superseded, e.g., the sand boats of Ballydehob, Achill Island sailing yawls and Waterford estuary prongs. Certain types of boats have been sustained through the dedication of associations such as the Galway Hooker and Húicéir na Gaillimhe, while others have continued to evolve as their builders incorporate modern materials, such as fibreglass and outboard engines – the Glass Reinforced Plastic (GRP) currach of the Galway coast is a case in point. However, many boats have been broken up or simply rotted away. Despite the general decline in the use of traditional boats, a county-wide survey of boats in County Galway found that south Connemara in particular has one of the most vibrant boat cultures in Europe. [Source: Robbie Murphy]

The distinctive elements of the coastal heritage of Ireland – remembering past events and spurring modern movements – are subject to evolving approaches to heritage management and appreciation. In conclusion, Ireland's coastal heritage plays a vital role in many aspects of life in Ireland, whether in culture, the environment or economic activities. So far, coastal and marine heritage has not received as much official attention as other aspects of Irish heritage. Despite this, individuals, voluntary groups and NGOs have worked on their own, or with the support of local authorities and other state organisations, to retain and promote many aspects of coastal heritage. The examples described in this chapter show the ways in which such groups have made use of their coastal heritage: as a force for constructive social engagement; as a mark of local distinctiveness;

as a focus in growing and reinforcing education and skills; or as a driver for alternative economic activities when traditional livelihoods are in decline.

Where local initiatives are the subject of recognition and support from state sources, these initiatives can bloom. The fruit of this work is starting to show signs of ripening, for example in the recognition of currach building in the National Inventory of Intangible Cultural Heritage and in success stories along the Wild Atlantic Way. Some initiatives have the potential to form the core of localised management programmes for the coast, or at the very least, to enlighten future coastal management programmes. Ireland's coastal heritage merits a higher level of official recognition of its potential value for coastal communities and as a stimulus for interesting, forward-thinking initiatives along these shores.

THE INHABITANTS OF IRELAND'S EARLY COASTAL LANDSCAPES

Peter Woodman and Robert Devoy

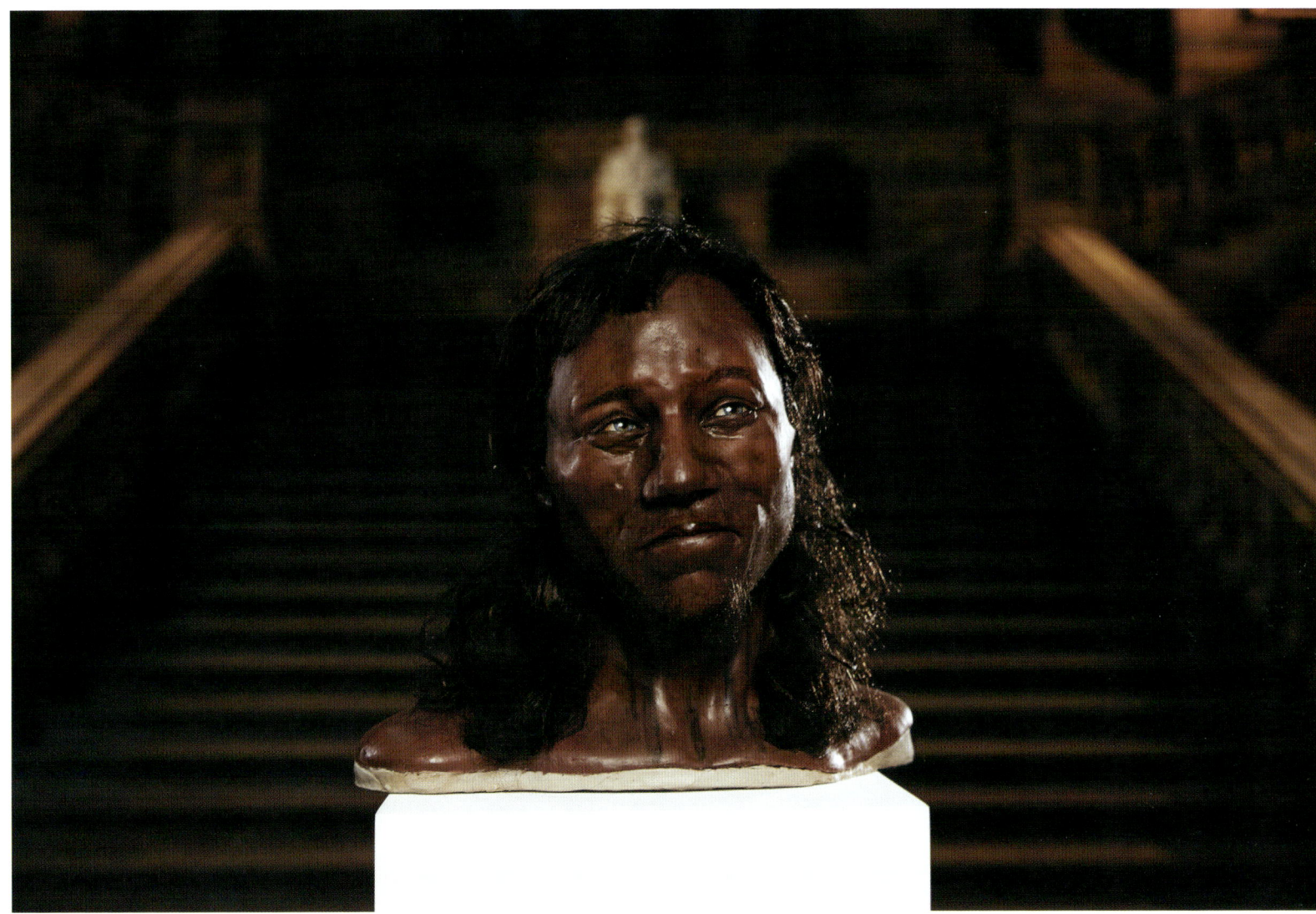

CHEDDAR MAN. The early Mesolithic settlers of Ireland were probably similar in appearance to Cheddar Man. The skeleton of a male found in Gough's Cave at Cheddar Gorge, Somerset, England, in 1903 has been radiocarbon-dated to having lived *c.*9,100 years ago in the early Mesolithic. Work on the DNA of the man shows that he probably had blue eyes, was dark- to very dark-skinned and had curly dark hair. Research of much later human remains of Mesolithic people in Ireland, dated to 6,000 years ago, shows that they had similar body characteristics, even at this later date. These features were typical of the European Mesolithic population of hunter-gatherers. The image shown here is a reconstruction of the Cheddar Man, which is held in the Natural History Museum in London. [Source: Tom Barnes]

One of the questions posed most commonly about prehistoric people's use of Ireland's coast involves the timing of their first arrival.[1] When did people first reach and settle on Ireland's shores? Also, when did people's permanent occupation of the island begin (Figure 16.1)?[2] This chapter seeks to explore these questions of the prehistoric settlement of Ireland (that is before approximately 2,000 years ago) and the associated changes in coastal landscapes (see Chapter 6: Glaciation and Ireland's Arctic Inheritance). It goes on to discuss the early settlers in the context of their use of the coast, including examination of what happened to coastal environments as farming was introduced from around 5,700 years ago (Figure 16.2). Finally, consideration is given to an understanding of how the in-built dynamic processes driving coastal systems affected prehistoric people (see Chapter 3: Coastal Waters: Marine biology and Ecology and Chapter 11: Beaches and Barriers. Figure 16.1 illustrates the principal coastal locations described in this chapter.

ENVIRONMENTS OF THE FIRST SETTLERS

There is good archaeological evidence of the presence of people in the neighbouring island of Britain for at least the last 700,000 years, although records of both the different population groups and their patterns of occupation are fragmentary and intermittent.[3] Given Ireland's close proximity to Britain, it might be expected that similar early traces of human activity would also occur in Ireland. To date, however, little substantial evidence of a human presence has been found until after c.30,000 years ago and more substantially after c.9,500 years ago, though this time picture is changing and becoming more controversial.[4]

BARRIERS TO THE FIRST PEOPLE

Extensive research in the Earth sciences indicates that before lateglacial time (Figure 16.2), Ireland was most likely inhospitable to colonisation and permanent habitation, not only by humans, but also by many animals and plants. The reasons why the island was an unattractive place for people to settle during much of the last 700,000 or more years are not hard to find: glaciations![5]

This is the period in Earth history known commonly as the Ice Age, covering the Quaternary period of geology (Figure 16.2). This lasted for over two million years and was a time of repeated global cycles of cold climate conditions. The last phase in glaciation ended only around 11,000 years ago. Ireland and Britain would have been impacted during the Ice Age, along with much of north-west Europe and North America, by long periods of glaciation and Arctic-type environments. These conditions would have lasted, with some short climate-warming interludes, for 60,000 to 80,000 years at a time. More than eight such major glaciation 'cold cycles' occurred during the Late Quaternary, alternating with intervening times of climate warmth, known as interglacials. These longer warm phases were commonly 15,000 to 20,000 years in duration, with temperatures and environments similar to those of the present day.[6]

It is quite possible, even probable, that early man, as groups of Palaeolithic hunter-gatherers from Britain and continental Europe, may well have lived in Ireland during the Late Quaternary, particularly during the interglacials. Either, their remains have not as yet been identified or, more likely, the evidence of a human presence (such as their human bones, tools and occupation sites) has been destroyed. Though this is less the case in neighbouring Britain and Europe, where the remains of Palaeolithic people do exist, why would this destruction of evidence have happened in Ireland? The cause is probably that the island is situated on the frontline boundary between continental land and the North Atlantic ocean. This setting provides the conditions for the establishment of a highly energetic Earth system, of the land, atmosphere and the ocean. Consequently, during times of glaciation the island has experienced repeated and rapid movements of ice, water and rocks. This created dynamic landscapes through the operation of geomorphological processes driven by the agents of ice and water, which were very energetic and extremely destructive.[7] Any remains of people on the land and coasts that may have existed would have been ground up and eroded away.

Environmental reconstructions of the last major glacial period (115,000–18,000 years ago) are well established and this time forms a good analogue for the sort of conditions with which people and other life would have had to cope earlier in the Quaternary. The period is still known in Ireland by many as the Midlandian, though referred to internationally now as Marine Isotope Stage 4 (MIS 4). The peak cold phase of this time ('the Last Glacial Maximum cold' or LGM) was probably one of the most environmentally intense, and the coldest, of the whole Ice Age. It lasted in Ireland and Britain from approximately 26,000 to 17,000 years ago (Figure 16.2). Palaeoenvironmental studies show that Ireland was covered by extensive areas of ice. A thick ice sheet lay over much of the island, with a separate regional ice cap over west Cork and Kerry. Numerous valley glaciers, scouring the mountain areas, also existed. During seasonal or longer periods of melting ice, broad and turbulent glacial river systems filled with rocks and sediments dissected the country (Figure 16.3). Regionally, relative sea levels were at least 100m below those of the present. Much of the inner continental shelf around Ireland, usually covered by the sea during times of warm conditions (the interglacials), lay exposed to the 'Arctic' climate.[8] Consequently, seasonally, and during other extensive phases of ice melt, the glacial meltwater systems were able to extend across the present shorelines onto these shelf areas. In the former Irish and Celtic seas, which separated Ireland from Britain and continental Europe, substantial glaciers remained until c.18,000 to 20,000 years ago.

These conditions were replaced almost immediately, under rapidly rising sea levels, by marine flooding of the ground surfaces as these emerged from beneath the retreating ice. Any areas of proto land would have been characteristically wet, unstable and cut by glacial meltwater rivers. Importantly, they would also have lacked any plants and animals. They may have served, however, as potential but short-lived landbridges. These would have formed routeways for plant and animal migration. Such routeways may have operated briefly across the Irish and northern Celtic Sea areas during the lateglacial, similar to the classic landbridges of the Quaternary. In many other parts of the world these surfaces have enabled the movement of people, plants and animals between formerly separated land areas at times of low sea level. A good example close to Ireland is that of Doggerland. This area, now under the North Sea, formerly connected Britain with continental Europe, until approximately 8,000 years ago, when rising sea levels and the Storegga tsunami caused its submergence (Figure 16.4). In Ireland, where the controlling Earth processes were more dynamic, any similar landbridge would have been ephemeral and gone before *c.*14,000 years ago.[9]

Further, environmental conditions on any landbridge would have been relatively arduous and probably a deterrent to the regular movements of people, though some possible landbridge positions in the Irish Sea (Figure 16.1) may just have happened to coincide with the routeways to Ireland taken by the first migrants. More importantly for these first settlers, these former glacial environments would have been real barriers to plants and animals attempting to cross into Ireland.

Most temperate-zone plants and animals had died out on the island during the glaciations, except perhaps from some

Fig. 16.1 PRINCIPAL COASTAL SITES OF THE MESOLITHIC PEOPLE AND SETTLEMENTS IN IRELAND. The probable three main routes of entry for people to Ireland from Britain are also shown.

'STONE AGE' ARCHAEOLOGY
Northern Europe

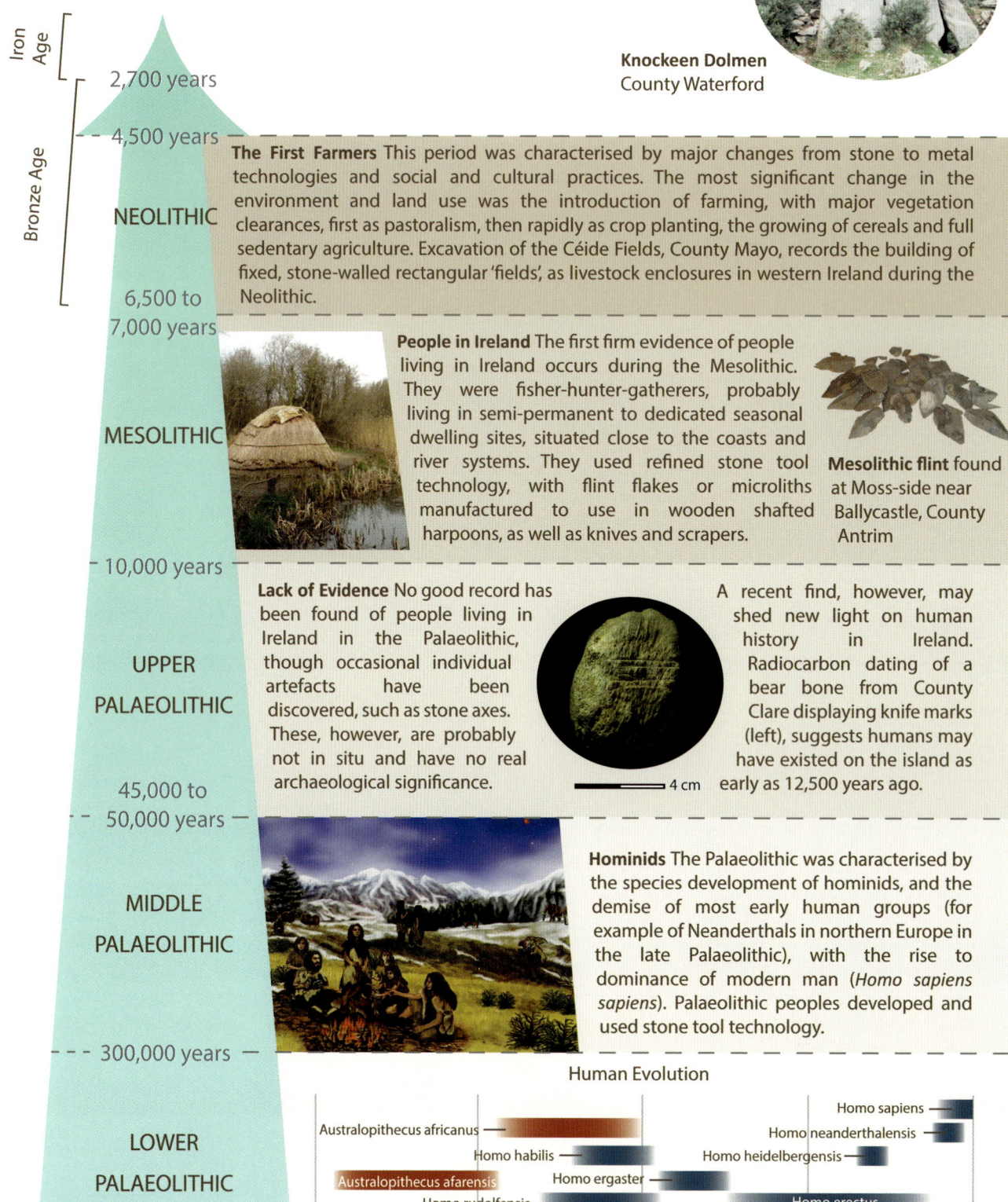

Knockeen Dolmen
County Waterford

Iron Age
2,700 years

Bronze Age
4,500 years

NEOLITHIC

The First Farmers This period was characterised by major changes from stone to metal technologies and social and cultural practices. The most significant change in the environment and land use was the introduction of farming, with major vegetation clearances, first as pastoralism, then rapidly as crop planting, the growing of cereals and full sedentary agriculture. Excavation of the Céide Fields, County Mayo, records the building of fixed, stone-walled rectangular 'fields', as livestock enclosures in western Ireland during the Neolithic.

6,500 to 7,000 years

MESOLITHIC

People in Ireland The first firm evidence of people living in Ireland occurs during the Mesolithic. They were fisher-hunter-gatherers, probably living in semi-permanent to dedicated seasonal dwelling sites, situated close to the coasts and river systems. They used refined stone tool technology, with flint flakes or microliths manufactured to use in wooden shafted harpoons, as well as knives and scrapers.

Mesolithic flint found at Moss-side near Ballycastle, County Antrim

10,000 years

UPPER PALAEOLITHIC

Lack of Evidence No good record has been found of people living in Ireland in the Palaeolithic, though occasional individual artefacts have been discovered, such as stone axes. These, however, are probably not in situ and have no real archaeological significance.

A recent find, however, may shed new light on human history in Ireland. Radiocarbon dating of a bear bone from County Clare displaying knife marks (left), suggests humans may have existed on the island as early as 12,500 years ago.

4 cm

45,000 to 50,000 years

MIDDLE PALAEOLITHIC

Hominids The Palaeolithic was characterised by the species development of hominids, and the demise of most early human groups (for example of Neanderthals in northern Europe in the late Palaeolithic), with the rise to dominance of modern man (*Homo sapiens sapiens*). Palaeolithic peoples developed and used stone tool technology.

300,000 years

Human Evolution

Australopithecus africanus
Homo habilis
Australopithecus afarensis
Homo rudolfensis
Homo ergaster
Homo sapiens
Homo neanderthalensis
Homo heidelbergensis
Homo erectus

LOWER PALAEOLITHIC

3.5 million years

4   3   2   1   0
MILLIONS OF YEARS AGO

Fig. 16.2 OUTLINE OF THE MAIN ARCHAEOLOGICAL TIME PERIODS FROM THE PALAEOLITHIC TO THE IRON AGE. These times of major population, cultural and technological change are linked in the diagram to the main phases of geological time and environmental changes in Ireland and north-west Europe. (Note: the suggested date for known people in Ireland has now been pushed even further back, to *c*.33,000 years ago.) [Photo sources: Knockeen Dolmen, County Waterford: Robert Devoy; reconstruction of Mesolithic hut: David Hawgoof, Wikimedia Commons, CC BY-SA 2.0 Unported; Mesolithic flint: Notafly, Wikimedia Commons, CC BY-SA 3.0 Unported; bear patella: Thorsten Kahlert courtesy of Marion Dowd; Palaeolithic hominid illustration: David Mark, Pixabay; Inishbofin: Alvaro, CC BY-SA 3.0 Unported; glacier: Jay Ruzesky, public domain]

'ICE AGE' EVENTS

'ICE AGE' EVENTS	AGE (Stage) In Ireland	EPOCH (Series)	PERIOD (System)
Postglacial (Warm)	LITTLETONIAN	HOLOCENE	
	— 11,700 years —		
Nahangian Stadial (Cold) Woodgrange Interstadial (Warm) Glenavy Stadial (Cold) } Late Glacial	MIDLANDIAN	Late	QUATERNARY
	— 120,000 years —	— 127,200 years —	
	Last Interglacial		
	— 132,000 years —		
	MUNSTERIAN		
	GORTIAN	Middle	PLEISTOCENE
	— 338,000 years —		
	PRE-GORTIAN Pre-MIS9		
	Glacial and interglacial climate cycles continue to occur, with over 100 of these recorded from ice and ocean cores dating back to 2.614 million years ago (MIS 104). Initially in the Quaternary, these cycles occurred every c.41,000 years. Later, over the last c.1 million years they have occurred at a frequency of approximately 100,000 years. Eight major glacial cycles have occurred over the past 740,000 years.	780,000 years / Early	
	— 2.6 million years —		
	PRETIGLIAN	PLIOCENE	TERTIARY
	— 2.3 million years —		
	REUVERIAN		
	5.3 million years		

Interglacial landscape

Palaeoclimates The Earth over the last 600 million years has been without large ice masses at its poles for longer than it has had any. The most recent development of ice in the Arctic region began in the middle Miocene, with ice sheets developing on Greenland *c.*18 million years ago. The development of full sea ice cover of the Arctic Ocean, with extreme glacial conditions beginning in the North Atlantic region, is more recent beginning some 3 million years ago. Ice sheet formation started much earlier in Antarctica, *c.*25 million years ago.

Glacial and Arctic landscape

Fig. 16.3 A COASTAL GLACIAL LANDSCAPE FROM EASTERN GREENLAND. This shows an Arctic tidewater, glaciated environment, which would have been similar to those that occurred around Ireland's mountains or uplands rim and its palaeocoasts at the end of the last glaciation, c.14,000–17, 000 years ago. [Source: Greenland Travel, flickr, CC BY 4.0]

bio refugia that existed close to today's coasts and on the inner shelf. The cold-adjusted plant communities, which had replaced those under glacial conditions, also began to suffer the same fate with the

end of glaciation and the return to a warm climate in the postglacial. By around 15,000 years ago, and before the landbridge routes closed during the lateglacial, many temperate European mammalian species had made their way back to Ireland. However, there are also many animals found in Britain that failed to migrate and colonise the island. The story for plants is similar, with over 30 per cent of plant species commonly found in Europe, some of which had existed in Ireland earlier in the Quaternary, failing to make the crossing.[10]

With regard to humans, radiocarbon dating evidence from animal bones, which retain tool cut marks, indicate that hunter-gather people were probably in Ireland at least c.30,000 years ago, though evidence of a more substantial presence (from human artefacts, bones, occupation sites) occurs only later in the postglacial (after c.11,700 years ago). The landbridge argument – as a means of delineating the timing of colonisation of Ireland by people, as well as by plants and animals – is uncertain. Potential short-lived land surfaces existed, but their viability remains questionable (see Chapter 7: Ancient Shorelines and Sea-level Changes). Plant and animal communities developing there would have been temporary and lacked species diversity. They would have provided a poor food resource for any hunter-gatherer peoples attempting to make the crossing. Given the inhospitable and dynamic characteristics of the region, people would have traversed the landbridge quickly. Boats, or rafts, would most likely have been needed to cross the rivers, developing sea embayments and wetlands characteristic of these surfaces. Realistically, any landbridge that existed in the region would have been short in length (<200km at maximum) and not a major barrier to people. In many locations along Irish Sea coasts, Ireland is visible from Britain, particularly Northern Ireland from Scotland and Cumbria. If people had wished to cross over they would have known where they were going.[11]

While glacial conditions and rising sea levels may have become less of a barrier to people crossing to Ireland with time, the absence of familiar foods and other resources could have remained a problem. By 8,000 years ago there would have been plenty of potential food resources on the island, with extensive woodlands of birch, hazel and willow, together with many other usable plants. These

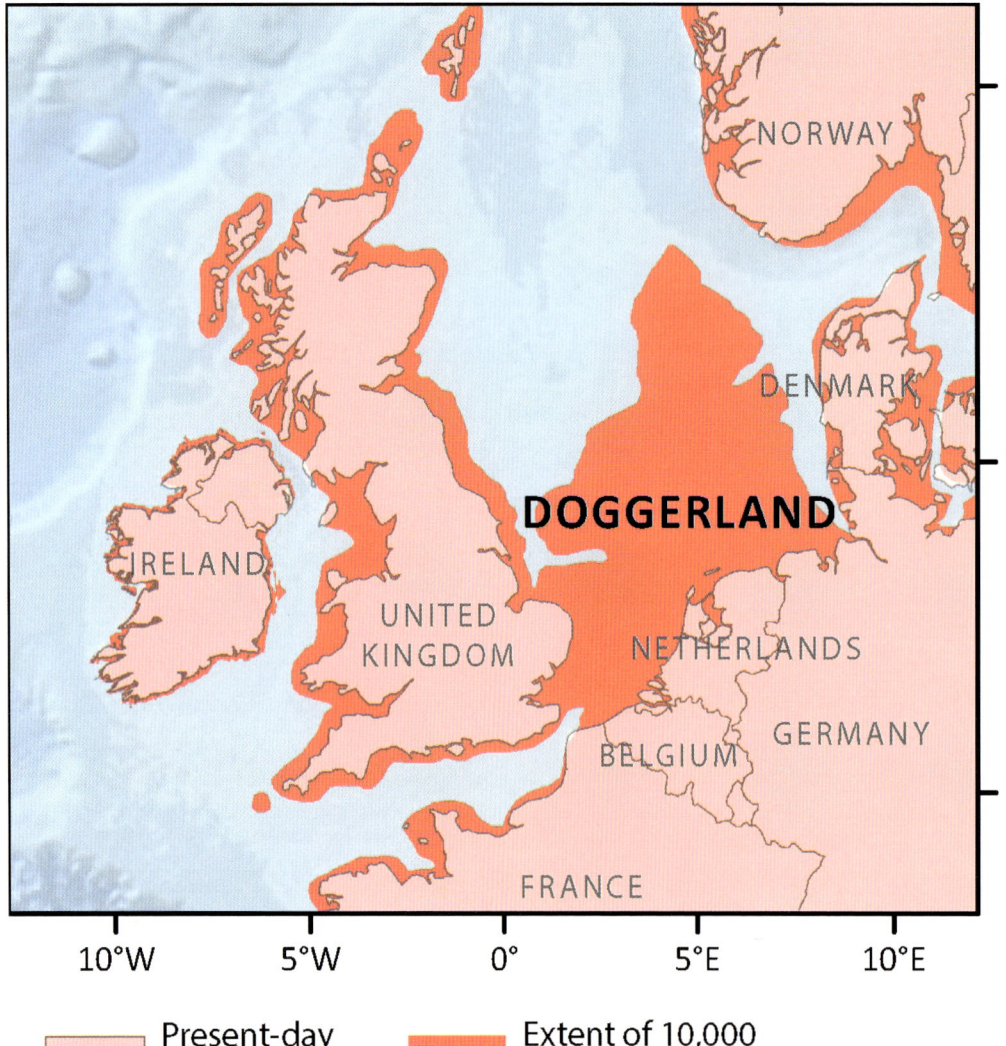

Fig. 16.4 CHANGES TO THE EUROPEAN COASTLINE FROM 10,000 YEARS AGO TO THE PRESENT DAY.

Present-day shoreline

Extent of 10,000 years BP shoreline

early peoples would have been used to a more diverse food and materials base in mainland Europe and Britain.[12] In Ireland, although a few animal and fish species had survived the lateglacial, many larger mammals and some fish species found elsewhere in Europe were missing. Wild cattle and elk were absent, as well as a range of freshwater fish, such as pike.[13] Consequently, colonists of the island would have had to cope with some changes to their diet. This likely unfamiliarity with the available foods, however, was perhaps much less of a deterrent to these hardy Mesolithic people than some commentators suggest! A few mammals, including wild boar, hare and wolf/dog were present by this time, but it is unclear how many of these animals, which become iconic food sources, were used by these early settlers.

THE FIRST PEOPLE

In the lateglacial, human settlement in western parts of Britain had been badly affected by the extreme low temperatures. After this time, people were able to re-settle these areas of western Britain. In Ireland, results from recent studies suggest possible seasonal human inhabitation from as early as 33,000 years ago. Later radiocarbon dating at 12,500 years ago of a bear bone (patella) displaying cut marks, found in County Clare, may indicate the existence of people in this area at this time (Figure 16.2).[14] This dating would fit with a timing for a

FIG. 16.5 PRINCIPAL COASTAL MESOLITHIC SITES. The concentration of sites in the north-east around Larne, County Antrim has led to these early Mesolithic people being known as the 'Larnians', a term that has now dropped out of academic usage, but is still in common use. The map may not show all known sites. [Data source: Woodman, P., 2015. *Ireland's First Settlers: Time and the Mesolithic*. Oxford: Oxbow Books]

lateglacial or, as is now suggested, an even earlier movement of people into Ireland. On the south coast in 2018, updated reporting of the finds of Mesolithic artefacts in the Creadan Head area of Dunmore East, County Waterford, occurred.[15] Some of these finds have been interpreted now as of early Mesolithic age, and may indicate that people arrived in this part of Ireland as early as the settlers in the north and west of the island.

At present, though, the view based on substantial archaeological site investigation and artefact data is that, accepting the likelihood of earlier seasonal migratory type visits, the main occupancy of the island by people did not take place until around 10,000 years ago. These settlers were 'late Stone-age' Mesolithic

hunter-gatherers. As discussed, they may have crossed the Irish Sea using boats, which may have been skin- or wood-bark-covered craft, similar to the umiaks or kayaks used by Arctic peoples today, or possibly by rafts as well (Figure 16.6). If this was the case, it would show that there was probably a practice of seafaring established in the early postglacial by the coastal people of Britain and Europe.

Conventional wisdom had been that they first arrived by crossing the narrow stretch of water between Scotland and the Antrim coast. There is now growing evidence that they arrived first on the Leinster coast and moved from there around Ireland's coastal zones as seasonal fisher-hunter-gatherers. It is more likely, however, that there was no single major first settlement point, but that multiple

Fig. 16.6 TRADITIONAL BOATS. a. The modern umiak, a skin-covered boat used by Yupik and Inuit Arctic peoples from Shishmaref, Alaska. The photograph shows the frame of a umiak. [Source: US National Parks Service]; b. A modern-day kayak from Shishmaref, Alaska. Using these types of boats, Inuit people are able to travel long distances over the Arctic ocean areas. [Source: US National Parks Service/Lia Nydes]

landfalls were made in the same broad timespan. Whichever routeways were taken, these early colonisers were most likely coming from the major Mesolithic cultural grouping of people in Britain, centred on Cumbria and Wales. The people here were well adapted to using the coastal environments of the region. The early colonisers of the Isle of Man were most likely part of this same cultural grouping and connected to the people colonising Ireland. As with Ireland though, rapid sea-level rise cut their landbridge route back to mainland Britain across the eastern Irish Sea.[16]

LOCATING IRELAND'S EARLIEST COASTAL SETTLERS

Given the difficult glacial and lateglacial terrains of Ireland's interior before 12,000–14,000 years ago, coupled with the potential issue of suitable food resources, the early colonists would have tended to stay and settle in coastal areas. Archaeological finds of their occupation sites show also that these Mesolithic people's colonisation extended into areas of connected rivers, sea loughs and lakes (Figure 16.7). These environments would have had an inherent diversity of more familiar and readily accessible resources.

Finding evidence of the first coastal settlement sites is problematic. Ireland's coastal zone at the macro-scale appears to have changed relatively little during the postglacial compared to other parts of north-west Europe. The palaeo-map of Ireland at 10,000 years ago would be easily recognisable as being the same island (Figure 16.8), but in parts of Britain and Europe significant land-area changes have occurred. For example, the former Doggerland area, roughly the same size as England, vanished under the North Sea about 8,000 years ago (Figure 16.4). Denmark has changed from a substantially large land area into a peninsula and an archipelago of islands. Land losses have also happened in Ireland, at lower scales. Rising sea levels have now drowned the locations of most early coastal settlement sites, as well as causing the complete loss of others through erosion (see Chapter 11: Beaches and Barriers).

Relative sea-level rise around Ireland has caused numerous river valleys to be inundated by marine flooding, submerged and even transformed into sea loughs (see Chapter 7: Ancient Shorelines and Sea-level Changes). This is particularly apparent in the

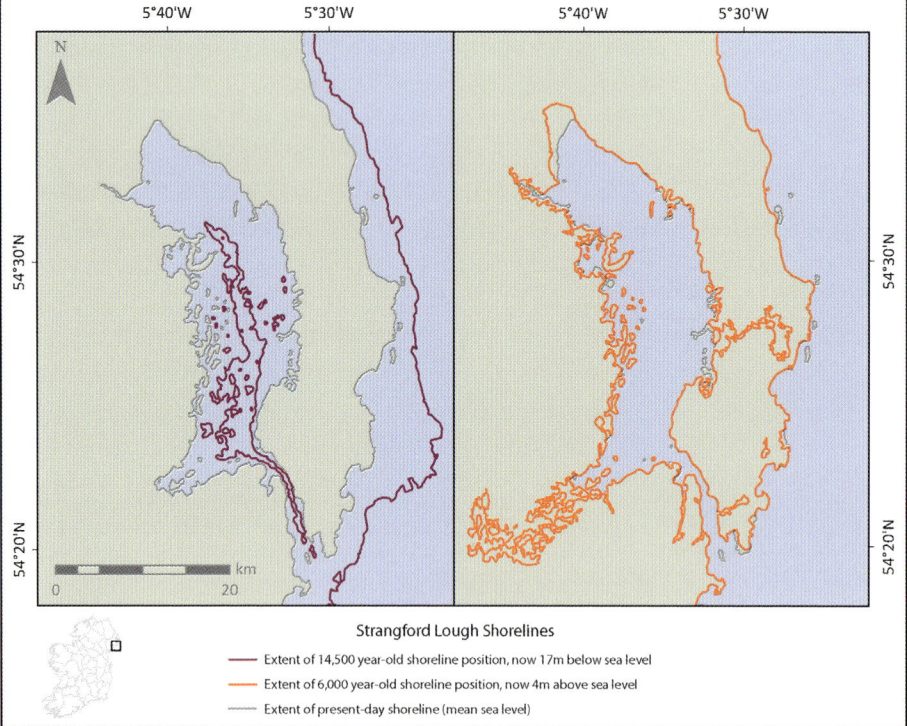

Strangford Lough Shorelines

—— Extent of 14,500 year-old shoreline position, now 17m below sea level
—— Extent of 6,000 year-old shoreline position, now 4m above sea level
—— Extent of present-day shoreline (mean sea level)

Fig. 16.7 FORMER SHORELINE POSITIONS AND LAND TO SEA-AREA CHANGES OF STRANGFORD LOUGH BETWEEN 14,500 AND 6,000 YEARS AGO. These maps show two snapshots of shoreline positions and the consequent losses in land areas since their formation, caused by postglacial relative sea-level changes, marine flooding and erosion. Note that the 6,000-year-old shoreline, along which Mesolithic people would probably have lived, is now elevated above the present sea level and forms a raised shoreline. This is due to regional land uplift, resulting from the rebound of the Earth's crust after former ice loading. [Data source: Edwards, R.J. and Brooks, A.J., 2008. The Island of Ireland: Drowning the myth of an Irish landbridge. In: Davenport, J.L., Sleeman, D.P. and Woodman, P.C., eds. *Mind the Gap: Postglacial colonisation of Ireland*. Special Supplement to the *Irish Naturalists' Journal*, pp. 19–34]

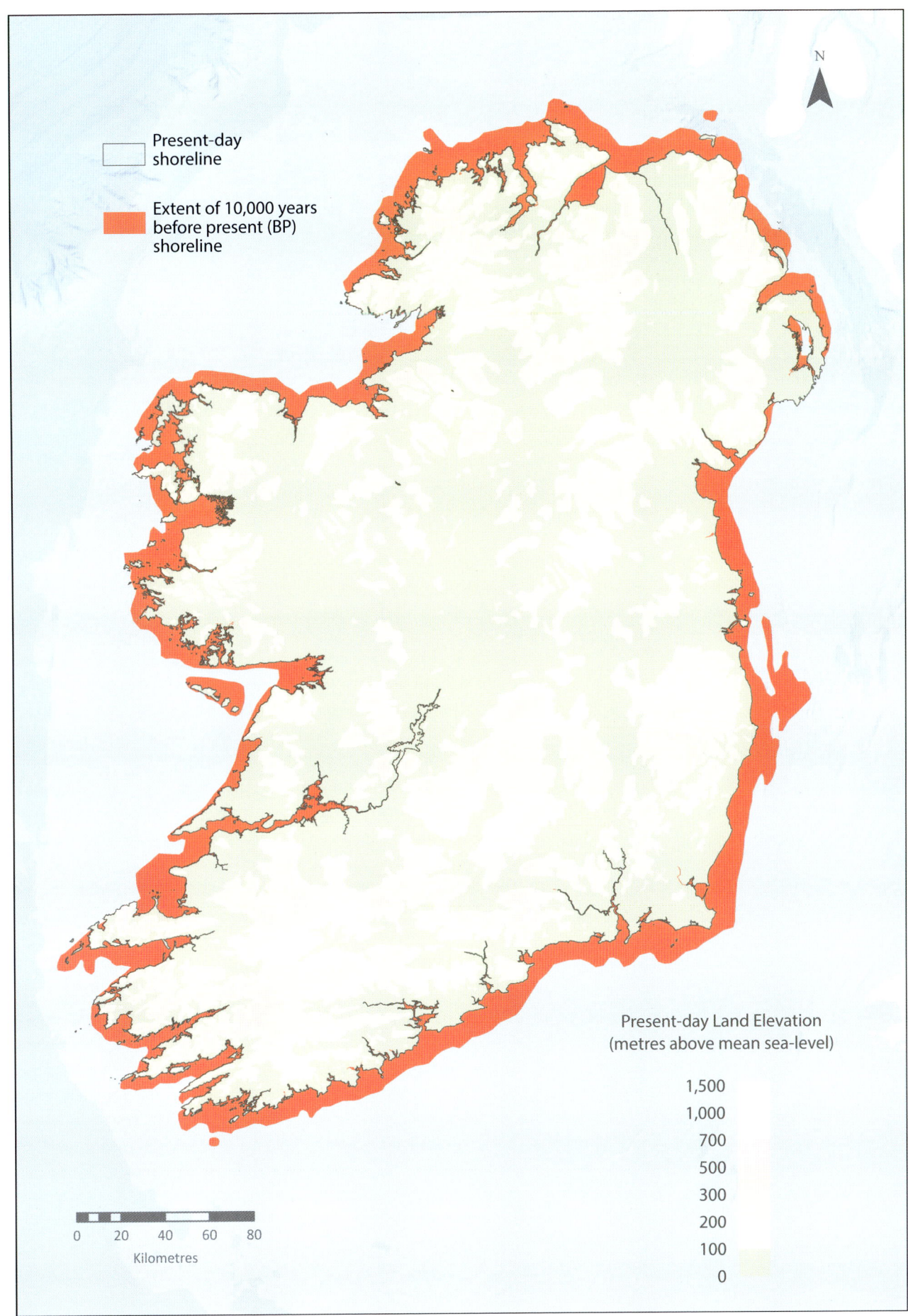

Present-day
shoreline

Extent of 10,000 years
before present (BP)
shoreline

N

Present-day Land Elevation
(metres above mean sea-level)

1,500
1,000
700
500
300
200
100
0

0   20   40   60   80
Kilometres

Fig. 16.8 IRELAND'S COASTLINE, AROUND 10,000 YEARS AGO, SHOWING THE EXTENT OF CHANGES IN SHORELINE POSITION AROUND THE ISLAND RELATIVE TO THE PRESENT DAY. [Adapted from Edwards, R.J. and Brooks, A.J., 2008. The Island of Ireland: Drowning the myth of an Irish landbridge. In: Davenport, J.L., Sleeman, D.P. and Woodman, P.C., eds. *Mind the Gap: Postglacial colonisation of Ireland.* Special Supplement to the *Irish Naturalists' Journal,* pp. 19–34]

Fig. 16.9 A. FLINT NODULES. This important stone material, together with chert and other fine-grained or 'glassy' rock materials, was in widespread use for making arrow-heads, harpoons and axes in the Mesolithic. Flint can be found commonly in the chalk cliffs and in linked bedrock outcrops on the coasts of north-west Ireland. It also occurs in the glacial sedimentary cliffs and beaches of Ireland's eastern and southern coasts, transported from the Irish Sea and Celtic Sea regions. Much of the flint found is flawed, but the nodules that survived wave and storm battering were more reliable for use; flint cobbles found in storm beach shingles were preferable to those from mined flint. It is likely that much of the flint found in Ireland at inland settlement sites – such as at Lough Gara in County Sligo, Killuragh cave in east Limerick, or those in Roscommon – was brought from adjacent coastal locations. Other stone materials also used – such as chert, mudstones and slates – can be found more widely in many locations. [Source: Robert Devoy]; b. Flint microliths and blades, used to make hunting equipment and other tools, based on archaeological finds of Mesolithic age. Microliths are created by the knapping of flint nodules, as shown in Figure 16.9a, or from other suitable rock materials. Flint microliths were mounted in wooden hafts, or handles, as appropriate, to make harpoons, axes and many other sharp bladed tools. [Source: © National Museums Northern Ireland, Ulster Museum]

BP coastal areas from central Ireland northwards (Figure 16.8). It was this inundation that initiated the drowning and development of the region's principal marine loughs, such as Strangford Lough (Figure 16.7).

As seen at Strangford Lough, many present areas of land uplift and emergent coasts in Northern Ireland have preserved the location of some Mesolithic settlement sites. Many others, however, have been lost to marine drowning and now lie (if they have survived at all) well offshore. Travelling southwards, away from the land uplift areas of Northern Ireland, the problem of marine inundation and the loss of pre-5,000 years ago coastal archaeology worsens. Earth crustal change and land submergence toward the south coast of Ireland has accentuated the losses of lateglacial-age coastal areas to the sea. Consequently, much of the archaeological record of the first settlers comes, either from artefacts washed up on the modern seashore, or from debris found in raised beaches that have been elevated above present-day sea level, as a result of land uplift caused by the removal of ice.

Known locations where the earliest settlements of Mesolithic people do survive below present-day sea level are rare. Flint tools in fresh, un-weathered condition have been found in intertidal peat deposits, as at Big Stone Bay in Strangford Lough and from an underwater site at Eleven Ballyboes near the entrance to Lough Foyle (Figure 16.9). Some of these coastal settlements, for example at Rough Island and other adjacent sites at the northern end of Strangford Lough, would have been elevated above sea level, allowing colonists to look out over the coastal plains (Figure 16.10).

In neighbouring areas, however, such as the outer coast of the Ards Peninsula, and along much of the Leinster coast, many early Mesolithic sites that must have existed have been eroded away. It is, of course, possible that some of these may be buried under contemporary areas of sand dunes, mudflats and estuary sediment. Elsewhere, settlement sites from the latter stages of the Mesolithic period (before 6,000 years ago) have been found in some present-day coastal areas, for example, at Fanore dunes, County Clare (Figure 16.11) and Ferriter's Cove, west County Kerry (Dún an Óir, Ballyferriter).[17]

Fig. 16.10 STRANGFORD LOUGH, SHOWING THE POSITION OF SOME OF THE MESOLITHIC OCCUPATION SITES DISCOVERED AROUND THIS LOUGH. These sites are clustered along today's coasts and would have been originally situated on elevated land, away from the contemporary shoreline but within view of the sea. [Data source: Davenport, J.L, Sleeman, D.P. and Woodman, P.C., eds., 2008. *Mind the Gap: Postglacial colonisation of Ireland.* Special Supplement to the *Irish Naturalists' Journal*]

5°40'W   5°30'W

Rough Island

54°30'N

54°20'N

● Early settlements on the coastal plain at Strangford Lough

—— Extent of 9,000 year-old shoreline position, now 6m below sea level

—— Extent of present-day shoreline (mean sea level)

Metres

0   125   250   375   500

Fig. 16.11 ARCHAEOLOGICAL EXCAVATION AT FANORE DUNES, COUNTY CLARE. The remains of a fisher-hunter-gatherer type occupation site survived here, situated on a terrace slightly above present-day sea level. This has been covered subsequently by Atlantic storm beach deposits and sand dunes. The excavations at Fanore and at Ferriter's Cove produced a wide range of stone types from the occupation surfaces, such as sandstone cobbles, metamorphosed mudstones (slates) and local volcanic rocks. These were probably used for tool-making, cooking, or for other domestic purposes. [Photo source: Peter Woodman; map data: Google, Digital Globe, Maxar Technologies]

SAND DUNES IN COASTAL ARCHAEOLOGY

Robert Devoy and Peter Woodman

Since the 1950s the experience of archaeology in Ireland has been that it is increasingly difficult to find prehistoric artefacts along these coasts. This results to a degree from contemporary coastal management practices, which have promoted the defence and stabilisation of dunes and linked coastal systems, as part of attempts to reduce the impacts of erosion. The difficulty has been compounded by the conservation and protection measures, enacted for sand dunes and other coastal areas, under broader-scale European environmental legislation (Figure 16.12).[18] Coasts where extensive quantities of archaeological artefacts have been discovered in the past, through excavations and other directed searches, are now protected. Examples of such sites can be found at Inch Spit, County Kerry, and Fanore, County Clare. Similarly, at Dundrum, County Down, the search for archaeological material in the sand dunes has been specifically prohibited as part of current coastal management.[19]

Fig. 16.12 VIEW TO THE SOUTH-EAST, OVER THE NORTHERN AREA OF THE SAND DUNES AT INCH SPIT, COUNTY KERRY: ADJOINING THE DINGLE PENINSULA, WHERE SHELL MIDDENS OF DIFFRENT AGES HAVE BEEN FOUND. Modern sand accretion and vegetation, as foredunes, can be seen in front of the main series of sand dune ridges. The dunes are aligned primarily north-to-south cand in places are over 20m in height. The formation of the dunes dates probably from *c.*5,000 years ago, with the dunes developing initially above glacial and mid-postglacial storm beach sediments. Many phases of dunes destabilisation and movement have occurred subsequently, with the most recent dunes dated to post-600 years ago. Between the dune ridges, shell middens of different time periods are exposed periodically as the dune sands move, and are associated with the shingle and cobble beach sediments. Some of these middens are of Neolithic age. The dune environment is a protected area under the European Natura 2000 legislation. The photograph also shows the use of Inch strand for beach recreational activities, such as surfing and walking, with cars parked on the upper-beach areas, a common scene still on many beaches in Ireland! [Source: Robert Devoy]

Fig. 16.13 COASTAL EROSION HAS EXPOSED THE STUMP OF A 7,500-YEAR-OLD PINE TREE PRESERVED WITHIN THE PEATS AND FOUND IN 2014 ON THE FORE-SHORE AT SPIDDAL, COUNTY GALWAY. This is one of many such pine, alder, hazel and oak trees comprising the remnant peats of drowned forests exposed now by storms along Ireland's coasts. These coastal lowland forests and wetlands, preserved now in coastal peats, commonly date to the period between *c.*6,000 and 4,500 years ago, before postglacial sea-level rise led to their submergence. [Source: Joe O'Shaughnessy].

Ironically, the impacts of increasing coastal storm magnitude and frequency under climate warming since *c.*2000 is causing an acceleration of coastal erosion rates.[20] This has led to the discovery of many new archaeological sites of all periods. An example comes from the northern shores of Galway Bay. Storm action has exposed large areas of peat between Furbo and Spiddal, which evidence the existence of former extensive forest and wetland environments, extending westwards across Galway Bay. The peats date from 8,000 to 3,500 years ago (Figures 16.13 and 16.14) and were developed under high, fresh groundwater tables, driven by postglacial sea-level rises and periods of increased climate wetness. Similar intertidal peats can be found around Ireland's coasts, particularly on the large bay and headland areas of the west (see Chapter 7: Ancient Shorelines and Sea-level

Fig. 16.14 (right) AT FURBO, GALWAY BAY, PART OF AN ANCIENT OAK WOOD TRACKWAY, OR POSSIBLY A 'CEREMONIAL PLATFORM', WAS DISCOVERED ON THE FORESHORE IN EXPOSURES OF THE COASTAL PEAT BEDS. The structure was built between 3,500 and 4,500 years ago. The peat containing the 'trackway' provides evidence of the existence of former forest and linked wetland environments, which became increasingly wet under the effects of sea-level rise and climate change. These areas were ideal for hunting and as a resource-base for Mesolithic and later Neolithic people. [Source: Joe O'Shaughnessy]

Changes and Chapter 12: Coastal Wetlands).

In Galway Bay, as elsewhere on these coasts, the sea finally broke through large off-shore beach and dune-barrier protection systems and inundated the forests and wetlands. The remaining peat deposits in the bay extended quite a way offshore in former times, but have been eroded away from present-day submarine and intertidal areas. Large storms since *c*.2012 are now providing many new exposures of these peats from beneath the overlying beach, coastal marsh and dunes sediments.

The finding of such sites, caused by the processes of coastal erosion, dates back many years. Collectors were often attracted particularly to areas of wind erosion (blowouts) formed within sand dune systems. Between the 1880s and the 1950s, collectors and natural historians flocked to such locations. They were known as 'treasure houses', where artefacts of all periods could be found. These areas became such an important source of information that, in the 1890s, the Royal Irish Academy set up a committee to investigate the archaeology associated with sand dunes.[21] Many artefacts have also been recovered from old soil horizons at the base of dune blowouts and their ages have sparked debates about human antiquity in Ireland. An example of this is that Stone-Age arrowheads (of different archaeological periods) and Elizabethan coins have often been found in the same dune hollows. In the minds of some, the presence of these stone tools and metal objects of a much later date was a strong argument in favour of Ireland's first people only arriving quite recently. Sand dunes would have been visited by people for their resources on a regular basis throughout much of prehistory. Today, sand dunes are often thought of as a constantly shifting coastal environment. Stratigraphic, sedimentary and archaeological evidence suggests, however, that these systems did experience periods of stability in the postglacial and may even have been settled during these times. Sand-dune destabilisation today is being caused by both the continuation of the impacts of people (from farming and beach recreation activity) as well as by the operation of natural coastal processes.

EARLY SETTLERS AND THEIR USE OF THE COAST

The idea that Stone-Age peoples, that is from the Palaeolithic to the Neolithic (Figure 16.2), relied only on stone as a material resource in their lifestyle is a myth. In Ireland, it is very likely that shorelines and the wider coastal environment – of estuaries, mudflats, back-beach dunes, lakes and wetlands – formed a key source of valuable raw materials. These included minerals (stone), animals, shellfish, fish and plant resources, such as berries and roots

as food, wood, bark and other plant fibres (Figure 16.15).

Much of these people's equipment and clothing would have been made from wood, bone, hides and possibly plant fibres. In Ireland, it is likely that the bones of sea mammals, for example seals and whales, as well as other cetaceans, played an important role in providing hunting and domestic equipment. Unfortunately, excavated prehistoric sites in Ireland, which have preserved significant quantities of animal bone and other faunal remains, are rare.

Fig. 16.15 ARTIST'S RECONSTRUCTION DRAWING OF A MESOLITHIC HUNTER-GATHERER COMMUNITY, *c*.7,000 YEARS AGO, ON THE SOUTH COAST OF BRITAIN. Many imaginings of Mesolithic people's living sites, clothing and lifestyles have been attempted in depictions from Ireland and Britain. These reconstructions are based mainly on the artefacts found at their living sites. Most show the significance of coastal waters, rivers or lakes for the site's location, together with the importance of hunting as a primary source of food, commonly from both the land and the sea. These people would have moved regularly from site to site with the seasons to follow different food sources. [Source: Illustration © Andy Gammon 2018 with kind permission of Wave Leisure, Newhaven Fort Prehistory Display, Sussex]

Mesolithic People and Ferriter's Cove

Peter Woodman

The 'later' Mesolithic site at Ballyferriter, County Kerry, is situated on the northern edge of Ferriter's Cove (Figure 16.16), a raised former interglacial shoreline, known as the South of Ireland Shore Platform (SISP). The former living surfaces of the people lie beneath a thin covering of blown sands. These form 1–3m thick sandhills and dunes, which thicken to the south-west, toward the centre of the coastal embayment. The dunes form the margin to extensive wetlands, which include large areas of grasses or sedges, particularly *Phragmites* grass. The wetland environments today lie directly behind the beach at Ferriter's Cove, and link northward into Smerwick Harbour in a series of shallow, sediment-filled basins formed by the bedrock. As shown by the organic sediments (peats) in these basins, the wetlands represent a continuation of the long-term development of open, freshwater environments in the area, such as bog, fen and swamp.[22]

When the Mesolithic hunter-gatherers used the site, dated to a period around 6,300 years ago, possibly even earlier, the wetlands were more extensive and occupied the whole of Ferriter's Cove. Peat from these environments can still be found outcropping from beneath the sands at low tides. A dune beach barrier formed here at the open coast in the early postglacial and provided protection to the embayment from marine flooding. The barrier moved progressively into the cove under sea-level rise and storm action. It has now been removed through coastal erosion. The sands covering the Mesolithic site are probably the remnants of this barrier.

The late Mesolithic occupation areas are situated in an elevated position on the old shoreline platform, approximately 3m above present high water (see Chapter 7: Ancient Shorelines and Sea-level Changes) (Figure 16.17). This would have given the people both a dry location, in this otherwise semi-water world, and a commanding view of the different embayment environments. Ferriter's Cove, with its easy access to the open coast, as well as to a wide range of contiguous inland and aquatic areas, formed an ideal biological, ecotone-type setting for Mesolithic people. It provided a diverse and rich range of food and material resources (such as wood, reeds and rocks). Excavation of the site has shown that the people collected hazelnuts and other plants and fruits, and hunted animals and birds, such as pig, cow, red deer, hare, guillemot and gannet. In addition, remains of fourteen fish species have been found, including those of wrasse,

Fig. 16.16 Ferriter's Cove, a coastal landscape preserving evidence of Mesolithic habitation sites in western Ireland. View north-west over the area of Ferriter's Cove, Dingle, County Kerry (in middle distance), and towards the coastal volcanic headlands of the Three Sisters and the water of Smerwick Harbour, with Ballydavid Head beyond. This area of Mesolithic-age occupation, with a coastal environment and landscape, has attracted people for millennia. [Source: Sarah Kandrot]

Fig. 16.17 ARCHAEOLOGICAL EXCAVATION OF THE FERRITER'S COVE HABITATION SITE. The satellite photograph shows the two main beaches and associated low dunes of the cove, separated by a small rock promontory. The Mesolithic seasonal living areas were discovered beneath the dune sands of the northernmost (upper) of these two beaches. The occupation surfaces are situated on the low, raised rock platform here, but above the level of daily wave activity in the cove. This rock surface is a former interglacial shoreline, known as the South of Ireland Shore Platform (SISP). Charcoal-rich shell midden deposits occur at the site (a. top-left inset photograph) composed of the local postglacial, temperate rock-shorelines shellfish species. These middens, and other parts of the living sites, are being removed today by coastal erosion. [Source: Ben Gearey] The excavations discovered a number of different camp sites (b. lower-right inset photograph), which were probably occupied seasonally and developed over several decades before the site was finally abandoned. [Source: Peter Woodman; map data: Google, Digital Globe, Maxar Technologies]

thornback ray, tope and haddock. The cow bones have been radiocarbon dated to *c.*6,300 years ago, making them the earliest evidence for cattle in Ireland.[23]

The site was probably used on a temporary basis over many years, with people travelling seasonally to Ferriter's Cove to exploit its resources. A particular attractor may have been the volcanic and other hard, fine-grained rocks from the area. These may have been of especial importance in tool-making, and slabs of rhyolite, slates and quartzites have been found scattered in site occupation debris. The spectacular Three Sisters coastal headlands to the north of the cove are a source of volcanic rhyolite rock (see Chapter 5: Geological Foundations).

THE HUNTER-GATHERER'S FOOD

The likely reduced range of food resources (animals and plants) available in Ireland, in comparison to mainland Europe, meant that in addition to coastal habitation sites, the linked estuary, river and lake environments also formed important environments for occupation by early Mesolithic settlers.[24] The remains of migratory salmon, eels and sea bass, radiocarbon dated to 9,700 years ago, have been found at Mount Sandel, County Londonderry (Figure 16.18). This site is situated today just above the limits of the Lower Bann Estuary (Figure 16.1), but would originally have been inland of the immediate coastline. The fish species found here, however, would have been caught in the open sea, which lay several kilometres to the north of the occupation site.

At present the earliest undisturbed coastal site adjacent to the sea can be found at the Port of Larne. Remains recovered there are about 1,000 years younger than those recovered at Mount Sandel.

The site itself was located originally on an island within a coastal embayment that was turning into a sea lough. Today, the area forms part of an open coastal setting and has been buried by storm-beach shingle. A range of sea-fish remains have been found there. These are roughly the same range of fish recovered commonly from Mesolithic period sites elsewhere in Ireland. Most of the fish remains found are of species available from inshore waters, such as whiting, pollock, cod and tope, together with wrasse and conger eels, which could be caught along rocky shores. There seems to be little indication of extensive deep-sea fishing by people, or of a reliance on a single fish species. Occasionally, however, evidence for a dependency on, or preference for, a single species does occur; the number of otoliths (ear bones) of whiting recovered from the Ferriter's Cove excavation shows that between twenty and forty individual fish may have been caught at one time, and that this fish was a very important food source. Other food resources, including

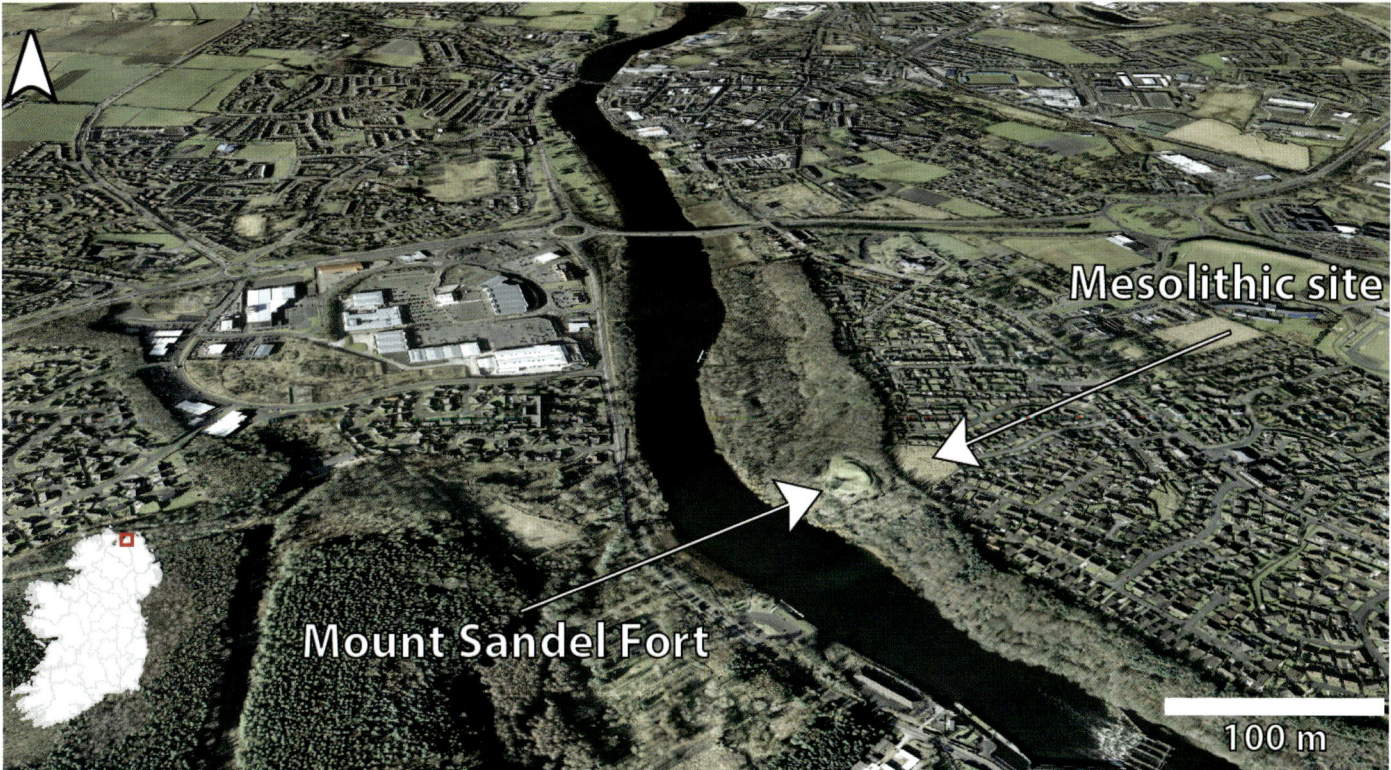

Fig. 16.18 THE SITE OF MOUNT SANDEL, COUNTY LONDONDERRY, LOCATED ON THE UPPER ESTUARY OF THE LOWER RIVER BANN. [Map data: Google, DigitalGlobe, Maxar Technologies]

shellfish and seabirds, seem to have been sought after by these hunter-gatherers, but seabirds would usually have been available only in the nesting season. Sea mammal bones such as those of grey seals have also been found, notably in significant quantities during excavations in Dalkey Island, County Dublin.

There is a tendency to assume that these earliest peoples were heavily reliant on shellfish and fish. This is perhaps because of assumptions that these were an obvious and easy source of food for hunter-gatherers, used to seasonal movements of habitation sites, particularly at coastal locations. Mesolithic peoples did, of course, collect shellfish. Oysters, for example, have been found in some locations, as at Strangford Lough and at Baylet on Lough Swilly. Evidence for the collection of periwinkles, whelks and limpets has also been found at the more rocky sites of Ferriter's Cove and Dalkey Island. Although there are somewhere around 500 known mounds of shells (shell middens) from Ireland's coasts,

relatively few of them, can be shown to date to the Mesolithic period.

There are two reasons why early shell middens are not that common. First, many of the older sites would have been destroyed by the rise in relative sea level and linked coastal erosion. Second, they can be very difficult to find, as they can be buried under larger middens of a much later date (that is, of Neolithic age to more recent ones from the nineteenth century) (see Chapter 20: The Great Famine). Further, although shellfish are plentiful in Ireland, very large quantities would be required if they were to be a dominant element in people's diet. For example, around 700 oysters would be required to feed one adult for one day. This would create a shell mound with a volume of approximately one cubic metre. It has been estimated that an adult would have to eat 52,000 oysters, or 31,000 limpets, to obtain the same food value as that from a red deer or an equivalent large mammal.

SHELL MIDDENS ON THE SOUTH COAST: PAST, PRESENT AND FUTURE

Peter Woodman

Shell middens, or shell dumps, have been discovered along many of Ireland's southern and western coasts. Due to the relatively late stabilisation of relative sea level in these areas, many of the shell middens preserved at the shoreline today were developed only within approximately the last 2,000 years. Most of these date from the period of the Great Famine and other famine times in the nineteenth century.

One relatively large group of shell midden sites, which are often almost entirely made up of oyster shells at each site, has long been documented in the Cork Harbour area. Several middens of moderate size have been found also in the outer harbour, at places such as

Fig. 16.19 LOCATION OF SHELL MIDDENS STUDIED IN CORK HARBOUR. The radiocarbon ages (in calibrated years before present) obtained from the shells found at some of the different shell middens in the area are shown.

Currabinny, Whitegate and Rostellan (Figure 16.19). The oysters found in these middens are the common European flat oyster (*Ostrea edulis*). This has come under strong environmental pressure since the late twentieth century, due to impacts of over-exploitation, pollution and disease.

The most notable concentration of sites occurs in the inner harbour, close to Little and Great islands, which is an area that may have become saline only in the last 1,000 years or so. Some of the middens here are particularly large, as at Ballintubrid, which was found to be up to 100m in length and in places up to 1.5m thick. Unfortunately, some of these sites are vulnerable and at risk of destruction. This is due not only to the impacts of coastal erosion and the linked effects of continuing sea-level rise, but from the current direct effects of human activity, such as housing development and road construction. In some cases, as at Brick Island, Rossmore (Figure 16.20), a midden of similar size to that at Ballintubrid has been partially bulldozed. Ironically, the site then had modern Pacific oyster shells dumped in the same area. Commercial oyster farming was formerly developed in this area, but ceased in the early 2000s.

Unlike Mesolithic and other prehistoric middens, these harbour sites have rarely produced evidence of settlement activity, such as fire places, faunal remains or discarded artefacts of a known date. Frequently, especially in local folklore, they were assumed to be associated with the Great Famine. In fact, the middens almost resemble industrial dumps. In many cases, the absence of charcoal, or animal bones, has meant that the age of these enigmatic dumps could only be dated using radiocarbon dating of the shells.

The radiocarbon dates from Ballintubrid, and from the spread of similar dates found elsewhere across the harbour area, indicate that as soon as oysters colonised the waters of the inner harbour their harvesting and hand-shelling began. It is likely that throughout the last centuries of the Iron Age, and possibly during all of the early medieval period, oysters played an important seasonal role in the diet of the local population. Oysters and other shellfish are also recorded as being important in the early modern period as well. There are historical references from the eighteenth and early nineteenth centuries stating that oysters were being taken from their shells, packed into kegs, and traded along the coast from Kenmare to Dungarvan.

A potential conundrum for these harbour midden sites exists, however. The extent of the shell mounds that can be associated with the medieval period is not as substantial as might be expected. The excavations in Cork city may provide the explanation, as these show that oyster shells were in common use as building material in the medieval city. In particular, they helped provide a firm footing along pathways through the soggy ground of Cork city's north and south islands. It is possible that, with the developing focus of a large town in the vicinity, it may have been more suitable to transport oyster shells to the city's markets. Rather than accumulating as rubbish, oyster shells were found to have other uses (such as a filler in pottery). Oyster shells have also been found in quantity adjacent to other medieval sites in the harbour area, such as at Carrigtohill.

Fig. 16.20 SHELL MIDDEN ON BRICK ISLAND, ROSSMORE TOWNLAND, CORK HARBOUR: SITE OF A LARGE OYSTER SHELL MIDDEN, WITH A MODERN DUMP ON THE FORESHORE OF PACIFIC OYSTER SHELLS. The County Cork archaeological inventory records twenty-five midden sites in the harbour area, though many more have disappeared through coastal erosion. This site was studied by the Reverend Professor Power in 1930, who observed the shell midden to be approximately 1.5m thick and composed almost completely of oyster shells, with the occasional find of mussel, cockle and whelk shells. Charcoal, formed as thin layers, was also found, together with large stones used to open the shells. Observation showed the deposit to extend along the low, cliffed shoreline for approximately 57m and to run inland for another 25m. The layer represents a volume of over 2,100 cubic metres of collected shells (that is, enough food for one adult for approximately six years). The age of the midden has been attributed to the Mesolithic period, though the evidence is uncertain. Finds of flint tools of possible late Mesolithic age have been made at other midden locations in the area, but so also have numerous pieces of pottery dating from the medieval period into the nineteenth century. [Source: Peter Woodman] Inset image: detail of the densely packed oyster shells, which comprise many of the middens found in Cork Harbour. [Source: Carolyn Howle]

From the scatter of distinctive stone tools found at habitation sites, as at Mount Sandel and Ferriter's Cove, it is clear that these Mesolithic people lived in a range of environments. This evidence has given rise to the question of whether life was one of constant migration between the coast and the interior? Some answer to this comes from studies of the amount of food in people's diet that came from a marine source. The evidence comes from differences in the carbon isotope $\delta^{13}C$ (pronounced 'delta c thirteen') of Mesolithic people's bones. Where people have a strongly terrestrial-based diet, the $\delta^{13}C$ concentration is usually between -20 and -23 parts per 1,000. The $\delta^{13}C$ evidence from human remains at Ferriter's Cove and from the bones of dog at Dalkey Island, for example, is approximately 14 parts per 1,000. This would be typical of communities that lived virtually all year round on the coast. Although human remains have not been recovered from the Strangford Lough sites, the narrow range of local raw materials used here in the manufacture of stone tools indicates that this and other groups also lived almost entirely on the coast. It may be, though, that some communities relied heavily on terrestrial and freshwater resources, while others moved between the coast and the interior.

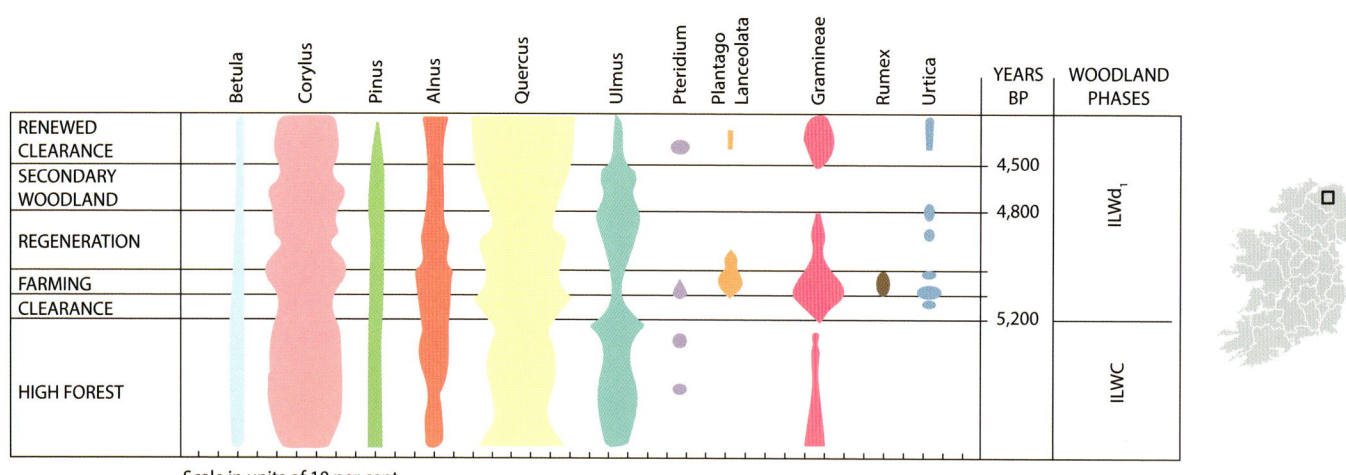

Fig. 16.21 A SCHEMATIC POLLEN DIAGRAM FROM FALLAHOGY BOG, COUNTY LONDONDERRY. This illustrates the sequence of changes in vegetation patterns in this region, and as also found elsewhere in Ireland during the Neolithic. Clearance of the forest and woodlands is indicated by the rapid expansion of light- and nutrient-demanding plants. Of importance in showing the first influences of people and agriculture is the increased frequency of the pollen of grasses and herb plant species, such as ribwort plantain, docks, nettles and bracken. A significant decline in elm (*ulmus*) pollen is also recorded in association with these vegetational changes. This was probably influenced strongly by the development of animal husbandry, with people beginning to keep livestock in enclosures and collecting elm and other trees' leaves as fodder. Issues of disease and climate change may also have influenced the timing of the elm decline in different places. The initial elm pollen decline shown (at *c*.5,200 years ago) is followed by the subsequent collapse of this important woodland tree in the vegetation after approximately 4,500 years ago. Climate and other environmental factors, for example, the nutrient status of soils, may also have been a contributory factor in these vegetational changes.

FIRST FARMERS

How farming was introduced to Ireland, or even the reason why this happened only shortly after its appearance in Britain, is beyond the scope of this chapter, but we can note simply that after *c*.5,700 years ago – taken conservatively as the beginning of the Neolithic (Figure 16.1) – evidence of agriculture in Ireland is apparent. This can be seen in the local to regional patterns of vegetation changes, evidenced in many pollen and spore palaeoenvironmental records (Figure 16.21). The subsequent development of farming activities, seen now as coming in quickly, possibly within 200 years or less, altered fundamentally people's uses of the land on the island and its different environments.[25]

Fig. 16.22 DISTRIBUTIONS OF NEOLITHIC, BRONZE AGE TO EARLY MEDIEVAL MONUMENTS IN IRELAND. Summary maps of different prehistoric monuments: a. known megalithic (large stone-built) structures, b. megalithic tombs (dolmens), c. stone circles and d. ringforts. These all have an Ireland-wide pattern of occurrence, which is not linked particularly to coastal settings. The construction of these monuments is probably associated more closely with people's agricultural and other uses of the land, as well as with their settlement locations. The distributions of these monuments evidence the extensive impacts of people on the landscapes of the island from at least Neolithic times and probably much earlier. [Source: © Government of Ireland. This dataset was created by the National Monuments Service, Department of Culture, Heritage and the Gaeltacht. This copyright material is licensed for re-use under CC BY 4.0]

Fig. 16.23 MONUMENTS OF THE NEOLITHIC PERIOD: POULNABRONE PORTAL TOMB, THE BURREN, COUNTY CLARE. The tomb produced human bones dated to *c.*5,600 years and excavation showed the remains of sixteen people, both adults and children, together with various artefacts. These structures (commonly called dolmen), which change in layout and construction over the Neolithic period, formed the stone interiors of graves. Some of these graves were single-chambered but many are multi-chambered in form and date probably from the late Neolithic onward. The uprights and cap stones of the tombs may have been covered originally by earth or stones, or part-supported, as appears to be the case at Poulnabrone, by mounds of rocks. The tombs were positioned in prominent, visible locations in the landscapes of the time. The larger and more complex structures, as found in the Boyne Valley (for example, Newgrange, Knowth and Dowth), were used for human burials well into the Bronze Age. Examples of other dolmen can be found throughout Ireland, including many in coastal locations, as at Ahaglaslin, Rosscarbery, County Cork, and at Arderawinny, near Schull, County Cork. [Source: Barry Brunt]

The first stages of farming are linked with the cutting down of mature forest and woodlands. At Fallahogy Bog, County Londonderry, for example, the start of this clearance has been radiocarbon dated to *c.*5,200 years ago, with similar times found elsewhere throughout Ireland for these changes (e.g., Lough Sheeauns, County Galway, or the Céide Fields, County Mayo). This indicator of human impact on the natural vegetation is probably linked initially with animal husbandry and is shown in the rapid expansion of light- and nutrient-demanding plants as the wooded areas were removed. A second significant change in Ireland's land use and linked landscapes was the appearance of built stone

Fig. 16.24 NEWGRANGE, BOYNE VALLEY, COUNTY MEATH. The construction of the Newgrange passage tomb dates to *c.*5,200 years ago. This monument forms part of the internationally famous Boyne Valley tombs, which include the similarly sized passage graves of Knowth and Dowth. These tombs are positioned inland, in the wide and agriculturally rich lands of the River Boyne Valley. This east–west orientated river enters the Irish Sea through its estuary close to Drogheda, some 30km to the east. The tombs form examples of the many such monument sites that are linked to the coastal zone. This may be just a matter of serendipity; alternatively, the sites, which remained of importance well into the Bronze Age and even to the present day, may indicate the significance for people of the environmental control of good communication routeways. Coastal lowlands, as found along this eastern central area of Ireland, in counties Dublin, Louth and Meath, provided ease of access to the resource-rich environments inland. Further, these tombs have important astronomical alignments as part of both their layout and internal construction, as found in many other such locations throughout western Europe. In particular, they were used to mark the occurrence of the winter solstice. This may have been of both religious and also agricultural land-use importance to people, with the consequent high community ceremonial significance of these tombs and other types of megalith, such as stone circles and standing stones. These structures may have formed major elements of the landscapes of 'the ancestors and the dead', as referred to in the archaeological work of Mike Parker-Pearson. The tomb sites would have most likely attracted many people to them on an annual basis for festival-type occasions. In the Boyne Valley examples, their location may also be associated with matters of trade and possibly ease of access even for international trade links. [Source: Dave Keeshan, flickr, CC BY SA 4.0; inset: Robert Devoy]

interior lowlands. Here, the vegetation had been composed primarily of woodlands dominated by trees of oak, hazel, alder, pine and elm. Landnam-style, 'slash and burn' agricultural practice was used to establish the woodland clearance. These land-use techniques are still found operating in tropical forest areas today and were widely practised in Europe in the past. In the transition from Mesolithic to Neolithic approaches to land use and later to fully sedentary agriculture practices, people probably continued to live initially in temporary or short-lived settlements. The hunting and gathering of local food resources would also have persisted. However, with the development of farming, together with the migration of different cultural groups into the island, changes in settlement locations occurred. In farming, animal husbandry was introduced first probably (dating of cereal grains, e.g., from Corbally, County Kildare, and cereal pollen records at Lough Sheeauns, to *c.*6,000 years ago may indicate some earlier use of crops). This was followed by the more extensive planting of cereal crops, mainly of barley and wheat. Field patterns became progressively well established in the landscape from the Neolithic, though this is contentious, and more definitely in Bronze Age times and later. This is shown nowhere better than in the internationally renowned archaeological site of the Céide Fields, County Mayo (Figure 16.25).

Fig. 16.25 THE CÉIDE FIELDS, BELDERG, COUNTY MAYO, WITH OVERVIEW OF THE AREA FROM THE COAST TO THE SITE OF THE VISITORS' CENTRE. The remains of stone walls (inset photograph), arranged in broadly rectangular patterns, can be seen emerging from beneath the remaining boglands. These 'fields' were probably used mainly as livestock enclosures and were developed possibly in the Neolithic, though this is disputed by some who argue for a later date in the Bronze Age. They provide evidence of significant human organisation of Ireland's coastal to interior landscapes, as forest and woodlands environments were first cleared. [Source: National Monuments Service, Department of Culture, Heritage and the Gaeltacht; inset image: Gretta Byrne]

The dominant way of life in Ireland from the later Neolithic onward was based on animal husbandry, particularly of cattle and, in time, with a growing reliance on pigs. Arable farming also became more extensive and well established. For coastal dwellers, this meant a continuation of gathering plants, shellfish and fishing. The collection of shellfish is evidenced in the large shell middens of Neolithic age found in many coastal areas, as on Dalkey Island, Inch Spit, the Kenmare River coasts, Lough Swilly and Strangford Lough. Most of the middens from this period, and later, are distinctive in that they consist almost entirely of shells and cover extensive areas. The shell midden at Culleenamore, County Sligo, for example, is several hundred metres in length and 1.5m thick.

structures (that is, megalithic tombs, now archaeological monuments). This began with the Neolithic age and evidences the widespread movements of people on the island. The consequences of these changes altered the earlier likely coastal and linked river-based dominance in settlement patterns (Figures 16.22, 16.23 and 16.24).

By 5,000 years ago, people had moved from the more lightly wooded coastal zones, and down from upland areas, into the

It would appear that after the Mesolithic, shore areas were used progressively by people in a different way. Visits to coastal areas became more likely as ones to obtain the meat in the shell for consumption elsewhere, or perhaps to acquire shells in exchange for other prized items. Stable isotope data from human remains found at Neolithic sites, for example, from Carrigdirty on the Shannon Estuary and from burial monuments at the southern end of Strangford Lough, support this view. These show that $\delta^{13}C$ concentrations in the remains lie between -20 and -23 per thousand. The values indicate a mainly land-based diet, in which coastal resources played only a limited role. In addition to the many shells accumulated in Neolithic middens, artefacts can also be found in some of them, notably at Dalkey Island and Strangford Lough. Unfortunately, relatively few bones of any mammals dating to this period are found in them, or the settlement sites. Fish and bird bones are also rare.

Following the Neolithic, the introduction of metal technology, for making copper and later bronze implements, first with the Copper Age and then the Bronze Age, allowed further radical changes in Ireland's environments and landscapes. There was a widespread expansion of settlement at this time. Many areas if not incorporated directly into farming were used in a kind of transhumance system for grazing. Bronze Age people were probably limited, however, in their settlements and linked agriculture and farming to the lighter and thinner soils of upland slopes, as well as to areas dominated by sands, as at the coast. Here, the cultivation of grounds cleared of their woodland vegetation cover would have been easier. Today, most Bronze-Age monuments are found in such locations (Figure 16.26), though many others from lowland interior areas may have been lost through subsequent agriculture practices, construction works and other impacts of people. There seems to be nothing particularly distinctive of coastal uses from these times, but this point remains to be proven!

The Iron Age, and the later early medieval period, in Ireland is associated

Fig. 16.26 BRONZE AGE MONUMENTS. a. Drombeg Stone Circle, Glandore, County Cork. Although the site was surveyed in the early 1900s, it was only formally excavated in 1957–1958. During this excavation the cremated human remains of a teenager were found in a pot in the circle's centre, together with broken pieces of pots and other artefacts. Radiocarbon dates from Drombeg indicate that it was in use from c.3,100 to 1,800 years ago. The site is comprised of seventeen closely spaced stones and measures 9.3m in diameter. Further, as a good example of a 'Cork–Kerry type' stone circle, it has a pair of 1.8m high outlying portal stones. These provide a south–west axis orientation to the monument, aligning it approximately with the setting sun during the mid-winter solstice. The ruins of two round stone-walled prehistoric huts and fulacht fiadh occur about 40m west of the monument. Bronze Age stone circles, as at this famous Drombeg site or at, for example, Knocknakilla, Millstreet, County Cork, are located particularly in the regions of the south-west and central north of Ireland. They are found commonly throughout Atlantic Europe and date from c.4,000 years ago onward. As with the Boyne Valley tombs, these structures invariably have an astronomical significance in their construction and in the alignment of the standing stones, as first proposed by Alexander Thom in the 1970s. They are not particularly coastal in their distribution and were built almost everywhere across the island; they are now preserved more in upland areas, as they have been progressively destroyed by farming activity. Their distribution and common occurrence indicates that by the Bronze Age people had occupied and were using the whole of Ireland. [Source: Lisa Fawkes]; b. Gallán (Standing Stone), with the Ardagh Martello tower visible in the distance on Bere Island, West Cork. These single-stone monuments, called menhirs (from the Breton), are common in Ireland but also widespread across western Europe, as well as being found in Africa and Asia. Dating typically from the middle Bronze Age in Europe, they were constructed in many different periods through prehistory and often occur in association with other megalithic structures. The standing stone on Bere Island is positioned in the exact centre of the island and is visible from Berehaven, as well as from many other points in the area, emphasising the significance of these stones to the people that erected them. [Source: Robbie Murphy]

Fig. 16.27 GREENAN STONEFORT (GRIANÁN AILIGH, GRIANAN OF AILEACH), INISHOWEN, COUNTY DONEGAL, WITH A VIEW NORTHWARDS OVER THE INISHOWEN PENINSULA, TOWARD LOUGH SWILLY. This well-known stonefort, or cashel, is situated prominently on top of Greenan Mountain at around 244m high, with clear views to both Lough Swilly and Lough Foyle. It was built in the early medieval period between the seventh and tenth centuries AD. The fort is constructed on the site of an earlier enclosure, probably of the Bronze Age from around 1100 BC, though these stratified levels at the site have yet to be excavated. Restoration of the stonefort was begun in 1874 by Dr Walter Bernard, a Derry antiquarian. The structure is over 23m in diameter, with stone walls 4.0–4.5m thick and 5m high. Inside, these are structured into three terraces, which are linked by steps and also contain two long passages within the walls. Buildings would have been situated inside the fort. There are many different types of such high-status stone-built forts situated throughout Ireland and of smaller, but similar structures of stone-walled and light-ly-fortified farms sites, named variably raths and cashels. Stonefort structures continued to be built well into the early Christian and early medieval periods in Ireland (post c.500 AD). Their functions varied both regionally and over time, but most would have served as the base for extended-family-unit living and farming. Many would also have been places to herd cattle and other farm animals for shelter and particularly defence during times of raiding. [Source: Mark McGaughey, Wikimedia CC BY SA 4.0]

Fig. 16.28 DISTRIBUTION OF COASTAL PROMONTORY HEADLAND FORTS IN IRELAND. [Data source: Lock, G. and Ralston, I., 2017. Atlas of Hillforts of Britain and Ireland. Available at: https://hillforts.arch.ox.ac.uk]

initially with the incursion of people from western Europe and Britain, beginning c.3,000 years ago. Periods of distinct expansion and contraction of population and land uses took place, a distinctly unsettled period, ending with a time of expansion in the fourth and fifth centuries AD prior to the beginning of the early medieval. This people pressure again brought land-use and occupation changes (Figure 16.27). Further, iron tool technology facilitated the cultivation of some of the wetter and heavier clay-based soils of lowland areas. Much of Ireland's interior, however, remained covered by the active growth of peat bogs, formed under the cool and wet climate of this late postglacial time. Consequently, the development and settlement of the extensive wet or fully flooded areas of the country, such as the huge Shannon lowland area, was delayed until well into historical times.

By the second century BC, farm settlements of wood and thatched round houses had become common in the landscape, together with many other different types of constructions associated with the emerging Iron Age kingdoms of Ireland, such as sites of ringed earth, stone-bank and palisaded enclosures, linear earthworks of banks and ditches that stretch over long distances (for example the 'Doon of Drumsna', situated close to the River Shannon, County Roscommon). Distinctive archaeological landscapes were established, such as Tara, County Meath, and Dun Ailinne, County Kildare. The hillforts of the earlier Bronze Age continued to be used during this period. Coastal promontory forts, such as that at Galley Head, County Cork, or Dun Aonghasa, Inish Mór belong to this time, with the coasts of south-west and western Ireland associated particularly with their development (Figure 16.28). The need for the defence of property was important in a culture where warfare and raiding

were commonplace. Promontory coastal headlands, as well as steep rock-cliffed coasts, became ideal locations for defensive-style use and possibly for settlements. Many of these distinctive coastal sites were certainly used in trading; progressively so with the Roman world of Britain and continental Europe, as identified for example in the artefacts from Drumanagh, Fingal, County Dublin, on the now important trading corridor of the Irish Sea. These links are indicated in the mapping work of Ireland by Ptolemy from the second century AD (see Chapter 8: Monitoring and Visualising the Coasts). The arrival of Christianity in Ireland in the fifth century BC again brought many new landscape changes and expansion of settlements, including the development of over 20,000 ringforts (raths) and linked embanked sites (such as cashels). By now the development role of coastal settlements in Ireland as the ports and trade centres of the later medieval period had begun to emerge (see Chapter 17: The Vikings and Normans: Coastal invaders and settlers).[26]

Case Study: Irish Promontory Forts

Muireann Ní Cheallacháin

Coastal promontory forts are found around the Irish coast wherever there are suitable headlands. The distribution of the forts appears to be determined mainly by the coast's topography and geology, together creating suitably indented, or crenulate, coastlines. There are about 500 known sites in Ireland, with County Mayo having the largest concentration – due to its steep rock-cliffed coast and many offshore islands – followed by County Cork.

Within Atlantic Europe, however, these coastal forts have a limited archaeological distribution. Apart from Ireland, they are found elsewhere only in coastal Scotland, Wales, north-west France, Cornwall and the Isle of Man. Their occurrence and distribution has led many

Fig. 16.29 THE PROMONTORY FORT OF DOONEENDERMOTMORE, NEAR TOE HEAD, COUNTY CORK, SHOWING THE ERODING COASTAL HEADLAND. The fort's walls are just visible on top of the cliffs of the headland (the low mounds), with the 'bridge' of land connecting the fort to the areas behind it. Coastal erosion has almost separated the site from the mainland (left of photograph), with the beginning of a sea arch forming beneath the bridge area as the sea breaks through. [Source: Muireann Ní Cheallacháin]

researchers to consider them an Atlantic tradition in themselves. Others argue that the definition of a coastal promontory fort is based on geographical and geological considerations, which do not necessarily require any relationship with the chronological or cultural features on the site.

A coastal promontory fort is defined as a headland area surrounded by natural cliffs, with the remaining landward access cut off by a human-constructed barrier (Figure 16.29). The fort may consist of a single line of defence, usually a thick stone wall, or a more complex, multivallate one with multiple defensive barrier constructions. These barriers can be constructed of earth, stone or a combination of both. The earth-barrier type is usually comprised of a ditch and corresponding earthen bank, whilst the stone type may consist of a simple stone wall or a wide stone rampart with rooms constructed within the walls, such as at Dunbeg, County Kerry (Figure 16.30).

In early Irish, a dún was the residence of a chief and, according to the early Irish laws regarding status, a king was required to have a dún. In modern Irish, the traditional place-name element *dún* refers to a fort or a

Fig. 16.30 THE REMAINS OF THE PROMONTORY HEADLAND FORT OF DUNBEG (*Dún Beag*), NEAR SLEA HEAD, DINGLE, COUNTY KERRY. The fort dates from *c*.1,200 years ago, the early Medieval. Major storms, particularly since 2012, have impacted this site significantly. The results of coastal erosion now threaten the complete loss of this structure in the coming decades. This is also the case with many other such forts and other coastal archaeological monuments in Ireland, particularly as the increasing repercussions of climate warming are felt. Many people interested in these sites say that the remaining monuments must be protected and these often form valuable sites for coastal tourism and local business interests. Issues of site heritage and conservation are not as easy as simply 'defending' them. There is much debate on the appropriate measures for protecting such monuments, led in Ireland by the Department of Culture, Heritage and the Gaeltacht, as part of people's learning to adapt to climate change in the twenty-first century. [Source: Bjørn Christian Tørrissen, Wikimedia Commons, CC BY-SA 3.0]

castle and is often associated with coastal promontory forts. Good examples of these can be seen at Dunbeg (small fort) in County Kerry and Doonagappul (fort of the horse) in County Mayo, or the now internationally known cliff-top fort site of Dun Aengus (Dún Aonghasa, or fort of Aonghas), Inish Mór, County Galway (see Chapter 21: Ireland's Islands).

Many structural and other built features are often found within promontory forts. These include stone castles, houses and huts of various shapes and sizes, underground storage or refuge tunnels (souterrains) and terraces, that is, steps built on the inner side of a stone rampart (Figure 16.31). There is limited knowledge, however, on what else may be found in the interior of these sites. Only twelve archaeological excavations have been undertaken on promontory forts in Ireland since 1935, when Gordon Childe partially excavated a simple ditch and bank, or 'univallate', fort at Larrybane in County Antrim.

Of the twelve sites investigated, seven have been dated by radiocarbon and related precise techniques. Larrybane, Duniney and Bruce's Castle in County Antrim, Dunmore in County Clare and Dalkey Island in County Dublin were dated to the early medieval

Fig. 16.31 (left) a. AERIAL VIEW OF DOOCAHER (*Dún Dúchathair*, OR BLACK FORT), KILLEANY, INISH MÓR, THE ARAN ISLANDS, COUNTY GALWAY. This wind-exposed fort is situated on the eroding rock-promontory's cliffs, some 100m above the Atlantic ocean, close to the site of Dun Aengus. Though the site has yet to be excavated, it is thought that the fort is contemporary with Dun Aengus, which has its origins in the Bronze Age. [Source: National Monuments Service, Department of Culture, Heritage, and the Gaeltacht]; b. Close-up view of part of the fort's interior. The fort's walls, which are about 6m high and 5m wide, show the three-tiered, or terraced, form of the structure. The ruins of some of the site's former internal stone houses, clocháns and other buildings are also visible. [Source: Lisa Harbin, Flickr CC BY SA 4.0]

period. Dooneendermotmore (little fort of big dermot) in County Cork appears to be late medieval in age, although it did have an undated earlier phase of activity.[27] These datings and accompanying excavations demonstrate multi-period construction promontory forts in Ireland, with many having even Bronze Age origins. In the cases of Dunbeg and Dooneendermotmore, amongst others, there is evidence for the later re-use of these sites. A similar broad chronology for these forts has been found in Wales, where promontory forts can be dated from the Late Bronze Age (c.2,500 years ago) up to the fourteenth century in the medieval period.[28] Scottish and Cornish promontory forts are dated generally to the Iron Age, though earlier and later examples are also present.

Various interpretations of function have been attributed to promontory and related coastal forts, based on parallels between countries, the number and style of their defences, and site excavation evidence. The most common interpretation is that of places of refuge, due to their highly defensive and sometimes dramatic location and design, as with Dun Aengus, Aran Islands. Other views have suggested that they were used as centres of local administration, trade and ritual, similar to the hill forts of England and Wales. Excavation of the fort at Dalkey Island provided a range of high-status imported artefacts, including Roman pottery. These discoveries led to interpretation of this fort as being an important early medieval trading centre or 'gateway community'.[29] The site is situated on a low headland and formed a perfect location for boats to pull up alongside.

The lack of occupation evidence from sites such as Carrigillihy and Portadoona, excavated in County Cork in the 1950s, supports the view that most promontory forts were indeed little-used refuges, or even simply lookout posts. Only two small hearths, or fire-places, were found at Portadoona. The people of the time may have lived inland in less-exposed areas, but were aware of the need to keep a lookout over the Atlantic for possible attack from farther afield. It is easy to imagine the poor souls picked for lookout duty in these forts at times of danger, sitting around the fires in the dark of night, with the waves crashing into the cliffs below them.

Many of the forts may have been inhabited year-round and could be the coastal equivalent of ringforts, with a primarily domestic function. The univallate fort at Larrybane was interpreted as the early medieval homestead of a small family group, based on the artefacts uncovered.[30] These included handmade 'souterrain ware' pottery, jewellery in the form of an amber bead and a bronze bracelet, a bone hair comb with ring and dot decoration and an iron sickle used for harvesting grain. Why a family group would choose to live in such an exposed and dangerous location might suggest a dire need for defence, in a period of possible unrest.

The second period of habitation at Dooneendermotmore in County Cork was in the post-medieval period of the sixteenth and seventeenth centuries (Figure 16.29). The living arrangements consisted of a fortified stone house surrounded by high cliffs and a deep ditch, which was crossed using a drawbridge. An interesting selection of artefacts uncovered from the site excavation shows that it may not have been all that bad living within the fort. These included soup bowls and tin glazed plates, a salt cellar, a sixteenth-century German wine bottle and stone game pieces from a board game. A quern stone, for grinding cereal grain into flour, and a pickaxe were also found; so, it wasn't all games, eating and wine drinking at Dooneendermotmore in this more recent time of occupation! The excavations showed that at least half the stone house had fallen into the sea during the time of use, emphasising the all-too-real hazard of erosion at these locations. It can only be hoped that the family using the fort were resettled safely before the ocean claimed their sea-front home.

There is no real current consensus on the core functions and dates of use of promontory forts in Ireland, or their links with those elsewhere in the Atlantic zone. More scientific excavations of individual sites are necessary to better understand initial periods of use and original function of these enigmatic archaeological sites.

As a final comment, this brief view of Ireland's prehistory shows that the island's coastal environments were very likely the locations for the first beginnings of its people. Shorelines and coasts provided a critical resource base and a home for the succession of different cultures that used them down the ages. In the post-archaeological world and up into the present day, as illustrated in the other chapters of this *Atlas*, the coasts continue to be of major significance in the life of the people on the island.

THE VIKINGS AND NORMANS: COASTAL INVADERS AND SETTLERS

John Sheehan and Michael Potterton

THE VIKINGS! The voyages of exploration and the progressive expansion of the Vikings into the North Atlantic region, which led to their impact on Ireland's coastal regions from the ninth century AD onwards, is surrounded in myth and legend (for example, as illustrated in the epic poems and writings of Norse mythology and early medieval Scandinavian history by Snorri Sturluson, an Icelandic historian and poet of the late twelfth to early thirteenth centuries). The actions of the Vikings in their famous longships is shown in a diversity of pictures and other imagery, from the romantic and dramatic in form, to the stark and violent. In their raids on Ireland, the Vikings would have experienced, and had to battle with, the regular storms and dangerous waters that affect its coasts. This oil on canvas painting by Carl Peter Lehmann (1826) – entitled *Frithiof Kills the Two Sea-Demons on the Sea* – conjures all the drama of a stormy ocean. The picture is a scene from a popular Romantic nineteenth-century Scandinavian poem, a re-telling of the legendary Icelandic Frithiof's Saga (published by Esaias Tegnér in 1825). [Source: Dag Fosse/KODE Art Museums]

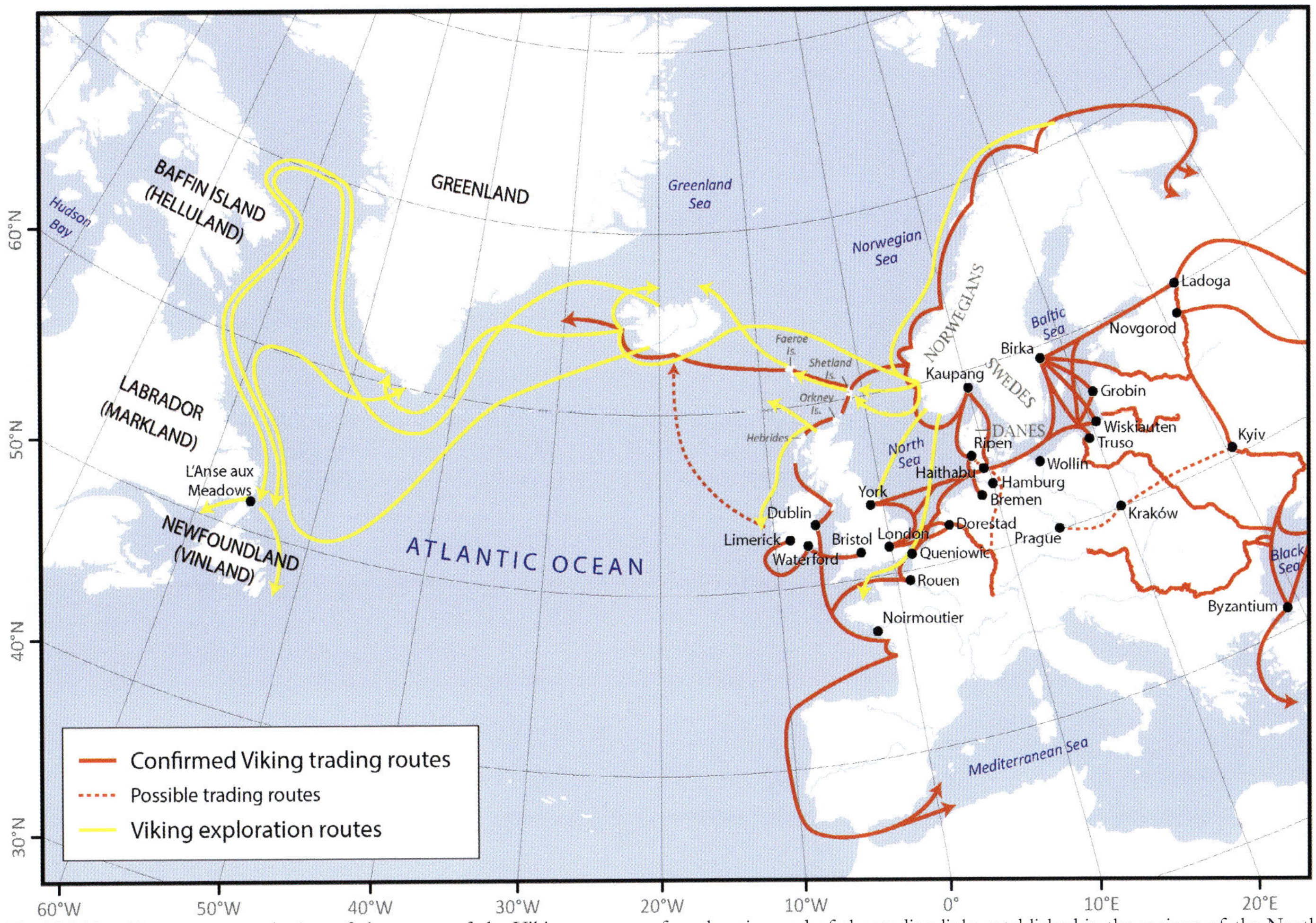

Fig. 17.1 THE VIKING WORLD. A view of the extent of the Viking voyages of exploration and of the trading links established in the regions of the North Atlantic, Europe and toward the Middle East from *c*.800–1050 AD. The Scandinavian peoples that first raided and subsequently settled progressively into the coastal margins of Ireland came originally, in the main, from the area of present-day Norway, though others from Denmark and wider in Scandinavia were also involved. Over time – and with inter-marriage into the Irish (Gaelic) population – the Viking family origins and allegiances in the developing Hiberno-Scandinavian society of Ireland became more complex and separated from their Scandinavian roots. Whatever about their origins, trade became of prime importance for the Hiberno-Scandinavians, with Dublin becoming a primary trading centre in the Viking world of the tenth and eleventh centuries AD. [Data sources: Clarke, H.B., Dooley S., Johnson, R., 2018. *Dublin and the Viking World*. Dublin: The O'Brien Press Ltd; Fitzhugh W.W., Ward, E.I., 2000. *Vikings: The North Atlantic saga*. Washington and London: Smithsonian Institution Press]

Fig 17.2 (right) THE *Sea Stallion from Glendalough* (FROM THE DANISH *Havhingsten fra Glendalough*), UNDER SAIL IN DUBLIN ON THE RIVER LIFFEY. This is a recent reconstruction of one of the largest Viking longships to have been found, the *Skuldelev 2*, one of five Viking ships excavated in 1962 from coastal waters north of Roskilde, Denmark. The reconstruction work was undertaken at the shipyard of the Viking Ship Museum, Roskilde, and completed in 2004. From dendro-provenancing evidence from the ship's timbers, it is known that the original vessel was first built near Dublin *c*.1042 AD. The reconstruction was done, as far as possible, using original shipbuilding techniques, materials and components. The ship is made from a variety of wood-timbers, which included oak for much of the ship's frame and planking, pine for the mast, yard and oars, together with lime and spruce for shields and willow for nails. The original ship was made with oak wood from the Glendalough area, County Wicklow, hence the vessel's name. It is 30m long and carried a rowing crew of sixty. The design of this light longship, which supported 112m² of sail, was for both strength and speed. It would have formed a versatile and formidable 'war machine' and undoubtedly would have been used in coastal raiding, as well as in long-distance travel. [Source: William Murphy, flickr, CC BY SA 4.0]

Inset (bottom left) is of the *Gokstad*, excavated in 1880 from a Viking chief's burial mound at Sandar, Sandefjord, Vestfold County, Norway, and now exhibited in the Viking Ship Museum, Oslo. This earlier ocean-going type of Viking vessel is similar to the *Sea Stallion*. The ship was built in 890 AD and, as with the Dublin ship, is made of oak, measuring 23.8m long and 5.1m wide. It had a square sail of *c*.110m², carried thirty-two oarsmen, had a speed of over 12 knots and could easily have sailed as far as Iceland, showing the speed, versatility and ability of these longships to connect the Vikings' maritime world. [Source: Barry Brunt]

THE VIKINGS IN IRELAND

The Vikings (referred frequently to as Norse, or Scandinavians) are renowned, first and foremost, as a seafaring people who used their ships to expand from their Scandinavian homelands on voyages of raiding, trading and settlement. This activity began in earnest towards the end of the eighth century and, by the beginning of the first millennium, they had discovered and settled many of the lands around the margins of the North Atlantic, from the Shetlands to Greenland, which created a new Viking world away from Scandinavia (Figure 17.1). The Viking expansion included the invasion and colonisation of England by the Danes. This was joined by the extensive settlement and trading activities of the Swedes, operating in the Baltic and along the great Russian rivers as far as the Black Sea and Caspian Sea, reaching southwards into the Islamic world. There are many factors that led to this Viking territorial expansion, including particularly those of environmental, demographic, economic, political and ideological determinisms, but most importantly, the extraordinary seafaring abilities of the Scandinavians (Figure 17.2).[1]

Case Study: The Brendan Voyage

Darius Bartlett

Fig. 17.3 TIM SEVERIN'S CURRACH *Brendan*, SEEN HERE UNDER SAIL IN THE NORTH ATLANTIC. In May 1976, the British explorer, Tim Severin, and a crew set sail from Brandon Creek in County Kerry, to cross the Atlantic in a 12m traditional Irish currach constructed of ox-hides lashed with leather thongs to a wooden frame. Their aim was to test the hypothesis that early medieval monks, under the leadership of the abbot, Brendan 'the Navigator', could have travelled from Ireland to North America and back in the sixth century, some 500 years before the Vikings. 'A truly awesome sight loomed up out of the dark just downwind of us – the white and serrated edge of a massive floe, twice the size of Brendan and glinting with malice' (Severin. T., 2005. *The Brendan Voyage: Across the Atlantic in a leather boat.* Dublin: Gill and Macmillan, p. 209). In June 1977, after a journey of over 7,000km, during which they had encountered ice floes, whales, fog and storms, and were almost mown down by a cargo ship passing them in the night, Severin and his crew landed at Peckford Island in Newfoundland. Even though academic debate continues to question whether or not sixth-century monks ever carried out such journeys in practice, Severin's reconstruction succeeded in establishing that such a journey was at least feasible in principle. [Source: © Tim Severin archives]

Who were the first Europeans to visit North America? Many people, if asked, would point to the Genoese adventurer, Christopher Columbus, who, 'in fourteen hundred and ninety two, sailed across the ocean blue' and landed in the Caribbean island of Hispaniola. Others might cite the Vikings: Norse sagas tell that around the year 985 AD, a fleet of some twenty-six ships led by Bjarni Herjólfsson, sailing from Iceland to Greenland, got blown off course and eventually sighted land further to the west (Figure 17.1). Some fifteen years later, the adventurer, Leifur Eiriksson, undertook an expedition to Vinland in North America, where he overwintered and established an ultimately short-lived settlement. The existence of this settlement was authenticated by archaeological excavations conducted at L'Anse aux Meadows in the northern tip of Newfoundland in Canada in 1960 and subsequently. However, could Irish seafarers have preceded the Norsemen and women, and crossed the Atlantic several centuries earlier?

An early medieval Latin manuscript, the *Navigatio Sancti Brendani Abbatis*, often referred to as the *Navigatio* or the *Voyage of Saint Brendan*, of which more than one hundred separate copies exist, relates how sometime towards the middle of the sixth century AD a priest visited Saint Brendan (an Irish monastic) and told him of a land far across the ocean to the west.[2] Inspired by this visitor's account, Brendan and a crew of seventeen monks embarked on a journey to visit this land. In a boat made of ox-hides lashed to a wooden frame, they sailed for seven years, during which they had numerous adventures, saw many marvellous and amazing things, and made landfall at various islands along the way. Eventually, Brendan arrived at this promised land and, having explored its coastal margins, he and his crew returned to Ireland.

Brendan himself was a significant figure in early Irish monastic history. He was born in Ireland, probably near Killarney in County Kerry, sometime around 489 AD. He was baptised into the church, rose to become an abbot and founded a number of monasteries, including at Ardfert in County Kerry, Inishdadroum in County Clare and Annadown and Clonfert, both in County Galway. Brendan also gained a reputation as a sailor, who travelled extensively along much of the Atlantic seaboard of northern Europe. He is known to have visited Saint Columba in the Scottish island of Iona, and also travelled to Wales, and perhaps to Brittany, Orkney and even the Faroes. Because of these journeys, he is often known as 'Brendan the Navigator'.

Many aspects of the *Navigatio*, and indeed the life of Brendan himself, are still the subject of much debate, some of it at times quite acrimonious.[3] In particular, scholars have been, and are still, divided over whether the *Navigatio* should be read purely as a Christian allegory,

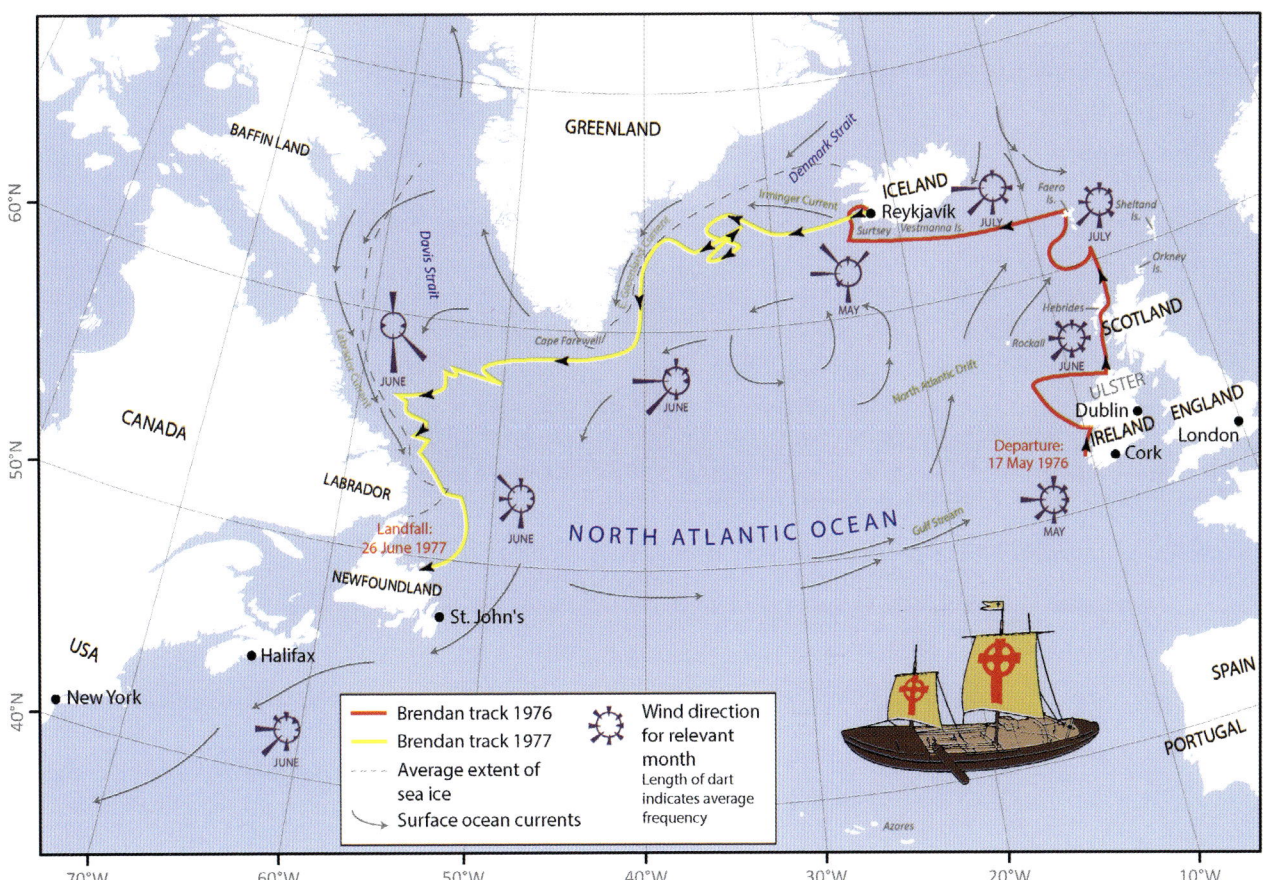

Fig. 17.4 RETRACING ST BRENDAN'S VOYAGE. The route taken by Tim Severin. [Data source: adapted from Severin, T., 2005. *The Brendan Voyage: Across the Atlantic in a leather boat*. Dublin: Gill and Macmillan]

or if it is in fact based on one or more historical accounts of actual journeys undertaken by medieval Irish monks. As John O'Meara wrote in the introduction to his translation from the Latin of the original *Navigatio* 'The *Voyage of Saint Brendan* hardly proves that he discovered America. Nevertheless, the tradition of Brendan's having not only voyaged, but discovered some western, fortunate, but lost island became in due course very persistent.'[4] At the same time, he suggests that it is also possible to see the *Voyage of Saint Brendan* in less literal terms, as a rewriting of the New Testament, with 'the journeyings, miraculous meals, fears and wonders that happened around the Sea of Galilee … transferred to the Atlantic and a strongly insular context'.

In 1976 the British explorer and adventurer, Tim Severin, decided to test, in a practical way, the hypothesis that Brendan could possibly have travelled to North America and back in a leather boat.[5] Drawing on the skills of traditional craftsmen in Ireland and England, Severin arranged for a 12m long traditional Irish currach to be built at Crosshaven Shipyard in Cork Harbour, using only materials and methods that would have been available to boatbuilders in Ireland at the end of the first millennium. This currach, later to be called the *Brendan* (Figure 17.3), had a frame of Irish oak and ash wood, lashed together with leather strips, and covered in hand-stitched, ox-hide leather. On 17 May 1976, Severin and his five-man crew set sail in the *Brendan*, from Brandon Creek on the north side of the Dingle Peninsula. Their course took them first northward along the Atlantic coast of Ireland, via the Aran Islands, to the Scottish Hebrides, which they reached in June. From there, they proceeded to the Faroe Islands and, two months after their initial departure, arrived in Reykjavik, on Iceland's south-west coast. Here the *Brendan* was laid up and stored for the remainder of that year. Early in 1977, Severin and his crew resumed their journey, sailing westward to Greenland, and then south-west across the Davis Strait. They eventually made landfall in Peckford Island, Newfoundland, on 26 June 1977, 7,200km from their departure point in Ireland (Figure 17.4).

By completing the journey from Ireland to Newfoundland in the *Brendan*, Severin did not prove that Irish monks had reached North America in the ninth and tenth centuries. However, he concluded that '*Brendan*'s success went a long way to vindicate the *Navigatio* itself'.[6] The expedition certainly demonstrated that voyages across the open Atlantic would have been feasible in the traditional Irish sea-going craft of the time. Severin also argued that many of the fantastic encounters and observations reported in the *Navigatio*, such as accounts of a pillar of crystal and an Island of the Fiery Mountain, can be seen plausibly as a medieval monk's attempted explanation for icebergs, volcanoes and other phenomena which the modern *Brendan* voyagers encountered along the way.

Tim Severin's voyage captured the public imagination. His account of the expedition became an international bestseller, while his 1978 film, based on footage shot during the voyage, was shown on RTÉ and subsequently worldwide. The composer, Shaun Davey, also wrote a major orchestral suite based on the journey, featuring the uilleann piping of Liam O'Flynn. Tim Severin's boat, the *Brendan*, was returned to Ireland and is now one of the attractions on display in the open-air museum at Craggaunowen in County Clare.

Fig. 17.5 THE IRISH OPPOSE THE LANDING OF THE VIKING FLEET, AS DEPICTED IN A MURAL IN THE ROTUNDA OF DUBLIN CITY HALL, BY THE BELFAST-BORN PAINTER, JAMES WARD (1851–1924). The painting is one of twelve frescoes showing scenes from the early history of Dublin and heraldic representations of Ireland's four provinces, commissioned by Dublin City Corporation from Ward, who was then headmaster of the Metropolitan School of Art, Dublin (now the National College of Art and Design). These works were begun in 1914 and completed in 1918. In this panel, Ward imagines the arrival of the Viking fleet off the Dublin coast, in 841 AD, which is being opposed by a somewhat anxious local population! Ireland's coasts, were being probed and plundered regularly from the end of the eighth century. The Vikings built a *longphort* (naval encampment) in the estuary of the River Liffey, probably in the area of the present-day Dublin Castle, overlooking the Black Pool (Dubh Linn), and overwintered there in 841–842 AD. [Source: Dublin City Council]

Ireland lay in this North Atlantic theatre of operation of the Vikings, but it was quite different from most other areas in this vast region. It was a prosperous and well-populated country, with a powerful configuration of local and regional kings (Figure 17.5). The lands were long settled, from the prehistoric period onward (see Chapter 16: The Inhabitants of Ireland's Early Coastal Landscapes). There was an internationally prominent Christian church, with many well-established and wealthy monastic foundations and churches (see Chapter 18: Era of Settlement: Trade, plantation and piracy and Chapter 21: Ireland's Islands) (Figure 17.6).

Unlike the remote Scottish Isles, or the unpopulated Faroe Islands and Iceland, Ireland could not easily be taken over and settled by the Scandinavians. For this reason,

Fig. 17.6 THE MONASTIC SITE AND THE RUINS OF ST COLMAN'S ABBEY ON THE ISLAND OF INISHBOFIN, COUNTY GALWAY, MID-WEST COAST OF IRELAND. (The name is anglicised from the Irish Inis Bó Finne, meaning 'Island of the White Cow'.) The abbey was founded *c.*665–668 AD by Saint Colman from Lindisfarne, north-east England, as recorded in the Annals of the Four Masters, which details some of the monastery's abbots until the early tenth century. The island is situated *c.*8km (five miles) from the Connemara mainland coast, opposite Cleggan and Ballinakill Harbour. The monastery was raided early in the first wave of the Vikings' incursions into Ireland in 795 AD, along with the monasteries of Iona (*Í Coluim Chille*), south-west Scotland; Rathlin (Rechru), Country Antrim; and Inishmurray (Inis Muiredaig), County Sligo. [Source: © Marie Coyne]

compared to elsewhere in western Europe, the activities of the Scandinavians in Ireland were different over the course of the Viking age (early to middle medieval period, *c.*800–*c.*1200 AD). Their activities were adapted to allow them to become a part of Ireland's ruling elite, rather than to conquer and settle regions of the country. As a result, the Vikings in Ireland, unlike those in England, remained landless for the most part.[7] Furthermore, their settlements, when they began to develop – both as *longphuirt* (singular *longphort*) in the ninth century (Figures 17.7 and 17.8) and later as urban centres from the tenth century onwards – were confined primarily to coastal areas. Here, access to sea-routes encouraged the development of settlements (Figure 17.9), while enhanced opportunities for trade could be considered to be at least the equivalent of beneficial land ownership.

The functions of these coastal settlements evolved to provide significant advantages for the allied Irish kingdoms, as much as for their own Scandinavian rulers. A distinguishing feature of such urban and linked port

Fig. 17.7 (above) THE OCCURRENCE OF *Longphuirt* AND RELATED BASES IN IRELAND. A *longphort* was established originally as a temporary, defended place of safety for the Vikings and their longships. As such, a short-lived *longphort*, perhaps established simply as a winter base, is likely to be different in form to ones that endured and developed over decades to become permanent and enduring bases, such as Dublin. It was common for these bases to be positioned on the boundaries of the Gaelic kingdoms and it seems likely that this strategy would have had the support of local rulers, who hoped to benefit from trading opportunities as well as the availability of mercenaries and potential allies.

Fig. 17.8 (right) AERIAL VIEW OF THE PROBABLE NINTH-CENTURY *longphort* ENCAMPMENT AT BALLYKEERAN, COUNTY WESTMEATH. Located on the edge of Lough Ree, this site is situated well inland, though it was accessible to ships and the sea via the River Shannon. Its embanked area, visible in the left side to centre of the photograph, is situated in a low, wet and muddy area of the lake margin. Such sites are open to different interpretations as to their origin and use, but the Viking ships would likely have been drawn-up in this protected area, and a temporary over-wintering camp established. [Source: Tom Fanning]

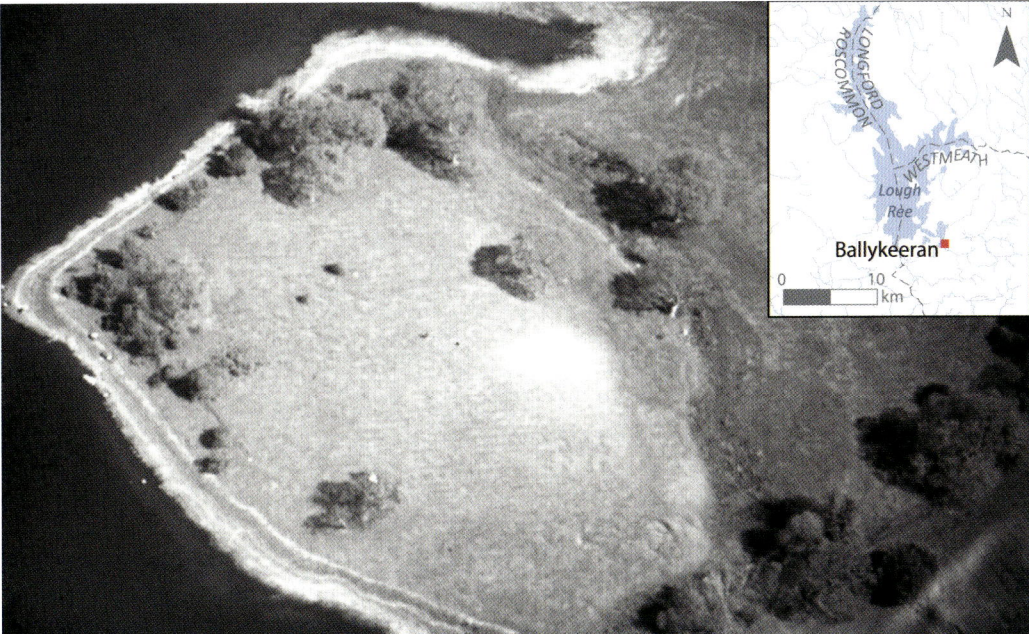

Fig. 17.9 WOOD QUAY, DUBLIN CITY. Excavations of a Viking coastal town that developed on the mudflats and banks of the estuary of the River Liffey. [Source: National Museum of Ireland]

settlements was their reliance on ships to engage in mercantile trade, just as the earlier *longphuirt* depended on ships to engage in raiding as well as trading. As a result, the occupants of these towns, together with their allies elsewhere in the Viking world, gained control of principal sea-routes and became part of a larger, international trading network. This allowed them, as Ireland's trading middlemen, to accumulate significant wealth and influence.

COASTAL TIDE MILLS

Colin Rynne

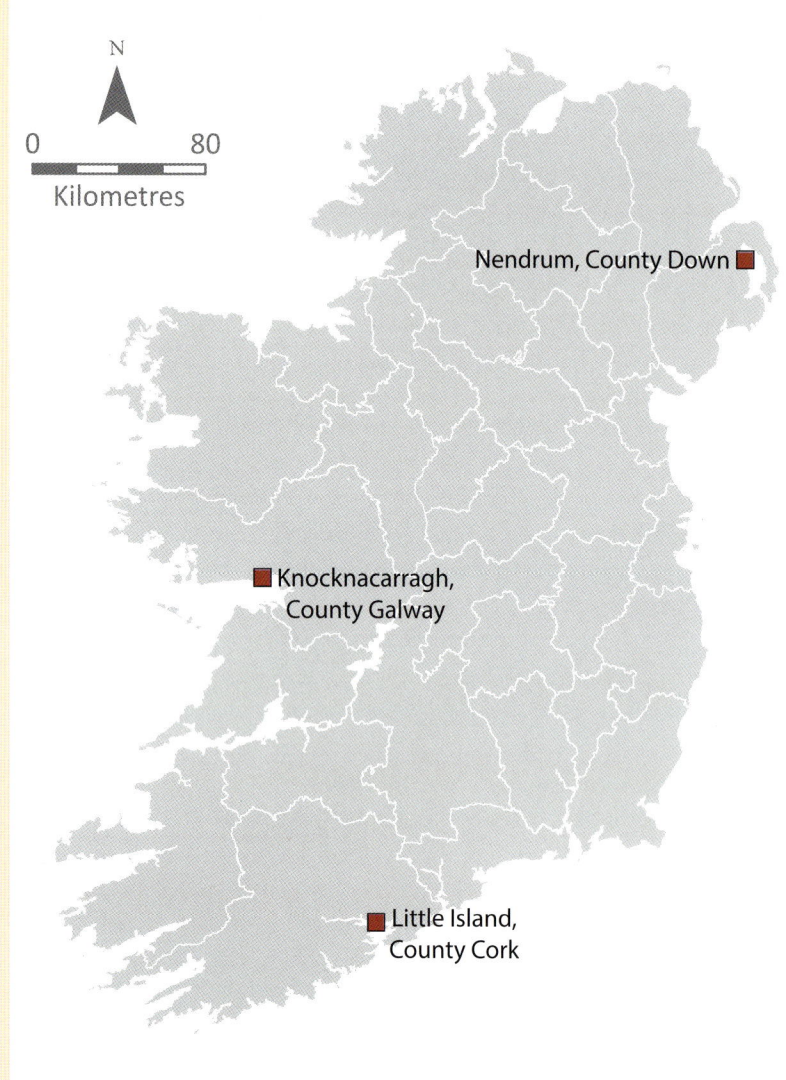

Fig. 17.10 LOCATION OF THE THREE EARLY MEDIEVAL TIDE MILLS STUDIED IN IRELAND. [Source: Colin Rynne]

Shorelines were an important source of raw materials for early settlers, who extracted value from resources, such as flint and shellfish (see Chapter 16: The Inhabitants of Ireland's Early Coastal Landscapes). Another valuable coastal resource was tidal energy. Harnessed from the rise and fall of the tide in coastal locations, tidal energy was used to drive early mills. These tide mills enhanced the development of agriculture, facilitated new trade and, consequently, influenced the growth of human settlements. They were common on the Atlantic coasts of Europe and the earliest examples of water-powered mills in Ireland were of this type.[8] To date, three early medieval coastal tidal mills have been investigated in Ireland: at Nendrum, County Down (Nendrum 1: 619 AD; Nendrum 2: 789 AD); Little Island, County Cork (630 AD), and Knocknacarragh, County Galway (973 AD) (Figure 17.10).[9]

It would appear that the tide mills at Nendrum and Little Island are the earliest known examples in the world. Despite the extraordinary evidence from these sites, we have no way of knowing how representative they were of early Irish or, more widely, early medieval tide mills in general. Only at Nendrum do we have a clear picture of how tidal mill ponds were created (Figure 17.11). Like its near contemporary at Little Island, it was situated on an island immediately adjacent to tidal, estuarine waters. Thus, we have good evidence for two early seventh-century tide mills, which, although at opposite ends of Ireland, appear to have been built in locations with similar coastal attributes that include a large tidal range (of between 3 and 4m), and protection from the full impacts of coastal storms. The construction details of the Little Island site would suggest, however, that this mill was sited

Fig. 17.11 PHILIP ARMSTRONG'S ARTISTIC IMPRESSION OF THE LATER PHASE OF THE NENDRUM TIDAL MILL AND ITS OPERATION. Different mills were built at this protected coastal and island site on Strangford Lough (Nendrums 1 and 2 are dated as spanning the period 600–800 AD). The drawing shows a later phase of the mill, with the area of coastal tidal inflow to the site on the right. The water outflow, mill and mill-race are on the left. [Source: © Crown DfC Historic Environment Division]

in a much more demanding operational environment. Less is known about the environs of the Knocknacarragh mill, although it may well have utilised a mixture of tidal and freshwater sources. For the present, we can only speculate as to why particular sites were chosen and why tidal sources were, in certain instances, preferred over freshwater ones.

A number of interesting wider questions are raised by archaeological evidence for the creation of mill landscapes in early medieval Ireland, and which have important implications for our understanding of similar developments elsewhere. In most of Europe, the tide mill is considered generally to be a later medieval development, yet Ireland's two earliest water mills appear at the beginning of the medieval period. In each case, considerable skill, effort and resources were involved in siting these mills, which were located often in the most demanding of positions, especially the example in Little Island. That such work was possible in the early decades of the seventh century raises the question of the possible transfer of technology between freshwater and tidal mill systems earlier than previously thought.

A second question concerns the association of the mill at Nendrum with an adjacent island monastery. Could this mill have been an afterthought, built sometime after the monastery had been founded? (Figure 17.12). Available evidence suggests strongly that the Nendrum water mills and their supply systems were designed, from the outset, to be integral to the monastery. There are numbers of instances of Irish monastic islands – such as Inishturk, County Mayo, and High Island, County Galway – where a freshwater supply was available for powering water mills. This option was clearly unavailable at Nendrum, where excavators of the mill believe that its location was an important consideration in the siting of the monastery itself. Careful planning, it appears, enabled the founders of the monastery to exploit a coastal site suited ideally to the creation of tidal mill ponds situated close to its outer enclosure wall.

The impounding of tidal waters facilitated a regularity of supply greatly in excess of many freshwater sources, which relied on the vagaries of annual rainfall. In choosing to construct a tide mill, and in accepting the risks of damage from winter storms, the builders of the Nendrum and Little Island mill complexes made a bold, but explicit, technological choice. Dependability of waterpower supply, whereby the mill could technically work all year round, was preferred over a potentially less consistent freshwater source. Such archaeological evidence serves as an important reminder of the strategic importance of the coastal zone to early settlers.

Fig. 17.12 AERIAL VIEW OF NENDRUM, COUNTY DOWN, SHOWING THE SITE OF THE TIDAL MILL AND MONASTERY. With the shallow coastal waters of Strangford Lough visible, the area circled in red contains the site and embankments developed in the different phases of the tidal mills (Nendrum 1: 619 AD and Nendrum 2: 789 AD). The approximately circular-plan walls of the monastery can be seen on the left of the photograph, only several hundred metres away from the mill site. [Map data: Google, Digital Globe, Maxar Technologies, 2019]

The first recorded Viking events in Ireland took place in 795 AD, when church sites were raided on the islands of Rathlin, County Antrim; Inishmurray, County Sligo and Inishbofin, County Galway (Figure 17.6). Over the following forty years or so, more sporadic attacks occurred around Ireland's coasts (Figure 17.13). These appear to have been mainly hit-and-run affairs, involving small forces targeting islands or coastal sites, though a small number of raids took place further inland. It is commonly understood that the Vikings returned to Scandinavia or bases in the Scottish Isles at the end of each season of these exploratory raids, but this may be a simplification of the situation. It is possible that some of the coastal islands, for instance, were used occasionally as temporary bases from which to launch further attacks over a season or two, or

that ships overwintered at anchor in sheltered harbours. If so, these Viking bases, which are undocumented in sources, pre-date the conventional foundation dates of overwintering *longphuirt*.[10]

Through exploratory raiding during the opening decades of the ninth century, the Vikings began to become familiar with the geography of Ireland's coastal landscapes – its islands, harbours, river estuaries and landing places – and began the process of applying Norse place names to strategic locations. This was also the period when the foundations of ties and alliances with the Irish began to be laid down, even though the first instance of formal military collaboration is not recorded until the early 840s AD. Communication with the Irish, and incipient bilingualism, was part of this process, and the Vikings started adding information

Fig. 17.13 NINTH-CENTURY VIKING RAIDS ON IRISH CHURCHES. This map shows the church sites that are recorded in the Irish annals as having been raided by the Vikings in the period between the first raids of 795 AD and 900 AD. The pre-840 AD raids were concentrated particularly on Ireland's coasts. During the decades following these initial incursions, more sporadic raids took place. These appear generally to have been small-scale affairs along the coastal zone (shown as black dots on the map). The Vikings also probably became better acquainted with Ireland's geographic, economic and political landscapes in these years, which facilitated later and more ambitious interventions. After around 840 AD, with the establishment of the *longphuirt*, which combined both military and economic activities, there was an escalation of raiding that began to penetrate into the heart of the country (shown as red dots on the map). Some of the more important church sites were raided several times. [Data source: based on Etchingham, C., 1996. *Viking Raids on Irish Church Settlements in the Ninth Century*, Appendix 1. Maynooth: St Patrick's College]

concerning the nature of Irish society and economy, and the character and settings of its kingdoms, to their geographical knowledge of its landscapes. This type of locally sourced knowledge, which must have been acquired mainly in the coastal regions, facilitated the development of strategic decisions about further raiding.

By the opening decades of the ninth century, the Vikings were in a position to identify sites that had particular links with local kingships, particularly church sites, and sometimes selected these for attack. These were not simple raids for booty, but rather were attacks on local dynasts that could lead to the payment of protection money and tribute, as well as ransom for the return of hostages. By the 840s, at the latest, the Vikings had accumulated more detailed knowledge of the geographical extent of kingdoms, and sometimes established *longphuirt* on their boundaries. These footholds positioned them to take advantage of rivalries that existed between bordering territories, sometimes siding with one kingdom against another, and to engage in trading. In some instances, a *longphort* may have been set up with the agreement of a local king.

NORSE PLACE NAMES

John Sheehan

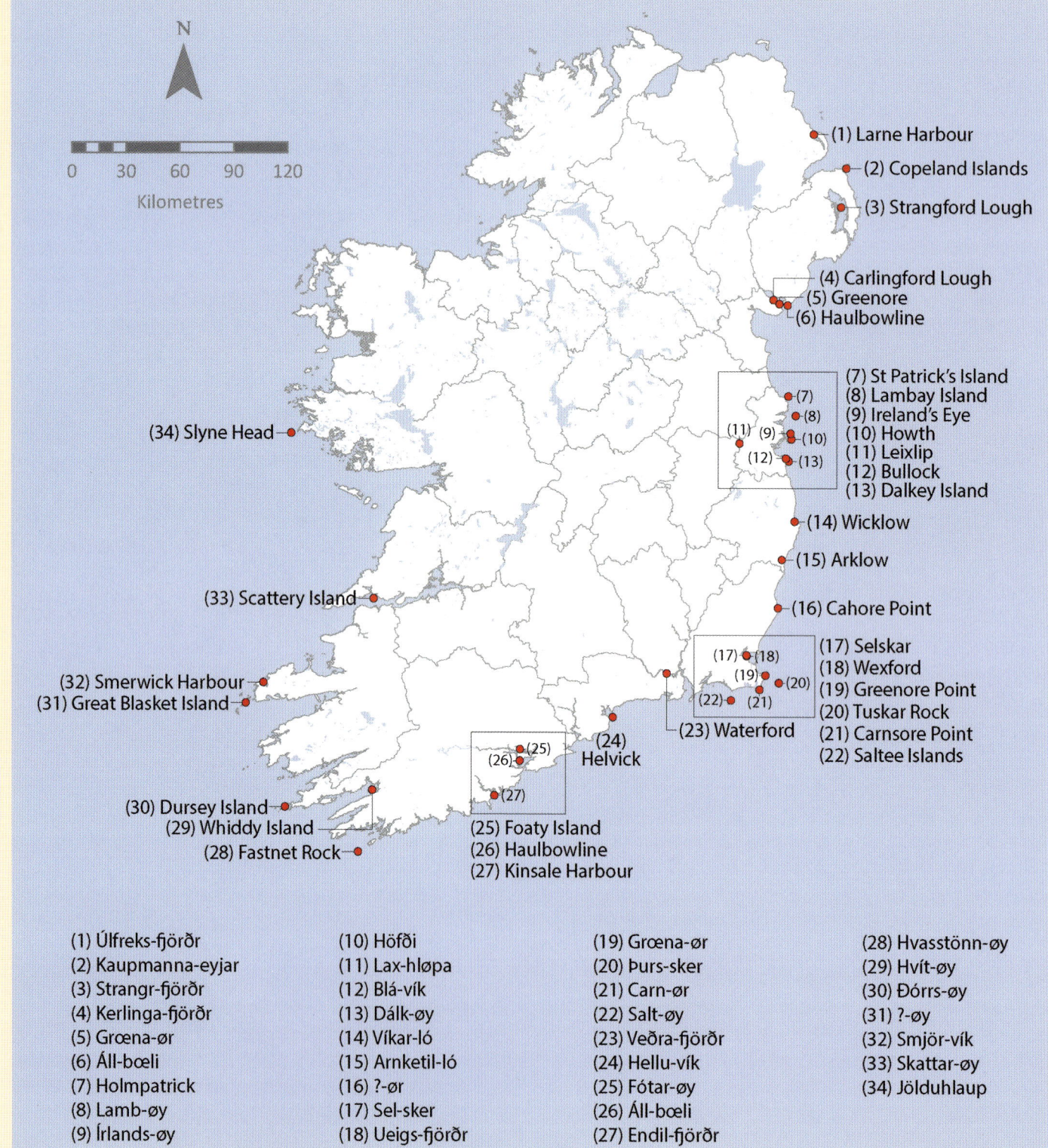

(1) Larne Harbour
(2) Copeland Islands
(3) Strangford Lough
(4) Carlingford Lough
(5) Greenore
(6) Haulbowline
(7) St Patrick's Island
(8) Lambay Island
(9) Ireland's Eye
(10) Howth
(11) Leixlip
(12) Bullock
(13) Dalkey Island
(14) Wicklow
(15) Arklow
(16) Cahore Point
(17) Selskar
(18) Wexford
(19) Greenore Point
(20) Tuskar Rock
(21) Carnsore Point
(22) Saltee Islands
(23) Waterford
(24) Helvick
(25) Foaty Island
(26) Haulbowline
(27) Kinsale Harbour
(28) Fastnet Rock
(29) Whiddy Island
(30) Dursey Island
(31) Great Blasket Island
(32) Smerwick Harbour
(33) Scattery Island
(34) Slyne Head

(1) Úlfreks-fjörðr
(2) Kaupmanna-eyjar
(3) Strangr-fjörðr
(4) Kerlinga-fjörðr
(5) Grœna-ør
(6) Áll-bœli
(7) Holmpatrick
(8) Lamb-øy
(9) Írlands-øy
(10) Höfði
(11) Lax-hløpa
(12) Blá-vík
(13) Dálk-øy
(14) Víkar-ló
(15) Arnketil-ló
(16) ?-ør
(17) Sel-sker
(18) Ueigs-fjörðr
(19) Grœna-ør
(20) Þurs-sker
(21) Carn-ør
(22) Salt-øy
(23) Veðra-fjörðr
(24) Hellu-vík
(25) Fótar-øy
(26) Áll-bœli
(27) Endil-fjörðr
(28) Hvasstönn-øy
(29) Hvít-øy
(30) Ðórrs-øy
(31) ?-øy
(32) Smjör-vík
(33) Skattar-øy
(34) Jölduhlaup

Fig. 17.14 NORSE PLACE NAMES. The Norse place names to be found in Ireland's coastal regions are, with one exception, found exclusively on the coastline and are generally topographical names attached to islands, headlands and the main harbour areas where the Scandinavians settled. Many of these names seem to have been given to the places that were used by Scandinavian seamen as strategic navigational landmarks, though a small number seem to indicate the use of some locations as settlements or trading places. [Data source: adapted from Mac Giolla Easpaig, D., 2002, L'influence Scandinave sur la toponymie irlandaise', *L'héritage maritime des Vikings en Europe de l'Ouvert*. Caen. (2002)]]

There are only limited numbers of Norse place names in Ireland, totalling just over thirty examples (Figure 17.14). Practically all of these are found on the coastline and are generally topographical names attached to islands, headlands and the main harbour areas in which the Scandinavians settled. In addition to these Norse place names, there are also numbers of hybrid Norse-Irish place names, although these tend to be found in the hinterlands of Hiberno-Scandinavian towns. Compared to parts of Britain, however, the Scandinavians did not leave a marked impression on the place names of Ireland.[11] The main reason for this is that, unlike in northern and eastern England, and in the Scottish Isles where extensive rural settlement took place, Scandinavian settlement in Ireland was more confined and had a significant urban focus. The study of Norse place names can yield some insight into the nature of Scandinavian activities in Ireland, as well as their linguistic interactions with the Irish.[12]

Most Norse place names in Ireland are to be found along the Irish Sea and Munster coastlines. Many of these seem to have been given to strategic navigational markers, such as headlands, that would have been utilised by the Scandinavian seamen who sailed the coastal routes (Figure 17.15). Examples include Howth, County Dublin; Carnsore Point, County Wexford, and Helvick, County Waterford.

The name Howth is derived from Norse *höfði*, meaning a headland connected by a neck of land to the mainland. This describes physically the form of Howth Head, and for Norse-speaking seafarers it would have been an accurate description of this landmark to the entrance of Dublin Bay. The name *höfði* was not, however, adopted by the Irish, who continued to use the headland's original name, Beann Éadair, but the Norse name was borrowed into English after the Anglo-Norman invasion. This trend is repeated in many other places

Fig. 17.15 COASTAL LANDSCAPES AROUND IRELAND THAT CAUGHT THE ATTENTION OF THE VIKINGS. a. Part of Strangford Lough [Source: Copyright Tourism Northern Ireland]; b. Great Blasket [Source: Noel O'Neill]; c. Slyne Head [Source: Irish Air Corps]; d. Ireland's Eye [Source: William Murphy, flickr, CC BY-SA 4.0]. From an early date, the Vikings began to become familiar with the geography of Ireland's coastal landscapes and started using Norse place names for strategic locations. Just over thirty examples are recorded, and these are generally topographical names attached to headlands, offshore islands and the harbour areas where the Scandinavians settled. Given the association of the Scandinavians with ships and sailing, it is not surprising that the names of several of Ireland's harbours bear the Norse words *fjörð* or *vík*, meaning 'bay' or 'harbour'. These include *Strangr-fjörðr* (Strangford Lough, County Down, (a.) and *Smjör-vík* (Smerwick Harbour, County Kerry), which also have historical and archaeological evidence for Scandinavian settlement and activity. The early forms of the name used in fourteenth-century maps for the Blasket Islands, County Kerry (b), seem to have the Old Norse termination *–øy*, meaning 'island'. Further up the west coast, the Old Norse name for Slyne Head, County Galway (c), *Jölduhlaup*, is a direct translation of its Irish name, Léim Lára, meaning 'mare's leap', which indicates close linguistic contact between Norse and Irish speakers. Linguistic exchanges sometimes, however, resulted in mistakes; for instance, the Norse name for a small island near Dublin Bay, Ireland's Eye, *Írlands-øy*, meaning 'Ireland's island' (Fig. 17.15d), is probably based on a mistranslation of its Irish name, Inis Ereann, meaning 'the island of Eria', resulting from confusion of this personal name with *Éire*, a name for Ireland.

in Ireland that have a Norse name; the Norse name is adapted into English, while a different name is used in Irish.

Norse island-names, as with headland-names, often seem to have been given to locations that served as coastal navigational points. Obvious examples of this include Tuskar Rock, County Wexford, and Fastnet Rock, County Cork. Other Norse island-names, however, seem to indicate their use as Scandinavian settlements or trading bases. An example of this may be found in the case of the Copeland Islands, County Down; the name is anglicised from Norse *Kaupmann-eyjar* – a compound of *kaupmann* (merchant) and the plural of *øy* (island) – signifying 'the merchant's islands'. These islands are referred to in a thirteenth-century Icelandic saga, *Hákonar saga Hákonarsonar*, as lying on the route between the Mull of Kintyre and the Isle of Man.

Dalkey Island, in Dublin Bay, was also a trading post in the centuries before its Norse name was adopted, and may have continued in this role during the Viking age. It is possible that the Saltee Islands, County Wexford, were the location of Scandinavian salt-manufacturing, leading to their place name, *Salt-øy* (salt island). In addition, islands were often used as places of refuge by the Scandinavians when necessary. For example, three of the islands that are noted in historic sources as being used in this way during the tenth century have Norse names: Dalkey, Ireland's Eye and Scattery.

The Norse name for Dalkey, *dálk-øy* (thorn island), may be a re-rendering of the first element of its Irish name, Deilginis Cualann (the thorn island of Cualu). This name can only have developed through a significant level of linguistic contact between the Irish and the Scandinavians, as it is based on a mutual understanding of the connection between the words *dálkr* and *delg*. The island, based on historical sources, was sometimes used as a slave-holding centre by the Dublin Vikings.

Linguistic contact between Norse and Irish speakers, evidenced in the case of the place name Dalkey, also explains how the Norse name for St Patrick's Island, off the coast of north Dublin, Holmpatrick, is a direct translation of its Irish name, Inis Pádraig. The same phenomenon is represented at Slyne Head, County Galway, which is referred to as *Jölduhlaup* (mare's leap) in Icelandic sagas (Figure 17.15c). It is a direct translation of the name of the headland in Irish, *Léim Lára*. Such linguistic exchanges sometimes resulted in mistakes; for instance, the Norse name for Ireland's Eye, *Írlands-øy* (Ireland's island) (Figure 17.15d), is probably based on a mistranslation of the Irish name for this small island, Inis Ereann (the island of Eria), resulting from confusion of this personal name with Éire, the name for Ireland.

Given the association of the Viking-age Scandinavians with ships and sailing, it is not surprising that the names of several of Ireland's harbours bear, or previously bore, the Norse words *fjörð* or *vík*. These include Larne, County Antrim, *Úlfreks-fjörðr*; Strangford Lough, County Down (Figure 17.15a), *Strangr-fjörðr*; Carlingford, County Louth, *Kerlinga-fjörðr*; Wexford, County Wexford, *Ueigs-fjörðr*; Waterford, *Veðra-fjörðr*; and Helvick, County Waterford, *Hellu-vík*; Kinsale Harbour, County Cork, *Endil-fjörðr*, and Smerwick Bay, County Kerry, *Smjör-vík*. Most of these are harbours that also have historical and/or archaeological evidence for Scandinavian activity, including the development of significant urban settlements at Waterford and Wexford.

The Hiberno-Scandinavians also exercised a degree of control over their urban hinterlands and this is reflected in some place names. The hinterland around Dublin, for example, was referred to as *Dyflinnarskíri* in the Icelandic sagas, and contains one inland Norse place name. This is *lax-hløpa* (salmon's leap), on the River Liffey, and has been anglicised as Leixlip, reflecting the importance of this particular food resource to the urban dwellers of Dublin.

Even though Irish historical sources focus almost exclusively on raiding in their coverage of ninth-century Viking activity in Ireland, it is more than likely that there were also peaceable contacts between the Scandinavians and the Irish. This was almost certainly based on trading, including slave-trading (Figures 17.16 and 17.17). Indeed, prior to the full-scale development of Viking-age mercantile trading, the distinctions between trading and raiding could be rather ambiguous; a trading voyage could easily transform into a raiding trip, and vice versa. Coastal landing places and offshore islands provide the most likely potential locations for initial trading contact, and it is worth noting that there were already precursors of this type of setting in both Scandinavia and Ireland. In Scandinavia, seasonal beach markets were commonplace from the eighth century onwards, while offshore trading locations were already a feature in Ireland from well before the arrival of the Vikings. Dalkey Island, for instance, was used as an importation and distribution base for high-status goods from the fifth century onwards, while Dunnyneill Island, in Strangford Lough, was both an importation and a manufacturing centre during the sixth and seventh centuries. At least two important

examples of Viking-age beach markets have been identified in the Irish Sea region; those at Meols, on the River Dee, near Chester, and Whithorn, at Galloway, south-west Scotland. No definite examples have yet been identified in Ireland. These transient beach markets by their nature, however, have left few archaeological traces and Viking-age landing places are very difficult to identify, given that ships could simply be hauled onshore.

It is clear that Viking-age beach markets existed around Ireland's shoreline, even if no definite example has yet been identified archaeologically, but there are some potential candidates. For instance, the discovery of a small Viking cemetery at Church Bay, Rathlin Island, suggests that this may have been the location of some sort of settlement, perhaps a beach market. The single graves at coastal locations elsewhere, such as Eyrephort, Connemara; Three-Mile-Water, near Arklow, and Ballyholm, near Bangor, might be interpreted in the same way. Dalkey Island, based on historical references, seems to have functioned for a time as a Viking slave market, while the Old Norse origins of the name of the Copeland Islands, County Down – *Kaupmanneyjar* – which means 'merchant's islands', is highly suggestive of them having served as a trading base.

It may be that beach markets are, in some respects, related to the concept of the *longphort*. The latter is often interpreted primarily as a raiding base, and to an extent this may be true, but archaeological investigations of the *longphort* at Woodstown, on the River Suir near Waterford Harbour, show that it was also a significant trading and manufacturing site.[13] Investigations of similar types of sites in England, associated with the campaign of the Danish 'Great Army' of 865–877 AD, have shown that they combined both military and economic activity. It is clear that *longphuirt*, or at least some of them, were more than just fortified military settlements.

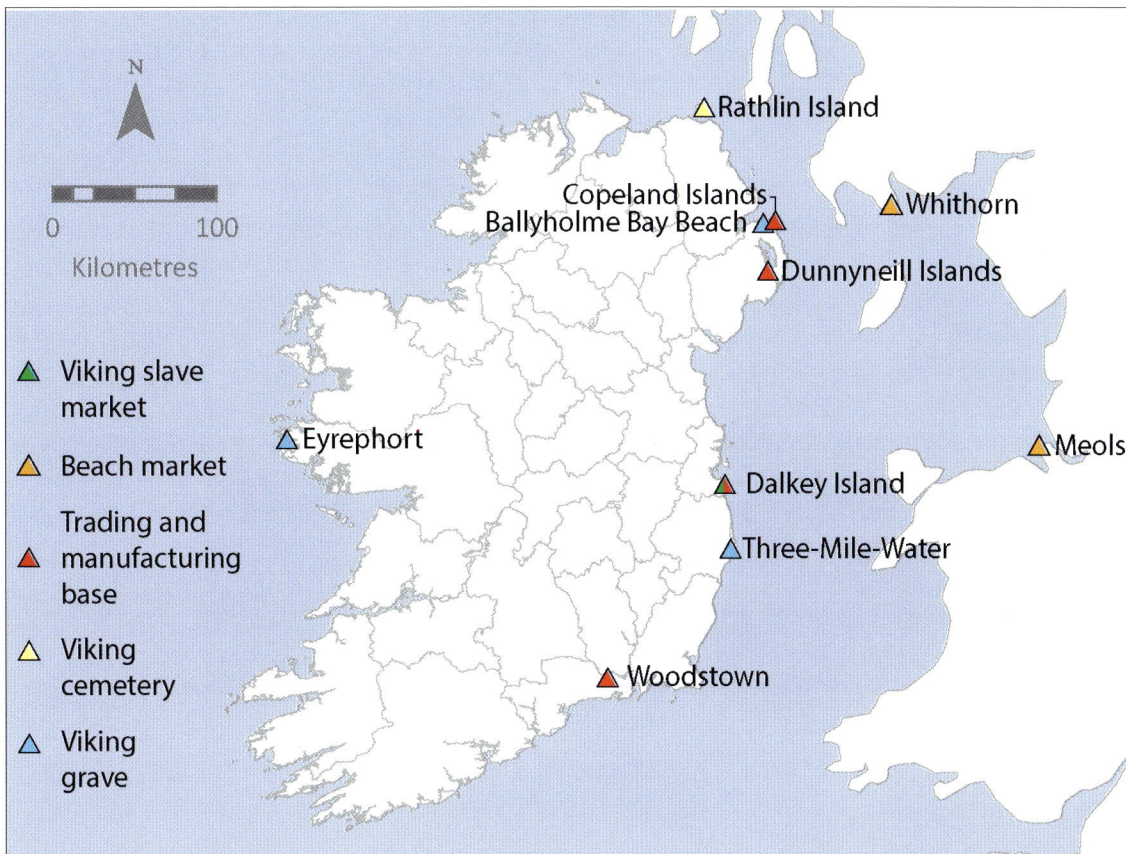

Fig. 17.16 TRADING, BEACH MARKETS AND OTHER SITES OF IMPORTANCE. In Scandinavia, seasonal beach markets were common during the Viking age, while offshore trading locations were a feature of Ireland from before this period. It is clear that Viking-age beach markets existed around Ireland's shoreline, even if no definite example has yet been identified archaeologically. However, there are some potential location candidates. For instance, the discovery of a small Viking cemetery at Church Bay, Rathlin Island, suggests that this may have served as some sort of settlement, perhaps a beach market. Further, the single graves at coastal locations elsewhere, such as Eyrephort, Connemara; Three-Mile-Water, near Arklow and Ballyholm, near Bangor, might also be interpreted in the same way.

The term *longphort* is first used in the Irish Annals in the early 840s. It was a new word, a compound based upon two loanwords from Latin that were borrowed into Irish at an earlier period: long from L. (*navis*) *longa* 'ship' and port from L *portus* 'port', 'landing place', 'shore'. It is interesting that the term itself has no explicit military connotations, even if it is normally used in the Annals within the context of raiding. This may suggest that those who coined the term 'longphort' understood that these sites had various functions, which included political, economic and military.[14] The appearance of the term, however, coincides with the beginnings of a phase from the late 830s to the early 840s in which Viking raiding would increase and intensify. It would seem that the decades of information gathering, based on various coastal contacts, prepared the Vikings to begin a new phase of more ambitious interventions in Ireland. From now on, the coastline would continue to be important as the base-zone, but the new raiding centres would be located on land rather than on offshore islands. These new bases included sites along the main rivers, and even on some lakes, with the focus of attacks no longer being confined mainly to the coastal zones.

At this time, the Irish Annals record the appearance of substantial fleets of Vikings and the occurrence of large-scale territorial raiding, and note that the raiders overwinter in their *longphuirt* (Figure 17.7). The first phase, in the 840s, involved the

establishment of several such sites along the coastal and inland waterways. These include Limerick; Lough Swilly, County Donegal; Cluain Andobair, probably on the River Barrow; Lough Neagh; Rossnaree, on the River Boyne; Lough Ree, on the River Shannon; Linn Dúachaill, County Louth, on the Irish Sea, and Dublin. By using river systems and the main lakes, the Vikings were now able to penetrate to the very heart of the country. For instance, the Annals of Ulster record that in 845, from their Lough Ree base, the Vikings plundered the kingdoms of Connacht and Meath and attacked, among other monasteries, the important foundations at Clonmacnoise, Clonfert, Terryglass and Lorra (Figure 17.18).

In total, the sources record the existence of around twenty-five *longphuirt* during the ninth century, and several more in the 920s and 930s, which were sometimes used in conflicts between the Scandinavians themselves. Further examples of *longphuirt*, not mentioned in historical sources, have been revealed by archaeology. In some cases, the foundation dates of such bases are provided in the sources, while for others only the dates of their destruction are given. What seems clear, however, is that while some *longphuirt* were only of a short-term duration, others achieved a longer-term status.

The initial phase of *longphort* foundation, accompanied by large-scale raiding, was to last for about twenty years. The Irish

Fig. 17.17 LOCATIONS OF POSSIBLE BEACH MARKET SITES IN IRELAND. a. Dalkey Island, County Dublin [Source: National Monuments Service, Department of Culture, Heritage, and the Gaeltacht] and b. Copeland Islands, County Down [Source: Commissioners of Irish Lights]. Dalkey Island, based on historical references, seems to have functioned for a time as a Viking slave market and slave-holding centre, while the Old Norse origins of the name of the Copeland Islands, County Down – *Kaupmanneyjar* – which means 'merchant's islands', is highly suggestive of having served as a trading base.

alliances by marriage. Marriage was also used, however, to maintain their links with Scandinavian kings in the Scottish Isles and elsewhere in the Irish Sea region. These developments, together with the beginnings of their conversion to Christianity as well as the expansion of bilingualism, began to belie impressions of the Scandinavians as out-and-out interlopers on the Irish scene.

By the tenth century, the Vikings in Ireland had as much in common with the Irish as they had with the populations of Scandinavia. They had undergone a partial cultural transformation, from being 'Vikings' to 'Hiberno-Scandinavians', and the locations of a number of their earlier settlements became, over time, the focus of enduring urban development (Figure 17.19).[15] These include the present-day cities of Dublin, Wexford, Waterford, Cork and Limerick.

It should be noted, however, that there was no real background to urbanism in Norway, the part of Scandinavia that was the origin for most of the Vikings involved in Ireland. Rather, the process of urbanisation and the emergence of towns in Ireland should be viewed as a gradual development that resulted from the transformation of Viking raiding and trading. These activities grew between the tenth and twelfth centuries into more sophisticated economic connections, based on conventional mercantile trading between the two elements of Irish society (the Irish and the Hiberno-Scandinavians), as well as expanding international maritime trade networks. The economic aspects of large Irish monastic establishments also had a role to play in this development, although the dominant economic orientation of early Irish towns was to the sea. These towns, therefore, should be considered to be Hiberno-Scandinavian

kings, however, were very successful in controlling the threat, often defeating the Vikings in battle, and ninth-century Viking settlement seems to have been confined to a small number of coastal bases and their immediate hinterlands. But the repercussions of this period would extend long after the mid-ninth century, since the military and economic importance of some of these bases began to accord the Vikings a place among Ireland's prevailing elites and allowed them to become increasingly integrated into the world of Irish politics. They frequently served as military allies for Irish kings, becoming embroiled in local power struggles and dynastic disputes, sometimes sealing their

rather than Viking in origin.

Over time, these urban power bases were to become accepted elements within the framework of local kingdoms that formed the political structure of early medieval Ireland. The towns became prosperous centres and developed important political and economic interests, both within Ireland and abroad. Some were of greater importance than others. During the tenth century, for example, Dublin developed into a commercial centre of international importance and was heavily involved in the political affairs of the Irish Sea region, including those of north-west England, York, the Isle of Man and Scotland. Limerick and

Waterford also prospered, while Cork and Wexford seem to have been of somewhat lesser importance.

Archaeological investigation has been able to illuminate a great deal of the nature of these Viking coastal settlements. Excavations of Viking Dublin, particularly since the 1980s, have revealed its defences, streets, property plots and houses, as well as valuable information on its living conditions, people's diet and their rich material culture (Figure 17.20).[16] Much of Dublin's trading links were clearly with England, but there were also trading contacts much further afield. Additionally, a historical account of Limerick in the late tenth century noted its 'gold and silver', its 'beautiful woven cloths of all colours and of all kinds' and its 'saddles beautiful and foreign'. Such records add further light to this picture of Ireland's prosperous Hiberno-Scandinavian towns, as do the finds of many other artefacts (Figure 17.21).

The towns, in order to ensure a stable supply of food, fuel, building materials and other everyday requirements, also developed strong relationships with their surrounding areas.[17] These urban–rural connections, which must have been mutually beneficial to both the Irish and the town dwellers, do not seem to have involved any significant level of displacement of the Irish from the hinterlands. Nevertheless, the towns do seem to have exercised a degree of control over their hinterlands. In the Icelandic sagas, for instance, the area around Dublin, which comprised a large part of the modern county and a coastal strip of County Wicklow, is referred to as *Dyflinnarskíri* (shire of Dublin), while in the Irish Annals its northern part is referred to as Fine Gall (territory of the foreigners). There is some archaeological evidence for Hiberno-Scandinavian settlement in the hinterland, where a presence is also indicated by place name evidence. Similar indications occur in the case of the other Hiberno-Scandinavian towns; the hinterland of Waterford, for instance, was named Gaultier, *Galltír* (land of the foreigners).

Fig. 17.18 THE EARLY MONASTERY SITE OF CLONMACNOISE, COUNTY OFFALY, ON THE RIVER SHANNON. During the ninth century AD, when Viking raids on Ireland's church sites were becoming commonplace, the high-ranking monastic foundation of Clonmacnoise was only raided on two occasions. This is probably because it was under the patronage of the powerful polity of the Clann Cholmáin kings, which produced two high kings of Ireland, Máel Sechnaill (845–862 AD) and Flann Sinna (879–916 AD). In fact, during their reigns not a single attack on Clonmacnoise was recorded in the different historical sources. [Source: National Monuments Service, Department of Culture, Heritage and the Gaeltacht]

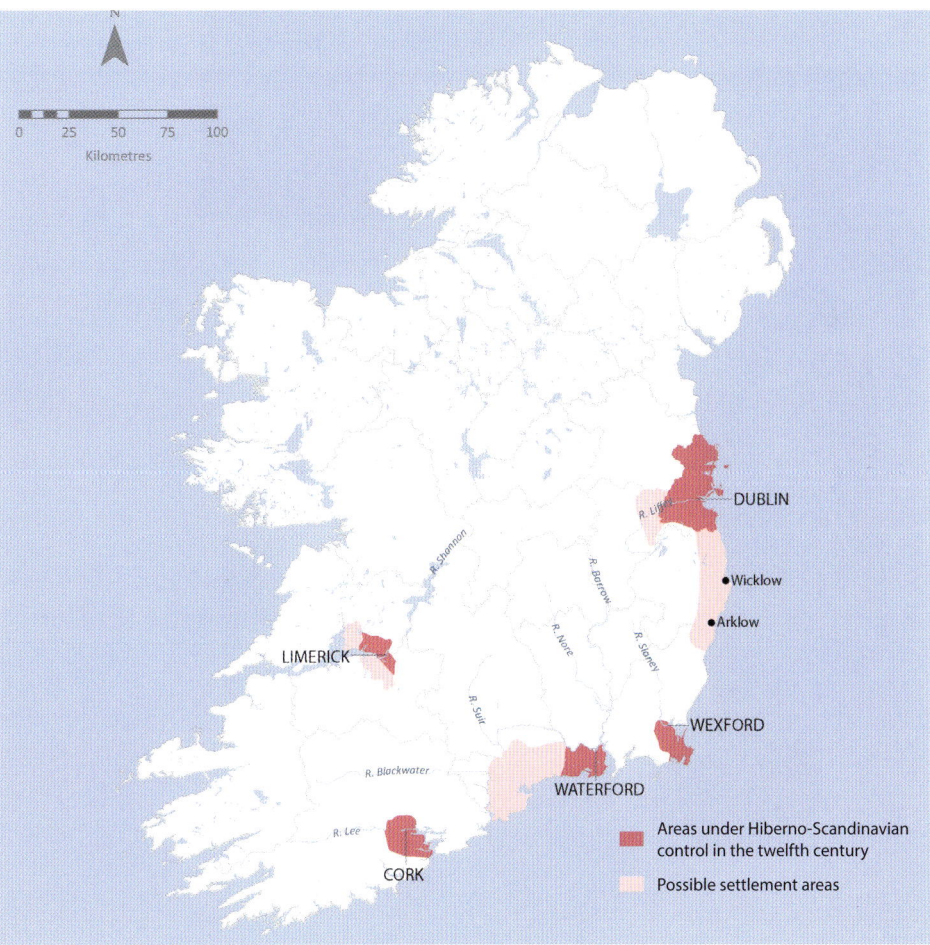

Fig. 17.19 THE DISTRIBUTION OF HIBERNO-SCANDINAVIAN SETTLEMENT IN IRELAND. During the tenth century a number of the earlier Viking bases began to become the focus of urban development. Five towns emerged on the coastlines of Leinster and Munster, each with an economic focus, and their inhabitants underwent a cultural transformation to become 'Hiberno-Scandinavians'. The towns had well-developed relationships with their surrounding hinterlands. These supplied the food for the urban centres, as well as other requirements and there is evidence that, in time, the towns came to exercise a degree of control over these areas. [Data source: adapted from Bradley, J., 1988. *Settlement and Society in Medieval Ireland: Studies presented to F.X. Martin O.S.A.* Kilkenny: Boethius Press]

Fig. 17.20 ARTIST'S IMPRESSION OF TENTH-CENTURY DUBLIN, BY SIMON DICK. The initial Viking settlement in Dublin took the form of a *longphort* centred around a pool close to where the Poddle enters the Liffey. This was known as Dubh Linn, meaning 'black pool', which gives its name to Dublin, and the *longphort*'s core fortified area was probably on its northern side, where Dublin Castle now stands. By the tenth century, Dublin had begun the process of becoming an important commercial centre and was heavily involved in the political affairs of the broad Irish Sea region. It was a wealthy settlement, with defences, streets, property plots and houses, and in the 990s AD it became the location of Ireland's first mint. A long process of conversion to Christianity among the Hiberno-Scandinavians culminated in the foundation, around 1028 AD, of Christ Church Cathedral. [Source: National Museum of Ireland]

The interrelationships between the Hiberno-Scandinavian towns were not always peaceful and were based largely on sea travel. As a result, there is some evidence for settlements in bays and harbours along the coastal routes, and these presumably served as maritime havens, or way-stations for passing ships, as well as trading posts. This is the best interpretation of the Hiberno-Scandinavian settlement on Beginish Island, in Valencia Harbour, County Kerry, which is located close to the royal centre of Ua Fáilbe, kings of Corcu Duibne, who would have benefited from its presence. A number of coastal middens in western Connemara have been radiocarbon dated to the Viking age, and some incorporate items associated with Hiberno-Scandinavian material culture and may represent fishing stations connected with Limerick (Figure 17.22). Some of these are located in the area of Slyne Head, which is referred to as *Jölduhlaup* (mare's leap), in the Icelandic saga, *Landnámabók*: 'From Reykjanes in southern Iceland there is a sailing of five days to Jölduhlaup in Ireland'. Old Norse *Jölduhlaup* is a direct translation of the headland's Old-Irish name, *Léim Lára*, and suggests

Fig. 17.21 EXAMPLES OF WOOD CARVING AND INCISED SHIP-GRAFFITI ON WOOD FROM VIKING DUBLIN. A wide range of crafts were practised in Dublin and the evidence for wood carving is especially interesting. Many of the surviving items are decorated in a distinctively Dublin version of the Scandinavian Ringerike art-style, as exemplified by an animal-headed crook (c.) which may have served as a whip-handle or the head of a walking stick. Popular art from Viking Dublin includes several examples of graffiti of ships on planks of wood, demonstrating the importance of the ship in Hiberno-Scandinavian culture (a. and b.). The ships, with their high stems and sterns, are clearly of Scandinavian type and they are depicted with mastheads, rigging ropes, hoisted yards and furled sails. On one of the ships a man is shown sitting on the ship's yard and he is presumably furling the sail. [Source: National Museum of Ireland]

Fig. 17.22 VIKING-AGE BEADS FOUND IN COASTAL MIDDENS IN DOG'S BAY AND MANNIN BAY, CONNEMARA, COUNTY GALWAY. There is some evidence for Hiberno-Scandinavian settlement in bays and harbours along the coastline, which presumably served as trading posts and way-stations for passing ships. Some coastal middens in western Connemara, radiocarbon dated to the Viking age, have produced Viking and Hiberno-Scandinavian artefacts. Some of the glass and amber beads illustrated above may be linked to Scandinavian manufacturing traditions by reference to their form and colour. The middens at these locations may indicate that there were fishing stations here, probably connected with Limerick. Some of them are located in the Slyne Head area, which is referred to in an Icelandic saga in a context that suggests the sailing route from Iceland to the west of Ireland was a familiar one. [Source: E.P. Kelly and National Museum of Ireland]

that there were direct voyages from Iceland to the west of Ireland. It may be that the Connemara fishing stations also served as way-stations for ships on this Iceland–Ireland route. Literary evidence also suggests there was a Hiberno-Scandinavian deep-sea fishing base in Bantry Bay, possibly centred in Whiddy Island, the name being derived from Old Norse *Hvít-øy* (white island). This too probably served as a way-station. Sites such as these, which must have been a fairly regular feature of the Irish Sea and Munster coastline, played an important role in supporting and facilitating the trade and commerce of the Hiberno-Scandinavian towns.

A HIBERNO-SCANDINAVIAN SETTLEMENT ON BEGINISH ISLAND, COUNTY KERRY

John Sheehan

A Scandinavian presence in Ireland's south-western coastal region has been revealed by archaeological excavations on Beginish Island, in Valencia Harbour, County Kerry. The eastern end of this island comprises a narrow sandy isthmus, with a beach on both sides, which terminates in a raised plateau known as Canroe (Figure 17.23). Here, the well-preserved remains of a settlement complex, consisting of house-sites and field-walls, were partially excavated in the early 1950s.[18] It is quite clear that the initial archaeological view was that the cultural background of the settlement was Irish, and that it and the adjacent ecclesiastical site on Church Island were related in purpose. The archaeologist involved does not seem to have been open to the possibility that Beginish was a Scandinavian settlement, even though the excavations resulted in the discovery of a rune-inscribed stone, probably the most distinctive of all Scandinavian monument-types. It should be noted, however, that until the 1980s the phenomenon of Scandinavian settlement in Ireland was regarded as being predominantly urban in nature. Therefore, the existence of such a settlement on an island off the Kerry coast may have seemed a rather implausible concept at the time of the excavations.

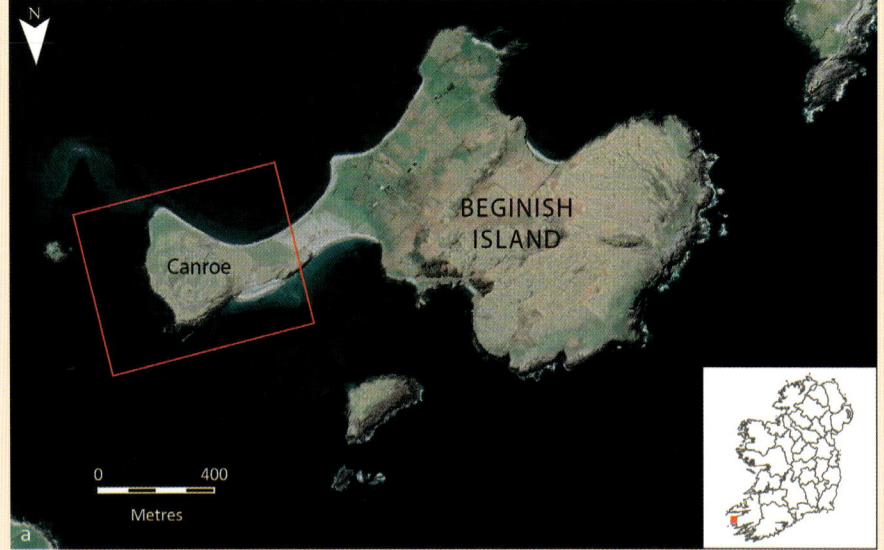

In the decades since these Beginish excavations, it has become apparent that a number of culturally diagnostic finds from the excavations are of Scandinavian or Hiberno-Scandinavian character. In addition, further finds of this type have emerged as stray artefacts located on the island. A review of these and a re-evaluation of the background of the best-preserved and largest structure on the island, House 1, formed the basis for a reassessment of the character of the settlement.[19] It was proposed that it should, at least in its second phase, be regarded as an enduring settlement of Hiberno-Scandinavian character that dates from the tenth century until the late eleventh or early twelfth century.

It has been suggested that the foundations of two unexcavated houses, which are situated adjacent to one another on the sheltered south-facing shore of the isthmus and now engulfed in sand, may represent a tenth-century settlement. The dating evidence comprises the stray finds that have come to light from the immediate vicinity of the houses. The most important of these is a round-bottomed schist or soapstone bowl, probably from Norway, which is the most complete example of its kind on record from Ireland. The overall temporal focus of these important artefacts is a tenth-century one. The fact that they are largely Scandinavian, or Hiberno-Scandinavia, in style, in terms of their cultural background and with no diagnostically Irish objects being represented, supports the likelihood that the occupants of these houses were of Hiberno-Scandinavian origin. The range and quality of such finds indicates that the inhabitants were probably as sophisticated, in terms of their material possessions, as were the occupants of Ireland's tenth-century Hiberno-Scandinavian towns, with the inhabitants of Beginish maintaining links with the mainland towns.

Beginish was not a short-lived settlement; the evidence of House 1 indicates that occupation of the island extended well into the eleventh century. This unusual house is a large, circular, stone structure of sunken-floored construction. It features a lintelled entrance passage, with a rune-inscribed stone serving as a lintel, approached through a sloping entranceway with retaining

Fig. 17.23 HIBERNO-SCANDINAVIAN SETTLEMENT SITES ON BEGINISH ISLAND AND CANROE. a. The boxed area of the photograph shows the area of Canroe and the eastern part of the narrow isthmus that joins it to the mainland of the island. This red-boxed area is where the archaeological excavations took place in the 1950s and where the evidence for a Viking settlement emerged. There are two beaches flanking the isthmus: the north-facing one is sandy, while the southern one is stony, but not rocky. Boats may land on either one. Elsewhere in Ireland and Scotland the Vikings often picked these types of locations where boats may land on the island in varying wind conditions; b. Higher-resolution air photograph of Canroe, showing the remains of numerous walls and other built structures visible across this part of the island. Many of these are thought now to be Hiberno-Scandinavian in origin: 1. In this area, above the stony beach, are the remains of a small number of sand-covered houses. From their immediate vicinity a number of finds of Viking-age artefacts of tenth-century date have emerged (including an almost complete soapstone bowl). 2. House 1, the largest Viking-age building on Canroe. It was here that excavation revealed a rune-stone and several diagnostic Viking finds. The house probably dates to the eleventh century. 3. A network of small fields, bounded by stone walls (many now collapsed), small houses and animal shelters. This may represent a farm associated with the Viking-age settlement. [Map data: Google, Digital Globe, Maxar Technologies and Department of Archaeology, University College Cork]

walls. In an Irish context, the Beginish house is very unusual on account of its sunken-floored form, which is a defining characteristic of the *grubenhaus* tradition of the Scandinavian, Germanic, Slavic and Anglo-Saxon worlds. In addition, an important feature of the house, its sunken entranceway, may only be paralleled in houses from Scandinavian contexts, including houses in Dublin and Waterford. In summary, it seems likely that the background of this Beginish house lies in a fusion of Irish and Scandinavian building traditions. Its distinctive entranceway and its sunken-floored form find strong parallels in Hiberno-Scandinavian contexts, while the stone construction and corbelled form of its walls most probably derive from the Irish building tradition.

A consideration of the purpose of the Beginish settlement raises a number of possibilities. While there is little doubt that its occupants engaged in farming, and probably fishing, this was hardly their primary motivation for settling there. It also appears doubtful that the settlement was established primarily as a trading station, as such markets are usually founded in locations of strategic economic importance. Another possibility – that the settlement functioned as some type of military base – seems unlikely. The idea that Beginish functioned as a maritime haven or way-station, controlled by the Hiberno-Scandinavians, has most to commend it. The island is situated within easy access of the open sea and in the shelter of one of the safest harbours on the route between the Irish Sea and the Shannon Estuary. It has also another important advantage; in poor weather boats could be hauled up on either of the shelving beaches of its isthmus. There can be little doubt about the necessity for way-stations along this dangerous sea-route. Places like Beginish may have played an important role in supporting and facilitating the trade and commerce that formed the raison d'être of the Hiberno-Scandinavian towns.

An enduring way-station, like Beginish, could not have been established without the agreement and support of local rulers. Research as part of UCC's Making Christian Landscapes project, has shown that the royal centre of one of the most important dynasties of the kingdom of Corcu Duibne was located on the mainland immediately adjacent to Beginish.[20] Corcu Duibne was divided into three sub-kingdoms; the largest of these, which covered much of the Iveragh Peninsula, was Áes Irruis Deiscirt and was ruled by a dynastic lineage, later known by the family name Ua Fáilbe (Falvey). It has been suggested that Ua Fáilbe's centre of power, or royal estate, was located on the northern side of the estuary of the Fertha River, at the western end of which lies Beginish.[21] Here, there is a remarkable grouping of three stone forts, which may be interpreted as the Ua Fáilbe royal residences. The juxtapositioning of Beginish and the Ua Fáilbe's royal estate may suggest that these Corcu Duibne rulers facilitated the establishment of this Hiberno-Scandinavian settlement and that both parties saw benefit to the arrangement.

One of the more interesting points concerning the Beginish settlement is how the material culture of its occupants remained identifiable as Hiberno-Scandinavian throughout a period of over 150 years. This is evidenced by the presence and re-use of the rune-inscribed stone, the form of the houses and the character of culturally diagnostic finds. One might have expected that the Beginish settlers would have become archaeologically indistinguishable from the Irish that surrounded them. However, this does not appear to have been the case. One explanation for this is that these people, like those in the Hiberno-Scandinavian towns, consciously used their material culture as a mark of ethnic identity.

In general terms, by the middle of the tenth century the Scandinavians in Ireland had been transformed from the Viking raiders and looters of the ninth century. They lived mainly in urban or proto-urban centres, engaged in commerce and trade, both nationally and internationally, and became increasingly integrated within Irish power structures. They also continued the process of adopting Christianity, and Scandinavian art-styles even began to appear on the metalwork of Christian reliquaries (Figure 17.24). The raiding and plundering aspects of the Vikings, however, had not entirely gone away, but had been replaced by a raiding policy that involved Hiberno-Scandinavian groups targeting political rivals and territories that were outside their own economic spheres. In this respect, they were no different than the Irish. Therefore, by the time of the invasion of 1169 AD, the Anglo-Normans seem to have regarded the Hiberno-Scandinavians as just another element of political and cultural mosaic that was medieval Ireland.

ANGLO-NORMAN IRELAND (c.1169–1400)

The Normans were a remarkable cultural and political phenomenon that emerged in the early tenth century when Scandinavian settlers mixed with local populations at the north-western edge of the Carolingian Empire, in what became West Frankia. The name Normandy means 'Land of the Northmen' and reflects the Viking origins of the people, rather than its geographical location as a region of France.

Within a generation, the Normans had more than doubled their territory and turned Normandy into one of the most powerful principalities in France. From there, they travelled widely, conquering England, southern Italy and Sicily, and establishing

Fig. 17.24 THE CROOK OF THE SUPERB CROZIER ASSOCIATED WITH THE ABBOTS OF CLONMACNOISE, COUNTY OFFALY. The crook dates to the eleventh century, with some later refurbishments. It is especially noteworthy for its figure-of-eight arrangements of snake-like animals, which are inlaid in silver and outlined in niello. This ornament is an Irish version of the Scandinavian transitional art-style known as Ringerike-Urnes, and similar designs occur on bone motif-pieces and wood carvings from Dublin. The crozier may have been produced in Dublin and exemplifies the degree to which the Scandinavians and their art-styles were integrated into Irish society. [Source: National Museum of Ireland]

outposts in northern Africa and the eastern Mediterranean. The Normans were unequalled at adaptation; wherever they went they intermarried, interbred and ultimately were subsumed into the local population. Depending on their surroundings, they became Norse invaders, Frankish crusaders, Byzantine monarchs or feudal overlords. They were masters of the high seas, ruthless warriors, skilled builders, pragmatic opportunists and experienced negotiators.

In 1014, Normans had already been involved alongside the Vikings at the Battle of Clontarf, and the discovery of pottery from Normandy in late eleventh-century contexts in Dublin is a reflection of the deepening contact between Ireland and the Norman world at that time (Figure 17.25). Once the Normans became established in England after the Battle of Hastings (1066), however, trade with Ireland grew and Norman England became a major influence on Ireland in terms of politics, urban development, economy, religion and art.[22] Commercial links with Chester and Bristol were particularly strong, as indicated by archaeologists' discoveries of West Country pottery imported into Hiberno-Norse Dublin and of West Country influences on Irish architecture. There were also important Irish/Anglo-Norman marriage alliances, such as those between the Montgomerys of Pembroke and the O'Briens of Munster.[23]

Three major, and interrelated, cultural developments in Ireland

in the first half of the twelfth century had a Norman or Anglo-Norman background: church reform; the arrival of continental religious orders, notably Cistercians and Augustinians, and the introduction of Romanesque architecture. The spread of Cistercian foundations across Ireland was part of a much larger phenomenon that saw hundreds of new monasteries being established across Europe. Ireland may have been geographically peripheral, but it was associated more closely than ever with developments on the continent. Romanesque was the first truly international movement of art and architecture. Emerging somewhere in western Europe in the first half of the eleventh century, Romanesque is so closely associated with the Normans that in England it is referred to simply as 'Norman architecture', or 'Norman style'. Ireland's finest Romanesque building is Cormac's Chapel on the Rock of Cashel. Cormac's Chapel is a classic example of Norman architecture influencing buildings in Ireland long before the Anglo-Normans themselves arrive. These developments can be interpreted as an element of the increasing Europeanisation of Ireland and, as contacts developed more and more with Britain and the continent, it was perhaps inevitable that people would soon follow; the coming of the Anglo-Normans to Ireland in the 1160s and 1170s could be considered a natural progression.

Diarmait Mac Murchada had been king of Leinster for over

Fig. 17.25 *The Battle of Clontarf*, SAMUEL WATSON'S OIL ON CANVAS PAINTING (1844) DEPICTING THIS SIGNIFICANT EVENT IN IRELAND'S HISTORY. The Battle of Clontarf occurred in 1014 in the then estuary area of the River Tolka, on the northern shorelands of Dublin Bay, at Clontarf, Dublin. Although not large in number, some Norman soldiers were part of the complex of Scandinavian-Viking alliances that joined with the Irish-Norse, principally from Leinster and Dublin, in battle against the Irish forces under Brian Boru (high king of Ireland). The battle was of significance in marking the continued waning of the Hiberno-Norse and wider Viking influences in Ireland and the greater importance of regional Gaelic kingdoms, which had now to contend with the invasion of the Anglo-Normans. [Source: painting: *The Battle of Clontarf*, Samuel Watson (1844), courtesy of The O'Brien Collection. Photography: Michael Tropea]

Legend:
- Anglo-Norman invasion routeways
- Robert fitz Stephen campaigns from May 1169
- Local campaigns with Maurice de Prendergast and Irish allies
- Raymond le Gros and Strongbow
- King Henry II of England
- Anglo-Norman settlers by c.1300
- *Sligo* Towns founded by the Anglo-Normans
- ó néill Gaelic lordship
- Stone castles
- Cistercian monasteries
- Benedictine monasteries
- Houses of Augustinian Canons Regular

Fig. 17.26 NORMAN INVASIONS OF IRELAND, TERRITORIAL EXPANSION AND SETTLEMENT. After the initial Wexford landings of the 1160s and 1170s, the Anglo-Normans spread north and west, establishing castles, towns and religious houses across the country. By c.1300 AD, they had settled more than half of the island and extended their influence through much of the rest of it. They controlled the coast from Cork, anti-clockwise, as far as Greencastle in County Donegal; as well as Sligo, Galway, Limerick and Tralee in the west. [Adapted from Killeen, R., 2003. *A Timeline of Irish History*. Dublin: Gill & Macmillan, p. 22]

thirty years when he was dethroned and exiled in 1166. Mac Murchada was in alliance with Henry II, king of England and great-grandson of William the Conqueror. Consequently, and after a number of false starts, in the early summer of 1169 three shiploads of mercenaries arrived at Bannow Bay on the Wexford coast, including fighting men and soldiers from England, Wales, France and Flanders. The following year, the main Anglo-Norman forces arrived with Richard fitz Gilbert (Strongbow), who captured Waterford (Figure 17.26). There was a fierce battle and Mac Murchada was restored as king of Leinster. He quickly became very

powerful with his new Norman soldiers, weaponry and organisation. He consolidated his hold on the kingdom of Leinster and the cities of Waterford and Dublin and then embarked on aggressive incursions into the bordering kingdom of Meath and on a bid to challenge for the high-kingship of Ireland.

The rise of Strongbow (Richard fitz Gilbert) caused something of a crisis for Henry II. In just a couple of years the Anglo-Normans had already conquered large areas of Ireland. Now, with Strongbow in a position of strength, Henry feared they could soon be prepared to attack England. Henry II, with an army of 4,000

Fig. 17.27 ANGLO-NORMAN WEAPONRY. In addition to new forms of fortification, such as motte-and-bailey castles and ringworks, the Anglo-Normans brought advanced weaponry to Ireland. Archery, using both longbows and crossbows, was a key component in the Anglo-Norman arsenal. A complete bow (of yew wood) from *c*.1220, found in Waterford, is one of the very few surviving examples from Britain or Ireland. [Source: Waterford Treasures – Three Museums in the Viking Triangle]

men, sailed to Waterford in October 1171 to reassert his authority over his Anglo-Norman subjects. He stayed for six months, but did not once engage his army. He received the backing of the church and the submission of a significant number of Irish kings. He secured recognition as overlord of Ireland, with Strongbow as lord of Leinster. In a move that is usually interpreted as an endeavour to keep Strongbow in check, Henry appointed Hugh de Lacy as bailiff of Dublin and granted him the ancient kingdom of Meath (a lordship that extended to almost one million acres and stretched from Clonmacnoise on the Shannon, eastwards to the Irish Sea). These arrangements transformed the nature of Anglo-Norman intervention in Ireland and forged a link between the English crown and Ireland, the repercussions of which are still reverberating. A new French-speaking aristocracy was installed in Ireland and the existing links with Normandy, England and the continent were broadened and crystallised.

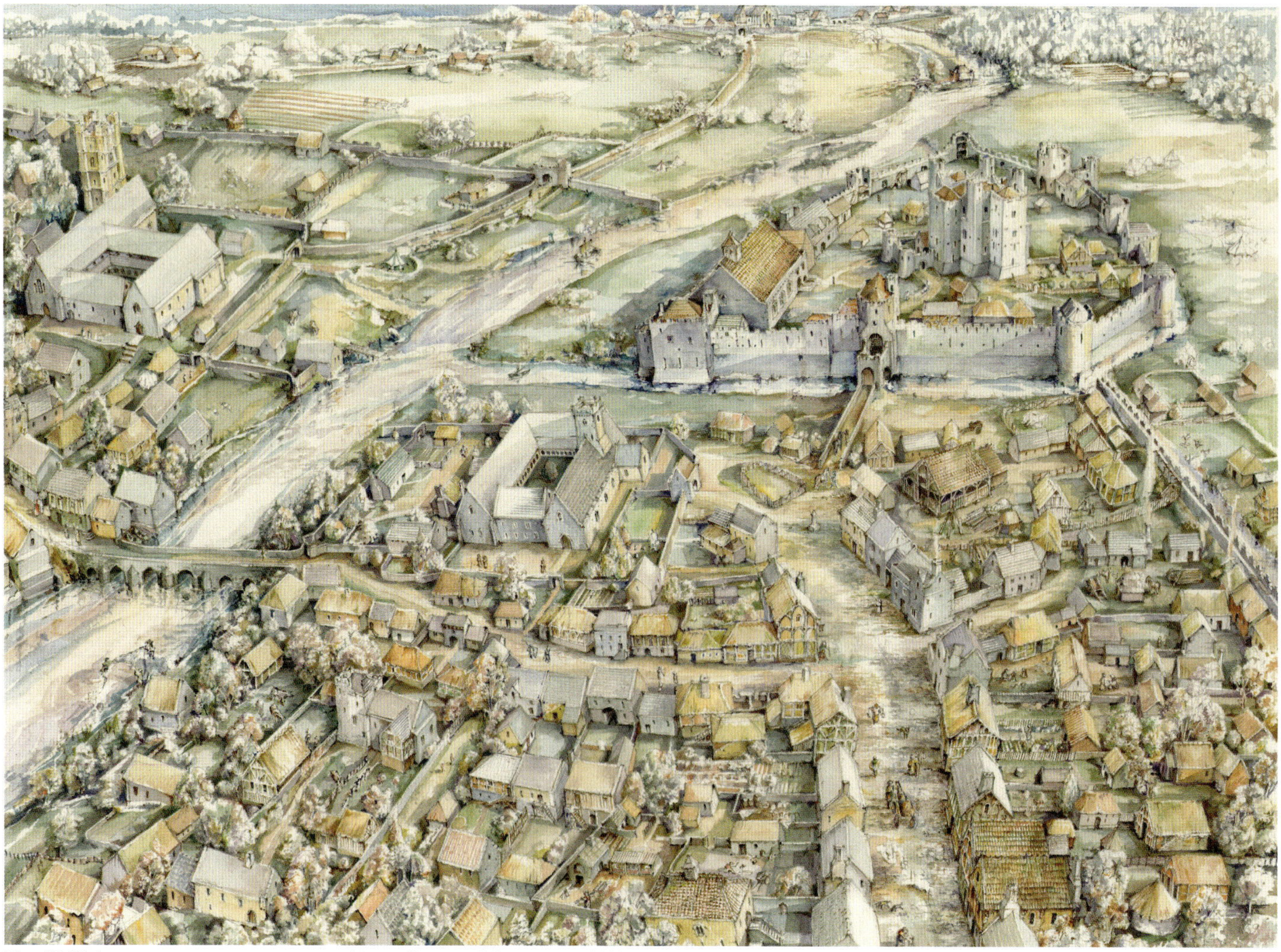

Fig. 17.28 ARTISTIC RECONSTRUCTION OF TRIM, COUNTY MEATH, IN *c*.1400 BY KEVIN O'BRIEN. Trim was a typical Anglo-Norman market centre, dominated by its castle overlooking the River Boyne and a fertile hinterland. Market Street was wide enough to facilitate crowds and temporary stalls once a week; religious houses played an important role; the houses of wealthy merchant families were built of stone, while those of most other residents were made of timber. The river, which was navigable from the Irish Sea 40km away, linked the town with the coast and onwards to international trade routes. [Source: courtesy of Kevin O'Brien]

ANGLO-NORMAN SETTLEMENT

The first fifty or sixty years after the arrival of the Anglo-Normans was a period of intensive settlement, expansion and consolidation. Capturing new territory was relatively easy for the newcomers, given their powerful weaponry, well-drilled soldiers and high levels of organisation (Figure 17.27). The challenge was to hold onto their gains. This they managed by both dividing conquered lands among loyal followers and colonists, in return for allegiance and military support (subinfeudation), and through the building of a systematic network of fortifications across the newly acquired areas (encastellation). Most of the new fortifications were earthwork castles in the form of mottes – with or without baileys – and ringworks. These were constructed at strategic locations, usually on the richest agricultural land, close to a river or the coast, and often next to the intersection of transport routes (Figures 17.28 and 17.29). This was a vast military operation. Under the de Lacys, for example, Meath became the most densely castellated part of Europe at this time.

Given the importance of trade and the need for revenue to bankroll the conquest, it was imperative for the Anglo-Norman settlers in Ireland to establish market centres and trade networks as rapidly as possible; consequently, new towns were

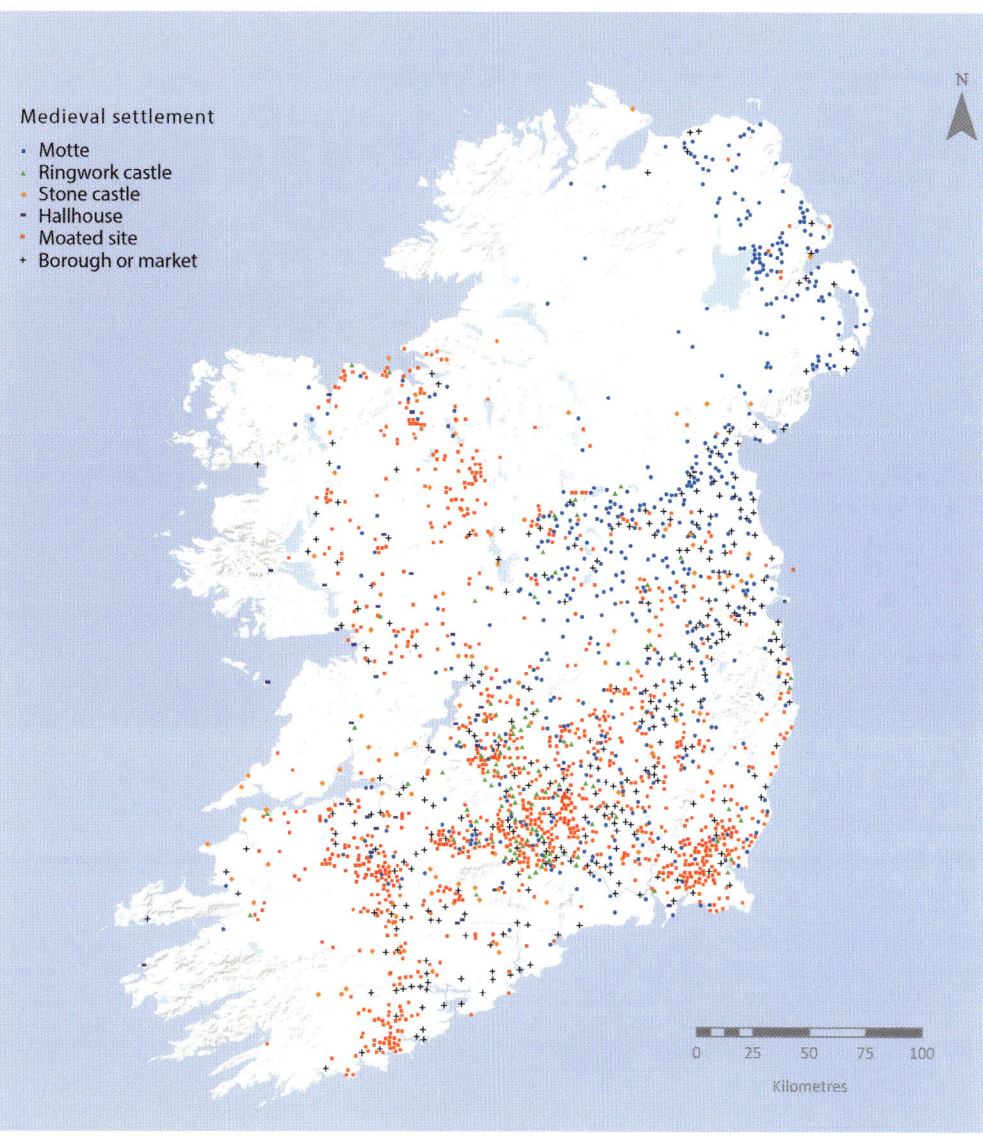

Fig 17.29 MEDIEVAL FORTIFICATIONS AND SETTLEMENTS IN IRELAND. The growth of the urban network in Ireland under the Anglo-Normans is a reflection of the importance of trade. Towns and ports (as boroughs) developed as market centres, where the produce of the rich hinterlands was taken for sale and often ended up on ships bound for Britain and the continent. Many of these were, at least initially, sited in coastal and linked areas. The majority of the earliest castles in Ireland (first wave of colonisation and settlement) were built of earth and timber, either as mottes or ringworks. Settlers of the second wave appear to have constructed moated sites in more peripheral areas. On closer inspection, many of the field monuments recorded by the Archaeological Survey of Ireland as 'rectangular enclosures' prove to be moated sites. [Data source: adapted from Stout, G. and Stout, M., 2011. Early Landscapes: From prehistory to plantation. In: Aalen, F.H.A. Whelan, K., and Stout, M., eds. *Atlas of the Irish Rural Landscape*. Cork: Cork University Press, p. 56]

developed. Their focus on urban development is epitomised by the speed at which they took over the Viking port towns of Waterford, Wexford and Dublin. They soon moved in on Cork and Limerick as well as a number of the larger monastic sites with urban attributes. The thirteenth century was, therefore, a boom-period for urban foundation and expansion in Ireland, with the Anglo-Normans establishing over fifty new towns, including many of twenty-first-century Ireland's largest and most successful urban centres. Many of the towns were enclosed by a defensive wall and about half of them were provided with extra protection in the form of a castle (Figure 17.28).

In addition to the crown-approved activities of Strongbow in Leinster and de Lacy in Meath, a number of disillusioned Anglo-Norman adventurers began to carve out their own areas of influence beyond the bounds of these lordships. John de Courcy, for

instance, made inroads into south-east Ulster in early 1177 with a force of fewer than 350. He established the headquarters of his earldom at Downpatrick and on the Antrim coast developed Carrickfergus as a strategic site for a castle, town and port (Figure 17.30). The new town of Dundalk provides an additional example of similar developments occurring in Ireland in areas under Anglo-Norman control. This town was founded to take advantage of the small harbour of Athlone (see Chapter 18: Era of Settlement: Trade, plataion and piracy), while transport inland from the new port was facilitated by a network of roads leading in various directions. By the 1220s, its annual November fair ran for three days, but was extended to eight days from 1230, possibly on account of its success.[24]

The development of seaports was of great strategic significance for the Norman invaders. The riverine location of most of Ireland's medieval towns is a reflection of the importance of

Fig. 17.30 CARRICKFERGUS CASTLE, COUNTY ANTRIM. Anglo-Norman adventurer John de Courcy from Somerset began the construction of Carrickfergus Castle in the late 1170s. The site was well chosen, dominating a natural coastal rock-promontory on the northern shore of Carrickfergus Bay (now Belfast Lough). As at Trim, the imposing stone-built castle at Carrickfergus provided shelter for a market town to develop in its shadow. When de Courcy was ousted by Hugh de Lacy in 1204, Carrickfergus Castle was seized by the crown, in whose hands it remained for most of the Middle Ages. [Source: © Tourism Northern Ireland]

transport and trade, while the coastal position of so many of the country's largest urban centres is due, in part, to the special emphasis placed on overseas trade in the Middle Ages (Figure 17.31). Understandably, the main Anglo-Norman ports in Ireland were on the east and south coasts (for example, Carrickfergus, Strangford, Drogheda, Dublin, Wexford, Waterford, New Ross, Youghal, Cork), although significant ports also existed elsewhere around the coastline (including Dingle, Limerick, Galway and Coleraine). Other smaller ports included Ardglass, Arklow, Carlingford, Dundalk, Kilclief, Kinsale, Portaferry, Sligo and Wicklow.[25] Crown customs officials were appointed at the larger ports to ensure that the appropriate taxes were collected on goods entering and leaving. The east-coast ports continued to trade especially with Bristol and Chester, as well as with Liverpool and some Welsh and Scottish ports. Irish ships also plied the waters between Ireland and France, Spain, Portugal, Belgium, the Netherlands, Denmark, Germany and Iceland.

It is worth looking at the development of several ports in a little more detail. Dublin witnessed great expansion under the Anglo-Normans. In addition to its new castle and bridge, it received several new churches and a piped water supply. The port appears to have been located at Wood Quay, where excavations have revealed evidence for major works beginning slightly before 1200.[26] Natural silting and active reclamation by the town's authorities combined to completely reshape the River Liffey quays by the mid-thirteenth century. New revetments using oak timbers and up-to-date carpentry techniques probably also functioned as docksides. By c.1300 Dublin had a new stone quay, but the continued silting up of the Liffey was a cause of considerable difficulty for merchants using

the port. This was probably part of the reason that Drogheda was able to expand so rapidly in the thirteenth century.

Situated on the Cooley Peninsula, roughly half-way between Dublin and Belfast, the town of Carlingford is sited on a naturally protected bay in Carlingford Lough that lends itself to use as a port and anchorage. The deepwater harbour is not prone to silting and in the Middle Ages was awash with herrings and oysters. The fertile and well-drained land in the area also made it especially attractive to Anglo-Norman settlers. This is reflected in the modern agricultural landscape of the peninsula, where the coastal lowlands are dotted with place names containing Norman elements while Gaelic names predominate in the uplands. Carlingford Castle, also known as King John's, was begun before 1200 (probably by Hugh de Lacy) on a narrow volcanic outcrop overlooking the town and harbour. The town thrived through the thirteenth and early fourteenth centuries, its reputation as a port indicated by its inclusion (as *Carenforda*) on an Italian portolan chart of 1339 by Angelino Dulcert.[27] The discovery in Carlingford of sherds of pottery from England and France reflect the town's medieval trading contacts, while a nineteenth-century record of a holy well dedicated to St James hints at a link with pilgrimage to Santiago de Compostela.

When Henry II permitted Strongbow to retain Leinster, he astutely held onto Waterford on the south coast as part of the royal demesne, recognising its advantageous location, rich hinterland, deep harbour and established population (Figure 17.32). Later, King John endeavoured to ensure the port's prosperity by proclaiming that all foreign ships entering Waterford Harbour must dock at Waterford. Here, they would be met by his customs officials, even if their destination was the port at New Ross situated some 40km inland. This led to three centuries of disputes between these two major ports, but enabled the port of Waterford to thrive as merchants from Wales, England, France, Italy and the Low Countries settled in the town.[28] The port facilitated the export of harvests from its rich inland areas and the import of necessities and luxuries from overseas, as well as from other parts of the island. Waterford became Ireland's largest importer of wine in the thirteenth century.

The riverine location of many inland towns allowed them and their hinterlands to benefit from overseas trade via navigable rivers: for example, Trim via the Boyne; Clonmel via the Suir; Inistioge and Thomastown via the Nore; Athy and Carlow via the Barrow and Lismore via the Blackwater. Even the town of Athlone, located furthest inland, was accessible from the sea via the Shannon.

SHIPS, PILGRIMAGE AND TRADE

New types of boats and ships began to appear in Irish waters – especially cogs, hulks, barges and galleys (and later caravels) – all involved in trade, transport, warfare and piracy.[29] A clearer picture of what these vessels looked like and how they were constructed is gradually emerging from research on ship timbers found on excavations in Dublin, as well as the analysis of medieval texts and illustrations. A thirteenth-century seal of Dublin, for example, depicts a double-ended, clinker-built galley reminiscent of Norse vessels of earlier centuries. Medieval images of ships on plaster in Ireland are found almost exclusively at ecclesiastical sites (Figure 17.33). They include the cog and several other vessels incised on the wall of Ennis Friary in County Clare. Another

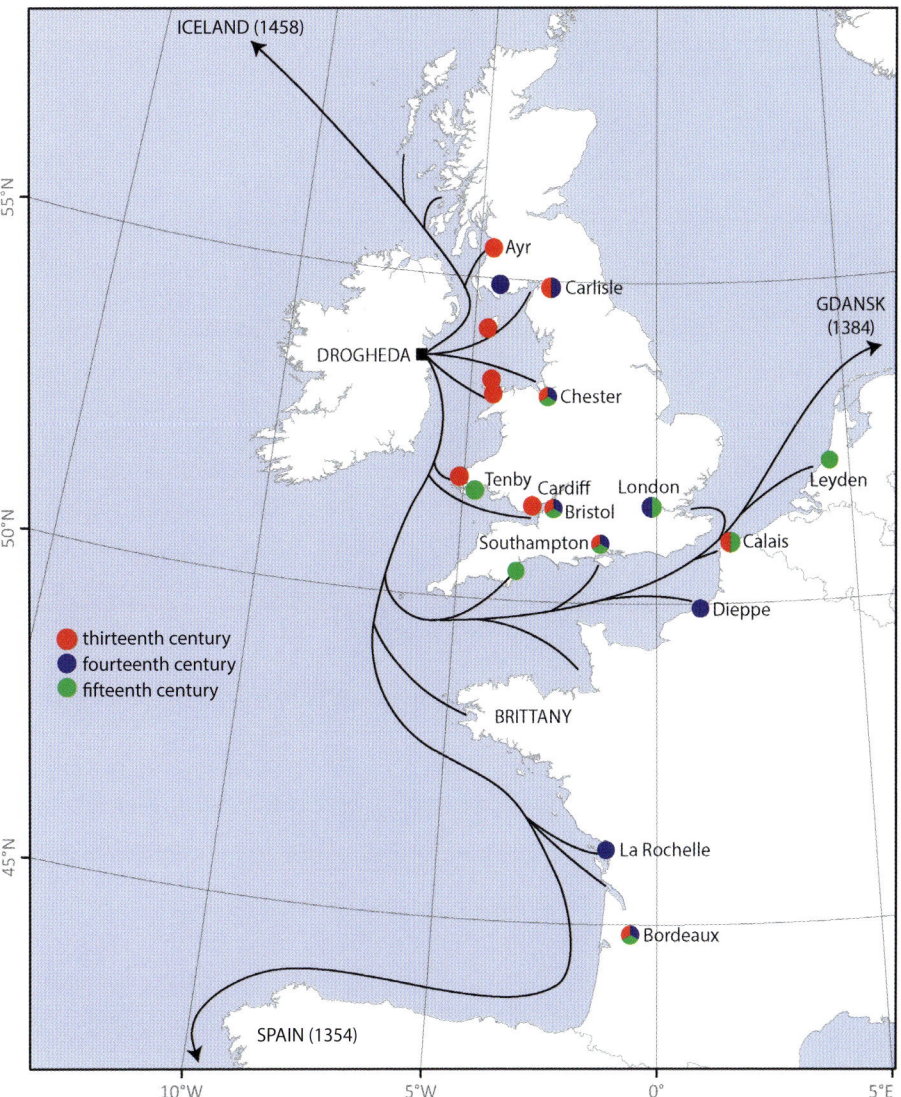

Fig. 17.31 THE ANGLO-NORMAN PORT TOWN OF DROGHEDA'S TRADE ROUTES WITH EUROPE, SHOWING THE BROAD PATTERNS OF TRADE OVER *c*.300 YEARS. The primary documentary sources for medieval Drogheda, at the mouth of the River Boyne, County Louth, indicate an extensive overseas trade, with Scotland, Wales and England, as well as with the north and west coasts of France. By the fourteenth century, the town's trade network extended southwards to Spain and northwards to the Hansa market at Gdansk. The following century saw the establishment of commercial contacts with Iceland and Portugal. Wine, iron and salt were major imports, while among the main exports were grain, hides, wool and fish. Hugh de Lacy first founded two towns at Drogheda *c*.1180 AD, one on each side of the River Boyne. Drogheda quickly overtook Dublin as the busiest port on the east coast, and by the late thirteenth century it was outranked only by the south-coast triumvirate of New Ross, Waterford and Cork. Boats and ships were built and repaired on the quayside at Drogheda, providing employment and requiring resources, especially timber, from the town's hinterland. [Data source: Bradley, J., 1978. The Topography and Layout of Medieval Drogheda. *Journal of the County Louth Archaeological and Historical Society*, 19(2), pp. 98–127]

simple stylised ship is found at St Mary's Church in New Ross, as well as the single-masted vessel (probably a hulk) on the north wall of the chancel at Cashel Cathedral, County Tipperary, where there is the depiction of a full-rigged three-masted ship.

Apart from merchants and mariners, pilgrims became one of the most important groups of foot-passengers passing through Ireland's ports.[30] Pilgrims from Ireland travelled to places such as Cologne Cathedral, Westminster Abbey, the Holy Land and Santiago de Compostela. Shops, hostels and taverns thrived at pilgrim destinations and along the well-trodden routeways leading to them. Pilgrims also made donations to churches and shrine-keepers as well as purchasing souvenirs, especially badges that they attached to their cloaks to demonstrate which shrines they had

visited. Pilgrimage had always been an important rite of passage for the Normans, with pilgrimage to Compostela from medieval Ireland being, for many decades, the preserve of the Anglo-Norman community. This is reflected in the number of churches dedicated to St James in the south and east of the country, as well as the use of the forename James among these communities. By the thirteenth century, pilgrim hostels were established at Drogheda, Dublin, Waterford and at other coastal locations. Among the discoveries made by archaeologists in Waterford are translucent amber prayer beads and badges brought back as souvenirs by pilgrims *c*.1250. Overseas visitors disembarked in Ireland to visit the famed island sanctuary of St Patrick's Purgatory on Lough Derg (County Donegal). This was Ireland's premier

Fig. 17.32 THE MEDIEVAL WALLED TOWN OF WATERFORD, BASED ON A DRAWING IN THE GREAT CHARTER ROLL OF WATERFORD, *c.*1372. The ships depicted are reminiscent of those portrayed in the Virgil Master's painting of the relief vessels arriving on the Irish coast in 1399 and of the ship on the thirteenth-century common seal of Dublin. The rabbits are significant! The rabbits shown, on the banks of the River Suir at Waterford, were probably descended from those introduced to the area by the Anglo-Normans in the late twelfth century. Before that, there were no rabbits in Ireland. Artificial warrens were constructed for them so they could be kept and bred for food and pelts. The easiest place to create such warrens was in the sandy soils and dunes in coastal and riverine locations, and this is where we find the earliest evidence for rabbits in Ireland (Murphy, P., 2016. Medieval Rabbit Farming and Bannow Island. In: Doyle, I., and Browne, B., eds. *Medieval Wexford: Essays in memory of Billy Colfer.* Dublin: Four Courts Press, pp. 292–311). They were often kept on small islands where they were relatively free from predators. Documentary evidence confirms that there were warrens at Portraine and on Lambay off the Dublin coast, for example, while place names indicate the presence of rabbits at Coney Island and at Balcummin near Skerries (coney = rabbit). A rabbit warren is recorded at Bunratty in County Clare in 1287. The Anglo-Normans were also responsible for the introduction to Ireland of a range of other wild and domesticated species, including black rats, hedgehogs, fallow deer and pike. [Source: courtesy of Sara Nylund]

pilgrimage site and drew pilgrims from as far afield as Catalonia and Hungary, England and France.

The most commonly encountered imported item on medieval excavations in Ireland is pottery (Figure 17.34).[31] It is found frequently in coastal towns, where it provides direct evidence for overseas trade and wine consumption. Predictably, sites on or near the east coast tend to turn up a higher proportion of English wares, while places closer to the south coast yield a greater range of material from France. Pottery from the Low Countries, northern Germany and the Rhineland and from the Mediterranean is also found, albeit in much smaller quantities. As much as 40 per cent of the late twelfth- to fourteenth-century pottery recovered on the Wood Quay excavations in Dublin was made up of English and continental wares. The vast majority of the continental pottery originated in France, but there were also pieces found from the Bruges area and from Paffrath near Cologne in Germany. Paffrath and other German wares have also been found in Cork, Limerick and Waterford, while sherds of Flemish pottery have been recovered in Cork, Dublin, Galway, Limerick and especially from Waterford.

Situated at the south-east corner of Ireland, County Wexford was ideally situated to benefit from overseas trade. It had as many as seven ports with international trading connections in the thirteenth and fourteenth centuries. This is reflected in the archaeology, architecture, landscape, place names, language, surnames and even the DNA of the people. These ports developed as part of an international trade network, exporting hides, boards, wool, wheat, oats, barley, beans, wine, meat and fish (especially herrings and salmon) and importing wine, salt and iron among other things. Sheep leather arrived from Spain, wine from St Emilion on the Dordogne, merchants from Florence and Lucca. Architectural stone, from the quarries at Dundry outside Bristol, can be seen at more than ten locations in south Wexford, all within a few kilometres of the coast.

Less obvious and less resilient than ceramics and stone, archaeological excavations have shown that other goods imported via Ireland's growing medieval trade hubs included figs, grapes, walnuts, hemp and serradella. Surviving historical sources indicate that Ireland's main imports included salt, spices, soap, hops, tools, weapons, nails, tin and lead. Iron was one of the most common imports, often arriving along with cargoes of salt in ships from Brittany, Spain and England.[32]

In terms of exports from Anglo-Norman Ireland, grain, wool, wax and livestock were significant, as were hides, furs, leather and cloth. Timber was an important raw material for construction, shipbuilding, barrel-making (coopering) and other crafts and purposes. Extensive forestry in the hinterland of Ireland's larger ports supplied the export market to England and France especially. Already in 1098 Muirchertach O'Brien allowed William II (Rufus) to take roofing timbers from the outskirts of Dublin for Westminster Hall, London. Similarly, modern analysis of the timbers used in the construction of the roof of Salisbury Cathedral revealed that roughly half of the wood was oak imported from Dublin in the first half of the thirteenth century.

In the late twelfth and thirteenth century, Irish ports became occupied increasingly in provisioning the king's armies overseas. Food, drink, fighting men, money, ships and other resources were sent in large quantities to Scotland, Wales, Flanders, Gascony (and France, by the start of the Hundred Years' War in 1337). It was not unusual for crown authorities to requisition ships in Irish ports to transport supplies to warzones.

FISH AND THE COAST AS A RESOURCE

The largest export from medieval Ireland was fish. Herring dominated, but cod, ling, pollock, salmon, hake and whiting were also shipped in large quantities. Despite this growing export market, the majority of the catch was destined for local consumers. There was always a high demand for fish, as all citizens were expected to eat fish on fast days, days of abstinence and throughout Lent. Fish was an especially important part of the monastic diet. The coastal location of many of Ireland's Anglo-Norman urban settlements and the well-established trade networks that existed on the island meant that most of the population had access to a wide variety of seafood.

Archaeological excavations in Dublin, for instance, have uncovered the remains of bream, cod, eel, gurnard, haddock, hake, herring, ling, plaice, ray and other fish, as well as cockles, mussels, oysters, periwinkles, scallops and crabs from medieval deposits (see Chapter 16: The Inhabitants of Ireland's Early Coastal Landscapes).[33] Excavations at Arran Quay have also revealed evidence for fish processing, while the remains of a possible herring fishery were excavated on the south of Dublin's Hammond Lane. Ling and cod were the main species represented, and although the cod may have been caught locally, ling are not normally found in waters shallower than 60m, indicating that fishermen were travelling quite some distance to catch them. The plaice and flounder caught were comparable in size to modern flatfish, at about 30–35cm (see Chapter 26: Coastal Fisheries and Agriculture).

Throughout the medieval period, most of the Irish coast was exploited for its fishing, both onshore and offshore. In Anglo-Norman territories, much of the coastal area was in ecclesiastical ownership; these, and other landholders, were keen to enforce their rights to fishing dues, customs and tithes. These revenues were sometimes collected in the form of a portion of the catch from boats that operated out of the havens and harbours on the lord's lands. Some landlords also insisted on being given preferential rates to purchase additional fish, or even being gifted the best fish of the catch.

Fish were typically caught by hook-and-line or in nets. Ponds in which fish could be stocked were also in use and sometimes constructed at coastal locations. Additionally, coastal fish-traps have been in use in Ireland since Mesolithic times and there is good

Fig. 17.33 A DRAWING OF 'RELIEF SHIPS' ARRIVING ON THE LEINSTER COAST. French historian and poet Jean Creton (fl. 1386–1420) accompanied English king Richard II to Ireland in 1399 and later wrote, *La prinse et mort du Roy Richart* ('The capture and death of King Richard') for the duke of Burgundy. Seven copies of this text survive, including one in Harley MS 1319 (fos 1r – 78v), which was illustrated in Paris in the early 1400s by the Virgil Master (fl. at the court of Jean, duc de Berry, 1390–1420). The text is accompanied by sixteen coloured and gold miniatures. One of these is shown here and depicts relief ships arriving on the Leinster coast, with provisions for Richard II and his retinue. [Source: © British Library Board. All Rights Reserved, Bridgeman Images]

Fig. 17.34 FLEMISH POTTERY FOUND IN WATERFORD. Wine was one of the most significant imports to medieval Ireland, arriving in the port towns from a range of continental markets. The discovery of overseas pottery on archaeological excavations is not uncommon in Ireland's medieval towns; the highly decorated glazed Flemish wine jugs shown here were found in Waterford in the 1990s. They were made near Bruges in the early thirteenth century and exported probably from its port at Sluys. Trade with this Hanseatic League port could have opened up a significant range of opportunities to merchants and mariners from Ireland. [Source: Waterford Treasures – Three Museums in the Viking Triangle]

Fig. 17.35 REMAINS OF FISH-TRAPS IN THE PRESENT INTERTIDAL AREA OF THE RIVER FERGUS' ESTUARY, ON THE SHORES OF THE SHANNON, COUNTY CLARE. Fish-traps were an important means of catching fish in the Middle Ages. Fragmentary examples of these have been found at several riverine and coastal locations around Ireland. Timbers tend to survive well in mudflats and are often exposed at low tide, such as in the estuary waters of the River Fergus. The examples seen here date to the fourteenth century and are situated today *c*.1.5km from dry land, below the port town of Clarecastle and close to Bunratty, in an area heavily populated by the Anglo-Normans from the thirteenth century onward. These traps are among the best examples known in north-western Europe; they are visible only during exceptionally low summer tides. Similarly, at Bunratty, on the Shannon Estuary mudflats in County Clare, a series of small but very well-preserved medieval fish-traps were discovered and recorded *c*.1km downriver from Bunratty Castle. They consisted of complex post-and-wattle fences and intricately woven baskets and were probably used for trapping salmon, eel, trout and flounder. There was an important Anglo-Norman borough at Bunratty between the mid-thirteenth and the early fourteenth century, with a population of about one thousand people. One of the Bunratty fish-traps dates to this period. In 1287 AD there was a castle, a court, a market, a mill, a rabbit warren and fish pond there. The pond provided an income of 20s. per year and was enlarged two years later. There is little discernible difference between fish-traps of the pre-Anglo-Norman and of the Anglo-Norman periods, which supports the belief that there was a large degree of social and economic continuity; some Gaelic Irish people worked as free tenants for the Anglo-Normans at Bunratty in the thirteenth and fourteenth centuries. [Source: courtesy of Aidan O'Sullivan, University College Dublin]

evidence for their widespread use around Irish coasts in the medieval period (Figure 17.35). A significant amount of labour and materials, mostly hazel and ash rods as well as oak posts, were required for the construction and maintenance of wooden fish-traps. The arrival of the Anglo-Normans, and especially the Cistercian monastic order, however, may have provided the impetus for a move from wooden to stone fish-traps.[34] Stone traps required less maintenance, were self-baiting (through molluscs attaching themselves to the stones), could be made larger than wooden ones, and were not dependent on a dwindling timber resource.

Their use was regulated either by the Anglo-Norman or Gaelic lords, or by the crown. Coastal fish-traps were built to catch fish moving with the ebbing and flooding tides (see Chapter 2: The Coastal Environment: Physical systems, processes and patterns). They were usually v-shaped structures of converging vertical post-and-wattle fences, or stone walls, at the apex or 'eye' of which was a basket to trap the fish that were funnelled in its direction. In the estuary of the River Deel, County Limerick, several small intertidal fish-traps of twelfth- to fourteenth-century date were recorded in the mid-1990s. The presence of the Anglo-

Norman castle and borough of Askeaton, just 3km to the south of the fish-traps, is significant and the fishmongers there would surely have been involved in the trade (see Chapter 18: Era of Settlement: Trade, plantation and piracy). Interestingly, the location of the fish-traps suggests that medieval sea levels were broadly similar to today's (see Chapter 7: Ancient Shorelines and Sea-level Changes).[35]

The use of fish-traps elsewhere in Ireland was also common, as in Strangford Lough. This area was a core component of the earldom of Ulster and, under the de Courcys, three Cistercian monastic houses were founded in the area (at Inch, Grey and Comber). That significant numbers of settlers arrived in this area is suggested by the Anglo-Norman townland names and the distribution of earthwork castles, of which there were at least seven on the shores of the lough. The monks at Inch operated a salmon fishery in the Quoile Estuary and a herring fishery on the outer Ards coast. A number of stone fish-traps (with walls up to 300m long) have been discovered along the north-east and north-west shores of the lough, which may have been used by the Cistercians at Grey Abbey and Comber.[36] There is also a concentration of fish-traps in Greyabbey Bay (more than half of all those recorded in Strangford Lough), including four wooden examples that probably date to the twelfth to thirteenth centuries, based on radiocarbon dates. The only dateable stone fish-traps in Strangford Lough are from the mid-thirteenth century. The traps in Greyabbey Bay were probably part of the monks' foreshore fishery and were worked as a commercial concern. The monks exported grain to England and elsewhere and in 1298 Edward I ordered 20,000 dried fish to be sent from Ireland to Skinburness in Scotland, this latter being the port of Holm Curtram, the mother house of Grey Abbey.

OTHER COASTAL RESOURCES

The coast provided more than just access to overseas markets and sea fisheries. Sand was collected on beaches and used to improve soil fertility and drainage, such as on the Earl of Norfolk's manors at Fenagh, Forth and Old Ross in the 1280s. The hospital of St John the Baptist without the New Gate, in Dublin, had access to a sandpit in 1312, the sand from which was used on its lands in the north of the county. Seaweed was also gathered for use as a fertiliser, which may explain why seaweed, seashells and the remains of insects associated with the seaside were found in garden soil deposits in Dublin. The availability of fertiliser, combined with sandy soils and a favourable coastal climate, made some sections of the coast ideal for producing vegetables, fruit and grain. This produce was used to supply the growing ports and towns in Ireland, as well as an expanding overseas market.[37]

Sand, gravel and stone were also quarried along rivers and the coast for use as building materials, road surfacing, millstones, ballast and other purposes. In Dublin, for example, the banks and bed of the River Dodder were quarried. When stone had to be transported over longer distances, it was mostly by river or along the coast. Grain was another expensive and relatively heavy commodity to transport where shipping by water had significant cost advantages, as was the case with timber and fuels. The riverine and coastal trade routes were the main arteries by which these commodities came to the port towns (see Chapter 19: Changing Coastal Landscapes). In the case of Dublin, as all of the country's well-wooded demesnes were situated along the south-east coastal fringe, timber from this area would have been shipped along the coast and unloaded at Wood Quay. The use of timber from coastal broad-leaved woodland in medieval Dublin is indicated by the presence of certain insect remains in wood samples from Back Lane (in the form of Coleoptera such as *Mesites tardii* and others).

Although trade around Ireland's coastline was both significant and grew strongly throughout the medieval period, its waters could be treacherous, due to the nature of the coastline and the frequency of often violent storms (see Chapter 2: The Coastal Environment: Physical systems, processes and piracy). Shipwrecks were quite common. In the medieval period, the right to collect flotsam and jetsam that washed up on the shore, particularly following a shipwreck, was known as the 'right of shipwreck'. Efforts, however, were made to reduce the dangers of seafaring and a range of beacons and lighthouses were operated around the coast of medieval Ireland. The best known example, Hook Tower, is a remarkable lighthouse built by the Marshal family at the tip of the Hook Peninsula in the mid-thirteenth century.[38] It is one of the oldest operational lighthouses in the world (see Chapter 23: Maritime Traditions and Institutions).

It is possible that salt-making was carried out on Ireland's foreshores in the Middle Ages (through the evaporation of seawater in specially constructed pans), but the records of imported salt suggest that it was certainly not widespread. In medieval Europe the first windmills were built generally along its shorelines in the late twelfth century. This pattern is also evident in Ireland, where onshore winds around the coast powered windmills in the same way that the tide powered tidal mills. As a result, the earliest documented windmills in the Dublin region were close to the coast, one in the city area at Oxmantown, the other to the north at Swords. The Knights Templar at Kilcloggan on the south Wexford coast operated a windmill to grind the grain produced from arable crops harvested on their estate, where they also had a brewery.

While proximity to the coast had many advantages, it was not without its risks. Flooding for these seafaring and coastal peoples would have been a regular occurrence and, together with other disasters, something to be coped with and overcome. In c.1285 Great Island in Waterford Harbour was abandoned after being flooded by seawater, while in 1325 several burgages in Wexford town were evacuated because they too had been inundated by the sea. Bannow and Clonmines suffered when the channels there silted up and it is likely that both were abandoned due to drifting sand. Farmland at Swords in County Dublin was flooded by the sea in 1326. The success of the ports became a drain on the resources of the towns and their hinterlands, with money, victuals, goods and people being shipped overseas to support royal war efforts. Ships from foreign lands brought rare treasures and luxury foodstuffs, but they also carried the Black Death. The plague spread rapidly through ports, along trade routes and within and between towns and so it affected the nucleated settler populace more than the dispersed native Irish. It is estimated that the main outbreak in 1348–1349, and subsequent aftershocks, depleted the colonial population by as much as 50 per cent.[39] The Black Death may have proved even more deadly in the towns and ports of continental Europe, thus shrinking drastically the market for Ireland's exports and adding to the decline of her overseas trade through the remainder of the fourteenth century.

Over the span of some six centuries the Vikings and Normans together transformed the earlier map, landscape and uses of coastal Ireland; they expanded the number and sizes of the existing towns, and of the other settlements, while fundamentally altering former commodities and trading patterns. For both the Hiberno-Norse and later the Anglo-Normans, the coastal zone provided the lifeblood base for commodities, trade and increasingly for living space. Both of these peoples founded major new coastal urban centres in Ireland, which by the eleventh and twelfth centuries had become of European significance. By the close of the medieval period, the ship-friendly coastal towns of the Scandinavians had been transformed into places of stone and architecture, whilst becoming absorbed into the new landownership and societal structures of their Anglo-Norman successors.

ERA OF SETTLEMENT: TRADE, PLANTATION AND PIRACY

James Lyttleton

ROCKFLEET CASTLE, A FOUR-STOREY TOWER HOUSE SITUATED ON THE NORTHERN FORESHORE OF CLEW BAY IN COUNTY MAYO, PROBABLY BUILT BY THE O'MALLEY CLAN IN THE LATE FIFTEENTH CENTURY. The construction of such iconic buildings along Ireland's coastline facilitated the control of local marine resources at a time when there was a significant expansion of the sea fisheries, with large fishing fleets from England and the European continent arriving off the Irish coastline. They also provided secure bases from which local lords could engage in acts of piracy to further enhance their prosperity and reputation. [Source: Christiaan Corlett, courtesy of Culture Stock]

Ireland experienced many profound changes during the sixteenth and seventeenth centuries. These were marked, in particular, by new ways of thinking, encapsulated by the Renaissance and the Reformation. This emerging modernity initiated new political, economic and social changes that eventually brought the later medieval world to an end. Particular events, such as the fall of the Geraldine House of Kildare in 1534 or the dissolution of the monasteries later in the same decade (Figure 18.1), are seen popularly as marking the beginning of the early modern era in Ireland. In reality, however, the process was a far more prolonged and convoluted affair that arguably did not reach a conclusion until the end of the seventeenth century.[1]

Politically, Ireland in 1500 was a mosaic of semi-autonomous, Gaelic-Irish and Anglo-Irish lordships, with crown control confined largely to a coastal region around the Port of Dublin (Figure 18.2). England, as a new Protestant power in Europe, was under threat from Catholic rivals on the continent – France and Spain – and there were concerns among English authorities that Ireland could be used as a base from which to launch an invasion of its larger neighbour. To forestall this, it was decided to renew and expand English control across the whole island. Native lords were encouraged to surrender their lands and have them re-granted under terms that favoured the spread of English legal norms and landholding practices. Another means was to provoke the native lords into rebellion, then confiscate their lands and redistribute them to New English landlords (typically Protestant, but not always) whose loyalty to the monarch was more assured. These schemes of confiscation and redistribution came to be known as plantations, as new landlords were encouraged to 'plant' or settle tenants from England, Wales and later Scotland. It was anticipated that these plantations would facilitate the creation of a new society that would ensure the continuation of English rule in Ireland (Figure 18.3).

While these plantations varied in success, the movement of new settlers into the country, as well as the appearance of new consumer goods and greater demands for agricultural and fishery products, brought about fundamental changes in material culture, architecture and landscape. These were to have profound

Fig. 18.1 THE RUINS OF TIMOLEAGUE FRIARY, COUNTY CORK. At the start of the sixteenth century, monasteries and friaries had become a conspicuous feature of the Irish landscape. The patronage of both Anglo-Norman and Gaelic-Irish lordships had allowed various religious orders to erect impressive buildings, while generous land grants enabled the orders to shape the surrounding land-use, often promoting quite productive agriculture. The Reformation and dissolution of the monasteries from the 1530s, however, resulted in the buildings and land being expropriated and ownership transferred usually to supporters of the crown. While dissolution would evolve over the following century or so, it resulted ultimately in the removal of their pivotal roles in shaping both the cultural and economic landscapes in their immediate vicinity. Despite experiencing damage, most of these impressive buildings have survived as iconic features in the landscape and are often important elements of the local heritage. Here, the ruins of Timoleague Friary are seen, located close to where the Argideen River enters into Courtmacsherry Bay. Founded probably before 1316 by either Donal Glas MacCarthy or William de Barry, the remains consist of the church with ranges of domestic buildings around a cloister. Edmund de Courci, made bishop of Ross in 1494 and who was buried in the friary in 1518, is credited with expanding the layout by building a steeple, dormitory, infirmary and library at Timoleague. The friary was damaged extensively following the Reformation. [Source: Sarah Kandrot]

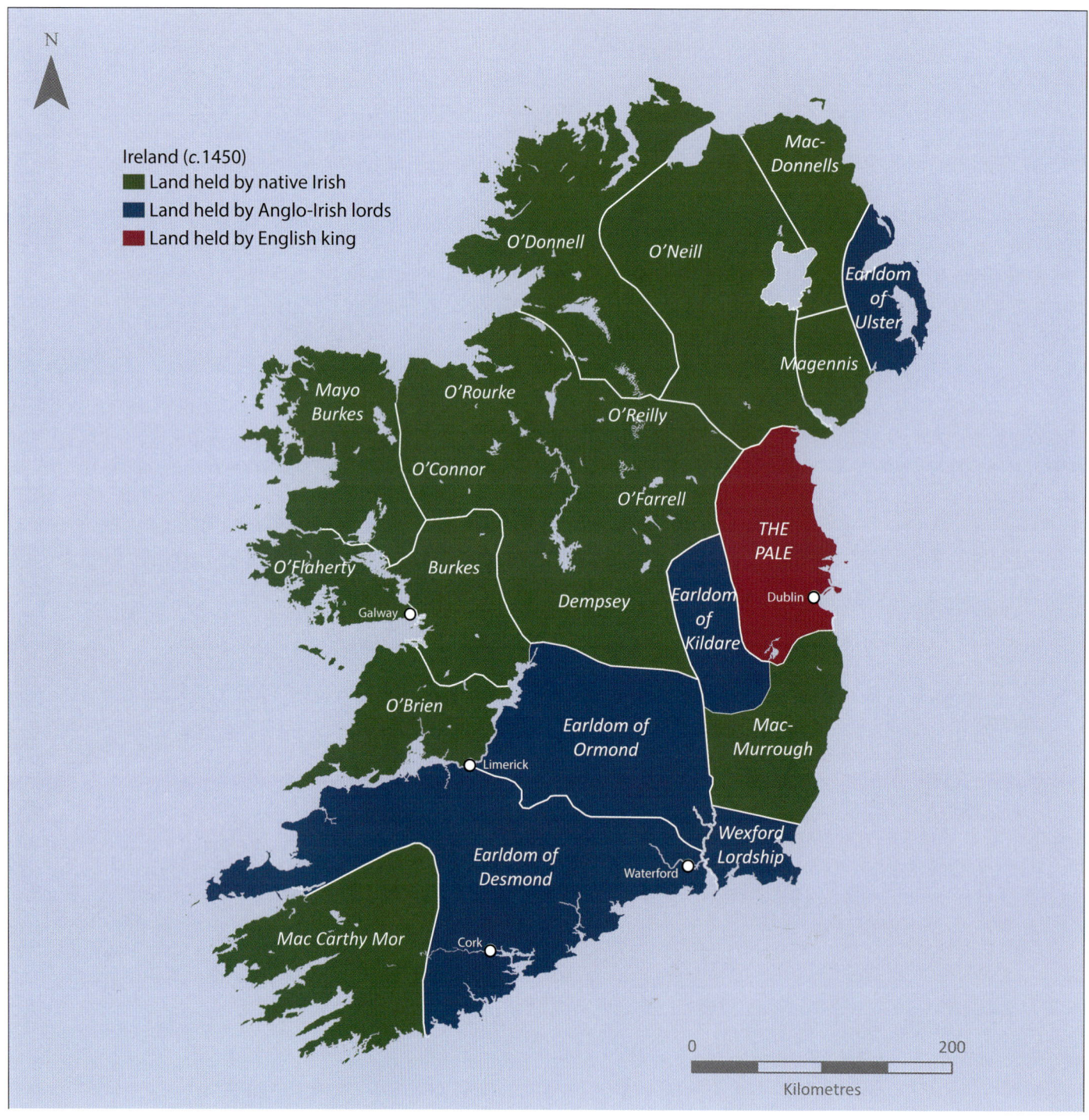

Fig. 18.2 THE POLITICAL GEOGRAPHY IN IRELAND (c.1450). At the start of the sixteenth century, the English crown's sphere of control in Ireland was limited essentially to a restricted area around Dublin, known as the Pale. Beyond the Pale were a series of semi-autonomous Anglo-Irish and Gaelic-Irish lordships. Major ports like Waterford, Cork, Limerick and Galway also enjoyed a degree of civic independence from government control, though agents for the crown did collect revenues from these ports. Along the western and northern coastlines, where Gaelic lords such as the O'Flahertys and the O'Donnells were dominant, a less formalised port structure was in place, with the emphasis more on controlling the exploitation of natural anchorages, convenient landing places and fishing grounds. [Data source: www.wesleyjohnson.com]

consequences for Ireland's coastal communities as they were forced to contend with new political, economic and social challenges. In spite of this, the nature of most coastal settlements in the early modern era would reflect as much continuity as change.

Port towns, for example, remained as significant foci of settlement, trade and activity. In addition, tower houses, constructed typically between the early fifteenth and early seventeenth centuries, were commonplace in urban and rural settings, along with thatched, mud-walled cabins, which were also derived from medieval building

traditions. Fishing remained a principal source of income for many coastal communities, as well as for Gaelic lordships that controlled extensive areas of the island's coastline. From strategic tower houses, many of these coastal lordships enhanced profits from legitimate trade by active engagement in piracy. Yet, changes were occurring, as can be seen with the rise of the country house and the arrival of new house styles, both stone-built and timber-framed, for the population at large. The onset of the Reformation was also to have profound consequences for coastal communities, with many churches and monasteries falling into ruin, as well as the break up of the surrounding estates once owned by the hierarchy and religious orders.

PORT TOWNS

Ireland's coastline was first mapped in detail on the portolan charts drawn by Italian and Catalan map-makers between the fourteenth and sixteenth centuries (see Chapter 8: Monitoring and Visualising the Coast). These depicted a rough outline of the island, with the principal ports and harbours marked along the coast. By this time, a network of ports was in place, situated mostly along the eastern and southern coastline, to take advantage of proximity to Britain and the European continent. These ports included a number of towns established originally by the Vikings in the ninth century, though most were founded by the Anglo-Normans in the later twelfth and thirteenth centuries. The towns were administered generally by civic authorities, led by a mayor and aldermen, drawn usually from the leading merchant families. Before the dissolution

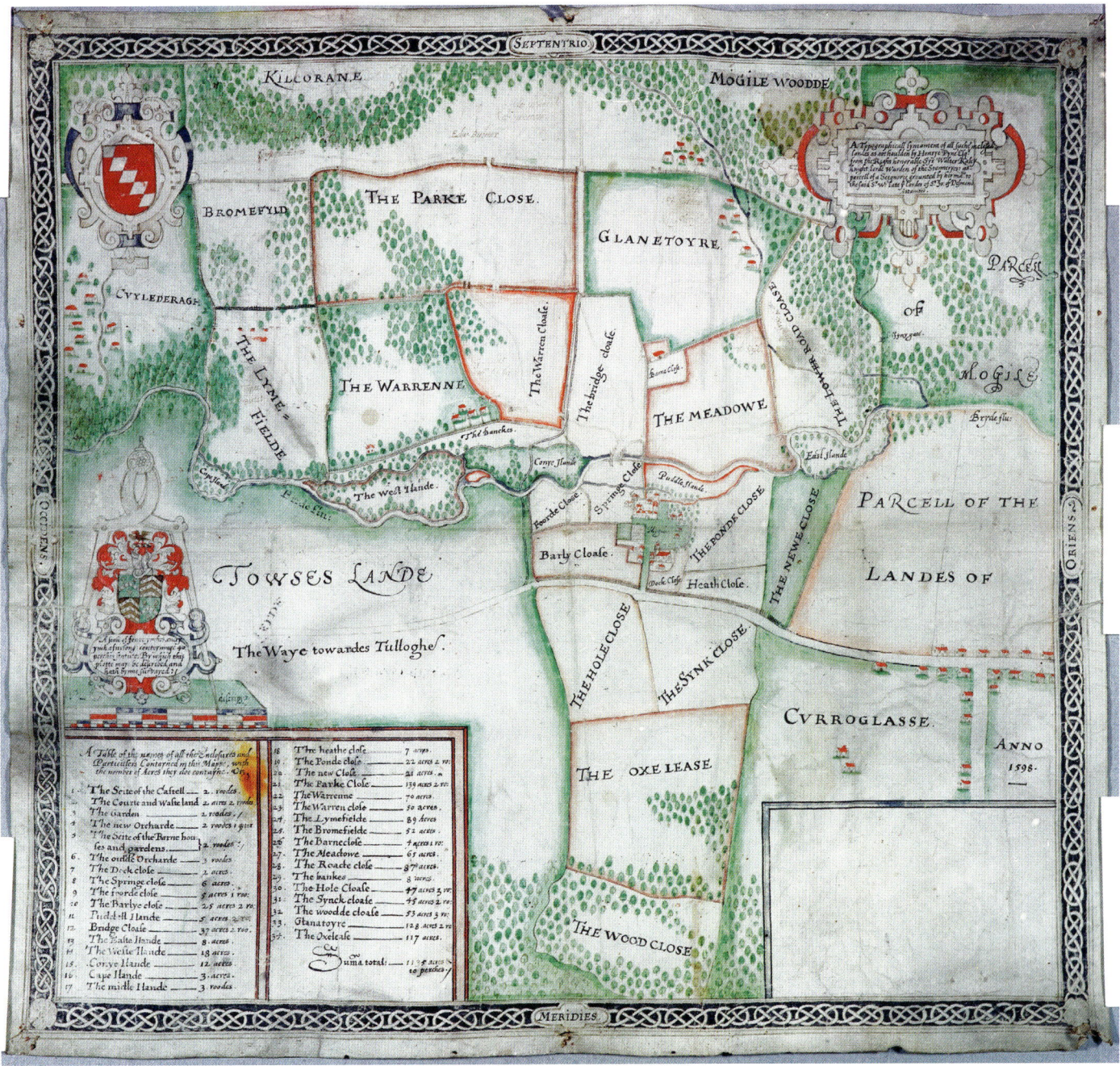

Fig. 18.3 MOGEELY ESTATE, EAST COUNTY CORK. This map of the Mogeely Estate, or plantation, in east County Cork, dated 1598, is held as part of the Lismore Estate Papers at the National Library of Ireland. It is the first known estate map produced of an Irish estate, possibly by the artist and cartographer John White, and gives an insight into the estate on the eve of the 1598 Rising. Estates such as this were part of the Munster Plantation scheme, which evolved from the 1580s in what was deemed to be a troublesome province, and pre-dates the more ambitious Ulster Plantation of the early seventeenth century. In essence, to reduce the threat of rebellion and consolidate English power, the crown found reasons to confiscate land, generally from Gaelic landlords, and to redistribute it to landlords and tenants who would be far more supportive of English cultural norms. This also consolidated England's political power in Ireland and began the process of reshaping the island's social, cultural and economic landscape. Much of the focus of this map is on the boundaries and natural resources of the estate, which was granted to Sir Walter Raleigh. These include the forests, which Raleigh and his tenant, Sir Henry Pyne, were to develop as a successful timber industry. The Blackwater River, seen winding through the centre of the estate, allowed easy transport of timber to the sea at Youghal. Development of this estate reflects an early attempt by the English both to exert political control and to develop the resources of this area in east Cork. [Source: National Library of Ireland, call no.: MS 22068]

of the monasteries, various monastic orders – such as the Augustinians, the Benedictines and the Franciscans – also owned a considerable amount of property in urban areas. Such land, however, would now pass into the hands of the civic authorities, including various merchant, gentry and aristocratic families considered deserving of royal largesse.

Fairs and markets were a fundamental part of a town's economy. The fairs were held annually, associated usually with the feast day of a particular saint, and were where both raw materials and finished products were sold. Most goods were produced locally, especially as town dwellers could be employed in numerous activities such as milling, tanning and leather-working, weaving and dying, metalworking, coopering, food processing and brewing. Markets were held weekly, and offered further opportunities for craftsmen and merchants to sell their wares, as well as for the townspeople to buy food provisions for the week ahead. Most

towns had an area designated as the market-place, which was usually located at a road junction of the principal street. In contrast, fairs were situated more commonly outside of the town walls, to better accommodate the large crowds attending such annual events.

Coastal towns, especially along the eastern and southern coastlines, were surrounded by relatively prosperous and populous hinterlands, and contrasted strongly with the lesser developed and rural landscapes that dominated inland Ireland. These provided a market for the importation of foreign goods, such as iron, salt and wine, and also the means of exporting locally produced goods like timber, wool and hides. These coastal towns were also of major strategic significance, facilitating both Anglo-Norman and Gaelic lords in the control of their territories. Major ports like Waterford, Cork, Limerick and Galway enjoyed a degree of civic independence from government control, though agents for the crown did collect revenues from these ports. Along the western and northern

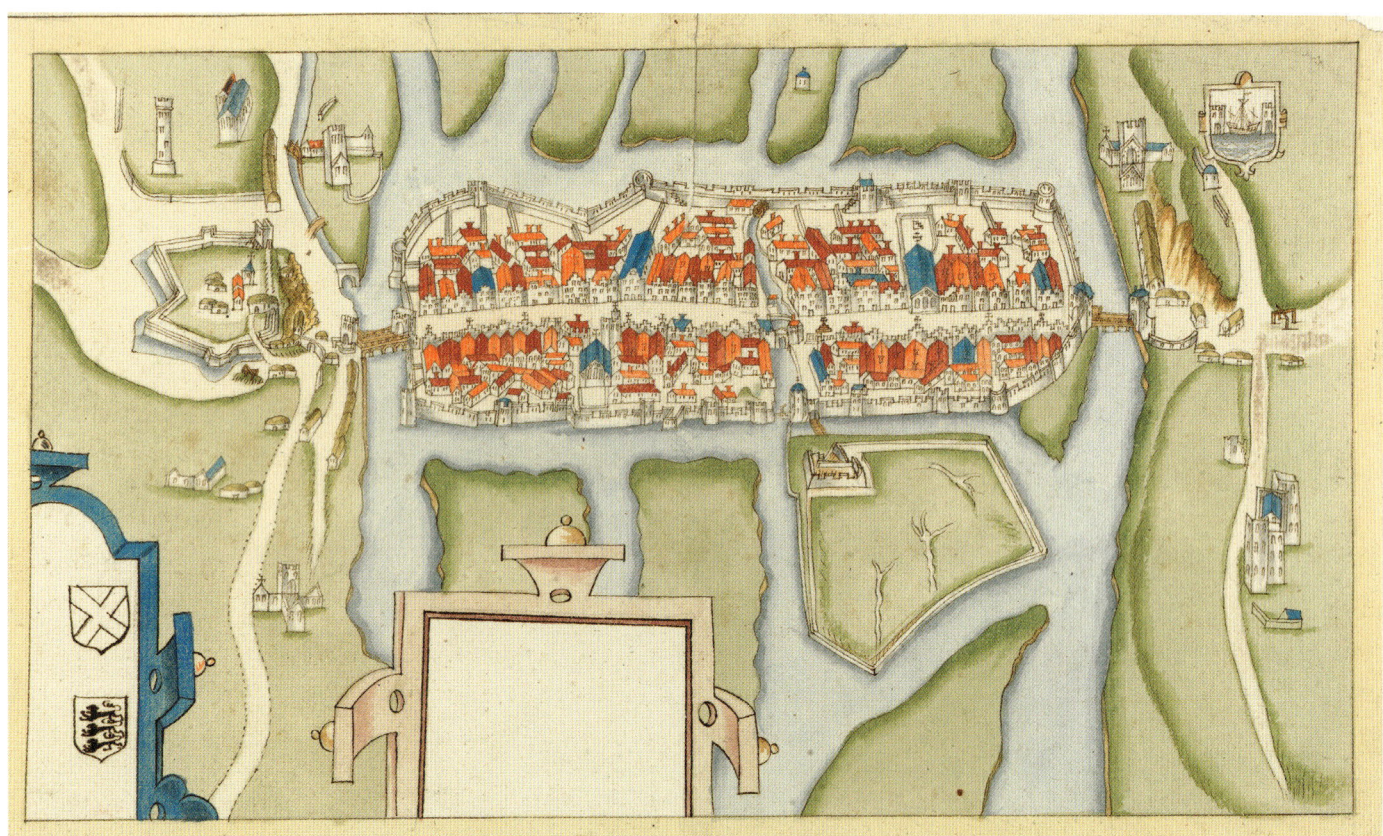

Fig. 18.4 THE TOWN OF CORK AND ITS STREETSCAPE (1602). While the foundation of Cork is often associated with the establishment of a monastery by St Finbarr in the seventh century, the urban origin of one of Ireland's oldest towns is linked more directly to its emergence as a Viking trading port in the mid-ninth century. Located on marshy islands where the River Lee enters Cork Harbour, the site was the lowest bridging point on this important river. As such, it was a site where land, river and maritime routes converged, offering access to both an extensive and rich hinterland as well as to an expanding European and global trade network. In 1177, the Anglo-Normans took control of the town. Recognising its strategic location, they drained and infilled several river channels to extend the area of dry land and developed the town's commercial activities. By the sixteenth century, however, Cork's population and trade remained less than that in the other Munster ports of Waterford, Youghal and Kinsale. The publication of detailed cartographic maps, paintings and plans from the early 1600s allows for an effective reconstruction of the urban morphology of port towns such as Cork. In 1602, Cork still retained much of its medieval character as a relatively small, compact and well-defended town. Surrounded by an impressive wall with eighteen towers, there were three principal entrances to the town. The North and South gates, with wooden bridges, gave access to a productive pastoral hinterland and defended the privileged community living within the walls. An east–west waterway, guarded to the east by a marine gate, gave direct access to the sea and also provided quays within the walled town for ships to berth in safety. The north–south spinal street dominated the layout and was fronted typically by three- or four-storey residential buildings with numerous windows and their gabled ends facing the street. Narrow laneways, orientated east–west, separated the burgage plots occupied by large residential units. Outside the walls evidence of an expanding population is apparent, as are a number of abbeys and friaries that had been built by new monastic orders attracted to Ireland and supported by the Anglo-Normans. Elizabeth Fort (1601) is shown on the southern ridge of the valley, with a commanding view of the town. Following the first decade of the seventeenth century, Cork was to be transformed into an early modern Atlantic port settlement. Its strategic location on Ireland's southern coast allowed it to become a key supply port, involved primarily in the provision of beef and butter for England's growing colonial trade with the West Indies and America. As a result of its economic strength, its population would increase eight-fold from 1600 to 1700, by which time it had acquired the status of one of Europe's key Atlantic ports, with a population five times greater than that of Liverpool. [Source: The Board of Trinity College Dublin]

coastlines, where Gaelic-Irish lordships were dominant, a less formalised port structure existed. Here, emphasis was placed more on controlling the exploitation of natural anchorages, convenient landing places and fishing grounds, through the siting of tower houses in strategic coastal locations.[2] While the number of vessels visiting Gaelic harbours was not necessarily fewer than those calling into ports on the eastern and southern coastline, they were less involved in handling large quantities of cargo; the principal focus related to offshore fishing and this did not require the provision of significant shore-based facilities.

By the beginning of the sixteenth century, the Anglo-Irish ports were still surrounded by the medieval town walls. These walls, with their gates and towers, controlled access to the quaysides, providing a level of security for the merchants, traders and town-dwellers. They also served to reinforce a sense of civic identity that was being cultivated by leading merchant families in order to create a greater sense of distinction between those who lived within the medieval town walls and those who lived outside. The various schemes of plantation that were introduced into Ireland between the mid-sixteenth and early seventeenth centuries also advocated the establishment of new towns in areas where, hitherto, there had been little urban development, such as Bandonbridge in the Munster Plantation and Londonderry in the Ulster Plantation.

Due to extensive rebuilding in later centuries, little now remains of the later medieval and early modern housing that would have dominated the streetscapes of these port towns (Figure 18.4). These would have consisted of stone and brick-built houses, as well as ones built of more perishable materials, such as clay, timber framing and even post-and-wattle. There are notable survivals of stone-built tower houses still to be seen in Galway (Lynch's Castle), Carlingford,

County Louth (Taaffe's Castle and The Mint) (Figure 18.5), and Youghal, County Cork (Tynte's Castle) . These urban tower houses were the homes of Anglo-Irish merchant families who, around 1600, also started to adopt more fashionable, non-castellated houses. A fine example is Rothe House in Kilkenny, a Jacobean town residence built around a number of rear courtyards by a family of the same name. Less obvious are buildings whose façades have been changed over time to render them indistinguishable from their modern neighbours, the townhouses of Galway being a notable example.

Archaeological excavation has revealed the foundations of timber-framed and stone-built houses of later medieval date in

Fig. 18.5 URBAN TOWER HOUSE 'THE MINT', CARLINGFORD, COUNTY LOUTH. Despite this building being constructed by an Anglo-Irish merchant sometime in the fifteenth century, the windows in the front façade are decorated with carvings of intricate knots and interlace of the Gaelic tradition, testimony to the close connections between the different ethnic groups. Similar carvings have been found in tower houses built by Gaelic-Irish lords elsewhere in the country. [Source: James Lyttleton]

various port towns, such as Dublin, Waterford, Dungarvan, Cork and Galway, but our knowledge of their early modern successors is more circumscribed. For example, the subsequent construction of Georgian basements in many locations has removed features and deposits of early modern date that lie over the medieval layers.[3] While most timber-framed buildings have disappeared, there are two remarkable survivals of timber-framed houses dating to the early seventeenth century in Ennis, County Clare, and Watergate

Street in Kilkenny. Notable examples may exist elsewhere but remain hidden under later renovation and rebuilding.

To some extent, contemporary cartographic and pictorial evidence has helped to ameliorate this lack of knowledge. For example, maps published in *Pacata Hibernia* in 1633 illustrate a number of Munster port towns as they would have appeared in the early seventeenth century.[4] Among these is Youghal, County Cork, a port of Anglo-Norman origin situated on the west bank of the

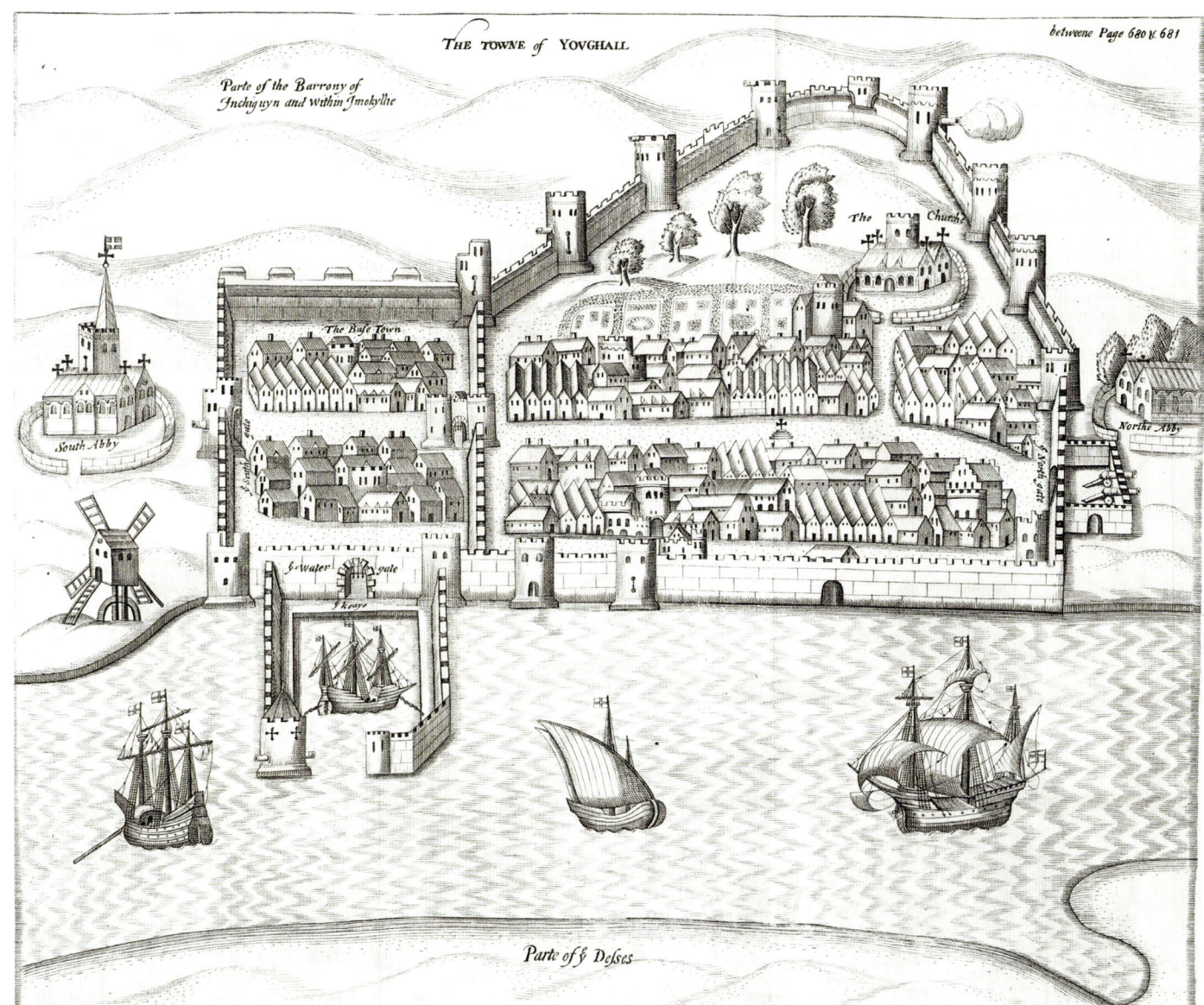

Fig. 18.6 MEDIEVAL TOWN OF YOUGHAL, COUNTY CORK. Although of Viking origin, Youghal would emerge as a significant Anglo-Norman town due to its strategic location on the western bank of the Blackwater River estuary. This provided direct access to an extensive and relatively prosperous hinterland, as the Blackwater was one of the longest and most navigable rivers along Ireland's southern coast. As a result, Youghal became a very successful port throughout the medieval period, and by the sixteenth century was the most important port on the south coast. Its prosperity and strategic location, however, also made the town a pivotal centre in the conflict between the earls of Desmond, who controlled much of Munster, and the English crown. Following the successful siege of Youghal by English forces in 1579, the trade of the port declined. The town would regain its prosperity largely under the influence of Richard Boyle, the first earl of Cork, who emerged as one of the most successful New-English planters in the late sixteenth and early seventeenth centuries. He was influential in securing a new charter for the town and stimulated economic development within the region. Trade flourished and Youghal re-emerged as a prosperous centre of trade. This map of Youghal in *c*.1600 (though not published till 1633 in a book called the *Pacata Hibernia*) illustrates the well-ordered layout of the town and the retention of much of its medieval character. It was enclosed by an impressive town wall with a number of guard towers and entrance gates, while the main street and market place are lined with gable-fronted houses on medieval burgage plots. The most dominant building in the town is St Mary's Collegiate Church (claimed to be the oldest church in Ireland, with continuous services from the thirteenth century) and, with the two abbeys located outside the north and south gates, this suggests the power of the church, especially prior to the Reformation. Outside the town walls, the harbour and quays were protected by their own walls, emphasising the importance of trade. The smaller walled enclosure to the left of the town was called 'Irish Town' and was separated from the main town by a guarded gate-tower. No longer a commercial port, Youghal emerged as an important tourist resort in the late nineteenth century, thanks largely to the now-discontinued rail connection with Cork city. Today, its historic waterfront area has attracted significant investment in refurbishment and construction, enhancing its role as both a tourist centre and commuter town for Cork city. [Source: by permission of the Royal Irish Academy © RIA]

Blackwater River estuary, which had established a major trading role along Ireland's southern coastline (Figure 18.6). The overall appearance of the town by the early seventeenth century was still medieval in character, being enclosed within a crenellated town wall, its circuit punctuated with mural towers and gate-towers.

A similar arrangement can be seen in other towns and much later in the seventeenth century. This is exemplified by a view of the port of Galway in 1684–1685 as provided by a Captain Thomas Phillips. It illustrates clearly the silhouette of an early modern town, with high pitched roofs incorporating dormer windows and tall chimney stacks, which rise above the town's medieval wall (Figure 18.7). Despite its outwards appearance of prosperity, however, Galway's role as a major port was in decline by the late seventeenth century. Effectively, Galway would remain, until the late twentieth century, a relatively small urban outlier along Ireland's western seaboard. In marked contrast, port towns along the eastern and southern coasts prospered in the late seventeenth century, benefiting from their more productive agricultural hinterlands and closer proximity to Britain. Dublin, in particular, would emerge as a dominant city and port, given its enhanced status as a major civic and military centre for the crown in Ireland.

Elsewhere along the coastline, besides the major ports, there were many smaller towns and villages whose focus was more on localised trade with their surrounding hinterlands. Archaeological surveys of foreshores have revealed the remains of quays, piers and slipways associated with these smaller settlements. One such survey of a small harbour just south of Bremore Head in north County Dublin, for instance, has uncovered the remains of a drystone-built pier. This same pier is recorded on a mid-seventeenth-century cartographic source, the Down Survey, in a harbour called 'Newhaven'.

TOWER HOUSES

Tower houses can be found at coastal locations around the country, and regional case studies have illustrated how these castles formed an integral component of the coastal landscape during the fifteenth and sixteenth centuries. This was a type of castle that was built by a wide spectrum of Irish society from *c*.1400 to *c*.1600 and was a settlement form adopted by both Gaelic-Irish and Anglo-Irish landowners across much of the country, as well as wealthy merchants who lived within the confines of walled towns (Figure 18.8). It has been estimated that some 3,000 tower houses were built, of which only 1,200 still remain in varying degrees of preservation.

Tower houses are typically rectangular in plan, rising to between three and five storeys in height, with the roofline marked by

Fig. 18.7 A VIEW OF THE PORT TOWN OF GALWAY (1684–1685). Following its Norman conquest in the thirteenth century, the walled coastal town of Galway emerged as a strategic stronghold for the English in the west of Ireland. Although surrounded by rebellious Gaelic lords and subjected to repeated attacks, the town remained a staunch supporter of the English crown in Ireland. This brought many privileges to the port, boosting its trade and prosperity. By the sixteenth century, Galway was conducting a lucrative international trade, especially with France and Spain. Wines, spices, silks and provisions were imported, while local produce of fish, hides, wool and tallow were exported. This enriched the merchant class and helped the town to enhance its role as one of England's few strategic ports along the western seaboard. Its historic allegiance to the English crown was highlighted by the fact that it would be the last major Irish town to be held by Catholic forces before its surrender to Cromwell after the siege of Galway in 1651–1652. This defeat, together with the marginalisation of the province of Connacht as a Gaelic outlier, resulted in the decline of the town's trade and prosperity. Captain Thomas Phillips' view of the town and port of Galway in 1684–1685 illustrates the silhouette of an early modern coastal town. High pitched roofs with dormer windows and tall chimney stacks peek out from behind the medieval town wall, while the church of St Nicholas dominates the skyline. Some trading vessels are berthed along the waterfront area in the shadow of the protective walls. Here is also located what is today known as the Spanish Arch, reflecting the city's historic trading links with Spain. Today, Galway has revitalised its role as the dominant urban centre in Connacht, and although its function as a commercial port has declined significantly, it has become a thriving retail, industrial, educational and tourist centre, as well as being one of the Republic's key gateway cities for the development of the west of Ireland. [Source: National Library of Ireland, call no.: MS 3137/29]

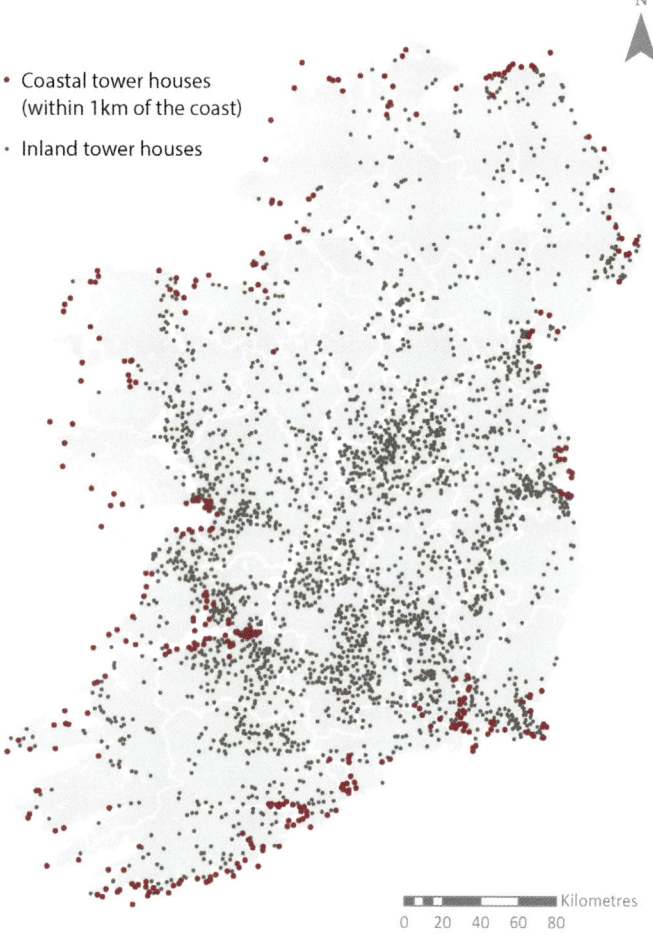

N

• Coastal tower houses
  (within 1km of the coast)
• Inland tower houses

Kilometres
0 20 40 60 80

Fig. 18.8 DISTRIBUTION OF TOWER HOUSES IN IRELAND, HIGHLIGHTING COASTAL LOCATIONS. Many tower houses are found along Ireland's coastline, where access to the sea facilitated trade and the exploitation of fisheries. Although the origins of the tower house lie in Anglo-Irish areas, the construction of such castles spread into Gaelic-Irish territories (though there is a notable absence in Ulster). Their distribution reflects political geographies and regional identities, with concentrations occurring especially in the earldoms of Ormond (counties Tipperary and Kilkenny), of Desmond (counties Limerick and Cork) and of Thomond (County Clare), and also in the Anglo-Irish/Gaelic lordships in County Westmeath. [Data source: Matthew Stout]

battlemented parapets. Despite much variation in design, the essence of tower houses is their verticality, with the chambers set on top of each other, instead of being laid out side-by-side. A number of defensive features were provided in most tower houses, including machicolation boxes and bartizans covering the immediate environs of these towers, while inside, a murder hole in the ceiling of the entrance lobby provided a last line of defence. They were also surrounded generally by a walled courtyard known as a bawn (from the Irish *bódhun* meaning 'cow fort'). These were normally rectangular in plan, and contained a number of ancillary buildings, including a hall that was usually placed right alongside the tower house (Figures 18.9 and 18.10).

Along the Atlantic littoral, Gaelic-Irish lords – such as the O'Driscolls and the O'Sullivan Beare in County Cork, the O'Flahertys in County Galway, the O'Malleys in County Mayo and the O'Donnells in County Donegal – were developing maritime lordships. In Anglo-Irish areas, tower houses can also be found in coastal locations, most notably in the Strangford Lough area of County Down or in south Wexford. Tower houses were built as the

Fig. 18.9 DUNMANUS CASTLE, LOCATED ON THE NORTHERN SHORE OF THE MIZEN PENINSULA, WEST CORK. From the mid-fourteenth to the seventeenth centuries, the waters off the south-west coast of Ireland attracted large numbers of fishing boats from Spain and England to work in these rich fishing grounds. While part of the catch would have been processed on the vessels and taken back directly to their country of origin, large quantities would have been taken ashore for processing. This, together with related activities – such as supplying provisions, as well as boat and net repairs – became an important element of the economies in these areas. To ensure that they benefited directly from such activities, as well as being able to tax vessels fishing in their offshore waters, landowners built coastal tower houses/castles to safeguard their interests. Such strongholds, therefore, occupied sites that held commanding views of their onshore and offshore territories. Such an example is Dunmanus Castle, built in the O'Mahony lordship on the northern shoreline of the Mizen Peninsula in west Cork. It stands on a rock outcrop, close to the shore and overlooking Dunmanus Harbour, a sheltered cove that allowed fishing and other vessels safe anchorage. It was also an easily defendable site, from which the O'Mahonys could control their territory. As with most coastal tower houses, Dunmanus Castle is now in ruins, but is still an impressive structure, a reminder of a time when houses such as this were key administrative and defensive centres for large stretches of Ireland's coastline. [Source: roaringwaterjournal.com]

Fig. 18.10 Illustration of a tower house, approximately 1500. The outstanding feature of a tower house was the stone-built tower itself. It formed a compact, vertical structure in which the rooms were placed on top of each other for three to five storeys, rather than creating a larger horizontal spread of rooms by locating them adjacent to each other. The tower itself was surrounded usually by a walled enclosure, called a 'bawn', within which other smaller buildings were located. Within this compound lived the lord and his family, together with servants and labourers, as well as a small armed retinue for protection. These buildings could include a hall sited adjacent to the tower, as well as a kitchen and brewhouse. Outside the walls were located gardens, orchards and other farm buildings necessary to work the land. In the more rugged terrain that typifies the western coastline, however, the extent of such productive farming activities around the bawn would have been quite restricted. Few of these 'bawns' survive, and now it is usually only the abandoned tower house that remains, surrounded by open fields or sometimes by more modern structures, especially if the tower house is located in, or adjacent to, an urban centre. [Source: Rhoda Cronin]

principal residences in these lordships, their form and layout acting as an architectural metaphor for the power that those families wielded within their localities. Furthermore, their construction and distribution were not dictated solely by defensive considerations, but were also governed by the spatial spread of different branches of ruling families and their supporters.

Construction of these buildings facilitated the control of local marine resources at a time when there was a significant expansion of the sea fisheries, with large fishing fleets from England and the European continent arriving off the Irish coastline. While these maritime lordships did not have the expertise, or interest, to engage directly in sea fishing to catch the likes of herring, hake and pilchards, they offered foreign fishermen permission to fish in their waters. In addition, they provided onshore facilities to enable visiting fishing vessels the opportunity to dry their catch, repair boats and ships, and the possibility to engage in trade. In return for such privileges and facilities, these lords would take a portion of the catch, thereby increasing their wealth. The construction of

these tower houses by the Gaelic-Irish and Anglo-Irish lords and their participation in the foreign fishing trade appears to have gone hand-in-hand.[5]

Given that maritime lordships were often located in mountainous and boggy terrain, and where good farmland was in short supply, an economic specialisation in fishing and trade provided minor lords like the O'Flahertys and the O'Driscolls with the necessary surpluses required for the building of tower houses. In addition, a significant number of maritime lordships used their tower houses as strongholds from which they could engage in acts of piracy and plunder against vessels passing through their territorial waters. This was often a lucrative, if dangerous, occupation, but helped to extend the power and wealth of some clan families, such as the O'Malleys. Of particular note was Grace O'Malley (Granuaile) who, as Ireland's 'Pirate Queen', menaced merchant shipping in the waters off Ireland's west coast from her fortified tower houses sited on the shoreline of south County Mayo.

Case Study: Piracy and Smuggling Around the Irish Coast

Connie Kelleher

The nature of much of the Irish coastline has lent itself to facilitating piracy and its willing bedfellow smuggling. From the earliest of times, references to unlawful maritime activities are to be found in both oral and written sources. These include the legend of St Patrick in the fifth century, recounting his capture in Wales by pirates and transport to Ireland as a slave, to the fourteenth century when the Irish Sea coast was reportedly infested with 'Irish' pirates operating out of the Hebridean Isles. This persisted into the fifteenth century, when Scottish men-of-war continued to extend their range of operations into the Irish Sea and along the length of the east coast of Ireland.[6]

Ruling Gaelic maritime lords who commanded ships – like those of the O'Donnells in County Donegal, O'Malleys in County Mayo, O'Sullivan-Beare and O'Driscolls in County Cork – engaged in acts of piracy as a means to exert control on those using their territorial waters. This also helped to extend their maritime power base and reap financial rewards from merchant vessels plying the waters of the Atlantic. Noted as pirates, the O'Driscolls, for example, were documented as being a 'menace to merchant shipping and who constantly remain upon the westerly ocean preying on passing ships' (Figure 18.11).[7]

Further north, the O'Malleys, who were based in Clew Bay, were also recognised as ruthless lords of the sea and, depending on who was commenting, as outright pirates, with perhaps the most famous member of the clan being Grace O'Malley (known also as Granuaile). She would become celebrated in lore and literary sources as Ireland's 'Pirate Queen', who operated out of a series of tower houses located strategically along the coastal fringe of Clew Bay and its adjacent islands. Her principal residence was Clare Island Castle, located on Clare Island, which dominated the approaches to the bay (Figure 18.12). Her galleys were involved in extensive maritime activity, including piracy along much of the western seaboard, for some fifty years or so until her death in 1603.

The Tudor period under Queen Elizabeth I, who famously met with Grace O'Malley when attempting to broker a maritime and political accord, saw a more organised system of piracy off the south-west coast of England and Ireland that went hand-in-hand with her 'seadogs'

Fig. 18.11 DEPICTION OF THE WATERFORD RAID ON BALTIMORE IN 1537. This raid was in retaliation for a piratical act of plunder by the O'Driscoll clan on a Portuguese merchantman named *Sancta Maria de Soci*, which was carrying a cargo of 100 tonnes of valuable Portuguese wine, bound for Waterford. The merchants of Waterford retaliated, with resulting devastation to the settlement and to the O'Driscolls of Baltimore. Such was the extent of Waterford's response, it was described by the London parliament as an act of treachery and revenge. [Source: Bernie McCarthy]

or privateers, the more famous among them being Sir Walter Raleigh and Sir Francis Drake. These individuals operated under commissions from the crown, and were thus officially sanctioned to attack traditional enemy ships, predominantly from the Iberian Peninsula. They regularly acted outside their remit and engaged in outright piracy and, in reality, who the pirates were at any given time depended on the viewpoints of the victim or perpetrator; the actuality of piracy in the age of sail was very much a subjective one. Though operating on a more localised and lesser scale than in the following century, piratical and smuggling operations began to grow to a level that became a serious menace to shipping and security at sea. Some local rulers, for instance, liaised with visiting pirates – like the 'partnership' of O'Sullivan Beare and Captain John Callice in the 1570s – to advance their mutual political and economic interests. While a specific alliance between ruling Irish lords and pirates did not develop until later, the assistance of local power-bases and willing natives ensured that the business of piracy flourished.

Such collaborations, however, were to become a more permanent and essential feature of piracy in the North Atlantic in the early seventeenth century. Referred to at the time as a 'confederacy', with areas along the Munster coast from Baltimore, Schull-Leamcon and westwards to Crookhaven described as 'the Nursery and Storehouse of Pirates', this was a well-organised and strategically orchestrated alliance (Figure 18.13). It can also be seen as part of the processes taking place in settlement patterns and maritime ventures under the Munster Plantation. Englishmen, many of whom were previously legitimate seamen but who had been forced to move from their original bases in the south-west of England to the south-west coast of Ireland, dominated within the alliance. In no small part, this was due to the actions of King James I of England (James IV of Scotland), who had outlawed privateering in 1603. A loophole in the law in Ireland, whereby a suspected pirate could not be tried in Ireland but had to be transported to England for trial, was an added enticement to those who wished to partake in the business of plunder around the Irish coast. While the south-west formed the primary location for piracy in the North Atlantic at that time, pirates operated all along the west coast, frequenting the harbours of Dingle, Limerick, Galway and Broadhaven in County Mayo, and as far north as County Donegal and Lough Foyle.

This alliance had a formalised hierarchy, with elected commanders and a chain of captains and crews that acted in hunting packs when the need arose, or operated as individuals when

Fig. 18.12 CLARE ISLAND CASTLE IN THE LATE NINETEENTH CENTURY. From the fifteenth to the seventeenth centuries, the O'Malley seafaring family used a line of fortified tower houses/castles located principally along the southern shoreline of Clew Bay and its islands to control an extensive territory in south-west County Mayo. These castles were sited strategically to monitor ships passing through their territorial waters. This allowed the O'Malleys to tax fishing boats working in these rich fishing grounds, but also to engage in acts of piracy against merchant vessels. The square-shaped Clare Island Castle, probably built by the O'Malleys in the fifteenth century, highlights some of the key attributes required for successful piracy operations. Sited on a rocky promontory and overlooking a sheltered harbour, a small fleet of galleys could sail quickly to intercept ships passing through Clew Bay and the seas off the west coast of County Mayo. It was also a site that could be defended against attackers. The castle became a principal residence and key base for the highly successful pirate operations of Grace O'Malley (Granuaile), Ireland's 'Pirate Queen'. In addition to the castle, the impressive ruins of the island's Cistercian abbey – Clare Abbey – is reputedly the site where Granuaile is buried, along with others of the O'Malley dynasty. [Source: National Library of Ireland, call no: L_ROY_06778]

opportunity presented itself. The success of the pirates at sea was bolstered by corrupt officials on land who participated actively in the black market trading of exotic goods, which could include pepper, cinnamon and nutmeg, along with Barbary gold and Spanish silver. Smuggling activity in remote bays and harbours was linked intimately with piracy ventures, and a network of carriers and traders waited on specific quaysides for the arrival of pirate ships and their rich cargoes. These goods were then transported overland to the major towns and cities in Ireland, where they could be sold on to willing buyers.

Many pirates and their families settled locally along Ireland's coastline and became integrated into the local communities. Such was the overt nature of the piracy in west Cork for the specific period in the early seventeenth century, that deep-sea plundering progressed into an integrated maritime and land-based operation. This amphibious industry evolved to combine outright piracy and coastal smuggling with the need for legitimate seasonal work, when the men involved in piracy could also be employed gainfully as tradesmen, such as shoemakers, labourers and carpenters, or in the flourishing fishing industry that existed there at the time.[8]

This alliance went into decline from 1614 when the law changed to allow trials for piracy to be held in Ireland. It was also influenced by a change in the form of predation that emerged from the 1620s onwards; a new type of piracy dominated by the Ottoman Turks and which included human captives among goods traded. The south-west coast of Ireland felt the impact of this change directly when, in 1631, north African pirates, led by a Dutch renegade who had converted to Islam, raided the coastal town of Baltimore. The attack resulted in the capture of over one hundred of its citizens, who were carried off by the pirates and sold in the slave markets of Algiers.

During the Irish Rebellion period in the 1640s, Irish Confederate privateers, acting in unison, wreaked havoc on parliamentary shipping and European merchant ships loyal to the Cromwellian cause. Although viewed essentially as 'Irish pirates' operating in the southern channel, their activities ranged far and wide such that trade to the main ports of the east and south-east coast of Ireland was brought to a virtual standstill. The scale of their maritime successes was tantamount to a war at sea.

While piracy continued in succeeding centuries, smuggling came into its own during the eighteenth century. This was due in part to rising excise duties on goods, such as tobacco, but also to the growth in the import of tea. Coastal areas along Ireland's east coast, but in particular the smaller ports of Rush and Skerries, became dens of smuggling activity. In this, their success was helped by their location along the island's

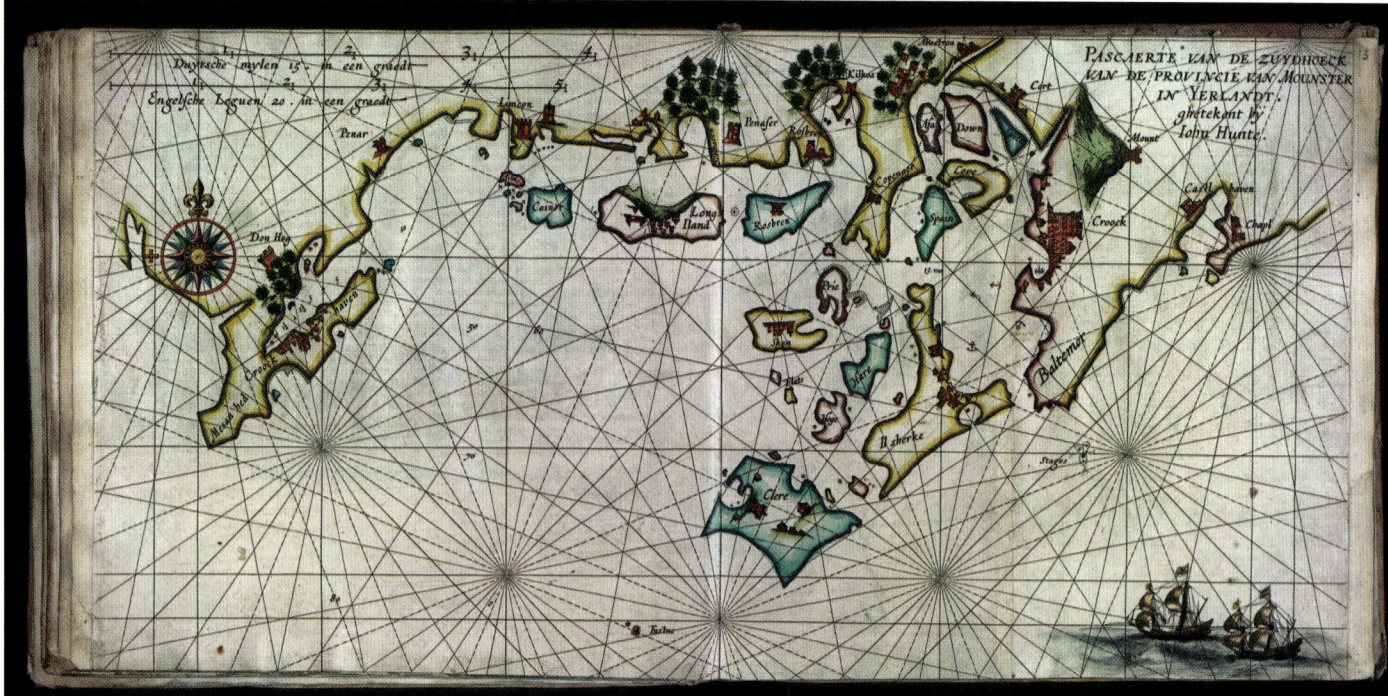

Fig. 18.13 THE GERRITSZOON-HUNT 'ANTI-PIRATE' CHART (1612). The publication in 1612 of the Gerritszoon-Hunt 'anti-pirate' chart shows the pirate havens of the south-west of Ireland, including Baltimore and Crookhaven. It was commissioned by the States-General (the Netherlands' legislature) to inform them of the pirate nests in Munster in advance of their attack in 1614 on the pirates in Crookhaven. The Dutch raid on the harbour inflicted major damage to the pirate base and was part of the reason why piracy of the time went into decline. The map is notable for the inclusion of straight-line sailing directions, sometimes known as windrose lines or portolani, a technique pioneered initially by Mediterranean seafarers in the fourteenth century and subsequently adopted by cartographers and chart-makers more widely. Their purpose was to assist in precise navigation, especially in coastal waters, and reliable identification of recognisable landmarks onshore. [Source: Georg-August-Universität, Göttingen, Germany SUB Göttingen, 4 H BRIT P III, 6 RARA]

Fig. 18.14 STEPS AND LIGHT NICHES AT DUTCHMAN'S COVE, NEAR CASTLEHAVEN IN WEST CORK. This photograph, which illustrates a series of rock-cut steps, a platform and light niches, provides tangible evidence for the former illicit activity of smuggling that occurred in Dutchman's Cove near Castlehaven. Such remote sites were ideal for the business of piracy and smuggling, though access points like this often required some human modifications if the shoreline was too difficult to transfer goods easily from boat to land. Steps and other features were therefore carved out of the sheer rock. The remote nature of the coastline of the west and south-west of Ireland presented many similar opportunities for smugglers and pirates. [Source: Connie Kelleher; by kind permission of Thomas and Jane Somerville, Castletownshend]

most populated coastline, as well as being proximate to Scotland and the Isle of Man. This ensured ready markets and a rapid movement of goods brought ashore by smugglers.[9]

During the eighteenth century, sources began to cite individual names associated with smuggling activities. Along the more remote coastal areas of County Kerry, for example, Maurice 'Hunting Cap' O'Connell, uncle of Liberator Daniel O'Connell, was noted as being among the most famous smugglers. He made his fortune through trading and smuggling goods from his base in Derrynane near Caherdaniel. In Castletownbere, long before the advent of the Wild Atlantic Way, Morty Oge O'Sullivan conducted similar illegal activities and proved to be the wildest character along that coast. Prior to finding notoriety as a pirate from the 1750s onwards, he had a distinguished military career in Spain. The tales of characters like these help to bring alive the clandestine activities of the time, within a landscape that was conducive to covert landings, surreptitious deals and secret rendezvous in the dead of night.

Recent archaeological discoveries, in tandem with historical and cartographic research, are also providing tangible proof for the use of parts of the Irish coastline in the activities of pirates and smugglers. Though sites within this landscape are beginning to

Fig. 18.15 STREEK HEAD SMUGGLER STEPS AND CAVERN, CROOKHAVEN, COUNTY CORK. Steps are visible at the very edge of Streek Head on Gokane Point, Crookhaven. Cut out of the sheer rock and only wide enough for one individual to access, the steps overhang a drop to the sea below, where a cavern runs through the headland and opens to the sea to the east. The cavern would have provided access for ships' boats or other small craft, while the use of a derrick or rope pulley system on the top of the rock face beside the steps would have facilitated the smuggling of goods in the age of sail. The cavern is still used today by kayakers who continue to enjoy and engage with this wild and remote part of Ireland's coast. [Source: Connie Kelleher]

reveal their secrets, applying a specific date to these can be difficult. The very nature of some sites, however, point unquestionably to their having been used by smugglers and pirates. Sites involving rock-cut steps, associated rock platforms and rock faces with carved-out lantern niches, provided coastal access points for the illicit transfer of goods (Figures 18.14 and 18.15). In effect, such sites were crucial to help guide pirates ashore to awaiting smugglers and present evidence of such activities in remote areas of Ireland's varied coastline.

Grace O'Malley

Barry Brunt

Grace O'Malley (c.1530–1603), more commonly referred to in Irish as Granuaile, was the daughter of the lord of the O'Malleys who controlled much of south-west County Mayo. From the fifteenth to seventeenth centuries, they developed a reputation as a powerful

Fig. 18.16 STATUE OF GRACE O'MALLEY AT WESTPORT HOUSE, COUNTY MAYO. This bronze statue by Michael Cooper of Grace O'Malley or, in the Irish language, Granuaile (*c*.1530–1603), stands in the gardens of Westport House, overlooking Clew Bay in County Mayo. The house was built in 1650, using for its foundations one of the castles that had been controlled by Granuaile and used to consolidate her power in this area. Although modified in later centuries, Westport House continues to promote its historic association with Ireland's 'Pirate Queen'. A permanent exhibition dedicated to her life and times is located, for example, in the basement of the house. More recently, the potential of the local folklore surrounding the exploits of Granuaile has been recognised with the opening of a Pirate Adventure Park to further boost tourist numbers visiting Westport House. [Source: Sue Mischyshyn]

seafaring family, as well as becoming ruthless pirates who terrorised coastal shipping entering and leaving the port of Galway. To protect and promote their interests, the family built an impressive line of castles on the shores of Clew Bay and on some of its many islands.

On the death of her father, Granuaile assumed leadership of the family and became an extraordinary seafaring leader. From her castle strongholds, such as Rockfleet Castle and her principal residence at Clare Island Castle on Clare Island (Figure 18.12), her fleet of galleys, other fighting vessels and armed crews were able to tax fishing activities and ships passing through her territorial waters. In addition, she extended her family's tradition of piracy and plunder to most of Ireland's western seaboard. In 1576, in a meeting with Granuaile, the lord deputy of Ireland, Sir Henry Sidney, acknowledged her fame by making the observation: 'There came to me also a most famous feminine sea captain called Granny Imallye … This was a notorious woman in all the coasts of Ireland.'[10]

Granuaile was often in conflict with the English as they sought to extend their influence and control over Gaelic Ireland as part of the expanding British empire. This is highlighted in a letter of the governor of Connacht in 1593 who declared her to be the 'nurse to all rebellions in the province for this forty years'.[11] In 1593, however, she was able to meet with Queen Elizabeth I in London, and petitioned for the release of members of her family who had been captured. In turn, Granuaile agreed to become a subject of the crown. Despite this, she continued with her pirate activities and supported rebellion against England's territorial expansion until her death in 1603. Tradition has it that she is buried in the Cistercian abbey on Clare Island.

Following her death, her exploits at sea and her leadership role against England in this turbulent period of Irish history became recognised increasingly (if embellished somewhat) by Irish nationalists. As a consequence, she has emerged as an Irish folk heroine of the sixteenth century and is celebrated as Ireland's 'Pirate Queen'. A bronze statue of Granuaile can be seen at Westport House, County Mayo (Figure 18.16), as well as an exhibition that celebrates her life and times.

THE SACK OF BALTIMORE

Bernie McCarthy

During the first half of the seventeenth century, Muslim corsairs of the Barbary Coast of north Africa were operating extensively throughout the Mediterranean and in the Atlantic. Often referred to as 'Turks', and comprised principally of European renegades, expelled Moriscos from Spain and the Ottomans, they operated large fleets and made repeated attacks on European coastal communities. Their main trade was in Christian slaves and, at this time, in order to fulfil the demand for slaves in cities like Algiers, Tripoli and Tunis, they began to venture beyond the Straits of Gibraltar towards northern Europe (Figure 18.17).

On 20 June 1631, one of these raids took place on the south-west coast of County Cork in Ireland at the small but successful English plantation town of Baltimore. By the 1620s, this had become a thriving maritime port based on its successful fishing industry, expansion of trade and long-term collusion between government officials, willing locals and Atlantic pirates. Better known as the 'Sack of Baltimore', the raid was a triumph for the north African corsairs, who reputedly captured 107 of the unsuspecting townspeople during the early morning attack.

FIG. 18.17 'BARBARY' GALLEY. A depiction of a 'Barbary' galley, typical of the type of ship used by the Ottomans as pirate vessels that plundered merchant ships, including those off the south-west coast of Ireland. Fully rigged ships were also used by the Ottoman pirates, including the favoured Dutch *fluyt* (a sailing vessel used for carrying cargo), when venturing on long sea journeys. [Source: Bibliothèque Nationale de France]

Fig. 18.18 STRAFFORD FOLIO MAP OF BALTIMORE TOWN AND HARBOUR IN THE EARLY SEVENTEENTH CENTURY. This map is of particular interest as it shows, in remarkable detail, the layout of Baltimore, with its fishing, naval and maritime activities, prior to the 'Barbary' raid in 1631. It depicts a prosperous community, which may have been one of the reasons for the attack. The town of Baltimore was in two parts. There was a group of houses in what was called the Cove area, denoting the original medieval settlement of Baltimore and separated by a narrow track from another group of houses in the main part of the town surrounding Dún na Séad castle, which at the time of the raid was the main plantation settlement. The north African corsairs would have anchored outside the harbour at a similar location to where two vessels are shown on the left of the map, waiting to enter the harbour. (The map forms part of the Strafford Papers series within the Wentworth Woodhouse Muniments at Sheffield Archives. The Wentworth Woodhouse Muniments have been accepted in lieu of Inheritance Tax by HM Government and allocated to Sheffield City Council.) [Source: Sheffield Archives]

One of the most detailed accounts of the pirate raid is contained in a letter written to Charles I in London, three days after the event. It stated that the leader of the attack was a Dutchman, Jan Janszoon, who, like many northern European seamen, had ventured south in search of greater fortune and joined the north African pirates, becoming one of the most successful corsairs of the Mediterranean. He converted to Islam, changed his name to Murat Reis (signifying that the Ottoman leaders accepted him as a worthy captain of their fleets) and prospered from a trade in human cargo. The Baltimore captives were taken to be sold in the slave markets of cities like Algiers.

In 1631, Murat Reis sailed towards Ireland with two ships, where he captured two Dungarvan fishing boats off the coast near Kinsale. He ordered their captain, John Hackett, to pilot him into Kinsale. Hackett, however, took them to Baltimore instead, claiming that it would be an easier port to take as the king's ships were stationed in Kinsale. The north African pirates arrived off Baltimore at dead of night and, hidden by high cliffs at the entrance, anchored outside the harbour's mouth. Under cover of darkness, they made a tour of the harbour noting the coastline and the layout of the town (Figure 18.18).

The raiders returned just before dawn and made a sudden attack into the Cove area, wreaking havoc and confusion. They made a ferocious spectacle, brandishing muskets and scimitars, long knives and canvas-soaked torches to set the town alight, and attacked twenty-six houses in the Cove, from where they claimed some one hundred captives. The houses that formed the main settlement around the castle of Dún na Séad at the other end of the town did not yield, as the noise and commotion of the initial assault on the Cove alerted the inhabitants to the danger. As a result, the alarm was raised in the main town which deterred the pirates from venturing further into the settlement of Baltimore.

The raiders returned to their ships with their captives and set sail swiftly for Algiers. The letter to the English king, cited above, stated that 107 captives were taken and listed the names of each of them. This was confirmed by the English ambassador in Algiers who reported the arrival of 107 captives from Baltimore five weeks later (Figure 18.19). In contrast, however, a French priest, Pierre Dan, who spent his life recording the arrival of captives into the port of Algiers, noted the arrival of 237 captives from Baltimore. One possible, though unproven, explanation for this discrepancy is that the larger number may have included Irish captives who had not been recorded by the English authorities, or may reflect captives taken from other coastal areas prior to the raid on Baltimore. He described the fate of the Baltimore captives as being a pitiful sight, with families separated in the slave markets.

The 'Sack of Baltimore' in 1631 was the only 'Barbary' raid to take place in Ireland and, whatever the actual number of captives taken, it had a devastating impact on the town. The raid has gone down in history as a significant event and, even to this very day, through tradition and local folklore, the north African raid on Baltimore is remembered and commemorated (Figure 18.20).

Fig. 18.19 SLAVES DISEMBARKING IN ALGERIA. This image, from Pierre Dan's *Histoire de Barbarie et de ses Corsaires* (1684), shows Christian slaves arriving into the port of Algiers to be sold in the open slave markets. Dan concludes that up to the year 1680, a million captured Christians were sold into slavery in Algiers. The trade during this period was the mainstay of the Algerian economy. [Source: Bibliothèque Nationale de France]

Fig. 18.20 MODERN AERIAL VIEW OF BALTIMORE CASTLE AND THE QUAYSIDE LAYOUT. Much has changed since the Strafford folio map of the early seventeenth century, due to extensive reclamation to facilitate the construction of a pier and quayside works. The castle of Dún na Séad, which was in a ruinous state since Cromwellian times, was restored between 1997 and 2005 by Patrick and Bernie McCarthy. [Source: Sarah Kandrot]

Due to their significance in terms of income generation, tower houses were located strategically at points along the coast where it was easier to control the movement of goods and traffic, as well as offering defensible positions. They were also situated in close proximity to fishing grounds, where levies could be collected from visiting fishing fleets. In 1572, for example, it was reported that over 600 Spanish vessels were fishing off the Cork and Kerry coasts, while in 1605 an exiled Domhnall O'Sullivan Beare wrote to Philip II of Spain 'that at least five hundred fishing boats come to his ports'.[12] Medium- to large-sized vessels could anchor in deeper waters nearby and be tendered by smaller boats from shore, with quays and jetties facilitating the movement of traffic and goods.

Fieldwork has revealed that the builders of many tower houses had a preference for sites possessing commanding views over sheltered anchorages and adjacent landing places. The O'Sullivan Beare on the Beara Peninsula in west Cork, for example, constructed tower houses at Dunboy, Ardea and Reenavanny (the latter on Whiddy Island) with this in mind. Elsewhere in west Cork, the O'Driscolls were active in controlling and exploiting coastal resources. The remains of a stone quay/slipway, for example, are associated with an O'Driscoll tower house at Inane on Ringarogy Island, to the north of Baltimore harbour, while another stone jetty on the island is possibly associated with the O'Driscoll fortification at Dunagall. Similar sites can be found on the west coast: the remains of a quay are found in association with an O'Flaherty tower house at Ard in south Connemara, while other tower houses were built by the same family overlooking landing places at Bunowen and Lettermullan. The O'Malleys built castles adjacent to harbours at Kildavnet on Achill Island, Clare Island and along the southern shoreline of Clew Bay in County Mayo (Figure 18.12), while the O'Donnells did the same at Rathmullan, Rathmelton and Doe in Lough Swilly in County Donegal.[13]

From the middle of the sixteenth century, as these coastal lords came under pressure from the English government, new fortifications appeared in Ireland whose origins lay not in medieval lordship but in the emergence of full-time professional armies with their cohort of trained officers and military engineers. These included star-shaped forts, which first emerged in Italy or France in the fifteenth century, and spread northwards into Germany, the Low Countries and England, before reaching Ireland. Examples of such star-shaped and bastioned forts occur along Ireland's coastline, the Cromwellian fort at Inishbofin Island, off the Connemara coast, being an impressive, well-preserved example. One of the finest examples of a star-shaped fort from the early modern period, however, is Charles Fort, constructed around 1680 and overlooking the entrance to Kinsale Harbour (Figure 18.21). It is situated directly across the harbour from an earlier star-shaped fort, James Fort, which had been built between 1602 and 1611.

The appearance of such fortifications marks the decline of the castle-building tradition, and indeed the declining fortunes of medieval lordship, as it contended with the expansion of crown authority across the country. No longer was there a need for

Fig. 18.21 CHARLES FORT. A. LOOKING TOWARDS THE TOWN OF KINSALE AND B. SEEN FROM ABOVE TO SHOW THE CLASSIC STAR-SHAPED PLAN. Ireland's exposed west and south-west coastline presented 'backdoor' opportunities for hostile powers, primarily France and Spain, to invade Britain. The landing of some 3,500 Spanish soldiers in Kinsale Harbour in 1601 to support the O'Neill rebellion against English rule highlighted this concern. Although defeated by the English forces at the Battle of Kinsale, the invasion emphasised the need for more effective coastal fortifications to be built to protect key harbour sites. The more traditional castle structures, including tower houses, however, were no longer adequate to confront new techniques in siege warfare, which involved major advances in artillery power. As a result, new designs in fortifications arrived in Ireland in the late sixteenth century from the European mainland. In particular, Sébastien de Vauban, Chief Engineer of Louis XIV, perfected the construction of massive defensive fortresses which took the form of a star-shaped design. One of the best examples of a star fort along the Irish coast is Charles Fort, constructed in the 1680s. Located on the western side of Kinsale Harbour and complementing an earlier, but much less impressive, star fort (James Fort) built in the first decade of the sixteenth century on the opposite side of the Bandon River estuary, it dominates the entrance to the harbour. Its imposing structure, many cannon emplacements and massive bastions provided defenders with an extensive field of fire to prevent hostile naval forces entering the harbour. It could also protect the large overseas trade that passed through Kinsale, which had become one of the most significant ports on Ireland's southern coast. Inland, however, the fort was overlooked by high ground which exposed it to artillery bombardment. This resulted in the defenders, who supported James II, having to surrender to Williamite forces following a siege in 1690. Unlike castles, which also functioned as centres of lordship, combining official, residential and defensive functions in one complex, star forts were designed solely for military defence and were occupied primarily by professional soldiers, as witnessed by the barrack buildings seen clearly at Charles Fort. The fort was taken over by the state in 1973 and declared a national monument. Extensive repair and conservation works have been undertaken to restore the fort to emphasise its seventeenth-century layout. It is now a significant tourist attraction. [Source: Shane O'Connor]

fortifications that combined the residential, judicial, economic and military roles of a lord. As a result, many of Ireland's traditional castles and tower houses went into gradual decline through the course of the early modern era. Britain's need to ensure a strong defensive capability in Ireland, however, remained of paramount importance. This was especially the case around Ireland's extensive coastline and at key harbour facilities, which could be exposed to invasion from forces hostile to Britain. Such defensive roles were allocated increasingly to these impressive star-shaped forts.

ORDINARY HOUSING OF THE NATIVE TRADITION

Cartographic and documentary sources from the period sometimes depict clusters of cabins or houses in the vicinity of tower houses in Gaelic-Irish and Anglo-Irish areas. Our knowledge of domestic housing is somewhat limited, however, compared with other monument types of the period, as are the daily activities of their inhabitants. Perishable materials, such as the timber, post-and-wattle, sods and mud that were used in the construction of these houses have conspired against the survival of upstanding structures. They also create ephemeral subsurface remains that are difficult to recognise during archaeological excavation.

Two traditions of domestic housing occur in Gaelic-Irish areas and have been identified as co-existing during this period. The first involved a relatively permanent style of building and the best pictorial evidence of these structures comes from Richard Bartlett's maps of military campaigns in Ulster c.1600 (Figure 18.22) and Thomas Raven's maps of the Londonderry plantation settlements in 1622 . These houses were typically single-storied and thatched, were mainly rectangular or sub-rectangular in plan and had a more or less centrallyplaced doorway. The walls were rather thick and were built of clay sods or post-and-wattle, covered in clay.

The second tradition focused on less permanent buildings, and

Fig. 18.22 DETAIL OF RICHARD BARTLETT'S MAP OF THE DISTRICT BETWEEN DUNDALK AND NEWRY (c.1602). This detail from Richard Bartlett's map shows a stretch of coast from Dundalk Bay to Carlingford Lough. The coastal settlements of Carlingford and Dundalk were depicted as walled towns, while elsewhere villages were centred around tower houses. The houses in both urban and rural settings appear typically to have been single-storeyed thatched cabins, mainly rectangular or sub-rectangular in plan and with a more or less centrally placed doorway. Also notable are the numbers of roofless churches – such as the one at Faughart, north of Dundalk – their ruined condition brought about by the impact of the various uprisings in Ulster, most notably the Nine Years' War. The Reformation also contributed to the lack of upkeep in many churches in this period. [Source: National Library of Ireland, call no.: MS 2656(1)]

Fig. 18.23 THE CASTLE AND WALLED TOWN OF CARRICKFERGUS (c.1560). A series of early maps dating from the Tudor era depict the port town of Carrickfergus, County Antrim. This was one of the few English strongholds in Ulster at the time, hence its strategic importance. This detailed map dating to c.1560 illustrates the various castellated tower houses that lined the principal streets. These would have been the residences of the more wealthy merchants and key administrators. Their concern with the defensive rather than residential attributes of these structures emphasises the frontier character of the town. The great tower or keep that dominates the castle was built by the Normans and, together with its defensive walls, overlooks the small port. The portrayal of boats highlights the importance of trade for the only sizeable town in Ulster in the mid-sixteenth century. Surprisingly, buildings that would have been more typical of rural settlements in Gaelic Irish lordships are also depicted. These are the oval or circular huts built of post-and-wattle, known as 'creats', and are located more to the periphery of the town but also occur along the water's edge. [Source: © British Library Board. All Rights Reserved/Bridgeman Images]

is widely attested to by mainly oval or circular structures, although some were rectangular in shape, and were built of post-and-wattle. As such, the walls were not load-bearing. Commonly known as 'creats', they were thatched, chimney-less and associated typically with the practice of transhumance. Cartographic evidence, however, suggests that such structures could also be found in urban environments like Carrickfergus, County Antrim (Figure 18.23).

Fortunately, a small number of low-status houses of late medieval/early modern date have been identified on the coastline. On the Beara Peninsula in west Cork, for example, a programme of fieldwork and excavation has revealed houses and hut sites, both nucleated and dispersed, at Blackrock, Canalough, Caheravart and Ballynacallagh.[14] Similarly, in the uplands of north Antrim, a small number of hut sites have been identified and excavated at Glenmakeeran, Gortin, Ballyutoag, Craigs and Goodland. In both regions, the houses were all generally of the same size, form and layout, and were defined by low foundation walls constructed of stone and clay. These supported the main walls of the huts, which were of a more ephemeral nature.[15]

Surviving field evidence suggests that the ordinary homes of rural populations living in coastal areas were modest in size and sub-rectangular in shape. Daily activities would have focused around a central hearth from which smoke escaped through a hole in the roof. People would have slept around this hearth and cooked their meals on it with hand-built ceramic pots. Additional structures appended onto such houses could contain farm animals or fishing equipment. Such houses tended to occur in small clusters from two or three to no more than a dozen in number, suggesting that extended family groups resided in such settlements. Little is understood of the lifeways of such ordinary families as the material evidence recovered from excavations of such houses is rather limited. Given their location on the coastline, however, the exploitation of marine resources was likely to have been a feature of their daily lives.

Case Study: Plantations

Annaleigh Margey

From the mid-sixteenth century, England's interaction with Ireland transformed with the development of plantation schemes across the Irish Midlands, Munster and Ulster. These schemes emerged in the aftermath of centuries of increasing English, Welsh and Scottish influence in Ireland. From Norman times, Anglo-Norman settlers and their descendants had taken up residence in areas along the eastern seaboard and south of Ireland. Moreover, Scottish settlers, such as the McDonnell family, had gained footholds along the north-east coast. This influence heightened with the emergence of an English-dominated region known as the Pale, roughly in the area of modern-day counties Dublin, Kildare, Meath and Louth (Figure 18.2). The Pale became the geographic centre of English administration in Ireland, with Dublin Castle serving as its headquarters. Outside the Pale and areas of Anglo-Norman settlement coinciding usually with coastal strongholds, Gaelic-Irish politics and settlement persisted. Here, Gaelic chieftains continued to hold sway over their own lordships and peoples. The mid-sixteenth century, however, heralded the beginning of a seachange in this political landscape.

So, why did this juncture become so significant? Historians have long argued that English attention to Ireland increased during Henry VIII's reign.[16] This was due to a number of factors, some of them external, such as the move towards overseas expansion in the Americas by many of England's European rivals, including Spain and Portugal. Factors internal to Ireland included Henry's increasing drive towards religious reform and his growing anxiety about the security of his existing possessions in Ireland, this conditioned by fear of a potential invasion of Ireland by his European enemies, as well as continuous, destructive raids by the Gaelic Irish into the Pale.

Located on the boundary of the Pale, the O'More and O'Connor Gaelic lordships were amongst the most problematic for this security. Since the 1520s, frequent rebellions had raised concerns, with attacks on the Pale quickly becoming part of the modus operandi of the two families.[17] By 1548, the English authorities sought to curb these incursions by undertaking a campaign in the region, which culminated in the placing of two forts, Fort Governor and Protector, within the lordships. Both became proto-urban sites, with spillover settlement emerging to house soldiers and their families. These soldiers became, in effect, the first of a new wave of English settlers in Ireland and, by November 1550, authorities were conducting surveys of the wider lands of both families to allow even further settlement of supportive tenants. In 1557, Mary I incorporated both forts, Protector and Governor, as the towns of Maryborough (present-day Portlaoise) and Phillipstown (Daingean), and created Queen's and King's counties (counties Offaly and Laois respectively). Throughout her reign, and that of Elizabeth I, renewed attempts were made to bring new settlers to the region.

In the 1580s, the establishment of plantations had emerged as a core English strategy in Munster. As a result, in the aftermath of the defeat of the Desmond Rebellion in 1583, plans emerged for the escheated lands of the earl and his followers. Commissioners conducted surveys in 1584 that identified lands across counties Limerick, Kerry, Cork and Waterford for settlement. Following these surveys, the

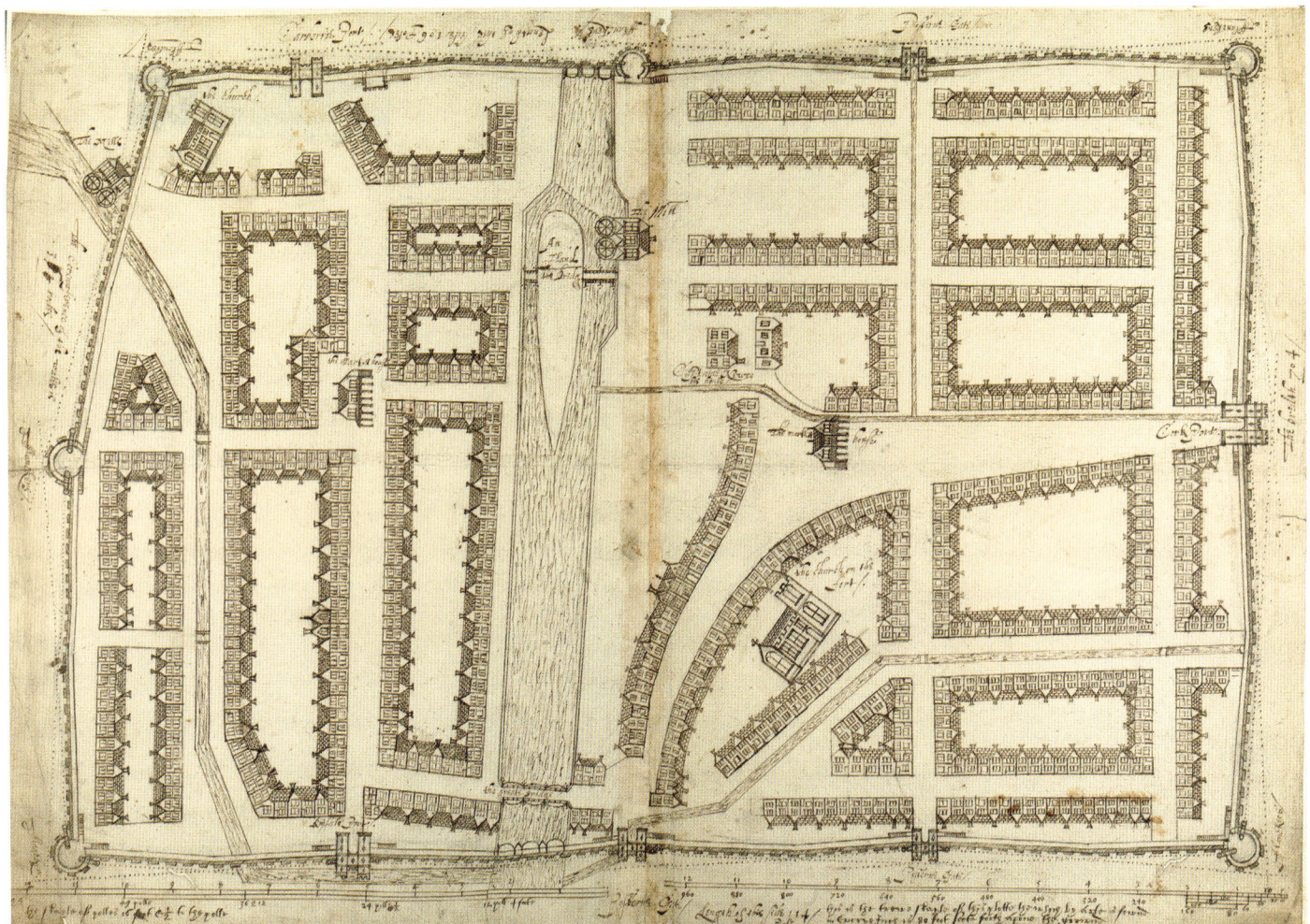

Fig. 18.24 'THE TOWNE OF BANDON BRIDGE AS IT IS NOW BUILT' (*c.*1620). The map depicts the town with its fortified town walls, a necessary feature for such a plantation settlement in what was considered to be a frontier area in west Cork, together with regulated streets and two market houses. Bandonbridge developed as part of the Munster Plantation under the direction of the two initial landowners in the area – Captain William Newce and John Shipward – who set the foundations of their own towns on either bank of the Bandon River. The town, however, entered its most significant phase of development under Sir Richard Boyle, who acquired both leases by 1618. As earl of Cork, Boyle's wealth and power allowed him to promote the settlement and establish it as the 'flagship' plantation town in Munster. Its site on the Bandon River, which is navigable for sea-going vessels upstream to Inishshannon, and from there by smaller river craft, provided the town's merchant classes with direct access to maritime trade through the thriving port of Kinsale, located at the mouth of the River Bandon. The town, now called Bandon, emerged quickly as an important regional market town for its rich agricultural hinterland, possessing a variety of functions, including a shambles area, significant church buildings and a garrison to consolidate English control in the area. While Bandon continues to act as a prosperous market town in Munster and a commuter centre for people working in Cork city, the role of the river has changed. It is no longer a trade artery, but is rather viewed as a recreational channel adding to the aesthetic value of the town. It is, however, now prone to frequent flooding of the town, a significant contemporary issue that current engineering and river management works are attempting to alleviate. [Source: The Board of Trinity College Dublin]

plantation conditions were laid down. These defined plantation estates or seignories of 12,000, 8,000, 6,000 and 4,000 acres, which were to be granted to 'undertakers', who would undertake to settle their estates with ninety-one Englishmen and embed English practices in agriculture, industry and settlement across the Munster lands.[18] These undertakers came from mainly coastal counties in England and Wales, and were drawn to Ireland by opportunities for improvements in their personal positions and due to the relative ease of access to home via the Irish Sea.[19]

Amongst the most entrepreneurial undertakers were those who built settlements and industries based around the natural landscape of Munster, but particularly along the province's major rivers and coastal lowlands. Sir Walter Raleigh, for example, who had been granted an estate in the Blackwater Valley, worked this plantation with his agent and tenant, Sir Henry Pyne, and developed a flourishing timber industry around Mogeely in County Cork (Figure 18.3).[20] The surrounding river network enabled goods to be transported directly to the thriving port of Youghal and, therefore, allowed sea access to and from the estate. Raleigh maintained his lands in Munster until the 1598 Rising.

In the renewed settlement after 1601, Raleigh´s lands were bought out by Richard Boyle, the future earl of Cork, who acquired seven seignories and the fledgling town of Bandonbridge in west County Cork (Figure 18.24). Bandonbridge straddled both sides of the Bandon River and emerged as the flagship plantation town in Munster, eventually attaining incorporation. Its strategic location on the Bandon River signified plans to utilise the river for trade and the transport of goods to and from the nearby port of Kinsale (see Chapter 19: Changing Coastal Landscapes). This is given added credence by the plans shown for two market houses, one on each bank of the river, a reflection of the relatively rich agricultural hinterland of the town and its prospects for profitable trade. While an unusual occurrence, it also reflected the town's initial growth under two separate landowners in the early years of the plantation.[21]

By the seventeenth century, English attention had been somewhat diverted to Ulster, which had been the least-known province and the proverbial thorn in their side. Up until this point, their interaction had been mainly with its eastern seaboard where English, and some Scottish, settlements had emerged from the twelfth century. The rest of the province, however, had remained outside their control and was, in effect, the last Gaelic outpost.

Concerned by the rebellion of Shane O'Neill in the 1560s, and the Nine Years' War of the 1590s and early 1600s, the English authorities attempted to extend order, control and 'civility' to the rest of the province. Following the Battle of Kinsale in 1601, the Lord Deputy Charles Blount, Lord Mountjoy, negotiated with Hugh O'Neill, giving rise to the Treaty of Mellifont. This offered O'Neill terms that allowed him to maintain both his life and his lands; however, in its aftermath, his position and that of the other Ulster lords declined continually under increasing influence from across the Irish Sea. By 1607, feeling their position to be untenable, O'Neill, Rory O'Donnell, earl of Tyrconnell, and many other Ulster lords and their families set sail from Rathmullan in County Donegal, in an event known as the Flight of the Earls. Contemporaries and historians often reflect on this event as the end of Gaelic order in Ireland as, following the flight, they were deemed to have committed treason and their lands confiscated.[22] In 1608, following Sir Cahir O'Doherty's rebellion, many of the smaller Ulster lords also had their lands confiscated. As a result, King James VI and I effectively became landowner of six Ulster counties: Armagh, Tyrone, Coleraine, Cavan, Fermanagh and Donegal.

Even before the departure of the earls, English administrators had discussed plans for a possible plantation in Ulster. This followed early anglicisation attempts, such as the shiring of the province from the mid-1580s. However, the idea of formal plantation only came to fruition in 1610 when, following two surveys – a written survey and Sir Josias Bodley's map survey – the authorities set out the plantation conditions. These determined that the plantation was to develop on the footprint of 'precinct' and 'proportion': precincts would equate roughly to a barony and proportions to a townland. Furthermore, the conditions established that plantation estates of 2,000, 1,500 and 1,000 acres would be granted to three categories of settler: undertakers, servitors and 'deserving' natives. They also tasked those in receipt of land grants with the building of variant combinations of stone houses and bawns, and required that twenty-four adult male settlers should be installed per 1,000 acres.[23]

Occurring in the aftermath of the Union of the English and Scottish crowns in 1603, settlers in the Ulster Plantation very much reflected the new British identity of the scheme. The undertakers comprised both English and Scottish settlers, the final list of names highlighting their relatively high status (for example, two of King James VI and I's own Scottish cousins received lands). Furthermore, the servitors also included men who had served in a military or administrative capacity in Ireland and again comprised many individuals of note, such as Sir Arthur Chichester.

The Ulster scheme differed significantly from the Munster scheme in one particular: the engagement of the City of London in the plans. In the seventeenth century, the City of London was often seen as a potential source of funding for settlement schemes around the Atlantic world. Regular petitions for funds to support the Virginia Company, as well as a potential settlement in Ireland, were directed towards the City's Common Council and the Twelve Great Livery Companies. In particular, the City of London was offered lands in north Tyrone and County Coleraine, to be developed under their auspices and using their resources. Within the scheme, the City established The Honourable The Irish Society, which was tasked with developing their assigned lands. These became the new plantation county of Londonderry, with two settlements – Londonderry and Coleraine – established to become part of a corporate town network across Ulster. Moreover, the Irish Society in turn divided County Londonderry into twelve lots, granting a lot to each of the Twelve Great Livery Companies to develop new plantation estates.[24]

The cities of Londonderry and Coleraine became the focal point of much of the early activity of the settlement. Londonderry developed on the bank of Lough Foyle (Figure 18.25). With its direct access to the sea, the city's location was to take advantage of a natural port and enabled a flourishing trade network to open between the cities and London. A network of smaller villages developed across the county lands. Many of these villages, such as New Buildings on the Goldsmiths' Proportion, developed on or near the river network, which in turn also allowed building materials to be easily transported to the new towns, as well as enabling them to participate in the early industries and markets of Ulster.[25]

By the mid-seventeenth century, plantation had become a significant tool of British expansion in Ireland. While the Leix-Offaly, Munster and Ulster plantations had formalised the structure across the country, the advent of smaller plantations across Wexford, the Irish Midlands, Longford and Leitrim consolidated it more widely. By the 1640s, only Connacht and other parts of the western seaboard remained outside formal plantation schemes. With the outbreak of the 1641 Rebellion, plantation was halted in its tracks, but by then the central purposes of plantation – anglicisation and control – had been well embedded across the country. The 1641 Depositions (the witness testimonies of mainly Protestant settlers to the events of the rising) for example, portray an extensive geographical pattern of British settlement. These settlements, in turn, adopted British practices in areas such as architecture, agriculture and commerce, creating more mixed economies and maximising production and trade.[26] The footprints of this plantation remained deeply connected to the motherland, with the commercial worlds of the coastal plantations conducting trade across the Irish Sea and further afield. These bonds continued to strengthen post-Cromwellian settlement, as a new wave of settlers set about re-developing and gentrifying the Irish landscape, connecting the commercial worlds of their Irish estates with those at home in Britain.

415

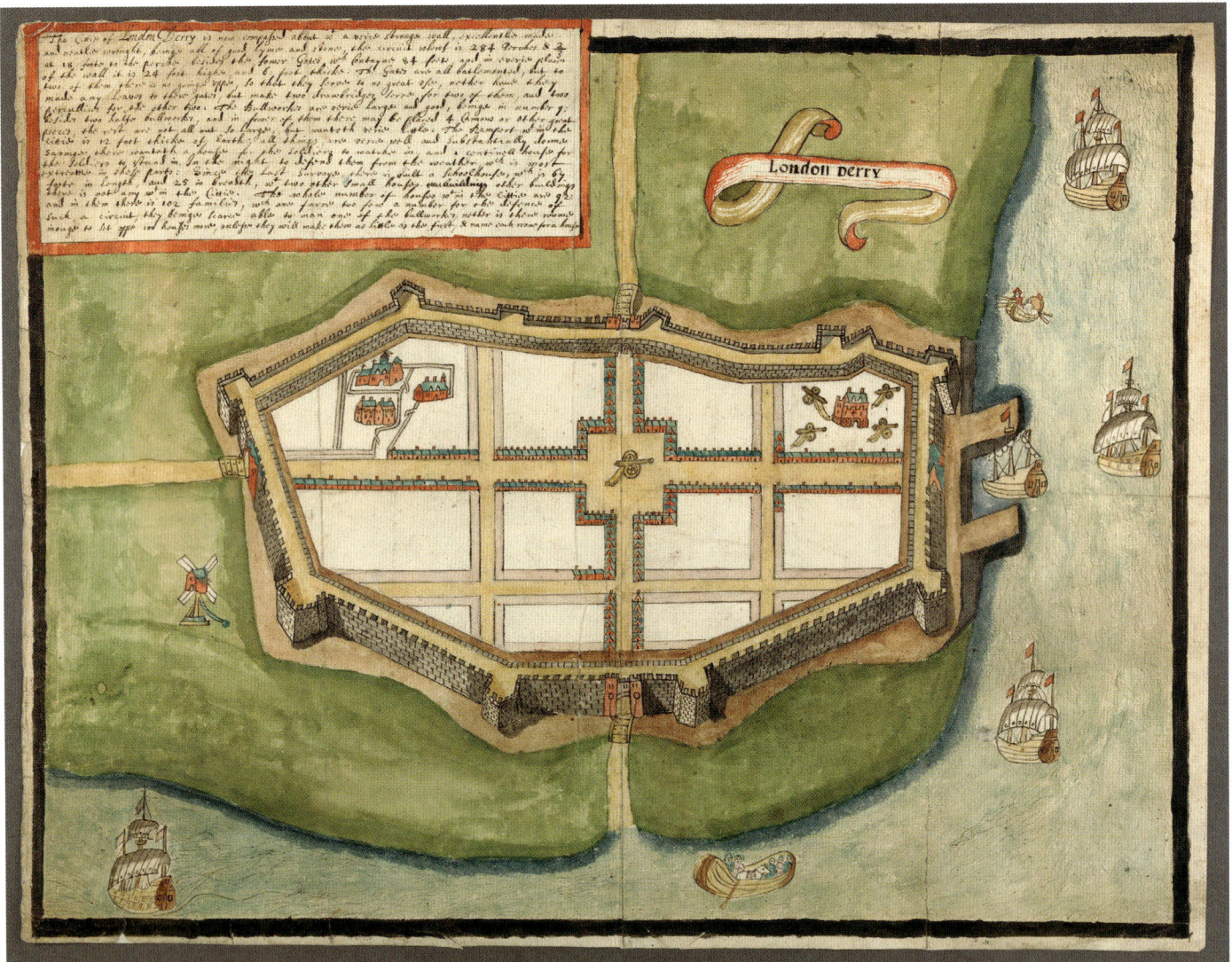

Fig. 18.25 'London Derrie', Nicholas Pynnar (1618). Londonderry became the flagship plantation town in Ulster. Developed on the site of Sir Henry Docwra's original fort, on the western bank of the River Foyle, it developed as a plantation citadel. With a circuit of fortified walls, complete with bastions and demi-bastions, the town espoused the plantation ideal of order, civility and security. Captain Nicholas Pynnar's map is the first to show the completed walls of the new town of Londonderry. The site comprised roughly 1.3ha, which could be considered modest in comparison to its medieval forerunners. The planners envisaged Londonderry and Coleraine as the two corporate towns within the Londonderry plantation. As the foremost plantation town in the north-west of Ireland, Londonderry was expected to become a settlement, commercial and defensive hub for the entire hinterland. As a result, the city comprised a well-ordered cruciform street structure, with many streets lined with houses, and others left empty in preparation for further expansion. Pynnar's map highlights the centre square, or Diamond, which became a core feature of many planned plantation towns across Ireland, and was intended to be a marketplace. The town also incorporated some pre-existing features, such as the castle, surrounded by cannon, which was used as a munitions store. Moreover, Pynnar's survey suggests the presence of a schoolhouse, close to a church, ensuring that the social norms of a plantation town emerged. A well-developed quay signalled the importance of the sea connections of the region, with ships having ease of access to trade between Londonderry, London and ports around Ireland. Moreover, the town's location on the banks of the River Foyle allowed it to have easy access to and within its hinterland and further afield. Londonderry proved a successful venture, eventually bursting the boundaries of its walls and expanding westwards across the boglands of the Donegal hinterland and eastwards to the Waterside. The plantation walls still remain in situ 400 years later, surviving the sieges of the seventeenth century and, more recently, the Troubles. They are now a tourist feature of the city, having been transformed into an accessible walkway. [Source: The Board of Trinity College Dublin]

THE RISE OF THE COUNTRY HOUSE

The early modern period witnessed significant cultural, economic, social and political change that transformed society in Ireland, including aspects of coastal settlement. New landlords and tenants arrived from England, Wales and Scotland, bringing with them novel ideas on agriculture, fashion and housing. This included the adoption of the country house, albeit in its earlier forms. These houses would have incorporated several defensive features, such as crenellated parapets, machicolation, flankers, gun loops and bawns that would not have looked out of place in tower houses. Examples

of these new country houses or manor houses are depicted on a number of pictorial maps produced by Thomas Raven in 1622 to record the progress made by the London Companies in establishing new settlements in County Londonderry following the plantation of that county as part of the Ulster Plantation.

This was the most ambitious of all the plantation schemes in terms of the amount of land redistributed, and the number of settlers established on the new estates. Notwithstanding their common association with the plantations, the construction of new manor houses crossed ethnic and religious lines, with many examples being built by Gaelic-Irish and Old-English families (as the Anglo-Irish came to be called at this time), as well as their New-English and Scottish counterparts. It has been suggested that these fortified examples bridged the division between castle and house, their form and layout being a compromise between the expectations of genteel living and the unstable political conditions on the ground.[27]

The plans and elevations of these manor houses showed great variation in design, with houses involving L-, T- and U-shaped plans constructed across the country. There were also houses of an oblong plan, which could be further extended with tower projections, themselves rectangular, circular or spear-shaped in plan. These buildings were provided with far more windows and fireplaces than was the case with earlier tower houses, allowing for the creation within of well-lit and well-heated spaces.

A number of impressive manor houses were built on Ireland's coastline, and include Mountlong in County Cork (built by John Long in 1631), Dunluce Castle in County Antrim (built by Randal MacDonnell, first earl of Antrim, by 1611) and Ballygally Castle in the same county (erected by James Shaw in 1625) (Figures 18.26, 18.27 and 18.28).[28] It has also been noted that, in contrast with the later medieval tower houses built in the lordship of the O'Sullivan Beare, a fortified house at Reenadisert in County Cork (erected by the same Gaelic family in the early seventeenth century) occupied a site that did not prioritise a dominant and strategic location with a commanding view over an extensive area of the coastline. Instead, the later structure was sited in a more inconspicuous location on the northern shores of an inlet, a placement possibly indicative of the reduced authority of that Gaelic-Irish family.[29]

The surrounding grounds of these evolving fortified houses were often occupied by landscaped gardens. Such a pleasure garden was created, for example, beside Dunluce Castle, where three broad terraces were laid out (Figures 18.27 and 18.28). Hedges, flowers and trees, all arranged in symmetrical plots (called parterres), were planted beside a wing of the castle that contained lodgings, so that the earl's guests could enjoy a view of the ornate garden. With its unusual and delicate plantings, placed against the backdrop of the wild Atlantic, they presented perhaps a suitable metaphor for civility (that is, British order) being wrestled from the power of unbridled nature (Gaelic-Irish order). In early modern elite society, gardens provided more than just pleasing views; they contained complex

Fig. 18.26 RUINS OF MOUNTLONG CASTLE, LOCATED NEAR KINSALE, COUNTY CORK. Mountlong Castle was built by John Long in 1631 on a site overlooking the Belgooly River, close to the estuary of Oysterhaven near Kinsale. It is a fine representation of a fortified house and a new building style that emerged in the early seventeenth century, and one that contrasts with the more defensive structures of the tower house which dominated in Ireland from the early fifteenth century. Builders of these new fortified houses often superimposed new architectural styles, imported by English settlers, onto the basic design plan of the more traditional tower house. In this, they would shift emphasis from more overt defensive structures involving, for example, slit windows and crenellated battlements, to designs that focused on larger, well-lit rooms and residential comfort. Some defensive features (such as gun loops) were retained, however, due to concerns to provide some protection against attack. In this example, four large towers project from each corner of the main building, which allowed for a more extensive layout of rooms. The more numerous and larger windows also provided better illumination while the inclusion of fireplaces meant the rooms could be heated for greater comfort. The emergence of these newer forms of fortified houses in Munster can be linked directly to the establishment of the Munster Plantation. [Source: Colin Rynne]

Fig. 18.27 PHILIP ARMSTRONG'S ARTISTIC IMPRESSION OF DUNLUCE CASTLE AND ITS SURROUNDS IN THE SEVENTEENTH CENTURY. In the seventeenth century, many fortified castles that had been built in an earlier period were redesigned, extended and also incorporated significant changes to land-use in their immediate vicinity to allow them to become more attractive residences for wealthy and powerful landowners. This is illustrated in the impression of the 'modernisation' of Dunluce Castle by the MacDonnell family in the early seventeenth century. Here, a new manor house was accompanied by a fashionable walled, terraced garden overlooking the waters of the Atlantic. At the same time as the remodelling of the later medieval castle, a new village was established by Sir Randal MacDonnell who became the first earl of Antrim. The focal point of the settlement was a marketplace located close to the entrance of the castle, reminding villagers and visitors alike that the village's existence was dependent upon MacDonnell patronage. The marketplace was called the Diamond, a term typically associated with plantation-period villages located elsewhere across Ulster. While this part of the province was not subject to plantation, it is unsurprising that large numbers of planters settled in the area following the end of the Nine Years' War in 1603, given its proximity to the south-west coastline of Scotland. [Source: © Crown DfC Historic Environment Division]

Fig. 18.28 AERIAL VIEW OF DUNLUCE CASTLE OVERLOOKING THE NORTH ANTRIM COASTLINE. While built initially around 1500 by the MacQuillan family, Dunluce Castle had passed into the hands of the MacDonnells by the mid-sixteenth century. The castle was of strategic importance to the latter family, who had interests in north-east Ulster and the Western Isles of Scotland. The original site of the castle gardens, dating from the early seventeenth century and consisting of three broad terraces and other earthworks, can be seen in the foreground. To the right of the gardens are the earthwork remains of part of the village or small town, founded by Sir Randal MacDonnell around 1608. This was an unusual location for a coastal settlement, given its exposed location close to north-facing cliffs and with a lack of direct access to the sea. [Source: © Crown DfC Historic Environment Division]

metaphors that expressed class, religious and cultural identities. In the case of Dunluce Castle, here was a garden placed in full view of the earl's esteemed guests, reinforcing the standing of the MacDonnell family in the new political environment of seventeenth-century Ireland.

Following the Restoration in 1660, defensive features disappeared from the façades of newly constructed and elite residences in Ireland. This heralded the appearance of the first 'true' country houses, although bawns would still have surrounded many of these buildings. As these places were the principal residential and functional centres of country estates, there was a need for additional buildings, such as dairies, wash-houses, brewhouses, malthouses as well as stables, storehouses and living quarters for servants. The household was generally busy, with many servants responsible for the upkeep and maintenance of the main residence, along with the comings and goings of various craftsmen and labourers who were employed in the surrounding estate. Like the ordinary dwellings of the planter tradition (discussed below) little of these additional buildings have survived. This has been due to either the perishable nature of their construction and/or the occupation and reuse of the bawns in subsequent centuries, which erased all traces of early modern usage.

ORDINARY HOUSING OF THE PLANTER TRADITION

Besides the fortified houses of landlords, the arrival of new settlers from varied social backgrounds also saw the introduction of more modest buildings. These new building types are also highlighted in Thomas Raven's detailed pictorial maps of the plantation of County Londonderry in 1622. They include, for example, the village of Ballykelly, established by the Fishmongers Company. A number of stone houses are depicted in such villages, which (apart from the manor houses) can be characterised generally as being of one storey in height or one-and-a-half storeys if including dormer windows. Their roofs were usually covered in slate or tile, although several examples were thatched. Few houses were equipped with gable hearths, with most having a chimney provided in a central position. Entry was also usually in a central position. The houses were gabled, with gable windows at a raised level indicating the presence of lofts. However, typifying the challenges posed to archaeologists working on this period, no upstanding remains of such houses recorded by Raven can be identified.

The English surveyor also mapped Killybegs Harbour in County Donegal, even though it lay outside the Londonderry plantation (Figure 18.29). In recent years, archaeological excavation has uncovered a number of houses mapped by Raven at Rough Point, close to Killybegs, where a cluster of five stone houses had been built near the shoreline. The interiors of these houses would have been heated by fireplaces placed at their gable ends, while evidence for stone paving and cobbling survived in places. Where the doorways survived, they were

placed centrally along the side walls. Four of the five houses shared the same orientation, suggesting a concern with exposure to prevailing winds.

The same series of maps also illustrate rather attractive depictions of timber-framed houses, which incorporate a central chimney stack, two dormer windows in the front elevation, and almost all roofs being slated or tiled (Figure 18.29). The timber box frames possessed both diagonal braces and horizontal ties, with no evidence of cruck construction. These timber-framed buildings with their central chimneys and exposed timbers were typical of vernacular building traditions at this time in both the north-west and south-east of England. As is the case with timber-framed houses in urban areas, such buildings have not survived, either as relic buildings or as part of the vernacular building tradition. Most would have been destroyed through the troubles of the seventeenth century. In addition, the disappearance of native woodlands, along with the shortage of suitably qualified English carpenters and craftsmen, contributed in no small part to the demise of timber framing.

Fig. 18.29 TRACE COPY OF THOMAS RAVEN'S MAP OF KILLYBEGS. Thomas Raven's map of Killybegs Harbour, County Donegal, in 1622 illustrates the nature of settlement that had grown up in the vicinity of the harbour following the establishment of the Ulster Plantation a decade earlier. Houses of different sizes dot the coastline, including sub-rectangular thatched cabins and stone-built gabled houses, testimony to the influence at the time of both native and planter architectural traditions in the built environment. The depiction of several fishing and trading vessels suggests the busy nature of this sheltered harbour and the prosperity of the town. In recent years, archaeological excavation has uncovered a number of the houses mapped by Raven at Rough Point (encircled), close to Killybegs, where a cluster of five stone houses had been built near the shoreline. The interiors of these houses would have been heated by fireplaces placed at their gable ends, while evidence for stone paving and cobbling survived in places. Where the doorways survived, they were placed centrally along the side walls. Four of the five houses shared the same orientation, suggesting a concern with exposure to prevailing winds. The original Thomas Raven map of Killybegs is held with Lambeth Library. [Source: Public Records Office of Northern Ireland and Lambeth Library]

CHURCHES AND MONASTERIES

Churches and monasteries were another feature of coastal settlement in both urban and rural areas in early modern Ireland. Most of the parish churches (particularly in rural areas) were simple, oblong buildings, though in the more populous urban parishes, churches had been provided with aisles, chantry chapels and belfry towers. During the Reformation in the sixteenth century, Ireland did not follow the general European experience of *cuius regio eius religio*, which saw the majority of the population following the religious tendencies of their rulers. Instead, most people in Gaelic-Irish and Anglo-Irish areas remained Catholic, with the country's official Protestant religion attracting only an elite minority; a situation in direct contrast to the Reformation as experienced in England.

In Ireland, the state-sponsored Church of Ireland retained ownership of all parish churches, even though their congregations were small. Consequently, there was little or no development of religious architecture, with most medieval church buildings falling into ruin due to a lack of support within the wider community. The situation was exacerbated by the displacement of the older landowning elite and fundamental changes taking place within established economic structures. Local economies that were once rooted in feudal concepts of clientship were now superseded by those centred primarily on market supply and demand, and a cash-based economy. It is not surprising then that Luke Gernon, travelling through east Limerick in 1620, reported that in every village there was a castle and a church, but that both were in ruins. This serves as a clear testimony to the profound and radical changes being experienced by local communities at the time.

From 1536 onwards, monasteries were closed as part of the Reformation, and their buildings and properties confiscated. While regarded as one of the chief events that marked the end of the later medieval era in Ireland, the process of dissolving the monasteries was less thorough in Ireland than was the case in Britain. For example, the earls of Thomond and Desmond were allowed to carry out their own campaigns of suppression in Munster without government interference. This meant that many of the religious houses that enjoyed their patronage continued to function until they were closed down finally following the Desmond Rebellion (which

Fig. 18.30 HISTORIC DRAWING OF ASKEATON ABBEY, COUNTY LIMERICK (1633). Maps published in *Pacata Hibernia* in 1633 illustrate a number of port towns in Munster as they would have appeared in the early seventeenth century. Among these is Askeaton in County Limerick, where one of the chief castles belonging to the earls of Desmond was located, situated on an island on the River Deel. The impressive castle was long perceived to be a significant threat to English interests and was destroyed by Cromwellian forces in 1652. A Franciscan friary ('the Abbye') and St Mary's parish church (with its ruined belfry) are situated near the castle. Oddly, the map omits the houses and streets that would have made up the town of Askeaton. The process of dissolving the monasteries was less thorough in Ireland than was the case in England. Many of the religious houses that enjoyed the patronage of the earls of Desmond continued to function until they were finally closed down following the Desmond Rebellion (1579–1583). [Source: National Library of Ireland, call no.: ET B206]

broke out in 1579 before being brought to a conclusion in 1583 with the killing of the earl of Desmond) (Figures 18.30 and 18.31). In the Gaelic areas of Connacht and Ulster, many of the monasteries also remained open until the 1580s when they too were suppressed. Even then, it was known in some instance for friars to regroup and re-establish themselves in certain localities, so that the dissolution of the monasteries was never complete. The suppression of these religious foundations, however, meant that the estates under their control were now appropriated by the crown and re-granted to lay landowners. This marked a significant change in patterns of land ownership in Ireland.

Monastic buildings themselves, however, were slow to disappear from the landscape, as the cost of demolishing such large structures simply outweighed the demand for the building materials. Only monasteries within or near to valuable urban sites, or located close to royal properties that required materials for repair, were subjected to demolition immediately following their dissolution. Consequently, many monasteries remained an integral part of the built environment along the coastline. A number of monasteries, for example, were re-used by the authorities to provide strongholds for garrisons. This is not surprising, as many of the religious houses had been established originally by Anglo-Norman lords four hundred years previously with a view to consolidating their control over newly gained territories. The various buildings arranged around cloisters and courtyards, along with stout precinct walls, lent themselves to be readily reused and transformed into garrison forts.

Another consequence of the closure of the monasteries was the shortfall in the provision of care for the sick and the poor. While this practice was continued in some foundations, such as St John's Newgate in Dublin, elsewhere new hospitals were founded. One such example was the Franciscan friary in Waterford that was acquired by a local Catholic merchant and converted into a hospice. Many monasteries had also fulfilled important education roles, but this was offset largely by the foundation of new grammar schools to educate the children of the gentry. In Dublin, the former buildings of the Augustinian priory of All-Hallows were modified to become the home of Trinity College.

In many instances, members of the New-English gentry converted the monasteries into residences, such as the Cistercian abbeys at Tintern and Dunbrody, County Wexford (Figure 18.32). Often, church towers and adjacent transepts were refurbished with the insertion of new windows, gun

Fig. 18.31 MODERN DRONE IMAGE OF ASKEATON ABBEY. The extensive and impressive remains of the Franciscan friary at Askeaton are situated on the east bank of the Deel, not far upstream from where the river enters the Shannon Estuary. This fourteenth- or fifteenth-century monastic complex consisted of a church and other buildings placed around a cloister. Later fifteenth-century additions include a sacristy and transept to the north of the church (in the background). Also added was a two-storeyed refectory, projecting from the south of the cloister (in the foreground). According to the map of Askeaton published in *Pacata Hibernia* (1633), a large belfry tower was present in the friary, but no trace of this now remains. The insertion of Renaissance-style doorways suggest repairs or refurbishments in the sixteenth century. The friary survived the dissolution in the 1540s and a provincial chapter was held here in 1564. During the Desmond Wars, the friary was captured by Nicholas Malby and was damaged extensively, with the friars being expelled. The friars returned in 1627, and were to occupy the site periodically until its permanent closure in 1740. [Source: Mike McCarthy, Wikimedia Commons, CC BY SA 4.0]

Fig. 18.32 TINTERN ABBEY, COUNTY WEXFORD. Europe in the twelfth century saw the establishment of many new religious orders anxious to spread the Christian faith. Among these were the Cistercians, founded in France, who became the first continental order to establish themselves in Ireland. This occurred in 1142 at Mellifont Abbey, located some 10km inland from Drogheda in County Louth. In the following century, some thirty-three additional Cistercian abbeys were founded on the island. One of the most important of these was Tintern Abbey, sited near the western shores of Bannow Bay on the Hook Peninsula in County Wexford. Founded by the powerful Norman knight, William Marshal, first earl of Pembroke, the abbey was colonised initially by monks brought to Ireland from the larger Tintern Abbey in south-east Wales. As with most religious institutions of that time, the abbey acquired large tracts of land. Tintern Abbey received an early endowment of some 9,000 acres. This, as in other Cistercian abbeys, allowed the order to play a significant role in developing agriculture. Their lands were generally divided into individual farms, or granges, each possessing their own buildings and labour force of lay monks. This encouraged productive use of the lands. At the time of the dissolution of Tintern Abbey in 1539, it was one of the richest and most powerful of the Cistercian abbeys in Ireland. Following its dissolution, the abbey and its confiscated lands were sold to the gentry and aristocracy who were supporters of the English crown. Many abbeys, as in the case of Tintern, were converted into new residences. Bought by the Colclough family, the church and adjacent transepts at Tintern were refurbished with the insertion of new windows, gun loops, stairwells and fireplaces. Such modifications allowed for the provision of comfortable residences and would have been comparable with those seen in the new, fashionable country houses being built elsewhere in the country. The Colclough family transferred ownership of Tintern Abbey to the state in 1963, after which it became a national monument. [Source: National Monuments Service, Department of Culture, Heritage and the Gaeltacht]

loops, stairwells and fireplaces. Such modifications allowed for the provision of comfortable lodgings and would have been comparable to those seen in the new country houses that were beginning to make their appearance at the time. Other buildings around the cloister, such as dormitories and kitchens, were also adapted with relative ease to suit the requirements of the new household. The remainder of the buildings within the former monastic precincts were used as agricultural buildings or just left to decay.[30] Overall, the radical transformation of the monasteries, including those dotted along the coastline, was reflective of a changing political, cultural and economic environment that ultimately brought about the birth of modern Ireland.

IRELAND AND SLAVERY: COASTAL CONNECTIONS THAT BECAME BITTERSWEET

Nini Rodgers

As Daniel O'Connell proudly declared, Ireland did not possess a slave-trading port. But as slavery boomed in the eighteenth century, Ireland's European maritime connections tied her into this lucrative and cruel commerce (Figure 18.33).[31] In England, France and Spain, Irish names can be found among the crews and captains who sailed to west Africa to exchange iron bars, firearms and alcohol for human cargo. The dream of those captaining such long, unhealthy voyages (Figure 18.34) was to make enough money to become merchants, which would allow them to stay in port cities such as Bristol, Liverpool, Nantes and Cadiz, and from where they could organise, outfit and take the major profits from such ventures. Among the successful personalities in these roles can be found a Callaghan in Bristol, Tuohy in Liverpool, Butler in Cadiz and, richest and most influential of all, the Walsh family of Nantes. Philip Walsh, a Dublin merchant who claimed to have conveyed the defeated King James II to France in 1690, settled in St Malo, where he entered the slave trade. His son Antoine, who retained close links with a rich, maternal uncle in Waterford, moved the family to Nantes, then emerging as France's premier slave-trading

Fig. 18.33 *Negres a fond de calle* (*Navio negreiro*) (NEGROES IN THE CELLAR OF A SLAVE BOAT) *c.*1830. This engraving, by Juan Mortiz Rugendas (the original of which is in the Museo Itaú Cultural in São Paulo, Brazil), depicts the Brazilian abolitionist poet Castro Alves's epic 'Navio Negreiro' (1880), which exalts the African people and the conditions in which they were brought across the Atlantic on slave ships to work in the colonies of the Americas. It is estimated that about 12 to 12.8 million Africans were transported across the Atlantic between the sixteenth and nineteenth centuries. During the eighteenth century, it typically took up to two and a half months to cross the Middle Passage from west Africa to the Caribbean, but one hundred years later the journey had reduced to two months, or even shorter. Particularly in the early years of the trade, it is estimated that as many as one in eight captured slaves died during the journey, victims of brutal treatment and the cramped, overcrowded and unhygienic conditions on board the slave ships. [Source: Wikimedia Commons, Public Domain Mark 1.0]

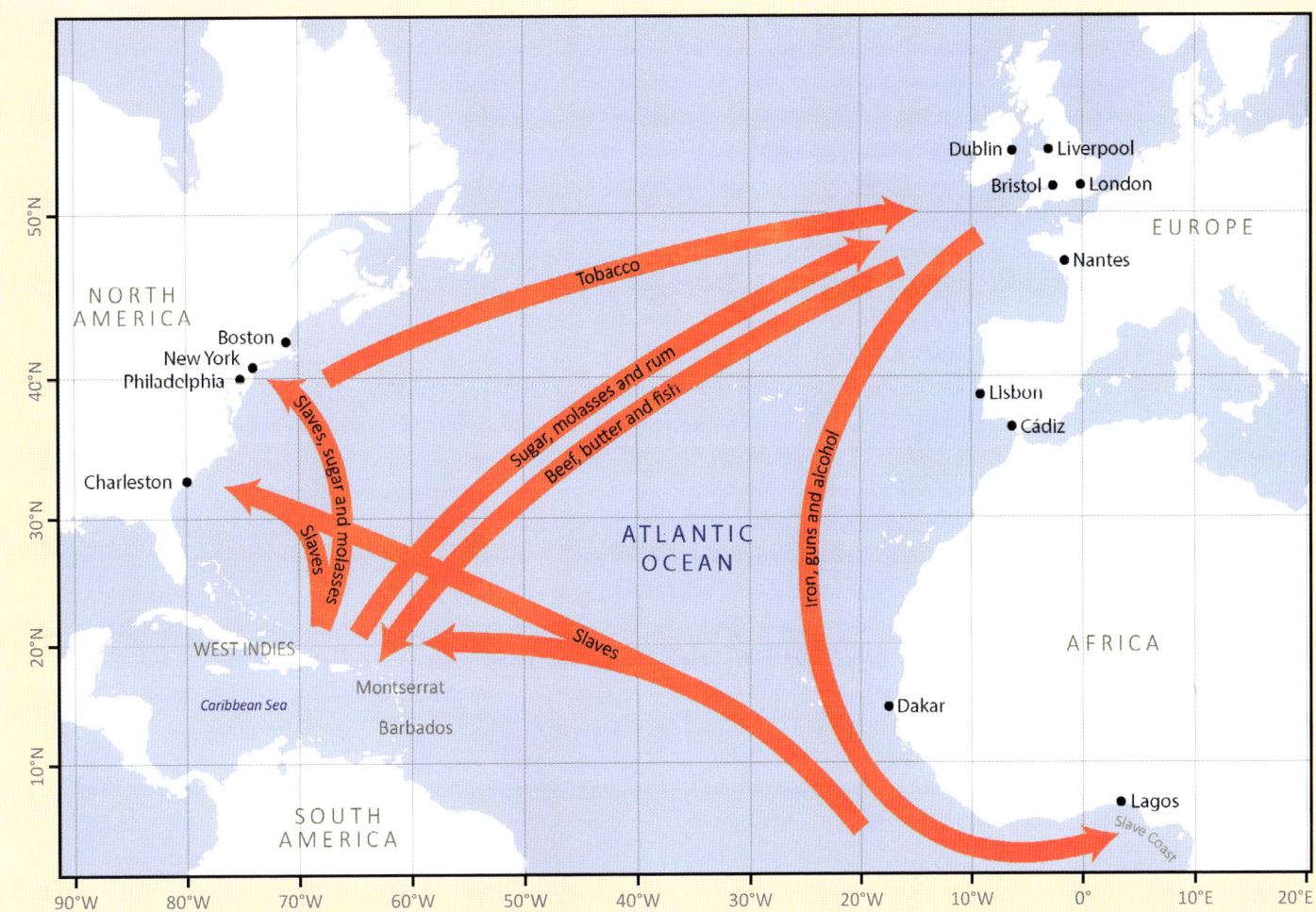

Fig. 18.34 NORTH ATLANTIC TRIANGULAR TRADE ROUTES. The transatlantic slave trade was initiated by the Portuguese who, in 1526, transported the first slaves from Africa to their colonies in Brazil. Other European countries, notably England, France, Spain and the Netherlands, soon followed suit. Taking advantage of the prevailing winds and currents of the North Atlantic, the triangular slave-trading routes were initiated by the English admiral, Sir John Hawkins, in 1562. The trade normally began at European ports, from where goods and commodities such as iron, cloth, guns and alcohol were carried south to slave-trading ports along the African coast, from Dakar in present-day Senegal, around the Gulf of Guinea, to the mouth of the Congo River and Benguela in Angola. At these ports, the goods were traded for slaves, mostly of central- and west-African origin and captured usually by other west Africans, but sometimes directly by the Europeans themselves. The slaves were taken, via the Middle Passage across the Atlantic, to be sold at markets in the Caribbean and New World colonies. Forming the third side of the triangle, the ships would then return to their European home ports, carrying cargoes of tobacco, cotton, rum, sugar, molasses and other produce, for sale to European customers. As well as slaves, mostly from Africa, numerous 'indentured servants', including several individuals and whole families from Ireland and indeed from England, were forcibly deported and transported to the colonies, particularly during the mid-seventeenth century. Although in practice the lives and working conditions of these servants often differed little from those of the slaves, in legal terms their status was somewhat different. Whereas slaves were treated as goods, whose status was perpetual, hereditary (that is, the children of slaves automatically became slaves at birth, and hence the property of their parents' owners) and viewed in law as sub-human, colonial servitude was temporary and non-hereditary, and the servants were accorded legal personhood. The main destinations for Irish indentured servants were Barbados and, especially, the island of Montserrat, where people of Irish origin represented both the colonised and the colonisers, and held positions from indentured servant up to governor.

port. There, over some thirty years, he organised voyages that transported some 12,000 slaves across the Atlantic. He used the wealth he amassed in a bid to dominate the French slave trade, and to convey Bonnie Prince Charlie to Scotland in 1745. Encountering failure on both fronts, he retired to his slave plantations on St Dominique in the Caribbean, then the most successful tropical colony in the world.

At home on Ireland's coast, the major consequence of developing Caribbean slave plantations was the growth of the provisions trade. The trend was set by Cork, such that its focus on this emergent and buoyant cross-Atlantic trade allowed it to outstrip quickly its historic port rivals of Kinsale, Bandon and Youghal. Special and secret recipes were devised for pickling and salting goods to survive in a hot climate; convenience food in an age before tin canning or refrigeration were invented. Thus, Cork would develop the most advanced meat-packing industry in Ireland during the eighteenth century, while its butter market in Mallow Lane carried an equally sophisticated product. All beef involved in the provision trade was graded; 'mess beef' was for the planters, 'cargo beef' for the white servants and 'cow beef' for the slaves. The latter was generally an unappetising item and was produced from elderly dairy cattle, fattened for slaughter in their last days and augmented by rejects from other carcasses. French planters depended heavily on Irish 'cow beef' as a source of protein for their slaves; the British favoured salt fish, which Ireland also supplied in the form of salted herrings. As slaughter yards, butchers, coopers, salters, butter buyers, weigh masters, porters and clerks proliferated, so too did industrial espionage. Spies from Denmark, the German Baltic states, the Netherlands, Newry, Belfast and Derry sought eagerly to unlock Cork's secrets. The demand for these provisions drove deep into the countryside, intensifying dairying and the keeping of beef cattle throughout Munster from the south coast of Cork to Limerick.

The most valuable product entering Ireland at this time was sugar from the Caribbean slave plantations. For much of the eighteenth century (1733–1780), it arrived by an eastern rather than a western route; so important was this commodity that imperial regulations laid down in Westminster demanded that it be conveyed direct from the British colonies to English ports, thereby requiring re-export to Ireland. Packed in barrels, the sugar arrived as 'muscovado'; heavy, wet and dark brown, requiring further refining. This meant that it was taken to a sugar-house, boiled and poured into cone-shaped pots. When it cooled, it emerged ready for sale as the classic, triangular-shaped 'sugar loaf', its status recorded in the multitude of hills and mountains around the world – including the renamed 'Holy Mountain' in Wicklow – known as the Great Sugar Loaf.

By the mid-eighteenth century, there were forty sugar-houses in Ireland (half of them in Dublin), capable of refining two-thirds of the country's supply. Dublin's silversmiths busied themselves producing sugar bowls, sugar dusters and sugar tongs. Once glittering on the tea tables of the well-to-do, they now twinkle from cases in the National Museum of Ireland, at Collins Barracks in Dublin. The punch bowl too was filled increasingly with another product of slave labour – West Indian rum, distilled from molasses (treacle), a by-product of sugar production. Less trammelled by Westminster restrictions than sugar itself, rum spread from the ports throughout the country. By 1769, over two million gallons were being consumed annually, seeping from the coast into the countryside and encouraging the spread of a money economy.

Tantalising and portable, sugar itself did not remain the monopoly of the rich. In Belfast the Committee of the Charitable Society struggled with a dilemma: if they issued their workers with a sugar allowance, they would be accused of supplying luxury to paupers; if they did not do so, their workers might steal the yarn they were spinning to pay for the desired commodity. In 1791, the Dublin parliament, Anglican and landed gentry, admitted that sugar was now seen 'as in some sort a necessary of life'.

Urban growth associated with the expansion and prosperity of the transatlantic provisions trade throughout the eighteenth century produced significant social change throughout Ireland, including the expansion of a Catholic middle class. In Dublin, for example, the wealthy sugar baker, Edward Byrne, emerged as the head of the Catholic Committee, employing Wolfe Tone as secretary and demanding the complete removal of the Penal Laws. Thus, Ireland's cries for liberty and independence in 1798 were fed and fuelled by the spread of African slavery in the West Indies.

The changes wrought upon coastal settlement in this period of Ireland's history are reflective of profound cultural, economic, political and social changes, which combined to transform the lives of most of the fishermen, sailors, merchants, craftsmen, farmers, gentry and aristocracy who lived in areas proximate to the shoreline. By the late seventeenth century, the political map had been redrawn, with the various semi-autonomous Gaelic-Irish and Anglo-Irish lordships that had dominated much of the island at the start of the fifteenth century now subsumed into a series of counties, governed by a unitary nation state centred on both Dublin and London. The crown's writ was no longer limited essentially to the Pale, larger coastal ports (including Dublin) and their surrounding hinterlands, but now extended across the whole island of Ireland.

The implementation of the various schemes of plantation across certain areas of Ireland since the mid-sixteenth century – most notably in Munster (1584) and later in Ulster (1609) – saw the arrival of the first substantial migration of settlers into the country since the first Anglo-Norman colonies in the late twelfth and thirteenth centuries. These new settlers from England, Wales and Scotland brought with them new ideas on housing, farming and material culture; they also spoke English and, in many cases (though not always), professed an adherence to Protestantism. Similar changes were also taking place in areas of the country outside the plantation schemes, suggesting that other transformative factors were at play, including the decline of native lordship, the Reformation and Counter-Reformation, the Renaissance, and nascent capitalism.

The political turmoil and conflict environments of the sixteenth century had depressed opportunities on the island of Ireland for sustaining prosperous trading links and facilitating significant economic growth. One important consequence of these unstable conditions was that while population totals elsewhere in Europe showed high levels of growth, Ireland's total population is reported as stagnating at c.0.75 million.[32] This absence of population growth would have been reflected in coastal areas.

The early seventeenth century, however, saw an expansion in trade, judging from the custom revenues of ports such as Bristol and Chester, which were involved in the Irish trade. Exports of hides, yarn, wool, timber, fish, butter and live cattle featured prominently in this trade. The country also benefited significantly from its strategic location, which allowed it to meet the high demands for provisions associated with England's growing trade with its Atlantic colonies.

Following military conflagrations, such as the Nine Years' War (1594–1603) and the Confederate War (1641–1653), the political environment also stabilised in the latter half of the seventeenth century. The net result was that the evolving economic and political environments in the country favoured population growth, particularly around port-centred communities and the more productive agricultural areas along the northern, eastern and southern coasts. By the end of the century, Ireland's population had almost doubled to 1.3 million.[33] While the outbreak of the Williamite War in the early 1690s did cause damage, the overall trajectory of Ireland's participation in global trading networks was to continue and indeed expand. As the country entered the eighteenth century, a society had emerged that was more recognisable to modern eyes, in comparison to the collection of feudal fiefdoms that had existed two centuries earlier.

CHANGING COASTAL LANDSCAPES

Patrick O'Flanagan

The majority of people in Ireland today live within approximately 50km of the coastline, making the coastal zone the most built upon and humanised of the island's landscapes. Most of Ireland's major settlements occur here and it is the locale for a dense network of infrastructures, including railways, roads, high-density housing, ports and harbours. Together, these have generated many polluted stretches of coastline. The transformation of this coastline that occurred between the early seventeenth century and the close of the nineteenth century, and the kinds of processes that prompted such dramatic alterations, will be explored in this chapter.

PORTS AND URBAN GROWTH

Urban expansion has been the most formidable non-physical driver of change on Ireland's coastline, as exemplified from the 1700s onwards by the intensification of commerce and trade. This development was facilitated by the building of roads, bridges, canals and railways, which allowed the island's principal port settlements to expand their hinterlands and encouraged new regionalisms to emerge. Over the same period, most of Ireland's ports also began to develop their handling facilities. Rivers were embanked and land was reclaimed, while docks, cranes and warehouses were constructed. Many ports expanded seawards as land was reclaimed

for commercial, industrial and residential purposes.

These developments were expressed especially in the dramatic growth of a series of port cities and towns, such as Dublin, Belfast, Cork, Limerick, Waterford and Londonderry (see Chapters 24: Ports and Shipping and 25: Urbanisation of Ireland's Coast). Such settlements became centres of innovation in commerce, engineering, industry and technology. Engineers, for example, were employed to deliver improvement works relating to defence, land reclamation, the building of new docks and shipping basins, embankment construction and shipbuilding. Important emblematic buildings were also completed, especially in Dublin, such as the Custom House (1791) (Figure 19.1), City Hall, known formerly as the Exchange (1779) and the Four Courts (1802). In effect, the major port cities developed differently from other large towns in Ireland and, by the end of the nineteenth century, were more cosmopolitan and functionally diverse than was typical on the island. This remains the case, particularly with regard to Dublin, Belfast and Cork.

The Port of Dublin has its origins on the banks of the River Liffey in the shadow of Christ Church.[1] Gradually, it moved out into the river, as land was reclaimed on the river's edge to provide for deeper water. From the sixteenth century the city quays began to extend downstream, though the centre of port operations

CUSTOM HOUSE, DUBLIN.

London Published July 1792 by Ja.ᵗ Malton and G. Owen, Dublin.

James Malton del. et fecit.

Fig. 19.1 *The Custom House, Dublin* (ENGRAVING), BY JAMES MALTON (1792). The view is taken from the River Liffey of the newly built Custom House, located on the waterfront of the then expanding old port of the city. The grandiose architectural style reflects the prosperity of Dublin as the second-largest city in Britain at the time. [Source: National Library of Ireland, call no.: PD 3181 TX81]

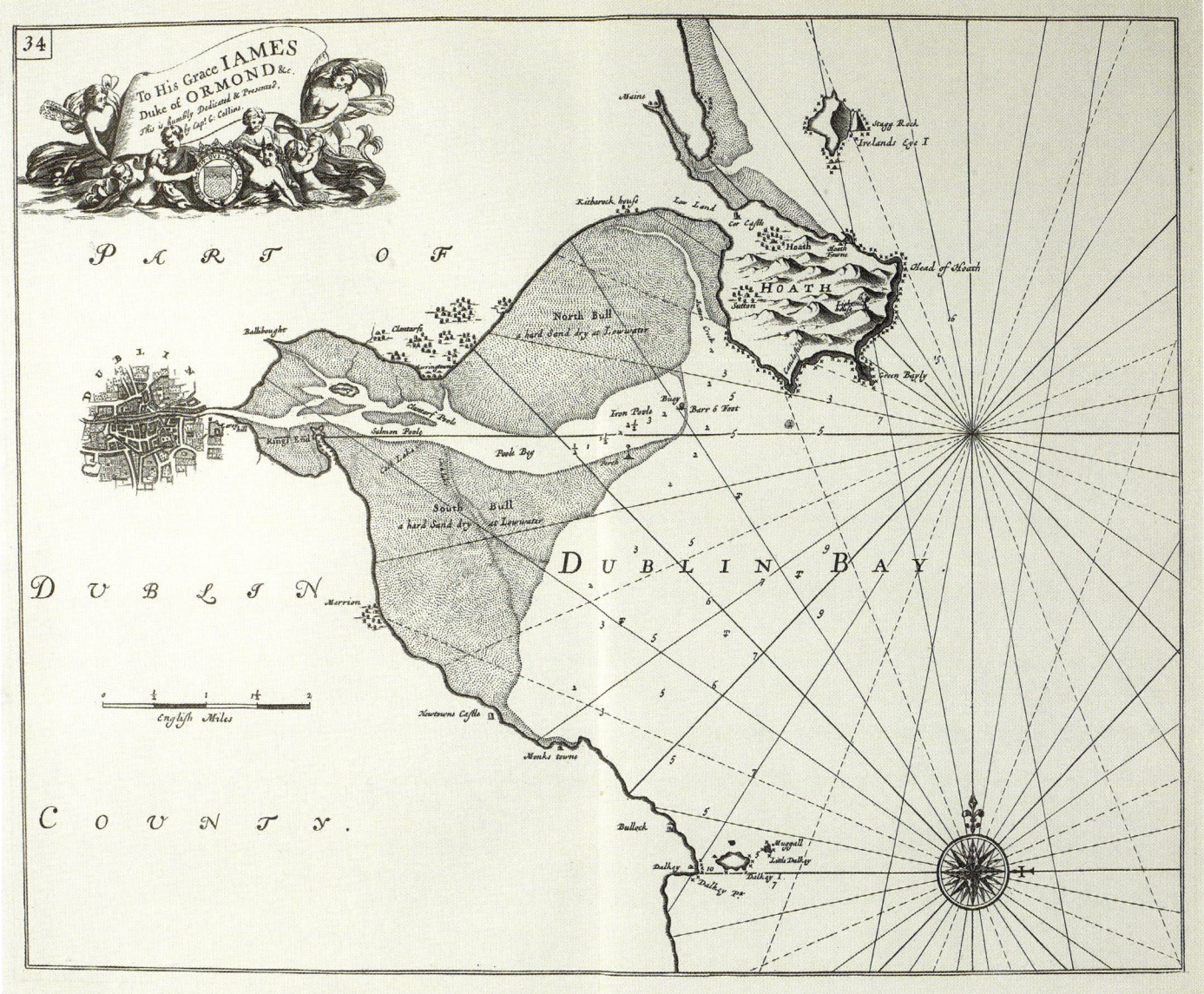

Fig. 19.2 COLLINS' MAP AND HYDROGRAPHIC CHART OF DUBLIN BAY (DATED TO 1693). Access to Dublin Port had long been restricted by wind and wave action, sediment, silting and wrecks. These factors resulted in many ships foundering while trying to access the port. The above map is probably the first to provide limited but valuable hydrographic detail for the entrance to Dublin Bay and Port, as an aid to help improve navigational safety to access the port. It shows, in particular, extensive areas of shallow water over mud and sand (shown as stippled shading), together with changing water depths marked in fathoms (1 fathom = 1.8288m). More detailed charts were developed later, for example, Brooking's 1723 map of Dublin Bay and Harbour, Gibson's map of 1756 and Rocque's 1757 map. [Source: National Library of Ireland, call no.: LBR 9142]

remained in the vicinity of the Custom House, then sited on what is now Wellington Quay. It was not until the early eighteenth century that Dublin Port began to encroach into the waters of Dublin Bay, with the construction of a series of breakwaters and walls to protect shipping.

Dublin, however, had two major disadvantages as a port. First, the river channels were constantly shifting as sands were carried into Dublin Bay by the tides, while second, a substantial sandbank or sandbar existed in the bay that blocked entry to the port, forcing ships to wait for high tide to enter or leave Dublin (Figure 19.2). To help address such problems, the first of a number of major projects was undertaken with the construction of the Great South Wall. This was completed in 1786, utilising blocks of granite from Dalkey Quarry. By this time, the Poolbeg Lighthouse had already been constructed (1767–1768) to replace the nearby 'Cassoon' light (Figures 19.3 and 19.4). Work on the building of the North Bull

Wall commenced in 1820, and was finished some five years later. These physical changes helped the harbour to de-silt, but it was also marked by the unexpected appearance of Bull Island, now a major public recreational resource.

The period between 1840 and 1860 witnessed railways being extended into Dublin docks, principally on the northern side of the River Liffey, to the newly constructed Grand Canal Dock. An extensive infrastructure was erected to facilitate the assembly, storage and shipping of a range of bulk goods – including lairages (animal pens), mills, railway sidings and warehouses– and consumed extensive spaces in and around the docklands.

Although the new harbour constructions assisted navigation into Dublin Bay and facilitated access to its port, local people and business interests looked to promote alternative sites and larger-scale developments. This was exemplified by attempts from 1807 to build an artificial harbour at Howth outside Dublin Bay. Initial

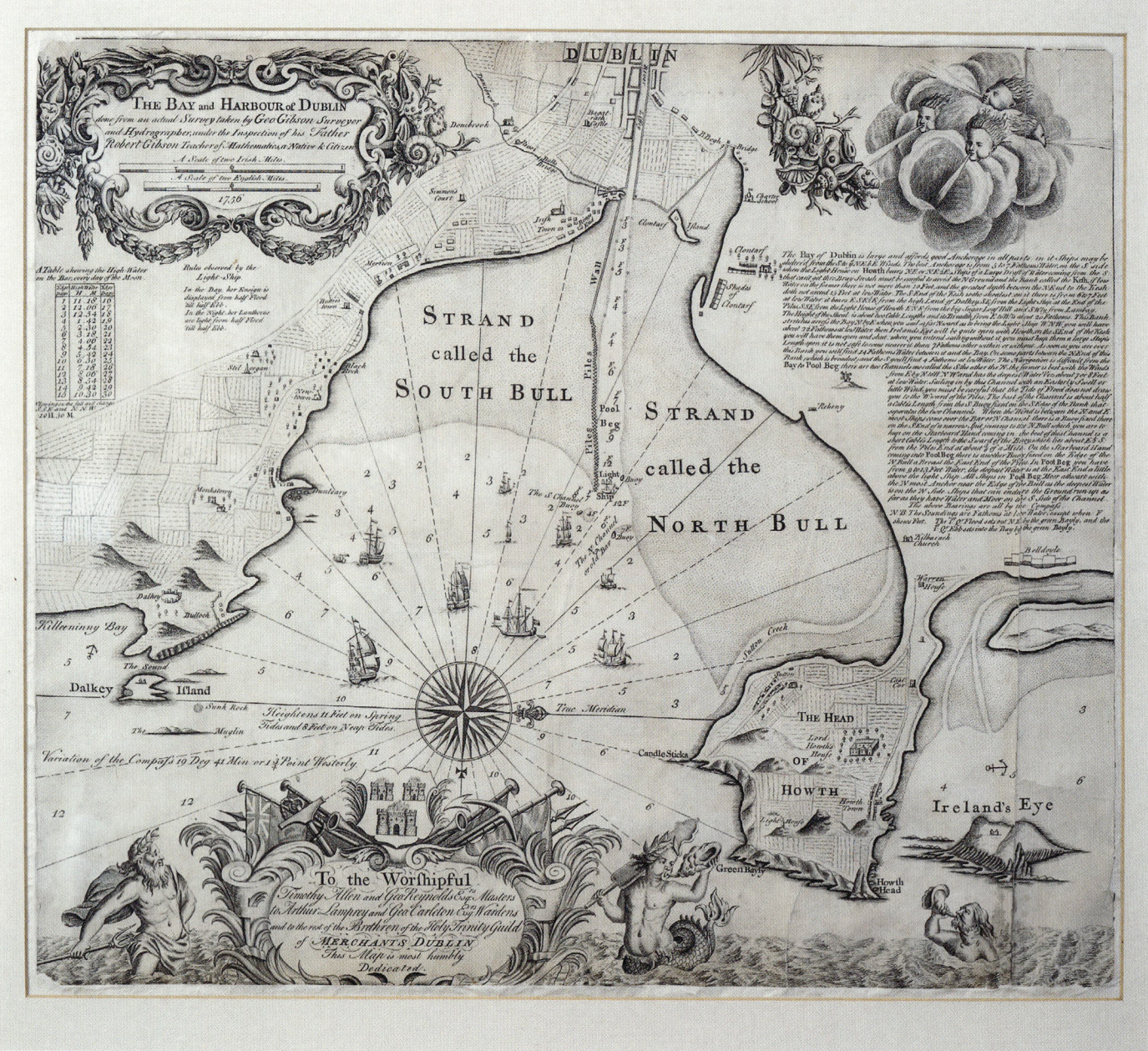

Fig. 19.3 MAP BY GEORGE GIBSON (1756). The map shows largely stylised key physical features of Dublin Bay, as well as some of the developing infrastructure of Dublin Port, such as Poolbeg Lighthouse and the channel embankment of the south 'Wall'. This wall was extended and developed further as a fully stone structure (the Great South Wall), completed in 1786. While an indication of the separation between the broad intertidal and subtidal areas of the bay is illustrated, and is similar to those of Collins' chart, detail of the many sandbanks, bars and shoals in the bay is only inferred, though the water depths (in feet) are shown. The map is a product of the developing eighteenth-century 'Age of Science', and the desire to communicate precise and practical information is indicated by the detailed notes, tables, scale bar and compass bearings on the map. [Source: Dublin Port]

successes in this venture were crowned by its capture of the Admiralty Mail Service functions. These achievements were short-lived, however, as the project lacked business vision and the harbour failed to provide enough secure shelter and berthage for vessels.

A series of marine tragedies in Dublin Bay at this time prompted the emergence of new thinking. A Norwegian engineer, Richard Toutcher (1758–1841), proposed the construction of an entirely new harbour, south of Dublin Port, as an alternative to the existing exposed anchorages and the limitations of Howth. Work commenced on the building of an ambitious and daring new port in 1817, but was completed only in 1859. Initially referred to as Asylum Harbour, and later called 'Dunleary Harbour', it was renamed Kingstown in 1821 after a visit by the then British

monarch. It reverted to its original Irish name of Dún Laoghaire in 1922.

The construction of Dún Laoghaire Harbour was an enormous civil engineering enterprise, involving some 1,000 quarrymen and stone masons. Most lived in the vicinity of Dalkey Commons beside Dalkey Quarry, which had been opened for the construction of Dublin's Great South Wall. Granite from this quarry was also transported by sea to other sites, the most notable being St Colman's Cathedral at Cobh, County Cork. Other workers lived closer to Dalkey town on the sites of what became Coliemore Road and Sorrento Road. This area is now one of the most sought-after residential locations on the island.

Technological innovation was associated with many aspects

of the construction of this vast harbour enterprise. Not least of these was the development of a funicular railway, functioning on gravity and running over a 4km track. This transported cut granite blocks from the quarries at Dalkey to the incipient harbour. In the 1840s, an even more ambitious project was superimposed on part of the funicular railway's site. Known as the Atmospheric Railway (Figure 19.5), it ran between Dalkey and Dún Laoghaire from 1844 to 1854, as an extension to the main railway line then connecting Dublin to 'Dunleary'.[2]

The completion of the Dublin–Bray railway line, begun in 1834 and finished in 1854, allowed many wealthy, inner-Dublin residents to build mansions and villas within the vicinity of the railway line. The area between Dalkey and Bray Head is dotted with many fine residences and emblematic gardens, often with private slipways. These belonged to bankers, lawyers and merchants who moved out from Dublin and built trophy houses on roads around Killiney Bay, between Vico Road and Sorrento Road (Figure 19.6) in Dalkey. They baptised them with exotic Italian names – such as Amalfi, Capri, Monte Alverno and Nerano – or names reflecting distant military experiences, such as Khyber Pass (now Sorrento Heights) and Fort Jamrud (now known as The Lodge). Today, they are the residences of well-known musicians, celebrities, sports people and lawyers. These road and house names reflect the extravagant perceptions of their owners, who aspired to Italian or even more exotic comparisons for their living contexts. In this, and other interlinked ways, the harbour and railway building and extensions transformed over three centuries this part of greater Dublin Bay, making it one of the most altered coastal regions on the island.

During this period, Dublin enhanced its role as Ireland's principal seaport in terms of goods throughput. The rapid increase in the size of ships trading internationally necessitated, however, ongoing development, which included substantial and innovative works to construct the Alexandra Basin as well as extensions to the

Fig. 19.4 A DRAWING BY WILLIAM HENRY BARTLETT FROM THE EARLY NINETEENTH CENTURY (ENGRAVED BY JOSEPH CLAYTON BENTLEY) SHOWING THE POOLBEG LIGHTHOUSE. The lighthouse was built at the end of the Great South Wall between 1767 and 1768 and replaced the earlier and nearby 'Cassoon' light. Apart from its significance to the shipping in Dublin Bay, and its role as a landmark, the lighthouse is now famous as the primary benchmark to fix the first Ordnance Datum position in Ireland. (The benchmark was cut at the position of the low water mark of spring tides against the base of the lighthouse in 1842.) The Great South Wall was, for many years, the longest sea wall in the world and today stretches from Ringsend to Poolbeg Lighthouse, a distance of approximately 6.4km. [Source: National Library of Ireland, call no.: ET B69a]

Fig. 19.5 DRAWING OF THE ATMOSPHERIC RAILWAY FROM DALKEY TO DÚN LAOGHAIRE. The novel railway was in service for a decade and provided public transport from Dalkey to Kingstown (Dún Laoghaire). [Source: *Illustrated London News* (1844)]

North Wall between 1871 and 1885. These constructions provided extended deepwater berths, vital to accommodate increased levels

429

Fig. 19.6 PAINTING OF SORRENTO TERRACE, WITH DALKEY ISLAND IN THE BACKGROUND, BY GERARD BYRNE. Reputedly the most expensive piece of terraced real estate in Ireland, Sorrento Terrace, completed in the 1850s, was built on top of a former lead mine. The building of the terrace coincided with the opening of the Dublin–Bray railway. Together, these projects facilitated the arrival of the rich from the city to live on this headland. Behind the terrace can be seen the Martello tower on Dalkey Island. It is one of some twenty-two towers built on the coast of County Dublin. Many more were erected elsewhere on the east and south coasts of Ireland in the early 1800s as part of an early warning defence system against a French seaborne invasion. They remain as a most emblematic testament to more turbulent times. [Source: Gerard Byrne, www.gerardbyrneartist.com]

of trade (see Chapter 24: Ports and Shipping, Figure 24.3). Additionally, Dublin emerged as the island's leading transport hub for roads and canals, extending subsequently to railways and, in the twentieth century, to air traffic. Such developments emphasised the centrality of Dublin and allowed it to consolidate its role as the administrative, colonial and military capital, as well as being the dominant commercial centre on the island. By 1911, the city of Dublin had a population of some 305,000, but was second to Belfast where rapid industrialisation had increased the population to almost 390,000.

THE PORT AND THE HARBOURS OF DUBLIN BAY

Rob Goodbody

Although developing strongly as a major Irish port from the eighteenth century, Dublin was not the only port in Dublin Bay (Figure 19.8).[3] From an early period, there were small fishing communities based at Clontarf on the northern shore, Dún Laoghaire and Bullock on the southern shore, while another operated from Ringsend, which was situated at the end of a sand spit projecting into the bay to the east of the mouth of the Dodder. At the south-eastern margin of the bay there was a port at Dalkey, which operated from the sound between Dalkey Island and the mainland. To the north, there was another fishing port at Howth, though this was outside the bay, on the northern side of Howth Head.

Fig. 19.7 (left) DUBLIN BAY AND KINGSTOWN HARBOUR (DÚN LAOGHAIRE) FROM KILLINEY QUARRIES (1842). A lithograph (originally published by Newman and Co., 48 Watling Street, London) depicting how the appearance of Dublin Bay was being radically and rapidly modified by the building of the new harbour at Kingstown. The enterprise was planned on a grand scale and witnessed the construction of a new town behind the port, also known now as Dún Laoghaire. The creation of this new harbour had been an enormous gamble, but it repaid its leading supporters handsomely. It was briefly Ireland's main mail port and was soon to become the principal passenger port linking to Holyhead in north Wales. The port was able to adapt successfully to the advent of steam shipping, but, while attracting some cargo functions, it did not displace Dublin Port as Ireland's largest cargo handling hub. [Source: National Library of Ireland, call no.: ET C180]

Fig. 19.8 SOME EARLY PORTS AND SETTLEMENTS AROUND DUBLIN BAY. This map depicts some of the early settlements, often medieval in origin, and other built and engineered structures around Dublin Bay. Some of these formed important early harbours, before the expansion of the main Port of Dublin and the nineteenth-century growth of Howth and Dún Laoghaire as key outer bay harbours. The settlement of Sutton, on the north side of the bay, grew up on the sands of the tombolo (coastal sandbar landform), which links the mainland to the former island of Howth.

Problems associated with the existence of a sandbar within the bay affected all of these ports. It was, for example, the principal reason for an anchorage at Dalkey in the Middle Ages, where ships unloaded their cargoes into lighters and could then enter the Dublin port more easily. It also restricted access to Ringsend, though passengers nonetheless used to disembark there and then proceed across the sands to the city. Howth was developed in the early years of the nineteenth century as a new, enlarged harbour to serve as a port for mail ships that also brought passengers from Holyhead. This enabled these ships to keep to the strict timetable that the mails required, free from the timing of the tides for entry to the Liffey. The fishing community at Clontarf, however, found itself within the enclosure created by the North Bull Wall in the 1820s, while the area beyond it, offshore from Raheny and Kilbarrack, was cut off from the open sea by the appearance of Bull Island. This resulted from deposits of sand caused by the change in tidal flow following the construction of the north and south walls. The harbours at Coliemore (near Dalkey) and at Bullock were both rebuilt on a significantly larger scale in the later nineteenth century.

The biggest change of all occurred at the harbour of Dún Laoghaire (Figure 19.7).[4] Developing as a fishing village from the medieval period, this small port had a population of about 300 people at the beginning of the nineteenth century. Its location had been used for a long time as the occasional landing place for passengers from Britain. By the eighteenth century, it had also developed a trade in coal from England, serving the south Dublin area, now free from the restrictions of the bar. In 1816, however, construction began on the substantial East Pier, designed to provide a shelter for ships caught in storms at a time when they could not cross the bar to Dublin Port. This facility arose from the substantial loss of life and ships that had occurred over the years due to the lack of a place of shelter within the bay. The decision in 1818 to construct the West Pier led to the provision of the great enclosed harbour of Kingstown, now Dún Laoghaire.

Rapid advances in technology in the nineteenth century changed the three principal harbours of Dublin, Kingstown and Howth. The latter, having been designed specifically for mail packet ships, became obsolete almost immediately with the introduction of steam ships in the very year that the harbour was completed. Once they appeared, the new ships increased quickly in size. The harbour at Howth was not deep enough for these larger ships and could not be dredged due to its rocky bottom. Kingstown Harbour could also have become obsolete, with the removal of the bar at Dublin Port obviating the need for a place of shelter for ships during storms. However, it was the natural replacement for Howth as a harbour for the mail and passenger ships and it thrived for the rest of the century.

The historic Port of Cork exploded onto the urban hierarchy as a major trading centre from the late seventeenth century and retained its status as Ireland's second city for more than two hundred years.[5] To facilitate trade, the city's inner harbour was extended by the fashioning of a navigation wall (Figure 19.9) begun around 1765. This was followed by a series of other improvements, many proposed and executed by the engineer, Alexander Nimmo. Such improvements allowed trade through Cork to flourish and it functioned as the island's major overseas provisions trading supply centre, sending vast quantities of barrelled beef, fish and butter to the New World, and especially to the Caribbean.[6] The butter trade was of particular significance and, in 1770, the Cork Butter Exchange was opened by the Cork Committee of Merchants to regulate this rapidly increasing trade. The volume of butter passing through Cork meant that for much of the nineteenth century, the Cork Butter Exchange was the largest butter market in the world. At peak trading in the 1880s, some 500,000 casks of butter passed through the Exchange, valued at some £1.5 million.[7]

The strength of the trade through the Port of Cork in the eighteenth century resulted in it becoming one of Atlantic Europe's leading ports, recording a resident population that equalled, or exceeded, centres such as Cadiz, Seville, Bilbao, Bordeaux and Cherbourg. In the city, its merchants' residences were amongst some of the most sumptuous urban buildings erected during the eighteenth and nineteenth centuries, and were located usually beside docks and warehouses, as on Patrick's and Bachelor's quays, or the nearby Cooper Penrose mansion at Woodhill (Figure 19.10). Many of these waterfront buildings have strong links to Dutch or Flemish architecture, reflecting well-established sea-trading partnerships. Similar influences can be seen in neighbouring small coastal towns, as in Youghal, County Cork, where its iconic 'Red House' also shows a Flemish cultural influence. Despite Cork's important trading function, the city's population in 1911 was only 77,000, significantly less than the emerging primate cities of Belfast and Dublin.

In other ports of significance in Ireland, management institutions and buildings to house them emerged in the late eighteenth to nineteenth centuries. Limerick, for example, enjoyed a boom period from the mid-eighteenth century up to the 1840s and here, as elsewhere, growing seaborne trade prompted urban and port expansion. Georgian Limerick was the outcome of this prosperity. By 1823, an agency known as Limerick Bridge Commissioners was set up and gained responsibility for the construction of new docks on both sides of the Shannon Estuary. These were completed by 1840. Renamed the Limerick Harbour Commissioners in 1847, this body managed all aspects of harbour operations and development.

A similar process operated in Belfast, where an act of 1785 established 'The Corporation for the preserving and improving the port of Belfast'. This was repealed in 1847 and a new statute constituted the Belfast Harbour Commissioners. It was this

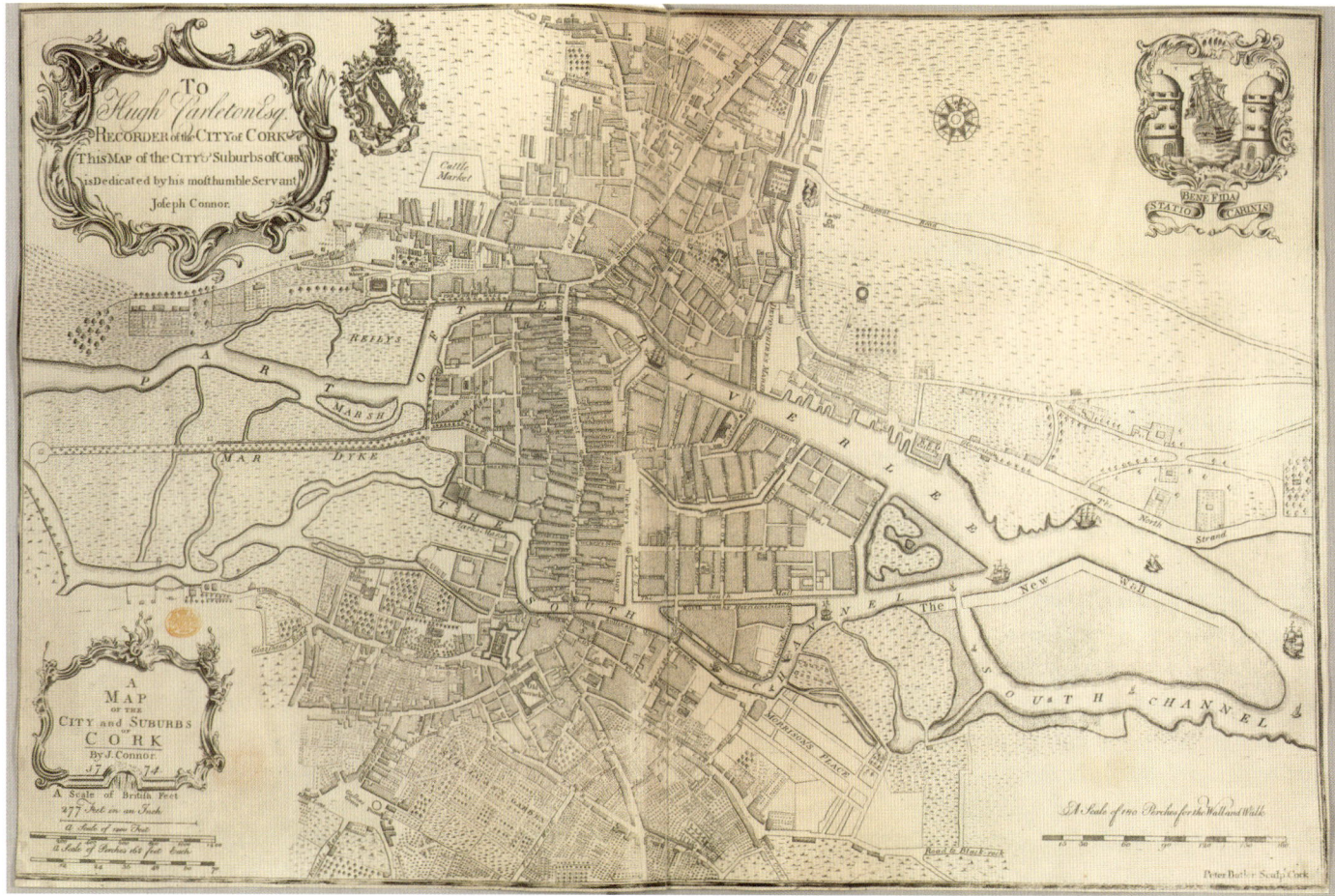

Fig. 19.9 THE 'NEW WALL', CORK CITY, FROM J. CONNOR'S MAP (1774). The erection of the wall in the 1760s went a long way towards regularising the various inner harbour and river channels facilitating ship access to the Port of Cork. [Source: © British Library Board. All Rights Reserved/Bridgeman Images]

Fig. 19.10 A VIEW OVER CORK CITY (MEZZOTINT), BY THOMAS ROBERTS (1760–1826). Substantial mansions, warehouses and other structures are shown along the waterfront areas and within the developing city core. The eastwards perspective, over the present-day inner Cork Harbour and river estuary, portrays this expanding coastal city and port, with its recently built navigation wall (the New Wall), begun in the late eighteenth century. [Source: Crawford Art Gallery, Cork]

agency that drove the initial modernisation of the Port of Belfast and later changes to land–use in adjacent coastal areas. These improvements in infrastructure and expansion of the port city were necessitated by the large-scale industrialisation of the city's hinterland. The result was that, by the mid-nineteenth century, Belfast replaced Cork as the island's most populous second city.

Other small seaports throughout the island also witnessed progressive investment from the early 1700s. Landlords at this time went on a spree of planned village and town building. Early examples of such initiatives include Courtmacsherry, Kilkee and Lehinch. The aspirations and scale of some of these developer-landlord planned settlements greatly exceeded the realities of commerce and trade. Thackeray's description of Westport stands out as a monumental example of this type of extraordinary utopian building excess, to be repeated often elsewhere in Ireland (Figure 19.11).

Fig. 19.11 WESTPORT QUAY, COUNTY MAYO (c.1900). Soon after completion, Westport Quay warehouses were a vast mausoleum of emptiness and a testament to overzealous investment during the nineteenth century. 'There was a handsome pier and one solitary cutter alongside; which may or may be not there now. There were about three boats lying near the cutter and six sailors lolling about the pier. As for the warehouses, they are enormous and might accommodate, I should think, not only the trade of Westport but that of Manchester too. There are huge streets of these houses, ten stories high with cranes, owner's names and marked with Wine Stores, Floor Stores, Bonded Tobacco Warehouses ...' (William M. Thackeray, 1844.) Today, many buildings here have been gentrified, having been converted into apartments, hotels and restaurants. [Source: National Library of Ireland, call no.: L_ROY_05061].

BELFAST PORT AND SHIPBUILDING

Stephen A. Royle

By 1700, Belfast, a seventeenth-century foundation, had developed some industry, including shipbuilding, but was most significant as a trading place. Belfast had replaced Carrickfergus as the dominant port on Belfast Lough, although its river, the Lagan, was not deep enough for large ships, which had to offload downstream. In the eighteenth century, the quaysides became rundown and only a few private improvements were made, such as George and Hanover quays in the 1710s. However, an act of 1785 established 'The Corporation for the preserving and improving the port of Belfast', known as the Ballast Board. The act was repealed in 1847 and a new statute constituted the Belfast Harbour Commissioners, which drove the initial modernisation of its port and subsequent significant changes to its contiguous coastal areas.[8] Under the earlier Ballast Board, the two Corporation Docks of 1796 and 1826 were built. They still exist as heritage features. Sugar and tobacco were significant imports to these; exports were foodstuffs and linen. The emigrant trade commenced at this time, principally to North America.

Dredging sand from the river helped reduce the shallowness of the early port, but in 1839 a channel was finally cut through a bend of the River Lagan, allowing substantial vessels to reach Belfast. Dredged material was tipped to the southern County Down side of the channel and Dargan's Island (after the engineer, William Dargan from Carlow) arose from the sloblands (Figure 19.12). This was renamed Queen's Island after Victoria's 1849 visit, when she progressed up a second cut forming Victoria Channel. This improved access helped Belfast become Ireland's largest port by 1852.

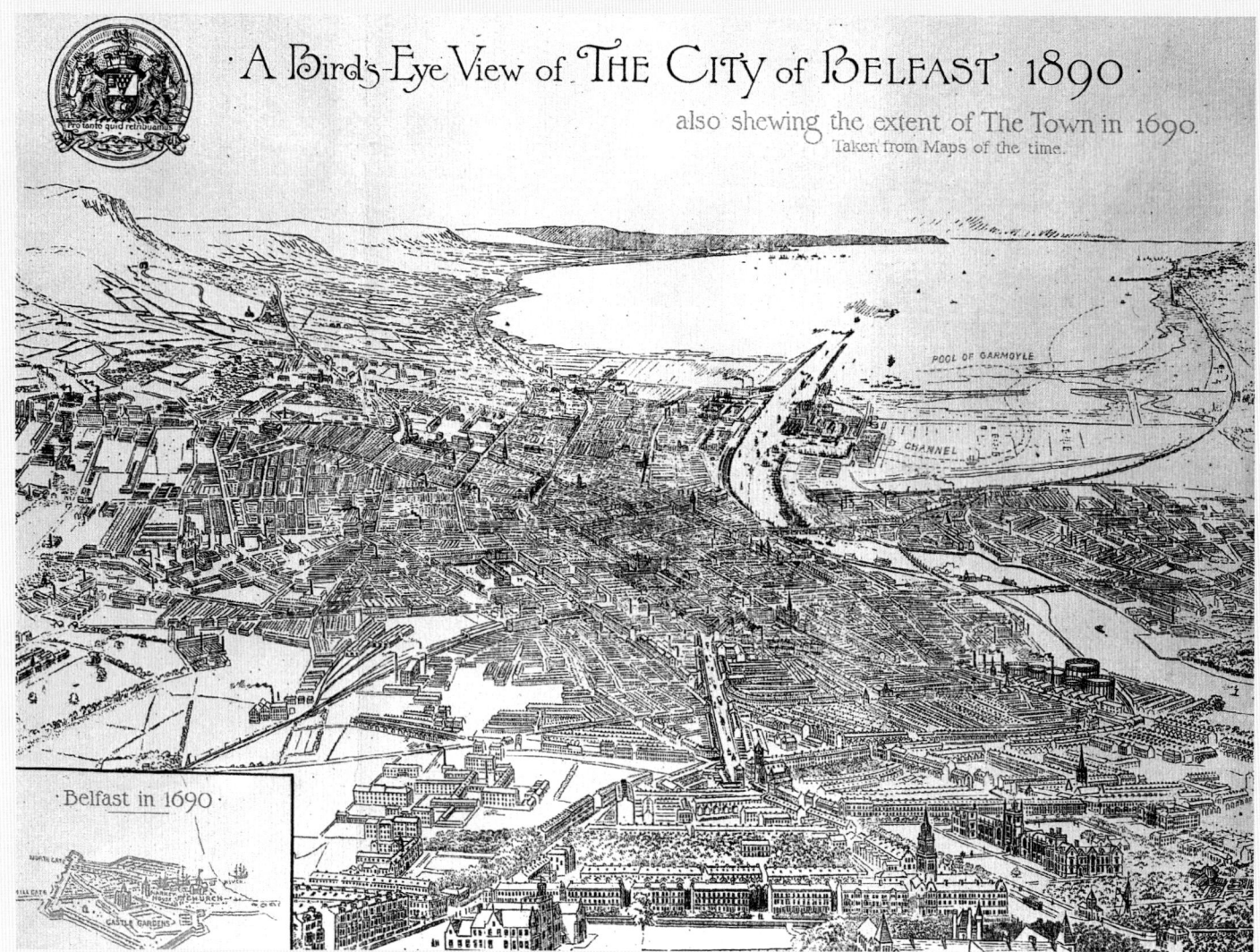

Fig. 19.12 AN 1890 BIRD'S-EYE VIEW OF THE DEVELOPING PORT AREA OF BELFAST, LOOKING NORTH-EASTWARDS OVER BELFAST LOUGH. The illustration shows the River Lagan, the main navigational channel to the port, and the expansion of the docklands over the sloblands on the right side of the channel. The large-scale expansion of the city, due to nineteenth-century industrialisation, can be seen in comparison to the relatively small and compact town of Belfast in 1690 (see inset).[Source: Public domain]

The number of ships entering Belfast rose 30 per cent from 1860 to 1914, and their increased capacity saw the tonnage grow from 800,000 to 3 million tonnes. This necessitated major changes in infrastructure. In 1872, the Spencer and Dufferin docks were built and the following year Albert Quay was extended and rebuilt. The Thompson Graving Dock, the world's largest dry dock, was built in 1911 and the new Musgrave Channel opened in 1903.[9]

In the early nineteenth century, shipyards on Belfast's County Antrim shore made wooden vessels. This was to change, however, when in 1820 the first Irish-built steam-powered ship was produced there. Shipbuilding became focused on Queen's Island. Robert Hickson developed a shipyard on Queen's Island, of which Edward Harland became manager in 1854. Harland bought out Hickson in 1858, making his assistant Gustav Wolff a partner in 1861, forming the internationally famous Harland and Wolff shipyard. Thomas Ismay began to order passenger ships from this shipyard in the 1870s for his White Star Line, initially to service the emigrant trade. Huge increases in this trade saw Harland's innovatively designed ships much in demand. Harland and Wolff employed 2,400 in 1870, rising to 14,000 in 1914.

Another significant shipyard and builder was the 'wee yard', or Workman Clark, established initially on the County Antrim shore before premises were bought on the opposite side. Several of the world's largest ships were built in Belfast, with the RMS *Titanic* being the most famous. The shipyards employed almost exclusively men, a counterpart to the female labour of the textile mills. Belfast's decent housing, cheap living costs and the availability of jobs for both genders helped foster its massive population growth, from 71,447 in 1841 to 386,946 in 1911.[10]

THE PORT OF LIMERICK

Des McCafferty

Limerick's foundation as an urban settlement dates from the establishment of a Viking port and trading centre in the early tenth century, and since then the city's development has been intertwined with the changing fortunes of its port. Standing at the head of the 100km-long Shannon Estuary, the settlement at Limerick acted as a gateway from the Atlantic ocean to the very heart of Ireland. The Viking port was located at the confluence of the Abbey River and the Shannon, in the centre of what became the medieval city of Limerick (Figure 19.14), and port activity remained focused in this area until the late eighteenth century.[11] Then, the development of an entirely new, planned town – Newtown Pery – led to a shift of port activity downstream.

From 1770 until 1840 an extensive programme of quay construction along the river in the new town reflected the growing confidence engendered in the city's merchants by the burgeoning commercial activity of the period. Indeed, such was the growth of trade in the early nineteenth century that quay development alone proved inadequate for the port's needs. Due to the large tidal range on the Shannon at Limerick, ships berthed at the quays 'took the bottom' at low tide and, as vessels increased in size, this became increasingly problematic. Consequently, under an act of parliament in 1847, Limerick Harbour Commissioners was established to oversee construction of a 'floating' dock, which would retain water at high tide level, thereby enabling ships to remain afloat at all times. The floating dock was completed in 1853 and facilities were further enhanced in 1873 by the construction of a graving dock designed to enable repairs to vessels arriving from the USA and Canada.[12] As the size and number of ships continued to increase through the early twentieth century, the area of the docks was extended in 1934. A new, western, entrance from the river was also planned at this time, though this was not provided until 1956.

Fig. 19.13 LIMERICK CITY QUAYS. This photograph was taken in the late nineteenth century, as steam was becoming more common in shipping and beginning to replace sail on the Shannon. The view is upstream (eastwards) towards the old port and the city and shows the quays, which were built in the first half of the nineteenth century. As well as the paddle steamer in the centre, the photograph captures several masted vessels berthed at the quays and the chimney stacks of a number of the city's factories. [Source: National Library of Ireland, call no.: L_ROY_05297]

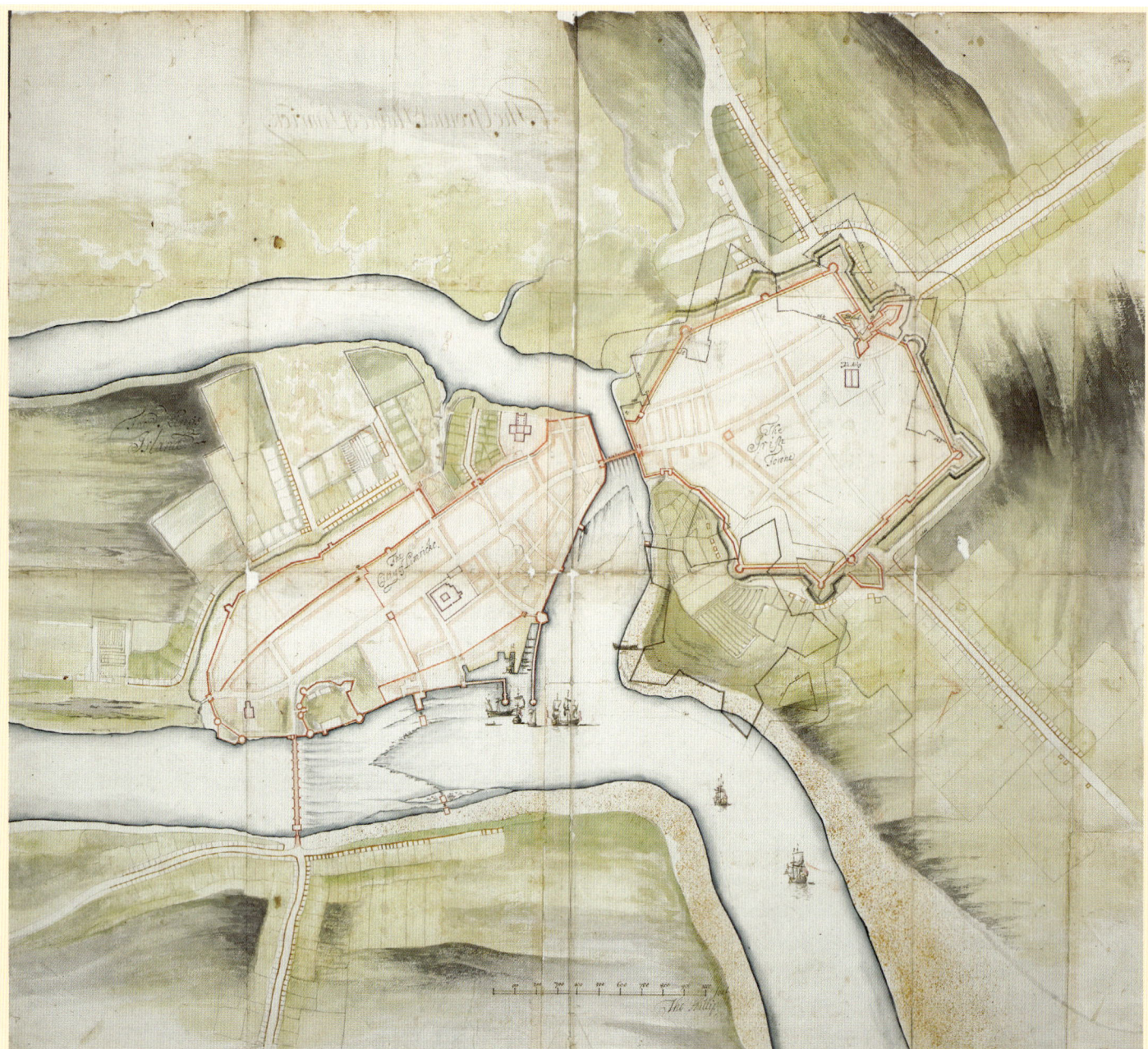

Fig. 19.14 THE EARLY CITY AND PORT OF LIMERICK. Thomas Phillips' 1685 map of Limerick shows the city just before the sieges of 1690–1691. The walled city consists of two parts. The area labelled 'City of Limericke', in the centre left, is the medieval settlement that developed at the southern end of King's Island – the land between the Abbey River (top of image) and the Shannon (bottom). Also referred to as the English Town, this area contained the original Viking harbour, which remained the focal point for the city's trade, as well as St Mary's Cathedral (centre) and the thirteenth-century King John's Castle. The 'Irish Towne', which was walled in the fifteenth century, lies to the south (right) of the English Town, to which it is connected by Baal's Bridge. The structures on the Shannon at the Curragower Falls, between Thomond Bridge and the harbour, are water mills. Suburban development is evident in Thomondgate (west of Thomond Bridge) and on the main roads leading from the Irish Town, though there is as yet no settlement in the area at the bottom right of the image which, from 1760, was developed as Newtown Pery. [Source: National Library of Ireland, call no.: MS 2557/21]

From the medieval period onwards, Limerick's economy depended on its port activities, which were oriented heavily towards exporting. Exports consisted of agricultural produce from the city's hinterland which, besides the whole of County Limerick, extended into counties Kerry, Tipperary and Clare. Rapid industrialisation and urbanisation in Britain in the early nineteenth century generated strong demand for, and a surge in exports of, products such as cereals, butter and pork. While industrialisation did not occur in the southern part of Ireland to anything like the same degree as in Britain, Limerick nevertheless diversified its economic base with the addition of significant industrial capacity in bacon curing, dairy processing and clothing. All of these industries were located close to the city centre. They served lucrative overseas markets and, initially, the port was essential to that development. By 1835, the value of exports through the port was estimated to be £800,000 per annum while that of imports was £325,000.

Events in the course of the nineteenth century affected the port in different ways. The advent of the railway in 1848 had an adverse impact, by allowing the city's food products to be conveyed by rail (rather than coastal vessel) to Dublin, Cork and Waterford for onward export to Britain. However, imports remained strong and, in the years immediately following the Great Famine, a considerable volume of wheat and corn was imported from North America through Limerick. This development can be attributed to two post-Famine trends in

Ireland: first, a greater dependence on bread and cereals for food and second, a shift in agriculture from tillage to pasture. While milling had long been an important enterprise in the city, it was now scaled up considerably and became a major contributor to port traffic. The Limerick millers were technologically innovative and, by the 1890s, Limerick surpassed Cork in its imports of grain.[13] As the industry grew, ownership became more concentrated and activity more centralised in the docks (Figure 19.13). This area also contained the timber and coal yards and nearby, the factories of Irish Wire Products and Irish Cement. The docks and the area adjacent to it remained a major employment hub in the city until the middle of the twentieth century. The port was particularly important as a source of employment for the city's working-class communities, though the majority of the jobs were relatively low-skilled, manual-labouring ones and dock work was highly casual in nature. In 1911, the city's population was 39,000.

RIVERS AND RAILWAYS

Prior to the emergence of canals and railways in Ireland during the nineteenth century, goods had to be delivered by road or river transport. Heavy goods could be carried only short distances overland. If, like live cattle, they could move themselves, the issue of distance was modified. Cattle, for example, could be marched, or driven, along roads from County Kerry to Cork docks. The movement of grain presented more difficulty. Only small quantities of this bulky commodity could move to the coast for warehousing and export.

Few Irish rivers were extensively navigable and bulky goods could not, therefore, be transported easily on the rivers. There were exceptions, and in this category was part of the course of Munster's River Blackwater and its tributary, the Bride. At high tide, the Blackwater was navigable from below Cappoquin and the Bride, for small lighters, after Tallowbridge.[14] Many farms on the River Bride possessed their own mini-quays, as at Ballyphilip Castle, County Waterford (Figure 19.15). From these minute quays, estate owners and tenant farmers loaded their grains onto lighters to be carried downriver to Youghal, where they were stored prior to being exported, chiefly to south-western England. Smaller and more occasional shipments of stock and passengers were also carried by small ferries and more than a dozen ferry quays were erected on the river's banks. From upstream, shipments of bricks, coal, sand and seaweed were delivered to the same quays.

By the mid-eighteenth century, Youghal's commerce expanded substantially, permitting merchants to modify the waterside by reclamation, the construction of new quays, warehouses and attractive residences. A constellation of different types of boats and lighters plied the river upwards from Youghal. Subsequently, competition from the railways later in the nineteenth century, and

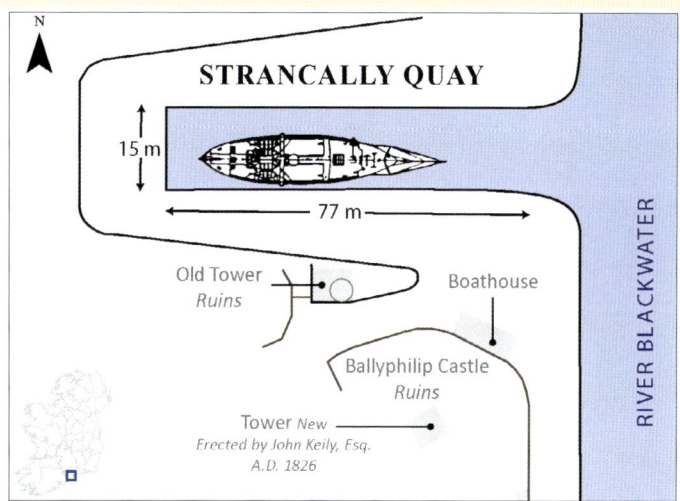

Fig. 19.15 PLAN OF STRANCALLY 'QUAY', COUNTY WATERFORD. Many boats plied the River Blackwater and its tributary, the Bride, during the eighteenth and nineteenth centuries. They carried grain downstream to Youghal and sand, seaweed and wood upriver. There were at least forty quays on these rivers, with Strancally Quay one of the largest and most elaborate, situated c.15km upriver from Youghal. [Data source: adapted from: Niall O'Brien, Marston, Tallow].

the building of bridges across these and many other rivers, deflected the transport of bulky goods to Cork. This allowed the city to emerge at this time as Ireland's premier grain-exporting port. The period marked a golden age for the port of Youghal and witnessed a widespread improvement in living conditions.[15]

Elsewhere in Munster, and wider in Ireland, significant improvement in roads and the building of new bridges– as at Mallow in County Cork – facilitated a re-ordering of port hinterlands. Cork became pre-eminent in Munster, heralding the gradual demise of now smaller rival ports, such as Kinsale and Youghal. The emergence of a railway network from the 1840s consolidated these trends and laid the foundations for the rise of Ireland's coastal seaside resorts.

Case Study: Coastal Railways

Ray O'Connor and Richard Scriven

The development of the railway network in the nineteenth and twentieth centuries transformed Ireland by connecting inland towns and villages with larger urban areas and ports on the coast (Figure 19.16). These ports were critical points of access and egress, with the emerging rail network shaping new patterns of movement of people, goods and services. The complex links between railways and the coastal social and economic geographies is an under-appreciated aspect of this age of steam.[16]

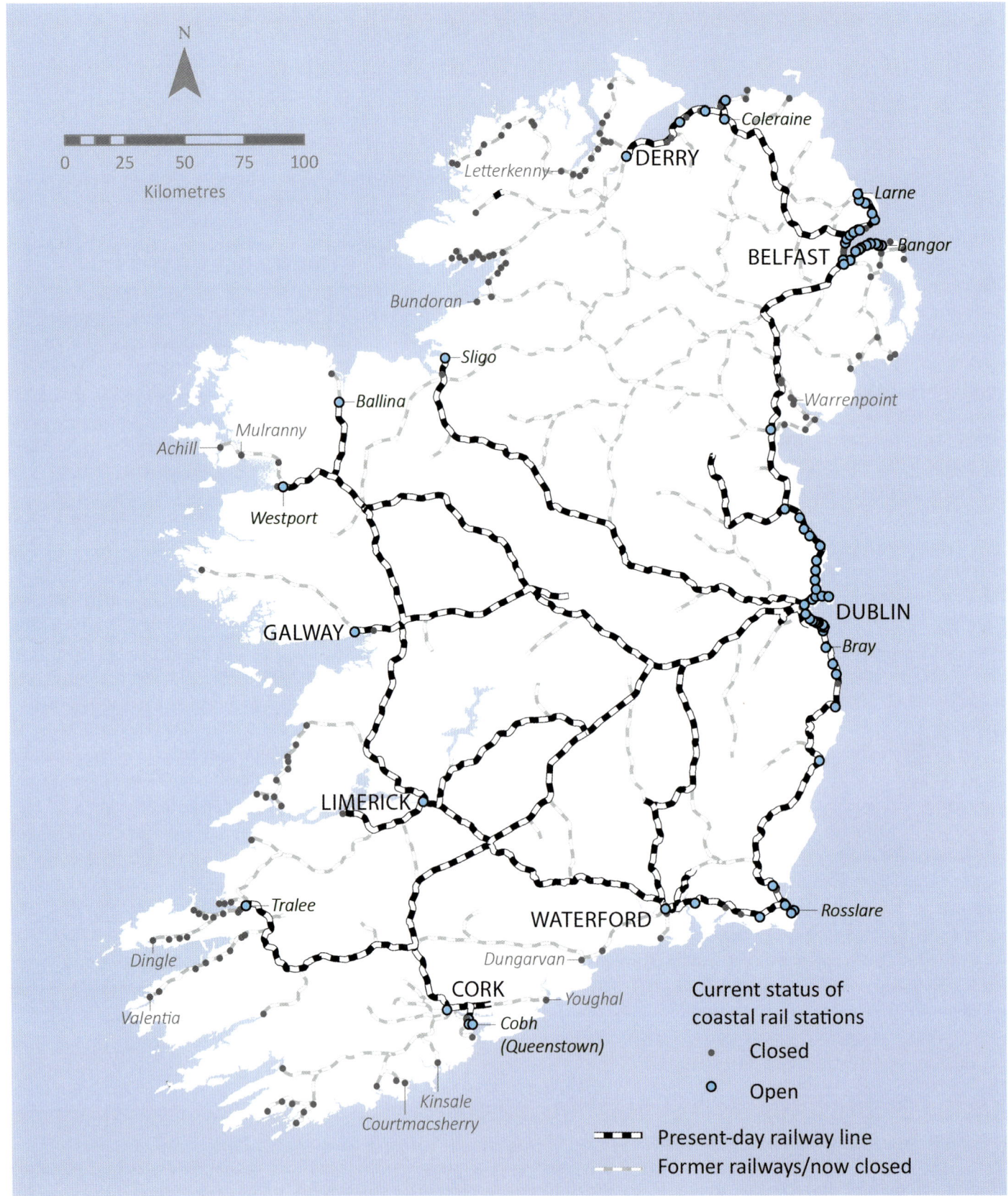

Fig. 19.16 THE CURRENT AND FORMER RAIL NETWORK IN IRELAND, SHOWING COASTAL RAILWAY STATIONS. Most of the island's coastal lines were closed and many had been dismantled by the 1970s. Many of the routeways are now used as part of the road system, as walking tracks and, more recently, as coastal 'greenways'. [Data source: modern network from Iarnród Éireann Irish Rail; historic network adapted from Viceregal Commission on Irish Railways, 1906]

HISTORICAL OVERVIEW

Ireland's railway experience differs from other countries in that the development of the rail network took place in the context of a major decline in population. While rail networks in other countries developed services for expanding passenger and freight markets, in Ireland the declining population acted as a significant constraint. This helps explain both the time lag between plans for lines that were made in the

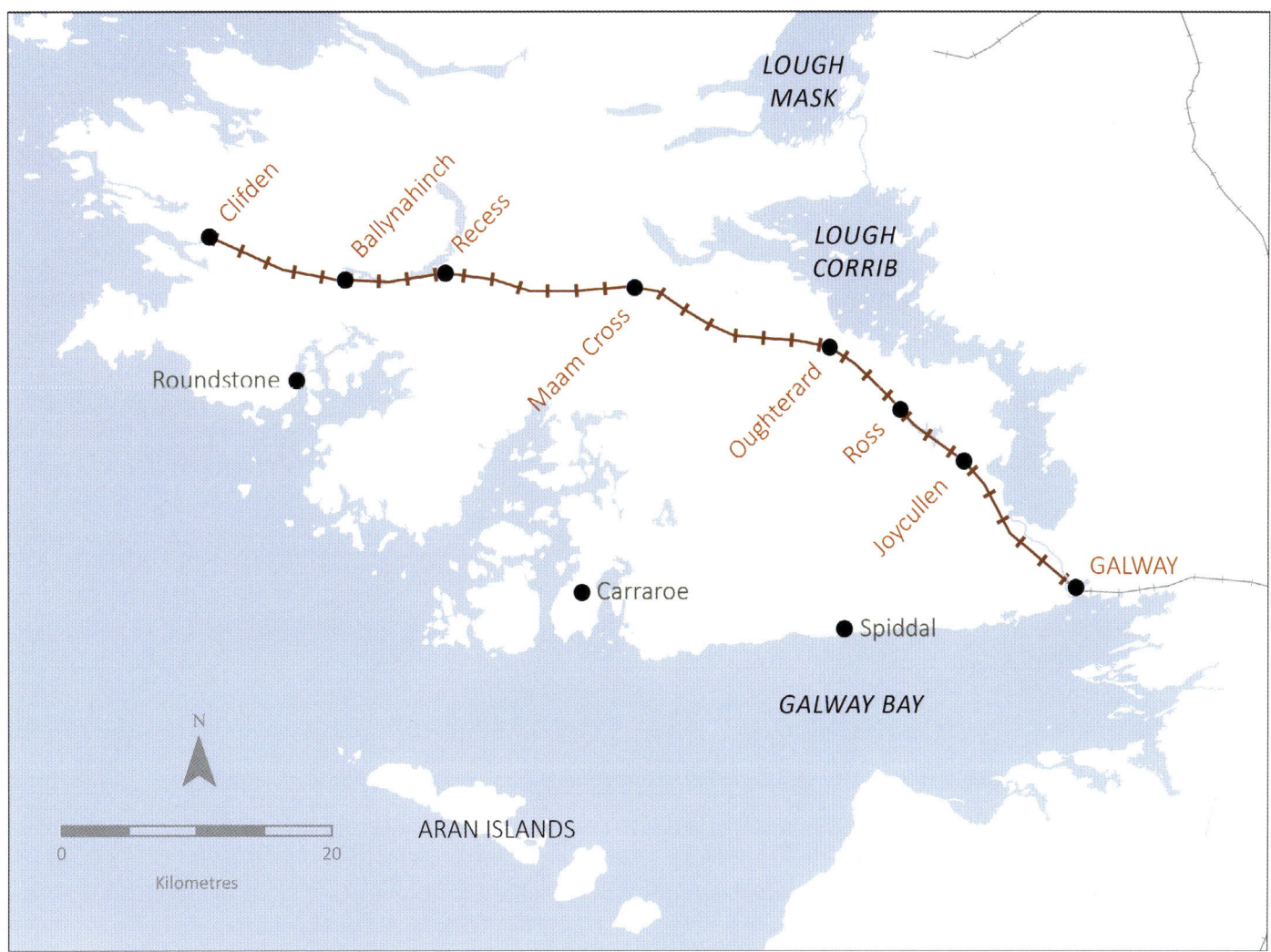

Fig. 19.17 CONNEMARA RAILWAY LINE. In response to a series of poor harvests, harsh winters and a decline in agricultural prices in the 1880s it was decided that, as a catalyst for immediate and longer-term economic development, a railway line would be built from Galway, through Connemara, to the coastal
village of Clifden. A grant of over £250,000 was made to the Midland Great Western Railway Company to organise and oversee the scheme. Its purpose
was to help in the development of fisheries and agriculture by providing overland access to markets. Prior to the construction of Balfour's Galway to Clifden line, all supplies to Clifden and surrounding villages were delivered by sea. A coastal route, favoured by locals and the Midland Great Western Railway
Company, through Spiddal, Carraroe and Roundstone, was not chosen by the decision-makers, the Royal Commission for Irish Public Works, and an inland
route was preferred. The line was completed in 1895. Not convinced that fisheries or agriculture could be developed to a sustainable level, the Midland Great
Western Railway Company promoted aggressively the Connemara region and this particular railway line. Such marketing, and an investment in a hotel in
Recess, eventually paid dividends and Connemara became popular as a tourist destination among the landed gentry. In spite of this, the railway never achieved
anything like the impact hoped for. [Data source: adapted from Viceregal Commission on Irish Railways, 1906]

1830s and early 1840s, and their continued delivery into the 1890s, as well as changes to the initial routes envisaged.

The railways, as originally constituted, were private companies and attracted a lot of speculative investment. Therefore, they were drawn
to areas with higher population densities and to the more prosperous regions where freight and passenger traffic were guaranteed. Ireland's
population declined by 3 million people between 1841 and 1881, with the western seaboard counties experiencing a disproportionate share
of this loss (see Chapter 4: People, Agriculture and the Coast, Figure 4.22). This uneven areal population decline explains the initial
concentration on developing connectivity between Ireland's cities, and why it took almost half a century for railways to reach out to the
rapidly depopulating coastal regions of western Ireland. For example, trains reached Dingle in 1891, over thirty years after the opening of
Tralee station in 1859, and Letterkenny, County Donegal, by 1909, fifty-seven years after establishment of the Londonderry Waterside
Station. In spite of this, in less than a century, from 1834 to 1914, the rail system expanded from a few fledgling lines to a complex network
that reached all corners of the island, including its coastal regions.

By 1845, there were roughly 160km of track in Ireland, this rose to 3,200km by 1865, and peaked at over 5,600km by 1920. With the
country still recovering from the effects of the Famine, there was ready availability of cheap land and labour, which made railway lines
relatively inexpensive to construct. The building of lines to relatively remote areas, especially along the western coast, was also facilitated
by government support, in particular the Light Railways (Ireland) Act 1889 championed by Arthur Balfour, and by related railway
development projects (Figure 19.17). An example of a 'Balfour' project funded entirely by the government was the Westport to Achill Island
rail extension in County Mayo in the 1890s. Poverty was endemic in this coastal region of Mayo and emigration rates high. During this time,
there was also a drive to extend routes to west-coast ports considered to possess good potential for trade, such as Clew Bay and Valencia,

Fig. 19.18 A CONTEMPORARY VIEW OF A SECTION OF THE FORMER CORK, BLACKROCK AND PASSAGE RAILWAY (1850–1932), LOOKING NORTH-EAST INTO CORK HARBOUR. This line had been a popular suburban and tourist service that ran originally to Passage West and was later extended to Carrigaline and Crosshaven in 1904. After closure of the line, large sections were developed as a public walkway, with remaining features, such as this tunnel, serving as points of industrial heritage interest. [Source: Kyle Fawkes, Sarah Kandrot and Zoë O'Hanlon]

County Kerry. The intention was that these places would become transatlantic hubs.

Since the 1920s, two independent companies (now Iarnród Éireann in the Republic and Northern Ireland Railways) manage the island's rail system. Until the 1980s, closures and declines were dominant trends as competition from road transport, reflected especially in the significant growth of car ownership levels, accelerated the loss of passengers and freight traffic on Ireland's rail network. Between 1950 and 1970, the length of the island's rail track was halved.

Since the 1980s, a process of consolidation has occurred as attempts were made to re-establish the role of railways in Ireland's economic and social life. This centred around the provision of a high-speed service for inter-city passenger traffic. In this, Dublin and Belfast have confirmed their roles as focal points of their state networks, as well as being terminals for a crucial cross-border service. The railways also acquired a growing role for commuter traffic. Much of this occurs within the expanding hinterlands of the capital cities, reflected in the multiplicity of stations along the coastal railway lines to the north and south of Dublin, as well as the coastline near Belfast (Figure 19.16). Despite the success of these functions, and curtailment of rail closures, Ireland's track length by 2020 was some 2,200km (c.340km in Northern Ireland); 40 per cent of its peak in 1920. Several former coastal railway lines have been revitalised as routes for walkers and cyclists (Figure 19:18). The Great Western Greenway, connecting Westport and Achill Island, and the Waterford Greenway, linking Dungarvan to Waterford city, are examples of what is a successful tourism and recreation model.

CONNECTIVITY

Railways offered a ground-breaking form of connectivity that linked inland regions to coastal areas, enabling new and faster flows of goods, people and services. By the late nineteenth century, development in the railway network had resulted in a general raising of living standards. Coastal and wider rural populations could now access new levels of service and consumable goods, while increased communications and movement allowed the spread of ideas and knowledge. The potential that railways offered to connect Ireland, both internally and to the wider world, was clearly realised by political and commercial figures. The *Royal Commission on General System of Railways in Ireland, 1836–1838* was formed to assess and prepare for the development of this new mode of transport across Ireland. Amongst the aims of the commission was the investigation of ports on the west and south coasts that could serve as sites for ships destined for North America, as well as for 'the construction of Lines of Railroad across Ireland, to such Port or Ports in connexion with the greatest possible collateral benefits to internal communication'.

In the 1840s, Daniel O'Connell had a vision for a railway that connected south Kerry to England via the packet station in Dublin and a transatlantic hub based in Valencia Harbour. He understood fully the transformative potential of railways, although when and where railway lines were built always represented a compromise based on economics, engineering and expediency. In the case of the Iveragh Peninsula, O'Connell's vision was realised only in part; south Kerry was connected to Dublin, but the line never made it all the way to Valencia Island. The transformative impact of a railway is also evident in the case of Rosslare, County Wexford. It was a relatively small port until 1906 when the Great Southern and Western Railway (GSW), in conjunction with the English (and Welsh) Great Western, established a mail boat service there connecting with Fishguard in Wales. This was part of a larger strategy of moving traffic across the Irish Sea to more southerly routes and thus away from competing companies linking to Dublin and Belfast. The Fishguard and Rosslare Railways and Harbour Company also provided a link from Rosslare and Waterford to connect with Cork. The GSW ran boat trains on these lines connecting Rosslare with the transatlantic port of Queenstown (Cobh).

As the rail network expanded, coastal settlements that had been focused on the sea and isolated by a poor road network were, for the first time, provided with a direct and reliable connection with Ireland's major cities. For areas bypassed by the rail network, new levels of isolation and even greater sets of challenges arose to threaten the existence of many coastal communities.

COMMERCE

Railways rapidly became a popular way to travel and sell goods, and there were huge commercial benefits to towns having railway stations. Farmers and local industries flourished, as their goods were brought to almost any destination in the country, both efficiently and cost-effectively. This encouraged Irish businesses to sell their products further afield. The transportation of livestock was an important feature of the Midland Great Western Railway, which carried cattle not only to fairs and marts across Ireland, but also to be exported to Britain. From the 1880s until the 1960s, wagons of cattle were brought to Dublin where they were taken either through Smithfield, or directly to awaiting steamers at the North Wall. Similarly, the rail line at Achill Island, County Mayo, allowed for the transportation and export of fish to Dublin and on to Birmingham, while in County Kerry, from 1920 to the early 1950s, the rail link between Renard and Cork city facilitated a thriving fishing industry. This was especially important during the Second World War when food shortages and rationing in Britain led to increased demand for salted mackerel (Figure 19.19).

TOURISM

The expansion of the rail network along with the road system opened up existing coastal spa towns, such as Buncrana, Salthill, Newcastle and Holywood.[17] Later, towns such as Bray, Ballybunion, Kilkee, Skerries, Youghal and Tramore were also transformed into significant seaside resorts. The running of day trips or excursions by both tour operators and railway companies to different parts of the country, along with the establishment of hotels to cater for train passengers, were central to the creation of Ireland's tourist sector and especially the emerging seaside resorts.

Fig. 19.19 THE RAILWAY BRIDGE APPROACHING CAHERSIVEEN FROM THE EAST IN THE LATE NINETEENTH CENTURY. The 'old' railway bridge spanning the River Fertha estuary connects Cnoc na dTobar (Hill of the Wells) with the town of Cahersiveen in County Kerry. At its peak, up to three trains daily left the town, carrying fish bound for Billingsgate in London. The fishing industry provided seasonal employment for nearly 500 people in an area where employment was scarce. On a less positive note, while the railways brought considerable economic benefit and much needed employment, they also facilitated emigration. The rail line was closed on 1 February 1960. Since the early 2000s, however, concerted efforts have been made to transform the section of the line that once connected Glenbeigh to Cahersiveen into a greenway. It is envisioned that this greenway would provide a cycleway that would allow visitors and locals to experience the stunning views and landscapes of south Kerry. [Source: National Library of Ireland, call no.: L_ROY_04573].

Fig. 19.20 CLOSED- AND OPEN-TOP CARRIAGES AT ROSTREVOR, ALONG THE WARRENPOINT–ROSTREVOR TRAMWAY (1877–1915). This scenic coastal location became a popular holiday destination with the arrival of the railway in 1849 and the addition of tourist infrastructure. This service connected Warrenpoint railway station to the Great Northern Railway's hotel at Rostrevor. It was over 5km long and was one of the first tramways in Ireland; it is an excellent example of an integrated tourist infrastructure developed to encourage passenger traffic. [Source: National Library of Ireland, call no.: L_CAB_02168]

Fig. 19.21 YOUGHAL EXCURSION: CROWDS ARRIVING AT YOUGHAL RAILWAY STATION IN 1908. The railway connected Cork city to Youghal in 1860, opening the way for the town to become a significant seaside resort, especially for family and school day-trips from Cork. The accessibility and affordability of the line enabled new types of leisure opportunities for the middle- and working-class populations of the city, who travelled in their thousands during the summer months. While the line was closed to regular traffic in 1963, it remained open for Sunday excursions and goods traffic until the late 1970s, illustrating the continuing popularity of the service. [Source: Cork County Library, © National Library of Ireland, call no.: L_ROY_11470].

In the early twentieth century, Achill Island used its rail link to develop its tourist sector by running special trains directly from Dublin during the summer months. Travel packages included stays in the rail company's hotel, located adjacent to the train station. Many railway companies adopted such a strategy and built their own hotels to cater for and attract passengers. For example, the Great Northern Railway Company had a series of 'Great Northern Hotels' – Bundoran, Warrenpoint and Rostrevor – which were extremely popular during the summer (Figure 19.20). At Mulranny, County Mayo, the Midland and Great Western Railway opened a luxury hotel in March 1897, and promoted a combined hotel and rail ticket that offered patrons the availability of a nine-hole links golf course overlooking Clew Bay, as well as access to boats and beaches.

The seasonal nature of tourism, however, affected railway companies, which also experienced growing competition from buses and cars from the 1940s. In spite of this, summer routes and special services, such as All-Ireland GAA trains and pilgrimage trains, kept many lines open even after much of their regular traffic had ceased. For example, the Bundoran Express running during the summer months brought tourists to the seaside town, as well as pilgrims to Pettigo, near Lough Derg.[18] The Cork–Youghal line also operated summer trains until the 1970s (Figure 19.21), while Courtmacsherry in west Cork, located on the Bandon–Timoleague railway, also remained a popular destination for Cork city residents until the 1960s.

The development of the tourism sector was a component of the larger modernisation process in Ireland during the late nineteenth century, in which coastal and wider rail networks were of critical importance. Railways transformed earlier seaside towns into planned, orderly venues, with promenades, modern hotels and well-designed terraces and villas. It was also a form of regional development, with the new infrastructures and services making previously depressed areas, especially along the western coast, tourist hotspots. The impact of these changes remain in many seaside locations, which continue to serve as tourist centres, long after the closure of the railways.

Case Study: Seaside Resorts

Patrick O'Flanagan

Ireland's coastline is dotted with a significant array of seaside resorts, the prime function of which is to provide leisure facilities for visitors. These resorts share a range of characteristics, which include particularly that of having short holiday seasons, confined mainly to the June to August period. Most resorts are small and few number more than 1,000 full-time residents. Some may even cluster around a single facility,

Fig. 19.22 MAIN STREET AT WATERVILLE, COUNTY KERRY (c.1900). This is a view of the small coastal resort of Waterville, looking southward toward the open bay and waterfront area. Interestingly, the houses and this view of Waterville have changed little since the photograph was taken. [Source: National Library of Ireland, call no.: L_ROY_08180]

Fig. 19.23 BRAY PROMENADE, COUNTY WICKLOW (*c.*1900). This depiction of the crowded Bray promenade illustrates the large numbers of tourists that travelled to Bray from Dublin utilising the regular rail services that connected the capital city to such seaside resorts. Bandstands, kiosks and other sources of recreation were often located on such promenades to help diversify their appearance and increase the attraction for visitors. [Source: National Library of Ireland, call no.: L_CAB_06647]

such as a hotel, as at Inchydoney, County Cork, or Mulranny, County Mayo (see Chapter 27: Tourism and Leisure).[19]

These resorts may be classified into different groups on the basis of their origins, size, functional complexity, type(s) of clientele, length of season and built fabric. Others may be defined simply by a main street, as at Waterville (Figure 19.22), Courtmacsherry and Lehinch. Most of Ireland's resort settlements developed during the late nineteenth century when mass transport became attainable as the island's railway network expanded and extended to remote areas, such as west County Clare and west County Kerry. Some centres provided

Fig. 19.24 SEASIDE RESIDENCES DEVELOPED SINCE THE 1950S IN THE CORK HARBOUR AND OPEN COASTAL AREAS BETWEEN MYRTLEVILLE AND CROSSHAVEN, COUNTY CORK. This is the area of the Ford box settlements near Crosshaven and Myrtleville. Many of these houses were built in the 1950s and 1960s from the remains of large wooden crates used to transport car parts to Cork from London for reassembly. The boxes were then re-utilised to serve as rudimentary summer homes south of Crosshaven, beside Cork Harbour. Many of these dwellings have been upgraded and re-developed subsequently and bear little resemblance to the original packing-case structures! [Source: Maxim Kozachenko]

additional attractions, as in Bushmills, County Antrim, which developed early rudimentary tourist facilities for those visiting the Giant's Causeway. The advent of the consumer society, especially since the 1960s, based on stable incomes and enhanced ownership of private cars, has further opened up many coastal settlements as holiday destinations. More recently, they have also become the locales for clusters of second homes, as at Courtmacsherry, Waterville, Strandhill, Rossnowlagh and Bundoran.

Beneath an apparent functional uniformity, many of Ireland's coastal resorts have diverse origins. Several began as fishing settlements, or as minor ports, and later accreted recreational dimensions, as at Bray, Kinsale, Youghal, Portrush and Portstewart. All of these resorts added their tourist functions and new residential infrastructures after they became connected to the national railway network. Cheap post-Second World War ship travel to Ireland from England, the Isle of Man, Scotland and Wales gave a new impetus to centres like Bray (Figure 19.23). In parts of Ulster, however, early and significant industrialisation prompted the emergence of a series of seaside resort facilities in centres such as Portstewart, Portrush, Newcastle, Rostrevor and Warrenpoint. These developed primarily to accommodate growing numbers of factory workers residing within the province.

Although emerging in part to meet the growing demands for tourism from Dublin's expanding urban population, Bray and Dún Laoghaire developed rapidly in the late 1950s as mass tourist

Fig. 19.25 EXAMPLES OF BATHING PLACES. a. Vico Men's Bathing Place, Dalkey, County Dublin. This bathing place on Vico Road, Dalkey, was opened in the late nineteenth century and served as a male-only facility for more than a century. It admitted women in the early 2000s. [Source: Patrick O'Flanagan]; b. Rudimentary bathing boxes on the beach at Enniscrone, County Sligo (c.1900). Minor settlements also grew up around seaweed baths in this resort and at similar locations, such as at Ballybunion. [Source: National Library of Ireland, call no.: L_CAB_07914]

destinations for well-paid English and Scottish industrial workers. This visitor bonanza melted away rapidly after the northern Mediterranean and the Canary Islands opened up to mass tourism, facilitated by cheap charter air travel. Other centres, such as Salthill, grew after a tram system opened, linking it to Galway city, while Strandhill in County Sligo emerged as a resort suburb for the nearby larger urban centre of Sligo town. Some scenically attractive centres – like Glengarriff in west County Cork with its Italianate gardens – became popular nodes on nineteenth-century grand journeys. A few decrepit clusters, such as Glenbeigh or Louisburgh, blossomed only after the 1970s, their development being linked to massive increases in private car ownership. At Myrtleville and the Crosshaven area, located at the mouth of Cork Harbour, an unusual development in tourism and second homes occurred. This was linked directly to the easy availability from the 1950s to the 1970s of large sturdy wooden boxes, which had contained car parts destined for the Ford assembly plant in Cork. These were taken by many craftspeople to build a kind of untidy, almost 'squatter' resort settlement. Many of these properties have now been extensively renovated (Figure 19.24).

Many larger resort settlements retain a distinctive suite of buildings, street furniture and other features to support leisure functions linked to sea-related activities. These facilities have often deeply modified the intertidal zone with, for example, the building of promenades (esplanades) at Cobh and Bray. The most extensive one at Salthill, and recently extended to Mutton Island, offers an unrivalled marine walkway around the inner sections of Galway Bay. At Dún Laoghaire, although its piers were never envisaged initially as a major urban leisure facility, they have now become adopted and adapted as one of south Dublin's leading urban walkways. Piers and breakwaters, acting as part of sea defences, have also been incorporated as parts of many promenades (see Chapter 30: Engineering for Vulnerable Coastlines). Separately, bathing places, often segregated until the late 1960s, or even later as at Vico Men's Bathing Place, Dalkey (Figure 19.25), were often placed in a picturesque setting. Bathing places such as this, and the even more famous 'Forty-Foot' at Sandycove, acquired bathing huts, sea-baths and pond facilities. In an effort to stimulate economic development at several remote coastal settlements, marinas have been built since the 1990s. This represents an effort to lure visiting mariners to use these locations, for example Cahersiveen, Courtmacsherry, Kilrush and Westport.

The extraordinary contemporary growth in mass tourism abroad, aided by low-cost airlines, presents a major challenge to seaside resorts in Ireland and their capacity to attract tourists. The meteoric growth in car ownership within Ireland has helped, in part, to compensate for the loss of the traditional family group as holidaymakers. Marketing devices, such as the promotion of certain coastal tourist routes – like the 'Ring of Kerry' and the 'Wild Atlantic Way'– have also enjoyed success in attracting many tourists to villages in more remote coastal areas, as at Waterville or Belmullet (see Chapter 27: Tourism and Leisure). Buttressed by investment in 'green' cycleways, some coastal resorts have managed to once again attract new family-based tourism, as along the Westport to Newport to Mulranny 'Greenway', which follows the route of a disused railway line in County Mayo. Interpretive centres, as at the Céide Fields and Ballycroy, County Mayo, have also drawn visitors into coastal areas otherwise seldom visited.

Nineteenth-century and earlier 'period' buildings have enriched several resort townscapes, especially those of some of the larger resorts, such as Dún Laoghaire and Youghal. In these settlements, local entrepreneurs have often invested in the construction or, more recently, in the refurbishment of large and very impressive hotels, as at Bray. Here, the extension of the Dublin to Dún Laoghaire railway to Bray in 1855 became a catalyst for expansion and ushered in a series of major urban developments. These included the erection of several impressive hotels, the laying out of a new chequerboard pattern of mainly residential roads behind the seafront and the building of an exotic and complex Turkish baths, opened in 1859 (Figure 19.26).

With the emergence of Ireland's Celtic Tiger economy from the mid-1990s, tax incentives made available to contractors for the construction of second homes and hotel development led to a frenzy of building at coastal resorts. With the 2008 economic crash and subsequent recession, many of these resorts became defined as possessing a glut of buildings, often becoming derelict and difficult to sell. Ireland's economic recovery since 2014, however, has led to a revival of interest in property development in coastal settlements and has contributed to a renewed prosperity for their communities.

Fig. 19.26 TURKISH BATHS, BRAY, COUNTY WICKLOW. This iconic building opened in 1859 and quickly became a major attraction in Bray, a new and impressive seaside resort the origins of which are linked to the opening of the Westland Row (now Pearse Street) train station in Dublin and the first train connection to Bray in 1854. The building was demolished in 1980. [Source: National Library of Ireland, call no.: STP_2915]

The recent advent of cruise-liner traffic to some ports has also thrown them a lifeline and opened up new economic possibilities for them and their hinterlands. A good example is Cobh, where some of the world's largest liners berth at this historic port and tower over Victorian crescents that lie adjacent to the quays (see Chapter 27: Tourism and Leisure). The waterside infrastructure in this small resort and many of the lodging houses used for earlier emigration, have been transformed into boutique hotels and restaurants. More specialised cruises in small luxury liners have brought similar new opportunities to smaller harbours, such as Glengarriff in west Cork and even Killybegs in County Donegal.

REMOTE COASTAL LIFE

Outside of the cities, towns and larger villages, how did coastal communities maintain themselves between approximately 1600 and 1900? In what kind of settlements did they reside, and to what extent were people able to modify pre-existing coastal cultural landscapes, or create entirely new ones? Such questions need to be answered when considering the elements of coastal landscape changes in this period.

A significant influence on Ireland's coasts in the early decades of the seventeenth century was the arrival to the south-west coast of Ireland of large numbers of settlers from England, as part of that country's colonial enterprise. Profound alterations to the coastal zone were propelled by this form of settler colonisation, including the injection of venture capital for the exploitation of indigenous resources – of the sea (fish), land (minerals), forests (wood) – and for the intensification of commercial agriculture. These ventures were undertaken by a cadre of entrepreneurial personalities, such as Richard Boyle, earl of Cork and his partner, William Hull.

In south-west County Cork, these New-English settlers built a series of fish 'palaces' or 'pallices'. The remains of these sites can still be found, for example, at Baltimore, Bantry and Clonakilty, or at Crookhaven, Goleen and Sherkin Island along the shores of Dunmanus Bay and on to Eyeries and Ardgroom (see Chapter 18: Era of Settlement: Trade, plantation and piracy). The fish palaces were designed to process mainly pilchards, involving the gutting, pressing, pickling and smoking of the fish, with the finished product being exported mostly to southern Europe to satisfy the Lenten trade. These enterprises employed an estimated 2,000 people, although it is not clear if the work was exclusively for the immigrant New English and what, if any, was the role of women in this business. They also spawned new settlements and promoted urban growth in existing ones, although in some instances – whether due to the aftermath of the 1641 Rebellion, or the failure of the fishery – several settlements disappeared subsequently.

During the late eighteenth century, linked to a boom in cereal growing and exports, a constellation of small milling settlements was founded along the County Cork coastline, such as at Ballinacurra, near Midleton, in Cork Harbour.[20] This was a region suited to cereal growing as plentiful sources of fertiliser were available locally from harvesting seaweed, to be carried subsequently by a combination of donkeys, or donkeys and carts, to adjacent fields.

Mining activities were developed from the late eighteenth century in a number of coastal environments – such as around Allihies, in west County Cork, and Bunmahon, County Waterford (see Chapter 29: Coastal Mining, Quarrying and Hydrocarbon Exploration) – and quarrying took place at Dún Laoghaire.[21]

Innovations in cable-laying and the decision to improve communication links between North America and Europe also had an important impact on Valencia Island, County Kerry. Knightstown became the eastern terminal for a new transatlantic cable link and, in 1866, the final stage of the 1,686 nautical miles of cable was pulled ashore in Newfoundland. These and other agri-rural businesses and linked economic stimuli were accompanied by the building of wharves and docks to facilitate the movement of goods. In aggregate, they encouraged the growth of many existing coastal settlements as well as the emergence of new ones.

The ingenuity and evolution of many coastal communities over these centuries is testified, in particular, by the fishing industry, and the range of boats, fishing methods and processing technologies utilised.[22] Throughout the eighteenth century, inshore fishing had intensified along large stretches of the Irish coast. In County Cork, for example, most small ports possessed many sand-lighters, small long-liners and seaweed harvesting craft (see Chapter 23: Maritime Traditions and Institutions). This enhanced fishing capacity, coupled with the availability offshore of rich resources of fish, allowed some coastal communities to supply distant markets with fish, such as the newly industrialised centres in Britain and on the continent.

In spite of the existence of nutritious and valuable marine resources in adjacent seas, for many coastal communities, and especially those located on the western seaboard, poverty and undernourishment were common descriptors of their living

Fig. 19.27 VIEW TO THE SEA FROM CILL RIALAIG, BOLUS HEAD, BALLINSKEL-LIGS, COUNTY KERRY. This and other adjacent clusters were diminutive settlement nodes in a lattice-like web of minute enclosures. Nestling on a hillside above the sea, this well-known house cluster, or clachan, was the home of the famous seanchaí Seán Ó Chonaill. His work was published in *Leabhar Sheáin Í Chonaill* (1948), edited by Séamus Ó Duilearga and translated subsequently into English by Máire MacNeill in 1981. [Source: Sarah Kandrot]

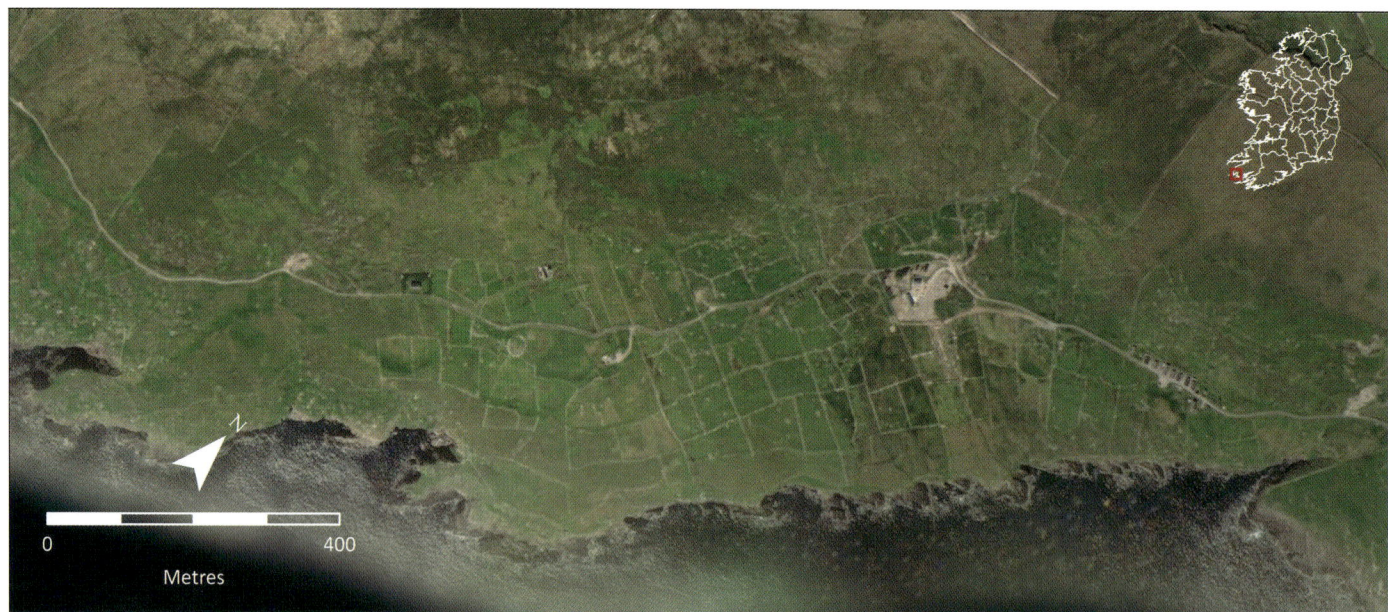

Fig. 19.28 PATTERN OF FIELD ENCLOSURES IN THE BOLUS HEAD AREA OF COUNTY KERRY. The web of small, walled enclosures underlines the struggle to sustain life on the edge of a steep Kerry hillside by the residents of Cill Rialaig and Bolus Head, Ballinskelligs, County Kerry. [Source: Google, Digital Globe]

standards. This can be explained partly by the lack of secure access to the sea for most small-scale farmer-fishermen. For example, some coastal settlement clusters, or clachans, even though located within a stone's-throw of the sea, often had no connections with it, as at Cill Rialaig, Ballinskelligs (Figure 19.27). This and other adjacent clusters were diminutive settlement nodes in a lattice-like web of minute enclosures (Figure 19.28). Críostóir Mac Cárthaigh, quoting a state report in 1836 regarding Seafield in County Clare, highlights this problem: 'Those without land have very little employment, and their families are left starving'.[23]

The lack of access was compounded further by the fact that in such contexts most fishermen only had access to minute 'gardens' and, if the fishing failed, the consequences were dire. In County Cork, Spillar, describing the fishing 'village' of Meelmore beside Courtmacsherry, noted: 'It stands on nine acres of ground and consists of fifty five houses of inferior description, mostly inhabited by fishermen and sailors. There are no shops in the town which is of regular and uninteresting appearance.'[24]

While offshore waters hold valuable resources, so too does Ireland's intertidal zone. Aquaculture practices – such as oyster and mussel cultivation, salmon farming and seaweed harvesting – have long been encouraged by environmental conditions found in the deeply indented bays and inlets that characterise much of the island's foreshore. Documentary evidence for such activities can be traced back to medieval times, and even further if shell middens are taken into account. Since the eighteenth century, the opportunities presented by such practices were taken up by thousands of part-time farmers as a means to improve their incomes and diets (See Chapter 4: People, Agriculture and the Coast). As a result, the intertidal zone of almost every coastal county has, to some extent, been modified.

Oyster farming has a well-established tradition, and ponds for oyster spat rearing have been laid out in diverse locations, including Ballysodare Bay, County Sligo; Kilrush, County Clare; Lough Swilly, County Donegal; Dundrum Bay, County Down, and Clew Bay, County Mayo.

The management of many of these enterprises, however, was often totally inadequate and the oyster beds over-exploited to generate immediate returns to owners. The result was the rapid degradation of the resource with immediate and negative consequences for coastal communities.

An example of an area that experienced a significant transformation of its foreshore to facilitate oyster cultivation and harvesting was the inner area of Galway Bay. Documentary evidence from nineteenth-century licensing applications confirms the earliest was awarded in 1864. Foreshore constructions included 'spatting ponds', where young oysters were raised, while other sites focused on the fattening or 'steeping' of the oysters in 'parcs'. These parcs formed low extensive stone enclosures (known still as bracai) and extended for many kilometres along the intertidal zone. An estimated twenty million oysters were delivered to Dublin by train from Galway to meet the demands of its restaurant trade.

Salmon hatcheries, mussel and razor fish farming and eel fisheries also contributed to alterations made within the coastal zone, although seaweed harvesting might have been the most transformative activity. Until recently, several seaweed 'farms' operated along the coast, with Mill Bay, County Down, being one of the largest. Here, the bay was divided into over 1,000 half-acre plots, which could be bought, leased, sold or lent, and were exploited primarily by local farmers. Other seaweed farms were located at, for example, Achill Sound and Murrisk, County Mayo, and Ardara and Carndonagh, County Donegal.

By the early nineteenth century most coastal areas, but primarily those along the western seaboard, could be characterised as being congested with people. Subsistence farming dominated the economic landscape and, although significant numbers were also involved in maritime activities and mining, this could not offset the poverty that characterised family life in most localities (see Chapter 4: People, Agriculture and the Coast).

448

In exceptional instances, however, enormous inequity existed in the wellbeing of some coastal communities. This is exemplified in and around mining settlements on the Beara Peninsula, for example, beside Allihies in west County Cork. Most miners lived in dire conditions in a constellation of house clusters within about a 15km radius of the mine, at centres such as Ballynacallagh, Firkeen and Kilmichael on Dursey Island. Literacy levels were low and housing consisted of 'class four houses' (as defined in the early census data in Ireland), in marked contrast to the elaborate and opulent residence of the mine owner, Mr Puxley, at Dunboy (Figure 19.29). More widely, some west-coast areas offered better environmental conditions and opportunities for settlement, such as Gweedore in County Donegal or the Mullet Peninsula, County Mayo, and became 'nurseries of humanity'. In addition, the more rapidly transforming and often 'genteel' urban settlements of the eastern coasts stood in stark contrast to the localised famines, distress, poor laws and emigration that typified most rural coastal communities along the western seaboard.

Although the impact of various nineteenth-century famines in coastal areas are well documented (see Chapter 20: The Great Famine), schemes designed to improve the wellbeing of deprived communities were a rarity. In some areas, however, enlightened landlords were responsible for the execution of diverse relief works, especially between 1822 and 1831, such as Lord George Hill in County Donegal. In addition, a number of engineers, but in particular Alexander Nimmo (1783–1832), stood out in promoting infrastructures designed to alleviate distress of communities living especially along the western seaboard.[25] Nimmo designed, for example, piers, causeways, towns and villages – such as Knightstown on Valencia Island in County Kerry (Figure 19.30) and Roundstone in County Galway – as well as connecting many inaccessible coastal communities with the national road network. He was one of a trio of engineers that included William Bald (1784–1857) and Richard Griffith (1784–1876), who made immensely positive and undervalued contributions to the lives of thousands. In particular, their activities helped improve the food gathering and production capacities of many isolated western coastal communities, chiefly by improving land connectivity to ameliorate the impact of food shortages. Their efforts were a prelude to the relief works of the Great Famine.

In 1831, the British government instituted the Office of Public Works (OPW) to manage funds providing grants and loans that could be used to promote a programme of public works, such as roads, piers, harbours and field drainage schemes. It quickly became the principal driver for the construction of public works and its role expanded significantly during the Great Famine (1845–1852) (see Chapter 30: Engineering for Vulnerable Coastlines). During these years, it had the almost impossible task of providing a works programme and employment to help relieve the destitution that confronted a large proportion of the population, but in particular along the west coast.

Restrictions on funding and the scale of the problem during the Famine and post-Famine years, however, limited the overall accomplishments of the OPW's public works programme. In spite

Figure 19.29 MR PUXLEY'S HOUSE, DUNBOY CASTLE, BEARA PENINSULA. The opulence of the castle and the wealth of its owner sets it apart from workers' settlements associated with the mines. In this instance, the skilled Cornish miners' 'village' was located beside one of the nearby mines in order to provide them with access to their place of work. The mine manager's house was sited separately, but also beside the mine. [Source: National Library of Ireland, call no.: L_CAB_02083]

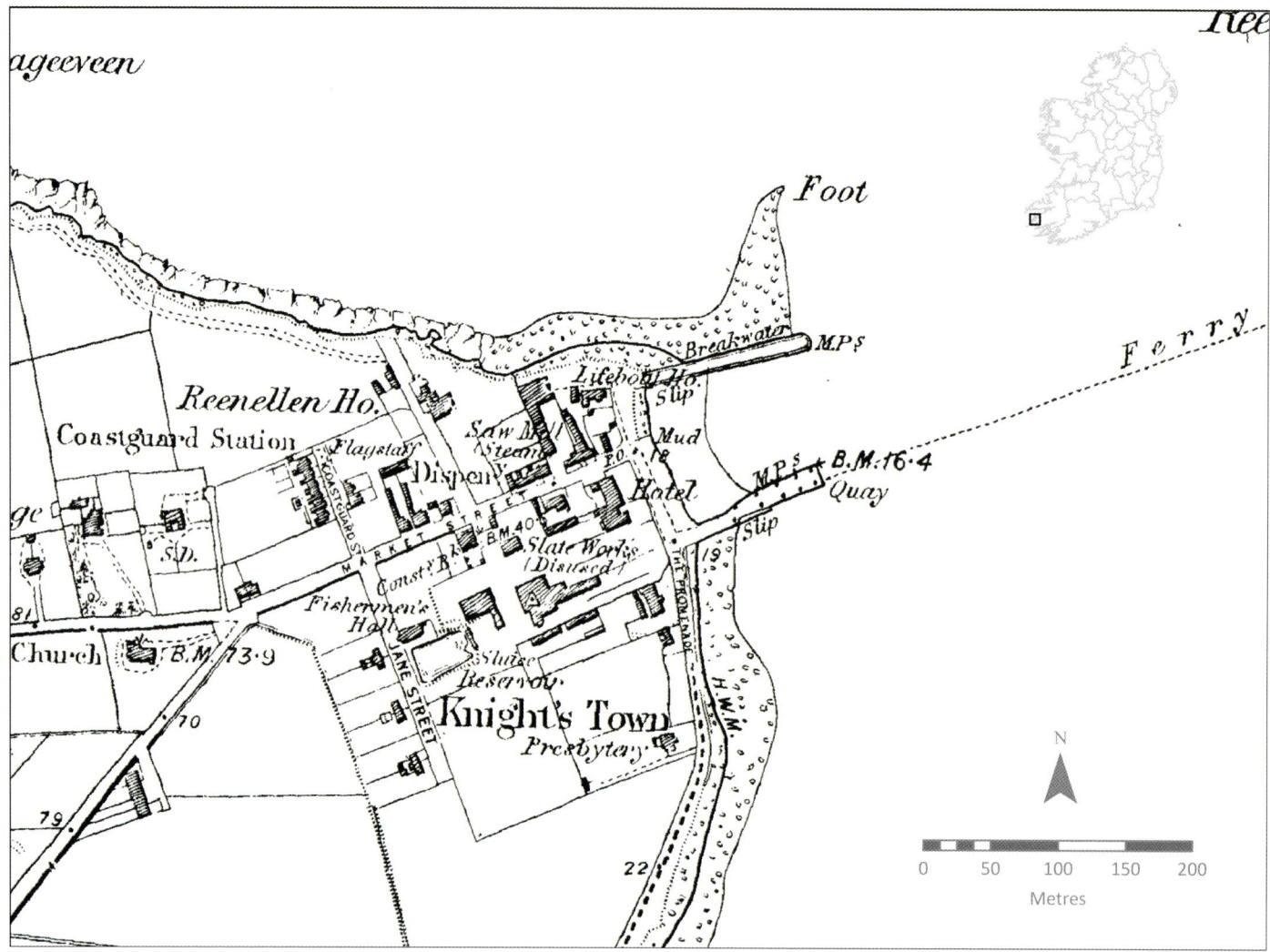

Fig. 19.30 KNIGHTSTOWN, VALENCIA ISLAND, COUNTY KERRY, AND NEIGHBOURING COASTAL AREAS. Knightstown was designed by the engineer, Alexander Nimmo, in the early 1830s and built during the 1840s. Known initially as the 'New Town of Valencia', it was a settlement that revolutionised life on Valencia Island, which had been previously bereft of towns. Nimmo also built harbours, piers, embankments and bridges, and designed other towns in Ireland. No other individual left such a legacy of shoreline modification, especially along the west coast of Ireland. [Source: Ordnance Survey Ireland, historic six-inch maps, *c.*1842]

of this, some notable successes were achieved. Between Black Head, County Clare, and Leenane, County Galway, for example, some 300 piers, jetties, wharves, breakwaters, small harbours and docks were built during the nineteenth century. Wilkins confirms that the majority of these were constructed expressly with public funds and that their presence contributed to the humanising of the foreshore, especially along the entire western seaboard.[26]

By the late nineteenth century, enormous numbers of people in the west of Ireland continued to live in poverty and relied almost exclusively on subsistence farming for their livelihood. In an attempt to address such poverty and living conditions, the Congested Districts Board (CDB) was set up in 1891 and focused on an extensive area along the western seaboard (Figure 19.31). To achieve its goals, the CDB offered financial incentives to encourage modernisation of farming through the amalgamation and consolidation of existing holdings, together with improvements to farm buildings and equipment (see Chapter 4: People, Agriculture and the Coast). The objective of the CDB was to develop greater competitiveness and self-sufficiency within the areas it covered; infrastructures were improved – such as transport systems and

harbours – while new business ventures were encouraged to help diversify the economic base. In this respect, manufacturing ventures were promoted, for example, boatbuilding, fish processing, net making and textiles, such as Aran knitwear on the Aran Islands.

In an evaluation of the CDB, Lee contends that its promises exceeded greatly its achievements in that it invested too heavily in uneconomic projects.[27] Once subsidies were removed, most projects collapsed. The result was that, despite the aim for managing migration by encouraging movement within Ireland from congested areas to those where resources were more readily available, as in eastern Ireland, the CDB continued to lose large numbers of people through emigration. Some minor success, however, was achieved under another government agency, the Irish Land Commission (1881), which succeeded in encouraging some people to move from the congested districts to newly acquired and dismantled landlord estates located in the east of Ireland, for instance in counties Carlow and Meath. With the establishment of the new Irish state, the CDB was dissolved in 1923.

From the early seventeenth to the late nineteenth centuries, the construction of new buildings, changing economic systems,

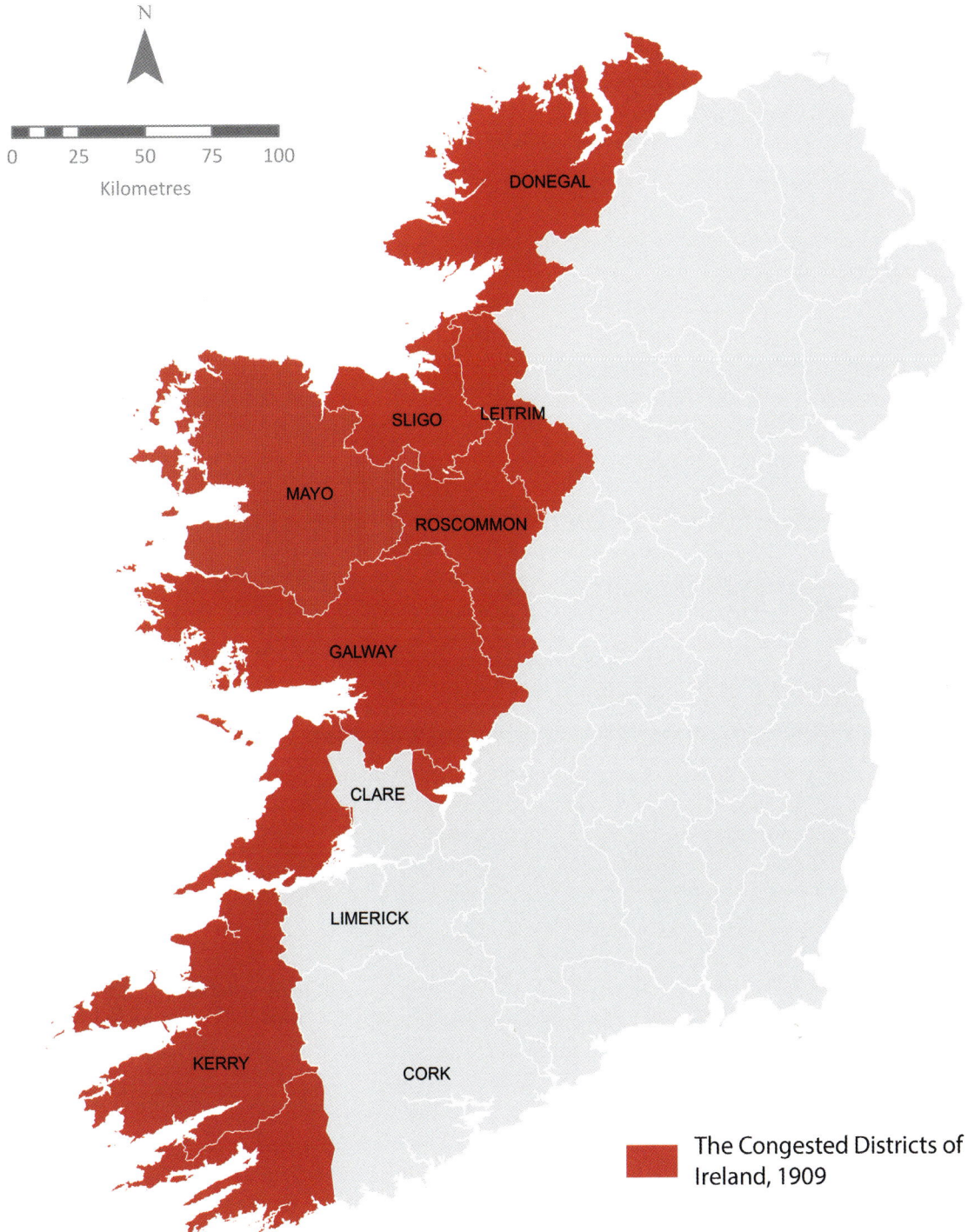

N

0   25   50   75   100
Kilometres

DONEGAL

SLIGO   LEITRIM

MAYO

ROSCOMMON

GALWAY

CLARE

LIMERICK

KERRY

CORK

The Congested Districts of
Ireland, 1909

Fig. 19.31 THE CONGESTED DISTRICTS OF IRELAND. The western and coastal orientation of these districts is apparent in the map. The Congested Districts Board (CDB) was established in 1891, resulting from a need for land reorganisation after famine times. It was replaced by the Land Commission of 1923. The remit of the CDB covered the lives of at least a half a million people, living mainly on Ireland's western coastal fringe. [Data source: adapted from Breathnach, C., 2005. *The Congested Districts Board of Ireland, 1891–1923. Poverty and development in the west of Ireland.* Dubin: Four Courts Press]

expanding urban and port environments, population movements and other processes combined to transform Ireland's coastal landscapes. Changes were especially intense along the eastern and north-eastern seaboard. Here, proximity to Britain and a coastal environment that was supportive to urban and port development encouraged urbanisation and some industrial growth, as well as a more productive agricultural base. This coastline included the core regions for both Northern Ireland and the new Irish Free State as the island moved into the post-First World War era.

In contrast, most areas located along the western seaboard displayed less significant cultural change and economic growth. Adjustment to external forces of modernisation proved far more difficult for these areas, with population decline, economic stagnation and poverty characterising most communities. As Ireland moved into the twentieth century, the contrasts between its eastern and western seaboards, and between the north and south of the island, were set to dominate the political, economic and cultural agendas of governments into the twenty-first century.

THE GREAT FAMINE

Marita Foster and Barry Brunt

THE FAMINE IN IRELAND, WHICH OCCURRED FROM 1845 TO 1851, HAD SEVERE, IMMEDIATE AND LONG-TERM CONSEQUENCES FOR THE COUNTRY. By 1851, repeated failures of the potato harvests resulted in the population on the island falling by some 20 per cent from its 1841 peak of 8.2 million. Communities along the western and southern seaboard were particularly affected. Here, the vast majority of people subsisted as tenant farmers on small plots of land or as cottiers and landless labourers. They often worked in difficult environments, and were almost totally dependent on the success of the potato crop for their survival. Failure of the crop was a catastrophe and one that transformed the culture and landscape of these coastal areas. Here, the desperate and starving people are shown to be scouring the shoreline to collect limpets, seaweed or anything to help offset hunger. [Source: World History Archive/Alamy Stock Photo]

The Great Irish Famine from 1845 to 1851 was caused primarily by repeated failures of the potato harvest.[1] Failure of the crop that had become the staple diet of at least three million people had a devastating impact. By 1851, Ireland's population had declined by some 20 per cent from its peak of 8.2 million in 1841. While no part of the island remained untouched, its impact was experienced most severely along western and southern coastal areas. Here, dependency on the potato was extremely high, land use was dominated by subsistence farming involving small-scale holdings, the physical environment was difficult and the vast majority of people lived in conditions of poverty and deprivation.

This chapter is concerned primarily with describing and assessing the impact of the Famine in Ireland's western and southern coastal areas. Following a brief introduction, it revolves around four sections. Attention is directed first to the establishment of food depots by the British government as an early initiative to combat the onset of famine. This is followed by a review of the public works programme, but especially those activities focused on coastal communities. The third section addresses the pivotal role of the poor law unions as the principal arbitrators of famine relief, while section four concentrates on some of the implications of famine on eviction rates and emigration. The chapter concludes with some images to highlight the relic landscapes of the Famine, which constitute an important part of Ireland's identity, culture and memory, especially in western and southern coastal areas.

Fig. 20.1 PREPARING LAZY BEDS FOR THE POTATO CROP. By the 1840s, the average daily consumption of potatoes by an adult male varied between 7lbs and 14lbs. This level of intake, especially to feed the large families that typified rural communities in the west of Ireland, demanded that every scrap of land be used. As a result, lazy beds became an integral part of the landscape in the marginal uplands and boglands along the western seaboard as tenant farmers eked out a living from the difficult environments. [Source: Maggie Blanck collection (http://www.maggieblanck.com), courtesy of Mayo County Library]

Fig. 20.2 OLD LAZY BEDS ON INISHKEA, SOUTH ISLAND, COUNTY MAYO. As Ireland's population increased, new land was required to provide additional food. For most farmers, the only available land was in marginal upland areas and boglands along the western seaboard. Here, soils were thin, of poor quality and had limited carrying capacity for crops. To facilitate the growth of potatoes in such environments, lazy beds were utilised. These involved digging low trenches by spade and piling up the dirt and sod between the trenches in raised (lazy) beds. Potatoes were planted in the ridges, with manure, rotted vegetation and seaweed added to facilitate growth. This is a labour-intensive process, especially in such environments, but productivity was high. These lazy beds, and the stone boundaries that defined the individual fields of tenants, were essential to ensure good crops and provide a supply of food for most of the year. [Source: Frank Fullard]

The potato was introduced to Europe by the Spanish from conquered territories in South America. It spread quickly through Europe and arrived in Ireland in the late 1580s. Over the following century, its cultivation spread throughout the island and became a key element in its agricultural economy. The high nutritional value and productivity of the potato proved vital for feeding Ireland's growing, but mainly impoverished, population.

New land to facilitate increased production of food and accommodate the expanding rural population was limited, apart from relatively large tracts of uplands and boglands, primarily along the island's western seaboard. Here, however, environmental conditions were generally unsuitable to support the intensive farming necessary to sustain high population densities. Weather systems coming off the Atlantic ocean give rise to many days dominated by cloud and rain. Furthermore, the relatively steep and rocky slopes along the coasts yield thin soils or bogland. Traditionally, such environments supported communities involved in extensive pastoral farming and low population densities.

The arrival of the potato changed this pattern of land use. Potatoes could be harvested successfully in such environments, but especially through the use of lazy beds, which increased levels of productivity. As a result, these areas became dominated by substantial numbers of small, subsistence holdings, occupied by impoverished tenant farmers, the cottier and labouring classes, and their large families. No other crop but the potato could have supported population levels that were well in excess of the carrying capacity of this natural environment (Figures. 20.1 and 20.2).

By the 1830s, one-third of Ireland's population depended on the potato for more than 90 per cent of their food requirements. This proportion was even higher along the western and southern coastline. Such levels of dependency posed problems, as was illustrated by the partial failures of the potato harvest that had occurred in 1822 and throughout the 1830s. The arrival of a new potato blight and the large-scale failure of the crop in 1845 found a society and government ill-prepared to address the crisis of famine.

In the autumn of 1845, the potato crop in Ireland was attacked by a new phenomenon, *phytophthora infestans*, a fungal disease or blight that originated in the north-east of the United States of America (Figure 20.3). About one-third of the crop was destroyed and that was followed by its near total failure in 1846. While the potato did not fail to any significant extent in 1847, little seed was available for planting and, as a result, it is estimated that the crop yield in that year was one-quarter of a normal pre-Famine harvest. The autumn of 1848 saw another almost total failure of the crop, a situation exacerbated by a very poor grain harvest. Blight was less severe in 1849 and 1850, although a substantial failure of the crop did occur in some counties in the latter year.

Relief Efforts – Establishment of Food Depots

In early November 1845, in response to the first reports of the failure of the potato crop, the prime minister, Sir Robert Peel, arranged for the secret purchase of £100,000 worth of Indian corn and corn meal, from the United States. This was not specifically to feed all those affected by the scarcity of food resulting from the blight, but instead was to be held in reserve and released into the market whenever prices rose to an unreasonable level. Local relief committees could then purchase the corn and sell it to the poor who were to pay for it with the wages they earned on public works, which were also a key element of the Peel government's relief policy. Gratuitous distribution could be considered only when the ability to pay was 'absolutely wanting'.

The first shipment of relief supplies arrived in Cork at the end of January 1846; the last was delivered in May. In February, as daily reports relating to the extent of the loss of the potato crop were being received by the Relief Commission, the commissariat branch of the army was ordered to establish provision for food depots around the country. Up to the autumn of 1846, the main

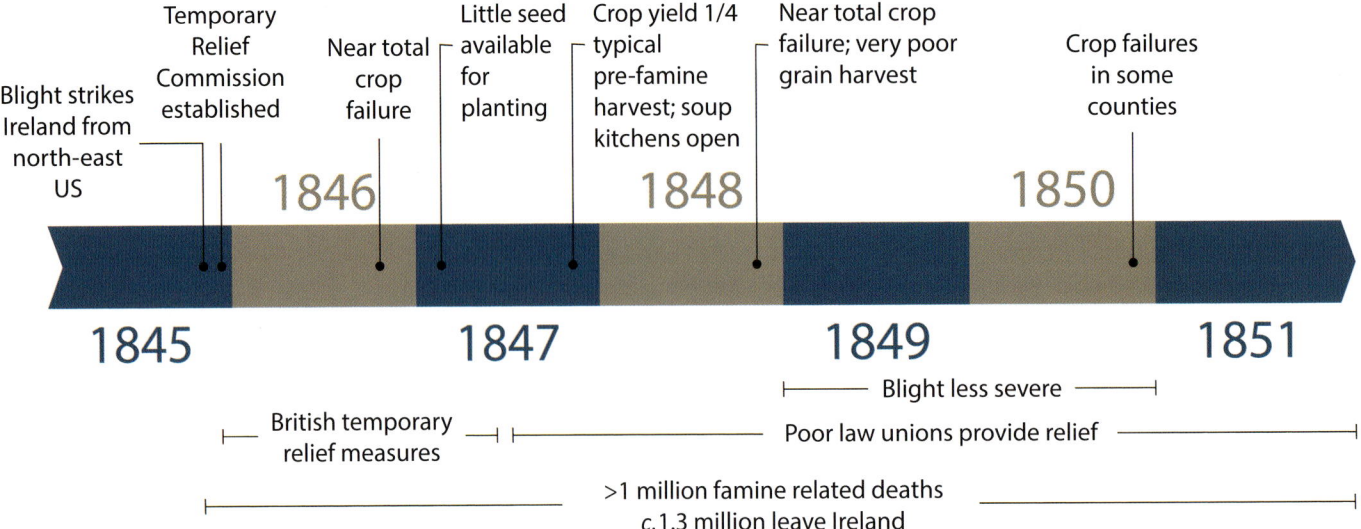

Fig. 20.3 TIMELINE OF MAJOR FEATURES ASSOCIATED WITH THE FAMINE. The repeated failure of a crop that was the staple diet of at least three million people had a devastating impact. It is now generally accepted that, at a minimum, one million died of famine-related causes between 1846 and 1851, while it is estimated that at least 1.3 million people emigrated to Britain and other destinations between 1846 and March 1851. The British government responded to this unprecedented crisis in Ireland by establishing a temporary Relief Commission, which introduced a series of short-term relief measures between 1845 and 1847. These included the importation of Indian corn, the introduction of public works and the establishment of soup kitchens.

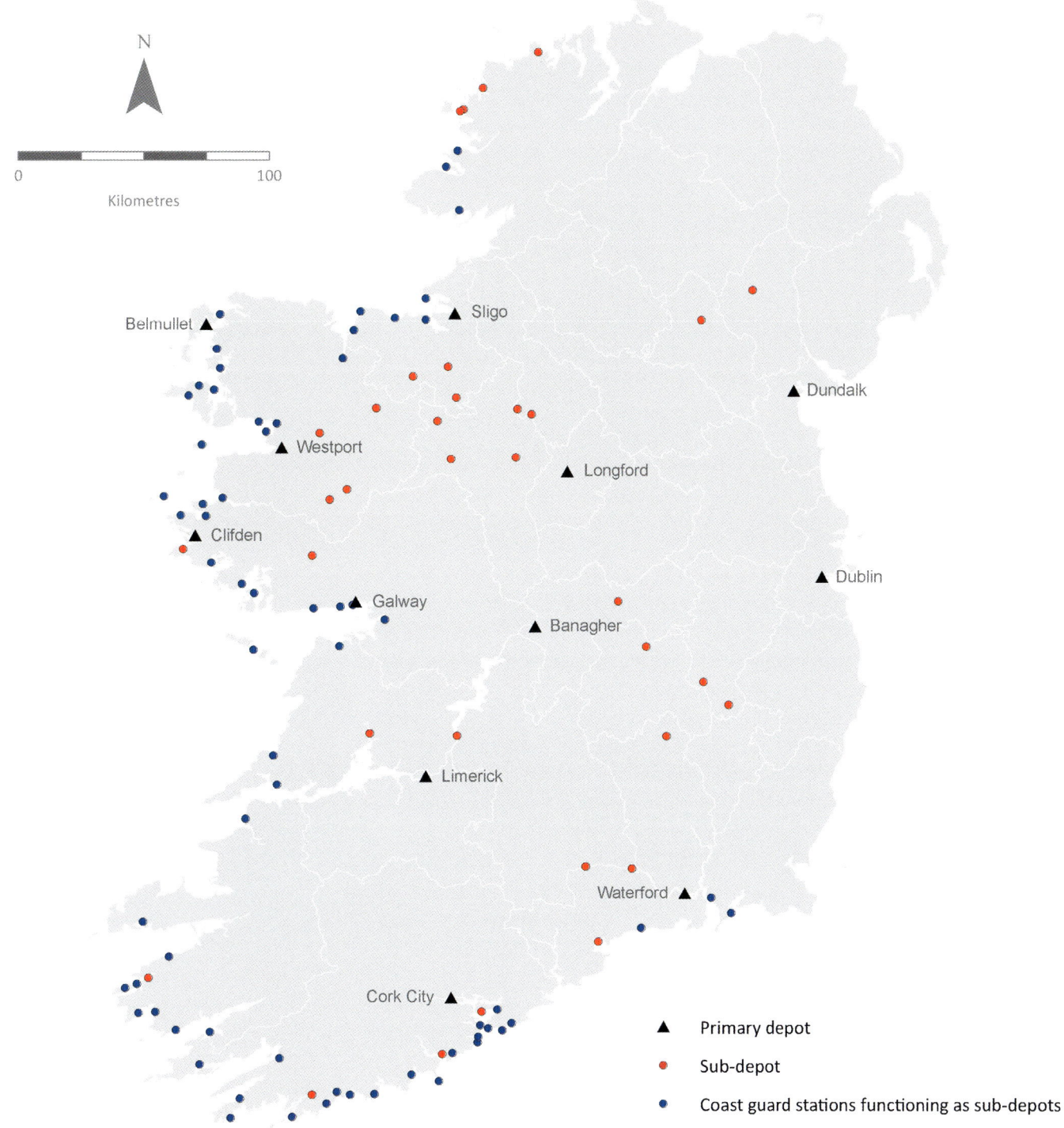

Fig. 20.4 LOCATION OF DEPOTS AND SUB-DEPOTS TO DISTRIBUTE FOOD RELIEF (1846). The map highlights the concentration of depots along the southern and western coastline. Primary depots were established initially at Limerick, Galway, Westport and Sligo, and were selected on the basis of their accessibility by sea from Cork. These acted as regional centres for the redistribution of supplies. Between February and September 1846, naval vessels made sixty-three voyages from Cork, distributing almost 48,000 tonnes of relief supplies. In addition, numerous sub-depots were established under the control of the coast guard and constabulary. Those under the charge of the coast guard stretched along the southern and western seaboard. Coast guard cutters played a vital role in complementing the role of the navy in that their smaller vessels were better equipped to supply the many destitute and more isolated coastal communities that lacked harbour facilities necessary for large vessels: 'When the rough and boisterous weather on this coast is considered, together with the minuteness of their issues, and the crowds of poor and destitute persons waiting to receive them, it is difficult to do justice to their exertions'. (Sir Randolph Routh. Letter to Charles Trevelyan, 31 July 1846. In *Correspondence explanatory of the measures adopted by Her Majesty's government for the relief of distress arising from the failure of the potato crop in Ireland* (Command Papers) p. 221.) [Data source: House of Commons Parliamentary Papers Online]

depot was in Cork, which was the centre for the importation, milling, storage and distribution of the government-imported corn. Much of the milled corn was held in the naval stores on Haulbowline in Cork Harbour prior to its distribution to depots around the country. First-order depots on the west coast were established initially in Limerick, Galway, Westport and Sligo, and were chosen on the basis of their accessibility by water from Cork. Access by sea was vital due to inadequate inland transportation, which demonstrates the isolation of communities along the western and southern coastline (Figure 20.4).

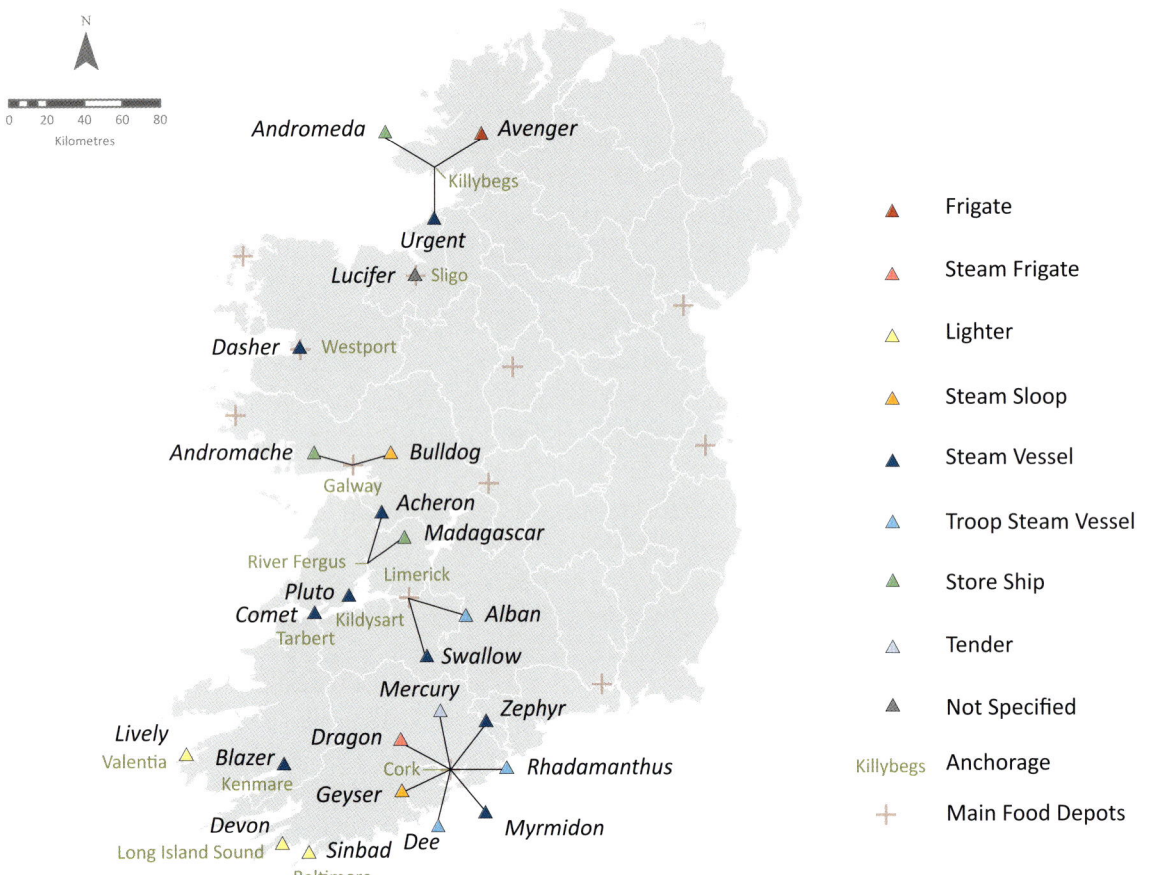

Ship	Under What Orders
▲ Acheron	To proceed to Black Sod Bay, and afford protection to the merchant vessels seeking shelter there, from acts of plunder.
△ Alban	To follow the instructions of Commissary-General Hewetson in the conveyance of supplies to the distressed districts.
▲ Andromache	As a depot for Indian meal and other provisions.
▲ Andromeda	As a depot for Indian meal and other provisions.
▲ Avenger	To land the meal now on board at Killybegs and return to Plymouth as soon as possible.
▲ Blazer	To land her cargo of meal for the Relief Committee at Kenmare, and return to Cork.
△ Bulldog	To dispose of the meal now on board as the Commissariat at that port may request.
▲ Comet	To follow the instructions of Commissary-General Hewetson in the conveyance of supplies to the distressed districts.
▲ Dasher	To follow the instructions of the Commissariat in the conveyance of supplies to distressed districts.
△ Dee	To take on board meal for the Relief Committees at Castle Island, Valentia and Cahirciveen.
△ Devon	To remain at this anchorage for the present as a floating depot.
△ Dragon	To land her cargo of meal shipped by the British Relief Association at any port that their agent may request.
△ Geyser	To take on board meal for the Relief Committees at West Cove (in Kenmare Water) and Dingle.
△ Lively	To deliver 30 tons of barley meal at Valentia, and proceed to Limerick to discharge the rest of her cargo.
▲ Lucifer	To follow the instructions of the Commissariat in the conveyance of supplies to distressed districts.
▲ Madagascar	As a depot for Indian meal and other provisions.
△ Mercury	To proceed to Castletownsend with supplies shipped by the British Relief Association.
▲ Myrmidon	Flag ship.
▲ Pluto	To afford protection to the meal vessels passing between Limerick and the Fergus.
△ Rhadamanthus	To return to Plymouth with 130 bags raised by the coastguard and to make good defects, etc.
△ Sinbad	Put into Baltimore to repair her mainsail, having been blown to leeward.
▲ Swallow	To be repaired at Limerick.
▲ Urgent	To proceed to any port, and deliver the supplies on board shipped by the British Relief Association that their agent may request.
▲ Zephyr	To proceed to Baltimore with supplies shipped by the British Relief Association, and return to Cork.

Fig. 20.5 LOCATION OF BRITISH NAVY VESSELS AROUND IRELAND'S SOUTH AND WEST COAST (4 FEBRUARY 1847). As in the previous year, the British navy, augmented by the coast guard service, continued to play a key role in bringing food and other supports to help relieve the hunger and other forms of deprivation that affected coastal communities. The map is a 'snapshot' of the strategic location of twenty-four naval vessels on one day at the height of the Famine, and the varied tasks assigned to them. In most cases, this involved the conveyance of meal and provisions to starving people, such as HMS *Avenger*, directed to Killybegs, and HMS *Devon*, which acted as a floating (food) depot for communities in west Cork. Note also the orders for HMS *Acheron* to counteract incidents of plundering as outlined in the text. [Data source: 'Correspondence from January to March 1847, relating to the measures adopted for the relief of distress in Ireland'. Commissariat series, part II, HC 1847 [796], pp. 72–74]

The sub-depots maintained by the coast guard were located strategically to provide relief to more isolated coastal communities along the western and southern seaboards, and could be accessed more easily by their smaller vessels. In addition to their role in the management of sub-depots, coast guard officers in stations all around the Irish coast played a key role in monitoring the advance of the blight. Coast guard personnel were very often the only official representatives of the state resident in these remote districts and, apart from local clergy, were the only other source of reliable information relating to conditions prevailing in their localities.

In July 1846, the blight reappeared. The result was that only two million tonnes of potatoes were harvested compared to almost fifteen million tonnes in the pre-Famine year of 1844. In spite of this, the government of Lord John Russell, which came to office in June, favoured a policy of non-intervention in the market to provide food and regulate prices. This translated into the decision to discontinue imports of Indian corn and to close food depots. The consequence was the aggravation of the crisis, with thousands dying from starvation and disease along the west and south coasts. Frustration and anger at the shortage of food led, almost inevitably, to public unrest in those areas most affected by famine. Among the most noteworthy were disturbances in Youghal in east Cork and Dungarvan in west Waterford. In Dungarvan, for example, the 1st Royal Dragoons were brought to the town to quell riots, while fishermen and their wives also blocked ships laden with grain for export from leaving the port until supplies of Indian corn meal were delivered to the town to alleviate food shortages.

Later in the year, unrest and the severity of famine led the government to reactivate the import of Indian corn. A limited number of food depots were also reopened in December along the western seaboard to assist in redistributing relief. As in the previous year, the navy and coast guard services were called upon to transfer supplies due to easier access by sea than land (Figure 20.5). The conditions observed by naval and coast guard personnel involved in such relief work were horrendous. As one naval commander noted, 'Never in my life have I seen such wholesale misery, nor could I have thought it so complete.'[2]

Another manifestation of the desperation of people in coastal areas were incidents where merchant vessels carrying cargoes of meal and other foodstuffs were targeted and plundered. This type of incident was prevalent particularly along the coast of County Mayo, around Erris and Blacksod Bay, from January to the autumn of 1847. One such episode occurred in April when the schooner, *Mavis of Dumfries*, which was sailing from Greenock for Galway with a cargo that included sixty-two tonness of wheat, was intercepted off Black Rock (near Achill Island). The mate described how two rowboats with fourteen men pulled along aside and boarded:

> The men got on board, seized the helm and brought the ship up in the wind, when eight more boats came alongside, and finding the vessel was laden with wheat, they broke open the hatches with axes of their own and filled the sacks which they had in their boats, about 30 in number, as well as the sacks which were on board the schooner. They then filled their

boats in bulk and pulled away in the direction of the Black Rocks. The boats' crew did not offer any personal violence but merely said that they wanted the wheat to eat.[3]

Generally, those involved were unarmed and many who engaged in such acts were apprehended by naval or coast guard patrols, which had been increased in this area. For example, plunderers who boarded the *Defiance of Cardigan*, Galway-bound from Belfast and off the coast of Belmullet, were apprehended by a coast guard vessel which seized ten currachs, loaded with Indian meal and crewed by thirty-four men. In October, 1848, a representative of the Commissioner of Customs visiting the north-west coast noted the cessation of all pirate activity. By that time, however, it is likely that those who engaged in such actions either no longer had the physical strength to continue to do so or had died, emigrated, been imprisoned or transported.

RELIEF EFFORTS – PUBLIC WORKS

While imports of Indian corn and the establishment of food depots had been a major component of the Peel government's response to the potato blight, public works were another. The provision of such works, principally the construction and repair of roads, had been a standard response during previous food crises in Ireland. Under bills presented to parliament in January 1846, this priority was continued (see Chapter 19: Changing Coastal Landscapes). Projects such as drainage works and improvement of navigation were also encouraged. Of particular significance to coastal communities, however, was the inclusion of a scheme to facilitate the development of harbours and piers to better promote fisheries, provide employment and diversify food supplies in coastal communities.

By October 1846, *c.*114,000 were employed on public works schemes, but this rose to a peak of over 700,000 by March 1847. These had the dual purpose of improving the country's physical infrastructure to facilitate development, but also to provide temporary relief through the payment of wages. This would allow workers to purchase food for themselves and their destitute families. While most areas benefited from the programme of public works, dependency on such projects was greatest in areas most affected by the Famine. This includes, in particular, the coastal counties of Connacht and Munster (Figure 20.6).

Despite good intentions, the programme was not successful. It was composed generally of one-off, palliative projects, which did little to enhance local development prospects or reduce significantly the isolation of most coastal communities, especially those along the western seaboard. While offering short-term relief through employment, wages were relatively low and did not keep pace with inflated food prices. Finally, increasing rates of starvation, death and emigration reduced significantly the numbers of able-bodied men capable of physical work, often undertaken in difficult environmental conditions. These failings, together with escalating costs of support (almost £5,000,000 by March 1847), resulted in the government abandoning the programme by mid-1847, although the scheme to improve piers and harbours was continued until 1851.

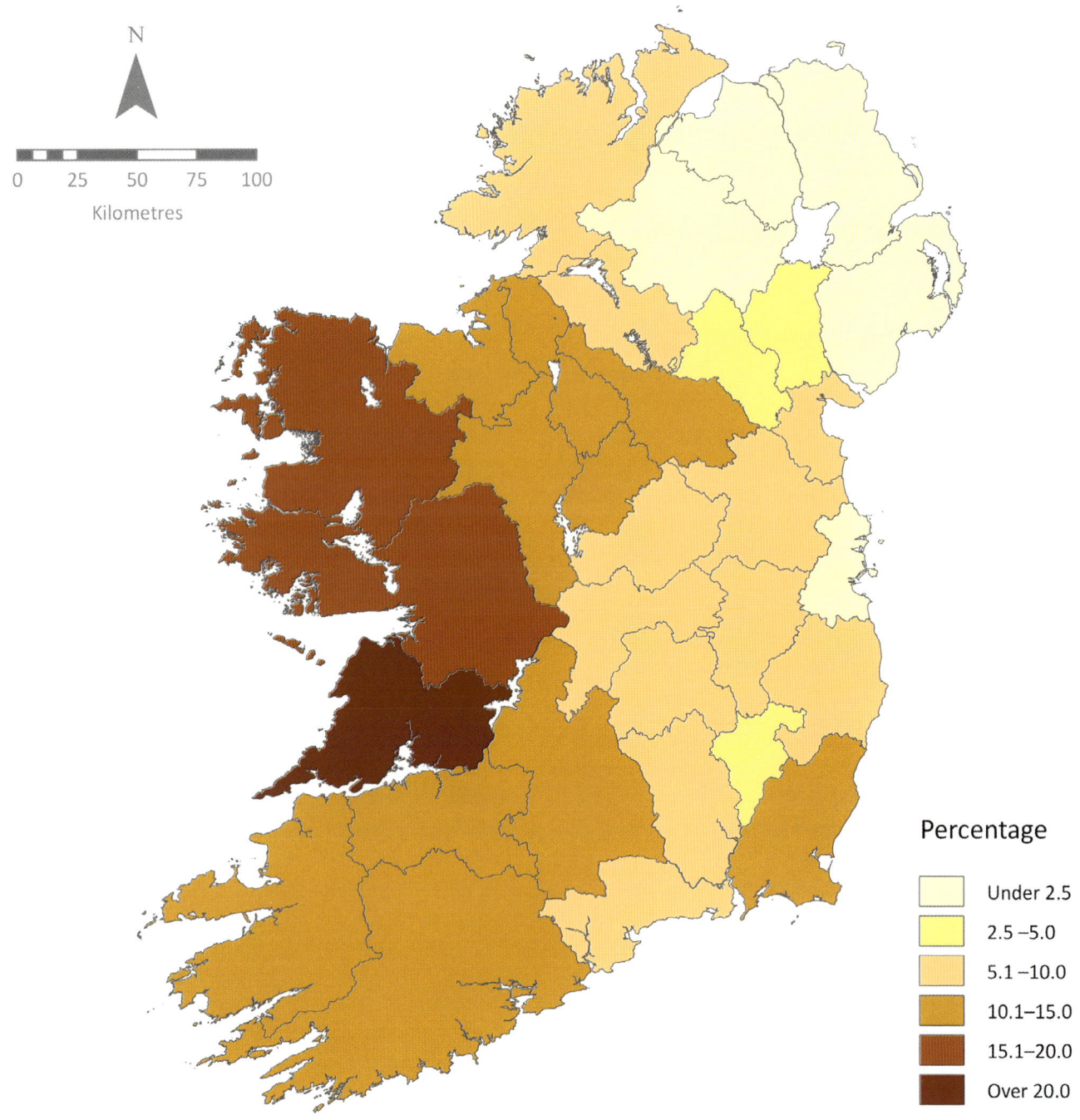

N

0   25   50   75   100
Kilometres

Percentage

Under 2.5

2.5 –5.0

5.1 –10.0

10.1–15.0

15.1–20.0

Over 20.0

Research by W.J. Smyth, Mapped by M. Murphy and S. Kandrot

Fig. 20.6 PERCENTAGE DISTRIBUTION OF POPULATION SUPPORTED BY PUBLIC WORKS PROJECTS IN EARLY 1847. The map shows the percentage of the 1841 population employed in public works in spring 1847. High percentages indicate communities most vulnerable to the consequences of severe famine. A core group of dependent counties is apparent. These are centred on Clare, where more than one in five of its population was engaged in public works, but extends into Galway and Mayo (15–20 per cent). Most remaining counties on the western and southern seaboards show dependency rates of 10–15 per cent. Lower dependencies occur around the northern and eastern coasts, linked to less severe famine conditions but, more especially, to better economic environments associated with the emerging industrial economy in Ulster and more successful commercial agriculture and trade in eastern Leinster. [Data source: adapted from Crowley, J., Smyth, W.J. and Murphy, M., eds, 2012, *Atlas of the Great Irish Famine,* Cork: Cork University Press]

Piers, Harbours and Fisheries

At the time of the Famine, despite possessing rich resources in its coastal waters, Ireland's fishing industry could best be described as backward and under-developed. Development was frustrated by factors such as the paucity of good infrastructure, especially piers and sheltered harbours, inadequate transport links to access markets and an absence of curing facilities. In addition, poor quality boats, nets, gear and tackle made deep-sea fishing extremely difficult, especially in the hazardous sea and weather conditions that occur frequently around the south and west coasts. At the heart of most problems, however, was the lack of investment capital.

The problem of capital availability was addressed partially through a government scheme that allocated £50,000 in 1846 for the improvement of piers and harbours. An additional £40,000 was sanctioned in the following year. Between 1846 and 1851, the Board of Works, which supervised the public works programme, received 180 applications for funding. By 1851, thirty-eight new and/or improved piers and harbours were completed (Figure 20.7).

458

Fig. 20.7 PIERS AND HARBOURS CONSTRUCTED OR IMPROVED (1846–1851). The thirty-eight projects were quite well distributed around the Irish coastline. Some clusters occur, however, in counties Cork, Kerry, Galway, Mayo and Donegal. These reflect the severity of the Famine in such coastal communities, together with difficult environmental conditions for fishing offshore. Most projects involved construction of new piers or the extension and repair of existing piers. This was to provide improved berthing and landing facilities for boats, while construction of breakwaters would allow easier and safer access to the piers and harbours. Total expenditure was $c.£106,000$. Much needed employment was also provided by these projects. Numbers employed varied considerably, but peaked in June 1848 when 6,296 workers were engaged. [Data source: Commissioners of the Public Works (Ireland)]

Curing stations

The other pillar of government policy relating to efforts to develop the fisheries during the Famine was the establishment of curing stations. From the autumn of 1846, reports were circulating from the west coast that fish were being left to rot on the shore, or were being used as manure, while thousands were dying for the want of food. Fisheries commissioners at the Board of Works were, therefore, encouraged to establish a number of curing stations on the west and south-west coasts. The immediate aim was to increase the stock of food and encourage extra employment in fisheries. A longer-term aim was to establish a more sustainable fishing industry by ensuring supplies of landed fish would not spoil and could be transported to more distant markets. To advance this project, the Irish Productive Loan Fund Society provided £5,000 to establish six curing stations at Killybegs, Belmullet, Roundstone, Valencia, Castletown Berehaven and Baltimore. Roundstone was closed shortly after its establishment in May 1848 and the government ordered the closure of the remaining stations by the end of that year, believing that their objectives had been accomplished. A more convincing reason was the government's preoccupation with costs and what were deemed to be unnecessary subsidies.

In any event, the curing stations were a failure. Many of the factors that had militated against the development of the fisheries in the pre-Famine period, including the need for proper boats and equipment, remoteness from markets and the poverty of the country, were all exacerbated by famine conditions. In addition, with the onset of famine, many fishermen had pawned their fishing gear and boats in order to buy food. The failure of the scheme to provide loans or grants, to enable the fishermen to redeem their equipment and allow them to get back to sea to resume fishing, contributed to their failure. Curing stations could not operate on an efficient or economical basis when the supply of fish was irregular.

The government, however, was opposed to the provision of small loans to help improve the boats and tackle of the fishermen, as they were concerned that this would create a culture of dependency. In contrast, the Society of Friends, or Quakers, took a more pragmatic approach. Reports received from their representatives within Ireland had made them well aware of the conditions confronting fishing communities around the coast, and especially the issue of pawning equipment to buy food. For example, on Achill Island it was observed that the waters could not be fished because nets and tackle had been pawned or sold to buy corn meal. There were similar reports from other fishing ports along the west and south coasts.

From January 1847, the Quakers provided loans to fishermen to help them redeem their tackle and boats. This, however, was just one aspect of their relief efforts. Others included a project making and repairing nets in Ballycotton in County Cork, establishing fishing stations at Newport on Achill Sound, Belmullet in Erris, Ballinakill near Clifden and Castletown Berehaven, and purchasing various types of fishing vessels, which they made available to fishing communities, but retained ownership of until the fishermen gradually cleared the debt. An example of a successful project occurred in Ring in County Waterford. In total, the Quakers invested over £5,300 in fisheries projects.

Case Study: Relief Efforts in Ring, County Waterford

Marita Foster

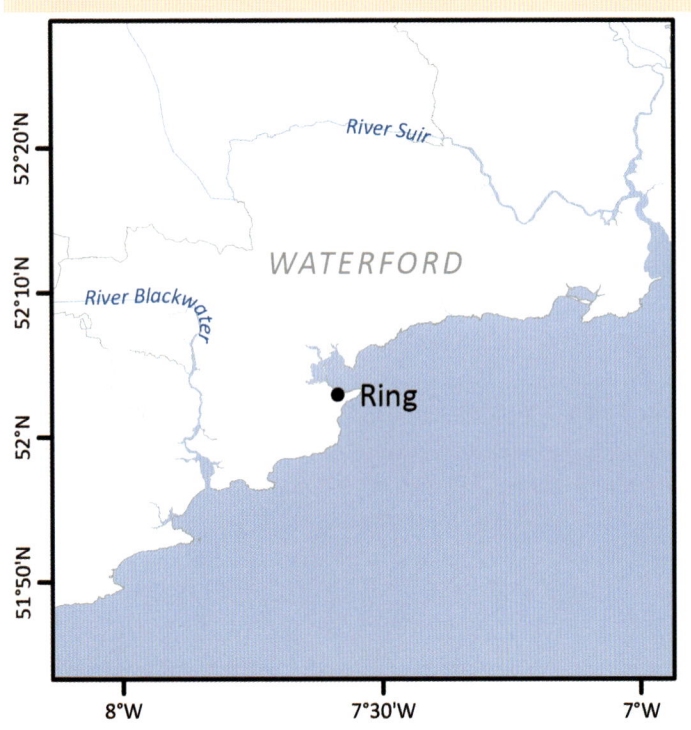

In the spring of 1847, as the Famine was having a devastating impact on communities in many areas along the coast, James Alcock, a Church of Ireland vicar in Ring, contacted the Waterford Auxiliary Committee of the Society of Friends (or the Quakers) seeking assistance for the local fishermen and their families: 'Famine was raging at its height, while fever and dysentery, whose virulence were much increased by the impurities so prevalent in fishing hamlets, was hurrying many to a premature grave'.[4] In one village alone, one hundred families were receiving public relief. Following an inspection of the local fishing boats, Alcock discovered that, as in other areas, much of the fishing gear had been pawned or sold for food, and the oars and boat linings used for fuel. In an effort to alleviate the situation, an initial grant of £57 was made to the vicar to allow small loans to be made to the fishermen. This was to enable them to repair their boats and resume fishing activities, making them producers of food instead of consumers. The crew of each boat also received two stone of Indian meal per week to ensure they could be independent of outdoor relief and give their full attention to fishing.

By the summer of 1847, Alcock reported that forty-nine boats had been assisted, totalling 150 crew members. In almost every case, the loans received by the fishermen had been repaid. By April of the following year, Alcock observed that there was no destitution in the area and the fishermen had a sufficient supply of gear to be constantly employed, whatever the weather. In addition, there was a noticeable improvement in the houses of the fishermen and their quality of life.

One of the main factors inhibiting the progress of the fisheries in Ring was the absence of a pier at which boats could discharge their catch. In November 1826, an application was made to the Commissioners of the Irish Fisheries by local landowners and fishermen to construct a pier at Ballynagaul. This was not progressed, but a renewed application in 1846 for a new pier and landing slip was sanctioned. Work commenced in August 1848 with an estimated cost of £1,800 (including a grant of £1,350). Savings on construction, however, allowed the project to be extended to include a breakwater, which ran parallel to the shore and ensured the harbour would be secure from all winds. Apart from providing a facility for the local fishing community, the pier was also used by schooners of heavy tonnage at which they could discharge their cargoes both securely and safely. This also increased trade within the community (Figure 20.8).

With the efforts to improve fishing boats and also the provision of better harbour facilities, the need for a curing house was proposed to process the anticipated larger catch. The Quakers were agreeable to supporting the construction of a curing house at Helvick, and the venture was initially successful. Produce, which included ling, cod and smoked haddock, found a market in England and provided additional employment for the community. The operators of the project, however, overextended themselves and the curing station closed. Alcock, with support from the Quakers, also made efforts to develop other aspects of the fisheries, including investing in and the commissioning of a model fishing boat to be built at Ring. This was designed to be more suitable for local conditions than pre-existing vessels. They also provided nets more suited to deep-water fishing than the trammel nets, which were used for shallower waters.

By September 1849, financial support to the fishing community in Ring had ceased. A number of individual Friends, however, continued to provide relief through other means, such as donations of clothes and the provision of hemp at wholesale price to make nets. Despite the ongoing scarcity of the potato, fishing operations continued, the people were able to support themselves without seeking poor-law relief and there was no significant diminution of the population in the locality. In January 1852, Alcock, in correspondence with the Society of Friends, reflected on what had been achieved:

When there has been no diminution of the population from evictions; no emigration, nor the abandonment of premises through distress; no dilapidation of houses; a reduced poor rate, and a large stock of cured fish on hands; it may fairly be inferred that much good has been effected in Ring through your means; and that, notwithstanding the heavy losses which the fishermen sustained on shore by the failure of their little crops, they have continued to support their families in comparative comfort for a considerable period, even after your favours were discontinued.[5]

Fig. 20.8 BALLYNAGAUL PIER AND FISHING BOATS (1886). The successful intervention of Quaker relief efforts not only had the immediate effect of reducing levels of starvation and emigration, but also laid the foundations for a longer-term impact. One vital development was the provision of funding for the construction of a new pier and landing slip at Ballynagaul to help promote and encourage the local fishing community. The legacy of such efforts played a key role in the preservation of a resilient fishing industry and an Irish-speaking community. Ring remains as a small outlier of the Gaeltacht along Ireland's southern coastline. [Source: National Library of Ireland, call no.: L_CAB_02083]

Historian Helen Hatton has argued that the relief efforts of the Quakers should be viewed positively, in that they provided hundreds of jobs in fishing and related activities, such as boatbuilding, making nets and clothing.[6] The Quakers, however, were more critical of their efforts, citing bad management and natural obstacles – such as severe weather and dangerous waters off the western and south-western coasts – as contributing to a relative failure. A more interesting observation made by the Quakers was that the main obstacle to the success of the fisheries was the absence of local demand. Although this seems astonishing in the context of the Famine, their rationale was that, with the exception of some very poor countries, fish was not a primary food or an absolute necessity of life. In a country like Ireland, fish was a luxury and was eaten with bread or potatoes: 'When the people have not means to purchase both, the fish is given up. On this account, the failure of the potato, so far from increasing the demand for fish, greatly diminished it, and thus presented an additional discouragement to fishing operations.'[7]

Despite the efforts of the Quakers and government relief schemes, the overall impact of the Famine on coastal communities was devastating. In their 1850 annual report for the twenty-eight fishing districts, into which the coast had been divided, inspecting commissioners highlighted 'the general disorganization which arose out of the unprecedented suffering and distress, which continued unabated for three years'.[8] Employment in fishing declined by one-third, while numbers of boats and vessels fell by 26 per cent when compared to pre-Famine conditions. Rather than helping to alleviate the consequences of famine, the underperforming fishing industry compounded the despair, deprivation and decline in most coastal communities.

POOR LAW UNIONS, WORKHOUSES AND SOUP KITCHENS

In 1838, poor law unions had been introduced in Ireland to act as new administration units to address the problems of the growing numbers of poor and destitute who needed relief. Each of the 130 unions (increased to 163 by 1850) had a workhouse located at its administrative centre (usually the main town), from which relief would be distributed. The cost of relief would be paid for by rates levied on landowning residents. Initially, pressures on the unions and workhouses were manageable.

However, the impact of famine from 1846 onwards generated a large-scale increase in demand for relief within the workhouses. This was particularly the case in western and south-western coastal areas, where population increases of more than 30 per cent had

Fig. 20.9 PEASANTS AT THE GATE OF A CROWDED WORKHOUSE (1847). The Poor Law (Relief) Act of 1847 made poor law unions the principal provider of famine relief. Their workhouses, in particular, symbolised the plight of the poor and starving masses seeking relief and were key institutional features in all coastal unions. Skibbereen in County Cork was one of the worst-affected unions and came to symbolise the desperation of the Famine years. Built in 1840–1841 to house 800 inmates, the onset of famine caused numbers to reach 1,449 in March 1847. Conditions deteriorated and, with such overcrowding, death rates rose to over sixty per week. However, with continuation of famine, more and more people clamoured for entry in the hope of some sustenance. By December 1848, the workhouse accommodated 4,221 inmates, living in desperate conditions. As a result of high death rates and emigration, the population of the Skibbereen Union fell by 36 per cent from 1845 to 1852, the highest rate of decline of all unions in Ireland. A mass Famine grave at Abbeystrewry, near Skibbereen town, contains the remains of between 8,000 and 10,000 people. [Source: Maggie Blanck collection (http://www.maggieblanck.com), courtesy of Mayo County Library]

been recorded between 1821 and 1841. Relatively high rates of marriage (and at a younger age) and inheritance laws, which facilitated sub-division of holdings into small tenant plots and colonisation of the poorer land with the potato because it could support large families, contributed to this growth. Failure of the potato crop was disastrous and found expression in high and increasing levels of starvation and destitution. By early 1847, most workhouses, especially those in the south of Ireland and Connacht, were severely overcrowded and experienced increased death rates of inmates (Figure 20.9). In many coastal land areas, particularly along Ireland's western coasts, people were forced to scavenge for food from the beaches, which provided the only free and readily available source of food. Shellfish were collected commonly from these coasts and the shell remains were dumped on the upper parts of the beaches and in adjacent sand dunes. Inch and Rossbehy, County Kerry, or on the Silver Strand to Carrownisky coast of County Mayo, provide excellent evidence of this activity (Figure 20.10).

Shell Middens

Robert Devoy

Extensive nineteenth-century, famine-related shell middens occur in the dunes and neighbouring coastal areas at Rossbehy and Inch, Castlemaine Harbour, County Kerry. An extensive midden layer is exposed here and visible along the coast at Reenanallagane, Rossbehy. In famine times, people in the area would have been driven to scour the foreshore and intertidal zones for any available food, such as shellfish. This would have been cooked on the spot, probably in large pots, consumed rapidly and the shell remains dumped in huge piles. Cockle and winkle shells are abundant in the area and dominate the midden layer. Similar famine-time middens can be found in many

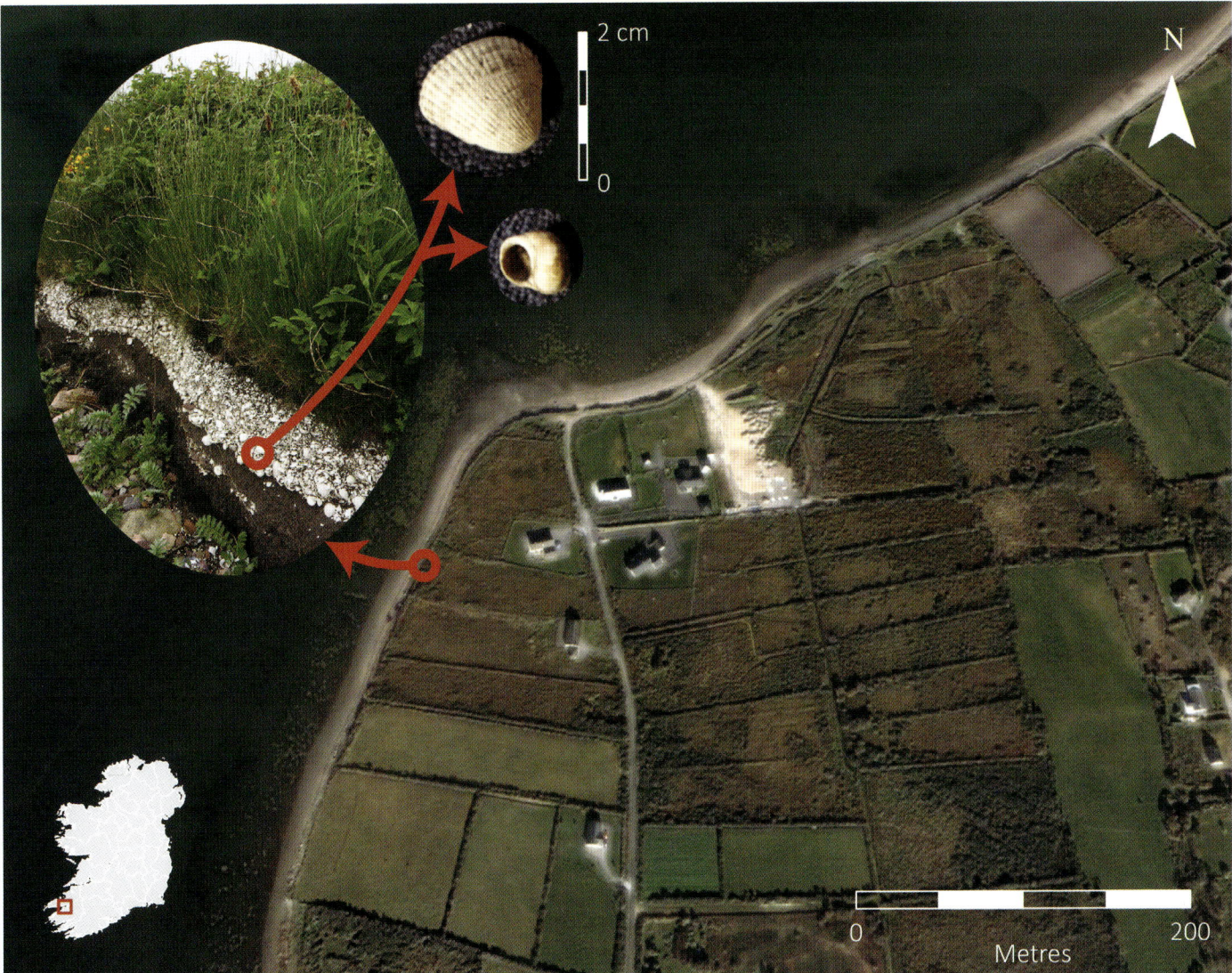

Fig. 20.10 SHELL MIDDEN AT REENANALLAGANE, COUNTY KERRY, AND AERIAL VIEW OF SURROUNDING AREA. [Aerial image: DigitalGlobe; shell midden: Aisling O'Grady; shell specimens: Robert Devoy]

locations along Ireland's western coasts.

The aerial image shows the wider area behind the coastal shell midden site of Reenanallagane today (Figure 20.10) showing the original settlement areas and cabin plots from which people would have come during the famines to collect shells and other foods from the adjacent beach and foreshore. Some of the nineteenth-century buildings (originally cabins) and associated land plots and strip fields still remain. Many of the buildings have been renovated, others remain abandoned or have disappeared completely.

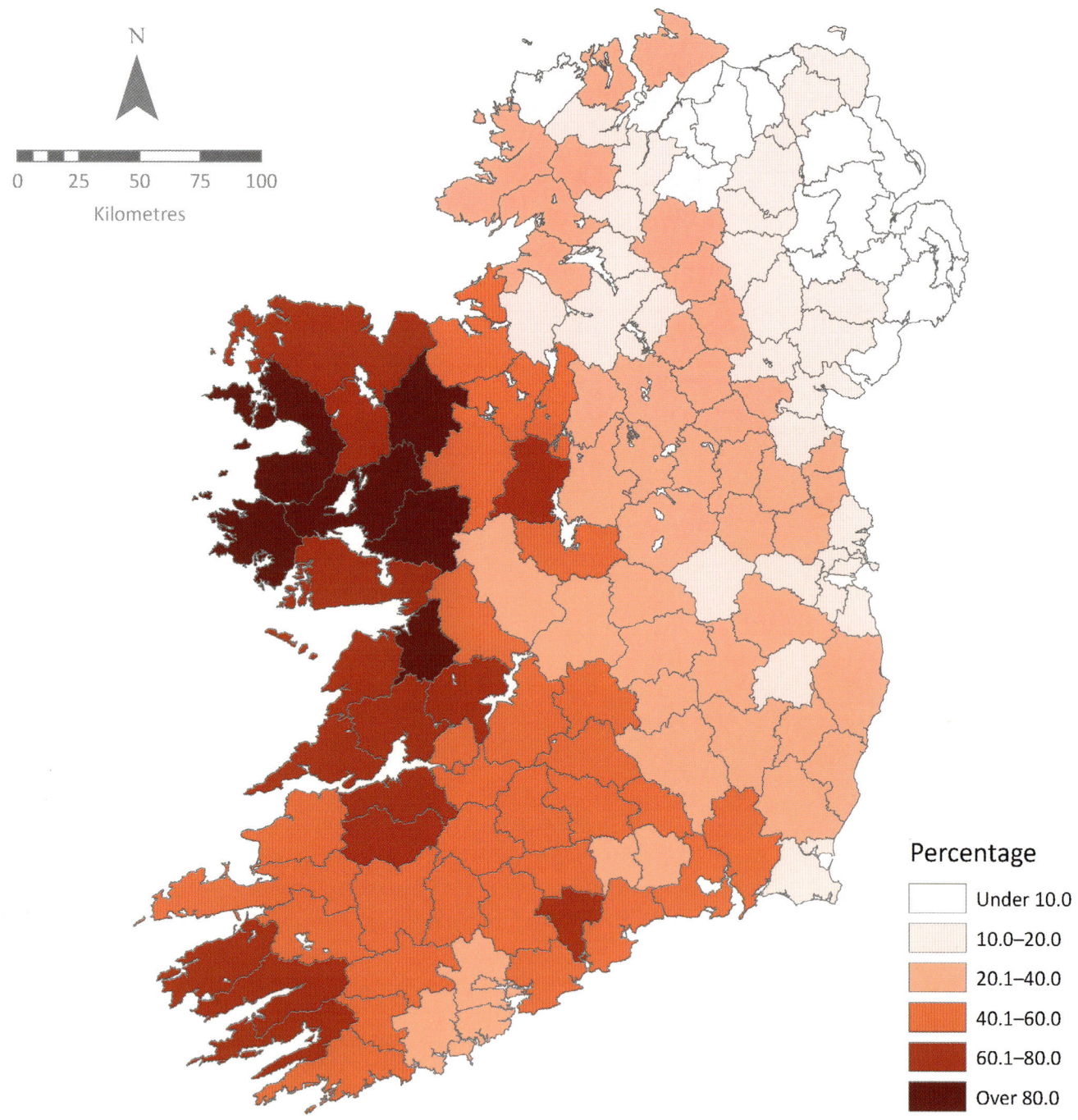

N

0 25 50 75 100
Kilometres

Percentage

Under 10.0
10.0–20.0
20.1–40.0
40.1–60.0
60.1–80.0
Over 80.0

Research by W.J. Smyth, mapped by M. Murphy and S. Kandrot

Fig. 20.11 PERCENTAGE DISTRIBUTION OF THE MAXIMUM POPULATION RECEIVING FOOD RELIEF UNDER THE 'SOUP KITCHEN' ACT (1847). Introduced in the spring of 1847, the establishment of soup kitchens to provide free, daily rations of food has been considered the most successful, albeit short-lived, intervention by government in the Famine years. In Ireland, almost one in three people depended on soup kitchens for survival. Dependency, however, was highest in the vulnerable coastal communities in the west and south-west. In several poor law unions in Connacht, more than four in every five people received relief from these kitchens. In counties Clare, Limerick and peninsulas of Cork and Kerry, dependency rates exceeded 60 per cent of the population. Without this intervention, death rates through starvation would have been much higher. In contrast, dependency on soup kitchens was low along the eastern coastline and, more especially, in Ulster. [Data source: adapted from Crowley, J., Smyth, W.J., and Murphy, M., eds, 2012, *Atlas of the Great Irish Famine*, Cork: Cork University Press]

In spite of overcrowded workhouses and widespread distress, the government pressed ahead with its decision to abandon its public works programme. To compensate for this, a Temporary Relief Act, known as the Soup Kitchen Act, was passed in February 1847, which allowed soup kitchens to be set up in each poor law union. These provided free daily rations of food to over three million starving and destitute people between spring and autumn 1847. Their impact was most pronounced in communities along the western and south-western coastline, which had become most vulnerable to the Famine (Figures 20.11 and 20.12).

In June 1847, the government passed the Poor Law (Relief) Act. Following closure of soup kitchens in September, this made the poor law unions the principal and only official means of relief for the poor in Ireland. This was despite the fact that the poor law system had been unable to deal with the demands for relief in the early months of 1847 as workhouses were filled to capacity. From its inception, the revised poor law system struggled to cope with the increasing scale and severity of the Famine, especially following the blighted harvest of 1848. Although accommodation in workhouses had been increased by one-third to 150,000 in the year to September 1848, and outdoor relief made available for some 800,000 people, the supply of relief failed to match the increasing rate of demand. Overcrowded spaces and appalling conditions facilitated the rapid spread of infectious disease, especially in many southern and western coast unions. By early 1849, mortality rates in workhouses in the west were as high as 2,500 per week. An Asiatic cholera outbreak in late 1848 caused almost 20,000 deaths in Ireland, with a high proportion occurring along the western seaboard, where the population was already weak from prolonged hunger (Figure 20.13). The historian, Christine Kinealy, has suggested that 'by the beginning of 1849, mortality in the western workhouses was proportionally as high as it had been in the winter of 1846–47, reaching 2,500 deaths per week'.[9]

The increasing severity of famine through the latter half of the 1840s exacerbated conditions of starvation and destitution. Despite the operation of relief measures, death rates soared with commentators agreeing that at least one million deaths can be attributed directly to the Famine. Mortality rates were highest in the western counties and in unions from Mayo to Cork (Figure 20.14).

EVICTIONS

A major influence on death rates (as well as rates of emigration) was the pattern and level of evictions of tenants by landlords during the Famine. Under poor law legislation, relief was funded by taxes placed on landowners, who were also responsible for all rates on holdings valued at £4 or less. With a high proportion of holdings around the western and southern coasts in this category, landlords were encouraged to reduce tenant numbers, especially if they were unable to pay their rent (Figure 20.15).

Evictions became a common 'tool' used by landowners to rationalise their estates and reduce costs by clearing tenants from their smallholdings. Smyth calculates from official records that at least 100,000 families (more than 500,000 people) were evicted in the Famine years.[10] Landlords also used other methods to persuade

Fig. 20.12 BUILDING HOUSING FORMER SOUP KITCHEN IN SKIBBEREEN, COUNTY CORK. Built in the 1780s as a mill adjacent to the Ilen River, the building became Ireland's first soup kitchen when it opened in November 1846. Operated by the Skibbereen Committee of Gratuitous Relief, it provided free soup or food to the thousands of destitute and starving people that flocked into the Skibbereen Union. [Source: Kyle Fawkes, Sarah Kandrot and Zoë O'Hanlon]

Fig. 20.13 IMPOVERISHED PEASANTS SCOUR THE LAND FOR FOOD. An impoverished boy and girl on the road to Cahera in the Skibbereen Union, turning up the ground to seek a potato to appease their hunger. This is one of the iconic sketches drawn by artist, James Mahony, for the *Illustrated London News* in 1847. Skibbereen in County Cork was identified as one of the worst-affected unions and Mahony's sketches helped make this town a symbol of the appalling consequences of the potato famine. [Source: Maggie Blanck collection (http://www.maggieblanck.com), courtesy of Mayo County Library]

Excess mortality (1846–1850) as a percentage of population

- Under 5.0
- 5.0–7.4
- 7.5–9.9
- 10.0–12.4
- 12.5–15.0
- Over 15.0

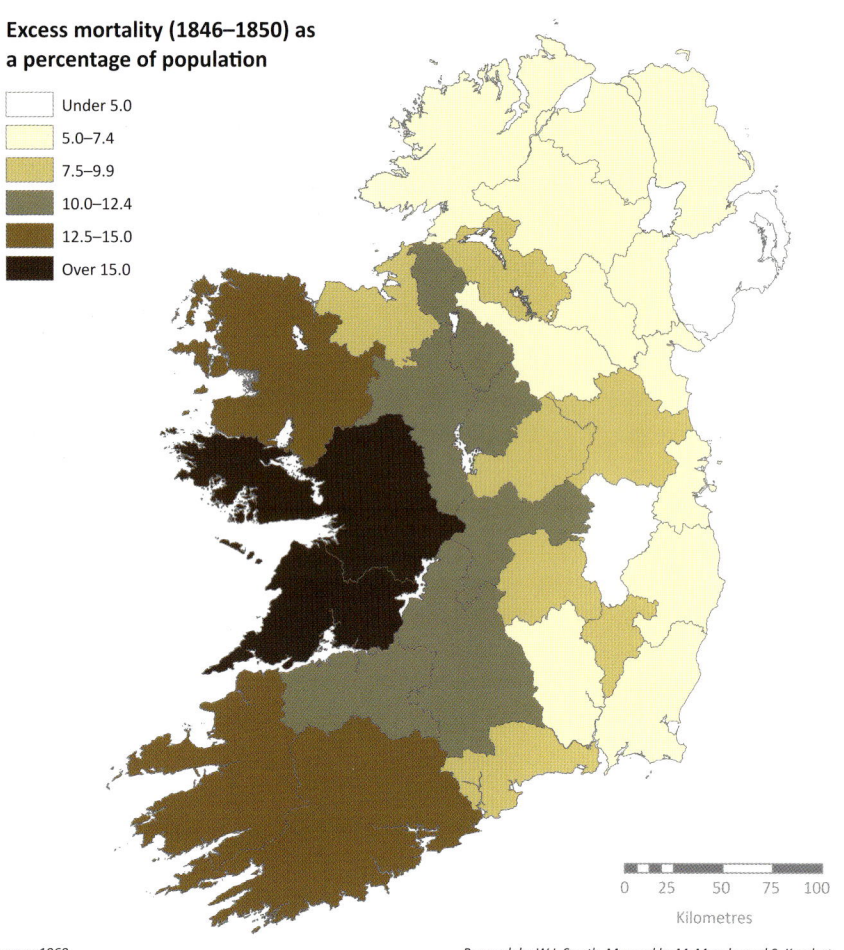

N

Cousens 1960

Research by W.J. Smyth, Mapped by M. Murphy and S. Kandrot

0   25   50   75   100

Kilometres

Fig. 20.14 (left) EXCESS MORTALITY (1846–1850) AS A PERCENTAGE OF THE 1841 POPULATION. In estimating excess deaths due to the Famine, the geographer, S.H. Cousens, suggests a figure of *c*.800,000. This is now considered an underestimate; the more realistic figure being at least one million. The map illustrates the distribution of excess deaths from 1846 to 1850 ascribed to the Famine. It shows clearly the relatively high figures from counties Mayo to Cork, where at least one in eight of the 1841 population was estimated to have died due to the Famine. This high vulnerability to the devastating consequences of famine in these coastal counties was ascribed to overdependence on the potato crop, subsistence farming on landholdings of generally less than twenty acres (many less than five acres), difficult physical environments and poverty. [Data source: adapted from Crowley, J., Smyth, W.J., and Murphy, M., eds, 2012, *Atlas of the Great Irish Famine*, Cork: Cork University Press]

Fig. 20.15 (below) THE EJECTMENT. The image displays the forcible removal of tenants and the destruction of their home. While the tenants plead for compassion, the landlord or his representative appears to urge on the process of destroying the premises, supported by the presence of armed soldiers. [Source: Maggie Blanck collection (http://www.maggieblanck.com), courtesy of Mayo County Library]

THE EJECTMENT.

tenants to 'vacate' their holdings, such as threats and cash inducements to emigrate. For large numbers, however, extreme poverty and starvation were sufficient to force families to move from their land, into the workhouses. In combination, Smyth estimates that between two-thirds and three-quarters of a million people lost their homes. Eviction rates were particularly high in coastal counties from Mayo to Cork.

EMIGRATION

Starvation, evictions and destitution made emigration the only option for people hoping to find a better life. Although emigration from Ireland had a long tradition, during the Famine it reached unprecedented levels (Figure 20.16). The spring 1847 exodus, for example, 'bore all the marks of panic and hysteria, down to the queues and fights for passage tickets at the embarkation points'.[11]

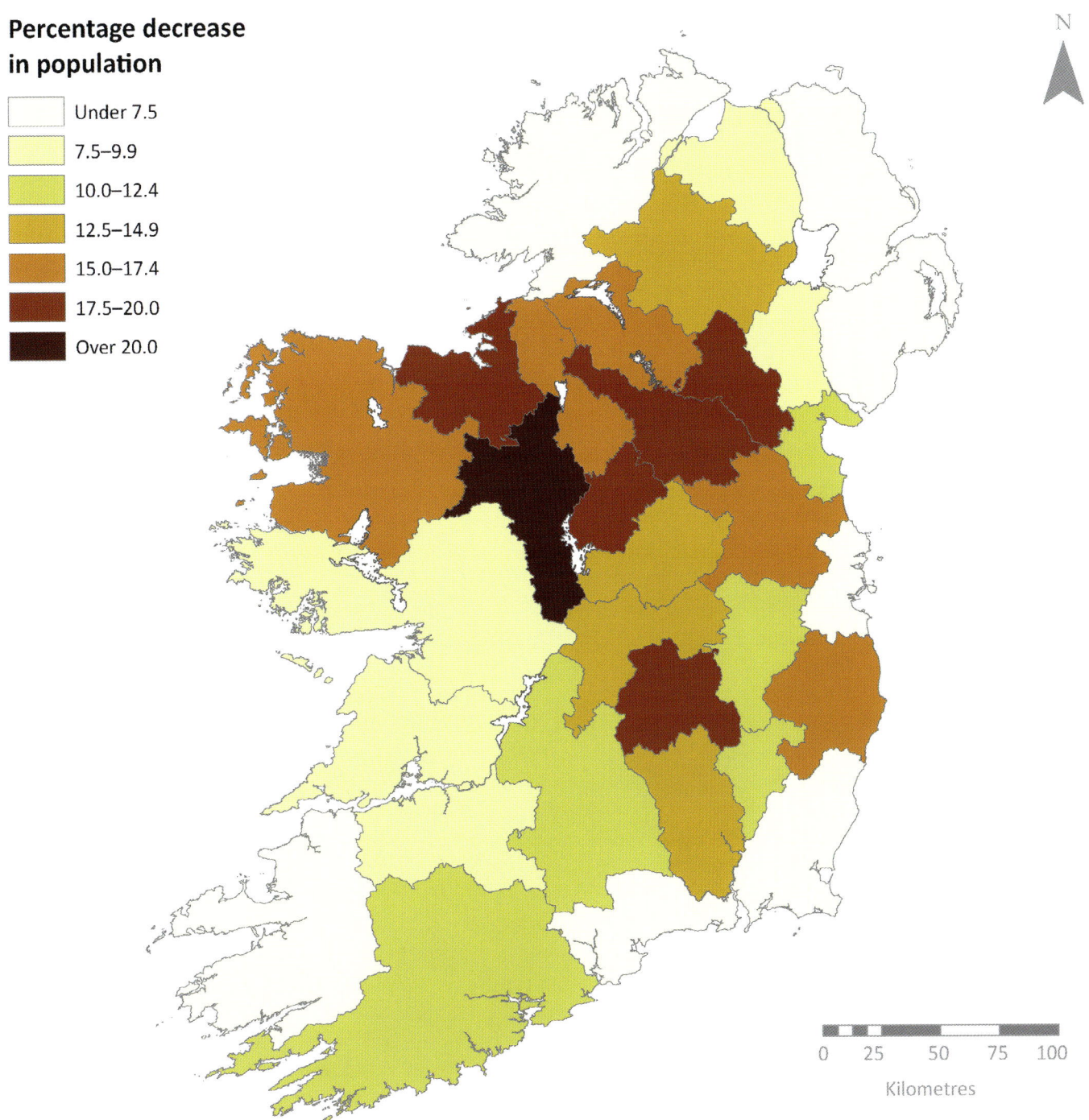

Percentage decrease in population

- Under 7.5
- 7.5–9.9
- 10.0–12.4
- 12.5–14.9
- 15.0–17.4
- 17.5–20.0
- Over 20.0

N

0   25   50   75   100
Kilometres

Research by W.J. Smyth, mapped by M. Murphy and S. Kandrot

Fig. 20.16 POPULATION DECLINE PER COUNTY DUE TO EMIGRATION (1841–1851). The map displays rates of emigration as a percentage of each county's 1841 population. Considerable variation exists, with highest rates occurring in the inland counties of the north-west centre of Ireland. Apart from Mayo, Sligo and Leitrim, counties along the western and southern coasts display rates of generally less than 10 per cent. Although most severely affected by the Famine, poverty levels here were so acute as to possibly prevent the purchase of tickets to avail of the option to emigrate. Coastal counties around the northern and eastern coastline possessed better employment prospects and were generally more prosperous, which reduced pressure to emigrate. In addition, major coastal cities, especially Belfast and Dublin, but also Cork, emerged as centres of immigration from the depressed rural hinterlands. This reduced levels of net emigration in such areas. [Data source: adapted from Crowley, J., Smyth, W.J., and Murphy, M., eds, 2012, *Atlas of the Great Irish Famine*, Cork: Cork University Press]

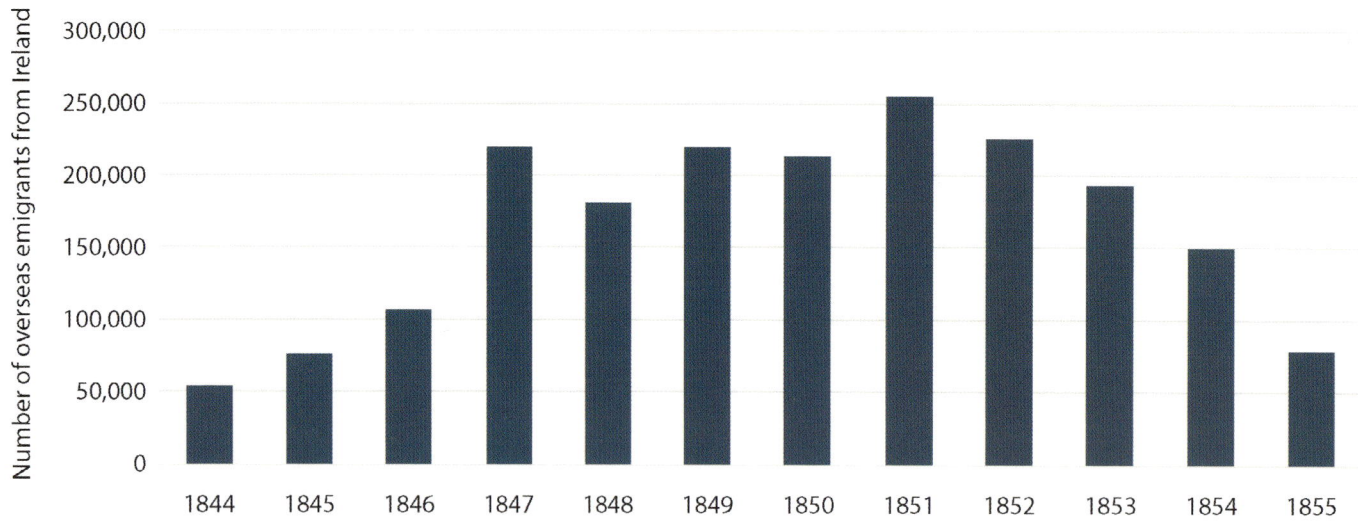

Fig. 20.17 NUMBERS OF OVERSEAS EMIGRANTS FROM IRELAND (1844–1855). In the pre-Famine year of 1844, just over 54,000 people left Ireland for overseas destinations (excluding emigration to Britain). With the arrival of the potato blight and Famine in 1845, numbers increased rapidly. Peak emigration was in 1851, when the cumulative impact of starvation and destitution saw almost 255,000 leave in that year. Although numbers then began to decline, the patterns and traditions of emigration had become established and have remained a key component in the evolving demographic and social structures of Ireland to this day. [Data source: Commission on Emigration and other Population Problems 1948–1954. Dublin, Stationery Office, 1954, p. 314]

Fig. 20.18 EMIGRANTS AT CORK QUAY. On 15 April 1847 the *Cork Examiner* reported that 'the quays are crowded every day with peasantry from all quarters of the country who are emigrating to America, both direct from this port and "cross channel" to Liverpool, as the agents here cannot produce enough ships to convey people from this unhappy country.' Between 1845 and 1851, it is estimated that 71,325 emigrated from Cork, almost exclusively to destinations in the United States and Canada. This large-scale pattern of emigration was continued throughout the second half of the nineteenth century. The demand for passage out of Ireland meant that almost every Irish port saw some share of the emigrant traffic during the Famine period, as vessels of every size and condition were pressed into service. For example, between 1846 and 1851, some 736 ships carrying emigrants into New York Harbour sailed directly from the large ports of Belfast, Cork (including Cobh/Queenstown), Dublin and Limerick, but also from a range of smaller ports that accommodated emigrants from their immediate hinterlands (such as New Ross, Youghal, Galway and Moville). Conditions on these vessels were often desperate, especially for the large numbers of poor classes. Death rates were often high on these 'coffin ships', but were almost accepted as part of the price paid to escape Famine Ireland. [Source: Maggie Blanck collection (http://maggieblanck.com), courtesy of Mayo County Library]

Given these conditions, those who departed during the Famine have been likened to refugees rather than considered emigrants.[12] Historian, Kerby Miller, has estimated that over 2.1 million Irish people emigrated over the period from 1845 to 1855, with more people leaving the island in those eleven years than in the preceding two centuries (Figure 20.17).[13]

Emigration rates varied within the country and around the coastline. Excluding a small group of counties in the north-west and Wicklow to the east, coastal counties experienced generally low rates of population loss through emigration. This suggests an inverse relationship between emigration and excess mortality, especially along the western and southern coastline (Figure 20.16 and Figure 20.14). A possible explanation for this is that poverty levels were so acute in such coastal communities that they could not afford the cost of tickets to escape the country. For such people, movement off the land involved the trek to the workhouse, or death from starvation and disease.

For those who did emigrate, the first step in their journey was from the home place to the port of embarkation on the coast (Figure 20.18).[14] These tragic journeys have been described by historian, Robert Scally, as involving 'the refugees of a thousand townlands' who, on their journey to the ports, could not have 'failed to notice the unburied corpses, the emaciated families, and the universal beggardom thronging the main roads and towns'.[15]

To conclude, while the whole of Ireland suffered during the Great Famine, communities along the western and southern coasts were decimated by its severity and the inadequacy of effective relief efforts. The scale of population losses in such areas transformed their physical and cultural landscapes, with implications that remain to the present day. The Famine reduced dramatically the labour-intensive use of land that typified these coastal areas, and which involved large numbers of small-scale tenant holdings capable of cultivating potato crops to sustain high densities of population. Much of the land reverted to more extensive forms of pastoral farming, although large areas of the more marginal upland slopes and bogland were abandoned. The result was a progressively depopulated landscape that incorporated large numbers of abandoned farm cottages (and some 'ghost villages') and the almost ubiquitous dry-stone walls that encircle the many disused fields and long-forgotten lazy beds. These all remain as relic features and poignant reminders of once vibrant communities and landscapes. They have also become central images in our perception of the west coast of Ireland.

In addition to the physical expressions of abandonment, the loss of so many people through death and emigration impacted the cultural landscape. Prior to the Famine, potato harvests had sustained large populations and vibrant communities along the western and southern seaboards. These were also Irish-speaking and

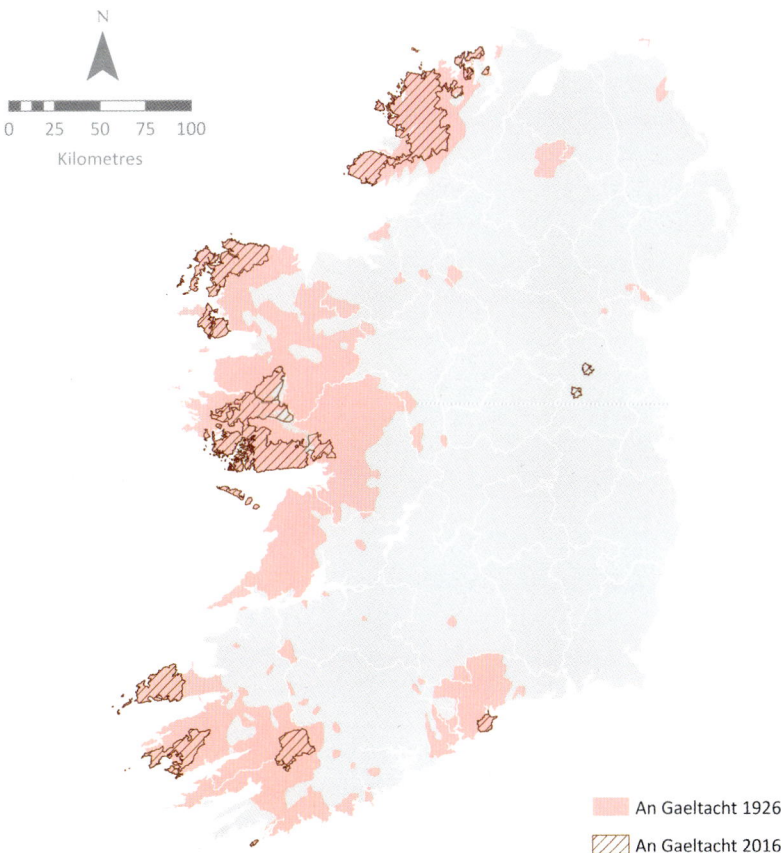

An Gaeltacht 1926

An Gaeltacht 2016

Fig. 20.19 IRELAND'S GAELTACHT (1926–2016). The Gaeltacht refers to those areas where the government recognises and supports the use of the Irish as the vernacular language. They originated in the 1920s with the foundation of the Free State and reflect the importance of the language for the new state's identity. Since their official designation in 1926, however, the Irish-speaking areas have declined significantly and the designation has failed to reverse the post-Famine trend of a collapse in numbers speaking Irish. The 2016 census, therefore, records only 20,586 people as speaking Irish on a daily basis, compared to some 247,000 in 1926. These are now concentrated in relatively small areas located along the western coast. Despite this, their communities continue to play a key role in the nation's cultural expressions and identity. For example, each year large numbers of school children travel to the Gaeltacht to enrol in residential Irish colleges for short-term courses to enhance their ability to speak Irish, as it remains a compulsory subject in the school syllabus. This increases the vitality of the Gaeltacht communities and also helps shape the perceptions of the western coast and its culture for many of Ireland's younger generations. [Source: Department of Culture, Heritage and the Gaeltacht]

could be considered to be a core of Gaelic culture as the process of anglicisation advanced in eastern and northern Ireland. Such communities were ravaged by the Famine and if the Irish language did not fall silent, it was reduced to a marginal state as English became the dominant language throughout the country. In the 1841 census, some four million Irish speakers were recorded on the island; by 1891 this had collapsed to 680,000.

The island retains a small number of Irish-speaking communities and these remain concentrated along the western coast (Figure 20.19). These are the Gaeltacht areas where the government supports and recognises Irish as the predominant spoken language, and music, song and folklore are nurtured as part of the country's identity and heritage. Here, and elsewhere along the western coast, these cultural expressions have also acquired a growing commercial value, as large numbers of tourists gravitate to them to immerse themselves in Ireland's unique cultural heritage, as well as absorbing the outstanding natural landscapes. After almost two centuries, the relic landscapes and memories of the Famine remain central to any understanding of these coastal areas (Figure 20.20).

Fig. 20.20 FAMINE AND EMIGRATION HAD DRASTIC IMPACTS ON POPULATIONS AND COMMUNITIES ALONG IRELAND'S WESTERN COAST, LEAVING MANY RELIC LANDSCAPES AS POIGNANT REMINDERS OF THIS TRAGIC EVENT. a. Ruined houses of the abandoned village of Slievemore, Achill Island, County Mayo. This is one example of many villages that became deserted following the Famine. Evidence of abandoned fields and relic lazy beds are also apparent. [Source: Frank Coyne]; b. Gort Na gCapall, Inis Mór, Aran Islands. A 'patchwork-quilt' of small, but now mostly abandoned, fields and enclosing stone walls bear testimony to a more densely populated past. Large-scale depopulation, linked to the Famine, has helped preserve the scenic and historic landscapes of extensive areas along the west coast. These have helped attract large numbers of tourists to this area. [Source: Mary Frances Beatty]; c. Abandoned cottage on the Dingle Peninsula, County Kerry, overlooking the Blasket Islands. These relic features highlight the consequences of the Famine. For most families located along the western seaboard, their problems were compounded by difficult environmental conditions for productive agriculture. Many of the sites of such houses commanded scenic views. Although this would have been largely irrelevant to poverty-stricken farmers, today such sites attract investors wishing to refurbish the ruins as holiday homes. [Source: Noel O'Neill]; d. The Famine Memorial, by the sculptor Robert Gillespie, was unveiled on Custom House Quay in the Dublin Docklands in 1997. The emaciated and desperate figures represent the vast number of people seeking to escape the ravages of poverty and starvation through emigration. The great irony is that its location is proximate to Dublin's highly successful docklands development, which has attracted significant overseas investment and immigration to avail of the city's and country's new prosperity. How time and spaces change! [Source: Bouchtrou, Wikimedia Commons, CC BY-SA 4.0]

IRELAND'S ISLANDS

Stephen A. Royle

INIS TUAISCEART (INISHTOOSKERT), COUNTY KERRY. Also known as *An Fear Marbh* (the dead man or the sleeping man). [Source: Noel O'Neill]

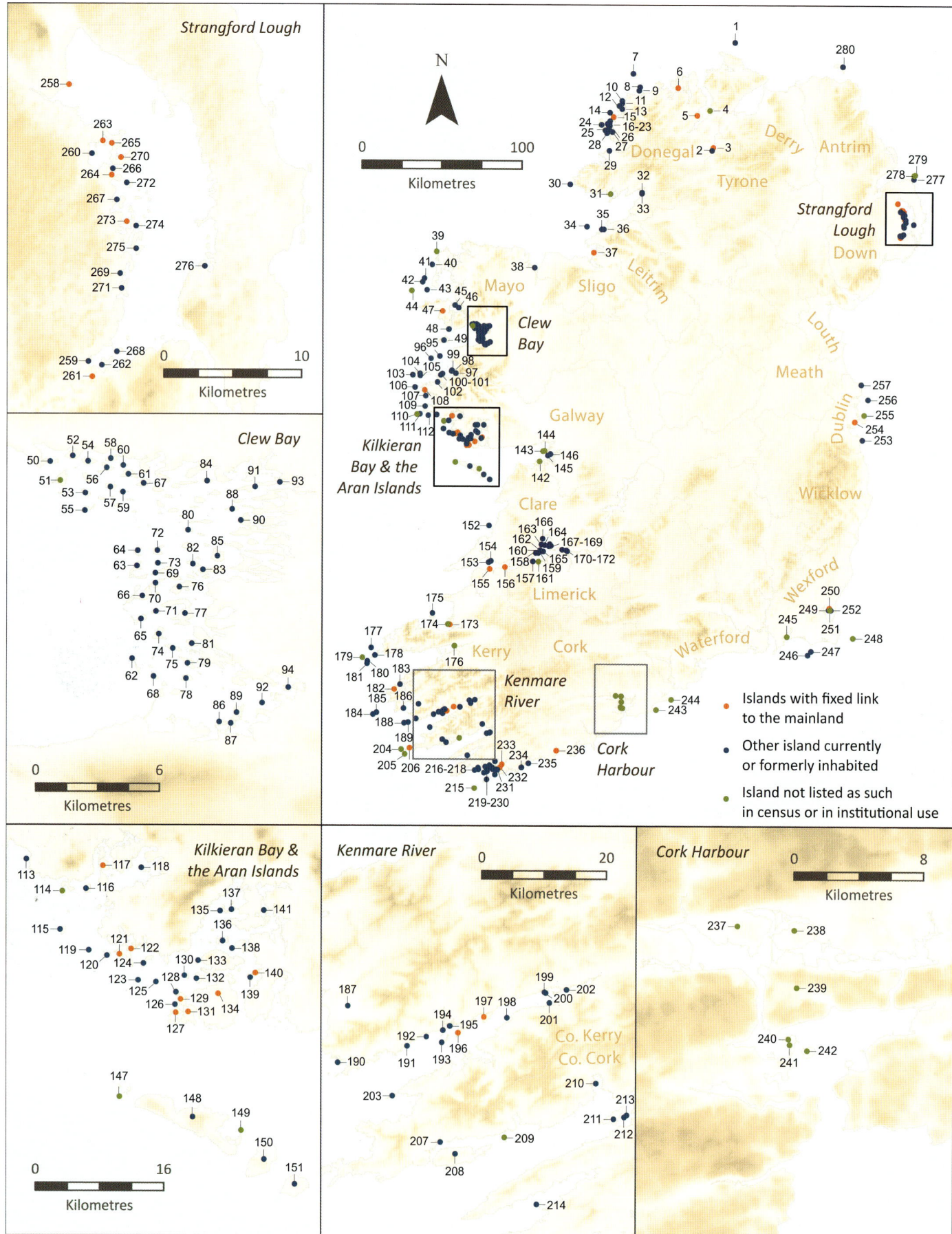

Fig. 21.1 IRELAND'S OFFSHORE ISLANDS. The vast majority of these islands lie off the south and west coasts, largely due to a combination of the prevailing east–west trending hard-rock geology, the erosive power of the Atlantic waves and factors of sea-level rise; however, in Clew Bay, and in Strangford Lough on the east coast of County Antrim, the islands are composed mostly of sediments deposited as drumlins by the retreating glaciers at the end of the last Ice Age.

1 ● Inishtrahull, Co. Donegal
2 ● Corkan, Co. Donegal
3 ● Island Beg, Co. Donegal
4 ● Inch, Co. Donegal
5 ● Aughnish, Co. Donegal
6 ● Island Roy, Co. Donegal
7 ● Tory, Co. Donegal
8 ● Inishdooey, Co. Donegal
9 ● Inishbofin, Co. Donegal
10 ● Inishsirrer, Co. Donegal
11 ● Inishmeane, Co. Donegal
12 ● Gola, Co. Donegal
13 ● Inishinny, Co. Donegal
14 ● Owey, Co. Donegal
15 ● Cruit, Co. Donegal
16 ● Inishinny, Co. Donegal
17 ● Lahan, Co. Donegal
18 ● RInralny, Co. Donegal
19 ● Eighter, Co. Donegal
20 ● Inishcoo, Co. Donegal
21 ● Duck, Co. Donegal
22 ● Edernish, Co. Donegal
23 ● Rutland, Co. Donegal
24 ● Arranmore, Co. Donegal
25 ● Inishkeeragh, Co. Donegal
26 ● Inishfree Upper, Co. Donegal
27 ● Inishal, Co. Donegal
28 ● Illancrone, Co. Donegal
29 ● Inishkeel, Co. Donegal
30 ● Rathlin O'Birne, Co. Donegal
31 ● Rotten Island, Co. Donegal
32 ● Rooney's, Co. Donegal
33 ● Inishpat, Co. Donegal
34 ● Inishmurray, Co. Sligo
35 ● Conor's, Co. Sligo
36 ● Dernish, Co. Sligo
37 ● Inishmulclohy (Coney), Co. Sligo
38 ● Bartragh, Co. Mayo
39 ● Eagle Island, Co. Mayo
40 ● Inishglora, Co. Mayo
41 ● Inishkea N., Co. Mayo
42 ● Inishkea S., Co. Mayo
43 ● Duvillaun More, Co. Mayo
44 ● Black Rock, Co. Mayo
45 ● Inishbiggle, Co. Mayo
46 ● Annagh Island, Co. Mayo
47 ● Achill Island, Co. Mayo
48 ● Achillbeg, Co. Mayo
49 ● Clare Island, Co. Mayo
50 ● Rosmurrevagh, Co. Mayo
51 ● Moynishmore, Co. Mayo
52 ● Rosturk, Co. Mayo
53 ● Inishcooa, Co. Mayo
54 ● Inishkeel, Co. Mayo
55 ● Roeillaun, Co. Mayo
56 ● Inishgowla, Co. Mayo
57 ● Inishnacross, Co. Mayo
58 ● Inishbobunnan, Co. Mayo
59 ● Inishilra, Co. Mayo
60 ● Inishlim, Co. Mayo
61 ● Inishtubbrid, Co. Mayo
62 ● Dorinish More, Co. Mayo
63 ● Inishbee, Co. Mayo
64 ● Inishoo, Co. Mayo
65 ● Inishgort, Co. Mayo
66 ● Island More, Co. Mayo
67 ● Inishquirk, Co. Mayo
68 ● Inishleague, Co. Mayo
69 ● Derinish, Co. Mayo
70 ● Knockycahillaun, Co. Mayo

71 ● Collan Beg, Co. Mayo
72 ● Inishgowla, Co. Mayo
73 ● Calf, Co. Mayo
74 ● Inishlyre, Co. Mayo
75 ● Crovinish, Co. Mayo
76 ● Clynish, Co. Mayo
77 ● Collan More, Co. Mayo
78 ● Inishraher, Co. Mayo
79 ● Inishgowla S., Co. Mayo
80 ● Inishfesh, Co. Mayo
81 ● Illanataggart, Co. Mayo
82 ● Inishcottle, Co. Mayo
83 ● Inishnakillew, Co. Mayo
84 ● Roslynagh Island, Co. Mayo
85 ● Inishturk, Co. Mayo
86 ● Annagh Island W., Co. Mayo
87 ● Annagh Island Middle, Co. Mayo
88 ● Inishturlin, Co. Mayo
89 ● Annagh Island E., Co. Mayo
90 ● Rosbarnagh Island, Co. Mayo
91 ● Rosmore, Co. Mayo
92 ● Illanroe, Co. Mayo
93 ● Lisduff, Co. Mayo
94 ● Roman Island, Co. Mayo
95 ● Caher Island, Co. Mayo
96 ● Inisturk, Co. Mayo
97 ● Inishbarna, Co. Galway
98 ● Inishdegil Beg, Co. Mayo
99 ● Inishdegil More, Co. Mayo
100 ● Crump Island, Co. Galway
101 ● Freaghillaun N., Co. Galway
102 ● Freaghillaun S., Co. Galway
103 ● Inishark, Co. Galway
104 ● Inisbofin, Co. Galway
105 ● Port Island, Co. Galway
106 ● High Island, Co. Galway
107 ● Omey Island, Co. Galway
108 ● Turbot Island, Co. Galway
109 ● Inishdugga, Co. Galway
110 ● Illaunamid, Co. Galway
111 ● Chapel Island, Co. Galway
112 ● Illaunurra, Co. Galway
113 ● Inishdawros, Co. Galway
114 ● Mutton, Co. Galway
115 ● Croaghnakeela Island, Co. Galway
116 ● Inishlackan, Co. Galway
117 ● Inishnee, Co. Galway
118 ● Illaungorm N., Co. Galway
119 ● St. Macdara's Island, Co. Galway
120 ● Mason Island, Co. Galway
121 ● Mweenish Island, Co. Galway
122 ● Rusheennacholla, Co. Galway
123 ● Inishmuskerry, Co. Galway
124 ● Finish Island, Co. Galway
125 ● Birmore Island, Co. Galway
126 ● Inisherk, Co. Galway
127 ● Crappagh, Co. Galway
128 ● Dinish, Co. Galway
129 ● Furnace, Co. Galway
130 ● Illauneeragh, Co. Galway
131 ● Lettermullan, Co. Galway
132 ● Inishbarra, Co. Galway
133 ● Inchaghaun, Co. Galway
134 ● Gorumna Island, Co. Galway
135 ● Illauneeragh W., Co. Galway
136 ● Inishtravin, Co. Galway
137 ● Illaunmore, Co. Galway
138 ● Inishlusk, Co. Galway
139 ● Inchamakinna, Co. Galway
140 ● Rossroe Island, Co. Galway

141 ● Inisheltia, Co. Galway
142 ● Aughinish, Co. Clare
143 ● Tawin W., Co. Galway
144 ● Tawin E., Co. Galway
145 ● Island Eddy, Co. Galway
146 ● Mweenish, Co. Galway
147 ● Eeragh, Co. Galway
148 ● Inishmore, Co. Galway
149 ● Straw Island, Co. Galway
150 ● Inishmaan, Co. Galway
151 ● Inisheer, Co. Galway
152 ● Mutton Island, Co. Clare
153 ● Scattery Island, Co. Clare
154 ● Inishbig (Hog Island), Co. Limerick
155 ● Carrig Island, Co. Kerry
156 ● Tarbert Island, Co. Kerry
157 ● Foynes Island, Co. Limerick
158 ● Inishcorker, Co. Clare
159 ● Inishtubbrid, Co. Clare
160 ● Inishmacowney, Co. Clare
161 ● Aughinish Island, Co. Limerick
162 ● Canon Island, Co. Clare
163 ● Inishmore, Co. Clare
164 ● Coney Island, Co. Clare
165 ● Inishloe, Co. Clare
166 ● Horse Island, Co. Clare
167 ● Feenish, Co. Clare
168 ● Deenish Island, Co. Clare
169 ● Inishmacnaghtan, Co. Clare
170 ● Saint's Island, Co. Clare
171 ● Quay Island, Co. Clare
172 ● Green Island, Co. Clare
173 ● Samphire Island, Co. Kerry
174 ● Little Samphire Island, Co. Kerry
175 ● Illauntannig, Co. Kerry
176 ● Rosculien Island, Co. Kerry
177 ● Inishtooskert, Co. Kerry
178 ● Great Blasket Island, Co. Kerry
179 ● Tearaght Island, Co. Kerry
180 ● Inishnabro, Co. Kerry
181 ● Inishvickillane, Co. Kerry
182 ● Valentia Island, Co. Kerry
183 ● Beginish Island, Co. Kerry
184 ● Skellig Michael, Co. Kerry
185 ● Little Skellig, Co. Kerry
186 ● Horse Island, Co. Kerry
187 ● Church Island, Co. Kerry
188 ● Scariff, Co. Kerry
189 ● Deenish, Co. Kerry
190 ● Abbey Island, Co. Kerry
191 ● Illaundrane, Co. Kerry
192 ● Illaunleagh, Co. Kerry
193 ● Sherky Island, Co. Kerry
194 ● Garinish, Co. Kerry
195 ● Illaunslea, Co. Kerry
196 ● Rossdohan Island, Co. Kerry
197 ● Rossmore Island, Co. Kerry
198 ● Ormond's Island, Co. Kerry
199 ● Greenane Islands, Co. Kerry
200 ● Cappanacush, Co. Kerry
201 ● Dinish Island, Co. Kerry
202 ● Dunkerron Island, Co. Kerry
203 ● Inishfarnard, Co. Cork
204 ● The Bull, Co. Cork
205 ● The Calf, Co. Cork
206 ● Dursey Island, Co. Cork
207 ● Dinish Island, Co. Cork
208 ● Bere Island, Co. Cork
209 ● Roancarrigmore, Co. Cork
210 ● Garnish Island, Co. Cork

211 ● Whiddy Island, Co. Cork
212 ● Little Chapel, Co. Cork
213 ● Chapel, Co. Cork
214 ● Carbery Island, Co. Cork
215 ● Fastnet Rock, Co. Cork
216 ● Goat Island, Co. Cork
217 ● Coney Island, Co. Cork
218 ● Long Island, Co. Cork
219 ● Calf Island W., Co. Cork
220 ● Castle Island, Co. Cork
221 ● Calf Island Middle, Co. Cork
222 ● Cape Clear Island, Co. Cork
223 ● Calf Island E., Co. Cork
224 ● Horse Island, Co. Cork
225 ● Skeam W., Co. Cork
226 ● Skeam E., Co. Cork
227 ● Inishodriscol (Hare), Co. Cork
228 ● Sherkin Island, Co. Cork
229 ● Manin Island, Co. Cork
230 ● Sandy Island, Co. Cork
231 ● Spanish Island, Co. Cork
232 ● Ringarogy Island, Co. Cork
233 ● Inishbeg, Co. Cork
234 ● Horse Island, Co. Cork
235 ● Rabbit Island, Co. Cork
236 ● Inchydoney Island, Co. Cork
237 ● Little Island, Co. Cork
238 ● Fota Island (Foaty), Co. Cork
239 ● Great Island, Co. Cork
240 ● Haulbowline, Co. Cork
241 ● Rocky Island, Co. Cork
242 ● Spike Island, Co. Cork
243 ● Ballycotton, Co. Cork
244 ● Capel Island, Co. Cork
245 ● Bannow, Co. Wexford
246 ● Great Saltee, Co. Wexford
247 ● Little Saltee, Co. Wexford
248 ● Tuskar Rock Lighthouse, Co. Wexford
249 ● The Ridge, Co. Wexford
250 ● Begerin, Co. Wexford
251 ● Big Island, Co. Wexford
252 ● Breast, Co. Wexford
253 ● Dalkey, Co. Dublin
254 ● North Bull, Co. Dublin
255 ● Ireland's Eye, Co. Dublin
256 ● Lambay, Co. Dublin
257 ● Shenick, Co. Dublin
258 ● Rough, Co. Down
259 ● Gibb's, Co. Down
260 ● Wood, Co. Down
261 ● Castle, Co. Down
262 ● Gores, Co. Down
263 ● Rolly, Co. Down
264 ● Sketrick, Co. Down
265 ● Reagh, Co. Down
266 ● Rainey, Co. Down
267 ● Conly, Co. Down
268 ● Salt, Co. Down
269 ● Simmy, Co. Down
270 ● Mahee, Co. Down
271 ● Taggart, Co. Down
272 ● Trasnagh, Co. Down
273 ● Castle, Co. Down
274 ● Dunsy, Co. Down
275 ● Pawle, Co. Down
276 ● Ballywallon Island, Co. Down
277 ● Copeland, Co. Down
278 ● Lighthouse, Co. Down
279 ● Mew, Co. Down
280 ● Rathlin, Co. Antrim

Ireland is fringed by offshore islands, especially along its west coast (Figure 21.2), many of which have played a significant role in Ireland's history and culture. How many islands there are depends on what is counted: how big does a piece of land surrounded by water have to be before it graduates from being a rock to an island? How far up a river can an island be before it is classed as riverine rather than offshore? Approximately 300 pieces of land off the coast are of a size and scale as to be accepted as islands (Figure 21.1).

From the medieval period onwards, many islands, particularly those on the exposed west coast, were religious retreats and refuges, as evidenced by the numerous monuments they contain. These range from the splendid beehive huts on Skellig Michael, a UNESCO World Heritage Site, to the single hermit cells visible on islands such as Bishop's Island off County Clare. Elsewhere, Tory and Scattery

Fig. 21.2 BALTIMORE ON THE COUNTY CORK MAINLAND, WITH SHERKIN ISLAND AND CAPE CLEAR BEHIND. The island of Ireland is itself large, the twentieth largest by area in the world. Off its coasts, especially to the west, lie at least 350 small islands; this number is imprecise because there is no generally accepted means of deciding when an offshore rock is big enough to be classified as an island. The Irish census returns from 1821 onwards show that 284 islands have had a resident population during at least one census. Other islands would have lost their populations before official records were kept. The considerable majority of these islands do not now have any residents, the challenges of island life having seen the islands become depopulated. All the others have smaller populations now than in the past. The islands shown here – Sherkin and Cape Clear in County Cork – have retained populations of 111 and 147 permanent residents respectively in 2016, but these totals are much reduced on their peaks of 1,131 for Sherkin and 1,052 for Cape Clear in 1841, the last census before the Famine. The two islands are serviced by a regular ferry from Baltimore harbour. In the decade up to 2018, this subsidised service allowed almost 410,000 visitors to access Sherkin and 250,000 to Cape Clear. This has been vital to the islands' ability to retain a vibrant tourist sector. [Source: Robbie Murphy]

Fig. 21.3 (ROUND TOWER ON SCATTERY ISLAND. Foundations like this were used widely as part of the early Christian church in Ireland. Life in the monastic settlements on remote islands was difficult and unforgiving at times, but was nonetheless well suited to the contemplative and spiritual needs of the religious communities that occupied them. The monastic site on Scattery dates from the late tenth or early eleventh century, and was founded by Saint Senan. The round tower on this site is approximately 26m high, with a circumference of just over 5m at its base. As alluded to by the Irish term for this type of tower, *cloigtheach*, its main function was as a bell tower, although it may also have served as a lookout and a storeroom. Like most Irish round towers, the door faces the west doorway of the church with which it is associated. However, in the case of the Scattery tower the entrance door is at the base, whereas in comparable structures elsewhere (most of which were built later), the door was usually placed some distance above ground level, to enhance the structural integrity of the tower as well as for possible security purposes. [Source: National Monuments Service, Department of Culture, Heritage and the Gaeltacht]

Fig. 21.4 DÚN AONGHASA ON INIS MÓR, ARAN ISLANDS. One of seven *dúns* on the Aran Islands, Dún Aonghasa is one of Ireland's iconic and most-visited attractions. Perched on the cliff-edge, 100m above the waves, the fort commands a view over the length of the island as well as far out to sea. The walls of Dún Aonghasa were partially reconstructed in 1884–1886, and again in the twentieth century. Unfortunately, only incomplete documentary records exist for this rebuilding and it is possible that important evidence relating to the origins of the fort was destroyed in the process. Nevertheless, the dún itself is thought to date to the early Irish Celtic period and a possible influx of people from Iberia in the *c*.200 AD, while archaeological excavations in the 1990s indicate that the wider site at Dún Aonghasa was occupied from about 1500 BC, demonstrating clearly that the offshore islands of Ireland have attracted human settlement since far back in antiquity. [Source: National Monuments Service, Department of Culture, Heritage and the Gaeltacht; Long, H. and Rynne E., 1992. Dún Aonghasa. *Journal of the Galway Archaeological and Historical Society*, 44, pp. 11–28.]

Islands each have ruined monastic round towers (Figure 21.3), while religious activity on the Aran Islands contributed in no small way to Ireland becoming known as the 'land of saints and scholars'.

Several islands around the coast have been important for strategic purposes, at many periods in Ireland's history. Forts have been constructed on many of them, from the magnificent stone dúns of the Aran Islands, County Galway (Figure 21.4), to the more modern Cromwellian structure on Inishbofin and the later eighteenth-century star fort on Spike Island, this last still in use for military and penal purposes until the latter half of the twentieth century. In another strategic context, operating out of her stronghold on Clare Island, Grace O'Malley, Granuaile, proved problematic to the mighty English queen, Elizabeth I.

More recently, Valencia and, to a lesser extent, Rathlin islands both played significant roles in the development of modern wireless communications because of their locations.

Case Study: Skellig Michael (Sceilg Mhicíl)

John Crowley

The magic of Ireland is very strong for me when I see a beehive dwelling. Did you ever make the pilgrimage to Skellig Michael? If not, you have not yet seen Ireland.[1]

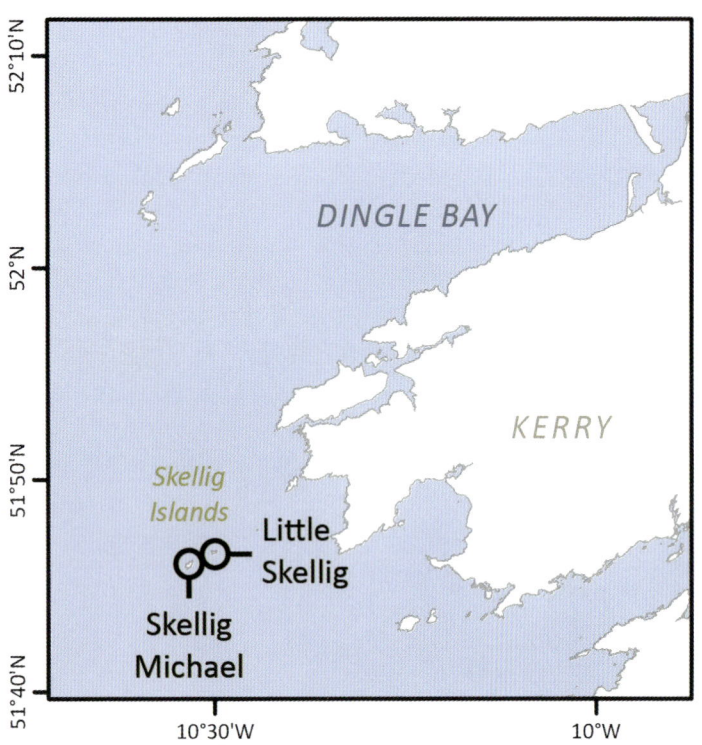

Lying 12km off the Kerry coast, Skellig Michael has inspired people over many centuries. Rising from the Atlantic, its bold, pyramidal-like structure has contributed to its mystique and otherworldly appearance (Figure 21.5). A small island, approximately 22ha in extent, it continues to stir the imagination, both spiritual and artistic. The writer and future Nobel laureate, George Bernard Shaw, was so taken with his visit there in 1910 that he spoke of the island in reverential terms: 'I tell you the thing does not belong to any world you and I have ever lived and worked in; it is part of our dream world'.[2] In contemplating the island, its monastic foundation and beehive dwellings, Shaw appreciated that Skellig Michael had a much wider and deeper cultural significance when it came to understanding Ireland and its contribution to the world. The monastery, located *c*.185m above sea level, is an architectural feat in every sense (Figure 21.6). Enthralled by the appearance of the beehive cells, Shaw was well aware of the powerful imagination that had willed them into existence. While Irish monks would produce magnificent manuscripts in scriptoria across Ireland and Europe, the monks on Skellig Michael created their own unique wonder of the world, a manuscript carved in stone that resonates just as much today as when it was built.

Early Christian psalms speak of God being a rock and a refuge. The monks who ventured across the sea to this remote island sanctuary in the wild Atlantic believed firmly that withdrawing from the material world would bring them closer to the God who was their refuge and their strength. These eremitical practices were inherited from the Eastern Church, where isolation was eagerly sought after. Both the monastery and the more elevated South Peak hermitage reveal the extent of their devotion (Figure 21.7). The initial steps down to the water's edge are hewn from the rock face while, further up, unique stairways (there are three on the island) lead to the hermitage and monastery. Located within

Fig. 21.5 A VIEW OF THE SKELLIGS FROM ST FINIAN'S BAY. Lying 12km off the Iveragh coast, the iconic islands can be viewed at intervals along the scenic Skellig Ring, which is part of the Wild Atlantic Way. [Source: John Crowley]

Fig. 21.6 THE DISTINCTIVE BEEHIVE CELLS AT SKELLIG MICHAEL (Sceilg Mhichíl) WITH THE LITTLE SKELLIG (An Sceilg Bheag) IN THE DISTANCE. Perched *c*.185m above sea level, the monastery's otherworldly appearance draws visitors from around the world. A UNESCO World Heritage Site since 1996, the monastic settlement is an integral part of the history of early Christianity. [Source: Valerie O'Sullivan]

the latter's confines are the distinctive beehive cells that are as iconic of early Irish Christianity as the high crosses and round towers found elsewhere.

According to tradition, the ecclesiastical settlement at Skellig Michael originated in the sixth century and is associated with a local Iveragh saint, Fionán. Documentary evidence for the settlement is scant, but an entry in the ninth-century *Martyrology of Tallaght* refers to the death of a monk '*Suibni in Scelig*',[3] while a Viking raid on the site in 824 is recorded in the *Annals of Inisfallen*:

Scelec do orgain do gentib *7 É do brith I mbrait co nerbailt gorta léo.*
[Scelec was plundered by the pagans and Étgal was carried off into captivity and he died of hunger on their hands]'[4]

A later Viking raid would again test the resilience of the monks, but the foundation survived until the thirteenth century, when the monks eventually retreated to the mainland and the site of the Augustinian priory at Ballinskelligs (Baile an Sceilg). The dedication of a church to St Michael at the monastery at some point between the ninth and tenth centuries would subsequently result in the naming of the island. The Augustinian friars at Ballinskelligs most probably oversaw Skellig Michael's later development as a place of pilgrimage, until the priory was dissolved in the sixteenth century. The importance of the island as a pilgrim and penitential site in this period is attested to by its inclusion on a list of fifteen pilgrimage sites in Ireland to be visited by 'Heneas MacNichaill in 1543 (diocese of Armagh) as penance for the killing of his son'.[5] In subsequent centuries, pilgrims continued to visit the site and were joined later by antiquarians such as John Windele, who made the journey to Skellig Michael in 1851 and recorded some of the island's antiquities for posterity.

By the time Windele had visited the island, two lighthouses designed by George Halpin, inspector of works and inspector of lighthouses, had been built and, just like the monastery before them, against great odds. Skellig Michael by then belonged to a Mr J. Butler of Waterville. Butler demanded an annual fee of £30 from the Board of Ballast, which was equivalent in value to the sixteen to eighteen stone (approximately 100–114kg) of puffin feathers he had previously received by way of rent (Figure 21.8). He eventually sold the island to the Board of Ballast for £780.[6] The lighthouses, which were first commissioned in 1820, took over five years to complete, the arduous work necessitating the blasting of rocks for the construction of roadways and paths. The upper lighthouse was closed in 1870, while the remaining lighthouse was automated in 1987, bringing to an end the era of keepers living on the island.

Skellig Michael was inscribed on the UNESCO list of World Heritage Sites in 1996, one of three sites on or around the island of Ireland to achieve such status. The designation recognises the site's unique contribution to world heritage and, in particular, its importance in the story of early Christianity: 'The monastery and hermitage on Sceilg Mhichíl represent a unique artistic achievement. They provide an outstanding example of a perfectly preserved early monastic settlement … The presence of the monks on the island for such a long period has imbued the place with a strong sense of spirituality.'[7]

Fig. 21.7 THE SMALL ENCLOSURE OR PRAYER STATION ON THE SOUTH PEAK. [Source: Valerie O'Sullivan]

Fig. 21.8 THE PUFFIN (*Fratercula artic*) IS SYNONYMOUS WITH SKELLIG MICHAEL AND IS A CONSPICUOUS PRESENCE ON THE ISLAND FROM APRIL TO JULY EACH YEAR. After wintering far out in the Atlantic they make the arduous return journey – frequently extending over thousands of kilometres – to their breeding grounds on Skellig Michael. Once hunted for their meat and feathers, they are now protected and are an integral part of the island experience. Increasing familiarity with visitors during the summer months contributes to their almost tame appearance. [Source: John Crowley]

The site is a jewel of Irish heritage and requires constant vigilance and stewardship. The filming of scenes from episodes of the *Star Wars* movie franchise on the island has created a new appeal, which is welcomed by local tourism interests; however, it also increases potential visitor pressure and risk of damage to the island. The state, as custodian, must manage the site so that present and future generations can encounter its magic and wonder.

Skellig Michael remains a profoundly important place to people of faith and of no faith. While shaped continually by the elemental forces of nature, the human imprint is also highly significant and impressive. Although the release of the recent *Star Wars* films has created another layer of meaning, the real story is carved in the rock and will be there for generations to come. In his poem 'Postscript', Seamus Heaney, another Irish Nobel laureate, speaks of special moments when the heart is caught off guard and blown open.[8] George Bernard Shaw must have felt such a sensation when he stood amongst the beehive cells for the first time. The power of the imagination was the key to understanding Skellig Michael's significance then, as it is today, and perhaps that is its ultimate gift to the world.

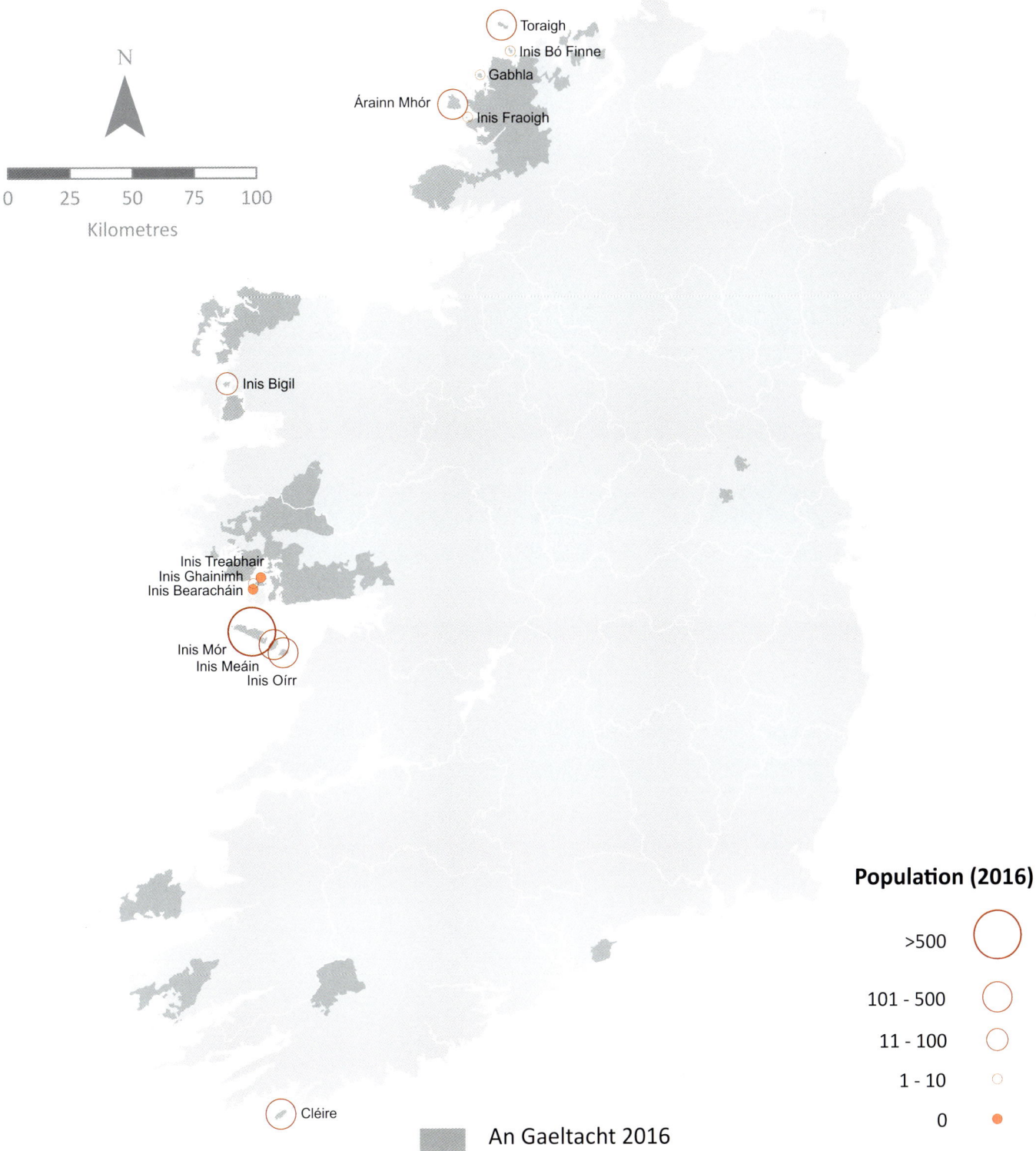

Fig. 21.9 GAELTACHT-DESIGNATED ISLANDS AROUND THE COAST OF IRELAND. Given their isolation and peripherality, many of the islands around the west coast retained daily use of the Irish language, long after it had largely given way to English elsewhere in Ireland. This significant cultural survival has been threatened by the population decline that has impacted Irish islands generally since the Famine and also, particularly since the 1960s, by the growing availability of English-language television broadcasts and the migration of young people to the mainland for third-level education. This is one reason why island life is now supported by the state and the European Union. Thirteen of the islands were part of the Gaeltacht in 2016. [Data source: Department of Culture, Heritage and the Gaeltacht]

The isolation of many islands, from Rathlin (County Antrim) and Tory (County Donegal) in the north, to Cape Clear (County Cork) in the south, has contributed to the perseverance of more traditional lifestyles and community values. This is seen, for example, in the preservation on several of these islands of their versions of the Gaelic language, which has been eroded severely by the domination of the English language throughout most of the Irish mainland. As a result, many islands have been designated as Gaeltacht areas (Figure 21.9), and the retention of the Irish language is now being actively promoted as a key element of these islands' identity and heritage (Figure 21.10).

There is a special affinity between Irish people and the

Fig. 21.10 CAPE CLEAR (*Cléire*) IS THE SOUTHERNMOST INHABITED PART OF IRELAND. Its people speak Irish and the island is part of the Gaeltacht. Its population of 147 at the 2016 census is only 14 per cent of its pre-Famine total, but shows a slight increase from the 124 people recorded in the previous census (2011). Cape Clear is noted as the setting for the autobiography of Conchúr Ó Síocháin, *Seanchas Chléire* (*The Man from Cape Clear*), which has been compared to the more famous Great Blasket autobiographies. [Source: Robbie Murphy]

country's offshore islands. From the mythological adventures of Oisín and Niamh to the Children of Lir, as well as from countless memoirs, poems and pieces of literature, the coastal landscape, the treacherous sea and the mysterious islands beyond have long been woven into Ireland's cultural identity. Numerous island writers, many but not all using Irish, have been celebrated for their immense contribution to Irish literature. One island famous for its literature, Great Blasket, even featured on the Irish £20 note prior to the adoption of the Euro (Figure 21.11).

Fig. 21.11 'YEATS' EDITION SERIES B £20 NOTE. The Central Bank of Ireland issued these banknotes from 1976 until 1992, to replace the earlier Series A notes. The £20 note featured the Great Blasket Island with, in the background, text from *An tOileánach* by Tomás Ó Criomhthain. The Series B notes were withdrawn from circulation in 1993 and replaced by Series C banknotes of different design, although the older ones remained legal tender until 1995. [Source: Courtesy of the Central Bank of Ireland Archives]

TRADITIONAL LIFE

Being valued for their heritage is only part of the narrative of Ireland's islands. There is also a bleaker story to tell. This is linked directly to the paucity of the local resource base and is compounded by difficult weather conditions resulting from the impacts of Atlantic storms that affect particularly the western seaboard of Ireland (see Chapter 3: Coastal Waters: Marine biology and ecology). These factors increase further the relative isolation of these communities, one result of which has been a long history of population decline on the islands.

A recurring theme painted by island writers is a portrait of hardship and frequent poverty, albeit leavened by a strong community spirit. This is evoked by Ó Síocháin for example, when he writes in *The Man from Cape Clear* that 'No one had the second pair of boots to put on his feet nor the second trousers to cover his bones.'[9] A more dispassionate, statistical picture of that life, at a time when its traditions had not been affected by either the Famine or by modernisation, can be gained from the rare survival of manuscript returns from the 1821 census for the Aran Islands (these documents were lost for almost every other place in Ireland, in the explosion at the Public Records Office in Dublin during the civil war). The manuscripts depict a subsistence, peasant economy where the few

Fig. 21.12 STONE WALLS AND SMALL FIELDS ON INISHMORE (INIS MÓR), ARAN ISLANDS. The three Aran Islands display a characteristic landscape of dry-stone walls surrounding tiny pocket hand-kerchief fields. The islands are composed mainly of limestone and little of the land area had natural soil. Over centuries much of the agricultural land has been created by filling in the cracks on the bare limestone pavement surface then covering it with layers of sand and sea-weed topped with soil, perhaps imported from the mainland. Loose rocks would be piled up into walls to take them off the surface and to act as windbreaks to protect the precious soil from erosion on these treeless islands. [Source: Stephen A. Royle]

resources available had to be exploited to their utmost.[10]

According to these census returns, most islanders worked in primary activities, within an economy that had little need for services or traders. Almost all Aran household heads were tenant farmers, some of whom also sold their time and effort as labourers. Apart from a few full-time fishermen, mainly in Killeany on Inish Mór, a number of landless men described just as labourers and a small number of widows who made fishing nets, few people had only one recorded occupation. Most people farmed; 60 per cent of household heads worked the land, but had other jobs too. This was necessary to maximise household income, mainly to meet the rental payments on their holdings. The necessity of occupational pluralism, highlighted on the Aran Islands, was replicated on most island communities. This was confirmed by Ó Síocháin, for Cape Clear, albeit for a later time:

The great majority of them had a small patch of land each, but even so, it was little they made out of it. In all truth but for the pennies they earned from the fishing they wouldn't be able to pay the rent, it was so big a burden on them.[11]

Agriculture on Aran was not easy. The mean holding size in 1821 has been calculated to have been 11.3ha, not that small for rural Ireland at that time, but the land itself was extremely difficult to work. Much land had been 'made' by creating soil on exposed limestone surfaces; this was achieved by spreading layers of sand and seaweed, topped with soil imported from the mainland. Loose rocks would be consumed by being crafted into the distinctive Aran dry-stone walls around the tiny fields, to protect the precious soil on the windy, virtually treeless islands (Figure 21.12). Crops such as potatoes were consumed for subsistence; cash came from selling on cattle. Kelp making (the gathering, drying and burning of seaweed, the ashes of which were exported for the production of chemicals, principally iodine) was an important component of the Aran economy, generating cash to help meet the rental payments. Many households were headed by a 'farmer and kelpmaker' (Figure 21.14).

Other activities known to have existed on Aran in the pre-Famine period, if not detailed in the 1821 census, included sealing, the taking of seabirds and the utilisation of 'wrack' – material washed ashore, particularly wood – in construction or for fuel. Another fuel was cow dung (*bualtrach*), which was dried on walls. In

481

Fig. 21.14 ARAN ISLANDS KELPMAKING. There is a long tradition of seaweed harvesting along the Atlantic seaboard of Ireland, with accounts of its use as food dating to before the twelfth century. More recently, and particularly from about 1700 onwards, the commercial harvesting of kelp (mainly *Laminaria hyperborea* and *Saccharina latissima*) was being undertaken on many Irish islands, including the Aran Islands, as seen here. The fronds and rods of the plant were collected from the seashore and burned in kelp kilns. The ash was exported for use in glassmaking and as a source of medicinal iodine. In 1947 a private company, Alginate Industries (Ireland) Limited, was opened in Connemara, County Galway, to buy storm-cast *Laminaria* from local suppliers for drying and exporting to Scotland for alginate extraction. Two years later, in 1949, the Irish government bought shares in the company, which was renamed Arramara Teoranta, and kelps were gradually replaced by the manual cutting and harvesting of another seaweed, a perennial wrack (*Ascophyllum nodosum*) found in the intertidal zone. Between 1964 and the first decade of the twenty-first century, up to 7,000 dry weight tonnes of the seaweed were cut in Ireland each year. In 2014, the government relinquished its stake in Arramara Teoranta, and the company was acquired by the international Acadian Seaplants group. Still based in Connemara, it continues to receive sustainably hand-cut seaweed from a network of harvesters along the west coast and fosters the production of a range of derived raw materials and premium-quality end-products for sale locally and for export. [Image source: National Museum of Ireland; Guiry, M.D. and Morrison, L., 2013. The Sustainable Harvesting of *Ascophyllum Nodosum* (*Fucaceae, Phaeophyceae*) in Ireland, with Notes on the Collection and Use of Some Other Brown Algae. *Journal of Applied Phycology*, 25(6), pp. 1823–1830.]

Fig. 21.13 (left) CLARE ISLAND. This map of Clare Island shows evidence of the work conducted there by the Congested Districts Board. This body was set up in 1891 to relieve poverty in the 'congested districts' of the west of Ireland, until it was dissolved in 1923. The board bought Clare Island in 1895 and replanned the island's agriculture and settlement. The farmers' then-scattered holdings were consolidated into strips in the south, each provided with a cottage. The north of the island was reserved for communal grazing (labelled Commonage) with a wall built across the island marking the division between the arable and pasture lands. *The Clare Island Survey of 1909–1911*, led by Robert Lloyd Praeger for the Royal Irish Academy, was an extensive investigation of its flora, fauna, geology and archaeology and was the first major multidisciplinary ecological survey of a specific area ever carried out. In 2016, the island's population was 159. *The New Clare Island Survey (1992–2009)* extended and updated the work.

short, the 1821 census shows an economy stretched to the utmost, with every feasible activity exploited. No wonder that in 1822, when bad weather caused the potato crop to fail, there was a crisis on these islands, serious enough to require assistance from outside.[12]

Such over-stressed island economies were rather primitive and were not capable of much expansion: 'Shovel, fork or spade were our only means of livelihood'.[13] The Congested Districts Board, established in 1891 to alleviate poverty in the west and parts of the north-west of Ireland, certainly tried to make improvements. For example, the landholding system in Clare Island, County Mayo, was remodelled from a fragmented pattern into seventy-four consolidated strip farms with new houses strung out along the flatter land to the south. A wall was built across the island with commonage being left to the north (Figure 21.13). However, even remodelled, island economies became increasingly unable to satisfy or retain people who, during the nineteenth and twentieth centuries, became aware of greater opportunities elsewhere.

DECLINE: ISLAND VOICES

The study of Irish islands in the late nineteenth and twentieth centuries is massively enriched by the accounts of visitors, such as Synge and Thomson,[14] but also by the islanders themselves, some of whom were encouraged to record their reflections by the visitors. Chief amongst these island voices were those of the three Great Blasket autobiographers: Tomás O'Crohan (Ó Criomhthain) (1856–1937), his great-nephew Maurice O'Sullivan (Ó Súilleabháin) (1904–1950), and Peig Sayers (1873–1958).[15]

O'Crohan details the way in which the Blasket islanders earned their living: farming, fishing, trading with the mainland, gathering wrack and cutting peat (Figure 21.15). He built his own house and was employed by the Congested Districts Board to build other properties. His book records bleak events, particularly regarding the deaths of children. It also describes a vibrant island community, successfully wrestling a livelihood from an unfavourable and unforgiving environment. The increasing impact of emigration on the island inspired O'Crohan to chronicle his small offshore world

Fig. 21.15 LIFE IN THE GREAT BLASKET (1940s). The picture shows the small harbour of Great Blasket, which is situated on its eastern shore, opposite the mainland of County Kerry. The harbour was vital for life on the island, not just for its fishing industry. As there were no services on Great Blasket, residents had to travel to Dunquin or further afield for everything, from obtaining goods to attending mass. The boats used were a local variant of the Irish currach. Typically, two men would propel the vessel using long, bladeless oars, a stack of which are shown propped against the rocks on the left. Currachs were light enough to be carried and were stored out of the water, upturned to keep them dry. Great Blasket was the residence of three famous autobiographers who vividly depicted the island's life and society before its depopulation in 1953. [Source: The Great Blasket Centre, Dunquin]

for posterity, 'for the like of us will never be again'. He describes how one man 'had a houseful of children then, [but] there are only three of them in my neighbourhood now, the rest of them are in America … like so many others', while many of his own immediate family left for America too, leaving Tomás 'to keep the little house' going for the sake of his parents.

Peig Sayers married into the island in 1892 (Figure 21.16). As a young woman, she had wanted to emigrate to America from her mainland home but her friend, who had gone before her, was unable to remit the money promised for Peig's passage. Her choices were then reduced to going into service or getting married. She opted for the latter, and took as her husband one

Pádraig Ó Gaoithín, from Great Blasket. She records her welcome to her new home, in 1892, when the island seemed 'black with people', but the population of the Great Blasket declined subsequently (Figure 21.17). The last census before its two-stage evacuation in 1953 and 1954 recorded a population of just twenty-seven people in 1951, compared with its maximum population of 160 in 1911.

Peig's island voice enables readers to understand why the Blasket islanders left: 'How lonely I am on this island in the ocean', she writes; 'I think this is a very confined place with the sea out there to terrorise me'; 'there's a great deal of pleasantry and hardship in the life of a person who lives on an island like this … going to bed at night with little food and rising again at the first chirp of the sparrow, then harrowing away at the world'. She details islanders, including family members, who migrated 'rather … than put down roots here'. One son 'hoisted his sail and went off to America'. Another 'never had any desire to leave Ireland. But … had to take the road like the others, his heart laden with sorrow.' As Peig had said to this son, Muiris, 'twould be a bad place that wouldn't be better for you than this dreadful rock'.

O'Sullivan's work is altogether lighter in tone. This is a young man detailing his upbringing in a lively style, although his island voice also considers emigration as an inevitability: 'The chief livelihood – that's the fishing – is gone underfoot, and when the

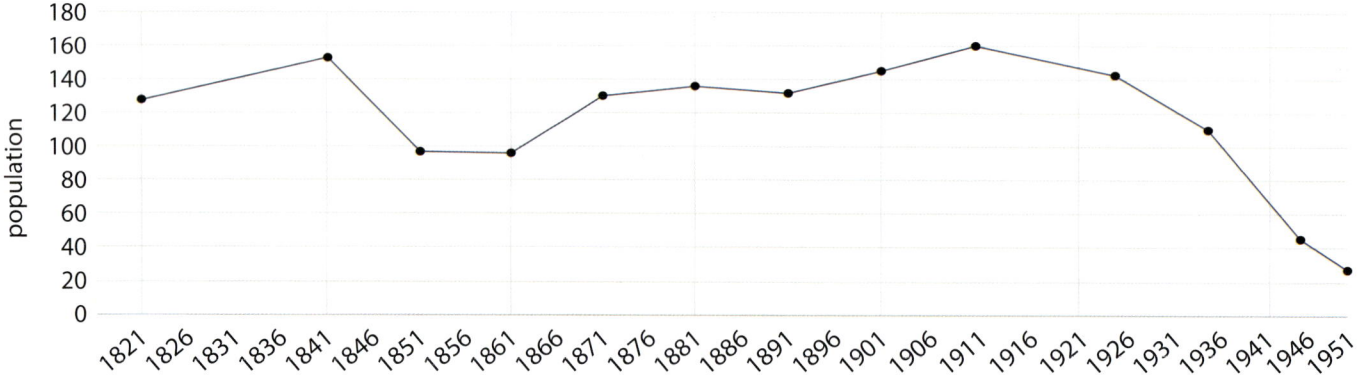

Fig. 21.17 FLUCTUATIONS IN THE POPULATION OF GREAT BLASKET ISLAND, 1821–1951. The graph shows that the population of Great Blasket rose between 1821 and 1841, but then fell precipitously during the Famine, with a loss of about a third of the total in ten years. The total rose in the 1860s and again in the early twentieth century to reach its peak recorded total in 1911, an unusually late date for an Irish island. However, population then fell sharply, a process captured in the famous autobiographies, especially the books by Peig Sayers. The fall between 1936 and 1946 was particularly steep; the resettlement programme then in operation saw Peig Sayers herself leave. The remaining islanders pressed the authorities to be taken off Great Blasket and the island was officially abandoned in 1953. [Data source: Central Statistics Office]

fishing is gone underfoot, the Blasket is gone underfoot, for all the boys and girls who have any vigour in them will go over the sea … our parents … will have to do without us'.

O'Sullivan's sister, Maura, departed for Springfield, Massachusetts, like many others. She told her father: 'don't you see everyone is going now and you will see me beyond like the rest of them'. The last Great Blasket child was born in 1947. His homecoming from the mainland hospital where he was born 'was deemed a great occasion; a new life had been injected into a dying community crushed by migration'. He was evacuated with the other residents when he was six.[16]

Not everybody abandoned island life. In *The Man from Cape Clear*, Conchúr Ó Síocháin wrote that, although conscious of better opportunities further afield,

If I were asked tomorrow to leave this wild, harsh place and go to live where I could have every advantage, convenience and opportunity I could wish for, this is what I would say: although I have lean cheeks and a stooped back from trying to make my living throughout my life in this isolated place, far from market or fair, still I and the patch of land I own, are so used to one another that I couldn't leave it.[17]

However, Ó Síocháin's determination to stay was not matched by many islanders. Patrick Heraughty was one of those evacuated from Inishmurray, an island off the coast of Sligo, when the last residents were taken off in December 1948. In Joe McGowan's book he records that by the 1920s it had become clear that 'life was better elsewhere' and the remaining islanders petitioned to be taken off the island. Another Inishmurray islander, Dan Brady, recollected the situation on the island prior to 1948:

The young people wanted to leave. The people of the island were all getting old. Old people were little good to you on an island. You needed young able-bodied men. If we needed a loaf of bread we couldn't walk down to the shop, we had to break our back rowing a boat all the way to Streedagh.[18]

Case Study: The Aran Islands

Piaras MacÉinrí

Considering the modest size of Ireland's offshore islands, the paucity of their populations, their relative economic insignificance and their lack of direct involvement in key historical events, some of them have nonetheless left a major imprint on the national imagination. Much of this can be explained by the role they have played in the cultural and folk tradition of a nation struggling to emerge and define itself in the late nineteenth and early twentieth centuries.

Of the various island groups, two in particular stand out: the Blaskets and the Aran Islands. The survival of the Irish language as a vernacular on these islands is certainly a significant reason, and the extraordinary flowering of writing and other creative endeavours that they catalysed, including music and song, captured an ancient cultural heritage and a people whose very existence was already threatened by the processes of modernisation and change, which began to affect even the most outlying parts of Europe by the middle of the nineteenth century.

While the Blaskets have been much studied and are famed for the work of a small group of key writers and visitors (discussed in the main text of this chapter), the case of the Aran Islands is equally interesting. A small group of three islands in Galway Bay, of no more than 46km[2], the total population was 3,521 in 1841, just before the Famine, while in 2016 it stood at just 1,226 persons.[19] Yet, they have come to occupy a central place in Irish cultural geography, history and anthropology. The reasons that brought this about will be explored briefly here.

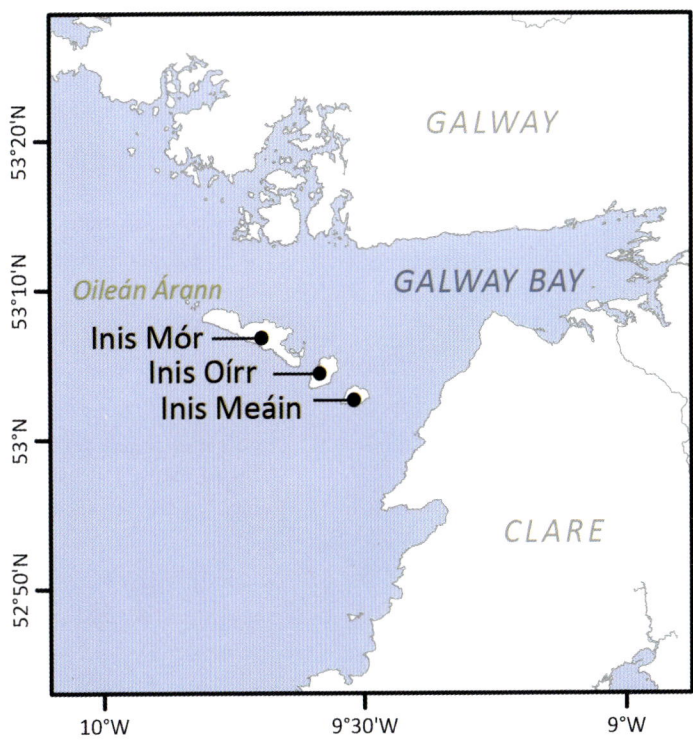

Fig. 21.18 INISHERE (*Inis Oírr*), 'FAMILY WALKING ON THE SHORE', ONE OF SEVERAL PHOTOGRAPHS TAKEN BY J. M. SYNGE DURING HIS STAYS ON INISHMAAN (INISMEÁIN) AT THE TURN OF THE TWENTIETH CENTURY. It shows a family walking on the shore of Inis Oírr, the middle of the Aran Islands. Both the parents (Micheál Ó Conaola, d. *c.*1932 and his wife Máire Uí Chonaola, née Ní Chualain, d. *c.*1934) are wearing homespun clothes, and the child standing partially hidden by the mother's skirts (Micheál Ó Conaola, d. 1965) is wearing a traditional Aran cap. The child in his mother's arms is Padraic Ó Conaola, d. 1969 (identification supplied by Micheál Ó Conaola's son Dara Ó Conaola). [Source: The Board of Trinity College Dublin]

AN IRISH THEME PARK? NATIVISM AND PRIMITIVISM, IRISH STYLE

Give up Paris ... Go to the Aran islands. Live there as if you were one of the people themselves; express a life that has never found expression.[20]

The well-known statement above is attributed to W.B. Yeats, who encouraged J.M. Synge to turn away from Europe's cosmopolitan, modern culture, symbolised by the city of Paris, and instead consider the lives of the people of the Aran Islands on the edge of the continent (Figures 21.18 and 21.19). The interest in 'primitive' societies, their culture and way of life, was a widespread phenomenon from the late nineteenth century onwards. In part this was due to the influence of such key figures as anthropologist, Franz Boas, for whom 'the anthropological concept of culture replaced race as the key to understanding human groups'.[21] It can also be seen as reflecting an interest on the part of those living in a rapidly industrialising, urbanised and fast-changing world in the notion of simplicity, a less stressful life, a return to roots and (sometimes) a rejection of the modern and the new.

There was a new interest throughout Europe in language, custom and tradition, in part arising out of German intellectual culture at that time, with its interest in Indo-European history and in the structural analysis of language. This search for the 'authentic' led to the idea that cultural values could be seen as core formative elements of 'the nation' in its purest form and, from there, to the celebration of identities and ways of life that were seen as untainted and original.

This earnest search for 'authenticity' in remote places by bookish men was not without its funny side and Synge himself has a little fun at their expense in *The Aran Islands*:

Fig. 21.19 SYNGE'S COTTAGE ON INIS MEÁIN. John Millington Synge (1871–1909) visited the Aran Islands in 1898, where he stayed in this cottage, owned at that time by Bríd and Páidín MacDonnchadha (MacDonagh). The house has been restored by Ms Theresa Ní Fhatharta, the great granddaughter of Synge's hosts and is now a museum containing photographs, drawings and letters related to the playwright, while a converted stone outhouse holds a small reference library of relevant publications by and about Synge, Yeats and Lady Gregory. [Source: Piaras Mac Éinrí]

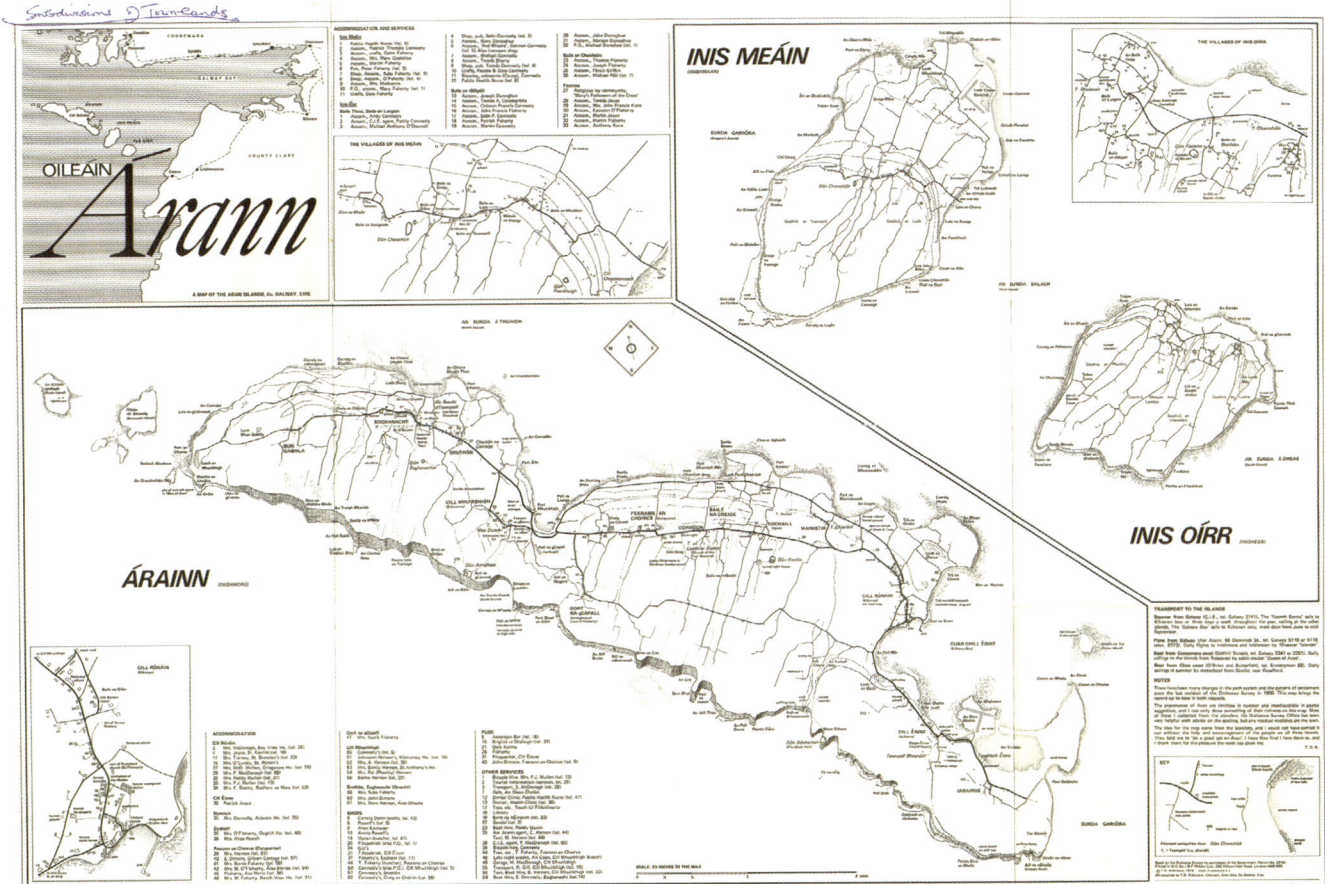

Fig. 21.20 TIM ROBINSON'S MAP OF THE ARAN ISLANDS. The English mathematician, writer and cartographer, Tim Robinson, lived in the west of Ireland. A much-respected and largely self-taught Gaeilgeoir, in 1972 he began a forty-year-long project to map and interpret the Galway Bay area, with specific studies of Aran Islands, the Burren and Connemara, for which he has won numerous awards nationally and internationally. His hand-drawn map of the Aran Islands shown here was followed by his much-acclaimed studies of the coastline (Robinson, T., 1986. *Stones of Aran: Pilgrimage*. Dublin: Lilliput Press) and the interior of the Aran Islands (Robinson, T., 1995. *Stones of Aran: Labyrinth*. Dublin: Lilliput Press). Taken together, the map and the two written volumes present a minutely detailed and comprehensive collection of observations on the language, landscapes, toponymy (place names), traditions, culture and folklore of Aran. The Library of the National University of Ireland, Galway (NUIG), acquired Tim Robinson's entire archive of work on the landscape of the west of Ireland in 2013, and is currently undertaking a major digitising project to make these materials available for further study and analysis. [Source: reproduced courtesy of the James Hardiman Library, NUI Galway]

Most of the strangers they see on the islands are philological students, and the people have been led to conclude that linguistic studies, particularly Gaelic studies, are the chief occupation of the outside world.

I have seen Frenchmen, and Danes, and Germans', said one man, 'and there does be a power of Irish books along with them, and they reading them better than ourselves. Believe me there are few rich men in the world now who are not studying the Gaelic.[22]

The Celtic Revival of the late nineteenth century soon merged with cultural debates of a more explicitly political nature. By the first decade of the twentieth century the Irish Ireland movement, most closely associated with the journalist, D.P. Moran, wanted an Ireland free of all foreign encumbrances, in which being Irish meant embracing an Irish-speaking, Catholic nation.[23] This led to the Aran Islands being presented as the archetypal example, seen as an eternal and never-changing landscape, where nature and humanity combined to embody the essential character and identity of the Irish people.

However, antiquarian, linguistic and scientific interest in the Aran Islands was not new even in the late nineteenth century. Ashley traces this to the 1860s and earlier:

Thereafter, philologists and enthusiasts had been drawn to the Aran Islands in pursuit of their linguistic riches. The relative isolation of the islands had maintained the Irish language there when it began to decline with precipitous speed elsewhere in Ireland over the course of the nineteenth century. And, inevitably, as the Aran Islands were being invented as bastions of the ancient sublime, so the islanders themselves were endowed with nationalist and racial significance. They were modern primitives, insulated from the deadening hand of progress and anglicization, true Irishmen and women, models for an Ireland freed from British dominion. They were a pure Gaelic stock, uncorrupted by infusions of degenerate blood from the mainland ... The influential Victorian scholar and popularizer on all matters Gaelic, Samuel Ferguson, had in 1852 written: 'If any portion of the existing population of Ireland can with propriety be termed Celts, they are this race'.[24]

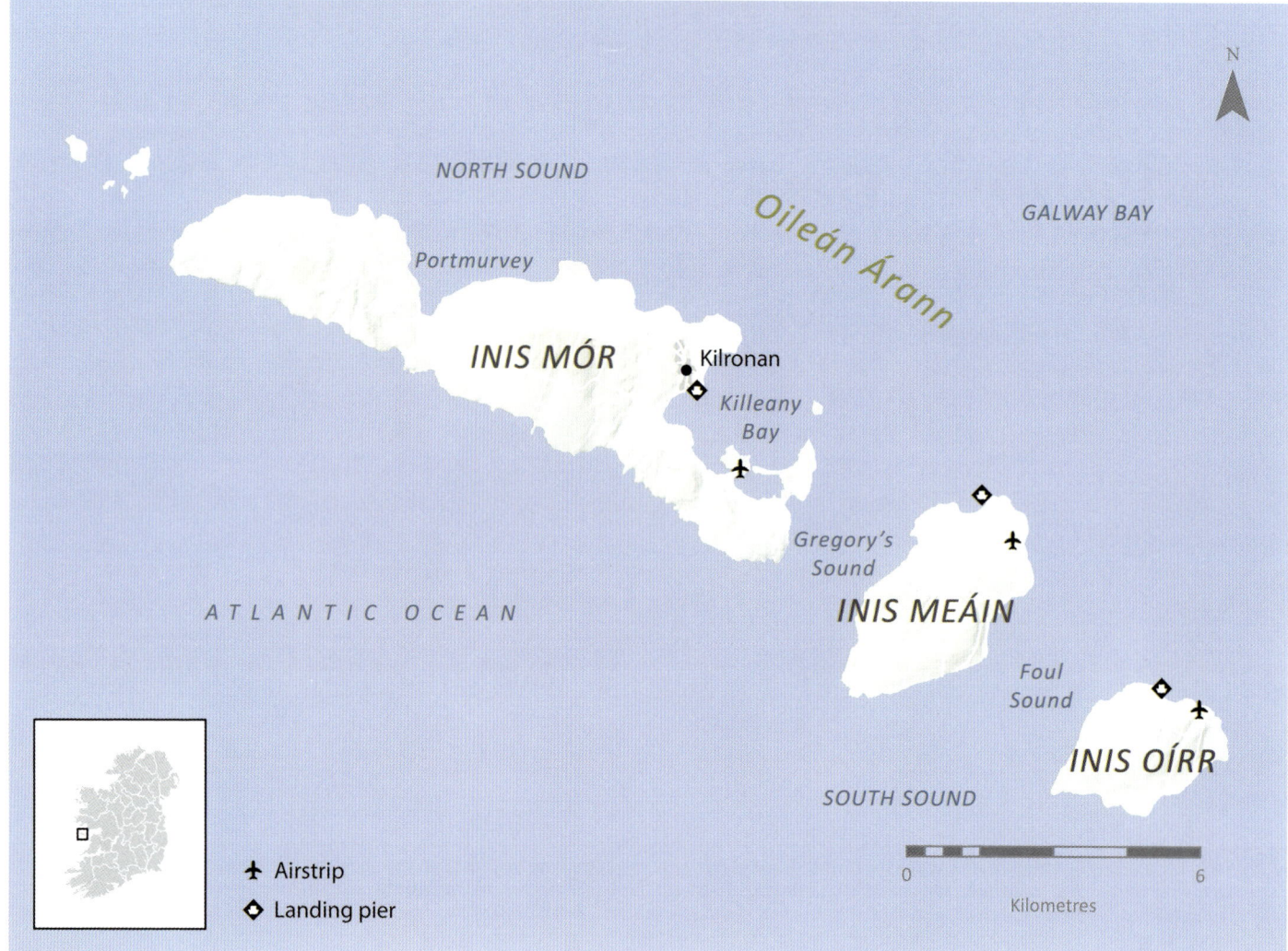

Fig. 21.21 MAP OF THE ARAN ISLANDS.

While Synge's visits in the 1890s as a kind of early amateur ethnographer are well known, it is worth noting that the same decade also saw a pioneering scientific study of the population of the islands by Alfred Cort Haddon and Charles Browne, regarded as founding fathers of British anthropology. To quote Ashley again:

… from the outset they [Haddon and Browne] signalled that there was to be more to the Aran study than mechanical measurement; like most of the emerging British anthropologists of the day, Haddon worked under the shadow of the Oxford sage, Edward Tylor, and his 1871 classic, *Primitive Culture*. Tylor had defined culture as 'that complex whole which includes knowledge, belief, art, morals, law, custom and any other capabilities and habits acquired by man as a member of society', thereby laying out a social anthropological programme that is still, in its essence, with us today.[25]

To these layers of multiple meanings and interpretations may be added those of writers and artists. In spite of their small size, the islands are famed, not just for the writers who grew up there, but for some distinguished visitors as well. Indigenous writers included Liam O'Flaherty and his brother Tom; his nephew, journalist and writer Breandán Ó hEithir; and distinguished poet, Máirtín Ó Direáin. Apart from Synge, already mentioned, whose classic *The Aran Islands* (1907) was and remains a hugely influential text, visitors possibly better known to the outside English-speaking world included American film-maker Robert Flaherty, whose *Man of Aran* (1934) became (although quasi-fictional) almost the definitive visual imagining of life on the islands; and Tim Robinson, whose extraordinary writing and map-making work includes *Stones of Aran*,[26] 'one of the most sustained, intensive and imaginative studies of a place that has ever been carried out'(Figure 21.20).[27] The Aran Islands continue to attract cultural and aesthetic interest and questioning, as in the work of the London-Irish playwright Martin McDonagh (with roots in Connemara himself), one of whose plays, *The Cripple of Inishmaan*,[28] is actually a radical re-reading of the reaction of the islanders when Flaherty came to make his film back in 1934.

THE REALITIES: ECONOMIC AND SOCIAL

So much for the mythification of the Aran Islands: what of the reality? For one thing, there was no 'pure' Celtic race, a myth to begin with, especially in the Aran Islands, where there had been an important English garrison in the seventeenth century. Like other parts of rural Ireland after the Famine, high mortality levels, widespread poverty and high levels of emigration were not short-term phenomena, but continued long after the 1850s. The situation was compounded by the presence, recorded on many occasions in the 1880s, of such diseases as scarlatina, dysentery, smallpox, typhus and even typhoid.[29] Population density was higher than on the mainland and islanders paid high rents for generally poor land, on which they had little by way of security of tenure. Infrastructure was almost entirely lacking, notably in the area of fisheries, which could at least have held out the prospect of a viable industry. All of this was very far from an idealised 'primitive' life, where the people were poor but contented.

The first official response to this challenge was the establishment of the Congested Districts Board in 1891. It is interesting that this initiative coincides, precisely, with the period when interest in the Islands was at its greatest in terms of the kinds of idealising and primitivist agendas touched upon earlier. The board was an advanced idea for its time and was intended to lessen poverty levels through public works, including the construction of piers, the modernisation of farming, the sponsoring of factories and the like. It had four principal powers: promotion of industry; amalgamation and improvement of holdings; resettlement of tenants to viable holdings and instruction in modern farming methods.

How successful were these efforts? In some respects the approach taken was startlingly modern. Thus, the establishment of a fisheries industry was accompanied by the setting of an intervention price, that is, the price paid to fishermen was guaranteed. Public works were also carried out by the board, housing was built and assistance was given for the establishment of local industries, such as knitwear, which has continued in one form or another down to the present day.

But there were tensions accompanying these development initiatives. The obvious ones concerned conflicts of interest between landlord interests and island inhabitants, including evictions, turf battles with other districts within the Board, continuing under-investment in infrastructural works and ongoing poverty and high emigration. In the end, a change in priorities – from economic and infrastructural development to land acquisition – put paid to any prospect of real improvements in the medium to long term. The Board itself was dissolved in 1923. It did not succeed in halting long-term population decline or in providing a sustainable economic infrastructure, although it did bring about some real improvements in people's lives.[30]

THE ARAN ISLANDS TODAY

Jumping forward by several decades, it would probably be fair to say that in spite of official lip-service, the Aran Islands remained in a relatively underdeveloped state for much of the twentieth century. Emigration continued at a high level, with a drop of almost 20 per cent in the population between 1911 and 1926. It continued to decline before reaching a degree of stability in the past two decades.

Change did come about, however, notably through the establishment of Údarás na Gaeltachta in 1980,[31] with an economic, social and cultural role reflecting but also expanding in some ways on the original mission of the Congested Districts Board nearly a century earlier, and promoting the development and implementation of more effective transport, tourism and industrial policies. Moreover, a generation of dedicated local leaders has emerged that is committed to conserving the unique heritage of the islands but is also driven by an enterprising, hard-headed approach, which is fully informed by economic planning and realities. Today, all three islands are served by pier

Fig. 21.22 INIS MEÁIN KNITTING COMPANY. Inis Meáin Knitting Company was founded in 1976 by an islander, Áine Ní Chonghaile, and her husband, Tarlach de Blácam. They decided to make their lives on the island and started the company to support themselves and also to provide employment for young islanders, which would help to stem emigration from the island. As befits a small factory whose output is limited, the company's products – many based on local traditions, including the Aran sweater – are small runs of high-quality garments and are not aimed at a mass market. The success of such localised ventures has helped to stabilise island populations and has maintained a functional connection with the mainland and further afield. In 1996 the population of Inis Meáin was 191, falling to 154 in 2006, but had increased to 183 in the 2016 census. [Source: Matthew Thompson]

and port facilities, as well as an air link (although the future viability of the latter has been called into question) (Figure 21.21). A modern knitwear factory operates in Inishmaan (Figure 21.22) and, on all three islands, accommodation and restaurants cater for the growing numbers of tourists. Perhaps most hopeful of all, the islands now have an innovative second-level school system located on Inishmaan, for the first time, using modern technologies and a hosting programme for students from the mainland to sustain a viable and successful operation.

Finally, the islanders themselves have never conformed to the stereotypical images painted earlier, of a kind of Irish 'Noble Savage' living an isolated existence and extolling the values of a primitive but authentic culture. A strong streak of anti-authoritarianism and even anti-clericalism may be recognised among an otherwise pious community. Isolation was always more apparent than real and the high rates of emigration, paradoxically, ensured a far-flung community and a constant engagement with the realities of the outside world. Perhaps nothing illustrates this better than the lives and careers of Inis Mór brothers, Tom and Liam O'Flaherty, mentioned earlier. Liam went on to be recognised as a great writer in two languages. Tom became passionately involved in revolutionary politics in the United States, writing a column every day for the daily newspaper of the Communist Party of the USA. Liam, for his part, was involved in the Irish Communist Party and Russian was one of the first foreign languages into which his work was translated.

In conclusion, the Aran Islands can no longer be seen as an under-developed, isolated place lost in the mist of time; indeed this was never, in the strict sense, true. Nevertheless, they face considerable challenges, with a demographic structure weighted towards an older population, total education levels at the lower end of the qualifications range and higher unemployment rates than the national average. That said, dramatic changes have taken place; in 2019 almost half of all households had broadband access, compared to less than 5 per cent in 2006.[32] Meanwhile, 'the residents of the three islands, are working towards becoming self-sufficient in locally generated renewable energy and free of dependence on oil, coal and gas by 2022'.[33] Modern planning and policy have at last begun to embrace the need both to conserve and protect a heritage and environment of outstanding cultural and natural importance, on the one hand, while working towards sustainable, resilient communities that can offer decent socio-economic conditions and a future, on the other.

DECLINE: CENSUS STATISTICS

The personal views of the individual islanders, who wrote about their experiences, can be scaled up to the national picture of island decline by reference to census statistics. In the modern census of Ireland there is a useful table where the populations of the inhabited islands are classified separately. In earlier census forms this was not the case. Working with these earlier records, data on island populations can be accessed by trawling through the substantial published decennial census volumes. This lengthy process was undertaken by the author, leading to the statistics presented in Figure 21.23. Before turning to the graph it is necessary to explain some protocols behind its construction.

First, the islands listed are those of the twenty-six counties that make up the Republic of Ireland. The three coastal counties in Northern Ireland do not have many offshore islands: Rathlin Island, off County Antrim, is the major one, while County Down has the Copeland Islands, as well as a number of islands in the nearly totally enclosed Strangford Lough (Figure 21.1). Northern Ireland's islands are excluded from this graph, not because they tell a different story (Rathlin's population has declined like those in the Republic, and the Copeland Islands were depopulated in 1953 when the last two inhabitants left), but because after the partition of Ireland the two jurisdictions undertook census recordings in different years, making it impossible to make direct comparisons.

Also not included in Figure 21.23 are lighthouse islands, such as Fastnet Rock and Tuskar Rock. Inishtrahull, County Donegal, was included only until 1926 since the last of its traditional population left in 1929 and after which it was peopled only by lighthouse-keepers, until 1987, when the light was automated. Other islands were excluded because they were in institutional use, for example, as a coast guard station; a prison (as in the case of Spike Island in Cork Harbour); or a military base (as in the naval base on

Haulbowline Island, also in Cork Harbour).

Great, Little and Fota islands, all in Cork Harbour, are likewise excluded from the graphs in Figure 21.23. This is because these islands have long been incorporated, physically and administratively, into the adjacent mainland. Functionally, they have not been treated as offshore islands for many decades; the census data does not record population totals for Great Island after 1861. Great Island is where the town of Cobh is located. It is a port, an urban place and is not functionally an island. Analysing Irish islands taking into account Great Island would be like studying offshore islands in the USA whilst incorporating Manhattan. Thus, the protocol for the inclusion of an island in this analysis means that these figures differ a little from those published for island populations by the Central Statistics Office, although, of course, the trends are the same.

The population of Ireland's islands was at its peak (36,352) in the pre-Famine 1841 census. Consequences of the Famine, however, caused a decline in this total of almost 9,000 (25 per cent) by 1851. Paradoxically, the number of inhabited islands increased slightly as distressed populations along the western seaboard sought out any land that could be cultivated and/or gave access to marine resources.

From 1851 to 1881, population totals fluctuated, with increases recorded between 1851 and 1861 and during the 1870s. The latter increase can be related directly to the role of the Congested Districts Board, which began to provide work and improved infrastructures. Particularly significant was the rise in population of some of the islands of south Connemara, including Gorumna, whose population increased from 1,417 to 1,798 between 1871 and 1881. By the latter years, Ireland's total island population (28,892) exceeded that recorded in 1851, although the number of inhabited islands had fallen slightly to 194 from a peak of 208 in 1861.

After 1881, island population fell at each census to 2002,

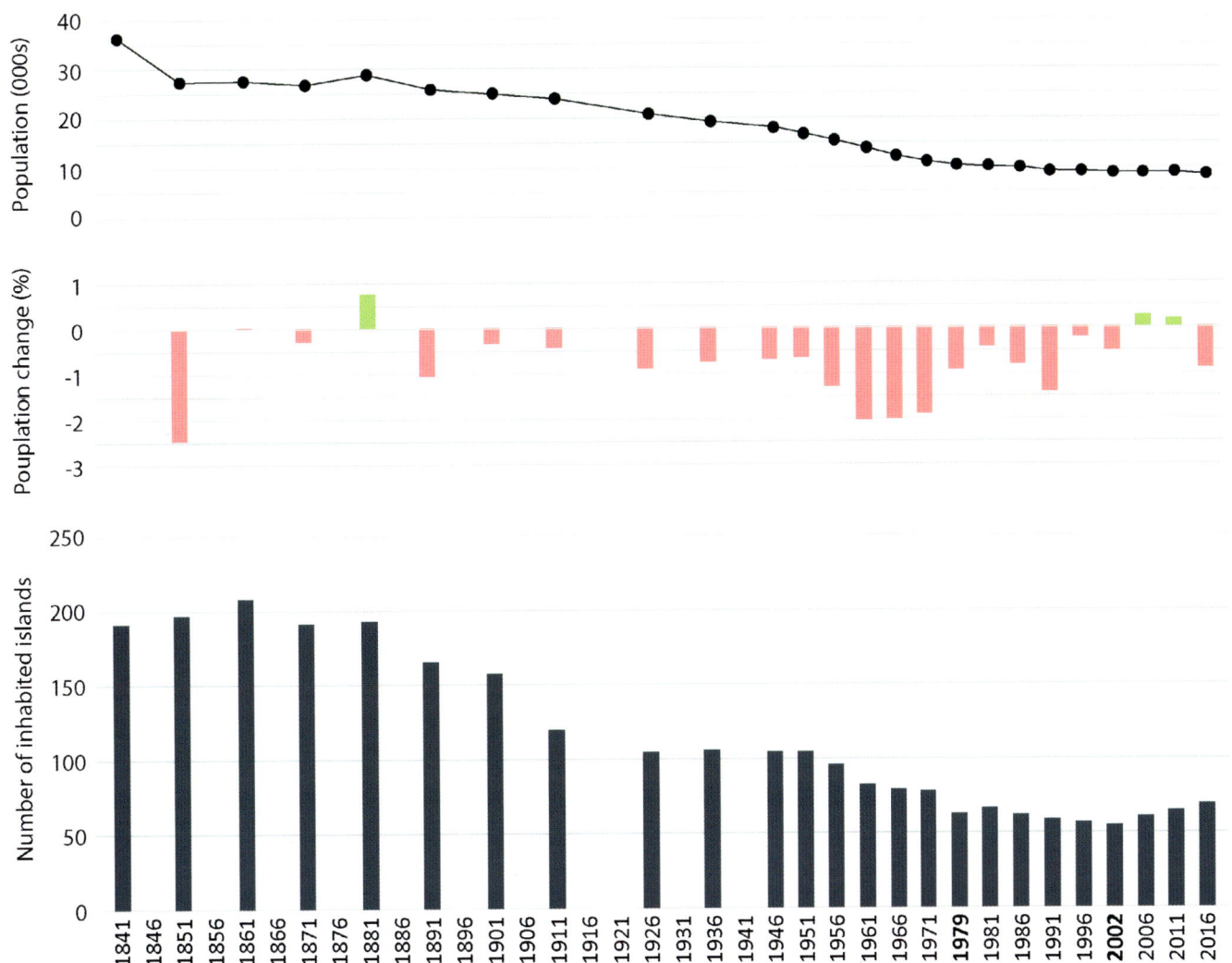

Fig. 21.23 THREE MEASURES OF IRISH ISLAND DEMOGRAPHICS. These graphs are based on figures presented in the official censuses from 1841 to 2016: total population size, population change between censuses and the total number of inhabited islands. The three measures show broadly the same trend, that of decline, which was associated initially with the Famine; thereafter with emigration as islanders sought better economic opportunities on the Irish mainland or overseas. The graphs for population totals and population change march in step, of course, illustrating the considerable losses in the Famine decade of the 1840s and steady, high losses in the decades following the Second World War. More recently, the decline has slowed, and in the first decade of the twenty-first century island populations rose for two successive censuses. The third graph of islands populated is a little different. Numbers actually rose after the Famine, presumably as a few people sought to exploit coastal and marine resources from an island base. The most dramatic fall was early in the twentieth century, while there has been a sustained increase in the number of islands inhabited since 2002. This can be associated with in-migration of people seeking the exclusivity and security of an island base, usually without having to make their living directly from island resources. In 2016, the total island population was recorded as 8,196. [Data source: Central Statistics Office]

representing an overall loss of some 70 per cent (20,146). Rates of decline were particularly high in the 1950s and 1960s as post-war reconstruction in Europe and the beginning of a phase of industrialisation in Ireland encouraged populations to move away from the islands. The number of inhabited islands also declined, if less steadily, reaching the low of fifty-five in 2002. Since that year, the number of inhabited islands rose, to reach seventy by 2016. This had a positive impact for total population between 2002 and 2011, and although the net gain of 176 was marginal, it was nevertheless a welcome reversal of over a century of decline. By 2016, however, total population (85,210) had again declined, especially on some of the larger islands, including Inishmore and Arranmore.

Spike Island, County Cork

Barra O'Donnabhain

With an area of just over 40ha, Spike Island (Figure 21.24) occupies a strategic position at the entrance to Cork Harbour. At the time of the creation of diocesan boundaries in the twelfth century, the bishops of both Cork and Cloyne laid claim to Cork Harbour. This may account for narratives of a seventh-century ecclesiastical settlement on the island. A seventeenth-century poet, Diarmaid Mac Sheáin

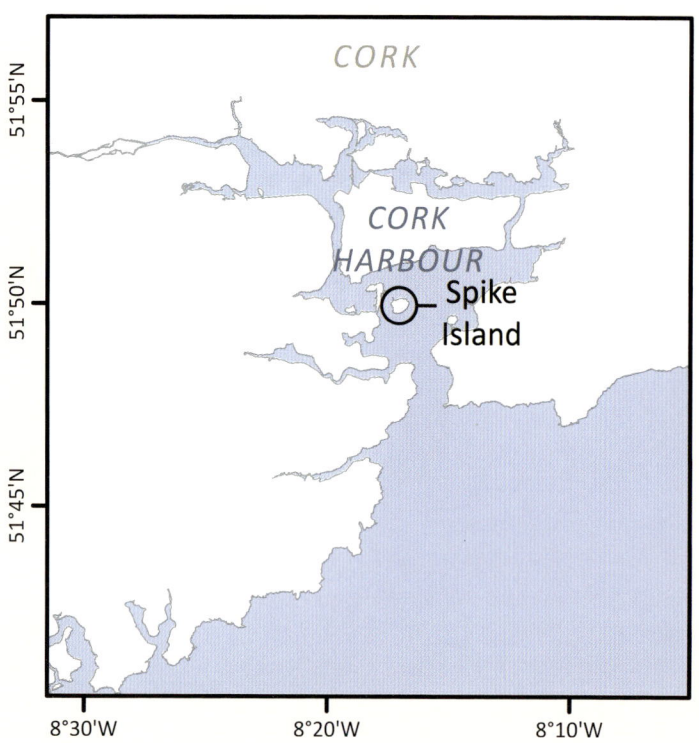

Bhuidhe Mac Cárrthaigh, mentioned the use of the island as a prison during the Cromwellian Wars, when it was a holding place for Irish prisoners prior to their transportation to Jamaica and other locations. The island's strategic location was recognised in the late eighteenth century when the first of a sequence of fortifications was established there. This coincided with the moving of the Royal Navy's provisioning station from Kinsale to Haulbowline Island, immediately west of Spike.

Early in the Napoleonic Wars, construction commenced on a massive new fortress (Fort Westmoreland, now known as Fort Mitchel), that was still largely incomplete at the time of the Battle of Waterloo over a decade later. An opportunity to finish the project arose in the late 1840s when the Irish prison system came under severe strain due to the perceived increase in criminality during the Great Famine. The unfinished fort became a civilian prison and within three years of its opening in 1847, was the largest prison in the then United Kingdom, holding over 2,300 men. Until 1853, the island was an embarkation point for the transportation of convicts overseas, with Bermuda, Gibraltar and Van Diemen's Land (now Tasmania) being the main destinations. Forced labour was an important element of the penal regime and

Fig. 21.24 SPIKE ISLAND. Sometimes referred to as 'Ireland's Alcatraz', Spike Island is located at the mouth of the large natural harbour at Cork. Originally a Napoleonic-era fortress named Fort Westmoreland (since Irish independence renamed Fort Mitchel), it was converted to a convict prison in 1847, the worst year of the Great Irish Famine (1845–1852), during which one million of Ireland's population of eight million starved to death. The prison was part of the British colonial government's response to the rise in public disorder that characterised the Famine and, in its early years, Spike Island was an important holding centre for convicts being transported to Australia and Bermuda. By the early 1850s, the island housed over 2,300 prisoners, many of whom died while awaiting shipment. When the prison closed in 1883, the island fort reverted to military use, and was to remain in such use until 1985, when it was turned over to the civilian prison service. The prison was finally closed in 2004 and, since then, the island has been redeveloped into a tourism and heritage centre. In 2017 it was named Europe's leading tourist attraction at the World Travel Awards. [Source: Cork County Council]

and the convicts completed the construction of the fortifications. Convict labour was also used in the construction of sister forts on either side of the entrance to the harbour, Forts Camden (now Fort Meagher) and Carlisle (now Fort Davis). From the mid-1860s, the focus of prison labour was on Haulbowline Island, the size of which was more than doubled by reclaiming some of the sandbar to the east of the island. The Victorian prison closed in 1883 when the island reverted to the military. By then, unfree labour had transformed the landscape of Cork Harbour. The island was again used as a civilian prison from 1985 to 2004, since when it has been developed to much international acclaim, as a heritage tourism destination.

RESURGENCE

The faltering resurgence in the sustainability of Irish islands since 2002 may be attributed to a number of factors, one of which is the emergence of co-operatives, which scale up demand for goods and services to the benefit of members (Figure 21.25).

In the mid-1980s, it was mainly the secretaries of island co-operatives who got together to establish the Comhdháil Oileáin na hÉireann (the Irish Islands Federation), which pressed for policies to be developed nationally for the provision of services required collectively by the islands. As well as lobbying for the provision of adequate health, education, waste management, childcare and social services, Comhdháil Oileáin na hÉireann campaign for improvements in access; attractive, sustainable employment; adequate, appropriate housing; and support for new, sustainable enterprises. The federation also pressed the government of Ireland to establish ministerial responsibility for island affairs. Since 1977 that responsibility has rested in the Department of Culture, Heritage and the Gaeltacht, which has the following mission statement:

> Our coastal islands are an integral part of the state's heritage. Around thirty of these islands are inhabited and hold a wealth of cultural heritage. A central objective of this Department is to ensure that sustainable vibrant communities continue to live on the islands.

Both anglophone and Gaeltacht islands are supported by the department, but islands with bridges or causeways, such as Achill Island and Valencia Island, do not come under its oversight. The European Union, especially through the Regional Development Fund, also supports the islands and many of them display blue plaques acknowledging EU support, especially for infrastructural projects, such as harbour improvements and provision of electricity and other services.

Improvements in accessibility are fundamental to ensure the sustainability of island communities. While islanders may aspire to self-sufficiency, the reality is that the majority have to commute to the mainland to access employment, schools, retail outlets, etc., and their economies are rarely viable without some external support. Regular, reliable and subsidised ferry services are, therefore, vital to ensure low-cost and efficient delivery and collection of goods and people travelling between islands and, more especially, to and from the mainland (Figure 21.26). Upgrading of the ferry boats and harbour facilities is also essential. The paucity of such services and infrastructures makes island living difficult and emphasises the sense of isolation.

While the provision of regular and reliable transport links is essential for the promotion of sustainable employment and enterprise on the islands (Figure 21.27), a recent key element towards resurgence has been the development of new economies. The tyrannies of distance and isolation, once so debilitating to traditional island enterprise, are not so significant in today's internet-connected world. Provided good broadband internet connections are available, it is nowadays possible to work, and even study, remotely from the islands. Modern developments have also

Fig. 21.25 THE ARRANMORE CO-OPERATIVE. Many of the Irish islands have co-operatives. These organisations co-ordinate the activities of island producers, scaling up the opportunities for goods and marketing to more efficient levels, as well as providing a pool of machinery for members to use. Co-operatives also act as agencies to promote their islands' economic, cultural and social wellbeing. The Arranmore Co-op (Comharchumann Oileán Árainn Mhóir) was founded in 1976 and has been involved in housing, tourism projects, the development of a heritage centre, a helipad and the provision of Irish language courses, as well as providing a machinery pool and the store shown. Despite such efforts, the island's population declined by over 20 per cent to 469 between 1996 and 2016. [Source: Stephen A. Royle]

influenced agriculture and fishing practices. The increase in aquaculture is notable (Figure 21.28) (see Chapter 26: Coastal Fisheries and Aquaculture), with fish farms emerging in locations such as Clare and Bere islands, although plans for a large-scale aquaculture development in Galway Bay, to be managed from the Aran Islands, proved very controversial in the mid-2010s and the

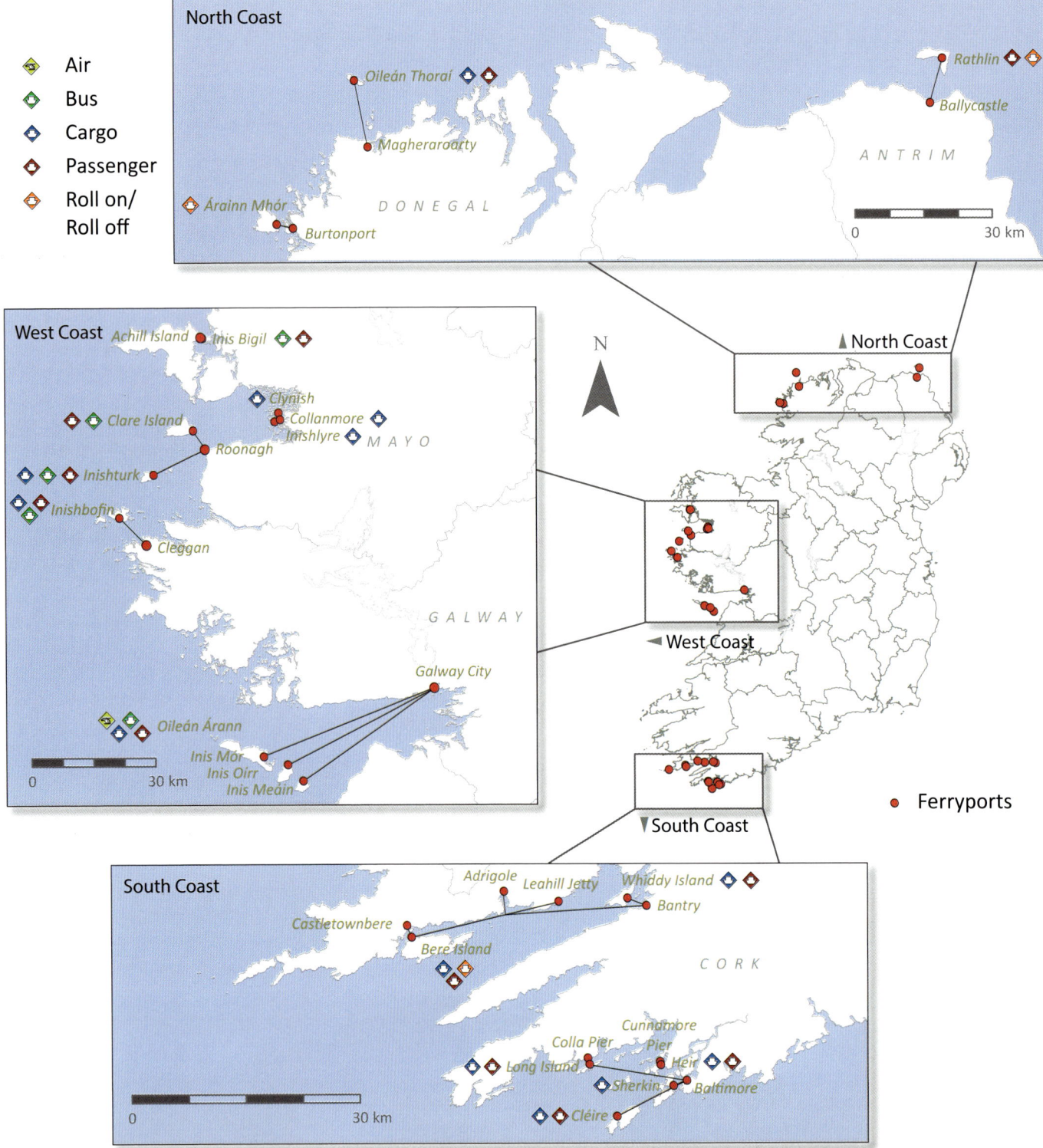

Fig. 21.26 GOVERNMENT-SUBSIDISED AIR AND FERRY LINKS TO THE ISLANDS. The Republic's National Planning Framework to 2040 recognises specifically the cultural significance of island communities, but equally highlights the fragility and isolation of island life. The Islands Programme of the Department of Culture, Heritage and the Gaeltacht, therefore, supports island communities by providing subsidised transport links between the islands and the mainland, to ensure the continued viability and sustainability of these communities. While air services are supported, as between Galway and Inishmore (Inis Mór), the most utilised links involve ferry services. The department thus provides subsidised services to islands that are not connected to the mainland, are permanently inhabited and not in private ownership. In 2018, these involved ferry and cargo links to fifteen offshore islands in counties Cork, Donegal, Galway and Mayo. Their importance can be appreciated in that between 2008 and 2017, there were over 410,000 sailings on subsidised services, which carried almost four million passengers (some 60 per cent being visitors to the islands) and 160,000 tonnes of cargo to meet the demands of island communities. While the services are all-year operations, sailings are in highest demand through the summer months to accommodate the large and growing numbers of tourists who travel to the islands. In 2017, for example, non-island passenger numbers to Árainn Mhór (the most-visited island) more than doubled to 46,634, compared to only 22,930 in 2008. The only ferry service in operation in Northern Ireland is to Rathlin Island. [Data source: Department of Culture, Heritage and the Gaeltacht]

Fig. 21.27 INISHTURK PIER AND FERRY BOAT. The small island of Inishturk lies some 15km off the coast of County Mayo and is located between the larger islands of Inishbofin and Clare Island. Like most inhabited islands around Ireland's coastline, Inishturk depends on regular, reliable and subsidised ferry services to the mainland, for the delivery and collection of goods and the maintenance of many other aspects of daily life. Most of these ferry services are provided by small, passenger-only boats, such as the one shown here, though some of the larger and more populous islands are also served by bigger vessels, able to carry vehicles and cargo as well as foot passengers. Prior to 1997, Inishturk had no ferry service and its population used local fishing boats to access the mainland. Since then, however, an all-year ferry service operates from Roonagh Pier, some 6km from Louisburgh in County Mayo. This docks on Inishturk at a pier built in the 1980s to provide shelter for local fishing vessels. Around this 'lifeline' to the mainland is located one of the island's small settlements, which offers accommodation and a pub and is a focal point for community activities. Tourism is a vital component in sustaining the island's resident population and in 2017 almost 5,900 visitors arrived via the ferry service, more than treble that in 2008. Despite this, Inishturk's permanent population continues to decline from a pre-Famine peak of 577 to the 2016 census total of fifty-one. [Source: Barry Brunt]

application was subsequently withdrawn. Efforts by Údarás na Gaeltachta to encourage manufacturing have led to the development of a small number of factories and industrial activities on some islands, of which the textile factory on Inishmaan, Galway Bay, is perhaps the most notable (Figure 21.22).

Of all the sectors, the modern tourist industry has become a major component of the economy on most islands. Modern tourists demand ease of access to their preferred destinations. If that is achieved, tourism can become a useful way of generating alternative income and employment, such as in the provision of bed-and-breakfast accommodation and local craft-based industries. Tory Island, for example, thanks to tourism, has benefited from improved public services, which provide employment and opportunities for local people.

In addition to the formal provision of accommodation, the tourist sector has become associated increasingly with holiday homes, which are now a feature of many islands. Once uninhabited islands, such as Gola (Figure 21.29), now have second homes, and ring with voices on occasion, even if they are not the voices of the departed traditional community. Some islands have even attracted back their emigrants, either to retire or to return with their families

so that their children can enjoy the safe and fulfilling childhoods possible in an insular setting.

Island tourism can also generate negative impacts. A key problem with the tourist industry, for example, is the highly seasonal nature of the business. This is exacerbated on islands, where access

Fig. 21.28 MOWI FISH FARM, DEENISH ISLAND, COUNTY KERRY. Deenish sea farm has produced Atlantic salmon since 1991 and currently supports seven full-time and three part-time employees from the counties of Cork and Kerry. The farm produces approximately 2,000 metric tonnes of salmon over a two-year cycle, and was the first in Ireland to be awarded the Aquaculture Steward-ship Council standard, established by the World Wide Fund for Nature (WWF). [Source: Mowi]

Fig. 21.29 GOLA HOLIDAY HOME. In recent years there have been signs of revival on some of the Irish islands with some return migration. The 1996 census records no people resident on Gola Island in County Donegal, for example, whereas six years later, in 2002, the population had risen to five and by 2011 had tripled to fifteen, before falling to five once more in 2016. However, many islands also attract people seeking quietude and tranquility, and these movements need not be for the whole year. In many cases, such as shown here on Gola, once-abandoned property has been restored and modernised for use as second homes or holiday lets. [Source: Stephen A. Royle]

is challenging during the harsh winter months. In addition, most tourists interact with islanders through English, which can have negative implications for the local Irish language and traditional culture. Habitat degradation, energy demands and the disposal of waste in an island context can also be challenging, particularly because the seasonal occurrence of large numbers of visitors can lead to significant peaks and troughs in the use of wastewater and other infrastructure. Careful planning and management are required to reduce the associated environmental impacts and to preserve the pristine habitats that make Ireland's islands so special. One example of innovative action to counter the negative impacts of tourism, and to improve progress towards sustainable development, is the effort on the Aran Islands to develop community-based solutions in the area of energy efficiency.

While innovations have undoubtedly made island life more comfortable and economically viable, the outlook into the future remains uncertain. As has been demonstrated, many of the traditional aspects of island living have declined and populations have fluctuated dramatically. While island communities, over generations, have proved to be resilient in addressing the numerous challenges that have confronted them, effective long-term planning and management will be key to ensuring their viability. Some examples of how this is being undertaken may be found in the accompanying Rathlin Island case study.

Case Study: Rathlin Island

Stephen A. Royle

Northern Ireland has few offshore islands, discounting those in the almost totally enclosed waters of Strangford Lough. Since the depopulation of the Copeland Islands in 1946, only Rathlin Island, off County Antrim in the north-east corner of Ireland, has been populated. Rathlin has the shape of 'an Irish stockin', the toe of which pointeth to the mainland', as William Petty put it in 1660.[34] It is about 6km from east to west along its 'leg' and 4km from north to south down the 'foot'. The island is composed largely of limestone and basalt forming rugged tableland with high cliffs to the west, reaching 144m at Slieveard. It is more gentle and rolling along the 'foot' and also at Church Bay (the 'instep' of the 'stockin"), where the principal harbour lies opposite the mainland harbour of Ballycastle, about 10km to the south, although it is only 3km from Rue Point at the 'toe' of the island to Fair Head.

Rathlin was significant in ancient times, being occupied from the Mesolithic period, and was the site of manufacture of stone axes made from a deposit of porcelenite, the axes being widely traded. Lying between Ireland and Scotland, Rathlin was claimed by both until a court case in 1617 assigned it to Ireland, partly because of its (Irish) lack of snakes. Due to its strategic position, Rathlin was often caught up in wider struggles. It was subject to Viking raids in 790, 973 and 1038, and later involved in conflicts between the English, Irish and Scots. Famously, Rathlin was the refuge of Robert the Bruce in 1306 who, inspired by viewing an indefatigable spider's many and finally successful attempts to extend its web, returned to Scotland and retook his crown. Massacres occurred on Rathlin, notably in 1557 by Henry Sydney, lord lieutenant of Ireland, and again in 1575, when the earl of Essex, assisted by Francis Drake, slaughtered the Macdonnells who held the island. Oliver Cromwell declared Rathlin to be forfeited because of support for the Royalist cause. In 1642 the island was 'swept bare of every living thing' by the Campbells, the family of the earl of Argyll, whose people were aligned with the Scottish MacDonald. Tradition has it that only two people survived.[35] Recovery was slow and a population estimate of 1657 shows only seventy-five inhabitants. Then Rathlin, according to its earliest historian in 1851, arrived 'at an epoch when the history of the island loses the character of romance … we no longer hear of its little community of Churchmen invaded and their property plundered by the merciless Danes … Distressed monarchs do not now seek for shelter within its seclusion, nor do we hear of ships of war with armed troops sent by the English government to storm its fortresses.'[36]

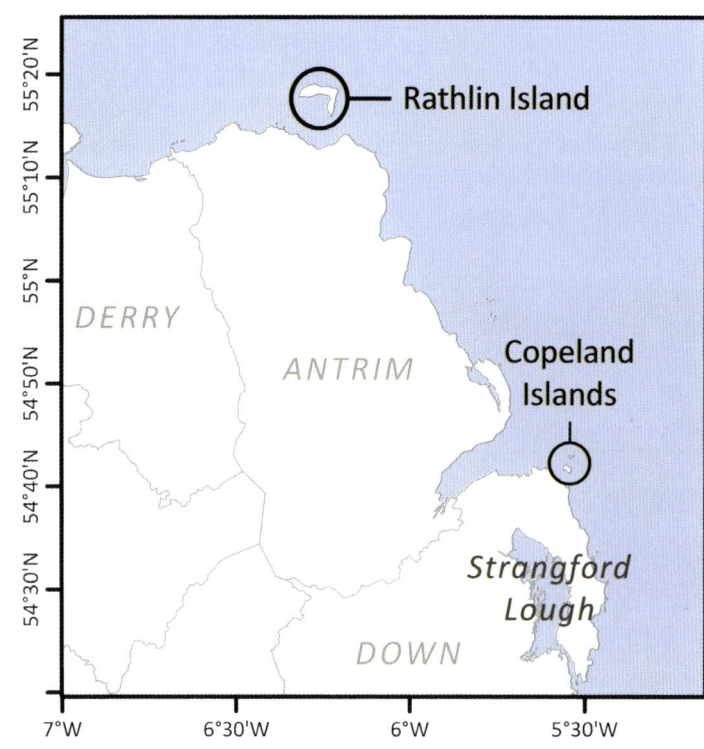

In 1746 Rathlin's ownership passed from the earl of Antrim (a Macdonnell) to Reverend John Gage, for the substantial sum of £1,750. Gage's family held the island under their benign sway until the 1925 Land Purchase Act system changed the tenure. The Gages built Rathlin's most prominent building, the Georgian Manor House at Church Bay (Figure 21.30) in the 1760s, and their reforms helped the population rise to an estimated 1,200 people by 1784. At this period Rathlin was relatively self-contained, sugar and tea imports excepted. Some fuel was brought in, too; there was little peat or wood, and cow dung was burnt. Few people emigrated or even travelled outside the island; mainland Ireland was regarded as a foreign kingdom. Cattle and sheep were kept; barley, oats and potatoes grown; fishing was carried out. There were ducks and hares to be caught and dulse was cut. Poitín

Fig. 21.30 CHURCH BAY, RATHLIN ISLAND. Since the depopulation of the Copeland Islands off County Down, Rathlin Island is the only inhabited offshore island of Northern Ireland, if one discounts some small islands in the almost totally enclosed Strangford Lough. Rathlin used to belong to the Gage family and the large buildings visible at the back of Church Bay include their eighteenth-century dwelling, Manor House, now used as a hotel, restaurant and event space, tourism having become a significant aspect of the island economy. The Manor House overlooks the harbour, from which Rathlin Island Ferry Ltd operates two boats to Ballycastle on the County Antrim mainland. Rathlin is noted for its extensive seabird populations, has a Royal Society for the Protection of Birds reserve and is a Special Area of Conservation. [Source: Douglas Cecil]

Fig. 21.31 RATHLIN ISLAND FERRY. Even in fine weather, the seas around many of Ireland's islands can be turbulent, making access uncomfortable at best and often completely disrupted. In this photograph, the *Spirit of Rathlin* passenger and vehicle ferry is seen crossing the Straits of Moyle, between Rathlin Island and Ballycastle on the County Antrim mainland. The cost of providing this ferry service is subsidised by the Northern Ireland administration to ensure its viability and fares are significantly lower for island residents than for visitors coming to the island. [Source: Douglas Cecil]

Fig. 21.32 LIGHTHOUSE AT RUE POINT, ON THE SOUTHERNMOST TIP OF RATHLIN ISLAND. Construction of a lighthouse at Rue Point was recommended in July 1914, sanction obtained from Trinity House in November of that year, and a temporary light installed towards the end of the following year. Eventually, in 1920–1921, the permanent six-sided concrete tower was built. Initially, two keepers were attached to Rue; they lived in a nearby wooden hut when on duty. The light was powered by a water-to-carbide-acetylene generator, housed inside the tower, and a fog gun was installed on the roof (although this was withdrawn in 1931). The acetylene-powered light was replaced by an electric light in 1965, which was run from batteries charged by a diesel engine and, in 2004, was further converted to mains electricity. The present light is 16m above mean high water spring tides (MHWS), and has a range of 14 nautical miles. [Source: Douglas Cecil]

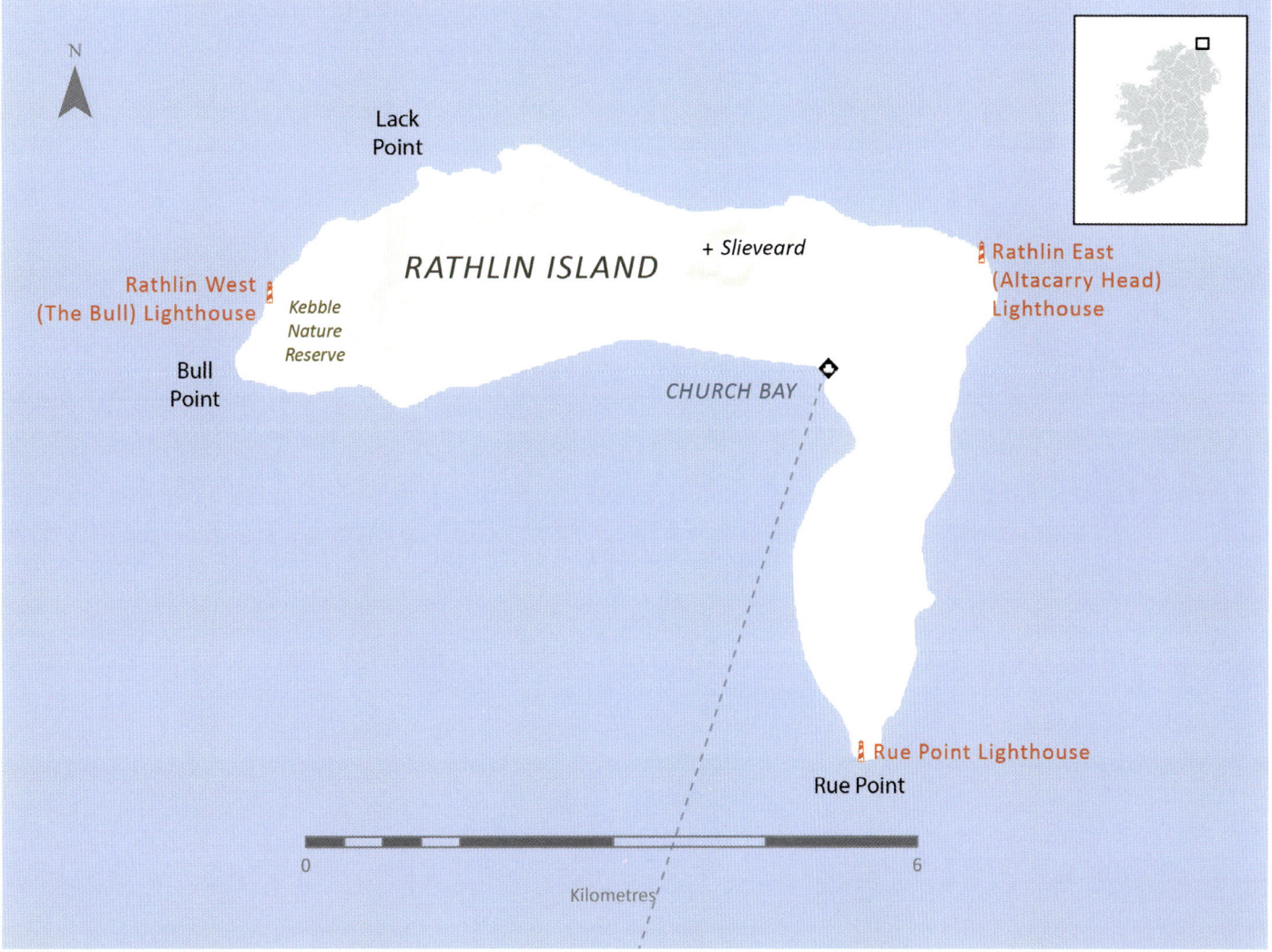

Fig. 21.33 MAP OF RATHLIN ISLAND. At the present day, Rathlin is the only populated island off the coast of Northern Ireland. Lying about 10km to the north of the mainland town of Ballycastle, Rathlin occupies an important strategic position in the often turbulent Straits of Moyle. It has been occupied since Mesolithic times, and was subject to raids by the Vikings as well as being caught up in wider conflicts between the Irish, the English and the Scots. It was claimed by both Scotland and Ireland until it was assigned to the latter in a court case in 1617.

was certainly distilled. Kelp burning became an important source of income after the industry started in 1774 and remains of kilns for burning and structures for the storage of kelp can still be seen.

In the early nineteenth century the head of the landowning Gage family was but a child and, under his unscrupulous agent, smuggling was facilitated. After achieving his majority, Robert Gage installed coast guards on the island in 1821, to halt the trade. He also rebuilt the Anglican church of 1722 in 1812 and revived, perhaps re-established, education on the island. Another notable nineteenth-century event was Rathlin becoming a site for early telegraphic communication when, in July 1898 under Guglielmo Marconi, a wireless telegraphy link to Ballycastle was established, to transmit news of shipping passing the island.

The population of Rathlin was 1,010 at the pre-Famine 1841 census, a high total ascribed to intensive cultivation under an active, resident landlord. The Famine, as in so many places in and off Ireland, brought about demographic change. The agricultural situation was accompanied by a lessening of fish stocks and, in contrast to the eighteenth century, emigration became established. Many Rathlin men went to Greenock in Scotland to work as shipwrights. The population declined to 753 by 1851 and more than halved to around 360 by the end of the nineteenth century. A century later, the population was under one hundred. J.H. Elwood wrote that by 1961 migration here, as elsewhere, had become 'a conditioned response of the native Irish in the face of adversity. The harshness of life on an offshore island, particularly for women has been increasingly rejected for the lure of the factory and a box-like urban existence.' He predicted the 'extinction' of the island in the 'near future'.[37]

Extinction has not happened; in fact the Rathlin Development and Community Association website states that the population total is currently 150 and growing.[38] Other estimates have it at a more modest one hundred at the winter low. A number of infrastructural improvements have helped, particularly harbour works in 1985. Mains electricity was established with an undersea link to the mainland. New houses have been built, many in the social sector, to try to ensure that they were available to local families, rather than people just retiring onto the island. The all-important ferry service was regularised after 1996. Before, it had been operated by islanders under licence but it was then taken over by the nationalised Scottish island ferry company, Caledonian MacBrayne, who ran the service with M.V. *Canna* until 2008.

In that year the contract was awarded to Rathlin Island Ferry Ltd, which runs a service several times a day with two boats. *Rathlin Express* is a fast catamaran that makes the crossing in twenty minutes and there is also a vehicle ferry; first it was the MV *Canna* until its replacement in 2017 by the new, purpose-built *Spirit of Rathlin*, which has vehicle and cargo capacity not available in the catamaran. The island harbour was upgraded to accommodate the *Spirit of Rathlin* (Figure 21.31). The ferry is subsidised by the Northern Ireland administration to guarantee this lifeline service to the island; islanders have 50 per cent of their ferry travel paid for and over-60s can use their 'smart passes' to travel free.

Another change was to the Gage Manor House, which came out of the family's possession in 1973 and has been remodelled, most recently as a hotel and restaurant with event spaces by the Rathlin Community and Development Association with the assistance of grant aid. Another contributor to Rathlin development was the entrepreneur, Richard Branson, who donated £25,000 as thanks for being rescued by an island boatman when his hot air balloon crashed into the sea off the island in 1987.

As with other islands off Ireland, the development of cyberspace has led to some people on Rathlin being able to work remotely. However, the island economy is still partly dependent upon seaweed production,[39] and also on agriculture, especially beef cattle. Farming is subsidised to help cover the extra costs involved in island working. Rathlin is a Special Area of Conservation and lies within the Antrim Coast and Glens Area of Outstanding Natural Beauty. These designations displeased some island farmers, who expressed concern that their ways of working would be challenged. However, they are important in conservation and in helping to attract tourists, tourism now being key to the island economy, facilitated by the regular ferry services. The Royal Society for the Protection of Birds is another conservation body active on Rathlin, which is notable for puffins, guillemots, kittiwakes, razorbills and fulmars, with a reserve around the West Light, one of three lighthouses on an island notorious for its shipwrecks (Figures 21.32 and 21.33). Rathlin has services and accommodation opportunities to facilitate its visitors and, of course, employment is provided thereby. In short, the extinction predicted in 1968 has not taken place; instead, as one tourist commented to Trip Advisor: 'This remote and quiet island is a preserved gem'.

FASTNET ISLAND, COUNTY CORK. This small rocky island, situated 6.5km south-west of Cape Clear Island (seen in the background) and 13km from the Irish mainland, is Ireland's most southerly point. It is sometimes known as 'Ireland's Teardrop', since it was the last part of Ireland that nineteenth-century emigrants saw when they departed for North America. The island is also important in competitive sailing and is the midpoint for the classic Fastnet yacht race, which runs from Cowes on the Isle of Wight in southern England, round the rock and back to Plymouth in Devon, an overall distance of 1,126km. The present-day lighthouse on the island entered service in 1904, replacing an earlier one constructed in the 1850s. [Source: Robbie Murphy]

Underwater Cultural Heritage

Karl Brady, Connie Kelleher and Fionnbarr Moore

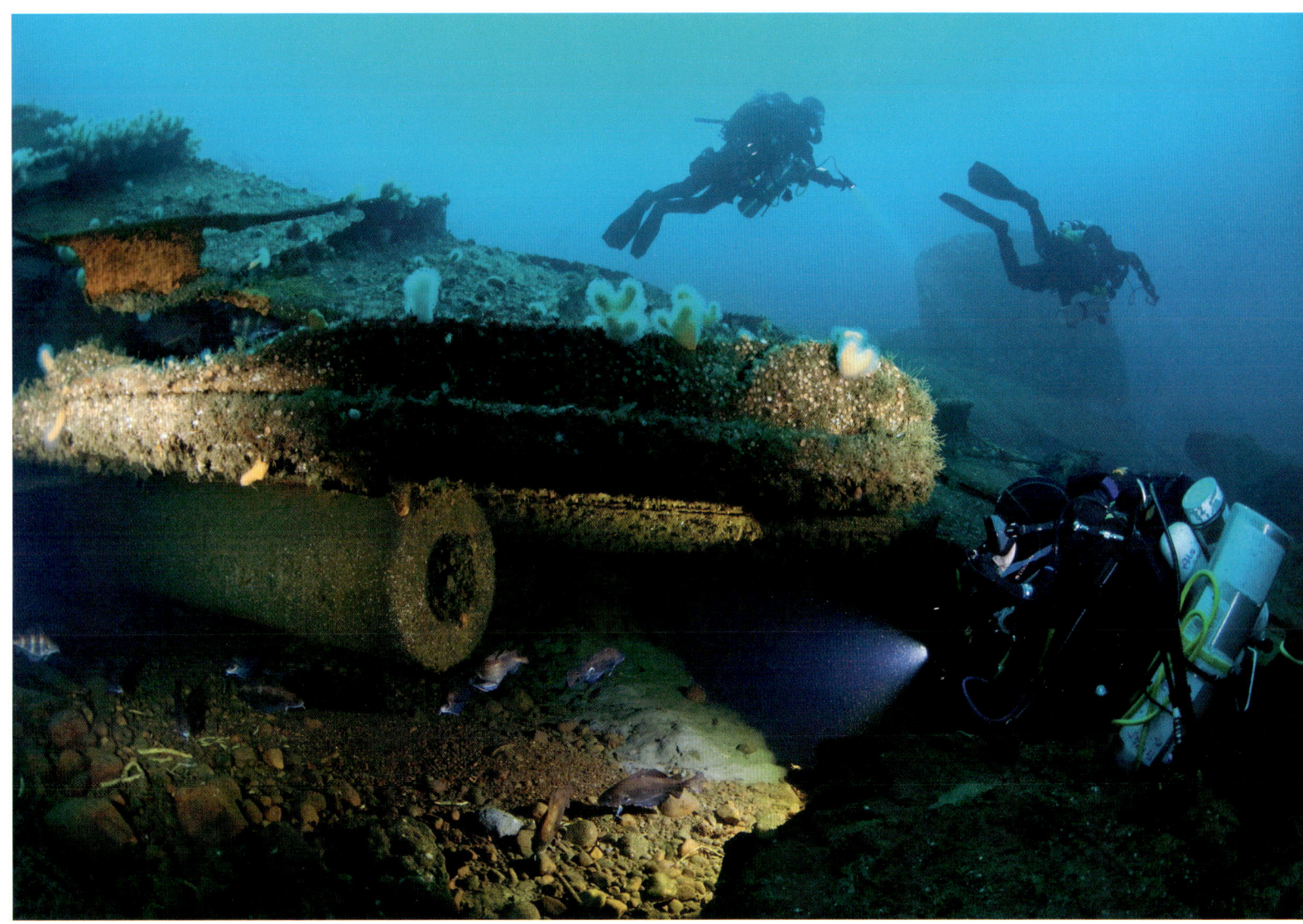

TECHNICAL DIVERS INVESTIGATING ONE OF THE 13.5" GUN BARRELS NEAR THE STERN OF THE WRECK OF HMS *Audacious*. HMS *Audacious* was a King George V-Class dreadnought battleship, which struck a mine laid by SS *Berlin* and sank on the 27 October 1914. It now lies at a depth of 64m off the north coast of Donegal. [Source: Barry McGill]

With a coastline that is more than 7,000km long and designated waters encompassing more than 900,000km² of seabed, it is not surprising that Ireland's coastal and underwater cultural heritage is both rich and diverse.[1] The richness of this archaeological resource demonstrates vividly the influence of the seas on the island throughout every period of its past.

Since the first settlers arrived by boat almost 10,000 years ago, Ireland's social, political, economic and cultural development has been linked with the sea, the interaction with coastal areas and the strategic usage of marine resources. The seas around Ireland are also maritime highways, linking people to a web of transport networks off the western seaboard of Europe and further afield. These highways have been used throughout time for fishing, trade, travel, exploration, migration, colonisation and warfare, all elements that need to be taken into account when considering the history and archaeology of Ireland.

Boats and ships are perhaps the first images that come to mind when contemplating the use of the sea, with the archaeological evidence for their ubiquity dotting the seabed and coastal fringes around Ireland. Remains of other aspects of maritime cultural heritage survive in a variety of forms – including evidence for submerged prehistoric landscapes, promontory forts, coastal settlement sites, later upstanding fortifications, industrial infrastructure relating to maritime activity, harbours and port facilities, shell middens, fish traps, salt production sites, historic reclamation works, jettisoned and lost cargoes from ships and isolated artefacts – all of which contribute to the creation of a fuller picture and understanding of Ireland's past.

This chapter captures the impact of Ireland's perilous storms and dangerous coastline on ships and shipping over the centuries (for example, the Great Storm of 1861 in Figure 22.1). It reviews the shipwreck heritage of Ireland's coastal waters and presents case-study accounts of key historical events in this context, namely the Spanish Armada of 1588 and the impact of the First World War on Ireland's shipwreck heritage.

SHIPWRECKS

Perhaps the most familiar and among the most fascinating of all underwater archaeological sites are shipwrecks. No site stirs the imagination more than the wreck of a long-lost ship, whether it is its loss in tragic circumstances, tales of rescue and bravery, the mystery surrounding its loss, the intrigue of search and discovery or the excitement of the recovery of ancient artefacts from the seabed. Some wrecks, such as the Tudor ship *Mary Rose* and the RMS *Titanic*, are globally renowned and their stories have been recounted many times and continue to have mass appeal for both the general public and professional alike. Ireland too has many wrecks that are known worldwide and which are of international importance, not least that of the RMS *Lusitania*, as well as other wrecks relating to the two World Wars or the twenty-six or so wrecks of the Spanish Armada lost on Ireland's western seaboard. The story of Ireland's maritime heritage would not be complete,

THE RECENT GALE.—WRECKS AT KINGSTOWN BAY OF DUBLIN.—SEE SUPPLEMENT, PAGE 177.

Fig. 22.1 ARTIST'S DEPICTION OF STORM WAVES AT DÚN LAOGHAIRE (FORMERLY KINGSTOWN) DURING THE GREAT STORM OF 1861. The Great Storm of 1861 began on 8 February and increased in severity throughout the next forty-eight hours. The wind changed from a north-easterly gale and increased to strong gale force, changing direction to north by the early hours of 9 February. The east coast of Ireland was battered, resulting in the loss of a great many ships in harbour. Harbours like that of Dún Laoghaire, pictured here, were washed by enormous waves and surges, with vessels of all sizes capsized and wrecked. [Source: *Illustrated London News*]

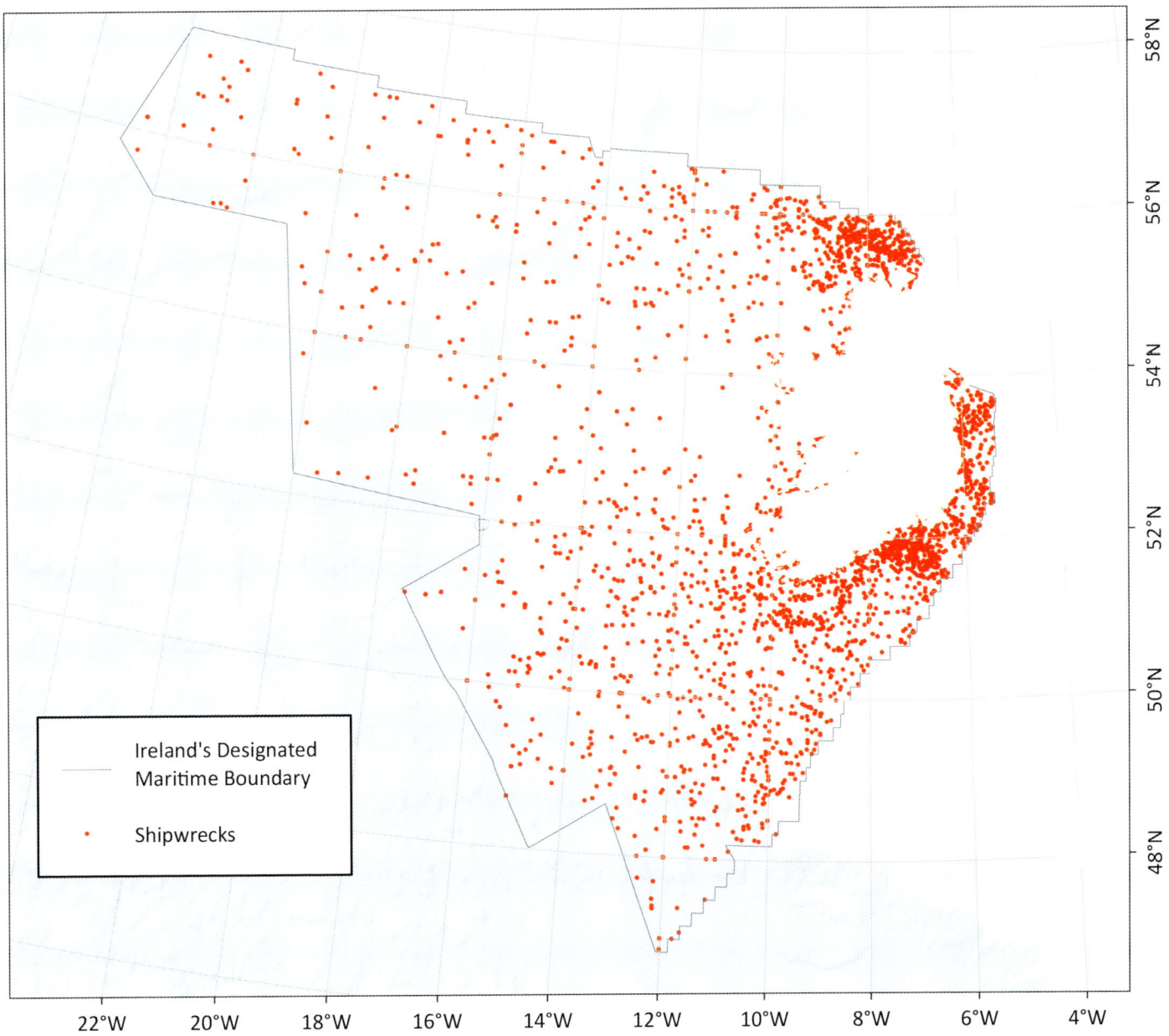

Fig. 22.2 KNOWN AND APPROXIMATE LOCATIONS OF WRECKS OFF THE IRISH COAST, WITHIN THE TERRITORIAL WATERS AND ON THE CONTINENTAL SHELF. Some 18,000 wrecks have been identified to date and this increases as new records come to light or new discoveries are made. The information illustrated here comes from the Wreck Inventory of Ireland Database (WIID). [Data source: WIID, National Monuments Service, Department of Culture, Heritage and the Gaeltacht]

however, without referencing the multitude of less well-known or smaller ships and boats lying wrecked around the coast, which play a key role in illuminating our understanding of the past.

Every wreck has an interesting story, be it a traditional vernacular craft – such as a fishing boat, coastal trader, hooker or barge – or a larger fully-rigged ship, ocean-going liner, U-boat or warship. Singular events involving the undocumented loss of ships and crew can be revealed when the remains of a wreck are discovered, while known wreck sites can provide unique insights into major events, with the Spanish Armada campaign of 1588, the attempted French invasion of 1797 or losses from the two World Wars among the better documented.[2] Wrecks associated with these events can assist in understanding the wider dynamics at work in the coastal environment over time and significantly expand the conventional interpretation of the archaeology and history of the littoral zone around the coast.

It is estimated that there are between three and four million shipwrecks lying in the seas and oceans of the world, and a percentage of these lie scattered on the seabed in the waters around Ireland. The Wreck Inventory of Ireland Database (WIID), held by the Department of Culture, Heritage and the Gaeltacht, lists over 18,000 records pertaining to wrecks in Irish territorial waters and in the designated area of the continental shelf (Figure 22.2). Of these, 15 per cent have precise locations. This figure, which is revised upwards regularly, is considered to represent only a fraction of the real number of wrecks that have occurred and it is estimated that the true figure could be as high as 30,000.[3]

Prehistoric logboats are some of the oldest vessels recorded and the evidence for their sudden sinking highlights the dangers the Irish coastline posed to seafarers from the earliest times, as well as the perils experienced by our ancestors when navigating the early waterways.

Early wrecks

The oldest logboat found in Ireland dates to *c*.5300 BC and was located at Brookend, Lough Neagh, in 1995.[4] The discovery of a *c*.5,000 year-old logboat at Greyabbey Bay, Strangford Lough, in 2000, however, also represents one of the earliest wrecks from a marine environment in Ireland.[5] The potential for further finds of logboats in maritime contexts is demonstrated by the discovery of a late Bronze Age logboat off Gormanston, County Meath, during pipe-laying works in 2002.[6]

Discovery of Roman burials in 1927 on the strand on Lambay Island, off the coast of Dublin, and a Roman amphora on the Porcupine Bank, 242km off the west coast of Ireland, in 1934, provide tantalising glimpses of possible ship losses dating to these early periods. While no wrecks have been found in association with or in the vicinity of these Roman discoveries to clearly suggest that they are related to wrecking events, it is the most likely scenario for such rare finds.[7]

Study of early Irish sources and the Irish Annals provides more direct historical evidence for ship losses from early medieval times onwards. Wreck entries or references during this earlier period, however, are not as numerous or as reliable as later sources. In general, when wrecks are recorded, the references are very often vague, containing limited information and often only in relation to an important naval engagement, a storm event or the drowning of an important ruler or ecclesiastical figure while voyaging at sea. An example of such an entry is an account in the Annals from 1528, which records that 'A great wind arose on the Friday before Christmas, which prostrated a great number of trees throughout Ireland … and swept away, sank, and wrecked many vessels'.[8] Another example is that of the *Nicholas*, a merchant ship from County Down, wrecked during a severe storm on Portmarnock Strand in 1306 with a cargo of wine, wax, coffers with jewels, copper pots, spices in barrels, tin, pitch and steel. When in full sail, the *Nicholas* would have been an impressive late medieval sailing ship and may be one of the nine wrecks listed in the Wreck Inventory of Ireland Database that becomes exposed on Portmarnock Strand when sands shift after stormy weather.[9]

Late medieval and post-medieval wrecks

With an increase in shipping activity due to the intensification of the fishing industry in the late-medieval period, increased cross-channel trade with England and the continent, more extensive global trade from the sixteenth century onwards, and a resultant increase in ships using the ports and harbours of Ireland for colonial expeditions, we see a gradual increase in the numbers of wrecks occurring in Irish waters.

This is well illustrated by the loss of ships like the *Emanuel* in Smerwick Harbour, near Dingle, County Kerry, in September 1578. The *Emanuel* was part of Martin Frobisher's final expedition, which attempted to find a navigable route through the Northwest Passage and on to the Far East. Just two years later (1580), during the Second Desmond War, the same site and harbour was the location of one of the greatest maritime massacres in Irish history. Dún an Óir fort was held by a papal force of 400–500 men, who had arrived in July 1579 by sea to assist the Irish loyal to James FitzMaurice, the earl of Desmond, who was engaged in a war for survival against the

Fig. 22.3 THE REMAINS OF A WOODEN BARREL FROM INSIDE THE HULL OF THE SEVENTEENTH-CENTURY RUTLAND ISLAND WRECK, BEING CAREFULLY EXCAVATED BY A DIVER FROM THE UNDERWATER ARCHAEOLOGY UNIT (UAU) OFF BURTONPORT, COUNTY DONEGAL. In view are barrel hoops and staves lodged within the hold of the wreck, where they would have been stored on board the ship during voyages. [Source: C. Kelleher, Underwater Archaeology Unit, National Monuments Service, Department of Culture, Heritage and the Gaeltacht]

Fig. 22.4 A DIVER FROM THE UNDERWATER ARCHAEOLOGY UNIT (UAU) EXCAVATING OUTSIDE THE STERN OF THE SEVENTEENTH-CENTURY RUTLAND ISLAND WRECK, OFF BURTONPORT, COUNTY DONEGAL. Sandbags have been placed within the stern to protect delicate structural timbers. In view are the transom timbers and remains of the rudder extending from the back of the stern timbers. [Source: C. Kelleher, Underwater Archaeology Unit, National Monuments Service, Department of Culture, Heritage and the Gaeltacht]

English crown. On the site too were many Irish, including women, encamped within the fort itself. In what became known as the Siege of Smerwick, Dún an Óir was blockaded by the combined land and sea forces of the English lord deputy, Arthur Grey and naval Commander, Admiral William Winter, whose warships commandeered the ships of the continental forces. The Lord deputy established siege works around the fort and on surrendering, the Irish, including the women who were present, were hanged and the continental forces put to the sword.[10] This wild and beautiful coastal site and the harbour of Smerwick itself links us to past conflicts, when Ireland's coast was the setting for many such tragic events, both on land and sea and to the wider European background.

Rather than being looked on as an island on the periphery of Europe, so often the perception in times past, sites like Dún an Óir demonstrate Ireland's strategic place within the wider North Atlantic and on the main transatlantic route for shipping. This pivotal location, in turn, helps explain why so many wrecks in Irish coastal waters are foreign in origin. A large number of ships involved in Atlantic voyages or en route to the continent never completed their intended journeys and were lost around the Irish coast (Figures 22.3 and 22.4). As a result, over 70 per cent of all wrecks documented in Irish waters are foreign-registered vessels.

The Dutch East Indiaman, the *Zeepaard*, is an example of one such wreck. Having left Weilingin in the Dutch Republic in October 1665, after the outbreak of the second Anglo-Dutch War, it was forced to take a circuitous route around Scotland and Ireland in order to avoid trouble in the English Channel. Reported as lost west of Scotland, and storm-battered, it eventually limped into Broadhaven Bay, on the Mayo coast, only to be wrecked at Inver Point. Seventy-four of the crew lost their lives and it remains unclear if all the cargo of silver, which was destined for the Far East, was ever recovered. The wreck was located in the late 1960s and since then has been dived a number of times. Several artefacts recovered by the Leinster Divers Sub-Aqua Club are now on display in the National Museum of Ireland, Collins Barracks, including several lead ingots with impressed merchant's marks.

Eighteenth century to the present day

While the inventory of recorded wrecks from the medieval and post-medieval periods, managed by the National Monuments Service, is growing as research continues, the majority of wrecks so far recorded in Ireland's territorial seas date to the eighteenth and nineteenth centuries. Contrary to what many people might expect, the heaviest concentrations of recorded wrecks occur along the east and south coast, rather than the west, despite the relentless pounding of the Atlantic ocean on the western seaboard. This is largely due to the high levels of marine traffic moving through the Irish Sea and St George's Channel, leading to higher numbers of recorded wrecks. A significant number of wrecks are also recorded off the north coast, many relating to losses from submarine warfare during the First and Second World Wars.

The nineteenth century was certainly the low point when it comes to the number of ships wrecked and, critically, the number of lives lost. The Wreck Inventory of Ireland Database has records for over 9,500 wrecks for this period, representing just over 50 per cent of known wrecks in Irish waters. The numbers of lives lost is correspondingly striking and, where records survive, over 8,400 crew and passengers are listed as drowned or missing as a result of shipwrecking in Irish waters. The true figure for such losses may be far higher as the sources do not always record the full number of lives lost. These high numbers are not surprising when one considers that in one storm in November 1807, two naval transport ships – the *Prince of Wales* and the *Rochdale* – were driven ashore in Dublin Bay with the loss of approximately 380 lives in that one incident. Another tragedy occurred in 1854 with the loss of between 290 and 400 lives, out of approximately 528 people on board, when the Liverpool clipper, *Tayleur*, on her maiden voyage carrying emigrants to Melbourne, foundered off Lambay Island.

Some years were worse than others. In 1852, 102 vessels were lost off the Irish coast; in 1855, 127 and in 1856, 155 vessels were lost. This averages at about one wreck every three days, reflecting massive economic losses as well as the tragic loss of human lives. With the gradual replacement of sail by steam power and improvements in shipping technology and safety in general, the numbers of shipwrecks began to decline later in the nineteenth century and during the early years of the twentieth century. This was to change with the outbreak of the First World War, one of the most ferocious periods of conflict and ship loss in the waters around Ireland. The large numbers of shipwrecks that lie scattered to the north and south of Ireland bear testimony to this violent period in Irish and European maritime history and highlight how the horrors of war were never far from the doorstep.

A notable loss during this conflict, and considered one of the worst Irish maritime tragedies, was the sinking of the RMS *Leinster*. The steamer had survived several close encounters with German submarines throughout the First World War, but with just weeks to go before the war ended, tragedy struck. On 10 October 1918, 8km east of the Kish Light Vessel and en route to Holyhead, the RMS *Leinster* was torpedoed and sunk by a German submarine, with the loss of 501 lives. This remains the largest recorded loss of life from an Irish vessel in Irish territorial waters.

As with the First World War, during the Second World War German U-boats patrolling the Atlantic approaches to the north and south of Ireland attacked merchant ships in an attempt to cut off Britain's vital supply of food, oil, weapons and raw materials, in a long drawn-out struggle (see Chapter 23: Maritime Traditions and Institutions, Case Study on Irish Shipping in the Second World War). This became known as the Battle of the Atlantic and it is estimated that there are over 600 wrecks in Ireland's designated waters of the continental shelf dating to this time of terrible conflict. Most of the wrecks are located in the deeper waters of the western approaches to Ireland.

With such a diverse and significant number of wrecks scattered around the Irish coast, there is a real need to ensure their effective management and protection and, where possible, their conservation. In situ preservation is the preferred option if possible, or if not, then by way of record through archaeological research and excavation. No other cultural site can tell us so much about a singular event in time as that of a lost ship; such research uncovers the story of the ship itself, as well as evidence for lives lived and lives lost, against the broader historical or archaeological context. Shipwrecks, when encountered or discovered, are timepieces that leave the imprint of past maritime endeavours on the cultural memory. It is this chance to peer back in time to a single event and to make sense of what happened that accounts for the fascination that shipwrecks hold. It is equally important to appreciate, however, that when undertaking such investigations we ensure they are carried out with due regard to best practice, in a dignified manner and with respect for those who were lost.

MAPPING AND PROTECTING SHIPWRECKS

There is strong national legislation in place to protect historic wrecks and archaeological monuments in Ireland's coastal waters and coastal environments. The National Monuments (Amendment) Act 1987 is the primary piece of legislation that affords automatic protection to all wrecks overone hundred years old and to all archaeological objects underwater. Monuments and archaeological sites within the coastal hinterland are also protected under the same legislation.[11] The National Monuments Service of the Department of Culture, Heritage and the Gaeltacht is active in assessing the known significant wrecks that have been identified in Irish coastal waters over the years, often with the assistance of local divers, dive clubs, local communities, historians and professional archaeologists.

A vast amount of new information relating to the number, location, extent and condition of shipwrecks in Irish waters has come to light in recent years as a result of work carried out during the Irish National Seabed Survey and its successor programme, INFOMAR (see Chapter 9: Underwater Surveys: The INFOMAR project). These comprehensive and far-reaching surveys, carried out under the direction of the Geological Survey of Ireland and the Marine Institute, have captured more than just seabed bathymetry. One of the significant by-products of the multibeam surveys undertaken in the waters off these shores has been the discovery of numerous shipwrecks, many of which came to grief as a result of hostile action during both World Wars. New imagery, with even more detail, has been captured during ongoing surveys by INFOMAR, including that of the SS *Manchester Merchant*, lost off Dingle in 1903 (Figure 22.5), the remains of the anti-submarine drifter, HMS *Guide Me II*, lost off the coast of Dublin in 1918, and the 1915 wreck of the ocean liner, RMS *Lusitania*, which lies in 93m of water, 11.5 nautical miles off the southern coast of Cork.

The collaborative efforts of a number of organisations – such as the National Monuments Service, the National Museum of

Fig. 22.5 3-D MULTIBEAM IMAGE OF THE SS *Manchester Merchant* LYING ON THE SEABED IN DINGLE BAY, COUNTY KERRY. The internal structure of the wreck is visible, with high points at the bow and amidships where the remains of the boilers stand over 5m above the seabed. Also visible is the large scour around the wreck in an otherwise flat, sandy seabed. [Source: INFOMAR Project, a joint seabed mapping project between the Geological Survey Ireland and the Marine Institute]

Ireland, the Geological Survey of Ireland, the Marine Institute, the Heritage Council, the Defence Forces, the Garda Síochána, the Receivers of Wreck, dive clubs, local interest groups, local communities and the public in general – have ensured that the state, as a whole, is now in a much better place to actively protect Ireland's underwater cultural heritage from present and future challenges and threats. The major threats in this regard are the potential effects of climate change, impacts from salvors and treasure hunters or the possibility of negative impacts from development-related works in the marine environment. The ongoing study of shipwrecks will continue to fascinate and intrigue both the enthusiast and professional alike, and many more fascinating wreck discoveries remain to be made in the coastal waters around the island of Ireland in the years to come.

THE SIXTEENTH-CENTURY DROGHEDA BOAT WRECK

Holger Schweitzer

In 2006, the remains of a wooden wreck were discovered in the River Boyne during dredging operations undertaken by Drogheda Port Company. Located approximately 1.5km downriver of the town of Drogheda in the southern half of the navigation channel, subsequent investigations and a full excavation in 2007 showed this to be a sixteenth-century, clinker-built, coastal trading vessel. Since the wreck was located in the main footprint of the dredging works, in situ preservation was not feasible and the wreck had to be fully archaeologically excavated and recovered from the riverbed to ensure it was not destroyed. The excavation and recovery was carried out by the Underwater Archaeology Unit, National Monuments Service, under the direction of the author.

Detailed recording was carried out on site in advance of the excavation, which at times was extremely difficult given the constant flow of maritime traffic and the wreck's location in the main navigation channel. Once the recording was completed, excavation began and the wreck was dismantled underwater, timber-by-timber, and safely recovered to the surface for further recording, research and reassembly. The wreck remains are currently undergoing conservation in the National Museum of Ireland and it is hoped that it will eventually go on full display in an appropriate museum setting in the County Louth area.

The wreck was lying on its starboard side, which was buried in the riverbed, while the upper portion, its port side, having been exposed through the intervening centuries since it sank, was largely eroded and for the most part did not survive. The boat measured

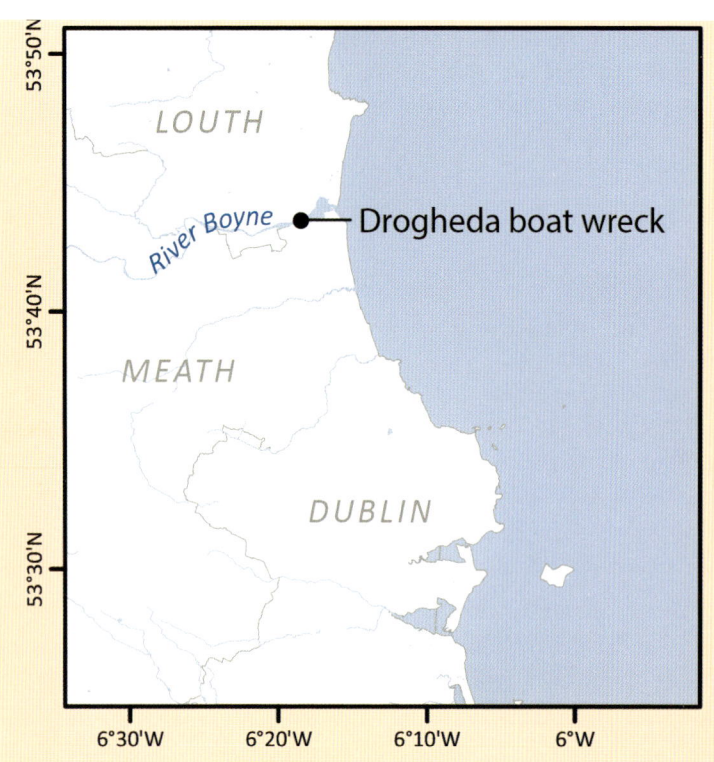

some 9m in length and 3m in width; the stem and sternposts were well preserved and were recovered during the excavation, along with the 6m-lon- keel. The floor and side timbers, the overlapping hull planking, fashioned clinker-style with iron clench nails and rectangular roves – used to fix the planking to the framing timbers – were all preserved intact and in situ. This provided exciting evidence for the continuity of boatbuilding techniques introduced into this part of the world by the Vikings in the ninth and tenth centuries (see Chapter 23: Maritime Traditions and Institutions).

The Drogheda boat had two masts when under full sail, as indicated by the presence of two mast steps. The form of the hull suggests it was designed to operate as a swift coastal trading vessel, with a good cargo capacity for its size. Strikingly, during the course of the excavation of the wreck, the remains of some of the actual cargo on board were discovered within the hold of the wreck in the form of wooden barrels. Although in a collapsed condition, the fourteen casks were well preserved, lying on top of the boat's floor timbers (Figure 22.6). The wooden staves, cants and even the well-preserved remains of the hoops with attendant withies were recovered. Post-excavation analysis has shown the barrels were reused to transport herring, but were originally used as wine casks.

Dendrochronology has provided a date between 1525 and 1535 for the construction of the vessel. These dates were obtained from a number of the floor timbers. There was a strong indication in the tree-ring chronology that the original timber was sourced in Northern Ireland, in the eastern part of County Antrim. Later repairs seem to have been carried out sometime between 1532 and 1560,

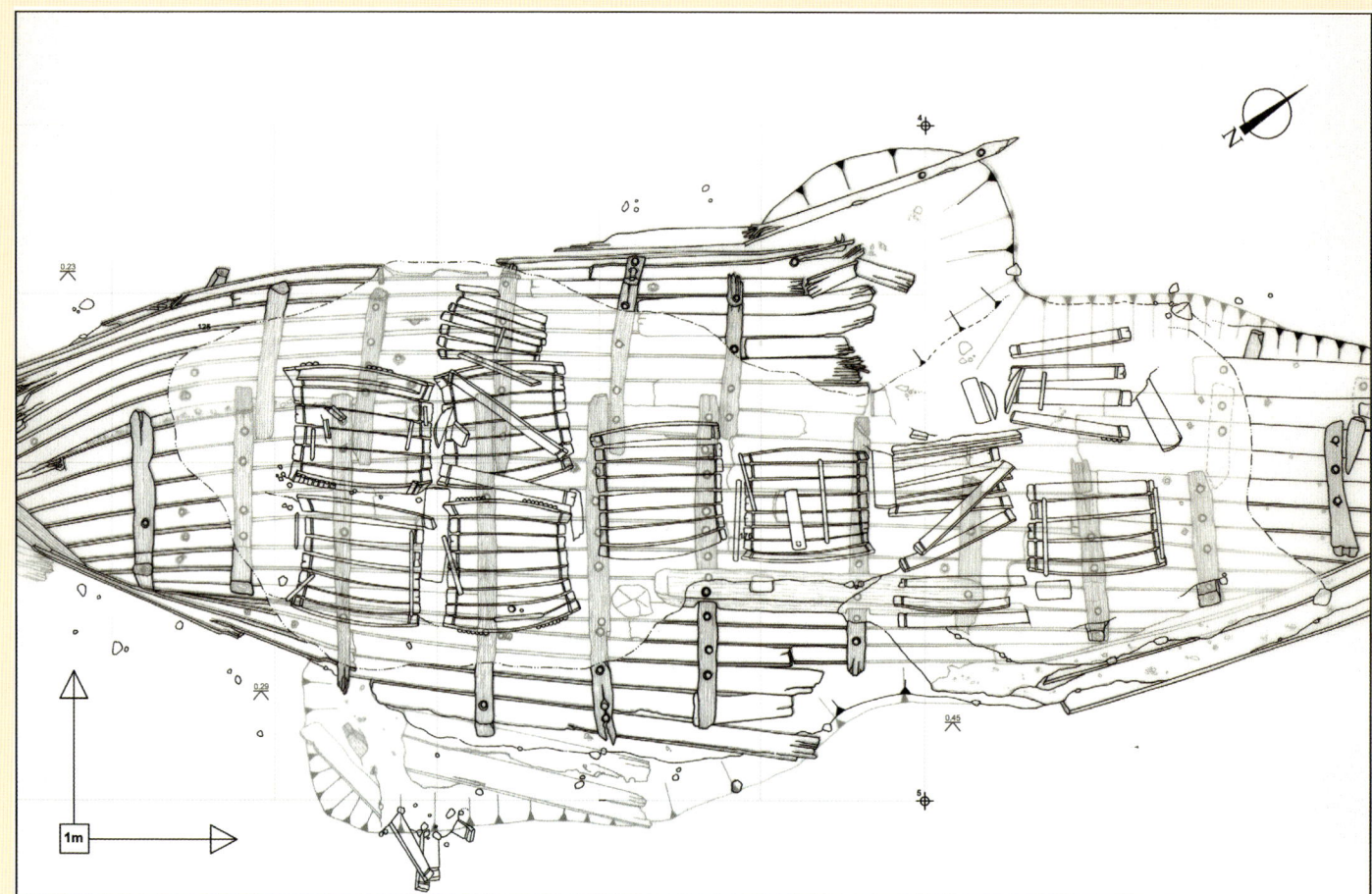

Fig. 22.6 SITE PLANS OVERLAID ON DROGHEDA BOAT WRECK. The image shows the wreck's structural timbers and main hold remains, with two areas of impact on either side caused by the dredger action that led to the wreck's discovery. Overlaid is the plan of the cargo of wooden barrels and their in situ positions within the hold of the wreck when discovered, and prior to excavation. [Source: National Monuments Service, Department of Culture, Heritage and the Gaeltacht]

giving us a good indication of the life span of the vessel. Analysis of the barrel timbers suggests a southern French origin, but this is not surprising considering that they originally contained wine. The Drogheda boat has provided us with evidence not only for boatbuilding technology from a period when we have very little information, but also for the fishing industry and trading connections of late medieval Drogheda around the coast of Ireland and as far afield as France during the sixteenth century.

GORMANSTON LOGBOAT

Niall Brady

While it has long been assumed that the earliest seafaring in Ireland was achieved using logboats and dugout canoes, the discovery of such craft in the marine environment eluded researchers until relatively recently. In 2002, archaeological monitoring undertaken by the Archaeological Diving Company (ADCO) of a subsea interconnector project successfully identified timbers at Gormanston, County Meath. Subsequent underwater inspection and excavation revealed this to be the remains of a logboat (Figure 22.7). Preliminary data suggested that the logboat originally measured *c*.7m in length and was located *c*.500m offshore. Dredging activities had an impact on the vessel, resulting in the recovery by the monitoring archaeologist of a series of timber fragments. At this point, dredging activity was suspended. Subsequently, ADCO investigated and excavated the remains of the wreck, exposing an intact portion of the logboat measuring 4.2m long and 1.1m wide. The boat was buried under 2m of sand and lay inverted atop marine till. The wooden vessel was made from a single oak bough, the interior of which was hollowed out to 70cm in depth. A defined bow section is preserved. Radiocarbon analysis dated the timber to 1,193–1,013 calendar years before present, which places it in the Later Bronze Age (*c*.1200–500 BC). There is no obvious indication of marine transgression along this stretch of coast during prehistory, suggesting that the shoreline since this time has remained unchanged.

The location of the find is revealing. Gormanston today is part of a long sandy beach that is partly used as a military coastal defence firing range. It is also immediately north of Bremore Point, on which is sited a series of megalithic burial tombs. A second cluster of passage tombs survives to the north in Gormanston,

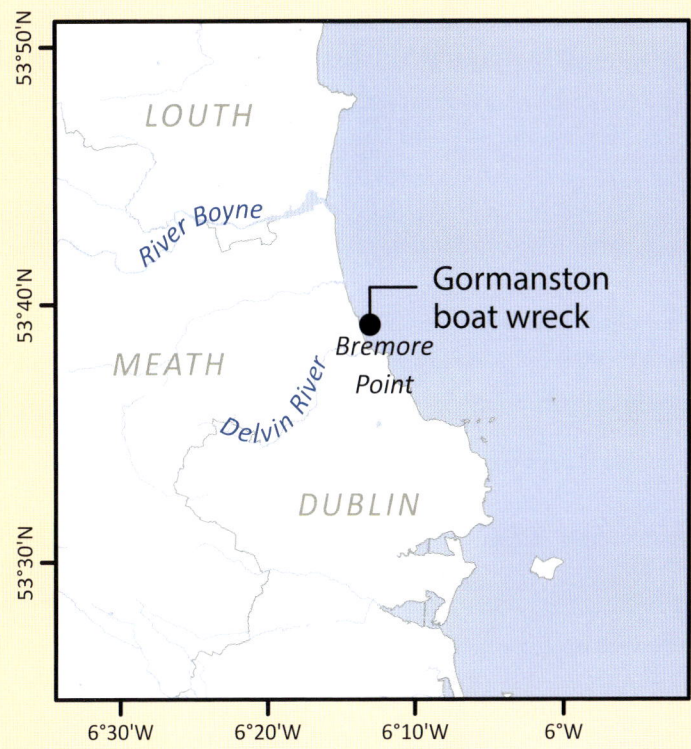

Fig. 22.7 (right) THE GORMANSTON LOGBOAT RECOVERY FOLLOWING ARCHAEOLOGICAL EXCAVATION BY THE ARCHAEOLOGICAL DIVING COMPANY (ADCO). The large wooden craft was discovered in 2002 during archaeological monitoring for the subsea interconnector project between Ireland and Scotland. The logboat has been dated to the Late Bronze Age. [Source: The Archaeological Diving Company Ltd]

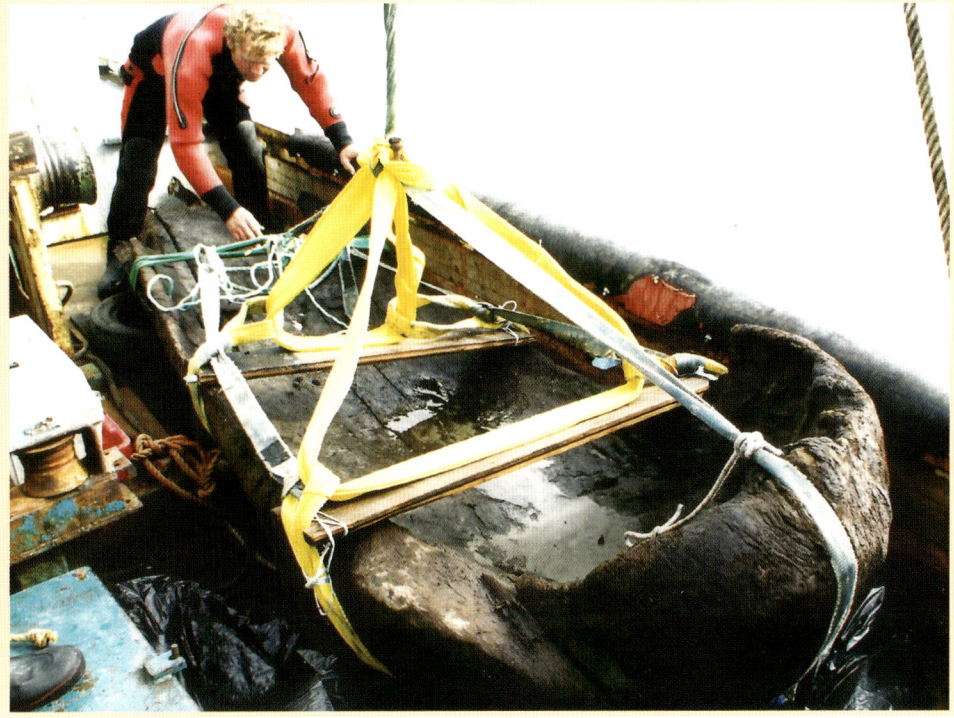

lying on the far side of the River Delvin, which forms the county boundary. Boulders on the beach may suggest the location of still other sites that have eroded into the sea. The extended complex of burial sites belongs to the Neolithic period (c.4000–3800 BC). The fact that early settlers chose a coastal promontory that straddles a river estuary, which is subsequently an important territorial boundary, calls to mind the entrance to the River Boyne further north. There, monolithic remains also stand watch over the river's entrance, as sentinels to the Boyne Valley complex, several kilometres upstream. These topographical associations highlight the importance of maritime and riverine resources during prehistory. Archaeological field-walking on Bremore has identified worked stone and other lithic tools of Neolithic date lying on the surfaces of present-day fields. Bremore remained a location of interest, and a series of low, circular, ditched enclosures were recorded south of the main tomb. Such small enclosures are characteristic of burial monuments of the Bronze Age. The discovery of the logboat just offshore fits the narrative of the strategic importance of the location and provides the first tangible proof for marine exploration and exploitation and the means by which early mariners traversed Ireland's east coast.

La Serveillante: 1797 WRECK OF A FRENCH ARMADA FRIGATE

Colin Breen

At the head of Bantry Bay, at a depth of 34m, lies one of the best-preserved historic shipwrecks in Irish coastal waters, the French frigate *La Surveillante*. Originally built at Lorient (north-west France) in 1778, the vessel was just under 44m long and 11m wide, had three masts and two main decks. It carried thirty-two guns and a crew of over 200 men. In 1779 it was one of only twelve French ships that were copper sheathed (in order to protect its hull from marine organisms), and remains of this copper sheathing are still visible on the wreck site today. The vessel played an important, but ultimately disastrous, role in an ill-fated French attempt to land an expeditionary force in Ireland to overthrow English rule. Theobald Wolfe Tone and the United Irishmen had persuaded the revolutionary government in France to launch an armada of forty-eight ships under the command of Morard de Galles from Brest, in December 1796. By the time the fleet reached the coast of Ireland it had been scattered by storms and only thirty-five ships made it to Bantry Bay, in County Cork. Further foul weather led to the loss of a number of vessels and the aborting of the invasion attempt. *La Surveillante* had arrived into Bantry Bay in such poor condition that the ship was abandoned and scuttled on 2 January 1797, to prevent it falling into English hands.

Following the 1979 Whiddy Island disaster, when the *Betelgeuse* oil tanker exploded at the island's terminal with the loss of fifty lives, the eighteenth-century shipwreck was rediscovered during seabed clean-up operations, just under 200 years since it was deliberately scuttled. The wreck is located 1.3km off the northern coastline of the island. Its partial burial in silt and the encasing of the vessel's hull in copper has ensured its extraordinary survival. Its bow stands to a height of 4.3m above the seabed. Within the wreck an assortment of ship's fittings, armaments and artefacts are partially visible, exposed within the fine silts. In the centre of the vessel lies a mound of brick and iron as well as a large section of anchor chain. An upturned anchor protrudes from the mound. This was likely the vessel's spare and was originally attached to the mast at this position. Fourteen guns have been identified across the site close to their contemporary position on the ship. A series of iron sections of the vessel's bilge drainage and pump system are scattered around the wreck. One of the ship's original bells was found underneath two of these flanges. It was recovered and is now housed in the National Museum of Ireland. Other artefacts include wooden musket stocks and rigging elements. At the eastern end of the wreck, the ship's stern is less well preserved and is partially separated.

The importance of the wreck of *La Surveillante* lies not just in the state of its preservation but also in the role it played in shaping succeeding events around Bantry Bay. As a direct consequence of this and subsequent invasion attempts, the landscape around the bay was heavily fortified with a series of Martello and signal towers in recognition of Bantry's strategic maritime importance.

Case Study: Encounter with the Irish Coast: The 1588 wrecks of the Spanish Armada

Connie Kelleher, Fionnbarr Moore and Karl Brady

… but take great heed lest you fall upon the Island of Ireland for fear of the harm that may happen unto you upon that coast.'[12]

Fig. 22.8 A CONTEMPORARY PAINTING OF ENGLISH SHIPS AND THE SPANISH ARMADA ENGAGING EACH OTHER IN THE ENGLISH CHANNEL, AUGUST 1588. The ships are shown under full sail, with some in distress and sinking. Of the 130 ships that set sail in the summer of 1588, just over half returned home. Of those lost, six wrecks have been found in Irish territorial waters, although there are likely to be others. Irish archaeological investigations have yielded artefacts that provide an insight into the wider world of the ships, their occupants and the theatre of conflict in which they played a part. [Source: BHC0262, National Maritime Museum, Greenwich, London]

HISTORY OF THE LOST SHIPS

The wrecking of the ships of the Spanish Armada is a tale of human tragedy, maritime loss and defeat (Figure 22.8). There is no better contemporary account to illustrate this than that written in Antwerp in 1589 by Spanish Captain, Francisco de Cuellar, who was on board *La Lavia* when it sank at Streedagh in County Sligo.[13] He provides a graphic description of the wrecking there of three ships, tells of his survival amid the carnage that took place on shore and recounts his experiences in Ireland and his subsequent escape, by way of Scotland, to the Spanish Netherlands and then home to Spain.

Of the 130 ships that set sail in the summer of 1588, just over half returned home, with the others scuttled, burned, captured or wrecked. Of those lost, as many as twenty-six may have been wrecked off the western coast of Scotland and the north, west and south-west coast of Ireland. The remains of just six in Irish territorial waters have been located and investigated, while the general location of some twelve others is known.

The shipwrecks so far identified and investigated include the galleass, *La Girona*, lost off the coast of Antrim in Northern Ireland, and the requisitioned Venetian merchantman, *La Trinidad Valencera*, discovered in 1971 in Kinnagoe Bay, County Donegal. In addition, in the 1980s three wrecks were located off Streedagh Strand, County Sligo. These were the Venetian-built *La Lavia*, the Ragusan ship, *Santa María de Visón* and the Catalan-built ship, *La Juliana*. The sixth wreck is the remains of the 945-tonne *Santa Maria de la Rosa* in the Blasket Sound, off the coast of Dingle in County Kerry.

Details of the large number of other vessels wrecked off the Irish coast are vague. Efforts to interpret the contemporary sources can be difficult, often leading to confusion and conflicting theories regarding the exact location and identity of many of the vessels wrecked. The first loss of ships is recorded in early September 1588, when two were presumed wrecked at sea north of Donegal. These were the

Fig. 22.9 A CONTEMPORARY VIEW OF KINNAGOE BAY, COUNTY DONEGAL, WHERE THE SPANISH ARMADA SHIP, *La Trinidad Valencera*, WAS WRECKED IN SEPTEMBER 1588. The ship was the fourth largest in the Armada, and formed part of the Levant Squadron. A requisitioned Venetian merchantman of 1,100 tonnes, *La Trinidad Valencera* ended its days in the shallows of the bay when trying to outrun one of the major storms of that year. Most of those on board survived the wrecking, but were captured and put to death upon reaching shore at Kinnagoe. [Source: K. Brady, Underwater Archaeology Unit, National Monuments Service, Department of Culture, Heritage and the Gaeltacht]

storm-battered 600-tonne *Barca de Amburg* with 263 men on board, and the 750-tonne *Castillo Negro*. The crew of the *Barca* transferred to *La Trinidad Valencera* and *El Gran Grifon*. Shortly afterwards, *La Trinidad Valencera*, which had been sailing with the previous two vessels, took refuge in Kinnagoe Bay (Figure 22.9), where it was subsequently wrecked, with the loss of forty people on board. The remaining 350 or so crew made it ashore but were later captured by the English soldiers stationed in the area and most were slaughtered. Some, however, managed to escape while the more important officers were transported to Drogheda, and later to London, for ransom.

Five Armada ships are believed lost off the Mayo coast, including one at the treacherous inlet of Inver in Broadhaven Bay. Attempts to retrieve cannons from the site in the 1630s seem to confirm there was an Armada vessel lost at this location but the wreck has yet to be found. Certain published accounts support the theory that the vessel was the 600-tonne *Santiago* with a crew of eighty-six men, while others argue that it was more likely to be the *San Nicholas Prodaneli*. Tradition also has it that another vessel was lost at the entrance to Broadhaven Bay, at Kid Island, but no wreck site has been found there to date.

Fig. 22.10 INIS TUAISCEART (INISHTOOSKERT), COUNTY KERRY. Also known as *An Fear Marbh* (the dead man or the sleeping man). Inis Tuaisceart is one of the Blasket Islands to the immediate west of Blasket Sound, off the Dingle coast in County Kerry. Here, in September 1588, two Spanish Armada ships were lost: *Santa Maria de la Rosa* and *San Juan*. There was only one survivor from the ships, a youth by the name of Giovanni de Genoa, who was captured when he made it ashore. After much interrogation he was put to death by the resident English agent, and is believed to be buried locally. [Source: C. Kelleher, Underwater Archaeology Unit, National Monuments Service, Department of Culture, Heritage and the Gaeltacht]

El Grand Grin, the 1,160-tonne vice-flagship of the Biscayan squadron, was reported lost in the environs of Clare Island in Clew Bay, a part of the coast where the sheer volume of islands alone acts as a natural hazard to ships, particularly those unfamiliar with the coastline on a stormy night. However, it is disputed whether the ship was lost off the island itself, off the Corraun Peninsula or at sea. To make matters more confusing, there is also lore of another Armada vessel lost in the area, possibly also in Clew Bay or off the Corraun Peninsula.

There is good documentation for the ship *La Rata Santa Maria Encoronada* recorded as having wrecked off Fahy Castle, Tullaghan Bay in County Mayo on 21 September 1588. After seeking shelter in Blacksod Bay, she dragged her anchors and grounded in the shallows below Fahy Castle. On hearing that another Armada ship, the *Dequesa Santa Ana*, was anchored nearby in Elly Bay, Captain Don Alonso De Leiva and his crew made the 40km trip with the hope of making it back to Spain safely. This proved to be disastrous, however, as the *Dequesa Santa Ana* was lost in late September, having made it as far north as Loughros More Bay, it ran ashore at Tramore Beach in County Donegal.

Off the Connemara coast, two vessels were reported lost. Over the years, scholars have similarly argued over their identities and exact places of loss, although it is generally accepted that the 300-tonne hulk, *Falcon Blanco Mediano*, was wrecked near Ballynakill Bay, possibly off Freaghillaun Island. A second vessel was reputedly wrecked at Mace Head, near Carna on a small islet called Duirling na Spáinneach. The likely candidate for this was *La Concepcion*, a 418-tonne ship of the Biscayan Squadron and captained by Juan Delcano.

The County Clare coast also took its toll on ships making their lonesome journey homeward. Both the 936-tonne *San Esteban* and the smaller 790-tonne *San Marcos* were sailing together when they were lost. The *San Marcos* went down somewhere between Mutton Island and Lurga Point, and the *San Esteban* near Doonbeg, at White Strand. Off Scattery Island, near the mouth of the Shannon, another Armada vessel, possibly the 703-tonne Ragusan ship, *La Anunciada*, considered unseaworthy, was scuttled and burnt.

Before departing the coast of Ireland, ships that made it to the south-west encountered the unforgiving coast of Kerry. Here, several ships are listed as having been lost. At Barrow Harbour, or perhaps Tralee Bay, one of the smaller *pataches* or *zabras* was wrecked, the identity of which is still disputed. Perhaps the most notable loss occurred within the Blasket Sound (Figure 22.10). Inspector general and paymaster of the galleons of Castile, Marcos de Aramburu, who was on board the *San Juan Bautista*, provided a vivid account. He made it back to Spain and recounted the probable loss of the smaller merchantman, *San Juan* (under commander Diego Flores and Captain Fernando Horra). He was also eyewitness to the sinking of the larger *Santa Maria de la Rosa*. Of the latter he wrote: '... she went down right away with every man on board, not a soul was saved, a most extraordinary and frightening thing'.[14]

THE ARCHAEOLOGY REVEALED

Though much has been written historically, a comprehensive treatment of the archaeology of the Armada shipwrecks remains to be compiled. Colin Martin's work has ensured that the investigation of Armada wrecks undertaken to date has been assessed scientifically and documented, and he has placed their story within the wider historical context of the period. Beginning with his work on the challenging site of *Santa Maria de la Rosa* in the 1960s, Martin went on to carry out extensive excavations on the site of *La Trinidad Valencera* in the 1970s. The majority of the material recovered from the latter wreck is now on view in the Tower Museum in Derry, while a selection of guns is on display in the National Museum of Ireland. Of the remaining wrecks discovered, the majority of the excavated material is housed in the Ulster Museum, Belfast.

Such material reflects that key moment in time when these objects, and the personal possessions of the mariners and soldiers on board, were sealed in their watery graves. These artefacts bring us back to those moments when the ships sank and provide an insight into the wider world of the ships, their occupants and the theatre of conflict in which they played a part. Stunning artefacts – like the exquisite gold and ruby-encrusted salamander from the *Girona* site or the gold Agnus Dei Reliquary and gold cross of the Order of Santiago – provide a glimpse of the wealth of some of the personnel on board and the

Fig. 22.11 THE FIGURE OF ST MATRONA EMBLAZONED ON ONE OF THE LARGE BRONZE GUNS RECOVERED FROM THE 1588 SPANISH ARMADA WRECK SITE OF *La Juliana* IN JULY OF 2015. The saint is shown holding a ship in one hand and a cross in the other. Matrona was a young saint of the third or fourth century and was martyred for her faith when caught administering to the sick. She is venerated particularly in Barcelona and Catalonia and by mariners. [Source: K. Brady, Underwater Archaeology Unit, National Monuments Service, Department of Culture, Heritage and the Gaeltacht]

Fig. 22.12 A SIEGE CARRIAGE WHEEL FOUND DURING AN ARCHAEOLOGICAL EXCAVATION IN JULY 2015 ON THE SITE OF THE SPANISH ARMADA WRECK, *La Juliana*, STREEDAGH, COUNTY SLIGO. The wheel is one of nine identified that were exposed on the seabed. They formed part of the large siege train carriages to carry the guns intended for use to conquer London and were stored in the hold of the ship during passage across the Atlantic. One wheel was raised to undergo conservation in the National Museum of Ireland. [Source: K. Brady, Underwater Archaeology Unit, National Monuments Service, Department of Culture, Heritage and the Gaeltacht]

strength of their religious beliefs. The beautifully ornate Remigy siege gun from *La Trinidad Valencera*, along with cannon recovered from the Sligo wrecks in the 1980s and most recently in 2015, with their ornate saintly figures, hint at the enormous power and significance of these lost ships (Figure 22.11). Equally, they are a source of fascination for the museum visitor, providing tactile evidence for tragic events in the past. The pewter Matute plate from *Santa Maria de la Rosa*, along with less familiar items from *La Trinidad Valencera* – including an intact olive jar, brass forks, candlesticks, buckles and remains of gaiters – tell the story of daily life on board these sailing ships. They evoke the tragic loss of life of the many who went down with the vessels, or were cruelly cut down when they crawled ashore on the lonely, remote beaches of Ireland in 1588.

RECENT DISCOVERIES AND RECOVERIES

For five weeks during late spring and over the summer of 2015, the Underwater Archaeology Unit (UAU) of the National Monuments Service carried out a survey and excavation at one of the three wreck sites off Streedagh Strand, County Sligo. The work focused on the remains of the Catalan-built ship, *La Juliana*. The remains of gun carriages and nine carriage wheels, four of which were large enough to support siege guns, lay exposed on the seabed, following winter storms (Figure 22.12).

A large articulated section of the hull of the ship itself was visible on the seabed, overlain in parts by sheets of concreted cannonballs and two large iron anchors that came to rest there as the ship broke up over 400 years ago. One of the cauldrons was recovered from the wreck site along with nine bronze guns of varying sizes and calibres. Seven of the guns bear the date of 1570, indicating their year of forging in an Italian foundry. They also retain symbols of saints, including Saintt Matrona, patron saint of Barcelona (Figure 22.13), Saint Rocho, shown with his dog, and Saint Sebastian in his martyred pose. The largest of the guns, a Sicilian-manufactured demi-cannon, displaying the symbol of Saint Peter holding the keys to heaven, retains superb floral motifs, sun symbols and embossed flame designs (Figure 22.14).

Fig. 22.13 ST MATRONA GUN DURING RECOVERY IN JULY 2015 FROM THE SITE OF SPANISH ARMADA WRECK, *La Juliana*. The gun is a large saker cannon, forged in the year 1570 by Dorina II Gioardi, the sixteenth-century Genoese gun founder. It formed part of the original gunnery of the ship. [Source: K. Brady, Underwater Archaeology Unit, National Monuments Service, Department of Culture, Heritage and the Gaeltacht]

Fig. 22.14 SAINT PETER CANNON DURING ARCHAEOLOGICAL EXAMINATION PRIOR TO RECOVERY IN JULY 2015. The bronze gun, a demi-cannon, displays floral and flame motifs and the saintly figure of Saint Peter holding the keys of heaven. It is a testament to the quality of work of the sixteenth-century Sicilian gun founder, Federico Musarra. [Source: K. Brady, Underwater Archaeology Unit, National Monuments Service, Department of Culture, Heritage and the Gaeltacht]

CONTINUED FASCINATION AND FUTURE WORK

The wrecks of the Spanish Armada continue to attract attention. Local initiatives underway in County Sligo and in County Clare are keeping the story alive and highlighting the need to protect these sites and to incorporate them and the Irish dimension to their story into the received narrative of history. More discoveries will undoubtedly be made in the future, given the number of Spanish Armada ships lost around the Irish coast. The scientific investigation of these sites will continue to add to our knowledge of the tragic events of that campaign, and to the wealth of archaeological material that such wreck sites retain. The story of the wrecking of the Spanish Armada ships is one underscored by the power of the natural environment and the unforgiving nature of Ireland's rugged coastline.

Case Study: Ireland and the First Battle of the Atlantic

Karl Brady

The First World War was a bloody and brutal conflict and led to the loss of up to 17 million lives, the destruction of numerous towns and cities around the world and long-lasting economic, social and political consequences that continue to reverberate to this day. The conflict was not just confined to the main theatres of war in Europe, Africa and the Middle East. A vicious campaign of naval conflict also occurred on the high seas between the warring nations, resulting in the loss of thousands of lives on both sides, with the sinking of over 5,000 Allied and neutral ships and hundreds of Axis vessels, including 180 U-boats. These wrecks lie dotted around the globe, with the highest concentrations in the Mediterranean, North Sea and North Atlantic.

The waters around Ireland and Britain did not escape this conflict, as both Britain and Germany attempted to starve each other into submission through naval blockades, including the use of a deadly submarine offensive by Germany. Such was the intensity of this naval conflict in the Atlantic and in the waters off Britain and Ireland that it became known as the First Battle of the Atlantic.

The high number of First World War wrecks off the Irish coast is a tangible reminder that much of this war took place within sight of these shores (Figure 22.15). When contemplating terms like the front line or Western Front, which are normally reserved for the war on land, we should perhaps consider that in reality the battle front lines extended to the waters and shorelines of both Ireland and Britain.

While only a small number of ships were sunk off Ireland in the

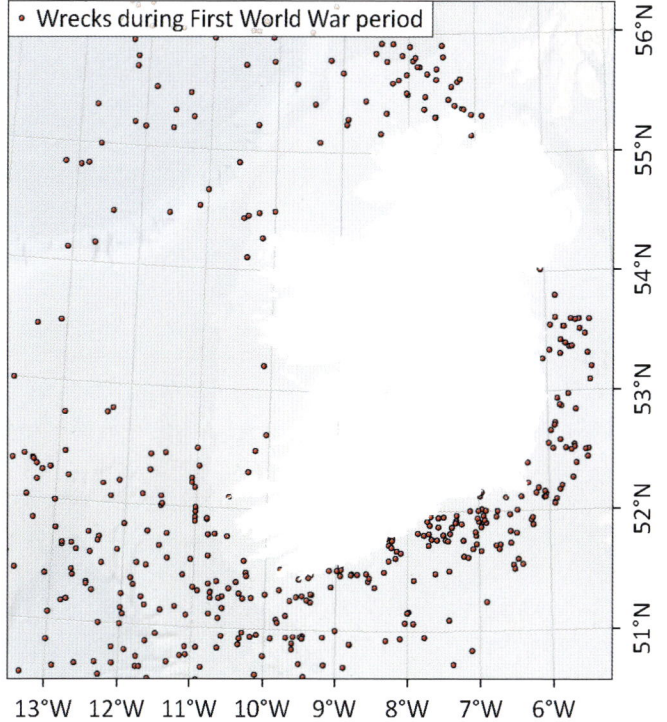

Fig. 22.15 FIRST WORLD WAR WRECKS WITH LOCATIONS (BOTH KNOWN AND APPROXIMATE) OFF THE IRISH COAST. Over 1,000 ships were lost during the First World War period, with huge loss of life. [Source: Wreck Inventory of Ireland Database, National Monuments Service, Department of Culture, Heritage and the Gaeltacht]

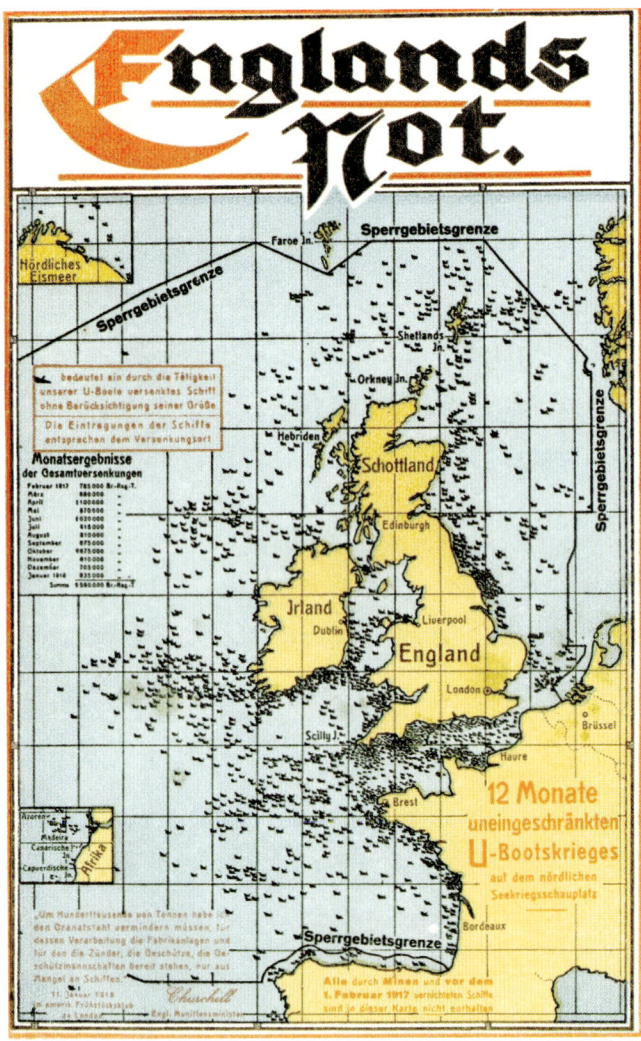

later months of 1914, they did include some significant losses, including the super dreadnought, HMS *Audacious,* and the SS *Manchester Commerce.* However, it was not until February 1915, when the Kaiser declared the waters surrounding Great Britain and Ireland a war zone, that there was an escalation in the number of ships attacked and sunk. The heaviest concentrations of wrecks lie to the north and south of Ireland. This was where the main shipping lanes and convoy routes from North America to Britain converged and, consequently, where U-boats centred much of their operations during the war. The devastating success rates of U-boat strikes resulted in these northern and southern approaches becoming known as 'killing lanes'. Three months after the Kaiser's war zone declaration, the RMS *Lusitania* was torpedoed and sunk by the *U-20* off the coast of Cork, with the loss of 1,198 lives. This incident – along with the sinking of the SS *Arabic* (forty lives lost), 81km south of the Old Head of Kinsale, and the sinking of the SS *Hesperian* (thirty-two lives lost), 137km south-west of Fastnet – made the Germans re-think their strategy and temporarily restrict their attacks on merchant or neutral shipping for fear of encouraging the US to enter the war.

Throughout 1916 the number of ship losses increased, with the Kaiserliche Marine inflicting enormous losses on Allied shipping. By the end of that year, some one hundred ships had been sunk in the coastal waters in or around Ireland. By far, the worst year for ship losses in Irish waters was 1917 (Figure 22.16), with a further 600 ships sunk as a result of unrestricted submarine warfare, which had resumed in February of that year. The patrol log of the *UC-65* illustrates the effectiveness of this strategy and the level of destruction that U-boats were inflicting on shipping in the Irish Sea and St George's Channel. In one month alone (March 1917), *UC-65* sank over twenty ships, destroying vital supplies for the British war effort and incurring the inevitable loss of many lives. An example of this devastating impact is illustrated by the activity of *UC-65* over four days in that month, when it sank five ships off Wicklow and Wexford: the SS *Ennistown,* torpedoed on 24 March 1917;

Fig. 22.16 FIRST WORLD WAR GERMAN PROPAGANDA POSTCARD. The postcard celebrates the sinking of almost 10 million tonnes of Allied shipping during a twelve-month period from February 1917 to January 1918 in the zone of unrestricted submarine warfare in waters around Ireland, Britain and France. [Source: the Ian Lawler Collection]

the *Adenwen,* torpedoed without warning on 25 March and the SS *Dagali,* SS *Wychwood* and LV *Guillemot,* all sunk on 28 March 1917. In all, thirteen lives were lost during these attacks.[15] The *UC-65* sank 115 ships in total during its ten months of service until it was itself torpedoed by British submarine *CI5* in November 1917. The highest number of losses for any period throughout the war occurred in April 1917, with 100 ships sunk and hundreds killed.

Fig. 22.17 THE RMS *Leinster.* Left: Postcard of the RMS *Leinster* departing Dún Laoghaire Harbour. [Source: National Monuments Service, Department of Arts, Heritage and Gaeltacht]; Right: The RMS *Leinster* in dazzle camouflage, painted by William Minshall Birchall (1884–1941). [Source: British Mercantile Marine Memorial Collection]

Fig. 22.18 MULTIBEAM IMAGE OF THE WRECK OF THE RMS *Leinster*. The RMS *Leinster* was torpedoed and sunk by a German submarine in October 1918. Inset: Pre-war photograph of the crew. [Source: INFOMAR project (a joint seabed mapping project between the Geological Survey Ireland and the Marine Institute) and University of Ulster; inset: courtesy of www.rmsleinster.com]

Over seventy Irish-based ships and boats were lost throughout the war, some as a consequence of bad weather, but most due to war action. One notable Irish loss was the SS *W.M. Barkley*. The Guinness-owned steamer was torpedoed without warning off the coast of Dublin in October 1917, with the loss of four lives, including the ship's master and all of its cargo of stout (see Chapter 9: Underwater Surveys: The INFOMAR project). Other Irish vessels lost were the fishing ketch the *Geraldine* (five lives lost), the City of Cork Steam Packet Company's steamships the *Innisfallen* (ten lives lost) and the *Inniscarra* (twenty-eight lives lost) and the City of Dublin Steam Packet Company's RMS *Leinster* (501 lives lost) (Figures 22.17 and 22.18). All highlight the human dimension to the deadly conflict, which had a lasting impact on Irish society and affected many Irish families throughout the country for generations to come.

The introduction of the convoy system, combined with vital support from the United States, which had entered the war in April 1917, helped stem the tide of destruction of Allied shipping, which had brought Britain to within weeks of defeat. Ship losses for 1918 (150 in total were far less than in 1917, and continued to reduce until peace was declared in November 1918. Overall, the seaborne conflict had devastating consequences. By the end of the war, over 1,000 ships had been lost in Irish waters and along the western approaches.

The role the First Battle of the Atlantic played in determining the outcome of the war has been recognised and, in this regard, wrecks of this period form an important element in the narrative of the First World War, as well as in the general maritime history and archaeology of Ireland. This is evident in the sheer number and distribution of wrecks on the seabed, which provide a visceral and tangible link to the dramatic events that were unfolding off the coast at that time.

The high numbers of lives lost illustrate that apart from wrecks as important underwater heritage sites in their own right, many of these sites are war graves and should be respected as such, and treated accordingly. Furthermore, apart from being an important aspect of Irish maritime heritage, First World War shipwrecks are also linked to the heritage of the many different countries involved, both directly and indirectly, in the global conflict. This fact is reflected in the high number of foreign vessels sunk in Irish waters. It is only through the preservation and study of such wrecks that the reality of the events of the time can be fully understood and appreciated. Tragedy and loss characterised the coastal waters of Ireland at that time, when the North Atlantic and the Irish territorial sea were a theatre of war.

THE PROTECTED WRECK OF THE RMS *Lusitania:* MANAGEMENT, PROTECTION AND PRESERVATION OF UNDERWATER CULTURAL HERITAGE

Fionnbarr Moore

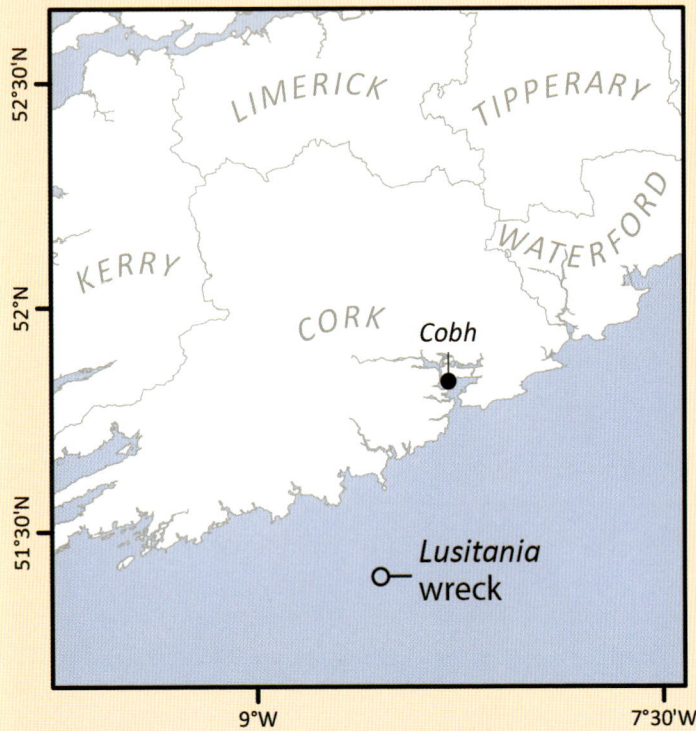

Fig. 22.19 ARTIST'S DEPICTION OF THE RMS *Lusitania* FOLLOWING IMPACT FROM THE GERMAN TORPEDO. The funnels were painted a grey-black, rather than their usual red and black, in an effort to make the liner less conspicuous during its Atlantic crossings, due to the threat of submarine activity in the area. A large explosion on the starboard side is also depicted, with smoke and ash billowing from the four funnels as the ship lists to starboard, before sinking beneath the waves. [Source: Brian Cleare]

In 2015, the centenary year of the sinking of the *Lusitania* (Figure 22.19), the president of Ireland, Michael D. Higgins, gave the keynote speech at the commemorative events held in early May in Cobh, County Cork. Today, the wreck lies on its starboard side at a depth of *c.*93m (Figure 22.20). The official death toll is given as 1,198 lives lost, including three stowaways. Only 289 bodies were recovered, most of which were interred in the Old Cemetery at Cobh. The remains of many others are entombed within the wreck itself, making it their final resting place.

The wreck of the *Lusitania* is significant in terms of Ireland's underwater cultural heritage for a number of reasons. While not designated as such, the enormous loss of life as outlined above would qualify it as a war grave. The passenger list shows the international impact of the tragedy. As well as those from the USA and Ireland, there were many listed as being from Canada, Belgium, Russia, Norway, Persia and Britain, in particular, who lost their lives. Many Irish were also listed as British, Canadian or American and many Poles as Russian. The ship's importance as a technical marvel of the age, being the largest and fastest liner ever built at the time, forms part of its allure for divers. The many unanswered questions that surround the cargo it was carrying and its role in influencing the eventual participation of the USA in the First World War continue to excite debate and inspire the production of numerous publications and documentaries.

The cause of a reported second explosion on the ship has attracted several explorative ventures over the years. Mr F. Gregg Bemis jnr, whose ownership of the wreck was upheld in the Admiralty Court of Norfolk, Virginia, in 1995 and in the High Court in Dublin in 1996, has promoted much of this research. Despite certain legal issues that arose between the state and the owner in relation to licensing since the Underwater Heritage Order (UHO) was placed on the wreck site in 1995, licences have been granted annually to the owner, his agents or other interested parties to undertake investigations that might help answer some of the research questions in relation to the vessel and its sinking. In 2011, project archaeologists and a conservator were engaged by *National Geographic* to draw up an archaeological and conservation methodology for an expedition and television documentary to be carried out under a five-year licence granted to Mr Bemis in 2007.

Fig. 22.20 MULTIBEAM IMAGE OF THE WRECK OF THE *Lusitania*. Acoustic image acquired by the INFOMAR crew on board the RV *Celtic Voyager* six hours after survey. [Source: INFOMAR project, a joint seabed mapping project between the Geological Survey Ireland and the Marine Institute]

This was the first time that an archaeological approach was brought to the investigation of the wreck. The benefits were clear to be seen, especially in the detail of the reporting and the subsequent analysis and conservation of the artefacts retrieved on the back of the expedition. In 2013, a memorandum of understanding was agreed between the National Monuments Service and the owner, Mr Bemis, with regard to present and future research projects on the wreck of RMS *Lusitania*. Mr Bemis was granted a three-year licence with a view to seeking funding for a new project aimed at finally resolving the mystery of the second explosion.

The National Monuments Service (NMS) maintains an ongoing programme of monitoring of the wreck site, and regulates diving and other activities on it through the licensing mechanism. In 2017, an archaeologically supervised and licensed expedition recovered the ship's telegraph, demonstrating the value of ongoing investigation (Figure 22.21). There is a need for further scientific analysis of the wreck's structural integrity with a view to developing predictive models for its potential to survive as a monument on the seabed into the foreseeable future. Recent seabed mapping of the wreck by the National Seabed Survey/INFOMAR, on behalf of the NMS, has enhanced our understanding of how the wreck is lying on the seabed and may form the basis for further targeted surveys. These will help build up knowledge of the wreck and perhaps also answer some of the outstanding questions that surround the rapid sinking of this once great liner following a single torpedo strike.

The RMS *Lusitania* has been a protected wreck site since 1995. An Underwater Heritage Order (UHO) was placed on it under Section 3 of the National Monuments 1987 (Amendment) Act by the then minister for arts, culture and the gaeltacht, Michael D. Higgins. The UHO allows for the protection of wrecks that are less than one hundred years old, based on their 'historical, archaeological or artistic importance'. A UHO stipulates that anyone seeking to undertake diving or other investigations aimed at the exploration of the site must apply to the NMS for a licence to do so. As the wreck is now over one hundred years old, the provisions of the National Monuments Act protecting such wrecks also apply to the RMS *Lusitania*.

Fig. 22.21 A DIVER EXAMINES THE TELEGRAPH AT THE WRECK OF THE RMS *Lusitania*. The telegraph was recovered by the dive team during an archaeologically supervised and licensed expedition to the wreck in July 2017. [Source: Barry McGill]

MARITIME TRADITIONS AND INSTITUTIONS

Daire Brunicardi

Asgard II. Sail-training vessel, brigantine *Asgard* II, was built by Jack Tyrrell in Arklow, County Wicklow, in 1981. Sailed by a permanent crew of five, and with twenty trainees on board for any given trip, *Asgard* II provided an experience of life at sea to thousands of young Irish people from all across the country. She was a familiar sight in Irish ports and harbours around the coast, until she sank in the Bay of Biscay in 2008. She was named after *Asgard*, a yacht used for gun running for Irish Volunteers in 1914. [Source: Brian Cleare]

Ireland is intrinsically linked to the sea, which has connected it, from earliest times, to neighbouring islands, to the European mainland and further afield. Ireland is shaped, not only by the physical features of the island and the seas and oceans surrounding it, but also by its people in their diversity and their connections with the world at large. The sea has provided the highway by which this connection was made and is maintained. The surrounding seas provide wealth and sustenance, contributing to people's wellbeing and to the lore and practices evoked by a maritime past. It is therefore essential that the seas, those who work on them and the traditions they have inspired be protected, nurtured and encouraged.

This chapter gives a flavour of how Irish people have interacted with the sea as individuals and through institutions, both past and present. It presents snapshots from the traditions that define the unique characteristics of coastal Ireland, from the picture-postcard images of traditional boatbuilding, to the contemporary nature of working in the Irish Naval Service. The chapter reveals how Irish maritime traditions are a response to the dramatic physical environment, whilst also being strongly influenced by socio-political elements, such as church and state, that impacted on cultural norms across the island, including the values of coastal communities.

IRISH MARITIME TRADITIONS

Irish maritime traditions, such as the Blessing of the Boats, have been preserved through generations and continue to feature prominently in the annual calendar of many ports and harbours around the country. Many traditions result from the need for safety at sea, such as pilotage, which has evolved to become a highly skilled practice in seaports in Ireland and all around the world. Some beliefs have become synonymous with modern-day marketing, such as the notion that Aran jumper patterns were used to identify the bodies of fishermen washed ashore after drowning at sea. Irrespective of the accuracy, it is fair to say that Irish maritime traditions and the nautical institutions that have evolved for the training of seafarers and the protection of life at sea evoke a sense of resilience and respect, used to cope with, and survive, the power of the ocean.

Case Study: Traditional Wooden Boats of Ireland

Críostóir Mac Cárthaigh

Much of Ireland's coast is enclosed by a rim of hills and mountains. In places, most notably in the west of Ireland, the land meets the sea in high cliffs interrupted occasionally by fjord-like arms of sea and sheltered bays. On the other hand, large stretches of the east coast offer relatively easy access to the interior of the country and also greater scope for contact and trade across the Irish Sea. It is significant that it was here, as well as in the natural harbours of the south coast, that the Vikings established their first footholds in Ireland. From the ninth century onwards, they followed the courses of the great, slow-moving rivers inland in their boats to explore and plunder. In the centuries that followed, increasingly large and varied types of vessels traded with Britain and continental Europe from these ports.

Where coastal topography was difficult and infrastructure poor, a greater variety of small boats capable of being hauled ashore and stored in safety above the tide evolved.[1] In a few favoured places, such as the Connemara coast, where sheltered bays are plenty and transport by road poor, a significant tradition of sail developed. Here, as elsewhere, water transport played a central role in people's lives. Boats were used for fishing and for carrying people and goods, including livestock, seaweed, sand and turf (Figure 23.1). Many offshore island

Fig. 23.1 GALWAY HOOKER NEAR INIS MÓR. The Galway Hooker, an iconic, traditional boat from the Connemara coast of Ireland, was typically used for fishing and for carrying cargo of turf, seaweed and even livestock. Hookers tend to have open or half open decks, and are gaff-rigged. The larger boats can be up to 40ft in length, and have a capacity to carry 7 to 8 tonnes of cargo, although there is considerable diversity in local, vernacular boats, which were adapted according to function and local traditions. Today, the fine craftsmanship in their design can be witnessed in the restored Galway Hookers frequently seen, with their distinctive red sails, in traditional boat festivals and regattas around the coast. [Source: Mary-Frances Beatty]

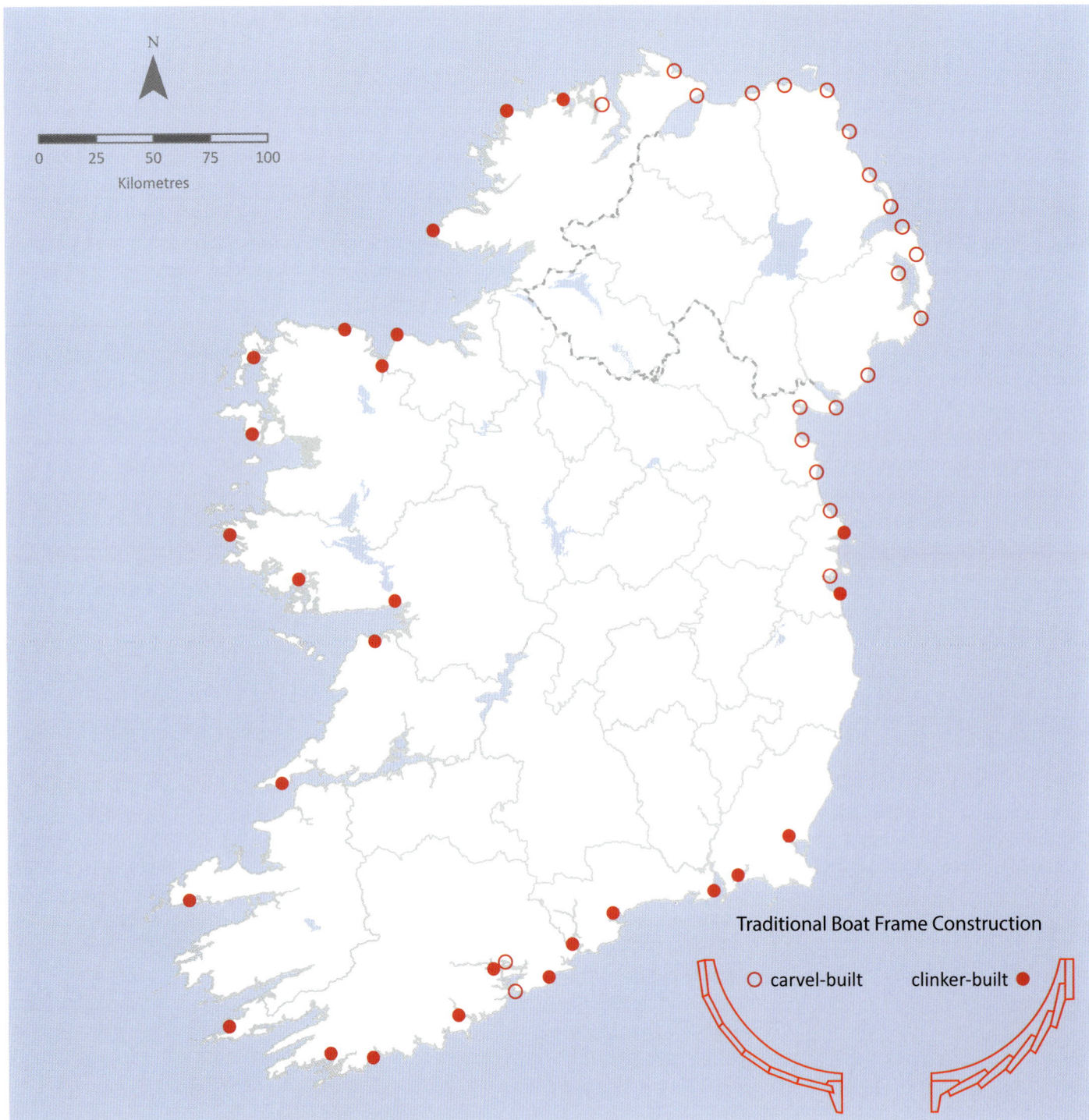

Fig. 23.2 THE EARLIEST EXAMPLE OF A WOODEN PLANK BOAT STRUCTURE YET DISCOVERED IN IRELAND DATES BACK TO THE THIRD OR FOURTH CENTURY AD. This boat (found in Lough Lene, County Westmeath) was built in the 'carvel' style, with edge-to-edge planking, the seams of which are caulked to seal the hull. This is in contrast to the 'clinker' method of boatbuilding associated with the Vikings, where planks overlap on their edges. Traditional wooden boatbuilding styles (clinker and carvel) are distributed according to an east/west divide, with carvel styles predominantly on the west coast, and clinker-built styles on the east coast. [Data source: Mac Cárthaigh, C., 2008. *Traditional Boats of Ireland: History, folklore and construction*. Cork: The Collins Press]

communities were maintained by virtue of boat transport (see Chapter 21: Ireland's Islands). Even today, most island communities of the west coast are wholly dependent on boats to carry them to and from the marketplace, school or church.

The finding of the remains of a small Roman-type boat at Lough Lene, County Westmeath, dating to perhaps the third or fourth century AD, is the earliest example of a wooden plank boat found in Ireland. The boat is almost certainly of Roman origin, built by someone familiar with Mediterranean boatbuilding techniques. However, the find may be exceptional, and the skills of the builder are unlikely to have been passed on to a native craftsman. Its function may well have been ceremonial, perhaps a status symbol possessed by a wealthy individual. The boat is built in the 'carvel' style, that is with edge-to-edge planking, the seams of which are caulked to seal the hull. This is in contrast with the 'clinker' method of boatbuilding associated with the Vikings, where the planks' edges overlap.

Early references in the Annals to seagoing craft are dominated by skin boats or 'currachs' (see Chapter 16: Inhabitants of Ireland's Early

522

Coastal Landscapes). Wooden boats feature only rarely. It is not until the tenth century that references to Irish-owned wooden plank boats, indeed fleets of boats on lake and sea, become frequent. Their appearance in the historical record coincides with the Viking period, during which the superior boat technology developed by the Scandinavians came to dominate small boat technology in northern Europe. Vikings penetrated Ireland's interior along the rivers and lakes, profoundly influencing boat design (see Chapter 17: Vikings and Normans: Coastal invaders and settlers). Present-day clinker-built lake boats may possibly represent a continuity of this tradition (Figure 23.2). The impact of the Scandinavians on Irish boat technology appears frequently in Irish-language boat terminology. For example *seas*, meaning 'thwart', is from Old Norse *sess*. *Stiúr,* and *stiúradh* meaning 'rudder' and 'to steer' respectively, are derived from the word *stýri*, while the term *ancaire*, meaning 'anchor', comes from the Old Norse *akkeri*.

There is also abundant evidence of carpentry and boatbuilding activity from the Wood Quay site in Dublin from the tenth-century through to the thirteenth-century (see Chapter 17: Vikings and Normans: Coastal invaders and settlers). Of the tools found in Dublin excavations, many would have formed part of the boatbuilder's tool chest. These include knives, spoon auger bits, gimlets, shaves and planes, all of which are still essential tools for modern wooden boatbuilders. The saw did not come into general use until the later Middle Ages. Before this, planks were cleft from logs with the aid of wedges and dressed with the axe or adze. The adze, still a key implement, is known from Roman times, though in its absence various hafted blades or axes would have served the purpose.

In the later Middle Ages large carvel-built vessels routinely visited Irish ports and documents of the period indicate that ships were also constructed by local shipwrights. Thus, the carvel method would have become increasingly familiar to all levels of Irish boatbuilding. As Michael McCaughan has shown, Ireland stands at a crossroads between a mainly northern European clinker zone and a southern European zone of carvel boatbuilding. There remains a strong preference for clinker-built boats on the north and east coasts and lakes, and for carvel-built boats on the south and west coasts. Of the former type, the double-ended 'Norway yawls' of the north and east coasts are the most

Fig. 23.3 TRADITIONAL BOAT BUILDING OF A CURRACH AT MEITHEAL MARA, COUNTY CORK. The currach is one of the best-known traditional boats of Ireland. Variations in the design of the currach depend on geographic origins in Donegal, Kerry, Clare, Mayo and Galway. The longevity of the currach extends back some *c*.2,000 years. This is a small, curved rowing boat, the hull of which is made by wooden slats covered by animal hide and sealed by tar. Today, canvas and resin replace these traditional materials. The tradition of currach construction is kept alive by enthusiasts and organisations such as Meitheal Mara, a community boatyard founded in 1993 in Cork, which supports the personal growth and progression of individuals by involving the public in boatbuilding and currach rowing activities. [Source: Meitheal Mara]

distinctive. These were introduced from Norway as recently as the eighteenth century, carried in unfinished form on the decks of ships delivering cargoes of timber and ice for the expanding towns and cities of the north of Ireland. The Norwegian origin of this boat type is reflected in the name 'Drontheim' (from Trondheim), which is still used in parts of the north coast. Native builders soon copied the type and distinctive local versions were produced, often acquiring the name of the harbour in which they were built and used. Examples of these include the Greencastle yawl of Inishowen and the Groomsport yawl of Belfast Lough. South of St John's Point in County Down, the Norway yawl was frequently termed a 'skiff', a term applied to a variety of small clinker-built sailing and pulling boats as far south as Wicklow.

Southwards of Arklow, one enters a mixed carvel/clinker zone. The Arklow yawl was a carvel-built sailing boat used by the large Arklow fleet that fished the Irish Sea. It too is now extinct but a fine replica, the *Ógra na Mara*, was built in 1988. When one rounds the Hook of Wexford, carvel boats like the Ballyhack yawl, associated with the harbours of Passage East, Fethard and Slade, predominate. Other examples, now extinct, include the Helvick hooker, the Rathcoursey hooker from Cork Harbour, and the Kinsale hooker. Along the west Cork coast a number of small sailing and pulling boats were used for fishing, piloting and carrying. The most widespread was known as the west Cork lugsail yawl. Other characteristic small carvel boats of the region include the Heir Island lobster boat and the Long Island mackerel boat. The carvel tradition is also represented by the seine boats of west Cork and Kerry, a boat type introduced in the early 1600s from Cornwall – a region of Britain noted for carvel building – to encourage the pilchard fishery of the region.

At the mouth of the Cashen River in north Kerry, and in the upper reaches of the Shannon Estuary, a most interesting flat-bottomed boat type is still used. This is the 'ganglo' or 'gandelow', a boat with a distinctive, pleasing sheer, and a bottom that is slightly 'rockered' to prevent it sticking to the mud of the estuary when it is being launched. At Coonagh, just outside Limerick, water reed is harvested in winter for the purpose of thatching, while in late spring the same boats were used until recently for salmon netting.

Further north, Galway and Mayo have strong traditions of small sailing and pulling boats too numerous to mention here. Most famous of these is the Galway hooker family, which includes several classes of sailing vessels, including the *bád mór*, the *púcán* and *gleoiteog*, and the *bád iomartha* or 'rowing boat'. Another example is the currach *adhmaid* or 'wooden currach', so-called because it mimics the shape of the canvas-covered currach. This small rowing boat, developed only in the last 120 years, is still widely used and has proved itself a successful general-purpose workboat (Figure 23.3). Galway hookers were also general along the west Mayo coast, where they continued in use to the Second World War.

As the tradition of pulling and sailing wooden working boats declines, replicas of many traditional boats are being made for recreational purposes. Increasingly, they are referred to as 'classic' boats – a recognition of their cultural importance. New uses for old boats are being found, and developments in aquaculture and in the shellfish industry have to some degree stimulated demand for new wooden boats, although the cheaper fiberglass (GRP) is an attractive alternative.

BLESSINGS OF THE BOATS

Elaine O'Driscoll-Adam

The connection between fishing families and the sea can be observed and experienced through the many rituals that occur in coastal communities. The 'Blessing of the Boats' is one such ritual. Understood to have originated in southern Europe, the ritual of blessing boats has been taking place for centuries. Wherever the origins, many fishing communities throughout the world continue to bless their fleets. Fishing communities in Ireland are no different. Ceremonies survive in fishing villages and towns in Ireland from Courtown, County Wexford, to Claddagh, County Galway. There is no distinct structure to these blessing ceremonies, revealing characteristics that are locally unique.

For example, in Castletownbere, west Cork, the boat blessing ritual has taken place for almost a century (Figure 23.4). It was traditionally held on the pier after Sunday mass on the August Bank Holiday weekend. At the turn of the twenty-first century, the organisation Mná na Mara (women of the sea) proposed a special Fishermen's Mass to coincide with the blessing, to be celebrated on

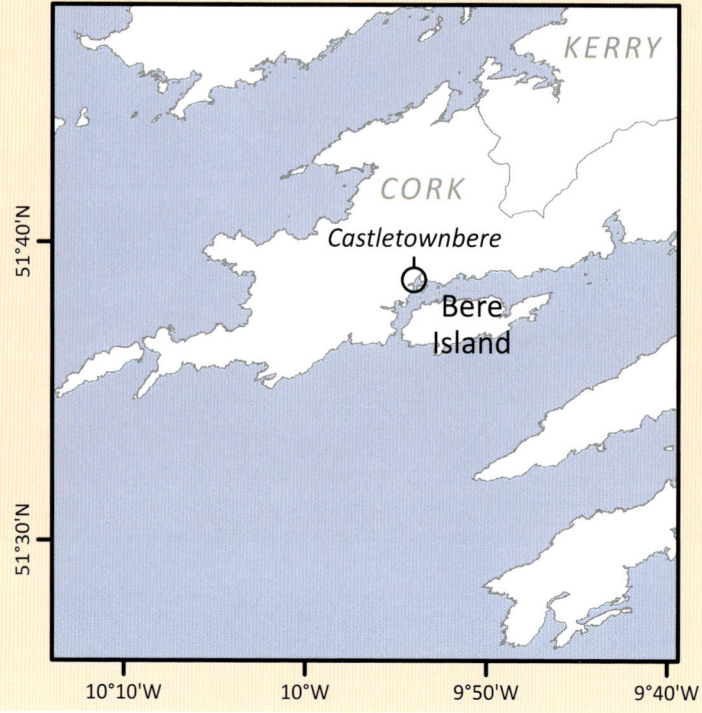

Fig. 23.4 BLESSING OF THE BOATS CEREMONY, CASTLETOWNBERE, COUNTY CORK. This scene from the annual Blessing of the Boats ceremony in Castletownbere is typical of scenes from fishing harbours all around the coast of Ireland. The blending of the Fishermen's Mass and the Blessing of the Boats highlights how 'the sacred has increasingly come to permeate ... the geographies of everyday life' (della, Dora, V., 2011. Engaging Sacred Space: Experiments in the field. *Journal of Geography in Higher Education*, 35(2). p..163) [Source: Elaine O'Driscoll-Adam].

the pier in the open air. The Castletownbere Fishermen's Mass, despite being held in early August every year, was frequently impacted by inclement weather conditions. As a result, in 2010, the mass was relocated to the former auction hall, with the Blessing of the Boats held on the pier immediately after. The occasion, which fuses past and present, and encourages reflection, now marks the start of the Castletownbere Regatta. Fishing artefacts used for the offertory procession – mending needles and twine, platters of seafood and lifejackets – reinforce the expression of a specific way of life. The combination of these two significant rituals – the Fishermen's Mass and Blessing of the Boats – can be regarded as a cultural identifier where both religion and a way of life converge through the performance of ritual.

Fishing is a way of life steeped in tradition, with fishing skills handed down through generations. While the arrival of modern technology has altered the way in which knowledge is acquired, through advances in communications systems and more accurate weather forecasting, certain traditions, such as the Fishermen's Mass or the Blessing of the Boats, prevail. These events are significant for many fishing and coastal villages and towns and are adapted to their locality.

THE ARAN JUMPER: A MARITIME TRADITION

Ken Cotter

The Aran jumper is undoubtedly an iconic symbol of Irish coastal identity. However, its profile might be more a triumph of marketing than the successful preservation of an ancient tradition. In today's more critical age, questions of authenticity arise. Is today's Aran jumper, available to purchase in so many tourist and craft shops across Ireland, a dubious 'invention of tradition', or the modern manifestation of an ancient coastal craft?

Irish knitwear has become an important and lucrative industry, and like most businesses that 'commodify tradition', the true provenance of the Aran sweater is conveniently left to remain nebulous and obscure. Most websites and shops selling Aran knitwear happily perpetuate the romantic mysticism that their products conjure up. Visions of simple, noble island communities living and working in harmony with their environment is appealing to a modern consumer tapping in to their real or imagined ancestry, or their desire to assert a national identity.

Historians and knitting experts have debated the Aran jumper for decades. In the 1960s Heinz Edgar Kiewe enthusiastically purported to have 'discovered' an Aran sweater in a shop in Dublin, knitted in 'Biblical white' and bearing all the hallmarks of an ancient custom rooted in early medieval Christianity. These symbols, he suggested, were associated with the Coptic-inspired monks that lived on Ireland's western seaboard. He went on to surmise that the motifs on Aran jumpers were a mixture of the religious and the secular. For instance, the rope patterns, or cables, could represent the fisherman's ropes, while the 'Ladder of Life' design could be considered a symbol of 'man's earthly climb to eternal happiness'.[2]

Further romance can be added to the tradition by recalling the tragic ritual, known to many coastal communities around Britain and Ireland, that drowned fishermen could be identified by the unique patterns on their jumpers, knitted by a loving mother, sister or wife. The playwright, J.M. Synge, borrowed this idea for his one-act play *Riders to the Sea*,[3] but, crucially, the garment in question was not an Aran jumper but, rather, a 'stocking' with some tell-tale stitches dropped.

However, the truth is almost certainly less extraordinary, but no less interesting. Where there is an abundance of photographic evidence from around the islands of Britain and Ireland of fishermen wearing sweaters or ganseys, there is precious little depiction of Aran men wearing anything similar. Films, photographs and various oral accounts by visitors throughout the nineteenth century mention Aran's traditional dress, but not the jumpers.

There was a strong tradition of knitting and weaving on the Aran Islands, but it was woollen stockings and hats, homespun trousers and hand-sewn leather shoes (known as pampouties) that gave the islanders the unique look lionized by filmmakers, poets and playwrights of the late nineteenth and early twentieth centuries. The robust, durable, multicoloured woollen belt worn by the menfolk, known as a *crios*, provided the colour.

By the end of the nineteenth century, the Congested Districts Board had been encouraging the production of local crafts, including knitting (see Chapter 19: Changing Coastal Landscapes). The income that could be derived from this endeavour was small, but nonetheless welcome to a community wrestling with poverty. The board brought instructors to the islands to coach the women in the finer details of producing garments to order. Outlets like the Country Shop on St Stephen's Green in Dublin, run by Dr Muriel Gahan, sold the products. It was in this shop that Heinz Edgar Kiewe 'discovered' his Aran jumper.

There is a distinct evolution in the specimens of gansey on display in the National Museum, according to Alice Starmore in her book, *Aran Knitting*.[4] Her belief is that the oldest example in the collection, knitted in the 1930s, was made by someone with an intimate knowledge of the Scottish coastal knitting tradition, and that this design unleashed a wave of creativity in the women of Aran.

A competition launched by the Country Shop in 1946, seeking Aran designs, received over fifty entries. The three winners were all from the western villages of Árainn (Aran Islands). Over the next thirty or so years, the Aran jumper evolved and a company called Galway Bay Products opened up new markets in the United States. This enterprise was no doubt helped by the appearance of Aran-clad folk stars, the Clancy Brothers and Tommy Makem, on the Ed Sullivan Show in 1961. Since then, locally significant knitwear producers

Fig. 23.5 THE ARAN JUMPER. Whether invention of tradition, or the modern manifestation of an ancient coastal craft, the Aran jumper, with its perceived root in coastal communities, is a highly regarded and popular garment to this day. This image, courtesy of the Inis Meáin Knitting Company archives, demonstrates the Aran style, which serves as a source of inspiration for modern collections. The Inis Meain Knitting Company on the Aran Islands, founded in 1976, provides local employment and derives inspiration from the heritage of the island to produce contemporary knitwear for worldwide export. [Source: Inis Meáin Knitting Co.]

have developed to provide important sources of indigenous enterprise in peripheral coastal communities (see Chapter 21: Ireland's Islands) (Figure 23.5).

What remains fascinating is how the women of Aran, when faced with the challenge of creating a piece of knitwear for the marketplaces of urban Ireland and beyond, created patterns bearing such unique ingenuity. What was it that released this rich vein of creativity? Did they indeed draw on the symbolism of their version of Christianity, with its strong medieval links, intertwined with traditional superstitions of a people living life on these 'three stepping stones out of Europe', in constant communion with the sea? The fact that the Aran jumper has become an instantly recognisable symbol of Irish identity and a channel for the creative spirit inspired by an ancient coastal tradition, real or perceived, is surely legacy enough for any item of clothing!

THE SEA AND THE SONGS

Ken Cotter

Many Irish people live adjacent to the sea and many more have worked on the sea or pondered its mystery from the shore, so it is no wonder that maritime motifs pepper the songs in the Irish folk songbook. Musicians and songwriters have been drawn to and inspired by the sea since the beginning of time. Coastal communities embraced the medium of song to tell their stories and to weave their narrative and lore. Local songs were crafted by the people to tell their own history and to bring to mind times of joy or misfortune, tales of work and weather, travels and travails, belonging and place, toil and tragedy.[5]

These expressions took many forms: shanties, *sean nós*, ballads, love songs, calls to arms and rousing folk songs (Figure 23.6). Musicians expressed the sentiments of their communities through the composition of slow airs, jigs, reels and polkas, and reflected their littoral roots in their titles: 'The Lady in the Boat', 'Cois Taoibh a Chuan', 'The Sailor Boy', 'The Bay and the Grey', 'Fuaim na dTonn'. Some songs are self-conscious appraisals of the sea, its beauty, its harshness, its life-giving and life-ending possibility. The lifestyle associated with sea embeds itself in the soundtrack of the coastal community, the fishing, the net-mending, boatbuilding, the journeying away and returning home. Some songs bask in the glorious physical beauty of the coast, the beaches and bays, or the mystic craft that ply the waves. Sometimes the sea and the coast are the setting for the songs, the stage scenery for an epic journey, an emotional recollection, a commemoration, a fight or flight. All the while, like in coastal life itself, the sea breathes in and out in the background.

Many Irish songs, or songs brought here from other coastal regions, derive their themes from the bustling activity of the fishing industry and celebrate the harvest of the sea, the shoals of herring, salmon or the ling, such as 'Amhrán na Scadán', a traditional song from Tory Island. These songs often delight in the return of the fishermen, with their holds bursting with the fruits of their labour, like 'The Boys of Killybegs Come Rolling Home'.

Shanties are a particularly familiar form of sea-related song. These songs were work songs used by sailors to keep time when rhythmic coordination was of the essence, for instance when the crew was weighing anchor or hoisting sails. Other shanties passed the time when working in the cramped conditions of the forecastle, perhaps stowing sails or mending nets. Their themes were often bawdy, but many had their sentimental side, evoking images of loved ones at home or lovers in faraway ports:

When I was a little lad
And so my mother told me
Way, haul away, we'll haul away, Joe!
That if I did not kiss the gals
Me lips would all grow moldy
Way, haul away, we'll haul away, Joe![6]

In the Irish context, of course, many songs and tunes from the sea centred on the national struggle, perhaps reflecting the desire for Ireland to strike out for liberty, taking her inspiration from the sea:

When Connacht lies in slumber deep …
… Sing, Oh! Let men learn liberty
From crashing wind and lashing sea.[7]

More songs of freedom reflect on the flight of the Wild Geese, a lover left alone, 'for he has gone, ochone mo chroí'. More sing

Fig. 23.6 ONE OF THE MOST WELL-KNOWN SEA SHANTIES 'WHAT SHALL WE DO WITH THE DRUNKEN SAILOR?'. It is sung to the traditional Irish folk song 'Óró sé do bheatha abhaile', although it is unknown which came first. [Source: Public Domain]

in tribute to the arrival of various French or Spanish fleets, destined to contribute their significant strength to the battle for Irish freedom.

Love of place is a strong and recurring motif. There are myriad songs in the Irish folk songbook dedicated to the love of the composer's own coastal town, townland, or village, or a longing to be back in a special home place:

So I'll wait for the wild rose that's waitin' for me –
Where the Mountains o' Mourne sweep down to the sea.[8]

There is often mentioned a favoured safe harbour where home-like comforts are afforded before an ineluctable return to the strenuous workload and spartan living conditions on board ship:

And now the storm it is over and we are safe on shore,
We will drink a toast to the Holy Ground and the girls that we adore.
We will drink strong ale and porter and make the rafters roar,
And when our money is all spent we'll go to sea once more.[9]

Traditions honouring place remain unbroken today through ballads like 'The Cliffs of Dooneen', or Jimmy MacCarthy's 'As I Leave Behind Neidín'. However, often these songs of place are tinged with melancholy. The pain and loss of emigration is never too far away. There's a sadness there, a feeling of loneliness and desolation, of being the last one of your community left. The forlorn writer of 'Bantry Bay', a song made famous by singer John McCormack (1884–1945), laments the passing of youth and the loss of friends and family through death and emigration.

Now I'm sitting alone in the gloaming
The shadows of the past draw near
And I see the loving faces all around me
That used to glad the old brown pier
Some are gone upon their last journey homing
Some are left, they're old and grey

And we're waiting for the tide now in the gloaming
To sail upon the Great Highway
To the land of rest that is unending
All peacefully from Bantry Bay.[10]

Historically, disaster is never far away from any coastal community. Bravery is often recounted like that of the Shannon pilots in the song 'The Kilbaha Pilots 1873', who ventured out from Loop Head into stormy seas to offer pilotage to an Austrian brig before being swamped by a wave:

They are gone away and away to death, as they pulled with a hearty will,
Whilst the watchers moved to an eager group, to the top of Dun Dahlen Hill.
They watched the pilots draw near the ship, and the sea give one fatal roll,
And they cried to the God of the Lord above, to have pity on those creatures' souls.[11]

Merchant shipping was no less hazardous than many other aspects of seafaring and many lives were lost at sea on vessels that sailed commercially around the world. A song called 'The Trader' recalls the sinking of a ship sailing from Galway to London, on which seven men perished:

The people there from everywhere come flocking that sad sight to see,
Seven heroes' corpses lying on the shore, the Trader's doleful company;
It is in Dunboe they're lying low where there you'll see their green green graves,
No friends were near but strangers dear, we buried them in sweet Articlave.[12]

The 'Arranmore Disaster', from Donegal, is typical of many songs commemorating heartbreaking loss of life in a coastal community. The song is particularly poignant because the men who died were on their way home, having completed seasonal work in Scotland. These tragedies ripped the heart out of small coastal communities with so many young lives lost, and songs were not only a way of recalling the tragedy and remembering the people who perished, but must have also provided succour to those left behind:

Twas in the year of thirty-five on a bleak November eve
This awful tragedy occurred, it caused us all to grieve;
Those cheerful lads returning from the Scottish harvest field
Unto the stormy ocean their lives were forced to yield.

What cheerful thoughts were in their mind when sailing up Lough Foyle
To view the hills of Inishowen, that land of Irish soil!
Their little boat came slowly on through Creeslough and Gweedore:
Oh God, who'd think they ne'er would reach their native Arranmore![13]

The lure of the sea and the passion for the age of sail burns strongly in modern songwriters too. Jimmy Crowley's 'My Love is a Tall Ship', speaks of the beauty of the Irish sail-training brigantine, *Asgard* II. The song is made all the more poignant since she no longer heaves to the wind; she sank in the Bay of Biscay in 2008.

THE TRADITION OF PILOTAGE: THE LIFE OF A PILOT

Michael Barry and Cormac Gebruers

Records of the practice of pilotage extend back to ancient Greek and Roman times. Pilotage involves the manoeuvring of ships through local waters, with the aid of the local knowledge of highly skilled pilots, who board visiting ships to ensure safe navigation and berthing. Before harbour boards were established in the early nineteenth century, pilots, known as hobblers, worked independently. Skilled seamen set out in skiffs to reach incoming ships, in order to receive payment for navigating incoming vessels safely to dock.

The UK Merchant Shipping Acts of the 1840s onwards allowed for the establishment of compulsory pilotage areas in ports. Liverpool, Dublin and the Clyde complied with the obligations, but many other ports did not. Arguments raged as to whether compulsory pilotage was a good idea. Many complained that it restricted the free market, created a monopoly and increased costs. Conversely, an unregulated system was claimed to put safety at risk. At the time, accounts abounded of exorbitant pilotage fees being charged in areas where pilotage was optional, especially in bad weather, or where a ship was in danger.

In 1868, Cork had not yet adopted compulsory pilotage. Changes in pilotage during that period are reflected in the biography of Cork Harbour pilot, Thomas Martin. At the age of twenty-four, Thomas Martin began an uncertain career as a freelance pilot. He appears to have worked with the Nash family out of Cobh in a cutter called *Mary Rose*. They would have plied their trade on the approaches to Cork Harbour as far west as the Fastnet Rock, competing for pilotage jobs against other freelance pilots from ports like Crookhaven. These south coast harbour pilots were often the first contact many ships would have had after months on long voyages from far-flung locations. In 1939, it was from a Cork pilot that the crew of the *Moshulu* learned that they had won what was to be the last of the Great Grain Races, beating their great rivals the *Padua* and *Passat*, having made the passage in ninety-one days from Port Victoria, South Australia, to Cork.

A pilot's job was a risky one. Before radio, ships' exact arrival times were unknown, and so Cork Harbour pilots maintained a watch at Roche's Point and launched their pilot cutter once a vessel was sighted. They would spend many uncomfortable and often dangerous hours at sea, waiting to board the ship. The hazardous boarding operation often took place in darkness and in poor weather. On four occasions Thomas Martin was carried away on board a vessel, being unable to get off due to bad weather.

Between 1868 and 1882, Thomas Martin established his reputation as a skilled pilot and ship handler. His good reputation saw him being appointed Cork Harbour pilot for the famous Cunard Line in 1882. Thomas' first Cunarder was the *Marathon*, a 336 feet, 2,400 tonne, three-masted auxiliary steam ship. In the years that followed he was to have charge of some of the greatest steam liners ever built, including the *Lusitania* and her sister ship, *Mauritania*. These ships were designed for the transatlantic route to New York, a route that was all about speed. By 1907 there was growing pressure to bypass Cork Harbour in order to shorten the passage time, and among the arguments put forward was a lack of water depth in Cork for these ocean giants. The water issue was, however, definitively settled on *Lusitania*'s third voyage when, as senior Cunard pilot, Thomas Martin brought the 787 feet, 61,550 tonne *Lusitania* into the inner harbour and anchored her safely off Whitegate.

Fig. 23.7 A PILOT CLIMBING ABOARD A COMMERCIAL SHIP IN CORK HARBOUR. This image demonstrates the technical, skilled and risky nature of the profession. The history of harbour pilotage is one that has evolved from a commercial, unregulated transaction to the formal, statutory basis of pilotage in support of navigation and shipping these days. The bravery of pilots of old is reflected in the tragedy that befell ports and harbours due to the perilous work of unlicensed pilotage. The term 'hobblers', extending back to the early nineteenth century, was given to the pilots in Kingstown (now Dún Laoghaire). These men (and sometimes boys) risked their lives in open skiffs in adverse weather conditions, to accompany ships into harbour. For example, in 1928 three hobblers lost their lives in a single incident involving the Dutch steamer *Hesbaye* in Dublin Bay. [Source: Captain Aidan Fleming]

A blazing headline in the *Cork Examiner* of 4 November 1907 announced 'Lusitania Voyage, Enters the Harbour at Low tide'. Thomas Martin recorded that he had six feet under his keel at the shallowest point, compared to just seven inches at the shallowest point in New York. In one act of superlative seamanship the argument for bypassing Cork Harbour on the grounds of depth was demolished.

This snapshot from the life of Thomas Martin, a pilot in Cork Harbour, provides a glimpse of the endeavours, skills and bravery that remain fundamental characteristics of the piloting profession to this day (Figure 23.7). Although technology has changed and progressed, risk management remains a fundamental component of the job, with larger ships and expanding operational weather windows forming part of the contemporary modus operandi of pilots working out of ports on the island of Ireland, from Cork to Belfast.

NAUTICAL INSTITUTIONS

There is a very interesting history behind the various agencies and organisations that epitomise Irish maritime and nautical institutions. This involves accounts of the evolution of the Naval Service, nautical education and training, lighthouses, lifeboats and the Coast Guard. The story is complex and, as will be seen, was very much affected by British power and influence over the centuries. Irish marine affairs were controlled by various laws and other restrictions, set in place to deter serious competition with English trade, naval development and fishing activities, mainly in the seventeenth and eighteenth centuries. In spite of these constraints, Irish trade and maritime interests managed not only to survive, but in some cases flourished. However, after Ireland became part of the United Kingdom, under the Act of Union in 1801, Irish marine affairs became 'British' by definition, and subordinate to the dominant maritime interests. This shaped Ireland's maritime development profoundly until recent years.

By the last decades of the nineteenth century, Irish-owned and operated shipping was minimal; with the exception of Belfast, shipbuilding and marine engineering were almost non-existent, and Irish fishing was, with a few exceptions, in serious decline (see Chapter 24: Ports and Shipping and Chapter 26: Coastal Fisheries and Aquaculture). Ireland had truly 'turned its back to the sea'. Even after the establishment of the Irish Free State in 1922 little was done to promote maritime affairs in the new jurisdiction until the modern era, when the institutional arrangements described below were established.

SALT IN OUR VEINS

In spite of this overshadowing by political events, there were sparks of leadership and marine enterprise glowing faintly in the ashes of maritime Ireland. One of the most persistent of these influences was the call of the sea to Irish men and boys, although much of this lure was to find an outlet in other countries. As a result, Irish names are found among the senior ranks in the histories of many of the navies of the world. For example, Commodore John Barry of Wexford (1745–1803) is considered 'father' of the United States

navy,[14] while Admiral William Brown of Foxford, County Mayo (1777–1857), is revered in Argentina as the principal hero of their war of independence (Figure 23.8).[15]

Many of Ireland's principal ports had established a long trading tradition with various regions of Europe. Irish trading houses and shipping enterprises were established in several continental ports, principally in the eighteenth century. For example, prominent merchant prince families from Cork traded with the Hanseatic League ports, such as Riga and Hamburg. Several Irish merchant families resided in ports such as Malaga, Seville and Cadiz in Spain. In addition to those who achieved success in the maritime world, a considerable cohort of Irishmen following the call of the sea did not achieve notable positions or amass wealth. One of the major attractions for many seamen was the British Royal Navy. At the time of the Battle of Trafalgar, about one-third of its manpower was Irish. Irish representation was second only to the number of Englishmen in the service, exceeding Welsh, Scottish and other nationalities combined. Officers of Irish birth were equally prominent in the Royal Navy, usually from the Irish ascendancy and unionist classes.

Fig. 23.8 a. PORTRAIT OF ADMIRAL WILLIAM BROWN (1777–1857) AND b. PORTRAIT OF COMMODORE JOHN BARRY (1745–1803) BY V. ZVEG FROM GILBERT STUART'S 1801 ORIGINAL. Originally from Foxford, County Mayo, William Brown emigrated to the United States with his family when he was ten years old. Following a period of hardship that saw him orphaned as a youth, Brown went on to become successful in the merchant marine and subsequently in the British Admiralty. He distinguished himself in numerous battles against the Spanish navy, and in the Napoleonic Wars against the French. Following a period of incarceration by the French, Brown escaped and reached England, where he resumed his military career and met his wife. Together they moved to South America, where in 1814, Brown was given the task of rebuilding the Argentinian naval fleet. His influence, skills as a military tactician and leadership led to him becoming regarded as the 'Father of the Argentine Navy' and as an Argentinian hero by the time he died there, in 1857. [Source: Henry Harvè, Wikimedia Commons, Public Domain]; b. Born in Tacumshane, County Wexford, John Barry spent his youth working on fishing boats in the Irish Sea. At sixteen, he emigrated to America, and subsequently he enlisted in the merchant marine. He fought as an officer in the Continental Navy during the American Revolution. Following the formation of the United States navy, he commanded several ships and went on to be appointed as senior captain of the frigate *United States*. He was distinguished by President George Washington with a commission in 1797 as the first flag officer and has since become known as the 'Father of the American Navy'. He is remembered as an exceptional navigator, mariner and leader. [Source: US Naval Historical Center, Wikimedia Commons PD-USGov]

Fig 23.9 a. *Muirchú*, b. LÉ *Gráinne*, c. LÉ *William Butler Yeats*. These three vessels represent different eras in the development of maritime defence and enforcement as the Irish state developed. *Muirchú* was a purpose-built (1908) research and patrol vessel for the Department of Agriculture and Technical Instruction of Ireland. She transferred to the Department of Agriculture and Fisheries of the new state in 1923, and was the only Irish Government vessel patrolling Irish waters until 1936 when she was joined by the trawler *Fort Rannoch*. Both of these vessels were taken over in January 1940 by the Defence Forces, with a flotilla of six motor torpedo boats, forming the wartime Marine Service. In 1947 this service was incorporated into the Permanent Defence Forces as the Naval Service. Three ex-British Navy Flower class anti-submarine vessels (corvettes) were purchased and in service until 1971 when replaced by LÉ *Gráinne*, two sister minesweepers, and a new purpose-built offshore patrol vessel LÉ *Deirdre*. In the 1970s and 80s three further similar offshore patrol vessels were constructed and a larger ship, LÉ *Eithne*, equipped to carry a helicopter. Also, the three minesweepers were replaced by two fast coastal patrol vessels in this era. In the 1990s a vessel replacement programme was started. With the commissioning of LÉ *Roisín* and LÉ *Niamh* in 1999 and 2001 respectively and the larger Beckett class, the Deirdre class were all disposed of over the following years. The fleet currently (2021) consists of LÉ *Eithne*, LÉ *Samuel Beckett*, LÉ *James Joyce*, LÉ *William Butler Yeats*, LÉ *George Bernard Shaw*, LÉ *Niamh*, LÉ *Roisín*, LÉ *Ciara* and LÉ *Órla*. Over the coming years the older vessels are due to be replaced. [Source: Irish Naval Service Archives]

at the time dominated the world. This was evidenced by the sheer numbers of ships registered under the British flag, leadership in shipbuilding and marine engineering and technology (for example, Belfast), and by the concentration of shipping markets and marine insurance based in London. The Great Depression of the 1930s decimated shipping worldwide, as well as what little Irish shipping enterprise existed at that time.

Equally, the management of commercial fisheries was not of prime importance in the new state. Fisheries issues were, however, raised at the time in the Dáil, with concerns centred around the promotion and development of the sector, protection from foreign incursion and the need to address the lack of appropriate regulation. The research and patrol ship, *Helga*, was acquired by the formative Department of Agriculture and Fisheries for policing Irish fisheries and for marine research. This purpose-built vessel, renamed *Muirchú*, was an excellent one, but was completely insufficient, on her own, for the task. This, coupled with inadequate regulation, made prosecuting offenders almost impossible. Nonetheless, early marine research was conducted on

THE EMERGENCE OF THE IRISH MARITIME INSTITUTIONS AND THE DEVELOPMENT OF IRISH MARINERS

Very little changed when the Irish Free State separated from the United Kingdom in 1922. Irish seafarers continued to serve mainly in British ships, both mercantile and navy. The need for a native and independent Irish shipping industry and for capacity to ensure the protection of fisheries was raised in parliament after Irish independence. However, shipping stakeholders showed little interest in moving away from the advantage of being part of the British mercantile marine. This link was important because British shipping

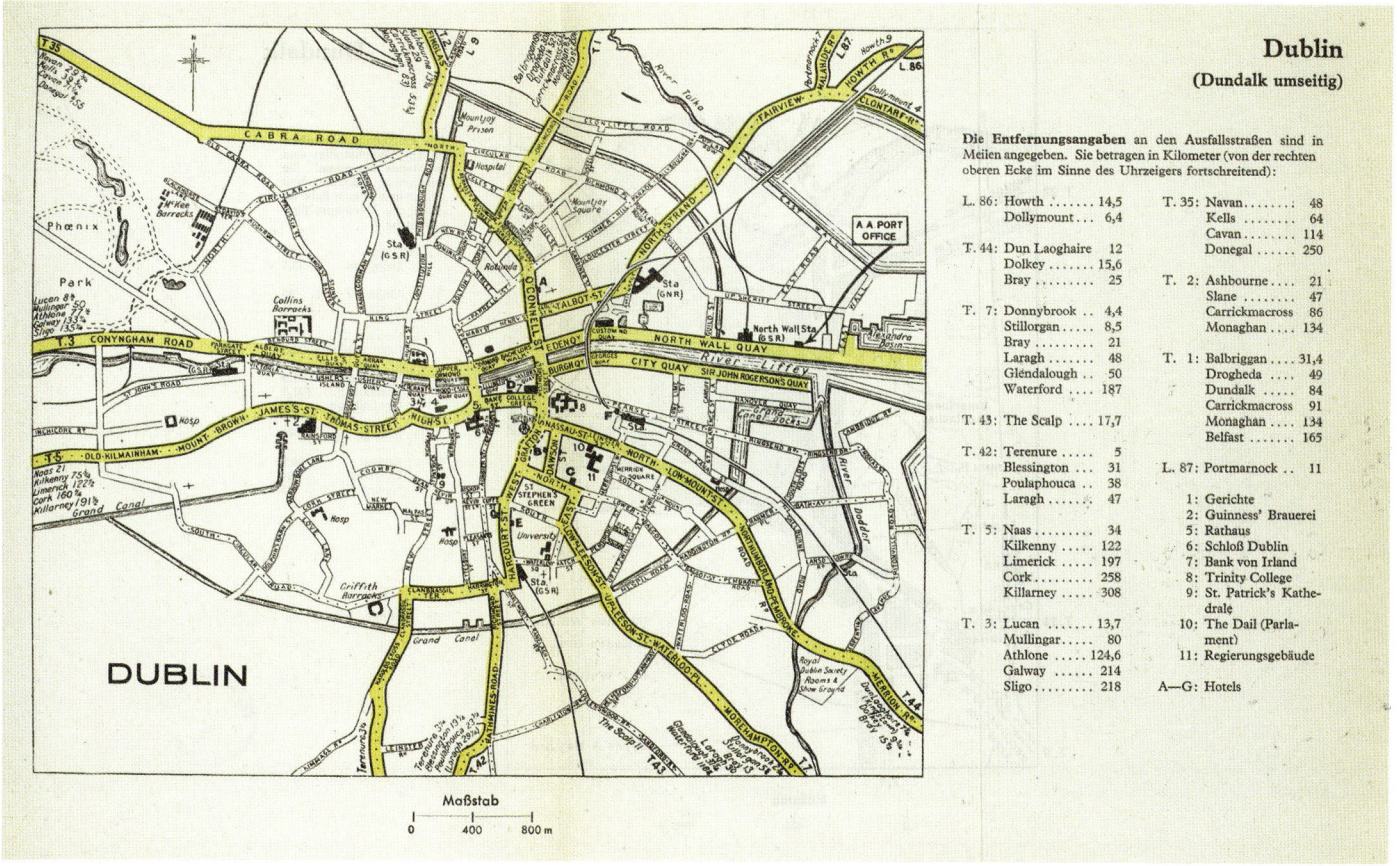

Fig. 23.10 MAP OF GERMAN INVASION PLANS IN THE SECOND WORLD WAR. As it became clear that Hitler was not going to win the Battle of Britain, which was fought largely in the air, Hitler switched strategies, with a bombing campaign, known as the 'Blitz', targeting cities across the UK. While the Battle of Britain and the Blitz are well-known keystones in the Second World War, there is little awareness of the vulnerability that came from Ireland's weak defence position. This weak status was exposed by detailed plans drawn up by the German High Command to invade Ireland via its key city ports. The map of Dublin above, shows how German military strategists analysed the main arteries into and around the city centre, with distance shown in kilometres from the city centre. Similar maps were produced for other port cities such as Cork and Galway. [Source: National Library of Ireland]

board the *Helga* by George Farren and Arthur Went. They were much respected internationally for their published works on fisheries of the eastern North Atlantic and in the International Council for the Exploration of the Sea (ICES). Their work was to be the forerunner of the modern Irish Marine Institute. In 1936, a deep-sea trawler, *Fort Rannoch*, was chartered as a patrol vessel to support *Muirchú* and new fisheries regulations were enacted. At last, something was being done to address the fisheries issues, whilst other maritime areas still required action and dedicated vessels, namely the mercantile marine; seaward defence and coastal surveillance; and search and rescue (Figure 23.9).

The Anglo-Irish Treaty of 1921 prohibited the establishment of an Irish navy until agreement had been reached with the United Kingdom. A conference took place in London five years after the signing of the treaty to discuss an Irish maritime defence force, but it came to nothing. The situation continued whereby the harbour defences of Cork, Berehaven and Lough Swilly were manned and maintained by the British army until 1938, while mooring buoys and a stock of fuel and other stores were maintained at Haulbowline Island in Cork Harbour. Two British destroyers were normally at these moorings, maintaining their presence in Irish waters. The naval dockyard on Haulbowline, however, had been handed over to the new Irish government in 1922. Despite attempts to keep the dockyard open as a viable commercial operation, the depressed state of international shipping and other factors mitigated against this. It

was shut down in 1929. Certain buildings and facilities were maintained by the Office of Public Works, mainly to comply with the terms of the treaty.[16]

The outbreak of the Second World War found Ireland completely unprepared. The Free State army was run down due to a lack of funding. The mercantile marine was similarly under-resourced, consisting of about fifty small ships suitable only for trading to Britain and the near continental ports. As a result, there was no seaward defence of any kind and Irish territorial waters were vulnerable. The Air Corps was in a similar position to the army, though it did have a limited offshore patrol capability.

The stories of the wartime measures are described in the case studies below, such as the Coast Watching Service, established in the summer of 1939 along the south and east coasts. Many accounts of Irish recent history point to the Second World War period as the Irish state's 'coming of age', demonstrating its ability to stand alone and pursue its own destiny and foreign policy. This was due largely to a combination of good luck and circumstances. The declaration of neutrality and the poor state of Irish defences hardly deterred the German High Command. In fact, plans were drawn up by the German army for an invasion of Ireland (Figure 23.10). The peripheral location of the island in Europe, and the strong presence of the British navy, together prevented an invasion by Germany.

The harsh lesson of the Second World War prompted two

Fig. 23.11 THE *Arklow Faith*, ONE OF THE ARKLOW SHIPPING FLEET. Arklow Shipping, the largest shipping company in Ireland, which dates back to 1966, currently operates a fleet of fifty-five ships. [Source: Arklow Shipping]

actions: the establishment of a naval branch in the permanent defence forces and a deep-sea merchant shipping company. The fortunes of both these organisations were to fluctuate over the years due to successive government policies.[17] The defence forces, and particularly the Naval Service, were easy targets for savings, as discussed further in the case study.

In the merchant shipping sector, Irish Shipping Ltd, the government-owned deep-sea shipping company, traded successfully for many years in the global shipping markets. The depression of the 1980s, however, saw its demise. The one shipping company that survived and thrives to this day is Arklow Shipping Ltd (Figure 23.11), the culmination of the seafaring and ship-owning tradition of that town. From a situation where it

operated a few small coastal trading vessels, it now comprises a substantial fleet of ships trading worldwide.

Today there are a range of maritime agencies devoted to protecting, serving and exploiting the seas around Ireland, and to protecting essential lines of communication by which the modern state and its economy exists and thrives. Each of these agencies has grown from a long tradition of trade, defence and seafaring. Together with other agencies of the state, such as the Marine Institute and the Geological Survey of Ireland, concerned with marine science and seabed mapping respectively (see Chapter 9: Underwater Surveys: The INFOMAR project), Ireland can boast a comprehensive array of shipping, navigation, defence and communications capabilities. No longer can it be said that Ireland is not a maritime nation.

THE COAST WATCHING SERVICE

Daire Brunicardi

Crudely constructed little concrete shelters, with glassless windows staring sightlessly over the ocean, can be found on various headlands around the Irish coast. These, and a few aerial signs, are all that remain of the wartime Coast Watching Service.[18] A depressed economy in the decades prior to the Second World War limited the development of coastal defences for Ireland. As the likelihood of war materialised in 1939, the government and the defence forces set about considering what action might be taken to protect the country. One course of action that prevailed, possibly because of its cost effectiveness, was the Coast Watching Service.

Established upon the declaration of war between Britain and Germany, the new service enlisted local men as volunteers in an army reserve unit. Some old coastal stations from the First World War were occupied initially, but in many cases crude Look Out Posts were constructed rapidly, many of which still exist (Figure 23.12). A twenty-four-hour watch was maintained, with two men on watch at any one

Fig. 23.12 REMNANTS OF THE COAST WATCHING LOOK OUT POST AT KNOCK-ADOON. a. The former Look Out Post (LOP) at Knockadoon Head is a good example of the type of shelters used to house coastwatchers who manned these stations, twenty-four hours a day, with pairs of observers working eight- or twelve-hour shifts. These formed part of a network of LOPs built all around the coast, following the hand-over of the Treaty Ports in 1938, as a measure to deal with the lack of defence and protection for Ireland; b. The view from the Knockadoon Coast Watching Service Look Out Post shows a strategic vantage point that would have been typical of the sites selected for LOPs, in order to report every activity observed in the air or on the sea. A number of historic log-book records have been preserved by the Irish Military Archives, and are available to view online on www.militaryarchives.ie. [Source: Darius Bartlett]

time. At first, any sighting reports required one of the men to leave the post and cycle to the nearest post office, often many miles away. Eventually, telephones were provided to all posts. After the invasion of France by the German armies, the number of posts was increased to cover the entire coastline, with each post within sight of at least one other (Figure 23.12).

Very soon, the reports from the Coast Watching Service showed that Ireland was indeed in the middle of a ferocious war. Belligerent aircraft made free with Irish air space; convoys were sighted on a daily basis; attacks on the convoys were frequently reported and casualties and survivors arrived or were washed up on the coasts. After the arrival of American forces in Britain and Northern Ireland in 1942, large signs were established at each Look Out Post, consisting of 15 foot letters spelling the word 'Eire' and the number of the post (Figure 23.13).

The Coast Watching Service continued until the end of the war in Europe. Its work, and the intelligence gathered, was of immense value in keeping the government and defence forces informed of the progress of the war and of dangers to the state and community (for example, floating mines were a continuous hazard). Much of the

Fig. 23.13. 'ÉIRE' SIGN AT BRAY HEAD. During the Second World War, the Coast Watching Service developed signage on headlands, usually near to Look Out Posts, to indicate to aircraft that they were flying over 'Eire' and thus neutral territory. Very quickly, the Americans published the positions and numbers of the signs on their aeronautical charts and they became an aid to navigation. A few of these signs are still visible, and some coastal communities have restored the signs as a historical feature of the locality. The image above, taken by the Irish Air Corps in summer 2018, shows the impact of gorse fires during a particularly dry summer. [Source: Irish Air Corps]

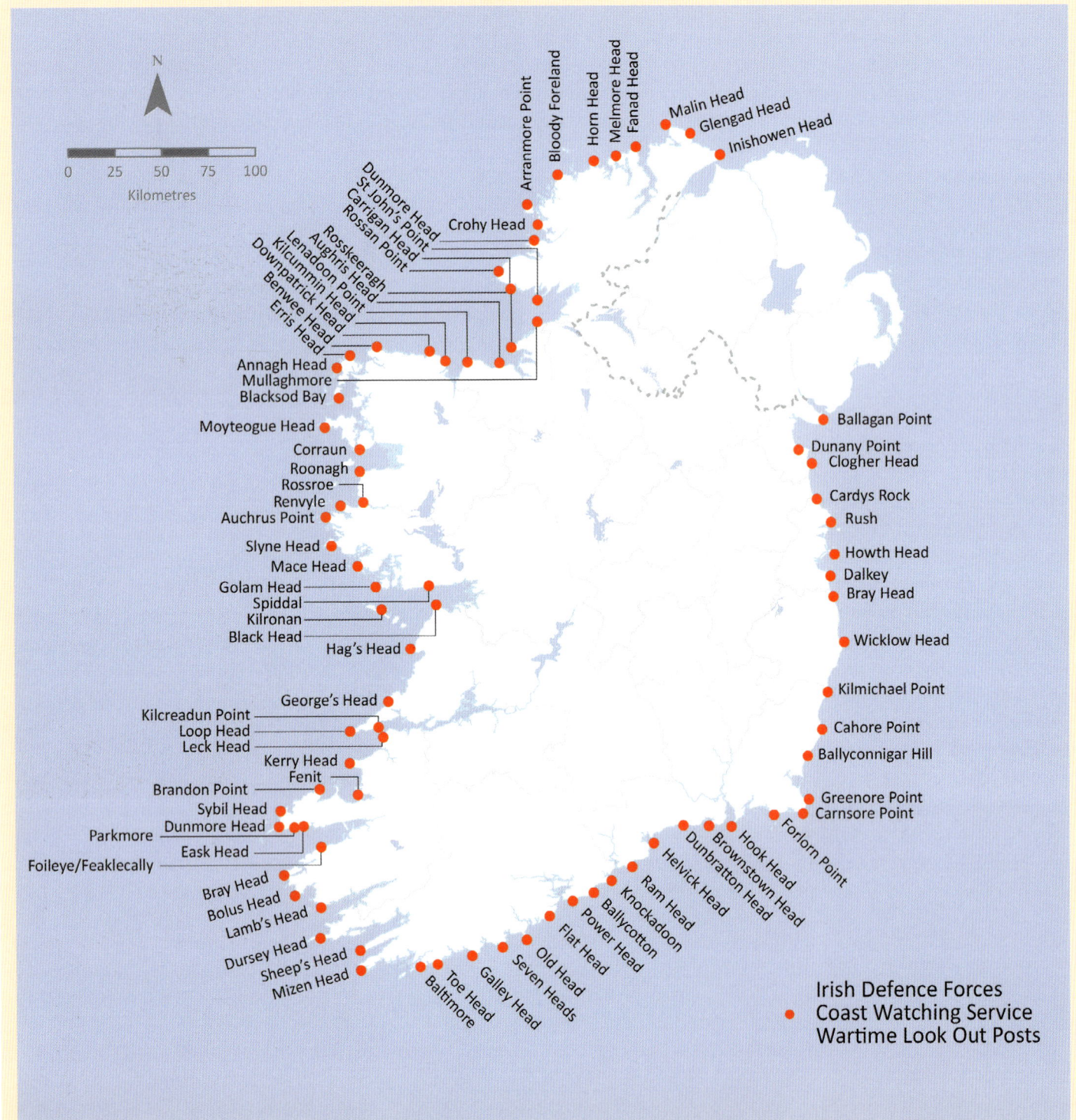

Fig. 23.14 DISTRIBUTION OF WARTIME LOOK OUT POSTS MAINTAINED BY THE COAST WATCHING SERVICE. Today, fifty-one Look Out Posts remain, while many of the eighty-three in total have been ravaged over time by the elements. [Data source: Irish Military Archives]

intelligence found its way to Allied interests, without publicity but with the agreement of government, a fact that has only recently been officially recognised. At the end of the war in Europe the Coast Watching Service was very quickly disbanded and the posts abandoned (Figure 23.14). The few remaining huts and 'Eire' signs remain as the only memorial to an important, even critical, service and are a historical legacy in Irish defence and maritime security.

The Daunt Rock Lightship Rescue

Ken Cotter

Fig. 23.15 THE DAUNT ROCK LIGHTSHIP RESCUE. [Source: Bernard Gribble/RNLI]

On the morning of 11 February 1936, following a severe south-easterly gale, an SOS was received by the Ballycotton lifeboat station. The lightship, *Comet*, had broken from its moorings and was drifting without power, posing a severe danger to the eight crew members onboard. Without waiting for orders, coxswain of the Ballycotton lifeboat, Patrick Sliney, took the decision to launch his vessel, the *Mary Stanford*. The *Comet* had drifted a quarter of a mile off-station overnight, and presented an extreme hazard to other shipping in the area. Initial attempts by the lifeboat crew to secure the vessel failed.

The *Innisfallen*, HMS *Toledo* and ILT *Isolda* stood by, allowing the *Mary Stanford* to return to Cobh a number of times to obtain essential supplies, including strong steel cables and fuel. Through strong seas, stormy winds and occasional fog, the lifeboat men eventually rescued all crew onboard (Figure 23.15). The extraordinary effort involved an especially daring rescue of the final two crewmen, who, due to fatigue, had to be physically grabbed from the guardrail of the stricken vessel. In total, the crew of the Ballycotton lifeboat spent forty-nine hours at sea, going twenty-five hours without food, and getting only three hours' sleep. By the time they returned to Ballycotton on Friday 14 February, they had been away from base for seventy-nine hours.

All of the crew of the *Mary Stanford* were awarded medals for their service, with Patrick Sliney receiving a gold medal for his 'bravery and fortitude'. The Daunt Rock rescue has entered the lore of the Royal National Lifeboat Institution (RNLI), who commemorated the achievement in a special stamp marking the institution's 150th anniversary.

Case Study: The Irish Naval Service

Daire Brunicardi

1940

Fig. 23.16 Aerial view of Haulbowline Island, headquarters of the Irish Naval Service, in 1940. Haulbowline provides a sheltered base within Cork Harbour, with easy and strategic access to the south coast of Ireland. To the left of the image (the western part of the island) is the main naval base, including stores and accommodation and ceremonial grounds. The eastern part of the island hosts the considerable basin that was built by the Admiralty, including a dry dock facility. This was developed on reclaimed land. Between 1938 and 2002, the Naval Service shared the island with Irish Steel (subsequently called ISPAT). The Irish Steel factory can be seen occupying the site towards the centre of the island, immediately to the west of the basin. Spoil from the factory was dumped over many years on the eastern flank of the island, giving rise to a large extended site of contaminated land. [Source: Irish Naval Service Archives]

As with many other aspects of Irish maritime history, the current naval service stems from a long tradition of Irish seafaring. Up to the Elizabethan era, the various maritime chieftains maintained fleets of galleys and longships that controlled and policed the seas in their areas. The most famous of these was Grace O'Malley, (Gráinne Ní Máille or Granuaile), the 'pirate queen' from Mayo (see Chapter 18: Era of Settlement: Trades, plantation and piracy).[19] After the decline of the Gaelic order in the early seventeenth century, the distinction between mercantile and naval service continued to be ill-defined; seamen traded in times of peace and manned warships during the frequent war years. This situation continued throughout the period of British maritime power into the twentieth-century.

Despite gaining independence in 1922, no concrete action was taken to form any sort of seagoing defence force. As a result, Ireland's territorial waters continued to be patrolled by the Royal Navy, as part of Britain's wider strategic view of European waters and the North Atlantic. In 1938, as war loomed in Europe, Britain's remaining naval bases in Ireland (Cork, Berehaven and Lough Swilly) were handed over to the Irish government. In early 1939 an order was placed by the Irish government, with J.L. Thornycroft in the United Kingdom, for two motor torpedo boats. The order was subsequently increased to six boats, but by the summer of 1940, only three of the torpedo boats had been provided. The two fisheries patrol vessels, the *Muirchú* and *Fort Rannoch,* were taken over by the defence forces, and each was armed with a twelve-pounder gun, some smaller weapons and a radio. An initial base for this emerging fleet was established in 1940 on the vacated Royal Navy base on Haulbowline Island in Cork Harbour (Figure 23.16). Haulbowline would become the headquarters of the Irish Naval Service and remains as its focal point to this day (Figure 23.17).[20] The ad hoc fleet was boosted in 1942 by the arrival of three additional torpedo boats and two other vessels, a small cargo steamer called *Shark* to be used as a 'mine planter', and a three-masted sailing schooner which would act as a training vessel. This was the fleet with which Ireland faced a world at war.

Two notable men were among several ex-Royal Navy personnel who formed this Marine and Coast Watching Service. Seamus Ó Muiris was from a distinguished west-coast family with a strong naval tradition. He took command of the service in 1941. Chief Petty Officer Jim Power, who had fought at the Battle of Jutland (as had Ó Muiris), was a mine and torpedo expert. These, and several other experienced Royal Navy officers and senior ratings, gave the new service its formative naval ethos.

Much of the Marine Service was disbanded after the Emergency (Second World War). The remainder was incorporated into a new service, the Naval Service, which became part of the Permanent Defence Forces. Three Flower-class corvettes were purchased from the United Kingdom, and recruitment started. The *Muirchú* and *Fort Rannoch* were decommissioned, one to the breaker's yard and the other to resume her role as a fishing trawler. Once again, officers and senior ratings from the Royal Navy were recruited, and others were sent to the Royal Navy for training. The purchase of three further corvettes was planned as well as a hydrographic survey ship and a training vessel.

A distinguished British naval officer, Captain H.J.A. Jerome, was employed to command the service for five years, a term that was subsequently extended to ten. As the memory of the crisis of 1939–1945 faded though, so did the fortunes of the Naval Service. No further ships were provided, and the motor torpedo boats were sold by 1950. Recruitment and retention of personnel became increasingly difficult. By the mid-1960s most of the now obsolete armament was stripped from the corvettes and they became solely fishery protection vessels. The ships were also showing their age and by 1969 only one was serviceable.

Moves were eventually made to modernise the Naval Service in the early 1970s.[21] An all-weather patrol ship was designed and laid down in the dockyard at Rushbrooke in Cork

Fig. 23.17 AERIAL VIEW OF HAULBOWLINE ISLAND IN 2018. Haulbowline Island has changed considerably from the 1940 image opposite. First, the Irish Steel factory, which closed in 2002, was demolished. The old factory site requires rejuvenation. However, priority has been given to the remediation of the contaminated slag heap to the east of the island, with over €40m invested by the state in a project to create a new landscaped environment, complete with a parkland for visitors to enjoy. The bridge connecting Haulbowline to Rocky Island and on to the mainland was constructed in 1966. The jetty in the foreground of the picture is part of the National Maritime Campus of Ireland (NMCI). The NMCI was developed adjacent to the Naval Service on Haulbowline Island, and is jointly run by the navy and Cork Institute of Technology. In 2015 University College Cork co-located the Beaufort Building, including the National Ocean Energy Test Tank Facility in Ringaskiddy, as part of the IMERC cluster initiative. The floating barge, also seen in the foreground, was deployed by the Port of Cork as a further development, involving a new public slipway and jetty, providing enhanced access to lower Cork Harbour for small boat users. [Source: SkyTec Ireland]

Fig. 23.18 FISHERIES REGULATORY ENFORCEMENT BY THE IRISH NAVAL SERVICE. The Naval Service acts as the official agency of the state for fisheries monitoring and enforcement. The Fisheries Monitoring Centre (FMC), headquartered on Haulbowline Island, Cork Harbour, co-ordinates monitoring and surveillance of all vessels with a Vessel Monitoring System (VMS) operating in the Irish Exclusive Fishery Zone. The FMC works in tandem with the Sea Fisheries Protection Authority and the Irish Air Corps, as well as co-ordinating with European agencies. Observations of the VMS data twenty-four hours a day every day, which involves satellite tracking, allow Naval Service personnel to identify any rogue behaviour, and to co-ordinate with Naval Service vessels at sea, involved in fisheries patrol and protection work. At-sea boardings are a fundamental component of the work of the Naval Service, necessary to enforce the quotas set by the Common Fisheries Policy (CFP) at the European level. This work requires deployment of Rigid Inflatable Boats (RIBs) from naval vessels, and boarding operations that are often conducted in challenging marine conditions, requiring high levels of training and skills of those involved. On average, the Naval Service will conduct *c.*2,000 boardings to inspect the catch and compliance of a fishing vessel, per year. [Source: David Jones, Irish Naval Service]

Fig. 23.19 IRISH NAVAL PASSING OUT PARADE. With a complement of *c.* 1,000 personnel from all across the island, the Irish Naval Service seeks to recruit and develop talent among officers, non-commissioned officers and able seamen. Challenges with pay and working conditions that can be less attractive than what is on offer by industry, hava made recruitment and retention of personnel an increasing issue for the defence forces. [Source: David Jones, Irish Naval Service]

Fig. 23.20 DRUG SMUGGLING BUST OFF THE IRISH COAST IN DUNLOUGH BAY. In 2007, €440 million of cocaine was recovered in Dunlough Bay off the west Cork coast. A small boat used by smugglers to bring illegal drugs ashore, capsized into the sea, leaving bales of cocaine floating in the bay. The Naval Service played a crucial role in the recovery of one of the biggest drugs consignments in the history of the state. This is one of many incidents requiring co-ordination of intelligence, also involving An Garda Síochána and Customs and Excise, to intervene in, and deter, criminal drug trafficking activity around the coast. [Source: David Jones, Irish Naval Service]

Harbour, and three minesweepers were purchased from the Royal Navy. A recruitment campaign was started and, by 1972, a reinvigorated Naval Service had four operational ships. In addition to the minesweepers, a naval diving unit was established, which became a major asset. This revitalisation of the navy was just in time. The worsening political situation in Northern Ireland had consequences for the Republic and the navy was able, for example, to arrest the gun-running ship *Claudia* in 1973. The event proved the necessity of a naval service for national security.

Since then, the fortunes of the navy have fluctuated. The accession to the European Economic Community (EEC) and regulations introduced under the Common Fisheries Policy created a need for new patrol ships, an improved version of the first 'all-weather' ship, and a larger helicopter-carrying patrol vessel. By the late 1980s, the minesweepers had been decommissioned and replaced by two fast coastal patrol ships. Today, fisheries protection is a key peacetime function of the Naval Service (Figure 23.18).

In 1996 the Irish Naval Service celebrated fifty years in existence. By this time the government was committed to maintaining a fleet of eight ships. The ships are underpinned by a complex of offices and logistical support facilities, still mostly located on Haulbowline Island. These consist of a small but effective dockyard, stores and supplies, the Naval College, the Naval Diving Unit and Naval Headquarters. The permanent force is retained at around 1,000 personnel. Today, recruitment and retention of talent to the navy is an important priority (Figure 23.19). There are also units of the Naval Reserve located in Dublin/Dún Laoghaire, Cork/Cobh and Limerick. These part-time naval personnel focus on the seaward defences of their particular coastal areas, and also train from time to time with the regular service.

The White Paper on Defence designates the Naval Service as the principal seagoing agency of the state, with a general responsibility to meet contingent and actual maritime defence requirements.[22] It provides an essential presence at sea, enforcing the state's entitlement to the vast maritime area to which Ireland lays claim. The presence of the Naval Service as a deterrent to gun-running, drug-smuggling and other illegal activities is key to the management of Ireland's territorial waters (Figure 23.20). The involvement of the Naval Service in humanitarian work in the Mediterranean emphasises the mobility and versatility of the service. Also, the appointment of a naval officer as chief of staff of the defence forces, with the rank of vice admiral, indicates a new recognition of the importance and status of the Naval Service within the defence organisation and within the state.

IRELAND AND THE HUMAN TRAFFICKING CRISIS IN THE MEDITERRANEAN

Brian FitzGerald

In April 2015, in the face of a great tragedy where 800 migrants were drowned in a single incident in the Mediterranean Sea south of Sicily, a European response was triggered. Ireland was one of the first countries to react with the commitment of a naval ship to participate in Search and Rescue (SAR) and humanitarian operations there. The Irish mission was named 'Operation Pontus' (after the Greek god of the sea) and comprised seven discrete ship deployments between 2015 and 2017, before subsequently joining the multinational EU Naval Force mission 'Operation Sofia' in late 2017, with a focus on solving the root causes of the crisis in addition to the humanitarian element. During

Fig. 23.21 'OPERATION PONTUS' DEALING WITH HUMAN MIGRATION IN THE MEDITERRANEAN. Irish Naval Service personnel in the co-ordination of the rescue of migrants in the humanitarian intervention under 'Operation Pontus' in the Mediterranean Sea. The professionalism and commitment of the men and women of the Naval Service in 'Operation Pontus' demonstrates how a small island nation can contribute significantly to international maritime situations. Although this scenario appears far removed from Irish shores, the global aspect of the migrant crisis demands preparedness and capability in monitoring and observing for human trafficking around the coast of Ireland, which is ongoing by the Naval Service on behalf of the state. [Source: David Jones, Irish Naval Service]

Fig. 23.22 NAVAL SERVICE PERSONNEL RETURNING HOME FROM A MISSION IN THE MEDITERRANEAN GREET THEIR FAMILY MEMBERS.[Source: David Jones, Irish Naval Service]

the period of slightly more than two years when Ireland participated in 'Operation Pontus', the Irish Naval Service rescued 17,404 migrants and coordinated or assisted in the rescue of several thousand more (Figure 23.21).

In 2017 the crisis was ongoing with wave after wave of migrants finding themselves adrift in the middle of the Mediterranean Sea as they endeavoured to cross from Africa and the Middle East to Europe in the pursuit of a better life. Irish sailors were faced with near overwhelming situations as they set about their tasks in saving lives, representing the very best traditions of the Naval Service and maritime Ireland.

For example, in a two-day period the seventy-strong crew of the LÉ *Eithne* directly rescued 1,417 migrants from unseaworthy craft, while simultaneously coordinating a number of non-governmental organisation vessels to rescue a total of 2,760. The lessons learned by Ireland in both the rescuing of migrants and the evolving system in the central Mediterranean are from the core of the maritime dimension of the crisis taking place there. The smuggling and trafficking of migrants by sea is a maritime security matter and dealing with it on such a scale requires a multi-faceted response.

The migrant flow depends heavily on trafficking, while its scale and geophysical nature makes it an international crisis. Dealing with the challenge is a substantial undertaking. The crisis has brought two key principles of international law into stark relief: first, the duty of mariners and coastal states to render assistance to those in peril at sea and second, the right of sovereign nations to control entry of non-nationals to their territories. Dealing with mass flows of migrants also presents overwhelming SAR challenges. States have therefore had to react in a much more hands-on and coordinated manner, with international cooperation emerging as key. Ireland can be proud of its response to date (Figure 23.22).

Case Study: Irish Shipping in the Second World War

Daire Brunicardi

At the time of the Second World War, the Irish Free State was a member of the British Commonwealth. More than any other Commonwealth member, the Irish Free State was strongly tied to Britain. Almost all external trade from Ireland was with Britain; the currency was sterling and controlled by the British Treasury; and the free movement of people between the two states emphasised the vague status of Irish independence.

Fig. 23.23 *The City of Limerick*, A KENNETH KING PAINTING OF AN IRISH MERCHANT SHIP IN THE SECOND WORLD WAR AFTER BEING TORPEDOED BY A GERMAN BOMBER. The painting depicts the vulnerability of Irish merchant shipping during the Second World War, and the risks and sacrifices made by seafarers to keep sea lines of communication and trade open. The neutral markings, such as tricolours and 'Eire' lettering on the side of ships, did little to protect them from attack, resulting in sinking and the loss of lives of many heroic sailors at sea. [Source: Kenneth King (1939–2019), on display in the National Maritime Museum of Ireland]

Declaring and maintaining neutrality during the Second World War was a defining event in the evolution of Irish statehood.[23] The Irish declaration of neutrality brought many aspects of Irish/British relations to the fore. The use of Irish bases for the defence of shipping was one issue; trade was another. Initially there was a reluctant acceptance of Irish neutrality on the part of the British government. The Irish government was pressed for the use of Irish bases which had only been vacated by British forces in 1938. This was refused, but an agreement was reached such that a certain percentage of shipping would provide for Irish needs. This was adhered to up to the summer of 1940. This agreement was vital to Ireland. It was far from self-sufficient in many commodities, including such basics as fertilisers, animal feedstuffs, wheat, coal, timber, tea, tobacco and many manufactured goods. Neither had it virtually any shipping of its own. The Irish-registered merchant fleet consisted of about fifty small ships intended only for voyages to Britain and the European continent. Many of these were sailing schooners, the majority of which were owned by various Arklow families.[24]

Irish–British relations deteriorated dramatically in the summer of 1940. France had capitulated as the German armies swept through the country and an invasion of Britain seemed imminent. The Germans now had the French Atlantic ports as submarine bases and the use of airfields there, and attacks on convoys of merchant ships intensified. When Winston Churchill became prime minister in the United Kingdom he demonstrated a visceral detestation for Irish neutrality. Churchill was intent on pressurising the Irish state. Why should British shipping carry Irish supplies through submarine and aircraft attacks, where huge numbers of ships were being sunk and seamen killed, when the Irish government refused the use of bases for the defence of that shipping?

By the winter of 1940 that pressure was certainly being felt in Ireland.[25] Severe food shortages were likely but, with strict censorship in force, the true extent of the crisis was kept from the public. The few small ships did their best, in spite of the risk of being attacked and sunk. The neutral markings (tricolours and 'Eire' in large letters painted on the ships' sides and flood lighting at night) did little to protect them (Figure 23.23). An taoiseach, Éamon de Valera, reported to the Dáil on 20 February 1941, concerning the attacks on Irish ships in 1940:

Of eight attacks, two of them twice, four were sunk, twenty lives lost and seven men injured. In addition to these four ships three others were sunk by mines and the cause of the loss of two others has not been ascertained. Seven of the ten attacks were from the air. In all but one case the attackers were identified as German. It is right to say however that I have also received reports of cases in which German planes have circled and examined Irish ships without attacking them.[26]

He went on to say that the loss of life and the sinkings had 'occasioned feelings of deep resentment here' and that the government had protested and lodged a claim for compensation with the German government.

Shipping operations became increasingly constrained. Ships trading to and from Ireland were required to adhere to a British imposition on shipping in the war zone, by presenting for inspection and a 'Navicert' at the port of Fishguard, on each return trip. Sailing in convoy meant delays and if sailing independently (to Lisbon almost exclusively), a long course, well to the west, had to be followed (Figure 23.24). Britain had almost exclusive control over chartering available vessels, and the United States had forbidden any of its shipping entering the 'war zone' around Britain and Ireland. Lisbon then became the transhipment port for any American cargoes, meaning further delays and restrictions on Irish trade and supplies (Figure 23.25).[27]

The supply situation became dire. Transport came to a halt, domestic gas almost ceased, tea and many normal foodstuffs were severely rationed. Faced with such a critical situation the Irish government embarked on a risky project in the spring of 1941 and established its own

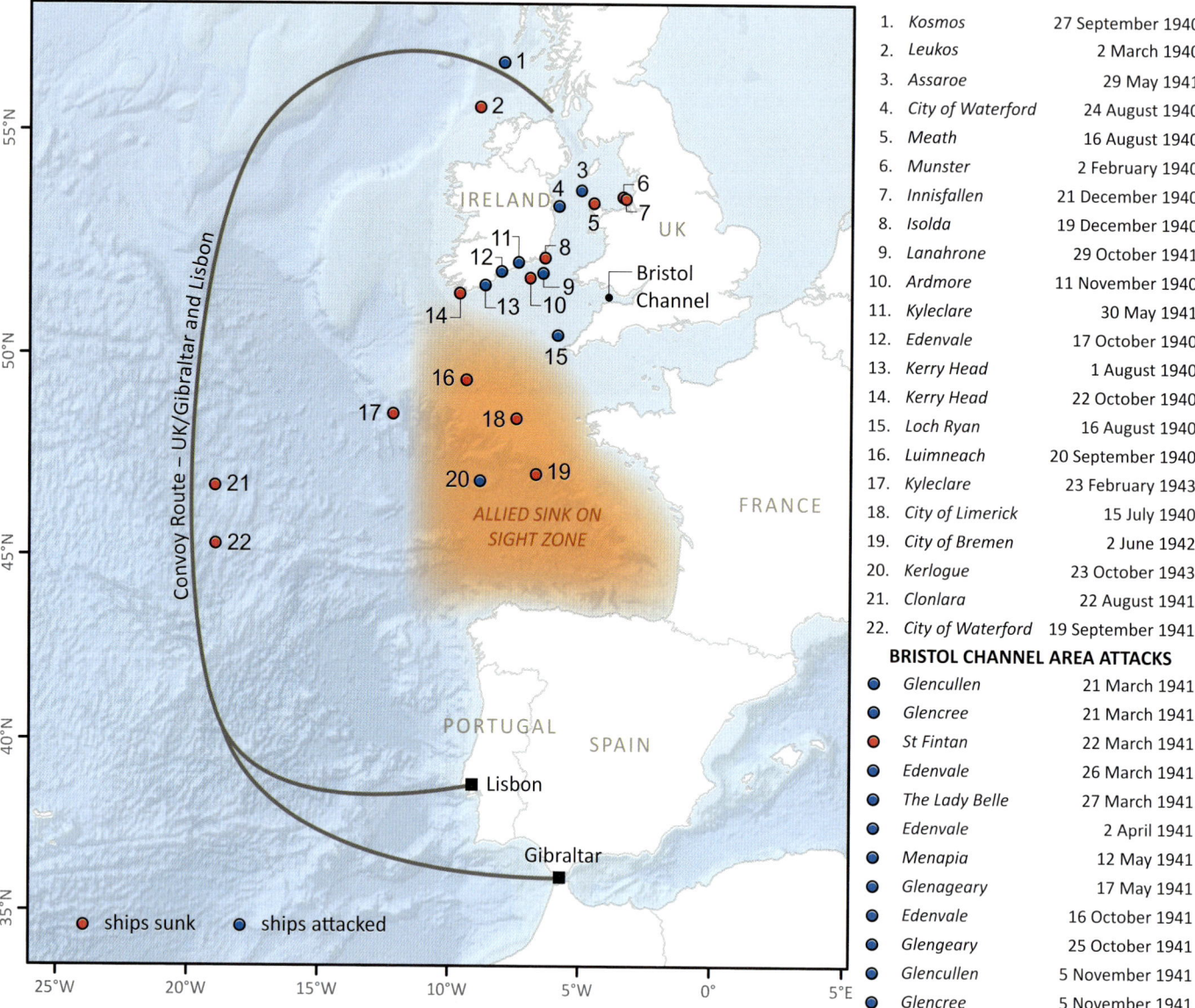

1.	Kosmos	27 September 1940
2.	Leukos	2 March 1940
3.	Assaroe	29 May 1941
4.	City of Waterford	24 August 1940
5.	Meath	16 August 1940
6.	Munster	2 February 1940
7.	Innisfallen	21 December 1940
8.	Isolda	19 December 1940
9.	Lanahrone	29 October 1941
10.	Ardmore	11 November 1940
11.	Kyleclare	30 May 1941
12.	Edenvale	17 October 1940
13.	Kerry Head	1 August 1940
14.	Kerry Head	22 October 1940
15.	Loch Ryan	16 August 1940
16.	Luimneach	20 September 1940
17.	Kyleclare	23 February 1943
18.	City of Limerick	15 July 1940
19.	City of Bremen	2 June 1942
20.	Kerlogue	23 October 1943
21.	Clonlara	22 August 1941
22.	City of Waterford	19 September 1941

BRISTOL CHANNEL AREA ATTACKS

●	Glencullen	21 March 1941
●	Glencree	21 March 1941
●	St Fintan	22 March 1941
●	Edenvale	26 March 1941
●	The Lady Belle	27 March 1941
●	Edenvale	2 April 1941
●	Menapia	12 May 1941
●	Glenageary	17 May 1941
●	Edenvale	16 October 1941
●	Glengeary	25 October 1941
●	Glencullen	5 November 1941
●	Glencree	5 November 1941

Fig. 23.24 MERCHANT SHIPPING ROUTES TO LISBON AND GIBRALTAR AND APPROXIMATE LOCATIONS OF SUNK AND ATTACKED IRISH MERCHANT SHIPS DURING THE SECOND WORLD WAR. Ireland's failure to develop the mercantile marine following the War of Independence resulted in a lack of vessels. Foreign ships, upon which Ireland depended, were less available from the outbreak of the Second World War. This factor, together with the inherent dangers to merchant shipping during the war, placed serious constraints on the movement of vital cargo, such as the import of fuel, fruits, cereals, fertiliser and tea, into the country, and the export of agricultural produce, primarily agricultural goods to Britain. Ireland suffered from severe hardship as a result of this disruption to trade. The majority of casualties occurred off the south and east coasts, with Irish merchant ships falling prone to attack as far as the Bay of Biscay. Trade across the Irish Sea was made increasingly difficult as Germany issued an ultimatum on Irish food exports to Britain. Merchant shipping routes were adapted, following Roosevelt's intervention preventing American ships from entering the war zone. As a result, Lisbon became an important transhipment hub. Irish ships on the 'Lisbon run' imported fruits and wheat from Iberia, as well as goods transhipped there from the Americas. This work was often conducted by small coasters, not originally designed for the function, which exacerbated the peril of going to sea during these times. Gibraltar also provided a transhipment function for Allied shipping. [Data source: Forde, F., 2000. *The Long Watch*. 2nd ed. Dublin: New Island Books.]

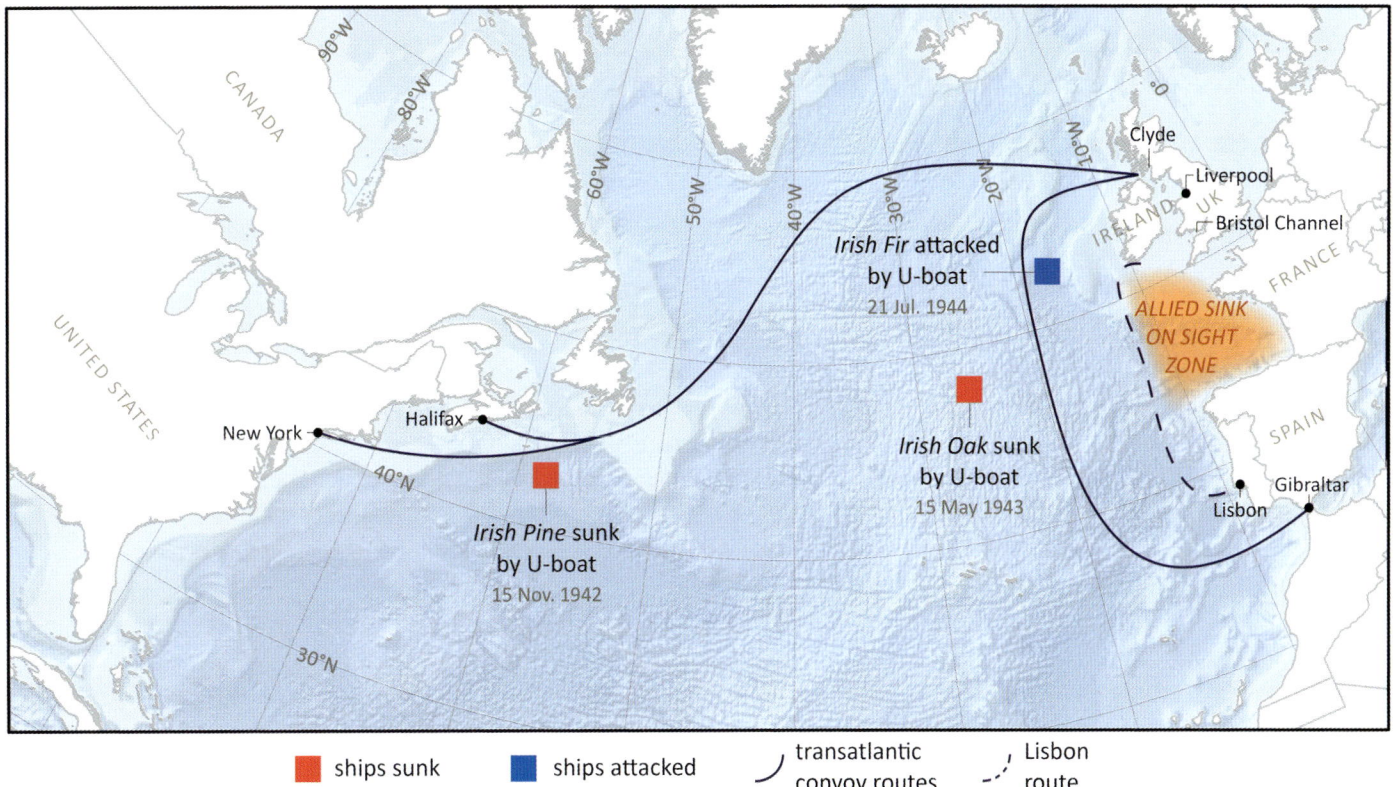

ships sunk ships attacked transatlantic convoy routes Lisbon route

Fig. 23.25 MAP SHOWING THE TRANSATLANTIC SHIPPING ROUTES USED BY IRISH MERCHANT SHIPS DURING THE SECOND WORLD WAR. Also highlighted are the locations of Irish merchant ships attacked and sunk in the broader Atlantic basin. [Data Source: Forde, F., 2000. *The Long Watch*. 2nd ed. Dublin: New Island Books]

merchant shipping fleet. Almost the entire US dollar holding of the state was used for this purpose. Fortunately, there was expertise in shipping operations and brokerage available in the existing Irish companies, principally Palgrave Murphy's Ltd, Wexford Steamship Company and Limerick Steamship Company. Also, the officials of the various government departments (often accused of lack of imagination and restrictive attitudes and practices) proved enthusiastic and imaginative, notably John Leydon, permanent secretary of the Department of Supplies and J.J. McElligott, secretary of the Department of Finance. They provide an instructive example of what can be done by Irish politicians and administrators in a crisis.[28]

The first difficulty was finding ships, as they were scarce and prices exorbitant, but several vessels were acquired through various means. Some ships were interned in Irish ports because of their states' involvement in the war. With great difficulty, two ships were chartered from the United States. Even though neutral itself, the Roosevelt government condemned Irish neutrality and, ironically, one of the hindrances in chartering the ships was the necessity of clearing the financial transaction with the British Treasury. Both of these ships were ultimately lost in Irish service due to belligerent action, although the fate of one, the *Irish Pine* and her entire crew, did not come to light until over twenty years later.[29]

Fifteen ships were eventually acquired. Initially, they sailed in convoy, but after a disastrous convoy battle in August 1941– in which one Irish ship was among two dozen sunk – Irish ships generally sailed independently. This increased productivity and although the Allies imposed circuitous routes, this was compensated for by the elimination of delays in convoy assembly. What of the men? The long tradition of Irish seafaring came to the nation's aid at this critical time. Whatever the difficulties there were in acquiring ships, no ship was held up for lack of crew. The sacrifice of those civilian seamen in the Second World War is no less heroic than that of their brothers in the British merchant navy, if less recognised by modern Ireland. The number of men killed due to belligerent action was 149, with thirty-two wounded. This was out of no more than about 800 at most. It is also worth mentioning that Irish ships rescued 521 men of all nationalities from ships attacked or sunk during the war. An impressive monument on City Quay in Dublin records the names of those lost on Irish ships during the war.

In all, Irish Shipping Ltd carried 712,000 tonnes of wheat, 178,000 tonnes of coal, 63,000 tonnes of phosphates, 24,000 tonness of tobacco, 19,000 tons of newsprint and 10,000 tons of timber to Ireland in those critical years, not forgetting the efforts of the fleet of smaller ships. This allowed the government to continue its policy of neutrality, which it would otherwise likely have been forced to abandon in 1941 or later.[30]

It should be noted that many Irish seafarers were also in employment in British ships, enduring all of the wartime dangers and horrors that this implied. The wartime sacrifices of the British merchant navy is one of the great examples of endurance and fortitude in which that nation justifiably takes immense pride. The magnificent monument to them is on Tower Hill, near the Tower of London, on which all the names of those lost during two world wars appear. There are dozens of Irish names there.

Case Study: Nautical Education in Ireland

Daire Brunicardi

Nautical education, as we know it today, has its roots in the introduction in 1851 of compulsory certification of masters and mates. Prior to this the approach to nautical training was piecemeal and, as a result, many mariners lacked even the most basic skills of seamanship and discipline. Conditions on British-registered ships were so appalling that in 1836 a commission was set up by parliament to investigate complaints. This found that the abnormal loss of life and ships was due to inferior construction, incompetence and drunkenness of masters and mates and the lack of harbours of refuge. As a result, the commission recommended the formation of a Mercantile Board to lay down a standard of professional knowledge by starting examinations and establishing nautical schools.

A bill was passed in 1845 providing for voluntary examinations for masters and mates. Since there was no government department with responsibility for shipping, the examinations were conducted by local pilotage authorities. For example, in Dublin, the Ballast Board (later the Dublin Port and Docks Board and now the Dublin Port Company) was charged with administering the examinations. Later, certificates were introduced for additional skills, such as engineering, and in 1901 radio officer certificates came into force. These were issued and administered by the Post Office.

As would be expected, the compulsion to pass examinations created the demand for schools to help candidates with their studies, even though there was no requirement in the examinations to attend a course of study. In Ireland, in the 1850s and 1860s, there were, as in the rest of the United Kingdom, public schools offering courses for masters, mates and apprentices. For example, in 1857 a school of navigation was established in the Waterford District Model School. Concurrent with these official schools, private academies were also set up by shipmasters in various ports. These were cramming establishments geared towards coaching candidates for successful examination outcomes.

In 1889, Captain H. Griffith Quirke opened his School of Navigation at 23 Eden Quay, Dublin, which operated until 1925. Its closure can be related to the disturbed state of the country during the War of Independence, which inhibited the attendance of many students from North Wales, who came traditionally to study in Ireland. During the same period a nautical academy operated in the Belfast Shipping District, where it continued until 1950. It was subsequently absorbed by the technical college.

THE IRISH NAUTICAL COLLEGE

The Irish Nautical College and Training School opened in 1925 in Dublin, with Captain R.E. Kellett as its first principal. During the fifty years of its existence, it specialised in preparing experienced seamen, apprentices and junior officers for their various statutory examinations. The college relocated from its initial site at Aston Quay in the 1930s, to Sir John Rogerson's Quay in 1953 and, finally, to bigger premises on the west pier in Dún Laoghaire. There, it extended its courses to include preparation for Lifeboatman's Certificate, Radar Observer and Simulator Certificates, and also introduced short induction courses for boys who had left school and were awaiting appointment to a ship. Block release courses were provided for groups of fishermen, and evening classes for yachtsmen were held during the winter.

By the late 1960s it was considered important to incorporate the various seagoing disciplines. By this time, however, there was no further room for expansion within the technical colleges in Dublin. In 1970, therefore, the Department of Education decided that Irish nautical education would be best served by integrating within the new Regional Technical College building, which was planned to open in Cork in 1974. Cork was chosen on the basis that its old technical college had provided a marine engineering course for cadets since 1955.

THE DEPARTMENT OF NAUTICAL STUDIES IN CORK REGIONAL TECHNICAL COLLEGE

In 1975, the Irish Nautical College and Training School in Dublin was closed. Its staff and several items of equipment were moved to Cork, to combine with the existing marine engineering facility to form a new Department of Nautical Studies. Even with this consolidation, marine training did not meet modern requirements. Three years after the move, issues with the provision of training were still being identified. These included, for example, the absence of training for deck cadets and radio navigational instrumentation technicians. In addition, the radar observers' course, formerly held in Dublin, was suspended due to the lack of a suitable site to position a live radar set in Cork.

On the positive side, a radar simulator had been transferred from Dublin and the tender, *Cill Airne*, was purchased as a training ship. Overall, however, it could be argued that Ireland's deficiencies in marine education and training at this time were due to national apathy regarding marine affairs, but things improved gradually. The training vessel, *Cill Airne*, was equipped with radar, which allowed radar observer training to resume. Lifeboat and Efficient Deck Hand courses were provided, and an extra full-time deck lecturer was recruited. The deck

Certificate of Competency courses became very popular with foreign students.

In 1984 the collapse of Irish Shipping Ltd was perceived to have had a negative impact on the employment opportunities within the country's maritime sector. This, of course, was not entirely the case, as several Irish shipping companies continued to operate a fleet of approximately fifty ships, which maintained a demand for well-trained seafarers. Many Irish mariners also continued to find employment in foreign shipping, mainly British.

The staff in the Nautical Studies Department of Cork Regional Technical College moved quickly to limit the damage caused by the demise of Irish Shipping Ltd. A programme for training school-leaving students to qualify as deck officers was proposed. The first students under this scheme were enrolled in 1985, the year after the demise of Irish Shipping Ltd, thus ensuring the continuation of career opportunities. Two years later the programme was accredited as a national diploma in science and it has continued with increasing success ever since. The marine engineering diploma course was also modified to meet the needs of graduates wishing to pursue shore-based employment opportunities.

THE NATIONAL MARITIME COLLEGE OF IRELAND (NMCI)

By the late 1990s the situation in the Department of Nautical Studies at Cork Institute of Technology (CIT) was becoming critical. Competition for space in the Bishopstown campus led to the consideration of moving to an alternative location. In parallel, the need to comply with the International Convention on Standards of Training, Certification and Watchkeeping for Seafarers in 1995 (STCW95), as specified by the International Maritime Organisation, created the urgent need for new facilities in the areas of sea survival and navigation. Even without these requirements the need for such re-equipping was imminent; the code made it critical.

Similarly, in the Naval Service, training accommodation and facilities were in need of major improvement. This was made more urgent by a management decision that naval officers should also be compliant with the provisions of the STCW95 code. Discussion between naval officers and CIT staff led to a proposal for a solution to the problems of both organisations. A proposal was made to government that a Department of Defence site at Ringaskiddy be developed to host a joint college, to derive value and synergies from shared facilities.

A government task force also recommended a subsidy for training and the establishment of a joint naval/mercantile marine college. A public–private partnership arrangement resulted in the construction of the present college in 2005. To this day, the NMCI is a world-class, state-of-the-art facility that uniquely trains naval officers, merchant marine cadets, ships watchkeeping, senior officers (both deck and marine engineers) and various maritime degree students under one roof. NMCI is a 14,000m² facility built on a ten-acre waterside campus (Figure 23.26).[31] Specialist spaces include survival pools, seamanship and shipwright workshops, firefighting and damage control units, a jetty and lifeboat facilities, as well as high-tech bridge and engine room simulators (Figure 23.27). It offers undergraduate programmes in nautical science and marine engineering to up to 800 undergraduates, as well as providing continued professional development to over 1,000 marine professionals from overseas on an annual basis.

Fig. 23.26 AERIAL IMAGE OF THE NATIONAL MARITIME COLLEGE OF IRELAND, RINGASKIDDY. The National Maritime College of Ireland (NMCI), was established in 2006 as a public–private partnership, to deliver training and education to the merchant marine and Naval Service personnel. The training of civil and military personnel under one roof is a unique aspect of NMCI. The college focuses training on nautical science and marine engineering. It is equipped with state of the art facilities such as a 360-degree bridge simulator and a sea survival pool. To the foreground is the Beaufort Building, which includes the National Ocean Energy Test Tank Facility. The Beaufort Building, owned by University College Cork, is part of a cluster of marine expertise in lower Cork Harbour. [Source: SkyTec Ireland]

Fig. 23.27 FACILITIES IN THE NATIONAL MARITIME COLLEGE OF IRELAND (NMCI) INCLUDE TRAINING POOLS TO PREPARE PERSONNEL WORKING OFFSHORE FOR SCENARIOS INVOLVING DITCHING AT SEA. The helicopter simulator is provided by Irish company Seftec. Images a.–d. show the rigorous Helicopter Underwater Escape Training (HUET) students are obliged to undertake in preparation for working on offshore platforms, helicopter flight crews or military operations. Trainees are gradually submerged in water, the helicopter simulator is rotated upside down, and students must swim through the open hatches to free themselves. Divers accompany the students to ensure their safety during training. [Source: David Jones, Irish Naval Service]

The Irish Coast Guard

Daire Brunicardi

The Coast Guard was established in 1822, with the principal purpose of aiding coastal defence, preventing smuggling and protecting revenue around the coast of Britain and Ireland.[32] During the nineteenth century, while its original functions remained, its remit was expanded gradually to incorporate the prevention of illegal fishing, Search and Rescue, and other emergency roles, such as assisting in famine relief.

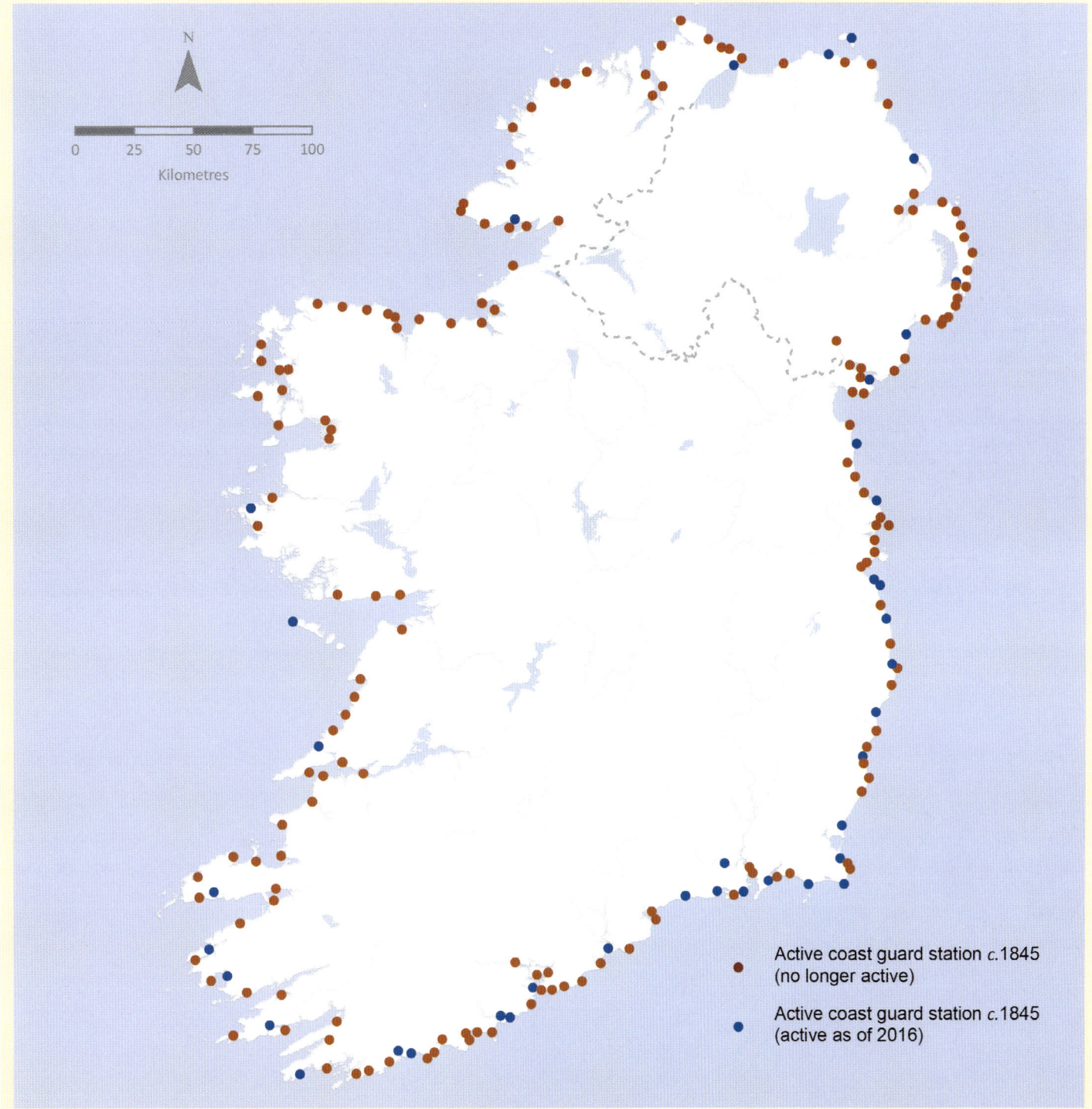

N

0   25   50   75   100
Kilometres

Active coast guard station *c.*1845
(no longer active)

Active coast guard station *c.*1845
(active as of 2016)

Fig. 23.28 PAST AND PRESENT COAST GUARD STATIONS. The Coast Guard established in 1822, evolved into the present-day organisation, with responsibility for Search and Rescue of life at sea, as well as other roles such as coordination of marine pollution efforts. Originally active in coastal defence its members formed part of the Royal Naval Reserve. Today, as part of the Department of Transport, coast guard operations cover the entire coastline of Ireland. While there are fewer stations than previously, the technology and capability is significantly enhanced. [Data source: Irish Coast Guard, coastguardsofyesteryear.org and Ordnance Survey Ireland historic maps]

In 1858, the Coast Guard came under the management of the Admiralty, and formed part of Britain's naval reserve. In Ireland, a number of station ships were located in ports around the coast, each ship being the headquarters for a coast guard district. Almost every harbour and creek had a coast guard station, amounting to over 150, and covering the whole coast of Ireland (Figure 23.28). These shore stations were supported by a number of revenue or coast guard cutters. Each year before the First World War, the coast guard men, who were mostly naval reservists, would present to the station ships. These would join the flagship of the 'Admiral Commanding on the Coast of Ireland' for training exercises, together with other units of the fleet from the United Kingdom.

Despite strong ties between the Coast Guard and the Royal Navy, the general public never developed the same level of antipathy towards the Coast Guard as they did, for example, towards the Royal Irish Constabulary. Nonetheless, they were regarded as agents of the state and, therefore, became a target for insurgents during the War of Independence. Coast guard stations were attacked in order to

Fig. 23.29 IRISH COAST GUARD RAPPELLING DOWN THE SIDE OF A CLIFF ON TRAINING MISSION NEAR HOWTH HEAD. Composed of about sixty full-time technical employees, the voluntary service members within the Coast Guard are a crucial resource. All coast guard personnel, whether full time or voluntary, carry out significant training in anticipation of a range of scenarios, from the shore, at sea, and from the air. Cliff rescue one of the functions of the coastguard is a highly skilled and dangerous operation. In 2016, coast guard volunteer Caitriona Lucas died in an incident at sea, the first time a Coastguard volunteer died during a search and rescue operation. Tragedy struck again in March 2017 when a coast guard helicopter crashed off the coast of Mayo and all four crew members on board were killed. Tributes were paid to the bravery of the men and women of the coast guard who sacrifice so much to ensure the safety of others. [Source: Irish Coast Guard]

commandeer the small amount of firearms and other equipment held in storage (see Chapter 15: Coastal Heritage). Generally, coast guard personnel were not harmed, and the presence of women and children in the stations usually meant that little resistance could be offered. All stations were evacuated in 1922.

On its foundation, the Irish Free State did not retain a coast guard organisation. Instead, it set up a minimal and voluntary Coast Life Saving Service (CLSS), later renamed the Coast and Cliff Rescue Service (CCRS). These were established to comply with international obligations, although some of the duties formerly under the remit of the Coast Guard were fulfilled by An Garda Síochána in the coastal villages. As time progressed, there was little development in maritime rescue services compared to other countries. In the 1960s, however, the Marine Rescue Coordination Centre was established at the naval base in Haulbowline, but this was transferred to air traffic controllers at Shannon airport in the 1970s. In spite of this, an inordinate reliance continued to be placed on

Fig. 23.30 IRISH COAST GUARD HELICOPTER NEAR KEEM BAY, ACHILL ISLAND. The Coast Guard contracts five medium-lift Sikorsky Search and Rescue helicopters, deployed from bases at Dublin, Shannon, Sligo and Waterford. Personnel are highly trained to deal with a range of terrestrial and marine-based missions, such as evacuation of survivors at sea, including rescue of personnel from ships facing imminent disaster. [Source: Irish Coast Guard]

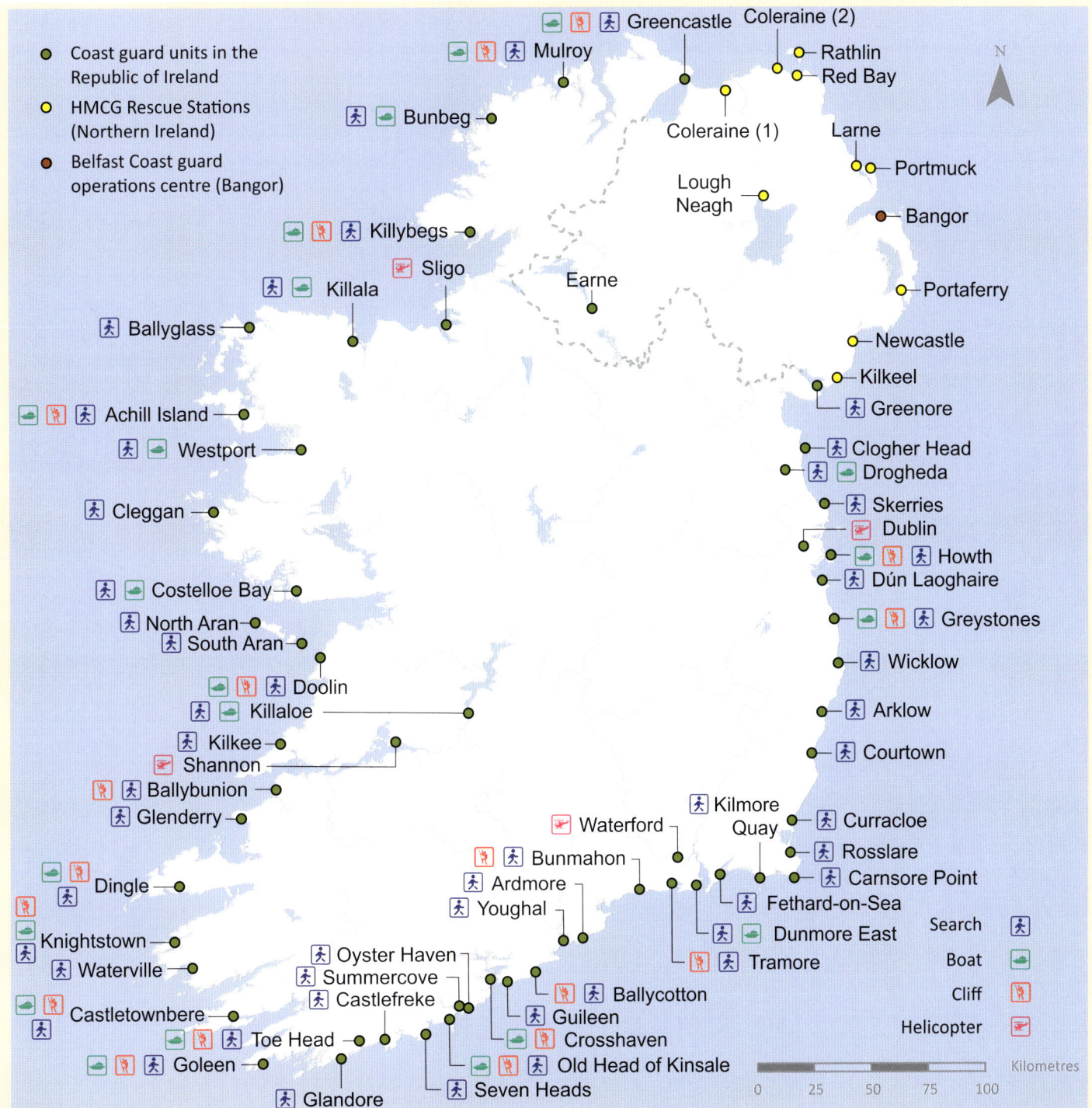

Fig. 23.31 CURRENT COAST GUARD OPERATIONS. The Irish Coast Guard is a civilian agency that calls upon the assets and support of the defence forces, including the Irish Naval Service and Air Corps, to respond to emergencies. Fifty-five coast guard units with one hundred volunteers are available around the clock, all year round. The community coast guard stations provide a range of services, including boat and cliff Search and Rescue capabilities, that can be mobilised in response to an incident at the coast. [Data source: Irish Coast Guard]

British services to provide helicopter coverage and other SAR (search and rescue) services in Irish waters.

In the 1980s two incidents brought matters to a head. One was a situation where a fisherman died as a result of an accident at sea when, had a rescue helicopter been available, he might have survived. This led to significant lobbying of the government to provide for a national air-sea rescue service. The other incident was the sinking of a huge bulk carrier. The *Kowloon Bridge* had been damaged at sea in heavy weather and was subsequently abandoned. She ultimately foundered on the Stag Rocks near Baltimore, in west Cork. There was no clearly designated government agency to take control of the situation with the power to make decisions or to take actions to prevent damage and pollution to the environment. As a result, the Irish Marine Emergency Service (IMES) was established in 1991, incorporating the CCRS. Its director was given powers to take charge in such situations and to set up and coordinate rescue services around the coast. A fleet of helicopters with a reasonable offshore range was provided; these were operated by the Air Corps from strategic locations around the coast.

The improved and expanded service changed its name in 2000 to become the Irish Coast Guard (IRCG). From its headquarters in Dublin, the IRCG continues to focus on maritime safety and SAR in the coastal waters around Ireland (Figure 23.29). To achieve this, it uses a variety of modern rescue vessels and other equipment located at its coastal stations. It has also modernised its helicopter fleet, using five larger and longer-range machines acquired under a contractual basis, and operated by a civilian crew. The red and white helicopters, stationed at Dublin, Sligo, Shannon and Waterford, are a familiar sight around the coast (Figure 23.30). Three communications stations are manned on a twenty-four hour basis, at Dublin, Valencia and Malin Head. Through a series of remotely operated radio stations, these monitor radio traffic and distress frequencies covering the Irish maritime area. In the event of a distress call, or other incident at sea, one of these stations will coordinate the response.

In addition to the IRCG stations, units consisting of volunteers located in various coastal communities are a vital element of the service (Figure 23.31). These units are equipped differently, depending on the nature of the coast and volunteers are trained accordingly. For example, areas with coastal cliffs will have cliff scaling equipment. Some units have Rigid Inflatable Boats (RIBs), but the emphasis is on inshore rescue, rather than offshore, which is generally covered by the RNLI.

Community Rescue Boats Ireland (CRBI) is also available to assist the IRCG. This is a nationwide group of independent, voluntary rescue boats and crew, trained and administered by Irish Water Safety. Currently, there are seventeen such units available in coastal communities that have been affected by local drowning tragedies.

THE COMMISSIONERS OF IRISH LIGHTS

Daire Brunicardi

The Commissioners of Irish Lights (CIL) carry out the international obligations of the British and Irish governments in relation to the provision of an aids to navigation (AtoN) service, including lighthouses, buoys, beacons and various electronic and radio navigation aids (Figures 23.32 and 23.33) around the coast of Ireland, for the safety of navigation under the international Safety of Life at Sea Convention (SOLAS).[33] The provision of such aids is commensurate with the volume of traffic and degree of risk in any particular area.

The provision of aids to navigation is said to go back to the fifth century, when the monks of Rinn Dubháin in County Wexford (now known as Hook Head) lit a beacon to warn shipping away from dangerous rocks. Around 1207, the lord of Leinster, William Marshal, ordered the building of a lighthouse on Hook Head to guide ships up river to his thriving town of New Ross. He enrolled the monks as the first lighthouse keepers and they remained as custodians for several centuries. Hook Head Lighthouse remains one of the oldest working lighthouses in the world.

Today, the organisation is based in Dún Laoghaire and employs 120 full-time staff, as well as many local part-time employees around the coast. An act passed by the British parliament in 1810 transferred control of all lighthouses to the Corporation for Preserving and Improving the Port of Dublin. The present Commissioners of Irish Lights came into being when the Merchant Shipping (Ireland) Act of 1867 separated the Corporation for Preserving and Improving the Port of Dublin into the Dublin Port and Docks Board and the Commissioners of Irish Lights, the latter with exclusive responsibility for the erection and maintenance of lighthouses, beacons and other aids to navigation.

CIL personnel have deployed a wide range of aids to navigation, which form an interlocking network around the coast. Their duties involve the provision and maintenance of lighthouses and marine aids to navigation to assist the safe and expeditious passage of all classes of mariners in general navigation; sanctioning the establishment, alteration or discontinuation of local aids to navigation in ports, harbours and coastlines within the jurisdiction of the local lighthouse authority; the inspection of local aids to navigation to ensure they comply with international standards, and the marking or removal of wreck that is a danger to navigation where no harbour or authority has the power to do so.

Each lighthouse or lighted buoy has its own unique characteristic sequence of flashes identified on nautical charts. Radio aids to navigation such as AIS (an automatic radio coding system that locates and identifies a vessel or an aid to navigation), RACONS (radar beacons that highlight an aid to navigation on a radar display) and differential GPS corrections (a system that adds accuracy and warning of failure to ships' GPS receivers) all play an important role in the mix of aids to navigation provided around the Irish coasts.

Traditionally, lighthouses were manned by keepers. At one time at certain lighthouses the keepers' families lived on the station. Automation overtook this service and the last Irish lighthouse to be unmanned was the Baily at Howth, County Dublin, in March 1997. All lighthouses around the Irish coast are now automated (Figure 23.35). Lighthouses and their associated equipment are centrally monitored on a twenty-four-hour basis, using remote controlled telemetry via radio and telephone. Similarly, manned light-vessels, warning of particular dangers, have been replaced by large light-buoys, which are also remotely monitored.

The Irish Lights vessel, *Granuaile*, maintains aids to navigation provided by the commissioners (Figure 23.34). The current vessel entered into service in February 2000 and is the third Irish Lights ship to be named after the pirate queen, Granuaile. She is the latest in a long line

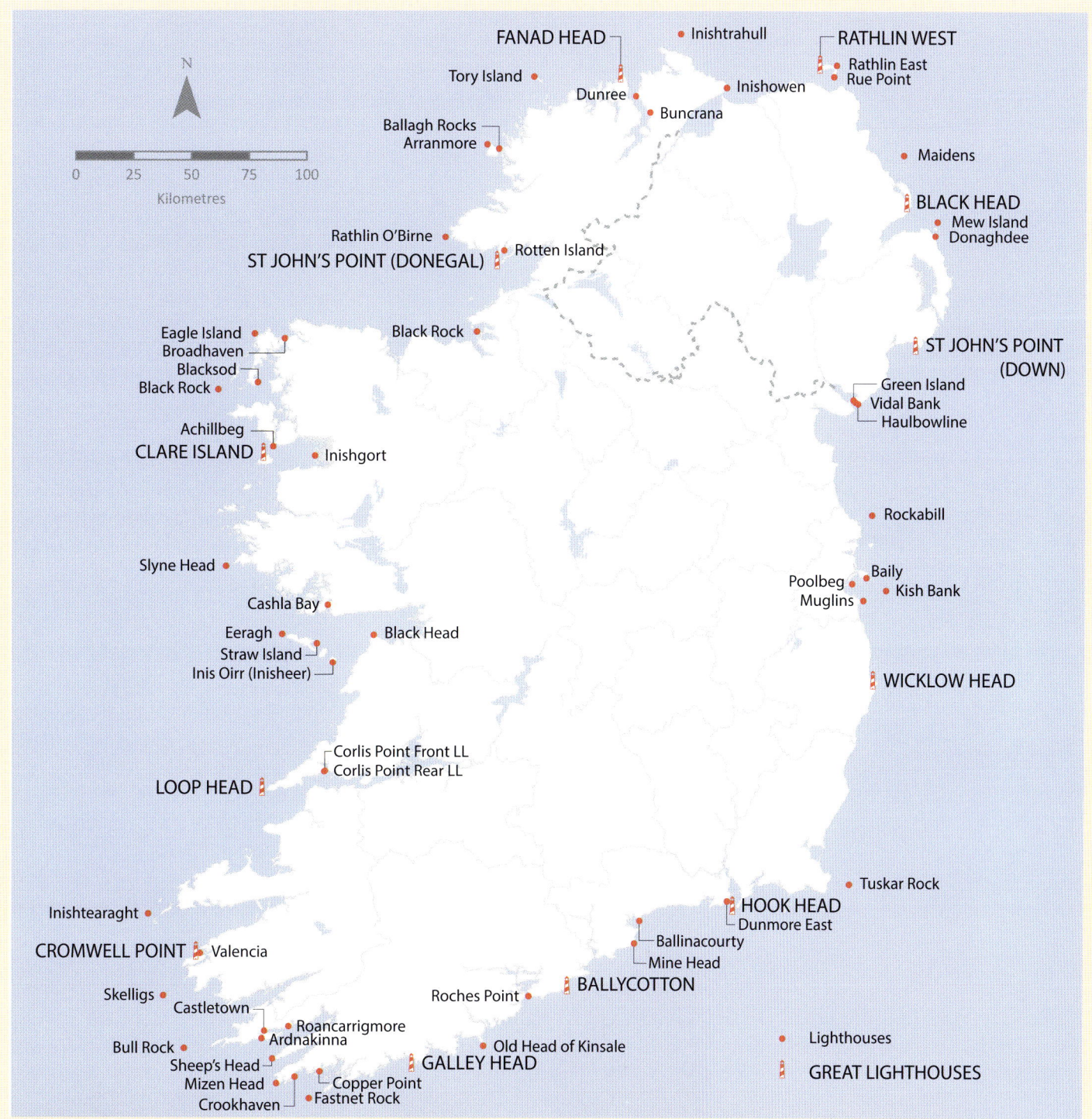

Fig. 23.32 LIGHTHOUSES AND GREAT LIGHTHOUSES OF IRELAND. For hundreds of years, the lighthouses along Ireland's coast, provided guidance for seafarers, especially in coastal areas hazardous for navigation. The Commissioners of Irish Lights (CIL) have overall responsibility for the maintenance and operations of the lighthouse system, as part of the overall marine navigation aid function. Technological advances resulted in the automation and remote monitoring of lighthouses in the 1980s. Given the scenic, heritage and remote aspect of many lighthouses, the CIL developed an all-island initiative, called The Great Lighthouses of Ireland, aimed at promoting access to lighthouse experiences to visitors. These Great Lighthouses are shown in the image, together with other operational lighthouses around the coast. [Data source: Commissioners of Irish Lights]

of ships servicing, refuelling and maintaining the lighthouses and other aids around the coast. *Granuaile* is a multifunction vessel, which can operate in challenging sea conditions. Fitted with dynamic positioning linked to satellite navigation systems, the vessel's primary function is to place and service buoys, replenish offshore lighthouses and serve as a helicopter platform for operational activities. *Granuaile* is also available to assist state agencies with Search and Rescue, emergency towing, oil pollution, surveying and offshore data collection.

Aids to navigation are a critical element in the transport chain, helping to ensure the safety of life, property, business and the coastal environment. Measures that keep the sea routes open, safe and running smoothly are essential both for the economy, particularly of an island country, and for safety of life and property. In recent years CIL have transformed how such services are delivered, putting the emphasis on efficiency, cost-effectiveness and sustainability, while exploiting new technology and new opportunities wherever possible.

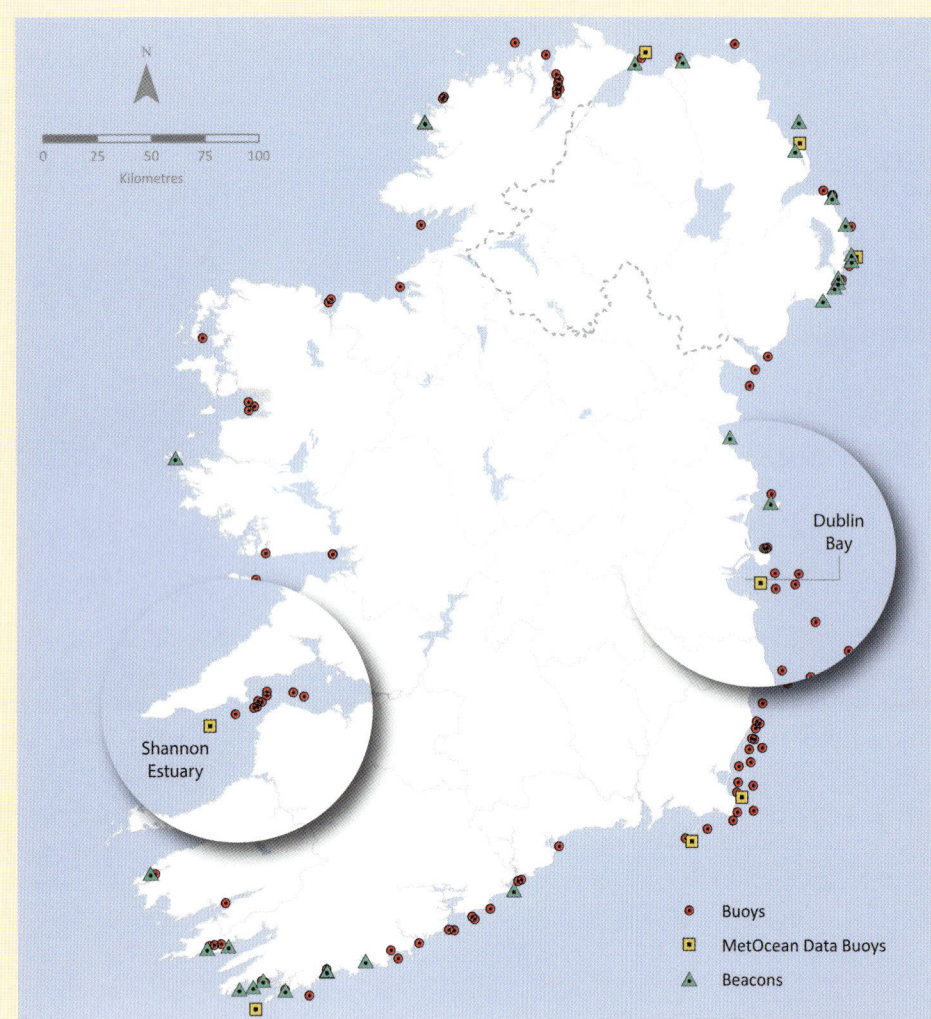

Fig. 23.33 BUOYS AND BEACONS AROUND THE IRISH COAST. As well as having responsibility for sixty-six lighthouses, the Commissioners of Irish Lights also maintain 116 buoys and twenty beacons as part of its aids to navigation (AtoN) system. Floating buoys include port and starboard markers denoting navigation channels; cardinal marks denoting danger features and special marks denoting activities such as fish farms. Beacons are fixed markers, such as the Baltimore Beacon high on the cliff, marking the entrance to Baltimore Harbour in west Cork. AtoN are essential for the safety of navigation in coastal waters, including for navigation in fog, in busy shipping lanes and in bad weather. The system is relied upon by all marine users from recreational boaters to fishers and shipping. [Data source: Commissioners of Irish Lights]

Irish Lights is one of three General Lighthouse Authorities (GLAs), the others being Trinity House, with responsibility for the coasts of England and Wales, and the Northern Lighthouse Board for Scotland. These three operate an integrated aids to navigation service throughout the coastal waters of Britain and Ireland. This service is delivered to recognised standards set by the International Association of Marine Aids to Navigation and Lighthouse Authorities (IALA) so as to meet the responsibilities of the Irish and British governments under the International Maritime Organisation (IMO) Safety of Life at Sea Convention.

Fig. 23.34 COMMISSIONERS OF IRISH LIGHTS ILV *Granuaile* SERVICING A NAVIGATIONAL AID. The ILV *Granuaile* was specifically designed as an aids to navigation support vessel. Commissioned in 2000, the vessel includes a large service deck for servicing buoys (such as the port-hand marker in the image), a Dynamic Global Positioning System (DGPS), and a helicopter platform for servicing offshore lighthouses, such as the infamous Fastnet Rock Lighthouse. [Source: Andrew Collins]

Fig. 23.35 IRISH LIGHTHOUSES. a. Valencia Lighthouse, County Kerry [Source: Kieran Minihane]; b. Fanad Head Lighthouse, County Donegal [Source: Malcolm Hough]; c. Old Head Lighthouse, County Cork [Source: Kieran Minihane]; d. Hook Head Lighthouse, County Wexford [Source: Alan O'Reilly]; e. Baily Lighthouse, County Dublin [Source: Pat Kehoe]; f. Poolbeg Lighthouse, County Dublin. [Source: Robert Linsdell, Wikimedia Commons, CC BY 4.0]

The Royal National Lifeboat Institution

Dick Robinson

The Royal National Lifeboat Institution (RNLI) is a voluntary organisation whose mission it is to end preventable loss of life at sea. The first RNLI lifeboat station in Ireland was opened at Arklow in 1826 (Figure 23.36). By 1852, the number of stations had increased to eight.

Fig 23.36 ARKLOW ROYAL NATIONAL LIFEBOAT STATION. The first RNLI station in Ireland was opened in 1826 in Arklow, County Wicklow, only two years after the founding of the RNLI service in Britain. This was due to the importance of providing such a service on the east coast of Ireland in response to the increasing number of vessels trading in the Irish Sea. Although closed in 1830, it was reopened in 1857 and continues to provide vital Search and Rescue services into the modern era. [Source: Nicholas Leach; inset source: Royal National Lifeboat Institution/Nigel Millard]

Most were located on the east coast where large numbers of wrecks and other distress calls placed high demands on RNLI services. This was due to local marine environmental conditions, but especially the large volume of maritime trade being funnelled through the Irish Sea, centred on Liverpool. In spite of this demand, the service was considered inefficient. As with the rest of the United Kingdom, however, the efficiency of the RNLI in Ireland increased significantly from the 1850s.

Following the creation of the Irish Free State in 1922, the government decided to retain the services of the RNLI, unlike other British agencies such as the Coast Guard. As a result, the twenty-four RNLI lifeboat stations located in Ireland in 1922 continued to function on an all-island basis. Since then, the role of the RNLI in Ireland has expanded. Its services have been modernised with more powerful and innovative boat designs, while training of volunteer crew and support services has been upgraded. The Irish RNLI is now headquartered in Skerries, County Dublin, and coordinates a range of key services. Its principal function, however, is to provide Search and Rescue (SAR) services for up to 100 nautical miles off Ireland's coastline.

In 2016, Ireland had forty-two lifeboat stations located around its coast, with a further three inland stations linked to river and lake rescue (Figure 23.37). Two principal types of lifeboats are launched from these stations, depending on local geographic and environmental conditions, as well as cover provided from neighbouring stations. The twenty-four all-weather vessels are designed for deeper-water SAR, while thirty-four inshore vessels focus on responding to problems within the country's increasing maritime leisure sector.

The RNLI continues to function primarily as a voluntary organisation. A growing majority of people living within coastal communities no longer have any direct relationship with the sea, creating a challenge in recruiting volunteers. Only one-in-ten volunteer crew originate from a maritime profession. Training is therefore vital for crew to be able to operate more sophisticated vessels and to function efficiently in extreme conditions. Much of the training has now been centralised in the headquarters of the British and Irish RNLI at Poole, Dorset, where the RNLI College was opened in 2004.

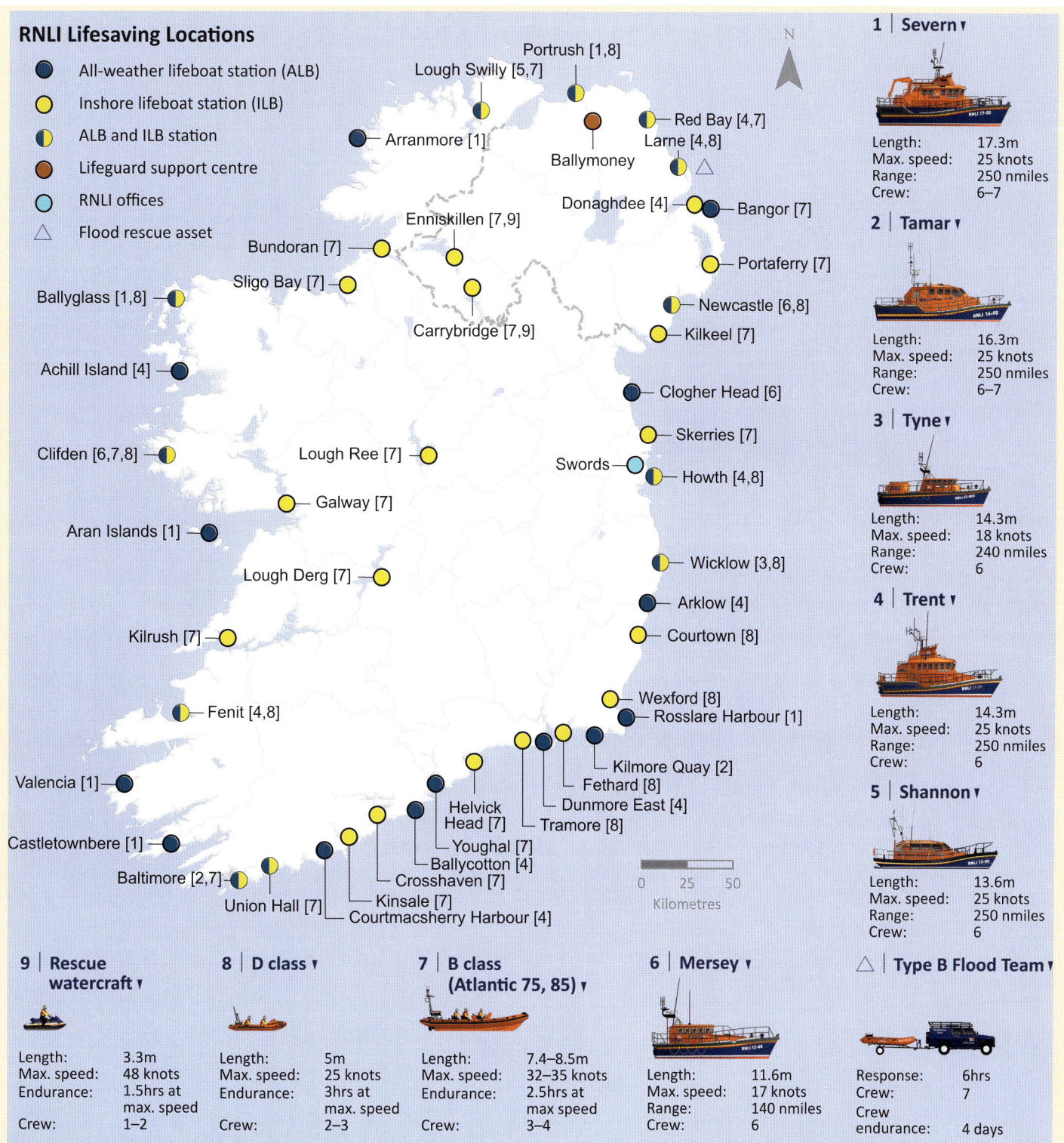

RNLI Lifesaving Locations

- ● All-weather lifeboat station (ALB)
- ○ Inshore lifeboat station (ILB)
- ◑ ALB and ILB station
- ● Lifeguard support centre
- ○ RNLI offices
- △ Flood rescue asset

Portrush [1,8]
Lough Swilly [5,7]
Arranmore [1]
Ballymoney
Red Bay [4,7]
Larne [4,8]
Donaghdee [4]
Bangor [7]
Enniskillen [7,9]
Bundoran [7]
Portaferry [7]
Sligo Bay [7]
Newcastle [6,8]
Ballyglass [1,8]
Kilkeel [7]
Carrybridge [7,9]
Achill Island [4]
Clogher Head [6]
Clifden [6,7,8]
Lough Ree [7]
Skerries [7]
Swords
Howth [4,8]
Galway [7]
Aran Islands [1]
Wicklow [3,8]
Lough Derg [7]
Arklow [4]
Kilrush [7]
Courtown [8]
Fenit [4,8]
Wexford [8]
Rosslare Harbour [1]
Valencia [1]
Kilmore Quay [2]
Fethard [8]
Helvick Head [7]
Dunmore East [4]
Castletownbere [1]
Tramore [8]
Youghal [7]
Baltimore [2,7]
Ballycotton [4]
Union Hall [7]
Crosshaven [7]
Kinsale [7]
Courtmacsherry Harbour [4]

0   25   50
Kilometres

1 | Severn ▾

Length: 17.3m
Max. speed: 25 knots
Range: 250 nmiles
Crew: 6–7

2 | Tamar ▾

Length: 16.3m
Max. speed: 25 knots
Range: 250 nmiles
Crew: 6–7

3 | Tyne ▾

Length: 14.3m
Max. speed: 18 knots
Range: 240 nmiles
Crew: 6

4 | Trent ▾

Length: 14.3m
Max. speed: 25 knots
Range: 250 nmiles
Crew: 6

5 | Shannon ▾

Length: 13.6m
Max. speed: 25 knots
Range: 250 nmiles
Crew: 6

9 | Rescue watercraft ▾

Length: 3.3m
Max. speed: 48 knots
Endurance: 1.5hrs at max. speed
Crew: 1–2

8 | D class ▾

Length: 5m
Max. speed: 25 knots
Endurance: 3hrs at max. speed
Crew: 2–3

7 | B class (Atlantic 75, 85) ▾

Length: 7.4–8.5m
Max. speed: 32–35 knots
Endurance: 2.5hrs at max speed
Crew: 3–4

6 | Mersey ▾

Length: 11.6m
Max. speed: 17 knots
Range: 140 nmiles
Crew: 6

△ | Type B Flood Team ▾

Response: 6hrs
Crew: 7
Crew endurance: 4 days

Fig. 23.37 COASTAL ROYAL NATIONAL LIFEBOAT STATIONS. There are forty-two lifeboat stations around the entire island of Ireland, together with three inland stations. The Republic host thirty-seven stations, the remaining eight are in Northern Ireland. The majority of these lifeboat stations, but especially those providing only inshore services, are located on the eastern and southern coastlines. This reflects the high volume of shipping in the Irish Sea and the trade from Ireland's principal ports of Dublin, Belfast, Cork and Waterford. Also, the high density of population and urban settlement along these coastlines generates much marine-related leisure activities and therefore the need for lifesaving services. In contrast, the west coast has fewer RNLI stations. Most of these are all-weather RNLI stations which provide SAR services for international shipping that follow deep-water trade routes off Ireland's west coast. The less populated coastal environment generates less demand for inshore services, although its growing tourist industry demands effective coverage along the coast. [Data source: Royal National Lifeboat Institution]

A twenty-four-hour, all-year service is available from the forty-five RNLI stations located around Ireland. In 2015 alone, vessels were launched some 1,100 times amounting to the rescue of almost 1,250 people. Today, as in the past, the RNLI continues to provide an essential service to coastal communities.

SECTION 4
RESOURCES:
COMMUNICATIONS AND
INDUSTRY

The port city of Galway [Source: Irish Air Corps]

PORTS AND SHIPPING

Barry Brunt

GENERAL CARGO VESSELS BELONGING TO IRISH SHIPPING LTD WAITING TO BE UNLOADED OR LOADED WITH GOODS AT THE INNER QUAYS, PORT OF CORK, IN THE EARLY 1950s. [Source: Brian Cleare, courtesy of Captain Michael McCarthy, Cork]

Fig. 24.1 IRISH PORTS. Ports account for over 95 per cent by weight of all merchandise imports and exports moved in and out of Ireland. They also facilitated the movement of over 4.7 million foot and car passengers in 2018. Due to this heavy reliance on maritime trade, an effective national port infrastructure is vital. It plays an essential role in the day-to-day functioning of the island's economy by enabling the efficient international movement of people and commodities. The photograph shows the container terminal at Tivoli, located on the River Lee downstream from Cork's historic city-centre port. Container traffic has increased greatly in significance for efficient, cost-effective, just-in-time delivery of goods. Such modern facilities, however, demand extensive areas adjacent to port infrastructure to store the containers. The Tivoli terminal accounted for over 170,000 loaded containers in 2018, about 10 per cent of the Republic's container traffic. Future plans for the the Port of Cork, however, include the relocation of its container traffic to a new, deep-water site in the Outer Harbour at Ringaskiddy, which can accommodate much larger container vessels. [Source: Maxim Kozachenko]

As an island economy located off the north-west coast of Europe, development within Ireland has been, and remains, associated with maritime trade. Ports are, therefore, a crucial element in the island's transport infrastructure. In particular, they operate to overcome its peripheral location and enhance connectivity within an expanding European Union and global trading system. As a result, the Republic of Ireland is the most trade-dependent economy, and most dependent on its port network, in the European Union (Figure 24.1).

For much of its history, Ireland functioned as part of Britain's colonial economy. The north-eastern counties, centred on Belfast and the Lagan Valley, emerged as an urban-industrial complex, specialising in the production and export of textiles, engineering goods and ships to Britain and its colonies. In contrast, the rest of Ireland experienced minimal industrialisation and was a provider of food and labour for Britain's urban-industrial system, and a market for that country's manufactured goods and raw materials.[1]

Two specialised regional economies had emerged on the island of Ireland in the nineteenth century. In effect, they were independent of each other, but dependent on Britain for trade and development. This dependency was facilitated by the multiplicity of ports, primarily along Ireland's east coast, which were the gateways for the flows of goods and people between the two islands.

Political independence in 1922 for twenty-six southern counties within Ireland aggravated socio-economic and political divisions on the island. While six counties of Northern Ireland opted to remain part of a United Kingdom, the new Irish Free State (Saorstát Éireann), which became a Republic in 1949, was determined to reduce its high levels of dependency on Britain. These political decisions had far-reaching consequences for development trends on the island and were reflected strongly in terms of port development and trading patterns.

PORT DEVELOPMENT AND MARITIME TRADE: 1922–1972

This fifty-year period, from the political division of Ireland to the eve of accession of both countries to the then European Community, involved fundamental changes to the island's development and trading patterns.[2] Northern Ireland continued to function in the context of Britain's historic commitment to free trade. As a result, inter-war recession and the drift to a more protectionist global trading environment exposed its specialised and export-orientated industrial base to significant decline. Despite some recovery in demand for its traditional products during the Second World War and post-war reconstruction, the emergence of severe de-industrialisation from the 1950s made it the most problematic regional economy in Britain. Although its maritime trade improved to 7.4 million tonnes by 1957, this was due primarily to imports rather than exports.

In contrast to Northern Ireland, the Republic's Fianna Fáil government in 1932 embraced protectionism and Import

Substitution Industrialisation (ISI) to kick-start the national economy. This strategy looked to replace imported industrial goods with those manufactured by indigenous producers, who would be protected by high tariffs. By the start of the Second World War, the Republic had one of the most protected economies in the world. This, together with recession and war, resulted in the value of the country's external trade falling by over 50 per cent from 1929 to 1944, and gave rise to significant trade deficits (Figure 24.2).

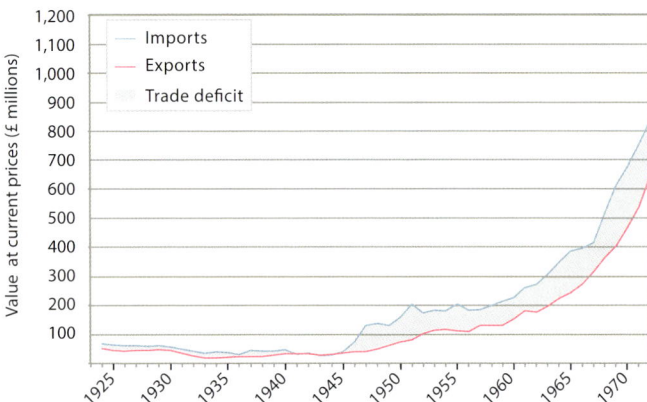

Fig. 24.2 VALUE OF THE REPUBLIC OF IRELAND'S MERCHANDISE TRADE (1924–1972). From independence to the end of the Second World War, protectionism, global recession and war limited opportunities to increase external trade. In addition, imports continued to exceed exports and imposed capital constraints on the new state. In the post-war period, Britain's demands to import low-cost food products boosted exports, although Ireland's growing reliance on imported industrial goods and energy resulted in a growing trade deficit. The 1960s mark clearly a turning point for trade. A commitment to open trade and the attraction of TNCs to establish branch plants in Ireland increased the range and volume of industrial exports. By 1972, the value of exports reached £648 million. However, the Republic's new assembly-line manufacturing economy necessitated the import of large amounts of relatively high-value goods, services and energy supplies. As a result, by 1972 the country had built up a trade deficit of £190 million, some 30 per cent of the value of total exports. This deficit had to be offset by government borrowings to ensure a continuation of national development. [Data source: Central Statistics Office, Annual External Trade]

The Republic's commitment to protectionism continued into the 1950s.[3] This emphasised its peripherality when most countries in Western Europe were enhancing their trading networks as members of the European Economic Community (EEC) or European Free Trade Association (EFTA). Furthermore, the country's small and poor consumer market prevented ISI from stimulating long-term, sustainable development, while protectionism encouraged a high-cost and low-productivity manufacturing sector, which could not compete in export markets. In addition, its port infrastructure was, in general, both underdeveloped and inefficient (Figure 24.3). By the late 1950s, therefore, the Republic had one of the least developed national economies in Western Europe. Despite this, the value of the country's external trade did increase between the end of the Second World War and 1957. As in Northern Ireland, however, this was due principally to imports of industrial goods, raw materials and energy supplies, rather than exports, which remained dominated by agricultural goods. Trade deficits continued to increase.

Going into the 1960s, both countries identified the need for a new development paradigm.[4] In the Republic, the first Programme for Economic Expansion was crucial in that it focused future

Fig. 24.3a. TRANSIT SHEDS AT PORT OF DUBLIN. Maritime trade through Irish ports in the inter-war and immediate post-war period was dominated by relatively small, general cargo vessels. They did not require deep-water berthage or much specialised port-side infrastructure. Cargo was generally loaded/off-loaded to or from the ships to transit sheds/warehouses located on the quayside and from where goods could be redistributed to markets within the port hinterlands. It was a labour-intensive process. In this image, many general cargo vessels are lined up alongside the inner-city quays and transit sheds that were located adjacent to the inner city, and highlight the close relationship between Ireland's major coastal cities and their port-based activities. [Source: Courtesy of Dublin City Library and Archive].

Fig. 24.3b. GRAND CANAL DOCKS AND SIR ROGERSON'S QUAY IN PORT OF DUBLIN (1926). On gaining political independence from Britain, the new Republic also sought to effect an economic independence through protectionism and import substitution. The policy was not successful, and economic and trade dependency on Britain remained high. This added to the significance of the Republic's principal ports, especially Dublin, due to its ease of access to Britain via the relatively short-distance shipping route across the Irish Sea. In addition, because the focus of much of the country's trade was on the capital city port, the optimal site for many industries in the state would be in or near to Dublin Port, due to its trading infrastructure and large domestic market. Here, merchant vessels line the quays at the Grand Canal Docks and Sir Rogerson's Quay. Numerous warehouses and industries are sited along the quays to provide easy access for traded goods. Interspersed between the industries and adjacent to the port are large housing areas that provided most of the workers required in these port-related activities. [Source: Courtesy of Dublin City Library and Archive]

development on Export Orientated Industrialisation (EOI). This embraced free trade and the attraction of foreign direct investment (FDI) from transnational corporations (TNCs), rather than protection of an underperforming indigenous sector. It also recognised key changes in the global context for trade including: the increasing liberalisation of trade, reflected in the success of the European Community (now the European Union); improvements in transportation, such as containerisation and bulk shipping,

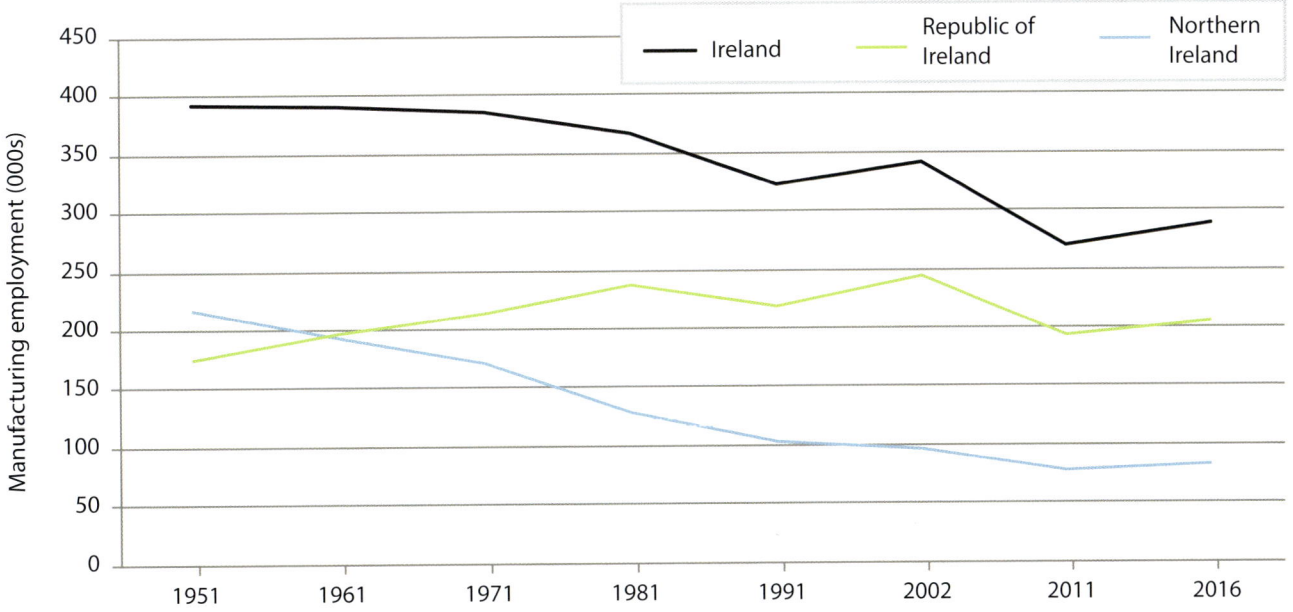

Fig. 24.4 CHANGES IN MANUFACTURING EMPLOYMENT (000s) (1951–2016). Emerging from the Second World War, Northern Ireland remained the island's dominant industrial economy. A loss of competitiveness in its traditional industries and failure to attract significant new investment resulted, however, in severe deindustrialisation and a net loss of some 140,000 jobs (65 per cent) over the period 1951 to 2011. Civil strife compounded economic problems within the state, although the 1990s' peace process succeeded in slowing down the continuing erosion of its manufacturing base. In the five years to 2016, Northern Ireland registered its first census increase in industrial employment in over sixty years. In contrast, the Republic's ability to attract large amounts of FDI resulted in an 'industrial revolution' in the 1960s and, by 1971, it had replaced Northern Ireland as the dominant industrial economy on the island. Its continued success in attracting FDI and the modernisation of its industrial economy saw employment continue to rise in the Republic, peaking at some 250,000 jobs in the Celtic Tiger. Although severe recession in the early 2000s caused a net loss in manufacturing employment, the economy recovered and, from the 2016 Census returns, accounted for three-quarters of the island's industrial workforce. [Data source: relevant census reports produced in the Republic and Northern Ireland]

which reduced frictional costs of distance; the global spread of TNCs and their search for low-cost production sites from which they could export to growing international markets.

Northern Ireland also emphasised FDI through incentives offered in the context of Britain's regional policies. The Republic, however, was more successful, as its independent government provided more generous and targeted incentives (especially tax) to TNCs to locate within the country. In effect, the state experienced its 'industrial revolution' and, by 1971, its industrial workforce exceeded that in Northern Ireland, where heavy losses in traditional

sectors failed to offset gains through FDI (Figure 24.4).

By 1972, the volume of trade through Ireland's port network had surged to almost 54 million tonnes. This, however, was influenced strongly by the 24 million tonnes of crude oil being transhipped through the new oil terminal opened officially at Bantry Bay in 1969. Discounting this specialist port function, maritime trade almost doubled between 1957 and 1972 to 29.8 million tonnes, with ports in the Republic accounting for 56 per cent of all-Ireland trade. This highlighted the success of both countries' moves to an open-trade policy.

WHIDDY ISLAND OIL TERMINAL: BANTRY BAY

Barry Brunt

The ria coastline of south-west Ireland is typified by a number of sheltered, deep-water bays. Bantry Bay, between the Beara and Sheep's Head Peninsulas, has used these natural advantages to provide a unique development on the Irish coastline.

From the 1960s, Europe became increasingly dependent on oil imports and this changed the logistics of the oil trade. Economies of scale provided by a new generation of supertankers greater than 250,000 Dead Weight Tonnes (DWT) became critical to deliver large volumes of crude oil to European refineries at low unit cost. Since most European ports could not accommodate such vessels, one strategy adopted was to construct large, deep-water terminals from which smaller tankers could tranship crude oil to coastal refineries.

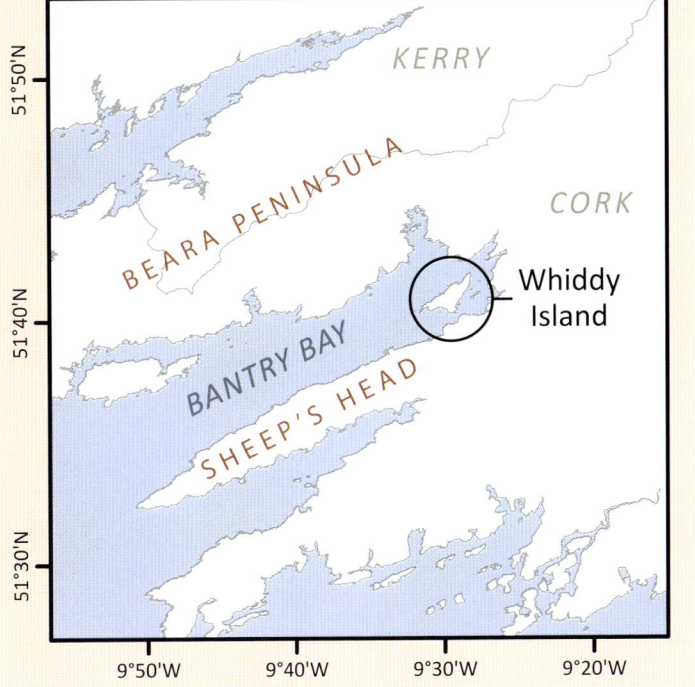

Fig. 24.5 THE WHIDDY ISLAND OIL TERMINAL, BANTRY BAY. Opened officially in 1969, this specialised, deep-water port facility dominated the volume of Ireland's maritime trade until the late 1970s. Currently, relatively small volumes of oil are traded through the terminal. The offshore berth is no longer used and the jetty that once linked it to the onshore storage tanks has been removed. Despite the significant decline in the trade of crude oil through the port, over 8 million barrels of oil can be stored in the tanks and are part of Ireland's strategic oil reserves. [Source: Zenith Energy]

In 1966, Gulf Oil selected Whiddy Island, at the head of Bantry Bay, as the optimal site for a crude oil terminal from which to service its European refineries.[5] Its low-lying, undulating topography and a small resident population (it had a population in 1966 of ninety-seven people) facilitated construction of extensive onshore facilities, which included twelve storage tanks with a capacity of 1.3 million tonnes of crude oil. These were connected by a 488m jetty and pipelines to a deep-water berth with the potential to accommodate supertankers of up to 500,000 DWT (Figure 24.5). The first delivery of crude oil was in 1968, from Gulf Oil's *Universe Ireland* (312,000 DWT), then the world's largest ship. By 1973, trade in oil through Bantry Bay was 34.2 million tonnes, two-thirds of all trade through ports in the Republic. Following the 1973–1974 oil crisis, however, declining demand and overcapacity in European refineries encouraged the adoption of a 'just-in-time' policy of using smaller tankers to supply crude oil directly to refineries, when demand occurred. This reduced the economy of scale advantages of major oil terminals and highlighted their disadvantage of having large amounts of capital tied up in storage. By 1978, oil trade through Bantry Bay had fallen to 11.9 million tonnes.

In 1979, a massive explosion on the French tanker *Betelgeuse*, while discharging oil at the deep-water berth, killed fifty people and effectively destroyed the jetty and berth. Gulf Oil did not re-open their facility after the 'Whiddy Disaster' and in 1986 sold the terminal to the Irish state. The terminal was operated commercially by the Irish National Petroleum Company (INPC) until it reverted to private ownership in 2001 (Phillips 66), transhipment of crude and refined oil products continued through a Single Point Mooring (SPM). The volume of its oil trade, however, declined to 0.55 million tonnes in 2018, less than half of that total representing imports to the facility. While the terminal's commercial viability is under review, it has acquired additional strategic value in that the National Oil Reserve Agency (NORA) has a contract to hold one-third of the Republic's strategic oil stocks (90-day reserve) at the terminal's storage tanks.

In 2015, Phillips 66 sold Whiddy terminal to Zenith Energy. The US company planned to invest in what it regarded as one of the most strategic fuel storage facilities in Europe. However, the evolution of EU energy policy to a low carbon footprint poses questions for the viability of this coastal facility. Early in 2020, reports suggested Zenith planned to sell the terminal.

Imports continued to dominate port trade in both countries, as most of the mass-produced and relatively low-cost goods manufactured on assembly lines in the new branch plant economy (primarily for export) depended on the importation of large quantities of component parts, raw materials and energy supplies. This import dependency found expression in the Republic's balance of trade figures which, by 1972, continued to exhibit high net deficits (Figure 24.2).

The direction of port traffic in 1972 illustrated the continued significance of Britain for Ireland's trade. This could be expected for Northern Ireland, given its political and economic integration as part of the United Kingdom. The Republic, however, had made a policy decision to complement its political independence from Britain with an economic independence. This proved difficult, given its geographic proximity to and strong economic ties with Britain. Although the growing presence of TNCs increased the

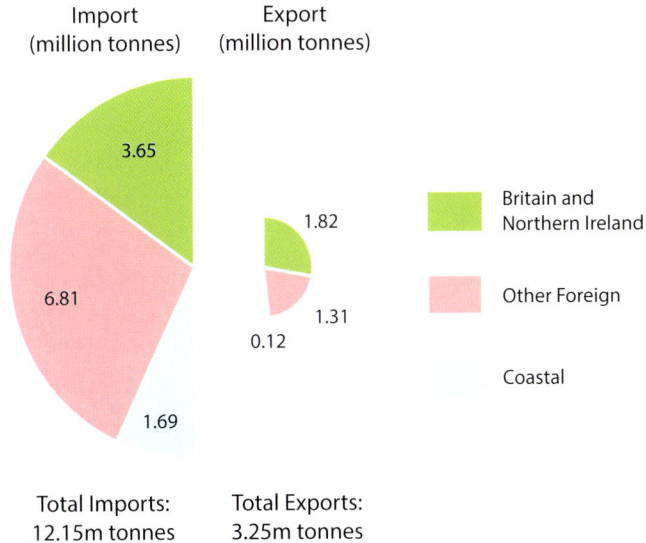

Import (million tonnes)   Export (million tonnes)

3.65

6.81

1.82

0.12

1.31

1.69

Britain and Northern Ireland

Other Foreign

Coastal

Total Imports: 12.15m tonnes

Total Exports: 3.25m tonnes

Fig. 24.6 SOURCES AND DESTINATIONS OF TRADE FOR REPUBLIC OF IRELAND, 1972. Despite a growing diversity in Ireland's geography of trade, by 1972 Britain was the destination of 56 per cent of the country's exports and the source of 30 per cent of all imports. Plans to join the EEC, and the growing number of American and European TNCs in Ireland, however, ensured an increasing volume of external trade with the rest of Europe and North America. Coastal trade refers, primarily, to the redistribution of goods between Cork and Dublin, especially refined oil products from the Whitegate oil refinery in Cork Harbour to storage depots at other Irish ports. The data excludes crude oil imports to the Whiddy oil terminal sourced from the Middle East. [Data source: Central Statistics Office, Statistics of Port Traffic, 1972]

ports were located along the eastern coastline, emphasising connections with Britain (Figure 24.7).

Six key ports in Northern Ireland controlled almost 90 per cent of its maritime trade. Belfast remained dominant (7.4 million tonnes), especially for imports, although its share of the country's increasing trade volume fell from 66 per cent in 1957 to 60 per cent in 1972. This was due, primarily, to the growth of container traffic at Larne, which complemented its passenger-ferry services. By 1972, this specialist roll-on, roll-off (ro-ro) port had doubled its share of Northern Ireland's freight traffic to 15 per cent (1.9 million tonnes).

The ten largest ports in the Republic (excluding Bantry Bay) monopolised maritime trade (95 per cent in 1972), with imports generally exceeding exports. Unlike Northern Ireland, however, the capital city did not possess the largest port. While Dublin remained important, especially for imports to meet the demands of the growing metropolitan region, and controlled 38 per cent (5.9 million tonnes) of all maritime trade, Cork had grown to become the Republic's largest port (6.1 million tonnes). This was due to the growth of bulk trade to serve the needs of a heavy industrial complex that was emerging in Cork Harbour and which included Irish Steel and the Verolme Shipyard, but above all reflected the scale of oil imports needed to meet the demands of Ireland's only refinery, opened at Whitegate in 1959.

diversity of foreign markets, Britain remained the single-most important country for Irish imports and, especially, exports (Figure 24.6).

By 1972, Ireland retained a well-distributed port network which involved sixteen ports (excluding Bantry Bay), each with annual trade exceeding 100,000 tonnes. These accounted for 93 per cent of the island's maritime trade. The vast majority of these

Fig. 24.7 IRELAND'S PORT NETWORK, 1972. In 1972, Ireland possessed seventeen ports trading in excess of 0.1 million tonnes. The specialised port of Bantry Bay dominated total volumes, transferring some 24 million tonnes of oil divided almost equally between imports and exports. Excluding Bantry Bay, Belfast (7.4 million tonnes), Cork (6.1) and Dublin (5.9) accounted for two-thirds of the island's external trade, and were the principal gateways linking Ireland to international markets. The dominance of the island's east and southern coastline is apparent. The peripherality and underdevelopment of the western coast is reflected in the fact that less than 10 per cent of the Republic's total maritime trade was shipped through Foynes, Limerick and Galway. [Data source: Central Statistics Office, Statistics of Port Traffic, and Northern Ireland Department of Economic Development, Trade at Principal Ports, 1972]

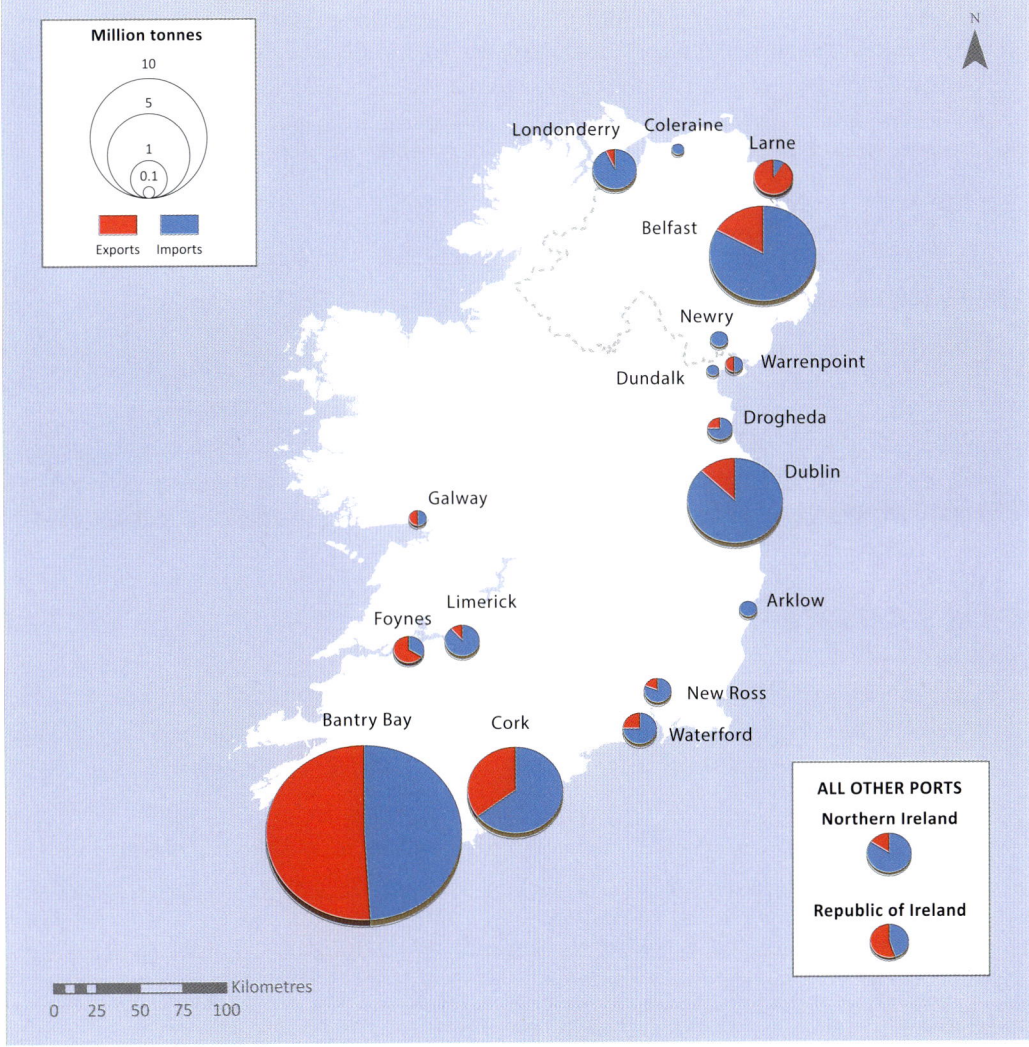

Furthermore, the beginning of some dispersal of branch plants to less-developed rural communities and aspirations for a more balanced regional development programme in the Republic in the 1960s created a demand to improve regional port infrastructures. The result was a more dispersed pattern of port infrastructure than in Northern Ireland, which would become an important locational attribute following accession to the European Community.

WHITEGATE OIL REFINERY, CORK HARBOUR

Barry Brunt

Whitegate, located on the eastern shore of Cork Outer Harbour, is the site of Ireland's only oil refinery. Opened in 1959 by the Irish Refining Company, the facility reflected the evolving strategies of national governments and oil corporations to locate new refining capacity on greenfield sites within deep-water ports. This encouraged efficient plant layout on sites distanced from major population centres, and allowed refineries to benefit from critical economies of scale via the bulk shipping of crude oil. In Ireland, it also reflected policy to decentralise development from the Dublin core region, while also meeting the growing demands for oil-based products from a national refinery.

The 131ha refinery site is accessible to a deep-water channel in Cork's Outer Harbour via a purpose-built marine terminal (Figure 24.8). This accommodates bulk tankers of some 70,000 DWT that discharge crude oil, via pipelines, to the refinery's storage tanks. The terminal is also used by smaller vessels to transfer refined products to distribution depots located around Ireland's coastline.

Since the 1960s, the refinery has played a strategic role in the modernisation and industrial take-off of the Republic. With a throughput of 2.2 million tonnes, it meets some 40 per cent of the country's fuel needs and contributes to security of energy supplies. The

Fig. 24.8 WHITEGATE REFINERY, CORK OUTER HARBOUR. The refinery has three components: the onshore refinery complex, crude oil storage tanks on Corkbeg Island and a jetty to the offshore, deep-water tanker terminal, which allows bulk carriers to discharge oil directly to the tanks. The refinery throughput of 75,000 barrels a day, some 300 full-time employees and an outlay of over €100 million a year to the local economy make it a key element of the region's industrial infrastructure. It is also an essential element in the Republic's energy infrastructure as it accounts for approximately 40 per cent of the country's fuel requirements. [Source: Port of Cork]

refinery is also a critical component of the port-related industrial complex that has evolved around Cork Outer Harbour. Furthermore, liquid bulk (primarily oil) accounted for 56 per cent of volume of trade through the Port of Cork in 2018 and is, therefore, central to its trading function.

In 2001, Phillips 66 bought the refinery from the Irish state. Company restructuring and concerns over increasing financial losses in operating the refinery, however, led to a decision by Phillips 66 not to renew its lease beyond 2016. This placed Whitegate in jeopardy, especially as a strong economic argument could be advanced to close what was seen as an uncompetitive facility and to supply Ireland's fuel needs from larger and more cost-efficient refineries located elsewhere in Europe – especially from Milford Haven in Wales. This would also conform to the European Union's strategy for a more unified and efficient European energy market. For Ireland, such a 'solution' was deemed unacceptable as it raised questions over security of energy supplies, a concern that was heightened further by the Brexit vote and its implications for the country's energy market. This course of action would also remove a key economic 'driver' for the Cork Harbour region. In October 2016, however, the Canadian company Irvine Oil bought Whitegate and has committed to invest in the facility to ensure its longer-term viability.

Recently, the possibility of the development of the nearby coastal Barryroe oil and gas field has re-emerged (See Chapter 29: Coastal Mining, Quarrying and Hydrocarbon Exploration). The developers of the oilfield, Providence Resources and their partners, give a recoverable production (mid-range projected values) of 311 million barrels of oil, and a potential tax revenue to the Republic of Ireland of the order of €3 billion in the period 2024–2040.[6] Whitegate would be a logical candidate for refining crude oil from this field. Furthermore, such a development could provide clear advantages for the country, as a large and indigenous source of hydrocarbon resources would help reduce the Republic's high level of dependency on foreign energy supplies, especially at a time of increasing international/geopolitical instability.

IRELAND'S MARITIME TRADE, 1973–2018

The Republic and Northern Ireland gained accession to the then European Community in 1973. This provided direct access to its large, prosperous market and enhanced the island's attraction for FDI. This was apparent especially in the Republic, where its low corporation tax (12.5 per cent), significant investments to upgrade physical infrastructure and labour market flexibility furthered its location as an optimal site for large numbers of export-orientated

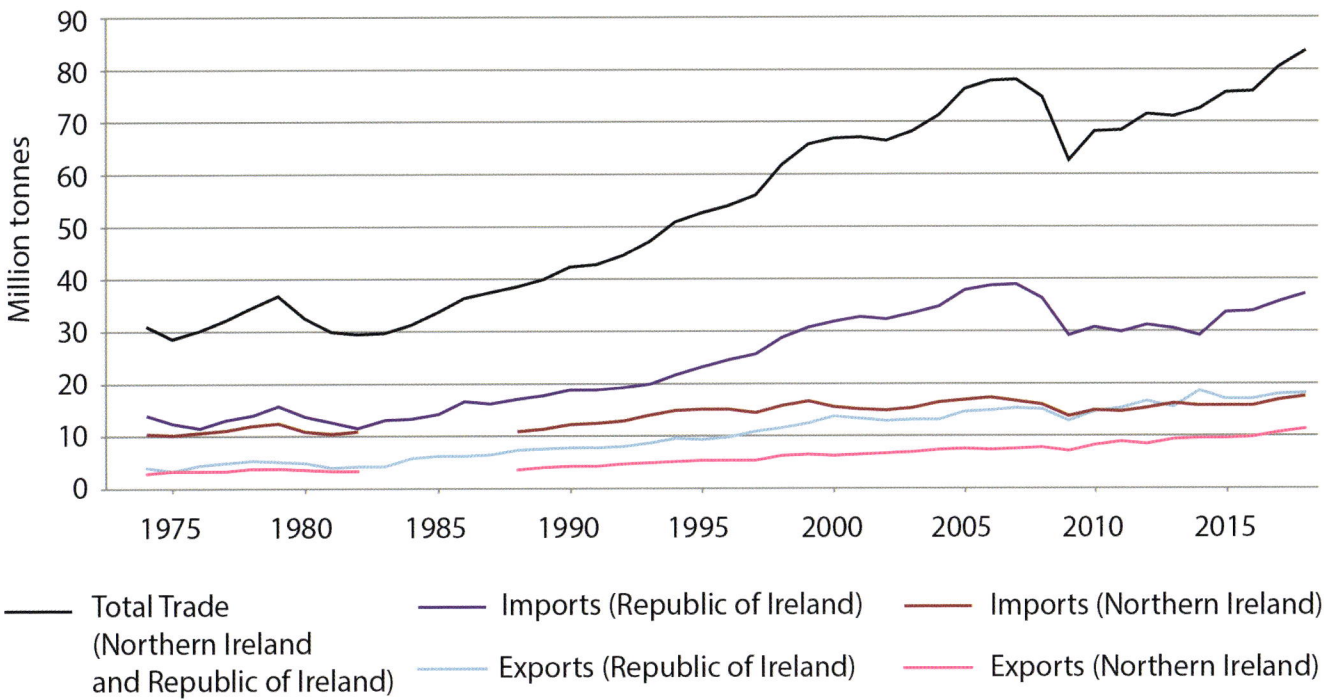

Fig. 24.9 TRENDS FOR TOTAL MARITIME TRADE IN IRELAND (MILLION TONNES) (1973–2018). For the first decade of membership of the European Union, both the Republic and Northern Ireland experienced difficulties in adjusting their economies to more competitive trading conditions. By 1982, total trade volume had fallen to 29.4 million tonnes. From this post-accession low point, Ireland's trade showed an almost uninterrupted growth to almost 80 million tonnes by 2007. Growth was particularly strong in the Republic, which accounted for almost 70 per cent of the island's total volume of trade, and was linked directly to its attractiveness for large numbers of TNCs. The volume of imports grew in particular, reflecting the country's dependency on external sources of bulk raw materials and energy supplies. The Celtic Tiger economy gave way, however, to a deep economic and financial crisis from 2007–2009, which impacted negatively on trade. Recovery has been pronounced, however, and 51.1 million tonnes were traded by the Republic in 2018. Growth in trade was initially less impressive in Northern Ireland, due to internal civil strife, which acted as a deterrent to FDI. As a result, by the 2007–2009 recession, Northern Ireland accounted for only 30 per cent of the island's trade. As in the Republic, trade revived as it was able to benefit from a more competitive economy and inward investment encouraged by the 'peace dividend' following the Good Friday Agreement (1998). Overall, trade on the island recovered to its pre-recession peak in 2017 and increased further to 83.5 million tonnes in 2018. [Data source: Central Statistics Office, Statistics of Port Traffic and Northern Ireland Department of Transport, Port Freight Statistics]

Fig. 24.10 VALUE OF REPUBLIC'S MERCHANDISE TRADE (1973–2018). On accession to the now European Union in 1973, the Republic's imports exceeded exports by €314 million. Direct access to the growing and prosperous European market boosted trade. Exports increased ten-fold to 1984, reflecting the success of the country as a low-cost production platform for TNCs to access the EU. The continued need to import large amounts of component parts, raw materials and energy supplies, however, meant the continuation of a trade deficit. In 1985, the Republic moved into a trade surplus, a characteristic that has grown in significance since that year. This reflects changes in development strategy to target high value-added goods and service sectors, rather than bulk production. As a result, Ireland has become one of Europe's most successful economies and one based on export-driven development. Although the growth rate for both exports and imports was reduced significantly for over a decade at the start of the twenty-first century, a positive and strong balance of trade was retained. Since 2012, economic recovery in the aftermath of the recession has been reflected in a renewed surge in the value of imports, but more especially exports. In 2015, exports exceeded €100 billion for the first time and, by 2018, reached €140.5 billion, as opposed to €92 billion in imports. This yielded a trade surplus equivalent to over one-third of the total value of exports. [Data source: Central Statistics Office, Annual External Trade]

TNCs.[7] Accession, however, also exposed Ireland to a more competitive trading system. Successful development would be conditioned strongly by its port infrastructure being able to provide an effective gateway for trade within the European Union and global markets.[8] In general, Ireland adapted well to the challenges of export-led growth within an open-trade system, and both the volume and value of trade increased strongly, especially from the 1980s. This was epitomised by the Republic's Celtic Tiger economy and its ability to recover from the 2007–2009 global recession.[9] (Figures 24.9, 24.10).

In terms of volume of merchandise trade through Irish ports, total tonnage showed an almost uninterrupted increase from 30 million in 1973 to 78 million tonnes, its pre-recession peak, in 2006. Although both states shared in the general increase, the trend was dominated by the success of the Republic's programme of industrial expansion, which was expressed in an almost four-fold increase in export trade. Economic recession (2007–2009) impacted, in particular, on this export trade although recovery has been impressive (Figure 24.9).

The benefit of the Republic's role as a production platform for TNCs wishing to access the European Union and the wider global market is apparent, with the value of imports and, more especially, exports rising steadily to 1985 (Figure 24.10). That year,

however, marked the start of a long-term trend that sees exports consistently exceed imports and provide the Republic with a growing and large-scale surplus of trade. Central to these trends was the re-orientation of the economy to higher-value production, based more on high-skill inputs rather than large volumes of low-cost inputs. Although recession, growing international competition, inflation and high operating costs linked to the success of the Celtic Tiger economy limited the rate of growth in the value of trade for the first decade or so of the new millennium, recovery has been pronounced. The Republic's surplus of trade in 2018 was €48.5 billion; almost one-third of the total value of its exports.

Growth in the volume and, especially, the value of Ireland's merchandise trade has been related to fundamental changes in the composition of its international trade.[10] In the early 1970s, the domestic economy of both states tended to be dominated by traditional, but declining, sectors such as agriculture and textiles, and low-value output from branch plants of TNCs. Changes in national development strategies from the 1980s (especially in the Republic) focused attention on attracting high value-added growth sectors, such as pharmaceuticals and data processing industries.[11] These depended more on the successful development of Ireland's higher labour market skills and productivity, and improved infrastructures, rather than on cheap, bulky imports. By 2018,

SITC*	Republic of Ireland (€ billion)		Northern Ireland (£ billion)	
	Import	Export	Import	Export
0 Food and live animals	7.8	11.1	1.4	1.5
5 Chemicals and related products	19.9	85.8	1.1	1.2
7 Machinery and transport equipment	39.8	20.0	2.0	3.4
8 Miscellaneous manufactured articles	8.8	15.8	1.3	1.4
Other	15.7	7.8	1.1	1.2
Total	92.0	140.5	6.9	8.7

Table 24.1 VALUE OF MERCHANDISE TRADE BY COMMODITY IN IRELAND (2018). By 2018, the development strategy designed to attract high-growth/high-value sectors to achieve sustainable growth had been particularly successful for the Republic. A trade surplus of €48.5 billion was achieved, compared to negative trade balances which had dominated until 1984. The Chemicals category was especially significant, amounting to over 60 per cent of the country's export value. These and most other growth sectors depend on efficient port systems – especially containerisation – to remain competitive in international markets. A surplus in balance of trade was also achieved within Northern Ireland. Here, Category 7 is most important, reflecting the state's traditions in engineering. [Data source: Central Statistics Office, Goods Exports and Imports by Commodity and H.M. Revenue and Customs, Commodity Analysis – Imports and Exports from Northern Ireland, 2018 *Standard Industrial Trade Classification]

therefore, the Republic's export trade in merchandise goods had become dominated by a range of multinational corporations manufacturing chemicals and related products. This sector accounted for over 60 per cent of the country's value of exports. Northern Ireland's value of exports, however, showed a more balanced distribution across the main sectors, and reflects the state's longer industrial tradition. The growing emphasis on international growth sectors has been fundamental in explaining

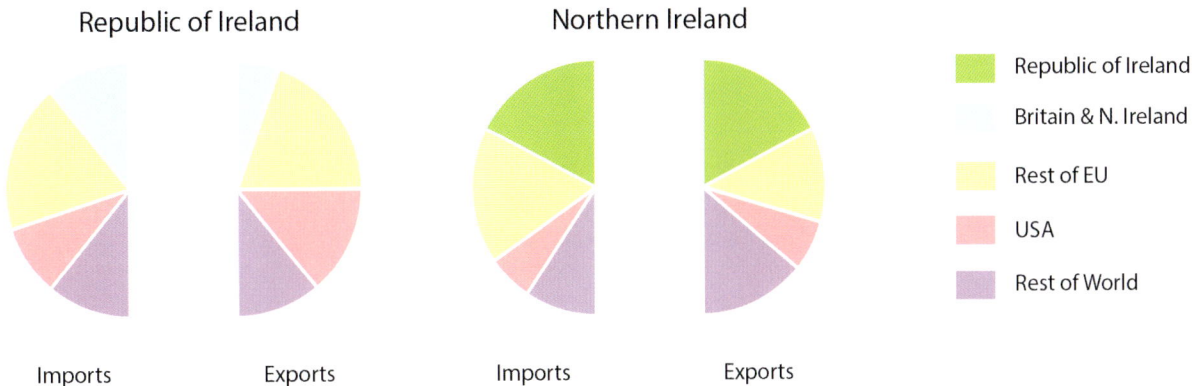

Fig. 24.11 VALUE OF IRELAND'S TRADE BY GEOGRAPHIC AREA (2018). Following accession to the European Union, the Republic's dependency on Britain for exports fell sharply and in 2018 amounted to only 11 per cent (€16.1 billion, of which €2 billion crossed the border to Northern Ireland). Despite this fall in the share of the British market for exports, it remains the single-most important source of imports (22 per cent; €19.8 billion). Failure to guarantee free access to this market post-Brexit would, therefore, pose major concerns for the Republic's economy. The rest of the European Union has become its most important export market (38 per cent; €54.9 billion), and provides a significant surplus for the country's balance of trade (€20 billion). Exports and imports to/from the rest of the world increased to 50 per cent and 40 per cent respectively, reflecting the Republic's commitment to globalisation. As part of this process, FDI and TNCs from the USA increased their presence in the Republic. The USA has, therefore, become the single-most important country for exports for the Republic (28 per cent; €39 billion), and provides a surplus for balance of trade of over €23 billion. For Northern Ireland, trade with Britain is excluded, since it forms part of Britain's single national market economy. An estimated 30 per cent of the value of Northern Ireland's trade is external to Britain. Of this, trade with the Republic is especially significant amounting to over one-third of its non-British trade. The rest of the EU is also responsible for some 30 per cent of this trade. This also highlights concerns within Northern Ireland over Brexit, and helps explain why the majority of its electorate voted in 2016 to remain in the European Union. The economic benefits of the Good Friday Agreement have been noticeable, especially in the increased globalisation of the state's trade. In 2018, over 40 per cent of its exports were with the USA and the rest of the world. [Data source: Central Statistics Office, Annual External Trade of Ireland and HM Revenue and Customs Imports and Exports from Northern Ireland, 2017]

the shift from deficits to growing surpluses in the balance of trade for both states (Table 24.1).

The direction of Ireland's trade also changed. In 1973, 51 per cent of the Republic's value of imports and 43 per cent of exports were with Britain.[12] A further 22 per cent of both imports and exports focused on the European Union. By 2018, however, the historic dependency on Britain had fallen significantly (Figure 24.11). The European Union had now become the Republic's dominant trading partner. Dependency on trade to the rest of the world also increased, especially with the USA.

As part of the United Kingdom, Northern Ireland's trade is dominated by this country (approximately two-thirds), although its trade has diversified, especially to the rest of the European Union. Apart from Britain, however, the Republic of Ireland is currently its single-most important trading partner. This highlights the concerns of both the Republic and Northern Ireland over Britain's decision to leave the European Union (Brexit) and its implications for free trade between both parts of the island.

IRELAND'S PORT NETWORK, 1973–2018

Official government statistics for trade in 1973 listed 21 commercial ports in the Republic and six within Northern Ireland. In the Republic, almost half of these ports showed annual trade flows of less than 100,000 tonnes (Figure 24.7). As a result, it could be argued that the small island economy was over-serviced by ports and reflected its long history of maritime trade. However, it was unlikely that all ports would be able to maintain, or grow, their trading functions in a more competitive, free-market environment. This highlighted the need to focus investment increasingly on a

select number of ports that would provide critical infrastructures to facilitate more efficient trade linkages – the key to success of an export-orientated development policy.

The evolution of Ireland's port network from the 1970s is linked directly to three key processes: government policies for economic development and the role of ports; Europeanisation and globalisation; innovation in shipping technologies.[13]

Government policies

Ireland's initial advantages for FDI centred on low-cost sites, generally in rural areas, to facilitate production of standardised goods for export.[14] This dispersal of industry, a specific objective for both governments, benefited many ports since poorly developed land transportation, especially in the Republic, added to the costs of linking factories to major ports, such as Dublin. Smaller regional ports were, in effect, able to benefit from servicing the needs of branch plants located within their limited hinterlands. Dispersal of trade was also encouraged by the absence of effective port policies to guide development. Ports were, therefore, able to pursue independent strategies to promote trade.

In 1973, over three-quarters of Ireland's maritime trade was focused on the top three ports in each state (Figure 24.12). Belfast (7.5 million tonnes) was the largest port and, with Dublin (6.2 million tonnes), controlled 44 per cent of the island's trade. In the twenty years to 1993, however, while the trend to polarise maritime trade on major ports continued in Northern Ireland, this did not occur in the Republic. Thus, despite impressive growth in bulk shipping through the Shannon Estuary, the loss of key industries in Cork in the 1980s (e.g. Verolme Shipyards, Fords, Dunlops) and

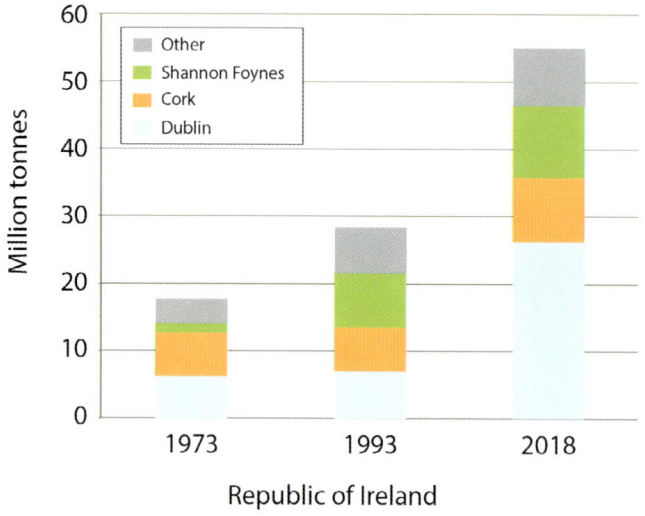

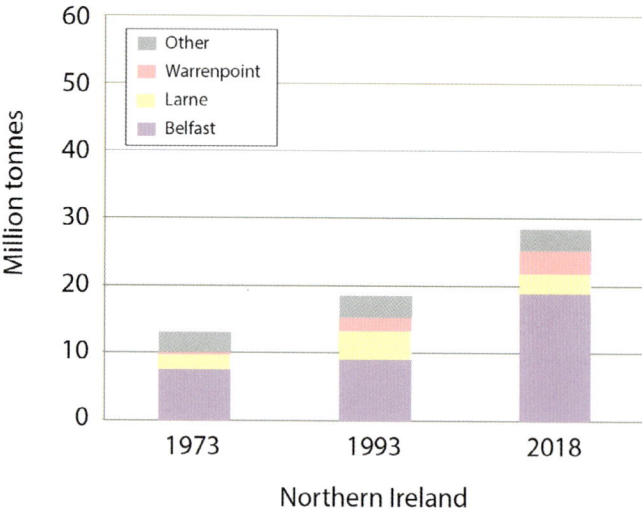

Fig. 24.12 TONNAGE OF TRADE THROUGH IRELAND'S SIX DOMINANT PORTS (1973–2018). In the Republic's 2013 Port Policy, Dublin, Cork and Shannon Foynes were designated as Tier 1 Ports of National Significance, highlighting their priority for development. Belfast, Larne and Warrenpoint are the three ports of strategic importance in Northern Ireland. From 1973 to 1993, Dublin's maritime trade grew weakly and it lost its status as the Republic's largest port to Shannon Foynes, where port-related industrial development stimulated over 8 million tonnes of (primarily bulk) trade. Inefficient infrastructure and high costs discouraged freight operators from using Dublin, with Culliton (1992) estimating that almost 40 per cent of the Republic's ro-ro trade used ports in Northern Ireland (Culliton, J., 1992. A Time for Change: Industrial policy for the 1990s. In: Report of the Industrial Policy Review Group. Dublin: Stationery Office). Larne captured most of this traffic and, by 1992, its ro-ro trade exceeded that at all ports in the Republic. Belfast also benefited and remained the island's largest port. By 2018, however, Dublin had become Ireland's premier port. Introduction in the 1990s of more efficient port management practices and infrastructures, as well as the export-led growth of the Celtic Tiger economy, saw its volume of trade triple to over 26 million tonnes (Power, J., 2011. Trends in Irish Merchandise Trade: A report for Dublin Port Company Limited. Dublin). Deflection of trade to Northern Ireland has been reduced considerably. This affected Larne, in particular, which also saw its Stena ferry service to Stranraer relocated to Belfast. In 2018, maritime trade through Belfast was 18.9 million tonnes. [Data source: Central Statistics Office, Statistics of Port Traffic and Northern Ireland Department of Transport, Port Freight Statistics, 2018]

Fig. 24.13 AERIAL VIEW OF BELFAST PORT. Since the industrial revolution of the nineteenth century stimulated trade through Belfast, its port became the largest on the island of Ireland. It retained this status until the early 1990s, when it was replaced by Dublin. Despite this, maritime trade through Belfast Port has continued to increase, and amounted to 18.9 million tonnes (two-thirds of Northern Ireland's trade) in 2018. Much of the recent growth is due to strong government support to modernise the port's infrastructure and reclaim land for further developments. In particular, the port focuses increasingly on containerisation and on improved ferry services across the Northern Corridor to Britain, primarily to Cairnryan, but also to Liverpool. Lo-lo and ro-ro infrastructure extend downstream from the historic port and dominate the left bank of the Victoria Channel and River Lagan. The former derelict, port-side land associated with the historic Harland and Wolff shipyards, located on the right bank of the Lagan, has been transformed into more urban-related developments (such as recreational space, apartments), and is centred on the Titanic Quarter. The iconic building adjacent to the green sward is Titanic Belfast, a major tourist centre. Downstream from this area there is evidence of recent diversification of port-related industry with operations given to offshore wind logistics and to the construction and repair of deep-water oil and natural gas platforms in a new and purpose-built deep-water quay. [Source: Belfast Harbour]

inefficiencies at Dublin Port saw a decline in polarisation on these ports from 79 per cent to 76 per cent of total maritime trade. In contrast, all three Northern Ireland ports grew strongly to control 83 per cent of that state's maritime trade. Belfast retained its position as the island's leading port (Figure 24.13).

During the 1990s, industrial policy changed to target high-value sectors.[15] These industries, however, have different locational requirements from branch plants, emphasising proximity to key services, flexible labour markets and ease of access to suppliers and customers. This favours larger urban centres, especially those with efficient ports. In the Republic, therefore, the National Spatial Strategy (2001) advocated nine urban centres to be gateways through which national development should be channelled. Eight of these are ports (see Chapter 25: Urbanisation of Ireland's Coast).[16]

A more focused strategy was also adopted in the development of ports. The Republic's National Port Policy (2013) established a hierarchy of designated ports (Figure 24.14). This would guide investment strategies to provide a more competitive and effective market for maritime transport services. In Northern Ireland, the strategic role of Belfast Port is also recognised for national development. By 2018, 72 million tonnes (86 per cent) of trade passed through Ireland's six dominant ports. The process of polarisation has been focused especially on the two capital city ports,

with Dublin controlling 31 per cent (26 million tonnes) and Belfast 23 per cent (19 million tonnes) of all maritime trade (Figure 24.12).

Europeanisation and globalisation

Membership of an enlarging European Union and a more liberalised global trading environment presented new opportunities and changed the geography of trade for many Irish ports. East coast ports, in particular, exploited their locational advantages to dominate increasing trade with Britain and Europe. This is highlighted by the fact that some two-thirds of the Republic's goods exporters use Britain as a 'landbridge' to connect Ireland to Europe. In addition, over one-half of Irish goods exports with the rest of the world (excluding Britain) pass through Britain. Efficient container services, especially on the Central and Southern Corridor routes across the Irish Sea are, therefore, pivotal for the success of Ireland's export-orientated economy. Apart from Shannon Foynes, ports on the west coast have become more marginal.

Dublin and Belfast dominate trade with Britain and the European Union (Figure 24.15). So important are these two market areas for the capital city ports, that 92 per cent of Dublin's total port traffic and some 85 per cent of Belfast's are linked directly to these markets. Most of the trade occurs through containers. Cork and Shannon also have important relationships with these areas,

Legend:

◉ Ports of National Significance (Tier 1)
◎ Ports of National Significance (Tier 2)
● Ports of Regional Significance
○ Major Northern Ireland port
┈ High speed rail line

Larne
Northern Corridor to Scotland
Belfast
M1
Warrenpoint
M1
Drogheda
Dublin
Central Corridor to Holyhead & Liverpool
Dún Laoghaire
Galway
M6
Wicklow
M18
M9
M11
Shannon Foynes
M20
M7
Bulk shipping to global markets
M8
New Ross
Waterford/ Belview
Rosslare
Cork
Bantry
Continental Corridor to France
Southern Corridor to South Wales

0 25 50 75 100
Kilometres

Fig. 24.14 IRELAND'S PORT NETWORK AND MAIN INLAND TRANSPORT ROUTES. Efficient management of Ireland's ports is of vital importance, given that 84 million tonnes of trade, valued at over €205 billion, was traded through these facilities in 2018. In the Republic, the government has established a hierarchy of ports to better manage and plan their development to serve both national and more regional/local interests. Five ports are designated to be of national significance, and are divided into two tiers. Five other ports are classified as ports of regional significance.

Tier 1 Ports of National Significance– Dublin, Cork and Shannon Foynes. These are each responsible for 15 to 20 per cent of tonnage through the Republic's ports and are considered to possess significant potential for further development for the state. Bantry Bay is operated as a subsidiary of the Port of Cork and has national importance, given its role for oil transhipment and as the location of the country's strategic oil reserves.

Tier 2 Ports of National Significance – Rosslare Europort and Waterford, although the port company's main centre of operations has been transferred 8km downstream to Belview). At least 2.5 per cent of tonnage pass through each of these two ports. They have been identified as having considerable potential to increase their handling of container traffic and have transport links to serve the national market.

Tier 3 Ports of Regional Significance. This category includes five ports that are judged to have limited national significance (3 per cent total tonnage), but have a role in promoting development within their hinterlands. Management is given over to local authorities to manage development in the interests of local and regional communities.

but the scale of the trade is much smaller. The 2.1 million and 2 million tonnes traded respectively through Rosslare and Waterford's new port at Belview (Figure 24.16) depend exclusively on British and European markets. Trade with the rest of Europe and the global market is dominated by the deep-water ports of Cork but, more especially, by the Shannon Estuary, which is capable of accommodating large volumes of bulk goods, such as coal (from Colombia) and bauxite (from Guinea).

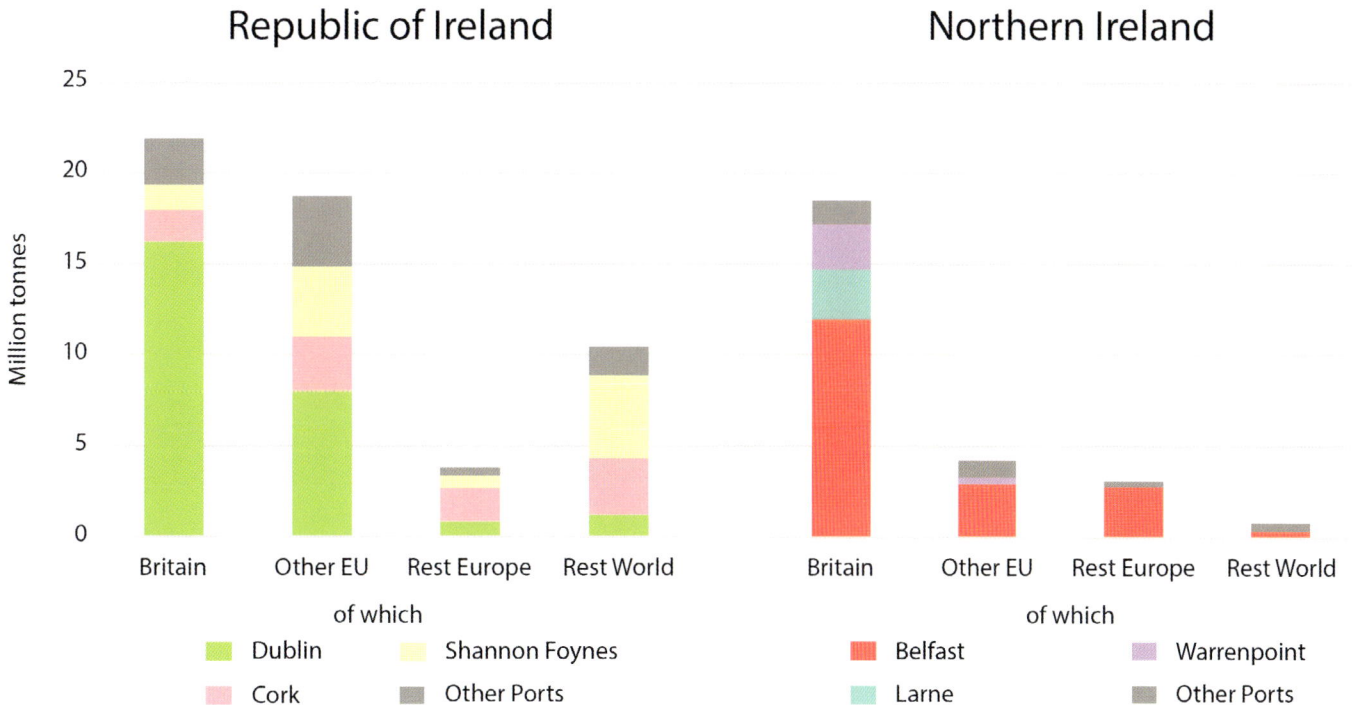

Fig. 24.15 IRELAND'S PORT TRADE BY GEOGRAPHIC AREA (2018). Dublin and Belfast dominate their countries' trade with Britain (74 per cent and 67 per cent respectively), benefitting from regular ferry services using the short sea crossings of the Central and Northern Corridors of the Irish Sea. Larne and Warrenpoint offer competition to Belfast for the British market, while Rosslare and Cork provide alternatives to Dublin in the Republic. The capital city ports also control much of the trade with the European Union, although to a lesser extent with the rest of Europe. This is focused mainly on regular container services to key European ports, but also involves some trade in bulk goods. Trade through bulk shipping, however, usually involves longer-distance movement, such as coal from Colombia and bauxite from Guinea (West Africa), and necessitates deep-water ports to accommodate large bulk carriers. Trade with the rest of the world is, therefore, focused primarily on Shannon Foynes and, to a lesser extent, Cork. [Data source: Central Statistics Office, Port Statistics of Ireland and Northern Ireland Department of Transport, Port Freight Statistics, 2018]

Fig. 24.16 THE NEW DOWNSTREAM PORT AT BELVIEW WATERFORD WITH LO-LO FACILITIES AND BULK STORAGE AREAS. Waterford is one of Ireland's historic towns and ports, serving an extensive area in the south-east of the island. Its historic port infrastructure occupied over eight hectares of land on the northern banks of the River Suir and near the core of the city. In 1992, 1.4 million tonnes of goods were traded through the port. Its congested site within the city and relatively shallow water acted as severe constraints to further development and to meet competition from other ports, especially Dublin. The decision was taken, therefore, to transfer port operations from the city to a new 265 hectare greenfield location at Belview on the River Suir, 8km from the city. The new downstream port emphasises its accessibility to British and European markets, provides modern lo-lo container cranes and ample storage areas for more efficient handling of both containers and bulk goods. The site is also adjacent to deeper water channels that facilitate the berthage of larger ships than was possible at Waterford. In effect, development at Belview has enabled the Port of Waterford to be categorised as a Tier 2 port in the Republic, with potential to increase its trade. The 2 million tonnes traded in 2018 is primarily with Europe and consists mainly of dry bulk (such as animal feed for the rich farming hinterland) and lo-lo containers. [Source: Port of Waterford]

TECHNOLOGY OF SHIPPING

Two innovations in shipping have particular significance for the differential development and changing geography of port activity in Ireland: bulk shipping and containerisation.

Bulk shipping

As the global economy expanded, demands increased for large volumes of low-cost raw materials and energy supplies. Sources of such inputs, however, are usually distant from regions of demand. This led to an increase in the size of ships, designed especially to transport large volumes at low unit costs – bulk shipping.

Industrial location theory suggests that a break-of-bulk site (where goods are transferred from one mode of transport to another, for example, a port) provides optimal conditions for industries to process inputs before trading them into the market. While port-related industries have a long history, their large-scale development occurred from the 1960s, and relates to the emergence of bulk shipping.

Bulk shipping and large-scale, port-related industrial development depend on the combination of three factors: deep water and sheltered harbours to accommodate large-scale vessels; extensive land banks adjacent to deep water berths for industrial development; good access to international shipping routes.

In 2018, 18 million tonnes of bulk cargo was traded through the Shannon Estuary and Cork's Outer Harbour, which can

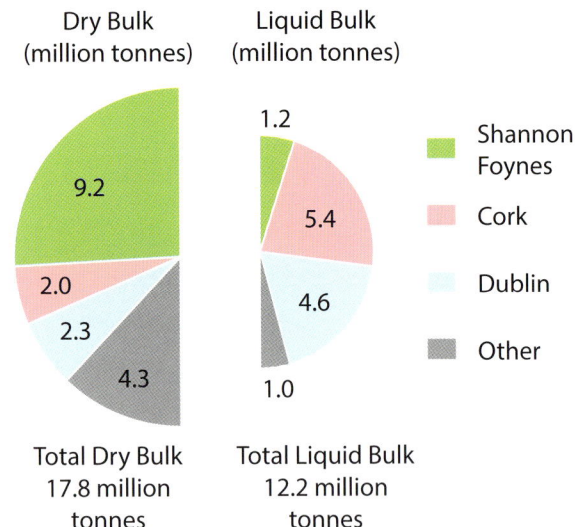

Dry Bulk (million tonnes)
9.2
2.0
2.3
4.3

Liquid Bulk (million tonnes)
1.2
5.4
4.6
1.0

Shannon Foynes
Cork
Dublin
Other

Total Dry Bulk 17.8 million tonnes

Total Liquid Bulk 12.2 million tonnes

Fig. 24.17 DISTRIBUTION OF BULK GOODS BY PORT IN THE REPUBLIC OF IRELAND (2018). Most of the Republic's trade in bulk goods involves imports of critical raw materials required by key sectors of the economy. Over 50 per cent of dry bulk goods are traded through specialised port facilities in the Shannon Estuary: coal at Moneypoint and bauxite for the Rusal Alumina plant at Aughinish. Cork and Dublin control a further 24 per cent of dry bulk trade, such as wood products and agricultural feed, although Dublin also has an important function for the export of zinc concentrate from the Tara Mine at Navan. Liquid bulk involves, almost exclusively, crude and refined oil. Cork is dominant in this bulk traffic, which is linked to its refinery at Whitegate. Demand for refined oil products in Dublin and its extensive hinterland ensures a significant role for the import of liquid bulk at the capital city port. [Data source: Central Statistics Office, Statistics of Port Traffic, 2017]

Fig. 24.18 THE EXTENSIVE WATERS OF THE SHANNON ESTUARY AND PORT OF FOYNES. The Shannon Foynes Port Company is the authority that covers port activities for the estuary which stretches 97km from Limerick city to the point where the River Shannon flows into the Atlantic ocean. Most shipping that once used the Limerick port infrastructure has gravitated downstream. The deep-water port at Foynes is now the centre of trade within the Shannon Estuary and can accommodate vessels of 200,000 tonnes. The port, shown in the foreground, possesses infrastructure to load/unload general cargo, but focuses on imports of animal feed, fertiliser, coal and oil for distribution within Ireland. Above all, however, the advantages of this port relate to bulk shipments. The vast majority of its traffic consists of shipments of dry bulk goods that utilise special jetties linked to large-scale, industrial complexes located on the shores of the Estuary. The photograph shows Rusal Alumina, the largest alumina refinery in Europe, which is sited near the port. Operational since 1983, it processes annually almost 2 million tonnes of alumina. A special jetty allows large bulk carriers to discharge the necessary raw material, bauxite, used in the process, of which over 4 million tonnes are imported each year primarily from Guinea, but also from Brazil. The storage area for the (red) residue left over from the refining process can be seen adjacent to the plant. Another major user of dry bulk goods in the Estuary is the Moneypoint Power Station. In effect, Shannon Estuary possesses all the necessary attributes to accommodate bulk shipping and large-scale, port-related industries, and future plans will focus on further developing these specialised functions. In 2018, 52 per cent of the Republic's dry bulk goods was traded through Shannon Foynes. [Source: Press 22]

Fig. 24.19 A LARGE, DRY BULK VESSEL CARRYING COAL BEING POSITIONED AT THE DEEP-WATER JETTY, WHICH SERVICES THE MONEYPOINT POWER STATION IN THE SHANNON ESTUARY. Moneypoint power station is the Republic's largest and only coal-burning electricity generating station. Commissioned in 1987, it has cost €1.2 billion to build and modernise, and meets some 20 per cent of the Republic's demand for electricity. It depends on imported coal to produce its output of 915MW. To ensure low-cost input of the 2 million tonnes of coal used annually, bulk shipping is essential. Most of the coal is sourced in Colombia and discharged at a special deep-water jetty, which accommodates vessels of up to 200,000 tonnes. To ensure continuity of supply, adjacent to the power plant is a 600,000 tonne coal storage area. Despite costly upgrades to improve its environmental standards, the large-scale burning of coal makes the plant the country's largest producer of harmful, greenhouse gases, in conflict with strict European Union policy for the environment. This concern with environmental standards and the projected end of its operational life in 2025 casts doubts on its future. Closure would have major implications for port traffic in the Shannon Estuary. [Source: Shannon Foynes Port Company]

accommodate vessels of some 200,000 and up to 90,000 tonnes respectively (Figure 24.17). Shannon Foynes depends almost exclusively on dry bulk goods (9.2 million tonnes), primarily meeting the needs of the Rusal plant at Aughinish, Europe's largest alumina refinery (Figure 24.18) and the coal-fired Moneypoint power station (Figure 24.19). Some 7.4 million tonnes of bulk goods were traded through Cork, which amounts to over three-quarters of the port's traffic. Most of this trade involves liquid bulk (oil) and relates to the Whitegate refinery in the Outer Harbour, but also includes dry bulk traffic unloaded mainly at the modern, deep-water terminal at Ringaskiddy. This is located adjacent to the 1,000ha IDA land bank on which is located the country's largest cluster of chemical industries (Figures 24.26, 25.13).

Containerisation

Containerisation involves handling traded goods in standardised units that can be transferred easily between different transport modes, such as from road to ship. Port trade using containers involves two systems: load-on, load-off (lo-lo) and roll-on, roll-off

(ro-ro). Lo-lo uses specialised cranes at the quayside to lift containers on/off a ship, while ro-ro involves containers being driven on/off a ship by accompanied or unaccompanied vehicles. Containerisation allows the uninterrupted flow of goods from origin to destination, thereby reducing time and cost of transport, which is vital for just-in-time production. It can also reduce the need for expensive stockpiling of goods and materials.

An increasing amount of European and global trade use containers. Therefore, the degree to which Irish ports have adopted this system has influenced strongly their ability to trade. For ports to offer effective container services, however, four key conditions have to be met: good access to the port from an extensive hinterland to ensure sufficient trade to justify investment; investment in expensive port infrastructure, such as container cranes, to allow for efficient turnaround of vessels; extensive port-side space to store containers; regular services to other container ports, but especially to key hub ports (such as Antwerp, Rotterdam) for connection into global trade.

In the Republic, four ports have been selected to provide container services: Cork, Dublin, Rosslare and Waterford (Figure

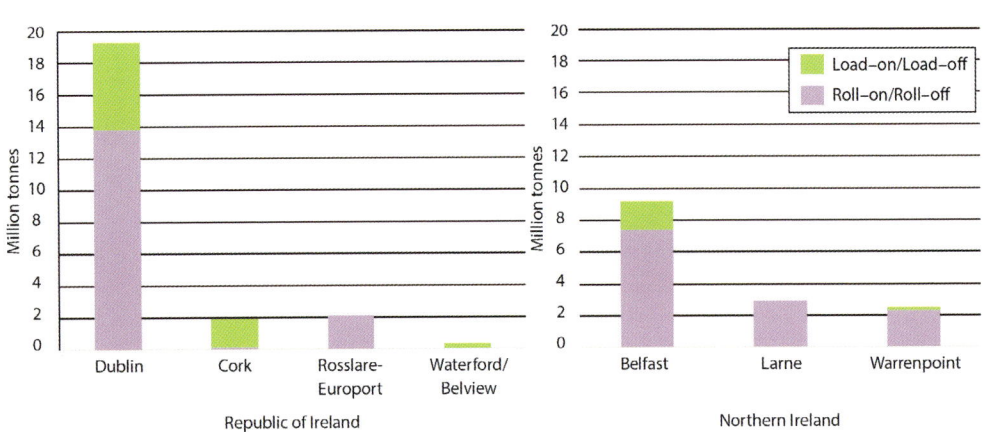

Fig. 24.20 RO-RO AND LO-LO TRADE THROUGH IRELAND'S PORTS (2018). Container traffic through Ireland's ports is dominated by Dublin and Belfast. In the Republic, 86 percent of ro-ro and 73 per cent of lo-lo traffic is routed through Dublin. Improvements to the national road network have extended the capital city's port hinterland to cover the entire country, and have encouraged freight hauliers to drive to Dublin to use its modern portside facilities and frequent container freight services to Britain and Europe. This is especially the case for ro-ro freight on the Central Corridor route to Britain, but also for the regular lo-lo services to mainland Europe. Rosslare provides some competition for ro-ro freight through its ferry services on the Southern Corridor to South Wales and the Continental Corridor to France, while lo-lo facilities at Tivoli and Ringaskiddy in Cork Harbour and Waterford's downstream development at Belview also offer alternatives to Dublin. In Northern Ireland, Belfast Port's dominance is more marked for lo-lo (90 per cent) than for ro-ro (59 per cent) due, in particular, to competition from Larne's ro-ro ferry service on the Northern Corridor route to Britain. Investments at Warrenpoint have also raised competition for Belfast, with improved freight ferry services to Heysham. [Data source: Central Statistics Office, Port Trade Statistics and Northern Ireland Statistics and Research Agency, 2018]

24.20). As a result, they have benefited from significant investment in infrastructure, while national transport policy (aided by Structural Funds) has improved their accessibility to larger hinterlands. Dublin, in particular, has benefited from its increasing specialisation on containers, and over 80 per cent of the Republic's container traffic now flows through this port (Figure 24.21). Northern Ireland's three main ports (Belfast, Larne, Warrenpoint) also show a high dependency on their container services (Figure 24.13). Central to the success of container ports has been their provision of frequent ferry services to Britain, but also to Europe.

Total container trade through these seven ports amounted to 38 million tonnes in 2018; approximately 45 per cent of all maritime trade.

Fig. 24.21 DUBLIN PORT. Dublin is Ireland's largest port (26.3 million tonnes in 2018). Like Belfast and Cork, it is also a multimodal port, in that it offers a full range of freight operations (dry bulk, liquid bulk, ro-ro, lo-lo and general cargo). The port authority has invested heavily in modernising infrastructure, especially for container traffic, which is a strong growth element in international trade. A specialised zone for lo-lo trade is seen on the left bank of the Liffey, where cranes and an extensive storage area for containers facilitate fast turnaround for container freight services to Britain and Europe. The right bank of the Liffey highlights the ro-ro terminals for ferries operating on the Central Channel route to Holyhead and Liverpool. Almost three quarters (19.3 million tonnes) of Dublin's total maritime trade occurs through container services and highlights its range of modern infrastructure and ease of access to a national hinterland and international shipping lanes. Liquid bulk (4.6 million tonnes), mainly oil, is also significant, as is dry bulk (2.3 million tonnes) involving mineral ores, coal and fertiliser. In recent years, Dublin has also become a significant destination for cruise liners. [Source: Dublin Port]

Dublin (19.3 million tonnes) and Belfast (9.1 million tonnes) provide both lo-lo and ro-ro services, and dominate the island's container trade (Figure 24.20). Larne (2.9 million tonnes), Rosslare (2.1) and Warrenpoint (2.5) focus mainly on ro-ro services to Britain, while Cork and Belview, Waterford's new out port, operate primarily as lo-lo ports serving European markets.

Case Study: Ireland's Passenger Ferry Ports

Barry Brunt

Although ports are associated generally with merchandise trade, they also perform a key function in enhancing connectivity between communities separated by relatively short-distance water divides, such as rivers, lakes and seas. These ferry ports provide terminal facilities for specially constructed ro-ro ferry boats to carry people, vehicles and goods across a body of water on a regular, scheduled route.

Ferry ports are of particular significance for the island of Ireland, given its long history of strong commercial and cultural connections with its neighbouring island of Britain. Furthermore, following independence for the Irish Republic in 1922, a Common Travel Area was agreed between the UK and Ireland to allow unrestricted travel between the two states for their citizens. As a result, several ports on Ireland's east coast have long-established ferry routes to facilitate passenger and trade movements with ports along the west coast of Scotland, England and Wales. Approximately 4.7 million passengers passed through Ireland's ferry ports in 2018, 2.7 million using routes centred on the Republic (Figure 24.22). Similar to general merchandise trade, ferry ports have defined four principal channels of connection to Britain and Europe: Northern, Central, Southern and Continental (Figure 24.23).

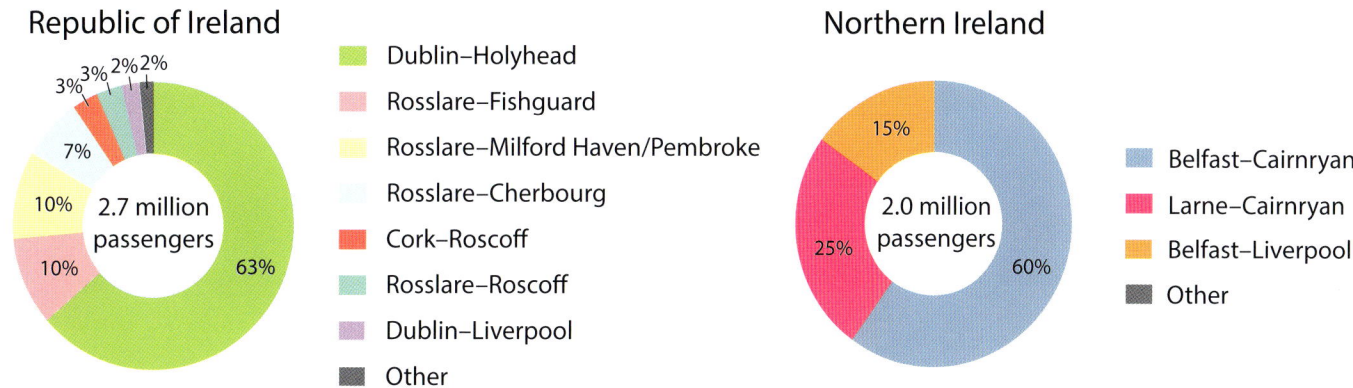

Fig. 24.22 SEAPORT PASSENGERS ON FERRY ROUTES FROM IRELAND (2018). In 2018, almost 4.4 million passengers used ferry services between Ireland and Britain. Two routes dominate – Dublin to Holyhead and Belfast to Cairnryan – and account for approximately 3 million passengers, two-thirds of all ferry traffic between the two islands. By focusing on key routes, the ferry ports and companies can concentrate their investment strategies to provide more efficient and faster services, as well as providing modern terminal facilities. This is vital to allow for effective competition with low-cost airlines. Over 0.5 million passengers used the Larne–Cairnryan service, while a similar number travelled on the two ferry services that operate from Rosslare to South Wales along the Southern Corridor. Approximately 0.35 million passengers used ferry links from Ireland to continental Europe. [Data source: Central Statistics Office and Northern Ireland Statistics and Research Agency, 2017]

THE NORTHERN CORRIDOR

Northern Ireland ferry ports catered for some 2 million people in 2018, representing over 40 per cent of all sea-ferry passenger traffic for the island of Ireland (Figure 24.22). This is significantly higher than Northern Ireland's relative share of the island's total population (30 per cent), and highlights the importance of such ferries in maintaining and promoting the deep historic connections between Northern Ireland and Britain, but especially with Scotland. As a result, almost 1.8 million passengers travel on Northern Corridor routes, which link Belfast and Larne to Cairnryan in Scotland.

The modern phase of regular ferry services between Northern Ireland and Scotland can be traced to the mid-nineteenth century, with a regular mail service being established between Larne and Stranraer in 1875. As a result, Stranraer became the principal Scottish ferry terminal for the Northern Corridor. In 1973, however, the P&O Line initiated a route from Larne to Cairnryan, a port located downstream from Stranraer on Loch Ryan. This became the shortest sea crossing on the Irish Sea, reducing costs and time connecting Northern Ireland to Scotland. Its development ultimately caused the Stena Line to move their Belfast–Stranraer service to a new terminal in Cairnryan in 2011 to compete more efficiently on the Northern Corridor. This ended the 150-year direct link between Stranraer and Northern Ireland.

Belfast is Northern Ireland's premier ferry port, providing some fifty sailings each week and catering for 1.5 million passengers in 2018. The Stena Line service to Cairnryan is dominant (1.2 million passengers), although Belfast also provides ferry links to Heysham and Liverpool in north-west England, and to Douglas in the Isle of Man.

Larne is the second-largest ferry port, with 0.5 million passengers carried on its Cairnryan route (Figure 24.24). From 2003–2015, Larne also offered a ferry link to Troon, but the P&O Line discontinued the service due to falling passenger traffic.

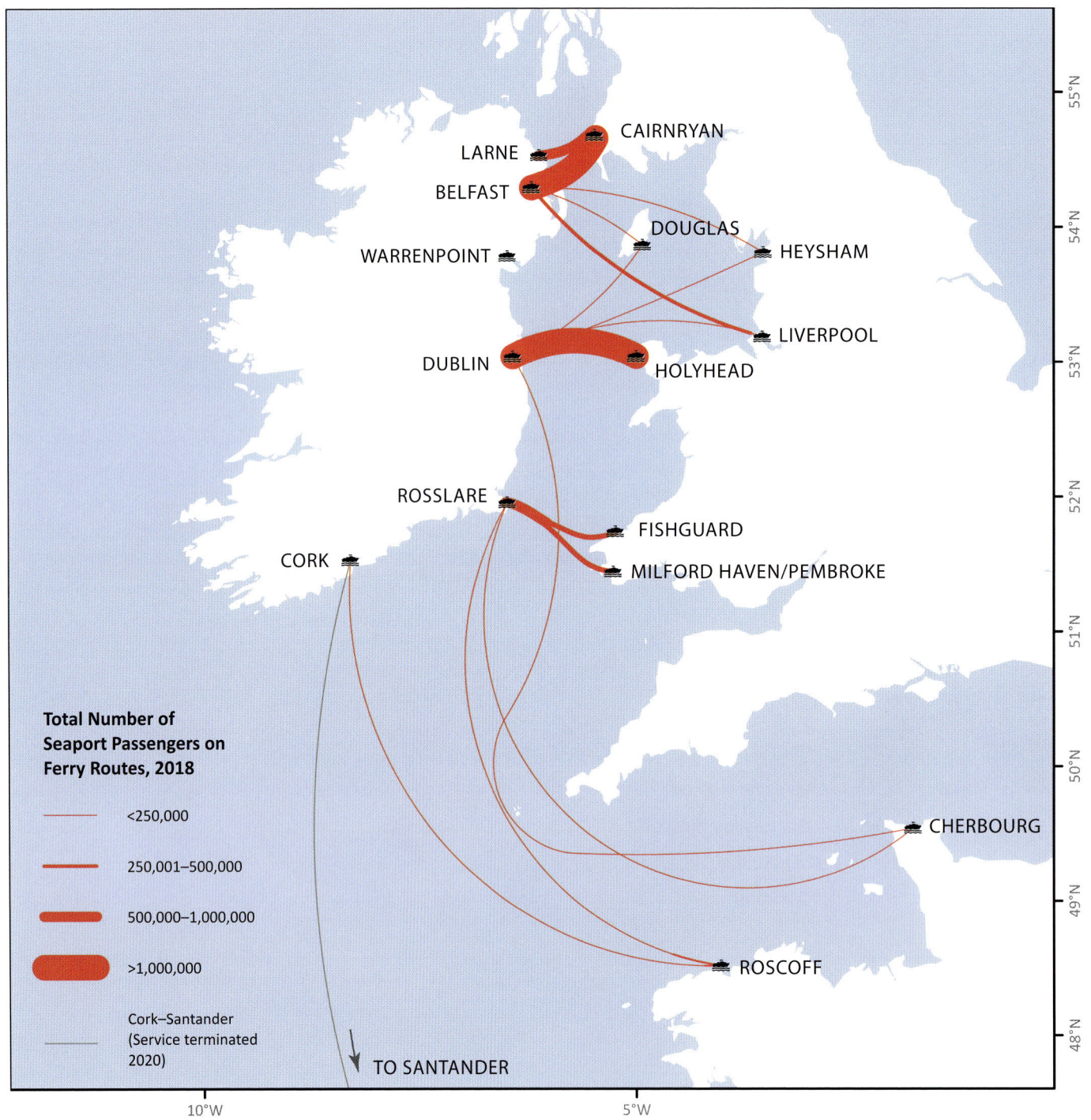

Fig. 24.23 FERRY ROUTES CONNECTING IRELAND TO BRITAIN AND EUROPE, 2018. Passenger ferry routes link Ireland to Britain and Europe through four principal corridors. The Central Corridor is the most utilised and involves competing ferry companies servicing primarily the Dublin to Holyhead route (1.7 million passengers). Strong cultural and economic links between Northern Ireland and Scotland are consolidated by frequent ferry services operating on the Northern Corridor, but especially Belfast to Cairnryan (1.2 million passengers). Both capital city ports also provide important links to Liverpool. Over 0.5 million passengers utilise the Southern Corridor from Rosslare to two Welsh ports. The Continental Corridor services to France are the least utilised, since competition from low- cost airlines has reduced the attraction of more lengthy trips between Ireland and Europe. A combination of Brexit and the Covid 19 pandemic (2020) is impacting ferry services. Passenger flows have declined on the Irish Sea, especially on the Central Corridor, although this is likely to be a short-term trend. Brexit also imposes additional costs/time delays at ports for road haulage companies operating between Britain and Ireland. Using Britain as a landbridge to Europe has therefore become less competitive. The result has been demands for additional direct ferry services on the Continental Corridor. Dublin, Cork, and Rosslare have all attracted new services, although the less congested port environments and shorter sea routes for ro-ro operators working through Rosslare and Cork could be more significant in the long term. [Data source: Central Statistics Office and Northern Ireland Statistics and Research Agency, 2018]

THE CENTRAL CORRIDOR

This corridor is focused on Dublin's ferry terminals, which link with Liverpool, Douglas (Isle of Man) and, above all, Holyhead in Anglesey, North Wales, and caters for the largest flow of passengers between Ireland and Britain. It is not surprising that the capital city's port should emerge as the Republic's largest ferry port. It is located strategically on a relatively short sea route to Britain and is the principal gateway into the country. In addition, the population of the Greater Dublin Area (1.9 million) provides a large catchment area for people wishing to travel to Britain

The Dublin to Holyhead services provided daily by Stena Line and Irish Ferries dominate this corridor. Regular ferry services can be traced to the mid-sixteenth century, but its prominence relates to developments in the nineteenth century. Following the Act of Union (1800) between Ireland and Britain, there emerged an increasing need for more effective and regular transport links, and especially between the dominant economic and political core regions. The Dublin–Holyhead sea route was identified as pivotal, and was confirmed by the building of road (1826) and rail (1859) bridges across the Menai Strait, which separates Anglesey from North Wales. These infrastructures, plus improved rail systems, allowed for large-scale increases in the movement of people, mail services and goods between both countries.

Although passenger flows have declined, due primarily to increased competition from airlines, the Dublin–Holyhead route catered for 1.7 million passengers and over 0.5 million passenger vehicles in 2018. As with the Belfast–Liverpool ferry service, ferries to Holyhead and Liverpool provide direct access to metropolitan centres of north-west Britain. Of greater importance, however, are the connections at Holyhead and Liverpool to Britain's extensive motorway and railway networks, which provide efficient links to all parts of Britain. In addition, many people and more especially freight vehicles use these links as a landbridge to access continental Europe via Britain's Channel ports. About 80 per cent of Irish goods exported to continental Europe, for example, go via these Channel routeways. Such high levels of dependency on using Britain and its ports as a through-route to access Europe and the rest of the world is now a source of growing concern, given Britain's decision to leave the European Union and the potential disruption this could have on Ireland's international trade.

THE SOUTHERN CORRIDOR

Until 2011, this corridor linking Ireland to South Wales was served by three routes: Cork to Swansea, Rosslare to Fishguard and Rosslare to Pembroke Dock.

The ferry link from Cork focused initially on Fishguard, but in 1969 a new Cork–Swansea route was inaugurated by the B&I Line. From 1987 to 2006, the service was provided by Swansea Cork Car Ferries and brought almost 3 million passengers to Cork over its period of operations. Rising costs, falling passenger numbers and increased competition from the shorter sea crossings from Rosslare (three hours compared to ten hours) resulted in the service being discontinued in 2006. Closure of the ferry route, however, had significant implications for counties Cork and Kerry; a 30 per cent decline in British tourists and the loss of €35 million in revenue. As a result, business interests in Cork tried to reactivate the service. This met with initial success in 2010, but momentum stalled and the ferry route was again discontinued in 2011.

Rosslare Harbour was opened in 1906 and was promoted by railway companies to service the increasing ferry traffic between Britain and Ireland. It is the closest port in the south of Ireland to Britain and the European mainland and has developed successfully as a major terminal for ro-ro passenger and freight traffic. Now called Rosslare Europort, it remains the only Irish port operating on the Southern Channel, with Irish Ferries and Stena Line carrying over 0.5 million passengers each year to Fishguard and Pembroke Dock.

THE CONTINENTAL CORRIDOR

Direct ferry services are provided from Cork to Roscoff (one sailing per week) and, more frequently, from Rosslare Europort, where two ferry operators offer services to Cherbourg (three sailings per week) and Roscoff (two per week). Although most international tourists travelling to and from Ireland use airlines, the attractions of an accompanying vehicle result in Rosslare terminal accommodating some 270,000 passengers on its France routes. Declining numbers using this route, however, caused Irish Ferries to announce the discontinuation of its Rosslare to France services from 2019. It remains to be seen whether the implementation of Brexit will lead to a reversal of this decision. At the time of writing, however, the company's French services will in future focus on its Dublin to Cherbourg route. Stena Line will continue to operate its French services from Rosslare, while over 90,000 passengers use the Ringaskiddy ferry terminal in Cork's Outer Harbour. In 2018, the first direct ferry service between Ireland and Spain was launched, with two sailings each week connecting Santander with the deep-water ferry terminal at Ringaskiddy. This was an encouragement for the Port of Cork and its commitment to focus future development in its Outer Harbour. Use of this link was, however, disappointing and in 2020 Brittany Ferries transferred the service to Rosslare Europort. A direct link to Spain is of strategic value as it provides the country with another direct route to the European Union and, therefore, reduces dependency on the British landbridge.

For Ireland, accession to the European Union and commitment to export-led development resulted in significant changes to the island's ports and patterns of maritime trade. Trade almost trebled (180 per cent) from 1973 to 2018 to reach 83.5 million tonnes, but occurred largely in the absence of coherent policies. This enabled the larger ports, but especially Dublin and Belfast, to benefit from demand-led growth. By the 1990s, Dublin had replaced Belfast as the largest port, and now accounts for almost one-third of the island's maritime trade, with Belfast contributing a further 22 per cent. Control of the rapidly increasing international trade through containerisation is even more apparent, with the two ports accommodating three-quarters of the island's maritime trade utilising this mode of transport. The result has been an increasing polarisation of maritime trade through the two capital city ports of Dublin and Belfast. In contrast, bulk shipping has

focused increasingly on the deep-water ports in the Shannon Estuary and Cork Harbour (Figure 24.25).

Competition between ports to attract trade has been intense, since increasing traffic generates considerable economic benefits for a port and its city region. Given the important strategic role of ports for Ireland, a laissez-faire approach to their development could not persist. Over the last decade, therefore, efforts to encourage more coordinated planning have emerged. This is essential, given trends in international trade, which emphasise the need for expensive, deep-water berths and efficient infrastructures to accommodate large vessels and provide effective access to both an extensive hinterland and foreland. Most Irish ports do not possess these critical site and situation advantages. In addition, the capital costs of providing modern infrastructures are very high. Together, this necessitates difficult political and business-orientated

Fig. 24.24 THE FERRY PORT OF LARNE. Larne is a thriving port and market town located on the western side of an inlet that links Lough Larne to the Irish Sea. Its location on the east coast of County Antrim is also only 40km from the Scottish mainland and this proximity to Scotland has been pivotal in influencing Larne's development. As one of Ireland's main ferry ports, Larne offers modern facilities to accommodate ro-ro freight and passenger traffic in the Northern Corridor linking Ireland to Britain. Although Belfast is Northern Ireland's dominant port, Larne is highly competitive for passenger traffic on the route to Cairnryan, advertised as the shortest and fastest ferry route across the Irish Sea. The service operated by the P&O Line provides frequent daily sailings and accommodated over 0.5 million passengers and 0.2 million tourist vehicles in 2018. Extensive areas of portside parking are provided adjacent to the terminal buildings to facilitate ease of access to the ferries. On the eastern side of the inlet is the Islandmagee Peninsula on which is located the Ballylumford power station. This is Northern Ireland's largest power staion, providing some 50 per cent of its energy requirement. [Source: Port of Larne]

decisions to focus investment on a small number of Irish ports that possess long-term, strategic advantages. Evidence that this reality has been appreciated by governments in both the Republic and Northern Ireland is reflected in the fact that almost 80 per cent of the island's maritime trade is now shipped through the four principal ports of Dublin, Belfast, Shannon Foynes and Cork.

The polarisation of maritime trade through these dominant ports is likely to be further enhanced by their natural advantages, which will ensure their selection and development as optimal sites for Ireland's growing offshore energy sector. While port congestion and limited quayside space adjacent to deep-water channels is likely to limit Dublin's future role, Belfast, Cork and Shannon Foynes are well placed to benefit. All three ports have locations that are strategic for current and projected offshore developments, such as wind farms and hydrocarbon resources. Furthermore, they have extensive land banks adjacent to deep-water berths on which to manufacture infrastructure (for example, oil rigs, assemblage of wind turbines) and to provide the range of maintenance services required by this international growth sector (see Chapter 28: Renewable Energies: Wind, wave and tidal power).[17]

As a result of this, together with the polarisation of trade, these four ports will continue to dominate. Apart from the two Tier 2 ports in the Republic, and Larne and Warrenpoint in Northern Ireland, all of which have specialised functions (for example, ferry ports), Ireland's smaller regional ports are likely to face a future of continuing decline in maritime traffic (Figure 24.24). The future of such minor ports and harbours located around the island will probably be linked to leisure and tourism, marinas and water-based recreation, together with the provision of local ferry services, especially along the coastline of western Ireland.

Belfast and all Tier 1 ports in the Republic have now replaced demand-led growth by a greater commitment to a supply-led strategy. In addition to national governments encouraging investment on these ports, their managing companies have also published plans to attract trade and achieve better commercial returns via economies of scale.

The Port of Dublin Masterplan 2012–2040, which was reviewed in 2018, illustrates the commitment to forward and more sustainable planning.[18] The significant increase in trade through the port since the launch of the initial plan (some 30 per cent in volume) has placed additional stress on infrastructure and has seen projected capacity at the port being increased from 60 to 77 million tonnes by 2040. Rather than undertaking expensive and environmentally contentious reclamation into Dublin Bay to provide additional land

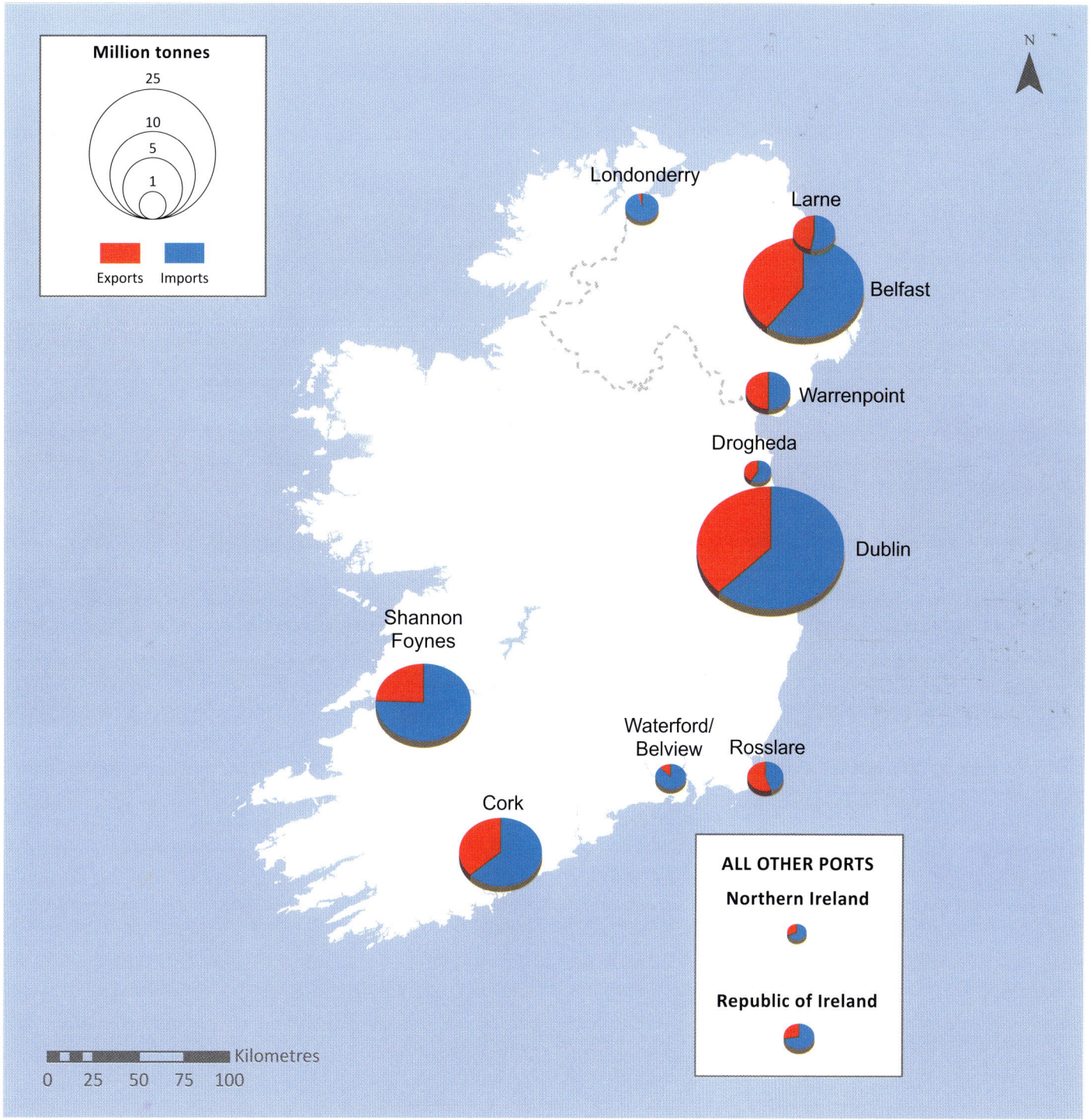

Fig. 24.25 IRELAND'S PORT NETWORK, 2018. In 2018, total trade through Ireland's port network was a little over 83.5 million tonnes. The advantages of Dublin for a multi-modal port is reflected in its fourfold increase in trade from 1972, to become the island's dominant port and responsible for 26.3 million tonnes of trade (almost one-half of the Republic's maritime traffic). Although growth in Belfast was less impressive (150 per cent), its 18.9 million tonnes accounted for two-third's of Northern Ireland's maritime trade. The need to focus maritime trade increasingly on a relatively small number of ports possessing site and situation advantages is reflected in the fact that by 2018, only ten ports shipped over 1 million tonnes of goods. Most, however, remain relatively small operations and depend for their success on specialised function, such as ro-ro ferry services at Larne. Growth has been particularly impressive at Shannon Foynes, which replaced Cork to become the third-largest port complex on the island (10.7 million tonnes). This is related, almost exclusively, to its role as an importer of dry bulk goods for major port-related industries located in the Shannon Estuary. Such development provides modern infrastructures to enhance the development potential of this part of the Atlantic seaboard. Apart from this port, the locational advantages of the east coast for success in maritime trade is apparent. Overall, the top four ports of Dublin, Belfast, Shannon Foynes and Cork account for almost 80 per cent (65.4 million tonnes) of Ireland's maritime trade. [Data source: Central Statistics Office, Statistics of Port Traffic, and Northern Ireland Department of Economic Development, Trade at Principal Ports, 2019]

for infrastructure, the strategy now includes moving non-core port activities to a new 44ha inland site near Dublin Airport. This will release land within the port for more effective use, such as the Alexandra Basin Redevelopment Project, and provide opportunities to enhance container facilities (especially ro-ro) which could account for 90 per cent of the port's future maritime trade. These developments will add to Dublin's locational advantages and its status as the premier port on the island of Ireland.

Fig. 24.26 THE PORT OF RINGASKIDDY IN CORK HARBOUR. Considered to be the second-largest natural harbour in the world, Cork Harbour possesses the attributes necessary for large-scale, port-related development. The port of Ringaskiddy is sited on the western side of the Outer Harbour, opposite to the town of Cobh, and is proximate to the naval base on Haulbowline Island and to Spike Island. Promotion of a new port at Ringaskiddy began in the 1970s as part of a long-term plan to move shipping from the congested and relatively shallow waters of Cork's inner-city quays, downstream to the Outer Harbour. Extensive land banks adjacent to deep-water channels offered optimal conditions to develop a new multi-modal port and site for large-scale, port-related industry. Significant development has occurred. Adjacent to the port are two large factories which form part of one of the largest clusters of chemical plants in Ireland. This includes Pfizer with its jetty extending into the Harbour to allow direct transfer of imported raw materials from ship to plant. The modern ro-ro ferry terminal can be seen adjacent to the deep-water, dry-bulk terminal, where a large vessel is berthed and awaiting unloading. The port is also one of the principal points of entry for large numbers of motor vehicles, which are stored on the extensive parking areas seen adjacent to the ro-ro terminal before being re-distributed to dealerships throughout the country. The extensive area of land adjacent to the ferry port is the site of a projected €80 million development of a new container terminal, which will replace the existing one sited upstream at Tivoli. This will add further to the significance of Ringaskiddy as one of the island's strategic ports. [Source: Port of Cork]

Dublin's principal rival port in Belfast also has plans to enhance infrastructure, especially its container and ferry operations. The plan envisages trade through the port increasing by 60 per cent by 2030.[19] This will confirm its status as the premier port and growth centre in Northern Ireland.

Elsewhere in the Republic, investments outlined in Shannon Foynes Vision 2041 (primarily at the deep-water port of Foynes) are projected to double trade through Shannon Estuary and strengthen its role as Ireland's leading bulk shipment port.[20] Finally, a new cargo container terminal is to be built at Ringaskiddy in Cork Harbour as part an €80 million investment by the Port of Cork (Figure 24.26). It forms the most recent and significant part of the port's process of downstream development taking maritime trade activities away from the more congested and shallower berths near the city. The existing container facility at Tivoli (Figure 24.1) will be phased out (allowing for urban-related developments) with trade focusing on a new and extensive deep-water berth, modern container-handling equipment and backed by a large cargo-

handling terminal. These developments will complement the existing deep-water terminal for bulk goods and confirm the central role of Ringaskiddy in Cork's status as a multi-modal port of national importance.[21]

Long-term strategic planning in key ports is essential for the continuing success of the island's economy.[22] Capital costs of creating and maintaining efficient and effective port infrastructure are very high and limited funding cannot be dispersed or duplicated over too many ports. This would, furthermore, reduce the benefits of economies of scale at preferred gateways for global trade and therefore limit their abilities to offer competitive, low-cost services to shipping companies and businesses. Polarisation of investment on a limited number of ports, however, creates political pressures as well as having the potential to create significant environmental stress, which will impact on communities living in and around major port complexes. Achieving an acceptable balance between competing economic, social and environmental interests will not be easy, but will have to be resolved.

URBANISATION OF IRELAND'S COAST

Barry Brunt

VIEW OF DUBLIN, IRELAND'S LARGEST CITY, LOOKING EASTWARDS ALONG THE RIVER LIFFEY TO ITS PORT, WHERE IT ENTERS THE IRISH SEA. [Source: Irish Air Corps]

The process of urban development is often associated closely with water, and intimate relationships usually emerge between the origin and development of urban areas and their strategic locations adjacent to rivers, canals, harbours, seas and oceans. Many urban places have their history, identity, morphology, culture and prosperity defined by their waterside location. As a relatively small island with a long history of dependency on external trade, Ireland is an excellent example of a country where urbanisation and coastal development are closely intertwined. Almost all of the island's major towns and cities evolved out of significant relationships that linked their dual roles as urban places and functioning ports. Many of Ireland's contemporary urban challenges and opportunities have also emerged in this context. By European standards, the spatial bias to coastal locations for the principal cities and most large towns is high, and this has been – and remains – crucial for patterns of population and employment.

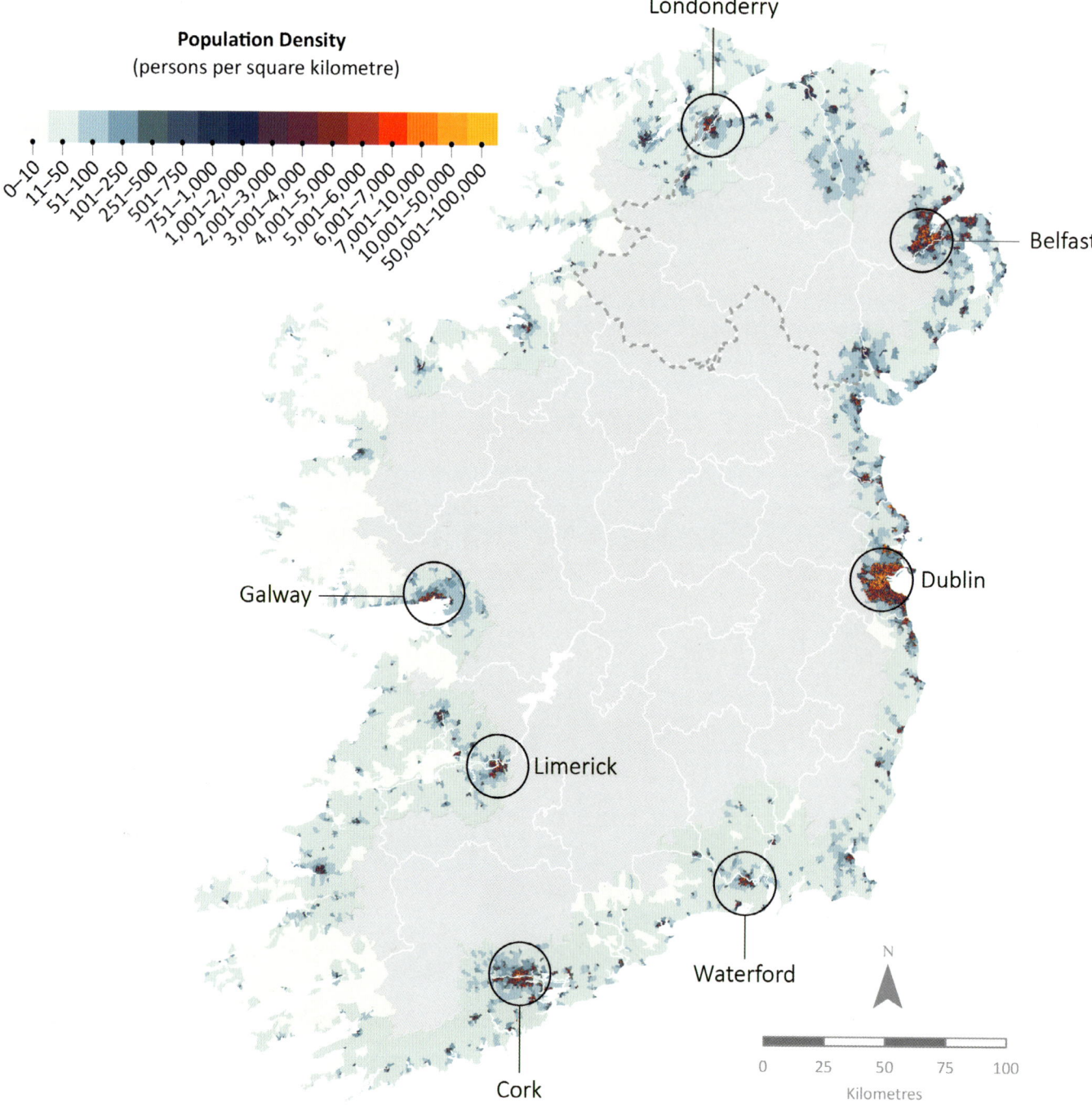

Fig. 25.1 POPULATION DENSITY PATTERNS FOR SMALL AREAS WITHIN 10KM OF IRELAND'S COASTLINES (2016). The population density of the Republic in 2016 was seventy persons per km². This is low by European standards and is approximately one-half of that in Northern Ireland. Since 2011, data has been available within Ireland for Small Areas, which allow for detailed mapping of population patterns. Using the data, concentration of people residing within 10km of the coast is focused primarily within and around the island's few cities and large towns, but especially for Dublin and Belfast. Extensive areas of the coastal zone, however, have density levels of fewer than fifty persons per km². While this suggests a rural landscape, numerous places can be identified, especially along the north, east and south coasts, where densities occur in excess of 100 persons per km². These are small towns and villages that have attracted a growing number of residents, which now form part of the expanding commuter zones that have developed around coastal cities and larger towns. As a result, they impose an increasing urban footprint along the coastline. Large areas along the western seaboard display density levels of less than ten persons per km². These are part of Ireland's empty landscapes, linked to their difficult physical environments and underdeveloped economies. Small tourist resorts and/or service centres occur intermittently along this peripheral coastline. [Data source: CSO and NISRA]

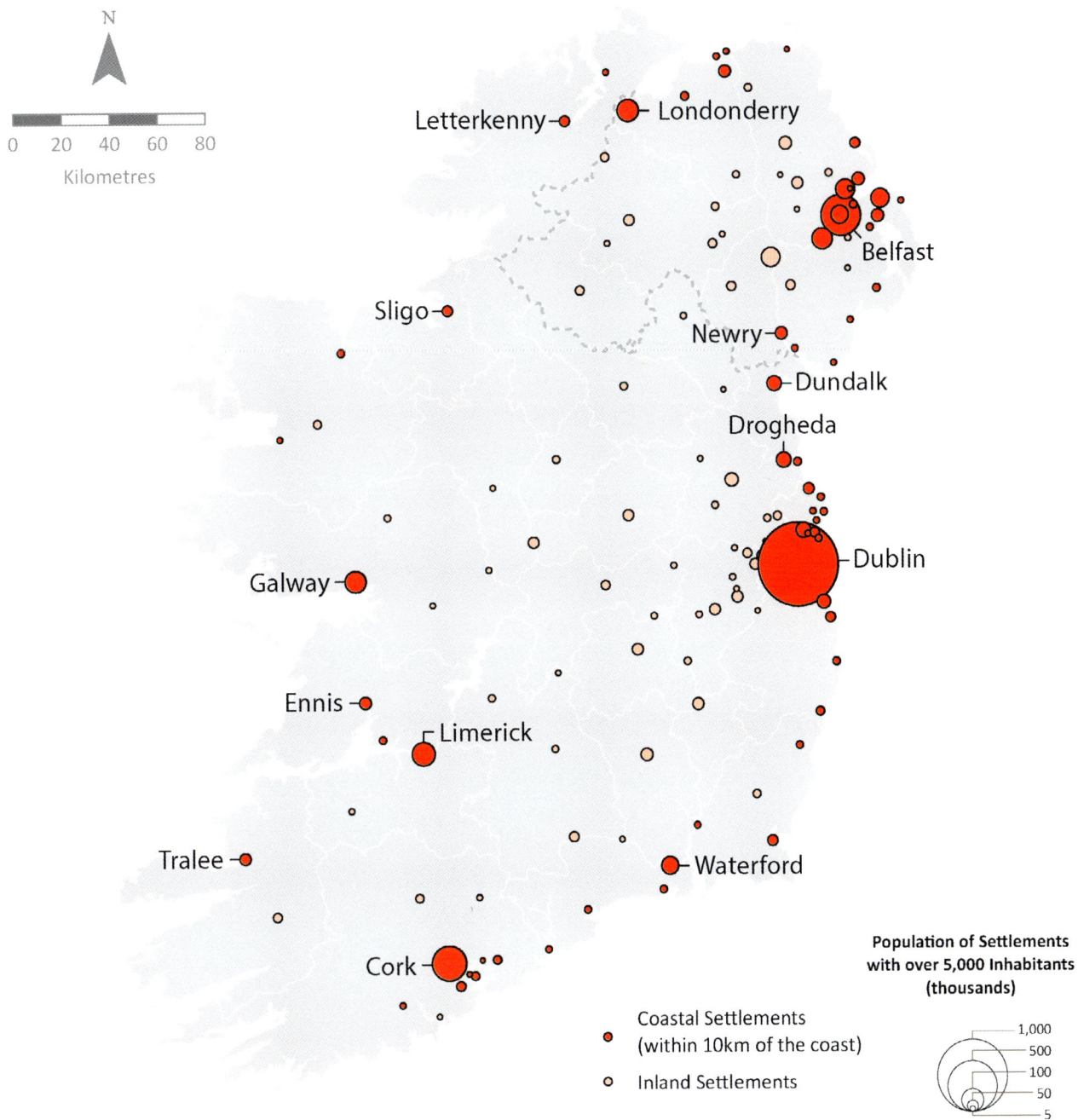

Fig. 25.2 LOCATION PATTERN OF URBAN CENTRES OF GREATER THAN 5,000 POPULATION (2016). Differences can be observed between east–west and north–south, but also between the island's coastal rim and interior. Most of Ireland's main towns and cities, including the capitals of Dublin and Belfast, are coastal sites, occurring within 10km of the coast. [Data source: CSO and NISRA]

In 2016, the population of Dublin and its environs was 1.17 million, the population of the Belfast Metropolitan Area approached 0.6 million.[1] Although these totals for capital cities are relatively small by international standards, they nevertheless dominate their urban hierarchies and much of the economic, social and political developments that occur within their state boundaries. Over one-in-four of the populations of the Republic of Ireland and Northern Ireland reside in these coastal capitals and their environs. Within the national context, they are disproportionately large and are excellent examples of primate cities. These primate cities and other large coastal towns have developed significantly since the 1960s, and have extended their influences over adjacent coastal areas, as well as inland. Figure 25.1 illustrates this strong orientation to coastal cities

and adjacent areas by mapping population densities for geographic units, defined as Small Areas, and located within 10km of Ireland's coast.[2] Such levels of development and consequent pressures on the environment suggest that Ireland's coastal zone has been influenced strongly by processes of urbanisation.[3]

For much of this chapter, a theoretical size of 5,000 people is used as a threshold for describing changes in Ireland's urban landscape. Settlements of fewer than 5,000 residents have recently lost, or are losing, many key functions such as schools, post offices and banks and, therefore, have become less attractive for inward investment. In contrast, larger settlements have gained in significance in shaping urban patterns, and these are strongly oriented towards coastal location. (Figure 25.2).

THE HISTORICAL CONTEXT

In general, the development of towns in Ireland was not an indigenous process. In effect, the first towns originated and evolved largely through external influences and invading cultures. As a result, Ireland's coastal zone has played a dominant role in shaping the distribution of towns and cities on the island of Ireland.

Continuous urban settlement can be traced initially to the Vikings who established fortified trading centres, primarily along Ireland's east and southern coasts, in the ninth and tenth centuries (Figure 25.3). Subsequently, these evolved to become principal cities of the present-day Republic, such as Dublin, Cork and Waterford (Figure 25.4). However, it was the Anglo-Norman invasion in 1169 that 'unequivocally introduced urban life into Ireland, as an integral part of the process of urbanisation'.[4] This was followed by a second wave of colonisation in the Tudor and Stuart times (sixteenth and seventeenth centuries), together with an extension of landlord estate towns in the eighteenth century. By 1800, a pattern of urban places had been established that would become the framework for Ireland's contemporary urban system (see also Chapters 18: Era of Settlement: Trade, Plantation and Piracy and 19: Changing Coastal Landscapes).

While towns spread inland to allow British colonists to better control and exploit the countryside, urban development remained focused along the eastern, southern and northern coastlines. These

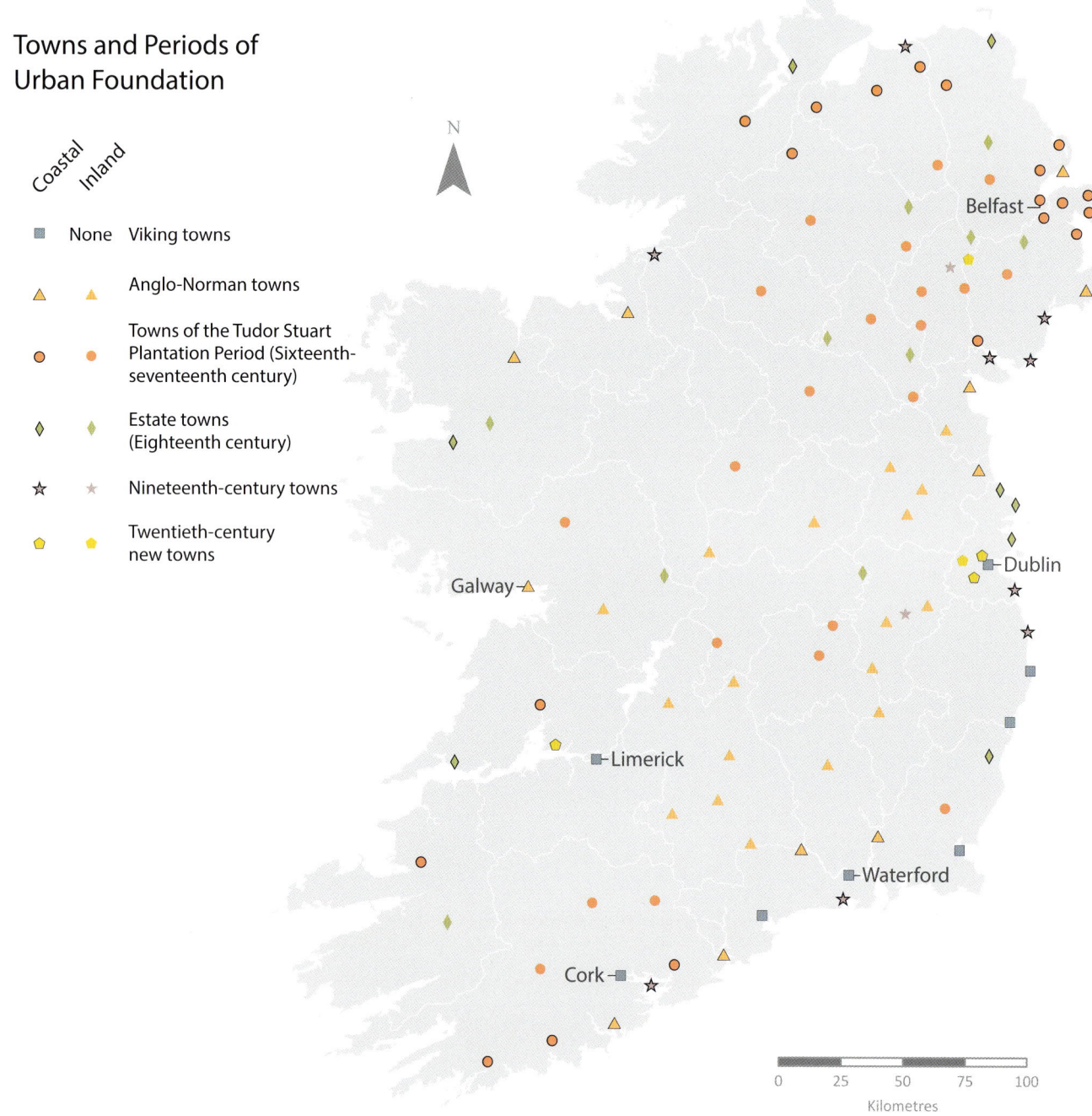

Towns and Periods of Urban Foundation

Coastal Inland

▪ None Viking towns

△ ▲ Anglo-Norman towns

◯ ● Towns of the Tudor Stuart Plantation Period (Sixteenth-seventeenth century)

◆ ◆ Estate towns (Eighteenth century)

☆ ★ Nineteenth-century towns

⬠ ⬠ Twentieth-century new towns

Galway
Belfast
Dublin
Limerick
Waterford
Cork

0 25 50 75 100
Kilometres

Fig. 25.3 PHASES OF URBAN FOUNDATION. Early Viking settlements, located primarily on the eastern and southern coastline, were consolidated by the Anglo-Normans. These extended their influence by establishing many inland towns throughout Munster and Leinster. In Northern Ireland, the importance of plantation settlements is apparent in the evolution of their urban system. While the industrial revolution increased the size and importance of many Northern Ireland towns, comparatively few new towns were established after 1800. The sparsity of towns in the western part of the island is a noticeable feature. [Data source: Johnson, J.H., 1994. *The Human Geography of Ireland*. Chichester: J. Wiley]

Fig. 25.4 VIEW OF WATERFORD (*c*.1736) BY ANGLO-DUTCH ARTIST, WILLEM VAN DER HAGEN. This is the first landscape in oils of an Irish city. Founded by the Vikings in the ninth century, Waterford became a key fortified city, port and administration centre to consolidate Anglo-Norman control over the south-east of Ireland. The painting illustrates the wealth and prosperity of the city in the eighteenth century when the former fortified walls adjacent to the River Suir had been replaced by elegant waterfront buildings. The city remains the principal urban centre for the south-east. From 1926 to 1961, population totals stagnated around 28,000. Despite some growth following the 1960s, and its selection as a growth centre and gateway for regional development, the city did not evolve as a counter pole to Dublin to the extent of Cork and Limerick. Recently, the city has experienced significant renewal, especially along its waterfront as the historic port has moved downstream to Belview. Under Project Ireland 2040, its role as a growth centre will be enhanced, and its population is projected to rise from 54,000 to over 80,000. [Source: Waterford Treasures – Three Museums in the Viking Triangle]

early and more prominent towns were located generally at the mouths of significant rivers, such as the Lagan, Lee, Liffey, Shannon and Suir. These sites provided sheltered harbours, while reclamation efforts increased the depth of water available to accommodate larger ships, and extended the area of waterfront land for port infrastructure and urban development. They also provided access to an extensive hinterland.

Ports located along the east coast in particular were also, and remain, well situated to provide relatively short-distance Irish Sea routes to Britain. As trade dependency with Britain increased, so the strategic gateway functions of these coastal towns were emphasised. This spatial preference of port-related urban growth reflected the advantages of site and situation. As a result, selected centres were able to extend critical functional relationships with both their hinterland, and Britain and a wider global market.

Following a long period of growth in population and urban development, Ireland's total population fell precipitously in the second half of the nineteenth century in the aftermath of the Great Famine (Figure 25.5; see also Chapter 20: The Great Famine). The southern counties that would form the Republic were affected most severely, given the dominance of a rural society

and an almost total dependency on the potato harvest to sustain large numbers of people. Here, the only areas to experience population increases were, in general, the larger coastal towns and their immediate hinterlands. These were able to prosper from their trading functions and attracted large-scale rural to urban migration. The advent of railways also allowed people to better access emerging coastal tourist resorts and encouraged the growth of towns such as Bray and Youghal.

In contrast, the north-east of Ireland had a more diversified agricultural economy but also experienced an industrial revolution in the nineteenth century. This stimulated significant urban growth and employment opportunities.[5] Dependency on Britain for critical industrial inputs and market outlets meant the focus of this new surge of urbanisation remained principally around Belfast and the Lagan Valley (see Chapter 19: Changing Coastal Landscapes, Case Study on Belfast). Despite high rates of emigration in the aftermath of the Famine, these enlarged, urban-based labour markets allowed for a lower rate of population loss compared with the rest of the island.

COASTAL URBANISATION, 1926–1961

In 1926, the first census of the recently created Irish Free State

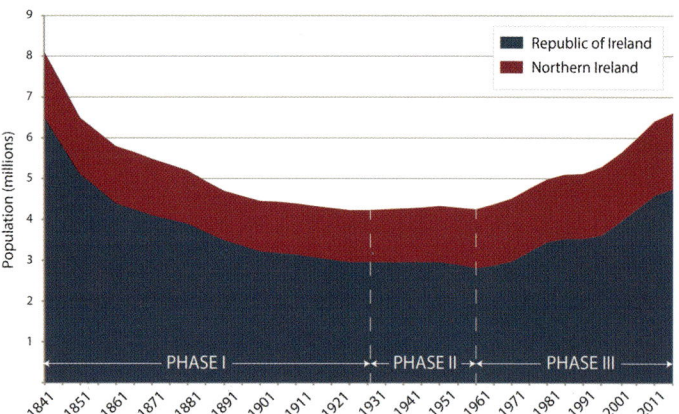

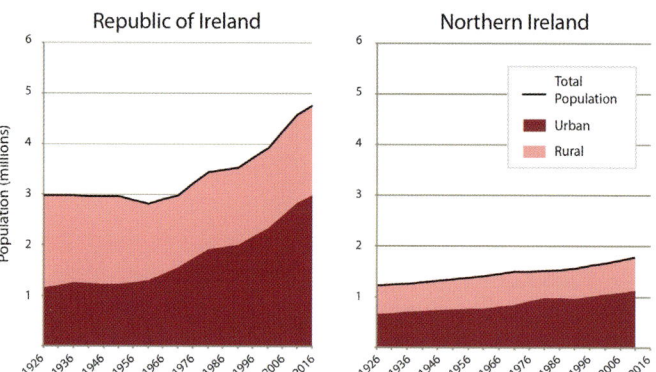

Fig. 25.5 POPULATION TRENDS FOR IRELAND (1841–2016). The trends can be divided into three phases. The first (1841–1926) shows a severe decline from the historic peak of 8.2 million to the lowest, all-island total (4.2 million) recorded in the twentieth century. This was due primarily to the consequences of the Great Famine. Rural southern counties were affected most, as a successful industrial revolution in the north-east helped offset the scale of losses through death and emigration in this region. From 1926 to 1961, the island's total population stabilised, but masked different trends between the new Republic and Northern Ireland. While the industrial economy of the North allowed for a slow growth, the Republic experienced a slight decline. By 1961, the Republic's total population of 2.8 million would be the lowest recorded in the twentieth century, and was less than one-half of its 1841 level. This continuous decline for 120 years is unique in Europe. The most recent phase (1961–2016) for the island's population has been more positive, increasing more than 50 per cent to 6.4 million. Growth, however, was now focused in the Republic, linked to its more buoyant economy. In contrast, civil unrest and deindustrialisation in Northern Ireland resulted in slow growth. By 2016, almost two-thirds of the island's population resided in the Republic. In spite of the more recent growth, the island's total population remains significantly less than that recorded for 1841. [Data source: CSO and NISRA, 1841 to 2016]

(Saorstát Éireann) revealed that almost two-thirds (1.9 million) of its population resided within 20 miles of the coast. This concentration can be linked to the relatively high carrying capacity for productive agriculture in coastal lowlands (especially in the east and south), together with opportunities presented by fishing and tourism. Above all, however, the dominant influence was the historic patterns of urbanisation that favoured coastal locations as strategic gateways for trade and administrative control under a colonial economy.

Despite experiencing some increases in its urban population in the period prior to its independence, the Irish Free State (Republic) remained predominantly a rural society.[6] By 1926, only 39 per cent of its people resided in defined urban places. This contrasted with 52 per cent in Northern Ireland (Figure 25.6). Apart from north–south differences, significant contrasts in levels of urbanisation occurred also around the island's coastline (Figure 25.7).

In the Republic, the ten largest towns in 1926 were coastal. Of these, only three (Limerick, Galway and Sligo) were located along the western seaboard. The few towns of greater than 5,000 residents in this area acted as outliers of urbanisation. In effect, a dispersed pattern of small towns and villages provided the limited range of services and market functions needed to meet the demands of a declining rural population and economy that typified the west of Ireland. Apart from Limerick, counties along the western seaboard showed levels of urbanisation that were significantly below an already low national average.

In contrast, the rest of the coastline, but especially the east,

Fig. 25.6 TRENDS IN URBANISATION IN IRELAND (1926–2016). Unlike Northern Ireland, the Republic entered the post-First World War period as a rural economy and society, with only 39 per cent of its population residing in settlements designated as towns. In Northern Ireland this figure was 52 per cent. From 1926 to 1961, the urban population in both states increased by over 100,000. A stagnant total population and large-scale rural depopulation due to both emigration and internal movement to urban centres, however, translated into higher growth rates for urbanisation in the Republic. The result was a convergent trend for the two states. In 1971, the Republic achieved the status of an urbanised country, when more than one-half (52 per cent) of its population resided in urban centres. The North had achieved this percentage by 1926. A transformation of the Republic's economy to embrace high-growth industry and services saw its rate of urbanisation increase significantly, as these sectors exhibit a marked preference to locate in, or near, large towns and cities. Urban growth was less impressive in Northern Ireland, where many of the negative multipliers of internal civil unrest were focused within urban environments. By 2016, the Republic's rate of urbanisation matched that of Northern Ireland (63 per cent). [Data source: CSO and NISRA, 1926–2016]

possessed a greater range of urban centres. These provided more diversified employment opportunities and services, while their roles as focal points for transport networks extended their hinterlands to facilitate long-established patterns of rural to urban migration. This was especially the case for the capital cities of Dublin and Belfast which dominated their urban hierarchies. In 1926, these two primate cities had a combined population of over 700,000. The location of the island's other cities was also pivotal in raising levels of urbanisation in their respective counties above the national average. In the case of Londonderry, however, the new political border between the Republic and Northern Ireland reduced significantly its natural hinterland and curtailed its growth potential (Figure 25.8).

From 1926 to 1961, three processes combined to influence urbanisation trends in Ireland and, particularly, its orientation to the coast. The first was the continued dependence on Britain of the island's trade and economy. Political independence had not reduced this to any significant extent for the Republic, while Northern Ireland functioned as a regional economy within the United Kingdom. Trade and economic development, therefore, continued to focus on large port cities and their hinterlands along the east coast.[7] It has been suggested that Ireland's urban system is unique in Europe in that it was 'one of the few countries of the continent whose urban network was closely tied to the needs of a single outside state over such a long period'.[8]

Second, government policy was influential, particularly in the Republic. Here, from the 1930s to the late 1950s, a policy of economic protection was adopted. Import substitution would become the mechanism to wean the new state from British dependency, and was designed to encourage growth of small,

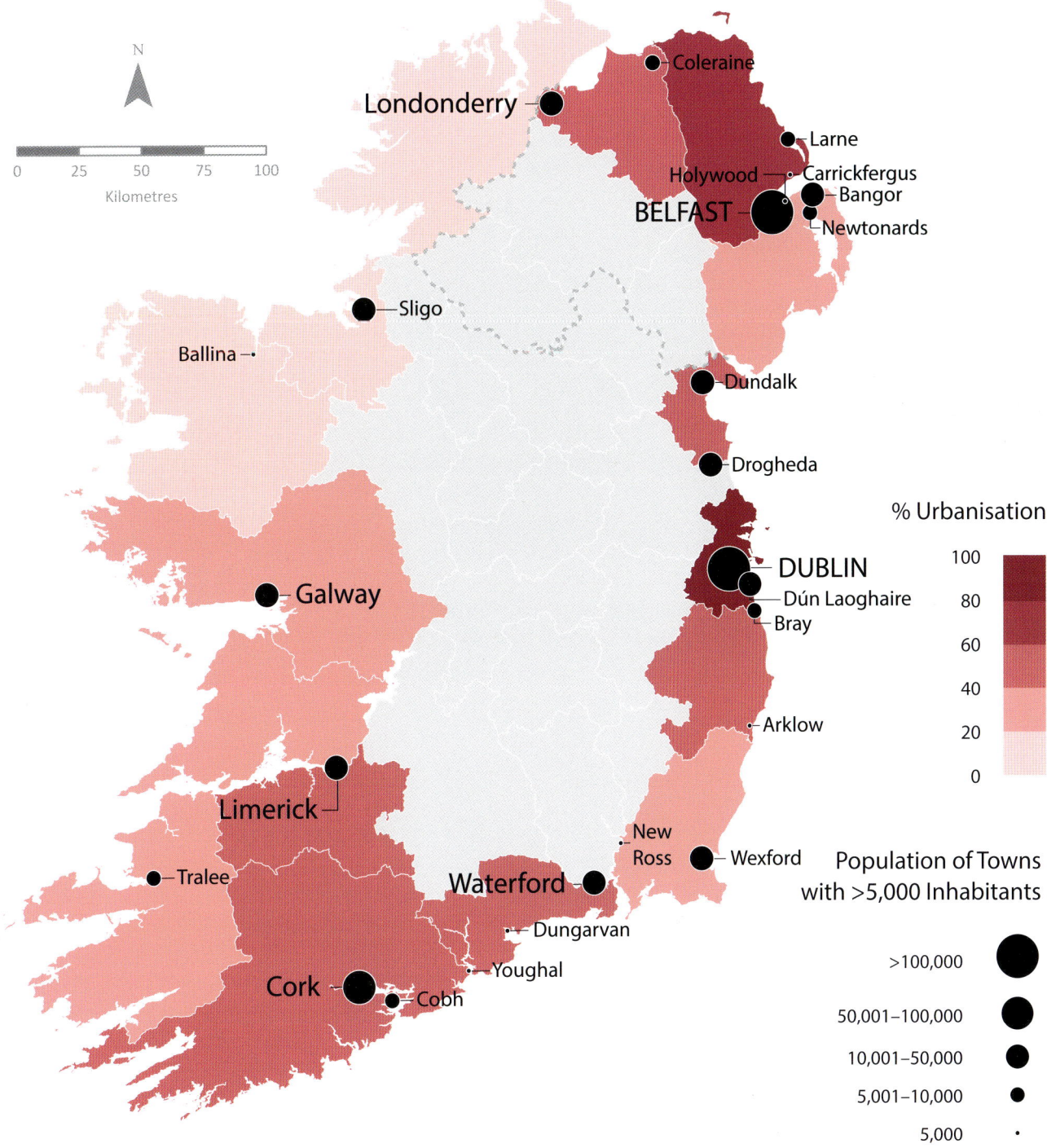

Fig. 25.7 URBANISATION RATES AND TOWNS OVER 5,000 POPULATION IN COASTAL COUNTIES OF IRELAND (1926). Following the partition of Ireland, the first census of 1926 illustrated the limited extent of urbanisation, especially in the Republic. Here, there were few towns (twenty-six) with a population of over 5,000, two-thirds of which (eighteen) were located along the coast. In both countries, the capitals and primate cities of Dublin and Belfast were pivotal in determining high rates of urbanisation in their respective counties. Counties Louth and Wicklow also benefited from dispersal of population from Dublin, as well as developing towns such as Drogheda, Dundalk and Bray. The cities of Cork, Limerick, Waterford and Londonderry, together with smaller market towns in their hinterlands, gave urbanisation rates of over 40 per cent for their counties. County Limerick appears as an outlier of urbanisation along the western coast, especially as Galway had not yet received city status and remained a small town of 14,000 people. Urbanisation was weakest in the north-west, where the combined population of towns in excess of 5,000 people in counties Mayo, Sligo and Donegal amounted to only 33,000. [Data source: CSO and NISRA]

indigenous enterprises to serve the national market. This would also facilitate a dispersal of industry from the large, coastal urban centres to smaller and inland rural communities, thereby nurturing the state's commitment to a rural Gaelic culture. However, rather than reducing what was perceived to be 'alien' urban influences, protectionism encouraged investment, as well as large numbers of migrants from rural areas, to gravitate increasingly to larger coastal cities. The net result was an increase in their dominance of national markets (Figure 25.9).

Furthermore, and contrary to a strategy of dispersal, both governments opted to centralise administrative functions within their capital cities. This acted as another catalyst for polarisation of development within the Dublin and Belfast urban areas, and was a significant restraint on development occurring in regional centres.

589

Fig. 25.8 THE PEACE BRIDGE CROSSING THE RIVER FOYLE IN LONDONDERRY. Northern Ireland's second-largest city (2011 Census: *c.*83,000, but with 105,000 in the Derry Urban Area) is located along the north-west coast of the island and traces its 'modern' phase of urban development to its selection as a strategic plantation town (1613–1619). The industrial revolution had a significant impact on the town as it became a major centre for textiles, and especially the manufacture of shirts, which employed 12,000 in 1920. The town's location close to the new border with the Republic caused the loss of much of its natural hinterland and this, together with its peripherality with respect to Belfast, impacted adversely on development trends. This was compounded further by its role as a centre for civil unrest during 'the Troubles'. The official ending of hostility in the 1990s, symbolised by the Peace Bridge (2011) designed to bring two communities together, raises hope for a more positive future for this historic coastal city. [Source: Mark McGaughey, Wikimedia Commons, CC BY-SA 4.0]

Fig. 25.9 THE DUNLOP TYRE FACTORY AND GRAIN SILOS ALONG CORK CITY QUAYS IN THE LATE 1960S. These port-related industries are sited upstream from the Ford motor vehicle plant, located at the city's Marina Quay. Ford began production of tractors at this site in 1919, but the factory was converted to car manufacture in 1921. Protectionism benefited the Republic's dominant port cities due to their more developed transport infrastructure, proximity and access to large domestic markets, and availability of a sizable and diversified labour market. By 1951, Cork had 12,000 jobs in manufacturing. Two companies, however, defined the city's industrial development: Ford and Dunlop. At their peak, some 8,000 jobs were provided directly and indirectly. Increasing competition within the European Union, and from other economies, impacted badly on these historic industries. As a result, both closed in 1983–1984 with the loss of over 2,500 jobs. This, plus the contraction of other economic activities that also failed to adapt to more competitive trading environments, transformed the vibrant waterfront area to a largely derelict site near the core of the city. Similar trends occurred in other port cities, but especially in Dublin and Limerick. [Source: Cork City and Council Archives]

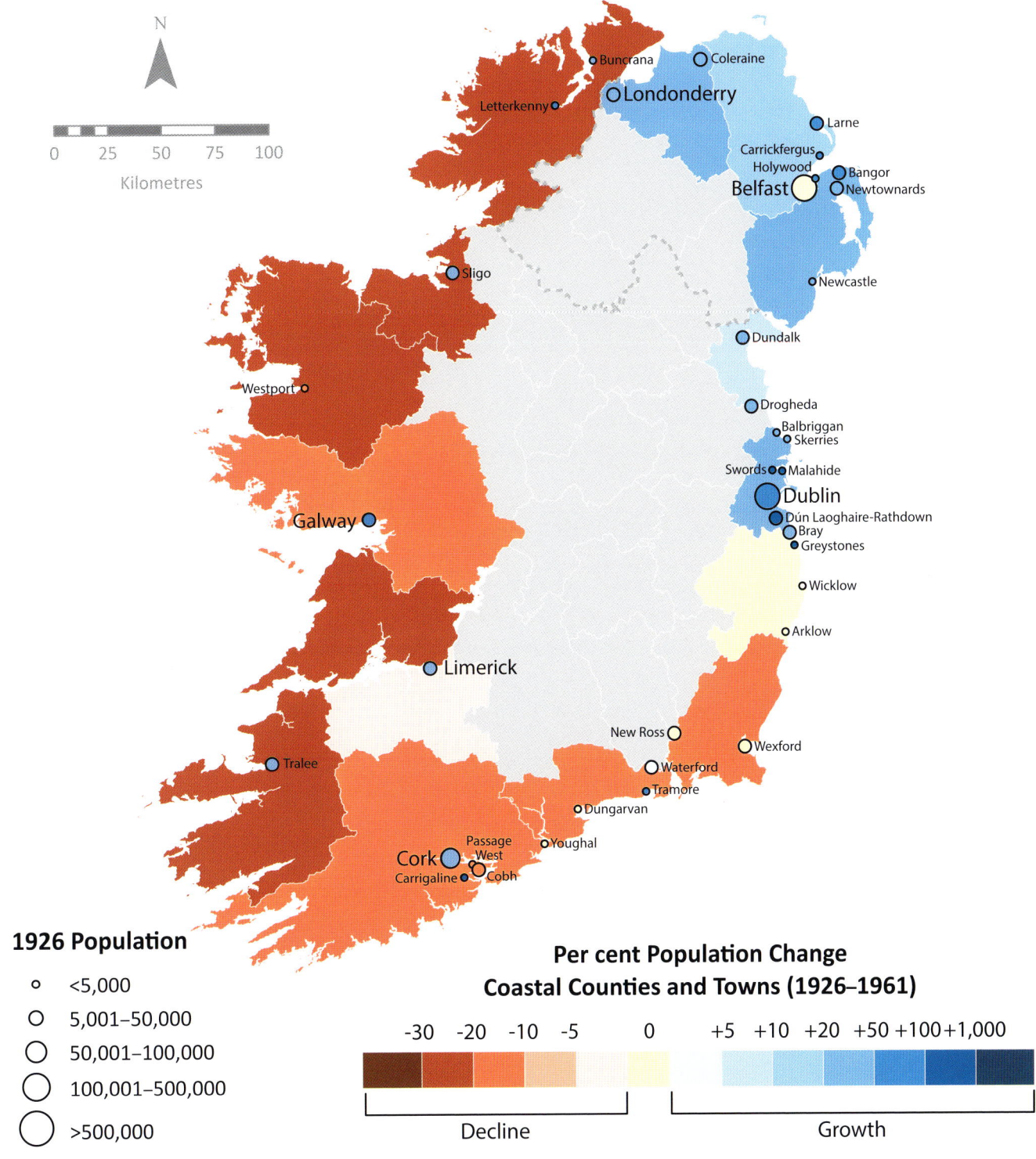

1926 Population

- ∘   <5,000
- ○   5,001–50,000
- ○   50,001–100,000
- ○   100,001–500,000
- ○   >500,000

Per cent Population Change
Coastal Counties and Towns (1926–1961)

−30   −20   −10   −5   0   +5   +10   +20   +50 +100 +1,000

Decline   Growth

Fig. 25.10 POPULATION CHANGE IN COASTAL COUNTIES AND LARGER TOWNS, 1926–1961. Most of the larger towns around Ireland's coastline experienced significant growth between 1926 and 1961. In the Republic, however, these increases usually failed to offset the high losses experienced in the large rural communities that typified most coastal counties. Along the western coast, apart from counties Limerick and Galway, declines of more than 20 per cent of the population were recorded. In contrast, the doubling of the population of Dublin city and its suburbs, and spill-over effects into nearby towns such as Dún Laoghaire, made County Dublin an 'island of growth'. Longer distance spill-over influences from the capital, together with the port and market functions of Dundalk and Drogheda, and the tourist/commuter role of Bray, enabled counties Louth and Wicklow to record some growth. Dispersal of population and employment from Belfast was pivotal in facilitating population growth in the three coastal counties of Northern Ireland, as was the importance of Londonderry as a regional outlier for development. (Data source: CSO and NISRA)

Third, the existing size distribution of towns was to be decisive, especially when combined with the first two factors. Larger towns and cities on the east and north coast were not only accessible to the British market, but were also of a size that allowed them to offer diversified labour markets, critical infrastructures and

a relatively large and prosperous domestic market for goods and services. Such factors were attractive for indigenous entrepreneurs searching for low-cost sites with good access to internal markets. Rather than dispersing to less accessible and smaller inland rural locations, developers invested in the larger coastal towns and cities located primarily along the eastern coastline. While Limerick and Galway experienced some growth, this did little to offset the marginalisation of western Ireland.

Between 1926 and 1961, while the population of the Republic fell by some 5 per cent to its lowest recorded total in the twentieth century (2.8 million), its urban population increased by 100,000. This had important implications for coastal cities and counties.

Apart from County Dublin and the adjacent Louth and Wicklow, all other coastal counties experienced population losses. Declines were particularly severe along the western seaboard, with five counties experiencing losses of between 20 and 30 per cent (Figure 25.10).

Dublin emerged as the focus of growth due to the increasing dominance of the capital city as the primary economic and political centre of the state. This continued to encourage large-scale internal

Fig. 25.11 AERIAL PHOTOGRAPH OF DOONAGH ON THE WEST COAST OF ACHILL ISLAND IN THE 1950S. The image portrays the landscape that typified much of the north-west coast of Ireland in the period prior to the 1960s. The exposure of the coastline to onshore weather systems and the rugged terrain combined to act as significant constraints for human settlement. A predominance of extensive pastoral farming, combined occasionally with offshore fishing, also offered few employment opportunities to sustain a high density of population. Migration rather than population growth was the dominant characteristic of these coastal areas, and there were few demands for services to meet the requirements of declining and depressed communities. The urban 'footprint' was, at best, marginal throughout most of these areas, with the few service centres being generally small and dispersed. Here, most of the rural housing can be seen to be orientated along the coastal road in a linear pattern. Where roads meet, as at Doonagh, small villages could be sustained and these provided basic functions, such as a church, primary school and some shops. While some tourists would be attracted to the beaches and coastal scenery of this area, the resident population of the village of Doonagh in 1961 was only 387. [Source: Independent Newspapers Ireland/National Library of Ireland]

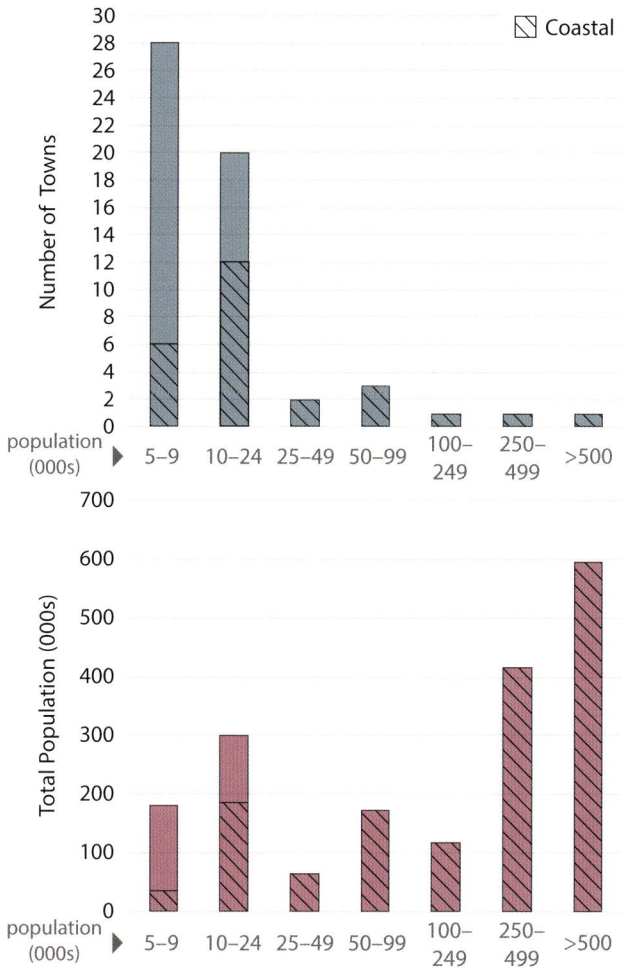

Fig. 25.12 IRELAND'S TOWNS AND CITIES BY SIZE CATEGORIES, 1961. In 1961, almost one-half (26:56) of Ireland's towns of greater than 5,000 population had a coastal location. The orientation to coastal sites, however, was biased to the larger centres; all cities and towns with resident populations of greater than 25,000 were coastal. Inland sites predominate only for towns with populations of 5,000–24,000. As a result, over 85 per cent of the island's urban population that resided in towns of greater than 5,000 occurred on the coast (1.6 of 1.8 million). This included both capital cities, with a combined population of just over 1 million, almost one-quarter of the island's total population. [Data source: CSO and NISRA].

migration, especially from depressed rural communities and, together with high birth rates, contributed to the significant growth of the capital city's total population. For many inner-city residents, however, housing conditions were appalling. Urban slums, especially in the form of high-density tenement buildings possessing totally inadequate provision of water supplies and effluent disposal, symbolised the grinding poverty and degradation of these people. A report in 1914, for example, had indicated that 30 per cent of Dublin's population lived in slums, with three-quarters of tenement households occupying a single room, and 60,000 people lived in housing unfit for human habitation. Such conditions did not improve with independence, especially given what could be perceived as an anti-urban bias of the government.

The large increases in inner-city populations and negative images associated with the tenements encouraged many of the middle classes to vacate Dublin's city centre for new suburban areas. These extended the city's built-up area beyond its legally defined administrative boundaries. Thus, between 1926 and 1961,

the population of Dublin city and its environs almost doubled to 0.7 million and accounted for one-in-four of the Republic's population. Positive spillover effects from Dublin's growing residential population were key factors in enlarging the urban communities and encouraging population growth in the adjacent counties of Louth and Wicklow.

Similar processes occurred in the Republic's other cities, but on a lesser scale. Population increases, and some expansion of suburban areas, occurred in Cork, Limerick and Waterford. Such growth helped, to some extent, to offset the large-scale rural depopulation that characterised most of the inland areas within these counties. The result was, therefore, a relatively low rate of overall population decline. For most counties along the western seaboard, however, the absence of larger towns to compensate for rural depopulation translated into high rates of population decline (Figure 25.11).

In Northern Ireland, the three coastal counties and their larger towns all experienced significant growth. This is linked to the state's stronger urban economy, but also to government efforts to reduce polarisation on the capital city by encouraging a dispersal of population and employment from Belfast to other urban communities. For Belfast city, therefore, total population stabilised over this period at some 0.42 million, and is in marked contrast to the trend experienced in Dublin

Between 1926 and 1961, the island's urban population increased to some 2 million. This trend, however, had been shaped mainly by the scale and growth of the larger towns and cities, which were located predominantly on the coast. By 1961, therefore, over 85 per cent of the island's population residing in towns of greater than 5,000 were located in coastal centres. This represented over one-third of the island's total population (Figure 25.12).

ECONOMIC GROWTH AND COASTAL URBANISATION, 1961–2016

Ireland entered the 1960s as a relatively underdeveloped economy with one of the lowest rates of urbanisation in Europe. This decade, however, marked a turning point, especially for the Republic. Here, an economic take-off occurred that initiated a phase of significant growth, culminating in the Celtic Tiger economy from the mid-1990s to early 2000s. In contrast, Northern Ireland began to experience early problems of deindustrialisation and social upheaval from the late 1960s associated with 'the Troubles'. The global financial crisis of 2007–2008 exposed both states to major problems, but there has been a significant recovery in recent years.

One consequence of the Republic's stronger economic performance has been higher rates of population growth and urbanisation than experienced in Northern Ireland (Figures 25.5 and 25.6). In the 1960s, the Republic recorded its first substantial increase in population for over a century. Since then, its population has continued to increase to reach 4.8 million in 2016, some 70 per cent higher than in 1961. In contrast, population in Northern Ireland over the same period grew by approximately one-third, to 1.9 million.

Ireland's increasing population has focused disproportionately

Fig. 25.13 AERIAL VIEW OF PHARMACEUTICALS CLUSTER AT RINGASKIDDY, CORK HARBOUR. The Pfizer pharmaceutical plant at Ringaskiddy in Cork's Outer Harbour is shown along the right margin of the photograph and adjacent to the modern, deep-water port. This large multinational commenced operation in 1969 and was the first of a number of similar industries to be based in the area, making the location one of the largest pharmaceutical clusters in Ireland. These industries, as with most TNCs attracted to the Republic since the 1960s, produce primarily for the export market. The harbour, regarded by many as the second-largest natural harbour in the world (after Sydney), is a significant attraction for export-oriented industries. The availability of large sites and the environmental quality of the landscape adjacent to modern port facilities, together with its close proximity to Cork city, have encouraged an increasing urban imprint. Across the water from Pfizer is Cobh, which has grown in size as a commuter town for Cork. It has also developed a tourist role centred on its heritage as an emigrant port linked to the Famine and the *Titanic*'s last port of call. It has a growing role for cruise liners. Other important and expanding harbour communities include Passage West, Crosshaven and Carrigaline. [Source: Irish Air Corps]

in urban centres, rather than in rural spaces. This enabled the Republic to attain the status of an urbanised country in 1971, when more than 50 per cent of its population resided in officially designated urban areas. Furthermore, its higher rates of urbanisation allowed the Republic to converge on levels experienced by Northern Ireland. By 2016, almost two-thirds of the population of both states resided in urban centres.

The 1960s in Ireland witnessed, therefore, the start of a process involving radical changes, as forces of modernisation impacted on the island's economy, society and environmental qualities. Changes were pronounced, especially within the coastal zone under the key driver of urban-led development. Reference can be made to five processes to explain the transformations, which were centred especially on the larger port cities and their hinterlands: export-led growth; changes to the economic system;

national spatial planning; changing urban layout; tourism.

Export-led growth

In the 1960s, the Republic changed its development strategy. This involved rejecting protectionism and import substitution and its replacement by a policy focused on free trade and the attraction of foreign direct investment (FDI) to stimulate economic take-off. Its success was aided by the country's low costs of land and labour, but above all by its generous tax incentives. The shift to export-led development was confirmed with accession to the European Union in 1973. This, together with an increasingly liberalised global trade environment, saw the Republic become a significant production platform for multinational enterprises wishing to export goods and services to the European Union and wider global markets (Figure 25.13).

Figure 25.14 RENEWAL OF DUBLIN DOCKLANDS. The development and dominance of the capital city is associated with its port and gateway functions. By the late 1980s, however, the inner-city port and some adjacent areas had become derelict and depressed, and people dispersed to suburban and hinterland locations. Since then, major physical, economic and social regeneration has occurred (see Dublin Case Study). Large inflows of international investment in financial services and other growth sectors are reflected in the modern office blocks and apartments that line the quays on either side of the River Liffey and which are linked by the new Samuel Beckett Bridge. Employment, growth and renewed prosperity are complemented by a return of population to modern apartment complexes and mixed community housing. Dublin has adapted well to forces of change and continues to influence the scale and nature of urbanisation in Ireland, and especially along the eastern coast. [Source: Daniel Vorndran, Wikimedia Commons, CC BY-SA 4.0]

Despite recessions (for example in the early 1980s and 2007–2013), the Republic has been remarkably successful in transforming itself from an underdeveloped periphery to a country experiencing some of the highest growth rates in the European Union. Export-led development, however, meant that most of the benefits of enhanced employment and prosperity gravitated to the country's large port cities, sited primarily along the east coast. This encouraged increases in their population and urban land uses, above all in Dublin, which enhanced its role as the Republic's principal gateway for international trade.

Northern Ireland also benefited from membership of the European Union. Internal political and social conflict, however, restricted its attraction for FDI and, therefore, growth in employment and population. In spite of this, but especially since the economic revival following the Good Friday Agreement (1998), the state's dependency on external markets and capital supports have encouraged a similar focus on coastal urbanisation.

Changes to the Economic System

Dependency on the primary sector declined dramatically from the 1960s as the importance of the industrial and services sectors increased. The initial surge in FDI generally involved branch plants that sought low-cost production sites in rural environments, rather than within established, but higher-cost urban centres.[9] One important consequence of rural industrialisation in the Republic was that it encouraged a trend towards greater convergence in levels of regional development, and also between urban and rural communities. Between 1961 and 1981, therefore, the share of manufacturing employment located in the five port cities fell from 53 per cent to 29 per cent. In part, this was due to the slow growth or decline of indigenous industries in the port cities as they suffered from lack of competitiveness once import duties and tariffs were relaxed in the post-protectionism era. Limerick provides an early example of such industrial decline.

Recession in the 1980s impacted severely on Ireland's branch-plant economy, which failed, in general, to meet increasing challenges in the international market and the relocation of production to even lower-cost sites. Many factories that had been the basis of a new sense of optimism in rural communities and other less developed areas were closed, or had their workforces reduced. In addition, large numbers of jobs were lost in Ireland's traditional, labour-intensive industries, which were located primarily within port cities and their environs. Cork city provides a significant example of industrial decline in this decade, with major industries such as Ford (vehicle assembly), Dunlop (tyres), Sunbeam-Wolsey (textiles) and Verolme (ship-building) all closing in the mid-1980s, with the loss of thousands of jobs (Figure 25.9). Losses of this magnitude would be a catalyst for Cork to plan the redevelopment of the city. Such levels of deindustrialisation at a national scale also demanded changes within the the Republic's economic structure.

In contrast to Northern Ireland, which continued to struggle to combat the consequences of civil conflict, the Republic moved quickly and successfully to modernise and promote a post-industrial economy. This involved prioritising the attraction of high-tech, high value-added industries and international services. For such growth sectors, low-cost sites are not critical. Instead, key considerations involve an educated, skilled and flexible labour market, high-quality infrastructure (including transport and communications, housing and education), and an attractive environment offering a range of recreational outlets (for example, retail, culture and sport).

These requirements are best found in large cities and their immediate hinterlands rather than in rural areas. The result was a reconfirmation of Ireland's urban base, and especially port cities, as optimal sites for contemporary development. Dublin and Belfast have both 'reinvented' themselves as international cities and are the focus of their national development (Figure 25.14). In addition, in the Republic, both Cork and Limerick have been active in transforming their urban images to support their designated roles as effective counter-poles to Dublin for economic and population growth.

Case Study: Dublin

Rob Goodbody

At the close of the Second World War, Dublin Port still retained vestiges of the old ways. Downstream from Butt Bridge, both sides of the River Liffey were in use as part of the port with bustling porters pushing hand trolleys, cranes loading and unloading ships and goods carried on horses and carts, which shared space with motor lorries until the 1960s. Railway wagons ran along tracks on the quayside, carrying coal to the gas works to the south of Sir John Rogerson's Quay, while on the opposite bank the railway brought cattle and other goods to the quay. Some efforts were made, however, to improve facilities, which included reclamation of land from the sea at East Wall and construction of new quays at Alexandra Quay East (Figure 25.15).

From the 1960s, more substantial changes were required if Dublin was to adapt successfully to new trends in global shipping (see Chapter 24: Ports and Shipping). In 1965, therefore, a five-year development plan proposed new quays and additional land reclamation at the eastern end of the port. This downstream development, involving new container facilities and ro-ro terminals, encouraged the relocation of car ferry and freight services from facilities along the inner quays. Various coal importers using the port were also encouraged to centralise their operations into a single downstream facility adjacent to deeper water and a storage area, thereby freeing up land along the North Wall Quay and Sir John Rogerson's Quay. In a similar way, and following the arrival of piped natural gas to the city in 1982, production of town gas ceased, which eliminated a major importer of coal and made available a substantial, but polluted, waterfront site near the inner city.[10]

While modernisation and downstream development allowed the port to remain competitive, and also made available land for development within the inner city, recession in the 1980s caused most of this land to remain derelict. In addition, loss of employment

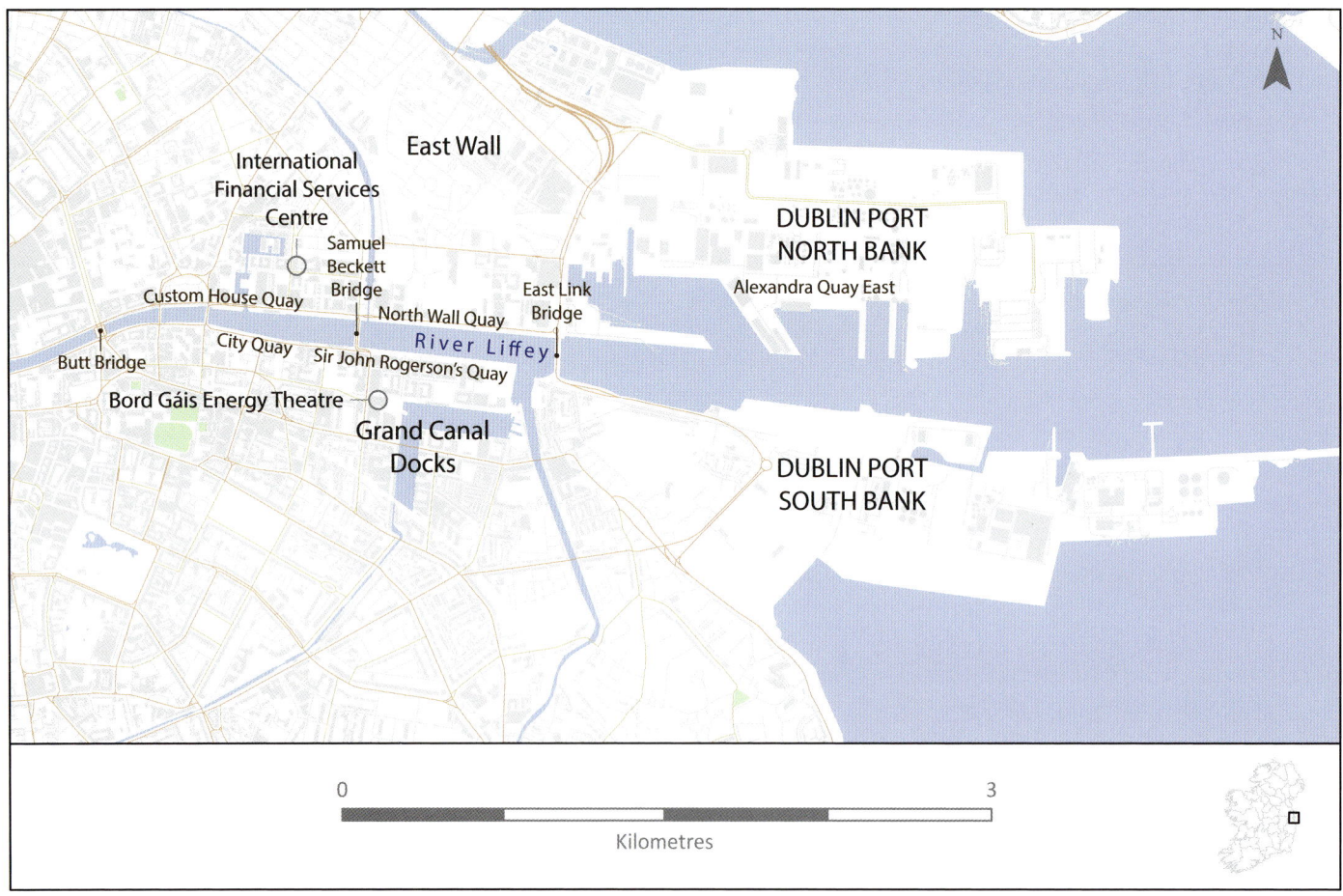

Fig. 25.15 DUBLIN CITY. The River Liffey flows through Dublin and widens as it progresses to the east of Butt Bridge. The land on either side of this part of the river is all reclaimed and, along with its widening channel, allowed for the berthing of larger ships and substantial trade to occur along the North Wall Quay and Sir John Rogerson's Quay. Most of this area has now lost its port-related functions and has been redeveloped as sites on which modern and large-scale urban-related functions are located, such as office blocks, other commercial activities and high-rise apartment units. Downstream from the East Link Bridge is the modern port, developed from the mid-nineteenth century onwards. It is now the focus of the city's maritime trade, and is the country's largest port.

Fig. 25.16 CUSTOM HOUSE AND ADJACENT QUAYS IN THE 1970s. At this time, ships were using, on a regular basis, the quays upriver to the Custom House. The photograph shows Custom House Quay on the left bank and George's Quay on the right bank. Both are still fitted out with bollards and mooring hooks to facilitate shipping and are open to the river, without the walls that now separate the quays from the Liffey. Beyond the Custom House is the land now occupied by the IFSC with the cranes testifying to the continuing port activities, as does the ship that is docked at City Quay. [Source: Courtesy of Dublin City Library and Archive]

(linked to deindustrialisation of this area) and dispersal of its population to suburban estates, caused Dublin's inner-city population to fall significantly between 1971 and 1991 (Figure 25.17).

The first significant move to capitalise on the downstream relocation of port activities occurred in 1980 when the Dublin Port and Docks Board applied for outline planning permission to redevelop the 11ha Custom House Dock area (Figure 25.16).[11] Despite a strategic location near the city centre, its development was delayed until 1986. In that year, and under the Urban Renewal Act, responsibility for the site was transferred to the Custom House Dock Development Authority (CHDDA). This, together with the government's new policy to attract high-tech, international investment to Ireland, saw the site cleared of disused warehouses along its redundant quays. The site became the location for the International Financial Services Centre (IFSC). With generous tax incentives, the IFSC has been remarkably successful and is the location for over 400 international companies, employing some 35,000 people (one-third in locations outside Dublin) and contributing an estimated 7 per cent of the country's GNP. This was the initial catalyst that led to the reimagining of Dublin as a global city. In 1994, a further 5ha was added to the authority's remit, allowing for the clearance of more disused port facilities and opening up much of the River Liffey's waterfront.[12]

This urban renewal and attraction of international growth sectors was pivotal for Dublin's continued primacy and leadership role in the country's Celtic Tiger economy. In 1997, the CHDDA was dissolved and its place taken by the Dublin Docklands Development Authority (DDDA). It was given responsibility for regeneration of 520ha of under-developed dockland property, stretching from the initial hub of the IFSC downstream on either side of the River Liffey. New infrastructures, such as the Samuel Beckett Bridge and the LUAS light rail system, have improved accessibility to the area, while new and iconic buildings, such as the Bord Gáis Energy Theatre and Convention Centre, emphasise its new urban status and break with the historic, port-centred activities. Pivotal to this image and function have been developments around the former Grand Canal Docks. As part of a highly successful regeneration scheme, an area known as 'Silicon Docks' has emerged. Here, some 7,000 technical professionals find employment in the offices and European headquarters of an array of high-tech global leaders including Google, Facebook, Twitter and LinkedIn. Such investments have confirmed Dublin as a successful and competitive post-industrial, global city and one that continues to define Ireland's contemporary, urbanised economy (Figure 25.18).

The initial dockland scheme envisaged a balanced development process encompassing economic, social and physical rejuvenation.

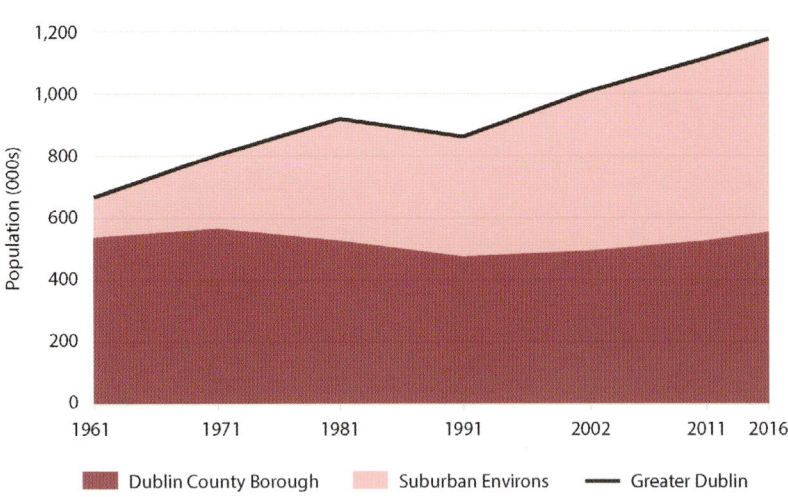

Fig. 25.17 (left) POPULATION TRENDS FOR DUBLIN CITY AND ITS EXTENDED SUBURBAN AREA (1961–2016). In 1961, 23 per cent of the Republic's population resided in Greater Dublin, which was comprised of the County Borough and its suburban area and which had a population of almost 0.65 million. Four out of every five people in this area lived within the city limits. While the Republic's population grew significantly to 2016, trends within Greater Dublin displayed some marked variation. The city, and more especially inner-city areas centred around the docklands, experienced population loss from 1971–1991 as economic problems and outdated urban infrastructure encouraged the dispersal of residents to more attractive suburban areas located beyond the city boundaries. Although gentrification, arising from renewal and economic success centred on the transformation of the former docklands, has helped to reverse population decline, total population within the city remained below the 1961 figure. In contrast, Dublin's suburban environs have almost quadrupled in population (+0.5 million) and now exceed numbers residing in the city. In 2016, Dublin and its suburban environs remained the domicile of almost one-in-four of the Republic's population, highlighting its continued attraction for both economic and residential development. [Data source: CSO]

Fig. 25.18 A MODERN VIEW OF THE DUBLIN DOCKLANDS. In the view downstream along the Liffey, Liberty Hall, just above centre and to the left, shows the point beyond which the port continued to operate until the latter part of the twentieth century. Both sides of the river past Samuel Beckett Bridge and up to the East Link Bridge in the distance are built up largely with substantial buildings that have no relationship to the port. Prominently visible are the IFSC, above Liberty Hall and the Convention Centre to the left of Samuel Beckett Bridge. The modern port of Dublin has moved downstream, beyond the view in the photograph. [Source: Joe Brady]

This did not occur, as emphasis was placed firmly on economic priorities, and many sites of traditional housing were sold to be redeveloped as private housing.[13] Despite this, the area continued to include extensive zones of working-class housing and flats, which accommodated significant numbers of dock workers and others employed in port-related activities. Although many residents were dispersed from the area, when the DDDA took over responsibility for regeneration in 1997, estimates suggest that some 17,500 people remained. By this time, however, the economic success of the Celtic Tiger allowed social considerations to receive a higher profile.[14] As a result, the DDDA embraced a stronger social input from community leaders within decision-making processes. The outcome has been a greater balance of land uses to incorporate, not only office blocks and high-priced apartment units for well-paid employees of high-tech industries, but also a mix of residential housing, including affordable social units. As a result, the population of Dublin's inner city area has shown an increase since 1991 (Figure 25.17).

After fifteen years of extensive change and redevelopment, the DDDA was wound up in 2012 and planning in the area reverted to Dublin City Council. It is now a Strategic Development Zone (SDZ), where the planning process is streamlined to avoid any costly delays in promoting new developments. Current plans for the SDZ include provision of 2,600 residential units and 305,000 m^2 of commercial floorspace, capable of sustaining a residential population of almost 6,000 and a workforce of some 23,000. The objective is clearly to develop not only a competitive location for international investment but also to make the rejuvenated city centre an attractive and liveable environment for residents and tourists.

In a relatively short period, Custom House Quay, North Wall Quay, City Quay, Sir John Rogerson's Quay and the Grand Canal Docks have been transformed from lively waterside port activities in low-rise buildings to a vibrant mass of corporate headquarters, apartments, cultural and leisure facilities, occupying buildings that are frequently eight storeys high or more. Only a few retained features – such as rail tracks, stone setts, mooring hooks and the occasional visiting ship – remain as reminders of the former port activities on these quays.

Case Study: Reimagining Cork as a Port City

William Brady

Cork is the second-largest city in the Republic of Ireland, with a 2016 census population of 125,000 and a wider metropolitan population of 305,000. Located on the River Lee and at the head of a large natural harbour, it is surrounded by a planned network of satellite towns and employment hubs, all within a high-quality landscape and coastal setting. As the state's second city, Cork fulfils an important function as a substantial economic hub with a strong international and national profile. Following a period of deindustrialisation in the 1980s, which affected in particular its portside industries, the city has recovered and performed strongly in economic terms over the past twenty years.[15]

As a trading port since perhaps the tenth century, Cork has enduring economic, cultural and social associations with its coastal context. Historically, and as an economic entity, the city can be understood as a series of islands, waterways and connected hinterlands, which function as a space where strategic commercial interactions between sea and land occur. As a seventeenth-century merchant trading centre, for example, Cork emerged as part of a European network of commercial ports, and the city's identity, heritage and morphology remain closely bound up in this relationship.

This functional and spatial relationship is continually evolving. Like many comparative European coastal cities, Cork is undergoing a significant transformation in the nature of the spatial relationship between port and city. This is manifested in a process of retreat and modernisation, as shipping activities abandon their urban locations in favour of deeper waters and less constrained sites. In Cork, the spatial intimacy between port and city is therefore being reduced, as shipping activity migrates from its historic core towards Tivoli and, more especially, to the lower harbour at Ringaskiddy.

Cork's evolution as a modern city port has not been dictated solely by economic and technological impulses; it is also part of careful and long-term planning for the future of the city, its harbour and the region.[16] In 1978, the Cork Land Use and Transportation Study laid the foundation for a continuous forty-year programme of integrated planning for the city and harbour. The result has been a template for a new phase of urban development, with port activities and industrial development being directed towards deeper waters and more accessible sites along the harbour, and the city's former docklands identified as the city's principal opportunity for development and expansion (Figure 25.19).

By the late 1990s, the scale of the opportunity for docklands redevelopment in the old urban port locations became clear and the city began to prepare actively for the wholesale redevelopment of former docklands sites. The Cork Docklands Development Strategy 2001 proposed that the former port area would accommodate 580,000m^2 of new non-residential uses (including offices, education, retail, culture and leisure facilities), 5,800 additional residential units as well as new recreational spaces, and river crossings to better integrate developments on either side of the River Lee.

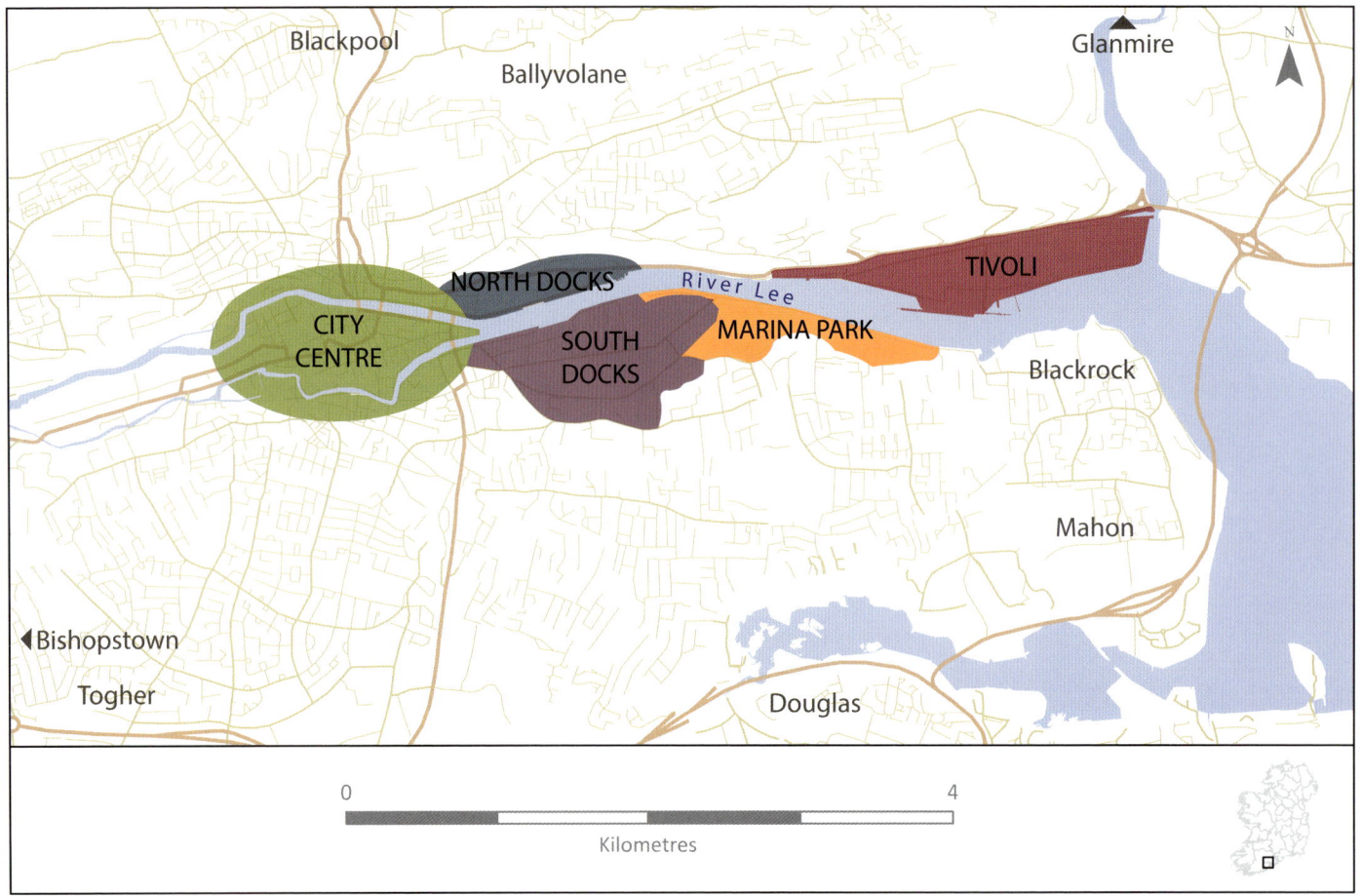

Fig. 25.19 CORK CITY CENTRE AND DOCKLAND AREAS. This diagram illustrates the scale and importance of former port locations (shown by colours) in the future expansion of Cork's urban centre. With increasing pressures to limit suburban expansion and promote more sustainable and compact forms of urban growth, old port locations have become critical components of the city's long-term growth strategy. The city's north and south docks are comparable in size to the established city centre island and present opportunities to accommodate an incremental expansion of the city centre core as part of a series of new mixed-use urban districts. Further downstream, Tivoli docks, where container traffic is being relocated to Ringaskiddy, is likely to become the site of Cork's most important urban residential expansion, with potential to accommodate at least 10,000 people in this extensive waterfront site. Collectively, these former port locations present Cork with an opportunity to fulfil its role as a driver of national development, as well as facilitating a balanced growth strategy for the city region. [Data source: City Docks Local Area Plan Issues Paper 2017]

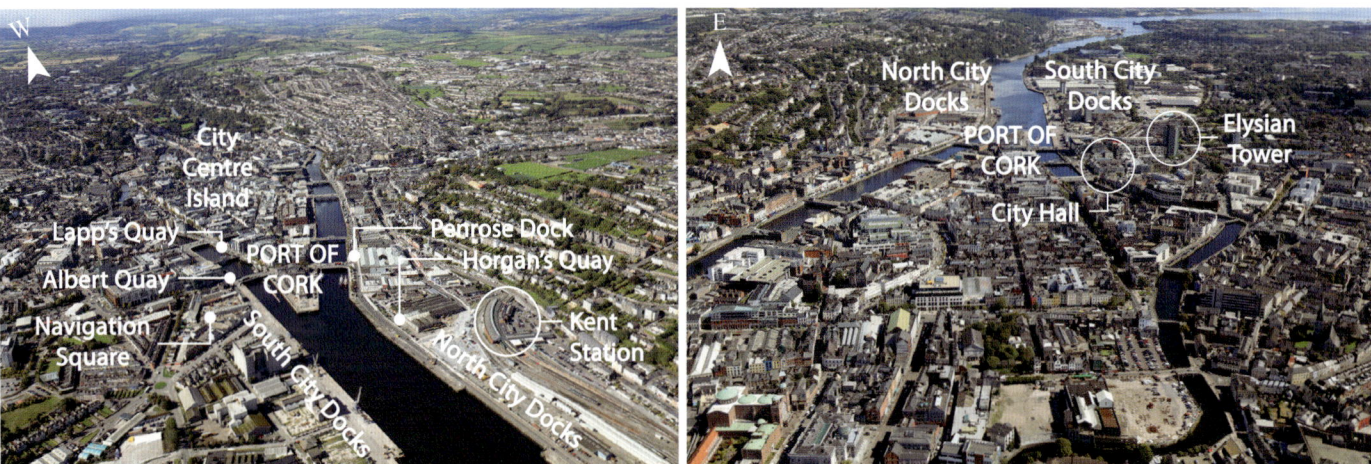

Figure 25.20 AERIAL PHOTOGRAPHS OF CORK CITY CENTRED ON ITS URBAN CORE AND DOCKLANDS. These westwards and eastwards views of Cork City show the city centre island located at the confluence of the River Lee, which includes Cork's commercial and retail core. The historic port extends eastwards downstream from this site on either side of the river before it transitions into the extensive waters of Cork Harbour. It also highlights the scale of brownfield redevelopment opportunities immediately downstream from the city's core related to historic port land uses at the north and south city docks. Since the 1990s, the urban core has been expanding gradually towards the port area; the civic quarter, for example, has consolidated around the City Hall, and also involved some substantial redevelopment of brownfield sites. This included mixed-use hotel/residential/office/restaurant activities, around Lapp's Quay and the Elysian Tower district. This ongoing transformation of the urban form and character at the interface between the established city and port areas highlights the aspirations of the planners to secure an organic and incremental expansion of the city's core, rather than encouraging isolated developments that could detract from the aim of preserving a dynamic city centre. Major ongoing developments are occurring at Albert Quay and Navigation Square at the south docks, while transformation of an extensive brownfield site adjacent to Kent railway station and Horgan's Quay/Penrose Dock will extend the city's urban footprint along the north bank of the River Lee. In combination, these schemes will involve a floorspace of 140,000m^2 and could provide over 12,000 jobs. The result will be the reimagining of the city's former port district to act as the catalyst for its future growth, and that of the entire city and wider metropolitan area. [Source: Dennis Horgan/Port of Cork]

Due to the ongoing presence of an active port, costs of developing brownfield lands, established patterns of land ownerships, the need for infrastructural investment and the lack of explicit government support, the process of renewal and development was delayed. Despite slow progress in the years following its publication – apart from a number of high-profile development projects at Lapp's Quay/City Quarter and Anglesea Street/Elysian in the 2000s – the Cork Docklands Development Strategy still provides a template for the city's expansion into the former port areas and the core planning vision remains intact (Figure 25.20).

In recent years there have been signs of renewed and active private-sector interest in docklands. As employment in commercial office space becomes increasingly urbanised, the latest phase of development in Cork's docklands reflects a desire for city-based locations and coincides with demands from a growing number of people for improved urban working and living environments (Figure 25.21).

Under Ireland's new National Planning

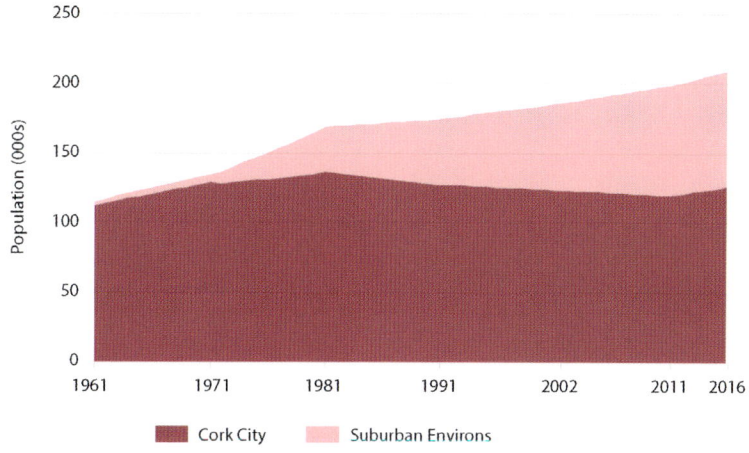

Fig. 25.21 POPULATION TRENDS FOR CORK CITY AND ITS SUBURBS (1961–2016). From 1961 to 1981, the population of Cork city continued to increase, reflecting the general wave of economic growth that typified the Republic. Suburban population also increased, especially in the decade following 1971, although the magnitude of growth was influenced by the wider definition presented by the 1981 census for suburban areas in all cities. The impact of economic recession and deindustrialisation in the city's dockland area caused its population to experience a continued fall to 2011. In contrast, Cork suburban areas grew by more than 50,000 (156 per cent). It appears that redevelopment of the city centre and docklands has reversed the city's trend of a declining population, with its first growth in thirty years between 2011 and 2016. By 2016, almost 210,000 people resided in Cork city and its suburbs, while the wider metropolitan hinterland (the Cork Metropolitan Area) recorded a population of 305,000 people. [Data source: CSO]

Framework, Cork has been elevated within the state's urban hierarchy and is expected to accommodate a substantial growth in population of between 105,000–125,000. Much of this will have to be located within its existing built-up area to reflect the need for densification and more compact growth patterns.[17] The former port facilities in the city's docklands and at Tivoli are expected to account for approximately half of the city's population growth targets and a substantial proportion of new urban employment; in this way, the old port becomes the new face of the city. The relationship between the city and the port of Cork continues to define and shape its economic and physical development. As the port migrates towards bluecoast locations in the harbour, the city advances to reclaim the brownfield terrains of former docksides and industrial heartlands. This is part of a continuing, synchronised set of spatial interactions which shape the geography of urban development.

Cork is now challenged with fulfilling a transformative urban planning agenda, reflecting the desire to develop as a European City region in scale and quality. The success of this is closely aligned with the extent to which the shared aspirations of city and port can be effectively managed.

National spatial planning

In the 1960s, in addition to promoting national development, governments introduced a spatial strategy to encourage balanced regional growth.[18] Under Matthews (1963) and Buchanan (1968), growth-centre strategies were proposed for both Northern Ireland and the Republic.[19] These were to limit the rate of growth in the capital cities and encourage investment in a small number of designated centres, which would act as catalysts for regional development. In the Republic, eight of nine designated growth centres were coastal (Figure 25.22). Of particular note was the inclusion of four regional centres along the western coastline (Limerick-Shannon, Galway, Sligo and Letterkenny). These would be promoted as catalysts for growth in an area of the Republic that remained underdeveloped. Many politicians, however, opposed the degree of bias given to large coastal towns and cities and demanded greater dispersal of investment to favour inland and rural locations. This complemented the preferences of the type of foreign companies being attracted in the 1970s to low-cost rural locations, and resulted in less growth than was projected for

coastal growth centres.[20]

The severity of the 1980s' recession, however, refocused government policy on national, rather than regional, priorities. A laissez-faire approach emerged which encouraged entrepreneurs to invest in their preferred locations, rather than having to be concerned with balanced regional development. This favoured the concentration of employment in larger coastal centres, but especially in and around the primate cities of Dublin and Belfast, and became a dominant feature of the Celtic Tiger economy.

Attempts to guide spatial development re-emerged in the 2000s, with governments reverting to a growth-centre approach.[21] Under the National Spatial Strategy (2002), investment would be prioritised in a small number of 'gateways', considered to have the potential for supporting sustainable growth. Development opportunities would 'trickle-down' from these gateways to designated smaller towns or 'hubs'. The competitive advantages of the gateways, the majority of which were coastal sites, would be enhanced by upgrading transport corridors that interconnected the gateways and hubs (Figure 25.23). This would be especially

Fig. 25.22 GALWAY, THE 'GATEWAY TO THE WEST'. This historic city is located where the River Corrib meets the Atlantic ocean in Galway Bay. It has long served as the principal market and services centre for Connacht, and is its provincial capital. In 1961, Galway attained city status, and was designated as one of eight growth centres within the Buchanan proposals of 1968, and a gateway for the Western Region in the Republic's plans for national development in 2002. The result has been rapid growth in its population (from 24,000 to 80,000 between 1961 and 2016), which is reflected in significant urban sprawl and, therefore, in the extension of its urban footprint around both Galway Bay and inland areas. Its success is based on numerous key assets including its national university, an array of high-tech industries, an all-year-round tourist sector and its attractive physical and cultural landscapes. The city's growth and coastal imprint is expected to continue, as Project Ireland 2040 identifies Galway as having a key strategic role on the Atlantic Economic Corridor and a future population of at least 120,000. [Source: Irish Air Corps]

important for the promotion of successful regional centres in Limerick and Galway along the west coast.

The 2002 strategy confirmed a concept that had first been introduced in the early 1990s, that of an 'economic corridor' linking the two largest cities on the island, Belfast and Dublin. It also suggested the need for a new 'Atlantic Corridor' from Cork to Letterkenny, which would enhance prospects for growth along the more peripheral western seaboard.

As with the Buchanan plan of 1968, the strategy to concentrate investment in designated larger centres again fell foul of political and other vested interests. Within the Republic, for example, a plan introduced in 2003 for the decentralisation of government departments and state agencies from Dublin identified over fifty different locations to which public servants were to be relocated. To a large extent, however, centres that had been designated in the spatial strategy were bypassed. Hence, a key opportunity to increase and diversify employment

opportunities and enhance the attraction of the gateways and hubs was lost at the outset. It appeared that the political agenda of government ministers and politicians ensured that decentralised jobs would be dispersed over a much greater number of urban centres located throughout the country. In spite of this attempted redirection of economic activities, development continued to polarise on major cities, especially Dublin and its environs. Regional and urban–rural divides have, therefore, been exacerbated, rather than reduced, since the Celtic Tiger. They show the strength of free-market forces and the attractive power of Ireland's coastal cities but, above all, the primate capital cities of Dublin and Belfast.

Project 2040 is the current plan for national development in the Republic.[22] While it again aspires to attain balanced regional development and a reduction in the urban-rural divide, it also recognises the reality that the country's continued economic success will depend on increasing the scale and competitiveness

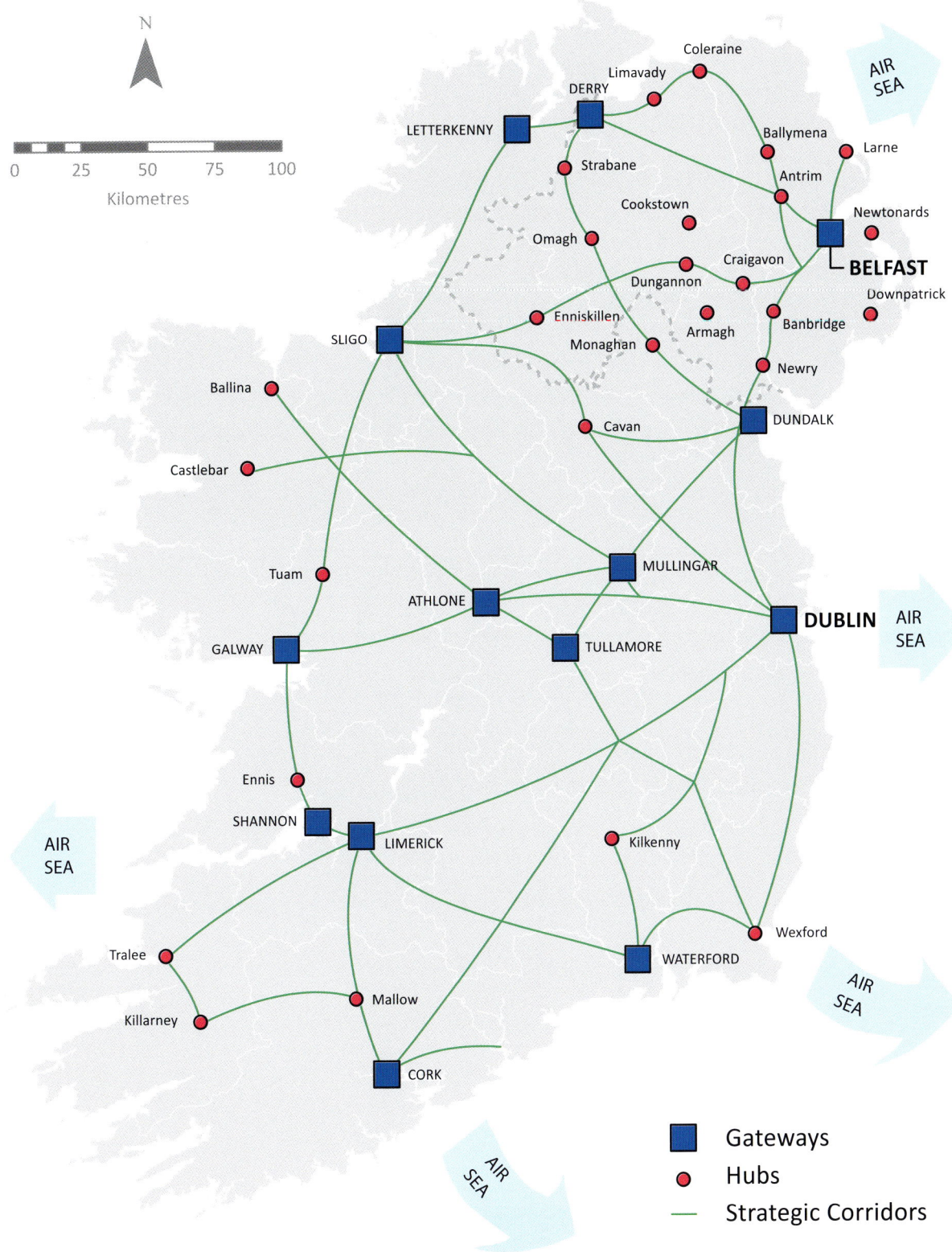

Fig. 25.23 GATEWAYS AND HUBS. In 2001, the Northern Ireland government published *Shaping our Future*, while in the Republic the *National Spatial Strategy* (2002) appeared. Both were adaptations of earlier growth centre strategies proposed in the 1960s for their respective states. Government investment would be prioritised for designated urban centres, or gateways. These were selected on the basis of their population size, quality of infrastructure and diversified labour markets, and were considered to possess the potential to attract and sustain economic growth. Smaller 'hubs' would benefit from a 'trickle-down' of prosperity from adjacent gateways, and would act as centres for regional development. The use of hubs is especially apparent in Northern Ireland to disperse growth opportunities from Belfast to relatively depressed inland towns. In the Republic, all gateways (except Athlone–Mullingar–Tullamore) are coastal, which confirmed the priority of the zone for national development. The cross-border gateway of Letterkenny–Londonderry (Derry) recognised the growing significance of trade between the two states, as did the corridor of growth between Belfast and Dublin. Strategic investment in transport emphasises further the pivotal locations of the main coastal towns and cities. This is especially apparent given the strategic need to promote an effective growth corridor along the the Atlantic seaboard by linking six gateway towns from Letterkenny to Cork. [Data sources: National Spatial Strategy, 2002; Regional Development Strategy, 2001]

of its major coastal cities for international investment. Thus, one-half of the Republic's projected increase of one million people by 2040 is expected to occur in its five coastal cities. The population of Dublin city and suburbs is expected to grow between 230,000 and 290,000 to at least 1.41 million, confirming its primacy and role as the country's major international city and gateway for global trade.

Case Study: The Port of Limerick Today

Des McCafferty

In spite of the earlier success of the port of Limerick, by the 1970s it was struggling, as were the city's other traditional industries (see Chapter 19: Changing Coastal Landscapes). This was the result of the loss of markets due to declining competitiveness in the new era of free trade.[23] Employment in the important port-based industry of flour milling was in steady decline, as were sectors such as bacon curing, dairy processing and clothing. Following a series of rationalisations, the last remaining miller, Ranks, closed in 1983. Among the reasons cited for closure was the relatively high cost of imports through Limerick port. This is linked specifically to the fact that Limerick is the furthest inland of Ireland's river ports, which presents difficulties for ever-larger sea-going vessels in navigating the tidal Shannon as far as the city (Figure 25.24). These considerations led to an increasing shift in activity to the deep-water port at Foynes, located approximately 40km closer to the mouth of the estuary. For example, coal imports were relocated to Foynes (and Cork) in 1992. The

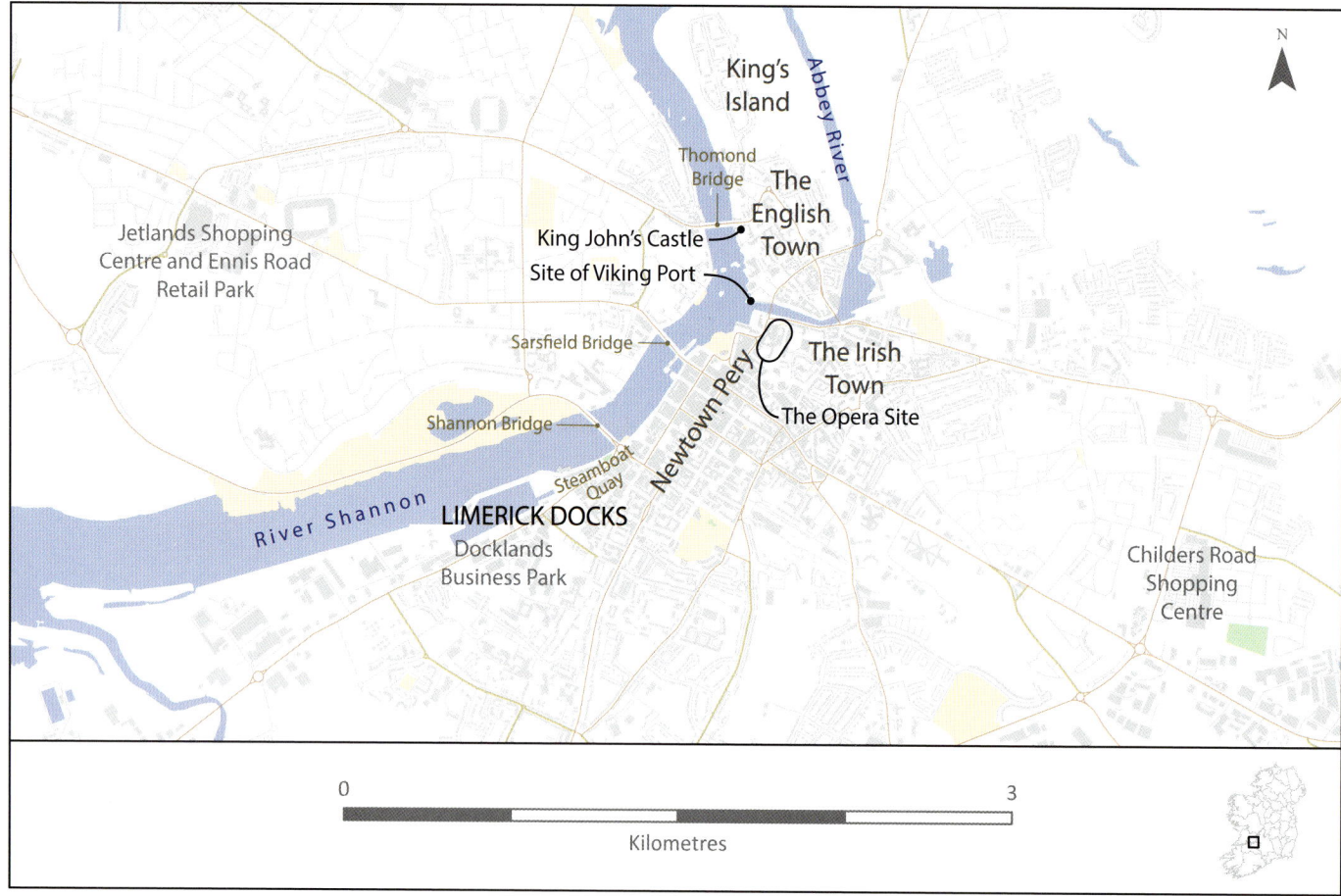

Fig. 25.24 LIMERICK CITY CENTRE: FROM MEDIEVAL TO MODERN. The walled medieval city consisted of the English Town on King's Island and the Irish Town on the southern side of the Abbey River. The modern city centre corresponds largely to Newtown Pery, which was built following the dismantling of the city walls in the late eighteenth century. This area is of considerable historical and architectural significance on account of its fine Georgian dwellings, which were laid out along wide straight streets arranged in a rectilinear ('grid-iron') pattern. Despite being the main retail hub of the city and the region, it has suffered from both the suburbanisation of population and, especially during the Celtic Tiger period, the large-scale development of out-of-town shopping centres, such as those at Jetland and the Childers Road. As a result, many of the Georgian buildings are now underutilised or, in some cases, vacant and derelict, and they present a considerable challenge for conservation. Limerick City and County Council's *Limerick 2030: An economic and spatial plan for Limerick* envisages the redevelopment of a large area in the oldest and most dilapidated part of Newtown Pery, known as the Opera site, as a mixed office/innovation/education space, with some elements of retailing and residential land use. The population of the city and its immediate suburbs in 2016 was 94,192, with 104,952 in the wider Limerick Metropolitan District, which includes a number of commuter villages, such as Annacotty, Castleconnell and Mungret.

decline of port-based economic activity contributed to a severe unemployment crisis in the city in the 1980s, which in turn gave rise to a complex array of social problems, especially in the city's local authority housing estates.[24]

By the mid-1980s, Limerick's quays and the area surrounding the docks had been in decline economically and demographically for several decades (Figure 25.25). As a result of widespread dilapidation, this part of the city was designated under the Irish government's 1986 and 1999 urban renewal schemes. The principal objective of the schemes was to stimulate property development through financial incentives. In this they were broadly successful, with investment in Limerick higher than in any city other than Dublin. Since the late 1980s, the cityscape along the southern bank of the Shannon River, from Sarsfield Bridge to the docks, has been transformed as derelict warehouses and former industrial buildings have been replaced with modern apartments, shops and offices. Early developments in the vicinity of the docks included apartments at Steamboat Quay and the Docklands Business Park, built on a site occupied previously by timber yards (Figure 25.26).

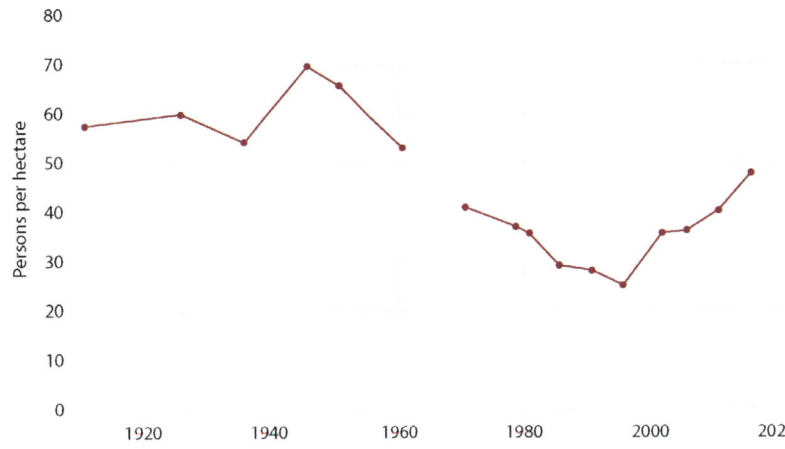

Fig. 25.25 POPULATION DENSITIES (1911–2016) IN LIMERICK'S RIVERSIDE WARDS. The graph depicts changes in the population density in the wards and electoral districts that adjoin the quays and docks in Limerick. Population decreased sharply after 1946 as part of a movement out of the city centre, reflecting the decline of traditional city centre employments, including dock work, and the process of suburbanisation. The graph also depicts the sharp turnaround brought about by the urban renewal schemes introduced from the late 1980s onwards. These schemes resulted in extensive construction of new city centre apartments close to the river, many of which have been occupied by immigrants. [Data source: CSO, 1911–2016. No ward-level data were reported for Limerick County Borough in the 1966 census, and in 1970 the eight original wards, including Shannon and Dock, were subdivided into thirty-seven new wards. To allow for changes in areas following this reorganisation, population densities rather than absolute populations are shown.]

There have also been changes in the governance of the port, which have implications for the redevelopment of Limerick. In 1996, the Limerick Harbour Commissioners were replaced by the Shannon Estuary Ports Company which, following amalgamation with Foynes Port Company, became the Shannon Foynes Port Company (SFPC) in 2000. In 2006, a call by SFPC for expressions of interest in Limerick docks from property developers led to proposals for redevelopment along similar lines to that carried out in Dublin's docklands. However, redevelopment was opposed vigorously by the remaining port users, as well as by the city councillors. Despite limited employment in, and traffic through, the city docks, this group, and others, argued that it would mean the end of Limerick as a 'working port', thereby bringing to an end the historically strong, but visibly weakening, relationship between the city and the port. There were also concerns about the loss to the city of an irreplaceable piece of infrastructure. In the face of this opposition, the redevelopment proposals were shelved.

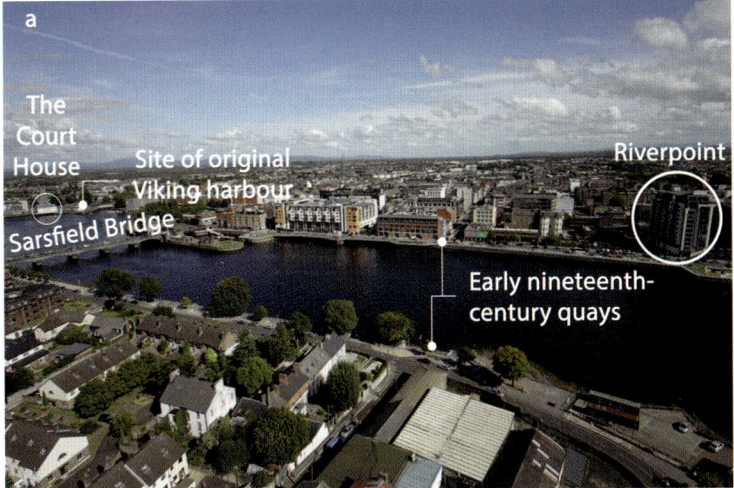

Fig. 25.26 CONTINUITY AND CHANGE: LIMERICK QUAYS AND DOCKS. Limerick's quays have undergone significant transformation in recent years, but the nineteenth-century docks maintain the city's links with its maritime past. The photograph on the left shows the quayside from Sarsfield Bridge (upstream, left of image) to Shannon Bridge (just out of view on the extreme right). The brightly coloured building facing the river above Sarsfield Bridge is the Court House, which stands close to the site of the original Viking port. The far bank of the river forms the western edge of the historical Newtown Pery district. The old stone warehouses and mills that once fronted the river in this area have been replaced with modern buildings dating from the early 1990s onwards. The tall building on the far right is part of the Riverpoint complex, a mixed office, retail and residential development on a site that was designated under the 1999 urban renewal scheme. Downstream of Shannon Bridge, Limerick's nineteenth-century docks are shown in the photograph on the right, as a bulk carrier prepares to exit the docks at high tide. Among the main exports through the docks are scrap metal (right foreground), which is shipped to steel mills throughout Europe, and cement produced at the nearby Irish Cement factory established in the late 1930s. On the left-centre of the photograph the former Ranks grain silo is visible, and beyond it the Bannatyne Mills, both relics of Limerick's flour-milling past and vacant for over thirty years. Current plans for the conversion of Bannatyne Mills envisage mixed-use as a 'flagship' office development, while Ranks silo may be adapted as a recreational space. [Sources: a. Limerick City and County Council and b. Shannon Foynes Port Company]

The need and potential for new development that incorporates the docklands area have, however, been recognised in recent plans developed at the local and regional level. These include Limerick City and County Council's *Limerick 2030: An economic and spatial plan* (2014), and SFPC's *Limerick Docklands Framework Strategy* (2018), both of which identify the docks as a site of strategic importance adjacent to the city centre. *Limerick 2030* recognises the need to connect the docks with the area already redeveloped along the quays, so as to complete the renaissance of the entire waterfront. While the SFPC Strategy is committed to maintaining cargo traffic through the docks, it also recognises that some two-thirds (31ha) of its dockland estate comprises 'non-core' assets (land and buildings not utilised in daily port operations). These include landmark buildings, such as the Bannatyne Mills and Ranks grain silo, both vacant since the demise of flour milling. The plan envisages the creation of a Limerick Docklands Economic Park, with a focus on attracting high-technology businesses and light manufacturing, particularly in the renewable energy sector. The overarching objective of the strategy, however, is again to reconnect the docklands with the city and revive the historic relationship between the city and its port.

The importance of the docks is also recognised in the Irish government's *Project Ireland 2040: The national planning framework* (2018), which identifies the area as a key element in Limerick assuming the role assigned to it under the strategy: a regional centre of national significance, because of its size and location mid-way between Cork and Galway on Ireland's 'Atlantic Economic Corridor'. The plan sets an ambitious target of between 50 and 60 per cent for population growth in the city to 2040. In keeping with the overarching commitment to more sustainable development, it is envisaged that 50 per cent of the new housing required to accommodate this growth will be located within the existing urban footprint, so that Limerick will become a more compact city. In ensuring this target is met, centrally located, publicly owned sites with untapped development potential will be key, and the area between the docks and the city centre is recognised as one such site. Thus, the work to enhance the city's national role is seen as beginning in the historic core of Limerick.

Changing urban layout

From the 1960s, Ireland's larger urban centres evolved rapidly from relatively compact and dense settlements to more expansive forms of land use. In particular, the physical distinction between urban and rural became increasingly blurred as built-up areas expanded beyond administrative boundaries and urbanisation of the countryside gathered momentum.[25]

Population losses began to occur in Ireland's inner-city neighbourhoods, especially from the 1960s, associated with a relocation of both working-class and middle-class residents.

Fig. 25.27 KILRUSH, COUNTY CLARE. Kilrush is one of a number of urban communities located on the shoreline of the Shannon Estuary. Its distance from Limerick city has limited its role as a commuter centre and, reflecting the general trend in south-west Clare, its population has decreased by 5.8 per cent since 1961. However, the attractive environment along the 100km-long estuary has encouraged its growth as a tourist resort and a location for second homes. Water-based leisure activities are emphasised and a modern marina and a lock were constructed in the early 1990s, providing protection from the tidal estuary. A nearby ferry service linking Killimer to Tarbert, County Kerry, encourages tourist flows through the town. [Source: Clare County Council]

Congested and inadequate housing conditions that typified many working-class communities were replaced frequently with large, public authority housing estates around the administrative boundaries of cities and larger towns. An increasingly negative perception of the quality of life within the cities, together with high costs, also encouraged growing numbers of middle-class residents to relocate to low-density, modern suburbs. The scale of relocation out of the two capital cities was encouraged further by decisions to build new towns to accommodate the overspill: Craigavon in Belfast and Blanchardstown, Lucan-Clondalkin and Tallaght around Dublin. In Belfast, therefore, the city's population fell from almost 416,000 to 236,000 between 1961 and 1991, while its outer suburbs increased from 163,000 to 244,000. During the same period, Dublin's inner-city population was halved.

Increasing levels of private car ownership and, especially in the Republic, a laissez-faire policy regarding housing development were also key factors in extending the urban footprint of coastal cities over a more extensive hinterland. Many small to medium-sized towns accessible to these cities became commuter centres, while linear patterns of settlement intensified along the roads that connected coastal centres with their hinterland.[26] Urban development was, in effect, 'leapfrogging' across the landscape around the coastal cities. Good examples occur around Belfast Lough, Dublin Bay, Cork Harbour, Galway Bay and the Shannon Estuary (Figure 25.27). In the case of Dublin, Hourihan noted that

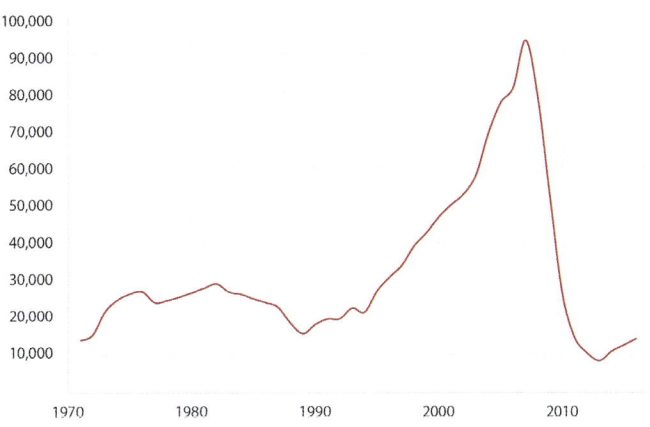

Fig. 25.28 ESTIMATED HOUSING COMPLETIONS IN THE REPUBLIC OF IRELAND (1971–2016). From 1971 to the early 1990s, the annual average construction of housing in the Republic was between 20,000–25,000 completions. The economic boom of the Celtic Tiger, easy access to cheap credit and rising personal prosperity translated into a construction boom. Annual completions increased dramatically, peaking at over 90,000 homes per year. A large number of new housing units were built in speculative housing estates around the edges of villages and small to medium-sized towns within the expanding commuter hinterland of coastal cities. With the financial crisis and recession of 2007–2009, many developments remained unfinished and became 'ghost estates'. One-off housing also predominated in rural areas, adding to concerns about 'bungalow blitz'. Economic recovery has increased demand for housing and the problem of ghost estates is receding. Now the problem is a shortage of housing to meet rising demand, especially in and around the cities. This raises concerns over 'boom-and bust' cycles that have characterised Ireland's construction industry since the Celtic Tiger [Data source: Housing Agency, National Statement of Housing Supply and Demand, 2016, Dublin, p. 8].

Fig. 25.29 COMMUTER TOWN OF GREYSTONES. Greystones is a coastal town located 24km south of Dublin, bordered by the Irish Sea and backed inland by the Wicklow Mountains. Its development from the mid-nineteenth century, initially as a tourist resort and later as a commuter centre for the capital city, was limited historically by its proximity to the larger county town of Bray, which is located 3.5km to the north of Greystones. Since the 1960s, however, a number of construction booms have occurred in the town and these have resulted in its population more than quadrupling, to over 18,000 by 2016. The modern growth has been due almost entirely to its role as a popular dormitory town within Dublin's expanding commuter zone. Its attractive environment and resort functions have also aided development. Reflective of its attractions for a relatively well-off resident population has been the upgrading of its harbour and promenade. In 2015, a redevelopment scheme costing well in excess of €300 million was finalised, which provided a new harbour and a 238-berth marina, together with a significant number of well-positioned apartments and townhouses forming the Marina Village, located along the seafront. The 2016 census indicated that Greystones was the Republic's most expensive town in terms of property values. [Source: Barrow Coakley Photo and Video]

'the speed with which [it] has been transformed into a sprawling, low-density city has been very rapid compared to developments elsewhere [in the world]'.[27]

The boom years of the Celtic Tiger accentuated these trends, but it also witnessed the first concerted attempts to reverse inner-city decline. A malfunctioning inner city, characterised by many derelict sites and buildings, became viewed by planners and government as an underutilisation of a valuable locational resource. In a post-industrial economy, well-serviced and modern city centres had become optimal sites for newly emerging and fast growing economic sectors. An Urban Renewal Act was therefore passed in the Republic in 1986, which incentivised private developers to regenerate the historic cores and waterfront areas of the five port cities.

The scheme has been successful in reimagining these cities, but especially in Dublin with the modernisation of its Custom House area. In Belfast, the Titanic Quarter illustrates a similar process in Northern Ireland. The rejuvenation of the city centre in Belfast has been fundamental for encouraging people to move back to the area after decades of decline. Its 2011 population (281,000) was almost 20 per cent higher than in 1991, although the outer suburbs increased even further (30 per cent) to 315,000, reflecting the continued attractions of Northern Ireland's primate city.

Modernisation of inner cities has accentuated their roles as concentrated centres of employment, based strongly on high-density office developments. Gentrification also occurred, which renewed their housing markets, principally through the construction of new apartment blocks. These attracted large numbers of relatively young and upwardly-mobile people seeking to access the well-paid employment that was gravitating to the coastal cities. While helping repopulate inner cities, these processes also inflated the costs of, and competition for, accommodation. The result has been increased levels of counterurbanisation, which involve a reverse migration of many young people and families from designated coastal gateway cities, as they seek access to a more affordable housing market. This push factor has been complemented by a pull factor involving speculative building of large new housing estates by private developers in and around most small to medium-sized towns within daily commuting distance of coastal cities (Figures 25.28 and 25.29).

The populations in most hinterland towns have grown faster than the built-up areas of the coastal cities. Many, however, were developed without adequate planning and lack effective social infrastructure to enhance quality of life. In addition, large numbers of unregulated, one-off houses have been built in rural areas by people seeking to combine the benefits of low-cost sites with more aesthetic environments. These, in effect, have become dormitory communities within the expanding daily urban systems of coastal cities, and are linked to them by congested routes facilitating large volumes of commuting. As a result, the contemporary imprint of urbanisation has spread over many parts of coastal Ireland, but especially along the northern, eastern and southern seaboards, as well as penetrating deeply inland.

Case Study: Belfast

Stephen A. Royle

By the early 1800s, Belfast had emerged as the dominant port in Northern Ireland. This was related to its gateway role for its rich hinterland in the Lagan Valley, which produced both agricultural goods and linen for export through the developing port infrastructure. The industrial revolution that occurred in Northern Ireland in the nineteenth century benefited the port-city in particular. The city attracted many linen factories, chemical plants, engineering works and, above all, a large shipbuilding industry focused on the Harland and Wolff Shipyards. By the outbreak of the First World War, this port-centred industrialisation caused the population of the city to increase more than five-fold from 1841, to almost 390,000. A contemporary report noted: 'there is no city in Ireland (if indeed in the United Kingdom) which has so rapidly developed from insignificance to vast importance as Belfast' (Figure 25.30).[28]

The sinking of the *Titanic* in 1912 was a major symbolic blow for Belfast, a situation made worse by the economic depression of the inter-war years and reduced demand for shipping. While Belfast recovered initially from the consequences of the Second World War, aided by an upsurge in demand for shipping, problems of deindustrialisation soon overtook the city. In the 1960s, falling world demand and competition from more innovative and cheaper producers eroded Belfast's significance as a global centre for shipbuilding. The decision was made to refocus on building large oil tankers. A massive building dock was, therefore, constructed and remains dominated to this day by the two huge and iconic yellow Harland and Wolff cranes: Samson (built 1969) and Goliath (1974). The strategy was not successful and the company has not built ships since 2003. It now concentrates on rig repair and offshore construction with a workforce of about 200, compared with its peak of some 35,000 during the Second World War.

The decline of shipbuilding was mirrored by the ending of linen production and many other jobs in manufacturing. The attraction of the aircraft manufacturing company, Shorts, to a large plant and airstrip constructed on land reclaimed from the Lough and adjacent to the port, did little to reduce the sense of depression and high unemployment experienced in the city. Another negative multiplier was

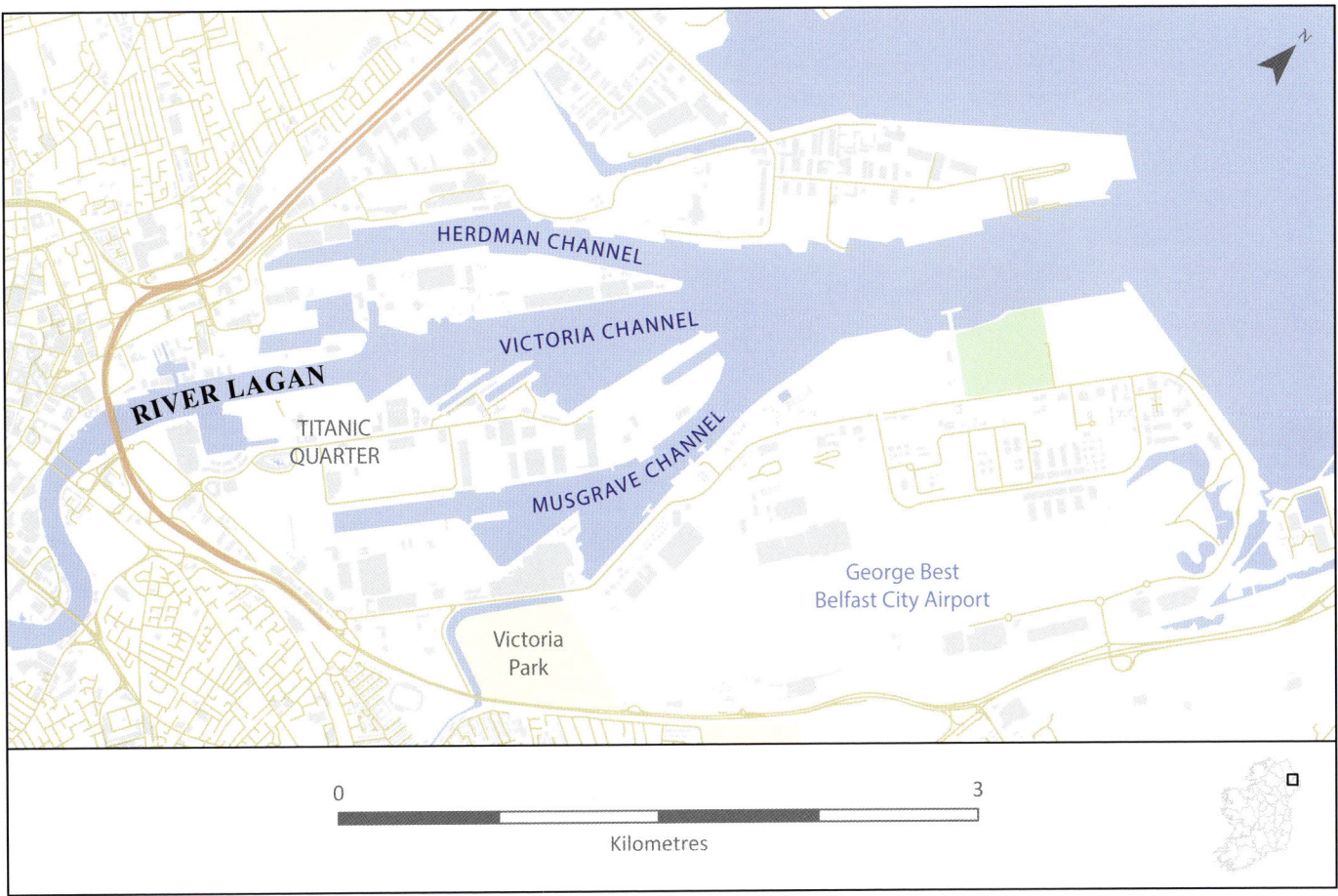

Fig. 25.30 BELFAST PORT AND PRINCIPAL CHANNELS. The city of Belfast has developed around the River Lagan, its port and harbour, and is situated at the landward end of Belfast Lough. The area has been described as a trident with, from the left, the Herdman Channel (1930–1933), the Victoria Channel (the Lagan itself, deepened and straightened from 1849) and the Musgrave Channel (1899–1903). The area of land between the Victoria and Musgrave Channels was dominated by the Harland and Wolff Shipyards. Today, much of this land is given over to the Titanic Quarter and is the catalyst for the recent transformation of Belfast city. The modern port infrastructures occur on the land between the Victoria and Herdman channels and downstream to Belfast Lough.

the thirty years of civil strife, or 'Troubles', from the 1960s.

Between 1951 and 2001, the population of the city fell from some 444,000 to only 238,000, as large numbers of residents dispersed to suburban estates. As a result, Belfast city's outer suburbs more than doubled to almost 266,000. This loss of inner-city population was

accompanied by the movement downstream of the port's principal activities to new, purpose-built structures on land reclaimed from the Lough. These changes in land-use have reduced greatly the intimate relationships that existed between port-based activities and the city. A new relationship, however, emerged from the 1990s, which has helped modernise the city and enhance its status as a dynamic and growth-orientated primate city. This is reflected in a return to growth in the population of the city of Belfast (Figure 25.31).

The relocation of port facilities and the decline of shipbuilding has released some 800ha (2,000 acres) of waterside land and this has been fundamental in the transformation of Belfast to a post-industrial city. The area is managed by the Belfast Harbour Commission and is now put to different land uses. Most of the initial regeneration was overseen by the Laganside Corporation (1989–2007), which held the mission to revive central Belfast's waterfront and act as a catalyst for urban renewal. The process has been successful, with over 700 firms located on the estate, providing employment to some 23,000 people.

In 1938, a runway was constructed adjacent to the port to serve the needs of the aerospace company, Shorts. This became a commercial airport in 1983, and was modernised further under plans

Fig. 25.31 POPULATION CHANGE, BELFAST (1911–2016). At the outbreak of the First World War, Belfast's population was approaching 0.4 million. Despite inter-war recession, the city continued to attract people, primarily from depressed rural communities and declining industrial towns located in its hinterland. Deindustrialisation, problems associated with the 'Troubles' and government policies to restrict growth in the capital city resulted in a decline in population of almost 50 per cent from 1951 to 1991. Economic revival linked to the benefits of the 'Peace Dividend' and transformation of the inner city and waterfront have encouraged the recent growth. In 2011, the city's population was over 280,000, while Belfast metropolitan area, which includes its commuter towns, approached 675,000. [Data source: NISRA]

609

Fig. 25.32 BELFAST CITY AND THE TITANIC QUARTER. Central to the transformation process of Belfast city is the Titanic Quarter, located around the dockland site where the iconic vessel was built and launched in 1912. Here, Queen's University Belfast has located its Institute for Electronics, Communications and Information Technology (ECIT). Belfast Metropolitan College has also moved to the Titanic Quarter, as did the Public Records Office of Northern Ireland. These now constitute a cluster of research-orientated activities. The former paint hall of the Harland and Wolff shipyard is a film studio while its drawing office became a boutique hotel. The most notable development in the Quarter, however, is Titanic Belfast, which is a key tourist facility opened in 2012 to mark the centenary of the sinking of the *Titanic*. This is sited beside the slipways where the *Olympic* and the *Titanic* were built. In 2016, Titanic Belfast was declared to be the World's Leading Tourist Attraction by the World Travel Awards. [Source: Belfast Harbour]

to transform the waterfront area. Since 2006, new terminal facilities were built and it was renamed the George Best Belfast City Airport. Shorts, now Bombardier Aerospace, remains Belfast's largest manufacturer. The strategic location and traditional skills of the historic port city have also attracted new growth sectors in manufacturing, but especially in offshore wind logistics at a purpose-built 200,000m² deepwater quay (Figure 25.32).

Most of the redevelopment of the docklands, which is central to the reimagining of Belfast, is focused on new services sectors. The SSE Arena, an entertainment complex, opened on a former scrap metal collection site as the Odyssey Complex in 2000. In addition, the modernisation process involved the opening of large retail outlets, construction of hotels and a number of new residential blocks and office buildings. A new boating marina and terminals to allow large cruise liners to berth have boosted the city's tourist trade and international image. As a result, Belfast has shown considerable success in detaching itself from the image of a 'troubled' and economically depressed city. It now competes successfully for international investment, while a more attractive urban environment is reflected in a growing population, both of which consolidate its status as Northern Ireland's primate city.

Tourism

This international growth industry has played a significant role in shaping urbanisation around coastal Ireland (see Chapter 27: Tourism and Leisure). Increasing tourist numbers have stimulated development of new accommodation, employment, retail outlets, cultural activities and improved roads to enhance accessibility to key sites. In combination, these increased the urban footprint of tourism along the coastline.

Since the 1960s, the west coast has experienced a large influx of both international and domestic tourists. This has allowed growth to replace the stagnation or decline that had characterised most of these coastal communities for many years. The tourists are

attracted by spectacular scenery, such as the peninsulas of counties Kerry and Donegal, and cultural traditions embedded in many towns and villages. Activity-centred tourism is also prominent, such as hiking, surfing and golf. These environmental and cultural attractions are now marketed under the Wild Atlantic Way (see Chapter 27: Tourism and Leisure).

Tourism has also surged around the rest of the island's coastline. Here, however, tourists often complement a large and growing volume of commuters who reside in coastal communities that are accessible to large urban centres and provide an added vitality to such places. Although perhaps less spectacular than the west coast, these other coastal areas have diverse and often

Fig. 25.33 PORTRUSH, COUNTY ANTRIM. The old town of Portrush and the core of its early tourist activities occur on the peninsula of Ramore Head. Here are located the railway station, large hotels and seafront developments constructed in the style of Victorian coastal resort towns. By the First World War it had established itself as one of Northern Ireland's most important resorts and it retains that status to the present day. Although the town experienced a period of relative stagnation from the 1930s to the 1960s, its population doubled to some 7,000 by 2011. This reflected a revival in its attraction for tourists, but also its increasing role as a commuter centre for Coleraine. Growth has extended its urban imprint to the west and inland from the historic core; to the east the Royal Portrush Golf Club occupies a considerable area of beachfront. Unlike in the Republic, where urban expansion into surrounding rural areas is often typified by an irregular, almost haphazard expression of low-density housing, here development is better controlled under stricter British planning regulations. The urban imprint is, therefore, characterised by high density and more compact housing estates. This intrudes less on the environmental qualities of surrounding areas. [Source: Aerial Vision NI]

picturesque landscapes, a wealth of historic sites and towns, and possess resources for leisure-based activities. The east coast, for example, is marketed to reflect its historic traditions as the 'Ancient East' (see Chapter 27: Tourism and Leisure). Domestic tourists are especially numerous, and include day-trippers wishing to 'escape' congested urban environments to access beaches, walking trails, golf, fishing and surfing (Figure 25.33).

Governments recognise the catalytic role that tourism can play in encouraging growth in coastal areas. In the Republic, therefore, government-approved programmes, such as the Seaside Resort Scheme (1995), have been used to incentivise the building of tourist accommodation, even though its impact has proved controversial in some coastal communities. Often, financial incentives encouraged construction of high-density, standardised groupings of cottages that now occur around the periphery of many small tourist resort towns and villages (Figure 25.34).

Since the 1960s, an increasingly prosperous society in Ireland took the decision to construct new, or reconstruct old, homes in attractive rural environments and especially along the coast. While some become permanent residences, large numbers are holiday or second homes. Many were built as detached bungalows strung out along roads leading to coastal towns and villages. Beginning in the 1970s, the scale of construction and the extent of intrusion into picturesque landscapes of stand-alone buildings gave rise to the term 'bungalow bliss' or, more appropriately, 'bungalow blitz' (Figure 25.35).

While providing a sense of modernisation and prosperity in coastal areas associated historically with declining populations, the improvement is superficial. For most of the year, many properties are vacant and the veneer of urbanisation adds little to a sense of community. In addition, during peak summer months, this urban footprint can place unsustainable pressures on the limited carrying capacity of such environments in terms of, for example, water supply and effluent disposal.

CHANGING PATTERNS, 1961–2016

Between 1961 and 2011, the number of towns in Ireland with resident populations in excess of 5,000 more than doubled to 128 (Figure 25.36). Coastal towns contributed strongly to this trend and, by 2011, had a combined population of over 2.6 million. The relative share of the island's population residing in such coastal centres declined, however, to 71 per cent. This reflects the above

Fig. 25.34 BALLYBUNION HOLIDAY COTTAGES AND CARAVAN PARK. Ballybunion is a small coastal resort in County Kerry with a resident population of some 1,500 people. This number inflates greatly in summer months when tourists arrive to access the beaches, surfing and its links golf course, one of Ireland's top-ten courses. Ballybunion was also one of seventeen places included in the 1995 Coastal Resorts Scheme, which aimed to promote and improve amenities in such resorts. Most were located along the western seaboard (for example, Kilkee, Achill and Bundoran). Incentives encouraged construction of relatively high-density groupings of rental cottages having a uniform design. In addition, Ballybunion has several areas designated as caravan parks on which large numbers of caravans are located on a permanent basis. Both holiday homes and caravan parks are located usually at the edge of a resort and represent an element of mass tourism, which fits poorly with the traditional character and culture of the resort. While providing some seasonal employment and income, the buildings and caravans are vacant for much of the year. Furthermore, high-intensity occupancy in the summer months places severe pressure on local environments (water supplies, effluent disposal) while inflating costs of housing and land for local people. This is not a good example for sustainable urbanisation of Ireland's coasts. [Source: Sarah Kandrot]

average growth of many small to medium-sized towns located inland from coastal cities, but forming part of their extensive commuting hinterlands. Dispersal trends to inland centres were generally more pronounced in Northern Ireland. Here, the operation of a stronger regional policy and more effective urban planning, which placed limitations on the growth of Belfast,

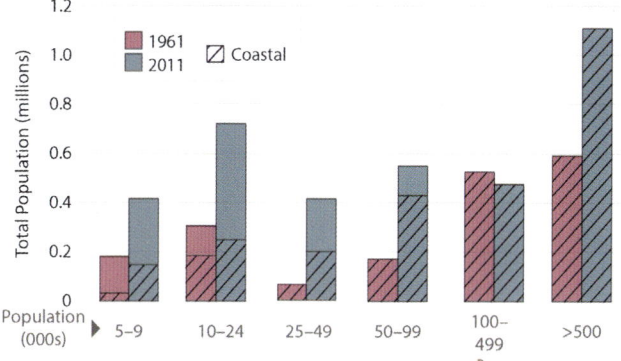

Fig. 25.35 'BUNGALOW BLITZ', BRINLACK, GWEEDORE BAY, COUNTY DONEGAL. The increased prosperity associated with the modernisation of Ireland's economy from the 1960s, combined with a laissez-faire approach by the planning system to new house construction in the Republic, resulted in the large-scale development of one-off houses in scenic rural and coastal environments. Frank McDonald, a long-time critic of unrestrained, one-off housing, described bungalows as spreading like a rash over the countryside, a phenomenon that became known as 'bungalow blitz'. In effect, it was a suburbanisation of the countryside but with minimal controls. One such landscape is depicted here in County Donegal, where we can see a large number of individual bungalows occupying sites in small fields. Many of these houses may have been built on sites acquired from family members, in a modern-day extension of the ancient 'clachan' form of settlement, which the geographer E. Estyn Evans identified as continuing into modern times in north-west Donegal. It is also likely that some of them are holiday homes, the development of which also surged with increasing prosperity. In many such landscapes, unprofitable farming has ceased and unused fields await a new crop of bungalows. [Source: Joseph Mischyshyn]

favoured the urban renewal of many inland industrial towns that had been adversely affected by deindustrialisation.

During this period, all of Ireland's coastal counties, apart from Antrim, recorded population growth (Figure 25.37). This exception was due to the loss of one-third of the population of Belfast over the period. Since 1991, however, the capital city has reversed this trend and has experienced some growth. This general pattern of growth is in marked contrast to the period 1926 to 1961, especially in the Republic where the majority of coastal counties experienced population losses (Figure 25.10). Above average increases were confined primarily to the eastern seaboard from County Wexford to County Down, where the primacy of Dublin and Belfast was pivotal in this trend. Population overspill from the two capitals extended their built-up areas, while dispersal of large numbers of people, together with improved local employment opportunities, contributed to significant growth in many coastal towns in adjacent counties, such as Swords and Bray in the Republic, and Bangor in Northern Ireland. Such developments confirmed further the pivotal role of the Belfast to Dublin corridor for the island's competitive economy.

Fig. 25.36 POPULATION TOTALS FOR COASTAL AND INLAND TOWNS BY SIZE CATEGORIES (1961–2011). In the fifty years to 2011, urban places with a resident population of over 5,000 more than doubled in number (56 to 128) and had a combined population of 3.7 million (60 per cent of the island's total population). Coastal centres continued to dominate, especially those with over 50,000 residents. By 2011, coastal cities and towns had gained more than a million population. The growth rate of inland centres, however, was even more impressive, increasing more than three-fold from 1961 to 2011. This was due primarily to a dispersal of people, primarily from the largest coastal cities, to an increasing number of small to medium-sized towns, which formed part of their expanding, inland commuter zone. The three size groupings from 5,000 to 49,000 illustrate this clearly; these inland towns gaining almost 0.7 million residents (+253 per cent). [Data source: CSO and NISRA]

Population growth in other coastal counties also correlated strongly to the presence and size of urban centres located along their coastlines. The regional capitals of Cork, Limerick, Galway, Waterford and Londonderry were, therefore, critical in enhancing their counties' population totals.

Perhaps the most significant changes occurred along the western seaboard, where growth replaced large-scale population decline, which had been endemic since the Great Famine (see Chapter 20: The Great Famine). County Galway, in particular, experienced above-average growth as its university city attracted

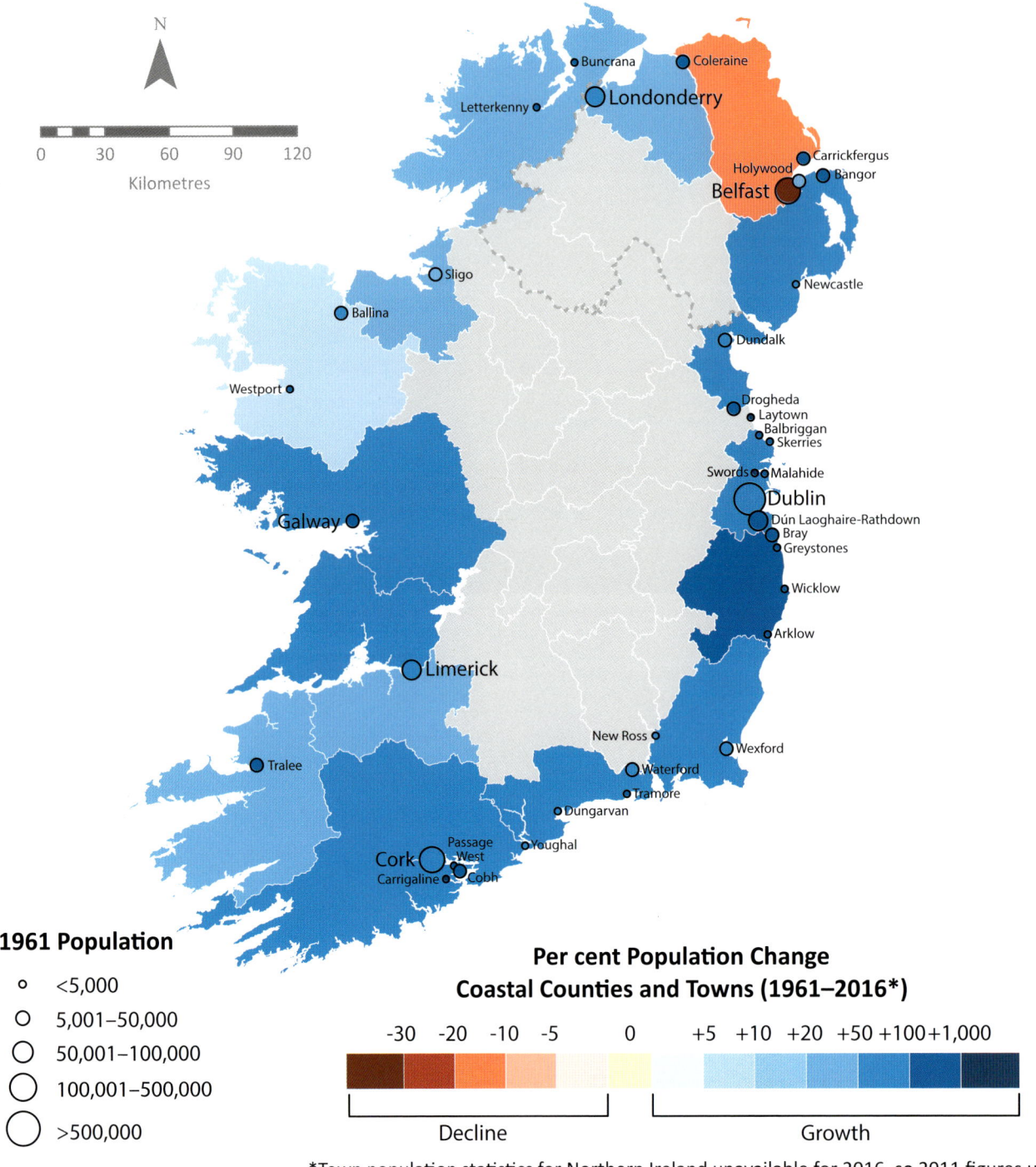

1961 Population

- o <5,000
- O 5,001–50,000
- O 50,001–100,000
- O 100,001–500,000
- O >500,000

Per cent Population Change
Coastal Counties and Towns (1961–2016*)

| -30 | -20 | -10 | -5 | | 0 | | +5 | +10 | +20 | +50 | +100 | +1,000 |

Decline — Growth

*Town population statistics for Northern Ireland unavailable for 2016, so 2011 figures used

Fig. 25.37 POPULATION CHANGES IN COASTAL COUNTIES AND LARGER TOWNS (1961–2016). Apart from Antrim, all coastal counties experienced population growth in this period. A committed policy to restrict the growth and spread of Belfast's contiguous built-up area, changes to the city's administrative boundaries and incentives to disperse its population contributed to this exception. In contrast, the adjacent County Down almost doubled in population, benefiting from dispersal out of Belfast and relocation within its towns. Bangor, for example, grew significantly to a town of over 60,000 as its traditional role as a coastal resort was overshadowed by its function as a major commuter centre for the Greater Belfast Area. In the Republic, Dublin and its county continued to dominate population growth with positive spill-over effects for adjacent counties. On the west coast, all counties reversed their long history of decline. Galway, which acquired city status in 1961, emerged as a strong growth centre for its county and region, while counties Clare and Limerick benefited from development at Shannon New Town and the port city of Limerick respectively. Despite this, the north-west counties remained poorly serviced with large urban centres and experienced below-average rates of growth. [Data source: CSO and NISRA]

significant investment in high-tech employment and new areas of residential housing. This consolidated its role as the 'gateway to the west'. In the adjacent County Clare, successful promotion of a new town and related airport complex at Shannon was pivotal for its population growth. These positive trends, centred on the two regional capitals, created prospects for a development zone that would link Limerick to Galway. This, furthermore, could anchor a much-needed western corridor of growth from Letterkenny to County Kerry, and possibly to Cork. It would provide a necessary counter to the concentration of Ireland's development along the eastern seaboard.

Coastal counties exhibiting the lowest growth rate remained in the north-west of the island. Here, the paucity of large towns and the region's peripherality continued to inhibit development that could sustain significant increases in population and job creation. County Mayo, for example, experienced only a 6 per cent increase

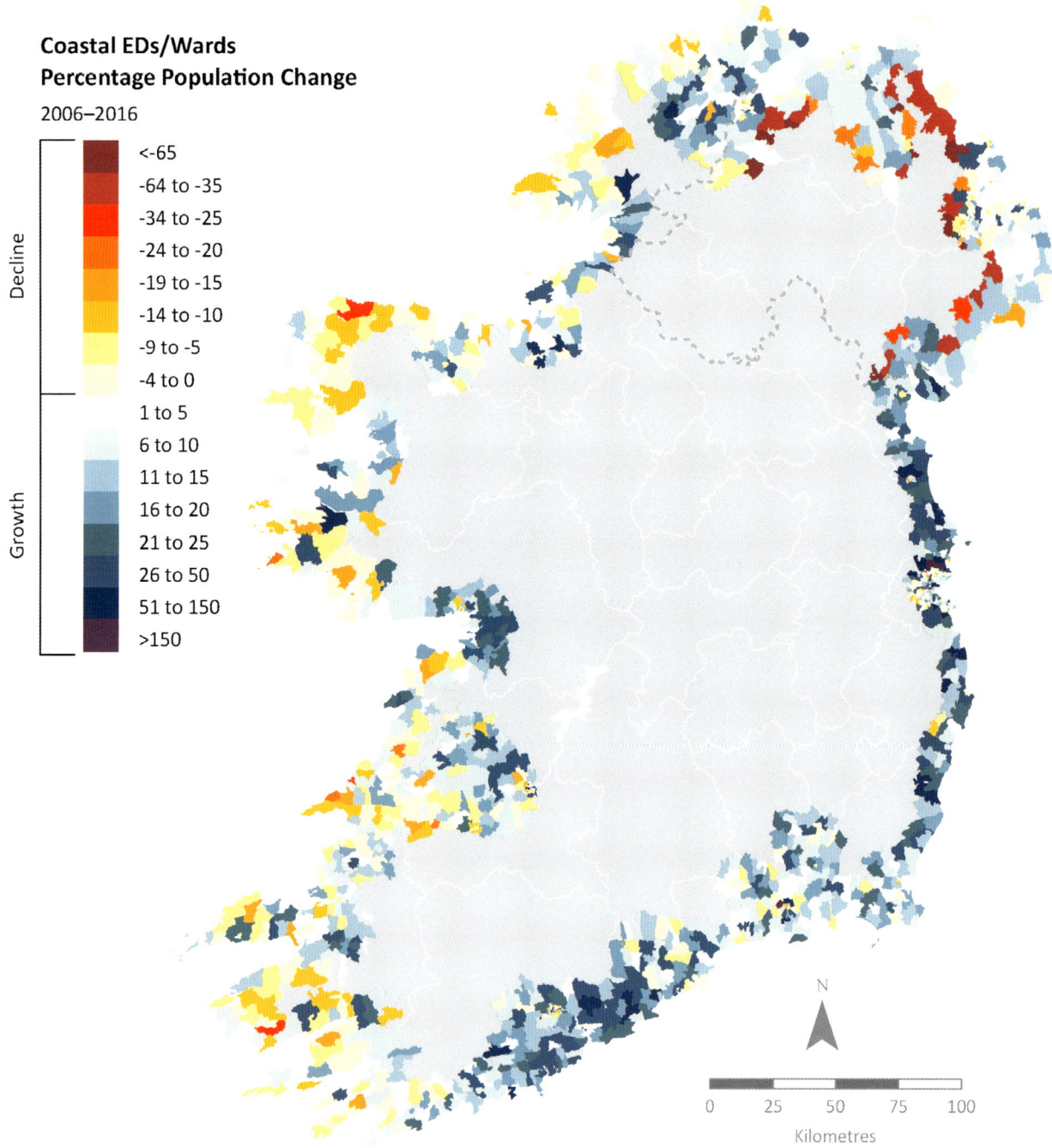

Coastal EDs/Wards
Percentage Population Change

2006–2016

Decline
	<-65
	-64 to -35
	-34 to -25
	-24 to -20
	-19 to -15
	-14 to -10
	-9 to -5
	-4 to 0

Growth
	1 to 5
	6 to 10
	11 to 15
	16 to 20
	21 to 25
	26 to 50
	51 to 150
	>150

N

0   25   50   75   100
Kilometres

Fig. 25.38 PERCENTAGE POPULATION CHANGES IN COASTAL ELECTORAL DIVISIONS AND WARDS (2006–2016). Significant growth in population occurred in almost all areas along the east and south coasts. Here, the many historic towns and villages, which acted as market centres for a relatively productive agriculture, have become cores around which modern housing developments have been built. These accommodate large and increasing numbers of commuters who travel to work in Cork, Waterford and, above all, Dublin. North of Dublin, a rapidly growing and self-sustaining coastal strip of urbanisation is apparent. It is anchored by expanding dormitory towns such as Swords, Skerries and Balbriggan, but also includes the historic port towns of Dundalk and Drogheda. Similar developments characterise the coastline south of Dublin to Wexford, and along the coast of County Cork and Cork Harbour. In contrast, most electoral districts along the western seaboard display population losses due to the lack of local employment opportunities and relatively few urban centres. Apart from growth in and around the urban gateways of Limerick, Galway, Sligo and Letterkenny, more isolated places of development focus on tourist resorts, such as Dingle, Westport and Bundoran. [Data source: CSO and NISRA]

in population over this fifty-year period.

Since the launch of the All-Island Research Observatory (AIRO) and Central Statistics Office (CSO) Census Mapping tools in 2011, population data have become available at a more granular spatial level across Ireland. This allows for more detailed analyses of the Irish coastal zone. In the Republic, data are publicly available online at the Electoral Division (ED) level from 2006 and at the Small Area (SA) level from 2011.[29] In Northern Ireland, equivalent datasets are available from 2001 (Census Area Statistics (CAS) wards) and after the 2011 census (Small Areas). Figure 25.38 shows population change from 2006 to 2016 for EDs/CAS wards located within 10km of the Irish coastline. This can be compared with the 2016 patterns for Small Area population density illustrated in Figure 25.1. The decade from 2006 to 2016 represents an important transitional period for the island's economy and society. It commences prior to the 2007–2009 global recession, incorporates the severe implications for Ireland's economy due to the global recession and fiscal crisis in the Eurozone, and ends at a time when it emerges successfully from recession.

In the Republic, EDs along the east coast from its border with Northern Ireland and extending to the peninsulas of West Cork experienced an almost universal growth in total populations from 2006 to 2016. Some exceptions occur, most notably in the cities of Dublin, Cork and Waterford, where changing patterns of land-use encouraged a dispersal of population from some of their long-established, built-up areas. In general, patterns of strongest growth reflect the continued expansion in the dormitory roles of many coastal towns and villages accessible to cities. These places accommodate the need for housing from a growing volume of increasingly mobile commuters whose employment opportunities remain within the cities.

Ireland's west coast presents a markedly different pattern. Large areas are dominated by electoral districts experiencing significant declines in population. These include the peninsulas of counties Cork, Kerry and Donegal together with much of the coastline of Clare and Mayo. Here, the difficult physical environments, peripherality, paucity of urban centres and limited employment prospects provide little incentive for population to grow. The limited areas of growth along the western seaboard are focused primarily around its relatively few cities and large towns, but also include small tourist centres dispersed along this coastline.

The wards around the coast of Northern Ireland illustrate growth, but generally at lower rates than the Republic's east coast. Furthermore, growth appears to be restricted to areas immediately adjacent to the coastline, with losses occurring inland. This is due, in part, to stronger planning controls relating to urban spread and the encouragement of more compact urban development. Population decline characterises some areas within Belfast, although dispersal from the city has benefited communities along the Ards Peninsula and south-west to the Republic's border.

Urbanisation of Ireland's coastal zone has increased dramatically from the 1960s. The 2016 census for the Republic, for example, indicated that 1.9 million people (40 per cent of total population) resided within 5km of the coast. This remarkable tendency for urban development to gravitate to the coast was not, however, evenly distributed or balanced in respect to its consequences for patterns and processes of development on the island. Two key characteristics emerged that defined urbanisation along Ireland's coast.

First is spatial bias along the the island's coastal zone. The focus of urbanisation remains orientated to the east coast as opposed to the western seaboard. This continues to limit opportunities for development for most locations in the west. In addition, and contributing to the east coast bias, is the strong preference for population and investment to gravitate to large coastal cities. Approximately, one-in-two of the island's urban population now reside in its seven coastal cities. Above all, Dublin and Belfast have confirmed their dominant roles as primate capitals, with almost 30 per cent of the population in both states residing in their extended built-up areas. This dominance is unlikely to change, given the need for Ireland to possess cities of sufficient scale to compete successfully for international investment. The government of the Republic recognises this reality. Although expressing the desirability of balanced development in its projections to 2040, at least 25 per cent of a one-million growth in total population will occur in Dublin. This underlines the importance of its capital city.

Second, rapid and large-scale urbanisation creates a 'footprint' that is unsustainable. Within the Republic, for example, what appears to be a laissez-faire approach to development has allowed employment opportunities to become centralised increasingly in the larger coastal cities. In contrast, it has encouraged a growing population and their housing needs to disperse into an expanding hinterland around the cities. This has given rise to urban sprawl involving the spread of uncontrolled and low-density housing over extensive areas of the countryside. These daily urban systems are typified by long-distance commuting, congested transport routes and both personal and environmental stress.

Projections in the Republic for 2040 suggest a population growth of 20 per cent (c.1 million people), 660,000 additional jobs and at least half a million new homes.[30] In Northern Ireland, population growth is projected at only 8 per cent (140,000 people). While both governments aspire to achieve a more balanced regional development and reduce the urban-rural divide, free market forces will be difficult to overcome. In contrast to political objectives, these will encourage further concentration in Ireland's larger coastal cities, which possess the scale and infrastructures to be competitive in the globalised economy.

The track record of government to follow through with the hard political decisions needed to achieve more balanced development on the island of Ireland has not been impressive. Failure in this case, however, will mean the vital resources provided by coastal communities and the coastal environment will come under even greater stress from unrestrained urbanisation.

Coastal Fisheries and Aquaculture

Mike Fitzpatrick, John Dennis, Donal Maguire, Emmet Jackson and Roy Griffin

Kilkeel, County Down [Source: Darius Bartlett]

Ireland has abundant fisheries resources arising from its extensive continental shelf and favourable hydrography. Its Exclusive Economic Zone (EEZ), one of the largest in Europe in comparison to its landmass, is characterised by productive waters mainly less than 200m in depth, and fishing grounds with strong species diversity. De Courcy Ireland's history of Irish fisheries states that 'by the fifteenth century, the Irish Sea fisheries were famous throughout western Europe and greedily coveted by foreigners'.[1] Despite this relative abundance, Ireland has not always prioritised nor maximised the value of its fisheries and we start this chapter by briefly setting out the historical context in which this paradoxical situation has evolved.

Our primary focus is on the contemporary commercial situation and to illustrate this, figures are provided showing overall quantities and values for both wild-caught fisheries and aquaculture, numbers employed in various segments of the seafood industry and vessel numbers. Maps of the main fishing grounds and principal landing ports are given, and the most important commercial fish species along with the fishing fleets that target them are described. Brief descriptions of the Irish aquaculture and the seaweed industries are also provided. The chapter addresses a number of important issues, including interactions between fisheries and other marine organisms, such as seals and jellyfish, and includes a detailed case study of the Celtic Sea Herring fishery. The chapter concludes with a brief analysis of future challenges, including Brexit.

HISTORICAL CONTEXT

Ireland, despite being an island with abundant fisheries resources, cannot be characterised as a major fishing state. The annual per capita consumption of fish of 22kg is very low when compared with 92kg for Iceland, 56kg for Portugal, 52kg for Norway and 45kg for Spain.[2] Seafood accounted for 0.4 per cent of Gross

Fig. 26.1 HISTORICAL ARDGLASS HARBOUR, SOMETIME BETWEEN 1865 AND 1914. Ardglass Harbour, on the Lecale Peninsula near Downpatrick in County Down, was a significant fishing port during its heyday in the nineteenth century. Up to four or five hundred sail-powered fishing boats fished for herring during the summer months, and for whiting and cod during the winter. The sheltered, deep-sea inlet, accessible at all stages of the tide, was popular for fishing boats from as far away as the east coast of Scotland. [Source: National Library of Ireland, L–RO411290]

Fig. 26.2 TRADITIONAL CURING OF FISH, CAHERSIVEEN, TAKEN SOMETIME BETWEEN 1865 AND 1914. The curing of herring was a common practice in Cahersiveen, County Kerry. The cured herring from boats fishing in Ballinskelligs Bay and Dingle Bay were despatched by train, which serviced Cahersiveen up until the late 1960s, to markets further afield. Curing was an important method of preserving fish at the time. Elsewhere, especially along the west coast, a lack of infrastructure for preserving fish had implications for the price fetched on the market. The poor price fishermen received for their products reduced their ability to invest in the industry. This had long-term implications for the relatively poor development of the sector in Ireland. [Source: National Library of Ireland, L–ROY–04566]

Domestic Product (GDP) in the Republic of Ireland in 2016, while in Iceland, it was 6.5 per cent, and in the same year, Norway's seafood exports were eighteen times greater than those from the Republic of Ireland. Some insights into the roots of the paradox of the Irish relationship with its fisheries resources are contained in MacLaughlin's *Troubled Waters: A social and cultural history of Ireland's sea fisheries*, which analyses the history of Ireland's sea fisheries up to the end of the nineteenth century.[3]

MacLaughlin describes evidence of reduced investment, incentives and export tariffs for Irish vessels, in comparison to English and Scottish fleets, whose vessels outnumbered their Irish counterparts in Irish ports in the nineteenth century.[4] At that time the east coast ports of Howth and Ardglass were the major fishing ports in the country, due to both having favourable access and transport links to the main markets in the UK (Figure 26.1). A lack of investment in infrastructure and curing facilities on the west coast meant that there were difficulties both with landing fish and with getting it to market before it spoiled (Figure 26.2). This resulted in fish regularly being used as pig feed or fertiliser, the poor price fishermen received for such products further reduced their ability to invest and compete with foreign fleets. This led to the fishing industry being described as 'backward and neglected' just prior to the Famine.[5]

One of MacLaughlin's key points is that the leadership of the emerging Irish independence movement also viewed fisheries as a backward element of Irish society. According to them, fishing lacked the relative success of farming and industry, thereby sowing the seeds of Irish disregard for the importance of the fisheries resource. MacLaughlin further attributes the marginalising of fishing communities to a Darwinian strand visible in the social progress theories that were in vogue at the time, which framed subsistence-level fishers, in common with Travellers and the

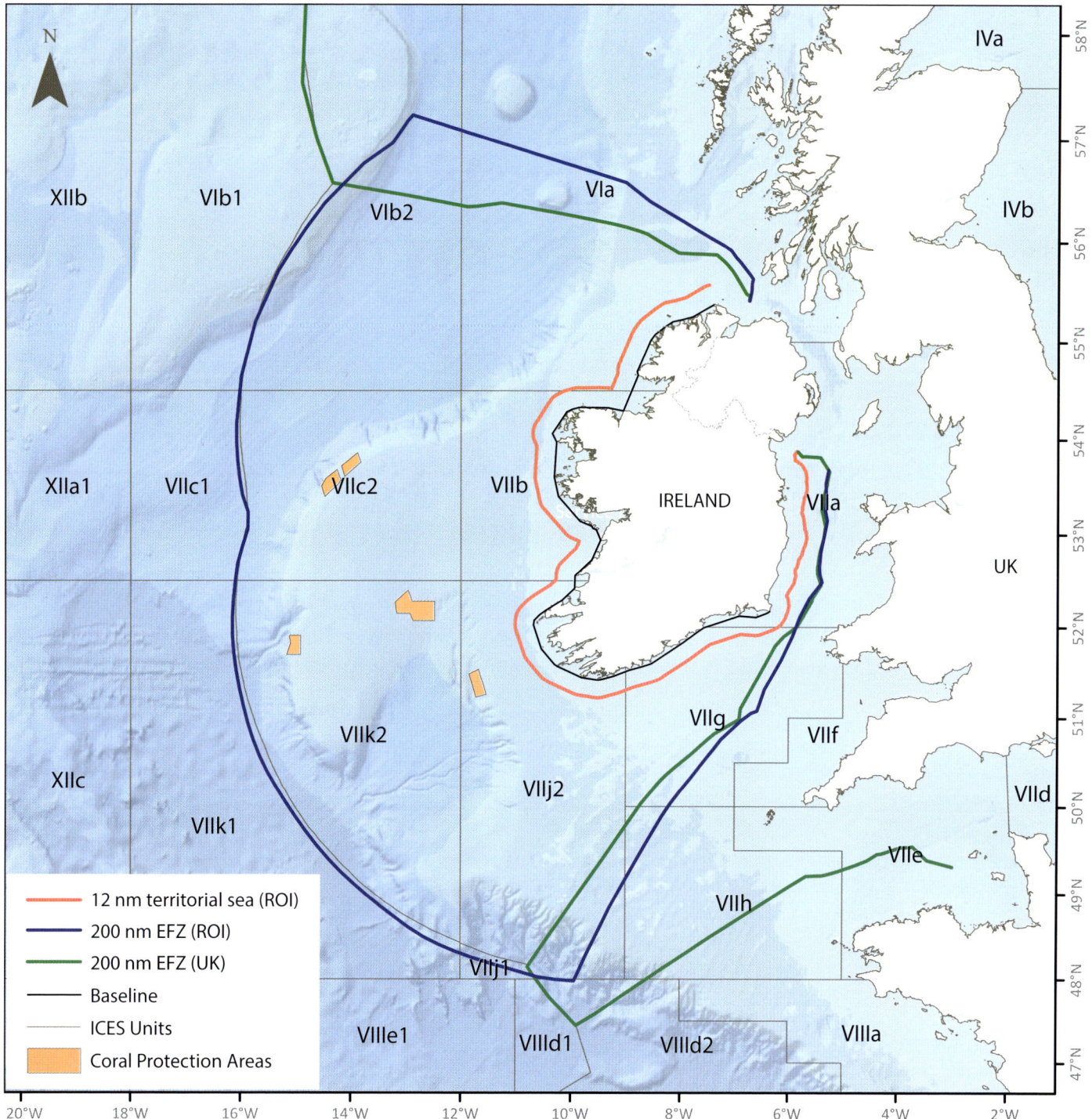

Fig. 26.3 EXCLUSIVE FISHERY ZONE LIMITS. These are based on the International Council for Exploration of the Sea (ICES) statistical area boundaries. The Exclusive Fishery Limits of the state comprise all seas that lie inside the outer limit of the Exclusive Economic Zone (EEZ) (200 nautical miles [nm]), so they include the 12nm territorial sea and the EEZ. The Exclusive Fishery Zone (EFZ) limits are specified in the Maritime Jurisdiction (Exclusive Fishery Limits) Order 1976 and extend to a distance of 200nm from the baselines, where there are no opposing or adjacent states. A baseline is the line from which the seaward limits of a state's territorial sea and certain other maritime zones of jurisdiction are measured. Where these limits cannot be applied because of a similar limit claimed by the UK and France, the Exclusive Fishery Limits is an equitable equidistant line as set out in the Schedule to the 1976 Order. None of these limits apply in Lough Foyle or in Carlingford Lough, where maritime boundaries are subject to a bilateral agreement between Ireland and the UK. These boundaries remain highly contested by local stakeholders. Fisheries management is usually conducted according to ICES Units. These units are set under the International Council for the Exploration of the Seas and ICES Units provide a framework for data collected by fisheries scientists, as well as a framework for fisheries closures or stock regulations. The Coral Protection Areas also provide for fisheries management by delineating cold water coral habitats, vulnerable to the practice of deep-sea trawling. [Data sources: Baseline (Department of Communications, Climate Action and Environment, 2016); 200nm EFZ, Republic of Ireland and United Kingdom (Irish Naval Service, 2004); ICES Units (Marine Institute, 2016)]

landless poor, as less socially evolved, or in the first stage of social development. The wildness of these communities was not in accordance either with the puritanical Catholic element of the independence movement. Accordingly, in contrast with farming,

no leaders of the Irish independence movement emerged from Irish fishing communities.

The marginalising of Irish fisheries continued into the twentieth century. Arthur Griffith stated in 1911 that 'the number

Fig. 26.4 NEWSPAPER HEADLINE (1977) CONCERNING IRELAND'S DEMANDS FOR MORE FAVOURABLE FISHING ZONATIONS FROM BRUSSELS. The Common Fisheries Policy (CFP) provides for shared access to the seas and resources within and between EU Member States. In the early stage of negotiations with the EEC, the government, in response to opposition from fishing communities, in particular on the equal access condition, sought a 50-mile Exclusive Economic Zone (EEZ) for Ireland. However, this was rejected in 1976 by the European Court of Justice. The issue of shared access to Ireland's fishing resources has remained a contentious point with the members of the Irish fishing community to this day. The grievance relates to the discrepancy between the volume of fish that Ireland contributes to the CFP, through its large and productive 200nm exclusive EEZ, and the share of fish stocks it has received through quotas allocated in the CFP. [Source: *Irish Examiner*]

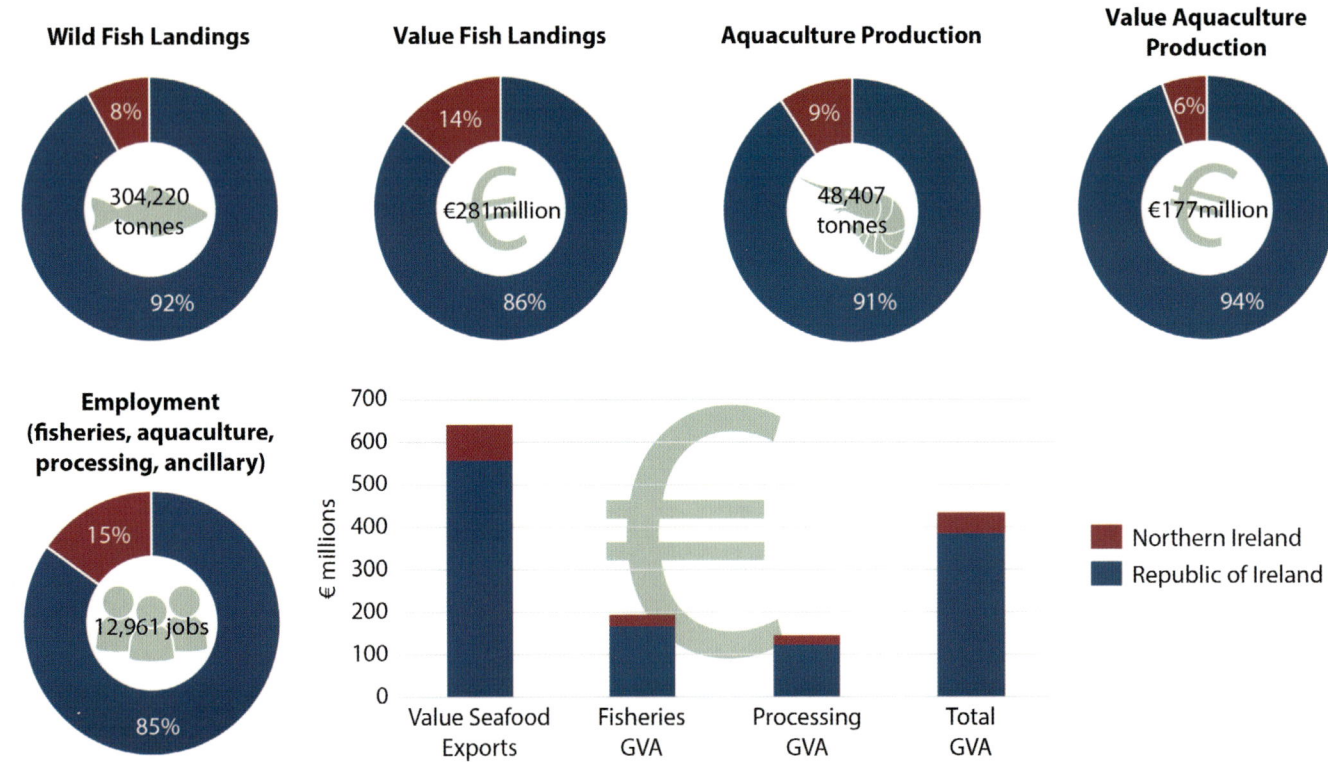

Fig. 26.5 KEY FISHERIES STATISTICS (2016). Key indicators, such as quantities of fish landed, levels of employment and value of catch and processing, give an insight into the fishing sectors in the Republic of Ireland and Northern Ireland. Overall, the figures for Northern Ireland make up an average of 10 per cent of the total figure for the island, for both fisheries and aquaculture. Landings data show the discrepancy between the levels of fish produced for market from capture fisheries versus aquaculture, around 304,000 tonnes versus 48,000 tonnes respectively. While aquaculture production appears to have a relatively high value compared to that of fish landings, in fact the latter is a measure of the return received by the producer, whereas the former value is a retail value. Overall figures for employment in 2016 show that fishing is a source of employment for almost 13,000 people on the island of Ireland. This is particularly significant considering the importance of fishing and ancillary activities as a source of employment in peripheral coastal communities. [Data source: The Business of Seafood, 2016; UK Sea Fisheries Annual Statistics Report, 2016]

Fig. 26.6 THE FISHING INDUSTRY REQUIRES A DIVERSITY OF LINKS. Illustration of the different links in the supply chain to provide for activities both off-shore and onshore, from boat builders and netmakers, to logistics and distributors. a. Brailer net full of herring trailing from *Aine* (February 2010) on the west coast of Norway. [Source: John Cunningham]; b. Mackerel-net making at Killybegs [Source: Val Cummins]; c. Inshore sprat being unloaded in Killybegs (2016) [Source: John Cunningham]; d. The *Sunbeam* as it accepts fish being passed over from the loaded-down *Aine*. Fish was passed over using a pump feed, January 2011. [Source: John Cunningham]

of public men in Ireland who realise that the sea fisheries of this country could be an industry second only to agriculture' might be 'counted on one hand' but died in 1922, having been president for less than a year, and never got to implement a vision that he had for a vibrant Irish fisheries industry.[6] From the foundation of the state, fisheries was seen as a minor responsibility and a stepping stone for junior politicians. The practice of the fisheries brief shuttling between different departments has arguably continued up to the present day. The early years of the Irish state also saw the cessation of fisheries and boat-building training, as well as the abandonment of funding for scientific research into fisheries (see Chapter 23: Maritime Traditions and Institutions). Government fisheries staff were employed in various departments and did not have a dedicated laboratory until 1950, while the numbers of full- and part-time fishers fell from 21,000 in 1910 to 7,351 in 1938. De Courcy Ireland characterised the first two decades of independence as ones of 'shameful neglect' with regard to fishing and claimed that the first generation of post-independence leaders 'almost unanimously turned their backs on the national maritime heritage and made it impossible for the ordinary citizen to remember we

even had one'.

Bord Iascaigh Mhara (BIM), the Irish Sea Fisheries Board, was established in 1952 with the remit of developing Irish fisheries and rectifying some of the past problems. A number of strategic and policy reviews and reports were commissioned over the next twenty years (Whitaker, 1952; Bjuke, 1959; MacArthur, 1960) and a more progressive approach seemed to be possible. During the 1960s, however, the new challenge of entry to the then European Economic Community (EEC) emerged. The principal issue was access to waters. The Common Fisheries Policy (CFP) was based on equal access to all waters by member states. Fishing communities around Ireland expressed their opposition to entry to the EEC and, in particular, to the equal access condition and held protest marches in Dublin. In order to address these concerns, the government minister in charge of BIM and Fisheries at the time, Paddy Donegan, sought a 50-mile EEZ for Ireland. However, in the end, the 50-mile limit sought by Ireland was rejected in 1976 by the European Court of Justice (see Figures 26.3 and 26.4).[7] During this period the familiar issue of a political disregard or ignorance of the value of Irish fisheries was evident in, for example, the sending

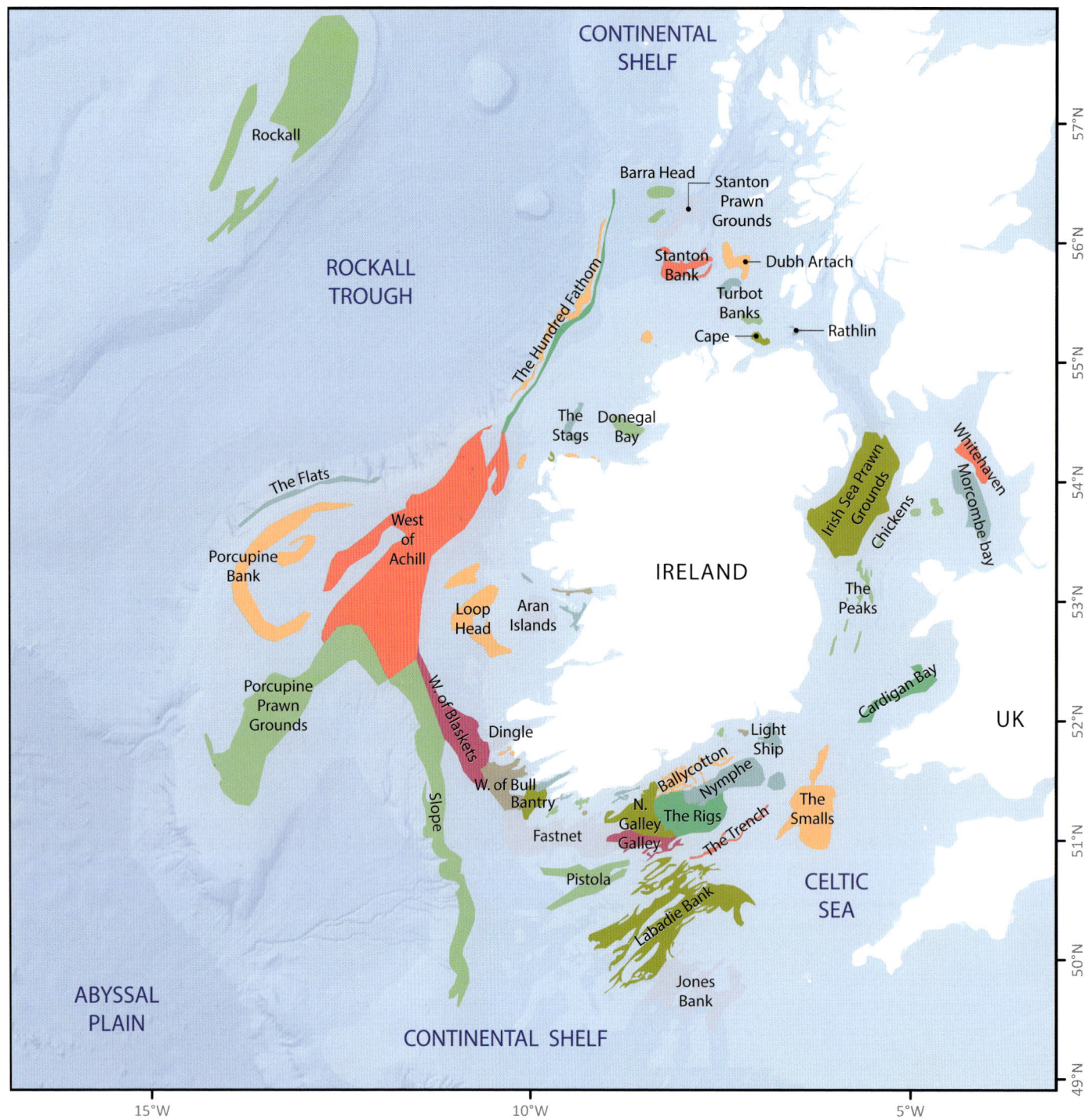

Fig. 26.7 THE MAIN IRISH FISHING GROUNDS TARGETED BY IRISH FLEETS, (the distribution includes neighbouring fishing areas such as Morecombe Bay, White-horn and Cardigan Bay, also visited by Irish vessels). Hotspots of fishing effort occur in the main prawn fishing grounds, namely the Irish Sea and Porcupine Prawn Grounds, the Smalls in the Celtic Sea, and the Aran Islands Grounds. The data on fishing grounds are derived from the Vessel Monitoring System (VMS), a satellite-based system used by the authorities for tracking fishing vessels. All vessels in excess of 12m in length are obliged by law to be fitted with a VMS. The map does not show the activity related to the smaller inshore fleet. [Data source: Marine Institute]

home of the chief executive of BIM, Brendan O'Kelly, in 1971, during accession negotiations, because Minister Patrick Hillery felt that he was too vociferous in his defence of Irish fishing interests.[8]

During the subsequent round of CFP negotiations in 1982, the then taoiseach, Garret Fitzgerald, said that his officials were unprepared and did not have the necessary information regarding a rational share of the quotas.[9] Despite this, these quotas were introduced and ratified according to the principle of 'relative stability' in the 1983 CFP. Dissatisfaction with the Irish share of fishing quotas has persisted and most recently was spelled out in

Ireland's official response to the 2010 CFP reform process: 'Irish fishermen and the wider Irish public remain deeply aggrieved at the discrepancy between the volume of fish which Ireland contributed to the CFP (through its large and productive 200 mile exclusive economic zone) [see Figure 26.3] and the share of fish stocks it has received through the CFP.'[10]

COASTAL FISHERIES AND AQUACULTURE TODAY

The volume and value of the seafood industry in 2016, for both the Republic and the North of Ireland, are shown in Figure 26.5.

These figures include landings from wild fisheries, aquaculture production, fish processing and ancillary industries such as boat building, net making, fisheries technology and others (Figure 26.6). Overall, Northern Ireland accounts for an average of 10 per cent of the total fishing output for the island, for both fisheries and aquaculture. The only Northern Irish fisheries comparable in volume to those in the Republic are the *Nephrops* (also variously known as Norway lobster, Dublin Bay prawn, langoustine or scampi) and scallop fisheries. This imbalance is reflected throughout the present chapter, resulting in a somewhat greater emphasis on the industry in the Republic of Ireland.

The data shown in Figure 26.5 might appear to suggest that the seafood industry is not significant on the island of Ireland as a whole, in either employment or economic terms. This would be misleading. The real significance is felt at a far more local level, where in many cases alternative industries and employment opportunities are few or absent altogether. It should also be noted that these figures are indicative only, and should be interpreted with caution. The value of the fisheries landings is a first-sale value (that is, the return received by the producer), whereas the aquaculture value is a retail value, and hence a direct comparison between the figures might lead to a misleading conclusion that the value of aquaculture in the Republic was almost 70 per cent that of wild fisheries.

In some cases it is difficult to find comparable figures for both the Republic and Northern Ireland due to differing ways in which the data are collated and reported in each jurisdiction. An example of this is the GDP figure of €1.1 billion for the Republic in 2016. GDP figures for the Northern Ireland seafood industry are not reported separately from other regions of the United Kingdom and so a comparable figure could not be provided. In order to provide a comparable figure for overall economic value, Gross Value Added (GVA) figures are presented.

Figure 26.7 shows the main fishing grounds in Irish waters, based on data from Vessel Monitoring Systems (VMS) on board the fishing vessels (see Chapter 8: Visualising, Mapping and Monitoring Coasts). These data indicate that the greatest intensity of fishing effort mostly occurs in the main prawn fishing grounds, namely the Irish Sea and Porcupine Prawn grounds, the Smalls in the Celtic Sea and the Aran Islands grounds.

Fisheries management goes beyond monitoring vessels and catch. Because fisheries have a direct impact on the marine ecosystem, which is also impacted by human activities, an ecosystem-based approach to fisheries management ensures that the complex relationships within dynamic social-ecological systems are considered holistically. This includes consideration of factors such as the impact of discarding on seabird foraging, the impacts of predation by seals on fish catches, and the need to deal with overfishing, as in the case of the Celtic Sea Herring fishery.

INTERACTIONS BETWEEN DISCARDS AND GANNETS

Mark Jessopp

Whilst seabird populations are in decline globally, gannets (*Morus bassanus*) are one of the few seabirds shown to be increasing in numbers. One possible reason for this is their ability to utilise fishery discards (Figure 26.8). Many fishing vessels will discard the unwanted portion of fish catches that are of little or no commercial value, throwing these overboard. Because this provides a food source that can be obtained without the energetic cost of diving, scavenging on fisheries discards has become a common foraging strategy for many seabirds. Gannets, in particular, compete vigorously for them. One study in 2014 tracked gannets from four colonies around the coast of Ireland, and showed they were more likely to feed in close proximity to fishing vessels.[11] Regional differences in such behaviour were also noted, reflecting the local availability of discards. The research found that gannets respond to fishing vessels within 11km of their location, highlighting the visual acuity of these birds. This might also explain how large aggregations of these birds occur at sea. The sight of fishing vessels, or other gannets diving to feed, is likely to indicate food and an ability to detect these visual cues over large distances is advantageous.

Gannets are known to compete aggressively against other seabirds for access to fishery discards and this has possibly led to them becoming dependent on this food source. While gannets seem to have benefited from the availability of discards, reform of the European Common Fisheries Policy post-2015 aims to reduce significantly the amount of fishery discards at sea. How changes in discarding rates in Irish waters will affect gannets is unknown. In light of this, close monitoring of both the diet and foraging behaviour of this species would be prudent.

Fig. 26.8 GANNETS AND GULLS ATTRACTED TO FEED ON FISH DISCARDS, UNWANTED CATCH, IN THE WAKE OF A FISHING VESSEL. Fisheries biologists are endeavouring to understand the benefits of a European ban on discards to fish stocks and also to the wider ecosystem. [Source: Mick Mackey]

INTERACTIONS BETWEEN SEALS AND FISHERIES

Michelle Cronin

With global fish stocks in decline, consumption of fish by marine mammals is a contentious issue. It has been estimated globally that marine mammals consume about four times as much fish as humans catch.[12] While commercial species only make up a relatively small proportion of the diet of marine predators, they are still considered competitors for this resource.

In Irish waters seals are a major cause of concern to small-scale coastal fishermen, where they remove or damage catch in static fishing gear, such as gill-nets, tangle nets and trammel nets (Figure 26.9). A study in Ireland estimated losses at nets attributed to seals to be as high as 59 per cent for certain fisheries like monkfish.[13] They have even rendered fisheries in some areas unviable, with significant economic impacts on small coastal communities along the south and west coasts in particular. Additionally, seals are often accidentally drowned in fishing gear as by-catch. Ireland's two species, the grey seal (*Halichoerus grypus*) and the harbour seal (*Phoca vitulina*), are both protected under the European Union Habitats Directive. This requires member states to monitor and minimise threats to the conservation status of protected species. As a result, the priorities of the fishing industry and conservation bodies come into conflict in what is a complex problem.

Simply reducing the seal population by culling, arguably unacceptable on ethical and conservation grounds, is also unjustifiable scientifically. Extensive research suggests that reducing a predator population will not necessarily make more prey available to humans. The ocean is a dynamic ecosystem and when one predator is removed, or reduced, others (for example, predatory fish, dolphins, seabirds) will move in to take its place. Furthermore, it is likely that the damage caused at nets is done by specific individual seals who have become 'specialised' in this type of behaviour. Culling a fraction of the population at the breeding grounds may not remove these individuals and problems would persist.

Modification of fishing practices and gear to minimise interactions with seals is considered to be the optimal solution and follows international best practice. Research was conducted in 2015 in waters off the west coast of Ireland to develop mitigation measures that might have potential to alleviate the problem of seals at nets.[14] This involved the use of an acoustic device that emits a short loud burst of sound in the hearing range of seals which produces a startle response, but which is harmless to seals and other marine organisms. Work is now underway to trial this approach and preliminary results suggest that the device is an effective deterrent to keep seals away from nets, as well as from fish farms where they can also be a pest. Further research is needed to determine the long-term implications of deterring seals from competing with commercial fisheries.

Fig. 26.9 A GREY SEAL CONSUMING A FRESHLY CAUGHT FISH. The interactions of seals with commercial fisheries needs to be better understood in order to deal with the impact of seals damaging fish caught in static gear. This is highly controversial in inshore fisheries, where losses at nets attributed to seals have been recorded as high as 59 per cent for monkfish. Seals are a protected species in Ireland, which means that measures to deter seals from preying on captured fish need to be sensitive to the conservation of seal populations. [Source: Pedro García Izquierdo]

The nature and distribution of fisheries effort around the coast of Ireland is shown in Figures 26.10 to 26.13. The first of these, Figure 26.10, shows landings for the top twenty fishing ports in Ireland. The port with the greatest volume of landings was Killybegs in County Donegal, which is the main landing port for pelagic species. Pelagic fish form large shoals and tend to be caught in large quantities in mid-water. Important commercial pelagic species in Ireland include mackerel, herring, horse mackerel and blue whiting.

The port with the highest value of landings is Castletownbere (Figure 26.11). This is because landings here are made up mainly of demersal or whitefish species, which have a higher average value than pelagic species. Demersal fish are usually caught on or near the sea bed, and include species such as *Nephrops*, haddock, whiting, cod and hake.

In some cases a port may only be used for a short period of the year for fisheries landings. This is the case with Ringaskiddy in Cork Harbour which would not typically be considered as a fishing port, but which is used as a landing port for the Celtic Sea Herring fishery between September and December. A number of ports, particularly those in the south and south-west, feature a significant proportion of landings by non-Irish fleets. Landings by Spanish and French vessels targeting demersal species on fishing grounds off the south and west coasts make up the majority of these (Figure 26.12).

624

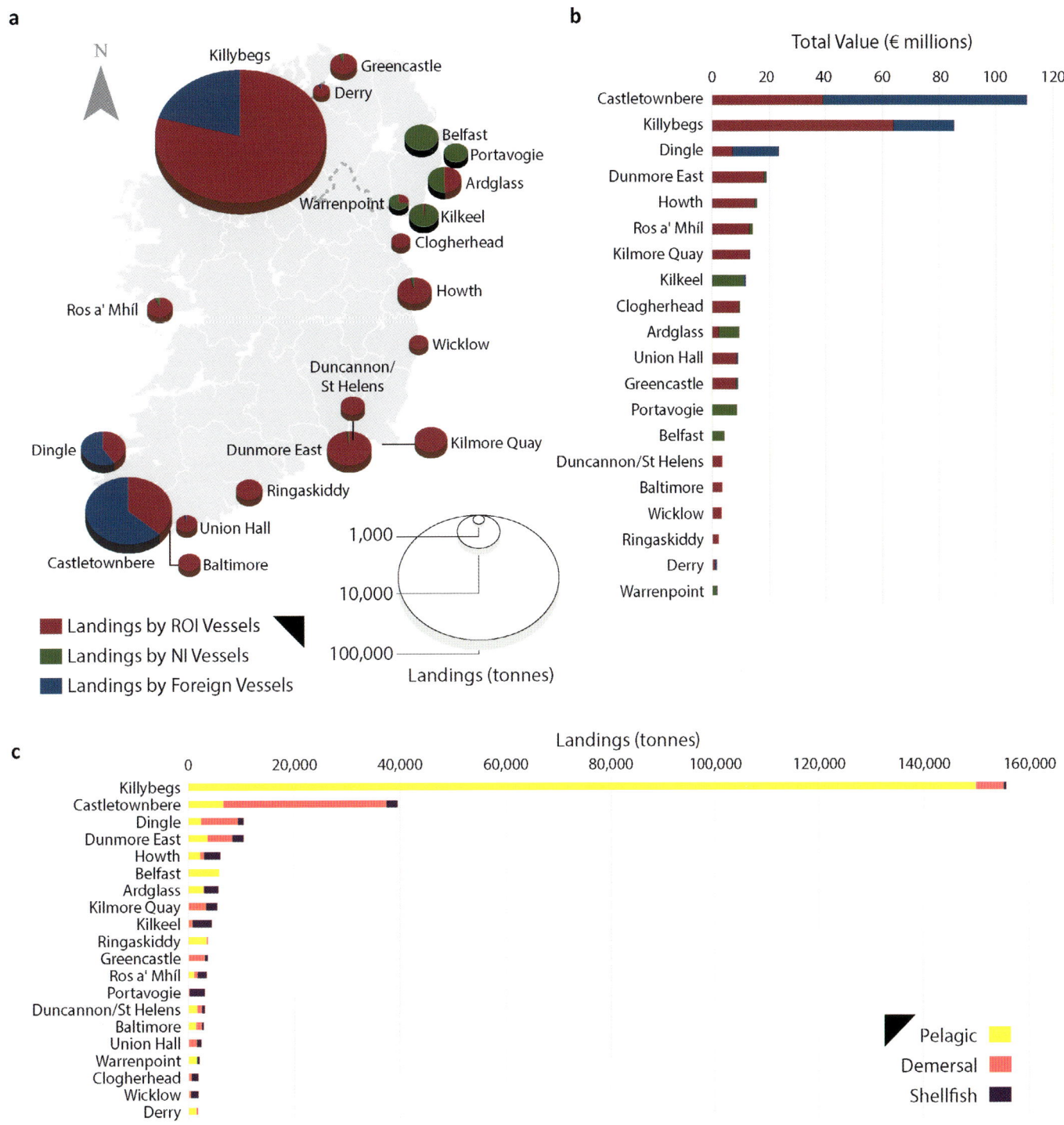

Fig. 26.10 TOP TWENTY FISHING PORTS IN IRELAND ACCORDING TO: a. LANDING SIZE AND IRISH VERSUS FOREIGN LANDINGS; b. LANDING VALUE FOR IRISH AND FOREIGN VESSELS; c. LANDINGS BY SPECIES GROUP. The ports of Killybegs, Castletownbere and Dingle present as the most significant ports on the west coast of Ireland, in terms of both landings by size and by value. These are also the principal locations for landings from foreign vessels. On the east coast, the primary ports are Dunmore East and Howth. The port of Killybegs handles the largest number of fish landings of any port on the island of Ireland, with approximately 150,000 tonnes of pelagic fish landed per year (including mackerel, herring and blue whiting). The contribution of pelagic fisheries to Killybegs is significantly greater than the 10,000 tonnes of demersal fish and shellfish combined that are landed on a yearly basis. In contrast, more demersal than pelagic fish are landed in Dingle and Castletownbere. Demersal fish (for example, haddock, hake, whiting) have a higher value than pelagic species, which makes Castletownbere the port with the highest value of landings. The different patterns of fishing around the coast are further evidenced by the predominance of shellfish over other fishing types in small ports on the east coast, including Kilkeel, Dunmore, Ardglass and Portavogie. Ringaskiddy in Cork Harbour is not usually thought of as a fishing port, but it serves as a base for landings for Celtic Sea Herring between September and December. [Data source: Sea Fisheries Protection Authority, 2016; DAERA, 2016]

REPUBLIC OF IRELAND FLEET STRUCTURE

There are four subsections that characterise the structure of the Irish fishing fleet: polyvalent, Refrigerated Sea Water pelagic, beam trawl and specific. The largest is the polyvalent sector with 1,726, out of a total of 2,044, registered vessels. Polyvalent vessels use a range of fishing gears and target a wide range of fisheries. A subsection within the polyvalent fleet comprises inshore vessels, which are only licensed to fish with pots (Figure 26.13). There are also larger vessels in this sector. These are increasingly focused on particular types of fisheries, rather than adapting gear to switch between fisheries on a seasonal basis, which would have previously been the case for most of the Irish fleet. Vessels targeting *Nephrops*

Fig. 26.11 FISHING BOATS IN CASTLETOWNBERE HARBOUR, WEST CORK. Castletownbere handles about 40,000 tonnes of fish landings per year. The majority of fish landed are high-value demersal species. The total annual value of landings is approximately €115 million per year to the local economy. The Irish (ROI and NI) fleet accounts for approximately €40 million of this, with the balance attributable to foreign landings, which indicates the importance of Castletownbere for both domestic and foreign vessels. [Source: Mike Fitzpatrick]

have undergone significant technical advancement over the past twenty to thirty years. In the 1990s Nephrops vessels started to switch from towing a single net behind the vessel to twin-rigging, which involved towing two nets. In the past ten years an increasing

Fig. 26.13 INSHORE FISHING WITH SHRIMP POTS, CASTLETOWNBERE, WITH THE PELAGIC TRAWLER, *Neptune*, ENTERING CASTLETOWNBERE IN THE BACK-GROUND (OCTOBER 2013). The structure of the Irish fishing fleet is characterised by the type of fishing activity and the size of the vessel. There is a total of 2,044 registered vessels in the Republic of Ireland fishing fleet, of which 86 per cent are inshore vessels of less than 12m in length. The boat in the foreground of the image is a small inshore vessel, licensed to fish for shrimp. This is an example of an extremely important source of income and diversity of employment in local coastal communities. Juxtaposed is the trawler entering the harbour, representing more substantial registered tonnage, which generates more significant landings, employment and economic value to Irish fishing as a whole. [Source: John Cunningham]

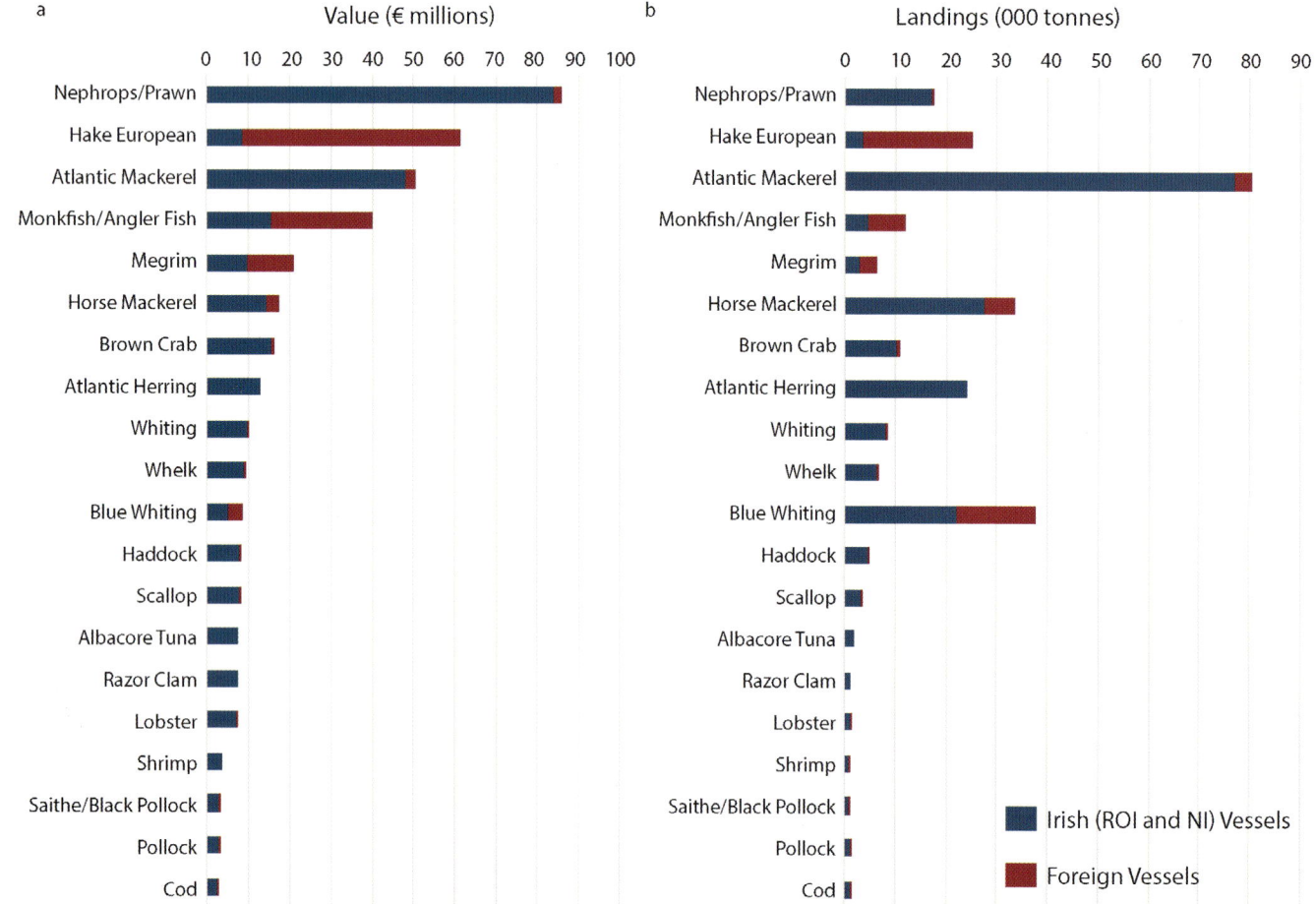

Fig. 26.12 TOP TWENTY SPECIES LANDED INTO IRISH PORTS, RANKED IN ORDER OF VALUE OF LANDINGS. a. Value breakdown by species; b. landings breakdown by species. [Data source: Sea Fisheries Protection Authority, 2016; DAERA, 2016]

Fig. 26.14 AERIAL VIEW OF KILLYBEGS HARBOUR, OCTOBER 2017. The use of Refrigerated Sea Water tanks by vessels targeting pelagic fish, as a means to preserve their catch in optimum condition, was first implemented in the 1970s on Irish vessels based in Killybegs. This led to the development of the RSW pelagic segment of the Irish fishing fleet. This segment has twenty-three vessels and contains the largest vessels in the Irish fleet, ranging up to 71m in length. The aerial view of Killybegs shows the quayside space and facilities necessary to accommodate vessels of this size. [Source: John Cunningham]

Fig. 26.15 (right) LANDINGS AND VESSEL NUMBERS BROKEN DOWN BY SIZE AND FISHING GEAR CATEGORIES (2016). The graphs show the relationship between the structural features of the Irish fishing fleet, and key variables of vessel size and gear type. Large vessels (over 12m) account for a relatively low number of vessels, but their catching capacity means they have a high impact in terms of landings by volume (large pelagic vessels) and by value (large demersal vessels). The structure of the fleet has four main determinants. Polyvalent vessels can target a range of fisheries as they deploy a range of different fishing gears. Refrigerated Sea Water pelagic vessels are the largest in the fleet, ranging up to 71m. Flatfish in the Irish and Celtic seas are targeted by beam trawl vessels. The specific segment, which fish for bivalve molluscs including scallops, make up the remainder of the Irish fishing fleet. [Data source: Scientific, Technical and Economic Committee for Fisheries, 2018]

number of vessels in this fleet have switched to quad-rigging, which enables them to tow four nets. Onboard freezing of the catch has also become increasingly common in this fleet as a means to preserve the catch and to add value.

Another major technical adaptation which has occurred in the Irish fleet over the past forty years is in the use of Refrigerated Sea Water (RSW) tanks by vessels targeting pelagic

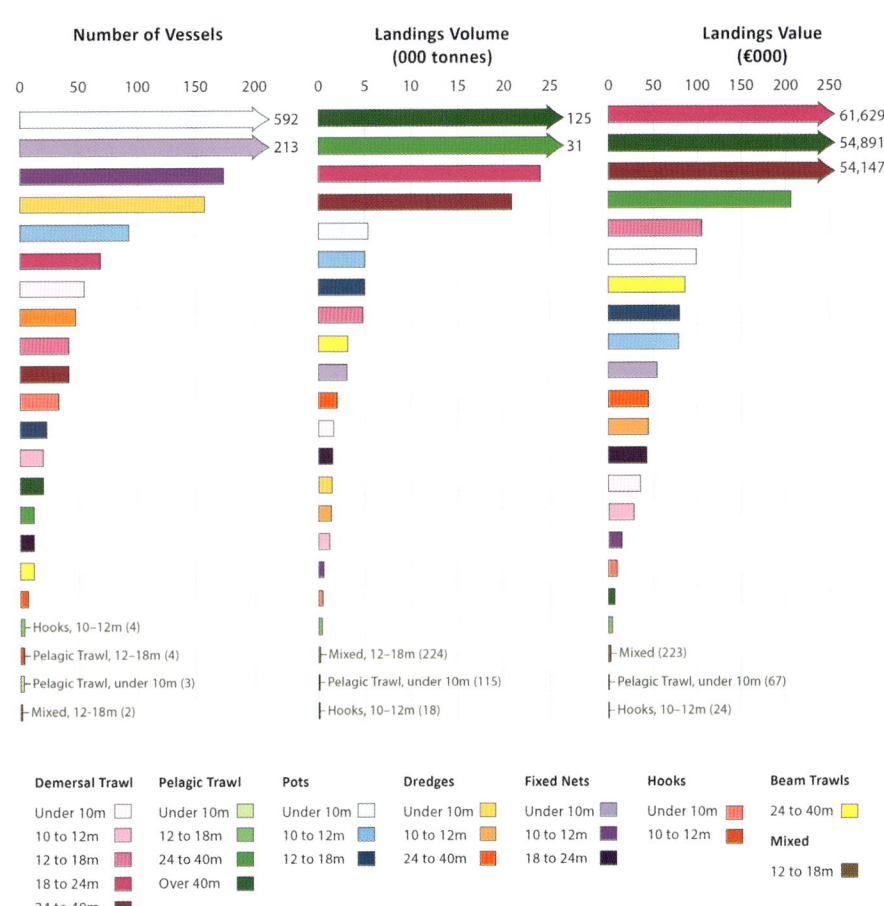

627

Fig. 26.16 LOADED FISHING VESSEL, *Neptune*, PULLING UP MACKEREL, WEST OF THE HEBRIDES (2013). [Source: John Cunningham]

fish, as a means to preserve their catch in optimum condition.

This technology was first used in the 1970s on Irish vessels based in Killybegs (Figure 26.14) and led to the creation of the RSW pelagic segment, which has twenty-three vessels and contains the largest vessels in the Irish fleet, ranging up to 71m in length. Some vessels in the polyvalent segment now also use RSW tanks and focus mainly on pelagic species.

The beam trawl segment (eight vessels), which fish mainly for flatfish in the Irish and Celtic Seas, and the Specific segment (140 vessels), which fish for bivalve molluscs including scallops, make up the remainder of the Irish fishing fleet.

Although the majority of vessels on the Republic's register are below 12m in length, the majority of fleet capacity (expressed as the registered tonnage of the vessel), landings, employment and economic value is generated by the large-scale fleet (above 12m) (Figures 26.15, 26.16).

CELTIC SEA HERRING FISHERY

Mike Fitzpatrick

The Celtic Sea Herring fishery is essentially a single-species pelagic fishery, predominantly targeted by Irish fishing vessels working in International Council for the Exploration of the Sea (ICES) areas VIIj, VIIg and the southern part of VIIa, see Figure 26.3 above (see also Figure 26.17).

It is conducted by a diverse fleet of vessels ranging from multi-purpose inshore vessels under 10m in length, up to modern pelagic vessels over 50m, equipped with refrigerated seawater tanks. It has traditionally been very important, both for the fleet and for the fish-processing sectors in the south of Ireland, although landings over the last thirty years have fluctuated widely and current landing levels are well below previous peaks. The fishery has in recent years been exploited almost entirely by Ireland, with small reported catches by other nations. The only other significant players involved are Dutch vessels and Dutch-owned vessels from France and Germany.

CURRENT MANAGEMENT INSTITUTIONS AND APPROACHES

The history of the fishery over the past fifty years has been one of alternating boom and bust cycles. This pattern is shown in Figure 26.18 which graphs the finshing landings and biomass trends since 1958.[15] In 2001 the ICES advice for the Celtic Sea Herring stock recommended a cut from the previous year's Total Allowable Catch (TAC) of 20,000 tonnes to a precautionary level of 6,000 tonnes for 2002. This was mainly based on a poor age profile for the stock, which showed an over-dependence on juvenile fish. That bleak outlook led to the establishment of the Celtic Sea Herring Management Advisory Committee (CSHMAC) in 2001 by concerned stakeholders in the fishery, including fishermen, processors, scientists,

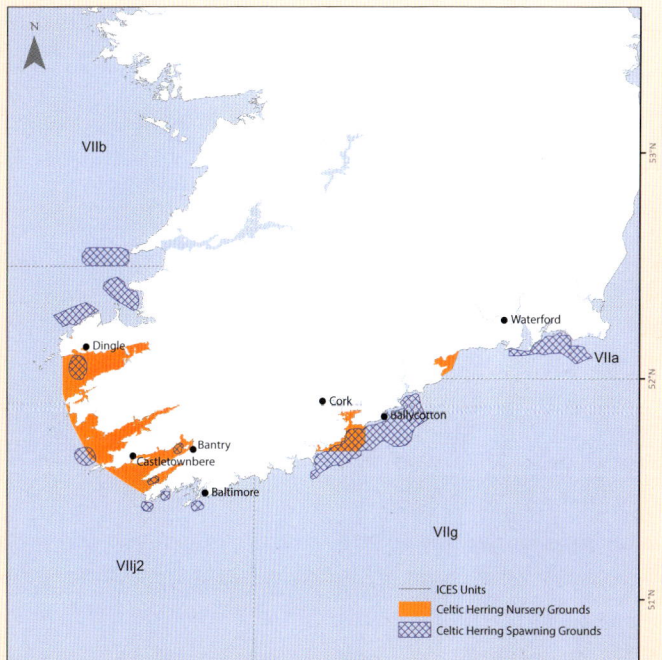

Fig. 26.18 (below) GRAPH OF a. LANDINGS AND b. BIOMASS TRENDS IN CELTIC SEA HERRING (1958–2017). The relationship between landings of herring and stock biomass levels is visible in the corresponding graphs, from 1958 to 2017. Overfishing of Celtic Sea Herring has been a cyclical problem for the last fifty years. Management interventions instigated in 2002 were brought in to address a critical imbalance in the age profile of the stock. Reductions in the Total Allowable Catches available to the sector had a negative impact on those dependent on the fishery in the short term, but the reductions were considered an essential measure to rebuild stocks to a sustainable level. Decisions on fisheries management, such as those applied to the Celtic Sea Herring fishery, depend on best available scientific data and information. Even with long-term records, the complex nature of fisheries science makes it challenging to understand the interplay between shifting natural and fishing dynamics, and to arrive at appropriate management measures. Most recent data for the Celtic Sea Herring (2018) indicate that the stock is under more pressure than perceived in recent years, despite almost twenty years of concerted effort by the Celtic Sea Herring Management Advisory Committee.

Fig. 26.17 (above) THE DISTRIBUTION OF CELTIC SEA HERRING SPAWNING AND NURSERY GROUNDS. Celtic Sea Herring are distributed around the south coast of Ireland, such that the fishery is targeted almost entirely by Irish fishers. The distribution and abundance of stocks are linked to these key spawning and nursery areas. The recruitment of herring has come under pressure from over-fishing in recent decades. Spawning and nursery areas need to be protected, as part of an ecosystems-based approach to fisheries management, to allow stocks to rejuvenate. Policy-makers need to consider the vulnerability of fish stocks in ICES Units, and in the context of other pressures, such as the need for marine aggregates, as marine spatial planning is progressed. ICES is an intergovernmental organisation aimed at advancing understanding of marine ecosystems, including fish stock distribution. [Data source: Marine Institute, 2017]

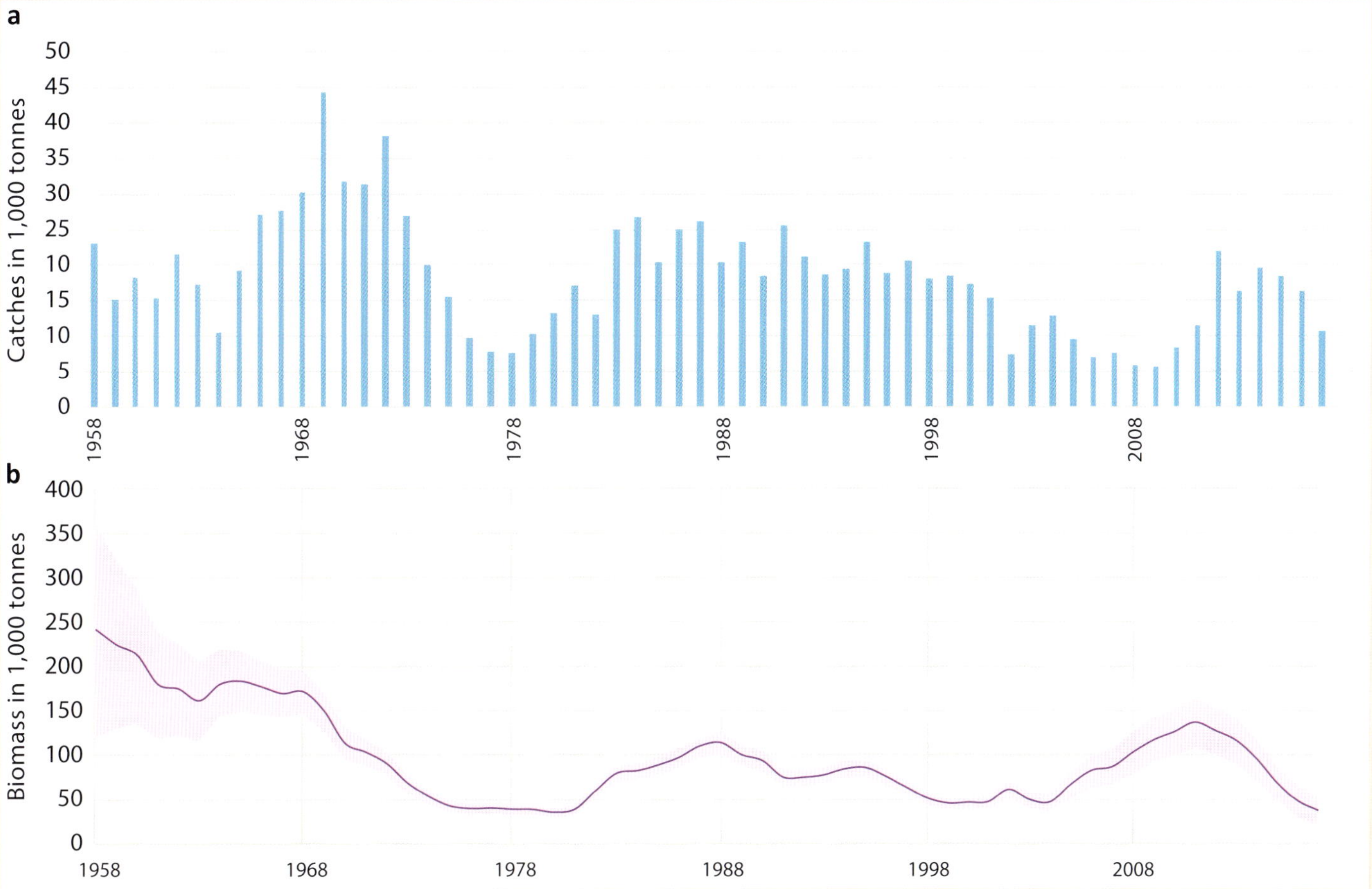

a

(Graph a: Catches in 1,000 tonnes, years 1958 to 2008+)

b

(Graph b: Biomass in 1,000 tonnes, years 1958 to 2008+)

Fig. 26.19 HAULING A BUMPER CATCH OF HERRING. [Source: John Cunningham]

NGOs and control authorities. The Committee was established with the goals of ensuring a sustainable level of catches, rebuilding the stock if necessary and improving the partnership between industry and scientists.

In 2005 the Committee was officially recognised as an advisory committee by the Irish fisheries minister and tasked with providing advice to the minister and managers from the fisheries department. Although officially only advisory, following ministerial recognition the Committee has found that more of its advice has been accepted and also partnership between industry and science has strengthened.

One of the most significant measures taken was the closure to larger vessels of a large coastal area off Dunmore East, known as the Dunmore Box, where herring spawning occurs and where fishing effort had previously been concentrated. This was aimed at reducing catches of small first time spawning herring. However, despite this initiative the TAC continued to decline, so in 2007 a rebuilding plan was developed by the CSHMAC in conjunction with scientists from the Marine Institute. The Rebuilding Plan set a very low fishing mortality level and strengthened the annual closure of the spawning area. The Rebuilding Plan also allowed for a small-scale or sentinel fishery to continue within the Dunmore Box. This sentinel fishery has the advantage of allowing smaller vessels, which do not have the option to move to another area to fish, to continue fishing. It also ensures that scientists have continued access to data on fish in the spawning area. The TAC in 2010 was increased by 70 per cent over the 2009 figure and in 2011 increased by a further 30 per cent (Figure 26.19). In 2011 the Rebuilding Plan achieved its aim of maintaining stock level above the precautionary biomass level for a third consecutive year and the parameters of a Long Term Management Plan (LTMP) were agreed and implemented. However, despite the existence of a LTMP, the efforts of the CSHMAC and the retention of the spawning area closure, the stock has continued to fluctuate and current scientific advice is for catch levels well below average.

In 2012 the Celtic Sea Herring fishery became certified by the Marine Stewardship Council (MSC) as a sustainably managed fishery. This certification is hugely beneficial nowadays in accessing the most valuable markets for fish. However, in February 2018, MSC certification was withdrawn due to new scientific advice that indicated a lower stock level than previously thought, and that fishing levels are unlikely to ensure that Maximum Sustainable Yield (MSY) can be achieved.

The most recent advice from the Marine Institute stresses that there has been an increase in marine anthropogenic activity in the

Celtic Sea area and, as herring are restricted to spawning on gravel beds, mining of marine aggregates (see Chapter 29: Coastal Mining: Quarrying and hydrocarbon exploration) or any other activity that may disrupt this spawning should be avoided.[16]

The history of exploitation of herring fisheries illustrates the difficulty of managing unpredictable fish stocks, even when good scientific data are available and a dedicated management committee is overseeing the industry. It remains to be seen whether yet another stock recovery can be achieved in this fishery and whether catch levels can be kept at sustainable levels into the future.

Case Study: Aquaculture in the Republic or Ireland

Herbie (John) Dennis, BIM

Ireland provides a harsh, demanding and unpredictable environment for the production of farmed fish and shellfish via aquaculture. Nevertheless, the marine environment is in general perceived to be clean and healthy, and this perception is an invaluable marketing tool for the Irish aquaculture sector. At the same time, problems of water quality do arise from time to time, relating to increased urbanisation and inadequate wastewater treatment in the coastal zone, while a range of further issues (including storm damage, red tides, floods, fluctuating water temperature and salinities, poor light conditions, the effects of parasites, and diseases old and new) can take their toll on production. Added to this are bureaucratic challenges in the form of red tape, legislative log-jams, rival stakeholder claims and public opposition to the industry in general.

Fig. 26.20 SALMON FARMING ON THE WEST COAST OF IRELAND. The salmon farming sector dominates national production of aquaculture in Ireland, with approximately 19,000 tonnes, worth €141 million, of farmed salmon produced in 2017. Salmon cages are located off the coasts of Cork, Kerry, Galway, Mayo and Donegal. Together, the fish farms in these locations account for about 210 full-time jobs. Irish salmon is grown to organic certification, making it a high-value product, suitable for export to Europe and the Far East. As global demands for high-quality protein are increasing, in tandem with a rapidly growing global population, opportunities exist to increase the production and market potential of farmed Irish fish. However, the expansion of the industry has been hampered by factors such as controversy over the impact of farmed salmon on wild salmon populations. Ireland underperforms in the production of farmed salmon, in comparison to market leaders in Scotland and Norway. [Source: Mowi]

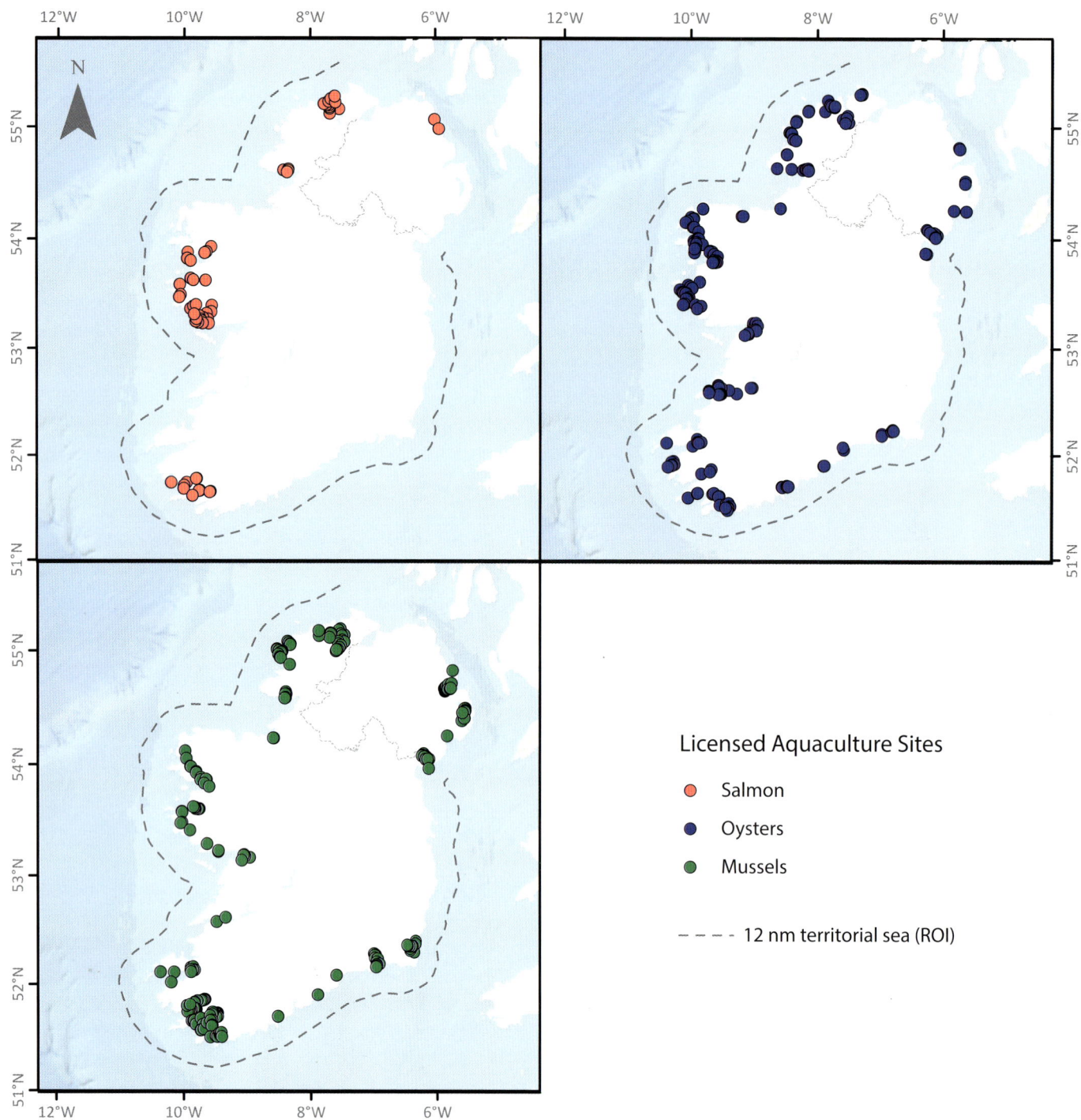

Fig. 26.21 THE LOCATION OF SHELLFISH SITES AND FISH FARM SITES AROUND THE COAST. More recent, higher-resolution data for the Republic of Ireland are available in *Ireland's Marine Atlas*. [Adapted from SEAI, 2018 and DAERA, 2017]

Compared to many other European countries, such as France and Holland, the practice of aquaculture in rural coastal areas has been poorly understood by the general public. More than 80 per cent of existing aquaculture sites lie within or adjacent to Natura sites (sites designated for special conservation measures under the EU Birds or Habitats Directives), which has made renewal or expansion of licensed sites a tortuously slow and complex process from the perspective of the fish or shellfish farmers. Those who make it successfully through a licensing process may endure further delays where an appeals process is instigated by objectors.

The public image and profile of aquaculture is undergoing a change for the better as the industry is finding ways to engage in a collaborative manner with coastal stakeholders on contentious issues, such as good husbandry and environmental stewardship. The marketing of the Wild Atlantic Way promotes wild and farmed Irish seafood to the public and to tourists alike. This helps to create an awareness and appreciation of aquaculture businesses in the public consciousness, and represents aquaculture as part of the local community and environment, rather than as an intrusion.

Irish aquaculture is made up principally of sea-reared salmon (Figure 26.20), bivalve molluscs, gigas oysters, blue mussels and a small number of inshore freshwater finfish units producing trout and salmon smolts. The majority of aquaculture is practised along the west coast, with approximately 250 family-owned micro-businesses managing *c*.280 production units (Figure 26.21). Shellfish farming is seasonal, but plays a vital role in sustaining communities in areas where alternative employment options are scarce. The large salmon farms tend to be owned by larger corporates. MOWI, previously known as Marine Harvest Ireland, is one such business. It owns the Irish Atlantic Salmon Company and is one of the world's largest suppliers of farmed Atlantic salmon.

TRENDS

Overall aquaculture output has varied from 30,000 to 50,000 tonnes per year over the past twenty years. Employment in the aquaculture sector as a whole, including fin and shellfish farming, has oscillated between 1,700 and just over 1,900. Male workers dominate, with female employment in the aquaculture sector at 7 per cent. The nationality of employment is shifting to reflect the growing proportion of non-nationals in the workforce.

OUTPUT

Irish aquaculture output in 2017 (which represents the most recent data available) was 47,000 tonnes overall, in line with the general trend, and was worth €208 million in retail value. Employment in the sector in 2017 stood at 1,913 people, with over 1,000 Full Time Equivalents (FTEs). Salmon production made up 41 per cent of volume and 68 per cent of value nationally. Output levels tend to follow cyclical trends in production. These patterns are primarily linked to productive and fallow output cycles from salmon farms, historically the most economically important aquaculture sub-sector. The most economically important shellfish aquaculture sector, farmed oysters, made up 21 per cent by volume and 21 per cent by value of national output. The oyster farming sector is also the biggest shellfish employer, with over 800 persons employed.

Fig. 26.22 MUSSEL HARVESTING, KILLARY HARBOUR. Images depict the view of a mussel farm in Killary Harbour, made up of rows of rope-grown mussel long-line systems. The mussel farming sector emerged around the coast of Ireland as a commercial activity in the 1980s. Since expansion of the sector in the 1990s, the mussel sector has faced challenges from increasing competition from low-cost suppliers, most notably Chile, as well as challenges associated with biotoxins in local water bodies, and licensing processes. Mechanical grading machines are used, either on a barge or on land, to sort mussels according to size. Mussels can take up to two years to achieve market size, at which point the mussels are typically harvested and distributed to one of the processing plants for packaging and distribution. Mussels are also sold locally and feature on seafood menus, especially throughout the tourism season. [Source: Andrew Downes, Xposure, courtesy of Killary Fjord Shellfish]

Fig. 26.23 OYSTER DEVELOPMENT PROCESS. a. Old oyster beds at Rostellan, on the east side of Cork Harbour. [Source: Darius Bartlett]; b. Oysterbags on trestles near Sherkin Island. [Source: Robbie Murphy]; c. Transfer of oysterbags from boat to tractor on Sherkin Island. [Source: Robbie Murphy]; d. Grading of Irish rock oysters. [Source: Robbie Murphy]; e. Fully grown oysters. [Source: Robbie Murphy]

MARKETS

Ireland's aquaculture produce is mainly for export, but its peripheral location relative to the centre of Europe makes it a challenge to supply the continental market in a consistent and cost-effective manner, especially as these are fresh products that have a short shelf-life. Up until now, the target markets in Europe have been competitive, with abundant home-produced supplies. The supply of fresh Irish mussels to the markets of France and Holland is a prime example of these challenges (Figure 26.22). Efforts are underway to develop the home market to counteract this. Furthermore, it is recognised that Irish aquaculture cannot compete on the basis of volume and therefore is moving away from bulk sales via wholesalers and distributors, to compete as a high-end, differentiated, Irish-branded product, supplied directly to retail. Oysters have traditionally been sold mainly to France but now markets as distant as the Far East are also being developed. Growing recognition of Irish oyster quality and the longer shelf life, ten days or more for fresh oysters, allows for greater market reach of fresh produce.

MAIN SPECIES/CULTURES

The salmon on-grown sector dominates national production and is up to 19,305 tonnes, worth €141 million in 2017. On-growing takes from nine to eighteen months depending on market needs. Employment in the salmon sector is mainly full-time with a total of 210 in direct employment at primary production sites in sea cages off the coasts of Donegal, Mayo, Galway, Kerry and Cork (Figure 26.21). The product, grown to exclusively organic certification standards, is exported to diverse markets including the EU, North America and the Near and Far East.

The farmed Pacific oyster (*Crassostrea gigas*) sector is the most valuable bivalve sector, producing 9,879 tonnes with a value of €43 million in 2017. Combined oyster employment (gigas and native oysters) is close to 1,300 people, mainly on gigas farms. Production is widespread along the coast, with concentrations in the south-east and north-west regions (Figure 21.23). Production is principally by bag and trestle, though other techniques are being introduced as knowledge of local site dynamics increases. Production is largely dependent on imported seed and the growth cycle is three years on average. France was traditionally the main market for Irish-grown oysters but, while this is still the main destination, sales to other European countries and to the Far East are expanding and, increasingly, home-branded product is also being sold directly to retail.

The rope mussel sector production in 2017 was 8,549 tonnes with a value of €6 million. The number of businesses operating and

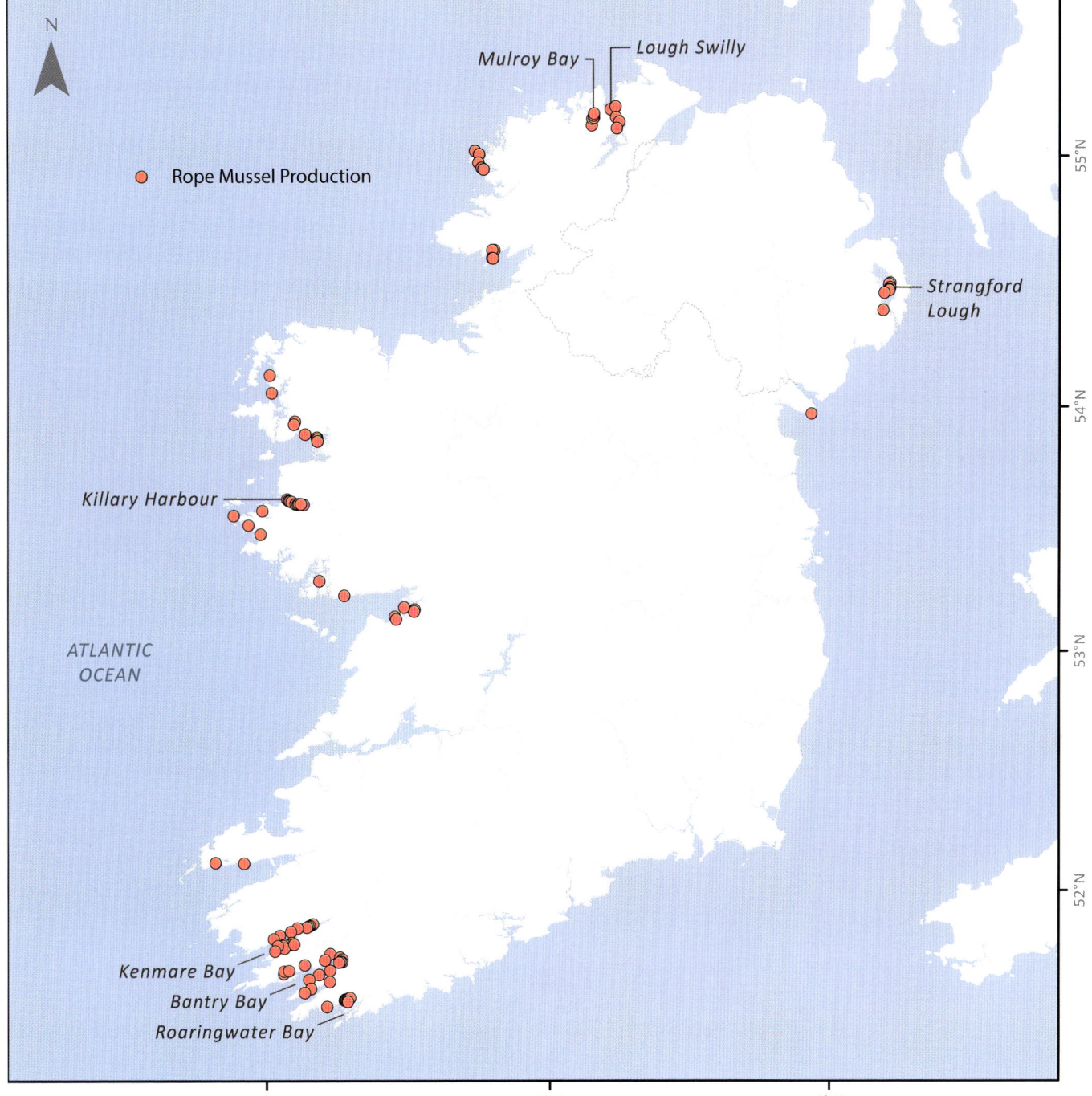

Fig. 26.24 MAIN ROPE MUSSEL GROWING AREAS. Rope mussel production is concentrated in the south-west in Cork and Kerry and to a lesser extent in the north-west, from Killary Harbour to Mulroy Bay. The distribution of rope mussel growing areas in the Republic of Ireland has remained relatively stable since 2009. [Adapted from: Dallaghan, 2011; NI Data source: DAERA, 2017]

the number of employees have both declined in recent years due to streamlining into larger units with specialist crews and equipment but there is still a large proportion of seasonal employment. Production is concentrated in Cork and Kerry in the south-west and to a lesser extent in the north-west, from Killary Harbour to Mulroy Bay (Figure 26.24). Production is supplied mainly by locally collected seed and the cycle is from one to two years depending on market requirements and site capacity. France and Holland remain the main market destinations. While red tide closures obstruct continuous production flow, the biggest impediment to sectoral growth remains the reliance on those markets that have a large home production stock, leading to periods of over-supply.

The bottom mussel sector production was 7,781 tonnes with a value of €9 million in 2017. The number of businesses in the sector has been in major decline in recent years, due to several poor seed settlement years and poor unit prices, and currently appears to have stabilised at twenty-four. These businesses are situated in Carlingford Lough, Wexford Harbour and Castlemaine Harbour with 114 mainly full-time employees. The industry has not, thus far, recovered in the north-west. The industry continues to depend almost exclusively on the level of annual wild seed settlement, while cost-effective alternative sources are pursued. Seed is fished by a fleet of mainly in-house dredgers and re-laid on licensed sites within suitable harbours. Harvesting occurs one to two years later, mainly for the fresh markets of Holland and France.

Research and Development in Aquaculture

Val Cummins

The rapid pace of innovation in the aquaculture industry has consequences for the operational aspects of fin and shellfish farming. Research and development into areas such as fish genetics, disease and feed have implications for survivability and productivity of farmed stocks. Other breakthroughs in R&D have a knock-on effect on engineering design, such as the use of bubble curtains generated by wave power, to mitigate against the impacts of jellyfish blooms on cages; or the development of closed-system aquaculture to create a barrier between farmed and wild fish, thus eliminating the most negative aspects of fish farming. Research into land-based grow-out of salmon suggests the use of closed containment conditions, which would see salmon farming migrate onshore. Research into different forms of aquaculture is undertaken in centres such as the Shellfish Research Centre in Carna, run by the National University of Ireland Galway; University College Cork's Aquaculture and Fishery Development Centre (AFDC), as well as in privately funded centres, such as the Bantry Marine Research Station (BMRS) in west Cork. Knowledge about lumpfish, which graze on sea lice, has led to the production of lumpfish for the salmon farming sector in BMRS (Figure 26.25).

Fig. 26.25 A LUMPFISH *(Cyclopterus lumpus)* PRODUCED AT A BESPOKE HATCHERY, FOR USE BY THE SALMON INDUSTRY AS A SUSTAINABLE SEA LICE MITIGATION TECHNIQUE. Lumpfish are 'cleaner fish', which means they naturally graze on parasites found on other fish species. Bantry Marine Research Station grows lumpfish until they reach approximately 15g, at which point they are sold to salmon farms. The salmon farms introduce the lumpfish to their salmon cages where they remove sea lice from the salmon. [Source: Laura Cannarozzi]

Case Study: Jellyfish and Aquaculture Interactions in Irish Coastal Waters

Damien Haberlin

WHAT IS A JELLYFISH?

Jellyfish are a common component of the plankton community around Irish shores. (see Chapter 3: Marine Biology and Ecology). They are particularly noticeable in summer and autumn, but some species persist year round, even throughout the cold winter months. Although jellyfish bodies are mostly composed of water (in some cases 98 per cent), they have nonetheless left a fossil record that stretches back

Fig. 26.26 INJURIES TO FARMED SALMON. These have been caused by dense aggregations of *Pelagia noctiluca* inundating a coastal salmon farm at Glenarm Bay, County Antrim, 2007. [Source: Hamish Rodger]

over 500 million years. These fossils show a remarkable degree of similarity to modern jellyfish, meaning they have not needed to adapt in any significant way to survive. All jellyfish species are simple animals, with no brain and only the most rudimentary senses, and yet the majority are voracious predators, capable of influencing and at times dominating entire ecosystems. The majority of species belong to the phylum Cnidaria and are united by a single trait, the possession of microscopic stinging capsules called cnidae. Jellyfish tentacles are packed with cnidae, some of which are like microscopic needles that are capable of piercing the skin or shell of prey and then injecting venom. Another characteristic of many species is the ability to 'bloom', or reproduce quickly when conditions are favourable. This can result in enormous numbers appearing suddenly. The conditions that bring about these blooms are not well understood at present and this lack of knowledge means such blooms are unpredictable. As a result these events have had severe negative impacts on a range of human activities, including fisheries, tourism and aquaculture. Often the scale of blooms has been unmanageable in any practical way, and on occasion, extreme blooms have caused the shut-down of nuclear power plants by clogging coolant water intakes.

SALMON AQUACULTURE AND JELLYFISH INTERACTIONS

In Ireland the industry most severely affected by jellyfish has been salmon aquaculture, with jellyfish contributing to annual mortality rates of approximately 10 per cent of the salmon stock, and periodically causing catastrophic losses. The worst single incident to date took place in Glenarm Bay, County Antrim, in 2007 when the mauve stinger jellyfish (*Pelagia noctiluca*) caused 100 per cent mortality of the fish being farmed, with over 120,000 fish lost at an estimated value of more than £1m (Figure 26.26).

While globally a range of jellyfish species contribute to aquaculture losses, in Ireland two species have been most problematic, the mauve stinger *Pelagia noctiluca*, and the small siphonophore *Muggiaea atlantica* (Figure 26.27). *M. atlantica* is a small transparent jelly, only a centimetre or so in length, and is common around Ireland from June through to November. *P. noctiluca* is much larger, reaching approximately 10cm in diameter. Despite the very different form of these two species, they present aquaculture with the same core problem, that they are both capable of inflicting injury to caged salmon. The small size of *M. atlantica* means it can easily enter salmon cages, where its tentacles make contact with the salmon. The delicate gill tissues of fish are particularly vulnerable to damage. *M. atlantica*

Fig. 26.27 JELLYFISH. a. Mauve stinger *Pelagia noctiluca*. [Source: Nigel Motyer]; b. The small siphonophore *Muggiaea atlantica*. [Source: Damien Haberlin]

is also small enough to be inhaled by the fish, passing directly over the gills as the fish respire. In contrast, the larger *P. noctiluca* gets fouled by the cage netting and broken into fragments before entering the cage. This fragmentation does not, however, diminish the stinging effect of the jellies. In fact, it may well make the problem worse by turning a discrete number of jellyfish into a diffuse cloud of stinging cells, which the fish cannot possibly avoid.

The density of the jellyfish present in the water determines the severity of the impact on the fish, with very high densities causing high mortalities and lower densities causing varying levels of injury and distress to the salmon. Even relatively low densities of jellyfish adversely affect the behaviour and feeding cycles of the fish, and these disruptions result in substantial economic costs for fish farms. Another concern is the risk of secondary bacterial infection after injury from jellyfish stings, and in a rather cruel twist, research is showing that the jellyfish can be carriers of the infecting bacteria, thus simultaneously injuring and infecting the fish. The fouling of *P. noctiluca* on cage netting can also cause further problems for fish farms, particularly when densities are very high. As jellies accumulate on the netting and clog up the net apertures, water starts to push against the cage rather than flowing through it. This deprives the fish of fresh oxygenated water and increases the physical load on the cages and their mooring systems, placing further stress on the fish and increasing greatly the risk of structural damage to the farm.

WHAT MAKES THESE SPECIES BLOOM?

The occurrence and distribution of both species in Irish waters is linked with permanent oceanic currents, with transient seasonal currents and fronts, and with local wind patterns. Local wind patterns can exert a similar influence over both species. Strong westerly winds, for example, can push surface aggregations into sheltered bays along the west coast and both species seem to peak in the late summer and autumn, often persisting into the early winter months. On a broader scale, the source of each species is quite different. *P. noctiluca* is an oceanic species that is often found in the deeper water of the western continental shelf. Recent evidence suggests it may reside in deep canyons off the western shelf, coming to the surface layers in the mid- to late summer; once on the surface the wind becomes the main driving factor.[17]

In contrast, *M. atlantica* is considered a more coastal species and is not generally found in deeper offshore waters. It appears likely that *M. atlantica* is brought into Irish coastal waters by a seasonal coastal current driven by the Celtic Sea gyre, a seasonal formation caused by stratification in the water column. It has been established that higher temperatures lead to enhanced reproduction in *M. atlantica*. However, the dynamic oceanography along the Irish coast means that sea temperatures are not simply a reflection of increased local sunshine; warm surface waters over the shelf can intrude into coastal areas and, equally, cold deep water can be upwelled into coastal areas. It is this complex interplay between weather and oceanography that makes forecasting blooms so difficult.

MANAGING THE IMPACT OF FUTURE BLOOMS

At the present time, there is little that fish farms can do to combat the impact of jellyfish blooms. Aeration of cages is thought to provide some relief for fish, but in the event of high-density blooms, farms will often opt to harvest early and remove fish out of harm's way. This is a costly exercise as farms lose substantial profits when fish are not at the normal market size. Current research is reviewing different aspects of jellyfish ecology, as well as technologies that might help the industry to manage blooms and mitigate against large-scale losses.

In order to understand the movement of jellyfish in shallow coastal areas, lion's mane jellyfish (*Cyanea capillata*) have been tracked using acoustic biotelemetry in Dublin Bay (see Chapter 3: Marine Biology and Ecology). The tracking research showed that jellyfish can adjust their vertical position in the water column quickly, but have far less control horizontally, moving with the prevailing currents.[18] This has implications for those salmon farms that are located in deep narrow bays, where tidal exchange occurs predominantly in the surface layers of the water column. By dropping out of the surface layers, jellyfish can accumulate in a bay, increasing the level of impact on a farm. A potential method of protecting farms from this threat is the use of a bubble curtain, which would entail pumping compressed air through a perforated tube on the seafloor. As the air expands and rises through the water, it creates an upward current that acts as a barrier to prevent jellyfish from encroaching on a farm. Further current research is combining knowledge of jellyfish ecology, oceanography and ecosystem modelling in order to predict the likelihood of blooms. This is a slow process as it requires years of data collection in order to make accurate predictions, but researchers in the Mediterranean have begun to make progress in predicting *Pelagia noctiluca* blooms and similar research is underway in Irish coastal waters.[19] In addition to understanding existing conditions, this type of knowledge will be important in understanding and anticipating future changes in our coastal ecology, particularly as a result of climate change. Enormous changes in zooplankton distribution are already evident in the north-east Atlantic, with some species of copepods having shifted north by over 1,000km during the last fifty years.[20] For jellyfish, increased temperatures may mean new species moving into Irish coastal waters and a widening of the reproductive season, potentially leading to an increase in blooms of both native and non-native species.[21] Equally, increased temperatures may be less favourable for native species and result in their decline. Notwithstanding which particular species wins or loses, the mid-latitudes where Ireland is located are likely to see the greatest changes.[22] Considering that increased aquaculture is a strategic goal nationally and at the European level, there is a pressing need to understand these changes in coastal ecology. Ireland was at one time a pioneer in jellyfish research, thanks to the tireless efforts of Maude Delap while living on Valentia Island in the late nineteenth and early twentieth centuries (see Chapter 3: Marine Biology and Ecolgy). Perhaps in addressing the challenges outlined above, Irish researchers can recapture some of that pioneering spirit.

Case Study: The Seaweed Harvesting Industry in Ireland

Niamh O'Donoghue and Sarah Kandrot

The island of Ireland has a culture of seaweed harvesting dating back centuries. Historically, seaweed has been collected to provide a source of food since the Middle Ages and was also used as fertiliser for potato growing, for animal fodder and to make glass and soap. It represents a valuable commodity and means of employment, particularly in remote or economically disadvantaged areas (see Chapter 21: Ireland's Islands). The west coast of the country is exceptionally rich in the resource, as its rocky bays and indentations along the open Atlantic ocean encourage seaweed settlement. This has led to a progressive shift, from exploitation of seaweed as a domestic enterprise to it becoming the basis for a commercial industry (Figure 26.28). This shift, however, has prompted impassioned stakeholder debate over controversial issues related to privatisation and regulation of harvesting activities, as well as a transition from traditional exploitation methods to modern mechanised ones. While the potential for continued growth in the industry is greater than ever, the challenges faced by industry stakeholders are significant.

Scientific studies of Irish seaweed species began in the eighteenth century and since then, 579 species have been identified.[23] These include 284 species of red algae (*Rhodophycota*), 155 of brown (*Phaeophycota*), eighty-six of green (*Chlorophycota*), forty-four of blue-green

(*Cyanophycota*), and ten species of yellow-green algae (*Vaucheria*).[24] An array of products are produced from these seaweeds, the most important of which, by both volume and value, being animal feeds, plant supplements, fertilisers and agricultural products. Other uses in Ireland are in sea vegetables such as carrageen moss and dulse, body care and cosmetics, thalassotherapy, biotechnology and biomedicine. The most widely harvested species, representing over 75 per cent of the harvest, is the brown seaweed, *Ascophyllum nodosum*, also known as knotted or egg wrack. *A. nodosum* is primarily harvested and processed in the west of Ireland. The species is most commonly used in the food processing, agriculture and horticulture sectors, although a shift in emphasis towards research into and development of value added products is trending.[25] Aside from *A. nodosum*, other Irish species of commercial interest include serrated wrack (*Fucus serratus*), dulse (*Palmaria palmata*), carrageen moss (*Chondrus crispus*), oarweed (*Laminaria digitata*) and to a lesser extent, *Laminaria hyperborea*, *Saccharina latissima*, *Fucus vesiculosus* and *Alaria esculenta* (Figure 26.29). Ireland still only exploits sixteen of its native species.[26] However, it is forecast that more will be utilised as we learn more about the practical uses of other species through research.

IRISH SEAWEED FACTS AT A GLANCE

Carrageen moss (*Chondrus crispus*) at Easky, County Sligo. Underwater, tips of Carrageen can have a violet iridescence.

Globally, less than **5%** of commercially produced seaweed comes from wild stock, and that figure has been declining. The majority of Irish seaweed is from wild resources.

The Irish seaweed production industry peaked in the 1970s, at which time **100,000 tonnes** of seaweed were harvested per annum. Today, only **25,000–40,000 tonnes** are harvested annually.

The main commercial species harvested in Ireland are **egg wrack (*Ascophyllum nodosum*)**, **kelp (*Laminaria hyperborea*)** and various types of red seaweeds, including **dulse (*Palmaria palmata*)** and **Carrageen Moss (*Chondrus crispus* and *Mastocarpus stellatus*)**.

Uses of wild Irish seaweed include:

Food   Fertilizer   Cosmetics

Animal Feed   Medicine   Industrial Uses

Knotted or egg wrack (*Ascophyllum nodosum*) (middle) interspersed with bladderwrack (*Fucus vesiculosus*) (right) and serrated wrack (*Fucus serratus*) (left) at Derk More, County Sligo.

Fig. 26.28 SEAWEED HARVESTING. [Data source: FAO, 2018; BIM; DHPLG, 2018; photo sources: carrageen moss: Sarah Kandrot; knotted wrack: Maxim Kozachenko]

Fig. 26.29 HARVESTING OF *Alaria esculenta* (WINGED KELP) FROM NEAR-SHORE LONG-LINES IN BANTRY BAY. Seeded lines are placed at sea between October and December, and harvested between March and May. Average yields are approximately 10kg per 1m of line (wet weight). At present, Bantry Marine Research Station is analysing the seaweed for fluctuations in bioactive compounds, and sells the bulk of the biomass to secondary processors for the production of food and feed products. [Source: Fiona Moejes]

Historically, Irish seaweed has been hand-harvested from naturally grown resources. In 2019, over 95 per cent was exploited from wild stock and only a fraction produced by aquaculture, which remains at a nascent stage. There is still ample room for development in the seaweed sector, which is distinguished by high-volume, low-value added processing activities.[27] However, a shift is evident with new entrants into the sector, such as the biotechnology and cosmetics markets, which could lead to significant growth. Issues related to regulation, monopolisation and management of the resource, however, may threaten this expansion.

One issue relates to foreshore licensing. In Ireland, the foreshore is state-owned and, according to the Foreshore Acts (1933–2011), a licence from the Department of Housing, Planning and Local Government (DHPLG) is required for the collection or removal of any beach material from the foreshore, effectively (but not explicitly) including seaweed. Despite this, it is estimated that 95 per cent of all seaweed harvested in Ireland is taken without a foreshore licence.[28] This did not emerge as an issue until 2014, when the largest seaweed processing company in Ireland, Arramara Teoranta (see Chapter 21: Ireland's Islands), was sold to a Canadian multinational, who made an application for an exclusive foreshore licence to harvest seaweed at twenty-one locations along the coast, ranging from County Mayo to County Clare. Prior to this, traditional harvesters believed that they were protected by seaweed-harvesting rights, sometimes dating back centuries, which were often written into land deeds and inherited as part of an individual's estate (similar to the 'turbary rights' that allow the cutting of peat for fuel). Indigenous enterprises that once relied on these traditional rights of the harvesters whom they contracted, subsequently submitted their own foreshore licence applications. It was the first time in the history of the Irish seaweed industry that the issue of foreshore licences had surfaced. On one hand, it was inevitable that this should happen, as industry growth threatens the sustainability of the practice and regulation will be necessary to prevent overharvesting. On the other hand, the rights and livelihoods of traditional harvesters were being called into question. In 2018, the DHPLG provided some clarity on the issue when they announced that those with seaweed harvesting rights would not be required to hold a foreshore licence to harvest seaweed.

In addition to bringing to light the issue of foreshore licensing, the sale of Arramara also raised concerns amongst harvesters and indigenous SMEs about the monopolisation of the industry, whereby harvesting activities might no longer be controlled by local communities – with vested interests in protecting the future viability of an Irish resource – but by a foreign multinational enterprise. Traditional harvesters have raised concerns that this could result in a lack of control over the price of seaweed. Industry representatives have promised publicly to work with local harvesters, though some harvesters remain sceptical.

A third issue relates to mechanical harvesting. The longevity and sustainability of the Irish seaweed industry stems from the hand-harvesting techniques employed, along with the knowledge that has been passed down through generations on the species' life cycles and regrowth periods.[29] Concerns have been expressed about the introduction to Irish waters of mechanical harvesting, which has been employed in countries such as Norway and France as intensive or over-harvesting can have negative consequences for both the sector and the environment. These include, for example, a reduction in the rate of seaweed regeneration, a reduction in biodiversity and increased exposure of coastal areas to wave action, potentially resulting in coastal erosion. More information from scientific research is required to adequately address these issues.

As of 2019, four foreshore applications have been submitted to mechanically harvest native kelp, from Kenmare Bay, Bantry Bay, Clew Bay and Galway Bay. The foreshore application to remove *Laminaria digitata* from Kenmare Bay was declined due to the bay's classification as a Special Area of Conservation (SAC 002158) under the EU Habitats Directive.[30] The applications for Clew Bay and Galway Bay are still open (2019). In 2014, a licence was granted for mechanical harvesting of *Laminaria hyperborea* and *Laminaria digitata* in Bantry Bay. The duration of this licence is ten years, permitting 1,860 acres of native kelp to be harvested from five zones in the bay, using a vessel fitted with a suction pump and cutter. As of 2019, however, harvesting in the bay has not commenced due to the formation of a campaign group that has organised petitions, protested and fundraised to have the licence revoked, until a thorough Environmental Impact Assessment has been carried out and confidently displays that no biodiversity will be threatened.

The seaweed harvesting industry in Ireland is in the midst of change. The challenge at present is to balance the interests of traditional harvesters with industry development. This should not be an unattainable goal. The industry will ultimately benefit from meaningful stakeholder collaboration on all sides – from harvesters to SMEs to multinationals – and such coordination could ultimately pave the way to a vibrant, thriving, Irish seaweed industry.

UNREGULATED HARVESTING: THE EDIBLE PERIWINKLE

Val Cummins

The edible periwinkle, *Littorina littorea,* is common around the coast of Ireland, mainly on rocky shores, where it can be found in the littoral zone between high and low water marks.[31] It is one of the most common, and one of the largest, shore gastropods. Sexual maturity occurs at twelve to eighteen months, once a shell height of approximately 11mm has been achieved, although shell heights of approximately 35mm can be attained. Periwinkles are capable of a long lifespan, with a record of one individual surviving for twenty years in an aquarium.

In general, the highest densities of periwinkles in Ireland are found on semi-exposed sites, where populations are influenced by multiple factors such as exposure to waves and the nature of the substratum (Figure 26.30).[32] Gravelly substrates are particularly attractive as they provide protection from dislodgement and predation. *L. littorea* is an omnivorous grazer, with a preference for green algae such as *Ulva lactuca* and *Enteromorpha intestinalis*. Feeding activity is influenced by the tidal cycle and the season. Populations on any one shore can fluctuate as a result of factors such as natural mortality, food availability and sub-tidal migration, harvesting can also introduce significant pressures.

One of the classic signs of imminent problems in a fishery is the proliferation of undersized individuals in a catch. Anecdotal reports of problems arising from over-picking on several shores led to the Marine Institute commissioning a dedicated study to acquire baseline data of the overall population.[33] The study included the impacts of harvesting and yielded insights into the unregulated nature of the sector.

Very few periwinkles are actually sold for consumption in Ireland. Periwinkles were traditionally sold to eat, wrapped in newspaper cones, on the promenade in places such as Kilkee in County Clare. However, they are considered a delicacy in overseas markets, such as France, where they are thought tender and flavoursome when cooked. This provides an export opportunity for wholesalers. Data from 2002 suggested that approximately 3,600 tonnes of periwinkles were harvested annually at that time, equating to about €9 million of value.[34]

Wholesalers collect periwinkles for distribution via refrigerated transport, from periwinkle harvesters around the coast. The volume and value of periwinkles for export is extremely low in comparison to shellfish such as mussels and oysters, but this fishery provides an important source of income and an alternative livelihood in many peripheral rural coastal areas. There were approximately 500 part-time pickers in 1998/1999. The concerns around overharvesting that triggered the Marine Institute baseline study were mitigated by findings that indicated a major decline in the age profile of periwinkle harvesters. Young people perceive periwinkle picking as difficult, labour-intensive work, with very little financial return for the effort involved. This trend, together with the lack of conclusive evidence to prove that shores were over-picked, meant that suggested interventions – such as the introduction of a closed season or minimum landing size – were not implemented.

The lengthy planktotrophic stage of *L. littorea* makes it less likely to be impacted by localised overexploitation, compared to species without such an ability to disperse. Where populations of large periwinkles are removed from one area, larvae can be recruited from more distant shores, and so, in time, over-picked areas may be re-seeded. The impact of changes in water quality and changing sea surface temperatures, from climate change, is unknown.

Fig. 26.30 PERIWINKLE HARVESTING ON EXPOSED ROCKS AT LOW TIDE, ARDNAHINCH STRAND NEAR BALLYCOTTON, COUNTY CORK. Archaeological records and the abundance of shell middens (piles of discarded shells) around the coasts of Ireland provide clear evidence that shellfish, collected from the seashore, has been an element in the diet of communities from the early Mesolithic period right up to recent times. Today, most shellfish consumed in Ireland comes from commercial aquaculture, but some wild collecting still takes place. Periwinkle harvesting, in particular, is often a family-based pastime activity, and is undertaken both for home consumption and occasionally as an additional source of domestic income. Provided due care and attention is given to the state of the tide and the potential slipperiness of the rocks, hunting for periwinkles and other sea shells hidden in rock pools and seaweed can also be a fun activity in introducing children to the ecology of the shore. [Source: Darius Bartlett]

The lack of any subsequent studies has also left a gap in information as to the current socio-economic status of the sector. Anecdotal information suggests that a new wave of harvesters entered the fishery, following inward migration through the 2000s from Eastern Europe. A group of thirteen periwinkle harvesters, working the shore in north Dublin, had to be rescued by the lifeboat when they became cut off by the tide in 2005. As long as the market demand is there, it is likely that periwinkles will continue to be harvested for export. Developments in in-growing and polyculture could provide new areas of opportunity for the sector.

FUTURE CHALLENGES

In recent years, an overall trend of improvement has been recorded in fish stock status in European seas.[35] In Irish waters, hake is a good example of this, as the stock was in very poor condition in the early 2000s but, following implementation of a recovery plan in 2002, the stock has increased greatly and landings have increased while remaining within sustainable levels. However, there are still some significant challenges, and this is evident in the fishing quotas for 2019 agreed by EU fisheries ministers in December 2018; Ireland's two most valuable fisheries, nephrops and mackerel, suffered cuts of 32 per cent and 20 per cent respectively.

The reasons for such quota cuts are difficult to clearly identify. In the case of nephrops, significant factors may include technological improvements and an increase in the number of vessels specialising in prawn fishing, both of which could be addressed by improved future management structures. In the case of mackerel however, the situation is less clear and, in addition to issues such as fishing pressure from modern fleets, climate change may also be playing an increasingly significant role on decisions relating to fisheries management. In recent years, there has been an ongoing dispute between Iceland and the Faroe Islands on the one hand, and EU member states (Ireland, the UK, the Netherlands and Denmark) plus Norway, on the other, regarding the allocation of mackerel quotas. It has been labelled a 'Mackerel war' similar to the 'Cod Wars' between Iceland and the UK from the 1950s to the 1970s.

Prior to 2006 Iceland caught almost no mackerel but the species began to appear in greater quantities in Icelandic waters in the late 2000s, and the country sought a greater share of the North East Atlantic Mackerel quota. When Iceland could not agree on a division of the quota with the EU and Norway, it decided unilaterally to take a share of the catch, which rose to a peak of 155,000 tonnes in 2011. This had significant implications for both the biological sustainability and the economic value of the fishery. The economic impact was due to the suspension of the certification of the fishery by the Marine Stewardship Council (MSC) in 2012, which meant that premium markets for products from this fishery were no longer accessible. The MSC certification was only restored in 2016, when scientific advice regarding the mackerel stock indicated that, despite higher catches by Iceland and the Faroe Islands, overall catches in the fishery were within sustainable limits. Now that more recent scientific advice yet again indicates a declining mackerel stock, there is potential for further disputes.

As can be seen from this example, the challenge of managing a single species, mackerel, is not insignificant. The changing migratory patterns of fish, the dynamic requirements of governance structures, and availability of scientific resources all represent significant challenges that will also become evident in other species in the future.[36] An emerging example of this is the claim by Irish fishermen for a share of the North Atlantic Bluefin Tuna (*Thunnus thynnus*) quota, since this species has been appearing in increasing numbers in Irish waters over the past few years.[37]

BREXIT

The other significant issue and challenge for fisheries, in both the Republic and Northern Ireland, is the British exit from the European Union. Overall 34 per cent of landings by Republic of Ireland vessels, based on 2011–2015 averages, are taken from UK waters.[38] For fleets from the Republic, the problem is highlighted again, by the two most valuable fisheries, nephrops and mackerel. A significant number of prawns and mackerel (60 per cent) catches by Republic of Ireland vessels are from UK waters. Although there are abundant mackerel in Irish waters, a large proportion of the mackerel catch is taken in Scottish waters, since this is where the fish tend to be in their optimum market condition in terms of fat content.

At the time of writing there is significant uncertainty about whether there will be a negotiated or a no-deal Brexit. In the case of a no-deal Brexit, it is possible that Irish vessels may not be able to enter UK waters and, even if they are, the share of the quota they receive may be less than when the UK was an EU member. There may also be a counterbalancing effect if UK vessels cannot enter Irish waters, as 16 per cent of landings from the Irish EEZ are from UK-registered vessels.

Transport is another significant issue for the seafood industry in a Brexit scenario, due to the highly perishable nature of most seafood products and high reliance on the UK and European export markets. Transport of fish by lorry through the UK would, in effect, represent an export out of the EU and then a re-import to the EU from the UK, unless an agreement can be reached for a 'bonded transit' system, whereby lorries transporting fish would be sealed upon entry to the UK, and the seal checked again on exiting the UK.

The situation for Northern Ireland fisheries also remains unclear. In common with fisheries in the Republic, a significant amount of Northern Ireland produce is exported to both the Republic and Europe, and transport logistics and import tariffs may become a significant issue post-Brexit. Additionally, issues such as the voisinage arrangements (which allow Republic of Ireland vessels smaller than 75 feet (23 metres) in length to fish in the territorial waters of Northern Ireland and vice versa) and territorial claims on Lough Foyle, which have not been major issues while the UK was an EU member, may have to be resolved post-Brexit. The future existence, or not, of a hard border on the island of Ireland, or a possible border in the Irish Sea instead, further complicates the issue for all fishing and aquaculture-related operations on the island.

In conclusion, like the oceans and coasts themselves, the Irish fishing and aquaculture industries have experienced many changes over the years. After several decades of neglect and decline, particularly during the early twentieth century, Irish fisheries are experiencing a revival, marked by iterative management interventions, diversification and investment. In part, this reflects wider trends at European and global levels, as well as changes driven by the availability of new technologies. One of the most significant of these changes is the transition from fishing as a form of hunter-gathering activity, to the highly sophisticated commercial operations today. The industry has also become more regulated. While some elements of specialised deep-sea fishing in international waters operate out of a few key fishing ports, such as Castletownbere and Killybegs, for the most part, Ireland's fishing industry is predominantly in inshore waters and coastal embayments, and is closely tied to localised onshore supporting infrastructures. The launch of the National Strategy for the Irish Inshore Fisheries Sector, 2019–2023, is a strong indication of the growing appreciation by government of the importance of this sector.

Nonetheless, despite these undoubted improvements, and growing appreciation of the role of the 'blue economy' in Ireland's future, many challenges remain. Chief among these are the political and economic uncertainties arising from the UK's departure from the European Union. The EU Common Fisheries Policy has been a key regulator for the industry in recent years, and is especially important in allocating national and regional shares in the European fisheries resource, as well as mediating and adjudicating between competing claims. With Britain withdrawing from the EU, and hence from the CFP, the direction of future relationships between the Irish and UK fishing fleets remains currently unknown. Furthermore, the ramifications of Brexit regarding customs, tariffs and cross-border trade may also impact negatively on Irish access to European markets, and transport to these of fish landed in Ireland.

In the longer-term, the equally or more profound potential impacts of climate change and ocean warming add further uncertainty regarding the future of the industry. Some species of current commercial importance may decline, or their distributions may change, while new ones – and new threats from unwelcome or harmful species such as jellyfish and algal blooms – may emerge or expand. It is likely also that the marine climate itself may change and, with it, the frequency and severity of extreme weather events that could also impact significantly upon the industry.

There are signs, however, that the urgency and seriousness of these challenges, particularly in the latter case, are being recognised by government and key stakeholders in the industry alike. The development of a Marine Spatial Plan for Ireland is likely to impact on the future direction of the Irish fishing and aquaculture industries, as regulators seek to reconcile the growing demands for activity in the context of Blue Growth.

COASTAL AND MARINE TOURISM: BUILDING OPPORTUNITIES IN BLUE GROWTH

Cathal O'Mahony and Stephen Conlon

CLIFFS OF MOHER, COUNTY CLARE. These spectacular coastal cliffs of the Burren Geopark are approximately 8km in length and are amongst the highest cliffs in western Europe, reaching elevations of 214m high. [Source: Elizabeth K. Joseph, flickr, CC BY 4.0]

Fig. 27.1 VENTRY BEACH, DINGLE, COUNTY KERRY. Over one-half of Ireland's *c.*7,000km of coastline is composed of sandy beaches and dune systems, as shown in this example at Ventry. These coastal types, and many others in Ireland, are recognised widely as having high scenic values that are comparable with, or exceed, those found in other European countries. Their environmental qualities provide superb raw conditions for many water-based activities, such as surfing, windsurfing, kite-surfing and sailing. In addition, the often-dramatic scenery associated with such coasts attracts a large and growing number of recreational walkers. In the above photograph, a coastal walking trail (signpost in foreground) directs people along this wide and sheltered beach and dunes area. Coastal housing and settlement can be seen at the northern end of the beach, nestling around Ventry Harbour. Much of this would be linked directly to providing facilities for tourists, or are second-homes used particularly during peak times of coastal tourism. [Source: Noel O'Neill]

Fig. 27.3 BALLYMASTOCKER BAY, THE FANAD PENINSULA, COUNTY DONEGAL. Also known as Portsalon Beach, the bay is located on the western side of Lough Swilly, between Rathmullan and Fanad Head, with stunning coastal views towards the Inishowen Peninsula. This spectacular Blue Flag beach has, in the past, been voted the second most beautiful in the world by readers of the *Observer* Magazine. [Source: Eskling, flickr, CC BY-ND 4.0]

Coastal locations are primary hubs for tourism in Ireland.[1] For many people this remains the traditional beach-based holiday, especially given the impressive scenic backdrops to the island's beaches (Figure 27.1).[2] A large number of these beaches now have the much valued European Blue Flag status.[3] Most coastal sites also enable access to many other types of land, beach and marine-based tourism activities. In addition, coasts provide rich offerings for those interested in

Fig. 27.2 THE GIANT'S CAUSEWAY, COUNTY ANTRIM. The Giant's Causeway is composed of over 40,000 polygonally shaped blocks of columnar basalt rock, some of these measuring *c.*12m in height and 1m across. Folklore has it that the legendary giant, Fionn Mac Cumhaill (Finn McCool), built the 'causeway'. These columnar blocks formed through the cooling of an ancient lava flow in this region and were created during the Paleogene period *c.*50–60 million years ago. The blocks are linked in age and structure to similar columnar basalts found in the Fingal's Cave area off the Isle of Staffa, in the Inner Hebrides. The Causeway was declared a UNESCO World Heritage Site in 1986 and was also voted the fourth-best natural wonder in the United Kingdom. Over one-million people visited the location in 2017, making it the most-visited tourist site in Northern Ireland. Many sea birds inhabit this scenically beautiful, historic and ecologically important coast, such as cormorants, razorbills and petrels. [Source: Patrick O'Hanlon]

Fig. 27.4 COASTAL AND MARINE TOURISM AND LEISURE ACTIVITIES ON IRELAND'S SHORES. a. The annual 'Sherkin Island to Baltimore, County Cork 2km swim' takes place in September. [Source: Robbie Murphy]; b. West Cork Islands five-a-side soccer tournament being played on Cow Strand (Trag Eoin Mhór), Sherkin Island. [Source: Robbie Murphy]; c. Portmagee Regatta, 29 August 2010, a classic seine boat race. [Source: Kevin Terrant]; d. Kite-sailing at Gary-lucas beach County Cork. [Source: Lisa Fawkes]; e. Lucy Hunt, Marine Biologist, Sea Synergy Marine Awareness and Activity centre, makes the shore on Scariff Island, south-west Kerry. [Source: Damian Foxall]

heritage and historical sites. This is due to the strong and long-established relationships that exist between settlement patterns and the coast, particularly in the development of transport routes (road and rail), and in the movement of goods and people. These routeways and functions have often developed national significance. In modern times, the role and use of second homes as holiday residences can now be added to the importance of these coastal locations, which form attractive sites for housing and personal investments.

Numerous examples of all of these different forms of use and tourism attraction exist, ranging from the beach holiday and coastal seaside towns, as at Youghal, County Cork, or Bundoran, County Donegal; the scenic 'hotspots' (such as the Cliffs of Moher, the Aran Islands); to the many adventure and activity leisure hubs dotted along the Wild Atlantic Way (WAW) and the historical and marine heritage-themed towns, such as those of

Cobh, County Cork, and Dún Laoghaire, County Dublin (Figures 27.2, 27.3). Our seaside resorts, beaches, harbours and sea access-ways, inland waterways and rivers are remarkably varied. Together, they provide the resourcing and facilities for a wide range of water-based recreational pursuits and these sports and other recreational activities attract large numbers of both domestic and overseas tourists (Figure 27.4). Similarly, in terms of visitor experience and leisure offerings to tourists, the south-west and western coasts of Ireland, in particular, are areas possessing long-established histories for boat-cruising. This more specialised activity around Ireland's coasts is considered exceptional in quality, and is acknowledged widely as being the best in Europe.

BLUE FLAG BEACHES

Cathal O'Mahony, Kathrin Kopke and Val Cummins

The Blue Flag seen flying over many of Ireland's popular beaches, from Bundoran in County Donegal to Barleycove in County Cork, is a recognisable symbol of environmental quality and a source of assurance to beach users that the water quality and facilities at the beach are of a high standard (Figure 27.5).[4] Indeed, the Blue Flag is a 'badge of honour', readily promoted by businesses and communities located around Ireland's coasts. The history of the Blue Flag award dates back to 1985, when local authorities in France developed an award to recognise compliance with bathing water regulations. Since that date, it has developed into a global initiative, administered by the Foundation for Environmental Education (FEE).[5] Almost 4,200 Blue Flags have now been granted worldwide to beaches and marinas. Ireland has a strong record in the Blue Flag programme; the island of Ireland was ranked eleventh in 2015, in an international table based on the number of awards issued (Figure 27.6).[6]

The level of local participation in promoting and maintaining beach environments that comply to high international standards is testimony to the value of the award to coastal economies. Many of these rely on beach tourism, and their ability to market a destination as one that delivers a quality visitor experience. In addition to the tourism advantage that a Blue Flag provides to a destination,[7] the process behind the programme has significant benefits, in terms of supporting the integrated management of our coasts (see Chapter 32:

Fig. 27.5 ROSSBEHY SPIT, DINGLE BAY, COUNTY KERRY, VIEW NORTH OVER ROSSBEHY STRAND TOWARD THE GLACIATED UPLANDS OF DINGLE. This wide, sandy beach area here faces due west into the high-energy wave conditions of the North Atlantic. The train of regular Atlantic swell waves can be seen approaching the beach and beginning to 'curl' and break. The dunes at the head of the beach are under active erosion, as part of the currently breaching Rossbehy Sand Spit. This shoreline is visited daily by many people, undertaking different activities, from dog-walking, to swimming, to kite-sailing. The inset image shows the Blue Flag signage, located at the car park end of the beach. [Source: Sarah Kandrot]

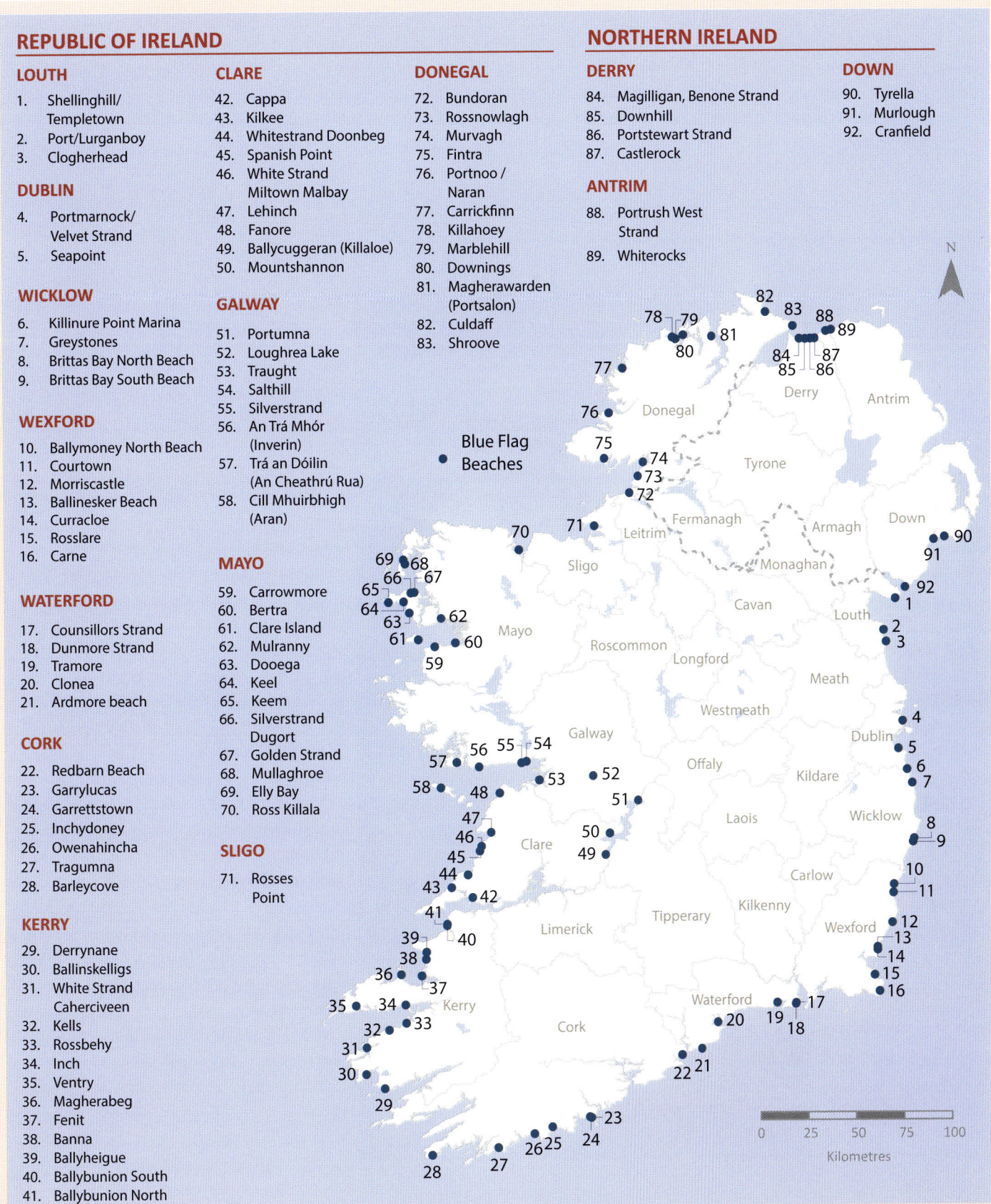

REPUBLIC OF IRELAND

LOUTH

1. Shellinghill/ Templetown
2. Port/Lurganboy
3. Clogherhead

DUBLIN

4. Portmarnock/ Velvet Strand
5. Seapoint

WICKLOW

6. Killinure Point Marina
7. Greystones
8. Brittas Bay North Beach
9. Brittas Bay South Beach

WEXFORD

10. Ballymoney North Beach
11. Courtown
12. Morriscastle
13. Ballinesker Beach
14. Curracloe
15. Rosslare
16. Carne

WATERFORD

17. Counsillors Strand
18. Dunmore Strand
19. Tramore
20. Clonea
21. Ardmore beach

CORK

22. Redbarn Beach
23. Garrylucas
24. Garrettstown
25. Inchydoney
26. Owenahincha
27. Tragumna
28. Barleycove

KERRY

29. Derrynane
30. Ballinskelligs
31. White Strand Caherciveen
32. Kells
33. Rossbehy
34. Inch
35. Ventry
36. Magherabeg
37. Fenit
38. Banna
39. Ballyheigue
40. Ballybunion South
41. Ballybunion North

CLARE

42. Cappa
43. Kilkee
44. Whitestrand Doonbeg
45. Spanish Point
46. White Strand Miltown Malbay
47. Lehinch
48. Fanore
49. Ballycuggeran (Killaloe)
50. Mountshannon

GALWAY

51. Portumna
52. Loughrea Lake
53. Traught
54. Salthill
55. Silverstrand
56. An Trá Mhór (Inverin)
57. Trá an Dóilin (An Cheathrú Rua)
58. Cill Mhuirbhigh (Aran)

MAYO

59. Carrowmore
60. Bertra
61. Clare Island
62. Mulranny
63. Dooega
64. Keel
65. Keem
66. Silverstrand Dugort
67. Golden Strand
68. Mullaghroe
69. Elly Bay
70. Ross Killala

SLIGO

71. Rosses Point

DONEGAL

72. Bundoran
73. Rossnowlagh
74. Murvagh
75. Fintra
76. Portnoo / Naran
77. Carrickfinn
78. Killahoey
79. Marblehill
80. Downings
81. Magherawarden (Portsalon)
82. Culdaff
83. Shroove

NORTHERN IRELAND

DERRY

84. Magilligan, Benone Strand
85. Downhill
86. Portstewart Strand
87. Castlerock

ANTRIM

88. Portrush West Strand
89. Whiterocks

DOWN

90. Tyrella
91. Murlough
92. Cranfield

Blue Flag Beaches

Fig. 27.6 BLUE FLAG beaches in Ireland, showing the location of these high-quality beach and bathing water environments. Ireland participated first in the Blue Flag programme in 1988, when nineteen beaches achieved the award in the Republic. In 2018, this number had risen throughout Ireland to 92 Blue Flag beaches. [Data Source: Blue Flag Programme]

Coastal Management and Planning).[8] As well as being an indicator of environmental quality, the educational element of the award has potential to further increase citizen awareness of the challenges linked to the protection of Ireland's marine ecosystems, particularly in encouraging people to act together to address emerging societal issues, such as marine litter and pollution by plastics [see Chapter 31: Pollution].[9]

SURFING IN IRELAND

Tristan MacCana

Although the western coast of Ireland is considered as prime territory for surfing, it is the north-western region especially (counties Donegal, Sligo, Mayo), that holds the majority of the most sought after of Ireland's waves (Figure 27.7). Here, a combination of factors come together to develop excellent wave conditions for surfing. First, much of this part of the coast faces northwards, which results in the island's predominantly south-westerly winds blowing offshore. This produces ideally shaped waves, in surfers' terminology, waves that are 'clean' and 'rideable'. Second, the coastline is well positioned to catch ocean swell waves, which are formed by the winter cyclones (low pressure systems) that track north-east across Ireland from the North Atlantic. These send wave-groundswell into the northward-facing bays and shores of this region, providing good surfing waves. Third, and most important, is the topography of the underwater coastal landscape in this region, which creates the wave conditions that attract surfers the world over to visit Ireland's shores (see Chapter 2: The Coastal Environment: Physical systems processes and patterns). Elsewhere, along much of the of the island's western coasts, similar and often comparable conditions are experienced.[10] Near-horizontal slabs of rock protrude into the sea, as with the limestone rock reefs of County Clare. These create what can be considered a perfect rock platform, which causes the Atlantic swell waves to break in the exact manner needed by surfers.

When such waves approach shallowing coastal waters they come into contact with the outer parts of these submerged rocks and reefs. Here, they are projected upwards and over the rocks, forming conditions of gradual wave-breaking. Put simply, waves move toward the shoreline at different speeds, depending primarily on the water depth. So, as a line of deep-water Atlantic swell comes into contact with a rock headland, or offshore reef, it slows as it reaches shallow water and 'bends' around the rock outcrops into the shape of the

Fig. 27.7 a. SURFERS IN TRAINING! INISHCRONE BEACH, COUNTY SLIGO. This glimpse of people from a local surfing school is now a common scene at Ireland's well-known surfing locations, where as many as upwards of a hundred would-be surfers can be seen learning the sport. [Source: 7th Wave Surf School]; b. 'Carving' the wave at Mullaghmore, County Sligo. Ireland's western coasts are rich in locations for good surfing. These are now attracting many visitors from abroad, particularly to the coasts of counties Kerry, Clare, Mayo and Donegal, where breaking Atlantic waves can be extreme! In County Clare, coastal sites, as at Doolin and Mullaghmore, frequently experience extreme wave conditions and attract inter-national surfing competitions, such as those sponsored by Red Bull. Here the participants can really show off their expertise by 'carving' and 'shredding' the waves. Noteworthy in this area of Mullaghmore and the Cliffs of Moher is Aileen's Wave, which can rise to 12m in height. Reputedly, it was named by a local surfer after the nearby headland of Aill na Searrach (the cliff of foals). [Source: Michal Czubala]

Fig. 27.8 LOCATIONS ALONG THE WILD ATLANTIC WAY. a. Keem Bay, Achill Island, County Mayo [Source: Vicuna R, CC BY-SA 4.0]; b. Coast view across Rossbeigh Strand and Dingle Bay toward the Dingle Peninsula, featuring the distinctive signage used to mark locations on the Wild Atlantic Way. [Source: Sarah Kandrot]; c. Doolin coast area, County Clare, cliffs view. [Source: Robert Devoy]; d. Slieve League, southwest County Donegal, valued as one of the most spectacular cliff coasts in Europe. [Source: Greg Clarke, Flickr CC BY 4.0].

coast (see Chapter 2: The Coastal Environment: Physical systems processes and patterns). When waves break like this (as technically 'plunging', or 'C-shaped' wave forms), they are called by surfers 'peeling' waves. The north-west of Ireland, and the central western coasts, experience many different types of such breaking wave-forms. For the surfer, these waves, especially in the more sheltered bay areas, can result in gloriously fun rides. These conditions allow surfers to 'carve' and even 'shred' the wall of an extreme wave, from top to bottom, for hundreds of metres.[11] In contrast, there are swell waves that impact and break directly onto the many underwater rock reefs that exist along coasts. These often create even bigger waves that can still be surfed, for those who have the combination of experience, skill and bravado to tackle them!

Surfing in Ireland is a unique experience. The combination of cold, wet, windy climate and the rugged coastline means that surfers are more likely to be climbing gates and traipsing through fields to check conditions, rather than hanging out on the beach getting a sun tan! Surfing here is sometimes a pursuit for the more eccentric. After all, it does take a certain type of character who is willing to brave the harsh winter elements for the fleeting thrills that surfing provides but, while surfing in Ireland might be less glamorous than other classic locations (such as Hawaii, New Zealand or Portugal), it does have a romance to it. You will find yourself exploring deserted parts of the most rugged coastlines, where one would otherwise have little reason to venture in winter. During times of torrential rain and freezing winds, little moments of perfection occur. When all the elements combine, they create the intoxicating feeling that comes with sliding across the wall of a fast-moving wave, hoping you have enough speed to outrun the crashing wave. When the wave ends, you paddle back hurriedly to the peak of the curling wave, getting the opportunity to soak up Ireland's uniquely beautiful coastline.

VALUING COASTAL AND MARINE TOURISM

Numerous definitions exist as to what constitutes coastal and marine tourism. A recent report prepared for the European Commission (EC) distinguishes between maritime and coastal tourism, with the former considered as being-water based and the latter beach based.[12] Tourism Development International defines marine tourism a little differently, as being the sub-sector of the tourism industry that is based on tourists taking part in leisure, holidays and journeys on coastal waters, or their shorelines.[13] Fáilte Ireland, however, uses a wider definition of marine tourism. This incorporates coastal and marine water-based activities and services adjacent to the coastline.[14]

This is the definition that is used in this chapter.

Within Europe, the importance of coastal and marine-based tourism to the social and economic well-being of its citizens is well established. This business sector's potential to foster a smart, sustainable and inclusive Europe is viewed as a key contribution to delivering the aims of Europe's Blue Growth Strategy into the twenty-first century.[15] Tourism, therefore, features within two of the four priorities of the Atlantic Action Plan, an initiative being implemented by the EU's Atlantic member states, as a means of supporting development within the marine and maritime economies of the Atlantic ocean region.[16]

THE WILD ATLANTIC WAY

Fáilte Ireland

The Wild Atlantic Way (WAW) has been promoted as a unifying and sustainable development concept for leisure and tourism along the western coasts of Ireland. The principal aim is to encourage increased numbers of international and home-based tourists to visit this stunning coastline (Figure 27.8). Currently, this is the longest-defined coastal touring route in the world, stretching over 2,600km along Ireland's highly indented coastline from the Inishowen Peninsula in County Donegal to Kinsale in West Cork (Figure 27.9).

The WAW route can be accessed at the northern end through the village of Muff in County Donegal and from the south through the town of Kinsale in County Cork, although visitors can also access the WAW at numerous entry points along its spinal route. The immediate catchment of the Wild Atlantic Way involves a corridor occupying land to both the east and west of the spinal route. In a busy and competitive tourism marketplace, the route focuses solely on the west coast of Ireland, and is intended to present a tourism experience and product at both a scale and uniqueness that will capture the imagination particularly of international visitors. Tourists, especially those arriving from outside the island, very often want an 'off the beaten track' experience that immerses them in multiple environments, so that they feel stimulated, energised and uplifted. The strategy is designed, therefore, to encourage all visitors to travel, explore and engage with tourism experiences and local communities resident along these coastlines and over as wide a geographical area as possible. The benefits of tourist expenditures can thus be dispersed to better facilitate development within these peripheral communities.

In 2014, a capital investment fund of €10 million was made available to bring the Wild Atlantic Way project to life.[17] It included the development of:

- a planned network of 3,850 road signs at 980 junctions along the route
- fifteen Signature Discovery Points and 173 Discovery and Embarkation Points
- a programme of ongoing trade and community engagement
- a suite of digital marketing platforms
- a rolling programme of domestic and international marketing and promotion
- a series of journey loops and spurs off the identified route, intended to encourage further tourist exploration and to increase the dwell time of international visitors.

The WAW covers an extensive area of c.40,000km^2 and, consequently, it is perhaps difficult for visitors to grasp initially its scale and the unique character of its diverse coastal environments. For this reason, six geographic zones have been identified to amplify the different

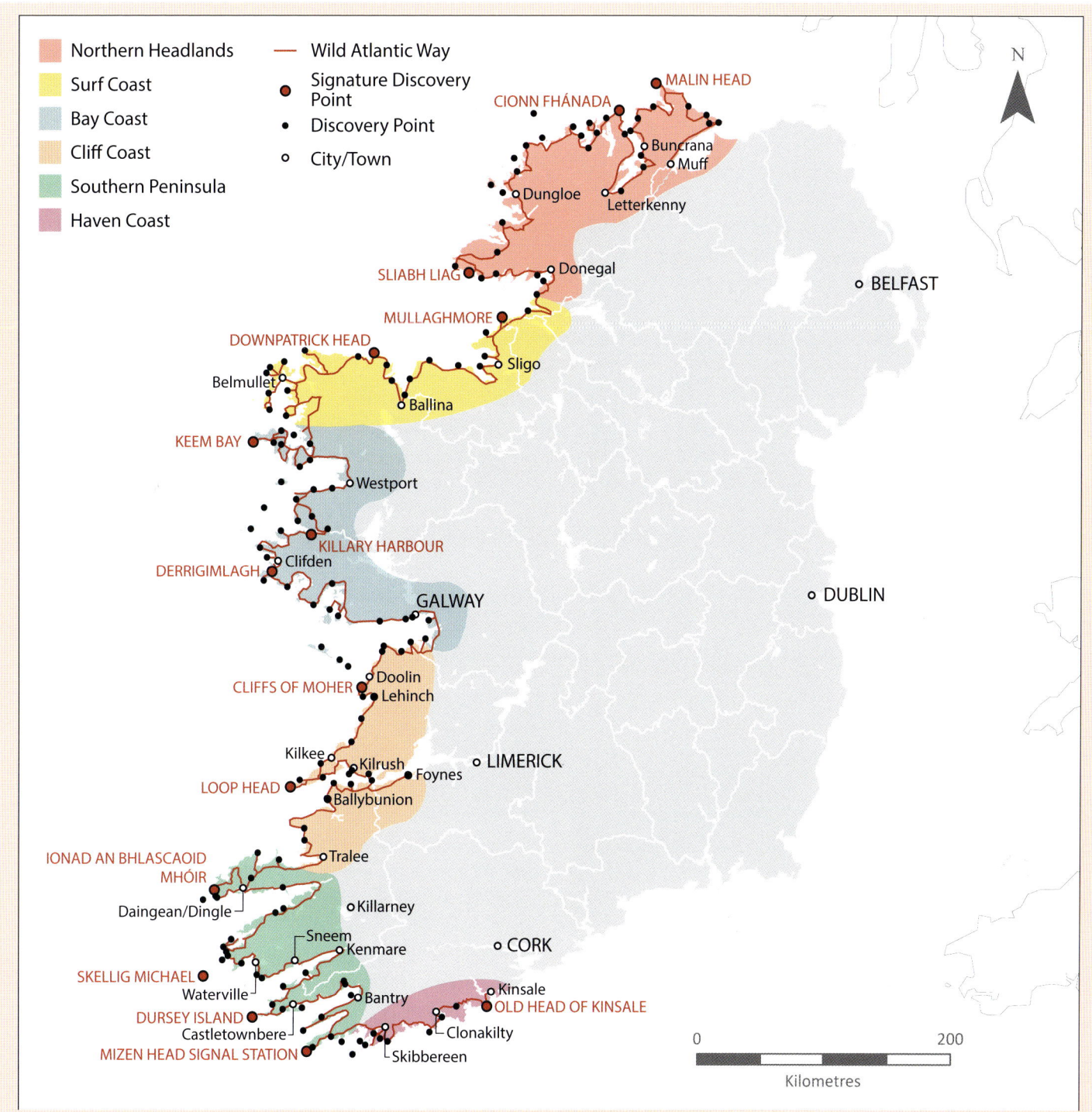

Fig. 27.9 THE WILD ATLANTIC WAY (WAW). The map shows the main regions, and their principal attraction designations, in Fáilte Ireland's flagship for coastal tourism development, as well as the route of the WAW. The route encompasses the coasts and hinterland of the nine counties of western Ireland: Donegal, Leitrim, Sligo, Mayo, Galway, Clare, Limerick, Kerry and Cork. There are 187 discovery points, 1,000 attractions and more than 2,500 activities along the route, which was officially launched in 2014 by minister of state for tourism and sport, Michael Ring. The success of the routeway has grown since it began, with, for example, the WAW becoming increasingly popular as a cycling route. More coastal areas in these regions now want to be put direct-ly on the route, in order to benefit from the positive multipliers gained from increased flows of tourists along the WAW. [Data source: Fáilte Ireland]

the unique character of its diverse coastal environments. For this reason, six geographic zones have been identified to amplify the different sections of the route and to make it simpler for consumers to orientate themselves, based on their motivations. The use of these zones allows, furthermore, the travel trade to better promote the destination brand of the Wild Atlantic Way through a series of distinctive itineraries of varying themes and duration.

It is important that the WAW is described and presented as a destination brand and not merely as 2,600km of tarmac. As such, the routeway is supporting currently over 2,500 tourism enterprises along Ireland's west coast. Most importantly, it provides opportunities for local businesses and communities to work together to develop tourism experiences that will increase the dwell time and expenditure of visitors in their particular locations. The branding also provides an umbrella marketing approach in key overseas markets for all businesses located along the route, thereby helping to establish a long-term, sustainable development project for this unique and fragile coastline.

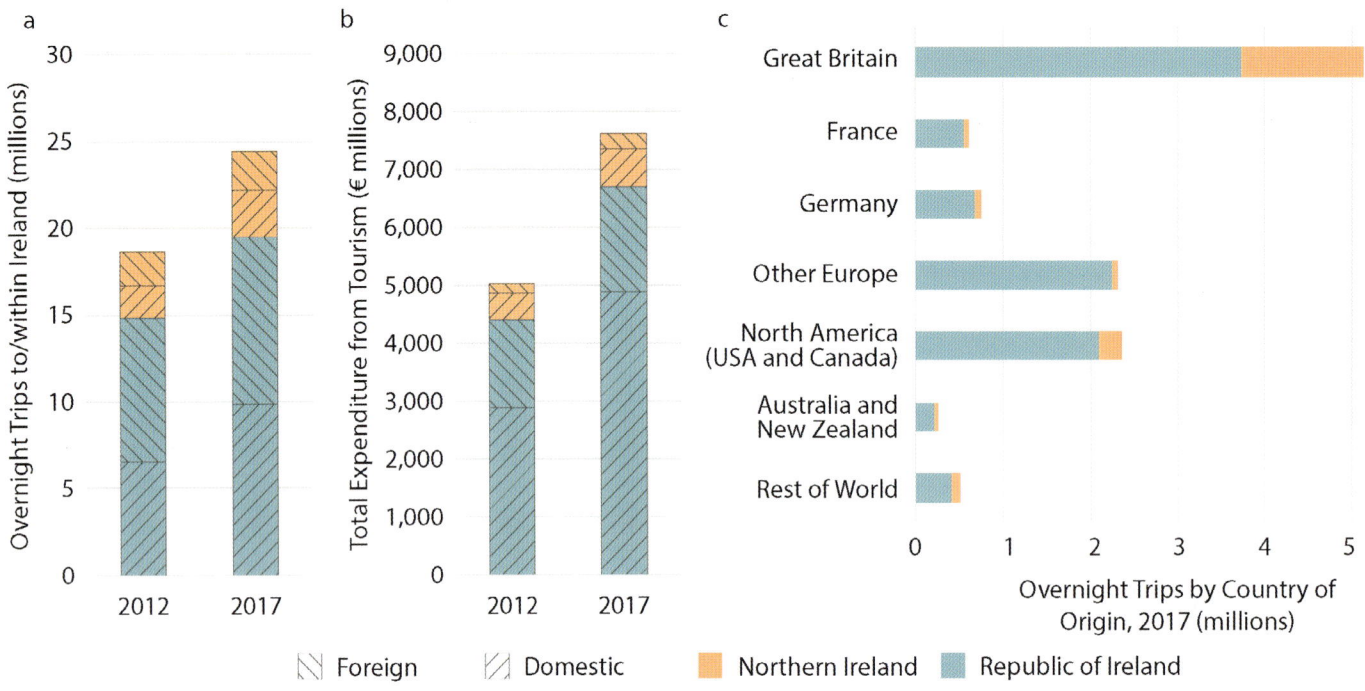

Fig. 27.10 TOURISM STATISTICS FOR IRELAND (2012–2017). Tourism is a major component of the island of Ireland's economy. Following an international recession, which reduced levels of global tourist traffic, the economic revival of recent years has encouraged an upsurge in this growth sector of world economies. Ireland has benefited from this trend and continues to be a magnet for international visitors, as well as being a dynamic market for domestic tourism. In 2017, 9.9 million overseas tourists visited the Republic and a further 2.7 million travelled to Northern Ireland. Britain is shown as the most important single country source of tourists, although in the Republic numbers from Europe (3.5 million) now almost equal those from Britain. The island of Ireland also generated almost 12 million overnight stays involving Irish residents. Approximately half of these stays were for the purpose of holidays. Overall, expenditure from tourism in Ireland amounted to some €7.7 billion. Most of this revenue is generated within the Republic and contributes significantly to the national well-being. Given that large numbers of these tourists are attracted to the more peripheral coastal and rural areas of the island, it is clear that this income is very significant. [Data source: CSO Statistical Year Book of Ireland, 2018; NISRA, 2018]

Fig. 27.11 MYRTLEVILLE BEACH, COUNTY CORK. Seaside holidays, day-trips and wider beach recreation, as seen in this summer beach tourism scene from Myrtleville, near Cork city, have become a well-established part of Ireland's coastal recreational activities since the end of the nineteenth century, and even more so from the 1950s. This is more popular amongst domestic tourists than international visitors and usually involves large numbers of people who reside in urban centres located in relatively close proximity to such beach environments. It is increasingly complemented in the present day by many other forms of leisure pursuits. These include, for example, sailing, surfing and other types of water sports; hang-gliding from Ireland's many rock cliffs and dunes areas and coastal walking. In 2017, domestic tourism involved almost 12 million people, and an expenditure of over €2 billion. It is, therefore, a vital component of the island's tourist economy. [Source: Sarah Kandrot]

At a national level, extensive references are made to coastal and marine tourism and to its potential contribution to Ireland's ocean economy, in *Harnessing Our Ocean Wealth: The integrated marine plan for Ireland*.[18] The importance of the coastal and marine tourism sector is evidenced by the fact that it is often seen as either a catalyst for economic recovery or a mainstay of economic performance. An example of this in the Republic of Ireland was the publication for County Cork of Marine Leisure Strategies. This strategy planning work was based at the municipal level on the different divisions of the Cork coastline, identified in the period 2008–2010. At the time of this stategy's release, these coastal divisions were dealing with contrasting economic conditions, but had the common objective of optimising the tourism potential of Cork's coast for the benefit of both the local and the regional economies.[19] Other local authorities in Ireland, such as Donegal County Council, have engaged with national and EU-funded initiatives to assist in the development of coastal and marine tourism. Similarly, the Irish Marine Federation and the Marine Institute have partnered in a number of European INTERREG programmes to develop aspects of the marine leisure and tourism sector within Ireland. These have included projects such as the Irish Sea Marine Leisure Knowledge Network (ISMLKN) and Marinas and Yachting in the North West Metropolitan Area (MAYA).

Estimates derived from surveys by Fáilte Ireland in 2014 suggested that coastal and marine tourism was worth

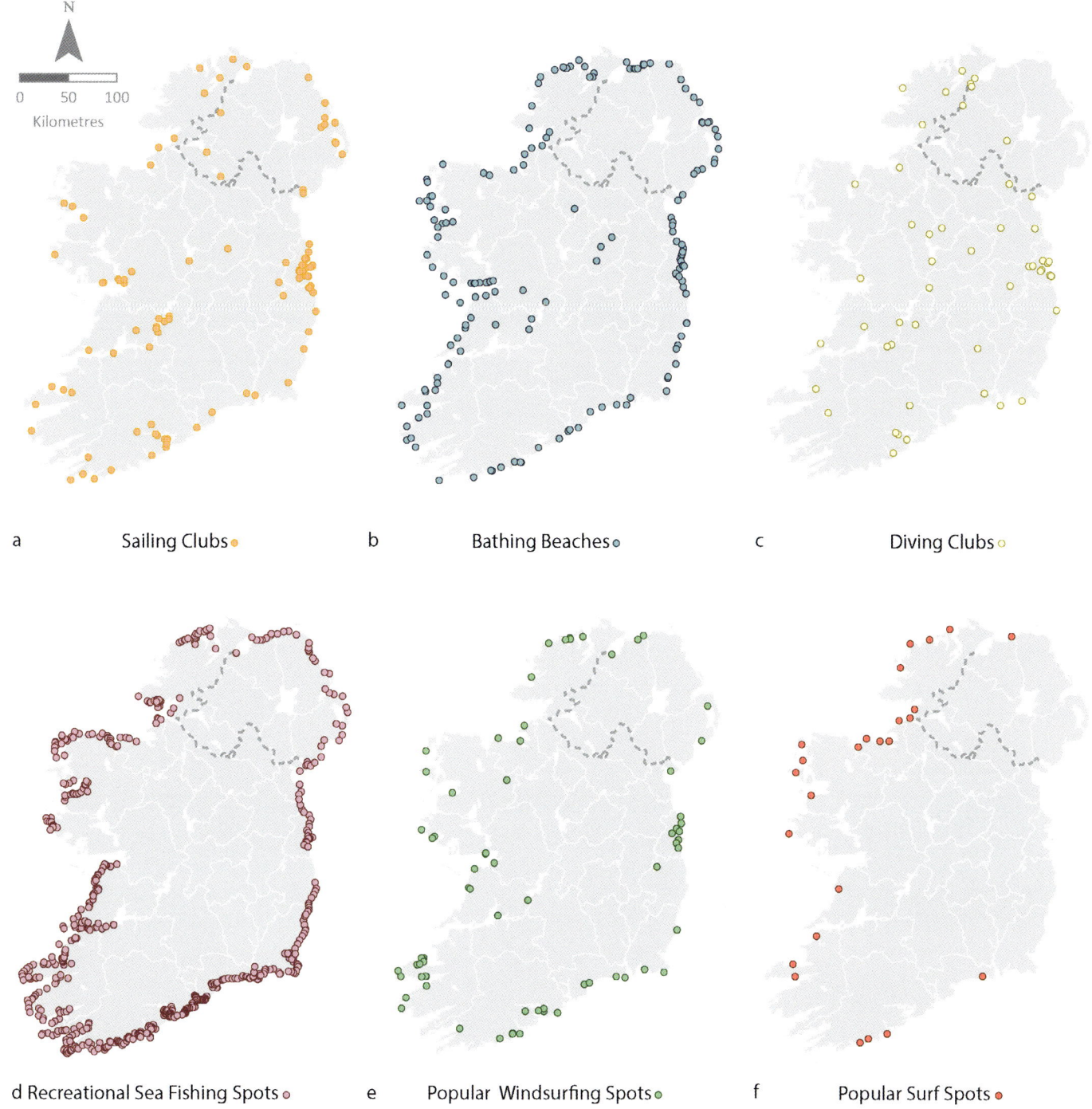

a   Sailing Clubs ●

b   Bathing Beaches ●

c   Diving Clubs ○

d Recreational Sea Fishing Spots ●

e   Popular Windsurfing Spots ●

f   Popular Surf Spots ●

Fig. 27.12 PRINCIPAL COASTAL TOURISM ACTIVITIES AROUND IRELAND'S COASTS. a. Sailing clubs. [Data Source: Irish Sailing Association]; b. Bathing beaches (2018). [Data source: ROI: EPA; NI: Department of Agriculture, Environment and Rural Affairs (DAERA)]; c. Diving clubs [Data source: Irish Underwater Council]; d. Recreational sea fishing. Data source: [Data source: ROI: Inland Fisheries Ireland (IFI); NI: DAERA]; e. Windsurfing. [Data source: Irish Windsurfing Association]; f. Surfing. [Data source: MIDA (2008) and Irish Surfing Association information].

approximately €2 billion annually to the Republic of Ireland's economy (Figure 27.10). Since that year, in which tourist expenditure from domestic and foreign visitors in the Republic amounted to some €5.3 billion, the tourist sector has continued to grow (Figure 27.11). By 2017, total expenditure levels from tourists reached almost €7 billion and would suggest clearly that the worth to coastal and marine tourism now greatly exceeds the value achieved in 2014.[20] Further, a survey conducted by the Economic and Social Research Institute (ESRI) for the Marine Institute showed that 1.48 million persons, representing 49 per cent of the

Republic's adult population, participated in some form of water-based activity (Figure 27.12).[21] In terms of specific events, the Volvo Ocean Race stopover in Galway was estimated to be worth €50 million to the economy.[22] A study conducted estimated the spend of each visiting boat to Irish marinas at €1,883.69 per boat, spent on visitor experiences in and around the marina area.[23] In addition, they found that the average per person per boat spend, excluding children, was €539.42, while the average including children was €513.73.

For Northern Ireland, comparable statistics for coastal and

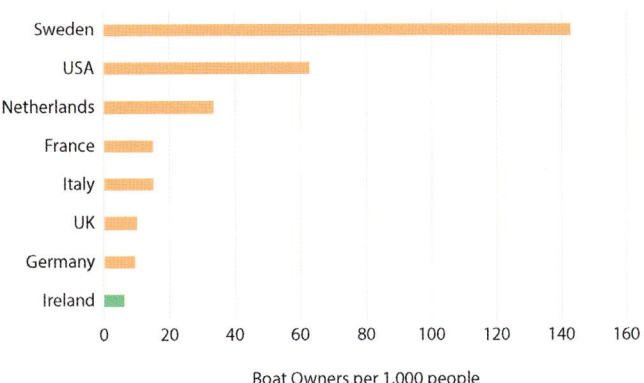

Boat Owners per 1,000 people

Fig. 27.13. BOAT OWNERSHIP RATES ACROSS EUROPE. These statistics for boat ownership have been compiled using widely accepted methodologies, developed by the International Council of Marine Industry Associations (ICOMIA) for estimating international boating activity. These data show that boat ownership in Ireland is only six for every 1,000 of the country's population. Most other developed countries have higher rates of ownership. This indicates a low level of engagement with water-based leisure pursuits on the island and may suggest that Ireland's population has developed, possibly through its history, an attitude to the sea that is centred more on its land-based activities, rather than on the seas around its coastline. In contrast, countries such as Sweden and the Netherlands have engaged far more positively with the sea, identifying the maritime environment as both a valuable resource to be exploited and an attractive outlet for recreation and leisure pursuits. Indicative of this is Sweden's boat ownership rate, which is over twenty times higher (143 per 1,000 population) than Ireland's. The low levels in Ireland do show, however, the potential for the growth of marine-based recreational activities in future. [Data source: Irish Sea Marine Sector Marketing and Business Development Programme, 2007]

marine tourism are not available, although information does indicate a steady growth and increased interest in coastal leisure pursuits. Those statistics show that the top two visitor attraction sites in Northern Ireland in 2017 were both associated directly with

aspects of coastal tourism. The World Heritage coastal site of the Giant's Causeway was the most-visited attraction, with just over one million people viewing this remarkable coastal formation (Figure 27.2). The second-most visited attraction (0.76 million visitors) was the Titanic Centre, Belfast. Here, attention is focused on the marine and historical interests that surround the historic shipyards of Harland and Wolff and the construction of the ill-fated *Titanic* ocean liner. Both locations continue to record significant increases in visitor numbers.[24]

The impact of marine tourism can also be seen internationally. For example, the British Marine Federation found that the economic impact of UK boating tourism is bigger than that of the tourism contribution from the London Olympics.[25] The statistic gives an indication of the scale of revenue that can be generated from such leisure activities. In addition to employment and income stimulated by coastal and marine tourism in Ireland, the marine tourism sector has provided significant contributions to a number of areas of national concern: addressing challenges of cleaner production and waste reduction;[26] advancing sustainable environmental practices;[27] and, contributing to human health and well-being.[28] Although the economic downturn experienced during Ireland's post-Celtic Tiger era affected the marine leisure tourism market, the natural asset that is Ireland's coastal and marine environment is still robust. It remains a critical resource and one that offers considerable potential if developed in the context of sustainable management. Not only should tourism experience a full economic recovery and future growth, but also, by acknowledging the importance of

Fig. 27.14 DÚN LAOGHAIRE MARINA, COUNTY DUBLIN, AN EXAMPLE OF A LARGE-SCALE MARINA DEVELOPMENT ON IRELAND'S COASTS. Many such developments are smaller than this, as at Ardglass, East Ferry and Rossaveal, but most have hopes of growing. There now exist many smaller developments of boatyard and jetty facilities, which are assuming informally the function of marina-style activities. [Source: Tim Wall/Dún Laoghaire Marina]

sustainability, it will be developed in a way that allows future generations to enjoy fully the diverse attractions of Ireland's coastal environments.

Given the value of the coastal and marine leisure experience on offer to both domestic and overseas visitors, it is surprising that the full potential of this economic sector remains unrealised in Ireland, an example being boating activities. Irish boat ownership per 1,000 population is quite low, in comparison with levels in other countries (Figure 27.13). In spite of this apparent detachment from boating, Ireland's two major sailing centres, at Dún Laoghaire (County Dublin) and Crosshaven (County Cork), both offer international competitive sailing events and regattas. Furthermore, they are acknowledged as being of such quality that allows them to compare well with other international events, such as Cowes Week (UK) and Kiel (Germany). There is now a significant economic opportunity to further develop these events. The race management standard is recognised as being world-class and Ireland is well-positioned to attract more international competitors. Clearly, Ireland has not reached its full potential in the marine leisure sector. Barriers exist that need to be addressed, particularly with regard to future national and regional planning policies, legislation and the provision of critical infrastructures.

MARINAS AND COASTAL TOURISM: THE CASE OF COBH

Liam Coakley

The number of marinas in Ireland has grown steadily since the 1980s and there are now over sixty marinas, or linked boat-berthing type facilities (such as pontoons and jetties), of varying size located around Ireland's coastline (Figure 27.14).[29] Most of the marinas are privately owned and managed. The permanent boat berths provided are generally leased to individuals and clubs, with some berthing spaces also available to visitors for a nightly fee. Estimates by the Irish Marina Operators Association suggest the value of marina activities to local communities, given that marina-based boat owners spend in excess of €34 million annually, with an average overseas visitor boat spend of €192 per night and an average length of stay of three nights.[30] In spite of their popularity, however, marinas may pose significant local or regional-scale development, community and economic challenges. Apart from needing to deal with issues of finding an environmentally suitable site, including meeting planning controls, there are also the longer-term impacts and costs of such developments. These may include, for example, roads and other built infrastructural improvements, new coastal protection works, increased demands on amenities, tourism accommodation, wastes disposal systems and boat maintenance and provisioning. Equally, such developments can generate local economic benefits, forming strong attractors for increased tourist use of coastal areas.

In Cobh and Cork Harbour, three areas for marinas have become established: East Ferry, Monkstown and Crosshaven. At present Cobh lacks such a facility in spite of a strong demand. Cobh (known formerly as Queenstown) is a small town of intense maritime activity. This is the largest town in Cork Harbour and one of Ireland's established heritage tourist destinations. By the end of the eighteenth century, Cobh was developing as a garrison town and transatlantic seaport. As the nineteenth and twentieth centuries progressed, industrial (for example, ship repairs) and residential urban satellite (to Cork) roles have been added to the earlier functions. In more recent years, Cobh has been re-imagined as a site for the localisation of Irish maritime heritage, and the town has forged a presence for itself in a number of key international tourist markets. Of particular significance has been the development of a dedicated cruise liner terminal, which is able to accommodate some of the largest liners afloat. In 2018, ninety-two cruise liners docked at Cobh with almost 160,000 passengers alighting to the benefit of the town and regional economy (Figure 27.15). Today, Cobh retains a very vibrant and palpable connection to the sea, with large numbers of tourist visitors experiencing this connection. Up to 50,000 tourists accessed Spike Island via ferry from Cobh in 2018.

Cobh people are also orientated towards the sea in their outdoor recreation and many residents participate in marine leisure pursuits. Sailing, sea-angling and motor-boating are popular. Some of these activities are facilitated through formal club structures, but most are not, and people in Cobh have historically used other maritime facilities existing in Cork Harbour (Figure 27.16). These are centred in the sailing hubs of East Ferry, situated about 8km to the east of the town; Crosshaven, on the Owenboy River estuary at the entrance to Cork Harbour; and Monkstown to the west. Cobh continues to lack much of the infrastructure needed to encourage local marine leisure activities, such as small boat marinas and yacht-club facilities. Instead, it has limited pontoon lengths. This gap has been recognised as forming an impediment to the continued development of Cobh as a maritime tourist town. Despite some setbacks to date, local stakeholders are active in seeking to develop a new mixed-leisure marina at Whitepoint, adjacent to Cobh's town centre.

This marina was first proposed by members of the Cobh Sailing Club in the late 1990s. Initial planning for development of the new facility began in 2010, when a group of local residents agreed to fund the planning and initial construction phases for the work. Regrettably, Ireland's Celtic Tiger period of prosperity was already coming to a close when the project started up. The subsequent catastrophic collapse of Ireland's economy in the period after 2008/2009 had a significant impact on the proposal to build a marina for Cobh. Ultimately, adequate financing for the project from external sources could not be secured by the small group of sailing enthusiasts and business interests that had been involved in initial seed-funding for the project. The development remains unfinished, but local stakeholders continue to work on behalf of the project with Cork County Council and local development organisations, such as the South and East Cork Area Development (SECAD). The Port of Cork is also in the process of seeking planning permission for a second berth for cruise liners in Cobh. The 300m pontoon, to be located at the eastern end of the town, will be available for boats, including yachts and superyachts, on the days when cruise ships are not visiting, which will equate to about 250 days per year.

The Cobh marina project is part of a wider structural engagement with marine leisure in Cork Harbour. Particular efforts are being made to create a new marine leisure infrastructural framework outside Cork's traditional sailing hubs. For example, new mixed leisure marinas have already been developed in the west of the harbour at Monkstown and in the east of the harbour, at Lower Aghada. It is hoped that these developments will do more than simply facilitate existing local marine leisure tourism, but will also help to stimulate increased commercial activity in the towns where they are situated. A key goal for this 'sea-to-shore transition' plan will be for the new marinas to be able to encourage the existing marine leisure enthusiasts to move beyond their traditional connections in the harbour region and create a wider range of local destination options for visiting marine leisure tourists from abroad. In pursuit of these aims, existing discretionary marine pursuits are being harnessed as catalysts in developing new businesses/commercial outlets onshore. It is hoped that the Cobh marina project, and projects such as the second cruise liner terminal, will have a similar impact on the Cobh area. In this way, this infrastructural development will serve to change the town from its historical position – as an outlier in people's thinking of the harbour's maritime leisure facilities – to a more central position in Cork.

Fig. 27.15 MSC *Meraviglia* LEAVING CORK HARBOUR. Tourism linked to a buoyant world market associated with cruise liners has developed significantly in the Republic. In the ten years to 2017, for example, the number of visitations by cruise liners increased from 130 to 234, while passengers alighting more than doubled from some 106,000 to almost 265,000. The ports of Dublin and Cork dominate this form of tourist traffic, with major benefits to the local and regional economies. This photograph of the cruise liner MSC *Meraviglia*, shown exiting Cork Harbour, illustrates the type and scale of such vessels visiting Ireland's ports. In 2018, this vessel of 168,000 tonnes, and able to accommodate 5,700 passengers, was the largest cruise liner to dock at Cobh's dedicated berth, which is located adjacent to the town's Heritage Centre. That year, ninety-two cruise ships visited Cobh, with more than 157,000 passengers and a further 69,000 crew alighting at the berth. Expenditure by these people in the town and at other tourist sites in the region (such as Blarney Castle and Midleton Distillery) amounted to about €12 million and has become a growing element of the local regional economy. In 2019 the Port of Cork received 103 cruise liners. [Source: Ryan O'Neill]

Fig 27.16 THE WATERFRONT AT COBH, COUNTY CORK. The view shows the town centre, St Colman's Cathedral and the waterfront quays. The sailing club is located further west along the waterfront at Whitepoint, which has been selected tentatively as the location for the Cobh marina development. [Source: Sarah Kandrot]

INFRASTRUCTURE

From the overall tourism perspective in Ireland, the ability to optimise coastal and marine resources is contingent 'on improved access to our shoreline, particularly using existing infrastructure'.[31] In order to advance the coastal leisure sector and to place it on a competitive footing with its international counterparts, it is important in future to both maintain and improve existing coastal and marine facilities and assets. An approach is required that blends investment in both current and future infrastructural needs, such as accommodation, transport and leisure outlets. Additionally, it should be recognised that these needs will vary greatly for different elements of the sector, for example, for yachting versus coastal walking. Of significance also would be the national to regional-level policies applied in meeting the development demands of the many different types of

settlement found on coasts, as characterised, for example, by the towns of Kinsale, Dingle, Kilrush, Kilkeel or Ballycastle.

The advent since 2000 of new tourism products in Ireland, such as the Wild Atlantic Way, or the more recent development of coastal heritage trails, should help in identifying the infrastructural investment needed. It is important to consider which locations will best support multiple service providers, or assist existing and start-up enterprises. In developing this thinking, the methods and approach taken by Cork County Council in dealing with these issues in recent years are worth examining. In County Cork, the auditing of coastal and marine infrastructures, products and facilities, to support strategic planning and investment in marine tourism, is a model that could be replicated elsewhere in Ireland.[32]

Dingle Town and Waterfront, County Kerry

Robert Devoy and Barry Brunt

This historic coastal town and harbour is a long-established centre for fishing and leisure boating activities in south-west Ireland, with a large marina facility developed in the 1990s within the harbour and fishing port (Figure 27.17). The town is also well known for Irish traditional music, which can be heard performed most nights in at least one or two of the town's pubs, as well as during music festival times. Dingle serves as a local and regional hub for tourists visiting the Dingle Peninsula and the wider coastal Kerry region. It provides

Fig. 27.17 DINGLE TOWN AND WATERFRONT, COUNTY KERRY. [Source: Barry Brunt]

many leisure attractions, including a large marine aquarium, many restaurants (offering varied cuisines including seafood specialities), marine excursions for fishing and 'whale watching', as well as the now-famous trips to spot Fungie, the dolphin, together with guided tours of the peninsula's coasts and its spectacular glaciated upland interior.

Within coastal locations, like Dingle, the development of a successful tourism and leisure industry must offer a range of essential infrastructures needed to satisfy an increasingly affluent tourism market. Central to this has been the growth and modernisation along Ireland's coastline of its small towns and villages. These urban places have become nodes where many tourists' demands can be satisfied. They offer commonly a range of accommodation options, from AirBnB and standard B&Bs to luxury hotels, restaurants, pubs and fast-food outlets, along with shops and centres for the retail of craft industries. Additionally, they often also provide other tourism and recreational needs, such as marinas and golf courses. In varied ways these small, rural urban centres have become important for local employment and income generation, which can spread into their immediate hinterlands.

Kinsale Harbour and Town, County Cork

Robert Devoy

This historic seaside town and port has its origins as a medieval fishing village, later serving from the sixteenth and seventeenth centuries onward as an important base for the British Navy and centre for shipbuilding. Following the First World War and the foundation of the

Fig 27.18 KINSALE HARBOUR AND TOWN, COUNTY CORK. [Source: Lisa Fawkes]

Irish Free State, the town went into severe decline as it lost its primary functions as a British naval port and garrison town. From 1911 to 1961, its population fell by 60 per cent, from 4,020 to 1,587, and many of its Georgian and Victorian buildings went into decay or were abandoned. Since the 1960s, however, Kinsale has been transformed, as it developed as a multi-faceted tourist centre (Figure 27.18). Key to this has been its emergence as an internationally acclaimed centre for cordon bleu-style cuisine and gourmet food. As a result, some people refer to Kinsale as the 'gourmet capital of Ireland'. Situated due south of the narrow rock headlands protecting the entrance to the Bandon River estuary, it is also now an established hub for yachts and other boating activities, with extensive marina and boat supply facilities. The town is linked into the Wild Atlantic Way and has developed a thriving regional to internationally scaled tourism and leisure industry. Amongst Kinsale's attractions are its history, with the town's narrow and winding streets and medieval street plan, old-style shops, ancient churches and the two prominent seventeenth-century, star-shaped fortifications protecting the harbour entrance of James Fort and the more spectacular Charles Fort. Apart from the marina and boating activities, other attractions include its many restaurants, shopping, sea angling, whale/cetacean watching, scuba diving, kayaking, climbing and abseiling on the areas rock-cliffed coasts. The town also forms a support service's centre to the nearby world-class golf course at the Old Head of Kinsale. Today, Kinsale is a thriving community with a population that has more than trebled from 1961 to its 2016 census total of 5,281.

COASTAL TRAILS AND IRELAND'S ANCIENT EAST

Cathal O'Mahony

Trails, or formalised walking routes, are recognised as a significant element of tourism. The development of trails in Ireland is being progressed rapidly in many coastal counties, particularly as a way of supporting tourism and recreation (Figure 27.19). These routeways often incorporate educational, well-being and environmental thematic features, such as heritage, crafts, food, wildlife biodiversity themes

Fig. 27.19 a. VIEW OF THE BEACH AT PORTMARNOCK, DUBLIN BAY. The beach here has been incorporated by Fingal County Council into a wider network of coast-centred trails and boardwalks. [Source: William Murphy, flickr, CC BY-SA 4.0]; b. Long-established coastal path in the Fingal area, north county Dublin, used regularly for local recreation (walking and dog-walking) and now part of a more extensive coastal walking route in this area. [Source: Jeremy Gault (MaREI)

and environmental interpretive information. The result has been the provision of a wide range of trail types and characters in Ireland's coastal areas. More recently, the popular development of Greenways has caught people's imaginations, and they exemplify the value placed on the infrastructure of trails in contributing to the portfolio of activities provided by the leisure and tourism economic sector.

Linked to the development of trails, which incorporate coastal landscapes and walkways, is Ireland's Ancient East (IAE). This new tourism promotional development by Fáilte Ireland is described as 'a visitor experience, encompassing the rich heritage and cultural assets that Ireland has to offer'.[33] Ireland's Ancient East is centred on the Midlands and Eastern regions of the country, covering areas outside of Dublin and east of the River Shannon (Figure 27.20). It extends from Carlingford to Cavan and south to Cork city, including east County Cork and east County Limerick. In many cases, trail development within IAE, and particularly the promotion of coastal walks, preceded its emergence. The launch of this initiative provides, however, a focal point for tourism in the southern and eastern regions of Ireland, in a similar manner to that of the Wild Atlantic Way on Ireland's western and southern seaboards.

Trails feature strongly within the narrative of the IAE and are promoted extensively within the marketing of the initiative. Focused primarily on heritage and history, this venture incorporates routeways that cover many other features designed to capture the interest of visitors (for example, food and brewing). The initiative also fosters a working approach, which is needed to facilitate the provision of access to many of the routes and distinctive human and landscape features that are critical to the attractiveness and success of trails. Coastal counties, such as Louth, Wexford, Waterford, Dublin, Wicklow and Cork, all offer trails within this marketing approach. These provide visitors the opportunity of exploring Ireland's coastline: whether in pursuit of artisanal foods along Cork's scenic coastline, tasting locally brewed beers while exploring the influence of Vikings in Waterford, or exploring wildlife along Wexford's award-winning 'Green and Clean Coast' designated beaches.

Given the heavy use of the many and varied trails established along Ireland's eastern and southern coasts, it is likely that the value of trails will continue to be a significant promotional element in Ireland's tourism industry. Further, by directing potential visitors to particular areas in need of economic support, the development of trail routes presents a number of distinctive benefits. These include increased levels of access to Ireland's natural and built heritage, supporting the tourism service sector and linked product suppliers, providing opportunities for environmental education and well-being, as well as fostering a partnership approach to support community engagement and rural enterprise.

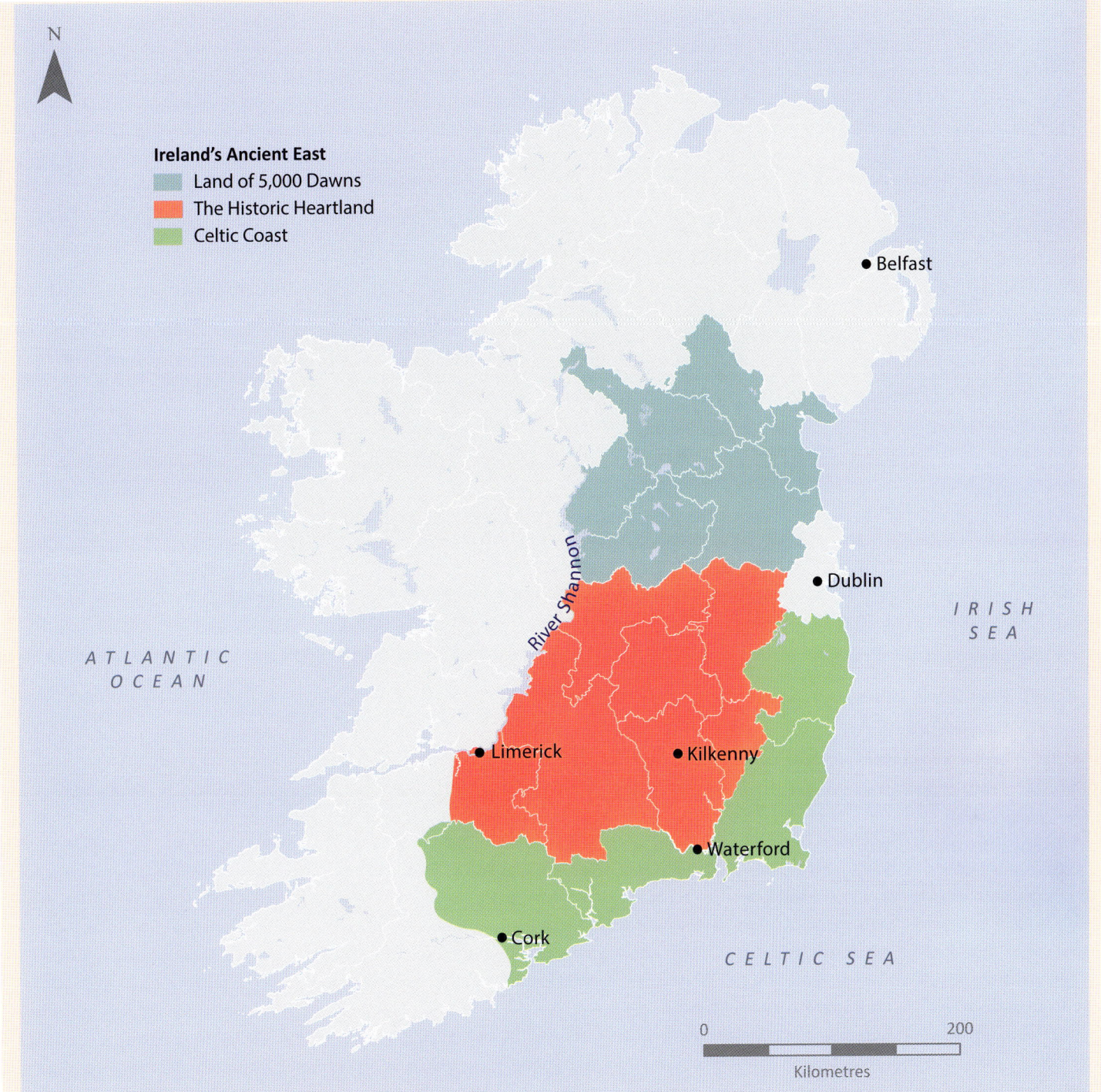

Fig. 27.20 THE REGIONS IN IRELAND INCLUDED IN FÁILTE IRELAND'S EAST COAST HERITAGE TOURISM DEVELOPMENT. [Adapted from: Fáilte Ireland, 2019]

THE CAUSEWAY AND MOURNES COASTAL ROUTEWAYS

Robert Devoy

Since the early 2000s, both the national and international tourism industries have become aware increasingly of the leisure resource value of Northern Ireland's coastal areas. The scenic and historic ingredients of its coastlines have been developed by tourism and other leisure interests as a range of designated way-marked routeways. The primary area promoted is that of the region's rugged northern coast, between Belfast and Londonderry (Derry), which has a great variety of spectacular sceneries, along with many historic sites to visit. This

Fig. 27.21 THE CAUSEWAY COASTAL ROUTE, NORTHERN IRELAND. References to the Wild Atlantic Way and Mourne Coastal Route are also shown.

route is known as both the Causeway Coast and also the Causeway Coastal Scenic Drive (Figure 27.21) and is linked in its advertising to the separate journey package to Rathlin Island (see Chapter 21: Ireland's Islands).

In its core section, the Causeway Coast covers over 130km of coastline, running across two counties and passing through three designated areas of outstanding natural beauty. The routeway is important in bringing visitors not only to these coasts, but also to neighbouring inland locations. It is comprised of more than forty sites of interest and is supported by widely promoted Causeway Coast tourism packages, which draw in many visitors from abroad.

The routeway has acquired quickly a significant international reputation, with Michael Palin describing the parts of the route accessible by rail as, 'one of the most beautiful rail journeys in the world' and, similarly, the road route is described by the planet D Blog, as 'one of the top road trips in the world. The Causeway Coast is centred on the UNESCO World Heritage geological attraction of the Giant's Causeway, County Antrim. This high-profile coastal site for international tourism forms a key part of the route. The Causeway Coast links up at its western end, at the village of Muff (County Donegal), with the final northern leg of the over 2,500km long Wild Atlantic Way, which is marketed separately by Fáilte Ireland.

Visitors can experience what the Causeway Coast has to offer through a wide variety of approaches and activities, including those of walking and heritage-type trails, package holiday tours, adventure and coastal leisure pursuits and from travelling parts of the coast by railway, but most commonly, by car and bus. Much of this coastal route uses A-class roads, with nine shorter loop-section drives on more minor roads. The routeway is promoted by Tourism Northern Ireland and linked agencies, together with many commercial travel and leisure companies. The main, northern Causeway route comprises a range of travel options, designed to appeal to people's different interests, either as visits to individual locations, or as part of different tourism packages involving a number of sites, or by travelling the whole routeway. Most people start this coastal tour in Belfast, reflecting the city's role as the region's capital and Northern Ireland's major international transport hub, with its two airports, railway and motorway road connections.

The landscapes of these northern coasts and of adjacent areas form the main attraction for most visitors, though these are best viewed in calm and sunny weather! High and often extreme-energy storm and wave action characterise these coasts, creating fast-eroding rock cliffs, as well as extensive sandy beaches and sand dunes (Figure 27.22). In addition, these areas of Northern Ireland are ones of ancient settlement, as at Mount Sandel and Fort, Coleraine, situated on the estuary shores of the River Bann, which date from the Mesolithic (Middle Stone Age), some 10,000 years ago (see Chapter 16: The Inhabitants of Ireland's Early Coastal Landscapes). Many historic towns, heritage-status buildings (for example, castles) and settlements are located either on or close to the route, together with artisanal-type industrial locations. These attractions include the Bushmills Irish Whiskey distillery, founded in 1784 and situated about 3km from the Giant's Causeway. This site forms a major draw to tourists in its own right, receiving over 120,000 visitors a year. An additional stimulus to people's awareness, and to the projection and marketing of the Causeway route and more widely of Northern Ireland's coasts, has been the filming of the acclaimed TV series, *Game of Thrones*. Much of the filming for this series has been on these northern coasts, such as the area of Ballintoy and Dunluce Castle (Figure 27.23).

The Causeway Coast has been added to more recently by two other scenic coastal routeways, on Northern Ireland's south-western coasts, of the Mourne Mountains region and the Ards Peninsula (Figure 27.21). The Mourne coastal scenic drive starts at its northern end at Strangford, the entrance to Strangford Lough, with Carlingford Lough forming the southern end of the route. The Mourne Mountains rise up sharply from this narrow coastal zone to heights of 850m (2,790ft), creating breathtaking views of the Irish Sea coasts. On a clear day it is possible to see across to south-west Scotland, or to the Isle of Man and beyond to Cumbria. At the heart of the drive are seaside towns, such as Ardglass, Newcastle and Kilkeel which, while picturesque, are busy fishing ports and boat harbours. The future of fishing in the region, though, is increasingly under review, particularly with the impact of Brexit (see Chapter 26: Coastal Fisheries and Aquaculture). Consequently, the expansion of the tourism and leisure-industry sector, centred on the use of its harbours for marinas and boating activities, as well as their seaside resort functions, will be of increasing importance for the local economy. A separate coastal drive of this region's northern Ards Peninsula begins at the seaside town of Bangor, close to Belfast, and progresses south along the 32km shores of the peninsula, from Donaghadee to Portaferry and on to Strangford Lough.

Fig. 27.22 CARRICK-A-REDE, CAUSEWAY COAST, NORTHERN IRELAND. A view west along this scenic, cliffed coastline and showing the rope bridge at this site, which connects the mainland to the former coastal rock promontory, now separated by coastal erosion to form the island of Carrickarede (right). Similar features can be seen in the distance, as lower elevation islands, reduced to rock pinnacles and sea stacks by the waves. A rope bridge was first built here by a local fisherman over 350 years ago, to allow ease of access to the island for salmon fishing. The modern bridge and the site is managed by the National Trust and forms a key visitor attraction on the Causeway Coastal route. [Source: Copyright Tourism Northern Ireland]

Fig. 27.23 *Game of Thrones* INFORMATION SIGNPOST, BALLINTOY HARBOUR, COUNTY MONAGHAN. This historic fishing harbour and the wider area is featured as the Iron Islands in the *Game of Thrones* television drama series. The harbour and its surrounds are part of the Causeway Coast. It is one of the many locations on the coast for fans of this famous television drama to visit, especially for dedicated *Game of Thrones* package tours, as well as for other tourists. The Fullerton Arms pub in nearby Ballintoy town is also a draw to visitors, as it has one of the ten *Game of Thrones* carved wooden doors, made from the Dark Hedges trees. [Source: Sonse, flickr, CC BY 4.0]

Additionally, in Ireland's ports sector, opportunities exist for small ports and harbours to transition their present infrastructures to service coastal and marine tourism, as a means of supporting their rejuvenation. This approach might provide much-needed facilities and also improve access to better support for coastal and marine tourism, as in the development of marinas. In the case of sailing and yachting, the ideal situation is one where marinas are not more than 20 nautical miles apart. This is to help provide safe havens for sailing boats during adverse weather conditions, particularly in circumstances where

Fig. 27.24 COASTEERING ON THE OLD HEAD OF KINSALE, COUNTY CORK. Coasteering is now a popular pursuit in the USA and Wales, for example, but is not so well established in Ireland, in spite of there being a physical coastal environment well suited to the activity. [Source: Bob Jones, CC BY-SA 2.0]

such conditions are unexpected, or atypical for that region and time of year. The distribution of marinas, jetties and slipways around Ireland's coasts, however, is patchy and regionally uneven (Figure 27.25).

On the west coast, there are no marinas between Fahan in Donegal and Kilrush Creek in the Shannon Estuary, although a new marina is being planned for Killybegs in Donegal. While some cruising boats do make the trip along the west coast of Ireland, many more would do so if the infrastructure was in place to support shorter intervals of travel along the route. The many natural harbours to be found along this coastline, together with the number of offshore islands, could encourage a sustainable tourism product, if only boats could land safely and securely. Facilities of this nature and scale can be provided through a series of 'yacht stations', designed to accommodate a small number of cruising boats but without providing a full marina. Kerry County Council have done this successfully in Portmagee, County Kerry, providing a safe haven between the locations of the Kenmare River and Dingle.

Apart from marina-type developments, the scale of investment required in the infrastructure for many leisure and tourism activities is generally much smaller. For example, all the surfer needs as minimum facilities is a safe place to park vehicles, access to the sea and a cold water shower. Further, recreation based on many other pursuits that occur in coastal areas – such as garden visiting, or restaurant and food tourism interests – are already well supported by existing infrastructures, like roads, signposting,

Fig. 27.25 DISTRIBUTION OF COASTAL MARINAS IN IRELAND. The development of coastal marinas in Ireland has a very patchy distribution, with a dominance of the facilities located on its eastern coasts. South-west Ireland, a primary region for boating leisure pursuits, has no large-sized marina between Kinsale, County Cork (apart from one at Lawrence Cove) and those of Dingle and Cahersiveen (County Kerry), 66 and 70 nautical miles away respectively. Although this coast has natural anchorages and small harbours in abundance, it has only a relatively small marina facility on Bere Island (Cork), a distance of 49 nautical miles from Kinsale. On western coasts even larger distances occur between marinas. [Data source: Afloat, 2013]

restaurants, publications and websites but there is always a case for the improvement of these. Similarly, the small-boat user needs only a slipway access to open water. In many places, however, particularly in urban coastal areas, such infrastructure does not exist, or is severely limited. Dublin city and region is a case in point, where access to public slipways is currently poor at best. Yet, Dublin as a major city on a large bay possesses extensive safe water areas suitable for recreational purposes. Although the major yacht clubs in the Dublin region all have access to open water, the general

public do not. This group, however, is from where much of the expansion in coastal and marine leisure and tourism will come. A relatively small public investment in such facilities could produce significant returns in revenue from such tourism.

In future, new tourism product developments that can thrive with only modest infrastructural requirements should be encouraged. These could involve, for example, rigid-hulled inflatable boat (RIB) touring; the development of opportunities to explore the natural and cultural heritage of the coastline, both water- and land-based; traditional activities such as sea angling, or more recent innovations such as coasteering. This latter pursuit is a physical activity that involves moving along the intertidal zone of a rocky coastline on foot, or by swimming, and without the aid of boats, surf boards or other craft (Figure 27.24). Clearly, there is room for the growth of the sport in Ireland.

Sea and Coastal Angling

Val Cummins

Fig. 27.26 SEA AND SHORE ANGLING ON THE COAST AT DOOAGHTRY, ARD NA MARA, COUNTY MAYO. [Source: Inland Fisheries Ireland]

Sea and shore angling, from beaches, rocky shores, sheltered harbours, or from a boat, is extremely popular all along Ireland's coasts (Figure 27.26). Whether in fishing for wrasse, pollock, bass, mackerel, flounder, a large ray, or for any of the multiple species found and targeted in this fashion by recreational fishers, the natural heritage of Ireland's coastal zone offers a unique experience to the angler all year round. Anglers can sometimes be spotted digging for bait on the beaches and estuary mud and sand flats, seeking species such as ragworm and lugworm, crab or razor clam. Saltwater fly fishing, and lure fishing using plastic lures, is also emerging as a popular new technique, alongside the traditional bait and hook mode. Both privately owned and charter boats are used for sea angling and this can be an important source of revenue in harbours around the coast. Inland Fisheries Ireland is the authority tasked with making these overall 'inland' fishery resources both sustainable and accessible to all users. In a review of the recreational angling sector in Ireland in 2012, Tourism Development International – on behalf of Inland Fisheries Ireland – found sea angling, along with salmon and brown trout angling, to be the most popular form of fishing for domestic anglers.

SAILING IN IRELAND

Val Cummins

Boating is an important tourism and recreational activity for marine users in Ireland. Despite the increased levels of participation and investment in yachting activity, little work has been progressed at the national level to promote yachting. More needs to be done to understand the carrying capacity of local harbours, or how to realise the opportunity to promote coastal areas, and enhance human well-being through sailing. An untapped opportunity exists to increase participation in sailing as a water-based sport, tourism and recreational activity.[34]

The Irish Sailing Association (ISA) is the national governing body representing sailing in Ireland, including all aspects of boating and sailing, including yachting, power boating and windsurfing. It has lobbied for more facilities and infrastructure around the coast, including critical marina and shore-based facilities. However, Ireland suffers from a paucity of marina coverage, especially along the exposed west and north-west coasts. In an era of recreational sailors keen to embrace new challenges – including sailing in harsher environments with more sophisticated sails and technology – the rugged aspects of the Wild Atlantic Way are appealing (Figure 27.27). However, opportunities to appreciate the scenic beauty of this stunning coastline from the sea are constrained by a lack of investment in marinas. The indented coast offers many favourable anchorages, but few inlets or harbours are accessible at all stages of the tide, or always provide the requisite shelter from wind and storms.

Cork Harbour is an excellent example of a safe anchorage. With naturally deep water channels, the harbour can easily be navigated year round. Founded in 1720 by Lord Inchiquin and friends, Cork has one

Fig. 27.27 CRUISING AROUND THE COAST OF IRELAND. Coastal cruising and offshore sailing activities are the pursuit of sailing enthusiasts on yachts that tend to be upwards of approximately 6m (20 feet). This image shows the renowned Irish sailor, Damian Foxall, preparing for a sail change in the Volvo Ocean Race, working on the bow as part of a crew. Irish sailors have also undertaken to sail in many single-handed races, such as La Solitaire du Figaro, or the Vendée Globe round the world race, which features 60-foot IMOCA class boats. Solo sailing requires extreme physical and psychological endurance. Closer to home, whether single-handed or part of a crew, Irish waters provide wonderful sailing grounds for gentle cruising, or for more adventurous racing. The Volvo Round Ireland Yacht Race, held every two years and hosted by Wicklow Sailing Club, sees a wide range of competitors from Ireland and further afield. As this race is non-stop, it does not rely on access to marina berths, except at the start and end of the race. However, the development of cruising in Ireland is hampered by a lack of marina facilities, especially along the west coast. [Source: Yann Riou]

Fig. 27.28 DOUBLE-HANDED SAILING. This type of sailing is becoming increasingly popular in both recreational and professional sailing. In this image, sailors representing Ireland are putting a catamaran through its paces. The boat is equipped with a trapeze harness, providing the crew with more leverage to counter the force of the wind in the sails. Dinghy boats are popular among novices and veterans alike. They range from high-performance dinghies, which are fast and powerful, to racing, cruising and classic dinghies. These boats can make sailing more accessible, as they can be relatively easy to tow and to launch. [Source: Damian Foxall]

of the oldest yacht clubs in the world. Originally known as the Water Club of the Harbour of Cork, the club is now known as the Royal Cork Yacht Club. Significant sailing events, such as the Cork300 in 2020 – celebrating the club's 300-year anniversary – serve to promote sailing, as well as generating a boost to local economies. Hubs for extreme ocean yacht racing such as Portsmouth in the UK, or Lorient in France, also serve as role models for how Irish harbours can develop in niche areas of the sport.

Apart from the promotion of racing regattas, a number of organisations promote cruising around the coast, including the Irish Cruising Club (ICC) and the Cruising Association (CA). The ICC produces the sailing almanac, a critical guide to navigating by sail around the coast. Sailing directions indicate hazardous features to avoid, such as rocks at the entrance to harbours. The Irish Sailing Association is the sport's governing body. As such, it has a different outlook and responsibilities. In early 2019, the ISA supported the opening of an elite sail training facility for the Tokyo Olympics in 2020. The success of Irish sailors in the competitive field can help to encourage the younger generation to take up sailing. In particular, Olympic sailor Analise Murphy and solo sailor Joan Mulloy have been an inspiration to aspiring female sailors (Figure 27.28).

One of the drawbacks in the development of the sport is the perception of sailing as the preserve of the well-off. As the barriers to entry to sailing come down, it is likely that recreational sailing will become more mainstream around the coast into the future. However, for sailing to become a viable source of alternative revenue for coastal communities, government investment will be required, to match a piecemeal, bottom-up approach from small boat owners of craft such as dinghies, which can be launched from a slipway or beach (Figure 27.28). Marinas are vital to derive the benefits from larger sail boats navigating along the coast, as demonstrated by the Sail Scotland initiative. Sail Scotland takes a strategic and collaborative approach, and reaps the benefits of the development and promotion of cruising there.

Case Study: Coastal Gardening

Verney Naylor

Travelling around Ireland's wild and glorious coastline, with its rocky cliffs, acres of sand dunes and marshy estuaries, you might be surprised to discover that, in spite of the very challenging conditions, many wonderful gardens have been created within a few kilometres of the sea. These range from large, historical ones such as Mount Stewart in County Antrim and Ilnacullin in County Cork, to smaller and more personal ones such as Caher Bridge in County Clare and The Glebe in County Cork. They include both public and private gardens with varying opening times. Many are featured in books, travel guides and on the internet. But, it should be remembered, gardens are ephemeral. They come and go, change ownership and, sadly, even become overgrown and derelict, so, it is advisable to check them out before struggling up too many muddy boreens! A good place to find the current status and opening times for coastal gardens, especially for the private gardens that have irregular openings, is online, or from one of the many available local garden trail leaflets, or on regularly updated websites for the different regions, such as those for west Cork, Connemara, Clew Bay, Donegal and Antrim (Figure 27.29).[35]

The Irish coastline is much indented, formally described as crenulate and exposed particularly to westerly moving storms. Whether gardens are in a fully exposed situation, facing directly into the open sea, or hidden away within a sheltered tidal inlet, they all experience, to some degree, similar conditions. It is the different ways in which gardeners have coped with these often challenging situations that makes their gardens so interesting to visit (Figure 27.30).

What are the challenges that seaside gardeners face? The main one is wind, regularly laden with salt. As a coastal gardener, even though I live 3km from the sea, I can still often find white speckles of salt

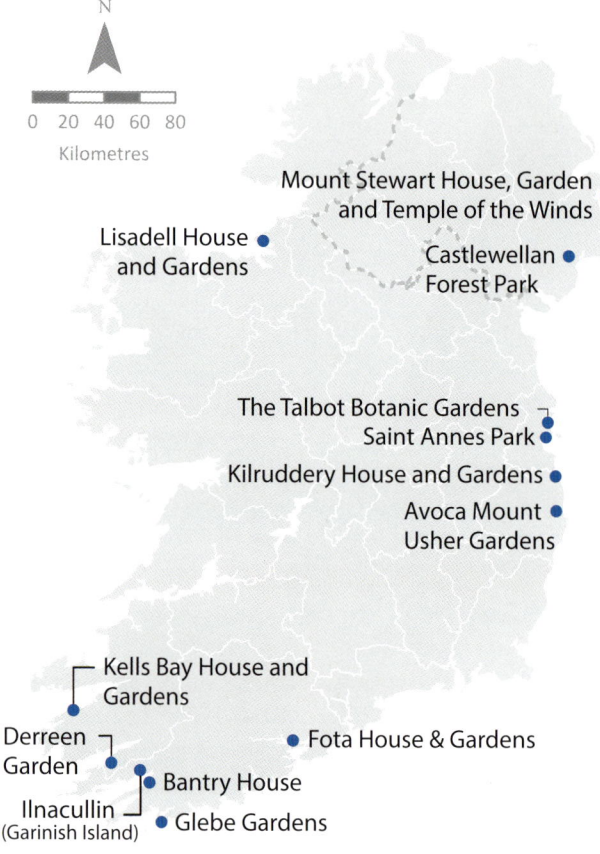

Fig. 27.29 LOCATION OF SOME OF THE ESTABLISHED GARDENS THAT HAVE DEVELOPED AROUND IRELAND'S COASTS. Many of these are adapted in their planting, character and displays to the local coastal landscapes and climates. [Source: Verney Naylor]

Fig. 27.30 COASTAL PLANTS AND GARDENS. a. Daisies (*Asteraceae*), an example of one of the many 'garden-escape' type plants now found growing abundantly and wild on Ireland's coasts; the view here is to Hook Light House, County Wexford. Other common examples of garden escapes on coasts include, fuchsia (*Fuchsia magellanica*), montbretia (*Crocosmia × crocosmiiflora*) and tree lupin (*Lupinus arboreus*). [Source: John Chessum, Wikimedia Commons, CC BY-SA 4.0]; b. Cliffs of Moher area, County Clare, ragwort or stinking willie (*Jacobaea vulgaris/Senecio jacobaea*), common in wild and untended places, as at the seaside. [Source: Rollant Pomaret, flickr, CC BY 4.0]; c. Heather (*Ericaceae spp.*) at Dunree, County Donegal, common on poor, peaty and acid soils of coastal rock cliffs. [Source: Greg Clark, flickr, CC BY 4.0]; d. Common fuschia (*Fuschia magellanica*) at Sherkin Island marina, and found on many coasts in southwest Ireland, particularly as hedges along the roadsides. [Source: Robbie Murphy]; e. Common gorse (*Ulex spp.*), at Ballycotton, County Cork. [Source: Darius Bartlett]; f. Thrift (*Armeria maritima*) in the Burren, County Clare, a plant common in rocky coastal areas throughout Ireland. [Source: Brendan Dunford]

appearing on the windows of my house after a westerly gale. The trees and hedges close to the sea have been bent over from the force of the prevailing wind. Some gardeners simply accept the wind and only grow low cushion-type plants that can take some salt. Others plant a shelterbelt of tough trees and shrubs around their garden, ideally well before any ornamentals are planted, so that the benefit of the shelter is already in place. Sometimes a hedge may be sufficient to stop the full force of the wind. Either way, it is wise to start with

Fig. 27.31 BANTRY HOUSE GARDEN. View of this formally laid-out and planted garden at the rear of the house toward Whiddy Island and of the low, 'rolling' glaciated drumlin landscapes of Bantry Bay. This coastally situated garden also contains many natural elements of the local flora. Built around 1700, Bantry House, called originally Blackrock house, was constructed on the south side of Bantry Bay. Since 1946 the house and gardens have been open to the public. [Source: Tony Lewis]

relatively small plants and allow them to adapt to the wind as they grow. Larger, top-heavy plants are more likely to be blown over before their root system is established. Shelter can be reinforced to protect new plants by erecting shelter netting as a temporary measure. These shelter methods work because they allow the wind to filter through but decrease its force. Walls and fences, on the other hand, completely protect the space immediately next to them. This protection, however, decreases gradually the further away from the wall you go and then the wind can cause turbulence, unless there are trees outside the wall. Despite this, walls can be really handsome and give a feeling of permanence to a garden.

As might be expected, most coastal gardens have spectacular views, whether to the open sea, a distant headland, a tidal river estuary or a rocky foreshore. These views change constantly, from high to low tide, from crashing waves to gentle, lapping wavelets, from winter storms to summer mists, sunsets and sunrise and moonlit paths on the ocean. A dilemma that many gardeners face is how to have both a view and shelter, because you can be sure the wind will always come straight out of the view but it is often possible (with careful planning) to have both. Breaking up a view with planting or built structures can provide the opportunity for a sitting area in different places according to the direction of the wind, or sun. The view can become even more interesting if seen from different angles (Figure 27.31).

Many coastal gardens are surrounded by rocky landscapes, so it makes sense to introduce local natural stone. Stone can be used to construct all sorts of walls from tall, neat estate walls, to lower, more old-fashioned walls made of irregular field stone. After a few years it is interesting to note how many plants will have seeded themselves into the cracks. Paving and gravel suit seaside gardens, especially in situations where lawns may get drenched in salt spray, or dry out and turn brown in drought conditions which can occur if the soil is thin and poor, maybe over sand or rock. Using natural stone – either the existing underlying rock or from a nearby local quarry – for paths and paved sitting areas gives a timeless look to a garden, as well as linking it to the surrounding geology and helping it to blend into its surroundings. Weathered boulders in various sizes, together with cobbles and gravel, accentuate the seaside nature of the location and, unlike trees and shrubs, stones and rocks are unlikely to be damaged in a storm! Gardens created in recent times are generally informal in their design, with today's gardeners appreciating a more relaxed mood and wanting to incorporate the nearby countryside and seascape into their view, for instance by introducing wild-flower meadows into their design. However, formality is easily managed in a walled garden with a regular shape. Walls provide shelter and privacy from neighbours.

Solid structures are used extensively in coastal gardens, as they are much less vulnerable to the prevailing, often difficult, climatic conditions. As well as using any existing rock outcrop and boulders, many man-made features help tie a design together. Pergolas and free-standing arches can be used to frame a view as well as acting as support for climbing plants. Some shelter from the wind can be found in a gazebo, summer house or bower. Having seats placed in several different locations makes a garden seem peaceful and restful, even if the gardener doesn't have time to sit down. These can be antique wrought iron, timber, logs or made of concrete, or merely a slab of rock. Ornaments such as sculptures and lovely pots, or containers, are often used to add interest and focus to the scene, but be careful in placing these, so that they do not 'argue' with an already existing beautiful view. They need to be placed where they make a 'picture' in their own right, maybe terminating the view at the end of a path, or seen through an arch or next to a seat.

The planting of a coastal garden, or indeed of any garden, forms the 'icing on the cake'. The boundaries, the lay-out, the paths and structures are all pretty permanent, but the plants themselves can change, often quite quickly, from season to season and year to year. This is where gardeners can have serious fun. If the soil is thin, then plants that suit these conditions can be chosen, or fresh top soil can be brought in to improve the growing medium. Despite the challenges of light soil, wind and salt, gardeners around the Irish coast have the major advantage of the ameliorating effect of the North Atlantic Drift lapping our shores (see Chapter 2: The Coastal Environment: Physical systems process and patterns). This allows Ireland to escape the extremes of winter cold and summer heat, so that the growing of many different tender plants is made possible, especially where there is shelter from the strongest winds. Look out for plants such as tree ferns (*Dicksonia antarctica*) and bottle brushes (*Callistemon spp.*) from Australia and New Zealand, red hot pokers (*Kniphofia*) and African lily (*Agapanthus* spp.) from South Africa, myrtles (*Myrtaceae*) from South America, Californian lilac (*Ceanothus* spp.) and tree lupins (*Lupinus spp.*) from the west coast, USA, and sunroses, or rock roses (*Cistaceae*) from the Mediterranean (Figure 27.32).

However, it can still be a 'hit or miss' situation. You need only one salty winter storm to turn evergreen leaves brown, or to blow trees right out of the ground, or at the very least to cause branches to die back. Luckily, there are some plants that appear dead after a hard

Fig. 27.32 DERREEN GARDENS, COUNTY KERRY. A well-known coastal garden, open to the public, which has made use of one of the many warmer micro-climatic locations of south-west Ireland, which are conditioned by the warm 'Gulf Stream' waters of the North Atlantic. Here, exotic plants of cool temperate forest and rainforest-type environments have thrived. These include plants originating in, for example, parts of South Australia, the Himalayas, New Zealand, western USA and Chile. Trees, understorey woody shrubs and other plants of these regions have been planted and have become established in this garden and in neighbouring coastal locations. Trees and other plants to be found at Derreen include eucalypts, tree ferns (*Dicksonia antarctica*), other ferns (for example, *Dryopteris spp.*), bamboos (*Bambusoideae*), sitka spruce (*Picea sitchensis*), hemlock (*Tsuga spp.*), other conifers and cypresses (*Cupressaceae*), together with elements of the native regional flora (such as oaks, alder ash, birch, holly). The vegetation is home to many wild birds and animals, including Sika deer, hares and red squirrel. Seals can often be seen from the shore, together with otters and Kerry's rarest mammal, the pine marten. Shoreline birds include cormorants, oystercatchers, gulls, great northern divers, guillemots and sea eagles. [Source: Verney Naylor]

winter, but will produce new growth from the base later in the summer, such as bottlebrush. Plants that have fine silvery hairs covering their leaves usually do well by the sea. For example, *Brachyglottis* (formerly *Senecio*) 'Sunshine' and the bush or golden daisy (*Euryops pectinatus*), both with yellow daisy flowers, the butterfly bush (*Buddleja davidii*) and the coastal daisy-bush (*Olearia* spp.) all generally survive rough weather. Shiny, waxy leaves also tend to tolerate seaside conditions, such as the African lily (*Agapanthus* spp.) and griselinia (*Griselinia* spp.). Herbs love the well-drained soils and the sun by the sea. Rosemary (*Rosmarinus officinalis*), in particular, always looks at home close to the seashore. Several cultivated plants have been so happy with our coastal conditions that they have escaped from our gardens into the wild. In July and August, the roadside hedges of County Kerry and County Cork are lined with the dripping crimson blooms of the common fuschia (*Fuschia magellanica*), originally introduced from South America, and the verges are set on fire by the vivid orange spikes of montbretia (*Crocosmia x crocosmiiflora*), hailing from South Africa.

The coastal gardens of Ireland are unique in their wonderfully romantic locations, the practical use of local stone and timber and especially in the incredible diversity of the plants that grow and thrive in them. Go, explore and discover them.

Case Study: The Burren and Cliffs of Moher Coastal Geopark – A model for sustainable tourism

Maria McNamara and Eamon Doyle

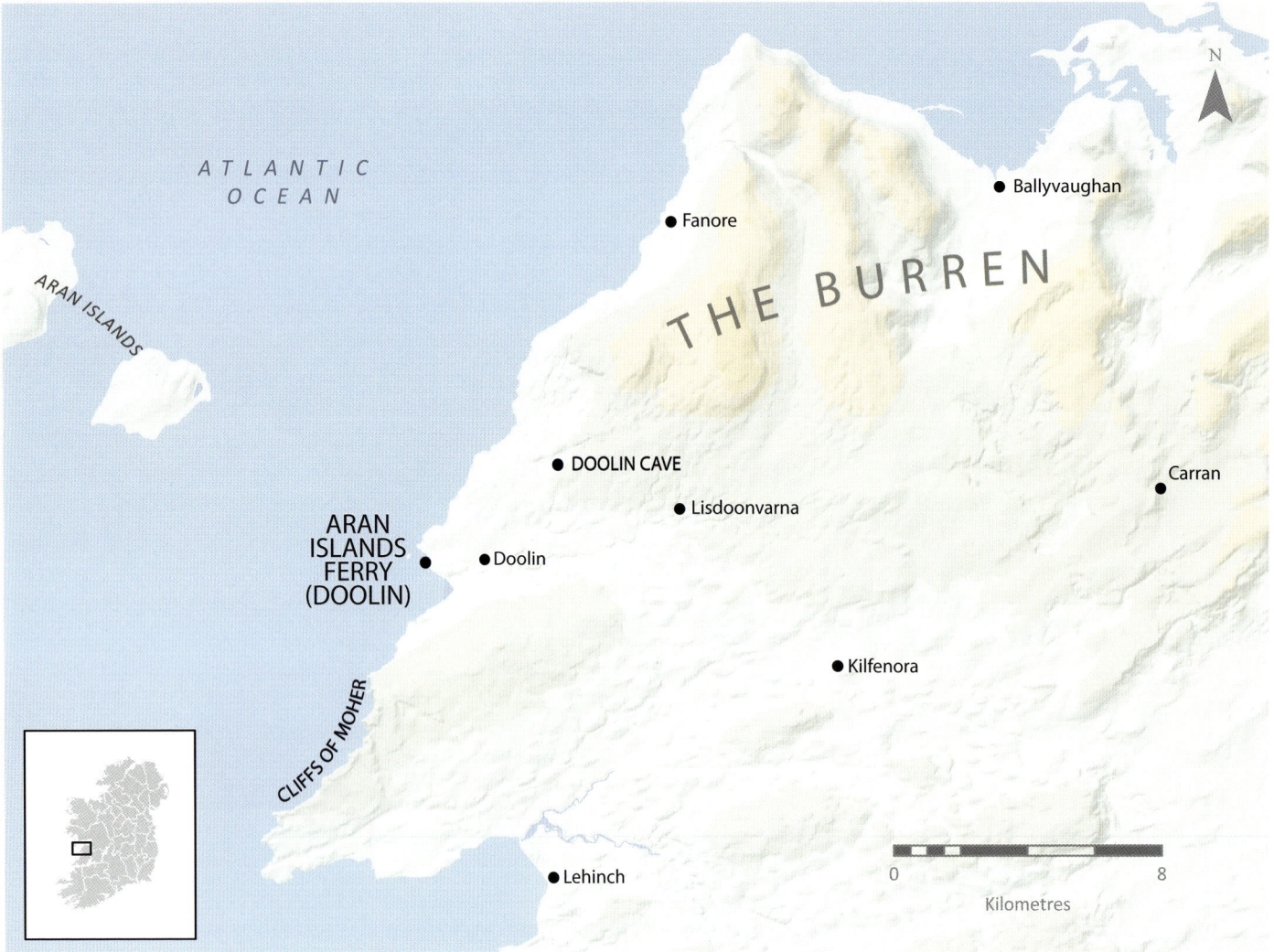

Fig. 27.33 THE BURREN, COUNTY CLARE. This area of the UNESCO Geopark and World Heritage site incorporates much of the northern part of the Burren region, including its coasts, from Doolin in the west, to Ballyvaughan in the north.

The Burren in north County Clare is renowned for its spectacular scenery and wild, haunting beauty (Figure 27.33). Its distinctive landscape is a rich tapestry of natural and cultural features. These include extensive exposures of bare rock, archaeological monuments spanning at least 5,000 years, a diverse and unique flora, several hundred kilometres of dry stone walls, and small towns and villages in which many traditional features of Irish rural life persist (Figure 27.34). The coastline of the Burren passes south into the Cliffs of Moher and together comprise one of the most dramatic coastlines in Ireland (Figure 27.35). It is, therefore, not surprising that this region is one of Ireland's primary tourist attractions, receiving over 1.5 million visitors each year. Tourism is a valuable source of income for many businesses and families in the Burren, but it also brings both opportunities and challenges. Managing visitors, while maintaining and protecting the landscape and traditional ways of life, have been major concerns for local people over the last couple of decades. This is also a central focus of activity for key stakeholders, notably Clare County Council, Fáilte Ireland, the Geological Survey of Ireland, the National Parks and Wildlife Service, the National Monuments Service, the Burren branch of the Irish Farmers Association (IFA), Burrenbeo Trust and the BurrenLIFE programme, now known as the Burren Programme.

Fig. 27.34 THE BURREN COAST, WITH A VIEW TOWARD SLIEVE ELVA AND THE BARE, LIMESTONE ROCK CENTRE OF THE GEOPARK. Inset image is of the spring gentian (*Gentiana verna*) (the blue flower), with *Lotus corniculatus* (bird's-foot-trefoil). [Source: Robert Devoy] The gentian is an example of the Burren's distinctive Lusitanian flora which, in part, is seen as inherited from the former Arctic environments of this area from the last Ice Age. [Source: David Hodgeson]

A UNESCO GLOBAL GEOPARK

Most of the Burren is designated as a Special Area of Conservation (SAC) and is protected under Irish law and EU Natura 2000, together with subsequent environmental legislation.[36] The development of sustainable tourism in the region has been accelerated and facilitated since 2011 when it achieved UNESCO Global Geopark status. A Geopark provides a holistic approach to tourism, supporting and encouraging people and organisations to work together to promote better understanding of local heritage, custodianship of the landscape, sustainable tourism and a vibrant community with strengthened livelihoods. Designation as a Geopark has ensured that the Burren has a cohesive management plan and integrated programmes for education, conservation and the sustainable development of the local economy. These provide a critical vehicle for managing tourism sustainably in one of Ireland's most dramatic natural landscapes. The Geopark also has high international visibility, marketed directly as a destination through the Global Geoparks Network, and via the inclusion of several of its specific attractions as Discovery Points on the Wild Atlantic Way. The entire Geopark is also located on the European Atlantic Geotourism Route, which supports the development of geotourism in dramatic landscapes along the European Atlantic coast.

Fig. 27.35 CLIFFS OF MOHER, COUNTY CLARE. The Atlantic cliff-top views attract large numbers of tourists to the area, who have been responsible for establishing the many informal paths at the cliff edges, which are highly dangerous! [Source: David Lee, flickr, CC BY-SA 4.0]

The Burren and Cliffs of Moher UNESCO Global Geopark is managed by Clare County Council, and also receives support and funding from the Geological Survey of Ireland and Fáilte Ireland. Management by the local authority, with support from key national organisations and the local population, is a criterion of UNESCO Geopark status. In particular, the role of the local authority is critical to the park's viability and to the success of linked initiatives. This is in both the provision at present of financial and logistical support and for its long-term planning and strategic development.

Today, the park is a dynamic and successful partnership at the destination scale, between local businesses, villagers and municipal authorities. As a result, it emphasises a community-based, collaborative approach to marketing the region's tourist facilities and activities, as well as encouraging the development and use of locally produced inputs and marketing materials to better support diverse aspects of local heritage. Furthermore, such a strategy facilitates improved connections between disparate tourism facilities, which allows for a more effective redistribution of the income generated by the sector.

Aside from the dramatic natural landscape, one of the key attractions of the Burren is the range of activities available in the region (Figure 27.36). In addition to eleven visitor centres, the Burren offers diverse opportunities for ecotourism through, for example, cycling, walking, heritage trails, maps and interactive apps, geosites, discovery points, cave tours, fishing and many different adventure sports. Several annual festivals have also been developed based around local geology, heritage and food, the latter encouraged, in particular, by a plethora of local food producers. In aggregate, the rich diversity of activities are all underpinned by a strong local culture and traditional music scene.

Central to the Geopark's strategy of community-led development is ensuring the effective distribution of educational information relating to the region's natural environment, heritage and the need for conservation. In this context, the Burren's different visitor centres are obvious hubs for learning and are utilised by local adults and schoolchildren, as well as tourists (Figure 27.37). In addition, the Geopark has developed, or has been involved in, a number of important environmental education initiatives in recent years. These include an evening course for local adults and a course in introductory geology, accredited by the National University of Ireland, Galway (NUIG), both of which are ongoing. Such initiatives dovetail with a programme of regular heritage-based public walks and evening lectures delivered by the Geopark and its associated partners throughout the Burren region. The provision of opportunities to learn about local heritage has been identified as an effective and important way to promote confidence and a sense of pride in the local natural environment. Critically, this also raises the profile of, and enhances public support for, the Geopark and other local educational organisations.

Much of the recent success of the Burren as a region is linked to its increasing commitment to ecotourism and its dedication to caring for the natural environment and balancing the needs of tourism and conservation. Through this approach, the park is able to provide a high-quality visitor experience while also improving local livelihoods and promoting conservation. Today, the Burren and Cliffs of Moher Geopark is recognised as a global leader in sustainable destination development, and has spearheaded various key ecotourism initiatives

Fig. 27.36 TRAILS AND WALKWAYS THROUGH THE RUGGED LIMESTONE TERRAINS OF THE BURREN GEOPARK. The horizontally bedded, massive rock-beds of the Burren's limestone geology can be seen out-cropping in the landscape above the walkers (centre foreground). The rock structures result often in the high coastal cliffs of the region, frequently reaching heights of above 30m. Large boulders of limestone, left behind from the last Ice Age, can also be seen here in the foreground. [Source: Maria McNamara]

Fig. 27.37 CLIFFS OF MOHER VISITOR CENTRE. This was set up in 2007 at a cost of €31.5 million and in 2017, for the fourth year running, received over one million tourists. The centre attracts the majority of the many visitors to the Geopark. It has extensive car parking, restaurant and linked facilities to cater for the numerous coach party groups and visitors' cars. Its design and construction, though perhaps controversial, is intended to fit into the existing landscape of the Cliffs of Moher area. Most importantly, the facility serves as an interpretative and educational centre that, using both fixed and interactive displays, provides information to bring alive the people, history, environment and marine life of the Burren. [Source: Barry Brunt]

that have proved critical in connecting the diverse number of stakeholders. Such actions have been recognised at the international scale via receipt of the prestigious Destination Leadership award for the 'tourism for conservation' project, GeoparkLIFE. This project is one of many highly successful ecotourism initiatives in the region, including the Burren Ecotourism Network and Burren Food Trail.

GeoparkLIFE was a project established specifically to develop an effective balance between the tourism interests and conservation needs within the Geopark region. The project was supported by the EU LIFE programme, together with various local stakeholders and operated from 2012 to 2017. It focused on developing a sustainable approach to local tourism, enhancing conservation actions by upskilling locals in conservation management and developing and testing new strategies for managing these activities at sites of high heritage value. As a result, the project delivered new management proposals for seven key sites in the Burren, linked with surveys undertaken on specific tourism pressures and their impacts on heritage sites, landscape and community. In addition, the project developed and promoted the incorporation of diverse conservation programmes into the local community, such as Leave-no-Trace, Meitheal, community-led Management of Monuments, Adopt a Hedgerow and environmental awareness initiatives with second-level students.

The Burren Ecotourism Network was created in 2008 by local businesses that had a shared vision of establishing the Burren region as a premier destination for sustainable tourism. Within the context of the Geopark, the work has grown substantially, to involve over fifty members who receive tangible benefits from their involvement in this network. These can include support with branding, marketing, packaging, information on funding, higher standards, enhanced training and skills, cost and energy savings, status, and better relationships within the local business community. Arising from the network was the creation of the award-winning Burren Food Trail (Figure 27.38) and the Burren Outdoor Adventure and Activity Trail. Together with the GeoparkLIFE project it has also developed support systems, such as a Code of Practice for Sustainable Tourism throughout the Burren.

Fig. 27.38 THE BURREN FOOD TRAIL. This is another highly successful, award-winning ecotourism initiative in the region. Created in 2013, the Food Trail highlights the importance of connections between growers, producers, chefs and retailers. The trail promotes the Burren region as a gastronomic destination, with a strong code of practice underlining the importance of environmental management. The Food Trail has won various awards, including the 2015 European Destination of Excellence for Tourism and Local Gastronomy, for developing a tourism offering based on local gastronomy that balances sustaining the local environment and promoting viable tourism. [Source: The Burren EcoTourism Network Company Limited]

677

ONGOING ISSUES AND FUTURE PLANS

Despite these successes, sustainable tourism in the Burren continues to face challenges. Coach tourism is one issue of increasing concern, both due to the volume of traffic and its distribution. A 2014 study commissioned by the Geopark revealed that in that calendar year, approximately 571,000 people visited the Burren region via coach. For 86 per cent of coach traffic, however, the Cliffs of Moher was the major destination. The total number of visitors to the Burren region is now well over one million. This bias towards the coast and, in particular, the Cliffs of Moher, leads to a highly unequal distribution of tourist-generated revenue within the Burren, which was valued at €6.9 million in 2014. It also causes the congestion of certain local roads during peak tourist season. The survey also revealed an annual total of

Fig. 27.39 THE COAST DUNES AT FANORE, WITH VIEW NORTH TOWARD COUNTY GALWAY. These dunes are part of the Wild Atlantic Way within the Burren Geopark. The area is serviced by large car parks and linked leisure facilities; these include facilities for surfers and wind-sailors, which are popular pursuits on this coast. At present, no tourism interpretive centre exits here, only information boards. At summer peak season, traffic congestion may lead to the complete blocking of access roads to this coast. The dunes have long attracted beach and other forms of coastal recreation, with established caravan parks and the development of some second homes in the Fanore area. These activities have put significant pressure on the dune environments, causing some erosion of the dunefield and have required significant investment in the coastal management of the beach and dunes. [Source: Robert Devoy]

approximately 20,000 coach stops occurring at attraction sites, which involve over 600,000 passengers alighting. Such traffic problems raise concerns about safety, access and environmental impact in the Geopark, albeit at a highly localised scale. Furthermore, this problem is likely to worsen under the strong emerging brand of the Wild Atlantic Way. Fanore forms a good coastal example of these and other linked contemporary environmental management issues (Figure 27.39).

Worryingly, there is evidence that some tourism sites are already close to capacity, with notices encouraging off-peak coach traffic. High levels of coach usage increases the risk of recreational damage, as large visitor numbers, even locally, are not consistent with the ethos of ecotourism. It also feeds into the broader issue of unequal tourism revenue across the region. The 'inner Burren' (that is the region surrounding villages such as Carran) typically attracts much less tourist traffic than the 'outer', coastal, areas. This is linked, in part, to the nature of the transport infrastructure, with the roads in the inner Burren being unable to accommodate large coaches. This problem is one of the fundamental driving forces behind the creation of the Burren Ecotourism Network and the Burren Food Trail as initiatives to raise the visibility of geographically diverse tourism sites and activities throughout the region.

Coastal Food

Regina Sexton

The work of Myrtle Allen, with Ballymaloe House and Restaurant (situated close to Ballycotton, County Cork), has been important nationally and internationally in popularising food and food-linked leisure in Ireland since the 1960s. Today she is joined by many others, whose restaurant and food-based activities are important attractions in bringing visitors into coastal areas and in stimulating other local businesses (Figure 27.40)

The innovative restaurants and cuisine that have been developed, as exemplified at Ballymaloe, attract significant numbers of visitors both from home and abroad each year into coastal areas. Coastal restaurants and food businesses are an important element of the restaurant industry sector, which employs around 72,000 people in Ireland and contributes €2 billion annually to the national economy.

The cuisine developed at Ballymaloe and elsewhere forms a new style of lighter seafood cookery with a distinctive Irish signature, one that challenges the former hegemony of French classic cuisine. The cookery style is built upon the supply of local produce, particularly fresh seafood and garden-grown food. This contemporary Irish approach characterises the work of a number of high-profile chefs, including Ross Lewis of Chapter One, Dublin; Billy Whitty of Aldridge Lodge, County Wexford; Marie-Thérèse and Ruairí de Blacam, Inis Meáin Restaurant, Aran Islands; J.P. McMahon, Aniar Restaurant and Tartare Café, Galway; and most recently Rob Krawczyk, Restaurant Chestnut, Ballydehob, and Robert Collender and James Ellis of Mews, Baltimore, County Cork.[37]

Fig. 27.40 BALLYMALOE HOUSE RESTAURANT. Wild salmon on the summer buffet landscape. [Source: Ballymaloe House]

GOLF TOURISM AND COASTAL GOLF COURSES

Barry Brunt and Robert Devoy

During the twentieth century, golf emerged as an increasingly popular sport and a significant component of a growing tourist trade. Pressures of golfing on existing courses and demands for the construction of new courses increased dramatically due to the growing numbers of both domestic and tourist golfers. In 2018, there were over 34,000 golf courses located throughout the world.

The development of golf in Ireland dates from the mid-nineteenth century. Its rapid growth, however, occurred principally from the 1960s, associated with the increasing ease of international travel and the commencement of strong economic growth that created a more affluent Irish population. In Ireland, almost 0.4 million people play golf with an additional 0.2 million golfing tourists attracted to play on the island's many golf courses (Figure 27.41).

In 2018, Ireland possessed over 300 golf courses, many of which offer a range of environmental conditions set to really test the golfer's skills. Just over one-half of the courses are located along the coast and many of them have become especially attractive for visiting golfers. This is due, in particular, to the fact that Ireland's coasts provide ideal conditions for the location of some of the finest links golf courses in the world. The term links became associated rapidly, in the early stages of the game's development, with coastal stretches of land. These lands have become almost useless in modern farming, but have proved attractive for the development of golf.

As the game evolved, links golf became identified with very specific environmental characteristics. These include, an undulating terrain, which is dominated commonly by sandy soils; few, if any, trees and the absence of inland water hazards; relatively narrow fairways, surrounded by extensive areas of vegetation. In Ireland, links courses are situated in areas of extensive coastal sand dunes, which are set generally within larger lowland bays and beach areas, as at Waterville, Lehinch, Ballyliffin and Portmarnock. These courses are comprised of sand dunes, parts of which remain active to coastal changes. They are covered semi-naturally with extensive areas of seaside grasses (for example, marram grass, *Ammophila* spp., and lyme grass, *Leymus arenarius* Hochst.), but with a conspicuous absence of trees. The environmental conditions presented to golfers by these links golf courses are very challenging. This is due especially to the exposed nature of these coastal courses, making wind an additional and major factor for concern. Judging the flight of the ball is a problem, as is finding any mishit ball that flies off the narrow fairways and into the deep 'rough'! This combination of different and such difficult conditions are rare at the global level, with almost all the world-class links courses being located in Britain and Ireland.

Although Ireland possesses less than 1 per cent of the world's golf courses, it currently has over 25 per cent of all links courses. This provides coastal Ireland with significant advantages in a very competitive market and in the business of attracting affluent international

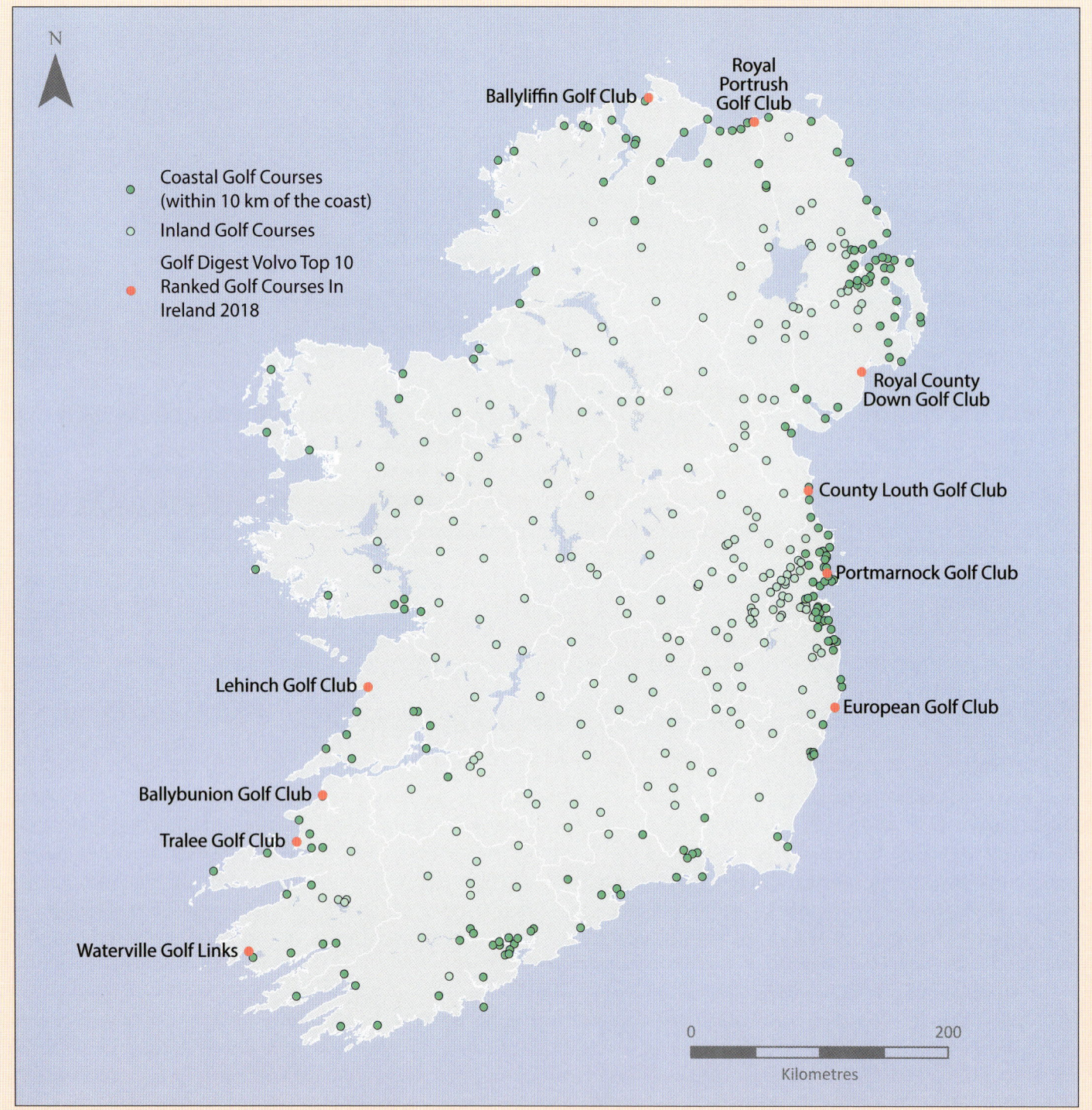

Fig. 27.41 GOLF COURSES IN IRELAND. Ireland has more than 300 golf courses and just over one-half of them are located within 10km of the coast. Furthermore, all of the island's top-ten courses, including the first-ranked course at Royal Portrush, occur on or near the coast (labelled red) and are known as links courses. Many of the links courses are able to use the diverse environmental resources of the island's coastline, set against the backdrop of oftentimes spectacular scenery. These conditions contribute to a range of highly attractive courses for domestic and international tourists. [Source: Brian Keogh, 2018. Golf Digest Volvo Top 100 Courses in Ireland. Irish Golf Desk. 8 March 2018]

golfers to visit the country to play on these recognised world-class links. Almost all of the island's top-ten courses are of the coastal links type, with Figure 27.42 providing examples of such courses. The economic benefits to communities, and to adjacent areas, of such courses are generally very positive. At the national level, consumer expenditure by domestic golfers amounts to some €540 million a year. A further €270 million is spent in Ireland by overseas golfing visitors. Between 80 and 90 per cent of such expenditure takes place off the courses and includes accommodation, food, drinks, craft industry purchases and travel. This also helps to create employment opportunities in a range of service industries, often in places where alternative jobs are not readily available. Given the disproportionate significance of coastal golf courses, and notably links courses, this segment of the tourist industry is of particular importance to the well-being of many of Ireland's coastal communities.

Due to the high levels of demand placed on Ireland's golf courses, and especially its links courses, pressures have increased to further develop their number, generally in order to raise additional revenue from this tourist sector. However, links courses, in particular, are defined by environmental qualities that are highly sensitive to change. Poorly controlled development in such circumstances could, therefore, erode the very attractions that development seeks to promote. The increasing popularity of links courses, and of their wider dunes areas for leisure activities, is creating tensions nationally for the sustainable use of these sand dune environments. Similar problems are being experienced internationally in such coastal sand dunes.

Coasts are naturally dynamic, changing continually under the impact of wind, wave and storm activity. Consequently, there is always a call from the users of these golf courses, and by neighbouring coastal landowners, for their protection from erosion. Engineered measures for such protection are expensive, with coastal defence projects easily costing over a quarter of a million euro and expenditure can reach levels of many millions. In an example from the County Clare coastline at Doonbeg, the local authority eventually agreed in 2017 to the defence of part of this prestigious golf course, which had also been a long-running cause for people's concern due its

Fig. 27.42 VIEW OF SIX OF IRELAND'S COASTAL LINKS GOLF COURSES. a. Waterville Golf Course, County Kerry. [Source: Lynne Connolly, courtesy of Waterville Golf Links, County Kerry]; b. The European Club, County Wicklow. [Source: Gerard Ruddy]; c. View of Louth Golf Club, which looks north with Clogher Head in view, and in the very far distance you see the mountains of Mourne. [Source: County Louth Golf Club]; d. The Irish Open 2015 at Royal County Down Golf Course with the Mourne Mountains in the background. [Source: © Tourism Northern Ireland]; e. Royal Portrush Golf Club. [Source: Royal Portrush Golf Club]; f. Ballybunion Golf Club. [Source: Evan Schiller, courtesy of Ballybunion Golf Club]

vulnerability to erosion. The scheme involved the construction of some 38,000 tonnes of rock armouring, in order to protect three of the golf links holes (including fairways and golf tee areas). This decision resulted from long-running pressures exerted by the golf course owners, other local landowners and the business community. The wisdom of such protection, however, continues to be contested strongly by environmental and coastal science pressure groups. It is argued by supporters of the defence works that the scheme will safeguard the jobs of the local people and the coastal economy, which is dependent on the continuance of the golf links. This view can be supported by the growing economic value of such courses in Ireland, but it will come, of course, at a significant public expense. This is seen by some as a legitimate and much-needed investment in coastal and rural economies; others see it as an unreasonable cost burden on scarce government financial resources.[38]

In future, the impact of climate warming will drive increased levels of coastal erosion along all of Ireland's coasts and, particularly, in these sensitive lowland, sands areas occupied by golf links. The problems of destruction of these courses will commonly become worse. Additionally, and this is where the real tension for these environments occurs, these original natural dunes and back-beach coastal systems are created by the operation of coastal processes as sand reservoirs. In time, such dune areas will re-supply the sands and nourish the beaches linked to them as they undergo change. Consequently, the conclusion reached by coastal scientists and managers from this observation and understanding, is that golf courses and dunes should, in many or even most cases, not be artificially protected and stabilised, a view that is supported internationally. This is particularly so as coastal environments collectively face into the impacts of climate warming. But, the argument about whether to defend or not is set to run for many years to come.

POLICY AND LEGISLATION

The report, *Harnessing Our Ocean Wealth: An integrated marine plan for Ireland*, represents an important step towards more joined-up planning and management for Ireland's coastal and marine domain. The plan makes reference to the steps needed to realise the potential of coastal and marine tourism in Ireland. Regrettably, an individual policy and strategy to drive investment in this tourism sector is still lacking. Similarly, plans aligned with discrete tourism products, such as the Wild Atlantic Way, are likely to result in consideration being given to the particular needs for tourism assets associated with the product, but may not necessarily fulfil the requirements of the entire sector, such as multiple activities, services and products.[39]

Local authorities are likely to continue to support the development of coastal and marine tourism within their functional areas. This is in the context primarily of supporting those local communities that are reliant on this sector for their livelihoods. Such local-level planning and support should continue and be encouraged. Coastal tourism would benefit, however, from larger-scale regional and national sector-specific frameworks, which could strategically guide investment and development. For coastal regions, the issue of second homes also arises. Houses in coastal areas are often vacant for much of the year and they can frequently outnumber the permanently occupied residences. This results in some coastal settlements having an air of being ghost centres, almost empty of residents for extended periods.

The Foreshore Acts, which govern coastal areas in investment, development and the use of Ireland's coastal resources, is well established. The current licensing and planning regime operating under these regulations, however, are viewed as constraints and barriers to growth (see Chapter 32: Coastal Management and Planning).[44] Revision of the foreshore legislation has been mooted for a number of years and is now long overdue. The existing legislation has led to much uncertainty and delays in terms of coastal developments. While these impediments are not specific to coastal and marine tourism, there is a reluctance on the part of investors, and communities, to devote resources (financial and human) in development planning that may not necessarily get to the first stage of an application.

To conclude, Ireland's coasts have much to offer the tourist, in terms of their natural and cultural resources and all estimates point to their significant socio-economic value. In addition, the sector demonstrates many examples of good practices in the delivery of services and products, and the ability of key stakeholders to secure regional development funding from EU programmes. Indications are that the contribution of marine and coastal tourism will continue to grow despite the challenges facing many national economies. In Ireland, elements of the sector have been characterised by sustained growth (for example, cruise tourism). Some key challenges remain, namely, the need for the development of a strategy that facilitates the implementation of planning and management frameworks for the sector, which in turn can guide investment in the maintenance and development of infrastructures. These should be appropriate to the demands and capacity of individual coastal locations. Such a policy would improve significantly the types of leisure activity offered and the economic return to the economy from coastal and marine tourism.

RENEWABLE ENERGIES: WIND, WAVE AND TIDAL POWER

Fiona Devoy McAuliffe and Tony Lewis

Ireland potentially has one of the best marine renewable energy resources in the world; its seas experience some of Europe's strongest winds and waves (Figures 28.1 and 28.2). Conditions are particularly favourable for a range of ocean renewable technologies, namely, offshore wind, wave and tidal energy. For example, potentially up to 5GW (5,000MW) of offshore wind could be sustainably developed in Ireland by 2030. It is estimated that a 100MW wave farm could power up to 47,000 homes.

Developing Ireland's offshore energy potential would enable the country to generate a significant proportion of its electricity from carbon-free, renewable sources, reduce greenhouse gas (GHG) emissions, develop an export market in green energy and enhance the security of supply, by reducing dependence on imported fuels. It could also have a significant economic impact. A study commissioned jointly by the Sustainable Energy Authority of Ireland and Invest Northern Ireland asserted that by 2030, a fully developed island of

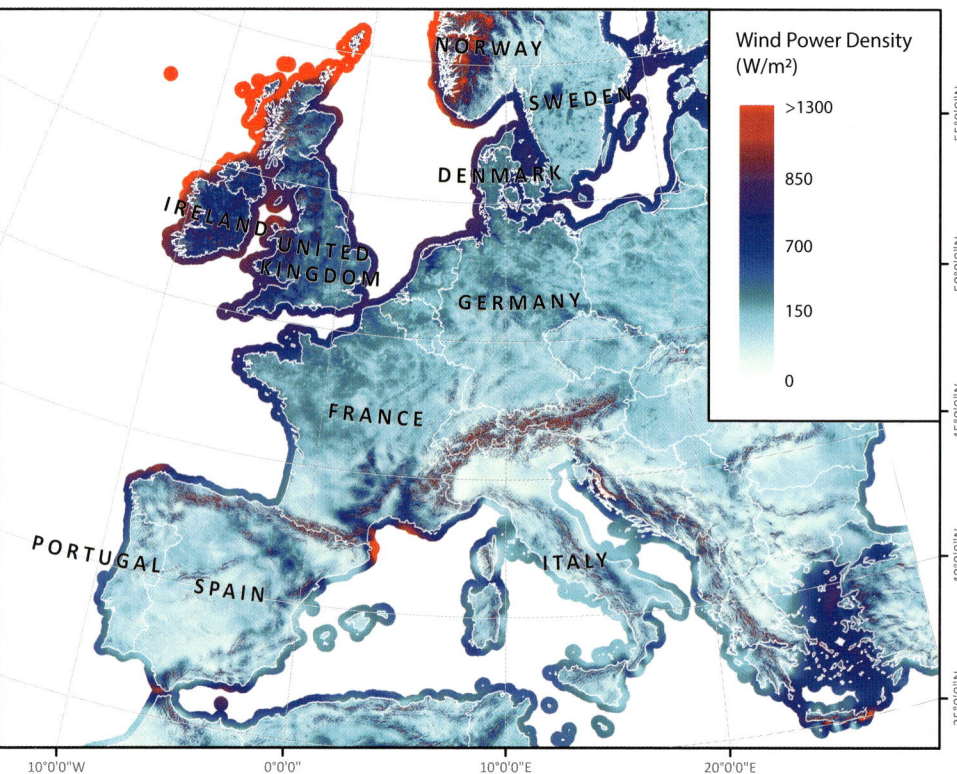

Fig. 28.1 EUROPEAN WIND RESOURCES. This shows wind power density (W/m²) across Europe, to *c.*30km offshore. Higher mean wind power densities indicate better wind resources. It is apparent from the map that Ireland has one of the best wind resources in Europe. This makes the seas around Ireland ideal for the development of offshore wind projects. This includes areas beyond 30km offshore, which are becoming viable for new offshore wind technology, linked to floating wind developments. [Data source: Global Wind Atlas, 2018]

Ireland ocean energy sector, serving a home and global market, could produce a total Net Present Value (NPV) of around €9 billion and thousands of jobs.[1]

The EU has also set targets to reduce greenhouse gas emissions by 55 per cent in 2030. Ireland aims to achieve net zero by 2050. Ocean renewable energy could significantly contribute to achieving Ireland's targets.

TECHNOLOGY: WIND (FIXED AND FLOATING), TIDAL AND WAVE

Different technologies are suited to different marine environments around the coast of Ireland. For example, the east coast is ideal for fixed offshore wind technology, with shallow waters near to shore. Offshore floating wind technology is being considered for the deeper waters off the south, west and north-west coasts. The south-east coast, the Shannon Estuary and the north-west coast are areas suitable for commercial tidal deployment (Figure 28.3). The best wave resources are along the west coast, where the deeper exposed waters of the Atlantic ocean have the greatest energy.

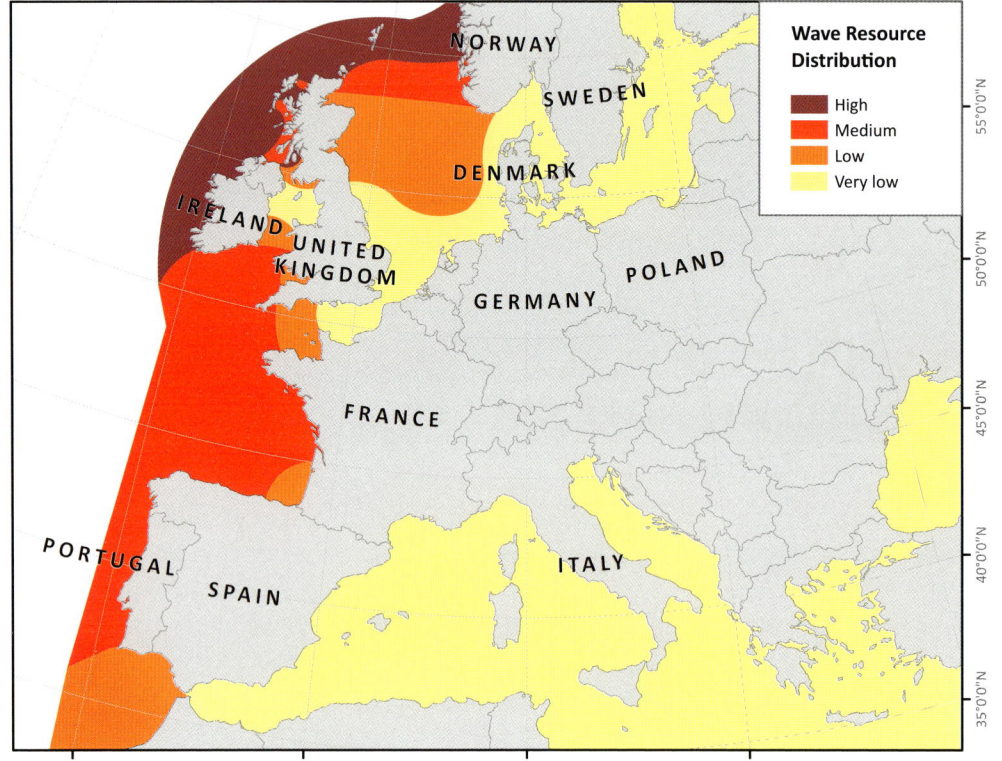

Fig 28.2 EUROPE'S WAVE ENERGY RESOURCE. The wave resource distribution map shows the geo-strategic location of Ireland in the context of wave energy. Traditionally, extreme wave conditions were regarded as a hindrance to the safety of navigation around the coast but, increasingly, these same wave conditions are regarded as an asset, because of the potential to convert them into alternative sources of renewable energy. [Data source: Aqua-RET Project, 2012]

Offshore wind energy

Wind turbines are a well-established commercial technology. Significant cost reductions and record bids for subsidy-free offshore wind farms in both Germany and the Netherlands have been achieved since 2017. However, in the Irish Sea, the Arklow Bank Wind Farm (Figure 28.4) has been the only wind farm to be developed off the coast of Ireland thus far. Although it received consent for 520MW capacity in 2002, only seven 3.6MW wind turbines (totaling 25MW) have been commissioned to date. Further expansion by the developer was put on hold due to a lack of government support for offshore wind and the need for the development of critical infrastructure for an export market. While the Codling Wind Farm, also in the Irish Sea, received

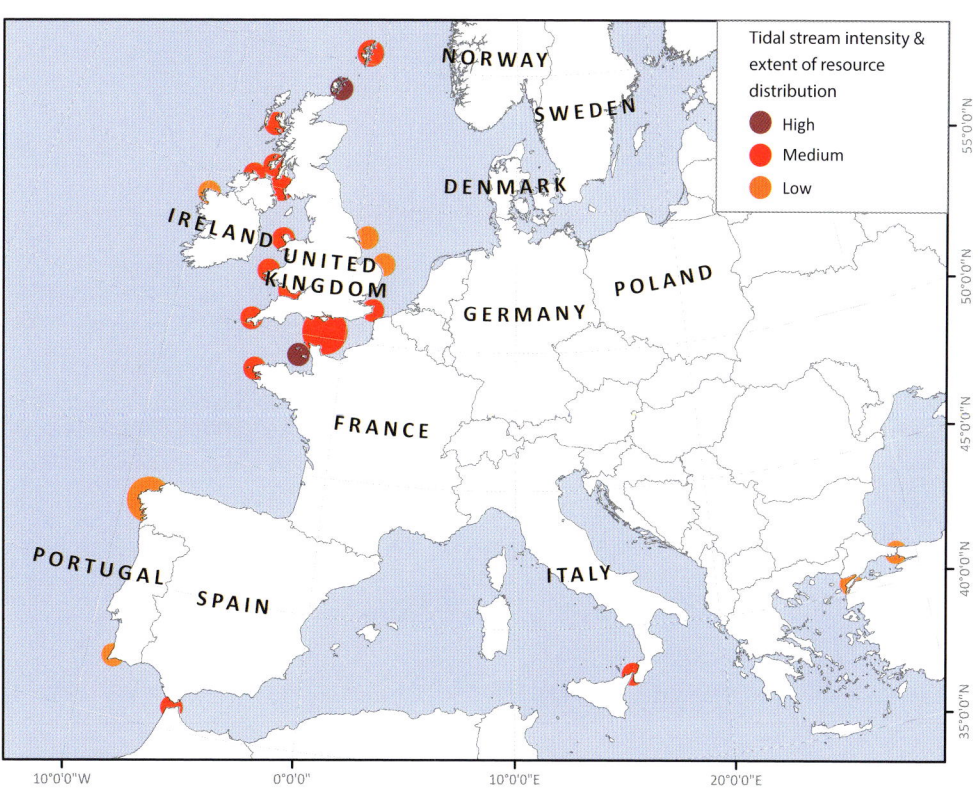

Fig 28.3 EUROPE'S TIDAL RESOURCE. Relative to other hotspots around Europe, Ireland has some limited potential for tidal energy projects. Key locations, primarily off the coast of Northern Ireland, are prime areas for the development of utility-scale tidal energy projects. These are areas where the tidal streams are most pronounced. [Data source: Aqua-RET Project, 2012]

Fig. 28.4 ARKLOW BANK OFFSHORE WIND FARM. The seven 3.6MW turbines standing on the Arklow Bank are a familiar sight to the coastal community who live nearby and to the mariners that navigate that part of the Irish Sea. The pace of development of offshore wind in Ireland has been extremely slow to date. However, sights like this may become more significant, as offshore wind is developed in Ireland. SSE are currently working on a project that will see the expansion of the Arklow array from a 25MW to a 520MW farm. This involves close engagement with stakeholders in the local area. [Source: GE Energy (Ireland)]

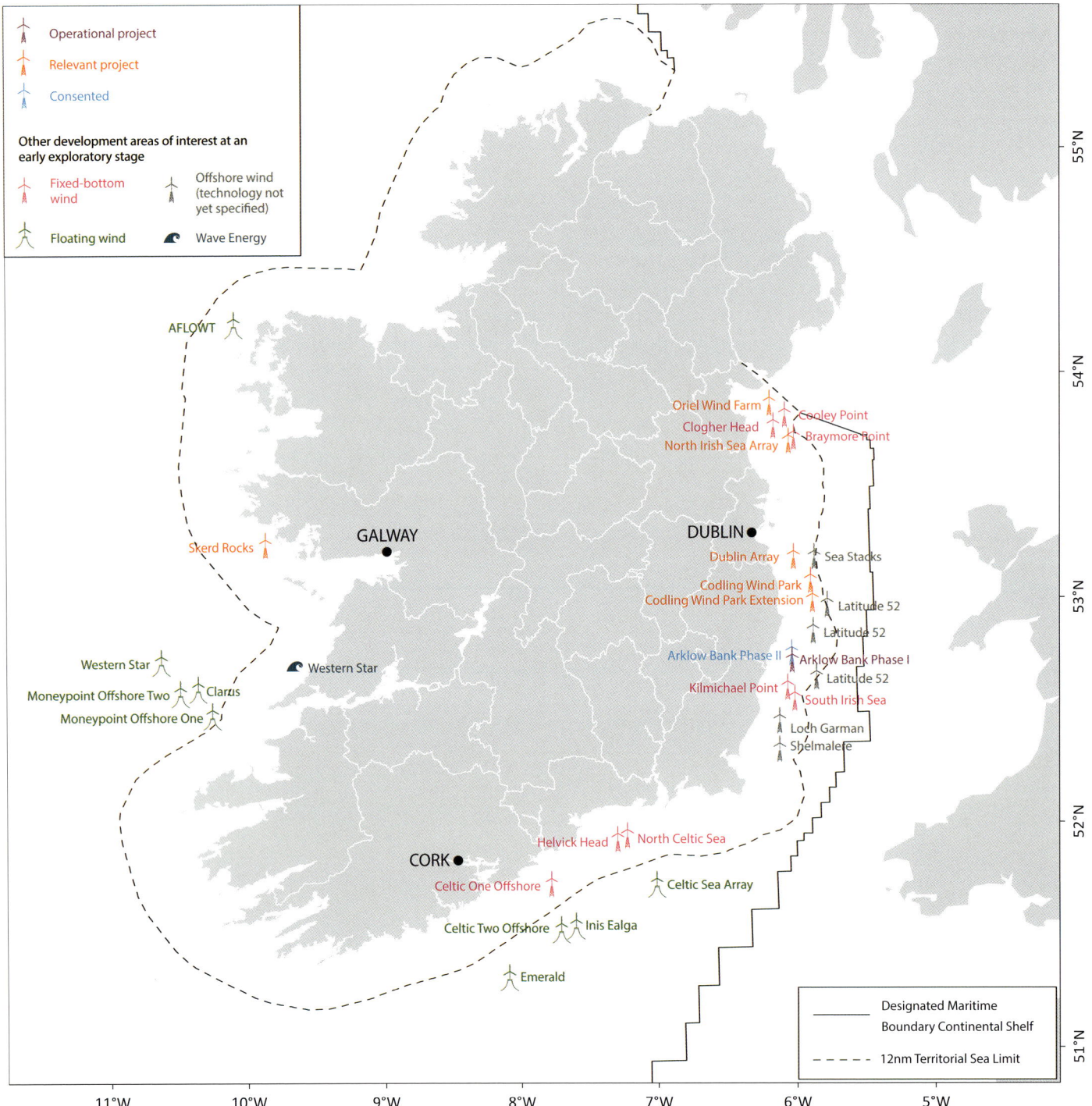

Fig. 28.5 a. EXISTING AND PLANNED OFFSHORE WIND FARMS. These offshore wind farms range from early stage plans to more mature, fully consented projects, such as the Oriel project. The Arklow Bank Wind Farm (Phase 1) is currently the only fully commissioned, operational wind farm developed in Ireland to date. Projects closer to shore are potentially more contentious, as a result of issues around visual impact. [Source: adapted from 4C Offshore, 2019]

consent for approximately 1GW capacity, with plans for 220 5MW turbines, developers also put the project on hold, preferring to wait for more favourable market conditions. Although the financial crisis of 2008 impeded development of the offshore wind sector, it was subsequently hoped that a proposed agreement would allow Ireland to export power to the UK and gain access to UK incentives. However, the bilateral government negotiations fell through. A proposal for a 300–400MW offshore wind farm, known as First Flight, was abandoned in 2014, following delays in the clarification of the subsidy system for Northern Ireland. The project had already been scaled back from a 600MW proposal, to

minimise impacts on shipping and commercial fisheries.

Traditional barriers to the development of offshore wind are rapidly being removed. For example, the Renewable Energy Support Scheme (RESS) provides a positive initiative from government to the sector in the Republic. The global technology advancement leading to the reduction in the Levelised Cost of Energy (LCOE), along with the natural advantages of offshore wind resources on the island of Ireland, the opportunities around the deployment of floating wind technology, and the rising demand in clean energy, all make Ireland an increasingly attractive emerging market for international and indigenous developers. Belfast

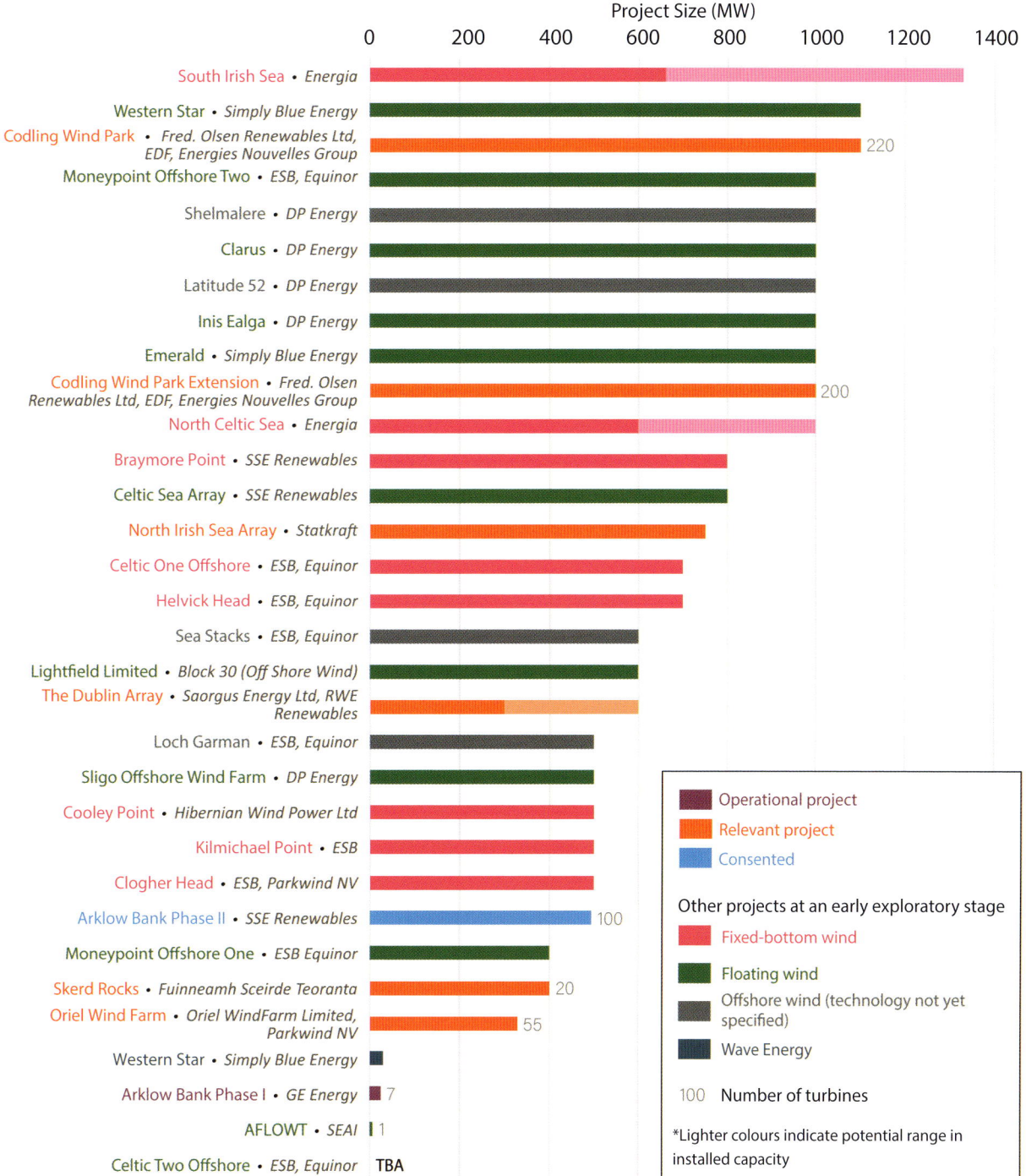

Project Size (MW)

| South Irish Sea • *Energia* |
| Western Star • *Simply Blue Energy* |
| Codling Wind Park • *Fred. Olsen Renewables Ltd, EDF, Energies Nouvelles Group* | 220 |
| Moneypoint Offshore Two • *ESB, Equinor* |
| Shelmalere • *DP Energy* |
| Clarus • *DP Energy* |
| Latitude 52 • *DP Energy* |
| Inis Ealga • *DP Energy* |
| Emerald • *Simply Blue Energy* |
| Codling Wind Park Extension • *Fred. Olsen Renewables Ltd, EDF, Energies Nouvelles Group* | 200 |
| North Celtic Sea • *Energia* |
| Braymore Point • *SSE Renewables* |
| Celtic Sea Array • *SSE Renewables* |
| North Irish Sea Array • *Statkraft* |
| Celtic One Offshore • *ESB, Equinor* |
| Helvick Head • *ESB, Equinor* |
| Sea Stacks • *ESB, Equinor* |
| Lightfield Limited • *Block 30 (Off Shore Wind)* |
| The Dublin Array • *Saorgus Energy Ltd, RWE Renewables* |
| Loch Garman • *ESB, Equinor* |
| Sligo Offshore Wind Farm • *DP Energy* |
| Cooley Point • *Hibernian Wind Power Ltd* |
| Kilmichael Point • *ESB* |
| Clogher Head • *ESB, Parkwind NV* |
| Arklow Bank Phase II • *SSE Renewables* | 100 |
| Moneypoint Offshore One • *ESB Equinor* |
| Skerd Rocks • *Fuinneamh Sceirde Teoranta* | 20 |
| Oriel Wind Farm • *Oriel WindFarm Limited, Parkwind NV* | 55 |
| Western Star • *Simply Blue Energy* |
Arklow Bank Phase I • *GE Energy*	7
AFLOWT • *SEAI*	1
Celtic Two Offshore • *ESB, Equinor*	TBA

Legend:
- Operational project
- Relevant project
- Consented

Other projects at an early exploratory stage
- Fixed-bottom wind
- Floating wind
- Offshore wind (technology not yet specified)
- Wave Energy

100 Number of turbines

*Lighter colours indicate potential range in installed capacity

Fig. 28.5 b. SIZES OF EXISTING AND PLANNED OFFSHORE WIND FARMS. Planned offshore wind farms typically range in scale from *c.*300MW to 1.1GW. Grid capacity needs to be enhanced to facilitate the delivery of this energy on to the grid and, ultimately, to enable export to other markets overseas. This is where the enterprise and export opportunity of this sector will be realised. The development of projects typically involves partnership between developers and financiers and the supply chain. This is a volatile industry and many more initiatives for offshore wind farms are emerging in 2021 and as the Climate Change COP26 summit approaches.

Harbour has developed considerable port infrastructure to support a number of offshore wind farms in the UK part of the Irish Sea, including the Dong Energy Offshore Wind Terminal. This critical port infrastructure will be a key enabler for other developments on the island of Ireland.

Extensive catch-up efforts are underway to progress the current pipeline of offshore wind developments. Figure 28.5b illustrates the locations of many of these developments, while Figure 28.5b summarises details of existing and planned wind farms. For example, in 2017, the Belgian offshore wind farm developer, Parkwind, invested in Oriel Wind Farm and subsequently partnered with the Electricity Supply Board (ESB) to progress plans to generate enough capacity to cover the needs of most of Louth and Meath. Subsequently, SSE (formerly Airtricity) began progressing plans to develop Arklow Bank Phase 2. At the same time, Innogy Renewables Ireland invested in the Dublin Array project; Statkraft acquired the North Irish Sea Array from Gaelectric; the ESB commenced plans to acquire and develop new projects (such as at Clogher Head and Kilmichael Point) and Equinor partnered with the ESB to develop the potential for fixed

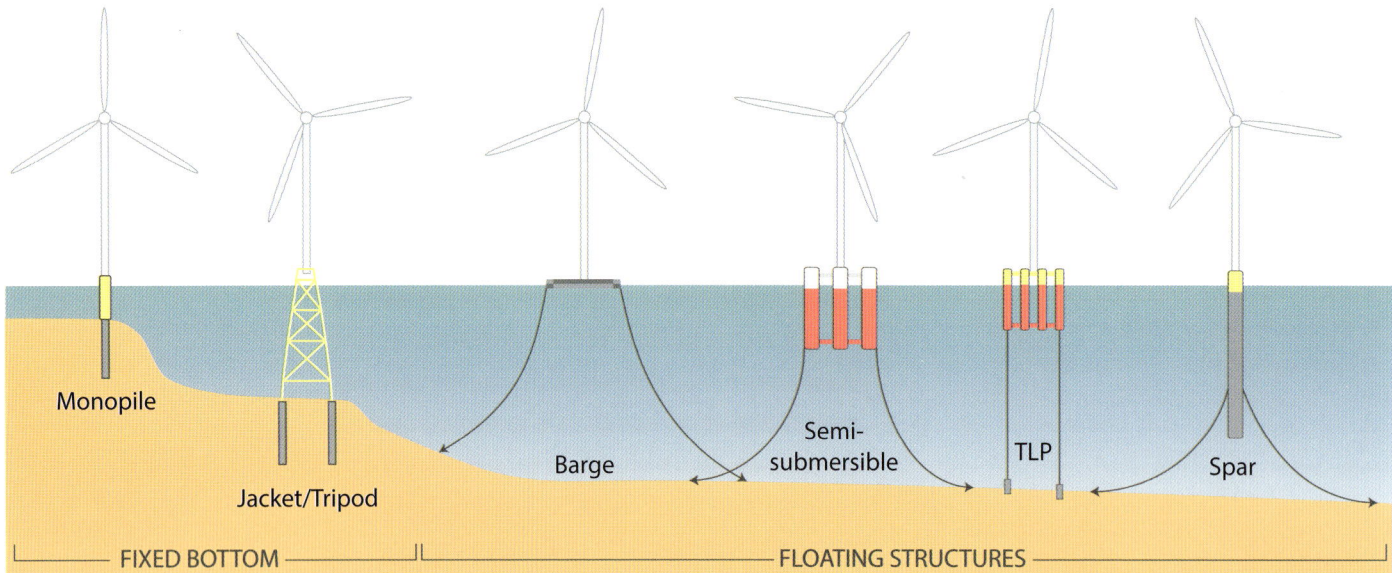

Fig. 28.6 OVERVIEW OF OFFSHORE WIND TECHNOLOGY PLATFORMS. Moving from left to right, the progression from fixed bottom (monopile and jacket/tripods to floating foundations (barge, semi-submersible, tension leg platform and spars) can be observed. Fixed bottom-fixed technology is enabled in shallower water depths, typically at *c.*30m, but up to 60m in more recent projects. For locations with over 60–80m water depths, fixed foundations are not viable, and floating wind turbines, moored to the seafloor, are required. Floating wind technology is a nascent sector. There are currently no floating wind farms in Ireland, but given the excellent wind resources offshore, floating technology is anticipated to be deployed off the south and west coasts this decade. [Source: adapted from Principle Power]

and floating farms. DP Energy Ireland Ltd has plans for a 1GW floating wind farm called Inis Ealga. Simply Blue Energy are also planning the Emerald floating wind farm (up to 1.3GW), with Shell, in the vicinity of the former Kinsale gas platform; and the Western Star floating wind and wave project off Clare. However, while floating wind will require support and subsidies, it is likely to mushroom as a result of the programme for government ambitions for 30GW of floating wind for Ireland. The first operational floating wind farm in the world was Hywind, a 30MW project built by Equinor off the coast of Scotland. Figure 28.6 shows the range of offshore wind technologies currently in play, including well-established 'bottom fixed' and emerging floating wind technology types.

Gannets and Offshore Windfarms

Mark Jessopp

The significant wind resource around the Irish coast has been identified as suitable for large-scale, offshore wind installations. However, evidence suggests that for some birds, both environmental disturbance and death due to collisions with turbines could have significant impacts on populations. Gannets (*Morus bassanus*) have been tracked using sophisticated Global Positioning System tags equipped with altimeters (Figure 28.7). These show that when foraging, they fly at heights within the sweep area of many turbines, creating risk of collision.[2] However, surveys around wind farms in the North Sea suggest that gannets exhibit strong avoidance tactics.[3] Since gannets can hunt over 500km from their colonies, and as wind farms take up more space, impacts may arise from the removal of available feeding grounds.[4] These issues need to be considered at the planning and consenting stages before developments are allowed to proceed.

Fig 28.7 GANNET IN FLIGHT. Gannets (*Morus bassanas*) are thought to be among the seabirds most vulnerable to collision risk with offshore wind turbines, as their flight patterns tend to place them at the level within the sweep of turbine blades. Their potential to avoid such structures, and to adapt to the presence of offshore wind farms, is a matter for ongoing research and scientific investigation by marine biologists. [Source: William Hunt]

Tidal energy

The tidal industry is in its infancy, with numerous technologies being considered. The first tidal energy farm was built by MeyGen in the Pentland Firth, Scotland. This consists of four 1.5MW turbines and generated over 8GW hours of energy by July 2018. On the island of Ireland, a number of single prototypes have been installed in demonstration sites, including Minesto's Deep Green device and Marine Current Turbines Ltd's SeaGen tidal turbine in Strangford Lough, Northern Ireland (Figure 28.8). The narrows between Strangford Lough and the Irish Sea have also been identified as an ideal test location for the PowerKite technology, involving a partnership with Queen's University Belfast.

Commercial deployments are still at the planning stages (2020), including the Fair Head Tidal Energy Park and the Torr Head project (Figure 28.9). DP Energy is an Irish company and one of the leading independent developers of tidal stream projects in the world. The company is currently developing projects in Scotland and Canada as well as the Fair Head Tidal Energy Park (100MW) in Northern Ireland in collaboration with Bluepower NV. If planning is approved, it would be capable of powering 70,000 homes. The Torr Head project, developed by Tidal Ventures Ltd., proposes a 100MW farm approximately 1km off the coast of Antrim. Approval is pending and it is unclear the impact the liquidation of co-developer OpenHydro will have on this project. The Dublin-based OpenHydro group were the first to deploy a tidal turbine at the European Marine Energy Centre (EMEC), Scotland, and to generate electricity from tidal streams onto the UK national grid. They developed operations in Ireland, Canada, France, Japan and Scotland. Early in 2013, Naval Energies took a majority stake in the company. However, in July 2018 they announced that they would no longer invest in the company due to the group's insolvency.

Irish device developer, Limerick-based company, GKinetic Energy Ltd. developed a technology involving two vertical axis turbines placed on either side of a buoyant vessel. The device can be scaled up to 1MW and installed in rivers, oceans or estuaries. It was tested (2015–2017) in a custom tow-testing facility in collaboration with Shannon Foynes Port Company.

Fig. 28.8 SeaGen Tidal Device, Strangford Lough (2011). The tidal regime in Strangford Lough is characterised by the Strangford Narrows, which generate very strong tidal flows, making it ideal for demonstrating the commercial viability of harnessing tidal energy, from a relatively accessible location. The SeaGen Tidal Device was the world's first commercial-scale tidal device when it was launched in Strangford Lough in 2008. The 1.2MW device featured two horizontal axis turbines, as shown in the image. When lowered into place, close to the seabed, the rotor blades were designed to operate on both the ebb and flow of the tide. Marine Current Turbines (MCT Ltd), the initial developers of the project, was acquired by Siemens, before the project was decommissioned in 2017, also making it the first commercial tidal project to have completed a full life-cycle. [Source: Siemens]

Wave energy

Given the massive challenges of installing, operating and maintaining devices in extreme open-sea conditions, the wave energy sector remains in the Research and Development (R&D) phase. There is a lack of convergence on the optimal technology,

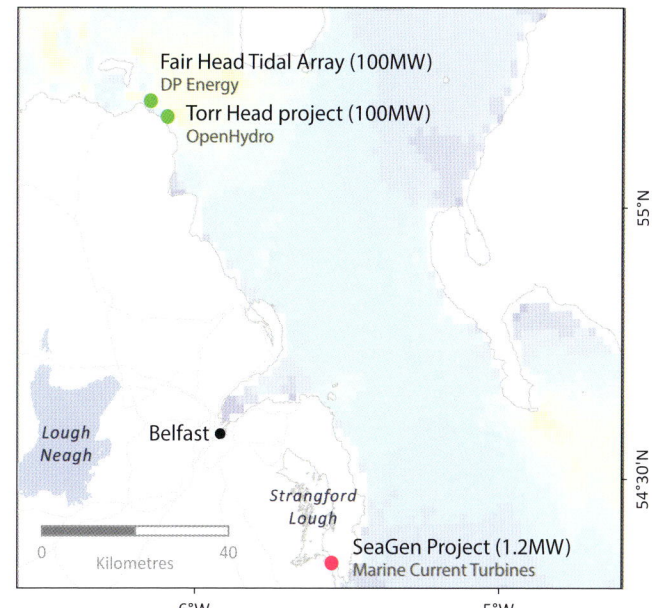

Tidal projects

- 🟢 Planned Projects
- 🔴 Demonstration Project

Peak currents speed of a Mean Spring Tide
Metres per second (m/s)

| 0.00–0.11 |
| 0.12–0.25 |
| 0.26–0.50 |
| 0.51–0.75 |
| 0.76–1.00 |
| 1.01–1.25 |
| 1.26–1.50 |
| 1.51–1.75 |
| 1.76–2.00 |
| 2.01–2.50 |
| 2.51–3.00 |
| 3.01–3.50 |
| 3.51–4.00 |

Fair Head Tidal Array (100MW)
DP Energy

Torr Head project (100MW)
OpenHydro

Lough Neagh

Belfast

Strangford Lough

0 Kilometres 40

SeaGen Project (1.2MW)
Marine Current Turbines

55°N
54°30'N
6°W
5°W

Fig. 28.9 Strategic sites earmarked for tidal development in Northern Ireland. Proposals for large-scale tidal energy deployments off the coast of Northern Ireland, taking advantage of the favourable tidal streams off two major headlands, have stalled due to the problems in the Northern Ireland Assembly (2021). If progressed, the plans will be significant, with each project delivering a potential installed capacity of 100MW per project. The Fair Head Tidal Array is led by Buttevant, County Cork-based developers, DP Energy. The Torr Head project has been further impacted by issues with the developer OpenHydro, which went into liquidation in 2018, following a decision by its French owner, Naval Energies. These setbacks, unprecedented at the time of project initiation, indicate the regulatory and corporate governance frameworks that need to be aligned to proceed with large offshore development projects. [Data source: UK Renewables Atlas, 2018]

with a range of Wave Energy Converter (WEC) concepts at various stages of development and testing. There are currently no commercial wave energy projects in the world. The ESBI's Ocean Energy Project, WestWave (located off the coast of Clare) would be one of the first wave energy pre-commercial demonstration trials globally. In collaboration with technological and research partners from Ireland and abroad, the project aims ultimately to trial five devices, but ESBI are still in the process of selecting a suitable technology.

Considering devices, Irish company Ocean Energy developed the 'OE Buoy' technology, based on the Oscillating Water Column (OWC) principle (Figure 28.10). It has one single moving part to improve reliability and survivability. Having tested the platform for over three years in Atlantic waters, a full-scale prototype (rated 1.25MW) has been fabricated in Oregon and deployed at the US Navy's Wave Energy Test Site off the coast of Hawaii.

Simply Blue Energy, an Irish-based blue economy development company, are seeking to develop a wave energy site off the coast of West Clare. The Saoirse project, is part of the Western Star co-developed floating wind and wave initiative. Initial plans for the wave element, are for a 5MW demonstration project by 2026, potentially using Corpower technology.

OCEAN ENERGY POLICY

As a general principle on the island of Ireland, ocean energy policy is nested in overarching energy and climate policies. The Republic's long-term energy policy framework is catered for in the 2015 Energy White Paper, Ireland's Transition to a Low Carbon Energy Future 2015–2030;[5] while the Climate Action Plan to Tackle Climate Breakdown provides government targets for 2030, and ambitions to 2050.[6] The National Energy and Climate Plan provides an integrated approach to decarbonisation and adaptation for the period 2021–2030. Ocean energy is specifically provided for in the Offshore Renewable Energy Development Plan (OREDP).[7] Published in 2014 and reviewed in 2018, the OREDP aimed originally to provide a framework for the sustainable development of the country's marine renewable resources, setting out key principles and policy actions. Renewable energy policy in Northern Ireland is guided by the Department of Enterprise, Trade and Investment's Strategic Energy Framework. The equivalent government initiative to the OREDP is the Offshore Renewable Energy Strategic Action Plan.[8]

Fig. 28.10 THE OE WAVE ENERGY CONVERTER (WEC) UNDER CONSTRUCTION IN THE US (2018). The Ocean Energy Wave Energy Converter was fabricated in Oregon in 2018, prior to being towed to Hawaii, where it is installed at the US Navy Wave Energy Test Site. The project is funded by the US Department of Energy, in collaboration with the Government of Ireland, through Enterprise Ireland and the Sustainable Energy Authority of Ireland. The 826 ton buoy is rated to produce 1.25MW in electrical power production when operational. The process of developing and testing the technology was supported through Ireland's marine energy R&D test-bed infrastructure, with testing in the Lir tanks and in Galway Bay, before full-scale production commenced with Oregon-based Vigor. Ocean Energy Ltd is an Irish company, headquartered in Cobh, Cork Harbour. [Source: Ocean Energy Ltd]

The Role of Ulva lactuca *in Biogas Production*

David Wall

Ulva lactuca is a species of green seaweed, commonly referred to as sea lettuce, that appears along the coastline of Ireland in shallow estuaries and beaches. The reason for the accumulation of sea lettuce is due to agricultural practices onshore, and more specifically eutrophication, a process whereby coastal waters become overly enriched with nutrients. Such conditions lead to the creation of 'green tides' or 'algal blooms'. This is not only a common occurrence in Ireland but evident worldwide, becoming more endemic in countries such as France, Denmark and Japan. Algal blooms can result in the closure of beaches and dangerous conditions due to the build-up of toxic gases, such as hydrogen sulphide, as the high-sulphur-containing seaweed rots.

However, in recent years the focus has shifted from seeing *Ulva lactuca* growth solely as a problem, and researchers are now looking into how this seaweed can be treated effectively, and utilised beneficially as a potential resource. In particular, *Ulva lactuca* can provide a source of third-generation biofuels. Such biofuels are considered advanced since they do not compete with food production and, furthermore, have no land requirements unlike traditional biofuels. One process that could be used in Ireland is anaerobic digestion, a process whereby the seaweed feedstock is biodegraded by a series of micro-organisms in a sealed tank in the absence of oxygen to produce biogas. *Ulva lactuca* could be combined with slurry and excess grass available from local farmers, or food waste from local supermarkets, to increase the biogas produced. Biogas typically comprises of 60 per cent methane and 40 per cent carbon dioxide. If the carbon dioxide is removed, the methane can

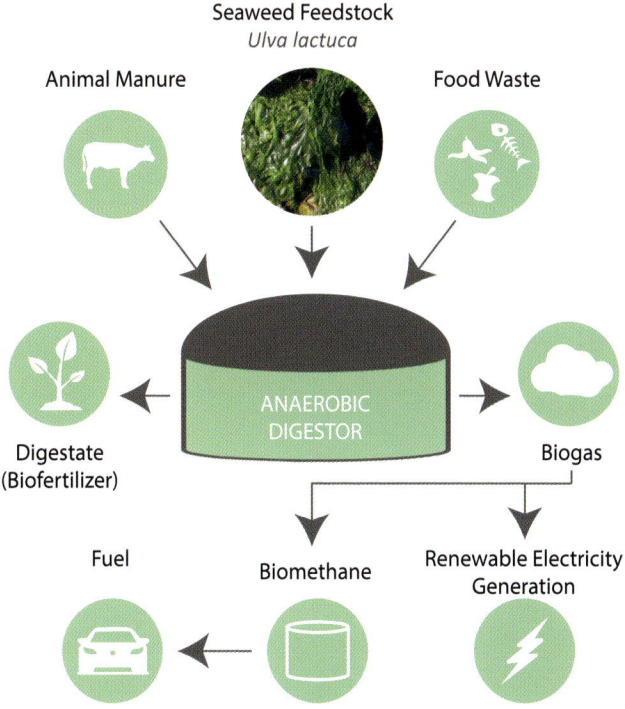

Figure 28.11 ANAEROBIC DIGESTION PROCESS. The potential to use seaweed as a biomass for biofuel production is an area of ongoing research and development. Seaweed is a wet biomass, which lends itself to anaerobic digestion for biomethane production. This involves a number of steps, whereby seaweed may be combined with food waste or animal manure. Brown seaweed such as *Laminaria* can be used, as well as *Ulva* (also known as sea lettuce), depicted in the image. Mass algal growths of *Ulva* can occur, especially when nutrients are loaded into a watershed. This abundance, together with its composition and degradability, make it a relatively good input to an anaerobic digester system. [Photo source: Robbie Murphy]

be used in the same manner as natural gas or even as a gaseous transport fuel. Utilising anaerobic digestion to treat *Ulva lactuca* would thus not only provide a source of indigenous energy in Ireland but also a means of reducing the detrimental effects caused by eutrophication to the amenity of the Irish coastline (Figure 28.11).

RESEARCH AND DEVELOPMENT

Third-level institutions across the island of Ireland have led a number of important research projects to help build capacity in the Ocean Renewable Energy (ORE) sector. These include initiatives such as MaRINET2, a Horizon 2020 European project that facilitates access to thirty-nine organisations representing the top ORE testing centres in Europe, and EirWind, a research-industry partnership aimed at the development of the offshore wind sector. Given the nascent state of the marine renewable energy sector, government policies in the Republic and in Northern Ireland have focused on building capacity for world-class test-bed R&D infrastructure, to encourage the development of ocean renewable energy technology (Figure 28.12 and Figure 28.13). Key elements of this infrastructure include:

Lir National Ocean Test Facility (NOTF)

Based in the Beaufort Building at University College Cork (UCC), Lir

NOTF is designed for the laboratory testing of offshore wind, wave and tidal energy devices. Facilities comprise four different wave tanks including a deep ocean wave basin, which is capable of producing waves of up to 1.1m high (*c.*1:15 scale testing); an ocean wave basin (*c.*1:50 scale testing); a wave and current flume with coastal/tidal testing capabilities (*c.*1:50 scale testing); and a wave demonstration flume. Lir NOTF also has a range of electrical and energy storage test infrastructures. It is estimated that in excess of seventy different wave energy technology types and five floating wind platforms have been tested at Lir NOTF.

SmartBay: Galway Bay Marine and Renewable Energy Test Site

A joint venture between the Marine Institute and the third-level sector, SmartBay Ireland includes a fully licensed test site for ocean energy devices aimed at testing small-scale devices in the early stages of development. It is located 4.5km east of Spiddal in Galway,

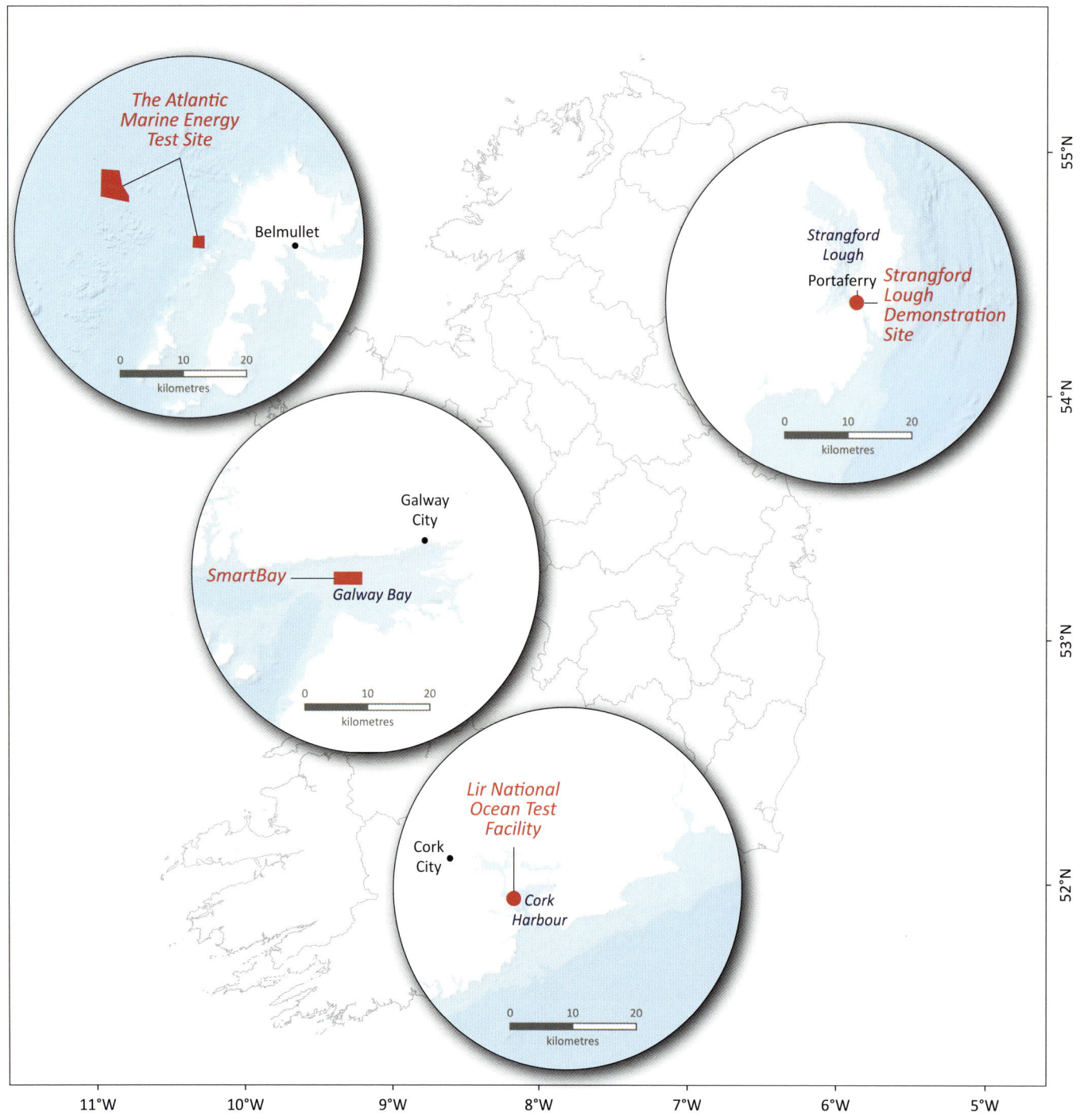

Fig. 28.12 MARINE RENEWABLE ENERGY RESEARCH AND DEVELOPMENT TEST FACILITY LOCATIONS, INCLUDING THE LABORATORY-SCALE TESTING AT THE LIR TANKS IN CORK HARBOUR QUARTER-SCALE TEST SITE IN GALWAY, DEMONSTRATION-SCALE TEST SITE IN STRANGFORD LOUGH AND FULL-SCALE TEST SITE, IN EXTREME CONDITIONS, OFF THE COAST OF BELMULLET, COUNTY MAYO. [Data source: Marine Institute]

1.5km offshore, in water depths of 20–25m. A comprehensive time-series of weather, wave and current data is available for the site, as well as an Information and Communication Technology (ICT) team to assist with data acquisition and transmission support. SmartBay Ireland also has a sub-sea cabled observatory in Galway Bay. This includes a fibre optic data and 400v power cable; high-speed communications via four pairs of optical fibres; and a sub-sea cabled sensor platform that hosts a variety of sensors and equipment, which can be tested in near real-time.

The Atlantic Marine Energy Test Site (AMETS)

AMETS is being developed to facilitate testing full-scale marine

renewable technology in an open ocean environment. It consists of two test areas located off Annagh Head, west of Belmullet in County Mayo: test Area A, 100m water depth, 16km out from Belderra Strand, 6.9km²; and test Area B, 50m water depth, 6km from the strand, 1.5km².

It will be connected to the national grid via a small onshore substation. Detailed information is available for interested developers including historical and live met-ocean data measurements from wave buoys, numerical wave modelling reports; and offshore site investigations including vibrocores and multi-beam surveys. The AFLOWT project is the first to utilise the AMETS test site for a floating offshore wind technology pilot.

Fig. 28.13 MARINE RENEWABLE ENERGY TEST SITE FACILITIES AROUND THE COAST. A. LIR National Ocean Test Facility [Source: Lir-NOTF]; b. Portaferry Coastal Wave Basin [Source: School of Natural and Built Environment, Queen's University Belfast, Northern Ireland]; c. Seapower Wave Energy Device at the Galway Bay Marine and Renewable Energy Test Site (SmartBay) [Source: SeaFever]; d. Waverider buoy deployment at the Atlantic Marine Energy Test Site. [Source: AMETS development team]

Queen's University Belfast (QUB)

QUB has a range of test facilities including the Belfast wave tank, which is 22 x 4.75m with an operating depth of 0.8m and an adjustable floor; the Queen's Marine Laboratory in Portaferry, including an 18 x 16m wave basin with an operating depth of 0.65m; and the Strangford Lough demonstration site where tidal devices can be tested in tidal flows from 2 to 4m/s. The Marlfield Bay wave test site is also close to Strangford Lough (Figure 28.14).

THE FUTURE: PROSPECTS AND CHALLENGES

At the time of writing, it is clear that the Republic of Ireland will not meet imminent targets for renewable energy (Figure 28.15). In both the Republic and Northern Ireland, an increased deployment rate for all renewable energy technologies will be required to fully comply with European Energy Directives and to achieve decarbonised economies by the middle of the

century. Failure to meet energy and emissions targets will lead to substantial fines. Failure to comply with the overall binding renewable energy target of 16 per cent could result in fines ranging from €65 to 130 million per percentage shortfall in the Republic of

Fig. 28.14 ARTIST'S IMPRESSION OF TWO-BLADED, FLOATING WIND TURBINES IN A COMMERCIAL ARRAY. The image is an example of the next generation of offshore wind technology under development. Innovative designs have been tested at reduced scales by Irish companies, in R&D test facilities such as the LIR tanks and QUB test facilities. Future plans for offshore wind development are likely to manifest in the roll-out of this technology in licensed zones off the coast. [Source: Bluwind Power Limited]

693

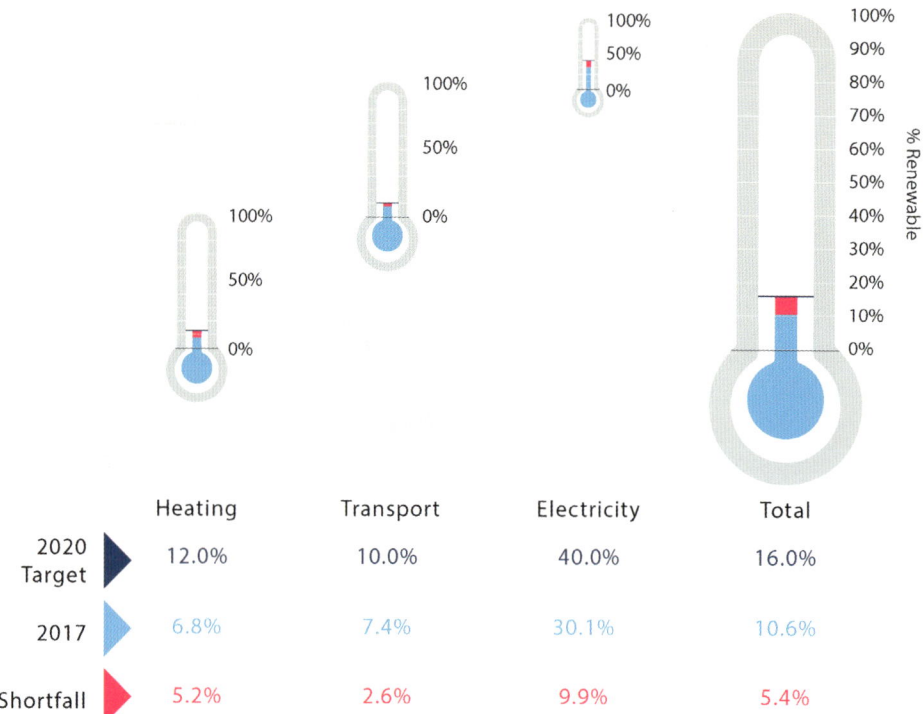

	Heating	Transport	Electricity	Total
2020 Target	12.0%	10.0%	40.0%	16.0%
2017	6.8%	7.4%	30.1%	10.6%
Shortfall	5.2%	2.6%	9.9%	5.4%

Fig. 28.15 CONTRIBUTION OF RENEWABLE ENERGY TO 2020 TARGETS. Government targets for 16 per cent of energy from renewable sources by 2020 have fallen short by approximately 5 per cent. The shortfalls are evident across specific targets established for renewable energy in heating, transport and electricity. Ireland is under increasing pressure politically to address this poor performance, as the urgency of decarbonising the economy by 2050 intensifies, in light of insights from climate science. This sharpens the focus on accelerating progress in the marine renewable energy sector, as part of a portfolio of investments required in clean, alternative sources of energy. [Data source: SEAI, 2018]

Ireland alone.[9]

In order to realise the opportunity presented by the island's abundance of marine renewable energy resources, a number of challenges need to be overcome. The Marine Renewable Industry Association (MRIA) and Wind Energy Ireland (WEI), both all-island industry representative organisations, advocate for clearer policy signals to indicate that Ireland is 'open for business' (Figure 28.16). While offshore wind is more advanced than tidal and wave energy, addressing barriers to progressing the former will also address the future needs of the latter. In order for government to address the combined challenges to the development of marine renewable energy, it is necessary to:

• provide technology-specific subsidies

• indicate clear planning and regulatory structures

Fig. 28.16 STAKEHOLDERS ATTENDING THE ANNUAL MARINE RENEWABLE INDUSTRY ASSOCIATION (MRIA) FORUM, (2019) IN DUBLIN. The annual MRIA stakeholder's forum is an important event in the annual calendar for the marine renewables sector. As an all-island association, it brings together industry, policy, research and civil society representatives from both sides of the border, to take stock of trends and issues in project development, policy, technology, supply chain research and demonstration initiatives. [Source: Anthony O'Reilly]

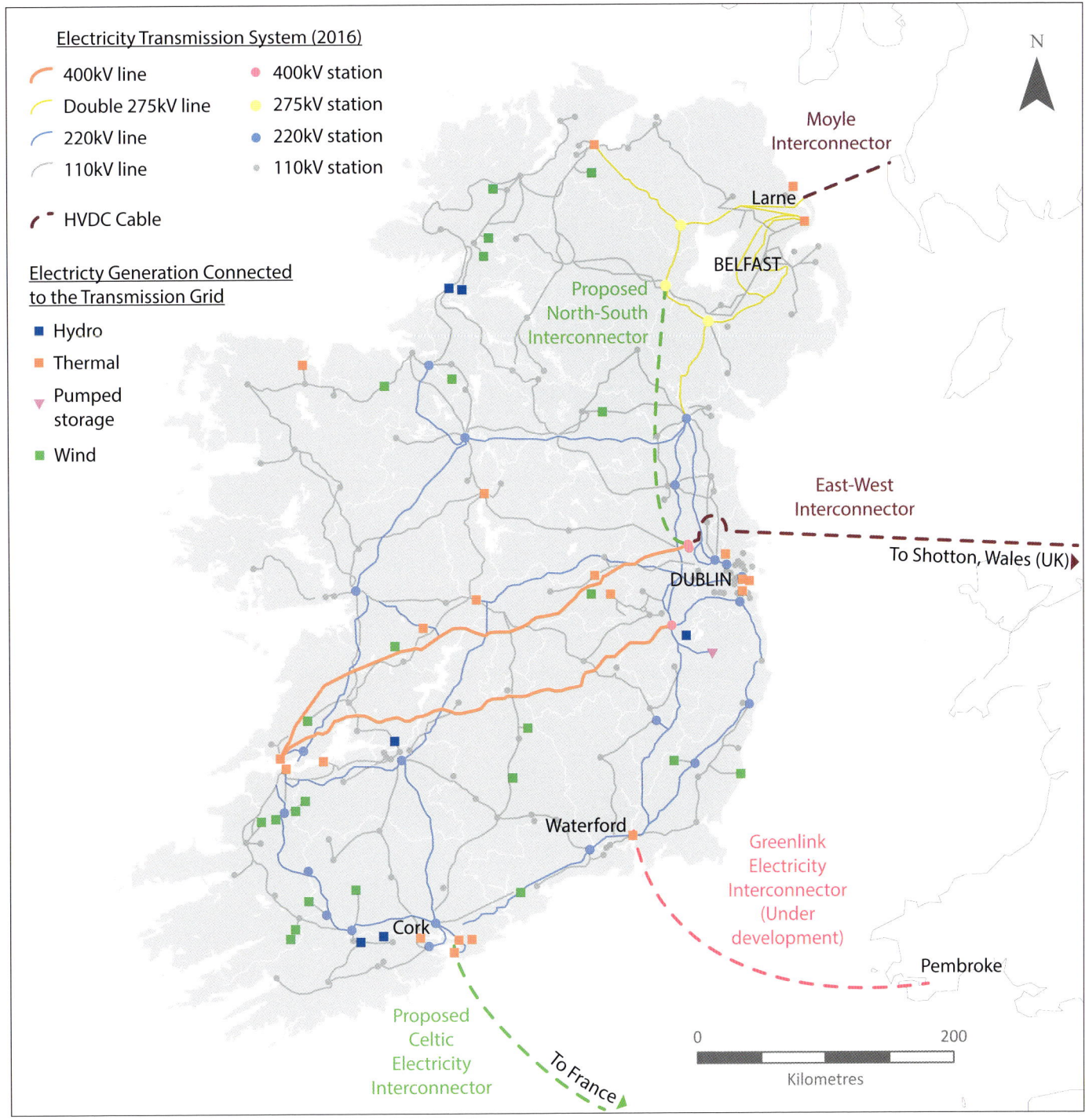

Electricity Transmission System (2016)

- 400kV line
- Double 275kV line
- 220kV line
- 110kV line
- HVDC Cable

- 400kV station
- 275kV station
- 220kV station
- 110kV station

Electricty Generation Connected to the Transmission Grid

- ■ Hydro
- ■ Thermal
- ▼ Pumped storage
- ■ Wind

Moyle Interconnector

Larne

BELFAST

Proposed North-South Interconnector

East-West Interconnector

To Shotton, Wales (UK) ▶

DUBLIN

Greenlink Electricity Interconnector (Under development)

Waterford

Pembroke

Cork

Proposed Celtic Electricity Interconnector

To France

0   200
Kilometres

N

Fig. 28.17 IRELAND'S ELECTRICITY GRID. The electricity grid, managed by Eirgrid in the Republic and by SONI in Northern Ireland, provides the capability to transmit electricity all around the island. As the transmission system operators, they play a critical role in developing and managing the infrastructure to provide current and future marine renewable energy developers with essential grid access. Eirgrid and SONI are working towards the delivery of the north–south interconnector by 2023, which necessitates dealing with legal challenges through the planning system. Eirgrid is also progressing plans for the Celtic interconnector between Ireland and France. Private developer Statkraft is developing a link to the UK, via the Greenlink connection. These developments are of vital importance to the marine renewable sector, as this opens much-needed links to export markets. Alternative routes to market, such as power to gas from offshore wind, are the subject of research and development in the Science Foundation Ireland, Eirwind project. [Adapted from Eirgrid, 2015]

• establish a strong export market through new routes to market (grid interconnections and green hydrogen production)

• develop the supply-chain required to support large-scale development (including investment in port infrastructure).

Ocean Renewable Energy (ORE) subsidies

A market support tariff is vital to incentivise investment. In the Republic, currently there are no ORE-specific subsidies but the Department of Environment, Climate and Communications

(formerly the DCCAE) proposed a Renewable Energy Support Scheme (RESS) in 2018. Based on a Floating Feed-in-Premium (FFiP) scheme, the RESS would determine a strike price for all renewable energy generated through 'technology-neutral' auctions. Depending on the lowest bid, all generators will receive the same strike price. The scheme then calculates the difference between the strike price and the reference market price, which varies over time. This method could create a level playing-field and increase competition, thereby lowering costs. 'Technology-specific' auctions

695

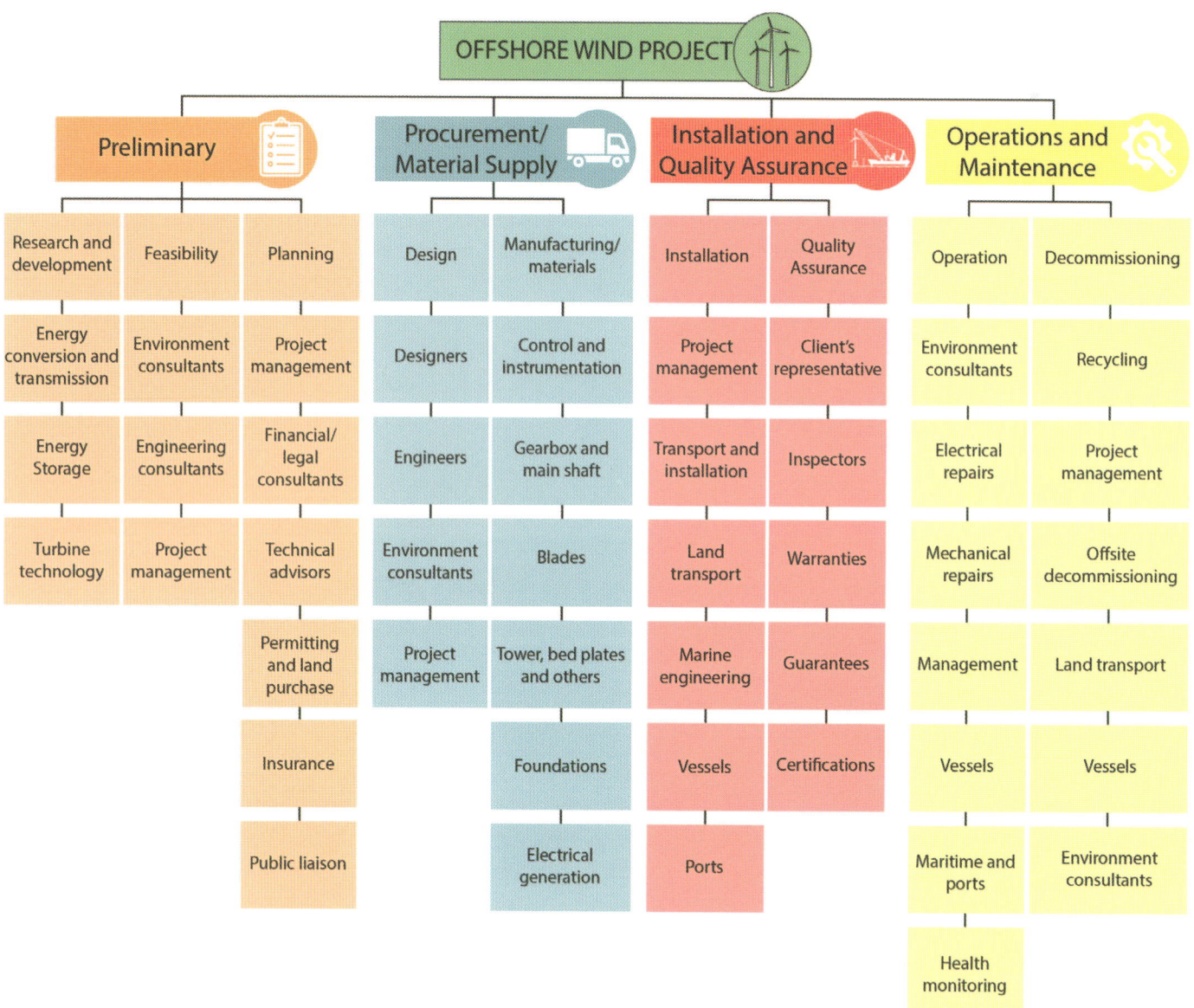

Fig. 28.18 EXAMPLE OF THE SUPPLY CHAIN FOR AN OFFSHORE WIND PROJECT. The development of a marine renewable industry in Ireland can facilitate a transition from a fossil-fuel, economy to a decarbonised economy. Significant additional value can be achieved from the sector in enterprise development. Enterprise Ireland has a long track record in supporting the growth of indigenous Irish companies to realise their export potential. In recognition of the increasing interest in Ireland as an emerging market for offshore wind, Enterprise Ireland has established an offshore wind cluster. The objective is to take stock of the supply chain opportunities, in terms of existing capacity and where capacity can be built, in support of all aspects of an offshore wind project – from the preliminary design stage to the operation and maintenance stage. Suppliers are required to deliver products and services in technically diverse areas from insurance and law to blade manufacturers and vessel supports, as shown in the diagram. [Adapted from Nguyen Dinh]

for offshore wind are being designed to encourage competition, but between comparable technologies. In addition, the OREDP proposed an initial market support scheme of €260/MWh for wave and tidal energy limited to 30MW. This amount does not compare to tariffs available in other European countries. The 30MW limit could be raised, with an amount ring-fenced for wave and tidal energy, and the balance available to floating wind and hybrid technologies.

The renewable energy market in Northern Ireland is based around Contracts for Difference (CfD). The replacement of the previous Renewables Obligation Scheme in 2015 with the proposed CfD auction-based subsidy was met with controversy, with concerns expressed over insufficient consultation, followed by challenges in making the new system effective in light of the collapse of the political power-sharing arrangements.

Planning and regulatory structures

The complexity and uncertainty of consenting processes increases risks and timelines, ultimately discouraging investors. In the Republic of Ireland, site investigation requires a foreshore licence and exclusive use of the foreshore for offshore electricity generation requires a foreshore lease. Beyond 12 nautical miles, there is no operational planning or consenting system. In addition to the requirements deriving from the Foreshore Acts (1933–2011), planning permission is required under the Planning and Development Acts (2000–2018) if there are any onshore facilities, such as sub-stations. Under the Electricity Regulation Act (1999), developers must have a Licence to Generate and Supply Electricity and an Authorisation to Construct/Re-construct a Generating Station. The latter are administered by the Commission for Regulation of Utilities (CRU). EirGrid is responsible for grid connection policy. Common stumbling blocks to development include the costly and time-consuming surveys to support Environmental Impact Assessment (EIA) and Appropriate Assessment (where applicable). Delays may also result from the lack of early and meaningful engagement by industry with local communities, increasing the risk of legal challenges to consenting applications.

Work is underway to introduce a new planning and consent system. The proposed Marine Area Planning Act seeks to integrate the foreshore consent process for ocean renewable energy projects into a streamlined process with planning provided to the extent of the EEZ. The proposed legislation would also provide for a single EIA process. It is anticipated that the new legislation will be enacted imminently. In addition, DCCAE (now the Department of Environment, Climate and Communications) published a guidance document on EIAs for ORE projects[10] and are developing approaches to community engagement.[11]

The Crown Estate, a UK-wide agency, owns and administers rights to the seabed in Northern Ireland, where developers must also ensure compliance with requirements for Environmental Impact Assessments to secure the necessary marine licence, and secure an electricity generating consent. The Marine Plan for Northern Ireland aims to ensure sustainable development of offshore resources. A consultation process on the spatial plan in 2018 facilitated viewpoints from diverse stakeholders on priority areas for development and conservation.

Spatial planning is critical in enabling development of the marine environment in a sustainable manner. The government in Dublin has also published a roadmap to develop Ireland's first marine spatial plan.[12] The National Marine Planning Framework (NMPF) aims to facilitate better management of marine resources, incorporating the need to encourage economic development of marine industries while ensuring environmental protection is a central consideration.

Upgrading the national grid and establishing an export market

There is an urgent need to upgrade the grid system and enable the integration of increasing amounts of renewable generation. Grid 25 sets out EirGrid's development strategy to achieve this by upgrading existing and building new transmission lines, essentially doubling the current size of Ireland's electricity grid by 2025.[13] In addition, EirGrid's DS3 Programme seeks to increase the share of variable renewable electricity that the all-island system can accommodate at any instant.

An export market would particularly encourage the development of offshore wind, given the advanced commercial state of bottom fixed wind energy technology. EU targets for electricity interconnection between member states is driving Ireland's policy response, with interconnectors between Ireland and Wales and between Northern Ireland and Scotland.[14] The north–south 400KV interconnector project would create a unified Irish electricity market, and has planning consent in the Republic, but is awaiting approval in Northern Ireland. Two additional interconnector projects have been designated Projects of Common Interest (PCIs) by the European Commission: the Celtic and Greenlink interconnectors, connecting Irish–French and Irish–British networks respectively (Figure 28.17). Greenlink is at the pre-

Fig. 28.19 TURBINES BEING LOADED ONTO SHIP IN BELFAST HARBOUR. Opened in 2013, Belfast Harbour's Offshore Wind Terminal was the first dedicated port facility in the UK for offshore wind. It operates as a preassembly site and load-out port for major offshore wind projects, such as the Dong Energy Burbo Bank Extension project, 8km off the coast of Liverpool Bay. The investment in the facility has paid off; the terminal handled 305,000 tonnes of wind-farm components in 2017. It is likely to be the port of choice for developments in the Irish part of the Irish Sea, particularly the installation process. [Source: Belfast Harbour]

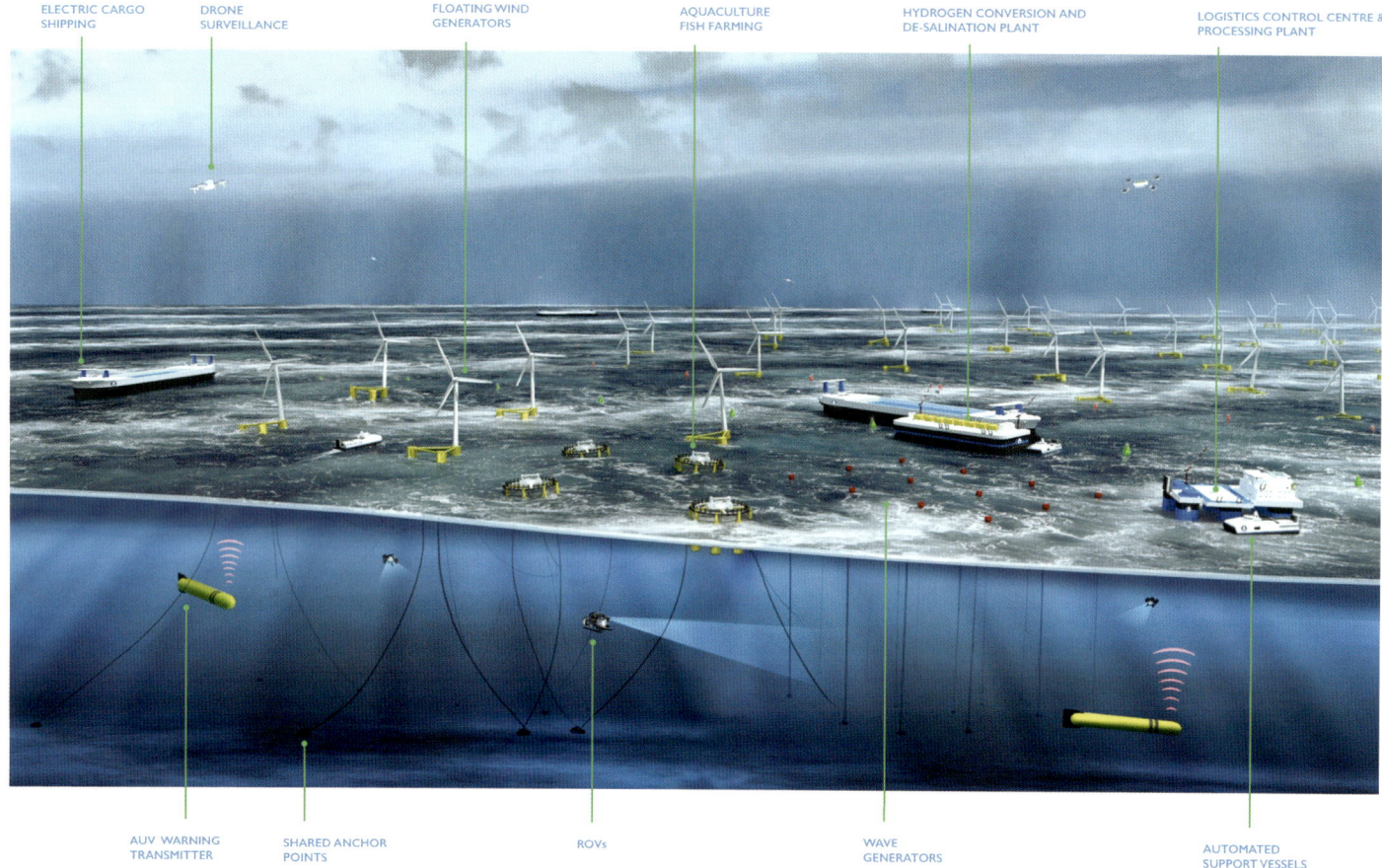

ELECTRIC CARGO SHIPPING   DRONE SURVEILLANCE   FLOATING WIND GENERATORS   AQUACULTURE FISH FARMING   HYDROGEN CONVERSION AND DE-SALINATION PLANT   LOGISTICS CONTROL CENTRE & PROCESSING PLANT

AUV WARNING TRANSMITTER   SHARED ANCHOR POINTS   ROVs   WAVE GENERATORS   AUTOMATED SUPPORT VESSELS

Fig. 28.20 A VISION OF THE FUTURE (2016). The coast of Ireland is set to become a busy place, especially if marine renewable energy enters the fray, adding to a considerable list of different users occupying Ireland's marine frontier, including fishers, mariners and hydrocarbon companies. The sea is relatively under-exploited and with vast resources, so it is imperative that good environmental governance be utilised, to ensure smart, sustainable use. An opportunity exists for marine renewable energy projects to be co-designed with fishers, aquaculture developers, scientists and conservationists, to seek optimum solutions for synergy, for example, through benefits from artificial reefs, or through co-location with multi-trophic aquaculture. [Source: Simply Blue Group]

planning stage but Brexit may jeopardise this as well as plans to join the Irish and Northern Irish electricity markets. Therefore, the government and EU have additionally placed their support behind the Celtic interconnector.

Developing a supply chain

A pressing concern is establishing an adequate supply chain and infrastructure to handle the requirements of future ocean renewable energy projects (for example, expanding Irish port capacity) (Figure 28.18). Failure to tackle gaps could result in reliance on overseas suppliers and foreign companies, resulting in a missed opportunity for Irish businesses and increasing energy prices. The SEAI and Enterprise Ireland have established the Marine Renewables Supply Chain Database, containing details of existing companies that provide services to the wind, wave and tidal energy sectors, including companies north of the border. Engineering consultants GDG completed a gap analysis of Irish port and harbour infrastructure for the ORE sector.[15] In Belfast Port, the Harland and Wolff shipyard is a significant hub for assembly and manufacturing

of turbines (Figure 28.19). However, while work is being done to identify opportunities and to address potential gaps in the supply chain, many companies are maintaining a watching brief on the industry until commercial-scale projects materialise.

In conclusion, while the building blocks are being put in place to support the ocean renewable energy sector, considerable work remains to be done. Progress is fundamentally linked to governments recognising their responsibility to develop sustainable, renewable, clean energy. The benefit of political support for marine renewables is evident in Scotland, which is now a world-leader with the first operational floating wind and tidal energy farms. Northern Ireland has made considerable progress in developing tidal energy projects, while the government of the Republic of Ireland has invested in world-class R&D and testing facilities. Improvements in consenting, subsidy supports, critical infrastructure and supply chain developments are required to ensure Ireland is poised to seize the opportunity offered by commercial ocean energy projects in the near future (Figure 28.20).

Coastal Mining, Quarrying and Hydrocarbon Exploration

David Naylor

REMAINS OF FORMER MINE BUILDINGS AT TANKARDSTOWN, NEAR BUNMAHON, IN COUNTY WATERFORD. Many locations around the coast of Ireland supported extensive mining industries, for copper, lead, zinc, silver and other minerals, particularly in the eighteenth and nineteenth centuries. Virtually all of this activity has now gone, but in many places it has left an important legacy of abundant industrial archaeology. On the coast of Waterford this led to the area being declared a European Geopark, known as the Copper Coast Geopark, in 2001, and a Global Geopark under the auspices of UNESCO in 2005. [Source: Copper Coast Geopark]

This chapter examines the impact that the search for a range of minerals, oil and gas has had, and continues to have, in Irish coastal and marine areas. There is abundant historic evidence of mining around the coast of Ireland, dating mainly from the eighteenth and nineteenth centuries. In a few locations, however, there is a much longer tradition of mining. On Rathlin Island off the north coast, and at Tievebulliagh in County Antrim, for example, the hard rock called porcellanite was quarried and worked to make axe heads and other implements by Neolithic peoples at least 7,000 years ago. On the other hand, the search for oil and gas began only in 1960. There are three main reasons for these contrasting timelines: the geology of Ireland and its offshore region, developing geological concepts,

and the development of new technologies.

A glance at the geology map (see Chapter 5: Geological Foundations) shows that onshore Ireland is composed mainly of Palaeozoic and older rocks. Like humans, rocks are subject to ageing processes and, over time, they become compressed, fractured, folded or heated. The ability rock sequences had to generate and store hydrocarbons is, therefore, diminished according to the severity of these processes. As a consequence, mineral deposits are found most frequently in older rocks and granites.

The geology map shows that the older rocks in Ireland are often located close to the coast, where they also tend to be well exposed. It is natural, therefore, that early searches for minerals

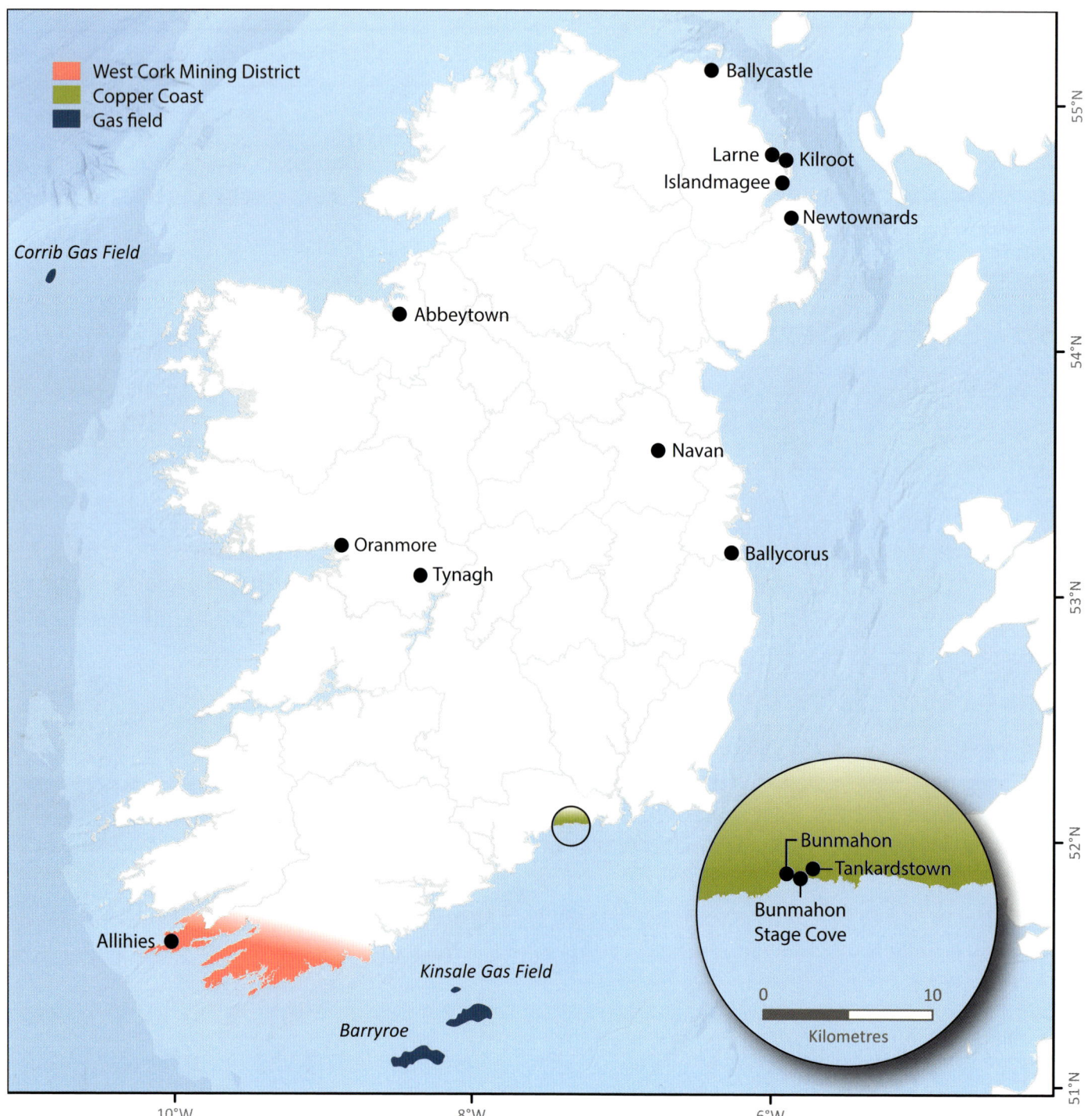

Fig. 29.1 IMPORTANT HYDROCARBON AND MINERAL LOCALITIES IN IRELAND.

were in coastal areas, rather than in the broad, central limestone plain with its lack of rock outcrop and cover of glacial drift. The point should also be made that the environmental footprint of mining operations is generally more marked and longer lasting than is the case with oil and gas operations.

Most of the world's oil and gas reserves, comprising around 86 per cent of the oil and 71 per cent of the gas, occur in younger Mesozoic and Cenozoic rocks. For Ireland, these younger rocks are found mostly offshore. The Republic of Ireland produces only 63 per cent of its natural gas needs from indigenous resources located at the Corrib and Kinsale gas fields (Figure 29.1). The balance of the natural gas requirement is imported from Britain. To facilitate that process, the island of Ireland taps into that supply through three undersea pipelines that connect directly to the British gas system at Moffat in Scotland. Production from Ireland's two existing gasfields (57 per cent of demand in 2018–19) is expected to decline with the move to renewable energy sources and unless further indigenous gas supplies are located, there will be increased dependency on imported supplies. There is no oil production in Ireland, so the total requirement for liquid hydrocarbons is imported.

MINING ACTIVITIES ON THE IRISH COAST

There are many signs of previous mining around the coasts of Ireland, and an early review by Cole (1998) of all Irish mineral localities is still useful and of interest.[1] At the present time, however, only one mining operation remains active on the island, the east Antrim salt mines. Most of the deposits that have been mined were generally small-scale operations and were either worked out quite quickly or became uneconomic in the face of competition from larger mines located elsewhere. Nevertheless, the physical evidence of historical mining remains visible at many places around the coastline (Figure 29.1). These are too numerous to mention individually, but the more important mining provinces and sites will be described.

West Cork

West Cork was a well-known mining district in the nineteenth century. A variety of minerals – copper, manganese, barytes, lead and silver – were produced, but the main product was copper (Figure 29.2). The copper ore occurred generally in quartz veins within the country rock and these could be from less than 1m to

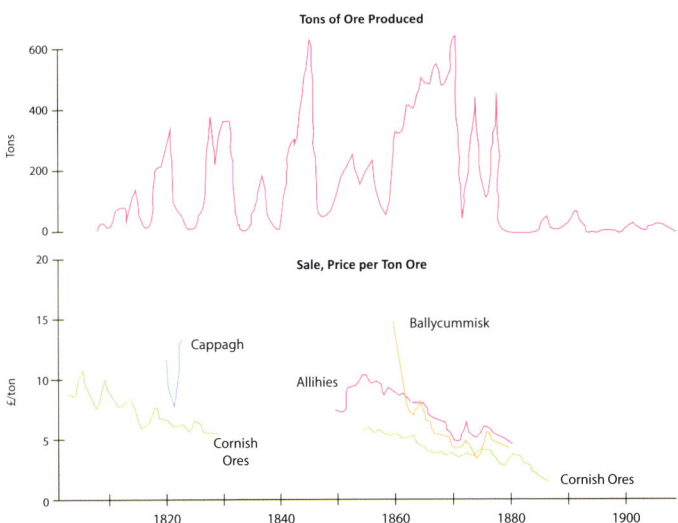

Fig. 29.2 COPPER ORE PRODUCTION, WEST CORK DISTRICT. The graph shows the fluctuating production from the copper mines of west Cork, and the price paid for the different ores compared with ore from Cornwall. It is notable that Irish ores fetched a higher price than the Cornish ones, perhaps generating greater revenues for the mine owners, but also putting them at a competitive disadvantage compared to their Cornish counterparts. The graph also shows the falling prices in the latter part of the nineteenth century that led to the inevitable closure of the working mines. [Data source: modified from Cowman, D., 1988. *The Abandoned Mines of West Carbery Promoters, Adventurers and Miners*. Dublin: Geological Survey of Ireland]

15m in width. There were many workings and mines within this district, a number of which were at, or close to, the coast (Figure 29.3). Reviews by Cowman and Reilly (1988) and Hodnett (2010) cover the historical development and geology of most of the mineral occurrences in west Cork.[2]

Allihies, located at the western end of the Beara Peninsula, was the most successful locality. Evidence of past mining may often be limited to old spoil heaps and small horizontal tunnels (adits) or shafts, but the ruins of old engine houses and dwellings are also preserved at some locations. Although the main phase of mining activity was in the mid- to late nineteenth century, evidence has been found in the mines of early phases of working, possibly dating from the period 1500–1800. Indeed, some mining for copper dates back to the Bronze Age, the best-known locality being the slopes of Mount Gabriel, near Schull, where diggings and spoil heaps are visible. Here, there was also small-scale working of quartz lodes in the mid-nineteenth century.

MOUNTAIN MINE, ALLIHIES, BEARA PENINSULA

David Naylor

The jewel in the crown of west Cork copper mining was the mines at Allihies, near the western extremity of the Beara peninsula (Figure 29.3) where the main ore mineral mined was chalcopyrite (copper-iron-sulphide).[3] Operations began in 1812 at the Dooneen mine, which followed a quartz lode with copper staining eastwards from its exposure on the coast. This mine had a number of phases and closures, and was worked to a depth of about 240m in 1878.[4] Ultimately, a total of six productive mines were worked in the Allihies area, and attracted significant numbers of miners from Cornwall and South Wales to supplement the local workforce. By the mid-nineteenth century,

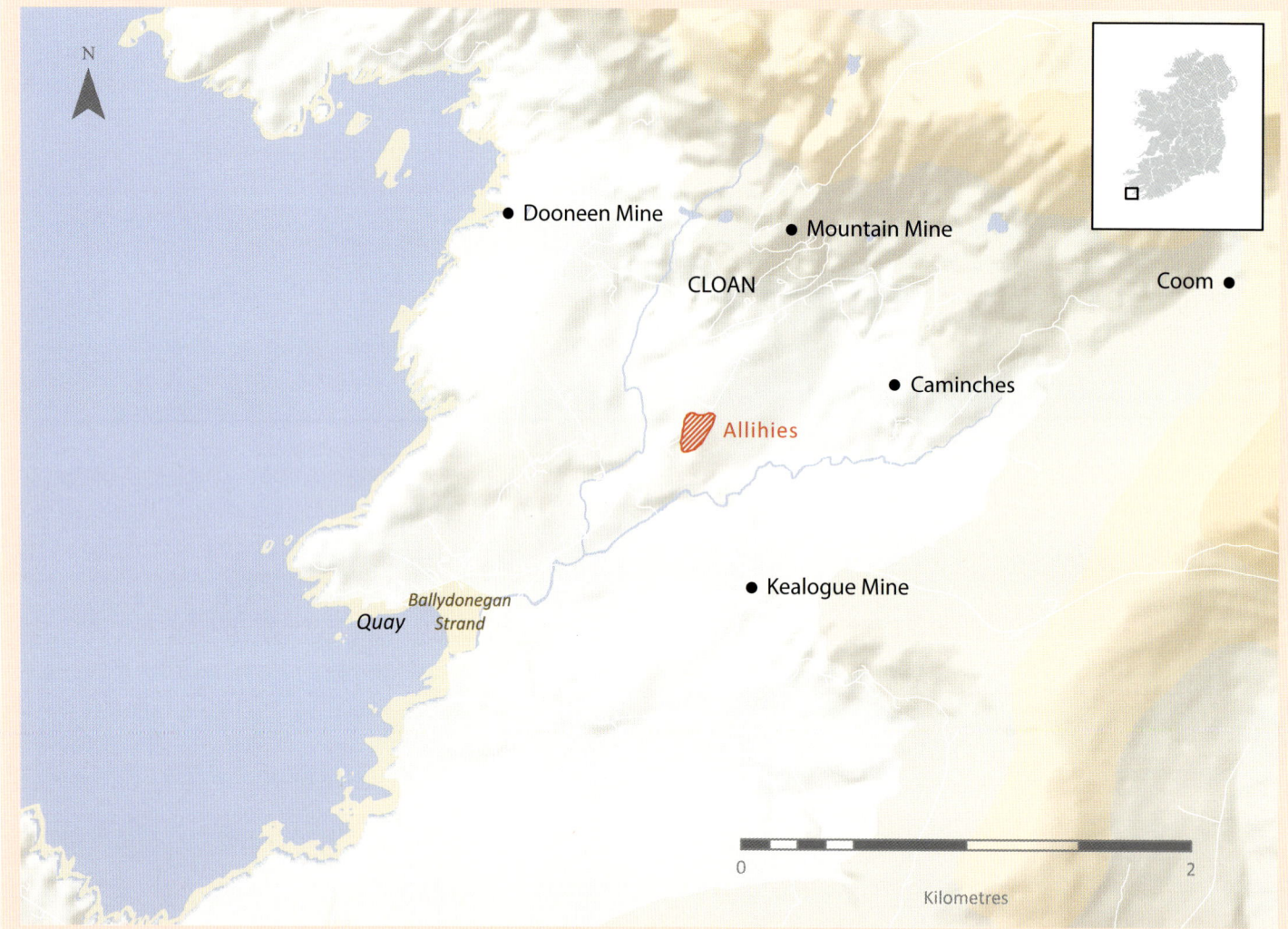

Fig. 29.3 IMPORTANT MINING LOCALITIES AROUND ALLIHIES. The village of Allihies grew up as a direct result of copper mining operations in the western part of the Beara Peninsula. The main ore mineral was chalcopyrite (copper-iron-sulphide). By the mid-nineteenth century, the success of the mines opened in the area had attracted mining families from South Wales and, more especially, from the copper-working areas of Cornwall. At this time, the mines employed around 1,600 miners and their families. Many lived in the village of Allihies, from where they walked to work in the nearby mines. The work was difficult and labour intensive. Following the manual extraction of the ore, it was brought to the surface for crushing and hand-sorting. This work, which was mainly carried out by women and children, was essential since it removed waste material and thereby not only improved the grade of the ore, but also reduced the weight of the ore, which had to be transported to local piers on the coast for shipment to Swansea in South Wales for sale and smelting. Between 1812 and 1912, some 297,000 tonnes of ore were shipped from the Allihies mines. Depressed copper prices and growing competition from other suppliers caused the run-down and eventual closure of mines in the area. Large numbers of miners emigrated, for example to the copper mines in Butte, Montana.

the Allihies mines employed around 1,600 people. Water wheels and steam engines were employed to crush ore and dewater the mines; remnants of these, including three engine houses, are still visible on the mountainside.

Mining began in 1813 at the Mountain Mine, which proved to be the most productive of the west Cork copper mines (Figure 29.4). A wide east–west surface quartz lode, an eastward extension of the zone worked at Dooneen, was mined by hand, initially as an opencast operation, then hauled to the surface in buckets where it was hand-crushed and sorted, mainly by women and children. The hand-sorting of the ore helped to improve its grade, prior to shipment to Swansea in South Wales from local piers located at the coast. Ore grades improved with depth, and the first steam engine house was erected in 1830 to pump water from the workings and allow deeper mining. Mining followed the inclined vein downwards and a Cornish Man Engine – the only example in Ireland – was installed in 1862 to facilitate the movement of workers to greater depth.[5] The mine was closed in 1882.

Subsequent attempts to resume operations at the Mountain Mine were made on several occasions to the 1920s, but to limited success. In 1959–1960, the mine was dewatered to help assess its potential for reopening. Drilling from the lower levels of the mine, however, failed to establish the existence of adequate reserves to allow commercial exploitation. Efforts to reopen the mine were, therefore, abandoned.

Other successful mining operations were concentrated on the Caminches lode, a mineralised quartz vein to the east of Allihies, extending from a point east of Cloan in the north to beyond Kealogue in the south (Figure 29.3). Little remains of the engine houses, stamping mills for ore crushing, and water mills that were constructed to work the northern part of the lode from 1818 onwards. The most productive and profitable mine on the lode was the Kealogue mine that operated from 1842 to 1882. The workings here mined the lode along a length of 2,000m and to a depth of 400m. Of the four original engine houses, only one – Puxley's engine house – still stands. Waste from the crushing floors was washed into the river to the north, and was carried down to Ballydonegan strand.

Fig. 29.4 MAN ENGINE HOUSE, MOUNTAIN MINE, ALLIHIES. This mine operated continuously until 1882 and was abandoned when it reached a depth of 421m below ground level (280m below sea level) due to deteriorating copper prices and overseas competition. These factors also led to the termination of other copper mining operations in Ireland. The ruins of the main engine house and associated chimney stack remain as reminders of the historic importance of mining in this locality. Below the site of the mine is the village of Allihies, where a significant mining community was located, comprised largely of miners and their families, who were brought into west Cork from Cornwall. With the decline of mining, the village reverted to being a small services centre which meets some of the needs of the small-scale farm communities that typify this pastoral landscape in West Cork. Today, it has sought to revive its mining heritage and potential for tourism, by opening a museum to highlight the history of mining in the area. This is housed in the former Methodist church, which was built to meet the religious needs of the Cornish miners. [Source: Tara Hanley]

The result of copper mining operations in the district gave rise to the growth of the large mining village of Allihies. With the collapse of mining, its population fell significantly, largely through emigration to other mining areas, such as Butte in Montana. In more recent times, however, the village has found a new focus. This has involved identifying and promoting the industrial archaeology and heritage value of the mine workings. To this end, the Man Engine House has been partially restored (Figure 29.4), while the village now also hosts a mining museum and a related Copper Trail that takes hikers to surviving mine-related buildings, powder houses and dressing floors of the various historical mine operations.

In addition to the mine workings around Allihies, evidence of mining activity can also be seen at a number of remote localities on the shores of Bantry and Dunmanus bays. On Sheep's Head, there are adits in the sea cliffs along the north coast at Glanalin and Caravilleen (Figure 29.5), and also further west at Killeen North (Glanroon) where the ruins of mine cottages still stand. These mines produced and sold only a small tonnage of ore.[6] Further west, beyond Cahergal, there is substantial evidence of mining at the isolated locality of Gortavallig. Initially, there was high expectation for Gotavallig, and development included the construction of two piers, adits and shafts, a dressing floor for the manual upgrading of the ore, and a row of miners' cottages (now in ruins). The mine was worked in the 1840s but shipped only modest amounts of ore, and closed in 1850.

A number of mining ventures operated along the south coast of Cork from Mizen Head eastwards.[7] At the western extremity of Mizen Head at Brow Head, there is extensive evidence of mining activity. Here, a series of adit entrances are visible in the sea cliffs. Small amounts of ore were mined in the 1850s, but the venture was

unprofitable and mining ceased in 1860. Attempts to re-open the mine in the early 1900s were unsuccessful. The ore was hauled to the cliff top where the ruins of mine houses and cottages, a cobbled dressing floor, a gunpowder house and mine waste tips are preserved.

At Crookhaven on Mizen Head, a lode was also worked in the 1840s and 1850s, employing a steam engine and pumps. The operation failed to produce significant amounts of ore and had a short working life.

The mine was located east of the village on the small peninsula where two circular mine powder houses and a ruined chimney and engine house can still be seen.[8]

Three copper mines, all with chequered histories, operated on the north shore of Roaringwater Bay. These were the Coosheen mine on the east side of Schull Harbour and, further east, the Ballycummisk and Cappagh mines near Rossbrin. Ballycummisk operated intermittently and on a small scale from 1814 until the 1850s, but was plagued by financial difficulties. However, a steam engine was introduced in 1857 that allowed mining to reach a depth

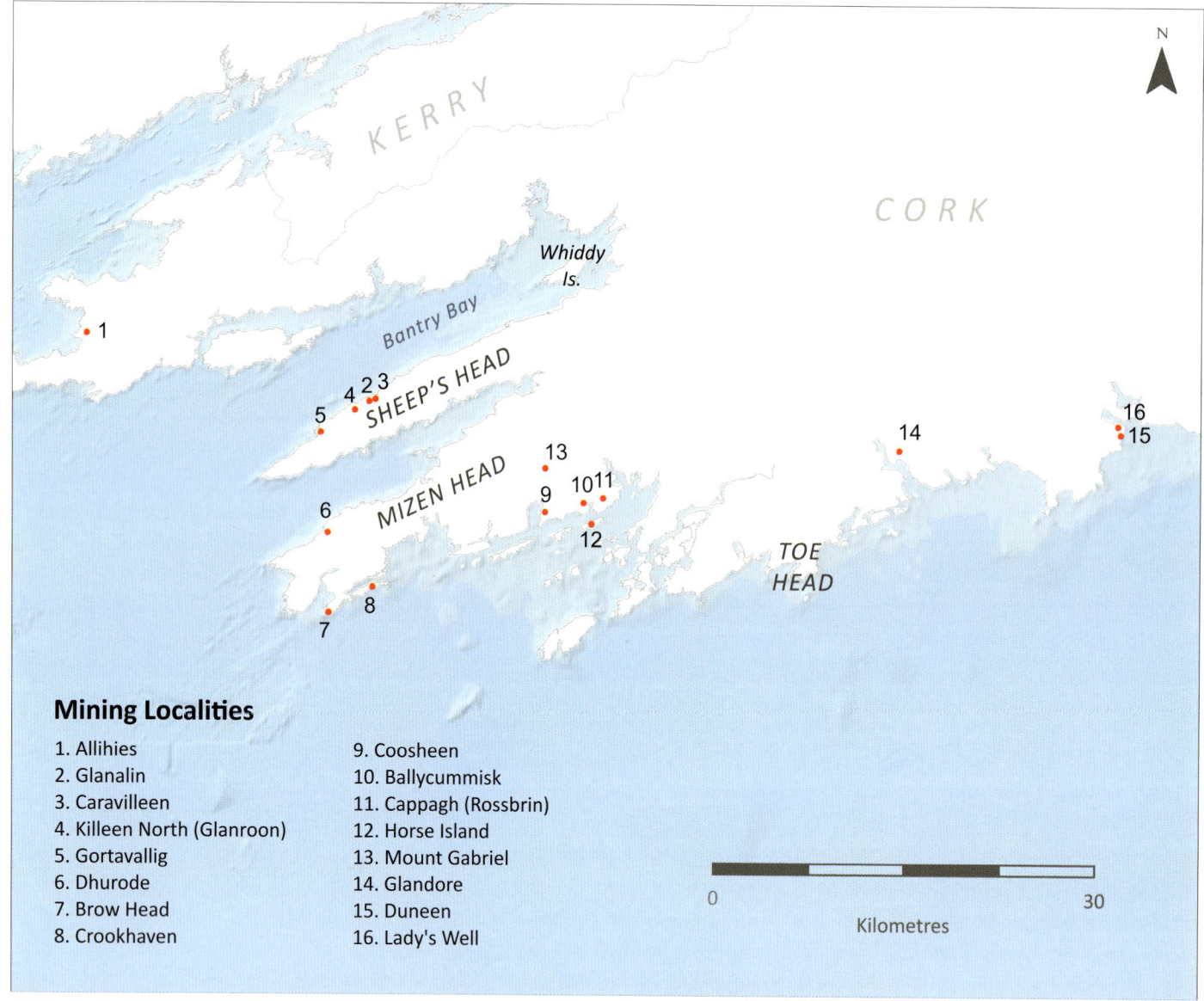

Mining Localities

1. Allihies
2. Glanalin
3. Caravilleen
4. Killeen North (Glanroon)
5. Gortavallig
6. Dhurode
7. Brow Head
8. Crookhaven
9. Coosheen
10. Ballycummisk
11. Cappagh (Rossbrin)
12. Horse Island
13. Mount Gabriel
14. Glandore
15. Duneen
16. Lady's Well

Fig. 29.5 MINING LOCALITIES IN WEST CORK. There were many attempts to replicate the commercial success of the Allihies copper mines elsewhere on the west Cork coastline. Some of these met with modest success, some were small or uneconomic, and yet others were used to float mining 'bubbles' or investment scams in London. Proximity to the coast was important for the commercial viability of these operations, since it allowed extracted ores to be loaded onto ships for transport and processing in South Wales and elsewhere. Many of the communities that developed around these mines were wholly dependent on the employment that mining operations provided. However, during the great Irish famines of the early and mid-1800s, many mine operators sought to offset the impacts of food shortages by bringing in provisions for their workers, thereby at least partially increasing the resilience of the communities in these otherwise precarious and marginal areas.

of about 360m. This helped increase output, and a substantial tonnage of ore was shipped to Swansea in the 1860s. Today, two derelict buildings are the only surface remains.

There were several attempts to mine the nearby lode at Cappagh, but only modest amounts of ore were mined. This operation ceased in 1873. The initial success at Ballycummisk also prompted renewed interest in a previously failed mining attempt at Coosheens which was restarted in 1860 with the installation of a steam engine. However, only a modest output of ore was achieved and, when falling copper prices forced the closure of Ballycummisk in 1877, Coosheen also ceased production. The ruin of the Coosheen Engine House still stands to the east of Schull Harbour, as does an ivy-covered gunpowder magazine.

A further mine was located on Horse Island, one of 'Carberry's Hundred Islands' within Roaringwater Bay itself (Figures 29.6 and

29.7). The earliest recorded mining on the island was by Colonel Robert Hall in 1814. In 1835, ancient workings were re-excavated by the West Cork Mining Company and found to contain a stone hammer, oak shovel and tub. By tradition, these old workings were ascribed to 'The Danes'. Throughout the nineteenth century there were phases of mining activity, which involved the driving in of a number of adits and shafts. This occurred mainly at the north-east end of the island, where ruined mine buildings still stand. Although the copper grades of the ore were high, the mined tonnage was relatively small and could not sustain a long-term viable enterprise. The last mining was in 1901–1902.[9]

Manganese/iron ore was mined at Glandore (on the east side of Glandore Harbour) in the 1830s and 1840s for use as a bleach. The venture underwent a further phase of activity between 1869 and 1881, but low prices for the ore forced its closure. Apart from a brief attempt in 1907–1908, this was the end of mining at this locality. The well-preserved remains of the mine engine house still stand.

While copper mining was declining, interest in barytes (used in paint, plaster of Paris, soap and other products), which had been encountered in various localities in west Cork, was revived in the 1870s.[10] This mineral was mined at a number of localities, but not in major amounts at any one mine. However, a successful mine was opened at Duneen, on the west side of Clonakilty Bay, which was to remain in continuous production for twenty-six years, and intermittently thereafter until 1923. Despite its closure, tests and trials continued in the area. In 1979, the mine was reopened as the Ladyswell Mine. The impetus for this was the use of barytes as an additive to provide necessary weight in the mud used as lubricant in oil-well-drilling operations. The operation, however, was short-lived and the mine closed in 1985.

Despite the profusion of mining attempts at different localities throughout the West Cork mineral district, there were few significant mines. The vein lodes, although rich in places, were relatively narrow. Success depended on good product prices and the availability of cheap manual labour, including women and girls, to manually sort the ore to a higher grade before shipment. In modern times, the availability of machinery, especially for use in extracting large volumes of ore from large-scale operations, together with the increasing cost of labour, have meant that small vein deposits can no longer compete with large and mechanised lower-grade mines.

Fig. 29.6 ABANDONED MINERS' DWELLINGS AT COOSHEEN MINE. The coastal landscapes of Roaringwater Bay today are predominantly rural and agricultural in character, and it is difficult to imagine how the area must have looked when mining was at its peak. During the course of the nineteenth century, a number of mining operations were established in and around Schull Harbour, and these mines supported a thriving community of miners and their families, all of whom required housing. It was always a precarious existence and, when production ceased and the mines closed, there were precious few employment opportunities to take their place in the area. This, combined with the catastrophic impact of the Great Famine in west Cork, led to massive depopulation from the mid-nineteenth century onward. Many of the cottages that the mine workers occupied now lie roofless, empty and derelict, as shown here, and the small gardens that once provided vegetables for their kitchens have been overtaken by bracken, brambles and hawthorn shrubs. However, since the 1960s, several of these abandoned cottages have been restored, modernised and put to new purpose as holiday homes. [Source: *Roaringwaterjournal.com*]

Fig. 29.7 HORSE ISLAND MINERS, IN FRONT OF WHAT APPEARS TO BE THE MINING CAPTAIN'S COTTAGE. The exact date of the photograph is unknown but a good estimate would be around 1900, which coincides with some of the later workings at the mine. It is interesting to note how many people the mine operation employed. Such employment, although often dangerous and poorly paid, was an important alternative occupation for communities dominated by low-income, pastoral farming. Today, Horse Island is in private ownership, and has a number of holiday homes. [Source: Barry Flannery]

Fig. 29.8 MINERS AT THE ENTRANCE TO A MINE, KNOCKMAHON, 28 MARCH 1906. A number of copper mines were developed along the Waterford coast at Knockmahon (Stage Cove) and Tankardstown, east of the River Mahon and Bunmahon village, especially during the early part of the nineteenth century. The miners' work was arduous, uncomfortable, labour-intensive, required long hours spent underground, and was almost entirely powered by human effort. It was also dangerous, with the *Waterford Mail*, the *Waterford Chronicle* and the *Dublin Morning Register*, among other local, national and even international newspapers of the time, carrying frequent reports of accidents occurring underground, many of which had fatal consequences. The Knockmahon mine workings eventually reached a depth of about 366m, and extended both inland and also south-eastwards under the sea. However, a recurring problem of flooding, combined with decreasing ore grades, eventually made the Knockmahon operations unsustainable, and the mine closed around 1855. There was an attempted revival of the mine in 1906, as shown in this photograph, but it was short-lived and ultimately unsuccessful. [Source: NLS POOLEWP1555a]

Waterford coast (Copper Coast)

In the nineteenth century, a stretch of the Waterford coast east of Bunmahon village (now known as the 'Copper Coast') was the site of intense mining activity. Between 1828 and 1878, significant volumes of copper ore were raised and exported from this area. Mining began in the 1820s in an area east of the River Mahon, where a number of veins or lodes had been located. Over the next two decades, a mine complex was developed on the cliff top at Knockmahon (Stage Cove), together with extensive ore-dressing facilities. Ore concentrate was shipped from a local cove. The mine eventually reached a depth of about 366m, with the workings extending underground and inland from Knockmahon, and also south-eastwards under the sea (Figure 29.8).[11] These latter lodes

Fig. 29.9 (left) THE COPPER WATERFALL IN TANKARDSTOWN MINE (left). The 'stope' is where an inclined quartz vein, bearing copper ore, has been systematically extracted. Even after mining operations have ceased, groundwater continues to percolate and flow through cavities in the rock. Here in the Tankardstown mine, this has led to the formation of an extensive jelly-like deposit of flowstone known as the 'Tankardstown Waterfall', or sometimes the 'Blue Waterfall'. The colours are generated by groundwater and sea air interacting with the remaining ore minerals to produce iron hydroxides (rusty colours) and, with the addition of chlorine derived from seawater, copper sulphate (the blues and greens). [Source: Sharon Schwartz, courtesy of Geological Survey Ireland]

Fig. 29.10 BALLYCORUS CHIMNEY. Chimneys such as this can be seen at many points around the coast of Ireland. Standing as modern counterparts to the round towers associated with medieval monastic settlements, they mark the location of abandoned mines, often long after most of the other buildings associated with the operations have gone. At Ballycorus in County Dublin, a vein of lead ore with associated barytes was worked between 1806 and 1860. Evidence of the initial opencast operation is still visible on the ground, although its output was never prolific. However, a smelting works with its associated chimney stack was built nearby, to receive both the locally extracted ore and also ores from mines located further afield in counties Wexford, Wicklow and elsewhere. The works continued to function until 1913. Today, the area is a popular hiking and tourist destination. [Source: Joe King, Wikimedia Commons, CC BY-SA 3.0]

eventually flooded and this problem, combined with decreasing ore grades, eventually resulted in the closure of the mine around 1855.

In the meantime, a new high-grade lode had been located about 1km east at Tankardstown (Figure 29.9) and work had begun there. The distance to the ore-dressing facilities, however, presented haulage problems and this led eventually to the construction of a mineral tramway to facilitate movement of the ore from Tankardstown to Knockmahon. Tankardstown mine remained open until about 1875, when decreasing grades of ore at depth forced its closure. Although much of the mining infrastructure has fallen into ruin, or was physically removed, a considerable effort has been made in recent years to preserve the remaining buildings and mine legacy of the area.

The 'Copper Coast' was adopted as a European Geopark in 2001 and as a Global Geopark, under the auspices of UNESCO, in 2005. Local voluntary effort, aided by the Mining Heritage Trust of Ireland, made this possible, and has allowed the remaining structures to be researched and preserved. There is now a Geopark Centre in Bunmahon and a Copper Coast trail, taking in most of the heritage sites (see the opening image for this chapter).

County Dublin

There were many mineral localities known around Dublin, and fifteen of these saw some degree of mining.[12] The principal mineral was lead-silver, occurring in relatively narrow veins, and operations were modest in size and involved only small tonnages of output. Much of the evidence for this activity has been obscured beneath urban expansion.

South of the city, at Dalkey and Killiney Hill, intermittent mining for lead with associated barytes took place in the period 1750–1850.[13] The barytes was shipped from Dublin to Liverpool for use in paint manufacture. There are many old adits and shafts at these locations, but they are difficult to access. The Ballycorus vein was worked between 1806 and 1860, and although the mine was never a prolific producer, the initial opencast work is still visible. The most obvious evidence of mining, however, is the smelter chimney at Ballycorus, south-west of Killiney, which overlooks much of south Dublin (Figure 29.10).

County Down

In the past, County Down was a mining district for silver-lead ore. Although this led to many trials and some short-lived mining ventures, only one or two successful mines were developed.[14] The Conlig and Whitespots mines near Newtownards, at the head of Strangford Lough, were the two most successful lead producers (Figure 29.11). In the decade 1848–1858, the Whitespots mine accounted for almost 40 per cent of the lead produced in Ireland. Thereafter, the mine experienced diminishing production until the 1880s. Some buildings remain standing, but are in a perilous state.

The Antrim coast

A different form of mining took place on the north Antrim coast, east of Ballycastle. Carboniferous strata with thin coal seams are exposed in the cliffs along this section of the coast, with the seams being accessible by adit. The diggings for coal can be traced to the medieval period, but their persistent development dates to the 1730s, when production rose to approximately 15,240 tonnes per annum.[15] Such levels of output, however, declined in the early nineteenth century, and although the seams were worked intermittently during the twentieth century, the last mine closed in 1967. In general, the seams were thin, the coal was of poor quality, and only the 1.3m thick main seam was considered to have been of commercial significance. Offsetting these disadvantages, however, was the fact that the seams outcropped directly at the coast, which meant that it was easy to transport the extracted coal by boat, using small jetties built from locally quarried stone (Figure 29.12). Being of comparatively low grade, most of the coal would have gone to local markets in Belfast and elsewhere in Northern Ireland, rather than being exported overseas. Signs of former mining activity – such as spoil heaps, remains of abandoned jetties and discarded equipment – are still visible along this stretch of the north Antrim coast.

In County Antrim, iron ore was also mined from the eighteenth century until after the Second World War. The main body was lateritic iron ore, which was developed above the aluminium ore bauxite, so that the two ores were often worked together by open cast or underground methods. The main development of these ores is in the Interbasaltic Formation that occurs between the Upper and Lower Basalts (see Chapter 5: Ireland's Geological Diversity). In the quiescent interval following the first phase of lava extrusions, there was a long period of tropical weathering, which produced a deep alteration on the surface of the Lower Basalts. This red alteration layer is seen clearly in the basalt cliffs above the Giant's Causeway (Figure 29.13). The mining of this ore, however, took place inland on the Antrim basalt plateau. Although the ore grades were generally good, the operations gradually waned against competition from large-scale overseas deposits.

North-west Ireland

There is a scattering of old mining locations from Donegal to Galway, dating mainly from the nineteenth century. Most of these were small, inland mines that yielded lead-silver ore, with the ore being transported to local piers on the coast for shipment. Another lead mine operated in the mid-nineteenth century at Rinville, Oranmore, on the east coast of the bay from Galway town. The mine took ore from an adit in Carboniferous limestone at the water's edge.

A more important operation that mined lead-zinc-silver ore from Carboniferous limestone was at Abbeytown, near Ballysadare, on the south coast of Sligo Bay. Here, a lode exposed at the shore has been worked intermittently and at a relatively small scale from at least the eighteenth century, and was continued into the mid-twentieth century. Further drilling at the site following World War Two led to the location of significant hidden reserves of ore. This stimulated a major mining operation between 1950 and 1961 which extracted 731,000 tonnes of ore.[16] Concentrates produced at the site were shipped from Sligo for smelting.

Fig. 29.11 THE CONLIG LEAD MINES, COUNTY DOWN. At its peak, the Conlig and the adjacent Whitespots lead mines, near Newtownards in County Down, constituted one of the largest underground mines in Ireland. Today, all that remains of this activity is an extensive site containing spoil heaps, along with the tailings impoundments, the capped mine shafts, and a number of buildings including the chimney seen here. The site is now used as a fieldwork location for training geology students from Queen's University Belfast and other colleges, as well as being popular with amateur rock and mineral collectors. A variety of minerals and crystals can still be picked up from the mined debris, including galena (lead ore), chalcopyrite, barite, dolomite, calcite and a rare mineral called harmotome. The Ulster Museum in Belfast has a number of specimens from this locality in its collections. [Source: Alastair Ruffell]

Figure 29.12 COAL BEING LOADED ONTO A BOAT NEAR FAIR HEAD, TO THE EAST OF BALLYCASTLE, SOMETIME BETWEEN 1865 AND 1914. Thin, but workable, seams of coal outcropped at many points in the cliffs between Ballycastle and Murlough Bay on the north Antrim coast. They could also be accessed further inland. A number of mines were developed to extract this coal from the early 1700s onward. Although the quality of the coal was generally poor, the coastal location meant that it was easy to take it in wheeled tubs, directly from the mine, and load it onto boats lying alongside jetties built from locally quarried sandstone. Around the middle of the nineteenth century, 10,000–15,000 tonnes of coal were being exported from the Ballycastle coalfields annually, most of it going to markets in Belfast and Dublin. However, from about 1865 onward the mining went into decline, and by the end of the century only four working mines remained. The Black Park mine closed in 1961 and the final mine, at Craigfad, ceased operation six years later. The cliffs of Fair Head, seen in the background of this photograph, are the highest cliffs in Northern Ireland, and some of the highest on the whole island of Ireland. Today they are a popular rock-climbing location, and are claimed to offer the biggest expanse of climbable rock-face in either Ireland or Britain. [Source: NLS L_ROY.02220]

For much of its productive life, the Abbeytown site was occupied by quarrying operations. However, the current mineral licence holder, Erris Resources, is involved in an active reappraisal (2019) of the ore body that involves six licensing areas covering 159km². Drilling from the surface has shown that mineralisation extends more than 500m south of the historic mine. Furthermore, in 2018 the old workings were re-entered, and the underground workings reconditioned and made safe. The company now plans to drill from underground locations to explore the ground between the original mine and the surface drill sites.

COASTAL MINING AT THE PRESENT DAY

To conclude, mining in the coastal areas of Ireland, particularly during the nineteenth century, provided local labour, albeit at low rates, but in the case of the main mines, led to the growth of mining communities, such as Allihies. In these areas, efforts are now being made to preserve the remaining

Fig. 29.13 THE RED ALTERATION LAYER IN THE BASALTS OF THE GIANT'S CAUSEWAY, COUNTY ANTRIM. The Antrim basalts were erupted during an extended period of volcanicity in the Tertiary geological period, at the time when the North Atlantic ocean was starting to open up and North America was separating from Europe. The red colour of the alteration layer is due to the presence of iron oxides (rust) that resulted when basalt lava weathered to create a soil, under tropical conditions during an extended quiet period between two episodes of volcanic activity. Despite their outcropping along the coast, the active mining for the iron ore has occurred only inland on the Antrim Plateau. [Source: Kieran Buckley, Pexels]

709

structures and mining heritage; little remains to show for the labours of generations of miners in many areas.

Since the 1960s, however, Ireland has become a major lead-zinc mining province based on the successful development of large-scale, inland mines. Important underground mines have been developed at Tynagh, County Galway; Navan, County Meath and at a number of other inland locations, designed to extract ore from the Carboniferous limestone. The main coastal impact of these enterprises occurs at the ports through which the ore is exported for smelting. The Tynagh mine operated for fifteen years and

contributed enormously to economic development in County Galway, with the Port of Galway benefiting from the maritime trade. Development of the Tara mine at Navan started in 1973, with production beginning four years later. It is currently the largest lead-zinc mine in Europe, with an ore reserve estimated at more than 100 million tonnes. The Navan ore is transported by rail to Dublin Port via Drogheda for shipment overseas (see Chapter 24: Ports and Shipping). Today, the only operational underground coastal mine in Ireland is the salt mine at Kilroot (Islandmagee) in County Antrim.

Case Study: East Antrim Salt Deposits

David Naylor

The only significant development of younger, post-Carboniferous rocks (see Chapter 5: Geological Foundations) in Ireland occurs in County Antrim. The main basin, termed the Larne–Lough Neagh Basin, extends along the east coast between Belfast Lough and Cushendall, and inland to the west side of Lough Neagh (Figure 29.14). Rock salt (sodium chloride), in the form of the mineral known as halite, is developed within the younger rock sequences in the coastal areas, and its presence is known from drilling and mining. This is the only development of salt onshore on the island of Ireland.

Immediately north of Belfast Lough, and for several miles along the coast towards Larne Lough, the salt-bearing beds outcrop at the surface. Due to faulting and folding, they are rarely deeper than 300m. Shallower sections in this area were exploited for salt intermittently and at different places between 1853 and 1958, by shaft, bell mining and, particularly, by solution mining to produce brine (Figure 29.15).

Today, the Irish Salt Mining and Exploration (ISME) Kilroot mine, located to the north-east of Carrickfergus, is the only active mine in operation around the Irish coastline. The operation commenced in 1965 and access is via an inclined adit, while mining is by the room-and-pillar technique (Figure 29.15). In 2010, Kilroot received planning approval for a major underground extension that would provide an additional thirty years of working life for the mine. Some sixty personnel are directly employed in the operation, while other companies and people in the locality are involved in supply and maintenance. Rock salt from the mine is crushed within the

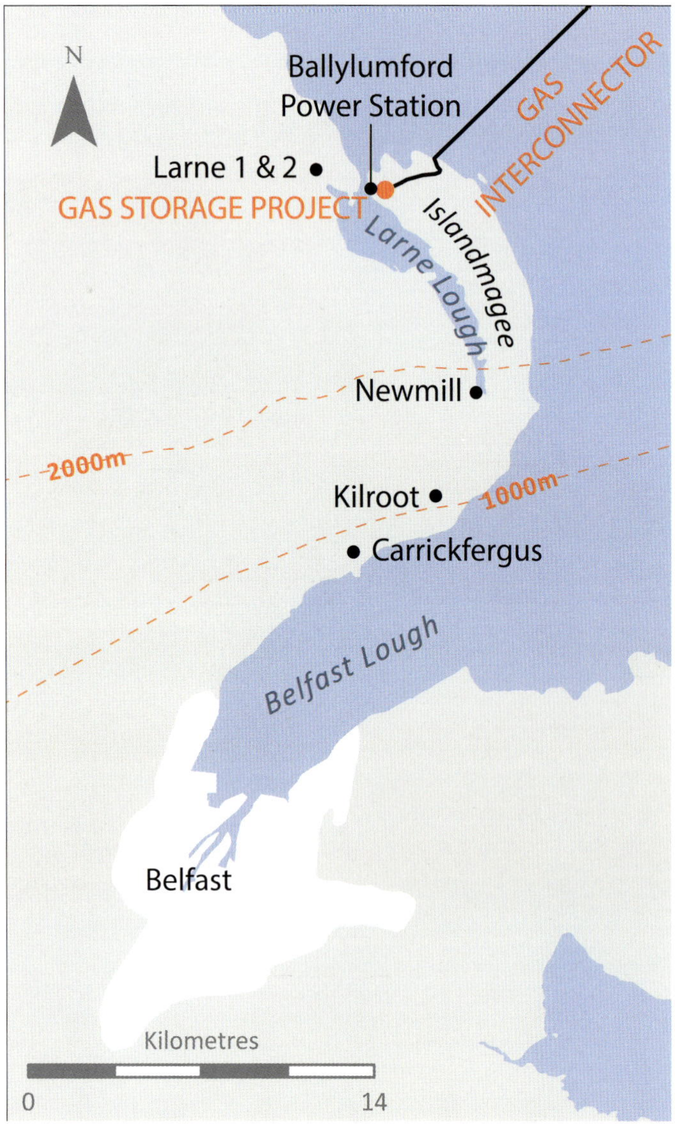

Fig. 29.14 ANTRIM SALT DEPOSITS (left). The Kilroot salt mine and the deep boreholes drilled at Larne and Newmill, together with the gas storage project on Islandmagee, are located along the coastline to the north of Belfast. The salt deposits lie on the southern margin of the Larne–Lough Neagh Basin. They were accumulated over long geological time, by the evaporation of seawater from a former shallow inland sea that extended in the Triassic geological period from present-day Northern Ireland, across Britain to Poland and Estonia. In the southern part of the basin the salt (halite) beds vary in thickness from 9m to 27m, but north from Belfast Lough the section thickens rapidly. The Larne No. 2 borehole revealed that 3,000m of younger rocks are developed in the basin centre at that point. The dashed red lines represent the thickness of rocks down to the base of the Permian, showing the southward thinning of the basin (only part of the Larne–Lough Neagh Basin is shown here. The full extent of the basin runs all the way to Cushendall and inland to Lough Neagh).

underground caverns, and is then carried to the surface by conveyor belt for storage. From here it is either taken away by lorry for redistribution within Ireland, or else conveyed to the company's deep-water jetty for loading onto bulk-carrier ships (Figures 29.16 and 29.17).

A petroleum exploration well, drilled by Shell/Marathon in 1971 at Newmill on the south-west shore of Larne Lough (Figure 29.14), intersected four major salt zones at depth, ranging from 25m to 60m in thickness.[17] A later geothermal test well, located further north at Larne, terminated at 2,907m. The thickest of the individual salt beds in this well ranged from 30m to 70m.[18] Deep wells drilled further inland, however, have lacked salt, demonstrating that the development of the salt deposits is restricted to a zone along the coast and immediately offshore.

The salt beds between Belfast and Larne have potential, other than mining, for the development of two other significant projects: onshore gas storage and compressed air storage for power generation.

ONSHORE STORAGE OF GAS

Salt caverns can provide a secure environment for the containment of substances that do not cause dissolution of the impermeable salt. Due to this attribute, they have been used in many countries for the storage of liquified petroleum gas (LPG) and natural gas. The Antrim salt deposit is particularly well positioned for this, since 90 per cent of natural gas imported into Ireland is through a single pipeline from Scotland to a terminal located on Islandmagee in Northern Ireland. Furthermore, this gas supply is sourced from the North Sea,

Fig. 29.15 MINERS SHOVELLING SALT INTO A WAGON IN THE KILROOT MINE, 700 FEET BELOW CARRICKFERGUS IN COUNTY ANTRIM, SOMETIME BETWEEN 1902–1939. The Antrim salt deposit was discovered sometime in the early years of the nineteenth century, by miners looking for coal. Initially, the salt was extracted at a number of mines in the area, mostly through manual excavation of rock salt in underground chambers, as shown here. Later, the mining was done by shaft, bell mining and, particularly, by solution mining to produce brine. A 'room-and-pillar' technique was also used, whereby pillars of rock were deliberately left standing to hold up the roof of the mine chambers. Most of the mines had ceased operation by the 1950s. Today, only the Kilroot mine remains active. It employs about sixty people directly, as well as supporting additional jobs in ancillary services, but much of the extraction work is now heavily mechanised. [Source: National Museums Northern Ireland]

which is a declining province and cannot guarantee the continuation of large-scale supplies. As a result, there is an obvious vulnerability of gas supplies for the island of Ireland due to the potential for disruption in supplies caused by external factors.

A number of studies of salt cavern potential have been carried out. In 1998, a French company, Sofregaz, acting on behalf of Bord Gáis, examined the feasibility of gas storage in salt caverns near Larne. They concluded that such a facility was too far from the Dublin market to be of practical value. Of particular interest, however, has been the evaluation of salt beds in Larne Lough by Islandmagee Energy Limited, a wholly owned subsidiary of UK company, Infrastrata PLC. The Islandmagee Gas Storage Project commenced in 2007 with the acquisition of 3D seismic data to image the Permian salt in the Larne Lough area. In some salt provinces throughout the world, the salt shows significant thickness variation over short distances, due to salt mobilisation and flow under pressure. There is little evidence of this problem in the Larne salt development and, therefore, the height of any planned cavern will be controlled by the normal thickness of the salt bed. Islandmageee Energy plans to create seven caverns, capable of storing a total of up to 500 million cubic metres of gas in Permian salt beds beneath Larne Lough (Figure 29.18).

To acquire more detail on the nature and thickness of the salt, a well (Islandmagee I) was drilled to a depth of 1,753m in May/June 2015. This showed that the layer of Permian salt has an average thickness of 200m over the area of the proposed gas storage caverns, at a depth of approximately 1,300m subsea. The salt layer has high purity and is suitable for cavern construction over its entire thickness. The project received EU grant aid and, having secured the necessary planning permissions and environmental permits, the company carried out the Front End Engineering Design (FEED) element of the project, completed in 2018. On completion of the construction phase, the company plans to initiate gas storage operations in 2023.

The Islandmagee Gas Storage Project has an estimated development and completion cost of £274 million. It has the potential to

Fig. 29.16 CRUSHED SALT BEING LOADED ONTO A BULK CARRIER VESSEL AT THE KILROOT MINE DEEP-WATER JETTY. The mined salt is used almost entirely for de-icing roads, so the quantity and value of salt shipped depends mainly on the severity of the winter in any one year. Approximately 20 per cent of the product goes to the Irish market, and the remainder is shipped to Great Britain. In previous years, salt would also have been shipped to the US, but the local Irish and UK markets now take all the salt that can be produced. [Source: Chris Heaney, courtesy of Irish Salt Mining and Exploration Company Ltd]

Fig 29.17 MODERN-DAY UNDERGROUND MINING OPERATIONS, KILROOT SALT MINE. As in earlier days, the mining is still undertaken by the 'room-and-pillar' method, but today is heavily mechanised. Mined salt is transferred underground by dump trucks to a crusher and converted into gritting salt for road usage. Crushed salt is then carried on conveyors to the surface, where it is stored under cover awaiting distribution. Up to 500,000 tonnes of salt, worth approximately £12 million, are produced each year and sold for road maintenance in Ireland and the UK. [Source: Chris Heaney, courtesy of Irish Salt Mining and Exploration Company Ltd]

create more than twenty high-quality permanent jobs, with the construction phase generating temporary employment for over 200 people, and knock-on indirect business for support services.

Gas storage caverns perform the obvious function of providing emergency supplies in case of any disruption to normal supplies. However, the storage facility will provide flexibility in being able to switch from injection to withdrawal to cater for short-term spikes in demand, such as a cold spell or a period of low wind generation. The proposed site is located close to two supporting infrastructural requirements: a source of power (adjacent to Ballylumford power station), and a connection to the main gas network (Ballylumford pressure reduction station at the connection point to the Scotland–Northern Ireland gas pipeline). Gas from the facility will also be able to feed into Ireland's onshore network via the south–north pipeline.

COMPRESSED AIR STORAGE

A second innovative use for salt caverns located in this area was researched by Gaelectric Energy Storage, beginning in 2006. The company planned to develop a compressed air energy storage facility, using a technique currently operating at only two other localities in the world. Energy is stored by compressing air into underground salt caverns using off-peak power, which can then be released, when needed, to drive turbines and generate power. As in the case of gas storage, this could provide Ireland's power system

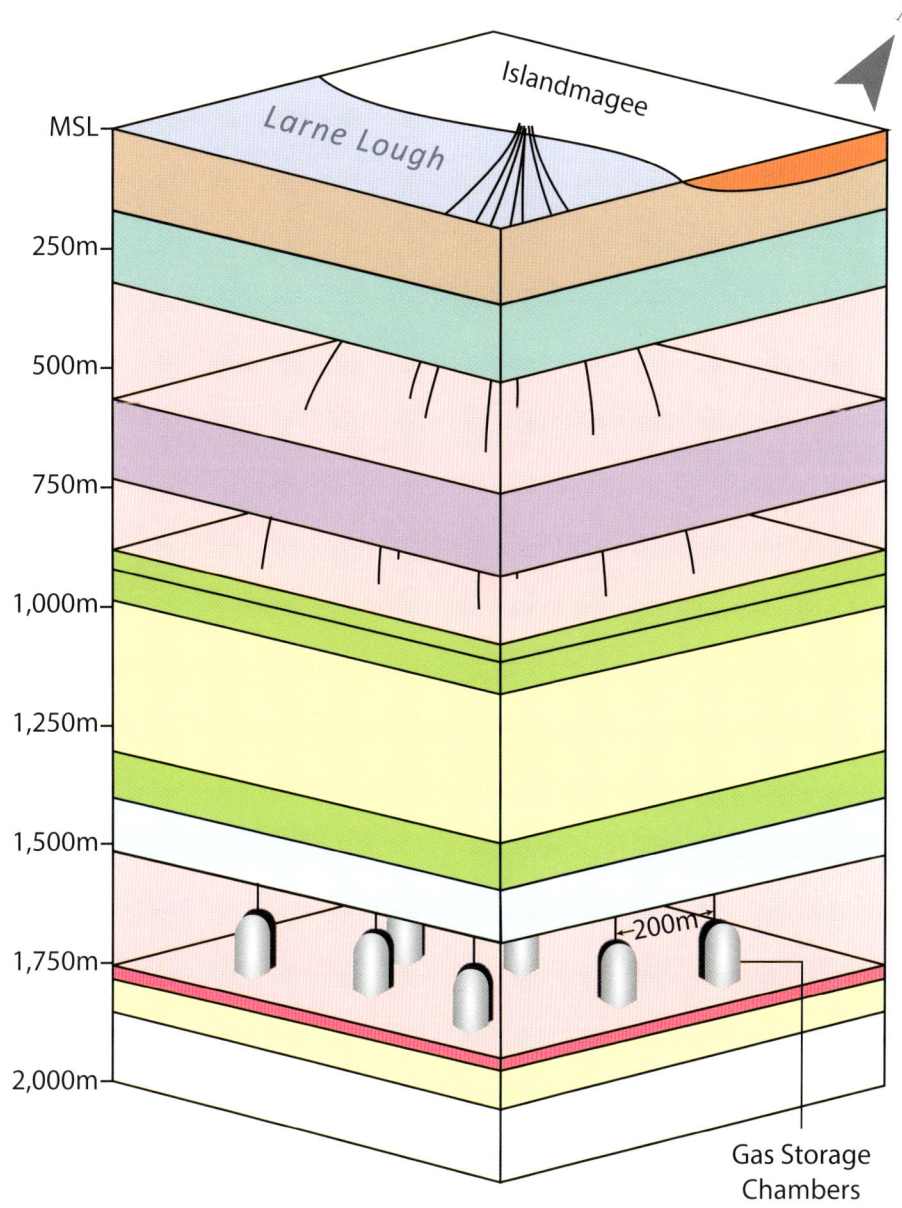

Fig. 29.18 PROPOSED GAS STORAGE CHAMBERS IN THE SALT DEPOSIT, LARNE LOUGH. The storage caverns will each be approximately 80m in diameter and 150m in height, and will be created by a technique known as 'solution mining'. This method dissolves the salt, under controlled conditions, and creates a cavern, deep underground in the salt layer. The facility will be designed to have an injection capability of 12 million cubic metres per day and a withdrawal capability of 22 million cubic metres of gas per day. The company has been granted planning permission for the above-ground facilities for a natural gas storage facility near the Ballylumford power station on Islandmagee. [Adapted from: Islandmagee Storage Ltd website]

some flexibility in dealing with periods of high demand or unexpected outages.

Initially, the company planned to develop a site in the relatively shallow salt beds to the south-east of Larne. Following exploratory drilling to assess the potential of the site, Gaelectric developed a strategic alliance with the engineering group Dresser-Rand in 2012. This facilitated the use of innovations in technology that allow for the use of deeper salt layers for the storage of compressed air. As a result, attention has now switched to deep salt layers (approximately 1,500m) at Islandmagee, which has the added advantage of being proximal to the Ballylumford power station. The outline plan included access to a power station that would be capable of generating up to 330 MW of electricity for periods up to six to eight hours (enough to meet the electricity needs of over 200,000 homes), and would be driven by compressed air from two underground salt caverns.

An initial grant of up to €6.47 million was received from the European Union to meet the costs of an environmental impact assessment, addressing planning issues, as well as front-ending engineering design for the project. In 2016 an additional €8.28 million was awarded for the drilling of an appraisal well, and detailed studies into the design and commercial structure of the project. Total project costs are estimated to be £300 million. However, in 2017 Gaelectric encountered severe financial problems and in the following year went into voluntary liquidation with the sale of its assets. There is, therefore, considerable doubt as to whether the Islandmagee project will proceed.

There are, of course, numerous local quarries for stone or sand and gravel around the Irish coastlines. The expansion and further exploitation of such sites, however, is often limited by conflicts with other land uses, and by costs associated with crushing, processing and transporting quarried bedrock. At the present time, most of these quarries tend to be relatively small, although their coastal location often allows the extracted rock to be loaded directly onto ships moored offshore, for export to markets in Britain or elsewhere. One specialist example of this was the quarrying of millstones at various coastal locations.

In some countries, notably in Norway and Scotland, coastal superquarries have been developed, where the entire scale of exploitation is very much larger. No superquarries exist, as yet, in Ireland. Furthermore, any such development would need a national debate to minimise conflicts since suitable rock types and locations for this type of activity are likely to be in areas of high scenic and nature conservation value.

Coastal Quarrying

Matthew Parkes and Alastair Lings

As well as being valued for the metallic ores or other minerals that they contain, the rocks of Ireland themselves have also been exploited as an economic resource. This includes the use of sandstone and other rock types as building stone for houses, boundary walls and other structures; limestone, for crushing and burning for fertiliser, building lime and other industrial uses; and slate for roofing and other building

Fig. 29.19 SLATE QUARRY ON VALENTIA ISLAND. Slate has been quarried on Valentia Island since 1816, initially for use as a roofing material and for gravestones used in local cemeteries. Valentia slate was also used in the construction of the Paris Opera House and in many buildings erected in London in the nineteenth century, including the Houses of Parliament, Westminster Abbey and Cathedral, St Paul's Cathedral and a number of the city's underground railway stations. A further use of Valentia slate in the mid-nineteenth century was for the beds of top-quality billiard tables. At its peak in the 1850s, the Valentia quarry employed up to 500 people, but it was unable to compete with the growing availability of cheaper slate from Wales, and the quarry closed in 1911. However, it reopened in 1999, and currently employs six people. Operated by the Valentia Slate Company, the quarry has contributed slate products to a number of notable civic and other buildings in Ireland, including the Marine Institute headquarters in Galway, the National State Laboratory in County Dublin, and the Gleneagle Hotel in Killarney. The company also produces a range of finished slate products for interior and exterior use, including floor tiles, fire surrounds, bar and shop counters, steps and staircases and funerary monuments. [Source: Lisa Fawkes]

purposes, as well as for funerary ornaments, interior house fittings, etc. (Figure 29.19).

The usual means of extracting this rock is via open pits, or quarries, rather than by means of mine shafts or adits. These pits can be small, in the case of quarries for domestic or farm use for example, or of considerable size when developed for commercial and industrial purposes.

Today, large volumes of rock material are still required frequently for civil engineering projects, many of which need rocks with specific characteristics or measurable physical properties, such as Polished Stone Value for skid-resistant road surfaces, for example. Yet the relative low value of the rock and the high cost of transport mean that most aggregate or bulk rock quarries have a quite restricted radius in which it is economic to supply the product. Locating a quarry on or near the coast offers distinct advantages over those further inland, since it is often possible to load the extracted rock by conveyor belt straight onto ships berthed in deep water alongside the source. Such bulk transport costs are then quite economic compared to road transport.

The idea is not entirely new, as Charles Stewart Parnell's quarry at Arklow was exporting stone for street paving setts to England from the 1880s onward. Today, Roadstone continue to operate this same quarry, and supply large blocks of the blue volcanic rock for rock armour against coastal erosion. Large rock blocks absorb the impact of waves and storms much better than the unconsolidated glacial sediments they may protect (see Chapter 30: Engineering for Vulnerable Coastlines). This technique has been used in many places around the Irish coast, for example, to protect the railway line at Killiney in County Dublin.

Another coastal quarry of note was the one at Brow Head near Crookhaven, County Cork, opened in 1925 by the Rowe Brothers and Company of Liverpool. The company exported crushed sandstone aggregate to England until 1939, employing up to seventy men. More recently, and also in County Cork, the Leahill Quarry on the northside of Bantry Bay provided material straight from quarry to ship. During its period of operation, between 1991 and 2009, this quarry provided crushed sandstone, which was shipped to asphalt plants in south-east England. A 60,000 tonne vessel could be loaded in less than twenty-four hours. However, although recent attempts have been made (2019) to reopen this quarry, these have not, so far, succeeded. As with all such operations, the broader economic conditions in society, as well as the qualities and specific properties of the resource, tend to dictate demand and consequent financial decisions as to investment or closure in a coastal quarry, similar to inland quarries.

The Coastal Millstone Quarries of Waterford Harbour

Niall Colfer

Three millstone quarries in the environs of Waterford Harbour exploited coastal outcrops of Devonian Old Red Sandstone in the manufacture of monolithic millstones during the post-medieval period. These include examples in the townlands of Graigue Great and Templetown (known locally as 'Millstone Hole' and Harrylock) on the Hook Peninsula, County Wexford, and a third on the southern side of Creadan Head in County Waterford (Figure 29.21). All millstones produced in Waterford Harbour were referred to by the generic name of 'Ballyhack Millstones', named after a further millstone quarry (a total of six were situated in the environs of Waterford Harbour), located on a hilltop at Ballyhack near the northern tip of the harbour.

The millstones were sold locally, nationally and internationally, with maritime transportation central to the commercial success of the quarries. This was not unique to Ballyhack millstones; further examples of economically successful millstone quarries located near the coast include Lough Eske, County Donegal, and Millstone Mountain in the Mourne Mountains, County Down, from where millstones were brought to Donegal town and Newcastle respectively to be loaded onto ships.

Fig. 29.20 UNCOLLECTED MILLSTONES AT HARRYLOCK MILLSTONE QUARRY. At many points along the shore, erosion and planing by the sea has formed extensive intertidal rock platforms, up to 50m or more in extent. On these platforms, exposed rock could easily be accessed at low tide and fashioned into millstones, as shown here at the Harrylock quarry on the western side of Hook Head. The red and white bands on the pole in the foreground of the photograph mark 10cm divisions, showing that the stone on the right was approximately 60cm in diameter. Once ready, the finished stones were attached to the bottom of boats and lifted on the rising tide, for removal and transport to market. [Source: Niall Colfer]

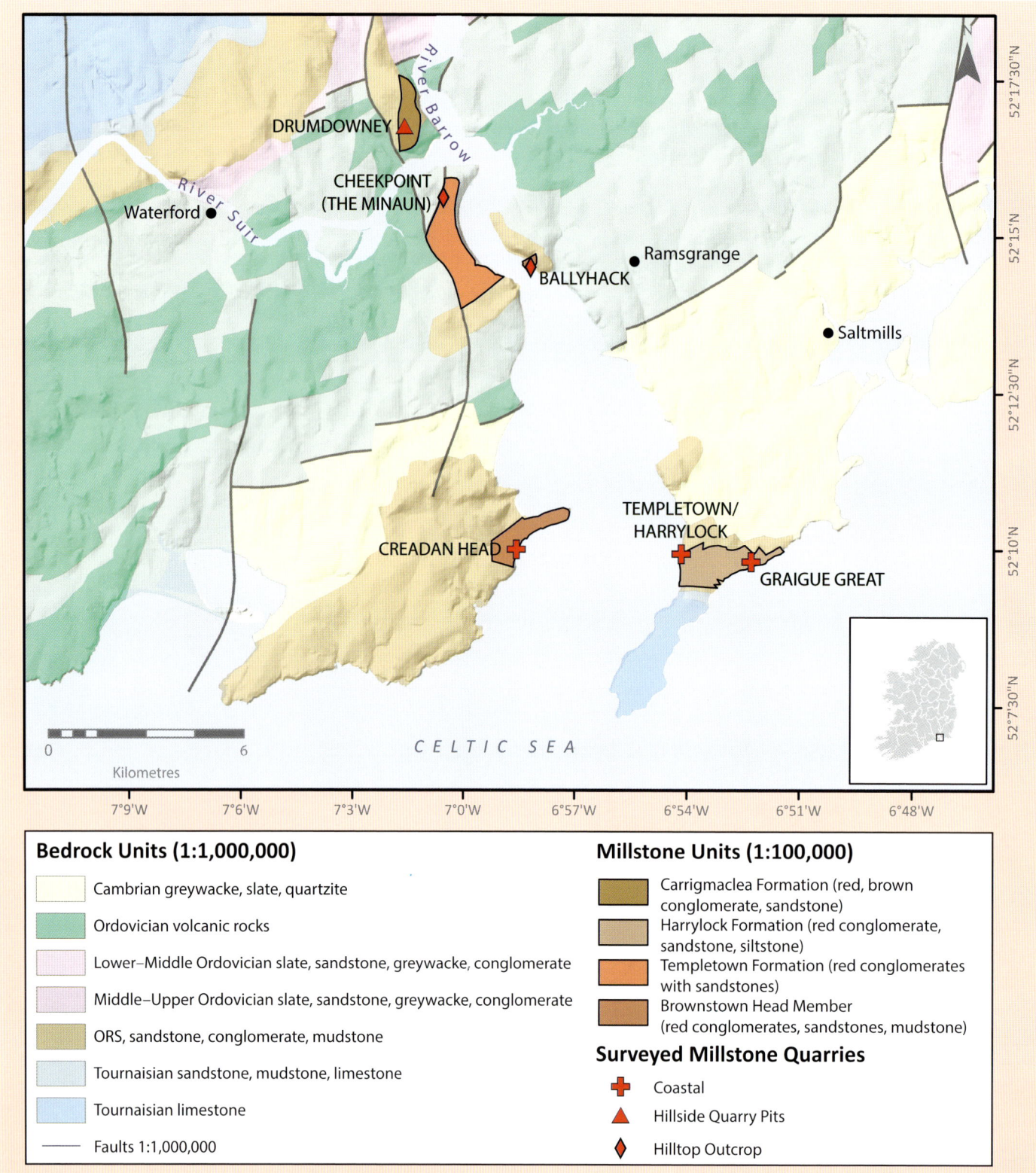

Bedrock Units (1:1,000,000)

- Cambrian greywacke, slate, quartzite
- Ordovician volcanic rocks
- Lower–Middle Ordovician slate, sandstone, greywacke, conglomerate
- Middle–Upper Ordovician slate, sandstone, greywacke, conglomerate
- ORS, sandstone, conglomerate, mudstone
- Tournaisian sandstone, mudstone, limestone
- Tournaisian limestone
- Faults 1:1,000,000

Millstone Units (1:100,000)

- Carrigmaclea Formation (red, brown conglomerate, sandstone)
- Harrylock Formation (red conglomerate, sandstone, siltstone)
- Templetown Formation (red conglomerates with sandstones)
- Brownstown Head Member (red conglomerates, sandstones, mudstone)

Surveyed Millstone Quarries

- ✚ Coastal
- ▲ Hillside Quarry Pits
- ◆ Hilltop Outcrop

Figure 29.21 MILLSTONE MINING LOCATIONS AROUND WATERFORD HARBOUR. The geology of counties Waterford and Wexford, around the estuary of the rivers Barrow and Suir, is dominated by a sequence of interleaved sandstones, mudstones and conglomerates. The coarse texture and strong durability of these rocks made them ideal raw materials from which to develop a specialist local industry based on the manufacture of millstones and grindstones. A number of quarries were developed in the region for this purpose, both inland and particularly along the coast.

An important distinction is that millstones from Waterford Harbour coastal quarries were transported by sea directly from the quarries where they were extracted. The millstones weighed over a tonne each and there were two known methods of transferring the product from the quarry. Firstly, at low tide the millstones were manoeuvred to the tidal zone of the quarry and on the rising tide they were strapped to the underside of a small boat (Figure 29.20).[19] Secondly, a series of tripod wooden cranes was utilised, as depicted on a painting of quarries at Graigue Great by George Victor Du Noyer in 1850 (Figure 29.22).

The suitability and popularity of Old Red Sandstone from the Waterford Harbour area for milling is evident in contemporary publications. An advertisement printed in *A Dictionary of Dublin* for the year 1738 directed prospective buyers to 'Nicholas Mellin-

Fig. 29.22 QUARRIED STONE BEING LOADED ONTO A BOAT. On the eastern side of Hook Head in County Wexford, millstones were carved from rock quarried from the coastal cliffs and intertidal rock platforms. As shown in this watercolour by the celebrated nineteenth-century artist and field geologist George Victor Du Noyer, cranes were used to load the quarried stone onto boats brought alongside the shore at high tide. [Source: Geological Survey Ireland]

merchant for Ballyhack millstones. George's Quay, next door to THE SUN'.[20] This is indicative of advertisements published elsewhere, such as in the *Belfast Newsletter*, where there are nine advertisements for 'Ballyhack Millstones' between the years 1760 and 1782.[21]

The millstone industry provided a viable, seasonal alternative source of employment to fishing and farming for the local population. In 1800, a stonemason in Waterford Harbour could receive up to six guineas for producing a pair of millstones, with a single stone taking a week to shape.[22] In the mid-nineteenth century, stonemasons on the Hook Peninsula lived in a farm cluster of seventeen houses at Harrylock, overlooking the millstone quarry of the same name.[23] The close proximity in which they lived encouraged cooperation, a pooling of resources and also the establishment of a tradition of stone working that was intrinsic to their social identity. The last stoneworker from Harrylock was Matthew Kelly. He produced the final millstone, at Millstone Hole, before he died in 1905 and left instructions that it should not be sold for less than one pound.[24]

Of potentially greater economic impact than quarrying could be the future offshore dredging of marine aggregates. At a European level, the extraction of marine aggregates from shallow seas is a mature and well-established industry, and is gradually extending its reach into deeper waters. In Ireland, marine aggregate operations to date have been mostly localised and comparatively small-scale. However, increasing demand for construction materials for housing and for energy and transport infrastructures is putting pressure on the availability of terrestrially sourced aggregate supplies. This, along with developments in seabed surveying and marine dredging technologies, is leading to expanding interest in offshore aggregates as a potential resource.

MARINE AGGREGATES

Gerry Sutton

The term 'marine aggregates' refers to sedimentary material – sand or gravel of various grain and class sizes (grades) – found on the seabed. In common with their terrestrial counterparts, marine sands and gravels are important economic resources, suitable for the development and maintenance of critical infrastructure, such as buildings, roads and bridges. Marine aggregates are also used routinely for beach nourishment and coastal defences. Demand for material for these latter purposes, in particular, is likely to increase as climate change, sea-level rise and associated effects on low-lying coastal areas drive investment in various adaptation measures.

The process of extracting marine aggregates typically involves raising them from the seabed using sophisticated, purpose-built

Fig. 29.23 THE TRAILING SUCTION HOPPER DREDGER *Sand Heron*, OPERATED BY CEMEX MARINE LTD, EXCAVATING SAND FROM AN AREA IN THE ENGLISH CHANNEL, 10KM SOUTH-WEST OF THE ISLE OF WIGHT. The commercial marine aggregates industry connects extraction activities at sea with supporting port-related infrastructure and a variety of service industries onshore. In countries where the industry is well established, including the United Kingdom, Belgium, the Netherlands and other European countries, dedicated wharves are located in most of the major ports to offload sands and gravels from dredging ships, and facilitate the storage and eventual onward carriage of the aggregates by road or rail to markets inland. Trailing suction hopper dredgers, such as the British ship, *Sand Heron*, are among the most commonly used vessels for extracting sands, gravels, silts and clays from the seabed. They are typically equipped with one or two suction tubes that are lowered to the seabed and trailed across the bottom. A pump system onboard the vessel then sucks up a mixture of sea water and the aggregate materials, and discharges it into hoppers in the hold of the vessel. The loaded vessel then sails under its own steam, either to a suitable receiving wharf for off-loading or else, if the dredged material is being used for landfill or beach replenishment, it can be discharged directly via a floating pipeline to where it is required onshore. [Source: British Marine Aggregate Producers Association]

dredging machinery (Figure 29.23). Recovered aggregates may be screened initially (sorted and cleaned) at sea before being transported to a wharf facility for unloading, storage or processing. Depending on the characteristics of the raw material at source, and regional demand for particular grades at a given time, additional processing is usually undertaken to remove salt and other contaminants prior to final screening/blending and delivery.

While marine aggregates are not yet exploited commercially in Ireland, many countries have sought to meet their demand for aggregates by utilising this seabed resource to replace, or complement, terrestrial material obtained from quarries and pits on land. Belgium, the Netherlands and the United Kingdom are among the major producers within Europe. These and other countries have long-established and mature practices of marine aggregate extraction. Furthermore, this has allowed for the evolution of sophisticated permission and environmental monitoring and evaluation procedures, to help ensure that the industry operates sustainably, and that environmental impacts are both minimal and localised.

Knowledge of how marine aggregates are distributed around Ireland's coast arises from a diversity of sources. Many localised marine mapping and geophysical surveys have been undertaken during the last forty or so years. When these results are pieced together and viewed in a regional geological context, they can provide a reasonable understanding of the size and potential of the overall resource. The bulk of aggregates found in Irish waters are derived from Quaternary (mainly Pleistocene) glacial deposits (from 2.6 million to approximately 11,700 years before present). These are supplemented by material derived from reworking of former deposits during the Holocene period (11,700 years ago to the present day) (see Chapter 6: Glaciation and Ireland's Arctic Inheritance) under the influence of rising sea levels, together with other more recent material eroded from the coast.

Aggregates are distributed widely around the entire coast of Ireland but, from an economic perspective, the practical focus of interest has been on the Irish Sea and key areas off the south coast (mainly in the vicinity of Cork, Youghal and Waterford harbours). Over the past three to four decades, some extraction of aggregates in limited volumes has been permitted from several nearshore coastal sites, specifically for localised beneficial purposes, such as beach nourishment, coastal protection, reclamation and backfill. Additionally, a number of targeted investigations undertaken with specialised survey equipment have enabled promising areas off the south and east coasts to be characterised, and large areas containing significant potential resources in the Irish Sea have been quantified. However, to date there has not been any commercial extraction of marine aggregates from Irish waters on a scale approaching that seen in the North Sea and elsewhere in Europe.

Between 2005 and 2008, collaborative work was carried out by teams of scientists from Ireland and Wales under the IMAGIN project, part-funded by the Ireland–Wales Interregional Fund.[25] This research identified and mapped key areas in the Irish Sea where the potential to extract aggregates of industrial quality was deemed highest. In total, researchers cited volumes in the range of between 2 and 6 billion cubic metres, occurring in water depths of between 20m and 60m, in the region between Dublin and Rosslare. Further areas of potential interest for aggregate extraction were identified off the North Wales coast (Figure 29.24). If all the high potential areas in the Irish waters of the Irish Sea are considered (that is, including areas identified on geophysical profiles but not quantified specifically by in-situ penetrative sampling), the potential maximum volume rises to 12 billion cubic metres. The future outlook for the commercial exploitation of these resources in Ireland is influenced strongly by two key factors: the balance of demand and available supply in key (mainly urban) markets; and the nature and stability of the planning and consenting regime.

Prior to the 2008 economic crisis, Ireland's Celtic Tiger economy and a construction boom caused demands for aggregate to run at thirty times the EU average. Such levels of demand suggested that a shift to exploiting marine aggregates appeared imminent. However, the onset of a fiscal crisis and deep economic recession depressed levels of demand. Ireland's exit from recessionary conditions, however, has reversed such depressed market conditions. In particular, the revival of strong economic growth, coupled with ambitious construction programmes and ongoing investment in coastal protective infrastructure, are leading to renewed interest in exploitation of marine aggregates.

With the current high demand for construction aggregates in Dublin and Cork, due to a resurgence in the building sector, securing access to certain grades of construction aggregate is becoming increasingly challenging. This is compounded by the fact that existing local supplies of these non-renewable resources remain depleted following the pre-2008 construction boom. There is, in particular, a noted scarcity of 'sharp' sand required for concrete products and, at present, available supplies of this are being supplemented by crushed rock. This has further negative implications, since concrete products made with crushed rock fines require a higher cement content than concrete made using natural sand. The additional energy required for crushing (together with higher CO_2 emissions) must also be taken into account. These factors are now starting to impact on both the availability and cost of supply of materials to the building sector.

Given the high entry costs associated with dredgers and processing plants, and the extended periods before an effective return can be made on investment (margins are relatively low), entry into the offshore aggregates market is likely to remain influenced strongly by the direction of government policy, and details of any regulatory consenting framework that might emerge. This latter will be especially important in the context of the ongoing Marine Spatial Planning (MSP) process. Marine aggregates have been recognised as an important sector of the economy, and are afforded a dedicated section within the National MSP Baseline Report.[26] This can be seen as echoing earlier recommendations by IMAGIN that called for a clear national policy to be developed in order to promote and facilitate the sustainable development and use of Irish marine aggregates.[27]

Any such provisions should also take account of the direct competition for space that now exists within Ireland's offshore areas. In particular, it is essential that an equitable balance must be struck between the need to safeguard aggregate resource areas from inadvertent, long-term sterilisation, especially given the legitimate interests in such areas of other marine uses, particularly those involved in offshore energy development, fishing, cables/pipelines, shipping lanes and nature conservation.

Bringing material from marine sources into the aggregates supply chain also raises some interesting and wider issues of environmental sustainability. Provided the direct physical and ecological impacts on the marine environment are monitored and managed carefully, there is strong evidence to suggest that a shift to marine sources can provide significant benefits in terms of reduced environmental cost. These are associated mainly with a dramatic reduction in vehicle movements and CO_2 emissions when compared to the exploitation of land-based aggregates.[28]

HYDROCARBON EXPLORATION IN IRELAND

While mineral exploration and extraction in Ireland has a long history, the island remained completely unexplored for hydrocarbons throughout the first half of the twentieth century. Major political events in Ireland during the first part of the twentieth century, the lack of onshore oil or gas seeps, and other negative geological factors, were all significant deterrents. However, in 1960, Ambassador Irish Oil Ltd was granted an

exploration licence for twenty years for the whole onshore area of the Republic of Ireland and the entire offshore area then under Irish jurisdiction. The apparent generosity of the terms of the licence has to be judged against the lack of previous exploration, as well as a general climate of low expectation prior to the first North Sea discoveries in 1965. Conoco and Marathon Petroleum Ireland Ltd joined Ambassador in the licence, and the consortium drilled six onshore exploration wells during 1962–1963, and a

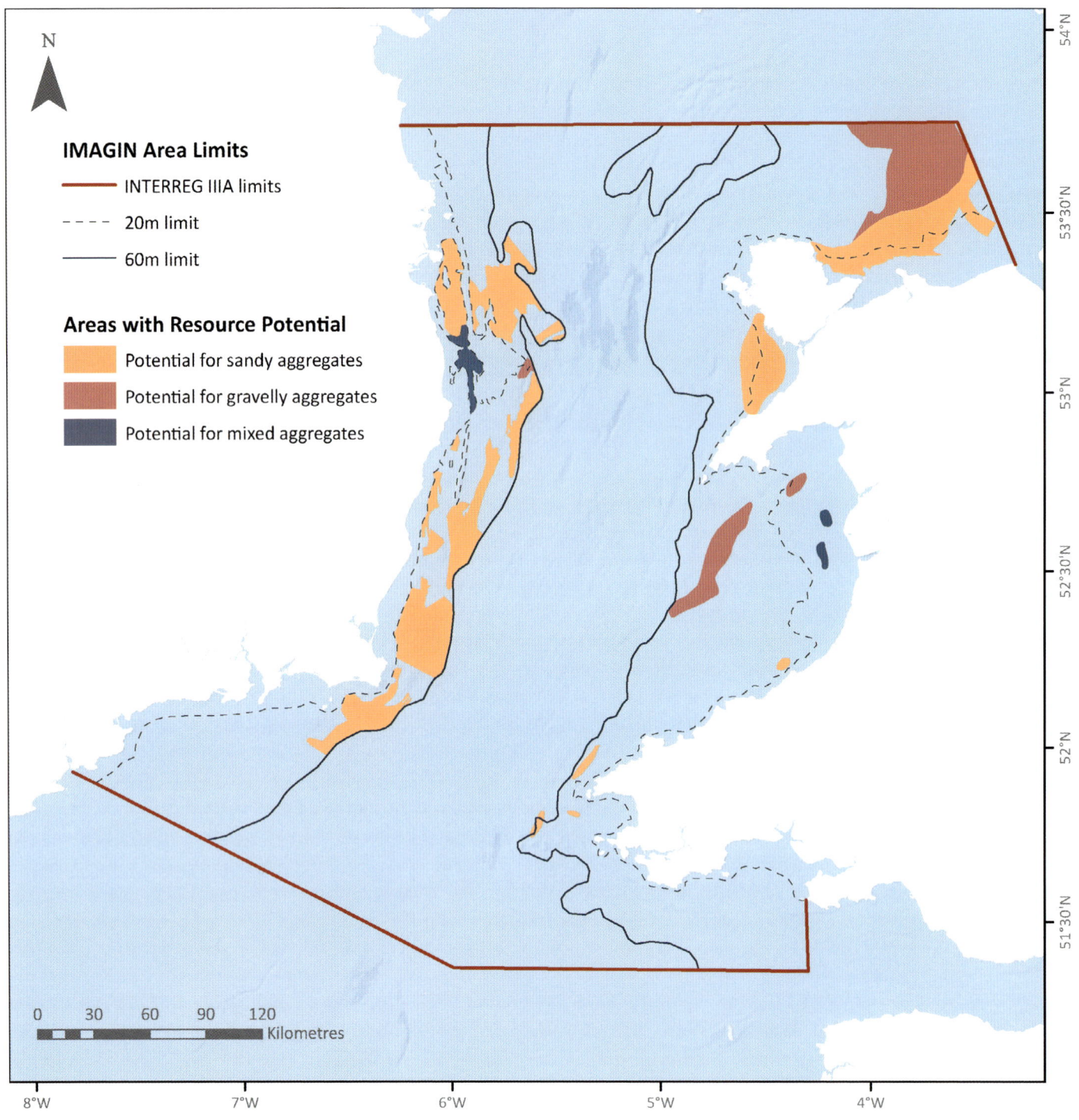

IMAGIN Area Limits

— INTERREG IIIA limits

- - - 20m limit

— 60m limit

Areas with Resource Potential

Potential for sandy aggregates

Potential for gravelly aggregates

Potential for mixed aggregates

Fig. 29.24 DISTRIBUTION OF MARINE AGGREGATE POTENTIAL IN THE IRISH SEA. Although currently not exploited in the Irish sector, many areas in the Irish Sea have been identified as having good potential to yield marine aggregates of commercial relevance. As land-based sources of sands and gravels for building and other purposes become progressively depleted, it is likely that interest in extracting marine aggregates will grow. However, seabed areas dominated by sands and gravels are also key habitats for commercially important fish species. In addition, the Irish and Celtic Sea areas contain busy shipping lanes, as well as being of interest for offshore wind farm development, among other uses. This leads to potential for many conflicts of interest, and future development of a sustainable marine aggregate industry will require very careful marine spatial planning, and the balancing of competing stakeholder demands. [Data source: Kozachenko, M., Fletcher, R., Sutton, G., Monteys, X., Landeghem, K. Van, Wheeler, A., Lassoued, Y., Cooper, A. and Nicoll, C., 2008. *A Geological Appraisal of Marine Aggregate Resources in the Southern Irish Sea. Technical report produced for the Irish Sea Marine Aggregates Initiative (IMAGIN) project funded under the INTERREG IIIA Programme.* Cork: University College Cork]

further three wells in Northern Ireland. Results were largely disappointing, except for some non-commercial flows of gas in wells located in County Cavan.

Almost all of the world's most economically important coal, oil and gas deposits are found in rocks of younger Mesozoic age (see Chapter 5 Geological Foundations). The only significant

development (up to 3,000m thick) of rocks of this age onshore in Ireland is in Antrim. Early deep drilling into these strata was in search of coal and halite (rock salt), with the latter leading to the establishment of the Kilroot salt mine discussed in the case study. The main hydrocarbon potential was considered to be gas, and four oil exploration wells have been drilled to date but without success.

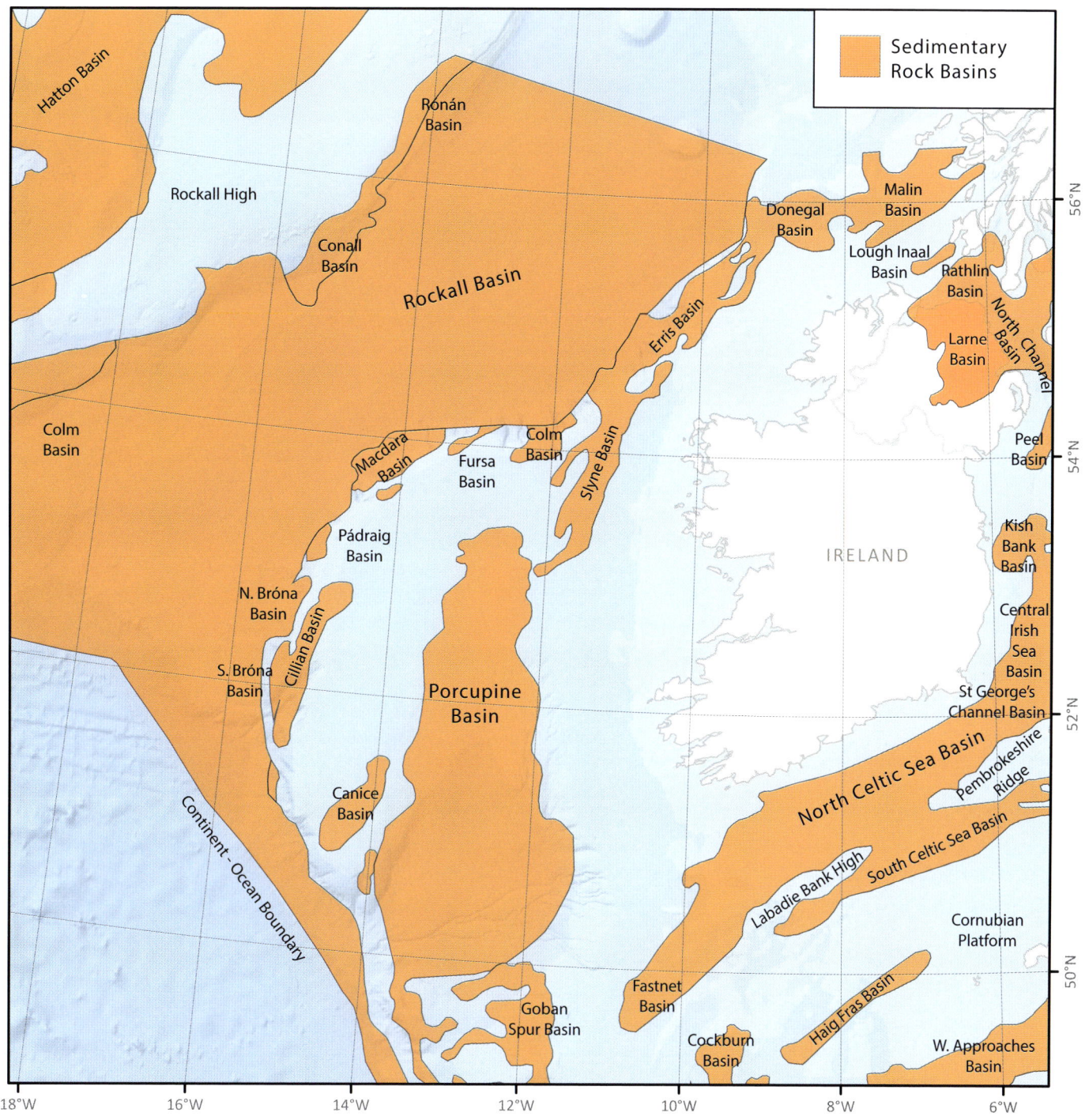

Fig. 29.25 Irish offshore basins. In the 1960s, geological and geophysical data derived from academic research and from the search for oil and gas revealed a pattern of linear troughs, filled with younger sedimentary rocks, on the continental shelves of western Europe. One system of interlinked troughs extended from the north Norwegian shelf southwards through the North Sea to the Netherlands, and provided the geological basis for the rapid expansion of the offshore oil industry in the North Sea, particularly from the oil crises of the 1970s onwards. To the west, another major trough system comprised the Irish Sea, Celtic Sea and Western Approaches basins, the Porcupine Basin, the Rockall–Hatton basins, and the Slyne–Erris–Donegal basins, together forming a ring around Ireland. Significantly, under the terms of the United Nations Convention on the Law of the Sea (UNCLOS) Part VI, most of the sea areas occupied by these latter troughs are deemed to lie within Ireland's territorial waters. This gave the state sovereign rights to explore and develop any natural resources, including oil and gas, that occur there. In the North Celtic Sea Basin, the Kinsale, Ballycotton and Seven Heads fields have been producing gas since the 1970s, and more recently the adjacent Barryroe oilfield has been assessed to contain potentially recoverable quantities of both oil and gas. The Corrib Gas Field, located in the Slyne Basin off the Sligo coast, has also come onstream and is contributing gas to the national distribution grid. It is likely that further deposits of both oil and gas await discovery in the basins of Ireland's western continental shelf but, given the often extreme harshness of the Atlantic ocean climate, accessing and harnessing these will pose major technological and other challenges. [Data source: modified from Naylor, D. and Shannon, P., 2011. *Petroleum Geology of Ireland*. Edinburgh: Dunedin Academic Press]

However, a deep geothermal well named Larne No. 2 was also drilled at the east coast, aiming to exploit the residual heat from the volcanism that produced the Antrim Basalts.

Following the initial and unsuccessful onshore drilling

campaign in the Republic, two consortium partners withdrew, and Marathon was left holding the concession. After further negotiations with government, Marathon was granted three offshore tracts covering a substantial portion of the limited offshore area claimed, at that time, by Ireland. It had become clear by the mid-1960s that linear basins with thick prospective sequences of younger rocks were present offshore (Figure 29.25). Furthermore, commercial discoveries of hydrocarbons in the North Sea gave rise to increasing interest in the Irish offshore. Five offshore designations in the period 1968–1977 extended the area claimed by Ireland westwards over the Porcupine and south Rockall regions. In 1970 Marathon began exploration drilling and, with their third well (1971) in the North Celtic Sea Basin, discovered the Kinsale Head Gasfield.

KINSALE HEAD GAS FIELD

David Naylor

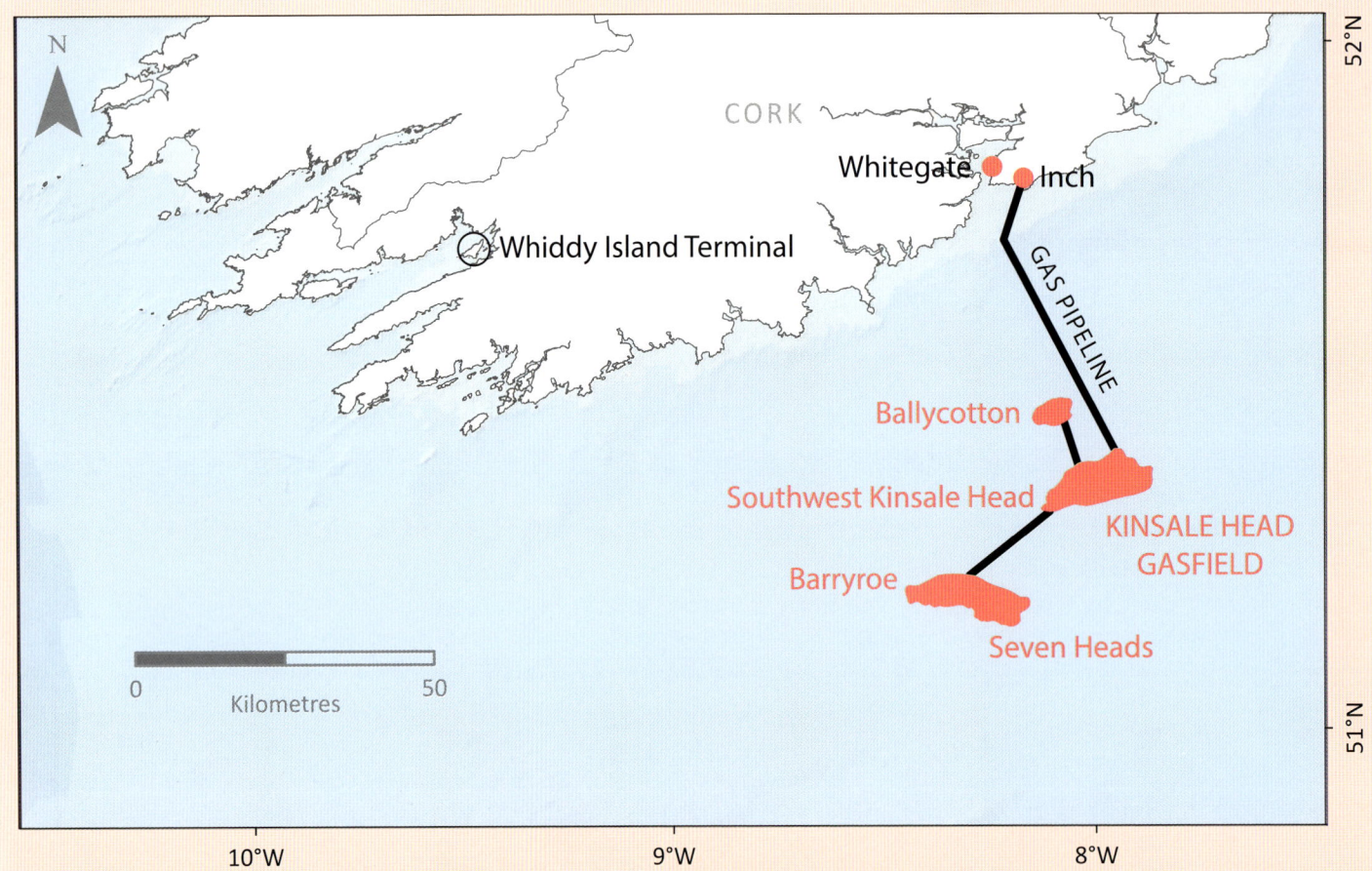

Fig. 29.26 LOCATION OF GAS FIELDS AND LINKED INFRASTRUCTURES IN THE CELTIC SEA AND ONSHORE. The Kinsale Head Gas Field was Ireland's first commercial discovery (1971) of natural gas, and is located in its offshore waters of the Celtic Sea. Production commenced in 1978 and, until 1995, it was the sole supplier of natural gas in Ireland. The subsequent development of three satellite fields of Ballycotton (1991), Southwest Kinsale (1999) and Seven Heads (2003) ensured the continuation of natural gas, following peak production at the Kinsale Head Field in 1995. In 2020 the field supplied about 5 per cent of the country's gas needs. The gas is brought ashore from two production platforms via an undersea pipeline to an onshore terminal at Inch. From here, it is connected to the country's gas grid for redistribution. Reserves of natural gas are declining and production terminated in 2021. From 2006, the Southwest Kinsale Head Field was redeveloped as Ireland's first offshore gas storage facility for natural gas piped from the UK. The present owner of the Kinsale complex of gasfields (Petronas), however, closed that facility in 2017. In spite of this, there remains interest in the possibilities of using the Kinsale fields, when finally decommissioned, as future storage chambers for natural gas and/or carbon dioxide. Discovery of commercially exploitable reserves of oil and gas at Barryroe may also offer some possibilities for retaining the undersea links to the Cork coastline, which could be used to feed into the oil refinery at Whitegate. Onshore storage of oil could also be provided at the Whiddy terminal in Bantry Bay. At the time of writing, no decisions have been made on these options. [Data source: Marine Institute]

The Kinsale Head Gas Field in the Celtic Sea, located approximately 50km off the Cork coast and in 90m of water, was discovered by Marathon in 1971. Production began in 1978, when it was estimated to contain ultimate recoverable reserves of 1.5 trillion cubic feet (42.3×10^9 m³) of gas.[29] The gas is contained within layers of porous sandstone rock (see Chapter 5: Geological Foundations) about 900m below the seabed. At this depth, the rocks are folded into a dome extending over 100km², with the gas trapped under layers of impermeable shale and chalk.

A number of smaller satellite accumulations of gas were also discovered: Ballycotton (1991), Southwest Kinsale Head (1999) and the Seven Heads accumulation (2003) (Figure 29.26). These were linked into the Kinsale Head production facility using subsea completion

Fig. 29.27 KINSALE HEAD ALPHA PRODUCTION PLATFORM. The natural gas produced from the Kinsale Head Gas Field is brought to the surface through two fixed steel production platforms – Alpha and Bravo – that were installed in 1977. Gas flows from all the satellite fields that make up the Kinsale Head gasfields are controlled remotely and brought from wellheads on the seafloor to one of these large and complex production platforms. Here, the gas from all the offshore fields is combined and compressed to raise its pressure to allow it to be transported ashore. The Bravo platform is linked to Alpha from which the gas is piped directly to the Inch gas terminal located near the mouth of Cork Harbour. With the cessation of gas production from the gasfields in 2020, the current owner, the Malaysian corporation, PETRONAS, has sought permission to plug and abandon the twenty-four gas wells. In addition, they plan to dismantle and remove the large production platforms of Alpha and Bravo, and other undersea infrastructures. This process of decommissioning is subject to regulatory approval and Environmental Impact Assessment to ensure the works do not cause any damage to sensitive marine ecosystems. [Source: PSE Kinsale Energy Limited]

technology. In 2012, an oil discovery at Barryroe was assessed to contain potentially recoverable quantities of both oil and gas but, to date, this site has not yet been brought into production.

Extraction of gas from the fields is controlled remotely from two production platforms, Alpha and Bravo, installed in 1977 (Figure 29.27), and the gas is brought ashore by submarine pipelines to a processing facility at Inch, east of Cork Harbour (see Chapter 24: Ports and Shipping). The terminal is linked directly to the island's system of gas pipelines and this allows for its distribution to consumers located throughout Ireland. Prior to this, gas had been produced from coal at a number of individual plants situated around the country. In addition to supplying the national grid, natural gas from the Kinsale fields proved to be instrumental in the decision to construct a large, new gas-fired electricity generating plant at Aghada, located proximate to the Inch terminal.

Peak production from these gasfields occurred in 1995, at 99 billion cubic feet (bcf) or approximately 2.8 billion cubic metres per year. They are now, however, in the final phases of decline, with the field and terminal now scheduled for decommissioning in 2021. In an effort to compensate for the decline and ultimate cessation of supplies of natural gas, the depleted Southwest Kinsale Head Gasfield was redeveloped as Ireland's first offshore gas storage facility. As such, it was designed to hold natural gas piped from the UK, and was licensed by the Commission for Energy Regulation in 2006. While this offered the prospect of continuity of gas supplies, a Malaysian company, Petronas, that bought the Kinsale assets (KinsaleEnergy Limited) in 2009 took the decision to close the Southwest Kinsale storage facility. The last of the storage gas was withdrawn from the reservoir in March 2017.

There is a strong argument for the subsurface, geological storage of natural gas, either as short-term storage for subsequent re-extraction as an energy resource, or the long-term storage of greenhouse gases (mainly carbon dioxide) as a contribution towards combating climate change. Suitable sites for these purposes could either be dissolution caverns in salt-bearing strata for small, typically onshore or shallow offshore sites, or larger offshore structures with suitable reservoirs. With a few exceptions, the onshore geology on the island of Ireland is not promising for either of these. In contrast, however, depleted offshore oil- and gasfields have been identified as having the potential for both short-term and long-term storage of gas.

A review of sites with the potential for long-term greenhouse gas storage around Ireland concluded that, on depletion, the Kinsale Head Gas Field offered the most promise for this purpose.[30] In this respect, the proven use of Southwest Kinsale Head for natural gas storage, to mitigate variations in demand, demonstrated the viability of this approach for short-term use. However, more stringent standards need to be applied before authorisation can be granted for its use in long-term storage of CO_2. The cost of accessing the Kinsale field for short-term storage was judged to be €15 million, although up to €80 million could be required to bring the field to a suitable level for longer-term storage. Although a gasfield may have stored natural gas without leakage for millions of years, the process of producing the gas and the subsequent depletion of the reservoir may have had an adverse effect on its storage potential. Also, any existing production wells are a source of potential leakage and may require further engineering work to secure them properly.

Since this chapter was written, the present government (ROI) (2020/21) has banned all new applications for offshore oil and gas exploration and there will be no more future licensing rounds. It is the end of an era for this hydrocarbons sector in Ireland. The focus in Ireland is now on developing clean, renewable sources of energy, including offshore wind, for the production of electricity.

In 1975, the Irish Department of Industry and Commerce held a first round of licensing, covering all blocks within Ireland's expanded area of designated offshore territory. The first Exploration Licences (in these waters) were granted in 1976.[31] This initiated a phase of drilling in the Celtic Sea basins (and subsequently in the Kish Bank Basin, east of Dublin), and also in the deeper waters of the Porcupine Basin off the west coast. Two further licensing rounds took place in the early 1980s, at a time when techniques for undertaking seismic surveys were improving. These enabled better definition of deep, previously unexplored geological structures within the North Celtic Sea and Porcupine basins. A total of 160 wells have been drilled to date in the Irish offshore, but with only limited success.

Of the more than eighty wells drilled in the Celtic Sea region, a number tested showed significant hydrocarbon flows. These were judged to be non-commercial at the time, although these have recently undergone reappraisal given the incentives of increased product prices and further improvements in technology. Currently, none have been brought to development.

Of particular interest is the Barryroe Oilfield, lying alongside the Seven Heads Field. This field, operated by Providence Resources, was discovered in the 1970s by Esso-Marathon. The early wells yielded a flow of oil, but with a wax content of 17–20 per cent which, at that time, constituted a severe technical obstacle. More recently, however, the drilling of additional appraisal wells, together with new seismic data, have yielded improved estimates of oil reserves. Furthermore, the availability of modern production techniques allow for successful production of the oil. In 2018 a Chinese company, Apec, joined the group and a further four wells are planned to assess the recoverable oil reserves of the field. Currently, these reserves are estimated at 311 million barrels and, should they become exploited, it is projected that the oil will be delivered by submarine pipeline to the Whitegate refinery in Cork. Given the country's total dependency on imported oil, and concerns over disruption to supplies post-Brexit, this development would be a boost to both the national and the Cork regional economy (see Chapter 24, Ports and Shipping).

By the early 1990s, drilling results from the Irish offshore had been largely disappointing. Some small, non-commercial discoveries had been made, but no major fields had been developed and drilling operations dropped to low levels. The decision as to whether a discovery is commercially viable, however, may depend as much on available technology and oil price as the volume of the estimated resource. In this context, there has been a marked improvement in the quality of seismic data which has helped to decrease the levels of risk associated with exploration. It also allows a more accurate estimate of potential reserves.

As a result of such improvements in offshore technology, it is also possible to drill in increasing depths of water. This has extended the range of possible drilling targets. Since 1993, the Irish licensing authorities have sought to promote more active exploration of the Atlantic margin. This involved the holding of a succession of Frontier Licensing Rounds, with different licensing terms being offered to interested parties. Although the number of wells drilled under the Frontier Rounds has been relatively small, such drilling did lead to the discovery, in 1996, of the Corrib Gas Field by a group operated by Enterprise Oil (now Shell) in the north Slyne Basin in 1996 (Figure 29.28). After many delays, production of the gas began in 2015.

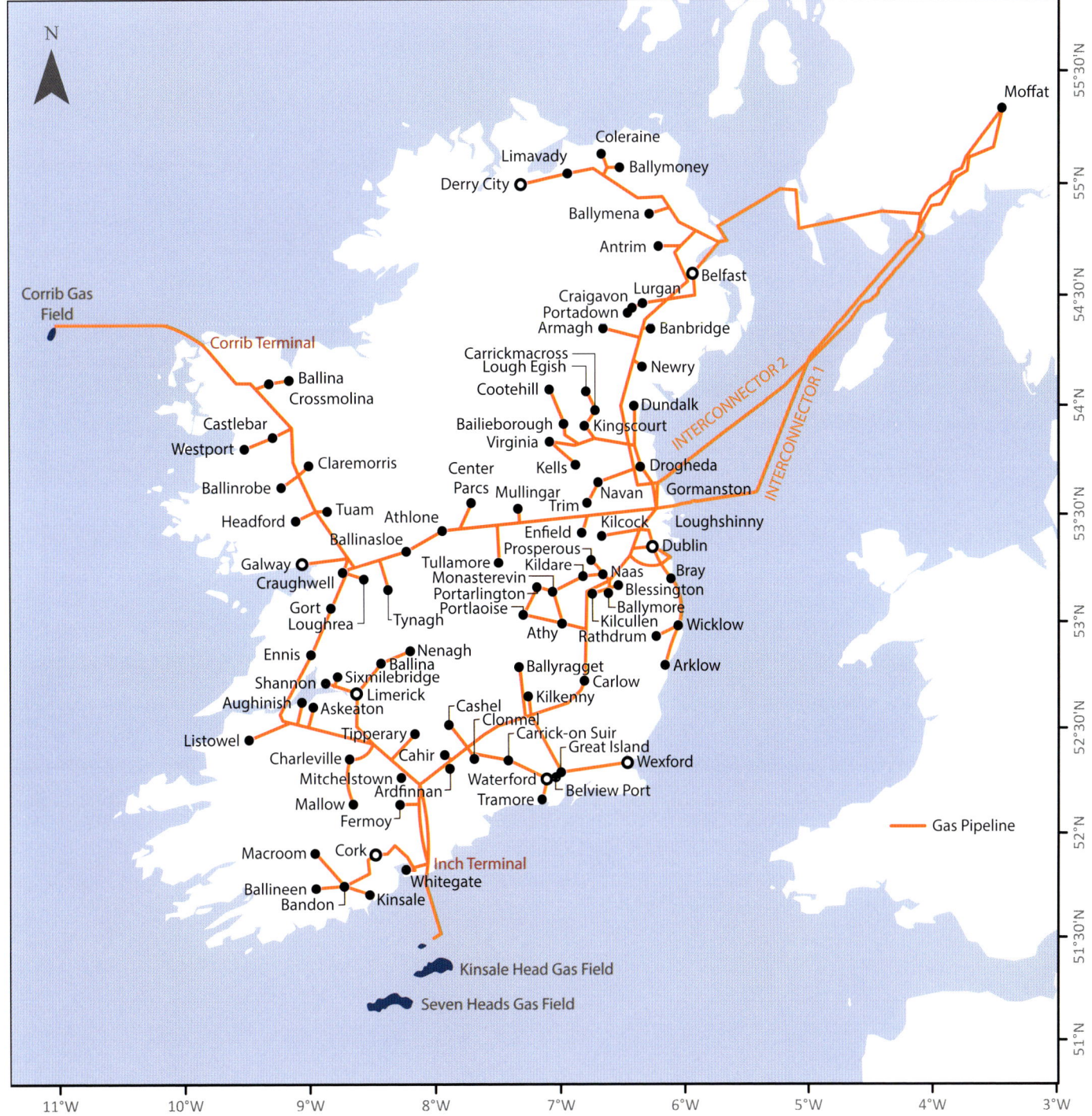

Fig. 29.28 IRELAND'S GAS NETWORK. The gas network serves to redistribute supplies of natural gas obtained from the Kinsale and Corrib gasfields in the Republic and via the interconnectors from Great Britain. [Data source: adapted from Gas Networks Ireland]

The Corrib Gas Field

Marcus Lange

The Corrib Gasfield is a small-to-medium natural gas resource, located within the North Slyne Basin, approximately 83km off Erris Head in County Mayo. It lies in Triassic Sandstones, at the margin of the Rockall Trough, at water depths of 355m. It was discovered in 1996 by a group operated by Enterprise Oil (now Shell). Construction work on infrastructure for extracting and processing the gas started in 2004, and the project was commissioned finally in 2015. Reserves in the field are believed to be about 1 trillion cubic feet (28×10^9 m^3), which

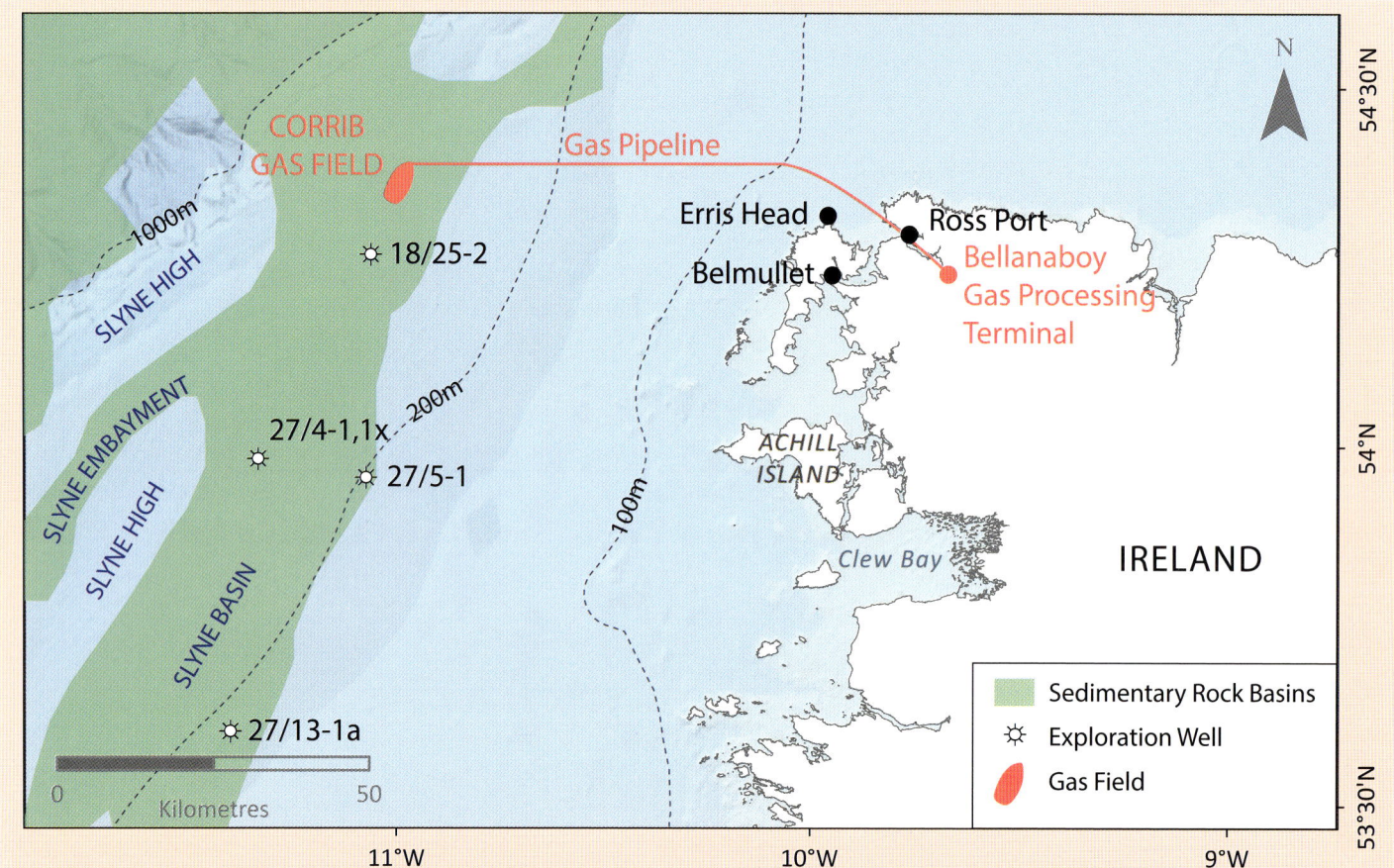

Fig. 29.29 SLYNE BASIN AND THE CORRIB GASFIELD. The Corrib gasfield is located in the Slyne Basin, one of the linear sedimentary troughs that occur around the coast of Ireland, and is located about 83km off Erris Head, County Mayo, in water depths of 355m. The gas is produced from a Triassic Sandstone reservoir 3,000m below the seabed. Reserves in the field are believed to be about 1 trillion cubic feet ($28\times10^9 m^3$). The gasfield was first discovered by Enterprise Energy Ireland Ltd in 1996, and in 2001 a government petroleum lease was granted to the company. The following year, the lease was taken over by a consortium led by Anglo-Dutch Shell, with Marathon and the Norwegian company Statoil (now Equinor) as partners, and work began on the construction of the onshore and offshore infrastructure needed to extract and process the gas. The gas is collected via five subsea wells located within the gasfield area, which feed in to a central, subsea manifold. This, in turn, is connected to the main offshore pipeline, which carries the gas from the manifold to a landfall at Broadhaven, near Rossport in County Mayo. From here, a third pipeline that runs under Sruwaddacon Bay delivers the gas to an onshore terminal at Bellanaboy Bridge. Almost from the outset, the project to develop the Corrib gas resource was mired in controversy, and led to several years of protests at the site. [Data Source: modified from Naylor, D. and Shannon, P., 2011. *Petroleum Geology of Ireland*. Edinburgh: Dunedin Academic Press]

equates to a production of 10 million standard cubic metres of purified gas per day, approximately 70 per cent of that at the Kinsale Head Gas Field off Ireland's south coast. Compared to Kinsale, the Corrib Gas Field offers a very pure form of gas, consisting of approximately 97 per cent methane and ethane. At peak production (5–10 years), and over a lifetime of 15–20 years, the project is capable of meeting up to 60 per cent of Ireland's gas needs. This is expected to bring down import dependency, particularly from the UK. At present, Ireland is connected to the UK via three gas and two electricity subsea interconnectors, with a new north–south electricity interconnector under development. However, the continuity of gas supplies from Britain, in the context of that country's exit from the European Union, together with the declining reserves of North Sea gas, is no longer guaranteed. This raises a very real concern on the island of Ireland regarding reconciling the future supply and demand of energy.

Extraction of gas from the Corrib fields is controlled remotely from a subsea facility which connects five subsea wells. An offshore pipeline connects the wells to a landfall in County Mayo, from where an onshore pipeline connects to a gas-processing terminal located 9km inland from the coast at Bellanaboy, near the town of Belmullet (Figure 29.29). This latter pipeline is routed through a tunnel for approximately 4.9km of its length, taking it underneath Sruwaddacon Bay. From the terminal, the gas enters the Irish gas transmission network for distribution via a 150km extension to Galway.

In terms of strategic importance, the Corrib project represents Ireland's largest-ever energy investment. During construction, more than 6,000 people worked on the project, with up to 175 full-time job equivalents being required to maintain the operation.[32] However, the Corrib Gas project was characterised by extreme conflict between the principal stakeholders. In particular, there was strong local opposition to the project, linked to a variety of issues. These included concerns over the safety and environmental impact of the pipeline, an insufficient flow of information and engagement with the local communities, limited opportunities to intervene in the planning process, and a perceived lack of benefit for the local Gaeltacht communities.

From 2004 until the summer of 2005 the situation escalated (Figure 29.30), particularly when development of the onshore pipeline in Rossport and construction of the terminal commenced. In 2005, five protesting farmers (known as the 'Rossport Five') from Rossport (Kilcommon) were jailed for ninety-four days. These individuals had failed to comply with a court injunction taken out by Shell to stop

Fig. 29.30 CAMPAIGNERS PROTESTING AGAINST THE CONSTRUCTION OF THE CORRIB GAS PIPELINE AND REFINERY. Royal Dutch Shell, commonly known simply as Shell, is an Anglo-Dutch company headquartered in Rotterdam, and is the sixth-largest company in the world as measured by its 2016 revenues. In the early years of the twenty-first century, the company became the lead developer, along with partners Statoil and Marathon, in the Corrib Gasfield off the Mayo coast, and sought to develop a terminal near Rossport to receive and process the extracted gas. A key element of the project involved routing a high-pressure raw gas pipeline through the small communities of Rossport, Glengad and Pullathomas. This proposal, in particular, was met with much local resistance, including this sit-down protest and attempt to block the entrance to the refinery site in September 2007. The response by Shell to the protests was considered generally to be heavy-handed, and there were widespread claims that the Gardaí and the government of the time were putting the commercial interests of a foreign multinational company ahead of those of the citizens of the state they were supposed to protect. These perceptions led to the Corrib controversy becoming a national and international cause célèbre, and people travelled from around Ireland and even from abroad to join the protest alongside local campaigners. Protests (including civil disobedience), a number of high-profile legal challenges to the planning permissions granted and the jailing of five local residents, who became known as the 'Rossport Five', delayed the project by more than ten years. The pipeline and terminal were eventually completed and gas began flowing in December 2015. [Source: William Hederman]

protesting at the site and attempting to disrupt the construction work. They, as well as other local opponents, had genuine safety concerns due to living in close proximity to a pipeline that pumped unprocessed natural gas under high pressure to the terminal.

Although the developer implemented a community benefit scheme to compensate the community financially, and the project itself was commissioned in 2015, community–developer relations, and even relations within and across the community, remained divided. The controversy generated over this project raised important questions in planning and development at all spatial scales, in particular, how to manage development projects (especially those of a large scale) which often involve the conflicting interests of local communities, multinational enterprises and the state.

The focus of Ireland's search for offshore hydrocarbon resources has turned increasingly to the deep-water Atlantic margin, and draws on geological insights provided by exploration successes offshore in West Africa and eastern Canada (Figure 29.31). This was reflected in increased levels of interest by international companies in the two Frontier Licensing Rounds held in 2011 and 2016. As a result, exploration licences now cover almost all of the Porcupine Basin (Figure 29.32). Such a positive sign of commitment from investors has been facilitated by technological advances made in detailed 3D seismic data. These have, in particular, allowed many licensees to obtain better images and understanding of the subsurface rocks and, therefore, contribute to lowering the risk levels in future expensive drilling operations.

Since 2015, significantly lower and variable oil prices have made planning for exploration and development in high-cost and long lead-time environments difficult for the oil industry. The two wells drilled in the Porcupine Basin in 2013 and 2017 yielded promising technical results, but no commercial hydrocarbons. The Atlantic margin awaits a new phase of drilling, which will have to evolve in line with new and emerging government policy dealing with climate

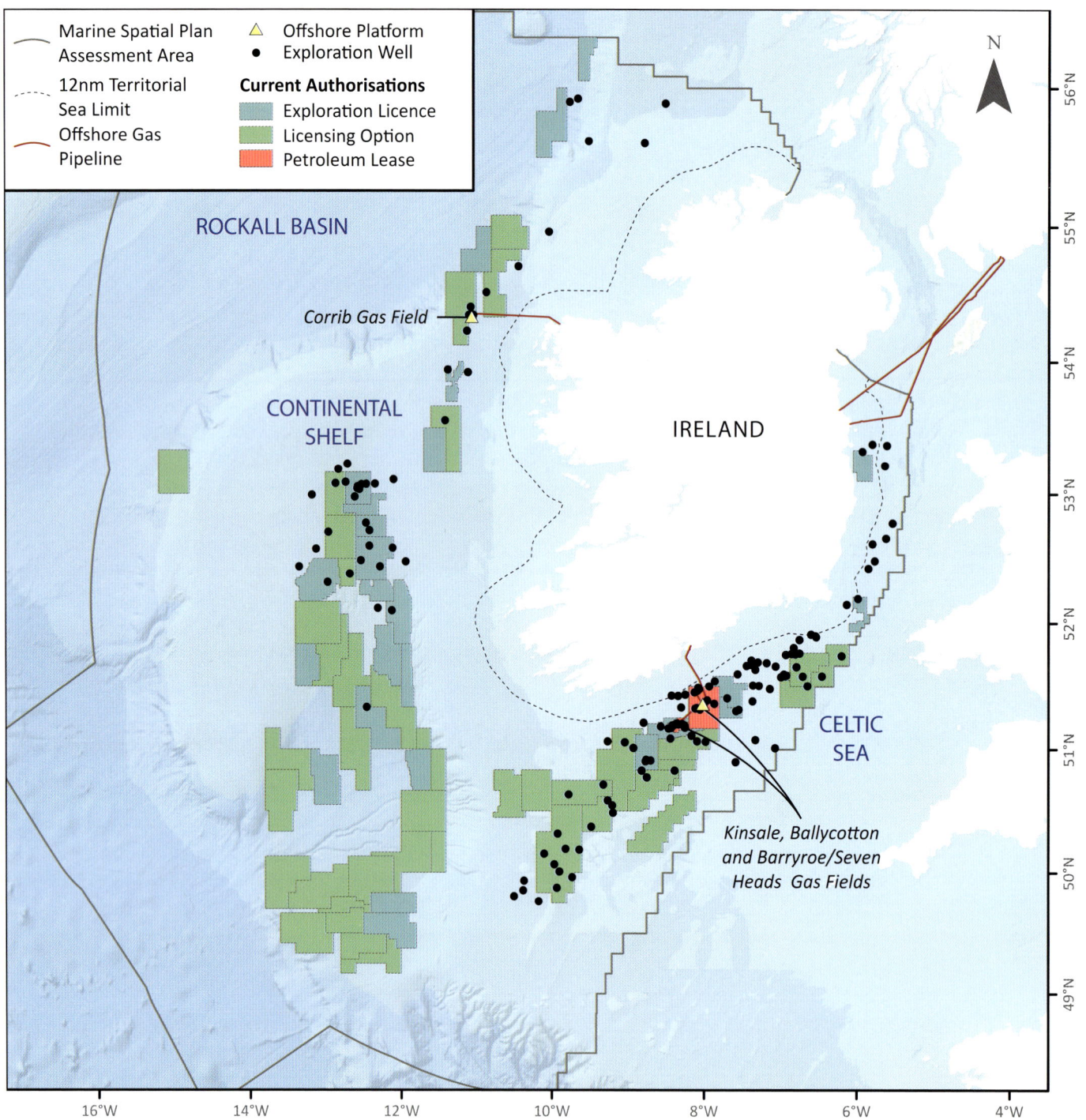

Fig. 29.31 OIL AND GAS LICENSING BLOCKS AND ASSOCIATED INFRASTRUCTURE IN IRISH COASTAL WATERS (ROI ONLY). Before the exploration for oil and gas, either onshore or offshore in Irish territorial waters, can commence, companies must obtain an authorisation to do so from the minister for the Department of Communications, Climate Action and Environment. These authorisations are issued under the Petroleum and Other Minerals Development Act, 1960, and are granted subject to the Licensing Terms for Offshore Oil and Gas Exploration, Development and Production. Often, two or more companies will come together in a consortium to bid for licences and share the subsequent costs of undertaking an exploration campaign. For licensing purposes, the Irish offshore is divided into 'Open Areas' and 'Closed Areas': in Open Areas, applications for exploration licences and licensing options may be made at any time; in Closed Areas, applications may only be made during a specified Licensing Round, at which time a particular area is declared open for licensing by the minister. Once obtained, a petroleum exploration licence gives the holder the exclusive right to carry out exploration for petroleum in all or part of a specific licensed offshore area, known as a licensing block. There are three types of licences available: a Standard Exploration Licence is issued for a period of six years, and for an area with water depths of up to 200m; a Deepwater Exploration Licence is issued for a period of nine years, and applies to an area with water depths exceeding 200m; while a Frontier Exploration Licence is issued in respect of an area with special difficulties related to physical environment, geology or technology, and is valid for a period of not less than twelve years, divided into a maximum of four exploration phases. Licensing Options give the holder the first right to an Exploration Licence over all or part of the area covered by the Option. [Data Source: Department of Communications, Climate Action and Environment]

change (see Chapter 33: Climate Change and Coastal Futures).

In conclusion, despite offshore exploration since 1969, most Irish offshore sedimentary basins remain lightly explored, particularly in the deeper water areas off the west coast. Initial

successes in both the Celtic Sea and Porcupine Basin have not been sustained. In some years, however, rigs or seismic boats were brought in, often from the North Sea, to develop a single well or conduct survey work. Despite this, drilling activity has

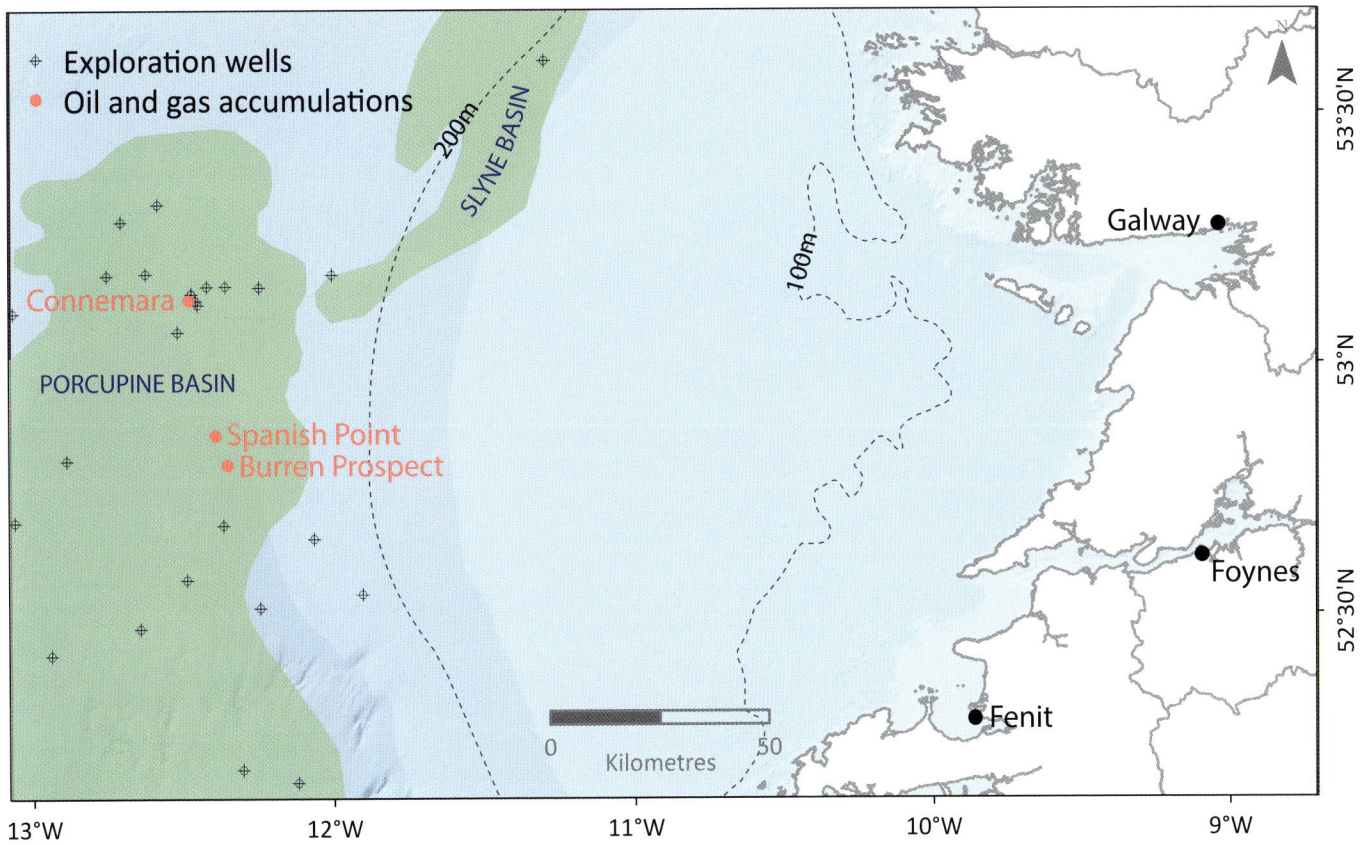

Fig. 29.32 HYDROCARBON EXPLORATIONS IN THE PORCUPINE BASIN. A number of hydrocarbon accumulations have been discovered in the northern part of the basin. A well drilled in 1978 flowed 730 barrels of oil per day (bopd) on test and this undeveloped discovery is now known as the Burren Prospect. During 1979, BP discovered the Connemara oil accumulation (undeveloped), with proven reserves in place of 120 million barrels of oil. Subsequent drilling showed the geology to be complex, and failed to establish the commerciality of the field. In 1981, Phillips Petroleum tested gas and condensate on a prospect that became known as 'Spanish Point', but at the time this was deemed non-commercial. The current licence holders have carried out a reappraisal, but no development plans had been announced at the time of writing.

remained low in Ireland's offshore waters since the early 1990s (Figure 29.33). With the high costs of deep-sea exploration and the new environmental policy of the European Union emphasising a shift from a hydrocarbon base for energy supply, prospects for major developments offshore are likely to remain limited.

A number of coastal facilities, however, have emerged as a result of developments in Ireland's offshore hydrocarbon sector. Of particular note are the important onshore gas terminals for the Kinsale and Corrib gasfields. Other structures related to the oil industry include the large Whiddy Island oil storage facility in Bantry Bay, which holds one-third of the country's strategic reserves of oil. Whitegate in Cork Harbour is the location of Ireland's only oil refinery, which generates a large trade in oil and oil-based products through the

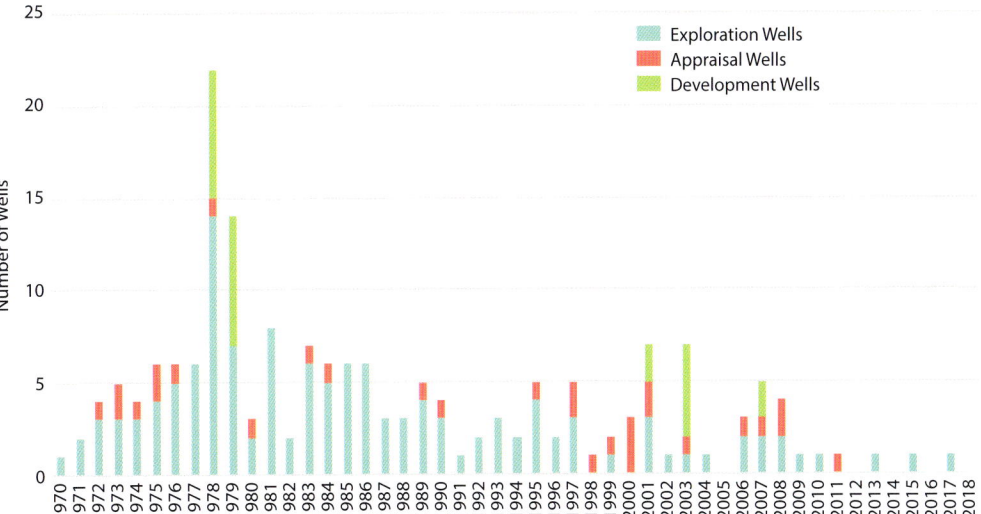

Fig. 29.33 OFFSHORE WELLS DRILLED EACH YEAR IN THE REPUBLIC OF IRELAND SINCE 1970. A total of just under 160 exploration and appraisal wells have been drilled offshore in Irish territorial waters since 1970, with a peak in 1978–9 and a gradual decline since then. Exploration wells (shown in blue on the graph) were drilled on potential untested targets; appraisal wells (orange) were drilled following promising exploration results to try to establish whether sufficient reserves existed for commercial development; and development wells (green) were drilled as part of the subsequent exploitation process. So far, these have served to access and produce gas from the two commercial gasfields – Kinsale Head (1978–1979) and Corrib (2001–2007) – while potentially exploitable reserves of oil have been found in the Porcupine Basin off the coast of Connemara, and in the Barryroe field adjacent to the Kinsale Head Gasfield. Running a drilling programme for oil or gas exploration requires much advance planning, and typically needs a lead-in time of at least a year before actual drilling can take place. The factors that can influence the number of offshore wells being drilled in any one year include the availability of equipment and personnel, political and commercial policies and priorities prevailing at the time, the state of the global economy, and sometimes even the weather. [Data source: Petroleum Affairs Division of the Department of Communications, Climate Action and Environment, modified from Naylor, D. and Shannon, P., 2011. *Petroleum Geology of Ireland*. Edinburgh: Dunedin Academic Press]

Port of Cork. In Northern Ireland, the import and storage of natural gas in the salt caverns at Kilroot is also of significance.

Although Ireland has not seen the development of a coastal hydrocarbon base comparable to Aberdeen or Stavanger, the ports of Cork and Foynes in the Shannon Estuary have benefited from this activity, albeit to a limited extent. Both are well placed to service any offshore activities. In addition, proposals to develop major terminals for the large-scale importation of Liquified Natural Gas (LNG) in both Cork Harbour and in the Shannon Estuary could, if enacted, further the role of these two port areas as dominant centres for energy supply within Ireland.

Case Study: Ireland and Liquified Natural Gas (LNG)

Barry Brunt

Despite the successful development of the Kinsale Gas Field and some growing commitment to renewable energy, the Republic of Ireland entered the twenty-first century with an import dependency of 85 per cent for its energy needs. As part of the UK's energy market, dependency levels in Northern Ireland were even higher. In the Republic, government was committed to reduce such levels of dependency, and especially on oil and gas from Britain's offshore fields located in the North Sea.

Although further development of its own offshore areas has yielded a positive result in the Corrib Gas Field, and development of renewable energy sources have increased to a level of meeting 30 per cent of the country's energy requirements, reports have indicated that Ireland will require new sources of natural gas in the 2020s to meet rising demand. One option of particular interest for government was to import liquified natural gas (LNG), however, plans for floating and land-based terminals have been shelved. The story of LNG in Ireland is an example of how the energy transition is changing plans for the utilisation of the coastal zone in the twenty-first century.

At the global level, reserves of natural gas exceed greatly those of oil, especially as new and relatively inexpensive sources of gas supply have become available, albeit through the controversial technique of fracking. However, while many of the large gas fields that have been developed have the potential to produce far in excess of their regional or even national requirements, they are often located at significant distances from import markets. In such circumstances, using pipelines to transport the gas is often neither economically nor technically feasible. The remaining option is to freeze the gas to a liquid state (LNG). This makes its transportation possible in very large, and specially designed, LNG tankers (Figure 29.34). On arriving at ports of importation, a regasification process occurs, which allows the gas to be stored and/or redistributed by pipelines into the designated market area.

In 2005, a development company, Shannon LNG Ltd, proposed the construction of a large LNG terminal and related infrastructure on a 281-acre (approximately 114 ha) site at Ballylongford, near Tarbert, in the Shannon Estuary. Planning permission was received in 2008 for a ten-year period to develop the facility. A US investment company (Hesse) backed the project, but withdrew in 2015. Problems linked to inflated costs and time delays due to the imposition of additional regulations and objections from environmental groups contributed to the failure.

As Ireland exited its economic recession, interest in the project was revived. This was linked to government encouragement, as highlighted in the 2018 *National Energy and Climate Plan for 2021 to 2030*. This recognised the critical role that LNG could play in improving national security of energy supplies if the country could be linked directly into the expanding global market for LNG. The endorsement for LNG and the Shannon project relates to a number of key issues, especially as Ireland remains the only country in NW Europe that does not possess a terminal for LNG supplies, and is now most likely to retain that status.

First, Ireland's indigenous supplies of offshore gas were expected to decline in the absence of major new discoveries. The Corrib and Kinsale fields supplied some 60 per cent and 5 per cent respectively of the country's gas needs in 2018, but Kinsale would terminate in 2021 while, by 2025, Corrib was likely to meet only 25 per cent of requirements.

Second, reserves of gas in Britain's North Sea fields are depleted. In addition, Brexit will remove its obligations under the EU Security of Supply Agreement, which are designed to ensure energy supplies for member states are not disrupted. These factors raise doubts over the continuity of supplies from Britain via the gas interconnector into Northern Ireland.

Third, while Ireland is committed to increase the role of renewable energy (which met one-third of electricity supply in 2018) this

Fig. 29.34 AN EXAMPLE OF A SPECIALLY DESIGNED LIQUIFIED NATURAL GAS TANKER. From the late twentieth century, liquified natural gas (LNG) has transformed the global market for natural gas by facilitating the long-distance movement of natural gas from fields that were considered previously to be uneconomic. This has been achieved by investments in new technologies, which allow the natural gas to be supercooled (-160°C) and which transform it to a liquid state. Not only does this enable the LNG to be moved more easily over considerable distances but also, in its liquid state, it is 600 times smaller in volume than in the gaseous state. To transport large volumes of LNG both efficiently and safely, specially designed LNG tankers (with an average capacity of approximately 3 billion cubic feet [approximately 85 million cubic metres]) are used to export the LNG from areas of surplus, such as the Middle East and USA, to importing regions, such as the European Union. With Ireland's need to diversify its energy market and reduce levels of carbon emissions, the import of LNG was regarded initially as being a viable option. If proposals to construct large-scale LNG terminals in the Shannon Estuary and Cork Harbour were to materialise, then vessels such as shown (LNG *Akwa Ibom* in the port of Las Palmas) could have been a frequent sight in these locations. Since 2020, however, government support of LNG has eroded as concerns over environmental implications have increased. In addition, there is now a growing recognition of the benefits of promoting clean, marine renewable energy from floating offshore wind farms relating to the abundance of offshore wind resources and deep water available off Ireland's south and west coasts. (Source: Vin Moore)

will not, especially in the near future, meet the increasing demands for energy anticipated on the island.

In such a context, the Shannon LNG project was reactivated in 2017, when An Bord Pleanála granted a five-year extension to the original planning permission. Arising from this, a US-based company(New Fortress Energy) bought out Shannon LNG Ltd. and, with initial government backing, anticipated that the project could come 'on stream' by 2020. Optimistic forecasts suggested that, on completion, Shannon LNG could supply 40 per cent of Ireland's requirements for gas. In particular, the project was 'slated' to fill the generation gap left by the anticipated closure in 2025 of the Moneypoint power station. In 2020, however, the project suffered a major setback when the High Court ruled to quash all development consents for the project.

The case for LNG in Ireland was further downplayed after a preliminary deal signed between the Port of Cork and the US energy company NextDecade expired in 2021. This project would have involved the location of a large Floating Storage and Regasification Unit (FSRU) in Cork Harbour. LNG would be shipped to the plant from the USA and, following regasification, could gain access to the national gas grid from an adjacent terminal at Inch, which had been built to receive natural gas from the Kinsale Gas Field. It was also anticipated that the gas-fired power plant at Aghada would provide a local market for the new supplies of gas and ensure continuation of energy supplies for an expanding industrial area around Cork Harbour.

Irish governments prior to 2020 had seen LNG as a viable option to ensure energy security of the national gas supply, given the country's dwindling offshore production. The prospects of LNG supplies to Ireland changed, however, with the formation of a new government in 2020 and its stated opposition to the import of LNG-backed fracked natural gas from the USA and elsewhere. A Joint Committee on Climate at the end of 2020 called for LNG import infrastructure to be banned due to environmental concerns. Furthermore, advocates of floating offshore wind made – and continue to make – a strong case to replace LNG options with clean, marine renewable energy from the abundance of offshore wind resources in the Celtic Sea and the Atlantic. The result has seen the folding of the NextDecade project and a growing political and public pressure which places the prospects of Shannon LNG in doubt. A final LNG project remains under consideration. This is the Predator Oil and Gas plan for an FSRU off the Cork coast in the vicinity of the former Kinsale Gas Field, and one which will not use shale gas as a feedstock. Uncertainty abounds in this sector but one thing remains certain; the energy transition is playing out in real time.

SECTION 5

MANAGEMENT OF THE COASTS AND THE MARINE ENVIRONMENT

Marina and coastal defences, Bangor, County Down. [Source: Geraldine Hennigan and Norman Kean]

ENGINEERING FOR VULNERABLE COASTLINES

Jimmy Murphy and Robert Devoy

COASTAL ENGINEERING: A VIEW OF THE DEEP OCEAN WAVE TANK AT THE LIR LABORATORY, IN THE SFI RESEARCH CENTRE FOR ENERGY, CLIMATE AND MARINE RESEARCH AND INNOVATION (MAREI), UNIVERSITY COLLEGE CORK (UCC). Lir is the National Ocean Testing Facility, for the development of built structures (for example, fixed coastal defences, or floating structures such as a wave-energy machines) that will be used in the water worlds of the open ocean and coast. The laboratory is involved particularly in the development of ocean wave energy 'machines' for deployment in inshore water areas and electricity production. It was commissioned in 2016, as the brainchild of Professor Tony Lewis, UCC. The earlier and smaller wave tank facility at UCC, known as the Hydraulics and Marine Research Centre (HMRC), was founded in the 1970s. MaREI, as the host institute (part of the ERI, UCC), forms a multidisciplinary combination of the earlier HMRC and the Coastal and Marine Research Centre (founded by Professor Robert Devoy), which are joined now by members of the Renewable Energies grouping, UCC. The new Lir facility is used both for the design, development and testing of engineered ocean and coastal structures. It is one of just a few such laboratories worldwide carrying out complex and multi-dimensional hydrodynamics work, based on both the numerical and physical modelling of coastal and ocean waters. Using the water tank as its main experimental tool, the facility is able to take the design concept for a structure (device) and subject it to the current and wave conditions it is likely to have to cope with. There are a number of differently shaped and sized water tanks in the Lir laboratory, built for different testing functions, and each capable of producing waves of different dimensions. The main tank (shown in the photograph), one of three large tanks in Lir, is able to generate waves of up to a metre high that can travel in multiple directions, much as occurs in the real ocean, or in the complex three-dimensional shapes of the coast. The waves are generated by an interconnected series of movable paddles, which can be seen arranged at the end of the tank behind the technician. [Source: Lir-NOTF]

Fig. 30.1 COASTAL FORTIFICATIONS, SCATTERY ISLAND, THE SHANNON ESTUARY. At first sight, this might appear to give a different interpretation to the expected use today of the term 'coastal protection', but the same principle is involved as with, for example, built sea walls: to defend and stop! Here, a gun platform and blockhouse are shown, which form part of a larger complex of coastal fortifications. They were built *c.*1805 to protect the Port of Limerick and shipping in the Shannon Estuary, serving as defence against Napoleon's expected invasion of Ireland at the beginning of the nineteenth century. Martello towers, and other less elaborate defence and observation post structures, were also built at this time around the whole of Ireland. Many of these can still be seen along the coastline today, some now converted to residences with splendid views! [Source: Robert Devoy]

Built and engineered coastal structures found today along Ireland's coasts date mainly from the late eighteenth century to the present, though older structures such as forts and castles do exist (Figure 30. 1). The construction of designed coastal works – for example, harbours, piers, jetties, sea walls and other types of sea defences – is, of course, much older. The evidence of these built structures can be found widely in the coastal archaeological and historical records of the island, and date from many different time periods (see Chapters 16: The Inhabitants of Ireland's Early Coastal Landscapes; 17: The Vikings and Normans: Coastal invaders and settlers; 18: Era of Settlement: Trade, plantation and piracy; 19: Changing Coastal Landscapes). The purpose behind these constructions has varied, but all focus on people's desire to use, control and manage the sea and its resources. In this context, the present chapter deals particularly with the intertwined themes of coastal erosion, flooding and the need for the planning and physical management of coastal areas, which includes their defence and protection (a term linked often to the idea of 'prevention'). The threats facing Ireland's coastline, primarily in relation to storminess and sea-level rise (SLR),

are considered along with potential coastal protection solutions.

Ireland's long and varied coastlines range from rocky cliffs, that can resist high-energy Atlantic waves and storms, to sandy beaches that seem to change with every tide (Figure 30.2). About half of this 7,000km or so length of coastline is considered to be hard in character, dominated by rock and not subject to significant change at decadal to century timescales (see Chapter 10: Rocky Coasts). The remaining coasts are developed from soft sediments. These are composed of glacial deposits, such as till and boulder clay, as well as unconsolidated gravels, sands and muds, that erode easily and form low-cliffed beaches, dunes and estuaries. These soft coasts are dynamic (see Chapters 6: Glaciation and Ireland's Arctic Inheritance and 11: Beaches and Barriers). Consequently, their positions can change rapidly, depending on the operation and impact of the different coastal process controls that produce them. These 'controls' include factors such as tides, sea-level change, waves, storm events, sediment supply and the impact of people (see Chapter 2: The Coastal Environment: Physical system, process and patterns).[1]

Fig. 30.2 DYNAMIC SAND AND 'SOFT' SEDIMENTARY COASTS SHOWING THE DUNE AND BEACH-BARRIER COASTAL SYSTEMS OF ROSSBEHY SPIT (BOTTOM LEFT) AND INCH SPIT (TOP CENTRE), TOGETHER WITH THE INTERTIDAL SANDBANKS AND SHOALS FORMED BEHIND CROMANE SPIT (TOP RIGHT), COUNTY KERRY. This photograph, taken from an orbiting satellite, shows clearly the pattern of beaches (bright, white coastal areas), as well as the inshore and sub- to intertidal sediments that link them, within Dingle Bay and Castlemaine Harbour, western Ireland. The image is taken at a time of low tide, exposing much of these coastal sediments. Central to the processes and functioning of the beaches is the water-driven (hydrodynamic) operation of the ebb-dominated tidal delta system, seen as sand plumes, shoals and bars offshore (visible centre, left). This delta forms a sediments store for the coast, from which sand moves out onto the beaches and back again over time and under different sized waves. It serves as a critical link to these otherwise separated coastal sand systems. These, and similar sand systems that have formed along these coasts, develop as part of the daily tidal, wave and current cycles. They can be adversely impacted and even destroyed by large storm events. Such events occur less frequently than these regular, daily wind and wave movements. As a rule, storms of progressively larger size occur inversely with time, the smaller events repeating annually and the biggest storms at decadal or even century timescales. In 2008 the dune barrier at Rossbehy was fully breached by a storm and again, partially, in the storms of 2014. These events have led to big changes in the size and position of the ebb-tidal delta system, together with reduction in the length of the spit and accelerated erosion of the coasts behind the barrier, in the Carragh River Estuary area (photograph centre) and also behind Cromane. [Source: Google, Digital Globe, 2019]

Globally, coastlines are coming under increased pressure from the impacts of climate change and increased human activity, with an estimated loss of land to the sea of over 100,000km² in the last thirty years, primarily from the impact of SLR. Even under the most conservative scenarios for projected SLR with future climate warming, the order of 1,000km² of coastal lands in Ireland may be lost from flooding and erosion by c.2030.[2] Since approximately 40 per cent of Ireland's population (1.9m people) live within 5km of the coast, and 40,000 people live within 100m (see Chapter 25: Urbanisation of Ireland's Coast), protecting the coastline from these threats is essential, not only for public safety but also for the sustainability of trade and business. In Ireland, flooding is generally regarded as being the more critical of these hazards at present. Significant national funding (in both Northern Ireland and the Republic) has been invested since the late 1990s in putting protection schemes in place for locations considered most at risk. In more populated coastal and estuarine areas, with most of the

island's urban centres located in the coastal zone, major flood protection and prevention plans have been established, for example, for Cork, Dublin and Belfast.[3]

Most coastal European countries today are proactive and innovative in dealing with the problems of erosion and flooding, as well as with the fundamental controlling issues of policy and management. Invariably though, the approach to coastal management and these linked matters in Ireland has been more traditional and reactive in character.[4] Past average coastal erosion

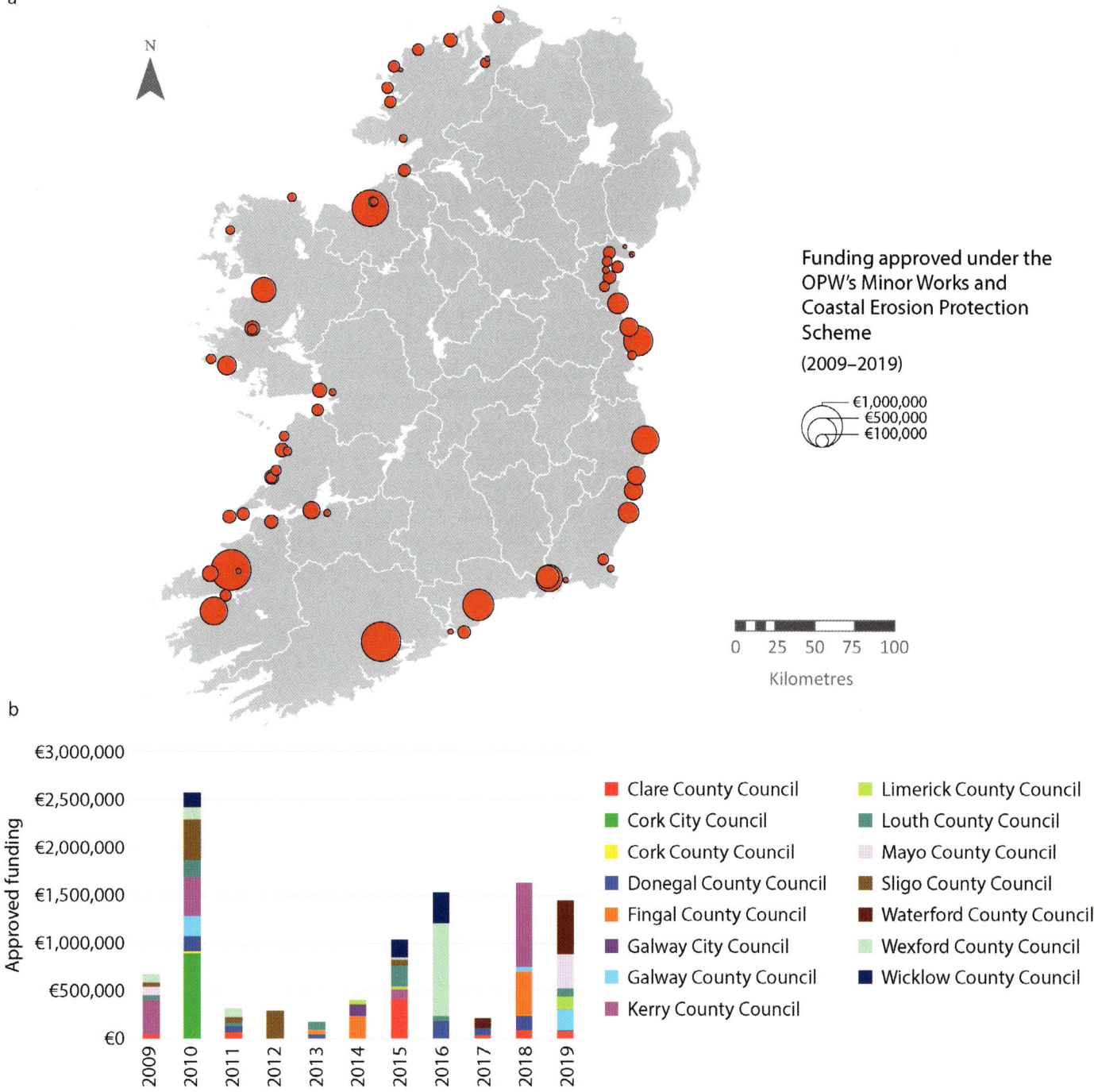

Fig. 30.3 ANNUAL FUNDING (2009–2019) SPENT ON PROTECTING COASTAL REGIONS, INCLUDING WORK ON EROSION, FLOODING AND LINKED PROJECTS WITHIN THE REPUBLIC OF IRELAND. The data show the level of funding provided, particularly through the Minor Flood Mitigation and Coastal Protection Scheme of the Office of Public Works (OPW) (https://www.opw.ie/en/flood-risk-management/operations/minorfloodworkscoastalprotectionscheme/listsofminorworksapprovals/). The map (a.) gives an approximate location and size of the projects funded; the bar graph (b.) charts the relative proportion of funding allocated regionally to the different coastal counties. Statistics relating to dedicated, large-scale coastal protection projects, of which there have been very few, are not included in this figure. The OPW Minor Works Scheme provides funding to local authorities to carry out minor flood mitigation, other works and studies costing less than €0.75 million each. Details of most small-scale works, not specifically part of this funding scheme (for example, culvert replacement, flood drains), are also not included in the spending calculations. [Data source: Office of Public Works, 2019]

rates have been of low-to-moderate levels, at <0.5m/year (in range of *c*.0.2–2.0m/year), especially on soft sections of Ireland's storm-dominated western coasts. Given that long sections of the coastline are largely unpopulated, or have low population density (see Chapter 4: People, Agriculture and the Coast), priority for protection works has inevitably been given to the more vulnerable areas, where public safety and/or critical infrastructure is most endangered. Built protection works have been installed mainly in response to well-defined, long-term and usually local coastal erosion problems and the hazards that they pose. These interventions have been managed mostly by engineers working with local authorities. Their approach has been aimed primarily at 'stabilising' and defending the line of the coast, using best 'cost-benefit' principles, though more holistic and environmentally friendly engineering solutions have occasionally come to the fore.[5] Further, and in spite of widespread government and public awareness of the probable impacts of future climate warming, there is still a tendency for coastal protection works and wider management planning to react and 'do something' only after particular storm events have occurred (see Chapter 33: Climate Change and Coastal Futures).

For many people, the coast doesn't really seem to change much from year to year, and because of this there is often poor understanding among the general public of the scale and long-term needs for coastal management, and of its complex of different approaches.[6] People's awareness of the truly dynamic character of coastal systems is limited. Further, increased familiarity with technology and science, together with perceived inaccuracies of forecasting environmental changes, can breed scepticism (such as the 'let-down' of people's expectations of the effects of Storm Lorenzo on Ireland, 2–3 October 2019). The experience of those working as coastal management practitioners is that many people's contact with the coast is through beach recreation and leisure pursuits (see Chapter 27: Coastal and Marine Tourism: Opportunities in Blue Growth). Unless people live in a seafront town, or travel to work by a coastal road or railway, as for example in Belfast, Dublin and east coast Ireland, then what happens at the coast doesn't really register in their daily life and can seem of minor importance.[7] This view may arise from the fact that the coastline, although being displaced and modified by erosion, does not ever disappear! So, when someone returns to the beaches and dunes of their last summer's holiday, it may not be obvious that these same dunes could have moved a metre or more landwards since the previous visit. Even with Ireland's relatively moderate overall rate of erosion, about 200ha of land (*c*.2km^2) are being lost along its coasts every year. Current projections of the changes to be expected under climate warming and associated sea-level rise show that rates of erosion of land and loss of wetlands will have increased significantly by as soon as 2030.[8] At some point in the future, therefore, there will need to be radical change in the approaches to managing Ireland's coastal environments.

One obvious stimulus to change will be dictated by the future frequency and magnitude of storms and associated marine flooding that attack the coastline.[9] For instance, Storm Ophelia (October 2017) generated huge waves off the south coast. A wave of 26.1m

was measured at the Kinsale energy platform, but coastal damage was limited because the peak of the storm did not coincide with the high tide. Storms of this magnitude (and also, in the future, probably hurricanes) will occur more frequently in Ireland, implying that a far more proactive approach to coastal management and protection will be required. The costs of undertaking protection schemes, however, could be prohibitive. This factor will require politicians and decision-makers to identify those areas that will receive priority for intervention, and those where nature will have to be allowed to take its course (Figure 30.3).[10]

COASTAL EROSION

Coastal erosion generally refers to the recession of the line of the coast, as defined by a sea-level position, such as the mean high water mark (MHWM), or mean sea level (MSL), or the seaward extent of land-based vegetation. The term 'erosion' indicates a displacement of rock and sediment from one location to another, a movement that is important as a critical part of the overall behaviour of coastal systems. Soft coastlines, as dynamic environments, try under natural conditions to stay in a state of equilibrium with the waves and tides that impact on them. When storm events occur there is a temporary excess of wave energy arriving at the shore, which results in sediment being pulled from, for example, beaches, dunes and cliffs into the nearshore area. Usually, some or all of this material will be returned to these environments in subsequent calmer conditions, and this cycle of behaviour is normal for many beaches. This, often seasonal, adjustment of the coastline through both erosion and accretion, in response to the amount of wave energy arriving at the shore, is a natural process. It is often disrupted by human interaction, however, through infrastructural developments (for example, the building of sea walls, harbour structures), as well as increasingly by the impacts of climate change. As a consequence, the natural cyclicity of coastal advance and retreat becomes a linear, one-directional process, which is frequently dominated by erosion. Where this impacts particularly on human livelihoods it will generate strong demand from affected coastal dwellers for the implementation of coastal protection measures.

Large storms, such as those of the winter of 2013–2014 (for example, Storm Darwin, 2014), or storms Ophelia (2017) and Lorenzo (2019), highlight the fragility of Ireland's coastline and its vulnerability to significant erosion, flooding and damage. They bring large waves and enhanced water levels to the coast and their impact is exacerbated when they occur at the top of a spring tide. An analysis of data from Irish tide gauges by the Marine Institute, from November 2013 to March 2014, showed that during storm events, water levels were frequently more than 1m higher than the predicted astronomical tide. On one occasion during this period, the Galway Bay tide gauge showed the actual water levels to be 1.6m higher than had been predicted.[11] As tide gauges are usually located in sheltered locations, the full extent of water-level rise during storm events may not be detected (Figure 30.4). These measurements in Ireland do not account for the full extent of storm surge as wave set-up (higher water levels due to new waves breaking on the shoreline), which can be significant on exposed coastlines. The direct impact of these

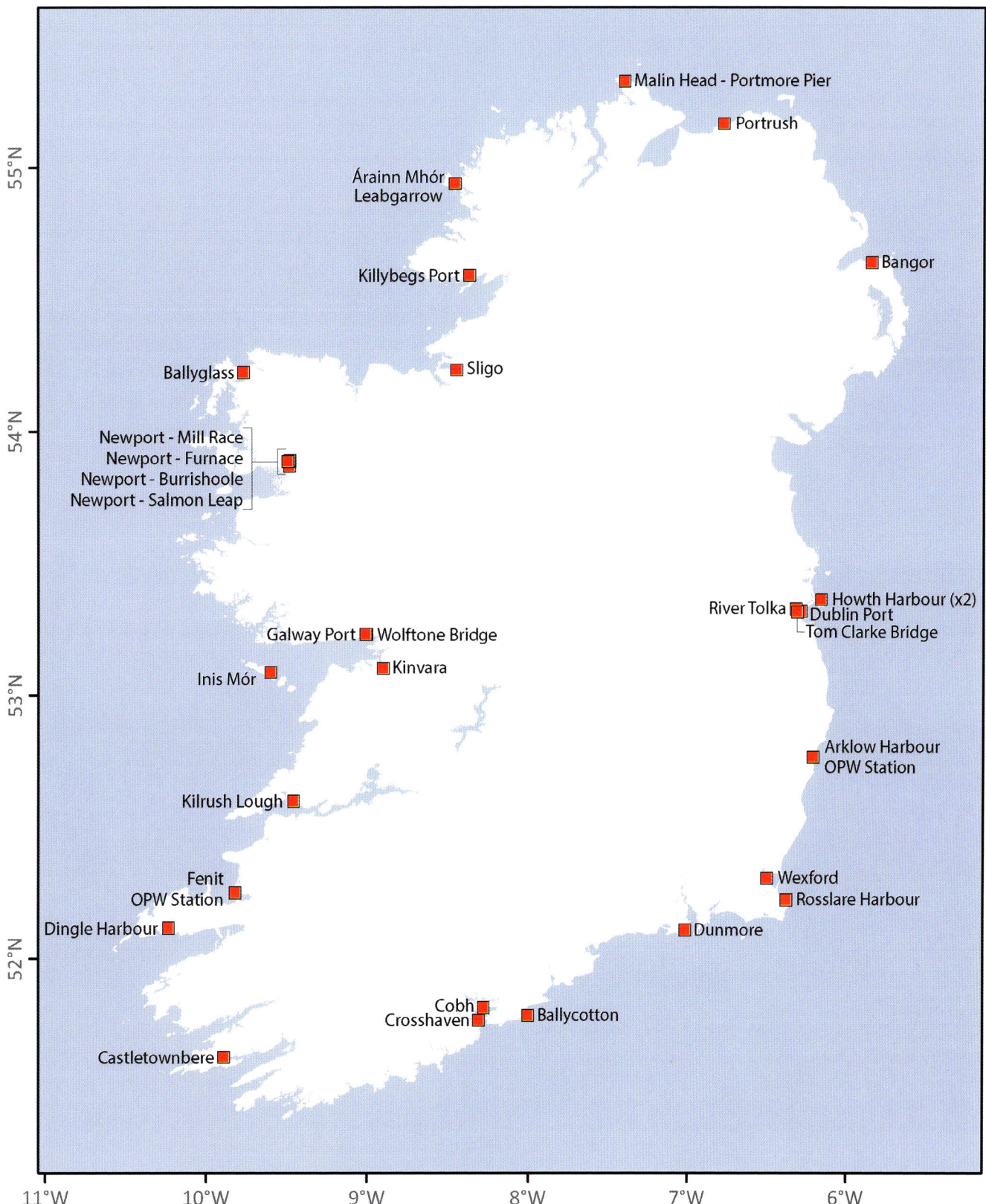

Fig. 30.4 LOCATION OF PRIMARY TIDE GAUGE MONITORING STATIONS AROUND THE COAST OF IRELAND. Different types and qualities of tide gauges, and resulting data output, now exist for coastal areas around the island. Many of these are used for short-term tidal monitoring only, that is tide observations that cover the daily changes over periods of weeks to months. These data are produced generally for specific port and harbour surveys, research projects, or for industrial purposes (for example, shipping activities, marine survey and coastal dredging), or for building schemes. For purposes of validation, these short-duration data need to be calibrated and linked to the longer-term network of higher-quality tidal monitoring stations (that is, annual records, which can run over many decades, as in the case of Newlyn, UK, and in some places for even centuries, such as in Amsterdam and Stockholm). This latter work is undertaken by groups such as port and harbour authorities, city and local governments (such as Cork, Galway, Belfast and Dublin) and the Marine Institute. The map doesn't show all the stations operating, but only those that are on the data network operated by the Marine Institute. The information from marine tide gauges are reported to the Lowest Astronomical Tide (LAT), which forms the lowest marine water level that can occur due to the astronomical tide. Tidal predictions for hourly to daily timescales are produced separately by a number of national and international oceanographic organisations, such as the United Kingdom Hydrographic Office, Taunton. This provides accurate tidal prediction for all standard ports in European waters and globally. Other predictions of tides are also readily available from many other web sources, though some of these can be of dubious quality and should not be seen as authoritative. [Data source: Marine Institute; British Oceanographic Data Centre]

Fig. 30.5 (left) EXTENSIVE COASTAL EROSION DAMAGE CAUSED BY STORM DARWIN (12 FEBRUARY 2014) TO THE CAR PARKING AREAS, ADJOINING ROADWAY AND RECREATIONAL FACILITIES AT DOOLIN, COUNTY CLARE. This was one of the largest storms to have affected Ireland since the 1960s, including the impacts of Hurricane Debbie and even possibly the now-legendary 'Big Wind' of 1839. Storm Darwin created waves of 25m in height at the Kinsale gas platform offshore, with sustained winds of 103–117km/h at coastal sites in western Ireland and a maximum wind-speed gust of 160km/h recorded at Shannon Airport. It was one of nine major storms to have affected coastal and other areas of Ireland during 2013–2014, which together caused hundreds of millions of Euros worth of damage nationally and at least five fatalities. Coastal erosion, accompanied by the destruction of coastal properties and infrastructure, occurs regularly with the autumn and winter storms on Irish coasts. Both meteorological records and projections from numerical climate models indicate that storm sizes, and consequently coastal damage and other impacts, are going to increase in the twenty-first century due to climate warming. [Source: Clare County Council]

higher water levels is flooding, mainly through greater wave overtopping of coastal structures, as well as by enabling larger waves to reach higher into the backshore dunes and cliff areas, and so cause increased coastal erosion.

Any wider-scale increase in sea levels (for example, through ocean water volumes, tides and wave-storm changes) also raises the point on the coast at which wave energy and water flooding potential can attack the shore, which may lead to a landward movement of the line of the coast. These changes in sea level can occur over a wide range of spatial and temporal scales, from those typical of a single storm event to the much longer-term elevation of local and global sea levels due to climate and other environmental changes (see Chapter 7: Ancient Shorelines and Sea-level Changes) (Figure 30.5). An SLR of 0.5m is generally specified as a design

Fig. 30.6 LOSS OF AGRICULTURAL LANDS, WITH ACCOMPANYING DAMAGE TO BUILDINGS, ROADS AND OTHER INFRASTRUCTURE DUE TO SEASONAL FLOODING OF THE SHANNON RIVER SYSTEM NEAR CARRICK-ON-SHANNON, COUNTY LEITRIM. This is now a common sight each year in low-lying coastal and adjoining inland districts across Ireland, particularly in western regions. Though many such areas were drained, with extensive land reclamation and improvement taking place from the period of c.1840–1920, this drainage work was not maintained adequately in the twentieth century. The problems of dealing with this flooding into the twenty-first century are being addressed within the Republic of Ireland under the national Catchment Flood Risk Assessment and Management Study (CFRAMS), and directed by the Office of Public Works. This flood management work is supported since 2018 by a ten-year programme of €1 billion investment in flood relief measures, as part of the Project Ireland 2040 National Development Plan. A similar approach is being taken in Northern Ireland through the Department of Infrastructure. The direct causes of flooding within the Shannon River catchment are the result of ground waterlogging coupled with high rainfall and water runoff to the rivers, particularly during the winter months. Less directly, Carrick-on-Shannon, in spite of being situated over 160km away from the direct influence of marine storms and tidal action, as well as being protected by weirs, barrages and locks, is still impacted by marine-induced increases in water levels. Long-term sea-level rise, coupled with storm surges, results in the progressive backing-up of river water discharges and a worsening of flooding, particularly in large river systems such as the Shannon and the River Bann, or fast-flowing and rain-responsive rivers such as the Lee. In the future it is possible that under these conditions and with increased rainfall, as projected with climate warming, much of the interior of the island will revert to a landscape of inland lakes. Some future reconstruction scenarios for Ireland under climate warming, carried out by the reinsurance industry, show that with a 1m sea-level rise, the Shannon catchment area could become an inland sea. For the actual outcome, much will depend on the engineered management and planning approaches taken to tackle the flood problem. [Source: Office of Public Works]

requirement in many Irish coastal protection projects and several have adopted the wider, international best practice of 1m or more. Irrespective of the timeframes involved, where the section of coast in question contains human settlements or resources, including in some instances agricultural land, this may lead to demands for civil engineering works to be undertaken to help remedy and minimise the adverse impacts of these adjustments, as in the Shannon River catchment and many other coastal areas of western Ireland (Figure 30.6).[12]

The impacts that SLR due to climate change might have on future coastal erosion are not easy to estimate. Even though values for averaged global mean SLR of the order of 3–4mm/year are often quoted (IPCC, 2019), the actual change at any particular location will depend on local factors (Figure 30.7). Longer-term vertical movements of the Earth's crust, for example, mean that some coastal lands are slowly sinking, whilst others are uplifting. These movements can add further variability to the impacts of shorter-term sea-level changes. For instance, water level measurements in Dublin Bay indicate an average sea-level rise since 2000 of 6–7mm/year, which is around twice the global mean value.[13] Climate change may also be altering the frequency, duration, location and intensity of storms on our coastlines. Most storms track towards Ireland from the Atlantic and any overall change in their paths will have different degrees of impact in coastal areas. If storms track further northwards across the island on a more frequent basis, then southern counties will benefit whilst northern counties could be harder hit. Also, if more storms are occurring overall, then the return period for particular types of events will reduce in number. As an example, the winter storms of 2013–2014 represented a cluster of severe events. In an analysis of these events, undertaken by the author for the Department of the Marine, it was found that the wave condition that would normally be expected only once per year was exceeded seven times in these storms, in a period of just greater than one month.[14]

MANAGEMENT AND TECHNIQUES

The late Professor Bill Carter (1946–1993) (University of Ulster at Coleraine) once remarked that humanity 'has an uneasy relationship with the coast. Throughout history, we have tried to ignore it, adjust to it, tame it or control it, more often than not, unsuccessfully.' In ideal circumstances and following 'best practice' as currently viewed, the coastline should be allowed to evolve naturally and humans would adapt to this change. However, in the modern world this

principle is rarely still seen as a feasible option, though it has been adopted increasingly in some other European countries and globally. More commonly, intervention is seen as necessary, to ensure public safety and to provide protection to important infrastructure and valuable coastal lands. Given sufficient resources, it is possible to 'engineer' a solution for any coastal erosion scenario that arises, but this is not always a practical option. Since the 1970s two main intervention approaches have been defined. These are characterised by the more traditional, 'hard' engineering-type measures, which are necessary in many situations but are in contrast to the innovation of a wide range of lower-cost and more flexible 'soft' measures (Figure 30.8).[15] Many of the lower-cost approaches have been designed to fit the constantly changing state of low-elevation, sand-dominated coasts, and also for rapid deployment in response to storms. They often supplement, or augment, an engineering solution when public safety and critical infrastructure is at risk.[16]

In many countries there is an overall coastal erosion management strategy, which dictates on a national, regional or local level the approach to be adopted should erosion occur. Such

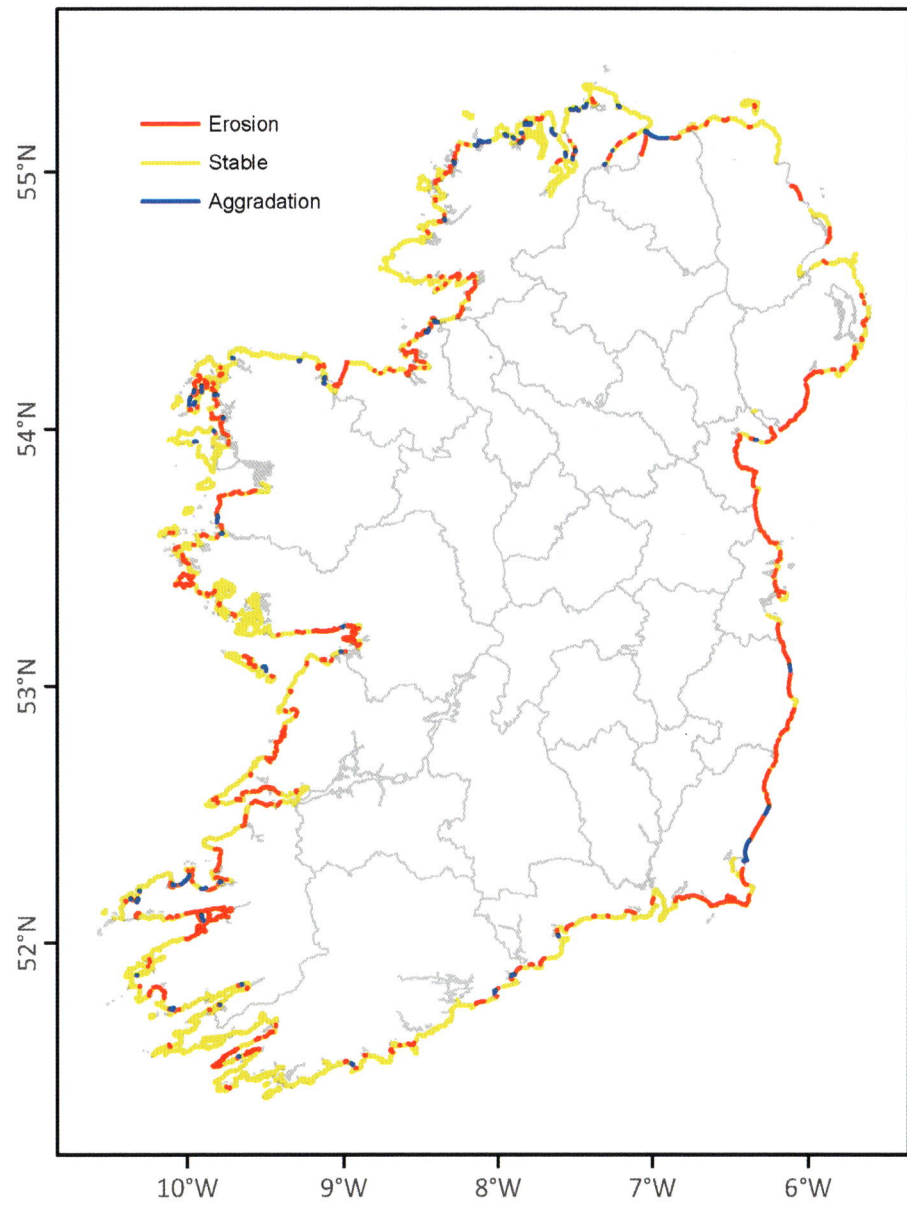

Fig. 30.7 COAST EROSION TRENDS IN IRELAND. [Source: Eurosion Project, 2004]

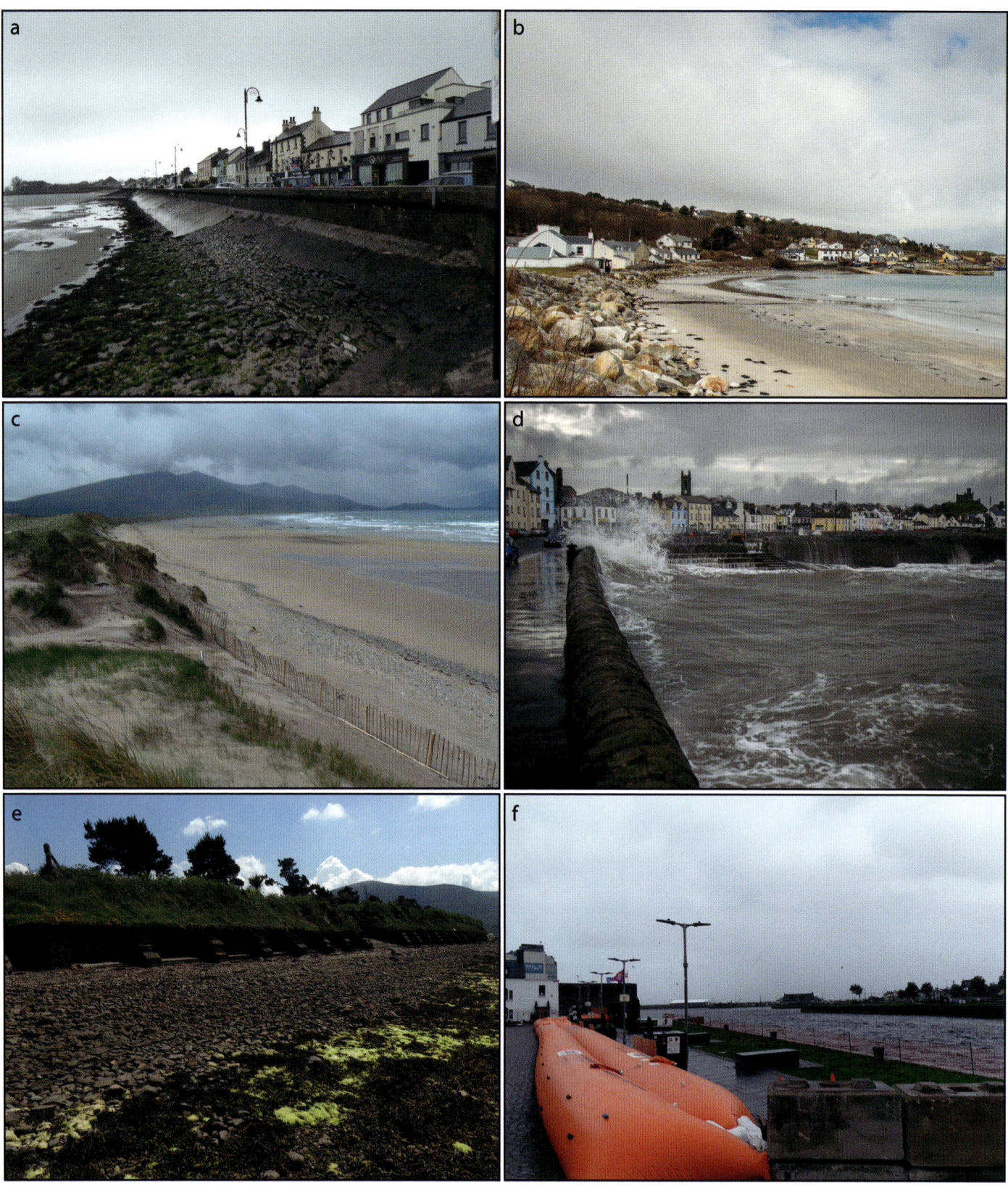

strategies are normally based on detailed knowledge of coastal processes and environmental risks. In Ireland, the Irish Coastal Protection Strategy Study (ICPSS) has been carried out in the Republic of Ireland. This has produced erosion risk maps based on past behaviour of the coastline (Figure 30.7). While this study has not yet led to a national erosion management strategy, it is now being used by local authorities to assess sensitive sections of coastline. Similar work has been undertaken in Northern Ireland,

through the Department of Infrastructure and the Department of Agriculture, Environment and Rural Affairs (DAERA), as part of the UK's strategic response to climate risks.[17]

As a general principle, in the context of coastal management practice, there are now four main strategies, or variants of these, to be chosen from when responding to coastal erosion and the loss of coastal lands:[18]

Fig. 30.8 (opposite, left) APPROACHES TAKEN IN COASTAL DEFENCE. a. Sea wall at Blackrock, County Louth, an example of hard engineering widely used internationally in coastal defence and a common approach taken in Ireland. Structures such as this are frequently used to protect the land behind from erosion. This is a classic technique used in the hold-the-line approach to coastal management. When storm waves hit a vertical sea wall structure, they are mostly stopped and their energy instantly released. Some are reflected back to sea again, rather than their energy being dissipated. When an outgoing wave meets incoming ones, the energy is amplified at their point of encounter, with the surplus energy still requiring work to be done. One common result of these processes of wave action created by the sea wall is that the sediment on the beach in front of the wall is scoured away, as seen here. The level of the beach is, therefore, lowered and the wall itself may be undermined, leading to its eventual collapse. The 'curve' design in the sea wall here is intended to help produce the optimal conditions for wave reflectance and energy dissipation. [Source: Darius Bartlett]; b. Use of rip-rap as a lower-cost approach in fixing a coastal position at An Leadhb Gharbh on Arranmore Island, County Donegal. Rip-rap (that is, an engineered arrangement of loose stone blocks) is a widely used technique, but a measure that needs constant repair and maintenance, as repeated wave and storm action destabilise the structure. This can be seen beginning to happen on the beach at An Leadhb Gharbh at Arranmore. [Source: Donegal County Council]; c. Wooden fencing in the sand dunes on the Maharees, Castlegregory, County Kerry, used as part of a series of low cost and 'soft' engineered techniques in efforts at dune system restoration. This technique for stabilising and encouraging dune regrowth, together with others – such as the use of boardwalks, dune grass planting (commonly of marram and lyme grasses on North Atlantic coasts) and of other types of sediment trapping – has long been used in Ireland. There is a problem with this approach, which can trap sands preferentially in these dune areas and cause sand-starvation elsewhere. Consequently, the use of such techniques needs careful management and monitoring of the along-coast effects. [Source: Eugene Farrell]; d. Donaghadee, County Down, with the sea walls under 'attack' by storm waves, a regular occurrence. Many such sea walls have been built around Ireland's coasts. Most had their origins in nineteenth-century constructions, such as here, or at Kilkee and 'thousands' of similar locations around the island. Such walls often have to be repaired after major storms using modern engineering methods, or reinstated completely. [Source: Ross McDonald]; e. Revetment wall at Glenbeigh beach, the 'Stooks', Caragh River area, County Kerry. The revetment is built of concrete blocks and buttress reinforcements. The buttresses used here add to the value of the revetment structure in helping retain and stabilise the slope behind the wall, as well as reducing the impact of wave erosion on the soft sediments cliffs. [Source: Sarah Kandrot]; f. Inflatable Flood Barrier, Galway city: a flexible Quick Defence Measure and Quick Accommodation Measure (QDM and QAM). This is one of a wide and flexible range of engineered techniques, as part of a new strategy internationally of rapid and relatively low-cost coastal protection. It is being used particularly in response to the increasing storm threat to the vulnerable coasts of urban areas, but has a much wider application. [Source: Ann Tracey, courtesy of Geoline Ltd]

- Hold the line involves maintaining the position of the shoreline, primarily through engineering intervention; an approach sometimes known as 'armouring the coast'. While it can be effective, a traditional type hold-the-line policy is expensive and is normally only applied when valuable infrastructure is at risk, for example, roads, rail, urban areas. In an Irish context, it is mostly relevant to the defence of coastal towns, such as Kilkee, Lehinch, Youghal or Waterville. The argument in such cases is that it is inconceivable that these towns will be allowed to be taken by the sea. This strategy is often linked in coastal settings needing flood protection measures as well to the use of similar hard-engineering construction techniques (Figure 30.9).

- Retreat/managed realignment is a strategy that allows

Fig. 30.9 (right) HOLD-THE-LINE! a. The use of fixed coastal position engineering techniques (for example, sea walls and revetments) in a hold-the-line approach at Lehinch, County Clare, showing the sea defences under attack in the storms of 2013/2014. [Source: Patrick Flynn]; b. The Thames Barrage, London – showing the use of barriers and barrages in flooding protection schemes – is related in approach to that of hold-the-line coastal protection. The Thames Barrage is an example of a large-scale and high-cost engineering scheme. The barrage has moveable gates to allow the passage of ships and forms a sophisticated flood prevention measure. It is part of an integrated system of embankments and sea walls that together provide protection to the centre of London, which is becoming more vulnerable to marine storms and the potential for flooding. The cost of the barrage when completed in 1982 was around £1 billion. It provides a defence against the long-established, and costly, problems of marine flooding for the infrastructures and high financial value areas of central London. In particular, it gives protection to the important business districts and urban living areas, with their iconic but vulnerable buildings. The barrier was designed to cope with the 1:100 year storm event but, in spite of its high cost and detailed engineering, was 'out of date' when it was commissioned, due to the rates of change in regional storm frequency and magnitude, as well as in sea-level rise. These factors of sea and coastal change present a major challenge in the design and application of flood barrages and of other large-scale defence projects, such as the Rhine-Meuse region, Venice or Mississippi barrage schemes. Such schemes, whilst having limitations including their high cost, may for some coasts be required to provide adequate protection for the people and the livelihoods of those who live there. In Ireland, the situations of Cork and Dublin could well require such an engineered measure in the future. In Belfast, the Lagan Weir (commissioned around 2000) already provides some upgraded barrier-type protection for future marine flooding threats to the city. [Source: UK Environment Agency; contains public sector information licensed under the Open Government Licence v3.0]

coastal recession to happen, but on a managed basis. Policy and planning decisions in this approach are based on taking the pragmatic view that it may, in the long term, be the only financially justifiable option. Politically, this can be a hard strategy to justify, since it often implies the need for some property, jobs and land to be sacrificed in the affected area (Figure 30.10).

A coast on which 'managed realignment' could potentially be applied is that of Rossbehy, County Kerry. The dune system of this beach barrier was breached by the sea in a storm event in 2008. The breach has since widened, resulting in a reduction

Fig. 30.10 THE EAST-COAST MAINLINE RAILWAY TO ROSSLARE EUROPORT, UNDER VISIBLE THREAT FROM THE SEA AND COAST EROSION AT ROSSLARE STRAND, COUNTY WEXFORD. This stretch of Ireland's east coast is continually eroding, with the buildings and infrastructures, such as rail and road, situated close to the coastline and needing continually to be evaluated as to their safety and future use. The main railway connection to Dublin provides an important transport link out of the busy ferry port of Rosslare. In many sections this railway runs along the coastline, which since the 1980s has been retreating at rates of *c*.0.5m–1m/year. In the Greystones area, County Wicklow, the railway has already had to be moved further back from the coast and realigned. Along the east coast – as elsewhere in Ireland, such as around Castlemaine Harbour, County Kerry; Belfast Lough, or Salthill, County Galway – coastal communities find that they are having to grapple with these impacts of 'coastal squeeze' as coastal lands increasingly flood or disappear. This problem of 'squeeze' is set to worsen under the impacts of future changes in coastal climates. The coastal railway example at Rosslare Strand illustrates well a significant issue for coastal engineering: what is the appropriate management and planning response to the rapid coastal changes now taking place in such low-height and soft sedimentary coastal environments? Is it one of planned retreat and shoreline use realignment, or the more traditional hold-the-line approach, or a combination of these? [Source: Iarnród Éireann Irish Rail]

of the whole barrier length by *c*.1.5km and the redistribution of large volumes of sand within the adjacent areas (Figure 30.2). This has exposed the land and people's properties on the lee side (Castlemaine Harbour side) of the dune barrier to increased flooding and cliff erosion. A further partial breaching occurred approximately midway along the barrier in 2014, at a different location to that of 2008. This has added to concerns that this coastal area is continuing to destabilise, and that it will soon reconfigure itself into a different type of coastal environment. Restoring or protecting Rossbehy beach is not a viable option, but lower-cost measures such as improvements to the dykes at the locations most at risk may be feasible.[19]

• Advance the line is a strategy that pushes the coastal position seaward and is usually undertaken on a local scale. Beach nourishment could be considered an example of this approach, although some people prefer to see it as a hold-the-line technique instead since it provides only a temporary progradation of the coast. Land reclamation is also an advance-the-line technique, though one normally employed for reasons other than protecting against erosion. Many examples can be seen around the coast of Ireland of land that has been reclaimed from the sea, as in Castlemaine Harbour, Wexford Harbour and in Lough Swilly, or in more urban contexts, such as those of Dublin Bay and Cork Harbour. Beach nourishment is another very common advance-the-line technique, used worldwide, and involves the pumping of dredged sand onto a beach to compensate for a deficit of sediment and to create a buffer against erosion (Figure 30.11). Over time this sand may be moved away by nearshore processes and the nourishment needs to be repeated. This recurrent cost is one of the reasons there are only a few significant applications of beach replenishment in Ireland (Rosslare and Bray). Many beaches, such as Lehinch, have sediment deficits and would benefit from periodic nourishment in front of the revetments.

• Do nothing allows only minimal intervention and perhaps best defines Ireland's past default approach to the problem of coastal erosion. However, in coastal management terms, implementing a 'do nothing' approach, and allowing the coastline to evolve naturally, is very much 'doing something'! It requires the use of strong planning restrictions, involving, amongst other measures, imposing minimum development setback distances from the shoreline. In these setback zones the construction of buildings and other high-value or vulnerable land uses are prohibited or severely limited. For instance, it can be specified that no new development is allowed on the seaward side of the road closest to the coastline, or a fixed distance can be provided (for example, no building within 100m of the shore). If and where this approach is adopted, it is important that it is well communicated to the public and discussions are held with affected property owners and other stakeholders (Figure 30.12). This approach, together with that of managed realignment, forms part of a much wider suite of adaptation responses to environmental changes, such as the impacts of sea-level rise due to climate warming. The focus in this type of management is more on people, on building their social and economic cohesion, resilience and capacity to change with the environment.[20]

The first three strategies are usually implemented by means of engineering intervention, through the application of various coastal protection techniques. It is important to emphasise that any intervention on the coastline should be undertaken based on a detailed knowledge of the local coastal processes involved, since there is a danger that otherwise the measures put in place could exacerbate, rather than remedy, the problem. High-cost measures,

involving the use of built structures, require engineering design to be undertaken and are used in the absence of any alternative, low-cost solutions. For instance, seafront towns are more likely to need built revetments or beach nourishment for their survival, whereas natural beach systems that are suffering local-scale erosion can often be rehabilitated by means of using lower-cost, non-built measures. Traditionally, the main approach and response to coastal erosion in Ireland has been the construction of sea walls, rock revetments and similar structures. Historically, however, because of ad hoc methods of design and construction, as well as an insufficient understanding of the coastal environment at the affected site, these have often been relatively short-lived structures. In the nineteenth century, particularly during and in the aftermath of the Great Famine, a large number of small piers and some coastal protection sea walls were built. Examples of these can be found at Donaghadee, Bangor, County Down, and Kilkee, County Clare (Figure 30.8). The wall at Kilkee is one of the longest-surviving structures in Ireland, even though part of it was destroyed in the storms of the winter of 2013/2014. Elsewhere in Ireland, sea walls can be found at the back of many beaches, particularly those near towns, or on beaches that have a high recreational value, such as those around Dublin Bay, at Salthill, County Galway, or Youghal, County Cork (see Chapter 19: Changing Coastal Landscapes). The use of old railway sleepers to build these vertical wall structures was also a common practice in Ireland in the 1950s and 1960s. Sea walls can be relatively effective

Fig. 30.11 THE 'SAND MOTOR' OR THE 'SAND ENGINE' NEAR KIJKDUIN, THE HAGUE, THE NETHERLANDS, AN EXAMPLE OF LARGE-SCALE COASTAL ENGINEER-ING. This image illustrates the pumping of dredged sands and gravels from offshore sediments sources into inshore areas close to Kijkduin, which is The Hague's second most popular beach resort. The work is part of a project involving the creation of new coastal lands and seabed reclamation, the Medvlaakt shoreline progradation scheme in the province of South Holland, which began in 2009. This is an advance-the-line and a linked beach-nourishment type large-scale engineered approach. The concept of the 'sand engine' was developed by Marcel Stive (professor of coastal engineering at the Delft University of Technology) and is now being used in coastal development and protection in other countries, such as the United States and China. The Kijkduin, Hague project cost about \$81 million and used approximately 21.4 million m^3 of sand dredged from the North Sea. In the next few decades the action of waves and tides will move these sands landwards to create new protective beach and dune-barriers stretching many kilometres along the coast. [Source: Rijkswaterstaat/Province of South Holland]

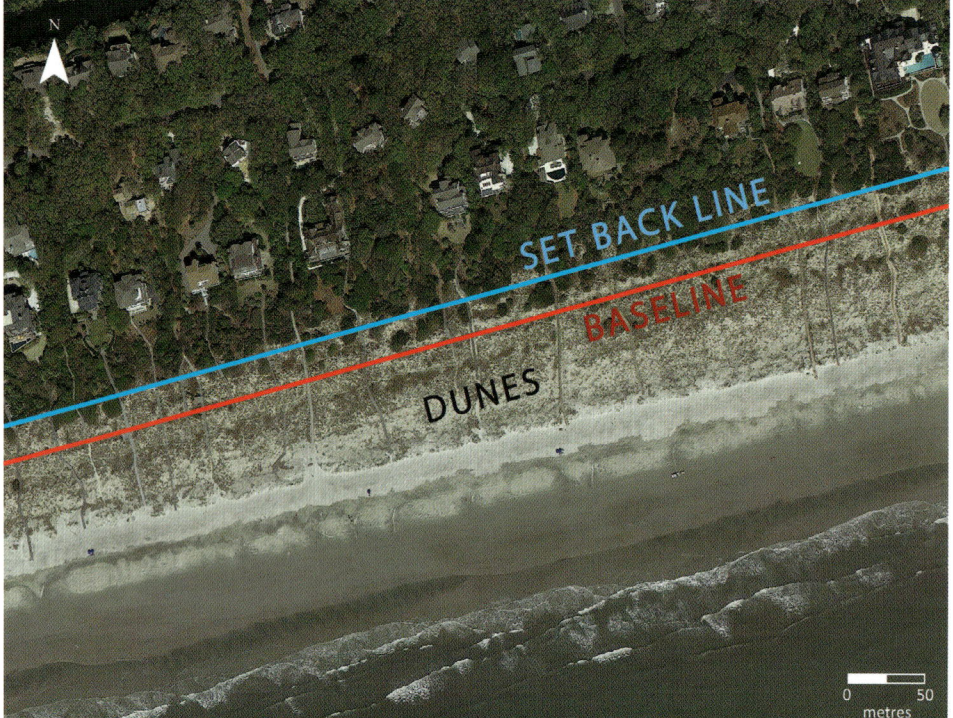

Fig. 30.12 (left) COASTAL SET-BACK LINES USED IN 'DO NOTHING' TYPE COASTAL HAZARDS MANAGEMENT AND PLANNING, KIAWAH ISLAND, SOUTH CAROLINA, USA. The idea of the measured distance 'set-back' line in coastal management was first developed in Florida in the 1960s. The approach formed a planned response to the increasing problems created by population pressures on coasts linked to coastal hazards and vulnerability from, for example, erosion, storms and flooding. As at Kiawah Island, set-back lines appear on planning maps and photographs as a simple line, which forms a measured linear distance from the active coastal position of a high water mark, or dune-crest (a baseline). The calculation of the distance in reality can be more complex and is based on the use of sophisticated coastal zone hazard mapping and numerical modelling techniques. Since its inception the concept has been applied commonly to vulnerable coasts worldwide and, increasingly, as part of low-cost shoreline realignment and retreat measures. The set-back line measure is used particularly in rural areas that would be too difficult and costly to defend with traditional built techniques. For whatever reason, the technique has not been used in Ireland, except perhaps less formally as part of local authorities' building controls . [Source: Google, Digital Globe, 2019]

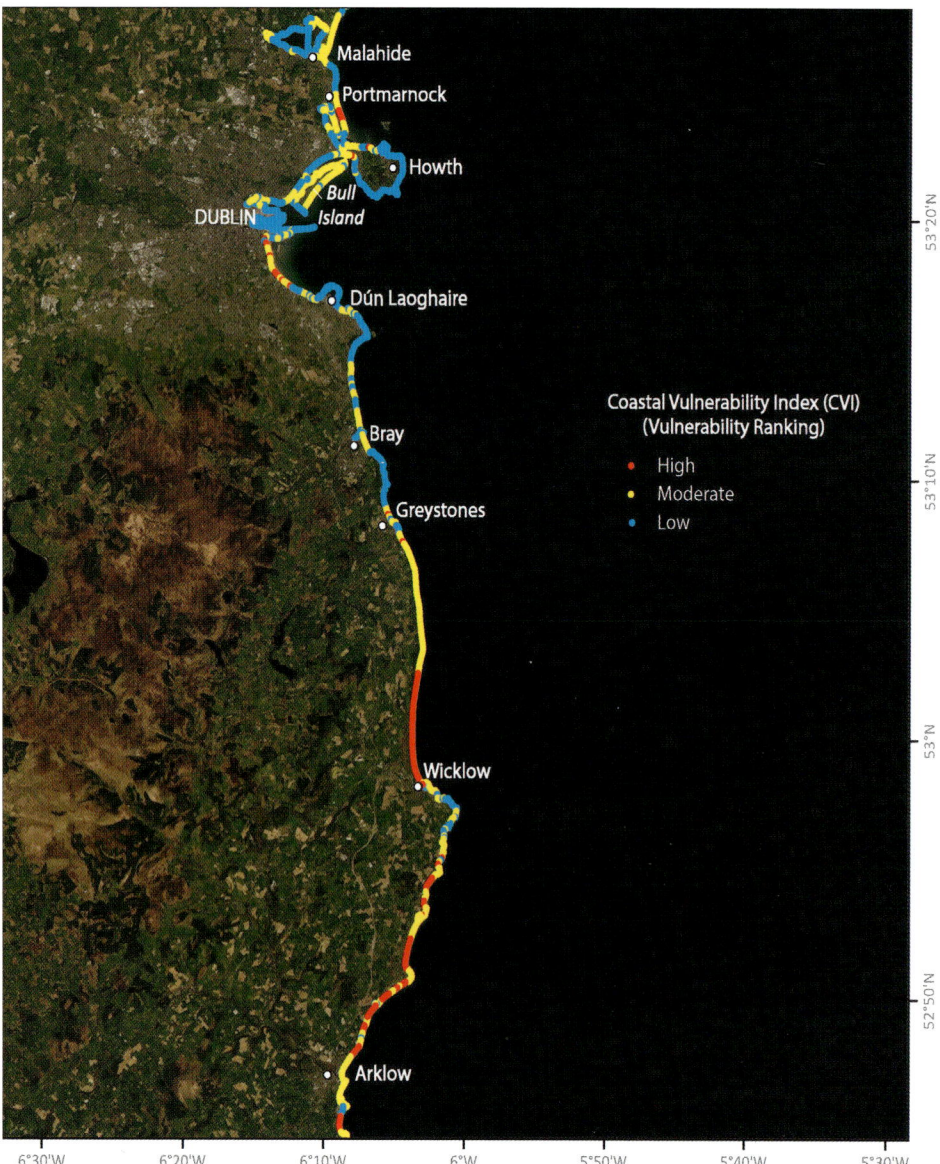

Coastal Vulnerability Index (CVI)
(Vulnerability Ranking)

- High
- Moderate
- Low

in reducing erosion, but in many cases they become undermined over time, due to falling beach levels, and fail.

Revetments are also constructed widely on coasts, to help retain and stabilise slopes, and provide protection from wave erosion. They may be built in the form of a shaped embankment (as a covering over an earth core), and are often sloped but can also be vertical structures. Constructed generally of rocks, concrete armour blocks or gabions (stone-filled wire cages), revetments of different types are found at the foot of dunes and cliffs on many Irish beaches. Specifically, it has been found that the angle of slope of the revetment is particularly important; if the face presented to the sea is too steep, then storm waves will reflect off the structure and the rebounding energy can lead to loss of sand from the beach. Because of this, it

Fig. 30.13 (left) VULNERABILITY RISK MAP OF CENTRAL EAST-COAST IRELAND. Coastal areas vulnerable to sea-level rise, and other physical environmental risk factors, identified in County Dublin and County Wicklow from Coastal Vulnerability Index (CVI) analysis, showing their vulnerability from high (red) to low (blue). Many examples of similar CVI-type mapping exist for Ireland, where this now widely used and low-cost technique in CZM was innovated at an early stage in its development internationally. The CVI approach is based upon the use of numerical methods in developing algorithms to establish the degrees of risk found along and within coastlines. [Source: Silvia da Caloca-Casao, unpublished PhD Thesis, National University of Ireland, Maynooth, 2018]

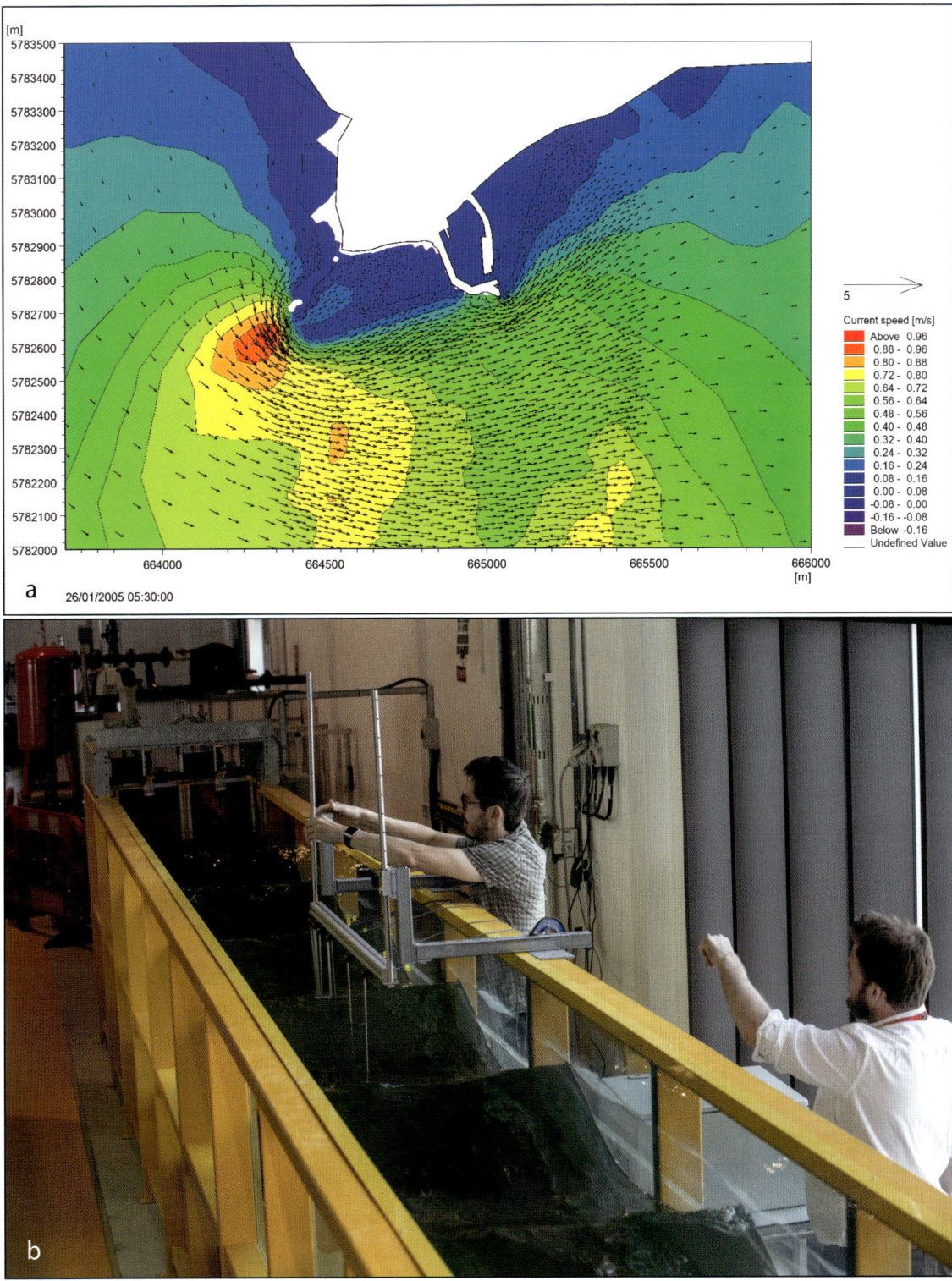

Fig. 30.14 NUMERICAL MODELLING. a. Coastal current projections from Kilmore Quay, County Wexford. Water flows showing sea current velocities and directions, based upon numerical modelling output data of water current velocities and direction: part of work carried out in 2008 for Kilmore Quay harbour redevelopment project, the modelling used the MIKE21 HDFM software. The projection forms part of a time series showing the modelled changes in the data over a period of time (generally for hours to days to weeks). These outputs are based on the input of known information from the area of the factors controlling water movements, such as seabed bathymetry and water depths, wave climate conditions, tides and mean water velocities. The outputs of the model (based on mathematical algorithms that link these controlling factors), or of any numerical modelling, are validated (checked) against observations from the area of the past conditions. The validation process is a way of checking the accuracy and the degree of uncertainty (or probability) of the projections. Such models form an invaluable tool in helping understand the functioning of Earth environments, the likely future changes in systems and in designing how built structures will work and fit into the environment. [Source: Jimmy Murphy]; b. Physical modelling of water and wave dynamics in a flume tank at the Lir test facility, MaREI, UCC. The mechanisms of water motions and their interaction with the seabed, and with coastal structures, can be observed and modelled in these types of experimental flume tank settings. Testing of the concepts of water motion and behaviour can also be undertaken. As shown here, such tanks can be many meters in length, with the capability of varying the internal shape, depth and bed surface roughness of the tank (using, for example., natural sediments, or fixed artificial moldable materials, such as fibreglass or cement) to provide analogue of real coastal, seabed or water channel conditions. Waves in the tank are produced by paddle mechanisms, generally fixed at one end. The tanks are instrumented to provide a digital record and data stream of measured change information and visual data (i.e., video film). Experiments, based on water height and velocity movements, are scaled to fit the reality of the world outside the tank. Similar, but bigger, water tanks are used to provide valuable means of testing the behaviour of designed structures, such as sea walls, or even full-scale harbour construction, such as Kilmore Quay (Fig. 30.13. B.). [Source: Lir-NOTF]

is important to design for shallow slopes (1:2.5–1:4 that is 1 vertical to 2.5 horizontal), as this leads to better sand retention.

There are numerous examples around the Irish coast where built attempts to combat coastal erosion have proved to be inappropriate to the site. The failure of sea defences and other unforeseen consequences have arisen, for example, where protection measures that have worked at one location have been transposed to another location, without adequate recognition that any solution must be tailored to site-specific conditions. There are many variables that influence nearshore sediment processes, and hence the causes of and vulnerability to erosion; these include wave exposure, nearshore bathymetry, tidal currents and the levels on backshore areas reached by tides, wind conditions, sediment properties and sediment supply (Figure 30.13) (see Chapters 2: The Coastal Environment: Physical systems, processes and patterns and 8: Monitoring and Visualising the Coast). Understanding these and the cumulative effects of their interaction, is very challenging. Numerical models that combine simulations of waves, tidal currents and sediment transport help bridge the knowledge gap, and are often used to design coastal structures (Figure 30.14).

It is also the case that just because a particular approach and built structure to combat erosion has been attempted in the past, does not mean that it should necessarily continue to be repaired if or when it is damaged. Nature has a way of telling us if something fits or not, so we should not be tied into repeating past mistakes! The main coastal engineering work done in Ireland following the storms of 2013/2014 was repair of damage to existing defences. These were rebuilt, and made stronger than before, with the expectation that they will be more capable of resisting future storms. The national repair costs (in the Republic of Ireland) of the significant coastal damage caused by the storms in 1989/1990, and again in 2013/2014, was over €100 million in each case and there have been many other major damaging storms. While such repairs may be necessary in the short term, they do not deal with the underlying factors that caused failure of coastal infrastructure in the first place. Deferring the provision of a real solution to coastal erosion and change through an integrated planning approach to some point in the future may prove more costly in the long run. Selecting the correct coastal protection approach and technique for a site, therefore, not only requires detailed knowledge of coastal processes, but also depends on having the scope and resources to be allowed to make the right decision.[21]

Case Study: A Recent History of Coastal Engineering in Waterville, County Kerry

Michael O'Shea

Waterville is located at the centre of Ballinskelligs Bay on the tip of the Iveragh Peninsula, County Kerry. The coastline is comprised of cobble-shingle-sand beaches, till (boulder-clay) sea cliffs and coastal dune systems. The village itself lies on a narrow, relatively low-lying isthmus between the freshwater Lough Currane to the east, Ballinskelligs Bay and the Atlantic ocean to the west, the Inny Estuary to the north and Finglas River to the south. The Finglas River links Lough Currane to the sea. It is at this location, where freshwater meets the sea, that Waterville House was established in 1775, from which the village grew (see Chapters 11: Beaches and Barriers and 19: Changing Coastal Landscapes) (Figure 30.15).

A study undertaken by engineers from University College Cork (UCC) on the coastal processes in Ballinskelligs Bay established that the wave climate is dominated by south-westerly and westerly swells, with waves being typically shore-parallel, due to the location of the village in line with the mouth of the bay.[22] The effect of wave refraction from Atlantic swells is at a minimum and the coastline has aligned itself with the incident wave direction. Due to this wave climate, the sediment transport regime in the vicinity of the village is dominated by onshore–offshore movement and alongshore sediment transport is minimal. These processes, together with the inherited glacial and coastal geomorphology of the bay, have a significant bearing on the type of built coastal structures that might be suitable for protecting Waterville's coastline. This knowledge was lacking when assessing the suitability of some of the coastal protection measures undertaken in the past at Waterville.

Considering its orientation and the soft nature of its coastline, Waterville is particularly vulnerable to coastal erosion. The need for protection was recognised early in the village's development, with numerous examples of attempts to defend the coastline against Atlantic storms still evident today. The historical development and protection of the coastline over the past two-and-a-half centuries can be seen on the Ordnance Survey's mapping of the area from the 1840s onward, and is related closely to the nature of economic activity in the County Kerry region. The first evidence of structural coastal development is the construction of a boat and watch house in the late 1700s by the revenue officer for Iveragh, who was also the local landlord. The primary function of these structures was to combat smuggling. A slipway at the southern end of the village was added later in the nineteenth century, along with the establishment of a coast guard station. This was situated in a row of houses, at the edge of a low cliff protected by a masonry sea wall. The remnants of the slipway and a dredged channel can still be seen at low tide today. The coast guard station was relocated to a site adjacent to the Commercial Cable Company in 1903, where

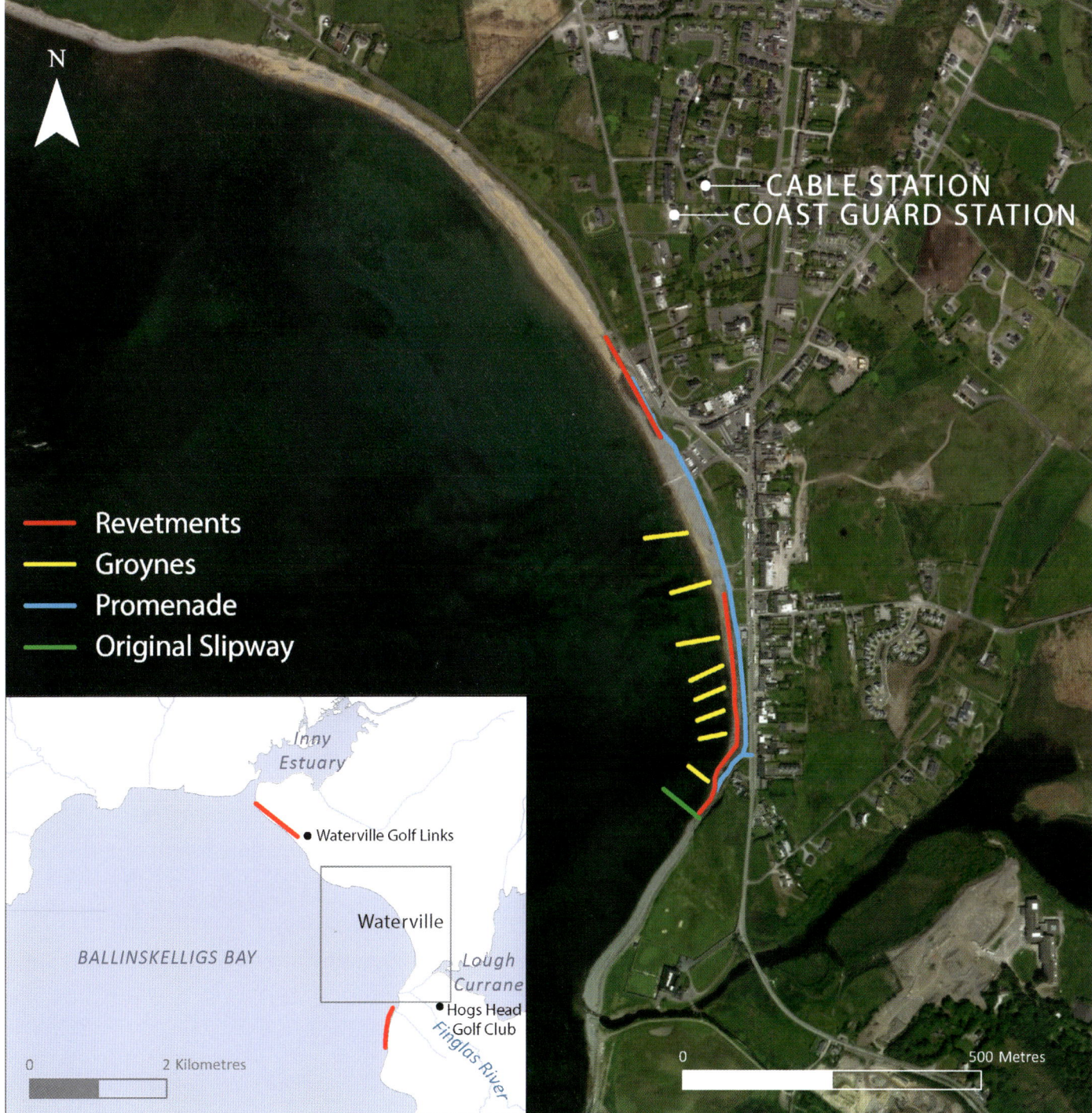

Fig. 30.15 THE WATERVILLE VILLAGE AND URBANISED COAST, BALLINSKELLIGS BAY, COUNTY KERRY, SHOWING THE POSITIONS OF KEY COASTAL PROTECTION INFRASTRUCTURE. [Data source: Michael O'Shea; satellite image: Google, Digital Globe, 2019]

it remains to the present day. A second slipway was constructed in the 1900s and extended in the 2000s but, due to the exposed nature of the bay and large tidal range, its functionality is limited. In hindsight, the original slipway was in the more suitable location, in terms of both the shelter it provided and the water depth.

In 1858 south-west Kerry became the focal point of a global development in telecommunications, with the establishment of the Transatlantic cables. This extraordinary engineering feat linked the American continent telegraphically for the first time with the then British Empire and continental Europe. As a key part of this technical innovation, the Commercial Cable Company developed a cable station in Waterville in 1884 and laid a comprehensive transatlantic cable network that came ashore at the village. The remnants of the landing structure were still evident on the foreshore at Waterville until the late 1990s, when the construction of coastal protection works obliterated them.[23]

The first photographic evidence of the early protection works at Waterville depicts a vertical masonry sea wall, presumably erected to protect the row of houses that included the original coast guard station (Figure 30.16). It is likely that erosion was becoming an issue at this location prior to the turn of the nineteenth century, as indicated by the subsequent relocation of the coast guard station to higher ground. Local anecdotal evidence suggests that a storm in October 1945 damaged the sea wall irreparably and rendered all but three of the houses

Fig. 30.16 THE WATERVILLE COAST, DATED TO THE PERIOD BETWEEN 1865 AND 1914. The first sea wall built in the village area of Waterville, which was built in the Victorian era and mostly destroyed by storm waves in 1945, is visible in the centre of the photograph, with the later promenade structure to the left of it. The soft and highly erodible glacial till cliffs, which were partly protected by the wall, can be seen at the back of the beach on the right-hand side of the photograph. [Source: National Library of Ireland, call no.: L_ROY_04619]

protected by it uninhabitable. An account provided by legendary Gaelic football player and manager, Micko O'Dwyer, whose family lived in one of the affected houses, tells that during this storm 'the shed and back garden disappeared overnight, along with our beagle hound'.[24] The O'Dwyers and five other families were forced to abandon their homes and relocate to safer accommodation. The construction of the sea wall is a classic example of the Victorian era's response to coastal erosion; it was a hard vertical structure, designed with little cognisance of the incident wave climate or the impact of wave reflection and scouring and it is likely that undermining and overtopping by waves ultimately led to the failure of the wall.

The promenade fronting the beach at Waterville is another Victorian engineering example, closely related to the sea wall. The original section was less than 200m long and consisted of a concrete vertical wall and rubble infill, with a masonry wall set back 3m from the crest. It was constructed in the late nineteenth century, possibly at the same time as the sea wall. Over the years, the promenade has been extended in both directions from this original structure and now runs for almost 800m along the shoreline. It is considered one of the village's most valuable amenities, providing an excellent panorama of the bay. The promenade's setback wall has also historically provided protection from overtopping during storms to the lands and road behind it. However, the original section of the promenade was constructed as a low structure, with a very low height or freeboard. In storms since the 2000s it has been overtopped regularly, with running repairs required after every winter season. The setback wall has also been rebuilt twice in the last twenty years, though the failure in this case cannot be completely laid at the feet of Victorian engineers; it is much more likely that increased storminess and a rise in sea levels in the 150 years since construction have contributed to the current dysfunctional nature of the promenade.

In the 1960s Kerry County Council (KCC) constructed eight rock groynes to combat the ongoing erosion problem that was threatening the main N70 Ring of Kerry road (Figure 30.17a). However, as was the case with the Victorian sea wall, this scheme was constructed with little knowledge of the primary factors driving the erosion. The function of a groyne system (that is, seawards projecting rock-built structures) is to capture sediments that are being transported alongshore and to build up the beach level. This further allows the effective wave height at the coast to be lowered and erosion and overtopping rates of shoreline areas are reduced. Because there is minimal alongshore transport in Waterville, and hence no significant supply of incoming sediment, the groynes were ineffective as a mechanism to build beach levels and they may even have caused localised funnelling effects that concentrated wave energy at high tide, thereby increasing the erosion at some points.

The most recent form of coastal defence attempted at Waterville is a series of rock armour revetments that have been constructed since the late 1990s (Figure 30.17b). One scheme protects the local road known as the Cliff Road, and another forms a replacement of the Victorian sea wall. Subsequent to these schemes, during the early 2000s coastal erosion threatened the Waterville Golf Links, north of the village, and the Hogs Head Golf Club (formerly Skelligs Bay) to the south. Two more revetment schemes were constructed to address these problems. Together, these four revetment schemes have had varying degrees of success.

There are several factors to consider when designing a revetment scheme, including incident wave height, extreme water levels, rock size related to wave forcing, beach-level range and ground conditions. A key parameter to consider when designing a revetment is the slope of the rock face; a shallower slope reduces the reflection of waves and encourages the greater dissipation of wave energy, leading to more effective protection. However, a shallow-sloped revetment has a greater land take, requires more material and is hence more expensive to build per unit length than a steeper, narrower-profile one.

Using best practice, a well-designed revetment has generally a design life of 100 years. In Waterville, the various schemes that have been put in place cover a range of rock armour designs. At lower Waterville the revetment is steep, with sections of the armour practically vertical. As the wave runs up the beach and meets the vertical section of rock armour, a large 'slamming' force is exerted on the revetment. This results in the loosening and displacement of the primary armour; the structure is already showing signs of distress. The revetment protecting the Cliff Road is a large structure with a shallow slope and has functioned well with little sign of degradation. The only negative is that due to its bulk it dominates the coastal landscape. The revetments protecting the large clay sea cliffs in Hogs Head Golf Course were constructed on an ad hoc basis

Fig. 30.17. COASTLINE PROTECTION AT WATERVILLE TOWN. a. Rock groynes on the main beach. The rock groynes were placed here first in the 1960s and by the date of the photograph in 2007/2008 much of their original shape had been changed, with the internal structure of the groynes exposed and degraded by wave action. The component rock boulders have been redistributed subsequently in the local beach areas, along the coast and even to the offshore. These structures have now largely gone following the major storm series of 2013/2014. [Source: Stephanie Desmond]; b. Waterville coastal revetment protection. The view is towards Waterville, showing the rock armour revetment works placed in front of the town, as well as adjacent to the promenade and beach front. The high wave energy causing the erosion problems is indicated by the relatively steep beach angle, together with the cobble and other large-sized beach sediments. The overall shape of Ballinskelligs Bay, coupled with the dominant wind-wave direction, mean that the town here will continually be under threat from erosion and storms, with the need for long-term renewal of any coast defence structures. [Source: William Hunt]

in the early 2000s. The revetment serves as a toe protection for the soft glacial sediments cliff, but the crest of the structure is not high enough to protect the cliffs from extreme storm-water levels. Erosion of the cliffs continues to be an issue.

The best example of revetment design in the area is probably that located at the Inny Strand, where it protects the Waterville Golf Links. Pitched at a slope of 1:2.5, the revetment has a deep toe to account for the variable beach level, and a large crest width to encourage dune establishment. The design blends structural stability with a sympathetic integration into the surrounding dune system. The performance of this revetment has proven effective at combating assaults from recent storms that damaged other dune systems in the area. Further, it has promoted dune progression over the crest of the revetment, helping the structure bed into the landscape.

A shoreline development plan of Waterville was commissioned in 1997.[25] This study outlined several measures that focused on enhancing the amenity value for the village and improving the level of coastal protection. To date, only a few of these measures have been implemented, with planning and management of this coastal zone remaining undecided. Two areas are of immediate concern: first, the cliffs to the south of the village; and second, the low-lying section of the original promenade. The cliffs are eroding rapidly, with cliff-toe protection urgently required to prevent undermining of the local road as well as several residences. There are also tentative plans, by the local development group, to raise and widen the original promenade. As sediment transport is cross-shore dominated in this area, rearranging the series of groynes here to create a system of offshore breakwaters is a solution that could contribute to beach build-up. If this is supplemented with a shore nourishment programme, then an amenity beach may even be possible. The scheme is, however, controversial.

What is clear from the analysis of the recent history of Waterville's coastline is that any proposed development requires a deep knowledge of coastal processes, coupled with an understanding of the impacts the development will have on these processes, to ensure that the mistakes of the past are not repeated.

ROSSLARE STRAND: EROSION AND PROTECTION

Jimmy Murphy

Rosslare Strand is located in south-east Ireland in a coastal embayment close to Wexford Harbour, bounded by Greenore Point to the south and by Raven Point to the north (Figure 30.18). It is a curved, concave-shaped beach, *c.*6.5km in length. It starts *c.*1.5km west of Rosslare Harbour and runs in a general north–north-west direction to the mouth of the inlet to Wexford Harbour. The southernmost 2km of the strand is backed by high cliffs composed of glacial sediments, which merge into low-level sand dunes for the remainder of the beach. Rosslare Strand is exposed to waves, which range in offshore directions from north–north-east to south–south-east, with the dominant wave directions being from the south, which are the result of Atlantic swell (that is, long-period ocean waves) being diffracted around Greenore Point. This results in a net south-to-north sediment transport and it is estimated, by using dredging records from Rosslare Harbour, that the annual northerly rate of sediment transport is of the order of 27,000m³.

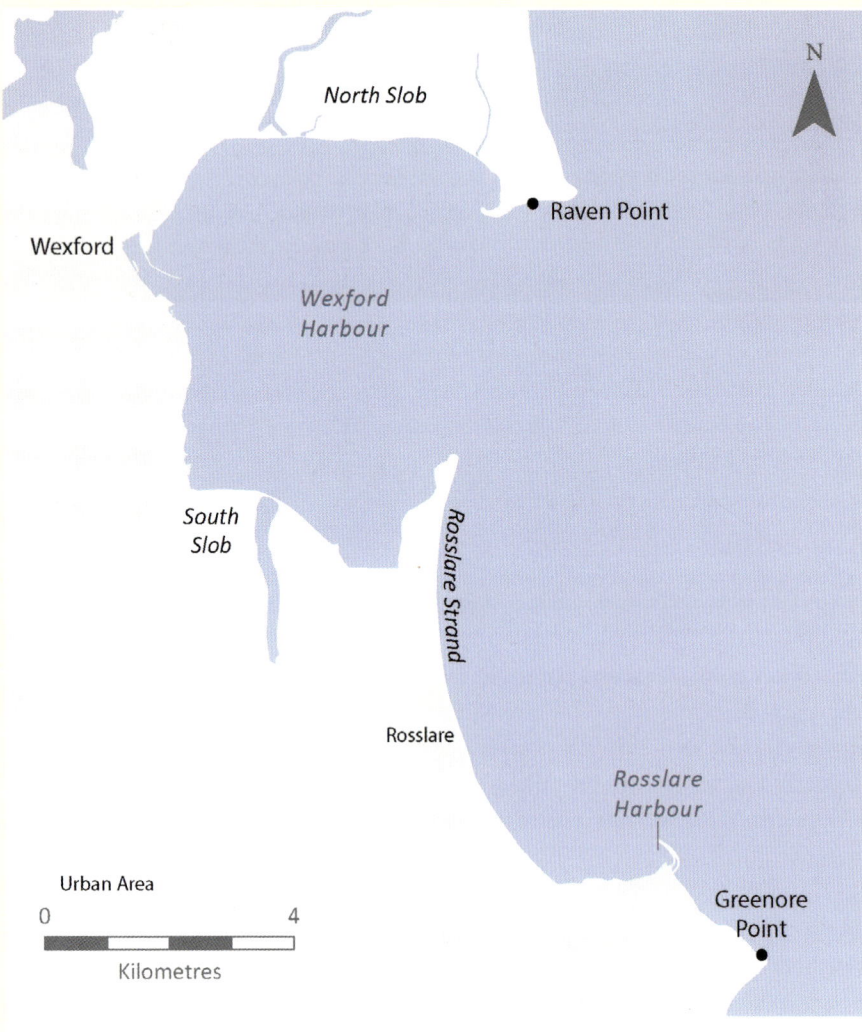

Fig. 30.18 ROSSLARE STRAND AND WEXFORD HARBOUR.

It was the northern motion of sediment along the Irish Sea coast that resulted in the formation of the sand spit at Rosslare Strand. This had grown and extended almost completely across the mouth of Wexford Harbour by the sixteenth century. At the end of the spit a small fort was built, called Rosslare Fort, to protect Wexford town. It is not known when exactly this fort was constructed, but it was first mapped in 1599 and underwent major extension in the 1640s. At its peak in the early 1800s about fifty families lived in and around the fort. It went into decline after the Great Famine in the 1840s and only a few families remained when, in the winter of 1924–1925, a severe south-easterly storm breached the spit. By 1940 the fort had gone, along with about 2.5km of the spit itself. The spit's beaches and dunes were replaced by various moving shoals and channels, together with low-lying to subtidal sand islands (Figure 30.19). The end of the spit was eventually protected from further erosion by the construction of a rock revetment, but at that stage almost 3km of it had been lost. Coastal erosion continues to be a critical issue on Rosslare Strand and, given its strategic importance to the south-east region, various protection schemes supported by studies of the coastal dynamics in this area have been undertaken. To the north, a shorter spit structure has also developed in the area of the Raven. This

Fig. 30.19. MAPS SHOWING THE DEVELOPED HISTORICAL POSITION OF ROSSLARE STRAND AND THE SAND SPIT. The main image depicts Wexford Harbour and Rosslare Strand in 1845, the inset satellite image shows the subsequent erosion and break up of Rosslare Spit. The area of the former spit is now being replaced largely by tidal channels, with sandbanks and shoals; much as would have been found in the medieval period. [Source: Wexford County Archives; inset: Google, Digital Globe, 2019]

forms a complex of beach and dune environments that are backed by glacial till cliffs. The sand spit has developed southward from here under the influence of southerly moving offshore sands.[26]

Given that Rosslare Strand had remained a very stable beach system for hundreds of years, as evidenced by the longevity of Rosslare Fort, the question arises as to why did such significant erosion take place over the space of a couple of decades? The answer lies in two very important changes that were made to this coastal system in the nineteenth century, that acted as a tipping mechanism in terms of changing the underlying processes and thus destabilising the coastline.

First, the construction of Rosslare Harbour at the southern end of Rosslare Strand. This work started in the 1860s, in order to provide easier access to open water than was afforded by Wexford Harbour. The initial design was quite innovative, as it included a viaduct to an offshore pier with the intention of allowing sediment to continue to flow along the coastline and onto Rosslare Strand. However, the leeside (western side) of the pier provided an area of calm water into which sediment was deposited and no longer moved northwards. To keep the harbour operational, regular dredging was undertaken and between 1913 and 1964 about 1.2 million m³ was dumped offshore, in deep water, where it was lost to the beach system. Following this, from 1964 to 1978, approximately 450,000m³ of further sediment was deposited on the southern part of the beach, which was a very positive action at that time.

In 1978, to reduce the dredging requirements, the viaduct was closed off, which effectively turned the harbour into a large impermeable barrier. This has resulted in a build-up of sand on the south side of Rosslare Harbour and added to the ongoing deficit of sediment on Rosslare Strand. In addition, sediment from any dredging that has been carried out since 1978 has largely been dumped offshore, due to licensing issues associated with using harbour sediment for beach nourishment. This action may also have further added to the developing sediment starvation of Rosslare Strand and spit.

Secondly, the reclamation of large areas of 'sloblands' within Wexford Harbour, up to 5,000 acres (2,023ha) from the north and south slobs combined, began in earnest during the Great Famine, with a resultant reduction in the intertidal area by half. This work has had a major influence on the tidal flows within the harbour area, leading in particular to a reduction in the ebb tide velocities. The general effect of this is that more sediment is transported into the harbour area during the flood tide than is transported out on the ebb tide. There is evidence of deposition within Wexford Harbour since completion of the reclamation works, through sedimentation of navigation channels. This has effectively put an end to commercial shipping entering Wexford town and contributed to the need for the building of Rosslare

Harbour. Studies of Rosslare Strand have attributed the increase of sediment volume within Wexford Harbour to the losses that have occurred along the strand's beaches.

The combined effect over time of these harbour construction and reclamation works has been a significant loss of the sand sediments along Rosslare Strand. The health of any beach is directly related to the availability of sufficient sediment to allow the beach to adapt to environmental conditions. Since the late nineteenth century Rosslare Strand has been in sediment deficit. This deficit first manifested itself in an obvious manner when the dunes were breached in 1924. The ongoing erosion problems on Rosslare Strand are testament to the unforeseen negative impacts that anthropogenic changes can have on a coastal system. It is also possible that other coastal and linked offshore sediments and wave dynamics factors are involved in the sediment losses experienced, but require further study to establish their influence on the strand.[27]

FIG. 30.20. ROSSLARE STRAND, COUNTY WEXFORD (SEPTEMBER 2019) SHOWING THE USE OF ROCK GROYNES TO HELP TRAP SANDS AND STABILISE THE POSITIONS OF THE BEACHES AND THE SPIT. [Source: Wexford County Council]

Protection works at Rosslare Strand have been ongoing since 1957, with many schemes being small-scale and localised, focused on stopping the erosion rather than addressing the underlying issue of a sediment deficit. Many of the works have been undertaken privately and include the use of timber groynes and revetments, plastic groynes, a rock breakwater (to stabilise the head of the spit), rock revetments and gabion mattresses. The majority of these works have been unsuccessful, but in the early 1990s the then Department of Marine and Natural Resources (DOMNR) initiated a major study of coastal processes, followed by construction works. The scheme included building twelve rock groynes, beach nourishment and various dune management measures, for example, fencing and grass planting, amongst others (Figure 30.20). This work was carried out on a phased basis over a period of five years, being completed finally in January 1995. The implementation of the beach nourishment element, involving the pumping of 276,788m³ of sand onto the beach was, and still is, very novel for Ireland and addressed directly the sediment deficit issue. It is estimated that approximately 27,000m³ of sand would move northward annually if not for Rosslare Harbour, therefore the added (nourished) material should have a maximum design life of ten years before erosion becomes a problem again. This coastal protection scheme was successful, but there has been no further sand nourishment since January 1995, and recent erosional problems in the back-beach area have led to rock revetments being constructed. The Rosslare coastal system continues to cause concern and is a focus for continued study. The aim of any future protection work would be to preserve Rosslare Strand, although it is unlikely that the Strand will ever return to its pre-1924 state.

COASTAL EROSION MANAGEMENT IN IRELAND

Historically there has been very little management of the Irish coastline, even though it is evident that erosion has been an issue for over a century or more. The earliest documented national study was by the Royal Commission in 1911, which set out the extent and causes of erosion. There is nothing further to report on a national level for nearly eighty years, until the severe storms of 1989–1990 caused extensive damage to many areas around Ireland's coasts. This motivated the government and local authorities to act, forming the National Coastal Erosion Committee. This reported that 1,500km of coastline was at risk of erosion, with 490km of this requiring urgent intervention. Over the quarter-century since 1992, many studies into the causes of and possible management responses to coastal erosion in Ireland have been carried out. These have included: Coastal Zone Management: A draft policy for Ireland; ECOPRO: Environmentally friendly coastal protection, code of practice; Coastal Zone Management; and the National Development Plan (2000–2006).[28]

Since 2003, the main focus in addressing the problems of protecting the coasts in the Republic has been through the Irish Coastal Protection Strategy Study (ICPSS). This national study took ten years to complete and involved detailed numerical modelling, data collection and analysis. The study enabled the mapping of flood risk and erosion, although the erosion mapping is based purely on historical trends so has a high level of uncertainty associated with it. This work has been complemented subsequently by the work of the

National Risk Assessment of the Impacts of Climate Change.[29] The main benefit of the ICPSS, although not an all-island project, is that it provides information where previously there has been none. However, it does not point to a national strategy in terms of managing coastal erosion. Instead, the ICPSS considers that extreme events, such as the storms of 2013–2014, will naturally identify the areas which are most at risk and they then can be managed using in-place procedures. This is seen as unsatisfactory and there are calls now, from local authorities as the primary coastal managers, as well as by many others from both the public and business sectors, for definite national-scale action on these issues. This comes, particularly in the face of increasing concerns about the risks to coasts under climate warming, with the expression of a need for an integrated approach to coastal protection and planning throughout the island of Ireland (see Chapter 33: Climate Change and Coastal Futures). A National Coastal Change Management Strategy Steering Group (Republic of Ireland) has begun work (September 2020) on dealing with these issues.[30]

The current management of the Republic of Ireland's coastline is undertaken by the Office of Public Works (OPW) and the local authorities, as the primary bodies responsible for planning and implementing coastal protection measures. Other agencies and departments relevant to coastal erosion include: the National Monuments Service; the National Parks and Wildlife Service; the National Inventory of Architectural Heritage; the Department of Tourism, Culture, Arts, Gaeltacht and Media; and the Department of Housing, Local Government and Heritage. In Northern Ireland, management is undertaken through the Department of Infrastructure and the Department of Agriculture, Environment and Rural Affairs (DAERA), together with input from other organisations, such as the National Trust, Northern Ireland.

The OPW is the government agency responsible for both coastal erosion and flood-risk management, and views these two functions as being linked closely in many coastline situations (that is, in the erosion of dunes resulting in increased and wider flood risk). Regarding coastal erosion, the OPW prioritises needs for action on the basis of risks to public safety and public infrastructure, with any works undertaken coming under the provisions of the Coastal Protection Act (1963). This provides the mechanism whereby local authorities apply to the OPW for funding. Applications for protection schemes are assessed and, where appropriate, finance is given for detailed initial design studies, in order to determine the best intervention to apply. On completion of the design study, a decision is made on whether capital funding will be provided. Whilst only low levels of national funding have been provided in the past for coastal protection, the OPW has been able to support some significant coastal protection and defences repair schemes (Figure 30.8).

Local authorities are at the frontline in terms of managing coastal erosion in Ireland. Their responsibilities include coastal maintenance and emergency works, identifying critical areas for action, applying for OPW funding, and managing coastal protection schemes. It seems that managing erosion is undertaken slightly differently in each of the nineteen coastal local authorities, depending perhaps on the level of importance given to this issue. Some counties have a dedicated engineer or planner who deals with coastal works; in others, it forms part of the remit of two or more engineers. What is certain is that coastal erosion is becoming more of a core issue in local authorities and for those living on the coast, perhaps with the realisation of the increasing threat that it poses to the public and to critical infrastructure.

Case Study: Maharees Conservation Association

Eugene Farrell

The Maharees Conservation Association (MCA) CLG (Company Limited by Guarantee) is a coastal community group that facilitates a wide range of local volunteer networks to make change locally, and works collaboratively with agencies and stakeholders, such as, Kerry County Council (KCC), National University of Ireland, Galway (NUIG), National Parks and Wildlife Service (NPWS), Clean Coasts, An Taisce and IT Tralee. Its work is devoted to protecting the coastline and natural heritage of the Maharees, raising awareness of the cultural and ecological importance of the area and ensuring the viability of the Maharees community.[31]

FROM LAND TO SEA

The Maharees (Na Machairí) Peninsula is a low-lying tombolo, composed of sand-dominated beaches and dunes. The tombolo separates Brandon Bay and Tralee Bay, and connects the two villages of Fahamore and Kilshanning to the mainland (total population 293) (Figure 30.21). It is an area of Outstanding Natural Beauty (ONB) and comprises part of the Tralee Bay and Maharees Peninsula, west to Cloghane Special Area of Conservation (SAC), due to the extensive coastal habitats and rare, diverse wildlife that it hosts. Three main types of sand dune dominate the southern section of the tombolo: fixed dunes, dune slacks and dunes with *Salix repens* ssp. *argentea* (creeping willow). The shoreline is delineated by a strip of mobile dunes vegetated with *Ammophila arenaria* (marram grass) and *Elytrigia juncea* (sand couch-grass). The only exceptions to sand dune cover are two caravan parks and a 10ha inland site of *Hippophae rhamnoides* (sea buckthorn) that was planted in the late 1970s by KCC to combat erosion (see Chapter 11: Beaches and Barriers, case study on Ecology of Sand Dune Habitats

Fig. 30.21. AERIAL VIEW OF THE MAHAREES ISLANDS AND PENINSULA (TOMBOLO), IDENTIFYING THE MAIN LOCAL COASTAL ISSUES AND ACCOMPANYING MANAGEMENT CHALLENGES. [Data source: Eugene Farrell; satellite image: Google, Digital Globe, 2019]

in Ireland). The northern section of the tombolo section of the tombolo lies on limestone rock-pavement. This land still retains fields with plough-drill patterns from a bygone era, when there was a thriving farming industry driving the local economy, the main crops being onions, carrots and potatoes. In the past three decades socio-economic forces have resulted in a transition away from farming and fishing, to focus upon the provision of tertiary goods and services for tourism and recreational visitors to the area (see Chapter 27: Coastal and Marine Tourism: Opportunities in Blue Growth). This social change has had a dramatic impact on the landscape and people have observed a migration of activities from land towards the sea (see Chapter 14: Imagining Coasts). The area has a new commercial pulse that hinges on water recreation activities, such as surfing, sea kayaking, windsurfing, paddle boarding, sea safari, waterparks and diving, and access to facilities that support these.

PRESSURES AND SOLUTIONS

The Maharees are increasingly subject to a number of pressures and concerns (that is, management challenges), that impact both on the natural environment of the peninsula and on the social fabric and quality of life of the resident community (Figures 30.21 and 30.22).

The title of 'water sports mecca' should be a reason to celebrate and a reason to continue to promote the area internationally, but there

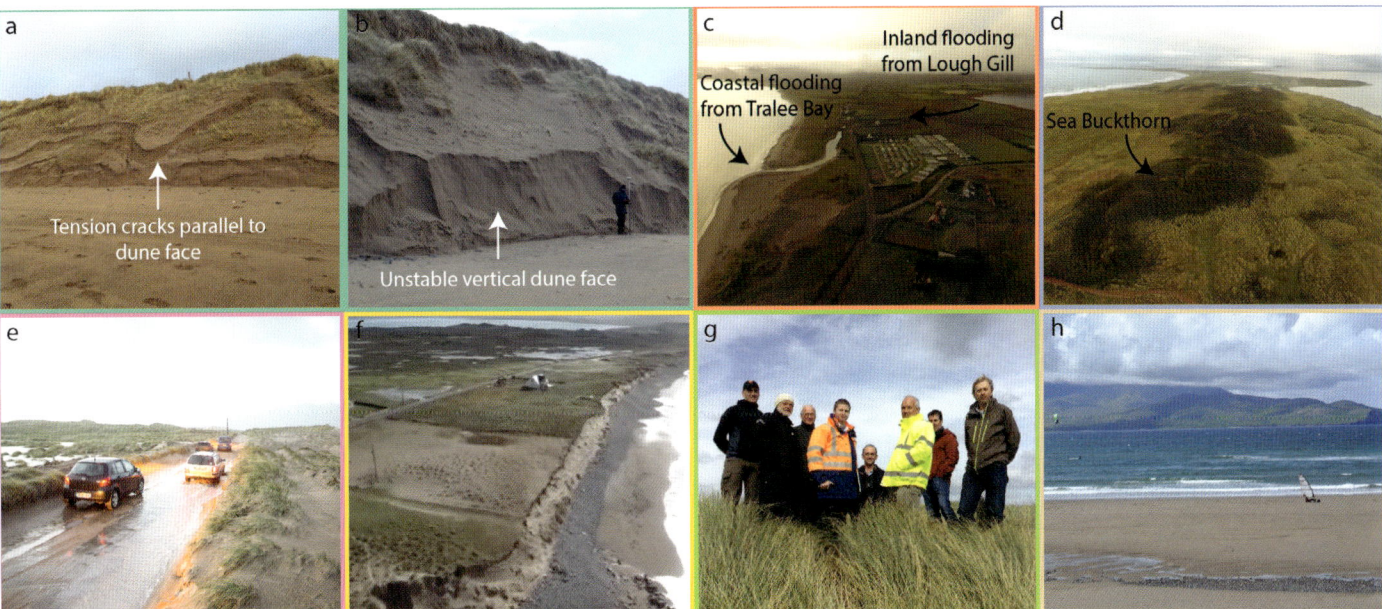

Fig. 30.22. THE CHALLENGES OF PROTECTING THE MAHAREES FROM THE IMPACTS OF PEOPLE AND CLIMATE CHANGE. Challenges 1 and 2: Dune and cliff erosion from storm waves. a. Dune erosion after Storm Eleanor (January 2018). Oversteepening (scarping) of the dune toe has led to undercutting and an increase in steepness of the dune face. The consequence of this is a state of tension in the upper part of the dune, with visible cracks appearing parallel to the dune face; b. Mass failure of the upper dune to dune toe, due to slumping processes. As wave attack continues, sand is removed. If wave attack ceases, slumped material provides temporary protection to the dune toe. Challenge 3: flooding; c. Kerry County Council carries out regular maintenance at the Trench Bridge and Lough Gill areas, in order to keep the stream outlet from blocking up and flooding this low-lying land. The ongoing management of this area is one of great concern to the local community, particularly in relation to climate change predicted sea-level rise. Challenge 4: Invasive Plant Species; d. Sea buckthorn (*Hippophae rhamnoides*) shown as dark green. Anecdotal evidence from local farmers and National Parks and Wildlife Service (NPWS) staff suggests that sea buckthorn was planted sometime in the 1970s by Kerry County Council to combat dune erosion. The plantations have been expanding ever since and are impacting adjacent native flora, as has also been the case on similar dune systems in east-coast Ireland. Recent study shows that the coverage by sea buckthorn in the Maharees, currently *c.*13.56ha, has been increasing at an average rate of 27 per cent, or 0.39ha/year (1995–2017). Challenge 5: sand deposition; e. The Maharees has one access road, which is frequently made impassable by sand deposition. This disruption has important implications for the servicing of the Maharees community, particularly the health and safety of residents. For example, the Maharabeg Cut location, shown here, was cleared seventeen times by Kerry County Council (KCC) in the winter of 2016. Sand fencing installed by the MCA has fixed this problem, at least for the present. Challenge 6: Human impacts; f. The extent of fixed dunes is rated as 'unfavourable-inadequate' by the EU Conservation Status Assessment of the Maharees. The decline in fixed dune area, as a result of erosion, is largely caused by human recreational activities, as well as by the overstocking of cattle that graze the dunes. Land tenure is an important constraint in managing the dunes. Other human impacts include pedestrian paths and horse trails to access beaches, car parking, camping, use of ATVs, dune scrambling, and fires on the dune and beach. Challenge 7, conservation and management, and the link to the stakeholders; g. Coastal protection designations (for example, EU legal and management instruments, such as Integrated Coastal Zone Management, 2002 and the Marine Strategy Framework Directive, 2008) can prevent local communities mobilising, or feeling empowered to carry out intervention at the local scale. The MCA has built a very strong relationship with the NPWS and Kerry County Council, as the environment managers, as well as with many universities and other technical groups. Challenge 8: valuing Ecosystem Goods and Services (EGS); h. These include the economic value of fisheries and agriculture, storm protection and erosion control, and of non-extractive uses like tourism. Many other values are not easily estimated. For EGS that are produced outside of an economic CBA market, economic assessments are lacking, or purposefully minimised as market externalities. It is critical, however, to include both market and non-market EGS to determine the full value of a functional system, often termed 'ecosystem services', to its users. [Source: Eugene Farell]

is a sense of foreboding that it has been a pyrrhic victory. The lack of basic amenities to facilitate the increased numbers of people accessing the shoreline has led to widespread dune degradation and conflict between landowners, residents and visitors. Every year, visitors and local communities in the Maharees highlight the paucity of designated parking, legal campsites, toilet facilities, water and showers, rubbish and dumping facilities, beach and dune access points, picnic areas, seating for visitors to appreciate the views and interpretative information on the local heritage that is of cultural and historical importance (see Chapter 21: Ireland's Islands). These issues are not only impacting the visitor experience, but have caused resentment amongst the local community, who observe their land being damaged every year by trespassing visitors and campers. Left unmanaged, the increased human activity has been accelerating the destabilisation of the fragile dune ecosystems that are already highly vulnerable to Atlantic storms and suffering chronic erosion. Historic shoreline analyses from a team based in the Discipline of Geography, NUIG, illustrate that the dunes have retreated over 75m in the past century, with increased erosion rates observed in recent years. The retreating mobile foredune system is the only line of defence from storm surges that frequently flood or block (with blowing sand deposits) the single access road on the peninsula (Figure 30.22e). For example, the road in the vicinity of the Maharabeg Cut had to be cleared seventeen times by KCC in the winter of 2016. This had serious implications for the daily commuting of the residents and to their health and safety, in terms of access for paramedics and emergency services. The long-term future of the tombolo is also precarious, considering climate change, and the associated projections of rising sea levels combined with potentially increased intensity of north-east Atlantic storms.

The local community mobilised during the Dingle Peninsula's inaugural Féile na Stoirme in February 2016, following a public lecture by Dr Eugene Farrell (NUIG), who presented results from an OPW-funded coastal monitoring project of coastal dunes in the Maharees. The scientific results supported anecdotal accounts of residents who had been observing the erosion and degradation of the dunes for generations. The talk provided a set of short- and long-term recommendations for future action, as a catalyst for the community to mobilise. Initially, the main aim of the MCA was to mitigate erosion and flood hazards that impact road access on the tombolo. This has since expanded to include raising awareness of the cultural and ecological importance of the area, and the rich human heritage that gives the location and people a unique identity. Since its inception in February 2016, the MCA has conducted over 500 work events and activities. These include, but are not limited to, beach clean-ups (25), dune vegetation planting (11), signage and access fencing (8), straw bales protection (4), dune fencing (6), cultural/heritage/biodiversity walks (13), public and committee meetings (27), fundraising events (17), media presentations (32), and partnership and outreach events (29).

The most vulnerable locations on the western seaboard of the peninsula are large sand blowouts, which also have the highest volume of traffic (for example, pedestrians, water sports, horses, vehicles). The community focused their attention on these priority sites in 2016. Restricted- and controlled-access measures were designed and installed using fencing. Multiple dune planting projects were co-ordinated with experts in An Taisce's Clean Coasts programme. Information signs on dune vulnerability and protection were prominently displayed on the dunes and back beach. Sand fences were installed on the fixed dunes adjacent to the road sections most prone to sand deposition. All these actions were led by the MCA in consultation with KCC, NPWS, An Taisce and coastal scientists from NUIG. The rapid recovery of the dunes in these treated areas gave extra impetus to the MCA to continue their efforts and expand their remit to include other priority sites on the tombolo. These include the area known locally as Trench Bridge, located on the very narrow eastern part of the tombolo that many consider to be the entrance to the Maharees. Kerry County Council regularly carries out maintenance work here, in order to prevent the stream outlet from Lough Gill blocking and flooding the adjacent low-lying area, including the private houses and the road.

EFFECTIVE COMMUNITY-LED CLIMATE ACTION

It is a remarkable testament to the MCA, and the people and organisations they represent, that their efforts are being used regionally and nationally as an exemplar for community engagement in climate action (Figure 30.23).[32] This national forum is an initiative to generate a country-wide discussion on climate action and, specifically, to engage with the public on climate change issues. The MCA is also included as a case study for An Taisce's Climate Ambassador Training Programme. The group has been the recipient of numerous local and national awards that focus on community engagement, including Clean Coasts Ocean Hero Award Group of the Year (2017), Pride of Place Special Award (2018) and the KCC Gold Community Award (2018).

Some valuable lessons and insights are also emerging from the Maharees that are informing other coastal communities by highlighting enablers and barriers to their success. The enablers include:

• building strategic partnerships and networks with key decision-makers within the different management agencies (KCC, NPWS, OPW, EPA, the Heritage Council, An Taisce);

• facilitating their local political representatives (TDs, councillors) to support their actions;

• collaborating with third-level institutes to avail of physical and social science expertise, for example, IT Tralee, NUIG, UCC and University of Limerick (UL);

• building key relationships with commonage users and individual landowners to allow the MCA to carry out coastal management projects (commonage is defined as land jointly owned by several individuals, who have shared grazing rights);

• good governance arrangements; the MCA committee comprises leaders with diverse backgrounds and skill sets;

• managing to build consensus despite internal conflict within the MCA group and the community regarding priority actions; attaining the long-term goals has always been the primary focus of the group'

• Ability of residents, visitors, landowners, and managers to see progress (no matter how small) from their actions (the 'feel good' factor);

• being granted Charitable Status by Charities Regulatory Authority;

• success at obtaining funding via multiple sources (online support request; table quizzes; Local Agenda 21 Fund; KCC Biodiversity Office; KCC Local Area Engineer; KCC Community Support Fund; Heritage Council; Clean Coasts; Tralee Chamber Alliance; LEADER);

• continuous engagement with the entire community (from the youngest to the oldest; from the farmers to the kitesurfers) to commit time to participate in volunteer projects;

• engagement with the local schoolchildren and teachers via field trips and class projects (for example, The Fishies Project: €5 keyrings made from marine litter by transition year Meánscoil Nua an Leith Triúigh students with all profits donated to the MCA; Marine Litter Mask

Fig. 30.23. VOLUNTEERS FROM THE MCA PLANTING MARRAM GRASS ON THE SURFACE OF AN ERODING DUNE FACE. Dune vegetation planted during 2017 survived the winter storms of December 2017–February 2018 and will reduce the risk of dune blowouts by wind action, and from moving landwards, as well as the storm breaching the dune system to impact essential roads. In other soft-type interventions by the MCA, dune restoration has been found to occur within twelve months of the works, such as the restriction of parking access, the deployment of straw bales, to help reduce dune toe erosion, education boards and warning signs. [Source: Ray Buckley]

Project: students made masks from marine litter collected by the MCA and displayed them at the Green Schools Marine Environment Conference (2017) organised by Clean Coasts; the Meánscoil Nua an Leith Triúigh students were featured on the front page of the *Examiner* wearing the masks);

• building a strong presence in print and broadcast media (local and national newspapers, radio and television) to highlight issues and/or achievements; using social media (Facebook) to document visually the achievements and maintain transparency in the process; social media serves as the community information hub for organising volunteer activities; MCA leaders have presented at national conferences (Sea, Land and Spirit (2017); Wild Mind Nature Festival (2018);

• hosting organisations to conduct field trips in the Maharees, to meet with the MCA to listen to their story and/or conduct complementary field research, for example, Irish Planning Institute, Rose of Tralee, CHERISH Project, KETB Surfing and Lifeguarding, NUIG, UL;

• embracing the rich heritage and identity of the area by broadening their 'protection' strategies, to include cultural and heritage awareness and education (for example, Maharees Heritage Trail; biodiversity walks led by students of the BSc in Wildlife Biology IT Tralee; species spotter sheets; dune ecology awareness signs).

Barriers include:

• competing values and priorities within the community;

• controlling access and signage perceived as militant;

• lack of recognition, or support at government level;

• perceived lack of expertise and under-resourcing within local authorities to make site-specific decisions for management strategies;

• the balance of interests between protecting the SAC, as required by EU legislation versus protecting the local communities;

• general lack of enforcement of local plans and county beach bye-laws due to limited Gardaí resources and ambiguity in the responsibilities of the local authorities;

• continued illegal sand mining from the foreshore by local farmers;

• two different commonage areas that require mitigation, the MCA has to obtain consent from thirteen landowners to carry out management projects.

THE FUTURE OF THE MAHAREES?

In March 2018 the OPW announced that they would be funding (award of €150,000) a 'Coastal Erosion and Flood Risk Management Study' for the Brandon Bay and Maharees 'priority cell'. This is to be undertaken by the MCA and is a major step forward in action, forming one of the milestone needs identified at the February 2016 Féile na Stoirme meeting. To have attained this goal in such a very short time underlines the real concern that exists for the future of the people of this coastal area, as well as the commitment (or fear) that the MCA and the Maharees community (that is, the landowners, residents and visitors) have. The aim of this OPW-funded study is to provide a suite of options, and assessments, of appropriate and alternative strategies to best manage the coastal erosion/accretion and flooding risks in the short (2025), medium (2050) and long term (2100). However, there is still no guarantee that the chosen options will receive capital funding. Further, the selected management approach and options may not be acceptable to the local community. This is especially so if the recommendation involves a 'do nothing' or a 'do minimum' approach, as may be the case in needing to comply with the Habitats Directive to protect the SAC, or with the developing national policy on coastal management. Ironically, the SAC comprises sand dunes that the MCA is trying to prevent from being washed away!

Provisioning Services	Cultural Services
• Freshwater	• Cultural heritage
• Food	• Recreation and tourism
• Fibre and fuel	• Aesthetic value
• Genetic resources	• Spiritual and religious value
• Mineral extraction	• Inspiration for art, folklore, architecture
• Landscape for industrial use	• Social relations
	• Educational resource

Regulating Services	Supporting Services
• Air quality regulation	• Soil formation
• Climate regulation	• Primary production
• Water regulation	• Nutrient cycling
• Natural hazard regulation	• Water recycling
• Pest regulation	• Photosynthesis
• Disease regulation	• Provision of habitat
• Erosion regulation	
• Water purification and waste treatment	
• Pollination	

Table 30.1 ECOSYSTEM GOODS AND SERVICES PROVIDED BY COASTAL SAND DUNES. Extensive sand dune systems are found at many locations around the coast of Ireland, and provide a wide range of services of actual or potential value. As the experience of the Maharees Conservation Association has shown, providing coastal communities with the tools and knowledge required to act as stewards and custodians of their own local dunes, and enabling 'bottom-up' participation in planning and decision-making, can greatly enhance the sustainable management of these often vulnerable coastal resources. [Source: Derived from Everard et al., 2010. Have We Neglected the Societal Importance of Sand Dunes? An ecosystem services perspective. *Aquatic Conservation, Marine and Freshwater Ecosystems*, pp. 480–483]

At the most fundamental level, the MCA has illustrated that there are many benefits to be derived from involving a coastal community in decision making (that is, bottom-up type governance). This approach is more likely to result in policy and planning decisions that are acceptable to all the parties concerned. Likewise, the future social and economic health of most coastal communities in Ireland, like the Maharees, depends on empowering people and groups such as the MCA to make changes locally (for example, to the environment, landscapes and infrastructure). This involves providing communities with the tools to adapt to the impacts of climate change, whilst developing new sustainable businesses and tourism opportunities. It is important to provide roadmaps to coastal communities that incentivise them to mobilise for change (that is, to minimise costs while maximising opportunities) and reach attainable goals. (Table 30.1).

In developing future national coastal policies, it is also imperative that Ireland builds on case studies such as that of the MCA, to deliver fit-for-purpose guidelines to local authorities to help build community resilience for climate action (see Chapter 33: Climate Change and Coastal Futures).[33]

Case Study: Buildings at the Coast: An architect's viewpoint

Anna Ryan

Good architecture facilitates the taken-for-granted 'operation' of the quotidian, offering new spatial experiences for social interaction. As an architect and cultural geographer, I believe that in the practice of architecture three elements, or concepts, are of particular importance. First, context: that is, understanding the nature of the site and how new structures, whether buildings or otherwise, respond and relate to the intrinsic characteristics and challenges of the site; second, materials: the appropriateness of the choice of materials, in terms of both their functional and their aesthetic performances and qualities; third, use: the opportunities for habitation offered by new architecture and its potential for enhancing day-to-day life. In this brief discussion of coastal architecture – the coast both as construction and construction (building) at the coast – it is useful to bear these three elements in mind.

The coast is an ongoing construction, taking place each time a wave sucks back from a sandy or shingly shore; each time a cliff stands in the way of the incoming tide, or each time a storm lifts and shifts whatever comes in its path. The pace of this construction happens over minutes, hours, tidal cycles, seasons, years, centuries and beyond. When we humans decide to build at the coast, we are adding other layers to this existing fluid construction. The layers of new construction that we add can take a number of approaches. These can include ones that show understanding and respect for the shifting physical and social characteristics of the coastal zone over a long time period, both permanent and temporary qualities. They can also range through a variety of other approaches, to ones that seek to maintain the status quo, rooted in ideas of preserving and protecting the 'here and now'.

The morphology and outlook of most Irish coastal settlements has changed significantly, particularly since the late eighteenth century, as can be seen in the case of Lehinch, County Clare (Figure 30.24) (see Chapters 4: People, Agriculture and the Coast and 19: Changing Coastal Landscapes). As a small fishing hamlet on the west coast of Clare, the village of Lehinch traditionally turned its gables to the sea. The grain of the village began as a number of parallel laneways running from the cliff-side to the main street. As it developed into a coastal tourism resort in the nineteenth century, and then with the advent of the West Clare railway, Lehinch altered its focus. Following the widespread fashion and demand for seaside 'improvements', a sea wall was built and a promenade constructed on top of the former shingle beach. Lehinch, as built place, re-orientated itself to effectively confront the sea. In this way, a precise line where the built edge of Lehinch meets the Atlantic was demarcated in 1893. Over a century on, we see now how wildly incompatible poured concrete and stone-and-mortar are with a shifting system of shingle beach and dunes. The golf club, situated close to the present-day town, has made efforts to protect its 'natural' golf links through placing a wall of stone gabions between the beach and the dunes. In effect, this has cut off the beach from its source of sand, the dunes. Following the power of the storms that shattered Lehinch during the winter of 2013–2014, this artificial coastal edge has been reinstated and strengthened with yet more concrete and rock armour. The sea here continues to meet these hard surfaces, but a flow or relationship of materials between land and sea is now no longer possible. Thus, there is a gradual but noticeable scouring out of the beach, and the sea seeks out and finds the nearest 'soft' points in the coastline to attack. Perhaps, following Edmund Burke's eighteenth-century treatise on the 'sublime and the beautiful', we still hold that softness is related to weakness,[34] but when it comes to the coast, our thinking needs to be fundamentally transformed. In terms of its interplay with the sea, the 'softness' of land is the essence of the coast. Our decision-making, based on an understanding of our own place in time, would be enhanced if we were to more fully embrace a positive conception of 'softness' in our approaches to coastal living, inhabitation and construction.

Our larger seaside towns hold the architectural legacy of nineteenth-century seaside holidays. Resorts such as Dún Laoghaire (Kingstown) and Bray present terraces of large houses with generous windows, a uniformity and elegance of proportion facing the horizon.

Fig. 30.24 LEHINCH, COUNTY CLARE, A VIEW CAPTURED SOMETIME BETWEEN 1865 AND 1914. [Source: National Library of Ireland, call no.: L_ROY_11334]

Piers, promenades, palm trees, bandstands, benches, railway lines, roof shelters and wrought ironwork together remain as an ongoing performative presence of the bourgeois seaside holiday of previous centuries (Figure 30.25). In contrast, our approaches to building at the coast in recent decades could be considered as less than holistic in nature. Where the nineteenth-century resort holds together and, even still, has resonance between its parts, the built form that is emerging from recent and current policies and culture is very piecemeal. As responses to tax incentives, cul-de-sacs of white holiday homes and complexes of large apartment blocks have appeared in coastal villages and towns. Their suburban and urban forms and arrangements are often very alien to their exposed semi-rural and rural contexts. Individually, the low-rise white bungalow-cum-holiday home is presumably thought to be sympathetic to the form of the traditional Irish cottage. However, when multiplied around curving tarmacadam roads, unconcerned with orientation in relation to sun and wind, reproduced from one holiday-home estate to the next and from the edge of one coastal settlement to another, the spatial starting point becomes entirely lost. Ireland's vernacular approach to grouping dwellings in clacháns could serve as a very useful precedent for these coastal developments, with their variety of volumes, their gathering around two-, three- and four-sided courtyards, their approach of sheltering from exposure.[35] The aggregative forms of these genuinely traditional cottages with outbuildings and their relationships with their neighbours, with trees, walls and hedging, formed a holistic whole. Such an approach to building made, and continues to makes sense to the particularities of the physical surroundings. It forms an embedded knowledge that we would do well to remember and draw from as people (government and communities) address the very complex challenges of our expanding and pressurised coastal settlements.

Appropriate built responses to these challenges of living and holidaying by the sea today are made more complex by the economic value of a view of the sea. Whether for holidaying, or for permanent living, we now want to be as close to the action of the coast as possible, whilst at the same time held at a 'safe' distance. The pressures on coastal habitation, both from our desires and fashions and from forces such as climate change and the sea itself, raise multiple and complex questions. Do we continue to permit the building status quo and continue to develop areas vulnerable to coastal inundation? Or, do we begin to really acknowledge the flowing materiality of the coast? Do we concede and then recede, allowing policy to develop a new coastline, a 'line' that perhaps doesn't actually meet the sea at all, that thickens the coastal zone to allow natural processes to take their course and thus recognise our short 'moment' of inhabitation within the wider flux of time?

When people see change happening, change that is part of natural coastal processes, our usual response is in terms of defending and protecting the land where we live. Our language and actions tend to emerge out of engineering and a perception of conflict: we talk of 'solutions' to the 'problems' of erosion. We try to 'protect' ourselves from the sea and the climate; we use 'armour' to 'defend' ourselves from the forces of nature. This language doesn't help us, as it implies that, at the coast, there is a winner and a loser. Something is strong,

Fig. 30.25 BRAY, COUNTY WICKLOW, 2016, AN EXAMPLE OF A NINETEENTH-CENTURY RESORT TOWN. Bray has retained much of its architecture and physical structure. Many such seaside towns occur around Ireland, for example, Portrush, Newcastle, Ballyheigue, Clonakilty and Tramore. As can be seen here at Bray, their appearance and character is being changed, and often rapidly in the twenty-first century, by the construction of new buildings, of glass, concrete and steel, together with the increase of road and communications infrastructure development. These transformations alter the architectural and living space feel, use and uniqueness of these coastal urban areas – often, some may think, not for the best. [Source: Mark and Dave Murphy]

something is weak, and natural processes are a threat. The human impulse is to protect that which is significant to us: our people, our homes, our familiar settings, our world the way we know it. It is extremely difficult to step out of this mindset, to think beyond our immediate lives and the needs of our own timeframe, to think instead of action and 'building' that is for a period much longer than we can conceive. This shift is needed to enable and allow worthwhile construction, an architecture of many scales and functions, that remains true to the complexity of the coast as site.

The notion of the coastal zone is understood by most people, and yet when it comes to construction at the coast, we build an 'edge'. Oftentimes, concrete, rock armour and gabions are placed in front of the high water mark, as an attempt to dissipate the energy of the seawater as it arrives to what is considered to be a vulnerable piece of land. Such measures can sometimes display a politics of preservation that focuses on an immediate infrastructure that is abstract to the nature of the coast. This infrastructure can often, with its imposition of foreign materials, ignore the wholeness and togetherness of land and sea at the coast, where one is of the other. As intended, rock armour assists the accretion of material, a reinforcing of a formerly natural process. In any case, the materials used for protection and defence often tend to be alien to the site, and are 'hard' in contrast to 'soft'. The visual incongruity of such imposed materials resonates with the artificiality of the challenge of coast as line, the challenge of building or maintaining an edge. Indeed, the mathematician, artist, cartographer and writer, Tim Robinson, has (using fractal geometry) demonstrated the artificiality of the concept of coast as a line, calculating the dimension of a typical coastline as 1.2.[36] In thinking back to the important aspects of architecture, when constructing and reconstructing the coast, there are alternatives to this opposition of materials. There are many wonderful examples where coastal protection measures move beyond a sense of defence and barrier and become 'growth'. The current work on the tombolo of the Maharees, County Kerry, provides a good illustration: a community-led project where marram grass and chestnut stakes are being used to re-nurture and re-grow the dunes. This project of 'saving the Maharees' resonates with the approach of good architecture, where the construction being undertaken respects the materiality of the site. The engineered responses to the needs of coastal protection are appealing, both visually and texturally; function and aesthetics are in collaboration.

We have many examples of coastal structures and coastal building types where function and aesthetics are also combined extremely successfully, leading to memorable spatial and material experiences of quality for those using and encountering them. Lighthouses are a coastal building type par excellence (see Chapter 23: Maritime Traditions and Institutions). The lighthouse presents a solidity of volume rising as a mediator between sea and land, either from a medium of constant movement, or perched on an edge. The lighthouse could be

Fig. 30.26 RUTLAND ISLAND, COUNTY DONEGAL. Concrete coastal structures, of jetty and boat slipway, have merged now with the rock and become a part of the 'natural' coastal landscape. [Source: Donegal County Council]

considered as the ultimate built encounter between human and more-than-human forces. Present in a three-dimensional rotation, the building form itself says much about its physical surroundings, from topography to climate to exposure. That its primary function is related to this physical form – a vertical beacon of light in a horizontal space – is its essence. It is through the simplicity and elegance of its beautiful form that the lighthouse building itself performs its functions, from safety to communication and beyond. From the Fastnet Rock to Hook Head to Loop Head, Ireland's lighthouses offer a very significant contribution to the country's architectural legacy.

Related architectural qualities are evident in the network of small-scale harbours and piers that weave around the nooks and crannies of our coast (Figure 30.26). Both formally designed and locally evolved, the locations of these structures are carefully considered. Whether sited at existing places of natural shelter, or having a particular orientation in relation to the roll of the swell, their physicality offers a method of negotiation from the world of the sea to the world of the land and vice versa. Of stone and concrete, weathered from time and exposure, these constructions have literally settled into the bedrock; their solidity sometimes appears almost 'natural' as they have close conversations with their context. Their simple lines form their horizontality, perform their function of shelter, and offer a stage-like presence for an open variety of uses.

Perhaps when responding, through construction, to the intensity of today's social and physical pressures on our coastal edge, we need to expand our scope. Holistic-design thinking has a role to play at the contemporary coast. Rather than remedy specific issues with particular solutions, we need to join the dots and think, design and act before such solutions are required. We would benefit from more examples of all-encompassing approaches and processes that involve more than the construction of individual structures. Perhaps there are alternative ways to conceive of coastal construction? When challenges arise, rather than think we need to build interventions that stop something, perhaps we need to reverse the approach and work to 'build' interventions. This approach would instead 'facilitate' something further, something that would consider both the way we live our lives now and how we might do so in the long-term future. People should ask now whether the design and construction of our coastal protection measures could serve society in multiple ways, by offering new spatial experiences of the coast. These measures could provide new spatial forms that combine use with beauty in ways paralleled by the longevity of the historic lighthouse, harbour, pier and promenade. In this approach, these new structures to 'protect' the coast can become equally, or more importantly, driven by the potential to holistically enhance an aesthetic and embodied relationship for all who engage and enjoy where land meets sea.

FUTURE ACTIONS

In terms of its coastal protection needs, Ireland just 'gets by'. Until now, coastal erosion has occurred primarily in relatively unpopulated locations and the frequency and magnitude of storms has not been great enough to significantly jeopardise public safety and infrastructure. There is uncertainty, however, as to whether the current approach of repairing damaged infrastructure, or protecting coastal assets, on the basis of 'as funding allows' will be sustainable into the future. For instance, if severe events limit themselves to every ten to twenty years, as has been the case in the past century, then a simple 'just repair it'-type policy may be the most feasible option. But, as the pattern of storminess has shown since c.2000, there is no guarantee that this timing in the frequency of major storms and big waves will continue. Observations of the sea state around Ireland, together with modelled projections of storm and wave climates, show that significant changes in coastal storms and waves have already begun under climate warming (see Chapter 33: Climate Change and Coastal Futures).

For the present, it is expected that the Republic of Ireland will continue with a minimal intervention strategy and will not prioritise coastal erosion in terms of national management plan; although this is under review (see endnote 30). A different set of UK-based strategic guidelines operates in Northern Ireland. Practitioners in coastal science and engineering, as well as those involved in the work of coastal management – both in local government and more widely within Ireland – are unhappy with this situation and also with the lack of integration in approaches on the island as a whole. Disagreement about the most appropriate future approach and policy in coastal protection does not mean, however, that Ireland should literally 'do nothing'. There are a number of measures, or actions, that should be adopted that will be of considerable benefit to future planning:

1. Develop a more consistent national approach to assessing erosion risk. Currently, each local authority develops its own methodology for the identification of sites to be included in submissions to the OPW for funding. It would be a useful initiative if local authorities established a formal collaborative mechanism to enable better co-operation, interaction and standardisation of methodologies. Work toward realising this is in progress (see endnote 30).

2. Monitoring programmes of coastal areas that are subject to erosion and may require future action, should be undertaken by local authorities and appropriate government agencies, in order to provide a good basis for any future coastal protection designs and works. Studies could include topographic and bathymetric surveys, as well as wave, tide and current measurements. Annual surveys would be relatively low-cost and would track the health of a coastal system over a period of time. Monitoring programmes could be specified at a national level, so that there would be consistency in the data collected. Further, all the data should be stored centrally and made available to the public.

3. In 1997 the Brady Shipman Martin Report stated that 'the legislative framework is out of date and inadequate to the task in hand' and that it was 'unwieldy and insufficient in scope'.[37] It would be worth examining the robustness of the 1963 Coastal Protection Act, which currently defines the legal framework for coastal control for much of the island. Such

work is in progress (see Chapter 32: Coastal Management and Planning). This review should determine if the act is now the most appropriate legislation to deal with future protection responses to the increased problems of coastal erosion.

4. Local authorities need to develop more open communication channels and offer solutions to coastal property owners to reduce the number of ad hoc private coastal protection works that are appearing around the coast. Group coastal protection schemes should be supported and, in cases where such an option is not feasible, relocation within the locality should be facilitated. Improved education and outreach to stakeholders should be offered, so that they can be better informed of the practical choices that must be made in relation to coastal erosion.

5. Examine more innovative options for managing and protecting the coastline. Coastal defences worldwide are becoming higher and bigger to take account of future climate uncertainty, which can have a negative local impact. It is likely that opposition to proposed solutions will become more common if engineers continue the traditional route in relation to coastal protection solutions. There are other potential solutions, which require research and development, that could provide viable alternatives to traditional approaches. These could include active and demountable systems that can deal with the severe events, but that need not be a dominant feature on the coastline.

6. There is a need to develop a critical mass of experts in Ireland, in a wide range of disciplines that involve the coastal environment.

It can be said, in conclusion, that the island of Ireland, as a whole, is entering a time of high uncertainty in relation to coastal behaviour, with climate change, sea-level rise and increased human activity at the coast all contributing to this uncertainty. There will be a tipping point in the short to medium term where the status given to the management of erosion will need to be upgraded. There is currently insufficient data and knowledge of coastal dynamics to make optimal management decisions and implementing some fundamental actions now can form the building blocks for the future. It is reassuring to know that coastal protection solutions exist that can fit any erosion scenario, but these often require significant resources to implement. In order to make the best decisions, we need to be more proactive, rather than allowing the repetitive pattern of 'the winter storms' dictate people's actions in the future.

POLLUTION

Evin McGovern and Shane O'Boyle

PLASTIC DEBRIS RECOVERED FROM THE FORESHORE AT SPANISH POINT, COUNTY CLARE. Particularly after a winter storm, discarded plastic items such as these can often be found among the seaweed, stones and sand of the foreshore. In many localities, teams of volunteers conduct regular community beach clean-ups and this, combined with increased public awareness of the problem and measures to combat it at source, means that this type of scene is now fortunately less common than it was towards the end of the last century. [Source: Alicia Mateos-Cárdenas]

As an island on the periphery of Europe, Ireland is surrounded by comparatively clean and healthy seas, is bathed by the waters of the North Atlantic Current, and has prevailing winds that come predominantly from the Atlantic. There are no major continental rivers discharging directly to the adjacent seas, and the island has no history of heavy industrial development. Because of these fortunate consequences of geography and history, the seas and coasts of Ireland are relatively unaffected by pollution in comparison to those of many other countries. Nonetheless, this is an island of 6.6 million inhabitants. Daily domestic life requires materials and chemicals, and waste is a by-product of both this and the sectors underpinning the economy, such as industry and agriculture. Some of this waste inevitably enters the sea, primarily into the coastal environment and shelf seas. Where it leads to harmful effects on habitats and resources, or where waste poisons organisms (including humans), it results in pollution. Moreover, some pollutants are now globally ubiquitous and certain human-originating substances can be detected in even the most remote parts of the planet, including the deepest reaches of our oceans. The impacts of some types of pollution, such as massive oil spills or extensive littering of beaches, are easily grasped by the public, particularly when images of oiled seabirds or entangled animals (Figure 31.1) appear in the media. However, other forms of pollution can be more insidious and much less obvious. This chapter examines the extent of marine pollution in Irish marine and coastal waters, and highlights initiatives, from local to global, to prevent marine pollution.

SOURCES AND PATHWAYS: HOW DO POLLUTANTS ENTER THE SEA?

The OSPAR Convention for the Protection of the North-East Atlantic (1992) defines marine pollution as 'the introduction by man [sic], directly or indirectly, of substances or energy into the maritime area which results, or is likely to result, in hazards to human health, harm to living resources and marine ecosystems, damage to amenities or interference with other legitimate uses of the sea.' The two key elements of this definition are: 1. something must be introduced to the environment, and 2. this must cause harm. This harm may be adverse impacts on marine life and ecosystems, but could also be impacts on the ecosystem services provided by our seas and coasts, such as fisheries and seafood production, or amenity uses. Fortunately, in many cases the use of the precautionary principle can help avoid and prevent pollution, rather than delaying action until harm has been shown.

Many substances can be introduced into the environment to cause pollution. They may be solids, liquids or gases such as, for example, untreated sewage discharging to coastal waters or oil

Fig. 31.1 A SEAL ENTANGLED IN A DISCARDED FISHING NET, KILMORE QUAY, COUNTY WEXFORD. The seal was rescued soon after it had become entangled, before it received any serious injuries or infection. It was released back into the sea immediately, as it was otherwise in very good condition. [Source: Seal Rescue Ireland]

spilled from ships; energy, including noise, heat and light/electromagnetic radiation, can also be a pollutant; while increased oceanic uptake of atmospheric carbon dioxide due to human emissions is leading to ocean acidification (see Chapter 33: Climate Change and Coastal Futures).

Marine pollutants originate primarily from terrestrial sources but can also arise from sea-based activities (Figure 31.2) such as shipping, fishing, exploration and extraction of oil and gas or other minerals, or the harnessing of offshore energy resources.[1] Pollutants may be discharged from a specific location ('point sources'), or can enter the environment over a widespread area ('diffuse' or 'non-point' sources). For example, marine oil spills or wastewater plant discharges are point sources, while nutrient runoff associated with agricultural fertiliser use is a diffuse source. Rivers and the atmosphere can transport pollutants from land-based sources to the sea, the latter potentially over great distances, such that anthropogenic substances can be detected in some of the remotest places on earth. Clearly, it is simpler to identify and control point-source pollution than diffuse sources. A range of key pollutants impacting on the Irish coastal zone are explored below, along with some of their effects. These include nutrients, pathogens and marine litter (including plastics, oil, heavy metals and noise).

THE STATE OF IRELAND'S COASTAL AND MARINE WATERS

Nutrient enrichment and ecological status

The enrichment of coastal waters with nutrients such as phosphorus and nitrogen, which can upset the natural balance of aquatic ecosystems, is a common problem in coastal waters around the world. In Ireland, the problem of nutrient enrichment and eutrophication – a term used to describe the excessive growth of algae fuelled by the presence of too many nutrients – is mostly restricted to inshore waters along the southern and eastern coasts.

Fig. 31.2 PATHWAYS FOR THE TRANSFER OF POLLUTANTS FROM AIR, LAND AND WATER TO THE OCEANS. Pollutants can enter the coastal and marine environment from both land- and sea-based activities, including industry, agriculture, transport, wastewater discharges and runoff, mining, and the extraction of hydrocarbons. Although dumping at sea is now illegal, legacy issues and accidental losses at sea can still arise. Rivers and airborne transport are important pathways that can deliver pollution to the sea from sources sometimes far inland. While some pollutants have a clearly definable point of origin, many others come from diffuse sources that can make them particularly difficult to address and manage. Furthermore, while many pollutants originate near their point of impact, some have important regional, or even global, transboundary causes and implications. [Data source: adapted from OSPAR quality status report 2010]

The most obvious example of eutrophication in Irish waters is the excessive growth of sea lettuce (*Ulva* spp.), a green seaweed or macroalgae that can grow quickly when there are excessive nutrients present in the environment.

SEA LETTUCE GROWTH IN RESPONSE TO HIGH NUTRIENT LEVELS

Robert Wilkes

The seaweeds commonly known as sea lettuce are made up of several species of the genus *Ulva* (Figure 31.3). These species are commonly found around Ireland's coasts and are a normal part of the marine flora. In undisturbed environments, they are found throughout the intertidal and subtidal zones and their distribution is related to their requirements regarding light, temperature and nutrients. However, sea lettuces are considered opportunistic species and when nutrient levels increase, they can utilise these resources to grow very quickly. When conditions are suitable the algae multiply, creating very large accumulations.

Algal growths are a feature of inshore coastal regions around the world, with notable problems occurring, for example, in western France and eastern China. The 2008 Olympic Games in China were famously impacted by the world's largest recorded seaweed bloom, with over one million tonnes of seaweed removed from the Qingdao region. Although problems of this scale are not seen in Ireland, the issue of too much sea lettuce growth, caused by nutrient pollution, is evident in many bays and estuaries around these coasts.

These seaweed blooms (Figure 31.4) can form extensive mats on mudflats and sandflats, which can lead to the accumulation of rotting algae when they begin to decay in mid-summer. These accumulations can affect the underlying sediments, causing changes in the structure and function of the animal communities that live on the seafloor. Mats can also affect the ability of other plant communities, such as seagrass, to survive by blanketing the substrate and smothering underlying organisms, which can lead to a decrease in dissolved oxygen as

Sea Lettuce (*Ulva*)

Fig. 31.3 SEA LETTUCE. Sea lettuce is the common name for seaweed of the genus *Ulva*. These green algae are a natural component of marine flora and are found in all Irish estuarine and coastal waters. Under the right conditions, and particularly when assisted by abundant supply of nutrients from terrestrial runoff, certain species of *Ulva* have the ability to grow rapidly and proliferate, creating large accumulations of biomass known as blooms. [Source: Robert Wilkes]

the organic matter contained in the bloom decays. This can result in low (hypoxic) or no oxygen (anoxic) conditions in the underlying sediments, as well as sulphide poisoning, which can be harmful to bottom-living animals like worms, shellfish and even some types of fish.

Fig. 31.4 SEAWEED BLOOM IN CLONAKILTY HARBOUR, COUNTY CORK. The large sheltered intertidal harbour at Clonakilty in west Cork has been affected by accumulations of green algae for many years, fed by nutrients from the surrounding hinterland. The seaweeds begin to bloom in late spring and reach peak biomass during the summer, forming extensive mats on the mud- and sandflats. As well as releasing unpleasant odours as they decay, these accumulations can affect the underlying sediments, causing changes in the structure and function of the animal communities that live on the seafloor. [Source: Dr Liam Morrison, National University of Ireland, Galway]

On some mudflats with overlying algae, the anoxic layer in the sediment can be at or near the surface, which can lead to odour issues when hydrogen sulphide (H_2S) gas is released (Figure 31.5).

In Ireland, macroalgal (seaweed) or microalgal (phytoplankton) blooms leading to the depletion of dissolved oxygen in the water column are thankfully rare. Low oxygen conditions near the seabed are apparent in some urbanised estuaries, but these are mostly linked to external inputs of organic matter (that is, from sewage inputs) rather than the growth and decay of algal blooms. However, while historically green tides were reported in Belfast Lough and Dublin Harbour, they have increased in frequency and abundance on many Irish coasts. Since the early 2000s recurring blooms of green algae have become a feature of summer conditions in some embayments and estuaries along the south coast. Areas such as Clonakilty Bay, the Argideen Estuary, Courtmacsherry Bay and the Colligan Estuary have large mats of green algae covering the intertidal zone, with accumulations of washed-up algae along the upper shore line, leading to a public nuisance. The local authorities have put in place beach clearing programmes to mitigate any potential public risk, but longer-term solutions are needed to reduce this problem. These can include actions such as improvements in sewage treatment and better agricultural practices, both of which can reduce the transfer of nutrients from both point sources and diffuse sources to the sea.

Fig. 31.5 OXYGEN-DEFICIENT BLACK MUD FOUND UNDER ROTTING LAYERS OF SEA LETTUCE ON THE FORE-SHORE. The blackness is caused by the presence of metal sulphides, which form as a result of the reaction between hydrogen sulphide and metal ions present in the mud. [Source: Robert Wilkes]

Nutrients can enter our coastal waters from a variety of sources; the main source of nitrogen is runoff from agricultural land (pasture) and the main sources of phosphorus are wastewater treatment and agriculture. Nationally, over 80 per cent of nitrogen and 50 per cent of phosphorus inputs to surface waters, and ultimately the sea, come from non-point sources.[2] The spatial distribution of nutrient inputs shows the relative importance of different sources in different regions (Figures 31.6 and 31.7).

A variety of methods are used to assess the extent of eutrophication in Irish waters. In nearshore areas, the Irish Environmental Protection Agency (EPA) uses an assessment scheme that looks at nutrient levels and the extent of algal growth to detect eutrophication in estuaries and coastal waters.[3] In a study carried out by the EPA in 2017, 18 per cent of waters examined were classified as either eutrophic or showing signs of becoming

eutrophic. Moving away from the coast, the OSPAR assessment method is designed to assess eutrophication in both nearshore and offshore marine waters. In 2017, the OSPAR assessment confirmed that the problem of eutrophication is confined to inshore areas, with little or no evidence of nutrient enrichment in the vast marine areas to the west and south of Ireland (Figures 31.8 and 31.9).

The broader ecological health of Ireland's estuaries and coastal waters is also assessed under the requirements of the EU Water Framework Directive (WFD).[4] In addition to looking at the impact of nutrient enrichment, these assessments also look at the potential impacts that might arise from other pressures, such as from chemical substances or from physical modifications to the morphology (that is, the shape and structure) and hydrology of these waters. For example, maintenance dredging to deepen a navigation channel may impact organisms living on the seabed adjacent to the channel. Water Framework Directive assessments

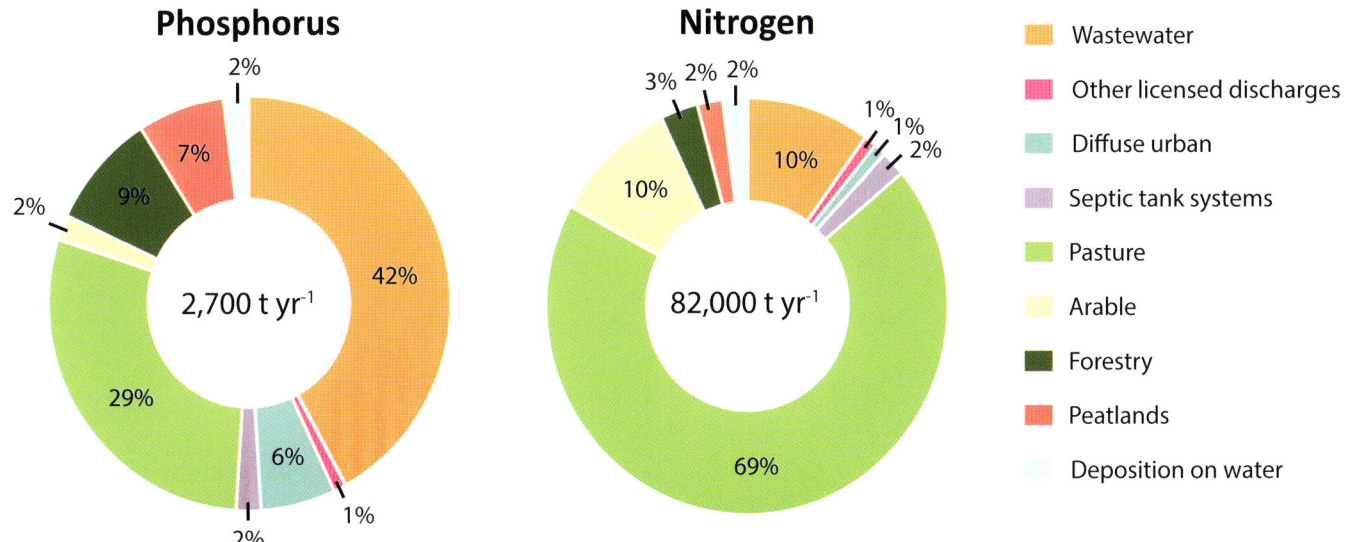

Fig. 31.6 RELATIVE CONTRIBUTION OF SOURCES OF PHOSPHORUS AND NITROGEN LOADS TO SURFACE WATERS IN THE REPUBLIC OF IRELAND. [Data source: EPA]

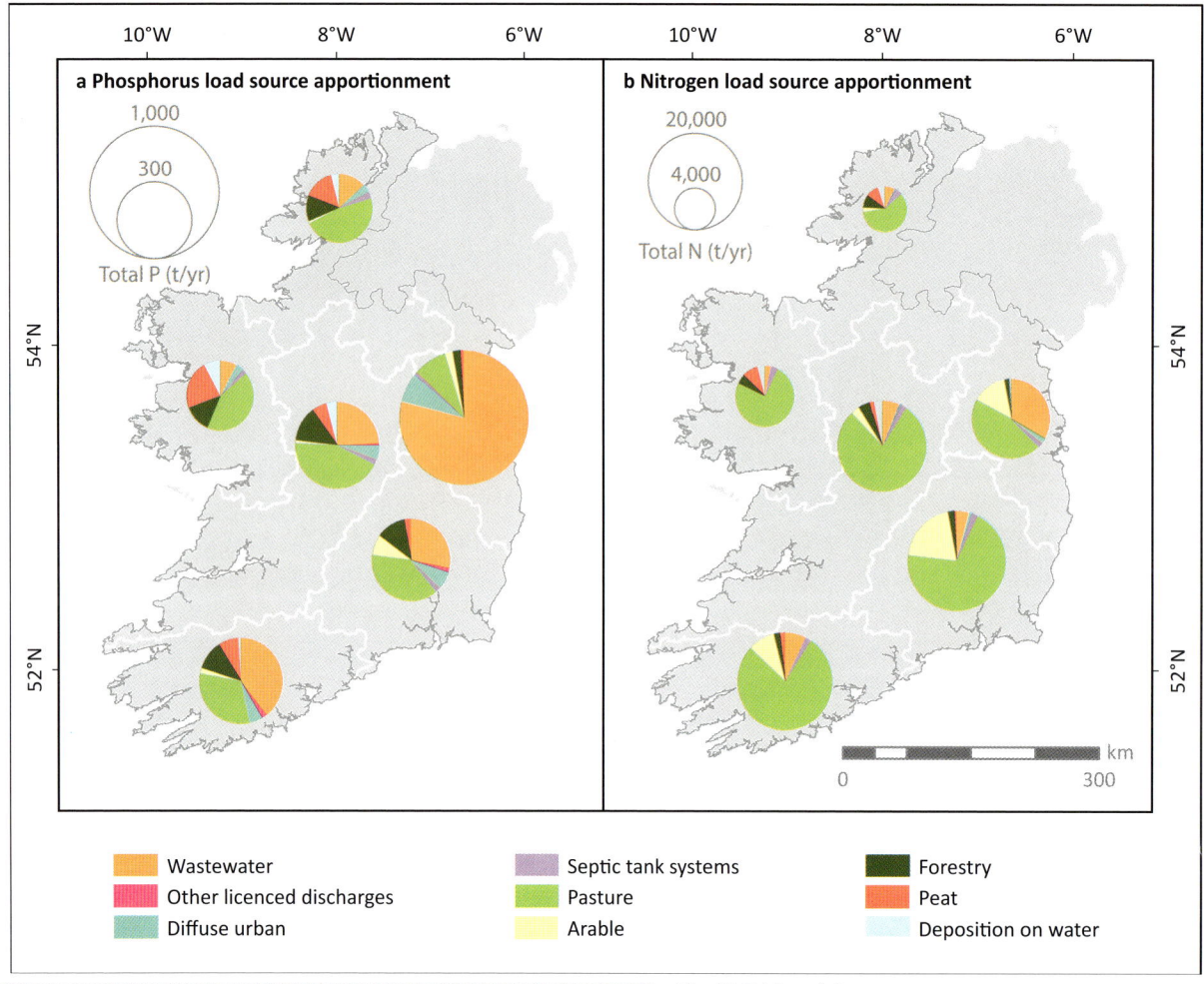

a Phosphorus load source apportionment

1,000
300
Total P (t/yr)

b Nitrogen load source apportionment

20,000
4,000
Total N (t/yr)

km
0   300

▉ Wastewater	▉ Septic tank systems	▉ Forestry
▉ Other licenced discharges	▉ Pasture	▉ Peat
▉ Diffuse urban	▉ Arable	▉ Deposition on water

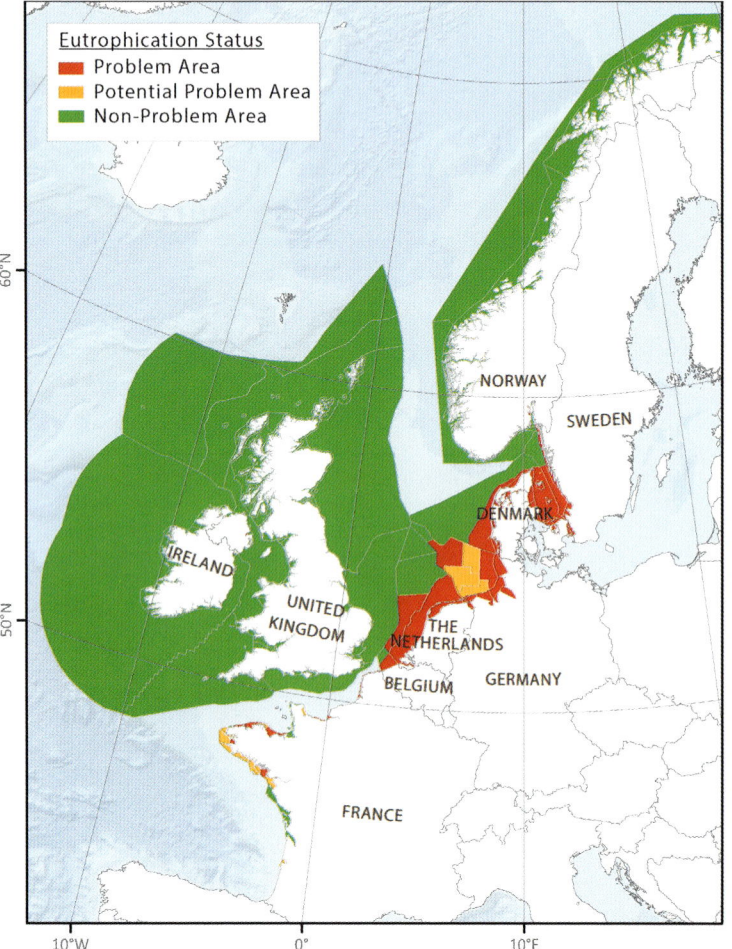

Eutrophication Status
▉ Problem Area
▉ Potential Problem Area
▉ Non-Problem Area

Fig. 31.7 (above) LOAD APPORTIONMENT OF a. PHOSPHORUS AND b. NITROGEN EMISSIONS TO WATER BY REGION IN THE REPUBLIC OF IRELAND. From the graphs it is noticeable that agriculture, and especially pastureland, is the main source of both phosphorus and nitrogen discharges throughout most of the island. The striking exception is the major urban area of Dublin, which is responsible for by far the greatest overall amount of phosphorous, more than three-quarters of which comes from domestic and industrial wastewater discharges. [Data source: Mockler, E.M., Deakin, J., Archbold, M., Gill, L., Daly, D. and Bruen, M., 2017. Sources of Nitrogen and Phosphorus Emissions to Irish Rivers and Coastal Waters: Estimates from a nutrient load apportionment framework. *Science of the Total Environment*, 601–602, pp. 326–339]

Fig. 31.8 (left) EUTROPHICATION AROUND THE COASTS OF NORTHERN EUROPE. In the regional context of northern Europe, the more open waters of the north Atlantic can be seen to enjoy generally good water quality. The most serious levels of over-enrichment by nutrients are mostly confined to the southern North Sea basin, where the Rhine, Elbe and other major European rivers carry agricultural and industrial pollutants from the population heartlands of the interior to the sea. This links into a second northern European eutrophication hot spot, in the Kattegat basin between Denmark and Sweden, where pollution-laden waters from eastern Germany, Poland and Russia leave the very enclosed and shallow Baltic Sea and discharge into the wider world ocean. Minor problem areas can also be seen along the Atlantic seaboard of France, primarily in the nearshore areas of Brittany, where abundant shellfish farming takes place and, further south, at the mouths of the Loire and Garonne rivers, both of which drain large, predominantly agricultural catchment areas. In the case of Ireland, Norway and the United Kingdom, eutrophication can and does also occur, but for these countries it is essentially a localised coastal/nearshore problem that does not show up on a map of this scale. [Data source: OSPAR, 2017]

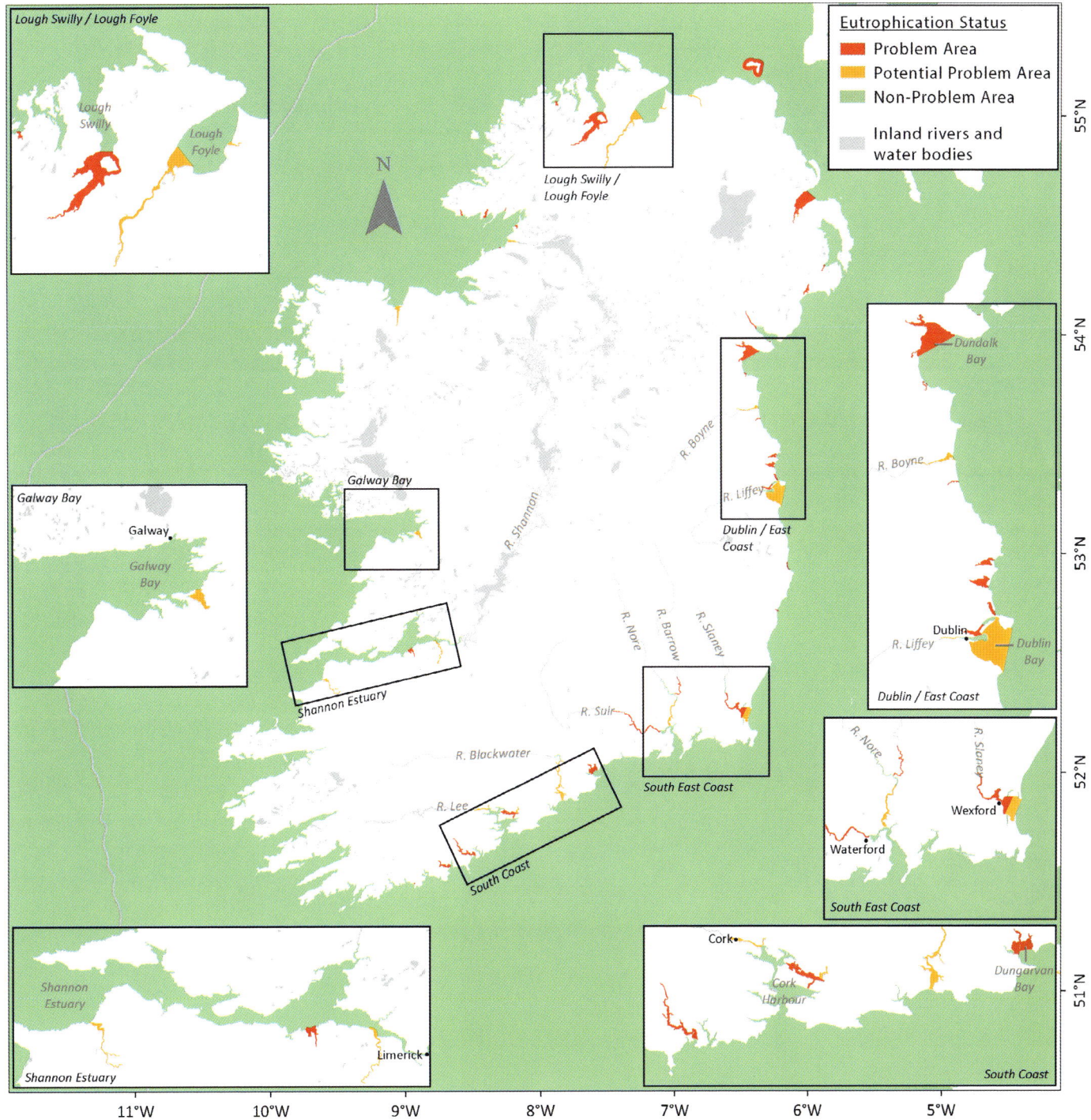

Fig. 31.9 THE EUTROPHICATION STATUS OF WATERS AROUND THE IRISH COAST. The nearshore and offshore waters surrounding the island of Ireland are free from any major eutrophication problems. Some issues may, however, be found in the more enclosed transitional waters of many estuaries and embayments, especially in Dublin Bay, inner Cork Harbour, Carlingford Lough, Belfast Lough, and the mouths of the Barrow, Suir and Slaney rivers on the south and east coasts. Here, the combination of shallow tidal waters, along with the arrival of fresh, sometimes nutrient-rich waters from the agricultural and urban areas inland, along with the presence of often abundant silt and clay sediments to act as nuclei for the accumulation of nitrates and phosphates, can lead to over-enrichment and can encourage the growth of algal blooms, particularly during the summer months. The apparent problem area around Rathlin Island appears to be an anomaly, but this is an artefact of the data as reported by OSPAR. [Data source: OSPAR Commission, 2014. Marine Litter Regional Action Plan]

also look at a broader range of biological indicators such as fish, benthic invertebrates (bottom-living organisms with no spine) seagrass and saltmarsh, in addition to the phytoplankton and macroalgae that are the main focus of the eutrophication assessments.

The WFD classification scheme for water quality is based on five status classes: high, good, moderate, poor and bad. High status

is achieved when the biological, chemical and morphological conditions are associated with no, or very limited, human pressures. In the most recent assessment, just under a third (31 per cent) of estuaries in Ireland were in 'good' ecological status according to the WFD criteria. This is not entirely surprising as many of Ireland's major towns and cities are located close to estuaries, which means these waters receive inputs directly from multiple point sources,

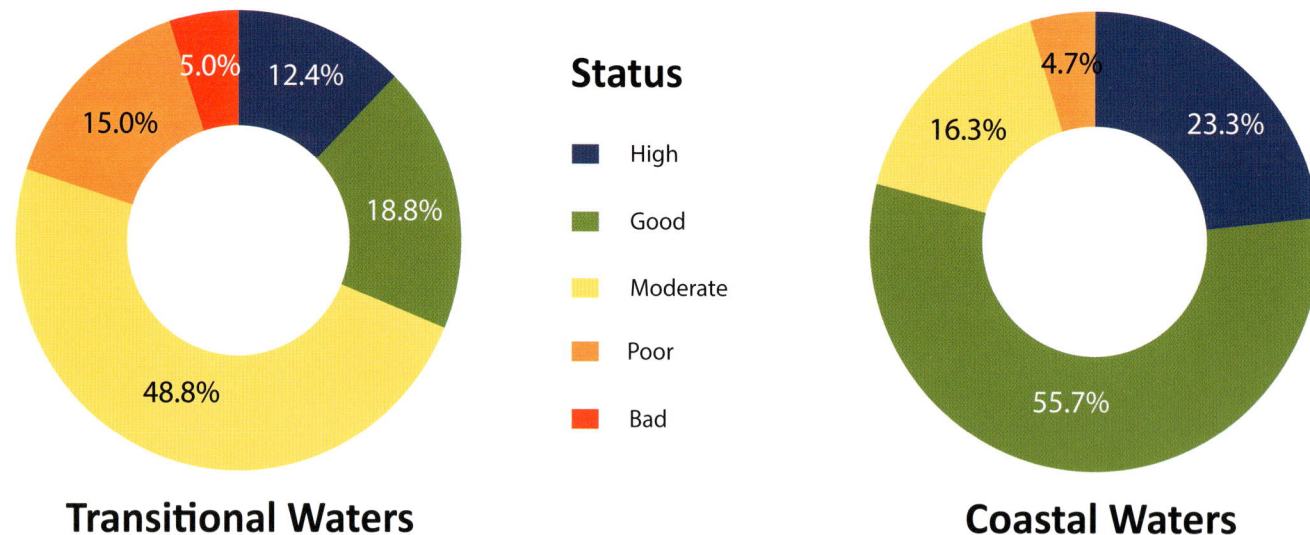

Status

■ High

■ Good

■ Moderate

■ Poor

■ Bad

Transitional Waters

Coastal Waters

Fig. 31.10 Percentage of transitional waters and coastal waters in each of the five Water Framework Directive quality classes in the Republic of Ireland. The Water Framework Directive (WFD) defines ecological status based on the assessment of general physico-chemical parameters (for example, salinity, temperature, nutrient concentrations, oxygenation) and biological quality elements (for example, phytoplankton, benthic invertebrates, seaweeds, angiosperms and fish). Water bodies are classified on a five-point scale from 'high' (generally unimpacted by anthropogenic influences) to 'bad' (several of the ecological parameters are in a degraded state). The WFD assigns the overall status based on the lowest class of the elements monitored in a water body. The key goals of the WFD are to ensure all Irish water meets at least 'good' status and that no deterioration occurs. [Data source: Environmental Protection Agency, 2017. *Water Quality in Ireland 2010–2015*]

such as sewage treatment plants and industrial discharges. Estuaries also receive inputs from the surrounding catchment, in many cases from a combination of multiple point and diffuse sources including forestry and agriculture. In contrast, the majority of coastal waters (79 per cent) are in 'good' or 'high' ecological status, which reflects the lower intensity of activities causing water pollution in Ireland's coastal environment, as well as the dynamic nature of the waters off Ireland's coast (Figure 31.10).

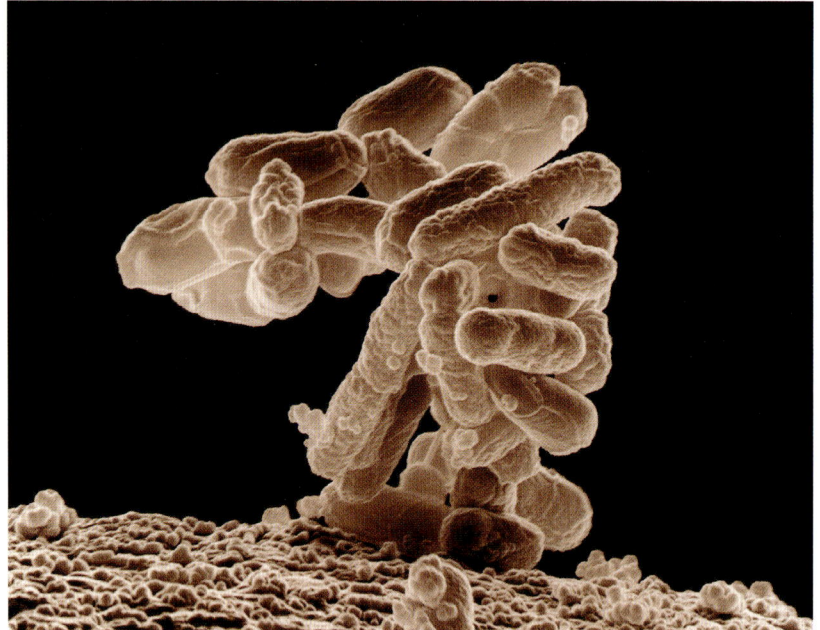

Fig. 31.11 A low-temperature electron micrograph of a cluster of *Escherichia coli* (*E. coli*) bacteria. Each individual bacterium is oblong shaped. *E. coli* and intestinal enterococci are found in the gastrointestinal tracts of all warm-blooded organisms. Most varieties of *E. coli* are harmless, but a few strains can lead to poisoning and disease in vulnerable people, including pregnant women, the very young, the elderly and people with weakened immune systems. Since 2011, local authorities in Ireland have been obliged to sample (or arrange for sampling of) bathing waters for *E. coli* and enterococci in late May, prior to the start of the bathing season, and then at regular periods from 1 June until 15 September. The results of this sampling are then submitted to the EPA, so that they may be assessed for compliance with relevant Irish and European water quality standards. Should any water sample from a bathing area fail to meet these standards, appropriate notifications to the public are then posted at the location, as well as occasionally via other media. [Source: Photograph by Eric Erbe, digital colourisation by Christopher Pooley. Photograph courtesy of the United States Department of Agriculture, Agricultural Research Service. https://www.ars.usda.gov/oc/images/photos/mar05/k11077-1/]

Microbial Pathogens: shellfish and bathing waters

Another possible source of pollution in Irish coastal waters, and of most concern for shellfish-growing waters and bathing waters, is microbiological contamination by bacteria and viruses. The dominant source of microbiological contamination in the marine environment is associated with inputs of human and animal faecal waste, which may cause human gastrointestinal illness. Bacteria found in the gastrointestinal tract of warm-blooded organisms, *Esherichia coli* (*E. coli*) (Figure 31.11) and intestinal enterococci, are used as indicators of faecal contamination in marine waters. Water-borne viral contamination, such as norovirus (NoV), can also lead to gastrointestinal illness, often called the winter vomiting bug.

Bivalve molluscs, such as mussels and oysters, are filter feeders and readily accumulate microbiological contaminants. The presence of *E. coli*, together with other indicators, is therefore also used to assess the quality of water used to cultivate shellfish. Its detection in environmental samples is an indication of the presence of faecal matter, and potentially the presence of disease-causing organisms. This information is used to classify waters to ensure that shellfish are appropriately treated before going on the market and to verify that the final product meets the appropriate food

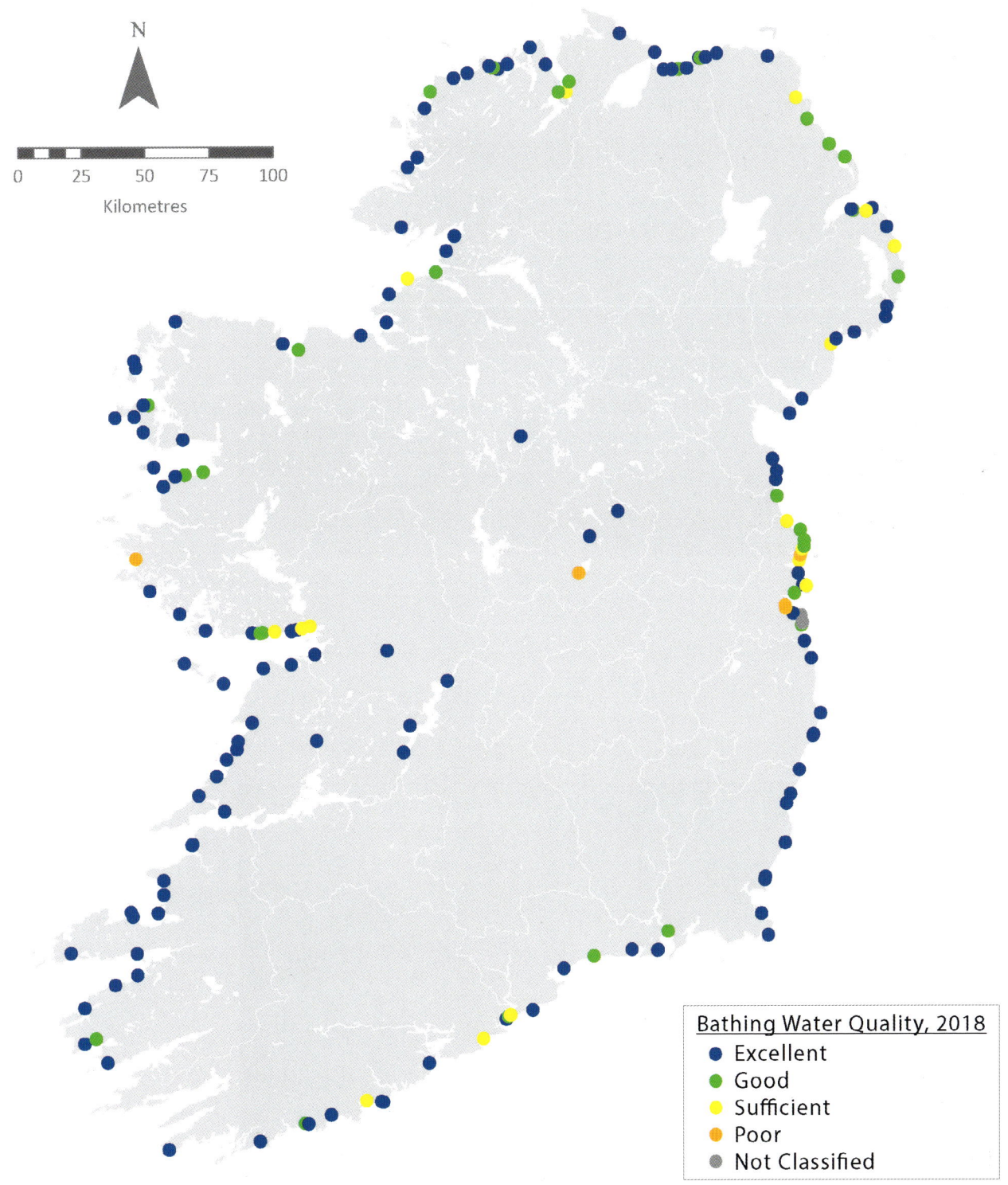

Fig. 31.12 BATHING WATER QUALITY STATUS IN IRELAND (2018). The EU Bathing Waters Directive requires member states to identify popular bathing places in fresh and coastal waters and monitor and assess them for presence of faecal bacteria (*E. coli* and intestinal enterococci) and other polluting substances at regular intervals between May and September of each year. The monitoring authorities are required to inform the public regarding the results (in Ireland this is done via a new bathing water information website, beaches.ie), and the member state must also submit an annual report of the national results to the European Environment Agency. Although both inland and coastal waters are monitored under the terms of the directive, the latter are subject to much stricter quality criteria. In 2018, Ireland identified and reported the results for 145 bathing waters overall, which is 0.7 per cent of all bathing waters in Europe. Of these, 71 per cent were deemed to be of 'excellent' quality, placing the country twenty-third in the rankings among all EU member states, while the United Kingdom (which includes results reported for Northern Ireland) was ranked twenty-five, with 63.2 per cent of its waters achieving this status. Although both of these results are below the EU-wide average of 85.1 per cent of 'excellent' bathing waters, most of the remaining bathing water areas around the island were of 'good' or 'sufficient' status, and very few (mostly around the much more urbanised coastline of Dublin Bay) were considered to be 'poor'. It is clear, however, that improvements still need to be made, in particular with regard to many small-scale and localised planning problems in coastal areas that can adversely affect overall results, and it is known that the quality of bathing waters in many places around the coast can also be impacted by heavy rainfall and other extreme weather events. [Data source: European Environment Agency]

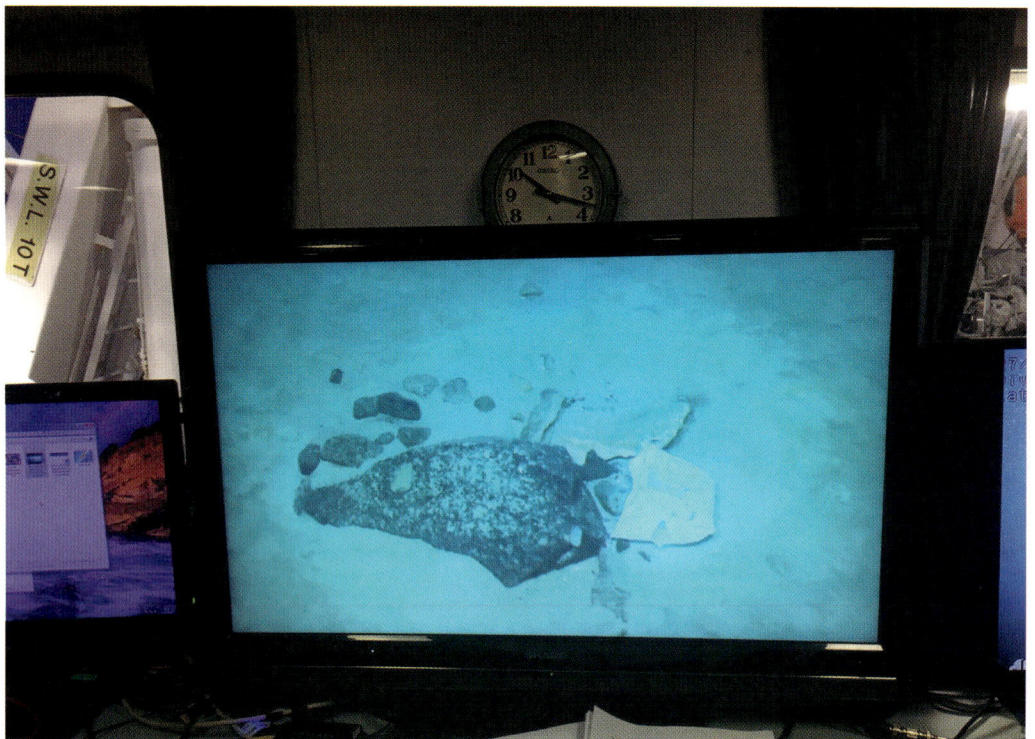

Fig. 31.13 A PLASTIC BAG FOUND AT 2KM WATER DEPTH IN THE PORCUPINE BANK CANYON, APPROXIMATELY 322KM DUE WEST OF IRELAND. Plastic waste was discovered on a research expedition led by Dr Aaron Lim of University College Cork, on board the Marine Institute's RV *Celtic Explorer*, in August 2019. During the recovery of lander systems used to monitor cold-water coral habitats in the Porcupine Bank Canyon, researchers discovered a large plastic bag at 2,125m water depth, caught between rocks within a submarine canyon. This image was captured via the Remotely Operated Vehicle (ROV) *Holland 1* and highlights the grim reality of human reach to the deep oceans. These cold-water coral habitats are vital biodiversity hotspots in otherwise barren parts of the deep seabed. The research team are using the acquired data to determine the controls on habitat variability and to investigate if microplastics are being fed to the corals from above. This may have serious implications for coral health and for the survival of habitats that have existed for thousands of years. [Source: Zoë O'Hanlon, courtesy of Aaron Lim]

Fig. 31.14 DISCARDED FISHING EQUIPMENT IN KILLARY HARBOUR, CONNEMARA, COUNTY GALWAY. Divers from the Netherlands and from Dive Centre in Galway, with a mussel boat to provide surface support, engaged in 'Operation Stone and Pots' in Killary Harbour in September 2018. The primary objective of the exercise was to recover abandoned lobster pots from the seabed (a.) However, during the operation, a 'ghost net' (b) was also discovered and was brought to the surface for recovery/recycling. [Source: Sarah Tallon, Dublin]

safety standards.

Bathing water quality in Ireland has been of a consistently high standard over several years (Figure 31.12). The 2018 EPA report on bathing water quality shows that 137 out of a total of 145 (94 per cent) of sampled sites for bathing waters met the EU minimum quality standards, with 125 of these classed as being either 'excellent' (103) or 'good' (22) quality.[5] Twelve were classed as 'sufficient' but at risk of pollution. Five locations were assessed to be in 'poor' quality: three coastal bathing areas in County Dublin and one in County Galway as well as an inland lake in County Westmeath. In many cases improved wastewater treatment and remedial action to reduce discharges from misconnected sewerage pipes will be required to improve the quality of these bathing areas.

Litter

Marine litter includes plastics, metal, wood, glass and textiles, although plastics are by far the predominant form. The traditional term 'flotsam' is used to refer to debris that enters the water accidentally, for example due to shipwreck, accident or by being washed overboard in a storm. This is in contrast to 'jetsam', which is defined as debris that is deliberately thrown overboard, originally by the crew of a ship in distress.

Massive volumes of discarded plastics end up in the sea or on our coasts and can persist for decades if not hundreds of years, although they can degrade and break down over time into smaller fragments. Most plastics float and are driven by surface currents and wind, as is evident in a walk along an Irish beach after a storm. However, denser pieces sink and plastics and other forms of litter are also found on the seafloor (Figure 31.13). Litter can present a risk to marine

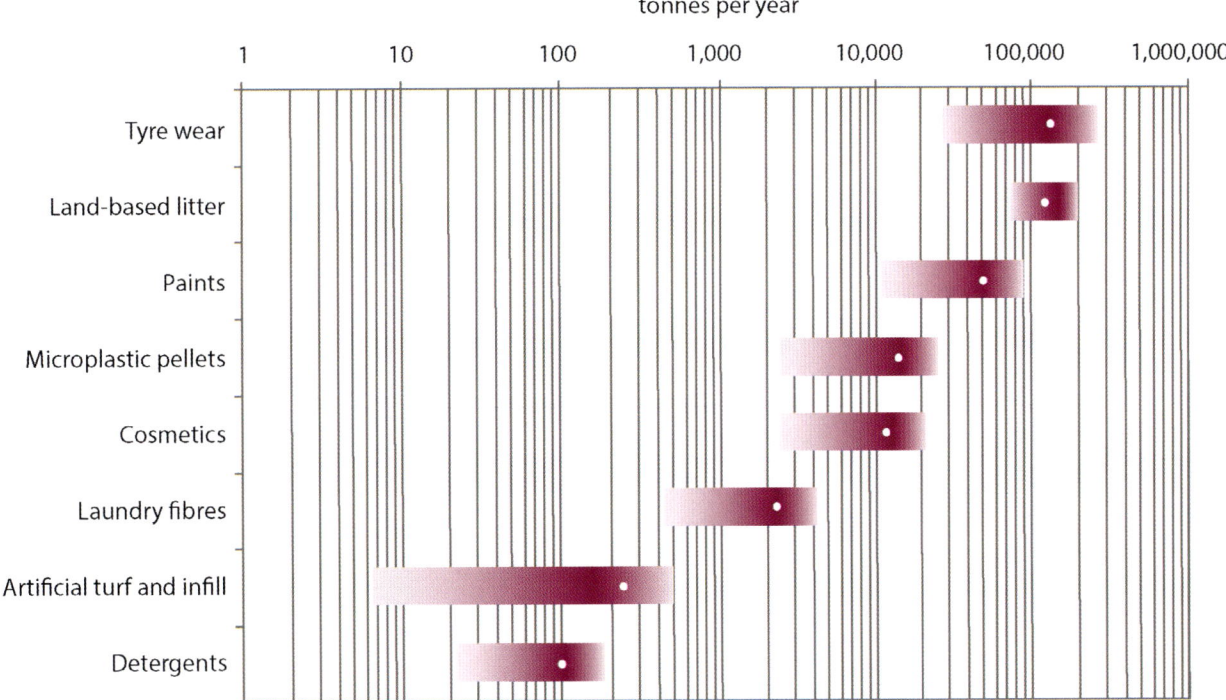

tonnes per year

Fig. 31.15 ESTIMATED SOURCES AND EMISSIONS OF MICROPLASTICS IN OSPAR CATCHMENTS (TONNES/YEAR). The bars represent the uncertainty margins of the emission, white dots represent the midpoint. [Data source: OSPAR Commission, 2017. *Assessment Document of Land-based Inputs of Microplastics in the Marine Environment.* Environmental Impact of Human Activities Series. London: OSPAR Commission]

animals, primarily through entanglement or ingestion. Lost fishing nets and crab or lobster pots, for example, have the potential to continue 'ghost' fishing for decades (Figure 31.14).

The term microplastics refers to plastic particles smaller than 5mm. They occur typically due to the breakdown of larger plastic materials, such as tyres, but may also be present due to household products, most notably exfoliating agents in personal care products, or fibres from clothing, washing into the sea (Figure 31.15).[6] The EU Marine Strategy Framework Directive identifies marine litter as one of eleven descriptors of 'Good Environmental Status',[7] and the OSPAR Convention has developed a Regional Action Plan with multiple actions to prevent litter entering the sea as well as its removal.[8] For example, the Fishing for Litter scheme serves both to remove litter from the sea and to engage with and educate members of the fishing community regarding their role and responsibilities as custodians of the sea. In addition to governmental and supranational initiatives (such as the banning of microbeads in personal-care and household products in Ireland and the UK), individual citizens can contribute significantly, by reducing the plastic waste they generate themselves, and/or engaging in community beach clean-ups, such as those organised through An Taisce's Clean Coast programme.

Case Study: Plastics in the Marine Environment

Róisín Nash, João Frias, Alicia Mateos-Cárdenas

Synthetic plastics were first mass-produced in the 1950s and were once seen as the material of the future – lightweight, strong and durable – the introduction of which into industrial processes and domestic products would increase productivity, reduce costs and enhance citizens' quality of life in a post-Second World War context. Instead, plastics have now become a global environmental issue of worldwide concern. The exponential growth in plastic production (Figure 31.16) is linked to the versatility and malleability of this material, which can be used in a wide range of applications in the medical, packaging, electric and electronic and multiple other industries.

The characteristics that made this material so unique – such as low density and resistance to water – are the same properties that have contributed to the widespread emergence of plastic pollution as an environmental issue. Despite records of plastic pollution being reported in scientific literature since the 1970s, it was only after Charles Moore discovered an extensive accumulation area of floating debris in the North Pacific Central Gyre in 1997 that the issue of plastic marine litter has entered public consciousness and become widely reported on in the mass media.

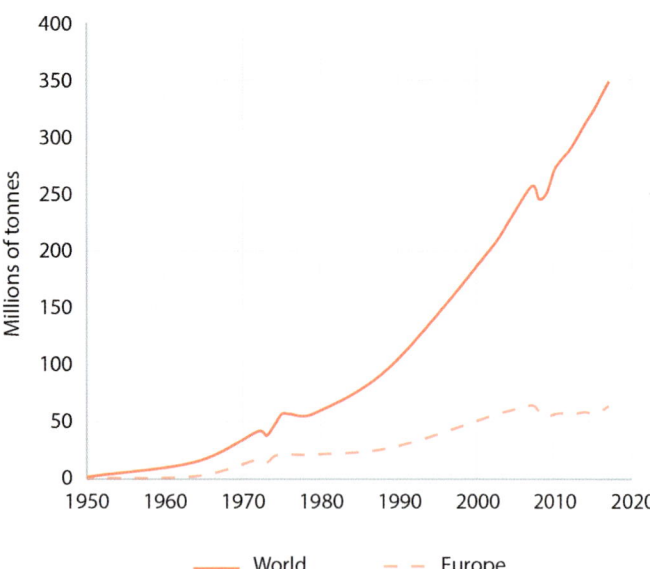

Fig. 31.16 WORLD AND EU PLASTIC PRODUCTION (1950–2017). The production and use of plastics worldwide has increased exponentially since these substances were first introduced in the mid-twentieth century. Hailed initially for their light weight, low cost of production and durability, plastics are now recognised as one of the major sources of persistent pollution, especially in the marine environment. Given current levels of public and scientific concern over the issue, as well as moves towards greater use of biodegradable and food-starch derived alternatives in many sectors, it will be interesting to see whether there will be a reversal of these trends in years to come. [Data source: PlasticsEurope Reports 2010–16]

Studies have estimated that the oceans could have between 7,000 and 35,000 tonnes of floating plastic[9] and more than five trillion pieces of plastic.[10] Numbers of synthetic fibres and fragments have been recovered from the water column and deep-sea sediments of the oceans. Regarding the Irish coast, an average plastic abundance of 2.5 particles/m³ was calculated for waters of the north-east Atlantic.[11]

In other research, marine fauna stranded on the coast of Ireland has also been studied to assess the impacts of plastic pollution. A quantification of plastic debris across sixteen species of seabirds concluded that 27 per cent of all species sampled had ingested plastic debris, with northern fulmars (*Fulmarus glacialis*) being the species showing highest presence (93 per cent) of plastic overall.[12] Ingestion of plastics by seabirds has been found to be increasing globally and it has been estimated that 99 per cent of all seabird species will have ingested plastics by 2050.[13] Furthermore, the digestive tracts of stranded and by-caught cetaceans between 1990 and 2015 in the Republic of Ireland and Northern Ireland were found to have ingested at least one plastic fibre or fragment.[14]

Macroplastic items fragment into smaller microscopic plastic particles, due to ultraviolet (UV) radiation and mechanical abrasion. However, the timescale of this process is unknown, which results in the persistence of microplastics in the marine environment.[15] Microplastics are currently the latest in a line of novel pollutants to appear on our shores and in our waterways. The reason why they have gone unnoticed for so long may be reflective of their size. Once pollution is visible it is easier for the general public to recognise the problem, particularly if it is on our doorstep. The upper size of what is termed a 'microplastic' is equivalent to the opening for the headphone cable on a mobile phone (about 3.5mm), while the lower range is generally considered to be about one micron. To put this in context, a human hair is about 75 microns thick, a red blood cell is 5 microns and the width of a strand of spider web silk is 3–8 microns. The unaided human eye cannot see anything smaller than forty microns, and this alone presents one of the difficulties when researching microplastics within this lower size range limit.

Microplastics can be further defined in accordance to their origin: 'primary' refers to microplastic particles that are produced deliberately and are <5mm, while 'secondary' microplastics result from fragmentation and/or degradation processes operating on larger plastic or macroplastic items that are already within the environment. There are six main types of plastics used in a wide range of commercial and industrial applications. These are often referred to as the 'big six' (Table 31.1), and they each have associated Resin Identification Codes (RIC), which indicate the type of plastic that an item is made from. These numbers are intended to help consumers know whether and how to recycle various plastic products and packages.

The most common physical forms in which these microplastics exist are pellets, fragments, fibres, films, rope and filaments, microbeads, sponge and/or foam and rubber (Figure 31.17). The type and shape of

Acronym	Type of plastic	Common applications	♺
PET	Polyethylene terephthalate	Shampoo bottles, body wash, fizzy drink bottles, clothing	♺1
HDPE	High density polyethylene	Milk, bleach and washing-up liquid bottles, jugs	♺2
PVC	Polyvinyl chloride	Furniture, food trays, buildings (guttering, windows, doors), vinyl discs	♺3
LDPE	Low-density polyethylene	Carrier bags, multi-can loop holders, plastic wrap, bin liners	♺4
PP	Polypropylene	Food packaging and car parts, film pouches, crates, container caps	♺5
PS	Polystyrene	Protection of electrical goods, cups, foam meat or fish trays, plastic cutlery, packaging, insulation, VHS tapes	♺6
O	Other	Any other complex plastics or plastic mixes that do not fall into any of the above categories	♺7

Table 31.1 THE SIX MOST COMMON TYPES OF PLASTIC AVAILABLE, SOME OF THE COMMON APPLICATIONS FOR THESE, AND THEIR ASSOCIATED RESIN IDENTIFICATION CODE. [Data source: adapted from Katz, D.A., 1998. *Identification of Polymers*. Online at https://www.scribd.com/document/394985958/Polymer-Identification (From original source: Kollman C.S. (1994) *Chem 13 News.*)]

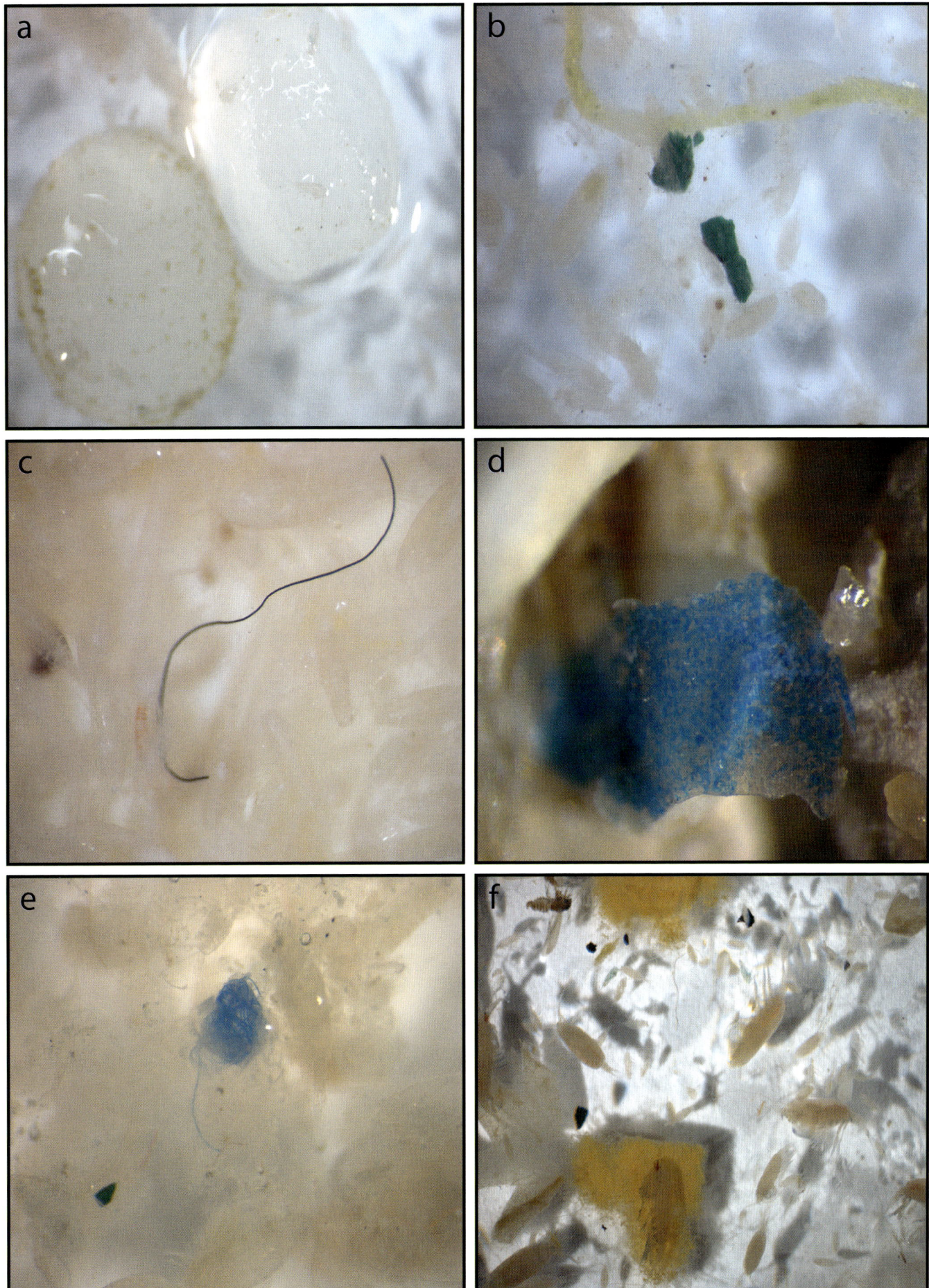

Fig. 31.17 Photomicrographs, showing different types of microplastics (particles ranging from about 3.5mm down to about *c.*1 micron in size) to be found in the environment. a. pellets; b. fragments; c. fibres; d. film; e. rope and filaments; f. sponge/foam. [Source: João Frias]

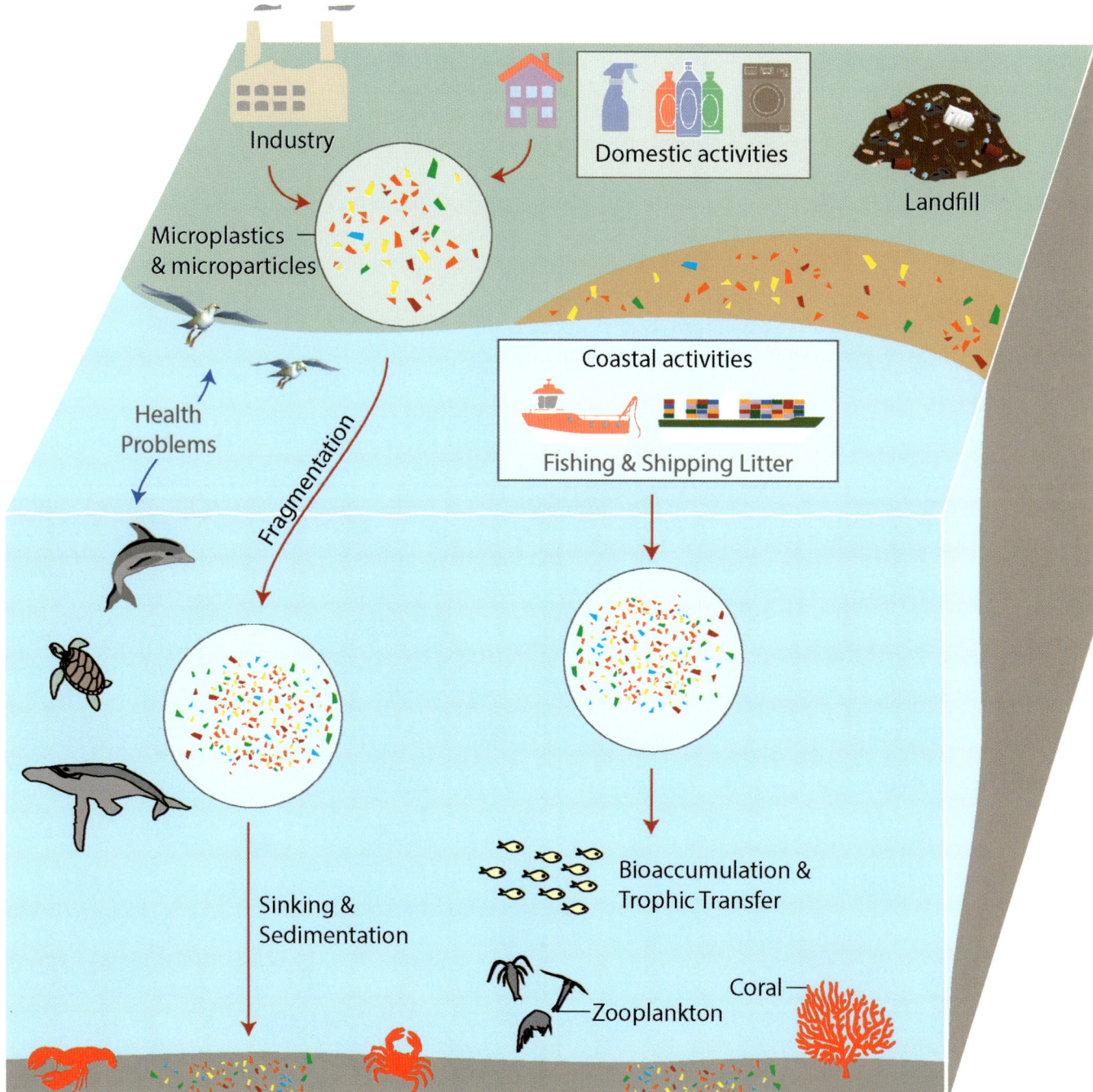

Fig. 31.18 SIMPLIFIED DIAGRAM, SHOWING PATHWAYS BY WHICH MICROPLASTICS CAN ENTER THE ENVIRONMENT. The scourge of plastic pollution can be seen worldwide, and in all environments, including in the seas around Ireland. Many microplastics are formed through fragmentation of larger plastic items, although primary microplastics can also be released into the environment from a variety of industrial and domestic activities, eventually making their way to the coast. While it is the larger macroplastics, and discarded plastic bags in particular, that tend to cause the most widely known health problems in marine organisms, the impacts of microplastics on environmental (and human) health are less well understood and are the subject of much current research. Landfill sites have been a problem in the past, but under current best practice are required to be capped and sealed to prevent leachate (and hence plastics) escaping. [Data sources: GRID-Arendal, 2016. Marine Litter Vital Graphics 2016. http://www.grida.no/resources/6933; Sharma, S. and Chatterjee, S., 2017. Microplastic Pollution, a Threat to Marine Ecosystem and Human Health: A short review. *Environmental Science and Pollution Research*, 24(27), pp. 21,530–21,547]

the microplastic will give an indication as to its original source: for example, fibres result largely from ropes and clothing; while microbeads are commonly found in cosmetic products and toothpastes, where they are used as scrubbers and/or exfoliants. However, recent legislation, driven largely by consumer pressure, has imposed limits and in some cases prohibition on the use of these beads, and the quantities of these entering the environment will likely reduce in the future.

To mitigate against pollution from plastics, including microplastics, one must first identify the primary source(s) from both land and sea. Marine or maritime-based sources are more easily identified since they most commonly relate to industries that operate directly at sea, including fishing, aquaculture, shipping, tourism, recreational activities, etc. Items of this type may be accidentally lost (flotsam) or deliberately discarded (jetsam) at sea, and they result from material/equipment dumping, paint chippings from abrasion or removal applications, nets and ropes, etc. However, most microplastics make their way to the sea via the waterways, from onshore sources that include runoff from roads through tyre wear, construction works, use of artificial grass, leaching from landfill, etc. Some of the current measures for mitigating against water pollution, such as wastewater treatment plants, have seen microplastics incorporated into effluent and becoming redistributed on to agricultural land through the spreading of sewage sludge.

As well as seeking better understanding of the extent and distribution of microplastic contamination within both the Irish terrestrial and marine environments, significant research effort is now also exploring the potential implications of microplastic contamination for seafood and food safety. While there is no doubt that Irish seafood webs are exposed to microplastics (Figure 31.18), the potential of microplastics to act as vectors for other environmental contaminants, and the potential risk of toxic chemicals leaching from plastic and entering the food chain, remains unknown. There is also an urgent need to find some effective mitigation measures and solutions to these problems.

OIL SPILLS

The image of oiled seabirds following oil spills is perhaps one of the most enduring images of marine pollution that the public generally sees (Figure 31.19). In fact, while catastrophic oil spills can cause massive environmental damage, most of the petroleum hydrocarbons that enter the marine environment are not from such spills but from terrestrial sources, natural seeps or

Fig. 31.19 (right) OILED BIRD, NORWAY. Cleaning an eider duck at a wildlife rehabilitation centre in Norway following a spill of bunker fuel from a grounded bulk carrier off Sastein Island, approximately 200m south of the Norwegian mainland, in July 2009. Oiled seabirds are among the most visible victims of oil spillages at sea, although scenes such as this are thankfully rare in Ireland. Contingency preparations for incidents of oiling include the training of specialist response teams and stockpiling of the detergents required for treating affected birds. [Source: International Tanker Owners Pollution Federation Limited]

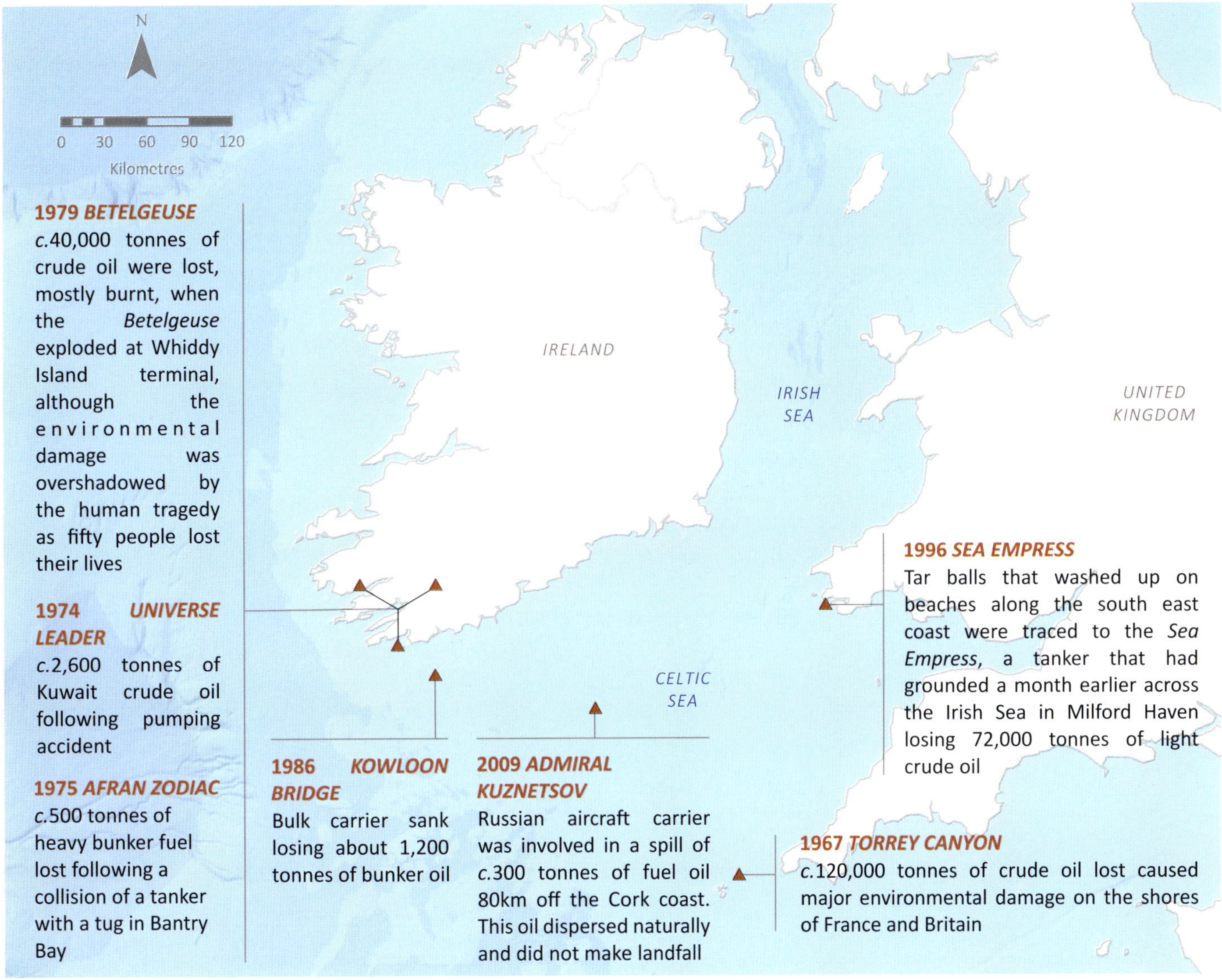

N

0   30   60   90   120
Kilometres

IRELAND

IRISH
SEA

UNITED
KINGDOM

CELTIC
SEA

1979 BETELGEUSE
c.40,000 tonnes of crude oil were lost, mostly burnt, when the *Betelgeuse* exploded at Whiddy Island terminal, although the environmental damage was overshadowed by the human tragedy as fifty people lost their lives

1974 UNIVERSE LEADER
c.2,600 tonnes of Kuwait crude oil following pumping accident

1975 AFRAN ZODIAC
c.500 tonnes of heavy bunker fuel lost following a collision of a tanker with a tug in Bantry Bay

1986 KOWLOON BRIDGE
Bulk carrier sank losing about 1,200 tonnes of bunker oil

2009 ADMIRAL KUZNETSOV
Russian aircraft carrier was involved in a spill of c.300 tonnes of fuel oil 80km off the Cork coast. This oil dispersed naturally and did not make landfall

1996 SEA EMPRESS
Tar balls that washed up on beaches along the south east coast were traced to the *Sea Empress*, a tanker that had grounded a month earlier across the Irish Sea in Milford Haven losing 72,000 tonnes of light crude oil

1967 TORREY CANYON
c.120,000 tonnes of crude oil lost caused major environmental damage on the shores of France and Britain

Fig. 31.20 LOCATIONS AND DETAILS OF SIGNIFICANT OIL SPILLS SINCE THE *Torrey canyon* SPILL, OFF THE COAST OF CORNWALL, UK (1967).

operational discharges.[16] Nonetheless, it is easy to appreciate the widespread damage that major spills associated with the production, transport or storage of oil can wreak on sensitive coastal environments, impacting on livelihoods as well as ecosystems.

The Irish coast was lucky to escape unharmed following the grounding of the supertanker, *Torrey Canyon*, off Cornwall in 1967.

The ensuing environmental disaster, the UK's and one of the world's worst, helped spark development of international oil-spill mitigation and response policy. However, while Ireland has been spared the worst impacts of catastrophic spills, there have been significant spills in, or close to, Irish waters (Figure 31.20), including oil lost as a result of the explosion of the carrier MV *Betelgeuse* in Bantry Bay in 1979.

The Betelgeuse *Disaster*

Darius Bartlett

The oil terminal at Whiddy Island, in Bantry Bay, was built by the Gulf Oil Corporation during the 1960s, to handle ultra large crude carrier (ULCC) oil tankers. These ships were too large to enter most older ports on the Atlantic, North Sea and English Channel coasts, but the deep water, westerly-facing aspect and sheltered anchorage of Bantry Bay was considered ideally suited for the purpose, especially since it was distant from any major urban areas or significant shipping lanes. The Whiddy Island terminal was planned to serve as a trans-shipment terminal, where oil could be offloaded from ULCCs, stored temporarily, and then transferred to refineries elsewhere in Europe via smaller vessels.

The 121,432 tonne oil tanker MV *Betelgeuse* had been built in 1968 and was approaching the end of its expected useful life. The ship was registered by Total S.A. at Le Havre in France. In late November 1978, the ship left the port of Ras Tanura in the Persian Gulf for

Fig. 31.21 THE MV *Betelgeuse* EXPLOSION AND FIRE AT THE WHIDDY ISLAND OIL TERMINAL, BANTRY BAY (JANUARY 1979). [Source: courtesy of *Irish Examiner*]

Leixões in Portugal, with a full cargo of crude oil. However, because of problems at Leixões, the *Betelgeuse* was instructed to divert to Bantry instead. She arrived on 4 January 1979, completed berthing two days later, and discharging operations commenced. Shortly after midnight on 8 January, fire broke out on the ship and spread rapidly, culminating in a massive explosion around 1.00a.m., followed by further explosions that caused the vessel to break in half (Figure 31.21). Much of the oil still on board the vessel ignited, generating temperatures estimated to have exceeded 1,000°C. About twelve hours after the explosion, *Betelgeuse* sank at her moorings in 40m (130ft) of water, with her stern becoming completely submerged. Fifty people died in the incident, including one salvage diver, and the subsequent costs of salvage, clean-up and compensation are believed to have totalled around US$120 million.

Gulf Oil never reopened the Whiddy Island terminal. In 1986 it was handed over to the Irish government, who sold it into private ownership in 2001. In the intervening years the terminal has seen significant investment and renewal of infrastructure and has currently a storage capacity of 7.5 million tonnes. As well as renewed commercial use, it also serves as one of five storage locations (along with the Whitegate refinery in Cork Harbour, Tarbert in County Kerry, Ringsend in Dublin and Kilroot in County Antrim) for the Irish national strategic oil reserve.

Under the National Contingency Plan,[17] the Irish Coast Guard is responsible for organising and directing marine spill response, and also directs and co-ordinates the functions of local authorities tasked with the clean-up of oil from beaches and nearshore waters. Fortunately, the Irish and European experience reflects global statistics, showing big reductions in the number of tanker spills since the 1970s, due to vastly improved mitigation measures – such as the phasing out of single-hull tankers – as well as the use of increasingly sophisticated electronic navigation aids and more accurate maritime weather forecasts (see Chapter 8: Monitoring and Visualising the Coasts).

Prevention and preparedness for significant pollution incidents is critically dependent on international co-operation. Ireland and the UK have both ratified the Oil Pollution Preparedness, Response and Co-operation Convention (OPRC) and both are contracting parties to the Bonn Agreement which aims to combat pollution from ships and offshore installations. The European Maritime Safety Agency (EMSA) also provides pollution response services to EU member

European Maritime Safety Agency
CleanSeaNet
ENVISAT 2009-02-14 10:53:38 UTC
© ESA (European Space Agency) / EMSA 2009

Fig. 31.22 OIL SPILL FROM THE RUSSIAN AIRCRAFT CARRIER, *Admiral Kuznetsov,* OFF THE SOUTH COAST OF IRELAND (FEBRUARY 2009). The 46,000 tonne Russian aircraft carrier *Admiral Kuznetsov* was refuelling in international waters, approximately 80km south-east of Fastnet Rock in February 2009, when an estimated 300 tonnes of light crude oil waste from the ship's bilges was accidentally discharged into the sea. Fortunately, the weather conditions at the time were good and the resulting slick was dispersed by waves and wind before it was able to reach shore. The spill was first identified by the European Maritime Safety Agency's CleanSeaNet satellite monitoring system. The CleanSeaNet service uses satellite imagery based on Synthetic Aperture Radar (SAR) technologies rather than visible light. Radar is able to penetrate fog and cloud cover and can also take images at night, giving worldwide twenty-four-hour coverage of maritime areas. Data from these satellites are processed and analysed for the presence, magnitude and possible sources of oil spills, along with information regarding meteorological and sea conditions in the affected area. [Source: European Maritime Safety Agency]

states, such as the ability to mobilise and co-ordinate response vessels and equipment in the event of an oil spill from a ship or offshore installation. A network of eighteen stand-by oil spill response vessels are maintained at different locations around Europe. EMSA also provides a satellite-based oil spill monitoring service known as CleanSeaNet (Figure 31.22). Once activated, the specific services come under the control and responsibility of the requesting party.

THE NATIONAL CONTINGENCY PLAN

David McMyler

To succeed as a nation and grow the economy, it is essential for Ireland to be able to trade with its near neighbours as well as globally, and the most effective and efficient way for the movement of most goods is by sea. The International Maritime Organisation estimates that over 80 per cent of all goods traded globally are carried by marine shipping. However, the sea, for all its riches and potential for development, can be one of the most hostile and dangerous environments on the planet.

Maritime crises, often unpredicted and unexpected, can arise out of a short chain of events. Once triggered, however, they can then develop and expand over much longer timeframes, and their consequences and the required responses may extend into days and months. The grounding of the MSC *Napoli* off the coast of England in 2007, and the MV *Rena* off New Zealand in 2011, highlighted the complexities involved when a major maritime incident occurs off a nation's coastline.

There is a significant risk to the state's international reputation from the mismanagement of a major maritime emergency, such as a ferry disaster or major oil tanker sinking. The likelihood of significant costs to both the economy and the environment will compound the difficulties faced by the emergency managers handling the event, while the effects of such an incident could have implications in terms of cost and loss of reputation that may continue to be felt for years after the event itself.

Ireland is a signatory to the UN Conventions on Oil Pollution at Sea.[18] This convention was set against a background of Regional Sea Agreements[19] and Regional Co-operation Agreements.[20] To support these, Ireland has enacted national legislation that designates the Irish Coast Guard as the competent authority tasked with the responsibility for preparedness for, and response to, any maritime emergency.[21] The Coast Guard has developed a National Plan, intended to promote a planned and nationally co-ordinated response to any marine pollution event or major shipping casualty. The need to deal with the mismatch between major pollution incidents and national capacity to respond is particularly challenging for small island states, including Ireland. A report into the grounding of the MV *Rena*, in New Zealand, found that '[a] major maritime casualty is both an industrial accident and a natural disaster. There will always be a gap between the worst-case risks presented by a major maritime casualty and the indigenous response capabilities of most sovereign states, especially smaller countries and particularly those with maritime domains as extensive as New Zealand's.'[22] In addition to the National Oil Spill Pollution Plan, Ireland's preparedness regime is supported by a number of international agreements and agencies. Since 2010 Ireland is a full signatory of the Bonn Agreement, a mechanism by which the governments of the Greater North Sea and its wider approaches co-operate in dealing with pollution of the North Sea by oil and other harmful substances. Ireland also has an agreement with the United Kingdom, to share personnel, expertise and resources in the event of a major oil spill within each other's EEZ.

Finally, Ireland can also draw on assistance from the European Maritime Safety Agency (EMSA), should this ever be required. Based in Lisbon, the EMSA is one of the EU's decentralised agencies, set up to provide technical assistance and support to the European Commission and its member states in the development and implementation of EU legislation on maritime safety, pollution by ships and maritime security. It has also been given operational tasks in the field of oil pollution response, vessel monitoring and in long-range identification and tracking of vessels, and has a number of support vessels and equipment stockpiles to assist member states should the need arise.

HAZARDOUS CHEMICALS AND RADIOACTIVE SUBSTANCES

The use of chemical substances has transformed society. Chemicals are used in domestic products, in manufacturing, in energy production, in medicine, in agriculture and food production and in almost every other aspect of modern life. The number of chemicals that are commercially produced and used globally is not precisely known, but it is likely that some tens of thousands of substances may be involved.

A somewhat cavalier approach was taken to the use of these chemicals by earlier generations, with little appreciation of the environmental consequences. The marine environment still retains the legacy of this disregard. Legacy substances and their breakdown products, such as the pesticide DDT, though banned for decades, are ubiquitous pollutants and may still be found in even the most remote regions of the Earth, such as Antarctica, due to long-range transport and their environmental persistence. However, attitudes have changed and strong European legislation is now in place to protect the environment and human health, including REACH,[23] which regulates the production and use of chemicals, and the Water Framework Directive,[24] which includes both inland and nearshore waters in its remit.

Tackling chemical pollution requires identifying the

Substance group	Examples	Some uses/sources
Metals	Mercury, cadmium, lead and other metals	Sources include metallurgic, mining, fossil fuel, biocides and incineration
PCBs and dioxins	Polychlorinated biphenyls (PCBs), dioxins and furans	PCBs were widely used in electrical products, especially transformers
		Dioxins and furans are unintentional by-products formed during combustion of certain materials
Pesticides, biocides	Pyrethroids, gyphosate, avermectins, chlorinated pesticides (e.g., DDT)	A wide range of substances used to control animal pests and weeds (herbicides), fungi, etc.
		Although persistentn chlorinated pesticides, such as DDT, are long banned, their persistence means we still have an environmental legacy associated with historical use
Hydrocarbons	Polyaromatic Hydrocarbons (PAH), crude oil and refined products	PAH are associated with incomplete combustion of fossil fuels Fuel spillages and releases
Brominated flame retardants	Polybrominated diphenyl ethers (PBDEs), Hexabromocyclododecane (HBCD)	Used in electrical products, textiles and other materials as flame retardants
Perfluorinated compounds	Perfluorooctanesulfonic acid (PFOS)	Wide variety of uses e.g., fire-fighting foam, fabric protector, metal plating
Organotins	Tributyltin (TBT)	Used in marine antifoulant paints (now banned)
		Organotins also have other uses including antifungals in textiles
Phthalates	Diethylhexylphthalate (DEHP)	Plasticizers
Other substances of emerging concern	Medicines Other endocrine-disrupting substances such as natural and synthetic oestrogens	Pharmaceutical and verterinary use Personal care products
Radioactive substances	caesium-137 (^{137}Cs), technetium-99 (^{99}Tc), tritium (^{3}H), iodine-131 (^{131}I)	The nuclear sector in some countries (associated with electricity generation)
		The non-nuclear sector (e.g., medical uses and offshore oil and gas industry)

Table 31.2 SOME OF THE BROAD SUBSTANCE GROUPS OF CONCERN FOR THE MARINE ENVIRONMENT WITH EXAMPLES AND USES. These groups are not mutually exclusive, and some substances could belong to multiple groups. International measures, such as EU Chemicals Policy and legislation, and the Stockholm Convention, are in place to protect human health and the environment. This can include phase-out of certain hazardous substances or restriction of use.

substances of most concern out of the vast number of chemicals that may be in use. The priority list focuses on toxic substances that accumulate in organisms over time and that are persistent, that is, very slow to break down to non-harmful substances. Some of these priority substances, such as trace metals, are naturally present in the marine environment, but anthropogenic sources can raise concentrations above their natural background, while others are synthetic and are only associated with human inputs (Table 31.2).

Hazardous substances can exhibit various, sometimes subtle, toxicological effects on organisms.[25] For instance, there is evidence that high concentrations of contaminants in marine mammals at the top of the food chain may increase susceptibility to disease by affecting the immune system.[26] Humans are also part of that food chain and while consumption of fish, especially oily fish, has very clear health benefits, seafood is also a potential source of contaminants such as mercury.[27]

WEIGHING THE HEALTH BENEFITS OF SEAFOOD CONSUMPTION AGAINST THE RISKS: A CASE STUDY ON MERCURY IN SEAFOOD

Evin McGovern and Christina Tlustos

Seafood is a highly nutritious food with clear health benefits to consumers, in particular due to the presence of long-chain n-3 polyunsaturated fatty acids in oily fish. These have a proven association with reduced heart disease and may also have beneficial effects in the neurodevelopment of children. Unfortunately, seafood can also be a dietary source of certain environmental contaminants and, most notably, is the primary dietary source of methylmercury. From a health policy perspective, it is desirable to encourage fish consumption, while at the same time ensuring intake of methylmercury is within safe levels.[28]

Mercury can enter the marine environment naturally but it is also introduced by human activities, such as coal burning, mining or as waste from other uses, although since the 1970s strict controls have dramatically reduced anthropogenic inputs of the chemical.[29] In the aquatic environment, inorganic mercury is transformed into more toxic methylmercury, primarily by microbes in sediments. This methylmercury can accumulate in marine organisms and can also magnify through the marine food chain, so that long-lived top predators – such as tuna and swordfish – can have relatively high levels of the chemical. Methylmercury is associated with adverse health effects, including impaired neurological development effects in infants and young children, a hazard that was forcefully brought home in Minamata, Japan, in the 1950s, where the occurrence of a deadly mystery disease, which led ultimately to the deaths of over 1,000 people, was traced to consumption of highly contaminated seafood following decades of industrial discharges of mercury into the bay.

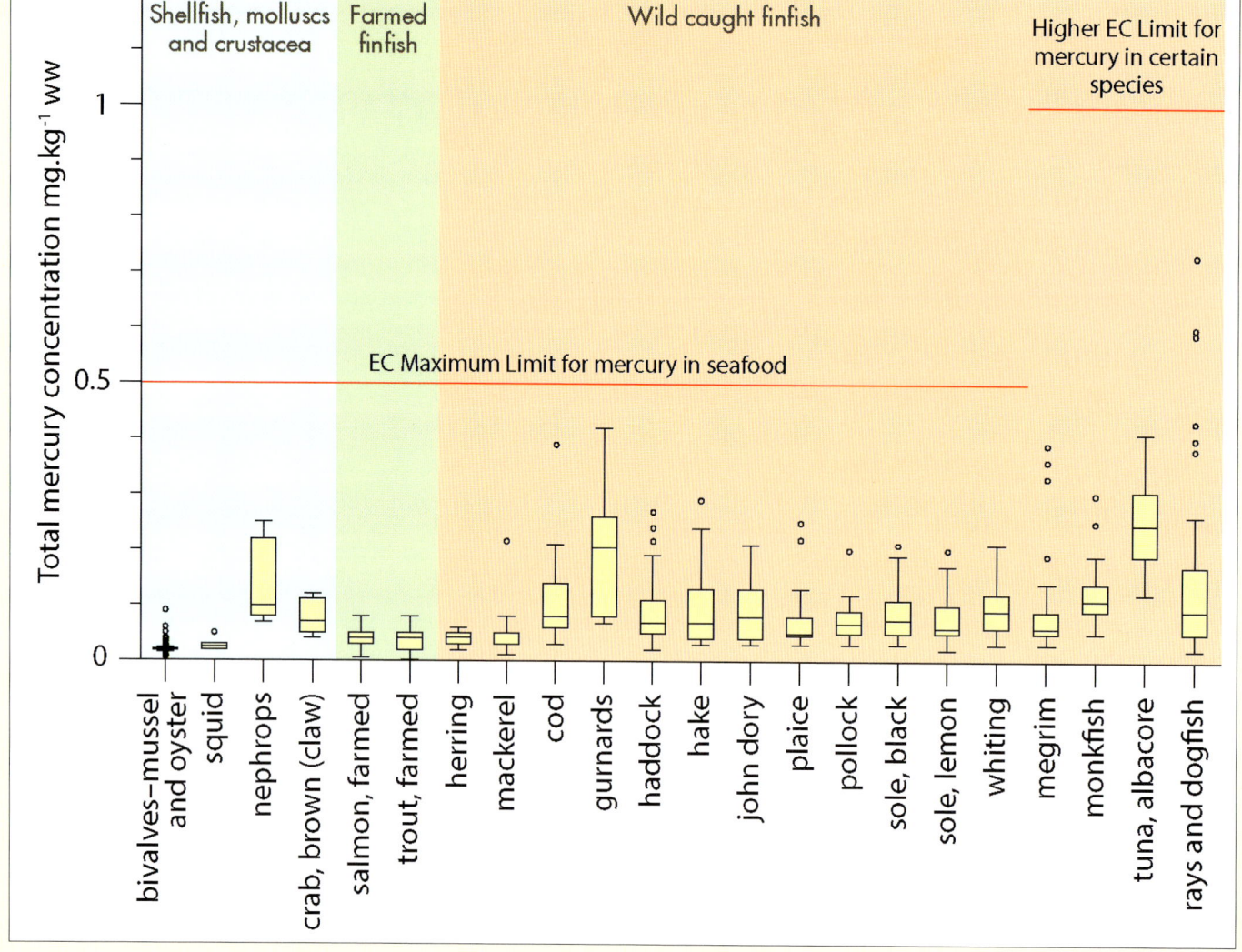

Fig. 31.23 DISTRIBUTION OF TOTAL MERCURY IN DIFFERENT SPECIES OF FISH AND SHELLFISH FROM IRISH WATERS. Regular monitoring shows that while mercury concentrations vary with species, wild fishery and aquaculture products harvested or landed in Ireland have levels that are consistently well below both the general limit for mercury in seafood and the higher species-specific maximum level set by the European Commission (red lines). [Data source: Marine Institute]

Policy makers use toxicologically based safe thresholds for dietary intakes of contaminants to protect human health and the European Food Safety Authority (EFSA) has established a tolerable-intake level of 1.3µg methylmercury, expressed as mercury per kg bodyweight per week, based on the prenatal neurodevelopmental toxicity. Taking account of this, the EC has set maximum limits for mercury in food, including fishery products. Seafood containing concentrations higher than these limits should not be placed on the market. Mercury levels measured routinely in fish and shellfish from Irish waters comply consistently with these maximum limits (Figure 31.23), and mercury intake for the typical Irish consumer of seafood is well within the safe levels.[30] The Food Safety Authority of Ireland encourages the consumption of fish as part of a balanced diet, but does advise that pregnant and breastfeeding women, women of childbearing age and young children should select fish from a wide range of species, but not eat predatory fish such as swordfish, marlin and shark, and limit consumption of tuna to one fresh tuna steak or two 8oz cans of tuna per week. Everyone should continue to eat one to two portions of fish per week, including one portion of oily fish such as salmon, to accrue the health benefits of seafood consumption. The FSAI is currently developing more detailed national advice to facilitate more frequent consumption of fish in line with recommendations of EFSA Scientific Committee.[31]

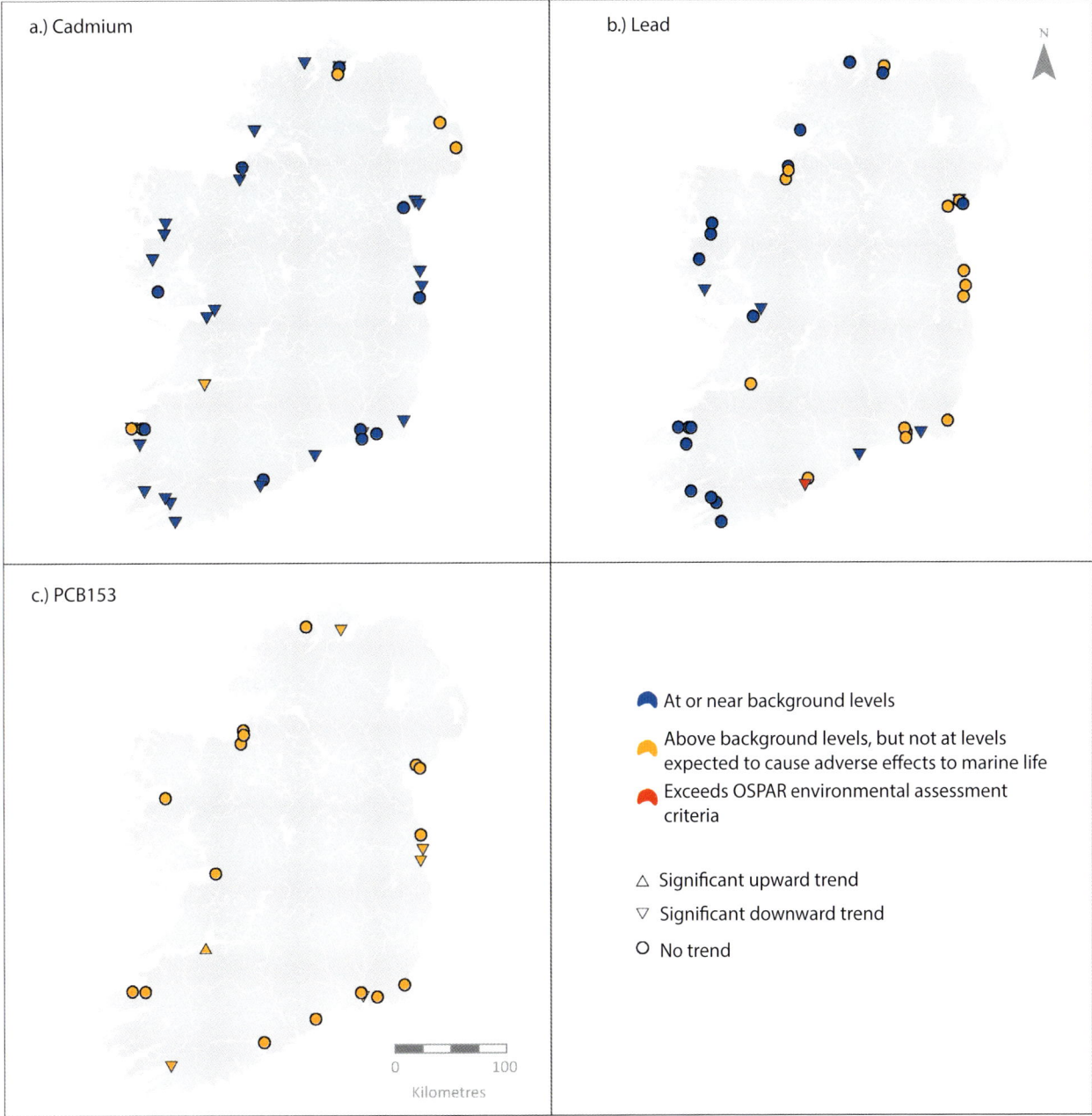

Fig. 31.24 OSPAR ASSESSMENT OF TRENDS AND STATUS OF CONTAMINANT CONCENTRATIONS IN SHELLFISH: a. CADMIUM, b. LEAD AND c. PCB153. The OSPAR assessments of contaminants in the marine environment of the north-east Atlantic are based on concentrations of contaminants in biota and sediments. Shellfish, such as mussels and oysters, act as good samplers, since they ingest and accumulate substances from the water column over time. Concentrations of contaminants in shellfish are thus excellent indicators of water pollution. The plots show OSPAR status and time-trend assessments for two heavy metals (cadmium and lead) and also for an indicator PCB (PCB153) in mussels and oysters. PCBs are persistent synthetic organochlorine chemicals that were widely used in electrical products and transformers in the past, but which are now banned. At most sites, the concentrations of these three contaminants are below levels deemed to be a problem, and are lowest in areas away from urban/industrial activities. At many sites the concentrations of lead and cadmium are 'near background'. The still-elevated but significantly decreasing concentrations of lead seen at a station in Cork Harbour reflects historical local pollution (near the now-closed Irish Steel plant); a downward trend is also evident for other metals at this site. [Data source: Marine Institute, OSPAR Coordinated Environmental Monitoring Programme. OSPAR contaminants assessment data are available at https://ocean.ices.dk/oat/]

Priority and other relevant substances in Irish coastal and transitional waters are monitored by the Marine Institute, to assess chemical and ecological status in accordance with the Water Framework Directive, and to ensure appropriate action is taken where good status is not achieved. Under the WFD, substances are normally measured in water samples to check that concentrations comply with toxicologically based Environmental Quality Standards (EQS) as set in European and/or Irish law. Substances found to be present at concentrations below an EQS are not considered a risk to aquatic organisms. Ireland has also an obligation under the OSPAR Convention to monitor the levels, trends and effects of hazardous substances and for this purpose concentrations of hazardous substances are normally measured in marine biota and sediments, where many of the substances of concern can accumulate. Table 31.2 lists some of the substances considered of potential concern for the marine environment and their uses/sources.

Many of these contaminants are often detected in Irish waters, but only rarely at concentrations that indicate a cause for concern (that is, they comply with standards set in European or Irish legislation). Coastal and transitional water bodies monitored during the 2012–2015 cycle of the WFD achieved good chemical status for priority substances in the water column, with the single exception being the Avoca Estuary, on the east coast, where the water EQS for a number of metals were found to be exceeded due to historical mining activities in the catchment.

Concentrations of metals and persistent organic pollutants in mussels and oysters are widely used as indicators of coastal contamination. Unsurprisingly, this monitoring shows that low levels of these pollutants occur along the west coast, where inputs are few, while higher levels are often found closer to urban agglomerations on the south and east coast, albeit rarely at problem levels. Maps of status and temporal trends for selected trace metals and PCBs in shellfish from Irish coastal waters are presented in Figure 31.24. It is reassuring that at some sites where higher concentrations of PCBs and metals have been recorded in the past, the levels have been shown to be decreasing. Another good news story is that the adverse effects of the antifoulant, tributyltin (TBT), on gastropod mollusc reproduction seem to be fading following the banning of this substance.

BOATS, PAINT AND TRANSGENDER SNAILS

Brendan McHugh and Michelle Giltrap

Fig. 31.25 THE DOGWHELK, *Nucella lapillus*, IN ITS NATURAL ENVIRONMENT. The dog whelk is a predatory marine sea snail, or gastropod, that is commonly found in the intertidal zone of rocky coasts and estuaries all around the island of Ireland, although it generally prefers more sheltered areas where wave action is less vigorous. It is carnivorous and preys on smaller herbivorous molluscs, such as barnacles, mussels and periwinkles and, in turn, is preyed on by crabs when the tide is in and seabirds when exposed at low water. In early Christian Ireland of the seventh and eighth centuries, dogwhelks were harvested in Ireland and Britain to extract a red-purple dye that was used to colour cloth, and also by monks to illustrate their illuminated manuscripts. In the 1950s a whelk-dyeing workshop dated to the seventh century was found by archaeologists on the island of Inishkea North in County Mayo, the only one to be found so far on these islands. [Source: Martin Talbot, courtesy of OSPAR]

The growth of barnacles and other organisms on ship hulls, a process known as biofouling, has always been a problem for mariners, with the resultant increase in friction drag reducing vessel speeds and greatly increasing fuel costs. Antifouling paints have been extensively used to prevent biological growth on marine vessels and structures.[32] By the 1980s it was realised that many of these paints were affecting certain non-target organisms.[33] The first indication of contamination by tributyltin (TBT) in Irish waters was reported in 1985, by Dan Minchin and Colm Duggan, scientists with the then Irish Fisheries Research Centre, who detected poor growth in oysters from Cork Harbour and Baltimore in County Cork,[34] and subsequently possible similar effects on the scallop population in Mulroy Bay in County Donegal.[35]

Each of these instances contributed to the evidence base against TBT, the antifouling agent in these paints, as being one of the most toxic contaminants then introduced into the marine environment. To address this, an Irish bye-law was passed in April 1987, prohibiting, except under special circumstances, the use of TBT on vessels under 25m, and on aquaculture netting and marine structures in Irish waters.[36] The International Convention on the Control of Harmful Anti-fouling Systems on Ships,[37] which prohibited the use of TBT on all shipping, came into force in 2008 and this was also transposed into EU legislation.[38] Any EU ship, or any ship entering EU ports, must either not bear TBT antifoulant, or else must have a hull coating that prevents leaching of TBT.

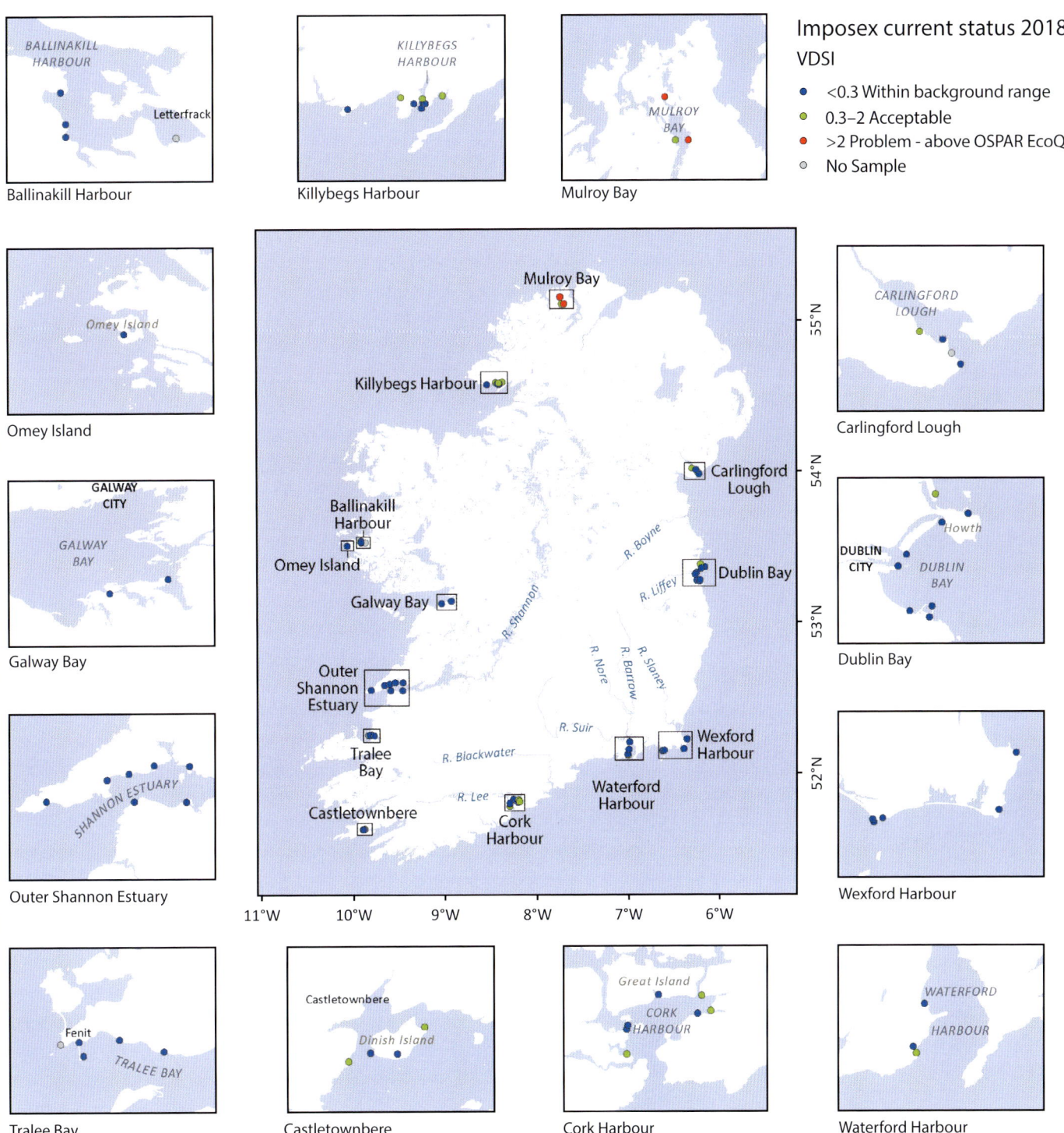

Fig. 31.26 STATUS OF IMPOSEX MEASURED AS VAS DEFERENS SEQUENCE INDEX (VDSI) IN DOGWHELKS, AN INDICATOR OF TBT POLLUTION, IN THE AREAS SURVEYED AROUND THE IRISH COAST IN 2018. High levels of imposex are attributable to TBT contamination. Levels of imposex in 2018 are typically low or within background range. [Data source: Marine Institute]

TBT is extremely toxic to many marine organisms, and in particular has been unequivocally linked to reproduction effects in several mollusc species. The most notorious effect of TBT exposure is now recognised as being the masculinisation of female gastropods. Dogwhelks, *Nucella lapillus* (Figure 31.25), are particularly sensitive to TBT even at very low levels of exposure, developing a condition known as imposex, whereby the females develop male sex organs and ultimately become sterile.

Since 1987, the extent of imposex in dogwhelks around the Irish coast has been monitored by various research initiatives[39] and from surveillance surveys undertaken by the Marine Institute (formerly, the Fisheries Research Centre) at approximately six-yearly intervals. The surveys put a particular emphasis on ports and their approaches.

The survey in 2018 measured the degree of imposex in female dogwhelks at fifty-nine sites from thirteen locations around the coasts of Ireland, from Carlingford Lough (County Louth) to Mulroy Bay (County Donegal) (Figure 31.26).[40] At three sites (North and South walls in Dublin Bay and Ahanesk Pier in Cork Harbour) small numbers of sterile females were found and reduced numbers of young individuals

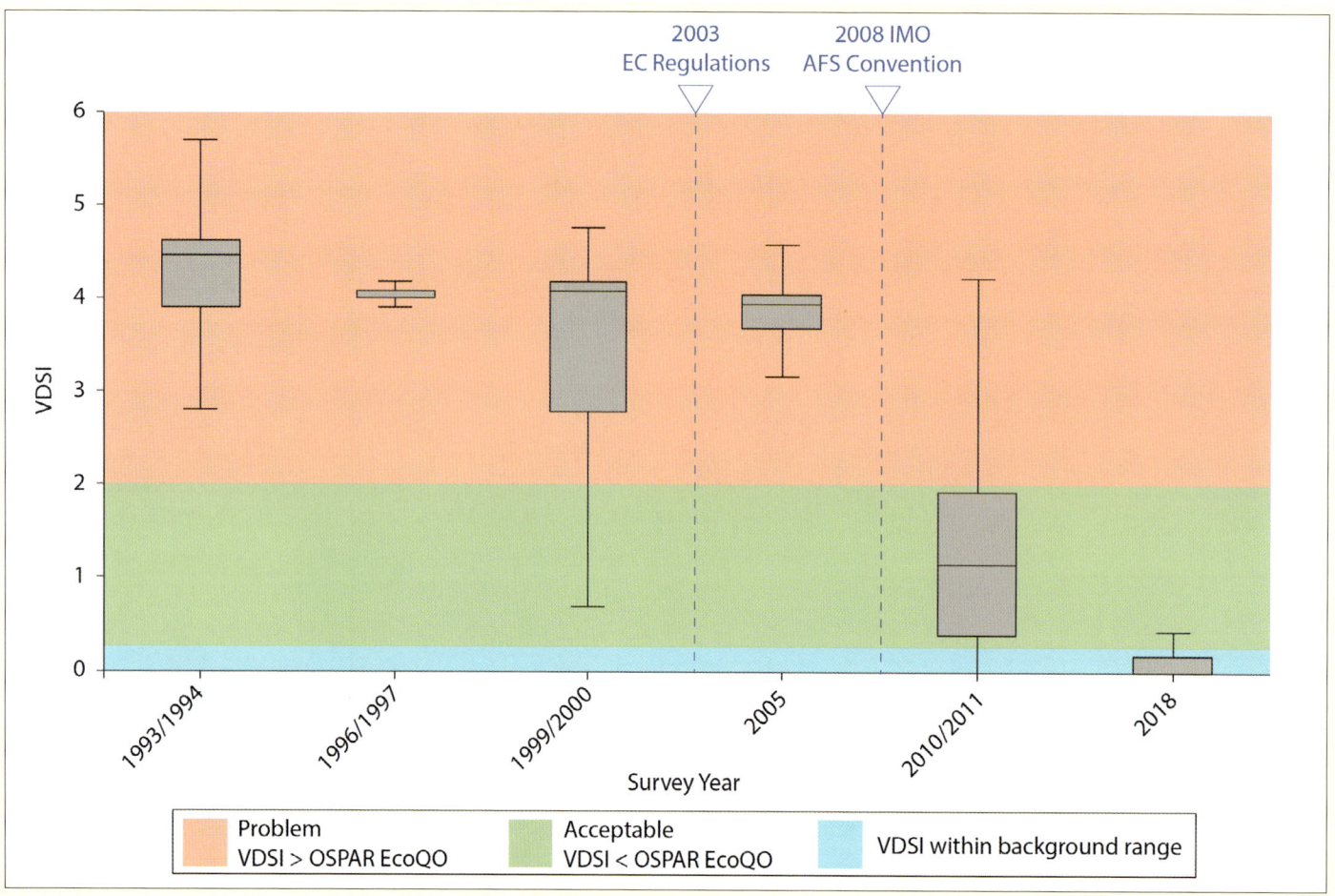

Fig. 31.27 Box-plot (outliers not shown) of all imposex measurements for periodic surveys undertaken since the early 1990s. Since 2005, a marked improvement in imposex status is evident. Dashed lines show the timings of adoption of EU regulations and ratification of the IMO Antifouling System Convention. While the locations sampled varied between surveys, the overall trend is clear for individual sites that previously showed high levels of contamination, such as the site at Aghada, Cork Harbour. [Data source: Marine Institute]

were noted, but at the majority of individual sampling sites there were no indications of significant TBT contamination. Trend assessments show that until the mid-2000s, levels of imposex at many sites remained consistently elevated. However, the 2010–2011 survey showed a notable improvement at many sites and in 2018 almost all locations were found to be at or close to background, exhibiting little evidence of imposex. While a few locations still exhibit some problem levels (for example, Mulroy Bay), it is clear that even at these the situation is improving (Figure 31.27).

Overall, the study determined that there has been a dramatic reduction of TBT contamination around the Irish coast, demonstrating that measures taken nationally and internationally to phase out known toxic substances can be very effective in reducing marine pollution.

The most significant source of artificial radionuclides in the Irish marine environment is the discharge of low-level liquid radioactive waste from the Sellafield Nuclear Fuel Reprocessing Plant in England to the eastern Irish Sea. However, EPA monitoring of fish and shellfish does not indicate a human health concern as doses to consumers are extremely low.[41]

Overall, the picture is of clean waters around the coast of Ireland. Nonetheless, vigilance is required to ensure new substances do not become an environmental threat and research is also needed to better understand the risk from real-world, and perhaps cumulative, exposure to complex mixtures of substances in the environment, as opposed to assessing the risk of contaminants on a substance-by-substance basis.

UNDERWATER NOISE

Many marine organisms, particularly mammals and many fish species, communicate using sound to find mates, to search for prey, to avoid predators and hazards, and for navigation. Human activities are adding to the background noise level in the sea. These anthropogenic sounds may be either impulsive (generated by a single trigger) or ambient (continuous). Impulsive sounds can arise, for example, due to pile-driving for offshore construction, or seismic surveys for oil and gas exploration; ambient sound is mainly associated with marine traffic. When the level of sound has the potential to cause adverse effects on marine life it is referred to as 'noise'. There is growing evidence that underwater noise can have negative effects for a range of marine taxa.

THE IMPACT OF ANTHROPOGENIC NOISE ON MARINE MAMMALS

Michelle Cronin and Mark Jessopp

Human-induced noise from shipping, seismic surveys and construction has increased significantly in the Irish marine environment. This raises particular concerns for marine mammals, as there is potential for impacts ranging from physical injury causing death or hearing damage, through to masking of acoustic signals that animals rely on for navigation, feeding and communication. While the noise impacts of shipping are considerable and widespread, seismic surveys are particularly problematic.

They involve firing arrays of airguns into a water column (Figure 31.28), with noise levels at the airgun source typically louder than a jet aircraft at take-off. Recorded effects in marine animals include temporary or permanent hearing damage, stress, and behavioral responses such as avoiding noisy areas and changing vocalisation rates.[42] Recent research work in Ireland has shown that the cumulative noise 'footprint' of seismic surveys can extend hundreds of kilometres from the source seismic vessels, at levels that cause hearing damage to some organisms.[43]

Perhaps less damaging is noise from coastal and marine construction, which includes activities such as pile driving, drilling and dredging. These activities are becoming more widespread, and can mask important signals used by marine animals for navigation, communication or feeding. A long-term marine mammal monitoring programme was established in north-west Ireland to assess impacts arising from the construction of a gas pipeline associated with the

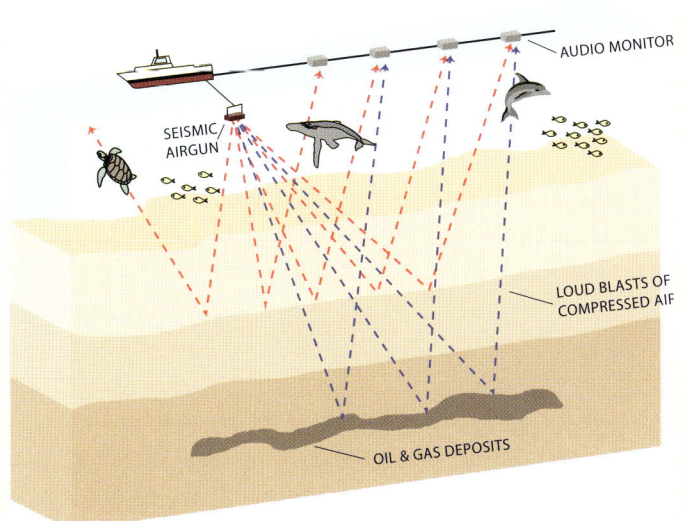

Fig. 31.28 SEISMIC SURVEYS ARE USED TO MAP THE SEABED AND ITS UNDERLYING GEOLOGY. Seismic surveys are commonly used for seabed mapping and hydrocarbon exploration and involve sending soundwaves to the seafloor and measuring the time it takes for the sound to bounce back to a receiver. However, the frequency of these sounds can overlap with the hearing sensitivity of marine animals and sometimes at volumes loud enough to cause impacts such as hearing damage. Dedicated marine mammal observers are required on seismic survey vessels to ensure marine mammals are not within the immediate area prior to airgun activity. [Source: adapted from Offshore Oil and Gas Exploration: Seismic Airgun Blasting. http://www.usa.oceana.org/sites/default/files/662/seismic_fact_sheet_long_final_7-25_0.pdf]

Fig. 31.29 COMMON DOLPHIN (*Delphinus delphis*), IRELAND. Charismatic species such as the common dolphin occur frequently in coastal waters where they are increasingly encountering noise from human activities that may affect their feeding, communication and habitat use. Studies in Ireland suggest that harbour porpoises, minke whales and bottlenose dolphins are all impacted by construction and vessel-related noise. [Source: William Hunt]

789

Corrib gas field (see Chapter 29: Coastal Mining Quarrying and Hydrocarbon Exploration). This found that construction-related activity reduced harbour porpoise (*Phocoena phocoena*) and minke whale (*Balaenoptera acutorostrata*) sightings, while an increase in vessel numbers impacted negatively on the presence of common dolphins (*Delphinus delphis*) (Figure 31.29). The research, however, suggests that any impact is likely to have been short-term.[44] Most marine mammal species in coastal waters are seasonally distributed, as they are influenced by factors such as prey availability and water temperature. As a result, it is quite challenging to establish the full impact of human activity on these species. Understanding of the potential impacts comes through maximising data collection, which will help to mitigate effectively against future impacts. Management responses will vary, depending on the activity involved (for example, pile-driving, dredging), the nature of the site (breeding or feeding area), and the characteristics of the species (behaviour and ecology). Under the Marine Strategy Framework Directive (MSFD), European Union member states, including Ireland, will need to take measures to ensure that underwater noise from human activities does not adversely affect the marine environment and ecosystem.[45]

ADDRESSING COASTAL AND MARINE POLLUTION: WHAT IS BEING DONE?

Thankfully, it is now recognised that the sea should no longer be seen as a dumping ground for waste. Tackling and preventing marine pollution requires both local policies and actions and also international co-operation. Ireland and the UK are both signatories to a plethora of global and regional treaties that contribute to addressing the issue. Some of these, such as the Stockholm Convention on Persistent Organic Pollutants, address the source of pollutants at a global scale; others address specific activities and pressures such as shipping, or dumping at sea. For maritime states of the north-east Atlantic, the most important regional instruments that protect the coastal and marine environment against pollution are the 1992 OSPAR Convention and various elements of European Union legislation (see Chapter 32: CoastalManagement and Planning).

The 1992 OSPAR Convention for the Protection of the

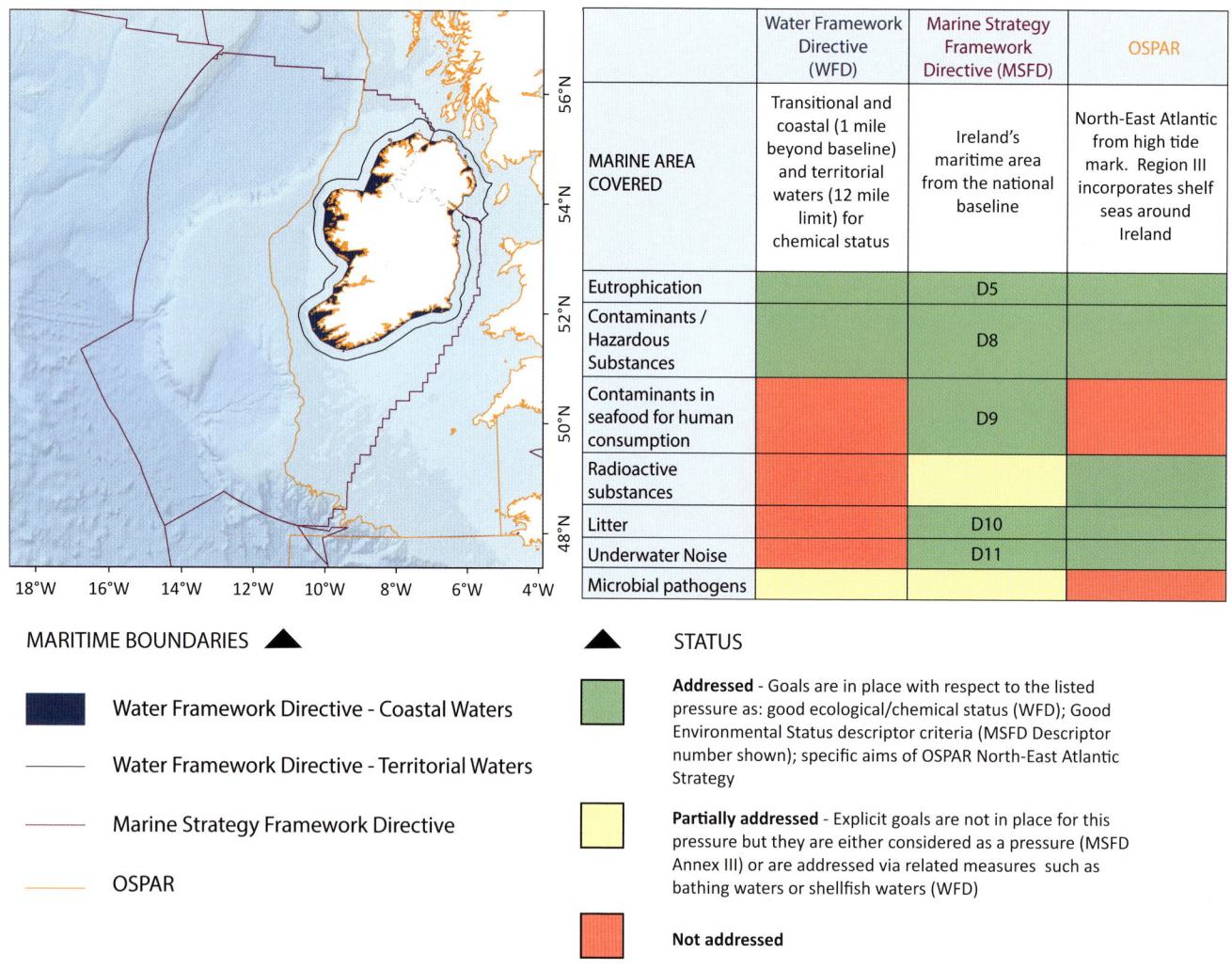

Fig. 31.30 THE GEOGRAPHIC SCOPE OF THE WATER FRAMEWORK DIRECTIVE (DIR. 2000/60, WFD), MARINE STRATEGY FRAMEWORK DIRECTIVE (DIR. 2008/56/EC, MSFD) AND 1992 OSPAR CONVENTION. The WFD requires Ireland to achieve good ecological status in coastal waters and good chemical status in territorial waters. The MSFD requires Ireland to assess various aspects of pollution within the nation's marine waters. The shelf seas to the west of Ireland and the UK and the Irish Sea form Region III of the OSPAR Convention for the Protection of the North-East Atlantic. [Data sources: Environmental Protection Agency; Northern Ireland Environment Agency; Department of Communications, Climate Action and Environment; United Kingdom Hydrographic Office Department of Housing, Planning, Community and Local Government; European Environment Agency; OSPAR]

Marine Environment of the North-East Atlantic stemmed from the 1972 Oslo and 1974 Paris conventions that aimed to prevent sea- and land-based marine pollution. In its north-east Atlantic strategy, OSPAR sets out thematic strategies on biodiversity and ecosystems, hazardous substances, eutrophication, human activities, radioactive substances and offshore industry, that together address the main threats to the marine environment.[46]

As mentioned above, the most important EU legislation to protect water quality in Europe is the Water Framework Directive, and this is implemented in Ireland by the Department of Housing, Planning and Local Government and by the Environmental Protection Agency, and in Northern Ireland by the Department of Agriculture, Environment and Rural Affairs.[47] Member states monitor and assess periodically the ecological and chemical status of water bodies, including coastal and transitional (estuarine) waters, and set out measures in River Basin Management Plans (RBMPs) to ensure at least 'good' status (as defined in the legislation) is achieved, and also to avoid deterioration of 'high'-status waters. The national River Basin Management Plan for Ireland 2018–2021 was published in 2018. The plan outlines the key measures that will be taken to protect and improve water quality over this period. Other EU legislation tackling pollution – such as Nitrates, Urban Waste Water Treatment and Bathing Waters directives – are also key to implementing RBMPs.

The 2008 Marine Strategy Framework Directive aims to protect the marine environment and applies an ecosystem-based approach to the management of human activities, enabling a sustainable use of marine goods and services.[48] The directive identifies eleven descriptors of Good Environmental Status (GES) that must be achieved. Five of these relate to pollution (Figure 31.30). Given the transboundary nature of marine issues, the directive envisages member states working together through the regional sea conventions (OSPAR for the north-east Atlantic) to deliver coherent programmes for the protection of the marine environment.

Thanks to a combination of its position adjacent to the Atlantic ocean, along with a growing range of policies, incentives and legislative measures at national, European and international levels as outlined above, Ireland's coastal and marine waters are relatively unpolluted, although it must be acknowledged that they are not actually pristine. Where it does exist, eutrophication is generally confined to a few inshore and estuarine water bodies, while hazardous substances, though detectable, rarely exceed levels of concern, and fish and shellfish from these waters are normally safe to eat. Public awareness of the impact of some pollution threats to the marine environment, especially marine litter, has grown in recent years and has encouraged welcome changes in behaviour and attitude, including the emergence in many coastal areas of volunteer beach clean-up and monitoring activities. These, along with a range of measures put in place in recent decades, driven primarily by European legislation, have delivered significant improvements in the cleanliness of Ireland's coasts and nearshore waters. Nonetheless, continued vigilance is required to protect the coastline and seas from pollution. Ongoing investment in scientific research is essential to assess the cumulative impacts of microplastics in marine ecosystems.

COASTAL MANAGEMENT AND PLANNING

Anne Marie O'Hagan and Val Cummins

THE PORT OF LARNE AT THE ENTRANCE TO LARNE LOUGH, COUNTY ANTRIM. [Source: Port of Larne]

The preceding chapters identify a range of issues arising from marine development, such as water quality, marine transport, energy, species and habitat conservation, fishing and other demands. Good coastal planning and management are essential prerequisites in dealing effectively with a range of issues. This chapter outlines the trends in planning and management for the Irish coast, covering legislation and policy, as well as recent developments in Maritime Spatial Planning (MSP). As policy and legislative provisions are constantly adapted and refined, this chapter has to be a snapshot of the instruments in effect at time of writing. First, it is necessary to reflect on drivers at the international and EU levels, given the huge influence these have had on environmental and coastal management in recent years. Insights are also provided on the management of the Northern Ireland coast, with a focus on the shared border bays of Lough Foyle and Carlingford Lough and consequent implications for transboundary marine management.

In both the Republic of Ireland and Northern Ireland, planning and management of coastal areas has proceeded largely in an ad hoc manner and has generally been based on sectoral legislation, although progress is being made towards more joined-up thinking through the advancement of policy for Maritime Spatial Planning, which will be discussed in more detail later. This EU legislation is enforced by a plethora of national and local bodies, such as government departments, agencies and local authorities, whose responsibilities are often divided according to the 'marine' and 'terrestrial' parts of the coast, leading to complex jurisdictional arrangements and, sometimes, unclear remits.

This chapter begins by providing some background on international instruments that influence the Irish approach to coastal management. This is followed by case studies that inform on the key principles of ownership and decision-making. The chapter then proceeds to delve more deeply into the institutional frameworks that influence the national response to coastal governance.

Before commencing, however, it is worth noting that the nature of this response can be viewed in the context of Blue Growth, which is one of the most significant drivers for development in the coastal zone today. The overarching goals of Blue Growth are to create jobs and stimulate growth, while driving marine innovation, research and technology, in a sustainable manner.[1] The Inter-departmental Marine Coordination Group (IMCG), established by the Irish government in 2009, was instrumental in developing Harnessing Our Ocean Wealth: An integrated marine plan for Ireland.[2] The targets of the plan are to double the value of Ireland's ocean wealth to 2.4 per cent of GDP by 2030 and to increase the turnover from Ireland's ocean economy.

THE CONSTITUTION FOR THE OCEANS: THE INTERNATIONAL LAW OF THE SEA

In the context of marine space, national legislation and management must fit within a much wider international legal framework. This is derived primarily from the United Nations Law of the Sea Convention (LOSC), which was ultimately agreed in 1982 and entered into force in 1994. The convention is known as the 'Constitution for the Oceans' and covers a huge range of significant issues, including maritime jurisdictional zones and their delimitation, navigation, transit regimes, continental shelf jurisdiction, deep seabed mining, protection of the marine environment, marine scientific research and settlement of disputes. The convention provides for the creation of a number of different jurisdictional zones, including Territorial Seas, the Contiguous Zone, the Exclusive Economic Zone (EEZ), the Continental Shelf and the High Seas (Figure 32.1). Under certain geological conditions coastal states can submit claims to extend their continental shelf beyond 200 nautical miles and thus gain additional marine territory and sovereignty rights. Ireland has made a number of such submissions, as extended continental shelf claims.

Areas Beyond National Jurisdiction (ABNJs) make up 61 per cent of the global ocean. These areas are coming under increasing pressure for exploitation, particularly for seabed mining. Only approximately 1 per cent of ABNJs have been established as Marine Protected Areas, exposing the vulnerability of this vast territory to maladaptive practices.[3] As a result, UN members are negotiating an international legally binding instrument under the UN Law of the Sea Convention (UNLOSC) on the conservation and sustainable use of marine biological diversity of ABJNs. Ireland has a role to play in adding its voice to these international deliberations. While the consequences of decisions pertaining to the High Seas may appear to be limited for coastal communities such as those living on the west coast of Ireland, the protection of the global ocean is fundamentally connected to the livelihoods and wellbeing of all coastal stakeholders.

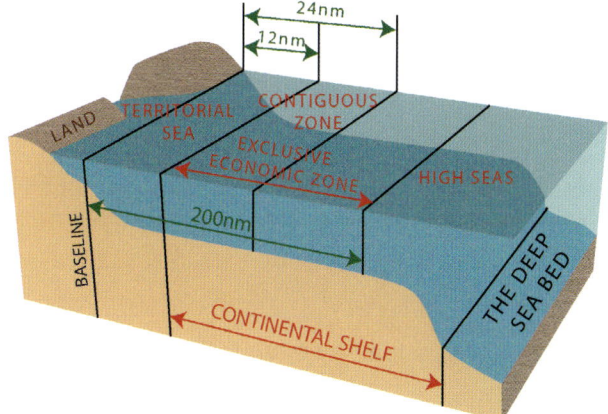

Fig. 32.1 JURISDICTIONAL BOUNDARIES EXTENDING OUTWARDS FROM THE COAST. The United Nations Law of the Sea Convention was developed to ensure freedom of navigation of the seas. The convention provides for the creation of a number of different jurisdictional zones including Territorial Seas, the Contiguous Zone, the Exclusive Economic Zone (EEZ), the Continental Shelf and the High Seas. First, the Territorial Sea is the area that extends up to 12 nautical miles from the baseline of a country's coast and is under the jurisdiction of that particular country. However, foreign ships have the right to innocent passage through those waters. The coastal state can intervene on matters such as drug smuggling or other illicit trafficking. The Contiguous Zone extends 12 nautical miles beyond the Territorial Sea limit. This is like a buffer zone where a coastal state can control activities that threaten regulations in the Territorial Sea. The Exclusive Economic Zone extends 200 nautical miles from the baseline. This bestows control to the coastal state over economic resources, including fishing, hydrocarbons and mining. The Contintental Shelf extends beyond the 200 nautical mile limit, up to 350 nautical miles from the baseline, or 100 nautical miles from the 2,500m isobath. Finally, the High Seas are the areas beyond the above. These are open to all states for freedom of navigation, fishing, research and exploration. The High Seas are the focus of major UN efforts to evolve the management of areas beyond national jurisdiction to mitigate against threats from human activities, including deep sea mining.

IRISH EXTENDED CONTINENTAL SHELF CLAIMS UNDER THE LAW OF THE SEA

David Naylor

Until the early twentieth century, the legal position regarding the marine areas of the world was simple but largely undefined. National offshore boundaries were limited to a narrow zone extending from the coastline – usually 3 nautical miles – beyond which the seas were considered to be international waters. However, the situation became more confused as the century progressed and countries began unilaterally to extend their maritime limits by different amounts. The Real Map of Ireland (Figure 32.2) denotes the results of decades of iterations of the United Nations Conferences on the Law of the Sea (UNCLOS) process to designate Irish maritime jurisdiction. The UNCLOS process is further outlined below.

However,, the United Nations held two Conferences on the Law of the Sea (UNCLOS), in 1956–1958 and 1960 respectively. Important among the provisions arising from the meetings was one allowing a state to exercise sovereignty over an area of the seabed up to a water depth of 200m. The continued development of improved offshore technology, increased fishing, and offshore oil and gas discoveries combined to pressure governments in the next decade to progressively designate extensions to their offshore areas.

The third United Nations Conference on the Law of the Sea (UNCLOS III, 1973–82) was convened to examine all elements of the Law of the Sea: the right of passage, fishing limits, continental shelf limits and deep seabed ownership. At the same time, countries continued

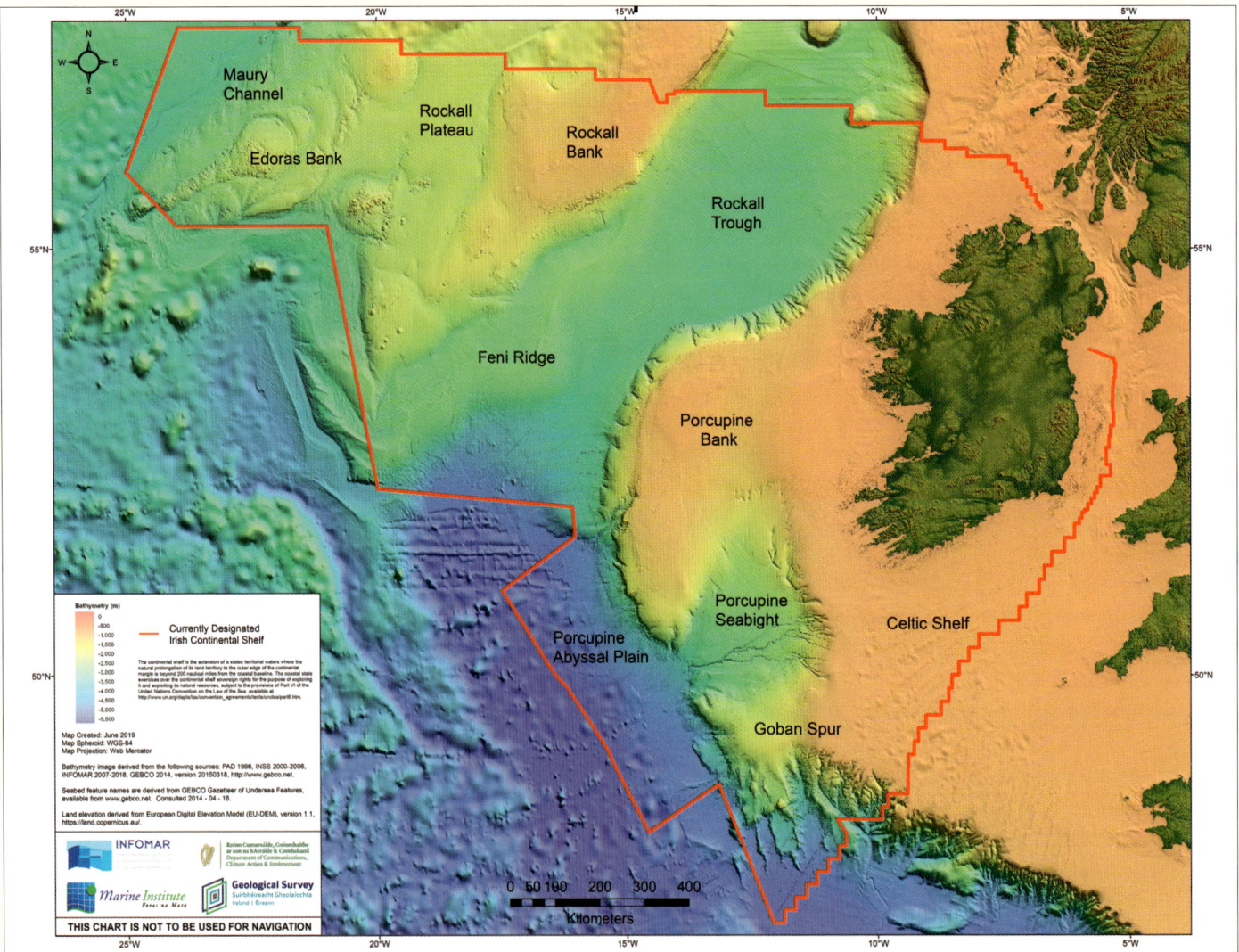

Fig. 32.2 THE REAL MAP OF IRELAND. Released in 2008, the map delineates the extent of Ireland's designated continental shelf. It was produced, in part, to emphasise the marine resource potential of Ireland and support its promotion. Ireland's marine environment comprises over 880,000km² (*c.*88 million hectares), making it one of the largest in the European Union. The Real Map of Ireland was developed by the Marine Institute and Geological Survey Ireland. It is based on data collected by the Irish National Seabed Survey and the Integrated Mapping for the Sustainable Development of Ireland's Marine Resource (INFOMAR). [Source: INFOMAR, Government of Ireland]

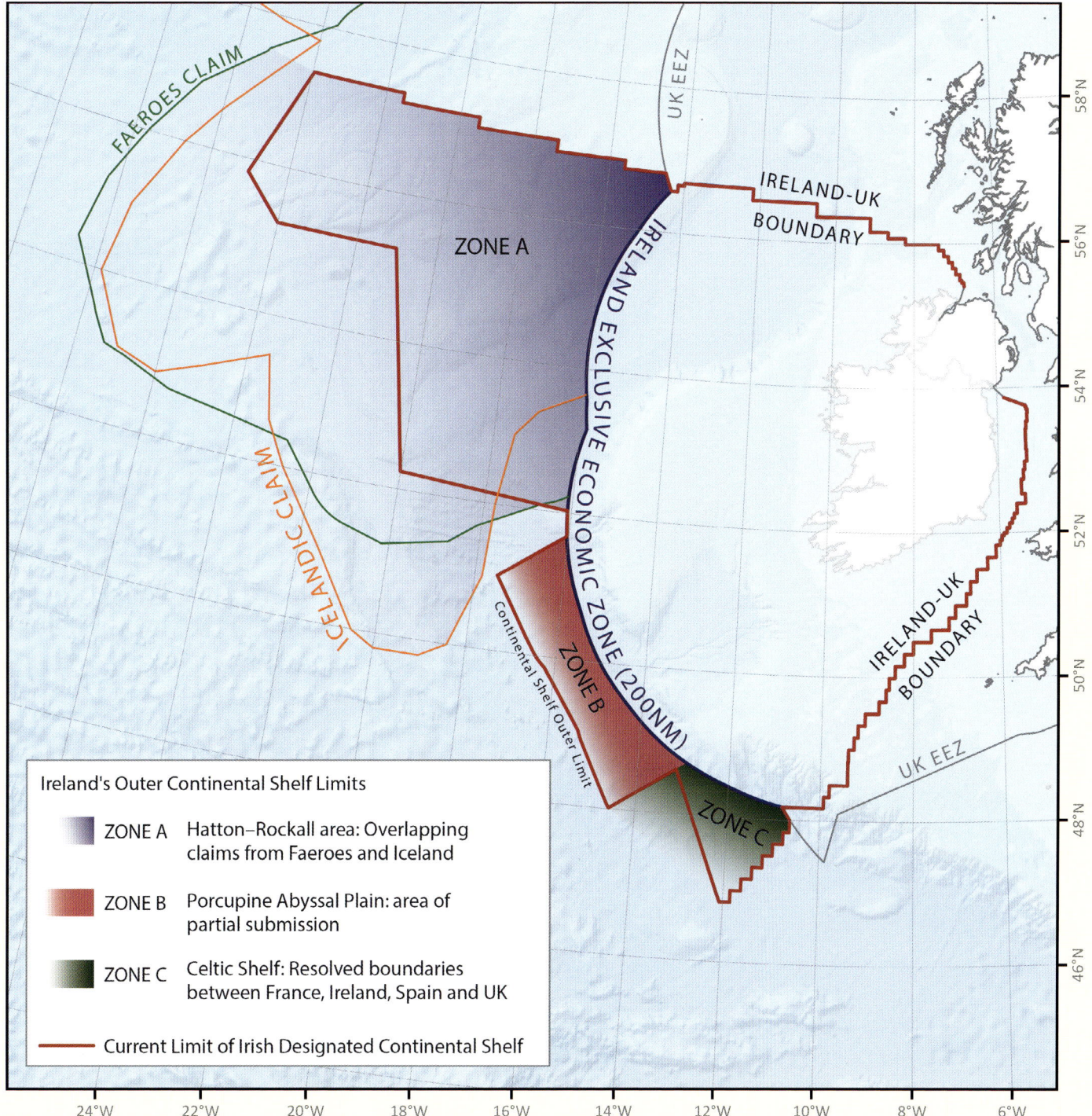

Fig. 32.3 CONTINENTAL SHELF CLAIMS ALONG THE IRISH ATLANTIC MARGIN. Due to overlapping claims by Ireland, Denmark, Iceland, France, Spain and the UK, the Irish submission to the Commission on the Limits of the Continental Shelf (CLCS) was divided into three parts: Zone A: the area is also claimed by Denmark (on behalf of the Faroe Islands), by Iceland and by the UK. Ireland lodged a sole submission for this area to the CLCS in 2009. Zone B: a central area on the Porcupine Abyssal Plain was judged not to prejudice the outcome of discussions to the north and south, and to concern only Ireland (submitted 2005). Recommendations were issued by the CLCS in April 2007 and the limits of the continental shelf established formally in 2009 by means of statutory order adding 39,000km² to the state's continental shelf. Zone C: the south-western extension of the Celtic Sea boundary agreed by Ireland and the United Kingdom enters into an area adjacent to the interests of both France and Spain. The four countries agreed to collaborate in a joint submission to the Commission (submitted 2006) and resolve the boundaries between themselves later. Recommendations on the shelf limits for this area were received from the CLCS in April 2009. [Data source: Petroleum Affairs Division of the Department of Communications, Climate Action and Environment; Irish Naval Service, 2015; Geological Survey of Denmark and Greenland, 2010; United Nations Division for Ocean Affairs and Law of the Sea, 1984]

to expand their offshore areas by designation. Ireland made a series of offshore designations in the 1960s and 1970s, extending westwards over waters denoted as the Porcupine and south Rockall regions. Arbitration cases between nations had set precedents regarding issues such as the drawing of baselines, status of fringing islands, etc., and these were largely incorporated into the UNCLOS III framework. This came into force in 1994, after the required sixtieth country ratified the treaty, and an International Seabed Authority (ISA) was established.[4]

From an Irish viewpoint, the convention contained two important elements. The first was the concept of Exclusive Economic Zones (EEZ) extending 200 nautical miles from the baseline, within which the coastal state has sole exploitation rights over all natural resources, including fishing rights. Second, the continental shelf is defined as the natural prolongation of the land territory to the outer edge of the continental margin, or 200 nautical miles from the baseline, whichever is greater (exclusive fishing rights do not extend beyond the EEZ). Rules for the definition of the continental margin were included in the text (Figure 32.3).

Where the continental shelf extends beyond 200 nautical miles, a state is required to make a submission to the Commission on the Limits of the Continental Shelf (CLCS). This submission sets out the co-ordinates of the outer limits of the shelf and is accompanied by technical and scientific data to support the claim. The commission assesses the limits and data submitted by the coastal state and makes recommendations, which are final and binding. The convention does not, however, provide authority to the United Nations to arbitrate in disputes between nations on delimitation of maritime boundaries.

Generally, coastal states exercise most control in areas closest to their shores, with rights and responsibilities changing in each zone, to the limits of national jurisdiction at 200 nautical miles. The majority of maritime jurisdictional zones are measured from baselines. The delineation of these baselines is a technical process that has significant consequences for maritime jurisdictions. The Irish baseline was remapped in 2014 and 2015. The technical aspects of this process, which have a knock-on effect on determining the different marine areas for management, are described below.

Case Study: Ireland's Baselines

Eoin V. Fannon

Territorial waters and the maritime jurisdiction of a state are measured from baselines. This case study explains the concept of defining the baseline, and the need for accurate charts to delineate marine boundaries.

Baselines consist of two types: normal baselines and straight baselines. The 'normal baseline' is the line of low water along the coast of a state. A 'straight baseline' is drawn up where the coastline is deeply indented and cut into, or if there is a fringe of islands along the coast. Ireland's topography along the east coast is largely flat with very few islands. In contrast, the west coast has a large number of islands

Fig. 32.4 BASELINE POINT NUMBER 49 AT BLACK ROCK, CARNSORE POINT, COUNTY WEXFORD. The 2014–2015 survey of the straight baseline fixed bronze bolts was carried out at each baseline point from Malin Head, County Donegal, to Carnsore Point, County Wexford. These points were then co-ordinated using modern Global Navigation Satellite System (GNSS)–Global Positioning System (GPS) technology. The baseline survey was led by the Department of Foreign Affairs and Trade, which is the functional department for legislation on baselines. Detailed preparation was carried out by the Geological Survey of Ireland. Ordnance Survey Ireland carried out the surveying in co-ordination with the Naval Service and Air Corps. Other contributors were the Office of the Attorney General, the Department of Defence, the Irish Coast Guard, the Commissioners of Irish Lights and the RNLI. [Source: Ray Healy]

along a deeply indented coast and an indented south coast.

Prior to international agreements under UNCLOS (1982), the government made two orders in 1959, under the Maritime Jurisdiction Act, 1959, which reflected international law as it then was. This established straight baselines from the Scart Rocks, Malin Head, in County Donegal, down the west coast and along the south coast to Carnsore Point in County Wexford (Figure 32.5). A normal baseline was applied to the east coast. In all, fifty baseline points were identified in the 1959 Baseline Order.

A separate order provided for the charts published by the British Admiralty to be used for the purpose of mapping internal waters, territorial seas, exclusive fishery limits and fishery conservation areas. As mentioned, a state's maritime jurisdiction extends from its baselines, the locations of which are found by reference to the Admiralty charts. Accurate charts are

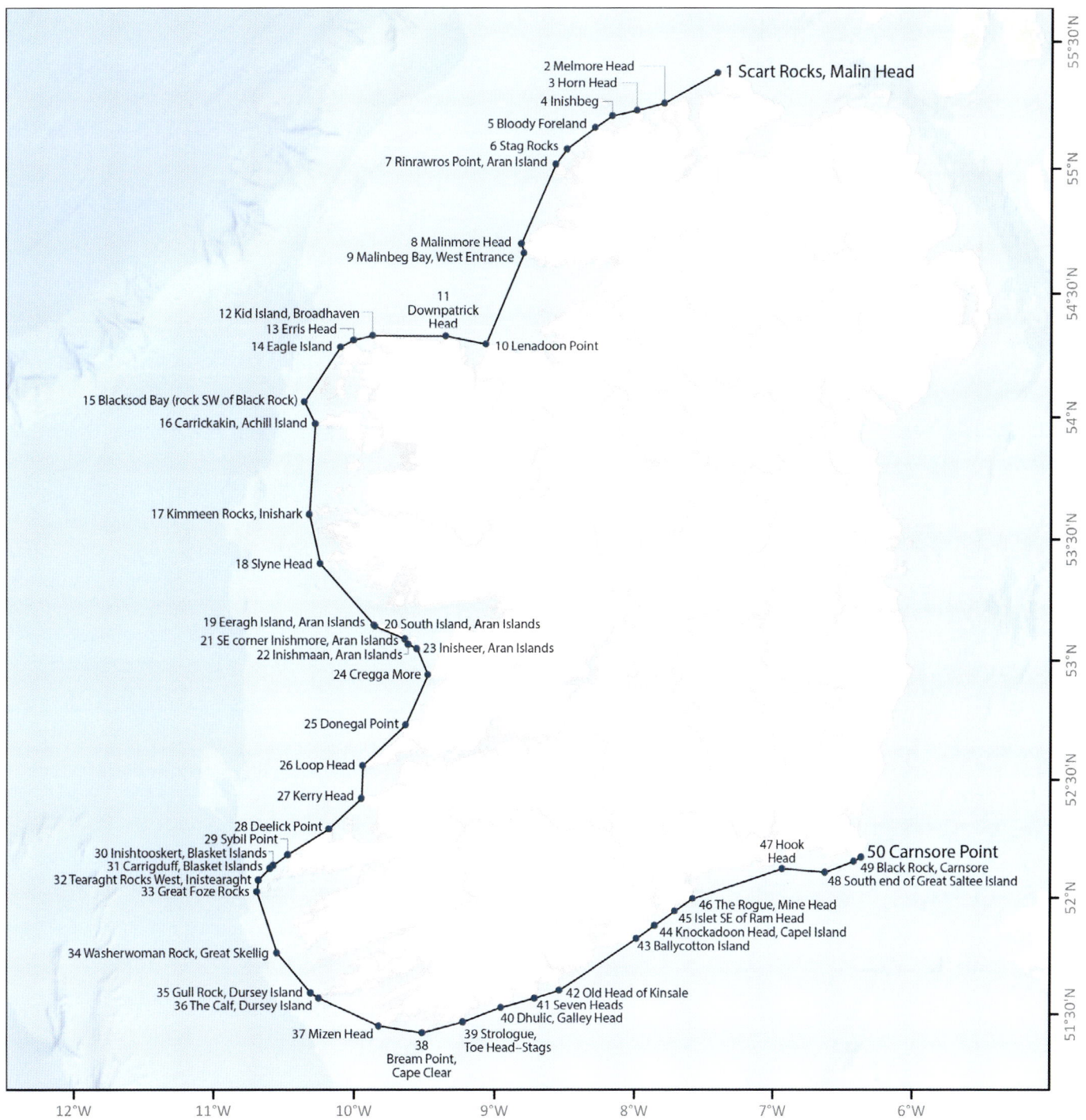

Fig. 32.5 IRELAND'S STRAIGHT BASELINE. Baselines consist of two types: normal baselines and straight baselines. The 'normal baseline' is the line of low water along the coast of a state. This is applied along Ireland's eastern coastline which has relatively few inundations or islands. The 'straight baseline', as shown on the map, can be used where the coastline is deeply indented or if there is a fringe of islands along the coast. This is the case on the south coast of Ireland, which is indented, and on the west coast, which has a large number of islands along a deeply indented coast. The fifty baseline points used to calculate this 'straight' baseline were resurveyed in 2014 and 2015 as part of a national effort to develop higher precision charts for navigation, delineation and other management purposes. [Data source: Department of Communications, Climate Action and Environment]

therefore necessary for establishing these legal imperatives.

Positions at sea are plotted with reference to a particular chart datum. Chart datum is the height of water at the lowest theoretical tide and it differs according to coastal locations. When Admiralty charts were first produced in the nineteenth century, they were drawn with reference to five different chart datums. The 1959 Baseline Order did not specify which chart datum to use as a standard. Without the correct chart datum, positional errors could arise. These could be amplified if small-scale charts are used. As Ireland's maritime jurisdiction exceeds 880,000km², a small error could translate into several hundred square kilometres of ocean. All EU member states extended their exclusive fishery limits to 200 nautical miles on 1 January 1977, using a European Datum 50 standard (ED 50). The standard was not generally used here. As Irish datums pre-existed ED 50, this could be a further source of error and the issue has been documented by legal experts.[5] An issue with the 1959 Baselines Order 'is that there are many datums in use and positions defined by reference to different datums

Fig. 32.6 SURVEYING THE BASELINES. The re-survey of the baseline points in 2014 and 2015 involved visiting some of the most exposed places on the Irish coast. Some inhospitable locations, like this rock near Achill Island, County Mayo, could only be accessed by Air Corps helicopter. Surveyors from the OSI had to undergo special training in advance of the project, which included sessions on sea survival, helicopter winching, rock climbing and abseiling. The team was outfitted with special boots for the steep and slippery terrain, as well as rock climbing equipment. [Source: Grace Fanning, courtesy of Irish Defence Forces Press Office]

can differ by several hundred metres'.[6]

As discussed, the 1959 order did not provide for international standards for chart datum, creating problems for accurate mapping of baselines, with variations in excess of 100m in some cases. Added to this was the need to take into consideration the coastal erosion that has occurred along various stretches of coast since 1959, and the need for clarity in the process of establishing baselines across most departments of state, not least in the office of the attorney general. This gave rise to a re-surveying of the baselines, undertaken by the state in 2014 and 2015. The purpose of the new survey was the accurate fixing of the baselines.

Present-day users of the sea, in the main, use electronic charts to determine positions (see Chapter 8: Monitoring and Visualising the Coast). Electronic charts need accurately surveyed baselines as a reference. The World Geodetic System (WGS84) is the standard used in satellite navigation and Global Positioning Systems (GPS). The re-surveying of all fifty baseline points in 2014 and 2015 was based on WGS84. Each baseline point is marked by a bronze bolt, numbered and set into the rock face (Figure 32.4). The survey was completed through the co-operation of several state and non-state agencies (Figure 32.6). Ireland's maritime zones can now be determined to an accuracy of better than 2cm.

This survey served as the precursor for the Maritime Jurisdiction (Straight Baselines) Order, 2016, which established the newly surveyed and more accurate co-ordinates for the baseline points and replaced the original Maritime Jurisdiction (Straight Baseline) Order, 1959.

EUROPE AS A DRIVER FOR HOW IRELAND MANAGES THE COAST

Membership of the EU has been a major driver of environmental legislation and policy, the impact of which has generally been positive, leading to an increased commitment to addressing issues such as water quality and enhanced levels of protection of natural habitats and species (for example through the Birds and Habitats Directives). There have been challenges regarding implementation of EU environment law, particularly in relation to how EU law is transposed into domestic law in both the Republic and in the UK.

It remains to be seen how Brexit will impact on coastal management and planning in Northern Ireland into the future.

In recent years there have been moves away from sectoral-based management towards more integrated or ecosystem-based management. In 2000, the European Union published the Water Framework Directive (WFD), which for the first time established a framework for the protection of rivers, lakes, estuaries, coastal waters (out to 1 nautical mile) and groundwater, and their dependent wildlife/habitats, in one legal instrument. A second

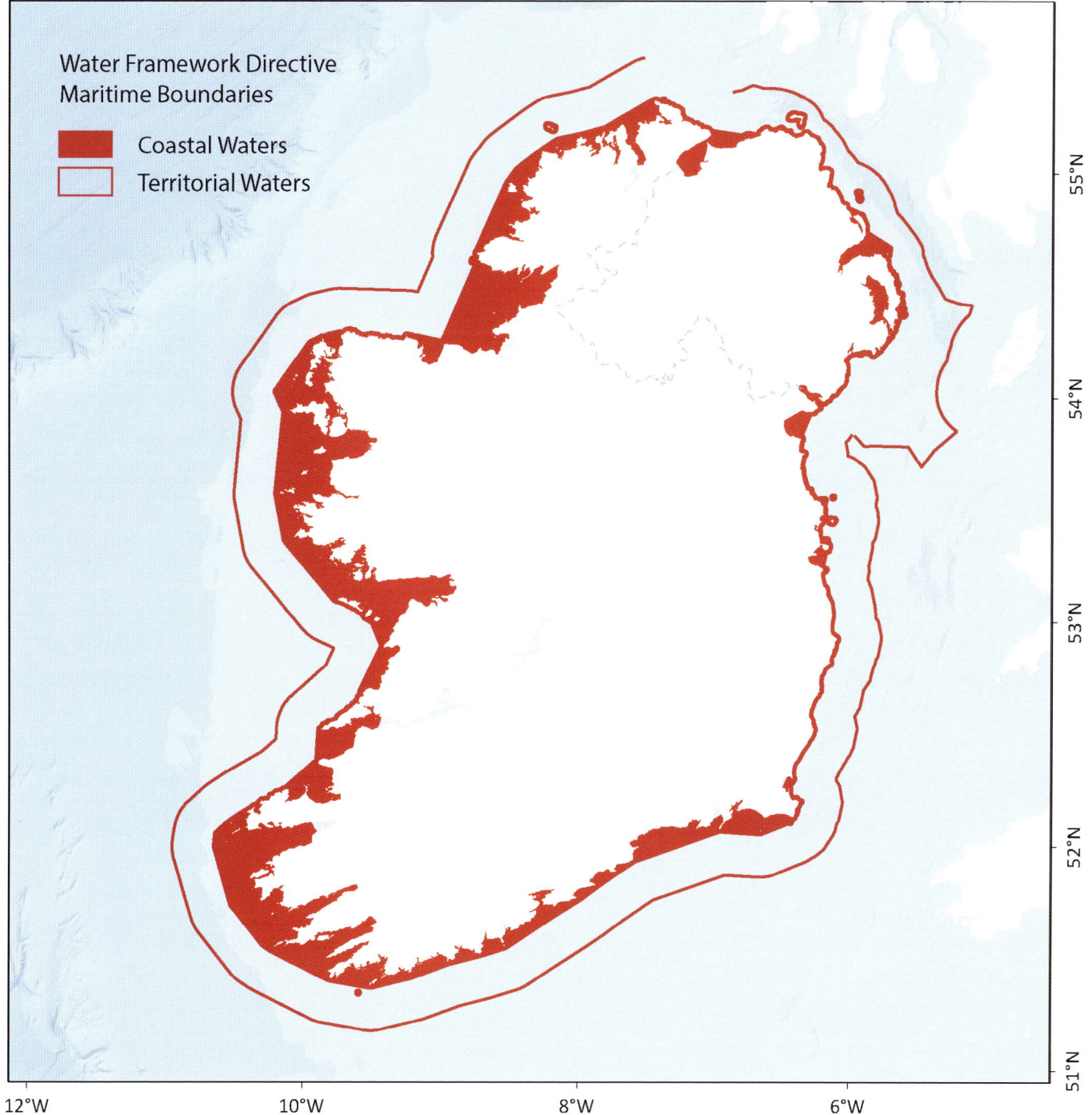

Fig. 32.7 THE GEOGRAPHIC SCOPE OF THE WATER FRAMEWORK DIRECTIVE. Coastal waters under Article 2(7) of the Water Framework Directive (WFD) are considered surface waters that extend from the land, or outer limit of transitional waters, to 1 nautical mile seaward of the baseline. Transitional waters are defined as those that exist near river mouths and may be characterised as partially saline but 'substantially influenced by freshwater flows'. In the Republic of Ireland, the Environmental Protection Agency is largely tasked with overseeing water quality monitoring activities for the WFD, while in Northern Ireland this responsibility rests with the Department of Agriculture, Environment and Rural Affairs. Coastal waters around Ireland are generally classified as having good or high ecological status under the WFD. [Data source: Department of Communications, Climate Action and Environment; UK Hydrographic Office; Department of Agriculture, Environment and Rural Affairs]

novel feature of this directive is that it moved away from planning and management based on administrative boundaries and focused on hydrographic units. It also links with other EU legal instruments, such as those relating to biodiversity (Birds and Habitats Directives), specific uses of waters (Drinking Water, Bathing Waters and Urban Waste Water Directives) and regulation of activities conducted in the environment (Environmental Impact Assessment Directive). The WFD aims at protecting/enhancing all waters (surface, ground and coastal waters) through achieving 'good status' for those waters

on a river basin or catchment basis whilst also involving the public (Figure 32.7) (see Chapter 31: Pollution). Similar to the WFD, the Marine Strategy Framework Directive (MSFD) takes a holistic, integrative approach. It aims to achieve Good Environmental Status (GES) of marine waters across Europe. For the purposes of the directive, marine waters extend effectively from the low water mark to the 200 nautical mile limit.

The provisions of the EU Maritime Spatial Planning (MSP) Directive also enable a more integrative approach to dealing with

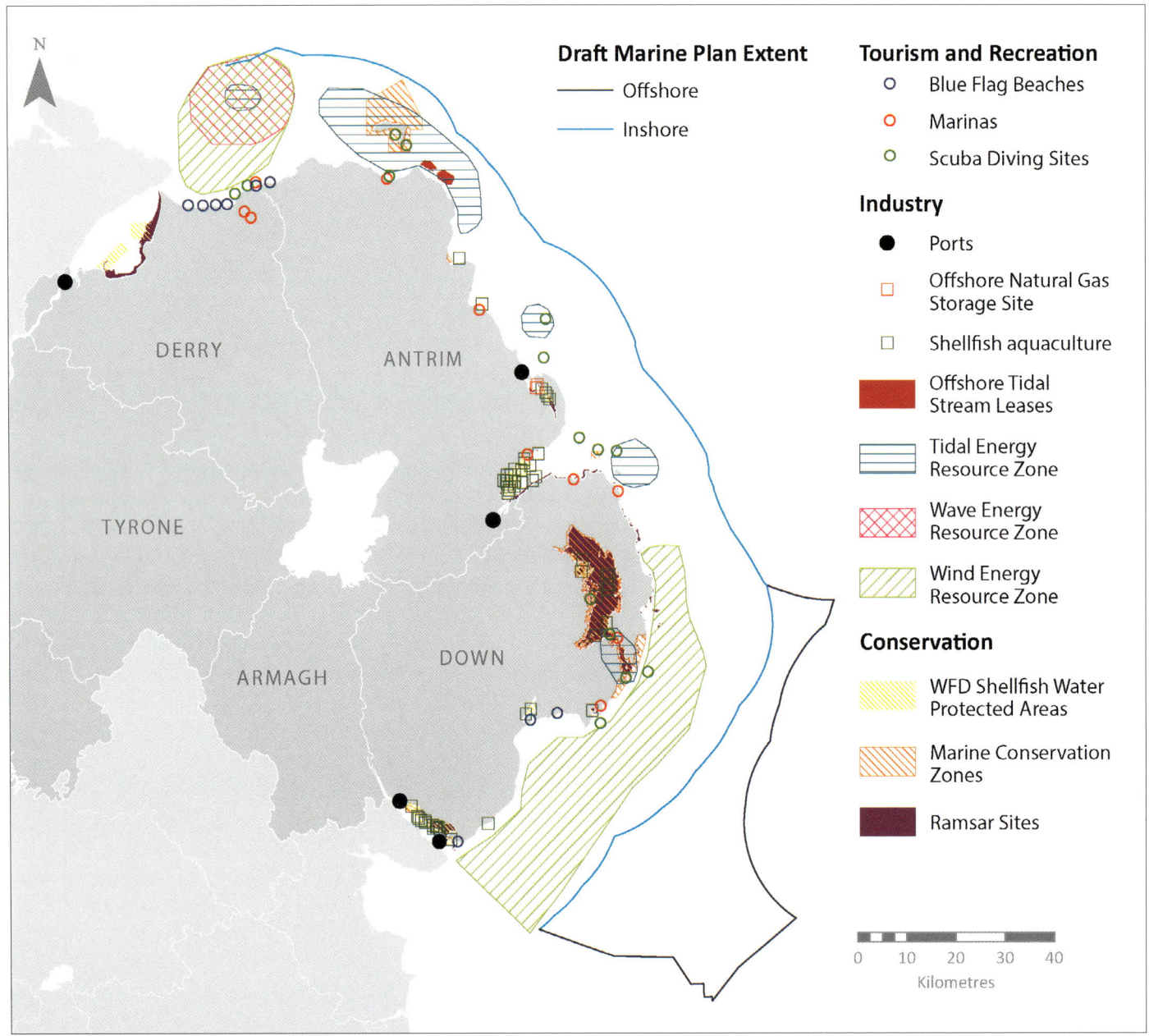

Fig. 32.8 A SELECTION OF COASTAL DESIGNATIONS AND ACTIVITIES THAT WERE CONSIDERED FOR MARINE SPATIAL PLANNING (MSP) IN THE DRAFT MARINE PLAN FOR NORTHERN IRELAND. The draft Marine Plan for Northern Ireland is on hold as a consequence of the collapse of the Northern Ireland Assembly in 2017. Prior to that, good progress had been made on issue identification. The map shows the range of issues that are typically covered and mapped in an MSP process. While the type, nature and scale of activities are contingent on geographic attributes, an MSP mapping process usually includes sectors of activity overlaid with conservation areas and planning zones. [Data source: Department of Agriculture, Environment and Rural Affairs, 2019; The Crown Estate, 2019; Afloat, 2013; Blue Flag Programme, 2018; Department for the Economy, 2019]

marine issues. As with all other directives, it must be transposed into the domestic law of the individual member states. In the Republic of Ireland, the provisions of the MSP Directive were incorporated into the Planning and Development (Amendment) Act, 2018. Prior to the MSP Directive there was a significant gap between terrestrial and marine planning. In short, the system for land-based planning is well developed, including comprehensive approaches to forward planning and development control, facilitated by local authority planning departments and An Bord Pleanála. However, there had been a complete absence of forward planning for the sea and the approach to consenting for marine space has been dysfunctional due to outdated and inadequate foreshore legislation. The advancement of MSP allows the minister to prepare one marine spatial plan for the entire maritime area or

different marine spatial plans for different parts of the maritime area. The plan is part of the National Marine Planning Framework, similar to the National Planning Framework that applies on land. The government has been building capacity for effective spatial planning at sea, by mapping all of the existing sectoral uses and consulting widely with the general public. MSP is a new way of planning for the marine area, by seeking to balance competing demands for space and resources with the imperative to protect the marine environment. Stakeholder engagement is a key principle of MSP.

According to the MSP Directive, MSP in European member states should follow the Ecosystem-Based approach to Management (EBM). This approach takes environmental, economic and societal aspects into consideration in a holistic way.

EBM, which has its roots in the principles of Integrated Coastal Zone Management (ICZM), was given credence by the UN Convention on Biological Diversity. It is one of the most significant paradigms to emerge in coastal management and is increasingly referred to as best practice. Despite this, the successful implementation of EBM can be challenging for planners and practitioners. To date, efforts to implement EBM have been limited to fisheries science.

There is no doubt that Maritime Spatial Planning represents a huge opportunity for the country in terms of planning and managing its extensive maritime area. In Northern Ireland, the Marine Plan remains in draft format as it requires ministerial approval (Figure 32.8).

Maritime Spatial Planning has become an important approach in Europe and elsewhere in promoting stakeholder dialogue, taking a long-term view and planning on the shared use of marine resources. However, it is not a panacea. MSP tends to focus on issues 'over the horizon' or at sea, rather than in the coastal zone where most conflicts tend to actually arise. ICZM is a management approach specific to the coast. ICZM seeks to deal with the land/marine divide that occurs typically in approaches to policy and planning. The US led the development of the concept of ICZM in the 1970s and since then it has been implemented in many countries around the world.[7] ICZM gained popularity as a concept in Europe in the 1990s. At this time ICZM was implemented in a limited fashion in different bays around the Irish coast on a project basis, for example in Bannow Bay, Cork Harbour and Clew Bay. The ICZM project in Bantry Bay was by far the most comprehensive and impactful of those initiatives.

Case Study: The Bantry Bay Charter

Val Cummins

The Bantry Bay Charter Project (BBCP) commenced as an EU-funded Demonstration Programme project on Integrated Coastal Zone Management (ICZM). It was one of thirty-five demonstration projects on ICZM funded and conducted simultaneously across Europe. These were initiated to inform a European view of how ICZM might work across member states. In 1997 the BBCP project leaders – Cork County Council in partnership with University College Cork and the Nautical Enterprise Centre – selected Bantry Bay as a pilot area to test a 'bottom-up' stakeholder-led approach to the management of the bay. The main aim was to establish consensus among all stakeholders to produce a sustainable management strategy for the rich coastal resources of Bantry Bay. The project was a unique and innovative approach to ICZM in Ireland at the time. It is further distinguished as the most concerted ICZM effort trialled in Ireland to date.[8] This case study summarises the key findings from a PhD thesis that examined the process and the outcomes of the BBCP as a model for participatory governance.[9]

From the outset, the BBCP was a challenging project, with a multitude of coastal management issues to be addressed. Bantry Bay is a large deep-water bay, with good water circulation, but incidents of localised impairment of water quality were a cause of concern. At the time, the Bantry Bay area hosted a population of c.12,000 people, dependent for a living on activities such as agriculture, tourism, fishing and aquaculture. Other activities included conservation, production from a super-quarry located at the coast (no longer operational) and the Whiddy Island oil trans-shipment terminal (Figure 32.9). Issues linked to the Whiddy Island oil terminal and the nascent aquaculture industry were the most contentious, deep-rooted and difficult issues to resolve. As a result, historic aspects of the development of the oil industry and aquaculture in Bantry are worth reflecting on.

Ireland's only oil terminal at Whiddy Island commenced operations in 1969. On 8 January 1979, when in the ownership of Gulf Oil, the oil tanker *Betelgeuse* exploded, killing fifty people and one salvage diver at the terminal (see Chapter 31: Pollution). Following the disaster, the terminal ceased to operate and 250 local employees lost their jobs. The disaster was also perceived by fishers to have had a detrimental impact on the local fishing industry.

In fact, the impact on the fishing industry was perceived by some actors to have extended back over the preceding decade. During that period, local anecdotal evidence suggested that up to approximately fifty oil spills had occurred as a consequence of terminal operations. A perceived lack of regulation was a cause for concern. In particular, the role and function of the Harbour Authority was called into question. When the Whiddy oil terminal was first developed, there was no harbour board in place to manage the terminal traffic. The Bantry Bay Harbour Commissioners were established under the Harbours Act, 1976, but they were initially set up without full harbour authority functions. This was one of the issues that set the context for conflict at the time of the Charter project.

The Whiddy oil terminal reopened for operations in 1998, around the time of the launch of the BBCP. From the terminal perspective, the BBCP offered terminal managers an opportunity to engage with the community in a new way. As a result, they embraced the project and gave it strong support. However, there was hostility to the terminal from the fishing and aquaculture sectors. There were concerns over potential impacts on the marine ecology of invasive species from the exchange of ballast water by tankers at sea. Ballast water had

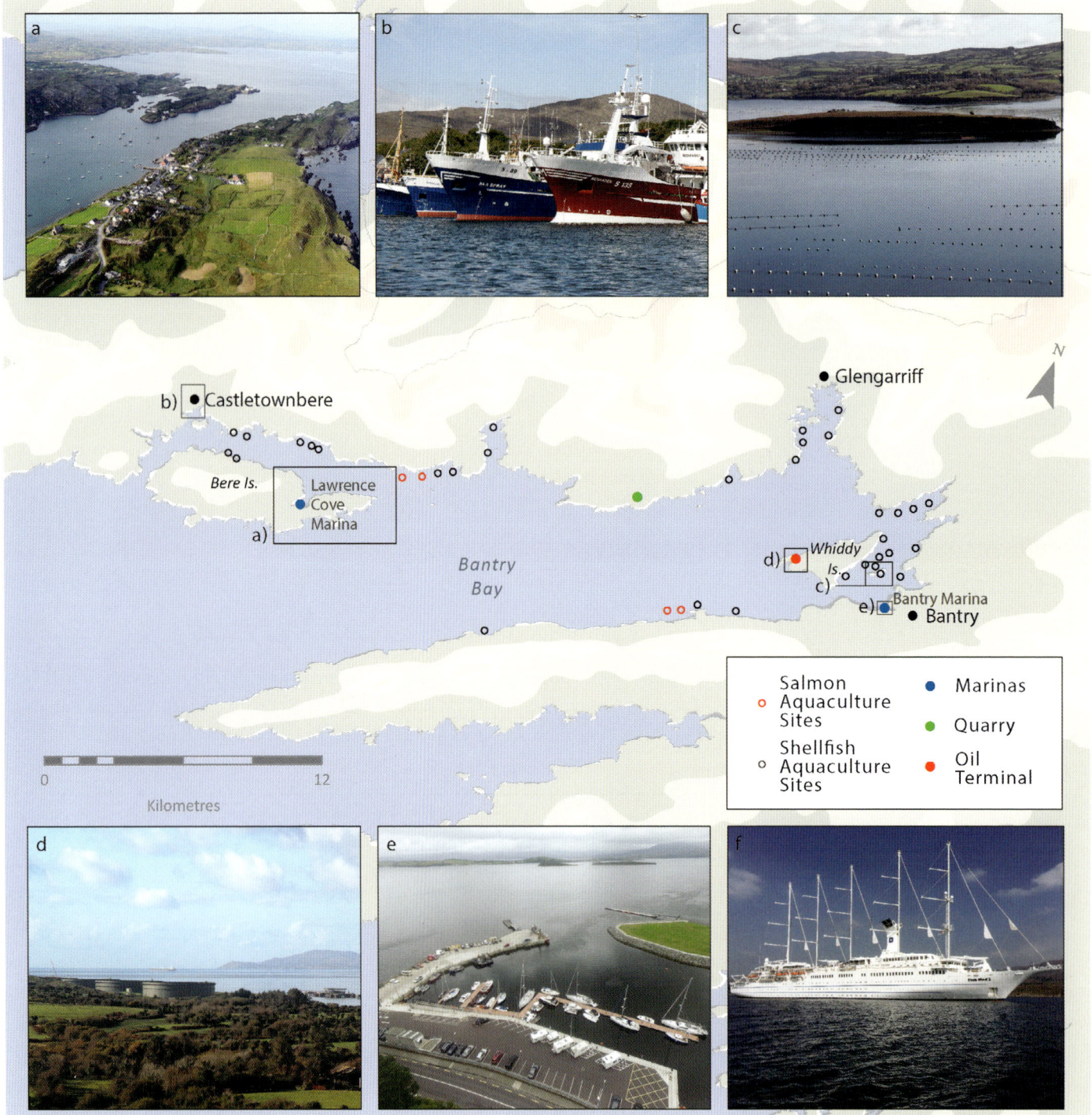

Fig. 32.9 OVERVIEW OF BANTRY BAY AS A COASTAL MANAGEMENT UNIT. Bantry Bay was selected as a focal point for a community-based experiment on public participation in coastal zone management as a result of the diversity of activities co-existing within the bay area. The photographs show a sense of this diversity, including developments that have occurred over the last decade, such as cruise tourism and enhanced shoreline access. The photographs show a. Lawrence Cove Marina and moored boats off Bere Island; b. fishing boats in Castletownbere, a fishing harbour of national significance; c. mussel lines in Bantry Bay, part of the shell and finfish farming activities; d. Whiddy Island oil terminal, the only oil terminal in the Republic of Ireland; e. Bantry Marina, opened by the Port of Cork company in 2017; and f. a cruise ship visiting Bantry Bay, where visitors enjoy popular destinations such as Glengarriff and Garnish Island. [Source: a: Irish Air Corps; b., c., d. Robbie Murphy; e., f. Bantry Bay Port Company DAC]

traditionally been pumped ashore, but new tanker design at the time of the reopening led to a change in practice. The fisheries sector was deeply concerned about this.

Opposition also came in particular from the aquaculture sector, which had developed extensive rope mussel farming in the intervening period. Mussel farmers were frustrated about water quality issues and maintained a negative perception of oil terminal operations. However, they were also deeply frustrated with the state agencies responsible for the development of fish farming. Mussel farming, first developed in Bantry Bay in 1981, grew quickly to become one of the most significant industries in the area. By 1999 the industry was producing 4,000 tonnes of mussels for export with a sales value of £2 million (€2.54m). It employed thirty people full-time and up to eighty people at peak. A further eighty to one hundred people were employed in processing and other downstream industries. Bord Iascaigh Mhara (the Irish Sea Fisheries Board) was actively implementing government policy to grow the nascent sector. Despite all of this, between 1982 and 1997 no licences had been issued for mussel farming in Bantry Bay. Numerous applications had been submitted to comply with the foreshore legislation but there was a systemic failure within the Coastal Zone Management Division (CZMD) of the Department of Marine and Natural Resources (DoMNR) – established with the primary purpose of issuing foreshore licences – to deal with the applications.

The lack of an efficient regulatory framework for aquaculture licensing led to the unplanned growth of the mussel farming industry within the bay. Issues arose as a result of mussel rafts encroaching into marine space used by other marine user groups, visual impact from barrels used to mark the mussel rafts, fishery closures due to harmful algal blooms and problems with associated marine litter and debris. Salmon farming in the bay produced a further £3m (€3.81m) worth of value in 1999. Both the salmon and mussel farming businesses also suffered from distribution issues due to poor transport infrastructure. In short, the tensions arising from the aquaculture and oil sectors were at breaking point when the BBCP was introduced to the local community.

BBCP STAKEHOLDER ENGAGEMENT OPPORTUNITY

Stakeholders were defined as 'anyone who has an immediate "stake" in the area, for example they live or work there'.[10] This term was also taken to include any public bodies with a direct executive local role. There was no formal stakeholder mapping process. Instead, a major public awareness campaign was initiated to invite participation from any interested stakeholders. Stakeholders had opportunities to engage through one-to-one interviews, roundtable processes and working groups. All of these mechanisms were facilitated by the project team over a three-year period. They were complemented by the production of technical reports and studies, as well as the development of a Geographical Information System, as a tool for planning. Careful consideration had to be given to facilitating meaningful engagement, particularly in a rural community that was widely distributed over considerable distances from one end of the bay to the other.

The outputs from the working groups and the consultation process were taken into consideration by a roundtable forum. The consensus-building process, which was gradual and iterative over the course of three years of trust building, nearly collapsed towards the latter stage of the negotiations, as a result of apparently irreconcilable differences stemming from the Mussel Cultivation Working Group. This group, besieged by difficulties with licensing, had little trust in officialdom and struggled to engage proactively with the BBCP. Finally, the charter, which consisted of over 200 recommendations, organised according to the headings of Government, Production, Protection and Infrastructure, was agreed by all stakeholders through consensus.[11]

The next step was to progress towards implementation and action. Despite an initial commitment of funding from Cork County Council, the implementation phase barely got off the ground. The project office was suddenly closed in 2002 due to budget cuts in the local authority. It was likely that the BBCP was positioned at a weak point within a system that was incapable of dealing with an initiative that would expose or threaten the politics and administration of the day. In other words, the participatory process perceptively threatened representative democracy. Politicians had deliberately been left out of the stakeholder engagement process, which meant that there were no political champions, and political representatives became worried about a process they could not seek to influence. The BBCP also threatened to expose the weakest link in the system, namely, the capacity of the DoMNR to deal with bottom-up approaches to conflict resolution.

Fig. 32.10 STAKEHOLDER CONFLICT IN BANTRY BAY. Despite widespread disappointment at the closure of the Bantry Bay charter Office in 2002, following a highly successful project, research has shown that many stakeholders are positively disposed towards revisiting such a project. A major benefit of the Bantry Bay Charter Project was perceived to have been the consensus-building process. While it cannot be seen as a panacea, these efforts in Integrated Coastal Zone Management can be helpful in navigating through contested issues, such as the dispute around kelp harvesting within the bay area. [Source: Niamh O Donoghue]

The PhD research revealed issues relating to leadership, power dynamics and the bureaucratic mentality across decision-making for coastal management in Ireland at the time. One interviewee referred to the 'tragic comedy of cycles' that earmarked the disintegration of marine affairs in three consecutive decades from 1980 to 2010, in which 'departmental structures and personalities had a role to play'.[12] In short, institutional analysis showed that leadership at the national level was weak and inconsistent when it came to marine affairs, both politically and administratively; the mentality was conservative and institutionalised; and there was a lack of power and credibility within the DoMNR. The Marine Institute, established in 1992, was curtailed in its ability to influence best practice in ICZM. This was added to by a number of other failings at the local level in the project structure, including a lack of experience in the project management team, a lack of a high-level champion and the overall lack of strategic engagement by senior managers in the local authority.

Despite the ultimate outcome, the majority of stakeholders interviewed for the PhD research, ten years after the project came to an end, said that they would re-engage in a similar process again, on the basis of a more simplified, focused and issue-oriented approach. Individual stakeholders were found to have derived considerable value from participation in the BBCP process in terms of knowledge and skills development at a personal level. Some even applied their learning about the consensus-building process to their professional work environments.

Over the last few years there have been notable improvements in issues such water quality and shore-based access due to infrastructure investments around the bay. However, issues around aquaculture and sustainable development prevail. A new conflict has arisen in relation to a licence granted to mechanically harvest kelp in Bantry Bay (see Chapter 26: Coastal Fisheries and Aquaculture and Figure 32.10). Locals have expressed concern about the potential environmental impact of kelp harvesting at scale, as well as a perceived lack of transparency in the decision-making process. Supporters of the proposal are keen to see an underutilised and abundant natural resource deliver jobs and economic benefits. Today, stakeholders are engaged in the consultation on Marine Spatial Planning, which, in due course, may provide a new opportunity for local engagement in the management of Bantry Bay.

The European Commission (EC) took the lessons from the Europe-wide Demonstration Programme on ICZM on board. Interest in a Europe-wide approach to ICZM was maintained through to 2013 when the EC proposed a draft directive, aimed at creating a framework for both ICZM and MSP. The aim was to improve planning and management of the land–sea interface. During the negotiation phase, however, ICZM was dropped from its contents. No official explanation is available but significant concerns were expressed by the Committee of the Regions, for example, as it was perceived that ICZM impinged substantially on

Fig. 32.11 COASTWATCH EVENT FOR WORLD OCEANS DAY (2018), COUNTY WEXFORD. The popularity and influence of bottom-up, community-based approaches to management, facilitated by organisations such as environmental non-governmental organisations (eNGOs), have grown substantially in the past couple of decades. eNGOs now undertake vital activities in efforts toward conservation and sustainable development, including awareness-raising, environmental data collection, scientific publishing and participation in decision-making processes. Due to the increasingly global nature of environmental issues, many of these eNGOs stimulate action locally, while maintaining an international perspective through umbrella initiatives with partners in other countries. The Irish Coastal Environment Group, or Coastwatch, is a national eNGO in the Republic of Ireland that works with local groups around the Irish coast to protect wetlands and other coastal habitats by raising awareness about their societal value and partaking in monitoring and clean-up activities. Their efforts are co-ordinated with the Ulster Wildlife Trust in Northern Ireland and other eNGOs in Europe through the Coastwatch Europe (CWE) network. Together, the CWE network aims to conserve and promote the sustainable use of coastal resources while also engaging the public in environmental management and planning. [Source: Coastwatch Ireland]

Fig. 32.12 GREEN COAST BEACHES (2019). The Green Coast award is a distinction granted to beaches for environmental excellence in Ireland and Wales. To be considered for the award, applicant sites need to be managed sensitively – in co-ordination with the local community and conservation organisations – swept carefully for litter (so as to avoid removing natural beach debris) and the visual appeal of signs and facilities maintained. Like Blue Flag beaches, Green Coast beaches must also maintain excellent water quality. The Green Coast award strongly supports community involvement by promoting the establishment of Clean Coasts groups. These groups develop goals and activities to maintain and protect a stretch of coastline in co-ordination with An Taisce's Coastal Programmes Officers. In 2019, a total of sixty-two sites were granted Green Coast awards across Ireland. While beaches along the south and west coasts were particularly successful, no awards were granted to beaches in Northern Ireland. The Green Coast award is run by An Taisce in the Republic of Ireland and Keep Northern Ireland Beautiful in Northern Ireland. [Data source: Green Coast Awards, 2019]

existing member state competences relating to spatial planning policy and practice at regional and/or local levels.[13]

Lessons learned from ICZM initiatives across Europe show that there is a need to develop a range of instruments for managing the coast. Government-led legislation and policy, referred to as top-down approaches to coastal management, need to be complemented with community-driven initiatives, or bottom-up approaches to management.[14] In both approaches, relevant and effective stakeholder engagement is important. The European Commission advocates for bottom-up approaches, such as citizen science, which is playing an increasingly important role in gathering the data and information necessary for effective planning and management (Figures 32.11 and 32.12). Finally, there is the issue of how we include societal norms and values in the development of coastal plans. This becomes a matter of increasing importance as society becomes more diverse and the potential for contestation over access to resources in the coastal zone increases. Inevitably, perspectives or worldviews shape decisions that are made, and these need to be understood. Planning for tourism development on the islands provides a good example of how stakeholder values and perspectives should be incorporated into decision making.

PLANNING FOR IRELAND'S ISLANDS: A MATTER OF PERSPECTIVE

Karen Ray and Brendan O'Sullivan

How we engage with an environmental management issue, such as the sustainability of the island of Ireland as a whole or individual island life, is often a question of the perspective from which we view it. All Irish islands – whether inhabited or not, whether inaccessible or accessible by bridge, causeway, air, cable car or scheduled ferry (Figure 32.13) – fall within the jurisdiction of a planning authority with legal powers concerning how land, habitats, buildings and landscapes can change or develop over time. Even in the most compact and well-connected counties, these local authority decision-making powers are exercised at a distance from the islands themselves. The question of perspective is particularly important when examining how decision-makers and politicians, along with the local people and visitors, engage with islands and island issues. Decisions about development proposals on the islands (whether by the local planning authority or by An Bord Pleanála on appeal) require a particular degree of sensitivity. But to what extent should this be seen in the wider context of the coastal zones, rural economies and landscapes within which the islands are located? How do discussions about these issues come about? Are all the important perspectives represented in these discussions? And are they as inclusive, wide ranging and rich as they need to be?

It is useful to look at this issue by using tourism as an example. As far back as 1936, in his reflections on Ireland's islands, Thomas H. Mason made the following observation:

Fig. 32.13 a. ISLAND ROY, COUNTY DONEGAL, AND b. TORY ISLAND, COUNTY DONEGAL. Island Roy and Tory Island in County Donegal are vastly different communities. While Island Roy lies within the protected waters of Drongawn Lough and features bridge access to the mainland, Tory Island lies 15km offshore, in the open north Atlantic and is only serviced by a ferry or helicopter in weather-permitting conditions. This diversity in the characteristics of Irish islands, coupled with their geographic distance from decision-making centres, present substantial challenges for local authorities to plan, develop and implement relevant policies for island communities. [Source: a. Donegal County Council; b. Geraldine Hennigan and Norman Kean]

More than one person has expressed to me the fear that the publicity given to the islands in recent years by books and films will make them popular as tourist resorts and destroy the finer characteristics of the people. Undoubtedly, excessive 'tourism' does tend in this direction, but in the case of islanders I do not think this will occur; their fundamentals are too deeply engrained, and the class of tourist who will go to the islands is of necessity a class with the finer instincts.[15]

The extract conveys an idyllic yet pragmatic perspective. On the one hand, Mason touches on an issue of sustainable tourism that remains as valid a concern today in any areas susceptible to losing authenticity or even environmental degradation. On the other, and perhaps more importantly, he implies an inconspicuous resilience among island cultures and landscapes that accommodates movement and certain levels of change.

With rural depopulation as one of the key contemporary concerns for Ireland, perspectives that eschew a purist or tokenistic approach ensure the islands are kept in focus. The reality is that on an ongoing basis, decisions about wind turbines, homes, tourism accommodation, rural enterprises and new infrastructure on the islands cannot be avoided. For fragile island economies it is essential that planning authorities get the balance just right; with such little room for error it is imperative that people engage robustly with these planning processes. While there is a perception among visitors that heritage is better preserved on the islands,[16] these incremental decisions can still have significant impacts in such environments, gradually eroding valued aspects of an island's integral character and heritage and in turn debunking the myth of islands as having an enduring 'other-worldly' stillness (Figure 32.14).

An interesting study on the marketing potential for tourism development on the islands, commissioned by Fáilte Ireland – particularly with regard to Gaeltacht islands – found that some of the main motivators for people to visit included 'the peace and quiet, slower pace of life, natural beauty of the landscape', and the 'real Irish cultural experience'.[17] Still-living places, often with vibrant communities, even the 'seemingly inhospitable' Inis Meáin of the Aran Islands has 'supported a community since prehistoric times'.[18] Islands have a timeless quality to them that makes them special.

Fig. 32.14 PERCEPTION OF MODERN ISLAND DEVELOPMENTS. Incremental changes to modernise, renovate and rebuild old residences can amount to substantial changes in the visual aesthetic of the landscape over time. Developments on a. Inis Oírr, County Galway, and b. Rutland Island, County Donegal, illustrate attempts to assimilate and conform with the surrounding environment. The degree to which this has been achieved is a matter of perspective. This question of perspective is an important consideration in coastal planning and management. Perspectives have a high degree of local specificity and are influenced by culture, norms and values. [Source: a. Stephen A. Royle; b. Donegal County Council]

We should also remember that, while the island angle on sustainable development is distinct from the mainland one in many ways, the two approaches are closely linked in planning terms especially when it comes to employment, economic activity and the provision of services; it can be said that an island will thrive if its adjacent mainland region is thriving as well.

INSTITUTIONALISATION OF COASTAL MANAGEMENT IN THE REPUBLIC

At the most fundamental level, the coast is simply defined as where land and sea meet. Such a simple definition cannot take account of the dynamic nature of coastal environments where active coastal processes ensure that the coast is not stable, but varies in both space and time. As can be seen throughout the *Atlas*, the landward and seaward influences of coastal processes are extensive. In addition to the legal measures that give rise to delineation of maritime boundaries, the legal instruments of particular relevance to coastal

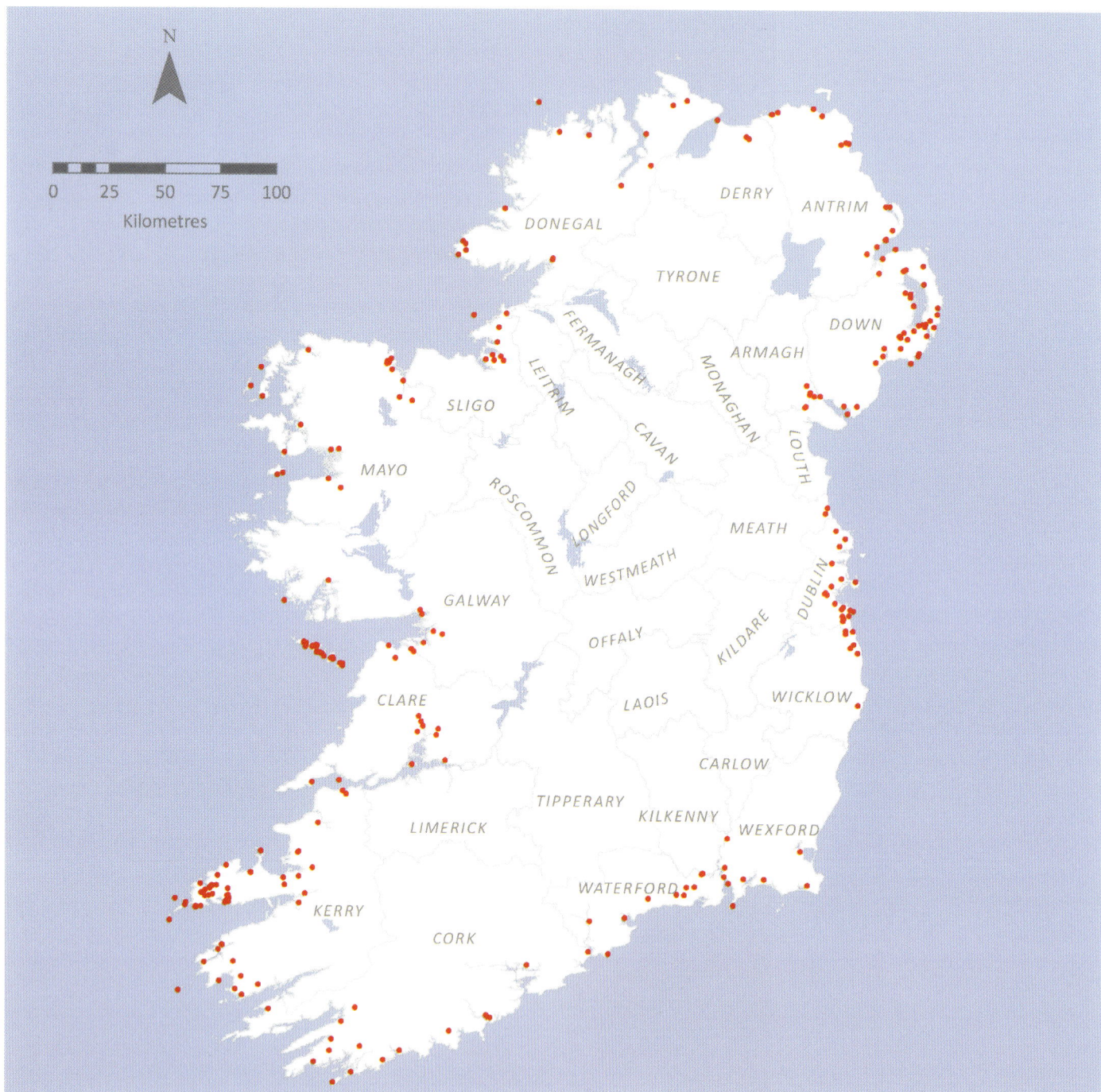

Fig. 32.15 NATIONAL MONUMENTS WITHIN 5KM OF THE COAST THAT ARE WITHIN STATE CARE. The National Monuments Acts, 1930–2004 strive to protect and preserve national monuments and archaeological objects. In this context, a 'national monument' means a monument 'the preservation of which is a matter of national importance by reason of the historical, architectural, traditional, artistic or archaeological interest attaching thereto' (Government of Ireland, National Monuments Act, 1930). Protection is also given to the means of access and any adjoining land necessary for its preservation. Under the provisions of the legislation, national monuments can be protected in one of four ways: 1. included in the Record of Monuments and Places (RMP); 2. registered in the Register of Historic Monuments (RHM); 3. subject to a preservation order and 4. under the ownership or guardianship of the Minister for Culture, Heritage and the Gaeltacht or a local authority. The level of protection varies according to how the monument is protected. The legislation also covers underwater archaeology. [Data sources: National Monuments Service; Department for Communities]

planning and management in Ireland are the Foreshore Acts, 1933–2011; the Planning and Development Acts, 2000–2018; the Harbours Acts, 1996–2015; the Wildlife Acts, 1976–2012; and the National Monuments Acts, 1930–2004 (Figure 32.15). It is important to emphasise that almost all coastal and marine activity will have some form of legislative basis attached to it and the aforementioned are only a sample of the key influencing instruments.

Marine consenting

At the time of writing, the government is developing a more integrated approach to the management of development and activities in the marine space. Given this transition, it is worth commenting on the approach taken to legislating for foreshore activities up to this point, before providing an overview of arrangements still in the pipeline.

Historically, in Irish law the term 'foreshore' has been used, rather than coastal zone. It was originally defined in the Foreshore Act, 1933 as 'the bed and shore, below the line of high water of ordinary or medium tides, of the sea and of every tidal river and tidal estuary and of every channel, creek, and bay of the sea or of any such river or estuary'. In Ireland an outer limit of the foreshore, corresponding to 12 nautical miles, was added for the first time by the Maritime Safety Act, 2005. The high water mark (HWM) generally separated the terrestrial planning jurisdiction of a local authority from the marine jurisdiction administered by central government departments, depending on the use of the marine area. At the time of writing, the Foreshore Acts, 1933–2011 allow for the granting of foreshore leases and licences utilised to convey rights to develop and utilise the foreshore, much the same as planning

permission operates on land. This legislation has been perceived as a major constraint on development to date, particularly among stakeholders concerned with the development of offshore wind.[19] The majority of foreshore consenting responsibility rests with the Department of Housing, Planning and Local Government.

In recent years the Department of Housing, Planning and Local Government has led a process to renew the foreshore legislation, to streamline the process of planning at sea and to align it with the requirements to build capacity to comply with the Maritime Spatial Planning (MSP) Directive. The outcome of this process will establish a new marine planning system under the auspices of the National Marine Planning Framework. The Marine Area Plan legislation envisages the granting of a Maritime Area Consent for occupying a maritime area. This replaces foreshore licences and leases. The new planning regime will extend beyond 12 nautical miles, as far as the outer limit of the state's continental shelf. Decisions on consents will be the concern of the minister for Housing, Planning and Local Government, with the exception of offshore renewable energy, which will be the concern of the minister for Communications, Climate Action and Environment. Aquaculture and fisheries are under the remit of the Department of Agriculture, Food and the Marine and will remain outside the scope of the new legislation awaiting enactment. An Bord Pleanála will be responsible for decisions for certain classes of strategic marine infrastructure. A formal public consultation process is pending to inform the final drafting of the legal text.

Planning and development

Planning law deals with forward planning and planning controls for the landward side of the coastal zone. The Planning and

Fig. 32.16 WATERFRONT DEVELOPMENTS IN SALTHILL, COUNTY GALWAY. Beginning as a quaint fishing community in the agricultural hinterland of Galway city, Salthill has transformed into a vibrant urban village at the heart of the Wild Atlantic Way. Despite its success, the 2017–2023 Galway City Council Development Plan identified that underutilisation of space, design inconsistencies, pedestrian access, connectivity to Galway city and environmental challenges would require attention in future developments. Moving forward, Galway City Council aims to promote Salthill's status as a coastal leisure and recreation hub, maintain quality and consistency in design, co-ordinate an environmental improvement scheme and develop a long-term management scheme. [Source: Chaosheng Zhang]

Development Acts, 2000–2020 operate on two levels: first, they provide for the making and implementation of structures regulating land use, and second they prohibit the development of land unless it is authorised and carried out in agreement with planning authority regulations. Under the act, every planning authority must make a development plan every six years (Figure 32.16). Development plans and local area plans operate as a framework within which planning applications are made on land, and planning permissions granted or refused. In general, land above the line of ordinary high water mark is within the administrative area of the appropriate county council which also acts as the planning authority. Historically, the foreshore lay largely outside the administrative area of the county council. The new marine planning framework, described above, considers an extension of the planning remit of local authorities over a limited distance from the shore.

Case Study: Planning for Ireland's Islands

Karen Ray and Brendan O'Sullivan

The aim here is to draw on examples of certain islands to examine how fine-tuned planning perspectives might deliver more sustainable futures for these special environments and places (see Chapter 21: Ireland's Islands). Sometimes, islands require unique planning interventions and the lessons observed can often be transferred or adapted, as appropriate, to other parts of the coast.

Without a co-ordinated coastal zone and islands strategy at national level, responsibility for the general co-ordination of the islands remains ill-defined. Islands can fall victim to administrative and policy gaps, often in the form of superimposed and often unsuitable mainland models. It is hoped that the ambitions of the National Planning Framework – in particular, its chapter 7: 'Realising our island and marine potential' – will go some way to addressing this during its lifetime (2018–2040).

To date, there has been a lack of centralised governmental co-ordination of planning and management of the islands. It has been twenty-three years since cross-government policy for the islands has been considered. Now, the voices of island communities are finally being taken into consideration, in a new plan to set out a national roadmap for the long-term sustainability and development of these offshore communities.

To date, planning responsibilities have predominantly devolved to local authority level, to where mainland models can overlook the nuances of natural and/or cultural components of island life (Figure 32.17). At this local scale, each island is – or should be – dealt with

Fig. 32.17 FISHERMAN AND HARBOUR ON GOLA, COUNTY DONEGAL. Gola is a small, moderately hilly island located approximately 1km off the coast of Gweedore. It hosted a population of several dozen fishing and farming families through the nineteenth and beginning of the twentieth century. While its population dwindled to zero in the late 1960s, people have begun to return to the island in the last few decades. Electricity connectivity was established in 2005 and a seasonal ferry service now operates out of Bunbeg Harbour. In the summer months a small community of fishermen and farmers inhabit the island, with hundreds of tourists making additional visits. In 2016 a crane was transported to the island to support fishing activities and vessel loading/unloading at the pier. The increased number of visitors on the island has also prompted plans for road upgrades. Despite these improvements, funding for infrastructure on Gola and many Irish islands remains contentious. The rural nature, context specificity and seasonality of many island communities can make planning decisions challenging for authorities. A new national roadmap for the islands is long overdue. This may prioritise some of the planning and management issues faced by island communities such as Gola in the future. [Source: Donegal County Council]

in the development plan for the corresponding authority. A small number of islands are accompanied by their own management or development plan, driven by particular issues or threats (usually nature-related) facing the sites. For example, the uninhabited North Bull Island in Dublin Bay was the first site to receive protection as a nature reserve under the 1930 Wild Birds Protection Act and as a result has benefited from a management plan implemented by the Parks and Landscapes Services of Dublin City Council since 2009. The North Bull is also an example of the paradox of sustainable tourism, whereby improved access by means of a causeway can enhance the appreciation of the island while simultaneously accelerating problems of environmental degradation (such as impacts on the sensitive dune habitats and rabbit population). The same paradox can equally apply to inhabited islands, where the economic and social vibrancy of the islands is increased with better access but is potentially accompanied by greater threats to the fragile physical and cultural island fabric.

Islands make up a significant proportion of the Gaeltacht and this presents another interesting perspective for planning in Ireland. The island of Árainn Mhór (Arranmore), off the coast of County Donegal, implemented its own development plan in 2008. Funded by Údarás na Gaeltachta and with a strong community co-operative involvement, the plan set out the strategic development of the island for a five-year period, with the steadying and possible reversal of population decline as a primary aim (see Chapter 21: Ireland's Islands). Whilst Gaeltacht islands such as Árainn Mhór are particularly vulnerable to such decline, planning policies and decisions promote development that sustains populations and jobs, thereby supporting Irish-speaking communities.

The Donegal County Development Plan, 2012–2018 states that 'while language is the foremost uniquely defining feature of this area, the Gaeltacht also maintains a rich social and cultural heritage'.[20] The Aran Islands, off the coast of County Galway, fall under a specific Local

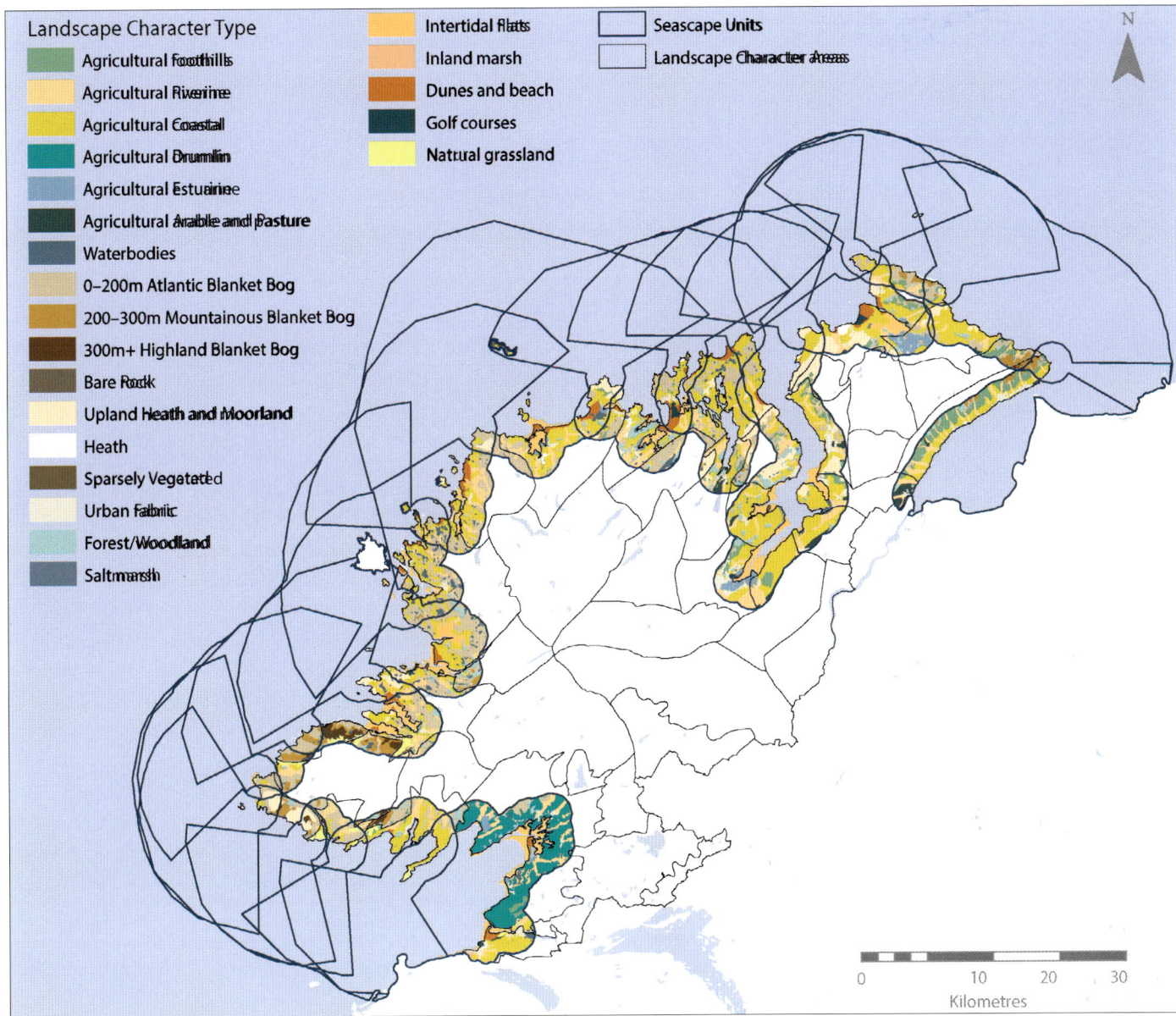

Fig. 32.18 INTEGRATED LANDSCAPE–SEASCAPE CHARACTER ASSESSMENT. Donegal County Council drew international attention for their work on the 2016 Donegal Seascape Character Assessment. The assessment combined traditional landscape and seascape character assessments with coastal geological classifications to develop a single integrated assessment for the coastal zone. Seascape units were classified according to physical type, use, environmental designations and sensitivities, visual and sensory qualities, key historic and cultural association and forces of change. The assessment process was also supported by public participation and stakeholder consultation sessions. These approaches to coastal planning and management can provide valuable visualisation aids, as well as being effective tools for deliberative democracy. [Data source: Donegal County Council]

Area Plan for the Gaeltacht while Cape Clear, which is also part of the Gaeltacht, falls within the West Cork Islands Integrated Development Strategy along with its neighbouring non Irish-speaking islands. Planners have learned that the best indicator for preserving linguistic heritage in the long run is the degree to which the island community is well served by access to jobs and services. It is not simply about people's individual fluency in the language; it is about the island's community life to sustain itself.

An interesting aspect of policy-making for islands comes from the perspective of landscape. It is arguably one of the most fundamental reasons why we value islands so highly. By law, all development plans must include objectives for preserving the character of the landscape. In order to do this properly, the distinctiveness and sensitivities of any given landscape must be thoroughly understood and meticulously recorded. Only then can we even begin to craft bespoke policies for dealing with change in that particular landscape.

Nowhere is the task of capturing distinctiveness more challenging than on these islands. This might seem surprising, given that islands are self-contained: it is usually quite clear where an island begins and ends. Yet when we refer to Ireland's islands, we are, in essence, referring to landscapes that are small – at least in comparison to adjacent mainland landscapes. In policy terms though, landscape appraisal is usually carried out at county or city level. Following the principles of the *European Landscape Convention*, every county has some form of landscape character assessment which divides the area into distinct landscape types. Island landscapes, which often bear different characteristics from those of the mainland, tend to be blended with the larger, more generic, landscape types of which they form part. The 'agricultural coast' type in the Donegal plan (Figure 32.18) is a good example of this, while all of the west Cork islands – which are devoid of many peninsular characteristics – fall into a landscape category called 'rugged ridge peninsula'.

In terms of perceptions then, the primary task is to look 'within' the islands themselves and to draw out their local character at an intimate scale. We then complement this with looking both 'at' the island from the mainland and visitor perspectives and 'from' them towards the coastal seascapes and the rest of the world. In some places, an island landscape is a continuation of the defining characteristics of the mainland, with identical sensitivities and with an identical capacity for change. In such circumstances perhaps we should not give undue attention to the islands alone; sometimes the adjacent mainland areas have sensitivities that require just as much close attention. As with economic and social considerations, a more refined landscape perspective for Ireland's islands – articulated at different scales – can balance individual fine-grained island character with that of the wider spatial environment.

In Ireland, there is a need to embrace the special characteristics that set islands apart and to appreciate that these often overlap with those of the wider coastal zones in which they sit. And all of this needs to be reflected in the political arenas and the quasi-judicial processes (such as An Bord Pleanála) where planning takes place. Indeed, planners are often in a position to affect small changes that can have significant long-term benefits for islands. For example, they can impose conditions on private coastal developments to ensure continued public rights of access to and from the coast. Or they can take a strong position on seemingly modest development proposals that might lead to incremental privatisation of some of the smaller islands. And because these often unseen approaches can meet strong resistance from individual interests, they require a degree of subtlety and care – not to mention community support – in order to be successful.

Strategic Infrastructure Development

The Planning and Development (Strategic Infrastructure) Act, 2006 was intended to 'fast-track' and streamline the consenting process applicable to developers of specified private and public strategic infrastructure developments, by allowing them to make applications directly to An Bord Pleanála rather than to the local planning authority. The projects to which the Strategic Infrastructure Development (SID) process applies cover major energy, transport, environmental and health infrastructure. Within the environmental category, for example, coastal protection works that exceed 1km are specified. Relevant proposals must be of strategic economic or social importance to the state, contribute to the objectives of the

Fig. 32.19 ORAL HEARING OF A COASTAL STRATEGIC INFRASTRUCTURE DEVEL-OPMENT PROJECT. Oral hearings are often convened for strategic infrastructure cases under consideration by An Bord Pleanála. While they are open to public attendance, participation in discussions may be limited. Hearings are organised to examine potential concerns in cases before An Bord Pleanála. Upon completion of a hearing, the An Bord Pleanála inspector issues a report to the board. This report is taken into consideration along with case documents before a decision is taken on the case. An Bord Pleanála maintains the authority to decide whether to convene an oral hearing. The use of the image at left does not imply that An Bord Pleanála approves or disapproves of the text that is around the image, the chapter in which it appears or the publication as a whole. [Source: An Bord Pleanála]

Fig. 32.20 ESTABLISHED AND PROPOSED NATURAL HERITAGE AREAS (WITHIN 5KM OF THE COAST) AS WELL AS AREAS OF OUTSTANDING NATURAL BEAUTY. In the Republic of Ireland, Natural Heritage Areas (NHAs) are legally protected under the Wildlife Amendment Act, 2000. NHAs are designated for nationally important habitats or regions containing organisms requiring protection. The proposed Natural Heritage Areas (pNHAs) are not subject to legal protection, but many of these will be considered for full NHA status in the coming years. Areas of Outstanding Natural Beauty (AONB) are designated for the national importance of their landscapes and scenic value. While AONBs are legally protected from certain types of development, much of their management is left to the discretion of local authorities. [Data source: National Parks and Wildlife Service, 2019; Northern Ireland Environment Agency, 2016]

National Spatial Strategy or have a significant effect on the area of more than one planning authority. A pre-application consultation stage is a mandatory part of the Strategic Infrastructure Development process (Figure 32.19).

Harbour management

Harbours and commercial seaport activities are governed by the Harbours Acts, 1996–2015. The 1996 Act provided for the establishment of state companies to manage and operate commercial ports in harbours such as Cork, Dublin, Foynes and Galway. The functions of a harbour company relate to the management of its harbour and approach channels; the provision of such facilities for ships, goods and passengers; the promotion of investment; the engagement in business activity advantageous to the harbour; and the utilisation of the resources available in a manner consistent with these functions. Under the act, a company is required to employ a harbour master, who also has significant powers. The Harbours Acts, 1996–2015 also cover safety of navigation and security in harbours in line with wider EU and international requirements. This complements the work of the Commissioners of Irish Lights (CIL), as General Lighthouse Authority for the whole island of Ireland.

Provisions for protecting wildlife

The Wildlife Acts, 1976–2012 seek to protect wildlife and habitats and to control some activities that may adversely affect wildlife, such as hunting and shooting. Nature and biodiversity are strongly protected in EU and national legislation. In Ireland these requirements tend to be implemented through a large number of statutory instruments, rather than one legal instrument, though the Wildlife Act has itself been amended to better reflect such obligations. The 1976 Act allowed for the conservation of a representative sample of important habitats through the designation of nature reserves, refuges for fauna or specific management agreements, though many aquatic species were not covered.

The 2000 Act sought to supplement this and address the 'aquatic' gap. It allowed the minister to designate Natural Heritage Areas (NHA), which can cover important geological and geomorphological sites, many of which are coastal (Figure 32.20). Whilst the acts reflect some of the features contained in the EU's Birds and Habitats Directives, there was still a need to enact specific statutory instruments for transposition purposes.

The Birds and Habitats Directives require the designation of certain habitats and species through Special Protection Areas (SPAs) for birds and Special Areas of Conservation (SACs) for habitats and species, which together form the Natura 2000 network (see Chapter 15: Coastal Heritage). Habitats occurring along the Irish coast include reefs, sea caves, sand dunes, machair, rivers, mudflats and sandflats, estuaries, sea inlets and bays (Figure 32.21). The conservation of common (harbour) seals and grey seals, along with all cetaceans, is required. Cetaceans are provided with strict

Fig. 32.21 DUNE SLACKS IN THE MAHAREES, COUNTY KERRY. The majority of coastal wet dune depressions (dune slacks) are protected habitats in Ireland. The slacks in the Maharees, which are perhaps the best conserved in Ireland, fall under the protection of the Tralee Bay and Maharees Peninsula Special Area of Conservation (SAC). These ponds are particularly important from an ecological standpoint, because they serve as breeding grounds for the endangered natterjack toad (*Bufo calamita*). Conservation designations such as these can be effective to a point. They can be most impactful when implemented together with an appropriate management plan, which is sometimes limited by the available resources. In 2008 the National Parks and Wildlife Service introduced a financial incentivisation scheme to farmers for digging suitable natterjack ponds on their properties. Additionally, the National Parks and Wildlife Service has started a breeding programme at Fota Wildlife Park to bolster survival rates. These schemes have helped to stabilise the natterjack population in Ireland over the last decade. [Source: Eugene Farrell]

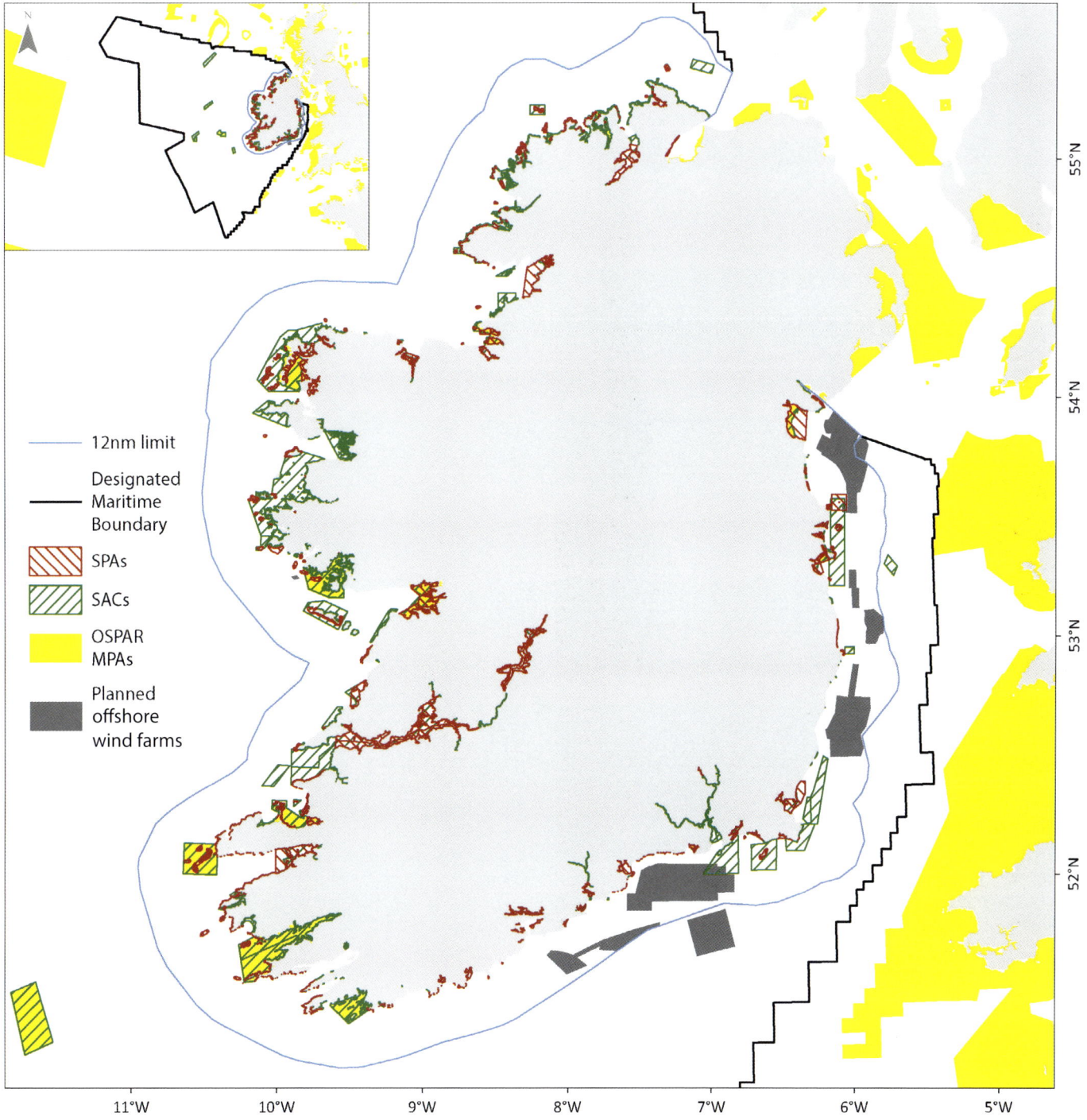

Fig. 32.22 DESIGNATING AREAS FOR THE PROTECTION OF HABITATS AND SPECIES. A network of protected sites around the coast, designated under European or national legislation, protects species and/or habitats prioritised as important. For example, 130 sites are designated as Special Areas of Conservation (SACs). However, not all of these would fall within the definition of Marine Protected Areas (MPAs) as defined by the International Union for Conservation of Nature (IUCN). This defines MPAs as 'any area of intertidal or subtidal terrain, together with its overlying water and associated flora, which has been reserved by law or other effective means to protect part or all of the enclosed environment' [Source: www.iucn.org]. This excludes terrestrial habitats, such as vegetated sea cliffs. The inset map shows some sites that have been designated for protection some distance offshore. These relate to areas considered important for biodiversity, such as cold water corals. OSPAR sites currently overlap with sites designated as SACs or Special Protected Areas (SPAs), which gives a legal mandate for intervention, as OSPAR designations in their own right do not confer any legislative protections. Nevertheless, the work of the parties to the OSPAR Convention, aimed at protecting the marine environment of the north-east Atlantic, is important for raising awareness of the need for Ireland to be more proactive in the designation of MPAs. [Data source: National Parks and Wildlife Service; OSPAR]

protection across their entire natural range, extending to the 200 nautical mile limit of the Exclusive Economic Zone. SACs for bottlenose dolphins, harbour porpoises, grey seals and common seals have also been designated.

The establishment of networks of Marine Protected Areas (MPAs) is increasingly recognised as a critical pathway to deal with the problems facing the marine environment. The current approach to MPA designation in Ireland is narrowly based on the parameters of the Birds and Habitats Directives. The exceptions are Lough Hyne, Ireland's only marine reserve, and the Irish Conservation Box (see Chapter 26: Coastal Fisheries and Aquaculture), which have specific conservation interventions. An opportunity exists for

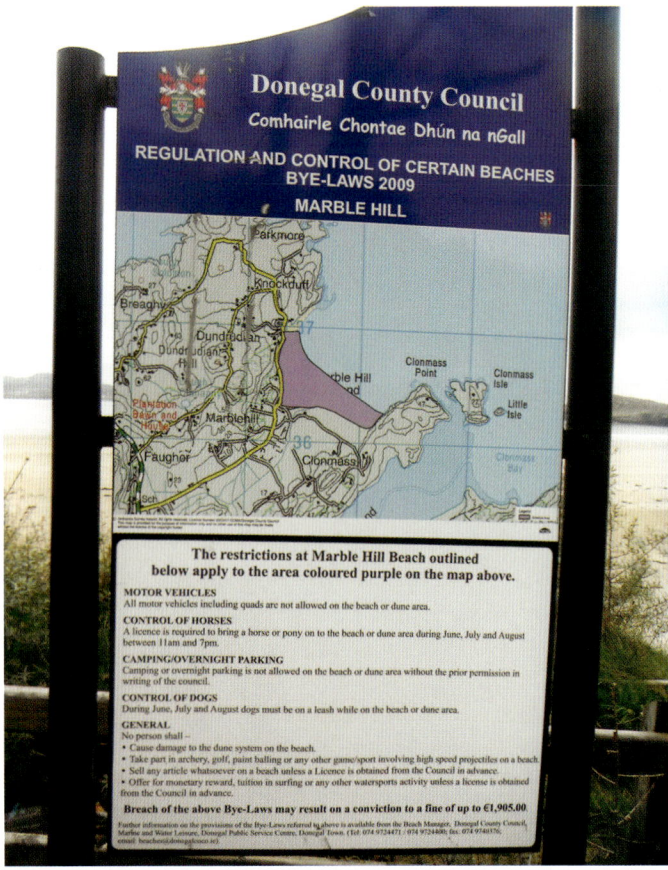

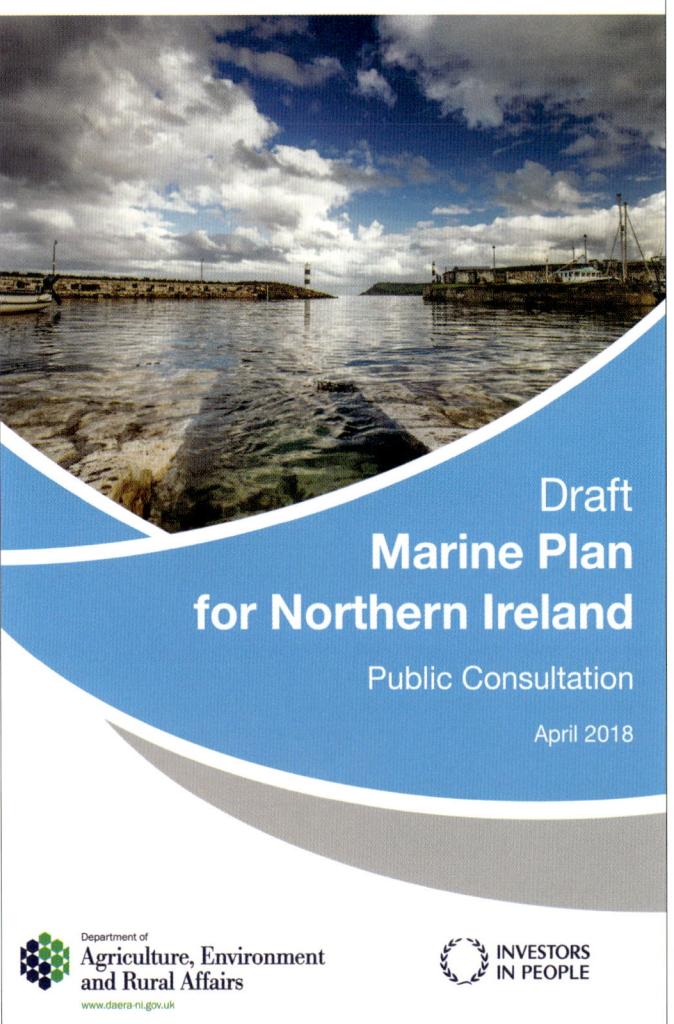

Fig. 32.23 (left) MARBLE HILL BYE-LAW SIGNAGE, COUNTY DONEGAL. Local authorities derive the power to make bye-laws from a number of legislative instruments. Most commonly, Part 19 of the Local Government Act, 2001 is used as the legal basis for beach bye-laws. This provides that a local authority can enact a bye-law where it is desirable in the interests of the common good of the local community. Specifically, this can apply to the foreshore and coastal waters adjoining the functional area of that, or an adjoining, local authority. [Source: Donegal County Council]

Ireland to step up its approach to MPA designation in line with international best practice and this is a goal of the National Biodiversity Action Plan, 2017–2021 (Figure 32.22).

Bye-laws

Many local authorities act as de facto coastal managers and carry out management functions through the creation of beach bye-laws. The bye-law section of the Local Government Act, 2001 is broad and enables local authorities to address matters such as prohibiting certain activities, regulating the conduct of persons at specified places and levying charges as they deem necessary (Figure 32.23). In recent years, the provisions of the Maritime Safety Act, 2005 in relation to beach bye-laws has increased across Ireland.[21] For the first time, this provided local authorities, harbour authorities and Waterways Ireland (which has a cross-border remit) with the power to make bye-laws to regulate or control certain types of recreational craft, such as jet-skis and other personal watercraft.

INSTITUTIONALISATION OF COASTAL MANAGEMENT IN NORTHERN IRELAND

In some respects, Northern Ireland could be considered as being further ahead than the Republic with respect to coastal and marine management. In 2006, the Department of the Environment Northern Ireland published an Integrated Coastal Zone Management (ICZM) strategy to run until 2026.[22] Though the ICZM strategy does not have a statutory basis, it seeks to improve the way the coast is managed by promoting integrated management and new approaches to management that bring regulators and stakeholders together to discuss, plan and resolve issues at a local level. Aside from the involvement of government departments, a Northern Ireland Coastal and Marine Forum was created with the purpose of facilitating stakeholder involvement, providing public information on ICZM, delivering expert advice and co-ordinating research in order to achieve the objectives of the strategy. This is no longer operational, although an Interdepartmental Marine Co-ordination Group (IMCG) has been established to facilitate joined-up thinking.

Marine planning in Northern Ireland was advanced more

Fig. 32.24 (left) DRAFT MARINE PLAN FOR NORTHERN IRELAND. Built upon the foundations of the UK Marine Policy Statement and in accordance with the Marine and Coastal Access Act, 2009 and the Marine Act (Northern Ireland), 2013, the draft Marine Plan for Northern Ireland emphasises support for the sustainable development of Northern Ireland's marine environment. This draft plan is supported by a publicly accessible marine mapviewer, which digitally illustrates activities that influence the marine environment in Northern Ireland. In anticipation of finalising a Marine Plan for Northern Ireland, the Department of Agriculture, Environment and Rural Affairs (DAERA) sought public consultation on the draft plan in 2018. The final plan will outline a new planning approach to be used by authorities to make decisions that will impact on Northern Ireland's marine area. [Source: Department of Agriculture, Environment and Rural Affairs]

rapidly than on the rest of the island. The Marine Act (Northern Ireland), 2013 gave practical effect to MSP in Northern Ireland and this fits within the overall framework established by the UK Marine and Coastal Access Act, 2009 (MCAA) and the UK Marine Policy Statement (HM Government, 2011). The UK framework for MSP was in place before the EU Directive on MSP. The draft Marine Plan for Northern Ireland was published in 2018 as a single document made up of two plans,[23] one for the inshore region under the Marine Act (Northern Ireland), 2013 and one for the offshore region under the MCAA (Figure 32.24).

As one of the devolved administrations of the United Kingdom, coastal and marine matters in Northern Ireland are governed by both UK and Northern Ireland legislation, depending on the subject matter. After the Good Friday Agreement, certain

powers were 'devolved' to the Northern Ireland Assembly, whilst the UK government and parliament retained responsibility for reserved and excepted matters. Legislative authority for reserved matters rests generally with Westminster, but the Northern Ireland Assembly can legislate for these with the consent of the secretary of state. Reserved matters include, for example, navigation and shipping, the foreshore and the seabed. Excepted matters, such as international relations and nuclear energy, are matters that are never expected to be considered for devolution. The enactment of the MCAA and subsequent Marine Act (Northern Ireland), 2013 could be said to provide a more integrated legal approach to the management of marine activities, with almost all licensing of activities and uses now done under this legislation, depending on whether it occurs in the inshore or offshore area. The Marine Act

Fig. 32.25 THE DEVELOPMENT OF GREENCASTLE FERRY TERMINAL, COUNTY DOWN. After nearly ten years of planning and consultation with authorities and local communities, Frazer Ferries Limited was granted a marine licence under the UK Marine and Coastal Access Act in 2016 for the development of a ferry terminal near Greencastle, County Down. Dock development was completed in 2017 and the ferry service commenced later that year. The Carlingford Lough Ferry now provides up to a dozen sailings daily between Greencastle and Greenore Point. This is the first ferry service to transport cars across Carlingford Lough and has become an important transit link for the east coast of Ireland, connecting the Republic's Ancient East to Northern Ireland's Mourne Mountains region. The 1.5km crossing now saves drivers the 50km journey around the rim of the lough. [Source: Courtesy of Carlingford Lough Ferry]

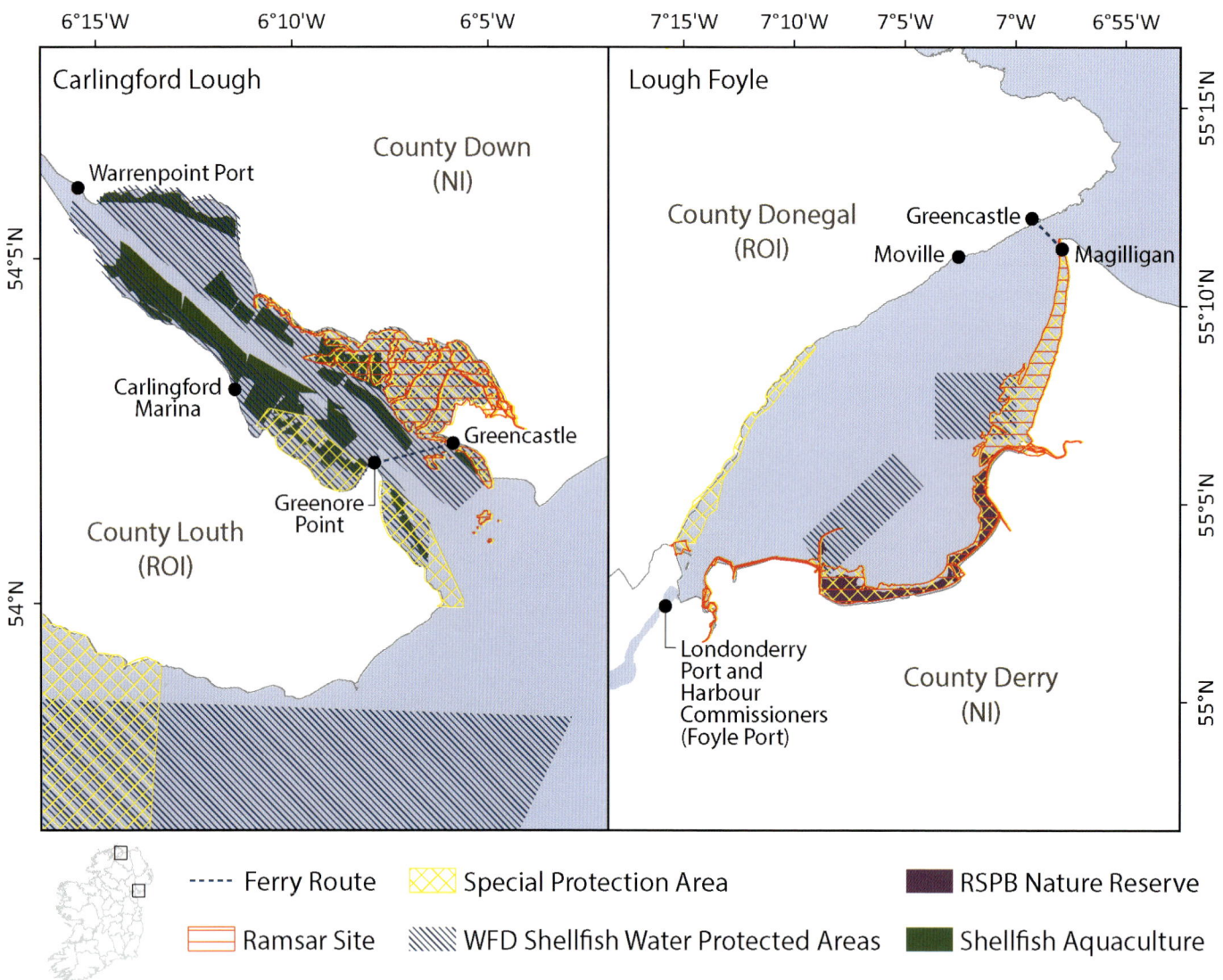

Fig. 32.26 COASTAL AND MARINE ACTIVITIES IN CARLINGFORD LOUGH AND LOUGH FOYLE. The loughs between the Republic of Ireland and Northern Ireland have become increasingly busy waterways in recent decades and play host to a wide variety of activities, including shipping, aquaculture, fishing, tourism and recreation. Several nature conservation designations have been established as well. Carlingford Lough, Warrenpoint Port and Greenore Port facilitate extensive shipping traffic, while Carlingford Marina supports pleasure cruising activity in the region. In Lough Foyle, the Foyle Port generates shipping and cruise traffic, while Moville and Greencastle ports host commercial fishing fleets. Both loughs support cross-waterway ferry services. Aquaculture is prominent in both loughs, with the industry focused particularly on Pacific oyster and blue mussel cultivation. Extensive areas of the loughs are designated as SPAs, Ramsar sites or Royal Society for the Protection of Birds (RSPB) Nature Reserves, which impacts on the level of development that can occur in these areas. Despite the scale and variety of ongoing activities, a lack of established maritime boundaries promotes ambiguity and thus affects management of the loughs. Concerns over the future management of the loughs are linked to uncertainties about Brexit. [Data source: Directorate-General for Environment, European Environment Agency; Department of Agriculture, Environment and Rural Affairs; Royal Society for the Protection of Birds]

(Norther Ireland) covers construction and removal activities, such as, harbours, marinas, beach replenishment, rock armouring, dredging and disposal of dredged material. The Marine and Fisheries Division of the Department of Agriculture, Environment and Rural Affairs (DAERA) is responsible for licensing and enforcement functions in the inshore area under the MCAA (Figure 32.25).

Transboundary marine and coastal management

The jurisdictional boundaries between the Republic and Northern Ireland in Lough Foyle and Carlingford Lough have been contested since the time of partition. As a result of partition, six parliamentary counties became governed by Britain. Parliamentary counties end at the low water mark so, arguably, territorial waters remained under the jurisdiction of (then) Southern Ireland. This took on a new

significance on the establishment of the Irish Free State. Ireland retained a constitutional claim to all waters around the island until the Good Friday Agreement (GFA) in 1998. In the associated referendum, the people of the Republic voted to change Articles 2 and 3 of the constitution, meaning the claim to territorial waters around Northern Ireland ceased. There is still no formal, legal delimitation of the maritime boundaries in the border bays between Northern Ireland and the Republic (Figure 32.26). For the purposes of marine renewable energy development, the Republic of Ireland and the UK adopted and signed a Memorandum of Understanding (MOU) in 2011, which states that they 'may each arrange for the lease of the seabed to facilitate the development of offshore renewable energy installations, and for the licensing of construction and operation of such installations, up to their respective sides of the two lines constituted by the lists of coordinates and depicted on

Fig. 32.27 GREENCASTLE, COUNTY DONEGAL, AND THE ENTRANCE TO LOUGH FOYLE. The Loughs Agency works to conserve, manage, promote and develop the marine resources of Carlingford Lough and Lough Foyle with the aim of contributing sustainable social, economic and environmental benefits to the regions. One of the primary responsibilities of the Loughs Agency is fisheries management. Greencastle is the largest commercial fishing port in Lough Foyle and one of the largest in the Republic of Ireland. While many of the vessels berthed at Greencastle focus on offshore pelagic fisheries, the port is also home to a thriving inshore fishery that engages in mussel dredging as well as crab and whelk fishing, in and around Lough Foyle. [Source: Cinematic Sky, courtesy of Scenic Lough Foyle Ferry]

illustrative maps'.[24] This MOU represents a political commitment, not a legal agreement on the boundaries.

The signing of the Good Friday Agreement in 1998 resulted in the creation of a number of transboundary institutions, including the North South Ministerial Council (NSMC); the British Irish Council (BIC); and six North South Implementation Bodies. Waterways Ireland is one of the North South Implementation Bodies and is responsible for managing, maintaining, developing and promoting over 1,000km of inland navigable waterways for recreational purposes. With respect to coastal and marine functions, the Foyle, Carlingford and Irish Lights Commission (FCILC) is the North South Implementation Body of most relevance (Figure 32.27). This consists of two agencies: the Loughs Agency and the Lights Agency. The Loughs Agency has responsibility for the promotion of development in Loughs Foyle and Carlingford Lough for commercial and recreational purposes in respect of marine, fishery and aquaculture matters. The Loughs Agency is also responsible for the transboundary management of fisheries and aquaculture development in the bays. The Foyle and Carlingford Fisheries (Northern Ireland) Order, 2007 and the Foyle and Carlingford Fisheries Act, 2007 (ROI) provided for a new regulatory system for aquaculture in the Foyle and Carlingford areas. To date those powers have not been enacted, in the form of a management agreement, so as to enable the Loughs Agency to manage, on behalf of both governments, the marine aquaculture in Lough Foyle. Uncertainty concerning the extent to which each side exercises jurisdiction within Lough Foyle has added to the practical difficulties in conducting activities there and in their regulation (Figure 32.28).

An assessment of the conditions for transboundary MSP in the shared waters of Northern Ireland and the Republic of Ireland was conducted in 2015.[25] This found that whilst some cross-border consultation takes place, it is often ad hoc with little or no evidence of joint planning.[26] There are often variations in approaches according to the topic being considered or managed. The electricity market, for example, is managed on an all-island basis. There are also arrangements in place for transboundary consultation on developments likely to have significant environmental impacts, deriving from the EU EIA and SEA Directives and the Espoo Convention on Environmental Impact Assessment in a Transboundary Context.

FUTURE DIRECTIONS

Irish legal and administrative frameworks for coastal planning and management are complex and remain in need of reform and improvement in a number of respects. The need to facilitate approaches for stakeholder engagement and for more integrated approaches to decision-making, at every level, emerges as a consistent message over many decades, in both Northern Ireland and the Republic of Ireland.

The national action arena for coastal planning and management in Ireland is nested in significant international and European efforts

819

Fig. 32.28 OYSTER TRESTLES NEAR QUIGLEY'S POINT IN LOUGH FOYLE, COUNTY DONEGAL. The boundary dispute between the Republic of Ireland and the UK over ownership of Lough Foyle has stifled efforts to manage aquaculture licensing in the region and created an administrative vacuum on the issue. The result has been rapid growth in the number of unlicensed oyster trestles in Lough Foyle, particularly along the Donegal coast. In 2016 it was estimated that 50,000 unlicensed oyster trestles had been established in Lough Foyle, a twenty-three-fold increase from 2010. [Source: Sarah Twomey]

to regulate activities in the marine environment. In particular, the proactive approach to the management of coastal, marine and maritime activity at EU level in recent years has stimulated action in Ireland on coastal and marine legislation and policy. Ambitions to develop the maritime economy through Harnessing Our Ocean Wealth serve as another driver for legislative and policy changes. The need for changes to key legislation for planning and development at sea, specifically the foreshore and aquaculture licensing legislation, has undoubtedly created tensions, such as those experienced in Bantry Bay. However, improvements in the form of marine spatial planning are on the horizon and much has been learned from the islands and other locations to help guide better engagement with communities in coastal planning and management going forward.

The need to integrate stakeholder voices into the decision-making process is one of the many forms of integration that needs to be pursued, to make coastal planning and management more effective. Community groups are increasingly taking initiatives to manage local beaches and engage in citizen science, etc. Integration across the land–sea divide may be served, to an extent, by MSP. It

remains to be seen how this will work out in practice. The EBM provides a framework for integrating environmental, social and economic management of the coast at multiple scales. This helps to address the three pillars of sustainability in decision-making, but there are large gaps, as yet, in implementation. Finally, the need for an integrated approach to deal with transboundary marine issues is essential, especially for communities in Lough Foyle and Carlingford Lough.

Whilst the situation in Northern Ireland could be said to be a little more advanced at a technical, policy-development level, moving towards integrated EBM represents challenges for both jurisdictions. Recent developments in areas such as foreshore legislation, marine spatial planning and transboundary management have the potential to significantly impact the future governance of coastal and marine environments (Figure 32.29). It is anticipated that the designation of Marine Protected Areas will come to the fore as another shortcoming to be addressed. Finally, the exit of the UK from the EU is expected to have significant implications for the island of Ireland, but the consequences are not currently known and cannot be explored definitively.

Fig. 32.29 BEACHES OF PORTRUSH, COUNTY ANTRIM. The sustainability of coastal communities in the coming years will greatly depend on the management and governance schemes implemented today. [Source: © Tourism Northern Ireland]

SECTION 6
FUTURE COASTS

SUNRISE OVER ESTUARY MUDFLATS, COUNTY DONEGAL. A new day: a new beginning. [Source: Pat Kehoe]

CLIMATE CHANGE AND COASTAL FUTURES

Val Cummins, Robert Devoy, Barry Brunt, Darius Bartlett and Sarah Kandrot

DUBLIN BAY: A PANORAMIC VIEW NORTH-EAST OVER THE HEART OF DUBLIN CITY. Extensive areas of Ireland's coastal lowlands, but especially those along its eastern seaboard, have been influenced by increasing pressures of urbanisation. The result is a coastal landscape that is dominated increasingly by urban land uses. This is evidenced particularly well in Dublin, Ireland's largest city, as well as the political capital and economic core of the Republic. The photograph shows the city's coastal zone, with its port facilities around the canalised channel of the River Liffey and then out over Dublin Bay and beyond to the Irish Sea. The sand dunes of North Bull Island and Dollymount Strand (visible, centre-left) illustrate important recreational spaces at the heart of the city and are the result of coastal engineering and reclamation works during the early nineteenth century. The advantages of the city for diverse business activities and housing have seen large-scale increases in its population and spatial expression. In 2016, over 0.55 million people resided within the 115km^2 that constitute the Dublin city authority's area and, although poorly regulated, urban sprawl extends this built-up area to cover almost 320km^2, which accommodates some 1.2 million people. This relatively small area is, therefore, domicile for some 25 per cent of the Republic's population (*c.*20 per cent for the island).With much of Dublin occupying land of low elevation near the coast, the threat to the city from SLR is a major concern. The terrain modelling work of Climate Central indicates that many of the city areas seen in the photograph – with the exception of the Raheny/St Anne's Park area and the core of Howth Island – are likely to be flooded by 2050 because of SLR and the effects of climate change on river water sources. To build capacity to cope with these changes is likely to require the construction of extensive and costly sea defence structures to protect the city and its vital infrastructures from marine flooding. An alternative and more innovative approach may be, as now being practised in the Netherlands, to prograde the coast seawards into Dublin Bay as protection, creating an entirely new coastal Dublin. [Source: Dublin Bay Biosphere Partnership]

Whether you have browsed random pages, or read large sections of the *Atlas* to reach this final chapter, we trust that you have garnered a greater appreciation of Ireland's beautiful and dynamic coastline. We also hope that you have gained an insight into how our coastal resources have provided, over many millennia, an abundance of ecosystem goods and services for coastal dwellers. The *Atlas* is intended as a lens to aid an appreciation and understanding of the many ways in which Ireland's coastline has altered over time. Some coastal changes have been a consequence of natural forces. These range from the moulding and shaping of glacial processes, incremental effects of coastal erosion and accretion, sudden impacts of storms, surges and even tsunami, to the many alterations of coastal biology and ecosystems. Others have arisen more as a consequence of human controls, such as the activities of coastal fishing and farming over generations, the physical, industrial and chemical impacts of people on coastal ecosystems and shifting patterns of contemporary urbanisation (see Chapters 1: Ireland's Coasts: Setting the scene and 4: People, Agriculture and the Coast).

People and society are grappling today with global-scale problems of climate and biodiversity changes, which form two of the existential crises of our time. Many other environmental issues of different proportions also exist for contemporary societies around the world, as well as, most recently, the coronavirus COVID-19 pandemic. It is argued, however, and agreed by most countries (albeit with great political rancour) that none of these threats are as fundamental as those of climate change, together with the interlinked losses in species from both terrestrial and marine environments.

Times of global cooling of, on average, 5°C during the last one million years allowed ice sheets, like those remaining in Antarctica, to spread from the poles almost to the Equator, lowering sea levels by 120m and together changing the face of the Earth. Today, however, the world environment is confronted by human-induced global warming, now of over 1.5°C and more likely over 3°C in the future. Such trends are set to impose fundamental changes to the planet, expanding deserts and areas of aridity throughout the continents and causing sea levels to rise by several metres in the next 150 years. The frontline in all these changes is our coasts. These factors fundamentally affect the Earth's marine and coastal environmental systems through which most life on the planet is sustained.[1]

This final chapter is an opportunity to take stock of these and other environmental issues that affect our coasts today and to consider what it means, from an Irish coastal perspective, to be living in this time of the Anthropocene (the current geological era, dominated by people) (see Chapters 5: Geological Foundations and 16: The Inhabitants of Ireland's Early Coastal Landscapes). The coastal zone provides a hugely important 'living laboratory' that will both compel and enable people to experiment with, and to develop, new ways of living in a greatly changed future. To help understand some of the key characteristics of the Anthropocene coast, this chapter provides an overview of the coastal megatrends of the twenty-first century and identifies the implications of these for the coastline of Ireland. We then proceed to reflect on the critical question of the viability of transformative change, which links to the sustainable development of our coasts. In the final section, we speculate on a vision for our coastal future and end by presenting reasons why we can be hopeful.

The underlying global metrics to these themes and issues are stark and deserve a brief review. Over 600 million people currently live in the coastal zone, most of them within 10m of sea level (c.10

Fig. 33.1 LÉ *William Butler Yeats* BEING USED AS A COVID-19 TEST CENTRE IN GALWAY. This Irish Naval Service offshore patrol vessel, designed to cope with the stormy conditions of the North East Atlantic, was commissioned in 2016 and has been employed mainly in maritime security and defence patrolling. This work is part of the Republic's response to help regulate and conserve marine resources. The ship has also been deployed to the Mediterranean to help in the rescue of migrants attempting to cross that sea to Europe. Many migrants are motivated by the poverty and instability of their homelands, which have resulted from adverse politics and economics, together with the growing impacts of the climate crisis. [Source: Irish Naval Service]

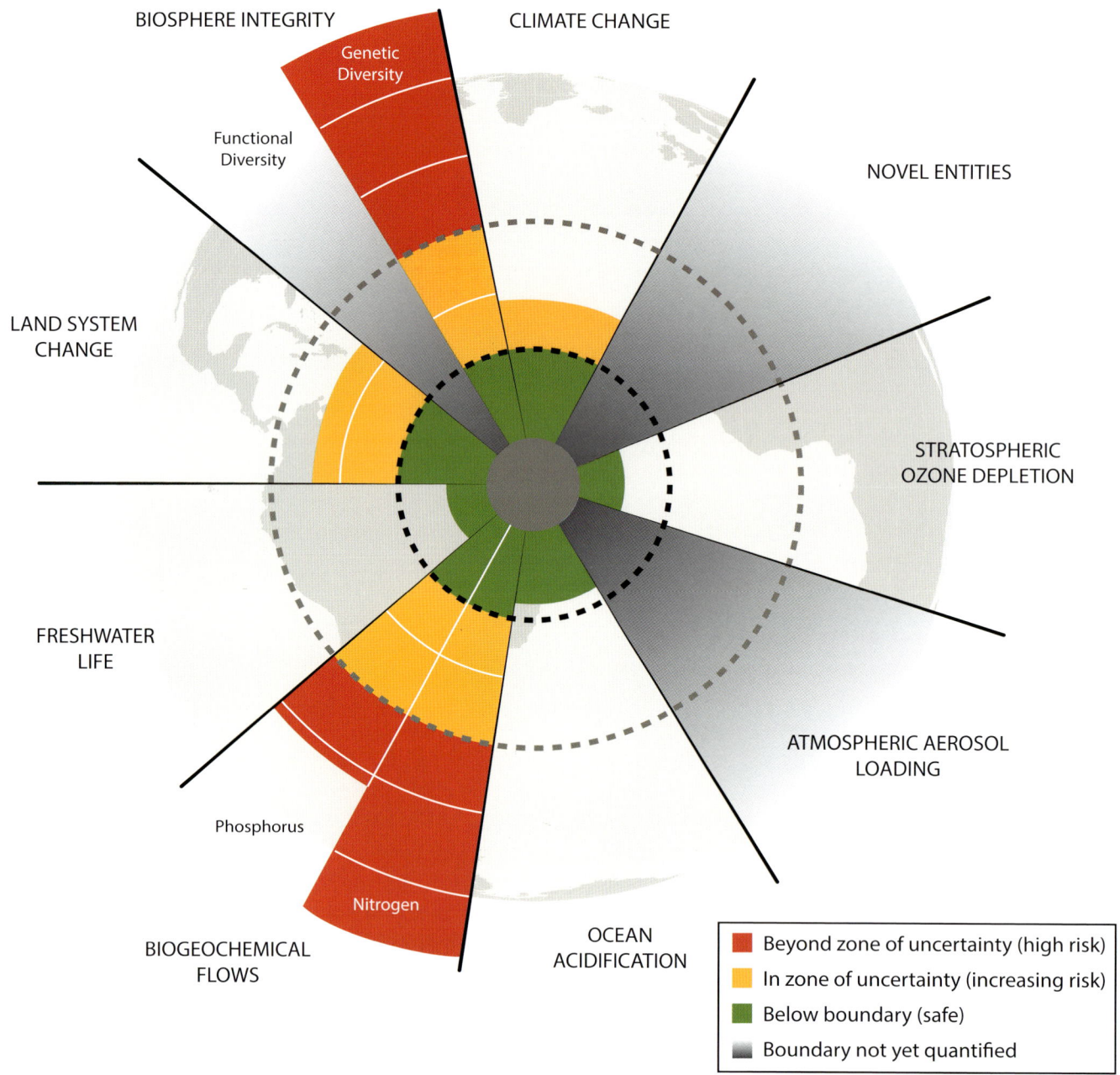

Fig. 33.2 THE ROCKSTRÖM PLANETARY BOUNDARIES (PB) CONCEPT. This illustrates how the different control variables for nine planetary boundaries have changed from 1950 (figure centre) to 2020 (figure edge) (concentric white lines as decadal timescale). The diagram shows large Earth system processes, incorporating planetary boundaries for environmental stability, as well as defining 'safe operating spaces' for sustainable development. The concept, developed in 2009, identifies factors (processes), boundaries and thresholds involved in the regulation and stability of the Earth system. The areas in red represent the estimated and current states of danger; the green areas show the estimated levels of safety; the dashed lines indicate the upper and lower boundaries and thresholds of uncertainty, between safe operation and danger in the different sectors. The original research identified processes, including climate change, that could push the Earth system into a new state, if crossed. These were also recognised as having a significant and pervasive influence on the others. In Ireland, as elsewhere, this approach can be applied also at regional and national scales in assessing environmental impacts and stocktaking. [Data source: adapted from Steffen et al., 2015. Planetary boundaries: Guiding human development on a changing planet. *Science*, 347, DOI: 10.1126/science.1259855]

per cent of the global population), a number that is increasing annually. As relative sea level (RSL) continues to rise now with climate change, many of these people face increasingly disastrous flooding or even the complete loss of their living space. Most significantly, the coastal zone contains many major cities and core centres of economic output, in agriculture, industry and services. In aggregate, the marine coastal economy, which includes ecosystem and cultural services provided by the oceans, is estimated at between US$3 and $6 trillion per year. Much of this, and the global economy in general, is also controlled through financial centres (such as London, New York and Tokyo) and associated communication

infrastructures that are situated on vulnerable coasts. With climate change, these areas will experience serious threats to their future as, by extension, will the rest of the world economy.[2]

The lack of progress in dealing with these core issues has already resulted in the significant degradation or unsustainable use of 60 per cent of the world's major marine ecosystems. These environments are vital for the biogeochemical functioning of the planet and in sustaining the world's population. Mean Earth sea surface temperatures have risen by *c.*1°C over the past 100 years and are likely to increase by over 3°C in some regions by the end of this century; oceans could be 150 per cent more acidic by 2100. The livelihoods and wellbeing of coastal communities that are dependent on healthy oceans, shelf seas and their coasts is in jeopardy.

The COVID-19 pandemic of 2020 has further sharpened international focus on the need to build resilience in human societies, in order to protect people from the worst health, social and economic consequences of the virus. At the same time, the pandemic is driving a new awareness of the links between the damage being done to the environment (of which coasts and the marine are a principal part), and the corresponding damage to global health, food, energy and climate security. This has included, importantly for the climate issue, an enhanced understanding that a 'business as usual' (BAU) approach to managing climate, health, economics and the environment is not sustainable in future (Figure 33.2).[3]

International concern for climate change, in particular, has established a number of large-scale, global climate-response actions, such as the Kyoto Protocol, an international treaty establishing mandatory limits on greenhouse gas emissions in industrialised nations. Similar concerns and international actions, but on smaller scales, have been shown for biodiversity losses, chemical pollution and the spread of plastics. In a more positive vein, the need to promote the use of marine resources and renewable ocean energies has also emerged as part of a push toward sustainable living (see Chapters 28: Renewable Energies: Wind, wave and tidal power and Chapter 31: Pollution). These large-scale, international, roadmap-type measures to address the causes of climate change have been slow to emerge. The Kyoto Protocol was first accepted nominally in 1997, but entered into force only in February 2005. It remains a contentious topic and a 'political football'. Many scientists, however, recognise that time is running out before a point of no-return for

the climate system is reached. To resolve such persistent societal problems, structural transformations (or transitions) are necessary to achieve a new dynamic equilibrium. This needs to occur within what may be described as a 'safe operating space for humanity', based on the intrinsic biophysical processes that regulate the stability of the Earth system (Figure 33.2; see Chapter 2: The Coastal Environment: Physical systems processes and patterns).[4]

GLOBAL COASTAL MEGATRENDS AND IRELAND'S COASTS
Megatrends that are reshaping our world at large have a cascading effect, often with direct consequences on the manner in which coastal communities live. These trends include those of demographics, food security, macroeconomics, climate change, biodiversity and the energy transition to renewable resources. All of these factors have implications for the shape of Ireland's coastal future, with an abundance of associated challenges and opportunities.

Economics and a post-statistical society
Different economic theories – such as the opposing approaches of Keynes and Friedman – have been proven to have fundamental weaknesses.[5] Most recently, economic theories of living with a 'finite planet' are starting to gain traction. Those who have added to such economics, based on planetary ecology and sustainability, envisage an alternative economic system reconciled with principles of equity and social justice. This alternative view is sometimes referred to as the Green New Deal, with its implied reference to President Roosevelt's response to the Great Depression in the USA. However, there is a long way to go before alternative value systems can be advanced to a point whereby this type of holistic view of environmental resources, including non-economic values, is

Fig. 33.3 HOUSING DEVELOPMENT AT BALLYDAVID, COUNTY KERRY, BUILT ON COASTAL FLOODPLAINS AND UNDERTAKEN DURING THE PERIOD OF THE CELTIC TIGER ECONOMIC BOOM. Many relatively low-lying lands around Ireland's coasts, especially those associated with river estuaries, former boglands and plains, have become increasingly subject to flooding since the 1980s, because of storms and heavy rainfall. Areas such as these have, however, provided convenient flat and accessible land for housing and industrial development. Inadequate planning systems, especially in a period of high demand for housing during the Celtic Tiger era, have given rise to serious problems linked with climate change. This is reflected in the risk to life and damage to property as flooding problems develop and become both more frequent and extensive. [Source: Óglaigh na hÉireann, flickr, CC BY 4.0]

factored into coastal planning and policy.[6]

The global economic system is one of the greatest influences in shaping people's activities in the coastal zone. The impact of globalisation on a small, open economy, such as the Republic of Ireland's, is outlined in the story of the Celtic Tiger.[7] Its initial knock-on effect on the housing market is part of a well-established narrative in Ireland, and involved stimuli to develop vulnerable, low-lying lands of the coast, particularly around estuaries, as in north County Dublin, or Oranmore and Galway city. This development is now recognised as a major error in planning (Figure 33.3). Lesser known are the impacts on Irish ports and their struggle to sustain business with the fall of the Celtic Tiger. Such economic system weakness and vulnerability was further exposed in 2020 by the COVID-19 pandemic and Brexit.

Another emerging global trend can best be described as the 'post-statistical society', which may appear, at first glance, to have little to do with the coast. This involves the business model that underpins today's digital world and which conditions every aspect of human communications, government and societal behaviour. An outcome of the digital world includes a new human behavioural futures market, where individual data on consumer preferences are the principal commodity.[8] Additionally, the domain of 'new knowledge', as shaped by multiple forms of digital social media, is part of this commodity, but is increasingly contentious and becoming the subject of political conflict. These factors – and their association with an apparent undermining of democracy, for example, in possible electoral manipulation – are seen as part of a scenario whereby trust in elected representatives, professionals, scientists and other leaders is being eroded.

Such influences and emerging conflicts with long-established patterns of human activity (for example, government) come, regrettably, at a time when coastal environments and communities have become highly vulnerable. This means that there is now a need for clear direction in decision-making for coasts, not conflict and uncertainty about how to develop policy and implement actions. Evidence-based policy-making is needed to steer a brave new path towards sustainability. The lessons of the coronavirus pandemic may, in hindsight, be seen to have had a seminal bearing on the development of such pressures of a post-statistical society, on its decision-making processes and the strength of government (see Chapter 32: Coastal Management and Planning).

Demographics

Population size and demographics together pose one of the most significant challenges for the planet. In this context, the coast has been a place where people and settlements have been concentrated over millennia and, arguably, is the birthplace of human civilisation. The complex, social-ecological systems that control this pattern of coastal community development can be seen as resulting from the product of three key factors: population size, affluence and technology $(I = P*A*T)$. The affluence variable represents the consumption of each person in the population and technology represents how resource-intensive the production of affluence is. Population size and density, however, are central to the scale of

physical impacts of people on coasts.[9]

Globally, the UN forecasts that world population will increase from c.7.7 billion (2019) to 9.8 billion in 2050 and about 11.2 billion in 2100, although the rate of this population growth is starting to slow and, by the end of the century, is likely to plateau. Overall, coasts (as within 100km of the sea) currently sustain c.40 per cent of the world's population.[10] In Ireland, the population of the Republic is set to increase from 4.7 million people in 2020 to 6.7 million by 2050, whereas in Northern Ireland the projected rate of population increase over this time will be much lower; from 1.9 million to c.2 million. The distribution of people on the island is a microcosm of the planetary picture, with the majority living on or near the coast, and being influenced in some way by the sea (see Chapters 2: The Coastal Environmental: Physical systems, processes and patterns, 4: People, Agriculture and the Coast, 19: Changing Coastal Landscapes and 27: Coastal and Marine Tourism: Opportunities in Blue Growth). Furthermore, the proportion of people living in urban areas is set to increase from 40 per cent in 1950 to 75 per cent by 2050, with almost all of the island's large urban areas located at the coast (see Chapter 25: Urbanisation of Ireland's Coast). These trends, and those related to spatial preferences in economic activities, will have broad-ranging implications for how people will impact on Ireland's coastal zone in the future. Coastal habitats, already stressed by increasing demands for space, resources, infrastructures and investment, will come under further pressure from climate change factors.[11]

People and food security

Global demographics are related closely to global food security.[12] To date, however, human ingenuity has proven to be largely ineffective in dealing with the problems of feeding people and controlling population growth. Evidence also suggests that sustaining world populations with current patterns of food and resources consumption is a losing battle. On the other hand, there are opportunities for new food products and services that have primary links to coasts. In particular, the contribution of aquaculture to economic expansion and alleviation of food poverty worldwide is growing. Global fish production peaked at about 171 million tonnes in 2016, with aquaculture representing 47 per cent of this total, and 53 per cent of the non-food uses of fish (including reduction to fishmeal and fish oil). Furthermore, the amount of input needed to produce a kilogram of fish protein is lower than that required for meat protein, which means that fish is the food livestock with perhaps the least impact on the environment.

The role and nature of fishing around Ireland's coasts altered radically in the twentieth century (see Chapter 26: Coastal Fisheries and Aquaculture). Changes are set to continue in this twenty-first century, as the marine environment around Ireland adjusts to ongoing pressures from ecosystem and climate changes, as well as to innovations in marine technology and evolving patterns of trade and political economies. A significant 'storm cloud' that remains far from being resolved for this industry is that of the outcome of Brexit negotiations between the United Kingdom and the EU, and the likely fallout for fishing activity within Ireland's waters, as well as for trade of marine products under new market arrangements.

The impact of changes, such as revisions to fishing boundaries and quotas, will be greater for some parts of the island than others, with significant effects likely for the inshore fishing industry of north and north-east Ireland.

Fishing has long been a core industry in Ireland that, together with farming, supported many rural coastal communities island-wide, often as part of a mixed and sustainable fisher-farmer economy (see Chapter 26: Coastal Fisheries and Aquaculture). Other countries – such as Portugal, Spain, France and the United Kingdom – had traditionally fished in Irish waters (then defined by the '12 nautical mile', or 22km territorial sea limit), which also supported a significant and sustainable inshore to mid-distance offshore Irish fishing industry. Markets within Ireland were often limited, however, as many people were not attracted to fish and shellfish as food.

Following the United Kingdom and Ireland's accession to the European Union, the operation of a Common Fisheries Policy (CFP) opened up the island's offshore waters to extensive fishing operations from other member states. This led to overfishing and the subsequent imposition of fish quotas to protect depleted stocks. Degradation of water quality and increasing incidence of fish diseases contributed further to unsustainable pressure on fish stocks in Irish offshore waters. The results of these influences on this former staple industry of coastal Ireland has been to concentrate offshore fishing operations in a small number of large ports, for example, Killybegs, Castletownbere and Belfast. The fishing fleets operating out of these specialised ports are now focused primarily on a relatively small number of large, deep-sea vessels, which are able to compete more effectively in international waters. While this has allowed the Republic's fishing fleet to take a larger catch, total employment has declined in this sector. The industry overall is now focused primarily on inshore fishing, as it is unable to compete with the international fleets.

Not all changes have been controversial, or negative, with several positive trends occurring, associated with fish becoming identified as part of an important and fast-growing global resource for food and development. The extent of Ireland's territorial and exclusive economic zone of control has now been extended to the '200 nautical mile' (370km) internationally accepted limit. This has helped stimulate a Blue Growth economy linked to development opportunities presented by Ireland's 880,000km² seabed and marine waters resource (see Chapter 32: Coastal Management and planning). The role of aquaculture has also expanded greatly; the many fish and shellfish farming sites now dominate the industry in terms of food output and employment numbers. Furthermore, although the traditional fishing industry is today smaller, in terms of direct employment and restricted to fewer locations, its fleet is composed of better-equipped inshore boats and offshore ships, with a more knowledgeable and science-based workforce. This has allowed the industry to respond more effectively to growing export opportunities for a diversity of marine foods, particularly shellfish,

SUSTAINABLE DEVELOPMENT GOALS

  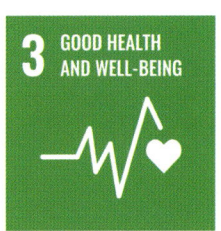   

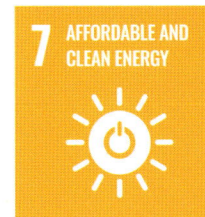 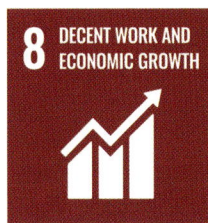    

Fig. 33.4 THE SEVENTEEN SUSTAINABLE DEVELOPMENT GOALS IDENTIFIED BY THE UN IN 2015. These form the blueprint, agreed by most countries, needed to achieve a more sustainable planetary future. They address the global challenges people face worldwide. Some – such as poverty, inequality, peace and quality education – are more applicable to particular regions and national settings than others. The goals as a whole are interconnected, so there is need for a concerted global effort to achieve them, with the agreed aim of the signatory countries to do so by 2030. For Ireland, as an island with significant dependency on the ocean in maintaining overall environmental health, Goal 14 has been seen as of particular importance. [Source: https://www.un.org/sustainabledevelopment/. The content of this publication has not been approved by the United Nations and does not reflect the views of the United Nations or its officials or Member States]

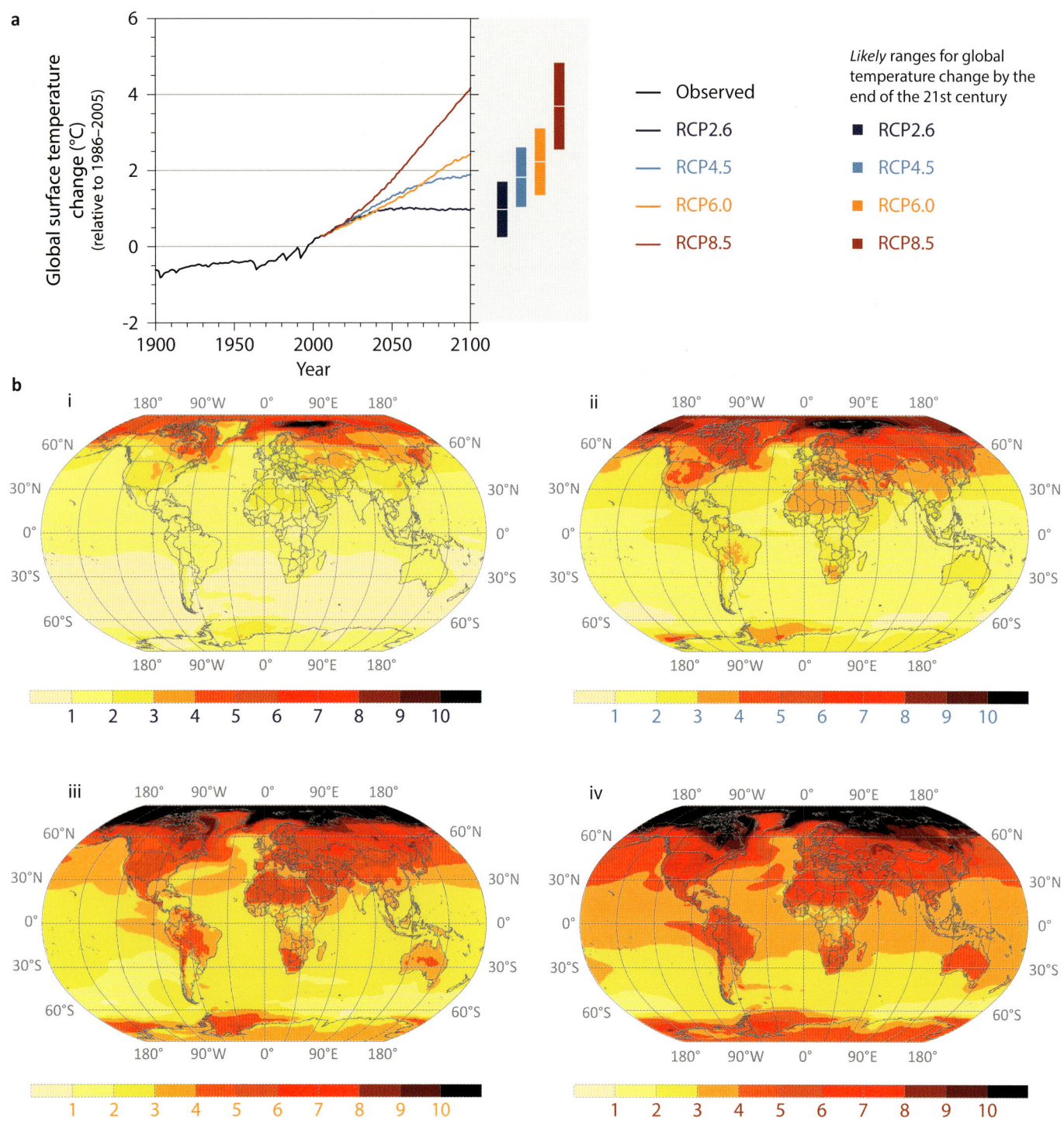

Fig. 33.5 Projections of different climate scenarios for global temperature changes. a. Climate ensemble model projections to 2100 for four established climate scenarios: Representative Concentration Pathways (RCPs) 2.6–8.5, for global mean temperature changes relative to 1986–2005. RCP2.6 forms the best-case scenario for limiting anthropogenic climate change. RCP8.5 – the nightmare scenario – of a 4°C global average temperature rise seems more plausible now (in 2020) than in the past five years. Likely ranges for global temperature change by the end of the twenty-first century are shown by the vertical bars on the right. These indicate the difference between two twenty-year means – 1986–2005 in relation to 2081–2100 – which accounts for the bars being centred at a smaller value than the end point of the annual trends. [Data source: adapted from: Collins et al., 2013. Long-term Climate Change: Projections, commitments and irreversibility. In: Stocker, T.F., D. Qin, G.-K. Plattner, M. Tignor, S.K. Allen, J. Boschung, A. Nauels, Y. Xia, V. Bex and P.M. Midgley, eds., *Climate Change 2013: The physical science basis. Contribution of Working Group I to the Fifth Assessment Report of the Intergovernmental Panel on Climate Change*. Cambridge/New York: Cambridge University Press; b. Four maps, (i) SSP1–2.6, (ii) SSP2–4.5, (iii) SSP3–7.0 (7 a new RCP) and (iv) SSP5–8.5, indicate the mean global spatial patterns of temperature changes in the twenty-first century. These are based on ensemble model outputs, which were produced in 2019/20, and use the Shared Socioeconomic Pathways (SSP) scenario approach, coupled with the RCPs. The maps show EC-Earth annual 2m temperature projections for 2071–2100 and 1981–2010, with the change presented in degrees Celsius. In each case, an average is taken of the five ensemble members used. Ireland's ocean margin. is shown to continue to act as a modifying and reducing influence on projected temperature rises. [Data and map source: Figure 3.2 in Nolan, P. and McKinstry A., 2020. EC-Earth Global Climate Simulations: Ireland's Contributions to CMIP6. EPA Research Report 310. Dublin: Environmental Protection Agency (EPA), Ireland]

as well as supplying a more vibrant, home-based market (see Chapter 27: Coastal and Marine Tourism: Opportunities in Blue Growth).

Ireland, with its abundance of clean coastal waters, improvements in environmental husbandry and well-established organic certification, has great opportunities to respond to future global market demands for protein. The island is also well-positioned to meet increasing consumer demands for organic products, grown with low-carbon footprints. There are concerns, however, with regard to the farming of caged salmon, particularly the possible impacts on wild salmon populations (see Chapter 21: Ireland's Islands, Figure 21.28). Current and projected trends in aquaculture underline the continued likely significance of the coastal zone in Ireland for the production of food. This will be both directly, through contributions to land-based farming and aquaculture production, but also indirectly, through the provision of the many support services needed in these industries, particularly in port infrastructures (see Chapters 4: People, Agriculture and the Coast and 26: Coastal Fisheries and Aquaculture).[13]

Sustainable development goals

Coastal systems and processes are all about the sustainability of environments, with the principles of sustainable development in western Europe extending back to at least medieval times. Fundamentally, sustainable development has been concerned with a balancing act between its three cornerstones: environment, economy and society. It is concerned essentially with ensuring the healthy self-regulating operation and maintenance of environments into the future.[14]

In 2015, all United Nations member states agreed to adopt what have become known as the UN Sustainable Development Goals (SDGs), following the success of the earlier UN Millennium Development Goals (MDGs), which between 2000 and 2014 had achieved a halving, globally, of the number of people living in poverty (Figure 33.4). The SDGs can be described as a blueprint for humanity. These are seen by some as an imperative set of objectives for everyone interested in progressing towards a more equitable world via a just transition; other commentators are more critical. In total, there are seventeen goals and 169 targets, many of which relate directly or indirectly to the oceans, seas and coasts of the world. The declared objective is to achieve these goals by 2030.[15]

Sustainable Development Goal 14, which concerns Life Below Water, is particularly relevant to the subject-matter of this *Atlas*. Appreciative of the fact that the oceans cover *c.*71 per cent of the planet and that over three billion people rely on marine and coastal biodiversity for their livelihoods, this goal focuses on a range of ocean conservation targets. These include, for example, the prevention and significant reduction of marine pollution; protection and restoration of marine and coastal ecosystems; implementation of the UN Convention on the Law of the Sea (UNCLOS) and effectively addressing issues such as ocean acidification, overfishing and many more.

In Ireland, the Republic published the SDGs Implementation Plan (2018–2020), which outlines a whole-of-government approach to implementing the SDGs.[16] To date, however, it is difficult to assess national progress towards Goal 14, due to the lack of an integrated, critical assessment of the current state of play in coastal affairs and of the future needs of the island's marine and coastal environments. One area of increasing debate, however – possibly motivated by the need to adhere to targets set out in the SDGs – is the question of the designation of Marine Protected Areas (MPAs).

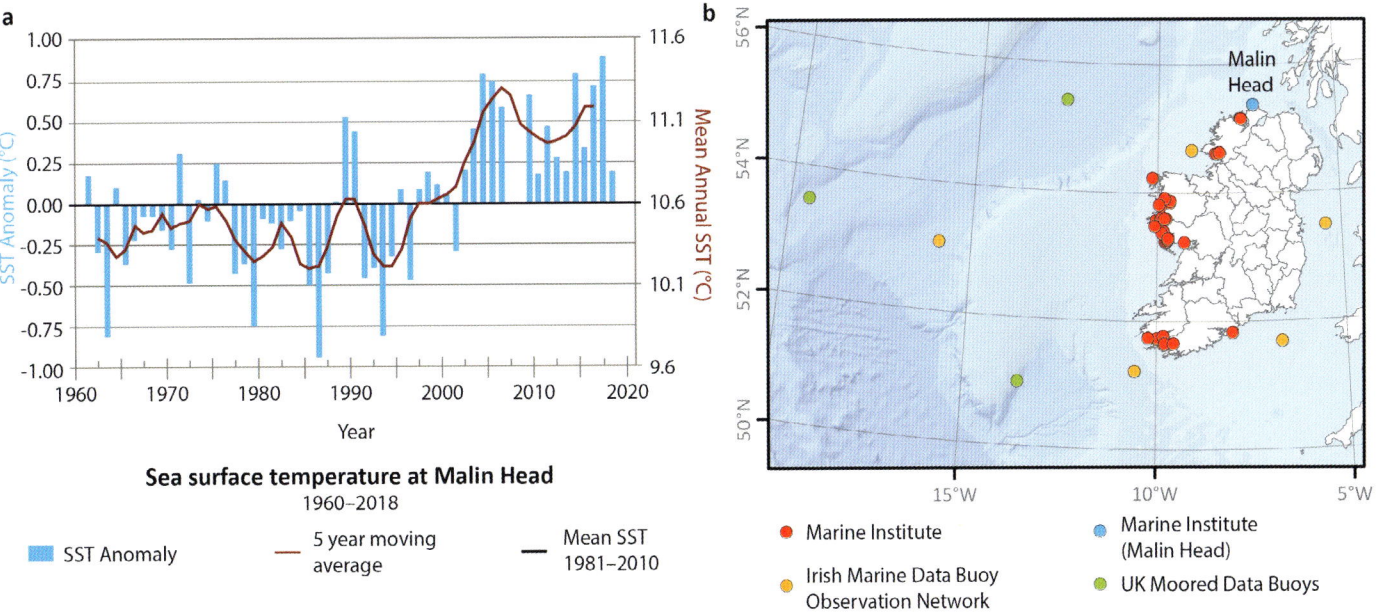

Fig. 33.6 SEA SURFACE TEMPERATURES CHANGE AROUND IRELAND. Sea surface temperature measurements a. that can be used for climate change studies, are made at multiple locations around Ireland (shown in 33.7b), including at the Irish Marine Data Buoy Observation Network and the UK moored data buoys.at Malin Head for the period 1960 to 2018 (scale is on the right Y-axis). The annual anomalies (the difference between the mean annual temperature and the 1981 to 2010 reference mean value) are presented in the same figure with the scale on the left Y-axis. The five-year moving average is also shown (red line). The overall trend from the mid-1990s to present is upward. Furthermore, all the annual means since the late 1990s are above the 1981–2010 mean. This behaviour is strongly linked to the natural cycle of variability in the North Atlantic known as the 'Atlantic Multi-decadal Oscillation or Variability', although approximately half of the recent warming is attributed to an underlying global warming trend. [Data source: Cannaby, H., and Husrevoglu, Y.S., 2009. The Influence of Low-Frequency Variability and Long-term Trends in North Atlantic Sea Surface Temperature on Irish Waters. ICES *Journal of Marine Science*, 66: 1480–1489; graph data: Marine Institute, Met Éireann; map data: Marine Institute, Met Éireann, UK Met Office]

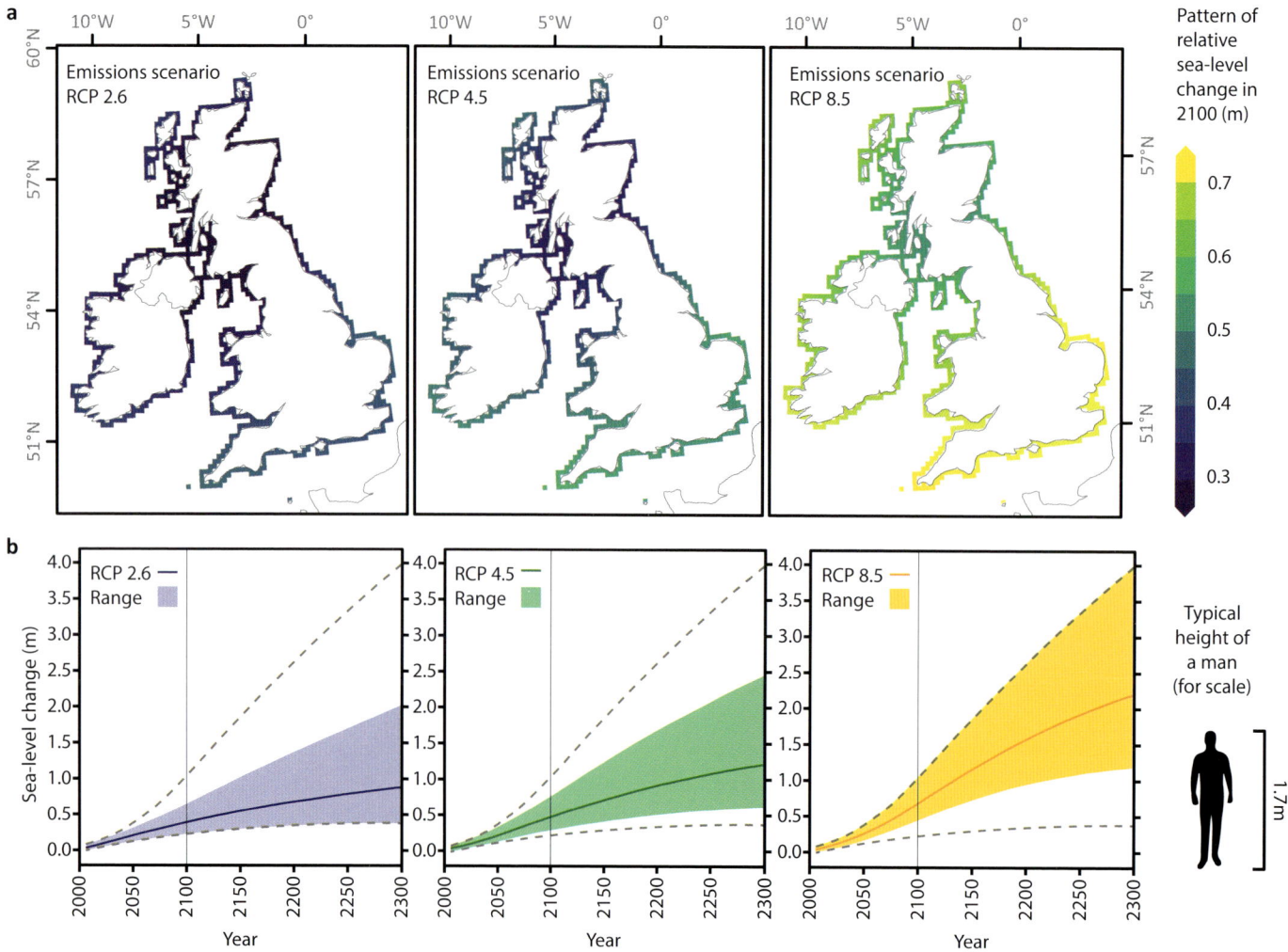

Fig 33.7 FUTURE REGIONAL PROJECTIONS FOR MEAN SEA-LEVEL CHANGES UNDER CLIMATE CHANGE. a. Relative sea-level changes for the twenty-first century around Ireland and Britain, based on UKCP18 Climate Risks Assessment, from the Hadley Centre. These show the results of ensemble model projection trends of averaged SLR for three climate change emissions scenarios, from RCP 2.6 as best-case, to 8.5 as worst-case scenario. These indicate a difference in the amount of relative SLR experienced north to south around Ireland, similar to that occurring in the present day. A lower impact of RSL rise is shown for the north versus southern regions, due primarily to continuance of the postglacial (Ice Age) control of gradual earth crustal uplift (north) and relative subsidence (south) resulting from glacio-isostasy. These values are consistent with IPCC model outputs globally and may be considered as conservative estimates of change. b. The longer-term trends of RSL changes for Ireland and Britain's coasts to 2300 under the three scenarios shown in a. [Data source: Met Office Hadley Centre, 2018. UKCP18 21st Century Time-mean Sea Level Projections around the UK for 2007–2100. Centre for Environmental Data Analysis. https://catalogue.ceda.ac.uk/uuid/0f8d27b1192f41088cd6983e98faa46e]

In the Republic, the government has pledged to meet targets of 10 per cent of marine areas to be designated as MPAs, with the aim of preserving biodiversity, by 2030. Work is underway to create a process specific to the identification, designation and governance of MPAs in the Republic, and wider, but it remains to be seen how fast this will progress.[17]

Climate change

The production and consumption of resources by communities living in the world's coastal zone will be affected increasingly by a broad range of climate-change impacts. The seas around Ireland have become warmer, more acidic and less productive in recent decades. Globally, melting glaciers and ice sheets are causing increased rates of sea-level rise (SLR), and coastal extreme events are becoming more severe. Climate-related events are having widespread impacts on human and natural systems. These have included losses of life, livestock and livelihoods due to extreme events, such as flooding. On coasts, intense storms – particularly in Bangladesh and other very low-lying countries, for example, Palau and other Pacific islands – have caused widespread damage.

The globally averaged, combined land and ocean surface temperature data now show a warming of over 1°C since 1880 (Figure 33.5). Sea surface temperature is also rising at an increasing rate (Figure 33.6). The rate at which this has occurred since 1994 (0.6°C per decade) is unprecedented in the 158-year observational record. This trend has implications for marine life, particularly the vital areas for biodiversity of coral reefs, ocean chemistry, storm intensity and SLR. Studies indicate that the climate of Ireland is changing in line with global patterns, although due to its ocean margin location climate responses over the island have lagged behind, but are now catching up. Temperature records show the clearest trends of this 'catch up'; over the last century, Ireland's air temperature has also increased by c.1°C (1880–2020).[18]

In the worst-case scenarios advanced by the Intergovernmental Panel on Climate Change (IPCC), global mean sea levels are projected to rise by a maximum of 0.8–1m by 2100 (Figure 33.7),

Research indicates that this may be a somewhat conservative view for many world coasts, including those in Ireland. Here, evidence from the past has shown they can experience large and regionally different variations in SLR. Satellite observations also indicate that sea levels around the island have already risen by approximately 0.1–0.12m since the early 1990s, and at increasing rates from *c*.0.2mm/yr to current values of *c*.3–3.5mm/yr (Figure 33.8). Tide gauges situated around these coasts, however, show more local to regional variations in these satellite statistics, confirming that differences and anomalies in SLR do occur at these scales. In addition to these more general rises in water levels, storm surges can have immediate and significant implications for coastal communities. Storm surges create the immediate SLR and associated marine flooding problems on coasts, through damage to infrastructures, loss of homes, as well as cliff and shoreline erosion (see Chapter 30: Engineering for Vulnerable Coastlines).

By mid-century, accelerating regional SLR – at projected rates of >5–8mm/yr – will become a more significant cause for concern as a control to the island's coasts. However, the present west to east pattern of high magnitude storm events across Ireland will remain as the most immediate coastal threat, in terms, for example, of flood damage, as well as the breaching and reorganisation of sand systems (for example, beaches, dunes and barriers). These projections are based on linear modelling approaches, which do not account for apparently randomly timed causes of SLR, such as catastrophic ice melts (for example, Greenland and Antarctic ice), or the lagged process, which refers to differential rates of response within environmental systems. The height of SLR on our coasts is likely, therefore, to be much higher, potentially >2m rise by early twenty-second century. Such developments, particularly as the result of rapid changes, are well established in long-term Earth records as forming the real pattern of SLR and coastal responses.

The repercussions of such large SLR events on unprepared coastal populations and environments around the world will probably put the COVID-19 pandemic crisis in the shade. The problem for government and coastal managers in Ireland, as elsewhere, is that such environmental systems' behaviour is difficult to model with certainty, and so patterns and rates of change are 'smoothed' and more conservative values of change accepted. Thus, the significance of the different approaches and outcomes of modelled projections for most elements of climate change, including SLR, remains controversial. It is important to note though, that the broad trends and patterns of climate change experienced to date have been predicted accurately in models.

For Ireland, it is well established that SLR will result in a range of increasingly significant coastal – and wider – environmental changes, many of which are already in progress. These include, accelerating coastal erosion, particularly of the long expanses of soft-sedimentary coasts (see Chapter 1: Ireland's Coasts: Setting the scene, Figure 1.3a); flooding of land areas, causing the squeeze in

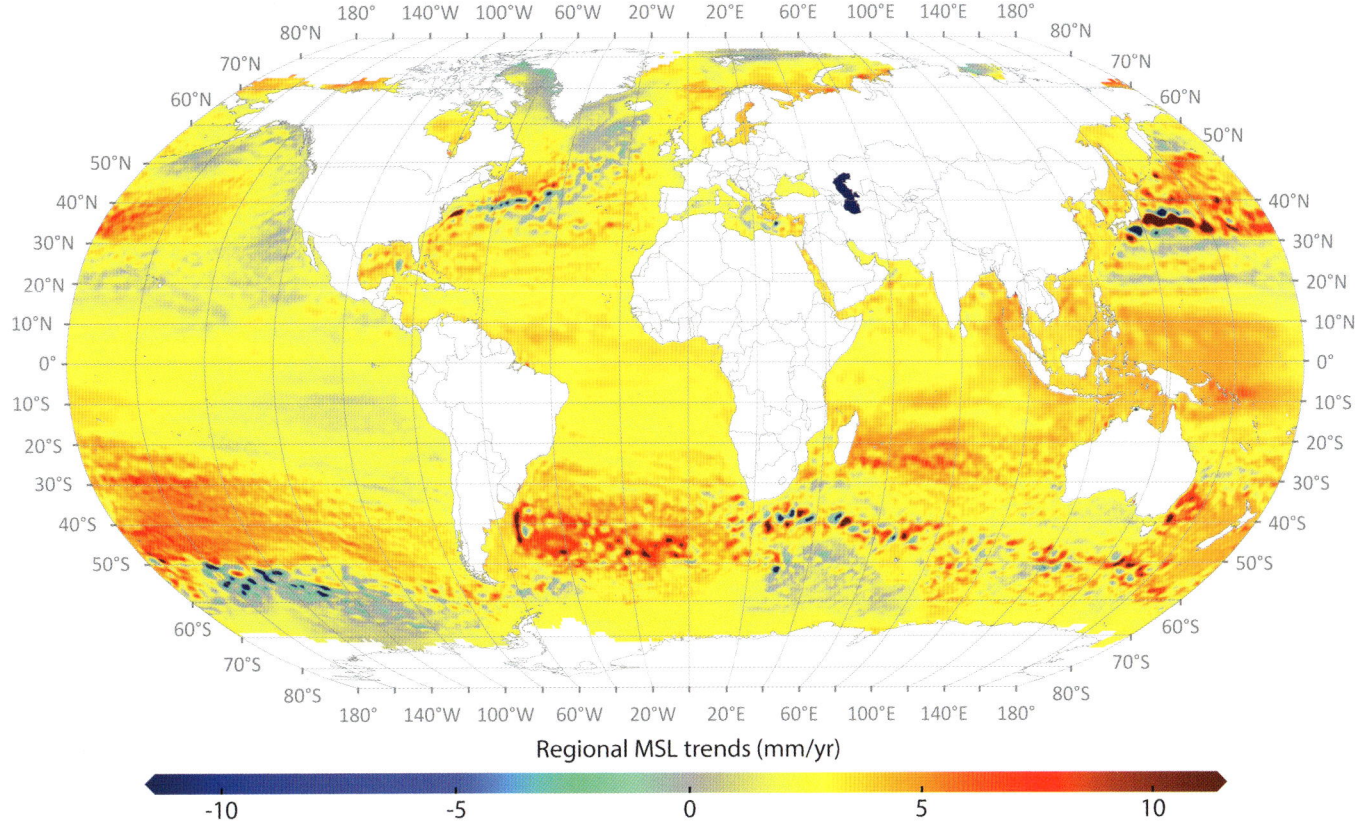

Fig. 33.8 THE PATTERN OF CONTEMPORARY GLOBAL RATES IN MEAN SEA-LEVEL (MSL) CHANGES OBSERVED FROM SATELLITE IMAGERY (SEPTEMBER 1992–MAY 2019). The data shown are based on a number of satellite observation platforms (for example, TOPEX, Poseidon) and indicate that sea levels around Ireland have risen at rates of *c*.3mm/yr (total SLR rise 0.04–0.06m) since the early 1990s. These rates are consistent with the regional pattern for North-west Europe and the North Atlantic. Elsewhere, globally, much larger rates of SLR have been experienced, as in the Pacific ocean, which contrast with areas in high latitudes (north and south) showing a fall in SLR rate. These differences result from the local to regional-scale accommodation and interaction of ocean water volumes (some as new ice meltwater) with earth crustal movements. [Map data: EU Copernicus Marine Service/CNES/LEGOS/CLS, 2020]

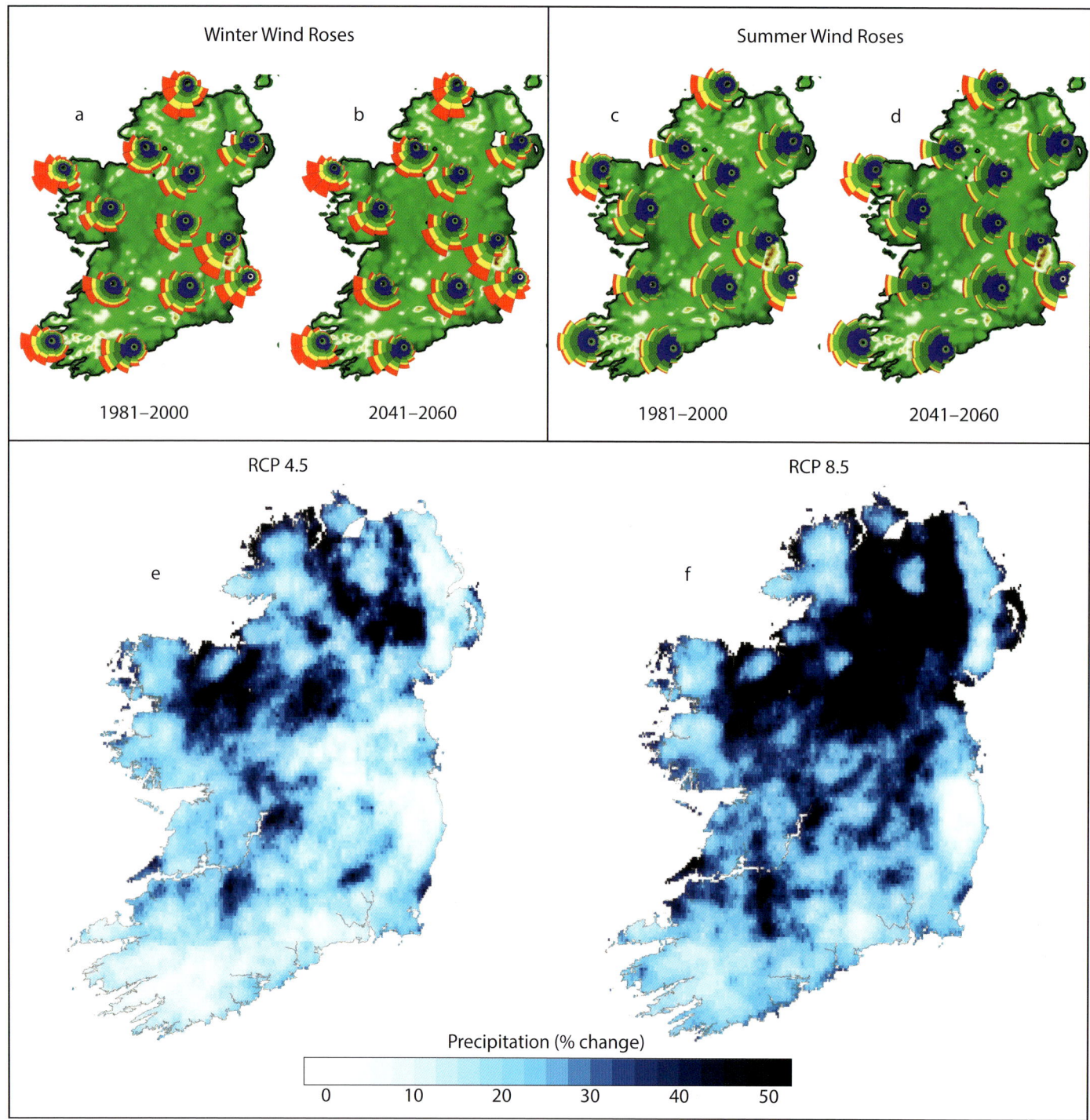

Fig. 33.9. EXAMPLES OF CLIMATE CHANGE PROJECTIONS FOR COASTAL IRELAND (WIND AND RAINFALL), AND NEIGHBOURING REGIONS, USING ESTABLISHED RCP SCENARIOS AND ENSEMBLE MODELLING. Maps a.–d. show projections of changes in wind speed and direction at 60m height (above ground surfaces) around Ireland, including for the offshore windfarm area of Arklow Bank, south-east Ireland, and Malin Head, north Donegal. The wind rose maps show the proportional (per cent) changes in wind speed (circular black–red colour ranges; red as highest wind speeds) plotted with wind direction. The four maps indicate: a. the winter ensemble of past simulations 1981–2000; b. winter ensemble of high emission future simulations (2041–2060); c. summer ensemble of past simulations (1981–2000); d. summer ensemble of high emission future simulations (2041–2060). Modelling these changes is valuable in assessing the likelihood of the occurrence of damaging winds (for example, storm surges, hurricanes), for impacts on, for example, coastal infrastructures, and the viability of wind farm operation. Projections indicate some increase in wind speeds in both winter and summer periods. Wind directions, however, remain substantially unchanged in the twenty-first century. For the winter months, a small increase in the frequency of south-westerly winds is noted, but in summer wind directions across Ireland show only minor changes. Given that the overall pattern of storms tracking across the island from the Atlantic shows only a little drift southward in future, it is not surprising that the projected changes in wind direction are small. (This image is from Met Éireann's publication, *Ireland's Climate: The road ahead* and has been reproduced here by kind permission of the editors.) [Source: Gleeson, E., McGrath, E., and Treanor, M, eds,. *Ireland's Climate: The road ahead*. Met Éireann, 2013. http://hdl.handle.net/2262/71304]; Maps e. and f. show two scenarios (RCPs 4.5, intermediate level and 8.5, worst-case) for future rainfall; as projected changes (per cent) by mid-century of the number of days with heavy precipitation > 30 mm/day. In each scenario, the future period 2041–2060 is compared with the past period 1981–2000. These data illustrate some of the possible extremes in rainfall that might be expected in future, though the exact values and patterns remain uncertain. [Data source: Nolan, P. and Flanagan J., 2020. High-Resolution Climate Projections for Ireland: A multi-model ensemble approach. EPA Research Report. Dublin: Environmental Protection Agency; Cartography: Sarah Kandrot and James Fitton]

living spaces of coastal lands; breaching of coastal defences; loss of wildlife habitat and other environmental amenities. The effects of these changes will be particularly detrimental for urbanised coasts, such as those of Belfast, Waterford, Galway and others. Both Cork and Dublin are extremely vulnerable to SLR and have a proven record of susceptibility to major flooding events, due their location on the former marshes of estuaries. Some 1,500km of Ireland's coastline is at risk from erosion, while some 490km in immediate danger. This will be particularly problematic for the soft-sedimentary cliffs and beaches of the east coast (see Chapter 30: Engineering for Vulnerable Coastlines).[19]

Research since the 1990s has shown that the Gulf Stream (and the North Atlantic Drift element of this that affects Ireland) is currently in its weakest state over a period of 1,600 years, including the time of the Little Ice Age in the fifteenth to seventeenth centuries. It is thought that this is linked, at least partially, to current climate changes.[20] While the Gulf Stream exerts a major positive temperature influence on Ireland's climate, scientists are still trying to establish the

risks and potential future implications of a sudden, versus a gradual, change in the flow of these warm Atlantic waters (see Chapters 2: The Coastal Environment: Physical systems, processes and patterns, 10: Rocky Coasts and 12: Coastal Wetlands). Other impacts of climate changes for Ireland and its coasts are much clearer and have been outlined in research since 1990 and increasingly since 2010. Illustrations of the model projections for wind, wave, temperature and rainfall changes in the twenty-first century are presented in this work, with frequent updates of these being published (Figure 33.9).[21] These show that the Republic and Northern Ireland, working in both international and national frameworks, have made significant progress in developing an understanding of the range of issues and the responses needed to address this major global problem. Continued monitoring and modelling of climate changes are required, together with the development of coherent government policies on actions, particularly for the coastal zone. This is taking place in the context of international and, particularly, EU support to address the issues effectively.[22]

Case Study: Ocean Acidification

Evin McGovern and Triona McGrath

As long ago as the 1860s, the Irish scientist John Tyndall published on the heat-trapping 'greenhouse' properties of carbon dioxide (CO_2). During the mid-twentieth century, the realisation began to dawn that ever-increasing amounts of CO_2 and other greenhouse gases being added to the atmosphere could cause significant perturbations to the Earth's climate system (Figure 33.10). However, in the opening years of the twenty-first century, scientists also began to appreciate fully another inevitable consequence of humanity's carbon dioxide emissions. Namely, the additional

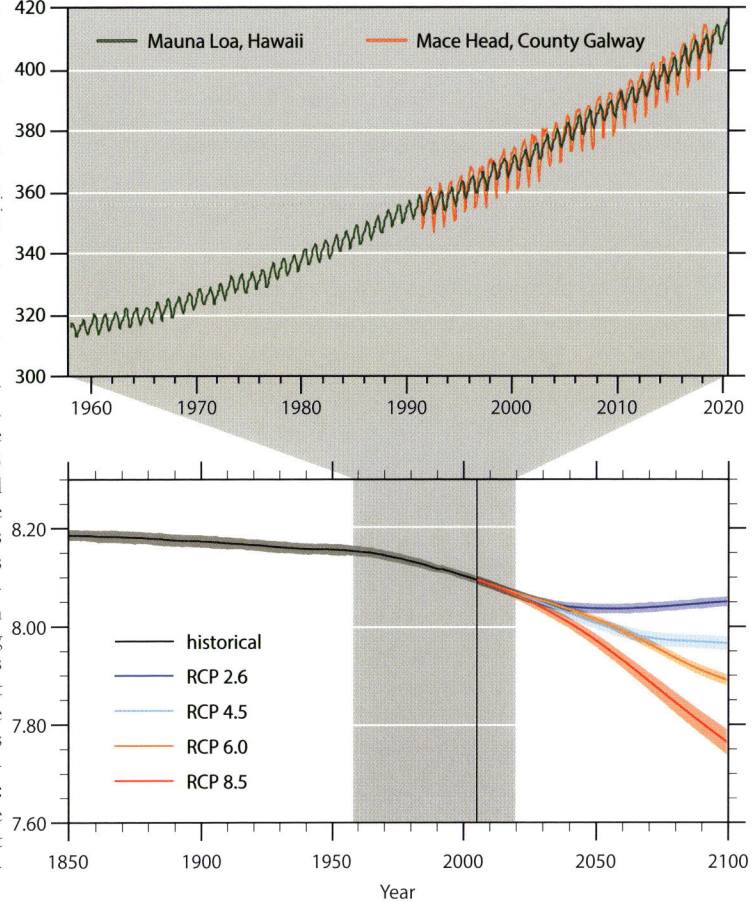

Fig. 33.10 (right) KEY SIGNATURES OF OBSERVED AND MODELLED CARBON EXCHANGES IN THE EARTH'S ATMOSPHERE OCEAN-SYSTEMS. a. Global monthly mean CO_2 gas levels in the Earth's atmosphere. The atmospheric CO_2 plot shows the Mauna Loa CO_2 time series (NOAA) as the longest available observation series, overlaid with the Mace Head (County Galway) time series. While the seasonal cycle is bigger for the Mace Head station, the match when the two series are superimposed is remarkable, despite being so remote from each other. This illustrates well the global nature of increasing CO_2. [Data sources: NOAA Global Monitoring Laboratory and Mace Head, Global Atmosphere Watch Station (GAWS)]; b. Modelled global surface ocean pH levels showing the outcome of the established IPCC RCP scenarios. These pH levels are believed to have decreased from values of 8.2 to 8.1 since the start of the industrial revolution, this is equivalent to $c.26$ per cent increase in acidity. By the end of the century, and under more of a 'business as usual' emission scenario (RCP 8.5), the increase in surface ocean acidity is expected to be $c.150$ per cent (pH 7.8). [Source: Stocker et al. [2013]: Technical Summary. In: Climate Change 2013: The physical science basis. Contribution of Working Group I to the Fifth Assessment Report of the Intergovernmental Panel on Climate Change. Cambridge/New York: Cambridge University Press]

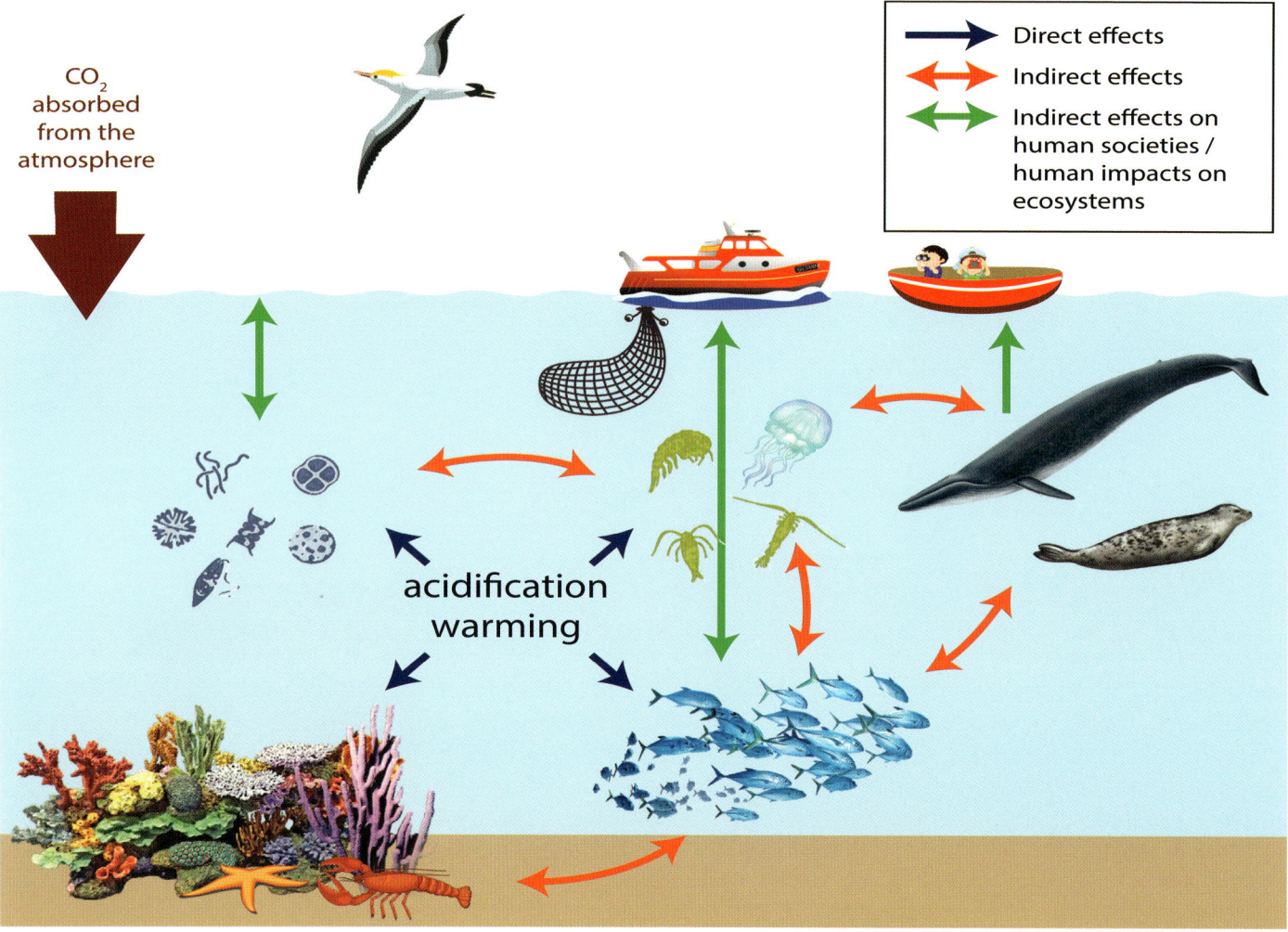

	Direct effects
	Indirect effects
	Indirect effects on human societies / human impacts on ecosystems

Figure 33.11 OCEAN ACIDIFICATION IN A NUTSHELL. The schematic shows atmospheric accumulation of CO_2 emissions and its uptake pathways in the oceans, which drive ocean acidification. Physical, chemical and biological processes control ocean acidification in coastal, shelf and open ocean waters. Rapid future ocean warming and acidification, due to unabated CO_2 emissions, may have serious consequences for marine ecosystems and the services provided to people, such as food security. Many organisms, for example, bivalve and gastropod molluscs, echinoderms, calcifying phytoplankton (coccolithophores) and corals, are likely to find it harder to construct and maintain their calcium carbonate shells and skeletons in the future ocean.

CO_2 absorbed by seawater is overwhelming the ocean's buffering capacity, and oceans are acidifying. Research in this area has intensified as scientists try to establish how marine life will respond to such a fundamental change in ocean chemistry. Of particular concern is the potential impact for calcifying organisms that use carbonate ions for making calcium carbonate shells and skeletons (Figure 33.12).

CARBON DIOXIDE EMISSIONS ARE CHANGING OCEAN CHEMISTRY

More than two trillion tonnes of carbon dioxide have been released into the environment as a result of human activities since the start of the industrial revolution, due to fossil fuel combustion, cement production and land-use change. In 2019, the average global concentration of CO_2 in the atmosphere was approximately 410 parts per million (ppm), well above the stable, long-maintained pre-industrial levels of about 280ppm (Figure 33.10). In fact, today's atmospheric CO_2 levels would have been much higher, and the climatic consequences much more severe, had not the oceans absorbed about a quarter of the net anthropogenic CO_2 emissions.[23]

The globally connected great ocean currents transport carbon around the planet and ultimately into the deep ocean, where it will remain for potentially hundreds of years. Carbon dioxide is taken up unequally across the oceans, with physical, chemical and biological processes driving spatial and temporal variability. For instance, cold polar waters are more effective at taking up CO_2, making them more vulnerable to acidification than warmer, tropical ones. In coastal waters, the picture is more complex, due to terrestrial influences, freshwater inputs and strong, biologically driven seasonality. Around the Irish coast, geology is a key factor in determining local water chemistry, as limestone river catchments can export large amounts of inorganic carbon and alkalinity; the former driving lower pH and the latter buffering changes of the carbonate system.[24]

HOW WILL OCEAN ACIDIFICATION AFFECT MARINE LIFE?

The current CO_2 emission trajectory has committed the oceans to increasing acidification which, once achieved, cannot be reversed in any practical way. Critically, this is occurring at a rate that is probably unprecedented over the last 300 million years, challenging the capacity of marine organisms to evolve and adapt to such a rapid change. Moreover, ocean acidification will not occur in isolation, but will operate in concert with sea surface warming and other environmental stressors. Research work is using a number of approaches to tease out the likely impacts of ocean acidification on marine ecosystems. Much of this involves experimental studies in the laboratory, or in the field, with simulated projected end-of-century CO_2 levels. On the west coast of the United States, ocean acidification superimposed on already low-pH water, periodically upwelled from depth, has been linked to recurrent failures in oyster hatcheries, providing an indication of how changes in ocean chemistry may lead to additional future impacts. Another approach to understanding the potential impact of ocean acidification involves the study of geological records, which can reveal insights into past high CO_2 events, sometimes associated with mass extinctions, although it is recognised that none of these events is fully equivalent to the current situation.

There will be winners and losers in a high CO_2 world and, while some organisms will cope well and possibly even thrive, many species are likely to be negatively impacted. Research shows complexity and variability in response to elevated CO_2, even between similar species. Some of the effects may be subtle and sensitivity can vary with different life stages, with early-life stages often most sensitive. Calcifying organisms include molluscs (fox example, mussels, oysters, sea snails), echinoderms (for example, urchins), coralline algae (for example, maërl), calcifying phytoplankton (coccolithophores) and cold and warm water corals (Figure 33.12 see Chapter 3: Coastal Waters: Marine biology and ecology).[25]

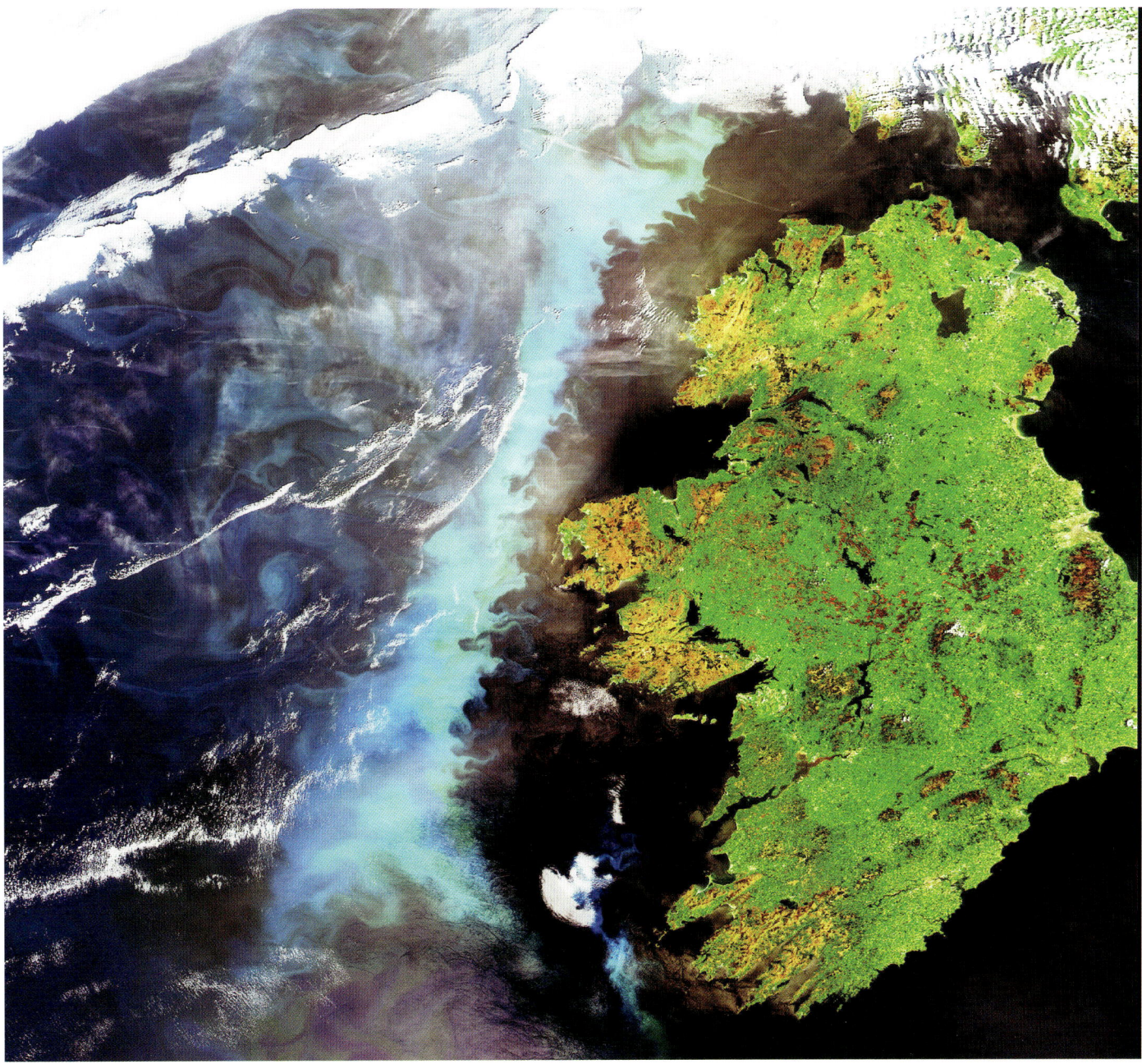

Fig. 33.12 SMALL ORGANISMS, BIG SCALE IMPACTS! AN ENVISAT image (2006) of a Coccolithophore bloom in the Atlantic Shelf waters west of Ireland. The bloom is shown by the broad (>480km length), pale-blue band running off Ireland's west coast, while the whiter shades to the north are mostly clouds. Coccolithophores are abundant unicellular marine algae that produce calcite plates and are so small they cannot generally be seen by the naked eye (coccolith sizes 5–30µm in diameter). They form an important element of the food chain as primary producers and are fed on by zooplankton. Coccolithophores also play a significant role in ocean biogeochemistry and they may be affected by rapid ocean acidification. [Source: © ESA, CC BY-SA 3.0 IGO]

Calcification has been shown to be impaired at reduced pH in some species and, even where there is no obvious impairment, diversion of energy resources to maintain calcification may be at a metabolic cost for organisms (Figure 33.13). This may impact many slow-growing, cold-water coral reef habitats that have been identified along the Irish continental shelf slopes (see Figure 2.7). There is some evidence, however, that living coral of *Lophelia pertusa* may be able to cope with ocean acidification. But if the ocean deep saturation horizon shoals (that is, the ocean depth below which $\Omega < 1$; where Ω is the saturation state for calcium carbonate minerals, such as aragonite and calcite) emerge as projected for the end of the twenty-first century, supporting reef structures may be exposed to corrosive waters. As a result, the aragonite mineral from which it is built may begin to dissolve (see Chapter 2: The Coastal Environment: Physical systems, processes and patterns). Experimental studies have also shown that ocean acidification can affect sensory systems and behaviour in fish and, possibly, in some invertebrates, altering responses to key environmental chemical cues.

Given the complexity of responses to increasing ocean acidity, it is difficult to generalise predictions of its future impacts in ocean environments. This is particularly challenging for coastal environments, given the variability of carbonate systems to which organisms are already exposed. Key research studies are focused currently on the resilience of organisms, acclimation, and adaptation over multiple generations, as well as the combined impact of multiple stresses. Nonetheless, there are already ample reasons for society to drastically reduce CO_2 emissions, given the certainty that global oceans will continue to acidify irreversibly, even beyond the moment when CO_2 emissions start to decline, as well as the ever-growing evidence-base regarding the effects of projected ocean acidification. The alternative could well be major global changes to marine ecosystems, affecting marine biodiversity and key services – such as fisheries and commercial shellfish production – bequeathing, ultimately, a diminished ocean to future generations.

Fig. 33.13 COASTAL AND MARINE ORGANISMS VULNERABLE TO OCEAN ACIDIFICATION FOUND IN IRELAND'S COASTAL WATERS. a. Mussels (*Mytilus edulis*) and barnacles (*Chthamalus* spp. and/or *Semibalanus balanoides*) on rocks off the coast of County Kerry. [Source: Maria Delaney, Wikimedia Commons, CC BY-SA 4.0]; b. Edible sea urchin (*Echinus esculentus*), Skellig Islands, County Kerry [Source: Derek Heasley, Wikimedia Commons, CC BY-SA 3.0]; c. Cold-water coral (*Lophelia pertusa*). [Source: University College Cork and the Marine Institute]

Cutting carbon emissions, both at global and at the critical national scales, is key to mitigating against the worst implications of climate change. Achieving carbon emission targets in Ireland requires strong political and governance leadership to stimulate markets for the decarbonisation of electricity, heat and transport, both in the Republic and Northern Ireland. Other initiatives include engineered infrastructures for flooding and the development of grassroot, community-based responses to the impacts of climate

change (see Chapters 11: Beaches and Barriers and 28: Renewable Energies: Wind, wave and tidal power). Studies show that the Irish coastal zone could become a hotspot for these initiatives – with development of marine renewable energy technologies, carbon capture and storage deployments – and for new shipping based on alternative fuels working out of, or passing through, Irish ports.[26]

In managing different response initiatives to climate change, leaders need to take a broad view of the linkages across sectors and

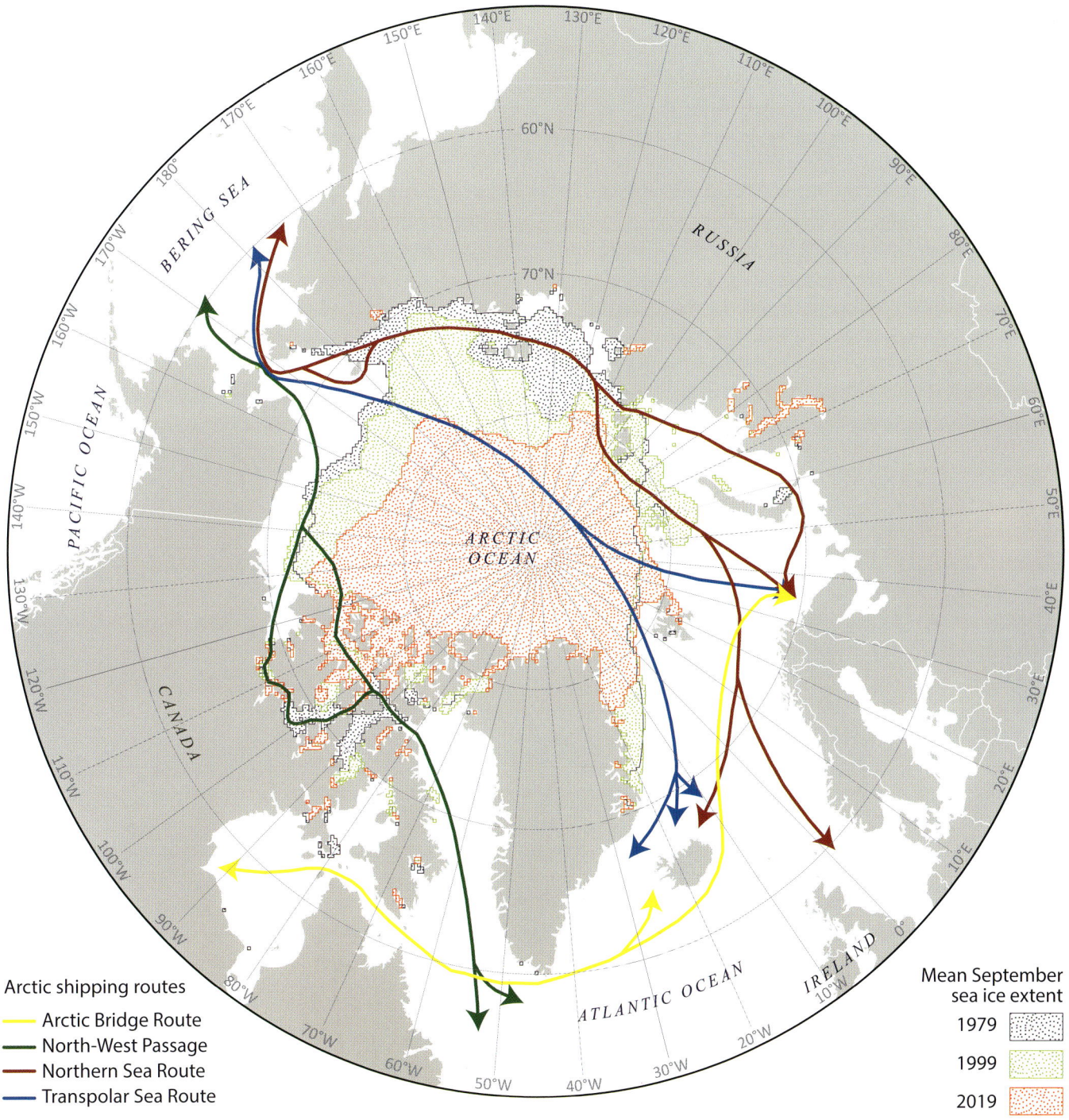

Fig. 33.14 ARCTIC SHIPPING ROUTES AND CHANGES IN SEA ICE EXTENT (1979–2019). The disappearance of Arctic ice since the 1980s, which will be complete by mid twenty-first century, means that new trans-polar passages will become a reality and will likely open the region to even greater development of its seabed and other ocean resources. The changing face of coastal futures in the Arctic has been recognised already by countries as far away as in south-east Asia. Singapore, even from its location near the Equator, has articulated its interest in Arctic governance and has requested observer status to the Arctic Council. New shipping routes from Europe to Asia, via the Arctic, would drastically increase shipping traffic off the Irish coast, perhaps creating potential for new trans-shipment hubs, or issues with pollution. It could well be that in future Ireland will be influenced by a much greater situational awareness of, and impacted by, Arctic issues. [Data source: shipping routes: Humpert, M. and Raspotnik, A., 2012. The Future of Arctic Shipping Along the Transpolar Sea Route. *Arctic Yearbook, 2012*(1), pp. 281–307; sea ice extent: Monthly sea ice extent September 1979, 1999, 2019 (G02135). Boulder, Colorado, USA: National Snow and Ice Data Center]

scales – from local, to national and global levels – not just in terms of the functioning of natural systems, but also in terms of socio-economic drivers. At first glance, the relevance of some global affairs may appear tenuous. This is demonstrated best by Ireland's links with the Arctic. At one level, the extreme environment of the Arctic appears unconnected to the temperate island of Ireland. However, there is an increasing understanding that the melting of ice in this region, which includes much of Greenland, has grave implications for climate change (in terms of, for example, temperature and ocean circulation). This will have significant knock-on effects for Ireland that include the risk of SLR and its impacts. Importantly, the opening up of the Arctic for new shipping routes and other resource uses, as has been happening in the last two decades, may also have consequences for Ireland's coastal future. In terms of changing port and transhipment facilities, for example, it could bring Ireland into the sphere of Arctic region trade and geopolitics (Figure 33.14).

Closer to home, the building of the coping capacity and resilience of people and environments to help adaptation to climate change is needed, and no more so than in the coastal zone. This includes measures concerned with the restoration of natural habitats, that can both help protect the coast from storms and also affect carbon capture at the same time. Other adaptation measures could include specifying design criteria for buildings and infrastracture that are adjusted for increased future flooding (Figure 33.15); avoiding development on flood plains; implementing early warning systems and helping to change stakeholder behaviour through education and awareness campaigns.

In Ireland, the Republic's National Climate Change Adaptation Framework vests government departments and local authorities with responsibility for the preparation of sectoral and local adaptation strategies respectively, subject to update and review every five years. Given the relatively early stage of its roll-out, benefits to coastal communities tend, at present, to be localised and confined to locations where pilot research and/or capacity-building projects have been conducted, or where a track record of coastal partnership exists. In future, the evolved ad hoc approach to integrated coastal zone management (ICZM) initiatives should be revisited, given the renewed interest in the value of coastal partnerships provided for in the National Marine Planning Framework and associated legislation (see Chapter 32: Coastal Management and Planning).[27]

Adaptation to climate change is very much a matter of the perspective, also, of stakeholder and other groups. People often hold very different views on what measures are needed to tackle the implications of climate change, while the ability to adapt can also be constrained by budget limitations. An example of such issues in

Fig 33.15 No. 1 ALBERT QUAY, CORK: DESIGNING FOR FUTURE FLOODING. Situated at the gateway to Cork Docklands in the River Lee Development Corridor, No. 1 Albert Quay is a fourth-generation, sustainable office development and is one of Cork city's most modern buildings. Completed in 2016, it lies within 10m of the south channel of the River Lee. Cork City Development Plan standards require all new city centre developments situated within the river's flood zone to have a finished ground floor level at a height of at least 3.8m OD, that is, some 1.2m above the adjacent and existing quays and street levels. The architect has designed the approach to this building to be via wide steps, with well-integrated ramped access; a modern palette of high-quality paving and façade materials, with appropriate building massing; a set-back distance from the quay and double-height colonnaded elements in the construction. These design features have successfully addressed ground-level differences, ensuring that the completed development enhances the streetscape while simultaneously masking the necessary flood building control measures. [Source: Ailbhe Cotter]

adaptation approaches for Ireland's coasts is provided in the discussion of the controversial €140m Lower Lee Flood Relief Scheme (LLFRS), developed by the Office of Public Works (OPW) and supported by Cork City Council. Designed to protect the city from flooding, the programme took a range of protection measures into consideration, including the use of an existing dam, direct flood wall type defences, the use of floodplain areas to the west of the city for water storage and the setting up of an early warning system. Following the voicing of community concerns during its review, elements of the plan were subsequently modified. This flood protection programme has experienced strong opposition from many sources in the community, as represented by the Save Cork City campaign (SCC), illustrating the diverging viewpoints that arise around such complex issues. The case study takes a particular perspective of the flooding problems in the economically important Cork city and harbour region, one that is influenced strongly by the desire to protect the city's valuable architectural heritage.

Architecturally, the core of Cork is very much an old city of the eighteenth and nineteenth centuries. By 1750, the streetscape that shapes Cork had been defined. Built on marshlands, this historic core is highly vulnerable to flooding, from both river and sea sources, which is exacerbated now by climate change. The SCC campaigners raised objections to the OPW recommendations for flood defences, which they argued would damage the aesthetic value of the city and much more. This view highlights the tough choices to be made, between preservation of built urban heritage, conservation of natural habitats and the protection of critical infrastructure in the short, medium and long terms. This is particularly challenging when resources are limited, and even more so when stakeholder perspectives are not aligned.

'SAVE CORK CITY': AN ARCHITECTURAL PERSPECTIVE

John Hegarty

During the late seventeenth and early eighteenth century, Cork city engaged increasing economic success thanks to its geographical position on the edge of the North Atlantic trade route and its proximity to Britain and the continent. It is within this context that the city grew to be a major port, with a population of over 80,000 people by 1841 (see Chapter 19: Changing Coastal Landscapes). An outcome of this prosperity was that, especially in the eighteenth century, the city experienced a golden age of architecture. New ideas, skills and influences passed between ports easily, with the sea representing the 'information superhighway' of the day. These ideas included concepts of building design, the origins of which were rooted firmly in the tradition of Greek and Roman urban design, principles that had been embraced during the Age of Enlightenment.

From around the time of the Norman invasion of the twelfth century, stone was adopted as a building material and as a basis of architectural expression (see Chapter 17: The Vikings and Normans: Coastal invaders and settlers). By the eighteenth century, the adaptation of minimalist Palladian design principles, combined with local building techniques and finishes, could be seen extensively in Cork (see Chapter 19: Changing Coastal Landscapes, Figure 19.10).[28] Along the riverfront and canals that traversed the city, the quays of Cork expanded from the city's medieval centre in a piecemeal fashion, often renewed by the building of new retaining walls in front of older structures. In the late eighteenth century, cut-stone quays were built in the tradition of great Georgian naval construction, representing a significant investment in the city (Figure 33.16).

Although the location of Cork gave it the trading advantage of being a safe port, it also left the city vulnerable to flooding. Historically, this flooding arose mostly from a combination of heavy rainfall in the river catchment upstream and high water levels caused by spring tides or tidal surges downstream, particularly when influenced by strong winds at sea. Today, these processes are exacerbated by the effects

Fig. 33.16 CITY OF CORK AND ITS EIGHTEENTH-CENTURY QUAYSIDES, AS SEEN AT SULLIVAN'S QUAY. Many of Cork's quays were reconstructed in the late eighteenth century with deep-cut stones, faced by hand and fitted together with precision by skilled masons, directed by architects. These remain in many locations along the city's north and south channels, forming an attractive streetscape, and have significant architectural and historic interest, as well as adding to the economic benefits to the city from sources such as tourism. [Source: Val Cummins]

of land drainage schemes, urbanisation upriver and, increasingly, the impacts of climate change and SLR.

The implications of these developments and the potential hazards to which they would give rise, were demonstrated to dramatic effect in 2009, and again in 2014, when Cork, and many parts of Ireland, were flooded extensively. In response, the Office of Public Works (OPW) presented proposals for a flood defence scheme for Cork city as part of the national Catchment Flood Risk Assessment and Management Study (CFRAMS) programme of flood alleviation.[29] Central to these proposals were plans to reinforce and raise the historic quay walls and streets of the city, to facilitate the passage of fast-moving water flowing downstream and out to sea, as well as measures to defend against tidal surge (see Figure 33.16).

A group called Save Cork City (SCC) was formed in January 2017 to promote discussion on the issue and to oppose the proposed OPW defences. It also aimed to increase awareness and appreciation of the architecture and urban landscape of the city as assets to be cherished for social and economic reasons. As the group noted, a survey by the Cork Town Planning Association and Chamber of Commerce, as early as 1925, had described the quays as 'one of Cork's assets which should be jealously guarded'. Also citing a 2012 study by the World Bank, the SCC argued that, '[a] city's conserved historic core can differentiate that city from competing locations, branding it nationally and internationally, thus helping the city to attract investment and talented people'.[30]

In 2017, SCC presented their own suggestions for an alternative approach aimed at protecting Cork from flooding, which they called the 'Save Cork City Solution'. This approach – drawn up by a group of professionals in architecture, engineering, coastal science, climate change and law – focused on the built environment and architectural heritage of the city centre as core assets to be preserved and protected. It argued for a flood protection scheme based on three main points: i. alternative plans for a tidal barrier; ii. slowing the flow of the river and iii. repairing the existing quayside walls.[31]

The tidal barrier proposed by SCC would be sited at Little Island, in shallow water at the mouth of Lough Mahon at the point where the River Lee transitions into the harbour beyond (Figure 33.17). It would temporarily store flood waters in the Lough Mahon area, and would also separate upstream floodwater from tidal surges. The group argued that installing a barrier at this location would limit adverse impacts on the essential character of the harbour, reduce environmental consequences and maintain maximum protection for the city, the docklands and the Douglas estuary from tidal surge. SCC also saw an advantage in the barrier being situated upriver from the city's main drainage processing plant at Little Island.

In contrast, the OPW have, post CFRAMS, posited a tidal barrier located further down the harbour and in deeper water, between Monkstown and Marlogue. Campaigners opposed this proposal on the grounds that it would be less effective with regard to environmental issues and protection of urban heritage and would cost more. From the perspective of SCC, there was no doubt that Cork, and many other coastal cities in Ireland, needed tidal barriers to manage flooding, and not in fifty years' time.

Regarding slowing the flow of the river, the group proposed changes in land management practices to help reduce the speed with which rainfall became floodwater. Better land management practices would also help increase the quality of the water supply and protect biodiversity. Another upstream measure perceived to be beneficial as part of an holistic approach, related to the Inniscarra and Carrigadrohid dams, built in the 1950s to generate electricity. The dams have often protected Cork city from flooding linked to fluvial water flows and/or

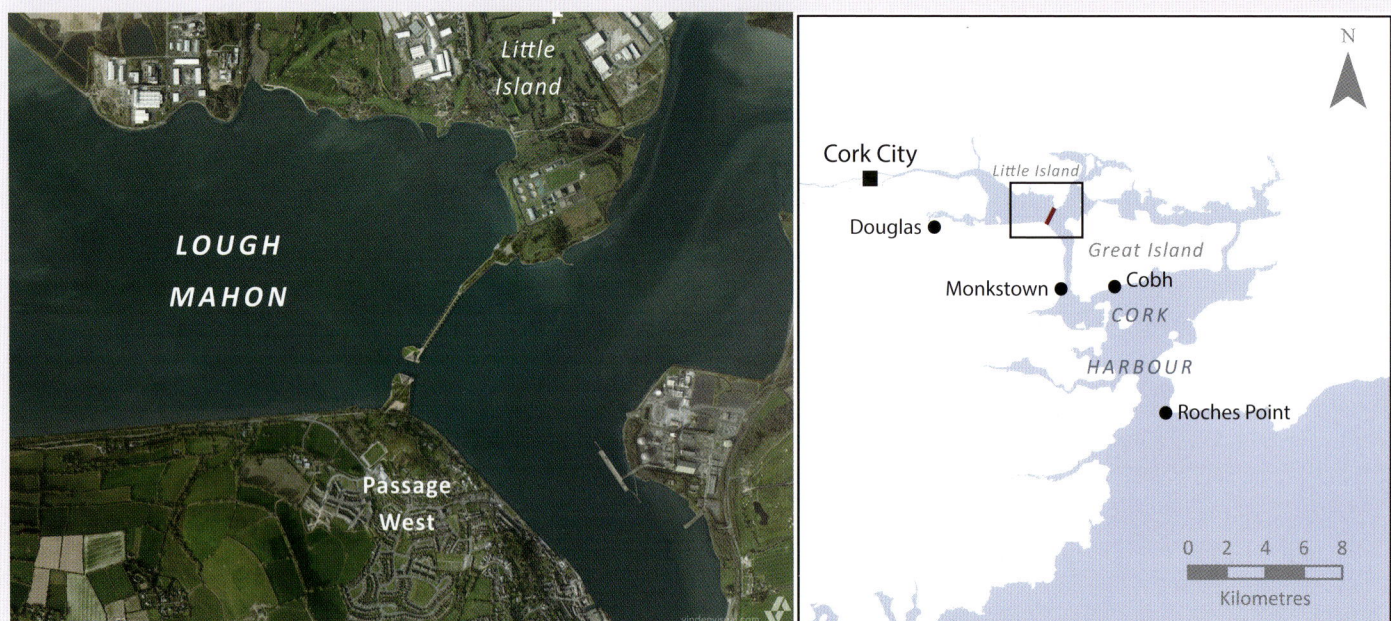

Fig. 33.17 A POTENTIAL TIDAL BARRIER LOCATION IN CORK HARBOUR AT CARRIGRENNAN POINT, LOUGH MAHON. A view of the coastal marine setting of Cork Harbour in the Little Island/Lough Mahon area is shown. The type of tidal barrier envisaged by SCC for this location would be of relatively low cost and provide for restricted access to shipping and allow the continued use of Cork's existing city port facilities. Others have suggested locations either closer to the city, within the outer channel of the River Lee, or, alternatively, further down estuary across the narrow, open coastal entrance to the harbour at Roches Point. These locations would require different types of engineering solutions for such a barrier, with both possible cost and resource use advantages and disadvantages. At present, construction of such a barrier remains a vision for the future, but, as SLR continues at increasing rates toward 2100 and beyond, a not too distant one. [Source: Save Cork City]

tidal surge. Both the OPW and SCC proposed the deployment of rainfall detectors in upstream catchments, to better inform the management of water levels behind the dams. Save Cork City advocated also, however, that the management and operation remit of the dams should be expanded to include, explicitly, flood prevention.

The third pillar of the campaign concerned the repair of the quayside landscape to increase amenity value and conserve the authenticity of Cork's urban heritage. One key benefit that arose in the short term from the choice of a tidal barrier over new flood defences – such as elevated sea walls in the city centre – was that it provided an opportunity to restore the existing city quays. This course of action would also provide the significant advantage of limiting the need for structural interference with complex groundwater conditions. (Groundwater levels need to be maintained in equilibrium to protect eighteenth- and nineteenth-century buildings supported by timber piles and historic archaeological remains.)

At the time of writing (2020), SCC felt that there was growing public support for the Save Cork City Solution and that their campaign was changing people's understanding, perceptions of the issues, and minds. A Love the Lee campaign also sought the protection of the city's historic core and the development of the tidal barrier at Lough Mahon.

HABITAT AND BIODIVERSITY LOSS

The first ever intergovernmental global assessment of biodiversity and ecosystem services, in 2019, issued a stark stocktake of global biodiversity issues. The figures published in the report, as relevant to coastal and marine environments, are vast and difficult to imagine. It showed that plastic pollution has increased tenfold since 1980; 300–400 million tonnes of heavy metals, solvents, toxic sludge and other wastes from industrial facilities are dumped into the world's waters annually. Fertilisers entering coastal ecosystems have produced more than 400 ocean 'dead zones', totalling more than 245,000km², a combined area about three times the size of the island of Ireland (see Chapter 31: Pollution, cover photograph). As a result, 66 per cent of the global marine environment has been altered significantly by human actions. The Food and Agriculture Organisation (FAO) of the United Nations, tasked with assessing the global state of fisheries and aquaculture, has determined that 30 per cent of global fish stocks are overfished, and 60 per cent are fished at their maximum sustainable yield threshold (see Chapter 31: Pollution).[32]

Threats to tropical coral reef ecosystems may appear a concern only for people living in far-off places. However, the global decline of marine biodiversity, as illustrated in the decline of coral reefs through ocean acidification, has both direct and indirect implications for coasts of the world, including Ireland's. Coral reefs cover 0.1 per cent of the entire ocean, yet support one quarter of all marine species and, therefore, contribute to the health of our oceans. Many Irish tourists visit locations such as the Great Barrier Reef to experience that wonder at first hand. An indirect consequence of reef decline may be that this kind of tourism will not be an option for future generations, with all the knock-on effects this will have on the economies of such regions. Closer to home, and considering more direct implications of marine biodiversity changes, habitats and species all along Ireland's coasts are at risk of a similar fate to those of reef corals. Major efforts in marine environmental management will be needed to halt these anthropogenic impacts of pollution and habitat destruction. An important environment for further study of these issues are the deep-water, cold-water coral reefs of the island's Atlantic shelf, designated a World Heritage site. These ancient reefs could also become badly affected as ocean acidification penetrates deeper into shelf-bottom waters, compounding damage inflicted already by intensive fishing through bottom-trawling (see Chapter 2: The Coastal Environment: Physical systems, processes and patterns).[33]

The National Parks and Wildlife Service (NPWS), in outlining the status of EU-listed habitats and species in the Republic, showed that most marine habitats are in an inadequate or bad condition. Non-indigenous species – or invasive species coming, for example, from southern waters as the north-eastern Atlantic has warmed – are now migrating northwards to Ireland's coasts (see Chapters 12: Coastal Wetlands and 13: Estuaries and Lagoons). These pose major threats to established native marine biodiversity and often generate significant economic costs. Examples of species that reproduce quickly and outcompete native species include the Pacific oyster (*Magallana gigas*), seagrass (*Spartina anglica*) and the Japanese skeleton shrimp (*Caprella mutica*). Other organisms, such as the carpet sea squirt (*Didemnum vexillum*), can alter the marine habitat in which they reside. Increasing globalisation of trade and movements of people have created new and multiple vectors and pathways by which non-native species are transported. Hull fouling on ships and discharge of ballast water are key vectors impacting on the marine environment.[34]

Issues such as these have been raised by many authors contributing to the *Atlas*. Clearly, a BAU approach cannot continue and public awareness of these issues has underscored the pressing need for transformative, societal change. Lessons as to how this might be achieved can potentially be learned from indigenous peoples and local communities in other parts of the world, where, in some cases at least, trends have been less severe or have been avoided altogether. Central to many of these successes has been a focus on bottom-up approaches to sustainability at the local level. This is already well established as an approach within ICZM, involving principles of community participation in capacity-building to respond to climate change and linked pressures on marine and coastal environments. Repercussions of the COVID-19 crisis will add to this need for change from the business as usual approach still operated by many communities in their relationships with marine environments.[35]

Case Study: Lessons from a Pristine Palau

Val Cummins

The United Nations refers to small island nations in the southern hemisphere, like Palau, as Small Island Developing States (SIDS). Many SIDS are based on low-elevation coral atolls and are exposed to the impacts of SLR. This may cause some of them to be abandoned in the future as marine flooding and storminess worsens in the twenty-first century. The key issue accompanying this development is that of climate refugees. Although not one that Ireland faces directly, this country will experience increasing problems of coastal squeeze associated with SLR and may also need to consider its role in accommodating such refugees in the future. Ireland has much to learn from SIDS and their understanding of the vital relationships between people and the sea, and from the visionary agendas for ocean conservation being driven by some political leaders of SIDS, in efforts to underpin thriving marine communities.

The Small Island Developing State of Palau is at the forefront of international best practice in ocean affairs. Palau is remote, situated east of the Philippines in the western North Pacific ocean (Figures 33.18 and 33.19). It has a population of 20,000 people distributed across twelve inhabited islands, though more were populated in the past. The land elevations of the main islands of Palau are mostly above the critical 2m level, which relates to future marine flooding under SLR. However, Palau's economy, as in other Pacific island states developed on low coral reefs and atolls, is heavily dependent on tourism and coastal infrastructures connected to such reef environments. Marine flooding could heavily impact these low-lying coastal areas in future.[36]

Promoted as 'Pristine Palau', the country boasts a marine environment of outstanding natural beauty. A barrier and fringing reef complex surrounds the archipelago, creating a vast lagoon and providing what are regarded as some of the best dive-spots in the world (Figure 33.18). Sharks, protected in Palau under the auspices of the shark sanctuary established in 2005, are a major attraction for divers. Tourism accounts for 40 per cent of employment and is a vital component of the economy. The government is now focused on attracting

Fig. 33.18 ROCK ISLANDS (ALSO CALLED CHELBACHEB), KOROR STATE, PALAU, SHOWING THE SHALLOW REEFS AND LAGOONS OF THE AREA. This collection of coral atolls and reefs is situated between Ngeruktabel and Mecherchar, in the chain of islands that make up the nation of Palau (see Figure 33.19). Formed when ancient corals were forced upwards by volcanic activity, these now uninhabited islands sustain a rich diversity of plants, birds and marine life. In addition, the islands and surrounding coastal lagoons form areas of outstanding natural beauty. In 2020, the government established the Palau National Marine Sanctuary here, which recognises the importance of protecting and sustaining such vulnerable environments (on which Palau's important tourist industry depends). This covers an area of 500,000km², some 80 per cent of Palau's Exclusive Economic Zone (EEZ), and in which no fishing, mining or extractive industries are allowed. It is acknowledged as one of the world's most significant conservation measures to protect and sustain strategic marine resources. The remaining 20 per cent of Palau's EEZ is categorised as a Domestic Fishing Zone, which accommodates the fishing rights of the indigenous population and, for some, tourist interests. Such an integrated and long-term vision for managing their valuable marine resources gives strong credibility to the government's use of 'Pristine Palau' to promote sustainable tourism. [Source: Charly W. Karl, flickr, CC BY-ND 4.0]

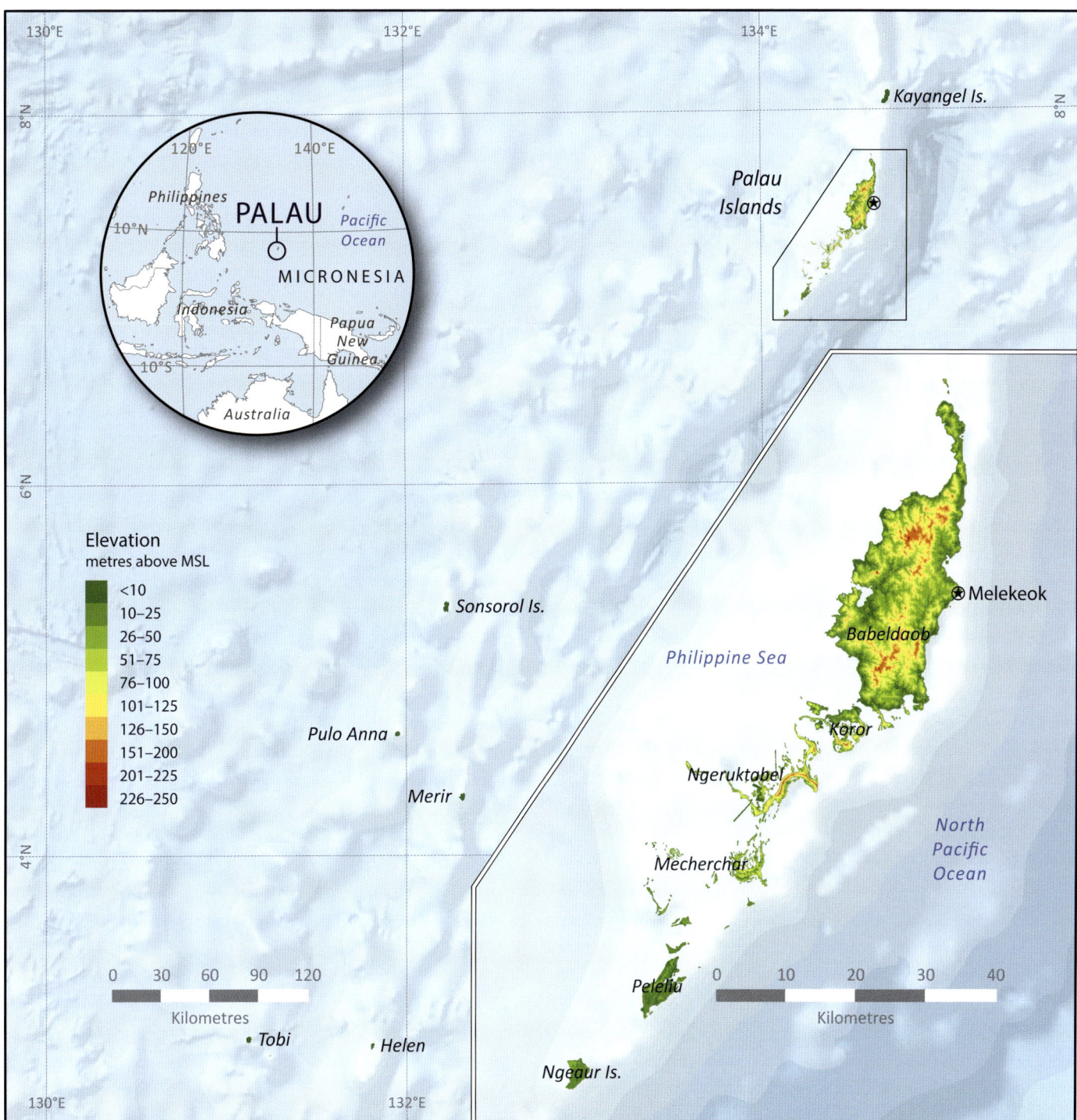

Fig. 33.19 PACIFIC OCEAN ISLANDS STATE OF PALAU. Palau is a relatively remote Small Island Developing State (SIDS), located in the western North Pacific. It forms an archipelago of over 500 islands, but only twelve of these mainly low-elevation coral atolls are inhabited, and are home to some 20,000 people. Despite its small land area, Palau controls a marine area of almost 0.7 million km². Although 'a world apart' from the island of Ireland (6.6 million population and surface area of 85,000 km²), Palau has much to offer in terms of how Ireland should manage its 0.88 million km² of territorial waters, and sustain viable coastal communities. Palau has emerged to take a leadership role in understanding the critical relationships that exist between the vitality of its coastal population and the need to conserve resources in surrounding marine areas. Currently, the United Nations estimates that only 7 per cent of the world's ocean area is protected, whereas a growing consensus in scientific bodies suggests this should be 30 per cent by 2030. Given the growing recognition of the value of its marine resource base, Ireland needs to support such conservation measures and become more proactive in applying sustainable development practices to its territorial waters. [Map elevation data source: NASA SRTM]

'high-end' ecotourists, with measures taken to deter large numbers of package tourists (for example, from China) from visiting and putting too much pressure on natural resources. It is expected that the short-term economic impact of a decline in tourist numbers resulting from this policy, from a peak of *c*.160,000 visitors in 2015, will be offset by longer-term returns from investment in ecotourism infrastructure, products and services. Reef fishery also provides for subsistence fishing among the local population, although the reefs are coming under increasing pressure from overfishing, exacerbated by seafood consumption within the tourist sector. Offshore, the domestic pelagic fisheries, including tuna, have been targeted by international fleets, facilitated under bilateral fisheries agreements with donor countries, such as the United States, China and Japan.

On the surface, there may be little in common between Ireland and Palau, although Ireland does share the challenge of developing sustainably a vast ocean territory and there is also much to be learnt from Palau's experience. Palau occupies a land area of only 500km², although its marine jurisdiction extends to almost 700,000km². In Ireland, the Republic's marine territory also extends far beyond its coastline and includes 880,000km² of ocean estate. A key difference is the scale of Palau's ambition when it comes to protecting the ocean resources upon which it depends. This is reflected in the nation being promoted as a Large Ocean State, rather than an SID.

Palau National Marine Sanctuary, which was created in 2020 incorporating areas such as Rock Islands, is one of the world's most significant ocean conservation initiatives. The work of the sanctuary is aimed at protecting the state's marine resources, as well as addressing the international issue of tuna stocks in its waters. Landmark legislation has provided for a no-take marine sanctuary, covering approximately 500,000km², and equating to *c.*80 per cent of Palau's Exclusive Economic Zone (EEZ). No fishing, mining or extractive activities will be permitted in this zone. It also creates a Domestic Fishing Zone (DFZ), covering approximately 20 per cent of the EEZ, where traditional and domestic fishing activities will be permitted.

This radical policy development is underpinned by three factors. The first is political leadership, which has endured local opposition and pressure from vested interests of donor countries. The second relates to norms and values at the heart of Palauan culture, which is deeply connected to the ocean and which appreciates the importance of effective ocean stewardship. The third concerns the role of science in developing meaningful policy outcomes. Palau has limited institutional capacity of its own, but has harnessed support from international scientific experts in conducting research on the impact of the sanctuary on food and climate security, and on marine resource sustainability.[37]

A high degree of uncertainty exists, however, regarding the impact that the marine sanctuary will have on the development of a domestic pelagic fishery, food security and societal wellbeing. Scientists are working on models to provide further insights into the implications of climate change, as well as developing capability to assess the responsiveness of integrated social-ecological systems to forces of change and disruption. Yet, the marine sanctuary policy has been progressed, without waiting for gaps in data and knowledge to be addressed. This is because of the perceived urgency of the need to protect Palau's vulnerable marine ecosystems and, in doing so, realise benefits for nature (a thriving ecosystem), as well as society (for example, population health) and the economy (for example, tourism and protecting 'Pristine Palau').

These observations regarding marine policy in Palau have some direct relevance for Ireland. It would also be wrong to think that the commonalities between these two states are only their shared status as islands and the possession of considerable ocean resources. For example, the implications of climate change, which will have different manifestations for each, particularly given Ireland's relative resilience and geographic location in the temperate belt of the North Atlantic, does provide common ground in the areas of developing marine environmental resilience and building adaptive capacity. The governments of Ireland need to take the lead given by SIDS like Palau and act with greater urgency to mitigate against climate change impacts on the marine, such as shifts in migration patterns of key commercial fish stocks. Like Palau, it should be taking an holistic view of the dynamic relationships between the large natural, social and economic systems. This is a complex endeavour, characterised by uncertainty, and with many pathways leading towards sustainability, based inevitably on trade-offs. In Palau's case, the stakes around developing the marine sanctuary were high, as it risked jeopardising finance from key donor countries in the region.

In Ireland, the analogous situation and question is whether we are prepared to set an agenda for the development of our maritime economies that remains firmly underpinned by adequate protection of the resources upon which that development is based, as opposed to swayed by political and economic experiences. The concept and process of Marine Spatial Planning – currently in its infancy in the Republic, but which has been discussed since the 2000s – provides an opportunity for the control and designation of areas for future use, including Marine Protected Areas (MPAs) (see Chapter 32: Coastal Management and Planning). The track record to date of the Republic in this regard – as the principal offshore resource stakeholder – has been, regrettably, poor. Policy progress may require consideration of complex co-location type opportunities. One potential solution is the co-location of MPAs together with offshore wind farms. These wind farms have the potential to act as de facto MPAs, since the use of commercial fishing gear may be constrained in the confined space between turbines. Such a co-location option has the potential to offer a win-win situation for all involved, if the sites can meet the mutual objectives of enhancing fish stocks for fisheries alongside nature conservation and socio-economic needs. Engagement with all stakeholders, including fishing communities, in planning for the optimisation of our marine space, will be critical to unlocking this potential. As in Palau, the stakes are high, the situation is urgent, and we have everything to win or lose.

THE GLOBAL ENERGY TRANSITION

The global energy transition describes a process through which nations, the corporate world and civilian society aim to transform patterns of energy production and consumption from fossil-based (non renewable resource) to zero carbon (renewable source), by a target date of 2050. The prime motivation is the need to decarbonise economies in response to climate change, and is centred around growing electrification, the expansion of the renewables sector and the globalisation of the natural gas market. In power markets, renewables have become the technology of

choice and will make up almost two-thirds of additions to global capacity by 2040, thanks to falling costs and supportive government policies. This is transforming the global power mix, with the share of renewables in generation expected to rise to over 40 per cent by 2040, from 25 per cent today, even though coal remains the largest source and gas the second-largest. Ireland, as an island, has a small developed offshore oil and gas industry and possesses some resource potential, dependent on economic market pricing, for future expansion in production. This could generate concomitant spin-offs to support and service industries located primarily in and

Fig. 33.20 BLIGH BANK 2, SOUTHERN NORTH SEA, BELGIUM, SHOWING AN OFFSHORE WIND TURBINE SUBSTATION AND PLATFORM WITH TURBINE ARRAYS IN THE AREA. Bligh Bank 2 is the second phase of a major wind park located 46km offshore from Zeebrugge, Belgium. Following installation of fifty-five wind turbine units under Phase 1, Phase 2 (2016–2017) saw development of an additional fifty turbines capable of generating 165MW of power. Undersea cables transfer the power to an onshore high voltage station for distribution through the national grid. About 10 per cent of Belgium's power supply is met by renewable wind, a third of which is from offshore developments. Ireland needs to progress in decarbonising its economy, reducing dependency levels on imported energy (2019: approximately 66 per cent) to meet its EU targets of increasing the use of renewable energy. To meet these goals Ireland will need to develop offshore wind farms, such as Bligh Bank 2, as these become technically and economically feasible. This is especially the case given the Republic's current failure to meet its EU targets (for example, 11 per cent compared to the 2020 target of 16 per cent for renewables in gross energy consumption), and a stated aim of the government to produce at least 70 per cent of electricity by renewables by 2030. Given the regular wind regime over the island, expansion of wind farms is a logical option, with considerable potential for offshore developments along the east coast, where waters are relatively shallow and high demand for electricity exists along the densely populated/urbanised coast. The west coast has considerable potential for floating offshore wind energy but remains problematic. [Source: Askjell Nicolas Raudøy, CC BY 4.0]

around Ireland's major ports and other coastal facilities.[38]

International progress towards increasing use of renewable energy sources has been incrementally slow since the 1990s, leading to the criticism of this being an inadequate response to environmental issues, such as those associated with pollution and climate change. As a small, open economy, the Republic is particularly vulnerable to global shifts in energy prices. Even though heavily dependent on imported fossil fuels, in 2019, and in line with its national climate action policy, the government decided to halt the development of offshore exploration for hydrocarbons. Domestic supplies of natural gas are finite, although at peak production the Corrib Gas Field supplies 60 per cent of the Republic's gas needs. This field is expected to supply gas for up to twenty years. The decommissioning of the Kinsale gas platforms, announced in early 2020, marks another milestone in the country's energy profile. Should proposals for the Liquified Natural Gas (LNG) terminal and Floating Production Storage and Offloading (FPSO) system, in Shannon and Cork Harbour respectively, meet a lack of political support, then the shift to renewables, including marine renewables, may be accelerated (see Chapter 29: Coastal Mining, Quarrying and Hydrocarbon Exploration).

In contrast, the benefits to Ireland's coastal communities associated with an energy transition to renewables have been identified by the EirWind research project, in the context of offshore wind (Figure 33.20). At a strategic level, development of this resource can facilitate the transition and strengthen security of energy supply. Furthermore, economic model outputs suggest that by 2030, 2.5–4.5GW of domestic offshore wind development could create between 11,424 and 20,563 supply chain jobs and generate between €763 million and €1.4 billion in gross value added. A demographic assessment of port areas with potential capabilities in the offshore wind supply chain, shows that east-, south- and west-coast ports could become hubs for employment. This provides opportunities to address the whole island's regional economic imbalances and issues such as rural depopulation (see Chapter 28: Renewable Energies: Wind, wave and tidal power).

Over a period of fifteen years, the Republic changed from having virtually no wind energy production, to number one in Europe for onshore wind. The question is whether the right ingredients (for example, economic, community, governance factors) are now aligned for the model to be repeated offshore? The International Energy Association (IEA) stated (2018) that, 'Over 70 per cent of global energy investments will be government driven and as such the message is clear: the world's energy destiny lies with decisions and policies made by governments.' If Ireland, the Republic and Northern Ireland together, is serious about meeting its climate targets, systematic preference for investment in marine and on-land renewable energy technologies is required. The country also needs to be much smarter about the ways in which the existing energy system is used, for example, by expanding the use of carbon capture and storage (CCS), green hydrogen production, improved energy efficiency and retiring capital stock, such as the planned decommissioning of the coal-fired Moneypoint power station, as early as possible.[39]

OVERCOMING MANAGEMENT CHALLENGES

Effecting change in attitudes in the governance and management of coasts requires a combination of both top-down and bottom-up approaches. Undoubtedly – both in the Republic and in Northern Ireland, despite many challenges and areas of uncertainty – political leadership increased the emphasis on these areas for maritime Ireland in the decade 2010–2020. As a consequence, there has been a seed change in the level of awareness of Ireland's marine opportunities. The publication of *Harnessing Our Ocean Wealth: An integrated marine plan for Ireland* (2012) was a key turning point in the Republic, as were studies such as the Strategic Energy Framework (2010) in Northern Ireland, both having been updated subsequently. This latter document sets out a long-term vision for energy policy in Northern Ireland until 2050, based on progressively increasing use of renewable resources, with wind playing a significantly large role.[40]

These forward-thinking policies adopted by government are now, only a decade later, under challenge. As the United Kingdom transitions into the post-Brexit era, the multiple implications for the island of Ireland, with regard to marine and coastal policies, is apparent and remains subject to many unknowns. The Northern Ireland Marine Task Force (NIMTF) pointed out, that in the exit of the UK from the EU, 'It is imperative that a common framework with regulatory alignment in marine management legislation is ensured in order to avoid a race to the bottom' or, a moving to the lowest common denominator in the management of these environments (see Chapter 32: Coastal Management and Planning).[41]

Investment in critical research and development infrastructure, such as the Marine Institute (MI) in Galway and MaREI in Cork, has provided a building block for knowledge generation on the island. It is clear that Ireland has advanced hugely in this area, compared to accounts of the neglected state of the marine sciences in the past.[42] Today's investment in scientific research needs to be harnessed to drive the commercialisation of research and new marine-based enterprise. However, this should not be at the cost of investment in traditional marine and coastal activities, such as fishing and marine recreation. These remain the mainstay of most rural coastal communities and can be a powerful force when organised through collective action. Under different actions for climate change in the Republic, such as the setting up of regional climate councils, government has given support to the development of local to regional-level community groups. These are to work practically on finding ways to adapt to the increasing impacts of

climate change and varying coastal risks (for example, those from erosion and marine flooding), and to contribute to the bottom-up development of governance (see Chapter 30: Engineering for Vulnerable Coastlines, Figures 30.22 and 30.23). Furthermore, stakeholder awareness and engagement can change attitudes and behaviours. Initiatives such as the Fisheries Local Action Groups (FLAGs) have been shown to have significant impact at the local level, through initiatives such as ecotourism and investment in local port and harbour facilities.[43]

Where principles of good governance have been neglected in the past, things have gone badly wrong. The breakdown of trust between government, industry and civil society was perhaps more pronounced in the Corrib gas project, County Mayo, than in any other controversy to affect communities around the coast (see Chapter 29: Coastal Mining, Quarrying and Hydrocarbon Exploration). Other notable examples include the applications for foreshore licences for the SmartBay marine instrumentation project in Galway and plans for seaweed harvesting in Bantry Bay, County Cork (see Chapter 26: Coastal Fisheries and Aquaculture). It is likely that as long as there are development proposals, there will be conflicts between competing interests, especially as demands intensify over access to, and the use of, coastal spaces and resources. However, unlike past instances, instituting and prioritising effective coastal planning and management can be achieved. By putting coastal issues at the heart of a national dialogue, and uniting around a vision for the future, coastal stakeholders have reasons to be hopeful.

Moving beyond immediate marine-related stakeholders, there is a need to inculcate a better awareness and appreciation of the importance of the marine and coastal environment in Irish society in general. Grassroots movements, such as the school strike over climate, have proven to be effective at triggering national dialogue about the future we want, and that our children deserve. Greta Thunberg has played a catalysing role, illustrating the importance of bottom-up approaches in influencing the future of policy. Deep concerns arising from the impacts of plastics and microplastics in marine environments are also motivating a groundswell of concerned citizens, including younger generations, towards active marine stewardship.

However, the way forward for coastal Ireland is not without challenges. In the Republic, successive governments have failed to prioritise Ireland's marine and coastal resources as strategic national assets. As a result, the starting point for better coastal governance stems from a low base, with research highlighting how bureaucracy, leadership and power dynamics negatively influenced coastal management practices in Ireland over preceding decades. Things have, however, progressed from the time when a marine ministerial post (combined with defence) was likened with having responsibility for the 'Department of Fish 'n Ships'. In recent years, the importance of oversight in marine issues in the Republic of Ireland was evident in it being elevated to the Department of An Taoiseach, which has facilitated better working of the important interdepartmental Marine Co-ordination Committee. This work has been added to more recently by the work role of the Climate

Change Advisory Council's Adaptation Committee, in highlighting policy changes required to meet coastal needs. Enabling mechanisms such as these have encouraged the move in government to better horizontal integration and joined-up thinking.[44]

Ireland's physical coastal systems will still be there (even in extreme scenarios) as climate changes. The soft coasts of dunes and beaches will just not be in the same places. There will be a lot more waterlands and flooding issues, unless dealt with proactively; flooding issues especially should be supported by sensitive approaches in coastal engineering. Biologically, coastal environments will also be different. New ecologies are developing, with new plant and animal species continuing to migrate to Ireland's coasts, with consequent serial shifts in plant communities developing and new diseases occurring. These will form some of the system responses as warmer and wetter/drier conditions develop locally to regionally under climate change. This means people in Ireland need to plan for the coasts of the whole island. The sea knows no boundaries! Coastal scientists, recognising the reality of joined-up coastal behaviour, have long advocated and promoted a single, statutory, unitary coastal authority for the island. Politics and inertia in the established governance systems have, so far, demurred, but the climate crisis is perhaps the time to make a change and a difference.

A VISION FOR IRELAND'S COASTAL FUTURE

Ireland's coasts will be places where people can continue the long-established traditions of deriving benefits from a healthy, thriving, secure and productive environment. We have reasons for hope in the realisation of this vision (Table 33.1), but this will involve a major reversal of the current unsustainable trajectory in people's uses of the environment and its development, both in Ireland and elsewhere (Figure 33.21). For the island's marine and coastal zones, this can be achieved by implementing nationally and internationally agreed sustainable development goals and targets.

At the time of writing (June 2020), with awareness of the scale of the climate, biodiversity and now COVID-19 coronavirus pandemic crises, and their attendant economic and political uncertainties, it is impossible to know what the 'new normal' world will look like. However, shaping our vision for Ireland's coasts can be a first, and a well-understood and planned, step in establishing a sustainable future. Society, particularly in Europe and the western world, has entered a point in history where there is a recognised need and, perhaps, a once-in-a-generation opportunity to re-set national and international scale economies. For Ireland, the island's coastal and marine resources should be central to deliberations on the Green Deal proposed by the European Commission, set up to help chart a course into this future. This policy has been promoted in Europe as an important contribution in overcoming climate change and environmental degradation. The United Kingdom's exit from the EU may add complications to the challenge of achieving cross-border compatibility in marine policy; this is part of the territory of uncertainty! Major opportunities do exist, however, to stimulate Ireland's joint economies, for example, through government investment in critical infrastructures, such as support for the development of offshore wind farms as part of the energy transition to renewable sources. Initiatives, such as plans to establish Marine Protected Areas in Irish territorial waters and to introduce policies aimed at habitat restoration, should deliver win-win situations; MPAs can enhance fish stocks, while habitat restoration projects contribute to biodiversity and blue carbon objectives.

The uncertainties of issues of economics, politics and policy planning are perplexing. The question, however, of what our future coasts will look like is clearer – they will be different! Under the impacts of climate change the island's coastlines will not be in the same place and some will look different. Many beaches and dunes, in particular, will have moved further onto the land, while some coastal territory will have been yielded to the sea. The plants and animals of Ireland's coast and marine zones will also be different, sometimes radically so, with a changed complex of species. People will have had to adapt and adjust to these changes of the coast, sometimes having to adopt very different living spaces. The question of 'at what cost?' remains. And what will people be doing on Ireland's shores in the future? Well, what they have always done! Coasts have and will always attract people; they give life to people, solace and encouragement. The coastal system will survive change. It is hardwired into its very fabric. We can, and must, manage coasts to better advantage for all, for people and the other species that share this precious space with us.

Table 33.1 REASONS FOR HOPE

1. We have the benefit of new knowledge, gained from investment in marine research, technology, development and innovation over the last twenty-five years.
2. There has been a seed change in our appreciation and understanding of the value of Ireland's coastal resources.
3. We understand the threats to our coastal zone and we know of solutions that can orientate us towards a more sustainable future.
4. We know that the choices we make today will provide us with a lifeline for tomorrow, and that we must stay within the thresholds of the planetary boundaries.
5. We can embrace new economic opportunities from green technology, marine renewable energy and the circular economy
6. Young people are motivated to fight for climate and sustainability issues, and understand the importance of protecting the coast and oceans.
7. We have evidence that nature-based solutions can enhance coastal resilience; such as the restoration of critical coastal habitats and also, allowing, where possible, coasts to naturally adjust their positions under SLR.
8. We have an opportunity to strengthen social cohesion through new coastal partnerships, and governance structures, that can learn from previous efforts in coastal zone management at home and abroad.
9. It is not too late in the coming decades to make a significant difference.

Fig. 33.21 THE MARCH FOR CLIMATE ACTION (MARCH 2019), WHICH WENT FROM STEPHEN'S GREEN TO THE DÁIL, DUBLIN. Protests such as this indicate that society and government are under notice from the future generations of Ireland that 'business as usual', in the way that we live in and manage the environment, needs to change. It is clear across society as a whole, but more especially in younger age groups, that concerns over the degradation of the Earth's environmental qualities have increased to, perhaps, a 'tipping point'. Changes in development strategies and lifestyles are being emphasised and a sustainable way forward is now recognised as being both desired and required, to protect the future of planet Earth. [Photo: Nick Bradshaw, for *The Irish Times*]

ENDNOTES

SECTION 1: THE PHYSICAL, BIOLOGICAL AND HUMAN SETTINGS

Ireland's Coasts: Setting the scene

1. Milliman, J. and B.U. Haq, 1996. *Sea-level Rise and Coastal Subsidence: Causes, consequences and strategies*. Netherlands: Springer; Pilkey, O.H., Neal, W.J., Kelley, J.T. and Cooper, J.A.G., 2011. *The World's Beaches*. Berkeley, University of California Press; Goudie, A., ed. 2013. *Encyclopedia of Geomorphology*. London. Routledge; Finkl, C.W. and Makowski, C., eds. 2019. *Encyclopedia of Coastal Science*. Netherlands: Springer.

The Coastal Environment: Physical systems, processes and patterns

1. Carter, R.W.G., 1988. *Coastal Environments: An Introduction to the physical, ecological and cultural systems of coastlines*. London: Academic Press; Woodroffe, C.D., 2002. *Coasts: Form, process and evolution*. Cambridge: Cambridge University Press.
2. Open University, 1989. *The Ocean Basins: Their structure and evolution*. Milton Keynes: OUP and Oxford: Pergamon Press.
3. Mienert, J. and Weaver, P., eds. 2003. *European Margin Sediment Dynamics: Side-scan sonar and seismic images*. Berlin, Heidelberg, New York: Springer-Verlag; Dorschel, B., Wheeler, A.J., Monteys, X. and Verbruggen, K., 2010. *Atlas of the Deep-water Seabed: Ireland*. Springer Science and Business Media.
4. Murphy, P. and Wheeler, A.J., 2017. A GIS-based Application of Drainage Basin Analysis and Geomorphometry in the Submarine Environment: The Gollum Canyon system, north-east Atlantic. In: Bartlett, D. and Celliers, L., eds. *Geoinformatics for Marine and Coastal Management*. Group Boca Raton, FL: Taylor and Francis, pp. 43–72.
5. Devoy, R.J.N., 1985. Holocene Sea-level Changes and Coastal Processes on the South Coast of Ireland: Corals and the problems of sea-level methodology in temperate waters. Proceedings of the Fifth International Coral Reef Congress, Tahiti, Vol. 3, 173–178; Roberts, J.M., Wheeler, A.J., Freiwald, A. and Cairns, S., 2009. *Cold-water Corals: The biology and geology of deep-sea coral habitats*. Cambridge: Cambridge University Press; Lim, A., Wheeler, A.J., Price, D.M., O'Reilly, L., Harris, K. and Conti, L., 2020. Influence of Benthic Currents on Cold-water Coral Habitats: A combined benthic monitoring and 3D photogrammetric investigation. *Nature Research: Scientific Reports*, 10:19433 [online] Available at: https://doi.org/10.1038/s41598-020-76446-y
6. Wheeler, A.J., Beyer, A., Freiwald, A., De Haas, H., Huvenne, V.A.I., Kozachenko, M., Olu-Le Roy, K. and Opderbecke, J., 2007. Morphology and Environment of Cold-water Coral Carbonate Mounds on the NW European Margin. *International Journal of Earth Sciences*, 96, pp. 37–56; Thierens, M., Pirlet, H., Colin, C., Latruwe, K., Vanhaecke, F., Lee, J., Stuut, J.-B., Titschack, J., Huvenne, V., Dorschel, B., Wheeler, A.J., Henriet, J. P., 2012. Ice-rafting from the British-Irish Ice Sheet since the Earliest Pleistocene (2.6 million years ago): Implications for long-term mid-latitudinal ice-sheet growth in the North Atlantic region. *Quaternary Science Reviews*, 44, pp. 229–240.
7. Cooper, J.A.G., Kelley, J.T., Belknap, D.F., Quinn, R. and McKenna, J., 2002. Inner Shelf Seismic Stratigraphy off the North Coast of Northern Ireland: New data on the depth of the Holocene lowstand. *Marine Geology*, 186, pp. 369–387; Peters, J.L., Benetti, S., Dunlop, P., Cofaigh, C.O., Moreton, S.G., Wheeler, A.J. and Clark, C.D., 2016. Sedimentology and Chronology of the Advance and Retreat of the Last British-Irish Ice Sheet on the Continental Shelf West of Ireland. *Quaternary Science Reviews*, 140, pp. 101–124.
8: Gallagher, C., 2002. The Morphology and Palaeohydrology of a Submerged Glaciofluvial Channel Emerging from Waterford Harbour on the Nearshore Continental Shelf of the Celtic Sea. *Irish Geography*, 35, pp. 111–132; Van Landeghem, K.J.J., Wheeler, A.J., Mitchell, N.C. and Sutton, G., 2009. Variations in Sediment Wave Dimensions Across the Tidally Dominated Irish Sea, NW Europe. *Marine Geology*, 263, pp. 108–119; Coughlan, M., Wheeler, A.J., Dorschel, B., Long, M., Doherty, P. and Mörz, T., 2019. Stratigraphical Model of the Quaternary Sediments of the Western Irish Sea Mud Belt from Core, Geotechnical and Acoustic Data. *Geomarine Letters*, 39, pp. 223–237.
9. Davies, G.L.H and Stephens, N., 1978. *Ireland*. London: Methuen; Lalor, B., 2003. *The Encyclopedia of Ireland*. New Haven, CT: Yale University Press; Devoy, R.J.N., 2009. Iveragh's Coasts and Mountains. In: Crowley, J. and Sheehan, J., eds. *The Iveragh Peninsula: A cultural atlas of the Ring of Kerry*. Cork: Cork University Press.
10. Hardisty, J., 2007. *Estuaries: Monitoring and modelling the physical system*. Oxford: Blackwell.

11. Von Bertalanffy, L. 1968, *General Systems Theory: Foundations, development, applications*. New York: Brazilier, p. 2.
12. Delaney, C.A., Devoy, R.J.N. and Jennings, S.A., 2012. Mid- to Late Holocene Relative Sea-level and Sedimentary Changes on European Atlantic Coasts: Evidence from Castlemaine Harbour, southwest Ireland. In: Duffy, P.J., Butler, D. and Nugent, P., eds. *Festschrift for Professor William J. Smyth*. Dublin: Geography Publications, pp. 697–746; Devoy, R.J.N., 2015. The Development and Management of the Dingle Bay Spit-barriers of Southwest Ireland. In: Randazzo, G., Cooper, J.A.G. and Jackson, D.W., eds. *Sand and Gravel Spits*. Coastal Research Library 12. Springer International Publishing, pp. 139–180. DOI 10.1007/978-3-319-13716-2_9.
13. Rohan, P.K., 1989. *The Climate of Ireland*, 2nd edition, Dublin: Meteorological Service; Dwyer, N., ed. 2012. *The Status of Ireland's Climate*. Dublin: EPA.
14. Met Éireann, 2020. Climate of Ireland. [online] Available at: https://www.met.ie/climate/climate-of-ireland.
15. Lozano, I., Devoy, R.J.N., May, W. and Andersen, U., 2004. Storminess and Vulnerability along the Atlantic Coastlines of Europe: Analysis of storm records and of a greenhouse gases induced climate scenario. *Marine Geology*, 210, pp. 205–225; Hickey, K.R., 2011. The Hourly Gale Record from Valentia Observatory, SW Ireland 1874–2008 and Some Observations on Extreme Wave Heights in the NE Atlantic. *Climatic Change*, 106, pp. 483–506.
16. Hickey, K., 2011. The Impact of Hurricanes on the Weather of Western Europe. In Lupo, A. ed. Recent Hurricane Research - Climate, Dynamics, and Societal Impacts. ISBN: 978-953-307-238-8, InTech [online] Available from: http://www.intechopen.com/books/recent-hurricane-research-climate-dynamics-andsocietal-impacts/the-impact-of-hurricanes-on-the-weather-of-west ern-europe.; Matthews, T., Murphy C., Wilby, R.L. and Harrigan, S., 2014. Stormiest Winter on Record for Ireland. *Nature Climate Change*, 4, pp. 738–740.
17. Gleeson, E., McGrath, R. and Treanor, M., 2013. *Ireland's Climate: The road ahead*. Dublin: Met Éireann; Nolan, P., 2015. *Ensemble of Regional Climate Model Projections for Ireland*. EPA Report Number 159. Dublin: EPA.
18. Tyrrell, J., 2003. A Tornado Climatology for Ireland. *Atmospheric Research*. 67–68, pp. 671–684; Met Éireann, 2018. *An Analysis of Storm Ophelia which Struck Ireland on the 16th October 2017*. Dublin: Met Éireann.
19. Carr, P., 1993. *The Night of the Big Wind*, 2nd ed. Belfast: White Row Press; Hickey, K.R., 2003. The Night of the Big Wind: The impact of the 1839 storm on Loughrea. In: Forde, J. et al., eds. *The District of Loughrea: Volume 1: History 1791–1918*. Loughrea: Loughrea History Project.
20. Carter, 1998. *Coastal Environments*; Haslett, S.K., 2009. *Coastal Systems*, 2nd ed. London: Routledge.
21. Orford, J.D., 1989. A Review of Tides, Currents and Waves in the Irish Sea. In: Sweeney, J. ed. 1989. *The Irish Sea: A resource at risk*6. Special Publication No. 3. Dublin: Geographical Society of Ireland; , pp. 18–4. Stone, G.W. and Orford, J.D., 2004. Storms and their Significance in Coastal Morpho-sedimentary Dynamics. *Marine Geology*, 210, pp. 1–5.
22. Pugh, D., 2004. *Changing Sea Levels: Effects of tides, weather and climate*. Cambridge: Cambridge University Press; Haigh, I.D., 2017. Tides and Water Levels. In: *Encyclopedia of Maritime and Offshore Engineering*. Wiley Online Library [online] Available at https://doi.org/10.1002/9781118476406.emoe122.
23. Finkl, C.W., 2004. Coastal Classification: Systematic approaches to consider in the development of a comprehensive scheme. *Journal of Coastal Research*, 20, pp. 166–213; Valiente, N.G., Masselink, G., Scott, T., Conley, D. and McCarroll, J.K., 2019. Role of Waves and Tides on Depth of Closure and Potential for Headland Bypassing. *Marine Geology*, 407, pp. 60–75.
24. Cowell, P.J. and Thom, B.G., 1994. Morphodynamics of Coastal Evolution. In: Carter, R.W.G and Woodroffe, C.D., eds. *Coastal Evolution: Late Quaternary Shoreline Dynamics*. Cambridge: Cambridge University Press. pp. 33–86 ; Slaymaker, O., Spencer, T. and Embleton-Hamann, C., 2009. *Geomorphology and Global Environmental Change*. Cambridge: Cambridge University Press.
25. Devoy, R.J.N., Delaney, C., Carter, R.W.G. and Jennings, S.C., 1996. Coastal Stratigraphies as Indicators of Environmental Changes upon European Atlantic Coasts in the Late Holocene. *Journal of Coastal Research*, 12 (3), pp. 564–588; Cooper, J.A.G., 2006. Geomorphology of Irish Estuaries: Inherited and dynamic controls. *Journal of Coastal Research*, Special Issue No. 39. Proceedings of the 8th International Coastal Symposium (ICS 2004), Vol. I, pp. 176–180.

26. National Parks and Wildlife Service, 2013. The Status of EU Protected Habitats and Species in Ireland. Habitat Assessments Volumes 1–3. Version 1.1. Unpublished Report. Dublin: Department of Arts, Heritage and the Gaeltacht.
27. Cooper, J.A.G., Jackson, D.W.T., Navas, F., McKenna, J. and Malvarez, G., 2004. Identifying Storm Impacts on an Embayed, High-energy Coastline: Examples from western Ireland. *Marine Geology*, 210, pp. 261–280; Williams, J., Esteves, L. and Rochford, L., 2015. Modelling Storm Responses on a High-energy Coastline with XBeach. Springer International Publishing.
28. O'Brien, L., Renzi, E., Dudley, J.M., Clancy, C. and Dias, F., 2018. Catalogue of Extreme Wave Events in Ireland: Revised and updated for 14 680 BP to 2017. *Nat. Hazards Earth Syst. Sci.*, 18, pp. 729–758; Cronin, A., Devoy, R.J.N., Bartlett, D., Nuyts, S. and O'Dwyer, B., 2018. Investigation of an Elevated Sands Unit at Tralispean Bay, South West Ireland: Potential high-energy marine event. *Irish Geography*, 51, pp. 229–260.
29. Farrell, E.J., Ellis, J.T. and Hickey, K.R., 2015. Tsunami Case Studies. In: Ellis, J.T. and Sherman, D.J., eds. *Coastal and Marine Disasters*. Amsterdam: Elsevier, pp. 93–128; Beese, A.P., 2017. Cork's Earthquake of 1755: interpreted as a seismic seiche. Special Report 707/2;. Carraigex. Hickey, Kieran and Beese, Anthony. 2018. The 1755 Lisbon Earthquake-Tsunami and the West Cork Coast. [http://www.deepmapscork.ie/past-to-present/climate/1755-lisbon-earthquake-tsunami-west-cork-coast/].
30. Highwave Project, 2020. European Research Council. [online] Available at: https://www.highwave-project.eu/index.php
31. Carter, R.W.G. and Devoy, R.J.N., 1987. The Hydrodynamic and Sedimentary Consequences of Sea-level change. *Progress in Oceanography*, 18, pp. 1–358; Swift, L.J., Devoy, R.J.N., Wheeler, A.J., Sutton, G.D. and Gault, J., 2006. Sedimentary Dynamics and Coastal Changes on the South Coast of Ireland. *Journal of Coastal Research*, SI 39, pp. 234–239.
32. Duffy, M. and Devoy, R.J.N., 1999. Contemporary Process Controls on the Evolution of Sedimentary Coasts under Low to High-energy Regimes: Western Ireland. *Geologie en Mijnbouw*, 77, pp. 333–349; Coughlan, M.J.C., 2015. Quaternary Geology of the German North Sea and Western Irish Sea Mud Belt: Revised stratigraphies, geotechnical properties, sedimentology and anthropogenic impacts. PhD Thesis. University College Cork.
33. Farrell, E., Bourke, M., Lynch, K., Morley, T., O'Dwyer, B., O'Sullivan, J. and Turner, J., 2020. *From Source to Sink: Responses of a coastal catchment to large scale changes: Golden Strand Catchment, Achill Island, Co. Mayo*. Research Report 2014-CCRP-MS.22. Dublin: EPA.
34. Cooper, J.A.G., McKenna, J., Jackson, D.W.T. and O'Connor, M., 2007. Mesoscale Coastal Behavior Related to Morphological Self-adjustment. *Geology*, 35, pp. 187–190; Masselink, G., Castelle, B., Scott, T., Dodet, G., Suanez, S., Jackson, D. and Floc'h, F., 2016. Extreme Wave Activity During 2013/2014 Winter and Morphological Impacts along the Atlantic Coast of Europe. *Geophysical Research Letters*, 43, pp. 2135–2143.
35. Randazzo, G., Cooper, J.A.G. and Jackson, D.W., eds. 2015. *Sand and Gravel Spits*. Coastal Research Library 12. Switzerland: Springer International Publishing; McClatchey, J., Devoy, R.J.N., Woolf, D., Bremner, B. and James, N., 2014. Climate Change and Adaptation in the Coastal Areas of Europe's Northern Periphery Region. *Ocean and Coastal Management*, 94, pp. 9–21.
36. Connolly, N., Buchanan, C., O'Connell, M., Cronin, M., O'Mahony, C., Sealy, H., Kay, D. and Buckley, S., 2001. *Assessment of Human Activity in the Coastal Zone: A research project linking Ireland and Wales*. Maritime Ireland/Wales INTERREG Report No. 9. Dublin: Marine Institute; Healy, M. and Hickey, K.R., 2002. Historic Land Reclamation in the Intertidal Wetlands of the Shannon Estuary. In: Cooper, J.A.G. and Jackson D.W.T., eds. Proceedings from the International Coastal Symposium (ICS) 2002 (Templepatrick, Northern Ireland). *Journal of Coastal Research*, Special Issue No. 36, pp. 365–373.
37. Falaleeva, M., Gray, S., O'Mahony, C. and Gault, J., 2013. Coastal Climate Adaptation in Ireland: Assessing current conditions and enhancing the capacity for climate resilience in local coastal management. Climate Change Research Programme (CCRP) 2007–2013. Report Series No. 28. Dublin: EPA; Gray, S., 2020. Making Sense of Changing Coastal Systems: Overcoming barriers to climate change adaptation using fuzzy cognitive mapping. PhD Thesis. National University of Ireland at Cork (NUIC), Cork.

Coastal Waters: Marine biology and ecology

1. Bouchet, P., 2006. The Magnitude of Marine Biodiversity. In: Duarte, C., ed. *The Exploration of Marine Biodiversity: Scientific and technological challenges*. Fundación BBVA, pp. 31–62.
2. Clarke, M., Farrell, E.D., Roche, W., Murray, T.E., Foster, S. and Marnell, F., 2016. *Ireland Red List No. 11: Cartilaginous fish [sharks, skates, rays and chimaeras]*. Dublin: National Parks and Wildlife Service, Department of Arts, Heritage, Regional, Rural and Gaeltacht Affairs.
3. Pollock, C.M., Reid, J.B., Webb, A. and Tasker, M.L., 1997. The Distribution of Seabirds and Cetaceans in the Waters around Ireland. *JNCC*, pp. 267–167.
4. Burke, B., Lewis, L.A., Fitzgerald, N., Frost, T., Austin, G., Tierney, D., 2018. Estimates of Waterbird Numbers Wintering in Ireland, 2011/12–2015/16. *Irish Birds*, 11, pp. 1–12.
5. Lewis, L.J., Austin, G., Boland, H., Frost, T., Crowe, O. and Tierney, T.D., 2017. Waterbird Populations on Non-estuarine Coasts of Ireland: Results of the 2015/16 Non-Estuarine Coastal Waterbird Survey (NEWS-III). *Irish Birds*, 10, pp. 511–522.
6. Merne, O.J. and Robinson, J.A., 2012. Estuaries and Coastal Lagoons. In: Nairn, R. and O'Halloran, J.K., eds. *Bird Habitats in Ireland*. Cork: The Collins Press, pp. 166–180
7. Curtis, T.G.F. and Sheehy Skeffington, M.J., 1998. The Salt Marshes of Ireland: An inventory and account of their geographical variation. *Biology and Environment: Proceedings of the Royal Irish Academy*, 98, pp. 87–104.
8. Merne, O.J. and Robinson, J.A., 2012, Estuaries and Coastal Lagoons. In: Nairn, R. and O'Halloran, J.K., eds. *Bird Habitats in Ireland*.
9. Burke, B., et al. Estimates of Waterbird Numbers Wintering in Ireland, 2011/12–2015/16, pp. 1–12.
10. Hughes, R.G., 2004. Climate Change and Loss of Saltmarshes: Consequences for birds. *Ibis*, 146, pp. 21–28.
11. Crowe, O., Lewis, L.J. and Anthony, S., 2013. Impacts of Sea-level Rise on the Birds and Biodiversity of Key Coastal Wetlands. Unpublished report to the Environmental Protection Agency. Wicklow: BirdWatch Ireland.
12. Lewis, L.J., Burke, B., Fitzgerald, N., Tierney, T.D. and Kelly, S., 2019. Irish Wetland Bird Survey: Waterbird status and distributions 2009/10–2015/16. *Irish Wildlife Manuals*, No. 106. Dublin: National Parks and Wildlife Service, Department of Culture, Heritage and the Gaeltacht.
13. Ellen Hutchins to Dawson Turner. Letter, 4 July 1809, Trinity College Cambridge.
14. Ellen Hutchins to Dawson Turner. Letter, 22 September 1810, Trinity College Cambridge.
15. Jessopp, M., Cronin, M., Doyle, T.K., Wilson, M., McQuatters-Gollop, A., Newton, S., Phillips, R.A., 2013. Transatlantic Migration by Post-breeding Puffins: A strategy to exploit a temporarily abundant food resource? *Marine Biology*, 160, pp. 2755–2762.
16. Cronin, M., Pomeroy, P., Jessopp, M., 2013. Size and Seasonal Influences on the Foraging Range of Female Grey Seals in the Northeast Atlantic. *Marine Biology*, 160, pp. 531–539.
17. Doyle, T.K., Bennison, A., Jessopp, M., Haberlin, D., Harman, L.A., 2015. A Dawn Peak in the Occurrence of 'Knifing Behaviour' in Blue Sharks. *Animal Biotelemetry*, 3, pp. 1–6.
18. Doyle, T.K., Haberlin, D., Clohessy, J., Bennison, A., Jessopp, M., 2017. Localised Residency and Inter-annual Fidelity to Coastal Foraging Areas May Place Sea Bass at Risk to Local Depletion. *Nature Scientific Reports*. 8:45841.doi:10.1038/srep45841.
19. Dale, B. and Murphy, M., 2014. A Retrospective Appraisal of the Importance of High Resolution Sampling for Harmful Algal Blooms: Lessons learned from long-term phytoplankton monitoring at Sherkin Island, S.W. Ireland. *Harmful Algae*, 40, pp. 23–33.
20. Bishop, G., 2003. *The Ecology of the Rocky Shores of Sherkin Island: A twenty year perspective*. Sherkin: Sherkin Island Marine Station. p. 305.
21. Cronin, M., Gerritsen, H., Reid, D. and Jessopp, M., 2016. Spatial Overlap of Grey Seals and Fisheries in Irish Waters, Some New Insights Using Telemetry Technology and VMS. *PLOS ONE*, 11(9), e0160564.
22. Russell, D.J., Brasseur, S.M., Thompson, D., Hastie, G.D., Janik, V.M., Aarts, G., McClintock, B.T., Matthiopoulos, J., Moss, S.E. and McConnell. B., 2014. Marine Mammals Trace Anthropogenic Structures at Sea. *Current Biology*, 24: R638–R639; Bodey, T.W., Jessopp, M., Votier, S.C., Gerritsen, H.D., Cleasby, I.R., Hamer, K.C., Patrick, S.C., Wakefield, E.D and Bearhop. S., 2014. Seabird Movement Reveals the Ecological Footprint of Fishing Vessels. *Current Biology*, 24: R514-R515.
23. SEMRU, 2019. *Ireland's Ocean Economy*. Report by the Socio-Economic Marine Research Unit, National University of Ireland, p. 84.

People, Agriculture and the Coast

1. Crowley, C., Walsh, J.A. and Meredith, D., 2008. *Irish Farming at the Millennium: A census atlas*. Maynooth: NIRSA.
2. NUI Maynooth, UCD, Teagasc, 2005. *Rural Ireland 2025: Foresight perspectives*. n.p. COFORD for NUI Maynooth, UCD and Teagasc.
3. Davies, G.L.H and Stephens, N. 1978. *Ireland*. London: Methuen and Co. Ltd.
4. Mitchell, F. and Ryan, M. 1997. *Reading the Irish Landscape*. Dublin: Town House and Country House Publishers.
5. Malone, S. and O'Connell, C., 2009. *Irish Peatlands Conservation Council Action Plan 2020: Halting the loss of peatland biodiversity*. Rathangan: IPCC.
6. Stout, G. and Stout, M., 1997. Early Landscapes: From prehistory to plantation. In Aalen, F.H.A. Whelan, K. and Stout, M. eds. *Atlas of the Irish Rural Landscape*. Cork: Cork University Press, pp. 31–63.
7. Feehan, J., 2003. *Farming in Ireland: History, heritage and environment*. Dublin: Faculty of Agriculture, UCD.
8. Canny, N., 2001. *Making Ireland British, 1580–1650*. Oxford: Oxford University Press.
9. Margey, A., 2018. Plantations, 1550–1641. In: Ohlmeyer, J. and Bartlett, T., eds. *The Cambridge History of Ireland, Volume II. 1550–1730*. Cambridge: Cambridge University Press.
10. Freeman, T. W., 1957. *Pre-Famine Ireland: A study in historical geography*. Manchester: Manchester University Press.
11. A fuller description of Ireland's agricultural regions at this time is provided in Whelan, K., 1997. The Modern Landscape: From plantation to present. In: Allen, F.H.A., Whelan, K. and Stout, M., eds. *Atlas of Irish Rural Landscape*. Cork: Cork University Press, pp. 67–105.
12. Aldecoa, I., 1958. *Gran Sol*. Madrid: Editorial Vitoria; Barros, J., 2005. *Os ausentes de Casteltón*. Santiago de Compostela: Editorial Medusa/Sotelo Blanco; Souto, X., 2016. *Cuentos del mar de Irlanda, de Gaicia al Gran Sol*. Cangas de Marrazo: Pulp Books.
13. Ó Cathain, S. and O'Flanagan, P., 1975. *The Living Landscape, Kilgalligan, Erris, County Mayo*. Dublin: Comhairle Bhéaloideas Éireann.
14. Ó Mongain, S. and Ní Ghearraigh, T., 2011. *Dordán Dúlra*. Béal an Átha: Comhar Dún Chaocháin Teo.
15. Mac Graith, U. agus Nic Gearraigh, T., 2009. *Logainmneacha agus Oidhreach Dhún Chaocháin i mBurúntacht, Iorrais, Condae Mhaigh Eo: The placenames and heritage of Dún Chaocháin in the Barony of Erris, County Mayo*. Béal an Átha: Comhar Dún Chaocháin.
16. Ó Gráda, C., Paping, R. and Vanhaute, E., eds. 2007. *When the Potato Failed: Causes and effects of the last European subsistence crisis, 1845–1850*. Turnhout: Brepols; Crowley, J., Smyth, W.S. and Murphy, M., eds. 2012. *Atlas of the Great Irish Famine*. Cork: Cork University Press.
17. Crotty, R., 1966. *Irish Agricultural Production: Its volume and structure*. Cork: Cork University Press.
18. Crowley, J., Ó Drisceoil, D., Murphy, M. and Borgonovo, J., 2017. *Atlas of the Irish Revolution*. Cork: Cork University Press.
19. Gillmor, D., 1965. The Agricultural Regions of the Republic of Ireland. *Irish Geography* 5(2), pp. 78–86

20. Bartley, B. and Kitchen, R., eds. 2007. *Understanding Contemporary Ireland*. Dublin: Pluto Press.

21. A number of atlases have been published since Ireland's accession to the European Union in 1973, and provide excellent coverage of the evolution of patterns and processes of change that have reshaped the Republic's agricultural landscapes: Horner, A.A., Walsh, J.A., and Williams, J.A., 1984. *Agriculture in Ireland: A census atlas*. Dublin: University College Dublin; Lafferty, S., Commins, P. and Walsh, J.A., 1999. I*rish Agriculture in Transition: A census atlas of agriculture in the Republic of Ireland*. Ashtown: Teagasc; Crowley, C., Walsh, J.A. and Meredith, D., 2008. *Irish Farming at the Millennium: A census atlas*. Maynooth: NIRSA.

22. Meredith, D. and Crowley, C., 2017. Continuity and Change: the geo-demographic structure of Ireland's population of farmers. *Irish Geography* 50(2), pp 111–136.

23. DEFRA, 2019. *The Agricultural Census for Northern Ireland, Results for June 2018*. National Statistics, UK.

24. Dillon, E., Moran, B., Lennon, J. and Donnellan, T., 2018. *Teagasc National Farm Survey Results, 2018*. Athenry: Teagasc.

25. Davidson, A. 1980. *North Atlantic Seafood*. London: Penguin, p. 468.

26. Pers. comm. JP McMahon, 6 February 2019.

27. McKenna, M. 1997. Putting Food on the Table: The story of the Blasket Islands 1850–1950. Dublin Institute for Advanced Studies, School of Celtic Studies, unpublished manuscript, 1997. (The author would like to express her gratitude and thanks to Professor Malachy McKenna for granting her access to his unpublished manuscript on Blasket Island food).

28. McKenna, 1997, Putting Food on the Table, p. 1.

29. McKenna, 1997, Putting Food on the Table, pp. 87–90. Lobster fishing began on the island *c.*1880 and catches were sold to British and French boats. When the market was steady the islanders made very good returns. The lobster fishery declined towards the end of the 1930s, even though the Islanders continued to fish for them until the last of the community moved to the mainland in the 1950s.

30. National Folklore Collection, UCD. Baile an tSagairt. Duchás, The Schools' Collection, vol. 1039, p. 76.

31. McKenna, 1997, Putting Food on the Table, p. 58.

32. *The EU Fish Market: 2018 edition*. Brussels: Directorate-General for Maritime Affairs and Fisheries. EUMOFA (European Market Observatory for Fisheries and Aquaculture Products), 2018.

33. Miller, D. and Mariani, S., 2013. Irish Fish, Irish People: Roles and responsibilities for an emptying ocean. *Environment Development and Sustainability* 15, pp. 529–546.

34. Safefood, 2006. Barriers to consumption of fish highlighted in new report [online] Available at: https://www.safefood.eu/News/2006/Barriers-to-consumption-of-fish-highlighted-in-new.aspx and safefood 'Barriers to eating fish still remain'; Safefood, 2012. Barriers to eating fish still remain. Available at: https://www.safefood.eu/News/2012/Barriers-to-eating-fish-still-remain.aspx

Geological Foundations

1. Playfair, J, 1802. *Illustration of the Huttonian Theory*. Edinburgh: Cadell and Davies.

2. Kearney, M., 2019. *Around Ireland in the Footsteps of George Du Noyer*, [online] Available at: https://www.bbc.com/news/uk-northern-ireland-31902629; Crawford Art Gallery, 2017. *Stones, Slabs and Seascapes: George Victor Du Noyer's images of Ireland*. Cork: Crawford Art Gallery.

3. Meere, P.A., MacCarthy, I.A.J., Reavy, R.J., Allen, A. and Higgs, K.T., 2013. *Geology of Ireland: A field guide*. Cork: Collins Press.

4. Geological Survey Ireland, 2019. *Geology of Ireland*. Available [online] Available at: https://www.gsi.ie/en-ie/geoscience-topics/geology/Pages/Geology-of-Ireland.aspx

5. Williams, E.A., Sergeev, S.A., Stössel, I., Ford, M. and Higgs, K.T., 2000. U-Pb Zircon Geochronology of Silicic Tuffs and Chronostratigraphy of the Earliest Old Red Sandstone in the Munster Basin, SW Ireland. *Geological Society*, Special Publications, 180, pp. 269–302, 1.

6. Stössel, I., Williams, E.A. and Higgs, K.T., 2016. Ichnology and depositional environment of the Middle Devonian Valentia Island tetrapod trackways, south-west Ireland. *Palaeogeography, Palaeoclimatology, Palaeoecology*, 462. pp. 16–40.

7. National Research Council, 1994. *Environmental Science in the Coastal Zone: Issues for further research*. Washington, DC: The National Academies Press, p. 44; Masselink, G., Hughes, M.G. and Knight, J., 2011. *Introduction to Coastal Processes and Geomorphology*. 2nd ed. Abingdon: Hodder Education.

8. Emery, K.O. and Kuhn, G.G., 1982. Sea Cliffs: Their processes, profiles and classification. *Geological Society of America Bulletin*, 93, pp. 644–654; Bird, E., 2008. *Coastal Geomorphology*. Chichester: Wiley.

9. White, I.D., Mottershead, D.N. and Harrison, S.J., 1993. *Environmental Systems: An introductory text*. London: Chapman and Hall.

10. Meere, et al., 2013. *Geology of Ireland: A field guide* ; Higgs, K., 2009, Geology of the Iveragh. In: Sheehan, J. and Crowley, J., eds. *Atlas of the Iveragh Peninsula*. Cork: Cork University Press.

11. Quinn, D., Meere, P.A. and Wartho, J.A., 2005. A Chronology of Foreland Deformation: Ultra-violet laser Ar-40/Ar-39 dating of syn/late-orogenic intrusions from the Variscides of southwest Ireland. *Journal of Structural Geology*, 27, pp. 1,413–1,425.

12. Geological Survey Ireland, 2019. Geology of Ireland. [online]

13. Geological Survey Ireland, 2019. Protection of County Geological Sites. [online] Available at: https://www.gsi.ie/en-ie/programmes-and-projects/geoheritage/activities/protection-of-county-geological-sites/Pages/default.aspx

14. UNESCO. Giant's Causeway and Causeway Coast. [online] Available at: https://whc.unesco.org/en/list/369

Glaciation and Ireland's Arctic Inheritance

1. Dunlop, P., Shannon, R., McCabe, M., Quinn, R. and Doyle, E., 2010. Marine Geophysical Evidence for Ice Sheet Extension and Recession on the Malin Shelf: New evidence for the western limits of the British Irish Ice Sheet. *Marine Geology*, 276, pp. 1–4; O'Cofaigh, C., Dunlop, P. and Benetti, S., 2012. Marine Geophysical Evidence for Late Pleistocene Ice Sheet Extent and Recession off Northwest Ireland. *Quaternary Science Reviews*, 44, pp. 147–159; Benetti, S., Dunlop, P. and O Cofaigh, C., 2010. Glacial and Glacially-related Features on the Continental Margin of Northwest Ireland Mapped from Marine Geophysical Data. *Journal of Maps*, 6(1), pp. 14–29.

2. Nesje, A. and Dahl, S.O., 2000. *Glaciers and Environmental Change*. London: Arnold; McCabe, A.M. and Dunlop, P., 2006. *The Last Glacial Termination in Northern Ireland*. Geological Survey of Northern Ireland; Coxon, P. and McCarron, S., 2009. Cenozoic: Tertiary and Quaternary (until 11,700 years before 2000). In: Holland, C.H. and Sanders, I.S. eds. *The Geology of Ireland*. 2nd Edition. Edinburgh: Dunedin Academic Press, pp. 355–396.

3. Dunlop et al. 2010. Marine Geophysical Evidence for Ice Sheet Extension and Recession on the Malin Shelf: New evidence for the western limits of the British Irish Ice Sheet; O'Cofaigh et al., 2012. Marine Geophysical Evidence for Late Pleistocene Ice Sheet Extent and Recession off Northwest Ireland.

4. Dunlop, P. and Clark, C.D., 2006. The Morphological Characteristics of Ribbed Moraine. *Quaternary Science Reviews*, 25, issues 13–14, pp. 1668–1691.

5. Close, M.H., 1867. Notes on the General Glaciation of Ireland. *Journal of the Royal Geographical Society of Ireland*, 1, pp. 207–242.

6. Greenwood, S.L. and Clark, C.D, 2009. Reconstructing the Last Irish Ice Sheet 2: A geomorphologically-driven model of ice sheet growth, retreat and dynamics. *Quaternary Science Reviews*, 28, pp. 3101–3123.

7. Dunlop et al., 2010. Marine Geophysical Evidence for Ice Sheet Extension and Recession on the Malin Shelf: New evidence for the western limits of the British Irish Ice Sheet; O'Cofaigh et al, 2012. Marine Geophysical Evidence for Late Pleistocene Ice Sheet Extent and Recession off Northwest Ireland; Benetti et al., 2010. Glacial and Glacially-related Features on the Continental Margin of Northwest Ireland Mapped from Marine Geophysical Data.

8. Dunlop et al., 2010. Marine Geophysical Evidence for Ice Sheet Extension and Recession on the Malin Shelf: New evidence for the western limits of the British Irish Ice Sheet.

9. Benetti et al., 2010. Glacial and glacially-related features on the continental margin of northwest Ireland mapped from marine geophysical data; McCabe, A.M., 2007. *Glacial Geology and Geomorphology: The landscapes of Ireland*. Edinburgh: Dunedin Academic Press.

10. Dunlop et al., 2010. Marine Geophysical Evidence for Ice Sheet Extension and Recession on the Malin Shelf: New evidence for the western limits of the British Irish Ice Sheet; O'Cofaigh et al, 2012. Marine Geophysical Evidence for Late Pleistocene Ice Sheet Extent and Recession off Northwest Ireland; McCabe, 2007. *Glacial Geology and Geomorphology: The Landscapes of Ireland*.

11. O'Cofaigh et al, 2012. Marine Geophysical Evidence for late Pleistocene Ice Sheet extent and Recession off Northwest Ireland; Benetti et al., 2010. Glacial and Glacially-related Features on the Continental Margin of Northwest Ireland Mapped from Marine Geophysical Data.

12. Olex, n.d. Maritime mapping and navigation. [online] Available at: www.olex.no

13. Peters, J.L., Benetti, S., Dunlop, P., O'Cofaigh, C., Moreton, S.G., Wheeler, A.J. and Clark, C.D., 2016. Sedimentology and Chronology of the Advance and Retreat of the last British-Irish Ice Sheet on the Continental Shelf West of Ireland. *Quaternary Science Reviews*, 140, pp. 101–124.

14. Thébaudeau, B., Monteys, X., McCarron, S., O'Toole, R. and Caloca, S., 2015. Seabed Geomorphology of the Porcupine Bank, West of Ireland. *Journal of Maps*, 12: 5, pp. 947–958.

15. Macayeal, D.R., 1993. Binge/purge Oscillations of the Laurentide Ice Sheet as a Cause of the North Atlantic's Heinrich Events. *Palaeoceanography*, 8(6), pp. 775–784; Bond, G.C. and Lotti, R., 1995. Ice Berg Discharges into the North Atlantic on Millennial Time Scales during the last Deglaciation. *Science*, 267, pp. 1005–1010.

16. Hughes, T.J., 1986. The Jakobshavns Effect. *Geophysical Research Letters*, 13, pp. 46–48; Shepherd, A., Wingham, D.J. and Mansley, J.A.D., 2002. Inland Thinning of the Amundsen Sea Sector, West Antarctica. *Geophysical Research Letters*, 29(10), pp. 21–24; Devoy, R.J.N., 2015. Sea-level Rise: Causes, impacts and scenarios for change. In: Ellis, J. and Sherman, D., eds. *Coastal and Marine Hazards, Risks and Disasters*. Amsterdam: Elsevier, pp. 197–242.

17. Powell, R.D., 1984. Glacimarine Processes and Inductive Lithofacies Modelling of Ice Shelf and Tidewater Glacier Sediments Based on Quaternary Examples. *Marine Geology*, 57 (1–4), pp. 1–52; Westley, K. and Edwards, R., 2017. Irish Sea and Atlantic Margin: Quaternary Paleoenvironments. In: Flemming, N., Harff, J., Moura, D., Burgess, A. and G. Bailey, eds. *Submerged Landscapes of the European Continental Shelf*. [online] Available at: https://doi.org/10.1002/9781118927823.

18. Eyles, N. and McCabe, A.M., 1989. The Late Devensian (<22,000 BP) Irish Sea Basin: The sedimentary record of a collapsed ice-sheet margin. *Quaternary Science Reviews*, 8(4), pp. 307–351; McCabe, A.M. and O'Cofaigh, C., 1996. Upper Pleistocene Facies Sequences and Relative Sea-level Trends Along the South Coast of Ireland. *Journal of Sedimentary Research*, 66, pp. 376–390.

19. McCabe, A.M., Knight, J., McCarron, S., 1998. Evidence for Heinrich Event 1 in the British Isles. *Journal of Quaternary Science*, 13(6), pp. 549–568; Forbes, D.L. and Hansom, J.D., 2012. Polar Coasts; McClusky, D., and Wolanski, E., eds. In: *Treatise on Estuarine and Coastal Science*. Volume 3. *Estuarine and Coastal Geology and Geomorphology*. Cambridge MA: Academic Press, pp. 245–283.

20. McCabe, A.M. and Clark, P.U., 1998. Ice-sheet Variability Around the North Atlantic Ocean During the Last Deglaciation. *Nature*, 392, pp. 373–377.

21. McCabe and Clark, 1998. Ice-sheet Variability Around the North Atlantic Ocean During the Last Deglaciation; McCabe et al, 1998. Evidence for Heinrich Event 1 in the British Isles; National Ocean and Atmosphere Agency (NOAA), 2019. Heinrich and Dansgaard–Oeschger Events. [online] Available at: https://www.ncdc.noaa.gov/abrupt-climate-change/

Heinrich%20and%20Dansgaard%E2%80%93Oeschger%20Events

Ancient Shorelines and Sea-level Changes

1. Devoy, R.J.N., 1987. *Sea Surface Studies: A global view*. London: Routledge-Croom Helm; Carter, R.W.G., Devoy, R.J.N. and Shaw, J., 1989. Holocene Sea-levels in Ireland. *Journal of Quaternary Science*, 4, pp. 7–24; Devoy, R.J.N., 2015. Sea-level Rise: Causes, impacts, and scenarios for change. In: Ellis, J.T. and Sherman, D., eds. *Coastal and Marine Hazards, Risks and Disasters*. Amsterdam: Elsevier, pp. 197–241; Carter, R.W.G. and Woodroffe, C.D., 1994. *Coastal Evolution: Late Quaternary shoreline morphodynamics*. Cambridge: Cambridge University Press; Plets, R.M.K., Callard, S.L., Cooper, J.A.G, Kelley, J.T., Belknap, D.F., Edwards, R.J., Long, A.J., Quinn, R.J. and Jackson, D.W.T., 2019. Late Quaternary Sea-level Change and Evolution of Belfast Lough, Northern Ireland: New offshore evidence and implications for sea-level reconstruction. *Journal of Quaternary Science*, 34, pp. 285–298.

2. Coxon, P., McCarron, S. and Mitchell, F., 2017. *Advances in Irish Quaternary Studies*. Dordrecht: Atlantis Press, Springer.

3. Devoy, R.J.N., 2015. The Development and Management of the Dingle Bay Spit-barriers of Southwest Ireland. In: Randazzo, G., Cooper, J.A.G. and Jackson, D.W., eds. *Sand and Gravel Spits*. Coastal Research Library 12, Springer International Publishing, pp. 139–180; Kandrot, S., 2016. The Monitoring and Modelling of the Impacts of Storms under Sea-level Rise on a Breached Coastal Dune-barrier System, Volumes 1 and 2. Unpublished PhD Thesis, National University of Ireland at Cork (NUIC).

4. Carter, R.W.G., 1982. Sea-level Changes in Northern Ireland. *Proceedings of the Geologists' Association*, 93, pp. 7–23; Carter, R.W.G., 1988. *Coastal Environments*. London: Academic Press; Edwards, R. and Craven, K., 2017. Relative Sea-level Changes around the Irish Coast. In: Coxon, P., McCarron, S. and Mitchell, F., eds. *Advances in Irish Quaternary Studies*. Dordrecht: Atlantis Press, Springer.

5. Orme, A.R., 1966. Quaternary Changes of Sea Level in Ireland. The Royal Geographical Society (with the Institute of British Geographers), 39, pp. 127–140; Devoy, R.J.N., 1983. Late Quaternary Shorelines in Ireland: An assessment of their implications for isostatic land movement and relative sea-level changes. In: Smith, D.E., and Dawson, A., eds. *Shorelines and Isostas* . Institute of British Geographers Special Publication No. 16. London: Academic Press, pp. 227–254; Devoy 2015. Sea-level Rise: Causes, impacts, and scenarios for change; Synge, F.M., 1985. Coastal Evolution. In: Edwards, K.J. and Warren, W.P., eds. *The Quaternary History of Ireland*. London: Academic Press.

6. Woodcock, N.H. and Strachan, R.A., 2012. *Geological History of Britain and Ireland* (2nd edition). Chichester: Wiley-Blackwell.

7. Devoy, R.J.N., 1987. Hydrocarbon Exploration and Biostratigraphy: The application of sea-level studies. In: Devoy, R.J.N., ed. *Sea Surface Studies: A global view*. London: Routledge, Croom Helm; Shannon P.M., Corcoran D.V. and Haughton P.D.W., 2001. The Petroleum Exploration of Ireland's Offshore Basins: Introduction. *Geological Society Special Publications*, 188, p. 480; Haq, B.U. and Schutter, S.R., 2008. A Chronology of Paleozoic Sea-level Changes. *Science*, 322, pp. 64–68.

8. Synge, F.M., 1980. A Morphometric Comparison of Raised Shorelines in Fennoscandinavia, Scotland and Ireland. *Geologiska Föreningens i Stockholm Förhandlingar*, 102, pp. 235–249; Carter, R.W.G., 1983. Raised Coastal Landforms as Products of Modern Process Variations and their Relevance in Eustatic Sea-level Studies: Examples from eastern Ireland. *Boreas*, 12, pp. 167–182; Edwards and Craven, 2017. Relative Sea-level Changes around the Irish Coast.

9. Devoy, 1983. Late Quaternary shorelines in Ireland; Carter, R.W.G., 1993. Age, Origin and Significance of the Raised Gravel Barrier at Church Bay, Rathlin Island, County Antrim. *Irish Geography*, 26, pp. 141–146; Edwards, R.J., 2013. Sedimentary Indicators of Relative Sea-level Change – Low Energy. In: Elias, S. and Mock, J., eds. *Encyclopedia of Quaternary Science*, Volume 4, 2nd edn. Amsterdam: Elsevier, pp. 396–408.

10. Sinnott, A.M. and Devoy, R.J.N., 1992. The Geomorphology of Ireland's Coastline: Patterns, processes and future prospects. *Hommes et Terres du Nord, Revue de l'Institut de Geographie de Lille, 1992–3*, pp. 145–153.

11. Devoy, R. J. N., Sinnott, A. M., Knudsen, K-L., Kristensen, P. and Peacock, J.D., 2000. Interglacial Relative Sea-level and Palaeoenvironmental Changes from Cork Harbour, Ireland. *Occasional Publication 02/ 99, Department of Geography*, NUIC; Devoy, R.J.N., 2005. Cork City and the Evolution of the Lee Valley. In: Crowley, J., Devoy, R.J.N., Linehan, D. and O'Flanagan, P., eds. *Atlas of Cork City*. Cork: Cork University Press, pp. 7–16; Coxon, P. and McCarron, S.G., 2009. Cenozoic: Tertiary and Quaternary (until 11,700 years before 2000). In: Holland, C.H. and Sanders, I.S., eds. *The Geology of Ireland*. Edinburgh: Dunedin Press, pp. 355–396.

12. Wright, W.B. and Muff, H.B., 1904. The Pre-glacial Raised Beach of the South Coast of Ireland. *Scientific Proceedings of the Royal Society of Ireland*, n.s. 10, part 2, pp. 250–324; Bryant, R.H., 1965. The 'Pre-glacial' Raised Beach in South-West Ireland, *Irish Geography*, 5: 2, pp. 188–203; O'Cofaigh, C., Telfer, M.W., Bailey, R.M. and Evans, D.J.A., 2012. Late Pleistocene Chronostratigraphy and Ice Sheet Limits, Southern Ireland. *Quaternary Science Reviews*, 44, pp. 160–179.

13. Edwards and Craven, 2017. Relative sea-level changes around the Irish coast.

14. McCabe, A.M. and O'Cofaigh, C., 1996. Upper Pleistocene Facies Sequences and Relative Sea-level Trends along the South Coast of Ireland. *Journal of Sedimentary Research*, 66, pp. 376–390; Devoy, R.J.N., Sinnott, A.M., O'Connell, Y., Dowling, L., Scourse, J.D. and Barlow, N., 2000. Quaternary Stratigraphic Reconstructions and the Marine Gortian Interglacial Sediments from Cork Harbour, Ireland. *Occasional Publication 01/99, Department of Geography*, NUIC; Devoy, 2005. Cork City and the Evolution of the Lee Valley, pp. 7–16.

15. Gallagher, C. and Thorp, M., 1997. The Age of the Pleistocene Raised Beach near Fethard, County Wexford, Using Infra-red Stimulated Luminescence (IRSL). *Irish Geography*, 30, pp. 68–89; Gallagher, C., Telfer, M.W. and O'Cofaigh, C., 2015. A Marine Isotope Stage 4 Age for Pleistocene Raised Beach Deposits near Fethard, Southern Ireland. *Journal of Quaternary Science*, 30, pp. 754–763; Coxon and McCarron, 2009. Cenozoic: Tertiary and Quaternary (until 11,700 years before 2000).

16. Carter, 1982. Sea-level Changes in Northern Ireland, 1983; Devoy. Late Quaternary Shorelines in Ireland; Synge, F.M., 1985. Coastal evolution. In: Edwards, K.J. and Warren, W.P., eds. *The Quaternary History of Ireland*. London: Academic Press; Plets et al, 2019. Late Quaternary Sea-level Change and Evolution of Belfast Lough, Northern Ireland.

17. Devoy, R.J.N., 1995. Deglaciation, Earth Crustal Behaviour and Sea-level Changes in the Determination of Insularity: A perspective from Ireland. In: Preece, R.C., ed. *Island Britain: A Quaternary perspective*. London: Geological Society of London, Special Publication, 96, pp. 181–208; Gallagher, C., 2002. The Morphology and Palaeohydrology of a Submerged Glaciofluvial Channel Emerging from Waterford Harbour onto the Nearshore Continental Shelf of the Celtic Sea. *Irish Geography*, 35, pp. 111–132; Kelley, J.T., Cooper, J.A.G., Jackson, D.W.T., Belknap, D.F. and Quinn, R.J., 2006. Sea-level Change and Inner Shelf Stratigraphy off Northern Ireland. *Marine Geology*, 232, pp. 1–15; Wilson, P. and Plunkett, G., 2010. Age and Palaeoenvironmental Significance of an Inter-tidal Peat Bed at Ballywoolen, Bann estuary, Co. Londonderry. *Irish Geography*, 43, pp. 265–275.

18. Devoy, R.J.N., Delaney, C., Carter, R.W.G. and Jennings, S.C., 1996. Coastal Stratigraphies as Indicators of Environmental Changes upon European Atlantic Coasts in the late Holocene. *Journal of Coastal Research*, 12 (3), pp. 564–588; Cooper, J.A.G., Kelley, J.T., Belknap, D.F., Quinn, R. and McKenna, J., 2002, Inner Shelf Seismic Stratigraphy off the North Coast of Northern Ireland: New data on the depth of the Holocene lowstand. *Marine Geology*, 186, pp. 369–387; Westley, K., Plets, R. and Quinn, R., 2014. Holocene Paleogeographic Reconstructions of the Ramore Head area, Northern Ireland, Using Geophysical and Geotechnical data: Paleo-landscape mapping and archaeological implications. *Geoarchaeology: An International Journal*, 29, pp. 411–430.

19. Delaney, C. and Devoy, R.J.N., 1995. Evidence from sites in western Ireland of late Holocene Changes in Coastal Environments. *Marine Geology*, 124, pp. 273–287; Delaney, C., Devoy, R.J.N. and Jennings, S., 2012. Mid- to Late Holocene Relative Sea-level and Sedimentary Changes in Southwest Ireland. In: Duffy, P.J. and Nolan, W., eds. *At the Anvil: Essays in honour of William J. Smyth*. Dublin: Geography Publications, pp. 697–746; O'Connell, M. and Molloy, K., 2017. Mid- and Late-Holocene Environmental Change in Western Ireland: New evidence from coastal peats and fossil timbers with particular reference to relative sea-level change; Shennan, I., Bradley, S.L. and Edwards,R., 2018. Relative sea-level changes and crustal movements in Britain and Ireland since the Last Glacial Maximum. *Quaternary Science Reviews*, 188, pp.143-159.

20. Plag, H.-P., Austin, W.E.N., Belknap, D.F., Devoy, R.J.N. et al., 1996. Late Quaternary Relative Sea-level Changes and the Role of Glaciation upon Continental Shelves. *Terra Nova*, 8, pp. 213–222; Finlayson, A., Fabel, D., Bradwell, T. and Sugden, D., 2014. Growth and Decay of a Marine Terminating Sector of the last British–Irish Ice Sheet. *Quaternary Science Reviews*, 83, pp. 28–45; Smedley, R.K., Chiverrell, R.C., Ballantyne, C.K., Burke, M.J., Clark, C.D., Duller, G.A.T., Fabel, D., MacCarroll, D., Scourse, J.D., Small, D. and Thomas, G.S.P., 2017. Internal Dynamics Condition Centennial Scale Oscillations in Marine-based Ice-stream Retreat. *Geology*, 45(9); Scourse, J.D., Ward, S.L., Wainright, A., Bradley, S.L. and Uehara, K., 2018. The Role of Megatides and Relative Sea Level in Controlling the Deglaciation of the British–Irish and Fennoscanadian Ice Sheets. *Journal of Quaternary Science*, 33, pp. 139–149.

21. Delaney, C. and Devoy, R.J.N., 1995. Evidence from Sites in Western Ireland of Late Holocene Changes in Coastal Environments. *Marine Geology*, 124, pp. 273–287; Brooks, A.J., Bradley, S.L., Edwards, R.J., Milne, G.A., Horton, B. and Shennan, I.J., 2008. Postglacial Relative Sea-level Observations from Ireland and their Role in Glacial Rebound Modelling. *Journal of Quaternary Science*, 23, pp. 175–192; Edwards, R.J. and Brooks, A.J., 2008. The Island of Ireland: Drowning the myth of an Irish landbridge? In: Davenport, J.J., Sleeman, D.P. and Woodman, P.C., eds. *Mind the Gap: Postglacial colonisation of Ireland*. Special Supplement to the *Irish Naturalists' Journal*, pp. 19–34; Delaney et al. Mid- to Late Holocene Relative Sea-level and Sedimentary Changes in Southwest Ireland.

22. Devoy, 2015. The Development and Management of the Dingle Bay Spit-barriers of Southwest Ireland; Edwards and Craven, 2017. Relative Sea-level Changes around the Irish Coast; Plets et al. Late Quaternary Sea-level Change and Evolution of Belfast Lough, Northern Ireland.

23. Edwards and Brooks, 2008. The Island of Ireland: Drowning the myth of an Irish landbridge?; Brooks, A.J., Bradley, S.L., Edwards, R.J. and Goodwyn, N., 2011. The Palaeogeography of Northwest Europe during the Last 20,000 Years. *Journal of Maps*, 7, pp. 573–587.

Monitoring and Visualising the Coast

1. Waddell, J., 1998. *The Prehistoric Archaeology of Ireland*. Dublin: Wordwell; Conditt, T. and Moore, F., 2003. Ireland in the Iron Age. Map of Ireland by Claudius Ptolemaeus. *Archaeology Ireland*, Heritage Guide No. 21.

2. Darcy, R. and Flynn, W., 2008. Ptolemy's Map of Ireland: A modern decoding. *Irish Geography* 41:1, pp. 49–69; Abshire C., Durham, A., Dmitri, A.G. and Stafeyev, S.K., 2018. Ptolemy's Britain and Ireland: A new digital reconstruction. *Proceedings of the International Cartographic Association*, 1.

3. Freeman, F., 2001. *Ireland and the Classical World*. Austin: University of Texas Press; Darcy and Flynn, Ptolemy's Map of Ireland; Cordero, M., 2019. A Critical Account of the Representation of Ireland in the Geography of Strabo and Ptolemy. *Atlas*, 17, pp. 11–19.

4. Wright, D. and Bartlett, D., 2000. *Marine and Coastal Geographical Information Systems*. London: Taylor and Francis, Ltd.

5. Bartlett, D. and Celliers, L., 2017. Geoinformatics for Applied Coastal and Marine Management. In: Bartlett, D. and Celliers, L., eds. *Geoinformatics for Marine and Coastal Management*. Boca Raton, FL: CRC Press, pp. 1–15.

6. O'Dea, L., Cummins, V., Wright, D., Dwyer, N. and Ameztoy, I., 2007. Report on Coastal Mapping and Informatics Trans-Atlantic Workshop 1: Potentials and limitations of coastal web atlases. Cork: University College Cork.

7. O'Dea et al. Report on Coastal Mapping and Informatics Trans-Atlantic Workshop 1.

8. Kopke, K. and Dwyer, N., eds. 2016. ICAN – Best Practice Guide to Engage your Coastal Web Atlas User Community. Paris: Intergovernmental Oceanographic Commission of UNESCO (IOC Manuals and Guides, 75), p. 35 (English) (IOC/2016/MG/75).

9. Wright, D.J., Dwyer, N. and Cummins, V., eds. 2011. *Coastal Informatics and Web Atlas Design and Implementation*. Hershey, PA: IGI Global, p. 321.

10. Kopke and Dwyer. ICAN – Best Practice Guide to Engage your Coastal Web Atlas User Community.

11. Maher, S.N., 2014. *Deep Map Country: Literary cartography of the Great Plains*. Lincoln NE: University of Nebraska Press.

12. Weintrit, A., 2017. Geoinformatics in Hydrography and Marine Navigation. In: Bartlett, D. and Celliers, L., eds. *Geoinformatics for Marine and Coastal Management*. Boca Raton, FL: CRC Press, pp. 323–348.

13. UK Hydrographic Office, 2019. *Admiralty Sailing Directions: Irish Coast Pilot*. 21st edition, NP40. Taunton: UKHO.

14. Towler, P. and Fishwick, M., 2020. *Reeds Nautical Almanac*. London: Bloomsbury.

15. Kean, N., ed. 2019. *South and West Coasts of Ireland Sailing Directions*. 15th ed. Kilbrittain: Irish Cruising Club Publications; Kean, N. ed, 2019. *East and North Coasts of Ireland Sailing Directions*, 13th ed. Kilbrittain: Irish Cruising Club Publications.

16. Wilcox, R., 2009. *Southern Ireland Cruising Companion*. London: Wiley Nautical; Swanson, G., 2005. Cruising Cork and Kerry. St Ives: Imray; Rainsbury, D., 2015. Irish Sea Pilot. St Ives: Imray.

17. Saunders, P., 2017. Navigating a Sea of Data: Geoinformatics for law enforcement at sea. In: Bartlett, D. and Celliers, L., eds. *Geoinformatics for Marine and Coastal Management*. Boca Raton, FL: CRC Press, pp. 203–223.

18. EMSA, no date. Vessel Tracking Globally. Understanding LRIT. European Maritime Safety Agency (EMSA). [online] Available at: emsa.europa.eu/operations/lrit/download/452/256/23.html

19. Kandrot, S., 2016. The Monitoring and Modelling of the Impacts of Storms under Sea-level Rise on a Breached Coastal Dune-barrier System. PhD Thesis. University College Cork. [online] Available at: https://cora.ucc.ie/handle/10468/3657

20. Heritage, G.L. and Large, A.R.G., 2009. *Laser Scanning for the Environmental Sciences*. Oxford: Blackwell.

21. Zhang, D., O'Connor, N.E. and Regan, F., 2017. Current and Future Information and Communication Technology (ICT) Trends in Coastal and Marine Management. In: Bartlett, D. and Celliers, L., eds. *Geoinformatics for Marine and Coastal Management*. Boca Raton, FL: CRC Press, pp. 97–127.

22. Delin, K.A., Jackson, S.P. and Some, R.R., 1999. Sensor Webs. *NASA Tech Briefs*, 23(80),p. 27.

23. Smartbay, 2019. [online] Available at: www.smartbay.ie

24. Argo, 2019. [online] Available at: www.argo.ucsd.edu

25. Zhang et al. Current and Future Information and Communication Technology (ICT) Trends in Coastal and Marine Management.

26. Ullgren, J.E. and White, M., 2010. Water Mass Interaction at Intermediate Depths in the Southern Rockall Trough, Northeastern North Atlantic. *Deep Sea Research Part I: Oceanographic Research Papers*, 57(2), pp. 248–257.

Underwater Surveys: The INFOMAR project

1. Kunzig, R., 2000. *Mapping the Deep: The extraordinary story of ocean science*. New York: Sort of Books, p. 2.

2. Dorschel, B., Wheeler, A.J., Monteys, X. and Verbruggen, K., 2010. *Atlas of the Deep-water Seabed: Ireland*. Heidelberg: Springer Science and Business Media.

3. European Parliament and Council, 2014. Directive 2014/89 of 23 July 2014 on Establishing a Framework for Maritime Spatial Planning.

4. Brissette, M.B. and Clarke, J.E., 1999. Side Scan versus Multibeam Echosounder Object Detection: A comparative analysis. *International Hydrographic Review*, Monaco, LXXVI(2).

5. Brissette and Clarke. Side Scan versus Multibeam Echosounder Object Detection.

6. Scott G., Monteys X., Hardy D., McKeon C. and Furey T., 2016. Mapping the Seabed. In Bartlett D., Celliers L., eds, 2016. *Geoinformatics for Marine and Coastal Management*. Boca Raton, FL: CRC Press, pp. 17–42.

7. International Hydrographic Organization, 2008. *IHO Standards for Hydrographic Surveys, Special Publication No. 44*. Monaco: International Hydrographic Bureau.

8. INFOMAR, 2020. [online] Available at: http://www.infomar.ie/data/DataAccess.php

9. Fealy, R.M., Green, S., Loftus, M., Meehan, R., Radford, T., Cronin, C. and Bulfin, M., 2009. *Teagasc EPA Soil and Subsoils Mapping Project-Final Report*. Volume I. Dublin: Teagasc.

10. Monteys, X., Harris, P., Caloca, S. and Cahalane, C., 2015. Spatial Prediction of Coastal Bathymetry Based on Multispectral Satellite Imagery and Multibeam Data. *Remote Sensing*, 7(10), pp. 13782–13806.

11. Brady, K., McKeon, C., Lyttleton, J. and Lawlor, I., 2012. *Warships, U-Boats and Liners: A guide to shipwrecks mapped in Irish waters*. Dubin: Department of the Arts, Heritage and the Gaeltacht, Geological Survey of Ireland and Stationery Office.

12. Molloy, J., 2006. *The Herring Fisheries of Ireland 1900–2005. Biology, research, development and assessment*. Galway: Marine Institute.

13. Breslin, J., 1998. The Location and Extent of the Main Herring (*Clupea harengus*) Spawning Grounds around the Irish Coast. Unpublished MSc. Thesis. University College Dublin.

14. Naylor, P., 2003. *Great British Marine Animals*. Sound Diving Publications.

15. Arroyo, I., ed. 2013. *Yearbook Maritime Law* (Vol. 1). Heidelberg: Springer Science and Business Media, p. 286.

16. PwC, 2008. Options Appraisal Report: Final Report. [online] Available at: http://www.infomar.ie/publications/Reports.php

SECTION 2: NATURAL COASTAL ENVIRONMENTS
Rocky Coasts

1. Davies, G.L.H. and Stephens, N., 1978. *The Geomorphology of the British Isles: Ireland*. London: Methuen; Devoy, R.J.N., 1983. Late Quaternary Shorelines in Ireland: An assessment of their implications for isotatic movement and relative sea-level changes. In: Smith, D. and Dawson, A., eds. *Shorelines and Isostasy*. London: Academic Press, pp. 227–254; Knight, J. and Harrison, S., 2018. Paraglacial Evolution of the Irish Landscape. *Irish Geography*, 51, pp. 171–186.

2. Davies and Stephens, *The Geomorphology of the British Isles: Ireland*; Devoy, R.J.N., 2005, Cork City and the Evolution of the Lee Valley. In: Crowley, J.S., Devoy, R.J.N., Linehan, D. and O'Flanagan, P., eds. 2011. *Atlas of Cork City*. Cork: Cork University Press; Moses, C. and Robinson, D., 2011. Chalk Coast Dynamics: Implications for understanding rock coast evolution. *Earth-Science Reviews*, 109, pp. 63–73.

3. Haslett, S.K., 2009. *Coastal Systems*. London: Routledge.

4. Trenhaile, A.S., 2017. *Coastal Erosion Processes and Landforms*. New York: Wiley; Geological Survey of Ireland and Heritage Council of Ireland, 2007. Ireland's Coastal Geology: The role of geology in Ireland's coastline. [online] Available at: https://www.heritagecouncil.ie/content/files/irelands_coastal_geology_2007_2mb.pdf

5. National Coastal Erosion Committee, 1992. *Coastal Management: A case for action*. Dublin: Eolas on behalf of County and City Engineers Association and the Irish Science and Technology Agency.

6. Bird, E., 2000. *Coastal Geomorphology: An introduction*. Chichester and New York: Wiley; Kennedy, D.M., Stephenson, W.J. and Naylor, L.A., eds. 2014. *Rock Coast Geomorphology: A global synthesis*. Memoir 40. London: Geological Society.

7. Devoy, R.J.N., 2008. Coastal Vulnerability and the Implications of Sea-level rise for Ireland. *Journal of Coastal Research*, 24 (2), pp. 325–341; Gault, J., Morrissey, C. and Devoy, R.J.N., 2007. *Dune Protection Plan for the Fingal Coast, County Dublin*. Report for Fingal County Council, Dublin. Cork: The Coastal and Marine Resources Centre, University College Cork; Cronin, A. and Kandrot, S., 2017. *Local Authority Coastal Erosion Policy and Practice Audit*. Report for Coastal Local Authorities in Ireland, Fingal County Council. Cork: Marine and Renewable Energy Institute (MaREI), University College Cork.

8. Bourke, M.C., Naylor, L.A., Flood, R., Nash, C., Cullen, N.D., Goffo, F. and Migge, K., 2016. *Investigation of Ireland's Shore Platforms: Location, type and coastal protection*. Geological Survey of Ireland Shortcall Reports. Dublin: Geological Survey of Ireland.

9. Trenhaile, A.S., 1987. *The Geomorphology of Rock Coasts*. Oxford University Press; USA; Sunamura, T., 1992. *Geomorphology of Rocky Coasts*, Chichester: John Wiley and Son Ltd.

10. Cruslock, E.M., Naylor, L.A., Foote, Y.L. and Swantesson, J.O., 2010. Geomorphologic Equifinality: A comparison between shore platforms in Höga Kusten and Färö, Sweden and the Vale of Glamorgan, South Wales, UK. *Geomorphology*, 114, pp. 78–88; Dasgupta, R., 2011. Whither Shore Platforms? *Progress in Physical Geography*, 35, pp. 183–209; Naylor, L.A., Stephenson, W.J., Smith, H.C., Way, O., Mendelssohn, J. and Crowley, A., 2016. Geomorphological Control on Boulder Transport and Coastal Erosion Before, During and After an Extreme Extra-tropical Cyclone. *Earth Surface Processes and Landforms*, 41, pp. 685–700.

11. Jackson, A. and McIlvenny, J., 2011. Coastal Squeeze on Rocky Shores in Northern Scotland and some Possible Ecological Impacts. *Journal of Experimental Marine Biology and Ecology*, 400, pp. 314–321; Naylor, L.A., Kennedy, D.M. and Stephenson, W.J., 2014. Synthesis and Conclusion to the Rock Coast Geomorphology of the World. In: Kennedy, D.M., Stephenson, W.J. and Naylor, L.A., eds. *Rock Coast Geomorphology: A global synthesis*. Memoirs 40. London: Geological Society, pp. 283–286.

12. Herterich, J.G., Cox, R. and Dias, F., 2018. How Does Wave Impact Generate Large Boulders? Modelling hydraulic fracture of cliffs and shore platforms. *Marine Geology*, 399, pp. 34–46.

13. Kinahan, G., Nolan, J., Leonard, H. and Cruise, R., 1878. Explanatory Memoir to Accompany Sheets 93 and 94, with the Adjoining Portions of Sheets 83, 84, and 103, of the Maps of the Geological Survey of Ireland, Memoir of the Geological Survey of Ireland. Dublin: HM Stationery Office.

14. Williams, D.M. and Hall, A.M., 2004, Cliff-top Megaclast Deposits of Ireland: A record of extreme waves in the North Atlantic—storms or tsunamis? *Marine Geology*, 206, pp. 101–117; Scheffers, A., Scheffers, S., Kelletat, D. and Browne, T., 2009. Wave-emplaced Coarse Debris and Megaclasts in Ireland and Scotland: Boulder transport in a high-energy littoral environment. *Journal of Geology*, 117, pp. 553–573; Cox, R., Jahn, K.L., Watkins, O.G., and Cox, P., 2018. Extraordinary Boulder Transport by Storm Waves, and Criteria for Analysing Coastal Boulder Deposits: *Earth-Science Reviews*, 177, pp. 623–636.

15. Haslett, *Coastal Systems*.

16. Neilson, B. and Costello, M.J., 1999. The Relative Length of Seashore Substrata around the Coastline of Ireland as Determined by Digital Methods in a Geographical Information System. *Estuarine and Coastal Shelf Science*, 149(4), pp. 501–508; Nairn, R., 2005. *Ireland's Coastline: Its nature and heritage*. Cork: Collins Press; Cruz-Motta, J.J., Miloslavich, P., Palomo, G., Iken, K., Konar, B., et al., 2010. Patterns of Spatial Variation of Assemblages Associated with Intertidal Rocky Shores: A global perspective. *PLOS ONE* 5(12): e14354.

17. Lozano, I., Devoy, R.J.N., May, W. and Andersen U., 2004. Storminess and Vulnerability along

the Atlantic Coastlines of Europe: Analysis of storm records and of a greenhouse gases induced climate scenario. *Marine Geology*, 210, pp. 205–225; Devoy, R.J.N., 2008. Coastal Vulnerability and the Implications of Sea-level Rise for Ireland. *Journal of Coastal Research*, 24 (2), pp. 325–341; Gallagher, S., Tiron, R., Whelan, E., Gleeson, E., Dias, F. and McGrath, R., 2016. The Nearshore Wind and Wave Energy Potential of Ireland: A high resolution assessment of availability and accessibility. *Renewable Energy*, 88.

18. Tepolt, C.K. and Somero, G.N., 2014. Master of all Trades: Thermal acclimation and adaptation of cardiac function in a broadly distributed marine invasive species, the European green crab, *Carcinus maenas*. *The Journal of Experimental Biology*, 217, pp. 1129–1138.

19. McGrath, D., 1984. The Seashore. In: de Buitléar, É., ed. *Wild Ireland*. Dublin: Amach Faoin Aer Publishing; Nairn, R., 2005. *Ireland's Coastline: Its nature and heritage*; O'Rourke, C., 2006. *Nature Guide to the Aran Islands*. Dublin: Liliput.

20. World Register of Marine Species, 2019 [online]. Available at: http://www.marinespecies.org/; Coughlan, J. and Crowe, T., 2019; A Brief Introduction to Rocky Shores and their Biota. Available at: http://www.ibioli.net/education/Rocky_shores_-_background_notes_for_IBIOLI_trip-1.pdf

21. Myers, A., ed. 2002. *New Survey of Clare Island*. Volume 3: *Marine intertidal ecology*. Dublin: Royal Irish Academy.

22. Buzer, J.S., 1975. Study of the Sediments from Lough Hyne. Unpublished PhD thesis. University of Bristol.

23. Davenport, J., Burnell, G., Cross, T., Emmerson, M., McAllen, R., Ramsay, R. and Rogan, E,. eds. 2008. *Challenges to Marine Ecosystems*. Proceedings of the 41st European Marine Biology Symposium. New York: Springer; School of Biological, Earth and Environmental Sciences, 2019. Lough Hyne. Available at: https://www.ucc.ie/en/bees/research/loughhyne/

24. Bell, J.J., 2002. Morphological Responses of a Cup Coral to Environmental Gradients. *Sarsia*, 87:4, pp. 319–330; Trowbridge, C.D., Kachmarik, K., Plowman, C.Q., Little, C. and McAllen, R., 2016. Biodiversity of Shallow Subtidal, Under-rock Invertebrates in Europe's First Marine Reserve: Effects of physical factors and scientific sampling. *Estuarine Coastal and Shelf Science*, 187, pp. 43–57.

25. Jessopp, M., McAllen, R., O'Halloran, J. and Kelly, T., 2011. *STRIVE Programme 2007–2013 Nutrient and Ecosystem Dynamics in Ireland's Only Marine Nature Reserve (NEIDIN)* (2007-FS-B-4-M5). Report Prepared for the Environmental Protection Agency by School of Biological, Earth and Environmental Sciences, University College Cork.

26. Johnson, M.P., Jessopp, M., Mulholland, O.R., McInerney, C., McAllen R., Allcock, A.L. and Crowe, T.P., 2008. What is the Future for Marine Protected Areas in Irish Waters? *Biology and Environment*, 108B(1), pp. 11–17.

27. McGrath, 1984. The Seashore, pp. 96–121, in *Wild Ireland*, edited by Éamon de Buitléar. Amach Faoin Aer Publishing; Trowbridge et al., Biodiversity of Shallow Subtidal, Under-rock Invertebrates in Europe's First Marine Reserve.

28. McGrath, 1984. The Seashore, pp. 96–121.

29. McGrath, 1984. The Seashore, pp. 96–121.

30. McGrath, 1984, The Seashore, pp. 96–121.

31. Myers, A. ed. 2002. *New Survey of Clare Island. Volume 3: Marine intertidal ecology*. Dublin: Royal Irish Academy, p. 225; Nairn, R., 2005. *Ireland's Coastline: Exploring its nature and heritage*.

32. Harrison, P.A., Berry, P.M. and Dawson, T.P., eds. 2001. Climate Change and Nature Conservation in Britain and Ireland: Modelling natural resource responses to climate change (the MONARCH project). UKCIP Technical Report, Oxford; Intergovernmental Panel on Climate Change, 2019. Summary for Policymakers. In: Pörtner, H.-O., Roberts, V., Masson-Delmotte, D.C., Zhai, P., Tignor, M., Poloczanska, E., Mintenbeck, K., Nicolai, M., Okem, A., Petzold, J., Rama, B. and Weyer N., eds. In Press. *Special Report on the Ocean and Cryosphere in a Changing Climate*; European Environment Agency, 2017. *Climate Change, Impacts and Vulnerability in Europe 2016: An Indicator-based Report*. Luxembourg: Publications Office of the European Union; Climate Change Advisory Council, 2019. Annual Review 2019. Dublin: Climate Council.

33. Bishop, G., 2003. *The Ecology of the Rocky Shores of Sherkin Island: A twenty-year perspective*. Sherkin Island Marine Station; Myers, A., ed. 2002. *New Survey of Clare Island. Marine Intertidal Ecology*. Dublin: Royal Irish Academy; Simkanin, C., 2004. Monitoring Intertidal Community Change in a Warming World. Unpublished MSc Thesis. Department of Life Sciences, Galway-Mayo Institute of Technology.

34. Marine Institute, 2019. Marine Biodiversity. Available at: https://emff.marine.ie/marine-biodiversity

Beaches and Barriers

1. Pilkey, O.H., Neal, W.J., Cooper, J.A.G. and Kelley, J.T., 2011. *The World's Beaches*. Oakland, CA: University of California Press; Bird, E. and Lewis, N., 2015. *Beach Renourishment*. Heidelberg: Springer, p. 24.

2. Carter, R.W.G. and Orford, J.D., 1984. Coarse Clastic Barrier Beaches: A discussion of their distinctive dynamic and morpho-sedimentary characteristics. *Marine Geology*, 60, pp. 377–389; Orford, J.D., Carter, R.W.G. and Jennings, S.C., 1991. Coarse Clastic Barrier Environments: Evolution and implications for Quaternary sea-level interpretation. *Quaternary International*, 9, pp. 87–104.

3. Clark, C.D., Hughes, A.L.C., Greenwood, S.L., Jordan, C.A. and Sejrup, H.P., 2012. Pattern and Timing of Retreat of the Last British-Irish Ice Sheet. *Quaternary Science Reviews*, 44, pp. 112–146.

4. Keary, R., 1967. Biogenic Carbonate in Beach Sediments of the West Coast of Ireland. *Science Proceedings of the Royal Dublin Society*, Series A3, pp. 75–85; Ryle, T., Murray, A., Connolly, C. and Swann, M., 2009. Coastal Monitoring Project 2004–2006. Unpublished Report. Dublin: National Parks and Wildlife Service.

5. De Grave, S., Fazakerley, H., Kelly, L., Guiry, M.D., Ryan, M. and Walshe, J., 2000. A Study of Selected Maërl Beds in Irish Waters and their Potential for Sustainable Extraction. Marine Resource Series. Galway: Marine Institute; Hall-Spencer, J.M., Kelly, J. and Maggs, C.A., 2010. Background Document for Maërl beds. *OSPAR Biodiversity Series 491/2010*.

6. Wilson, S., Blake, C., Berges, J.A. and Maggs, C.A., 2004. Environmental Tolerances of Free-living Coralline Algae (maërl): Implications for European marine conservation. *Biological Conservation* 120, pp. 283–293; OSPAR Commission, 2008. *OSPAR List of Threatened and/or Declining Species and Habitats*. Agreement 2008-6.

7. Betts, N., Orford, J.D., White, D. and Graham, C., 2004. Storminess and Surges in the South-Western Approaches of the Eastern North Atlantic: The synoptic climatology of recent extreme coastal storms. *Marine Geology*, 210, pp. 227–246; Marine Institute, 2015. Tidal Observations [online] Available at: http://www.marine.ie/Home/site-area/data-services/real-time-observations/tidal-flows-around-ireland?language=ga

8. Carter, R.W.G., 1988. *Coastal Environments: An introduction to the physical, ecological and cultural systems of coastlines*. London Academic Press. pp. 305–320; Smith, E.R., Ebersole, B.A. and Wang, P., 2004. Dependence of Total Longshore Sediment Transport Rates on Incident Wave Parameters and Breaker Type. US Army Corps of Engineers. ERDC/CHL CHETN-IV-62

9. Thomas, T., Phillips, M.R. and Williams, A.T., 2013. A Centurial Record of Beach Rotation *Journal of Coastal Research*, Special Issue No. 65, pp. 594–599; George, D.A., Largier, J.L., Storlazzi, C.D and Barnard, P.L., 2015. Classification of Rocky Headlands in California with Relevance to Littoral Cell Boundary Delineation. *Marine Geology*, 369, pp. 137–152.

10. Jackson, D.W.T., Cooper, J.A.G. and del Rio, L., 2005. Geological Control of Beach Morphodynamic State. *Marine Geology*, 216, no. 4, pp. 297–314.

11. Pilkey, O.H. and Cooper, J.A.G., 2014. *The Last Beach*. Durham, NC: Duke University Press.

12. Carter, R.W.G., 1988. *Coastal Environments*. London: Academic Press.

13. Duffy, M. and Devoy, R.J.N., 1999. Contemporary Process Controls on the Evolution of Sedimentary Coasts Under Low to High-energy Regimes: Western Ireland. *Geologie en Mijnbouw*, 77, pp. 333–349; Clarke, M.L and Rendell, H.M., 2009. The Impact of North Atlantic Storminess on Western European Coasts: A review. *Quaternary International*, 195, no. 1–2, pp. 31–41.

14. Wintle, A.G., Clarke, M.L., Musson, F.M., Orford, J. and Devoy, R.J.N., 1998. Luminescence Dating of Recent Dune Formation on Inch Spit, Dingle Bay, Southwest Ireland. *The Holocene*, vol. 8, pp. 331–339.

15. Ryle, T., Murray, A., Connolly, K. and Swann, M., 2004. Coastal Monitoring Project 2004–2006. Dublin: National Parks and Wildlife Service; Delaney, A., Devaney, F.M. Martin, J.M. and Barron, S.J., 2013. Monitoring Survey of Annex I Sand Dune Habitats in Ireland. *Irish Wildlife Manuals*, No. 75. Dublin: National Parks and Wildlife Service.

16. Sweeney, P., Sweeney, N. and Hurley C., 2013. Natterjack Toad Monitoring Project, 2011–2012. *Irish Wildlife Manuals*, No. 67. Dublin: National Parks and Wildlife Service.

17. Cooper, J.A.G., Jackson, D.W.T., Navas, F., McKenna, J. and Malvarez, G., 2004. Identifying Storm Impacts on an Embayed, High-energy Coastline: Examples from western Ireland. *Marine Geology*, 210, pp. 261–280.

18. Orford, J.D., Carter, R.W.G. and Jennings, S.C., 1996. Control Domains and Morphological Phases in Gravel-dominated Coastal Barrier. *Journal of Coastal Research*, 12, pp. 589–605; Kandrot, S., 2018. Monitoring and Modelling the Impacts of Storms Under Sea-level Rise on a Breached Barrier System. Vols 1 and 2. Unpublished Thesis, National University of Ireland Cork.

19. George, D.A., Largier, J.L., Storlazzi, C.D. and Barnard, P.L., 2015. Classification of Rocky Headlands in California with Relevance to Littoral Cell Boundary Delineation. *Marine Geology*, doi: 10.1016/j.margeo.2015.08.010

20. Short, A.D. and Jackson D.W.T., 2013. Beach Morphodynamics. In: ed. Shroder, J.F., *Treatise on Geomorphology*, San Diego: Academic Press, pp. 106–129.

21. Orford, J.D., Forbes, D.L. and Jennings, S.C., 2002. Organisational Controls, Typologies and Timescales of Paraglacial Gravel-dominated Coastal Systems. *Geomorphology*, 48, pp. 51–85.

22. Orford, J.D., 1986. Coasts: Environments and landforms. In: Fookes, P. and Vaughen, P., eds *Handbook of Engineering Geomorphology*. Glasgow: Thomson, pp. 203–217.

23. Carter, R.W.G. and Orford, J.D., 1988. Conceptual Model of Coarse Clastic Barrier Formation from Multiple Sediment Sources. *Geographical Review*, vol. 78, pp. 221–239; Greenwood, R.O. and Orford, J.D., 2007. Factors Controlling the Retreat of Drumlin Coastal Cliffs in a Low Energy Marine Environment – Strangford Lough, Northern Ireland. *Journal of Coastal Research*, 23, pp. 285–297.

24. Carter, R.W.G. and Orford, J.D., 1993. The Morphodynamics of Coarse Clastic Beaches and Barriers: A short and long-term perspective. *Journal of Coastal Research*, Special Issue No. 15, pp. 158–179.

25. Taylor, M. and Stone, G.W., 1996. Beach Ridges: A review. *Journal of Coastal Research*, 12 (3), pp. 612–621; O'Brien, L., Dudley, J.M. and Dias, F., 2013. Extreme Wave Events in Ireland: 14, 680 BP–2012. *Natural Hazards and Earth System Sciences*, 13, pp. 625–648.

26. Orford, J.D., Jennings, S.C. and Pethick, J., 2003. Extreme Storm Effect on Gravel-dominated Barriers. In: Davis, R.A. ed. *Coastal Sediments '03*, Proceedings of the International Conference on Coastal Sediments 2003. CD-ROM published by World Scientific Publishing Corp. and East Meets West Productions, Corpus Christi, Texas, USA.

27. Orford, J.D., Carter, R.W.G., Jennings, S.C. and Hinton, A.C., 1995. Processes and Timescales by which a Coastal Gravel-dominated Barrier Responds Geomorphologically to Sea-level rise: Story Head barrier, Nova Scotia. *Earth Surface Processes and Landforms*, 20, pp. 21–37.

28. Orford, J.D. and Carter, R.W.G., 1984. Mechanisms to Account for the Longshore Spacing of Overwash on a Coarse Clastic Dominated Barrier Beach in Southeast Ireland. *Marine Geology*, 56, pp. 207–226.

29. Delaney, C. and Devoy, R.J.N., 1995. Evidence from Sites in Western Ireland of Late Holocene Changes in Coastal Environments. *Journal of Coastal Research*, 124, (1–4), pp. 273–287; Nordstrom,

K.F., Bauer, B.O., Davidson-Arnott, R.G.D., Gares, P.A., Carter, R.W.G., Jackson, D.W.T. and Sherman, D.J., 1996. Offshore Aeolian Transport Across a Beach: Carrick Finn Strand, Ireland. *Journal of Coastal Research*, 12 (3), pp. 664–672.

30. Orford, J.D., Murdy, J. and Wintle, A., 2003. Prograded Holocene Beach-ridges with Superimposed Dunes in North-east Ireland: Mechanisms and timescales of fine and coarse beach sediment decoupling and deposition. *Marine Geology*, 194, pp. 47–64.

31. Devoy, R.J.N., 2015. The Development and Management of the Dingle Bay Spit-barriers of Southwest Ireland. In: Randazzo, G., Jackson, D.W.T. and Cooper, J.A.G., eds. *Sand and Gravel Spits*, Coastal Research Library no. 12. Switzerland: Springer International Publishing, pp. 139–180.

32. Carter, R.W.G., 1979. Recent Progradation of the Magilligan Foreland, County Londonderry, Northern Ireland. In: *Les côtes atlantiques d'Europe, évolution, aménagement, protection*, Publications du CNEXO, Brest, France, Actes de Colloques No. 9, pp. 17–28.

33. Daly, C.J., Winter, C. and Bryan, K.R., 2015. On the Morphological Development of Embayed Beaches. *Geomorphology*, doi:10.1016/j.geomorph.2015.07.040.

34. Devoy, R.J.N., 2015. The Development and Management of the Dingle Bay Spit-barriers of Southwest Ireland.

35. Mulrennan, M., 1992. Ridge and Runnel Beach Morphodynamics: An Example from the Central East Coast of Ireland. *Journal of Coastal Research*, 8, (4), pp. 906–918; Navas, F., Cooper, J.A.G., Malvarez, G.C. and Jackson, D.W.T., 2001. Theoretical Approach to the Investigation of Ridge and Runnel Topography of a Macrotidal Beach: Dundrum Bay, Northern Ireland. *Journal of Coastal Research*, Special Issue 34, International Coastal Symposium (ICS 2000), pp. 183–193.

36. Cooper, J.A.G. and Pilkey, O., 2012. *Pitfalls of Shoreline Stabilization: Selected case studies*, Coastal Research Library 3. Dordrecht: Springer, pp. 93–104; Pilkey, O.H. and Cooper, J.A.G., 2015. *The Last Beach*.

Coastal Wetlands

1. Doody, J.P., 2001. *Coastal Conservation and Management: An ecological perspective*. Berlin: Springer; Haslett, S., 2009. *Coastal Systems*. London: Routledge.

2. IWWR (Interagency Workgroup on Wetland Restoration), 2003. *An Introduction and User's Guide to Wetland Restoration, Creation and Enhancement*. Washington DC: US Environmental Protection Agency, Office of Water; Irish Ramsar Wetlands Committee, 2018. *Irish Wetland Types: An identification guide and field survey manual*. Johnstown Castle: EPA.

3. Curtis, T.G.F. and Sheehy Skeffington, M.J., 1998. The Salt Marshes of Ireland: An inventory and account of their geographical variation. *Biology and Environment: Proceedings of the Royal Irish Academy*, 98(B): pp. 87–104; Ryle, T., Murray, A., Connolly, K. and Swan, M., 2009. *Coastal Monitoring Project Report 2004–2006*. National Parks and Wildlife Service. Dublin: Department of Arts, Heritage and the Gaeltacht; Department of Culture, Heritage and the Gaeltacht, 2007. Inventory of Irish Coastal Lagoons 2007.

4. Cott, G.M., Jansen, M.A.K. and Chapman, D.V., 2012. Salt-marshes on Peat Substrate: Where blanket bogs encounter the marine environment. *Journal of Coastal Research* 28(3), pp. 700–706.

5. IWWR. *An Introduction and User's Guide to Wetland Restoration, Creation and Enhancement*.

6. Ramsar Convention on Wetlands, 2018. *Global Wetland Outlook: State of the world's wetlands and their services to people*. Gland: Ramsar Convention Secretariat.

7. Curtis and Sheehy Skeffington, *The Salt Marshes of Ireland*.

8. Delaney, C., Devoy, R.J.N. and Jennings, S., 2012. Mid to Late Holocene Relative Sea-level and Sedimentary Changes in Southwest Ireland. In: Duffy, P.J. and Nolan, W., eds. *At the Anvil: Essays in honour of William J. Smith*. Dublin: Geography Publications, pp. 697–746.

9. McCorry, M. and Ryle, T., 2009. *Saltmarsh Monitoring Project 2007–2008: Final report*. Dublin: National Parks and Wildlife Service.

10. Nellemann, C., Corcoran, E., Duarte, C., Valdés, L., De Young, C., Fonseca, L. and Grimsditch, G., 2009. *Blue Carbon: A rapid response assessment*. United Nations Environment Programme, GRID-Arendal, 80.

11. Duarte, C.M., Losada, I.J., Hendriks, I. and Mazarrasa, I., 2013. The Role of Coastal Plant Communities for Climate Change Mitigation and Adaptation. *Nature of Climate Change*, 3, pp. 961–968; Pendleton, L., Donato, D.C., Murray, B.C., Crooks, S., Jenkins, W.A., Sifleet, S., Craft, C., Fourqurean, J.W., Kauffman, J.B., Marba, N., Megonigal, J.P., Pidgeon, E., Herr, D., Gordon, D. and Baldera, A., 2012. Estimating Global 'Blue Carbon' Emissions from Conversion and Degradation of Vegetated Coastal Ecosystems. *PLOS ONE*, 7(9), e43542.

12. Adam, P., 1990. *Saltmarsh Ecology*. Cambridge UK: Cambridge University Press.

13. Cummins, H.A., 1930. Experiments on the Establishment of Rice Grass (*Spartina townsendii*) in the Estuary of the Lee. *Economic Proceedings (Royal Dublin Society)* 2, pp. 419–422.

14. Cott, G.M., Chapman, D.V. and Jansen, M.A.K., 2013. Salt Marshes on Substrate Enriched in Organic Matter: The case of ombrogenic Atlantic salt marshes. *Estuaries and Coasts*, 36: pp. 595–609.

15. McCorry, M., Curtis T. and Otte, M., 2003. Spartina in Ireland. In: Otte, M., ed. *Wetlands of Ireland: Distribution, ecology, uses and economic value*. Dublin: University College Dublin Press.

16. Austin, G.E. and Rehfisch, M.M., 2003. The Likely Impact of Sea Level Rise on Waders (*Charadrii*) Wintering on Estuaries. *Journal of Nature Conservation*, 11, pp. 43–58.

17. Devoy, R.J.N., 2008. Coastal Vulnerability and the Implications of Sea-level Rise for Ireland. *Journal of Coastal Research*, 24 (2), pp. 325–341; Devoy, R.J.N., 2015. Sea-level Rise, Causes, Impacts and Scenarios for Change. In: Ellis, J.T. and Sherman, D.J., eds. *Coastal and Marine Hazards and Disasters*. New York: Elsevier; IPCC, 2014; *Contribution of Working Group II to the Fifth Assessment Report of the Intergovernmental Panel on Climate Change 2014: Impacts, adaptation and vulnerability*. Cambridge: Cambridge University Press.

18. European Commission, 2003. *Common Implementation Strategy for the Water Framework Directive (2000/60/EC) Guidance Document No 12, Horizontal Guidance on the Role of Wetlands in the Water Framework Directive*. Luxembourg: Office for Official Publications of the European Communities.

19. Government of Ireland, 2019. *Climate Action Plan to Tackle Climate Breakdown*. Dublin.

20. Irish Ramsar Wetlands Committee, n.d. Ireland's Wetlands. [online] Available at: http://irishwetlands.ie/irelands-wetlands.

21. BirdWatch Ireland, 2011. *Action Plan for Shore and Lagoon Birds in Ireland 2011–2020: BirdWatch Ireland's group action plans for Irish birds*. Kilcoole: BirdWatch Ireland.

22. NPWS, 2019. The Status of Protected EU Habitats and Species in Ireland. Overview Volume 1, Lynn D. and O'Neill, F., eds. Unpublished Report, National Parks and Wildlife Services. Dublin: Department of Arts, Heritage and the Gaeltacht.

Estuaries and Lagoons

1. Dyer, K.R., 1997. *Estuaries: A physical introduction*. 2nd Edition. New York: John Wiley.

2. Sinnott, A.M. and Devoy, R.J.N., 1992. The Geomorphology of Ireland's Coastline: Patterns, processes and future prospects. In: *Hommes et Terres du Nord* (1992), Les littoraux, pp. 145–153; Cooper, J.A.G., 2006. Geomorphology of Irish Estuaries: Inherited and dynamic controls. *Journal of Coastal Research*, Special Issue No. 39. Proceedings of the 8th International Coastal Symposium (ICS 2004), I (Winter 2006), pp. 176–180.

3. Carter, R.W.G. and Woodroffe, C.D., 1994. *Coastal Evolution: Late Quaternary shoreline morphodynamics*. Cambridge: Cambridge University Press.

4. Devoy, R.J.N., 2005. Cork City and the Evolution of the Lee Valley. In: Crowley, J., Devoy, R.J.N., Linehan, D. and O'Flanagan, eds. *Atlas of Cork City*. Cork: Cork University Press.

5. Dyer, *Estuaries: A physical introduction*; Cooper, Geomorphology of Irish Estuaries: Inherited and dynamic controls, pp. 176–180.

6. Ní Longphuirt, S., O'Boyle, S., Wilkes, R., Dabrowski, T. and Stengel, D.B., 2015. Influence of Hydrological Regime in Determining the Response of Macroalgal Blooms to Nutrient Loadings in Two Irish Estuaries. *Estuaries and Coasts*, 39(2), pp. 478–494; Ní Longphuirt, S., Mockler, E.M., O'Boyle, S., Wynne, C. and Stengel, D.B., 2016. Linking Changes in Nutrient Source Load to Estuarine Responses: An Irish perspective. *Biology and Environment*. Proceedings of the Royal Irish Academy 2016.

7. Open University, 1989. *Waves, Tides and Shallow Water Processes*. Milton Keynes: The Open University and Pergamon Press; McLusky, D.S. and Elliott, M., 2004. *The Estuarine Ecosystem*. Oxford: Oxford University Press.

8. Orford, J.D., 1988. Coastal Processes: The coastal response to sea-level variation. In: Devoy, R.J.N., ed. *Sea Surface Studies*. Kent: Croom Helm, pp. 415–463.

9. Ní Longphuirt et al., 2016. Linking Changes in Nutrient Source Load to Estuarine Responses: An Irish perspective; Ní Longphuirt, S., McDermott, G., O'Boyle, S., Wilkes, R. and Stengel, D.B., 2019. Decoupling Abundance and Biomass of Phytoplankton Communities Under Different Environmental Controls: A new multi-metric index. *Front. Mar. Sci.* 6:312; Environmental Protection Agency, 2019. *Water Quality in Ireland 2013–2018*. Wexford: EPA.

10. Ducrotoy, J., Elliott., M., Cutts, N., Franco, A., Little, S., Mazik, K. and Wilkinson, M., 2019. Temperate Estuaries: Their ecology under future environmental changes. In: Wolanski, E., Day, J., Elliott, M., and Ramesh, R., eds. *Coasts and Estuaries: The future*. Amsterdam: Elsevier.

11. McAllen, R., Davenport, J., Bredendieck, K. and Dunne, D., 2009. Seasonal Structuring of a Benthic Community Exposed to Regular Hypoxic Events. *Journal of Experimental Marine Biology and Ecology*, 368, pp. 67–74.

12. Environmental Protection Agency. *Water Quality in Ireland 2013–2018*.

13. Lewis, L.J., Kelly, T.C. and Davenport, J., 2014. Black-tailed Godwits (*Limosa limosa islandica*) and Redshanks (*Tringa totanus*) Respond Differently to Macroalgal Mats in their Foraging Areas. *Wader Study Group Bulletin*, 121, pp. 21–29; Wan, A.H.L., Wilkes, R.J., Heesch, S., Bermejo, R., Johnson, M.P. and Morrison, L., 2017. Assessment and Characterisation of Ireland's Green Tides (Ulva Species). *PLOS ONE* 12(1).

14. Environmental Protection Agency, *Water Quality in Ireland 2013–2018*.

15. Wolanski et al. *Coasts and Estuaries: The future*; Climate Change Advisory Council, 2019. *Annual Review 2019*. Dublin: Climate Change Advisory Council.

16. Chapman, P., 2012. Management of Coastal Lagoons under Climate Change. *Estuarine, Coastal and Shelf Science*, 110, pp. 32–35.

17. Kjerve, B., 1994. *Coastal Lagoon Processes*. Elsevier Oceanography Series, 60. Elsevier Science Publishers BV; CEC, 1992. European Habitats Directive (92/43/EEC).

18. Oliver, G.A., 2007. Conservation Status Report: Coastal Lagoons (1150). Unpublished report to the National Parks and Wildlife Service, Dublin; Oliver, G.A., 2013. *Report on the Main Results of the Surveillance under Article 17 for Annex I Habitat Types (Annex D). Coastal lagoons*, 1150. Report to the National Parks and Wildlife Service, Department of the Arts, Heritage and the Gaeltacht, Ireland; Beer, N.A. and Joyce, C.B., 2013. North Atlantic Coastal Lagoons: Conservation, management and research challenges in the twenty-first century. *Hydrobiologia*, 701, 1–11; 25.

19. Oliver, G.A., 2005. Seasonal Changes and Biological Classification of Irish Coastal Lagoons. PhD Thesis. University College Dublin; Lettice, S., 2014. Environmental and Biological Characteristics of Lagoons on the Southwest Coast of Ireland. PhD Thesis. University College Cork.

20. Lettice, S., 2014. Environmental and Biological Characteristics of Lagoons on the Southwest Coast of Ireland.

21. Healy, B., Oliver, G. A., Hatch, P., and Good, J.A., 1997. *Coastal Lagoons in the Republic of Ireland. Volume 1. Background, outline and summary of the survey*. Report to the National Parks and Wildlife Service, Dublin.

22. Oliver, G.A., 2007. Conservation Status Report.

SECTION 3: PEOPLE AND THE COAST

Imagining Coasts

1. Foley, R., 2010. *Healing Waters: Therapeutic landscapes in historic and contemporary Ireland*. Aldershot: Ashgate; Ryan, A., 2012. *Where Land Meets Sea: Coastal explorations of landscape, representation and spatial experience*. Farnham: Ashgate.
2. Brown, M. and Humberstone, B., eds. 2014. *Seascapes: Shaped by the sea*. Farnham: Ashgate; Foley, R., Kearns, R., Kistemann, T. and Wheeler, B., eds. 2019. *Blue Space, Health and Wellbeing: Hydrophilia unbounded*. Abingdon and New York: Routledge.
3. Wylie, J.W., 2017. Vanishing Points: An essay on landscape, memory and belonging. *Irish Geography*, 50(1), pp. 3–18.
4. Kearns, R., and Collins, D., 2012. Feeling for the Coast: The place of emotion in resistance to residential development. *Social and Cultural Geography*, 13(8), pp. 937–955.
5. Massey, D., 2005. *For Space*. London: Sage Publications Ltd.
6. Löffler, S., 2015. Glas Journal: Deep mappings of a harbour or the charting of fragments, traces and possibilities. *Humanities*, 4, pp. 457–475.
7. Tóibín, C., 2011. *A Guest at the Feast: A memoir*. London: Penguin.
8. Lenček, L. and Bosker, G., 1998. *The Beach: The history of paradise on earth*. London: Secker and Warburg.
9. Shields, R., 1991. *Places on the Margin: Alternative geographies of modernity*. London: Routledge.
10. Ryan, *Where Land Meets Sea*.
11. Kearns and Collins, Feeling for the Coast.
12. Foley, *Healing Waters*.
13. Wood-Martin, C., 1892. *A History of Sligo: County and town from the close of the revolution of 1688 to the present time*. Dublin: Hodges, Figgis and Co., p. 410.
14. Walton, J., 1983. *The English Seaside Resort: A social history 1750–1914*. Leicester: Leicester University Press.
15. Houlihan, D., 2011. *Ballybunion: An illustrated history*. Dublin: The History Press.
16. Ó Dálaigh, B., ed. 1998. *The Strangers Gaze: Travels in County Clare, 1534–1950*. Ennis: CLASP Press.
17. Davies, K.M., 2007. *That Favourite Resort: The story of Bray, Co. Wicklow*. Bray: Wordwell.
18. Taylor, A., 1990. *Tramore: Echoes from a seashell*. Tramore: No publisher.
19. Department of Health, 2012. *Healthy Ireland: A framework for improved health and wellbeing, 2013–2025*. Dublin: Department of Health.
20. Foley, R. and Kistemann, T., 2015. Blue Space Geographies: Enabling health in place. *Health and Place*, 35, p. 158.
21. Bell, S.L., Foley, R., Houghton, F., Maddrell, A. and Williams, A.M., 2018. From Therapeutic Landscapes to Healthy Spaces, Places and Practices: A scoping review. *Social Science and Medicine*, 196, pp. 123–130.
22. Wheeler, B., White, M., Stahl-Timmins, W., Depledge, M., 2012. Does Living by the Coast Improve Health and Wellbeing? *Health and Place*, 18, pp. 1198–1201; White, M., Pahl, S., Ashbullby, K. and Herbert, S., 2013. Feelings of Restoration from Recent Nature Visits. *Journal of Environmental Psychology*, 35, pp. 40–51.
23. MacEvilly, B., and O'Reilly, M., 2016. *At Swim: A book about the sea*. Cork: The Collins Press.
24. Foley, R., 2017. Swimming as an Accretive Practice in Healthy Blue Space. *Emotion, Space and Society*, 22, pp. 43–51.
25. Fitzmaurice, R., 2017. *I Found my Tribe*. London: Chatto and Windus.
26. Olsberg SPI and Nordicity, 2017. *Economic Analysis of the Audiovisual Sector in the Republic of Ireland*. London: Olsberg SPI and Nordicity.
27. Sweeney, T., 2011. The Quiet Man – the movie that put Ireland on the tourist map. *Film Ireland*, 1 November.
28. *The Quiet Man*. 1952. [Film] Produced by John Ford. Studio City: Republic Pictures.
29. *Ryan's Daughter*. 1970. [Film] Directed by David Lean. Beverly Hills: Metro-Goldwyn-Mayer Studios Inc.
30. *Saving Private Ryan*. 1998. [Film] Directed by Steven Spielberg. Universal City: DreamWorks Pictures.
31. *Brooklyn*. 2015. [Film] Directed by John Crowley. Century City: 20th Century Fox.
32. *Star Wars: The Force Awakens*. 2015. [Film] Directed by J.J. Abrams. Burbank: Walt Disney Studios Motion Pictures.
33. *Star Wars: The Last Jedi*. 2017. [Film] Directed by Rian Johnson. Burbank: Walt Disney Studios Motion Pictures.
34. *Game of Thrones*. 2011–2019. [TV Programme] Created by David Benioff and D.B. Weiss. USA: HBO.
35. *Northern Ireland Screen*, 2017. Adding Value. Belfast: Northern Ireland Screen.
36. Ryan, *Where Land Meets Sea*.
37. Obrador-Pons, P., 2009. Building Castles in the Sand: Repositioning touch on the beach. *The Senses and Society*, 4(2), pp. 195–210.
38. Fiscina, J.E., Pakpour, M., Fall, A., Vandewalle, N., Wagner, C. and Bonn D., 2012. Dissipation in Quasistatically Sheared Wet and Dry Sand Under Confinement. *Physical Review E*, 86(2), pp. 020103(1)–020103(4).

Coastal Heritage

1. European Commission, 2018. *European Framework for Action on Cultural Heritage*. Commission staff working document.
2. This shift is expressed clearly in the Framework Convention on the Value of Cultural Heritage for Society (Faro Convention, 2005) from the Council of Europe.
3. Ireland's National Inventory of Intangible Cultural Heritage. Sea Currach Making. [online] Available at: https://nationalinventoryich.chg.gov.ie/sea-currach-making/
4. Colfer, B., 2004. *The Hook Peninsula, County Wexford*. Cork: Cork University Press.
5. Lankford, E., 2012. *An Logainmníocht in Oileán Cléire*. Cape Clear: Cape Clear Museum Society.
6. Camden Fort Meagher, 2019. Rescue Camden. [Online] Available at: http://www.camdenfortmeagher.ie/rescue-camden/
7. The Heritage Council, 2002. Bere Island County Cork Conservation Plan. Dublin: The Heritage Council.
8. Bardon, J., 2005. *A History of Ulster*. Newtownards: Blackstaff Press Ltd, pp. 431–445.
9. Nelson, C., 2014. 'Murderous Renegade' or Agent of the Crown? The riddle of Erskine Childers. *History Ireland*, 22(3), p. 34.
10. Brunicardi, D., 2012. *Haulbowline: The naval base and ships of Cork Harbour*. Cheltenham: The History Press; Nolan, L. and Nolan, J.E., 2009. *Secret Victory – Ireland and the war at sea 1914–1918*. Cork: Mercier Press.
11. Halpern, P.G., 1994. *A Naval History of World War I*. Annapolis: Naval Institute Press.
12. Minutes and Proceedings of the First Dáil of the Republic of Ireland 1919–1921. Message to the free nations of the World. 21 January 1919. Documents on Irish Foreign Policy. Reprint Dublin 1994.
13. Daly, M., Commission of Inquiry into Resources and Industries. In: Crowley, J., Ó Drisceoil, D., Murphy, M. and Borgonovo, J., 2017. *Atlas of the Irish Revolution*. Cork: Cork University Press, pp. 340–343.
14. Crowley et al., *Atlas of the Irish Revolution*, p. 339.
15. Chief Secretary for Ireland Weekly Survey of the State of Ireland, 18 July 1921. CAB 24/126/52. National Archives Kew.
16. For an example from the Mizen Head lighthouse, see Maurice Donegan, Bureau of Military History Statement 639, Irish Military Archives.
17. Cook, H.G., 1920. Explosives at Lighthouses. *Irish Independent*, 30 June, p. 6; Cook, H.G., 1920. Raids on Lighthouse Stations. *Irish Examiner*, 2 August, p. 3; Phelps, J.H., 1921. Raids Upon Lighthouses. *Freeman's Journal*, 5 July, p. 8.
18. For examples, see Bureau of Military History Statements 453, 779, 820, 986, 1018, 1179, 1309, 1493, 1495, 1518, 1524, 1528, 1570, 1611, Irish Military Archives.
19. Sheehan, W., 2009. *Hearts and Mines: The British 5th division, Ireland, 1920–1922*. Cork: The Collins Press, p. 67.
20. Sheehan, W., 2007. *Fighting for Dublin: The British battle for Dublin, 1919–1921*. Cork: The Collins Press, p. 50.
21. Sheehan, *Hearts and Mines*, pp. 66, 117–118, 304.
22. Kautt, W.H., 2014. *Ground Truths: British army operations in the Irish war of independence*. Dublin: Irish Academic Press, pp. 106, 145, 160–161.
23. Lynch, R., 2006. *The Northern IRA and the Early Years of Partition, 1920–1922*. Dublin: Irish Academic Press.
24. Borgonovo, J., 2011. *Military History of the Irish Civil War: The battle for Cork, July–August 1922*. Cork: Mercier Press.
25. Hopkinson, M., 2004. *Green Against Green: The Irish civil war*. Rev. ed. Dublin: Gill and Macmillan, pp. 142–171.
26. Linge, J., 1988. The Royal Navy and the Irish Civil War. *Irish Historical Studies*, 31(121), pp. 60–71; Crowley et al., *Atlas of the Irish Revolution*, p. 709.
27. Ireland, J.C., 1966. *The Sea and the Easter Rising*. Dublin: Maritime Institute of Ireland.
28. Sistermans, P. and Nieuwenhuis, O. Eurosion Case Study: Rosslare Wexford County (Ireland). Amersfoort: Royal Haskoning DHV. http://copranet.projects.cucc-d.de/files/000125_EUROSION_Rosslare.pdf
29. The National Archives, 1915. CAB37/132/5, report by the British ambassador, Washington, 21 July 1915.
30. Natural England, 2012. An Approach to Seascape Character Assessment: Natural England commissioned report NECR 105.
31. This is being rectified from a seaward perspective by the INSS and INFOMAR seabed mapping programme (2005–present).
32. There are some notable exceptions for example Molloy, J., 2004. *The Irish Mackerel Fishery and the Making of an Industry*. The Killybegs Fishermens' Organisation and the Marine Institute.
33. Bolton, J., 2008. *The Coastal Architecture of Dún Laoghaire Rathdown*. Dún Laoghaire: Dún Laoghaire Rathdown County Council.
34. Tully, D., 2006. *Audit of Maritime Collections*. Kilkenny: The Heritage Council.
35. Kelly, B. and Stack, M., eds. 2009. *Climate Change, Heritage and Tourism: Implications for Ireland's coasts and inland waterways*. Kilkenny: The Heritage Council.

The Inhabitants of Ireland's Early Coastal Landscapes

1. Woodman, P., 2015. *Ireland's First Settlers: Time and the Mesolithic*. Oxford: Oxbow Books.
2. Brace, S., Diekmann, Y., Booth, T.J., Faltyskova, Z., Rohland, N., Mallick, S., Ferry, M., Michel, M., Oppenheimer, J., Broomandkhoshbacht, N., Stewardson, K., Walsh, S., Kayser, M., Schulting, R., Craig, O.E., Sheridan, A., Pearson, M.P., Stringer, C., Reich, D., Thomas, M.G. and Barnes, I., 2018. Population replacement in early Neolithic Britain. bioRxiv.org/content/10.1101/267443v1
3. Jones, R.L. and Keen, D.H., 1993. *Pleistocene Environments in the British Isles*. London: Chapman and Hall.
4. Dowd, M. and Carden, R.F., 2016. First Evidence of a Late Upper Palaeolithic Human Presence in Ireland. *Quaternary Science Reviews*, 139, pp. 158–163; Dalton, E., 2018. Artefacts Found in Waterford Could Reveal 10,000-year-old Settlement. *The Irish Times*, 22 August; O'Keeffe, A. 2021. First humans came here 33,000 years ago, reindeer bones show. Independent Newspapers, https://www.independent.ie/irish-news/news/first-humans-came-here33000-years-ago-reindeer-

bones-show-40326319.html.

5. Devoy, R.J., 1985. The Problem of a Late Quaternary Landbridge between Britain and Ireland. *Quaternary Science Reviews*, 4(1), pp. 43–58; Devoy, R.J.N., 2005. Cork City and the Evolution of the Lee Valley. In: Crowley, J., Devoy, R.J.N., Linehan, D. O'Flanagan, P., and Murphy, M. eds. *Atlas of Cork City*. Cork: Cork University Press.

6. Nesje, A. and Dahl, S.O., 2000. *Glaciers and Environmental Change*. London: Arnold.

7. Coxon, P., McCarron, S. and Mitchell, F.J.G., 2017. *Advances in Irish Quaternary Studies*. Paris: Atlantis Press.

8. Devoy, R.J.N., 1995. Deglaciation, Earth Crustal Behaviour and Sea-level Changes in the Determination of Insularity: A perspective from Ireland. *The Geological Society*, 96(1), pp. 181–208; Edwards, R. and Craven, K., 2017. Relative Sea-level Change around the Irish Coast. In: P. Coxon, S. McCarron and F.J.G. Mitchell eds. *Advances in Irish Quaternary Studies*. Paris: Atlantis Press, pp. 181–215.

9. Edwards, R.J. and Brooks, A.J., 2008. The Island of Ireland: Drowning the myth of an Irish landbridge. In: Davenport, J.L., Sleeman, D.P. and Woodman, P.C. eds. *Mind the Gap: Postglacial colonisation of Ireland*. Special Supplement to the I*rish Naturalists' Journal*, pp. 19–34.

10. Woodman, P.C., McCarthy, M. and Monaghan, N., 1997. The Irish Quaternary Faunas Project. *Quaternary Science Reviews*, 16, pp. 129–159; Montgomery, W.I., Provan, J., McCabe, A.M. and Yalden, D.W., 2014. Origin of British and Irish Mammals: Disparate postglacial colonisation and species introductions. *Quaternary Science Reviews*, 98, pp. 144–165.

11. Sleeman, D.P., Devoy, R.J.N. and Woodman, P.C., 1986. Proceedings of the Postglacial Colonisation Conference. *Occasional Publication of the Irish Biogeographical Society*, 1, pp. 1–88; Davenport, J.L., Sleeman, D.P. and Woodman, P.C., eds. 2008. *Mind the Gap*.

12. Birks, H.H., Birks, H.J.B., Kaland, P.E. and Moe, D. 1988. *The Cultural Landscape: Past, present and future*. Cambridge: Cambridge University Press.

13. Sleeman, D.P., Devoy, R.J.N. and Woodman, P.C., 1986. Proceedings of the Postglacial Colonisation Conference, pp. 1–88.

14. Dowd and Carden, First Evidence of a Late Upper Palaeolithic Human Presence in Ireland, pp. 158–163; O'Keeffe, First humans came here 33,000 years ago, reindeer bones show, https://www.independent.ie/irish-news/.html

15. Dalton, Artefacts Found in Waterford Could Reveal 10,000-year-old Settlement.

16. Zong, Y. and Tooley, M.J., 1996. Holocene Sea–level Changes and Crustal Movements in Morecombe Bay, Northwest England. *Journal of Quaternary Science*, 11, pp. 43–58.

17. Woodman, *Ireland's First Settlers: Time and the Mesolithic*.

18. Devoy, R.J.N., 2016. *Fanore Beach and Dune Management Report: Current problems and planning for the future*. Marine and Renewable Energy Institute (MaREI), University College Cork and Clare County Council.

19. Collins, A.E.P., 1952. Excavations in the Sandhills at Dundrum, Co. Down 1950–1951, *Ulster Journal of Archaeology*, 15, pp. 91–123.

20. Devoy, R.J.N., 2015. The Development and Management of the Dingle Bay Spit–barriers of Southwest Ireland. In: Randazzo, G., Cooper, J.A.G. and D.W. Jackson eds. *Sand and Gravel Spits*. Cham: Springer International Publishing, pp. 139–180.

21. Woodman, P.C., 2004. The Exploitation of Ireland's Coastal Resources: A marginal resource through time. In: Gonzalez, M.R. and Clark, G.A eds. *The Mesolithic of the Atlantic Façade*. Tempe: Arizona State University, p. 37.

22. Devoy, R.J.N. and Woodman, P.C., 1986. Site 13, Dún an Óir. In: Warren, W.P., ed. *Irish Quaternary Association Field Guide Number 9: Corca Dhuibhne*. Dublin: Irish Association for Quaternary Studies.

23. Woodman, *Ireland's First Settlers*.

24. Woodman, *Ireland's First Settlers*.

25. O'Kelly, M.J., 2008. *Early Ireland: An introduction to Irish prehistory*. Cambridge: Cambridge University Press; O'Connell, M., Molloy, K. and Jennings, E., 2020. Long-term human impact and environmental change in mid-western Ireland, with particular reference to Céide Fields: An overview. E&G – *Quaternary Science Journal*, 69(1), pp. 1–32, https://doi.org/10.5194/egqsj-69-1-2020.

26. Cunliffe, B., 2001. *Facing the Ocean: The Atlantic and its peoples 8000BC–AD1500*. Oxford and New York: Oxford University Press.

27. O'Kelly, M.J., 1952. Three Promontory Forts in Co. Cork. *Proceedings of the Royal Irish Academy. Section C: Archaeology*, 55, pp. 25–59.

28. Barker, L. and Driver, T., 2011. Close to the Edge: New perspectives on the architecture, function and regional geographies of the coastal promontory forts of the Castlemartin Peninsula, south Pembrokeshire, Wales. *Proceedings of the Prehistoric Society*, 77, pp. 65–87.

29. Mytum, H., 1992. *The Origins of Early Christian Ireland*. London: Routledge.

30. Proudfoot, V. B. and Wilson, B.C.S., 1961. Further Excavations at Larrybane Promontory Fort, Co. Antrim. *Ulster Journal of Archaeology*, 24/25, pp. 91–115.

The Vikings and Normans: Coastal invaders and settlers

1. Barrett, J.H., 2008. What Caused the Viking Age? *Antiquity*, 82(317), pp. 671–685.

2. O'Meara, J.J., 1978. *The Voyage of Saint Brendan: Journey to the Promised Land*. Navigatio Sancti Brendani Abbatis / Translated with an introduction by John J. O'Meara. Dublin: Dolmen Press.

3. Wooding, J.M., 2011. The date of Navigatio S. Brendani Abbatis. *Studia Hibernica*, 37, pp. 9–26.

4. O'Meara, *The Voyage of Saint Brendan*, pp. xi–xv.

5. Severin. T., 1979. *The Brendan Voyage: Across the Atlantic in a leather boat*. Reprint 2005. Dublin: Gill and Macmillan.

6. Severin, *The Brendan Voyage*, p. 234.

7. Clarke, H.B. and Johnson R. eds. 2015. *The Vikings in Ireland and Beyond: Before and after the battle of Clontarf*. Dublin: Four Courts Press.

8. Rynne, C., 2009. Water-Power as a Factor of Industrial Location in Early Medieval Ireland: The environment of the early Irish water mill. *Industrial Archaeology Review*, 31, pp. 85–95.

9. McErlean, T. and Crothers, N., 2007. *Harnessing the Tides: The early medieval tide mills at Nendrum Monastery, Strangford Lough*. Norwich: The Stationery Office.

10. Purcell, E., 2015. The First Generation in Ireland, 795–812: Viking raids and Viking bases? In: Clarke and Johnson, eds. *The Vikings in Ireland and Beyond*, pp. 41–54.

11. Fellows-Jensen, G., 2015. Through a Glass Darkly: Some sidelights on Viking influence on personal names and place-names in Ireland. In: Clarke and Johnson, eds. *The Vikings in Ireland and Beyon*d, pp. 268–283.

12. Mac Giolla Easpaig, D., 2002. L'influence Scandinave sur la Toponymie Irlandaise. In: Ridel, É, ed, *L'Héritage Maritime Des Vikings en Europe de l'Ouest: Colloque International de la Hague (Flottemanville-Hague, 30 Septembre-3 Octobre 1999)*. Caen: Presses Universitaires de Caen, pp. 441–482.

13. Russell, I. and Hurley M.F., 2014. *Woodstown: A Viking-age settlement in Co. Waterford*. Dublin: Four Courts Press.

14. Sheehan, J., 2008. The *Longphort* in Viking Age Ireland. *Acta Archaeologica*, 79, pp. 282–295.

15. Wallace, P.F., 1992. The Archaeological Identity of the Hiberno-Norse town. *Journal of the Royal Society of Antiquaries of Ireland*, 122, pp. 35–66.

16. Wallace, P.F., 2015. *Viking Dublin: The Wood Quay excavations*. Dublin: Irish Academic Press.

17. Bradley, J., 1988. The Interpretation of Scandinavian Settlement in Ireland. In: Bradley, J. ed, *Settlement and Society in Medieval Ireland: Studies presented to F.X. Martin OSA*. Kilkenny: Boethius Press, pp. 49–78.

18. O'Kelly, M.J., 1956. An Island Settlement at Beginish, Co. Kerry. *Proceedings of the Royal Irish Academy*, 57, pp. 159–194.

19. Sheehan, J., Stummann Hansen, S. and Ó Corráin, D., 2001. A Viking-age Maritime Haven: A re-assessment of the island settlement at Beginish, Co. Kerry. *Journal of Irish Archaeology*, 10, pp. 93–119.

20. MacCotter, P., Ecclesiastical and Royal Estates in Corcu Duibne. In: Ó Carragáin, T. *Churches in the Irish Landscape*, pp. 400–1100.

21. Ó Carragáin, T., Mac Cotter, P. and Sheehan J., 2011. Making Christian Landscapes in Corcu Duibne. *Group for the Study of Irish Historic Settlement Newsletter*, pp. 1–5.

22. O'Keeffe, T., 2000. *Medieval Ireland: An archaeology*. Stroud: Tempus Publishing.

23. Duffy, S. ed. 2005. *Medieval Ireland: An encyclopedia*. New York and London: Routledge.

24. O'Sullivan, H., 2006. I*rish Historic Towns Atlas: Dundalk*. Dublin: Royal Irish Academy.

25. Bradley, J., 1978. The Topography and Layout of Medieval Drogheda. *Journal of the County Louth Archaeological and Historical Society*, 19(2), pp. 98–127.

26. Halpin, A., 2000. *The Port of Medieval Dublin: Archaeological excavations at the Civic Offices, Winetavern Street, Dublin, 1993*. Dublin: Four Courts Press.

27. O'Neill, T., 1987. *Merchants and Mariners in Medieval Ireland*. Dublin: Irish Academic Press.

28. McEneaney, E. and Ryan, R. eds. 2004. *Waterford Treasures: A guide to the historical and archaeological treasures of Waterford city*. Waterford: Waterford Museum of Treasures.

29. Brady, K., 2008. *Shipwreck Inventory of Ireland: Louth, Meath, Dublin and Wicklow*. Dublin: Wordwell Books.

30. Cunningham, B., 2018. *Medieval Irish Pilgrims to Santiago de Compostela*. Dublin: Four Courts Press.

31. Hurst, J., 1988. Medieval Pottery Imported into Ireland. In: Mac Niocaill, G. and Wallace, P. eds. *Keimelia: Studies in medieval archaeology and history in memory of Tom Delaney*. Galway: Galway University Press, pp. 229–253.

32. Childs, W. and O'Neill, T., 1987. Overseas Trade. In: Cosgrove, A. ed, *A New History of Ireland, vol. ii: Medieval Ireland, 1169–1534*. Oxford: Oxford University Press, pp. 492–524.

33. Murphy, M. and Potterton, M., 2010. *The Dublin Region in the Middle Ages: Settlement, land-use and economy*. Dublin: Four Courts Press.

34. McErlean, T., McConkey, R. and Forsythe, W., 2002. *Strangford Lough: An archaeological survey of the maritime cultural landscape*. Newtownards: Blackstaff Press Ltd.

35. O'Sullivan, A., 2001. *Foragers, Farmers and Fishers in a Coastal Landscape: An intertidal archaeological survey of the Shannon estuary*. Dublin: Royal Irish Academy.

36. O'Sullivan, A., McErlean, T., McConkey, R. and McCooey, P., 1997. Medieval Fishtraps in Strangford Lough, Co. Down. *Archaeology Ireland*, 39, pp. 36–38.

37. Murphy and Potterton, *The Dublin Region in the Middle Ages*.

38. Colfer, *The Hook Peninsula*.

39. Colfer, B., 2013. *Wexford Castles: Landscape, context and settlement*. Cork: Cork University Press.

Era of Settlement: Trade, plantation and piracy

1. Smyth, W.J., 2006. *Map-Making, Landscapes and Memory: A geography of colonial and early modern Ireland c.1530–1750*. Cork: Cork University Press. This text provides a comprehensive review of the major changes that occurred over these two hundred years and which helped transform the country and its coastal areas into the early modern era.

2. Breen, C., 2001. The Maritime Cultural Landscape in Gaelic Medieval Ireland. In: Duffy, P.J., Edwards, D. and FitzPatrick, E. eds. *Gaelic Ireland c.1250–c.1650: Land, lordship and settlement*. Dublin: Four Courts Press, pp. 418–430.

3. For evidence of pre-1700 housing in Ireland's coastal cities, a number of useful chapters relating to the post-medieval archaeology of Dublin, Carrickfergus, Belfast, Derry and Galway can be found in Horning, A., Ó Baoill, R., Donnelly, C. and Logue, P., 2007. *The Post-medieval Archaeology of Ireland, 1550–1850*. Dublin: Wordwell Press.

4. O'Grady, S., 1896. *Pacata Hibernia: A history of the wars in Ireland during the reign of Queen Elizabeth*. London: Downey and Co.

5. Breen, C., 2005. *The Gaelic Lordship of the O'Sullivan Beare: A landscape cultural history*. Dublin: Four Courts Press, pp. 204–206.

6. Appleby, J.C., 2009. *Under the Bloody Flag: Pirates of the Tudor age*. Stroud: History Press, pp. 146–147.

7. O'Neill, T., 1987. *Merchants and Mariners in Medieval Ireland*. Newbridge: Irish Academic Press, p. 126.

8. Kelleher, C., 2014. Depicting a Pirate Landscape: The anti-pirate chart from 1612 and archaeological footprints on the ground. *Journal of Irish Archaeology*, 22, pp. 77–92; Gerritszoon, H. and Hunt, J., 1612. Beschrijvinghe van de Zeecusten ende Havenen van Yerlandt/Description of the Seacoasts and Ports of Ireland. Göttingen: Georg–August–Universität, SUB Göttingen. 4 H BRIT P III, 6 RARA: 23 pages, including 4 charts of Ireland.

9. Cullen, L.M., 1968. The Smuggling Trade in Ireland in the Eighteenth Century. *Proceedings of the Royal Irish Academy*, 67(5), pp. 149–175.

10. Chambers, A., 2009. *Granuaile – Grace O'Malley – Ireland's Pirate Queen*. Dublin: Gill and Macmillan, p. 85.

11. Chambers, A., 2003. *Pirate Queen of Ireland: The true story of Grace O'Malley*. New York: MJF Books, p. 52.

12. Breen, *The Maritime Cultural Landscape in Gaelic Medieval Ireland*, p. 426.

13. Doran, L. and Lyttleton, J., 2007. *Lordship in Medieval Ireland: Image and reality*. Dublin: Four Courts Press.

14. Breen, *The Gaelic lordship of the O'Sullivan Beare*, pp. 90–98, 102–106 and 142–150.

15. Horning, A., 2013. *Ireland in the Virginian Sea: Colonialism in the British Atlantic*. Chapel Hill: University of North Carolina Press, pp. 36–42 and 56–60.

16. Maginn, C., 2018. Continuity and Change, 1470–1550. In: Smith, B. ed. *The Cambridge History of Ireland, Vol. 1, 600–1550*. Cambridge: Cambridge University Press, pp. 315–320.

17. Margey, A., 2018. Plantations, 1550–1641. In: Ohlmeyer, J. and Bartlett, T. eds. *The Cambridge History of Ireland, Vol. II. 1550–1730*. Cambridge: Cambridge University Press, p. 562.

18. Lennon, C., 1994. *Sixteenth-century Ireland: The incomplete conquest*. Dublin: Gill and Macmillan.

19. Canny, N., 2001. *Making Ireland British, 1580–1650*. Oxford: Oxford University Press, pp. 134–159.

20. Quinn, D.B., 1966. The Munster Plantation: Problems and opportunities. *Journal of the Cork Historical and Archaeological Society*, 71. pp. 19–40.

21. O'Flanagan, P., 1996. Bandon. In: Simms, A. and Andrews, J., eds. *Irish Historic Towns Atlas. Vol. 1, Kildare, Carrickfergus, Bandon, Kells, Mullingar, Athlone*. Dublin: Royal Irish Academy, pp. 1–2.

22. Robinson, P.S., 1994. *The Plantation of Ulster: British settlement in an Irish landscape, 1600–1670*. Belfast: Ulster Historical Foundation, p. 38.

23. Robinson, *The Plantation of Ulster*, pp. 63–65.

24. Margey, Building Early Modern Property Portfolios, pp. 70–71.

25. Margey, Building Early Modern Property Portfolios, pp. 69–81.

26. Margey, *The Cambridge History of Ireland, Vol. II. 1550–1730*, pp. 578–583.

17. Margey, A., 2018. Plantations, 1550–1641. In: Ohlmeyer, J. and Bartlett, T. eds. *The Cambridge History of Ireland, Vol. II. 1550–1730*. Cambridge: Cambridge University Press, p. 562.

27. Craig, M.J., 1982. *The Architecture of Ireland: From the earliest times to 1880*. Portrane: Lambay Books, p. 123.

28. Breen, C., 2012. *Dunluce Castle: History and archaeology*. Dublin: Four Courts Press.

29. Breen, The Gaelic Lordship of the O'Sullivan Beare, p. 160.

30. Moss, R., 2012. Reduce, Reuse, Recycle: Irish monastic architecture *c*.1540–1640. In: Stalley, R.A. ed. I*rish, Gothic Architecture: Construction, decay and reinvention*. Dublin: Wordwell Press, pp. 115–159.

31. Rodgers, N., 2007. *Ireland, Slavery and Anti-Slavery 1612–1865*. Basingstoke: Palgrave MacMillan. This work serves as a general reference for Ireland and Slavery: Coastal Connections that Became Bittersweet.

32. Canny, N., 1989. Early Modern Ireland, *c*.1500–1700. In: Foster, R.F. ed. *Ireland*. Oxford, Oxford University Press, pp. 104–160.

33. Canny, *Ireland*, p. 155.

Changing Coastal Landscapes

1. For an evolution of Dublin Port and city see: Clarke, H.B., 2002. *Irish Historic Towns Atlas. Number 11. Dublin. Part I, to 1610*. Dublin: Royal Irish Academy; Lennon, C., 2008. *Irish Historic Towns Atlas. Number 19. Dublin. Part II, 1610 to 1756*. Dublin: Royal Irish Academy; Goodbody, R., 2014. *Irish Historic Towns Atlas. Number 26. Dublin. Part III, 1756 to 1847*. Dublin: Royal Irish Academy.

2. Murray, K., 1954. The Atmospheric Railway Episode. *Journal of the Irish Railway Record Society*, 14(3), pp. 169–181; O'Flanagan, F. M., 1941–1942. Glimpses of Old Dalkey. *Dublin Historical Record*, IV, pp. 41–57.

3. Nairn, R., Jeffrey, D. and Goodbody, R., 2017. *Dublin Bay: Nature and history*. Cork: The Collins Press; Gilligan, H.A., 1988. *A History of the Port of Dublin*. Dublin: Gill and Macmillan.

4. Horner, A. 2013. Dún Laoghaire's Great Harbour. *History Ireland*, 21(5), pp. 24–27.

5. O'Flanagan, P., 1998. 'The Cork Region': Cork and County Cork, c.1600–c.1900. In: Brunt, B.M. and Hourihan, K. eds *Perspectives on Cork*. Dublin: Geographical Society of Ireland, pp. 1–18.

6. O'Flanagan, P., 2005. Beef, Butter Provisions and Prosperity in a Golden Eighteenth Century. In: Crowley, J., Robert, D., Linehan, D., O'Flanagan, P. and Murphy, M. eds. *Atlas of Cork City*. Cork: Cork University Press, pp. 149–159.

7. Rynne, C., 1998. *At the Sign of the Cow: The Cork Butter Market, 1770–1924*. Cork: The Collins Press.

8. Owen, D.J., 1917. *A Short History of the Port of Belfast*. 1st ed. Belfast: Mayne Boyd.

9. Sweetnam, R. and Nimmons, C., 1985. *Port of Belfast 1785–1985: An historical review*. Belfast: Belfast Harbour Commissioners.

10. Royle, S.A., 2011. *Portrait of an Industrial City: 'Clanging Belfast', 1750–1914*. Belfast: Ulster Historical Foundation.

11. O'Flaherty, E., 2010. Limerick. In: *Irish Historical Towns Atlas No. 21*. Dublin: Royal Irish Academy.

12. Hoctor, M., 2003. The Contribution of Limerick Docks to the Commercial Development of Limerick. In: Lee, D. and Jacobs, D. eds. *Made in Limerick. Vol. 1, History of industries, trade and commerce*. Limerick: Limerick Civic Trust in association with FÁS and Shannon Development.

13. Fenton, L., 2011. Economy of Limerick in the Aftermath of the Great Famine. *North Munster Antiquarian Journal*, 51, pp. 93–103.

14. O'Brien, N., 2008. *Blackwater and Bride: Navigation and trade, 7,000 BC to 2007*. Ballyduff Upper: Niall O'Brien Publishing.

15. Kelly, D. and O'Keeffe, T., 2015. *Irish Historic Towns Atlas*. Dublin: Royal Irish Academy.

16. Ferris, T., 2008. *Irish Railways: A new history*. Dublin: Gill and Macmillan; Winchester, C., 2014. Railways of Ireland: 180th Anniversary 1834–2014. Stroud: Amberley Publishing; O'Connor, K., 1999. *Ironing the Land: The coming of the railways to Ireland*. Dublin: Gill and Macmillan.

17. James, K.J., 2014. *Tourism, Land, and Landscape in Ireland: The commodification of culture*. New York: Routledge.

18. McCutcheon, W.A., 1970. *Railway History in Pictures: Ireland*. Newton Abbot: David and Charles.

19. O'Connor, B. and Cronin, M. eds. 1993. *Tourism in Ireland: A critical analysis*. Cork: Cork University Press; Cusack, T., 2010. 'Enlightened Protestants': The improved shorescape, order and liminality at early seaside resorts in Victorian Ireland. *Journal of Tourism History* 2(3), pp. 165–185.

20. Burke, T., 1974. County Cork in the Eighteenth Century. *Geographical Review*, 64(1), pp. 61–81.

21. O'Flanagan, P. and Buttimer, C.G., 1993. *Cork: History and society*. Dublin: Geography Publications.

22. Mac Cárthaigh, C., 2008. *Traditional Boats of Ireland: History, folklore and construction*. Cork: The Collins Press.

23. Mac Cárthaigh. *Traditional Boats of Ireland*, p. 526.

24. O'Flanagan and Buttimer. *Cork*, p. 437.

25. Wilkins, N P., 2009. *Alexander Nimmo, Master Engineer: Public works and civil surveys*. Newbridge: Irish Academic Press.

26. Wilkins, N.P., 2017. *Humble Works for Humble People. A history of the fishing piers of County Galway, 1800–1922*. Newbridge: Irish Academic Press.

27. Lee, J.J., 1973. *The Modernisation of Irish Society, 1848–1918*. Dublin: Gill and Macmillan.

The Great Famine

1. For an excellent, comprehensive and well-illustrated account of the Famine, and one that serves as the key reference for this chapter, see Crowley, J., Smyth, W.J. and Murphy, M., eds. 2012. *Atlas of the Great Irish Famine*. Cork: Cork University Press. See also Donnelly, J.S. Jr, 2001. *The Great Irish Potato Famine*. Stroud: Sutton Publishing and Ó Gráda, C., Paping, R. and Vanhaute, E. eds. 2007. *When the Potato Failed: Causes and effects of the last european subsistence crisis, 1845–1850*. Turnhout: Brepols.

2. Commander Caffin. Letter to Commander Hamilton, February 1847. In: *Correspondence from January to March 1847 Relating to the Measures Adopted for the Relief of the Distress in Ireland. Commissariat Series. Second Part*. London: W. Clowes and Sons, p. 163.

3. Swords, L., 1999. *In Their Own Words: The Famine in north Connacht 1845–49*. Dublin: The Columba Press.

4. Johnson, J., 1995. The Quaker Relief Effort in Waterford. In: Cowman, D. and Brady, D. eds. *The Famine in Waterford 1845–1850: Teacht na bprátaí dubha*. Dublin: Geography Publications in association with Waterford County Council, p. 225.

5. Central Relief Committee of the Society of Friends (Dublin, Ireland), 1852. *Transactions of the Central Relief Committee of the Society of Friends during the Famine in Ireland in 1846 and 1847*. (Reprints from the collection of the University of Michigan Library). Dublin: Hodges and Smith.

6. Hatton, H.E., 1993. *The Largest Amount of Good: Quaker relief in Ireland, 1654–1921*. Kingston and Montreal: McGill-Queen's University Press.

7. Society of Friends. *Transactions of the Central Relief Committee*, p. 109.

8. Commissioners of the Public Works (Ireland), 1850. *Eighteenth Annual Report from the Board of Public Works Ireland with Appendices*, p. 190.

9. Kinealy, C., 2015. *A Death-dealing Famine: The great hunger in Ireland*. London and Chicago: Pluto Press, p. 136.

10. Smyth, W., 2012. Exodus from Ireland: Patterns of emigration. In: Crowley, Smyth, and Murphy, eds. *Atlas of the Great Irish Famine*, p. 498.

11. MacDonagh, O., 1976. Irish Famine Emigration to the United States. In: *Perspectives in American History*, Volume 10. Charles Warren Center for Studies in American History, p. 408.

12. Gray, P., 1995. *The Irish Famine*. London: Thames and Hudson.

13. Miller, K.A., 2012. Emigration to North America in the Era of the Great Famine, 1845–55. In: Crowley, Smyth, and Murphy, eds. *Atlas of the Great Irish Famine*, pp. 214–227.

14. Irish ports played an important role in providing emigrant ships offering direct routes to North America and elsewhere throughout the Famine period. However, the vast majority of Irish emigrants seeking passage to the United States or Canada first travelled by boat from east- and south-coast ports to Liverpool.

15. Scally, R.J., 1995. *The End of Hidden Ireland: Rebellion, famine and emigration*. Oxford: Oxford University Press, p. 176.

Ireland's Islands

1. George Bernard Shaw. Letter to Mabel Fitzgerald, 1 December 1914. In: O'Toole, F., 2017. *Judging Shaw: The radicalism of GBS*. Dublin: Royal Irish Academy, p. 123.

2. O'Toole, 2017. *Judging Shaw*.

3. Rourke, G.D., 2009. Skellig Michael: Monastic island retreat in the Atlantic. In: Crowley, J. and Sheehan, J., eds. *The Iveragh Peninsula: A cultural atlas of the Ring of Kerry*. Cork: Cork University Press, pp. 129–135.

4. Ó Corráin, D., 2009. The Vikings and Iveragh. In: Crowley, J. and Sheehan, J., eds. *The Iveragh Peninsula*, p. 142.

5. Cunningham, B., 2018. *Medieval Irish Pilgrims to Santiago de Compostela*. Dublin: Four Courts Press, p. 22.

6. Long, B., 1993. *Bright Light, Still Water*. Dublin: New Island Books, p. 119.

7. UNESCO, n.d. *Sceilg Mhichíl–UNESCO World Heritage Centre*. [online] Available at: https://whc.unesco.org/en/list/757

8. Heaney, S., 2018. *100 Poems*. London: Faber and Faber, p. 135.

9. Ó Síocháin, C., 1975. *The Man from Cape Clear*. Dublin: Mercier Press.

10. Royle, S.A., 1983. The Economy and Society of the Aran Islands, Co. Galway, in the Early Nineteenth Century. *Irish Geography*, 26(1), pp. 36–54.

11. Ó Síocháin, *The Man from Cape Clear*.

12. Royle, 1983. The Economy and Society of the Aran Islands.

13. Scanlan, D., 2003. *Memories of an Islander: A life on Scattery and beyond*. Ennis: Clasp Press.

14. Synge, J.M., 1907. *The Aran Islands*. Dublin: Maunsell; Thomson, G., 1982. *The Blasket That Was: The story of a deserted village*. Dublin: An Sagart.

15. Ó Criomhthain, T., 1929. *The Islandman*. Dublin: Muintir C.S. Ó Fallamhain, Teo; Ó Súilleabháin, M., 1933. *Fiche Blian ag Fás (Twenty Years A'growing)*. Dublin: Clólucht agus Talbóidig; Sayers, P., 1936. *Peig*. Dublin: Clólucht agus Talbóidigh; Sayers, P., 1939. *Machtnamh Seana-Mhná (An Old Woman's Reflections)*. Dublin: Oifig an tSoláthair.

16. Ó Catháin, G.C., 2014. *The Loneliest Boy in the World: The last child of the Great Blasket*. Cork: The Collins Press.

17. Ó Síocháin, *The Man from Cape Clear*.

18. McGowan, J., 2004. *Inishmurray: Island voices*. Sligo: Aeolus, p. 170.

19. Central Statistics Office, 2017. Population of Inhabited Islands off the Coast 2011 to 2016 by Sex, Islands and Census Year.

20. Yeats, W.B., 1961. *Essays and Introductions*. London: Macmillan, p. 299.

21. Gleason, P., 1991. Americans All: World War 2 and the shaping of American identity. In: Pozzetta, G.E., *Americanization, Social Control, and Philanthropy*. New York: Garland Publishing.

22. Synge, J.M., 1911. *The Aran Islands*. Boston: J.W. Luce and Company, p. 39.

23. Moran, D.P., 2007. *The Philosophy of Irish Ireland*. Dublin: University College Dublin Press.

24. Ashley, S., 2001. The Poetics of Race in 1890s' Ireland: An ethnography of the Aran Islands. *Patterns of Prejudice*, 35(2), pp. 5–18.

25. Ashley, The Poetics of Race in 1890s' Ireland.

26. Robinson, T., 1986. *Stones of Aran: Pilgrimage*. Dublin: Lilliput Press.

27. MacFarlane, R., 1986. Introduction. In: Robinson, T. *Stones of Aran*, p. 2.

28. McDonagh, M., 1998. *The Cripple of Inishmaan*. London: Heinemann.

29. Harvey, B., 1991. Changing Fortunes on the Aran Islands in the 1890s. *Irish Historical Studies*, 27(107), pp. 237–249.

30. Harvey, Changing fortunes on the Aran Islands in the 1890s.

31. Údarás na Gaeltachta, 2019. *Our Role*. [online] Available at: http://www.udaras.ie/en/faoin-udaras/ar-rol [Accessed 11 Jan. 2019]

32. Ernst and Young, 2014. *Review of the PSO Air Service for the Aran Islands: Report to the Department of Arts, Heritage and the Gaeltacht*. Dublin: Ernst and Young.

33. Aran Islands Energy Co-op, 2019. *Fuinnimh Oileáin Árann: Aran Islands Energy*. [online] Available at: http://www.aranislandsenergycoop.ie/

34. Clark, W., 1988. *Rathlin: Its island story*. 2nd ed. Limavady: North-West Books.

35. Elwood, J.H., 1968. The Population of Rathlin Island. *The Ulster Medical Journal*, 37(1), pp. 64–70.

36. Gage, C. and Dickson, J.M., 1995. *A History of the Island of Rathlin*. Ulster Local History Trust and North Eastern Education and Library Board. Ballymena: J.M. Dickson.

37. Gage and Dickson. *A History of the Island of Rathlin*.

38. Rathlin Community and Development Association, 2018. *Rathlin Community: Another world of sea, sky and shore*. [online] Available at: rathlincommunity.org

39. BBC Radio 4, 2015. Rathlin Island seaweed. *On Your Farm*. 23 August.

Underwater Cultural Heritage

1. Shields, Y., O'Connor, J. and O'Leary, J., 2005. *Ireland's Ocean Economy and Resources*. Galway: Marine Institute.

2. Brady, K., McKeon, C., Lyttleton, J. and Lawler, I., 2012. *Warships, U-boats and Liners: A guide to shipwrecks mapped in Irish waters*. Dublin: Stationery Office.

3. Breen, C. and Forsythe, W., 2004. *Boats and Shipwrecks of Ireland*. Stroud: Tempus Publishing.

4. Brady, K., 2008. *Shipwreck Inventory of Ireland. Louth, Meath, Dublin and Wicklow*. Dublin: Stationery Office.

5. Brady, K. and Moore, F., 2012. The Underwater Archaeology Unit and the Protection of Ireland's Shipwrecks. In: Brady, McKeon, Lyttleton, and Lawler, eds. *Warships, U-Boats and Liners*.

6. Fry, M., 2000. *Coití: Logboats from Northern Ireland*. Antrim: Greystone Press on behalf of the Environment and Heritage Service, p. 116; Forsythe, W.E.S. and Gregory, N., 2007. A Neolithic logboat from Greyabbey Bay, Co. Down. *Ulster Journal of Archaeology*, 66, pp. 6–13; Brady, N., 2002.

Gormanston Boat Discovery. *Archaeology Ireland*, 16(2), p. 6.

7. Bateson, J.D., 1973. Roman Material from Ireland: A reconsideration. *Proceedings of the Royal Irish Academy*, 73C, pp. 21–97.

8. O'Donovan, J., 1856. *Annals of the Kingdom of Ireland by the Four Masters*. Year 1528, Entry 1395. Dublin: Hodges, Smith and Co.

9. Brady, N., 2008. *Shipwreck Inventory of Ireland*, p. 31.

10. Shiels, D., 2007. Dún an Óir 1580: The potential for intact siege archaeology. *Journal of the Kerry Archaeological and Historical Society*, 7, pp. 59–69.

11. Kelleher, C., 2011. Ireland's Treasure Hunting Past: The case for underwater archaeology. In: Castro. F. and Thomas, L. eds. *ACUA Archaeological Proceedings*. Advisory Council on Underwater Archaeology Publication, pp. 74–83.

12. Sidonia, D. of M., 1588. *The Defeat of the Spanish Armada*. Vol. II.

13. De Cuellar, F., 1588. *Captain Cuellar's Adventures in Connacht and Ulster*. Translated by Robert Crawford in 1897. Cork: CELT, The Corpus of Electronic Texts.

14. Douglas, K., 2009. 21 September 1588: The great gale. In: Douglas, K., ed. *The Downfall of the Spanish Armada in Ireland: The Grand Armada Lost on the Irish Coast in 1588*. [eBook] Dublin: Gill Books, p. 244.

15. Brady, K., 2008. *Shipwreck Inventory of Ireland. Louth, Meath, Dublin and Wicklow*. Dublin: Stationery Office.

Maritime and Nautical Traditions and Institutions

1. Mac Carthaigh, C., 2008. *Traditional Boats of Ireland: History, folklore and construction*. Cork: The Collins Press, p. 647.

2. Kiewe, H.E., 1971. *The Sacred History of Knitting* 2nd ed. Oxford: Art Needleworth Industries.

3. Synge, J.M., 1916. *Riders to the Sea*. Boston: John W. Luce and Co.

4. Starmore, A., 2010. *Aran Knitting*. Revised and enlarged edition. New York: Dover Publications.

5. Munnelly, T., 1980. Songs of the Sea: A general description with special reference to recent oral tradition in Ireland. *Béaloideas* 48/49, pp. 30–58.

6. 'Haul Away Joe', traditional.

7. Thomas Davis (1814–1845). 'The West's Asleep'. Public domain.

8. Percy French (1896). 'The Mountains of Mourne'. Public domain.

9. The Holy Ground, traditional.

10. John McCormack and Gerald Moore (1941). 'Bantry Bay'. Public domain.

11. Amy Griffin, 1873. The Kilbaha Pilots. The full song lyrics can be found in the National Folklore Collection, UCD and dúchas.

12. The full song lyrics may be found in Shields, H., 1981. *Shamrock, Rose and Thistle: Folksinging in North Derry*. Belfast: Blackstaff Press. No. 68 'The Trader', pp. 150–151. Singer: Eddie Butcher. Online version available at: https://www.itma.ie/features/discover/shamrock-rose-thistle

13. The full song lyrics may be found in Shields, *Shamrock, Rose and Thistle*, The Arranmore disaster, p. 43. Singer: John Butcher junior. Online version available at: https://www.itma.ie/features/discover/shamrock-rose-thistle

14. Clark, W.B., 1938. *Gallant John Barry 1745–1803: The story of a naval hero of two wars*. New York: Macmillan.

15. Hudson, T., 2004. *Admiral William Brown: The master of the River Plate*. Buenos Aires: Libris.

16. Brunicairdi, D., 2012. *Haulbowline: The naval base and ships of Cork Harbour*. Cheltenham: The History Press.

17. De Courcy Ireland, J., 1985. *Ireland and the Irish in Maritime History*. Dún Laoghaire: Glendale Publishing.

18. Kennedy, M., 2008. *Guarding Neutral Ireland: The Coast Watching Service and military intelligence, 1939–1945*. Dublin: Four Courts Press.

19. Chambers, A., 2014. *Pirate Queen of Ireland: The true story of Grace O'Malley*. The Collins Press.

20. Brunicairdi, *Haulbowline*.

21. McIvor, A., 1994. *A History of the Irish Naval Service*. Dublin: Irish Academic Press.

22. Government of Ireland, 2015. White Paper on Defence.

23. Fisk, R., 1983. *In Time of War: Ireland, Ulster and the price of neutrality, 1939–45*. Dublin: Gill and Macmillan.

24. De Courcy Ireland, *Ireland and the Irish in Maritime History*.

25. O'Hailpin, E., 2002. Irish Neutrality in the Second World War. In: Wylie, N., ed. *European Neutrals and Non-belligerents during the Second World War*. Cambridge: Cambridge University Press.

26. Houses of the Oireachtas, 1941. Dáil Debates .[online] Available at: https://www.oireachtas.ie/en/debates/debate/dail/1941-02-20/22/

27. Forde, F., 2000. *The Long Watch*. Dublin: New Island Books.

28. Raymond, J., 1984. World War II and the Foundation of Irish Shipping Ltd., 1941–1945, *Éire-Ireland*, vol. XIX, number 3.

29. Forde, *The Long Watch*.

30. Raymond, *World War II and the Foundation of Irish Shipping Ltd*.

31. National Maritime College of Ireland, 2019. Facilities. [online] Available at: https://www.nmci.ie/facilities

32. Department of Transport, Tourism and Sport, 2019. About the Irish Coast Guard. [online] Available at: http://www.dttas.gov.ie/maritime/english/about-irish-coast-guard-0

33. Commissioner of Irish Lights, 2019. What We Do – Our Role. [online] Available at: https://www.irishlights.ie/who-we-are/what-we-do.aspx

SECTION 4: RESOURCES, COMMUNICATIONS AND INDUSTRY
Ports and Shipping

1. Haughton, J., 2011. The Historical Background. In: Hagan, J.W., and Newman, C., eds. *The Economy of Ireland: National and sectoral policy issues*, 11th ed. Dublin: Gill and Macmillan.

2. Haughton. The Historical Background.

3. Brunt, B., 1988. Ireland in the Changing Postwar World. In: Brunt, B., *The Republic of Ireland*. London: Paul Chapman Publishing. pp. 1–55.

4. Brunt, B., 1989. The New Industrialisation of Ireland. In: Carter B., and Parker, A.J., eds. *Ireland: A contemporary geographical perspective*. London: Routledge, pp. 201–236; Bartley, B., and Kitchin, R., 2007. *Understanding Contemporary Ireland*. London: Pluto Press.

5. Butlin, R.A., 1965. The Bantry Bay Crude Oil Terminal. *Irish Geography*, 5(2), pp.481–484.

6. Daly, G., 2019. Barryroe oil field to raise a slick $3bn in taxes for exchequer. *Sunday Times*, 6 January. [online] Available at: https://www.thetimes.co.uk/article/barryroe-oil-field-to-raise-a-slick-3bn-in-taxes-for-exchequer-9306rfx2r

7. Brunt, The New Industrialisation of Ireland; Van Egerat, C., and Breathnach, P., 2007. The Manufacturing Sector. In: Bartley B., and Kitchin R., eds. *Understanding Contemporary Ireland*. London: Pluto Press, pp. 128–145.

8. Brunt, B., 2000. Ireland's Seaport System. *Tijdschrift voor Economische en Sociale Geografie*, 91(2), pp. 159–175.

9. Ruane, F., 2016. Ireland – A Remarkable Economic Recovery? *Australian Economic Review*, 49(3), pp. 241–250.

10. Brunt. Ireland's Seaport System; Power, J., 2011. *Trends in Irish Merchandise Trade: A report for Dublin Port Company Limited*. Dublin.

11. Van Egerat and Breathnach. The Manufacturing Sector.

12. Brunt. Ireland's Seaport System.

13. Brunt. Ireland's Seaport System.

14. Brunt. The New Industrialisation of Ireland.

15. Van Egerat and Breathnach. The Manufacturing Sector.

16. Department of the Environment and Local Government, 2001. *The National Spatial Strategy: Indications for the way ahead*. Dublin.

17. Government of Ireland/Marine Coordination Group, 2012. *Harnessing our Ocean Wealth: An integrated marine plan for Ireland*. Dublin: Department of the Taoiseach.

18. Dublin Port Company, 2012. *Masterplan, 2012–2040*. Dublin: Dublin Port Company.

19. Belfast Harbour Commissioners, 2019. Belfast Harbour 2035 Outlook. Belfast: Belfast Harbour.

20. Shannon Foynes Port Company, 2014. *Vision 2041*. Limerick: HRA Planning.

21. Port of Cork Company, 2017. *Strategic Development Plan, 2010*. Cork: Port of Cork.

22. Government of Ireland, 2018. In: Project Ireland 2040: National Planning Framework. Dublin, Stationery Office, pp. 98–105.

Urbanisation of Ireland's Coast

1. At the time of publication, the most recent census for the Republic was 2016; for Northern Ireland it was 2011, although estimates for 2016 can be derived from NISRA. As a result, contemporary maps and population data are provided for 2016.

2. Small Areas are areas of population comprising between 50 and 200 dwellings created by the National Institute of Regional and Spatial Analysis (NIRSA) on behalf of the Ordnance Survey Ireland (OSi) in consultation with CSO. Small Areas were designed as the lowest level of geography for the compilation of statistics in line with data protection and generally comprise either complete or part of townlands or neighbourhoods. The equivalent geographical units for Northern Ireland are also called Small Areas and were introduced after the 2011 census.

3. Urbanisation can be described as a process involving the movement of population from rural to urban areas. It implies a transition from dependency on the primary sector to that on manufacturing and services, which have preferences for urban spaces. Measurement of urbanisation usually involves the percentage of an area's total population that reside in officially designated urban places. However, it is exceptionally difficult to define 'urban', and considerable international variation exists regarding population totals necessary for a place to be classified as 'urban'. In the Republic of Ireland, a minimum of 1,500 people living in a continuously built-up area and exhibiting a higher population density than surrounding areas is used as a threshold for an urban place. In Northern Ireland, however, groupings or bands of population are used to separate rural from urban, although a settlement size of 4,500 is often used as a reference. Using such definitions, approximately two-thirds of the population of both states are currently urbanised.

4. Johnson, J.H., 1994. *The Human Geography of Ireland*. Chichester: J. Wiley, p. 96.

5. Ó Gráda, C., 1995. *Ireland: A new economic history 1780–1939*. Oxford: Oxford University Press.

6. The new Irish Free State (Saorstát Éireann) emerged in 1922 as a self-governing dominion of the British Commonwealth, but in 1937 its constitution changed the state's name to Ireland. In 1949, however, it left the Commonwealth and became a Republic. To avoid confusion and to distinguish between the two political entities on the island, the following titles will be used, irrespective of the time period under review. Republic refers to the independent state that has existed since 1922, while Northern Ireland to the six counties that remain part of the United Kingdom. The term Ireland refers to the whole island.

7. Brunt, B., 1988. Urban Ireland. In: Brunt, B., *The Republic of Ireland*. London: Paul Chapman Publishing, pp. 133–168.

8. Huff, D.L. and Lutz, J.M., 1979. Ireland's Urban System. *Economic Geography*, 55(3), p. 198.

9. Brunt, B., 1989. The New Industrialisation of Ireland. In: Carter, R.W.G., and Parker, A.J., eds. *Ireland: A contemporary geographical perspective*. London: Routledge, pp. 201–236; Van Egeraat, C. and

10. Gilligan, H.J., 1988. *History of the Port of Dublin*. Dublin: Gill and Macmillan.

11. Gilligan. *History of the Port of Dublin*, p. 212.

12. Bunbury, T., 2009. *Dublin Docklands – an Urban Voyage*. Dublin: Montague Publications Group, pp. 73–74.

13. Moore, N., 2008. *Dublin Docklands Reinvented: The post-industrial regeneration of a European city quarter*. Dublin: Four Court Press, p. 111.

14. Moore. *Dublin Docklands Reinvented*, p. 150.

15. Brunt, 2005. Industry and Employment. In: Crowley, J., Devoy, R., Linehan D. and Murphy, M., eds. *Atlas of Cork City*. Cork: Cork University Press, pp. 369–376.

16. O'Sullivan, B., Brady, W., Ray, K., Sikora, E. and Murphy, E., 2014. Scale, Governance, Urban Form and Landscape: Exploring the scope for an integrated approach to metropolitan spatial planning. *Planning, Practice & Research*, 29, pp. 302–316.

17. Government of Ireland, 2018. *Project Ireland 2040: National Planning Framework*. Dublin: Stationery Office.

18. Walsh, J., 2007. *Regional Development: Trends, policies and strategies*. In: Bartley B. and Kitchin, R., eds. *Understanding Contemporary Ireland*. London: Pluto Press, pp. 44–56; Murray, M.R. and Murtagh, B., 2007. Strategic Spatial Planning in Northern Ireland. In: B. Bartley and R. Kitchin, eds. *Understanding Contemporary Ireland*. London: Pluto Press, pp. 112–124.

19. Matthews, R., 1963. *Belfast Regional Survey and Plan: Recommendations and Conclusions*. Belfast: HMSO; Buchanan, C., 1968. *Regional Studies in Ireland*. Dublin: An Foras Forbartha.

20 Breathnach, P., 1982. The Demise of Growth Centre Policy: The case of the Republic of Ireland. In: Hudson, R. and Lewis, J.R., eds. *Regional Planning in Europe*. London: Pion Ltd, pp. 35–56.

21. Government of Ireland, 2002. National Spatial Strategy for Ireland, 2002–2020, Dublin: Stationery Office; Department for Regional Development, 2001. Shaping Our Future. Regional Development Strategy for Northern Ireland, 2025. Belfast: HMSO.

22. Government of Ireland, 2018. *National Planning Framework*. Dublin.

23. McCafferty, D., 2009. Aspects of Socio-economic Development in Limerick City since 1970: A geographer's perspective. In: Irwin, L., Ó Tuathaigh, G. and Potter, M., eds. *Limerick: history and society*. Dublin: Geography Publications, pp. 593–614.

24. McCafferty, D., 2011. Divided City: The social geography of post-Celtic Tiger Limerick. In: Hourigan, N., ed. *Understanding Limerick: Social exclusion and change*. Cork: Cork University Press, pp. 3–22.

25. McCafferty, D., 2007. Recent Development of the Irish Urban Systems. In: Bartley, B. and Kitchin,R., eds. *Understanding contemporary Ireland*. London: Pluto Press, pp. 57–70.

26. Horner, A.A., 1999. The Tiger Stirring: Aspects of commuting in the Republic of Ireland, 1981–1996. *Irish Geography*, 32(2), pp. 99–111.

27. Hourihan, K., 1982. Urban Population Density Patterns and Change in Ireland, 1901–1979. *The Economic and Social Review*, 13(2), p. 143.

28. O'Neill, H., 1901. The Progress of Sanitary Science in Belfast. *Journal of the Statistical and Social Inquiry Society of Ireland*, 11, pp. 35–45.

29. Electoral districts are the smallest legally defined administrative areas in the Republic of Ireland for which Small Area Population Statistics (SAPS) are published from the census. The equivalent geographical units for Northern Ireland are called Census Area Statistics (CAS) wards.

30. Government of Ireland, 2018. *Project Ireland 2040: National Planning Framework*. Dublin: Stationary Office.

Coastal Fisheries and Aquaculture

1. de Courcy, J., 1981. *Ireland's Sea Fisheries: A history*. Dublin: The Glendale Press.

2. European Commission, 2017. *The EU Fish Market*. European Market Observatory for Fisheries and Aquaculture Products.

3. Mac Laughlin, J., 2010. *Troubled Waters: A social and cultural history of Ireland's sea fisheries*. Dublin: Four Courts Press.

4. Mac Laughlin. *Troubled Waters: A social and cultural history of Ireland's sea fisheries*.

5. Woodham-Smith, C., 1962. *The Great Hunger – Ireland 1845–9*. London: Hamish Hamilton.

6. de Courcy, *Ireland's Sea Fisheries*.

7. *The Irish Times*, 1978. 'Give EEC Fish Plan a Chance' plea rejected. *The Irish Times*, 18 February. p. 5.

8. Siggins, L., 1995. Government was Warned against Deal. *The Irish Times*. 30 December. p 8.

9. Browne, H., 2008. Where Will they Get the Fish? *The Dublin Review*, 33. [online] Available at: https://thedublinreview.com/article/where-will-they-get-the-fish/

10. Department of Agriculture, Food and Fisheries, 2010. *Ireland's Response to the Commission's Green Paper on the Reform of the Common Fisheries Policy*. Dublin: Department of Agriculture and Food, p. 44.

11. Bodey, T.W., Jessopp, M.J., Votier, S.C., Gerritsen, H.D., Cleasby, I.R., Hamer, K.C., Patrick, S.C., Wakefield, E.D. and Bearhop, S., 2014. Seabird Movement Reveals the Ecological Footprint of Fishing Vessels. *Current Biology*, 24(11), pp. R514–R515.

12. Kaschner, K., Karpouzi, V., Watson, R. and Pauly, D., 2006. Forage Fish Consumption by Marine Mammals and Seabirds. In: Alde, J. and Pauly, D., eds. *On the Multiple Uses of Forage Fish: From ecosystems to markets*. Vancouver: Fisheries Centre. University of British Columbia.

13. Cosgrove, R., Gosch, M., Reid, D., Sheridan, M., Chopin, N., Jessopp, M. and Cronin, M., 2015. Seal Depredation in Bottom-set Gillnet and Entangling Net Fisheries in Irish Waters. *Fisheries Research*, 172, pp. 335–344.

14. Cosgrove et al., *Seal Depredation in Bottom-set Gillnet and Entangling Net Fisheries in Irish Waters*.

15. Marine Institute, 2018. *The Stock Book 2018: Annual review of fish stocks in 2018 with management advice for 2019*. Galway: Marine Institute.

16. Marine Institute, 2017. *The Stock Book 2017: Annual review of fish stocks in 2017 with management advice for 2018*. Galway: Marine Institute.

17. Canepa, A., Fuentes, V., Sabatés, A., Piraino, S., Boero, F., and Gili, J.M., 2014. *Pelagia noctiluca* in the Mediterranean Sea. In: Pitt, K.A. and Lucas, C.H., eds. *Jellyfish Blooms*. Dordrecht: Springer Netherlands.

18. Hays, G.C., Bastian, T., Doyle, T.K., Fossette, S., Gleiss, A.C., Gravenor, M.B., Hobson, V.J., Humphries, N.E., Lilley, M.K.S., Pade, N.G. and Sims, D.W., 2012. High Activity and Levy Searches: Jellyfish can search the water column like fish. *Proceedings of the Royal Society B: Biological Sciences*, 279(1728), pp. 465–473.

19. Canepa et al. *Pelagia noctiluca* in the Mediterranean Sea.

20. Edwards, M., 2016. Impacts and Effects of Ocean Warming on Plankton. In: Laffoley, D., and Baxter, J.M. eds. 2016. *Explaining Ocean Warming: Causes, Scale, effects and consequences*. Gland, Switzerland: IUCN. pp. 75–86.

21. Boero, F., Brotz, L., Gibbons, M.J., Piraino, S. and Zampardi, S, 2016. Impacts and Effects of Ocean Warming on Jellyfish. In: Laffoley, D., and Baxter, J.M. eds. 2016. *Explaining Ocean Warming: Causes, scale, effects and consequences*. Gland, Switzerland: IUCN, pp. 213-237.

22. Boero et al. Impacts and Effects of Ocean Warming on Jellyfish.

23. Morrissey, J. Kraan, S. Guiry, M.D., 2001. *A Guide to Commercially Important Seaweeds on the Irish Coast*. Dublin: Bord Iascaigh Mhara/Irish Sea Fisheries Board.

24. Nic Dhonncha, E. and Guiry, M.D., 2002. Algaebase: Documenting seaweed biodiversity in Ireland and the world. *Biology and Environment: Proceedings of the Royal Irish Academy*. 102B(3), pp. 185–188.

25. Walsh, M. and Watson, L. 2013. *A Market Analysis towards the Further Development of Seaweed Aquaculture in Ireland*. Irish Sea Fisheries Board. Dublin: Bord Iascaigh Mhara.

26. National Seaweed Forum, 2000. *National Seaweed Forum Report*. Galway.

27. Walsh and Watson, *A Market Analysis towards the Further Development of Seaweed Aquaculture in Ireland*.

28. Houses of the Oireachtas, 2014. *Licensing and Harvesting of Seaweed in Ireland: Discussion*. Joint Committee on Environment, Culture and the Gaeltacht Debate.

29. O'Toole, E. and Hynes, S., 2009. *An Economic Analysis of the Seaweed Industry in Ireland*. Socio-Economic Marine Research Unit (SEMRU), NUIG.

30. European Council, 1992. Council Directive 92/43/EEC on the Conservation of Natural Habitats and of Wild Fauna and Flora. *Official Journal of the European Communities*.

31. Cummins, V., Coughlan, S., McClean, O., Connolly, N., Mercer, J. and Burnell, G., 2002. *An Assessment of the Potential for the Sustainable Development of the Edible Periwinkle*, Littorina littorea, *Industry in Ireland*. Galway: Marine Institute.

32. Cummins et al. *An Assessment of the Potential for the Sustainable Development of the Edible Periwinkle*.

33. Cummins et al. *An Assessment of the Potential for the Sustainable Development of the Edible Periwinkle*.

34. Cummins et al., *An Assessment of the Potential for the Sustainable Development of the Edible Periwinkle*.

35. European Commission, 2018. *Communication from the Commission to the European Parliament and the Council on the State of Play of the Common Fisheries Policy and Consultation on the Fishing Opportunities 2019*. Brussels, 11.6.2018.

36. Burden, M., Kleisner, K., Landman, J., Priddle, E., and Ryan, K. eds. 2017. Workshop report: *Climate-related impacts on fisheries management and governance in the North East Atlantic*. New York: Environmental Defence Find.

37. Irish Skipper, 2018, No Equal Access to a Common Resource. *Irish Skipper*. November 2018, pp. 2–3.

38. Government of Ireland, 2018. *Brexit and the Irish Fishing Industry Factsheet*. Dublin: Department of Agriculture, Food and the Marine.

Coastal and Marine Tourism: Building opportunities in Blue Growth

1. Jennings, S., 2004. Coastal Tourism and Shoreline Management. *Annals of Tourism Research*, 31(4), pp. 89–922; Hall, M.C., 2001. Trends in Ocean and Coastal Tourism: The end of the last frontier? *Ocean and Coastal Management*, 44(9), pp. 601–618.

2. Heuston, J., 1993. Kilkee – the Origins and Development of a West Coast Resort. In O'Connor, B. and Cronin, M., eds. *Tourism in Ireland: A critical analysis, Cork*: Cork University Press, pp. 13–28.

3. O'Sullivan, K., 2018. Blue Flags 2018: Record number of Irish beaches given awards. [online] *The Irish Times*. 21 May. [online] Available at: https://www.irishtimes.com/news/environment/blue-flags-2018-record-number-of-irish-beaches-given-awards-1.3502795

4. Morton, A., 2012. Blue Flags. *Marine Pollution Bulletin*, 64, pp. 1983–1984.

5. O'Sullivan. Blue Flags 2018.

6. Nelson, C., Morgan, R., Williams, A.T. and Wood, J. 2000. Beach Awards and Management. *Ocean and Coastal Management*, 43(1), pp. 87–98.

7. An Taisce, 2019. All Blue Flag Awarded Sites Per Country. The Results of the Blue Flag International Jury 2018. *Blue Flag Programme*. [online] Available at: http://www.blueflag.global/all-bf-sites/

8. McKenna, J., Williams, A.T. and Cooper, J.A.G. 2011. Blue Flag or Red Herring: Do beach awards encourage the public to visit beaches? *Tourism Management*, 32(3), pp. 576–588.

9. O'Hagan, A.M. and Cooper, J.A.G., 2001. Extant Legal and Jurisdictional Constraints on Irish Coastal Management. *Coastal Management*, 29(2), pp. 73–90.

10. Department of Environment, Community and Local Government. 2016. Marine Strategy Framework Directive Programme of Measures for Ireland – Summary Report. Dublin: Department of Environment, Community and Local Government.

11. Fáilte Ireland, 2019. Mullaghmore Surf on the Wild Atlantic Way. [online] Wild Atlantic Way. [online] Available at: https://www.wildatlanticway.com/things-to-do/features/surfing/making-waves-on-mullaghmore-head

12. Degree 33 Surfboards, 2019. *10 Surf Terms You ABSOLUTELY Need To Know*. [online] Available at: https://www.degree33surfboards.com/blogs/gettin-pitted/14071217-10-surf-terms-you-absolutely-need-to-know

13. ECORYS, 2013. *Study in Support of Policy Measures for Maritime and Coastal Tourism at EU Level*. Report prepared for Directorate-General for Maritime Affairs and Fisheries under Specific Contract FWC MARE/2012/06 - SC D1/2013/01-SI2.648530.

14. Tourism Development International, 2007. *A Strategy and Action Plan for the Development of Marine Tourism and Leisure in Lough Foyle and Carlingford Lough Areas*. Report prepared for The Loughs Agency, East Border Region Committee and North West Region Cross Border Group.

15. Vega, A., Corless, R. and Hynes, S., 2013. *Ireland's Ocean Economy, Reference Year: 2010*. Socio-Economic Marine Research Unit, Galway: NUIG.

16. European Commission, 2012. *Blue Growth Opportunities for Marine and Maritime Sustainable Growth. Communication from the Commission to the European Parliament, the Council, the European Economic and Social Committee and the Committee of the Regions, COM(2012)* 494 final.

17. European Commission, 2013. *Action Plan for a Maritime Strategy in the Atlantic Area. Communication from the Commission to the European Parliament, the Council, the European Economic and Social Committee and the Committee of the Regions, COM(2013)* 279 final.

18. Fáilte Ireland, 2015. *Wild Atlantic Way Operational Programme 2015-2019*. Dublin: Fáilte Ireland.

19. Marine Co-ordination Group, 2012. *Harnessing our Ocean Wealth: An integrated marine plan for Ireland roadmap: new ways, new approaches, new thinking*. Dublin: Department of the Taoiseach..

20. O'Donnell, V. and O'Mahony, C., 2011. Maintaining a Marine Leisure Industry in a Recession. *Journal of Coastal Research*, SI 61, pp. 133–139.

21. CSO Statistical Yearbook of Ireland 2017. [online] Available at http://www.cso.ie/en/releasesandpublications/ep/p-syi/psyi2018/ Dublin: NISRA.

22. Marine Institute, 2003. *A National Survey of Water-Based Leisure Activities in Ireland 2003*. Galway: Marine Institute.

23. Marine Co-ordination Group. Harnessing our Ocean Wealth.

24. Egan, M. and Hynes, S., 2014. *Results from a Short Survey of Marina Visitors and Operators in Ireland*. Socio-Economic Marine Research Unit Working Paper 14-WP-SEMRU-08.

25. Northern Ireland Statistics and Research Agency, Northern Ireland Annual Tourism Statistics, 2017. [online] Available at http://www.nisra.gov.uk/news/northern-ireland-annual-tourism-statistics-2017

26. Egan and Hynes. *Results from a Short Survey of Marina Visitors and Operators in Ireland*.

27. Tepelus, C.M. and Cordoba, R.C., 2005. Recognition Schemes in Tourism – From 'Eco' to 'Sustainability'? *Journal of Cleaner Production*, 13(2), pp. 135–140.

28. O'Mahony, C., Gault, J., Cummins, V., Köpke, K. and O'Suilleabhain, D., 2009. Assessment of Recreation Activity and its Application to Integrated Management and Spatial Planning for Cork Harbour, Ireland. *Marine Policy*, 33(6), pp. 930–937; Budeanu, A., 2005. Impacts and Responsibilities for Sustainable Tourism: A tour operator's perspective. *Journal of Cleaner Production*, 13(2), pp. 89–97.

29. Fleming, L.E., McDonough, N., Austen, M., Mee, L., Moore, M., Hess, P., Depledge, M.H., White, M., Philippart, K., Bradbrook, P. and Smalley, A., 2014. Oceans and Human Health: A rising tide of challenges and opportunities for Europe. *Marine Environmental Research*, 99, pp. 16–19.

30. Egan and Hynes. *Results from a Short Survey of Marina Visitors and Operators in Ireland*; Baily Publications Ltd., 2019. *Irish Coastal Marinas, Pontoons and Jetties*. AFLOAT. [online] Available at: https://www.housing.gov.ie/sites/default/files/public-consultation/files/responses/095_irish_marina_operators.pdf; Quinn, S., 2014. Growing Tourism in Marine and Coastal Areas. Presentation given at Harnessing Our Ocean Wealth Conference, Dublin, 18 June; Coastal and Marine Research Centre, 2010. A Marine Leisure Infrastructure Strategy for the Southern Division of Cork County Council. Cork: Cork County Council.

31. Irish Marina Operators Association, 2018. Irish Marina Operators Association: National Marine Planning Framework Baseline Report Submission Introduction. [online] Available at: h t t p s : / / w w w . h o u s i n g . g o v . i e / s i t e s / d e f a u l t / f i l e s / p u b l i c - consultation/files/responses/095_irish_marina_operators.pdf; Quinn, S., 2014. Growing Tourism in Marine and Coastal Areas. Presentation given at Harnessing Our Ocean Wealth Conference, Dublin, 18 June; Coastal and Marine Research Centre, 2010. A Marine Leisure Infrastructure Strategy for the Southern Division of Cork County Council. Cork: Cork County Council.

32. Timothy, D.J. and Boyd, S.W., 2014. Tourism and Trails: Cultural, ecological and management issues. Bristol: Channel View Publications; Ordnance Survey Ireland, 2018. Ireland's Greenways and Trails. [online] Available at: https://www.osi.ie/blog/irelands-greenways-and-trails/; Thompson S., 2016. What's So New about Ireland's Ancient East? *The Irish Times*. 6 August. [online] Available at: https://www.irishtimes.com/life-and-style/travel/ireland/what-s-so-newabout- ireland-s-ancient-east-1.2746682.

33. Tourism Ireland 2018. New Tourism Campaign Promotes Causeway Coast and Northern Ireland to 7 million Europeans. [online] Available at: https://www.tourismireland.com/Press-Releases/2018/May/New-tourism-campaign-promotes-Causeway-Coast-and-N; Causeway Coast and Glens Heritage Trust, 2019. Causeway Coastal Route Alive. [online] Available at: http://ccght.org/about-the-area/ccr-alive/; Independent, 2017. Travel Northern Ireland: Take the Causeway Coastal Route. [online] Available at: https://www.independent.ie/life/travel/northern-ireland/take-the-causeway-coastal-route-

35479175.html; Causeway Coast Holiday, 2019. Mountsandel Fort. [online] Available at: https://www.causewaycoast.holiday/locations/mountsandel-fort; Norah, L., 2019. Highlights of the Causeway Coastal Route in Northern Ireland. [online] Available at: https://www.findingtheuniverse.com/causeway-coastal-route-northern-ireland/; Toner, N., 2017. Northern Highlights, Four Great Coastal Drives in Northern Ireland. Coast Monkey. [online] Available at: http://coastmonkey.ie/four-great-coastal-drives-northern-ireland/

34. Kopke, K., O Mahony, C., Cummins, V., and Gault, J., 2007. Assessment of Coastal Recreational Activity and Capacity for Increased Boating in Cork Harbour. Corepoint project technical report; St Leger, A., 2005. A History of the Royal Cork Yacht Club. Crosshaven: RCYC.

35. Garden.ie, 2018. Gardens Open. [online] Available at: https://www.garden.ie/gardens-open/

36. Burren Connect, 2013. Burren Ecotourism Network. [online] Available at: http://www.burrenecotourism.com; National Parks and Wildlife Service, 2014. Black Head-Poulsallagh Complex SAC (Site Code 20): Conservation objectives supporting document – coastal habitats. Dublin: NPWS, Department of Arts, Heritage and the Gaeltacht; Burren Connect. Burren Ecotourism Network; Christ, C., 2016. World Legacy Awards. National Geographic Traveler Magazine, April/May Issue, pp. 89–90; Devoy, R.J.N., 2016. Fanore Coast and Dune Management: Current problems and planning for the future. Report for Clare County Council. Clare County Council and Marine and Renewable Energy Centre (MaREI), University College Cork.

37. Restaurants Association of Ireland, 2016. Key Action Points for the Restaurant Industry. [online] Available at: https://www.rai.ie/key-issues/

38. 'A gift from Scotland': Golf 's Early Days in Ireland. History Ireland Features, Issue 5 (Sep/Oct 2006), Volume 14. [online] Available at: https://www.historyireland.com/20th-centurycontemporary-history/a-gift-from-scotlandgolfs-early-days-in-ireland-2/; Deegan, G., 2017. Trump Doonbeg Golf Links Gets Go-Ahead for 38,000-tonne Wall. The Irish Times. 21 December. [online] Available at: https://www.irishtimes.com/news/environment/trump-doonbeg-golf-links-gets-go-ahead-for-38-000-tonne-wall-1.3335412.

39. Marine Co-ordination Group. Harnessing our Ocean Wealth; O'Hagan and Cooper. Extant legal and jurisdictional constraints on Irish coastal management; Department of Enterprise, Trade and Employment, 2009. Developing the Green Economy in Ireland. Dublin: Department of Enterprise, Trade and Employment/for Fás; ECORYS. Study in Support of Policy Measures for Maritime and Coastal Tourism at EU Level.

Renewable Energies: Wind, wave and tidal power

1. MRIA, 2016. Funding the Development of the Ocean Energy Industry in Ireland. [online] Available at: http://www.mria.ie/documents/a524d9555b5a399944adbd8d7c.pdf.

2. Cleasby, I.R., Wakefield, E.D., Bearhop, S., Bodey, T.W., Votier, S.C. and Hamer, K.C., 2015. Three-dimensional Tracking of a Wide-ranging Marine Predator: Flight Heights and Vulnerability to Offshore Wind Farms. Journal of Applied Ecology, 52(6), pp. 1471–1482.

3. Krijgsveld, K.L., Flijn, R.C., Japink, M., van Horssen, P.W., Heunks, C., Collier, M.P., Poot, M.J.M., Beuker, D. and Dirksen, S., 2011. Effect Studies Offshore Wind Farm Egmond aan Zee: Final report on fluxes, flight altitudes and behaviour of flying birds. Commissioned by NoordzeeWind. Bureau Waardenburg Report.

4. Busch, M.A., Garthe, S. and Jessopp, M., 2013. Consequences of a Cumulative Perspective on Marine Environmental Impacts: Offshore wind farming and seabirds at North Sea scale in context of the EU Marine Strategy Framework Directive. Ocean and Coastal Management, 71, pp. 213–224.

5. Department of Communications, Energy and Natural Resources, 2015. Ireland's Transition to a Low Carbon Energy Future 2015–2030. Dublin.

6. Government of Ireland, 2019. Climate Action Plan 2019 to Tackle Climate Breakdown. Dublin.

7. Department of Communications, Energy and Natural Resources, 2014. Offshore Renewable Energy Development Plan. A Framework for the Sustainable Development of Ireland's Offshore Renewable Energy Resource. Dublin.

8. Department of Enterprise, Trade and Investment, 2012. Offshore Renewable Energy Strategic Action Plan 2012–2020. Dublin.

9. Sustainable Energy Authority of Ireland, 2017. Ireland's Energy Projections – Progress to Targets, Challenges and Impacts. Dublin.

10. Department of Communications, Climate Action and Environment, 2017. Guidance on EIS and NIS Preparation for Offshore Renewable Energy Projects. Dublin.

11. Department of Communications, Climate Action and Environment, 2016. Code of Practice for Wind Energy Development in Ireland Guidelines for Community Engagement. Dublin.

12. Department of Housing, Planning and Local Government, 2017. Towards a Marine Spatial Plan for Ireland – A Roadmap for the Delivery of the National Marine Spatial Plan. Dublin: Department of Housing, Planning and Local Government, 2018. National Marine Planning Framework Baseline Report. Dublin.

13. Eirgrid, 2008. GRID25 – A Strategy for the Development of Ireland's Electricity Grid for a Sustainable and Competitive Future. Dublin.

14. European Commission, 2015. Energy Union Package – Communication from the Commission to the European Parliament and the Council – Achieving the 10% Electricity Interconnection Target, Making Europe's Electricity Grid Fit for 2020.

15. Gavin and Doherty Geosolutions (on behalf of Sustainable Energy Authority of Ireland), 2018. Gap Analysis for the use of Irish Port & Harbour Infrastructure for Renewable Energy. Dublin: SEAI.

Coastal Mining, Quarrying and Hydrocarbon Exploration

1. Cole, G.A.J., 1998. Memoir of Localities of Minerals of Economic Importance and Metalliferous Mines in Ireland. 3rd ed. Dublin: Mining Heritage Society of Ireland.

2. Cowman, D. and Reilly, T.A., 1988. The Abandoned Mines of West Carbery Promoters, Adventurers and Miners. Dublin: Geological Survey of Ireland; Hodnett, D., 2010. The Metal Mines of West Cork. Camborne: Trevithick Society.

3. Williams, R.A. 1991. The Berehaven Copper Mines. British Mining No. 42. Reprint by A.B. O'Connor & Co. Ltd, 1993.

4. Cole. Memoir of Localities of Minerals of Economic Importance and Metalliferous Mines in Ireland.

5. Morris, J.H., 2001. Man Engine House, Allihies, Co. Cork. Journal of the Mining Heritage Trust of Ireland, 1, pp. 39–48.

6. Hodnett. The Metal Mines of West Cork.

7. O'Sullivan, P., 2006. The Mines of Sheep's Head and Mizen Head Peninsulas, County Cork. Journal of the Mining Heritage Trust of Ireland, 6, pp. 23–36.

8. Hodnett. The Metal Mines of West Cork.

9. Hodnett. The Metal Mines of West Cork.

10. Cowman, D., 2010. The Baryte Mines of West Cork. Journal of the Mining Heritage Trust of Ireland, 10, pp. 3–10.

11. Morris, J.H., Tietzsch-Tyler, D. and Scanlon, R., 2005. The Knockmahon–Tankardstown Mineral Tramway, Bunmahon, Co. Waterford. Journal of the Mining Heritage Trust of Ireland, 5, pp. 53–74.

12. Cowman, D., 2001. The Metal Mines of Dublin City and County, c.1740–1825. Journal of the Mining Heritage Trust of Ireland, 1, pp.61–66.

13. Barnett, J., 2006. Quarries, Mines and Railways of Dalkey. Journal of the Mining Heritage Trust of Ireland, 6, pp. 17–21.

14. Schwartz, S.P. and Critchley, M.F., 2013. A History of the Silver-lead Mines of County Down, Northern Ireland. Journal of the Mining Heritage Trust of Ireland, 13, pp. 23–95.

15. Arthurs, J.W. and Earls, G., 2004. Minerals. In: Mitchell, W., ed. The Geology of Northern Ireland: Our Natural Foundation. Belfast: Geological Survey of Northern Ireland, p. 318.

16. Kelly, J., 2007. A History of Zn-Pb-Ag mining at Abbeytown, Co. Sligo. Journal of the Mining Heritage Trust of Ireland, 7, pp. 9–18.

17. Naylor, D. and Shannon, P., 2011. Petroleum Geology of Ireland. Edinburgh: Dunedin Academic Press.

18. Penn, I.E., 1981. Larne No. 2 Geological Well Completion Report. London: Institute of Geological Sciences. 81/6.

19. Colfer, B., 2004. The Hook Peninsula, County Wexford. Cork: Cork University Press.

20. Dublin City Public Libraries, 2000. A Directory of Dublin for the Year 1738. Dublin: Dublin City Public Libraries.

21. Greene, J.C. 2004. The Belfast Newsletter Index, 1737–1800. [online] Available at: http://www.ucs.louisiana.edu/bnl/

22. Tighe, W., 1998. William Tighe's Statistical Observations Relative to the County of Kilkenny Made in the Years 1800 and 1801. Kilkenny: Grangesilvia Publications.

23. Colfer, N., 2016. Turning Stone into Bread: The millstone quarries of medieval and post-medieval Ireland. Ph.D., University College Dublin.

24. Colfer. The Hook Peninsula, County Wexford.

25. Kozachenko, M., Fletcher, R., Sutton, G., Monteys, X., Landeghem, K. Van, Wheeler, A., Lassoued, Y., Cooper, A. and Nicoll, C., 2008. A Geological Appraisal of Marine Aggregate Resources in the Southern Irish Sea. Technical Report Produced for the Irish Sea Marine Aggregates Initiative (IMAGIN) Project Funded Under the INTERREG IIIA Programme. Cork: University College Cork.

26. Department of Housing Planning and Local Government, 2018. National Marine Planning Framework Baseline Report. Dublin.

27. Sutton, G., O'Mahony, C., McMahon, T., Ó Cinnéide, M. and Nixon, E., 2008. Policy Report – Issues and Recommendations for the Development and Regulation of Marine Aggregate Extraction in the Irish Sea. Marine Environment & Health Series, 32.

28. Sutton et al. Policy Report – Issues and Recommendations for the Development and Regulation of Marine Aggregate Extraction in the Irish Sea.

29. Winn, R.D., 1994. Shelf Sheet-sand Reservoir of the Lower Cretaceous Greensand, North Celtic Sea Basin, Offshore Ireland. AAPG Bulletin, 78, pp.1775–1789.

30. Lewis, D., 2009. Geological Storage of CO_2 in the Island of Ireland: A review. European Geologist, 27, pp. 8–11.

31. Naylor and Shannon. Petroleum Geology of Ireland.

32. Shell E&P Ireland Limited, 2014. Corrib Development Biodiversity Action Plan 2014–2019. [online] Available at https://so8.static.shell.com/context/dam/shellnew/local/country/irl/downloads/pdf/corrib-development-biodiversity-aaction-plan-2014-2019.pdf

SECTION 5: MANAGEMENT OF THE COASTS AND THE MARINE ENVIRONMENT

Engineering for Vulnerable Coastlines

1. Swift, L.J., Devoy, R.J.N., Wheeler, A.J., Sutton, G.D. and Gault, J., 2006. Sedimentary Dynamics and Coastal Changes on the South Coast of Ireland. Journal of Coastal Research, SI 39 (Proceedings of the 8th International Coastal Symposium), pp. 227–232; Devoy, R.J.N., 2015. Sea-level Rise: Causes, Impacts, and Scenarios. In: Ellis, J.T. and Sherman, D.J., eds. Coastal and Marine Hazards, Risks and Disasters. Amsterdam: Elsevier. pp. 197–270.

2. Devoy, R.J.N., 2008. Coastal Vulnerability and the Implications of Sea-level Rise for Ireland. Journal of Coastal Research, 24, pp. 325–321; Neumann, B., Vafeidis, A.T., Zimmermann, J. and Nicholls, R.J., 2021. Future Coastal Population Growth and Exposure to Sea-level Rise and Coastal Flooding: A global assessment. PLOS ONE, 10(3), e0118571; Intergovernmental Panel

on Climate Change, 2014. Summary for Policymakers. In: Edenhofer, O., Pichs-Madruga, R., Sokona, Y., Farahani, E., Kadner, S., Seyboth, K., Adler, A., Baum, I., Brunner, S., Eickemeier, P., Kriemann, B., Savolainen, J., Schlömer, S., von Stechow, C., Zwickel, T. and Minx, J.C., eds. *Climate Change 2014: Mitigation of climate change. Contribution of working group III to the fifth assessment report of the intergovernmental panel on climate change.* Cambridge and New York: Cambridge University Press; Intergovernmental Panel on Climate Change, 2018. Summary for Policymakers. In: Masson-Delmotte, V., Zhai, P., Pörtner, H.O., Roberts, D., Skea, J., Shukla, P.R., Pirani, A., Moufouma-Okia, W., Péan, C., Pidcock, R., Connors, S., Matthews, J.B.R., Chen, Y., Zhou, X., Gomis, M.I., Lonnoy, E., Maycock, T., Tignor, M. and Waterfield T. eds. Global *Warming of 1.5°C. An IPCC special report on the impacts of global warming of 1.5°C above pre-industrial levels and related global greenhouse gas emission pathways, in the context of strengthening the global response to the threat of climate change, sustainable development, and efforts to eradicate poverty.* Geneva: World Meteorological Organization.

3. Office of Public Works, 2019. Catchment-based Flood Risk Assessment and Management (CFRAM) Studies; UK Committee on Climate Change, 2017. *UK Climate Change Risk Assessment 2017 Synthesis Report: Priorities for the next five years.* London: Her Majesty's Stationery Office.

4. McKenna, J., Cooper, A. and O'Hagan, A.M., 2008. Managing by Principle: A critical analysis of the European principles of integrated coastal zone management (ICZM). *Marine Policy*, 32, pp. 941–955; Cooper, J.A.G. and Cummins, V., 2009. Coastal Management in Northwest Europe. *Marine Policy*, 33, pp. 869–937; Coastal Areas, 2019. Climate Adapt. [online] Available at: https://climate-adapt.eea.europa.eu/eu-adaptation-policy/sector-policies/coastal-areas

5. Ecological Production in a Post-Growth Society (ECOPRO), Department of the Marine Great Britain, Department of the Environment Northern Ireland and Forbairt and L'Instrument Financier pour l'Environnement (LIFE) programme, 1996. *Environmentally Friendly Coastal Protection: Code of practice.* Dublin: Stationery Office.

6. Gray, S., O'Mahony, C., Hills, J., O'Dwyer, B., Devoy, R. and Gault, J., 2016. Strengthening Coastal Adaptation Planning through Scenario Analysis: A beneficial but incomplete solution. *Marine Policy*, DOI: 10.1016/j.marpol.2016.04.031.

7. National Coastal Erosion Committee, 1992. *Coastal Management: A case for action.* Dublin: Eolas on behalf of County and City Engineers Association and the Irish Science and Technology Agency.

8. Carter, R.W.G., 1991. Sea-level Changes. In: McWilliams, B.E., ed. *Climate Change: Studies on the implications for Ireland.* Department of the Environment. Dublin: Stationery Office, pp. 125–171; Devoy, R.J.N., 2008. Coastal Vulnerability and the Implications of Sea-level Rise for Ireland. *Journal of Coastal Research*, 24, pp. 325–321; Climate Ireland, 2019. [online] Available at https://www.climateireland.ie

9. Nolan, P., 2016. *Report No. 59: Ensemble of regional climate model projections for Ireland.* Johnstown Castle Estate: Environmental Protection Agency; Intergovernmental Panel on Climate Change, 2019. Summary for Policymakers. In: Pörtner, H.-O., Roberts, V., Masson-Delmotte, D.C., Zhai, P., Tignor, M., Poloczanska, E., Mintenbeck, K., Nicolai, M., Okem, A., Petzold, J., Rama, B. and Weyer N. eds. In Press. *Special Report on the Ocean and Cryosphere in a Changing Climate.*

10. Cronin, A. and Kandrot, S., 2017. *Local Authority Coastal Erosion Policy and Practice Audit.* Dublin: Fingal County Council.

11. Murphy, J., 2014. Coastal Erosion around Ireland and Engineering Solutions. *Engineers Journal*, 5 June 2015, https://www.engineersireland.ie/Engineers-Journal/.

12. Devoy, R.J.N., 2015. Sea-level Rise: Causes, impacts, and scenarios. In: Ellis, J.T. and Sherman, D.J., eds. *Coastal and Marine Hazards, Risks and Disasters.* Amsterdam: Elsevier, pp. 197–270; Intergovernmental Panel on Climate Change, 2019. Summary for Policymakers; European Environment Agency, 2017. *Climate Change, Impacts and Vulnerability in Europe 2016: An indicator-based report.* Luxembourg: Publications Office of the European Union.

13. O'Doherty, C., 2019. Coastal Homes at Risk from Catastrophic Sea-level Rise. *Independent*, 26 September; Devoy, R.J.N., 2021. Sea-level Changes. In, D.J. Sherman, ed. *Treatise in Geomorphology.* Elsevier, New York.

14. Gallagher, S., Tiron, R., Whelan, E., Gleeson, E., Dias, F. and McGrath, R., 2016. The Nearshore Wind and Wave Energy Potential of Ireland: A high resolution assessment of availability and accessibility. *Renewable Energy*, 88, pp. 494–516.

15. Carter, R.W.G. and Woodroffe, C.D., 1994. *Coastal Evolution.* Cambridge: Cambridge University Press; Devoy, R.J.N., 2015. The Development and Management of the Dingle Bay Spit-Barriers of Southwest Ireland. In: Randazzo, G., Jackson, D.W.T. and Cooper, J.A.G., eds. *Sand and Gravel Spits.* New York: Springer, pp. 139–180; Intergovernmental Panel on Climate Change, 2018. Summary for Policymakers. In: Masson-Delmotte, V., Zhai, P., Pörtner, H.O., Roberts, D., Skea, J., Shukla, P.R., Pirani, A., Moufouma-Okia, W., Péan, C., Pidcock, R., Connors, S., Matthews, J.B.R., Chen, Y., Zhou, X., Gomis, M.I., Lonnoy, E., Maycock, T., Tignor, M. and Waterfield T. eds. *Global Warming of 1.5°C. An IPCC special report on the impacts of global warming of 1.5°C above pre-industrial levels and related global greenhouse gas emission pathways, in the context of strengthening the global response to the threat of climate change, sustainable development, and efforts to eradicate poverty.* Geneva: World Meteorological Organization.

16. Sánchez-Arcilla, A., García-León, M., Gracia, V., Devoy, R., Stanica, A. and Gault, J., 2016. Managing Coastal Environments under Climate Change: Pathways to adaptation. *Science of the Total Environment*, 572, pp. 1336–1352.

17. Office of Public Works, 2013. *Irish Coastal Protection Strategy Study (ICPSS)*; Cooper, J.A.G. and Cummins, V., 2009. Coastal Management in Northwest Europe. *Marine Policy*, 33, pp. 869–937; McKibbin, D., 2016. *Legislative and Policy Response to the Risk of Coastal Erosion and Flooding in the UK and Ireland.* Northern Ireland Assembly, Research and Information Service Research Paper NIAR, pp. 274–316; Government, 2020. National Coastal Change Management Strategy Steering Group meets for the first time. Department of Housing, Local Government and Heritage. https://www.gov.ie/en/press-release/7b418-national-coastal-change-management-strategy-steering-group-meets-for-first-time/ .

18. Meur-Ferec, C., Flanquart, H., Deboudt, Ph., Morel, V., Hellequin, A-P. and Longuepee, J., eds. 2009. The Littoral: Facing constraints, initiating dialogue, taking action. *Journal of Coastal Conservation*, 13, pp. 49–183; Cooper, J.A.G., Muir, D. and Pétursdóttir, G., eds. 2014. Coastal Climate Change Adaptation in the Northern Periphery of Europe. *Ocean and Coastal Management*, 94, pp. 1–98.

19. Kandrot, S., 2017. The Monitoring and Modelling of the Impacts of Storms Under Sea-level Rise on a Breached Coastal Dune-barrier System. PhD University College Cork; Institute of Civil Engineers (ICE), 2015. Managed realignment at Medmerry, Sussex. [online] Available at https://www.ice.org.uk/knowledge-and-resources/case-studies/managed-realignment-at-medmerry-sussex

20. Sabatier, F., Samat, O., Brunel, C., Heurtefeux, H. and Delanghe-Sabatier, D., 2009. Determination of Set-back Lines on Eroding Coasts. Example of the beaches of the Gulf of Lions (French Mediterranean Coast). *Journal of Coastal Conservation*, 13, pp. 57–64; Falaleeva, M., Gray, S., O'Mahony, C. and Gault, J., 2013. *Coastal Climate Adaptation in Ireland: Assessing current conditions and enhancing the capacity for climate resilience in local coastal management.* Johnstown Castle: Environmental Protection Agency.

21. Cooper, J.A.G. and Pilkey, O.H., eds. 2012. *Pitfalls of Shoreline Stabilization.* Dordrecht: Springer; Climate Change Advisory Council, 2019. *Annual Review 2019.* Dublin: Climate Change Advisory Council; Joint Committee on Climate Action, 2019. *Report of the Joint Committee on Climate Action Climate Change: A cross-party consensus for action.* Dublin: Houses of the Oireachtas; Farrell, E., Bourke, M., Lynch, K., Morley, T., O'Dwyer, B., O'Sullivan, J. and Turner, J., 2021. From Source to Sink: Responses of a coastal catchment to large scale changes: Golden Strand Catchment, Achill Island, Co. Mayo. Research Report 2014-CCRP-MS.22. Dublin: EPA.

22. Hydraulics and Maritime Research Centre, 2001. *Inny Strand Coastal Erosion and Protection Study.* Cork: University College Cork.

23. Engineering and Technology History Wiki, 2019. Milestones: County Kerry Transatlantic Cable Stations, 1866. [online] Available at https://ethw.org/Milestones:County_Kerry_Transatlantic_Cable_Stations,_1866

24. O'Dwyer, M., 2019. Personal interview by Michael O'Shea. 19 January.

25. Murphy, J., 1997. *Waterville Shoreline Development Study.* Cork: Hydraulics and Maritime Research Centre, University College Cork.

26. Sinnott, A., 1999. Sea-level and Related Coastal Change in South and Southeast Ireland. PhD University College Cork.

27. Carter, R.W.G., 1988. *Coastal Environments.* London: Academic Press; Ecological Production in a Post-Growth Society (ECOPRO), Department of the Marine Great Britain, Department of the Environment Northern Ireland and Forbairt and L'Instrument Financier pour l'Environnement (LIFE) programme, 1996. *Environmentally Friendly Coastal Protection: Code of practice.*

28. National Coastal Erosion Committee, *Coastal Management*; Brady Shipman Martin, 1997. *Coastal Zone Management: A draft policy for Ireland.* Dublin: Department of the Marine and Natural Resources; Ecological Production in a Post-Growth Society (ECOPRO), Department of the Marine Great Britain, Department of the Environment Northern Ireland and Forbairt and L'Instrument Financier pour l'Environnement (LIFE) programme, 1996. *Environmentally Friendly Coastal Protection: Code of practice*; Department of the Environment and Local Government, 2001. *Coastal Zone Management*; Department of the Taoiseach, 2007. *National Development Plan 2007–2013*: Transforming Ireland. Dublin: Stationery Office; Cronin and Kandrot, Local Authority Coastal Erosion Policy and Practice Audit; Department of Communications, Climate Action and Environment, 2018. *National Adaptation Framework: Planning for a climate resilient Ireland.*

29. Office of Public Works, *Irish Coastal Protection Strategy Study (ICPSS)*; In 2018, the Marine and Renewable Energy Institute (MaREI) with support from the Environmental Protection Agency, began work on a national risk assessment of the impacts of climate change for Ireland. This research involves broad scoped expert reviews of all data and associated literature to establish the risks and impacts of climate change in Ireland today as well as into the future.

30. Climate Change Advisory Council, *Annual Review 2019*; Government, 2020. A National Coastal Change Management Strategy Steering Group. [online] Available at https://merrionstreet.ie/en/newsroom/releases/national_coastal_change_management_strategy_steering_group_meets_for_first_time.html; Farrell, E., Bourke, M., Lynch, K., Morley, T., O'Dwyer, B., O'Sullivan, J. and Turner, J., 2020. From Source to Sink: Responses of a coastal catchment to large scale changes: Golden Strand Catchment, Achill Island, Co. Mayo. Research Report 2014-CCRP-MS.22. Dublin: EPA.

31. IPB (Irish Public Bodies Mutual Insurance) Pride of Place, 2018. *Maharees Coastal Area.* [online] Available at https://prideofplace.ie/portfolio-item/special-award-maharees-coastal-area/

32. Environmental Protection Agency, 2019. *Innovative Methods of Community Engagement: Towards a low carbon climate resilient future workshop proceedings.*

33. The work of the Maharees Conservation Association can be viewed at: https://www.mahareesconservation.com

34. Thompson, I., 2009. *Rethinking Landscape: A critical reader.* Oxford: Routledge. pp. 41–44.

35. McCullough, N. and Mulvin, V., 1987. *A Lost Tradition: The nature of architecture in Ireland.* Dublin: Gandon Editions.

36. Robinson T., 1996. *Setting Foot on the Shores of Connemara and Other Writings.* Dublin: The Lilliput Press.

37. Brady, Shipman, Martin, *Coastal Zone Management: A draft policy for Ireland.*

Pollution

1. Tornero, V. and Hanke, G., 2016. Chemical Contaminants Entering the Marine Environment from Sea-based Sources: A review with a focus on European Seas. *Marine Pollution Bulletin*, 112(1–

2), pp. 17–38.

2. Mockler, E.M., Deakin, J., Archbold, M., Gill, L., Daly, D. and Bruen, M., 2017. Sources of Nitrogen and Phosphorus Emissions to Irish Rivers and Coastal Waters: Estimates from a nutrient load apportionment framework. *Science of the Total Environment*, 601–602, pp. 326–339.

3. Toner, P., Bowman, J., Clabby, K., Lucey, J., McGarrigle, M., Concannon, C., Clenaghan, C., Cunningham, P., Delaney, J., O'Boyle, S., MacCárthaigh, M., Craig, M. and Quinn, R., 2005. Water Quality in Ireland 2001–2003. Wexford: Environmental Protection Agency.

4. European Parliament and Council, 2000. Directive 2000/60/EC of 23 October 2000 Establishing a Framework for the Community Action in the Field of Water Policy.

5. Environmental Protection Agency, 2019. Bathing Water Quality in Ireland–A Report for the Year 2018. Wexford: Environmental Protection Agency.

6. OSPAR Commission, 2017. Assessment Document of Land-based Inputs of Microplastics in the Marine Environment. Environmental Impact of Human Activities Series. London: OSPAR Commission.

7. European Parliament and Council, 2008. Directive 2008/56/EC of 17 June 2008 on Establishing a Framework for Community Action in the Field of Marine Environmental Policy (Marine Strategy Framework Directive).

8. OSPAR Commission, 2014. Marine Litter Regional Action Plan. London: OSPAR Commission.

9. Cózar, A., Echevarría, F., González-Gordillo, J.I., Irigoien, X., Úbeda, B., Hernández-León, S., and Fernández-de-Puelles, M.L., 2014. Plastic Debris in the Open Ocean. *Proceedings of the National Academy of Sciences*, 111(28), pp. 10,239–10,244.

10. Eriksen, M., Lebreton, L.C.M., Carson, H.S., Thiel, M., Moore, C.J., Borerro, J.C., Galgani, F., Ryan, P.G. and Reisser, J., 2014. Plastic Pollution in the World's Oceans: More than 5 trillion plastic pieces weighing over 250,000 tons afloat at sea. *PLOS ONE*, 9(12), p.e111913.

11. Lusher, A.L., Burke, A., O'Connor, I. and Officer, R., 2014. Microplastic Pollution in the Northeast Atlantic Ocean: Validated and opportunistic sampling. *Marine Pollution Bulletin*, 88(1–2), pp. 325–333.

12. Acampora, H., Lyashevska, O., Van Franeker, J.A. and O'Connor, I., 2016. The Use of Beached Bird Surveys for Marine Plastic Litter Monitoring in Ireland. *Marine Environmental Research*, 120, pp. 122–129.

13. Wilcox, C., Van Sebille, E. and Hardesty, B.D., 2015. Threat of Plastic Pollution to Seabirds is Global, Pervasive and Increasing. *Proceedings of the National Academy of Sciences*, 112(38), pp. 11,899–11,904

14. Lusher, A.L., Hernandez-Milian, G., Berrow, S., Rogan, E. and O'Connor, I., 2018. Incidence of Marine Debris in Cetaceans Stranded and Bycaught in Ireland: Recent findings and a review of historical knowledge. *Environmental Pollution*, 232, pp. 467–476.

15. Thompson, R.C., Olsen, Y., Mitchell, R.P., Davis, A., Rowland, S.J., John, A.W. and Russell, A.E., 2004. Lost at Sea: Where is all the plastic? *Science*, 304(5672), p. 838.

16. National Research Council, 2003. Oil in the Sea III: Inputs, fates and effects. Washington, DC: The National Academies Press.

17. Department of Transport, Tourism and Sport, 2013. Irish National Contingency Plan. Dublin.

18. International Maritime Organization, 1990. International Convention on Oil Pollution Preparedness, Response and Co-operation.

19. Ireland became a member of the Convention for the Protection of the Marine Environment of the North-East Atlantic – OSPAR Convention – in 1998.

20. After fifteen years with Observer Status, Ireland became a full member of the Bonn Agreement in 2010; Irish Coast Guard and Her Majesty's Coastguard, 2010. Memorandum of Understanding on the Conduct of Operations between Her Majesty's Coastguard (HMCG) and the Irish Coastguard for Search and Rescue.

21. Government of Ireland, 1991–2006. Sea Pollution Acts.

22. Murdoch, S., 2013. Independent Review of Maritime New Zealand's Response to the MV *Rena* Incident on 5 October 2011. Wellington: Maritime New Zealand.

23. European Parliament and Council, 2006. Regulation (EC) No 1907/2006 of 18 December 2006 Concerning the Registration, Evaluation, Authorisation and Restriction of Chemicals (REACH).

24. European Parliament and Council, Directive 2000/60/EC of 23 October 2000 on Establishing a Framework for the Community Action in the Field of Water Policy.

25. Giltrap, M., McHugh, B., Ronan, J., Wilson, J. and McGovern, E., 2014. Biological Effects and Chemical Measurements in Irish Marine Waters. Galway: Marine Institute.

26. Jepson, P.D., Deaville, R., Barber, J.L., Aguilar, A., Borrell, A., Murphy, S., Barry, J. and Al, E., 2016. PCB Pollution Continues to Impact Populations of Orcas and other Dolphins in European Waters. *Scientific Reports,* 6: p. 185–73.

27. McGovern, E., McHugh, B., O'Hea, L., Joyce, E., Tlustos, C. and Denise, G., 2011. Assuring Seafood Safety: Contaminants and residues in Irish seafood 2004–2008. Galway: Marine Institute.

28. European Food Safety Authority, 2014. Scientific Opinion on Health Benefits of Seafood (Fish and Shellfish) Consumption in Relation to Health Risks Associated with Exposure to Methylmercury. *EFSA Journal*, 12(7), p. 3,761; Food Safety Authority of Ireland, 2016. Report on a Total Diet Study Carried out by the Food Safety Authority of Ireland in the Period 2012–2014. Monitoring and Surveillance Series. Dublin: Food Safety Authority of Ireland.

29. United Nations Environmental Programme, 2013. The Global Mercury Assessment 2013: Sources, emissions, releases and environmental transport. Geneva: United Nations Environmental Programme Chemicals Branch.

30. McGovern, et al. Assuring Seafood Safety: Contaminants and residues in Irish Seafood 2004–2008.

31. European Food Safety Authority Scientific Committee, 2015. Statement on the Benefits of Fish/seafood Consumption Compared to the Risks of Methylmercury in Fish/seafood. *European Food Safety Authority Journal.* 13(1):3982, p. 36.

32. Dafforn, K.A., Lewis, J.A. and Johnston, E.L., 2011. Antifouling Strategies: History and regulation, ecological impacts and mitigation. *Marine Pollution Bulletin*, 62, pp. 453–465.

33. Alzieu, C., Heral, M., Thibaud, Y., Dardignac, K.J. and Feuillet, M., 1982. Influences des peintures antisalissures a base d'organostanniques sur la calcification de Ja coquille de l'huitre. *Revue des Travaux de l'Institut des Pêches Maritimes*, 45(2), pp. 101–106.

34. Minchin D. and Duggan, C.B., 1986. Organotin Contamination in Irish Waters. International Council for the Exploration of the Sea, CM 1986/F:48.

35. Minchin, D., Duggan, C.B. and King, W., 1987. Possible Influence of Organotins on Scallop Recruitment. *Marine Pollution Bulletin*, 18(11), pp. 604–608.

36 Department of the Marine, 1987. Restriction of Use of Organotin Antifouling Compounds Bye-law No. 657. Dublin: Stationery Office.

37. International Maritime Organization, 2001. International Convention on the Control of Harmful Anti-fouling Systems on Ships.

38. European Parliament and Council, 2003. Regulation (EC) No 782/2003 of 14 April 2003 on the Prohibition of Organotin Compounds on Ships; European Parliament and Council, 2008. Commission Regulation (EC) No 536/2008 of 13 June 2008 Giving Effect to Article 6(3) and Article 7 of Regulation (EC) No 782/2003 of the European Parliament and of the Council on the Prohibition of Organotin Compounds on Ships and Amending that Regulation.

39. Wilson J., Minchin, D., McHugh, B., McGovern, E., Tanner, C.J. and Giltrap, M., 2015. Declines in TBT Contamination in Irish Coastal Waters 1987–2011, Using the Dogwhelk (*Nucella lapillus*) as a Biological Indicator. *Marine Pollution Bulletin*, 100(1), pp. 289–296.

40. Giltrap, M., Kennedy, R., McGovern, E., Joyce, E., Brophy, L., Parker, M., McDonnell, O., Fryer, R., Conway, A. and McHugh, B., In prep. Improving Status of Imposex in Dogwhelk (*Nucella lapillus*) in Irish Marine Waters. Galway: Marine Institute.

41. Environmental Protection Agency, 2017. Ionising Radiation Regulation: Key Findings from the 2016 Inspection and Enforcement Programme.

42. Gordon, J., Gillespie, D., Potter, J., Frantzis, A., Simmonds, M.P., Swift, R. and Thompson, D., 2003. A Review of the Effects of Seismic Surveys on Marine Mammals. *Marine Technology Society Journal*, 37(4), pp. 16–34.

43. Folegot, T., Clorennec, D., Sutton, G. and Jessopp, M., 2016. Seismic Survey Footprints in Irish Waters: A starting point for effective mitigation. In: Popper A., Hawkins A., eds. *The Effects of Noise on Aquatic Life II. Advances in Experimental Medicine and Biology*, vol 875. SpringerNew York: Springer. [online] Available at: https://doi.org/10.1007/978-1-4939-2981-8_37

44. Culloch, R., Anderwald, P., Brandecker, A., Haberlin, D., McGovern, B., Pinfield, R., Visser, F., Jessopp, M. and Cronin, M., 2016. The Effect of Construction-related Activities and Vessel Traffic on Marine Mammals. *Marine Ecology Progress Series*, 549, pp. 231–242.

45. European Parliament and Council, Directive 2008/56/EC of 17 June 2008 on Establishing a Framework for Community Action in the Field of Marine Environmental Policy (Marine Strategy Framework Directive).

46. OSPAR Commission, 2010. The North-East Atlantic Environment Strategy: Strategy of the OSPAR Commission for the Protection of the Marine Environment of the North-East Atlantic 2010–2020.

47. European Parliament and Council, Directive 2000/60/EC of 23 October 2000 Establishing a Framework for the Community Action in the Field of Water Policy.

48. European Parliament and Council, Directive 2008/56/EC of 17 June 2008 on Establishing a Framework for Community Action in the Field of Marine Environmental Policy (Marine Strategy Framework Directive).

Coastal Management and Planning

1. European Commission, 2012. Communication from the commission to the European parliament, the council, the European economic and social committee and the committee of the regions. Blue Growth Opportunities for Marine and Maritime Sustainable Growth. Brussels. COM(2012) 494 final.

2. Inter-Departmental Marine Coordination Group, 2012. Harnessing Our Ocean Wealth – An Integrated Marine Plan for Ireland. Dublin: Inter-departmental Marine Coordination Group.

3. UNEP-WCMC, IUCN, 2019. Marine Protected Planet. Marine Protected Areas, [online] Available at: www.protectedplanet.net

4. Naylor, D. and Shannon, P., 2011. *Petroleum Geology of Ireland*. Edinburgh: Dunedin Academic Press.

5. Symmons, C., 2000. *Ireland and the Law of the Sea*. 2nd ed. Dublin: Round Hall Sweet & Maxwell.

6. Edwards, J. and Mellett, M., 1999. Ireland's Maritime Boundaries and the Prosecution of Offences within the Territorial Seas of the State. *University of Limerick Law Review*, pp. 99–104.

7. Cummins, V., Glavovic, B. and Page, G. (in review). Our Coastal Futures: Advancing global progress Towards the sustainable development of the coastal zone. Special Issue Coastal Management.

8. The Bantry Bay Charter was awarded the National Planning Achievement Award by the Irish Planning Institute in 2000 as well as a Special Merit Award by the European Council of Town Planners in 2002.

9. Cummins, V., 2011. Organisational Tools for Integrated Coastal Zone Management: Public Participation and Sustainability Science in Coastal Decision Making. PhD Thesis, University College Cork, p. 479.

10. Bantry Bay Charter, 2001. Bantry Bay Coastal Zone Charter – Process 1. [online] Available at https://bantrybaycharter.ucc.ie/pages/introduction/introprocess.htm

11. Bantry Bay Charter, 2001. Government in the Coastal Zone – Co-ordination of Policies and Plans. [online] Available at: https://bantrybaycharter.ucc.ie/pages/government/govcoord.htm

12. Cummins, V. 2011. Organisational Tools for Integrated Coastal Zone Management, p. 123.
13. European Committee of the Regions, 2013. Opinion of the Committee of the Regions on Proposed Directive for Maritime Spatial Planning and Integrated Coastal Management (2013/C 356/18).
14. Cummins, V., O'Mahony, C. and Connolly, N., 2004. Review of Integrated Coastal Zone Management and Principles of Best Practice. Kilkenny: The Heritage Council.
15. Mason, T.H., 1936. *The Islands of Ireland*. London: B.T. Batsford, p. 12.
16. Ipsos MORI, 2008. Development and Marketing of the Islands of the Ireland Product – Qualitative Research. Dublin: Fáilte Ireland.
17. Ipsos MORI, 2008. Development and Marketing of the Islands of the Ireland Product – Qualitative Research, p. 2.
18. Ipsos MORI, 2008. Development and Marketing of the Islands of the Ireland Product – Qualitative Research.
19. O'Hanlon, Z. and Cummins, V. (in review). A Comparative Insight of Irish and Scottish Regulatory Frameworks for Offshore Wind Energy – An expert perspective. Submitted to *Marine Policy*.
20. Donegal County Council, 2012. County Donegal Development Plan 2012–2018. p. 124.
21. O'Mahony, C., O'Hagan, A.M. and Meaney, E., 2012. A Review of Beach Bye Law Usage in Supporting Coastal Management in Ireland. *Coastal Management*, 40(5), pp. 461–483.
22. Department of the Environment Northern Ireland, 2006. Towards an Integrated Coastal Zone Management Strategy for Northern Ireland 2006–2026.
23. Department of Agriculture, Environment and Rural Affairs, 2018. Draft Marine Plan for Northern Ireland.
24. The Governments of the United Kingdom or Great Britain and Northern Ireland and of Ireland, 2011. Memorandum of Understanding between the United Kingdom and Ireland Governments for Lease of the Seabed for Development of Offshore Renewable Energy Installations.
25. Flannery, W., O'Hagan, A.M., O'Mahony, C., Ritchie, H. and Twomey, S., 2015. Evaluating Conditions for Transboundary Marine Spatial Planning: Challenges and opportunities on the island of Ireland. *Marine Policy*, 51, pp. 86–95.
26. Payne, I., Tindall, C., Hodgson, S. and Harris, C., 2011. Comparative analysis of Maritime Spatial Planning (MSP) Regimes, Barriers and Obstacles, Good Practices and National Policy Recommendations. Brussels: European Wind Energy Association.

SECTION 6 FUTURE COASTS

Climate Change and Coastal Futures

1. United Nations Framework Convention on Climate Change, 2019. Information Update on Chile COP25, to be Held in Madrid on 2–13 December 2019. [press release] 5 November. [online] Available at: https://unfccc.int/news/information-update-on-chile-cop25-to-be-held-in-madrid-on-2-13-december-2019; UNESCO, 2020. Facts and Figures on Marine Biodiversity. [online] Available at: http://www.unesco.org/new/en/natural-sciences/ioc-oceans/focus-areas/rio-20-ocean/blueprint-for-the-future-we-want/marine-biodiversity/facts-and-figures-on-marine-biodiversity/
2. United Nations, 2017. The Ocean Conference, United Nations, New York, 5–9 June 2017 - Factsheet: people and oceans. United Nations, New York. [online] Available at: https://www.un.org/sustainabledevelopment/wp-content/uploads/2017/05/Ocean-fact-sheet-package.pd
3. IPBES, 2019. *Summary for policymakers of the global assessment report on biodiversity and ecosystem services of the Intergovernmental Science-Policy Platform on Biodiversity and Ecosystem Services*. IPBES Secretariat, Bonn, Germany, pp. 56; IPCC, 2014. Intergovernmental Panel on Climate Change: Synthesis Report. Contribution of Working Groups I, II and III to the Fifth Assessment Report of the Intergovernmental Panel on Climate Change. IPCC, Geneva, Switzerland, 151 pp.
4. Klein, N., 2014. *This Changes Everything: Capitalism vs. the climate*. Allen Lane/Penguin, UK; Steffan, W., Richardson, K., Rockstrom, J., Cornell, S.E., Fetzer, I., Bennett, E.M., Biggs, R., Carpenter, S.R. et al., 2015. Planetary boundaries: Guiding human development on a changing planet. *Science*, 347, Issue 6223. DOI: 10.1126/science.1259855; Environmental Protection Agency, 2020. The Kyoto Protocol. [online] Available at: https://www.epa.ie/climate/thekyotoprotocol/
5. Keynes, J.M., 1936. *The General Theory of Employment, Interest and Money*. Macmillan, London; Friedman, M., 1962, *Capitalism and Freedom* (fortieth anniversary edition). University of Chicago Press: Chicago. [Online] available at: https://press.uchicago.edu/ucp/books/book/chicago/C/bo68666099.html
6. Dasgupta, P., 2008. Nature in Economics. *Environ Resource Econ.*, 39, 1–7; Jackson, T., 2009. Prosperity without Growth: Economics for a finite planet. Earthscan, London.
7. Donovan, D. and Murphy, A.E., 2013. *The Fall of the Celtic Tiger: Ireland and the Euro Debt Crisis*. Oxford Scholarship Online.
8. Zuboff, S., 2019. *The Age of Surveillance Capitalism. The fight for a human future at the new frontier of power*. Profile Books Ltd, London.
9. Haggett, P., 2001 (3rd ed.). *Geography: A modern synthesis*. Harper and Row, London.
10. United Nations, Department of Economic and Social Affairs, Population Division, 2019. World Population Prospects 2019: Highlights (ST/ESA/SER.A/423); ICAN, 2020. ICAN: International Coastal Atlas Network. [online] Available at: https://www.oceandocs.org/handle/1834/5926
11. Department of Housing, Planning and Local Government, 2002. National Spatial Strategy for Ireland 2002–2020: People, places and potential. Government of Ireland, Dublin; Central Statistics Office, 2020. [online] Available at: https://www.cso.ie/en/index.html; Northern Ireland Statistics and Research Agency, 2020. Statistics. [online] Available at: https://www.nisra.gov.uk/statistics
12. Ehrlich, P.R. and Ehrlich, A.H., 2009. The Population Bomb Revisited. *The Electronic Journal of Sustainable Development*, 1(3), pp. 63–71.
13. FAO, 2018. The State of World Fisheries and Aquaculture 2018: Meeting the sustainable development goals. Food and Agriculture Organization of the United Nations, Rome.
14. Brundtland, G., 1987. Report of the World Commission on Environment and Development: Our Common Future. United Nations General Assembly document A/42/427
15. Lomazzi, M., Laaser, U., Theisling, M., Tapia, L. and Borisch, B., 2014. Millennium Development Goals: How public health professionals perceive the achievement of MDGs. *Global Health Action*, 7: 10.3402/gha.v7.24352; Kotter, R., 2018. Understanding sustainable development, *International Journal of Environmental Studies*, DOI. 10.1080/00207233.2018.1453385
16. Department of Communication, Climate Action and Environment, 2018. The Sustainable Development Goals National Implementation Plan 2018-2020. Department of Communications Climate Action and Environment, Government of Ireland, Dublin.
17. Glenn, H., Wattage, P.. Mardle, S., van Rensburg, T., Grehan, A. and Foley, N., 2010. Marine Protected Areas: Substantiating their worth. *Marine Policy*, 34, 421–430. https://doi.org/10.1016/j.marpol.2009.09.007; NPWS, 2020. Marine Habitats. [online] Available at: https://www.npws.ie/marine/marine-habitats
18. Desmond, M., O'Brien, P. and McGovern, F., 2017. A Summary of the State of Knowledge on Climate Change impacts for Ireland. EPA Research Report 11 Environmental Protection Agency, Government of Ireland; IPCC, 2019a. Summary for Policymakers. In: *Climate Change and Land: an IPCC special report on climate change, desertification, land degradation, sustainable land management, food security, and greenhouse gas fluxes in terrestrial ecosystems* [Shukla, P.R., Skea, J., Calvo Buendia, E., Masson-Delmotte, V., Pörtner, H.-O., Roberts, D.C., Zhai, P., Slade, R., Connors, S., van Diemen, R., Ferrat, M., Haughey, E., Luz, S., Neogi, S., Pathak, M., Petzold, J., Portugal Pereira, J., Vyas, P., Huntley, E., Kissick, K., Belkacemi, M., Malley, J., (eds.)]. In press.; IPCC, 2019b: Summary for Policymakers. In: *IPCC Special Report on the Ocean and Cryosphere in a Changing Climate* [Pörtner, H.-O., Roberts, D.C., VMasson-Delmotte, V., Zhai, P., Tignor, M., Poloczanska, E., Mintenbeck, K., Alegría, A., Nicolai, M., Okem, A., Petzold, J., Rama, B., Weyer N.M., (eds.)]. In press.
19. Devoy, R.J.N., 2008. Coastal Vulnerability and the Implications of Sea-level Rise for Ireland. *Journal of Coastal Research*, 24 (2), 325–341; Devoy, R.J.N., 2015a. Sea-level Rise: Causes, Impacts and Scenarios for Change. In, Ellis, J. & Sherman, D. (eds.), *Coastal and Marine Hazards, Risks and Disasters*. Elsevier, Amsterdam, pp. 197–242; Devoy, R.J.N., 2015b. The Development and Management of the Dingle Bay Spit-barriers of Southwest Ireland. In, Randazzo, G., Cooper, J.A.G. and Jackson, D.W. (eds.), *Sand and Gravel Spits. Coastal Research Library 12, Springer* International Publishing, Switzerland, pp. 139–180. DOI 10.1007/978-3-319-13716-2_9.
20. Caesar, L., Rahmstorf, S., Robinson, A., Feulner, G. and Saba, V. 2018. Observed fingerprint of a Weakening Atlantic Ocean Overturning Circulation. *Nature* 556, pp. 191–196. https://doi.org/10.1038/s41586-018-0006-5; Thornalley, D.J.R., Oppo, D.W., Ortega, P., Robson, J.I., Brierley, C.M., Davis, R., Hall, I.R., Moffa-Sanchez, P., Rose, N.L., Spooner, P.T., Yashayaev, I. and Keigwin, L.D. 2018. Anomalously weak Labrador Sea convection and Atlantic overturning during the past 150 years. *Nature* 556, pp. 227–230. https://doi.org/10.1038/s41586-018-0007-4
21. Gleeson, E., McGrath R. and Treanor M., 2013. *Ireland's Climate: The road ahead*. Dublin: Met Éireann, Dublin; Nolan, P. 2015. Ensemble of regional climate model projections for Ireland (2008-FS-CC-m). EPA Research Report 159. EPA; Nolan, P. and Alastair McKinstry, 2015. EC-Earth Global Climate Simulations: Ireland's Contributions to CMIP6 (2015-CCRP-FS.23) EPA Research Report Number 310. Environmental Protection Agency, Johnstown Castle, Co. Wexford, Ireland; Gallagher, S., Gleeson, S., Tiron, R., McGrath, R. and Dias, F., 2016. Twenty-first century wave climate projections for Ireland and surface winds in the North Atlantic Ocean. *Advances in Science and Res.*, 13, pp. 75–80. doi:10.5194/asr-13-75-2016.
22. Falaleeva, M., O'Mahony, C., Gray, S. and Desmond, M., 2011. Towards Climate Adaptation and Coastal Governance in Ireland: Integrated architecture for effective management? *Marine Policy*, 35(6): pp. 784–793; Sweeney, J., Albanito, F., Brereton, A., Caffarra, A., Charlton, R., Donnelly, A., Fealy, R., Fitzgerald, J., Holden, N., Jones, M and Murphy, C., 2014. Climate Change: Refining the impacts for Ireland. Strive Report, 2001-CD-C3-M1. Environmental Protection Agency, Johnstown Castle, Wexford; CCAC, 2020. Progress Towards a Climate-Resilient Ireland. Climate Change Advisory Council Report 2020. Government of Ireland, Dublin.
23. Ciais, P., Sabine, C. Bala, G., Bopp, L., Brovkin, V., Canadell, J., Chhabra, A., DeFries, R., Galloway, J., Heimann, M., Jones, C., Le Quéré, C., Myneni, R.B., Piao, S., and Thornton, P., 2013. Carbon and Other Biogeochemical Cycles. In: *Climate Change 2013: The Physical Science Basis. Contribution of Working Group I to the Fifth Assessment Report of the Intergovernmental Panel on Climate Change* [Stocker, T.F., Qin, D., Plattner, G.-K., Tignor, M., Allen, S.K., Boschung, J., Nauels, A., Xia, Y., Bex V., and Midgley, P.M., (eds.)]. Cambridge University Press, Cambridge, United Kingdom and New York, NY, USA; IPCC, 2019. IPCC Special Report on the Ocean and Cryosphere in a Changing Climate [Pörtner, H.-O., Roberts, D.C., Masson-Delmotte, V., Zhai, P., Tignor, M., Poloczanska, E., Mintenbeck, K., Alegría,A., Nicolai, M., Okem, A., Petzold, J. Rama, B., Weyer, N.M., eds.]. In press.
24. Bates, N.R., Astor, Y.M., Church, M.J., Currie, K., Dore, J.E., González-Dávila, M., Lorenzoni, L., Muller-Karger, F., Olafsson, J. and Santana-Casiano, J.M. 2014. A Time-Series View of Changing Ocean Chemistry Due to Ocean Uptake of Anthropogenic CO_2 and Ocean Acidification. *Oceanography*, 27: pp. 126–141; McGrath, T., McGovern, E., Cave, R. R., and Kivimäe, C. 2015. The Inorganic Carbon Chemistry in Coastal and Shelf Waters Around Ireland. Estuaries and Coasts, 1–13.
25. Secretariat of the Convention on Biological Diversity, 2014. Global Biodiversity Outlook 4.

Montréal, pp. 155; Secretariat of the Convention on Biological Diversity, 2014. An Updated Synthesis of the Impacts of Ocean Acidification on Marine Biodiversity (eds.: S. Hennige, J.M. Roberts & P. Williamson). Montreal, Technical Series No. 75, pp. 99; IPCC, 2019: Summary for Policymakers. In: IPCC Special Report on the Ocean and Cryosphere in a Changing Climate, Pörtner, H.-O., Roberts, D.C., Masson-Delmotte, V., Zhai, P., Tignor, M., Poloczanska, E., Mintenbeck, K., Alegría, A., Nicolai, M., AOkem, A., Petzold, J., Rama, B., Weyer N.M., (eds.)]. In press.

26. Department of Communication, Climate Action and Environment, 2019. Climate action plan 2019: to tackle climate breakdown. Government of Ireland, 2019; Cummins, V. and McKeogh, E., 2020. EirWind Blueprint, [online] Available at https://www.marei.ie/project/eirwind/

27. Cooper, J.A.G. and Cummins, V., 2009. Coastal Management in Northwest Europe. *Marine Policy*, 33, pp. 869–938; Falaleeva, M., Gray, S., O'Mahony, C. and Gault, J., 2013. Coastal climate Adaptation in Ireland (CLAD): assessing current conditions and enhancing the capacity for climate resilience in local coastal management. EPA CCRP Report 3.6. Environmental Protection Agency, Johnstown Castle, Wexford; Shine, T., 2018. Climate-Resilient Ireland. EPA Report 252. Government of Ireland, Dublin; Climate Change Advisory Council, 2020. Annual review report 2020. McCummiskey House, Government of Ireland, Dublin.

28. Crowley, J., Devoy, R.J.N., Linehan, D. and O'Flanagan, P., eds. 2005. *Atlas of Cork City*. Cork University Press, Cork.

29. Office of Public Works, 2016. The Lower Lee (Cork City) Flood Relief Scheme (Drainage Scheme). Exhibition Report. Office of Public Works, Dublin.

30. Abercrombie, P., 1925. Architectural Character: City Streets and quays. In: *Cork Town Planning Association*, Cork: A civic survey, chapter 11. Cork Town Planning Association, Cork City; Licciardi, G., Amirtahmasebi, R. 2012. *The Economics of Uniqueness: Investing in historic city cores and cultural heritage assets for sustainable development*. Urban Development;. Washington, DC: World Bank.

31. Save Cork City, 2017. Potential Cork: the save Cork city solution. Save Cork City Campaign, Cork. [online] Available at: http://savecorkcity.org/docs/Potential-Cork-Save-Cork-City-Solution.pdf

32. Intergovernmental Panel on Climate Change, 2014. *Climate Change 2014: Synthesis Report. Contribution of Working Groups I, II and III to the Fifth Assessment Report of the Intergovernmental Panel on Climate Change*. Intergovernmental Panel on Climate Change, Geneva, Switzerland, 151 pp.; FAO, 2018. The State of World Fisheries and Aquaculture 2018: Meeting the sustainable development goals. Rome. Licence: CC BY-NC-SA 3.0 IGO; IPBES, 2019. Summary for policymakers of the global assessment report on biodiversity and ecosystem services of the Intergovernmental Science-Policy Platform on Biodiversity and Ecosystem Services, IPBES Secretariat, Bonn, Germany.

33. Ragnarsson S.Á. et al., 2017. The Impact of Anthropogenic Activity on Cold-water Corals. In: Rossi, S., Bramanti L., Gori A., Orejas C., eds. Marine Animal Forests. Springer, Cham., New York; UNEP, 2020. Protecting Coral Reefs. [online] Available at: https://www.unenvironment.org/explore-topics/oceans-seas/what-we-do/protecting-coral-reefs

34. National Parks and Wildlife Service, 2019. The Status of EU Protected Habitats and Species in Ireland. Volume 1: Summary Overview (Edited by: Deirdre Lynn and Fionnuala O'Neill). Unpublished NPWS report. Dublin.

35. Klein, N., 2014. *This Changes Everything: Capitalism vs. the climate*. Allen Lane/Penguin, UK; Gray, S., 2020. Making Sense of Changing Coastal Systems: Overcoming barriers to climate change adaptation using fuzzy cognitive mapping. PhD Thesis. National University of Ireland at Cork (NUIC), Cork.

36. COE, 2018. Republic of Palau Ministry of Finance – Localizing the Agenda in SIDS: Palau's Aspirations with the SDGs- 2017. [online] Available at: http://www.sustainablesids.org/kb/sids-pacific/palau

37. Stanford Center for Ocean Solutions, 2020. Palau's National Marine Sanctuary: Managing Ocean Change and Supporting Food Security. [online] Available at: https://oceansolutions.stanford.edu/pnms-report

38. Lange, M., 2018. Development of a conceptual model towards an innovative solution for Marine Energy decision-making in Ireland. PhD Thesis. National University of Ireland at Cork (NUIC), Cork.

39. International Energy Association, 2018. World Energy Outlook 2018 examines future patterns of the global energy system at a time of increasing uncertainties. [press release] 13 November 2018. [online] Available at: https://www.iea.org/news/world-energy-outlook-2018-examines-future-patterns-of-global-energy-system-at-a-time-of-increasing-uncertainties; Cummins, V. and McKeogh, E., 2020. EirWind Blueprint, [online] Available at https://www.marei.ie/project/eirwind/

40. Cummins, V., 2011. Organisational tools for integrated coastal zone management: public participation and sustainability science in coastal decision making. PhD Thesis. National University of Ireland at Cork (NUIC), Cork; Department of Enterprise, Trade and Investment, 2010. A Strategic Framework for Northern Ireland, September 2010. Department of Enterprise, Trade and Investment. Belfast; Marine Institute, 2012. Harnessing Our Ocean Wealth: An Integrated Marine Plan for Ireland. Interdepartmental Marine Coordination Group. Government of Ireland.

41. Northern Ireland Marine Task Force, 2018. Northern Ireland Marine Task Force Response to Department of Housing Planning and Local Government – National Marine Planning Framework Baseline Report. [online] Available at: https://www.housing.gov.ie/sites/default/files/public-consultation/files/responses/068_northern_ireland_marine_task_force.pdf

42. McK Bary, B., 1986. Marine Science and the Development of Resources in Ireland. Sherkin Island Marine Station, *Bulletin No. 3*.

43. CRCG, 2019. Climate Research Coordination Group: First Report on Activities: June 2017–December 2018. Government of Ireland. Dublin; CCAC, 2020. Progress Towards a Climate-Resilient Ireland. Climate Change Advisory Council Report 2020. Government of Ireland, Dublin.

44. Cummins, V., 2011. Organisational Tools for Integrated Coastal Zone Management: Public participation and sustainability science in coastal decision making. PhD Thesis. University College Cork (NUIC), Cork; Gray, S., 2020. Making Sense of Changing Coastal Systems: Overcoming barriers to climate change adaptation using fuzzy cognitive mapping. PhD Thesis. University College Cork (NUIC), Cork; CCAC, 2020. Progress Towards a Climate-resilient Ireland. Climate Change Advisory Council Report 2020. Government of Ireland, Dublin.

INDEX

Seachange

Theo Dorgan

Sheep at night,
picking their way across
the face of the mountain.

Stars on a deep sky
crisp and clear
far out to sea.

Salt on the wind,
the sheep farmer
fears for the sailor,
the wind rising.

Rain on the wind,
the sailor observes
the stars hesitate,
then huddle for shelter.

for Francis Harvey